HANDBOOK OF PHARMACEUTICAL BIOTECHNOLOGY

BICENTENNIAL
1807
WILEY
2007
BICENTENNIAL

THE WILEY BICENTENNIAL—KNOWLEDGE FOR GENERATIONS

$\mathcal{E}$ach generation has its unique needs and aspirations. When Charles Wiley first opened his small printing shop in lower Manhattan in 1807, it was a generation of boundless potential searching for an identity. And we were there, helping to define a new American literary tradition. Over half a century later, in the midst of the Second Industrial Revolution, it was a generation focused on building the future. Once again, we were there, supplying the critical scientific, technical, and engineering knowledge that helped frame the world. Throughout the 20th Century, and into the new millennium, nations began to reach out beyond their own borders and a new international community was born. Wiley was there, expanding its operations around the world to enable a global exchange of ideas, opinions, and know-how.

For 200 years, Wiley has been an integral part of each generation's journey, enabling the flow of information and understanding necessary to meet their needs and fulfill their aspirations. Today, bold new technologies are changing the way we live and learn. Wiley will be there, providing you the must-have knowledge you need to imagine new worlds, new possibilities, and new opportunities.

Generations come and go, but you can always count on Wiley to provide you the knowledge you need, when and where you need it!

WILLIAM J. PESCE
PRESIDENT AND CHIEF EXECUTIVE OFFICER

PETER BOOTH WILEY
CHAIRMAN OF THE BOARD

HANDBOOK OF PHARMACEUTICAL BIOTECHNOLOGY

Edited by

SHAYNE COX GAD, PH.D., D.A.B.T.
Gad Consulting Services
Cary, North Carolina

WILEY-INTERSCIENCE
A John Wiley & Sons, Inc., Publication

Published by John Wiley & Sons, Inc., Hoboken, New Jersey
Published simultaneously in Canada

Library of Congress Cataloging-in-Publication Data:

Gad, Shayne C., 1948–
    Handbook of pharmaceutical biotechnology / Shayne Cox Gad.
       p. ; cm.
    Includes bibliographical references and index.
    ISBN: 978-0-471-21386-4
    1. Pharmaceutical biotechnology—Handbooks, manuals, etc. I. Title.
    [DNLM: 1. Drug Design. 2. Biotechnology. 3. Chemistry, Pharmaceutical.
QV 744 G123h 2007]
RS380.G33 2007
615′.19—dc22
                                                                   2006030898

Printed in the United States of America
10 9 8 7 6 5 4 3 2 1

CONTRIBUTORS

Toshihiro Akaike, Department of Biomolecular Engineering, Graduate School of Bioscience and Biotechnology, Tokyo Institute of Technology, Yokohama, Japan, *Case Studies—Development of Oligonucleotides*

Thomas Anchordoquy, School of Pharmacy, University of Colorado at Denver and Health Sciences Center, Denver, Colorado, *Basic Issues in the Manufacture of Macromolecules*

Aravind Asokan, Division of Molecular Pharmaceutics, School of Pharmacy, and Gene Therapy Center, University of North Carolina at Chapel Hill, Chapel Hill, North Carolina, *Strategies for the Cytosolic Delivery of Macromolecules: An Overview*

Joseph P. Balthasar, University at Buffalo, The State University of New York, Buffalo, New York, *Development and Characterization of High-Affinity Anti-Toptecan IgG and Fab Fragments*

Sathy V. Balu-Iyer, University at Buffalo, State University of New York, Amherst, New York, *Formulation and Delivery Issues of Therapeutic Proteins*

Randal W. Berg, Cancer Research Laboratory Program, London Regional Cancer Program, London, Ontario, Canada; Department of Oncology, The University of Western Ontario, London, Ontario, Canada; and Lawson Health Research Institute, London Health Sciences Centre, London, Ontario, Canada, *Pharmacokinetics of Nucleic-Acid-Based Therapeutics*

Isabelle Bertholon, Faculty of Pharmacy, University of Paris XI, Chatenay-Malabry, France, *Integrated Development of Glycobiologics: From Discovery to Applications in the Design of Nanoparticular Drug Delivery Systems*

Günter Blaich, Abbott Bioresearch Center, Worcester, Massachusetts, *Overview: Differentiating Issues in the Development of Macromolecules Compared with Small Molecules*

Jeanine I. Boulter-Bitzer, Department of Environmental Biology, University of Guelph, Guelph, Ontario, Canada, *Recombinant Antibodies for Pathogen Detection and Immunotherapy*

Tania Bubela, Department of Marketing, University of Alberta, Edmonton, Alberta, Canada, *Intellectual Property and Biotechnolgy*

Brian E. Cairns, University of British Columbia, Vancouver, British Columbia, Canada, *Growth Factors and Cytokines*

Heping Cao, Peace Technology Development, North Potomac, Maryland, *Growth Factors, Cytokines, and Chemokines: Formulation, Delivery, and Pharmacokinetics*

María de los Angeles Cortés Castillo, Technology and Health Service Delivery, Pan American Health Organization, Washington, DC, *Regulation of Small-Molecule Drugs versus Biologicals versus Biotech Products*

Jin Chen, Department of Pharmaceutical Sciences, University at Buffalo, The State University of New York, Buffalo, New York, *Development and Characterization of High-Affinity Anti-Toptecan IgG and Fab Fragments*

Roland Cheung, Division of Molecular Pharmaceutics, School of Pharmacy, and Gene Therapy Center, University of North Carolina at Chapel Hill, Chapel Hill, North Carolina, *Strategies for the Cytosolic Delivery of Macromolecules: An Overview*

Moo J. Cho, Division of Molecular Pharmaceutics, School of Pharmacy, and Gene Therapy Center, University of North Carolina at Chapel Hill, Chapel Hill, North Carolina, *Strategies for the Cytosolic Delivery of Macromolecules: An Overview*

Yong Woo Cho, Asan Institute for Life Sciences, University of Ulsan College of Medicine, Seoul, Korea, *PEGylation: Camouflage of Proteins, Cells, and Nanoparticles Against Recognition by the Body's Defense Mechanism*

Albert H.L. Chow, School of Pharmacy, The Chinese University of Hong Kong, Shatin, N.T., Hong Kong, China, *Stability Assessment and Formulation Characterization*

Ezharul Hoque Chowdhury, Department of Biomolecular Engineering, Graduate School of Bioscience and Biotechnology, Tokyo Institute of Technology, Yokohama, Japan, *Case Studies—Development of Oligonucleotides*

Daan J.A. Crommelin, Department of Pharmaceutics, Utrecht Institute for Pharmaceutical Sciences (UIPS), Utrecht University, The Netherlands, *Immunogenicity of Therapeutic Proteins; Biosimilars*

Mary Jane Cunningham, Houston Advanced Research Center (HARC), The Woodlands, Texas, *Toxicogenomics*

Vincenzo De Filippis, Department of Pharmaceutical Sciences, University of Padova, Padova, Italy, *Protein Engineering with Noncoded Amino Acids: Applications to Hirudin*

Pascal Delépine, EFS Bretagne—Site de Brest, INSERM U613, Brest, France, *Assessing Gene Therapy by Molecular Imaging*

Binodh DeSilva, Amgen, Inc., Thousand Oaks, California, *Analytical Considerations for Immunoassays for Macromolecules*

José Luis Di Fabio, Technology and Health Service Delivery, Pan American Health Organization, Washington, DC, *Regulation of Small-Molecule Drugs versus Biologicals versus Biotech Products*

Karen Lynn Durell, Center for Intellectual Property Policy, Faculty of Law, McGill University, Montreal, Quebec, Canada, *Intellectual Property and Biotechnolgy*

Andrew Emili, Banting and Best Department of Medical Research, Department of Medical Genetics and Microbiology, Donnelly Centre for Cellular and Biomolecular Research, University of Toronto, Toronto, Ontario, Canada, *Enhanced Proteomic Analysis by HPLC Prefractionation*

Claude Férec, EFS Bretagne—Site de Brest, INSERM U613, Brest, France, *Assessing Gene Therapy by Molecular Imaging*

Zoltan Gombos, National Institute for Nanotechnology, University of Alberta, Edmonton, Alberta, Canada, *Proteins: Hormones, Enzymes, and Monoclonal Antibodies—Background*

Matthew D. Gray, MDG Associates, Inc., Seattle, Washington, *RNA Interference: The Next Gene-Targeted Medicine*

J. Chris Hall, Department of Environmental Biology, University of Guelph, Guelph, Ontario, Canada, *Recombinant Antibodies for Pathogen Detection and Immunotherapy*

Pierre C. Havugimana, Banting and Best Department of Medical Research, Department of Medical Genetics and Microbiology, Donnelly Centre for Cellular and Biomolecular Research, University of Toronto, Toronto, Ontario, Canada, *Enhanced Proteomic Analysis by HPLC Prefractionation*

Suzanne Hermeling, Department of Pharmaceutics, Utrecht Institute for Pharmaceutical Sciences (UIPS); Central Laboratory Animal Institute, Utrecht University, The Netherlands, *Immunogenicity of Therapeutic Proteins*

M.D. Mostaqul Huq, Department of Pharmacology, University of Minnesota Medical School, Minneapolis, Minnesota, *Protein Posttranslational Modification: A Potential Target in Pharmaceutical Development*

Yukako Ito, Department of Pharmacokinetics, Kyoto Pharmaceutical University, Kyoto, Japan, *Pharmacokinetics*

Bernd Janssen, Abbott GmbH & Co. KG, Ludwigshafen, Germany, *Overview: Differentiating Issues in the Development of Macromolecules Compared with Small Molecules*

Wim Jiskoot, Department of Pharmaceutics, Utrecht Institute for Pharmaceutical Sciences (UIPS), Utrecht University, The Netherlands; Division of Drug Delivery Technology, Leiden/Amsterdam Center for Drug Research (LACDR), Leiden University, Leiden, The Netherlands, *Immunogenicity of Therapeutic Proteins; Biosimilars*

Beth Junker, Merck Research Laboratories, Rahway, New Jersey, *Process Validation for Biopharmaceuticals*

David Keast, Site Chief of Family Medicine, Parkwood Hospital, St. Joseph's Health Care, London, Ontario, Canada; Clinical Adjunct Professor of Family Medicine, University of Western Ontario, London, Ontario, Canada, *Unexpected Benefits of a Formulation: Case Study with Erythropoetin*

Marian Kelley, Centocor R&D, Inc., Randor, Pennsylvania, *Analytical Considerations for Immunoassays for Macromolecules*

Naoya Kobayashi, Department of Surgery, Okayama University Graduate School of Medicine and Dentistry, Okayama, Japan, *Overview of Stem and Artificial Cells*

James Koropatnick, Cancer Research Laboratory Program, London Regional Cancer Program, London, Ontario, Canada; The University of Western Ontario (Departments of Microbiology and Immunology; Oncology; Physiology and Pharmacology; and Pathology), London, Ontario, Canada; and Lawson Health Research Institute, London Health Sciences Centre, London, Ontario, Canada, *Pharmacokinetics of Nucleic-Acid-Based Therapeutics*

Krishnanand D. Kumble, Genesis Research, Parnell, Auckland, New Zealand, *Microarrays in Drug Discovery and Development*

Sarita Kumble, Pictor Limited, Glendowie, Auckland, New Zealand, *Microarrays in Drug Discovery and Development*

Anne E. Kwitek, Human and Molecular Genetics Center, Department of Physiology, Medical College of Wisconsin, Milwaukee, Wisconsin, *Genetic Markers and Genotyping Analyses for Genetic Disease Studies*

Denis Labarre, Faculty of Pharmacy, University of Paris XI, Chatenay-Malabry, France, *Integrated Development of Glycobiologics: From Discovery to Applications in the Design of Nanoparticular Drug Delivery Systems*

Hung Lee, Department of Environmental Biology, University of Guelph, Guelph, Ontario, Canada, *Recombinant Antibodies for Pathogen Detection and Immunotherapy*

Kang Choon Lee, Drug Targeting Laboratory, College of Pharmacy, SungKyunKwan University, Jangan-ku, Suwon, Korea, *Capillary Separation Techniques*

Corinne Lengsfeld, Department of Engineering, University of Denver, Denver, Colorado, *Basic Issues in the Manufacture of Macromolecules*

Jun Li, School of Chemical Biology and Pharmaceutical Studies, Capital Medical University of Medical Sciences, Beijing, China, *Pharmaceutical Bioassay*

Rui Lin, Peace Technology Development, North Potomac, Maryland, *Growth Factors, Cytokines, and Chemokines: Formulation, Delivery, and Pharmacokinetics*

John C. Lindon, Department of Biomolecular Medicine, Faculty of Medicine, Imperial College, South Kensington, London, United Kingdom, *An Overview of Metabonomics Techniques and Applications*

Donald E. Mager, Department of Pharmaceutical Sciences, University at Buffalo, The State University of New York, Buffalo, New York, *Preclinical Pharmacokinetics*

Krishan Maggon, Pharma Biotech R & D Consultant, Geneva, Switzerland, *R&D Paradigm Shift and Billion-Dollar Biologics*

Mandeep K. Mann, University of British Columbia, Vancouver, British Columbia, Canada, *Growth Factors and Cytokines*

Wayne Materi, National Institute for Nanotechnology, University of Alberta, Edmonton, Alberta, Canada, *Proteins: Hormones, Enzymes, and Monoclonal Antibodies—Background*

Razvan D. Miclea, Roswell Park Cancer Institute, Buffalo, New York, *Formulation and Delivery Issues of Therapeutic Proteins*

Jan Moebius, Rudolf-Virchow-Center for Experimental Biomedicine, Würzburg, Germany, *Chromatography-Based Separation of Proteins, Peptides, and Amino Acids*

Dong Hee Na, College of Pharmacy, Kyungsung University, Nam-ku, Busan, Korea, *Capillary Separation Techniques*

Nalu Navarro-Alvarez, Department of Surgery, Okayama University Graduate School of Medicine and Dentistry, Okayama, Japan, *Overview of Stem and Artificial Cells*

Michael Oettel, Prof. med. Vet. habil., Jena, Germany, *The Promise of Individualized Therapy*

Andrew V. Oleinikov, Seattle Biomedical Research Group, Seattle, Washington, *RNA Interference: The Next Gene-Targeted Medicine*

Michael Olivier, Human and Molecular Genetics Center, Department of Physiology, Medical College of Wisconsin, Milwaukee, Wisconsin, *Genetic Markers and Genotyping Analyses for Genetic Disease Studies*

Aleksandra Pandyra, Cancer Research Laboratory Program, London Regional Cancer Program, London, Ontario, Canada; Department of Microbiology and Immunology, The University of Western Ontario, London, Ontario, Canada, *Pharmacokinetics of Nucleic-Acid-Based Therapeutics*

Jae Hyung Park, Department of Advanced Polymer and Fiber Materials, College of Environment and Applied Chemistry, Kyung Hee University, Gyeonggi-do, Korea, *PEGylation: Camouflage of Proteins, Cells, and Nanoparticles Against Recognition by the Body's Defense Mechanism*

Ji Sun Park, Asan Institute for Life Sciences, University of Ulsan College of Medicine, Seoul, Korea, *PEGylation: Camouflage of Proteins, Cells, and Nanoparticles Against Recognition by the Body's Defense Mechanism*

Kinam Park, Departments of Pharmaceutics and Biomedical Engineering, Purdue University, West Lafayette, Indiana, *PEGylation: Camouflage of Proteins, Cells, and Nanoparticles Against Recognition by the Body's Defense Mechanism*

Steve Pascolo, Institute for Cell Biology, Department of Immunology, University of Tübingen, Tübingen, Germany, *Plasmid DNA and Messenger RNA for Therapy*

Shiqi Peng, School of Chemical Biology and Pharmaceutical Studies, Capital Medical University of Medical Sciences, Beijing, China, *Pharmaceutical Bioassay*

Nicholas J. Pokorny, Department of Environmental Biology, University of Guelph, Guelph, Ontario, Canada, *Recombinant Antibodies for Pathogen Detection and Immunotherapy*

Vivek S. Purohit, R & D, Eurand, Inc., Vandalia, Ohio, *Formulation and Delivery Issues of Therapeutic Proteins*

D.M.F. Prazeres, Institute for Biotechnology and Bioengineering, Center for Biological and Chemical Engineering, Instituto Superior Técnico, Lisbon, Portugal, *Production and Purification of Adenovirus Vectors for Gene Therapy*

Murali Ramanathan, Department of Pharmaceutical Sciences, University at Buffalo, The State University of New York, Buffalo, New York, *Preclinical Pharmacokinetics*

Raymond M. Reilly, Departments of Pharmaceutical Sciences and Medical Imaging, University of Toronto, Toronto, Ontario, Cananda, *The Radiopharmaceutical Science of Monoclonal Antibodies and Peptides for Imaging and Targeted* in situ *Radiotherapy of Malignancies*

Jorge David Rivas-Carillo, Department of Surgery, Okayama University Graduate School of Medicine and Dentistry, Okayama, Japan, *Overview of Stem and Artificial Cells*

Gregory Roth, Abbott GmbH & Co. KG, Ludwigshafen, Germany, *Overview: Differentiating Issues in the Development of Macromolecules Compared with Small Molecules*

Gabor M. Rubanyi, Cardium Therapeutics, Inc., San Diego, California, *Gene Therapy—Basic Principles and the Road from Bench to Bedside*

Jochen Salfeld, Abbott GmbH & Co. KG, Ludwigshafen, Germany, *Overview: Differentiating Issues in the Development of Macromolecules Compared with Small Molecules*

J.A.L. Santos, Institute for Biotechnology and Bioengineering, Center for Biological and Chemical Engineering, Instituto Superior Técnico, Lisbon, Portugal, *Production and Purification of Adenovirus Vectors for Gene Therapy*

Huub Schellekens, Department of Pharmaceutics, Utrecht Institute for Pharmaceutical Sciences (UIPS); Central Laboratory Animal Institute, Utrecht University, The Netherlands, *Immunogenicity of Therapeutic Proteins; Biosimilars*

Frank-Ranier Schmidt, Sanofi-Aventis Deutschland, Frankfurt am Main, Germany, *From Gene to Product: The Advantage of Integrative Biotechnology*

John C. Schmitz, VACT Healthcare System, VA Cancer Center, and Yale Cancer Center, Yale University School of Medicine, West Haven, Connecticut, *Pharmacokinetics of Nucleic-Acid-Based Therapeutics*

Herbert Schott, Institute of Organic Chemistry, University of Tübingen, Tübingen, Germany, *Delivery Systems for Peptides/Oligonucleotides and Lipophilic Nucleoside Analogs*

R.A. Schwendener, Institute of Molecular Cancer Research, University of Zurich, Zurich, Switzerland, *Delivery Systems for Peptides/Oligonucleotides and Lipophilic Nucleoside Analogs*

Gerhard K.E. Scriba, Friedrich-Schiller-University Jena, School of Pharmacy, Jena, Germany, *Bioanalytical Method Validation for Macromolecules*

Tatiana Segura, Chemical and Biomolecular Engineering Department, University of California, Los Angeles, Los Angeles, California, *Formulations and Delivery Limitations of Nucleic-Acid-Based Therapies*

Mrinal Shah, Houston Advanced Research Center (HARC), The Woodlands, Texas, *Toxicogenomics*

Nobuhito Shibata, Department of Biopharmaceutics, Faculty of Pharmaceutical Science, Doshisha Women's College of Liberal Arts, Kyoto, Japan, *Pharmacokinetics*

Dany Shoham, Begin–Sadat Center for Strategic Studies, Bar Ilan University, Israel, *Bioterrorism*

Albert Sickmann, Rudolf-Virchow-Center for Experimental Biomedicine, Würzburg, Germany, *Chromatography-Based Separation of Proteins, Peptides, and Amino Acids*

Alejandro Soto-Gutierrez, Department of Surgery, Okayama University Graduate School of Medicine and Dentistry, Okayama, Japan, *Overview of Stem and Artificial Cells*

Patrick A. Stewart, Department of Political Science, Arkansas State University, State University, Arkansas, *Comparability Studies for Later-Generation Products—Plant-Made Pharmaceuticals*

Remco Swart, LC-Packings—A Dionex Company, Amsterdam, The Netherlands, *Chromatography-Based Separation of Proteins, Peptides, and Amino Acids*

Kanji Takada, Department of Pharmacokinetics, Kyoto Pharmaceutical University, Kyoto, Japan, *Pharmacokinetics*

Henry H.Y. Tong, School of Health Sciences, Macao Polytechnic Institute, Macao, China, *Stability Assessment and Formulation Characterization*

Jack T. Trevors, Department of Environmental Biology, University of Guelph, Guelph, Ontario, Canada, *Recombinant Antibodies for Pathogen Detection and Immunotherapy*

Christine Vauthier, Faculty of Pharmacy, University of Paris XI, Chatenay-Malabry, France, *Integrated Development of Glycobiologics: From Discovery to Applications in the Design of Nanoparticular Drug Delivery Systems*

Ioannis S. Vizirianakis, Laboratory of Pharmacology, Department of Pharmaceutical Sciences, Aristotle University of Thessaloniki, Thessaloniki, Greece, *From Defining Bioinformatics and Pharmacogenomics to Developing Information-Based Medicine and Pharmacotyping in Health Care*

Li-Na Wei, Department of Pharmacology, University of Minnesota Medical School, Minneapolis, Minnesota, *Protein Posttranslational Modification: A Potential Target in Pharmaceutical Development*

David S. Wishart, National Institute of Nanotechnology, Departments of Biological Science, Computing Science, and Pharmaceutical Research, University of Alberta, Edmonton, Alberta, Canada, *Proteins: Hormones, Enzymes, and Monoclonal Antibodies—Background*

Peter Wong, Banting and Best Department of Medical Research, Department of Medical Genetics and Microbiology, Donnelly Centre for Cellular and Biomolecular Research, University of Toronto, Toronto, Ontario, Canada, *Enhanced Proteomic Analysis by HPLC Prefractionation*

Eugene Zabarovsky, Microbiology and Tumor Biology Center, Karolinska Institute, Stockholm, Sweden, *Sequencing the Human Genome: Was It Worth It?*

Ming Zhao, School of Chemical Biology and Pharmaceutical Studies, Capital Medical University of Medical Sciences, Beijing, China, *Pharmaceutical Bioassay*

Ying Zheng, Institute of Chinese Medical Sciences, University of Macao, Macao, China, *Stability Assessment and Formulation Characterization*

CONTENTS

Preface   xix

1.1   **From Gene to Product: The Advantage of Integrative
      Biotechnology**   1
      Frank-Ranier Schmidt

1.2   **Sequencing the Human Genome: Was It Worth It?**   53
      Eugene Zabarovsky

1.3   **Overview: Differentiating Issues in the Development of
      Macromolecules Compared with Small Molecules**   89
      Günther Blaich, Bernd Janssen, Gregory Roth, and Jochen Salfeld

1.4   **Integrated Development of Glycobiologics: From Discovery
      to Applications in the Design of Nanoparticular Drug
      Delivery Systems**   125
      Christine Vauthier, Isabelle Bertholon, and Denis Labarre

1.5   **R&D Paradigm Shift and Billion-Dollar Biologics**   161
      Krishan Maggon

2   **From Defining Bioinformatics and Pharmacogenomics to
      Developing Information-Based Medicine and Pharmacotyping
      in Health Care**   201
      Ioannis S. Vizirianakis

3.1   **Toxicogenomics**   229
      Mary Jane Cunningham and Mrinal Shah

                                                                           xiii

3.2 **Preclinical Pharmacokinetics**   253
     Donald E. Mager and Murali Ramanthan

3.3 **Strategies for the Cytosolic Delivery of Macromolecules:
     An Overview**   279
     Aravind Asokan, Roland Cheung, and Moo J. Cho

4.1 **Basic Issues in the Manufacture of Macromoleucles**   297
     Corinne Lengsfeld and Thomas Anchordoquy

4.2 **Process Validation for Biopharmaceuticals**   319
     Beth Junker

4.3 **Stability Assessment and Formulation Characterization**   371
     Albert H.L. Chow, Henry H.Y. Tong, and Ying Zheng

4.4 **Protein Posttranslational Modification: A Potential Target
     in Pharmaceutical Development**   417
     M.D. Mostaqul Huq and Li-Na Wei

4.5 **PEGylation: Camouflage of Proteins, Cells, and Nanoparticles
     Against Recognition by the Body's Defense Mechanism**   443
     Yong Woo Cho, Jae Hyung Park, Ji Sun Park, and Kinam Park

4.6 **Unexpected Benefits of a Formulation: Case Study with
     Erythropoetin**   463
     David Keast

5.1 **Capillary Separation Techniques**   469
     Dong Hee Na and Kang Choon Lee

5.2 **Pharmaceutical Bioassay**   511
     Jun Li, Ming Zhao, and Shiqi Peng

5.3 **Analytical Considerations for Immunoassays for
     Macromolecules**   573
     Marian Kelley and Binodh DeSilva

5.4 **Chromatography-Based Separation of Proteins, Peptides,
     and Amino Acids**   585
     Jan Moebius, Remco Swart, and Albert Sickman

5.5 **Bioanalytical Method Validation for Macromolecules**   611
     Gerhard K.E. Scriba

5.6 **Microarrays in Drug Discovery and Development**   633
     Krishnanand D. Kumble and Sarita Kumble

5.7 **Genetic Markers and Genotyping Analyses for Genetic Disease Studies**   661
Anne E. Kwitek and Michael Olivier

6.1 **Proteins: Hormones, Enzymes, and Monoclonal\Antibodies— Background**   691
Wayne Materi, Zoltan Gombos, and David S. Wishart

6.2 **Formulation and Delivery Issues of Therapeutic Proteins**   737
Sathy V. Balu-Iyer, Razvan D. Miclea, and Vivek S. Purohit

6.3 **Pharmacokinetics**   757
Nobuhito Shibata, Yukato Ito, and Kanji Takada

6.4 **Immunogenicity of Therapeutic Proteins**   815
Suzanne Hermeling, Daan J.A. Crommelin, Huub Schellekens, and Wim Jiskoot

6.5 **Development and Characterization of High-Affinity Anti-Topotecan IgG and Fab Fragments**   835
Jin Chen and Joseph P. Balthasar

6.6 **Recombinant Antibodies for Pathogen Detection and Immunotherapy**   851
Nicholas J. Pokorny, Jeanine I. Boulter-Bitzer, J. Chris Hall, Jack T. Trevors, and Hung Lee

6.7 **The Radiopharmaceutical Science of Monoclonal Antibodies and Peptides for Imaging and Targeted *in situ* Radiotherapy of Malignancies**   883
Raymond M. Reilly

7.1 **Gene Therapy—Basic Principles and the Road from Bench to Bedside**   943
Gabor M. Rubanyi

7.2 **Plasmid DNA and Messenger RNA for Therapy**   971
Steve Pascolo

7.3 **Formulations and Delivery Limitations of Nucleic-Acid-Based Therapies**   1013
Tatiana Segura

7.4 **Pharmacokinetics of Nucleic-Acid-Based Therapeutics**   1061
John C. Schmitz, Aleksandra Pandyra, James Koropatnick, and Randal. W. Berg

7.5   **Case Studies—Development of Oligonucleotides**   1087
Ezharul Hoque Chowdhury and Toshihiro Akaike

7.6   **RNA Interference: The Next Gene-Targeted Medicine**   1109
Andrew V. Oleinikov and Matthew D. Gray

7.7   **Delivery Systems for Peptides/Oligonucleotides and Lipophilic Nucleoside Analogs**   1149
R.A. Schwendener and Herbert Schott

8.1   **Growth Factors and Cytokines**   1173
Mandeep K. Mann and Brian E. Cairns

8.2   **Growth Factors, Cytokines, and Chemokines: Formulation, Delivery, and Pharmacokinetics**   1197
Heping Cao and Rui Lin

9   **Protein Engineering with Noncoded Amino Acids: Applications to Hirudin**   1225
Vincenzo De Filippis

10.1   **Production and Purification of Adenovirus Vectors for Gene Therapy**   1261
D.M.F. Prazeres and J.A.L. Santos

10.2   **Assessing Gene Therapy by Molecular Imaging**   1297
Pascal Delepine and Claude Férec

11   **Overview of Stem and Artificial Cells**   1313
Alejandro Soto-Gutierrez, Nalu Navarro-Alvarez, Jorge David Rivas-Carrillo, and Naoya Kobayashi

12.1   **Regulation of Small-Molecule Drugs Versus Biologicals Versus Biotech Products**   1373
María de los Angeles Cortés Castillo and José Luis Di Fabio

12.2   **Intellectual Property and Biotechnology**   1391
Tania Bubela and Karen Lynne Durell

12.3   **Comparability Studies for Later-Generation Products— Plant-Made Pharmaceuticals**   1433
Patrick A. Stewart

12.4   **Biosimilars**   1453
H. Schellekens, W. Jiskoot, and D.J.A. Crommelin

13.1   The Promise of Individualized Therapy   **1463**
Michael Oettel

13.2   Enhanced Proteomic Analysis by HPLC Prefractionation   **1491**
Pierre C. Havugimana, Peter Wong, and Andrew Emili

13.3   An Overview of Metabonomics Techniques and Applications   **1503**
John C. Lindon

13.4   Bioterrorism   **1525**
Dany Shoham

Index   **1653**

PREFACE

This *Handbook of Pharmaceutical Biotechnology* represents a unique attempt to overview the full range of approaches to discovering, selecting, and producing potentially new therapeutic moieties resulting from biological process. Such moieties are the backbone of both the pharmaceutical industry and the prime axis for the advancement of medical science.

The volume is unique in that it seeks to cover possible approaches to the biotechnology drug process as broadly as possible while not just doing so in a superficial manner. Thanks to the persistent efforts of Gladys Mok, these 50 chapters cover all major approaches to the problem of identifying, producing, and formulating new biologically derived therapeutics and were written by leading practitioners in each of these areas.

I hope that this second course of our banquet is satisfying and useful to all those working in or entering the field.

Select figures of this title are available in full color at ftp://ftp.wiley.com/ public/sci_tech_med/pharmaceutical_biotech/.

S.C. GAD

1.1

FROM GENE TO PRODUCT: THE ADVANTAGE OF INTEGRATIVE BIOTECHNOLOGY

FRANK-RANIER SCHMIDT
Sanofi-Aventis Deutschland, Frankfurt am Main, Germany

Chapter Contents

1.1.1 Introduction   2
1.1.2 Production of Organisms and Expression Systems   3
     1.1.2.1 Industrially Established Recombinant Expression Systems   3
     1.1.2.2 Evaluation of Secretory Expression Systems for Pharmaceutical Purposes   5
     1.1.2.3 Criteria for the Choice of Recombinant Expression Systems   11
1.1.3 Enhancement of Productivity   13
     1.1.3.1 Secretory Recombinant Expression Systems   13
     1.1.3.2 Natural Products   13
     1.1.3.3 Approaches and Goals for Further Strain Improvement   16
1.1.4 Biosynthetic Structure Modification   19
     1.1.4.1 Combinatorial Biosynthesis   19
     1.1.4.2 Precursor Directed Biosynthesis   20
1.1.5 Fermentation Optimization and Scale-Up   22
     1.1.5.1 Reduced Mixing Quality and Enhanced Stress Exposure   22
     1.1.5.2 Process Characterization   26
     1.1.5.3 Process Optimization   28
     1.1.5.4 Physical Scale-Up Parameters   29
     1.1.5.5 Development of Fermentation Models and Strategies   31
1.1.6 Downstream Processing   31
     1.1.6.1 Product Recovery and Purification   31
     1.1.6.2 Downstream Processing Optimization and Economization   32
     1.1.6.3 Downstream Processing Scale-Up   33
     1.1.6.4 Downstream Processing of β-Lactam Compounds   34

Handbook of Pharmaceutical Biotechnology, Edited by Shayne Cox Gad.
Copyright © 2007 John Wiley & Sons, Inc.

1.1.7   Postsynthetic Structure Modification   34
    1.1.7.1   β-Lactam Side Chain Cleavage   35
1.1.8   Quality Issues   37
    1.1.8.1   Virus and Endotoxin Removal   38
1.1.9   Conclusion   38
    References   38

1.1.1   INTRODUCTION

Biotechnology and biotechnology-based methods are increasing in importance in medical therapies and diagnostics as well as in the discovery, development, and manufacture of pharmaceuticals. Biotechnologically manufactured pharmaceuticals will soon reach a market volume of more than $100 billion USD and, thus, some 20% of the total pharmaceutical market. The key step in their manufacture is the conversion of the genetic information into a product with the desired pharmacological activities by an appropriate selection, design, and cultivation of cells and microorganisms harboring the corresponding biosynthetic pathways and physiological properties.

The intention of this chapter is to give insights into the typical issues and problems encountered in the manufacture of biopharmaceuticals, to mediate general ideas and current strategies on how to proceed in the design and development of biotechnological processes, and to deliver the immediate theoretical backgrounds necessary for comprehension rather than to give detailed experimental instructions like a manual does. In focusing on gene recombinant proteins and peptidic antibiotics, the biotechnologically produced pharmaceuticals with the highest market share, representative aspects will be discussed (1) for process development and optimization approaches to increase product yield and process rentability and to ensure a consistent product quality, (2) for experimental approaches to design and to modify the molecular structure of compounds to meet specific medical needs, (3) for the replacement of chemical procedures by economically and ecologically advantageous biotechnological processes, (4) for critical issues of product purification, and (5) for specific demands in pharmaceutical production to conform to regulatory requirements. Finally, the advantage of an integrative biotechnology is emphasized, which designs the biosynthetic steps of the product in accordance with the requirements of product purification procedures already during early development stages.[1]

For further reviews comprehensively illustrating issues of biotechnological production processes, the reader is referred to further review articles [1–8]. To look up basic subjects of molecular and cellular biology, the reader is referred to textbooks [9, 10].

---

[1] Parts of this chapter were excerpted and modified from previously published reviews of the author [16, 86, 98, 113, 273] by courtesy of Springer-Verlag, Heidelberg.

1.1.2 PRODUCTION ORGANISMS AND EXPRESSION SYSTEMS

Design and development of all microbial production processes start with the selection of appropriate organisms, strains, and expression systems enabling high yields and high quality of a desired product with defined pharmacological properties.

1.1.2.1 Industrially Established Recombinant Expression Systems

Industrially established expressions systems for production of the marketed compounds are, besides inclusion, body-forming *Escherichia coli* strains, the yeast *Saccharomyces cerevisiae* and mammalian cells like CHO- and BHK-cells (Table 1.1-1). These systems were the genetically and physiologically most advanced and therefore mostly applied when recombinant production processes were starting to be developed in the mid-1980s and are now widely accepted by regulatory bodies. *E. coli* and *S. cerevisiae* can be grown cheaply and rapidly, are amenable to high cell density fermentations with biomasses of up to 130 g/L, possess short generation times, have high capacities to accumulate foreign proteins, are easy to handle, and are established fermentation organisms.

However, because gene recombinant pharmaceuticals continuously gain an increasing importance in medicine and are expected to help curing diseases that are not yet treatable today, new expression systems have to be exploited enabling the production of such pharmaceuticals with innovative properties that simultaneously meet key criteria like consistent product quality and cost effectiveness. Of particular interest in this regard are expression systems enabling the secretion of

TABLE 1.1-1. Industrially Used Recombinant Expression Systems

Product	Company	System
Blood coagulation factors (VII, VIII, IX)	Novo-Nordisk/Bayer/ Centeon Genetics Baxter/ Centeon/Wyeth	BHK-Cells CHO-Cells
Calcitonin	Unigene	*E. coli*/CHO-Cells
DNase (cystic fibrosis)	Roche	CHO-Cells
Erythropoetin	Janssen-Cilag/Amgen/ Boehringer	CHO-Cells
Darbepoetin	Amgen	CHO-Cells
Follicle stimulating hormone (follitropin)	Serono/Organon	CHO-Cells
Luteinisation hormone	Serono	CHO-Cells
Gonadotropin	Serono	CHO-Cells
Glucagon	Novo-Nordisk	*S. cerevisiae*
Glucocerebrosidase (Gaucher-disease)	Genzyme	CHO-Cells
Growth hormones (somatotropines)	Pharmacia & Upjohn/Lilly/ Novo-Nordisk/Ferring/ Genentech	*E. coli*
	Serono	Mouse Cell Line
	Serono/Bio-Technology General Corp	CHO-Cells

TABLE 1.1-1. *Continued*

Product	Company	System
Eutropin (Human growth hormone derivative)	LG Chemical	*S. cerevisiae*
Growth factors (GCSF u. GMCSF)	Novartis/Essex/Amgen/ Roche	*E. coli*
	Chugai Pharmaceuticals	CHO-Cells
Platelet-derived growth factor (PDGF)	Janssen-Cilag	*S. cerevisiae*
PDGF-Agonist	ZymoGenetics	*S. cerevisiae*
Hepatitis B vaccine	GlaxoSmithKline	*S. cerevisiae*
	Rhein Biotech	*H. polymorpha*
Hirudin	Sanofi-Aventis/Novartis	*S. cerevisiae*
Insulin and muteins	Sanofi-Aventis/Lilly/Berlin- Chemie	*E. coli*
Insulin	Bio-Technology General Corp	*E. coli*
	Novo-Nordisk	*S. cerevisiae*
Interferon alpha and muteins	Roche/Essex/Yamanouchi	*E. coli*
Interferon beta	Schering	*E. coli*
	Biogen/Serono	CHO-Cells
Interferon gamma (mutein)	Amgen/Boehringer	*E. coli*
Interleukin 2	Chiron	*E. coli*
Oprelvekin (interleukin 11-agonist)	Wyeth	Human Cell Line ROMI 8866
OP-1 (osteogenic, neuroprotective factor)	Curis/Striker	*E. coli*
Tissue plasminogen-activator	Genentech/Roche/Boehringer	CHO-Cells
Recombinant plasminogen- activator	Genentech/Roche/Boehringer	*E. coli*
Stem cell factor	Amgen	CHO-Cells
Tumor necrosis factor	Boehringer	*E. coli*

Note: Overview on the currently worldwide commercialized recombinant pharmaceuticals and the expression systems employed for their production. The substances are not listed strictly alphabetically but are partially grouped according to therapeutic areas. Antibodies, which are mostly manufactured by hybridoma cell line systems, are not considered. Data were extracted from the European patent database Esp@cenet (http://de.espacenet.com) and the IDdb3-database (http://www.iddb3.com). BHK = baby hamster kidney; CHO = chinese hamster ovary. (Taken from Ref. 16 © Springer-Verlag, Heidelberg)

correctly glycosylated and folded proteins into the culture broth. Such secretory systems offer advantages in terms of simple and fast product purification procedures and the avoidance of costly cell rupture, denaturation, and refolding processes (see Section 1.1.4) and thus conform to the requirements of an integrated production process.

Even though animal and plant systems (molecular pharming, [11]) and secretory plant cell culture systems have received a great deal of attention, their commercial feasibility is still under investigation, particularly with respect to their slightly different posttranslational modification modus leading to an altered pharmacological behavior and to allergenic properties [12, 13]. Established in the pharmaceutical industry as production organisms are, besides the above-mentioned systems, further prokaryotic and yeast species as well as filamentous fungi, which are already employed for the manufacture of natural products (see Section 1.1.2.2).

The suitability of the most prominent secretory systems among these organisms from the viewpoint of an integrative process design for the manufacture of recombinant proteins will be evaluated in the next section by discussing their potential productivity and their physiological properties.

1.1.2.2   Evaluation of Secretory Expression Systems for Pharmaceutical Purposes

Escherichia Coli.   As *E. coli* lacks fundamental prerequisites for efficient secretion, the marketed pharmaceuticals (Table 1.1-1) manufactured by *E. coli*-systems are mostly produced as inclusion bodies. Due to the membrane structure, the low chaperone and foldase level and the high periplasmatic protease concentration *E. coli*-secretion systems allow only comparably low product yields, making them suitable only for compounds marketed in small quantities like orphan drugs. Genentech (San Francisco, CA), for instance, has patented a secretory *E. coli*-system for the preparation of human growth hormone [14]. The secretory potential of *E. coli* is indicated by exceptional high product titers in the range of several grams per liter, which were reached in a system developed for secretion of hirudin using the alpha-cyclodextringlykosysl-transferase signal sequence as a leader und secretor mutants deficient in their membrane structure [15]. Titers of human-insulin-like-growth-factor or human-epidermal-growth-factor were reported to be as high as 900 mg/L and 325 mg/L, respectively [references in 16]. Most of the reached and published data, however, refers to processes leading to a periplasmatic product concentration (e.g., 2 g/L of a human antibody fragment, 700 mg/L of a monoclonal antibody) or stays below 100 mg/L, a value that generally is not considered to be in a competitive and economic range. Efforts are thus undertaken to condition *E. coli*-strains to efficient secreters. The main strategies to enhance secretion efficiency [17–19] comprise (1) employment of well-characterized secretion pathways like the alpha-hemolysin system [20] or components of such pathways like efficient signal sequences from efflux proteins [21] or outer membrane proteins [22], for instance, the maltose binding protein [23] or the TolC-protein [24]; (2) variation of the signal peptide; (3) cocloning of and coexpression of chaperones and foldases [25–27]; (4) enhancement of gene expression by employment of strong promotors and efficient transcription termination sequences; (5) generation of protease deficient mutants; (6) generation of cell wall lacking or cell wall deficient mutants [28, 29]; and (7) modulation of the protein primary structure that was found to exert a strong influence on productivity and secretion efficiency by influencing protease resistance, folding efficiency, and the tendency to form inclusion bodies. Details of the strategies for the design and development of secretory *E. coli* strains as well as for the controlled soluble cytoplasmatic expression of recombinant proteins may be taken from general reviews [17, 30–32].

Alternative Prokaryotic Expression Systems.   In addition to conditioning *E. coli*-strains to efficient secreters, alternative species, which are considered to inherently possess a superior secretion capacity, are tried to be established as expression systems. Comparably high product yields of 2 g/L and 1 g/L were reported for production of human calcitonin by *Staphylococcus carnosus* [33] and of proinsulin by *Bacillus subtilis* [34], an organism that is continuously characterized and improved as a cell factory for pharmaceutical proteins [8]. *Bacillus megaterium*, which is thought to be as efficient as *B. subtilis*, is currently developed as a secretory expression system by a Collaborative Research Center (SFB) of the German Research Community (DFG). For *Ralstonia eutropha* (formerly *Alcaligenes eutrophus*), employed at ICI and Monsanto for polyhydroxyalkanoate production at a scale of several 100 m^3 and genomically completely sequenced, 1,2 g/L of secreted organophosphohydrolase, a model enzyme proned to form inclusion bodies in *E. coli*, were reported [35]. *R. eutropha* displays a more efficient carbohydrate metabolism than *E. coli* and is easily amenable to high cell density fermentations with biomass concentration of more than 150 g/L dry weight. This permits a lower specific productivity that in turn reduces the inclinement to form inclusion bodies and thus enables a more efficient secretion. *Rhodococcus, Corynebacterium, Mycobacterium*, actinomycetes, and streptomycetes [36] are also considered to be potentially suitable for the development of efficient secretion systems. A comparative study with recombinant alpha-amylase demonstrated that final yields as well as enzyme activity were considerably higher when produced by *Streptomyces lividans*, by which it was completely secreted than by *E. coli* in which it was concentrated periplasmatically [37]. The yields reported so far, however, are still below cost-efficient ranges. A system developed by Hoechst/Aventis for insulin production yielded around 100 mg/L, and the yields of correctly folded human CD4-receptor sites are in the range of 200 mg/L. Attached as signal proteins were the prepeptide of the alpha-amylase inhibitor from *S. tendae* (tendamistat) and the signal sequence of a protease inhibitor (LTI) from *S. longisporus*. In the course of these studies, it was found that the choice of the linker and its length strongly influences secretion efficiency. The comparably low yields, however, demonstrate that still a lot of fundamental research is necessary to render streptomyces systems competitive. (Strategies and examples for enhancement of recombinant protein expression in *S. lividans* and the current status of the genetic and physiological development are given in Refs. 38 and 39.) A general focus of research will be the detailed exploration of the twin arginine translocation (TAT) pathway, which has been recently discovered in addition to the conventional prokaryotic secretory (sec) pathways and enables the export of proteins with cofactors in a fully folded conformation [40, 41]. It evidently plays a more important role in *Streptomyces* species [42] but might also be useable in other species.

As the potential and capacity of prokaryotes for posttranslational modification appear to be quite limited and the knowledge about the pathways is quite scarce (reviews on bacterial protein glycosylation see Refs. 43 and 44), the employability of most of the known prokaryotes usually is restricted to the preparation of proteins that are naturally not glycosylated, such as insulin, hirudins, or somatotropins, or to natively glycosylated proteins that are pharmacologically also active without glycosylation, like various cytokines (tumor necrosis factor, interleukines, interferones). For production of proteins that are pharmacologically active only with an appropriate modification pattern, eukaryotic cell systems are more suitable.

Yeasts. Besides possessing complex posttranslational modification pathways, they offer the advantage to be neither pyrogenic nor pathogenic and to secrete more efficiently. Species established in industrial production procedures are *Saccharomyces cerevisiae*, *Kluyveromyces lactis*, *Pichia pastoris*, and *Hansenula polymorpha*, which will be dealt with more in detail in this section. Whereas *S. cerevisiae* is the best genetically characterized eukaryotic organism at all and still is the prevalent yeast species in pharmaceutical production processes (Table 1.1-1), *P. pastoris*, first employed by Phillips Petroleum for single-cell-protein production, is currently the most frequently used yeast species for heterologous protein expression in general. Whereas only just a few proteins were expressed by *Pichia* species at the beginning of the last decade [45], the expression of more than 400 proteins have been meanwhile reported now [46, 47]. *P. pastoris* is considered to be superior to any other known yeast species with respect to its secretion efficiency and permits the production of recombinant proteins without intense process development. The highest yields were reported for murine collagen (15 g/L), tetanus toxin fragment C (12 g/L compared with 1 g/L in *S. cerevisiae*), human serum albumin (10 g/L compared with 3 g/L in *K. lactis* and to 90 and 150 mg/L in *S. cerevisiae*), and human interleukin 2 (10 g/L). The highest reported yields for *H. polymorpha* relate to phytase (13,5 g/L) and to hirudin (g/L range) and for *K. lactis* to human serum albumin (3 g/L, see above). Even though *S. cerevisiae* offers a high secretory potential as evidenced by some 9-g/L secreted *Aspergillus niger* glucose oxidase [references in 16], such data document a general inferior secretory capacity, the reasons for which are numerous. For the methylotrophic species *Hansenula* and *Pichia* and the lactose using *K. lactis*, natively strong promoters are available that derive from the methanol and lactose assimilating pathways and their enzymes (e.g., alcohol- and methanol oxidase, lactose permease, galactosidase). As the enzymes of these pathways account for up to 30% of the total protein content, the metabolic efficiency with respect to the secreted protein is significantly higher as documented by Buckholz and Gleeson [48]: Only one or a few gene copies are sufficient in *P. pastoris* to gain the same yields as with 50 gene copies in *S. cerevisiae*. Furthermore, proteins with a molecular mass of above 30 kD are retained in the cytoplasma of *S. cerevisiae*, whereas *H. polymorpha* efficiently secretes proteins with a molecular mass of up to 150 kD, like the glucoamylase of *Aspergillus niger*. [Detailed overviews on the physiological properties of methylotrophic yeasts and *K. lactis* with respect to their use for recombinant protein production is given in Refs. 46, 47, and 49–53]. Further reasons for the differing secretion rates among the species are the specific proteolytic activities and the specific degrees and patterns of glycosylation. Besides having an impact on the protein's final pharmacological activity, glycosylation also exercises an influence on the folding and secretion efficiency. Among the discussed species, *S. cerevisiae* was shown to possess, besides a higher enzymatic activity in the secretion vesicles that leads to a reduced portion of intact secreted proteins [54], also the highest glycosylation capacity leading to a hyperglycosylation of the protein and a reduced secretion rate. Both the degree and the pattern of glycosylation are dependent on the genetic background of the species and strains employed as well as on the sequences of the expressed protein and adjacent regions. By employment of the natively highly glycosylated alpha-mating type factor as a secretion signal, the extent of the glycosylation of the product can be diminished or completely avoided as shown for human interleukin 6 [55]. NovoNordisk reported leader sequence-dependent insulin yields in *S. cerevisiae* [56] and in *S. cerevisiae* and *P. pastoris* [57]: The sequence and therewith the degree of glycosylation of the leader influences the efficiency of the multistage cleavage and folding processes as well as the insulin glycosylation rate and secretability. Further enhancement of the secretion efficiency can be achieved by (1) mutating secretion enhancer genes [58], (2) suppressing secretion blocking functions, and (3) reducing proteolytic activities in secretion vesicles [54]. So-called supersecreter strains of *S. cerevisiae* have, for instance, been generated by inactivation of the PMR1 (SSC1) function and suppression of the secretion blocking ypt1-1 gene: the yields of non-glycosylated human pro-urokinase [59], of human serum albumin, and of human plasminogen-activator have been augmented to a factor of up to 10 [60]. The traditional approaches pursued for enhancement of gene expression are gene amplification, employment of strong promoters, and enhancement of the transcription and translation rate. A high amplification of the gene copy number [45, 46, 61] usually is achievable with episomal vectors, which however do not reach the mitotic stability of integrative systems like the transposon (e.g., Ty-element) mediated embedment of reiterative, dispers repetitive sequences in *S. cerevisae*. Transcription rates were reported to be enhanced up to 100 times through cotransformation with transcription activators and enhancers, which evidently are limiting factors for overexpression of foreign proteins [62, 63]. Translation efficiency can be enhanced by preventing an accelerated degradation of transcripts and the yeast typical random transcription termination through modulation of the recognition sequences. Prevention of the random transcription termination led to an increase of tetanus toxin fragment C yields in *S. cerevisiae* by a factor 2000–3000 to 1 g/L and 3% of the total soluble protein fraction [64].

Despite their physiologically advantageous properties and natively high expression and secretion capacity, just one industrial application is reported for each of the alternative yeast species: *H. polymorpha* is employed for hepatitis B vaccine production at Rhein-Biotech (Düsseldorf, Germany), *K. lactis* for bovine prochymosin production in a 40-m^3 scale at Gist-Brocades (Delft, Netherlands), and *P. pastoris* for production of recombinant carboxypeptidase B and trypsin at Roche (Basel, Switzerland). For pharmaceutical application, it has to be considered that the methylotrophic yeasts in contrast to *S. cerevisiae* and *K. lactis* are not used in the production of foodstuffs and therewith have no GRAS (generally regarded as safe) status according to the U.S. Food and Drug Administration (FDA) criteria and have to be grown in expensive explosion-proofed equipments when the above-mentioned native induction systems are used. The discussed properties of the mentioned yeast species are compiled in Table 1.1-2 for comparison. The employ-

TABLE 1.1-2. Typical Sequence of Biotechnological Production Process Steps

Step	Method/Approach
Selection/design/engineering/ development of an appropriate species/strain/ expression system	**Criteria:** pharmacological activity and properties of the compound, productivity, process behavior, suitability for downstream processing steps, spectrum and pharmacological activity of side products to be removed, experimental experience with the respective system, biological and medical safety, acceptance by regulatory bodies

TABLE 1.1-2. *Continued*

Step	Method/Approach
Strain improvement	Mutation/selection, strain recombination (e.g., breeding/protoplast fusion), directed genomic alteration/metabolic engineering/enhancement of gene expression rates
Biosynthetic product structure modification	Amino acid exchange/combinatorial biosynthesis/precursor directed biosynthesis
Fermentation optimization	Empirical optimization of culture conditions (media components, pH, oxygen supply), clarification of the influence of process parameters on growth, productivity and side product formation
Fermentation scale-up	Detailed process characterization for reduction of stress exposure, insurance of homogenized reaction conditions, and identification of suitable scale-up parameters (e.g., with the help of chemometric modeling, computational fluid dynamics)
Downstream processing	
Cell separation and harvest, removal of particulate matters	Centrifugation (decanter, disk-stack separator, (semi)continuous centrifuges), filtration (dead-end, tangential flow filtration)
Cell rupture	High-pressure homogenization, bead mills, sonication, enzymatic treatments
Product capture	Filtration (micro-, ultra-, nano-filtration), precipitation, solvent extraction, ion exchange, size exclusion, affinity chromatography
Product purification/polishing	Hydrophobic interaction/reversed-phase chromatography
Clearance of contaminant agents (e.g., viruses, endotoxins)	Nano-filtration, heat, pH, chemical inactivation, ultraviolet, gamma irradiation
Drying	Heat, freezing, vacuum
Galenic preparation/filling	Addition of galenic excipients, supplements, stabilizers, and adjuvants

Notes: Comprehensive overview illustrating the typical steps of a biotechnological production process. Production starts with the selection of an appropriate species and strain and the engineering of an expression system that enable high product yields and permit easy handling during strain development, fermentation, and downstream processing. Classic measures for improvement of the selected strains and expression systems are mutation and selection runs, breeding, directed genomic alterations, metabolic engineering, and measures aimed at the enhancement of expression rates of genes involved in product biosynthesis. Cultivation of the cells occurs under suitable conditions permitting strong growth, high product yield, and high product quality and facilitating product recovery and purification during downstream processing, which usually starts with separation of the product from the biomass and the culture broth by centrifugation or filtration. Before the separation of intracellular occurring products, cells have to be ruptured by high-pressure homogenization (e.g., in a homogenizer, a Dyno mill, or a French press). For high product purity, high recovery rates and simultaneous maintenance of the pharmacological activity, product purification and polishing usually occurs by solvent and solid-phase extraction. The choice of suitable resins and solvents and their combinations is highly dependent on the chemical and physical properties of the product and the side metabolites to be removed.

The final step consists of product concentration and drying, e.g., by heat, freezing, and/or vacuum before the galenic preparation. The discrimination between product capture and polishing is somewhat arbitrary as is the assignment of the respective methods. Depending on the product and the individual product properties, the transitions between the steps are gliding and the methods can change accordingly.

ability of yeasts in some cases however might reach a limit, particularly when the pharmacological activity of the product is impaired by the glycosylation pattern. In *K. lactis*, for instance, which usually does not hyperglycosylate, an exceptional high glycosylation of human interleukin 1β has been observed, reducing the biological activity to 5% [65, 66]. In such cases, a postsynthetic chemical modification has to be considered or the employment of higher developed organisms.

Filamentous Fungi. Filamentous fungi are higher organized than yeasts and consequently have a more complex posttranslational modification apparatus more similar to mammals. Some proteins like t-PA in *Aspergillus nidulans* are produced with the natural human glycosylation pattern. For recombinant protein production, species are prefered, which are broadly employed in industry for production of enzymes, acids and antibiotics and thus possess GRAS-status: *A. nidulans, A. niger, A. sydowii, A. awamori*, various *Fusarium* and *Trichoderma* species, *Penicillium chrysogenum*, and *Acremonium chrysogenum*. Their productivity and secretion potential, which is in the range of 30–40 g/L for homologous enzymes like cellulases and amylases, is considered to be superior to any other system, but unfortunately could not be converted into corresponding yields of recombinant products, even when these were fused to such homologous enzymes. The highest yields are still obtained with heterologous fungal enzymes: 4-g/L *Fusarium* protease in *A. chrysogenum* and 4.6-g/L *A. niger* glucoamylase in *A. awamori*. The highest yield of a mammalian protein was reported for human interleukin 6 in *A. sojae* in a range of 300 mg/L [references in 16]. The yields of most human proteins like t-PA and various interferons, however, were reported to be below 1 mg/L [67–71]. Possible reasons for these incompetitive yields are restrictions in posttranslational metabolic steps like intracellular transport, folding, and processing. Even though filamentous fungi are industrially used now for decades, they are not adequately characterized on the physiological and genetic level. Little is known about details of the modification and secretion metabolism, and efficient gene transformation is hampered by degradation of foreign DNA, low transformation rate, and therefore, low copy numbers of the transferred genes and random genomic integration. Expression rates are restricted by a high RNA turnover, incorrect processing of the foreign messenger, and incomplete folding and secretion, which are, as in yeasts, both influenced by the glycosylation pattern. Proteins not completely or incorrectly glycosylated and excreted are rapidly degraded as shown for human interleukin 6 [71]. Currently, filamentous fungi cannot be regarded as a serious alternative for the production of pharmaceuticals. To fully exploit the potential of filamentous fungi, their physiology, particularly the glycosylation metabolism, thus has to be investigated and clarified in more detail.

Insect and Mammalian Cell Cultures. Animal cell cultures are the systems with highest similarity to human cells with respect to the pattern and capacity of posttranslational modifications. However, their cultivation is more complicated and costly and usually yields lower product titers. Among the known systems, insect cells transformed by baculovirus vectors have reached a comparable popularity as *Pichia* among yeasts because they are considered to be more stress resistant, easier to handle, and more productive compared with mammalian systems and are thus frequently employed for high-throughput protein expression. The highest reported yields refer to human collagenase IV in *Trichoplusia* (300 mg/L). Yields reported for non-insect cultures were 80 mg/L for human apolipoprotein A1 in chinese hamster ovary (CHO)-cells and 1 mg/L for human laminin in human embryonal kidney cells [references in 16]. For commercial application scale-up, related questions have to be clarified, particularly concerning oxygen supply [72] and carbon dioxide accumulation [73], stability of the cell line, and the bioreactor type to be employed [74]. Further deficiencies of insect cells are observed in (1) an inefficient processing and an impairment of the folding and secretion capacity due to the baculovirus infection [75]; (2) the high, in part baculovirus encoded, protease activity and the resulting necessity to routinely employ protease inhibitors in the culture media or to develop protease deficient vectors [73]; (3) an insufficient expression strength; and (4) deviations of the posttranslational modification pattern, which could act immunogenic. For optimization, the construction of new innovative vectors and the coexpression of chaperones, foldases, and folding factors such as canexin [76] have been suggested as has been the broader development and application of alternative systems like *Drosophila* [75]. Preferably applied in pharmaceutical production processes are mammalian systems like CHO- and baby hamster kidney (BHK)-cells (Table 1.1-1). These systems are generally considered to be genetically more stable and easier to transform and to handle in scale-up processes, grow faster in adherent and submerged cultures, and are more similar to human cells and more consistent in their complete spectrum of modification [77], which minimizes the risk of the formation of structurally altered compounds with immunogenic properties. In some cases, mammalian systems can be the only choice for the preparation of correctly modified proteins. Studies of Tate et al. [78] comparing the expression of a rat serotonin transporter in various expression systems demonstrate that biologically active transporter was only synthesized in mammalian cells, whereas it was partially degraded in *E. coli*, not correctly folded in *Pichia* and not correctly glycosylated in insect cells. CHO- and BHK-cells further have the advantage to be recognized as safe regarding infectious and pathogenic agents and therefore to have a higher acceptance by regulatory bodies, which accelerates or at least does not delay approval procedures. [Issues related to the physiology and the technical handling of mammalian cell cultures in the manufacture of human therapeutics are discussed in detail in Refs. 74 and 79–81].

1.1.2.3   Criteria for the Choice of Recombinant Expression Systems

In summary, the criteria for the choice of an expression system in pharmaceutical production [82] are the existing expertise, the available physiological and genetic know-how and tools, the patent situation, and to avoid delays of product launch and commercialization, regulatory aspects like the acceptance by the approving authorities.

The overall decision criteria, however, is the pharmacological activity profile of the yielded protein in context with the posttranslational modification pattern followed by rentability.

For production of non-glycosylated proteins and proteins that are natively glycosylated but pharmacologically active also without glycosylation, prokaryotes, which usually lack metabolic pathways for glycosylation, theoretically are the most suitable organisms, offering two alternatives: Either *E. coli*-strains are conditioned to efficient secreters, or efficient native secreters like *Bacillus*-species are accordingly developed. (The implications of the glycosylation pattern for the choice of expression systems and the physiology of glycosylation is reviewed and discussed in Ref. 83.) To fully exploit the secretion capacity of fungal species, a deeper understanding of their posttranslational modification physiology will be necessary to steer the degree and pattern of glycosylation, which influences both folding and secretion efficiency. Insect and mammalian cells display posttranslational modification patterns more similar or identical to humans, but in view of the entailed expenditures, their employment can only be justified if their modification machinery is required to ensure a desired pharmacological activity.

E. coli, *P. pastoris*, and Baculovirus-based systems are currently preferred in fundamental research for structural and functional analysis of proteins and are employed as high-throughput-expression systems but will certainly find their ways into production processes in the near future. None of the systems, however, can be considered to be generally superior to any other. For each product, the most suitable expression system has to be identified and optimized individually both on the genetic and on the fermentative level by taking into account the properties of the product, the organism, and the expression cassettes, which suggests to always have a set of accordingly developed expression systems available. Some key properties of the established expression systems are compiled in Table 1.1-3. A review on properties and prospects of further alternative systems is given in Ref. 84.

Once an expression system has been selected, the respective strains are continuously submitted to improvement programs to render fermentation processes more efficient by increasing strain productivity and by modifying physiological properties and process behavior to enhance process economy. This does not only apply to the fermentation process, but also to parts of downstream processing, thus requiring an integrated strain selection and process design. Strategies to improve productivity of secretory recombinant systems and of strains employed for the manufacture of natural products will be outlined in the next section.

TABLE 1.1-3. Features and Characteristics of Expression Systems

Property	Bacteria	Yeast	Insect and Mammalian Cell Cultures
Growth	Fast	Fast	Slow
Nutrient demand	Minimal	Minimal	Complex
Costs of media	Low	Low	High
Possible product yield	High	High	Low
Secretory capacity	Limtited	High	Medium
Glycosylation capacity	Limited	High	High
Modification capacity	Limited	High	High
Risk of retroviral Contamination	Low	Low	High
Risk of pyrogens	High	—	—
Scalability	Good	Good	Low
Process robustness	Good	Good	Low

1.1.3 ENHANCEMENT OF PRODUCTIVITY

1.1.3.1 Secretory Recombinant Expression Systems

In terms of downstream processing efficiency, secretory expression systems indeed offer potential advantages for production of recombinant proteins compared with inclusion body-forming cytosolic systems, but most of the potentially available secretory systems are not yet fully competitive for high-volume therapeutics like insulin and therefore still require intensive improvement efforts.

Current strategies to improve productivity and secretion efficiency comprise (1) enhancement of gene expression rates, (2) optimization of secretion signal sequences, (3) coexpression of chaperones and foldases, (4) creation of protease deficient mutants to avoid premature product degradation, and (5) subsequent breeding and mutagenesis.

1.1.3.2 Natural Products

Completely different is the situation faced in the production of the so-called natural products, which are mostly secondary metabolites and therewith the endproducts of complex biosynthetic pathways of filamentous bacteria and fungi and thus require different approaches. With few exceptions like the antidiabetic drug acarbose (Glucobay) [5] or the chosterol-lowering drug lovastatin (Mevicor) [85], such fermented pharmaceuticals of natural origin currently belong for the most part to the class of antibiotics [86].

Enhancement of the productivity is achieved at first by rounds of random mutation and selection and, if possible, by breeding, and/or by DNA injection or fusion techniques like protoplasting and at further stages, after a comprehensive characterization of the usually complex biosynthetic pathways, directed genomic alterations and metabolic engineering.

These approaches, developments, and states in industrial strain improvement and pathway characterization will be illustrated for some of the best characterized organisms, namely the two fungal species mainly employed in industrial β-lactam antibiotic production, the penicillin producing *Peni-cillium chrysogenum*, and the cephalosporin C producer *Acremonium chrysogenum*.

Characterization of Biosynthetic Pathways. The productivity of *Penicillium chrysogenum* could have been augmented impressively during the last decades: Penicillin titers have been increased by a factor of 50,000 from a very few milligrams/ Liter in the 1940s to more than several 10 g/L by now and also some several 10 g/L cephalosporin are currently gained.

This development, however, has come to a halt during the last couple of years. As these titers have been reached through conventional mutation and selection [87–90] and no further significant increase in productivity could be noted, the tools of genetic engineering were more and more included into the improvement programs [91]. To gain a profound basis for directed genomic alterations, studies were conducted at various corporations and universities to elucidate the mechanisms and genes involved in β-lactam biosynthesis to characterize the respective biosynthetic pathways. As the results and achievements of these studies have already been reviewed extensively and in detail elsewhere [92–97], the current knowledge will only shortly be summarized.

Cephalosporin C Synthesis in Acremonium Chrysogenum. Biosynthesis (Figure 1.1-1) starts with the polymerization of L-α-aminoadipic acid, L-cysteine, and L-valine to the linear tripeptide L-α-aminoadipyl-L-cysteinyl-D-valine (ACV-peptide). This reaction is catalyzed by the ACV-synthase (MW about 420 kD) through the following steps: (1) the ATP-dependent activation of these amino acids to bind them as thiolesters, (2) the epimerization of L-valine, and finally (3) the condensation by a thiotemplate mechanism [99].

Cyclization of the ACV-peptide to the bicyclic isopenicillin N (IPN) occurs under oxygen-, Fe^{2+}-, ascorbate-, and α-ketoglutarate-dependent action of the IPN-synthase (IPNS), which has an MW of about 38 kD. Inhibitory to IPNS activity are cobalt ions and glutathione.

Further pathway reactions are as follows:

- IPN-epimerization to penicillin N (IPN-epimerase).
- Penicillin N conversion to deacetoxycephalosporin C (DAOC) by expansion of the five-membered thiazolidine ring to a 6 C-dihydrothiazine-ring (DAOC-expandase).
- Formation of deactylcephalosporin C (DAC) by dehydroxylation and oxidation of the methylgroup in C3-position (DAC-hydroxylase).
- The acylation of DAC to cephalosporin C (DAC acetyltransferase).

DAOC-expandase and DAC-hydroxylase activities in *A. chrysogenum* are exerted by the same enzyme (MW 41 kD), which like the IPNS belongs to the group of α-ketoglutarate-dependent dioxygenases.

The first two enzymes, ACV- and IPN-synthase, are encoded by the *pcbAB*- and the *pcbC* gene, respectively. Both genes are linked to each other on chromosome VI by a 1.2-kb intergenic region carrying the putative promotor sequences from which they are divergently transcribed. The *pcbC*-promotor seems to be about five times stronger than the one of the *pcbAB*-gene [100]. The *cefEF*-gene and the *cefG*-gene encoding for the bifunctional expandase/hydroxylase and the DAC-acetyltransferase, respectively, are located adjacent to each other on chromosome II. Again the genes are separated by an intergenic region of 938 bp, which is supposed to harbor the promotors from which they are transcribed in opposite directions. The *cefG*-gene has been proven to contain two introns [101]. Due to an extreme *in vitro* lability, no information is yet available on the epimerase converting IPN to penicillin N as well as its putative *cefD* gene. Until recently, the structure of the *cefD* gene could have not been elucidated. Meanwhile, data could have been obtained indicating the existence of two reading frames (*cefD1* and *cefD2*) containing all characteristic motifs of mammalian acyl-CoA ligases and α-methyl-acyl-CoA racemases [93].

Regulatory studies demonstrated that the early functions are expressed simultaneously, whereas the later pathway genes *cefD* and *cefEF* seem to be induced sequentially.

In contrast, the penicillin biosynthesis genes in *P. chrysogenum* are considered to be expressed completely concomitantly.

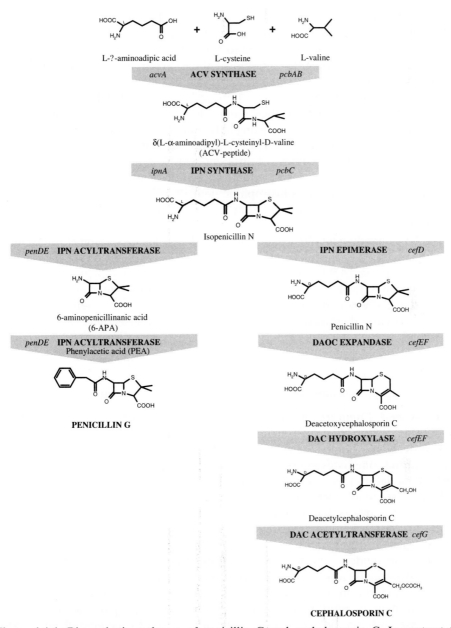

Figure 1.1-1. Biosynthetic pathways of penicillin G and cephalosporin C. In contrast to recombinant proteins, the structure and production of which can easily be modified by alterations of the encoding DNA sequence and the use of stronger promoters, respectively, structure modification and enhancement of expression rates of natural products by directed genomic alteration and metabolic engineering require a detailed characterization of the biosynthetic pathways as it has been performed for penicillin and cephalosporin biosynthesis in the filamentous fungi *Penicillium chrysogenum* (left) and *Cephalosporium acremonium* (right). Enzymes (bold) and genes are underlined by gray arrows. The first two reaction steps are common to both pathways. Cleavage of isopenicillin N to 6-APA and the subsequent acylation to penicllin G in *P. chrysogenum* and the expandase/hydroxylase activity in *A. chrsyogenum* are exerted by bifunctional enzymes. For details, see text. (Graph taken from Ref. 98, © Springer-Verlag, Heidelberg.)

Penicillin Formation by Penicillium Chrysogenum. The first reactions of the penicillin biosynthetic pathway are identical to the ones in *A. chrysogenum* (Figure
1.1-1). IPN, however, is not epimerized to penicillin N; instead it is converted to 6-
aminopenicillanic acid (6-APA) by removal of the L-α-aminoadipic acid side chain,
which is substituted by a hydrophobic acyl group. Both steps are catalyzed by the
same enzyme, the acyl coenzyme A: IPN acyltransferase (IAT). The enzymatic
activity of IAT is believed to be the result of the processing of a 40-kD monomeric
precursor into a dimeric form consisting of two subunits with MWs of 11 and 29 kD.
Due to the broad substrate specifity of IAT, various penicillin derivatives are synthesized naturally by attachment of different acyl-CoA derivatives to the 6-APA-
core. For industrial purposes, to facilitate extraction by organic solvents, synthesis
usually is directed to the less hydrophilic penicillin V or penicillin G. This is by
addition of phenoxyacetic acid or phenylacetic acid, respectively, as precursors to
the culture broth.

   The IAT encoding *penDE*-gene (also named *aatA*) is located on chromosome I
and organized as a cluster together with the ACV-synthase and IPN-synthase
encoding genes *pcbAB* (also named *acvA* in *Penicillium*) and *pcbC* (*ipnA*).

   The enzymes involved in penicillin biosynthesis are distributed at different sites
of the cell: ACV-activity was found to be bound to vacuole membranes, IPN-
synthase occurs dissolved in the cytoplasm and IAT-activity is microbody
associated.

1.1.3.3   Approaches and Goals for Further Strain Improvement

Analysis and Comparison of Strains. To get hints for more rational strain
improvement approaches, (1) highly mutated production strains were genetically
and physiologically compared with their less-productive ances-tors, (2) concentrations
of pathway intermediates were determined to identify potential pathway bottlenecks,
and (3) regulatory mechanisms were investigated.

   In the course of such studies, high-performance strains of *P. chrysogenum*
turned out to possess amplified copies of single genes like the *pcbC* gene or even
copies of the whole cluster as well as increased steady-state transcript levels of
pathway genes [102–104]. In some strains, the amplifications were shown to be
organized in tandem repeats, which presumably were generated by a hot-spot
TTTACA hexanukleotide [105]. Comparison of promotor strengths of these genes
from high and low productive strains did not reveal any differences. This indicates
the involvement of additional unknown trans-acting factors [106], as the amounts
of increased mRNA did not correlate with the degree of gene amplification. Also
a high specific activity of IPN synthase was reported in a more evolved *Penicillium*
strain, which was independent from transcript amounts and probably due to a
higher enzyme stability [102, 107].

   In *A. chrysogenum*, no amplification of the relevant genes could be detected.
Nevertheless, transcript amounts in production strains are significantly increased
[108].

   Also chromosome rearrangements could be detected in high titer strains [109–
111], but their causative influence on productivity remains unclarified for the
moment.

Directed Genomic Alterations. *Enhancement of Gene Expression.* Con-sidering these findings, the experiments conducted so far to improve productivity mainly concentrated on enhancing the pathway gene dosages and on enhancing promotor strengths to remove presumed pathway bottlenecks. For instance, the *cefEF*-activity was generally believed to constitute a potential rate-limiting step in view of the high accumulation of penicillin N in various strains of *A. chrysogenum* and accordingly amplified.

Such approaches resulted in partially significant increases of productivity in low and medium titer strains: Improvements of the final yields by 50% have been reported [96].

However, all improvements achieved so far with high-performance produc-tion strains remained with a 5% maximum in the range of normal statistical deviations.

Also more exotic approaches like the (presumed) improvement of the intracel-lular oxygen supply through cloning and expressing of a bacterial hemoglobin gene from *Vitreoscilla* [112] failed to increase productivity in high titer strains.

The reasons for these failures are manifold:

1. As both direction and site of integration of the imported DNA cannot be controlled sufficiently and transformation efficiency is still quite low, the probability of finding new, higher producing mutants even in a large-scale screening is rare.

2. Most of the data available are from studies performed at academic institu-tions with original strains. Even when sharing the same ancestral strain, the high-performance strains employed in industries have an individual geneal-ogy and mutation-selection history, so that their physiological behavior and properties differ drastically. Data and knowledge obtained with a particular strain thus cannot always be generalized and transfered to other strains. Even among industrial strains, significant differences have been revealed. For instance, the expression of the *cefG*-gene and the acetyl transferase activity was reported to constitute a possible rate-limiting step in Panlabs-strains [113]. Studies conducted on other industrial strains, however, could not confirm this observation.

3. The genetic instability and drift rises with the degree of genomic alterations, particularly during long-term vegetative propagation with numerous genera-tion cycles in industrial large-scale fermentations.

4. The complex interdependence with other metabolic areas is not yet com-pletely investigated. Consequently, after removal of an obvious β-lactam pathway bottleneck, unknown reaction steps of the linked and preceding pathways might become flux limiting. To overcome such flux limitations, one of the most exciting recent discoveries in molecular biology, namely RNA interference (RNAi), might play a major role in the future.

The Possible Impact of RNAi in Strain Improvement. The mechanisms of p osttranscriptional gene silencing, generally summed up as RNAi [114] offer potential approaches for the specific shut down of any gene of interest: Native genes can be silenced by triggering the RNAi cascade through introduction of homologous genetic sequences. As a result of these possibilities, RNAi has already revolutionized fundamental research in molecular biology, particularly in functional genomics, and will substantially contribute to novel therapeutic approaches in medicine [115, 116]. For fungi, which evidently were among the first organisms in which RNAi and associated phenomena were observed [117–122], this RNAi triggered cosuppression phenomenon has been termed "quelling" and bears an enormous potential with respect to the study of biosynthetic pathways and the improvement of metabolic productivity.

On the one hand, the implication of RNAi may supply an explanation for observations made in the course of strain improvement programs of *Penicillium chrysogenum* and *Acremonium chrysogenum*, according to which antibiotic productivity can decrease drastically upon amplification of genes involved in antibiotic biosynthesis.

The characterization and inactivation of these cosuppression mechanisms thus might help to overcome such limitations as does an application of fungal RNAi tools for silencing of pathway blocking or metabolic flux limiting genes, e.g., by the use of a novel vector system consisting of dsRNA viruses [114].

To fully realize the potential for further enhancement of productivity, which is generally assumed to be in the range of some 300% and to fully exploit the potential of RNAi, current and future studies thus aim at filling the actual gaps of knowledge and will be focused on

- The exploration of regulatory mechanisms and circuits on the transcriptional, translational, and posttranslational level.
- The determination of the specific intermediate turnover rates resulting from specific enzyme activities, enzyme titers, and enzyme stabilities.
- The detailed investigation of linked and preceding pathways.
- The investigation of product secretion mechanisms.

However, as no great leaps in productivity are currently reached, strains emerging from the development programs are also selected for beneficial alterations of genetic, physiological, and morphological properties that contribute to an enhanced process economy.

Improvement of Process Behaviour. Among the desired alterations are:

- A decrease of side products, which (1) consume metabolic energy at the expense of the desired main product, (2) hamper the final product purification, or (3) even can be inhibitory to the production organism. For instance, a significant reduction of DCPC concentrations has been reached through knocking out esterases that hydrolyze cephalosporin C to DCPC during fermentation. Esterase inactivation was by conventional mutagenesis as well as through gene-disruption and by introduction of anti-sense genes.
- Increase of the tolerance toward toxic side products and fed precursors like phenyl- or phenoxyacetic acid (see above).
- Further reduction of feedback inhibition by endproducts and metabolites (see below).
- Further reduction of catabolite repression (see below).

- Enhancement of genetic stability.
- Enhancement of viability, stress resistance, and life span.
- Accelerated product formation and enhancement of time-specific productivity to shorten fermentation time.
- Enhancement of the strain-specific productivity to reduce the amount of biomass to be processed and disposed.
- Increase of product secretion rate.
- Improvement of the fermentation behavior like (1) higher efficiency of substrate/precursor consumption, (2) diminished oxygen demand, and (3) diminished shear sensitivity.
- Improvement of filterability.

The in-depth characterization of physiological properties and biosynthetic pathways not only aims at improving process behavior and process economy, but also it relates to modifications of the inherent biosynthetic pathways (pathway engineering) to create innovative, novel products with improved properties and enhanced therapeutic values, e.g., to overcome the antibiotic resistance problem, a major global health-care problem of today [86].

1.1.4   BIOSYNTHETIC STRUCTURE MODIFICATION

Although the metabolic engineering of beta-lactams by combining β-lactam encoding genes from various sources [95, 123] is still in an explorative state, the study of polyketide pathways and the creation of polyketide antibiotic derivatives is more advanced.

1.1.4.1   Combinatorial Biosynthesis

Compounds emerging from the polyketide pathways are the classic subject of metabolic engineering [124] and for creation of novel structures. To create novel antibiotics, the natural biodiversity of polyketides is trying to be enhanced artificially by modifying and newly combining the biosynthetic pathway steps and enzymes involved [125, 126]. Polyketide synthases [127, 128] are multienzyme systems, which due to their modular structure and the occurrence of their encoding genes as clusters, are considered to be most amenable for directed alterations, genetic intervention, and heterologous expression in foreign hosts with different metabolic pathways. Concepts and strategies that are currently being pursued for creating novel and un-natural polyketides [129–132] comprise (1) creating block mutants at various biosynthetic levels to accumulate intermediate metabolites; (2) elimination of unwanted groups; (3) modification of polyketide synthases to further broaden substrate range, eventually combined with a precursor-directed biosynthesis (see below); (4) formation of new structures by deleting or introducing additional modules; (5) modification, addition, or elimination of postsynthetic modification steps like glycosylation, lactonization, and amidation; (6) the cloning and combination of various polyketide synthase genes in heterologous hosts to construct hybrid pathways and antibiotics [130]; and (7) linking and combining routes of polyketides with pathways of nonribosomal peptide synthesis to create hybrid peptide–polyketide compounds [131–135]. As the microbial nonribosomal peptide synthases share many structural and catalytic properties [136] with polyketide synthases like their organization into catalytic modules and domains [137], they are also envisaged as ideal subjects for metabolic engineering by creating large sets of modules and subsequently combining them artificially [95, 138, 139]. The rapidly growing body of patents makes evident that an increasing number of companies are working on the field of combinatorial biosynthesis. However, it also becomes evident that, despite their promising prospects, these strategies of combinatorial biosynthesis are still in an initial state and need further intense research to elucidate the required genetic and physiological details.

This is just in contrast to another approach of modifying the molecular structure of peptidic compounds on the biosynthetic level, namely the directed biosynthesis of the desired compound structures by feeding of appropriate biosynthetic precursors or stimulating agents during fermentation, as it is routinely performed for production of penicillin V und G by the addition of phenoxyacetic acid or phenylacetic acid, respectively, as precursors to facilitate penicillin extraction from the culture broth.

1.1.4.2 Precursor Directed Biosynthesis

Microbial peptidic compounds can be modified not only in their peptide structure by direct incorporation of amino acids supplemented during fermentation but also, if present, in their fatty acid moiety [140]. The possibilities of precursor-directed derivatization of cyclic peptides cores and fatty acid moieties by amino acids will be illustrated for three compound complexes (Figure 1.1-2) with unique modes of action and outstanding spectra of activity against multidrug resistant germs, which were developed by Aventis, Eli Lilly, and Cubist Pharmaceuticals. The complex A1437 of Aventis is synthesized by *Actinoplanes friulensis* [141], which originally produced a mixture of altogether eight lipopeptides (A–H), classified according to the exocyclic amino acid position and the type of their fatty acid chains [142] (Figure 1.1-2). Through addition of L-valine and asparagines, the biosynthesis could be directed toward the production of the desired D-peptide [143], which exhibited superior pharmacological properties in preclinical trials and is characterized by an exocyclic asparagine and an iso-C14 fatty acid moiety (Figure 1.1-2). The A54145 complex from Eli Lilly [144] exhibits the noteworthy characteristic, that a built-up resistance was immediately lost in the absence of this compound [145]. It consists of eight lipopeptides produced by *Streptomyces fradiae* containing four similar peptide nuclei in combination with three different fatty acid acyl side chains. The nuclei differ in their valine/isoleucine and glutamate/3-methyl-glutamate substitutions at one or both of two locations of the peptide ring. The composition of the peptide nucleus as well as the percentage of branched-chain fatty acid acyl substituents could be directed by supplementation of either valine or isoleucine [144] (Figure 1.1-2). Likewise, the complex A21978 from *Streptomyces roseosporus*, also developed at Eli Lilly, can be influenced in its composition by modification of the fatty acid chain. Precursing with valine results in an enhanced formation of the compound C2, whereas the compound C1 was preferably built upon feeding of isoleucine [146] (Figure 1.1-2). Compound C1 was inlicensed as

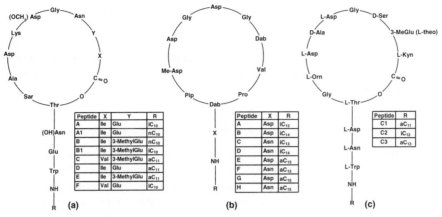

Peptide	X	Y	R
A	Ile	Glu	iC$_{10}$
A1	Ile	Glu	nC$_{10}$
B	Ile	3-MethylGlu	nC$_{10}$
B1	Ile	3-MethylGlu	iC$_{10}$
C	Val	3-MethylGlu	aC$_{11}$
D	Ile	Glu	aC$_{11}$
E	Ile	3-MethylGlu	aC$_{11}$
F	Val	Glu	iC$_{10}$

(a)

Peptide	X	R
A	Asp	iC$_{13}$
B	Asp	iC$_{14}$
C	Asn	iC$_{13}$
D	Asn	iC$_{14}$
E	Asp	aC$_{13}$
F	Asn	aC$_{13}$
G	Asp	aC$_{15}$
H	Asn	aC$_{15}$

(b)

Peptide	R
C1	aC$_{11}$
C2	iC$_{12}$
C3	aC$_{13}$

(c)

Figure 1.1-2. Structure modification of peptide antibiotics by precursor directed biosynthesis. Biosynthetic structure modification can be performed by metabolic engineering requiring an in-deph pathway characterization or by feeding of appropriate precursor compounds incorporated into the product. Examples are the lipopeptide complexes A 54145 from Eli Lilly (a), A 1437 from Aventis (b), and A 21978 developed by Eli Lilly and Cubist Pharmaceuticals (c) synthesized by *Streptomyces fradiae, Actinoplanes friulensis,* and *S. roseosporus,* respectively. The composition of both the peptide cores as well as the fatty acid moieties can be modulated *in vivo* by supplementation of amino acids like valine, leucine, isoleucine, and asparagine. The addition of valine leads to the preferable formation of peptides A 54145 C and F. The biosynthetic pathways of *A. friulensis* could be directed toward a more specific production of the desired peptide A 1437 D by feeding of valine and asparagines, whereas the preferred compound A 21978 C1 (daptomycin) could be stimulated by the addition of isoleucine. For the further details and references, see text. (Graph taken from Ref. 84, © Springer-Verlag, Heidelberg.)

daptomycin by Cubist Pharmaceuticals for clinical development. However, due to rhabdomyolytic activities and a slight haemolytic potential [138, 147, 148], it has been approved by the U.S. Food and Drug Administration in 2003 as Cubicin® salve for topical use only. Despite such disadvantageous properties, peptides in general, including those that are from nonmicrobial sources and ribosomally synthesized, are currently considered to be among the most promising compound classes for development of novel antibacterials [86, 149, 150].

However, it has to be taken into account that amino acid and precursor incorporation as well as regulation mechanisms of biosynthetic pathways also depend on growth phases, on environmental influences, and thus on the conduction of the fermentation process during which the cells are propagated.

A prerequisite for a successful fermentation are controlled and homogenous environmental conditions for the cellular reactions. However, the ensurance of controlled reaction conditions, particularly in large production scales, has been turned out to be one of the most critical issues of industrial biotechnology, as will be illustrated in the next section. For a detailed discussion of problems specifically related to the fermentation of β-lactam antibiotics, the reader is referred to the review by Schmidt [98].

1.1.5   FERMENTATION OPTIMIZATION AND SCALE-UP

The scale-up of fermentation processes as a central problem in biotechnology was first recognized and described during industrial penicillin production at the beginning of the 1940s and has been studied in more detail in *E. coli* and recombinant protein production.

1.1.5.1   Reduced Mixing Quality and Enhanced Stress Exposure

Fermentation scale-up is aimed at the manufacture of larger product quantities, if at all possible, with a simultaneous increase or at least consistency of specific yields and product quality. The changed geometric and physical conditions in larger scales, however, has lead to a less favorable mixing behavior and to impaired physiological reaction conditions, which in turn may lead to a decreased process constance and reproducibility, to reduced specific yields, to an increase of unwanted side products, and thus ultimately, to a diminished batch-to-batch consistency and product quality, which are all key issues in industrial production processes.

The problem of reduced mixing quality in larger scales is aggravated with increasing vessel sizes: The opposite substrate and oxygen gradients along the vessel height that are formed as a result of the conventional fermenter design, according to which substrate feed usually occurs from the top and aeration from the bottom, are more pronounced in larger reactors due to (1) longer distances to be covered leading to larger substrate and oxygen depletion zones, (2) larger volumes of culture broth to be stirred and therewith longer mixing times, and (3) stronger hydraulic pressure gradients influencing the oxygen transfer rate (OTR; Equations 14 and 15, Table 1.1-4) [151–155]. Cells at the fermenter top are exposed to excess glucose concentrations and simultaneously suffer from oxygen limitations, whereas those at the bottom are exposed to glucose starvation. Excess glucose concentrations (threshold value for *E. coli* at 30 mg/L) result in acetate overproduction (overflow-metabolism), and a simultaneous oxygen limitation further induces the formation of ethanol, hydrogen, formiate, lactate, and succinate (mixed acid fermentation: Refs. 154, 156, and 157). The produced acids can become reassimilated in oxygen-rich zones but in any case first lead to a temporary acidification of the microenvironment and later eventually of larger regions. Combined with a decreased transportation and elimination of carbon dioxide, detrimental metabolites, and surplus heat generated by agitation and metabolic processes, resulting in zonal overheating, the lower mixing rates in large scales thus lead to the formation of zones with enhanced stress conditions. The subsequent activation of stress genes (a survey on bacterial stress proteins is given in Ref. 158) only partially protects the cell against detrimental stress effects. For instance, despite activation of heat stress genes like *E. coli* dnaK and clpB, reducing misfolding and aggregation of heat-sensitive proteins [159], metabolic changes and damages such as translocation of proteins and membranes have been observed upon heat exposure [160]. The evidently unavoidable detrimental effects of the repeated and cyclic passing of different stress zones and the subsequent continuous activation and shut down of the corresponding stress genes are believed to lead to a completely altered physiology [155, 161] with metabolic shifts, which ultimately reduce growth and productivity and increase byproduct formation.

TABLE 1.1-4. Fermentation parameters, coefficients and terms, implicated in mixing, aeration, oxygen, and heat transfer, suitable as scale up variables to be kept constant alone or combined with each other or other process relevant variables, preferably but not necessarily, as dimensionless groups as described and discussed e.g., by Hubbard [228] and Wang and Cooney [230]. For instance, the k_La—value as the most frequently applied physical scale up variable has been combined with [1] the vessel backpressure p by maintaining the product of $k_La \times p$ constant through variation of pressure, agitation, and aeration for scale up of *Bacillus thuringiensis* fermentations (Flores et al. [232]) or [2] the aeration rate *vvm* under variation of power input, working volume and aeration according to the Wang-Cooney-equation [217]. For details and further explanations see text (taken from Ref 273 © Springer-Verlag, Heidelberg).

Parameter / coefficient	Mathematical characterization	Symbol explanation
Power input (P); volumetric power input (P/V)	(1) $P = 2 \pi n M = N_{Po} \rho n^3 d_I^5$ [$kg\,m^2\,s^{-2} = W$]	P = power input; n = stirrer speed; M = momentum; N_{Po} = dimensionless power number (impeller specific);
Dimensionless power number (N_{Po})	(2) $N_{Po} = P / \rho\, n^3\, d_I^5$	ρ = density of the medium; n = stirrer speed; d_I = impeller diameter
Impeller tip speed (v_{tip})	(3) $v_{tip} = 2 \pi n d_I$ [m / sec]	n = stirrer speed; d_I = impeller diameter
Reynolds number (Re)	(4) $Re = n\, d_I^2\, \rho / \eta$	Re = Reynolds number; n = stirrer speed; d_I = impeller diameter; ρ = density of the medium; η = dynamic viscosity
Mixing time (T_m)	(5) $T_m = f\,(n, d_I, \upsilon) = V / Q = V / N_{fl}\, n\, d_I^3$ [sec]	T_m = mixing time; n = stirrer speed; d_I = impeller diameter; υ = kinematic viscosity; V = volume; Q = volumetric flow rate; N_{fl} = pumping number
Dimensionless mixing time ($T_m n$)	(6) $T_m n = f\,(Re, N_{Po})$	T_m = mixing time; n = stirrer speed; $T_m n$ = dimensionless mixing time; Re = Reynolds number N_{Po} = dimensionless power number
Modified dimensionless power number	(7) $N'_{Po} = N_{Po}\, Re^3\, d_F / d_I = P\, d_F\, \rho^2 / \eta^3$	N'_{Po} = modified dimensionless power number $T'_m n$ = modified dimensionless mixing time
Modified dimensionless mixing time	(8) $T'_m n = n\, T_m / [(d_F / d_I)^2\, Re] = T_m\, \eta / d_F^2\, \rho$	N_{Po} = dimensionless power number; Re = Reynolds number; d_F = inside vessel diameter; d_I = impeller diameter; P = power input; n = stirrer speed; t_m = mixing time; ρ = fluid density; η = dynamic viscosity

23

TABLE 1.1-4. *Continued*

Parameter / coefficient	Mathematical characterisation	Symbol explanation
Aeration rate (volume per volume per minute, vvm)	(9) $A_R = F_G / V_R$ [m³/ m³ min]	A_R = aeration rate; F_G = volumetric gas flow rate; V_R = fermenter reaction volume
Superficial gas velocity (v_s)	(10) $v_s = F_G / A$ [m / sec]	v_s = superficial gas velocity; F_G = volumetric gas flow rate; A = fermenter cross section
Gas hold up (τ)	(11) $\tau = V_R / F_G$	τ = gas hold up; V_R = fermenter reaction volume; F_G = volumetric gas flow rate; ε = fractional gas hold up; V_G = dispersed gas volume;
Fractional gas hold up (ε) Gassing number (N_{Og})	(12) $\varepsilon = V_G / V_R$ (13) $N_{Og} = f\,(Fr, Re, S, (\rho_g / \rho_0), (\eta_g / \eta_0)) = q_g / n\, d_l^3$	N_{Og} = gassing number; Fr = Froude number; Re = Reynolds number; S = fluid constant ρ = density; η = dynamic viscosity; g = aerated; 0 = not aerated; q_g = gas throughput; n = stirrer speed; d_l = impeller diameter
Oxygen transfer rate (OTR)	(14) $OTR = k_L a\,(C_G - C_L) = k_L a\, L_{O2}\,(pO_{2G} - pO_{2L})$ [kg O₂/m³h] with (15) $C_G = 0{,}526\, p_i / 36 + T$ [mg / l]	OTR = oxygen transfer rate from gas to liquid phase k_L = mass transfer coefficient; a = specific interfacial surface area; C_G = oxygen saturation concentration in the gas phase; C_L = measured oxygen saturation concentration in the liquid phase; L_{O2} = oxygen solubility in the liquid phase; pO_{2G} = partial pressure of oxygen in the gas phase; pO_{2L} = partial pressure of oxygen in the liquid phase; p_i = vessel back pressure [bar]; T = temperature [°C]
Volumetric oxygen mass transfer coefficient ($k_L a$)	(16) $k_L a = a'\,(P / V_R)^b\, v_s^c$ [s⁻¹]	$k_L a$ = oxygen transfer coefficient; P = power input; V_R = fermenter reaction volume; v_s = superficial gas velocity; a', b and c as specific fluid constants to be determined experimentally

k_La scale dependent according to Wang-Cooney	(17) $k_La = k'(P_g/V_R)^a (v_s)^b (B/6)^{0.8} (j/d_1)^{0.3}$ [s^{-1}]	k', a, b = vessel specific coefficients; P_g = power input in aerated vessel; V = volume of culture broth; V_s = superficial gas velocity; B = number of stirrers j = baffle width; d_1 = impeller diameter
Respiratory quotient (RQ)	(18) $RQ = CER / OUR = pCO_{2E} - pCO_{2i} / pO_{2i} - pO_{2E}$ [mol CO_2 / mol O_2]	RQ = respiratory quotient; CER = carbon dioxide emission rate; OUR = oxygen uptake rate; pCO_2 = partial pressure of carbon dioxide; pO_2 = partial pressure of oxygen; $_i$ = air feed; $_E$ = exhaust air
Energy balance (Q$_{ges}$)	(19) $Q_t = Q_{met} + Q_{ag} + Q_{aer} - Q_{evap} - Q_{hxch}$ [kWh = kJ] with (20) $Q_{met} = Y_{ho}\,R_o$ [kWh = kJ]	Q_t = total energy; Q_{met} = energy generated by metabolic activities; Q_{ag} = energy generated by agitation; Q_{aer} = energy generated by aeration; Q_{evap} = energy losses through evaporation; Q_{hxch} = energy losses through cooling; Y_{ho} = approximate metabolic constant (460 kJ/mol O_2); R_o = molar oxygen uptake rate (mol O_2 / s)
Heat transfer coefficient	(21) $\alpha = (dQ/dt) / A\,dT$	α = heat transfer coefficient; dQ/dt = heat transfer; A = surface area; dT = temperature difference

Altered physiology. Changes in respiratory states like the switch from aerobic to anaerobic conditions and the imposed stress conditions, increased by induction and overproduction of the recombinant protein, lead to shifts in the coupled amino acid biosynthesis, an altered composition of the amino acid pools and changes in the protein biosynthesis machinery (162, 162a, b, c). Even though translation processes are considered to be generally performed precisely [163], the resulting shifts in the amino acid pool in turn lead, due to the inherent inaccuracy of the translation machinery (unspecificity of aminoacyl-tRNA-synthases, "wobbling" of tRNA), to aminoacid misincorporations in freshly synthesized proteins, particularly in phases of high protein production rates, which are characterized by amino acid shortages. Muramatsu et al. [164] report the incorporation of β-methylnorleucin instead of isoleucin into recombinant *E. coli* hirudin in positions 29 and 59. A misincorporation of norvalin instead of leucin and methionine into human recombinant hemoglobin in *E. coli* has been published by Apostol et al. [165] and Kiick et al. [166]. Amino acid misincorporations are thus a cause for an increase of byproducts at the expense of the desired main product yields.

Reduced Plasmid Stability. An essential prerequisite for high product yields particularly in larger scales, in which cultures pass a higher number of generations due to larger culture broth volumes and longer inoculation chains from the cell bank to the production stage, is the stable propagation of plasmids to daughter cells. Plasmid stability is influenced by the plasmid properties, including size and nucleotide sequence; by the genetic background of the host [167, 168]; as well as by process parameters like temperature, growth rates, and substrate concentrations. Lin and Neubauer [169] show that rapid glucose oscillations favor plasmid stability and recombinant protein production rate, whereas high glucose concentrations diminish plasmid stability [170]. Plasmid stability and plasmid numbers are thus negatively influenced, more difficult to control, and less easy to be maintained in larger scales. A concept to render plasmid stability and expression rates more independent from such physiological influences is the application of runaway-plasmids, the replication of which can be induced separately from growth in the desired fermentation phase [171, 172], enabling copy numbers of up to 1000 in the expression phase and, due to the separation of their replication and expression from the growth phase, simultaneously a more precise replication and amino acid incorporation. Further possible measures to enhance the accuracy of metabolic processes are a reduction of the generation time and numbers throughout the process and a deceleration of the metabolic speed, e.g., through reduction of temperature, and the development of more stress-resistant microbial strains.

A prioritized goal of process optimization and scale-up thus consists of an appropriate process design that improves the physiological conditions and the metabolic accuracy by minimizing microbial stress exposure.

1.1.5.2   Process Characterization

To identify process-specific stress factors and to understand the physiological responses to the vessel-specific physical conditions, the mutual influences and interactions of the various physical and physiological parameters have to be analyzed in detail. An overview on the analytical methods currently applied and yet in development is given in the following.

Analytical Methods. Among the physiologically most relevant parameters are, besides the pH as discussed, biomass, cell viability, the concentrations of substrates, metabolites and products, the partial oxygen and carbon dioxide pressures in culture broth, and the composition of the exhaust gas, giving information about respiratory states as indicated, e.g., by the respiratory quotient (RQ; Equation 18, Table 1.1-2). Established analytical methods and tools are colorimetric procedures, chromatography [173], mass spectroscopy [174, 175], enzymatic and electrophoretic methods [176–178], hybridizing techniques, biochips [155, 161, 179], or flow cytometry. Multiparameter flow cytometry permits, in addition to the analysis of usual metabolites, also the analysis of the cellular DNA-, RNA-, and protein content, cell viability, membrane potential, intracellular pH, cell size, and cell development stages [180–183]. In general, the data are, if possible, preferably captured as real-time values by in situ online measurements as occurring for physical parameters [184] to avoid falsified or gappy data due to time differences between sampling and analysis. *Ex situ* methods can be designed as real-time procedures in combination with robust automated sampling and sample preparation methods like ultra-filtration [173, 175, 185] and flow injection analysis (FIA) [173, 186–190]. An online FIA glucose analysis method of the native culture broth without prior sampling and filtration has been presented by Arndt and Hitzmann [191] and Kleist et al. [192]. For their robustness, *in situ* measurements of biological parameters are preferably performed by optical probes (optodes), working either on a physical base, e.g., by refractional or spectroscopic measurements, chemically as by ion exchange reactions [193] or enzymatically by gel-embedded biocatalysts, the activity of which can be measured by means of pH-changes or by fluorophores (chemo- and biosensors) [194, 195]. The hitherto employed enzymes have restricted their application mostly to the analysis of sugar compounds [190], whereby the most frequent applications are reported for glucose [187]. However, despite intensive research, only a few concepts practically applicable for industrial purposes could have been developed. Broader applications are physical optodes enabling the simultaneous measurement and quantification of several parameters and metabolites through continuous scans or excitations and subsequent absorption and scattering measurements within defined wavelength ranges like (1) 2-D-fluorescence spectroscopy [196–198] as shown for the determination of NAD(P)H concentrations through excitation at 350 nm and the fluorescence measurements at 450 nm [190]; (2) near-infrared-spectroscopy (NIR) [199–201], which has been employed as a 700–2500-nm scan in the antibiotic fermentation at Pfizer [202, 203] for the measurement of nutrient and product as well as side product concentrations; or (3) measurements in infrared [204], ultraviolet, or visible wavelength spectra [190]. By means of capacitance measuring electrodes (Aber Instruments Ltd., U.K.), viable cell counts can be performed by seizing the electric capacitance and conductivity of cells with intact membranes in a generated electric field. To gain a picture as comprehensive as possible by enhancing the amount of analyzable substances, multichannel arrays with parallel employment of various analytical procedures monitoring several parameters simultaneously have been established and new concepts like artificial noses [205] and electronic tongues [206] currently being integrated into fermentation technology, allowing the analysis of a large number of analytes by cross-talking semi-specific sensors that act in analogy to the human olphactory sensoric system. The realization of this concept will be facilitated by a further miniaturization of the sensors and chips employed [207] as it has been achieved with biochips developed for the pharmaceutical natural product screening, onto which the needed number of reactants is fixed in smallest space or is brought together in microchannels and the analytes are separated in microcapillars (lab-on-a-chip-technology) [208–210].

1.1.5.3 Process Optimization

For analysis, interpretation and correlation of the obtained signals and data through chemometric modeling [195, 197], neural network—tools (e.g., Unscrambler 9.0, Camo, Norway)—seem to have gained an increasingly important role [211]. Despite inherent limitations [212, 213], adequately trained neural networks are capable of recognizing and revealing nonlinear, highly complex, and even nonobvious, hidden relations among a multitude of various physical, biochemical, and physiological process parameters and to design the according process models. Hooked up to expert systems, they allow immediate and short-term reactions to process deviations by anticipating physiological drifts and their influences on product yield and quality and to change the according set points and process profiles. In this way, neural network tools help to optimize process control and to facilitate the identification of suitable strategies for process optimization and scale-up, e.g., with respect to an optimized glucose–oxygen equilibrium.

As the maximum glucose and oxygen uptake rates are not constant but depend on growth phases and rates and get reduced by induction of product synthesis [170], Lin et al. [214] established an integrative kinetic model combining these parameters with the aid of the simulation program SCILAB (Inria, F) enabling the determination of the maximum glucose and oxygen uptake rates through determination of the time that passes between glucose pulses and subsequent changes of partial oxygen pressure (pO_2) and biomass concentration. Such a model thus allows the alignment of the glucose feed meeting the maximum uptake capacity and therefore avoiding overflow metabolism. The observed pattern of glucose oscillation even can serve as a key parameter condition in later scale-up stages [169]. The equilibration of carbohydrate and oxygen supply thus constitutes a key component of process control and optimization. Further strategies aiming at the avoidance of oxygen limitation and glucose overflow metabolism consist of the employment of alternative carbon sources like glycerol [215] or galactose [216] and of a drastic glucose restriction. Wong et al. [217] report that minimal concentrations of glucose and yeast extract yielded the highest concentrations of 0.5 and 1 g/L of a recombinant K99 antigen. Yim et al. [218] achieved constant yields of 4.4 g/L of human granulocyte colony stimulating factor during scale-up from 2.5 to 30 L by limiting glucose feed and keeping the growth rate at the minimum ($\mu = 0.116 h^{-1}$). Besides the above-mentioned parameters like the maximum uptake capacities, the various different physiological or directly measurable physical process parameters like pH, ammonium consumption, pO_2, OTR, or the growth parameters can be applied as a reference parameter for glucose feed in a closed-loop design. Dantigny et al. [219] describe a biomass controlling feed depending on the ethanol formation rate and RQ for *S. cerevisiae*. To achieve a more uniform glucose distribution in large-scale tanks, Larsson et al. [151] and Bylund et al. [152] suggested a glucose feed at the dynamic zones of the fermenter bottom together with the injected air. A step ahead in this regard would be a further enhancement of the mixing quality by a reactor and process design permitting a multilevel injection of both air and substrates into high turbulence zones.

A supportive approach for comprehension of the large-scale hydrodynamic and reaction conditions to ensure homogenous reaction conditions and to reduce both the size of stress zones and the zonal residence times is the depiction of these zones through transfer into small reactors (scale-down) or through high-performance computing computational fluid dynamics (CFD) [220]. Whereas the sort and amount of parameters that can be simulated by the scale-down approach appear to be restricted, CFD is meanwhile broadly applied, e.g., the flow modeling software tools of Fluent Inc., Lebanon, NH (Fluent 4,5,6 and MixSim). A simulation of the trajectories and distributions of gas bubbles and mass transports enables the determination of zonal pO_2 values and oxygen transfer rates. The combination with parameters like substrate concentrations and gradients, the residence times in the respective zones, population dynamics, and metabolic fluxes (structured metabolic models) leads to integrative models (integrated fluid dynamics, IFD), which permit the prediction of physiological effects and reactions. Even though it is emphasized that the integration of CFD and structured biokinetics currently requires a deeper understanding of the dynamics of metabolic and regulatory networks and cascades of signal transduction triggered by microenvironmental fluctuations and thus further research [221], it will most likely contribute long term to a realistic modeling of the interplay between physics and physiology and thus will facilitate the identification of key parameters influencing product yield and quality the most and therewith of suitable scale-up parameters and strategies. Problems related to the scale-up of physical parameters are illustrated in the next section.

1.1.5.4   Physical Scale-Up Parameters

Suitable for employment as physical scale-up parameters are all known process parameters and coefficients exerting known physiological effects, particularly those affecting oxygen supply; heat transportation and mixing, such as power input (Equation 1; Table 1.1-4), aeration, and agitation rate (Equations 3, 9, and 10; [222]); mixing time (Equation 5; [223]); pO_2; OTR (Equations 14, 15; [224, 225]); oxygen mass transfer coefficient (k_La, Equations 16, 17); and biocalorimetric variables like heat fluxes and heat transfer coefficients (Equations 19, 20, and 21). Biocalorimetric measurements were found to be in definite correlation with metabolic activities and with OTR and are employable as growth phase indicators [226, 227]; the calculations, however, require a precise measurement and knowledge of heat fluxes and sources, energy inputs, and the temperature distribution in the vessel. In most cases, however, it is, depending on the scale-up factor, not, or only restrictedly, possible from the physical viewpoint to keep physical parameters constant throughout the scale-up. Constance of even a single specific parameter mostly leads to an uncontrolled and unpredictable change of other variables into dimensions that are technically not realizable. Classic examples are (1) the mixing time, which inevitably increases in larger vessels due to the larger volumes to be stirred and that cannot be unlimitedly compensated and kept constant by increasing stirrer speeds and energy inputs, and (2) the volumetric energy input P/V. A constant volumetric power input has indeed been successfully applied as a scale-up parameter for the early industrial penicillin fermentations (1 hp per gallon, equivalent to $1.8 kW$ per $1 m^3$) and in fermentations with low energy inputs [216], but it is limited in fermentations requiring high energy inputs, like recombinant *E. coli* cultures.

For this reason, mathematically driven approaches are pursued for both process and reactor scale-up by forming dimensionless coefficients (e.g., by the Buckingham-Pi method as cited by Hubbard [228]), which are kept constant by an appropriate choice and adjustment of the relevant influencing process and fermenter parameters. Well-established examples are, among many others, the dimensionless power number (Equation 2), Reynolds number (Equation 4), gassing number (Equation 13), and the modified dimensionless power number (Equation 7) and the modified dimensionless mixing time (Equation 8). The comparison of the latter two coefficients and of the corresponding curves enables the identification of appropriate types of stirrers capable of exerting the desired mixing performance at a given stirrer speed with a minimum of energy consumption and therewith to compensate the above-mentioned limitations of volumetric energy inputs at larger scales. (A quantum leap in stirrer technology in this regard seems to be the Visco-Jet of Inotec, which possesses cone-formed short tubes instead of blades and pretends to exhibit the best mixing performance with a minimum of power input and shear stress without foam generation.) The performance of further various types and designs of stirrers is presented and discussed by Junker et al. [229]. Stirrer performance coefficients and curves are thus crucial scale-up aids.

Furthermore, any of these parameters and coefficients can be combined with other variables to set up and create new, process-specific parameter correlations, coefficients, terms, groups, and characteristic curves (examples given, e.g., by Wang and Cooney [230]).

This relates also to the $k_L a$—value (Equation 16), which currently is the most applied physical scale-up variable because it includes the relevant parameters influencing oxygen supply like agitation (via energy input) and aeration as superficial gas velocity (Equation 10) and that, as a component of dimensionless terms, is also frequently employed for reactor scale-up [231].

Taking into account the limitation of a $k_L a$–oriented scale-up in the form of a limited power input, OTR, and tolerable shear stress, Flores et al. integrated the backpressure p by maintaining the product of $k_L a \times p$ constant through a variation of pressure, agitation, and aeration for scale-up of *Bacillus thuringiensis* fermentations [232]. Although the specific growth was reduced and the biomass production and sporulation efficiency remained constant, fermentation time could be shortened and toxin yields were increased. Wong et al. successfully scaled-up *E. coli* fermentations from 5 L to 200 L by keeping constant the product of $k_L a$ and aeration rate (vvm) under variation of power input, working volume, and aeration according to the Wang–Cooney equation (Equation 17; [230]), which reflects these parameters in dependence of the fermentation scale [217]. Diaz and Acevedo [224] argue that the oxygen transfer capacity, indicated by the $k_L a$ value, is not the most process-relevant parameter, but the effective oxygen transfer rate as a product of $k_L a$ and mass transfer potential is, i.e., the difference between the pO_2-values in the gas and the liquid phase, and suggest an OTR-based scale-up strategy (Equations 14 and 15).

A common, simple and robust method is the maintenance of a constant pO_2 in the culture broth by variation of stirrer speed and aeration rate. To avoid shear stress and to keep the energy input (P) at the lowest possible level, the pO_2 is steered at the minimum limit. Riesenberg et al. employ a stirrer speed and glucose-feed steered pO_2 of 20% for interferon α–*E. coli*-fermentations in 30-L and 450-L scales [233]. These examples demonstrate that, for each process, appropriate variables and coefficients and therewith scale-up strategies have to be identified individually.

1.1.5.5 Development of Fermentation Models and Strategies

Despite the central role of the scale-up issue in biotechnology and the comparably large body of literature, no common, generally applicable strategy seems to be established. For each product, process, and facility, a suitable scale-up strategy has to be elaborated.

A wholistic scale-up strategy consists of a comprehensive and detailed process characterization to identify key stress factors and key parameters influencing product yield and quality the most, and of an appropriate process control and process design ensuring optimum mixing and reaction conditions, supported by appropriate knowledge and data-driven models [234, 235] as well as computational tools.

It should be kept in mind, however, that any approach and any model will always be approximative and that a compromise in the process-related knowledge will always be gappy and that the known mathematical methods and relations cannot completely reflect the highly complex interactions and relationships of the physical conditions governing the fermentation process [236].

As a matter of fact, it seems, that, in view of the high complexicity of the fermentation parameters influencing each other and the only rudimentary and fragmentary reflection of the reality any model can deliver and finally the different layouts of vessels and facilities that rarely are designed according to strict scale-up criteria, successful scale-up in most cases will not be the result of a conclusive and straight-lined experimental strategy but will be the outcome of an independent optimization on each scale that highly depends on the experience, skill, and last, but not least, intuition of the experimentalist.

To a lesser extent, this also holds true for the design of procedures aimed at the isolation and purification of the compounds from the culture broth. As these purification steps follow upon the biosynthetic steps and the upstream processing in the bioreactor, they are usually referred as "downstream processing."

1.1.6 DOWNSTREAM PROCESSING

1.1.6.1 Product Recovery and Purification

Biopharmaceutical products consist of a multitude of compounds and structures with most having different physical and chemical properties and derive from a large variety of sources like human and animal tissues, body fluids, plant material [237], and as illustrated above, microbial fermentations. Accordingly, purification strategies have to be developed individually and empirically that reflect the physicochemical properties of the product, of the product source, and of potential contaminants by finding appropriate sorts, sequences, combinations, and operation modes of the respective downstream processing steps to finally achieve high purities and high recovery rates while maintaining the pharmaceutical activity of the molecule.

Traditionally, downstream processing steps are roughly subgrouped into operations related to cell harvest, cell rupture in case of intracellularly occurring products, product capture, product purification, and product polishing to manufacture a drug ready for galenic preparation and consist of a sequence of solid–liquid separation and solvent and solid-phase extraction steps. Large-scale cell harvest usually occurs by preparative centrifugation, e.g., in a decanter or in disk stack separator or by filtration to simultaneously separate existing suspended particles from the surrounding broth. Fragile mammalian cells are usually separated by filtration methods only.

If the value product occurs intracellularly, the slurry is taken up in washing buffers of appropriate ionic strength for preparation of the cell rupture, which occurs, e.g., by high-pressure homogenization, sonification, bead milling, or enymatic procedures.

Product capture is defined as the first product extraction and purification step aimed at a product concentration by volume reduction and partial purification, whereas polishing constitutes the final purification step, which removes persistent residual minor impurities like denatured, aggregated, or nonfunctional isoforms of the product, and which follows the preceding (intermediate) purification steps aimed at the removal of contaminating solutes like host cell proteins, DNA, and media components. Product capture usually starts with precipitation, solvent extraction, and/or a cascade of ultra-filtration and nano-filtration steps before the first ion exchange or size exclusion chromatography. Antibodies and antibody derivatives, which account for about 20% of the biopharmaceutical products currently in development, are preferably bound to affinity matrices [238] by natural immunoglobulin-binding ligands such as the *Staphylococcus aureus* membrane and cell wall protein A [239] or the streptococci surface protein G. To broaden the availability of specific structures with suitable binding properties, synthetic ligands with enhanced stability and resistance to chemical and biochemical degradation are continuously attempting to be developed, supported, e.g., by computer-aided design. The development of synthetic ligands with enhanced selectivity and stability, tailored to specific biotechnological needs and product structures, will lead to more efficient, less expensive, and safer procedures not only for purification of antibodies but also for other proteins at manufacturing scales. Also, antibodies can be employed as a means for protein purification in the form of immunoaffinity chromatography [240]. A novel emerging technology is the use of synthetic single-stranded nucleic acid molecules (aptamers) as affinity ligands [241], which fold up into unique three-dimensional structures specifically binding to the desired target structure as it has been shown for the purification of thyroid transcription factor or selectin receptor globulin [242, 243].

The subsequent purification steps including polishing are mostly performed on hydrophobic interaction resins and/or on reversed-phase matrices.

1.1.6.2 Downstream Processing Optimization and Economization

Mostly due to the tremendous costs of sometimes several thousand € per liter for chromatographic resins and production scales of up to several hundred liters column volume and due to the prices for the columns themselves of up to hundreds of thousands of U.S. dollars, downstream processing expenses account for the largest part (50–90%) of the total production costs and are therewith a focal area for process economization attempts. One of the many criteria for the choice of a resin is its shelf life. Bearing in mind that every downstream processing step coincides with a minimum product loss of 5–10%, the removal of a particular step not only helps to save the respective capital investment, but concomitantly to increase the final product yields, and thus to render the whole production process more economical in several respects. To reduce the number of required steps, a selection has to be made for resins exhibiting excellent separation, resolution, and yield with respect to a given product and the contaminants, i.e., for resins exhibiting excellent product binding selectivity and capacity. Likewise, the operation mode of the column has to be optimized with respect to the specific product load, adjustment of the physico-chemical operational conditions like pressure, temperature, pH(-shift), ionic strength, and sort of buffers and eluents (e.g., evaluation of continuous vs. step gradients).

A further approach is the combination of the solid–liquid separation and product recovery into a single step, e.g., by expanded bed adsorption [244, 245], which permits the direct initial purification from culture broths without the need of prior removal of suspended solids. In contrast to the conventional column operation, the mobile phase is pumped upward through the column bed from beneath. The bed thus starts to expand at a liquid flow rate above a critical value, and the gaps between the sorbent beads expand. The controlled distribution of bead size and bead weight results in a stable expanded bed in which the beads oscillate around a steady position and thus avoid clogging of the column. Whereas preparative liquid chromatography methods [246] have become a well-established separation and purification method in the pharmaceutical industry, techniques like two-phase system partitioning [247] and its variant reverse micellar extraction [248], which also potentially offer advantages like the possibility of a direct extraction from the culture broth through partitioning of the compounds into two immiscible phases, need further developmental activities for economic large-scale applications in the production of biopharmaceuticals and are less common.

A further criterion for the choice and design of a downstream processing procedure is thus its scalability.

1.1.6.3 Downstream Processing Scale-Up

Comparable with the scale-up of fermenters and fermentation processes, the scale-up of downstream processes, particularly of column and column operation, is not simply a matter of increasing size. Increasing column diameters, reducing the stabilizing wall effects, and increasing resin bed heights may lead to an altered settling behavior of the beads, to channel formation through shrinking and swelling of the packed matrix, to column clogging with a concomitant reduction of the flow rates and an increase of the column back pressure, to altered hydrodynamic behavior and residence times of the process fluids, and to an altered pattern for the product/contaminants adsorption/desorption pattern [249]. To avoid the resulting quality impairments and deviations, resins with a suitable scale-up behavior have to be selected and the resin bed height is tried to be kept constant as far as possible as is the flow rate as key scale-up parameters, which requires an appropriate experimental design in early laboratory development stages. As this, however, is only possible to a limited extent, downstream processing scale-up also is the result of an optimization of the respective operation conditions (see above) on each scale.

A procedure that has been optimized for decades and scaled-up to dimensions of several cubed meters is the commercial β-lactam purification.

(For further general articles dealing with problems and issues of downstream processing, the reader is referred to the reviews in Refs. 249–255).

1.1.6.4   Downstream Processing of β-Lactam Compounds

As most of the product is secreted and is thus concentrated in the culture broth, the recovery process starts with filtration, usually with a rotary vacuum filter, followed by a cascade of solvent and solid-phase extraction steps.

Common penicillin extraction solvents are amyl- and butyl-acetate or methylisobutylketone. As penicillin is extracted as a free acid at pH 2–2.5 and the molecule is instable at this pH, extraction occurs in a counter-flow using Podbielniak- and Luwesta-centrifugal extractors to shorten the contact time with the solvent and to prevent product decomposition.

Purification is through subsequent and repeated crystallization from an aqueous solution after alkalization.

In contrast to penicillin, the hydrophilic cephalosporin is more suitable for solid-phase extraction. For high yield and purity extraction, a combination of several, different chromatographic steps is used.

Hydrophobic interaction chromatography on neutral polyaromatic resins like Amberlite XAD 4, 16, 1180, or Diaion HP20 is widely used in combination with weak basic anion exchangers like Diaion WA-30 or strong acidic cation exchangers like Amberlite XAD 2000 and Diaion SK-1B [256, 257]. The mentioned resins are recognized for their high sorption capacity and their long shelf life.

Whereas the improvements on the extraction level are less spectacular, significant progress over the last few years can be noticed with respect to side-chain cleavage to 6-aminopenicillanic acid (6-APA) and 7-aminocephalosporanic acid (7-ACA) in switching from chemical to enzymatic procedures.

1.1.7   POSTSYNTHETIC STRUCTURE MODIFICATION

Most antibiotics in therapeutic use are synthesized or modified exclusively by the means of chemistry and are derived from the established compound classes, which have been known for decades. Chemical derivatization thus has also been applied to substances that are of microbial origin and that therefore are termed semi-synthetic.

Among the most recent semi-synthetic antimicrobials are Aventis' streptogramin derivative Synercid and the erythromycin derivatives clarithromycin (Klacid) from Abbott (Abbott Park, Illinois) and roxithromycin (Rulid) and the brand new ketolide telithromycin (Ketek), both from Aventis (Paris, France) [86].

Despite these recent successes and increasing efforts, the yields of therapeutically useful entities emerging from such chemical derivation programs are continuously decreasing.

For this reason, large-scale derivation methods (combinatorial chemistry) are integrated in the search for innovative antibiotics as are biotechnological approaches for structure modification. Biocatalytic procedures, which have been shown to be useful for generating novel antibiotic structures like Loracarbef, a novel β-lactam compound from Eli Lilly (Indianapolis, Indiana) highly active against various β-lactamase producing species [258, 259], are also about to replace continuously chemical production processes. At several universities and corporations, screening studies have been initiated to find appropriate enzymes for compound conversions, as it has been successfully established in production processes for penicillin and cephalosporin. (A review discussing criteria like resource and energy consumption, emissions, health risk potential, area use, and environmental effects for the evaluation of the eco-efficiency of biotechnological processes is given by Saling [260]).

1.1.7.1 β-Lactam Side-Chain Cleavage

6-APA and 7-ACA as the key intermediates for the production of semisynthetic penicillins and cephalosporins (see Section 1.1.3.2) are obtained by removal of the acyl side chains.

There is a large body of patents existing for chemical and enzymatic splitting procedures, but enzymatic processes have been more successful for economical as well as for ecological reasons.

Enzymatic 7-ACA splitting procedures [for general review, see 261] have been developed and commercialized by companies like Asahi Chemical, Hoechst, and Novartis. The replacement of the hitherto employed chemical deacylation processes like the imino ether (Figure 1.1-3) or the nitrosyl chloride method [262] resulted in a cost reduction of 80% and a decrease of the waste volume by a factor 100 from 31 t to 0.3 tons per 1-ton 7-ACA. Chlorinated hydrocarbons like dimethyl aniline and methylene cloride as well as heavy metal ions can be completely avoided. Instead of zinc salt formation, multiple silylation, formation of the imino chloride, imino ether, and finally an imino ether hydrolysis, the side chain is removed in two enzymatic steps (Figure 1.1-3).

Cephalosporin C is first oxidized and deaminated by a D-amino acid oxidase (DAO), which can be obtained from various fungal species, like the yeasts *Trigonopsis variabilis* and *Rhodotorula gracilis* or the ascomycete *Fusarium solani*. The resulting α-keto-adipyl-7-ACA, upon decarboxylation, is converted into glutaryl-7-ACA (G-7-ACA). DAO is a flavoenzyme containing flavin adenine dinucleotide as the prosthetic group and catalyzes oxidation of D-amino acids to their corresponding keto acids. In a second step, the glutaryl side chain of G-7-ACA is deacylated by a glutarylamidase from *Pseudomonas diminuta* [263]. The molecular data of other potentially suitable enzymes and genes from various sources are given by Isogai [264]. It is noteworthy that the enzymatic splitting process could have only been rendered economical and therefore commercially employable through a significant increase of glutarylamidase yield on the fermentation level by using a gene-recombinant *E. coli*-strain.

A recombinant amidase from *E. coli* is also the most commonly employed enzyme for 6-APA production by deacylation of penicillin G [265], a process that has been established now for decades [266]. With an annual turnover of 30 tons, the *E. coli* penicillin amidase is one of the most widely used biocatalysts, despite

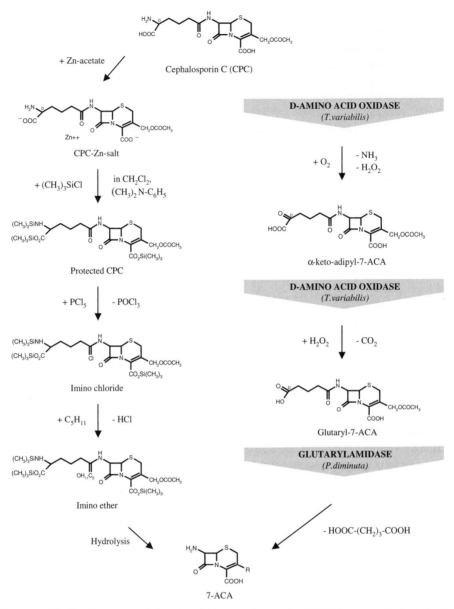

Figure 1.1-3. Replacement of the chemical cephalosporin C splitting procedure by a bio-technological process. Biotechnological production and derivatization procedures are continuously being developed to replace chemical processes for both economical and ecological reasons. An established industrial process is the enzymatic splitting of cephalosporin C to 7-ACA by D-amino acid oxidase of the yeast *Trigonopsis variabilis* and a glutarylamidase of *Pseudomonas diminuta* (right) replacing the chemical imino ether method (left). For further details, see text. (Graph taken from Ref. 98, © Springer-Verlag.)

the discovery of further similar acting enzymes from various microbial sources, including penicillin V acylases [267] from the basidiomycetes *Bovista plumbea* [268, 269] and *Pleurotus ostreatus* [270]; from the ascomycete *Fusarium* [271]; or from the yeast *Cryptococcus*.

A significant milestone in economization of enzymatic β-lactam production has been reached by enzyme immobilization permitting a preservation and multiple use of the cleavage enzymes. The currently most employed resins in industries for this purpose are epoxyacrylic acids (Eupergit) and silica gel derivatives (Deloxan). The general economic criterion for the preparation of a biocatalyst is production costs in relation to yield, turnover rate, storage stability, and operational/mechanical stability. Further criteria are outlined in the *Guidelines for the Characterization of Immobilized Biocatalysts* by the European Federation of Biotechnology. Application criteria are filterability, sedimentation velocity, and particle firmness. Operational and storage stability as well as the activity of this biocatalyst and the quality of 6-APA with respect to color are improved by sulfur reducing compounds [272]. As pointed out, the ensurance of product quality and drug safety is a key issue in pharmaceutical production processes.

1.1.8 QUALITY ISSUES

According to regulatory requirements, pharmaceutical production facilities and processes have to be proven to function properly across the entire range of process critical parameters by a qualification of the facilities and equipment, which consists of the steps design qualification (DQ), installation qualification (IQ), operational qualification (OQ), and performance qualification (PQ) and which results in a validation of the production process to substantiate that the complete system is capable of manufacturing the respective product reproducibly in a consistent quality as expected within preset specifications and in compliance with all laws and guidelines, like the cGMP (current good manufacturing practice) regulations, among others, issued by regulating authorities like the FDA and European Medicines Agency (EMEA).

The products have to be sterile and free of any contaminations that might derive from process material, residues of preceding production campaigns (cross-contaminations) in the case of product changes, and of the agents used for facility cleaning. To restrict the introduction of such contaminations into the repective production steps, the production facilities, equipments, and solutions are usually cleaned according to validated cleaning procedures (e.g., with 1 M NaOH) and sterilized by heat or filtration before their use. Product sterility is achieved by aseptical performance of the last manufacturing steps and by micro-filtration (0.2 μm) or, if possible, by terminal heating of the product before filling. Residues of chemical cleaning agents are removed according to validated procedures by rinsing with purified water.

As biopharmaceuticals are potentially contaminanted with possibly harmful viruses and allergenic and pyrogenic acting endotoxins as residual components of their host cell, guidelines and directions have been issued by the FDA's Center for Biologics Evaluation and Research (CBER) requesting validated procedures for virus and endotoxin removal and inactivation.

1.1.8.1 Virus and Endotoxin Removal

Endotoxins, which consist of the lipopolysaccharide fraction present in the cell wall of gram negative germs and which tend to adhere to equipment surfaces and to persist in products and product solutions, as do viral contaminants deriving from cell culture processes or from starting plasma or tissue material, can also be efficiently removed by final filtration steps. Whereas ultra-filtration with a cut-off of or below 10 kDa is sufficient for endotoxin removal, viruses need to be removed by nano-filtration steps complemented by a variety of inactivation measures like pasteurization, pH inactivation, solvent/detergent treatment, or ultraviolet and gamma ray irradiation. Usually, endotoxins and viruses are already deconcentrated to a significant extent by the chromatographic steps and the solvent exposure in the course of the normal purification procedures [251].

1.1.9 CONCLUSION

Marching in step with the technical and methodological progresses, biotechnology is gaining in increasing importance in pharmaceutical production processes by replacing chemical production procedures for economical and ecological reasons and in the development and commercialization of novel therapeutic principles.

To fully exploit the potential of biotechnological production methods, an integrated process design will be necessary considering the downstream processing requirement during the design of upstream operations and vice versa (A comprehensive overview illustrating the complete biotechnological production chain is given in Table 1.1-3.) As the fate of a compound as a pharmaceutical also depends on the prospective production costs already estimated in the development phase and production processes cannot be significantly changed after approval without running the risk of having to perform ad-ditional clinical trials, it is of utmost importance to choose and design production organisms, strains, vectors, expression cassettes, and fermentation procedures as early as possible in compliance with the prospective harvest and purification methods also to be chosen and designed according to cost criteria. Integrative biotechnology thus is not only a key prerequisite for the development of competitive and economical production processes to relieve the downward price pressure applied by generic drug makers after patent expiration and for the acceleration of market approval for new drugs in the highly competitive environment of the biopharmaceutical industry, but also it is a key technology to save and liberate capital for the development of novel drugs for the benefit of mankind.

REFERENCES

1. Choi J H, Lee S Y (2004). Secretory and extracellular production of recombinant proteins using *Escherichia coli*. *Appl. Microbiol. Biotechnol.* (online).
2. Li Y, Wei G, Chen J (2004). Glutathione: a review on biotechnological production. *Appl. Microbiol. Biotechnol.* 66:233–242.
3. Leathers T D (2003). Biotechnological production and application of pullulan. *Appl. Microbiol. Biotechnol.* 62:468–473.

4. Hermann T (2003). Industrial production of amino acids by coryneform bacteria. *J. Biotechnol.* 104:155–172.

5. Wehmeier U F, Piepersberg W (2004). Biotechnology and molecular biology of the alpha-glucosidase inhibitor acarbose. *Appl. Microbial. Biotechnol.* 63:613–625.

6. Pulz O, Gross W (2004). Valuable products from biotechnology of microalgae. *Appl. Microbiol. Biotechnol.* 65:635–648.

7. Lee P C, Schmidt-Dannert C (2002). Metabolic engineering towards biotechnological production of carotenoids in microorganisms. *Appl. Microbiol. Biotechnol.* 60:1–11.

8. Westers L, Westers H, Quax W J (2004). Bacillus subtilis as cell factory for pharmaceutical proteins: A biotechnological approach to optimize the host organism. *Biochim. Biophys. Acta.* 1694:299–310.

9. Miesfeld R (1999). *Applied Molecular Genetics*, Wiley, New York.

10. Weaver R F (1994). *Molecular Biology of the Cell*, Garland Publishing, New York.

11. Faye L, Boulaflous A, Benchabane M, Gomord V, Michaud D (2005). Protein modifications in the plant secretory pathway: Current status and practical implications in molecular pharming. *Vaccine* 23:1770–1778.

12. Hellwig S, Drossard J, Twyman R M, Fischer R (2004). Plant cell cultures for the production of recombinant proteins. *Nat. Biotechnol.* 22:1415–1422.

13. Goldstein D A, Thomas J A (2004). Biopharmaceuticals derived from genetically modified plants. *QJM* 97:705–716.

14. Gregory G, Heyneker H L (1988). Secretion of correctly processed human growth hormone in *E. coli* and *Pseudomonas*. U.S. Patent 4755465.

15. Habermann P, Bender R (2001). Signal sequences for the production of leu-hirudin via secretion by *E. coli* in a culture medium. International Patent WO 01/21662.

16. Schmidt F R (2004a). Recombinant expression systems in the pharmaceutical industry. *Appl. Microbiol. Biotechnol.* 65(4):363–372.

17. Choi J H, Ryu Y W, Seo J H (2005). Biotechnological production and applications of coenzyme Q_{10}. *Appl. Microbiol. Biotechnol.* (epub ahead of print).

18. Weickert M J, Doherty D H, Best E A, Olins P O (1996). Optimization of heterologous protein production in *Escherichia coli*. *Curr. Opin. Biotechnol.* 7:494–499.

19. Makrides S C (1996). Strategies for achieving high-level expression of genes in *Escherichia coli*. *Microbiol. Rev.* 60:512–538.

20. Gentschev I, Dietrich G, Goebel W (2002). The *E. coli* alpha-hemolysin secretion system and its use in vaccine development. *Trends Microbiol.* 10:39–45.

21. Ward A, Hoyle C, Palmer S, O'Reilly J, Griffith J, Pos M, Morrison S, Poolman B, Gwynne M, Henderson P (2001). Prokaryote multidrug efflux proteins of the major facilitator superfamily: Amplified expression, purification and characterization. *J. Mol. Biotechnol.* 3:193–200.

22. Pines O, Inouye M (1999). Expression and secretion of proteins in *E. coli*. *Mol. Biotechnol.* 12:25–34.

23. Bassford P J, Silhavy T J, Beckwith J R (1979). Use of gene fusion to study secretion of maltose-binding protein into *Escherichia coli* periplasm. *J. Bact.* 139:19–31.

24. Yamanaka H, Nomura T, Fujii Y, Okamoto K (1998). Need for TolC, an *Escherichia coli* outer membrane protein, in the secretion of heat-stable enterotoxin I across the outer membrane. *Microb. Pathogen.* 25:111–120.

25. Lund P A (2001). Microbial molecular chaperones. *Adv. Microb. Physiol.* 44:93–140.

26. Hartl F U (1996). Molecular chaperones in cellular protein folding. *Nature* 381:571–580.

27. Collier D N (1993). SecB: A molecular chaperone of *Escherichia coli* protein secretion pathway. *Adv. Protein. Chem.* 44:151–193.

28. Kujau M J, Hoischen C, Riesenberg D, Gumpert J (1998). Expression and secretion of functional miniantibodies McPC603scFvDhlx in cell-wall-less L-form strains of *Proteus mirabilis* and *Escherichia coli*: A comparison of the synthesis capacities of L-form strains with an *E. coli* producer strain. *Appl. Microbiol. Biotechnol.* 49:51–58.

29. Rippmann J F, Klein M, Hoischen C, Brocks B, Rettig W J, Gumpert J, Pfizenmaier K, Mattes R, Moosmayer D (1998). Procaryotic expression of single-chain variable-fragment (scFv) antibodies: Secretion in L-form cells of *Proteus mirabilis* leads to active product and overcomes the limitations of periplasmic expression in *Escherichia coli*. *Appl. Environ. Microbiol.* 64:4862–4869.

30. Mergulhao F J, Summers D K, Monteiro G H (2005). Recombinant protein secretion in *Escherichia coli*. *Biotechnol. Adv.* 23:177–202.

31. Swartz J R (2001). Advances in *Escherichia coli* production of therapeutic proteins. *Curr. Opin. Biotechnol.* 12:195–201.

32. Sorensen H P, Mortensen K K (2005). Soluble expression of recombinant proteins in the cytoplasm of *Escherichia coli*. *Microb. Cell. Fact.* 4:1–8.

33. Dilsen S, Paul W, Sandgathe A, Tippe D, Freudl R, Thömmes J, Kula M R, Takors R, Wandrey C, Weuster-Botz D (2000). Fed-batch production of recombinant human calcitonin precursor fusion protein using *Staphylococcus carnosus* as an expression-secretion system. *Appl. Microbiol. Biotechnol.* 54:361–369.

34. Olmos-Soto J, Contreras-Flores R (2003). Genetic system to overproduce and secrete proinsulin in *Bacillus subtilis*. *Appl. Microbiol. Biotechnol.* 62:369–373.

35. Srinivasan S, Barnard G C, Gerngross T U (2002). A novel high-cell-density protein expression system based on *Ralstonia eutropha*. *Appl. Env. Microbiol.* 68:5925–5932.

36. Nakashima N, Mitani Y, Tamura T (2005). Actinomyctes as host cells for production of recombinant proteins. *Microb. Cell. Fact.* 4:7.

37. Pierce J J, Robinson S C, Ward J M, Keshavarz-Moore E, Dunnill P (2002). A comparison of the process issues in expressing the same recombinant enzyme periplasmatically in *Escherichia coli* and extracellularly in *Streptomyces lividans*. *J. Biotechnol.* 92:205–215.

38. Anné J, van Mellaert L (1993). *Streptomyces lividans* as host for heterologous protein production. *FEMS Microbiol. Lett.* 114:121–128.

39. Paradkar A, Trefzer A, Chakraburtty R, Stassi D (2003). *Streptomyces* genetics: a genomic perspective. *Crit. Rev. Biotechnol.* 23:1–27.

40. Berks B C, Sargent F, Palmer T (2000). The Tat protein export pathway. *Molec. Microbiol.* 35:260–274.

41. Teter S A, Klionsky D J (1999). How to get a folded protein across a membrane. *Trends Cell. Biol.* 9:428–431.

42. Schaerlaekens K, van Mellaert L, Lammertyn E, Geukens N, Anne J (2004). The importance of the Tat-dependent protein secretion pathway in Streptomyces as revealed by phenotypic changes in tat deletion mutants and genome analysis. *Microbiol.* 150:21–31.

43. Upreti R K, Kumar M, Shankar V (2003). Bacterial glycoproteins: functions, biosynthesis and applications. *Proteomics* 3:363–379.

44. Benz I, Schmidt A (2002). Never say never again: Protein glycosylation in pathogenic bacteria. *Molec. Microbiol.* 45:267–276.

45. Russel C, Mawson J, Yu P L (1991). Production of recombinant products in yeasts: A review. *Aust. J. Biotechnol.* 5:48–55.

46. Lin Cereghino J L, Cregg J M (2000). Heterologous protein expression in the methylotrophic yeast *Pichia pastoris*. *FEMS Microbiol. Rev.* 24:45–66.

47. Lin Cereghino G P, Lin Cereghino J, Ilgen C, Cregg J M (2002). Production of recombinant proteins in fermenter cultures of the yeast *Pichia pastoris*. *Curr. Opin. Biotechnol.* 13:329–332.

48. Buckholz R G, Gleeson M A (1991). Yeast systems for the commcercial production of heterologous proteins. *Bio/technol.* 9:1067–1072.

49. Gellissen G (2000). Heterologous protein production in methylotrophic yeasts. *Appl. Microbiol. Biotechnol.* 54:741–750.

50. Gellissen G, Weydemann U, Strasser A W M, Piontek M, Janowicz Z A, Hollenberg C P (1992). Progress in developing methylotrophic yeasts as expression systems. *Trends Biotechnol.* 10:413–417.

51. Cregg J M, Lin Cereghino J L, Shi J, Higgins D R (2000). Recombinant protein expression in *Pichia pastoris*. *Mol. Biotechnol.* 16:23–52.

52. Swinkels B W, van Ooyen A J J, Bonekamp F L (1993). The yeast *Kluyveromyces lactis* as an efficient host for heterologous gene expression. *Antonie van Leeuwenhoek* 64:187–201.

53. Schaffrath R, Breuning K D (2000). Genetics and molecular physiology of the yeast *Kluyveromyces lactis*. *Fungal Genet. Biol.* 30:173–190.

54. Gellissen G, Hollenberg C P (1997). Application of yeasts in gene expression studies: A comparison of *Saccharomyces cerevisiae*, *Hansenula polymorpha* and *Kluyveromyces lactis*—a review. *Gene* 190:87–97.

55. Guisez Y, Tison B, Vandekerckhove J, Demolder J, Bauw G, Haegeman G, Fiers W, Contreras R (1991). Production and purification of recombinant human interleukin-6 secreted by the yeast *Saccharomyces cerevisiae*. *Eur. J. Biochem.* 198:217–222.

56. Kjeldsen T (2000). Yeast secretory expression of insulin precursors. *Appl. Microbiol. Biotechnol.* 54:277–286.

57. Kjeldsen T, FrostPettersson A, Hach M (1999). Secretory expression and characterization of insulin in *Pichia pastoris*. *Biotechnol. Appl. Biochem.* 29:79–86.

58. Chow T Y, Ash J J, Dignard D, Thomas D Y (1992). Screening and identification of a gene, PSE-1, that affects protein secretion in *Saccharomyces cerevisiae*. *J. Cell Sci.* 101:709–719.

59. Turner B G, Avgerinos G C, Melnick L M, Moir D T (1991). Optimization of pro-urokinase secretion from recombinant *Saccharomyces cerevisiae*. *Biotechnol. Bioengin.* 37:869–875.

60. Sleep D, Belfield G P, Ballance D J, Steven J, Jones S, Evans L R, Moir P D, Goodey A R (1991). *Saccharomyces cerevisiae* strains that overexpress heterologous proteins. *Bio/technol.* 9:183–187.

61. Fleer R (1992). Engineering yeast for high level expression. *Curr. Opin. Biotechnol.* 3:486–496.

62. Poletti A, Weigel N L, McDonnell D P, Schrader W T, O'Malley B W, Conneely O M (1992). A novel, highly regulated, rapidly inducible system for the expression of chicken progesterone receptor, cPR_A, in *Saccharomyces cerevisiae*. *Gene* 114:51–58.

63. Purvis I J, Chotai D, Dykes C W, Lubahn D B, French F S, Wilson E M, Hobden A N (1991). An androgen-inducible expression system for *Saccharomyces cerevisiae*. *Gene* 106:35–42.

64. Romanos M A, Makoff A J, Fairweather N F, Beesley K M, Slater D E, Rayment F B, Payne M M, Clare J J (1991). Expression of tetanus toxin fragment C in yeast: Gene

synthesis is required to eliminate fortuitous polyadenylation sites in AT-rich DNA. *Nucl. Ac. Res.* 19:1461–1467.

65. Fleer R, Chen X J, Amellal N, Yeh P, Fournier A, Guinet F, Gault N, Faucher D, Folliard F, Fukuhara H, Mayaux J F (1991a). High-level secretion of correctly processed recombinant human interleukin-1β in *Kluyveromyces lactis. Gene* 107:285–295.

66. Fleer R, Yeh P, Amellal N, Maury I, Fournier A, Bacchetta F, Baduel P, Jung G, L'Hôte H, Becquart J (1991b). Stable multicopy vectors for high level secretion of recombinant human serum albumin by *Kluyveromyces lactis. Bio/technol.* 9:968–975.

67. Jeenes D J, Mackenzie D A, Roberts I N, Archer D B (1991). Heterologous protein production by filamentous fungi. *Biotechnol. Genet. Eng. Rev.* 90:327–367.

68. Radzio R, Kück U (1997). Synthesis of biotechnologically relevant heterologous proteins in filamentous fungi. *Proc. Biochem.* 32:529–539.

69. Gouka R J, Punt P J, van den Hondel C (1997). Efficient production of secreted proteins by *Aspergillus*: Progress limitations and prospects. *Appl. Microbiol. Biotechnol.* 47:1–11.

70. Conesa A, Punt P J, van Luijk N, van den Hondel C (2001). The secretion pathway in filamentous fungi: A biotechnological view. *Fungal Genet. Biol.* 33:155–171.

71. Punt P J, van Biezen N, Conesa A, Albers A, Mangnus J, van den Hondel C (2002). Filamentous fungi as cell factories for heterologous protein production. *Trends Biotechnol.* 20:200–206.

72. Reiter M, Blüml G (1994). Large-scale mammalian cell culture. *Curr. Opin. Biotechnol.* 5:175–179.

73. Ikonomou L, Schneider Y J, Agathos S N (2003). Insect culture for industrial production of recombinant proteins. *Appl. Microbiol. Biotechnol.* 62:1–20.

74. Birch J R, Froud S J (1994). Mammalian cell culture system for recombinant protein production. *Biologicals* 22:127–133.

75. McCarroll L, King L A (1997). Stable insect cultures for recombinant protein production. *Curr. Opin. Biotechnol.* 8:590–594.

76. Betenbaugh M J, Ailor E, Whiteley E, Hinderliter P, Hsu T A (1996). Chaperone and foldase coexpression in the baculovirus-insect cell expression system. *Cytotechnology* 20:149–159.

77. Gervais A, Hammel Y A, Pelloux S, Lepage P, Baer G, Carte N, Sorokine O, Strub J M, Koerner R, Leize E, Van Dorsselaer A (2003). Glycosylation of human recombinant gonadotropins: Characterization and batch-to-batch consistency. *Glycobiology* 13:179–189.

78. Tate C G, Haase J, Baker C, Boorsma M, Magnani F, Vallis Y, Williams D C (2003). Comparison of seven different heterologous protein expression systems for the production of the serotonin transporter. *Biochim. Biophys. Acta.* 1610:141–153.

79. Grampp G E, Sambanis A, Stephanopoulos G N (1992). Use of regulated secretion in protein production from animal cells: An overview. *Adv. Biochem. Engin. Biotechnol.* 46:35–62.

80. Wurm F, Bernard A (1999). Large-scale transient expression in mammalian cells for recombinant protein production. *Curr. Opin. Biotechnol.* 10:156–159.

81. Hesse F, Wagner R (2000). Developments and improvements in the manufacturing of human therapeutics with mammalian cell cultures. *Trends Biotechnol.* 18:173–180.

82. Vapnek P (1991). Choosing a production system for recombinant proteins. *Biol. Rec. Microorg. Anim. Cells* 34:1–14.

83. Brooks S A (2004). Appropriate glycosylation of recombinant proteins for human use: implications of choice of expressions system. *Mol. Biotechnol.* 28:241–255.

84. Sodoyer R (2004). Expression systems for the production of recombinant pharmaceuticals. *BioDrugs* 18:51–62.

85. Sutherland A, Auclair K, Vederas J C (2001). Recent advances in the biosynthetic studies of lovastatin. *Curr. Opin. Drug Discov. Devel.* 4:229–236.

86. Schmidt F R (2004b). The challenge of multidrug resistance: Actual strategies in the development of novel antibacterials. *Appl. Microbiol. Biotechnol.* 63:335–343.

87. Parekh S, Vinci V A, Strobel R J (2000). Improvement of microbial strains and fermentation processes. *Appl. Microbiol. Biotechnol.* 54:287–301.

88. Penalva M A, Rowlands R T, Turner G (1998). The optimization of penicillin biosynthesis in fungi. *TIBTECH* 16:483–489.

89. Rowlands R T (1992). Strain improvement and strain stability. In: Finkelstein D B, Ball C (eds.), *Biotechnology of Filamentous Fungi, Technology and Products*, CRC Press, Boca Raton, FL, pp. 41–64.

90. Rowlands R T (1984). Industrial strain improvement: mutagenesis and random screening procedures. *Enzyme Microb. Technol.* 6:3–10.

91. Lal R, Khanna R, Kaur H, Khanna M, Dhingra N, Lal S, Gartemann K H, Eichenlaub R, Ghosh P K (1996). Engineering antibiotic producers to overcome the limitations of classical strain improvement programs. *Crit. Rev. Microbiol.* 22:201–255.

92. Brakhage A A, Sprote P, Al-Abdallah Q, Gehrke A, Plattner H, Tuncher A (2004). Regulation of penicillin biosynthesis in filamentous fungi. *Adv. Biochem. Eng. Biotechnol.* 88:45–90.

93. Schmitt E K, Hoff B, Kück U (2004). Regulation of cephalosporin biosynthesis. *Adv. Biochem. Eng. Biotechnol.* 88:1–43.

94. Brakhage A A (1998). Molecular Regulation of β-lactam biosynthesis in filamentous fungi. *Microbiol. Mol. Biol. Rev.* 62:547–585.

95. Martin J F (1998). New aspects of genes and enzymes for β-lactam antibiotic synthesis. *Appl. Microbiol. Biotechnol.* 50:1–15.

96. Skatrud P L (1992). Genetic engineering of β-lactam antibiotic biosynthetic pathways in filamentous fungi. *TIBTECH* 10:324–329.

97. Brakhage A A, Turner G (1995). Biotechnical genetics of antibiotic synthesis. In: Kück U (ed.), *The Mycota II, Genetics and Biotechnology*, Springer, Berlin, pp. 263–285.

98. Schmidt F R (2002). Beta-Lactam Antibiotics: Aspects of Manufacture and Therapy. In Osiewacz H D (ed.), *The Mycota X; Industrial Applications*, Springer, Berlin, pp. 69–91.

99. Zhang J, Wolfe S, Demain A L (1988). Phosphate repressible and inhibitible β-lactam synthetases in *Cephalosporium acremonium* strain C-10. *Appl. Microbiol. Biotechnol.* 29:242–247.

100. Menne S, Walz M, Kück U (1994). Expression studies with the bidirectional pcbAB-pcbC promotor region from *Acremonium chrysogenum* using reporter gene fusions. *Appl. Microbiol. Biotechnol.* 42:57–66.

101. Guiterrez S, Velasco J, Fernandez F J, Martin J F (1992). The cefG gene of *Cephalosporium acremonium* is linked to the *cefEF* gene and encodes a deacetylcephalosporin C acetyltransferase closely related to homoserin O-acetyltransferase. *J. Bacteriol.* 174:3056–3064.

102. Smith D J, Bull J H, Edwards J, Turner G (1989). Amplification of the isopenicillin N synthase gene in a strain of *Penicillium chrysogenum* producing high levels of penicillin. *Mol. Gen. Genet.* 216:492–497.

103. Smith D J, Burnham M R K, Bull J H, Hodgson J E, Ward J M, Browne P, Brown J, Barton B, Earl A J, Turner G (1990). β-lactam antibiotic biosynthetic genes have been conserved in clusters in prokaryotes and eukaryotes. *EMBO J.* 9:741–747.

104. Barredo J L, Diez B, Alvarez E, Martin J F (1989). Large amplification of 35-kb DNA fragment carrying two penicillin biosynthetic genes in high penicillin producing strains of *P. chrysogenum. Curr. Genet.* 16:453–459.

105. Fierro F, Barredo J L, Diez B, Guiterrez S, Fernandez J, Martin J F (1995). The penicillin cluster is amplified in tandem repeats linked by conserved hexanucleotide sequences. *Proc. Natl. Acad. Sci. USA* 92:6200–6204.

106. Newbert R W, Barton B, Greaves P, Harper J, Turner G (1997). Analysis of a commercially improved *Penicillium chrysogenum* strain series: involvement of recombinogenic regions in amplification and deletion of the penicillin biosynthesis gene cluster. *J. Ind. Microbiol. Biotechnol.* 19:18–27.

107. Brakhage A A, Browne P, Turner G (1994). Analysis of penicillin biosynthesis and the expression of penicillin biosynthesis genes of *Aspergillus nidulans* by targeted disruption of the acvA gene. *Mol. Gen. Genet.* 242:57–64.

108. Kück U, Walz M, Mohr G, Mracek M (1989). The 5′-sequence of the isopenicillin N-synthetase gene (pcbC) from *Cephalosporium acremonium* directs the expression of the prokaryotic hygromycin B phosphotransferase gene (hph) in *Aspergillus niger. Appl. Microbiol. Biotechnol.* 31:358–365.

109. Walz M, Kück U (1991). Polymorphic karyotypes in related *Acremonium* strains. *Curr. Genet.* 19:73–76.

110. Skatrud P L, Queener S W (1989). An electrophoretic molecular karyotype for an industrial strain of *Cephalosporium acremonium. Gene* 79:331–338.

111. Smith A W, Collis K, Ramsden M, Fox H M, Peberdy J F (1991). Chromosome rearrangements in improved cephalosporin C producing strains of *Acremoinium chrysogenum. Curr. Genet.* 18:235–237.

112. DeModena J A, Gutierrez S, Velasco J, Fernandez F J, Fachini R A, Galazzo J L, Hughes D E, Martin J E (1993). The production of cephalosporin C by *Acremonium chrysogenum* is improved by the intracellular expression of a bacterial hemoglobin. *Bio/Technology* 11:926–929.

113. Mathison L, Soliday C H, Stepan T, Aldrich T, Rambosek J (1993). Cloning, characterization, and use in strain improvement of the *Cephalosporium acremonium* gene cefG encoding acetyl transferase. *Curr. Genet.* 23:33–41.

114. Schmidt F R (2005a). About the nature of RNA interference. *Appl. Microbiol. Biotechnol.* 67:429–435.

115. Hannon G J, Rossi J J (2004). Unlocking the potential of the human genome with RNA interference. *Nature* 431:371–378.

116. Wassenegger M (2002). Gene silencing-based disease resistance. *Transgenic. Res.* 11:639–653.

117. Schmidt F R, Davis N D, Diener U L, Lemke P A (1983). Cycloheximide induction of aflatoxin synthesis in a nontoxigenic strain of *Aspergillus flavus. Bio/technology (Nature)* 1:794–795.

118. Schmidt F R, Lemke P A, Esser K (1986). Viral influences on aflatoxin formation by *Aspergillus flavus. Appl. Microbiol. Biotechnol.* 24:248–252.

119. Schmidt F R (2004c). RNA interference detected 20 years ago? *Nat. Biotechnol.* 22:267–268.

120. Pandit N N, Russo V E (1992). Reversible inactivation of a foreign gene, hph, during the asexual cycle in *Neurospora crassa* transformants. *Mol. Gen. Genet.* 234:412–422.

121. Romano N, Macino G (1992). Quelling: transient inactivation of gene expression in *Neurospora crassa* by transformation with homologous sequences. *Mol. Microbiol.* 6:3343–3353.

122. Pickford A S, Catalanotto C, Cogoni C, Macino G (2002). Quelling in *Neurospora crassa. Adv. Genet.* 46:277–303.

123. Liras P, Rodriguez-Garcia A, Martin J F (1998). Evolution of the clusters of genes for beta-lactam antibiotics: A model for evolutive combinatorial assembly of new beta-lactams. *Int. Microbiol.* 1:271–278.

124. Nielsen J (2001). Metabolic engineering. *Appl. Microbiol. Biotechnol.* 55:263–285.

125. Weber T, Welzel K, Pelzer S, Vente A, Wohhleben W (2003). Exploiting the genetic potential of polyketide producing streptomycetes. *J. Biotechnol.* 106:221–232.

126. Kantola J, Kunnari T, Mantsala P, Ylihonkoa K (2003). Expanding the scope of aromatic polyketides by combinatorial biosynthesis. *Comb. Chem. High Throughput Screen* 6:501–512.

127. Hutchinson C R, Fujii I (1995). Polyketide synthase gene manipulation: A structure-function approach in engineering novel antibiotics. *Annu. Rev. Microbiol.* 49:201.

128. Katz L, Donadio S (1993). Polyketide synthesis: prospects for hybrid antibiotics. *Annu. Rev. Microbiol.* 47:875–912.

129. Floss H G (2001). Antibiotic synthesis: from natural to unnatural products. *J. Ind. Microbiol. Biotechnol.* 27:183–194.

130. Pfeifer B A, Khosla C (2001). Biosynthesis of polyketides in heterologous hosts. *Microbiol. Mol. Biol. Rev.* 3:106–118.

131. Staunton J, Wilkinson B (2001). Combinatorial biosynthesis of polyketides and non-ribosomal peptides. *Curr. Opin. Chem. Biol.* 5:159–164.

132. McDaniel R, Katz L (2001). Genetic engineering of novel macrolide antibiotics. In: Lohner K (ed.), *Development of Novel Antimicrobial Agents: Emerging Strategies.* Horizon Scientific Press Norwich, U.K., pp. 45–60.

133. Du L, Shen B (2001). Biosynthesis of hybrid peptide–polyketide natural products. *Curr. Opin. Drug. Discov. Dev.* 4:215–228.

134. Du L, Sanchez C, Shen B (2001). Hybrid peptide—polyketide natural products: Biosynthesis and prospects towards engineering novel molecules. *Metab. Eng.* 3:78–95.

135. Kleinkauf H, von Döhren H (1995). Linking peptide and polyketide biosynthesis. *J. Antibiot.* 48:563–567.

136. Cane D E, Walsh C T (1999). The parallel and convergent universes of polyketide synthases and non-ribosomal peptide synthases. *Chem. Biol.* 6:R319–325.

137. Marahiel M A, Stachelhaus T, Mootz H D (1997). Modular peptide synthases involved in nonribosomal peptide synthesis. *Chem. Rev.* 97:2651–2673.

138. Mankelow D P, Neilan B A (2000). Non-ribosomal peptide antibiotics. *Expert Opin. Ther. Pat.* 10:1583–1591.

139. Trauger J W, Walsh C T (2000). Heterologous expression in *Escherichia coli* of the first module of nonribosomal peptide synthase for chloroeremomycin, a vancomycin-type glycopeptide antibiotic. *Proc. Natl. Acad.* 97:3112–3117.

140. Kaneda T (1977). Fatty acids of the genus Bacillus: an example of branched chain preference. *Bacteriol. Rev.* 41:391–418.

141. Aretz W, Meiwes J, Seibert G, Vobis G, Wink J (2000). Friulimicins: Novel lipopetide antibiotics with peptidoglycan synthesis inhibiting activity from *Actinoplanes friulensis* sp. Nov. I. Taxonomic studies of the producing microorganism and fermentation. *J. Antibiotics* 53:807–815.

142. Vertesy L, Ehlers E, Kogler H, Kurz M, Meiwes J, Seibert G, Vogel M, Hammann P (2000). Friulimicins: Novel lipopetide antibiotics with peptidoglycan synthesis inhibiting activity from *Actinoplanes friulensis* sp. nov. II. Isolation and structural characterization. *J. Antibiotics* 53:816–827.

143. Vertesy L, Aretz W, Decker H, Ehlers E, Kurz M, Schmidt F R, Knauf M, Kogler H (1999). Lipopetide calcium salts: Process for their preparation and their use. *EP* 1068223.

144. Boeck L D, Wetzel R W (1990). A54145, A new lipopeptide antibiotic complex: factor control through precursor directed biosynthesis. *J. Antibiotics* 43:607–615.

145. Counter F T, Allen N E, Fukuda D S, Hobbs J N, Ott J, Ensminger P W, Mynderse J S, Preston D A, Wu C Y E (1990). A54145 A new lipopetide antibiotic complex: microbiological evaluation. *J. Antibiotics* 43:616–622.

146. Zmijewski M J, Briggs B, Occolowitz J (1986). Role of branched chain fatty acid precursosrs in regulating factor profile in the biosynthesis of A21978 complex. *J. Antibiot* 39:1483–1485.

147. Stephenson J (2001). Researchers describe latest strategies to combat antibiotic resistant microbes. *JAMA* 285:2317–2318.

148. Tally F P, DeBriun F (2000). Development of daptomycin for Gram-positive infections. *J. Antimicrob. Chemo.* 46:523–526.

149. Hancock R E W, Patrzykat A (2002). Clinical development of cationic antimicrobial peptides: From natural to novel antibiotics. *Curr. Drug Targets Infect. Disord.* 2:79–83.

150. Mor A (2000). Peptide–based antibiotics: a potential answer to raging antimicrobial resistance. *Drug Dev. Res.* 50:440–447.

151. Larsson G, Törnkvist M, Wernersson E S, Trägard C, Noorman H, Enfors S O (1996). Substrate gradients in bioreactors: origin and consequences. *Bioproc. Engin.* 14:281–289.

152. Bylund F, Collet E, Enfors E O, Larsson G (1998). Substrate gradient formation in the large scale lowers cell yield and increases byproduct formation. *Bioproc. Eng.* 18:171–180.

153. Bylund F, Guillard F, Enfors S O, Trägardh C, Larsson G (1999). Scale down of recombinant production: a comparative study of scaling performance. *Bioproc. Engin.* 20:377–289.

154. Bylund F, Castan A, Mikkola R, Veide A, Larsson G (2000). Influence of scale-up on the quality of recombinant human growth hormone. *Biotechnol. Bioengin.* 69:119–128.

155. Enfors S O, Jahic M, Rozkov A, Xu B, Hecker M, Jürgen B, Krüger B, Schweder T, Hamer G, O'Beirne D, Noisommit-Rizzi N, Reuss M, Boone L, Hewitt C, McFarlane C, Nienow A, Kovacs T, Trägardh C, Fuchs L, Revstedt J, Friberg P C, Hjertager B, Blomsten G, Skogman H, Hjort S, Hoeks F, Lin H J, Neubauer P, van der Lans R, Luyben K, Vrabel P, Manelius A (2001). Physiological responses to mixing in large scale bioreactors. *J. Biotechnol.* 85:175–185.

156. Xu B, Jahic M, Bomsten G, Enfors S O (1999). Glucose overflow metabolism and mixed acid fermentation in aerobic large-scale fed-batch processes with *Echerichia coli. Appl. Microbiol. Biotechnol.* 51:564–571.

157. Castan A, Enfors S O (2001). Formate accumulation due to DNA release in aerobic cultivations of *Escherichia coli. Biotechnol. Bioengin.* 77:324–328.

158. Kwint K, Nachin L, Diez A, Nyström T (2003). The bacterial universal stress protein: Function and regulation. *Curr. Opin. Microbiol.* 6:140–145.

159. Mogk A, Tomoyasu T, Goloubinoff P, Rudiger S, Roder D, Langen H, Bukau B (1999). Identification of thermolabile *Escherichia coli* proteins: Prevention and reversion of aggregation by DnaK and ClpB. *EMBO J.* 18:6934–6949.

160. Umakoshi H, Kuboi R, Komaswa I, Tsuchido T, Matsumura Y (1998). Heat-induced translocation of cytoplasmic beta-galactosidase across inner membrane of *Escherichia coli. Biotechnol. Prog.* 14:210–217.

161. Schweder T, Krüger E, Xu B, Jürgen B, Blomsten G, Enfors S O, Hecker M (1999). Monitoring genes that respond to process-related stress in large-scale bioprocesses. *Biotechnol. Bioengin.* 65:151–159.

162. Fenton D, Lai P H, Lu H, Mann M, Tsai L (1997). Control of norleucine incorporation into recombinant proteins. U.S. Patent 5599690.

162a. Martinez-Force E, Benitez T (1992). Changes in yeast amino acid pool with respiratory versus fermentative metabolism. *Biotechnol. Bioengi.n* 40:843–849.

162b. Hondorp ER, Matthews RG (2004). Oxidative stress inactivates cobalamin-independent methionine synthase (MetE) in Escherichia coli. *PLoS Biology* 2:1138–1158.

162c. Hoffmann F, Rinas U (2004). Stress induced by recombinant protein production in *Escherichia coli. Adv. Biochem. Eng. Biotechnol.* 89:73–92.

163. Weickert M J, Apostol I (1998). High-fidelity translation of recombinant human hemoglobin in *Escherichia coli. Appl. Env. Microbial.* 64:1589–1593.

164. Muramatsu R, Negishi T, Mimoto T, Miura A, Miosawa S, Hayashi H (2002). Existence of β-methylnorleucine in recombinant hirudin produced by *Escherichia coli. J. Biotechnol.* 93:131–142.

165. Apostol I, Levine J, Lippincott J, Leach J, Hess E, Glascock C B, Weickert M J, Blackmore R (1997). Incorporation of norvaline at leucine positions in recombinant human hemoglobin expressed in *Escherichia coli. J. Biol. Chem.* 272:28980–28988.

166. Kiick K L, Weberskirch R, Tirell D A (2001). Identification of an expanded set of translationally active methionine analogues in *E. coli. FEBS Lett.* 502:25–30.

167. Friehs K (2003). Plasmid copy number and plasmid stability. *Adv. Biochem. Engin. Biotechnol.* 86:47–82.

168. Thiry M, Cingolani D (2002). Optimizing scale-up fermentation processes. *Trends Biotechnol.* 20:103–105.

169. Lin H Y, Neubauer P (2000). Influence of controlled glucose oscillations on a fed-batch process of recombinant *Escherichia coli. J. Biotechnol.* 79:27–37.

170. Neubauer P, Lin H Y, Mathiszik B (2003). Metabolic load of recombinant protein production: Inhibition of cellular capacities for glucose uptake and respiration after induction of a heterologous gene in *Escherichia coli. Biotechnol. Bioengin.* 83:53–64.

171. Ansorge M B, Kula M R (2000). Production of recombinant L-leucine dehydrogenase from Bacillus cereus in pilot scale using the runaway replication system *E. coli* (pIET98). *Biotechnol. Bioengin.* 68:557–562.

172. Nordstrom K, Uhlin B E (1992). Runaway-replication plasmids as tools to produce large quantities of proteins from cloned genes in bacteria. *Biotechnology* 6:661–666.

173. van de Merbel N C (1997). The use of ultra-filtration and column liquid chromatography for an on-line fermentation monitoring. *Trends. Anal. Chem.* 16:162–173.

174. Buchholz A, Takors R, Wandrey C (2001). Quantification of intracellular metabolites in *Escherichia coli* K12 using liquid chromatographic electrospray ionization tandem mass spectrometric techniques. *Anal. Biochem.* 295:129–137.

175. Matz G, Lennemann F (1996). On-line monitoring of biotechnological processes by gas chromatographic—mass spectrometric analysis of fermentation suspensions. *J. Chromatogr.* 750:141–149.

176. Klyushnichenko V (2004). Capillary electrophoresis in the analysis and monitoring of biotechnological processes. *Methods Mol. Biol.* 276:77–120.

177. Nishi H, Tsumagari N, Kakimoto T, Terabe S (1989). Separation of β-lactam antibiotics by micellar electro-kinezic chromatography. *J. Chromatogr.* 477:259–270.

178. Nishi H, Fukuyama T, Matsuo M (1990). Separation and determination of aspoxicillin in human plasma by micellar electrokinetic chromatography with direct sample injection. *J. Chromatogr.* 515:245–255.

179. Gabig-Ciminska M, Holmgren A, Andresen H, Bundvik Barken K, Wumpelmann M, Albers J, Hintsche R, Breitenstein A, Neubauer P, Los M, Czyz A, Wegrzyn G, Silfversparre G, Juregn B, Schweder T, Enfors S O (2004). Electric chips for rapid detection and quantification of nucleic acids. *Biosens. Bioelectron.* 15:537–546.

180. Hewitt C J, Nebe-von Caron G (2001). An industrial application of multiparameter flow cytometry: Assessment of cell physiological state and its application to the study of microbial fermentations. *Cytometry* 44:179–187.

181. Hewitt C J, Nebe-Von-Caron G (2004). The application of multi-parameter flow cytometry to monitor individual microbial cell physiological state. *Adv. Biochem. Eng. Biotechnol.* 89:197–223.

182. Hewitt C J, Nebe-von Caron G, Axelsson B, McFarlane C M, Nienow A W (2000). Studies related to the Scale-up of high-cell-density *E. coli* fed-batch fermentations using multiparameter flow cytometry: Effect of a changing microenvironment with respect to glucose and dissolved oxygen concentration. *Biotechnol. Bioengin.* 70:381–390.

183. Kell D B, Ryder H M, Kaprelyants A S, Westerhoff H V (1991). Quantifying heterogenity—flow cytometry of bacterial cultures. *Anton. van Leeuwen. Int. J. Gen. Mol. Microbiol.* 60:145–158.

184. Harms P, Kostov Y, Rao G (2002). Bioprocess monitoring. *Curr. Opin. Biotechnol.* 13:124–127.

185. Schaefer U, Boos W, Takors R, Weuster-Bootz D (1999). Automated sample device for monitoring intracellular metabolite dynamics. *Anal. Biochem.* 270:88–96.

186. Rocha I, Ferreira E C (2002). On-line simultaneous monitoring of glucose and acetate with FIA during high cell density fermentation of recombinant *E. coli*. *Bioelectrochem.* 56:127–129.

187. Tothill I E, Newman J D, White S F, Turner A P F (1997). Monitoring of the glucose concentration during microbial fermentation using a novel mass-producible biosensor suitable for on-line use. *Enzyme Microb. Technol.* 20:590–596.

188. Schügerl K, Brandes T, Dullau K, Holzhauer-Rieger K, Hotop S, Hübner U, Wu X, Zhou W (1991). Fermentation monitoring and control by online flow injection and liquid chromatography. *Anal. Chim. Acta.* 249:87–100.

189. Huang Y L, Foellmer T J, Ang K C, Khoo S B, Yap M G S (1995). Characterization and application of an on-line flow injection analysis/wall-jet electrode system for glucose monitoring during fermentation. *Anal. Chim. Acta.* 317:223–232.

190. Vaidyanathan S, Macaloney G, Vaughan J, McNeil B, Harvey L M (1999). Monitoring of submerged bioprocesses. *Crit. Rev. Biotechnol.* 19:277–316.

191. Arndt M, Hitzmann B (2004). Kalman filter based glucose control at small set points during fed-batch cultivation of Saccharomyces cerevisiae. *Biotechnol. Prog.* 1:377–383.

192. Kleist S, Miksch G, Hitzmann B, Arndt M, Friehs K, Flaschel E (2003). Optimization of the extracellular production of bacterial phytase with *Escherichia coli* by using different fed-batch fermentation strategies. *Appl. Microbiol. Biotechnol.* 61:456–462.

193. Spichiger-Keller U E (1997). Ion- and substrate specific optode membranes and optical detection modes. *Sensors and Actuators, B: Chemical* B38:68–77.

194. Wolfbeis O S (2002). Fiber-optic chemical sensors and biosensors. *Anal. Chem.* 74:2663–2678.

195. Griffiths D, Hall G (1993). Biosensors—what real progress is being made. *Trends Biotechnol.* 11:122–130.

196. Boehl D, Solle D, Hitzmann B, Scheper T (2003). Chemometric modeling with two-dimensional fluorescence data for Claviceps purpurea bioprocess characterization. *J. Biotechnol.* 105:179–188.

197. Solle D, Geissler D, Stark E, Scheper T, Hitzmann B (2003). Chemometric modeling based on 2D-fluorescence spectra without a calibration measurement. *Bioinformatics* 19:173–177.

198. Stark E, Hitzmann B, Schügerl K, Scheper T, Fuchs C, Koster D, Märkl H (2002). In-situ-fluorescence probes: a useful tool for non-invasive bioprocess monitoring. *Adv. Biochem. Eng. Biotechnol.* 74:21–38.

199. Hall J W, McNeil B, Rollins M J, Draper I, Thompson B G, Macaloney G (1996). Near-infrared spectroscopic determination of acetate, ammonium, biomass and glycerol in an industrial *Escherichia coli* fermentation. *Appl. Spectroscopy* 50:102–108.

200. Vaccari G, Dosi E, Campi A L, Gonzalez-Vara A, Matteuzi D, Mantovani G (1994). A near infrared spectroscopy technique for the control of fermentation processes: an application to lactic acid fermentation. *Biotechnol. Bioengin.* 43:913–917.

201. Macaloney G, Hall J W, Rollins M J, Draper I, Thompson B G, McNeil B (1994). Monitoring biomass and glycerol in an *Escherichia coli* fermentation using near-infrared spectroscopy. *Biotechnol. Techn.* 8:281–286.

202. Hammond S V (1992). NIR analysis of antibiotic fermentations. In: Murray I, Cowe I A (eds.), *Making Light Work: Advances in Near Infrared Spectroscopy*, vol 1. VCH Publishers, New York, pp. 584–589.

203. Hammond S V, Brookes I K (1992). Near infrared spectroscopy—a powerful technique for at-line and on-line analysis of fermentations. In: Ladisch M R, Bose A (eds.), *Harnessing Biotechnology for the 21st Century: Proceedings of the 9th International Symposium and Exhibition*, Am. Chem. Soc., Washington, DC, pp. 325–333.

204. Pollard D, Buccino R, Connors N, Kirschner T, Olewinski R, Saini K, Slamon P (2001). Real-time analyte monitoring of a fungal fermentation, at a pilot scale, using in mid-infrared spectroscopy. *Bioproc. Biosyst. Engin.* 24:13–24.

205. Dickinson T A, White J, Kauer J S, Walt D R (1998). Current trends in artificial nose technology. *TIBTECH* 16:250–258.

206. Turner C, Rudnitskaya A, Legin A (2003). Monitoring batch fermentations with an electronic tongue. *J. Biotechnol.* 103:87–91.

207. Moser I, Jobst G, Urban G A (2002). Biosensor array for simultaneous measurement of glucose, lactate, glutamate, and glutamine. *Biosens. Bioelectron.* 17:297–302.

208. Collins G E, Wu P, Lu Q, Ramsey J D, Bromund R H (2004). Compact, high voltage power supply for the lab-on-chip. *Lab. Chip.* 4:408–411.

209. Neils C, Tyree Z, Finlayson B, Folch A (2004). Combinatorial mixing of microfluidic streams. *Lab. Chip.* 4:342–350.

210. Farinas J, Chow A W, Wada H G (2001). A microfluidic device for measuring cellular membrane potential. *Anal. Biochem.* 295:138–142.

211. Ferreira L S, de Souza M B, Folly R O M (2001). Development of an alcohol fermentation control system based on biosensor measurements interpreted by neural networks. *Sens. Act B: Chem.* 75:166–171.

212. Chen V C P, Rollins D K (2000). Issues regarding artificial neural network modeling for reactors and fermenters. *Bioproc. Biosyst. Engin.* 22:85–93.

213. Trelea I C, Titica M, Landaud S, Latrille E, Corrieu G, Cheruy A (2001). Predictive modeling of brewing fermentation: From knowledge-based to black box models. *Math Comp. Sim.* 56:405–424.

214. Lin H Y, Mathiszik B, Xu B, Enfors S O, Neubauer P (2001). Determinantion of the maximum specific uptake capacities for glucose and oxygen in glucose-limited fed-batch cultivations of *Escherichia coli*. *Biotechnol. Bioengin.* 73:347–357.

215. Lee S Y (1996). High-cell density culture of *Escherichia coli*. *Trends Biotechnol.* 14:98–105.

216. Kim C H, Rao K J, Youn D J, Rhee S K (2003). Scale-up of recombinant hirudin production from *Saccharomyces cerevisiae*. *Biotechnol. Bioproc. Engin.* 8:303–305.

217. Wong I, Hernandez A, Garcia M A, Segura R, Rodriguez I (2002). Fermentation scale-up for recombinant K99 antigen production cloned in *Escherichia coli* MC1061. *Proc. Biochem.* 37:1195–1199.

218. Yim S C, Jeong K J, Chang H N, Lee S Y (2001). High-level secretory production of human granulocyte-colony stimulating factor by fed-batch culture of recombiant *E. coli*. *Bioproc. Biosyst. Engin.* 24:249–254.

219. Dantigny P, Ninow J L, Lakrori M (1991). A new control strategy for yeast production based on the L/A approach. *Appl. Microbiol. Biotechnol.* 36:352–357.

220. Cant S (2002). High-performance computing in computational fluid dynamics: Progress and challenges. *Philosoph. Transact.: Mathem. Phys. Engin. Sci.* 360:1211–1225.

221. Schmalzriedt S, Jenne M, Mauch K, Reuss M (2003). Integration of physiology and fluid dynamics. *Adv. Biochem. Eng. Biotechnol.* 80:19–68.

222. Alves S S, Vasconcelos J M T (1996). Optimization of agitation and aeration in fermenters. *Bioproc. Biosyst. Engin.* 14:119–123.

223. Oniscu C, Galaction A I, Cascaval D, Ungureanu F (2001). Modeling of mixing in stirred bioreactors 1. Mixing time for non-aerated simulated broths. *Roum. Biotechnol. Lett.* 6:119–129.

224. Diaz A, Acevedo F (1999). Scale-up strategy for bioreactors with Newtonian and non-Newtonian broths. *Bioproc. Engin.* 21:21–23.

225. Russel A B, Thomas C R, Lilly M D (1995). Oxygen transfer measurements during yeast fermentations in a pilot scale airlift fermenter. *Bioproc. Biosyst. Engin.* 12:71–79.

226. Türker M (2004). Development of biocalorimetry as a technique for process monitoring and control in technical scale fermentations. *Thermochimica Acta.*, In Press.

227. Anderson R K I, Jayaraman K, Voisard D, Marison I W, von Stockar U (2002). Heat flux as an on-line indicator of metabolic activity in pilot scale bioreactor during the production of *Bacillus thuringiensis* var *galleriae*-based pesticides. *Thermochimica Acta.* 386:127–138.

228. Hubbard D W (1987). Scaleup strategies for bioreactors containing non-newtonian broths. *Ann. NY Acad. Sci.* 506:600–607.

229. Junker B H, Stanik M, Barna C, Salmon P, Buckland B C (1998). Influence of impeller type on mass transfer in fermentation vessels. *Bioproc. Biosyst. Engin.* 19:403–413.

230. Wang D I C, Cooney C L (1997). Translation of laboratory, pilot, and plant scale data. In: Wang D I C, et al. (eds.), *Fermentation and enzyme technology*. Wiley, New York, pp. 194–211.

231. Yawalkar A A, Heesink A B M, Versteeg G F, Vishwas G P (2002). Gas-liquid mass transfer coefficient in stirred tank reactors. *Can. J. Chem. Engin.* 80:840–848.

232. Flores E R, Perez F, de la Torre M (1997). Scale-up of *Bacillus thuringiensis* fermentation based on oxygen transfer. *J. Ferm. Bioengin.* 83:561–564.

233. Riesenberg D, Menzel K, Schulz V, Schumann K, Veith G, Zuber G, Knorre W A (1990). High cell density fermentation of recombinant *Escherichia coli* expressing human interferon alpha 1. *Appl. Microbiol. Biotechnol.* 34:77–82.

234. Berkholz R, Guthke R (2001). Model based sequential experimental design for bioprocess optimization—an overview. *Focus Biotechnol.* 4:129–141.

235. Berkholz R, Rohlig D, Guthke R (2000). Data and knowledge based experimental design for fermentation process optimization. *Enzyme Microb. Technol.* 27:784–788.

236. Liden G (2002). Understanding the bioreactor. *Bioproc. Biosyst. Engin.* 24:273–279.

237. Drossard J (2004). Downstream processing of plant-derived recombinant therapeutic proteins. In: Fischer R, Schillberg S (eds.), *Molecular Farming.* Wiley-VCH, Weinheim, Germany.

238. Roque A C, Lowe C R, Taipa M A (2004). Antibodies and genetically engineered related molceules: Production and purification. *Biotechnol. Prog.* 20:639–554.

239. Jungbauer A, Hahn R (2004). Enegineering protein A affinity chromatography. *Curr. Opin. Drug Discov. Devel.* 7:248–256.

240. Burgess R R, Thompson N E (2002). Advances in gentle immunoaffinity chromatography. *Curr. Opin. Biotechnol.* 13:304–308.

241. Hage D S (1999). Affinity chromatography: A review of clinical applications. *Clin. Chem.* 45:593–615.

242. Murphy M B, Fuller S T, Richardson P M, Doyle S A (2003). An improved method for the in vitro evolution of aptamers and applications in protein detection and purification. *Nucl. Ac. Res.* 31:18–110.

243. Romig T S, Bell C, Drolet D W (1999). Aptamer affinity chromatography: combinatorial chemistry applied to protein purification. *J. Chromatogr. B Biomed. Sci. Appl.* 731:275–284.

244. Hubbuch J, Thommes J, Kula M R (2005). Biochemical engineering aspects of expanded bed adsorption. *Adv. Biochem. Eng. Biotechnol.* 92:101–123.

245. Anspach F B, Curbelo D, Hartmann R, Garke G, Deckwer W D (1999). Expanded bed chromatography in primary protein purification. *J. Chromatogr. A* 865:129–144.

246. Guiochon G (2002). Preparative liquid chromatography. *J. Chromatogr. A* 965:129–161.

247. Cunha T, Aires-Barros R (2002). Large-scale extraction of proteins. *Mol. Biotechnol.* 20:29–40.

248. Krishna S H, Srinivas N D, Raghavarao K S, Karanth N G (2002). Reverse micellar extraction for downstream processing of proteins/enzymes. *Adv. Biochem. Eng. Biotechnol.* 75:119–183.

249. Walter J (2000). Scale-up of downstream processing. *J. Chromatogr. Libr.* 61:765–783.

250. Kadir F, Hamers M (2002). Production and downstream processing of biotech compounds. In: Crommelin D J A, Sindelar R D (eds.), *Pharmaceutical Biotechnology* (2nd ed.). Taylor and Francis Ltd, London, U.K., pp. 53–71.

251. Kalyanpur M (2002). Downstream processing in the biotechnology industry. *Mol. Biotechnol.* 22:87–98.

252. Chisti Y (1998). Strategies in downstream processing. In: Subramanian G (ed.), *Bioseparation and Bioprocessing* (2nd ed.). Wiley-VCH, Weinheim, Germany, pp. 3–30.

253. Desai M A (2002). *Downstream Processing of Proteins.* Humana Press, Totowa, NJ.

254. Bhat S T, et al. (2005). *Chemistry of natural products.* Springer, Berlin, Germany.

255. Verrall M S (1996). Downstream processing of natural products. A practical handbook. Wiley, New York.

256. Gosh A C, Bora M M, Dutta N N (1996). Developments in liquid membrane separation of beta-lactam antibiotics. *Bioseparation* 6:91–105.

257. Gosh A C, Mathur R K, Dutta N N (1997). Extraction and purification of cephalosporin antibiotics. In: Scheper T (ed.), *Advances in Biochemical Engineering/ Biotechnology*, vol 56. Springer-Verlag, Berlin, pp. 111–145.

258. Bandak S I, Tumak M R, Allen B S, Bolzon L D, Preston D A (2000). Assessment of the susceptibility of Streptococcus pneumoniae to cefaclor and loracarbef in 13 countries. *J. Chemother.* 12:299–305.

259. Troxler R, von Grävenitz A, Funke G, Wiedemann B, Stock I (2000). Natural antibiotic susceptibility by *Listeria* species: *L. grayi*, *L. innocua*, *L. ivanovii*, *L. monocytogenes*, *L. seeligeri* and *L. welshimeri* strains. *Clin. Microbiol. Infect.* 6:525–535.

260. Saling P (2005). Eco-efficiency analysis of biotechnological processes. *Appl. Microbiol. Biotechnol.* 68:1–8.

261. Parmar A, Kumar H, Marwaha S S, Kennedy J F (1998). Recent trends in enzymatic conversion of cephalosporin C to 7-aminocephalosporanic acid (7-ACA). *Cri. Rev. Biotechnol.* 18:1–12.

262. Bianchi D, Bortolo R, Golini P, Cesti P (1998). Application of immobilized enzymes in the manufacture of beta-lactam antibiotics. *Chim. In. (Milan)* 80:875–885.

263. Binder R G, Numata K, Lowe D A, Murakami T, Brown J L (1993). Isolation and characterization of a *Pseudomonas* strain producing glutaryl-7-aminocephalosporanic acid acylase. *Appl. Env. Microbiol.* 59:3321–3326.

264. Isogai T (1997). New processes for production of 7-aminocephalosporanic acid from cephalosporin. *Drugs Pharm. Sci.* 82:733–751.

265. Gomez A, Rodriguez M, Ospina S, Zamora R, Merino E, Bolivar F, Ramirez O T, Quintero R, Lopez-Munguia A (1994). Strategies in the design of a penicillin acylase process. In: Galindo R, Ramirez OT (eds.), *Advances in Bioprocess Engineering.* Kluwer, Dodrecht, pp. 29–40.

266. Bruggink A, Roos E C, de Vroom E (1998). Penicillin acylase in the industrial production of beta-lactam antibiotics. *Org. Process Res. Dev.* 2:128–133.

267. Shewale J G, Sudhakaran V K (1997). Penicillin V acylase: Its potential in the production of 6-aminopenicillanic acid. *Enzyme Microb. Technol.* 20:402–410.

268. Stoppok E, Wagner F, Zadrazil F (1981). Identification of a penicillin V acylase processing fungus. *Eur. J. Appl. Microbiol. Biotechnol.* 13:60–61.

269. Schneider W J, Roehr M (1976). Purification and properties of penicillin acylase of *Bovista plumbea. Biochimica. Biophysica. Acta.* 452:177–185.

270. Brandl D, Kleiber W (1972). Verfahren zur Herstellung von 6-Aminopenicillansäure. Austrian Patent no. 297923, Vienna, Austria.

271. Singh K, Sehgal S N, Vezina C (1967). Conversion of penicillin V to 6-aminopenicillanic acid by the use of spores. U.S. Patent no. 3.305.453.

272. Wilms J, de Vroom E (1998). Preparation of β-lactam antibiotics. European Patent no. 0201688.

273. Schmidt F R (2005b). Optimization and scale-up of industrial fermentation processes. *Appl. Microbiol. Biotechnol.* 68:425–435.

1.2

SEQUENCING THE HUMAN GENOME: WAS IT WORTH IT?

Eugene Zabarovsky

Microbiology and Tumor Biology Center, Karolinska Institute, Stockholm, Sweden

Chapter Contents

1.2.1	Summary of the Chapter	54
1.2.2	Definitions of Keywords	54
1.2.3	Introduction. Short History of the Human Genome Project	55
1.2.4	Vectors. Description of the Main Types of Vectors Used in the HGP	57
	1.2.4.1 Lambda-Based Vectors and Main Approaches to Construct Genomic Libraries	57
	1.2.4.2 YACs	61
	1.2.4.3 Bacterial (BAC) and P1-Derived (PAC) Artificial Chromosome Vectors	63
1.2.5	Mapping of the Human Genome	65
1.2.6	Main Approaches to Sequence Human Genome	69
	1.2.6.1 Hierarchical and the Whole Genome Shotgun Sequencing Scheme (WGS)	69
	1.2.6.2 Massively Parallel Sequencing Approaches	71
	1.2.6.3 Slalom Library: A Novel Approach to Genome Mapping and Sequencing	73
1.2.7	Identification of Genes. How Many Genes Are in the Human Genome?	77
	1.2.7.1 Many Approaches Exist for Gene Identification	77
	1.2.7.2 CpG Islands	78
	1.2.7.3 SAGE and CAGE Approaches to Analyze Human Transcriptome and to Estimate the Number of Human Genes	80
1.2.8	Was it Worth to it Sequence Human Genome?	82
	Acknowledgment	83
	References	83

Handbook of Pharmaceutical Biotechnology, Edited by Shayne Cox Gad.
Copyright © 2007 John Wiley & Sons, Inc.

1.2.1 SUMMARY OF THE CHAPTER

An almost complete sequence of the human genome is available now, and progress in the sequencing of other model organisms is impressive. This sequence information provides enormous possibility for identification of disease genes. The first step for sequencing of large genomes usually is the construction of a high-resolution, long-range physical map based on sequence tagged site (STS) markers. Using this map, the contig of overlapping clones is constructed and sequenced. This strategy (hierarchical) was used by the International Human Genome Sequencing Consortium to sequence the human genome. The second approach, the whole genome sequencing (WGS) strategy, was used by researchers from Celera Genomics. This strategy, in principle, doesn't need construction of high-resolution physical map and is based mainly on shotgun sequencing of large and small insert clones.

Unfortunately, both strategies for mapping and sequencing are not appropriate for the challenge of high-throughput comparative genomics. Alternative and complementary strategies need to be developed, and the imperative now is to find cost-effective and convenient methods that allow comparative genomics projects to be performed by a wide range of smaller laboratories. Recently developed novel techniques open new perspectives for nonexpensive and fast sequencing. These novel developments can be grouped into two classes. The first type of improvement is enormous, increasing the speed and throughput of producing sequences. These techniques are based on massively parallel sequencing approaches (MPS and "494" techniques) where billions of individual DNA molecules could be sequenced simultaneously. Another type of improvements is a recently suggested completely different and efficient strategy for simultaneous genome mapping and sequencing. The approach is based on physically oriented, overlapping restriction fragment libraries called slalom libraries. Both techniques allow increasing the speed and decreasing the cost of sequencing 10–100 times. Importantly, they can be combined to increase efficiency even more.

1.2.2 DEFINITIONS OF KEYWORDS

Adaptor—an oligonucleotide that after ligation to a DNA molecule confers to it specific features.

Allele—is one of several alternative forms of a gene (or in the wider sense of any region of the chromosomal DNA) that occupies a given locus of chromosome.

Cap—the chemical modification at the 5′-end of most eukaryotic mRNA molecules.

Centromere—the region of a chromosome where the pairs of chromatids are held together.

Clone contig—a collection of clones (i.e., vectors or cells containing the same recombinant molecule) whose DNA inserts overlap.

Coding RNA—an RNA molecule that codes for a protein (mRNA).

DNA sequencing—establishing the order of nucleotides in a DNA molecule.

Episome—a vector or DNA sequence that can exist as an autonomous extra-chromosome replicating DNA molecule.

Exon—a coding region within a discontinuous gene.

Genetic selection—in cloning, this usually means selection against parental, nonrecombinant molecules in favor of recombinant ones. SupF selection exploits vectors carrying *amber* (nonsense mutation UAG) mutations (*ochre* is a nonsense mutation UAA). These vectors cannot replicate without the supF gene, which must be present either in the host or in the cloned insert. If the insert carries the supF gene, only recombinant phages will be able to replicate in an *Esherichia coli* host without the suppressor gene.

Intron—a noncoding region within a discontinuous gene.

Kilobase pairs (kb)—1000 base pairs.

Ligation—formation of a 3′–5′ phosphodiester bond between nucleotides at the end of the same or two different DNA or RNA molecules.

Marker—a gene or just DNA fragment that was located on a genome map.

Megabase pairs (Mb)—1000 kb.

Polylinker—a short DNA fragment (in the vector) that contains recognition sites for restriction enzymes that can be used for cloning DNA fragments into this vector.

Restriction enzyme—an enzyme that recognizes a specific sequence in DNA and can cut at or near this sequence. In cloning procedures, the most commonly used enzymes produce specific protruding (sticky) ends at the ends of a DNA molecule. Each enzyme produces unique sticky ends. The DNA molecules possessing the same sticky ends can be efficiently joined with the aid of DNA ligase (see **Ligation**).

STS—sequence tagged site. This is a short (200–500 bp) sequenced fragment of genomic DNA that can be specifically amplified using polymerase chain reaction (PCR) and represent or is linked to some kind of marker (i.e., it is mapped to a specific locus on a chromosome).

Shot-gun approach—a sequencing strategy for when a large DNA molecule is randomly broken, cloned, and sequenced. The final sequence of a DNA molecule is assembled by a computer program.

Telomere—the end of a eukaryotic chromosome.

1.2.3 INTRODUCTION. SHORT HISTORY OF THE HUMAN GENOME PROJECT

The Human Genome Program (HGP) was suggested and widely discussed within the scientific community and public press during the mid-1980s and over the last half of that decade [1–4]. In 1985, Dr. Charles DeLisi, then Associate Director for Health and Environmental Research at the Department of Energy (DOE) began to discuss an unprecedented biology project—to sequence the complete human genome. DOE funding began in 1987. The National Institutes of Health (NIH) established the Office of Human Genome Research in September 1988. The Director of this Office (renamed the National Center for Human Genome Research, NCHGR) was Dr. James Watson. The Human Genome Initiative was proposed to the Congress of the United States in 1988 by James Watson, initiating a worldwide coordinated research activity to sequence the DNA of humans and several other test organisms. In the United States, the (DOE) initially, and the NIH soon thereafter, were the main research agencies within the U.S. government responsible for developing and planning the project. By 1988, the two agencies were working together, a relationship that was formalized by the signing of a Memorandum of Understanding to "coordinate research and technical activities related to the human genome." The initial planning process culminated in 1990 with the publication of a joint research plan. The HGP started in 1990 as a 15-year program. One of its main coordinators was Dr. Francis Collins, Director of the National Human Genome Research Institute (NHGRI) at the NIH. Soon, the Human Genome Organization (HUGO) spread to Europe (HUGO Europe) and Japan (HUGO Pacific) to coordinate the research performed in other countries outside of the United States. The HGP was an international research program designed to construct detailed genetic and physical maps of the human genome, to determine the complete nucleotide sequence of human DNA, and to localize the estimated 50,000–100,000 genes within the human genome. The similar analyses on the genomes of several model organisms were also planned to be performed. The scientific products of the HGP were suggested to comprise a resource of detailed information about the structure, organization, and function of human DNA, information that constitutes the basic set of inherited "instructions" for the development and functioning of a human being. Achievement of the ambitious aims of HGP would demand the development of new technologies and necessitate advanced means of disseminating information widely to scientists, physicians, and others in order that the results may be rapidly used for the public good. Important for the success of the HGP was the statement that a fraction of the project money should be devoted to the ethical, legal, and social implications of human genetic research. It was clearly recognized that acquisition and use of such genetic knowledge would have momentous implications for both individuals and society and would pose several policy choices for public and professional deliberation. The HUGO program aims at a detailed understanding of the organization of the human genome, to enable a molecular definition of the basis for normal function as well as genetic disorders affecting human cells and the organism as a whole. An intermediate aim in this program was the construction of a high-resolution physical map for the human genome. At the beginning of this work, with sequencing still slow and expensive, the HGP adopted a "map–first, sequence later strategy." In the early 1990s, two Parisian laboratories, the Centre d'Etude du Polymorphisme Humain and Genethon, had an integral role in mapping of the human genome. The laboratories' driving forces were Drs. Daniel Cohen and Jean Weissenbach. Later the genome project constructed a higher resolution map that was used to sequence and assemble the human genome. In July 1995, a team led by Dr. J. Craig Venter published the first sequence of a free-living organism, *Haemophilis influenzae* (1.8 Mb) [5]. In October 1996, an international consortium publicly released the complete genome sequence of the yeast *Saccharomyces cerevisiae* (12 Mb) [6]. In May 1998, Venter announced a new company named Celera and declared that it will sequence the human genome within 3 years for $300 million. In December 1998, Drs. John Sulston, Robert Waterston, and colleagues completed the genomic sequence of *Caenorhabditis elegans* (100 Mb) [7]. The first complete human chromosome sequence—number 22 (33 Mb)—was published in December 1999 [8], and in June 2000, leaders of the public project and Celera announced completion of a working draft of the human genome sequence. Finally, in February 2001, the HGP consortium published its working draft of the human genome sequence in *Nature* and Celera published its draft in *Science* (2.9 billion bp) [9, 10].

1.2.4   VECTORS. DESCRIPTION OF THE MAIN TYPES OF VECTORS USED IN THE HGP

In this section, common vectors used for everyday cloning and subcloning like plasmid and M13 vectors will not be discussed. Only vectors suitable for cloning relatively large DNA inserts will be mentioned with the exception of lambda phage-based vectors specially designed to clone cDNA.

1.2.4.1   Lambda-Based Vectors and Main Approaches to Construct Genomic Libraries

Numerous modifications have been made to bacteriophage lambda-based vectors to facilitate their handling and to extend their use for new types of biological experiments [11–15]. The application of each vector is usually limited to a specific task: the construction of general genomic libraries that contain all genomic DNA fragments or special genomic libraries that contain only a particular subset of genomic DNA fragments. Among these special libraries, NotI linking and jumping libraries have particular value for physical and genetic mapping of the human genome.

Despite their obvious advantages for special purposes, novel vectors (YACs, BACs, PACs; see below) cannot compete with lambda-based vectors for ease of handling, screening, amplification, and biological flexibility, due to the almost complete knowledge about phage lambda genes and biology. The easiest way to obtain a maximal proportion of recombinant molecules is to perform ligation at a high concentration of vector and insert (genomic) DNA because an elevated concentration of DNA facilitates intermolecular ligation instead of self-ligation of the vector's molecules. In this case, the main product is long DNA chains (>200 kbp) containing many copies of vector and genomic DNA fragments. Extremely efficient *in vitro* systems for packaging such DNA into lambda phage particles (10^9 plaque-forming units per microgram of DNA) to produce viable phages were developed and are commercially available.

Extensive modifications of lambda phage vectors were developed that combine the features of different vector systems (see Figures 1.2-1 and 1.2-2, [13–15]).

Cosmids are essentially plasmids that contain the *cos* region of phage lambda responsible for packaging of DNA into the phage particle. Any DNA molecule, containing this region and having a size between 37.7 kb and 52.9 kb, can be packaged into phage particles and introduced into the *E. coli*. The advantages of cosmids are their easy handling (as with plasmids) and big cloning capacity. As the plasmid body is usually small (3–6 kb), large DNA molecules (46–49 kb) can be cloned in these vectors.

Phasmids are lambda phages that have an inserted plasmid. They have the same basic features as lambda phage vectors, but inserted DNA fragment can be separated from the phage arms DNA and converted into a plasmid form. There are two main ways for such conversion: biological and enzymatic. In the first case, the

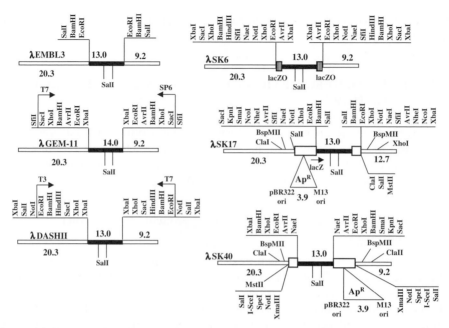

Figure 1.2-1. Examples of vectors used for construction of genomic libraries (maximum capacity 24 kb). Standard λ vectors (λEMBL3, λGEM11, λDASH, λEMBL3, and λSK6) and diphasmid vectors (λSK17 and λSK40) are shown. Sizes are in kilobases and are not shown in the scale. Not all restriction sites are shown. Heavy thick lines represent stuffer fragments; open boxes mark plasmid and M13 sequences; lacZO is the lac operator sequence. T3, T7, SP6-promoters for T3, T7, and SP6 RNA polymerases.

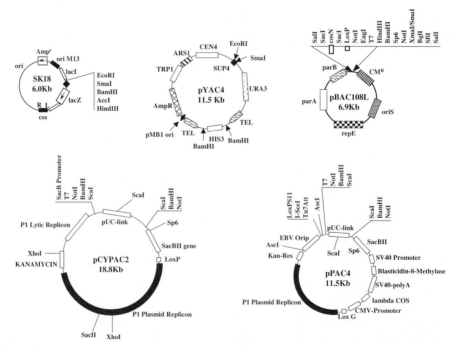

Figure 1.2-2. Examples of vectors used for construction of genomic libraries (capacity more than 45 kb). Sizes are in kilobases and are not shown to scale. Not all restriction sites are shown. For more details, see the text.

phasmid vector contains signal sequences bordering the insert and plasmid body in the phage vehicle. These signal sequences (e.g., from P1, lambda, M13 phages) can be recognized by specific proteins (e.g., *cre*-recombinase), and the cloned DNA fragment together with plasmid body will be cut out [16, 17].

The enzymatic conversion (see, for example, SK17 and SK40 vectors in Figure 1.2-1) can occur if the inserted DNA fragment and plasmid sequences are placed between two recognition sequences of some rare-cutting restriction endonuclease (e.g., I-SceI enzyme that recognize 18 bp and probably does not have any recognition site in the human genomic DNA). This enzyme can be used to separate the cloned DNA fragment together with the plasmid sequences from phage arms. Self-ligation of the molecules with insert and plasmid sequences will result in the production of a recombinant plasmid that can be introduced into the *E. coli* host.

Diphasmids combine the advantages of lambda and M13 phages, and plasmids, i.e., the three main types of vectors used in everyday molecular cloning. They can be divided into two classes: (1) diphasmids that can replicate as phage lambda (a further improvement of phasmids); see SK17 in Figure 1.2-1; and (2) diphasmids that are not able to replicate as phage lambda, i.e., a cosmid that can be packaged into phage M13 particles, e.g., SK18 in Figure 1.2-2.

In some cases, it is more convenient to work with a genomic library in plasmid than in lambda phage form. The construction of representative genomic library directly in a plasmid vector has several drawbacks and difficulties. However, these problems can be easily solved with the help of phasmid and diphasmid vectors. In this case, a genomic library is constructed in lambda phage (e.g., SK40), and then the whole library is converted to plasmid form.

Although the work with different lambda-based vectors is very similar, vectors for construction of genomic and cDNA libraries still have some special features. The major difference between cDNA and genomic cloning is the size of insert: short for cDNA and long for genomic fragments. Due to this difference, it is obvious that the enzymatic way of transferring inserts into the plasmid form is more dangerous (with the respect to the representativity of the library) for genomic libraries than for cDNA, because the probability that the insert contains a recognition site for the particular restriction enzyme increases with the length of the DNA insert. Most of cDNA vectors are "insertion" vectors, and genomic vectors are "substitution" vectors, which means that inserted foreign DNA fragments replace the stuffer vector fragment that does not have any important replicative function.

Full-length cDNAs are the most important material for the identification of all human genes. Cloning of full-length cDNA inserts has been hampered by problems related to both the preparation and the cloning of long cDNAs. Part of the difficulty associated with the preparation of long cDNAs has been overcome with the introduction of a thermostabilized and thermoactivated reverse transcriptase and the development of cap-based full-length cDNA selecting techniques. In contrast to standard cloning techniques, full-length cDNA cloning has the inherent risk of the under-representation or absence of clones corresponding to long mRNAs in the libraries, because the truncated cDNAs are usually not cloned. However, even in a perfect full-length cDNA library, cDNAs deriving from very long mRNAs will not be cloned if the capacity of the vector is insufficient. The most available cloning vectors show bias for short cDNAs: Shorter fragments are cloned more efficiently than the longer because of the intermolecular competition that occurs at the ligation and library amplification steps. To solve this problem, substitution vectors for cDNA cloning were designed.

For conversion into the plasmid form, the biochemical approach for SK-vectors was used [14]. Some vectors, like SK16 and SK17, have a cloning capacity of 0.2–15.4 kb and others, for example SK23, could be used for cloning of 3.7–18.9-kb inserts only. These cDNA vectors retain the most important advantages of the genomic vectors, e.g., the possibility of biochemical selection when producing vector arms and genetic selection against nonrecombinant phages. Importantly, the libraries made in some of these SK-vectors can be used as both expressed and nonexpressed libraries. For instance, libraries constructed in SK16 would be expressed only in *E. coli* strains containing *ochre*-codon mutation, and SK17 has *amber*-codon-controlled expression of cDNA.

In lambda FLC (full-length cDNA cloning), cloning capacity is either 0.2–15.4 kb or 5.7–20.9 kb. In these vectors, *cre*-recombinase mediated excision was used [17]. The average size of the inserts from excised plasmid cDNA libraries was 2.9 kb for standard and 6.9 kb for size-selected cDNA. The average insert size of the full-length cDNA libraries was correlated to the rate of new gene discovery, suggesting that effectively cloning rarely expressed mRNAs requires vectors that can accommodate large inserts from a variety of sources without bias for cloning short cDNA inserts.

One of the most important features of a genomic library is a representativity. A representative genomic library means that every genomic DNA fragments will be present at least in one of the recombinant phages of the library. In practice, however, this is difficult to achieve. Some genomic fragments are not clonable because of the strategy used for construction of the genomic library. In other cases, genomic DNA fragments (contain poison sequences) can suppress the growth of the vector or the host cell so its cloning can be restricted to specific vector systems. The important reason for the decreased representativity is different replication potential of different recombinant vectors. Thus, amplified libraries have significantly worse representativity. Usually a library is considered to be representative if after the first plating (before amplification) it contains several recombinant clones together containing genomic DNA fragments equal to a 7–10 genome equivalent. The way in which the genomic DNA fragments are produced for cloning is also important. The more randomly the genomic DNA is broken, the more the representative library can be obtained. Clearly, the EcoRI enzyme (6-bp recognition site) will cut genomic DNA less randomly than Sau3AI (4 bp recognition). Probably, the shearing of DNA molecules using physical methods (sonication, shearing using syringe) is the most reliable way to obtain randomly broken DNA molecules.

There are many different modifications to construct libraries; however, the three ways to construct genomic libraries are the most commonly used and will be exemplified below using genomic lambda-based vectors (Figure 1.2-3, [12–14]).

The classic method includes generation of sheared genomic DNA fragments using physical or enzymatic fragmentation followed by the physical separation of fragments of a particular size using, for example, ultra-centrifugation or gel-electrophoresis. The vector DNA is digested with two (or even three) restriction enzymes; those recognition sites are located in the polylinker. The arms and the stuffer fragment are purified, and on the further steps, arms/stuffer would be not able to ligate because they have different sticky ends, preventing recreation of the original vector molecules during subsequent ligation with genomic DNA fragments.

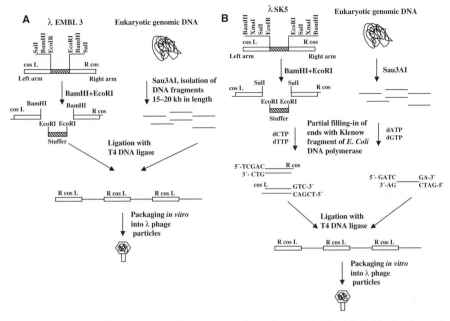

Figure 1.2-3. Two main approaches for construction of genomic libraries: (a) classic method and (b) partial filling-in method. cosL and Rcos—left and right parts of the *cos* site, respectively.

In the "dephosphorylation" approach [18, 19], the phage arms are prepared as described above. Genomic DNA is partially digested to obtain DNA fragments with sizes in the desired range; i.e., for the lambda vectors FIXII and SK6, it is 15–22 kb. These DNA fragments are dephosphorylated (which prevents their ligation to each other) and then ligated to the vector arms.

The third, "partial filling in" method [20], also avoids fractionation steps (Figure 1.2-3B). Phage arms are prepared by double digestion as described before (in this particular case, SalI and EcoRI are shown, but many other combinations can be used), and the sticky-ends produced after digestion are partially filled-in with the Klenow fragment of DNA polymerase I (or other DNA polymerase) in the presence of dTTP and dCTP. Genomic DNA partially digested with Sau3AI is also partially filled-in but in the presence dATP and dGTP. Under these conditions, self-ligation of vector arms or genomic DNA is impossible.

1.2.4.2   YACs

The mapping and analysis of complex genomes require vectors that allow cloning and work with large DNA fragments. Lambda-based vectors have many advantages as described above; however, a comparatively small size of inserts (maximum 48–49 kb) complicates the use of these vectors for the mapping of complex genomes.

Technological improvements made the cloning of large DNA pieces possible using artificially constructed chromosome vectors that carry human DNA fragments as large as 1Mb. Such vectors are maintained in yeast cells as artificial chromosomes (YACs). YAC methodology drastically reduces the number of clones to be ordered; many YACs span entire human genes. A more detailed map of a large YAC insert can be produced by subcloning, a process in which fragments of the original insert are cloned into smaller insert vectors. YAC cloning was widespread, and many laboratories used this technique to obtain large cloned DNA fragments from different eukaryotic genomes.

One of the most widely used YAC vectors was pYAC4 (Figure 1.2-2), and it has the main features of the YAC vectors. This vector can replicate in *E. coli* and in yeast (*S. cerevisiae*). It consists of five important genetic elements inserted into *E. coli* pBR322-like plasmid:

1. Centromeric *CEN4* sequences required for centromere function.
2. Autonomous replicating sequences *ARS1* that are necessary for replication in yeast.
3. Telomere sequences from *Tetrahymena* ribosomal DNA (rDNA) that seed the formation of functional telomeres at high efficiency.
4. Selective marker for cloning, *SUP4* (*ochre*-suppressing allele of tyrosine tRNA gene). As cloning EcoRI and SmaI sites are located inside *SUP4*, insertion of any large DNA fragment into the cloning sites will inactivate *ochre* suppression function.
5. Two yeast selectable markers: wild-type genes *TRP1* and *URA3*, which are located on opposite sides of the cloning sites.

The yeast *HIS3* gene is just a stuffer fragment to separate two arms. The general scheme of the library construction is as follows (Figure 1.2-4).

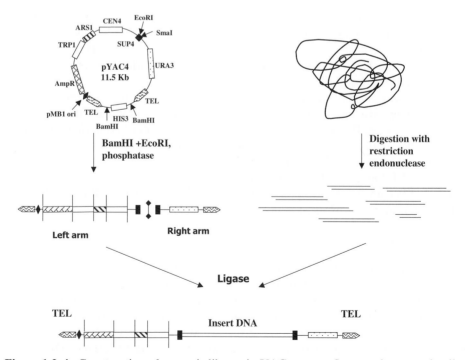

Figure 1.2-4. Construction of genomic library in YAC vectors. See text for more details.

The pYAC4 DNA is double digested with BamHI and SmaI (or EcoRI) to generate three fragments: left chromosomal arm containing *TEL*, *TRP1*, *ARS1*, and *CEN4* and right chromosomal arm with *TEL* and *URA3*. *HIS3* gene is discarded during the cloning procedure. The two arms are treated with alkaline phosphatase to prevent self-ligation and ligated with genomic DNA (usually) partially digested. The ligation product is transformed into yeast and selected for complementation of a mutated *ura3* gene present in yeast strain by the *URA3* gene in the vector. The grown transformants are selected for complementation of a host *trp1* mutated gene by the wild-type *TRP1* gene from the vector.

This double selection is necessary to ensure the presence in a recombinant YAC of both left and right arms and inactivation of the *SUP4* gene. *SUP4* is an especially useful cloning marker because in an *ade2-ochre* host, the cells with the functional *SUP4* gene are white and those in which the suppressor was inactivated form red colonies.

Using YAC vectors, several important genes were identified and cloned, e.g., *VHL*, *CDPX1*, *GART, and SON* [21–23]. Moreover, the first physical map covering the whole human genome was constructed with YAC contigs [24].

Although YACs greatly contributed to the progress in long-range mapping of the human genome, the construction of YAC contigs is not free from problems—certain regions of the human genome are difficult to clone in YACs. Moreover, repeat rich human sequences cloned in YACs are frequently unstable in yeast. Low transformation efficiency, the presence of an abundance of chimeric clones in nearly all YAC libraries, insert instability, and difficulties in DNA manipulation relative to bacterial systems are other important limitations.

Efforts to overcome the limitations of cosmids and YACs resulted in alternative large-insert cloning approaches using bacteriophage P1-based vectors and vectors based on the *E. coli* fertility plasmid (F-factor).

1.2.4.3   Bacterial (BAC) and P1-derived (PAC) Artificial Chromosome Vectors

BACs represent the state-of-the-art technology for such large-insert DNA library development. It has been demonstrated that BAC libraries are invaluable and desirable genetic resources for all kinds of modern structural, functional, and evolutionary genomics research. Genome-wide, as well as regional, physical maps of the human genome from BACs have been developed in different laboratories.

The BAC system is based on the well-studied *E. coli* F factor. Replication of the F factor in *E. coli* is strictly controlled. The F plasmid is maintained in low copy number (one or two copies per cell), thus reducing the potential for recombination between DNA fragments carried by the plasmid. Furthermore, F factors carrying inserted bacterial DNA are capable of maintaining fragments as large as 1 megabase pair, suggesting that the F factor is suitable for cloning of large DNA fragments. Individual clones of human DNA seem to be maintained with a high degree of structural stability in the host, even after 100 generations of serial growth.

The F factor not only codes for genes that are essential to regulate its own replication but also controls its copy number. The regulatory genes include *oriS*, *repE*, *parA*, and *parB*. The *oriS* and *repE* genes mediate the unidirectional replication of the F factor, whereas *parA* and *parB* maintain a copy number at a level of one or two per *E. coli* genome. The BAC vector (pBAC108L) incorporates these essential genes as well as a chloramphenicol resistance marker and a cloning segment (Figure 1.2-2). The cloning segment includes (1) the bacteriophage lambda cosN and phage P1 loxP sites, (2) two cloning sites (HindIII and BamHI), and (3) several CG-rich restriction enzyme sites (Not I, Eag I, Xma I, Sma I, Bgl I, and Sfi I) for potential excision of the inserts. The cloning site is flanked by T7 and SP6 promoters for generating RNA probes for chromosome walking and for DNA sequencing of the inserted segment at the vector–insert junction. The cosN site provides a fixed position for specific cleavage with the bacteriophage lambda terminase. The loxP site can be used similarly. In this case, P1 Cre protein catalyzes the cleavage reaction in the presence of the loxP oligonucleotide. These sites (cosN and loxP) allow convenient generation of ends that can be used for restriction-site mapping to arrange the clones in an ordered array. Restriction maps of the individual clones can be determined by indirect end-labeling and subsequent partial digestion.

PAC vectors combine the best features of the P1 and BAC systems. One of the most commonly used PAC vectors is pCYPAC2 (Figure 1.2-2) that was constructed for the cloning of large DNA fragments using electroporation. Previously developed P1-based vectors (like pAd10SacBII) employed bacteriophages P1 or T4 *in vitro* packaging systems, enabling the cloning of recombinants with inserts in the 70–120-kb range.

The pCYPAC2 vector was constructed by removing the stuffer fragment from the pAd10SacBII vector and by inserting a pUC19 plasmid into the BamHI cloning site. During the cloning process, the pUC19 sequences are removed through a double-digestion scheme using BamHI and ScaI. Religation of the pUC19 and producing of nonrecombinant clones is prevented at three levels, as follows:

1. pUC19 is cleaved into two ScaI fragments. ScaI sticky ends are not compatible with BamHI.
2. Oligonucleotides BamHI–ScaI connecting vector sequences with the stuffer fragment are physically removed from the vector and stuffer fragments.
3. pCYPAC2 fragments are treated with alkaline phosphatase to inhibit self-ligation.

NotI sites are flanking inserts that make it possible to obtain an insert that is fragment free from the vector sequences.

Different modifications of original BAC and PAC vectors were constructed, e.g., capable of replicating in human cells too [25]. These vectors, e.g., pPAC4 (Figure 1.2-2) and pBACe4, facilitate the use of large-insert bacterial clones for functional analysis and contain two additional genetic elements that enable stable maintenance of the clones in mammalian cells:

(1) The Epstein–Barr virus replicon, *oriP*. It was included to ensure stable episomal propagation of the large insert clones upon transfection into mammalian cells.
(2) The blasticidin deaminase gene is placed in a eukaryotic expression cassette to enable selection for the desired mammalian clones by using the nucleoside antibiotic blasticidin.

Sequences important to select for loxP-specific genome targeting in mammalian chromosomes were also inserted into the vectors. In addition, the attTn7 sequence present on the vectors permits specific addition of selected features to the library clones. Unique sites have also been included in the vector to enable linearization of the large-insert clones, e.g., for optical mapping studies. The pPAC4 vector has been used to generate libraries from the human, mouse, and rat genomes.

Libraries constructed in lambda-based, BAC, and PAC vectors are widely used now in different laboratories, and almost nobody is currently working with libraries constructed in YAC vectors.

1.2.5 MAPPING OF THE HUMAN GENOME

As mentioned in the Introduction, mapping of the human genome was a major step in achieving the complete human genome sequence. Human genome maps were also important by themselves because they led to the isolation of many disease genes.

The 3 billion bp in the human genome are organized into 24 distinct, physically separate microscopic units called chromosomes. All genes are arranged linearly along the chromosomes. The nucleus of most human cells contains two sets of chromosomes, one set given by each parent. Each set has 23 single chromosomes— 22 autosomes and an X or Y sex chromosome. A normal female will have a pair of X chromosomes; a male will have an X and Y pair. Chromosomes contain roughly equal parts of protein and DNA, and chromosomal DNA contains on average 150 million bases.

In general, all maps can be divided in two main classes: genetic and physical [see Refs. 3 and 4]. The genetic map is based on the frequency of recombination between two genetic markers on the chromosome. In fact, such a map just shows how these two markers are "linked" in respect to the frequency of recombination. Physical maps are based on physical distances between markers. Recombination frequencies define a genetic distance that is not the same as a physical distance. Two loci that show 1% recombination are defined as being 1 centimorgan (cM) apart on a genetic map. However, for distances above about 5 cM, human genetic map distances are not simple statements of the recombination fraction between pairs of loci. Loci that are 40 cM apart will show less than 40% recombination. This result reflects that recombination fractions never exceed 50% no matter how far apart are the loci. The female genetic map is larger than male due to a larger frequency of recombination. In general, for the whole human genome, the sex-averaged figure is 1 cM = 0.9 Mb, but the actual correspondence varies very significantly in different chromosomal regions. Usually there are more recombinations toward the telomeres of chromosomes and less toward the centromeres.

The value of the genetic map is that an the inherited disease can be located to the particular chromosomal region exploiting the inheritance of a DNA marker present in affected individuals (but absent in unaffected individuals), even though the molecular basis of the disease may not yet be understood nor the responsible gene identified. Genetic maps have been used to find the exact chromosomal location of several important disease genes, including cystic fibrosis, sickle cell disease, Tay-Sachs disease, fragile X syndrome, and myotonic dystrophy. Currently, some genetic linkage maps contain more than 100,000 polymorphic markers (see, for example, http://www.ncbi.nlm.nih.gov/SNP/).

Different types of physical maps are based on different techniques. For instance, the cytogenetic map is based on detection chromosomal bands, and an average band has from one to several Megabase of DNA. Chromosomal breakpoint maps can be based on somatic cell hybrid panels containing human chromosome fragments derived from natural translocation or deletion chromosomes. The resolution for such maps is usually several Megabases. Monochromosomal radiation hybrid (RH) maps have a distance between breakpoints that is often several Megabases long. Whole genome RH maps have much higher resolution and can be as high as 0.5 Mb. The RH maps were very important for joining genetic and physical maps. They are based on *in vitro* radiation-induced chromosome fragmentation and cell fusion (merging human with other cells from other species to form hybrid cells) to create panels of cells with specific and varied human chromosomal components. Assessing the frequency of marker sites remaining together after radiation-induced DNA fragmentation can establish the order and distance between the markers.

Very useful were clone contig maps that could be constructed with YAC (or BAC, PAC) clones, and in this case, the average DNA insert has several hundred kilobases of DNA [24, 26]. If such a clone contig map was created using overlapping cosmid clones (average insert is 40–45 kb), the quality of such a map is significantly higher. A restriction map created with rare-cutting restriction enzymes like NotI can be very useful to check RH or clone contig maps [27].

A sequence-tagged site (STS) map requires prior sequence information and is based on mapping short known sequences using PCR amplification.

An expressed sequence tag (EST) map is based on STSs that were generated from cDNA.

DNA fragments are called DNA markers after they were mapped to particular chromosomal regions. Such DNA markers are prerequisites for physical and genetic mapping of the genome. DNA markers are also important in the diagnosis of genetic diseases. DNA markers can be divided into several different classes depending on the way in which the markers were selected among the fragments of genomic DNA. Examples of such classes are anonymous, micro- and minisatellites, restriction fragment length polymorphism (RFLP) markers, *NotI* linking clones, ESTs, STSs, and so on.

Another strategy of mapping is based on the completely different principle of NotI jumping and linking libraries [28, 29].

A big advantage to using chromosome jumping and linking clones is that they are small-insert vectors and mapping with these clones can be completely automated.

The basic principle of jumping is to clone only the ends of large DNA fragments rather than a continuous DNA segment (Figure 1.2-5). The DNA between two ends is deleted with the different biochemical techniques, and clones containing only the ends of a large DNA molecule are enriched by the mean of genetic or biochemical selection.

In general, jumping clones contain DNA sequences adjacent to neighboring NotI restriction sites, and linking clones contain DNA sequences surrounding the same NotI restriction site.

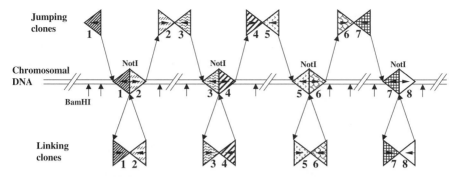

Figure 1.2-5. General scheme of long-range mapping using jumping and linking libraries. Small vertical arrows designate BamHI, and horizontal arrowheads show short sequences flanking NotI sites. Triangles indicate NotI–BamHI boundary fragments. See text for details.

A NotI jumping library is a collection of cloned DNA fragments that contains the ends of large NotI genomic DNA fragments. A NotI linking library is a collection of cloned genomic DNA fragments flanking the same NotI site. Basically, a linking library in combination with pulsed field gel electrophoresis (PFGE) is sufficient to construct a physical chromosomal map. When linking clones are used to probe a PFGE genomic blot, each clone should reveal two DNA fragments, and they must be adjacent in the genome. In principle, with just a single library and digest, one should be able to order the rare cutting sites, but in reality, it is not possible because many fragments could have the same size. Using these two types of libraries, it is possible to move fast along the chromosome because the human genome contains a limited number of recognition sites for this enzyme (10,000–15,000). The easiest way to establish the order of NotI sites is a shotgun sequencing of NotI linking and jumping clones (see below and Figure 1.2-5).

The most well-known and efficient way (integrated approach) to construct NotI linking and jumping libraries is shown in Figure 1.2-6 [30, 31].

To construct linking libraries (Figure 1.2-6B), genomic DNA is completely digested with BamHI. (Libraries can also be constructed with BglII, EcoRI, etc., but not with frequently cutting enzymes like MboI or infrequently cutting enzymes like SalI.) Subsequently, the DNA is self-ligated at a low concentration of DNA (without a *supF* marker), to yield circular molecules as the main product. To eliminate any remaining linear molecules, the sticky ends are partly filled-in with the Klenow fragment in the presence of dATP and dGTP (this is for the genomic DNA digested with BamHI; for other enzymes, partial filling-in with other nucleotides should be used). This way, all BamHI sticky ends will be neutralized and nearly all ends originated from linear molecules will be unavailable for ligation. Resulting DNA fragments are digested with NotI and cloned into λSK4, λSK17, and λSK22 vectors. These vectors permit the cloning of DNA fragments from 0.2 to 24 kb. The resulting phage particles are used to infect *E. coli* cells in which only recombinant phages could grow (spi- selection, [12]).

To construct the NotI jumping library, the same DNA and vectors can be used. However, the first steps are different. High-molecular-weight DNA is digested with

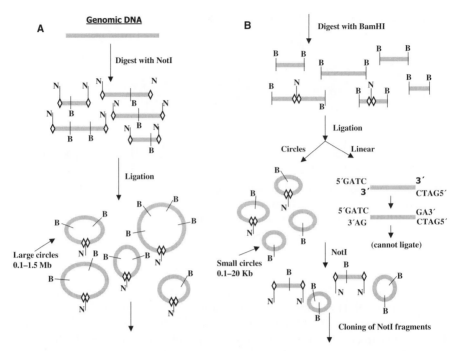

Figure 1.2-6. Construction of jumping (a and b) and linking libraries (b). B represents BamHI, and N denotes NotI sites. Digestion of genomic DNA with BamHI is a first step in construction of a linking library.

NotI and circularized at very low DNA concentration (0.1 μg/μL). Other steps were the same (Figure 1.2-6A).

The libraries can be converted into plasmid form, and sequences flanking NotI sites can be generated with standard sequencing primers. This sequence information can be used for generation of sequence tagged sites (STS). It was also demonstrated that the sequencing of these clones is a very efficient method for gene isolation because even short nucleotide sequences (approximately 500 bp from each side) flanking NotI sites are sufficient to detect genes [32, 33]. This approach was used for isolation of numerous new genes. The reason for this linkage of NotI sites and genes is the localization of practically all NotI sites in the CpG islands, which are tightly associated with genes.

A shotgun sequencing approach useful for the whole genome mapping was proposed for NotI linking and jumping libraries constructed in λSK17 and λSK22 vectors using an integrated approach [14, 29, 34]. This strategy was based on the fact that NotI sites in both libraries were available for the sequencing with standard sequencing primers, and sequence information about 800–1000 bp surrounding each NotI site could be easily obtained by automated sequencing. Plasmid DNA could be isolated automatically, and thousands of clones could be sequenced. Subsequently the linear order of the NotI clones can be established using a computer program. This approach is schematically illustrated in Figure 1.2-5.

NotI linking and jumping libraries were very important for the mapping of two regions in the short arm of chromosome 3 and isolation of tumor suppressor genes from the AP20 and LUCA regions (*RASSF1, G21, SEMA4B, RBSP3*, etc.).

1.2.6   MAIN APPROACHES TO SEQUENCE HUMAN GENOME

1.2.6.1   Hierarchical and the Whole Genome Shotgun Sequencing Scheme (WGS)

The international public (HGP) and American private (Celera) efforts to determine the sequence of the human genome exploited different approaches [9, 10, 35]. It is still not clear whether Celera's whole genome shotgun sequencing (WGC) approach was really efficient as they used publicly available data and it is not known to what extent.

The major points in the hierarchical or HGP approach [9] (Figure 1.2-7A) are as follows:

1. To construct a precise high-density STS map of the human genome. This map was integral and included not only STS but also genetic, microsatellite, FISH, and all other available mapping information.
2. To construct the clone contig of overlapping BAC and PAC clones. This contig was created and checked using a map constructed at the first step.
3. Each BAC or PAC clone was sequenced using a shotgun sequencing strategy. These sequences were assembled into the draft human genome sequence.

The shotgun sequencing approach for sequencing relatively small inserts was developed a long time ago [3, 4]. This method is based on the building of a complete

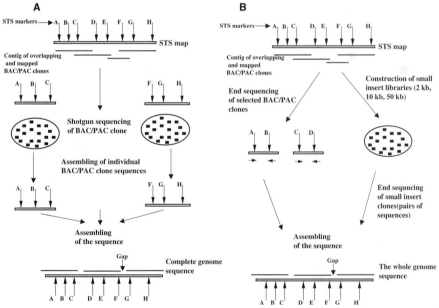

Figure 1.2-7. Two main approaches used to sequence the human genome. The hierarchical approach is based on creation of clone contigs covering the whole genome and shotgun sequencing of individual cosmid/BAC/PAC clones (a). The whole genome shotgun sequencing scheme (WGS) shown in (b) is based on pair-end sequencing of small-insert and selected BAC/PAC clones. See text for details.

insert sequence by randomly sequencing clones with overlapping inserts. To do this the insert DNA is subjected to partial digestion with a 4-bp cutter (like MboI) and the partially overlapping fragments are cloned at random into a suitable M13 or plasmid vector and sequenced. The sequence data obtained are fed into a computer programmed to detect overlaps between sequences and to assemble a composite sequence. Inevitably, this means a waste of efforts: The same sequence will be obtained over and over again.

Celera used a variant of this strategy that it named WGS [10] (Figure 1.2-7B). Central to the WGS strategy is preparation of high-quality plasmid libraries in a variety of insert sizes so that pairs of sequence reads (mates) are obtained; one is read from both ends of each plasmid insert. High-quality libraries have an equal representation of all parts of the genome and are constructed in three different sizes: 2 kb, 10 kb, and 50 kb. The method also involved end sequencing of some PAC clones that were carefully mapped. DNA fragments in these small-insert clones were generated by physical shearing of whole genomic DNA. According to the original predictions, this method requires the generation of sequences covering the whole genome not more than seven times.

The HGP and Celera announced that they have obtained a sequence of 92–97% of the human genome sequence and that a 7–10-fold genome sequence coverage is enough to generate an almost complete sequence for any human or mammalian genome. After 5 years, the scientific community understood that these estimations were overstated (see below), and even now when more than 10-fold genome sequence coverage was achieved, the human genome sequence is far from being complete.

An important point of both approaches is that the genome should be covered by sequences at least 10 times to yield contigs covering a significant part of the genome. If the coverage is not extensive enough, then sequences and clones will represent unconnected and unordered islands. Thus, despite impressive progress, mapping and sequencing even of small genomes, like from one bacterial strain, is still expensive and laborious. As a result, genome mapping and sequencing efforts need to be concentrated to big centers, and independent work by smaller groups is seriously hampered.

Partly in view of these logistic and economic limitations, HGP and Celera approaches may not be the optimal solution, especially for several experiments addressing biocomplexity. After completion of one sequence from an organism, in many cases there will be a great demand for comparison with other individuals, related species, pathogenic and nonpathogenic strains, and so on, in the growing field of comparative genomics. Such comparisons will be highly relevant for the better understanding of health, evolution, and ecology questions but must be performed by laboratories receiving considerably less funding than those involved in the recent high-profile sequencing efforts. Another area is a sequencing of related bacterial strains and species in order to identify the genomic basis of their biological differences and their interactions within the human intestinal flora. What is a difference between pathogenic and nonpathogenic members of the gut flora?

The whole genome sequencing will be performed only for the selected organisms. However, analysis of disease genes in other species may be extremely useful for the understanding of fundamental processes, leading to development of disease.

As the body of knowledge concerning the genome structure and genes in different organisms grows, it also becomes unnecessary to collect total sequences with 10–15-fold coverage for all organisms that would be interesting to study.

One of the most perspective solutions for this problem is sequencing using microarrays [36]. However, this approach will be not discussed here as it cannot be a strategy for the *de novo* sequencing. Two other alternative approaches that could be used for both the *de novo* and the comparative sequencing will be described below.

1.2.6.2 Massively Parallel Sequencing Approaches

The first publication using beads for massively parallel sequencing was published by Dr. Sydney Brenner et al. (2000) [37]. They described a novel sequencing approach that combined non-gel-based signature sequencing with *in vitro* cloning of millions of templates on separate 5-μM-diameter microbeads. After constructing a microbead library of DNA templates by *in vitro* cloning, a planar array of a million template-containing microbeads in a flow cell at a density greater than 3×10^6 microbeads/cm^2 was assembled. Sequences of the free ends of the cloned templates on each microbead were then simultaneously analyzed using a fluorescence-based signature sequencing method that does not require DNA fragment separation. Signature sequences of 16–20 bases were obtained by repeated cycles of enzymatic cleavage with a type IIs restriction endonuclease (BbvI), adaptor ligation, and sequence interrogation by encoded hybridization probes. The approach was validated by sequencing over 269,000 signatures from two cDNA libraries constructed from a fully sequenced strain of *S. cerevisiae*, and by measuring gene expression levels in the human cell line THP-1. This method is complicated and expensive and was used mainly for generation of SAGE tags (see below).

Recently MPSS principles were applied for long-range sequencing. Dr. George Church and his colleagues introduced a sequencer that uses a microscope and other off-the-shelf equipment [38]. With this technology, his team sequenced a strain of *E. coli* and could detect single-base-pair changes from an almost identical *E. coli* genome. The approach reduces sequencing costs by 90%. Dr. Jonathan Rothberg has demonstrated the power of another cost-cutting technology called "454" [39]. Using this approach, *Mycoplasma genitalium* was sequenced.

Both groups save money by eliminating the need for bacteria and miniaturizing the process wherever possible. Instead of bacteria, they attach DNA to aqueous beads encased in oil where PCR amplification was performed to produce the necessary amount of DNA. That change alone could reduce by two thirds the costs associated with space and personnel. Moreover, both perform many thousands of these sequencing reactions at once in miniature "reactors," decreasing the need for expensive chemicals. Once the DNA is ready, the two technologies diverge: The "454" technique uses pyrosequencing [40] to identify the bases and Church's technique (MPS, multiplex polony sequencing) uses bursts of different fluorescent colors, one each to a particular base, to distinguish the bases (sequencing by ligation, [41]). Both use high-speed charge-coupled device cameras to record the labeled bases.

The "454" technique uses a novel fiber-optic slide of individual wells and can sequence 25 million bases, at 99% or better accuracy, in one 4-hour run. To achieve an approximately 100-fold increase in throughput over current Sanger sequencing technology, an emulsion method for DNA amplification and an instrument for sequencing by synthesis using a pyrosequencing protocol optimized for solid support and picolitre-scale volumes was developed. In a pilot experiment, 96% of the *M. genitalium* genome was covered by sequence contigs at 99.96% accuracy in one run of the machine. The method could potentially allow one individual to prepare and sequence an entire genome in a few days. The sequencer itself, equipped with a simple detection device and liquid delivery system, and housed in a casing roughly the size of a microwave oven, is actually relatively low-tech. The complexity of the system lies primarily in the sample preparation and in the microfabricated, massively parallel platform, which contains 1.6 million picoliter-sized reactors in a 6.4-cm^2 slide.

Sample preparation starts with fragmentation of the genomic DNA, followed by the attachment of adaptor sequences to the ends of the DNA pieces. The adaptors allow the DNA fragments to bind to tiny beads (around 28µm in diameter). This is done under conditions that allow only one piece of DNA to bind to each bead. The beads are encased in droplets of oil that contain all reagents needed to amplify the DNA using a standard tool called the PCR. The oil droplets form part of an emulsion so that each bead is kept apart from its neighbor, ensuring the amplification is uncontaminated. Each bead ends up with roughly 10 million copies of its initial DNA fragment.

To perform the sequencing reaction, the DNA-template-carrying beads are loaded into the picoliter reactor wells—each well having space for just one bead. The technique uses a sequencing-by-synthesis method developed by Dr. Uhlen and colleagues [40], in which DNA complementary to each template strand is synthesized. The nucleotide bases used for sequencing release a chemical group as the base forms a bond with the growing DNA chain, and this group drives a light-emitting reaction in the presence of specific enzymes and luciferin. Sequential washes of each of the four possible nucleotides are run over the plate, and a detector senses which of the wells emit light with each wash to determine the sequence of the growing strand.

This new system shows great promise in several sequencing applications, including resequencing and *de novo* sequencing of smaller bacterial and viral genomes. It could potentially allow research groups with limited resources to enter the field of large-scale DNA sequencing and genomic research, as it provides a technology that is inexpensive and easy to implement and maintain. However, this technology cannot yet replace the Sanger sequencing approach for some of the more demanding applications, such as sequencing a mammalian genome, as it has several limitations [42].

First, the technique can only read comparatively short lengths of DNA, averaging 80–120 bases per read, which is approximately a tenth of the read-lengths possible using Sanger sequencing. This means not only that more reads must be done to cover the same sequence, but also that assembling short reads into longer genomic sequences is a lot more complicated. This is particularly true when dealing with genomes containing long repetitive sequences.

Second, the accuracy of each read is not as good as with Sanger sequencing—particularly in genomic regions in which single bases are constantly repeated. Third, because the DNA "library" is currently prepared in a single-stranded format, unlike the double-stranded inserts of DNA libraries used for Sanger sequencing, the technique cannot generate paired-end reads for each DNA fragment. The paired-end information is crucial for assembling and orientating the individual sequence reads into a complete genomic map for *de novo* sequencing applications.

Finally, the sample preparation and amplification processes are still complex and will require automation and/or simplification.

In the MPS technique, sheared, size-selected genomic fragments (approximately 1 kb in size) were circularised with a linker bearing MmeI (type IIs restriction enzyme) recognition site (TCCRAC). Then all noncircularized material was destroyed with exonuclease and circular molecules were amplified using rolling circle amplification with random hexamers. Afterward, resulting material was digested with MmeI. MmeI digests at a distance of 18/20 or 19/21 from its recognition site. In the MPS protocol, MmeI digestion results in 17-bp or 18-bp tags of unique genomic sequence. A sharp band at approximately 70 bp was purified with 6% PAGE, and DNA ends were blunted and ligated to two adaptors of different lengths. The molecules with two different adaptors at their ends were again purified using 6% PAGE and PCR amplified. Then sharp bands at approximately 135 bp were purified on a 6% PAGE gel and emulsion PCR with 1-μm magnetic beads containing biotin labeled primer was performed. Millions of beads with amplified DNA fragments (paired tags) were immobilized in acrylamid-based gel system and a four-color sequencing by ligation was made [41]. This method allowed to obtain 13-bp sequencing information per tag separated by a 4- to 5-bp gap (sequencing per each tag was done from both ends). Thus, 26-bp information per amplicon was obtained (2 tags × 13 bp). In a pilot experiment, this technology was applied to resequence an evolved strain of *E. coli*, and according to the estimation, this was done on less than a one-error-per-million consensus bases. In this pilot experiment, two deletion genomic fragments and several 1-bp substitutions were detected in the analyzed *E. coli* strain.

Still, neither "454" nor the MPS method is up to speed yet. The accuracy of both should be improved by at least one order of magnitude. Also, to sequence mammalian genomes the length of sequence generated should be about 700 bases, but reads reported from these new approaches are between 26 and 110 bases.

1.2.6.3 Slalom Library: A Novel Approach to Genome Mapping and Sequencing

All described above sequencing approaches were based on sequencing of randomly generated DNA fragments.

We recently suggested a completely different and efficient strategy for simultaneous genome mapping and sequencing. The approach was based on physically oriented, overlapping restriction fragment libraries called slalom libraries. Slalom libraries combined features of general genomic, jumping, and linking libraries. Slalom libraries could be adapted to different applications, and two main types of slalom libraries will be discussed below. This approach was used to map and sequence (with ~46% coverage) two human PAC clones, each of ~100 kb. This model experiment demonstrates the feasibility of the approach and shows that the efficiency (cost-effectiveness and speed) of existing mapping/sequencing methods could be improved at least 5–10-fold. Furthermore, as the efficiency of contig assembly in the slalom approach is virtually independent of length of sequence reads, even short sequences of 19–20 bp produced by rapid, high-throughput sequencing techniques would suffice to complete a physical map and a sequence scan of a small genome.

Two slalom libraries are used in the first type of approach. This approach allows us to construct contigs of plasmid clones covering a whole genome. However, these contigs will contain some gaps. Comparatively modest sequence information can be generated (20–25% of a genome). Using this technique, two (or more) bacterial strains (e.g., pathogenic and nonpathogenic) could be quickly compared and pathogenic islands could be identified.

The main principle of the type I slalom libraries is shown in Figure 1.2-8 and looks similar to the computer-assisted long-range mapping with NotI linking and jumping clones (Figure 1.2-5). In this case, the role of NotI jumping library plays a standard EcoRI genomic (slalom "R") library that is produced by complete digestion of genomic DNA with EcoRI, and the role of a NotI linking library–EcoRI linking (slalom "BR") library. Slalom "BR" library is constructed in exactly the same way as a NotI linking library using BamHI as a second enzyme and fulfills the connecting function joining EcoRI fragments. Shortly, DNA is digested with BamHI, circularized and cut with EcoRI ("BR" library). Thus, EcoRI in this case plays the role of a NotI enzyme.

Sequences from "BR" and "R" libraries are produced using standard reverse and forward sequencing primers and overlapping clones are found using a computer program. The homologies found will join the ends of EcoRI fragments in the "BR" library with the ends of EcoRI fragments in the "R" library to yield an ordered set of BamHI–EcoRI clones/STSs distributed along the genome.

Figure 1.2-8 shows the scheme where EcoRI and BamHI sites alternate. In reality, EcoRI sites and BamHI sites do not always alternate, and this will lead

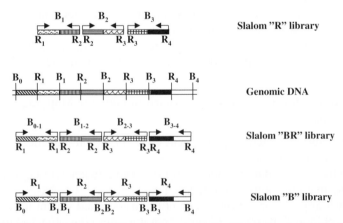

Figure 1.2-8. The main idea of the slalom approach. Numbers designate identical end sequences in different libraries that can be joined by a computer program in a contig of overlapping clones. B represents BamHI, and R denotes EcoRI sites. Small horizontal arrows show short sequences flanking EcoRI sites. The "BR" library plays a role of a linking library where EcoRI replaces NotI. The ordinary "R" library plays the role of a jumping library where EcoRI replaces NotI.

to the gaps in this set of overlapping clones, if several EcoRI or BamHI sites lie together. Thus, the genome will not be completely covered by clones because the information between some EcoRI and BamHI sites will be missing.

The problem with the gaps can be solved using the second variant of the slalom libraries (type II) where three libraries are used:

1. *Standard, completely EcoRI digested (slalom "R") library*
2. *Standard, completely BamHI digested (slalom "B") library*
3. *EcoRI jumping (slalom "RBR" or connecting) library*

This approach results in generation of plasmid clone contigs covering a whole genome without gaps. A large volume of sequencing information can be obtained (45–55% of a genome). The connecting library in this case is similar to a NotI jumping library. If the ultimate aim is to sequence the complete genome, then sequencing gaps could be filled-in using standard methods such as primer walking or transposon-mediated sequencing.

An "RBR" slalom library or EcoRI jumping library is prepared in the same way as a NotI jumping library using as a second enzyme BamHI. The only difference is that circularization after BamHI digestion is performed in the presence of a selective marker (e.g., Kan^R gene or oligonucleotide adaptor, etc.). The circular constructs are opened with EcoRI and cloned [43].

The "RBR" library can be constructed in a simpler way. Plasmid DNA isolated *en masse* from a slalom "R" library is digested with BamHI and circularized in the presence of Kan^R. Then *E. coli* cells are transformed with ligated DNA and plated on agar with kanamycin. The clones obtained in this manner will be identical in structure to the clones from an EcoRI jumping library prepared using the classic method.

By comparing the end sequences of the "B" library clones with the internal BamHI (from the marker fragment) sequences of the slalom "RBR" library clones, the BamHI clones can be positioned in relation to each other. After the comparison of end sequences in "R" and "RBR" libraries, EcoRI clones will be positioned relative to each other. Finally, EcoRI and BamHI clones will be assembled into a contig representing their organization in complete genome.

The EcoRI jumping library provides the connecting function in type II slalom libraries.

The major difference between the slalom library mapping/sequencing approach and other sequencing strategies is that the clones are generated according to a specific scheme and using complete digestion. As a result, the number of variants required to cover the whole genome decreases significantly. The preparation of libraries for the slalom approach is remarkably simple: Only complete digestion with EcoRI or BamHI is used. There is no need for size separation, agarose gel purification, or establishing conditions for partial shearing/digestion. There is no need to develop a new sequenator as any existing instrument can be used. Importantly, it is not necessary to keep all slalom clones because sequencing information can be used to design PCR primers and even large fragments (up to 40–50 kb) can be amplified by long-range PCR.

The slalom library approach differs fundamentally from the shotgun sequencing approach with respect to the efficiency of assembly (EOA). The EOA for the latter method is strictly dependent on the length of sequencing reads. The longer the reads, the higher the EOA. As the slalom approach uses nonrandom fractionation of the DNA and each start site is tightly linked with the recognition site for the restriction enzyme, even very short sequences will, in principle, be enough to create a contig of the overlapping clones. The EOA of the slalom library approach is, therefore, essentially independent of read-length. Even the short sequences generated by pyrosequencing, MPS, or MPSS [37–40] should be sufficient to completely cover a genome. As one person can generate thousands of sequences a day using a pyrosequencer, the minimal set of overlapping clones covering a 4-Mb genome can be completed in a couple of days. Repeat sequences are in fact less of a problem for the slalom approach than for the shotgun sequencing approach for several reasons [43].

It is important to stress that the benefits of the slalom approach are most obvious in comparative sequencing experiments in combination with high-throughput techniques like pyrosequencing or MPSS (which are not compatible with the shotgun sequencing approach). Our particular experiments demonstrated that approximately 4% coverage would be enough to construct the contig of the overlapping clones and subsequently generate >20% ordered sequences with almost 100% efficiency, i.e., with 0.2-fold coverage.

Of course, closing of the gaps will be done with significantly lower efficiency. However, the finishing stages of the shotgun sequencing approach are also the most expensive and time-consuming part of the process. Lander et al. [9] distinguished three types of gaps: gaps within unfinished sequenced clones; and gaps between sequenced clone contigs, but within fingerprint clone contigs; and gaps between fingerprint clone contigs. The first type is the simplest, and the third is the most complicated to close because constructing a contig of overlapping clones is the most difficult procedure. With the slalom approach, we already have a contig of overlapping clones, and thus, we will only suffer from the first, simplest type of gaps. It is important to mention another difference in the finishing stages of these two approaches. With the shotgun sequencing approach, sequences from different clones must be connected, and here highly related repeats, gene families, and polymorphisms will represent a major problem. These problems are nonexistent in the slalom approach, where a single insert should be sequenced.

The slalom strategy can increase several orders of magnitude the efficiency of mapping and sequencing, i.e., decrease the cost and labor, and increase the speed of sequencing project fullfillment (see Table 1.2-1). This strategy will allow the establishment of a physical map with a minimal set of overlapping clones that will pinpoint differences in genome organization between organisms. At the same time, a considerable sequence coverage (about 50%) of the genome at a less than onefold coverage will be achieved. This will make it possible to locate virtually every gene in a genome for more detailed study. The method makes it economically and logistically possible for a wide range of laboratories to gather the detailed information about large numbers of genomes. It offers a way to use genomics techniques for the study of biocomplexity, evolution, and taxonomy in the close future.

TABLE 1.2-1. Slalom Strategy and Shotgun Sequencing Approach

	Slalom	Shotgun
Sequence coverage 1.2-fold	A. Minimal set of overlapping clones; restriction map B. 20–25% of sequences are ordered into STS islands covering the whole genome (+appr. 30% of unordered sequences)	A. Random clones, no map B. Practically no ordered sequences (appr. 50–70% of unordered sequences)
Pyrosequencing 4% coverage	Minimal set of overlapping clones; restriction map	No use
Pyrosequencing plus sequencing unique clones (25–30% coverage)	20–25% of sequences are ordered into STS islands covering the whole genome	No use
Sequence coverage needed to achieve 95% of complete sequence	1	10–15
Sequence coverage needed to compare two genomes and isolate clones/genes completely covering regions of divergence	4% (0.04-fold)	At least 6–8-fold (600%–800%)

1.2.7   IDENTIFICATION OF GENES. HOW MANY GENES ARE IN THE HUMAN GENOME?

1.2.7.1   Many Approaches Exist for Gene Identification

The most important for the majority of research tasks is identification of genes in a particular region of the human genome. The different tissues in the body express information across a wide range, between less than 2000 (in specialized and terminally differentiated white blood cells) and more than 20,000 (in the placenta). Several methods were suggested previously for gene identification [3, 4]. Genomic DNA can be used as a probe to screen cDNA libraries. Different approaches were developed to capture cDNA clones using genomic DNA attached to a solid substrate (e.g., magnetic beads, nylon filters) as a target to that cDNA clones can hybridize. To identify genomic sequences that could be expressed as a special approach, exon trapping was suggested. For example, in the method of Church et al. [44], the DNA is subcloned into a plasmid expression vector pSPL3 that contains an artificial minigene that can be expressed in a suitable host cell. The minigene consist of the following:

1) Segment of the simian virus 40 (SV40) genome that contains an origin of replication plus a powerful promoter sequence.

2) Two splicing-competent exons separated by an intron that contains a multiple cloning site (MCS).

3) SV40 polyadenylation site.

The recombinant DNA is transfected into a strain of monkey cells, known as COS cells. COS cells were derived from monkey CV-1 cells by artificial manipulation, leading to integration of a segment of the SV40 genome containing a defective origin of replication (*ori*). The integrated SV40 segment in COS cells allows any circular DNA that contains a functional *ori*SV40 to replicate independently of the cellular DNA. Transcription from the SV40 promoter results in an RNA transcript that normally splices to include the two exons of the minigene. If the DNA cloned into the MCS contains a functional exon, then the additional exons can be spliced to the minigene exons. After isolation of RNA and preparing cDNA, PCR reactions with primers specific for the minigene exons is made. Agarose gel electrophoresis can easily distinguish between normal splicing and splicing involving exons in the insert DNA.

Another method to identify expressed sequences is called zoo-blotting. In this method, genomic DNA fragments are hybridized at reduced hybridization stringency against a Southern blot of genomic DNA samples isolated from different species. This method is based on a fact that expressed sequences are usually well conserved, and for example, human DNA fragment with exons from *RB1* gene will hybridize to mouse and pig genomic DNA fragments containing *RB1* gene.

However, all of these methods are laborious, time consuming, and inefficient.

Currently, three approaches are the most popular for the identification of expressed sequences/genes in the genomic DNA: computer analysis of DNA sequences for possible exons, identification of CpG islands that can be done both experimentally and using bioinformatics, and large-scale sequence analysis of transcribed sequences, like sequencing of ESTs, SAGE, and CAGE.

Bioinformatic methods to identify genes will not be discussed here as special journal issues are focused on this topic, and many new programs appeared every year. In principle, these programs could be divided into two classes. One type of program predicts genes using only genomic sequences. These programs can also compare genomic DNA sequences from different organisms and search for conserved sequences. The second type of program compares different cDNA, EST, and genomic sequences to find the most possible gene sequence. This method is more reliable; however, it is important to remember that all computer programs frequently predict nonexisting genes and exons and miss really existing genes and exons. Thus, all computer-predicted genes/exons need experimental confirmation. In cases when a gene has alternative 5′ and 3′ ends and different splicing forms computer programs are not very much helpful.

1.2.7.2 CpG Islands

The dinucleotide CpG is notable because it is greatly under-represented in human DNA, occurring at only about one fifth of the roughly 4% frequency that would be expected by simply multiplying the typical fraction of Cs and Gs (0.21×0.21). The deficit occurs because most CpG dinucleotides are methylated on the cytosine base, and spontaneous deamination of methyl-C residues gives rise to T residues.

Spontaneous deamination of ordinary cytosine residues gives rise to uracil residues that are readily recognized and repaired by the cell. As a result, methyl-CpG dinucleotides steadily mutate to TpG dinucleotides. However, the genome contains many CpG islands that represent stretches of unmethylated DNA with a higher frequency of CpG dinucleotides when compared with the entire genome [9, 10, 13, 32, 33, 45]. CpG islands are believed to preferentially occur at the transcriptional start of genes, and it has been observed that most housekeeping genes have CpG islands at the 5′ end of the transcript. In addition, experimental evidence indicates that CpG island methylation is correlated with gene inactivation and has been shown to be important during gene imprinting and tissue-specific gene expression. Thus, identification of CpG methylation is crucial for isolation of tumor suppressor genes and early diagnosis of cancer [46, 47].

CpG islands are usually 1–2 kb long and are dispersed in the genome. They are called CpG (rich) islands because the %(G+C) of CpG island sequences usually exceed 60% and the human genome contains on average about 40% of C+G content. The combination of G+C richness and lack of CpG suppression means that CpG islands contain 10–20 times more CpGs than an equivalent length of non-island DNA. Altogether, there are about 30,000 to 45,000 islands in the haploid genome (the average spacing is about 1 per 100 kb). It is now clear that the majority (if not all) of CpG islands are associated with genes.

As one of the main goals of the human genome project is the isolation of all genes and the construction of a transcriptional gene map, it is clear that markers located in the CpG islands have an additional value for physical mapping. It has been shown that recognition sites for many of the rare cutting enzymes are closely associated with CpG islands. For example, at least 82% of all NotI and 76% of all XmaIII sites are located in the CpG islands. More than 20% of CpG island-containing genes have at least one NotI site in their sequence, whereas about 65% of those genes have XmaIII site(s). Summarizing the data for all genes (with or without CpG islands), we can conclude that approximately 12% of all well-characterized human genes contain NotI sites, and 43% of them have XmaIII sites. For human genome mapping this means that by sequencing DNA fragments containing a NotI site, it is possible to tag up to one fifth of all expressed genes. As discussed, the recombinant clone containing a NotI (or other rare cutting enzymes) recognition site is called a linking clone.

How many genes contain the human genome? This question is difficult, and the answer is very much dependent on the definition of a gene. According to Celera and HGP, the human genome contains approximately 30,000–35,000 genes.

However, sequencing of NotI linking clones suggested that the human genome contains 15,000–20,000 *Not*I sites, of which 6000–9000 are unmethylated in any particular cell. It means that the human genome contains 45,000–60,000 genes [33].

Comparison of our database containing experimentally obtained NotI linking sequences with complete chromosome 21 and 22 sequences revealed several interesting features. First, it was shown that chromosome 22 contains greater than twofold more genes than chromosome 21 (545:225 = 2.4), and we see the same ratio within the NotI flanking sequences (119:49 = 2.4). We have demonstrated that nearly all our NotI clones contained genes and suggested that 12.5–20% of all genes contain NotI sites [32]. This correlates well with the number of genes on chromosomes 21 and 22 (168/770 = 22%). Second, according to sequencing, the two chromosomes contain 390 NotI sites. Therefore, if we assume that each NotI site is associated with a gene, then almost half of the genes contain NotI sites. This estimate seems excessive.

The human genome sequence data cannot discriminate between methylated and unmethylated cytosines. There are several algorithms for the identification of CpG islands on the basis of primary sequence. One quantitative definition holds that CpG islands are regions of DNA >200 bp with C+G content of >50% and a ratio of the "observed vs. expected" frequency of CG dinucleotides, which exceeds 0.6 [48, 49]. The ratio for the entire genome is approximately 0.2 [9]. According to the previous data, 82% of NotI sites are located in CpG islands (see above). It is important to note that these data were obtained either using computational methods or limited experimental datasets. Using the NotI cloning method, only unmethylated NotI sites can be isolated. Comparing the experimental NotI cloning method and computer-based identification of CpG islands [9, 33], two main features are apparent: The fraction of sequences with >80% CG content is nine times higher in the NotI collection, i.e., 142 vs. 22 sequences. Another striking finding is that even NotI flanking sequences with CG content less than 50% have a very high ratio of observed versus expected frequency of CG dinucleotide: 0.71. This suggests that essentially all NotI flanking sequences generated in the study are located in CpG islands, and therefore, the computational method misses at least 8.7% of CpG islands associated with NotI sites. Thus, existing tools for computational determination of CpG islands fail to identify a significant fraction of functional CpG islands, and unmethylated DNA stretches with a high frequency of CpG dinucleotides can be found even in regions with low CG content.

1.2.7.3 SAGE and CAGE Approaches to Analyze Human Transcriptome and to Estimate the Number of Human Genes

Another approach to identify genes and analyze their expression pattern in human/mammalian cell is large-scale sequencing of transcribed sequences. Sequencing of ESTs and isolation of full-length cDNA clones was known a long time ago and will not be discussed here. Recently two novel high-throughput methods for the analysis of human transcriptome were developed: serial analysis of gene expression (SAGE, [50]) and cap analysis gene expression (CAGE, [51]). Both methods are based on the generation of short sequence tags. Each of these tags is assumed to identify a gene transcript.

SAGE is based on two principles. First, a specific adaptor is ligated to cDNA close to the 3′ end of the mRNA (polyA tail). This adaptor contains a recognition site for a type IIs restriction aendonuclease. Originally, it was FokI, and in a recent modification (LongSAGE), MmeI was used [50, 52]. FokI generates 13–14-bp, and MmeI generates 20–21-bp tags. The 21-bp tag consists of a constant 4-bp sequence representing the restriction site at which the transcript was cleaved, followed by a unique 17-bp sequence derived from an adjacent sequence in each transcript. Theoretical calculations show that >99.8% of 21-bp tags are expected to occur only once in genomes the size of the human genome. Similar analyses based on actual sequence information from ~16,000 genes suggest that >75% of 21-bp tags would be expected to occur only once in the human genome, with the remaining tags matching duplicated genes or repeated sequences. Second, to optimize the quantification of transcripts, tags are ligated together to form "ditags," which are then concatenated and cloned. Sequencing tag concatemers in parallel allows the identification of up to ~30 tag sequences in each sequencing reaction. Matching tags to genome sequences identifies the gene corresponding to each tag, and the number of times a particular tag is observed provides a quantitative measure of transcript abundance in the RNA population.

CAGE is similar to SAGE in principle; however, it is based on preparation and sequencing of concatamers of DNA tags deriving from the initial 20 nucleotides from 5′ end mRNAs.

The method essentially uses cap-trapper full-length cDNAs to the 5′ ends of which linkers are attached [53, 54]. This is followed by the cleavage of the first 20 base pairs by class IIs restriction enzymes (Mme) and then another linker is ligated. After it, PCR amplification, excision of tags, their concatamerization, and cloning of the CAGE tags was performed. CAGE tags derived by sequencing these clones are mapped to the genome and used for expression analysis, as well as for the determination of the 5′ end borders of new transcriptional units. Thus, in contrast to SAGE, CAGE allows high-throughout gene expression analysis and the profiling of transcriptional start points, including promoter usage analysis.

SAGE and CAGE concatamer sequencing is more cost-effective than a full-length cDNA library sequencing because of the much higher throughput of identified tags.

In a recent study (FANTOM 3 project), full-length cDNA isolation and 5′- and 3′-end sequencing of cloned cDNAs was combined with CAGE, gene identification signature (GIS), and gene signature cloning (GSC) ditag technologies for the identification of RNA and mRNA sequences corresponding to transcription initiation and termination sites [55]. In this study, paired initiation and termination sites were identified and the boundaries for 181,047 independent transcripts in the mouse genome were established. In total, 1.32 5′ start sites for each 3′ end and 1.83 3′ ends for each 5′ end are found. Based on these data, the number of transcripts is at least one order of magnitude larger than the estimated 22,000 "genes" in the mouse genome, and the large majority of transcriptional units have alternative promoters and polyadenylation sites. To extend the mouse data, two human CAGE libraries, one constructed with random primers and the other with oligo-dT primers, were combined to produce 1,000,000 CAGE tags. Mapping of these tags to the human genome identified the likely promoters and transcriptional starting site of many genes and clearly indicates that the same level of transcriptional diversity occurs in humans as in mice and is at least 10 times as great as the number of "genes."

Of the 102,281 FANTOM3 cDNAs, 34,030 lack any protein-coding sequence (CDS) and are annotated as non-protein-coding RNA (ncRNA). The function of ncRNAs is a matter of debate [56], but it is clear that some of them are very important for transcription regulation. Some ncRNAs are highly conserved even in distant species. According to the previous data, only 1–1.5% of the human genome is spanned by exons and 75% of the genome is non-transcribed intergenic DNA. However, FANTOM 3 data demonstrated that the majority (more than 60%) of the mammalian genome is transcribed and commonly from both strands.

Analysis of the output of FANTOM 3 suggested that many more transcripts were still to be discovered and only the future can reveal the actual complexity in the mammalian transcriptome. Very similar conclusions were made in another detailed study of human transcriptome [57, 58].

1.2.8   WAS IT WORTH IT TO SEQUENCE HUMAN GENOME?

The Human Genome Project was controversial from the beginning. Some criticism concerns scientific strategy and the ways in which it was executed [1–4]. Is it important to sequence the whole human genome or only coding DNA sequences (genes)? Now it is clear that DNA sequences recognized previously as a "junk" in fact are very important for understanding how human genome functions. This "junk" DNA contains important regulatory sequences, ncRNAs, and probably the most intriguing discoveries will come from areas considered as wastelands. Many ethical questions arose during the HGP. For instance, sample collection from individuals was often done without proper consultation or explanation of the possible future use. As the HGP moved forward, commercial concerns were increasingly taking a stake in the research. Large financial investments from big companies can monopolize certain research areas, for example, sequencing of cDNA and gene identification. By subsequently seeking patents to protect their financial investments, they raised a question: Who owns the human genome?

Any major scientific advance carries with it the fear of exploitation, and HGP is not an exception. For example, knowledge of disease-associated mutations can be used for disease prevention, and at the same time, it can be used for discrimination. The comprehensive knowledge of human genome and genes can lead to biological determinism and revival of eugenics with all its negative consequences.

Another objection was that funding of the HGP would be done at the expense of other scientific directions.

The human genome map and sequence still contain many uncertainties, and only careful examination by some research group focusing on a particular genomic region can produce a reliable map and sequence. For instance, we have cloned all *Not*I sites and constructed a physical map for two chromosome 3 regions containing tumor suppressor genes [26, 59, 61]. In the course of these studies, it became apparent that large-insert vectors from these regions were unstable, sensitive to deletions and rearrangements, and the original maps produced by HGP were erroneous [26, 59, 60].

It should be emphasized that the enormous efforts deployed on sequencing the human genome are extremely important; however, a critical role remains for verified, integrated maps. In sequencing, the short and long repeats spread throughout the genome are sources of numerous errors. These errors are difficult to identify with the shotgun strategy, but they become evident when mapping information is combined with the sequence. Furthermore, difficulties in sequence assembly caused by the existence of large families of recently duplicated genes and pseudogenes are easier to resolve using integrated maps.

In many cases, sequence and mapping information is duplicated, overlapping, or contradictory. One must always keep in mind that even absolutely correct and long nucleotide sequences may be localized incorrectly along the chromosomal DNA if the appropriate accompanying mapping information is ignored. For this reason, despite the vast amount of information currently available, there is an urgent need to reconcile this information in a unified framework, to generate an integrated noncontroversial map for each chromosome.

The draft human genome sequence produced by HGP was estimated to cover at least 92% of the whole human genome. However, it contained only 55.7% of the *Not*I flanking sequences and Celera's database contained matches to 57.2% of the clones. The data suggest that the shotgun sequencing approach used to generate the draft human genome sequence resulted in a bias against cloning and sequencing of *Not*I flanks. Even in 2005 up to 5% of our NotI flanking sequences (that are also incomplete!) are not present in the human genome sequence available in public databases.

Several explanations can be offered to account for the low representation of *Not*I flanking sequences in the draft human genome sequences. One potential explanation is that the cloning of some *Not*I containing regions may be selected against experiments with large-insert cloning vectors. Our experience has also proven that even in small-insert plasmid vectors, some human sequences are more easy to clone than other. In our procedure, we directly selected clones containing *Not*I sites, and in a shotgun sequencing approach, such sequences could be under-represented. An alternative explanation, based on the observation that some *Not*I flanking sequences can have 100% identity over long DNA stretches [62], is that some *Not*I sites were incorrectly fused in the assembly process. Furthermore, our experience demonstrated that sometimes it is very difficult to read *Not*I flanking sequences because of extremely high CG content. During human genome assembling, such sequences would be eliminated as possessing low-quality data. Further experimental analysis is needed to conclusively identify the cause(s) of the bias.

HGP resulted in enormous progress in automation of sequencing and other molecular biology methods.

It gives us new tools for disease-gene discovery. In fact, the information obtained during the HGP helps rather than hinders small research groups. It is still important to remember that only the euchromatin portion of the genome was sequenced, and it is not clear to what extent it was done. Many heterochromatin regions are practically not sequenced, and they most probably contain many unknown genes and other functionally important sequences. Thus, the draft human genome sequence was made, but it is a very long way to obtain a really complete human genome sequence and to understand how many genes it contains and how it functions.

ACKNOWLEDGEMENT

This work was supported by research grants from the Swedish Cancer Society, the Swedish Research Council, the Swedish Foundation for International Cooperation in Research and Higher Education (STINT), the Swedish Institute, the Royal Swedish Academy of Sciences, INTAS, and the Karolinska Institute.

REFERENCES

1. The Human Genome. Special Issue. (2001). *Nature*, *409*, No.6822.
2. The Human Genome. Special Issue. (2001). *Science*, *291*, No.5507.

3. Brown T A (1999). *Genomes*, Wiley, New York co-published with BIOS Scientific Publishers, Oxford, UK.

4. Strachan T, Read A (1999). *Human Molecular Genetics*, Wiley, New York co-published with BIOS Scientific Publishers, Oxford, UK.

5. Fleischmann R D, Adams M D, White O, et al. (1995). Whole-genome random sequencing and assembly of Haemophilus influenzae Rd. *Science*. 269:496–512.

6. Goffeau A, Barrell B G, Bussey H, et al. (1996). Life with 6000 genes. *Science*. 274:546–567.

7. The *C. elegans* Sequencing Consortium. (1998). Genome sequence of the nematode C. elegans: A platform for investigating biology. *Science*. 282:2012–2018.

8. Dunham I, Hunt A R, Collins J E, et al. (1999). The DNA sequence of human chromosome 22. *Nature*. 402:489–495.

9. Lander E S, Rogers J, Waterston R H, et al. (2001). Initial sequencing and analysis of the human genome. *Nature*. 409:860–921.

10. Venter G J, Adams M D, Myers E W, et al. (2001). The sequence of the human genome. *Science*. 291:1304–1351.

11. Ausubel F M, Kingston R E, Brent R, et al. (1987–2004). *Current Topics in Molecular Biology*, Wiley, New York.

12. Sambrook J, Fritsch E F, Maniatis T (1989). *Molecular Cloning: A Laboratory Manual* (2nd ed). Cold Spring Harbour Laboratory Press, Cold Spring Harbour; N.Y.

13. Zabarovsky E R (2004) Genomic DNA Libraries, Construction and Applications. In R A Meyers (ed.), *Encyclopedia of Molecular Cell Biology and Molecular Medicine*, Vol. 5 Wiley-VCH, Weinheim, Germany, pp. 441–468.

14. Zabarovsky E R, Winberg G, Klein G (1993). The SK-diphasmids—vectors for genomic, jumping and cDNA libraries. *Gene*. 127:1–14.

15. Zabarovsky E R, Turina O V, Nurbekov M K, et al. (1989). SK18—diphasmid for constructing genomic libraries with blue/white selection, multiple cloning sites and with the M13 and lambda ori. *Nucl. Acids Res*. 17:3309.

16. Short J M, Fernandez J M, Sorge J A, et al. (1988). Lambda ZAP: A bacteriophage lambda expression vector with in vivo excision properties. *Nucleic Acids Res*. 16:7583–7600.

17. Carninci P, Shibata Y, Hayatsu N, et al. (2001). Balanced-size and long-size cloning of full-length, cap-trapped cDNAs into vectors of the novel lambda-FLC family allows enhanced gene discovery rate and functional analysis. *Genomics*. 77:79–90.

18. Frischauf A M, Lehrach H, Poustka A, et al. (1983). Lambda replacement vectors carrying polylinker sequences. *J Mol Biol*. 170:827–842.

19. Frischauf A M, Murray N, Lehrach H (1987). Lambda phage vectors–EMBL series. *Methods Enzymol*. 153:103–115.

20. Zabarovsky E R, Allikmets R L (1986). An improved technique for the efficient construction of gene library by partial filling-in of cohesive ends. *Gene*. 42:119–123.

21. Latif F, Tory K, Gnarra J, et al. (1993). Identification of the von Hippel-Lindau disease tumor suppressor gene. *Science*. 260:1317–1320.

22. Wang I, Franco B, Ferrero G B, et al. (1995). High-density physical mapping of a 3-Mb region in Xp22.3 and refined localization of the gene for X-linked recessive chondrodysplasia punctata (CDPX1). *Genomics*. 26:229–238.

23. Cheng S, Lutfalla G, Uze G, et al. (1993). GART, SON, IFNAR, and CRF2–4 genes cluster on human chromosome 21 and mouse chromosome 16. *Mamm Genome*. 4:338–342.

24. Cohen D, Chumakov I, Weissenbach J (1993). A first-generation physical map of the human genome. *Nature*. 366:698–701.

25. Frengen E, Zhao B, Howe S, et al. (2000). Modular bacterial artificial chromosome vectors for transfer of large inserts into mammalian cells. *Genomics*. 68:118–126.

26. Protopopov A, Kashuba V, Zabarovska V I, et al. (2003). An integrated physical and gene map of the 3.5-Mb chromosome 3p21.3 (AP20) region implicated in major human epithelial malignancies. *Cancer Res*. 63:404–412.

27. Ichikawa H, Hosoda F, Arai Y, et al. (1993). A NotI restriction map of the entire long arm of human chromosome 21. *Nat Genet*. 4:361–366.

28. Poustka A, Lehrach H (1988). Chromosome Jumping: A Long Range Cloning Technique. In Setlow J K (ed.), *Genetic Engineering Principles and Methods*, Vol. 10, Brookhaven National Laboratory, Upton, N.Y. and Plenum Press, New York, pp. 169–193.

29. Zabarovsky E R, Kashuba V I, Gizatullin R Z, et al. (1996). NotI jumping and linking clones as a tool for genome mapping and analysis of chromosome rearrangements in different tumors. *Cancer Detect. Prevention*. 20:1–10.

30. Zabarovsky E R, Boldog F, Thompson T, et al. (1990). Construction of a human chromosome 3 specific *Not*I linking library using a novel cloning procedure. *Nucl. Acids Res*. 18:6319–6324.

31. Zabarovsky E R, Boldog F, Erlandsson R, et al. (1991). A new strategy for mapping the human genome based on a novel procedure for constructing jumping libraries. *Genomics*. 11:1030–1039.

32. Allikmets R L, Kashuba V I, Bannikov V M, et al. (1994). NotI linking clones as tools to join physical and genetic mapping of the human genome. *Genomics*. 19:303–309.

33. Kutsenko A, Gizatullin R, Al-Amin A N, et al. (2002). NotI flanking sequences: A tool for gene discovery and verification of the human genome. *Nucleic Acids Res*. 30:3163–3170.

34. Zabarovsky E R, Kashuba V I, Zakharyev V M, et al. (1994). Shot-gun sequencing strategy for long range genome mapping: First results. *Genomics*. 21:495–500.

35. Myers E W, Sutton G G, Delcher A L, et al. (2000). A whole-genome assembly of Drosophila. *Science*. 287:2196–2204.

36. Drmanac R, Drmanac S, Chui G, et al. (2002). Sequencing by hybridization (SBH): Advantages, achievements, and opportunities. *Adv. Biochem. Eng. Biotechnol*. 77:75–101.

37. Brenner S, Johnson M, Bridgham J, et al. (2000). Gene expression analysis by massively parallel signature sequencing (MPSS) on microbead arrays. *Nature Biotechnol*. 18:630–634.

38. Shendure J, Porreca G J, Reppas N B, et al. (2005). Accurate multiplex polony sequencing of an evolved bacterial genome. *Science*. 309:1728–1732.

39. Margulies M, Egholm M, Altman W E, et al. (2005). Genome sequencing in microfabricated high-density picolitre reactors. *Nature*. 437:376–380.

40. Ronaghi M, Pettersson B, Uhlen M, et al. (1998). A sequencing method based on real-time pyrophosphate. *Science*. 281:363–365.

41. Jarvius J, Nilsson M, Landegren U (2003). Oligonucleotide ligation assay. *Methods Mol. Biol*. 212:215–228.

42. Rogers Y H, Venter JC (2005). Genomics: Massively parallel sequencing. *Nature*. 437:326–327.

43. Zabarovska V I, Gizatullin R G, Al-Amin A N, et al. (2002). Slalom libraries: A new approach to genome mapping and sequencing. *Nucleic Acids Res*. 30:1–8.

44. Church D M, Stotler C J, Rutter J L, et al. (1994). Isolation of genes from complex sources of mammalian genomic DNA using exon amplification. *Nat. Genet.* 6:98–105.

45. Bird A (2002). DNA methylation patterns and epigenetic memory. *Genes Dev.* 16:6–21.

46. Palmisano W A, Divine K K, Saccomanno G, et al. (2000). Predicting lung cancer by detecting aberrant promoter methylation in sputum. *Cancer Res.* 60:5954–5958.

47. Sugimura T, Ushijima T (2000). Genetic and epigenetic alterations in carcinogenesis. *Mutat. Res.* 462:235–246.

48. Gardiner-Garden M, Frommer M (1987). CpG islands in vertebrate genomes. *J. Mol. Biol.* 196:261–282.

49. Larsen F, Gundersen G, Lopez R, et al. (1992). CpG islands as gene markers in the human genome. *Genomics.* 13:1095–1107.

50. Velculescu V E, Zhang L, Vogelstein B, et al. (1995). Serial analysis of gene expression. *Science.* 270:484–487.

51. Shiraki T, Kondo S, Katayama S, et al. (2003). Cap analysis gene expression for high-throughput analysis of transcriptional starting point and identification of promoter usage. *Proc. Natl. Acad. Sci. U.S.A.* 100:15776–15781.

52. Saha S, Sparks A B, Rago C, et al. (2002). Using the transcriptome to annotate the genome. *Nat. Biotechnol.* 20:508–512.

53. Carninci P, Hayashizaki Y (1999). High-efficiency full-length cDNA cloning. *Methods Enzymol.* 303:19–44.

54. Shibata Y, Carninci P, Watahiki A, et al. (2001). Cloning full-length, cap-trapper-selected cDNAs by using the single-strand linker ligation method. *BioTechniques.* 30:1250–1254.

55. Carninci P, Kasukawa T, Katayama S, et al. (2005). The transcriptional landscape of the mammalian genome. *Science.* 309:1559–1563.

56. Szymanski M, Barciszewska M Z, Erdmann V A, et al. (2005). A new frontier for molecular medicine: Noncoding RNAs. *Biochim. Biophys. Acta.* 1756:65–75.

57. Kapranov P, Drenkow J, Cheng J, et al. (2005). Examples of the complex architecture of the human transcriptome revealed by RACE and high-density tiling arrays. *Genome Res.* 15:987–997.

58. Cheng J, Kapranov P, Drenkow J, et al. (2005). Transcriptional maps of 10 human chromosomes at 5-nucleotide resolution. *Science.* 308:1149–1154.

59. Wei M-H, Latif F, Bader F, et al. (1996). Construction of a 600 Kilobase cosmid clone contig and generation of a transcriptional map surrounding the lung cancer tumor suppressor gene (TSG) locus on human chromosome 3p21.3.: Progress toward the isolation of a lung cancer TSG. *Cancer Research.* 56:1487–1492.

60. Lerman M I, Minna J D. et al. (2000) The International Lung Cancer Chromosome 3p21.3 Tumor Suppressor Gene Consortium. The 630-kb lung cancer homozygous deletion region on human chromosome 3p21.3: Identification and evaluation of the resident candidate tumor suppressor genes. *Cancer Research.* 60:6116–6133.

61. Imreh S, Klein G, Zabarovsky E (2003). Search for unknown tumor antagonizing genes. *Genes Chromos. & Cancer.* 38:307–321.

62. Kashuba V I, Protopopov A I, Kvasha S M, et al. (2002). hUNC93B1: A novel human gene representing a new gene family and encoding an unc-93-like protein. *Gene.* 283:209–217.

FURTHER READING

Burke D T, Carle G F, Olson M V (1987). Cloning of large segments of exogenous DNA into yeast by means of artificial chromosome vectors. *Science*. 236:806–812.

Burke D T, Olson M V (1991). Preparation of clone libraries in yeast artificial-chromosome vectors. *Methods Enzymol*. 194:251–270.

Shizuya H, Birren B, Kim U J, et al. (1992). Cloning and stable maintenance of 300-kilobase-pair fragments of human DNA in Escherichia coli using an F-factor-based vector. *Proc Natl Acad Sci USA*. 89:8794–8797.

Ioannou P A, Amemiya C T, Garnes J, et al. (1994). A new bacteriophage P1-derived vector for the propagation of large human DNA fragments. *Nat Genet*. 6:84–89.

FURTHER READING

1.3

OVERVIEW: DIFFERENTIATING ISSUES IN THE DEVELOPMENT OF MACROMOLECULES COMPARED WITH SMALL MOLECULES

GÜNTER BLAICH,[1] BERND JANSSEN,[2] GREGORY ROTH,[2] AND JOCHEN SALFELD[2]

[1] Abbott Bioresearch Center, Worcester, Massachusetts
[2] Abbott GmbH & Co. KG, Ludwigshafen, Germany

Chapter Contents

1.3.1	Introduction	90
1.3.2	Target Selection and Drug Discovery Approaches	91
1.3.3	Target Product Profile	94
1.3.4	Lead Identification	94
	1.3.4.1 Small Molecules	98
	1.3.4.2 Macromolecules	100
1.3.5	Lead Optimization	100
	1.3.5.1 Small Molecules	100
	1.3.5.2 Macromolecules	102
1.3.6	Selection of Clinical Candidates and Preclinical Development	102
1.3.7	Production of Active Pharmaceutical Ingredients	104
	1.3.7.1 Small Molecules	108
	1.3.7.2 Macromolecules	108
1.3.8	Formulation and Delivery Devices	110
	1.3.8.1 Formulation Support in Discovery	110
	1.3.8.2 Formulation Activities Until Clinical Proof of Concept	112
	1.3.8.3 Formulation of Activities in the Commercial Phase (Clincal Phase IIB until Market)	112
1.3.9	Clinical Development	113
1.3.10	Intellectual Property	114
1.3.11	Generics, "Biosimilar Products," or "Follow-on Biologics"	115

Handbook of Pharmaceutical Biotechnology, Edited by Shayne Cox Gad.
Copyright © 2007 John Wiley & Sons, Inc.

1.3.12   Summary   116

    1.3.12.1   Step 1: Target Selection and Validation   116

    1.3.12.2   Step 2: Choice Between Small-Molecule and

                 Macromolecule Platforms   116

    1.3.12.3   Step 3: Definition of Target Product Profile   117

    1.3.12.4   Step 4: Drug Discovery and Development   117

    Acknowledgment   118

    References   118

1.3.1   INTRODUCTION

The probability of success in drug discovery and development is highly dependent on good planning right from the beginning, a rational approach to each phase in the light of the ultimate goal, and—unsurprisingly—good science. There are many possible reasons for the recent dearth of new molecular entity or new biological entity submissions to regulatory agencies [1, 2]. One possible reason is suboptimal discovery and development strategies for novel targets with little known biology [3]. In the late 1980s and early 1990s, drug discovery was specialized to an extent, with small-molecule drug discovery the domain of pharmaceutical companies and macromolecule discovery practiced by biotechnology companies. Today, drug discovery in both areas is practiced widely in both arenas. Large pharmaceutical companies have embraced biological therapies, in particular therapeutic antibodies, and large biotechnology companies have added small-molecule discovery to complement their macromolecular portfolio with orally available medicines. Pharmaceutical companies realize this portfolio diversification either by expanding or readjusting internal research or by collaborations with biotechnology companies. In either case, preclinical development is often conducted by the larger of the partners, and a comprehensive understanding about the different approaches to either drug class is critical.

Small molecules and macromolecules differ in much more than size, and this overview will discuss the consequences for each stage of drug discovery and development.

With a few exceptions, small molecules are synthesized chemically and are smaller than 600 daltons. For the purpose of this overview, macromolecules are defined as protein molecules of approximately 150 Kd, designed to bind to and functionally affect drug targets. The majority of these macromolecules are antibodies and antibody derivatives, which constitute approximately 25% of therapeutic entities currently in development [4]. Another highly represented class includes receptor fusion proteins. This overview will not describe other large molecules, including therapeutic proteins (naturally existing human proteins used for therapeutic purposes), antisense or siRNA, gene therapy, and others. However, many of the general concepts discussed here can be applied to such drug candidates as well.

The molecular weight and domain structure of antibodies or antibody derivatives are the basis for important limitations and potential advantages when

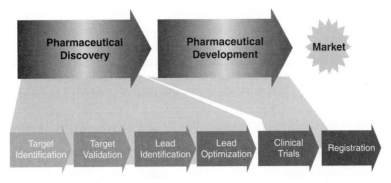

Figure 1.3-1. Critical path in the drug discovery process.

compared with the traditional drug discovery process. Although antibodies are at somewhat of a disadvantage because they are limited to soluble, cell-surface, and tissue matrix targets, their domain structure leads to very significant time-based advantages in the lead finding and the lead optimization phase (Figure 1.3-1).

Drug discovery today relies on the selection of specific and disease-relevant drug targets. Traditional (pre-1990) drug screening, based purely on observed pharmacological effects *in vivo*, has fallen out of favor due to regulatory pressures to conduct discovery research on well-characterized and validated targets. New screening technologies, coupled with the concept of using small molecules to complement gene-based methods of perturbing protein function, has generated renewed interest in this past paradigm and allows for the successful combination of pharmacological drug screening with the identification of specific and potentially new and novel targets in the "druggable genome" [5, 6]. The current general focus toward drug candidates with known mechanisms of action has important consequences for preclinical development. Whereas more standardized safety assessments were appropriate for drug candidates with unknown mechanisms, the understood biology of the drug action today allows and requires an appropriate adjustment at each stage. However, caution about any claim of a "known mechanism" is warranted. Indeed, although the industry worked previously on drug targets with, in some cases, more than 20 years of academic research into target biology, the Human Genome Project has led to work and focus on poorly understood targets. Therefore, drug discovery and development of the future has to undertake target validation work in parallel with drug discovery, consider the assumed mechanism in all formal pharmacology and safety assessments, and maintain the safety net of traditional drug safety studies.

1.3.2 TARGET SELECTION AND DRUG DISCOVERY APPROACHES

The term "target" describes a specific molecular entity, most often a protein, presumed to play a role in the pathogenesis of a disease state. A target can then be defined precisely with molecular biology tools and can be affected by the action of a specific drug in a manner beneficial to the patient. Estimates place the number of targets modulated by currently marketed drugs at between 120 and 500, with only about 7% of medicines not having a clearly defined target [7]. Target selection starts typically with an analysis of met or unmet medical need and a comprehensive understanding of the pathophysiogical processes underlying the human disease of interest. Companies or academic researchers are typically focused on a limited number of key indications (such as cardiovascular disease, diabetes, autoimmune disease, and cancer) and have significant and relevant internal discovery expertise. This includes, to some extent, novel target discovery and will in most cases include target identification and validation [8]. Indeed, some level of confidence in mechanism-based target validation is a key component of any discovery project considering that, industry-wide, 30–40% of experimental drugs fail because an inappropriate or non-physiologically relevant biological target was pursued [9].

Once a preferred set of targets is identified, the viability and tractability of a small molecule or macromolecule approach can be assessed. In general, differences in physicochemical properties between small and macromolecules may help drive the approach strategy. For example, intracellular and nuclear proteins are often best targeted by small molecules that would more easily overcome cellular permeability barriers in order to reach the desired target. The molecular weight of approximately 150 Kd precludes antibodies from entering cells. Some initial progress has been made to overcome this limitation through the use of viral peptides as vectors for macromolecular delivery into cells [10–15]. Commercially viable therapeutic applications in this direction are not expected to materialize in the short term, because the research efforts are still in the early discovery phase. Another approach under investigation is the use of antibodies directed toward specific internalizing receptors, which can then carry their "payloads" into cells [16]. Cell-surface receptors are often ideal candidates for both small molecule and macromolecule approaches, whereas soluble or circulating proteins are most often targeted by high-affinity macromolecules. Typical selection criteria are summarized in Table 1.3-1 and may be used to determine which technology could be applicable. Additional factors that can influence target selection include those listed in Table 1.3-2. The key point is that the decision of whether to use a small molecule or macromolecule approach within a discovery project needs to be made early in the process, because intrinsic molecular differences will subsequently affect program strategy all the way through to the clinical setting.

The size and structure of small molecules and macromolecules also have very important consequences for some key steps in drug discovery. It is readily apparent that the molecular weight of a drug determines the "footprint" of the drug on the target and related molecules. Macromolecules, including antibodies and receptor-based fusion proteins, have evolved to be able to accurately distinguish between closely related target structures. They can accomplish this by using structural differences on a larger surface area than is typically accessible for small-molecule affinity. Indeed, the footprint of antibodies is estimated to be on the order of about 5 times larger than that of a typical small molecule. However, high specificity requires careful antibody selection and optimization. Desired specificity has to be accomplished while maintaining orthologue (similar gene family member in another species) binding to enable preclinical development.

TABLE 1.3-1. Criteria for Selecting a Small Molecule or Macromolecule Approach

Target Features	Small Molecule	Macromolecule
Extracellular Location		
Soluble nonenzymatic protein/peptide	−	+
Soluble enzyme	+	(+)[1]
Soluble hormones and mediators	−	+
	−	+
Receptor for protein/peptide (proteinergic GPCR)	−	+
Receptor for nonpeptide hormone (aminergic GPCR)	+	(+)[2]
Pathogen external protein, nonenzymatic	−	+
Pathogen, external protein, enzymatic	+	(+)[1]
Intracellular Location		
Receptor associated enzymatic proteins	+	(+)[3]
Cytosolic enzymatic proteins	+	−
Nonenzymatic proteins	+	−
Enzymes of pathogenic organisms	+	−

[1] Enzyme inhibitory antibodies are theoretically possible, but evidence is sparse [17].
[2] Agonistic or agonistic antibodies are theoretically possible, but evidence is sparse [18].
[3] Indirect effect by interaction with extracellular receptor portion feasible; no evidence for enzyme inhibitory antibodies [19].
Abbreviation: GPCR = G-protein-coupled receptor.

TABLE 1.3-2. Additional Target Selection Criteria

Target Selection Criteria	Small Molecule	Macromolecule
CNS location	Requires specific compound characteristics for BBB penetration	Required dose to achieve effect *in vivo* may be too high
Homology within protein family	Specificity may be difficult to achieve	Specificity achievable, but orthologue binding required
Druggability	Assessed after high-throughput library screen, consider macromolecule if screen fails	Consider different approaches if the first fails (*in vivo*/*in vitro*)
Protein/protein interaction	Difficult to achieve, but examples available [20]	Often achievable

Abbreviation: CNS = central nervous system.

Another key difference related to size is the difference between small molecules and macromolecules in the degree of interdependence of domains. Although careful evaluation of a structure-activity-relationship (SAR) can readily determine domains in small molecules affecting metabolism, specificity, and other alternative domains, the optimization of these domains will often influence other parameters in a positive or, more often, in a negative manner. In macromolecules, the protein domains are for the most part independent, and optimization of a single parameter, such as specificity in the antigen binding region, will typically not affect half-life, which is determined by a different domain [21–23]. Optimization of small molecules is therefore "multidimensional," whereas macromolecule optimization is, to the most extent, "one-dimensional." For this, and other reasons, drug discovery using macromolecules tends to be more rapid than lead identification and optimization of small molecules.

1.3.3   TARGET PRODUCT PROFILE

The critical role of an early agreement on a target product profile (TPP) for any new discovery project has been reviewed recently [3]. Briefly, a target product profile describes the exact expectations for the drug candidate in specific quantitative terms. The parameters will vary depending on whether the desired molecule would be "first in class" or "best in class" for a given target. For best & in class, the TPP is often more rigorous because the new therapeutic must be demonstrated to be superior to accepted standard-of-care entities (gold standard) currently on the market. In general, the TPP describes the main development objectives in a precise, focused manner. It can serve as a tool for planning and decision making through the course of the research and development program. Ideally it anticipates product claims that can be realistically expected based on preclinical results and can be translated to and confirmed during development. For small-molecule drugs, this includes the proposed mechanism of drug action, specificity, affinity, *in vitro* and *in vivo* potency, pharmacokinetics, tissue distribution, solubility (LogP [less hydrophobic], $LogS_w$ [more soluble]), metabolism, chemical accessibility, acceptable safety profile (both mechanistic and nonmechanistic), and biomarker requirements. For macromolecules, the criteria include mechanism of action, epitope, specificity (including reactivity with relevant preclinical toxicology species), affinity, potency, constant region isotype, effector function readout, expression rate in mammalian cells, stability in typical formulations, side effects, pharmacokinetics from relevant models, and biomarker requirements. Target product profiles should be agreed to by all functions who will be involved in the project during its lifetime, including discovery, preclinical safety, clinical, manufacturing, and others. Modifications of a target product profile should be considered carefully and only based on new data.

1.3.4   LEAD IDENTIFICATION

Table 1.3-3 summarizes the most prevalent key technologies available today for lead identification of small molecules and macromolecules. This landscape is ever-

TABLE 1.3-3. Lead Identification Technologies

Technology	Small Molecule		Macromolecule (Antibody)		
	Description	Reference	Technology	Description	Reference
High-throughput screening	An instrumentation-driven, rapid, high-volume approach to assay single compounds or compound mixtures against a biochemical target	Archer [24]; Bajorath [25]; Kariv [26]	Humanization of murine antibodies	*In vitro* technologies CDR grafting with and without framework mutations	Lo [27]
Affinity screening by mass spectrometry	High-throughput assay technology based on ligand–protein affinity as detected by mass spectrometry	Markara [28]; Comess [29]	Resurfacing of murine antibodies	Selective replacement of surface residues of mouse antibodies	Roguska [30]; Zhang [31]
			Phage display	*De novo* antibody selection from very large and diverse phage libraries	Hoogenboom [32]; Marks [33]
Directed and focused combinatorial libraries	Knowledge-driven combinatorial library design and synthesis around target class-related structures	Keri [34]; Miller [35]	EBV transformed human B cells	*In vivo* technologies Useful for natural or anti-pathogen antibodies, can be combined with phage display	Maeda [36]

TABLE 1.3-3. *Continued*

Small Molecule			Macromolecule (Antibody)		
Technology	Description	Reference	Technology	Description	Reference
Fragment-based screening Example: NMR screening/SAR by NMR techniques	A technique for identifying one or more low-molecular-weight, weak affinity molecules for a given target with synthetic optimization driven toward increasing affinity	Mitchell [37]; Verdonk [38]; Petros [39]; Sillerud [40]	Trioma technology	Selection of human antibodies to pathogen or cancer targets by fusion of patient lymphocytes with a human heteromyeoloma line and subsequent selection in microtiter plates	Vollmers [41]
Virtual screening	The use of knowledge-driven computational selection and scoring methods to identify molecules that have a high probability for interaction with a biochemical target	Alvarez [42]; Anderson [43]	Transgenic mice	Immunization of mice transgenic for portions of the human Ig repertoire followed by traditional hybridoma technology	Lonberg [44]

Pharmacophore modeling	A complementary virtual selection or design exercise based on specific compound class SAR knowledge; uses three-dimensional conformation of molecules and spatial requirements	Guener [45]
QSAR	Mathematical relationships linking chemical structure and pharmacological activity in a quantitative manner for a series of compounds. Methods that can be used in QSAR include various regression and pattern recognition techniques	Lill [46]

Abbreviations: CDR = complementary determining region; EBV = Epstein–Barr virus; NMR = nuclear magnetic resonance; QSAR = quantitative structure-activity relationship; SAR = structure-activity relationship.

changing, and new technological innovations occur frequently. As a caveat, fusion proteins will often use preexisting binding domains (typically receptors) and are not considered in this table.

1.3.4.1   Small Molecules

Collections and Screening. Frequently, a small-molecule drug discovery approach identifies the initial lead molecules primarily by high-throughput screening of vast corporate compound collections against a biochemical target that is associated with a disease state. In general, most pharmaceutical companies will screen between 750,000 to up to 2 million chemical entities in search of a hit set. Although many corporate collections can exceed the 2 million sample mark, it becomes an economic exercise to reduce this, as well as a case of data management to effectively process and act on the results. A variety of innovative high-throughput screening assay formats have been developed that render this process efficient for surveying a reduced set of 500,000 to about 1.5 million entities as either single compounds or mixtures [25]. With increasing pressure to identify and optimize molecules faster, this process has been supplemented with alternative lead identification approaches such as nuclear magnetic resonance (NMR) [39, 40], fragment-based screening [37, 38], and others (*see* Table 1.3-3). Outside of the traditional microtiter plate-based biochemical assays, significant utilization of affinity selection mass spectrometry (ASMS) has become a primary screening paradigm in the past several years. Mass spectrometry-based measurement of small-molecule–protein interaction has subsequently produced results for challenging targets and has provided a cost-effective, rapid method to screen large compound collections while validating hits using plate-based biochemical techniques on smaller subsets [28, 29]. Various *in silico* hit selection techniques have been developed and applied as complementary approaches [47] to minimize the physical number of entities to be tested as well as to optimize the chances, based on appropriate training sets, to identify active chemical space within a large collection. The most common techniques involve a target-based virtual screening process whereby compounds with molecular similarity to known active small molecules are estimated to have affinity toward a biological target binding site as estimated through dock-ing and scoring functions [48, 49]. Selection techniques can be supplemented through the use of pharmacophore modeling [45] and quantitative structure-activity relationship (QSAR) screening [46]. Results with this approach can be limited because it does not initially involve direct biological assay of diverse compounds but a knowledge-based selection of a "high probability" subset of compounds to be screened. A high level of success remains to be achieved for this technique. More frequently, a diverse subset of compounds that should represent "hit-like" and "lead-like" molecules is selected based on calculated physicochemical properties [26]. These compounds are expected to provide ideal starting points for a medicinal chemistry optimization program because they are preselected to have the probability of acceptable solubility, perme-ability, and properties such as molecular weight where they can be subsequently elaborated on and optimized.

To further enhance the chances for success in the screening arena, most phar-maceutical companies have expended considerable effort in acquiring molecules with hit-like and lead-like properties to improve the quality of their collections [50]. Besides general acquisitions of diverse arrays of molecules, many have also focused on the generation of gene family or target class directed libraries [34], and a slight resurgence in natural products and natural product-like libraries has been seen [51]. Currently on the horizon are the emerging technologies using the concepts of chemical genomics and high content screening, and a variety of industry and academic laboratories are paving the path forward [52].

Hit to Lead. The overall probability that a high-throughput screening (HTS)-derived lead will reach launched drug status is about 1 in a million, shifting the focus of initial post-HTS research efforts toward identifying and characterizing good quality leads [53]. The distinction between hit (compound meeting initial screening criteria) and lead (verified on target activity and selectivity with initial exploration of SAR) is critical [54–56]. It is felt that a rigorous hit to the lead selection process will enhance the probability of overall success. Typical evaluation metrics that a compound series should meet, in order to proceed to lead status, are listed in the points below in Table 1.3-4.

TABLE 1.3-4. Typical Assessment Parameters for Start of Lead Optimization of a Small Molecule

Attributes	Hit-to-Lead Assessment Parameters
Chemical	Reproducible activity between HTS assay and local laboratory assay
	Dose responsive activity
	Confirmed structural identity
	Purity established
	No evidence of compound class instability
	Tractable synthetic route established
	Favorable intellectual property and competitive assessment for compound class
	Demonstrated exploitable SAR
Biochemical	Support for interaction with molecular target (SAR due to desired mechanism)
	Selectivity profile established
Pharmacology	Assessment of druggability and potential attributes that will need optimization
	Secondary assay paradigm established (cell assay, mechanism assay, *in vivo* model, etc.)

Abbreviations: HTS = high-throughput screening; SAR = structure-activity relationship.

With this, it is no surprise that hit to lead has emerged as a unique discipline within many companies, typically built or grown at the expense of the former technology-driven combinatorial chemistry groups. Some companies even have devoted entire departments to the activity so that consistent methods and quality metrics are applied in a rapid time frame [57].

1.3.4.2 Macromolecules

The lead identification process of macromolecules, and especially therapeutic anti-body leads, differs for the *in vivo* and *in vitro* approaches (*see* Table 1.3-3). In the case of the *in vivo* approaches, the screening of initial leads occurs through an enzyme-linked immunosorbent assay or through functional assays drawn parallel to the HTS approaches discussed above. However, the number of antibody candidates typically screened is significantly lower because target-specific antibodies have been enriched by the immunization of transgenic mice [44] or preexist in patients. For the *in vitro* approaches, the selection is template based. As an example, the antibodies from phage or yeast libraries are isolated by a process similar to affinity chromatography [32, 33].

1.3.5 LEAD OPTIMIZATION

The biggest difference between small-molecule drugs and therapeutic antibodies lies in the area of lead optimization; however, the overall concepts are very similar. The original leads are optimized in iterative cycles to achieve candidates meeting the pre-agreed target product profile. The drug structure and size makes the lead optimization of small molecules multidimensional (change in one parameter often affects another) and therefore lengthy, whereas the domain structure of antibodies allows for rapid, primarily one-dimensional optimization of the binding and eff-ector regions, where optimization of one has little or no effect on the other.

Although opportunities for macromolecule optimization were limited in the past, antibody technologies today allow one to dial in, more or less, any affinity, potency, or specificity as directed by the target product profile.

1.3.5.1 Small Molecules

In traditional small-molecule drug discovery, hit-to-lead selection is followed by careful, systematic exploration of SAR and *in vivo* compound properties. Iterative cycles of design and data gathering are used, often supported by generating small-molecule–target binding hypotheses using computational resources. For small mole-cules, this represents the research stage where chemical series with demonstrated affinity of the receptor, enzyme, or protein target of interest are optimized to build in drug-like properties. More properly defined, lead optimization is the synthetic modification of a biologically active compound, to fulfill all stereoelectronic, physi-cochemical, pharmacokinetic, and toxicologic properties required for clinical usefulness.

This is a joint potency, selectivity, and property optimization process that builds on data (design/analysis cycle) generated during the hit-to-lead prioritization phase and is heavily driven by absorption, distribution, metabolism, excretion, and toxi-cology (ADMET) and physicochemical property profiling as early as practical. Combinatorial or parallel synthesis techniques resulting in the generation of scaffold-targeted libraries are often used to understand and develop both the chemical and the intellectual property (IP) space. If amenable, specific analog design guided by QSAR or homology modeling hypotheses is incorporated. It is preferable to have more discrete structural information either from protein NMR studies or ligand/target x-ray crystallography if tractable for the target.

In the effort to choose and accelerate programs that offer the best possible chance to move forward, a value and risk assessment at the start of lead optimization typically occurs along with project and resource prioritization (Table 1.3-5).

TABLE 1.3-5. Typical Assessment Parameters for Start of Lead Optimization

Priority	Lead Optimization Assessment Parameters	Small Molecule	Macromolecule
1st	Potential for therapeutic advance	+	+
	Commercial potential/market size of indication area	+	+
	Competitive environment	+	+
	Intellectual property position	+	+
	Relevance of target for disease	+	+
	Feasibility of biological predictive preclinical model	+	+
	Quality of lead candidates (several dimensions, *see* hit-to-lead criteria)	Specific	Specific
2nd	Target related tolerability	+	+
	Fit to therapeutic area strategy/portfolio	+	+
	Clinical trial design feasibility	+	+
	Amenability of rational or structural drug design	+	−

The chemistry strategy involves defining critical regions of the molecule that can be identified as interacting with the specific amino acid residues of the targets in a three-dimensional model and thus drives potency and selectivity. Areas of the molecule that are SAR "neutral" can be found, which can be used as positions to alter the net physicochemical properties. Small modifications will often affect the *in vivo* bioavailability, the metabolism profile, and the absorption properties of the molecule. Optimization of chemical entities is therefore always multidimensional, and a compromise must be reached before preclinical candidate selection can occur. Any improvement in specificity may be gained with the loss in absorption or half-life, a correction of which will again have effects on specificity. Because of the many iterations of changes and retesting, traditional drug discovery projects are typically long and very costly. Not infrequently, non-mechanism-related toxicity is observed later in development and is not predicted based on the assays available preclinically. To address some of these complications, new methods for predictions of half-life and toxicity are being developed [58]. Introduction of early experimental ADME testing has been credited with significant improvements in the probability of successful lead optimization [59, 60] and is now a widely used practice. To date, the crucial component to small-molecule discovery is the ability to assess, as early as possible, the potential for ADMET failure. The key is to "fail fast" in order to minimize the substantial total research and development investment and costs incurred in the process of learning the shortcomings of a candidate at the clinical trial stage.

1.3.5.2 Macromolecules

Macromolecules, like antibodies, have a defined domain structure separating the domains for specificity and affinity (Fv) from the constant region domains responsible for half-life (Fc receptor [FcRn] binding domains) [61] and effector functions (antibody-dependent cellular cytotoxicity [ADCC] and complement-dependent cytotoxicity [CDC]) [62]. This separation is important because optimization can be done in the first instance in "one dimension." Extensive variation of the complementary determining region 3 (CDR3) regions of the heavy or light chain to increase affinity will in most or all cases have no affect whatsoever on half-life or effector functions of the antibody. Similarly, changes in the half-life or effector functions typically would not affect the affinity. However, examples of interaction between the Fc domain and the Fv antigen binding site have been reported [63, 64] and can be monitored and ruled out by appropriate testing at each stage of optimization. It is easy to see that this one-dimensional optimization should be more straightforward.

Antibody optimization of the antigen binding portion should focus on hypervariable regions, primarily both the heavy- and the light-chain CDR3 regions and preferably the heavy-chain CDR3, as not to introduce undesirable epitopes. Older methods for antibody optimization include a variety of mutation and selection techniques [65], most of which are tedious. The optimal method is yeast display, combining rapid library creation and rapid selection using cell sorters, with a wide range of selection options (selection for optimal "on-rate," "off-rate," "affinity," etc.) [66, 67]. Other methods include ribosome display [68]. Fc optimization includes optimization of effector functions [62] and half-life [21]. Interestingly, the differential effects of Fc mutants on either Fc effector functions (ADCC or CDC) or half-life is further proof that the overall concept of one-dimensional antibody optimization remains correct. This notwithstanding, careful analytical follow-up at the end of the project or at appropriate intervals with meaningful and well-characterized assays will ascertain that indeed no change in functions of another domain has occurred. Because of the specificity and lack of accessibility to intracellular targets, antibodies typically are devoid of nonmechanistic toxicity, which further reduces the discovery and development risks. However, meaningful and appropriate safety testing still will be necessary to assure the safety of therapeutic antibodies [69].

1.3.6 SELECTION OF CLINICAL CANDIDATES AND PRECLINICAL DEVELOPMENT

Selection of clinical candidates and the identification of new chemical entities or new biological entities that best match the TPP is a multidisciplinary task that relies heavily on *in vitro–in vivo* correlations. For first-in-class targets, mechanism of action validation (target validation) research typically parallels the lead identification and lead optimization phases of any drug discovery project. For best-in-class targets, the preferred initial small-molecule validation reagents are competitor's leads or gold standard therapeutics currently on the market, their own initial screening leads, or monoclonal antibodies to the target orthologue. As the project progresses, target validation continues with the drug leads until the *in vivo* data in relevant disease models meets the expectation set out in the TPP. For macromolecule discovery projects, target validation requires either surrogate antibodies or drug leads that cross-react with the orthologue of the preferred pharmacology model species. Usually, a small series of candidate molecules that best match the TPP set at the initiation of the project are subjects of consideration. Outliers are also considered if they can provide unique insight or datasets that might challenge the original TPP, because potential opportunities may not have been realized at the outset of the program. As the project progresses, target validation continues with the selected drug leads until the *in vivo* data in relevant disease models meets the expectations set out in the TPP. For macromolecule discovery projects, target validation requires either surrogate antibodies or drug leads that cross-react with the orthologue of the preferred pharmacology model species.

The topics in Table 1.3-6 are frequently evaluated as part of the risk assessment before starting development activities. Comprehensive *in vivo* efficacy and preliminary safety testing, preferably in multiple species, with simultaneous pharmacokinetic assessment is typically the last dataset collected before making a decision on a development candidate and possible back-up compounds. The Core Battery for safety pharmacology, which can be found in the International Conference on Harmonisation (ICH) Guideline S7A (Good Laboratory Practice [GLP]) document series [70], governs safety assessment and data generation guidelines. The suitable second (nonrodent) toxicity species should be identified in combination with drug metabolism, pharmacokinetics, and pharmacology studies in that species. An overall evaluation of the toxicity and toxicokinetic data should be undertaken to understand the risk for adverse effects, target organ issues, and systemic exposure liability. This should allow researchers to set a preliminary limit dose (mg/kg in relation to C_{max}) for toxicity studies and give recommendations for further

TABLE 1.3-6. Topics Evaluated as Part of the Risk Assessment Prior to Starting Development Activities

- CMC developability drug substance
- CMC physicochemical properties drug substance
- CMC developability formulated drug product
- Relevance of target for disease/indication
- Efficacy in experimental models
- Feasibility of clinical proof of concept
- Exposure and half-life
- Linearity of pharmacokinetics
- Drug–drug interaction potential
- Safety pharmacology
- General pharmacology
- Toxicology
- Potential for therapeutic advance
- Commercial potential/indication area/market size
- Fit to therapeutic area strategy/portfolio
- Patent status/freedom to operate

Abbreviation: CMC = chemical manufacturing and control.

toxicity assessments. The assessment also should include data from predevelopment lead optimization work. For small molecules, the 2-week toxicology study in rodents, including toxicokinetic data, should be suitable as dose-range finding for further repeat-dose (4 week) studies. If performed according to GLP, the safety data from the 2-week study could support phase I. Optional toxicity studies at this stage can include, but are not limited to, single-dose toxicity (GLP, rat) and genotoxicity tests (GLP, Ames, mouse lymphoma) [71, 72].

Early interaction with process research and development teams allows for intermediate-scale synthesis and production of material required for these advanced studies and will enable an assessment of any possible issues around scale-up, cost of goods, stereochemical issues, and compound solid-state crystallinity and stability. Selection criteria are often rigorous, and given the multidisciplinary nature of the profiling, collaborative, tightly scheduled efforts are required across research and development department structures. The preferred criteria set for candidate molecules include those listed in Table 1.3-7.

A discussion of regulatory requirements for the preclinical safety assessment of drug candidates intended for use in humans would require significant detail. The reader is referred to excellent resources and websites (www.ich.org, www.fda.gov, and www.emea.com) for detail and additional information.

Although the safety assessment of small-molecule drug candidates is more standardized, the preclinical program for a macromolecular biological medicine requires adaptation (i.e., case-by-case and highly tailor-made approach), simultaneously meeting the basic requirements for types of studies, duration, and exposure [73–75]. The key principles are outlined in ICH guidance S6 and Center of Biologics Evaluation and Research (CBER) 1997, and some key highlights are summarized in Table 1.3-8 [76, 77].

1.3.7   PRODUCTION OF ACTIVE PHARMACEUTICAL INGREDIENTS

Differences in structural complexity, continuous versus stepwise production, and the overall production techniques and equipment results in significant differences between the manufacturing of small and macromolecule bulk drug substances or active pharmaceutical ingredients (APIs). In addition, continuous improvement of analytical tools is changing the landscape of both process types over time. Small-molecule synthesis consists typically of 6 to 10 (or more) separate synthetic steps, whereas macromolecule production consists of 15 to 20 connected steps. Each step in small-molecule synthesis carries its own risk, but because they are separate or can be separated from each other, low efficiency or failed steps can be repeated, thus minimizing loss. On the macromolecule side, the process is continuous and all steps are directly linked. Failure at any step will typically result in complete batch failure. Another key difference is that small-molecule process research can and often begins during the lead optimization phase of research. This "front-loaded" model [78] is in place in many companies, and initial research toward the efficient production of a promising chemical series is initiated, at risk, early on. With a macromolecule, typically the final sequence or composition of the candidate molecule must be clearly identified so that appropriate constructs can be generated and then optimized for fermentation processes.

TABLE 1.3-7. Criteria for Development Candidates for Internal Decision Making

Criteria	Small Molecule	Macromolecule
Chemical Manufacturing and Control		
Synthetic accessibility	Synthetic steps, chiral centers, reagent cost of goods	—
LogP	$-4 \leq logP \leq 4.2$	—
LogSw	≥ -5	—
Formulation	Project specific	≥ 50 mg/mL
Production yields	Project specific (overall cost of goods)	50–100 pg/cell/day
In Vitro Pharmacology		
Cell assay potency with and without serum	Project specific	Project specific
Protein binding	Availability required	—
Target specificity	Specificity within close and distant target family members	ELISA specificity within diverse set of related proteins
Affinity	Project specific	BiaCore $>10^{-10}$ M
Tissue cross-reactivity	—	Binding to tissues with known target expression
In Vivo Pharmacology/PK/PD/ADMET		
Efficacy in relevant model at appropriate dose, route, and schedule	Project specific	Project specific
Metabolites	Identified and characterized	—
PK/PD relationship	Project specific	Availability desired
Biomarker for PD effect	Availability desired	Availability desired
No overt toxicity in repeat dose (1 to 4 week); toxicity studies in a rodent and nonrodent species	Availability required	As follow-up for unexpected pharmacological observations
Drug–drug interaction	No significant interactions	—
Human CYP inhibition	No relevant interactions; also consider comedications	—

Abbreviations: ADMET = absorption, distribution, metabolism, excretion, toxicology; CYP = cytochrome P450; ELISA = enzyme-linked immunosorbent assay; PD = pharmacodynamics; PK = pharmacokinetics; — = not applicable.

TABLE 1.3-8. Key Issues Around the Designs of Safety Studies [76, 77]

Issue	Small Molecule	Macromolecules	Guidance for Macromolecules
Selection of relevant animal species	Typically 1 rodent and 1 nonrodent species	Studies in irrelevant species are of no value or might even be misleading. Criteria for species selection include: • Tissue cross-reactivity studies comparing human and animal tissues • Pharmacologically active (i.e., expressing the appropriate epitope or receptor)	• Closest pattern to human tissues defines animal species • One species (often the monkey) might be sufficient; if no species available, use of a surrogate and/or transgenic animal might be appropriate • Early and close interaction with agencies highly recommended
Dose level selection	Based on the anticipated highest clinical dose, high dose should show toxicity		Three dose groups plus a control group with multiples of anticipated highest clinical dose (e.g., 2, 10, and 50 times clinical dose); high dose selection often not driven by toxicity
Treatment frequency	Should parallel the treatment regimen in humans		Depending on PK and immunogenicity, a higher dose frequency might be needed to achieve comparable exposure
Local tolerability	With oral dosing not an issue	Application route predominantly IV and/or SC; it is important to test the clinical formulation	Visual inspection and histopathology of injection site
Recovery period	Long enough to allow clearance of the drug		At least 5 half-lives; can be up to 3 months
Immunogenicity	Not an issue	Induction of ADA, characterization (i.e., neutralization, impact on drug exposure because of increased clearance) At high dose level immune response often suppressed	• If exposure decreases due to ADA formation, no adequate long-term treatment might be possible • In general, animals are not predictive of immunogenic response in humans

Genotoxicity studies	Standard battery	Not applicable	Interactions with DNA or chromosomes is not expected; large amounts might even lead to uninterpretable results
Safety pharmacology	Standard core battery	For mAbs, generally not necessary. If cause of concern (e.g., cross-reactivity with CNS, CV, or respiratory tissues), specific study might be indicated	As many parameters as possible to be included in toxicology study. A study in telemetrized animal to address CV and respiratory endpoints might be an alternative approach
Carcinogenicity	Required	Not applicable	The standard carcinogenicity bioassays in rat and mouse are generally inappropriate
Reproductive toxicology	Standard package (Fertility, embryo-fetal toxicity and peripost natal toxicity)	Dependent on product, intended use and patient population. Study must be conducted in relevant species. Expectations from authorities still different for U.S., EU, and Japan regarding timing of the studies	The primate is often the only relevant species; however, because of the small numbers of animals available and the lack of a broad dataset, testing of a rodent homologue might be an alternative U.S. not required to support phase I/II
ADME	Standard battery *in vitro* and *in vivo*	Not applicable	Not relevant as the drug enters the normal proteolytic degradation pathway
Biomarkers	Should be included whenever possible to demonstrate the intended pharmacology of the drug		PK/PD relationship helps to define safety margins as well as selection of dose level and regimen

Abbreviations: ADA = anti-drug antibodies; ADME = absorption, distribution, metabolism, excretion; CNS = central nervous system; CV = cardiovascular; EU = European Union; IV = intravenous; mAbs = monoclonal antibodies; PD = pharmacodynamics; PK = pharmacokinetics; SC = subcutaneous; U.S. = United States.

1.3.7.1   Small Molecules

As discussed, pressure to move new medicines to the market faster has compressed many aspects of the research and development cycle. Process research initiates their search for scalable routes early in the discovery programs in the hope of having an efficient or semi-efficient route to prepare up to 100 kg of API as a program nears the final preclinical stages and prepares for phase I studies in humans [79].

Although any synthetic transformation or use of a specialized reagent can be scaled up, cost and safety are two primary driving factors in the manufacturing process. In a recent survey of GMP bulk reactions run in the last several decades, a predominant subset of transformations were shown to fall into favor [80], with less than a 2% change within the categories. This result indicates that significant process research efforts are often required to advance an initial medicinal chemistry analog synthesis route to one that is amenable for use in large-scale production. Once route selection occurs at phase I, synthetic efficiency is critical and reactions must be controlled so that consistency of yield and purity is ensured, especially as a program nears phase II in the clinical setting. This is the point at which the process for generating the API must be finalized and approval packages initiated for the regulatory agencies. To control costs, many manufacturers will partner with specialty firms to transform key synthetic intermediates in the stepwise production process toward the final API (which is almost always prepared within the company's facility to meet regulatory requirements). Such external, contract synthesis often involves sensitive or specific technology-based transformations or separations for purification that would have significant cost impact if a production facility needed substantial capital improvement and/or steep licensing fees to conduct. Environmental aspects of production also have become increasingly important. Considerable steps are taken to minimize waste streams, and the advent of "green chemistry," a move to decrease usage of organic solvents in favor of water, when appropriate, is actively under investigation [81]. In addition, energy-efficient methods and technologies such as microwave-driven processes are being explored to accelerate cycle times [82] within synthesis routes.

Small-molecule manufacturing is also a system of checks and balances in which any small increase or decrease in reaction step performance can have consequences downstream. As always, patient safety is of utmost concern. Any manufacturing process changes that alter an impurity level or introduce a new impurity, even as low as 0.1%, may necessitate additional toxicity studies and documentation for review and registration. In addition, an improvement in reaction efficiency may alter bulk product crystallinity or polymorph composition that can affect formulation and human pharmacokinetics. Once process parameters are finalized, the ultimate manufacturing step involves selection of a manufacturing site, transfer of the process, and preparation of a demonstration batch followed by a minimum of three consecutive validation batches of API to demonstrate that the synthesis of material can be controlled within analytical specifications and reproducibility.

1.3.7.2   Macromolecules

Macromolecules, including therapeutic antibodies and fusion proteins, are complex molecules requiring extensive analytical characterization. Native antibodies comprise a heterodimer of two light and two heavy chains with a single N-glycosylation site in the heavy-chain constant region. Proper assembly and glycosylation is important for antibody function. Fusion proteins often consist of natural binding proteins (such as receptors) fused to the nFc (CH2 and CH3 IgG domains) region. Due to interchain disulphide bridges, Fc-fusion proteins assemble into homodimers. The complexity of fusion proteins depends on the structure of the ligand binding domain and the number of N-glycosylation sites. The single N-glycosylation site in antibodies is "buried" into the opposite heavy chain constant region, whereas the carbohydrate residues on ligand binding portions of Fc-fusion proteins are exposed and may have a pronounced effect on the *in vivo* behavior of the molecule. For these reasons, full-length antibodies and fusion proteins should be expressed in mammalian cells. Modified yeast systems capable of appropriate glycosylation and fungal systems have been proposed [83]. By contrast, antibody fragments have been expressed successfully in *Escherichia coli* and yeast. Decisions for an expression system should be based on the complexity of the molecule, the specific biological activity achieved with a given system, and the appropriate technical, IP, and regulatory considerations for manufacturing at scale. Expression of therapeutic antibodies in a variety of systems has been reviewed recently [84, 85]. Methods include transgenic animals [86], expression in plants [87], or the more standard myeloma systems [88] or CHO cell expression [89]. The recovery and purification of therapeutic antibodies also has been reviewed recently [90]. For the reasons discussed, when designing purification schemes, attention should be paid to reduction of aggregation and DNA content, especially if bacterial systems are involved, to minimize any immunogenicity risks. In each case, a defined set of assays needs to be applied to the purified final product to ascertain the purity and specific activity of the expressed antibody. This is especially important if reconstitution and refolding is part of the process [91].

Historically, macromolecule products were characterized with a limited set of tools and additional methods beyond that were defined by their specific process. This led to complications when changes in the process were desired. A chance event changed this landscape. Briefly, biotechnology company Biogen, Inc. (today Biogen Idec, Inc., Cambridge, Massachusetts) and the Rentschler daughter company, Bioferon GmbH (Laupheim, Germany), formed a joint venture in 1984 for the clinical development of interferon-beta, called BG9015. Rentschler was responsible for the creation of the cell line and production and Biogen for clinical development, which commenced in 1990 with a U.S. clinical study in patients with multiple sclerosis. When this joint venture failed in 1993, Biogen completed the initial trial with the remaining BG9015 and switched to BG9216 from their own process. After the U.S. Food and Drug Administration (FDA) determined that these products were not identical, a new process for what is today called Avonex (interferon beta-1a; Biogen Idec, Inc.) was approved and was considered sufficiently identical with the original BG9015 for the original study results to be considered as part of the Avonex biologic license application. Extensive biological, biochemical, biophysical, and human pharmacokinetic data were the basis for the FDA conclusion that BG9015 and Avonex were "biochemically and functionally equivalent," leading to Avonex approval on May 17, 1996 [92]. This decision created a path for others to follow, leading to a fundamental change in macromolecule production and ushering in the concept of "well characterized biologic" [93]. Increasingly sophisticated analytical techniques have facilitated process changes even further, and today, macromolecules require very detailed characterization, resembling more and more the requirements for small-molecule pharmaceutical ingredients [94]. Indeed, analytical tools have become so sophisticated that much of the current dispute on biogenerics (or "follow-on biologics") focuses on the true equivalence of macromolecule products [95].

1.3.8 FORMULATION AND DELIVERY DEVICES

Drug product development takes place in a highly regulated environment with a clear expectation for quality, safety, and efficacy in the final product. From a formulation perspective, quality primarily translates into a defined composition, assurance of purity, potency, an acceptable stability, and availability of a robust and consistent large-scale manufacturing process. Thus, the ultimate goal of formulation activities is to achieve the systemic delivery of a drug product that can be successfully commercialized. Product life-cycle management strategies are considered at the beginning of development and often include evaluation of specialized administration devices for parenteral (injectable) products, especially for macromolecules. If these options are considered early in development, a product family can often be designed to minimize subsequent development efforts by using a common platform (e.g., prefilled syringe-based autoinjectors). One challenge in the development of devices is a thorough understanding of the limitations of the device, limitations of the formulation, and finding the right compromise between both. Also, one has to keep in mind that the regulatory path for devices and fixed-device drug combination products is different, and significant differences exist among the major markets such as the United States, the European Union, and Japan.

The scope of formulation activities is dependent on the development stage of the drug substance. In general, the development process can be divided into three major phases, namely formulation support of discovery, formulation activities until clinical proof of concept (end of clinical phase IIa), and formulation activities in the commercial phase (phase IIb until market).

1.3.8.1 Formulation Support in Discovery

Formulation support in discovery usually starts at the lead optimization phase with a focus on key properties of molecules to support selection of the best candidates. The preferred route of administration for small molecules is oral, which requires the development of a solid dosage form. Because only dissolved molecules can be absorbed, assessment of solubility is of prime interest. In general, macromolecules are not orally absorbed and a parenteral dosage form is required. Some proteins, like monoclonal antibodies, often require high doses, and therefore, it is important to understand the solubility range for the protein in question. Compared with small molecules in which established strategies exist for tackling solubility problems on the basis of chemical modifications, solubility issues with proteins are mechanistically poorly understood and very difficult to address on a molecular basis. Most small-molecule drug candidates are crystalline in nature, whereas macromolecules are usually dissolved in aqueous media. The composition/formulation of the solution determines the properties of the protein, and therefore, a basic understanding about suitable pH ranges, temperature sensitivity, buffer composition, and so on, is important. These parameters have to be defined and controlled early in the discovery phase. The complex molecular structure of proteins poses a significant challenge to the analytical techniques and the ability to detect the instabilities of the molecule. Characterization of degradation products is a prerequisite to understand the major degradation pathways of the protein, which will enable a rational design of the formulation. Degradation pathways of proteins include deamidation, oxidization, hydrolysis, fragmentation, and aggregation. Formulation strategies have been developed to address these instabilities and can act as guiding principles in early development phases. Once the selection of clinical candidates has been initiated, the number of drug molecules will be narrowed down significantly and profiling of the proposed candidate for development can take place. For oral small molecules, solid-state characterization is a very relevant topic for a solid dosage form as summarized in Table 1.3-9. For a protein delivered parenterally, it is

TABLE 1.3-9. Formulation Attributes at the Development Project Stages

Stage	Small Molecule	Macromolecule
	Discovery Stage	
Common dosage route (all stages)	Oral	Parenteral
Critical factors to optimize	Bioavailability, solubility, logP, permeability, absorption	Bioavailability, solubility, pH stability, freeze/thaw behavior
Physicochemical attributes	Crystalline solids or salts	Solution stability, aggregation
	Early Development Stage	
Common dosage form	Powder in bottle, tablets	Lyophilized powder in bottle
Critical factors to optimize	Dosage form	Temperature sensitivity and storage conditions, dose per unit and dosage form
Physicochemical attributes	Crystal polymorphism, solid-state stability, optimized salt selection	Conformational stability, buffer formulation, surfactant or excipient stability, aggregation
Device	Tablet, capsule, oral suspension or solution	Prefilled syringe
	Commercial Stage	
Develop commercial process	Yes	Yes
Batch scale-up	Yes	Yes
Batch stability	Yes	Yes
Process validation	Yes	Yes
Establish commercial supply chain	Yes	Yes

important to define the composition of the drug substance. Ideally, the drug substance composition should be as simple as possible to allow for flexibility in later development phases. In the most simple situation, the drug substance composition consists of a protein (concentration at least half of the highest single dose for first in human studies), a buffer system (suitable for parenteral administration) in a pH range of 5 to 8, and a surfactant, if necessary.

1.3.8.2 Formulation Activities Until Clinical Proof of Concept

For early clinical trials, preliminary dose formulations or "clinical service forms" are developed. Small-molecule oral products can be either powder in a bottle or oral solutions/suspensions. Proteins are usually initially formulated as lyophilized powders that maximize long-term stability at the expense of a lesser convenient dosage form due to the necessity of the reconstitution step. Parenteral dosage form manufacturing requires a highly technical and sophisticated infrastructure, including highly trained personnel and well-established GMP systems in order to assure sterility of the products. These requirements limit flexibility in the development of parenterals compared with oral products. Temperature stability studies have to be conducted to justify use of material that has experienced temperature variations on storage. This is usually not an issue for small molecules, but sometimes specific safety considerations are more important and require special handling and shipment conditions.

During the course of the early clinical development program, a more detailed definition of the commercial product evolves. The most important information is the dose per unit and the specific dosage form. Oral products often require modified release dosage forms in which extensive pharmacokinetic modeling/testing of prototype formulation have to be performed. For parenteral products, self-administration and convenience is becoming more and more an expectation. Prefilled syringes that allow subcutaneous administration of liquids satisfy this need. The main issues involved with development of a liquid protein formulation in a prefilled syringe are using volumes less than 1 mL (consequently, protein concentration can be as high as 100–200 mg/mL), aggregation of proteins due to the high protein concentration, viscosity issues, interaction of the protein with silicone oil present in the prefilled syringe, and the higher chance of chemical degradation in a liquid solution. Aggregation of proteins and conformational stability is a major obstacle because reliable detection of small amounts of aggregates is analytically challenging and aggregates or conformationally altered protein molecules are suspected to be involved in undesired immunogenic responses to therapeutic proteins. Conformational changes can be induced by small changes in drug substance/drug product manufacturing. Therefore, changes in manufacturing have to be carefully evaluated on product impact and often require additional stability testing because the impact is not predictable.

1.3.8.3 Formulation Activities in the Commercial Phase (Clinical Phase IIb until Market)

The main activities are similar for small molecules and macromolecules at this stage. The primary focus is in the areas of scale-up, generation of primary batches for stability testing, and development and validation of the commercial manufacturing process.

For solid dosage forms, scale-up is a major challenge and includes ensuring the bioequivalency of the clinical product to the commercial product. For proteins, shear-induced aggregation can be an issue, as well as establishment of a commercial supply chain that is robust, reliable, and able to deal with large amounts of temperature-controlled materials.

1.3.9   CLINICAL DEVELOPMENT

There are differences in the pharmacokinetic properties of typical small molecules and monoclonal antibodies that will affect the clinical development path. A few of the differences are listed in Table 1.3-10. Small molecules are usually administered orally, which is desired because of convenience for patients. On the other hand, the administration of monoclonal antibodies is limited to the parenteral route, i.e., intravenously (*i.v.*), intramuscularly (*i.m.*), or subcutaneously (*s.c.*). Because of the availability of liquid formulation or reconstituted lyophilized powder to support parenteral administration, measurement of absolute bioavailability after *s.c.* or *i.m.* administration is necessary for monoclonal antibodies. The relative bioavailability is usually obtained for small molecules because of the expense of developing liquid formulations. Both renal and hepatic clearance pathways are important for elimination of small molecules, and an understanding of drug disposition in patients with renal or hepatic impairment is important to support the registration documentation. Small molecules are often metabolized in the liver by cytochrome P450 enzymes; therefore, drug–drug interaction studies are an important component of the clinical data package. Therapeutic monoclonal antibodies follow the typical pathway of elimination for immunoglobulins (IgGs), involving the neonatal (FcRns). Briefly, endothelial cells lining the blood vessel are thought to be the major site of IgG homeostasis, with the FcRns acting as salvage receptors,

TABLE 1.3-10. Differences Between Small Molecules and Therapeutic Monoclonal Antibodies that Can Affect Clinical Development

Small Molecules	Monoclonal Antibodies
Usually administered orally	Given parenterally
Transporters often important for disposition	Transporters are not important for disposition
Metabolism by hepatic cytochrome P450	Proteolytic degradation
Renal clearance often important	Renal clearance is not important for molecular weight higher than 20 Kd
Half-life is shorter (usually <24 h)	Half-life is long (1–3 weeks)
PD usually does not affect PK	PK can be dependent on PD
Immunogenicity is a not an issue	Immunogenicity is an important issue

Abbreviations: PD = pharmacodynamics; PK = pharmacokinetics.

protecting IgGs against lysosomal degradation. The IgGs are taken up by endothelial cells by nonspecific pinocytosis and then enter acidic endosomes. If the IgG is bound to the FcRn, then it is protected and released back into the blood stream; otherwise, it is degraded. Thus, the FcRns serve the dual function of transcytotic shuttle (i.e., placental transfer, etc.,) and homeostat [22, 61]. As long as therapeutic antibodies bind to the FcRns, they are thought to participate in this pathway. Whether modulation of that interaction by mutating the IgG Fc portion can predictably extend or shorten IgG half-life is the subject of ongoing research. Metabolic interactions such as inhibition or induction of cytochrome P450 enzymes are usually not a consideration. Transporters must be a consideration for the understanding of the disposition of small molecules, whereas except for FcRns transporters, transporters are of lesser concern for monoclonal antibodies.

Allergic reactions are a potential issue for both small and large molecules. The potential for development of significant levels of antidrug antibodies needs to be considered for any recombinant protein, including fusion proteins and monoclonal antibodies, regardless of whether the therapeutic antibody is chimeric, humanized, or fully human. Assessment of immunogenicity begins in phase I and continues throughout clinical development. Incidence of immunogenicity can differ for each patient population and should be assessed for each indication. The development of assays to assess immunogenicity begins before filing the investigational new drug application. Validated assays are necessary by phase III. In the case of fusion proteins, the cross-reactivity of antidrug antibodies with the endogenous human protein in the patient needs to be considered.

1.3.10 INTELLECTUAL PROPERTY

Farnley and colleagues [96] have recently reviewed the complexity of intellectual property around biotechnology and biotechnology products. The authors note that the IP issues for small molecules and macromolecules, with a few exceptions, may be more similar than commonly assumed. The most important difference relevant to this overview is that each step in the discovery process may potentially be subject to different intellectual property considerations, which the authors note as unusual. The major differences between small-molecule and macromolecule patenting are shown in Table 1.3-11.

TABLE 1.3-11. Intellectual Property Considerations for Small Molecules and Macromolecules

Patents	Small Molecule	Macromolecule
Target	Yes	Yes
Drug class ("antibody to target X")	No	Yes
Substance	Yes	Yes
Drug screening	Yes	Yes
Drug synthesis steps in discovery stage	No	Yes+

1.3.11   GENERICS, "BIOSIMILAR PRODUCTS," OR "FOLLOW-ON BIOLOGICS"

Although a full assessment of generics is beyond the scope of this introduction, the differences between small-molecule and macromolecule generics are highlighted briefly in Table 1.3-12. Importantly, the terminology "generics" should not be used for noninnovator developed off-patent macromolecules. They are not "generic" in the true sense, and "biosimilar products" or "follow-on biologics" are the more appropriate terms. Reasons include the independently developed manufacturing process, the resulting possible differences in product heterogeneity, and the need for preclinical and clinical development. Many companies around the world are working toward registration of off-patent recombinant proteins. Interferons, human growth hormone, and erythropoietin are some of the more popular therapeutics that are targeted.

The development path for small-molecule generics has been clear for many years. In contrast, the path for biosimilar products or follow-on biologics has only recently been clarified in Europe (October 2005), and regulatory uncertainty remains in the United States. The new European guideline 2004/27/EC clearly

TABLE 1.3-12. Regulatory Criteria for Biosimilar Products or Follow-On Biologics

	Small Molecule[1]		Macromolecule	
Terminology	Generics		Biosimilar products or Follow-on biologics	
	U.S.	EU	U.S.	EU
Regulatory framework	21 CFR Part 314 Subpart C.	Article 10 (1) of Directive 2001/83/EC	TBD	Annex I to the Directive 2001/83/EC and 2004/27/EC
Application and review	ANDA in CTD format FDA moving to QbR by 1/07	CTD format	TBD	CTD Agency consultation by "scientific advice request" strongly recommended
Nonclinical studies required	No	No	TBD	Yes
Clinical studies required	No	No	TBD	Yes

[1] Including some small proteins/peptides in the United States (e.g., insulin, human growth hormone).

Abbreviations: ANDA = Abbreviated new drug application; CFR = Code of Federal Regulations; CTD = Common technical document; EC = European Commission; EU = European Union; FDA = Food and Drug Administration; QbR = question-based review; TBD = to be determined; U.S. = United States.

anticipates a more or less full development program for biosimilar products or follow-on biologics, with material from the sponsor's manufacturing process compared with the innovator reference product.

Importantly, for registration in countries of the European Union, the sponsor of a follow-on biologic has to develop a process generating a protein sufficiently similar to the reference product without access to the typically trade-secret-protected innovator's manufacturing process or the manufacturing cell banks. In addition, a significant nonclinical safety and clinical program has to be conducted.

No biosimilar products/follow-on biologics have been approved in the United States, European Union, or Japan to date. Products confusingly referred to as "generic interferon alpha-2b" or "generic human growth hormone" available in some countries in Asia, Eastern Europe, and Mexico were registered after completion of clinical studies and not through the process used for small-molecule generics. In January 2006, the European Medicines Agency (EMEA) issued a positive opinion for the Novartis subsidiary, Sandoz GmbH (Holzkirchen, Germany), application for the recombinant human growth hormone Omnitrope (somatropin) as biosimilar to the already marketed reference product Genotropin.

1.3.12 SUMMARY

Biologics-based medicines have become an increasingly important part of the armamentarium available for physicians to therapeutically manage disease. Many promising new entities are in various stages of discovery and development. Although some conceptual differences between small-molecule and macromolecule drug discovery have decreased over recent years, the fundamental overall structural differences dictate unique approaches for each.

Previously, macromolecules were used with little or no optimization, whereas small-molecule and macromolecule discovery today are both geared toward developing the optimal molecule that meets the precise needs of the underlying biology. In mimicking natural processes, biologics, and especially antibodies, can be optimized in all binding characteristics without limitation. The overall discovery/development process can be divided into four convenient steps:

1.3.12.1 Step 1: Target Selection and Validation

Successful drug discovery generally starts with the definition of a well-defined and biologically understood target. Importantly, the role of the target in disease needs to be demonstrated with unequivocal target validation studies and the role in normal physiology and existence of variants or closely related family members needs to be investigated. Drug discovery today often fails because targets with uncertain biology are being pushed into drug discovery too early.

1.3.12.2 Step 2: Choice Between Small-molecule and Macromolecule Platforms

At this phase of the discovery process, the biology and location of the target, the disease, and the commercial goal should drive the decision of whether a small-molecule or biologics discovery approach is optimal. Overall, the management of a mixed portfolio using both drug discovery platforms needs to be considered in light of managing a risk-balanced portfolio. Biologics benefit from a more rapid discovery process and a higher probability of success, with one factor being the lack of nonmechanistic toxicity, often leading to first-in-class commercial opportunities.

1.3.12.3   Step 3: Definition of Target Product Profile

Once the target is defined and the technology platform selected, the TPP needs to be defined. The TPP is an agreement between all stakeholders in drug discovery, development, manufacturing, commercial, and others on the specific goal of the project considering all phases of development. The TPP lays down in detail the expected characteristics of the development candidate considering the needs of a formal preclinical program. As an example, if the biologics discovery project does not consider cross-reactivity with an orthologue in the planned toxicology as important, the final candidate may bind to the human target only and face huge hurdles in development.

1.3.12.4   Step 4: Drug Discovery and Development

With key decisions from Steps 1–3 complete, drug discovery using either platform can proceed. It is critical that target validation work continues throughout this period. The TPP planning discussed under Step 3 helps to anticipate reagents needed much later, but it may be time consuming to prepare (e.g., transgenic mice with human target proteins for pharmacology studies).

In contrast to the previous decade, biologics-based medicines can be finely tuned to meet challenging affinity, potency, or specificity goals, and large advances have been made toward dialing in appropriate half-life and effector functions. Antibody affinity optimization, for example, if attempted at all, may have taken a year or more previously, but it can be accomplished in 3 months today.

The predominant message within this introductory chapter is held in the concept that the drug discovery process involves a rational path of critical steps that a pharmaceutical scientist must think through (*see* Figure 1.3-1). Decisions are made early in a program, often in the conceptual stage, that are forward thinking and with an eye toward the clinical setting and meeting crucial medical needs in the general population. Although these steps appear identical when comparing macromolecular biomolecules and small-molecule paths, many key differences exist. The stepwise nomenclature tends to be consistent, and the two processes can be compared as waves in which there are distinct nodes of intersection yet many distinct divergences between phases. Many of these divergences involve small steps that can be technology driven. For example, the exploratory phases hold similar target identification and validation steps, whereas the subsequent lead identification step may greatly differ in assay development strategies and high-throughput screening with hit-to-lead follow-up (small molecule) versus surrogate selection and characterization (macromolecule). Surrogate antibody characterization (a lead identification function) can often run simultaneously with human antibody generation (a lead optimization function). Even at this early stage, a macromolecule most likely will be subject to *in vivo* experiments early in the lead identification phase, whereas it may take many months of even preliminary property optimization of a small molecule to probe *in vivo* pharmacology, especially for intracellular targets. As emphasized early in this overview, a great divergence is the one-dimensional versus multidimensional lead optimization dependencies when comparing small molecules and macromolecules and that much discovery speed can be gained given the distinct macromolecule domains. On the side of the drug development process, developing a large or small molecule that satisfies the TPP parameters is similar, and this can be extended to its measurement and performance in clinical trials. Not only do manufacturing processes diverge greatly, but also the ability to make small or incremental process improvements without affecting the composition and integrity of the bulk pharmaceutical product. In a small-molecule manufacturing process, a solvent or reagent can potentially be substituted to improve yield but must have no effect on the overall final product impurity profile, whereas the addition of an alternative feedstock or genetic optimization of a cell culture line can drastically alter a macromolecule's characteristics. In summary, a great appreciation can be gained for the strategic aspects of advancing small-molecule and macromolecular entities and how each step of a rational process is interdependent on the next to efficiently and successfully advance a therapeutic concept to the pharmacy shelves. Considering all aspects of small-molecule and biologics research and development, including the differences in attrition and speed [1], optimal integration of both approaches into the research and development portfolio may be the best strategy for the future.

ACKNOWLEDGMENT

We thank our colleage, Dr. Hans-Jürgen Krause, for very helpful contributions to and discussion of the Formulation and Drug Device sections and Lori Lush, PharmD, for helpful assistance with the manuscript.

REFERENCES

1. Kola I, Landis J (2004). Can the pharmaceutical industry reduce attrition rates? *Nat. Rev. Drug Discov.* 3:711–715.
2. Kubinyi H (2003). Drug research: myths, hype and reality. *Nat. Rev. Drug Discov.* 2:665–668.
3. Salfeld J G (2004). Use of a new biotechnology to design rational drugs against newly defined targets. *Best Pract. Res. Clin. Rheumatol.* 18:81–95.
4. Glennie M J, van de Winkel J G (2003). Renaissance of cancer therapeutic antibodies. *Drug Discov. Today* 8:503–510.
5. Mayer T U (2003). Chemical genetics: Tailoring tools for cell biology. *Trends Cell. Biol.* 13:270–277.
6. Stockwell B R (2004). Exploring biology with small organic molecules. *Nature.* 432:846–854.
7. Zambrowicz B P, Sanda A T (2003). Knockouts model the 100 best-selling drugs—Will they model the next 100? *Nat. Rev. Drug Discov.* 2:38–51.

8. Bagowski C P (2005). Target validation: An important early step in the development of novel biopharmaceuticals in the post-genomic era. In: Knäblein J (ed.), *Modern Biopharmaceuticals: Design, Development, and Optimization.* Vol. 2. Wiley-VCH, Weinheim, Germany. pp. 621–647.

9. Butcher S P (2003). Target discovery and validation in the post-genomic era. *Neurochem. Res.* 28:367–371.

10. Chen B X, Erlanger B F (2002). Intracellular delivery of monoclonal antibodies. *Immunol. Lett.* 84:63–67.

11. Hong G, Chappey O, Neil E, et al. (2000). Enhanced cellular update and transport of polyclonal immunoglobulin G and fab after their cationization. *J. Drug Target.* 8:67–77.

12. Kabouridis P S, Hasan M, Newson J, et al. (2002). Inhibition of NF-kappa B activity by a membrane-transducting mutant of 1 kappa B alpha. *J. Immunol.* 169:2587–2593.

13. Mie M, Takahashi F, Funabashi H, et al. (2003). Intracellular delivery of antibodies using TAT fusion protein A. *Biochem. Biophys. Res. Commun.* 310:730–734.

14. Temsamani J, Vidal P (2004). The use of cell-penetrating peptides for drug delivery. *Drug Discov. Today* 9:1012–1019.

15. Zhao Y, Lou D, Burkett J, et al. (2001). Chemical engineering of cell penetrating antibodies. *J. Immunol. Methods* 254:137–145.

16. Marks J D (2004). Selection of internalizing antibodies for drug delivery. *Methods Mol. Biol.* 248:201–208.

17. Prabbu R, Khalap N, Burioni R, et al. (2004). Inhibition of hepatitis C virus nonstructural protein, helicase activity, and viral replication by a recombinant human antibody clone. *Am. J. Pathol.* 165:1163–1173.

18. Goetzl E J (2004). An IgM-kappa rat monoclonal antibody specific for the type 1 sphingosine 1-phosphate G protein-coupled receptor with antagonist and agonist activities. *Immunol. Lett.* 93:63–69.

19. Bardelli C, Sala M, Cavallazzi U, et al. (2005). Agonist Met antibodies define the signalling threshold required for a full mitogenic and invasive program of Kaposi's Sarcoma cells. *Biochem. Biophys. Res. Commun.* 334:1172–1179.

20. Fry D C, Vassiley L T (2005). Targeting protein-protein interactions for cancer therapy. *J. Mol. Med.* 83:955–963.

21. Hinton P R, Xiong J M, Johlfs M G, et al. (2006). An engineered human IgG1 antibody with longer serum half-life. *J. Immunol.* 176:346–356.

22. Lencer W I, Blumberg R S (2005). A passionate kiss, then run: Exocytosis and recycling of IgG by FcRn. *Trends Cell. Biol.* 15:5–9.

23. Vaccaro C, Zhou J, Ober R J, et al. (2005). Engineering the Fc region of immunoglobulin G to modulate in vivo antibody levels. *Nat. Biotechnol.* 23:1283–1288.

24. Archer J R (2004). History, evolution, and trends in compound management for high throughput screening. *Assay Drug Dev. Technol.* 2:675–681.

25. Bajorath J (2002). Integration of virtual and high-throughput screening. *Nat. Rev. Drug Discov.* 1:882–894.

26. Kariv I, Rourick R A, Kassel D B, et al. (2002). Improvement of "hit-to-lead" optimization by integration of in vitro HTS experimental models for early determination of pharmacokinetic properties. *Comb. Chem. High Throughput Screen.* 5:459–472.

27. Lo B K C (2004). Antibody humanization by CDR grafting. *Methods Mol. Biol.* 248:135–159.

28. Makara G M, Athanasopoulos J (2005). Improving success rates for lead generation using affinity binding techniques. *Curr. Opin. Biotechnol.* 16:1–8.

29. Comess K M, Schurdak M E (2004). Affinity-based screening techniques for enhancing lead discovery. *Curr. Opin. Drug Discov. Dev.* 7:411–416.

30. Roguska M A, Pedersen J T, Henry A H, et al. (1996). A comparison of two murine monoclonal antibodies humanized by CDR-grafting and variable domain resurfacing. *Protein Eng.* 9:895–904.

31. Zhang W, Feng J, Li Y, et al. (2005). Humanization of an anti-human TNF-α antibody by variable region resurfacing with the aid of molecular modeling. *Mol. Immunol.* 42:1445–1451.

32. Hoogenboom H R (2005). Selecting and screening recombinant antibody libraries. *Nat. Biotechnol.* 23:1105–1116.

33. Marks J D, Bradbury A (2004a). Selection of human antibodies from phage display libraries. *Methods Mol. Biol.* 248:161–176.

34. Keri G, Szekelyhidi Z, Banhegyi P, et al. (2005). Drug discovery in the kinase inhibitory field using the Nested Chemical Library technology. *Assay Drug Devel. Technol.* 3:543–551.

35. Miller J L (2006). Recent developments in focused library design: Targeting gene-families. *Curr. Topics Med. Chem.* 6:19–29.

36. Maeda F, Takekoshi M, Nagatsuka Y, et al. (2005). Production and characterization of recombinant human anti-HBs Fab antibodies. *J. Virol. Methods* 127:141–147.

37. Mitchell T, Cherry M (2005). Fragment-based drug design: Fragments offer the prospect of a more efficient approach to drug discovery—resulting in the generation of high-quality leads with better chance of success in clinical development. *Innov. Pharm. Technol.* May:34–36.

38. Verdonk M L, Hartshorn M J (2004). Structure-guided fragment screening for lead discovery. *Curr. Opin. Drug Discov. Devel.* 7:404–410.

39. Petros A M, Dinges J, Augeri D J, et al. (2005). Discovery of a potent inhibitor of the antiapoptotic protein Bcl-x_L from NMR and parallel synthesis. *J. Med. Chem. Web.* Release Date: December 20, 2005.

40. Sillerud L O, Larson R S (2005). Nuclear magnetic resonance-based screening methods for drug discovery. *Methods Mol. Biol.* 316:227–289.

41. Vollmers H P, Braendlein S (2005). The early birds: Natural IgM antibodies and immune surveillance. *Histol. Histopathol.* 20:927–937.

42. Alvarez J, Shoichet B, eds. (2005). *Virtual Screening in Drug Discovery.* Taylor & Francis; Boca Raton, Florida.

43. Anderson A C, Wright D L (2005). The design and docking of virtual compound libraries to structures of drug targets. *Current Computer-Aided Drug Design* 1:103–127.

44. Lonberg N (2005). Human antibodies from transgenic animals. *Nat. Biotechnol.* 23:1117–1125.

45. Guener O F (2005). The impact of pharmacophore modeling in drug design. *IDrugs* 8:567–572.

46. Lill M A, Dobler M, Vedani A (2005). Multi-dimensional QSAR in drug discovery: Probing ligand alignment and induced fit—application to GPCRs and nuclear receptors. *Curr. Comp. Aided Drug Design* 1:307–324.

47. Lang P T, Kuntz I D, Maggiora G M, et al. (2005). Evaluating the high-throughput screening computations. *J. Biomol. Screen.* 10:649–652.

48. Jain A N (2003). Surflex: Fully automatic flexible molecular docking using a molecular similarity-based search engine. *J. Med. Chem.* 46:499–511.

49. Pirard B (2005). Knowledge-driven lead discovery. *Mini Rev. Med. Chem.* 5:1045–1052.

50. Schuffenhauer A, Popov M, Schopfer U, et al. (2004). Molecular diversity management strategies for building and enhancement of diverse and focused lead discovery compound screening collections. *Comb. Chem. High Throughput Screen.* 7:771–781.

51. Newman D J, Cragg G M, Snader K M (2003). Natural products as sources of new drugs over the period 1981–2002. *J. Nat. Prod.* 66:1022–1037.

52. Kubinyi H, Müller G, eds. (2004). *Chemogenomics in Drug Discovery: A Medicinal Chemistry Perspective.* Vol. 22. Wiley-VCH, Weinheim, Germany.

53. Oprea T I (2002). Current trends in lead discovery: Are we looking for the appropriate properties? *J. Comput. Aided Mol. Des.* 16:325–334.

54. Davis A M, Keeling D J, Steele J, et al. (2005). Components of successful lead generation. *Curr. Top Med. Chem.* 5:421–439.

55. Deprez-Poulain R, Deprez B (2004). Facts, figures and trends in lead generation. *Curr. Top. Med. Chem.* 4:569–580.

56. Olah M M, Bologa C G, Oprea T I (2004). Strategies for compound selection. *Curr. Drug Discov. Technol.* 1:211–220.

57. Alanine A, Nettekoven M, Roberts E, et al. (2003). Lead generation—enhancing the success of drug discovery by investing in the hit to lead process. *Comb. Chem. High. Throughput Screen.* 6:51–66.

58. Van De Waterbeemd H, Gifford E (2003). ADMET in silico modeling towards prediction paradise? *Nat. Rev. Drug Discov.* 2:192–204.

59. Roberts S A (2003). Drug metabolism and pharmacokinetics in drug discovery. *Curr. Opin. Drug Discov. Devel.* 6:66–80.

60. Yu H, Adedoyin A (2003). ADME-Tox in drug discovery: Integration of experimental and computational technologies. *Drug Discov. Today* 8:852–861.

61. Ghetie V, Ward E S (2000). Multiple roles for the major histocompatibility complex class I-related receptor FcRn. *Ann. Rev. Immunol.* 18:739–766.

62. Presta L G (2005). Selection, design, and engineering of therapeutic antibodies. *J. Allergy Clin. Immunol.* 116:731–736.

63. Pandley J, Astemborski J, Thomas D L (2004). Epistatic effects of immunoglobulin GM and KM allotypes on outcome of infection with hepatitis C virus. *J. Virol.* 78:4561–4565.

64. Williams R C Jr, Malone C C (1992). Heteroclitic polyclonal and monoclonal anti-Gm(a) and anti-Gm(g) human rheumatoid factors react with epitopes induced in Gm(a), Gm(g-) IgG by interaction with antigen or by nonspecific aggregation. A possible mechanism for the in vivo generation of rheumatoid factors. *J. Immunol.* 149:1817–1824.

65. Balint R, Larrick J W (1993). Antibody engineering by parsimonious mutagenesis. *Gene* 137:109–118.

66. Chlewicki L K, Holler P D, Monti B C, et al. (2005). High-affinity, peptide-specific T cell receptors can be generated by mutations in CDR1, CDR2 or CDR3. *J. Mol. Biol.* 346:223–239.

67. Feldhaus M J, Siegel R W, Opresko L K, et al. (2003). Flow-cytometric isolation of human antibodies from a nonimmune Saccharomyces cerevisiae surface display library. *Nat. Biotechnol.* 21(2):163–170.

68. Schaffitzel C, Zahnd C, Amstutz P, et al. (2005). In vitro selection and evolution of protein-ligand interactions by ribosome display. In: Golemis E A, Adams P D (eds.), *Protein-Protein Interactions, 2nd ed.* pp. 517–548.

69. Cavagnaro J A (2002). Preclinical safety evaluation of biotechnology-derived pharmaceuticals. *Nat. Rev. Drug Discov.* 1:469–475.

70. ICH Steering Committee. Safety pharmacology studies for human pharmaceuticals S7A. Available: http://www.ich.org/LOB/media/MEDIA504.pdf.

71. Gollapudi B, Krishna G (2000). Practical aspects of mutagenicity testing strategy: An industrial perspective. *Mutat. Res.* 455:21–28.

72. Krishna G, Urda G, Theiss J (1998). Principles and practices of integrating genotoxicity evaluation into routine toxicology studies: A pharmaceutical industry perspective. *Environ. Mol. Mutagen.* 32:115–120.

73. Brennan F R, Shaw L, Wing M G, et al. (2004). Preclinical safety testing of biotechnology-derived pharmaceuticals. *Mol. Biotechnol.* 27:59–74.

74. Green J D, Black L E (2000). Status of preclinical safety assessment for immunomodulatory biopharmaceuticals. *Hum. Exp. Toxicol.* 19:208–212.

75. Pilling A M (1999). The role of the toxicologic pathologist in the preclinical safety evaluation of biotechnology-derived pharmaceuticals. *Toxicol. Pathol.* 27:678–688.

76. Food and Drug Administration, Center of Biologics Evaluation and Research (1997). Points to consider in the manufacture and testing of monoclonal antibody products for human use. Fed Reg 62:9196.

77. ICH Steering Committee. Preclinical safety evaluation of biotechnology-derived pharmaceuticals S6. Available: http://www.ich.org/LOB/media/MEDIA503.pdf. Accessed January 3, 2006.

78. Federsel H J (2000). Building bridges from process R&D: From a customer-supplier relationship to full partnership. *Pharm. Sci. Technol. Today* 3:265–272.

79. Federsel H J (2003). Logistics of process R&D: Transforming laboratory methods to manufacturing scale. *Nat. Rev.* 2:654–664.

80. Dugger R W, Ragan J A, Brown Ripin D H (2005). Survey of GMP bulk reactions run in a research facility between 1985 and 2002. *Org. Process Res. Dev.* 9:253–258.

81. Andraos J (2005). Unification of reaction metrics for green chemistry II: Evaluation of named organic reactions and application to reaction discovery. *Org. Process Res. Dev.* 9:404–431.

82. Gronnow M J, White R J, Clark J H, et al. (2005). Energy efficiency in chemical reactions: A comparative study of different reaction techniques. *Org. Process. Res. Dev.* 9:516–518.

83. Lehmbeck J, Wahlbom F. Expression systems comprising heterokaryon fungus host cell and nucleic acids encoding heterologous signal peptide and cellulose-binding domain for industrial scale production of human antibodies. PCT Int. Appl. WO 2005070962 A1. Filing date: 20 January 2005.

84. Humphreys D P, Glover D J (2001). Therapeutic antibody production technologies: Molecules, applications, expression and purification. *Curr. Opin. Drug Discov. Devel.* 4:172–185.

85. Sanna P P (2002). Expression of antibody Fab fragments and whole immunoglobulin in mammalian cells. *Methods Mol. Biol.* 178:389–395.

86. Houdebine L M (2002). Antibody manufacture in transgenic animals and comparisons with other systems. *Curr. Opin. Biotechnol.* 13:625–629.

87. Schillberg S, Twyman R M, Fischer R (2005). Opportunities for recombinant antigen and antibody expression in transgenic plants-technology. *Vaccine* 23:1764–1769.

88. Yoo E M, Chintalacharuvu K R, Penichet M L, et al. (2002). Myeloma expression systems. *J. Immunol. Methods* 261:1–20.

89. Miescher S, Zahan-Zabal M, DeJesus M, et al. (2000). CHO expression of a novel human recombinant IgG1 anti-RhD antibody isolated by phage display. *Br. J. Haematol.* 111:157–166.

90. Fahrner F L, Knudsen J L, Basey C D, et al. (2001). Industrial purification of pharmaceutical antibodies: Development, operation, and validation of chromatography processes. *Biotechnol. Genet. Eng. Rev.* 18:301–327.

91. Lee M H, Kwak J W (2003). Expression and functional reconstitution of a recombinant antibody (Fab') specific for human apolipoprotein B-100. *J. Biotechnol.* 101:189–198.

92. *Berlex Laboratories Inc., Plaintiff, v. Food And Drug Administration, et al., Defendants.* Civil Action No. 96-0971 (JR), United States District Court for the District of Columbia, 942 F. Supp. 19, 1996 U.S. Dist. LEXIS 15169. Available: http://www.law.berkeley.edu/institutes/bclt/courses/fall99/berlex.html. Accessed March 15, 2006.

93. Henry C (1996). FDA, reform, and the well-characterized biologic. *Anal. Chem.* 68:674A–677A.

94. Lubarskaya Y, Houde D, Woodard J, et al. (2005). Analysis of recombinant monoclonal antibody isoforms by electrospray ionization mass spectrometry as a strategy for streamlining characterization of recombinant monoclonal antibody charge heterogeneity. *Anal. Biochem.* Available: sciencedirect.com.

95. Herrera S (2004). Biogenerics stand-off. *Nat. Biotechnol.* 22:1343–1346.

96. Farnley S, Morey-Nase P, Sternfeld D (2004). Biotechnology—a challenge to the patent system. *Curr. Opin. in Biotechnol.* 15:254–257.

1.4

INTEGRATED DEVELOPMENT OF GLYCOBIOLOGICS: FROM DISCOVERY TO APPLICATIONS IN THE DESIGN OF NANOPARTICULAR DRUG DELIVERY SYSTEMS

CHRISTINE VAUTHIER, ISABELLE BERTHOLON, AND DENIS LABARRE

University of Paris XI, Chatenay-Malabry, France

Chapter Contents

1.4.1 Introduction   126
1.4.2 Carbohydrate Research: Main Steps of Development and Discovery of Their Functionalities   126
1.4.3 Development of Nanoparticular Drug Carriers: Current States and Limitations   128
1.4.4 Integration of Carbohydrate in the Design of Nanoparticle Drug Carriers With New Functionalities   129
     1.4.4.1 Carbohydrates to Improve Mucoadhesion of Nanoparticle Drug Carriers   130
     1.4.4.2 Use of Carbohydrates to Control the Fate of Drug Carrier After Intravenous Administration   131
     1.4.4.3 Carbohydrates and Drug Carrier Targeting Perspectives   134
1.4.5 Enhancement of Carbohydrate Therapeutic Activity by Association With Nanoparticle Delivery Systems   139
     1.4.5.1 Anti-infectious Agents   139
     1.4.5.2 Anticancer Activity   141
     1.4.5.3 Control of Vascular Smooth Muscle Cell Proliferation   142
     1.4.5.4 Activity on the Immune System   142
     1.4.5.5 Control of Blood Coagulation and Complement Activation   144
1.4.6 Nanoparticle Drug Carriers Made of Carbohydrates for the Delivery of Fragile Molecules   145
1.4.7 Conclusion   147
     References   148

Handbook of Pharmaceutical Biotechnology, Edited by Shayne Cox Gad.
Copyright © 2007 John Wiley & Sons, Inc.

1.4.1   INTRODUCTION

From the three major classes of biomolecules including proteins, nucleic acids, and carbohydrates, the carbohydrates entering the constitution of glycans and polysaccharides are, at the moment, the least exploited in biology. The recent development of operating methods for the analysis of carbohydrates and for the synthesis of oligosaccharides support the efforts made to investigate the roles of carbohydrates in biological phenomena [1–4]. Different already identified functions of carbohydrates may be interesting to be integrated within drug delivery systems conferring new properties and opening avenues for innovative developments. In this view, carbohydrates may be used either as a drug due to their own biological activities or as a molecule to design nanoparticles with new functionalities [5–8]. Developing drug delivery systems, the main challenge is to design systems able to carry and target a drug to the diseased organs or cells after administration in the body by the appropriate route [6, 7, 9, 10]. This assumes that the biodistribution of the carrier system is perfectly controlled after *in vivo* administration. In the case of a soluble carrier, this can be achieved by grafting define types of carbohydrate ligands on the polymer backbone [10, 11]. Using colloidal particles like liposomes and polymer nanoparticles, it is now well accepted that their biodistribution is mainly controlled by their surface properties [9, 12–15]. Several carbohydrates and polysaccharides were identified at the surface of cells, bacteria, and viruses as having a key role in biological recognition and signaling events. Therefore, the potential of their use as coating material or as ligands to promote specific recognition between carrier and target cells was carefully considered as a possible way to develop biomimetic drug delivery systems having different functionalities [16–24]. In this chapter, we will summarize the main stages of the development of carbohydrates and polysaccharides as components of interest to be used either as a targeted drug or as a component entering the design of drug delivery systems. The first part of the chapter will focus on the principal milestones that allowed discovery of most of the carbohydrates and polysaccharides known today in terms of their chemical analysis and biological activity investigations. This part will also present status about the chemical synthesis of these compounds. The second part of the chapter will briefly discuss the current state and limitations to the developments of nanoparticle drug carriers. Integration of carbohydrate and polysaccharide developments in the design of drug delivery systems with new functionalities will constitute the third part of the chapter. The fourth and fifth parts will, respectively, review the current research about applications of nanoparticles as delivery systems for biologically active carbohydrates and the use of nanoparticle drug carriers made of polysaccharides for the delivery of fragile bioactive molecules.

1.4.2   CARBOHYDRATE RESEARCH: MAIN STEPS OF DEVELOPMENT AND DISCOVERY OF THEIR FUNCTIONALITIES

Forty years ago, Eylar [25] suggested for the first time that the carbohydrate units found in biologically active glycoproteins could act as a kind of "chemical label" or "chemical passport" to promote specific interactions of glycoproteins with corresponding cell-surface receptors to achieve their transportation across the cell membrane. However, progress in the understanding of the biological role of glycan moieties included in biological structures was hampered for a long time, because glycans and polysaccharides were much more complex structures to study than were proteins and nucleic acids. Thus, it was only 30 years ago, while full chemical synthesis of peptides was already possible and that of nucleic acid was in its infant age, that various carbohydrate containing compounds came into light as important elements in the number of biological activities [26–30]. For instance, the years 1976 to 1983 were critical in the discovery of the antithrombine binding site of heparin, which now can be used to confer anticlotting and anticomplement activation properties to drug carrier systems [30–32]. The study of lectines isolated from animals that are now known to be involved in many biological recognition molecules involving carbohydrates started to grow rapidly in the 1980s thanks to the development of recombination techniques to produce them [28]. Identification of the role of these molecules in recognition phenomena as well as understanding their functionality are interesting to be able to promote specific recognition between target cells and drug delivery systems achieving drug targeting. It is also noteworthy to point out the discovery of carbohydrates and carbohydrate derivatives that are currently used as major anticancer drugs and antibiotics in clinics [33–35]. Because of the high toxicity of these compounds for the healthy tissues, they are good drug candidates to be incorporated in nanoparticle drug carrier systems improving their specific biodistribution.

For long, progresses in glycobiology were hampered due to technological difficulties in analytical order. The appearance of the high-performance anion–exchange chromatography (HPAEC) in the mid-1990s revolutionized the analytical capacity and considerably improved the accuracy of determining the carbohydrate sequences and arrangements in glycans isolated from glycosylated biological compounds [2, 4]. This was a significant milestone that boosted the resolution of glycan structures and in parallel opened new areas to further investigate their biological activity. The analytical techniques were constantly improved by the subsequent introduction of pulsed amperometry detection (PAD) and more recently by online coupling the HPAEC to mass spectrometry devices and nuclear magnetic resonance (NMR) methods using a nano-probe [3, 4, 36–38]. New methodologies continue to be developed improving performance of the existing analytical methods [39]. The introduction of screening methods using microarray and chip technologies appeared within the last 3 years accelerating discovery and progress toward the understanding of the role of carbohydrates in biology [29, 40, 41].

Research into the biological roles of carbohydrates may be possible because sufficient quantities may be isolated from natural extract. Because of their structural complexity, it is extremely difficult to isolate pure carbohydrates from natural sources. Progress in the analytical chromatographic methods has helped to improve preparative techniques providing purified fractions of defined glycans from various natural extracts [36, 42]. However, the only way to access very pure carbohydrates relies on their chemical or enzymatic synthesis, which also constitutes the more attractive route for the production of the sufficient quantities for biological studies designed to identify biological roles and eventually medicinal interests. The synthesis of polysaccharides and glycans was an important technological challenge. The intensive efforts based on innovative strategies led to major advances, making the synthesis of complex carbohydrates now possible [1, 43–45]. Stereoregular polysaccharides, including linear, branched, amino, and deoxy polysaccharides, can be synthesized by ring opening polymerization of anhydro-sugar derivatives [1]. The technically extremely difficult synthesis of highly branched carbohydrates was made possible with the development of solid-phase synthesis strategies [44]. Another important milestone was completed thanks to the recent development of an automated oligosaccharide synthesizer leading to accelerated access to many highly branched carbohydrates for identification of their biological function [44]. Even if this methodology provides just enough glycan to study the biological role of the product, this really opens avenues for the identification and creation of important compounds for future biochemical and medicinal applications [44]. For instance, after years of research, many synthetic heparins proposed as anticoagulant drugs can be synthesized on a production scale. They are currently at different stages of clinical development [46]. The chemistry remains complicated, and some synthetic routes that have been scaled up comprised up to 65 steps.

1.4.3   DEVELOPMENT OF NANOPARTICULAR DRUG CARRIERS: CURRENT STATES AND LIMITATIONS

Despite several advances in chemotherapy, the real therapy of cancer and major infections remains a challenge. One problem comes from the nonspecific drug distribution resulting in low tumor or infected tissue concentrations and systemic toxicity. Another problem comes from the lack of stability of drugs in biological media and their incapacity to cross biological barriers. To improve drug concentration in the targeted tissues, several approaches have been suggested, including the association of the drug to carrier systems. A large piece of work has been done using liposomes as carriers, which has led to several marketed formulations so far. For instance, the liposomal formulation of Amphotericin B, which is the leading compound in the treatment of leishmaniasis is today the more efficient treatment against this parasite and other fungal infections [47]. The toxicity of Amphotericin B was reduced by a factor of 50- to 70-fold, which is among the most obvious benefit for the use of this liposome formulation in clinics [48]. The main drawback of liposomes remains the cost of the phospholipids entering their composition. To reduce cost, the biggest challenge is believed to come from polymer nanoparticles. Indeed polymers are a much cheaper material compared with phospholipids, and methods for producing nanoparticles are generally simpler. Other advantages of the polymer systems are their higher stability in biological media especially for the administration of drugs by mucosal routes and upon slight changes in the formulation. Additionally, they should be theoretically tailor-made thanks to the progress in colloid and polymer technology [9, 49]. However, only a couple of polymers are suitable as main components of the polymer nanoparticles designed as drug carrier systems [6, 7, 9]. As for liposomes, polymer nanoparticles were developed as drug carriers to improve the efficacy of the associated drug thanks to a better control of its biodistribution toward the diseased organ or cells and to their faculty to overcome biological barriers [9, 50, 51]. These colloidal particles have demonstrated enhanced efficacy for numerous drugs compared with conventional formulations. Indeed, the biological activity of the drug associated with the carrier is generally improved, whereas the side effects due to their toxicity are reduced thanks to a better control of the biodistribution [52–57]. The improved efficacy can also be explained by a higher stability of the drug in biological media and ability to overcome biological barriers. However, the main limitation of these systems remains the full control of their biodistribution because they still lack precise targeting specificities. This is the main challenge that remains to be addressed. According to our current knowledge, a better control of the interactions between the drug delivery system and surrounded biological media should improve targeting efficacy as well as promote the occurrence of highly specific interactions. Thus, it was postulated that the control of the biodistribution of drug carriers could be possible by modifying and by adjusting the surface characteristics of the drug delivery systems. The modification of surface properties of drug delivery systems by coating with poly(ethylene glycol) actually rerouted the carrier from a very efficient recognition and uptake by macrophages of the mononuclear phagocyte system (MPS) toward long circulating particles in the blood stream [58, 59]. It is now admitted that the biodistribution of a drug carrier greatly depends on the nonspecific and specific interactions between its exposed surface and the biological components of the surrounded medium found either in blood, on the cell surface, or at the level of biological barriers, i.e., endothelium or mucosa. Thus, efforts are now carried on tailoring the design of the carrier surface [12].

To be used as drug carriers, different drugs should be incorporated into the nanoparticles. Most nanoparticles made of synthetic polymers, including polyesters or polyanhydrides and poly(aminoacids), can incorporate enough lipophilic drugs. However, they are totally inefficient to incorporate hydrophilic molecules such as peptides, proteins, or nucleic acids. These molecules with a high therapeutic potential hardly cross biological barriers. They need to be associated with a drug carrier enhancing their transport across the biological barriers and preserving their biological activity because they are highly un-stable in biological media. This problem could be solved using nanopar-ticles made of poly(alkylcyanoacrylate) [60]. Polysaccharides extracted from natural compound could also be used to formulate new types of drug delivery systems in which macromolecules from biotechnology could be incorporated [6–8, 61, 62].

1.4.4 INTEGRATION OF CARBOHYDRATE IN THE DESIGN OF NANOPARTICLE DRUG CARRIERS WITH NEW FUNCTIONALITIES

As mentioned, the interactions of nanoparticle drug carriers with the surrounded biological environment are key factors controlling the *in vivo* fate of the transported drug. Three levels of interactions may be distinguished, but all will involve surface phenomena between nanoparticle surface and compounds of the surrounding environment. Interactions occurring at the level of a mucosa may be critical to promote the absorption of the drug. They will be discussed in the first paragraph. The interactions taking place in the blood compartment that greatly affect the biodistribution of the carrier due to interactions with blood proteins are considered in the second paragraph. Finally, the more highly specific interactions between drug carrier and cells in the view of targeting and allowing delivery of the drug to precise cells will be considered in the last paragraph of this section. As will be discussed, carbohydrates seem, in all of these cases, as versatile tools to control the interactions between the drug carriers and the biological surrounding media.

1.4.4.1   Carbohydrates to Improve Mucoadhesion of Nanoparticle Drug Carriers

Mucosal routes, especially the oral route, are the preferred methods for the administration of drugs. However, many drugs remain poorly available when they are administered by these routes. Among other reasons, this can be due to low mucosal permeability for the drug, permeability restricted to a region of the gastrointestinal tract, low or very low solubility of the compound that results in a low dissolution rate in the mucosal fluids and in elimination of a fraction of the drug from the alimentary canal prior to absorption, and lack of stability in the gastrointestinal environment, resulting in a degradation of the compound before its absorption (e.g., peptides and oligonucleotides).

To reduce problems of low permeability and low solubility, it was proposed to retain the drug at the surface of the mucosa using bioadhesive formulations increasing the contact time between the drug and the epithelium and in turn chance to the drug to be absorbed. In this aim several authors have suggested to use polysaccharides to design bioadhesive nanoparticle drug delivery carriers. Chitosan obtained from deacetylation of chitin is the most commonly used bioadhesive polysaccharide. At the level of mucosa, chitosan is known to enhance drug absorption thanks to it bioadhesive properties and to its capacity to induce transient opening of the tight junctions without damaging the cells [63, 64]. The bioadhesive properties of chitosan were described for the first time by Lehr et al. [65]. They depend on the molecular weight [66, 67] and on the pH of the surrounded media, which affect the polysaccharide solubility. Chitosan is soluble at pH below 6, and its bioadhesive properties require that the molecule is either soluble or strongly swollen [68]. Thus, to improve the solubility properties of chitosan especially at pH higher than 6 encountered at the level of several mucosa, modified chitosan were synthesized. For instance, the partial substitution of the amino function of chitosan by methyl-5-pyrrolidone led to derivatives enhancing the absorption of drugs at the level of the bucal and vaginal mucosa [68–70]. Enhancement of the bioadhesive properties and of drug absorption was also reported after introduction of thiol groups in the structure of chitosan. In this case, the thiol groups are assumed to form disulfide bonds with the cystein-rich domains of the mucus glycoproteins standing on the gastrointestinal mucosa [71, 72]. Drug carriers that have been coated with chitosan also showed enhanced bioadhesive properties at the level of different mucosa, including the ocular, nasal, gastrointestinal, and pulmonary mucosa [8, 73–76]. However, it is worth pointing out that a simple adsorption of chitosan at the nanoparticle surface is not stable enough to resist desorption when the nanoparticles are in contact with the mucus [77]. Thus, several authors have suggested preparing nanoparticles directly with chitosan obtaining plain chitosan nanoparticles [7, 75] or using copolymers of chitosan and the polymer constituting the nanoparticle, i.e., poly(lactic acid) [61] or poly(alkylcyanoacrylate) [19, 21, 62, 78].

According to Prego et al. [8], the mucoadhesive properties of the carrier and its interaction with the mucus occurring *in vivo* are determinant to provide an efficient transport of the nanodevice carrying the drug across the epithelium, resulting in an increase of the bioavailability. The mucoadhesive properties were shown to depend on the coating material but also on the size of the nanoparticles [8, 79]. Generally, the smallest nanoparticles accumulate more because of a higher diffusion in the mucus layer compared with bigger particles. However, according to a recent study, Bertholon et al. showed that the conformation of the chitosan chains grafted at the surface of poly(alkylcyanoacrylate) nanoparticles also influenced the bioadhesive properties of the nanoparticles on the intestinal mucosa [21]. Indeed, these authors reported differences in bioadhesion with nanoparticles onto which chitosan was grafted either on a "side-on" conformation or on the "end-on" conformation. Adhesion of the nanoparticles to the intestinal mucus layer can be viewed as a complex process, including the diffusion of the particles in the hydrogel, depending on their size, followed by their interactions with the mucus glycoproteins depending on both the nature of the polysaccharide and its conformation at the nanoparticle surface. It is likely that the molecular interactions played a key role in this latter process. Interestingly these results suggest that the bioadhesive properties of nanoparticulate systems can be modulated by a fine tuning of the surface properties of the particles.

Several other polysaccharides were used to obtain bioadhesive formulations of drugs [80, 81]. Hyaluronic acid, which is applied in ocular surgery to protect the corneal endothelium and to manipulate intraocular tissue is also a remarkable bioadhesive. It was suggested as coating material at the surface of nanoparticles made of poly(epsilon-caprolactone) for the development of bioadhesive ocular formulation of bioactive molecules to increase the retention of the drug at the level of the precorneal zone [82]. The other polysaccharides found in pharmaceutical formulations that may be interesting to use as bioadhesive coated material at the surface of nanoparticle drug carriers are alginate, dextran, starch, and cellulose derivatives, including carboxymethylcellulose and hydroxypropyl cellulose. These polysaccharides have in common high swelling properties in an aqueous environment.

1.4.4.2 Use of Carbohydrates to Control the Fate of Drug Carrier After Intravenous Administration

The fate of nanoparticles administered by the intravenous route is controlled by the interactions between the nanoparticle surface and the serum proteins. In the blood, nanoparticles are normally considered foreign bodies and are rapidly recognized as such by the host defense system. The primary event of the recognition phenomena is basically nonspecific and results in a massive adsorption of serum proteins at the nanoparticle surface [14, 58, 83]. Consequently, the complement system is activated, leading to the labeling of the nanoparticle surface with specific activated proteins of the complement systems. Once the nanoparticles are labeled with activated complement proteins, they are efficiently recognized by macrophages of the mononuclear phagocyte system (MPS) due to the presence of the corresponding receptors at the cell surface. The overall process is so efficient that less than 5 minutes are generally necessary to remove from the blood 80% to 90% of the injected dose of the drug carrier. After these events, the drug and its carrier are found in macrophages of the liver and of the spleen [9, 12, 84–87]. This phenomenom is highly beneficial to target drugs in the liver and spleen but hamper the distribution of the drug toward other biological territories.

Targeting territories, organs, and tumors located outside the MPS organs can be achieved by triggering the host defense systems thanks to the modification of blood–protein interactions with the nanoparticle surface in order to obtain long circulating nanoparticles. The more general approach consists of grafting poly(ethylene glycol) chains at the nanoparticle surface to reduce both opsonization of the carrier surface by blood proteins and complement activation [14, 15, 58, 88]. An alternative route is based on a biomimetic approach using carbohydrates as material to coat the nanoparticle surface in order to modify interactions with blood proteins [19, 20, 31, 32, 89–92]. The design of biomimetic drug delivery systems using carbohydrate as surface coating material is almost infinite as much of the diversity of carbohydrates appears to be in Mother Nature as exposed at the surface of the eukaryotic cells, bacteria, and viruses [5, 40, 44]. So far, only simple polysaccharides were used to modify the surface properties of nanoparticle drug carriers. The simplest polysaccharide was dextran, which is an alpha-1-6-poly(glucose). Measurements of complement activation given by different dextran-coated nanoparticles provided contradictory results depending on the method of the nanoparticle preparation. Indeed, dextran-coated poly(alkylcyanoacrylate) nanoparticles prepared by redox radical polymerization induced only a low complement activation [19, 20, 31, 32], whereas dextran-coated nanoparticles prepared by other methods induced a strong activation of complement, as does sephadex, which is a reticulated dextran [19, 20, 32, 90, 93, 94]. The contradictory complement activation measured could not be correlated to a marked difference in the interaction of the nanoparticles with blood proteins. Indeed, each type of the nanoparticles showed a proper pattern of blood protein adsorption and no general behavior could be drawn [32]. The hypothesis suggested that explains the results of complement activation is that it may depend on the conformation of the dextran chains at the nanoparticle surface, which can be defined by the method of nanoparticle preparation. This hypothesis is supported by the results of a very recent work that showed that nanoparticles with dextran chains grafted at the nanoparticle surface on the "end-on" conformation are nonactivators when the molecular weight of dextran passes above a certain limit (40,000 g/mol), whereas the nanoparticles on which dextran are arranged on a "side-on" conformation activated complement whatever was the molecular weight of dextran [20]. The effect of the polysaccharide chain conformation was more evident with a series of chitosan-coated nanoparticles. Indeed, nanoparticles with chitosan on the "end-on" conformation were frankly nonactivators, whereas the nanoparticles with chitosan on the "side-on" conformation were strong activators [20]. Although the conformation of the polysaccharide chains at the nanoparticle surface seemed to be a key parameter to control complement activation, several authors showed that the presence of dextran at the nanoparticle surface reduced the adsorption of albumin in favor of a reduction of opsonization phenomena [92, 95]. In terms of biodistribution of the corresponding nanoparticles, the nanoparticles having dextran on the "side-on" conformation were recognized by macrophages of MPS and concentrated in the liver and in the spleen [13], whereas the nanoparticles with dextran on the "end-on" conformation remained in the blood circulation for a longer period [89].

   Although chitosan-coated nanoparticles showed contradictory results in complement activation depending on the conformation of the chains at the nanoparticle surface [20], the level of complement activation induced by chitosan was also influenced by the physico-chemical properties of the polysaccharide, including its solubility, degree of deacetylation, and molecular weight [20, 96–98]. Using different chitosan and chitosan derivatives, nanoparticles with a prolonged circulation time in the blood can be obtained. Additionally, their distribution can be modulated in a certain way because they were found to accumulate in defined tissues outside the MPS [99–104].

In Mother Nature, sialic acids play a key role in the biological mask hiding epitopes at the cell surface, which are immediately recognized by macrophages when the sialic acid is removed from the cell surface [103–107]. The most typical example is given by the erythrocytes, which are rapidly destroyed when sialic acids are removed from their surface [106, 108, 109]. Highly virulent bacteria have also adopted sialic acids as coating material to escape host defense mechanisms [110–112]. Sialic acid, which seems to inhibit complement activation, allows cells and particular material to stay in the blood without being recognized by macrophages [110]. Thus, it was suggested to coat nanoparticles with sialic acids to mask their surface and avoid recognition by macrophages. Two different approaches were followed. At first, the ad-sorption of a sialic acid-rich glycoprotein, orosomucoid, at the surface of poly(alkylcyanoacrylate) nanoparticles reduced opsonization of the nanoparticles and complement activation. However, the orosomucoide that was simply adsorbed at the nanoparticle surface was rapidly displaced from the surface by other serum proteins due to the Vroman effect, and the effect of the sialic acids could be shown on short duration [17, 113]. Second, the nanoparticles coated with sialic acid were obtained from a copolymer of poly(sialic acid) and poly(lactic acid). These nanoparticles coated with a polysialic acid remained in the blood circulation longer than the uncoated nanoparticles after intravenous administration to mice [114]. Poly(sialic acids) have the advantage of being biodegradable, and their catabolic products (i.e., neuraminic acids) are not known to be toxic. Another important advantage of poly(sialic acids) is their T-independent antigens, meaning that they do not induce immunological memory. Because of these properties, poly(sialic acids) are very promising molecules to design biomimetic drug carrier systems with an increased half-life in the blood stream of the drug carrier [16, 115]. The main limitation of the development of such drug carriers so far may be explained by the poor availability of poly(sialic acids), which remain a problem because they are produced by highly pathogenic bacteria strengths at a very high cost.

The other bioactive polysaccharide that seemed interesting to be used to produce surface-modified nanoparticles reducing their recognition by the host defense was heparin. Heparin is used as a drug for its anticoagulation properties. Additionally, it is an inhibitor of the complement activation phenomenon [116–118]. It was demonstrated that heparin-coated nanoparticles did not activate the complement system [19, 31, 32] and remained in the blood stream for a longer time compared with nanoparticles, which do not show heparin on the nanoparticle surface [89]. Other polysaccharides extracted from mushrooms were found to inhibit the activation process of the complement. They could be alternative polysaccharides to produce nanoparticles with a reduced capacity to activate the complement, such as heparin [119].

According to the current experiences, several polysaccharides can be used to modulate the interactions of nanoparticle drug carriers, with blood proteins modifying their blood clearance and consequently their biodistribution. Two categories of polysaccharides giving nanoparticles an ability to escape recognition from macrophages of the MPS can be identified. The first category, including heparin and sialic acids, specifically interact with serum protein according to a well-defined mechanism, thanks to their specific biological activity being inhibitors of the complement activation phenomenom. The polysaccharide of the other category, including dextran, interacts with serum protein through nonspecific interactions. In addition to the specific and nonspecific interactions with blood proteins, it was clearly demonstrated that the conformation of the chains grafted at the nanoparticle surface is one key parameter that controls the complement activation phenomena and the opsonization of the nanoparticle surface. The nature of the polysaccharide and the molecular weight took on the second level of importance [20].

1.4.4.3 Carbohydrates and Drug Carrier Targeting Perspectives

Colloidal carriers have proved to be interesting to deliver drugs by the intravenous route because of their potential to improve the therapeutic index of the carried drug while reducing their side effects. Although with the current systems available it can be possible to enhance accumulation of the drug in diseased tissue, especially in tumors, the interactions of the carriers with diseased cells and their uptake remain insufficient. The main strategy that was proposed to further enhance drug delivery and retention at the level of diseased cells is based on the active targeting of the drug [12, 120–122]. This supposes that drugs reached target cells thanks to a highly specific biodistribution of the carrier. Our experience with drug carriers suggests that such a degree of specificity will only be obtained if the drug carrier can recognize the target cells through specific interactions.

One way to promote recognition between drug carriers and target cells is to attach ligands at the carrier surface that can bind specifically to target cells [12, 121, 122]. For this purpose, more or less complex carbohydrates can be used as ligands of endogenous receptors for carbohydrates known as lectins at the cell surfaces. Lectines are universally found in the microbial world, on vegetal and animal cells [5, 28, 123]. They can bind various types of carbohydrates from simple monosaccharides to oligo- or polysaccharides of complex structures with a very high specificity of recognition resulting in strong interactions. The binding is a very fast process, and the recognition between the carbohydrate and the corresponding lectin can occur whether the carbohydrate is free or bound on a macromolecule or on the surface of cells and of drug carriers [123, 124]. So far, the majority of the work done in the field of active targeting was performed with liposomes and soluble polymers. Currently, only a few studies were carried out on nanoparticles [12, 121, 125]. Microbial and vegetal lectins were bound to drug carriers as targeting moiety to find complementary carbohydrate epitopes at the surface of target cells [126–128]. However, with the recent progress in glycobiology, it seems that a better option would consist in grafting the carbohydrates on the drug carrier instead of lectins, which often lose activity during the grafting process. Another advantage of using carbohydrates as the targeting moieties is that the drug carrier would be targeted toward cell receptors with expression levels that are subjected to dramatic modifications upon physiological conditions of the cells. For instance, in several cancer cells, some receptors are specifically over-expressed, whereas other are downregulated and almost disappear from the cell surface. Thus, it seems better to use the ligand receptor, i.e., carbohydrate, as the targeted moiety attached on the drug carrier to take advantage of differences in the level of receptor expression in the targeting strategy. This method, which was applied with folic acid, actually showed high targeting potential [129]. Additionally, it would probably be easier to find universal methods for grafting various types of carbohydrate moiety on nanoparticle surface in comparison with the coupling methods of proteins, which need to be finely adjusted for each protein receptor to preserve the functionality of their active site [44].

Many endogenous lectins in animals might serve as target receptors for drug carriers exhibiting complementary carbohydrates at the surface. Examples of animal lectins and principal characteristics, locations, and specificities are given in Table 1.4-1, whereas Table 1.4-2 summarizes variations of the level of expression of lectins in several diseases that were already identified and reported in the literature.

This last table may be especially useful for identifing over-expressed lectins on diseased cells and may be used as a target to direct drug delivery carriers, bearing the corresponding carbohydrate ligand specifically toward them. Such a targeting strategy was successful using soluble drug carriers [126, 137, 141, 142], and in the case of targeting DNA/polymer complexes [130, 143–147] bearing simple carbohydrate or polysaccharides. Most of the work done so far with nanoparticles included mono- and di-saccharides as the targeting moiety. Lectins can recognize galactose and mannose residues after chemical grafting at the surface of polymer nanoparticles [148, 149]. Galactose was used to target paclitaxel-loaded nanoparticles toward hepatic cells. The lectins of the hepatic cells specifically recognized the galactose residues at the surface of the nanoparticles, improving the targeting of the drug in these cells. Thus, the nanoparticles were found to be more toxic toward the cells expressing the lectin at the surface compared with cells devoid of the corresponding lectin [147]. It also seems that lactose bearing nanoparticles that are recognized by asialoglycoprotein receptors of hepatic cells of mice are internalized by endocytosis [148]. A carbohydrate of Lewis was suggested as targeting moiety to target poly(lactide-co-glycolide) nanoparticles toward cells expressing E selectins on endothelial cells or in inflammatory sites. It was found that, for this more complex carbohydrate corresponding to an oligosaccharide, the affinity of the E-selectins for the Lewis carbohydrate exposed at the nano-particle surface depended on the density of grafting [121].

According to these data, targeting strategies that involve carbohydrates are principally directed toward cell receptors involved in the mechanisms of cell regulation and communication either between cells or with the surrounding medium. They are based on specific interactions with high affinity between the ligand and the corresponding receptor. So far, they have been applied with simple carbohydrates, including monosaccharides, di-saccharides, and some Lewis carbohydrates. It provided more than satisfactory results regarding the level of targeting specificity as evaluated *in vitro* using relevant cell models. Only a few studies that went over the *in vitro* stage were performed with liposomes and soluble drug carriers, providing extremely convincing perspectives [122, 141, 152]. There is no doubt that similar targeting efficiency may be obtained with success using nanoparticles. It can also be expected that current investigations aiming to identify structures and the biological functionalities of carbohydrates at a large scale will contribute to amplify our knowledge of cell communication and therefore highlight new targets and target carbohydrate-based materials [5, 35, 40, 44, 153]. This promises very exciting

TABLE 1.4-1. Principal Characteristics of Some Animal Lectins

Class	Subclass	General Properties	Localization	Specific Ligand	Non-Specific Ligand	Ref.
C–Lectins Calcium dependent	Endocytic receptors	Asialoglycoprotein receptors	Hepatocyte	Ending β D galactose Ending N Acetylgalactosamine		[120, 130]
		Involved in the clearance of desialylated proteins	Liver endothelial cells	Ending N Acetylgalactosamine		
			Kuppfer cells	Ending N Acetylgalactosamine		
			Macrophages	Mannose		
			Parenchymal liver cells	Arabinogalactane Ending Galactose	Dextran	[131, 132]
			Non-parenchymal liver cells	Pullulan Manan	Dextran Pullulan Arabinogalactan	[133]
	Selectins	Transmembrane glycoproteins involved in cell traffic to injury and inflammation sites	Platelets (P selectin) + endothelial cells	Heparin		[121, 125]
			Endothelial cells (E selectin)			[121, 125, 134]
			Leukocyte (L selectin)	Sulfated and sialylated ligands Heparin		[132, 134]
	Collectins	Involved in innate immune phenomena		Saccharides on pathogens or allergens		[132]

136

	Name	Function	Receptor/Cells	Sugar	References
C—Lectins **Calcium dependent**	Proteoglycans	Involved in regulation mechanisms of cell growth, differentiation, and lipid metabolism	*Hyaluronate receptor:* CD44: hematopoietic cells, Fibroblast; RHAMM: Fibroblast, Smooth muscle cells, Macrophages, Lymphocytes; ICMA 1: Leukocytes, Macrophages	Hyaluronic acid and sodium salts	[135–138] [139]
			Non hyaluronate receptors: Skeletal muscle	Fructose, Galactose	[132]
S-Lectins **Calcium independent**	Galectins 1	Cell adhesion and proliferation, regulation of immune infections	Skeletal muscle	B Galactoside Glucose	[132, 140]
	Galectins 3	Allergy, inflammatory process	Eosinophils Neutrophils Macrophages Epithelial cells	β Galactoside Glucose	[132–140]

137

TABLE 1.4-2. Modification of Expression Level of Different Lectins in Several Pathologies

Pathologies	Over-Expressed Lectins	Under-Expressed Lectins	Ref.
Cancer	P Selectins in plasma Hyaluronate receptor CD44 and RHAMM Galectins		[121, 125, 140]
Metastasis	E Selectins Hyaluronate receptor CD44 and RHAMM		[132] [135–138]
Colon cancer	Galectins 1 et 3	Galectin 8	[126]
Meningeal leukaemia	L Selectins		[121]
Malignant B-cells	L Selectins		[121]
Angiogenesis	E Selectins		[132]
Connective tissue disease	P Selectins in plasma		[121, 125]
Rheumatoid arthritis	P Selectins in plasma E Selectins		[121, 125]
Asthma, allergy	P Selectins in plasma E Selectins	L Selectins	[121, 125]
Multiple sclerosis	L Selectins		[121]
Kawasaki disease	E Selectins	L Selectins	[121]
Guillain—Barre syndrome	E Selectins		[121]
Intestine disease	P Selectins in plasma		[121, 125]
Sever infection (malaria)	P Selectins in plasma E Selectins		[121, 125]
Neonatal bacterial infection		L Selectins	[121]
HIV	L Selectins		[121]
Insulin-dependent diabetes	L Selectins		[121]
Risk factors of stroke		L Selectins	[121]
Cardiopulmonary bypass surgery		L Selectins	[121]

developments in the future. However, as stressed by Nobs et al., one has to keep in mind that the ability to achieve targeting *in vivo* depends directly on whether the target is accessible from the vascular compartment [12]. In the first case, particles bearing targeting moiety and having long circulating properties have a high probability to reach the target cells. On the contrary, when the target is extravascular, for instance, in solid tumors, penetration of the carrier in the tumor tissue is more difficult because it needs to extravasate. In this case, the size of carriers needs also to be adjusted so that they can cross the vessel endothelial through the pores of the local discontinuous tumor microvasculature, which may vary between 100 and 780 nm. Another requirement would probably concern the orientation of the carbohydrate moiety at the nanoparticle surface. Indeed, biological recognition systems are so sensitive that they are, for example, capable of differentiating small changes in the conformation of epitopes at the surface of pathogens.

1.4.5 ENHANCEMENT OF CARBOHYDRATE THERAPEUTIC ACTIVITY BY ASSOCIATION WITH NANOPARTICLE DELIVERY SYSTEMS

Research to understand biological functions of carbohydrates is currently extremely active providing new data almost everyday. The literature is growing exponentially in this very fast moving area. The purpose of this chapter was not to review all the different biological activity known for carbohydrates today but was to focus on carbohydrates whose biological activity is interesting for therapeutic purposes and whose therapeutic efficacy may be improved by association with a nanoparticle drug carrier.

1.4.5.1 Anti-Infectious Agents

The emergence of antibiotic resistant strengths of bacteria, parasites, fungus viruses, and new pathogens are very demanding for the discovery of new anti-infectious agents. If the discovery of new molecules is undoubtedly the leading strategy in finding effective treatments, all active molecules will need to reach sites of infections where pathogens are located at sufficient concentration. Several efficient carbohydrate antibiotics were extracted from various strengths of *Streptomyces*. They are simple aminoglycoside molecules known as streptomycins, including neomycin, gentamicin, or amikacin. Other recently discovered carbohydrate antibiotics, including the family of the orthosomycins, are more complex molecules. These carbohydrate antibiotics are very attractive because they are efficient against pathogen bacteria, including those resistant to antibiotics used as a last resort (penicillin or vancomycin). Their interest is so huge that several methods for their synthesis were already suggested [154–156]. However, they are generally very toxic for healthy tissues limiting the maximal dose of antibiotic that can be administered in patients and consequently the efficacy of treatment. In general, the efficacy of antibiotics can be considerably increased when they are associated with drug carriers thanks to a reduction of the toxicity and a higher accumulation in the infected cells that, in many infections, are cells of the mononuclear phagocytes system. The improvement of the safety profile of the targeted drug allowed administration of higher doses of the antibiotic when requested to promote treatment efficacy [48, 157]. In other cases, the therapeutic index of the antibiotic was markedly increased by a factor ranging from 6 to up to 120; thus, the dose of antibiotic required to kill infectious agents can be considerably reduced [158–164]. Complete sterilization of organs can be obtained with the antibiotic targeted with nanoparticles, whereas the corresponding liposomal formulation can not completely sterilized the organs in the same experimental condition [165].

Currently, there is only one formulation of Amikacin in liposomes under clinical trials for treatment of tuberculosis. The antibiotic was found to be twofold to sixfold more active than the free drug in an acute experimental model of murine tuberculosis in which bacteria are located in macrophages. The count of viable bacteria found in the liver and spleen was reduced by a factor of 3-$\log_{10}$ compared with the untreated mice [158]. However, in human, this formulation was not as active as expected. To explain the disappointed results obtained, it was suggested that the liposomes could target the antibiotics in the macrophages but this was not enough to reach extracellular bacilli, which are clustered in cavity caseum in the human infection and can be responsible for reinfection. It was suggested that these residual bacteria may be reached using drug carriers with long circulating properties.

So far, the more advanced studies on the targeting of antibiotics that reach the clinic stages were performed with liposomes. The only formulation marketed so far, AmBisome (Gilead, Foster City, CA), is the more efficient treatment against leishmaniasis and other fungal infections [47]. This demonstrates the high potential of targeted formulations of antibiotics to challenge devastating pathogens with toxic antibiotics for healthy tissue provided that they are targeted to the infected cells. It may be expected that research will continue considering targeting strategies of the most interesting carbohydrate antibiotics to improve the efficacy of anti-infectious treatments. In this view, it will also be interesting to pay attention to nanoparticle formulations that seemed more efficient than liposomes when they were tested in the same conditions probably because of their higher stability in the blood compartment *in vivo* [157, 161].

Some polysaccharides showed either antibacterial and antifungal activities or antiviral activity. For instance, the antibacterial and antifungal activity of chitosan was studied for a long time both *in vitro* and *in vivo*. They are used to control bacterial infections during the wound healing process [166]. The activity of chitosan against pathogens is maximal at acidic pH [166, 167]. The degree of acetylation has no influence on the activity, but high-molecular-weight chitosan was reported to be active against the Gram-negative bacteria *Escherichia coli* [168]. Depending on the authors, either high- or low-molecular-weight chitosan were reported to be active on the Gram-positive bacteria *Staphylococcus aureus* [168, 169]. To explain activity of chitosan, it is assumed that high-molecular-weight chitosan may form a film around the bacterial preventing the exchange of nutriment, whereas low-molecular-weight chitosan may enter bacteria cells to disturb metabolism [168]. Other mechanisms were suggested such as chelation of metal ions required for the synthesis of the bacterial toxin or interaction with cell membrane components inducing agglutination of bacteria or increasing of the permeability of the cell membrane [167, 170]. Chemically modified derivatives of chitosan such as diethy–methyl chitosane are more active than chitosan [171]. Chitosan has not yet been used in a targeted formulation for *in vivo* application. However, it was associated at the surface of nanoparticles to obtain fabrics with antibacterial activity.

Finally, antiviral activities were reported for polysaccharides bearing sulfonate groups. Most studies considered either the activity against the human herpes virus simplex 1 and 2 or against the Human Immunodeficiency Virus (HIV). Generally, the antiviral activity depended on the molecular weight, the amount and the distribution of the sulfonate groups, and the conformation and flexibility of the polysaccharide chains [172]. It was suggested that the anti-HIV activity can result from electrostatic interactions occurring between the negatively charged sulfonate groups of the polysaccharide and the positively charged amino-groups of the glycoprotein GP120 of the virus envelope [170]. Polysaccharides with antiviral activity can be obtained by grafting sulfonate or sulfamide groups [172, 174]. Modified dextranes containing carboxymethyl benzylamine and carboxylmethyl benzylaline sulfonates groups showed anti-HIV activity [175], whereas the anti-herpes virus activity was described for carraghennan corresponding to sulfated polygalactanes extracted from a red algae [172]. More interestingly, a sulphated polysaccharide extracted from the algae *Caulerpa racemosa* is active against strengths of the herpes virus resistant toward acyclovir, which is one of the leading antiviral compound used in clinics [176]. These polysaccharides showing antiviral activity may be interesting compounds to target against infected cells.

1.4.5.2 Anticancer Activity

Several small carbohydrates derivatives showed anticancer activity and are used in clinics as antitumoral drugs [33–35]. Pharmacologically active concentrations of such an anticancer drug in the tumor tissue are often reached at the expense of massive contamination of the rest of the body. As it was discussed with antibiotics, this poor specificity creates a toxicological problem that represents a serious obstacle to effective antitumor therapy. Another obstacle to their efficient delivery is the occurrence of multidrug resistance (MDR), which may appear either as a lack of tumor size reduction or as a clinical relapse after an initial positive response to the tumor tissue; it can be either directly linked to specific mechanisms developed by the tumor cells or connected to the physiology of the tumor tissue, including a poor vasculature and unsuitable physico-chemical conditions [177]. Outside the tumor tissue, the resistance to chemotherapy can be due to the more general problem of the distribution of a drug relative to its targeted tissue [15]. To overcome drug resistance, many attempts have been made using strategies that consider the more general problem of the control of the drug biodistribution either at the cellular level or at the tissue level [178, 179]. These reasons make small carbohydrate antitumoral drugs relevant candidates to be formulated in drug delivery systems allowing their targeting in tumors as it was already demonstrated with the anthracyclin doxorubicin [50, 180–183].

Antitumor activity was also reported for several polysaccharides [35]. Nanoparticles made of chitosan either loaded with copper or unloaded showed a higher cytotoxicity against tumoral cells, whereas their cytotoxicity remained low toward healthy cells of human liver L-02. The antitumoral efficacy depended on the nanoparticle diameter being the highest with the smaller nanoparticles (40 nm) [184]. Heparin showed interesting antitumoral activity especially when it formed a complex with poly(beta-aminoesters), a positively charged polymer, which enhanced the intracellular penetration of heparin by endocytosis into cancer cells. The mechanism of the anticancer activity of heparin was explained by an induction of apoptosis in cells in which the polysaccharide was internalized. The specificity of the activity against cancer cells was explained by a favorable competition of the endocytosis toward these cells compared with the lower endocytosis activity of healthy cells. However, it may be suggested that the specificity of the delivery may further be increased by a specific delivery of the conjugate at the level of the tumor [185, 186]. Many other carbohydrates and polysaccharides have recently been isolated showing antitumoral activity. The characterization of their structure and mechanism of action are still not fully elucidated, and this is part of a very active field of investigation. Among the very recently isolated compounds showing an antitumoral activity, several were extracted from plants, mushrooms, and microorganisms [143, 187–191] or were collected from their culture medium [192]. It was shown that some extracted polysaccharides may interact with the host defense mechanisms to induce a production of interleukin-2, which in turn can stimulate either cytotoxic T lymphocytes, Natural Killer lymphocytes, or the production of an antibody against tumoral cells [184]. Another hypothesis suggested that the polysaccharides can stimulate macrophages resulted in a production of nitrogen oxide (NO) and of tumor necrosis factor-alpha (TNF-α) [143, 187, 192–194]. A direct activity against tumoral cells was also assumed [188, 189]. Finally, certain active compounds may act on the angiogenesis reducing the formation of neovessels required for the irrigation of the tumor tissue [190, 191].

1.4.5.3   Control of Vascular Smooth Muscle Cell Proliferation

Several sulfonated polysaccharides showed an ability to control proliferation of vascular smooth muscle cells. Certain compounds may act on the angiogenesis reducing the formation of neovessels required for the irrigation of tumor tissues [190, 191]. They can contribute to reduce tumor size growth. Others can inhibit the proliferation of the vascular smooth muscle cells on a damaged artery to prevent and control artherosclerosis [195]. The inhibitor activity of the polysaccharides seemed to increase with their molecular weight because it is assumed that they are better internalized by cells. Among sulfonated polysaccharides, fucan showed a higher activity compared with heparin [196, 197]. With modified dextran, the inhibition of cell growth depended on the degree of substitution by sulfonate groups and on the presence of benzylamide groups. The production of collagen by the vascular smooth muscle cells was influenced by modified dextran and heparin [198]. Complexes among sulfonated polysaccharides, such as heparin, hyaluronate, and chondroitine sulphate with chitosan, were also reported to control proliferation of vascular smooth muscle cells [199].

Sulfonated polysaccharides present different biological activities. To take advantage of their antiproliferative activity toward vascular smooth muscle cells, it will be necessary to associate the polysaccharides with targeted drug carriers such as nanoparticles. The nanoparticles may also enhance the cellular endocytosis of the polysaccharides to further promote its antiproliferative activity on vascular cells.

1.4.5.4   Activity on the Immune System

Many antigens located at the surface of pathogens show more or less complex carbohydrate epitopes. The immune response induced by these antigens results in the production of antibodies, which is relevant in the aim of vaccination [200–203]. Thus, a few vaccines were developed using polysaccharide antigens against meningitis [203, 204], pneumonia [202], and against *Staphylococcus aureus*, which is responsible for nocosomial infections [201]. Usually, an efficient vaccine requires formulation of the antigen with an adjuvant. The more recent formulation strategy that was followed for vaccine development was based on the design of antigen presenting devices using nanotechnology [205–207]. Systems proposed so far are, on the one hand, derived from the liposome technology based on phospholipid bilayer reconstitution with incorporation of antigenic protein or glycoprotein, i.e., virosomes [205, 206], and, from the other hand, on the reconstitution of empty viral capside with viral recombinant proteins, i.e., virus like particles [207]. Finally, the immune stimulating complex (ISCOM) corresponds to a nanosized aggregate (40 nm in diameter) of saponin, lipids, and antigen held together by hydrophobic interactions occurring among the three components [208]. These systems formed spontaneously upon self-molecular assembly mechanisms. They were mainly developed as vaccines against viral infections and are mimicking viruses showing the corresponding antigenic epitopes at the surface. Although they are efficient for parenteral vaccinations, their application in mucosal vaccines is subjected to controversy because of possible stability problems in the biological environment of mucosa. Thus, there is a need for development of antigen presenting systems for bacterial antigens including those of a polysaccharide nature and for antigen presenting devices stable enough in biological media encountered at the level of mucosa to develop mucosal vaccination. Several types of polymer nanoparticles were proposed as carriers for antigens to be used as vaccine adjuvants [209, 210]. In general, in these systems, the antigen is mostly encapsulated within the nanoparticles and is not "visible" from the outside of the carrier as it is the case of systems obtained by self-assembly molecules, i.e., in virosomes and virus-like particles. Polysaccharide-coated nanoparticles would be more suitable as possible antigen presenting devices. To this purpose, nanoparticles were specifically designed to exhibit polysaccharides at the nanoparticle surface and can easily be obtained with various polysaccharides [211]. Their versatility should offer interesting developments for new vaccines as a polysaccharide antigen presenting device.

Although carbohydrate antigens were first described a long time ago, their interactions with the immune system are not limited to the production of antibodies in the frame of vaccination. Indeed, some polysaccharide can modulate the immune system activity in a specific manner. For instance, several polysaccharides interact with interleukin 10 (IL-10) [212]. Human IL-10 is known to stimulate CD16 and CD64 receptor expression on the monocyte and macrophage population within peripheral blood mononuclear cells. Soluble heparin, heparan sulfate, chondroitin sulfate, and dermatan sulfate inhibit the human IL-10-induced expression of CD16 and CD64 in a concentration-dependent manner. The antagonistic effect of heparin on human IL-10 activity depends on N-sulfation, the de-N-sulfated heparin had little or no inhibitory effect on the IL-10- induced expression of CD16, and the effect of de-O-sulfated heparin was comparable with that of unmodified heparin. Furthermore, the inhibition of cell-bound proteoglycan sulfation reduced the human IL-10-mediated expression of CD16 molecules on monocytes and macrophages. Taken together, these findings support the hypothesis that soluble and cell-surface glucosaminoglycans and, in particular, their sulfate groups are important in binding and modulation of human IL-10 activity and, therefore, promote or inhibit several pathways of the immune system. It may be suggested that these molecules grafted at the surface of a polymer nanoparticles may regulate the expression of the receptors at the cell surface modifying the immune system response in a very definitive and precise manner.

Small oligosaccharides can inhibit immune response by reducing production of oxidative species in immune cells. These components may be interesting to control the inflammatory process and for treatment of autoimmune diseases. Activity depended on both length and conformation of the oligosaccharides. It seemed that oligosaccharides longer than six residues and adopting a helical conformation enhanced interactions with immune system components [213]. The immuno-modulation properties shown by such oligosaccharides can be combined with anti-inflammatory properties of other polysaccharides extracted from *Costus spicatus* [214]. Although several oligosaccharides can inhibit immune response, others stimulate the immune system. Polysaccharides extracted from *Lycium barabrum* promoted expression of IL-2, TNF-alpha, in the peripheral mononuclear cells in human [215]. Those extracted from *Discirea opposita* induced proliferation of T lymphocytes [216]. It was suggested that stimulation of macrophages resulted from an interaction occurring between the n-acetylglucosamine groups of the oligosaccharides with the mannose calcium-independent receptors of the macrophages [217, 218].

Presenting under the form of nanoparticles, these polysaccharides may be better targeted toward macrophages and monocytes to focus the activity of the polysaccharide in macrophages and to avoid reaching other cells that may be affected by unspecific action on metabolism or physiology. Indeed, it will be very important to target precisely polysaccharides like heparin because it has other biological activity. The various activities found for polysaccharides on the immune system may be used to finely tune the activity of the very complex defense system of the host. This could lead to the emergence of new therapeutic strategies, provided the polysaccharides or the oligosaccharides will be precisely targeted against the right cells of the immune systems.

1.4.5.5 Control of Blood Coagulation and Complement Activation

Polysaccharides and oligosaccharides may induce or inhibit blood coagulation depending on their structure. Chitosan and polysaccharides extracted from the Chinese laquer showed biological activity promoting the blood coagulation [173, 219]. In general, negatively charged polysaccharides, including sulphate, sulfonate, carboxylic, and sulphonamide groups, show an opposite activity having anticoagulant properties. Heparin, which is used in clinics and for storage of blood in the blood banks, is for sure the more generally known natural anticoagulant agent [134, 220]. For the other polysaccharides, including sulphated and carboxylated derivatives of dextran, galactomannan, pullulan, and chitosan, it was shown that the amount of functional groups, their distribution along the chain, and the molecular weight of the polysaccharide chains are all parameters that influence the anticoagulant activity of the polysaccharide [174, 220–226]. In clinics, the main problem encountered with treatments based on the use of heparin is the rapid elimination of the polysaccharides from the blood stream. Long circulating formulations of heparin with longer biological and therapeutic efficacy may consist of heparin associated with long circulating drug carriers. It was reported that some nanoparticles coated with heparin displayed long circulating properties and that the anticoagulant activity of the heparin was preserved after grafting on the nanoparticle surface, making these nanoparticles interesting as a potential long-acting anticoagulant agent [19, 89]. Heparin that is used in clinics must be administered via the parenteral route, which is expensive, inconvenient, unsafe, and limits use by outpatients. The development of an oral form of heparin is more than warranted because the oral route remains the preferable route of administration of drugs for patient. It is also the safer and less-expensive route for the administration of drugs so far. However, heparin is a polyanionic macromolecule and is unstable under acidic conditions of the stomach, thus exhibiting poor oral bioavailability [227]. Nanoparticles were found to enhance absorption of hydrophilic macromolecules across the intestinal epithelium [51]. They seemed to be an interesting alternative for the development of an oral formulation of heparin to promote its bioavailability by this challenging route of administration.

1.4.6 NANOPARTICLE DRUG CARRIERS MADE OF CARBOHYDRATES FOR THE DELIVERY OF FRAGILE MOLECULES

Among the different delivery approaches explored so far to deliver complex molecules such as peptides, proteins, and nucleic acids by mucosal routes, those based on the use of polysaccharide nanoparticles were found to be interesting alternatives toward this ambitious goal thanks to physico-chemical and biopharmaceutical properties of these polymers [6–8, 51]. Indeed, the mucosal delivery of these molecules is very challenging, and despite their increasing market value, their clinical use was hampered due to their poor natural transport across biological barriers and their extremely rapid degradation in biological media. Elaboration of the polysaccharide nanoparticles requires neither an organic solvent nor a powerful homogenization procedure, possibly damaging macromolecular drug activity [75, 228]. This is an advantage regarding all other preparation methods of nanoparticles. Indeed, the mild conditions used to prepare polysaccharide nanoparticles suits with those required to preserve the biological activity of peptides, proteins, and nucleic acids in the nanoparticle formulations [9]. So far, nanoparticles made of chitosan have opened promising alternatives for the delivery of peptides, proteins, and nucleic acids by the mucosal route, including oral and nasal administration and for topical delivery of drugs in the eye [229–231]. The system that is stable in the mucosal environment, including in the harsh conditions of the gastrointestinal tract, can interact with the negatively charged glycoproteins of the mucus, which cover the epithelium retaining the nanoparticles on the mucosa surface. This retention effect due to the bioadhesive properties of chitosan to mucus increases the chance for the drug to be absorbed. The drug-loaded nanoparticles either made of chitosan or bearing a chitosan coating enhanced drug penetration and absorption after administration by different mucosal routes. For instance, the oral absorption of salmon calcitonin was improved when it was associated with chitosan nanoparticles. Interestingly, this system elicited long-lasting hypocalcemia levels, whereas calcitonin emulsions used as control led only to a negligible response [8]. To explain the effect, it was suggested that the nanosystems were able to enter the epithelia and provided a continuous delivery of the peptide to the blood stream [8, 232, 233]. Insulin-loaded chitosan nanoparticles also improved the delivery of insulin by the nasal route [234]. This formulation may be interesting for applications requiring a rapid supply of insulin in the blood. Indeed, the hormone is absorbed in about 30 minutes, but the duration of the effect was quite short, lasting for 1.5 to 2 hours.

Chitosan nanoparticles as well as chitosan-coated nanoparticles were also found interesting for the delivery of antigens to the Nasal Associated Lymphoid tissue. The results indicated that the nanoparticles could facilitate the transport of the associated antigen across the nasal epithelium leading to an efficient antigen presentation to the immune system. This elicited both an IgA and an IgG response increasing over time. For the immune response, it was suggested that not only the size and surface properties of the nanoparticles may influence the immune response but also the polymer composition and the structural architecture of the nanosystems, which were found critical for the optimization of such antigen carriers [210]. For the delivery of antigen by the oral route, it has very recently been proposed to coat chitosan nanoparticles with alginate in order to prevent a burst released of the antigen in the intestinal medium and to further improve the stability of the chitosan nanoparticles in this very aggressive medium [235].

Chitosan nanoparticles and chitosan-coated nanoparticles showed interesting bioadhesiveness on the cornea and good ocular tolerance. Their potential for the delivery of drugs at the ocular level was investigated. Both systems showed suitable delivery properties to improve topical administration of drugs at the level of the ocular mucosa [231]. Chitosan-coated nanoparticles, which are very versatile regarding their drug-loading capacity and release properties, increased the drug levels in cornea and aqueous humor to a significantly greater extent compared with the commercial drug preparation or the drug-loaded uncoated nanoparticles. These nanoparticles were clearly able to enhance transportation of the drug across the cornea to reach the inner part of the eye. In contrast, chitosan nanoparticles accumulated into the corneal and conjunctival epithelia, where an increased concentration of drugs can be maintained for a prolonged duration. Initial experiments have indicated the adequate tolerance and low toxicity of these ocular drug carriers [73].

Several polysaccharides can form a complex with nucleic acids. This property was exploited to formulate DNA loaded nanoparticles to deliver genes and antisens oligonucleotides *in vivo*. Most polysaccharide-based carriers were obtained from chitosan and alginate [7, 236]. *In vitro* expression of gene transfected into cells using chitosan nanoparticles reached a similar level than the expression level obtained with reference transfection agents, including Lipofectin (Invitrogen, Cergy-Pontoise, France) and Surperfect (Qiagen S.A., Coutaboef, France) [237]. The nanoparticles presented high encapsulation efficiency and good protection of DNA from DNase digestion [238]. To further improve the transfection properties of chitosan nanoparticles and enhance the specificity of the carrier toward the target cells, Manssouri et al. suggested to attach folic acid residues to the nanoparticle surface [239]. Nanoparticles made of polyesters or silica coated with chitosan or modified chitosan may also serve as carriers for DNA plasmids or for antisense oligonucleotides [7, 62, 240–243]. In these nanoparticles, the nucleic acid is adsorbed onto the carrier surface thanks to the formation of a polyion complex with the polycationic polysaccharide standing at the nanoparticle surface. Another system made of alginate nanoparticles was reported to associate antisense oligonucleotides [244]. The stability of the associated oligonucleotides in the presence of serum was considerably improved, indicating that the carrier system may be suitable for the *in vivo* delivery of antisense oligonucleotides. The main objective of gene therapy is to obtain successful *in vivo* transfer of the genetic material to the targeted tissue. However, the growing potential of gene therapy will not achieve this goal until the issue of gene delivery has been resolved [245]. Naked DNA cannot be delivered efficiently. Consequently, good gene delivery formulations are needed requiring good encapsulation efficiency of DNA, protection from degradation, and specific targeting to desired cells. The optimal system is still not found. The polysaccharide nanoparticles presenting so far good complexing and transfection efficiency are interesting potential systems. It may be expected that research on these systems will continue to grow and expand in the close future.

Polyanionic polysaccharides including heparin were shown to interact strongly with hemoglobin preserving its functionality as oxygen carrier [246]. This was exploited to load functional hemoglobin on heparin-coated nanoparticles to be used as a possible oxygen carrier for *in vivo* applications [247]. The nanoparticles having a diameter of 80 nm showed the highest hemoglobin-loaded capacity compared with other carrier systems. Indeed, they can associate up to 40-mg hemoglobin per gram of nanoparticles, whereas other nanoparticles showed one third of this loading capacity [247, 248]. The functionality of hemoglobin standing in the heparin corona at the nanoparticle surface was preserved as well as the anticoagulant activity of heparin. This system in which a drug may additionally be incorporated in the polymer core can constitute a new generation of multifunctional drug delivery system.

1.4.7   CONCLUSION

In the recent years, there has been a resurgence of interest in the biological role of carbohydrates. Active research that started a couple of years ago already highlighted the many biological roles and the diverse functions of carbohydrates in physiological and diseased processes. Several are useful for application as powerful drugs providing they are specifically delivered to the diseased cells. Their combination with nanoparticle drug carriers may be required improving their efficacy and specificity at the target site. From another point of view, several polysaccharides seemed to be interesting constituents of a drug carrier having remarkable mucoadhesive properties and promoting drug delivery by mucosal routes. The wide applications of carbohydrates in the design of drug delivery systems are illustrated in Figure 1.4-1.

Carbohydrate-based nanoparticles seemed to be well tolerated so far. However, full toxicological studies have not been performed yet and safety of the formulated

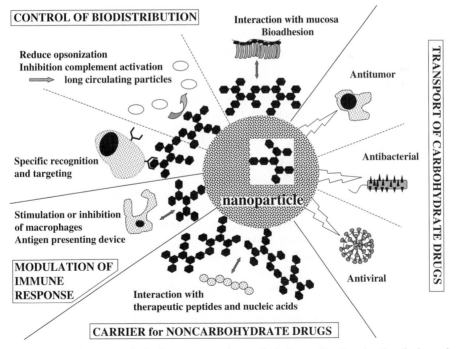

Figure 1.4-1. Diverse carbohydrate properties and their application in the design of a nanoparticle drug delivery system.

delivery systems will be an important point to clarify in the near future for further development for *in vivo* use. Finally, because carbohydrates are implicated in many cell recognition mechanisms and cell signaling phenomena based on high specific interactions with corresponding cell receptors, they may provide very useful tools to increase the targeting specificity of drug delivery systems toward defined cells *in vivo*. Considerable work remains to be done before the potential of carbohydrates will be depicted and will be ready to be integrated into the design of better defined drug delivery systems with more precise functions. This will require the understanding of the role of biological functions of more carbohydrates found in Mother Nature and production or synthesis of sufficient amounts of the defined carbohydrate to clarify their functionalities. From the first results of the experiences done so far on quite simple systems, it may be expected that carbohydrates will allow us to step over a new milestone in the development of drug delivery systems integrating the most recent developments in glycobiology.

REFERENCES

1. Yoshida T (2001). Synthesis of polysaccharides having specific biological activities. *Progress in Polymer Sci.* 26:379–441.

2. Venkataraman G, Shriver Z, Raman R, et al. (1999). Sequencing complex polysaccharides. *Science.* 286:537–542.

3. Toukach F V, Shashkov A S (2001). Computer-assisted structural analysis of regular glycopolymers on the basis of 13C NMR data. *Carbohydrate Research.* 335:101–114.

4. Lee Y C (1996). Carbohydrate analyses with high-performance anion-exchange chromatography. *J. Chromato. A.* 720:137–149.

5. Gabius H-J (2004). The sugar code in drug delivery. *Adv. Drug Del. Rev.* 56:421–424.

6. Vauthier C, Couvreur P (2000). Development of Polysaccharide Nanoparticles as Novel Drug Carrier Systems. In D L Wise (ed.), *Handbook of Pharmaceutical Controlled Release Technology.* Marcel Dekker, New York, pp. 413–429.

7. Janes K A, Calvo P, Alonso M J (2001). Polysaccharide colloidal particles as delivery systems for macromolecules. *Adv. Drug Del. Rev.* 47:83–97.

8. Prego C, Garcia M, Alonso M J (2005). Transmucosal macromolecular drug delivery. *J. Control. Rel.* 101:151–162.

9. Vauthier C, Fattal E, Labarre D (2004). From polymer chemistry and physicochemistry to nanoparticulate drug carrier design and applications. In Yaszemski M J, Trantolo D J, Lewandrowski K V, et al. (eds.), *Tissue Engineering and Novel Delivery System*: Marcel Dekker, New York, pp. 562–598.

10. Nori A, Kopecek J (2005). Intracellular targeting of polymer-bound drugs for cancer chemotherapy. *Adv. Drug Del. Rev.* 57:609–636.

11. Duncan R (2003). The dawning era of polymer therapeutics. Nature Reviews. *Drug Disc.* 2:347–360.

12. Nobs L, Buchegger F, Gurny R, et al. (2004). Current methods for attaching targeting ligands to liposomes and nanoparticles. *J. Phar. Sci.* 93:1980–1992.

13. Torchilin V P (2005). Recent advances with liposomes as pharmaceutical carriers. *Nat. Rev. Drug Disc.* 4:145–160.

14. Moghimi S M, Szebeni J (2003). Stealth liposomes and long circulating nanoparticles: Critical issues in pharmacokinetics, opsonization and protein-binding properties. *Progress Lipid Res.* 42:463–478.

15. Moghimi S M, Hunter A C, Murray J C (2001). Long-circulating and target-specific nanoparticles: Theory to practice. *Pharmacol. Rev.* 53:283–318.

16. Gregoriadis G, McCormack B, Wang Z, Lifely R (1993). Polysialic acids: Potential in drug delivery. *FEBS Lett.* 315:271–276.

17. Olivier J-C, Vauthier C, Taverna M, et al. (1996). Stability of orosomucoid-coated polyisobutylcyanoacrylate nanoparticles in the presence of serum. *J. Control. Rel.* 40:157–168.

18. Passirani C, Ferrarini L, Barratt G, et al. (1999). Preparation and characterization of nanoparticles bearing heparin or dextran covalently-linked to poly(methyl methacrylate). *J. Biomater. Sci. Polym. Ed.* 10:47–62.

19. Chauvierre C, Labarre D, Couvreur P, et al. (2003). Novel polysaccharide-decorated poly(isobutyl cyanoacrylate) nanoparticles. *Pharm. Res.* 20:1786–1793.

20. Bertholon I, Vauthier C, Labarre D (2006). Core-shell poly(isobutylcyanoacrylate)-polysaccharide nanoparticles: Influences of surface morphology, length and type of polysaccharide on complement activation in vitro. *Pharm. Res.* 23:1313–1323.

21. Bertholon I, Ponchel G, Labarre D (2006). Bioadhesive properties of poly(alkyl-cyanoacrylate) nanoparticles coated with polysaccharide. *J. Nanosci. Nanotechnol.* 6:3102–3109.

22. Sihorkar V, Vyas S P (2001). Potential of polysaccharide anchored liposomes in drug delivery, targeting and immunization. *J. Pharm. Pharmceut. Sci.* 4:138–158.

23. Drotleff S, Lungwitz U, Breunig M, et al. (2004). Biomimetic polymers in pharmaceutical and biomedical sciences. *Euro. J. Pharmaceutics and Biopharmaceutics.* 58:385–407.

24. Bravo-Osuna I, Schmith T, Bernkop-Schnurch A (2006). Elaboration and characterization of thiolated chitosan-coated acrylic nanoparticles. *Int. J. Pharm.* 316:170–175.

25. Eylar E H (1966). On the biological role of glycoproteins. *J. Theoret. Bio.* 10:89–113.

26. Kimmerlin T, Seebach D (2005). "100 years of peptide synthesis": Ligation methods for peptide and protein synthesis with applications to beta-peptide assemblies. *J. Peptide Res.: Official J. Amer. Peptide Soc.* 65:229–260.

27. Zamecnik P (2005). From protein synthesis to genetic insertion. *Ann. Rev. Biochem.* 74:1–28.

28. Sharon N, Lis H (2004). History of lectins: From hemagglutinins to biological recognition molecules. *Glycobiology.* 14:53R–62R.

29. Khersonsky S M, Ho C M, Garcia M-A F, et al. (2003). Recent advances in glycomics and glycogenetics. *Curr. Topics Medic. Chem.* 3:617–643.

30. Petitou M, Casu B, Lindahl U (2003). 1976–1983, A critical period in the history of heparin: the discovery of the antithrombin binding site. *Biochimie.* 85:83–89.

31. Passirani C, Barratt G, Devissaguet J-P, et al. (1998). Interactions of nanoparticles bearing heparin or dextran covalently bound to poly(methyl methacrylate) with the complement system. *Life Sci.* 62:775–785.

32. Labarre D, Vauthier C, Chauvierre C, et al. (2005). Interactions of blood proteins with poly(isobutylcyanoacrylate) nanoparticles decorated with a polysaccharidic brush. *Biomaterials.* 26:5075–5084.

33. Di Marco A, Cassinelli G, Arcamone F (1981). The discovery of daunorubicin. *Cancer Treat. Rep.* 65:3–8.

34. Osborn H M, Evans P G, Gemmel N, et al. (2004). Carbohydrate-based therapeutics. *J. Pharm. Pharmacol.* 56:691–702.

35. Barchi J J (2000). Emerging roles of carbohydrates and glycomimetics in anticancer drug design. *Curr. Pharm. Des.* 6:485–501.

36. Kabel M A, Schols H A, Voragen A G J (2001). Mass determination of oligosaccharides by matrix-assisted laser desorption/ionization time-of-flight mass spectrometry following HPLC, assisted by on-line desalting and automated sample handling. *Carboh. Polym.* 44:161–165.

37. Broberg A, Thomsen K K, Duus J O (2000). Application of nano-probe NMR for structure determination of low nanomole amounts of arabinoxylan oligosaccharides fractionated by analytical HPAEC-PAD. *Carboh. Res.* 328:375–382.

38. Morelle W, Michalski J-C (2005). Glycomics and mass spectrometry. *Curr. Pharm. Des.* 11:2615–2645.

39. Iwafune M, Kakizaki I, Nakazawa H, et al. (2004). A glycomic approach to proteoglycan with a two-dimensional polysaccharide chain map. *Analyt. Biochem.* 325:35–40.

40. Ratner D M, Adams E W, Disney M D, et al. (2004). Tools for glycomics: Mapping interactions of carbohydrates in biological systems. *Chembiochem: A Euro. J. Chem. Bio.* 5:1375–1383.

41. Howbrook D N, van der Valk A M, O'Shaughnessy M C, et al. (2003). Developments in microarray technologies. *Drug Dis. Today.* 8:642–651.

42. Vives R R, Goodger S, Pye D A (2001). Combined strong anion-exchange HPLC and PAGE approach for the purification of heparan sulphate oligosaccharides. *Biochem. J.* 354:141–147.

43. Plante O J (2005). Combinatorial chemistry in glycobiology. *Combinatorial Chem. & High Throughput Screening.* 8:153–159.

44. Holemann A, Seeberger P H (2004). Carbohydrate diversity: Synthesis of glycoconjugates and complex carbohydrates. *Curr. Opin. Biotech.* 15:615–622.

45. Köhn M, Breinbauer R (2004). The Staudinger ligation-a gift to chemical biology. *Angew. Chem. Int. Ed. Engl.* 43:3106–3116.

46. de Kort M, Buijsman R C, van Boeckel C A A (2005). Synthetic heparin derivatives as new anticoagulant drugs. *Drug Dis. Today.* 10:769–779.

47. Croft S L, Coombs G H (2003). Leishmaniasis–current chemotherapy and recent advances in the search for novel drugs. *Trends Parasitol.* 19:502–508.

48. Adler-Moore J P, Proffitt R T (1993). Development, characterization, efficacy and mode of action of ambisome®. *J. Liposome Res.* 3:429–450.

49. Couvreur P, Barratt G, Fattal E, et al. (2002). Nanocapsule technology: A review. *Crit. Rev. Ther. Drug Carrier Syst.* 19:99–134.

50. Vauthier C, Dubernet C, Chauvierre C, et al. (2003). Drug delivery to resistant tumors: The potential of poly(alkyl cyanoacrylate) nanoparticles. *J. Control. Rel.* 93:151–160.

51. Alonso M J (2004). Nanomedicines for overcoming biological barriers. *Biomed. Pharmacother.* 58:168–172.

52. Fouarge M, Dewulf M, Couvreur P, et al. (1989). Development of dehydroemetine nanoparticles for the treatment of visceral leishmaniasis. *J. Microencapsulation.* 6:29–34.

53. Gaspar R, Preat V, Opperdoes F R, et al. (1992). Macrophage activation by polymeric nanoparticles of polyalkylcyanoacrylates: activity against intracellular Leishmania donovani associated with hydrogen peroxide production. *Pharm. Res.* 9:782–787.

54. Kattan J, Droz J P, Couvreur P, et al. (1992). Phase I clinical trial and pharmacokinetic evaluation of doxorubicin carried by polyisohexylcyanoacrylate nanoparticles. *Investigat. New Drugs.* 10:191–199.

55. Barraud L, Merle P, Soma E, et al. (2005). Increase of doxorubicin sensitivity by doxorubicin-loading into nanoparticles for hepatocellular carcinoma cells in vitro and in vivo. *J. Hepatol.* 42:736–743.

56. Sangro B (2005). Refined tools for the treatment of hepatocellular carcinoma. *J. Hepatol.* 42:629–631.

57. Tyagi R, Lala S, Verma A K, et al. (2005). Targeted delivery of arjunglucoside I using surface hydrophilic and hydrophobic nanocarriers to combat experimental leishmaniasis. *J. Drug. Target.* 13:161–171.

58. Stolnik S, Dunn S E, Garnett M C, et al. (1994). Surface modification of poly(lactide-co-glycolide) nanospheres by biodegradable poly(lactide)-poly(ethylene glycol) copolymers. *Pharm. Res.* 11:1800–1808.

59. Gref R, Minamitake Y, Peracchia M T, et al. (1994). Biodegradable long-circulating polymeric nanospheres. *Science.* 263:1600–1603.

60. Vauthier C, Couvreur P (2002). Biodegradation of poly(alkylcyanoacrylates) in Matsumara J P and Steinbuchel A (eds.), *Handbook of Biopolymers*, Wiley-VHC, Weinheim, Germany, pp. 457–490.

61. Ravi Kumar M N V, Bakowsky U, Lehr C M (2004). Preparation and characterization of cationic PLGA nanospheres as DNA carriers. *Biomaterials.* 25:1771–1777.

62. Bertholon-Rajot I, Labarre D, Vauthier C (2005). Influence of the initiator system, cerium-polysaccharide, on the surface properties of poly(isobutylcyanoacrylate) nanoparticles. *Polymer.* 46:1407–1415.

63. Schipper N G M, Olsson S, Hoogstraate J A, et al. (1997). Chitosans as Absorption Enhancers for Poorly Absorbable Drugs 2: Mechanism of Absorption Enhancement. *Pharm. Res.* 14:923–929.

64. Smith J M, Dornish M, Wood E J (2005). Involvement of protein kinase C in chitosan glutamate-mediated tight junction disruption. *Biomaterials.* 26:3269–3276.

65. Lehr C-M, Bouwstra J A, Schacht E H, et al. (1992). In vitro evaluation of mucoadhesive properties of chitosan and some other natural polymers. *Int. J. Pharm.* 78:43–48.

66. Schipper N G, Varum K M, Artursson P (1996). Chitosans as absorption enhancers for poorly absorbable drugs. 1: Influence of molecular weight and degree of acetylation on drug transport across human intestinal epithelial (Caco-2) cells. *Pharm. Res.* 13:1686–1692.

67. Chae S Y, Jang M-K, Nah J-W (2005). Influence of molecular weight on oral absorption of water soluble chitosans. *J. Control. Rel.* 102:383–394.

68. Thanou M, Verhoef J C, Junginger H E (2001). Oral drug absorption enhancement by chitosan and its derivatives. *Adv. Drug Deliv. Rev.* 52:117–126.

69. Avadi M R, Jalali A, Sadeghi A M M, et al. (2005). Diethyl methyl chitosan as an intestinal paracellular enhancer: ex vivo and in vivo studies. *Int. J. Pharm.* 293:83–89.

70. Sandri G, Rossi S, Ferrari F, et al. (2004). Assessment of chitosan derivatives as buccal and vaginal penetration enhancers. *Eur. J. Pharm. Sci.* 21:351–359.

71. Kafedjiiski K, Föger F, Werle M, et al. (2005). Synthesis and in Vitro Evaluation of a Novel Chitosan-Glutathione Conjugate. *Pharm. Res.* 22:1480–1488.

72. Bernkop-Schnurch A, Pinter Y, Guggi D, et al. (2005). The use of thiolated polymers as carrier matrix in oral peptide delivery–Proof of concept. *J. Control. Rel.* 106:26–33.

73. De Campos A M, Sanchez A, Gref R, et al. (2003). The effect of a PEG versus a chitosan coating on the interaction of drug colloidal carriers with the ocular mucosa. *Eur. J. Pharm. Sci.* 20:73–81.

74. Yamamoto H, Takeuchi H, Hino T, et al. (2000). Mucoadhesive liposomes: Physicochemical properties and release behavior of water-soluble drugs from chitosan-coated liposomes. *S.T.P. Pharma. Sci.* 10:63–68.

75. Calvo P, Vila-Jato J L, Alonso M J (1997). Evaluation of cationic polymer-coated nanocapsules as ocular drug carriers. *Int. J. Pharm.* 153:41–50.

76. Yamamoto H, Kuno Y, Sugimoto S, et al. (2005). Surface-modified PLGA nanosphere with chitosan improved pulmonary delivery of calcitonin by mucoadhesion and opening of the intercellular tight junctions. *J. Control. Rel.* 102:373–381.

77. Messai I, Delair T (2005). Adsorption of chitosan onto poly(D,L-lactic acid) particles: A physico-chemical investigation. *Macromolecular Chem. Phys.* 206:1665–1674.

78. Yang S C, Ge H X, Hu Y, et al. (2000). Formation of positively charged poly(butyl cyanoacrylate) nanoparticles stabilized with chitosan. *Colloid Polym. Sci.* 278:285–292.

79. Ponchel G, Irache J (1998). Specific and non-specific bioadhesive particulate systems for oral delivery to the gastrointestinal tract. *Adv. Drug Del. Rev.* 34:191–219.

80. Valenta C (2005). The use of mucoadhesive polymers in vaginal delivery. *Adv. Drug Del. Rev.* 57:1692–1712.

81. Song Y, Wang Y, Thakur R, et al. (2004). Mucosal drug delivery: Membranes, methodologies, and applications. *Crit. Rev. Ther. Drug Carrier Syst.* 21:195–256.

82. Barbault-Foucher S, Gref R, Russo P, et al. (2002). Design of poly-e-caprolactone nanosphere coated with bioadhesive hyaluronic acid for ocular delivery. *J. Control. Rel.* 83:365–375.

83. Leroux J-C, De Jaeghere F, Anner B, et al. (1995). An investigation on the role of plasma and serum opsonins on the evternalization of biodegradable poly(D,L-lactic acid) nanoparticles by human monocytes. *Life Sciences.* 57:695–703.

84. Lenaerts V, Nagelkerke J F, Van Berkel T J, et al. (1984). In vivo uptake of polyisobutyl cyanoacrylate nanoparticles by rat liver Kupffer, endothelial, and parenchymal cells. *J. Pharm. Sci.* 73:980–982.

85. Lemarchand C, Gref R, Couvreur P (2004). Polysaccharide-decorated nanoparticles. *Eur. J. Pharm. Biopharm.* 58:327–341.

86. Torchilin V P (2000). Drug targeting. *Eur. J. Pharm. Sci.* 11:S81–S91.

87. Owens III D E, Peppas N A. Opsonization, biodistribution, and pharmacokinetics of polymeric nanoparticles. *Int. J. Pharm.* In Press, Corrected Proof.

88. Torchilin V P, Trubetskoy V S (1995). Which polymers can make nanoparticulate drug carriers long-circulating? *Adv. Drug Del. Rev.* 16:141–155.

89. Passirani C, Barratt G, Devissaguet J P, et al. (1998). Long-circulating nanoparticles bearing heparin or dextran covalently bound to poly(methyl methacrylate). *Pharm. Res.* 15:1046–1050.

90. Lemarchand C, Gref R, Passirani C, et al. (2006). Influence of polysaccharide coating on the interactions of nanoparticles with biological systems. *Biomaterials.* 27:108–118.

91. Jaulin N, Appel M, Passirani C, et al. (2000). Reduction of the uptake by a macrophagic cell line of nanoparticles bearing heparin or dextran covalently bound to poly(methyl methacrylate). *J. Drug Targeting.* 8:165–172.

92. Rouzes C, Gref R, Leonard M, et al. (2000). Surface modification of poly(lactic acid) nanospheres using hydrophobically modified dextrans as stabilizers in an o/w emulsion/evaporation technique. *J. Biomed. Mat. Res.* 50:557–565.

93. Carreno M P, Maillet F, Labarre D, et al. (1988). Specific antibodies enhance Sephadex-induced activation to the alternative pathway in human serum. *Biomaterials.* 9:514–518.

94. Olivier J-C (1995). Développement d'un vecteur biomimétique, PhD Dissertation, Université Paris 11, Faculté de Pharmacie, Chatenay-Malabry. France, p. 133.

95. Maruyama A, Ishihara T, Kim J S, et al. (1997). Nanoparticle DNA carrier with poly (L-lysine) grafted polysaccharide copolymer and poly(D,L-lactic acid). *Bioconjug. Chem.* 8:735–742.

96. Benesch J, Tengvall P (2002). Blood protein adsorption onto chitosan. *Biomaterials.* 23:2561–2568.

97. Minami S, Suzuki H, Okamoto Y (1998). Chitin and chitosan activate complement via the alternative pathway. *Carbohyd. Polym.* 36:151–155.

98. Suzuki Y, Okamoto Y, Morimoto M, et al. (2000). Influence of physico-chemical properties of chitin and chitosan on complement activation. *Carbohyd. Polym.* 42:307–310.

99. Mitra S, Gaur U, Ghosh P C, et al. (2001). Tumour targeted delivery of encapsulated dextran-doxorubicin conjugate using chitosan nanoparticles as carrier. *J. Control. Rel.* 74:317–323.

100. Banerjee T, Mitra S, Kumar Singh A, et al. (2002). Preparation, characterization and biodistribution of ultrafine chitosan nanoparticles. *Int. J. Pharm.* 243:93–105.

101. Banerjee T, Singh A K, Sharma R K, et al. (2005). Labeling efficiency and biodistribution of Technetium-99m labeled nanoparticles: Interference by colloidal tin oxide particles. *Int. J. Pharm.* 289:189–195.

102. Son Y J, Jang J-S, Cho Y W, et al. (2003). Biodistribution and anti-tumor efficacy of doxorubicin loaded glycol-chitosan nanoaggregates by EPR effect. *J. Control Rel.* 91:135–145.

103. Kato Y, Onishi H, Machida Y (2000). Evaluation of N-succinyl-chitosan as a systemic long-circulating polymer. *Biomaterials.* 21:1579–1585.

104. Hyung Park J, Kwon S, Lee M (2006). Self-assembled nanoparticles based on glycol chitosan bearing hydrophobic moieties as carriers for doxorubicin: In vivo biodistribution and anti-tumor activity. *Biomaterials.* 27:119–126.

105. Schauer R (1985). Sialic acids and their role as biological masks. *Trends Biochem. Sci.* 10:357–360.

106. Durocher J R, Payne R C, Conrad M E (1975). Role of sialic acid in erythrocyte survival. *Blood.* 45:11–20.

107. Schauer R (1992). Sialic acids regulate cellular and molecular recognition. In Ogura H, Hasegawa A, Suami T, Carbohydrates, Kodansha Ltd., Tokyo, pp. 340–354.

108. Jancik J, Schauer R (1974). Sialic acid–a determinant of the life-time of rabbit erythrocytes. *Hoppe Seylers Z Physiol. Chem.* 355:395–400.

109. Jancik J, Schauer R, Anders K H, et al. (1978). Sequestration of neuraminidase-treated erythrocytes. *Cell Tiss. Res.* 186:209–226.

110. Ourth D D, Bachinski L M (1987). Bacterial sialic acid modulates activation of the alternative complement pathway of channel catfish (Ictalurus punctatus). *Devel. Compar. Immunol.* 11:551–564.

111. Sakarya S, Oncu S (2003). Bacterial adhesins and the role of sialic acid in bacterial adhesion. *Med. Sci. Monitor: Internat. Med. J. Exper. Clin. Res.* 9:RA76–RA82.

112. Vimr E R, Kalivoda K A, Deszo E L, et al. (2004). Diversity of microbial sialic acid metabolism. *Microbiol. Mol. Biol. Rev.* 68:132–153.

113. Olivier J C, Taverna M, Vauthier C, et al. (1994). Capillary electrophoresis monitoring of the competitive adsorption of albumin onto the orosomucoid-coated polyisobutyl-cyanoacrylate nanoparticles. *Electrophoresis.* 15:234–239.

114. Huve P (1994). Comprendre et éviter la capture des nanoparticules de poly(acide lactique) par le système des phagocytes mononucléaires, PhD Dissertation, Université Paris Sud, Faculé de Pharmacie, Chatenay-Malabry, France, p. 193.

115. Yamauchi H, Yano T, Kato T, et al. (1995). Effects of sialic acid derivative on long circulation time and tumor concentration of liposomes. *Int. J. Pharm.* 113:141–148.

116. Kazatchkine M D, Fearon D T, Silbert J E, et al. (1979). Surface-associated heparin inhibits zymosan-induced activation of the human alternative complement pathway by augmenting the regulatory action of the control proteins on particle-bound C3b. *J. Exp. Med.* 150:1202–1215.

117. Keuren J F, Wielders S J, Willems G M (2003). Thrombogenicity of polysaccharide-coated surfaces. *Biomaterials.* 24:1917–1924.

118. Weber N, Wendel H P, Ziemer G (2002). Hemocompatibility of heparin-coated surfaces and the role of selective plasma protein adsorption. *Biomaterials.* 23:429–439.

119. Jeong S C, Yang B K, Ra K S (2004). Characteristics of anti-complementary biopolymer extracted from Coriolus versicolor. *Carbohyd. Polym.* 55:255–263.

120. Yamazaki N, Kojima S, Bovin N V, et al. (2000). Endogenous lectins as targets for drug delivery. *Adv. Drug Del. Rev.* 43:225–244.

121. Ehrhardt C, Kneuer C, Bakowsky U (2004). Selectins-an emerging target for drug delivery. *Adv. Drug Del. Rev.* 56:527–549.

122. Yamazaki N, Kojima S, Yokoyama H (2005). Biomedical nanotechnology for active drug delivery systems by applying sugar-chain molecular functions. *Curr. Appl. Phys.* 5:112–117.

123. Bies C, Lehr C-M, Woodley J F (2004). Lectin-mediated drug targeting: History and applications. *Adv. Drug Del. Rev.* 56:425–435.

124. Faivre V, Costa M d. L, Boullanger P, et al. (2003). Specific interaction of lectins with liposomes and monolayers bearing neoglycolipids. *Chem. and Phys. of Lipids.* 125:147–159.

125. Forssen E, Willis M (1998). Ligand-targeted liposomes. *Adv. Drug Del. Rev.* 29:249–271.

126. Minko T (2004). Drug targeting to the colon with lectins and neoglycoconjugates. *Adv. Drug Del. Rev.* 56:491–509.

127. Lehr C-M, Gabor F (2004). Lectins and glycoconjugates in drug delivery and targeting. *Adv. Drug Del. Rev.* 56:419–420.

128. Montisci M J, Giovannuci G, Duchene D, et al. (2001). Covalent coupling of asparagus pea and tomato lectins to poly(lactide) microspheres. *Int. J. Pharm.* 215:153–161.

129. Reddy J A, Allagadda V M, Leamon C P (2005). Targeting therapeutic and imaging agents to folate receptor positive tumors. *Curr. Pharm. Biotechnol.* 6:131–150.

130. Gao S, Chen J, Dong L, et al. (2005). Targeting delivery of oligonucleotide and plasmid DNA to hepatocyte via galactosylated chitosan vector. *European J. of Pharmaceutics and Biopharmaceutics.* 60:327–334.

131. Hubbard A L, Wilson G, Ashwell G, et al. (1979). An electron microscope autoradiographic study of the carbohydrate recognition systems in rat liver. Distribution of 125I-ligands among the liver cell types. *J. Cell. Biol.* 83:47–64.

132. Kishore U, Eggleton P, Reid K B (1997). Modular organization of carbohydrate recognition domains in animal lectins. *Matrix Biol.* 15:583–592.

133. Tanaka T, Fujishima Y, Hanano S, et al. (2004). Intracellular disposition of polysaccharides in rat liver parenchymal and nonparenchymal cells. *Int. J. Pharm.* 286:9–17.

134. Blondin C, Bataille L, Letourneur D (2000). Polysaccharides for vascular cell targeting. *Crit. Rev. Ther. Drug Carrier Syst.* 17:327–375.

135. Vercruysse K P, Prestwich G D (1998). Hyaluronate derivatives in drug delivery. *Crit. Rev. Ther. Drug Carrier Syst.* 15:513–555.

136. Sy M S, Guo Y J, Stamenkovic I (1991). Distinct effects of two CD44 isoforms on tumor growth in vivo. *J. Exp. Med.* 174:859–866.

137. Luo Y, Bernshaw N J, Lu Z-R, et al. (2002). Targeted Delivery of Doxorubicin by HPMA Copolymer-Hyaluronan Bioconjugates. *Pharm. Res.* 19:396–402.

138. Gunthert U, Hofmann M, Rudy W, et al. (1991). A new variant of glycoprotein CD44 confers metastatic potential to rat carcinoma cells. *Cell.* 65:13–24.

139. Entwistle J, Hall C L, Turley E A (1996). HA receptors: Regulators of signalling to the cytoskeleton. *J. Cell Biochem.* 61:569–577.

140. Danguy A, Camby I, Kiss R (2002). Galectins and cancer. *Biochim. Biophys. Acta.* 1572:285–293.

141. Julyan P J, Seymour L W, Ferry D R, et al. (1999). Preliminary clinical study of the distribution of HPMA copolymers bearing doxorubicin and galactosamine. *J. Control. Rel.* 57:281–290.

142. Jensen K D, Kopeckova P, Bridge J H, et al. (2001). The cytoplasmic escape and nuclear accumulation of endocytosed and microinjected HPMA copolymers and a basic kinetic study in Hep G2 cells. *AAPS PharmSci.* 3:E32.

143. Kim G-Y, Choi G-S, Lee S-H, et al. (2004). Acidic polysaccharide isolated from Phellinus linteus enhances through the up-regulation of nitric oxide and tumor necrosis factor-[alpha] from peritoneal macrophages. *J. Ethnopharm.* 95:69–76.

144. Zhang X Q, Wang X L, Zhang P C, et al. (2005). Galactosylated ternary DNA/polyphosphoramidate nanoparticles mediate high gene transfection efficiency in hepatocytes. *J. Control Rel.* 102:749–763.

145. Wu C, Wilson J, Wu G (1989). Targeting genes: delivery and persistent expression of a foreign gene driven by mammalian regulatory elements in vivo. *J. Biol. Chem.* 264:16985–16987.

146. Wilson J, Grossman M, Wu C, et al. (1992). Hepatocyte-directed gene transfer in vivo leads to transient improvement of hypercholesterolemia in low density lipoprotein receptor-deficient rabbits. *J. Biol. Chem.* 267:963–967.

147. Fajac I, Thévenot G, Bédouet L, et al. (2003). Uptake of plasmid/glycosylated polymer complexes and gene transfer efficiency in differentiated airway epithelial cells. *J. Gene Med.* 5:38–48.

148. Cade D, Ramus E, Rinaudo M, et al. (2004). Tailoring of bioresorbable polymers for elaboration of sugar-functionalized nanoparticles. *Biomacromol.* 5:922–927.

149. Racles C, Hamaide T (2005). Synthesis and Characterization of Water Soluble Saccharide Functionalized Polysiloxanes and Their Use as Polymer Surfactants for the Stabilization of Polycaprolactone Nanoparticles. *Macromol. Chem. and Phys.* 206:1757–1768.

150. Jeong Y-I, Seo S-J, Park I-K, et al. (2005). Cellular recognition of paclitaxel-loaded polymeric nanoparticles composed of poly([gamma]-benzyl l-glutamate) and poly(ethylene glycol) diblock copolymer endcapped with galactose moiety. *Int. J. Pharm.* 296:151–161.

151. Uchida T, Serizawa T, Ise H, et al. (2001). Graft copolymer having hydrophobic backbone and hydrophilic branches. 33. Interaction of hepatocytes and polystyrene nanospheres having lactose-immobilized hydrophilic polymers on their surfaces. *Biomacromol.* 2:1343–1346.

152. Sinha V R, Kumria R (2001). Polysaccharides in colon-specific drug delivery. *Int. J. Pharm.* 224:19–38.

153. Jozefowicz M, Jozefonvicz J (1997). Randomness and biospecificity: random copolymers are capable of biospecific molecular recognition in living systems. *Biomaterials.* 18:1633–1644.

154. Walker J B (2002). Enzymatic synthesis of aminoglycoside antibiotics: novel adenosylmethionine:2-deoxystreptamine N-methyltransferase activities in hygromycin B- and spectinomycin-producing Streptomyces spp. and uses of the methylated products. *Appl. Environ. Microbiol.* 68:2404–2410.

155. Macmillan D, Daines A M (2003). Recent developments in the synthesis and discovery of oligosaccharides and glycoconjugates for the treatment of disease. *Curr. Med. Chem.* 10:2733–2773.

156. Wong C H, Bryan M C, Nyffeler P T, et al. (2003). Synthesis of carbohydrate-based antibiotics. *Pure Appl. Chem.* 75:179–186.

157. Basu M K, Lala S (2004). Macrophage specific drug delivery in experimental leishmaniasis. *Curr. Mol. Med.* 4:681–689.

158. Donald P R, Sirgel F A, Venter A, et al. (2001). The early bactericidal activity of a low-clearance liposomal amikacin in pulmonary tuberculosis. *J. Antimicrob. Chemother.* 48:877–880.

159. Fielding R M, Lewis R O, Moon-McDermott L (1998). Altered tissue distribution and elimination of amikacin encapsulated in unilamellar, low-clearance liposomes (MiKasome). *Pharm. Res.* 15:1775–1781.

160. Fielding R M, Moon-McDermott L, Lewis R O, et al. (1999). Pharmacokinetics and urinary excretion of amikacin in low-clearance unilamellar liposomes after a single or repeated intravenous administration in the rhesus monkey. *Antimicrob. Agents Chemother.* 43:503–509.

161. Fattal E, Youssef M, Couvreur P, et al. (1989). Treatment of experimental salmonellosis in mice with ampicillin-bound nanoparticles. *Antimicrob. Agents Chemother.* 33:1540–1543.

162. Pinto-Alphandary H, Andremont A, Couvreur P (2000). Targeted delivery of antibiotics using liposomes and nanoparticles: research and applications. *Int. J. Antimicrob. Agents.* 13:155–168.

163. Schiffelers R, Storm G, Bakker-Woudenberg I (2001). Liposome-encapsulated aminoglycosides in pre-clinical and clinical studies. *J. Antimicrob. Chemother.* 48:333–344.

164. Barratt G (2003). Colloidal drug carriers: achievements and perspectives. *Cell. Mol. Life Sci.* (CMLS). 60:21–37.

165. Fattal E, Balland O, Alphandary H, et al. (1993). Colloidal carriers of antibiotics as an alternative approach for the treatment of intracellular infections. In Raoult D (ed.), *Antimicrobial agents and intracellular pathogens*: CRC Press, Boca Raton, F.L., pp. 63–72.

166. Senel S, McClure S J (2004). Potential applications of chitosan in veterinary medicine. *Adv. Drug Del. Rev.* 56:1467–1480.

167. Rabea E I, Badawy M E-T, Stevens C V, et al. (2003). Chitosan as antimicrobial agent: applications and mode of action. *Biomacromol.* 4:1457–1465.

168. Zheng L Y, Zhu J F (2003). Study on antimicrobial activity of chitosan with different molecular weights. *Carbohyd. Polym.* 54:527–530.

169. Jeon Y J, Park P J, Kim S K (2001). Antimicrobial effect of chitooligosaccharides produced by bioreactor. *Carbohyd. Polym.* 44:71–76.

170. Liu H, Du Y, Wang X, et al. (2004). Chitosan kills bacteria through cell membrane damage. *Intern. J. Food Microbiol.* 95:147–155.

171. Avadi M R, Sadeghi A M M, Tahzibi A, et al. (2004). Diethylmethyl chitosan as an antimicrobial agent: Synthesis, characterization and antibacterial effects. *Euro. Polym. J.* 40:1355–1361.

172. Carlucci M J, Pujol C A, Ciancia M, et al. (1997). Antiherpetic and anticoagulant properties of carrageenans from the red seaweed Gigartina skottsbergii and their cyclized derivatives: correlation between structure and biological activity. *Int. J. Biol. Macromol.* 20:97–105.

173. Yoshida R L, Nakashima H, Premanathan M, et al. (2000). Specific biological activities of Chinese lacquer polysaccharides. *Carbohyd. Polym.* 43:47–54.

174. Hattori K, Yoshida T, Nakashima H, et al. (1998). Synthesis of sulfonated amino-polysaccharides having anti-VIH and blood anticoagulant activities. *Carbohyd. Res.* 312:1–8.

175. Seddiki N, Mbemba E, Letourneur D, et al. (1997). Antiviral activity of derivatized dextrans on HIV-1 infection of primary macrophages and blood lymphocytes. *Biochim. Biophys. Acta.* 1362:47–55.

176. Ghosh P, Adhikari U, Ghosal P K (2004). In vitro anti-herpetic activity of sulfated polysaccharide fractions from Caulerpa racemosa. *Phytochem.* 65:3151–3157.

177. Hobbs S K, Monsky W L, Yuan F, et al. (1998). Regulation of transport pathways in tumor vessels: role of tumor type and microenvironment. *Proc. Natl. Acad. Sci. USA*, 95:4607–4612.

178. Krishna R, Mayer L D (2000). Multidrug resistance (MDR) in cancer. Mechanisms, reversal using modulators of MDR and the role of MDR modulators in influencing the pharmacokinetics of anticancer drugs. *Eur. J. Pharm. Sci.* 11:265–283.

179. Jain K K (2005). Nanotechnology-based Drug Delivery for Cancer. *Technol. Cancer Res. Treat.* 4:407–416.

180. Jain R K (2001). Delivery of molecular medicine to solid tumors: lessons from in vivo imaging of gene expression and function. *J. Control. Rel.* 74:7–25.

181. Brigger I, Dubernet C, Couvreur P (2002). Nanoparticles in cancer therapy and diagnosis. *Adv. Drug Del. Rev.* 54:631–651.

182. Minko T, Dharap S S, Pakunlu R I, et al. (2004). Molecular targeting of drug delivery systems to cancer. *Curr. Drug Targets.* 5:389–406.

183. Couvreur P, Vauthier C. Nanotechnology in pharmacology. Submitted.

184. Qi L, Xu Z, Jiang X, et al. (2005). Cytotoxic activities of chitosan nanoparticles and copper-loaded nanoparticles. *Bioorg. Medic. Chem. Lett.* 15:1397–1399.

185. Berry D, Lynn D M, Sasisekharan R (2004). Poly(b-amino ester)s promote cellular uptake of heparin and cancer cell death. *Chem. Bio.* 11:487–498.

186. Linhardt R J (2004). Heparin-induced cancer cell death. *Chem. Bio.* 11:420–422.

187. Peng Y, Zhang L, Zeng F, et al. (2005). Structure and antitumor activities of the water-soluble polysaccharides from Ganoderma tsugae mycelium. *Carbohyd. Polym.* 59:385–392.

188. Zhang M, Zhang L, Cheung P C K, et al. (2004). Molecular weight and anti-tumor activity of the water-soluble polysaccharides isolated by hot water and ultrasonic treatment from the sclerotia and mycelia of Pleurotus tuber-regium. *Carbohyd. Polym.* 56:123–128.

189. Yoo S-H, Yoon E J, Cha J, et al. (2004). Antitumor activity of levan polysaccharides from selected microorganisms. *Internat. J. Biol. Macromol.* 34:37–41.

190. Ho J C K, Konerding M A, Gaumann A, et al. (2004). Fungal polysaccharopeptide inhibits tumor angiogenesis and tumor growth in mice. *Life Sci.* 75:1343–1356.

191. Cheng J-J, Huang N-K, Chang T-T, et al. (2005). Study for anti-angiogenic activities of polysaccharides isolated from Antrodia cinnamomea in endothelial cells. *Life Sci.* 76:3029–3042.

192. Peng Y, Zhang L, Zeng F, et al. (2003). Structure and antitumor activity of extracellular polysaccharides from mycellium. *Carbohyd. Polym.* 54:297–303.

193. Kabanov A V, Batrakova E V, Miller D W (2003). Pluronic block copolymers as modulators of drug efflux transporter activity in the blood-brain barrier. *Adv. Drug Del. Rev.* 55:151–164.

194. Saima Y, Sarkar K K, Sen Sr A K, et al. (2000). An antitumor pectic polysaccharide from Feronia limonia. *Int. J. Biol. Macromol.* 27:333–335.

195. Clowes A W, Karnowsky M J (1977). Supression by heparin of smooth muscle cell proliferation in injured arteries. *Nature.* 265:625–626.

196. Logeart D, Prigent-Richard S, Boisson-Vidal C, et al. (1997). Fucans, sulfated polysaccharides extracted from brown seaweeds, inhibit vascular smooth muscle cell porliferation. II. Degradation and molecular weight effect. *Euro. J. Cell Bio.* 74:385–390.

197. Logeart D, Prigent-Richard S, Jozefonvicz J, et al. (1997). Fucans, sulfated polysaccharides extracted from brown seaweeds, inhibit vascular smooth muscle cell porliferation. I. Comparison with heparin for antiproliferative activity, binding and internalization. *Euro. J. Cell Bio.* 74:376–384.

198. Mestries P, Borchiellini C, Barbaud C, et al. (1998). Chemicaly modified dextrans modulate expression of collagen phenotype by cultured smooth muscle cells in relation to the degree of carboxymethyl, benzylamide, and sulfation substitution. *J. Biomed. Mater. Res.* 42:286–294.

199. Chupa J M, Foster A M, Summer S R, et al. (2000). Vascular cell response to polysaccharide materials: in vitro and in vivo evaluations. *Biomaterials.* 21:2315–2322.

200. Sato N, Nakazawa F, Ito T (2003). The structure of the antigenic polysaccharide produced by Eubactrium saburreum T15. *Carbohyd. Res.* 338:923–930.

201. Fattom A I, Horwith G, Fuller S (2004). Development of StaphVAX(TM), a polysaccharide conjugate vaccine against S. aureus infection: From the lab bench to phase III clinical trials. *Vaccine.* 22:880–887.

202. Menzel M, Muellinger B, Weber N, et al. (2005). Inhalative vaccination with pneumococcal polysaccharide in healthy volunteers. *Vaccine.* 23:5113–5119.

203. Ruben F L, Froeschle J E, Meschievitz C, et al. (2001). Choosing a route of administration for quadrivalent menigococcal polysaccharide vaccine: intramuscular versus subcutaneous. *Clin. Infec. Dis.* 32:170–172.

204. Moe G R, Tan S, Granoff D M (1999). Molecular mimetics of polysaccharide epitopes as vaccine candidates for prevention of Neisseria meningitidis serogroup B disease. *FEMS Immunol. Med. Microbiol.* 26:209–226.

205. Westerfeld N, Zurbriggen R (2005). Peptides delivered by immunostimulating reconstituted influenza virosomes. *J. Pept. Sci.* 11:707–712.

206. Moser C, Metcalfe I C, Viret J F (2003). Virosomal adjuvanted antigen delivery systems. *Expert Rev. Vacc.* 2:189–196.

207. Garcea R L, Gissmann L (2004). Virus-like particles as vaccines and vessels for the delivery of small molecules. *Curr. Opin. Biotechnol.* 15:513–517.

208. Morein B, Hu K F, Abusugra I (2004). Current status and potential application of ISCOMs in veterinary medicine. *Adv. Drug Del. Rev.* 56:1367–1382.

209. Kreuter J (1995). Nanoparticles as adjuvants for vaccines. *Pharm. Biotechnol.* 6:463–472.

210. Koping-Hoggard M, Sanchez A, Alonso M J (2005). Nanoparticles as carriers for nasal vaccine delivery. *Expert Rev. Vacc.* 4:185–196.

211. Chauvierre C, Couvreur P, Labarre D, et al. (2002). Copolymères à structure séquencée composé d'un segment saccharidique lié à au moins un segment hydrophobe bioérodable et particules correspondantes, Patent n°WO 02/399-79.

212. Salek-Ardakani S, Arrand J R, Shaw D, et al. (2000). Heparin and heparan sulfate bind interleukin-10 and modulate its activity. *Blood.* 96:1879–1888.

213. Bland E J, Keshavarz T, Bucke C (2004). The influence of small oligosaccharides on the immune system. *Carbohyd. Res.* 339:1673–1678.

214. da Silva B P, Parente J P (2003). Bioactive polysaccharides from Costus spicatus. *Carbohyd. Polym.* 51:239–242.

215. Gan L, Zhang S H, Liu Q, et al. (2003). A polysaccahride-protein complex from Lyciul barbarum upregukates cytokine expression in human peripheral blood mononuclear cells. *Euro. J. Pharmac.* 471:217–222.

216. Zhao G, Kan J, Li Z, et al. (2005). Structural features and immunological activity of a polysaccharide from Dioscorea opposita Thunb roots. *Carbohyd. Polym.* 61:125–131.

217. Peluso G, Petillo O, Ranieri M, et al. (1994). Chitosan-mediated stimulation of macrophage function. *Biomaterials.* 15:1215–1220.

218. Han Y, Zhao L, Yu Z, et al. (2005). Role of mannose receptor in oligochitosan-mediated stimulation of macrophage function. *Int. Immunopharmacol.* 5:1533–1542.

219. Klokkevold P R, Fukayama H, Sung E C, et al. (1999). The effect of chitosan (poly-N-acetyl glucosamine) on lingual hemostasis in heparinized rabbits. *J. Oral Maxillofac. Surg.* 57:49–52.

220. Jordan R, Beeler D, Rosenberg R (1979). Fractionation of low molecular weight heparin species and their interaction with antithrombin. *J. Biolog. Chem.* 254:2902–2913.

221. Zeerleder S, Mauron T, Lämmle B, et al. (2002). Effect of low molecular weight dextran sulfate on coagulation and platelet function tests. *Thrombosis Res.* 105:441–446.

222. Mauzac M, Jozefonvicz J (1984). Anticoagulant activity of dextra derivatives. Part I: synthesis and characterization. *Biomaterials.* 5:301–304.

223. Magel-Din Hussein D, Helmy W A, Salem H M (1998). Biological activities of some galactomannans and their sulfated derivatives. *Phytochem.* 48:479–484.

224. Alban S, Schauerte A, Franz G (2002). Anticoagulant sulfated polysaccharides: Part I. Synthesis and structure-activity relationship of new pullulan sulfates. *Carbohyd. Polym.* 47:267–276.

225. Krentsel L, Chaubet F, Rebrov A, et al. (1997). Anticoagulant activity of functionalized dextrans. Structure analyses of carboxymethylated dextran and first Monte Carlo simulations. *Carbohyd. Polym.* 33:63–71.

226. Ronghua H, Yumin D, Jianhong Y (2003). Preparation and anticoagulant activity of carboxybutyrylated hydroxyethyl chitosan sulfates. *Carbohyd. Polym.* 51:431–438.

227. Siddhanta A K, Shanmugam M, Mody K H, et al. (1999). Sulphated polysaccharides of Codium dwarkense Boergs. from the west coast of India: Chemical composition and blood anticoagulant activity. *Int. J. Biol. Macromol.* 26:151–154.

228. Ross B P, Toth I (2005). Gastrointestinal absorption of heparin by lipidization or coadministration with penetration enhancers. *Curr. Drug Deliv.* 2:277–287.

229. Rajaonarivony M, Vauthier C, Couarraze G, et al. (1993). Development of a new drug carrier made from alginate. *J. Pharm. Sci.* 82:912–917.

230. Vila A, Sanchez A, Tobio M, et al. (2002). Design of biodegradable particles for protein delivery. *J. Control Rel.* 78:15–24.

231. Sakuma S, Hayashi M, Akashi M (2001). Design of nanoparticles composed of graft copolymers for oral peptide delivery. *Adv. Drug Del. Rev.* 47:21–37.

232. Alonso M J, Sanchez A (2003). The potential of chitosan in ocular drug delivery. *J. Pharm. Pharmacol.* 55:1451–1463.

233. Behrens I, Pena A I V, Alonso M J, et al. (2002). Comparative uptake studies of bio-adhesive and non-bioadhesive nanoparticles in human intestinal cell lines and rats: the effect of mucus on particle adsorption and transport. *Pharm. Res.* 19:1185–1193.

234. Mao S, Germershaus O, Fischer D, et al. (2005). Uptake and Transport of PEG-Graft-Trimethyl-Chitosan Copolymer-Insulin Nanocomplexes by Epithelial Cells. *Pharmac. Res.*, In Press.

235. Fernandez-Urrusuno R, Calvo P, Remunan-Lopez C, et al. (1999). Enhancement of Nasal Absorption of Insulin Using Chitosan Nanoparticles. *Pharm. Res.* 16:1576–1581.

236. Borges O, Borchard G, Verhoef J C, et al. (2005). Preparation of coated nanoparticles for a new mucosal vaccine delivery system. *Int. J. Pharm.* 299:155–166.

237. Douglas K L, Tabrizian M (2005). Effect of experimental parameters on the formation of alginate-chitosan nanoparticles and evaluation of their potential application as DNA carrier. *J. Biomater. Sci. Polym. Ed.* 16:43–56.

238. Li X W, Lee D K L, Chan A S C, Alpar H O (2003). Sustained expression in mammalian cells with DNA complexed with chitosan nanoparticles. *Biochimica et Biophysica Acta (BBA)—Gene Struc. Expres.* 1630:7–18.

239. Peng J, Xing X, Wang K, et al. (2005). Influence of anions on the formation and properties of chitosan-DNA nanoparticles. *J. Nanosci. Nanotechnol.* 5:713–717.

240. Mansouri S, Cuie Y, Winnik F, et al. (2006). Characterization of folate-chitosan-DNA nanoparticles for gene therapy. *Biomaterials.* 27:2060–2065.

241. Ravi Kumar M N V, Samedt M, Mohapatra S S, et al. (2004). Cationic silica nanoparticles as gene carriers: synthesis, characterization and transfection efficiency in vitro and in vivo. *J. Nanosci. Nanotechnol.* 4:876–881.

242. Messai I, Lamalle D, Munier S, et al. (2005). Poly(D,L-lactic acid) and chitosan complexes: interactions with plasmid DNA. *Colloids and Surfaces A: Physicochem. Engin. Aspects.* 255:65–72.

243. Zimmer A (1999). Antisense Oligonucleotide Delivery with Polyhexylcyanoacrylate Nanoparticles as Carriers. *Methods.* 18:286–295.

244. Zobel H P, Kreuter J, Werner D, et al. (1997). Cationic polyhexylcyanoacrylate nanoparticles as carriers for antisense oligonucleotides. *Antisense & Nucleic Acid Drug Dev.* 7:483–493.

245. Aynie I, Vauthier C, Chacun H, et al. (1999). Spongelike alginate nanoparticles as a new potential system for the delivery of antisense oligonucleotides. *Antisense Nucleic Acid Drug Dev.* 9:301–312.

246. El-Aneed A (2004). An overview of current delivery systems in cancer gene therapy. *J. Control. Rel.* 94:1–14.

247. Riggs A (1981). Preparation of blood hemoglobins of vertebrates. *Methods Enzymol.* 76:5–29.

248. Chauvierre C, Marden M C, Vauthier C (2004). Heparin coated poly(alkylcyanoacrylate) nanoparticles coupled to hemoglobin: a new oxygen carrier. *Biomaterials.* 24:3081–3086.

1.5

R&D PARADIGM SHIFT AND BILLION-DOLLAR BIOLOGICS

KRISHAN MAGGON

Pharma Biotech R&D Consultant, Geneva, Switzerland

Contents

1.5.1	Introduction	162
1.5.2	Regulatory Approvals	163
1.5.3	Blockbuster Human Medicines	168
1.5.4	Biologics Gold Rush	173
1.5.5	R&D Overview	176
	1.5.5.1 R&D Productivity	178
1.5.6	R&D Success	180
	1.5.6.1 Erythropoietin	180
	1.5.6.2 Tumor Necrosis Factor Inhibitors	181
	1.5.6.3 Psoriasis	182
	1.5.6.4 Insulin	182
	1.5.6.5 Interferon	183
1.5.7	R&D Failures	184
	1.5.7.1 Alzheimer's Disease	184
	1.5.7.2 Anti-infective Exodus and U-turn	185
	1.5.7.3 RSV	186
	1.5.7.4 Sepsis	186
1.5.8	Cost Constraints	187
1.5.9	Public–Private Partnerships	189
1.5.10	Information Resources	191
1.5.11	Industry and Medicinal Brands in 2006	192
1.5.12	Conclusion	194
	Acknowledgment	195
	References	195

Handbook of Pharmaceutical Biotechnology, Edited by Shayne Cox Gad.
Copyright © 2007 John Wiley & Sons, Inc.

1.5.1 INTRODUCTION

There is no good measure of R&D productivity in the pharmaceutical and biotechnology industry because of the long maturation time needed to bring drugs to the market. The R&D investment has a long span of 7–15 years of payback time before drugs are approved. It takes even longer to market in major markets and turn the product into a profitable blockbuster [1–4]. Interim measures of R&D productivity like the number of patents filed and granted, scientific publications (variable impact factor), R&D expenses, and new drug approvals provide only partial measures of success. The real measure of R&D success is the market and profits from blockbuster sales after recovery of development costs. The success of a drug is based on its safety and efficacy established in clinical studies and favorable benefit-to-cost ratio and advantages over current therapy. A marketed product with limited sales and low growth, which barely recovers development costs, is an R&D failure. New projects with potential medicine fail due to lack of safety and/or efficacy. The major objective of the industrial R&D is to plan and execute the shortest development path to gain regulatory approval, strong patent position, and market the drug on a global basis in a short time frame and make it a blockbuster drug by expanding to related indications or new markets [4–8].

Company R&D productivity and efficiency can be measured by its introduction of new billion- and multibillion-dollar medicines through in-house R&D or extramural R&D, licensing, partnership, or marketing rights. Pharmaceutical and biotechnology R&D concentrates essentially on existing therapeutic areas of billion- and multibillion-dollar medicines and/or of first-in-class breakthrough products in new disease areas with double-digit growth. The R&D goal is to create, protect, and extend a company's franchise, to compete with market leaders with products with enhanced safety and efficacy or me-too follow-up for a breakthrough drug. Most current models measure past R&D performance, provide conflicting and variable results, and do not predict current or future outcome.

The modern blockbuster products era started four decades ago with Valium (Roche, diazepam), which was the most prescribed global medicine between 1969 and 1982 with 2.3 billion doses taken in 1978 and the first medicine to achieve peak sales of $200 million. It was later joined by benzodiazapine, antibiotics, and nonstroidal analgesics and by several innovative medicines like Captopril, Capoten, and Mevacor in the 1980–1990 period [9–13]. The period of billion-dollar pharmaceutical brands started 10 years ago and of biotechnology brands about 5 years ago. A commercial report estimated the total blockbuster medicine sales in 2005 at $145 billion and forecasted strong growth in biologics [14].

The profitability of the top pharmaceutical companies and their market value has increased over the years despite a decline in new drug approvals since 1990. Mergers and acquisitions, cost cutting, job elimination, and rationalization of operations achieved short-term profits. The higher R&D budgets failed to increase the number of new regulatory approvals. Moreover, patent expirations on existing blockbuster drugs, frequent and lengthy patent challenges, high prices of medications, drug supplies to poor countries and patients, product injury litigation, drug withdrawals, activists' pressure to shift R&D resources to tropical and neglected diseases, and tough new regulatory requirements for safer drugs and black box warnings hit the industry hard [15–19]. The price multiples or price earnings ratio of innovative pharma was flat or declined over the past 2 years and was converging with that of generic companies like Forrest and Teva. It is ironic that the industry that took pride in development of new drugs to combat human diseases is now compared with big tobacco and big oil firms [1–4, 15–19].

Profits from billion-dollar (blockbuster) products invariably cover R&D and market failures, fund future R&D investments, pay financial incentives to executives and dividends to shareholders, and fund philanthropy and charity. Fast track regulatory approval, medical need, media coverage, and stories of responding patients contribute to the success of a new medicine. Reports of links to serious adverse reactions, high prices with questionable benefits or low response rates, and black box warnings or regulatory delays have negative effects on the marketing of a new or existing medicine.

The Gold Rush of biotechnology-derived blockbuster bands and the paradigm shift in innovation from pharmaceutical to biotechnology R&D, within biologics towards monoclonal antibodies and from Europe to the United States will be discussed. The successes and failures of biotechnology R&D, criteria for blockbuster brands, global sales of biologics blockbuster brands, changing regulations, cost constraints, public–private partnership to deliver and develop low-cost treatments for tropical diseases, and information resources for biologics will be discussed in this chapter.

1.5.2 REGULATORY APPROVALS

Regulatory approval in the United States, Europe, and Japan is the first step before marketing of a new/future blockbuster; this aspect is first covered before discussions of blockbuster drugs, global sales, and R&D performance.

Regulatory approval of a new medical product is based on the safety and efficacy of a treatment established in open-label and double-blind comparative clinical trials of an adequate number of patients. For life-saving medicines, the number of patients is low, whereas for chronic diseases, the number of patients may go up to 25,000 to define rare and unexpected toxicities or adverse drug reactions. The major markets are in the United States [Food and Drug Administration (FDA)], Europe [European Medicinal Evaluation Agency (EMEA)], and Japan, with common registration requirements for innovative and biotechnology products (International Conference of Harmonization, ICH). Regulatory approval is used to measure the success of a company R&D, but the product may fail to make an impact in the market, only generate modest sale leading to a loss, or barely recover costs of R&D. The R&D success of a company should only be linked to the profitable new blockbuster drugs introduced each year and to a sales increase of $1 billion for existing products.

Biotechnology companies contributed over 60% of the new approvals by the FDA in 2005. Despite the increasing R&D costs [2, 5, 6] and the increasing numbers of active investigational new drugs (INDs) (41% increase from 2001 to 2004), the number of new drug and new molecular entity and first-in-class FDA approvals from top pharmaceutical companies has declined (Table 1.5-1). Since 2001, over 40% of the approved products by top pharmaceutical companies were licensed from smaller biotechnology companies. FDA approvals for new drugs in 2005 included 32 products including Byetta (exenatide, Lilly), Levemir (insulin

TABLE 1.5-1. FDA Drug Approvals 2001–2006

Categories	2006	2005	2004	2003	2002	2001
New drugs	26	75	118	72	78	66
New molecular entities (NMEs)	18	20	38	21	17	24
Priority review	10	20	29	14	11	10
Generics	NA	361	384	263	321	234
Biologics	4	5	7	22	20	7
Active INDs	5445	5029	4827	4544	4158	3883

Data based on information available at www.fda.gov. The European Medicinal Evaluation Agency (EMEA) provides an annual review of drug approvals on its web site, *www.emea.eu.int.*

detemir, Novo Nordisk), Hydrase (hyaluronidase, Prima Pharm), Naglazyme (galsulfase, BioMarin), Copagus (peginterferon alpha-2a+ribavarin, Roche), Revlimid (lenalidomide, Celgene), Nexavar (sorafenib, Bayer, Onyx), and Orencia (abatacept, BMS). The FDA and the EMEA approved breakthrough medicines like Macugen (pegaptanib, Pfizer, OSI) for age-related macular degeneration or Avastin (bevacizumab, Roche, Genentech) and Erbitux (cetuximab, BMS, ImClone) for cancer in 2004 [4]. In 2005, several big pharma companies failed to gain approval for any compounds discovered in their laboratories. Most of the approved drugs were for rare diseases. FDA approved 26 new medicines in 2006 including Januvia (Sitagliptin, Merck) for type 2 diabetes, Sutent (sunitinib, Pfizer), Sprycel (desatinib, BMY), and Zolinga (vorinostat) for cancer. The 8 new biologics approved in 2006 by FDA included Rotateq (rotavirus vaccine, Merck) for gastroenteritis, Gardasil (human papillomas virus vaccine, Merck) for cervical cancer, Zostavax (Zoster vaccine live, Merck) for herpes Zoster in adults, Vectibix (penitumumab, Amgen) for colon cancer, Lucentis (ranibizumab, Genentech, Novartis) for age related wet macular degeneration, inhaled insulin Exubera (Pfizer), Myozyme (alglucosidase alpha, Genzyme) for Pompe disease, and Elaprase (idursulfase, Shire) for Hunter syndrome. The European regulatory agency EMEA approved 51 products in 2006, including first approvals for NCE like Accomplia (rimonabant, Sanofi Aventis) for obesity, Nexavar (sorafanib, Bayer-Schering) for renal cell carcinoma, Thelin (sitaxentan, Encysive) for pulmonary arterial hypertension, Baraclude, Sutent, Sprycel, and Tygacil. Biologics approved included 2 biosimilar human growth hormones, Atryn (r antithrombin alpha, Genzyme) for deep vein thrombosis, Byetta, Elaprase, Lucentis, and Tysabri. EMEA gave first approvals for the new vaccine Daronix (H5N1 whole inactivated antigen, GSK) for avian influenza pandemic and Proquad (measles, mumps, rubella, and varicella, Sanofi Aventis), in additional to Gardasil, Rotateq, and Zostavax. Rotarix (rotavirus vaccine, GSK) was approved in late 2005 and launched in early 2006.

In the past 5 years, at least six or seven products a year, each with peak sales potential of more than $1 billion annually, were terminated in late-stage phase III development or at the new drug application (NDA) filing stage, resulted in market value loss of tens of billion dollars for the affected companies. All big pharma companies like Merck, Lilly, BMS, GSK, and Novartis had their share of phase III failures, regulatory delays, NDA rejections, and requests for additional data [16, 17]. During the past 5 years, 15 drugs with peak sales potentials of a combined $11 billion a year were withdrawn from the market for safety reasons, some of which had received fast-track priority approval. The resulting product injury litigation cost for just one of the withdrawn drugs, Vioxx, may reach $15 billion (Table 1.5-2) [20, 21]. Black box warnings, media coverage, and links to serious adverse effects resulted in significant sales decline for affected drugs in 2005 (Table 1.5-3). Patient safety and comfort was given priority by the black box warning for increased cardiovascular risks with COX II inhibitors and antidepressants link to suicidal behavior in children. The FDA rightly imposed the requirements for long-term safety

TABLE 1.5-2. Major Drug Withdrawals

Drug	Company	Year	Adverse Effect	Patients Claims	Litigation $
Benoxaprofen	Lilly	1982	Liver failure	2,000	$10 million
Redux/Pondimin (Fen-Phen)	Wyeth	1997	Heart valve problems	130 deaths	16 billion paid
			Pulmonary hypertension	62,000 cases	5 billion reserve
Rezulin (Troglitazone)	Pfizer	2000	Liver failure	400 deaths 4,000 cases	15 billion
Baycol/Lipobay	Bayer	2001	Rhabdomyolysis Kidney failure	100 deaths 1,500 cases	1 billion paid
Vioxx	Merck	2004	Heart attacks Heart failure	10,000 cases	15–25 billion

Two drugs initially withdrawn were allowed back by the FDA in U.S. markets. Lotronex (Alosetron, Glaxo Smith Kline) for irritable bowel syndrome was approved in early 2000 and withdrawn 9 months later due to serious gastrointestinal problems. It was reapproved with restrictions in 2002. Tysabri (Natalizumab, Biogen Idec, Elan) for multiple sclerosis was approved in late 2004 and withdrawn in 2005 due to rare fatal brain lesions and was allowed back in mid-2006.

TABLE 1.5.3 Significant Sales Decline in 2005–2006

Generic Name	Brands	Companies	Indications	Sales $ billion		
				2004	2005	2006
Celecoxib*	Celebrex	Pfizer	Pain	3.3	1.73	2.04
Paroxitine*	Paxil	Glaxo Smith Kline	Depression	3.9	2.2	1.15
Azithromycin**	Zithromax	Pfizer	Antibiotic	1.8	2.02	0.64
Pravastatin**	Pravachol	BMS, Daiichi Sankyo	Cholesterol	5.5	5.0	4.0
Sertraline*	Zoloft	Pfizer	Depression	3.4	3.25	2.11
Simvastatin**	Zocor	Merck	Cholesterol	5.2	4.4	2.8

Sales of top brands in 2005 were down due to safety concern about increased suicidal tendencies in younger children linked to Paxil and Zoloft and increased risk of cardiovascular events (heart attacks) due to COX II inhibitors and withdrawal of Vioxx and added black box warnings imposed by FDA. Medicinal brands with highest sales decrease in 2006 were mainly NCE synthetic products and the loss was due to patent expiry and generic competition.
*Black box warning **Patent Expiry and generics

data for new coxibs and carcinogenicity data before phase III trials for Peroxisome Proliferator activator receptor (PPAR) agonists [20–22].

The current regulatory requirements mandate extensive testing of all biological products even with small differences. Demonstrating that a biosimilar product is as safe and effective as the originator is a difficult task [18, 20]. Europe has taken the lead in biosimilar products by establishing a legal framework for authorization of biogenerics, and EMEA issued several final and draft guidelines in 2006 to cover erythropoietin, insulin, interferon, somatotropin, and granulocyte colony stimulating factor. Sandoz generic human growth hormone was approved in USA and Europe (EU) in early 2006 after a 3-year delay. The recent episode of an unexpected severe inflammation and multiple organ failure adverse reaction in six healthy subjects in phase I trials in England, with an immunomodulatory humanized agonistic anti-CD28 monoclonal antibody for leukemia, multiple sclerosis, and arthritis, and similar reactions in cancer patients in the United States with another anti CD28 antibody, after investigations, may result in new safety regulations for testing biologics and monoclonal antibodies.

The history of modern drug regulations is closely linked to the incidence of drug-induced injury, organ failure, and deaths. Each tragedy linked to marketed drugs resulted in increased regulations and additional testing to prevent future episodes with new drugs of the same class (Table 1.5-4). The Elixir of Sulfanilamide episode resulted in the Food, Drug and Cosmetic Act of 1938 and Thalidomide tragedy in early 1960 resulted in requirements for safety and efficacy testing for drugs in animals and humans.

The withdrawal of benoxaprofen resulted in a 10-fold increase in the number of patients (from 200 to 400 patients/year exposure to 2500 to 5000 patients/year exposure to the study drug) required for newer nonsteroidal anti-inflammatory drugs (NSAIDs) and trials in the elderly, hepatic and renal impairment, and drug interactions. Astra Zeneca total patient exposure in clinical trials submitted for approval of rosuvastatin (12,500) was considerably greater than the 2000–3000 patients submitted for most other marketed statins. Since the withdrawal of troglitazone (Rezulin), no other PPAR agonist has been approved because of the concerns about the class carcinogenicity, hepatotoxicity, and cardiotoxicity [19, 22]. Merck was the dominant R&D-driven and most admired company during the 1980s and 1990s and had one of the highest market valuations in the industry. Thus, a sale loss of $2.5 billion of rofecoxib resulted in a loss of $27 billion market value in one day and another $20 billion within the next month. Analysts have provided estimates of up to $20 billion for rofecoxib litigation. Similarly Pfizer lost $20 billion in one day in market value after discontinuation of its torcetrapib-Lipitor Phase III trials and termination of all trocetrapib (HDL promoter) development projects due to higher number of deaths in the treated group.

The cost of drug development has reached astronomic proportions. Considering the projections of previously contested estimates from 2005 and adding in marketing and manufacturing costs, a new drug for a chronic disease requires $2 billion in investment [2, 23, 24]. Industry critics have doubted earlier estimates, which included interest payments on blocked R&D funds and excluded R&D tax credits. A new GMP facility for a biological product costs an additional $500 million [2, 20]. With the new FDA safety focus, an average drug NDA file for registration for chronic diseases includes data on over 10,000–20,000 patients enrolled in 100–200

TABLE 1.5-4. Regulatory Impact of Health Crisis/Drug Withdrawals

Drug	Year	Regulatory Action
Elixir Sulfanilamide	1937	Food Drug and Cosmetic Act 1938
Thalidomide	1962	Safety and efficacy testing
Contraceptives Risks	1970	Package inserts
Benoxaprofen	1982	Studies in special populations, elderly, renal & hepatic insufficiency, genetic, gender differences
Rare diseases	1983	Orphan drug act
Life-threatening diseases	1987	Treatment IND for life saving drugs
Global regulations	1990	International Conference on Harmonization ICH 1-ICH 7(2006): GMP; GLP, GCP
High costs	1992	Generic drug act (1984 ANDA)
Drug safety	1993	MedWatch
AIDS	1995	Fast-track approval
Fen-Phen	1997	Orlistat NDA 5,000 patients, Safety for new obesity drugs
Rezulin	2000	Carcinogenicity prior to Phase III for PPAR agonists, CV safety for Murgaglitazar in 2005
Tropical/neglected diseases	2000	Public–private partnerships: MMV, GAVI,
Baycol	2001	Safety for new statins, Rosuvastatin NDA 12,500 patients
Vaccines	2002	NDA >20,000–60,000 patients safety, Rotavirus
Vioxx	2004	CV safety in phase III for new COX II inhibitors Etoricoxib & Lumiracoxib NDA >25,000 patients
Reduced new drug approvals	2004	Critical Path Initiative: 0 Phase Clinical trials, biomarkers, and clinical trial
Low R&D productivity		design to speed up development
Black box warnings	2005	Reduced sales of affected medicines
Biogeneric guidelines	2006	Europe (EMEA), FDA 2008?

NDA-New drug application; ANDA-Abbreviated new drug application; CV-cardiovascular; PPAR-Peroxisome Proliferator Activator Receptor; MMV-Medicines for Malaria Venture; GAVI-Global Alliance for Vaccines and Immunization; EMEA-European Medicinal Evaluation Agency

clinical studies over a 7–10-year period. The requirements for life-threatening diseases require only a few thousand patients. The development of a new drug typically takes 7–12 years, and clinical trials for a new vaccine require 25,000–100,000 individuals enrolled in a 2:1 ratio in the treatment and placebo groups. The recently approved rotavirus vaccines [Glaxo Smith Kline, Rotarix (in Europe) and Merck, Rotateq (in the USA and Europe)] to protect children from severe gastroenteritis were studied in 60,000–90,000 patients. These two NDAs established a historic record number of patients monitored for safety. With such progression both in the duration of treatment and in the number of patients, new drug development is going to be even more costly (Table 1.5-4).

The FDA's current twin objectives are fast-track approvals of life-saving drugs and long-term safety and efficacy data for chronic conditions. The FDA has taken several initiatives to speed up drug development by introducing 0 phase trials at low doses, improving clinical trials study design and finding validated biomarkers. The objective is to improve R&D productivity [4]. Fast-track approvals contribute greatly to the success of the breakthrough drugs or drugs with significant therapeutic advantage and bring along media interest and coverage of the medical meetings, expert opinion, and raised demand, hope, and awareness of health-care professionals and patients for the new medicine. The FDA and EMEA fast-track approvals have contributed to the blockbuster sales of several new anticancer and antiviral products for AIDS and RSV and new vaccines. The first-year sales for Neulasta, Pegasys, and Avastin were $1300, $760, and $550 million, respectively, due to favorable regulatory and media buzz. Flu Mist (MedImmune) failed to benefit from the U.S. shortage of flu vaccines in 2004, and despite the surge in demand due to Avian Influenza in 2005, because the FDA had limited its use in adults (who do not require flu vaccination) and excluded its use in the elderly and infants.

1.5.3   BLOCKBUSTER HUMAN MEDICINES

The "Gold Rush" of billion- and multibillion-dollar human medicine brands started in 1995 for the pharmaceuticals and in 2000 for the biotechnology-derived brands and will continue to expand in the future. Despite the proclaimed and much heralded "End of the Blockbuster Model" for the pharmaceutical industry, the number of blockbuster drugs, their market share, and contribution to profits for companies has steadily increased (Table 1.5-5).

TABLE 1.5-5 Number of Blockbuster Drugs and Market Share

	$ billion	
	2005	**2006***
Global Pharma Market	603 IMS	640
Global Biotech Market	63 Ernst & Young	75
Global Generic Market	60 IMS	65
R&D Expenses	50 PhRMA	55
Biotech R&D	20 Ernst & Young	25

Year	Billion dollar drug brands		Total share of global market %
	NCE	NME	
1991	4	0	6
1998	29	3	16
2000	55	5	25
2003	65	13	32
2004	79	18	38
2005	109	29	42
2006	125	35	45

NCE-New Chemical Entity (Pharmaceuticals); NME-New Molecular Entity (Biologics)
*Global 2006 market estimates by author.

BOX 1.5-1

The success criteria for blockbuster biologics are as follows:

- Life saving, reduction in hospitalization days, critical care, and life support systems
- Improve, maintain, restore, or reverse declining vital/organ functions
- Complete or partial disease remission
- Lifestyle changes, reverse or slow aging process, obesity, diabetes, and hypertension
- High safety and efficacy profile
- Long-term benefits
- Fast-track regulatory approvals
- Strong patent protection and repeated extension for new indications
- Global sales, marketing, and manufacture to meet any surge in demand

From 1991 to 2006, the number of billion-dollar medicines increased 30-fold and their market share by 7-fold and accelerated for biologics brands in 2004 and 2005. The share of biotech sales of the total pharma sales increased in the United States from 4% in 1993 to 15% in 2005. The total number of blockbuster medicinal brands increased to 125 in 2006 and included 35 biologic brands with global sales of biologics reaching $75 billion (Table 1.5-5).

The criteria for blockbuster drugs and diseases with unmet medical need and market success to $5 billion, $10 billion, and $15 billion are as described (Box 1.5-1) [15].

The first effective and safe treatments for Alzheimer's, Parkinson's, chronic graft rejection, permanent graft acceptance, obesity, chronic obstructive pulmonary disease (COPD), adult respiratory distress syndrome (ARDS), sepsis, and vaccines [common cold, respiratory syncytial virus (RSV), severe acute respiratory syndrome (SARS), AIDS, malaria, Avian Influenza, radiation, contraception, emerging new infections] will be multibillion-dollar products [15, 16].

Table 1.5-6 lists the R&D budgets, total sales, % contribution of blockbuster brand sales to human pharmaceutical sales and the list of blockbuster brands by different companies. Only companies with either R&D budgets of over $1 billion or with at least one blockbuster medicine in 2006 are included. The IMS sales figures for the same brand often differed from company figures by more than $500 million for several drugs. The contribution of blockbuster brand sales for top pharmaceutical companies ranged from 37% (Roche) to 74% (Johnson & Johnson). For biotechnology companies, the contribution of a blockbuster brand was even greater and ranged from 66% (Serono) to 98% (Amgen). The contribution of these drugs to the profitability and the market value of the company was even greater. Profits from billion-dollar (blockbuster) products cover R&D and marketing failures in other projects, fund future R&D investments, increase dividend payments to shareholders, fund philanthropy and charity, provide access to medicines for needy patients, and bonuses and motivation to executives. This aspect has not been included in R&D productivity evaluation by commercial reports.

TABLE 1.5-6. Multibillion-Dollar Brand Medicines in 2006

Company	Sales $ Billion (% Blockbuster Sales)		R&D Budget $ Billion		>2 Billion	>$ 1 billion
	2005	2006	2005	2006		
J&J	50(74)	53.3	6.3	7.1	Remicade Risperdal Procrit ***Topamax***	Aciphex, Duragesic, Floxin,
Pfizer	51(55)	48(55)	7.4	7.6	Lipitor, Norvasc, Zoloft, *Celebrex* ***Zithromax***	Viagra, Xalatan, Zyrtec, *Lyrica*
Glaxo Smith Kline	34(70)	43	5.7	6.4	Advair, Avandia ***Vaccines***	Augmentin, Coreg, Flovent, Imigran, Lamictal, Paxil, Valtrex, Wellbutrin, Zofran Combivir
Novartis	32(45)	37	4.8	5.3	Diovan, Glivec	Lotrel, Zometa Neoral, Lamisil
Sanofi Aventis	32(49)	36.8	4.8	5.76	Plavix, **Lovenox** Ambien *Aprovel, Eloxatine,* ***Lantus, Taxotere Vaccines***	Copaxone, Tritace, Allegra, ***Fluzone***
Roche	28(37)	33.6	4.6	5.2	***Avastin***, *Tamiflu,* ***Herceptin, Rituxan***	Cell Cept, **NeoRecormon, Pegasys**
Astra Zeneca	24(73)	26.5	3.4	3.9	Nexium, Seroquel, *Crestor*	Arimidex, *Atacand,* Casodex, Prilosec, Pulmicort, Symbicort, Seloken, Zoladex
Merck	22(61)	22.6	3.8	4.8	Fosamax, Cozaar, Singulair, Zocor	*Vytorin, Zetia,* ***Vaccines***
Wyeth	18(57)	20.4	2.7	3.1	Effexor, **Enbrel**	**Prevenar**, Protonix; *Premarin*
Abbott	22(71)	22.5	1.8	2.2	Prevacid ***Humira***	Depakote, Kaltera, Mobic *TriCor,* Biaxin
Bristol Myers Squibb	19(40)	18	2.7	3.0	Plavix, Pravachol *Avapro*	*Abilify,* ***Erbitux*** *Pravachol*
Bayer Schering	10.4	15.2	2.2	2.8		**Betaferon, *Kogenate,*** *Yasmin*

Company						
Lilly	14(52)	15.6	2.7	3.1	Zyprexa	Evista, Zemzar, **Humalog**, **<u>Humulin</u>**, *Cymbalta*
Amgen	12.4(98)	14.3	2.3	3.3	**Aranesp, Epogen, Enbrel, Neulasta**	**Neupogen**
Boehringer Ingelheim	12.3	13.8	1.5	2.0	*Spiriva*	Flomax, Micardis, <u>Mobic</u>
Takeda	10	10.5	1.3	1.5	Prevacid, *Actos*, *Blopress*	**Lupron**
Schering Plough	9	10.6	1.8	2.2		Zetia, *Vytorin*, Remicade
Baxter	10	10.4	0.5	0.6		**Advate rAHF-PFM (Factor VIII)**
Merck-Serono	6.9	9.84	1.2	1.36		**Rebif**, ***Erbitux***
Astellas	8	9	1.2	1.3		Prograf, Harnal
Teva	4.7	8.4	0.37	0.46	*Copaxone*	
DaiichiSankyo	8	8	1.4	1.45	Benicar	Levofloxacin, Mevalotin, Pravastatin
Novo Nordisk	5	7.2	0.6	1.0	**Insulins**	
Eisai	5	5.7	0.83	0.92		Aricept, Aciphex
Nycomed Atlanta	4	4.4	0.6	0.65	Protonix	
Otsuka	3.0	3.7	0.49	0.5	Abilify	
Allergan	2.3	3.45	0.39	1.0		***Botox***
UCB Schwarz	3.0	3.3	0.66	0.78		*Keppra*
Genzyme	2	3.2				***Cerezyme***
Gilead	2.03	3.03	0.28	0.38		*Truvada*
Forest	3.5	2.96	0.3	0.4		Lexapro
Biogen Idec	2.4	2.7	0.74	0.72		**Avonex**
MedImmune	1.2	1.3	0.4	0.42		**Synagis**
Purdue	NA	NA	NA	NA		OxyContin

NA-Not Available

Companies with one blockbuster brand or R&D budget of >$1 billion are listed. Marketing joint ventures like Takeda-Abbott Pharmaceutical, Merck-Schering Plough are not listed.

The % contribution of blockbuster human medicinal brands to the total human pharmaceutical 2005 sales of the company which increased even more in 2006. Biologics are in **bold**, new blockbuster drugs are in *italics*, brands going off blockbuster status in 2006 are <u>underlined</u>.

In 2006, there was 1 brand with over $12 billion dollar, 3 with over $5 billion, 9 with over $4 billion and additional 8 brands crossed $3 billion annual sales (Table 1.5-7). Although Merck has only four blockbuster drugs, all four had sales over $3 billion each, which is equal to 12 blockbuster drugs. Lipitor alone with sales of over $12 billion is equivalent to 12 billion-dollar drugs. Table 1.5-7 lists all new chemical entities with annual sales over $3 billion in 2006. Lipitor, Plavix, and Advair each had explosive sales growth during the 2002–2005 periods. An increase of over $1 billion a year was recorded with Lipitor during 2000–2005 and for Plavix during the 2001–2005 periods. For Pfizer and Sanofi Aventis, it is like adding a new block-buster drug each year during the above period.

The U.S. patents on 35 drugs with global sales totaling more than $22 billion expired in the year 2006, resulting in the loss of market to generics. Blockbuster products coming off patent are valued at $27 billion in 2007 and $29 billion in 2008 [17]. Blockbuster brands nearing patent expiration in therapeutic areas like ulcers, hypertension, lipid lowering, depression, schizophrenia, and cancer are replaced by new molecules under patent cover with improved safety and efficacy. Thus, Esmoprazole and two other proton pump inhibitors have replaced Prilosec (omeprazole). Similarly the angiotensin converting enzyme inhibitors class has been taken over by angiotensin receptors blockers with four brands in the multibillion-dollar sales category.

TABLE 1.5-7. Top Brand Medicines with Over $ 3 Billion Sales in 2006

Generic Name	Brands	Companies	Indications	Sales $ billion		
				2004	2005	2006
Atorvastatin	Lipitor	Pfizer	Cholesterol	10.8	12.2	12.9
Fluticasone Salmetrol	Advair	Glaxo Smith Kline	Asthma	4.5	5.5	6.13
Clopidrogel	Plavix	Bristol Myers Squibb, Sanofi Aventis	Atherosclerosis	5.2	6.2	5.55
Esomaprazole	Nexium	AstraZeneca	Ulcers	3.88	4.63	5.2
Amlodipine	Norvasc	Pfizer	Hypertension	4.46	4.76	4.85
Glanzapine	Zyprexa	Lilly	Schizophrenia	4.42	4.2	4.36
Valsartan	Diovan	Novartis	Hypertension	3.1	3.67	4.22
Risperidone	Risperdal	J&J	Schizophrenia	3.0	3.55	4.18
Pravastatin	Pravachol	BMS, Daiichi Sankyo	Cholesterol	5.7	5.0	4.0
Lansoprazole	Prevacid	Takeda, Abbott	Ulcers	3.1	3.8	3.8
Venlafaxine	Effexor	Wyeth	Depression	3.3	3.5	3.7
Montelukast	Singulair	Merck	Asthma	2.6	3.0	3.6
Glatiramer	Copaxone	Teva, Sanofi Aventis	Multiple Sclerosis	1.8	2.4	3.6
Pentoprazole	Protonix	Atlanta, Wyeth	Ulcers	3.3	4.0	3.4
Losartan	Cozaar	Merck	Hypertension	2.8	3.0	3.2
Alendronate	Fosamax	Merck	Osteoporosis	3.1	3.2	3.1

Sales as reported by companies.
Zocor which was 5th in 2005 failed to make the list in 2006.

New-patented products in the same class replaced blockbuster anticancer products in the class taxol, camptothecin, and platinum after patent expiry. In biotechnology-derived biologics, second- and third-generation products with improved properties, like improved insulin, erythropoietin, and peg-interferon, replace the first-generation products. Some companies failed to switch to new patent-covered formulations or analoges like Lilly with Prozac and BMS with Glucophage (no follow-up), and Schering Plough from Claritin to Clarinex. The pharmaceutical industry is very creative in producing me-too blockbuster drugs with extended patent life. There is no reason to believe that this trend will stop in the next 5 years due to looming patent expiration of several blockbuster drugs as some commercial reports have suggested.

1.5.4 BIOLOGICS GOLD RUSH

IMS [12] data estimated the global pharmaceutical market at $600 billion and biotechnology products at $52 billion in 2005. Ernst and Young [1] estimated the total biotechnology product sales at $54.6 billion in 2004 and $63 billion in 2005. Datamonitor estimated the 2005 global biotechnology market at only $40 billion. All sales forecasts are on the low side as the total sales of the listed biotechnology products in Tables 1.5-8 and 1.5-9 alone were $60.4 billion in 2005 and $71.25 billion in 2006. There are 300 approved biotech products in the United States and Europe, 100 under regulatory review, and another 400 in advanced clinical trials for 200 diseases [1].

Several brands of EPO, TNF inhibitors, interferon, insulin, and GCSF had sales of over $1, $2, and $3 billion each in 2004 and 2005. The period 2002 to 2005 recorded a sales increase of over $1 billion per year for TNF inhibitors. The top five best-selling human medications were all biotechnology drugs by sales in 2005, apart from Lipitor. There were five biologic brands with over $3 billion and six brands with over $2 billion in sales in 2005. As several companies market erythropoietins, interferons, insulins, GCSF, and human growth hormone under different brand names, these proteins rarely make bestsellers lists in commercial databases, unless taken together (Tables 1.5-8, 1.5-9) [22, 25].

A review of several therapeutic areas and brands indicates intense market competition for market leadership and change of market leadership among different brands and fast growth areas (Tables 1.5-8, 1.5-9). This has been previously observed with me-too or follow-on pharmaceuticals. Losec overtook Zantac, which had overtaken the first-in-class Tagamet; Enalapril over Captotril (Capoten); and Lipitor over Zocor and the first-in-class Mevacor. Lipitor has retained market dominance despite the arrival of superstatin Crestor and newer combination products. Diovan overtook the first-in-class Cozaar in 2004, for the market leadership of the angiotensin II receptor blocker ARB antihypertensive agents [22]. Astra Zeneca retained its dominance of the antiulcer market by successful switching patients to Nexium and limiting the decline in Losec sales after patent expiration. The success of Atlanta to launch a blockbuster antiulcer Protonix shows that follow-on patented me-too drugs carry low risk and high rewards (Table 1.5-7). Monoclonal antibodies for cancer and immunoinflammatory conditions had global sales of $20.5 billion in 2006 with tumor necrosis factor blockers accounting for

TABLE 1.5-8. Best Selling Biologics (Top 1–5)

Protein	Indication	Total sales $ Billion 2004	2005	2006	Company	Brands	Sales $ Billion 2004	2005	2006
Erythropoietin PEG, α, β, Darbepoetin	Anemia	11.8	12.3	12.1	Amgen	Epogen	2.6	2.45	2.51
						Epogin			
						Aranesp	2.4	3.3	4.12
					J&J	Procrit	3.6	3.3	3.2
						Eprex			
					Roche	NeoRecormon	1.8	1.7	1.8
						Epogin			
					Kirin	ESPO	0.50	0.55	0.50
Tumor Necrosis Factor Inhibitors	RA, JRA, Ps, PsA, AS	5.55	8.6	10.6	Amgen Wyeth	Enbrel	2.6	3.7	4.4
	RA, UC, CD, Ps, PsA, AS				J&J Schering Plough	Remicade	2.1	3.5	4.2
	RA, AS, PsA, CD				Abbott	Humira	0.85	1.4	2.0
Human Insulin	Diabetes	6.5	7.2	9.05	Novo Nordisk	Novolin	2.1	2.4	2.5
						Analogs	0.9	1.2	1.7
					Lilly	Humulin	1.0	1.12	0.92
						Humalog	1.1	1.0	1.3
						Byetta		0.43	
					Sanofi Aventis	Lantus	0.96	1.4	2.2
						Apidra			
Interferons α, β	Hepatitis C	6.8	6.8	6.7	Schering Plough	PegIntron	1.8	1.4	1.04
	Multiple sclerosis				Roche	Pegasys	1.4	1.4	1.2
					Biogen Idec	Avonex	1.4	1.5	1.7
					Merck Serono	Rebif	1.1	1.3	1.45
					Bayer Schering	Betaseron	1.0	1.2	1.3
Granulocytes-Colony Stimulating Factor, G-CSF	Granulocytes stimulator	2.9	3.5	3.9	Amgen	Neulasta	1.7	2.3	2.7
						Neupogen	1.2	1.2	1.2

RA-Rheumatoid Arthritis; JRA-Juvenile Rheumatoid Arthritis; Ps-Psoriasis; PsA-Psoriatic arthritis; CD-Crohn's Disease; UC-Ulcerative Colitis; AS-Ankylosing Spondylitis.

TABLE 1.5-9. Best Selling Biologics (6–20)

Generic Name	Brands	Companies	Indications	Sales $ billion		
				2004	2005	2006
Rituximab	Rituxan	Roche, Biogen Idec	Leukemia, Lymphoma Rheumatoid Arthritis	2.8	3.2	4.7
Trastuzumab	Herceptin	Roche	Breast Cancer	1.3	1.65	3.14
Factor VIII	Advate, Helixate, Kogenate, ReFacto Recombinate	Baxter, Bayer Schering Sanofi Aventis Wyeth	Hemophilia	1.9	2.1	2.5
Bevacizumab	Avastin	Genentech, Roche	Colon cancer	0.55	1.3	2.4
Human Growth Hormone	Serostim, Saizen, Humatrope, Protopin, Neutropin, Genotropin	Merck Serono, Biogen Idec, Roche, Novo Nordisk, Akzo Nobel, Lilly, Pfizer	Dwarfism	1.8	2.2	2.4
Influenza Vaccines	Fluvin, Flumist, Fluzone, Fluvirin, Fluarix, Influvac	Novartis, GSK, Merck, Sanofi Aventis, Crucell Berna, Tanabe	Influenza	1.2	1.9	2.3
Pneumococcal Conjugate Vaccine	Prevanar	Wyeth	Pneumococcal Dis	1.2	1.5	1.9
Luprorelin	Lupron	Takeda, Abbott	Prostate Cancer	1.2	1.5	1.7
FSH	Gonal F, Follistim	Merck-Serono, Akzo Nobel	Infertility	0.95	0.93	1.3
Hepatitis Vaccines		GSK, Merck, SanofiAventis	Hepatitis A, B	1.0	1.1	1.2
Pediatric Vaccines		GSK, Sanofi Aventis Merck	Measles, Mumps, Rubella Chickenpox, tetanus, diphtheria, acellular pertussis	1.0	1.1	1.2
Palivizumab	Synagis	MedImmune	RSV	0.95	1.1	1.1
Botulin toxin	Botox	Allergan	Wrinkles	0.70	0.83	1.1
Glucocerebrosidase	Cerezyme, Ceradase	Genzyme	Gaucher's disease	0.84	0.93	1.0
Factor VII	Novo Seven	Novo Nordisk	Hemophilia	0.76	0.81	0.95

Sales Growth of blockbuster brands

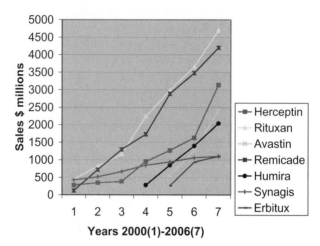

Figure 1.5-1. Blockbuster monoclonal antibodies sales growth.

$10.6 billion. Vaccines were second with global sales reaching $15 billion. The other major categories of biologic brands with global sales were erythropoietins for anemia with $12 billion, traditional and newer insulins like long acting at $9 billion and interferon for hepatitis and multiple scelrosis at $7 billion.

Erythropoietin has remained the best-selling human medicine for the past 3 years, followed by Lipitor (Atorvastatin), and both of these had over $12 billion in sales [20–22]. The sales of erythropoietin in Japan were over $1 billion. Tumor necrosis factor inhibitors, interferons, and insulins with, respectively, sales of $8.6 billion, $7.3 billion, and $6.5 billion in 2005 were the third, fourth, and fifth best-selling products after EPO and Lipitor (Table 1.5-8) [20–22, 24, 25]. The time taken to reach $1 billion annual sales for monoclonal antibodies has varied and was much faster for follow up and cancer products (Fig. 1.5-1). Rituxan is the leading antibody since 2003 by sales. The growth of Avastin and Erbitux has been much faster as compared to Herceptin and took only two years to reach $1 billion sales.

1.5.5   R&D OVERVIEW

An overview of the blockbuster medications during the past few years and a comparison of R&D productivity clearly show an ongoing paradigm shift of innovation from Europe to the United States and from pharmaceutical to biotechnology R&D. A comparative analysis of R&D productivity of pharmaceutical and biotechnology firms is provided.

Pharmaprojects lists over 7300 drugs in active R&D, whereas the PHRMA site using Kluwer Health Adis R&D Insight lists over 10,000 active R&D projects. The number of new drugs in R&D has almost doubled to 1970 drugs in 2004 from 1010 drugs in 1995 [1]. There are over 300 approved biotech products in Europe and the United States, 100 under regulatory review, 200 in late phase II–III clinical trials, and 800 in early clinical trials [1, 16, 20]. Monoclonal antibodies outnumber all other biologics combined in all stages of preclinical and clinical development. The biotechnology industry spent over $23 billion on R&D and generated over 7000 patents in the year 2005, with patents in cancer leading over other therapeutic categories. The highest number of patents was issued for cancer, and the success of cancer vaccines is attracting a lot of R&D funds. Only a strong R&D pipeline contributes to long-term growth of the company.

Several pharmaceutical companies spent over $3 billion on R&D in 2005 for a portfolio of 80–150 projects, of which 50–80 were in clinical development. In general, only about 30% NCE are truly innovative, the rest are me-too products or line extensions. About 30–40% of the projects of big pharma are licensed in from other sources [2, 3]. The share of the top 10 global R&D companies in the innovative first-in-class type of new molecular entities has steadily decreased over the past decade. It now takes 10–15 years to bring a product from discovery to markets. The success rate has decreased over the years, and out of a million compounds screened in highly automated computerized systems High Throughput Screening (HTS), only 250 advanced to become development candidates for additional testing [16].

All major pharmaceutical companies run their own discovery, development, licensing, partnership, manufacturing, marketing, and sales for the majority of their product pipelines and portfolios with a relatively high risk and investment [1, 3]. Strategic review, analysis, and consulting firms have prompted big pharmaceutical firms to shift focus to targeting specific populations or disease areas like the biotechnology companies and shift R&D resources to vaccines, biologics, and monoclonal antibodies that have higher rates of success, form partnerships in all stages of drug discovery and development to reduce risks and share rewards and focus on disease and needs of the patients, and operate as strategic business units [5, 9].

There has been a paradigm shift in innovation from the pharmaceutical to biotechnology R&D based on the number of new billion-dollar biologics introduced. There is a clear indication from Table 1.5-6 that biotechnology companies with smaller R&D budgets were more efficient in creating new blockbuster brands. [10, 11, 14, 16]. With the large-scale production of antibiotics, vaccines, and insulin during the post-World War II period, the pharmaceutical industry acquired and developed skills to manufacture, market, and supply medicines worldwide and retains its edge over biotechnology companies. An earlier paradigm shift from academic to industrial R&D happened in the 1970–1980 period. Pharmaceutical firms' R&D laboratory scientists won Nobel prizes and made important discoveries leading to new drug classes like ACE inhibitors, proton pump inhibitors, NSAIDs, and immunosuppressive agents for organ transplantation. The recent paradigm shift during the past 5 years has been the transfer of innovative R&D from Europe to the United States in parallel with the shift from pharmaceutical to the biotechnology R&D [25–28]. The strength of the biotechnology industry, university industry relationship, groundbreaking discoveries, National Institutes of Health (NIH), absence of price controls, and rapid uptake of high-price new treatments to avoid potential litigation from patients have attracted R&D investment from European and Japanese companies.

1.5.5.1 R&D Productivity

There was no clear-cut relationship between a higher R&D budget and creation of new blockbuster medicines. If one only goes by the number of blockbuster medicines, the top four companies did not increase the number of such drugs in the year 2005 from 2004. Several companies like Pfizer, Merck, BMS, and Wyeth had decreased numbers, whereas Sanofi Aventis, Astra, Roche, and Amgen had increased numbers. The international R&D expansion activities, research capabilities, and patent output of 65 Japanese pharmaceutical firms from 1980 to 1991 were studied. It was observed that firms benefited from international R&D only when they had in-house existing research capabilities in the underlying technologies [29]. Another study of the Japanese pharmaceutical industry innovative capabilities from 1975 to 1995 found that the unattractive home market pushed these firms into trivial innovations [30]. The global expansion of the Japanese companies and their creation of blockbuster drugs do not support the above conclusion. German majors like Bayer, Schering, and mostly generic Merck failed to create any new blockbuster medicine during the past few years, indicating the decline of the German industry from global to regional levels. Boehringer Ingelheim and mid-sized Atlanta were able to add new billion dollar molecules. The German industry moved strongly in 2006 towards consolidation and mergers. Bayer purchased Schering and Merck made a move into biotechnology by acquiring Swiss Serono. Atlanta was overtaken by Danish Nycomed and another mid-size, Schwartz, was absorbed by Belgian UCB. These moves reflect a desire to consolidate marketing and R&D.

Drews [31, 32] raised concerns about the low R&D productivity gap despite the mergers in the early and late 1990s. These mergers in fact resulted in reduced R&D productivity to produce new approvals and blockbusters. Glaxo Smith Kline of today was formed from Glaxo, Wellcome, Smith Kline French, Beecham, Beckman, Affymatrix, Sterling, and a host of other smaller companies. Similarly Sanofi-Aventis was a merger of the following component companies: Hoechst, Rhone Poulenc, Marion Merell Dow, Roussel, Rorer, Sanofi, Connaught Labs, Merieux, and Synthelabo. Pfizer acquired Pharmacia, Werner Lambert, Parke Davis, Searle, Upjohn, and a host of other smaller companies. This applies to most top pharmaceutical companies except Merck. Mergers have started within the biotechnology industry, like Amgen taking over Immunex and Tularik, and Biogen-Idec and pharma companies like Roche taking over Genentech, Novartis-Chiron, and J&J—Alza, Centocor. Astellas was formed from the merger of Fujisawa and Yamanouchi, and Daichi and Sankyo merged in 2005.

As several competitive pharma R&D units were merged, several projects were terminated or given low priority and funds, and R&D staff was reduced or shifted, resulting in high turnover and low morale. The bigger company centralized R&D units became more risk averse. Several once promising areas of research like combinatorial chemistry produced huge chemical libraries with minor structural variations; HTS, gene therapy, proteomics, antisense, vaccines for AIDS, sepsis, RSV, and malaria, Alzheimer's disease, and Parkinson's disease have increased our knowledge and understanding and the number of targets. However, no breakthrough blockbuster medicine has emerged out of these new high-speed and costly technologies. Demain [33] attributes such failure to the elimination of natural products and extracts from the HTS screening and recommends including natural products in HTS screens. The partnerships between different companies for R&D and marketing alliances/joint ventures to develop products have a higher probability of blockbuster success [17, 34]. Historically, blockbuster drugs were aimed at the larger markets of depression and ulcers. Other large-market indications like cholesterol and lipid lowering agents, antihypertensives, osteoporosis, asthma, and schizophrenia emerged due to the availability of effective and safe drugs (Table 1.5-7). These therapeutic areas have dominated the pharmaceutical markets for the past two decades. Several of these drugs were discovered and initially developed by smaller companies and licensed to big pharma for marketing and late development.

The successful biological products were developed to treat anemia, cancers, infections, and inflammatory joint diseases. The biologic blockbuster medicines have significantly contributed to the recent increase in the number of such drugs and made several previously ignored indications like multiple sclerosis, pulmonary arterial hypertension, anemia, cystic fibrosis, vaccines, and age-related macular degeneration commercially attractive and highly competitive areas for new drug development (Tables 1.5-6, 1.5-7) [1–4, 16, 20, 34–37]. The demand for new and better medications remains strong as currently only about one third of the over 35,000 human diseases have some treatment options available.

Based on new product approvals and creation of blockbuster brands, the success rates of biotech products are higher as compared with traditional chemically derived small-molecule new chemical entities. The biotechnology industry growth and market valuation has outpaced the pharmaceutical industry during the past 2 years. The increasing dependence of big pharma on small biotech is evident by the increasing number of licensing and R&D funding deals. The bulk of new medications approved in the years 2003–2005 originated in small biotech companies [35–39].

The higher R&D productivity of biotech companies is linked to their close relationship with academic biomedical research, higher risk taking by boldly testing first-in-class, and new concepts and designs in clinical trials, which are driven by unmet medical needs even for diseases for which the pharmaceutical industry has shown no interest. The R&D failure rate of biotech companies is much higher as compared with big pharma testing of follow-on or me-too drugs [1–3, 16, 20, 32–35]. For blockbuster drugs, the biotechnology companies have followed the standard pharmaceutical company strategy. They have followed fast-track approval for the first indication in an orphan or niche disease with high unmet medical need, market the drug at high prices, and expand approvals in other related diseases and new markets.

The risk associated with the development of human antibodies and biologics is lower than that with an NCE with a probability of success at 8% in phase I. Chimeric or humanized antibodies have a much higher success rate at the phase I stage (25%), due to recent advances in target selection and recombinant technology. New developments include techniques for converting murine monoclonal antibodies to mouse–human chimeric antibodies, for humanizing the antibodies, and producing human antibodies using transgenic mice. These approaches minimize or eliminate the immune responses and rejection associated with the earlier murine antibodies [37–40]. The strong sales and demand from cancer and arthritis patients have outdated almost all commercial market research reports and sales forecasts for monoclonal antibodies. The higher success rate, low cost of entry and IND filing, and use of human/humanized monoclonal antibodies—with 6 blockbuster brands, half of which arrived in 2005, and the potential for more billion-dollar monoclonal antibodies to reach the market—has fueled the current Monoclonal Antibody Gold Rush. The bandwagon has been joined by 200 companies with hundreds of new projects and targets and has attracted billions of dollars in R&D investment. Factors contributing to the success are unmet medical needs of patients, lower cost of entry into R&D and production, the overcoming of technical hurdles in large scale manufacturing, and availability of humanized or fully human monoclonal antibodies.

The R&D productivity of the big pharma with biotech companies has been compared for the period 1998–2004 in commercial reports [1–3]. The R&D expenses and the FDA approvals of the new molecular entities were linked [16, 20, 23, 24]. The biotech approvals were less than 10, and the pharma NCE approvals declined from 25 to 15 from 1998 to 2005. The decline of the first-in-class medications from big pharma was even more dramatic. The biotech approvals were more numerous than pharma approvals in 2003 and 2004, and the trend reversed in 2005. The big pharma R&D expenses were $50 billion in 2004, resulting in 10 NME approvals, whereas biotech approvals were 20 for an R&D budget of $20 billion. Tables 1.5-6 and 1.5-7 list the top-selling major therapeutic proteins and biological brands created by the biotechnology industry since 1997. The list provides therapeutic areas of blockbuster brands, marketing wars between different brands for market leadership, and shift to newer brands with improved properties.

1.5.6   R&D SUCCESS

The biotechnology industry had its share of major R&D and clinical trial success. The big gains for biotech products are mainly in anemia; cancer; TNF alpha inhibitors; human insulin with short, mixed, and long duration of action; interferon for multiple sclerosis and hepatitis C; monoclonal antibodies for cancer; RSV; and immuno-inflammatory diseases. This success has resulted in attracting top pharmaceutical companies and R&D funds for biosimilar products with improved properties. It is easier for start-up companies in the above areas to attract capital and angel investors or to conclude licensing deals and find marketing partners.

As EPO, insulin, interferon, and monoclonal antibodies are covered in other chapters, only a brief resume will follow of these major success stories of biotechnology R&D.

1.5.6.1   Erythropoietin

Erythropoietin, Lipitor, and TNF inhibitors are likely to cross the $15 billion mark within the next few years. Amgen commercial success with its EPO brands has attracted several competitors with which Amgen has been involved in costly and protracted patent and marketing litigation to protect its franchise. Erythropoietin is the top litigated biotech product. In 2005, over 20 erythropoietins were in development either as follow-on or improved biosimilar products [20, 41]. Numerous players are striving to dominate this market, including, among others, Amgen, J&J, Roche, Chugai, Shire, Kirin and Elanex, Teva, Sandoz, Stada, Cangene, Microbix and Gene Medix. Amgen patents expired in Europe in 2004 but are valid in the United States until 2015, and it has shifted patients to its new Aranesp brand. In Europe, two forms of rEPO are available: Eprex (J&J) and NeoRecormon (Roche). Two other forms Dynepo (Shire, Sanofi Aventis) and CERA (Roche, PEG-EPO) are likely to be approved in 2007. Amgen has filed a preemptive patent infringement against Roche for its PEG-EPO (CERA), thereby delaying its market launch in the United States. The FDA's recent alert about erythropoietin (EPO) use in cancer patients with over 12 gm/dL of hemoglobin and the black box warning linking EPO agents to increased mortality, cardiovascular events, tumor growth, and hyptertension are likely to result in market and sales decline in 2007. The FDA advised doctors to use the lowest dose of EPO for patients to achieve hemoglobin levels of 12 gm/dL.

1.5.6.2 Tumor Necrosis Factor Inhibitors

The commercial success of antitumor necrosis factor monoclonal antibodies like Remicade (Infliximab), Humira (Adalimumab), and Enbrel (Etanercept) in rheumatoid arthritis is due to better tolerance and superior efficacy as compared with older disease modifying antirheumatic drugs (DMARDs) with variable and short response rates but higher toxicity leading to either their low clinical use or market withdrawal (Table 1.5-6) [42–44]. TNF inhibitors with methotrexate provide a response rate of 50–60% by the FDA, American and European Rheumatology remission criteria. Risks associated with TNF-blockers are serious infections, congestive heart failure, demyelinating diseases, and systemic lupus erythematosus, but in most cases, they can be identified and managed. Biological agents have revolutionized the treatment of autoimmune/inflammatory diseases like rheumatoid arthritis and Crohn's Disease (CD). Enbrel was the first TNF inhibitors approved for treatment of rheumatoid arthritis patients in 1998. The FDA (fast track) initially approved Remicade for Crohn's Disease the same year. Both agents were tried and failed to show any efficacy in early trials in cancer, sepsis, and congestive heart failure. Both Enbrel and Remicade are approved for rheumatoid arthritis (RA), psoriatic arthritis (PsA), and ankylosing spondylitis (AS). There is some differentiation as Enbrel is effective in juvenile rheumatoid arthritis (JRA) and psoriasis (Ps). Remicade is approved for CD and ulcerative colitis (UC) and has shown efficacy in sarcoidosis and Wagener's granulomatosis. The third agent, Humira, is approved for use in RA, PsA, AS and CD.

After the launch of Enbrel, the demand for patients during the first 3 years of marketing was so strong that Immunex and its marketing partner Wyeth (American Home Product) were unable to increase good manufacturing practice (GMP) production to meet the surge in demand. There was a 30% shortfall, and these patients drifted to Remicade. Only Amgen's takeover of Immunex in 2002 resolved the supply problems to meet demand. Remicade, with its approval in UC and CD, was the market leader and overtook Enbrel in sales in the 2002–2003 period. Enbrel has been the current market leader in its class since 2004. Kineret (anakinra, Amgen) is a recombinant, nonglycosylated synthetic form of the human interleukin-1 receptor antagonist (IL-1Ra) that mimics the body's endogenous mechanism for regulating IL-1 and is approved for treatment of RA patients not responding to DMARDs and TNF inhibitors. Kineret (anakinra, Amgen) has failed to gain a significant market share of the RA patients because it cannot be used in combination with TNF inhibitors. Orencia (Bristol-Myers Squibb abatacept) is the first T-cell costimulation modulator approved for the treatment of RA and was marketed in early 2006. Orencia in combination with TNF inhibitors or other biological agents like Anakirna is not allowed. Roche Rituxan, a monoclonal B-cell-specific antibody to CD20 (MabThera), was approved for use in RA patients along with methotrexate.

Recent R&D failures include Biogen Idec, Lilly-ICOS, and Alexion products in clinical trials. The Roche monoclonal antibody to the IL-6 receptor for RA (MRA) has shown efficacy in RA patients not responding to methotrexate and TNF inhibitors. If long-term safety, efficacy, and remission of new biologics are established in osteoarthritis trials, it will be an important therapeutic advance and create a large market.

1.5.6.3 Psoriasis

Alefacept (Amevive, Biogen Idec) and Efalizumab (Reptiva, Xoma, Genentech, Serono), the two monoclonal antibodies, which block the activation of T cells, were hailed as major advances in the treatment of psoriasis. The sales of Reptiva were $112 million and $160 million and of Amevive were $48 and $15 million in 2005 and 2006 below initial sales forecasts mainly due to competition from TNF inhibitors [45]. The technical success of R&D of non TNF biologicals for treatment of psoriasis has not translated into commercial success due to the large clinical data base and dominance of TNF antagonist products.

1.5.6.4 Insulin

A review of the insulin blockbuster brands in Table 1.5-10 indicated the market entry of a new company Sanofi Aventis to challenge the market dominance by Lilly and Novo Nordisk and the rapid growth of the newer insulin analogs with improved properties. The aim of the R&D is to move away from several daily insulin injections. There has been intense R&D activity during the past few years to provide patients with improved insulin that is fast acting, of a long duration, and dual action. This approval and launch of the first inhaled insulin Exubera (Pfizer) should expand the market, and initial peak sales projections are $3 billion per year depending on its long-term effects on lung functions. The UK cost effectiveness agency has determined that the high costs of Exubera are not justified as it does not offer any advantages over insulin injections, and concerns about long term toxicity to lung cells has delayed marketing by Pfizer in USA despite FDA approval. Table 1.5-8 lists new improved marketed and approved insulin analogs and those in development [46–48].

TABLE 1.5-10. Improved Insulin Analogs

Company	Product	Description	2006 Sales Status $ Million
Sanofi Aventis	Lantus Apidra	Insulin glargine Insulin glulisine	2200
Novo Nordisk	Insulin Aspart Levemir	Rapid insulin Insulin detemir	1700
Lilly/Amylin	Byetta	Exenatide incretin mimetics	430
Pfizer	Exubera	Inhaled insulin powder	Approved
Lilly/Alkermes	AIR Insulin	Insulin powder	Phase III
Mannkind	Technosphere Insulin	Powder liquifies in lungs	Phase III
Novo Nordisk Aradigm	NN 1998	Aerosol Mist	Phase III
Kos	KI 02 212	Liquid suspension	Phase II
Novo Nordisk	NN 344 NN 5401	Long-acting insulin, improved analog	Phase I

1.5.6.5 Interferon

Interferon alpha in combination with ribavarin for the treatment of hepatitis C is a commercial success. The biotech industry had its once promising drugs failing in clinical trials, and drug withdrawal by Biogen and Elan of Tysbari (natalizumab) for multiple sclerosis due to deaths of some patients from a rare brain infection was a setback. Multiple sclerosis (MS) is a neurological disorder affecting 2 million patients worldwide. Teva/Sanofi-Aventis Copaxone joined the three interferon blockbuster brands for the multiple sclerosis market led by Avonex in 2005. The total MS market sales reached $7 billion in 2006. Copaxone and return to market of Tysbari will expand the market and replace older interferon brands. Interferon sales slowed in 2005 after 2 years of rapid growth due to the new use of a poly ethylene glycol (PEG)–interferon combination with ribavarin in hepatitis C and withdrawal of Tysabri.

Biologics are important for new emerging threats, bioterrorism, and infectious diseases such as pandemic influenza, West Nile Virus (WNV), SARS, antibiotic-resistant tuberculosis, and the prion agent that causes the human form of mad cow disease, variant CJD. New vaccines to protect against HIV, RSV, anthrax, small-pox, hepatitis, malaria, WNV, and bird (avian) flu are in clinical testing [49–51]. Monoclonal antibodies and peptides have made a great contribution to the treatment of cancer patients, and several antibodies and peptides have become block-buster drugs. Many of these antibodies carry black box warnings. Several new anticancer agent brands are blockbuster drugs, and the list grows each year [52–54].

The advent of Macugen and Lucentis to treat age-related macular degeneration and endothelin antagonists (Tracleer, Thelin, Ambrisentan) and prostacyclin analogs (Epoprostenol, Treprostinil, Beraprost, Iloprost) to treat pulmonary arterial hypertension are important treatment advances. Both of these therapeutic areas have several active R&D projects at the approval and the clinical trial stage.

Roche has been the most efficient and productive industrial R&D during the past 2 years. The association of Roche with 5 of the top 10 biotech products is a tribute to its success in acquiring Genentech, an investment in biotechnology and its alliance with biotech companies. Amgen and Genentech market cap was in the $90–$100 billion range in early 2006 but has come down recently, and was still higher than Wyeth, BMS, or Schering Plough. Other big companies like J&J and Lilly have a strong pipeline of biological products in development.

1.5.7   R&D FAILURES

The biotechnology industry had its share of major R&D and clinical trial failures. The big late-stage setbacks for biotech products are mainly in Alzheimer's disease, sepsis, and vaccines for RSV, malaria, and AIDS. Despite such failures, the R&D funding by big pharma continues in Alzheimer's and sepsis but has bailed out in vaccines and infectious diseases. The long history of failures has made it difficult for a biotechnology company working in the above areas to attract venture capital and angel investors or to conclude licensing deals or find marketing partners.

1.5.7.1   Alzheimer's disease

Alzheimer's disease (AD), the most common cause of cognitive impairment in older patients, is characterized by the development of senile plaques and neurofibrillary tangles, which are associated with neuronal loss affecting to a greater extent cholinergic neurons. A cascade of pathophysiological events is triggered in AD that ultimately involves common cellular signaling pathways and leads to cellular and neural networks disfunction, failure of neurotransmission, cell death, and a common clinical outcome. The viable neurons remain an important target for therapeutic intervention at each stage of disease evolution. Currently, symptomatic drugs inhibiting the degradation of acetylcholine within synapses and more recently glutamate receptor antagonists represent the mainstay of therapy. However, interventions able to halt or slow disease progression (i.e., disease-modifying agents) are necessary. Although much progress has been made in this area, there are currently no clinically approved interventions for AD classed as disease-modifying or neuroprotective drugs [55].

The first four approved and marketed drugs Aricept (Pfizer, Eisai, donepezil), Reminyl (J&J, galantamine), Exelon (Novartis, Rivastigmine), and Cognex (Pfizer, Tacrine) are reversible cholinesterase inhibitors [55]. Memantine is an *N*-methyl-D-aspartate (NMDA)-type glutamate receptors antagonist, which is the first in a new class of AD medications. Glutamate is the most common excitatory neurotransmitter in the central nervous system, and the modulation of glutamatergic neurotransmission is a major target for the treatment of Alzheimer's disease. Memantine was developed by Merz and licensed to Forest for the U.S. and Lundbeck for the European and international markets. Memantine is marketed under the brands Axura and Akatinol by Merz, Namenda by Forest, and Ebixa by Lundbeck. Clinical studies of the safety and efficacy of memantine for other neurological disorders, including glaucoma and other forms of dementia, are currently ongoing [56]. A series of second-generation memantine derivatives are currently in development and may have better neuroprotective properties than memantine. Recent failures have included Axonyx, Forest Labs, Elan, Myriad, and Neurocrine. Elan and Wyeth (American Home Products) ended trials of their Alzheimer's vaccine (AN-1792) in 2002, after patients suffered swelling of the central nervous system or inflammation of the brain and spinal cord. The vaccine activated the patient's own immune system to attack a protein called beta-amyloid that is implicated in Alzheimer's. Several tested drugs provided good results in phase II but failed in phase III. Some drugs showed a biphasic response curve. Several muscarinic receptors agonists failed in phase II–III trials.

1.5.7.2   Anti-infective Exodus and U-turn

During the 2001–2003 period, several companies, namely Aventis, Bristol Myers Squibb, Eli Lilly, Roche, GlaxoSmithKline, Proctor & Gamble, and Wyeth, eliminated or greatly curtailed their anti-infective R&D [57]. The reasons for pharmaceutical companies leaving the anti-infective area were slow growth in the West, unpredictable and seasonal demand, pricing pressure to donate supply or at cost to poor countries and patients, the vaccine unprofitability, and low investment in vaccine companies. Vaccine companies were scared and hesitated to launch new vaccines with sales in the tens to hundreds of millions of dollars because of the threat of billion-dollar punitive damage settlements. The liability issues and current regulatory requirements for new vaccines have a negative impact on vaccine R&D. Public–private partnerships, along with NGO and public initiatives, have resulted in increased funding of newer vaccines for tropical, orphan, and neglected diseases like malaria, tuberculosis, and leishmaniasis.

Vaccines were back in favor and were highlighted in R&D portfolios of major companies (Sanofi Aventis, Glaxo Smith Kline, Merck, Novartis, and Wyeth). Global sales in 2006 were $15 billion because of scare about avian influenza, bioterror organisms, and new emerging infections like SARS. Introduction of new cancer and rotavirus vaccines will greatly expand the vaccine market in the next 2 years. During the past few years, the spotlight on SARS, Avian Influenza, new emerging infections, and bioterror agents have seen a few big companies like GSK, Merck, Novartis, and Sanofi Aventis declaring vaccine production as a high priority area of R&D. It is ironic that Roche, which had closed its anti-infectives R&D unit, was in the news to supply its antiviral Tamiflu to stop Avian Influenza. The success of Prevenar (Wyeth), the explosive demand for influenza vaccines, and the approval of two new Rotavirus vaccines with huge safety databases have brought back the interest of the industry in vaccines. Pediatric vaccines, influenza, and hepatitis B vaccines have been the leading categories. Although the demand for influenza vaccine was strong, MedImmune FluMist 2005 sales of $21 million remained below forecast, due to high cost and its use limited to younger patients (5 to 17 years of age, and healthy adults, 18 to 49 years of age). A new refrigerator-stable formulation was 55% more effective in children in the age range of 6 and 59 months. New vaccines for cancer, RSV, AIDS, and malaria will expand the market [57–61]. Vaccines have been the most cost-effective treatment and probably were undervalued, but the higher prices of newer vaccines in the $100–$400 range are likely to keep even these out of the reach of poor patients [49–51].

1.5.7.3   RSV

The commercial success of a new product in most indications or additional use in new indications results in much higher R&D activity and funding by competitors. However, the success of Synagis (Palivizumab) in RSV and Drotrecogin alpha (activated protein C) in sepsis has resulted in industry bail-out due to the failure of several promising projects during the past decades [57, 62, 63]. The new drugs in development and currently approved drugs for RSV were reviewed recently and are listed in an updated Table 1.5-11 [57], which confirms a significant decline from over 25 to only 4 active RSV projects. Several RSV vaccines and antivirals, which were in phase III testing during the last review, are considered terminated either due to low efficacy or safety concerns due to adverse effects. These projects are not listed in the companies' current R&D pipeline, but many commercial databases still list these as active projects.

1.5.7.4   Sepsis

Severe sepsis or septic shock occurs when an infection (bacterial, viral, fungal, or parasitic)—often the result of trauma, surgery, burns, or cancer—triggers a cascade of immune system responses that can lead to acute organ disfunction and often death [62]. The death rate from severe sepsis ranges from 30% to 50%. Every day 1400 people worldwide die from severe sepsis. The International Sepsis Forum estimates that more than 750,000 individuals develop severe sepsis in North America each year, with similar estimates for Europe, and all need to be actively treated in the hospital. Sepsis kills 215,000 people annually in the United States and is the most common cause of death in noncoronary intensive care units. The annual costs associated with the treatment of patients with severe sepsis are estimated to be US$17 billion. The incidence of sepsis is expected to increase to 2.2 million per annum in the leading markets over the next 10 years as the elderly population grows. Thus, a safe and effective treatment, which saves lives, can be a blockbuster drug [62, 63].

Activated protein C (Drotrecogin alpha, Xigris) is the first and only drug shown to cure those suffering from sepsis. It had sales of $214 million and $192 million

TABLE 1.5-11. Marketed and R&D Development Products for Prevention of RSV

Product	Characteristics	Company	Sales 2006 Status
Palivizumab Synagis	hAnti F-glycoprotein	MedImmune	$1100 million
RespiGam	Immunoglobulin	MedImmune	negligible
Ribavarin	Antiviral	Valeant	Generics
Motavizumab Numax	Monoclonal antibody	Medimmune	Phase III
A 60444	Antiviral	Arrow	II
Live attenuated	RB/HPIV3-RSV-A/B	MedImmune	I
ALN RSV01	RNAi	Alnylam	I

Notes: Several commercial databases still list RSV vaccines and antiviral in clinical development by companies, e.g., Wyeth, Sanofi Aventis, and Glaxo Smith Kline. These projects are considered discontinued as companies' websites do not list any active RSV projects or there has been no new information for the past 3 years after completion of phase II or III trials.

in 2005 and 2006, respectively. In Lilly's PROWESS clinical trial, drotrecogin alpha only reduced the absolute risk of death by 6%. It seems that, with a $6800 price tag, this was insufficient to persuade physicians to prescribe the drug outside the most severe cases. In 3.5% of drotrecogin alpha-treated patients (compared with 2% of placebo treated patients), the drug also increased the risk of serious bleeding events. Another obstacle to drotrecogin alpha's widespread adoption is that it can be hard to identify candidates for treatment [62–64].

Drug development for sepsis and septic shock has been disappointing: Numerous agents have demonstrated promising activity in early clinical phase I/II and then failed to show significant efficacy in phase III trials. In recent years, several biotechnology firms have gone bankrupt or were taken over trying to develop a drug to combat sepsis. Billions of dollars have been wiped out in market values of the biotechnology companies due to failures of their drugs in sepsis.

In the early 1990s, two promising drugs named Centotoxin and E5 from Centocor and Xoma were in a race to be the first effective treatment. The two drugs had shown efficacy in early phase II clinical studies and were rushed into large-scale phase III clinical studies. Both drugs failed to show improvements in mortality in phase III and were rejected by the FDA and its advisory committee in 1992 [62].

Randomized trials of anti-inflammatory therapies agents, including NSAIDs and corticosteroids, failed to show improvement in survival from sepsis and septic shock. Three monoclonal antibodies to endotoxin were tested in clinical trials, and all failed to improve survival from sepsis and septic shock. The Centocor project went down in 1992, and drugs from Chiron, Xoma, and Synergen failed in 1994. Many anticytokine therapies have not improved survival from septic shock. These trials include the use of naturally occurring antagonists to tumor necrosis factor and interleukin-1, antibodies to tumor necrosis factor, bradykinin antagonists, ibuprofen, and a platelet-activating-factor antagonist. Additional endotoxin-directed therapies being studied were the use of bactericidal permeability-increasing factor, soluble CD14 receptors, and reconstituted high-density lipoproteins [62].

Anti-inflammatory interventions under study include efforts to modulate nitric-oxide activity, therapies directed at neutrophil adhesion molecules, the infusion of antithrombin III, tissue-factor pathway inhibitors, and pentoxyfylline. Alternatively, GCSF is being studied in an effort to enhance phagocytic-cell function in patients with severe sepsis [62]. During the past 5 years, Chiron and Pharmacia's tifacogin failed late-stage clinical trials at the end of 2001 and endotoxin from Baxter/Xoma (Neuprex) has failed to show clear benefits in large-scale studies and requires extensive additional studies (2002). ICOS and Suntory, Pafase, a platelet-activating factor acetylhydrolase; Lilly's LY-315920 phospholipase A2 inhibitor; and Knoll-Abbott's monoclonal murine anti-tumor necrosis factor-alpha antibody, Afelimomab, are some recent phase III trial failures.

Despite past failures, there are still several active R&D projects in late-stage clinical trials (Box 1.5-2).

1.5.8   COST CONSTRAINTS

The start-up cost of development, manufacture, and distribution of biological products is very high, and only a few companies can manufacture and test the product to current regulatory standards of GMP, GLP, and GCP (Good Manufacturing,

---

BOX 1.5-2

Phase III

- Eisai Eritroran, an antagonist of lipopolysaccharide
- Takeda TAK 242, which inhibits signal transduction through Toll Like receptor 4 Phase
- Prosthetics' CytoTAb target anti-tumor necrosis factor alpha polyclonal antibody (licensed in late 2005 to Astra Zeneca)

Phase II

- Novartis (PMX 622) covalent conjugate of Polymyxin B and Dextran-70
- GSK 270773 phospholipid anti-endotoxin emulsion
- Behring Complement CI inhibitor

Phase I

- Idun (acquired by Pfizer in 2005) IDN 6556 caspase inhibitor

---

TABLE 1.5-12. High-priced Biologic Medications 2005–2006

Company	Trade Name	Generic Name	Indication	Price per Month $
Biogen Idec	Zevalin®	Ibritumomab	Lymphoma NHL	24,000
BMS	Erbitux®	Cetuximab	Colon Cancer	17,000
Genzyme	Fabrazyme®	Agalsidase Beta	Fabry Disease	15,000
Genzyme	Aldurazyme®	Laronidase	MPS 1	16,600
Genzyme	Cerezyme®	Imiglucerase	Gaucher disease	16,600
Serono	Serostim	HGH	Dwarfism	7,000
Lilly	Xigris®	Activated Protein C	Sepsis	6,800
MedImmune	Synagis® J	Palivizumab	RSV	5,600
Genentech	Avastin®	Bevacizumab	Colon Cancer	4,400
Roche	Herceptin	Trastuzumab	Breast Cancer	

Laboratory, and Clinical Practice). The biologics for cancer and rare and other life-threatening diseases also rank as the most expensive treatments, and these products achieve blockbuster sales sooner than traditional pharmaceuticals (Table 1.5-12, Fig. 1.5-1). The price of biosimilar or follow-on products will be higher as compared with traditional generic drugs as very few companies can invest in manufacturing technology.

With the introduction of biologics for the treatment of rheumatoid arthritis and psoriasis, the annual cost of treatment has multiplied 10-fold to $25,000. Only about 10% of the eligible RA patients in the United States and Europe are currently on TNFα therapy. If all RA patients were to receive TNFα therapy, the total cost of therapy would reach $100 billion just for one disease only. The current or future health-care systems are not likely to fund such cost increases for anemia, cancers, infectious, cardiac, metabolic, neurological, and rare diseases as well as increasing acute care, diagnostic, and device costs [65, 66].

Treatment of psoriasis with new biologics costs in the range of $25,000–$45,000 per year as compared with $2200 for methotrexate and $3000–$5000 per year for phototherapy. The cost of erythropoietin for renal dialysis is $10,000 per year and for cancer patients $1000 per month of treatment. New vaccines are priced in the $200–$400 range. Xigris is priced at $6800 per injection and drug-coated stents at over $3000. The European government-funded healthcare systems and insurance companies try to delay and limit access to costly new medicines and try to negotiate lower prices.

The average price of new cancer treatments and for rare diseases has increased from the $20,000–$25,000 range to the $100,000–$280,000 range per year of treatment [51–54, 66]. If two or three new drugs were to be combined for treatment of colon cancer like Eloxatin, Erbitux, and Avastin, the yearly cost could easily reach $0.5–1 million per patient. If Erbitux was used for all eligible colon cancer patients in the United States, it would cost $1.2 billion. Several European health insurance systems, oncologists, and patient organizations have raised questions about the cost–benefit ratio and called for regulatory rejection of anticancer medications with short-term limited benefits. In the future, about 80% of cancer patients will be from poor countries or without insurance coverage. Colon cancer survival doubled in the past decade, but the costs multiplied by 340 for the first 2 months of treatment [66].

The clinical testing of these high-priced drugs in poor and uneducated patients in the developing countries is problematic due to ethical and moral grounds. These drugs are not likely to be marketed in poor countries or made available to responding patients at the end of the trial. The high cost of these life-saving medicines raises several critical issues concerning access, payment, and the legal, moral, and ethical aspects of supply and demand to poor countries and patients [67, 68].

Initial reluctance and resistance by industry to reduce the prices of AIDS medications for poor countries/patients, let aside patent protection to increase supplies, or reduce prices in national emergencies like the Anthrax scare in the United States (Bayer Ciprofloxacin), SARS, and Avian Influenza (Roche Tamiflu) has eroded the image of the research-based innovative pharmaceutical industry.

Biotechnology products have so far escaped pricing pressure to reduce costs as the treatments were for life-saving or rare diseases and were marketed by big pharmaceutical firms. As biological medications prove effective in treating chronic diseases like arthritis, psoriasis, macular diseases, and diabetes, big biotechnology firms will come under pressure to reduce costs and face class-action litigation for product injury. Biotechnology industry leaders should find an appropriate balance between the current "high cost high profits, low volume" model to "affordable cost, high volume, reasonable profits" model to increase the availability of life-saving medications to poor patients and countries.

1.5.9 PUBLIC–PRIVATE PARTNERSHIPS

Several public–private partnerships support development of drugs, vaccines, and diagnostics to address diseases that predominantly afflict the poor or to make drugs available to patients in poor countries and operate on the virtual company model [57, 69, 70]. As most pharmaceutical and biotech R&D has been directed toward diseases of the Western World, there has been no interest in developing vaccines and treatments for the tropical and neglected diseases like AIDS, Chagas' Disease, contraception, dengue, diarrhea, leprosy, leishmaniasis, lymphatic filariasis, malaria, onchocerciasis, schistosomiasis, trypanosomiasis, tuberculosis, and sexually transmitted diseases. None of the major companies has an R&D laboratory in Africa, and only two, Glaxo Smith Kline and Astra Zeneca, have R&D laboratories in Asia (outside Japan). International and national organizations like the World Health Organization (WHO) failed to develop new medications for tropical diseases and lacked sufficient resources. Universities or research institutes were not able to attract interest from the industry for promising leads, and companies terminated their own R&D leads because of lack of profits or commercial success.

Developments in the area of neglected diseases have opened up new opportunities for licensing by universities and companies. Over the past decade, private foundations and donors have provided social venture capital to launch several nonprofit "companies" that have now collectively raised over a billion dollars from philanthropic and government donors to support product development.

The price of AIDS drug tritherapy is $10,000 a year per patient in the United States and Europe. Indian companies have reduced the cost to $130 per patient per year and introduced a single tritherapy combination. When Roche was initially reluctant to increase production or reduce the price of its Tamiflu to treat Avian Influenza for developing countries claiming a difficult synthetic process, some Indian companies made the active ingredient from Shikimic acid within one week

The Foundation for the National Institutes of Health identifies and develops opportunities for innovative public–private partnerships involving industry, academia, and the philanthropic community (www.fnih.org). There were over 100 public–private partnerships funded with over $1 billion, with 60 active R&D projects for vaccines and drugs for tropical and neglected diseases by the end of 2005 (Table 1.5-13).

TABLE 1.5-13. Public–Private Partnership for Tropical and Neglected Diseases

Aeras, Global TB Vaccine Foundation (Aeras) (http://aeras.org)
Bill and Melinda Gates Foundation (www.gatesfoundation.org/GlobalHealth/InfectiousDiseases/)
Drugs for Neglected Diseases Initiative (www.dndi.org)
Foundation for Innovative New Diagnostics http://www.finddiagnostics.org
Global Alliance for TB Drug Development (www.tballiance.org)
Global Alliance for Vaccines and Immunization GAVI at WHO (www.who.int)
Global Forum for Health Research (www.globalforumhealth.org)
Global Fund to fight AIDS, Tuberculosis and Malaria (www.theglobalfund.org)
Institute for One World Health http://www.oneworldhealth.org
International AIDS Vaccine Initiative (www.iavi.org)
International Partnership for Microbicides http://www.ipm-microbicides.org
Malaria Vaccine Initiative (MVI). (http://www.malariavaccine.org)
Medicine for Malaria Venture (www.mmv.org).
Pediatric Dengue Vaccine Initiative http://www.pdvi.org
Program for Appropriate Technology in Health http://www.path.org

1.5.10 INFORMATION RESOURCES

Traditional and new media play a vital role in projecting breakthrough and block-buster medications and raising awareness among the general population (Table 1.5-14). The FDA/EMEA website provides detailed summaries and assessments of the actual data submitted in support of the NDAs. The transcripts of the FDA Advisory Committee meetings are valuable sources of information for approved

TABLE 1.5-14. Web Information Resources for Biotechnology Products

FDA	http://www.fda.gov
EMEA	http://www.emea.eu.int
WHO	http://www.who.int/en/
Canada	http://www.hc-sc.gc.ca/
UK	http://www.mhra.gov.uk/
NICE	http://www.nice.org.uk/
NIH	http://www.nih.gov/
CDC	http://www.cdc.gov
BIO	http://www.bio.org/
EuropaBio	http://www.europabio.org/
PHRMA	http://www.phrma.org/
IFPMA	http://www.phrma.org/
GPHA	http://www.gphaonline.org/
DIA	http://www.diahome.org/
IMS	http://www.ims-global.com/
DataMonitor	http://www.datamonitor.com/
Prous	http://www.prous.com/
ScienceDirect	http://www.sciencedirect.com/
Thompson	http://www.thompsonpharma.com/
PharmaProjects	http://www.pjbpubs.com/pharmaprojects/
BioCentury	http://www.biocentury.com/
BioSpace	http://www.biospace.com/
Mayo	http://www.mayoclinic.com/
CenterWatch	http://www.centerwatch.com/
ClinTrialsGov	http://www.clinicaltrials.gov/
Cochrane	http://www.cochrane.org/index0.htm

http://www.blackwell-synergy.com
http://www.springerlink.com/
http://www3.interscience.wiley.com/
http://www.biomedcentral.com/
http://gateway-di.ovid.com/
http://www.isinet.com/
http://www.centerwatch.com/

Accenture http://www.accenture.com/
Bains, http://www.bain.com/bainweb/home.asp
Boston Consultancy Group http://www.bcg.com/
IBM. http://www.ibm.com/us/
Frost & Sullivan. http://www.frost.com/prod/servlet/frost-home.pag
Reuters. http://www.reutershealth.com/en/index.html
Ernst & Young http://www.ey.com/global/content.nsf/International/Home

and new drugs under review, concerns about safety and efficacy and about prescribing information, and requirements for additional studies (http://www.fda.gov/foi/electrr.htm). Similar information is available on the Internet for public scrutiny, by the FDA/EMEA for the accepted IND files of new medications. The NIH (NIAID), Centers for Disease Control, Institute of Medicine, Mayo Clinic, Center Watch, WHO, and Public Citizen websites offer valuable information about the disease and its treatment and the list of ongoing trials. Company websites offer prescription information and other details about the currently approved drugs, clinical trials results, sales data, and R&D portfolio, patent litigation, and expenses. The commercial websites of Pharma Projects, Prous Science, Thompson, and BioCentury are other important sources for drug R&D. BiomedCentral and PubMed offer free access to many full text papers and abstracts. The ISI web of Science, Science Direct and BiomedCentral, and PubMed were consulted for reviews. To keep the number of references reasonable, original research papers have often been replaced by very recent reviews. Cochrane Reviews is a good source of new drug clinical safety and efficacy. For commercial reports, only information available in the public domain was used and cited. Negative media stories about high prices, profitability, quality/ethical problems, and links to serious adverse reactions may lead to declining sales or regulatory action.

1.5.11   INDUSTRY AND MEDICINAL BRANDS IN 2006

Most of the top companies had reported 2006 annual earnings and sales of top selling brands within the first two months of the new year. The Japanese companies had reported annual sales or forecasts based on actual results of 1–3 Q 2006, which was used. The new data was incorporated in the tables and text at the last-minute stage.

A review of the 2006 sales, market, earnings, and R&D showed several new trends and surprises in the top ten company listing as well as best selling medicinal brands [71,72]. Pfizer, which has been the most valuable company during the past five years, was overtaken by Johnson & Johnson in sales last year and briefly for a few months in market value but has now regained its top position in market valuation. Pfizer retained the top spot as the most profitable and with the highest R&D budget. Glaxo Smith Kline (GSK) retained the third position by total sales, market cap, and profitability, and topped the list with highest sales increase in 2006. Roche consolidated its return to the top ten groups by overtaking Novartis in market cap and closing the gap with Amgen in sales of biologics. Roche and Sanofi Aventis ranked second and third by sales of biologics as in 2005. Overall, increased R&D spending by the pharmaceutical and biotechnology industry failed to increase the number of new drugs approved. Only Merck and Amgen increased their R&D budget in 2006 by $1 billion over previous year. Roche, with no blockbuster going off patent in 2006, came out as the most efficient R&D company. Sanofi Aventis, despite loss of 3 blockbuster drugs to generic competition, came out second with creation and growth of several blockbuster brands. Merck and BMS ranked in the third position, while Takeda and Astra Zeneca took the fourth and fifth positions (Tables 1.5.8, 1.5.15) [71,72].

TABLE 1.5-15. Pharmaceutical Biotechnology Industry Performance Rankings 2006

Sales Increase $ Billion		Biologics Sales $ Billion		Profits $ Billion		Market Cap $ billion 7 March 2007		R&D Efficiency*
GSK	9	Amgen	14.3	Pfizer	19.3	J&J	179	Roche
Roche	5.6	Roche	13.5	J&J	14.5	Pfizer	176	Sanofi Aventis
Novartis	5	Sanofi Aventis	8.7	GSK	14.43	GSK	151	BMS, Merck
Sanofi Aventis	4.8	Novo Nordisk	7.25	Sanofi Aventis	12.6	Roche (Genentech 86)	142	Takeda
Teva	3.7	J&J	6.2	Astra Zeneca	8.54	Novartis	128	Astra Zeneca
J&J	3.3	Baxter	4.4	Roche	7.34	Sanofi Aventis	114	Abbott
Wyeth	2.4	Allergan	3.5	Novartis	7.20	Merck	96	J&J, Teva
Novo Nordisk	2.2	Wyeth	3.46	Merck	6.22	Astra Zeneca	82	Daiichi Sankyo Pfizer
Amgen	1.9	Genzyme	3.2	Takeda	5.4	Abbott	81	Genzyme, Merck-Serono,
Lilly	1.6	Merck Serono	3.2	Lilly	3.42	Amgen	72	Wyeth Lilly
Schering Plough								

*R&D productivity is linked to the introduction of new blockbuster medicinal brands and maintenance or sales growth of existing billion dollar brands and R&D resources shift to biologics and monoclonal antibodies (Maggon K. 2007).

TABLE 1.5-16. Top Selling Medicinal Brands in 2006 ($ billion)

NCE		NME		Sales Increase		Sales Decline	
Lipitor	12.9	Rituxan	4.7	Herceptin	1.59	Zocor	1.6
Advair	6.13	Enbrel	4.4	Rituxan	1.5	Zithromax	1.38
Plavix	5.55	Remicade	4.2	Copaxone	1.2	Zoloft	1.14
Nexium	5.2	Aranesp	4.12	Avastin	1.1	Paxil	1.05
Norvasc	4.8	Procrit	3.2	Tamiflu	0.9	Pravachol	1.0
Zyprexa	4.36	Herceptin	3.14	Aranesp	0.82	Allegra	0.87
Diovan	4.22	Neulasta	2.7	Lantus	0.80	Plavix	0.6
Risperdal	4.18	Epogen	2.5	Enbrel	0.70	Protonix	0.6
Pravachol	4.0	Novulin	2.5	Lipitor	0.70	Losec	0.3
Lansoprazole	3.8	Factor VIII	2.5	Remicade	0.70	Amaryl	0.29

NCE-New chemical entity; NME-New molecular entity

Several pharmaceutical companies highlighted their R&D in biologic, deals and links with biotechnology companies, and sales of biologic brands. Roche is likely to overtake Amgen in biologic sales in 2007 to emerge as the top biotechnology company. One of Roche's unit, Genentech, was the most valued biotechnology company in 2006. The big pharmaceutical companies made significant strides in biologic sales and Roche, Sanofi Aventis, Merck-Serono, J&J, and Wyeth made the list of the top ten biotechnology companies, leaving only 5 pure biotech companies in the top ten [71,72]. Teva, the world's top generic company, took the fourth position in increased net sales (Table 1.5.15).

Lipitor remains the top selling medicinal brand with $12.9 billion in sales, followed by Advair and Plavix. Rituxan, with sales of $4.7 billion, was the best selling biologic brand followed by Enbrel, Remicade, and Darbepoetin. Medicinal brands with the highest sales increase in 2006 were mostly biologic and all the brands with highest sales decline due to patent expiry and introduction of generics were synthetic chemicals. Anticancer, tumor necrosis factor inhibitors, monoclonal antibodies, and vaccines had significant sales growth. The monoclonal antibody Rituxan for cancer and arthritis emerged as the top selling biologic brand followed by TNF inhibitors (Enbrel, Remicade) and Darbepoetin, each with sales above $4 billion. All the top sales decliners in 2006 were new chemical entities in patent expiry phase in major markets. There was no biologic brand in this list. Sanofi Aventis with 3 blockbuster brands lost to patent expiry was followed by Bristol Myers Squibb and Pfizer with two each. Five top products each had sales drops of over $1 billion each (Table 1.5.16) [71,72].

1.5.12 CONCLUSION

R&D success has been clearly linked to the development and creation of billion-dollar human medication. Current regulations favor fast-track approval of life-saving medication and long-term safety and efficacy data for chronic diseases. The blockbuster drugs are the main driver of industrial R&D and contribute a major share of the total pharmaceutical sales, profits, and market value of the successful companies. The criteria for blockbuster drugs are defined, and output of blockbuster drugs by pharmaceutical and biotech R&D are compared. Biotech R&D was more productive and successful than pharmaceutical R&D in creating new blockbuster brands. The current Gold Rush of biologics indicates a paradigm shift in innovation from pharmaceutical to biotechnology R&D. Another perceptible shift is the industrial R&D bail-out from Europe toward the United States, once again based on the success of the U.S. biomedical science and biotechnology firms. Biotechnology R&D success includes several blockbuster brands of erythropoietin, tumor necrosis factor alpha blockers, insulin, interferon, and GCSF, monoclonal antibodies, vaccines, and anticancer, which had sales of over $1 billion during the 2004–2006 period. Biotechnology R&D failure includes Alzheimer's Disease, sepsis, and vaccines for RSV, malaria, and AIDS. Several public–private partnerships, operating on the virtual drug development company model, have taken up the void left by industry bail-out of tropical and neglected diseases. The high cost of biologics and link to serious adverse reactions will become important as biologics are used to treat chronic conditions like rheumatoid arthritis and psoriasis. New and old media have played an important role in raising awareness about new blockbuster biologics brands.

There were no indications or signs during the past 3 years from the big consultancy firms predicting the "End of the Blockbuster Era." On the contrary, the number of blockbuster brands has shown steady and continuous growth and now accounts for a major share of market, sales, and profits of the pharmaceutical and biotechnology industry. Pharmaceutical companies' outright acquisition of biotechnology companies and the licensing of technology/late stage projects in development increased significantly. The market and sales data once again provide strong support for the R&D paradigm shift to biologics and within biologics towards human monoclonal antibodies with a message: "It's biologics, Stupid."

ACKNOWLEDGMENT

The author wishes to thank the following persons for their review and critical comments on the draft: Dr. Daniel Brandt, Debiopharm, Lausanne; Mr. Mohammad Akram, Business Consultant, Geneva; and my wife Dr. Sunita Maggon for editing, data collection, and literature search. I wish to thank Mr. Robert Posey (PDP, Geneva) for providing office facilities and administrative support.

REFERENCES

1. Ernst & Young. (2005 & 2006). Global Biotech Report and Global Pharmaceutical Trends. Levinson AD (2005). What distinguishes biotech from big pharma? Global Industry Perspectives.
2. Bains & Company Reports. Rebuilding Big Pharma's Business Model. (2003) Addressing the Innovation Divide: Imbalanced innovation (2004).
3. Boston Consultancy Group. The gentle art of licensing. (2004) Rising to the productivity challenge. (2004).
4. FDA. (2004, 2006). Innovation or stagnation: Challenge and opportunity on the critical path to new medicinal products. CDER (2004). Report to the nation. Improving public health through human drugs.

5. Cockburn I M (2004). The changing structure of the pharmaceutical industry, *Health Affairs*. 23:10–22.

6. Ng R (2003). *Drugs: From Discovery to Approval*. John Wiley, New York, pp. 368.

7. Pfizer (2005). Annual Report. Available: www.pfizer.com.

8. Trombetta B (2005). Industry Audit, *Pharm. Exec.* 10:68–80.

9. Sellers L J (2004). Top 50 Pharma Companies. *Pharm. Exec.* 24:60–70. Available: www.pharmexec.com.

10. Parker M G, Amar D (2005). Building blockbusters. *Pharm. Exec.* 10:82–88.

11. Forbes (2004). Potential blockbuster drugs. Available: www.forbes.com/2003/10/08/cz_fmc_1008sf.html.

12. IMS Top Line Global Industry Data (2005). Available: www.imshealth.com, http://www.ims-global.com/insight/insight.htm.

13. Reuters Health. Reports (2004). The Pharmaceutical Market Outlook to 2010. Available: http://www.reutersbusinessinsight.com/content/rbhc0091t.pdf.

14. Datamonitor Potential blockbuster drugs (2004). Available: www.datamonitor.com.

15. Maggon K (2003). The ten billion dollar molecule. *Pharm. Exec.* 23:60–68.

16. Maggon K (2005). Best Selling Human Medicines 2002–2004. *Drug Disc. Today.* 10:738–742.

17. Maggon K (2004). The future of the research based pharmaceutical industry. *Express Pharma. Pulse.* January 29, 2004. Available: http://www.expresspharmapulse.com/20040129/oped01.shtml.

18. Maggon K (2004). Impact of Vioxx withdrawal on pharma R&D. *Medicine and Biotech. com*. November, 15, 2004. Available: http://www.medicineandbiotech.com/therapy.html1.html.

19. Maggon K (2002). Ups and down in drug development. *Express Pharma. Pulse.* Mumbai. October 3, 2002. Express Healthcare Management Mumbai, October 16, 2002. http://www.expresshealthcaremgmt.com/20021031/edit2.shtml.

20. Niazi S K (2005). *Handbook of Biogeneric Therapeutic Proteins: Regulatory, Manufacturing, Testing, and Patent Issues*. Taylor & Francis CRC Press, Boca Raton, F.L., p. 584.

21. Rader R A (2005). *Biopharmaceutical Products in the USA and European Markets*. Bioplan Associates, Rockville, M. D., p. 1234.

22. Maggon K (2004). Setting new standards about drug safety. *Express Pharma. Pulse.* 22 May 2004. Available: http://www.expresspharmapulse.com/20030522/edit2.shtml.

23. DiMasi J A, Hansen R W, Grabowski H G (2003). The price of innovation: New estimates for drug development costs. *J. Health Econ.* 22:151–185.

24. Grabowski H, Vernon J (2000). The determinants of pharmaceutical research and development expenditures. *J. Evo. Econ.* 10:201–215.

25. Biobusiness (2005). Sales of recombinant drugs. Available: www.i-s-b.net/business/recombinant_f.htm.

26. Danzon P M, Nicholson S, Pereira N S (2005). Productivity in pharmaceutical–biotechnology R&D: The role of experience and alliances. *J. Health Econ.* 24:317–339.

27. Higgins M J, Rodriguez D (2005). The outsourcing of R&D through acquisitions in the pharmaceutical industry. *J. Financial Econ.*, In Press, Corrected Proof, Available online 2 November.

28. Schmid E F, Smith D A (2004). Is pharmaceutical R&D just a game of chance or can strategy make a difference? *Drug Disc. Today*, 9:18–26.

29. Thomas L G (2004). Are we all global now? Local vs. foreign sources of corporate competence: the case of the Japanese pharmaceutical industry. *Strat. Manag. J.* 25:865–886.

30. Penner-Hahn J, Shaver J M (2005). Does international research and development increase patent output? An analysis of Japanese pharmaceutical firms. *Strat. Manag. J.* 26:121–140.

31. Drews J, Ryser S (1996). Innovation deficit in the pharmaceutical industry. *Drug Inform. J.* 30:97–107.

32. Drews J (1998). Innovation deficit revisited: Reflections on the productivity of pharmaceutical R&D. *Drug Disc. Today.* 11:491–494.

33. Demain A L (2002). Prescription for an ailing pharmaceutical industry. *Nature Biotechn.* 20:331–334.

34. Ho R J Y, Gibaldi M (2003). *Biotechnology and Biopharmaceuticals: Transforming Proteins and Genes into Drugs.* Wiley, New York.

35. Maggon K (2004). Rise of biologicals. *Express Pharma. Pulse.* February 5, 2004. Available: http://www.expresspharmapulse.com/20040205/editorial02.shtml.

36. Maggon K (2005). Biotech segment: Moving ahead of pharmaceutical industry. *Express Pharma. Pulse.* April 28, 2005. Available: http://www.expresspharmapulse.com/20050428/edit02.shtml.

37. Collins S W (2004). *The Race to Commercialize Biotechnology: Molecules, Market and the State in Japan and the US.* p. 224 Routledge, Oxford, U.K., p. 224.

38. Liossis S N C, Tsokos G C (2005). Monoclonal antibodies and fusion proteins in medicine. *J. Allergy Clin. Immunol.* In Press, Corrected Proof, Available online 1 September 2005.

39. Qu Z, Griffiths G L, Wegener W A, et al. (2005). Development of humanized antibodies as cancer therapeutics. *Methods.* 36:84–95.

40. Pavlou A K, Belsey M J (2005). The therapeutic antibodies market to 2008. *Eur. J. Pharmaceut. Biopharmaceut.* 59:389–396.

41. Charles S A (2005). Super Generics: A better alternative for biogenerics. *Drug Disc. Today.* 10:533–535.

42. Sesin C A, Bingham C O (2005). Remission in rheumatoid arthritis: Wishful thinking or clinical reality? *Seminars in Arthritis and Rheumatism.* 35(3):185–196.

43. Hochberg M C, Lebwohl M G, Plevy S E, et al. (2005). The benefit/risk profile of TNF-Blocking agents: Findings of a consensus panel. *Seminars in Arthritis and Rheumatism.* 34:819–836.

44. Miossec P (2005). Therapeutic targets in rheumatoid arthritis: More to come but which one(s) to select? *Drug Disc. Today: Disease Mech.* 2:327–330.

45. Thomas V D, Yang F C, Kvedar J C (2005). Biologics in psoriasis: A quick reference guide. *J. Amer. Acad. Dermatol.* 53:346–351.

46. Gómez-Pérez F J, Rull J A (2005). Insulin therapy: Current alternatives. *Arch. Med. Res.* 36:258–276.

47. Gerich J E (2002). Novel insulins: Expanding options in diabetes management. *American J. Med.* 113:308–316.

48. Feemester R (2006). Holding their breath: Pipeline watch. *Pharmceut. Exec.* 26:58–62.

49. Girard M P, Cherian T, Pervikov Y, et al. (2005). A review of vaccine research and development: Human acute respiratory infections. *Vaccine.* 23:5708–5724.

50. Manoj S, Babiuk L A, van Drunen Littel-van den Hurk S (2004). Approaches to enhance the efficacy of DNA vaccines. *Critical Reviews in Clinical Laboratory Sciences*. 41:1–39.

51. Plotkin S A (2005). Vaccines: Past, present and future. *Nature Med. Suppl.* 11:s5–s11.

52. Nahta R, Esteva F J (2005). Herceptin: Mechanisms of action and resistance. *Cancer Letters*. In Press, Corrected Proof, Available online 30 March 2005.

53. Ferrara N, Hillan K J, Novotny W (2005). Bevacizumab (Avastin), A humanized anti-VEGF monoclonal antibody for cancer therapy. *Biochem. Biophy. Res. Commun.* 333:328–335.

54. Erlichman C, Sargent D J (2004). New Treatment Options for Colorectal Cancer. *N. Engl. J. Med.* 351:391–392.

55. Silvestrelli G, Lanari A, Parnetti L, et al. (2006). Treatment of Alzheimer's disease: From pharmacology to a better understanding of disease pathophysiology. *Mech. of Ageing and Devel.* 127:148–157.

56. Lipton S A (2004). Failures and successes of NMDA receptor antagonists: Molecular basis for the use of open-channel blockers like memantine in the treatment of acute and chronic neurologic insults. *NeuroRX*. 1:101–110.

57. Maggon K, Barik S (2004). New Drugs and treatment for respiratory syncytial virus. *Rev. Molec. Virology*. 14:149–168.

58. Tongren J E, Zavala F, Roos D S, et al. (2004). Malaria vaccines: If at first you don't succeed . . . *Trends in Parasitology*. 20:604–610.

59. Girard M P, Osmanov S K, Kieny M P (2006). A review of vaccine research and development: The human immunodeficiency virus (HIV). *Vaccine*. In Press, Corrected Proof, Available online 28 February 2006.

60. Lowrie D B (2006). DNA vaccines for therapy of tuberculosis: Where are we now? *Vaccine*. 24:1983–1989.

61. Okaji Y, Tsuno N H, Saito S, et al. (2006). Vaccines targeting tumour angiogenesis—A novel strategy for cancer immunotherapy. *Euro. J. Surgical Oncology*. In Press, Corrected Proof.

62. Maggon K (2003). Activated Protein C in treating sepsis. *Express Pharma. Pulse*. Mumbai February 27, 2003, Express Healthcare Management Mumbai, March 1, 2003.

63. Maggon K (2003). Risk benefit assessment of APC. *Express Pharma. Pulse*. Mumbai. March 6, 2003. Express Healthcare Management Mumbai, April 1, 2003.

64. Griffin J H, Fernández J A, Mosnier L O, et al. (2006). The promise of protein C. *Blood Cells, Mol. Dis*. In Press, Corrected Proof, Available online 7 February 2006.

65. Cacciotti J, Shew B (2006). Pharma's next top model. *Pharm. Exec.* 26:82–86.

66. Schrag D (2004). The price tag on progress—Chemotherapy for colorectal cancer. *N. Engl. J. Med.* 351:317–319.

67. Angell M (2004). The truth about drug companies. Random House, New York, p. 336.

68. Maggon K (2004). Regulatory reforms and GCP Clinical Trials with New Drugs in India. *Clinical Trials*. 1:1–7.

69. Gardner C, Gardner C (2004). Technology licensing to nontraditional partners: Nonprofit health product development organizations for better global health. Available: http://www.mihr.org/?q=taxonomy_menu/1/6.

70. Moran M (2005). The new landscape of neglected disease drug development. *Wellcome Trust, LSE and OECD Report*. Available: http://www.wellcome.ac.V

71. Maggon K (2007). Market intelligence and pharma data mining. http://www.
medicineandbiotech.com. e-Published, Feb 1, 2007.

72. Maggon K, Maggon S (2007). Overview of the pharmaceutical and biotechnology
industry performance in 2006. http://www.medicineandbiotech.com. e-Published,
March 1, 2007.

2

FROM DEFINING BIOINFORMATICS AND PHARMACOGENOMICS TO DEVELOPING INFORMATION-BASED MEDICINE AND PHARMACOTYPING IN HEALTH CARE

IOANNIS S. VIZIRIANAKIS

Aristotle University of Thessaloniki, Thessaloniki, Greece

Chapter Contents

2.1 Introduction   201
2.2 Pharmacogenomics and Pharmacotyping for Complex and Polygenic Diseases   203
2.3 Pharmacotyping Concepts and CYP-mediated Metabolism of Drugs   207
2.4 Pharmacotyping Concepts for Modulating the Pharmacodynamics and
     Pharmacokinetics of Drugs   210
2.5 Cancer Pharmacogenomics and Pharmacotyping   211
2.6 The Challenges of Pharmaceutical and Medical Education to Meet
     the Central Dogma of Pharmacotyping as well as of Personalized and
     Information-Based Medicine   215
2.7 The Development of Information-Based Platforms to Ensure Major Benefits
     for Pharmacotyping in Clinical Practice   217
2.8 Concluding Remarks   222
     References   223

2.1 INTRODUCTION

Over the last three decades, the scientific efforts mainly focused on molecular biology, genetics, recombinant DNA (rDNA) technology, and lastly genomics and bioinformatics have had a huge impact on our understanding of disease pathophysiology and of drug actions at the molecular level. In addition, the accumulated genomic knowledge has provided the ability to predict drug–drug interactions and the emergence of adverse drug reactions (ADRs) in clinical practice, a fact of great importance in modern pharmacology and therapeutics. The elucidation of specific clinical- as well as genetic-related factors predisposing to individual drug–drug interactions, ADRs, or drug disposition also gives new dimensions to new drug development and pharmacotherapy [1–5]. Furthermore, the evolution of genomic technologies offered new tools for extracting drug-related information especially through the emergence of pharmacogenomics in recent years [6–9] (for term definition in pharmacogenetics, pharmacogenomics, and other drug-related areas, see the website: http://www.genomicglossaries.com/). Pharmacogenomics aims to strengthen drug efficacy and to prevent most, if not all, ADRs from emerging during pharmacotherapy. Furthermore, pharmacogenomics provides the methodology and the knowledge to medical practitioners to genetically test each patient before the administration of the drug of interest, a practice that obviously can lead to personalized drug therapy. To this end, improved pharmacotherapy outcomes are expected by anticipating and minimizing toxicity related to drug delivery in everyday patient care. By integrating the genomic drug-related data into drug delivery, pharmacogenomics is moving toward the application of pharmacotyping in drug prescriptions, where clinical and genomic information will be used to ensure maximum efficacy and safety in clinical practice, as it has been recently proposed [10].

Interestingly, this is happening in parallel with the efforts focusing on the creation and establishment of unified information-based platforms in medicine, a fact that obviously will accelerate the technological advancement for the integration of genomic data in clinical practice. The latter has been recently facilitated through the technological developments that now allow the application of molecular diagnostics and devices at the nanoscale level. Such a direction, however, stresses the need for a multidisciplinary approach in future health care by permitting nanomedicine to apply molecular diagnostics and new innovative drugs in a way that is beneficial in routine medical practice both in disease diagnosis and in therapeutics. This is a major advantage, because the development of nanotools is in high demand in information-based medicine and the application of suitable platforms for the exploitation of genomic information in medicine and pharmacy. Thus, the intercorrelation of the molecular knowledge extracted from the clinical, genomic, and technological health-related areas as well as the transformation into a form readily applicable in health care can be achieved. As a matter of fact, the evolvement of nanomedine seems to go in parallel with the capability of genomic technologies to transform medicine and pharmacy, and in this regard, their use depends on well- and properly educated practitioners able to work in a well-organized and information-based clinical infrastructure. Collectively, this approach in pharmacotherapy belongs to the broader concept of personalized medicine because its major goals are considered the anticipation and minimization of the individual risk of disease onset and progression as well as the individualized drug therapy for improved patient care [11].

Overall, by using the principles of pharmacogenomics, personalized medicine, and pharmacotyping in patient care, and by monitoring their impact on current drug development and delivery, a major benefit is expected upon co-evaluation of genomic with clinical data to improve pharmacotherapy outcomes. In this regard, the clinical validation of genomic data for routine drug prescription, the training of future health-care professionals, and the integration of molecular medicine into clinical trials for new drug development have to be defined and clearly addressed. Also, the need for extensive discussions about the ethical, societal, and economic impacts arising from the application and use of genomic data in patient care is now, more than ever, a necessity. The latter stresses the need for the proper education of health-care professionals in order for them to actively be involved in, and positively influence, all these issues in health care. The experience already gained has shown that this can be achieved in the near future, because the first pharmacogenomics test (AmpliChip Cytochrome P450) approved by the U.S. Food and Drug Administration (FDA), recently, to improve the delivery of drugs that are substrates or inhibitors of cytochrome P450 (CYP)2D6 and/or CYP2C19 has been introduced in everyday clinical use. Also, recently released on March 22, 2005 by the FDA, the guidelines for industry concerning the submission of pharmacogenomics data upon the process of application of new drug development, a fact expected to have a major impact on the flow and generation of data in clinical pharmacogenomics and the application of pharmacotyping in drug prescription. This means that the detection of single nucleotide polymorphisms (SNPs) or other genetic variations and the establishment of specific genotypes and haplotypes for several genes must be first defined and clinically validated at such a level as to explain why one patient responds well in drug therapy, another does not, or even another experiences serious ADRs. In fact, this will be a major transition in pharmacotherapy because pharmacotyping will be finally achieved, and, to this end, the medical practitioner in collaboration with other specialties will also be based on patient's genotyping–haplotyping analysis data for initiating and maintaining drug dosage therapeutic profiles for individual patients [5, 10]. However, in order for these new directions in therapeutics to succeed with maximum benefits to patients and society, the future advances in pharmacogenomics and bioinformatics have to first allow the smooth digital integration and second the easy end-user utilization of genomic and clinical data in unified information-based platforms within the healthcare system.

2.2   PHARMACOGENOMICS AND PHARMACOTYPING FOR COMPLEX AND POLYGENIC DISEASES

The concept of ensuring maximum efficacy and safety upon drug delivery is the long-desirable target of pharmaceutical care that implies improved pharmacotherapy profiles and outcomes in clinical practice and coincides chronologically with the establishment of pharmacology as a basic and clinical discipline. Also, the ability to lower the incidence of drug toxicity and the emergence of ADRs is a fundamental issue for the empowerment of health care in terms of patient's quality of life and cost (Box 2.1). Nowadays, this situation in pharmacotherapy has been, more or less, closer to reality. This happens due to the rapid development of genomic and information-based technologies and their application in pharmacotherapy and clinical practice (Figure 2-1). For example, it is now well accepted that, in terms of drug toxicity, many ADRs arise because of inter-individual genetic differences in drug metabolizing enzymes, drug transporters, ion channels,

BOX 2.1 MAJOR POINTS RAISED IN HEALTH CARE UPON THE INCIDENCE AND THE SEVERITY OF ADRS IN CLINICAL PRACTICE

- In a meta-analysis from 39 prospective studies in U.S. hospitals, it has been proposed that 6.7% of hospitalized patients developed ADRs during pharmacotherapy and 0.32% showed fatal ADRs, the latter causing about 100,000 deaths per year in the United States [12].
- The report released from the Institute of Medicine in December 1999 has shown that nearly 98,000 deaths in the United States annually were attributed to medication errors including ADRs [13].
- Furthermore, 10–17% of patient hospitalizations are directly related to the emergence of ADRs, an effect that poses patient pharmaceutical care and safety as a fundamental issue in each healthcare system [14].
- It has been estimated in the United States that the cost attributed to the emergence of ADRs is approximately $100 billion, a fact that further stresses for the improvement of drug delivery in clinical practice [15].
- In a 2-year period of study conducted in by the Department of Internal Medicine in a Norwegian hospital and from 732 deaths reported amongt the 13,993 in-patients, 133 of them were directly attributed to ADRs observed during pharmacotherapy [16]. This result gives almost 10 deaths per 1000 hospitalized patients, a number that urgently demands an improvement of drug delivery by strengthening drug efficacy and minimizing toxicity in the healthcare system.
- It has also been estimated that about 7% of all hospital admissions in the United Kingdom and Sweden are due to ADRs developed in clinical practice [15].

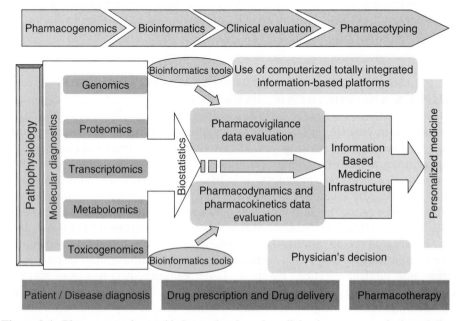

Figure 2-1. Pharmacotyping and information-based medicine in postgenomic drug delivery and health care (see the text for details).

receptors, and other drug targets [1, 2, 5, 10, 17, 18]. Furthermore, several pharmacogenomics studies have revealed various examples of clinically relevant genetic polymorphisms that are associated with altered drug response in patient care. Representative examples of such polymorphic gene variants of pharmacogenomics relevance are shown in Box 2.2. Although these intriguing data are very interesting in trying to understand the correlation between genetic make-up and various drug effects in the body, much work is needed for this type of drug-related genomic information to be transformed, integrated in everyday clinical practice, and, finally be available on a broader basis in health care.

Through the sophisticated technological advances achieved thus far, the genomic knowledge is entering clinical practice and being used to analyze complex diseases including cardiovascular diseases, asthma, cancer, and degenerated disorders [38–42]. Furthermore, the broad use of automated DNA sequencing techniques and the application of DNA microarray (DNA chip) systems permit the rapid and reliable assessment of a large number of genes implicated in drug response and/or disease phathogenesis [43–45]. Alternatively, and of equal importance, specific gene variants have been found either to predispose for, to contribute to, or even to increase the susceptibility of individual persons for disease development and progression. Unanimously, the availability of DNA microarray technology helped genomics research to achieve widespread interest particularly in clinical sciences and pharmacotherapy. As a matter of fact, it is a valuable technique for the improvement of clinical diagnosis and the understanding of molecular etiology and pathophysiology of polygenic and multifactorial conditions, e.g., cardiovascular diseases and degenerating disorders, because it gives researchers the ability to assess and evaluate each time the expression of nearly all genes presented in the human genome. For example, there have been several attempts, in the past few years, to establish specific polymorphic gene variants at pharmacodynamic loci predisposing to cardiovascular diseases or to affect drug response such as angiotensin-converting enzyme (ACE), angiotensin II type-1 receptor (AT-1), apolipoprotein E (APOE), α-adducin, and adrenergic receptors that could alternatively be used to improve the pharmacotherapy profiles of drugs exerting their effects through binding to these receptors [46–49]. It is also very interesting for anyone to see now, that genomic studies, throughout the last period of 10–15 years, have revealed several chromosomal loci and genes implicated for example in cardiomyopathies, arrhythmias, heart failure, and hypertension. The identification of several genes whose function is associated with diseases like hypertrophic or dilated cardiomyopathy, human long-QT syndrome, and essential hypertension give new insights on a molecular basis to heart disease progression and may help in orienting the proper drug therapies in a given population. Among them are included genes encoded for β-myosin heavy-chain, cardiac troponins and α-tropomyosin in hypertrophic cardiomyopathy; dystrophin, tafazzin, and actin in dilated cardiomyopathy; cardiac sodium channel (SCN5A) and potassium channel (HERG) in long-QT syndrome; as well as ACE, α-adducin, angiotensinogen, and β_2-adrenergic receptor (β_2AR) in essential hypertension [33, 39, 50]. Furthermore, even more impressive is the fact that, in the case of cardiomyopathies, genomic analysis has been able to establish specific molecular gene expression profiles that clearly define different molecular classes of cardiomyopathy such as of sarcomyopathy, cytoskeletalopathy, and cytokinopathy [51–54]. However, although these represent intriguing results in genomic

BOX 2.2 GENETIC POLYMORPHISMS IN SPECIFIC GENES ASSOCIATED WITH VARIOUS DRUG RESPONSE AND EMERGENCE OF ADRS IN CLINICAL PRACTICE*

Genetic polymorphisms of pharmacogenomics importance that can modulate the pharmacological response of specific drugs implicate the following genes:

1) *CYP2C9* that has been implicated in warfarin, phenytoin, tolbutamide, and/or glipizide has had various clinical responses [1].

2) *CYP2C19* that has been involved in proton pump inhibitors (e.g., omeprazole) has had an altered clinical response [19].

3) *ABCB1* gene that has been associated with the development of molecular resistance of anti-epileptic drugs [20, 21].

4) UDP-glucuronosyltransferase 1A1 (*UGT1A1*) that has been involved in irinotecan toxicity [20, 22].

5) Human leukocyte antigen (*HLA-B*) that has been associated with abacavir hypersensitivity [23, 24].

6) Glutathione-S-transferases (*GSTs*) that have been correlated with platinum chemotherapeutic agents and D-penicillamine altered response [25, 26].

7) N-acetyltransferases (*NATs*) that have been implicated with the emergence of sulfonamides hypersensitivity and the toxicity observed in some cases of isoniazid, procainamide, and hydralazine administration [27].

8) HERG potassium channel (*KCNH2*) that has been found to modulate quinidine and/or cisapride safety in clinical practice as well as KvLQT1 potassium channel (*KCNQ1*) that causes a similar effect on terfenadine and disopyramide response [28, 29].

9) Thiopurine methyltransferase (TPMT) activity has been well correlated with 6-mercaptopurine, 6-thioguanine, and azathioprine toxicity seen in some patients [1, 30, 31].

10) Several genetic polymorphisms of serotonin *5-HT2A* and *2C* receptors have been implicated with various pharmacological responses to clozapin delivery [32].

11) The angiotensin II type I-receptor (*AT-1*) gene has been associated with the altered drug response of losartan seen in certain patients [33].

12) The apolipoprotein E4 (*APOE4*) allele has been correlated with the variability of tacrine pharmacotherapy [34].

13) The gene of 5-lipoxygenase (*ALOX5*) has been studied in relation to zileuton and montelukast altered drug response [28].

14) Several genetic polymorphisms of β_2-adrenergic receptors (β_2AR) have been shown to modulate the β_2-agonist (e.g., albuterol) effect in clinical practice [35, 36].

15) The ABCG2 transporter gene, a member of the ATP-binding cassette transporters, in which an Arg to Gly mutation at amino acid 482 (G482 mutation) has been shown to confer high-level resistance to various antifolate chemotherapeutic agents [37].

* Modified from Ref. 5.

medicine, the recognition of the complexity of such diseases clearly demands the development and clinical validation of suitable genetic diagnostic markers; the application and use of more sophisticated, clinically validated genomic and bioinformatic tools; as well as the design of prospective pharmacogenomics studies, in order to further advance the molecular understanding of the disease pathophysiology and to improve the therapeutic intervention in clinical practice.

Regarding the degenerating diseases, the structural and functional genomics analysis of patients with Alzheimer's disease (AD) has recently identified several genetic loci that potentially contribute to disease pathophysiology in cooperation with both environmental and epigenetic factors [40]. In particular, genetic predisposition to AD development is demonstrated either for genes whose variants exhibit a mendelian inheritance pattern, e.g., the amyloid precursor protein (APP), and the presenilin-1 (PS1) and -2 (PS2), or for those considered to be potential susceptibility genes contributing to AD predisposition, e.g., APOE, ACE, interleukin-1α (IL1α), nitric oxide syntase-3 (NOS3), low-density lipoprotein-related protein 1 (LRP1), and α-2-macroglobulin (A2M). Furthermore, recent results have suggested that specific genotypes and haplotypes may cause different drug response rates in individual AD patients receiving cholinesterase inhibitors and/or non-cholinergic agents used for their therapeutic intervention [34, 40]. This progress in AD genomics clearly shows that pharmacogenomics analysis of complex and polygenic disorders has started to give valuable clinical results. As a matter of fact, the broader application of pharmacogenomics can be an alternative and suitable platform for further assessing the pathophysiology and molecular diagnosis of complex diseases, as well as for the identification of new drug targets to improve healthcare outcomes and patient's quality of life. To this end, however, as in the case of cardiovascular diseases, the establishment of clinically validated specific molecular diagnostic biomarkers and pharmacotyping profiles for individual patients is still a demand, in order for these efforts to be finally achieved in clinical practice.

2.3 PHARMACOTYPING CONCEPTS AND CYP-MEDIATED METABOLISM OF DRUGS

By considering the example of CYPs that are involved in the phase I drug metabolism in the body, the importance of pharmacogenomics studies and their clinical relevance in drug disposition will be further defined. Nowadays, 57 putative functional genes encoding different CYP enzymes and 58 pseudogenes have been characterized [55, 56] (see also the website: http://drnelson.utmem.edu/CytochromeP450.html). The major human CYP enzymes involved in drug metabolism include CYP1A2, CYP2A6, CYP2B6, CYP2C8, CYP2C9, CYP2C19, CYP2D6, CYP2E1, CYP3A4, and CYP3A5 that exhibit several polymorphic sites of clinical significance (see the website: http://www.imm.ki.se/cypalleles). Indeed, several experimental studies of applied pharmacogenomics research have already been published that correlate specific CYP gene variants with altered drug efficacy and safety and clearly shed light on the elucidation of the molecular mechanisms involved in inter-individual drug response observed among patients suffering from the same disease. The existed complexity at the genomic level in drug metabolism is additionally complicated by considering the high number of polymorphic gene variants encoding CYP enzymes discovered, especially those belonging into CYP families 1, 2, and 3. It has been estimated that enzymes belonging to these families mediate about 70–80% of all phase I-dependent metabolism of clinically used drugs. Interestingly, polymorphic enzymes that mainly account for ~40% of CYP-mediated drug metabolism are those of CYP2C9, CYP2C19, and CYP2D6 and, for this reason, are considered very important in pharmacotherapy by making problematic the application of a general drug dosing scheme for the drugs whose disposition is determined by these enzymes [15]. Similarly, by evaluating the incidence of ADRs, it has been shown that ~56% of drugs cited in ADR-related studies are metabolized by polymorphic enzymes of phase I, in which ~86% account for the CYP-mediated metabolism. On the contrary, only 20% of drugs that are substrates of non-polymorphic enzymes are cited in the ADRs reports [15, 57].

It is now well known that the genetic variability observed in CYP enzymes corresponds into four phenotypes of clinical relevance in drug delivery. The definitions used to identify these four phenotypes of CYP-related drug metabolism are as follows:

1. Poor metabolizers (PMs) for those individuals that have shown no enzymatic activity in diagnostic tests. Therefore, this is a phenotype that refers to people who lack the functional CYP enzyme and slowly metabolize the drug substrates, allowing their concentration to reach high plasma levels.

2. Intermediary metabolizers (IMs) for those individuals with reduced enzymatic activity. These people are heterozygous for one deficient allele or carry two alleles that cause lower CYP enzyme activity.

3. Extensive metabolizers (EMs) for individuals with physiological enzymatic activity, a fact that suggests the existence of two wild-type (normal) alleles allowing normal drug metabolism.

4. Ultrarapid metabolizers (UMs) for individuals having higher than normal enzymatic activity. Therefore, this is the phenotype applied to people carrying multiple gene copies of the respected enzyme. That trait is dominantly inherited and allows these individuals to rapidly metabolize the drug substrates and to achieve low, and sometimes inefficient, plasma levels.

Several examples of drugs whose dosage is related to CYP phenotypes have already been reported that further envisage on the importance of CYP polymorphisms for routine drug prescription and the modulation of pharmacotherapy outcomes. For example, CYP2D6 and CYP2C19 are responsible for the metabolism of most psychoactive drugs, including antidepressants. In particular, specific dosage recommendations for patients with EM, IM, and/or PM phenotypes of either CYP2D6 or CYP2C19 have been suggested for 14 antidepressant drugs [58]. Furthermore, the kinetics of nortriptyline is dependent on the number of active CYP2D6 genes, and the dosage required to reach the same plasma levels varies from 30–50 mg in CYP2D6 PMs to 500 mg in UMs [15]. Another example applies to sertraline, a substrate of CYP2C19 (and an inhibitor of CYP2D6), because CYP2C19 PMs develop ADRs (nausea and dizziness), a fact that might be attributed to increased sertraline plasma concentrations. In addition, the metabolism of valproate (commonly used to treat bipolar disorders) has been found to be influenced by CYP2C9 polymorphism, an effect of clinical relevance. Thus, the broader application of CYP-related pharmacogenomics knowledge to implement clinical experience can improve pharmacotherapy outcomes through the establishment of individualized dosage-regimens that take the difference in drug metabolic capacity into account.

Similarly, the fundamental role of CYP-related knowledge in minimizing the incidence of drug–drug interactions in clinical pharmacology and the application of such information in improving pharmacotherapy outcomes is now well appreciated in drug delivery and can be further defined by the example of cholinesterase inhibitors (rivastigmine, tacrine, donepezil, and galantamine), which are agents used to treat symptoms of patients suffering from AD. By assessing the drug–drug interactions of this group of agents published in several clinical studies, it can be suggested that the individual CYP-isoform implicated in the metabolism of each cholinesterase inhibitor also specifies the type of interacting drugs referred to modulate its pharmacokinetic parameters and clinical outcome [59, 60]. In particular, because the metabolism of rivastigmine is mediated predominantly by esterases rather than by hepatic CYP enzymes, clinically relevant drug interactions implicated by this phenomenon are unlikely. This fact for rivastigmine was indeed verified by *in vitro* assays and pharmacokinetic studies conducted in humans. To this regard, in a retrospective analysis including four major clinical trials with 2459 patients, it has been shown that no increase in adverse events occurred among those patients who received 22 different therapeutic classes of drugs commonly used in elderly populations, including antihypertensive, anti-histamine, anti-inflammatory, or anxiolytic agents [59].

Tacrine is primarily metabolized by the hepatic CYP1A2 (and to some extent by CYP2D6), so substrates, inhibitors, or inducers mainly of CYP1A2 are expected to alter its bioavailability. This is indeed the case: (1) with fluvoxamine (50 or 100 mg/day), which is a potent CYP1A2 inhibitor that caused ~85% reduction of tacrine metabolism; (2) with cimetidine, also a CYP1A2 inhibitor, where it has been observed that plasma concentrations of tacrine were increased by ~30%, so a reduction of its dosage has been advised when co-administered with cimetidine; (3) with estradiol also metabolized by CYP1A2 that caused an increase of ~60% in AUC and of ~46% in the mean C_{max} of tacrine. Following these data, it has also been reported that tacrine decreased by ~50% the clearance of theophylline, which is also a substrate of CYP1A2, so reduction of theophylline dosage has been suggested upon its co-administration with tacrine. In addition, tobacco smoking that induces CYP1A2 metabolism significantly accelerates tacrine metabolism, a result indicating that clinicians should be aware that smokers may require higher doses of tacrine than that would be given to nonsmokers [60].

Similarly, in the case of donepezil that is primarily metabolized by the hepatic CYP3A4 and CYP2D6, it is expected that substrates, inhibitors, or inducers of CYP3A4 and of CYP2D6 can alter its bioavailability. This prediction is indeed verified in certain pharmacotherapy dosage regimens: (1) with ketoconazole (200 mg/day), a potent inhibitor of CYP3A4, where a significant increase (23–30% at steady state) in the plasma concentrations of donepezil was observed; and (2) with paroxetine and sertraline, which are both of them CYP2D6 inhibitors, in which a careful patient monitoring has been recommended upon co-administration, due to possible interaction of donepezil with these drugs through inhibition of CYP2D6. On the contrary, upon the delivery of donepezil with risperidone and/or thioridazine, the latter two are substrates, but not inhibitors, of CYP2D6, and no significant interaction was observed. The same has been shown with digoxin (no significant CYP metabolism), theophylline (substrate of CYP1A2), warfarin (substrate of CYP2C9), and/or cimetidine (inhibitor of CYP1A2 and CYP2C19) [60].

By proceeding through the analysis to galantamine, similar results have been obtained. Galantamine is also metabolized by CYP3A4 and CYP2D6, so again the potent inhibitors of these isoenzymes are expected to result in significant drug–drug interactions. Such interaction was indeed observed: (1) with paroxetine (an inhibitor of CYP2D6 and CYP2C19) that caused an increase in the bioavailability of galantamine by 40%, as it has been reported upon their co-administration; and (2) with ketoconazole (an inhibitor of CYP3A4 and CYP2C19) that exerted a significant increase in the bioavailability of galantamine by 40%. However, no significant interaction of galantamine with warfarin and digoxin has been observed in clinical studies [60].

The above-mentioned example of cholinesterase inhibitors clearly shows the predictive value of CYP-related information during everyday drug prescription, because the avoidance of co-administration of drugs implicated with the same CYP isoform can ultimately improve pharmacotherapy outcomes, at least for drugs whose metabolism is the rate-limiting step upon the establishment of their therapeutic concentrations in the body. However, and more importantly, this knowledge has already started to implement routine clinical practice, since the first pharmacogenomics test was approved by the FDA on December 23, 2004. The "AmpliChip Cytochrome P450 Genotyping test" permits the identification of 31 different polymorphisms of CYP2D6 (29 polymorphisms) and CYP2C19 (2 polymorphisms), and it is used along with clinical evaluation and other tools to determine the best treatment options for patients taking drugs that are either substrates or inhibitors of these metabolizing enzymes. The development of this test is of great importance in pharmacotherapy, because it represents the first step toward the broader application of pharmacotyping concepts in drug prescription.

2.4 PHARMACOTYPING CONCEPTS FOR MODULATING THE PHARMACODYNAMICS AND PHARMACOKINETICS OF DRUGS

Interestingly, and in addition to CYP gene variants already mentioned in the previous section, specific genetic polymorphisms were also analyzed in genes encoding P-glycoprotein (P-gp), G-protein coupled receptors (GPCRs), thiopurine methyltransferase (TPMT), ACE, arylamine N-acetyltransferases (NATs), and/or UDP-glucuronosyltransferases (UGTs), just to mention a few, which have also been shown to correlate with altered drug response and incidence of ADRs in humans [20,22, 27, 30, 35, 46, 61–63], (see also Box 2.2). To better clarify the correlation of such gene variants with the pharmacotherapy outcome, experimental data related to either the ABC-type transporters function or to the variability of anti-asthmatic drug delivery will be discussed. P-gp is an ATP-dependent transporter (efflux pump) belonging to the superfamily of ABC transporters that includes 48 already identified members in humans. The role of P-gp in limiting intestinal, brain, and placental transport, as well as in biliary and urinary excretion of its substrates, is now well recognized. Thus, variations in its gene (*ABCB1* or *MDR1*) that existed in several individuals may influence drug disposition and alter drug response [20, 63]. These genetic variations can affect the function of P-gp as transporter by altering its binding with drugs and other xenobiotics [20]. Confirmatory to this notion is the mutant variant of *ABCB1* designated as ABCB1-C3435T, which has been characterized recently and has been shown to be well associated with the variability of antiepileptic drug response [21]. At the same time, further functional characterization of specific P-gp gene variants, or other ABC-type transporters, and elucidation of the molecular mechanisms that underlined their involvement in interindividual drug variability are expected to give new insights into the selective modulation of the P-gp function in the blood-brain barrier for optimizing therapeutic intervention and improving patient pharmaceutical care [64]. The clinical relevance of the mutant alleles of ABC-type transporters in pharmacotherapy was further confirmed in another member of this class. Recent studies have established that an Arg to Gly mutation at amino acid 482 (G482 mutation) in the ABCG2 transporter confers high-level resistance to various antifolate chemotherapeutic agents (e.g., methotrexate) [37].

Nowadays, specific polymorphisms in individual genes have been studied to explain the variability seen upon anti-asthmatic drug delivery, like β_2-agonists (see also Box 2.2). For example, it has been estimated that as much as 60% of interindividual variability in anti-asthmatic drug response observed with β_2-agonist delivery might involve genetic factors modulating the pharmacotherapy outcome [35]. Thirteen SNPs identified in the promoter and the coding region of β_2AR gene were found to be organized into 12 haplotypes. Some of these haplotypes have shown significant divergence in different ethnic populations, and this fact is associated with the bronchodilator response to β_2-agonists in patients with asthma [36]. As asthma is considered a complex and polygenic disease, several other gene variants have also been shown to contribute to the pharmacotherapy outcome in asthmatics in addition to β_2AR polymorphisms. Such genes include those of 5-lipoxygenase (*ALOX5*) for zileuton, muscarinic receptor-2 (M_2) and 3 (M_3) for muscarinic antagonists (e.g., ipratropium bromide), glucocorticoids receptor for glucocordicoids (e.g., prednisolone, beclomethasone), and phosphodiesterase-4A (*PDE4A*) and 4D (*PDE4D*) for theophylline delivery [65]. However, although these correlations add valuable knowledge to basic and clinical pharmacology, more systematic efforts at the molecular level must be undertaken toward thorough elucidation of the underlined mechanisms of genetic drug response variability, in order for the anti-asthmatic therapy and outcome to be finally enriched and improved.

2.5 CANCER PHARMACOGENOMICS AND PHARMACOTYPING

Recent progress in cancer pharmacogenomics suggests that improvement of chemotherapy outcomes for specific anticancer drugs, like 5-fluorouracil (5-FU), irinotecan, platinum agents, and thiopurine-like drugs, can be achieved. In particular, specific polymorphisms in genes encoding the enzymes dihydropyrimidine dehydrogenase and thymidylate synthase have been shown to influence the efficacy and toxicity of 5-FU in certain individuals. The applicability, however, of such an approach by selecting patients who are likely to tolerate and respond to 5-FU therapy remains very complicated, and it is not easily attainable for everyday care [20, 66]. Furthermore, it has been proposed that polymorphisms in the promoter region of the gene encoded UDP-glucuronosyltransferase 1A1 (*UGT1A1*) may be clinically useful for predicting severe patients' toxicity to irinotecan, thus making this drug an excellent candidate for individualized therapy, as it has been recently proposed [67]. Irinotecan belongs to the camptothecin class of topoisomerase I inhibitors and has been shown to have potent activity against many types of solid human tumors, in particular, gastrointestinal and pulmonary malignancies [22, 67]. It is a prodrug that is converted in the body by carboxyesterase-2 to SN-38 metabolite with more than 1000-fold enhancement of cytotoxic activity. The rate-limiting step of SN-38 pharmacokinetics is considered the UGT glucuronidation reaction *via* hepatic UGT1A1, whereas the presence of the homozygous *UCT1A1*28* genotype in patients as a molecular predictor of toxicity risk (the promoter region contains seven TA repeats instead of six in the normal wild-type genotype) seems to be associated with high risk of grade 4 neutropenia in clinical practice [67]. The predictive value of the *UCT1A1*28* genotype, in order clinicians to address more definitive guidelines regarding the effective and safe dose of irinotecan in these patients, however, remains to be established.

Also, in the case of platinum chemotherapeutic agents (cisplatin, carboplatin, and oxaliplatin), several polymorphic genes have been implicated in their efficacy and toxicity. These genes are the *XPD* and *XRCC1* that encode proteins involved in the cellular DNA excision repair system and that of glutathione-S-transferases (GSTs) [66]. Although these results can be con-sidered as positive indicators for the improvement of cancer patient care, additional work must be done in order for the general applicability and acceptance of genetic information related to platinum agents response to be established in clinical practice.

However, one of the most studied examples of applied pharmacogenomics that clearly shows how the translation of genomic information is being transferred to guide patient therapeutics is related to TPMT gene polymorphisms. Interestingly, specific TPMT gene variants have been isolated and clinically validated in studies showing that their presence is associated with high-risk toxicity in individuals taking thiopurine drugs (6-mercaptopurine, 6-thioguanine, azathioprine) [30, 31, 68]. Currently, eight TPMT alleles have been identified, including three variant alleles designated as TPMT*2 (with G238C mutation), TPMT*3A (with G460A and A719G mutations), and TPMT*3C (with A719G mutation) that account for about 80–95% of intermediate or low TPMT enzymatic activity, and for this reason, they are analyzed during genotyping and haplotyping population studies [30]. The pharmacogenomics data regarding the TPMT deficiency correlated well with thiopurine drug toxicity and were so clear and convincing that a TPMT genetic test was developed for clinical use before the initiation of 6-mercaptopurine delivery to children suffering from acute lymphoblastic leukemia (ALL) [69, 70]. The clinical usefulness of this genetic test was based on pharmacogenomics data suggesting that the calculation of a drug dosage scheme could be achieved though the identification of the specific TPMT genotype in each child. However, the difficulty of applying the genetic test in clinical practice was further demonstrated by the discussions after the attempts to apply TPMT pharmacogenomics data upon thiopurine drug prescription. Such discussions have been extended even to the need and usefulness of the genetic testing as compared with the traditional biochemical tests, e.g., the use of a red-cell assay for the detection of TPMT enzyme deficiency [11, 70]. The example of a TPMT genetic test clearly shows the difficulties that these new concepts of drug delivery will meet before their general acceptance in routine everyday use. Undoubtedly, to ascertain for applied pharmacogenomics a final success, more systematic work and advancement in information-based technologies are needed. In turn, this will allow better transformation of genomic data into clinical forms to finally lead into the wide use of genetic testing with high-quality and validity outcomes, thus ensuring specific dosage recommendations of individual drugs in routine pharmaceutical care [71, 72].

Of special interest, however, related to the application of pharmacotyping in anticancer therapeutics are issues concerning the development of drug resistance, as this phenomenon consistently affects the pharmacotherapy outcomes and patient's quality of life. Over the past several years, novel exploitable targets for cancer drug development have been revealed that recently resulted in the development of some innovative drugs already applied to clinical practice. Such targets include enzymes like tyrosine kinases involved in signal transduction pathways, genes regulating apoptosis and differentiation of malignant cells, cell-lineage transcriptional factors, angiogenesis factors, and unique proteins driving the cell cycle machinery [73–78].

The development of imatinib (Gleevec) as an effective agent capable of inducing cellular apoptosis for the treatment of patients suffering from chronic myelogenous leukemia (CML) represents such a bright example of molecularly targeted anticancer therapeutics [79]. Gleevec is a selective inhibitor of Bcr-Abl fusion tyrosine kinase specifically present in CML patients that results from a reciprocal translocation between chromosomes 9 and 22. This translocation causes the genetic fusion of *bcr* with exon-1 of *c-abl* resulting in a defect known as Philadelphia chromosome. Although the use of imatinib give very impressive and promising results for the therapy of CML as well as for other tumors by also inhibiting the c-kit kinase, the development of drug resistance causes a real restriction point into its clinical use for all CML patients [80]. This restriction is partially attributed to the occurrence of specific mutations in the gene encoding the Bcr-Abl fusion tyrosine kinase, a result that poses the drug resistance development of major concern in modern cancer pharmacotherapy.

The difficulties in reaching improved pharmacotherapy outcomes have also been reported upon the clinical use of gefitinib (Iressa) developed for solid tumor therapeutic intervention, which has had limited, success. This agent, indeed, has been shown to be effective and beneficial only in non-small cell lung cancer (NSCLC) patients whose tumor cells exhibit mutations in the epidermal growth factor receptor (EGFR) gene leading to activated forms of the extracellular domain of EGFR kinase [81–84]. Furthermore, the failure of gefitinib to prolong patient's lives limited its use in the United States during 2005, whereas at the same time caused a withdrawal of its marketing application in Europe [85]. This effect, however, has not been seen in the case of erlotinib (Tarceva), which is another EGFR tyrosine kinase inhibitor in clinical use, where a survival benefit in lung cancer patients has been measured in prospective clinical studies. At the same time, discussions regarding the development of multitargeted tyrosine kinase inhibitors to overcome the limited efficiency of gefitinib, like zactima that inhibits both tyrosine kinases of EGFR and vascular endothelial growth factor (VEGF) receptor-2, have sparked new research efforts in the field of developing tyrosine kinase inhibitors as new anticancer drugs. Overall, the experience already gained from the development and use in clinical practice of drugs of the new class of tyrosine kinase inhibitors (imatinib/Gleevec; gefitinib/Iressa; erlotinib/Tarceva) has clearly shown that the path from the initial enthusiasm of innovative drug discovery to improving pharmacotherapy outcomes *via* the identification of the molecular basis of responders and nonresponders has not been straightforward and easily attainable [77].

The problems confronted in clinical practice upon the delivery of tyrosine kinase inhibitors as drugs obviously support the notion for the exploitation of alternative and more sophisticated ways that must be undertaken in order to better cure cancer patients. Also, it stresses the need for thorough understanding of the molecular mechanisms that underline vital cellular processes before pharmacotyping can be generally applied in clinical practice. To this end, and to avoid such undesirable effects, the development and use of drugs that can induce simultaneously differentiation and apoptosis of leukemia cells have also been suggested [78]. Both processes result in restricted cell growth and thus represent a desirable target for the development of new anticancer therapeutics. To this end, the process of cell renewal in leukemia versus that of cell differentiation and/or apoptosis seems to be linked somehow to each other, thus giving hope that both differentiation and apoptosis can be simultaneously activated in order to serve as potential platforms for developing combined therapeutic approaches [78, 86, 87]. Toward this direction, the use of model cellular systems for more detailed analysis of the function of genes and the elucidation of their specific interactions in normal and disease states represents a useful alternative experimental design of clinical significance, as it has been recently shown [78, 87].

Interestingly, great progress has recently occurred in breast cancer biology and pharmacotherapy by revealing the structural interactions of the drug Herceptin and its receptor HER2. Herceptin is a humanized monoclonal antibody (known as trastuzumab) that was approved by the FDA in 1998 for breast cancer treatment and was the first genomic-research-based targeted anticancer therapeutic entered into clinical practice [77]. Its effects are achieved through the binding in the HER2 receptor (also known as Neu or ErbB2), a member of the epidermal growth factor receptor (EGFR; also known as ErbB) family of tyrosine kinases, and its prescription is based on the expression level of HER2 in the patient's tumor tissue (individualized therapy) [88]. The recent elucidation of the crystal structure of the entire extracellular regions of HER2 complexed with the Herceptin antigen-binding fragment (Fas) gives new information on the drug-receptor structural interactions, and this knowledge may lead to better drug design and development of new therapeutics of this type [89]. Furthermore, such studies envisage a better understanding of cancer biology at the molecular level and permit the design of more specific pharmacogenomics studies to help establish pharmacotyping in cancer chemotherapy. In particular, such an interesting step related to human breast cancer progression and metastasis is considered in the recently published work on the gene expression profiles of the premalignant, pre-invasive, and invasive stages of this type of tumor, as well as the identification of a "gene signature" for breast cancer metastasis to lungs [90, 91]. Through this approach of genomic analysis, the better understanding of disease pathogenesis and metastatic potential can be achieved and the identification of specific genomic targets for better drug development could also be attained. Moreover, by analyzing gene expression profiles in cancer pharmacotherapy treatments, it would be possible to characterize molecular gene networks that in turn might have important clinical cosequences for improving patient care. This happened in the case of ALL where 124 genes were identified to discriminate between different treatments of chemotherapeutic agents (methotrexate and mercaptopurine), a result that can be reversely applied afterward in order for the individualized therapy to be based on the expression level of such genes. Such a direction implies the application of genomic analysis of the identified genes for each ALL patient before initiating therapy with the referred drugs [92]. This procedure that uses transcriptional profiling analysis has been recently proposed as one way for applying pharmacogenomics in cancer pharmacotherapy [93]. Overall, however, the accumulated knowledge at the molecular level is expected to help toward elucidating the underlined complexity of both cancer biology and anticancer drug response variability. In any case, however enriching our knowledge of the molecular biology of cancer will ensure that better pharmacological interventions will be achieved in the near future.

2.6 THE CHALLENGES IN PHARMACEUTICAL AND MEDICAL EDUCATION TO MEET THE CENTRAL DOGMA OF PHARMACOTYPING AS WELL AS OF PERSONALIZED AND INFORMATION-BASED MEDICINE

The achievements and challenges in pharmaceutical research that have already been discussed in previous sections have added new and innovative knowledge in basic pharmacology and therapeutics and now stress for the appropriate adjustments in education and curricula in both medicine and pharmacy faculties [94–96]. Also, new roles for healthcare providers have emerged in the changing environment of pharmacotherapy and clinical practice that clearly coincide with their need for better education in pharmacogenomics and personalized medicine. It is evident that better training will be achieved by the development of new curricula based on the integration of drug-related genomic knowledge and bioinformatics technologies into the teaching process. However, healthcare educators already face difficulties in teaching pharmacogenomics and personalized medicine concepts in order to give students the skills and the knowledge to keep them up to date with recent advances in relevant drug-related issues [94]. The classic background bridge of physiology with chemistry and pharmacology needed in order for students to understand drug-related actions and effects in the body has now been expanded to include the knowledge coming from biochemistry, molecular biology, genetics, and finally genomics (Figure 2-2). The latter is now considered crucial for more thorough understanding of several aspects of drug delivery in order for future health and pharmaceutical care practitioners to improve their skills for better drug selection, dosage regimens, and co-administration of drugs in individual patients. Undoubtedly, this need for better training in pharmacogenomics, for both educators and healthcare providers, will be in increasing demands in the years to come, because developments are achieved by genomic technologies in all drug-related areas and rapidly introduced into health care and clinical practice [5, 10, 25, 94]. However, the successful integration of such knowledge in the educational process

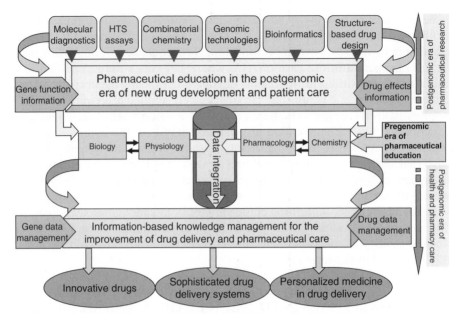

Figure 2-2. Convergence of genomic and clinical knowledge in postgenomic new drug development and health care (see the text for details).

is a difficult task. At first, this means the development and use of new interactive computer-assisted learning methods capable of combining drug-related information with the function of genes and proteins being expanded through the genome and proteome. Second, it implies that the proper introduction of these programs into the teaching process organized in a tight timetable for completing both theory and clinical practice is needed [25, 97–100]. Another challenging issue in the design of new curricula has to deal with students' future demands to keep pace with the new trends in clinical and pharmacy practice rapidly entering and enriching the drug-related era, even after their graduation, because such profound changes in pharmacotherapy are continuously entering into the clinics, thus affecting the practice of modern medicine and pharmacy.

Pharmaceutical education also has to address some ethical aspects related to the application of genomic technologies in pharmacotherapy. In other words, by integrating genomics data into the clinical practice and the drug delivery process, several societal- and clinical-related issues are raised and must be taken into account during the educational process. It is evident that the application of pharmacotyping requires the health and pharmaceutical care practitioner to look into the patient's genetic makeup and to correlate the extracted information with clinical knowledge in standardized pharmacogenomics protocols in order to validate the compatibility of the drug dosage profile for the individual patient examined. In addition, pharmacogenomics data impinge on people's genetic variations and traits implicated in altered drug response that are usually different among populations from various parts of the world [15, 101, 102]. So, an obvious question is raised: How can pharmacogenomics-based molecular diagnostic tests be clinically tested and validated in order to be generally applied to all populations around the world, by taking apart the patient's ethnic or racial origin? The same question seems to apply to the case of the drug developmental process and the criticism developed regarding the use of pharmacogenomics criteria for the stratification of patients recruited in clinical trials in order to gain better benefits for the efficacy and safety of the drug being tested [103–106]. As recently claimed, at least 29 drugs (or combination of drugs) have been shown to produce different effects in terms of efficacy and safety among ethnic or racial groups [102]. Although there is enough skepticism on these issues, one cannot easily preclude the discussion of such phenomena from the drug delivery process and must group patients according to the existed drug-related information that correlates with variability observed in pharmacological response. The latter is strengthened by the fact that several individuals show a genetic etiology implicated in their altered pharmacotherapy outcome. In any case, the categorization in pharmacotherapy profiling, i.e., pharmacotyping, must be focused and always be referred to specific drugs and not to specific ethnic populations [5, 25]. In addition, the development of DNA-based molecular diagnostics by no mean must dismiss the utilization of classic biochemical tests such as measurement of enzymatic activity, but they have to work in parallel. As molecular diagnosis is getting simpler, cheaper, and readily automated, then the application of pharmacogenomics will become more easily integrated into the clinical practice [107–110]. Undoubtedly, as the field of pharmacogenomics and personalized medicine continues to expand toward the development of relevant clinical research data, more issues will be raised for society that will need to be addressed. However, until pharmacogenomics research is generally accepted in patient care, its beneficial clinical impact together with the ethical and societal aspects raised must be taken into consideration by all scientists involved in its application to clinical practice. In particular, clinicians along with pharmacists and bioinformaticians must work together in order for pharmacogenomics clinical use to be compatible with good pharmaceutical and clinical practice guidelines [5, 25]. Furthermore, issues related either to quality or validation of the genetic tests that could be applied in pharmacotherapy and therapeutics must be clearly addressed, before the usage of genomic information in routine patient care. In any case, however, the positive consequences in the clinical practice will be greatly enhanced by imperatively incorporating the fundamental principles of pharmacogenomics, personalized medicine, and pharmacotyping into the core curricula of pharmacy and medical schools. And for sure, this will be beneficial for both health and pharmaceutical care, as well as for the society and the public in general.

2.7 THE DEVELOPMENT OF INFORMATION-BASED PLATFORMS TO ENSURE MAJOR BENEFITS FOR PHARMACOTYPING IN CLINICAL PRACTICE

As mentioned, genetic variations detected in genes encoding proteins implicated in drug action could affect their function and thus might result in an altered drug response (e.g., decreased metabolism, inadequate intracellular transport, impaired function on target site, etc.). Also, genomic technologies provide valuable techniques for uncovering the molecular mechanisms leading to pathogenesis and phenotype of specific human diseases. The accumulated data are organized into specific databases; most of them can be accessed through the Web. Some of these functional genomics databases related to genetic disorders, gene polymorphisms, ADRs, and drug response variability are shown in Box 2.3. In this frame of genetic research, new genes, proteins, and genetic variations of clinical importance can be detected in individuals, because the new technology allows the screening of thou-

---

BOX 2.3 FUNCTIONAL GENOMICS DATABASES RELATED TO GENETIC DISORDERS, GENE POLYMORPHISMS, ADVERSE DRUG REACTION, AND DRUG RESPONSE VARIABILITY (MODIFIED FROM REF. 25)

1. David Nelson's Cytochrome P450 Homepage
   http://drnelson.utmem.edu/CytochromeP450.html
2. Human Cytochrome P450 Nomenclature Website
   http://www.imm.ki.se/CYPalleles
3. The SNP Consortium-High-density maps of SNPs in the human genome
   http://snp.cshl.org
4. Adverse Drug Reactions Unit (Australia)
   http://www.health.gov.au/tga/adr/index.htm
5. Australian Adverse Drug Reactions Bulletin
   http://www.health.gov.au/tga/adr/aadrb.htm
6. Database of cytochrome P450-mediated drug interactions
   http://www.drug-interactions.com
7. Database of drug-induced arrhythmia
   http://www.torsades.org
8. Entrez SNP
   http://www.ncbi.nlm.nih.gov/entrez/query.fcgi?db=snp
9. Pharmacological Targets Database (PTbase)
   http://research.bmn.com/ptbase
10. The "Online Mendelian Inheritance in Man" (OMIM) database (catalog of human genes and genetic disorders)
    http://www.ncbi.nlm.nih.gov/entrez/query.fcgi?db=OMIM
11. The International HapMap Research Project with haplotyping data of the human genome
    http://www.hapmap.org/
12. The official database of the International Union of Pharmacology (IUPHAR) on Receptor Nomenclature and Drug Classification
    http://www.iuphar-db.org/iuphar-rd/index.html
13. The Pharmacogenetics and Pharmacogenomics Knowledge Base (PharmGKB) (It is an integrated resource database that contains genomic data and molecular, cellular, as well as clinical phenotype data related to drug response.)
    http://www.pharmgkb.org/

14. The Endogenous GPCR List
   http://www.tumor-gene.org/GPCR/gpcr.html

15. GPCRDB: An Information System for G Protein-Coupled Receptors (GPCRs) (Version 8)
   http://www.gpcr.org/

16. NucleaRDB: An Information System for Nuclear Receptors (Version 4)
   http://receptors.ucsf.edu/NR/

17. The Wnt Gene Homepage (Wnt proteins form a family of highly conserved secreted signaling molecules that regulate cell-to-cell interactions during embryogenesis. Wnt genes and Wnt signaling are also implicated in cancer.)
   http://www.stanford.edu/~rnusse/wntwindow.html

18. Genomic glossaries and Taxonomies—Evolving Terminology for Emerging Technologies (Cambridge Healthtech Institute)
   http://www.genomicglossaries.com/

19. Genome Programs of the U.S. Department of Energy Office of Research
   http://www.doegenomes.org/

20. Human Genome Project Information
   http://www.ornl.gov/sci/techresources/Human_Genome/home.shtml

21. The "GeneTests" website is a publicly funded medical genetics information resource developed for physicians, other healthcare providers, and researchers in order to provide authoritative information on genetic testing and its use in diagnosis, management, and genetic counseling.
   http://www.geneclinics.org

sands of genes simultaneously. Genotyping and haplotyping analysis of individuals can identify specific gene variations and correlate them with drug effects, an approach that can obviously lead to individualized drug therapy. Furthermore, the conventional concept applied in drug therapy of considering patients suffering from the same disease as the homogenous population and using a similar drug therapy of all patients is changing in the genomic era. Under the new bioinformatic and genomic technologies used in personalized medicine, each patient is being handled as an individual, a practice that applies differentiation in pharmacotherapy [5, 25]. As a matter of fact, the detection of SNPs or other variations and the establishment of specific genotypes and haplotypes for several genes could explain why one patient responds well in drug therapy, another does not, or even why some people experience serious ADRs. Overall, if our understanding of these idiosyncratic pharmacological effects is finally achieved, then drug prescription would be greatly influenced. In fact, this will be a major transition in pharmacotherapy because pharmacotyping will be finally achieved, meaning that the physician could be based on the patient's genotyping–haplotyping analysis data for initiating and maintaining drug dosage therapeutic regimens for individual patients [5, 10, 25]. As the future prescription process is expected to be done by the physician in a fully computerized environment with the aid of genomic and bioinformatic information concerning patient status, pharmacotyping could better ensure drug efficacy and safety. Such a case also means a transition from a drug-selection process mainly based on the physician's own experience into a more, highly integrated, information and computer-aided pharmacotherapy, thus making drug delivery digitized, more efficient, and safer. It is evident, however, that the genetic information derived from pharmacogenomics studies must be incorporated into clinical practice in such a way to assess simultaneously, besides drug efficacy, the risk of ADRs attributed to specific drug specimens. This fact can be further exploited through the recent scientific achievements of nanotechology and nanomedicine that obviously facilitate the development, validation, and establishment of information-based infrastructure in health care. As the advances in health sciences, either at a basic or a clinical level, are incorporated into clinical practice, nanotechnology can create the suitable diagnostic and drug delivery platforms in the healthcare system for better exploitation of such information at the molecular level. However, the path toward this goal is difficult enough, especially considering the existing complexity in physiology of individual cells and even more in the whole organism. The latter coincides with the high-level degree of cellular dynamics and homeostatic capability to respond to intrinsic or environmental factors by extremely complex structural and functional diversity. Toward this direction, the successful development of suitable nanodevices and the construction of biocompatible nanomaterials in nanomedicine implies that several factors ranging from the level of physiology and molecular biology to that of physics and chemistry must be taken into consideration. Furthermore, the identification of functional SNPs and the establishment of specific haplotypes for genes implicated in drug response, or ADRs, could also be used as a marker for drug prediction effects in a given individual, or even for a specific group of patients [25]. In this way, a standard framework for information-based medicine must first be developed, in order for this later approach to support an infrastructure for personalized medicine to achieve pharmacotyping in routine pharmaceutical care.

Another crucial aspect toward the successful application of pharmacogenomics and personalized medicine is related to the way in which the unprecedented load of the genomic information available at the molecular level has to be easily and rapidly transformed into a type-form applicable for everyday use by the medical practitioner or other specialties in health care. Indeed, the fast growing accumulation of genomic data concerning drug action is a real challenge that needs to be clarified in terms of the development of suitable tools capable of analyzing such information in a manner to be readily compatible for clinical use. This is of great importance in current drug delivery and biomedicine. In such a case, the validation of the genomic data mining processes will totally be based on the development of suitable computerized and data integration systems in order for these then to support the clinical transformation of the gathered genotyping and haplotyping profiling data from individual patients, before making such information applicable for routine patient care. Moving toward personalized medicine also means the existence of tools and molecular diagnostics capable of assessing genome-related clinical information in laboratory medicine to make the extracted information easily applicable by the physician. This is task very difficult, because before the transfer of techniques from genomic-related research laboratories to those used for the routine analysis of clinical samples in diagnostic laboratories, these techniques must be extensively assessed for their ethical, social, and cost–benefit consequences [104]. As a consequence, the gradual integration of technologies transferring genomic information into the clinic needs the development of carefully selected and evaluated specific genetic biomarkers for the diagnosis of disease and prediction of drug response, as well as the use and application of genome-wide linkage analysis capable of genotyping, gene array, proteomics, transcriptomics, and metabolomics profiling [5, 105–107]. These developments of laboratory medicine should also highlight the way that must be undertaken in order to assure the level of quality and validity needed in genetic and molecular diagnostics tests. If this is confirmed in practice, then the individualized drug delivery profiling based on a patient's clinical, genomic, and bioinformatics data will be achieved in pharmacotherapy, e.g., pharmacotyping [5, 10]. Furthermore, the advent of specialized techniques coming from the advances of functional genomics will greatly influence the application of genomic information in laboratory and clinical practice. This also means that unified platforms must be developed to permit compatibility in handling different data gathered from unrelated sources like those of drug databases, clinical trials, DNA sequencing, and functional genomic analysis, in a way to ultimately support the clinical application of pharmacogenomics, personalized medicine, and pharmacotyping in health care.

The use of computational and bioinformatics approaches to predict the pharmacokinetics (absortion, distribution, metabolism, excretion; ADME) and pharmacodynamics properties of a drug are well appreciated throughout the process of drug development and delivery. Moving forward, the application of *in silico* methods and technology to evaluate safety and efficacy issues in pharmacotherapy by predicting mainly the emergence of drug interactions and ADRs in clinical practice are now considered major advancements to ultimately improve the success of new drug discovery and delivery outcomes [111]. Unfortunately, the data and the information generated up to now through the application of genomic and high-throughput technologies are impressive in scale but limited in clinical usefulness due to different database system formats and organization. For such an effort to finally succeed, information-based platforms in the drug development era and health care have to be developed to evaluate and integrate knowledge from different genomic and clinical sources in a manner that is simple for the end user. To this end, the application of semantic technologies with ontologies able to integrate the proper knowledge in a way that will be reusable by several applications and in different scientific areas can be a more beneficial approach toward better and quicker exploitation of genomic information in health care and therapeutics [112]. In any case, however, a crucial aspect in health care and pharmacotherapy to achieve major benefits from the technological and scientific advances in genomics and bioinformatics has been to create infrastructure and utilities that easily integrate each time the knowledge generated from different disciplines and facilities.

The clinical integration of genomic data in information-based platforms will also be greatly advanced by careful design and experimentation of pharmacogenomic studies, by ensuring genetic test quality, and evaluating and validating the data in routine patient care. This direction in medical and pharmacy practice will positively affect the rate by which genotyping data are transformed into specialized types applicable in personalized medicine, thus allowing pharmacotyping concepts to be generally accepted as the new dimension in drug delivery to ensure drug efficacy and safety. The future advancement of patient care is directly related to the ability of medical practitioners to collaborate with other specialties to apply genomic technologies in clinical practice through the development of suitable tools and information-based platforms compliant with the good pharmaceutical and clinical practice guidelines [5]. Although this direction needs a multidisciplinary approach in order to finally be achieved, a recently published study has clearly shown that it is an affordable task. In particular, the maintenance warfarin dose was estimated through an algorithm generated by evaluating at the same time several clinical, pharmacogenetic, and demographic factors that were obtained before the warfarin initiation therapy was applied to patients [113]. The development of such a pharmacogenomics-based algorithm to predict an individual's response to coumarin therapy and the patient's maintenance dose has been recently discussed, showing that the better elucidation of genetic variants that can affect the coumarin drug response is an efficient and beneficial way to improve pharmacotherapy outcomes [114]. However, before the generalized application of such pharmacotyping/pharmacogenomics concepts in drug delivery, the complexity of personalized prescribing in terms of raised ethnicity issues has clearly to be addressed and dismissed in clinical practice [115]. By means of suitable, cost-affordable, and precious pharmacogenomics methodology, the *in vivo* selection, structural genomics analysis, functional verification, and clinical validation of biological samples can be safely performed in laboratory medicine in a manner that is compatible with information-based platforms. Overall, pharmacogenomics- and informatics-based approaches are expected to be finally used, giving hope, in the years to come, that the association of specific genetic markers with drug delivery dosage regimens, even for the therapy of complex diseases, is a realistic approach and can be achieved in everyday health care.

2.8   CONCLUDING REMARKS

By knowing how to better use drugs in clinical practice with enhanced efficacy and safety through the exploitation of genomic knowledge, the drug delivery era will certainly be advanced and medical practitioners will gain further roles in this challenging health and pharmaceutical care environment. But in order for this attempt to finally succeed, the appropriate education must be given to healthcare professionals through the development of new curricula and educational approaches. The training has to be focused on pharmacogenomics, personalized medicine, and pharmacotyping concepts as well as on bioinformatics and information-based medical practice. Also, the differences and peculiarities in health care and education found among several countries all over the world must be seriously taken into account, when the profound changes in post-genomic drug delivery and clinical practice are organized through the development of information-based platforms to implement and improve patient care. The trends in patient care-related issues or even the changes happening in one part of the world must be carefully examined and then adjusted before their use in other regions and vice versa. And for sure, this will be beneficial for both health and pharmaceutical care, as well as for the society and the public in general.

REFERENCES

1. Evans W E, McLeod H L (2003). Pharmacogenomics—drug disposition, drug targets, and side effects. *N. Engl. J. Med.* 348:538–549.
2. Pirmohamed M, Park B K (2001). Genetic susceptibility to adverse drug reactions. *Trends Pharmacol. Sci.* 22:298–305.
3. Müller M (2003). Pharmacogenomics and the complexity of drug response. *Int. J. Clin. Pharmacol. Ther.* 41:231–240.
4. Ginsburg G S, McCarthy J J (2001). Personalized medicine: Revolutionizing drug discovery and patient care. *Trends Biotechnol.* 19:491–496.
5. Vizirianakis I S (2005). Improving pharmacotherapy outcomes by pharmacogenomics: From expectation to reality? *Pharmacogenomics*, In Press.
6. Housman D, Ledley F D (1998). Why pharmacogenomics? Why now? *Nature Biotechnol.* 16:2–3.
7. Evans W E, Relling M V (1999). Pharmacogenomics: Translating functional genomics into rational therapeutics. *Science.* 286:487–491.
8. McLeod H L, Evans W E (2001). Pharmacogenomics: Unlocking the human genome for better drug therapy. *Annu. Rev. Pharmacol. Toxicol.* 41;101–121.
9. Roses A D (2000). Pharmacogenetics and the practice of medicine. *Nature.* 405: 857–865.
10. Vizirianakis I S (2004). Challenges in current drug delivery from the potential application of pharmacogenomics and personalized medicine in clinical practice. *Curr. Drug Deliv.* 1:73–80.
11. Nebert D W, Nebert L J, Vesell E S (2003). Pharmacogenomics and "individualized drug therapy": High expectations and disappointing achievements. *Am. J. Pharmacogenomics.* 3:361–370.
12. Lazarou J, Pomeranz B H, Corey P N (1998). Incidence of adverse drug reactions in hospitalized patients: A meta-analysis of prospective studies. *J. Am. Med. Assoc.* 279:1200–1205.
13. Kohn L T, Corrigan J M, Donaldson M S (Eds.) (2000). *To Err is human: Building a safer health system. Committee on Quality of Health Care in America.* Institute of Medicine, The National Academy Press, Washington, D.C.
14. Flockhart D A, Tanus-Santos J E (2002). Implications of cytochrome P450 interactions when prescribing medication for hypertension. *Arch. Intern. Med.* 162:405–412.
15. Ingelman-Sundberg M (2004). Pharmacogenetics of cytochrome P450 and its applications in drug therapy: The past, present and future. *Trends Pharmacol. Sci.* 25: 193–200.
16. Ebbesen J, Buajordet I, Erikssen J, et al. (2001). Drug-related deaths in a department of internal medicine. *Arch. Intern. Med.* 161:2317–2323.
17. Ingelman-Saundberg M, Oscarson M, McLellan R A (1999). Polymorphic human cytochrome P450 enzymes: An opportunity for individualized drug treatment. *Trends Pharmacol. Sci.* 20:342–349.
18. Meyer U A (2000). Pharmacogenetics and adverse drug reactions. *Lancet.* 356:1667–1671.
19. Furuta T, Shirai N, Sugimoto M, et al. (2004). Pharmacogenomics of proton pump inhibitors. *Pharmacogenomics.* 5:181–202.
20. Schwab M, Eichelbaum M, Fromm M F (2003). Genetic polymorphisms of the human MDR1 drug transporter. *Annu. Rev. Pharmacol. Toxicol.* 43:285–307.

21. Siddiqui A, Kerb R, Weale M E, et al. (2003). Association of multidrug resistance in epilepsy with a polymorphism in the drug-transporter gene ABCB1. *N. Engl. J. Med.* 348:1442–1448.

22. Guillemette C (2003) Pharmacogenomics of human UDP-glucuronosyltransferase enzymes. *Pharmacogenomics J.* 3:136–158.

23. Hetherington S, Hughes A R, Mosteller M, et al. (2002). Genetic variations in HLA-B region and hypersensitivity reactions to abacavir. *Lancet.* 359:1121–1122.

24. Mallal S, Nolan D, Witt C, et al. (2002). Association between presence of HLA-B*5701, HLA-DR7, and HLA-DQ3 and hypersensitivity to HIV-1 reverse-transcriptase inhibitor abacavir. *Lancet.* 359:727–732.

25. Vizirianakis I S (2006). The transformation of pharmacogenetics into pharmacogenomics reinforces personalized medicine towards pharmacotyping for improved patient care. In Barnes L P (ed.), *New Research on Pharmacogenetics*: Nova Science Publishers, Inc., New York., In Press.

26. Watters J W, McLeod H L (2003). Cancer pharmacogenomics: Current and future applications. *Biochim. Biophys. Acta.* 1603:99–111.

27. Meisel P (2002). Arylamine N-acetyltransferases and drug response. *Pharmacogenomics.* 3:349–366.

28. Anantharam A, Markowitz S M, Abbott G W (2003). Pharmacogenetic considerations in diseases of cardiac ion channels. *J. Pharmacol. Exp. Ther.* 307:831–838.

29. Tsai Y J, Hoyme H E (2002). Pharmacogenomics: The future of drug therapy. *Clin. Genet.* 62:257–264.

30. McLeod H L, Siva C (2002). The thiopurine S-methyltransferase gene locus— implications for clinical pharmacogenomics. *Pharmacogenomics.* 3:89–98.

31. Coulthard S A, Hall A G (2001). Recent advances in the pharmacogenomics of thio-purine methyltransferase. *Pharmacogenomics J.* 1:254–261.

32. Arranz M J, Munro J, Birkett J, et al. (2000). Pharmacogenetic prediction of clozapine response. *Lancet.* 355:1615–1616.

33. Cadman P E, O'Connor D T (2003). Pharmacogenomics of hypertension. *Curr. Opin. Nephrol. Hypertens.* 12:61–70.

34. Manasco P K, Rieser P, Renegar G (2002). Pharmacogenetics and the genetic basis of ADRs. In: Mann, R D, Andrews, E B (eds.), *Pharmacovigilance*: Wiley, West Sussex, pp. 516–553.

35. Small K M, Tanguay D A, Nandabalan K, et al. (2003). Gene and protein domain-specific patterns of genetic variability within the G-protein coupled receptor super-family. *Am. J. Pharmacogenomics.* 3:65–71.

36. Drysdale C M, McGraw D W, Stack C B, et al. (2000). Complex promoter and coding region beta 2-adrenergic receptor haplotypes alter receptor expression and predict in vivo responsiveness. *Proc. Natl. Acad. Sci. U.S.A.* 97:10483–10488.

37. Shafran A, Ifergan I, Bram E, et al. (2005). ABCG2 harboring the Gly482 mutation confers high-level resistance to various hydrophilic antifolates. *Cancer Res.* 65:8414–8422.

38. Loktionov A (2004). Common gene polymorphisms, cancer progression and prognosis. *Cancer Let.* 208:1–33.

39. Anderson J L, Carlquist J F, Horne B D, et al. (2003). Cardiovascular pharmacogenomics: Current status, future prospects. *J. Cardiovasc. Pharmacol. Ther.* 8:71–83.

40. Cacabelos R (2004). The application of functional genomics to Alzheimer's disease. *Pharmacogenomics.* 4:597–621.

41. Ulrich C M, Robien K, Mcleod H L (2003). Cancer pharmacogenetics: Polymorphisms, pathways and beyond. *Nat. Rev. Cancer.* 3:912–920.

42. Sayers I, Hall I P (2005). Pharmacogenetic approaches in the treatment of asthma. *Curr. Allergy Asthma Rep.* 5:101–108.

43. Rusnak J M, Kisabeth R M, Herbert D P, et al. (2001). Pharmacogenomics: A clinician's primer on emerging technologies for improved patient care. *Mayo Clin. Proc.* 76:299–309.

44. McCarthy J J, Hilfiker R (2000). The use of single-nucleotide polymorphism in pharmacogenomics. *Nature Biotechnol.* 18:505–508.

45. Carpenter A E, Sabatini D M (2004). Systematic genome-wide screens of gene function. *Nature Rev. Genet.* 5:11–22.

46. Rudnicki M, Mayer G (2003). Pharmacogenomics of angiotensin converting enzyme inhibitors in renal disease—pathophysiological considerations. *Pharmacogenomics.* 4:153–162.

47. Humma L M, Terra S G (2002). Pharmacogenetics and cardiovascular disease: Impact on drug response and applications to disease management. *Am. J. Health-Syst. Pharm.* 59:1241–1252.

48. Ferrari P, Bianchi G (2000). The genomics of cardiovascular disorders. *Drugs.* 59:1025–1042.

49. Nakagawa K, Ishizaki T (2000). Therapeutic relevance of pharmacogenetic factors in cardiovascular medicine. *Pharmacol. Ther.* 86:1–28.

50. Ferrari P (1998). Pharmacogenomics: A new approach to individual therapy of hypertension? *Curr. Opin. Nephrol. Hypertens.* 7:217–222.

51. Liew C-C, Dzau V J (2004). Molecular genetics and genomics of heart failure. *Nat. Rev. Genet.* 5:811–825.

52. Roberts R, Brugada R (2003). Genetics and arrhythmias. *Annu. Rev. Med.* 54:257–267.

53. Saavedra J M (2005). Studies on genes and hypertension: A daunting task. *J. Hypertension.* 23:929–932.

54. Kurland L, Lind L, Melhus H (2005). Using genotyping to predict responses to antihypertensive treatment. *Trends Pharmacol. Sci.* 26:443–447.

55. Wrighton S A, Schuetz E G, Thummel K E, et al. (2000). The human CYP3A subfamily: Practical considerations. *Drug Metab. Rev.* 32:339–361.

56. Nelson D R, Zeldin D C, Hoffman S M, et al. (2004). Comparison of cytochrome P450 (CYP) genes from the mouse and human genomes, including nomenclature recommendations for genes, pseudogenes and alternative-splice variants. *Pharmacogenetics.* 14:1–18.

57. Phillips K A, Veenstra D L, Oren E, et al. (2001). Potential role of pharmacogenomics in reducing adverse drug reactions: A systematic review. *J. Am. Med. Assoc.* 286:2270–2279.

58. Kirchheiner J, Brosen K, Dahl M L, et al. (2001). CYP2D6 and CYP2C19 genotype-based dose recommendations for antidepressants: A first step towards subpopulation-specific dosages. *Acta Psychiatr. Scand.* 104:173–192.

59. Gabelli C (2003). Rivastigmine: An update on therapeutic efficacy in Alzheimer's disease and other conditions. Curr. Med. Res. Opin. 19:69–82.

60. Bentue-Ferrer D, Tribut O, Polard E, et al. (2003). Clinically significant drug interactions with cholinesterase inhibitors: A guide for neurologists. *CNS Drugs.* 17:947–963.

61. Sadee W, Hoeg E, Lukas J, et al. (2001). Genetic variations in human G protein-coupled receptors: Implications for drug therapy. *AAPS PharmSci.* 3(3):E22. Available: http://www.pharmsci.org.

62. Desta Z, Zhao X, Shin J-G, et al. (2002). Clinical significance of the cytochrome P450 2C19 genetic polymorphism. *Clin. Pharmacokinet.* 41:913–958.

63. Sakaeda T, Nakamura T, Okumura K (2003). Pharmacogenetics of MDR1 and its impact on the pharmacokinetics and pharmacodynamics of drugs. *Pharmacogenomics.* 4:397–410.

64. Bredel M, Zentner J (2002). Brain-tumour drug resistance: The bare essentials. *Lancet Oncol.* 3:397–406.

65. Hall I P (2002). Pharmacogenetics, pharmacogenomics and airway disease. *Respir. Res.* 3:10.

66. Watters J W, McLeod H L (2003). Cancer pharmacogenomics: Current and future applications. *Biochim. Biophys. Acta.* 1603:99–111.

67. McLeod H L, Watters J W (2004). Irinotecan pharmacogenetics: Is it time to intervene? *J. Clin. Oncol.* 22:1356–1359.

68. Krynetski E Y, Evans W E (2000). Genetic polymorphism of thiopurine S-methyltransferase: Molecular mechanisms and clinical importance. *Pharmacology.* 61:136–146.

69. O'Kane D J, Weinshilboum R M, Moyer T P (2003). Pharmacogenomics and reducing the frequency of adverse drug events. *Pharmacogenomics.* 4:1–4.

70. Marshall E (2003). Preventing toxicity with a gene test. *Science.* 302:588–590.

71. Coulthard S A, Hogarth L A, Little M, et al. (2002). The effect of thiopurine methyltransferase expression on sensitivity to thiopurine drugs. *Mol. Pharmacol.* 62:102–109.

72. Goldstein D B (2003). Pharmacogenetics in the laboratory and the clinic. *N. Engl. J. Med.* 348:553–556.

73. Traxler P (2003). Tyrosine kinases as targets in cancer therapy—successes and failures. *Expert Opin. Ther. Targets.* 7:215–234.

74. Tenen D G (2003). Disruption of differentiation in human cancer: AML shows the way. *Nature Rev. Cancer.* 3:89–101.

75. Graf T (2002). Differentiation plasticity of hematopoietic cells. *Blood.* 99:3089–3101.

76. Hockenbery D M, Giedt C D, O'Neill J W, et al. (2002). Mitochondria and apoptosis: New therapeutic targets. *Adv. Cancer Res.* 85:203–242.

77. Fischer O M, Streit S, Hart S, et al. (2003). Beyond Herceptin and Gleevec. *Curr. Opin. Chem. Biol.* 7:490–495.

78. Tsiftsoglou A S, Pappas I S, Vizirianakis I S (2003). Mechanisms involved in the induced differentiation of leukemia cells. *Pharmacol Ther.* 100:257–290.

79. Capdeville R, Buchdunger E, Zimmermann J, et al. (2002). Clivec (STI-571, imatinib), a rationally developed, targeted anticancer drug. *Nature Rev. Drug Discov.* 1:493–502.

80. Cowan-Jacob S W, Guez V, Fendrich G, et al. (2004). Imatinib (STI571) resistance in chronic myelogenous leukemia: molecular basis of the underlying mechanisms and potential strategies for treatment. *Mini Rev. Med. Chem.* 4:285–299.

81. Suzuki T, Mitsudomi T, Hida T (2004). The impact of *EGFR* mutations on gefitinib sensitivity in non-small cell lung cancer. *Personalized Med.* 1:27–34.

82. Golsteyn R M (2004). The story of gefitinib, an EGFR kinase that works in lung cancer. *Drug Discov. Today.* 9:587.

83. Paez J G, Janne P A, Lee J C, et al. (2004). EGFR mutations in lung cancer: Correlation with clinical response to gefitinib therapy. *Science.* 304:1497–1500.

84. Lynch T J, Bell D W, Sordella R, et al. (2004). Activating mutations in the epidermal growth factor receptor underlying responsiveness of non-small-cell lung cancer to gefitinib. *N. Engl. J. Med.* 350:2129–2139.

85. Golsteyn R M (2005). Gefitinib does not increase survival in lung cancer patients. *Drug Discov. Today.* 10:381

86. Domen J (2001). The role of apoptosis in regulating hematopoietic stem cell numbers. *Apoptosis.* 6:239–252.

87. Tsiftsoglou A S, Pappas I S, Vizirianakis I S (2003). The developmental program of murine erythroleukemia cells. *Oncol. Res.* 13:339–346.

88. Lindpaintner K (2002). The impact of pharmacogenetics and pharmacoge-nomics on drug discovery. *Nature Rev. Drug Discov.* 1:463–469.

89. Cho H-S, Mason K, Ramyar K X, et al. (2003). Structure of the extracellular region of HER2 alone and in complex with the Herceptin Fab. *Nature.* 421:756–760.

90. Ma X J, Salunga R, Tuggle J T, et al. (2003). Gene expression profiles of human breast cancer progression. *Proc. Natl. Acad. Sci. U.S.A.* 100:5974–5979.

91. Minn A J, Gupta G P, Siegel P M, et al. (2005). Genes that mediate breast cancer metastasis to lung. *Nature.* 436:518–524.

92. Cheok M H, Yang W, Pui C H, et al. (2003). Treatment-specific changes in gene expression discriminate in vivo drug response in human leukemia cells. *Nature Genet.* 34:85–90.

93. Slonim D K (2001). Transcriptional profiling in cancer: The path to clinical pharmacogenomics. *Pharmacogenomics.* 2:123–136.

94. Vizirianakis I S (2002). Pharmaceutical education in the wake of genomic technologies for drug development and personalized medicine. *Eur. J. Pharm. Sci.* 15:243–250.

95. Gurwitz D, Weizman A, Rehavi M (2003). Education: Teaching pharmacogenomics to prepare future physicians and researchers for personalized medicine. *Trends Pharmacol. Sci.* 24:122–125.

96. Brock T P, Valgus J M, Smith S R, et al. (2003). Pharmacogenomics: Implications and considerations for pharmacists. *Pharmacogenomics.* 4:321–330.

97. Wieczorek S J, Tsongalis G J (2001). Pharmacogenomics: Will it change the field of medicine? *Clin. Chim. Acta.* 308:1–8.

98. Hughes I, Hollingsworth M, Jones S J, et al. (1997). Knowledge and skills needs of pharmacology graduates in first employment: How do pharmacology courses measure up? *Trends Pharmacol. Sci.* 18:111–116.

99. Landro J A, Taylor I C, Stirtan W G, et al. (2000). HTS in the new millennium: The role of pharmacology and flexibility. *J. Pharmacol. Toxicol. Methods.* 44:273–289.

100. Rodriguez R, Vidrio H, Lopez-Martinez E, et al. (1997). Changing the countenance of pharmacology courses in medical schools. *Trends Pharmacol. Sci.* 18:314–318.

101. Varmus H (2002). Getting ready for gene-based medicine. *N. Engl. J. Med.* 347:1526–1527.

102. Tate S K, Goldstein D B (2004). Will tomorrow's medicines work for everyone? *Nat. Genet.* 36:S34-S42.

103. Lindpaintner K (2002). The impact of pharmacogenetics and pharmacogenomics on drug discovery. *Nature Rev. Drug Discov.* 1:463–469.

104. Issa A M (2000). Ethical considerations in clinical pharmacogenomics research. *Trends Pharmacol. Sci.* 21:247–250.

105. Issa A M (2002). Ethical perspectives on pharmacogenomic profiling in the drug development process. *Nat. Rev. Drug Discov.* 1:300–308.

106. Weber W W, Caldwell M D, Kurth J H (2003). Edging toward personalized medicine. *Curr. Pharmacogenomics.* 1:193–202.

107. Debouck C, Goodfellow P N (1999). DNA microarrays in drug discovery and development. *Nature Genet.* 21(Suppl):48–50.

108. Sander C (2000). Genomic medicine and the future of health care. *Science.* 287: 1977–1978.

109. Schmitz G, Aslanidis C, Lackner K J (2001). Pharmacogenomics: Implications for laboratory medicine. *Clin. Chim. Acta.* 308:43–53.

110. Hardiman G (2004). Microarray platforms-comparisons and contrasts. *Pharmacogenomics.* 5:487–502.

111. Ekins S (2004). Predicting undesirable drug interactions with promiscuous proteins in silico. *Drug Discov. Today.* 9:276–285.

112. Gardner S P (2005). Ontologies and semantic data integration. *Drug Discov. Today.* 10:1001–1007.

113. Gage B F, Eby C, Milligan P E, et al. (2004). Use of pharmacogenetics and clinical factors to predict the maintenance dose of warfarin. *Thromb. Haemost.* 91:87–94.

114. Voora D, McLeod H L, Eby C, et al. (2005). The pharmacogenetics of coumarin therapy. *Pharmacogenomics.* 6:503–513.

115. Hall I P (2005). Pharmacogenetics and ethnicity: More complexities of personalized prescribing. *Am. J. Respir. Crit. Care Med.* 171:535–536.

3.1

TOXICOGENOMICS

MARY JANE CUNNINGHAM AND MRINAL SHAH
Houston Advanced Research Center (HARC), The Woodlands, Texas

Chapter Contents

3.1.1 Introduction   229
3.1.2 Genomics in Toxicology   230
3.1.3 Proteomics in Toxicology   234
3.1.4 Metabonomics in Toxicology   237
3.1.5 Pharmacogenomics in Toxicology   238
3.1.6 Systems Biology   239
3.1.7 Future Technologies in Toxicogenomics   240
      References   242

3.1.1 INTRODUCTION

Toxicogenomics is a relatively new discipline within the field of toxicology. The phrase was first coined in 1998 at the first Toxicogenomics Workshop held as part of the U.S. and European Community Consortium on Molecular Toxicology in Palo Alto, California. In its broadest sense, it is defined as the use of OMICS technologies to investigate issues of toxicity [1]. In its narrowest sense, it is defined as investigating the safety of compounds by using only cutting-edge gene expression technologies [2, 3]. In this chapter, the broader definition will be applied. Toxicogenomics is the use of OMICS technologies to assess the safety of new chemical entities or other compounds used in diagnostics or therapeutics [4–6].

OMICS technologies encompass genomics, proteomics, metabonomics, and pharmacogenomics [5]. Other descriptive terms (for example, transcriptomics, toxicoproteomics, and toxicogenetics) have been used intermittently, but most

Handbook of Pharmaceutical Biotechnology, Edited by Shayne Cox Gad.

references have centered on those listed above. Genomics is the study of gene expression through the use of high-throughput screening techniques, such as gene expression microarrays. Other gene expression techniques have been added to this category, and these methods include gene reporter assays, branched DNA amplification assay, scintillation proximity assay (SPA), rapid analysis of gene expression (RAGE), serial analysis of gene expression (SAGE), and various polymerase chain reaction PCR-based assays (e.g., real-time, quantitative, representational difference analysis, differential display) as has already been reviewed [7]. However, in some cases, it may be a stretch to qualify the method as "high-throughput" because several of these technologies either address gene expression more indirectly, such as SAGE, or can only address a few genes at a time, such as PCR. Proteomics, on the other hand, is the study of protein expression using either two-dimensional polyacrylamide gel electrophoresis (2DE) annotated by mass spectrometry (MS) or protein expression microarrays [7, 8]. Metabonomics is the use of high-resolution combinations of nuclear magnetic resonance (NMR), chromatography, or mass spectroscopy to evaluate metabolite profiles of body fluids or cells [9]. Finally, pharmacogenomics is the study of genetic variability to explain the adverse effects caused by compound–cell interactions [10–12]. Most pharmacogenomics studies have focused on single nucleotide polymorphisms (SNPs). This chapter will describe all of these OMICS methods and show how they can be applied to studying toxicity in pharmaceutical development. Case studies will be presented from very early citations to examples of current applications. Data analysis methods will not be discussed but are discussed in Chapter 2. This chapter will conclude with the authors' views on what the future may hold.

3.1.2 GENOMICS IN TOXICOLOGY

The applications of genomics to toxicology will be discussed with the focus on gene expression microarrays and how they have been used to investigate toxicity. Microarrays are covered in detail in Chapter 5.6. Briefly, they are manufactured by attaching pieces of DNA or RNA molecules to a nitrocellulose filter, a glass slide, or a silicon wafer [13]. Although early forms of arrays used nitrocellulose filters (now referred to as "macroarrays"), the most common substrate used today is a glass slide. Once the arrays have been printed with portion(s) of DNA or RNA (referred to as "probes"), they can be stored for further processing. The next step is to make the "targets" or isolated RNAs from both control cells and treated cells. These RNAs are labeled with a tag (most commonly fluorescent) and hybridized to complementary probes on the printed arrays. The microarrays are washed, scanned, and image analyzed to derive quantitative values for each probe signal. The remaining steps include data analysis of this raw information yielding (1) statistical analysis of replicate arrays and a measure of the variability in the processing steps, (2) similarities between gene expression patterns of control and treated samples, (3) identification of significantly expressed genes that may be biomarkers, and (4) patterns of gene–gene and gene–cell interactions.

Microarrays have been used as a screening tool in drug discovery and development. However, the focus and application of arrays in toxicology differs from discovery in several aspects. The emphasis in discovery is feasibility: to ascertain the interaction of the compound and drug target rapidly. The emphasis in toxicology is development and validation: to understand this interaction more fully and, more importantly, to assess if any adverse effects are occurring. The focus here is on (1) validation of the compound–target interaction, (2) the prediction of adverse effects, and (3) the discovery of any alternative mechanisms of action resulting in "off-target" effects and aiding in the redesign of the candidate to a less toxic substance.

One of the most significant factors in microarray experiments is the experimental design. The overall design is dictated by whether a one-color (one label) or two-color (two labels) array system is being used [14, 15]. A one-color system requires that the control and treated samples are run on separate arrays, and therefore, intra-array as well as interarray variability needs to be considered. A two-color system requires that both the control target and the treated target compete for a complementary probe molecule on the same array. In this system, the importance of the intra-array variability is lessened. This aspect is the same whether a discovery approach or a toxicology approach is being used.

However, several aspects of the experimental design do differ between microarray screens used in discovery or toxicology. In toxicity assessments using microarrays, the emphasis on validation is an essential requirement of the experimental design. Particular attention needs to be paid to (1) types of species; (2) gender; (3) cell or tissue system; (4) treatment scheduling, dose, and route; (5) appropriate controls; (6) numbers of replicates, and (7) correlation with independent morphological and pathological toxicity assays [4, 5, 14]. Making these choices is critical for minimizing the process variability in order to maximize and visualize the biological variability. In 2001, The Microarray Gene Expression Data (MGED) Society proposed a list of guidelines for microarray work, known as the Minimum Information About a Microarray Experiment (MIAME) guidelines [16]. These guidelines are recommended requirements for how experiments are designed and reported. A set of guidelines specifically for toxicogenomics applications was drafted, but this draft has not yet been agreed to [17, 18].

Another aspect unique to toxicology research is the emphasis on analyzing the activity of annotated genes as opposed to unannotated genes. Annotated genes are genes whose sequence and function is already known. Unannotated genes are genes whose sequence is known but have not been assigned a function. As the focus for toxicology is validation, one priority is correlating significant expression of annotated genes observed in the data with their previously published values. A secondary priority is discovering which annotated genes or proteins are involved in any adverse or "off-target" effects. The final priority is discovering whether any unannotated genes are involved in these adverse effects and what their function or role may be. Therefore, when selecting microarrays to use, arrays that have the most annotated genes are preferred. Two exceptions to this recommendation are (1) whole genome arrays and (2) focus arrays. The availability of whole genome arrays, or arrays containing the entire complement of genes for that organism, make the selection of arrays biased toward annotated genes a mute point. Focus arrays are sets of a smaller number of genes printed several times on one substrate. These arrays allow for profiling different samples using one array and, conceivably, hundreds of compounds could be screened in a high-throughput manner.

A third aspect of the experimental design that differs is the use of control or untreated samples. Discovery-focused experiments have emphasized the use of pooled samples to maximize consistency and minimize cost [19–25]. Pooled samples are individual samples from a cell or tissue system within a species and from the same treatment group that are pooled together to make one sample. In some cases, the control samples (from normal, non-diseased individuals) are pooled to make a "universal" reference [19]. However, the concerns are that (1) the resulting gene expression is an average of the group of samples, (2) the whole range of expression for the individuals within the group is not observed, and (3) pooling can only take place if the biological variation is greater than the process variation. [21–24]. One proposed method to address some of these concerns is "sub-pooling" where only subsets of samples are pooled and there are still replicates run for each group [25].

Toxicology-focused experiments may also include different sets of control samples. One set of control samples includes untreated samples both at the initial time point as well as at subsequent time points. If sufficient activity is found in the untreated control curve, this activity may need to be subtracted from the activity of the test curve. Another set of control samples includes samples from treatments using both positive and negative control compounds. Investigating the gene expression of the test compound is done by comparing it with the gene expression of the known control compounds. All of these controls need to be woven into the data analysis scheme.

As the focus is on validation, part of the evaluation is to determine how much the individual samples from a group vary in their gene expression. If the samples are pooled, the differences between responders and nonresponders will not be seen [24].

To understand toxicogenomics, it may be important to understand how microarrays evolved. An ongoing effort for developing more efficient sequencing by hybridization methods led to various initial microarray efforts where large numbers of genes could be screened at a time [26–30]. The first use of a microarray was in 1987 by Augenlicht et al., where macroarrays were used to differentiate a disease state [31]. In this case, cDNAs from a reference polyA mRNA library of the human colon carcinoma cell line, HT-29, were inserted into bacterial plasmids. Over 4000 clones were isolated and gridded onto several nitrocellulose filters. Radiolabeled cDNA probes from several biopsy samples (ranging from patients at low risk for colon cancer to familial adenomatous polyposis [FAP] patients to colon cancer patients) were then hybridized to the filters. The amount of radiolabel was scanned and analyzed. The amazing fact from this experiment was that a high percentage (20%) of FAP biopsies (in which the cells had not yet accumulated into adenomas) was upregulated as compared with the low-risk biopsies. These results suggested that increased gene expression seemed to correlate with early stages of the disease.

Initial toxicology studies using microarrays focused on screening and prioritizing lead compounds. Gray et al. designed a combinatorial library of protein kinase inhibitors used in cancer therapy and initially screened the compounds through an *in vitro* toxicity activity assay [32]. Then, three compounds were selected, and their interactions with yeast were investigated with oligonucleotide microarrays. The microarray results confirmed the diminished activity of one compound observed in the *in vitro* assay. Another collaborative study between Incyte and Tularik screened several lead compounds for efficacy and toxicity using two-color microarrays [33]. The result was that the optimized lead compound had a similar profile to a known toxin, which led it to be redesigned to a better nontoxic lead compound.

One early study focused on discovering alternative modes of action. Karpf et al. used microarrays to broadly screen for genes expressed when a colon adenocarcinoma cell line was exposed to 5-aza-2′-deoxycytidine [34]. Several genes were significantly expressed. When investigated further, an alternative pathway was discovered and linked to the signal transducer and activator of transcription (STAT) genes.

The first comprehensive gene expression profiling time-course study using microarrays was first cited in 1998 by Cunningham et al. [35–38]. Using an acute short-term exposure regimen, rats were treated with toxic doses of three known hepatotoxins: benzo(a)pyrene (BP), acetaminophen (APAP), and clofibrate (CLO). mRNA isolated from the livers were analyzed using a cDNA microarray containing 7400 rat genes. Significant gene activity was observed at the early time points (12 hours, 1 day, and 3 days), and less activity was observed at the later time points (7, 14, and 28 days). All three compounds resulted in different expression profiles. Several cytochrome P450 genes and genes involved in phase II reactions were expressed over all six time points; however, the genes were induced at different times depending on the compound used (Figure 3.1-1). Three different data analysis methods were compared. It was hypothesized that both nongenotoxic (causing damage by a non-DNA mechanism) APAP and CLO would have the most significantly expressed genes in common compared with genotoxic (damage caused by a DNA mechanism) BP. However, with all three analysis methods, APAP and BP showed the most overlap of significantly expressed genes. Interestingly, both APAP and BP have their primary metabolic pathway involving cytochrome P450, whereas the metabolism of CLO involves both cytochrome P450 and β-oxidation.

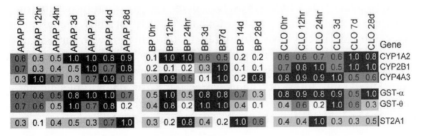

Figure 3.1-1. Normalized differential expression values for each gene across all time points (0 h, 12 h, 24 h, day 3, day 7, day 14, and day 28) for the following treatments: acetaminophen (APAP), benzo(a)pyrene (BP), and clofibrate (CLO). The highest relative expression is 1.0, and the lowest is 0.1. Three genes from the cytochrome P450 family (CYP1A2, CYP2B1, and CYP4A3) and two genes from the glutathione-S-transferase family (GST-α and GST-θ) give expression profiles that are similar to other members of the family but not identical. Gene expression profiles differ among all three gene families: CYP, GST, and sulfotransferase (ST).

Since these early citations, the use of gene expression microarrays as a tool for predictive toxicology has progressed. An early emphasis was to use microarrays to classify compounds based on their resulting gene expression profiles [39–46]. More recently, specific "sets of genes" giving rise to a gene expression profile have been used to distinguish classes of compounds [47, 48]. For example, Hu et al. showed through hierarchical clustering that distinct differences were observed between the gene expression profiles of direct- and indirect-acting genotoxins. When an array of over 9900 probes was used, a set of 58 genes could distinguish between these classes of compounds.

Another emphasis was to use gene expression microarrays to investigate mixtures. The profiles of the cellular interactions of mixtures were compared to the profiles of each component of the mixtures [49, 50]. Nadadur et al. compared the profile of residual oil fly ash (ROFA) in rat lung to the single profiles of vanadium sulfate and nickel sulfate, which are known components of ROFA. ROFA is derived from the exhaust plume of a power plant and is used as a model mixture to investigate the effects of air pollution. Using a focus macroarray, 12 genes were found to be significantly expressed in all three treatments. In a more recent study, arrays were shown to be able to distinguish toxic components of another type of mixture, wastewater effluent. From this study, Wang et al. found 11 genes that may ascertain the presence of endocrine-disrupting compounds in the environment [51].

A third emphasis was to match gene expression profiles to phenotypic changes. This correlation is the current goal for toxicogenomics. It will help toxicogenomics to be established as a tool in the regulatory arena. Early investigations focused on correlating gene expression from *in vivo* biological systems to gene expression from *in vitro* systems [52, 53]. Taking this concept further, attempts were made to correlate gene expression with specific organ regions exhibiting toxicity [54]. More recent studies have focused on anchoring specific cellular histopathological changes to gene expression [55–61]. To identify these changes, investigators have used laser capture microdissection (LCM) to extract specific cells exhibiting the phenotypic changes and then comparing the gene expression profiles of these cell populations with "normal" cell populations [62–66].

Finally, the most pressing issue in toxicogenomics may be the relevance of gene expression to toxicity. In other words, how does both the qualitative and the quantitative expression of a gene relate to toxicity? Several studies, including most of the papers already cited, have attempted to find this answer. Investigations by Heinloth et al. and Zhou et al. reported seeing gene expression changes before the observance of the overt toxicity [63, 67]. In each paper, *subchronic* doses of a toxicant were used and links of their expression with phenotypic changes were attempted. Today, research is ongoing, with the ultimate goal, to define a set of biomarkers that can predict toxicity earlier than the occurrence of phenotypic changes.

3.1.3 PROTEOMICS IN TOXICOLOGY

Two new words, *proteome* and *proteomics*, were coined a decade after classic terms such as *genome* and *genomics*. Here *proteome*, the complement of a *genome*, refers to the expression of all proteins in an organism, organ, or cell line at any given time. The study of the proteome, its interactions, and its post-translational modifications (such as glycosylation, phosphorylation, etc.) that may be associated with a specific disease or toxicant is defined as *proteomics* [68, 69]. One advantage of proteomics is the ability to profile several thousands of proteins on a single platform and provide insights into post-translational modifications. Another advantage is that because proteins govern the normal physiology and disease processes, a more accessible endpoint of any changes in gene expression can be determined by direct analysis of proteins. Also, unlike genomics, proteomics offers the advantage of readily assaying both tissues and body fluids (e.g., plasma and serum) to investigate the molecules correlating with disease and drug action [70]. The proteomic analysis of body fluids is of particular interest as it can provide a noninvasive method for biomarker identification [71].

In the realm of toxicology, proteomics is one of many new technologies being used today to identify novel protein biomarkers and signature patterns in protein profiles that measure sensitive cellular changes in response to xenobiotic exposure [72–75]. The most widely used technique to date for proteomic science is high-resolution 2DE coupled with MS. First developed in 1975, it separates proteins in the first dimension based on their isoelectric points and in the second dimension by molecular weight [76]. Typically, 2DE can resolve hundreds to thousands of proteins from a single sample [77, 78]. To avoid overcrowding on a single gel, researchers now run multiple "zoom" gels that cover narrow pH ranges (e.g., covering 1 pH unit) [79]. In the mid-1990s, 2DE was coupled to MS allowing for annotation of the resolved spots. Proteins resolved using 2DE are analyzed by MS to give mass spectra. Subsequently, individual spots are excised from the gel and trypsin-digested to give shorter peptide fragments for sequencing. With the help of software programs and sequence databases, the peptide mass spectrum is compared with known partial gene and protein sequences and is then assigned an annotation. Recent developments in MS techniques, such as "tandem MS" (MS/MS), allow for better analysis of protein sequences, by avoiding ambiguity arising when MS data match more than one protein sequence. A more detailed description of proteomic methods is found in Chapter 14.2.

The use of proteomics in toxicology can be divided into two classes: (1) investigative studies and (2) screening/predictive toxicology [71]. The mechanism of toxic damage in several model systems has been elucidated with proteomics [72, 79, 80]. But proteomics offers the prospect of identifying new toxic mechanisms along with traditional toxicology methods [74, 79, 81]. Such insights may provide valuable information about specific effects caused by a specific group or class of compounds. In addition to supporting investigative studies, proteomics promises to provide insight into screening and predictive toxicology. The screening of new compounds for toxicity can be possible by following specific signatures of protein expression, thus identifying them as predictive biomarkers of toxicity [82]. The sensitive nature of proteomics offers the detection of toxic effects at low dosage levels, which escape conventional screening methods, such as histopathology and clinical chemistry [83]. In addition, proteomics provides a method for the early detection of potential toxicity and the opportunity of ranking compounds during drug development. A recent review lists publications that use proteomics in toxicology studies, with the most popular application being biomarker identification [84]. The issues surrounding the experimental design and the emphasis on development and validation covered in the previous section for genomics applies to proteomics as well.

As an investigative tool, proteomics has been used to determine the failure of several drug candidates due to adverse drug reactions [72, 74]. In an early study, Anderson et al. developed a rodent liver proteomic toxicity database [85]. This database, containing profiles of effects of 43 compounds on liver, was among the first available for the detection, classification, and characterization of a wide range of hepatotoxins. The development of such a database has considerable implications for predictive toxicology. The potential hepatotoxicity of a new and unknown compound can be determined by comparing its effects on the proteome against known toxins from the database [74, 85].

New biomarker discovery from such studies can lead to early detection of deleterious effects. In a more recent study, researchers validated toxicological protein markers from an *in vivo* system (rat liver) as well as from an *in vitro* system (human HepG2 cell line) [86]. They reported a total of 11 protein markers with reactivity toward multiple toxic compounds and no reactivity toward nontoxic compounds. An important conclusion from this work is that cells in culture can be used as an *in vitro* toxicity testing system to assess hepatotoxicity. However, in the future, a much more extensive study may be required to identify a larger group of toxicology markers to detect more diverse types of toxic reactions.

Toward understanding the mechanistic action of toxicology using 2DE, Fountoulakis et al. showed the hepatotoxic effects of paracetamol (acetaminophen) overdose in human and rodent liver [74]. They found that the expression of 35 proteins was altered after treatment with paracetamol or its nontoxic regioisomer 3-acetamidophenol. Paracetamol selectively increases or decreases the phosphorylation state of proteins. This effect translates to a decrease in protein phosphatase activity [80]. The control of cellular functions of cells may be lost as a result of the dephosphorylation of certain regulatory proteins due to paracetamol overdose.

A dose- and time-dependent proteomics study with gentamicin confirmed the histopathological findings of renal toxicity at high doses and renal regeneration during the recovery phase [72]. Gentamicin, a known renal toxicant, represents a class of aminoglycosides. The effects of gentamicin treatment were observed after proteomic evaluation of rat kidney cortex samples. Examination of rat serum samples exposed to varying doses of gentamicin identified a protein being consistently overexpressed. Surprisingly, this particular protein marker was present both at the low treatment dose as well as at early time points before changes were observerd by routine clinical pathology. Intriguingly, during the recovery phase, the marker returned to control levels, thereby highlighting the sensitive nature of proteomics. Such protein markers are of great interest as they provide a noninvasive means of monitoring the onset of toxicity before any evident cellular damage.

Although 2DE offers a high-quality approach to proteomic research, it has certain limitations: Only major components of protein mixtures are visualized, the detection of low and high molecular mass of basic and hydrophobic proteins is inefficient, and it is a highly laborious technique. Alternative technologies, such as the protein microarrays, allows investigators to control conditions, such as pH, temperature, ionic strength, and different stages of protein modification, while monitoring the protein–protein interactions of thousands of proteins spotted on an array [87]. Protein microarrays not only provide a technique for high-throughput screening but also a method to study interactions of proteins with non-protein-aceous molecules. Zhu et al. have developed a protein microarray of the yeast proteome from 5800 open reading frames [88]. The expressed proteins are purified and printed on glass slides with high spatial density to form a yeast proteome microarray. They were used to screen for interactions with proteins and phospholipids. The researchers identified many new calmodulin- and phospholipid-interacting proteins. Thus, protein microarrays can serve to screen for several hundreds to thousands of biochemical activities on a single chip. In toxicology, these arrays can be used to screen for adverse protein–drug interactions, to detect post-translational modifications leading to disease, and to identify biomarkers for drug targets and early disease detection [89].

A recent advancement is the coupling of MS with LCM for identification of tumor markers specific for cancer [89]. Specific cells of pathological interest are selected with LCM under direct microscopic visualization, laser captured, and removed from the tissues. The extracted lysate from the cells is applied directly to spots on the substrate of a surface-enhanced laser desorption/ionization (SELDI) MS, ionized, and desorbed from the surface, and a time-of-flight (ToF) proteomic profile of the entire cellular system is observed. This method has been recently applied to toxicology [90].

Several efforts are attempting to improve proteomic technologies to map and measure proteomes and subproteomes. However, no single proteomic platform seems to ideally suit and quantify the broad range of protein expression in a given cell/tissue system. More than one proteomic platform may likely be needed to distinguish the multiple forms and post-translational modifications of proteins, to address the inadequate annotation of proteomes, or to accomplish integration of proteomic data with genomic and metabonomic data. Additionally, advances in genomics data through investigations of pathways and subcellular structures induced by toxicity may guide studies in proteomics.

3.1.4 METABONOMICS IN TOXICOLOGY

The term metabonomics was derived from early work in Dr. Jeremy Nicholson's laboratory [9]. This term along with metabolomics has been used in several citations interchangeably [91–94]. Attempts have been made to clarify the definition of each term [95–99]. It seems that agreement has been reached where metabonomics is the broader encompassing term of metabolite profiling and metabolomics refers specifically to profiling in cells [9, 95, 98].

This technology is covered in detail in Chapter 14.3, but briefly it uses combinations of NMR, MS, and chromatography to profile multiple components in biofluids, tissues, or cells. The main advantage of metabonomics is the serial sampling of a biofluid noninvasively. The earliest papers used high-resolution ^{1}H-NMR alone for profiling [100, 101]. The technique became more sophisticated when bioinformatics tools, such as pattern recognition and expert systems, were applied to the data [102, 103]. Today, the following combinations of techniques are used: liquid chromatography–nuclear magnetic resonance (LC–NMR), magic angle spinning–nuclear magnetic resonance (MAS–NMR), gas chromatography–mass spectroscopy (GC–MS), LC–MS, capillary electrophoresis–mass spectroscopy (CE–MS), and

LC–NMR–MS [97, 98]. The first technique listed in the combinations above refers to the method used to separate the multiple components of the sample, and these components are identified using the second technique [96]. The advantages and limitations for each combination of techniques have already been reviewed [94, 97].

Two of the earliest papers using metabonomics in toxicology were studies by Nicholson et al., which investigated mercury and cadmium toxicity in the rat [101, 104]. Each of these papers used an experimental design that reiteratively sampled a biological system over either dose or time using high-resolution ^{1}H NMR. The changes in the urinary excretion patterns correlated with the expected histological changes. However, an interesting note was that in the case of cadmium, a urinary metabolite related to nephrotoxicity was found that was undetectable using traditional methods. A more extensive study that included a-naphthylisothiocyanate and 2-bromoethylamine showed changes in the metabolite profiles ahead of the time point where the clinical chemistry or microscopic changes would be detected [105].

As with the previously described OMICS technologies, one emphasis has been to profile groups of compounds using metabonomics [103, 105, 106]. Each compound gave a unique metabonomic profile and was successfully classified using pattern recognition techniques. A second emphasis was to be able to monitor not only biofluids, such as urine or plasma, but also tissues and cells. Moka et al. performed the first study to classify carcinoma biopsy samples by MAS–NMR [107]. Since then, two studies have used metabonomics to investigate toxicity using a wider approach where urine, plasma, and tissue samples are all analyzed [108, 109].

To collaborate and share information, a proposal for a users group in metabonomics was put forth [110]. Since then, the Consortium for Metabonomic Toxicology (COMET) was formed and the first report from this group was given in 2003 [111]. The final report was just published [112]. This report describes what methodologies have been assessed, the development of a curated database for metabonomic profiles, and computer-based experts systems for data analysis.

3.1.5 PHARMACOGENOMICS IN TOXICOLOGY

Pharmacogenomics and pharmacogenetics are two terms that have created confusion because they have been used interchangeably and defined differently by several authors [113]. The most widely accepted term, pharmacogenetics, is the study of an individual's response to a drug as determined by their genetic makeup [114]. Pharmacogenetics as a field of study has been around for several decades [113]. Early studies focused on observations of patient variability in metabolizing various drugs such as primaquine, succinylcholine, and dibucaine [115–118]. Comprehensive reviews have been written, including one by Werner Kalow, a pioneer in this field, and more details are covered in Chapter 5.7 [11, 113, 114, 119, 120]. In this chapter, the term "pharmacogenomics" will be used to refer to the use of high-throughput screening techniques for the detection of the genetic variation of an individual and its role in toxicity, especially in causing adverse effects.

In toxicology, most pharmacogenomics studies have focused on variations within the drug metabolizing enzymes (DMEs) [11]. These DMEs catalyze the phase I and phase II metabolic reactions. Several examples have been cited for arylamine N-acetyltransferase (NAT1 and NAT2), thiopurine-S-methyltransferase (TPMT), glutathione-S-transferase (GST), as well as the various cytochrome P450 enzymes (e.g., CYP2A6, CYP2C8, CYP2C9, CYP2C19, CYP2D6, and CYP3A4) [10, 11, 120–122]. Early studies showed that individuals with these genetic variations can be classified as "poor," "extensive," and "ultrarapid" metabolizers [114, 121, 123]. From a toxicity standpoint, this variability is extremely important in that poor metabolizers could accumulate the drug within their tissues causing a toxic response, whereas in extensive metabolizers, the drug has an extremely short half-life and may not stay around the tissues long enough to exert its desired effect. For example, an individual with the gene variant CYP2D6*2, which represents an "ultrarapid metabolizer," can possess a much higher activity for cytochrome P450 isozyme 2D6, whereas an individual with the gene variant CYP2D6*10 possesses a defective enzyme and therefore is a "poor metabolizer" [121, 123]. Individuals who have slow or fast metabolism rates have been observed in several mammalian species [124–128].

Early examples of polymorphisms being detected in a high-throughput manner are studies that used oligonucleotide microarrays containing perfect-match and mismatch sequences. These studies investigated polymorphic changes in breast cancer, HIV infection, and cystic fibrosis [129–131]. A more recent study used a chip containing over 11,000 SNPs to detect genetic variabilities in over 100 DNA samples—a high-throughput variation on association studies [132]. Another array format blended nanotechnology by using multiplexed nanoparticle probes with printed oligonucleotides to distinguish genetic polymorphic variants [133].

3.1.6 SYSTEMS BIOLOGY

Systems biology is a recent term that encompasses looking at the entire biological system from the molecular, cellular, tissue, organism, population, and finally, ecosystem levels. As Witkamp describes it, "systems biology is a realization that organisms do not consist of isolated subsets of genes, proteins and metabolites" [134]. It is also the application of network biology or comprehensive computational methods to analyze the data produced by the combination of OMICS technologies [134–137]. The ultimate goal of toxicogenomics is to predict compound–cell and cell–cell interactions *in silico*, and the application of systems biology brings this one step closer [5, 138, 139].

In attempts to obtain this goal, several studies have correlated datasets from combined genomics and proteomic studies and combined genomics and metabonomics studies. An early toxicogenomics study by Cunningham et al. combined the datasets from the rat toxicity study with three hepatotoxins evaluated by gene expression microarrays described earlier with a matched proteomics arm (Figure 3.1-2) [8, 35–38, 140, 141]. In this study, a portion of the gene and protein expression profiles matched but the remaining did not. Four other studies also tried correlating gene and protein expression in evaluating the effects of zinc, carbon tetrachloride, lithium, and bromobenzene [142–145]. The experimental design for these studies used only one or two time points with one dose, and in all three studies, only a portion of overlap between gene and protein expression was observed.

DMSO   Clofibrate

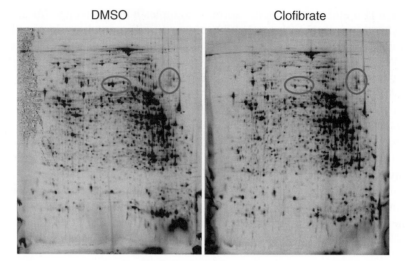

Figure 3.1-2. Proteome changes in livers at day 3 from rats treated with the vehicle DMSO (left) and clofibrate (right). The dose of clofibrate was a toxic dose at 250 mg/kg i.p. Two sets of specific changes are observed within the outlined areas.

Finally, Coen et al. attempted to correlate metabonomics with genomics [146]. They observed some overlap between the two technologies and, hence, concluded that these technologies are complementary in providing a view into toxicity.

A fully integrated approach using genomics, proteomics, and metabonomics was attempted by two groups of investigators. Schnackenberg et al. studied the hepatotoxicity effects of valproic acid [147]. Correlative changes in glucose metabolism were shown by both metabonomics and proteomics methods. However, no significant changes were observed with the gene expression method. Kleno et al. used genomics, proteomics, and metabonomics to investigate the hepatotoxicity of hydrazine [148]. Disruption in both glucose and lipid metabolism were present in all three datasets, and therefore, once again, these three technologies were shown to be complementary.

3.1.7 FUTURE TECHNOLOGIES IN TOXICOGENOMICS

The field of toxicogenomics continues to evolve and develop. For all the technologies listed above, research is continuing to develop more efficient, more high-throughput versions. The field of nanotechnology may aid in this effort. The various engineered nanomaterials may enable the current microarray format to be redeveloped on a nanoscale level and provide more rapid and possibly instaneous results. Some of these engineered nanomaterials, such as single-walled carbon nanotubes, are being developed for applications in the communications and information technologies field and may lead to increased capacity and advances in bioinformatics. The entire field of biosensors is being redeveloped on the nanoscale to nanosensors—from microelectronic membranes (MEMs) to nanoelectronic membranes (NEMs). A preliminary study done by Cunningham et al. assesses the toxicity of these nanomaterials using genomics (Figure 3.1-3) [5, 149, 150]. In this study, the gene expression

Figure 3.1-3. K-means clustering of gene expression profiles from human skin cells treated with TiO_2 (Compound A: A1, A2, A3), carbon black (Compound B: B1, B2, B3), and SiO_2 (Compound C: C1, C2, C3). Columns are the top 100 significantly expressed genes; rows are the samples representing triplicate arrays for each biological sample. The red color denotes upregulation of the gene, and the blue color denotes downregulation of the gene. Insignificantly-expressed genes are denoted by the black color. The expression profiles for TiO_2 and SiO_2 are more similar than the profile for carbon black. (This figure is available in full color at ftp://ftp.wiley.com/public/sci_tech_med/pharmaceutical_biotech/.)

profiles of three known nanoscale materials were compared and two were found to be more similar with each other. This similarity correlated with the similarity in their chemical structures. As more and more engineered nanomaterials are being made, the applications for these materials in the life sciences will only grow. As a result, their safety will also need to be assessed.

In addition to the microarray formats discussed above, different substances are now being printed as reviewed by Cunningham [13]. Arrays are being manufactured with chemical ligands [151, 152], whole cells [153–155], tissue slices [156–158], peptide nucleic acids [159], and nanomaterials [160–166]. These new formats will increase the ability to investigate cell and tissue interactions more extensively than previously known.

Facing the future, other new technological fields are making their way to the forefront of research and could have pertinent applications for investigating toxicity. RNA interference (RNAi) is a burgeoning area (as covered in Chapter 7.6). This field was first thought to include only investigations of naturally-occurring microRNA (miRNA) molecules and synthesized short-interfering RNA (siRNA) molecules. An early paper used siRNAs to the aryl hydrocarbon receptor (AhR) and the AhR nuclear translocator (ARNT) to look at gene silencing [167]. Since this 2003 paper, many references have now reported using siRNA molecules as knockout molecules to investigate mechanisms of toxicity. miRNA molecules, on the other hand, have been sequenced and printed on macroarrays and microarrays [168–169]. These array tools may allow the expression of these molecules, and consequently, their function to be investigated. Recently, other types of short non-coding RNA molecules have been reported as well as naturally-occurring siRNAs [170–172]. It will be interesting to see what roles, if any, each of these new classes of short RNA molecules play in toxicity.

Another technological field is the exciting area of stem cells. Although work has been ongoing with adult stem cells for many years, research involving embryonic stem cells continues to hold great promise. Cell lines for mouse and rat have been used in several research areas to develop new animal models and devise new therapies [173, 174]. The next front is human embryonic stem (HES) cells. Currently, this tool is in the midst of an ethics and morality debate [175]. However, the same debate was waged when fetal tissue was first used in the laboratory setting over two decades ago. If an investigator could direct the differentiation pathway of HES to more mature tissue-specific cells, the need for primary human tissue, which is already difficult to get, would be lessened. This application would increase dramatically the capability of *in vitro* toxicity assays.

Toxicogenomics is a growing and ever-changing field within toxicology. All of these advances in research and development could completely change the safety assessment landscape within just a few years. As the regulatory assays become more predictable, the time from discovery of a pharmaceutical agent to its market release will be lessened.

REFERENCES

1. Vojta P J, Kazmer V P, Kier L D (2001/2). Toxicogenomics: Insights into the present and future. *Drug Disc. World.* 3:18–22.
2. Farr S, Dunn R T (1999). Concise review: Gene expression applied to toxicology. *Toxicol. Sci.* 50:1–9.
3. Nuwaysir E F, Bittner M, Trent J, et al. (1999). Microarrays and toxicology: The advent of toxicogenomics. *Molecular Carcinogen.* 24:153–159.
4. Cunningham M J, Zacharewski T, Somogyi R, et al. (2002). Two-stepping through toxicogenomics: A basic primer. *Toxicol. Sci.* 66:1.
5. Nadadur S, Cunningham M J (2006). Predictive power of novel technologies (cells to "Omics"): Promises, pitfalls and potential applications. *Toxicol. Sci.* 90:1.
6. Mohrenweiser H W (2004). Genetic variation and exposure related risk estimation: will toxicology enter a new era? DNA repair and cancer as a paradigm. *Toxicol. Pathol.* 32:(1):136–145.
7. Cunningham M J (2000). Genomics and proteomics: the new millennium of drug development and discovery. *Journal of Pharmacol. Toxicolog. Methods.* 44:291–300.
8. Page M J, Moyses C, Cunningham M J, et al. (2003). Proteomics. *EXS.* 93:19–30.
9. Nicholson J K, Lindon J C, Holmes E (1999). "Metabonomics": Understanding the metabolic responses of living systems to pathophysiological stimuli via multivariate statistical analysis of biological NMR spectroscopic data. *Xenobiotica.* 29:1181–1189.
10. Sengupta L K, Sengupta S, Sarkar M (2002). Pharmacogenetic applications of the post genomic era. *Current Pharmac. Biotechnol.* 3:141–150.
11. Regnstrom K, Burgess D J (2005). Pharmacogenomics and its potential impact on drug and formulation development. *Critical Rev. Therap. Drug Carrier Sys.* 22:465–492.
12. Severino G, Del Zompo M (2004). Adverse drug reactions: Role of pharmacogenomics. *Pharmacological Res.* 49:363–373.
13. Cunningham M J, Dat D D (2005). Microarrays-Fabricating, Fuchs J and Podda M, (ed.), Marcel Dekker, *Encyclopedia of Diagnostic Genomics and Proteomics.* New York, pp. 819–823.
14. Lee K M, Kim J H, Kang D (2005). Design issues in toxicogenomics using DNA microarray experiment. *Toxicol. Appl. Pharm.* 207:S200–S208.

15. Yauk C L, Berndt M L, Williams A, et al. (2004). Comprehensive comparison of six microarray technologies. *Nucleic Acids Res.* 32:e124.

16. Brazma A, Hingamp P, Quackenbush J, et al. (2001). Minimum information about a microarray experiment (MIAME)-toward standards for microarray data. *Nature Genetics.* 29:365–371.

17. Microarray Gene Expression Data (MGED) Society. A MIAME for Toxicogenomics-MIAME/Tox, Based on Tox 1.1., Draft Publication, August, 2003. Available: http://www.mged.org.

18. Mattes W B, Pettit S D, Sansone S A, et al. (2004). Database development in toxicogenomics: Issues and efforts. *Environ. Health Persp.* 112:495–505.

19. Novoradovskaya N, Whitfield M L, Basehore L S, et al. (2004). Universal reference RNA as a standard for microarray experiments. *BMC Genomics.* 5:20.

20. Shih J H, Michalowska A M, Dobbin K, et al. (2004). Effects of pooling mRNA in microarray class comparisons. *Bioinformatics.* 20:3318–3325.

21. Zhang S D, Gant T W (2005). Effect of pooling samples on the efficiency of comparative studies using microarrays. *Bioinformatics.* 21:4378–4383.

22. Glass A, Henning J, Karopka T, et al. (2005). Representation of individual gene expression in completely pooled mRNA samples. *Biosci. Biotechnol. Biochem.* 69:1098–1103.

23. Kendziorski C M, Zhang Y, Lan H, et al. (2003). The efficiency of pooling mRNA in microarray experiments. *Biostatistics.* 4:465–477.

24. Jolly R A, Goldstein K M, Wei T, et al. (2005). Pooling samples within microarray studies: A comparative analysis of rat liver transcription response to prototypical toxicants. *Physiol. Genomics.* 22:346–355.

25. Peng X, Wood C L, Blalock E M, et al. (2003). Statisical implications of pooling RNA samples for microarray experiments. *BMC Bioinformatics.* 4:26.

26. Khrapko K R, Lysov Y P, Khorlyn A A, et al. (1989). An oligonucleotide hybridization approach to DNA sequencing. *FEBS Lett.* 256:118–122.

27. Southern E M, Maskos U, Elder J K (1992). Analyzing and comparing nucleic acid sequences by hybridization to arrays of oligonucleotides: Evaluation using experimental models. *Genomics.* 13:1008–1017.

28. Maskos U, Southern E M (1992). Oligonucleotide hybridisations on glass supports: A novel linker for oligonucleotide synthesis and hybridisation properties of oligonucleotides synthesised *in situ*. *Nucleic Acids Res.* 20:1679–1684.

29. Drmanac R, Drmanac S, Strezoska Z, et al. (1993). DNA sequence determination by hybridization: A strategy for efficient large-scale sequencing. *Science.* 260:1649–1652.

30. Beattie W G, Meng L, Turner S L, et al. (1995). Hybridization of DNA targets to glass-tethered oligonucleotide probes. *Molecular Biotechnol.* 4:213–225.

31. Augenlicht L H, Wahrman M Z, Halsey H, et al. (1987). Expression of cloned sequences in biopsies of human colonic tissue and in colonic carcinoma cells induced to differentiate *in vitro*. *Cancer Res.* 47:6017–6021.

32. Gray N S, Wodicka L, Thunnissen A M W H, et al. (1998). Exploiting chemical libraries, structure and genomics in the search for kinase inhibitors. *Science.* 281:533–538.

33. Braxton S, Bedilion T (1998). The integration of microarray information in the drug development process. *Curr. Opin. Biotech.* 9:643–649.

34. Karpf A R, Peterson P W, Rawlins J T, et al. (1999). Inhibition of DNA methyltransferase stimulates the expression of signal transducer and activator of transcription 1, 2, and 3 genes in colon tumor cells. *Proc. of the National Academy of Sciences USA.* 96:14007–14012.

35. Cunningham M J, Zweiger G, Bailey D, et al. (1998). Application of a genome-based screen. *Archives of Pharmacol.* 358:R564.

36. Cunningham M J, Liang S, Fuhrman S, et al. (2000). Gene expression microarray data analysis for toxicity profiling. *Annals of the New York Acad. Sci.* 919:52–67.

37. Fuhrman S, Cunningham M J, Wen X, et al. (2000). The application of Shannon entropy in the identification of putative drug targets. *BioSys.* 55:5–14.

38. Fuhrman S, Cunningham M J, Liang S, et al. (2000). Making sense of large-scale gene expression data with simple computational techniques. *Amer. Biotechnol. Lab.* 18:68–70.

39. Burczynski M E, McMillian M, Ciervo J, et al. (2000). Toxicogenomics-based discrimination of toxic mechanism in HepG2 human hepatoma cells. *Toxicol. Sci.* 58:399–415.

40. Waring J F, Jolly R A, Ciurlionis R, et al. (2001). Clustering of hepatotoxins based on mechanism of toxicity using gene expression profiles. *Toxicol. Appl. Pharm.* 175:28–42.

41. Thomas R S, Rank D R, Penn S G, et al. (2001). Identification of toxicologically predictive gene sets using cDNA microarrays. *Molec. Pharmacol.* 60:1189–1194.

42. Hamadeh H K, Bushel P R, Jayadev S, et al. (2002). Gene expression analysis reveals chemical-specific profiles. *Toxicol. Sci.* 67:219–231.

43. Hamadeh H K, Bushel P R, Jayadev S, et al. (2002). Prediction of compound signature using high density gene expression profiling. *Toxicol. Sci.* 67:232–240.

44. de Longueville F, Atienzar F A, Marcq L, et al. (2003). Use of a low-density microarray for studying gene expression patterns induced by hepatotoxicants on primary cultures of rat hepatocytes. *Toxicol. Sci.* 75:378–392.

45. Ishida S, Shigemoto-Mogami Y, Kagechika H, et al. (2003). Clinically potential subclasses of retinoid synergists revealed by gene expression profiling. *Molec. Cancer Therapeutics.* 2:49–58.

46. Kier L D, Neft R, Tang L, et al. (2004). Applications of microarrays with toxicologically relevant genes (tox genes) for the evaluation of chemical toxicants in Sprague Dawley rats in vivo and human hepatocytes in vitro. *Mutation Res.* 549:101–113.

47. Hu T, Gibson D P, Carr G J, et al. (2004). Identification of a gene expression profile that discriminates indirect-acting genotoxins from direct-acting genotoxins. *Mutation Res.* 549:5–27.

48. McMillian M, Nie A Y, Parker J B, et al. (2004). A gene expression signature for oxidant stress/reactive metabolites in rat liver. *Biochem. Pharmacol.* 68:2249–2261.

49. Nadadur S S, Schladweiler M C J, Kodavanti U P (2000). A pulmonary rat gene array for screening altered expression profiles in air pollutant-induced lung injury. *Inhalation Toxicol.* 12:1239–1254.

50. Bae D S, Hanneman W H, Yang R S H, et al. (2002). Characterization of gene expression changes associated with MNNG, arsenic, or metal mixture treatment in human keratinocytes: Application of cDNA microarray technology. *Environ. Health Persp.* 110(Suppl. 6):931–941.

51. Wang D Y, McKague B, Liss S N, et al. (2004). Gene expression profiles for detecting and distinguishing potential endocrine-disrupting compounds in environmental samples. *Environ. Sci. Technol.* 38:6396–6406.

52. Jessen B A, Mullins J S, de Peyster A, et al. (2003). Assessment of hepatocytes and liver slices as *in vitro* test systems to predict *in vivo* gene expression. *Toxicol. Sci.* 75:208–222.

53. Boess F, Kamber M, Romer S, et al. (2003). Gene expression in two hepatic cell lines, cultured primary hepatocytes, and liver slices compared to the *in vivo* liver gene

expression in rats: Possible implications for toxicogenomics use of *in vitro* systems. *Toxicol. Sci.* 73:386–402.

54. Konu O, Kane J K, Barrett T, et al. (2001). Region-specific transcriptional response to chronic nicotine in rat brain. *Brain Res.* 909:194–203.

55. Hamadeh H K, Knight B L, Haugen A C, et al. (2002). Methapyrilene toxicity: Anchorage of pathologic observations to gene expression alterations. *Toxicol. Pathol.* 30:470–482.

56. Kriete A, Anderson M K, Love B, et al. (2003). Combined histomorphometric and gene-expression profiling applied to toxicology. *Genome Biol.* 4:R32.1–R32.9.

57. Sadlier D M, Connolly S B, Kieran N E, et al. (2004). Sequential extracellular matrix-focused and baited-global cluster analysis of serial transcriptomic profiles identifies candidate modulators of renal tubulointerstitial fibrosis in murine adriamycin-induced nephropathy. *J. Biolog. Chem.* 279:29670–29680.

58. Kisby G E, Standley M, Lu X, et al. (2005). Molecular networks perturbed in a developmental animal model of brain injury. *Neurobiol. Dis.* 19:108–118.

59. Luo W, Fan W, Xie H, et al. (2005). Phenotypic anchoring of global gene expression profiles induced by N-hydroxy-4-acetylaminobiphenyl and benzo[*a*]pyrene diol epoxide reveals correlations between expression profiles and mechanism of toxicity. *Chem. Res. Toxicol.* 18:619–629.

60. Boverhof D R, Burgoon L D, Tashiro C, et al. (2005). Temporal and dose-dependent hepatic gene expression patterns in mice provide new insights into TCDD-mediated hepatotoxicity. *Toxicol. Sci.* 85:1048–1063.

61. Huang Q, Jin X, Gaillard E T, et al. (2004). Gene expression profiling reveals multiple toxicity endpoints induced by hepatotoxicants. *Mutation Res.* 549:147–168.

62. Patel S K, Ma N, Monks T J, et al. (2003). Changes in gene expression during chemical-induced nephrocarcinogenicity in the Eker rat. *Molecular Carcinogen.* 38:141–154.

63. Heinloth A N, Irwin R D, Boorman G A, et al. (2004). Gene expression profiling of rat livers reveals indicators of potential adverse effects. *Toxicol. Sci.* 80:193–202.

64. Ogawa K, Asamoto M, Suzuki S, et al. (2005). Downregulation of apoptosis revealed by laser microdissection and cDNA microarray analysis of related genes in rat liver preneoplastic lesions. *Medical Molec. Morphol.* 38:23–29.

65. Chung C Y, Seo H, Sonntag K C, et al. (2005). Cell type-specific gene expression of midbrain dopaminergic neurons reveals molecules involved in their vulnerability and protection. *Human Molec. Genetics.* 14:1709–1725.

66. Lefebvre d'Hellencourt C, Harry G J (2005). Molecular profiles of mRNA levels in laser capture microdissected murine hippocampal regions differentially responsive to TMT-induced cell death. *J. Neurochem.* 93:206–220.

67. Zhou T, Jia X, Chapin R E, et al. (2004). Cadmium at a non-toxic dose alters gene expression in mouse testes. *Toxicol. Lett.* 154:191–200.

68. Anderson N L, Anderson N G (1998). Proteome and proteomics: New technologies, new concepts, and new words. *Electrophoresis.* 19:1853–1861.

69. Yarmush M L, Jayaraman A (2002). Advances in proteomic technologies. *Ann. Rev. Biomed. Engin.* 4:349–373.

70. Anderson N L, Anderson N G (1991). A two-dimensional gel database of human plasma proteins. *Electrophoresis.* 12:883–906.

71. Kennedy S (2002). The role of proteomics in toxicology: Identification of biomarkers of toxicity by protein expression analysis. *Biomarkers.* 7:269–290.

72. Kennedy S (2001). Proteomic profiling from human samples: The body fluid alternative. *Toxicol. Lett.* 120:379–384.

73. Fountoulakis M (2001). Proteomics: Current technologies and applications in neurological disorders and toxicology. *Amino Acids.* 21:363–381.

74. Fountoulakis M, Berndt P, Boelsterli U A, et al. (2000). Two-dimensional database of mouse liver proteins: Changes in hepatic protein levels following treatment with acetaminophen or its nontoxic regioisomer 3-acetamidophenol. *Electrophoresis.* 21:2148–2161.

75. Ruepp S U, Tonge R P, Shaw J, et al. (2002). Genomics and proteomics analysis of acetaminophen toxicity in mouse liver. *Toxicol. Sci.* 65:135–150.

76. O'Farrell P H (1975). High resolution two-dimensional electrophoresis of proteins. *J. Biolog. Chem.* 250:4007–4021.

77. McGregor E, Dunn M J (2003). Proteomics of heart disease. *Human Molec. Genetics.* 12:R135–R144.

78. James P (2001). Protein expression analysis: From "tip of the iceberg" to a global method. *Disease Markers.* 17:235–246.

79. Anderson N L, Esquer-Blasco R, Richardson F, et al. (1996). The effects of peroxisome proliferators on protein abundances in mouse liver. *Toxicol. Appl. Pharm.* 137:75–89.

80. Bruno M K, Khairallah E A, Cohen S D (1998). Inhibition of protein phosphatase activity and changes in protein phosphorylation following acetaminophen exposure in cultured mouse hepatocytes. *Toxicol. Appl. Pharm.* 153:119–132.

81. Witzmann F A, Carpenter R L, Ritchie G D, et al. (2000). Toxicity of chemical mixtures: Proteomic analysis of persisting liver and kidney protein alterations induced by repeated exposure of rats to JP-8 jet fuel vapor. *Electrophoresis.* 21:2138–2147.

82. Steiner S, Aicher L, Raymackers J, et al. (1996). Cyclosporine A decreases the protein level of the calcium-binding protein calbindin-D 28kDa in rat kidney. *Biochem. Pharmacol.* 51:253–258.

83. Bandara L R, Kelly M D, Lock E A, et al. (2003). A correlation between a proteomic evaluation and conventional measurements in the assessment of renal proximal tubular toxicity. *Toxicol. Sci.* 73:195–206.

84. Wetmore B A, Merrick B A (2004). Toxicoproteomics: Proteomics applied to toxicology and pathology. *Toxicol. Pathol.* 32:619–642.

85. Anderson N L, Taylor J, Hofmann H-P, et al. (1996). Simultaneous measurement of hundreds of liver proteins: Application in assessment of liver function. *Toxicol. Pathol.* 24:72–76.

86. Thome-Kromer B, Bonk I, Klatt M, et al. (2003). Toward the identification of liver toxicity markers: A proteome study in human cell culture and rats. *Proteomics.* 3:1835–1862.

87. MacBeath G, Schreiber S L (2000). Printing proteins as microarrays for high-throughput function determination. *Science.* 289:1760–1763.

88. Zhu H, Bilgin M, Bangham R, et al. (2001). Global analysis of protein activities using proteome chips. *Science.* 293:2101–2105.

89. Petricoin E F, Zoon K C, Kohn E C, et al. (2002). Clinical Proteomics: Translating benchside promise into bedside reality. *Nature Rev. Drug Disc.* 1:683–695.

90. Petricoin E F, Rajapaske V, Herman E, et al. (2004). Toxicoproteomics: Serum proteomic pattern diagnostics for early detection of drug induced cardiac toxicities and cardioprotection. *Toxicol. Pathol.* 32:122–130.

91. Kell D B (2002). Metabolomics and machine learning: Explanatory analysis of complex metabolome data using genetic programming to produce simple, robust rules. *Molecular Biol. Reports.* 29:237–241.

92. Watkins S M, German J B (2002). Toward the implementation of metabolomics assessments of human health and nutrition. *Curr. Opin. Biotech.* 13:512–516.

93. Griffin J L, Bollard M E (2004). Metabonomics: Its potential as a tool in toxicology for safety assessment and data integration. *Curr. Drug Metabol.* 5:389–398.

94. Robertson D G (2005). Metabonomics in toxicology: A review. *Toxicol. Sci.* 85:809–822.

95. Adams A (2003). Metabolomics: Small molecule "Omics". *The Scientist.* 17:38–40.

96. Griffin J L (2004). Metabolic profiles to define the genome: Can we hear the phenotypes? *Philos. T. Roy. Soc. B.* 35, 857–871.

97. Weckwerth W, Morganthal K (2005). Metabolomics: From pattern recognition to biological interpretation. *Drug Disc. Today.* 10:1551–1558.

98. Keun H C (2006). Metabonomic modeling of drug toxicity. *Pharmacol. Therap.* 109:92–106.

99. Reo N V (2002). NMR-based metabolomics. *Drug and Chem. Toxicol.* 25:375–382.

100. Nicholson J K, Buckingham M J, Sadler P J (1983). High resolution ^{1}H NMR studies of vertebrate blood and plasma. *Biochem. J.* 211:605–615.

101. Nicholson J K, Timbrell J A, Sadler P J (1985). Proton NMR spectra of urine as indicators of renal damage: Mercury-induced nephrotoxicity in rats. *Molec. Pharmacol.* 27:644–651.

102. Gartland K P R, Beddell C R, Lindon J C, et al. (1991). Application of pattern recognition methods to the analysis and classification of toxicological data derived from proton nuclear magnetic resonance spectroscopy of urine. *Molec. Pharmacol.* 39:629–642.

103. Holmes E, Nicholls A W, Lindon J C, et al. (1998). Development of a model for classification of toxin-induced lesions using 1H NMR spectroscopy of urine combined with pattern recognition. *NMR in Biomedicine.* 11:235–244.

104. Nicholson J K, Higham D P, Timbrell J A, et al. (1989). Quantitative high resolution ^{1}H NMR urinalysis studies on the biochemical effects of cadmium in the rat. *Molec. Pharmacol.* 36:398–404.

105. Robertson D G, Reily M D, Sigler R E, et al. (2000). Metabonomics: Evaluation of nuclear magnetic resonance (NMR) and pattern recognition technology for rapid *in vivo* screening of liver and kidney toxicants. *Toxicol. Sci.* 57:326–337.

106. Beckwith-Hall B M, Nicholson J K, Nicholls A W, et al. (1998). Nuclear magnetic resonance spectroscopic and principal components analysis investigations into biochemical effects of three model hepatotoxins. *Chem. Res. Toxicol.* 11:260–272.

107. Moka D, Vorreuther R, Schicha H, et al. (1998). Biochemical classification of kidney carcinoma biopsy samples using magic-angle-spinning ^{1}H nuclear magnetic resonance spectroscopy. *J. Pharmac. Biomed. Anal.* 17:125–132.

108. Waters N J, Holmes E, Williams A, et al. (2001). NMR and pattern recognition studies on the time-related metabolic effects of α-naphthylisothiocyanate on liver, urine, and plasma in the rat: An integrative metabonomic approach. *Chem. Res. Toxicol.* 14:1401–1412.

109. Coen M, Lenz E M, Nicholson J K, et al. (2003). An integrated metabonomic investigation of acetaminophen toxicity in the mouse using NMR spectroscopy. *Chem. Res. Toxicol.* 16:295–303.

110. Car B D, Robertson R T (1999). Commentary: Discovery toxicology-a nascent science. *Toxicol. Pathol.* 27:481–483.

111. Lindon J C, Nicholson J K, Holmes E, et al. (2003). Contemporary issues in toxicology: The role of metabonomics in toxicology and its evaluation by the COMET project. *Toxicol. Appl. Pharm.* 187:137–146.

112. Lindon J C (2005). The consortium for metabonomic toxicology (COMET): Aims, activities and achievements. *Pharmacogenomics.* 6:691–699.

113. Goldstein D B, Tate S K, Sisodiya S M (2003). Pharmacogenetics goes genomic. *Nature Rev. Genetics.* 4:937–947.

114. Shastry B S (2005). Genetic diversity and new therapeutic concepts. *J. Human Genetics.* 50:321–328.

115. Kalow W (1956). Familial incidence of low pseudocholinesterase level. *Lancet.* 2:576–577.

116. Carsen P R, Flanagan C L, Iokes C E (1956). Enzymatic deficiency in primaquine-sensitive erythrocytes. *Science.* 124:484–485.

117. Kalow W, Staron N (1957). On distribution and inheritance of atypical forms of human serum cholinesterase, as indicated by dibucaine numbers. *Canadian J. Biochem. Physiol.* 35:1305–1320.

118. Motulsky A G (1957). Drug reactions, enzymes and biochemical genetics. *J. Amer. Med. Assoc.* 165:835–837.

119. Kalow W (1997). Pharmacogenetics in biological perspective. *Pharmacol. Rev.* 49:369–379.

120. Ingelman-Sundberg M, Rodriguez-Antona C (2005). Pharmacogenetics of drug-metabolizing enzymes: Implications for a safer and more effective drug therapy. *Philos. T. Roy. Soc. B.* 360:1563–1570.

121. Danielson P B (2002). The cytochrome P450 superfamily: Biochemistry, evolution and drug metabolism in humans. *Curr. Drug Metabol.* 3:561–597.

122. Cascorbi I (2006). Genetic basis of toxic reactions to drugs and chemicals. *Toxicol. Lett.* 162:16–28.

123. Ingelman-Sundberg M (2005). Genetic polymorphisms of cytochrome P450 2D6 (CYP2D6): Clinical consequences, evolutionary aspects and functional diversity. *Pharmacogenomics J.* 5:6–13.

124. Barham H M, Lennard M S, Tucker G T (1994). An evaluation of cytochrome P450 isoform activities in the female dark agouti (DA) rat: Relevance to its use as a model of the CYP2D6 poor metaboliser phenotype. *Biochem. Pharmacol.* 47:1295–1307.

125. Vorhees C V, Reed T M, Schilling M A, et al. (1998). CYP2D1 polymorphism in methamphetamine-treated rats: Genetic differences in neonatal mortality and effects on spatial learning and acoustic startle. *Neurotoxicol. Teratol.* 20:265–273.

126. Paulson S K, Zhang J Y, Jessen S M, et al. (2000). Comparison of celecoxib metabolism and excretion in mouse, rabbit, dog, cynomolgus monkey and rhesus monkey. *Xenobiotica.* 30:731–744.

127. Mise M, Yadera S, Matsuda M, et al. (2004). Polymorphic expression of CYP1A2 leading to interindividual variability in metabolism of a novel benzodiazepine receptor partial inverse agonist in dogs. *Drug Metabol. Dispos.* 32:240–245.

128. Kusano K, Seko T, Tanaka S, et al. (1996). Purfication and characterization of Rhesus monkey liver amido hydrolases and their roles in the metabolic polymorphism for E6123, a platelet-activating factor receptor antagonist. *Drug Metabol. Dispos.* 24:1186–1191.

129. Cronin M T, Fucini R V, Kim S M, et al. (1996). Cystic fibrosis mutation detection by hybridization to light-generated DNA probe arrays. *Human Mutation.* 7:244–255.

130. Kozal M J, Shah N, Shen N, et al. (1996). Extensive polymorphisms observed in HIV-1 clade B protease gene using high-density oligonucleotide arrays. *Nature Med.* 2:753–759.

131. Hacia J G, Brody L C, Chee M S, et al. (1996). Detection of heterozygous mutations in BRCA1 using high density oligonucleotide arrays and two-colour fluorescence analysis. *Nature Genetics.* 14:441–447.

132. Kulle B, Schirmer M, Toller M R, et al. (2005). Application of genomewide SNP arrays for detection of simulated susceptibility loci. *Human Mutation.* 25:557–565.

133. Bao Y P, Huber M, Wei T F, et al. (2005). SNP identification in unamplified human genomic DNA with gold nanoparticle probes. *Nucleic Acids Res.* 33:e15.

134. Witkamp R F (2005). Genomics and systems biology—how relevant are the developments to veterinary pharmacology, toxicology and therapeutics? *J. Veterinary Pharmacol. Therapeut.* 28:235–245.

135. Barabasi A L, Oltvai Z N (2004). Network biology: Understanding the cell's functional organization. *Nature Rev. Genetics.* 5:101–113.

136. Heijne W H, Kienhuis A S, van Ommen B, et al. (2005). Systems toxicology: Applications of toxicogenomics, transcriptomics, proteomics and metabolomics in toxicology. *Expert Rev. Proteomics.* 2:767–780.

137. Ekins S, Nikolsky Y, Nikolskaya T (2005). Techniques: application of systems biology to absorption, distribution, metabolism, excretion and toxicity. *Trends Pharmaceut. Sci.* 26:202–209.

138. Cunningham M J (2000). Toxicological database mining in the twenty-first century. *Toxicol. Sci.* 54(1-S):412.

139. Tomita M (2001). Whole-cell simulation: A grand challenge of the 21st century. *Trends in Biotechnol.* 19:205–210.

140. Cunningham M J, Holt G, Moyes C, et al. (2000). The use of proteomics in molecular toxicology. *Toxicol. Sci.* 54(1-S):62.

141. Keitt S K, Fagan T F, Marts S A (2004). Understanding sex differences in environmental health: A thought leaders' roundtable. *Environ. Health Perspec.* 112:604–609.

142. Hogstrand C, Balesaria S, Glover C N (2002). Application of genomics and proteomics for study of the integrated response to zinc exposure in a non-model fish species, the rainbow trout. *Compar. Biochem. Physiol. Part B, Biochem. Molec. Biol.* 133:523–535.

143. Fountoulakis M, de Vera M C, Crameri F, et al. (2002). Modulation of gene and protein expression by carbon tetrachloride in the rat liver. *Toxicol. Applied Pharmacol.* 183:71–80.

144. Bro C, Regenberg B, Lagniel G, et al. (2003). Transcriptional, proteomic and metabolic response to lithium in galactose-grown yeast cells. *J. Biolog. Chem.* 278:32141–32149.

145. Heijne W H M, Stierum R H, Slijper M, et al. (2003). Toxicogenomics of bromobenzene hepatotoxicity: A combined transcriptomics and proteomics approach. *Biochem. Pharmacol.* 65:857–875.

146. Coen M, Ruepp S U, Lindon J C, et al. (2004). Integrated application of transcriptomics and metabonomics yields new insight into the toxicity due to paracetamol in the mouse. *J. Pharmac. Biomed. Anal.* 35:93–105.

147. Schnackenberg L, Jones R C, Thyparambil S, et al. (2006). An integrated study of acute effects of valproic acid in the liver using metabonomics, proteomics, and transcriptomics platforms. *OMICS: A J. Integ. Boil.* 10:1–14.

148. Kleno T G, Kiehr B, Baunsgaard D, et al. (2004). Combination of "omics" data to investigate the mechanism(s) of hydrazine-induced hepatotoxicity in rats and to identify potential biomarkers. *Biomarkers.* 9:116–138.

149. Cunningham M J, Magnuson S R, Falduto M T (2005). Gene expression profiling of nanoscale materials using a systems biology approach. *Toxicol. Sci.* 84(1-S):9.

150. Monteiro-Riviere N, Cunningham M J (2006). Screening methods for assessing skin toxicity of nanomaterials. *Toxicol. Sci.* 90(1-S):5.

151. You A J, Jackman R J, Whitesides G M, et al. (1997). A miniaturized arrayed assay format for detecting small molecule-protein interactions in cells. *Chem. & Biol.* 4:969–975.

152. Kuruvilla F G, Shamji A F, Sternson S M, et al. (2002). Dissecting glucose signalling with diversity-oriented synthesis and small-molecule microarrays. *Nature.* 416:653–657.

153. Xia Y, Whitesides G M (1998). Soft lithography. *Angewandte Chemie (International Edition in English).* 37:550–575.

154. Kane R S, Takayama S, Ostuni E, et al. (1999). Patterning proteins and cells using soft lithography. *Biomaterials.* 20:2363–2376.

155. Ziauddin J, Sabatini D M (2001). Microarrays of cells expressing defined cDNAs. *Nature.* 411:107–110.

156. Kononen J, Bubendorf L, Kallioniemi A, et al. (1998). Tissue microarrays for high-throughput molecular profiling of tumor specimens. *Nature Med.* 4:844–847.

157. Moch H, Schraml P, Bubendorf L, et al. (1999). High-throughput tissue microarray analysis to evaluate genes uncovered by cDNA microarray screening in renal cell carcinoma. *Amer. J. Pathol.* 154:981–986.

158. Bubendorf L, Kolmer M, Kononen J, et al. (1999). Hormone therapy failure in human prostate cancer: Analysis by complementary DNA and tissue microarrays. *J. National Cancer Institute.* 91:1758–1764.

159. Matysiak S, Reuthner F, Hoheisel J D (2001). Automating parallel peptide synthesis for the production of PNA library arrays. *Biotechniques.* 31:896–904.

160. Fritz J, Baller M K, Lang H P, et al. (2000). Translating biomolecular recognition into nanomechanics. *Science.* 288:316–318.

161. McKendry R, Zhang J, Arntz Y, et al. (2002). Multiple label-free biodetection and quantitative DNA-binding assays on a nanomechanical cantilever array. *Proc. of the National Academy of Sciences.* 99:9783–9788.

162. Choi J W, Park K W, Lee D B, et al. (2005). Cell immobilization using self-assembled synthetic oligopeptide and its application to biological toxicity detection using surface plasmon resonance. *Biosens. and Bioelectron.* 20:2300–2305.

163. How S E, Yingyongnarongkul B, Fara M A, et al. (2004). Polyplexes and lipoplexes for mammalian gene delivery: From traditional to microarray screening. *Combinat. Chem. & High Throughput Screen.* 7:423–430.

164. Piner R D, Zhu J, Xu F, et al. (1999). "Dip-pen" nanolithography. *Science.* 283:661–663.

165. Park S J, Taton A T, Mirkin C A (2002). Array-based electrical detection of DNA with nanoparticle probes. *Science.* 295:1503–1506.

166. Taton A T, Mirkin C A, Letsinger R L (2000). Scanometric DNA array detection with nanoparticle probes. *Science.* 289:1757–1760.

167. Abdelrahim M, Smith R, Safe S (2003). Aryl hydrocarbon receptor gene silencing with small inhibitory RNA differentially modulates Ah-responsiveness in MCF-7 and HepG2 cancer cells. *Mol. Pharmacol.* 63:1373–1381.

168. Krichevsky A M, King K S, Donahue C P, Khrapko K, Kosik K S (2003). A microRNA array reveals extensive regulation of microRNAs during brain development. *RNA* 9:1274–1281.

169. Miska E A, Alvarez-Saavedra E, Townsend M, Yoshii A, Sestan N, Rakic P, Constantine-Paton M, Horvitz H R (2004). Microarray analysis of microRNA expression in the developing mammalian brain. *Genome Biol.* 5:R68.

170. Dalmay T (2006). Short RNAs in environmental adaptation. *Proc. Biol. Soc.* 273:1579–1585.

171. Yang Z, Wu J (2006). Small RNAs and development. *Med. Sci. Monit.* 12:RA125–RA129.

172. Ying S Y, Chang D C, Miller J D, Lin S L (2006). The microRNA: Overview of the RNA gene that modulates gene function. *Methods Mol. Biol.* 342:1–18.

173. Downing G J, Battey J F (2004). Technical assessment of the first 20 years of research using mouse embryonic stem cell lines. *Stem Cells.* 22:1168–1180.

174. Ruhnke M, Ungefroren H, Zehle G, et al. (2003). Long-term culture and differentiation of rat embryonic stem cell-like cells into neuronal, glial, endothelial, and hepatic lineages. *Stem Cells.* 21:428–436.

175. Bremer S, Hartung T (2004). The use of embryonic stem cells for regulatory developmental toxicity testing in vitro—the current status of test development. *Curr. Pharmac. Design.* 10:2733–2747.

3.2

PRECLINICAL PHARMACOKINETICS

Donald E. Mager and Murali Ramanathan

University at Buffalo, The State University of New York, Buffalo, New York

Chapter Contents

3.2.1 Introduction   253
3.2.2 Disposition of Macromolecules   254
    3.2.2.1 Absorption   254
    3.2.2.2 Distribution   255
    3.2.2.3 Metabolism   258
    3.2.2.4 Excretion   259
3.2.3 Pharmacokinetic Data Analysis   261
    3.2.3.1 Noncompartmental Analysis   262
    3.2.3.2 Pharmacokinetic Models   265
3.2.4 Predicting Macromolecule Pharmacokinetics in Humans   268
    3.2.4.1 Allometric Scaling   268
    3.2.4.2 Physiologically Based Pharmacokinetic Modeling   271
3.2.5 Conclusions   273
    Acknowledgment   273
    References   273

3.2.1 INTRODUCTION

The preclinical pharmacokinetics of protein pharmaceuticals is distinctively different from small-molecule drugs. The phase I and phase II metabolizing enzymes, for example, the cytochrome P450 and glucoronosyl transferases, which play dominant roles in the metabolism of small-molecule drugs, are not significantly involved in protein metabolism. Instead, because all protein pharmaceuticals invariably share a common polypeptide backbone, they are susceptible to proteolysis as a clearance mechanism. However, the disposition, pharmacokinetics,

Handbook of Pharmaceutical Biotechnology, Edited by Shayne Cox Gad.
Copyright © 2007 John Wiley & Sons, Inc.

and pharmacodynamics of proteins can be modulated by a variety of additional factors, including glycosylation and receptor binding, binding proteins, and neutralizing antibodies.

3.2.2 DISPOSITION OF MACROMOLECULES

3.2.2.1 Absorption

During drug development, the oral route is often targeted as a preferred route of administration for small molecules. However, the oral route is not usually an option for protein pharmaceuticals. Each organ in the gastrointestinal tract contains a variety of digestive proteases that enable it to break down nutritional proteins to small peptides and amino acids that can be absorbed in the intestine via secondarily active transport mechanisms. These digestive proteases also cause the extensive hydrolysis of the polypeptide bond in protein drugs in the gastrointestinal tract resulting in poor bioavailability. Protein pharmaceuticals are usually, therefore, administered via injection, such as the intravenous, subcutaneous, or intramuscular routes.

The absorption of protein drugs from extravascular sites is dependent on the size of the molecule. Upon subcutaneous administration, small molecules with molecular weights less than 1 kD predominantly enter the systemic circulation via blood capillaries, whereas subcutaneously administered proteins with molecular weights greater than 16 kD enter the systemic circulation predominantly via the lymphatic system. The extent of the lymphatic recovery after subcutaneous administration is a linear function of molecular weight (Figure 3.2-1) [1–5].

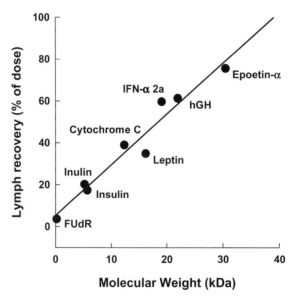

Figure 3.2-1. Linear relationship between cumulative lymphatic recovery and molecular weight of various therapeutic peptides and proteins. Data were extracted from the literature [1–5].

Nonparenteral routes of administration for protein drugs in general are not extensively used; however, recombinant human deoxyribonuclease I solution (dornase alfa or PULMOZYME) and, more recently, insulin powder by inhalation (EXUBERA) have been approved for clinical use in the United States [6]. The approval of inhaled insulin may provide the stimulus for greater interest in the inhalation route for other protein drugs in the near future. The lung presents a large surface area for transport and high permeability compared with other barriers. Inhaled insulin is administered as a novel powder formulation that is metered from a propellant-free metered dose inhaler. However, delivery to the lungs, particularly to the alveolar regions deep in the lung, presents several technological challenges and the overall system bioavailability (which is based on the amount of drug loaded into the inhaler device) of inhaled insulin is 9–22% that of the subcutaneous route. The reduced bioavailability is managed by dose adjustment and compared with subcutaneously administered regular insulin, the serum insulin concentrations after inhalation peak earlier (at 7–80 minutes compared with 42–274 minutes for subcutaneous regular insulin) and decay more rapidly; the pharmacodynamics of glucose changes are also correspondingly more rapid [7]. The duration of inhaled insulin action are comparable with regular insulin [6].

3.2.2.2 Distribution

The distribution of protein drugs is limited compared with that of small-molecule drugs and is primarily a function of the physico-chemical properties of the macromolecule and the physiological characteristics of the vasculature and cellular membranes [8]. Primary molecular descriptors, such as size, charge, and lipophilicity, will determine or influence the extent of specific and nonspecific protein binding and passage through various types of capillary endothelia [8]. The relatively large size of peptide and protein drugs alone can often account for the limited volume of distribution of these compounds, which is generally restricted to the extracellular space. Following intravenous injection, the initial volume of distribution for most macromolecules is confined approximately to plasma volume (3–8 L), and the total volume at steady state is typically only up to two fold greater.

Active transport and receptor-mediated uptake also may influence the biodistribution of macromolecules. For example, the internalization of monoclonal antibodies into cells may occur via Fcγ receptors, found on various immune and hematopoietic cells [9], or binding to membrane-localized antigens [10]. Thus, active uptake processes can serve to increase the volume of distribution to a large extent relative to that which would be expected based only on physico-chemical properties. Receptor-mediated endocytosis may represent a significant elimination pathway as well (see the Receptor-mediated elimination subsection) and is recognized as a major process influencing the overall disposition of macromolecules [11].

One must carefully interpret the volumes of distribution of peptides and proteins reported in the literature. Most studies rely on a so-called noncompartmental analysis to estimate primary pharmacokinetic parameters (see Section 3.2.3.1). However, this method is only valid for linear systems, assuming that the site of drug elimination is in rapid equilibrium with the sampling site (plasma). The former condition is often assumed for compartmental models as well, including those with nonlinear mechanisms of elimination. Therefore, uptake processes and relatively slow or complex tissue catabolism may confound reported distribution volumes of select macromolecules estimated by these common methods.

Binding Proteins. For some therapeutic proteins, specific and nonspecific protein interactions can modulate distribution and activity. Soluble receptors are the best-characterized specific binding proteins and α_2 macroglobulin, an acute phase protein, has been implicated in the binding of several cytokines. Binding and neutralizing antibodies are another class of binding protein that will be discussed separately in the section on Immunogenicity, Binding, and Neutralizing Antibodies.

Soluble cytokine receptors (SR) occur naturally in human serum and urine and play a role in determining the disposition and biological activities of cytokines in health and in pathological states. Additionally, bioengineered soluble receptors and their immunoglobulin derivatives have been investigated as specific cytokine-inhibiting therapeutic agents [12, 13]. However, the physiological role(s) of naturally occurring SR is controversial and not well understood.

As expected, the SR can inhibit cytokine bioactivity by competing with the receptor and, when these antagonist-like effects are observed, the dose-response curve of the cytokine effect as a function of SR dose usually decreases monotonically, and almost complete inhibition is observed at the highest SR doses. In animal models, soluble receptors for interleukin-4 [14], interleukin-7 [15], interleukin-6 [15, 16], and tumor necrosis factor-α [17] have been shown to enhance rather than inhibit cytokine responses under certain conditions [17–19]. When these agonist-like effects are observed, the dose-response curve of the cytokine effect as a function of SR dose is bell-shaped, with no inhibition of cytokine bioactivity at low SR doses, maximal cytokine bioactivity at intermediate SR doses, followed by increasing inhibition of cytokine activity at higher SR doses. As the SR effects of the cytokines biological response is dependent on experimental conditions and can be either agonist-like or antagonist-like, the physiological roles of SR have been difficult to assess. Cytokines generally have short half-lives consistent with their physiological roles in cell-cell communication and the apparent SR-mediated agonism of cytokine activity occurs because the SR stabilizes the cytokine in vivo either by reducing the rate of spontaneous activity decay [17] or by increasing the half-life of cytokine [20]. The formation of SR-cytokine complex reduces cytokine-free concentrations, but the increases in circulation time can offset the effect of reduced receptor occupancy and contribute to overall increases in exposure [20]. The binding half-life of insulin-like growth factor-1 (IGF-1) in rats deficient in IGF-1 binding protein is 20–30 min; but in rats with normal levels of the binding protein, the half-life of IGF-1 increases 8–12-fold to 4 hours [21]. In the plasma, nearly 40% of human growth hormone (GH) is also bound to binding proteins; the high-affinity component of the GH binding activity consists of the extracellular domain of the GH receptor [22, 23]. The metabolic clearance of the SR-bound GH fraction is 10-fold slower than free GH because the bound GH is not filtered by the kidney and the competitive inhibition of receptor binding reduces elimination via receptor-mediated pathways [24]. Indirect evidence for the physiological role of SR binding is provided by the data from short-stature Mountain Ok people of Papua, New

Guinea who have SR levels 50% that of controls of normal stature; the levels of GH, IGF-1, and low-affinity GH-binding protein levels are comparable in the two groups [25].

Immunogenicity, Binding, and Neutralizing Antibodies. The bioactivity and distribution of therapeutically administered proteins can be modulated by their immunogenicity; humoral responses to therapeutic proteins can result in binding and neutralizing antibodies. Generally, chronic or intermittently administered proteins are more likely to be affected by these pathways than proteins administered for acute conditions. The underlying protein structure, aggregation state, and protein formulation can contribute to the risk of binding and neutralizing antibodies. Factors related to the dosing regimen such as dose, frequency, and route of administration are also important determinants. The role of disease-related and individual-specific variables to the development of antibodies is poorly understood. The time to development of antibody depends on the dosing regimen and immunogenicity of the preparation. For some proteins, the development of antibodies to endogenous proteins can cause serious adverse events (e.g., red cell aplastic anemia can occur with anti-erythropoietin antibodies). The impact of antibodies on drug efficacy in clinical trials should be assessed during the period the antibodies are present.

As a case study, consider interferon-β, an endogenous cytokine, recombinant forms of which are used chronically in multiple sclerosis (MS). Two types of IFN-β exist: IFN-β-1a and IFN-β-1b. IFN-β-1a is glycosylated and has the same amino acid sequence as endogenous IFN-β. IFN-β-1b is not glycosylated, lacks the N-terminal methionine, and is engineered with cysteine replaced by serine. Two approved IFN-β-1a products exist that are dosed by the once-weekly, intramuscular, 30-μg and three-times-weekly, subcutaneous, 44-μg dosing regimens; the IFN-β-1b product is an every-other-day, subcutaneous, 250-μg dosing regimen. As with other protein drugs, some patients develop neutralizing antibodies (NAB) upon chronic administration of all IFN-β products. The proportion of patients developing NAB ranges from 25% for three-times-weekly, subcutaneous IFN-β-1a regimen to 2% for the once-weekly, IM IFN-β-1a regimen [26]. In a clinical trial where the weekly IM dose in the once-weekly, IM IFN-β-1a regimen was increased from 30 μg to 60 μg, the percentage of patients developing NAB titers greater than 20 after 3 years of treatment increased from 2.2% to 5.8% and suggests that, for a given route of administration, the frequency of IFN-β-1a administration is probably the major factor influencing the immunogenicity of an IFN-β product and the increase in dosage has relatively less impact. Published data from large randomized clinical trials demonstrate that disease activity in NAB-positive patients became similar to placebo-treated patients [27]. Interestingly, skin necrosis, an adverse effect that occurs with SC IFN-β administration, occurred only in the NAB-negative patients [27]. Despite the evidence from large clinical trials, some controversy exists regarding the appropriate clinical decisions for treatment of NAB-positive MS patients, especially those with a clinically stable disease, because (1) many NAB-positive patients test negative over the course of treatment [28], (2) no accepted IFN-β treatment biomarkers exist that are MS-relevant [29, 30] and, (3) market and competitive factors occur because of the availability of multiple IFN-β products for MS.

3.2.2.3   Metabolism

The cytochrome P450 enzymes, expressed in the endoplasmic reticulum of the liver, kidney, and intestine, are the predominant metabolizing pathways for small-molecule drugs. After the phase I oxidation reactions, small molecules typically are further modified by phase II conjugation reactions mediated by glucoronosyl transferases that render them hydrophilic for renal elimination. However, as noted earlier, the metabolism of protein drugs does not involve these small-molecule pathways: Endogenous pathways involved in protein homeostasis mediate protein drug metabolism.

As indicated earlier, proteolysis is the predominant mechanism for protein drug metabolism, but diverse endogenous proteolytic pathways exist in humans. Further, the half-lives of protein *in vivo* are, not surprisingly, regulated by a variety of intracellular mechanisms. The half-life of intracellular proteins is regulated via the ubiquitin pathway, which targets proteins to degradation in a large multimeric protease complex referred to as the proteasome. Ubiquitin is a widely expressed globular protein with a protruding carboxyl-terminus that tags proteins for degradation. Ubiquitin-activating enzymes activate through a process that requires ATP, and the activated ubiquitin is then transferred to a lysine on the target protein by another enzyme, the ubiquitin ligase. Multiple ubiquitin chains can be ligated on a single target protein lysine by conjugating glycine-lysine bonds between successive ubiquitin chains. Some target proteins are also tagged by binding proteins referred to as "recognin" that bind sequence elements called "degrons" that specify half-life. Sufficiently ubquitinated proteins are recognized and degraded in the proteasome. Although the ubiquitin pathway is critical for cellular protein homeostasis, its involvement in the metabolism of exogenously administered proteins is probably minor.

Some circulating proteins, most notably albumin and immunoglobulin G (IgG), have anomalously long half-lives because they are protected from degradation by specific mechanisms after cellular uptake. The best characterized of these protein-stabilizing mechanisms is the neonatal Fc receptor or FcRn, which is present on a variety of cell types and tissues including the kidney, liver, and lung endothelial cells, hepatocytes, monocytes, and intestinal macrophases. It confers a long half-life to IgG (mean half-life in humans = 23 days) relative to other types of immunoglobulins (half-lives for IgA, IgD, IgE, and IgM are 6, 3, 2.5, and 5 days, respectively). Most proteins generally undergo degradation in lysosome after pinocytosis and receptor-mediated endocytosis from the cell surface. However, the constant Fc region of immunoglobulin G and antibody-based drugs are internalized by pinocytosis or, when they bind, the cell surface antibody-binding Fcγ receptors. However, upon internalization into the endosomes, immunoglobulin G is bound by FcRn and the complex, upon endocytosis, is protected from proteolysis and recycled to the cell surface instead of being processed for degradation in lysosomes. The FcRn receptor for IgG is often referred to as the Brambell receptor after the authors [31] who postulated their existence to explain the relatively long half-life and the increased rates of IgG degradation at higher IgG concentration. Mice that are β_2 microglobulin-deficient do not express FcRn and have IgG half-lives that are 10–15-fold shorter than mice with intact FcRn [32–34].

Overall, 3 major mechanisms for clearing protein drugs exist: (1) renal, (2) hepatic, and (3) receptor-modulated.

3.2.2.4   Excretion

The rate of elimination of proteins is strongly dependent on the size, structure, and, in the case of chronically administered protein drugs, the presence or absence of antidrug antibodies. Renal clearance is relatively uncommon for protein drugs with molecular weights greater than 50 kD. Specific primary structural features such as the presence of a humanized immunoglobulin constant (Fc) region can confer prolonged stability via endogenous recycling pathways. Likewise, glycosylation and the addition of polyethylene glycol moieties (a synthetic modification often referred to as PEGylation) can also prolong circulation times.

Protein drugs with molecular weights less than 20 kD are filtered out of circulation in the Bowman's capsule of glomerulus of the kidney and enter the luminal space of the nephron. Even under pathological conditions, urinary excretion of intact protein is unusual. The filtered protein drugs are reabsorbed from the lumen and released into the circulation either as intact drugs or degraded into peptides. The proximal tubule is the predominant site for reabsorption, which transfers proteins into apical vacuoles that fuse with lysosomes. Lysosomal proteases degrade proteins into amino acids that re-enter the circulation. Small peptides can also be directly degraded by proteases on the luminal side of the renal tubule and the resulting amino acids can be reabsorbed by transport mechanisms.

Hepatic metabolism is important to the clearance of therapeutic proteins with molecular weights greater than 20 kD, although renal clearance can contribute to proteins as large as 50 kD. The Kupfer cell and hepatocytes of the liver are involved in clearing protein drugs.

Pegylation. Pegylation refers to the covalent modification of proteins by polyethylene glycol (PEG) moieties. As PEG is a hydrophilic, degradation-resistant polymer, pegylation can increase the circulating half-life of proteins, reduce renal elimination, alter the distribution, and reduce the immunogenicity of proteins [35]. The addition of PEG increases the molecular weight of the protein, which is instrumental in reducing renal filtration, and it provides a hydrophilic shield containing tightly associated water molecules that protects against proteolysis and some nonspecific interactions that mitigate therapeutic protein immunogenicity. However, pegylation can reduce receptor-binding affinity. When engineering pegylation during drug development, the effects of reduced receptor binding can be offset by the increased drug exposures; the improvements in circulating half-life and protein disposition effects and reduced immunogenicity benefit of this modification. The chemistry of PEG conjugation involves chemical activation of PEG followed by covalent attachment at the reactive amino groups of lysine or the N-terminal amino acid or the sulfhydryl groups of cysteine. As a rule of thumb, the molecular weight of the PEG moiety has to be comparable with that of the therapeutic protein for pegylation to be effective. Pegylation chemistry is also capable of conjugating branched PEG polymers that can increase the mass of PEG added at a each site and can better protect the protein therapeutic from degradation and immune recognition [36]. The pegylation approach is quite general and has also been used for liposomes: It prolongs the circulation times of liposomes by reducing the rapid elimination by the reticuloendothelial system.

Interferon-α-2b (IFN-α-2b, PEGINTRON), which is used for treating hepatitis, is a case study on the impact possible with the pegylation approach because considerable preclinical and clinical data are also available for the IFN-α-2b unmodified by pegylation (INTRON-A). As a result of pegylation, the molecular weight of IFN-α-2b is increased from 19 kilodaltons to approximately 31 kilodaltons and the half-life increases from 3–8 hours to 65 hours. Other pharmacokinetic variables altered include a decreased volume of distribution (from 31–73 L to 8–12 L) and decreased systemic clearance (from 6.6–29.2 L/hour to 0.06–0.1 L/hour). The pharmacodynamic effects of antiviral and antitumor activities of pegylated IFN-α-2b are 12–136-fold and 18-fold, respectively, greater than the unpegylated form. The net result of these pharmacokinetic and pharmacodynamic improvements is that the frequency of pegylated IFN-α-2b administration for treating chronic hepatitis C is once weekly compared with three times weekly for the unpegylated form.

Hyperglycosylation. Hyperglycosylation is the addition of sugar moieties to proteins and can be engineered using molecular biology techniques to specific locations on a protein. Many cellular proteins have N-linked or O-linked glycan modification that are added in posttranslational and cotranslational processes in the endoplasmic reticulum and in the Golgi apparatus [37–39]. The N-linked glycosylations occur at amino groups of the asparagine in Asn-X-Ser/Thr sequences, although not all potential glycosylation sites may be modified and heterogeneity can occur within given protein molecules. The O-linked glycosylations occur at the hydroxyl groups of serine and threonine.

Receptor-Mediated Elimination. For some proteins, pharmacokinetic non-linearities can occur because of receptor-mediated elimination. When therapeutic proteins bind their cognate cell-surface targets, the bound therapeutic protein can be internalized by receptor-mediated or fluid-phase endocytosis and traffic to the lysosomes where it may be degraded (Figure 3.2-2) [40]. The capacity-limited nature of these systems can manifest as saturable dose-dependent drug clearance and may be susceptible to further modulation through functional adaptation mechanisms such as receptor upregulation or downregulation. The saturable internalization of erythropoietin (EPO) in erythroid progenitor cells is a classic example [41]. Chapel et al. have shown that the bone marrow, a major site for erythropoietic cells, represents a significant EPO elimination pathway [42]. Busulfan or 5-fluorouracil-induced bone marrow ablation in sheep was associated with a decrease in EPO clearance, revealing the *in vivo* significance of target-mediated uptake and elimination for this therapeutic protein.

Complex target-mediated clearance mechanisms suggest that both pharmacokinetics and pharmacodynamics might have to be considered concurrently to understand the disposition of select macromolecules. Peptide and protein drugs that alter the expression of target cells may have the ability to modulate their pharmacokinetic properties. For example, thrombopoietin (TPO) is an endogenous hematopoietic growth factor that stimulates the proliferation and differentiation of megakaryocytes and thus the production of new platelets, which, in turn, would yield a net increase in the density of available TPO receptors (c-Mpl) and reconcile

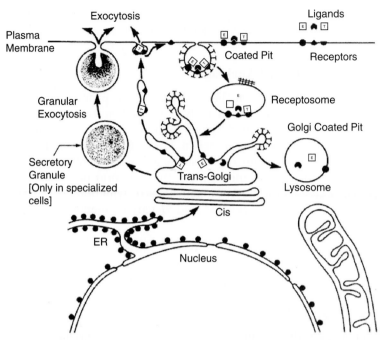

Figure 3.2-2. Schematic for the pathway of receptor-mediated endocytosis. Example ligands include EGF (E), transferrin (T), and α_2-macrogbuilin (●), and they are shown along with their respective receptors (●, ◄, and ▲). These ligands are examples of systems in which both ligand and receptor are transferred to lysosomes (EGF), both ligand and receptor are recycled to the cell surface (transferrin), and the ligand is delivered to lysosomes but the receptor is efficiently recycled (α_2-macroglobuilin). For some systems, receptors may undergo significant turnover in the absence of ligand. (Figure and legend reproduced with permission [from Ref. 40].)

the reciprocal relationship between platelet count and circulating plasma TPO concentrations [43].

3.2.3   PHARMACOKINETIC DATA ANALYSIS

The primary objective of pharmacokinetic data analysis is to characterize quantitatively the drug disposition processes discussed in Section 3.2.2. Sufficient knowledge of the preclinical pharmacokinetics of macromolecules, which can only be obtained via mathematical formalisms, may be used to anticipate relevant pharmacokinetic properties in humans (Section 3.2.4), thereby serving to guide dose selection in subsequent clinical trials. The two most commonly used methods for characterizing pharmacokinetic data are noncompartmental analysis and compartmental modeling. Basic assumptions and techniques are presented in this section and are placed within the context of biotechnology pharmaceuticals, which may provide guidance for selecting the most appropriate and desirable method for a given dataset. It is important to recognize that no method of data analysis can offset shortcomings in experimental designs and assays, and that careful consideration of such details is a sine qua non in pharmacokinetics [44].

3.2.3.1 Noncompartmental Analysis

Intravenous Drug Disposition. The estimation of primary pharmacokinetic parameters using noncompartmental analysis is based on statistical moment theory [45, 46]. The relationships defined by this theory are valid under the assumption that the system is linear and time-invariant. For simplicity, we further assume that drug is irreversibly removed only from a single accessible pool (e.g., plasma space). Regardless of the route of administration, the temporal profile of plasma drug concentrations, $C_p(t)$, can represent a statistical distribution curve. As such, the zeroth and first statistical moments (M_0 and M_1) are defined as:

$$M_0 = \int_0^\infty C_p(t)\,dt = AUC \tag{3.2-1}$$

$$M_1 = \int_0^\infty t \cdot C_p(t)\,dt = AUMC \tag{3.2-2}$$

where AUC and $AUMC$ are the so-called area under the plasma concentration-time curve and area under the first-moment curve. Several methods have been described for estimating the values of these moments, some of which will be presented in this section, and may be used to calculate relevant pharmacokinetic parameters.

Under linear conditions, the time-course of plasma drug concentrations following a single IV bolus dose can be characterized by a sum of exponentials:

$$C_p(t) = \sum C_i \cdot e^{-\lambda_i t} \tag{3.2-3}$$

where the coefficients (C_i) and exponentials (λ_i) may be obtained via curve-stripping, or more appropriately, from curve-fitting using nonlinear regression analysis. One advantage of fitting Equation (3.2-3) to pharmacokinetic data is the relative simplicity in calculating AUC and $AUMC$ values [47]:

$$AUC = \sum C_i / \lambda_i \tag{3.2-4}$$

$$AUMC = \sum C_i / \lambda_i^2 \tag{3.2-5}$$

Primary pharmacokinetic parameters such as total systemic clearance (CL) and the volume of distribution at steady state (V_{ss}) can be defined as:

$$CL = D_{iv} / AUC \tag{3.2-6}$$

and

$$V_{ss} = \frac{D_{iv} \cdot AUMC}{AUC^2} = CL \cdot MRT_{iv} \tag{3.2-7}$$

where D_{iv} represents the IV bolus dose and MRT_{iv}, or mean residence time, is the average time a drug molecule resides in the system before being irreversibly removed. Also, the central volume of distribution (V_c) can be calculated as:

$$V_c = D_{iv} \Big/ \sum C_i \qquad (3.2\text{-}8)$$

Additional advantages to resolving AUC and $AUMC$ from curve-fitting the sums of exponentials are the stability of the estimates and that nonlinear regression procedures provide measures of error about the pharmacokinetic parameter estimates.

Peptide and protein drugs requiring an IV route of administration may also be given as a constant rate infusion. Pharmacokinetic data from experiments designed to achieve a steady-state plasma concentration (C_{ss}) allow for the direct calculation of drug clearance according to the equation:

$$CL = k_0 / C_{ss} \qquad (3.2\text{-}9)$$

where k_0 is the constant zero-order rate of drug infusion. The steady-state volume of distribution can be obtained from:

$$V_{ss} = \frac{k_0 \cdot T - CL \cdot AUC_0^T}{C_{ss}} \qquad (3.2\text{-}10)$$

where T is the duration of the IV infusion. For short-term constant rate infusions, when plasma concentrations do not attain C_{ss}, Equation (3.2-6) is valid for calculating clearance by substituting $k_0 \cdot T$ for D_{iv}. Moment analysis also may be used to generate V_{ss} under these conditions from the equation:

$$V_{ss} = \frac{k_0 \cdot T \cdot AUMC}{AUC^2} - \frac{k_0 \cdot T^2}{2 \cdot AUC} \qquad (3.2\text{-}11)$$

Extravascular Drug Disposition. In general, peptide and protein drugs exhibit poor systemic bioavailability following oral administration owing primarily to gastrointestinal enzymatic degradation. On the other hand, IV drug administration may not result in a desirable time-course of drug exposure and certainly lacks appeal for patients requiring chronic therapy. One of the most common routes of extravascular peptide and protein drug administration is subcutaneous injection, with additional routes, such as inhalation, intranasal, and transdermal delivery representing the targets of current investigation [48].

The time-course of plasma drug concentrations following extravascular administration is often irregular, and simple analytical solutions for calculating AUC and $AUMC$, such as Equations (3.2-4) and (3.2-5), are not readily available. For these cases, numerical integration methods may be used to evaluate the integrals in Equations (3.2-1) and (3.2-2), such that:

$$AUC = \int_0^{t^*} C_p(t)\,dt + \frac{C_p^*}{\lambda_z} \qquad (3.2\text{-}12)$$

and

$$AUMC = \int_0^{t^*} t \cdot C_p(t)dt + \frac{t^* \cdot C_p^*}{\lambda_z} + \frac{C_p^*}{\lambda_z^2} \qquad (3.2\text{-}13)$$

where C_p^* and t^* are the last observed plasma concentration and time values, respectively, λ_z is the terminal slope of the curve, and the quotients provide estimates of the area extrapolating from t^* to infinity. The trapezoidal and log-trapezoidal rules are the most common methods for numerically evaluating the integrals in Equations (3.2-12) and (3.2-13) [49]. The trapezoidal rule for calculating the area under a curve within an interval t_i and t_{i-1} is given as:

$$\int_{i-1}^i y \cdot dy = 0.5 \cdot [(y_i + y_{i-1})(t_i - t_{i-1})] \qquad (3.2\text{-}14)$$

where y is $C_p(t)$ for AUC and $t \cdot C_p(t)$ for $AUMC$. For the log-trapezoidal rule, the equation is:

$$\int_{i-1}^i y \cdot dy = \frac{1}{\ln\left(\dfrac{y_i}{y_{i-1}}\right)} \cdot [(y_i + y_{i-1})(t_i - t_{i-1})] \qquad (3.2\text{-}15)$$

The cumulative sum of the interval areas thus provides an estimate of the AUC and $AUMC$ from time zero to t^*. Yu and Tse have evaluated the performance of several numerical integration algorithms for calculating AUC [49]. In practice, it is most common to employ the trapezoidal rule for ascending ($y_i > y_{i-1}$) and plateau ($y_i \approx y_{i-1}$) regions of the curve and the log-trapezoidal rule when measured values are decreasing ($y_i < y_{i-1}$). Such an approach is used for noncompartmental analysis in the computer program WinNonlin (Pharsight, Mountain View, CA), a nonlinear regression analysis program specifically designed for pharmacokinetic and pharmacodynamic data analysis.

The noncompartmental analysis of pharmacokinetic data after extravascular drug administration, when coupled with that of IV dosing, can yield additional relevant pharmacokinetic parameters, particularly regarding absorption processes. For example, the systemic availability (F), which represents the net fraction of the drug dose reaching the systemic circulation after extravascular administration, is defined as:

$$F = \frac{D_{iv} \cdot AUC_{ev}}{D_{ev} \cdot AUC_{iv}} \qquad (3.2\text{-}16)$$

where catabolism at the extravascular site of administration and other processes may result in incomplete drug absorption. Also, the mean absorption time (MAT), or the average time a drug molecule remains unabsorbed, can be calculated from the relationship:

$$MAT = MRT_{ev} - MRT_{iv} \qquad (3.2\text{-}17)$$

where the mean residence times derive from the ratios of *AUMC* and *AUC* [e.g., Equation (3.2-7)].

Overall, the relative simplicity of noncompartmental analysis, as compared with the methodological challenges of developing and validating structural mathematical models [50], forms the basis for it being the most common method for ascertaining the major pharmacokinetic properties of drugs. It is beyond the scope of this chapter to contrast the merits of these two methods, and the reader is referred to more complete treatments and the references therein [44, 51]. However, as discussed in Section 3.2.2, various pharmacokinetic processes of many peptide and protein pharmaceuticals are inherently nonlinear, which violate the fundamental assumptions of linearity and time-invariance in statistical moment analysis. For such compounds, the construction of mechanism-based mathematical models is required and essential for characterizing the *in vivo* pharmacokinetics of the drug (Section 3.2.3.2). In any event, noncompartmental analysis represents a highly useful starting point for any pharmacokinetic data analysis [44]. For simple linear systems, this technique may be sufficient for calculating primary pharmacokinetic parameters from preclinical experiments and anticipating dose-concentration relationships in humans (Section 3.2.4.1). Additionally, moment analysis of plasma concentration-time profiles resulting from a suitable range of dose levels may be used initially to identify whether nonlinearities exist, such as disproportionate *AUC* values (relative to dose) indicating dose-dependent clearance or bioavailability, thus necessitating the implementation of more sophisticated methods of pharmacokinetic systems analysis.

3.2.3.2 Pharmacokinetic Models

Compartmental modeling involves the specification of a structural mathematical model (commonly using either explicit or ordinary differential equations) and system parameters are estimated from fitting the model to pharmacokinetic data via nonlinear regression analysis or population mixed effects modeling. The methodological issues involved in the development, application, and interpretation of pharmacokinetic models are beyond the scope of this chapter and are discussed in detail elsewhere [50, 52]. The most popular structural model applied to macromolecule pharmacokinetic data is the open two-compartment model shown in Figure 3.2-3. This system is described, in general, by the following differential equations:

$$\frac{dA_p}{dt} = \frac{Input}{V_c} - \left(\frac{CL}{V_c} + \frac{CL_D}{V_c}\right) \cdot A_p + \frac{CL_D}{V_t} \cdot A_t \tag{3.2-18}$$

$$\frac{dA_t}{dt} = \frac{CL_D}{V_c} \cdot A_p - \frac{CL_D}{V_t} \cdot A_t \tag{3.2-19}$$

where A_i represents the amount of drug in the plasma ($i = p$) or tissue ($i = t$) compartments, *Input* represents a function describing the appearance of drug based on the route of drug administration, CL_D is the distributional clearance between compartments, and V_t is the volume of the second or tissue compartment. Plasma drug concentrations are thus specified as $C_p = A_p/V_c$.

Input

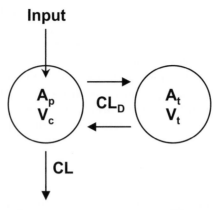

Figure 3.2-3. Schematic of a two-compartment mammillary model commonly used to characterize macromolecule pharmacokinetics. Drug input and elimination (*CL*) occur to and from a central compartment (A_p) and may distribute (CL_D) to a peripheral site (A_t, V_t). Plasma drug concentrations reflect the amount in the central compartment relative the volume of distribution of that site (V_c).

TABLE 3.2-1. Input Functions for Specific Pharmacokinetic Applications

D, dose; *F*, bioavailability; *fr*, fraction of dose undergoing first-order input in a dual absorption model; K_0, zero-order input rate; k_a and k'_a, first-order absorption rate constants; *MAT*, mean absorption time; NV^2, normalized variance of Gaussian density function, *t*, nominal time; t_{lag}, duration of time-lag; τ, duration of rapid input in dual absorption model; T_{inf}, duration of zero-order IV infusion; *T*, modulus time.

Condition	Input Function		Equation
Rapid intravenous injection	0		—
Intravenous infusion	K_0	$t \le T_{inf}$	—
	0	$t > T_{inf}$	
Linear first-order absorption with a time-lag	0	$t \le t_{lag}$	3.2-20
	$F \cdot D \cdot k_a \cdot e[-k_a \cdot (t - t_{lag})]$	$t > t_{lag}$	
Two consecutive first-order processes	$F \cdot D \cdot \dfrac{k_a \cdot k'_a}{(k'_a - k_a)} \cdot [e(-k_a \cdot t) - e(-k'_a \cdot t)]$		3.2-21
Rapid zero-order input followed by linear first-order absorption	$F \cdot D \cdot (1 - fr)/\tau$	$t \le \tau$	3.2-22
	$F \cdot D \cdot fr \cdot k_a \cdot e[-k_a \cdot (t - \tau)]$	$t > \tau$	3.2-23
Inverse Gaussian	$F \cdot D \cdot \sqrt{\dfrac{MAT}{2\pi \cdot NV^2 \cdot T^3}} \cdot e\left[-\dfrac{(T - MAT)^2}{2 \cdot NV^2 \cdot MAT \cdot T} \right]$		3.2-24

The drug input function in Equation (3.2-18) will depend on the experimental design and several common functions are listed in Table 3.2-1. For rapid IV injection, the input function is set equal to zero and the dose (*D*) is reflected in the initial condition for Equation (3.2-18) [$A_p(0) = D$]. A constant rate IV infusion also is easily handled by setting *Input* equal to the zero-order infusion rate (K_0) for time

(t) less than or equal to the infusion time (T_{inf}), otherwise *Input* = 0. The remaining equations in Table 3.2-1 typically are used to describe extravascular drug administration. The simplest function [Equation (3.2-20)] assumes a linear first-order absorption process (k_a) from an extravascular site following a short time-lag (t_{lag}). If a time-lag is not observed, t_{lag} may be set to zero, resulting in a simple linear absorption model, as exemplified in a pharmacokinetic model for recombinant human growth hormone following SC dosing in monkeys [53]. As mentioned in the previous section, IV pharmacokinetic data must be analyzed simultaneously with SC data to estimate drug bioavailability (F).

Owing to the complexities of drug absorption from extravascular sites (see Section 3.2.2.1), peptide and protein drugs may require an additional delay compartment [Equation (3.2-21)] comprising two consecutive first-order processes [54, 55], which may be accompanied by an initial rapid input of a fraction of the dose as defined by Equations (3.2-22) and (3.2-23). Such a dual absorption model was used to describe the input of recombinant erythropoietin after SC administration in monkeys [56]. Interestingly, the bioavailability was shown to be dose-dependent and ranged from 0.268 to 1.0, increasing with increased dose level. This property may be caused by saturation of catabolism at the site of drug administration; however, further research is needed to elucidate the specific processes involved in drug disposition following extravascular dosing. Many other complex functions are available, such as an inverse Gaussian model [Equation (3.2-24)], which may be substituted into pharmacokinetic models to describe the absorption of macromolecules.

The clearance of several macromolecules may be characterized using simple linear first-order elimination as shown in Equation (3.2-18) [53, 57], but may depend on the animal species and range of dose levels being evaluated. In contrast, many peptide and protein drugs demonstrate saturable or capacity-limited clearance, and CL in Equation (3.2-18) may be defined with a concentration-dependent function, such as:

$$CL = \frac{V_{\max} \cdot V_c}{K_m \cdot V_c + A_p} + CL_{ns} \qquad (3.2\text{-}25)$$

where $V_{\max}$ and K_m are the traditional Michaelis–Menten parameters and CL_{ns} represents a linear nonsaturable clearance pathway. The nonlinear Michaelis–Menten function [quotient in Equation (3.2-25)] may be specified alone [56] or in parallel with CL_{ns} [58, 59]. Thus, at high plasma drug concentrations (i.e., $C_p \gg K_m$), the saturable clearance function approaches a limiting value ($V_{\max}$), and at relatively low concentrations will represent a first-order elimination rate ($V_{\max}/K_m$).

Receptor-mediated clearance is an example of target-mediated drug disposition (TMDD), where binding of a compound to a pharmacological target (e.g., enzyme, receptor, or other proteins) influences the pharmacokinetics of the drug [60]. Such systems often exhibit a dose-dependent decrease in the apparent volume of distribution (approaching a limiting value). The general pharmacological expectations of TMDD and techniques for characterizing this phenomenon have been reviewed recently [61]. A general pharmacokinetic model of TMDD has been described [62], and the operative equations defining the rate of change of plasma drug concentrations typically include:

$$\frac{dC_p}{dt} = -k_{on} \cdot R_f \cdot C_p + k_{off} \cdot RC \tag{3.2-26}$$

$$\frac{dRC}{dt} = k_{on} \cdot R_f \cdot C_p - (k_{off} + k_{int}) \cdot RC \tag{3.2-27}$$

where k_{on} and k_{off} are second- and first-order rates of association and dissociation, R_f and RC represent free and bound receptor (or pharmacological target) concentrations, and k_{int} is a first-order internalization rate constant of RC. Under certain conditions, the total receptor concentration (R_{tot}) may be assumed to be time-invariant [11, 62], and thus $R_f = R_{tot} - RC$. Therefore, at high drug concentrations, $RC \to R_{tot}$ and effectively limits the distribution and elimination of drug from the central compartment. This approach has been used to characterize the complex nonlinear pharmacokinetics and pharmacodynamics of IFN-β 1a in monkeys [54]. Alternatively, an additional equation may be introduced that describes the rate of change of R_f directly [55]. Also, an equilibrium solution to the general TMDD model is available, where the drug-binding microconstants (k_{on} and k_{off}), which are often difficult to estimate from routine pharmacokinetic data, are replaced with an equilibrium dissociation constant ($K_D = k_{off}/k_{on}$) [63].

3.2.4 PREDICTING MACROMOLECULE PHARMACOKINETICS IN HUMANS

One of the major goals of preclinical pharmacokinetics is to obtain relevant information on the *in vivo* disposition of drugs early in the drug development process and to apply such knowledge to anticipate probable pharmacokinetic properties in humans for initial dose selection in phase I clinical trials. Allometric scaling and physiologically based pharmacokinetic (PBPK) models are the most popular techniques for predicting pharmacokinetic properties of drugs in one species from data collected in one or more other species. Although these methods have been applied to small-molecular-weight compounds for decades with variable and often limited success, it is hypothesized that these techniques are more accurate for peptide and protein drugs owing to the relative species conservation of mechanisms that control the biodistribution and elimination of such compounds [64, 65].

3.2.4.1 Allometric Scaling

Many physiological processes and organ sizes scale across species according to the well-known power-law relationship or allometric equation [66]:

$$Y = a \cdot W^b \tag{3.2-28}$$

where Y is a physiological parameter of interest, W represents body weight, and a and b are the allometric coefficient and exponent. The linearized form of Equation (3.2-28) is obtained by taking its logarithm:

$$\log Y = \log a + b \cdot \log W \tag{3.2-29}$$

where log a and b represent the intercept and slope of the physiological data presented on a log-log plot. The units of a will depend on the units of Y and W, and its absolute value will depend on the specific property of interest. In contrast, b is a unitless parameter characterizing the rate of change of Y with respect to W. According to the power-law relationship, Y may increase more rapidly than W ($b > 1$), less rapidly than W ($0 < b < 1$), or be directly proportional ($b = 1$). For rates of metabolism and clearance processes, the allometric exponent tends to be around 0.75, whereas organ sizes or physiological volumes tend to be proportional to body weight [66]. Physiological times or the duration of physiological events (e.g., heartbeat and breath duration, lifespan, or turnover times of endogenous substances or processes) scale across species with b values around 0.25 [66, 67]. These interspecies relationships appear to manifest from the fractal nature of biological systems [68, 69].

Retrospective analysis suggests that primary pharmacokinetic parameters of many small synthetic molecules scale across species according to allometric principles, primarily reflecting the physiological processes controlling such characteristics [67, 70]. However, the use of allometric scaling for prospective predictions of human pharmacokinetic parameters and dose selection of these compounds is controversial and is the topic of considerable debate [71, 72]. In addition, several correction factors and techniques have been suggested for improving the prediction of drug clearance in humans from preclinical values [73, 74]. Few studies have been conducted that specifically address the role of parameter uncertainty and species selection; however, those that have been reported suggest careful study designs and caution for prospective allometric scaling predictions [75, 76]. Research directed toward improving the understanding of interspecies pharmacokinetic relationships is likely to continue and necessary for resolving key methodological issues.

Notwithstanding concerns associated with prospective allometric scaling, primary pharmacokinetic parameters of select macromolecules appear to follow allometric relationships [65, 77–81], possibly resulting from a conservation of processes controlling peptide and protein disposition [64, 65]. Mahmood reviewed the prediction of total clearance of 15 therapeutic proteins from interspecies scaling, confirming the need for a suitable number of animals and that simple allometry may be used for most proteins [82]. As an example, Figure 3.2-4 shows log-log plots of total systemic clearance and the volume of distribution at steady state for interferon-α in mice, rats, rabbits, dogs, monkeys, and humans [78]. The lines of regression were generated without considering human values and the relationships were as follows: $CL = 3.7W^{0.71}$ ($r^2 = 0.98$) and $V_{ss} = 0.2W^{0.94}$ ($r^2 = 0.99$). Typical values in man (open symbols) were slightly underestimated using these equations, and Lave et al. suggest that transformations of plasma concentration-time profiles make more use of preclinical data and may provide improved predictions of human pharmacokinetic parameters [78]. Building on this concept, Cosson et al. described a method of fitting preclinical pharmacokinetic profiles and estimating allometric relationships simultaneously using a population mixed effects modeling technique [83]. Such an approach has been applied to characterize the interspecies pharmacokinetic relationships of pegylated human erythropoietin [84]. The final model was used to predict typical pharmacokinetic parameters in humans, as well as to simulate expected concentration-time profiles from IV and SC administration.

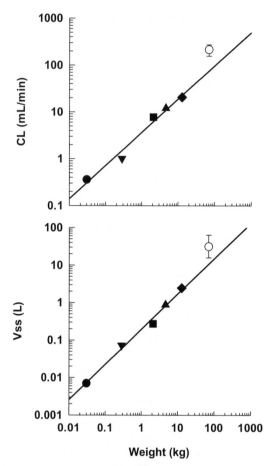

Figure 3.2-4. Interspecies scaling of total systemic clearance (CL; top panel) and steady-state volume of distribution (V_{ss}; bottom panel) of interferon-α as a function of total body weight. Species include mouse (●), rat (▼), rabbit (■), African green monkey (▲), dog (◆), and human (○). Human values were not included in the line of regression, and data were obtained from Lave et al. [78].

Despite successful applications of interspecies scaling to macromolecule pharmacokinetics, potential complicating factors exist, the most significant of which relate to saturable or dose-dependent pharmacokinetic processes. As discussed in Section 3.2.2, nonlinear behavior may manifest from several sources including receptor capacity and binding kinetics, macromolecule physico-chemical properties, and major mechanisms of peptide and protein transport and elimination. Thus, evidence of nonlinear pharmacokinetics in preclinical experiments would raise concerns as to the utility of allometric scaling for predicting such properties in humans. It remains to be determined whether a combination of *in vivo* preclinical data, compartmental modeling, and *in vitro* uptake studies might reliably anticipate macromolecule pharmacokinetics in humans under such nonlinear conditions [85].

3.2.4.2 Physiologically Based Pharmacokinetic Modeling

Classic pharmacokinetic modeling attempts to characterize drug disposition with systems comprised of a relatively small number of compartments (see Section 3.2.3.2). In contrast, PBPK models seek to mimic physiological pathways and processes controlling the time-course of plasma drug concentrations and represent the state-of-the-art in advanced pharmacokinetic systems analysis. As Dedrick has noted, "Physiologic modeling enables us to examine the joint effect of a number of complex inter-related processes and assess the relative significance of each" [86]. Basic concepts of PBPK modeling and applications to macromolecule pharmacokinetics are described in this section, and several reviews may be consulted for further details regarding the specification, implementation, and evaluation of PBPK models [87, 88].

The compartments in PBPK models represent organs and tissues of interest and are arranged and connected according to anatomical and physiological relationships (Fig. 3.2-5). Most PBPK models contain principal components, such as arterial and venous blood pools, liver, and kidney, along with additional tissues depending on its role in specific disposition processes (e.g., absorption, transport, and elimination), whether it represents a potential site of action (biophase) or is likely to account for a significant proportion of the administered dose [88]. A series of mass-balance equations are specified that define the time-course of drug concentrations within each compartment. In the simplest case, it is assumed that mass

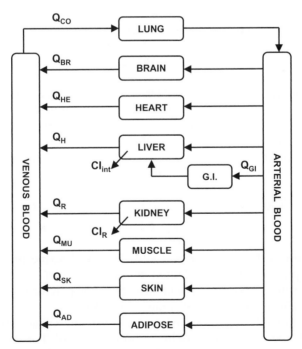

Figure 3.2-5. Model structure of a hypothetical physiologically based pharmacokinetic model. Tissue blood flows are shown (Q) along with renal (Cl_R) and nonrenal (Cl_{int}) drug clearance pathways.

transfer is blood-flow or perfusion-rate limited and compartments are well-stirred spaces. Under this assumption, the rate of change of drug concentration (C) in a noneliminating tissue (T) may be described by the following ordinary differential equation:

$$V_T \cdot dC_T / dt = Q_T \cdot (C_A - C_T / K_{pT})$$ (3.2-30)

where V and Q represent the volume and blood flow associated with a particular tissue, K_p is the tissue:plasma partition coefficient, and C_A is arterial drug concentration. Drug elimination (via excretion or metabolism) in specific organs often is described using a first-order approximation or a Michaelis–Menten function, separately or in parallel [Equation (3.2-25)]. Physiological parameters (e.g., V and Q) are frequently fixed to experimentally-measured values or literature-reported estimates [89, 90]; however, these terms have also been estimated during model fitting with a Maximum *a Posteriori* Bayesian method [91]. Several *in vitro* and *in vivo* experimental methods have been described for determining drug-specific parameters [92, 93]. Again, most contemporary applications of PBPK modeling involve estimating such terms from fitting the model to pharmacokinetic data.

Sugiyama and Hanano constructed a simple "minimal" PBPK model, consisting of plasma, liver, and kidney compartments, to evaluate the role of receptor-mediated clearance in the overall disposition of epidermal growth factor (EGF) in rats [11]. A remainder compartment was included to account for the rest of the body of the animal. The kinetics of the liver compartment incorporated receptor turnover processes (i.e., synthesis and degradation), binding of drug to free cell-surface receptors, and internalization of the drug-receptor complex (endocytosis and degradation). Computer simulations were conducted to evaluate this model, revealing that nonlinear kinetics of EGF may result from capacity limitation produced by either a saturation of free receptors or dose-dependent receptor downregulation.

The simple assumptions that constitute blood-flow-limited PBPK models often are inadequate for characterizing the pharmacokinetics of macromolecules. Instead, a membrane- or permeability-rate-limited model is more common, where it is assumed that mass transfer across the cell membrane is rate-limiting. For these models, organ compartments are subdivided into at least two well-stirred spaces representing vascular ($C_{T,V}$) and extravascular ($C_{T,EV}$) compartments. Such a system might be described by the following equations for a noneliminating tissue:

$$V_{T,V} \, dC_{T,V} / dt = Q_T \cdot (C_A - C_{T,V}) - f_{uB} \cdot PS_T \cdot (C_{T,V} - C_{T,EV} / K_{pT})$$ (3.2-31)

$$V_{T,EV} \, dC_{T,EV} / dt = f_{uB} \cdot PS_T \cdot (C_{T,V} - C_{T,EV} / K_{pT})$$ (3.2-32)

where f_{uB} is the free fraction in blood and PS_T is the permeability surface area product. Examples of membrane-limited models developed for protein molecules include β-endorphin, a 31 amino acid straight-chain polypeptide [94], and ISIS 1082, a 21-mer phosphorothioate oligonucleotide [95].

Monoclonal antibodies represent a diverse group of therapeutic proteins typically presenting with complex pharmacokinetic properties [10]. For these compounds, more comprehensive, mechanism-based PBPK models have been described [96, 97]. Baxter et al. developed and evaluated a bifunctional antibody PBPK model in mice and scaled the model to predict its pharmacokinetics in humans [96]. A membrane-limited model was used; however, additional equations were required to describe up to nine molecular species in each tissue (resulting from specific and nonspecific binding), and mass transport across the membrane included convective and diffusive components (so-called two-pore system). The latter model, by Friedrich et al., further incorporated lymphatic circulation and identified critical properties requiring optimization for this treatment modality to be successful [97].

Although the PBPK approach has existed for several decades, reports of its use in characterizing macromolecule pharmacokinetics have been relatively sparse. The technological requirements for PBPK modeling are often seen as formidable; however, advances in computer hardware and software are bringing this methodology within common reach. The significant potential for scaling preclinical models to anticipate human pharmacokinetics in various relevant tissues (including target sites) represents a strategic advantage of PBPK modeling in drug development. Furthermore, this systems approach provides a means of understanding the pharmacological implications of the complexities associated with the disposition of macromolecules.

3.2.5 CONCLUSIONS

The pharmacokinetic properties of biotechnology pharmaceuticals are complex relative to small-molecule drugs, and preclinical studies are critical to the discovery and development of these therapeutic agents [98]. Understanding the mechanisms and implications of *in vivo* disposition of macromolecules provides a means for investigating and designing novel drug candidates and delivery systems to optimize exposure-response relationships or therapeutic efficacy. Mathematical modeling of preclinical pharmacokinetic data is necessary for characterizing such properties, and techniques like allometric scaling and PBPK modeling may be used to anticipate human pharmacokinetics under certain conditions. The utility of these approaches may be further enhanced by coupling pharmacokinetic data with mechanism-based pharmacodynamic modeling of biomarkers and surrogates of drug effects [8, 99, 100].

ACKNOWLEDGMENT

Support from the National Science Foundation (Research Grant 0234895) and the National Institutes of Health (P20-GM 067650) is gratefully acknowledged.

REFERENCES

1. Supersaxo A, Hein W, Steffen H (1990). Effect of molecular weight on the lymphatic absorption of water-soluble compounds following subcutaneous administration. *Pharm. Res.* 7:167–169.
2. Charman S A, McLennan D N, Edwards G A, et al. (2001). Lymphatic absorption is a significant contributor to the subcutaneous bioavailability of insulin in a sheep model. *Pharm. Res.* 18:1620–1626.

3. Charman S A, Segrave A M, Edwards G A, et al. (2000). Systemic availability and lymphatic transport of human growth hormone administered by subcutaneous injection. *J. Pharm. Sci.* 89:168–177.

4. McLennan D N, Porter C J, Edwards, et al. (2003). Pharmacokinetic model to describe the lymphatic absorption of r-metHu-leptin after subcutaneous injection to sheep. *Pharm Res.* 20:1156–1162.

5. McLennan D N, Porter C J, Edwards, et al. (2005). Lymphatic absorption is the primary contributor to the systemic availability of epoetin Alfa following subcutaneous administration to sheep. *J. Pharmacol. Exp. Ther.* 313:345–351.

6. Odegard P S, Capoccia K L (2005). Inhaled insulin: Exubera. *Ann. Pharmacother.* 39:843–853.

7. Patton J S, Bukar J G, Eldon M A (2004). Clinical pharmacokinetics and pharmacodynamics of inhaled insulin. *Clin. Pharmacokinet.* 43:781–801.

8. Braeckman R (1999). Pharmacokinetics and Pharmacodynamics of Protein Therapeutics. In: Rereid (ed.), *Peptide and Protein Drug Analysis*, University of British Columbia, Vancouver, B.C., pp. 633–639.

9. Ravetch J V, Kinet J P (1991). Fc receptors. *Annu. Rev. Immunol.* 9:457–492.

10. Lobo E D, Hansen R J, Balthasar J P (2004). Antibody pharmacokinetics and pharmacodynamics. *J. Pharm. Sci.* 93:2645–2668.

11. Sugiyama Y, Hanano M (1989). Receptor-mediated transport of peptide hormones and its importance in the overall hormone disposition in the body. *Pharm. Res.* 6:192–202.

12. Kurschner C, Ozmen L, Garotta G, et al. (1992). IFN-gamma receptor-Ig fusion proteins: Half-life, immunogenicity, and in vivo activity. *J. Immunol.* 149:4096–4100.

13. Teng M N, Turksen K, Jacobs C A, et al. (1993). Prevention of runting and cachexia by a chimeric TNF receptor-Fc protein. *Clin. Immunol. Immunopathol.* 69:215–222.

14. Sato T A, Widmer M B, Finkelman F D, et al. (1993). Recombinant soluble murine IL-4 receptor can inhibit or enhance IgE responses in vivo. *J. Immunol.* 150:2717–2723.

15. Finkelman F D, Madden K B, Morris S C, et al. (1993). Anti-cytokine antibodies as carrier proteins. Prolongation of in vivo effects of exogenous cytokines by injection of cytokine-anti-cytokine antibody complexes. *J. Immunol.* 151:1235–1244.

16. Schobitz B, Pezeshki G, Pohl T, et al. (1995). Soluble interleukin-6 (IL-6) receptor augments central effects of IL-6 in vivo. *Faseb. J.* 9:659–664.

17. Aderka D, Engelmann H, Maor Y, et al. (1992). Stabilization of the bioactivity of tumor necrosis factor by its soluble receptors. *J. Exp. Med.* 175:323–329.

18. Debets R, Savelkoul H F (1994). Cytokine antagonists and their potential therapeutic use. *Immunol. Today.* 15:455–458.

19. Klein B, Brailly H (1995). Cytokine-binding proteins: stimulating antagonists. *Immunol. Today.* 16:216–220.

20. Ramanathan M (1997). A physicochemical modelling approach for estimating the stability of soluble receptor-bound tumour necrosis factor-alpha. *Cytokine.* 9:19–26.

21. Zapf J, Hauri C, Waldvogel M, et al. (1986). Acute metabolic effects and half-lives of intravenously administered insulinlike growth factors I and II in normal and hypophysectomized rats. *J. Clin. Invest.* 77:1768–1775.

22. Baumann G, Shaw M A, Amburn K (1994). Circulating growth hormone binding proteins. *J. Endocrinol. Invest.* 17:67–81.

23. Baumann G, Vance M L, Shaw M A, et al. (1990). Plasma transport of human growth hormone in vivo. *J. Clin. Endocrinol. Metab.* 71:470–473.

24. Baumann G, Amburn K D, Buchanan T A (1987). The effect of circulating growth hormone-binding protein on metabolic clearance, distribution, and degradation of human growth hormone. *J. Clin. Endocrinol. Metab.* 64:657–660.

25. Baumann G, Shaw M A, Brumbaugh R C, et al. (1991). Short stature and decreased serum growth hormone-binding protein in the Mountain Ok people of Papua New Guinea. *J. Clin. Endocrinol. Metab.* 72:1346–1349.

26. Panitch H, Goodin D S, Francis G, et al. (2002). Randomized, comparative study of interferon beta-1a treatment regimens in MS: The EVIDENCE trial. *Neurology.* 59:1496–1506.

27. The IFNB Multiple Sclerosis Study Group and the University of British Columbia MS/MRI Analysis Group (1996). Neutralizing antibodies during treatment of multiple sclerosis with interferon beta-1b: experience during the first three years. *Neurology.* 47:889–894.

28. Rice G P, Paszner B, Oger J, et al. (1999). The evolution of neutralizing antibodies in multiple sclerosis patients treated with interferon beta-1b. *Neurology.* 52:1277–1279.

29. Pachner A R (2001). Measurement of antibodies to interferon beta in patients with multiple sclerosis. *Arch. Neurol.* 58:1299–1300.

30. Rice G (2001). The significance of neutralizing antibodies in patients with multiple sclerosis treated with interferon beta. *Arch. Neurol.* 58:1297–1298.

31. Brambell F W, Hemmings W A, Morris I G (1964). A theoretical model of Gamma-Globulin Catabolism. *Nature.* 203:1352–1354.

32. Ghetie V, Hubbard J G, Kim J K, et al. (1996). Abnormally short serum half-lives of IgG in beta 2-microglobulin-deficient mice. *Eur. J. Immunol.* 26:690–696.

33. Israel E J, Wilsker D F, Hayes K C, et al. (1996). Increased clearance of IgG in mice that lack beta 2-microglobulin: Possible protective role of FcRn. *Immunology.* 89:573–578.

34. Junghans R P, Anderson C L (1996). The protection receptor for IgG catabolism is the beta2-microglobulin-containing neonatal intestinal transport receptor. *Proc. Natl. Acad. Sci. USA* 93:5512–5516.

35. Harris J M, Chess R B (2003). Effect of pegylation on pharmaceuticals. *Nat. Rev. Drug Discov.* 2:214–221.

36. Monfardini C, Schiavon O, Caliceti P, et al. (1995). A branched mono-methoxypoly(ethylene glycol) for protein modification. *Bioconjug. Chem.* 6:62–69.

37. Hart G W (1992). Glycosylation. *Curr. Opin. Cell Biol.* 4:1017–1023.

38. Jenkins N, Parekh R B, James D C (1996). Getting the glycosylation right: Implications for the biotechnology industry. *Nat. Biotechnol.* 14:975–981.

39. Jones J, Krag S S, Betenbaugh M J (2005). Controlling N-linked glycan site occupancy. *Biochim. Biophys. Acta.* 1726:121–137.

40. Pastan I, Willingham M C (1985). The pathway of endocytosis. In I Pastan, M C Willingham (eds.), *Endocytosis*: Plenum Press, New York, pp. 1–44.

41. Mufson R A, Gesner T G (1987). Binding and internalization of recombinant human erythropoietin in murine erythroid precursor cells. *Blood.* 69:1485–1490.

42. Chapel S, Veng-Pedersen P, Hohl R J, et al. (2001). Changes in erythropoietin pharmacokinetics following busulfan-induced bone marrow ablation in sheep: Evidence for bone marrow as a major erythropoietin elimination pathway. *J. Pharmacol. Exp. Ther.* 298:820–824.

43. Kuter D J, Rosenberg R D (1995). The reciprocal relationship of thrombopoietin (c-Mpl ligand) to changes in the platelet mass during busulfan-induced thrombocytopenia in the rabbit. *Blood.* 85:2720–2730.

44. Jusko W J (2005). Guidelines for collection and analysis of pharmacokinetic data. In M Leslie, J J Schentag, W E Evans, et al. (eds.), *Applied Pharmacokinetics & Pharmacodynamics: Principles of Therapeutic Drug Monitoring*: Applied Therapeutics, Inc., Vancouver, Canada, Chapter 2.

45. Yamaoka K, Nakagawa T, Uno T (1978). Statistical moments in pharmacokinetics. *J. Pharmacokinet. Biopharm.* 6:547–558.

46. Gibaldi M, Perrier D (1982). *Pharmacokinetics*, 2nd ed.: Marcel Dekker, Inc., New York.

47. Caprani O, Sveinsdottir E, Lassen N (1975). SHAM, A method for biexponential curve resolution using initial slope, height, area and moment of the experimental decay type curve. *J. Theor. Biol.* 52:299–315.

48. Tang L, Persky A M, Hochhaus G, et al. (2004). Pharmacokinetic aspects of biotechnology products. *J. Pharm. Sci.* 93:2184–2204.

49. Yu Z, Tse F L (1995). An evaluation of numerical integration algorithms for the estimation of the area under the curve (AUC) in pharmacokinetic studies. *Biopharm. Drug Dispos.* 16:37–58.

50. Bellissant E, Sebille V, Paintaud G (1998). Methodological issues in pharmacokinetic-pharmacodynamic modeling. *Clin. Pharmacokinet.* 35:151–166.

51. Foster D M (2001). Noncompartmental vs. compartmental approaches to pharmacokinetic analysis. In A J Atkinson, C E Daniels, R L Dedrick, et al. (eds.), *Principles of Clinical Pharmacology*: Academic Press, New York, pp. 75–92.

52. Gabrielsson J, Weiner D (2000). *Pharmacokinetic and Pharmacodynamic Data Analysis: Concepts and Applications*, 3rd ed.: Swedish Pharmaceutical Press, Stockholm, Sweden.

53. Sun Y N, Lee H J, Almon R R, et al. (1999). A pharmacokinetic/pharmacodynamic model for recombinant human growth hormone effects on induction of insulin-like growth factor I in monkeys. *J. Pharmacol. Exp. Ther.* 289:1523–1532.

54. Mager D E, Neuteboom B, Efthymiopoulos C, et al. (2003). Receptor-mediated pharmacokinetics and pharmacodynamics of interferon-beta1a in monkeys. *J. Pharmacol. Exp. Ther.* 306:262–270.

55. Segrave A M, Mager D E, Charman S A, et al. (2004). Pharmacokinetics of recombinant human leukemia inhibitory factor in sheep. *J. Pharmacol. Exp. Ther.* 309:1085–1092.

56. Ramakrishnan R, Cheung W K, Farrell F, et al. (2003). Pharmacokinetic and pharmacodynamic modeling of recombinant human erythropoietin after intravenous and subcutaneous dose administration in cynomolgus monkeys. *J. Pharmacol. Exp. Ther.* 306:324–331.

57. Benincosa L J, Chow F S, Tobia L P, et al. (2000). Pharmacokinetics and pharmacodynamics of a humanized monoclonal antibody to factor IX in cynomolgus monkeys. *J. Pharmacol. Exp. Ther.* 292:810–816.

58. Bauer R J, Dedrick R L, White M L, et al. (1999). Population pharmacokinetics and pharmacodynamics of the anti-CD11a antibody hu1124 in human subjects with psoriasis. *J. Pharmacokinet. Biopharm.* 27:397–420.

59. Veng-Pedersen P, Chapel S, Al-Huniti, et al. (2003). A differential pharmacokinetic analysis of the erythropoietin receptor population in newborn and adult sheep. *J. Pharmacol. Exp. Ther.* 306:532–537.

60. Levy G (1994). Pharmacologic target-mediated drug disposition. *Clin. Pharmacol. Ther.* 56:248–252.

61. Mager D E (2006). Target-mediated drug disposition and dynamics. *Biochem. Pharmacol.* 72:1–10.

62. Mager D E, Jusko W J (2001). General pharmacokinetic model for drugs exhibiting target-mediated drug disposition. *J. Pharmacokinet. Pharmacodyn.* 28:507–532.

63. Mager D E, Krzyzanski W (2005). Quasi-equilibrium pharmacokinetic model for drugs exhibiting target-mediated drug disposition. *Pharm. Res.* 22:1589–1596.

64. Ferraiolo B L, Mohler M A, Gloff C A (1992). *Protein Pharmacokinetics and Metabolism.* Plenum Press, New York.

65. Mordenti J, Chen S A, Moore J A, et al. (1991). Interspecies scaling of clearance and volume of distribution data for five therapeutic proteins. *Pharm. Res.* 8:1351–1359.

66. Adolph E F (1949). Quantitative relations in the physiological constitutions of mammals. *Science.* 109:579–585.

67. Boxenbaum H (1982). Interspecies scaling, allometry, physiological time, and the ground plan of pharmacokinetics. *J. Pharmacokinet. Biopharm.* 10:201–227.

68. West G B, Brown J H (2005). The origin of allometric scaling laws in biology from genomes to ecosystems: Towards a quantitative unifying theory of biological structure and organization. *J. Exp. Biol.* 208:1575–1592.

69. West G B, Brown J H, Enquist B J (1997). A general model for the origin of allometric scaling laws in biology. *Science.* 276:122–126.

70. Boxenbaum H, Ronfeld R (1983). Interspecies pharmacokinetic scaling and the Dedrick plots. *Am. J. Physiol.* 245:R768–R775.

71. Bonate P L, Howard D (2000). Prospective allometric scaling: Does the emperor have clothes? *J. Clin. Pharmacol.* 40:335–340.

72. Mahmood I (2000). Critique of prospective allometric scaling: does the emperor have clothes? *J. Clin. Pharmacol.* 40:341–346.

73. Tang H, Mayersohn M (2005). A mathematical description of the functionality of correction factors used in allometry for predicting human drug clearance. *Drug Metab. Dispos.* 33:1294–1296.

74. Tang H, Mayersohn M (2005). A novel model for prediction of human drug clearance by allometric scaling. *Drug Metab. Dispos.* 33:1297–1303.

75. Hu, T M, Hayton W L (2001). Allometric scaling of xenobiotic clearance: Uncertainty versus universality. *AAPS PharmSci.* 3:E29.

76. Tang H, Mayersohn M (2005). Accuracy of allometrically predicted pharmacokinetic parameters in humans: role of species selection. *Drug Metab. Dispos.* 33:1288–1293.

77. Mordenti J, Osaka G, Garcia K, et al. (1996). Pharmacokinetics and interspecies scaling of recombinant human factor VIII. *Toxicol. Appl. Pharmacol.* 136:75–78.

78. Lave T, Levet-Trafit B, Schmitt-Hoffmann A H, et al. (1995). Interspecies scal-ing of interferon disposition and comparison of allometric scaling with concentration-time transformations. *J. Pharm. Sci.* 84:1285–1290.

79. Bazin-Redureau M, Pepin S, Hong G, et al. (1998). Interspecies scaling of clearance and volume of distribution for horse antivenom F(ab')2. *Toxicol. Appl. Pharmacol.* 150:295–300.

80. Richter W F, Gallati H, Schiller C D (1999). Animal pharmacokinetics of the tumor necrosis factor receptor-immunoglobulin fusion protein lenercept and their extrapolation to humans. *Drug Metab. Dispos.* 27:21–25.

81. Lin Y S, Nguyen C, Mendoza J L, et al. (1999). Preclinical pharmacokinetics, interspecies scaling, and tissue distribution of a humanized monoclonal antibody against vascular endothelial growth factor. *J. Pharmacol. Exp. Ther.* 288:371–378.

82. Mahmood I (2004). Interspecies scaling of protein drugs: prediction of clearance from animals to humans. *J. Pharm. Sci.* 93:177–185.

83. Cosson V F, Fuseau E, Efthymiopoulos C, et al. (1997). Mixed effect modeling of sumatriptan pharmacokinetics during drug development. I: Interspecies allometric scaling. *J. Pharmacokinet. Biopharm.* 25:149–167.

84. Jolling K, Perez Ruixo J J, Hemeryck A, et al. (2005). Mixed-effects modelling of the interspecies pharmacokinetic scaling of pegylated human erythropoietin. *Eur. J. Pharm. Sci.* 24:465–475.

85. Proost J H, Beljaars L, Olinga P, et al. (2006). Prediction of the pharmacokinetics of succinylated human serum albumin in man from in vivo disposition data in animals and in vitro liver slice incubations. *Eur. J. Pharm. Sci.* 27:123–132.

86. Dedrick R L (1973). Animal scale-up. *J. Pharmacokinet. Biopharm.* 1:435–461.

87. Gerlowski L E, Jain R K (1983). Physiologically based pharmacokinetic modeling: Principles and applications. *J. Pharm. Sci.* 72:1103–1127.

88. Nestorov I (2003). Whole body pharmacokinetic models. *Clin. Pharmacokinet.* 42:883–908.

89. Davies B, Morris T (1993). Physiological parameters in laboratory animals and humans. *Pharm. Res.* 10:1093–1095.

90. Brown R P, Delp M D, Lindstedt S L, et al. (1997). Physiological parameter values for physiologically based pharmacokinetic models. *Toxicol. Ind. Health.* 13:407–484.

91. Xu L, Eiseman J L, Egorin M J, et al. (2003). Physiologically-based pharmacokinetics and molecular pharmacodynamics of 17-(allylamino)-17-demethoxygeldanamycin and its active metabolite in tumor-bearing mice. *J. Pharmacokinet. Pharmacodyn.* 30:185–219.

92. Lin J H, Sugiyama Y, Awazu S, et al. (1982). In vitro and in vivo evaluation of the tissue-to-blood partition coefficient for physiological pharmacokinetic models. *J. Pharmacokinet. Biopharm.* 10:637–647.

93. Houston J B (1994). Utility of in vitro drug metabolism data in predicting in vivo metabolic clearance. *Biochem. Pharmacol.* 47:1469–1479.

94. Sato H, Sugiyama Y, Sawada Y, et al. (1987). Physiologically based pharmacokinetics of radioiodinated human beta-endorphin in rats: An application of the capillary membrane-limited model. *Drug Metab. Dispos.* 15:540–550.

95. Peng B, Andrews J, Nestorov I, et al. (2001). Tissue distribution and physiologically based pharmacokinetics of antisense phosphorothioate oligonucleotide ISIS 1082 in rat. *Antisense Nucleic Acid Drug Dev.* 11:15–27.

96. Baxter L T, Zhu H, Mackensen D G, et al. (1995). Biodistribution of monoclonal antibodies: Scale-up from mouse to human using a physiologically based pharmacokinetic model. *Cancer Res.* 55:4611–4622.

97. Friedrich S W, Lin S C, Stoll B R, et al. (2002). Antibody-directed effector cell therapy of tumors: Analysis and optimization using a physiologically based pharmacokinetic model. *Neoplasia.* 4:449–463.

98. Galluppi G R, Rogge M C, Roskos L K, et al. (2001). Integration of pharmacokinetic and pharmacodynamic studies in the discovery, development, and review of protein therapeutic agents: A conference report. *Clin. Pharmacol. Ther.* 69:387–399.

99. Toutain P L (2002). Pharmacokinetic/pharmacodynamic integration in drug development and dosage-regimen optimization for veterinary medicine. *AAPS PharmSci.* 4: E38.

100. Mager D E, Wyska E, Jusko W J (2003). Diversity of mechanism-based pharmacodynamic models. *Drug Metab. Dispos.* 31:510–518.

3.3

STRATEGIES FOR THE CYTOSOLIC DELIVERY OF MACROMOLECULES: AN OVERVIEW

Aravind Asokan, Roland Cheung, and Moo J. Cho

University of North Carolina at Chapel Hill, Chapel Hill, North Carolina

Chapter contents

3.3.1	Introduction	279
3.3.2	Endosome-to-Cytosol Transfer	280
	3.3.2.1 pH-Sensitive Drug Carriers	281
	3.3.2.2 Acid-Labile Linkers and Drug Carriers	284
	3.3.2.3 Redox/Enzyme-Triggered Systems	285
	3.3.2.4 Caveolae-Mediated Cytosolic Delivery	286
	3.3.2.5 Cell-Penetrating Peptides: HIV-TAT	286
3.3.3	Direct Cytosolic Delivery Across the Plasma Membrane	287
	3.3.3.1 Cell-Penetrating Peptides: Penetratin and Other Peptides	288
	3.3.3.2 Molecular Umbrellas	289
3.3.4	Assays for Monitoring Cytosolic Delivery	290
	References	291

3.3.1 INTRODUCTION

The advent of molecular biology and genetics has identified numerous biotechnology-derived macromolecular drugs as potential targets for development. The factors required for the development of this new class of drugs are, however, significantly different from those required for conventional drug compounds with molecular weight less than 1 kDa or so [1, 2]. Efforts directed toward the delivery of small molecules have largely sought a compromise between two major factors, i.e., aqueous solubility and membrane partition properties. On the other hand, not only does the primary structure of proteins and genes require protection from enzymatic degradation but also higher order structures from denaturation [3]. In addition, for many macromolecules with therapeutic potential, be it peptide or protein-based drugs or nucleic acid-based drugs such as antisense oligonucleotides, ribozymes, siRNA, aptamers, and plasmid DNA, delivery to the cytosol becomes a necessary, if not sufficient, condition for their full pharmacological activity. For example, antisense oligomers or siRNA, which are extensively sequestered in endosomes after cellular uptake, require cytosolic delivery to bind ribosomal mRNA/nuclear pre-mRNA in order to inhibit protein translation or correct splicing [4–6]. As a result, several strategies to promote the cytosolic delivery of macromolecular therapeutics have been developed over the past few years. We have classified these diverse strategies into (1) those that exploit the biochemical aspects of the endosomal lumen to promote endosome-to-cytosol transfer and (2) approaches that bypass the endocytic pathway by facilitating direct transport of macromolecules across the plasma membrane.

3.3.2   ENDOSOME-TO-CYTOSOL TRANSFER

The plasma membrane is a mosaic network of lipids and proteins decorated with complex carbohydrates. This 5–10-nm-thick lipid bilayer is burdened with the dual role of serving as a barrier to external agents, while regulating the entry of nutrients into the cell. Conventional small molecules, such as amino acids, sugars, and ions, generally traverse the plasma membrane through a con-certed function of integral membrane protein carriers or channels. However, macromolecules, by virtue of their polarity and large hydrodynamic diameter, are taken up in membrane-bound vesicles derived by the invagination and pinching-off of sections of the plasma membrane in a process termed endocytosis. Endocytosis can occur via multiple mechanisms that can be broadly classified into (1) phagocytosis or the uptake of large particles and (2) pinocytosis or the uptake of fluid and solutes. Phagocytosis is typically restricted to specialized mammalian cells, such as macrophages and other antigen presenting cells, whereas pinocytosis occurs in all cells by at least four known mechanisms: macropinocytosis, clathrin-mediated endocytosis, caveolae-mediated endocytosis, and clathrin- and caveolae-independent endocytosis [7].

Internalization of macromolecules and a variety of pathogens via caveolae is another endocytic mechanism alternative to the clathrin-mediated process. Originally discovered in the 1950s, caveolae are uncoated, 50–100-nm flask-shaped invaginations of the plasma membrane. The process shares several similarities with clathrin-mediated typical endocytosis in that docking and fusion proteins involved in the latter are also found in caveolae. However, an important difference, especially from the perspective of delivery of intact therapeutic macromolecules, is the finding that the resulting "caveosomes" can avoid the degradative route to lysosomes [8]. Mechanistic aspects of such diverse and highly regulated endocytic pathways are beyond the scope of this review and discussed elsewhere. Nevertheless, three biochemical traits of endocytic vesicles or endosomes are particularly relevant in the context of macromolecular drug delivery.

First and foremost, it is well documented that the endosomal compartments maintain an acidic internal pH ranging from pH 5.5 to 6.5 and seem to mature or fuse with lysosomes, which maintain a lower pH (~5.0 to 5.5). This is vital for sorting of ligand-receptor complexes and the activation of hydrolytic enzymes such as cathepsins. Although the H^+-ATPase pumps protons into the endosome, the Na^+/K^+-ATPase acts to limit proton pumping into the endosomal compartment. In its normal orientation, the pump generates a positive interior membrane potential by pumping three Na^+ ions into the endosome for every two K^+ ions pumped out, thereby regulating the proton-translocating activity of the membrane bound H^+-ATPase. Evidence supporting these events has been provided by the treatment of isolated endosomes with inhibitors of Na^+/K^+-ATPase such as ouabain, which causes a fall in the pH within the endosomal compartments. Lysosomes, on the other hand, seem to lack Na^+/K^+-ATPase, and their luminal acidification is mainly regulated by H^+-ATPases.

Second, of several proteases that are known to participate in peptide/protein hydrolysis, the cysteine and aspartic proteases, also known as cathepsins, constitute a majority of the endosomal/lysosomal proteolytic machinery. These enzymes are often synthesized in an inactive form requiring excision of the pro-domain facilitated by other proteases or autocatalytic mechanisms activated by acidic pH. Cathepsins seem to have originally evolved to catabolize both internalized and endogenous proteins crucial for cellular homeostasis, autophagy, apoptosis, and antigen presentation. The optimal pH range for enzymatic activity of cathepsins ranges from 5.0 to 6.5, as is the case with endosomes/lysosomes.

Last, several redox pathways are thought to function within endosomes and lysosomes, although the exact mechanism(s) of disulfide reduction within the endosomal lumen are not well understood. Primary evidence stems from antigen processing within the endocytic pathway, which involves partial denaturation of internalized proteins facilitated in part by the acidic pH. Disulfide bonds, however, are not susceptible to lysosomal proteolysis and remain chemically stable at acidic pH. Disulfide reduction within endosomes/lysosomes has been attributed to active transport of cysteine (free thiol form) into lysosomes and redox enzymes such as γ-Interferon-inducible lysosomal thiol reductase (GILT) that function at an acidic pH optimum [9]. However, a recent report disputes the long-held view that endosomes and lysosomes are reductive environment. Rather, the experimental data support that they may exhibit an oxidizing environment suggesting that disulfide-based systems may function via a nonreductive mechanism or that they may be reduced in the cytosol after discharged from the endosome [10]. Overall, strategies for the endosome-to-cytosol transfer of macromolecular drugs can broadly be classified on the basis of the three aforementioned characteristics of the endosomal lumen, namely, acidic pH, proteolytic activity of cathepsins, and disulfide bond reduction.

3.3.2.1 pH-Sensitive Drug Carriers

Viruses have evolved proteins that perturb the host cell membrane by fusion. Stemming from this knowledge is the creation and use of pH-dependent fusogenic peptides. For example, diINF-7 is one such peptide that resembles the amino-terminal domain of the influenza virus hemagglutinin HA-2 subunit. At an endosomal pH of 5.0, the peptide undergoes a conformational change allowing it to penetrate deeper into the endosomal membrane resulting in membrane destabilization [11]. The pH-sensitive behavior of viral peptides and bacterial proteins originates from that of weak acids with pKa values in the range of 4.5 to 7.0, primarily Glu residues with a pKa of approximately 5.0. Thus, the pH-dependent membrane partitioning or the partition coefficient of weak acids (with elevated membrane-bound pKa values) depends critically on the endosomal pH. Although the negatively charged side chains of the Glu residues in viral peptides such as INF-7, initially repel each other, they collapse into a hydrophobic α-helix due to protonation of the side-chain carboxylate ions at a pH close to their pKa. Subsequently, the helical peptide inserts itself into the endosomal membrane and leads to fusion or pore formation and eventually endosomal release. Based on these principles, Li et al. designed the GALA peptide, which consists of repeats of the peptide motif Glu-Ala-Leu-Ala and adapts a α-helical conformation at acidic pH [12]. The peptide was subsequently found to efficiently release calcein from phosphatidylcholine liposomes in a pH-specific manner, presumably via formation of pores or perturbation-mediated flip-flop of lipids.

Weak acid-based homopolymers such as poly(Glu) and poly(alkylacrylic acid) have been shown to disrupt membranes in a pH-dependent manner [13, 14]. Poly(ethylacrylic acid) is thought to form cation-selective aqueous pores in lipid bilayers at relatively low concentrations, providing a potential application in promoting endosome-to-cytosol transfer of macromolecules. Mechanistically, the carboxylate head groups become neutral at a pH < pKa triggering polymer contact with endosomal membranes, resulting in membrane permeabilization. Polycationic molecules have been extensively used to promote the cellular delivery of macromolecules, due to their potent membrane disruptive, destabilizing, or fusogenic activity. A proton gradient such as the one found in endosomes and lysosomes can be exploited to protonate weakly basic drug carriers with pKa values between 5.0 and 7.0. The cationic charge thus generated can induce dramatic changes in the physico-chemical properties of the carrier.

Poly(His) was perhaps the first pH-sensitive polymer shown to promote fusion between lipid bilayers [15]. The imidazole functionality in the side chain of the His residue is protonated at low pH, and it interacts with negatively charged head groups of lipids. Although proximity-driven lipid and content mixing is observed at lower concentrations, a higher concentration of the polymer causes disruption of the bilayer and leakage of contents. Imidazole-containing polymers and their endosomolytic properties have received much attention lately, as exemplified by a pH-sensitive His-containing peptide analogous to the influenza HA-2. Histidylated polylysine and gluconic acid modified poly(His) have also been shown to promote efficient gene transfer [16]. Lee et al. constructed a core-shell-type micelle from the two block copolymer components poly(L-histidine)-*b*-poly(ethylene glycol) and poly(L-lactic acid)-*b*-PEG-*b*-polyHis-biotin and showed that after biotin receptor-mediated endocytosis, the micelle exhibited pH-dependent dissociation, thus allowing for the release of the model drug doxorubicin to be released in the early endosomal pH [17]. Furthermore, the collapsed micellar components could subsequently destabilize the endosomal membrane allowing for doxorubicin to enter the cytosol.

Polyethyleneimine (PEI) and starburst dendrimers are other examples of pH-sensitive cationic polymers, which become protonated incrementally over a wide pH range [18, 19]. The application of these pH-sensitive cationic polymers to oligonucleotide and gene delivery has been studied extensively. Other examples of the use of pH-sensitive materials are hydrogels made of chitosan and copolymers of chitosan [20]. These represent a class of polymeric delivery systems used for controlled drug release. Hydrodynamic swelling or shrinking of these gel-like polymers at low pH (due to protonation of the polymer side-chain moieties) is thought to release the entrapped contents. The proton sponge hypothesis suggests a mechanism by which the polymers such as PEI containing protonatable moieties lead to chloride influx from the cytosol and subsequent osmosis into the endosome, resulting in endosomal swelling and lysis. Endosomal lysis allows the polymer and its cargo to be released into the cytosol [21].

Acid-sensitive liposomes have been extensively used to deliver macromolecular therapeutics agents. The first example of pH-sensitive liposomes was outlined by Yatvin et al. [22]. The mechanism of acid-induced fusion between liposomes composed of DOPE and palmitoylhomocysteine was further established by Connor et al. [23]. As outlined in the review by Simões et al. [24], three hypothetical mechanisms of cytosolic delivery from liposomes have been proposed: (1) Destabilization of pH-sensitive liposomes destabilizes the endosomal membrane possibly through pore formation, (2) destabilization of pH-sensitive liposomes allows cargo to diffuse through the endosomal membrane, and (3) liposomal membrane fusion with the endosomal membrane (illustrated in Figure 3.3-1). Several pH-sensitive lipids have been designed to promote endosome-to-cytosol transfer of macromolecules using liposomal systems. Fusogenic lipids such as oleic acid and cholesteryl hemisuccinate contain weak acids as head groups. These lipids, when incorporated into liposomes, induce pH-dependent fusion. Wang and Huang demonstrated the utility of oleic acid-containing pH-sensitive liposomes in mediating targeted delivery of transgenes to lymphoma tumors *in vivo* [25]. Other examples for weak acid-based pH-sensitive lipids include the titratable, double-chain amphiphiles 1,2-dipalmitoyl-sn-3-succinylglycerol (1,2-DPSG), 1,2-dioleoyl-sn-3-succinylglycerol (1,2-DOSG), and 1,3-dipalmitoylsuccinylglycerol (1,3-DPSG) developed by Collins et al. [26]. The fusogenicity of these lipids in the liposomal formulation is greatly enhanced by the incorporation of dioleoylphosphatidyl ethanolamine (DOPE), a neutral fusogenic lipid.

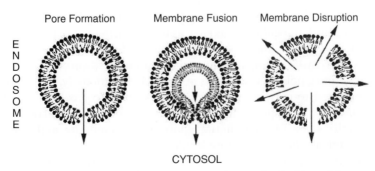

Figure 3.3-1. Mechanisms of endosome-to-cytosol transfer facilitated by various drug carriers.

The neutral dioleoyldimethylaminopropane (DODAP) free base tends to be fusogenic; however, protonation of the tertiary amino head group results in a bilayer-forming cationic lipid. It has been shown that interaction of the cationic lipid head group with anionic membrane-associated moieties such as proteoglycans results in a charge neutralization, which in turn, triggers the lamellar-to-hexagonal phase transition of such cationic lipid-based carriers. This nonbilayer phase transition is thought to promote extensive fusion between the liposomal formulation and the cellular membrane. Another example of weak base carriers is a series of amphiphilic imidazole-containing lipids. These lipids have been shown to efficiently promote the delivery of plasmid DNA to the cytoplasm [27]. As observed in the case of weak acid lipids, the fusogenic activity of the weak base lipids is potentiated by the presence of DOPE, a "helper" lipid that prefers the hexagonal phase above a temperature of 10°C.

Chen et al. synthesized a series of acyloxyalkyl imidazole lipids, with membrane-bound pKa values in the range of 5.0 to 5.5 [28]. It was expected that these molecules, once protonated in the acidic environment of the endosome, would become surface active and permeabilize the endosomal membrane. Eventually these lipids are degraded by cytosolic esterases into imidazole, formaldehyde, and the corresponding fatty acid. However, a serious limitation that precludes utilization of single-tailed, pH-sensitive detergents for the cytosolic delivery of macromolecules is their low limit of incorporation in stable liposomal formulations. Limited incorporation of detergent molecules resulted in a lack of liposomal collapse that leads to minimal endosomal membrane destabilization [29]. However, over-incorporation of such detergents destabilizes the liposomal membrane and would result in premature leakage of entrapped cargo. To address this issue, two Gemini surfactants or "bis-detergents" have been prepared by cross-linking the headgroups of single-tailed, tertiary amine detergents through oxyethylene (BD1) or acid-labile acetal (BD2) moieties [30]. The ability of BD2-containing liposomes to promote efficient cytosolic delivery of antisense oligonucleotides was confirmed by (1) their diffuse intracellular distribution observed in fluorescence micrographs, and (2) the upregulation of luciferase in an antisense functional assay. The low pH-responsive, bis-detergent constructs described herein are suitable for triggered release strategies targeted to acidic intracellular or interstitial environments.

3.3.2.2   Acid-Labile Linkers and Drug Carriers

Acid-labile linkers that are cleaved off in the endosomal milieu have been used in conjugating chemotherapeutic agents such as doxorubicin to targeting moieties such as folic acid, antibodies, and antibody fragments. Some common examples are acid-labile ketals, hydrazones, diorthoesters, and the *cis*-dicarboxylic acid linkers, which undergo intramolecular catalysis at low pH. Additionally, such moieties can also be conjugated to carrier molecules such as lipids to create an inactive pro-carrier. Hydrolysis of the pro-moiety at a low pH activates the carrier, which disrupts, destabilizes, or fuses with the endosomal membrane and subsequently promotes the release of entrapped molecules.

Recently, Murthy et al. have synthesized an acid-sensitive microgel material for the development of protein-based vaccines. The chemical design of these microgels is such that they degrade under the mildly acidic conditions found in the phagosomes of antigen-presenting cells (APCs). The rapid cleavage of the microgels leads to phagosomal disruption through a colloid osmotic mechanism, releasing protein antigens into the APC cytoplasm for class I antigen presentation [31, 32]. Other examples of such acid-labile hydrogels include poly(orthoester) microspheres [33] and poly(ketal) nanoparticles [34] that undergo acid-catalyzed hydrolysis into low-molecular-weight hydrophilic compounds and release encapsulated therapeutics at an accelerated rate in acidic environments.

With respect to lipid-based drug carriers, an attractive approach for triggered release is to design cleavable surfactants with uncharged functional groups whose hydrolysis is catalyzed by acidic conditions. The chemical principles behind this approach have been described in an outstanding review published almost three decades ago [35]. For example, Boomer and Thompson reported several mono- and diplasmenyl lipids with an acid-sensitive vinyl ether linkage between the headgroup and one or both of the hydrocarbon side chains [36]. Upon exposure to low pH or photo-oxidation, the vinyl ether side chains are cleaved from the lipids, leading to structure defects in the bilayer and the release of liposomal contents. In another example, Guo and Szoka used a diorthoester bridge to link PEG to distearoylglyc-erol. This lipid efficiently stabilizes DOPE bilayer vesicles at a total lipid concentration as low as 10%. Release of the PEG by cleavage of the acid-labile diorthoester linker disrupts the lamellar organization and promotes the fusogenic lipid configuration of DOPE within the endosome. Subsequent fusion with the endosomal membrane resulted in cytoplasmic delivery of liposomal contents. These strategies have been reviewed extensively by Guo and Szoka [37, 38].

3.3.2.3 Redox/Enzyme-Triggered Drug Carriers

In addition to lower pH, the endosome also exhibits other characteristics that can be used in cytosolic delivery. The endosome houses many enzymes, including redox enzymes [9], and may exhibit a slightly reductive environment, although the latter portion of this statement currently remains a subject of controversy [10]. Using a listeriolysin O-protamine conjugate, the reducing potential of the endosome was exploited for plasmid DNA delivery into the cytosol [39]. The conjugate contained a redox-labile disulfide bond between the cysteine of listeriolysin O and the poly-cationic peptide protamine, which could be reduced in the reductive environment of the endosome. Upon disulfide cleavage, listeriolysin O exerted its endosomolytic activity, resulting in the cytosolic release of condensed DNA polyplexes.

Liposomes prepared from DOPE and a disulfide-containing cationic lipid showed a greater release of plasmid DNA than a non-disulfide-containing analog [40]. The disulfide bond of the lipid, 1-2-dioleoyl-sn-glycero-3-succinyl-2-hydroxy-ethyl disulfide (DOGSDSO), was shown to be cleaved in the reductive environment leading to the destabilization of the liposome-DNA complex (lipoplex) and increased transfection activity. A novel set of cationic lipids was designed to confer reduction-sensitivity for lipopolyamines [41]. These lipids contain a disulfide bond at different sites on the lipid backbone. Compared with nondegradable lipopoly-amine analogs, increased transfection efficiency is observed when inserting the disulfide between one lipid chain and the rest of the molecule. However, loss or a substantial decrease in transfection activity occurs when the disulfide bridge is placed in the linker position between the polyamine and the lipid chains.

Chittimalla et al. have designed cationic thiol-based detergents [42]. Subsequent detergent polymerization by formation of intermolecular disulfide bonds within the condensed plasmid DNA leads to 32-nm-large neutral particles that promote efficient gene delivery.

Enzymatic cleavage of a synthetic peptide-lipid conjugate has been shown to trigger liposomal fusion to cellular membranes, thus allowing liposomal cargo to be distributed to the cytosol [43, 44]. The conjugate, formed from the covalent linkage of dioleoyl phosphatidylethanolamine (DOPE) with an elastase substrate N-acetyl-ala-ala-, having no intrinsic fusogenic activity, required cleavage by elastase or proteinase K and subsequent conversion to the fusogenic lipid DOPE. Only a few examples of enzymatic triggering have been described here although many enzymatic triggers have been studied. A few of these have explored the use of alkaline phosphatase [45], matrix metalloproteinases [46], and elastases [47].

3.3.2.4 Caveolae-Mediated Cytosolic Delivery

A variety of lipid-anchored membrane proteins are associated with caveolae that exist as a subdomain in a biochemically defined glycolipid raft. Not surprisingly, the intracellular trafficking of a given ligand is largely determined by what protein in caveolae they bind. It ranges from ultimate transcytosis as often found in endothelial cells, direct discharge to cytosol of folates, or endoplasmic reticulum and Golgi as in the case of simian virus 40 and cholera toxin, respectively [48–50]. The glycosylphosphatidylinositol (GPI)-anchored folate receptor-α (FR-α) internalizes and subsequently discharges 5-methyltetrahydrofolate and other folates directly to cytoplasm upon dissociation with FR-α in response to an acidic environment [48]. Although clustering of GPI-anchored proteins such as FR-α in lipid raft remains controversial [51], the current literature is rich in its application to cellular delivery of macromolecules and submicron drug carriers [52, 53]. Another unresolved issue is whether chemical modification of α-Glu of folate affects recognition of the conjugate by FR-α. A recent study reported no difference in liposomal delivery of antisense oligonucleotides when the surface was grafted with α- or γ-modified of folic acid [54].

3.3.2.5 Cell-Penetrating Peptides: HIV-TAT

The HIV TAT peptide is a short polybasic peptide sequence derived from the HIV-1 TAT protein, which plays a critical role in viral replication. The 11 amino acid TAT peptide is capable of penetrating cell membranes without causing membrane disruption. The TAT peptide was originally thought to enter cells directly at the plasma membrane via a nonsaturable, nonenergy-dependent mechanism [55, 56], possibly via transient formation of non-bilayer structures during the membrane penetration [57]. Recently, however, this picture has changed dramatically to a view that CPPs enter cells via saturable, energy-dependent processes involving endocytosis. Some early studies indicating nonendocytic uptake are now believed to reflect fixation artifacts [58]. Furthermore, cell-surface receptors for TAT protein chimeras were identified as being heparan sulfate proteoglycans [59, 60]. Additional studies showed that TAT and several other cationic CPPs clearly entered the cell by endocytosis, possibly via lipid raft [61, 62], and their release into the cytoplasm involves endosome acidification [58, 63–65].

3.3.3 DIRECT CYTOSOLIC DELIVERY ACROSS THE PLASMA MEMBRANE

Until recently, strategies for cytosolic delivery of macromolecules have largely been formulated by mimicking the cellular entry pathways of viruses and bacteria. Among these strategies, most have focused on promoting endosomal escape of macromolecules into the cytosol exploiting features such as acidic pH and redox/enzyme activity within endocytic vesicles. Nevertheless, simpler methods to deliver large macromolecules directly across the plasma membrane, while bypassing the degradative endocytic pathway, have been pursued. In particular, the seminal discoveries that peptides derived from glycoproteins embedded in the lipid envelope of viruses such as HIV and Herpes virus as well as those derived from the homeodomain of certain transcription factors can directly translocate across the plasma membrane of mammalian cells have opened up new avenues in macromolecular drug delivery. In general, these "cell-penetrating peptides (CPPs)" are approximately 30-amino-acid-long amphipathic peptides with a net positive charge [66]. The most prominent examples of CPPs are the HIV-TAT, discussed in the previous section, penetratins, and synthetic peptides derived thereof. Less-studied examples such as the peptides derived from HIV gp41 and the Herpes VP22 are also discussed (Table 3.3-1).

TABLE 3.3-1. Examples of Peptide Sequences of Common Cell-Penetrating Peptides

HIV-TAT	H_2N-**GRKKRRQRRRPPQ**-CONH$_2$
PENETRATIN	H_2N-**RQIKIWFQNRRMKWKK**-CONH$_2$
TRANSPORTAN	H_2N-**GWTLNSAGYLLGKINLKALAALAKISIL**-CONH$_2$
SN50	H_2N-**AAVALLPAVLLALLAP**-CONH$_2$

The main application of CPPs is the attachment of biologically active cargo for translocation into cells. Conjugating cargo to CPP can be achieved through several means. When the cargo is a peptide or protein, CPP and cargo are most often synthesized or expressed in tandem as a fusion protein. Alternatively, a suitable amino acid side-chain or bifunctional spacer molecule can be used to couple cargo to CPPs through maleimide-thiol, disulfide, or amine-carboxylate chemistry. Attachment of cargo to CPP can also be achieved via noncovalent bonds, employing for instance the interaction between streptavidin and a biotinylated CPP conjugate. When chemically conjugated, TAT has been shown to enhance intracellular delivery of oligonucleotides [67–71], peptides and proteins [72, 73], and particulates [74, 75]. The strategy has met with moderate success *in vivo* [76–79]. Several strategies involving target proteins fused to the TAT peptide have also been described [80, 81]. The technology requires the synthesis of a fusion protein, linking the TAT transduction domain to the molecule of interest using a bacterial expression vector, followed by the purification of this fusion protein under either soluble or denaturing conditions. The original 11-residue TAT sequence (47–57) has been extensively modified to improved versions of TAT [82], poly-arginine peptides [83, 84], and guanidinium-rich peptoids [84].

3.3.3.1   Cell-Penetrating Peptides: Penetratin and Other Peptides

Penetratins are a class of CPPs derived from transcription factor proteins of Antennapedia (a *Drosophila* homeoprotein). Using site-directed mutagenesis, it was discovered that the third helix (amino acids 43–58) of the latter homeodomain was necessary and sufficient for translocation. Structure–function studies of penetratin yielded several active and inactive analogs that allowed formulation of the inverted micelle internalization model. The cellular internalization of penetratins occurs at 37°C and 4°C through a nonsaturable mechanism. The demonstration that retro-inverso penetratin analogs and similar peptides derived from other homeodomains are capable of membrane translocation suggests that the number of member peptides comprising the penetratin family is high.

Transportans are 27-amino-acid-long chimeric peptides designed and synthesized from other unrelated peptides. The first 13 amino acids are derived from the highly conserved amino-terminal part of galanin, whereas the 14-residue C-terminus is derived from the wasp venom peptide toxin, mastoparan. Although the N-terminus portion is the smallest highly active galanin receptor ligand with agonist properties, the C-terminus end is thought to penetrate the cell membrane by creating transient pores by translocating into the inner leaflet of the membrane bilayer. Applications of transportans have been similar to that of HIV-TAT and penetratins and have been discussed in detail elsewhere. Morris et al. designed a synthetic CPP that contains a hydrophobic domain derived from the HIV fusion protein, gp41, and a polycationic sequence derived from the nuclear localization sequence of the SV40 T-antigen [85]. The peptide vector exhibits a high affinity for single- and double-stranded DNA in the nanomolar range and has been used for the nuclear delivery of oligonucleotides, siRNA, and plasmid DNA. Although unclear, the peptide vector seems to facilitate direct cytosolic and nuclear delivery of cargo through nonendocytic means.

Elliott and O'Hare demonstrated that the Herpes simplex virus protein VP22 exhibits the remarkable property of intercellular trafficking, whereby the protein spreads from the cell in which it is synthesized to many surrounding cells [86]. This function of VP22 has been exploited to construct fusion proteins similar to aforementioned strategies that retain their ability to spread between cells and accumulate in recipient cell nuclei. The extreme C-terminus (residues 267–301) of VP22 is thought to be required for membrane translocation. However, further studies by Aints et al. indicate that although this region facilitates VP22 transport, a core region in the protein (residues 81–195) also contributes to transport activity [87]. VP22 has effectively delivered proteins such as GFP, p53, and the SV40 T antigen in both actively dividing and terminally differentiated cell lines. It has been successfully employed to enhance the effectiveness of thymidine kinase and cytosine deamidase suicide gene therapy systems for cancer therapy both *in vitro* and *in vivo*. VP22 fusion proteins have also been successful in enhancing the efficacy of lentivirus and adenovirus-based gene therapy systems. Additionally, when combined with small oligonucleotides, VP22 forms spherical particles, named Vectosomes, that are taken up by an energy-dependent process when applied to cells in culture.

Hawiger has developed CPPs derived from the membrane-translocating hydrophobic sequence of the h-region of the signal peptide sequence of K-FGF [88].

These alanine and leucine-rich peptides adopt a predominantly alpha-helical conformation in solution. A particularly striking characteristic of these CPPs (SN50) is the presence of a kink in the center of the alpha-helix due to a single proline, which is thought to play a key role in the membrane translocating activity of these peptides. Several studies testing the ability of these CPPs to carry functional cargo have been applied to control signal transduction-dependent subcellular traffic of transcription factors such as NF kappa-B mediating the cellular responses to different agonists. More recently, the authors have used the SN50 peptide sequence for intracellular protein therapy with SOCS3 to inhibit inflammation and apopotosis induced by bacterial enterotoxins [89].

3.3.3.2   Molecular Umbrellas

A conventional strategy to enhance transbilayer transport of several drugs, at least in the case of small molecules, is to simply modify the lipophilicity of the molecule by chemical conjugation to hydrophobic species such as fatty acids or cholesterol. However, such strategies seem merely to facilitate the adsorptive pinocytosis of macromolecules owing to their highly polar nature. Steven Regen et al. have extended this principle beyond simply altering the hydrophilic/hydrophobic balance of macromolecules. Their strategy hinges on an "umbrella" mechanism involving chemical conjugates that adopt a shielded conformation to traverse the plasma membrane (Illustrated in Figure 3.3-2, [90]).

Molecular umbrellas are a unique class of amphiphiles that are capable of shielding an attached agent from an incompatible environment. Typically, such molecules are composed of a central scaffold that contains two or more facially amphiphilic units. When a hydrophilic agent is attached to an umbrella molecule, contact with water favors an exposed conformation such that intramolecular hydrophobic interactions are maximized. Alternatively, when immersed in a hydrophobic environment, such as the lipid bilayer, the molecular umbrella can shield the agent by providing a hydrophobic exterior. Recent mechanistic studies have provided strong evidence that facial amphiphilicity plays a major role in umbrella transport across lipid bilayers. The fact that facial amphiphilicity is more important than hydrophobic/hydrophilic balance in promoting transbilayer transport lends strong support for

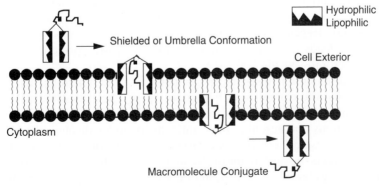

Figure 3.3-2. Proposed mechanism of cystolic delivery of macromolecules conjugated to molecular umbrellas.

an umbrella mechanism. In essence, the permeation pathway involves monolayer insertion of a shielded conformer, transbilayer diffusion, and entry to the opposite membrane/water interface. These molecular umbrellas were demonstrated to promote transport and release of small peptides by using a detachable handle based on disulfide cleavage. More recently, a series of molecular umbrellas derived from cholic acid, deoxycholic acid, spermidine, lysine, and cleavable disulfide linkages were found capable of transporting an attached 16-mer oligonucleotide across liposomal membranes. The ability of these novel agents in promoting the membrane translocation of macromolecules in a cellular setting remains to be seen [91].

3.3.4   ASSAYS FOR MONITORING CYTOSOLIC DELIVERY

In principle, the drug carrier or cargo molecules or both could be labeled with a variety of different radionuclides such as [^{125}I] or [^{3}H] for quantitative assessment. However, the disadvantage associated with such assays is the need for separating the cytosolic fraction from organelles such as endosomes, lysosomes, and nuclei using differential centrifugation techniques. Zaro and Shen have developed a novel method to quantitate the membrane transduction efficiency of CPPs and other drug carriers [92]. In essence, the amount of oligopeptide located in the cytoplasmic compartment was separated from the amount located in the intracellular vesicles using size exclusion chromatography. This separation allowed for the measurement of the membrane transduction efficiency of various [^{125}I]-labeled oligopeptides.

Most assays to determine the efficiency of cellular uptake and cytosolic delivery of macromolecules have involved the use of fluorescence microscopy or flow cytometry techniques. However, a major drawback associated with such qualitative fluorescence-based assays is their propensity for generating false-positives. For example, in the abundant literature dealing with CPPs, optimistic reviews highlighting their promising properties coexist with articles questioning their actual efficacy. The tight binding of most CPPs to negatively charged cell surfaces is responsible for microscopy misinterpretations. Methods such as methanol fixation of cells before microscopic observations have been shown to redistribute CPPs. Thus, the experimental assessment of the cytosolic delivery of proteins and other macromolecules is hindered by these artifacts. In fact such artifacts, which affect conventional as well as confocal microscopy analyses, has led to the false conclusion that CPPs, such as TAT and Penetratin, enter cells through an endocytosis-independent fashion [58, 93]. Recent studies suggest that the interaction of HIV-TAT with the cell surface leads to internalization of HIV-TAT fusion proteins by lipid raft-dependent macropinocytosis.

Owing to aforementioned issues, assays that reflect the functional activity of the cargo rather than the carrier have been developed. These assays can broadly be categorized on the basis of protein-based drugs and nucleic acid therapeutics. Prerequisite conditions for presentation of antigenic peptides through the MHC class I pathway are endosomal escape, followed by ubiquitin conjugation in the cytosol, which in turn targets the protein for proteasomal degradation. Conventional antigen presentation assays based on activation of cytotoxic T lymphocytes by monitoring cytokine levels and [^{51}Cr] release from target cells have been described elsewhere. Loison et al. have devised an ubiquitin-based assay for monitoring the cytosolic delivery using CPPs [94]. The authors developed a stringent enzymatic assay based on the processing of a CPP-Ubiquitin-Cargo fusion protein by exclusively cytosolic deubiquitinating enzymes. Based on this strategy, insertion of a ubiquitin moiety between the CPP and the protein cargo serves as a stringent test of cytosolic delivery by ensuring the cytosolic release of the cargo, followed by its proper folding and localization, which is required for interaction with intracellular targets.

Another relatively simple approach to assess cytosolic delivery is to monitor direct enzymatic activity of the protein cargo, such as β-galactosidase or horse radish peroxidase. However, as such enzyme activity could be recovered from within endosomes after fixation and staining, it is not possible to quantitate the cytosolic fraction exclusively. To address this question, a functional assay using a conjugate of TAT-RNase A was tested for its ability to inhibit protein synthesis through the nonspecific degradation of cellular RNA. In this model, if TAT-RNase A were entering the cell and remaining within endosomes, there should be no effect on protein synthesis. However, cytosolic entry of TAT-RNase A was found to be sufficient to decrease cellular protein synthesis as monitored by $[^{35}S]$-methionine incorporation. Another functional assay involved the use of TAT-Cre-Recombinase to mediate recombination of a *lox*P-STOP-*lox*P enhanced green fluorescent protein (EGFP) reporter gene in live mouse reporter T cells (tex.loxP.EG) as a measure of cellular uptake [81]. This system requires that exogenous TAT-Cre protein enter the cytosol, translocate to the nucleus, and excise the transcriptional STOP DNA seg-ment in live cells in a nontoxic fashion before scoring positive for EGFP expression.

Nucleic acid-based therapeutics such as antisense oligonucleotides, siRNA, and plasmid DNA act by interaction with translational or transcriptional protein machinery to downregulate a specific protein or promote transgene expression. In contrast to protein cargo, the interpretation and quantitation of cytosolic delivery of RNA/DNA-based therapeutics is relatively straightforward. Within this framework, reporter proteins such as luciferase and EGFP have been commonly used. Among several functional assays, the upregulation of reporter proteins by correction of aberrant splicing using antisense oligomers is notable [95]. This assay has recently been extended into an *in vivo* mouse model. In this model, the expressed gene (*EGFP-654*) encoding enhanced green fluorescence protein is interrupted by an aberrantly spliced mutated intron of the human β-globin gene. Aberrant splicing of this intron prevents expression of EGFP in all tissues. However, cytosolic entry of the splice site-targeted antisense oligomer restores corrective splicing and results in upregulation of EGFP in several tissues and organs.

REFERENCES

1. Cho M J, Juliano R (1996). Macromolecular versus small-molecule therapeutics: Drug discovery, development and clinical considerations. *Trends Biotechnol.* 14:153–158.

2. Jain R K (1998). The next frontier of molecular medicine: delivery of therapeutics. *Nat. Med.* 4:655–657.

3. Brockman S, Murphy R (1993). Endosomal and lysosomal hydrolases. In T J Raub, K Audus (eds.), *Biological Barriers to Protein Delivery*: Plenum Press, New York, pp. 51–65.

4. Gray G D, Basu S, Wickstrom E (1997). Transformed and immortalized cellular uptake of oligodeoxynucleoside phosphorothioates, 3′-alkylamino oligonucleotides, 2′-O-methyl oligoribonucleotides, oligodeoxynucleoside methylphosphonates, and peptide nucleic acids. *Biochem. Pharmacol.* 53:1465–1476.

5. Juliano R L, Alahari S, Yoo H, et al. (1999). Antisense pharmacodynamics: Critical issues in the transport and delivery of antisense oligonucleotides. *Pharmaceut. Res.* 16:494–502.

6. Braasch D A, Corey D R (2002). Novel antisense and peptide nucleic acid strategies for controlling gene expression. *Biochem.* 41:4503–4510.

7. Conner S D, Schmid S L (2003). Regulated portals of entry into the cell. *Nature.* 422:37–44.

8. Cohen A W, Hnasko R, Schubert W, et al. (2004). Role of caveolae and caveolins in health and disease. *Physiol. Rev.* 84:1341–1379.

9. Saito G, Swanson J A, Lee K D (2003). Drug delivery strategy utilizing conjugation via reversible disulfide linkages: role and site of cellular reducing activities. *Adv. Drug Del. Rev.* 55:199–215.

10. Austin C D, Wen X, Gazzard L, et al. (2005). Oxidizing potential of endosomes and lysosomes limits intracellular cleavage of disulfide-based antibody-drug conjugates, *Proc. of the National Academy of Sciences of the United States of America.* 102: 17987–17992.

11. Mastrobattista E, Koning G A, van Bloois L, et al. (2002). Functional characterization of an endosome-disruptive peptide and its application in cytosolic delivery of immuno-liposome-entrapped proteins. *J. Bioli. Chem.* 277:27135–27143.

12. Li W, Nicol F, Szoka F C, Jr (2004). GALA: A designed synthetic pH-responsive amphipathic peptide with applications in drug and gene delivery. *Adv. Drug Deli. Rev.* 56:967–985.

13. Murthy N, Robichaud J R, Tirrell D A, et al. (1999). The design and synthesis of polymers for eukaryotic membrane disruption. *J. Control. Rel.* 61:137–143.

14. Stayton P S, El-Sayed M E, Murthy N, et al. (2005). "Smart" delivery systems for biomolecular therapeutics. *Orthodn. Craniofacial Res.* 8:219–225.

15. Midoux P, LeCam E, Coulaud D, et al. (2002). Histidine containing peptides and polypeptides as nucleic acid vectors. *Somat Cell Molec. Genetics.* 27:27–47.

16. Pack D W, Putnam D, Langer R (2000). Design of imidazole-containing endosomolytic biopolymers for gene delivery. *Biotechnol. Bioengineer.* 67:217–223.

17. Lee E S, Na K, Bae Y H (2005). Super pH-sensitive multifunctional polymeric micelle. *Nano Letters.* 5:325–329.

18. Boussif O, Lezoualc'h F, Zanta M A, et al. (1995). A versatile vector for gene and oligonucleotide transfer into cells in culture and in vivo: polyethylenimine, *Proc. of the National Academy of Sciences of the United States of America.* 92:7297–7301.

19. Kukowska-Latallo J F, Bielinska A U, Johnson J, et al. (1996). Efficient transfer of genetic material into mammalian cells using starburst poolyamidoamine dendrimers, *Proc. of the National Academy of Sciences of the United States of America.* 93:4897–4902.

20. Shi L, Yang L, Chen J, et al. (2004). Preparation and characterization of pH-sensitive hydrogel of chitosan/poly(acrylic acid) co-polymer. *J. Biomat. Sci. Polymer Ed.* 15:465–474.

21. Sonawane N D, Thiagarajah J R, Verkman A S (2002). Chloride concentration in endosomes measured using a ratioable fluorescent Cl- indicator: Evidence for chloride accumulation during acidification. *J. Biolog. Chem.* 277:5506–5513.

22. Yatvin M B, Kreutz W, Horwitz B A, et al. (1980). pH-sensitive liposomes: Possible clinical implications. *Science.* 210:1253–1255.

23. Connor J M, Yatvin B, Huang L (1984). pH-sensitive liposomes: Acid-induced liposome fusion, *Proc. of the National Academy of Sciences of the United States of America.* 81:1715–1718.

24. Simões S, Slepushkin V, Duzgunes N, et al. (2001). On the mechanisms of internalization and intracellular delivery mediated by pH-sensitive liposomes. *Biochim. Biophys. Acta.* 1515:23–37.

25. Wang C Y, Huang L (1987). pH-sensitive immunoliposomes mediate target-cell-specific delivery and controlled expression of a foreign gene in mouse, *Proc. of the National Academy of Sciences of the United States of America.* 84:7851–7855.

26. Collins D, Litzinger D C, Huang L (1990). Structural and functional comparisons of pH-sensitive liposomes composed of phosphatidylethanolamine and three different diacylsuccinylglycerols. *Biochim. Biophys. Acta.* 1025:234–242.

27. Semple S C, Klimuk S K, Harasym T O, et al. (2001). Efficient encapsulation of antisense oligonucleotides in lipid vesicles using ionizable aminolipids: formation of novel small multilamellar vesicle structures. *Biochim. Biophys. Acta.* 1510:152–166.

28. Chen F J, Asokan A, Cho M J (2003). Cytosolic delivery of macromolecules: I. Synthesis and characterization of pH-sensitive acyloxyalkylimidazoles. *Biochim. Biophys. Acta.* 1611:140–150.

29. Asokan A, Cho M J (2003). Cytosolic delivery of macromolecules. II. Mechanistic studies with pH-sensitive morpholine lipids. *Biochim. Biophys. Acta.* 1611:151–160.

30. Asokan A, Cho M J (2004). Cytosolic delivery of macromolecules. 3. Synthesis and characterization of acid-sensitive bis-detergents. *Bioconjug. Chem.* 15:1166–1173.

31. Murthy, N, Thng Y X, Schuck S, et al. (2002). A novel strategy for encapsulation and release of proteins: Hydrogels and microgels with acid-labile acetal cross-linkers. *J. Amer. Chem. Soc.* 124:12398–12399.

32. Murthy N, Xu M, Schuck S, et al. (2003). A macromolecular delivery vehicle for protein-based vaccines: acid-degradable protein-loaded microgels, *Proc. of the National Academy of Sciences of the United States of America.* 100:4995–5000.

33. Wang C, Ge Q, Ting D, et al. (2004). Molecularly engineered poly(ortho ester) microspheres for enhanced delivery of DNA vaccines. *Nat. Mat.* 3:190–196.

34. Heffernan M J, Murthy N (2005). Polyketal nanoparticles: A new pH-sensitive biodegradable drug delivery vehicle. *Bioconjug. Chem.* 16:1340–1342.

35. Cordes E H, Bull H G (1974). Mechanism and catalysis for hydrolysis of acetals, acetals and orthoesters. *Chem. Rev.* 74:581–603.

36. Boomer J A, Thompson D H (1999). Synthesis of acid-labile diplasmenyl lipids for drug and gene delivery applications. *Chem. Phys. Lipids.* 99:145–153.

37. Guo X, Szoka F C, Jr (2001). Steric stabilization of fusogenic liposomes by a low-pH sensitive PEG–diortho ester–lipid conjugate. *Bioconjug. Chem.* 12:291–300.

38. Guo X, Szoka F C, Jr (2003). Chemical approaches to triggerable lipid vesicles for drug and gene delivery. *Accounts of Chem. Res.* 36:335–341.

39. Saito G, Amidon G L, Lee K D (2003). Enhanced cytosolic delivery of plasmid DNA by a sulfhydryl-activatable listeriolysin O/protamine conjugate utilizing cellular reducing potential. *Gene Ther.* 10:72–83.

40. Tang F, Hughes J A (1998). Introduction of a disulfide bond into a cationic lipid enhances transgene expression of plasmid DNA. *Biochem. Biophys. Res. Communic.* 242:141–145.

41. Byk G, Wetzer B, Frederic M, et al. (2000). Reduction-sensitive lipopolyamines as a novel nonviral gene delivery system for modulated release of DNA with improved transgene expression. *J. Med. Chem.* 43:4377–4387.

42. Chittimalla C, Zammut-ltaliano L, Zuber G, et al. (2005). Monomolecular DNA nanoparticles for intravenous delivery of genes. *J. Amer. Chem. Soc.* 127:11436–11441.

43. Pak C C, Ali S, Janoff A S, et al. (1998). Triggerable liposomal fusion by enzyme cleavage of a novel peptide-lipid conjugate. *Biochim. Biophys. Acta.* 1372:13–27.

44. Pak C C, Erukulla R K, Ahl P L, et al. (1999). Elastase activated liposomal delivery to nucleated cells. *Biochim. Biophys. Acta.* 1419:111–126.

45. Davis S C, Szoka F C, Jr (1998). Cholesterol phosphate derivatives: Synthesis and incorporation into a phosphatase and calcium-sensitive triggered release liposome. *Bioconjug. Chem.* 9:783–792.

46. Sarkar N R, Rosendahl T, Krueger A B, et al. (2005). "Uncorking" of liposomes by matrix metalloproteinase-9. *Chem. Communicat.* 999–1001.

47. Meers P (2001). Enzyme-activated targeting of liposomes. *Adv. Drug Del. Rev.* 53:265–272.

48. Anderson R G (1998). The caveolae membrane system. *Ann. Rev. Biochem.* 67:199–225.

49. Nabi I R, Le P U (2003). Caveolae/raft-dependent endocytosis. *J. Cell Biol.* 161:673–677.

50. Pelkmans L, Helenius A (2002). Endocytosis via caveolae. *Traffic.* 3:311–320.

51. Sabharanjak S, Mayor S (2004). Folate receptor endocytosis and trafficking. *Adv. Drug Del. Rev.* 56:1099–1109.

52. Drummond D C, Hong K, Park J W, et al. (2000). Liposome targeting to tumors using vitamin and growth factor receptors. *Vit. Hormones.* 60:285–332.

53. Low P S, Antony A C (2004). Folate receptor-targeted drugs for cancer and inflammatory diseases. *Adv. Drug Del. Rev.* 56:1055–1058.

54. Leamon C P, Cooper S R, Hardee G E (2003). Folate-liposome-mediated antisense oligodeoxynucleotide targeting to cancer cells: Evaluation in vitro and in vivo. *Bioconjug. Chem.* 14:738–747.

55. Derossi D, Calvet S, Trembleau A, et al. (1996). Cell internalization of the third helix of the Antennapedia homeodomain is receptor-independent. *J. Biolog. Chem.* 271:18188–18193.

56. Vives E, Brodin P, Lebleu B (1997). A truncated HIV-1 Tat protein basic domain rapidly translocates through the plasma membrane and accumulates in the cell nucleus. *J. Biolog. Chem.* 272:16010–16017.

57. Lindgren M, Hallbrink M, Prochiantz A, et al. (2000). Cell-penetrating peptides. *Trends Pharmacol. Sci.* 21:99–103.

58. Richard J P, Melikov K, Vives E, et al. (2003). Cell-penetrating peptides: A reevaluation of the mechanism of cellular uptake. *J. Biolog. Chem.* 278:585–590.

59. Ghibaudi E, Boscolo B, Inserra G, et al. (2004). The interaction of the cell-penetrating peptide penetratin with heparin, heparansulfates and phospholipid vesices investigated by ESR spectroscopy. *J. Peptide Sci.* 11:401–409.

60. Tyagi M, Rusnati M, Presta M, et al. (2001). Internalization of HIV-1 tat requires cell surface heparan sulfate proteoglycans. *J. Biolog. Chem.* 276:3254–3261.

61. Foerg C, Ziegler U, Fernandez-Carneado J, et al. (2005). Decoding the entry of two novel cell-penetrating peptides in HeLa cells: lipid raft-mediated endocytosis and endosomal escape. *Biochem.* 44:72–81.

62. Jones S W, Christison R, Bundell K, et al. (2005). Characterisation of cell-penetrating peptide-mediated peptide delivery. *British J. Pharmacol.* 145:1093–1102.

63. Fischer R, Kohler K, Fotin-Mleczek M, et al. (2004). A stepwise dissection of the intracellular fate of cationic cell-penetrating peptides. *J. Biolog. Chem.* 279:12625–12635.

64. Potocky T B, Menon A K, Gellman S H (2003). Cytoplasmic and nuclear delivery of TAT-derived peptide and a beta-peptide after endocytic uptake into HeLa cells. *J. Biolog. Chem.* 278:50188–50194.

65. Wadia J S, Stan R V, Dowdy S F (2004). Transducible TAT-HA fusogenic peptide enhances escape of TAT-fusion proteins after lipid raft macropinocytosis. *Nat. Med.* 10:310–315.

66. Langel U (2005). Cell-penetrating peptides, mechanisms and applications. *Curr. Pharmaceut. Des.* 11:3595.

67. Astriab-Fisher A, Sergueev D, Fisher M, et al. (2002). Conjugates of antisense oligonucleotides with the Tat and antennapedia cell-penetrating peptides: Effects on cellular uptake, binding to target sequences, and biologic actions. *Pharmaceut. Res.* 19:744–754.

68. Koppelhus U, Awasthi S K, Zachar V, et al. (2002). Cell-dependent differential cellular uptake of PNA, peptides, and PNA-peptide conjugates. *Antisense and Nucleic Acid Drug Devel.* 12:51–63.

69. Moulton H M, Hase M C, Smith K M, et al. (2003). HIV Tat peptide enhances cellular delivery of antisense morpholino oligomers. *Antisense Nucleic Acid Drug Dev.* 13:31–43.

70. Oehlke J, Birth P, Klauschenz E, et al. (2002). Cellular uptake of antisense oligonucleotides after complexing or conjugation with cell-penetrating model peptides. *Euro. J. Biochem.* 269:4025–4032.

71. Xia H, Mao Q, Davidson, B L (2001). The HIV Tat Protein Transduction Domain Improves the Biodistribution of β-Glucuronidase Expressed from Recombinant Viral Vectors. *Nat. Biotechnol.* 19:640–644.

72. Bucci M, Gratton J P, Rudic R D, et al. (2000). In vivo delivery of the caveolin-1 scaffolding domain inhibits nitric oxide synthesis and reduces inflammation. *Nat. Med.* 6:1362–1367.

73. May M J, D'Acquisto F, Madge L A, et al. (2000). Selective inhibition of NF-kB activation by a peptide that blocks the interaction of NEMO with the IkB kinase complex. *Science.* 289:1550–1554.

74. Kaufman C L, Williams M, Ryle L M, et al. (2003). Superparamagnetic iron oxide particles transactivator protein-fluorescein isothiocyanate particle labeling for in vivo magnetic resonance imaging detection of cell migration: uptake and durability. *Transplantation.* 76:1043–1046.

75. Torchilin V P, Levchenko T S, Rammohan R, et al. (2003). Cell transfection in vitro and in vivo with nontoxic TAT peptide-liposome-DNA complexes, *Proc. of the National Academy of Sciences of the United States of America.* 100:1972–1977.

76. Lee H J, Pardridge WM (2001). Pharmacokinetics and delivery of tat and tat-protein conjugates to tissues in vivo. *Bioconjug. Chem.* 12:995–999.

77. Pooga M, Soomets U, Hallbrink M, et al. (1998). Cell penetrating PNA constructs regulate galanin receptor levels and modify pain transmission in vivo. *Nat. Biotechnol.* 16:857–861.

78. Leifert J A, Harkins S, Whitton JL (2002). Full-length proteins attached to the HIV tat protein transduction domain are neither transduced between cells, nor exhibit enhanced immunogenicity. *Gene Therapy.* 9:1422–1428.

79. Ye D, Xu D, Singer A U, et al. (2002). Evaluation of strategies for the intracellular delivery of proteins. *Pharmac. Res.* 19:1302–1309.

80. Becker-Hapak M, McAllister S S, Dowdy S F (2001). TAT-mediated protein transduction into mammalian cells. *Methods.* 24:247–256.

81. Wadia J S, Stan R V, Dowdy S F (2004). Transducible TAT-HA fusogenic peptide enhances escape of TAT-fusion proteins after lipid raft macropinocytosis. *Nat. Med.* 10:310–315.

82. Ho A, Schwarze S R, Mermelstein S J, et al. (2001). Synthetic protein transduction domains: enhanced transduction potential in vitro and in vivo. *Cancer Res.* 61:474–477.

83. Futaki S (2002). Arginine-Rich Peptides: Potential for Intracellular Delivery of Macromolecules and the Mystery of the Translocation Mechanisms. *Internat. J. Pharmac.* 245:1–7.

84. Wender P A, Mitchell D J, Pattabiraman K, et al. (2000). The design, synthesis, and evaluation of molecules that enable or enhance cellular uptake: Peptoid molecular transporters, *Proc. of the National Academy of Sciences of the United States of America.* 97:13003–13008.

85. Morris M C, Vidal P, Chaloin L, et al. (1997). A new peptide vector for efficient delivery of oligonucleotides into mammalian cells. *Nucleic Acids Res.* 25:2730–2736.

86. Elliott G, O'Hare P (1997). Intercellular trafficking and protein delivery by a herpesvirus structural protein. *Cell.* 88:223–233.

87. Aints A, Guven H, Gahrton G, et al. (2001). Mapping of herpes simplex virus-1 VP22 functional domains for inter- and subcellular protein targeting. *Gene Therapy.* 8:1051–1056.

88. Hawiger J (1999). Noninvasive intracellular delivery of functional peptides and proteins. *Curr. Opin. Chem. Biol.* 3:89–94.

89. Jo D, Liu D, Yao S, et al. (2005). Intracellular protein therapy with SOCS3 inhibits inflammation and apoptosis. *Nat. Med.* 11:892–898.

90. Janout V, Staina I V, Bandyopadhyay P, Regen S L (2001). Evidence for an umbrella mechanism of bilayer transport. *J. Amer. Chem. Soc.* 123:9926–9927.

91. Janout V, Jing B, Regen S L (2005). Molecular Umbrella-Assisted Transport of an Oligonucleotide across Cholesterol-Rich Phospholipid Bilayers. *J. Amer. Chem. Soc.* 127:15862–15870.

92. Zaro J L, Shen W C (2003). Quantitative comparison of membrane transduction and endocytosis of oligopeptides. *Biochem. Biophys. Res. Communicat.* 307:241–247.

93. Lundberg M, Johansson M (2001). Is VP22 nuclear homing an artifact? *Nat. Biotechnol.* 19:713–714.

94. Loison F, Nizard P, Sourisseau T, et al. (2005). A ubiquitin-based assay for the cytosolic uptake of protein transduction domains. *Molec. Ther.* 11:205–214.

95. Kang S H, Cho M J, Kole R (1998). Up-regulation of luciferase gene expression with antisense oligonucleotides: Implications and applications in functional assay development. *Biochem.* 37:6235–6239.

4.1

BASIC ISSUES IN THE MANUFACTURE OF MACROMOLECULES

CORINNE LENGSFELD AND THOMAS ANCHORDOQUY

University of Denver, Denver, Colorado

Chapter Contents

4.1.1 Introduction to Bioprocessing 298
    4.1.1.1 General Method for Manufacturing Therapeutic Proteins 298
    4.1.1.2 General Methods for Manufacturing Plasmid DNA 298

4.1.2 Review of Molecular Structure 299
    4.1.2.1 Comparison of Protein and DNA Primary Structure 299
    4.1.2.2 Comparison of Protein and DNA Higher Order Structure 299

4.1.3 Manufacturing Concerns 300
    4.1.3.1 pH Sensitivity 301
    4.1.3.2 Metal Catalyzed and Enzymatic Degradation 301
    4.1.3.3 Hydrodynamic Degradation 301
    4.1.3.4 Cavitation 304
    4.1.3.5 Surface Interaction 305
    4.1.3.6 Freeze–Thaw Cycling 307

4.1.4 Example Process—Atomization 307
    4.1.4.1 Common Techniques Employed in Processing and Delivery 307
    4.1.4.2 Degradation Behavior of Linear Genomic DNA 309
    4.1.4.3 Degradation Behavior of Plasmid DNA 310
    4.1.4.4 Applications to siRNA Molecules 310
    4.1.4.5 Degradation Behavior of Proteins 311

4.1.5 Separation Techniques 312
    4.1.5.1 Classic Approaches 312
    4.1.5.2 Novel Molecular-Scale Probing Concepts 313

4.1.6 Chapter Summary 314
    References 314

Handbook of Pharmaceutical Biotechnology, Edited by Shayne Cox Gad.
Copyright © 2007 John Wiley & Sons, Inc.

4.1.1   INTRODUCTION TO BIOPROCESSING

In general the manufacturing methods for proteins and plasmid DNA require a cellular medium to either express a protein or replicate genetic material, which is later harvested and separated from undesirable cellular material. The type of cell culture will play a large role in the efficiencies of macromolecular production as well as in the separation process. Often forgotten are the manufacturing steps that bridge the raw purified product and that which is delivered. Lyophilization, encapsulation, delivery, and formulation steps suffer many of the same issues as those in the initial growth and harvesting steps. In the following sections, we briefly describe the initial manufacturing steps involved in the production of therapeutic proteins and plasmid DNA so that later sections specific manufacturing issues can be placed in context. However, it is in these later sections we will draw connections to the later manufacturing stages.

4.1.1.1   General Method for Manufacturing Therapeutic Proteins

The production of therapeutic protein starts with the design of a recombinant DNA that is typically transfected into mammalian cells, *Escherichia coli*, or yeast, although some transgenic plants and animals are being employed. Conditions, especially in the medium, that elicit the excretion of the desired proteins into extracellular medium are typically preferred over harvesting through cell disruption because of the reduction in contaminants and subsequent purification steps required. The recovery of the product in active form is complicated by protein aggregation that can be promoted by separation and purification processes. It should be recognized that the medium that generated the best protein extrusion may now hinder recovery because of high viscosity or a tendency to promote colloids due to the elevated protein concentrations. The isolation stage normally involves high-pressure homogenization, bead milling, centrifugation, filtration, precipitation, and/or ultrafilration. This stage is followed by a purification stage involving chromatographic sequence (i.e., polishing, desalting, buffer exchange, ion exchange, and reverse phase) that balances yield and purity. An in-depth discussion of all unit-based application can be found in Ref. 1.

4.1.1.2   General Methods for Manufacturing Plasmid DNA

The production of plasmid DNA begins with the replication of the desired therapeutic gene by bacterial cells, most often *E. coli* [2]. The cells are allowed to grow in a nutritious broth before they are harvested, typically by centrifugation. In a process called cell lysis, the harvested cells are broken open either mechanically (e.g., high-pressure homogenizer), enzymatically, or chemically (e.g., alkaline lysis). During this process the contents of the cells, including plasmid and genomic DNA, are most susceptible to shearing forces as they are no longer protected by the bacterial membrane. There are several ways of removing and purifying the plasmid DNA from the rest of the cell debris. These ways include centrifugation, filtration, and chromatography methods, which are briefly reviewed by Durland and Eastman [3].

4.1.2   REVIEW OF MOLECULAR STRUCTURE

Like all polymers, proteins and DNA are composed of a sequence of repeating units that interact at various levels to produce the active structure. Although the primary structure of DNA does largely determine its physical and biological properties, that of proteins is of minimal use in determining the desired active structure. This section is not intended to replace a biochemistry textbook, but it provides the basic information necessary to understand how the molecular structure of proteins and plasmid DNA differ, and how this impacts macromolecular stability in a general sense.

4.1.2.1   Comparison of Protein and DNA Primary Structure

Proteins, peptides, and polypeptides are single-stranded polymer chains composed of repeating amino acids (20 different amino acids are available) and only differ in the number of amino acids that comprise the polymer chain. Although the definitions of these terms are not strictly defined, peptides typically have less than 50 amino acids, whereas polypeptides possess up to 100 amino acids, and proteins generally have more than 100 amino acids covalently linked (through "peptide" bonds) into a polymer chain. Analogous terms for small polymers of nucleic acids, "oligonucleotide" and "polynucleotide," generally refer to chains of <50 and >50 nucleotides (adenosine, guanosine, cytosine, or thymidine), respectively. There is no term for nucleic acids that is analogous to "protein," but long strands of nucleic acids are typically produced as a plasmid, and they are generally >3000 nucleic acids in length. A critical difference between proteins and plasmid DNA is that proteins are a single polymer chain, whereas plasmid DNA consists of two strands that are associated in the classic double helix. Although DNA used for antisense therapy is typically single stranded, thereby designed to associate with a specific sequence to exert a therapeutic effect, these molecules are typically 10–30 nucleotides in length (oligonucleotides) and are chemically synthesized instead of being produced in bacteria as described above. Although the chemical synthesis of oligonucleotides presents its own set of purification issues that are reviewed by Sanghvi and Schulte [4], our discussion will focus on issues involved with the bioprocessing of plasmid DNA.

Simply based on the number of atoms participating in the primary structure of these biopolymers, proteins have a smaller hydrodynamic diameter (<30 nm), whereas typical plasmid DNA molecules range from 100 to 400 nm. As a result, proteins are less susceptible to mechanical forces, especially hydrodynamic forces. This resistance is simply due to a size range that is well below the continuum limit of the fluid. In other words, when particle sizes are well below 100 nm (Figure 4.1-1) in aqueous solutions, they "slip" between water molecules rather than having a drag force applied continuously over the particle surface.

4.1.2.2   Comparison of Protein and DNA Higher Order Structure

Unlike most polymers, these biopolymers have extraordinary secondary and tertiary structures that can dramatically affect therapeutic efficacy. For example, proteins form complex geometries based on hydrophobic interactions and hydrogen

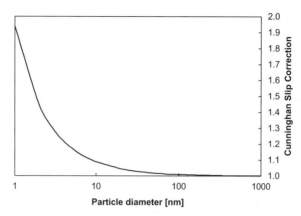

Figure 4.1-1. Cunningham slip correction vs. particle diameter for water fluid flow.

bonding, which results in well-defined structures, e.g., beta sheet and alpha helix. Because these structures rely on relatively weak bonding interactions (i.e., noncovalent), protein molecules are acutely sensitive to changes in pH and hydrophobic/hydrophilic interfaces. In contrast, plasmid DNA is a double-stranded helix that is produced as a circular form. In addition, this circular form is coiled in a direction opposite that of the double helix, resulting in "negative supercoils" that create torsional stress in the supercoiled form of the molecule. The added stress due to supercoiling is stored in the plasmid and cannot be relieved unless a break occurs in the covalent bonds comprising the phosphate backbone. However, the cleavage of a single covalent bond in the backbone of either strand composing the double helix results in the release of this torsional stress and conversion of the supercoiled form to the relaxed, "open circle" form. Another cleavage in the phosphate backbone of the opposite strand that is in the same vicinity (within a few nucleotides) of the first break will cause the circular integrity to be broken, resulting in a linear form of the plasmid. Because the supercoiled form requires an intact backbone in both DNA strands, supercoil content (the percentage of plasmids that are in the supercoiled form) is used as a measure of quality for a plasmid preparation. Typically, plasmid preparations used in clinical trials have supercoil contents of >90%. Not surprisingly, it is critical that stresses during bioprocessing be low enough to ensure that the integrity of the phosphate backbone is retained so that preparations with sufficiently high supercoil contents are produced.

4.1.3 MANUFACTURING CONCERNS

Basic manufacturing concerns fall into three categories: those associated with chemical interaction, mechanical phenomena, or thermodynamic energy states. In the following sections, we attempt to provide a general overview of the most common problems as well as the most commonly perceived problems: whether real or imaginary. A clear understanding of the source of degradation during a specific manufacturing process is essential to developing successful mitigation strategies. Numerous mitigation techniques have been described in literature, but it should be recognized that the best option is dependent on the ultimate form in which the molecule will be administered, e.g., free or in a delivery system.

4.1.3.1   pH Sensitivity

Extreme pH has been shown to have a role in the break down of macromolecular structure in aqueous solution via hydrolysis [5–7] and aggregation. For example, DNA can disintegrate by depurination reactions resulting in the loss of a purine base and the creation of apurinic (AP) sites. This absence of a base weakens the molecular chain, thereby leading to an increased susceptibility to hydrolytic attack [6]. The destabilized DNA can then undergo beta-elimination of a phosphodiester bond in the backbone, resulting in strand breaks and loss of the supercoiled form. These reactions can be accelerated if pH levels fall below 7.5 or rise above 9.0 [8].

In contrast, proteins will not denature in the same way when exposed to varying pH solutions. Therapeutic proteins are often only stable over a narrow pH range; outside these limits, the aggregation rate is strongly influenced by pH. A solution pH governs the total charge on a molecule; thus, it can affect intra- and intermolecular interactions. Increasing the acidity or alkalinity of the solution beyond the stable range will highly charge nonspecific groups within the protein generating strong repulsive forces, thus destabilizing the protein [9].

pH sensitivity is well known by process engineers, and solution pH is usually carefully controlled, but manufacturing steps that include a gas–liquid interface can unintentionally lower pH if solutions are not properly buffered. The diffusion of carbon dioxide into the aqueous solution could be facilitated by the use of a driving gas that produces a large gas–liquid interface across which substantial gas exchange can occur. Localized nonequilibrium supersaturation of carbon dioxide can lead to large transient changes in pH because carbon dioxide in water can form carbonic acid that can deprotonate and acidify the solution [10].

$$CO_2 + H_2O \rightarrow H_2CO_3$$
$$H_2CO_3 \rightarrow H^+ + HCO_3^-$$

4.1.3.2   Metal Catalyzed and Enzymatic Degradation

Trace amounts of reduced transition metal ions such as iron and copper are known to generate hydroxyl radicals resulting in oxidized nucleotide bases and strand breaks by means of Fenton-type reactions [11–15]. Diethylenetriaminepentaacetic acid (DTPA) and ethylenediamine tetraacetic acid (EDTA) are compounds that chelate metal ions present in solution or bound to macromolecules. These chelators have also been shown to inhibit DNA degradation via DNase by binding various divalent cations (e.g., Mg^{2+}) necessary for enzymatic activity [16].

4.1.3.3   Hydrodynamic Degradation

Observations of hydrodynamic shear-induced degradation of long polymer molecules first appeared in the literature approximately 60 years ago when Frenkel [17] mathematically demonstrated that hydrodynamic forces along a long polymer molecule could be sufficient to break covalent bonds. Early experimental investigations into the degradation of long linear DNA molecules blamed hydrodynamic shear forces for the observed fragmentation [18, 19]. These studies varied shear rates through needles and measured degradation through sedimentation rates. This technique did not allow for the clear distinction between molecular scission and structural changes, but these early experiments demonstrated the sensitivity of DNA to hydrodynamic shear stress encountered during stirring and pipetting. But as experimental techniques evolved, it become clear that the shear *rate* instead of the shear *force* was the dominant contributor. In their classic paper, Bowman and Davidson [20] provided the groundwork for relating the probability of fracture to the strain rate. These authors realized that the experimentally measured intact DNA concentration divided by the original concentration represented the total probability ($\langle P \rangle$); an integrated average over each radial position depending on the tube radius (R), volumetric flow rate (Q), tube length (l), and rate constant (k), which depends on strain rate (G). A complete review of the classic theory is available in Lengsfeld and Anchordoquy [21]:

$$\langle P \rangle = \frac{2\pi l}{Q} \int_0^R k r \, dr \qquad (4.1\text{-}1)$$

When even more conditions (i.e., length and flexibility of the molecule, strain rates, and shear forces) where explored, it became clear that the original classic theory needed modification to be applicable to bioprocessing. Recent work indicates that the degradation of very large biopolymers is similar to cell fragmentation, which has been linked to unsteady flow and turbulent intensity [22, 23]. When energy is transferred to a fluid, some is imparted to the motion of the bulk in a downstream direction, whereas some is consumed by turbulent mixing. This energy is transmitted from eddies on the order of the characteristic length through a cascade of sizes that terminate at the Kolmogorov length scale. At this length scale, energy is dissipated via viscosity.

Several studies investigating the dynamics of individual polymer molecules in flow demonstrated the transition of polymers in their natural coiled state to a stretched configuration in extensional flows [24–27]. Investigations that focus on simple shearing flow have demonstrated that the tumbling of the molecules prevents the elongation of the molecule [27–29]. In these cases, simple shear flow is defined as flow in which the strain rate is equal to the magnitude of the vorticity [28]. Random flows, however, have demonstrated that polymer molecules such as DNA can be effectively stretched [30–32]. In these cases, where the viscous scale of the flow exceeds the length of the polymer, constant strain is applied locally by the turbulent flow [32].

Theoretical models suggest that the transition from a coil-stretch transition within random flows depends on the principle Lyapunov exponent $\lambda 1$ and the longest relaxation time for the molecule τ_R. To stretch the molecules, $\lambda_1 \tau_R$ must be >1 [31, 32]. The Lyapunov exponent can be approximated by taking the fluctuating velocity u', divided by the characteristic length scale, which in turbulence, is determined by the eddies at the viscous scale; i.e., $\lambda = u'/\eta$ [31].

Yim and Shamlou [22] provide a comprehensive list of energy dissipation rates in each unit process used in the manufacture of biomolecules (Table 4.1-1). Worden [23] showed that degradation commences when the size of this small-scale turbulent structure becomes equal to or less than the molecule's hydrodynamic diameter (Figure 4.1-2). The scale of these smallest eddies η can be calculated based on the energy dissipation rate ε and the diffusivity v using the relationship derived by Kolmogorov in 1941 [33]:

$$\eta = \left(\frac{v^3}{\varepsilon}\right)^{1/4}$$ (4.1-2)

Estimating the size of the smallest length scale is relatively simple. One could use computational fluid dynamic modeling techniques or estimate them based on the power input to the system (head loss) and the mass of the fluid being powered. For example, in pipe flow, the energy dissipation rate is a function of the total head loss in the flow h_{lt}, the volumetric flow rate Q, the density of the solution ρ, and the mass of the solution m, which in this case is the mass of fluid contained within the pipe [Equation (4.1-3)]

TABLE 4.1-1. Energy Dissipation Rate Estimates by Yim and Shamlou for Common Bioprocessing Equipment [22]

System	Energy Dissipation Rates [W kg^{-1}]
Homogenizer	10^7–10^9
Liquid whistle	10^7
Centrifuge	10^6–10^7
Collid mills	10^5
Pumps, jets	10^3–10^5
Agitated vessels	10^0–10^2
Static mixers	10^0–10^1
Bubble columns	$<10^1$

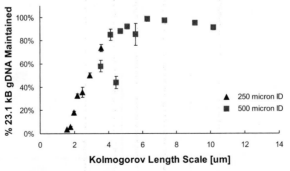

Figure 4.1-2. Degradation behavior of 23.1-kb genomic DNA during simple pipe flow with an initial contraction. Kolmogorov length scales were varied through changes in contraction ratio (i.e., pipe diameter) and pressure drop.

$$\varepsilon = \frac{h_{lt}Q\rho}{m} \qquad (4.1\text{-}3)$$

Note that to create length scales lower than those of proteins in water requires energy inputs of 1012 W/kg, which is well above that achieved in bioprocessing. Thus, hydrodynamic degradation is only an issue in the production of large macromolecules, e.g., plasmid DNA.

4.1.3.4 Cavitation

A phenomena present in most pumping systems, but until recently ignored by process engineers is the impact of cavitation on macromolecular stability. For example, the pressure waves induced by the piezoelectric element in an ultrasonic nebulizer have compressive crests and tensile troughs (Figure 4.1-3). If sufficient tensile stress is applied at the fluid during the low-pressure cycle, a void of vapor will be formed. Dissolved gases can diffuse to this surface causing the bubble to grow. In stable cavitaion, the bubble oscillates in size about a mean diameter with the passing of the pressure waves. However, in transient cavitation, once the bubble has reached a critical size, the compressive pressure at the crest is sufficient to cause implosion. The bubble collapse releases significant energy in the form of a shock exhibiting high temperatures and pressures, which can generate free radicals.

The factors influencing cavitation are the solution's viscosity, surface tension, vapor pressure, and the presence of contaminants. In addition, the applied forcing frequency and amplitude can contribute to the efficiency of cavitation.

Manufacturing engineers can easily track transient cavitation using a well-established method. When water molecules are exposed to a shock wave (intense temperature and pressures), it becomes ionized into hydrogen and hydroxyl radicals, and two hydroxyl radicals can combine to form hydrogen peroxide. If potassium iodide is present in the solution, the peroxide will react and form iodide and potassium hydroxide. If the solution contains excess potassium iodide, the iodide will form triiodide, which can be measured spectroscopically using ultraviolet absorption at 350 nm. This is essential for plasmid DNA systems where we have determined that the DNA molecule serves as a nucleation site for cavitation. Lentz et al. [34] observed a reduction in nucleation efficiency with size, suggesting that proteins are too small (<30 nm) to nucleate cavitation:

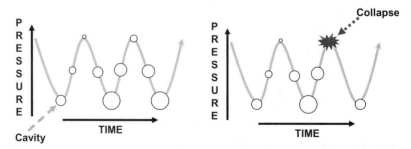

Figure 4.1-3. Schematic for the evolution of cavitation in (a) stable and (b) transient processes.

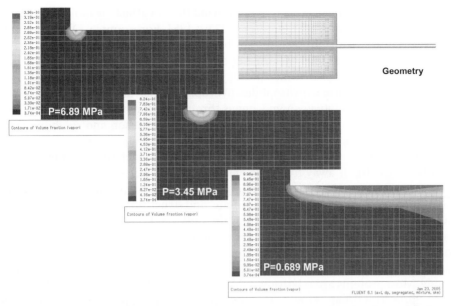

Figure 4.1-4. Computational fluid dynamics simulation of vapor fraction formation from a tube contraction as a function of tube length (diameter = 500 μm, pressure drop = 3.45 MPa). Red areas indicate 100% vapor regions within the pipe. (This figure is available in full color at ftp://ftp.wiley.com/public/sci_tech_med/pharmaceutical_biotech/.)

$$H-O-H \xrightarrow{\text{Cavitaton}} H^{\bullet} + H^{\bullet}$$

$$2OH^{\bullet} \longrightarrow H_2O_2$$

$$KI + H_2O_2 \longrightarrow I_2 + 2KOH$$

$$I_2 \xrightarrow{\text{Excess KI}} I_3^{\bullet}$$

However, unlike DNA, proteins will degrade at the bubble interface. Therefore, modeling stable cavitation may be important in the design phase of protein bioprocessing systems. Figure 4.1-4 demonstrates this process using a commercial computational fluid dynamics (CFD) package to observe the formation of vapor phases in low-pressure regions around a tube contraction. As anticipated, the volume of vapor formed increases with decreasing tube length.

4.1.3.5 Surface Interaction

Gas–liquid interfaces are problematic when processing macromolecules [35, 36]. Proteins, due to their amphipathic nature, undergo conformational changes by adsorbing to the gas–liquid interface [36, 37]. The Gibbs free energy available at the interface allows the protein to rearrange such that hydrophilic components remain in solution while hydrophobic moieties associate with the gas phase. The addition of mild shear stresses can expedite this unfolding process. It was Maa and Hsu [37] who clearly showed the correlation between protein denaturation and surface-to-volume ratio under hydrodynamic shear conditions.

In addition, some research has hypothesized that an attack by molecular oxygen could accelerate the degradation rates of DNA [38]. For example, DNA bases and sugars would be oxidized, resulting in base alterations or strand breaks [5, 39]. However, the investigations by Lentz et al. [40] into the aerosolization of genetic therapeutics showed not evidence of this effect. Although proteins act as surface-active agents and diffuse to hydrophilic–hydrophobic interfaces, DNA is not surface active and remains relatively homogenous throughout the bulk of the fluid (Figure 4.1-5). As a result, oxidative attack is controlled by diffusion and solubility limits in aqueous solutions.

It is important to note that these molecules are highly charged and may preferentially diffuse to, or away from, the electric double layer formed at the interface of a metal surface and water. Adsorption to solid surfaces is controlled by electrostatic interaction, dispersion, hydration state, and structural rearrangement because the molecule must be transported from the bulk to the surface, attach, and then spread (complete review in Ref. 41). For example, in rigid proteins (e.g., those in poor solvents), the surface energy increases necessary to spread along the surface must be overcome. Flexible proteins (e.g., those in good solvents) will spread more readily. Most proteins are relatively stable in solution when processed under appropriate buffer conditions. However, at high concentrations, proteins are more likely to adhere to solid surfaces, especially glass and plastics.

What can be done to inhibit molecular adsorption to solid surfaces? Several techniques exist for the protection against surface-induced damage during fermentation, purification, freeze-drying, shipping, and storage. Surfactants and osmolytes (e.g., small organic solutes like sugars, methylamines, and amino acids) all have demonstrated protection against protein aggregation. Surfactants can enhance

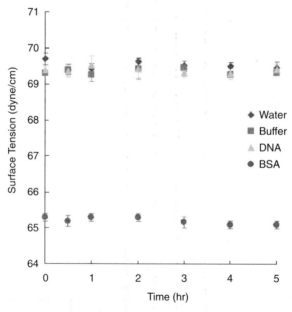

Figure 4.1-5. Magnitude of surface tension as a function of buffer, DNA, and protein (BSA) concentration [40].

protein stability in four ways by competing with proteins for adsorption sites on surfaces, binding to hydrophobic regions inhibiting intermolecular interactions, increasing the free energy of the unfolding pathway, and acting as a simple chaperone [42]. On the other hand, osmolytes, such as sucrose, are preferentially excluded from the protein surface, thereby increasing the potential energy and subsequently compacting the protein [43, 44]. Instead of manipulating the solution formulation, one can also alter the free energy of the surface. Coating the surfaces with materials that are not porous (i.e., large surface area) or those that are more hydrophilic will reduce adhesion. Exposure to gamma radiation or pulsed plasma [45] can alter the hydrophilic/hydrophobic state of a polymer surface by changing the functionalization of the surface molecules. Other groups have demonstrated that using polyDuramid or poly-N-hydroxyethacrylamid (PHEA) inhibits adhesion, whereas others in the microfluidic industry use polyethylene glycol (PEG) (plastics evaluation in Ref. 46).

4.1.3.6   Freeze–Thaw Cycling

The formation of discrete ice crystals during freezing causes boundaries to form between different crystal orientations. These boundaries have high surface energies that are known to contribute to fragmentation. As crystal size decreases, the probability that protein or DNA molecules will span one of these high-energy interfaces increases. When storing the therapeutic (plasmid DNA or protein) before use in secondary processing, it is essential to consider the freezing rate because high rates lead to smaller crystal sizes. These small differences in crystal morphology can cause significant batch-to-batch variability.

Embedding a macromolecular therapeutic in a highly crystalline polymer may experience a similar decrease in activity with decreasing particle size. Control over particle size during product formation may therefore be extremely important during fabrication steps.

4.1.4   EXAMPLE PROCESS—ATOMIZATION

Many issues that plague bioprocessing steps make atomization of biologically active macromolecules difficult. This section will begin with a review of nearly all the classic atomization techniques invented; we will compare and contrast these methods in regard to their ability to sustain molecular conformation from large micron-sized linear polymers down to less than 30 nm for proteins. The goal here is to establish the mechanism for degradation in each technique and to suggest a variety of procedures to help mitigate these issues.

4.1.4.1   Common Techniques Employed in Processing and Delivery

The generation of an aerosol is as varied as the methods for manufacturing macromolecules. All employ some sort of external force to divide the liquid bulk into individual droplets. Some methods employ aerodynamic forces, mechanical vibration, or pressure oscillations, whereas others harness electrical attraction and repulsion.

Figure 4.1-6. Photo of a Spraying Systems, Inc. hollow cone misting nozzle.

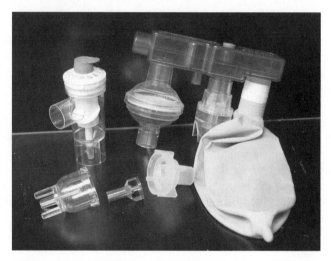

Figure 4.1-7. Photo of various commercial jet nebulizers.

Pressure-driven devices (Figure 4.1-6), such as pressure swirl atomizers, use aerodynamic drag at the gas–liquid interface to amplify natural disturbance to pinch droplets off liquid sheets and columns. Finer mists can be generated by adding a centrifugal force to the process that must overcome viscous damping and surface tension.

Jet nebulization (Figure 4.1-7), the cheapest and most common aerosol device in the pulmonary delivery arena, uses driving gas to accelerate liquid from a reservoir and achieve primary atomization. These primary droplets then hit an obstruction that causes splashing and spreading, which generates secondary droplets in the 1- to 10-μm range.

Ultrasonic nozzles use a metal horn geometry to amplify a small peizoelectric vibration. These vibrations drive surface instabilities along a thin film and generate droplets of very uniform size, but at low flow rates.

The details of how ultrasonic nebulizers create an aerosol is limited. A piezoelectric ceramic drives a pressure disturbance within the fluid at a rate of 15 to

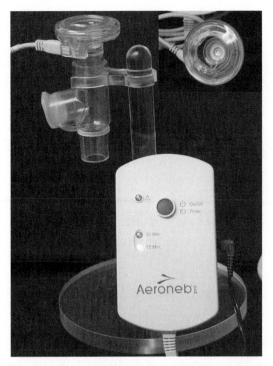

Figure 4.1-8. Photo of an Aeroneb Pro vibrating mesh atomizer.

19,000 kHz. These pressure waves can form small capillary wave crests at the surface of the liquid or on a central fountain that ejects small liquid droplets from the crest. A third hypothesis is that cavitation events within the liquid cause large pressure singularities that eject liquid mass out from the liquid pool. Most likely, combinations of all three of these phenomena work to generate a fine aerosol suitable for inhalation.

Vibrating mesh technology (Figure 4.1-8) uses microfluidic manufacturing techniques to create arrays of small orifices that act as nozzles. Liquid enters these pores via capillary action but are ejected when a high-frequency vibration is imparted to the mesh using a piezoelectric crystal. The droplet uniformity is high and velocity low, making them excellent for respiratory applications.

Fine aerosols can also be generated at low flow rates by creating a potential force between a charged fluid and a ground plate creating a charged jet stream. The charged jet stream breaks into droplets when electrical stresses overcome the surface tension of the fluid. The final size of these droplets are a function of fluid properties (i.e., electrical conductivity, surface tension, density, viscosity, and dielectric constant), fluid flow rate, and applied voltage.

4.1.4.2   Degradation Behavior of Linear Genomic DNA

The degradation behavior of large molecules in atomizers can be correlated with turbulent length scales when it is the only mechanism acting on the system. Figure 4.1-9 demonstrates that when large DNA (7 µm) is atomized by various system; only those with very large turbulent length scale maintain structural integrity.

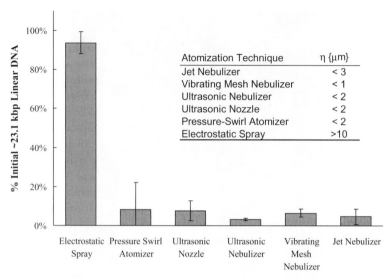

Figure 4.1-9. Degradation behavior of 23.1 kb genomic DNA after atomization. [23].

4.1.4.3 Degradation Behavior of Plasmid DNA

The structural degradation behavior of 5-, 10-, and 38-kbp plasmid/cosmid DNA in all atomizers presented so far has also been thoroughly studied. Lentz et al. [47] found that the jet nebulizer, vibrating mesh, and ultrasonic nebulizer are ineffective for all sizes. This is due to the extremely small turbulent length scales featured in the jet and vibrating mesh nebulization, and due to cavitation events nucleated at the DNA during ultrasonic nebulization. The ultrasonic nozzle is only suitable for very small molecular sizes based on length scales, whereas the pressure atomizer and electro spray seem to perform well over all size ranges.

4.1.4.4 Applications to siRNA Molecules

This chapter is focused on the effect of bioprocessing on proteins and plasmid DNA, but the use of small interfering RNA molecules to silence specific genes is gaining much attention and is currently being tested in clinical trials. Therefore, we thought it would be useful to briefly discuss how the principles described above for DNA might apply to small ($\approx$21 bp), double-stranded RNA molecules. Our laboratory recently undertook an investigation to determine whether these small biopolymers would be degraded during aerosolization. Previous work in DNA indicated that jet nebulization is one of the worst processing techniques; thus, the degradation behavior of a 21-bp siRNA with a calculated melting point of 46°C was subjected to aerosolization by a jet nebulizer (in contrast, plasmid DNA typically has melting temperatures above 100°C). Because these biopolymers are much too small to experience significant shear stress, they are more likely to denature (separation of the two strands in the double helix) than break a phosphodiester bond in their backbone. Degradation can be observed by using molecular probes that bind specifically to double-stranded oligonucleotides (not single-stranded) and

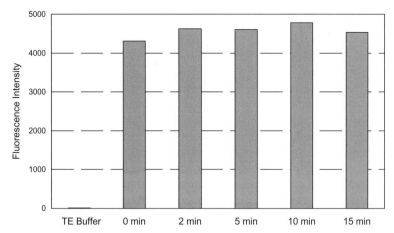

Figure 4.1-10. Changes in fluorescence intensity during nebulization of 21 bp siRNA with Pico Green fluorescent probes [Zelles-Hahn, Anchordoqury, and Lengsfeld, unpublished data].

fluoresce upon binding. In our experiments, we verified that Pico Green fluorescence was practically eliminated when the siRNA molecule was incubated above its melting point. Upon cooling to temperatures that allow annealing, Pico Green fluorescence was restored. When the siRNA was aerosolized for progressively longer times at room temperature, no change in fluorescence was observed (Figure 4.1-10). This indicates that even siRNA molecules with relatively low melting points remain double-stranded upon aerosolization with a jet nebulizer. Although it is possible that siRNA molecules with lower melting points might exhibit denaturation, sufficient stability can usually be achieved by increasing salinity to enhance the double-stranded form.

4.1.4.5   Degradation Behavior of Proteins

Work by Maa and Hsu [37] and Maa et al. [36] definitively demonstrated that protein degradation during atomization processing steps is due to the large interface to which the molecule readily diffuses (Figure 4.1-11). Here the application of a small shear stress is all that is needed to facilitate the unfolding of the amphipathic molecule to preferentially expose hydrophilic moieties to the gas phase (i.e., remove them from water). Thus, increased protein degradation is observed at smaller droplet sizes (Figure 4.1-11). In less-stable therapeutic proteins, we have seen batch-to-batch variability, whereas others have had success with using excipients, e.g., sugars and buffers. However, our group and others (Gomez et al. [48], Dunn-Rankin) have observed little to no degradation of protein during electrostatic spraying. This maybe because we all used relatively stable proteins or because our droplet sizes were not well controlled at sizes below 10 μm, but we have not experienced the severe difficulty found with numerous other techniques. Electrosprays are difficult to control with an aqueous solution, so if droplet uniformity and performance is needed over a broad range of conditions (or worse variable conditions), this is not the technique for you.

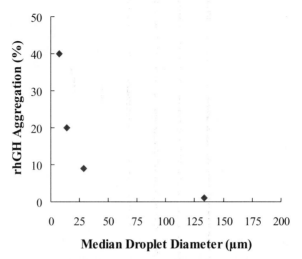

Figure 4.1-11. Data from Maa demonstrating protein aggregation of rhGH was linked to the surface-to-volume ratio of the liquid droplet produced during atomization [36, 37].

4.1.5 SEPARATION TECHNIQUES

The separation and purification steps in macromolecular bioprocessing typically cause the largest loss of product. The sources for molecular degradation presented above apply to these techniques, but it is informative to understand the classic approaches used today as well as methods under development for the future. To this end, the following sections provide a general overview of the subject followed by a brief review of molecular scale technologies currently under investigation.

4.1.5.1 Classic Approaches

Structural changes and contamination must be limited for macromolecular therapies to become a clinical reality. Current techniques for assessment, detection, and separation continue to struggle to meet industrial needs; a novel high-throughput, integrated system capable of qualifying and separating contamination from the bulk is highly desirable.

Conventional DNA purification techniques have been adapted from principles governing protein purification, but the copurification of *E. coli* chromosomal DNA with the desired plasmid presents a unique challenge. Cell lysis, differential precipitation, enzymatic digestion, column chromatography, desalting, buffer exchange, and sterile filtration represent the typical processing steps. The large size of genomic DNA makes it especially susceptible to shearing, which results in fragments that are similar in size to the desired plasmid. Due to their similar size and virtually identical chemical composition, it is difficult to separate the fragmented genomic DNA from the plasmid product. As a result, contemporary plasmid purification protocols sacrifice approximately 30–50% of the therapeutic plasmid in an attempt to minimize genomic DNA contamination. The most common large-scale purification method is column chromatography where separation is achieved largely based on molecule size using anion exchange, size exclusion, and reverse phase. Although a single pass is relatively effective for proteins (i.e., >95% purity), plasmid purification usually requires more than one of these columns to achieve acceptable levels of purity. In-process quality assessment is essential to biopharmaceutical development as the quality of the product is dependent on the initial stages of growth through every step of purification. Currently, in-process quality assurance tests are conducted off-line using electrophoresis and high-preformance liquid chromatography (HPLC) to assess plasmid concentration, hybridization to measure *E. coli* genomic DNA contamination levels, immunological assays for protein contaminations, and a standard *Limulus* amebocyte lysate assay to assess endotoxin levels. However, the major challenge is the detection of small quantities of bacterial genomic DNA in the large bulk quantity of plasmid DNA. This can be achieved with amplification techniques, but as with many of the other assessment techniques, this must be conducted off-line in a time-consuming manner. As a result, it is not determined whether batches are satisfactory until processing is completed. Clearly, it would be advantageous to develop techniques that allow process modification in real time to improve quality essential to sustaining the batch. For example, these sensors could be used to optimize production rates. High throughput usually requires higher flow velocities, which are known to fragment large macromolecules, but conservatively low velocities lead to long batch production times and revenue limitations. Real-time detection of contaminating DNA fragments could signal the operating system to reduce the energy per mass input to the system, lowering turbulence-induced fragmentation and thereby continuously optimizing throughput and product quality regardless of processing scale. A technique that probes individual molecules would provide an opportunity to overcome this multimillion-dollar problem. Because probing individual molecules requires a device with length scales at or below the molecular length scale, the answer must lie in a nanotechnology-based solution.

4.1.5.2 Novel Molecular-Scale Probing Concepts

Some work already exists in the area of using micron or submicron fluidic devices to enhance DNA separation techniques. Most of these techniques use electric fields or electrosmotic flow [49] to induce movement in these macromolecules. Electric field-induced diffusion through cylindrical or rectangular obstacles set at an angle of 45° to the field has demonstrated that small molecules move faster than large ones, making continuous sorting possible [50, 51]. Switching the electric field in 10-s intervals can provide enhanced separation rates by slowing the larger molecules to a greater degree [52, 53]. Instead of rectangular post arrays, as Han and Craighead [54] demonstrated, requiring the macromolecule to change conformation by altering the channel dimension from thick to thin, can create entropic traps that collect a particular size range. These confinement channels (post array limited or trap geometries) separate DNA fragments of a different size using the fact that as channel walls prohibit some molecular conformations as channel size approaches the molecular scale or smaller [55–57]. These limitations result in excess free energy (or an entropy gradient), which may impart a measurable force. For example, Turner et al. [55] report forces on the order of femto-Newtons. If entropy is proportional to molecular length, matrices can be designed such that only a specific length scale will have the necessary inherent force to move through the system. Yi and Lengsfeld [58] have built off these concepts and computationally explored the development of converging channels to alter the speed and compression of molecules at the molecular scale. Their new approach to analyzing molecular scale behavior is computationally efficient, and it will enable extension to circular and supercoiled DNA conformations where the excess energy generated by confinement might not only be governed by molecular length, but also by structurally restricted degrees of freedom inherent to the supercoiled and open circle conformation.

4.1.6   CHAPTER SUMMARY

Remember when dealing with proteins that (1) the small physical size of the molecule is below the continuum limit and thus the molecule is not susceptible to shear induced degradation; (2) structures that control activity are a result of relatively weak hydrophobic interactions and hydrogen bonds that result in an unstable active structure; and finally (3) proteins are surface active due to the amphipathic nature, and thus large surface-to-volume ratios should be avoided.

Double-stranded RNA molecules, although only discussed in a limited fashion, hold the best traits of DNA and proteins, i.e., siRNA's structure does not lend itself to high surface activity, and thus degradation by absorption to solid– or gas–liquid interfaces is not an issue. Furthermore, siRNA's small size is well below the continuum limit, and therefore, the molecule is not susceptible to shear-induced degradation.

When dealing with plasmid DNA, it is important to remember that the size straddles the continuum limit; thus, the molecule may be susceptible to hydrodynamic forces, especially turbulence. Reductions in power inputs to the fluid volume must be regulated to keep the Kolmogorov length scale larger than the molecule; alternatively, mitigation techniques using condensing cationic molecules must be used. Overall, the best-performing aerosolization devices are those with low residence times, such as pressure swirl devices, and low turbulence systems free of cavitation, e.g., electrostatic sprays.

REFERENCES

1. Hatti-Kaul R, Mattiasson B, eds. (2003). Isolation and Purification of Protiens. 1st ed.: Marcel Dekker Inc, New York, pp. ix–xvii.
2. Prazeres D M, Ferreira G N, Monteiro G A, et al. (1999). Large-scale production of pharmaceutical-grade plasmid DNA for gene therapy: problems and bottlenecks. *Trends in Biotechnol.* 17:169–174.
3. Durland R H, Eastman E M (1998). Manufacturing and quality control of plasmid-based gene expression systems. *Adv. Drug. Deliv. Rev.* 30:33–48.
4. Sanghvi Y S, Schulte M (2004). Therapeutic oligonucleotides: The state-of-the-art in purification technologies. *Curr. Opin. Drug. Discov. Devel.* 7(6):765–776.
5. Lindahl T (1993). Instability and decay of the primary structure of DNA. *Nature.* 362:709–715.
6. Lindahl T, Andersson A (1972). Rate of chain breakage at apurinic sites in double-stranded deoxyribonucleic acid. *Biochem.* 11:3618–3623.

7. Lindahl T, Nyberg B (1972). Rate of depurination of native deoxyribonucleic acid. *Biochem.* 11:3610–3618.

8. Evans R K, Xu Z, Bohannon K E, et al. (2002). Evaluation of degradation pathways for plasmid DNA in pharmaceutical formulations via accelerated stability studies. *J. Pharmac. Sci.* 89:76–87.

9. Chi E Y, Krishnan S, Randolph T W, et al. (2003). Physical stability of proteins in aqueous solution: Mechanism and driving forces in nonnative protein aggregation. *Pharmac. Res.* 20(9):1325–1336.

10. Tservistas M, Levy M S, Lo-Yim M Y A, et al. (2001). The formation of plasmid DNA loaded pharmaceutical powders using supercritical fluid technology. *Biotechnol. Bioengin.* 72:12–18.

11. Breen A P, Murphy J A (1995). Reactions of oxyl radicals with DNA. *Free Radical Biol. Med.* 18:1033–1077.

12. Graf E, Mahoney J R, Bryant R G, et al. (1984). Iron-catalyzed hydroxyl radical formation. Stringent requirement for free iron coordination site. *J. Biol. Chem.* 259:3620–3624.

13. Keyer K, Imlay J A (1996). Superoxide accelerates DNA damage by elevating free-iron levels, *Proc. of the National Academy of Sciences USA.* 93:13635–13640.

14. Luo Y, Han Z, Chin S M, et al. (1994). Three chemically distinct types of oxidants formed by iron-mediated Fenton reactions in the presence of DNA, *Proc. of the National Academy of Sciences USA.* 91:12438–12442.

15. Middaugh C R, Evans R K, Montgomery D L, et al. (1999). Analysis of plasmid DNA from a pharmaceutical perspective. *J. Pharmac. Sci.* 87:130–146.

16. Cowan J A (1998). Metal activation of enzymes in nucleic acid biochemistry. *Chem. Rev.* 98:1067–1087.

17. Frenkel J (1994). Orientation and rupture of linear macromolecules in dilute solutions under the influence of viscous flow. *Acta Physicochim. USSR.* 19:51–76.

18. Davison P F (1959). The effect of hydrodynamic shear on the deoxyribonucleic acid from T2 and T4 bacteriophages, *Proc. of the National Academy of Science.* 45:1560–1568.

19. Levinthal C, Davison P F (1961). Degradation of deoxyribonucleic acid hydrodynamic shearing forces. *J. Molec. Biol.* 3:674–683.

20. Bowman R D, Davidson N (1972). Hydrodynamic shear breakage of DNA. *Biopolymers.* 11:2601–2624.

21. Lengsfeld C S, Anchordoquy T J (2002). Shear-induced degradation of plasmid DNA. *J. Pharmac. Sci.* 91:1581–1589.

22. Yim S S, Shamlou P A (2000). The engineering effects of fluids flow on freely suspended biological macro-materials and macromolecules. Adv. Biochem. *Engin./Biotechnol.* 67:83–122.

23. Worden L (2005). Degradation behavior of genomic DNA during bioprocessing. Masters Thesis, University of Denver, Denver, C.O., pp. 69–73.

24. De Gennes P G (1974). Coil-stretch transition of dilute flexible polymers under ultra-high velocity gradients. *J. Chem. Phys.* 60(12):5030–5042.

25. Larson R G, Magda J J (1989). Coil-stretch transitions in mixed shear and extensional flows of dilute polymer solutions. *Macromo.* 22(7):3004–3010.

26. Perkins T T, Smith D E, Chu S (1977). Single polymer dynamics in an elongation flow. *Science.* 276:2016–2021.

27. Smith D E, Chu S (1998). Response of flexible polymers to a sudden elongation flow. *Science.* 281:1335–1340.

28. LeDuc P, Haber C, Bao G, et al. (1999). Dynamics of individual flexible polymers in a shear flow. *Nature.* 399:564–566.

29. Hur J S, Shaqfeh E S G, Babcock H P, et al. (2002). Dynamics and configurational fluctuations of single DNA molecules in linear mixed flows. *Phys. Rev. E.* 66(1):Pt 1, 011915.

30. Lumley J L (1973). Drag reduction in turbulent flow by polymer additives. *J. Polymer Sci.* Part D: Macromolecular Review, 7(1):263–290.

31. Balkovsky E, Fouxon A, Lebedev V (2001). Turbulence of polymer solutions. *Phys. Rev. E.* 64:056301:1–14.

32. Thiffeault J (2003). Finite extension of polymers in turbulent flow. *Phys. Lett.* A:308:445–450.

33. Kundu P K, Cohen I M (2002). Fluid Mechanics, 2nd ed.: Academic Press, San Diego, C.A., pp. 438–440.

34. Lentz Y, Anchordoquy T A, Lengsfeld C S (2006). DNA acts as a nucleation site for transient cavitation in ultrasonic nebulizers. *J. Pharmac. Sci.* 95(3):607–619.

35. Levy M S, O'Kennedy R D, Ayazi-Shamlou P, et al. (2000). Biochemical engineering approaches to the challenges of producing pure plasmid DNA. *Trends in Biotechnol.* 18:296–305.

36. Maa Y F, Nguyen P A, Hsu S W (1998). Spray-drying of air-liquid interface sensitive recombinant human growth hormone. *J. Pharmac. Sci.* 87:152–159.

37. Maa Y F, Hsu C C (1997). Protein denaturation by combined effect of shear and air-liquid interface. *Biotechnol. Bioengin.* 54:503–512.

38. Levy M S, Collins I J, Yim S S, et al. (1999). Effect of shear on plasmid DNA in solution. *Bioproc. Engin.* 20:7–13.

39. Imlay J A, Linn S (1988). DNA damage and oxygen radical toxicity. *Science.* 240:1302–1309.

40. Lentz Y, Worden L, Anchordoquy T A, et al. (2005). Impact of jet nebulization of plasmid DNA: degradation mechanism and mitigation methods. *J. Aerosol Sci.* 36:973–990.

41. Malmsten M ed. (2003). Biopolymers at Interfaces, 2nd ed.: Marcel Dekker Inc, New York.

42. Chou D K, Krishnamurthy R, Randolph T W, et al. (2005). Effects of Tween 20 and Tween 80 on the stability of albutropin during agitation. *J. Pharmac. Sci.* 94(6): 1368–1381.

43. Kim Y S, Jones L S, Dong A, et al. (2003). Effects of sucrose on conformational equilibria and fluctuations within the native-state ensemble of proteins. *Protein Sci.* 12:1251–1261.

44. Kendrick B S, Carpenter J P, Cleland J L, et al. (1998). A transient expansion of the native state precedes aggregation of recombinant human interferon-γ, *Proc. of the National Academy of Science USA.* 95:14142–14146.

45. Teare D O H, Barwick D C, Schofield W C E (2005). Functionalization of Solid Surfaces with Thermoresponsive Protein-Resistant Films. *J. Phys. Chem. B.* 109:22407–22412.

46. DeFife K M, Hagen K M, Clapper D L, et al. (1999). Photochemically immobilized polymer coatings: effects on protein adsorption, cell adhesion, and leukocyte activation. Journal of Biomaterials Science. *Polymer Edition.* 10(10):1063–1074.

47. Lentz Y, Anchordoquy T J, Lengsfeld C S (2006). Impact of Aerosolization Devices on plasmid DNA. *J. Aerosol Med.* In Press.

48. Gomez A, Bingham D, de Juan L, et al. (1998). Production of protein nanoparticles by electrospray drying. *J. Aerosol Sci.* 29(5–6):561–574.

49. Huang M F, Kuo Y C, Huang C C, et al. (2004). Separation of long double-stranded DNA by nanoparticle filled capillary electrophoresis. Anal. Chem. 76(1):192–196.

50. Chou C F, Bakajin O, Turner S W, et al. (1999). Sorting by diffusion: An asymmetric obstacle course for continuous molecular separation. *Proc. Natl. Acad. Sci. USA.* 96(24):13762–13765.

51. Kaji N, Tezuka Y, Takamura Y, et al. (2004). Separation of long DNA molecules by quartz nanopillar chips under a direct current electric field. *Anal. Chem.* 76(1): 15–22.

52. Cole K D, Tellez C M (2002). Separation of large circular DNA by electrophoresis in agarose gels. *Biotechnol. Prog.* 18(1):82–87.

53. Bakajin O, Duke T A I, Tegenfeldt J, et al. (2001). Separation of 100-kilobase DNA molecules in 10 seconds. *Anal. Chem.* 73(24):6053–6056.

54. Han J, Craighead H G (2000). Separation of long DNA molecules in a microfabricated entropic trap array. *Science.* 288(5468):1026–1029.

55. Turner S W P, Cabodi M, Craighead H G (2002). Confinement-induced entropic recoil of single DNA molecules in nanofluidic structure. *Phys. Rev. Let.* 88(12):128103.

56. Cabodi M, Turner S W P, Craighead H G (2002). Entropic recoil separation of long DNA molecules. *Anal. Chem.* 74:5169–5174.

57. Foquet M, Korlach J, Zipfel W, et al. (2002). DNA fragment sizing by single molecule detection in submicrometer sized closed fluidic channels. *Anal. Chem.* 74:1415–1422.

58. Yi Y B, Lengsfeld C S (2006). Mechanical modeling of a CA molecule motion in simple channels. *J. Applied Phy.* In Press.

4.2

PROCESS VALIDATION FOR BIOPHARMACEUTICALS

BETH JUNKER

Merck Research Laboratories, Rahway, New Jersey

Chapter Contents

4.2.1 Biopharmaceuticals 320

4.2.2 Process Validation Overview 320

    4.2.2.1 Definitions 320

    4.2.2.2 Purpose and Benefits 321

    4.2.2.3 Scope 322

    4.2.2.4 Regulatory 323

4.2.3 Evaluating Measures of Product Quality and Consistent Process Performance 326

    4.2.3.1 Critical Process Input Variables/Parameters 326

    4.2.3.2 Noncritical Process Input Variables/Parameters 329

    4.2.3.3 Process Control/Input Parameters (Xs) 329

    4.2.3.4 Performance Uncontrolled Output/Variables/Attributes (Ys) 330

4.2.4 Background and Requirements 331

    4.2.4.1 Types of Process Validation 331

    4.2.4.2 Aspects of Process Validations 332

    4.2.4.3 Process Validation Execution Timing 339

4.2.5 Development Support and Technology Transfer Activities 341

    4.2.5.1 Process Development 341

    4.2.5.2 Assessment of Process Risks 344

    4.2.5.3 Validation of Assays 347

    4.2.5.4 Change Control and Ongoing Process Validation 350

    4.2.5.5 Process Validation Documentation 351

4.2.6 Process Validation Strategies 352

    4.2.6.1 Historical Data Review 352

    4.2.6.2 Statistical Design of Experiments (DOE) and Analysis for Process Characterization 355

    4.2.6.3 Family/Matrix Validation and Platform Technologies 356

Handbook of Pharmaceutical Biotechnology, Edited by Shayne Cox Gad.
Copyright © 2007 John Wiley & Sons, Inc.

	4.2.6.4 Scale-Down Models	357
	4.2.6.5 Well-characterized Biotechnology Products	359
4.2.7	Future Trends	359
	4.2.7.1 Process Analytical Technologies (PAT)	360
	4.2.7.2 Process Reliability	361
	4.2.7.3 Biogenerics	362
	4.2.7.4 Outsourcing	363
	4.2.7.5 Risk-Based GMPs	363
4.2.8	Summary	364
	Nomenclature	364
	References	365

4.2.1   BIOPHARMACEUTICALS

Biopharmaceuticals are drugs, vaccines, and pharmaceutical agents (typically therapeutic proteins or polypeptides) developed or produced using techniques of biotechnology [1]. In many cases they are complex proteins exhibiting varying levels of heterogeneity [2], derived from recombinant gene expres-sion or hybridoma production systems [3]. These products of biotechnology generally are classified as biologicals by regulatory agencies [4]. Specific categories of products include monoclonal antibodies and other recombinant proteins produced in cell culture, enzymes, hormones, soluble receptors, growth factors, imunogens, blood and plasma products, and animal-derived products [3, 5]. Biopharmaceuticals range in molecular weight from a low value of 22,000 for human growth hormone to about 150,000 for an intact immunoglobulin [6]. These agents are among the most expensive products to manufacture, with typical inventory costs of hundreds to thousands of dollars per gram [3]. There is also intense pressure to introduce a product to the market in a timely manner [2, 7] to recover research and facil-ity investments. Conse-quently, the reliable production of quality biopharmaceuticals is of the highest importance, translating into an intense focus on the creation of an effective and economical production process. Process validation is a key mechanism to success-fully accomplish the task of process creation.

4.2.2   PROCESS VALIDATION OVERVIEW

4.2.2.1   Definitions

Validation derives from the Latin word *valere* meaning "to have the power" [8], presumably referring to the power to prevent an unexpected and undesirable occurrence. It establishes that the process used is reliable and reproducible [9]. The level of control implemented is supported by scientific knowledge of the culture and product, along with equipment capabilities [10]. The U.S. Food and Drug Administration (FDA) definition of process validation (PV) is documented evidence to provide a high degree of assurance that a specific process consistently produces a product meeting predetermined specifications and quality characteristics [11]. The International Conference on Harmonization (ICH) definition is similar, stating that process validation is documented evidence that the process, when operated within established parameters, performs reproducibly and effectively to produce an intermediate or active pharmaceutical ingredient (API) meeting predetermined specifications and quality attributes [11]. Most simply stated, process validation provides a high degree of assurance that process will routinely achieve the stated goals [12]. It accomplishes this goal by verifying product and process specifications to determine their weaknesses and reproducibility [13] and by demonstrating at which steps contaminants are removed and the product stream achieves acceptance criteria [14]. A verification of process consistency is conducted, highlighting the ability to produce essentially the same product (with the same safety and efficacy) over time, despite normal variability in operating conditions [12]. Consistency of multiple batches is demonstrated at full scale, with specific focus on the process operating in a consistent manner, reproducible removal of contaminants to acceptable levels, and monitoring parameters that demonstrate this consistency [14].

Process validation for biopharmaceuticals is somewhat different in application from traditional pharmaceuticals because the technology is different, but the theory and principles are similar [15]. Biotechnology quality control is defined by the individual production process and the product itself [16] because quality cannot be tested into the product [4]. The manner in which the biopharmaceutical is produced is part of its description; thus, biopharmaceuticals are characterized by chemical structure as well as by operational definition [17]. Few doubt the need for process validation, but there is disagreement and confusion about what needs to be validated and how to perform it [18]. Operational definitions of process validation have been developed to suit the philosophies of individual companies [3]. Thus, there is increased interest in developing industry guidelines for the design and performance of validation studies [19].

The term *validation* has different implications under different circumstances, and thus, its use can be confusing [19]. Furthermore, the term *validation* has been used interchangeably with other terms such as *performance qualification* and *process characterization*, when discussing process as well as equipment testing. Recognizing that individual firms distinguish among these terms differently, there has not been an attempt in this chapter to create strict definitions that artificially introduce restrictions where none exist currently. Rather, the terminology employed has remained flexible regardless of the specific terminology used in the source material, such that it reads as internally consistent. In addition, the focus of this chapter primarily is on process validation for steps required to produce bulk biological substances, but the principles discussed also can be applied to the final processing steps involved in formulation and packaging.

4.2.2.2 Purpose and Benefits

Validation has been a critical success factor in product approval and ongoing commercialization [14]. It serves as a formal link of the process to the product [10], focusing on quality but often concurrently addressing critical business concerns. Each Good Manufacturing Practice (GMP) step must be controlled to maximize the probability that the finished product meets all quality and design specifications [14]. In addition, each step must be evaluated for its function and effect on product identity, purity, potency, safety, or stability [12, 20]. An unvalidatable process is unacceptable and unreliable from both regulatory and manufacturing (i.e., business) perspectives [12].

There are two goals of process validation: (1) to support the safety, efficacy, and quality of the product; and (2) to identify major sources of process variability so appropriate controls can be implemented to provide consistency [10]. Achievement of these goals demonstrates that potential risks have been adequately reduced and assures products consistently possess the required established quality attributes [12]. Weaker points of the manufacturing process are uncovered and strengthened during the validation effort [12]. Thus, process validation deepens process understanding, which in turn decreases the risk of processing problems/failures, defect costs, lot rejections, reworking/reprocessing, product shortages, or regulatory noncompliance [3, 21]. Assurance of process consistency maximizes productivity and increases cost efficiency. It influences process economics in many ways; one example is through the reduction of in-process control and endproduct testing [21] by reducing the number and scope of quality control tests per lot [12].

4.2.2.3 Scope

Process validation is the culmination of all qualification and validation activities, representing process testing in its entirety [22]. It is a dynamic process varying in depth according to the product life-cycle timeline, spanning from initial process design through ongoing commercial operation [14]. Its execution can be envisioned as creating a "design space," a multidimensional space that defines how different variables interact [23]. This design space (1) establishes the process ranges demonstrated to provide quality assurance, and (2) defines areas in which process adjustments and refinements may be undertaken without regulatory involvement. The dimensions of this design space are based on the scientific understanding of critical process parameters that has been created through development studies [23, 24].

For a product, the overall list of subcomponent systems to be validated includes equipment, facilities, utilities, computers (including software and controls), processing environment, analytical methods, and operating procedures (e.g., cleaning and sterilization), as well as the process itself [14, 25]. Before process validation begins, these subcomponent systems are qualified [11]. As process validation rests on the fidelity of the qualification of the subcomponent systems required for process operation, the undetected shortcomings of these efforts can compromise subsequent process validation [10]. As with equipment qualification, process validation is most effective when conducted in combination with effective process and facility design efforts [12]. Also, as with equipment qualification, predefined acceptance criteria and action plans in case of test failure are devised [26]. The use of predetermined specifications serves to reduce the goals of a validation program to actual practice [12], assisting in implementation.

Process validation for the production process is only one means to assure product consistency. Other factors include quality control for the final product, in-process assays, raw material controls, appropriately designed facilities, personnel monitoring and training, and environmental monitoring [4, 10, 27]. Additional factors that become concerns for multiproduct biopharmaceutical facilities include (1) segregation of materials, personnel, and equipment; (2) cleaning and product changeover; and (3) product cross-contamination [4]. Characteristics of biopharmaceutical processing that generate specific areas of quality control concern are large, complex product molecules, complicated manufacturing systems, long production times, and pooling of fermentation lots before isolation [4]. Other product quality concerns include genetic stability, product yield and stability, host and nonhost cell component contamination, and posttranslational processing, including glycosylation and folding [16]. As biopharmaceutical manufacturing processes are complicated with multiple steps and extended production times, adequate in-process testing, thorough process validation, and appropriate measures to eliminate adventitious agents are necessary [28].

4.2.2.4 Regulatory

The completion of process validation represents the point at which the science of the process is ready to be explained to regulatory agencies [14]. Process validation data are initially presented to the regulatory agency during application submission [11]. These data set the bar by which the process is to be judged during future inspections, and consequently, one of most common FDA 483 observations is the lack of process validation [14]. As manufacturing is a key factor in achieving the required quality and safety of biotechnology products [27], process validation is a regulatory requirement for product licensure, and limited process validation is considered essential even for product destined for clinical trials [3]. Process validation is required in license submissions for all products regulated by either CBER or CDER [14]. There are added challenges for biopharmaceuticals, however, as process consistency often replaces detailed product analysis as a key method of quality assurance, and biologically based production systems contribute biological contaminants that pose safety risks [29].

It has been stated that validation should be written as if the FDA were the customer [14], and this approach still is valid because worldwide harmonization efforts are well underway [24]. Several guidance documents have been published in draft and finalized forms from various regulatory agencies and industry associations worldwide. Key documents are listed in Table 4.2-1. New guidance, as well as interpretations of prior guidance, is continuously published, suggesting that there is not yet one consistent approach to process validation. Additional guidance leading to the improved definition of process validation requirements and reduction in marginally helpful studies is welcomed by industry [3]. Several comprehensive publications describing the interpretation and application of process validation by industry are available, but terminology and definitions have become blurred and overlapping. A summary of various terms used is shown in Table 4.2-2. Consequently, the use of the term selected must be fully and consistently understood by authors, reviewers, approvers, executors, and auditors in the context of its specific application.

Pitfalls. Validation (consisting of process, equipment, and facility systems) is a costly and time-consuming investment with more resources believed devoted to its

TABLE 4.2-1. List of Selected Relevant Guidance Documents

Topic	Year of Publication	Authoring Organization	Title
Process validation	1987, revised 1996	FDA	Guideline on the general principles for process validation
	2005 (draft)	PDA	Process validation of protein manufacturing
Clinical materials	2001	EMEA	EU GMP Guide: Manufacture of investigational medicinal products
Inspections/ manufacturing	1985	FDA	Points to consider in the production and testing of new drugs and biologicals produced by recombinant DNA technology
	1991	FDA	Biotechnology inspection guide
	1987, 1997	FDA	Points to consider in manufacture and testing of monoclonal antibody products for human use
	2000, 2003	ICH	Q7A Good Manufacturing Practice guidance for active pharmaceutical ingredients
	2005 (draft)	ICH	Q8, Pharmaceutical development
Cell substrates	1997	ICH	Derivation and characterization of cell substrates used for production of biotechnological/ biological products
	1998	ICH	Q5A Viral safety evaluation of biotechnology products derived from cell lines of human or animal origin
	1985, 1993	FDA	Points to consider in the characterization of cell lines used to produce biologicals
Assay validation	1995	ICH	Guideline on validation of analytical procedures
	1996	ICH	Q2B Validation of analytical procedures: Methodology
	1999	ICH	Specifications: Test procedures and acceptance criteria for biotechnology/biological products
	2001	FDA	Guidance for Industry— Bioanalytical method validation
PAT	2003 (draft), 2004 (final)	FDA	Guidance for Industry—PAT— A framework for innovative pharmaceutical manufacture and quality assurance
Risk	2003	WHO	Application of hazard analysis and critical control point (HACCP) methodology to pharmaceuticals
	2005 (draft)	ICH	Q9, Risk management guidance

TABLE 4.2-2. Selected Process Validation Terminology

Aspect	Classification Level		
	A	B	C
Process	Validation	Variability	Definition
	(Performance) qualification	Reproducibility	Understanding
	Characterization	Optimization	Capability
	Robustness	Stability	
	Consistency		
Product	Biologicals	API	Protein
	Biopharmaceuticals	Bulk	
	Biotechnology products	Drug substance	
	Recombinant DNA products		
Scale	Manufacturing	Clinical	Small scale
	Production	Development	Laboratory scale
	Commercial	Pilot scale	
	Marketable		
	Full scale		
	Large scale		
Exercise	Test	Protocol	Variable
	Study	Procedure	Parameter
	Run	Plan	Factors
	Experiment	Document	Inputs/outputs
	Evaluation		
	Assessment		
	Investigation		
Result	Quality	Change	Fault
	Specifications	Amendment	Failure
	Requirements	Corrective	Hazard
	Standards	action	Nonconformance
	Acceptance criteria		Excursion
			Deviation
			Atypical

completion than to the initial process development and production of clinical supplies [3]. The actual cost of process validation is thought by some to be uncontrolled and excessive because there is no consensus on how much is enough [30]. One key risk associated with biopharmaceutical product manufacturing is the failure to gather sufficient information to support process decisions when faced with issues arising in process validation, failure investigations, product specifications, analytical methods, and equipment qualification [30].

Common pitfalls associated with process validation are numerous. They include a lack of overall strategy, insufficient planning/management, delay in early consultation with regulatory agencies, inadequate product definition, failure to observe current GMPs (cGMPs) and follow standard operating procedures (SOPs), poorly defined cell bank genealogy, inadequate analytical procedures, confusion over participant responsibilities, excessive process changes after process validation completion, insufficient prior process characterization or qualification, starting process validation too early and before process finalization, poor validation study design/

definition, disagreement on evaluation/interpretation of results, and inappropriate acceptance criteria [8, 14]. When overall validation activities begin too late, the work required to validate methods, facilities, and/or processes typically has been underestimated [4], placing product launch at risk. When processes perform erratically during process validation, further investigation to uncover inadequate development or lack of control of one or more process variables is required, followed by additional validation studies [12].

Validation must be science based [31]. Sound science during process validation is the primary vehicle to reduce manufacturing costs while remaining compliant with regulatory expectations [30]. Overly complex plans contradict desires to be efficient, lean, and cost-effective. More intensely planned projects produce more efficient, lower cost, and higher quality products in a faster time frame [30]. Resulting risks for biopharmaceutical companies from ineffective process validation include patient (safety and efficacy), operational (safety, contamination, variability), financial (product loss, reputation, legal costs), and regulatory (FDA 483s, warning letters, recalls) factors [11].

4.2.3 EVALUATING MEASURES OF PRODUCT QUALITY AND CONSISTENT PROCESS PERFORMANCE

A streamlined, integrated, and comprehensive validation process is desired [32]. The key to achieving a manageable process validation design is determining which process variables can be relegated to a noncritical status or maintained under tight control. Only when absolutely necessary should parameter ranges be validated. Thus development studies are required to identify critical parameters, limits to their operating ranges, and those process characteristics and equipment that enable their tight control [32]. It is necessary to distinguish between measures of product quality and process performance (consistency) and to differentiate between release specifications (such as critical quality attributes, CQA) that are measures of product quality and process attributes that demonstrate consistent process performance. Validation studies are required to examine process parameter variations that cause yield losses by affecting product quality (i.e., critical parameters); they are not needed for process parameter variations that cause yield losses that do not affect product quality even though these losses can cause economic/business concerns [10]. A comparison of process validation terminology for variables is shown by Table 4.2-3. A recent document has attempted to clarify these definitions further [33].

4.2.3.1 Critical Process Input Variables/Parameters

The basis of all process validation studies is the demonstration of control of critical process input parameters for reproducible operation of a commercial-scale production process [11, 32]. Critical parameters vary among processes so they must be separately assessed for each new process [10]. As a typical process has hundreds of variables, critical process input parameters must be identified first, and then the extended range and normal (target) operating range for each critical parameter must be determined during process development and/or using historical data [11,

TABLE 4.2-3. Comparison of Process Validation Terminology for Variables (bold entries according to Ref. 33)

Term	Criticality		Process Position		Control	
	Critical	Noncritical	Input	Output	Controlled	Uncontrolled
Operational parameters	X	X	X		X	
Input variables/ parameters	X	X	X		X	X
Output variables/ parameters	X	X		X		X
Control parameters	X	X	X		X	
Critical variables/ parameters	X		X		X	
Critical control points, CCP	X		X		X	
Key process input variables, KPIV	X		X		X	
Key/non-key operational parameter		X	X		X	
Xs (dependent)	X	X	X		X	
Ys (independent)	X	X		X		X
Quality/performance attributes, CQA	X			X		X
Acceptance criteria	X			X		X
Performance parameters	X	X		X		X
Process attributes		X		X		X

34]. Critical variables are identified using defined procedures combining elements of data analysis and scientific judgment and making maximum use of available data gathered during process development [20]. Various companies have developed differing decision matrices to determine critical parameters, some of which are based on formal risk analysis exercises. A critical variable is defined as a parameter whose operating range lies near the edge of failure [34]. Using this definition, in practice, most identified critical variables can be engineered out of the study. Critical steps are those steps that are difficult to control because they usually contain at least one or more critical variables that cannot be removed [34].

Critical control points (CCPs) are those locations within a process where control can be applied to protect product quality, specifically to prevent and eliminate quality hazards or to reduce them to acceptable levels [11]. For controlling hazards, the appropriate CCP may be positioned in the sub-sequent step from the actual hazard location. In addition, more than one CCP can be used to control a single hazard and more than one hazard can be controlled by a single CCP [11].

Critical parameters are studied next to determine the effects of their outer limit values on important process characteristics [10]. These critical outer limits must be attainable, accurate, robust, and scientifically based [11]. The control parameter range is the span of values that lies between two outer limits or control levels of a parameter, and it represents the highest and/or lowest values of a given control parameter that actually are evaluated during validation [25]. It is not usually necessary to test both ends of the control parameter range for all parameters because scientific judgment can help determine which end of the range (high or low) is more likely to generate an adverse effect [10]. In other cases, testing of both the upper and the lower control limits for a large number of variables is required to identify worst-case conditions [35]. Acceptable ranges are narrowed, and experiments are repeated until the acceptance criteria are met using an iterative process that requires substantial execution [3]. The edge of failure is reached when an exceeded control parameter value adversely affects the process or the product in that process performance degrades or product fails [20]; however, it is not necessary to validate the edge of failure when validating a process [25].

One strategy is to select worst-case conditions that encompass the upper and lower processing limits of critical input parameters, along with situations and circumstances over the entire process (rather than only a single step) that together result in the greatest chance of process or product failure compared with the ideal or target conditions [31, 34]. Worst-case runs are long and complex because the entire process must be run completely, usually at large scale or full scale [10, 20]. Because the worst-case approach comprehensively in-cludes the direct effects and actions of all variables, if all experiments are successful, only a few studies are required [20]. To mitigate the high likelihood of failure of a worst-case study that results in repeating lengthy runs using revised conditions, worst-case runs are performed after factorial experiments, usually conducted at a smaller scale and of a narrower scope, have confirmed a satisfactory outcome under the proposed worst-case conditions [20]. A key philosophical weak point of the worst-case run strategy is that it is less probable in practice that all critical parameters experience simultaneous excursions to outer limits [10]. However, because all critical parameters are tested at their extreme range, the worst-case approach can identify interactions that design of experiment (DOE) methods cannot detect [10].

Regardless of the strategy employed, worst-case conditions for critical parameters are established and process validation assures that the process yields an acceptable product when challenged under these worst-case conditions [30], even though full-scale validation runs may not have been conducted at many of these worst-case limits. Then operating ranges and control limits are set to establish a process robust enough to produce acceptable product [14, 35], and full-scale validation runs are executed to demonstrate performance consistency. The assumption is that if the product is of acceptable quality when parameters are controlled at the outer limits of control ranges, then the quality also will be acceptable when these parameters are held within those control ranges. This strategy is analogous to the approach of using fractional cycles for validation and full cycles for actual operation when testing at cleaning and sterilization procedures [36].

For a controlled variable, several types of ranges exist around the set point (Table 4.2-4). The normal operating range (NOR), or the alert limit, is established by trending performance during clinical material or qualification lot production, and then assigning usually two or three standard deviations to the set point [20]. The proven acceptable range (PAR) encompasses all values of a given control parameter that fall between proven high and low worst-case conditions [25] and constitutes the proposed operating range [3]. The PAR should be greater than or equal to the MOR [3]. The maximum operating range (MOR), or the action limit,

TABLE 4.2-4. Ranges of Controlled Variables/Parameters in Order of Increasing Breadth

Controlled Variable/Parameter	Definition
Set point	Target
Normal operating range (NOR)	Alert limit, variations typically +/− 2 or 3 SDs from set point
Maximum operating range (MOR)	Action limit, product quality acceptable
Proven acceptable range (PAR)	Upper/lower worst-case conditions; outer or control limits/control parameter range; characterization, tolerance, or design space range; product quality acceptable
Limit/edge of failure	Unacceptable product quality, critical limit

is the range of the controlled variable within which product quality is acceptable, but not necessarily at the limit of failure; thus, if the variable value falls outside the MOR, it is not necessarily a failure [20]. As MORs can lead to a decline in process performance even if product quality remains acceptable, their frequent occurrence can become an economic or operational issue. The MOR should be sufficiently broader than the NOR to reduce the frequency of unplanned deviations [20]. The MOR can be extended based on manufacturing experience when excursions outside the MOR occur without adverse quality effects [20]. Overall, a demonstrated state of control is the condition when all operating variables that can affect performance remain within such ranges that performance is consistent and as intended [25].

4.2.3.2 Noncritical Process Input Variables/Parameters

Noncritical or operational parameters are process control set points that define the process recipe and show that the process can be executed consistently [14], but for which there is no established evidence of product quality impact [11]. If the process performance is consistent when noncritical variables are controlled within their ranges and if minor deviations (e.g., instrumentation inaccuracy) from these ranges do not affect product quality, then the variable or step is classified as noncritical [34]. Typical noncritical control ranges are wide and/or the variable/parameter is easy to control within the range [33]. Although noncritical parameters do not require further study [10], they also are controlled and monitored during process validation to reduce variability and operator error [11]. Recently, noncritical process parameters have been designated as key (ensuring operational reliability and desired process performance when maintained within a narrow range, significant impact but not failure within range) and nonkey (well-controlled within a wide range but a potential process or quality impact when outside range, no significant impact within range) [33].

4.2.3.3 Process Control/Input Parameters (Xs)

Process control parameters are operating variables that have assigned values to be used for control [25]. They usually are readily maintainable away from a point of failure and within a tested manufacturing range by a feedback control mechanism [15, 35]. Control parameters may also be thought of interchangeably as input parameters, defined as operational parameters that can be controlled [32]. Control/input variables are those operating parameters with set points or ranges that define process execution, e.g., pH, flowrate, temperature, raw material amounts and specifications, properties, and hold times for in-process streams of prior steps [3, 14]; thus, these parameters can be validated as a range or as set points [32]. They are a subset of operating variables, all factors that potentially affect process control or product quality [25], that reasonably are expected to affect operation of a process step [15]. Control/input parameters also are referred to as Xs, analogous to independent variables [37].

All input parameters must be identified and demonstrated to be in control for the product to be in a validated state and the more tightly the range is controlled, the less critical the parameter is considered [32]. Validation steps are (1) identification; (2) classification as critical or noncritical, subclassification of critical parameters into those to be validated over a broad range and those to be validated as a set point with a narrow tolerance range; (3) validation execution for those critical input parameters with a broad range; and (4) validation execution using target set points for all process input parameters both critical and noncritical. The complexity of process validation is dictated by several critical process input parameters requiring either a broad range or a broad set point. As fewer range testing studies are required if more parameters can be tightly controlled, one strategy is to design tight control set points for as many critical process input parameters as possible (e.g., pH and temperature) [32].

4.2.3.4 Performance Uncontrolled Output Variables/Attributes (Ys)

Performance parameters reflect the outcome of a given step and indicate that the process gave the desired result [14] or quality attribute. They are uncontrolled performance variables [15] without a control action [35]. Their natural variation is defined by operating history; specifically, their variability is characterized from known historical data or estimated based on similar process performance [35]. Similarly, output variables reflect the step outcome and indicate performance was acceptable in terms of performance attributes for the step (e.g., titer and yield) or properties of the product stream (e.g., product homogeneity, purity, contaminant levels, and chromatography peak shape) [3, 14]. Still another term used is critical Ys (analogous to dependent variables), defined as product and process output variables that relate to critical quality attributes (CQAs), which are measurable outputs of each process step that are used to provide evidence that the step performed correctly [37].

An input parameter value is directly related to an output parameter or quality attribute; specifically, the output parameter of one unit operation often is the input parameter for the subsequent unit operation [32]. Equivalently stated, critical Ys (outputs or responses) from one stage often are inputs for next the stage [37]. Input parameters are assessed as to whether they significantly impact critical output parameters (e.g., product safety, identity, and efficacy), although the definition of significant varies among applications [32]. As all product quality attributes are not routinely measured during process development, some output variables are selected based on a hypothesis that they are linked to desirable product characteristics [20]. An input parameter is determined as critical based on whether its operating point is located near the edge of failure and how well it can be controlled [32]. Input parameters then are prioritized based on the (1) output sensitivity to input variations, (2) step proximity to the final product, and (3) technical difficulty controlling the input within the specified range [32].

A more formal approach involves listing all key process input variables (KPIVs), factors [both process (physical) and product (chemistry)] that could potentially affect the mean and standard deviation of critical Ys (outputs) [37]. KPIVs are identified using brainstorming to generate initial information. Then the identified factors are reviewed to select those that should be maintained constant based on scientific judgment. Next, a review of process development data and additional experiments using DOE are used to identify those KPIVs which influence critical Ys [37]. Experimental results are inserted into a cause and effect diagram which statistically correlates input and output variables by constructing scatter plots [20].

4.2.4   BACKGROUND AND REQUIREMENTS

4.2.4.1   Types of Process Validation

All three types of validation, prospective, concurrent, and retrospective, are part of most comprehensive process validation programs that occur throughout a product's lifetime [3]. As a process approaches commercial status, validation approaches completion [25].

Prospective. Prospective validation is validation conducted before the distribution of either a new product or a product made under a revised manufacturing process where revisions may affect the product characteristics [37]. Prospective validation assures that process quality attributes are met before manufacturing operations have commenced [26]. It is based on a preplanned protocol [25]. Usually process validation is prospective [18] because a science-based, prospective validation is integrated as a continuous part of the development process [36]. Validation performed prospectively is relatively less dependent on in-process or endproduct testing [25]. Manufacturing lots can be released faster if prospective rather than concurrent validation is performed [26].

Concurrent. Concurrent validation is based on information generated during the actual implementation of the process [25], at the same time the marketed product is being manufactured [26]. Concurrent validation reduces the load at the time of regulatory submission because experiments are delayed until the manufacturing process is operating, but it places marketable lots at risk if a failure is detected. Until process validation (e.g., useful resin lifetime) can be established at the manufacturing scale via concurrent validation, a thorough performance analysis of the step (e.g., chromatography) must be undertaken for each manufactured lot, which is quarantined until the results are evaluated [26]. Thus, concurrent validation is based on batchwise process control and requires the use of validated in-process sampling and methods [25].

Retrospective. Retrospective validation is based on the review and analysis of historical information (e.g., batch records, in-process control testing, and stability testing) to reconfirm formally that control parameter ranges are appropriate [25]. Validation protocols sometimes are not needed, but the final results are formally approved. As with concurrent validation, retrospective validation requires batchwise control of the process and usually depends on in-process testing [25]. Retrospective validation has been found useful to augment (but not replace) premarket prospective validation for new products or changed processes [3]. The proven acceptable range (PAR) can be developed using the principles of retrospective validation to link acceptable product lots to normal/permissible values for process operating parameters [12].

Retrospective analysis applies to a large number of batches that have been prepared similarly by evaluating input/output parameter variation and relating it directly to batch product quality [10]. Specifically, it is most effective when many similar batches (e.g., tens to hundreds) exist. However, the observed range of some input parameters may be insufficient if they were tightly controlled during processing to effectively determine the impact of wider variations [10]. In cases where disparate historical data sources may not be directly comparable or missing, a screening DOE might be appropriate to establish that existing specifications maintain CQAs at desirable levels [37].

4.2.4.2 Aspects of Process Validations

Process validation starts with the identification of product quality attributes and justification of acceptance criteria, followed by a review of the risk analysis, execution of process development runs, and compilation of clinical material manufacturing data to set specifications considering process variability [11]. There is a greater focus on process validation for downstream steps rather than for upstream steps because downstream steps are associated with virus removal. Process validation is just one approach used to control virus contamination, however; others include cell bank characterization, in-process testing, inactivation procedures, control of raw materials, containment, and postmarket surveillance [6].

One significant source of variability in downstream process steps is changes in input streams between research/pilot-scale to full-scale [38]. Thus, input streams of high and low quality are used when evaluating the performance of the subsequent step in a pair-wise validation strategy [3]. Specifically, forward linkage studies are performed to understand downstream consequences for critical parameters [10]. When process steps are studied individually in a factorial approach, forward linkage variables are the output variables of given step that affect the performance of the next step [20].

Fermentation. A validatable fermentation process demonstrates a controllable method of growing cells that reliably express the biopharmaceutical product that can be reproducibly recovered if broth is harvested within a set of specified conditions that allow the product to meet its quality specifications [10]. The goals of fermentation process validation are to provide documented evidence that all aspects of the process perform as intended to generate consistent fermentation broths at harvest; this consistency in turn permits downstream steps to yield a purified product consistently meeting its predetermined specifications [39]. Allowable ranges for culture conditions must be justified with scientific data [30]. Typical components of fermentation process validation include the fermenter sterile medium hold, inoculum train, fermenter inoculation, sampling, feeding, fermenter growth and production phases, and harvesting [9]. Specifically, the trigger for fermentation harvest (e.g., cultivation time, cell density, and viability) should be demonstrated to yield acceptable product [30]. CQAs defined for fermentation include product yield, cell density, and culture purity; critical process parameters include inoculum state, nutrient feeding scheme, and harvest time [40]. To ensure consistent broth quality, critical process parameters are best defined based on a culture physiological event rather than on elapsed cultivation time and are preferably those that exhibit distinctive online monitoring characteristics [40].

To link fermentation process input parameters to output parameters, specifically behavior in downstream isolation steps and ultimate product quality, usually a partial or total purification is necessary, which requires substantial labor to process samples [10]. Sample evaluation also is limited depending on the level of product characterization. Consequently, fermentation process validation has been viewed as less rigorous compared with other forms of process validation [10].

Viral Clearance Studies

Requirements. Any biopharmaceutical product using animal-derived materials during the manufacturing process has the potential for virus contamination [5]. The risk varies depending on the animal species and country of origin of the raw material [5] and the safety concerns associated with expression and amplification of the production system [13]. Process validation is conducted to determine the process capacity to remove and/or inactivate a virus using a wide variety of viruses in virus clearance studies. These studies complement analytical viral testing of product (at various stages of production, including the final product), starting materials (e.g., cell banks), and raw materials [5, 6]. Each process validation viral clearance study is reviewed individually to determine the log reduction performance versus requirements, which are set based on experimental limitations as well as on risk factors (i.e., number of patients and dosage) [5, 41]. In practice, viral inactivation testing is less rigorous than microbial sterilization testing, viral test methods generally are cumbersome with variable results (although PCR technology has improved technical robustness), and viral inactivating agents less robust than saturated steam [10]. Consequently, viral clearance studies are a substantial component of process validation.

Viral clearance can be demonstrated by two distinct mechanisms that both generate high clearance values. One mechanism is inactivation to reduce infectivity by chemical or physical means without affecting protein product stability. Methods include using pH (acidic, low pH conditions for pH inactivation for protein A, basic, high pH conditions for ion exchange resins), chaotropic agents (urea), ultraviolet irradiation, and solvent, detergent, or heat treatment [42–45]. The second mechanism is removal or partitioning to effect a physical separation of virus from the product. Methods include using chromatography (adsorbing protein and leaving virus in the column flowthrough or vice versa) and filtration (depth filters with a 20–40-nm pore size or ultra-filtration membranes with a cutoff of <300 kDa) [5, 7, 46–49].

Processing conditions significantly affect virus inactivation/removal performance. Specifically pH, buffer composition, protein concentration, flow rate, and product binding strength to resin influence chromatography steps [5]. Membrane chemistry and pore size, geometry, velocity and pressure of process streams, transmembrane pressure, and product load per area (volume/area, concentration/area) influence filtration steps [47]. When virus-inactivating buffers are used in chromatography, virus reduction is effected in the eluted product from a combination of removal and inactivation [46]. Analytical methods that can evaluate infectious as well as inactivated virus distribution in process streams can assist in understanding virus removal mechanisms [46].

Steps to be Validated for Viral Clearance. Process validation for viral clearance is conducted only on robust steps that can (1) be scaled down accurately and (2) reproducibly and effectively remove and/or inactivate a wide variety of potential viral contaminants under a wide variety of process conditions [5, 41]. The number of steps selected for validation depends on estimated viral clearance effectiveness based on historical data and target clearance values [5]. The FDA demands at least two different steps for virus reduction to guarantee safety and efficacy [7]. Potentially only two steps are required for antibody processes that use serum-free medium, but additional steps might be required if viral contamination risk is increased by using serum-containing medium [7]. Due to the use of live viruses to perform clearance studies, this work usually is outsourced to reduce cross-contamination issues [14].

The viral clearance reduction factor, the common logarithm of the ratio between the total virus loads before and after clearance, is established for viruses known to contaminate the production process [50]. Individual step clearances are combined to obtain the total clearance reduction factor. This reduction factor is used in combination with an assessment of step robustness to classify the step as effective (≥ 4 reduction factor and unaffected by small changes in process variables), moderately effective ($4 >$ reduction factor > 1), or ineffective (≤ 1 reduction factor) with respect to virus clearance [50]. Clearance factors are usually multiplied if the mechanism is different for two separate steps and sometimes are added if the mechanism is same [51]. In other cases, if two independent steps have similar mechanisms of clearance, only one step is included in the summation because virus particles removed via that mechanism would only be expected to be removed in the first step [3, 5]. A total clearance of 12–15 logs is desired for lipid-enveloped viruses and fewer logs for nonenveloped viruses (e.g., polio) [30].

Virus loads in unprocessed, prepurified bulk (harvest samples) are quantified, typically via transmission by electron microscopy or infectivity, to estimate actual virus load versus the expected virus removal capacity of the process [47, 50]. Total virus clearance should exceed measured levels by at least 3–5 logs [47]. An example of a typical virus load is 10^5 to 10^7 RVLPs (retrovirus-like particles) per milliliter of unprocessed bulk from CHO cells [3]. Clearance factors of 15–20 logs are typical for murine retroviruses [3] and slightly more than for lipid-enveloped viruses. For most viruses besides endogenous retroviruses, however, there is no reasonable way to establish virus load in the bulk because these types of viruses should not normally be present. In addition direct testing methods are limited inherently because they are designed to detect only known and not known contaminants [5]. Thus, the capacity of the downstream purification process to remove/inactivate known viruses should they be inadvertently introduced is defined for an extra degree of assurance [3, 5].

Spiking Studies. Often, if viruses are present in process streams, they are present at levels below analytical detection, so spiking-clearance studies are required for purification steps [3]. Spiking studies are repeated so that their reproducibility and variability can be assessed [5]. There are concerns that large spikes may not be representative of actual low-level virus behavior, however. Virus spikes are desired to contain as little protein as possible to reduce the impact of their addition on purification procedures [5]. Addition of a virus spike can alter protein concentration and viscosity, which can change elution characteristics and prevent the small-scale model from accurately predicting full-scale performance [52]. In practice, this goal is more readily achievable for non-lipid-enveloped viruses because enveloped viruses need a certain amount of protein to maintain stability [5]. Maintaining the salt concentration constant or evaluating product resolution and purity are two ways to demonstrate that the virus spike addition had no significant effect [52]. Consequently, there is a compromise between adding the highest amount of virus possible to more rigorously challenge and maximize the step clearance and maintaining the spike volume low (<10%) to avoid appreciable alteration of product composition and thus purification performance [5, 50].

Virus Selection. The goal of model virus selection is to demonstrate inactivation or removal of as many types of viruses as possible without compromising product activity and quality [50]. One virus can act as a model for a group of viruses with similar physio-chemical characteristics [5]. Model viruses should be possible to grow in cell culture to high titers and detectable using a simple but sensitive assay [47, 53]. Some viruses that are known to potentially contaminate starting materials grow quite poorly in cell culture; thus, some model viruses have been tissue-culture adapted and consequently may not be truly representative of wild-type viruses [41]. Another limitation of using model virus clearance studies is that model viruses may not behave similarly to actual viruses [5], but in many cases, model viruses are the only practical alternative.

Viral selection is based on (1) relevant viruses that are actual viruses (or of the same species as actual viruses and relevant to the host cell) that have been identified as contaminants (or potential contaminants) of the process, (2) specific model viruses that are closely related to actual viruses (e.g., same genus or family) and have similar physico-chemical properties, and (3) nonspecific model viruses believed to be representative of the spectrum of different virus physio-chemical characteristics [5, 50]. Nonspecific model viruses are used to show inactivation/removal of viruses in general and to characterize purification robustness [50]. Virus clearance studies should cover emerging viruses and viruses currently believed to be absent in raw materials. These concerns are not addressed when relying on direct testing to ensure safety, specifically consideration of future virus removal requirements in anticipation of future regulatory changes [5, 41].

Model viruses are selected that cover the range of physico-chemical properties of different virus species such as size (i.e., diameter and geometry), enveloped or nonenveloped, RNA or DNA, and single- or double-stranded genome [47, 51, 53].

Examples of model viruses selected for various characteristics are medium-to-large DNA/enveloped (Herpes Simplex I, pseudorabies), small/DNA/nonenveloped (Simian virus SV-40), small/RNA/nonenveloped (Sabin Type I Polio, animal parvovirus), medium-to-large RNA/enveloped (Influenza Type A, parainfluenza virus, Sinbis virus 1), and retrovirus/RNA/enveloped for murine hybridomas (Moloney murine leukemia) [3, 51]. Preference in the selection of specific model viruses is given to those viruses with significant resistance to physical removal/chemical agents [47, 49, 51].

The number and type of viruses selected for spiking studies depends on the process materials (e.g., host cell) and product development stage [5]. Before phase I, virus validation must be conducted. Typical viruses studied are the murine leukemia virus as a model virus for retrovirus-like particles produced by murine cell expression systems (if applicable to the process) and parvovirus [50]. Before phase II, two to four additional viruses are selected that depend on the host cell and media origin [50]. It is advised to solicit feedback on virus selection from the relevant regulatory agencies depending on where the trials are to be conducted before conducting viral clearance studies for product destined for initial clinical trials [14].

Starting and Raw Materials. Starting and raw materials are chemical, biochemical, or biological components used in a biopharmaceutical manu-facturing process [53]. Quality control strategies for these materials have been developed based on the assumption that these materials may be contaminated with a virus [54]. Direct testing is limited by the fact that viral contamination can be nonhomogenously distributed in a raw material lot; thus, additional sampling has little value. In addition to being effective only if contaminants are homogeneously distributed, direct viral testing also only is effective if the contaminant is present at high titer, false positives can be investigated promptly, a broad spectrum of viruses are detectable by available assays, assays are adequately sensitive, and raw materials do not inhibit assay performance [54].

Cell Substrate. Since the genetically altered cell substrate is considered the main component in the production process, complete characterization plus evaluation of genetic stability are primary concerns [27]. Cell bank characterization in conjunction with product characterization (e.g., peptide mapping and amino acid sequencing) demonstrates the stability of the production process [55]. Cell bank characterization in conjunction with process validation and endproduct testing demonstrate product safety [55]. The most likely source of viral contaminations in a large-scale cell culture is the cell substrate [54]. Process validation plus cell bank validation provides assurance that final product is free of contaminating viruses and other adventitious agents [55]. There is little concern about viruses that do not replicate in the cell substrate [54], although nonreplication can be hard to demonstrate.

As one of the most critical raw materials in production is the cell bank, defined acceptance/rejection criteria and parameters under which the bank will be used are required [55]. Assays are performed on master and working cell banks to demonstrate stability and freedom from adventitious agents. Cells are analyzed both from master/working bank vials as well as from the end of the production run or a few doublings later [55].

During cell bank characterization, phenotypic characteristics of the master cell bank are established such as morphology, doubling time, product expression rates, karyology (for diploid cell lines), and isoenzyme analysis [55]. Genotypic characteristics also are established, including a description of the vector and inserted gene, copy number, restriction maps, and the nucleotide sequence of a cloned gene. Tests are conducted for adventitious agents such as bacteria, fungi, mycoplasmas, and a wide range of potential viruses (e.g., retroviruses) [55]. Routine stability programs are undertaken to monitor genetic stability with respect to phenotype, expression, and nucleic acid integrity [4]. A passage history (e.g., split ratio, storage conditions, and media used) is documented as well as the cell line origin [55]. As a result of these regulatory requirements, cell banks must be prepared by those well versed in the technical processing steps, but also well aware of regulatory requirements.

Media. Basal medium and other supplemental components are tested in a scaled-down mimic of the production process to determine cell growth and product yield, and the product characteristics are compared against reference materials [16]. It is necessary to assure that multiple lots of raw materials meet user specifications for performance and quality before selecting a vendor [52] and that at least one back-up vendor is available.

Step Hold Times. Documentation of bioburden and endotoxin control at the manufacturing scale over the entire hold time is accomplished via monitoring of intermediates using in-process samples [33]. In addition, small-scale studies can provide relevant worst-case data. The cumulative worst-case hold times are obtainable from the summation of the maximum individual unit operation hold times. However, this worst case is not required to be validated because such a series of unexpected events is likely to cause adverse drug quality as well as high endotoxin or bioburden. Instead, hold time studies are conducted for process solutions over the maximum durations that they will be held in tanks used for manufacturing [33].

Real-time stability data are required for reliably assigning expiry periods [4]. Stability data are obtained on the drug product stored in the proposed container under recommended storage temperatures [56], and this strategy has been applied for in-process streams. Although elevated temperature stability studies are used to accelerate product degradation rate to predict shelf-life [56], this acceleration methodology is usually not used to determine step hold times, which typically are a few days, weeks, or in rarer cases months. Additional stability and degradation concerns arise from the possible interaction of protein with stoppers and container closure systems and interaction of proteins with residual cleaning agents [4].

Contaminant/Impurity Clearance. Classes of contaminant removal include process-related, host-cell related, and product-related (e.g., aggregates, deaminations, and oxidations) substances [14]. Contaminant clearance studies can be used to avoid lot-to-lot acceptance criteria because safety margins of several logs of excess clearance are demonstrated [33]. In these studies, a contaminant is added to the input feed stream at the small scale, its recovery is measured at each stage of the process step (e.g., column flowthrough, product pool, and regeneration fractions), and mass balance measurements are performed [51]. In a similar manner as viral clearance studies, each step's impurity clearance is challenged separately; then overall clearance is calculated by the product of individual clearances [51]. Impurity clearance studies are similar in concept to viral clearance studies without many of the virus handling and personnel contact safety issues.

Process-related contaminants include cell substrate-derived (e.g., host cell protein, DNA, retroviral-like particles, and endotoxin for *E. coli* cultures), cell culture-derived (e.g., inducers, antibiotics, and serum/media components such as antifoam), downstream-derived (e.g., enzymes, processing reagents such as glycerol and guanidine, inorganic salts, solvents, leachables, and ligands), and adventitious agents (e.g., virus, bioburden, and endotoxin) [33]. Cell substrate-derived and adventitious agents have been discussed. For many process-related contaminants (e.g., antifoams and enzymes), the key to tracking and demonstrating their removal is the development of a robust, sensitive, and yet straightforward assay.

A documented methodology is required for how product-contact materials are selected and purchased [57] for specific process use. Leachate process validation demonstrates that leachates are below limits of detection in each eluent pool; insignificant leaching occurs under severe chemical, time, and temperature exposure, and if significant leaching does occur, it is removed by the first few bed volumes of equilibration buffer [58]. Product extractable testing evaluates product contact surfaces in terms of materials of construction, process solutions or solvents, duration of product contact with various surfaces, product contact surface areas, and potential for extractables at various process stages [57, 59]. Materials are prioritized to direct studies at the most critical, high-risk materials. High-risk materials are based on the proximity to the final API, extraction capability of solvent, length of contact, product contact surface area, cytotoxicity of extractables, temperature, and inherent material resistance to extraction [57]. Example high-priority materials to assess are containers (e.g., stainless steel cans and plastic bags) for unfrozen and frozen storage of fermentation media/purification buffers/product intermediate streams, chromatography resins, valve diaphragms/transfer hoses, elastomers/pumps especially on the final purification skid, filtration membranes, final sterilizing filter, gaskets/o-rings especially on the final bulk tank, and silicone boot/tubing for the filling machine [57]. Material extractable data can be obtained in-house, or often it is available from the vendor.

Product-related impurities include truncated product forms, modified product forms (e.g., deaminated, isomerized, disulfide-linked, and oxidized), and product aggregates [33]. One way these undesired forms arise is during freezing and thawing owing to product concentration and pH gradients [60]; thus, the number of freeze/thaw cycles is minimized. Another way is from mixing liquids and dissoluting powders [33]. The minimum mixing time until homogeneity is determined for which results vary only within assay variability; maximum mixing times are established to avoid shear, oxidation, and/or product degradation.

Reuse of Chromatography Resin. It is necessary to demonstrate that chromatographic media can be reused, cleaned, and sanitized [61]. Used resins receive additional analysis not required with new resins such as the titration of small ion binding capacity, measurement of total protein capacity, comparison of flow versus pressure plots to indicate particle size (and attrition), particle size distribution, total organic carbon (TOC) removal by cleaning solutions, and microbial contamination/endotoxin (LAL) analysis [58]. Exposure to cleaning/regeneration solutions, rather than contact with mild buffers and protein solutions during normal processing, is most likely to cause chemical degradation [58]. Three types of studies thus apply to resins: Column reuse studies verify continued resin performance after multiple uses followed by small-scale carry-over runs to confirm cleanability with multiple uses, characterization studies prove the process is robust and defines acceptable operational limits, and another third study type demonstrates that regeneration sufficiently sanitizes the resin and that the storage solution is bacteriostatic and fungistatic [61].

Columns reused for many cycles may not provide the same virus clearance as new columns [41]. Thus, virus spiking studies can be performed on new and aged resins to ensure clearance does not vary [5, 26]. Alternatively, virus clearance can be studied for new resins only, and performance attributes expected to decay before virus clearance ability can be monitored subsequently during actual reuse [62, 63]. Validation of the column resin cleaning and sanitization regime for virus inactivation also is necessary to demonstrate that bound virus does not accumulate and subsequently elute [41]. The appropriate number of reuses is selected by balancing resin cost with the time required to conduct the reuse study and the expected annual number of production runs [61]. Using resin to its functional lifetime maximizes its cost basis, which helps economics if the resin cost is high or the number of cycles is low [26]. Virus removal filters [64] also have the potential to be reused, and although less, if anything, has been published about their reuse, the performance characteristics necessary to demonstrate are expected to be similar to resin reuse.

4.2.4.3 Process Validation Execution Timing

The execution timing of key process validation elements varies according to the clinical phase of the product. As validation starts well before the last three lots before actual production [31], it is essential to consider the question of "how much process validation to do and when to do it" as early as possible [14]. Creating the validation package too early might result in redoing substantial portions [14]; creating it too late may delay product launch owing to an incomplete regulatory package. Process validation is a continual process and should be monitored closely for adherence to timelines, scope, and cost [4]. It evolves along with the development of the process and should be complete at the end of phase III [14].

At the preclinical product phase, critical and noncritical classification of process input parameters should be initiated [32]. Critical components of facility subsystem validation need to be essentially complete before phase I product manufacture [15]. For phase I, it is necessary to validate aspects of the process related to product safety (e.g., sterility, mycoplasma, viral clearance, impurity removal, and stability) [14]. Abbreviated viral clearance studies for model viruses/retroviruses and impurity clearance studies for host cell DNA often are acceptable, resulting in fewer downstream steps validated at this product stage [3, 5]. If viral clearance results are available in sufficient time, the results can be applied to developing the phase I process steps. All assays do not have to be validated at this stage, but some (especially product-specific ones) should be at least qualified [14].

For phase II, no specific process validation activity is required unless process improvement changes potentially affect prior phase I validation work [14]. Assay validation efforts should continue, however, so assays are ready for phase III validation and process development should be finalized before phase III [14]. Most process and assay validation activities begin in phase II and often extend far into phase III [4], occurring in parallel with phase II and phase III clinical material production [3]. Intensive process characterization often is delayed until after completion of phase II studies to conserve resources [65]. About 12–15 months are allowed for its completion before the process validation runs [65]. To minimize risk, if all phase II clinical data are not available, these activities can be ramped up slowly over the first few months. Raw material vendor audits should be performed between phases II and III before the manufacturing process is fixed [47].

Before phase III and process finalization, process robustness is evaluated via several scaled-down runs to quantify process variability [14]. These studies form the basis for the process validation program if they are well documented, use operational range extremes, and test different resin/raw material lots. Based on the resulting data, additional optimization to improve robustness is conducted. Fully developed analytical assays are employed during these pre-phase III scale-down runs to ensure the process meets the quality acceptance criteria. Then during phase III, any remaining assays are validated [14].

The heavy-duty process validation occurs in phase III when the process is better understood, a commercial process is in place, and there are likely to be fewer changes, adjustments, and mistakes that can cause deviations [14]. First equipment and other subsystem validation are completed and followed by process validation during three or more consecutive batches [14]. The master validation plan usually is approved during phase III clinical material manufacture [32], and it sometimes includes manufacturing equipment validation in addition to process validation [47]. Viral clearance studies are conducted with a greater number of model viruses possessing a greater range of virus characteristics, and the full range of impurity clearance and operating range validations is undertaken [3]. Mass balances are attempted, duplicate runs are tested, and resin lifetime studies are conducted [14]. Range testing validations typically are completed during phase III clinical production before the manufacture of qualification lots, except for viral clearance studies that are done earlier [32]. Thus, the pivotal phase III clinical studies are conducted using a fixed process with refinements already incorporated to achieve a robust and cost-effective process [15]. At this stage, operational parameters are defined and performance parameters are specified within wide limits until more extensive data are accumulated [15].

Process performance qualification typically is conducted during the production of qualification lots using target set points for all process input parameters to demonstrate consistency and reproducibility of input and output parameters [32]. This process consistency is demonstrated by multiple full-scale batches, for example, three-to-five consecutive purification lots from three consecutive fermentation lots, during which yield and product concentration are monitored at each step [14]. A failed lot can occur without resetting the count if it is due to a known assignable cause that is not process-related (e.g., operator error or equipment failure) [33]. The cause of noncompliance must be determined and corrected, the noncompliant product disposition designated, and any corrective actions taken recorded [11]. At this stage, all assays are validated, all equipment and support systems are qualified, all personnel are trained, all critical input parameters with ranges are validated, and all critical input parameters with set points and noncritical input parameters are validated as a by-product of range-finding studies [32].

4.2.5 DEVELOPMENT SUPPORT AND TECHNOLOGY TRANSFER ACTIVITIES

4.2.5.1 Process Development

Role, Stages, and Plans. Process validation is an integral part of process development [13]. At the early stage of process development, the process is basic and unpolished [14] with purity ill defined and poorly controlled [2]. At the mid-stage, an optimized and more robust process exists, and by the end stage, the process is fully characterized, validated, and qualified [14]. During process development and scale-up, critical variables either are removed or else highly understood to develop a robust process [34]. Identification should start early and be based on literature knowledge, prior experience, or specific product information [32]. Some critical variables are identified via close in-process monitoring and thorough investigation when unexpected results are obtained [34]. Early, reliable, and accurate measurement of process variability increases the probability of operating and maintaining a robust, well-understood, and controlled process [66].

Successful process development establishes a clear logic path that streamlines subsequent validation design and testing [22, 32]. Process development serves to establish evidence that all key process control parameters and all control parameter ranges are validated and optimized to meet acceptance criteria before process validation [14, 25]. These acceptance criteria, specifically the critical process output parameters for scalability studies and range testing validations, in turn are derived from data obtained during development, preclinical, and clinical material production [32].

The greatest validation problems surface when insufficient resources and time are budgeted for understanding and optimizing production [14]. Development personnel require sufficient time and financial support for thorough validation and technical transfer activities, specifically remaining involved in process validation activities along with manufacturing personnel [2]. Consequently, development staff should understand compliance issues [13] to accurately anticipate the appropriate level of quality assurance applicable to process development studies [30]. Actual formal product development procedures have been recommended to ensure consistency [13], as well as capture best practices.

Process Development Documentation. As a considerable amount of relevant data is generated during process development, its documentation must be sufficient to support the selection of manufacturing steps [2]. Development work and decisions made during development are captured in development reports that support future regulatory inquiries [14] and demonstrate that regulatory requirements have been considered and addressed [30]. Development reports should highlight potential weaknesses in the product and process design with the goal of suggesting which ones should undergo subsequent development and validation [13]. They also include an analysis of critical process parameters [13]. The first reports can be generated as early as before the initial product transfer to the process development group [32]. Additional reports are generated after the end of process development and before the start of process validation [32]. All of these reports need to be compiled and indexed so they are readily retrievable [13], perhaps stored in a database for easy access [14]. They serve as a resource for the product's process development history if an employee leaves (or is reassigned) or as training for new employees [14].

An upfront listing of the titles and scope of expected reports felt to be necessary to support process scale-up, validation, and regulatory filings is created and reviewed within the multidisciplinary validation team [30]. This listing helps identify required studies and form a process development plan that can facilitate future tracking of executed studies. Once the required component tasks and their execution order are established, this plan should be reviewed semiannually, adding accuracy and sophistication over time. Formal evaluation of the status of the process development plan demonstrates whether the project is on track and identifies deviations in time and cost. The interim goals, resources, and milestones listed in the basic plan focus efforts on priorities, permit staff to see how tasks are integrated, and reduce the likelihood of running out of time [30].

Process Characterization. The terms *process characterization*, *robustness*, and *validation* are sometimes used interchangeably, and consequently, the distinction among these terms consequently is blurred. Process characteriza-tion is more encompassing than process validation because process validation uses only a subset of the data generated from process characterization [15]. Characterization studies are similar in nature to validation studies except with less formal compliance to regulatory guidelines [12]. It is the part of the process development examining manufacturing ranges, robustness, and the edge of failure for a limited number of critical parameters [67]. For characterization studies, established scientific and engineering principles are used, along with methods that are not validated in the regulatory sense but are qualified scientifically with appropriate controls and written procedures. [12]. Typical causes of unexpected results in process characterization are inadequate procedures, or inadequate following of adequate procedures, and uncontrolled variables [37]. Process robustness provides assurance that the process will not fail when operating within defined control limits for key process variables [35]. This definition of robustness has been adapted from the ICH definition of analytical assay robustness (ICH-Validation of analytical methods, 1993), not from the guidelines for process validation [35].

Process characterization defines process capability and facilitates prospective process validation at the production scale [40]. Full process characterization is valuable in maintaining smooth manufacturing operations and minimizing lost batches, and it provides supporting information for lot release justification for atypical batches [15]. Its goals are to identify key operating and performance parameters, define control limits for key process parameters, demonstrate robust-ness of the commercial process, and provide technical information about the process [68]. The steps involved in process characterization include risk assessments, scale-down model qualification, single-parameter ranging studies, parameter interaction studies, and worst-case studies [68]. Process characterization is conducted initially during the preclinical through phase III product stages and continues during the ongoing commercial manufacturing product stage [65].

The overall goals of process characterization are to (1) understand the roles of each process step, (2) become aware of the effect of process inputs (operating parameters) on process outputs (performance parameters) and identify key operating and performance parameters, (3) assure that the process delivers consistent product yields and purity in all operating ranges, and (4) develop acceptance criteria for in-process performance parameters [65]. Example operating and performance parameters for various process steps are shown in Table 4.2-5. Before beginning process characterization activities, advance preparation consists of examining available data and assessing risk, qualifying scale-down models, and developing process characterization plans. Execution involves four steps to achieving these process characterization goals: (1) characterize process performance ("process fingerprinting data") using screening experiments to eliminate additional parameters from further characterization studies, (2) identify interactions between key parameters, (3) establish process redundancy by determining the effect of feed quality on each unit operation, and (4) develop unit operation reports to document key operating and performance parameters and their respective ranges [65].

One recommended strategy for process characterization is to (1) measure as much as possible early in process development, (2) determine if changes in the

TABLE 4.2-5. Example Operating and Performance Parameters for Various Process Steps

Process Stage	Operating Parameter	Performance Parameter
Seed (shake flasks or fermenters)	Agitation rate Working volume Vial thawing conditions Growth temperature Number of generations	Time to reach transfer cell density
Production fermentation	DO/dCO_2 Broth pH Growth temperature Agitation rate Aeration rate Back-pressure Working volume Percentage inoculum Transfer cell density for inoculum	Titer Specific productivity Plasmid loss/genetic stability Growth rate/doubling time Peak growth cell density Harvest cell density/viability Product quality (glycoform and/or sialylation distribution) Bioburden/adventitious agents
	Broth pH Point of nutrient feed initiation Nutrient feeding rate Osmolarity Point of induction	Feed/acid/base usage Nutrient and waste by-product metabolism Time to induction

TABLE 4.2-5. *Continued*

Process Stage	Operating Parameter	Performance Parameter
	Concentration of inducer	
Filtration	Transmembrane pressure	Process flux and water flux
	Feed/recyle/permeate flowrate	recovery
	Membrane area/liquid volume	Cell lysis
	or mass	Permeate turbidity
	Membrane geometry	
	Volume concentration	
	factor/diafiltration volumes	
	Protein concentration	Yield
	Cell density/viability	Purity
Centrifugation	Centrifugal force	Cell lysis
	Feed/centrate flowrate	Impurity clearance (host
	Feed/centrate solids content	cell proteins, nucleic acids,
		endotoxins)
	Back-pressure	Centrate turbidity
Chromatography	Column pressure	Yield
	Column flowrate	Product concentration
	Column loading level	Purity
	Column load/wash/elution	Impurity clearance (host
	conditions (pH, conductivity,	cell proteins, nucleic acids,
	temperature, gradient slope,	endotoxins)
	volume)	Viral clearance
	Buffer volumes	Ligand leakage
		Column HETP
		Peak shape
Precipitation	pH	Yield
	Conductivity	Product concentration
	Precipitating agent concentration	Purity
	Precipitate agent addition rate	Impurity
	Temperature	Precipitate yield
	Protein concentration	Supernatant turbidity
General	Pre- and/or poststerilization	Growth rate/doubling time
	media/solution pH	Specific productivity
	Media/solution sterilization	
	conditions (time, temperature,	
	filtration)	

magnitude of each variable/parameter are relevant to process (or process model) performance, (3) select the relevant variables to be controlled and documented, and (4) collect, organize, and archive all raw data [69]. As the total number of factors to be tested can be very large, factors are combined and treated as single variables based on the experience gained during process development [60]. Variables expected to have similar effects are linked so that their cumulative effect will be additive [60].

4.2.5.2 Assessment of Process Risks

Quality assurance (QA) groups often desire a 100% guarantee of quality, leading to risk avoidance and promoting "over-validation" in the form of additional testing and repetitive documentation that increases production costs [31]. Although absolute safety is a theoretical concept just like absolute sterility or product purity, it can be approximated using established scientific and engineering principles [12]. Risk analysis identifies the steps, critical factors, and parameters that affect product quality [11]. A documented risk analysis determines the extent of process robustness, summarizes results in a form readily presentable to internal and external regulatory personnel, meets regulatory requirements, identifies critical parameters for monitoring, increases process understanding, assists in troubleshooting, and, most appropriately, focuses process validation efforts where they are most warranted and beneficial [11]. It is first necessary to determine what level of each putative risk factor is considered safe, considering the intended clinical use of the biopharmaceutical (e.g., administration route, dose size and frequency, and total amount consumed) [12].

The process design must include steps that reliably remove or inactivate potential risk factors [12]. Rare, significant, and real risk factors must be distinguished from frequent, less significant, theoretical, or putative risk factors. In some cases, targets are more stringent for real and present risks than for putative ones, and reliance on removal/inactivation steps is less important if risks are purely theoretical. The list of putative risk factors is developed early in the process development cycle and then updated periodically as additional information and experience are obtained. Each potential risk is defined in experimental terms by assessing its biological, biophysical, and biochemical aspects. Two categories of risks are (1) potential pharmacological and toxicological effects of the product's active ingredient(s), and (2) other product components arising from the preparation process. Examples of specific risk factors are the presence of intact cells, adventitious agents (e.g., bacteria, fungi, mycoplasma, and viruses), endogenous retroviruses, residual cellular nucleic acid/proteins, microbial contaminants (e.g., endotoxin and proteins), and process chemicals (e.g., antibiotics, ligands, solvents, cleaning chemicals, inducers, and nutrients) [12].

Risk assessment provides a systematic estimation and evaluation of the risk potential for all components used in the manufacturing process, including the origin and fate of each raw material [53]. Risk assessment also estimates potential risks associated with the application of certain procedures during the manufacturing process (e.g., isolation, validation, and testing) by quantifying a procedure's robustness and susceptibility to errors. It assesses which contaminants have the highest probability of being present and then provides a basis to establish a reasonable and economical testing scheme [53].

Each risk factor must be identified and quantified in terms of its likelihood of occurrence and its severity [12]. After the initial first key steps of risk identification and quantification, the greatest risks then can be reduced [29]. Reducing the probability of a risk's occurrence is achieved via implementations such as process automation, tighter controls, or removal of open processing steps [70]. Reducing a risk's severity factor is achieved by process changes or product redesign [70]. In contrast, raising the probability of failure detection by increasing sampling and monitoring activities has been found to be the least productive and most time-consuming solution with little opportunity for automation [70]. If a significant risk cannot be reduced sufficiently, then the associated process step can be validated to provide additional performance assurances.

Risk Assessment Tools.   Risk assessment tools (FTA, FMEA, and HACCP) can be used alone or in combination [11]. They are tools for problem prevention, requiring failure anticipation rather than analyzing failures that have already occurred [71]. After process characterization is completed, formal risk analysis exercises can be repeated so future focus can shift to newly uncovered unit operations and operating parameters requiring attention [65].

FTA.   Fault tree analysis (FTA) is a deductive, top-down approach to failure mode analysis [11]. Its open-ended structure is readily expandable, making it helpful for complex problems [71]. The failure or hazard is assumed; then the combination of conditions or probable causes necessary for that event to occur is systematically defined by identifying how that high-level failure is caused by lower level, primary failures, specifically the failure of an individual component or subsystem. This information is presented in the form of a fault tree diagram with the assumed failure/hazard listed first at the top and the associated conditions/probable causes listed as branches with successive levels listed as further branches [11, 71]. Events at different levels of the tree are listed below the main failure event and connected by event statements and logic gates defined by Boolean logic [11, 71]. "And" is used if events need to occur together to cause the failure/hazard; "or" is used if the event alone would cause the failure/hazard [11]. As a fully developed FTA can be complex, special symbols are used to describe events and gates to simplify graphical representation of casual relationships between branches [71].

FMEA.   Failure mode and effect [criticality] analysis (FME[C]A) is a preventative, bottom-up approach to identifying potential failures [11] that has been recently adapted to quantifying GMP risks [70]. FMEA has been applied to audit preparation, batch record review, production validation, change control evaluation, retrospective validation, trend analysis, and validation master plan development [72]. Although it includes identifying the effects or consequences of failure modes, it is a more practical tool and a structured guide to analyze major areas identified by the FTA [11, 71]. However, it is not considered as effective as FTA in identifying root causes for intractable process failures [71]. FMEA uses its structured methodology to facilitate identification of critical variables, specifically reducing 20 or more to 2–3 to permit efficient further optimization [72]. FMEA progresses over time, starting in the early stages of process development [11], and it is most successful if the process to be examined is clearly defined at the outset [71]. Overall, FMEA serves to (1) readily identify potential problem areas where operators must be especially alert and where intensified test procedures may be required, (2) define operational constraints and preventative maintenance actions for equipment, (3) provide information for planning performance qualification, (4) prioritize corrective actions and improvements, (5) foster interdisciplinary teamwork and communication, and (5) facilitate decision making throughout the development process [71].

The first step in FMEA is to develop a working matrix that identifies all information categories that are intended to be studied [67, 71]. A flowchart describing process steps serves as the left column of the matrix. An accurate description of each step helps stimulate thinking about how the step might fail. A list of reasons is generated describing why each step might fail to complete its intended function (potential failure modes) and what would happen if failure occurs (potential downstream effects) [73]. All possible causes, however improbable, are identified that might result in failure; then the risk attached to each failure is evaluated using an established scale, for example, values from 1 to 5 [71, 73].

The next step is to rate the importance of the identified failure in terms of its probability of occurrence based on the controls in place (P), severity (S), and ability to detected (D), resulting in a risk priority number (RPN) [11, 70]. It asks the following questions: What could go wrong, what is the probability of its going wrong, what are the consequences of it going wrong, and can in-process or final product testing detect it when it does go wrong [72]? RPNs, also known as the criticality index, then are used to sort the identified causes/effects of failures, which enables FMEA to identify those failures posing the greatest threat to the manufacturing process [71]. P, S, and D are all rated between the low and high values; P and S directly increase with numerical rating, and D inversely increases with numerical rating. The RPN value is calculated by the product, $P \times S \times D$ [71–73]. If P, S, or D is completely unknown, then a high value rating (e.g., 5) is assigned [72]. Thus, in the worst case RPN = 125, and in the best case RPN = 1. In practice, RPNs only are calculated for the worst aspects of each failure to reduce workload and to permit identification of the most critical failures first without requiring detailed calculations for each identified cause [71]. Frequently, the highest RPN steps are those steps involving human intervention because P is typically 3 or more [72].

The output of an FMEA study is the list of preventative and corrective actions necessary to improve the manufacturing process that now can be prioritized based on RPN values [71]. An acceptable RPN should depend on the severity rating of the failure, specifically if S = 5 (high value), then the RPN should be ≤5, but if S is below 5, the acceptable RPN can be higher [72]. The RPN then can be recalculated after corrective actions have been implemented providing measurable proof of impact [71].

HACCP. Hazard analysis and critical control points (HACCP) is a systematic approach that is system-based to determine high-risk steps [11]. The definition of hazard includes both safety and quality concerns for biopharmaceuticals. An effective HACCP system reduces endproduct testing because sufficient validated safeguards are instituted [11]. HACCP provides detail and documentation to show process/product understanding through identifying parameters to control and monitor. Its stages are to (1) conduct the hazard analysis, (2) determine critical control points, (3) establish critical limits, (4) establish monitoring procedures, (5) establish corrective actions, (6) establish verification procedures, and (7) establish record-keeping and documentation procedures [11, 74].

4.2.5.3 Validation of Assays

Purpose. As biopharmaceutical products are large and complex molecules and no single test method is sufficient, product analysis uses multiple, different analytical methods strategically designed to be complementary with respect to selectivity and specificity [4, 27]. More sensitive test methods allow proof of greater removal of a putative risk factor [12], and the more key parameters that can be measured assure that process variability is understood and controlled [66]. However, it is important to avoid setting product specifications at the assay's limit of detection [2]. Assays provide adequate in-process testing via process validation and a comprehensive assessment of the final product using a variety of analytical methodologies for identity, purity, potency, strength, safety, and stability [4].

Assay validation characterizes the assay performance so that the significance of the measured assay values obtained is readily understood [75]. Test methods should be validated when important decisions are to be based on the data generated [30]. Thus, the extent of method validation depends on the stage of clinical supply manufacture. The key elements of assay validation for a method are to establish reliability, the intra- and interlaboratory test variation, and relevance, the meaning of the results for a specific purpose [19]. The robustness of an analytical procedure (according to the ICH-Validation of Analytical Methods, 1993) is its measured capacity to be unaffected by small variations in controlled parameters and reliability under normal usage [35].

Scope. The scope of assay validation is to assess the essential test method performance characteristics of accuracy, reproducibility, repeatability, linear-ity, and limit of quantitation/detection [41]. A test method is evaluated for its readiness for assay validation against the following criteria: (1) description of the test basis and scientific purpose, (2) case for relevance, (3) proposed practical application, (4) statement of limitations, and (5) acceptable intra- and interlaboratory precision [8]. Before actual validation, the test method should be optimized and standardized [8], specifically controls necessary to invalidate inappropriate out-of-specification results should be incorporated [30]. Assays must be validated for each matrix in which samples will be run [75] and for its capability on a protein-by-protein basis [56].

Assay ruggedness also is measured in terms of the similarity of results from different laboratories, manufacturer's equipment, and analysts [2]. Interoperator, interday, and intraassay variations can be quantified by comprehensive validation [5]. Interassay (intermediate) precision is the precision of multiple determinations of a single sample analyzed in various runs [75]. Intra-assay precision or repeatability is the precision of multiple determinations of a single sample within one assay run. Interlot precision tests analyze values of aliquots from a single sample run using lots of different assay components. Finally, lot-to-lot precision is the precision of multiple determinations of a single sample analyzed using different lots of assay components [75]. These multiple types of variations must be considered so that release criteria ranges are not set too tight.

The limit of quantitation (LOQ) is the lowest concentration that produces a signal 10-fold above background, whereas the limit of detection (LOD) is the lowest concentration that produces a signal threefold above background [75]. The assay sensitivity and achievable concentration limits of additives define a "window of clearance," which is the difference (on a log scale) between the highest attainable initial contaminant/impurity concentration and the lowest detectable concentration (LOD) of that additive [12]. This difference is the amount that can be measured and thus the amount that can be claimed to have been removed by the process [12].

A low-level sensitivity is required for viral test methods to maximize reduction factors [5]. However, the ability to detect low virus concentrations is limited by statistical sampling [5]. The probability of detection (POD) for low-level virus concentrations depends on test volumes, batch volumes, and test sensitivity [40]. The POD is high when sample virus concentrations are well above the test method sensitivity, the POD is lower when concentrations are at or below the test method sensitivity, and the POD is even lower for very low concentrations due to the low probability of obtaining a virus in the sample as described by the Poisson distribution. Test method sensitivity should be based on a POD ≥95% [40].

In-process and release (endproduct) parameters, both critical and operational controls, are controlled and monitored during production using assays [11]. In-process controls generate data at critical process steps, which serve as a basis for specification development for product intermediates [2]. Specific in-process controls, routine sampling, and monitoring at specific process steps are established to monitor key process outputs [32]. Information only tests are used postvalidation without preestablished acceptance criteria until sufficient data have been collected to statistically establish clear limits or else omit the testing [32]. Thus, sampling and obtaining data for information only is an important part of process validation that must be recognized.

Assay Types

Potency. Binding assays for quantitation use the binding of at least two molecules to form an interaction that withstands multiple washing steps, and they are the least variable [75]. Cell-based bioassays for potency use a cell line mixed with a sample to generate a cellular response that is quantified [75]. They have more variability [75] and are tricky to use to measure potency because the CV often is >30% [2]. Whole animal assays for potency are time-consuming and highly variable [75], and they have the typical disadvantages of *in vivo* tests in that they are expensive, inaccurate, slow, and raise animal ethics issues [19].

The development and validation of *in vitro* tests that correlate with potency, typically desirable for vaccines, is tedious and rarely successful, requiring thorough knowledge of the immune response after infection or vaccination [19]. An alternative approach is to demonstrate batch comparability and then assume identical properties *in vivo* [19]. For an increasingly greater number of vaccines, no routine animal potency testing is being conducted because the product is well characterized, critical production process steps are controlled, immunology knowledge permits functional test development, analytical knowledge permits antigen characterization, and test procedures are becoming harmonized and validated [19].

Impurities. Analytical methods for impurity clearance require sensitivities capable of demonstrating sufficient clearance to well below safety margins [33]. rDNA technology has stimulated the development of a wide variety of assay types [16]. Specifically, PCR is used to understand how manufacturing steps impact retroviral clearance due to its high precision, high throughput, and low cost [76]. Typical testing for in-process impurities includes endotoxin, host cell and other proteins, DNA, viruses, and altered protein forms (e.g., proteolytic clips, deamination, oxidized methionines, and amino acid substitutions) [16]. Clearance of host cell proteins is measured using immunoassay [50]. Host-cell DNA is radiolabeled by nick-translation and then measured in column fractions (also radiolabeled host-cell proteins), and it is measured directly in production streams [51]. Cytotoxicity and interference testing using non-virus-containing samples are performed to ensure that sample matrices in a viral clearance study do not adversely affect the ability to titrate virus [5]. These tests are performed well in advance of the actual validation so that conditions may be found for which dilution is not required, and thus, assay sensitivity is not reduced by dilution [5].

There is a lack of practical laboratory assays to detect bovine spongiform encephalopathy (BSE)-like agents so the potential introduction of these agents into the process must be controlled by avoiding bovine-based materials, using materials from low-risk bovine tissues, and/or sourcing materials from countries with good BSE control measures [21].

Typical testing on final purified bulk includes those tests conducted with in-process testing plus total protein content and potency [16]. The final product (sterile filtered, filled and sometimes lyophilized, then packaged into vials and ampoules) is tested for sterility, pyrogenicity, particulates, content uniformity, identity, excipient chemical content, potency, and protein content [16].

Standards. Key to the success of any assay method is the availability of a reliable standard. A primary standard is one recognized by a national or international agency, whose characteristics have been established by collaborative effort [77]. Secondary standards are calibrated against primary standards. Reference material is used for the evaluation of other materials to generate comparative data. Primary reference material has been thoroughly characterized, and it is the purest material available in quantities large enough for use. Biochemical reference standards are used for all product tests and for process-specific tests both for material release as well as for upstream/downstream process monitoring and assay development. They are unique to a specific product in its amino acid sequence and glycosylation but different for each manufacturer. To ensure compatibility with the matrix of the intended assay, different standards can be required for different process steps. It can be challenging to provide sufficient quantities of the various types of reference material early in the product development cycle [77]. A working reference material is required to validate analytical procedures [2]. It should be stable for at least 2–4 years, and its stability should be monitored regularly [77].

4.2.5.4 Change Control and Ongoing Process Validation

Validation maintenance, an ongoing activity for manufacturing processes [11], continues through the life cycle of the process with a changing focus as the process matures [20]. There is a need to address the process life cycle as a whole and not to suspend process validation after three production scale runs are completed [14]. To support this life-cycle approach, process expert teams are created to rapidly resolve process deviations, determine trends toward loss of control, comply with regulatory requirements, assess process change impact, and identify areas for process improvement [11].

A validation program is specific for a product manufactured by a given process [12]. A strong change control program needs to be established once the process is validated [15]. Changes to process equipment, operation, or facilities should be reviewed carefully before the change is approved to determine the potential validation impact and if repeating validation studies are required [12, 25]. Such changes include process improvements, raw materials changes, step substitutions, and scale-up [3]. Changes that are expected to shift the value of a performance parameter require revalidation [15]. Specifically, if the purity is decreased at a specific step, then recovery of this purity must be demonstrated downstream [15]. In general the potential quality impact of process changes is higher with glycosylated rather than with nonglycosylated proteins [13].

The process change is documented in a proposal with its justification, along with the number of batches necessary and strategy (e.g., samples and data needed, extent of validation testing, and acceptance criteria) to evaluate the change [15]. In some cases, the original risk assessment (e.g., FMEA) is revisited or a new one conducted to ensure an unintentional new failure mode has not been created [71]. A formal change control report is written and approved [15].

Revalidation is the repetition of the validation process or specific part of it [25], and it has some conceptual overlap with retrospective validation. Periodic revalidation has been conducted in some cases even if nothing has changed [25]. Specific elements of most validation programs are repeated (at least partially) at regular, appropriate intervals (e.g., every 6–24 months) to verify that the original parameters are still effective [3, 12]. In other cases, if there have been no changes, process revalidation can consist of a formal review of operating and performance parameters, along with the nature and frequency of excursions, which can be executed as part of the annual product review [15]. When existing processes are reassessed, some input parameters may be more tightly controlled than previously permitted or a noncritical parameter may be found to be critical [32]. Product failures occurring within validated ranges of a process might be due to simultaneous excursions of several variables not studied during the initial process validation or from the use of an inadequate scale-down model [20].

Trending of process variables is used to refine range limits [3]. Manufacturing data can lead to superseded NORs and/or MORs, which are justified by the associated operational trends and product quality achieved [20]. Extreme parameter values subsequently can become acceptable in process validation if the acceptability of final product has been confirmed by multiple observations [12].

4.2.5.5 Process Validation Documentation

Documentation of process validation serves to establish that its goals are reasonable and achievable, as well as to ensure that the study design is based on solid scientific and engineering principles [12]. The process validation master plan defines the scope and rationale of process validation, giving an overview of the validation, its philosophy, and general testing strategy [33]. It describes how the validation is to be done, including the schedule, studies to be performed, systems to be validated, responsibilities, and approvals required [3, 15, 33].

The process validation protocol is a preapproved written plan stating how the validation will be conducted and identifying acceptance criteria as well as sampling and assay requirements [33] and other testing details [3]. It is a prospective experimental plan that when executed produces documented evidence that the system has been validated [25]. The protocol defines the system to be validated, identifies operating variables and probable control parameters, and indicates the number of replications required to provide statistical significance [25]. There is a major savings of cost and time realized by adopting potential streamlining measures, for example, combining resin reuse and chromatography step characterization studies into a single protocol [61]. The entire process validation study protocol is reviewed and approved by manufacturing and quality personnel [12] as well as by process development representatives. The process validation report includes the validation protocol, deviations from protocol, validation results, and conclusions about validation status of the step [3]. Copies of most critical validation reports and summaries of other reports are included in the license application [3], so their careful assembly, clarity, and accuracy is critical. Common validation submission deficiencies include the lack of data or protocols, inadequate data summaries, missing acceptance criteria, and inadequate monitoring or sampling [78].

4.2.6 PROCESS VALIDATION STRATEGIES

Several value-added, science-based, best practices are evolving for process validation, which are appropriate from both business and regulatory viewpoints. Selected industrial applications of process validation are summarized in Table 4.2-6. For a biopharmaceutical process, the potential number of variables and possible interactions among variables is large [12]. One possible approach is to conduct a worst-case test to evaluate each known critical variable separately, and if necessary to run steps more than once if they possess more than one critical variable [34]. Although the use of simple experimental designs to develop the relationship and describe the link between input and output parameters is straightforward, it can result in high numbers of experimental runs, which are impractical [10]. Potential streamlined approaches include retrospective analysis, qualified scale-down studies to identify critical parameters and important interactions, and worst-case runs [10]. In addition, more than one variable can be coupled together into a single variable if they are believed, based on science, to have a predictable effect on output variables [35]. Scouting studies then are used if process development data are limited regarding one or more variables or if the ranges previously studied were insufficiently large [20].

4.2.6.1 Historical Data Review

Retrospective analysis of historical data can be used to assess process reliability [88]. The initial process capability specifications are established based on a few runs (usually less than around 5–10) with acceptable results [30]. This number of runs often is too small to accurately define and characterize actual process variability [30, 31]. It can be difficult to claim that the previously set process capability specifications were flawed after validation runs fail to meet them [30]. Thus, the selection of acceptance criteria must consider the benefits of broad versus narrow tolerance ranges.

It is best to develop acceptance limits for parameters such as yield and purity using statistics applied to historical data [89]. The difficulty in setting acceptance criteria depends on the availability and quality of historical data. Small datasets tend to underestimate process variance because they contain only a limited number

TABLE 4.2-6. Selected Industrial Applications of Process Validation

Process Stage	Specific Description	Strategy	Reference (Company)
Fermentation	Recombinant *E. coli* protein production (defined medium, induced fed-batch process)	Bridged process changes to former process used for earlier clinical material	[38] (Amgen)
	Monoclonal NSO (GS) humanized antibody production (fed-batch process with concentrated medium additions)	Statistical DOE used random combinations of selected control parameters at high/low values in scale-down model	[79] (GlaxoWellcome)
	Recombinant yeast vaccine production (defined medium, fed-batch process)	Scale-down model to identify critical parameters	[40] (Merck)
Chromatography	Chromatography of recombinant *E. coli* protein	Parameters varied over narrow range and effect on step yield/pool purity determined	[80, 81] (Amgen)
	Reuse of column with affinity ligand for chromatography of antibody	Automation used to perform repeat cycles to evaluate elution profiles, yield, and purity	[82] (Celltech)
	Affinity chromatography of recombinant protein produced by CHO cells	Validated scale-down model used to test robustness (via fractional factorial design), reuse and virus removal	[83] (Wyeth/Genetics Institute)

353

TABLE 4.2-6. *Continued*

Process Stage	Specific Description	Strategy	Reference (Company)
	Chromatography steps for purification of monoclonal antibody produced by CHO cells	Scale-down studies for resin use and viral clearance fully predictive of larger scale	[26] (Genentech)
	Cation exchange chromatography of nerve growth factor produced by CHO cells	Process variables combined and linked, then statistically analyzed	[60, 84] (Genentech)
	Retroviral clearance in chromatography of antibody produced by hydrodomas	Moloney murine leukemia virus (model retrovirus) used in scaled-down chromatography column	[85] (Celltech)
	Viral reduction during manufacture of Factor IX	Scale-down models used to demonstrate log reduction.	[86] (CSL)
	Purification of a multivalent bacterial vaccine	Matrix/family approach to reduce number of full scale validation lots from 69 to 32	[36] (Merck)
Overall purification	Purification of a therapeutic monoclonal antibody	Scale-down characterization studies for various filtration and chromatography steps (including viral clearance) to predetermine acceptance criteria	[87] (Biogen Idec)

of historical production runs [89]. If historical data are used, excessively large ranges are difficult to defend.

Process variability is determined using available tools such as preparing historical trends of the mean ±3 standard deviations SDs to demonstrate agreement of manufacturing scale runs with clinical manufacturing/development runs [11]. After 15 or preferably 30 data points are obtained, a control chart based on a moving range is the best tool for monitoring process stability [89]. Statistical process control (SPC) and multivariate analysis can be applied to historical data to identify strategies for improving yield through investigation of cause-and-effect relationships using correlation tools and process knowledge [88].

4.2.6.2 Statistical Design of Experiments (DOE) and Analysis for Process Characterization

Experimental designs are used to efficiently identify those input variables that significantly affect output [20]. The use of DOE to easily screen a variety of operating parameters provides a framework for the design and interpretation of experiments, identifies those factors with the greatest effects, segregates key from nonkey parameters, identifies interactions and synergistic effects, and models how outputs relate to inputs [37, 65]. DOE is a powerful, prospective method compared with retrospective analysis [10]. Its ability to determine interactions between key parameters leads to the identification of other weak spots in the process not observed during initial screen experiments [65]. DOE is limited to assessing interactions among input parameters that can be tested within the same experiment [10]. Usually there are more variables than can be tested in single experiment, however [10]. To prepare for successful characterization via DOE, there is a need to design for simplicity by removing non-value-added experiments and to remove as much experimental variation as possible [37].

DOE modeling can be used early in process characterization to set guard bands, ranges for key process input variables where no statistical change to output is observed [31]. These guard bands are set to locate optimal settings that maximize output and minimize variation. Subsequent validation consists first of experimental design to screen key process input variables set at the guard bands previously determined and then to confirm that the output is within specification [31]. Modeling DOE helps understand the relationship between key input factors and critical Ys, relates input factors to responses (critical Ys) via a model (transfer function) that essentially is process characterization, determines input factor settings that optimize responses, and uses a model to set a limit on input variation such that no statistical changes in KPIVs are observed [37]. This approach is consistent with the PAT initiative.

The maximum information per experiment can be obtained by combining studies and varying factors (i.e., variables) using fractional factorial design [61]. A set of fractional factorial runs provides an estimate of each variable's effect and interactions, but no single run can be compared with a control run [35]. This design permits a large number of variables to be tested with lower number of experiments. Often variables are tested at a low and a high level that permits only linear interpolation (i.e., assumes a linear process response) [35]. If center points (i.e., midpoint of the tested range) are added to the experimental design, nonlinearity or curvature can be detected if present. Replication of low- and high-level test conditions increases the precision of the estimate of that factor's effect. Two-way variable interactions are tested with factorial design, but additional experiments may be necessary to deconvolute the effects of two factors that are confounded (i.e., indistinguishable effects from one another). Fractional fac-torial designs test the extremes of all variables and then generate the best-fit model with an evaluation of the statistical confidence of the model response predictions. If no process failures are predicted by the best-fit model, then the process step is considered robust over the ranges of control parameters tested [35].

Statistical analysis demonstrates which input parameters are important contributors to output parameters [10]. JMP statistical software (SAS, Cary, NC) can be used to analyze data and generate separate analyses for each response variable of interest [61]. A difference of 3 SDs often is selected as the criterion for significance corresponding to a 95% confidence level [61], and the test acceptance criteria range typically are set as the mean plus/minus 3 SDs (industry standard) [91].

4.2.6.3   Family/Matrix Validation and Platform Technologies

Common validation studies are required across different processes to demonstrate capabilities for viral clearance, resin and membrane reuse/sanitization/storage, buffer stability, filter extractables, and resin leachables [14]. Furthermore, many biopharmaceuticals are being developed using common platform technologies (e.g., cell line, medium, unit operations) to reduce development time [7]. Thus, templates can be developed that identify key input and output parameters for a certain unit operation and then customized to address unique process or product specifics [14]. Generic validated assays can be used for impurity removal, if a platform cell line and similar culture conditions are used that then are qualified for each new product [14]. In addition, generic validation can be developed in which the supplier and potential industry users define generic operating conditions under which the technology is to be used, and the equipment manufacturer conducts much of the testing for every industry user [90]. This reduces additional user validation testing to only product- or user-specific applications (e.g., product compatibility) [90].

To significantly reduce the number and extent of new studies associated with new products processed using similar or identical process operations (i.e., platform technologies), family and matrix validation approaches are applied to reduce repetitive work [33]. In a family approach, one piece of equipment is validated by three consecutive runs and other similar equipment is validated by one confirming run in each unit [33]. This approach can be applied to the process validation for a product with several distinct, different but similar, components (e.g., antigens for a multivalent vaccine) by (1) treating different product components as a product family, (2) classifying product components into groups based on their similarities, and (3) selecting one component from each group to validate, possibly based on worst-case experience during process development [36]. This strategy is facilitated when (1) the final product can be fully analytically characterized, (2) there is significant prior manufacturing experience with similar production processes and significant process development experience with the new production process, and (3) clinical trial data show comparable safety and efficacy profiles for some product components [36].

In a matrix approach, validation is conducted at the full range or extremes of process parameters, and then values in between are assumed validated [33]. This generic strategy is used to support a wide range of bracketed conditions that include the most common operating parameters for more than one product, as long as process streams and key step components (e.g., resin type) are similar [14, 91]. The matrix approach reduces the extent of process validation required for successive products, at the expense of only a limited impact on the initial product's process validation, as long as the platform technology is reasonably maintained. Example applications have been retrovirus inactivation via low pH treatment [42] and viral clearance using anion exchange chromatography [91].

4.2.6.4 Scale-Down Models

Although the most reliable data are obtained from actual production runs [12], determination of operational ranges at the manufacturing scale is laborious, time-consuming, and costly [40]. Thus, typical manufacturing scale validation is limited to impurity clearance (e.g., proteins, DNA, and small molecules) and resin/membrane cleaning [14]. If the process step cannot be scaled down successfully, then range definition validation studies are required at the large scale [32].

Small-scale process models are an essential and valuable tool for process validation and ongoing troubleshooting [32]. Scale-down factors ranging from 10- to 10,000-fold are employed [3] with the extent of scale-down de-pending on the actual production scale and smallest scale that can reliably reproduce the process [51]. Small-scale equipment is permissible only if the scalability of the unit operation is demonstrated [12]. The same technical principles for scale-down are used as for scale-up for the step of interest [3].

The scale-down model should represent the large-scale model with respect to input parameters and typical values for output parameters [10]. Data are collected under conditions relevant to full-scale manufacturing GMP process [12]. Qualified scale-down models are used in combination with strong statistical tools like DOE [10], but sample handling at the small scale is easier than during GMP manufacturing runs [26]. Multiple validation experiments can be conducted in parallel employing several identical scale-down units, thereby shortening the experimental time expended [3]. Also a hybrid approach can be instituted where samples from full-scale process streams are used as feed streams to scaled-down steps [3]. Overall the amount of re-sources used in small-scale validation is likely to be less than that needed for manufacturing scale validation, although in manufacturing scale validation, resource utilization is spread over a longer period of time [26].

Scale-down studies have been used for a wide variety of process validation studies, including resin lifetimes, in-process stream hold times, buffer stability, virus clearance, harvest criteria, filter extractables, resin leachables, and cell age at harvest [14, 91, 92]. The ease of scale-down differs depending on step and should be considered in selecting those steps to be validated [5]. In fact, certain validation issues can be addressed only via small-scale models (e.g., virus clearance evaluation, nucleic acid and other impurities/additives removal, cleaning and storage procedure evaluation, and column lifetime estimation) because their use increases worker safety, reduces costs, and permits use of higher titer samples for improved accuracy of prediction [52]. Specifically, viral clearance studies can only be performed at the small scale because virus introduction into cGMP manufacturing scale equipment is impractical and inappropriate [5, 26]. For viral clearance studies, the ideal scale-down factor is 10- to 100-fold [6].

Although scale-down models have been used largely in purification, they also have been used in fermentation. Fermenter scale-down models use the same set points for volume-independent operating parameters (e.g., temperature, pH, and DO) and adjust the set points for volume-dependent parameters such as mixing and gassing (e.g., dCO_2 stripping) [68]. The inoculum split ratio for cell culture scale-down is maintained constant [3, 68]. Characterization studies are performed in small-scale fermenters, and then it is demonstrated that the ranges established are insensitive to scale-up [40]. Only the critical process parameters found to be sensitive to scale-up require range studies at the manufacturing scale [40].

There is increased regulatory concern over the accuracy of results obtained from small-scale models [52]. Consequently, the scale-down model must be validated to be consistent with full-scale manufacturing before validation studies begin. For chromatography steps, it is necessary to determine that product purity and yield are the same [52] or demonstrate comparable HETP, peak asymmetry, and retention time [5]. Any product streams tested should be representative of a commercial-scale process feed stream [52].

For chromatography steps, the scale-down model consists of a column with system and auxiliary components (e.g., distributor, monitors, and fraction collectors) [52]. Scale equivalence for chromatography columns maintains constant column bed height and linear flow rate, while varying column diameter, buffer, and load amounts, which are scaled to column volume [61]. Specifically, the column diameter is decreased, but the bed height is maintained to maintain residence time, and the volumetric flow rate is reduced, but the linear flow rate is maintained [5, 52]. Constant residence time is important if viral clearance is due to inactivation rather than removal [52]. Configuration, transport distances (tubing diameters as well as tubing lengths), and materials of construction are similar to the commercial-scale system. The same methods of preparing buffers are used, including the same quality of water, buffer, and salts. The chromatography medium has the same base matrix, functional groups, and degree of substitution, as well as the lot used should meet approved specifications [52]. Finally, the product mass loading and relative buffer/load volumes between scale-down and production systems are similar [33].

Although chromatography scale-down is extensively studied owing to its use for viral clearance validation, the scale-down of other purification steps also is conducted. Virus filtration scale-down maintains the same linear or filtration pressures, product feed concentration, ratio of feed volume to filter area, and temperature as the production system [33]. Other steps such as solution inactivation (e.g., pH and heat treatment) are relatively easy to scale down because it is simply necessary only to maintain buffer composition, pH, protein concentration, and temperature consistent with production conditions [5]. The time course of the inactivation is quantified along with the upper tolerable range of inactivation agent concentration and minimum exposure time. Residence time differences between the production and small scales might change parameters such as buffer temperature [52], which might affect solution as well as chromatography inactivation steps.

4.2.6.5 Well-Characterized Biotechnology Products

A well-characterized biological (biopharmaceutical) is a chemical entity whose identity purity, impurity, safety, potency, and quality can be determined and controlled [19, 93]. To be well characterized, the drug substance must be >95% of the main component and/or related isoforms [93]. Well-characterized molecules are evaluating using sensitive and discriminating tests that are quantitative and relevant *in vivo* and *in vitro* potency assays [93]. The recent explosion in the availability of sophisticated, reliable, analytical instrumentation, combined with extensive, sensitive product analysis methods has resulted in great advancements in defining "well-characterized" biopharmaceuticals [28].

Although for a well-characterized biological the FDA trend is to rely on reduced final product testing, process validation must be present to assure product quality and consistency [15]. There has been an elimination of regulatory agency lot release for specific well-characterized biopharmaceuticals [19, 93]. In addition, process-related impurities (e.g., host cell protein and DNA) are controlled by process validation *in lieu* of lot-release testing [27]. Furthermore, many process changes are justifiable using analytical data and process validation studies without a clinical trial [3]. The following principles can be applied to the manufacturing and testing of a well-characterized product: (1) The manufacturing process should be robust, reproducible, validated, and designed to produce active drug substance; (2) the safety of the final product is assured through process validation using validated analytical testing to demonstrate removal of impurities to safe levels; (3) tests for identity, purity, and potency should be sensitive, quantitative, and validated; and (4) specifications should be quantitative and based on historical manufacturing data and clinical experience [27].

The more characterizable the biopharmaceutical product, the more emphasis is placed on validating that the correct molecule is being produced than assuring that the cell biology is being controlled [10]. However, the difficulty of characterization increases in direct proportion to the complexity of molecule [93]. When less is known about the product, a greater reliance is placed on product and process consistency [10]. Most vaccines used today are not well-characterized biologicals, so batch release must be performed by the manufacturer and by the appropriate national control laboratory [19]. The quality of vaccines is increasingly guaranteed by the use of robust and reproducible production processes [19], and thus, a high reliance is placed on process validation.

4.2.7 FUTURE TRENDS

Future developments in the biopharmaceutical industry are likely to impact the scope and execution of process validation activities. The nature of process validation is continuously changing to meet the requirements of biopharmaceutical development and manufacturing. The understanding and implementation of new strategies based on future developments is key to maintaining an effective process validation methodology.

4.2.7.1 Process Analytical Technologies (PATs)

PAT is a system for designing, analyzing, and controlling manufacturing through timely in-process measurements of critical quality and performance attributes, leading to the goal of ensuring final product quality [94]. The implementation of PAT potentially minimizes the number of tests that need to be performed at the time of lot release, or in the future for certain products, it might eliminate lot-release testing by the manufacturer entirely. Its precedent was the dropping of lot-release testing for certain impurities when validation of their removal was able to be assured [94].

Online and automatic methods have been demonstrated to be more accurate and precise compared with manual methods because (1) sampling and sample pretreatment artifacts are eliminated, minimized, or reproduced; (2) measurement frequency is increased; and (3) results are independent of personnel availability [69]. Continuous, real-time monitoring is preferred when feasible [11]. PAT is used to implement in-line controls to increase productivity and reduce costs [31]. In addition, accumulated data from numerous commercial batches are used to generate trend analysis profiles showing high and low control limits for process parameters [25]. These limits then are used to predict when processes may be deviating from their validated control ranges, although they are currently operating within them [15].

Relationships among manipulated (controlled) variables, online measured variables, and product (uncontrolled) variables in most biosystems are nonlinear to some extent [95]. A forward model is when parameters, starting conditions, and relevant equations governing behavior are known, readily measurable inputs and the outputs are variables; an inverse model is when the inputs are readily measurable variables and the outputs are difficult to measure parameters [69]. The forward model is most applicable to process validation, whereas the inverse model is most applicable to metabolic pathway analysis. Modeling systems such as neural networks have been used to describe the characteristics of extremely complex bioprocess systems [95].

Mathematical models can be used to simulate the controlled process variables and thus predict the effect of variations on the uncontrolled process variables [35]. Linking online sensing to modeling is effective when (1) there is product yield and biomass variation among optimal values, (3) optimal values may vary over time for processes that are not stationary, (4) parameters can be measured directly or otherwise, (5) the cost of online control is a small fraction of capital and expense costs to run a process, and (5) historical data combined with mathematical methods provide identification of key parameters and their optimal values to achieve rational process improvement [69]. Achieving good modeling is not always easy and involves correct identification of parameters and variables, and selection of the type of model (forward or inverse) along with key inputs and outputs, showing the relationship between predictions and actual data to assess model effectiveness [69].

A solid understanding of a product's critical quality attributes and what aspects of the manufacturing process control them is necessary to fully gain the benefits of PAT [94]. Furthermore, the inherent complexity of protein-based biopharmaceutical products makes it difficult to use PAT to assess product characteristics critical to safety, efficacy, and stability. In addition, it is difficult, compared with small molecules, to directly monitor the desired product concentration during upstream/early downstream processes in a background of protein impurities deriving from medium components or host cells [94]. However, PAT can be used for process monitoring and control that then indirectly results in improved productivity, process consistency, and product quality.

The pharmaceutical industry lags behind other automated industries in manufacturing quality analysis because it is mostly lab-oriented with little closed-loop, real-time control, and limited enterprise-wide data availability [96]. There is a fear of potential regulatory agency reprisals if PAT is implemented for existing processes, and its use detects behavior that would not have been detected using conventional process monitoring [96]. Before PAT is widely implemented, the risks associated with data variability need to be evaluated and reliable methods need to be instituted to distinguish among signal, process, and product variability [94].

4.2.7.2 Process Reliability

More increasingly, the supply capability of a commercial organization depends not only on manufacturing but also on process reliability, GMP compliance, and operational metrics [79]. Process reliability lies at the heart of lean manufacturing and Six Sigma programs because it ensures that the manufacturing process can be maintained in a compliant, validated state [97]. Performance factors are measured and monitored on an ongoing basis so that improvement efforts can be targeted at the most significant opportunities leading to the removal of variation from the manufacturing process [97]. The greatest threats to process reliability are production variances and associated investigations [79]. Variance analysis identifies systems prone to problems, and then it uncovers the root causes of these variances, strengthening the mechanical and operational aspects of the process [79].

Process reliability is the ability of the process to consistently produce required results as measured through the primary dimensions of uptime, dependability, and first run yield [97]. Uptime is the time the process is in operation compared with the available time the process is scheduled to be in operation. Dependability is the repeatability of a process' actual run rate compared with its scheduled run rate. First run yield is the ability to produce quality outputs the first time through production run without any rework. Reliability losses are measured in terms of unscheduled downtime, run rate losses, yield losses, and reprocessing [97].

Process reliability is improved most effectively using cross-functional teams by the following process: (1) identification of the opportunity; (2) collection of reliability loss data regarding uptime, dependability, and first run yield; (3) analysis of loss data and determination of the root cause of the problem; (4) brainstorming of solution alternatives; (5) selection and implementation of solutions considering solutions that are safe, feasible, implementable, and cost-effective; and (6) monitoring of results and repetitions to assess the impact of solutions and to prioritize remaining opportunities [97]. Six Sigma states these concepts for continuous process improvement as DMAIC: define, measure, analyze, improve, and control [37], which are readily applicable to process validation. Measurable benefits of improved process reliability include increased customer service (e.g., timely delivery), increased schedule adherence, reduced lead time, reduced inventory levels, increased effective capacity (i.e., greater process uptime and reduced unplanned downtime), decreased cost, creation of time available for predictive maintenance, incremental capacity for new products or increased volume without additional capital investment, increased process dependability, improved quality and first run yield, and reductions in setup and changeover time [97].

Process flexibility is needed for rapid turnover and reduced inventories that are key to lean manufacturing, which in turn are crucial to making more customized drug products and remaining competitive [97]. Flexibility losses are when time, people, and material waste are consumed by product changeovers. Flexibility is improved by mapping the changeover process, analyzing steps, and then eliminating, resequencing, or improving activities. In addition, standardizing procedures and setup locations and advance preparation for changeovers improve efficiency [97].

4.2.7.3 Biogenerics

Biogenerics, also known as "follow-on" biologics [98], are pharmaceutical preparations, using a generic name, that are based on drug substances arising from recombinant technology (complex structure and high molecular weight) that have been demonstrated to be essentially similar to an original pharmaceutical with an expired patent [99]. The main features of biogenerics processes (such as reproducibility, validation, and controls) need to be maintained similar to the patented biopharmaceutical product and should include quality, safety, and efficacy [100]. Key scientific issues for biogenerics are product characterization, demonstration of bioequivalence, and manufacturing issues (e.g., lack of ownership or access to a contracted facility, incomplete disclosure of patented product processes, and difficulty obtaining a suitable cell line and raw materials) [100]. The process as well as the strain for biogenerics are different from those of the original product, and the impact of even small changes on biopharmaceuticals is not clear [99].

The potential world market for biogenerics has been estimated at $5.4 billion/ year by 2008, assuming the evolution of a favorable regulatory environment [101], compared with a total biopharmaceutical market of $30 billion/year [99]. These estimates may be too high for the market potential for biogenerics because patients often are switched to the next-generation versions of branded products due to rapid innovations in these products [101]. In fact, the main barriers to entry are the fast pace of existing product enhancement or new product introduction, and an ill-defined regulatory framework [99]. In addition, due to the expense of biomanufacturing and developing a biomanufacturing process, the cost reduction of biogenerics may not be as significant as with non-biopharmaceuticals [1]. One estimate is that biogenerics will sell for 10–20% less than branded counterparts rather than 40–80% less as with small-molecule drugs [101].

Generic biopharmaceutical products, possessing a similar complex structure and high molecular weight as the original product, are harder to characterize than small-molecule generic drugs; thus, they are harder to prove equivalent [101]. To show bioequivalence, that the biogeneric is "essentially similar" to the patented biopharmaceutical, sophisticated analytical analysis is required. Comparability studies are required to prove the similar nature of the generic product, and the nature of these studies is product-class specific [101]. Prior experience is that even small changes in the product can result in significant safety or efficacy differences.

Biogenerics are currently in a state of "regulatory purgatory" [98]. Key regulatory issues are that the approval process is not well defined and is uncertain, particularly in the United States [100]. The FDA says creating a regulatory pathway for biogenerics is a top priority (Europe already has a biogenerics regulatory framework), but little resolution has been reached [98].

Regulatory authorities are requesting extensive demonstrations of biological activity, which translate to lengthy and costly clinical trials [99]. Proof of identity is difficult for biopharmaceuticals because they are difficult to analytically characterize. "The process defines the product" notion still exists when there is limited knowledge [100] and experience regarding cause-and-effect relationships [99]. Extensive oversight of biogenerics manufacturers is expected because the entire manufacturing process is as important as the final product and needs to be carefully regulated [1]. These requirements make development timelines and costs substantially higher than for small-molecule generics [99].

4.2.7.4 Outsourcing

As biopharmaceutical companies redefine their goals and priorities, outsourcing of workload can maximize investment [28] and assure an ability to meet capacity demands [102]. Previously, outsourcing has been done in clinical research development and fill/finish manufacturing steps [28, 103]. Now it is being done for QC/QA functions because every quality function except final release responsibility for biopharmaceutical drug product for human clinical trials can be outsourced [28]. Outsourcing also is occurring in many areas such as process development, clinical material production, and marketable production [102, 104]. Specific areas such as cell line development (high producer screening) and scale-up also can be outsourced [104]. Risk management systems are used to proactively identify the risks and liability between the organizations [105].

Contract companies can provide more than simply additional capacity. They can provide translation of bench-scale operations to cGMP compliant manufacturing, technology transfer, and process validation [104]. Contract services also are provided by equipment suppliers to optimize/develop processes using their equipment [90]. These services can be more cost-effective than third-party laboratories, contract manufacturing, or in-house.

4.2.7.5 Risk-Based GMPs

In the future, when manufacturing processes are developed and understood, there will be increased innovation and regulatory oversight proportional to risk [23]. The major goals of risk-based GMPs are to focus GMP attention on potential risk areas, ensure regulations do not impede innovations, enhance GMP inspection consistency, and encourage the use of the latest scientific advances in manufacturing [106]. To accomplish these goals, work is pro-gressing based on several major principles, including risk-based orientation, science-based policies/standards, integrated quality systems orientation, international cooperation, and strong public health protection. The advent of risk-based GMPs provides a timely opportunity for industry to prospectively evaluate ways to streamline the complexity of existing and proposed processes using sound and defendable methodologies for risk assessment [106].

4.2.8 SUMMARY

Overall, process validation is a methodology to permit organizations to take advantage of process improvement opportunities presenting themselves over the course of a product's lifetime. Process validation continues to be a formidable undertaking, requiring extensive resources and time for full execution. Improvements in efficiency are likely to arise as technologies for high-throughput scale-down units and fast-turnaround, sensitive analytical assays become more readily available. Focused guidance on process development, characterization, and validation also should be beneficial to evaluate and align varied industry practices.

NOMENCLATURE

API—active pharmaceutical ingredient

BSE—bovine spongiform encephalopathy

CBER—Center for Biologics Evaluation and Research, US FDA

CCP—critical control points

CQA—critical quality attributes

CDER—Center for Drug Evaluation and Research, US FDA

CV—coefficient of variation

DO—dissolved oxygen

dCO2—dissolved carbon dioxide

DOE—design of experiments

EMEA—European Medicines Evaluation Agency

EU—European Union

FDA—(United States) Food and Drug Administration

FME(C)A—failure mode and (critical) effect analysis

FTA—fault tree analysis

(c)GMP—(current) Good Manufacturing Practice

HACCP—Hazard analysis and critical control point

ICH—International Conference on Harmonization

KPIV—key process input variables

LAL—limulus-amoebocytes-lysate

LOD—limit of detection

LOQ—limit of quantitation

MOR—maximum operating range

NOR—normal operating range

PAR—proven acceptable range

PAT—process analytical technology

PCR—polymerase chain reaction

PDA—Parenteral Drug Association

POD—probability of detection

PV—process validation

QA—Quality Assurance
QC—Quality Control
RPN—risk priority number
SD(s)—standard deviation(s)
SOP(s)—standard operating procedure(s)
SPC—statistical process control
TOC—total organic carbon
WHO—World Heath Organization

REFERENCES

1. Dibner M D (2005). Biopharmaceuticals in 2005 and beyond. *Bioexec. Internat.* 1: 2–8.
2. Akers J, McEntire J, Sofer G (1994). Biotechnology product validation, part 1: Identifying the pitfalls. *BioPharm.* 7:40–43.
3. Zabriskie D W, Smith T M, Gardner A R (1999). Process validation. In M C Flickinger, S W Drew (eds.), *Encyclopedia of Bioprocess Technology: Vol. 4, Fermentation, Biocatalysis and Bioseparation*: Wiley, New York, pp. 2070–2079.
4. Federici M M (1994). The quality control of biotechnology products. *Biologicals.* 22:151–159.
5. Darling A (2002). Validation of biopharmaceutical purification processes for virus clearance evaluation. *Molec. Biotechnol.* 21(1):57–83.
6. Hill C R (1990). The manufacture of proteins for human therapeutic use. In D L Pyle (ed.), *Separations For Biotechnology Vol. 2*, Elsevier Applied Science, London, pp. 431–443.
7. Sommerfeld S, Strube J (2005). Challenges in biotechnology production- generic processes and process optimization for monoclonal antibodies. *Chem. Engin. Process.* 44:1123–1137.
8. Hendriksen C F M (1999). Validation of tests methods in the quality control of biologicals. In F Brown, C F M Hendriksen, D Sesardic (eds.), *Alternatives to Animals in the Development and Control of Biological Products for Human and Veterinary Use. Dev. Biol. Stand.* Basel Switzerland, Karger, 101:217–221.
9. Naglak T J, Keith M G, Omstead D R (1994). Validation of Fermentation Processes. *BioPharm.* 7:28–36.
10. Lubiniecki A S, Gardner A R, Smith T M, et al. (2003). Validation of fermentation processes. In F Brown, A S Lubineck (eds.), *Process Validation for Manufacturing of Biologics and Biotechnology Products, Devel. Biolog.* pp. 37–44, 111–112.
11. Mollah A H (2004). Risk analysis and process validation. *BioProcess. Internat.* 2(9):28–35.
12. Lubiniecki A S, Wiebe M E, Builder S E (1990). Process validation for cell culture-derived pharmaceutical proteins. *Bioproc. Technol.* 10:515–541.
13. Bliem R F (1995). Impact of research and development on validation, GMP and registration of biopharmaceuticals. *Pharmac. Engin.* 15(3):48–54.
14. Rathore A S, Noferi J F, Arling E R, et al. (2002). Process validation: How much to do and when to do it. *BioPharm.* 15:18–28.
15. Murphy R, Seeley R J (2001). Validation of biotechnology active pharmaceutical ingredients. In I R Berry, D Harpaz (eds.), *Validation of Active Pharmaceuticals Ingredients*: IHS Health Group, Englewood C. O., pp. 451–470.

16. Garnick R L, Solli N J, Papa P A (1988). The role of quality control in biotechnology: An analytical perspective. *Anal. Chem.* 60:2456–2557.

17. Jungbauer A, Boschetti E (1994). Manufacture of recombinant proteins with safe and validated chromatographic sorbents. *J. Chromatog. B: Biomed. Applic.* 662:143–179.

18. Ferreira J, Cooper D (1996). Process validation and drug cGMPs: A detailed analysis of the proposed revisions. *J. Valid. Technol.* 3:80–86.

19. Metz B, Hendriksen C F M, Jiskoot W, Kersten G F A (2002). Reduction of animal use in human vaccine quality control: Opportunities and problems. *Vaccine.* 20:2411–2430.

20. Gardner A R, Smith T M (2000). Identification and establishment of operating ranges of critical process variables. In G Sofer, D W Zabriskie (eds.), *Biopharmaceutical Process Validation*, Marcel Dekker, Inc.: Basel Switzerland, pp. 61–76.

21. Dellepiane N, Griffiths E, Milstien J B (2000). New challenges in assuring vaccine quality. *Bulletin of the W.H.O.*, 78(2):155–162.

22. Martin-Moe S, Kelsey W H, Ellis J, et al. (2000). Process validation in biopharmaceutical manufacturing. In G Sofer, D W Zabriskie (eds.), *Biopharmaceutical Process Validation*, Marcel Dekker, Inc.: Basel Switzerland, pp. 287–298.

23. FDC Reports (2005). Design space to facilitate quality initiative implementation. *The Gold Sheet.* 39(4):1.

24. Graffner C (2005). PAT-European regulatory perspective. *Proc. Analyt. Technol.* 2(26):8–11.

25. Chapman K G (1993). Validation terminology. In I R Berry, R A Nash (eds.), Pharmaceutical Process Validation, 2nd ed.: Marcel Dekker, Inc., New York, pp. 587–597.

26. O'Leary R M, Feuerhelm D, Peers D, et al. (2001). Determining the useful lifetime of chromatography resins. *BioPharm.* 14:10–18.

27. Zoon K, Garnick R (1998). Definition of a well-characterized biotechnology product. In F Brown, A Lubiniecki, G Murano (eds.), *Characterization of Biotechnology Pharmaceutical Products, Vol. 96, Dev. Biol. Stand.* Karger, Basel Switzerland, pp. 191–197.

28. Geigert J (1997). The challenge of managing quality control/quality assurance for loosely structured and virtual biopharmaceutical companies. *Drug Inform. J.* 31(1):97–100.

29. Lubiniecki A S (1987). Safety considerations for cell culture-derived biologicals. In B K Lydersen (ed.), *Large Scale Cell Culture Technology*: Hanser Publishers, New York, pp. 232–247.

30. Bobrowicz G (1999). The compliance costs of hasty process development. *BioPharm.* 12:35–38.

31. Welch K A (2003). Validation of checkweighers: A case study. *J. Validat. Technol.* 9(3):246–252.

32. Martin-Moe S, Ellis J, Coan M, et al. (2000). Validation of critical process input parameters in the production of protein pharmaceutical products: A strategy for validating new processes or revalidating existing processes. *J. Pharmac. Sci. Technol.* 54(4):315–319 and 54(5):423.

33. PDA (2005b). Process validation of protein manufacturing. PDA Technical Report No. 52, *PDA J. Pharmac. Sci. Technol.* 59(S-4):1–28.

34. Seely R J, Hutchins H V, Luscher M P, et al. (1999). Defining critical variables in well-characterized biotechnology processes. *BioPharm.* 12(4):33–36.

35. Kelley B D (2000). Establishing process robustness using designed experiments. In G Sofer, D W Zabriskie (eds.), *Biopharmaceutical Process Validation*: Marcel Dekker, New York, pp. 29–59.

36. Pujar N S, Gayton M G, Herber W, et al. (2005). Process validation of a multivalent bacterial vaccine: A novel matrix approach. In A S Rathore, G Sofer (eds.), *Process Validation*, Taylor and Francis, New York, pp. 529–550.

37. Welch K A, Fung C A, Schmidt S R (2004). Risk/science-based approach to validation: A win-win-win for patients, regulators, and industry. *PDA J. Pharmac. Sci. Technol.* 58(1):15–23.

38. Sofer G (1994), Validation of chromatography processes. In L Alberghina, L Frontali, P Sensi (eds.), *Proc. of the 6th European Congress on Biotechnology*, Elsevier Science B. V., Florence, Italy, pp. 977–980.

39. Sniff K S, Garcia L A, Hassler R A, et al. (1996). Process development, scale-up and validation of a recombinant *E. coli*-based fermentation process. *ASTM Spec. Tech. Publ., STP 1260*: 97–106.

40. Zhang J, Reddy J, Salmon P, et al. (1998). Process characterization studies to facilitate validation of a recombinant protein fermentation. In B D Kelley, R A Ramelmeier (eds.), *Validation of Biopharmaceutical Manufacturing Processes. ACS Symposium Series.* 698:12–27.

41. Darling A, Spaltro J J (1996). Process validation for virus removal: Considerations for design of process studies and viral assays. *BioPharm.* 9(9):42–50.

42. Brorson K, Krejci S, Lee K, et al. (2003b). Bracketed generic inactivation of rodent retroviruses by low pH treatment for monoclonal antibodies and recombinant proteins. *Biotechnol. Bioeng.* 82:321–329.

43. Sofer G (2002a). Virus inactivation in the 1990s- and into the 21st century, Part 2: Red blood cells and platelets. *BioPharm.* 15:42–49.

44. Sofer G (2002b). Virus inactivation in the 1990s- and into the 21st century, Part 3a: Plasma and plasma products (heat and solvent detergent treatments). *BioPharm.* 15:28–42.

45. Sofer G (2003b). Virus inactivation in the 1990s- and into the 21st century, Part 4: Culture media, biotechnology products, and vaccines. *BioPharm.* 16:50–57.

46. Lau A S L, Lie Y S, Norling L A, et al. (1999). Quantitative competitive reverse transcription-PCR as a method to evaluate retrovirus removal during chromatography procedures. *J. Biotechnol.* 75:105–115.

47. Walter J, Allgaier H (1997). Validation of downstream processes. In R Wagner, H Hauserv (eds.), *Mammalian Cell Biotechnology in Protein Production*: Walter De Gruyter Inc., New York, pp. 453–482.

48. Hirasaki T, Tsuboi T, Noda T, (1992). Removability of virus particles and permeability of protein from cell culture medium using cuprammonium regenerated cellulose hollow fiber (BMM™). In H Murakami, S Shirahata, H Tachibana (eds.), *Animal Cell Technology: Basic & Applied Aspects*, Kluwer Academic Publishers, The Netherlands, pp. 49–55.

49. PDA (2005a). Virus filtration- Introduction. *PDA J. Pharm. Sci. Technol.* 59 (S-2): 8–42.

50. Siklosi T (2005). Manufacturing of biopharmaceutical proteins. *Genetic Engin. News.* 25(17):58–59.

51. Levine H L, Tarnowski S J, Dosmar M, (1992). Industry perspective on the validation of column-based separation processes for the purification of proteins. *J. Parental Sci. Tech.* 46(3):87–97.

52. Sofer G (1996) Validation: Ensuring the accuracy of scaled-down chromatography models. *BioPharm.* 9(9):51–55.

53. Hesse F, Wagner R (2000). Developments and improvements in the manufacturing of human therapeutics with mammalian cell cultures. *Tibtech.* 18:173–180.

54. Garnick R L (1998). Raw materials as a source of contamination in large-scale cell culture. In F Brown, E Griffiths, F Horaud, J C Petricciani (eds.), *Safety of Biological Products Prepared from Mammalian Cell Culture, Vol. 93, Dev. Biol. Stand.* Karger, Basel Switzerland, pp. 21–29.

55. Facklam T J, Geyer S. The preparation and validation of stock cultures of mammalian cells. In Y H Chui, J L Gueriguian (eds.), Drug Biotechnology Regulation: Scientific Basis and Practices. Marcel Dekker, Inc., New York, 1991, pp. 54–85.

56. Geigert J (1989). Overview of the stability and handling of recombinant protein drugs. *J. Parenteral Sci. Technol.* 43(5):220–224.

57. Bennan J, Bing F, Boone H, et al. (2002). Evaluation of extractables from product-contact surfaces. *BioPharm Internat.* 22–34.

58. Seely R J, Wight H D, Fry H H, et al. (1994). Biotechnology product validation, Part 7: Validation of chromatography resin useful life. *BioPharm.* 15:7:41–48.

59. Depalma A (2006). Bioprocessing: Bright sky of single use bioprocess products. *GEN.* 26(3):1.

60. Webb S D, Webb J N, Hughes T G, et al. (2002). Freezing biopharmaceuticals using common techniques—and the magnitude of bulk-scale freeze-concentration. *BioPharm.* 15:22–34.

61. Breece T N, Gilkerson E, Schmelzer C (2002a). Validation of large-scale chromatographic processes, Part 1: Case study of neuleze capture on macroprep high-S. *BioPharm.* 15:16–20.

62. Brorson K, Brown J, Hamilton E, et al. (2003a). Identification of protein A media performance attributes that can be monitored as surrogates for retroviral clearance during extended use. *J. Chrom. A.* 989:155–163.

63. Norling L, Lute S, Emery R, et al. (2005). Impact of multiple re-use of anion exchange chromatography media on virus removal. *J. Chrom. A.* 1069:79–89.

64. Brough H, Antoniou C, Carter J, et al. (2002). Performance of a novel Viresolve NFR virus filter. *Biotechnol. Prog.* 18:782–795.

65. Seely J E, Seely R J (2003). A rationale, step-wise approach to process characterization. *BioPharm Internat.* 16: 24–34.

66. Krause S O (2005). How to maintain suitable analytical test methods: Tools for ensuring a validation continuum. *BioPharm Internat.* 18(10):52–60.

67. Seely R J, Haury J (2005). Applications of failure modes and effects analysis to biotechnology manufacturing processes. In A S Rathore, G Sofer (eds.), *Process Validation in Manufacturing of Biopharmaceuticals*: Taylor and Francis, New York, pp. 13–29.

68. Seely J E (2005). Process characterization. In A S Rathore, G Sofer (eds.), *Process Validation in Manufacturing of Biopharmaceuticals*: Taylor and Francis, New York, pp. 31–67.

69. Kell D B, Sonnleitner B (1995). GMP- good modelling practice: An essential component of good manufacturing practice. *Tibtech.* 13:481–492.

70. Noble P T (2001). Reduction of risk and the evaluation of quality assurance. *PDA J. Pharmac. Sci. Technol.* 55(4):235–239.

71. Sahni A (1993). Using failure mode and effects analysis to improve manufacturing processes. *Med. Dev. Diagn. Ind.* 15:47–51.

72. Kieffer R G, Bureau S, Borgmann A (1997). Applications of failure mode effect analysis in the pharmaceutical industry. *Pharma. Technol. Europe.* 10:36–49.

73. Mollah A H (2005). Application of failure made and effect analysis (FMEA) for process risk assessment. *BioProcess Internat.* 3:12–21.

74. Armbruster D, Feldsien T (2000). Applying HACCP to pharmaceutical process validation. *Pharmac. Technol.* 13:170–178.

75. Little L E (1995). Validation of immunological and biological assays. *BioPharm.* 36–42.

76. Xu Y, Brorson K (2003). An overview of quantitative PCR assays for biologicals: Quality and safety evaluation. In F Brown, A S Lubiniecki (eds.), *Process Validation for Manufacturing of Biologics and Biotechnology Products*, Basel, Karger Switzerland, pp. 89–98.

77. Federici M M, Garnick R L, Geigert J (1991). A perspective on reference standard and reference material requirements in biotechnology-derived pharmaceutical protein products. *Pharmacopeial Forum.* 17:2683–2687.

78. Bozzo T (1995). Biotechnology product validation, Part 9: A former FDAer's view. *BioPharm.* 8:35–38.

79. Moran E B, McGowan S T, McGuire J M, et al. (2000). A systematic approach to the validation of process control parameters for monoclonal antibody production in fed-batch culture of a murine myeloma. *Biotechnol. Bioengin.* 69(3):242–255.

80. Rathore A S, Johnson G V, Buckley J J, et al. (2003). Process characterization of the chromatographic steps in the purification process of a recombinant *Escherichia coli*-expressed protein. *Biotechnol. Appl. Biochem.* 37:51–61.

81. Rathore A S (2005). Process optimization and characterization studies for purification of an *E.coli*-expressed protein product. In A S Rathore, G Sofer (eds.). *Process Validation in Manufacturing of Biomanufacturing of Biopharmaceuticals*, Taylor and Francis, New York, pp. 451–468.

82. Francis R, Bonnerjea J, Hill C R (1990). Validation of the re-use of protein a sepharose for the purification of monoclonal antibodies. In D L Pyle (ed.), *Separations for Biotechnology 2*, Elsevier Applied Science, New York, pp. 491–498.

83. Kelley B D, Tannatt M, Magnusson R, et al. (2004). Development and validation of an affinity chromatography step using a peptide ligand for cGMP production of factor VIII. *Biotechnol. Bioengin.* 87(3):400–412.

84. Breece T N, Gilkerson E, Schmelzer C (2002b). Validation of large-scale chromatographic processes, Part 2: Results from the case study of neuleze capture on macroprep high-S. *BioPharm.* 15:35–42.

85. Brady D, Bonnerjea J, Hill C R (1990). Purification of monoclonal antibodies for human clinical use: Validation of DNA and retroviral clearance. In D L Pyle (ed.), *Separations for Biotechnology 2*, Elsevier Applied Science, New York, pp. 472–480.

86. Johnston A, MacGregor A, Borovec S, et al. (2000). Inactivation and clearance of viruses during the manufacture of high purity factor IX. *Biologicals.* 28:129–136.

87. Conley L, McPherson J, Thommes J (2005). Validation of the ZEVALIN® purification process- A case study. In A S Rathore, G Sofer (eds.), *Process Validation*, Taylor and Francis, New York, pp. 469–521.

88. McGurk T L (2004). Ramping up and ensuring supply capability for biopharmaceuticals. *BioPharm. Internat.* 17(1):38–44.

89. Seely R J, Munyakazi L, Haury J (2001). Statistical tools for setting in-process acceptance criteria. *BioPharm.* 14:28–34.

90. Martin J (2005). Turning compliance challenges into competitive advantages. *Chem. Engin.* 112(8):23–26.

91. Curtis S, Lee K, Blank G S, et al. (2003). Generic/matrix evaluation of SV40 clearance by anion exchange chromatography in flow-through mode. *Biotechnol. Bioengin.* 84(2):179–186.

92. Godarvarti R, Petrone J, Robinson J, et al. (2005). Scale-down models for purification processes: Approaches and applications. In A S Rathore, G Sofer (eds.), *Process Validation*: Taylor and Francis, New York, pp. 69–142.

93. Zoon K C (1998). Well-characterized biotechnology products: Evolving to meet the needs of the 21st century. In F Brown, A Lubiniecki, G Murano (eds.). *Characterization of Biotechnology Pharmaceutical Products, Dev. Biol. Stand.*, Basel Switzerland, Karger, 96:3–8.

94. Webber K (2005). FDA Update: Process analytical technology for biotechnology products. *J. Process Analytical Technol.* 2(4):12–14.

95. Montague G, Morris J (1994). Neural-network contributions in biotechnology. *Tibtech.* 12:312–324.

96. Rockwell Automation (2004). PAT initiative expected to invigorate pharmaceutical industry with improved quality, better efficiency, and improved profits. White Paper. Publication LIFE-WP001A-EN-P, Nov.

97. Geismar D, White G (2004). Process reliability and flexibility—A tool to improve pharmaceutical plan floor operations. *Pharm-Ind*, 11a:1399–1403.

98. Katsnelson A (2005). Teva to acquire Chinese biogenerics manufacturer. *Nat. Biotechnol.* 23(7):765.

99. Polastro E T, Tulcinsky S (2004). Are biopharmaceuticals immune to the generics threat? *J. Gene. Med.* 1(3):259–265.

100. Henry Steward Publications (2005). An overview of the scientific and regulatory issues facing biogenerics in the USA today. *J. Gener. Med.* 2(2):169–172.

101. Ainsworth S J (2005). Biopharmaceuticals: Patent expirations are beckoning generic drug companies, but numerous hurdles remain to a profitable business. *Chem. Engin. News.* 83(23):21–29.

102. Hempel J C, Hess P N (2005). Contract manufacturing of biopharmaceuticals including antibodies or antibody fragments. In J Knablein (ed.), *Modern Biopharmaceuticals*, Wiley-VCH, Weinheim, Germany, 1083–1107.

103. Kulkarni M (2005). Outsourcing QA: Perspectives on future strategies. *Contract Pharma.* 7(9):58–61.

104. Fox S (2004). Maximizing outsourced biopharma production. *Contract Pharma.* 6: 72–78.

105. Ryan C S, Wan M (2000). Risk management and liability for biologics. *BioPharm.* 13:40–42.

106. Shenoy P (2004). Risk based cGMP—A new initiative by USFDA *PharmaTimes.* 36(2):23–24.

FURTHER READING

Brown F, Lubiniecki A S (eds.) (2003). *Process Validation for Manufacturing of Biologics and Biotechnology Products*, Basel, Karger Switzerland.

Kelley B D, Ramelmeier R A (eds.) (1998) Validation of Biopharmaceutical Manufacturing Processes. ACS Symposium Series 698, San Francisco, California.

Rathore A S, Sofer G (eds.), (2005). *Process Validation in Manufacturing of Biopharmaceuticals*, Taylor and Francis, New York.

Sofer G, Zabriskie D W (eds.) (2000). *Biopharmaceutical Process Validation*, Marcel Dekker, Inc., Basel Switzerland.

4.3

STABILITY ASSESSMENT AND FORMULATION CHARACTERIZATION

ALBERT H.L. CHOW,[1] HENRY H.Y. TONG,[2] AND YING ZHENG[3]

[1] The Chinese University of Hong Kong, Shatin, N.T., Hong Kong, China
[2] Macao Polytechnic Institute, Macao, China
[3] University of Macao, Macao, China

Chapter Contents

4.3.1 Chemical Instability and Potential Degradation Pathways 372
    4.3.1.1 Hydrolytic Reactions in Proteins and Peptides 372
    4.3.1.2 Hydrolytic Reactions in Nucleic Acids 378
    4.3.1.3 Oxidation 381
    4.3.1.4 β-Elimination in Nucleic Acids, Proteins and Peptides 384
    4.3.1.5 Maillard Reaction in Nucleic Acids, Proteins, and Peptides 385
    4.3.1.6 Covalent Dimerization and Polymerization in Proteins 386
4.3.2 Physical Instability 387
    4.3.2.1 Denaturation 387
    4.3.2.2 Aggregation and Precipitation 388
    4.3.2.3 Adsorption 390
4.3.3 Considerations in the Stability Testing of Biopharmaceutcals 391
    4.3.3.1 Temperature 391
    4.3.3.2 Moisture 392
    4.3.3.3 Light and Ionizing Radiation 392
    4.3.3.4 Physical Stress 393
    4.3.3.5 Freeze-Drying and Spray-Drying 394
    4.3.3.6 Excipients 394
4.3.4 Regulatory Requirements in the Stability Testing of Biopharmaceuticals 395

Handbook of Pharmaceutical Biotechnology, Edited by Shayne Cox Gad.
Copyright © 2007 John Wiley & Sons, Inc.

4.3.5   Formulation and Characterization of Biopharmaceuticals   395
    4.3.5.1   Protein and Peptide Drugs   396
    4.3.5.2   Nucleic Acid-Based Drugs   401
4.3.6   Summary and Conclusions   404
    References   404

A drug must be formulated in a safe, consistent, and stable dosage form with proven clinical efficacy before it can be approved for specific therapeutic indications by drug regulatory authorities. Although biotechnological/biological products or biopharmaceuticals (therapeutic proteins, peptides, and nucleic acids) are no exception to this requirement, the inherently complex primary structures of this important class of therapeutic agents coupled with their ability to form higher order structures have posed additional challenges and constraints in their formulation and assessment.

The current chapter is intended to provide an overview of the stability problems of biopharmaceuticals and their assessment as well as the formulation approaches that can be used to circumvent these problems. Presented next is a literature review on the known degradation/denaturation mechanisms of various biopharmaceuticals, followed by a general discussion on the formulation strategies using specific excipients/additives for improving the product shelf-life.

4.3.1   CHEMICAL INSTABILITY AND POTENTIAL DEGRADATION PATHWAYS

4.3.1.1   Hydrolytic Reactions in Proteins and Peptides

Deamidation. Deamidation is a common degradation pathway for proteins and peptides, involving the conversion of the side chains of the asparagine and glutamine residues to the respective carboxylates of aspartate and glutamate. The reaction is highly pH dependent. Under acidic conditions, the reaction is initiated by the direct intermolecular attack of water at the side-chain amide linkage, yielding ammonia and the corresponding acids (i.e., aspartic acid for asparagine residue or glutamic acid for glutamine residue) [1–3]. At neutral or basic pH, the degradation is predominated by intramolecular deamidation with a relatively minor contribution from direct hydrolysis [2].

Mechanistically, deamidation of asparagine proceeds via the formation of a succinimide intermediate, involving the intramolecular attack by the nitrogen in the peptide bond on the C-terminus (Scheme 4.3-1) [2, 4]. The resulting cyclicimide tends to racemize at α-carbon. Hydrolysis of the cyclicimide yields a mixture of peptide products differing in the type of amide linkage in the polypeptide backbone; for the α-carboxyl linkage, the hydrolytic products are a mixture of *d*- and *l*-normal aspartyl peptides, whereas the β-carboxyl linkage affords a mixture of *d*- and *l*-isoaspartyl transpeptidation products [3, 5]. Spontaneous deamidation of the polypeptide chain at the asparagine residue can also occur via the attack by the

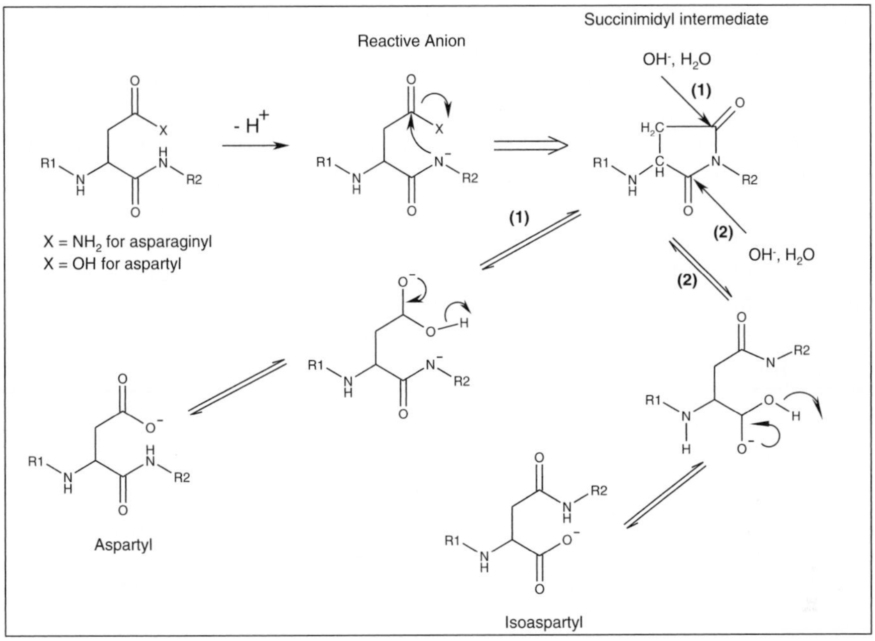

Scheme 4.3-1.

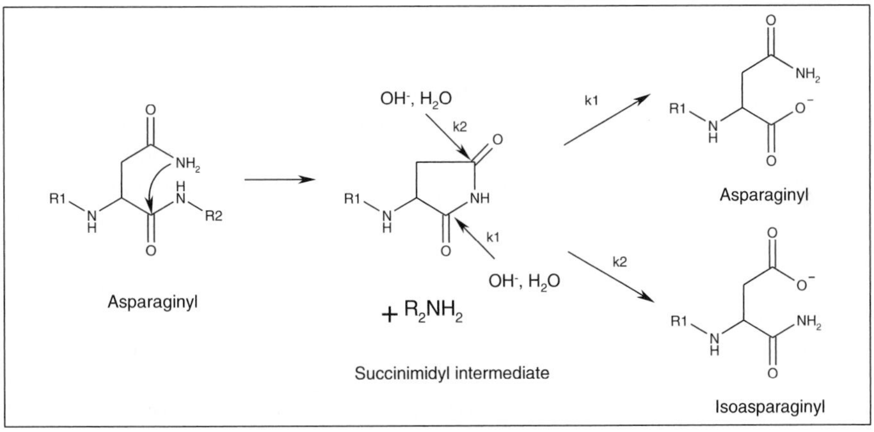

Scheme 4.3-2.

nitrogen in the side chain of the amide on the following peptide-bond carbonyl to form two polypeptides, one of which contains a C-terminal succinimide residue (Scheme 4.3-2) [3]. Most biopharmaceuticals are formulated at physiological pH (7.4) for parenteral administration and thus are prone to decomposition by deamidation via the aforementioned mechanisms. These products encompass recombinant human parathyroid hormone [6], orthoclone OKT3 [7], human epidermal growth factor [8], insulin [9], antiflammin 2 [10], recombinant tissue plasminogen activator [11], monoclonal immunoglobulin h1B4 [12], and murine monoclonal antibody MMA383 [13]. Adenovirus, a therapeutic viral gene carrier, is also known to degrade by deamidation and isoaspartate formation at four susceptible asparagine residues, and the stability of a recombinant adenovirus AV3760 has demonstrated considerable improvement after site-directed mutagenesis of the key asparagine residues in the hexon protein [14].

Similar to the asparagine residue (Scheme 4.3-1), glutamine residue undergoes deamidation at neutral or alkaline pH, where intramolecular attack results in glutarimide formation. Having one more methylene group than the asparagine residue, the glutamine residue yields a six-membered ring intermediate structure, which is formed at a much slower rate than that of the five-membered succinimide intermediate for the asparagine residue because of entropic factors [1, 11]. Although the asparagine residue is widely known to be more susceptible to deamidation than the glutamine residue, more facile deamidation for the glutamine residue under acidic conditions has been observed with the glucagon fragment 22–29 [15]. The lower reactivity of the asparagine residue in this case has been attributed to a decreased electrophilicity of its side-chain carbonyl carbon imparted by a parallel cleavage pathway at this residue.

Primary sequence has an important bearing on the spontaneous deamidation in peptides. It has been demonstrated using model peptides that the structure of amino acid residue immediately following the asparagine residue N + 1 exerts a significant impact on the overall rate of succinimide formation. For example, glycine residue in the N + 1 position results in a deamidation rate 70- to 100-fold faster than that found for amino acid residues with branched hydrophobic side chains [2]. In another study, the bulky side chain of N + 1 histidine residue has been found to limit the flexibility of the model peptide around the reaction site, thereby reducing the reaction rate [3]. As isoaspartate formation depends on both primary sequence and chain flexibility, two important generalizations can be made [11]. First, isoaspartate tends to form preferentially at the asparagine-glycine, asparagine-serine, aspartic acid-glycine, and possibly asparagine-histidine sites. Second, isoaspartate formation is favored in the highly flexible region of a polypeptide chain. However, it must be noted that these generalizations, which are based on solution-state data, may not be applicable to solid peptides because the mechanisms of deamidation involved in the solution and solid states are intrinsically different. As has been demonstrated in the solution state, a glutamine residue in the N + 1 position can either hinder asparagine deamidation by electrostatic and inductive effects or enhance the reaction though general acid/base catalysis, whereas in the solid form, a glutamine residue in the N + 1 position can increase the hydration of the peptide and afford asparagine deamidation rates similar to those in solution [16].

Higher order structures, viz. secondary, tertiary, and quaternary structures, also influence deamidation in proteins by positioning the asparagine or glutamine residue in spatial proximity to a functional group elsewhere in the primary sequence, which is capable of accelerating or inhibiting the deamidation. In addition, the conformational structure of the asparagine or glutamine residue and the nearby peptide region containing it may be strongly conditioned by the nature of the structural environment that incorporates the residue [17]. In general, α-helices, β-turns, and possibly β-sheets can stabilize asparagine residues against deamidation and drastically reduce the rate of succinimide formation for a given sequence relative to the same sequence in a random coil [11]. However, the effect is relatively weak with α-helices and destabilization of the asparagine residues may even occur

[17]. Conformational restrictions and reduced nucleophilic reactivity of the backbone NH centers have been suggested as probable stabilizing mechanisms with the higher order structures. In recombinant human lymphotoxin, asparagine deamidation can be inhibited by secondary and tertiary structures due to reduced polypeptide flexibility [18].

Apart from protein sequence and structure, temperature, pH, and the type of buffers can all influence the deamidation and isoaspartate formation. The effect of temperature on the deamidation reaction rate generally follows the Arrhenius law. Deamidation activation energies around 21~22 kcal mol^{-1} have been reported for two model peptides under alkaline conditions [19, 20]. The deamidation is also subject to general acid/base catalysis, as evidenced by an increase in deamidation rate with an increase in buffer concentration [2].

In stability assessment, chromatographic methods of high sensitivity are normally required to analyze the degradation products formed from parent drugs in trace amounts. Biological potency is not a sensitive marker for detecting deamidation and isoaspartate formation, which may or may not influence the biological activities of protein and peptides. If the asparagine and glutamine residues are not involved in receptor binding, there may not be any effect on biological activity upon deamidation, as in the case of human growth hormone [21], insulin [22, 23], recombinant human flial cell line derived neurotrophic factor [24], interleukin-1α [25], and recombinant human interleukin-2 [26]. However, activity loss associated with deamidation is probably the norm rather than the exception, particularly if the asparagine or glutamine residues take part in receptor binding, as exemplified by calmodulin, CD4, human and murine epidermal growth factors, growth hormone releasing factor, methionyl murine interleukin-1β, parathyroid hormone, and bovine pancreatic ribonucleases A [25]. It is important to note that in addition to the direct effect of receptor binding on activity, modification of a site outside the binding region may affect the biological activity indirectly through a local conformational change, and isoaspartate formation can also elicit autoimmunity against the proteins/peptides [11].

Peptide Bond Cleavage at Aspartic Acid Residues. The breakdown of proteins and peptides may occur by cleavage at the amide bonds. Although all amide bonds are theoretically susceptible to hydrolysis, they normally exhibit different hydrolytic rates. For instance, the rate of acid-catalyzed hydrolysis in polypeptides containing the aspartic acid residues are at least 100 times higher than those of the other peptide bonds [3, 4, 27]. The cleavage proceeds via either the *N*- or *C*-terminal bond adjacent to the aspartic acid residue (Scheme 4.3-3) [4]. For example, in acidic solutions, antiflammin 2 undergoes both *C*- and *N*-terminal hydrolyses across the aspartyl peptide bonds, and the peptide bonds at C-termini of the aspartyl residues are most susceptible to hydrolysis [28]. In addition, isoaspartate formation with aspartic acid residue is possible via the cyclicimide intermediate, as shown in Schemes 4.3-1 and 4.3-2. The influence of primary sequence on aspartic acid degradation is significant, but probably not as obvious as that observed for the asparagine residue deamidation [3]. The N + 1 glycine, which confers the greatest flexibility on the polypeptide chain, has been shown to be most vulnerable to hydrolysis at the aspartic acid residue in model peptides, which are structurally analogous to adrenocorticotrophin fragment 22–27 [29]. In a similar series of

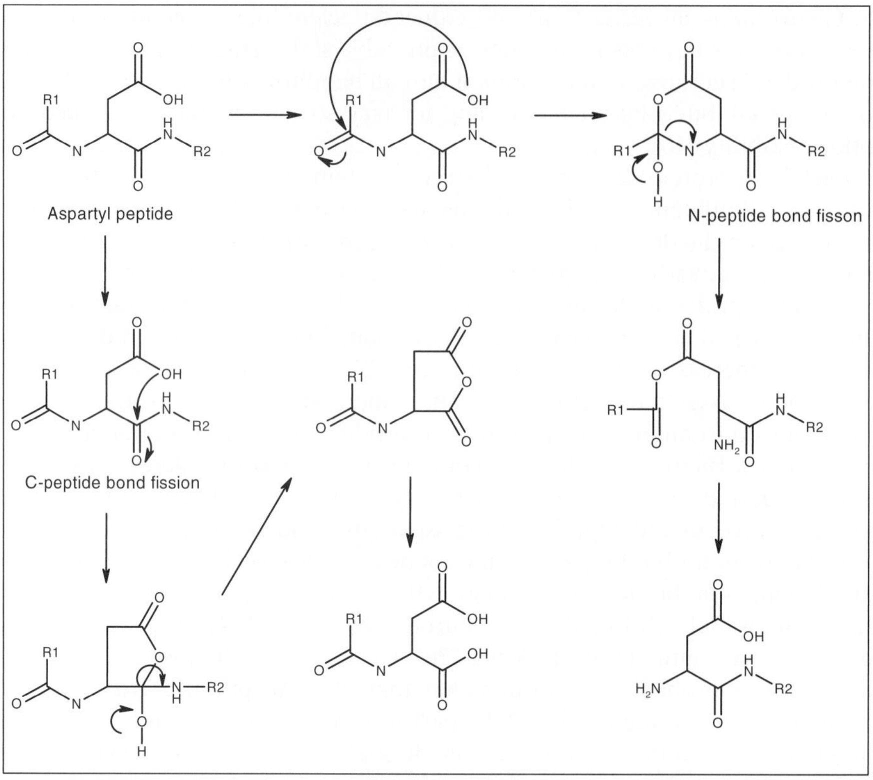

Scheme 4.3-3.

hexapeptides, it has been observed that the rate of hydrolysis decreases with an increase in steric hindrance of the N + 1 residue [30].

It is well established that aspartic acid-proline peptide bonds are particularly labile and undergo hydrolysis under conditions where other aspartic acid peptide bonds are stable, e.g., at low pHs [4, 31]. The hydrolysis proceeds via intramolecular catalysis by carboxylate anion displacement of the protonated nitrogen of the peptide bond [31]. This mechanism has been demonstrated for recombinant human interleukin 11, which undergoes peptide cleavage specifically at aspartic acid[133]-proline[134] residues in acidic solution [32]. The increase in reaction rate, by approximately 8- to 20-fold, may be ascribed to the weakened amide bond resulting from the increased electronegativity of the protonated proline nitrogen [4, 27].

As with aspartic acid containing amide bonds, peptide bonds carrying serine or threonine are generally more labile than other amide bonds. The degradation involves the nucleophilic attack of the serine or threonine hydroxyl group on the neighboring amide bond via the formation of a cyclic tetrahedral intermediate [27]. This mechanism exhibits predominance in luteinizing hormone-releasing hormone antagonist RS-26306 at pH ~5 [33].

It is worth noting that hydrolysis of the peptide bond can also occur with other amino acids. For example, salmon calcitonin undergoes hydrolysis at the cysteine[1]-serine[2] peptide bond where the cysteine remains linked to the peptide by a disulfide bond [34].

N-*terminal Degradation via Diketopiperazine and Pyroglutamic Acid Formation.* If the proteins and peptides possess a penultimate proline residue in an *N*-terminal sequence, the first amino acid in the primary sequence may be lysed from the polypeptide skeleton via the formation of diketopiperazine (DKP). This reaction is commonly termed nonenzymatic aminolysis, which results from the intramolecular nucleophilic attack of the *N*-terminal nitrogen on the carbonyl carbon of the peptide bond between the second and the third amino acid residues (Scheme 4.3-4) [3]. Pharmaceutical materials known to follow this degradation pathway include substance P [35], recombinant human growth hormone [36], histrelin [37], RMP-7 (a bradykinin analog) [38], as well as aspartame and aspartylphenylalanine in the solid state [39].

The rate of DKP formation generally increases with increasing temperature, as shown for RMP-7 [38] and phenylalanine-proline-*p*-nitroaniline (Phe-Pro-*p*-NA) [40]. Avoidance of general base catalysis or decreasing the buffer concentration at high pHs can minimize this cyclization reaction [38, 40]. The degree of ionization of the *N*-terminal group, which is governed by the pK_a of the amino group and the pH of the medium, also affects the rate of DKP formation because the first protonated amino acid in the sequence will not be available to effect base-catalyzed reaction. However, with Phe-Pro-*p*-NA and its analog, X-Pro-*p*-NA (where X is the other amino acid), the influence of primary sequence on the rate of DKP formation appears to be mediated through the ability of the X-Pro peptide to undergo *cis-trans* isomerization rather than through the difference in extent of ionization or in steric hindrance of the *N*-terminal amino acids [3, 40].

Pyroglutamic acid formation is a result of the condensation of either the amino group with γ–carboxylic acid group of glutamic acids or the amino group with γ–carboxamide group of glutamines at the *N*-terminal positions (Scheme 4.3-5) [27].

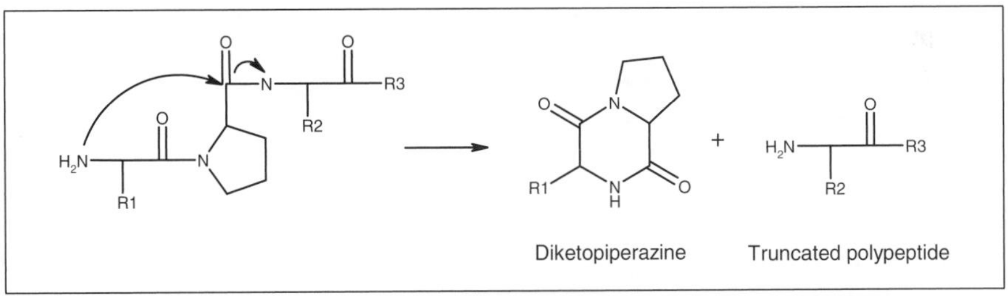

Diketopiperazine   Truncated polypeptide

Scheme 4.3-4.

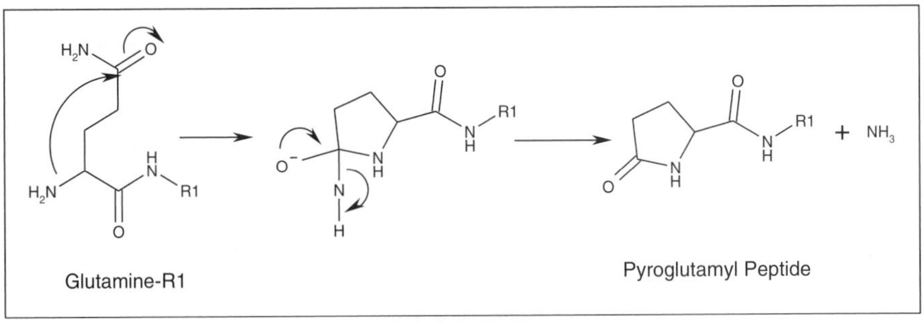

Glutamine-R1   Pyroglutamyl Peptide

Scheme 4.3-5.

It has been shown with model peptides that the rate of pyroglutamic acid formation is the fastest when the N + 1 amino acid residue next to glutamine is asparagine, leucine, or glycine [41]. Pyroglutamic acid formation in proteins and peptides is usually accompanied by a loss of biological activity. For example, decapeptide ELA, a Melan-A/MART-1 antigen immunodominant peptide analog, exhibits reduced binding to the specific class I major histocompatibility complex and a loss of cytotoxic T-lymphocyte activity after conversion of its *N*-terminal glutamic acid to pyroglutamic acid [42]. However, several peptides of pharmaceutical interest (including buserelin and gonadorelin) have been developed with the chemically more stable pyroglutamic acid instead of N-terminal glutamine or glutamic acid without loss of pharmacological activities.

4.3.1.2 Hydrolytic Reactions in Nucleic Acids

Depurination and Depyrimidation. Depurination, which refers to the hydrolysis of the *N*-glycosidic bond, is an important degradation pathway for nuclei-acid-derived drugs. By inhibiting free radical oxidation, depurination and β-elimination (Section 4.3.1.4) represent the most significant pathways for the degradation of DNA [43]. Due to the inductive effect of the 2′-OH group, deoxyribonucleosides and deoxyribonucleotides in DNA are more susceptible to depurination than are the ribonucleosides and ribonucleotides in RNA. The hydrolysis of *N*-glycosidic bond in DNA can be acid- (Scheme 4.3-6) or base-catalyzed (Scheme 4.3-7) [44]. The acid-catalyzed reaction pathway involves the protonation of the respective

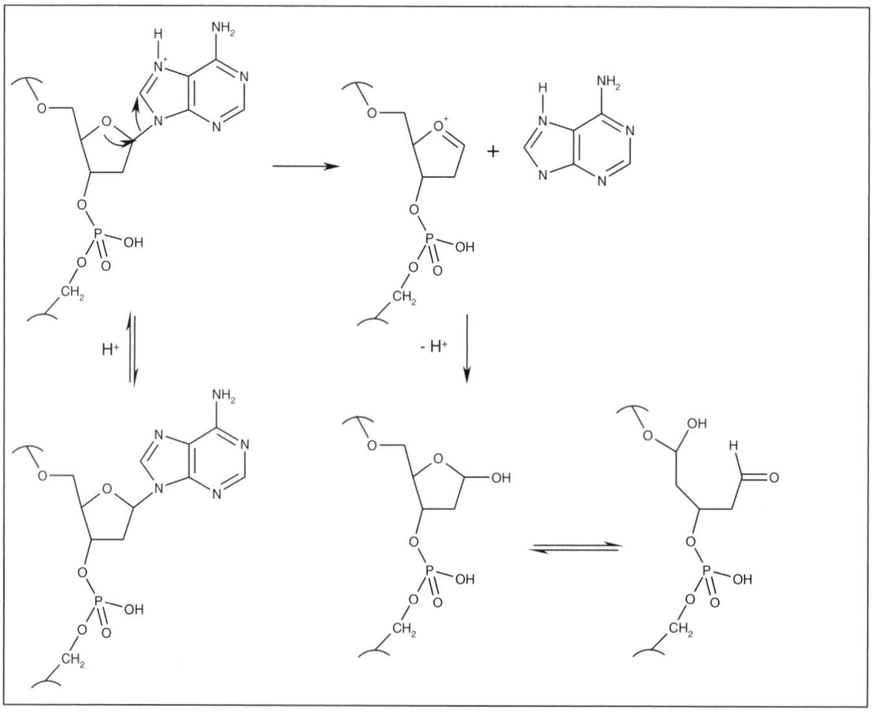

Scheme 4.3-6.

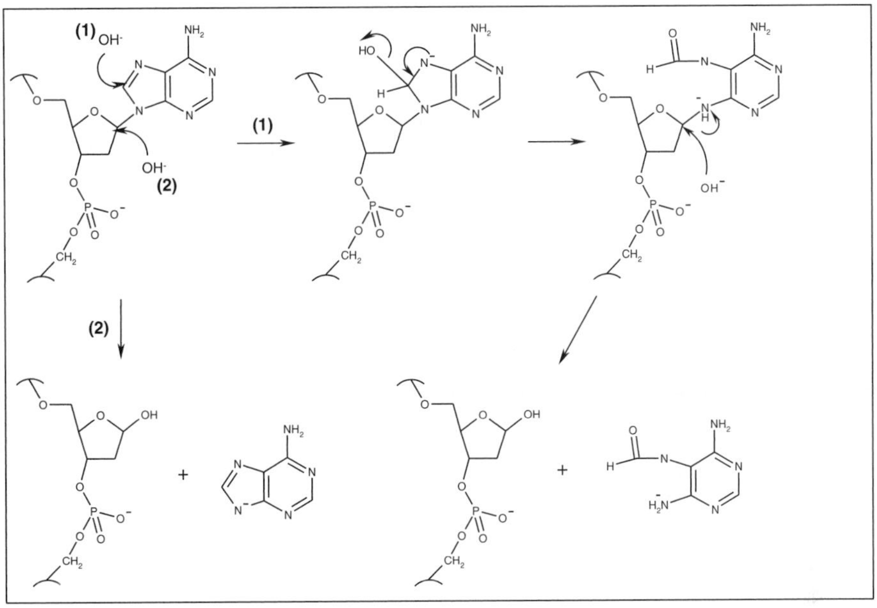

Scheme 4.3-7.

base, followed by a rate-limiting cleavage step to yield the C-1' carbonium ion (oxocarbonium ion) (Scheme 4.3-6). Hence, the stability of the N-glycosidic bond can be enhanced or reduced by destabilizing or stabilizing the C-1' carbonium ion being formed. The base-catalyzed hydrolytic reaction pathway (Scheme 4.3-7) proceeds by two S_N2 mechanisms at either C-8 or C-1', which are dependent on temperature and the hydroxide ion concentration [44]. Depurination is more rapid with guanine and adenine, whereas the loss of cytosine and thymine (depyrimidation) proceeds at 5% of the depurination rate [45]. The double helical structure in DNA can offer only limited protection against depurination, as suggested by the mere 4-fold difference in depurination velocity observed between single-stranded and double-stranded DNAs. Introduction of a single break in the DNA backbone is capable of converting supercoiled plasmid DNA to the open circular form, and thus monitoring of this conversion affords a convenient and sensitive means of determining DNA stability [46]. For instance, V1JpHA plasmid DNA at 50°C and pH 7.4 underwent depurination and β-elimination with respective rate constants of $5.6 \times 10^{-11} s^{-1}$ and $1.4 \times 10^{-6} s^{-1}$, reflecting an accumulation of ~0.5 apurinic sites per plasmid after 2 years of storage at room temperature and a corresponding decrease of ~20% in supercoiled DNA content [43].

Hydrolysis of the Phosphoester Bond. The hydrophilic surface of multistranded oligonucleotides on the sugar-phosphate backbones is a prominent target for hydrolysis. The nucleophilic phosphodiester cleavage can be mediated through both intermolecular and intramolecular reactions. DNA, being devoid of the 2'-OH group, is much less susceptible to base-catalyzed hydrolysis than RNA [44]. As shown in Scheme 4.3-9, the intramolecular displacement of the 5'-linked nucleoside by the 2'-OH group results in the phosphodiester cleavage. In acid-catalyzed

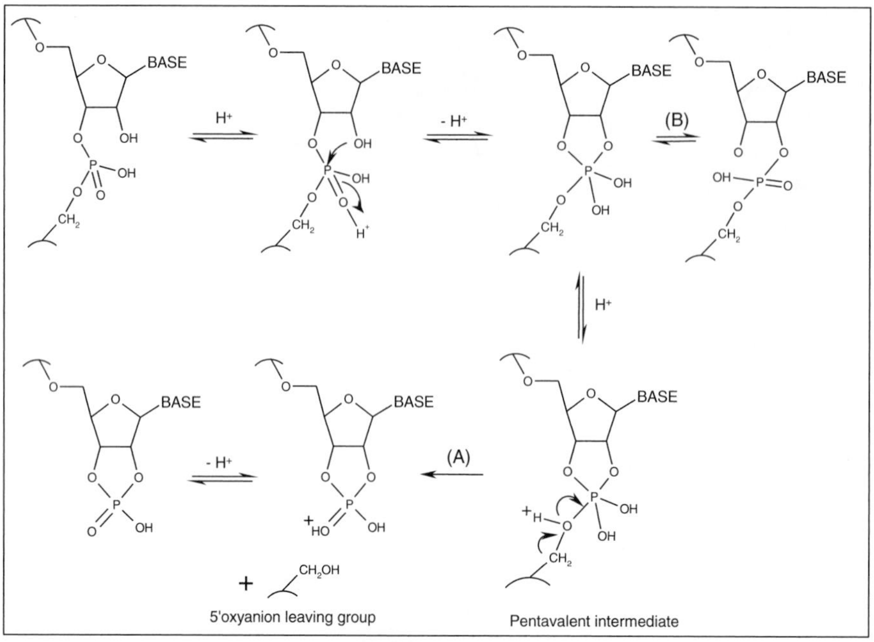

Scheme 4.3-8.

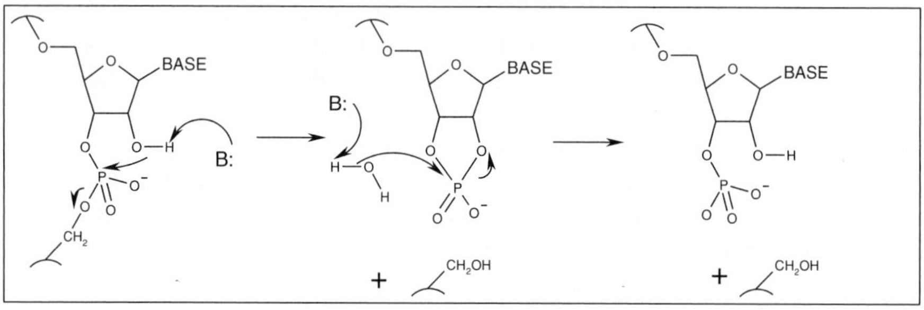

Scheme 4.3-9.

hydrolysis, isomeri-zation may also be involved (Scheme 4.3-8) [44]. Generally, pyrimidine nucleoside 3′-phosphodiesters react faster than their purine counterparts.

The hydrolysis of phosphoester bond can be accelerated by several catalytic factors, including enhancement of the nucleophilicity of the 2′-oxygen (Scheme 4.3-9), stabilization of the leaving 5′ oxyanion group (Scheme 4.3-8), stabilization of the pentavalent intermediate (Scheme 4.3-8), and the presence of buffers (e.g., imidazole, morpholine, and carboxylates) and divalent or trivalent metal ions (e.g., Mg^{2+}, Zn^{2+}, Ca^{2+}, Cu^{2+}, and Fe^{3+}) [44].

Deamination of Pyrimidine and Purine Nucleosides. Deamination of pyrimidine nucleosides into uridines is a common hydrolytic degradation for nucleic acid-derived drugs under both acidic and basic conditions. The main targets of this hydrolytic reaction are cytosine and its homologue 5-methylcytosine [45], whereas adenine and guanine are more prone to undergo deamination via the dediazoniation pathways [47]. In DNA, cytosine can be converted to uracil via direct deamination by base-catalyzed hydrolysis and parallel attack of a water molecule on the protonated base [44]. Consequently, the hydrolysis shows no strong pH dependence around pH 7.4. The reaction for single-stranded DNA is 150- to 200-fold more rapid than that for double-stranded DNA, reflecting the considerably better protection offered by the double helical structure [44, 47].

Purine nucleosides behave similarly to their pyrimidine counterparts, but their contribution to the overall degradation is insignificant [44]. For instance, adenine is converted to hypoxanthine in single-stranded DNA at only 2–3% of the rate of cytosine deamination [48].

4.3.1.3 Oxidation

Oxidation is one major chemical degradation pathway for biopharmaceuticals. The reaction involves reactive oxygen species such as hydroxyl radical (OH), hydrogen peroxide (H_2O_2), superoxide (O_2), and singlet oxygen ($^1O_2$), which can be generated by autooxidation, photoactivation, and metal catalyzed oxidation. Autooxidation refers to the oxidation in the sheer absence of oxidants. True autooxidation, being characterized by extremely slow reaction rate, is normally not regarded as an important degradation pathway for proteins [5, 49]. Oxidation can be catalyzed by various external factors, such as the presence of transition metal ions and ultraviolet (UV) irradiation [5, 49]. Among them, Fenton-type metal catalyzed oxidation is the most relevant generating source of reactive oxygen species in pharmaceutical formulations where contaminating peroxides may react with traces of redox-active transition metals such as iron or copper.

Metal Catalyzed Oxidation. Metal catalyzed oxidation is known to adversely influence the stability of proteins, peptides, plasmid DNAs, and nucleic acid-derived drugs by inducing conformational and functional changes in these biopharmaceuticals [43, 44, 49, 50]. In proteins and peptides where metal catalyzed oxidation is extremely common, the presence of trace metal ions can catalyze oxidation by either directly reacting with the side chains of certain amino acid residues to produce free radicals or complexing with oxygen to produce various reactive oxygen species. Due to site-specificity, a faster oxidation rate is commonly observed for amino acid residue having closer proximity to the metal-binding sites [51]. However, site-specific oxidation can be greatly reduced by replacement of the existing amino acid residues with their larger, bulkier, or charged counterparts, as in the case of condon-222 mutant subtilisin [52].

Generally, an electron donor or reducing agent known as prooxidant is also required to reduce the transition metal ions. The prooxidant may be present as contaminant in buffers or formulations. For instance, the presence of reducing sugar impurities in mannitol has been suggested as a possible cause for the oxidative degradation of L-367073, a potent fibrinogen receptor antagonist, which involves the formation of a Schiff base intermediate [53]. The role of prooxidants in the oxidative degradation of proteins and peptides has been experimentally verified under controlled conditions. For instance, in the presence of oxygen, Fe^{3+} and an appropriate electron donor (e.g., ascorbic acid and dithiothreitol), the methionine residues in proteins are rapidly oxidized to methionine sulfoxides [49].

Although fewer stability studies have been reported for plasmid DNA and other nucleic acids than for proteins and peptides, the ability of trace metal ions to catalyze many oxidative processes (including Fenton reaction) even in the absence of hydrogen peroxide has been clearly demonstrated with the nucleic-acid-derived drugs [46]. Free radical oxidation is the major degradation process exhibited by plasmid DNA in pharmaceutical formulations in the absence of free radical scavengers and/or specific metal ion chelators [43]. Plasmid DNA formulations with demetalated phosphate buffers have significantly higher stability than those with non-demetalated buffers, strongly reflecting the important role of trace metal ions in the oxidative degradation of DNA [43].

Specific Oxidation Reactions in Proteins and Peptides. In general, the amino acid residues exposed on the surface of proteins are more susceptible to oxidation than those buried within the hydrophobic core. For instance, different methionine residues in proteins can exhibit substantial differences in oxidation rate, as observed for human growth hormone [54], granulocyte colony-stimulating factor [55], and human parathyroid hormone [56]. Such differences in methionine reactivity are consistent with the location and solvent accessibility of the methionine residues.

Oxidation in proteins and peptides are highly amino acid specific. The amino acid residues that are susceptible to oxidation generally fall into two groups, viz. those containing a sulphur atom (methionine, cysteine) and those with an aromatic side chain (histidine, tryptophan and tyrosine) [49].

Cysteine. The free thiol groups present in cysteine residue can undergo spontaneous oxidation to form either intra-/intermolecular disulfide bonds and cross-links or monomolecular by-products such as sulfenic acid, sulfinic acid and sulfonic acid [49]. The mechanisms of cysteine oxidation by hydrogen peroxide have been investigated in detail. In the absence of metal ions and at low concentration of H_2O_2, the oxidation of cysteine is a two-step nucleophilic reaction that generates cystine as the final product, the rate-determining step being the reaction of thiolate anion with the neutral H_2O_2 to form cysteine sulfenic acid as an intermediate [57]. At high concentration of H_2O_2, the percentage of disulfide formation declines due to cysteine depletion or operation of alternative reaction pathways for consumption of cysteine sulfenic acid, leading to increased formation of cysteine sulfonic acid [57].

The oxidation of cysteine to cystine disulfide is strongly catalyzed by transition metal ions. As the formation of disulfide bonds and cross-links can bring about not only changes in tertiary structure but also covalent dimerization and aggregation, oxidation of cysteine has been of particular concern with many biopharmaceuticals, including interleukin-1B and interleukin-2 [54]. The spatial orientation of thiol groups is an important factor affecting the oxidation of cysteine to cystine. As two cysteine residues are required for the reaction, the oxidation rate is usually inversely related to the distance between the thiol groups [49]. In 28 hen lysozyme variants with random coil structure, the extent of disulfide bond formation in each variant has been shown to be proportional to the distance between the cysteine residues [58]. As cysteine oxidation is favored by deprotonation of the thiol group, the local structure governing the ionization or pK_a of cysteine residue will also influence the cysteine oxidation rate [49, 59]. As demonstrated using seminal ribonucleases segment 29–34 [60] and α-lactalbumin [61], introduction of an electronegative environment into the structures raises the pK_a of the thiol groups and reduces the cysteine oxidation rate. In contrast, cysteine[25] residue in papain, which displays a lower pK_a due to the presence of a nearby ionizable histidine group, is chemically more reactive than expected [59].

Methionine. The major oxidative degradation product of methionine is methionine sulfoxide. In acidic medium, methionine can be readily oxidized to methionine sulfoxide through a nucleophilic substitution reaction by alkyl hydrogen peroxides present in a solvent complex intermediate form [49]. In the presence of peracid, methionine can be oxidized to its sulfone derivative [54].

Light exposure, storage temperature, and oxygen level can all significantly influence the oxidation rates of protein pharmaceuticals [62, 63]. Local structure in the immediate vicinity of an amino acid residue in peptides may also affect the oxidation pattern of the residue. For instance, oxidation of threonine-methionine promotes the cleavage of the side chain of threonine residue [64]. Histidine facilitates the oxidation of neighboring methionine, and *C*-terminal methionine is oxidized more readily than the *N*-terminal or intra-chain methionine [54]. In addition, the formation of methionine sulfoxide is pH dependent, being favored at low pHs [54] and different methionine residues in the same protein molecule exhibit different oxidation rates, which also vary with pH [55, 56]. In human parathyroid hormone, the oxidation of the two methionine residues is virtually independent of pH between 4 and 8, but the rate of oxidation increases at pH 2 [56]. In contrast, the four methionine residues in granulocyte colony-stimulating factor show much larger differences in oxidation rate with H_2O_2 at various pHs [55]. Such pH dependence of the methonine oxidation has been attributed to the changes in local environment of the methonine residues in response to pH changes, i.e., the protonation or deprotonation of the nitrogen atoms in the neighboring histidine groups.

Aromatic Amino Acids (Histidine, Tryptophan and Tyrosine). Histidine residues are highly susceptible to oxidation, as shown for human growth hormone [65] and relaxin [66]. The resulting degradation products are aspartic acid, asparagines, and 2-oxo-histidine [65, 67]. Metal catalyzed oxidation of histidine may alter the secondary/tertiary structures of proteins. As has been demonstrated, oxidation of the human relaxin histidine[A(12)], which exists in an extended loop that joins two α-helices, alters the protein conformation, resulting in pH-dependent protein aggregation and precipitation [66, 68].

Tryptophan is also sensitive to oxidation, yielding mostly N′-formylkynurenine and 3-hydroxykynurenine [69]. The formation of N′-formylkynurenine can be activated by photoionization of tryptophan residues arising from the formation of solvated electrons [70]. The reaction is also catalyzed by the presence of metal ions [69]. Tryptophan residue can also mediate photooxidation of the disulfide bond in bovine somatotropin without being oxidized itself [70].

Tyrosine is also prone to oxidation, yielding 3,4-dihydrophenylalanine and dityrosine cross-link as the major reaction products [69]. For example, dityrosine-linked dimer was formed upon photoactivation of γB-crystallin at 445 nm [71].

Other Amino Acid Residues. N-terminal glycine residue is also a potential site for oxidation. Oxidation of Pexiganan has been reported to generate N-glyoxylyl-des-gly$_1$-pexiganan, which can be further decomposed to yield other products, as verified by high-performance liquid chromatography (HPLC) and mass spectrometry [72].

Specific Oxidation Reactions in Nucleic Acids. Nucleobases and sugar moieties are the oxidation sites of nucleic acids. Similar to metal catalyzed oxidation in proteins and peptides, DNA damage through oxidative stress can also be induced by Fenton-type reactions involving transition metal ions, e.g., Fe^{2+} and Cu^{2+} [73]. The electron-rich purines and pyrimidines are the prime targets for the reaction with electrophilic oxidizing and photooxidizing agents [74, 75], and hydrogen abstraction and radical addition are the two known mechanisms responsible for the oxidation of aromatic nucleobases [44]. Such oxidation reactions will result in base cleavage and subsequent strand scission [76, 78]. As nucleobase oxidation removes electron density from the heterocycle, the oxidized nucleobase is a better leaving group, leading to an enhancement of hydrolytic depurination and depyrimidation [78]. Guanine, being the most readily oxidized site among the nucleobases, can be activated by light and catalyzed by trace metal ions to yield a wide range of oxidation products such as 8-hydroxyguanine, xanthine [76], 8-oxoguanine (8-oxoG), and 2,6-diamino-4-hydroxy-5-formamidopyrimidine [78]. Of all oxidation products of guanine, 8-oxoG is the most common oxidative lesion observed in duplex DNA, and is generally regarded as a key lesion in toxicological assessment [78]. Another well-characterized guanine lesion is 7,8-dihydro-8-oxo2'-deoxyguanosine (8-OH-dG) [79], which has been used as a degradation marker for a plasmid DNA in pharmaceutical formulation [43]. Adenine can also be oxidized readily, although significantly fewer oxidative lesions are formed because of its higher one-electron redox potential [78].

For the sugar moieties, the only oxidation mechanism is radical-mediated hydrogen abstraction, which can be induced by Fenton- or radiation-generated hydroxyl radicals [80]. Various hydrogen atoms in sugar moieties, such as H-1', H-2', H-3', H-4', and H-5', are all potential sites for abstraction, although not all hydrogen atoms of deoxyribose in B-DNA have equal probability of being abstracted from duplex DNA [80]. The reaction depends on the helical structure of the nucleic acid, the orientation of the oxidant relative to the sugar as well as the stability of the carbon-centered radical being formed [44, 80]. As resonance stabilization is more difficult with sugar moieties than with nucleobases, the sugar moieties are less-favorable targets than nucleobases for radical attacks. The accessibility of the C–H bond also affects the rate and selectivity of hydrogen abstraction. For instance, oxidation of sugar moieties in the B-form of DNA mainly occurs at C-4' due to its accessibility [73, 78, 80].

4.3.1.4 β-Elimination in Nucleic Acids, Proteins and Peptides

The two-step process of depurination (see the "Depurination and depyrimidation" Subsection in Section 4.3.1.2) and β-elimination is an important DNA degradation pathway in aqueous medium [45]. Upon depurination, the apurinic site contains a chemically altered sugar that alternates between a cyclic furanose form and an

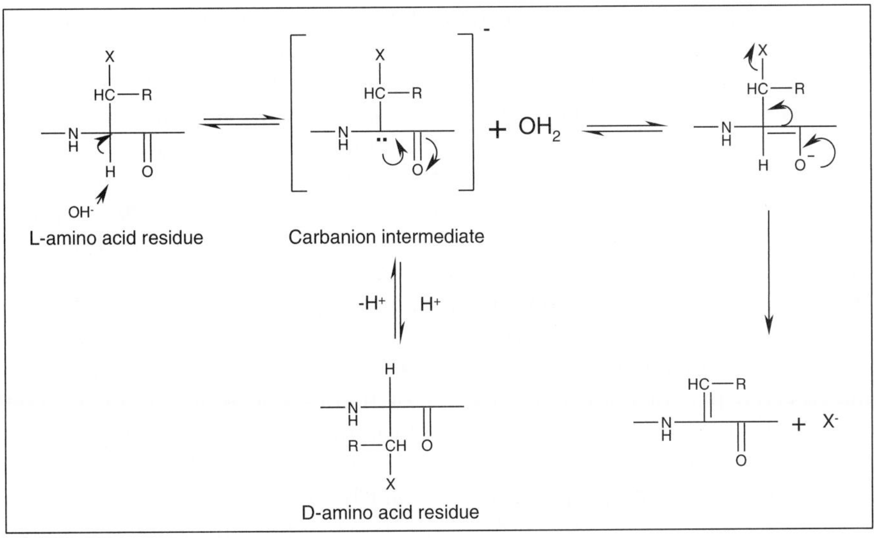

Scheme 4.3-10.

acyclic form containing an aldehyde functional group on the 1' carbon. The aldehyde form of the sugar then undergoes β-elimination, which commences with the abstraction of a proton from the 2' carbon by the OH⁻, leading to strand breakage on the 3' carbon of the abasic site [81].

In proteins and peptides, the β-elimination step involves the abstraction of a proton from the α-carbon of an amino acid residue in the peptide chain to form a carbanion intermediate (Scheme 4.3-10) [3]. Racemization can then occur via the addition of a proton to the opposite plane of the intermediate. Alternatively, the carbanion intermediate undergoes further reaction to form a dehydroalanine residue (Scheme 4.3-10) [3]. Cysteine, serine, threonine, phenylalanine, and lysine are the amino acid residues known to undergo β-elimination in peptides and proteins. Protein pharmaceuticals that can decompose by this reaction route include leuprolide [82, 83], salmon calcitonin [34], and insulin [84]. Leuprolide, which is formulated as a highly concentrated solution in DMSO for delivery in DUROS osmotic implant, shows increased β-elimination for the serine⁴ residue in nonaqueous media [82, 83]. For insulin, the covalent and noncovalent aggregation of lyophilized preparation has been attributed to the intermolecular thiol-catalyzed disulfide interchange following β-elimination of an intact disulfide bridge in the insulin molecule [84]. This degradation process can be accelerated by elevated temperature; increased moisture content of the insulin, lyophilization, and/or dissolution of the insulin in alkaline media; as well as the presence of Cu^{2+} (catalyst for the oxidation of free thiols).

4.3.1.5 Maillard Reaction in Nucleic Acids, Proteins, and Peptides

Maillard reaction, which is responsible for the nonenzymatic browning of materials, has been extensively investigated with food materials, but less so with pharmaceutical substances. The reaction consists of an initial reaction of reducing sugars with amino or free amine groups in proteins, peptides, or nucleic acids, followed by Schiff base conversion and the formation of brown pigments via Amadori and Heyns rearrangement [85, 86]. For proteins and peptides, the amino acid residue involved in the reaction is usually lysine because of its free ε-amino group, but other bases such as arginine, asparagine, and glutamine can also react with the sugars [86]. In nucleic acids, the reaction may occur between the aldehyde of the sugar at the apurinic sites and the amino groups on the DNA bases in another plasmid molecule to produce concatamers [46]. Biopharmaceuticals documented to undergo Maillard reaction include glucagon [85], human relaxin [87], β-lactoglobulin variant A [88], lysozyme [89, 90], recombinant human serum albumin [90], calf thymus DNA [91], and plasmid pBR322 DNA [92].

Maillard reaction can give rise to significant changes/losses in physico-chemical properties and pharmacological activities of therapeutic proteins. Of particular concern in this regard is the modification of protein antigenicity, which is associated with the formation of advanced glycation endproduct (AGE) [93]. Glucose, a monosaccharide with reducing property, is particularly reactive toward the amino acid residues in proteins and peptides [87–90], whereas other sugars such as trehalose and mannitol have very little effect, as attested by the findings of the formulation studies with these sugars on glucagons [85] and relaxin [87]. Sucrose can also cause glycation when it is hydrolyzed to glucose and fructose by heat treatment, as observed for β-lactoglobulin variant A, lysozyme, and rHSA during antiviral heat bioprocessing [88–90].

As with proteins and peptides, nucleic acids can suffer substantial losses of biological activities as a result of Maillard reaction upon aging [91, 92]. Age-related dysfunction in gene expression has been demonstrated in DNA modified by reducing agents [91]. In addition, the amino groups of DNA have been shown to react nonenzymatically with reducing sugars (i.e., glucose) to afford a yellow-brown coloration, and the adducts formed *in vitro* have been found to decrease the transfection ability of f1 phage DNA [91, 92].

4.3.1.6   Covalent Dimerization and Polymerization in Proteins

Precipitation is a macroscopic process characterized by a visible change of an otherwise clear solution in the form of an increase in solution viscosity, clouding of solution, or phase separation of solid materials. The formation of precipitates may be due to covalent polymerization and/or noncovalent aggregation (Section 4.3.2.2). In covalent polymerization, dimer formation usually precedes the precipitation, and the phenomenon is best illustrated by insulin, which has been extensively reported to form covalent dimers.

Covalent dimer is the main degradation product of insulin upon storage [94]. Covalent trimer, tetramer, oligo-, and polymer formation is also possible with insulin when stored at ambient or higher temperatures. In general, the rate of insulin polymerization is virtually independent of the protein source (e.g., bovine, porcine, and human), but it varies with the composition and formulation of the protein and, for the isophane preparation, with the strength of preparation [94]. The rate of polymerization is one order of magnitude slower than insulin hydrolysis except for the NPH preparations [94, 95]. The covalent aggregation of insulin can be inhibited by self-association at high insulin concentration [9]. Both deamidation and covalent dimer formation in human insulin solution at low pH seem to originate from a common cyclic anhydride intermediate [96]. Covalent dimer formation of human insulin is also evident in lyophilized solid state, accounting for no more than 15% of the total degradation [97]. Similar to the effect observed in solution, an increase in pH favors dimerization in lyophilized insulin, which can be ascribed to the increased nucleophilicity of the terminal amino groups after deprotonation. However, covalent dimerization can be almost completely suppressed by incorporating trehalose into the lyophilized matrix, which restricts the molecular mobility of insulin and keeps the protein molecules in a fully dispersed state.

Besides insulin, other biopharmaceuticals known to exhibit covalent polymerization include recombinant tumor necrosis factor-alpha, which forms nonreducible dimers and oligomers [98], and human insulin-like growth factor I, which is prone to covalent aggregation [99].

4.3.2 PHYSICAL INSTABILITY

4.3.2.1 Denaturation

Denaturation refers to an alteration of the global fold of a biopharmaceutical with higher order structures without an accompanying change in the primary structure. For proteins, these higher order structures refer to the secondary, tertiary, and quaternary structures with different levels of complexity. Although nucleic acids are structurally less complex than proteins, more elaborate structures such as double helix and supercoiling exist in DNAs, and many RNAs also have well-defined tertiary structures. Any conformational changes in these structures can lead to denaturation of these biopharmaceuticals.

The conformational integrity of native protein is maintained by a proper balance of weak nonbonding forces. Conceptually, an equilibrium exists between the native-folded and the denatured-unfolded forms of the protein, with or without the involvement of unfolding intermediate(s), or molten globule(s), in definable thermodynamic state(s). As the free energy of protein unfolding ΔG_{unf}, is typically in the range of 10–20 kcal/mol, the native structure of the protein is readily disrupted by a change of environment (e.g., application of heat, change of pH, and addition of detergents or chaotropes) [100]. Such structural perturbation, which is either permanent or temporary, leads to a change of conformationally sensitive physical properties (e.g., optical rotation, viscosity, and UV absorption) and, most seriously, to a concomitant loss of biological activities. If the denaturation process is reversible, the loss of native structure can be recovered once the stress is removed. However, no such structure recovery is possible with irreversible denaturation involving covalent bond formation such as disulfide exchange or cross-linking. Irreversible denaturation also leads to aggregation and precipitation with accompanying changes in the secondary and/or tertiary structures of the proteins. The soluble aggregates derived from denatured-unfolded protein and unfolding protein intermediates are prone to aggregation and precipitation in pharmaceutical formulation, resulting in a severe activity loss, as illustrated by recombinant human granulocyte colony stimulating factor [101]. Addition of sucrose to this protein brings about an increase in ΔG_{unf} and a consequential decrease in aggregation and an increase in conformational stability.

Almost all water-soluble proteins are denatured by heat treatment. Although some denaturation processes follow the Arrhenius kinetics, the ΔG_{unf} in most cases follow a parabolic dependence on temperature, as depicted by the following equation [100]:

$$\Delta G_{unf} = \Delta H_m \left(1 - T/T_m\right) - \Delta C_p \left(T_m - T\right) + T \ln\left(T/T_m\right) \qquad (4.3\text{-}1)$$

where ΔH_m is enthalpy of melting, ΔC_p is the specific heat capacity, and T_m is the melting temperature. In an actual situation, once the temperature is raised and a significant fraction of the protein ensemble is unfolded, aggregation and/or precipitation can proceed rapidly, resulting in irreversible denaturation. In addition to heat denaturation, cold denaturation of proteins can occur during freeze-drying. In general, freeze-drying causes considerable conformational changes and aggregation in protein molecules, thereby promoting the formation of the β-sheet structure at the expense of the α-helix and random structures [102].

Protein stability is highly sensitive to pH changes and decreases sharply at acidic or basic pH. Because of the differences in pK_a of certain amino acid groups between the native and unfolded proteins, protons released in response to pH changes can play an important role in the unfolding process [100]. It has been shown that protein conformational fluctuations generally agree with the predictions based on the pK_a values of the titrating groups in the protein [103]. Aside from pH manipulation, protein unfolding can be achieved with chaotropic agents (e.g., urea and guanidine hydrochloride). However, in contrast to the less-than-complete protein unfolding induced by thermal denaturation, the denatured state produced by such chaotropes is virtually devoid of structure, resembling a random coil conformation [104].

Subjection of duplex DNA to various stress factors such as heat, pH, and ionic strength can lead to a collapse of the native DNA structure. The two complementary strands separate and assume a random coil conformation with concomitant changes in physical properties (e.g., viscosity and UV absorbance) and biological activities. The thermodynamic stability of a nucleic acid duplex is a complex function of temperature and pressure, being strongly dependent on the denaturation temperature T_m [105, 106]. Interestingly, the relationship between the free energy associated with helix-to-coil transitions of nucleic acid duplexes ΔG_m and the corresponding ΔH_m, T_m, and ΔC_p conforms to that predicted by Equation (4.3-1) for the protein unfolding process, where a parabolic curve is expected [106]. Cold denaturation of nucleic acid duplex can also occur when the temperature is held below $-120°C$ at atmospheric pressure [105].

Duplex stability is dependent on DNA conformation. Classic B conformation has a lower free energy of helix-to-coil transition and lower T_m than B′ conformation, whereas B′ conformation has a higher thermal stability than A conformation [106]. Base pairs also influence duplex stability.

4.3.2.2   Aggregation and Precipitation

Aggregation and precipitation are common problems of instability with protein biopharmaceuticals, e.g., keratinocyte growth factor [107], human immunoglobulin [108], and human serum albumin [109]. In clinical practice, an aggregation level of as low as 1% over a period of 2 years is deemed unacceptable [110]. Aggregation differs from self-association in that it is confined only to non-native proteins, and it enamates from a conformational intermediate produced during folding or denaturation of the proteins, whereas self-association occurs in native proteins in response to deliberate changes of solvent environment (e.g., pH, ionic strength, and solvent composition) to effect crystallization or isoelectric precipitation.

Protein aggregation can be described as a two-step process. The first step entails protein unfolding and exposure of buried hydrophobic residues to the aqueous solvent. The second step involves intermolecular interactions among the exposed hydrophobic residues of the unfolded protein intermediates, which are thermodynamically favored by the tendency of the exposed hydrophobic amino acid residues to minimize their contact with the aqueous medium. The reaction is site specific, as the initial stage of aggregation involves the interaction of one protein molecule with another at specific sites. Two such sites per protein molecule are sufficient for aggregation to proceed longitudinally, leading to the formation of long fibers. As more of these stable aggregates comprising a few molecules are formed (nucleation), they will develop into larger aggregates (growth) and eventually into floccules, which will precipitate from the solution once their solubility limit is exceeded.

In aqueous solution, non-native protein aggregation is dependent on various factors, such as temperature, solution pH, salt type, and concentration as well as the presence of ligands, cosolutes, preservatives, and surfactants [111]. Intrinsic conformational stability of native proteins as well as the colloidal stability between protein molecules may play a crucial role in the aggregation process. As aggregation requires the presence of two or more protein molecules, the aggregation reaction rate is expected to be higher because of increased probability of molecular collision [112]. However, the problem of protein aggregation in solution cannot be solved simply by preparing the protein in solid form alone, as solid phase aggregation is also common with lyophilized or freeze-dried formulations. There are a wide variety of mechanisms responsible for moisture-mediated solid-phase aggregation, encompassing thiol-disulfide interchange, thiol-catalyzed disulfide exchange after β-elimination, non-covalent aggregation, and non-disulfide covalent aggregation [113].

Aggregation can lower protein potency simply by decreasing the effective protein concentration. Moreover, it can increase the immunogenicity of proteins, as observed for insulin, human growth hormone, and recombinant human interferon-α2a [114]. Various low-molecular-weight excipients have been investigated for their potential to minimize protein aggregation or stabilize protein conformation [112]. Simple sugars, polyhydric alcohols, and amino acids have all been shown to effectively prevent freeze/thaw-induced aggregation in chimeric L6, a mouse-human monoclonal antibody [115]. Sucrose is known to inhibit the aggregation of recombinant human granulocyte colony stimulating factor [116]. Sorbitol has also been demonstrated to reduce moisture-induced aggregation in tetanus and diphtheria toxoids [117]. β-Cyclodextrins have also been found to inhibit the aggregation of recombinant human growth hormone [118]. However, there are also excipients that can adversely impact the protein aggregation and precipitation. For example, benzyl alcohol can induce precipitation of human insulin-like growth factor I from solution [119] as well as aggregation of human interleukin-1 receptor antagonist (a predominantly β-sheet protein) in aqueous solution [120] and reconstituted lyophilized formulation [121].

4.3.2.3   Adsorption

Interfaces between two separate phases, such as air/water, oil/water, and solid/water, are potential adsorption sites of biopharmaceuticals. Adsorption often involves simple diffusion of surface-active solute molecules in the bulk to the interface, and hence the rate of adsorption is generally dependent on the solute concentration. At saturation, a close packed monolayer of protein molecules corresponding to 0.1 to $0.5\,\mu g/cm^2$ is normally formed at the interface [122], and this adsorption behavior is of particular concern for high-potency therapeutic proteins. However, certain proteins do not conform to such saturation-limited adsorption behavior and tend to show increased adsorption with increasing protein concentration, attaining a local protein concentration at the interface 1000 times higher than the initial concentration in the bulk solution [123].

The surface activity resulting from different amino acid compositions of polypeptides has an important bearing on the extent of surface adsorption. As amino acids can be hydrophobic (e.g., tryptophan, phenylalanine, and isoleucine), hydrophilic (e.g., serine and threonine), negatively charged (e.g., aspartic acid and glutamic acid), or positively charged (e.g., histidine), the proteins involved can potentially adsorb on to solids such as plastics and glasses with different affinities. Owing to their intrinsic polyelectrolyte nature, the adsorption of proteins on solids is highly pH dependent, reaching a maximum at their respective isoelectric points (pIs) [122]. It has been demonstrated that recombinant human interleukin 11, which is monomeric and highly basic (pI > 10.5), displays a nonspecific loss to container in alkaline solution [32]. Surface adsorption alone can cause more than 40% activity reduction of interleukin 11 in solution after 3 hours of storage at room temperature [124].

Upon adsorption to container, conformational changes can occur in the higher order structures of proteins. As hydrophobic amino acid residues tend to stay inside a protein (i.e., away from the aqueous environment), surface interaction may unfold the protein to expose the inside structure, thereby leading to structural rearrangement of the protein. For instance, it has been shown that the native structure of α-helices and extended loop bands of recombinant interleukin 2 disappeared, whereas the β-sheet appeared upon contact with a pump-based delivery system (borosilicate glass pump reservoir and catheter tubing), resulting in irreversible structural changes and major biological activity loss of the protein [125].

Apart from solid surfaces, liquid surfaces are potential sites for protein adsorption. By creating a large surface area, agitation may induce protein adsorption at the liquid surface where the area occupied per unit mass of protein in the monolayer region is high ($\sim 1\,m^2/mg$) compared with that at the solid–liquid interface ($\sim 0.1\,m^2/mg$) [122]. The adsorption process coupled with interfacial and shear forces may lead to protein denaturation through unfolding as well as aggregation and precipitation (Section 4.3.2.2), as observed for recombinant factor VIII SQ [126].

Addition of surface-active agents (e.g., polysorbates) can prevent protein adsorption and stabilize the protein preparation. Numerous reports have documented the stabilizing effect of surface-active agents against adsorption of biotechnological products. For example, polysorbate-80 exerts a primary stabilizing effect on advenovirus type-5 based vaccine by inhibiting adsorption of the vaccine to glass vials

[127]. The presence of 0.12% polysorbate-20 reduces the adsorption of recombinant human interferon-γ to both air–liquid and ice–liquid interfaces [128]. Similarly, the presence of polysorbate-80 or polysorbate-20 reduces surface adsorption of recombinant factor VIII SQ and, hence, the agitation-induced denaturation of the protein [126]. Inclusion of albumin, which minimizes the contact of proteins with glass surface, may also help to stabilize the proteins against glass adsorption, as demonstrated for salmon calcitonin [129].

Paradoxically, the adsorption process can be used to stabilize biopharmaceuticals, particularly vaccines [130]. Aluminum salts have been widely used as adjuvants in intramuscular vaccine injections. Novel vaccine formulations containing aluminum are available; for example, alum-adsorbed hepatitis B surface antigen and diphtheria-tetanus toxoid vaccines have been optimized for epidermal immunization in powder form [131].

4.3.3 CONSIDERATIONS IN THE STABILITY TESTING OF BIOPHARMACEUTICALS

4.3.3.1 Temperature

The stability of pharmaceuticals, including biopharmaceuticals, is highly temperature dependent. For most small organic molecules, thermal degradation generally follows Arrhenius kinetics. However, many biomaterials, e.g., proteins, do not exhibit Arrhenius degradation kinetics even over a narrow temperature range, which can be attributed to phase transitions, pH shifts, poor relative humidity control at elevated or subambient temperature, and involvement of complex reaction mechanisms [132]. Multistep reaction pathways are particularly common in biopharmaceuticals because of the existence of higher order structures. For instance, with recombinant bovine granulocyte colony stimulating factor, a reversible equilibrium between the native protein and an intermediate state is established, followed by irreversible aggregation [133]. Although individual denaturation steps seem to obey the Arrhenius law, the overall kinetic behavior for product formation may not [132, 133]. Thus, for protein pharmaceuticals, extrapolated shelf-life prediction from accelerated stability test must be viewed with caution and real-time stability monitoring should also be conducted whenever feasible.

As many therapeutic proteins are formulated in aqueous solution and stored at 2–8°C, they are liable to cold denaturation at such low temperatures and possibly to freezing/thawing inactivation arising from temperature fluctuation during storage or handling. When a protein solution freezes, water crystallizes out as solid ice and the solution becomes more concentrated in the protein and salt components present, causing dramatic changes in pH and ionic strength and hence protein inactivation [134]. For instance, during freezing of sodium phosphate solutions, crystallization of the disodium salt can reduce the pH by as much as 3 units [135]. The effects of freezing and thawing rates on the stability of model proteins have been investigated in the absence of cryoprotectants, and higher activity recovery at a slower freezing rate and a faster thawing rate has been observed [136]. During fast freezing, small ice crystals are preferentially formed and the associated increase in total surface area of the ice–liquid interface will enhance the exposure of protein molecules to the interface and increase the protein damage. Upon thawing, additional damage to proteins may be caused by recrystallization [136]. Many biopharmaceuticals, including recombinant human interferon-γ [135], tetanus toxoid conjugate vaccines [137], and pCMVβ-gal plamid DNA/lipid complexes [138] show a reduction in activity upon freezing/thawing. To circumvent the aforementioned protein inactivation problems, various excipients have been used in biopharmaceutical formulations. For example, Tween 80, a polysorbate surfactant, has proved useful for reducing surface-induced protein denaturation during freezing [139], and glycine can also stabilize proteins through the preferential exclusion mechanism [135].

4.3.3.2   Moisture

Moisture is well documented to adversely affect the chemical stability of solid biopharmaceuticals. The presence of water enhances the mobility and flexibility of the bioactive macromolecules in solid excipient matrix, thereby facilitating their rearrangement and increasing their susceptibility to chemical degradation. For instance, the deamidation rate of an asparagine containing hexapeptide in lyophilized poly(vinyl alcohol) (PVA) and poly(vinyl pyrrolidone) (PVP) increases with increasing moisture uptake and water activity of the system [140]. The asparagine deamidation seems to correlate closely with the extent of water-induced plasticization of PVA and PVP matrices (as determined by the respective glass transition temperature T_g), suggesting that the chemical stability of the peptide may be predicted by the physical state of the formulation [140]. Similarly, aspartate isomerization in humanized monoclonal antibody, which can be accelerated at elevated temperatures, has been observed to follow the Arrhenius law above the T_g of a lyophilized formulation [141].

The presence of moisture can also reduce physical stability. The mechanisms of moisture-induced aggregation are well documented and have been exemplified by a good number of solid biopharmaceuticals, including salmon calcitonin spray-dried powders for inhalation [142], human serum albumin [143], and recombinant human albumin [109]. To prevent such protein aggregation, excipients are often employed in protein formulation. The excipients, which comprise mostly sugars such as mannitol, lactose, trehalose, and cellobiose, prevent protein aggregation by occupying the water-binding sites of protein in the dried state [144]. However, these sugar excipients may also present formulation problems, notably sugar crystallization at high excipient-to-protein ratios [144] and phase separation due to prevalent sugar–sugar interactions [145].

The conventional wisdom of moisture content control for biopharmaceuticals is "the drier, the better." As has been demonstrated with a lyophilized humanized monoclonal antibody formulation, moist cakes tend to have higher aggregation rates than drier samples if stored above their T_g [141]. However, excessive moisture removal can destabilize the protein, as observed with bovine immunoglobulin [146].

4.3.3.3   Light and Ionizing Radiation

Eletromagnetic radiations of different wavelengths, i.e., UV-visible light and x-ray, are known to induce degradation in proteins [147]. Radiation, which is commonly employed for sterilization purpose, can generate free radicals and lead to peptide chain cleavage and aggregation in solid protein pharmaceuticals [147]. In aqueous solution, the protein may also degrade through the destruction of the amino acid residues by the hydroxyl radicals and electrons produced from water molecules [147].

As biopharmaceuticals are formulated mostly for parenteral administration, complete shielding of light is impossible. Although amber glass containers can be used for shielding against UV light, their utility in light protection is often limited by the glass thickness. Light-induced degradation or photodegradation can be minimized by formulation into solid dosage forms. By formulating into a lyophilized form, recombinant human factor VIII showed no significant loss of activity after accelerated photostability testing, whereas the reconstituted preparation displayed partial activity loss after similar light exposure, and the observed photodegradation could be effectively prevented by packaging the protein in tinfoil wrap [148].

Nucleic acids are also susceptible to UV light. DNA molecules absorb photon energy at wavelengths below 320 nm, which gives rise to the widely reported mutagenic DNA damage. DNA photocleavage can occur at both deoxyriboses and nucleobases [75]. The presence of photosensitizers, such as psoralen, chlorpromazine, and fluoroquinolones, may promote the light-induced degradation by generating reactive oxygen radicals or singlet oxygens, which then interact with DNA to produce damage [149].

4.3.3.4 Physical Stress

Biopharmaceuticals may be inactivated by physical stress associated with vial filling, shipping, storage, and handling. Insulin is a well-known example of agitation-induced instability, and patients are advised to avoid vigorous shaking of the insulin preparation. Shaking is known to accelerate the degradation of insulin by way of covalent dimerization [150]. To assess the adverse impact of agitation on the stability of insulin preparations, two automated physical stress tests, viz., the Temperature Cycling and Resuspension Test (TCRT) and the High Temperature and Extreme Agitation Test (HTEAT), have been applied to a commercial portable insulin pen-cartridge device [151]. TCRT involves temperature cycling (25–37°C) in an incubator unit combined with resuspension (three sets of 10 rolls plus 10 inversions) conducted twice daily, whereas HTEAT requires continuous high-temperature (37°C) exposure in an incubator unit combined with periodic daily agitation for 4 hours at 30 rpm. These two tests afford a rapid means for screening potential aggregation during the formulation development of proteins [151].

Apart from insulin, monomeric recombinant human growth hormone has been shown to aggregate rapidly within 10 hours of shaking [152]. The presence of Tween 20, a nonionic surfactant, at an excipient:protein molar ratio larger than 4 effectively inhibited this aggregation. Similar aggregation was observed with IgG1-antibody upon the application of physical stress. However, differences in aggregation mechanism exist among different mechanical stress methods (e.g., shaking and stirring) [153]. Moreover, the aggregation kinetics is dependent on the shear rate applied, and trimer formation is particularly evident at a high shear rate, as observed for human serum albumin [154]. Even simple vortexing for just 1 minute can cause substantial aggregation and precipitation, as shown for recombinant human growth hormone [155].

Agitation-induced instability is not limited to protein pharmaceuticals. For instance, agitation has been shown to significantly reduce transfection rates in pCMVβ-gal, a plasmid DNA, and in complexes prepared with three different commercially available lipid formulations, namely, DMRIE-C, lipofectAMINE, and DOTAP:DOPE [138].

4.3.3.5 Freeze-Drying and Spray-Drying

Lyophilization, the most widely used process for preparing solid proteins, consists of two major steps: freezing of a protein solution and drying of the frozen solid under vacuum. The drying step can be further divided into two phases, namely, primary drying (removal of frozen water) and secondary drying (removal of the non-frozen bound water) [156]. As alluded to, freezing can cause cold denaturation and introduce additional stresses to the proteins through ice crystallization (Section 4.3.3.1). Likewise, drying can bring about protein degradation. By removing the hydration shell covering the protein surface, the native state of protein may be disrupted, resulting in denaturation [157]. Water molecules may also be an integral part of protein active sites. Removal of these functional water molecules can inactivate proteins during dehydration [157]. Thus, it is perhaps not surprising that freeze-drying-induced degradation is common with protein pharmaceuticals, e.g., *Humicola lanuginose* lipase [158], lysozyme [159], and recombinant human factor XIII [160], as well as nucleic-acid-derived drugs, e.g., poly((2-dimethylamino)ethyl methacrylate)-based pCMV-lacZ plasmid gene delivery system [161] and cationic lipid-protamine-DNA complexes [162]. Aggregate formation is also apparent during storage of freeze-dried biotechnological products, e.g., ribonuclease A [163], and sugars, polyols, and polymers are the excipients commonly used for stabilizing them [157].

Spray-drying, a single step operation that converts a liquid feed to a dried particulate form, has proved useful for producing powders within the respirable range for pulmonary delivery of biopharmaceuticals. Many of these products, including recombinant human growth hormone [164], recombinant human deoxyribonucleases [165], and recombinant humanized anti-immunoglobulin E monoclonal antibody (rhuMAbE25), have been prepared by the spray-drying process [166]. As spray-drying of these proteins alone in solutions can result in unfolding, aggregation, and inactivation because of temperature and interfacial effects, stabilizers such as polysorbates [164, 167], mannitol [166], sucrose, and trehalose [168], are usually added to improve the processing and storage stability of these preparations [169].

4.3.3.6 Excipients

The inclusion of chemical excipients/additives can inhibit or slow degradation reaction in biotechnological products. In the case of metal catalyzed oxidation, chelating agents such as ethylene diamine tetra acetic acid (EDTA) can suppress the reaction by removing trace transition metals ions from the bulk solution. Some sugars and polyols (mannitol, glucose, glycerol, ethylene glycol) can inhibit metal catalyzed oxidation by complexation with the transition metal ions, whereas polymeric additives such as dextran may function as a hydroxyl radical scavenger [170]. Sodium thiosulfate has been reported to reduce the formation of a disulfide-linked high-molecular-weight species of OKT3 antibody in solution [171]. However, precautions must be taken when using these excipients for stabilization purposes. As has been reported, inclusion of aliphatic alcohols as inhibitors of metal catalyzed oxidation can give rise to cosolvent-induced perturbation of the metal binding site in recombinant human growth hormone [172].

Although several excipients have been made available for prolonging the shelf-life of biopharmaceuticals, potential reactions between excipients and actives may adversely influence the overall stability of the formulations. For instance, excipients containing carbonyl functional groups, such as PVP, can form a covalent adduct with peptides and proteins containing primary amines (i.e., N-terminus and lysine side chain) at high temperatures (70°C) and low RH (~0%) [173]. The presence of sucrose can promote methionine oxidation in recombinant factor VIIa, possibly due to the increase of chemical potential or a small conformational change in microenvironment [174]. Furthermore, excipients may be contaminated with traces of impurities, such as transition metals and reducing agents, which can compromise the stability of the protein products. For instance, both food-grade and pharmaceutical-grade polysorbates often contain traces of alkylhydroperoxide and hydrogen peroxide arising from the bleaching step in the synthetic process [175]. During storage, polysorbates may also decompose to peroxides, and the decomposition is accelerated by increased exposure to air and light, elevated temperature, and the trace presence of metal catalysts [54, 175].

4.3.4 REGULATORY REQUIREMENTS IN THE STABILITY TESTING OF BIOPHARMACEUTICALS

In the development of stability testing protocols for biotechnological/biological products, all relevant regulatory guidelines must be closely followed. The International Conference on Harmonization (ICH) guidelines for stability testing are widely adopted and are available online (http://www.ich.org). It is important to check the logistics and requirements of the appropriate stability programs before initiation of the stability testing. These guidelines have defined the storage conditions for long-term, intermediate, and accelerated stability testing for different climatic zones [Q1A(R2); Q1C; Q1F]. Photostability is also considered under the relevant section [Q1B]. Experimental design and data analysis with specific requirements also constitute an important part of the stability testing programs [Q1D; Q1E]. Given the complexity of biotechnological products and their sensitivity to environmental factors, such as temperature, humidity, light, oxygen, ionic strength, and shear stress, additional guidelines are available for assuring their quality and Q5B aims to regulate the stability testing of these products. Apart from the method of batch selection for the testing, Q5B guidelines have a particular emphasis on the establishment of stability-indicating profiles for the active drug substances. Potency testing should be included in the stability assessment, and the purity of biotechnological products should normally be determined by more than one method.

4.3.5 FORMULATION AND CHARACTERIZATION OF BIOPHARMACEUTICALS

As discussed, biopharmaceuticals are characterized by large molecular sizes and various levels of structural complexity (i.e., primary, secondary, tertiary, or quaternary), which render them particularly vulnerable to various forms of degradation or denaturation. Most current strategies for formulating biopharmaceuticals into appropriate dosage forms or delivery systems are aimed specifically at preserving the structural integrity and biological activities of these complex macromolecules as well as at improving their bioavailability or targeting them to specific sites in the body. Presented below is a general discussion on the formulation of biopharmaceuticals with particular emphasis on the techniques using pharmaceutical excipients or additives to achieve these aims. Biotechnological techniques that use viral vectors to deliver genes or nucleic acid-based drugs are outside the scope of the current discussion.

4.3.5.1 Protein and Peptide Drugs

With the aid of newer production techniques, such as recombinant DNA technology and transgenic protein production, coupled with cost-efficient protein purification techniques, proteins of high purity can now be mass-produced for therapeutic uses.

In protein production, selection of the host cells (e.g., bacterial, yeast, or mammalian cells) for the expression of recombinant proteins is critical, as this will determine the chemical properties (e.g., amino acid sequence/composition and glycosylation) as well as the pharmacokinetic behavior, biological activities, and potential toxicities of the resulting proteins [176]. For instance, granulocyte macrophage colony stimulating factor (GM-CSF) expressed in *Escherichia coli* bacteria is a glycosylated form with six fewer amino acids and one more methionine residue than the non-glycosylated native protein [177, 178]. Although both yeast and mammalian expression systems can produce glycosylated proteins, only the mammalian system can generate the more complex carbohydrate moieties, which may be associated with higher immunogenicity. In addition, various types of contaminants (e.g., viruses, toxins, and nucleic acids) may arise during the production, and thus establishment of the protein purity at the initial stage of formulation development is of paramount importance [179, 180].

A challenge to develop a high-concentration protein formulation is perhaps the limited solubility of the protein in aqueous solution. The aqueous solubility of a protein is governed by the interaction of its polar residues with water, whereas the non-polar side chains and peptide groups are normally buried in the folded protein to maintain the native state. In general, minimum protein solubility occurs at the isoelectric point (*p*I), where the protein molecule acquires a net-zero charge [181]. Thus, to ensure adequate solubility in water, proteins should not be buffered at or near their *p*Is. However, when selecting an appropriate pH for the vehicle, considerations should also be given to the chemical integrity and activity of the proteins, as these are also pH-dependent.

Proteins are amino acid chains that fold into unique three-dimensional structures. The primary structure of proteins is the peptide sequence of the whole molecule, which can be characterized by the amino acid composition, N-terminal amino acid sequence, C-terminal amino acid sequence, and peptide mapping.

The secondary structure refers to the general three-dimensional local regions of the protein, encompassing regions of α-helices, β-sheets, and supersecondary structures. Such local structures can be analyzed by circular dichroism, Fourier transform infrared spectroscopy, and fluorescence spectroscopy.

The tertiary structure refers to the overall three-dimensional structure of the polypeptide units of a given protein. Just as the secondary structure may be conceptually linked to distinct domains within a protein, the tertiary structure can be associated with the way by which these domains are related to one another. Interactions among such domains are governed by various physical forces, including hydrogen bonding as well as hydrophobic, electrostatic, and van der Waals interactions. The tertiary structure of proteins can be determined using a combination of analytical and complementary computer-aided modeling techniques, including solution nuclear magnetic resonance (NMR) spectroscopy, single-crystal x-ray crystallography, neutron diffraction, and molecular dynamics simulation.

In general, no single analytical technique can be used to fully characterize biopharmaceuticals. Aside from structural characterization, functional analyses, including immunoassay and bioactivity assay, are also essential for ascertaining the quality of biopharmaceuticals. Although immunoassay is more specific than bioassay in the presence of interfering factors, the data derived from the two techniques may not show any direct correlation with each other, mainly because the former detects not only biologically active proteins, but also denatured proteins and fragments as well as inactive protein-receptor complexes [182]. Some commonly used protein characterization techniques are summarized in Table 4.3-1. More specific reviews on protein and peptide analysis are presented elsewhere [183, 184].

An appropriate choice of excipients is also an important consideration in protein/peptide formulation, as it can profoundly influence the shelf-life, pharmacokinetic behavior, and efficacy of the final products. Table 4.3-2 shows a list of excipients commonly used in protein/peptide formulation [185]. Extensive reviews on excipient selection are available in literature [186, 187]. In general, excipients are experimentally selected by screening [188, 189]. The selection is guided by their abilities to produce the desired outcomes for the final dosage forms. For example, the stability of hybrid interferon-α in solution, which is sensitive to both pH and buffer composition, can be improved by adding mannitol or lactose at pH 4.0 [190]. However, at pH 7.6, lactose exerts an accelerating effect on the degradation of interferon-α, suggesting that an optimal pH exists for the excipient to function as a stabilizer. In another study, hydroxypropyl-β-cyclodextrin has been shown to effectively protect β-galactosidase against degradation during spray-drying [191]. Apart from the type of excipients used, the physical form of the excipients is also critical for the stabilization of proteins. For instance, mannitol in the amorphous state can preserve the structure and prevent the aggregation of recombinant humanized anti-IgE monoclonal antibody [166].

As many excipients are capable of stabilizing proteins under different conditions, it is unlikely that the stabilization effect is excipient-specific. It is widely believed that the stabilizing mechanism of excipients involves the preferential exclusion of the added excipient from the protein in aqueous solution and during freeze–thawing [192]. Briefly, a protein in an aqueous environment is in dynamic equilibrium between the native and the denatured (unfolded) forms. As the exclusion of excipient from the unfolded form is greater than that from the native form, the equilibrium will shift toward increasing concentration of the native form. Thus, the unfolding of protein in the presence of excipient is a thermodynamically

TABLE 4.3-1. Summary of the Commonly Used Analytical Techniques for Protein/ Peptide Characterization

Technique	Principle	Application	References
Colorimetric analysis	Based on specific chemical reaction and associated color change	Quantitative analysis; BCA assay	[207]
High pressure liquid chromatography (HPLC)			[226]
Reversed-phase chromatography (RPC)	Separation by hydrophobicity	Quantitative, stability (aggregation), purity analysis	
Size exclusion chromatography (SEC)	Separation by size (larger molecule elutes faster)		
Ion-exchange chromatography (IEC)	Separation by electrostatic interaction		
Electrophoresis			[227]
SDS-PAGE	Separation based on molecular sieving by gel matrix	Size, purity, stability analysis	
Capillary electrophoresis (CE)	Electrophoretic separation on capillary	Peptide analysis, purity, stability analysis	
Isoelectric focusing (IEF)	Separation based on continuous pH gradient (proteins migrate according to charge properties until the point of zero charge, i.e., isoelectric point or pI, is reached.)	Purity analysis, pI determination	
Mass spectrometry (MalDI-TOF)	Separation based on mass-to-charge ratio (m/z) in magnetic or electric field	Quantitative analysis (high sensitivity) Identification and integrity	[228, 229]
Immunochemical analyses			
Enzyme-linked immunosorbent assay (ELISA)	The first antibody is coated onto plastic surface and conjugated with antigen, whereas the second antibody is coupled to an enzyme; detection by color change of a chromogen when it is cleaved by the enzyme	Quantitative analysis; epitope quality	[207]
Bioassay	Based on biological effects of proteins (e.g., cell proliferation, cytotoxicity, and antiviral activity.)	Quantitative analysis; potency test	[230–231]

TABLE 4.3-2. Excipients Commonly Used to Stabilize Biopharmaceuticals (from Ref. 185)

Excipient Type	Examples
Amino acids	Glycine, arginine, alanine, proline, aspartic acid, glutamic acid, lysine
Sugars	Trehalose, sucrose, maltose, fructose, raffinose, lactose, glucose
Surfactants	Poloxamer 407, Poloxamer 188, polysorbate 80, polysorbate 20, octoxynol-9, polyoxyethylene-(23) lauryl alcohol, polyxyethylene-(20) oleyl alcohol, sodium lauryl sulphate
Salts	Sodium sulphate, ammonium sulphate, magnesium sulphate, sodium acetate, sodium lactate, sodium succinate, sodium proprionate, potassium phosphate
Polyols	Cyclodextrins, mannitol, sorbitol, glycerol, xylitol, inositol
Antioxidants	Ascorbic acid, glutathione
Polymers	Polyethylene glycol, dextran, polyvinylpyrrolidone
Chelating agents	EDTA, tris(hydroxymethyl)aminomethane (TRIS), diethylenetriaminepentaacetic acid, inositol, hexaphosphate, ethylenediaminebis (O-hydroxyphenylacetic acid), desferal

unfavorable process. This is to be expected because the excipient would denature the protein if it were to bind favorably to the exposed hydrophobic sites of the unfolded protein.

Two nonexclusive mechanisms, namely, glass dynamics hypothesis and water substitution concept, have also been proposed to explain the stabilization effects of excipients during the dehydration stage in freeze-drying [193]. According to the glass dynamics hypothesis, freeze-dried excipients form a rigid, amorphous, inert matrix where the dispersed protein molecules have limited mobility, being restricted from undergoing arrangement and eventual phase separation. Therefore, stabilization is via a kinetic mechanism and would be expected to correlate with the molecular mobility in the rigid matrix. The water substitution theory suggests that the stabilization effect is due to the substitution by the excipient for water to maintain hydrogen bonds with surface-accessible polar sites on the protein. Such interaction facilitates the preservation of the native-like solid-state protein structure and may help reduce the tendency of the protein molecules to form aggregates during storage. These stabilization mechanisms have been reviewed in depth by Pikal [194].

Most commercial protein/peptide products are administered by the parenteral route (e.g., subcutaneous, intramuscular, or intravenous) in the form of injections. The major disadvantages of parenteral injections are the inconvenience to the patient, high cost, and poor patient compliance. To overcome these problems, alternative administration routes such as peroral [195, 196], buccal [197], pulmonary, nasal [198], and transdermal [199] routes have been considered. However, nonparenteral protein delivery also suffers certain drawbacks, most of which are related to the inherent physicochemical characteristics of the proteins, including their large molecular size, low lipid membrane permeability, susceptibility to enzymatic degradation, potential immunogenicity, and bio-incompatibility as well as the propensity to undergo aggregation, adsorption, and denaturation. Not all nonparenteral routes share these limitatons. For instance, absorption via the buccal mucosa is free from degradation by protease activity and allows the absorbed protein to bypass the liver to gain direct entry into the systemic circulation, thus avoiding the first-pass hepatic metabolism. Pulmonary delivery also offers a similar advantage of bypassing the first-pass effect.

The feasibility of protein delivery via different routes of administration is best illustrated by insulin, a well-characterized and extensively studied therapeutic protein. To date, attempts to exploit the nasal, oral, and transdermal routes for insulin delivery have not been very successful for reasons related mostly to the stability and efficiency of delivery [200]. For instance, oral insulin delivery is subject to activity loss due to the acidity in the stomach and the presence of proteases and peptidases in the gut. Moreover, the intestinal absorption of the protein is limited by its large molecular size and hydrophilic nature. Various specific strategies to overcome these delivery barriers have been attempted, including the use of permeation enhancers, protease inhibitors, and enteric-coated microsphere formulations [201]. As an alternative solution to these problems, the pulmonary route has also been extensively investigated for its utility in protein delivery [202]. As has been well documented, lungs have a large surface area, which allows rapid drug absorption due to the close contact between the alveoli in the deep lung and the systemic circulation. As a result of intense formulation research, a novel inhalation insulin product by the commercial name, Exubera, was developed and subsequently approved by the Food and Drug Administration (FDA) Advisory Committee Panel in September 2005 for the treatment of adult type 1 and type 2 diabetes. The product is a rapid-acting, dry powder form of insulin derived through rDNA. The main concern with inhaled insulin products is the potential long-term adverse effects resulting from the intra-alveolar deposition of insulin within the lung, because insulin has growth-promoting properties. It can be envisaged that once long-term safety and efficacy is established, inhalation may become the first nonparenteral route used clinically for insulin delivery.

Apart from the aforementioned problems with oral delivery, certain undesirable pharmacokinetic properties of proteins, notably the short biological half-life, have presented additional hurdles in their formulation development. To lengthen the biological half-life and prolong the action of proteins in the body, several novel drug delivery strategies have been developed through chemical modification of the protein molecule (e.g., pegylation) or new formulation development (e.g., controlled or sustained release formulation).

Pegylation, defined as the chemical process by which polyethylene glycol chains are attached to protein and peptide drugs, have been successfully applied to produce peginterferon alfa-2a (PEGASYS) and peginterferon alfa-2b (PEG-Intron), both of which have been approved by the FDA for the treatment of chronic hepatitis C [203].

Another approach to extend the action of proteins is to develop biodegradable polymeric microspheres. Owing to their excellent biocompatibility, the biodegradable polyesters, poly(lactic acid) (PLA) and poly(lactic-co-glycolic acid) (PLGA), are the most frequently used biomaterials to achieve sustained action [204, 205]. Polymeric microspheres are commonly prepared by the solvent extraction/evaporation methods [206]. In brief, protein in a solid or liquid form is mixed with a polymer (dissolved in an organic solvent, e.g., dichloromethane) to prepare a solid-in-oil or water-in-oil emulsion. The resulting emulsion is further mixed with water containing an emulsifier (e.g., polyvinyl alcohol) to effect protein loading into the PLGA microspheres [207]. Release of protein from such microspheres usually occurs in three phases, namely, initial burst release, diffusion-controlled release, and erosion-controlled release.

The intrinsic drawbacks with these microencapsulation methods are that the organic solvents used and the low pH generated inside the matrix due to PLGA degradation may denature the protein. As an alternative for these biodegradable polyesters, hydrogels, which are prepared from natural or synthetic hydrophilic polymers, have been evaluated for use as protein-releasing matrices either alone or in combination with PLGA (i.e., as composite microspheres). Using the latter approach, a novel sustained-release insulin formulation has been developed by incorporating insulin into the hydrogel to yield microparticles, followed by encapsulation in a PLGA matrix as microspheres [208]. The swelling behavior of hydrogels, which governs drug release, is dependent on the external environmental conditions, such as temperature and pH. Hydrogel-based delivery devices can be used for oral, rectal, ocular, epidermal, and subcutaneous application. Excellent reviews on this topic can be found in literature [209, 210].

In general, proteins are more stable in the solid state than in the liquid state. A variety of methods are available for preparation of protein powders, including precipitation, lyophilization, and spray-drying. The main concern with these methods is the potential protein denaturation or inactivation during the preparation. More specialized techniques for preparing protein materials with specific applications have also been developed, such as supercritical fluid processing for human growth hormone [211], spray-drying for producing inhaled insulin [212], spray–freeze-drying into liquid to produce stable bovine serum albumin (BSA) nanostructured microparticles [213], supercritical fluid-based coating technology to produce lipid coated BSA particles [214], and emulsion precipitation [207]. Maa and Prestrelski [215] have presented a systematic review on protein powder production techniques. Table 4.3-3 shows a comparison of the advantages and disadvantages of these powder preparation methods. For assessment of the utility of these techniques, it is important that the resulting powders be subjected to rigorous physical and chemical characterization, including crystal form, particle morphology, particle size distribution, surface properties, and moisture content [212].

4.3.5.2 Nucleic Acid-based Drugs

Nucleic acid-based drugs also represent an important group of biopharmaceuticals that are still at their very early stage of development. Nucleic acids are complex, and high-molecular-weight macromolecules composed of nucleotide chains that convey genetic information. The most common nucleic acids are DNA and RNA. RNA is usually single stranded, whereas DNA is normally double stranded and can form a double helix via hydrogen bonding. Single-stranded RNA molecules tend to associate with one another via hydrogen bonding to attain the minimum energy state. DNA and RNA can be characterized for primary sequence by biochemical techniques [e.g., Southern blotting and polymerase chain reaction (PCR) for DNA and Northern blotting and reverse transcription-PCR (RT-PCR) for RNA]. They can also be analyzed in very much the same way as proteins by spectroscopic and

TABLE 4.3-3. Advantages and Disadvantages of Different Powder Preparation Methods (from Ref. 215)

Method	Advantage	Disadvantage
Freeze-drying	High yield; convenient operation; aseptic process; expensive process	Poor particle size control; irregular-shaped particles; broad size distribution
Spray-drying	Good particle size control; spherical-shaped particles; easy and convenient operation	Yield dependent on formulation; heat inactivation; surface denaturation
Spray–freeze-drying	Good particle size control; spherical-shaped particles; high yield; good aerosol properties	Poor density control
Supercritical fluid	Good particle size control; spherical-shaped particles	Complex process; limited solubility in organic solvents
Pulverization	Easy operation; high yield	Poor particle size control; irregular-shaped particles; broad size distribution
Precipitation	Simple and convenient operation; high yield	Poor particle size control; irregular-shaped particles; broad size distribution; protein denaturation; difficult to handle multicomponents
Emulsification followed by precipitation	Easy to control; particle size <3 μm; spherical-shaped particles; applicable to a variety of materials	Yield limited by solubility; protein denaturation by organic solvent; difficult to handle multicomponents

chromatographic techniques, including ultraviolet, infrared, Raman, fluorescence spectroscopy, circular dichroism, NMR, mass spectrometry, and HPLC [46].

Nucleic acid-based therapeutics aims to deliver nucleic acid-derived products into the nuclear compartment, which may act at the molecular genetic level either to replace a defective gene or to modulate a specific genetic function. Six types of nucleic acid-based therapeutics have been proposed, namely, antisense (molecules that interact with complementary strands of nucleic acids and modify expression of genes), ribonucleic acid inhibition (RNAi), gene therapy (insertion of genes into an individual's cells and tissues to treat a disease, particularly hereditary diseases), nucleoside analogs, ribozymes (RNA enzymes or catalytic RNAs that catalyze chemical reactions), and aptamers (DNAs or RNAs that are selected from random pools based on their ability to bind nucleic acids, proteins, or small organic compounds). Only one antisense drug named fomivirsen (Vitravene® developed by Isis Pharmaceuticals, Carlsbad, California) has been approved by the FDA for the treatment of cytomegalovirus retinitis and is currently available in the clinic, whereas others are still under development.

Compared with other organic drugs, nucleic acids are relatively large, hydrophilic, and negatively charged molecules. Systemic delivery of nucleic acids targeted at the nuclear compartment is limited by the hostile extracellular environment, including extreme pH, degradation enzymes (proteases and nucleases), as well as the immune defense and scavenger systems. Lechardeur et al. [216] have summarized the major cellular, metabolic, and physico-chemical barriers to the delivery of plasmid DNA into nucleus. A series of strategies employing physical forces (e. g., microinjection, naked DNA injection, electroporation, and gene gun), nonviral gene carriers (e.g., cationic liposome or cationic polymers), and/or biological carriers (e.g., viral gene carrier) have been developed to deliver nucleic-acid-derived drugs into cells [217]. Viral gene carriers, which rely on cell infection by live (nonpathogenic) viruses (e.g., adenovirus) to achieve specific site targeting, may cause severe immune response in the patient, and their delivery is limited by the gene size, which can be accommodated by the viral genome. Nonviral gene medicine is composed of three elements: a gene encoding a therapeutic protein, a plasmid-based gene expression system that controls the function of a gene within a target cell, and a synthetic gene carrier that controls the stability and gene delivery within the body. The most extensively investigated synthetic gene carriers involve the combination of plasmid DNA with either cationic lipids (lipoplexes) or polymers (polyplexes), or in association with encapsulation into microspheres or preparation of nanoparticulate systems.

Lipoplexes are prepared by the interaction of anionic nucleic acids with the surface of cationic lipid to afford multilamellar lipid–nucleic acid complexes. Cationic liposomes can protect nucleic acids from serum nucleases and facilitate the cellular uptake and release of nucleic acids into the cytosol [218]. Pedroso et al. [219] have extensively discussed the structure-activity relationships of cationic liposome/DNA complexes and the key formulation parameters influencing the properties of lipoplexes. In addition, optimization of the cationic liposomal complexes for *in vivo* application has been reviewed by Smyth [220].

Polyplexes are formulated by condensation of negatively charged nucleic acids via electrostatic interactions with polycationic condensing polymers to yield compact particles, which can then protect the nucleic acids and mask the negative DNA charges. The polycations used include natural DNA binding proteins such as histones, synthetic polymers such as poly[dimethylaminoethyl-methacrylates], linear or branched polyethylenimine, diethylaminoethyl modified dextran, or modified chitosan. The physico-chemical and biopharmaceutical behaviors of these polymers in nucleic acid delivery have been extensively reviewed elsewhere [221, 222]. Besides these polymers, polyethylene glycol-conjugated copolymers have also been introduced as an alternative drug carrier material [223]. However, these polyplexes are incapable of affording sustained release for nucleic acids and prolonging gene transfer. To achieve prolonged action of nucleic-acid-based drugs, the feasibility of encapsulating polyplexes into certain matrices to obtain sustained-release nanoparticles or microparticles is currently under investigation. Yun et al. [224] have prepared microspheres by physically combining poly(ethylene glycol)-grafted chitosan with poly(lactide-co-glycolide) using a modified conventional in-emulsion solvent evaporation method to achieve prolonged delivery of DNA/chitosan polyplexes.

Nanoparticles are submicronic (less than 1 μm) colloidal systems, which can be classified into two main categories depending on whether the formation of nanoparticles requires a polymerization reaction. Nanoparticulate systems have been developed to deliver antisense oligonucleotides with enhanced stability and delivery efficiency [225].

4.3.6 SUMMARY AND CONCLUSIONS

Biopharmaceuticals constitute a unique class of therapeutic agents that are difficult to handle and process because of their structural complexity and inherent chemical/physical instability. Depending on the modes of chemical degradation or physical denaturation, a wide variety of formulation excipients or additives, including buffers, antioxidants, sugars, and polysaccharides, can be employed to preserve the structural integrity and biological activity of such vulnerable macromolecules. In addition, an appropriate choice of administration route is critical for maximizing their bioavailability and efficacy. Sustained or prolonged action for specific therapeutic indications has also been made possible by formulating them in synthetic polymeric carriers, including biodegradable polymers, hydrogels, and polycationic polymers. It can be envisioned that as more is known about the relationship between the structures and the bioactivities of these vital macromolecules, the existing formulation strategies can be further improved or fine-tuned for particular biopharmaceuticals to achieve more promising therapeutic outcomes.

REFERENCES

1. Clarke S, Stephenson R C, Lowenson J D (1992). Lability of asparagine and aspartic acid residues in proteins and peptides: Spontaneous deamidation and isomerization reactions. In T J Ahern, M C Manning, (eds.), *Stability of Protein Pharmaceuticals: Part A Chemical and Physical Pathways of Protein Degradation*, Plenum Press, New York, pp. 1–29.

2. Brennan T V, Clarke S (1995). Deamidation and isoaspartate formation in model synthetic peptides: The effects of sequence and solution environment. In D W Aswad, (eds.), *Deamidation and Isoaspartate Formation in Peptides and Proteins*, CRC Press, Boca Raton F.L., pp. 65–90.

3. Goolcharran C, Khossravi M, Borchardt R T (2000). Chemical pathways of peptide and protein degradation. In S Frokjaer, L Hovgaard, (eds.), *Pharmaceutical Formulation Development of Peptides and Proteins*, Taylor & Francis, London, pp. 70–88.

4. Manning M C, Patel K, Borchardt R T (1989). Stability of protein pharmaceuticals. *Pharm. Res.* 6:903–918.

5. Schoneich C, Hageman M J, Borchardt R T (1997). Stability of peptides and proteins. In K Park, (ed.), *Controlled Drug Delivery*, American Chemical Society, Washington, D.C., pp. 205–228.

6. Nabuchi Y, Fujiwara E, Kuboniwa H, et al. (1997). The stability and degradation pathway of recombinant human parathyroid hormone: Deamidation of asparaginyl residue and peptide bond cleavage at aspartyl and asparaginyl residues. *Pharm. Res.* 14:1685–1690.

7. Kroon D J, Baldwin-Ferro A, Lalan P (1992). Identification of sites of degradation in a therapeutic monoclonal antibody by peptide mapping. *Pharm. Res.* 9:1386–1393.

8. Son K, Kwon C (1995). Stabilization of human epidermal growth factor (hEGF) in aqueous formulation. *Pharm. Res.* 12:451–454.

9. Darrington R T, Anderson B D (1995). Effects of insulin concentration and self-association on the partitioning of its A-21 cyclic anhydride intermediate to desamido insulin and covalent dimer. *Pharm. Res.* 12:1077–1084.

10. Wolfe J L, Lee G E, Potti G K, et al. (1994). Degradation of antiflammin 2 in aqueous solution. *J. Pharm. Sci.* 83:1762–1764.

11. Aswad D W, Paranandi M V, Schurter B T (2000). Isoaspartate in peptides and proteins: Formation, significance and analysis. *J. Pharm. Biomed. Anal.* 21:1129–1136.

12. Tsai P K, Bruner M W, Irwin J I, et al. (1993). Origin of the isoelectric heterogeneity of monoclonal immunoglobulin h1B4. *Pharm. Res.* 10:1580–1586.

13. Perkins M, Theiler R, Lunte S, et al. (2000). Determination of the origin of charge heterogeneity in a murine monoclonal antibody. *Pharm. Res.* 17:1110–1117.

14. Blanche F, Cameron B, Somarriba S, et al. (2001). Stabilization of recombinant adenovirus: Site-directed mutagenesis of key asparagine residues in the hexon protein. *Anal. Biochem.* 297:1–9.

15. Joshi A B, Kirsch L E (2002). The relative rates of glutamine and asparagine deamidation in glucagons fragment 22–29 under acidic conditions. *J. Pharm. Sci.* 91:2332–2345.

16. Li B, Gorman E M, Moore K D, et al. (2005). Effects of acidic N + 1 residues on asparagine deamidation rates in solution and in the solid state. *J. Pharm. Sci.* 94:666–675.

17. Xie M, Schowen R L (1999). Secondary structure and protein deamidation. *J. Pharm. Sci.* 88:8–13.

18. Xie M, Shahrokh Z, Kadkhodayan M, et al. (2003). Asparagine deamidation in recombinant human lymphotoxin: Hindrance by three-dimensional structures. *J. Pharm. Sci.* 92:869–880.

19. Geiger T, Clarke S (1987). Deamidation, isomerization, and racemization at asparaginyl and aspartyl residues in peptides. *J. Biol. Chem.* 262:785–794.

20. Patel K, Borchardt R T (1990). Chemical pathways of peptide degradation. III. Effect of primary sequence on the pathways of deamidation of asparaginyl residues in hexapeptides. *Pharm. Res.* 7:787–793.

21. Pearlman R, Bewley T A (1993). Stability and characterization of human growth hormone. In Y J Wang, R Pearlman, (eds.), *Stability and Characterization of Protein and Peptide Drugs: Case Histories*, Plenum Press, New York, pp. 1–58.

22. Brange J (1994). *Stability of Insulin: Studies on the Physical and Chemical Stability of Insulin in Pharmaceutical Formulation*, Kluwer Academic Publishers, Dordrecht, pp. 6–59.

23. Jars M U, Hvass A, Waaben D (2002). Insulin aspart (Asp[B28] human insulin) derivatives formed in pharmaceutical solutions. *Pharm. Res.* 19:621–628.

24. Markell D, Hui J, Narhi L, et al. (2001). Pharmaceutical significance of the cyclic imide form of recombinant human glial cell line derived neurotrophic factor. *Pharm. Res.* 18:1361–1366.

25. Teshima G, Hancock W S, Canova-Davis E (1995). Effect of deamidation and isoaspartate formation on the activity of proteins. In D W Aswad, (ed.), *Deamidation and Isoaspartate Formation in Peptides and Proteins*, CRC Press, Boca Raton F.L., pp. 167–191.

26. Sasaoki K, Hiroshima T, Kusumoto S, et al. (1992). Deamidation at asparagine-88 in recombinant human interleukin 2. *Chem. Pharm. Bull.* 40:976–980.

27. Powell M F (1994). Peptide stability in aqueous parenteral formulations. In J L Cleland, R Langer, (eds.), *Formulation and Delivery of Proteins and Peptides*, American Chemical Society, Washington D.C., pp. 100–117.

28. Ye J M, Lee G E, Potti G K, et al. (1996). Degradation of antiflammin 2 under acidic conditions. *J. Pharm. Sci.* 85:695–699.

29. Stephenson R C, Clarke S (1989). Succinimide formation from aspartyl and asparaginyl peptides as a model for the spontaneous degradation of proteins. *J. Biol. Chem.* 264:6164–6170.

30. Oliyai C, Borchardt R T (1994). Chemical pathways of peptide degradation. VI. Effect of primary sequence on the pathways of degradation of aspartyl residues in model hexapeptides. *Pharm. Res.* 11:751–758.

31. Piszkiewicz D, Landon M, Smith E L (1970). Anomalous cleavage of aspartyl-proline peptide bonds during amino acid sequence determinations. *Biochem. Biophys. Res. Commun.* 40:1173–1178.

32. Kenley R A, Warne N W (1994). Acid-catalyzed peptide bond hydrolysis of recombinant human interleukin 11. *Pharm. Res.* 11:72–76.

33. Strickley R G, Brandl M, Chan K W, et al. (1990). High-performance liquid chromatographic (HPLC) and HPLC-mass spectrometric (MS) analysis of the degradation of the luteinizing hormone-releasing hormone (LH-RH) antagonist RS-26306 in aqueous solution. *Pharm. Res.* 7:530–536.

34. Windisch V, DeLuccia F, Duhau L, et al. (1997). Degradation pathways of salmon calcitonin in aqueous solution. *J. Pharm. Sci.* 86:359–364.

35. Kertscher U, Bienert M, Krause E, et al. (1993). Spontaneous chemical degradation of substance P in the solid phase and in solution. *Int. J. Peptide Protein Res.* 41:207–211.

36. Battersby J E, Hancock W S, Canova-Davis E, et al. (1994). Diketopiperazine formation and *N*-terminal degradation in recombinant human growth hormone. *Int. J. Peptide Protein Res.* 44:215–222.

37. Oyler A R, Naldi R E, Lloyd J R, et al. (1991). Characterization of the solution degradation products of histrelin, a gonadotrophin releasing hormone (LH/RH) agonist. *J. Pharm. Sci.* 80:271–275.

38. Straub J A, Akiyama A, Parmar P, et al. (1995). Chemical pathways of degradation of the bradykinin analog, RMP-7. *Pharm. Res.* 12:305–308.

39. Leung S S, Grant D J W (1997). Solid state stability studies of model dipeptides: Aspartame and aspartylphenylalanine. *J. Pharm. Sci.* 86:64–71.

40. Goolcharran C, Borchardt R T (1998). Kinetics of diketopiperazine formation using model peptides. *J. Pharm. Sci.* 87:283–288.

41. Orlowska A, Witkowska E, Izdebski J (1987). Sequence dependence in the formation of pyroglutamyl peptides in solid phase peptide synthesis. *Int. J. Peptide Protein Res.* 30:141–144.

42. Beck A, Bussat M C, Klinguer-Hamour C, et al. (2001). Stability and CTL activity of *N*-terminal glutamic acid containing peptides. *J. Pept. Res.* 57:528–538.

43. Evans R K, Xu Z, Bohannon K E, et al. (2000). Evaluation of degradation pathways for plasmid DNA in pharmaceutical formulations via accelerated stability studies. *J. Pharm. Sci.* 89:76–87.

44. Pogocki D, Schoneich C (2000). Chemical stability of nucleic acid-derived drugs. *J. Pharm. Sci.* 89:443–456.

45. Lindahl T (1993). Instability and decay of the primary structure of DNA. *Nature.* 362:709–715.

46. Middaugh C R, Evans R K, Montgomery D L, et al. (1998). Analysis of plasmid DNA from a pharmaceutical perspective. *J. Pharm. Sci.* 87:130–146.

47. Glaser R, Rayat S, Lewis M, et al. (1999). Theoretical studies of DNA base deamination. 2. Ab initio study of DNA base diazonium ions and of their linear, unimolecular dediazoniation paths. *J. Am. Chem. Soc.* 121:6108–6119.

48. Karran P, Lindahl T (1980). Hypoxanthine in deoxyribonucleic acid: Generation by heat-induced hydrolysis of adenine residues and release in free form by a deoxyribonucleic acid glycosylase from calf thymus. *Biochem.* 19:6005–6011.

49. Li S, Schoneich C, Borchardt R T (1995). Chemical instability of protein pharmaceuticals: Mechanisms of oxidation and strategies for stabilization. *Biotechnol. Bioeng.* 48:490–500.

50. Meucci E, Mordente A, Martorana G E (1991). Metal-catalyzed oxidation of human serum albumin: Conformational and functional changes. *J. Biol. Chem.* 266: 4692–4699.

51. Li S, Nguyen T H, Schoneich C, et al. (1995). Aggregation and precipitation of human relaxin induced by metal-catalyzed oxidation. *Biochem.* 34:5762–5772.

52. Wells J A, Powers D B, Bott R R (1987). Protein engineering of subtilisin. In D L Oxender, C F Fox, (eds.), *Protein Engineering*: Alan R. Liss, Inc. New York, pp. 279–287.

53. Dubost D C, Kaufman M J, Zimmerman J A, et al. (1996). Characterization of a solid state reaction product from a lyophilized formulation of a cyclic heptapeptide. A novel example of an excipient-induced oxidation. *Pharm. Res.* 13:1811–1814.

54. Nguyen T H (1994). Oxidation degradation of protein pharmaceuticals. In J L Cleland, R Langer, (eds.), *Formulation and Delivery of Proteins and Peptides*, American Chemical Society, Washington, D.C., pp. 59–71.

55. Chu J W, Yin J, Wang D I C, et al. (2004a). Molecular dynamics simulations and oxidation rates of methionine residues of granulocyte colony-stimulating factor at different pH values. *Biochem.* 43:1019–1029.

56. Chu J W, Yin J, Wang D I C, et al. (2004b). A structural and mechanistic study of the oxidation of methionine residues in hPTH(1-34) via experiments and simulations. *Biochem.* 43:14139–14148.

57. Luo D, Smith S W, Anderson B D (2005). Kinetics and mechanism of the reaction of cysteine and hydrogen peroxide in aqueous solution. *J. Pharm. Sci.* 94:304–316.

58. Shioi S, Imoto T, Ueda T (2004). Analysis of the early stage of the folding process of reduced lysozyme using all lysozyme variants containing a pair of cysteines. *Biochem.* 43:5488–5493.

59. Shaked Z, Szajewski R P, Whitesides G M (1980). Rates of thiol-disulfide interchange reactions involving proteins and kinetic measurements of thiol pKa values. *Biochem.* 19:4156–4166.

60. Parente A, Merrifield B, Geraci G, et al. (1985). Molecular basis of superreactivity of cysteine residues 31 and 32 of seminal ribonucleases. *Biochem.* 24:1098–1104.

61. Kuwajima K, Ikeguchi M, Sugawara T, et al. (1990). Kinetics of disulfide bond reduction in α-lactalbumin by dithiothreitol and molecular basis of superreactivity of the cys6–cys120 disulfide bond. *Biochem.* 29:8240–8249.

62. Fransson J, Florin-Robertsson E, Axelsson K, et al. (1996). Oxidation of human insulin-like growth factor I in formulation studies: Kinetics of methionine oxidation in aqueous solution and in solid state. *Pharm. Res.* 13:1252–1257.

63. Fransson J, Hagman A (1996). Oxidation of human insulin-like growth factor I in formulation studies, II. Effects of oxygen, visible light, and phosphate on methionine oxidation in aqueous solution and evaluation of possible mechanisms. *Pharm. Res.* 13:1476–1481.

64. Schoneich C, Zhao F, Yang J (1997). Mechanisms of methionine oxidation in peptides. In Z Shahrokh, V Sluzky, J L Cleland, et al. (eds.), *Therapeutic Protein and Peptide Formulation and Delivery, ACS Symp. Ser. 675*, American Chemical Society, Washington, D.C., pp. 79–89.

65. Schoneich C (2000). Mechanisms of metal-catalyzed oxidation of histidine to 2-oxo-histidine in peptides and proteins. *J. Pharm. Biomed. Anal.* 21:1093–1097.

66. Khossravi M, Shire S J, Borchardt R T (2000). Evidence for the involvement of histidine A(12) in the aggregation and precipitation of human relaxin induced by metal-catalyzed oxidation. *Biochem.* 39:5876–5885.

67. Tomita M, Irie M, Ukita T (1969). Sensitized photooxidation of histidine and its derivatives. Products and mechanism of the reaction. *Biochem.* 8:5149–5160.

68. Khossravi M, Borchardt R T (2000). Chemical pathways of peptide degradation. X: Effect of metal catalyzed oxidation on the solution structure of a histidine-containing peptide fragment of human relaxin. *Pharm. Res.* 17:851–858.

69. Bummer P M, Koppenol S (2000). Chemical and physical considerations in protein and peptide stability. In E J McNally, (eds.), *Protein Formulation and Delivery*, Marcel Dekker, Inc. New York, pp. 5–69.

70. Miller B L, Hageman M J, Thamann T J, et al. (2003). Solid-state photodegradation of bovine somatotropin (bovine growth hormone): evidence for tryptophan-mediated photooxidation of disulfide bonds. *J. Pharm. Sci.* 92:1698–1709.

71. Kanwar R, Balasubramanian D (1999). Structure and stability of the dityrosine-linked dimer of γB-crystallin. *Exper. Eye Res.* 68:773–784.

72. Feibush B, Snyder B (2000). Oxidation of the N-terminal Gly-residue of peptides: Stress study of pexiganan acetate in a drug formulation. *Pharm. Res.* 17:197–204.

73. Meneghini R, Martins E A L, Calderaro M (1993). DNA damage by reactive oxygen species. The role of metals. In G Poli, E Albano, M U Dianzani, (eds.), *Free Radicals: From Basic Science to Medicine*, Basel, Birlhauser Verlag, pp.102–113.

74. Beckman K B, Ames B N (1997). Oxidative decay of DNA. *J. Biol. Chem.* 272: 19633–19636.

75. Armitage B (1998). Photocleavage of nucleic acids. *Chem. Rev.* 98:1171–1200.

76. Jaruga P, Dizdaroglu M (1996). Repair of products of oxidative DNA base damage in human cells. *Nucleic Acids Res.* 24:1389–1394.

77. Henle E S, Linn S (1997). Formation, prevention and repair of DNA damage by iron/hydrogen peroxide. *J. Biol. Chem.* 272:19095–19098.

78. Burrows C J, Muller J G (1998). Oxidative nucleobase modifications leading to strand scission. *Chem. Rev.* 98:1109–1151.

79. Evans M D, Dizdaroglu M, Cooke M S (2004). Oxidative DNA damage and disease: induction, repair and significance. *Mutation Res.* 567:1–61.

80. Pogozelski W K, Tullius T D (1998). Oxidative strand scission of nucleic acids: Routes initiated by hydrogen abstraction from the sugar moiety. *Chem. Rev.* 98:1089–1107.

81. Suzuki T, Ohsumi S, Makino K (1994). Mechanistic studies on depurination and apurinic site chain breakage in oligodeoxyribonucleotides. *Nucleic Acids Res.* 22:4997–5003.

82. Hall S C, Tan M M, Leonard J J, et al. (1998). Characterization and comparison of leuprolide degradation profiles in water and dimethyl sulfoxide. *J. Pept. Res.* 53:432–441.

83. Stevenson C L, Leonard J J, Hall S C (1999). Effect of peptide concentration and temperature on leuprolide stability in dimethyl sulfoxide. *Int. J. Pharm.* 191:115–129.

84. Costantino H R, Langer R, Klibanov A M (1994). Moisture-induced aggregation of lyophilized insulin. *Pharm. Res.* 11:21–29.

85. Colaco C A L S, Smith C J S, Sen S, et al. (1994). Chemistry of protein stabilization by trehalose. In J L Cleland, R Langer, (eds.), *Formulation and Delivery of Proteins and Peptides*, American Chemical Society, Washington, D.C., pp. 222–240.

86. Lai M C, Topp E M (1999). Solid-state chemical stability of proteins and peptides. *J. Pharm. Sci.* 88:489–500.

87. Li S, Patapoff T W, Overcashier D, et al. (1996). Effects of reducing sugars on the chemical stability of human relaxin in the lyophilized state. *J. Pharm. Sci.* 85: 873–877.

88. Smales C M, Pepper D S, James D C (2000a). Mechanisms of protein modification during model anti-viral heat-treatment bioprocessing of β-lactoglobulin variant A in the presence of sucrose. *Biotechnol. Appl. Biochem.* 32:109–119.

89. Smales C M, Pepper D S, James D C (2000b). Protein modification during antiviral heat bioprocessing. *Biotechnol. Bioeng.* 67:177–188.

90. Smales C M, Pepper D S, James D C (2002). Protein modification during anti-viral heat-treatment bioprocessing of factor VIII concentrates, factor IX concentrates, and model proteins in the presence of sucrose. *Biotechnol. Bioeng.* 77:37–48.

91. Bucala R, Model P, Cerami A (1984). Modification of DNA by reducing sugars: A possible mechanism for nucleic acid aging and age-related dysfunction in gene expression. *Proc. of the National Academy of Sciences of the United States of America*, 81:105–109.

92. Lee A T, Cerami A (1987). Elevated glucose 6-phosphate levels are associated with plasmid mutations *in vivo*. *Proc. of the National Academy of Sciences of the United States of America*, 84:8311–8314.

93. Davis P J, Smales C M, James D C (2001). How can thermal processing modify the antigenicity of proteins? *Allergy.* 56(Suppl 67):56–60.

94. Brange J, Havelund S, Hougaard P (1992). Chemical stability of insulin. 2. Formation of higher molecular weight transformation products during storage of pharmaceutical preparations. *Pharm. Res.* 9:727–734.

95. Brange J, Langkjaer L, Havelund S, et al. (1992). Chemical stability of insulin. 1. Hydrolytic degradation during storage of pharmaceutical preparations. *Pharm. Res.* 9:715–726.

96. Darrington R T, Anderson B D (1995). Evidence for a common intermediate in insulin deamidation and covalent dimer formation: Effects of pH and aniline trapping in dilute acidic solutions. *J. Pharm. Sci.* 84:275–282.

97. Strickley R G, Anderson B D (1996). Solid-state stability of human insulin I. Mechanism and the effect of water on the kinetics of degradation in lyophiles from pH 2–5 solutions. *Pharm. Res.* 13:1142–1153.

98. Hora M S, Rana R K, Smith F W (1992). Lyophilized formulations of recombinant tumor necrosis factor. *Pharm. Res.* 9:33–36.

99. Fransson J R (1997). Oxidation of human insulin-like growth factor I in formulation studies. 3. Factorial experiments of the effects of ferric ions, EDTA, and visible light on methionine oxidation and covalent aggregation in aqueous solution. *J. Pharm. Sci.* 86:1046–1050.

100. Baldwin R L, Eisenberg D (1987). Protein stability. In D L Oxender, C F Fox, (eds.), *Protein Engineering*: Alan R. Liss, Inc., New York, pp. 127–148.

101. Chi E Y, Krishnan S, Kendrick B S, et al. (2003). Roles of conformational stability and colloidal stability in the aggregation of recombinant human granulocyte colony-stimulating factor. *Protein Sci.* 12:903–913.

102. Roy I, Gupta M N (2004). Freeze-drying of proteins: Some emerging concerns. *Biotechnol. Appl. Biochem.* 39:165–177.

103. Zhou H X, Vijayakumar M (1997). Modeling of protein conformational fluctuations in pK_a predictions. *J. Mol. Biol.* 267:1002–1011.

104. Shirley B A (1992). Protein conformational stability estimated from urea, guanidine hydrochloride, and thermal denaturation curves. In T J Ahern, M C Manning, (eds.), *Stability of Protein Pharmaceuticals: Part A: Chemical and Physical Pathways of Protein Degradation*: Plenum Press, New York, pp. 167–194.

105. Dubin D N, Lee A, Macgregor R B, et al. (2001). On the stability of double stranded nucleic acids. *J. Amer. Chem. Soc.* 123:9254–9259.

106. Tikhomirova A, Taulier N, Chalikian T V (2004). Energetics of nucleic acid stability: The effect of ΔC_P. *J. Amer. Chem. Soc.* 126:16387–16394.

107. Chen B L, Arakawa T, Morris C F, et al. (1994). Aggregation pathway of recombinant human keratinocyte growth factor and its stabilization. *Pharm. Res.* 11:1581–1587.

108. Bermudez O, Forciniti D (2004). Aggregation and denaturation of antibodies: a capillary electrophoresis, dynamic light scattering, and aqueous two-phase partitioning study. *J. Chromatogr. B.* 807:17–24.

109. Klibanov A M, Schefiliti J A (2004). On the relationship between conformation and stability in solid pharmaceutical protein formulations. *Biotechnol. Lett.* 26:1103–1106.

110. Kendrick B S, Carpenter J F, Cleland J L, et al. (1998). A transient expansion of the native state precedes aggregation of recombinant human interferon-γ. *Proc. of the National Academy of Sciences of the United States of America*, 95:14142–14146.

111. Chi E Y, Krishnan S, Randolph T W, et al. (2003). Physical stability of proteins in aqueous solution: Mechanism and driving forces in nonnative protein aggregation. *Pharm. Res.* 20:1325–1336.

112. Krishnamurthy R, Manning M C (2002). The stability factor: Importance in formulation development. *Curr. Pharm. Biotechnol.* 3:361–371.

113. Costantino H R, Langer R, Klibanov A M (1994). Solid-phase aggregation of proteins under pharmaceutically relevant conditions. *J. Pharm. Sci.* 83:1662–1669.

114. Hermeling S, Crommelin D J A, Schellekens H, et al. (2004). Structure-immunogenicity relationships of therapeutic proteins. *Pharm. Res.* 21:897–903.

115. Paborji M, Pochopin N L, Coppola W P, et al. (1994). Chemical and physical stability of chimeric L6, a mouse-human monoclonal antibody. *Pharm. Res.* 11:764–771.

116. Krishnan S, Chi E Y, Webb J N, et al. (2002). Aggregation of granulocyte colony stimulating factor under physiological conditions: characterization and thermodynamic inhibition. *Biochem.* 41:6422–6431.

117. Schwendeman S P, Costantino H R, Gupta R K, et al. (1995). Stabilization of tetanus and diphtheria toxoids against moisture-induced aggregation. *Proc. of the National Academy of Sciences of the United States of America.* 92:11234–11238.

118. Tavornvipas S, Tajiri S, Hirayama F, et al. (2004). Effects of hydrophilic cyclodextrins on aggregation of recombinant human growth hormone. *Pharm. Res.* 21:2369–2376.

119. Fransson J, Hallen D, Florin-Robertsson E (1997). Solvent effects on the solubility and physical stability of human insulin-like growth factor I. *Pharm. Res.* 14:606–612.

120. Zhang Y, Roy S, Jones L S, et al. (2004). Mechanism for benzyl alcohol-induced aggregation of recombinant human interleukin-1 receptor antagonist in aqueous solution. *J. Pharm. Sci.* 93:3076–3089.

121. Roy S, Jung R, Kerwin B A, et al. (2005). Effects of benzyl alcohol on aggregation of recombinant human interleukin-1-receptor antagonist in reconstituted lyophilized formulations. *J. Pharm. Sci.* 94:382–396.

122. Horbett T A (1992). Adsorption of proteins and peptides at interfaces. In T J Ahern, M C Manning, (eds.), *Stability of Protein Pharmaceuticals: Part A: Chemical and Physical Pathways of Protein Degradation*: Plenum Publishing Corporation, New York, pp. 195–214.

123. Brange J (2000). Physical stability of proteins. In S Frokjaer, L Hovgaard, (eds.), *Pharmaceutical Formulation Development of Peptides and Proteins*: Taylor & Francis, London, pp. 89–112.

124. Page C, Dawson P, Woollacott D, et al. (2000). Development of a lyophilization formulation that preserves the biological activity of the platelet-inducing cytokine interleukin-11 at low concentrations. *J. Pharm. Pharmacol.* 52:19–26.

125. Tzannis S T, Prestrelski S J (1999). Moisture effects on protein-excipient interactions in spray-dried powders. Nature of destabilizing effects of sucrose. *J. Pharm. Sci.* 88:360–370.

126. Fatouros A, Sjostrom B (2000). Recombinant factor VIII SQ—the influence of formulation parameters on structure and surface adsorption. *Int. J. Pharm.* 194:69–79.

127. Evans R K, Nawrocki D K, Isopi L A, et al. (2004). Development of stable liquid formulations for adenovirus-based vaccines. *J. Pharm. Sci.* 93:2458–2475.

128. Webb S D, Golledge S L, Cleland J L, et al. (2002). Surface adsorption of recombinant human interferon-γ in lyophilized and spray-lyophilized formulations. *J. Pharm. Sci.* 91:1474–1487.

129. Chang S L, Hofmann G A, Zhang L, et al. (2003). Stability of a transdermal salmon calcitonin formulation. *Drug Del.* 10:41–45.

130. Matheis W, Zott A, Schwanig M (2001). The role of the adsorption process for production and control combined adsorbed vaccines. *Vaccine.* 20:67–73.

131. Maa Y F, Shu C, Ameri M, et al. (2003). Optimization of an alum-adsorbed vaccine power formulation for epidermal powder immunization. *Pharm. Res.* 20:969–977.

132. Waterman K C, Adami R C (2005). Accelerated aging: Prediction of chemical stability of pharmaceuticals. *Int. J. Pharm.* 293:101–125.

133. Roberts C, Darrington R T, Whitley M B (2003). Irreversible aggregation of recombinant bovine granulocyte-colony stimulating factor (bG-CSF) and implications for predicting protein shelf life. *J. Pharm. Sci.* 92:1095–1111.

134. Volkin D B, Middaugh C R (1992). The effect of temperature on protein structure. In T J Ahern, M C Manning, *Stability of Protein Pharmaceuticals: Part A: Chemical and Physical Pathways of Protein Degradation*: Plenum Press, New York, pp. 215–247.

135. Pikal-Cleland K A, Cleland J L, Anchordoquy T J, et al. (2002). Effect of glycine on pH changes and protein stability during freeze-thawing in phosphate buffer systems. *J. Pharm. Sci.* 91:1969–1979.

136. Cao E, Chen Y, Cui Z, et al. (2003). Effect of freezing and thawing rates on denaturation of proteins in aqueous solutions. *Biotechnol. Bioeng.* 82:684–690.

137. Ho M M, Mawas F, Bolgiano B, et al. (2002). Physico-chemical and immunological examination of the thermal stability of tetanus toxoid conjugate vaccines. *Vaccines.* 20:3509–3522.

138. Anchordoquy T J, Girouard L G, Carpenter J F, et al. (1998). Stability of lipid/DNA complexes during agitation and freeze-thawing. *J. Pharm. Sci.* 87:1046–1051.

139. Chang B S, Kendrick B S, Carpenter J F (1996). Surface-induced denaturation of proteins during freezing and its inhibition by surfactants. *J. Pharm. Sci.* 85:1325–1330.

140. Lai M C, Hageman M J, Schowen R L, et al. (1999). Chemical stability of peptides in polymers. 1. Effect of water on peptide deamidation in poly(vinyl alcohol) and poly(vinyl pyrrolidone) matrices. *J. Pharm. Sci.* 88:1073–1080.

141. Breen E D, Curley J G, Overcashier D E, et al. (2001). Effect of moisture on the stability of a lyophilized humanized monoclonal antibody formulation. *Pharm. Res.* 18:1345–1353.

142. Chan H K, Clark A R, Feeley J C, et al. (2004). Physical stability of salmon calcitonin spray-dried powders for inhalation. *J. Pharm. Sci.* 93:792–804.

143. Costantino H R, Langer R, Klibanov A M (1995). Aggregation of a lyophilized pharmaceutical protein, recombinant human albumin: Effect of moisture and stabilization by excipients. *Biotechnol.* 13:493–496.

144. Costantino H R, Carrasquillo K G, Cordero R A, et al. (1998). Effect of excipients on the stability and structure of lyophilized recombinant human growth hormone. *J. Pharm. Sci.* 87:1412–1420.

145. Tzannis S T, Hrushesky W J M, Wood P A, et al. (1996). Irreversible inactivation of interleukin 2 in a pump-based delivery environment. *Proc. of the National Academy of Sciences of the United States of America.* 93:5460–5465.

146. Sarciaux J M, Mansour S, Hageman M J, et al. (1999). Effects of buffer composition and processing conditions on aggregation of bovine IgG during freeze-drying. *J. Pharm. Sci.* 88:1354–1361.

147. Yamamoto O (1992). Effect of radiation on protein stability. In T J Ahern, M C Manning, *Stability of Protein Pharmaceuticals: Part A: Chemical and Physical Pathways for Protein Degradation*: Plenum Press, New York, pp. 361–421.

148. Parti R, Ardosa J, Yang L, et al. (2000). *In vitro* stability of recombinant human factor VIII (Recombinate™). *Haemophilia.* 6:513–522.

149. Gocke E (2001). Photochemical mutagenesis: Examples and toxicological relevance. *J. Environ. Pathol. Toxicol. Oncol.* 20:285–292.

150. Oliva A, Farina J B, Llabres M (1996). Influence of temperature and shaking on stability of insulin preparations: Degradation kinetics. *Int. J. Pharm.* 143:163–170.

151. Shnek D R, Hostettler D L, Bell M A, et al. (1998). Physical stress testing of insulin suspensions and solutions. *J. Pharm. Sci.* 87:1459–1465.

152. Bam N B, Cleland J L, Yang J, et al. (1998). Tween protects recombinant human growth hormone against agitation-induced damage via hydrophobic interactions. *J. Pharm. Sci.* 87:1554–1559.

153. Mahler H C, Muller R, Frieb W, et al. (2005). Induction and analysis of aggregates in a liquid IgG1-antibody formulation. *Euro. J. Pharm. Biopharm.* 59:407–417.

154. Oliva A, Santovena A, Farina J, et al. (2003). Effect of high shear rate on stability of proteins: kinetic study. *J. Pharm. Biomed. Anal.* 33:145–155.

155. Katakam M, Bell L N, Banga A K (1995). Effect of surfactants on the physical stability of recombinant human growth hormone. *J. Pharm. Sci.* 84:713–716.

156. Nail S L, Jiang S, Chongprasert S, et al. (2002). Fundamentals of freeze-drying. In S L Nail, M J Akers, (eds.), *Development and Manufacture of Protein Pharmaceuticals*: Kluwer Academic, New York, pp. 281–360.

157. Wang W (2000). Lyophilization and development of solid protein pharmaceuticals. *Int. J. Pharm.* 203:1–60.

158. Kreilgaard L, Frokjaer S, Flink J M, et al. (1999). Effect of additives on the stability of *Humicola lanuginose* lipase during freeze-drying and storage in the dried solid. *J. Pharm. Sci.* 88:281–290.

159. Liao Y H, Brown M B, Martin G P (2004). Investigation of the stabilization of freeze-dried lysozyme and the physical properties of the formulations. *Euro. J. Pharm. Biopharm.* 58:15–24.

160. Kreilgaard L, Frokjaer S, Flink J M, et al. (1998). Effects of additives on the stability of recombinant human factor XIII during freeze-drying and storage in the dried solid. *Arch. Biochem. Biophys.* 360:121–134.

161. Cherng J Y, Talsma H, Crommelin D J A, et al. (1999). Long term stability of poly((2-dimethylamino)ethyl methacrylate)-based gene delivery systems. *Pharm. Res.* 16:1417–1423.

162. Li B, Li S, Tan Y, et al. (2000). Lyophilization of cationic lipid-protamine-DNA (LPD) complexes. *J. Pharm. Sci.* 89:355–364.

163. Townsend M W, Deluca P P (1991). Nature of aggregates formed during storage of freeze-dried ribonuclease A. *J. Pharm. Sci.* 80:63–66.

164. Maa Y F, Nguyen P A T, Hsu S W (1998). Spray-drying of air-liquid interface sensitive recombinant human growth hormone. *J. Pharm. Sci.* 87:152–159.

165. Maa Y F, Nugyen P A, Andya J D, et al. (1998). Effect of spray drying and subsequent processing conditions on residual moisture content and physical/biochemical stability of protein inhalation powders. *Pharm. Res.* 15:768–775.

166. Costantino H R, Andya J D, Nguyen P A, et al. (1998). Effect of mannitol crystallization on the stability and aerosol performance of a spray-dried pharmaceutical protein, recombinant humanized anti-IgE monocolonal antibody. *J. Pharm. Sci.* 87:1406–1411.

167. Adler M, Lee G (1999). Stability and surface activity of lactate dehydrogenase in spray-dried trehalose. *J. Pharm. Sci.* 88:199–208.

168. Liao Y H, Brown M B, Nazir T, et al. (2002). Effects of sucrose and trehalose on the preservation of the native structure of spray-dried lysozyme. *Pharm. Res.* 19:1847–1853.

169. Lee G (2002). Spray-drying of proteins. In J F Carpenter, M C Manning, (eds.), *Rational Design of Stable Protein Formulations: Theory and Practice*: Kluwer Academic, New York, pp. 137–158.

170. Li S, Patapoff T W, Nguyen T H, Borchardt R T (1996). Inhibitory effect of sugars and polyols on the metal-catalyzed oxidation of human relaxin. *J. Pharm. Sci.* 85: 868–872.

171. Rao P E, Kroon D J (1993). Orthoclone OKT3: Chemical mechanisms and functional effects of degradation of a therapeutic monoclonal antibody. In Y J Wang, R Pearlman, (eds.), *Stability and Characterization of Protein and Peptide Drugs: Case Histories*: Plenum Publishing Corporation, New York, pp. 135–158.

172. Hovorka S W, Hong J Y, Cleland J L, et al. (2001). Metal-catalyzed oxidation of human growth hormone: Modulation by solvent-induced changes of protein conformation. *J. Pharm. Sci.* 90:58–69.

173. D'Souza A J M, Schowen R L, Borchardt R T, et al. (2003). Reaction of a peptide with polyvinylpyrrolidone in the solid state. *J. Pharm. Sci.* 92:585–593.

174. Soenderkaer S, Carpenter J F, van de Weert M, et al. (2004). Effects of sucrose on rFVIIa aggregation and methionine oxidation. *Euro. J. Pharm. Sci.* 21:597–606.

175. Donbrow M, Azaz E, Pillersdorf A (1978). Autooxidation of polysorbates. *J. Pharm. Sci.* 67:1676–1681.

176. Robert T (1993). Clinical properties of yeast-derived versus *Escherichia coli*-derived granulocyte-macrophage colony-stimulating factor. *Clin. Therap.* 15:19–29.

177. Schrimsher J L, Rose K, Simona G, et al. (1987). Characterization of human and mouse granulocyte-macrophage colony-stimulating factors derived from *Escherichia coli. Biochem. J.* 247:195–199.

178. Burgess A W, Begley C G, Johnson G R, et al. (1987). Purification and properties of bacterially synthesized human granulocyte-macrophage colony-stimulating factor. *Blood.* 69:43–51.

179. Briggs J, Panfili P R (1991). Quantitation of DNA and protein impurities in biopharmaceuticals. *Anal. Chem.* 63:850–859.

180. Berthold W, Walter J (1994). Protein purification: Aspects of processes for pharmaceutical products. *Biologicals.* 22:135–150.

181. Tan J, Leung W, Moradian-Oldak J, et al. (1998). The pH dependent amelogenin solubility and its biological significance. *Conn. Tiss. Res.* 38:215–221.

182. Piscitelli S C, Reiss W G, Figg W D, et al. (1997). Pharmacokinetic studies with recombinant cytokines. Scientific issues and practical considerations. *Clin. Pharmacokinet.* 32:368–381.

183. Metz B, Hendriksen C F M, Jiskoot W, et al. (2002). Reduction of animal use in human vaccine quality control: opportunities and problems. *Vaccine.* 20:2411–2430.

184. Reid R E (2000). *Peptide and Protein Drug Analysis*: Marcel Dekker. New York, pp. 309–859.

185. Parkins D A, Lashmar U T (2000). The formulation of biopharmaceutical products. *Pharm. Sci. Technol. Today.* 3:129–137.

186. Liao Y H, Brown M B, Quader A, et al. (2002). Protective mechanism of stabilizing excipients against dehydration in the freeze-drying of proteins. *Pharm. Res.* 19:1854–1861.

187. Bilati U, Allémann E, Doelker E (2005). Strategic approaches for overcoming peptide and protein instability within biodegradable nano- and microparticles. *Euro. J. Pharm. Biopharm.* 59:375–388.

188. Ruiz L, Reyes N, Duany L, et al. (2003). Long-term stabilization of recombinant human interferon α-2b in aqueous solution without serum albumin. *Int. J. Pharm.* 264:57–72.

189. Broadhead J, Rouan S K E, Hau I, et al. (1994). The effect of process and formulation variables on the properties of spray-dried β-galactosidase. *J. Pharm. Pharmacol.* 46:458–467.

190. Allen J D, Bentley D, Stringer R A, et al. (1999). Hybrid (BDBB) interferon-α: preformulation studies. *Int. J. Pharm.* 187:259–272.

191. Branchu S, Forbes R T, York P, et al. (1999). Hydroxypropyl-β-cyclodextrin inhibits spray-drying-induced inactivation of β-galactosidase. *J. Pharm. Sci.* 88:905–911.

192. Arakawa T, Kita Y, Carpenter J F (1991). Protein-solvent interactions in pharmaceutical formulations. *Pharm. Res.* 8:285–291.

193. Chang L Q, Shepherd D, Sun J, et al. (2005). Mechanism of protein stabilization by sugars during freeze-drying and storage: Native structure preservation, specific interaction, and/or immobilization in a glassy matrix? *J. Pharm. Sci.* 94:1427–1444.

194. Pikal M J (2004). Mechanisms of protein stabilization during freeze-drying and storage: The relative importance of thermodynamic stabilization and glassy state relaxation dynamics. In L Rey, J C May, (eds.), *Freeze-drying/Lyophilization of Pharmaceutical and Biological Products*: Marcel Dekker. New York, pp. 63–108.

195. Shah R B, Ahsan F, Khan M A (2002). Oral delivery of proteins: progress and prognostication. *Crit. Rev. Therapeut. Drug Carr. Sys.* 19:135–169.

196. Mahato R I, Narang A S, Thoma L, et al. (2003). Emerging trends in oral delivery of peptide and protein drugs. *Crit. Rev. Therapeut. Drug Carr. Sys.* 20:153–214.

197. Senel S, Kremer M, Nagy K, et al. (2001). Delivery of bioactive peptides and proteins across oral (buccal) mucosa. *Curr. Pharm. Biotechnol.* 2:175–186.

198. Johnson K A (1997). Preparation of peptide and protein powders for inhalation. *Adv. Drug Del. Rev.* 26:3–15.

199. Banga A K, Prausnitz M R (1998). Assessing the potential of skin electroporation for the delivery of protein- and gene-based drugs. *Trends Biotechnol.* 16:408–412.

200. Owens D R, Zinman B, Bolli G (2003). Alternative routes of insulin delivery. *Diabetes Med.* 20:886–898.

201. Carino G P, Mathiowitz E (1999). Oral insulin delivery. *Adv. Drug Del. Rev.* 35:249–257.

202. Mandal T K (2005). Inhaled insulin for diabetes mellitus. *Am. J. Health-Sys. Pharm.* 62:1359–1364.

203. Pedder S C (2003). Pegylation of interferon alfa: structural and pharmacokinetic properties. *Seminars in Liver Disease.* 23(suppl 1):19–22.

204. Jiang W, Cupta R K, Deshpande M C, et al. (2005). Biodegradable poly(lactic-co-glycolic acid) microparticles for injectable delivery of vaccine antigens. *Adv. Drug Del. Rev.* 57:391–410.

205. Schwendeman S P (2002). Recent advances in the stabilization of proteins encapsulated in injectable PLGA delivery systems. *Adv. Drug Del. Rev.* 19:73–98.

206. Freitas S, Merkle H P, Gander B (2005). Microencapsulation by solvent extraction/evaporation: reviewing the state of the art of microsphere preparation process technology. *J. Control. Rel.* 102:313–332.

207. Yang J, Cleland J L (1997). Factors affecting the *in vitro* release of recombinant human interferon-γ (rhIFN-γ) from PLGA microspheres. *J. Pharm. Sci.* 86:908–914.

208. Jiang G, Qiu W, DeLuca P P (2003). Preparation and *in vitro/in vivo* evaluation of insulin-loaded poly(acryloyl-hydroxyethyl starch)-PLGA composite microspheres. *Pharm. Res.* 20:452–459.

209. Peppas N A, Bures P, Leobandung W, Ichikawa H (2000). Hydrogels in pharmaceutical formulations. *Euro. J. Pharm. Biopharm.* 50:27–46.

210. Hennink W E, De Jong S J, Bos G W, et al. (2004). Biodegradable dextran hydrogels crosslinked by stereocomplex formation for the controlled release of pharmaceutical proteins. *Int. J. Pharm.* 277:99–104.

211. Velega S P, Carlfors J (2005). Supercritical fluids processing of recombinant human growth hormone. *Drug Dev. Ind. Pharm.* 31:135–149.

212. Stahl K, Claesson M, Lilliehorn P, et al. (2002). The effect of process variables on the degradation and physical properties of spray dried insulin intended for inhalation. *Int. J. Pharm.* 233:227–237.

213. Yu Z, Garcia A S, Johnston K P, et al. (2004). Spray freezing into liquid nitrogen for highly stable protein nanostructured microparticles. *Euro. J. Pharm. Biopharm.* 58:529–537.

214. Ribeiro D S I, Richard J, Thies C, et al. (2003). A supercritical fluid-based coating technology. 3: Preparation and characterization of bovine serum albumin particles coated with lipids. *J. Microencap.* 20:110–128.

215. Maa Y F, Prestrelski S J (2000). Biopharmaceutical powders: particle formulation and formulation considerations. *Curr. Pharm. Biotechnol.* 1:283–302.

216. Lechardeur D, Verkman A S, Lukacs G I (2005). Intracellular routing of plasmid DNA during non-viral gene transfer. *Adv. Drug Del. Rev.* 57:755–767.

217. Wagner E, Kircheis R, Walker G F (2004). Targeted nucleic acid delivery into tumors: new avenues for cancer therapy. *Biomed. Pharmacother.* 58:152–161.

218. Tranchant I, Thompson B, Nicolazzi C, et al. (2004). Physicochemical optimisation of plasmid delivery by cationic lipids. *J. Gene Med.* 6(suppl 1):S24–S35.

219. Pedroso de L M C, Neves S, Filipe A, et al. (2003). Cationic liposomes for gene delivery: From biophysics to biological applications. *Curr. Med. Chem.* 10:1221–1231.

220. Smyth T N (2003). Cationic liposomes as *in vivo* delivery vehicles. *Curr. Med. Chem.* 10:1279–1287.

221. Smedt S C, Demeester J, Hennink W E (2000). Cationic polymer based gene delivery systems. *Pharm. Res.* 17:113–126.

222. Prabaharan M, Mano J F (2005). Chitosan-based particles as controlled drug delivery systems. *Drug Del.* 12:41–57.

223. Lee M, Kim S W (2005). Polyethylene glycol-conjugated copolymers for plasmid DNA delivery. *Pharm. Res.* 22:1–10.

224. Yun Y H, Jiang H, Chan R, et al. (2005). Sustained release of PEG-g-chitosan complexed DNA from poly(lactide-co-glycolide). *J. Biomat. Sci., Polym. Ed.* 16:1359–1378.

225. Lambert G, Fattal E, Couvreur P (2001). Nanoparticulate systems for the delivery of antisense oligonucleotides. *Adv. Drug Del. Rev.* 47:99–112.

226. Luykx D M A M, Goerdayal S S, Dingemanse P J (2005). High-performance anion-exchange chromatography combined with intrinsic fluorescence detection to determine erythropoietin in pharmaceutical products. *J. Chromatogr. B.* 1078:113–119.

227. Runkel L, Meier W, Pepinsky R B, et al. (1998). Structural and functional differences between glycosylated and non-glycosylated forms of human interferon-beta (IFN-beta). *Pharm. Res.* 15:641–649.

228. Chan J H, Timperman A T, Qin D, et al. (1999). Microfabricated polymer devices for automated sample delivery of peptides for analysis by electrospray ionization tandem mass spectrometry. *Anal. Chem.* 71:4437–4444.

229. Diwan M, Park T G (2003). Stabilization of recombinant interferon-alpha by pegylation for encapsulation in PLGA microspheres. *Int. J. Pharm.* 252:111–122.

230. Mire-Sluis A R, Thorpe R (1998). Laboratory protocols for the quantitation of cytokines by bioassay using cytokine responsive cell lines. *J. Immunol. Meth.* 211:199–210.

231. Rosendahl M S, Doherty D H, Smith D J, et al. (2005). A long-acting, highly potent interferon α-2 conjugate created using site-specific PEGylation. *Bioconjug. Chem.* 16:200–207.

4.4

PROTEIN POSTTRANSLATIONAL MODIFICATION: A POTENTIAL TARGET IN PHARMACEUTICAL DEVELOPMENT

M.D. MOSTAQUL HUQ AND LI-NA WEI

University of Minnesota Medical School, Minneapolis, Minnesota

Chapter Contents

4.4.1   Introduction   418
4.4.2   What Is Protein Posttranslational Modification?   418
4.4.3   Overview of Protein Posttranslational Modification   419
4.4.4   Why Proteins Need to be Modified?   419
4.4.5   Prokaryotic versus Eukaryotic PTM   420
4.4.6   Studies of PTM   420
      4.4.6.1   Mass Spectrometry-Based PTMs   420
4.4.7   Mapping of PTM Sites on RIP140 by LC–ESI–MS/MS Analysis   427
4.4.8   The role of PTM in Cellular Function and Signaling   433
4.4.9   Protein Posttranslational Modification and Diseases   434
4.4.10   PTM as a Target for Therapeutic Development   436
4.4.11   Conclusion   438
       Acknowledgment   438
       References   438

4.4.1 INTRODUCTION

Protein posttranslational modification (PTM) is a hallmark of signal transduction and it allows the cells to rapidly respond to extracellular signals. In most cells, only a fraction of the genes are turned on at any given time, producing no more than 6000 primary proteins in a process called translation. However, several hundred types of PTMs could occur, greatly amplifying the diversity of proteins present in the cells [1]. These modifications, which involve stable and, sometimes, irreversible additions of nonamino acid chemical groups to primary translation products, occur in a large proportion of the cellular protein pool. In some instances, the types of PTMs of a protein can be predicted from its primary sequence. In many cases, however, such predictions are not possible. These modifications can result in an enormous amplification of the diversity of a protein pool. For example, glycosylation, the addition of sugar chains of varying lengths and compositions of one unmodified protein at three sites, can generate 11,520 protein variants. PTM of a protein is predicted to generate a combinatorial effect on protein structure and functions [2, 3]. The known PTMs include phosphorylation, acetylation, methylation, glycosylation, oxidation, sumoylation, ubiquitination, and nitration [4–10]. PTM of core histone proteins has been widely studied in the context of chromatin structure and gene transcription [11]. However, PTM of nonhistone proteins has only recently begun to receive attention [12–15].

This chapter will be focusing on several important issues in PTM studies such as:

- Why the proteins need to be modified and how they are modified.
- The relationship between prokaryotic and eukaryotic PTMs.
- Basic tools for studying PTM and the recent advancement in facilities and resources.
- PTM and signal transduction by nuclear receptors, co-regulators, and transcription factors.
- PTM as a target in rational design and development of therapeutic agents.

4.4.2 WHAT IS PROTEIN POSTTRANSLATIONAL MODIFICATION?

Simple Definition. The primary structure of a protein is dictated by its genetic code. All the proteins from the genome constitute the "proteome." After translation, the proteome is covalently modified by a variety of small molecules of exogenous or endogenous origin, and also by other proteins or peptides. The event is defined as the "protein posttranslational modification."

Studies of proteins present in a cell revealed a rich repertoire of expressed proteins way beyond what is expected from direct translation of the messages produced by a genome. Proteins can be modified posttranslationally by covalent attachment of one or more of several classes of molecules, by the formation of intramolecular or intermolecular linkages, by proteolytic processing of the newly synthesized polypeptide chain, or by any combination of these events. These modifications endow a protein with various properties that may be specifically required under a particular condition.

4.4.3 OVERVIEW OF PROTEIN POSTTRANSLATIONAL MODIFICATION

The most common PTMs include:

- *Glycosylation*—the addition of one or more sugar molecules to asparagine (N-linked) and Ser/Thr (O-linked) residues, which may involve more than one type of sugar, as shown in estrogen receptor (ER) [16] and androgen receptor (AR) [17].
- *Phosphorylation*—the addition of phosphate groups to Ser/Thr and Tyr residues, as shown in receptor interacting protein 140 (RIP140) [4], orphan testicular receptor 2 (TR2) [18], retinoic acid receptor (RAR) [19], and retinoid X receptor (RXR) [19].
- *Myristoylation and Prenylation*—the addition of certain fatty acids as shown in G-protein coupled receptor (GPCR) [20] and estrogen receptor (ER) [21].
- *Acetylation*—the addition of acetate groups to lysine residue as shown in nuclear receptors [22, 23], p53 [6], NF-κB [24], and histone proteins.
- *Methylation*—the addition of methyl groups to Arg/Lys/Gln/Asn (N-methyl) and Ser/Thr (O-methyl) as shown in p53 [25], PGC-1α [13], CBP/p300 [26], and YY1 [27].
- *Ubiquitination*—the addition of one or more ubiquitin molecules to proteins, which marks the protein for degradation as shown in p53 [28].
- Addition of a prosthetic group (e.g., heme in hemeproteins, such as hemoglobin), which is required for the protein's function).
- Addition of a certain chemical bond (i.e., a disulfide bond) between two sulfur-containing amino acids.
- Addition of a target leader sequence (a small removable peptide) at the beginning of the protein chain to allow the protein to be targeted to or from subcellular organelles (e.g., nuclei and mitochondria).

4.4.4 WHY THE PROTEINS NEED TO BE MODIFIED?

Proteins sometimes need to be modified to become functional. Secondly, the functional diversity of a protein can be greatly amplified by different posttranslational modifications. Modification of protein can dramatically alter its physico-chemical properties (i.e., hydrophobicity, stereochemical structure and conformation, and protein stability). Phosphorylation and glycosylation make a protein more hydrophilic, whereas acetylation, methylation, and prenylation are generally known to increase hydrophobicity of the protein. Ubiquitination typically functions as a tag for the proteins, which could be recognized by proteases (proteosomes) for degradation. Reactive oxygen species can oxidize cystein and methionine, whereas nitric oxide can react with cystein and tyrosine residues. These modifications can also modulate the function of the protein by influencing processes such as protein–protein interaction, transport, and stability.

4.4.5 PROKARYOTIC VERSUS EUKARYOTIC PTM

It is generally believed that proteins are not posttranslationally modified in pro-karyotic cells with some exception for glycosylation of the membrane protein, which is essential for the membrane integrity and virulence property of many bac-teria. However, recent investigation showed that prokaryotic proteins could also be modified, such as by phosphorylation. Some prokaryotic protein kinases, which were evolutionarily related to kinases of the eukaryotes, were discovered [29]. It was also speculated that bacterial protein could be modified by methylation, as shown by metabolic labeling using radioactive S-adenosylmethionine. Methylation was shown to enhance the chemotaxic response of the bacteria as well as the sensory transduction in *E. coli* [30]. Thus, prokaryotic proteins could also be modi-fied to enhance their functional diversity. Our recent investigation using mass spectrometry-based proteome analyses of nuclear co-regulator RIP140 expressed from bacteria also revealed that proteins expressed in bacteria can be modified by methylation at asparagine (N) and arginine (R) (Huq and Wei, unpublished result).

It has become increasingly clear that PTM can also occur in *Archaea*. *Archaea* express proteins that enable them to strive in harsh habitats. Archaeal proteins are able to remain properly folded and functional under extremely harsh conditions such as high salinity and temperature, and other adverse physical conditions that would normally cause protein denaturation, loss of solubility, and aggregation [31].

4.4.6 STUDIES OF PTM

Studies of PTM have been benefited from the recent advancements in mass spec-trometry, the introduction of new software and Internet-based MS data search facilities, computer-assisted topology prediction for a variety of PTMs (visit http://ca.expasy.org/), chemical synthesis of modified peptides and proteins, development of modified peptide specific antibody, *in vitro* modification techniques, exploitation of other eukaryotic cells such as insect cells for protein expression [4, 5, 16, 32], and progress in affinity purification of modified proteins.

4.4.6.1 Mass Spectrometry-based PTMs

Mass spectrometry has revolutionized the idetinification of PTM. In the past, the metabolic radio-labeling technique was the principal methodology used to deter-mine whether the protein is modified. However, identification of a specific residue that is modified by PTM using metabolic labeling relies on point mutation of the protein, which is practically time- and effort-consuming and sometimes unreliable because of existance of other similar sites of PTM [33]. The mass spectrometric analysis can identify all kinds of modification on a protein in a single experiment. However, sometimes it can be a laborious and lengthy process because of the need to purify the protein. Data analysis itself can also be time-consuming. We will use several MS-based PTM analyses in our own experiences as examples just to illus-trate the design and execute the entire experiment.

- *Sample.* The type of protein (i.e., membarne, cytosolic or neuclear protein) the pI value, the molecular weight, the solubility and stability. The amino acid sequence of the protein needs to be considered, which is necessary for the selection of proper proteases for proteolytic digestion.

- *Sample Preparation.* Sample preparation is the most critical step in mass analysis. The prurifacation procedure and the presence of potentially interfering substance for MS should be evaluated. For instance, the type and the concentartion of detergent used for solubilizing the protein may not be compatibe with mass spectrometry.

- *Type of Modifications.* The specific chemistry of the type of modification and its stability during the procedure.

- *Sites of Modification (residues).* The residues (amino acid) where the modification could occur, such as Arg/Lys for methylation, Ser/Thr or Tyr for phosphorylation, Tyr or Cys for nitration. The sites of modification is also important because it may interefer with protease digestion of the protein. Selection of the protease sometimes depends on the location of the sites of modification.

- *Mass Spectrometry Facilities.* The facilites of mass spectrometry (i.e., MALDI–TOF, MALDI–TOF–MS/MS, LC–MS, LC–ESI–MS/MS) and other resources for data search and management. In most cases, each machine is designed for a particular purpose.

- *Knowledge and Training.* It is crucial to acquire adequate working knowledge about mass spectrometry in addition to understanding the principle. It is also important to receive hands-on training in the operation of MS database search and manual data interpretation skills to verify the PTM and PTM sites, which is particularly critical to accurately interpret the data and to avoid any false-positive or fasle-negative result.

These points are presented in Scheme 4.4-1.

A practical, step-by-step guideline for successful studies of PTM is provided based on our own experience in studying PTMs of several transcription regulatory proteins including RIP140, RAR, RXR, TR2, TR4, and NF-κB.

Step 1. Determine the types of modification to be examined. It would be useful to gather information on whether this type of modification can be enriched. For example, enzymatic induction, *in vitro* chemical reaction, and expression of protein in suitable eukaryotic (i.e., insect cells) or prokaryotic systems. Some commercial products for enrichment of modified peptides such as phosphopeptides are currently available (visit Ionsource website at http://www.ionsource.com/ for more information about this option).

Step 2. Purify the protein in a system in which the modification is stable. Avoid any kind of external contamination during purification, for example, keratin, which is the most common impurity present in MS analysis. If purification is not possible, it can be enriched, separated, and recovered from an SDS–PAGE. The amount of sample is also important. A very sensitive machine can identify the protein at the famtomole level.

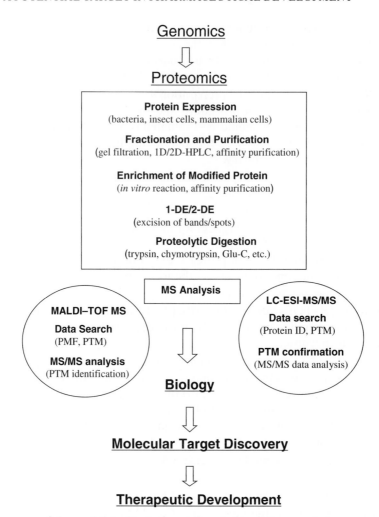

Scheme 4.4-1. Strategies and goal of proteome analysis.

Step 3. Carry out theoretical digestion of the protein to select proteases (i.e., trypsin, chymotrypsin, and Glu-C) using software available online (http://ca. expasy.org/ and MS–DIGEST at http://prospector.ucsf.edu/). Analyze the theoretical mass of each peptide and select the proper protease for digestion that will give you the maximum coverage of the protein with the molecular weight of each peptide no more than 3000 Da. However, most of the researchers select trypsin initially.

Step 4. Run a 1-DE/2-DE gel and stain the gel by commassie blue. Silver-stained gels need extra steps in destaining. All operation should be carried out in a dust-free envrionment to avoid contaminants like keratin. Staining or destaining solution and PAGE running buffer should not be recycled. Excise the protein band of interest and conduct in gel digestion. In-solution tryptic digestion can also be done. However, most investigators prefer in-gel digestion. Several protocols for in-gel digestion on commassie-stained and silver-stained gels are available online at the

University of Minnesota proteomics core facilities website (http://www.cbs.umn.edu/msp/).

Step 5. Desalt the tryptic peptides using ZipTip (Millipore). The ZipTip protocol is also available online (http://www.cbs.umn.edu/msp/) and spotted on a MALDI target over the crystal of matrix molecule (i.e., α-cyano-4-hydroxycinnamic acid). The full-scan spectrum is recorded and the peak lists are generated by software. The experimental mass value (monoisotopic) from the peak lists could be copied and pasted to the Mascot peptide mass fingerprint data search (http://www.matrix-science.com/) or MS-fit data serach (http://prospector.ucsf.edu/) for peptide mass fingerprinting (PMF) to identify the protein of interest or verify the protein identity. The search result will show the identified protein according to the match of mass of a number of peptides, their scores, and the sequence coverage of the total protein. An overview about PMF is described below, which is also available at Ionsource website as tutorials (http://www.ionsource.com/). When the protein is properly identified, the investigator can look at the original mass spectrum and compare with the mass data of theoretical digestion and subtract the peak manually from the full-scan MALDI spectrum one by one. The remaing peaks in the spectrum could originate from the PTM-modified trypsin autolysis peaks or contaminants. If the spectrum is of high quality, the investigator can analyze each individual peak and the neighboring peaks to determine the mass shift between the neighboring peaks, providing some insights into whether the protein is modified. An important consideration is that the modification should be stable or suitable for ionization by MALDI. If the mass shift between the two neighboring peaks give a mass shift equal to the mass shift caused by modification (i.e., 80 amu for phosphorylation and 42 amu for acetylation), that particular ion peak is likely to be originated from the modified peptide. To confirm this modification, the MS/MS spectrum of that parent ion can be recorded from the same sample spotted on a different target.

Peptide Mass Fingerprinting. PMF is an important tool to identify proteins by matching their constituent fragment ions (peptide masses) to the theoretical peptide masses generated from a protein or DNA database. The conceptual basis of PMF is that every unique protein generates a unique set of peptides and hence unique peptide masses. An intact unknown protein is first cleaved with a proteolytic enzyme to generate peptides. A PMF database search is conducted following MALDI–TOF mass analysis of the protein sample. Identification is accomplished by matching the observed peptide masses to the theoretical masses derived from a sequence database. PMF identification relies on observing a large number of peptides (5 or more) from the same protein at high mass accuracy. This technique is particularly suited for 1D/2D gel spots where the protein purity is high. However, it is less efficient for complex mixtures of proteins.

Although this technique was introduced in early 1990s, it was the introduction of a MALDI–TOF instrument capable of 50 ppm mass accuracy that made PMF a routine procedure. In MALDI–TOF mass spectrometry, peptides appear as singly charged species in the mass spectrum (Figure 4.4-1). Unlike an electrospray (ESI) mass spectrum, which displays multiply charged species, the MALDI–TOF spectrum is simple to interpret. PMF can also be used to identify proteins in ESI spectra, but it is seldom used because the peptide masses would need to be deconvoluted for each search.

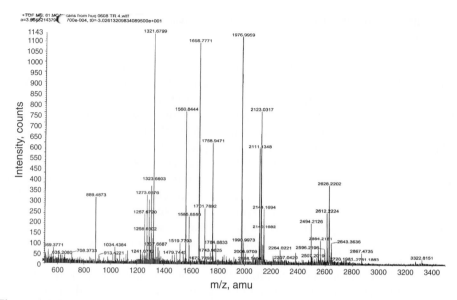

Figure 4.4-1. MALDI–TOF mass spectrum of orphan nuclear receptor TR4. This was collected on a QSTAR-XL MALDI–TOF mass spectrometer and was an average of 240 scans; peptide peaks appear as $[M+H]^{1+}$ ions.

LC–ESI–MS/MS analysis. To examine the potential PTM of a protein after MADLI–TOF mass, a liquid chromatography-tandom mass spectrometry (LC–ESI–MS/MS) is run to obtain information-dependent acquisition (IDA) data. Alternatively, an LC–ESI–MS/MS analysis can be performed without prior MALDI–TOF. However, it is the most common practice to first conduct a MALDI–TOF mass to determine the amount of sample to be loaded on a subsequent LC–ESI–MS/MS. The IDA data could then be searched at Mascot MS/MS data search (available online at http://www.matrixscience.com/) by selecting several parameters, such as protein data bank (i.e., NCBI and MSDB), the species, the fixed modification (i.e., carbamidomethyl cystein), the variable modifications, and the mass tolerance. The search result will generate information about the protein. The reconstructed MS/MS spectrum along with the calculated mass value of product ions matched in the original MS/MS spectrum can be obtained, which will show the residue modified by PTM. To confirm the modification, manual analysis of the MS/MS data is essential.

MS/MS Fragmentation and Nomenclature. The fragmentation pattern and the nomenclature commonly used in MS/MS studies are discussed below. This information was taken from Mascot website (http://www.matrixscience.com/).

Sequence Ions. The types of fragment ions observed in an MS/MS spectrum depend on many factors including primary sequence, the amount of internal energy, how the energy was introduced, and the charge state. The MS/MS fragmentation pattern is shown in Chart 4.4-1.

Fragments will only be detected if they carry at least one charge. If this charge is carried by the N-terminal fragment, the ion is called either *a*, *b*, or *c*. If the charge

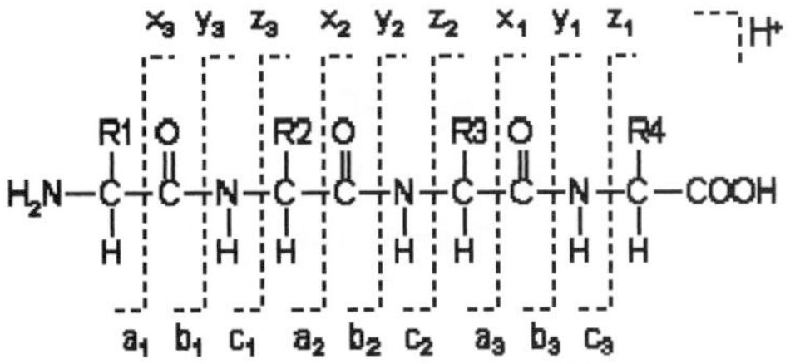

$$x_3\ y_3\ z_3\quad x_2\ y_2\ z_2\quad x_1\ y_1\ z_1\qquad H^+$$

$$H_2N-\underset{\underset{H}{|}}{\overset{\overset{R1}{|}}{C}}-\overset{\overset{O}{\|}}{C}-N-\underset{\underset{H}{|}}{\overset{\overset{R2}{|}}{C}}-\overset{\overset{O}{\|}}{C}-N-\underset{\underset{H}{|}}{\overset{\overset{R3}{|}}{C}}-\overset{\overset{O}{\|}}{C}-N-\underset{\underset{H}{|}}{\overset{\overset{R4}{|}}{C}}-COOH$$

$$a_1\ b_1\ c_1\quad a_2\ b_2\ c_2\quad a_3\ b_3\ c_3$$

Chart 4.4-1.

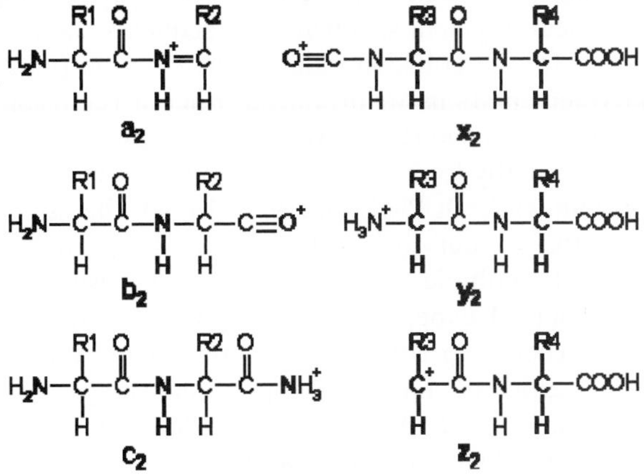

Chart 4.4-2.

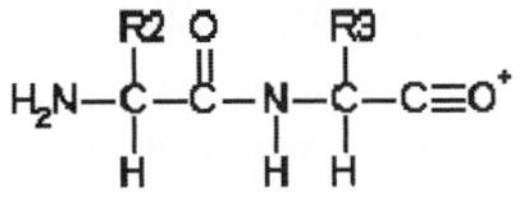

a_2, x_2, b_2, y_2, c_2, z_2

Chart 4.4-3.

is retained on the C-terminal fragment, the ion is classified as either x, y, or z. A subscript indicates the number of residues in the fragment. In addition to the proton(s) carrying the charge, c ions and y ions abstract an additional proton from the precursor peptide. The structures of the six singly charged sequence ions are shown in Chart 4.4-2.

The above structures show only a single charge carrying proton. However, in electrospray ionization, tryptic peptides generally carry two or more charges, so fragment ions may carry more than one proton.

Internal Cleavage Ions. Double-backbone cleavage gives rise to internal fragments (Chart 4.4-3), which are usually formed by a combination of b-type and y-type cleavage to produce the illustrated structure, an amino-acylium ion. Sometimes, internal cleavage ions can form by a combination of a-type and y-type cleavage, an amino-immonium ion.

Immonium Ions. An internal fragment with a single side chain formed by a combination of a-type and y-type cleavage is called an immonium ion (Chart 4.4-4).

Chart 4.4-4.

Protein Identification with MS/MS Data. The previously described PMF uses the intact masses of the peptides to fit a protein in a sequence database. The MS/MS spectral matching employs the uninterpreted peaks in a peptide fragment spectrum to match to a theoretical fragment spectrum in a sequence database. This MS/MS spectrum is acquired from a collision-induced dissociation (CID) within a mass spectrometer. The fragmentations are produced either in a collision cell in a tandem mass spectrometer or within an ion trap. In an MS/MS-based protein ID experiment, multiple peptides are usually found and all of their fragment spectrums are used to identify the protein. In both PMF and MS/MS ID, the larger the number of peptides identified, the greater the confidence in the protein correlation. The correlation made with multiple MS/MS spectra are usually superior to the identification made in a PMF experiment. It is impossible to rely on a single peptide mass in a PMF experiment to correlate a protein. However, the heterogeneity imparted to a single peptide in an MS/MS experiment can generate enough amino acid sequence to correlate a protein. Tutorials from Ionsource website (http://www.ionsource.com/) for a step-by-step simplification of how sequence database search programs accomplish peptide and protein ID are provided in the following.

A Step-by-Step General Scheme for MS/MS Protein ID

Step 1. Enzyme specificity constraint: Some programs preindex the sequence database based on the enzyme specificity, which facilitates the search much faster because tryptic peptides can be indexed and mass lists can be premade. The downside is the necessity of a separate database to be indexed for each enzyme or each time a potential modification is changed.

Step 2. Peptide matching: It involves matching the parent mass of the intact peptide to the peptides in the database. Generally, the narrower parent mass constraint, the faster the search will proceed because fewer peptides will need to be correlated in the next step.

Step 3. Peptide listing and comparison: It takes the list of peptides identified by parent mass in step two and compares the theoretical fragment masses of these peptides to the experimentally derived fragment spectra.

Step 4. Peptide ranking: Hits are ranked by how many of the fragment masses match the theoretical fragment masses in the sequence database.

Step 5. Significance of hit: If more than one peptide is searched, all the peptides found are correlated to their prospective proteins. The protein with the greatest number of well-correlated peptides is usually the most significant hit.

Step 6. Probability factor and validation: Some programs show the probability number to back-up the proposed match, which gives the users some comfort in the designation. Select the most important individual protein hit and manually validate the spectral match.

In the following, we describe data management of PTM analysis of RIP140, where we detected 11 phosphorylation and eight acetylation sites.

4.4.7 MAPPING OF PTM SITES ON RIP140 BY LC–ESI–MS/MS ANALYSIS

Expression and Purification of RIP140 from Sf21 Insect Cells. Sf21 insect cells (1×10^6) were infected with recombinant Baculovirus vector. Using affinity chromatography under a denaturing condition, we were able to purify the recombinant protein to over 95% homogeneity [4]. The eluted protein from the affinity column was further resolved by a SDS–PAGE and a distinctly separated single band of RIP140 was obtained. The gel bands were excised for trypsin digestion and the tryptic peptides were analyzed by MALDI–TOF and LC–ESI–MS/MS.

SDS-PAGE, In-gel Tryptic Digestion of RIP 140. The purified His6-RIP140 protein (500 ng) from insect cells and GST-RIP140 (1 mg) from *E. coli* were resolved by an 8% SDS–PAGE. The bands were visualized by commassie staining. Gel slices containing RIP140 were subjected to overnight in-gel tryptic digestion The tryptic peptides were extracted from the gel with 5% acetic acid, followed by 5% acetic acid in 50% ACN. The digests and the extracted solution were combined and dried. The sample was re-dissolved in 5% acetic acid and desalted using ZipTip C18 reverse-phase desalting Eppendorf tips (Millipore). The peptides were eluted with 2% ACN containing 0.1% TFA to a volume of 50 µL.

Mass Spectrometric Analysis of RIP140. For identification of the proteins, the samples were analyzed by MALDI–TOF MS in a positive ion reflection mode (Qstar XL, Applied Biosystems) using α-cyano-4-hydroxycinnamic acid as a matrix. For LC-MS, the tryptic peptides were subjected to a 20-fold dilution using 2% acetonitrile in water containing 0.1% trifluoroacetic acid. An LC packings (LCP, a Dionex Company, Sunnyvale, CA) Famos autosampler aspirated 27.5 µL of the peptide solution into a 100-µL sample loop using the Famos µl-pick-up injection mode and 98:2 water:ACN, 0.1% formic acid (load buffer) as the transfer reagent. An LCP Switchos pump was used to concentrate and desalt the sample on an LCP C18 nano-precolumn (0.3 mm internal diameter × 5 mm length). The precolumn was switched in-line with a capillary column and peptides were eluted at 350 nL/min using an LCP Ultimate LC system. The capillary column (100 mm internal diameter) was packed in-house to 12 cm length with 5-µm, 200-Å pore size C18 particles. Peptides were eluted with a linear gradient of solvent B (5:95 water:ACN, 0.1% formic acid) from 0% (100% solvent A: 95:5, water:ACN, 0.1% formic acid) to 40% over 60 minutes and 40% to 90% for 5 minutes. The LC system was online with an Applied Biosystems, Inc. (ABI, Foster City, CA) QSTAR Pulsar quadrupole time-of-flight (TOF) mass spectrometer (MS), which was equipped with Protana's nano-electrospray source. An electrospray voltage of 2250 V was applied distal to the analytical column. The TOF region acceleration voltage was 4 kV and the injection pulse repetition rate was 6.0 kHz. The $[M + 3H]^{3+}$ monoisotopic peak at 586.9830 m/z and $[M + 2H]^{2+}$ monoisotopic peak at 879.9705 m/z from human renin substrate tetradecapeptide (Sigma-Alrdich, St. Louis, MO) were used for external calibration. As peptides eluted from the column, they were focused into the MS. The IDA was used to acquire MS/MS data with experiments designed such that the three most abundant peptides were subjected to collision-induced dissociation, using argon as the collision gas every 15 seconds. Collision energies were varied as a function of the m/z and the charge state of each peptide. To avoid continued MS/MS of peptides that had already undergone collision-induced dissociation, a dynamic exclusion was incorporated for a further 45 seconds. IDA mode settings included continuous cycles of three full-scan TOF MS of 400–550 m/z, 550–750 m/z, and 750–1200 m/z (1.5 seconds) plus three product ion scans of 50–4000 m/z (3 seconds each). Precursor m/z values were selected from a peak list automatically generated by Analyst QS software (ABI) from the TOF MS scans during acquisition, starting with the most intense ion. Peptide mass software MS–Digest from the ProteinProspector (available online http://prospector.ucsf.edu) was used to generate a theoreticaltrypticdigestofRIP140byconsideringserine-,threonine-,andtyrosine-containing peptides to account for phosphorylation and lysine for acetylation. From the LCMS data, the molecular weights of the detected peptides were calculated using the LCMS reconstruct feature of Analyst QS. Experimentally measured peptide masses were compared with the theoretical digest. To confirm the sequence of the peptides and the sites of modification, MS/MS spectra were examined. Peaks with a minimum height of 3% relative to the base peak were considered and a 100-ppm tolerance was used to establish matches with the theoretical b and y ions that were predicted using Bioanalyst software (Applied Biosystems). In addition, in all cases, the data from IDA experiments were searched using MASCOT (http://www.matrixscience.com) MS/MS data search.

Mapping of Phosphorylation and Acetylation Sites on RIP140. The tryptic digested samples were first analyzed by MALDI–TOF mass to identify the RIP140 and posttranslational modification by phosphorylation at serine, threonine, and tyrosine residues and acetylation sites on lysine residues. The MALDI–TOF MS data were subjected to a MASCOT search at the NCBI data bank. The mass tolerance of both precursor ions was set at 175 ppm, carbamidomethyl cystein was specified as a static modification, and phosphorylated serine, threonine, and tyrosine and acetylation on lysine residues were specified as variable modifications. The search result confirmed the identification of the protein with significant sequence coverage to over 60% of the total protein [4]. However, only a few phosphopeptides were predicted from the MALDI–TOF mass data. The MS/MS analysis was carried out for the precursor ions of the predicted phosphopeptides and acetylated peptides to identify the peptide sequence and phosphorylation/acetylation sites. Unfortunately, in most cases, high-quality MS/MS spectra for the identification of the phosphorylation sites by MALDI–TOF mass were not obtained. However, sequences of the unmodified peptides were identified properly from the MS/MS analysis. Based on this information, we speculated that the poor quality of MS/MS spectra of the phosphopeptides/acetylated peptides could be caused by improper ionization of the RIP140 tryptic peptides by MALDI. Therefore, LC–ESI–MS/MS was applied using nanoelectrospray source for ionization to solve this problem. We recorded three independent full-scan ion chromatograms of 400–$550\,m/z$, 550–$750\,m/z$, and 750–$1200\,m/z$. The IDA was used to acquire MS/MS data. IDA analyses were performed on the tryptic digests of RIP140 expressed in *E. coli* and insect cells. The data from IDA experiments were searched using a MASCOT search. The mass tolerance of both precursor ion and the MS/MS fragment ions was set at $\pm 0.1\,Da$ and carbamidomethyl cystein was specified as a static modification. Phosphorylated serine, threonine, and tyrosine and acetyaled lysine were specified as variable modifications. The result revealed 11 tryptic phosphopeptides (Table 4.4-1) and eight acetylated peptides (Table 4.4-2) from RIP140 expressed in insect cells. For a control, no phosphorylated/acetylated peptides were detected from RIP140 expressed in *E. coli* [4, 5, 34].

Mapping of Phosphorylation Sites. The MS/MS data of the phosphopeptides in comparison with the unmodified form were analyzed manually to map the phosphorylation sites on RIP 140. As an example, the MS/MS spectrum of tryptic peptide spanning residues (100–111) is presented to explain how phosphorylation was assigned (Figure 4.4-2). The MS/MS fragmentation patterns for both the precursor ions of the unmodified (top panel) and modified peptide (bottom panel) residues 100–111 were identical, especially from $b1$ to $b4$ ions and $y1$ to $y4$ ions. The $b5$ ion of the unmodified form appeared as a singly charged ion at $559.26\,m/z$, whereas the $b5$ ion in the phosphopeptide appeared at $541.26\,m/z$ because of a – 18 amu delta-mass shift. This delta-mass shift was considered to be the β-eliminated product of $b5$ ion caused by loss of either H_3PO_4 or H_2O from Ser-104. Similar β-elimination was also shown at $b6$ ion of the modified peptide. The $b8$ ion of the modified peptide showed an 80-amu (H_3PO_4) delta-mass shift ($965.39\,m/z$) along with its β-eliminated ion signal at $867.41\,m/z$ due to loss of H_3PO_4 (98 amu). These data asserted that the β-eliminated $b5$ and $b6$ ions were from the loss of H_3PO_4 from Ser-104. Thus, the phosphorylation site was assigned to Ser-104. This phosphorylation site was also confirmed by an independent MS/MS analysis (data not shown) of the phosphopeptide residues 101–111 and 101–112 (Table 4.4-1).

Mapping of Acetylation Sites on RIP140. The MS/MS data of the acetylated peptides were also analyzed manually to identify the acetylation sites on RIP140. To identify the acetylated peptides, a 42-amu mass shift was considered as an indication of acetylation. To distinguish acetylation from trimethylation, both rendering the same integral mass shift, some marker ion signals, such as an immonium ion at $126\,m/z$ specific for acetylated lysine was monitored [35, 36]. In addition, the absence of a marker ion for trimethylated lysine generated from loss of trimethylamine from b or y ions as $[M-59]^+$ was also confirmed. Significant differences in mass shift were caused by loss of acetyl moiety ($42\,m/z$) and trimethylamine ($59\,m/z$). Therefore, a 42-amu net loss for acetyl group from y or b ions was also accounted to assign the acetylation sites [33, 34]. Initial analysis of the mass data showed some peptides displaying a 43-unit mass shift instead of 42 units for acetylation, which raised a possibility of carbamylation of lysine residue rather than acetylation in the modified peptides. But careful analysis of MS/MS data finally confirmed that this 1-unit positive delta mass shift was originated from the deamidation of either glutamine or asparagine residue present in the acetylated peptides. All of these

TABLE 4.4-1. LC–ESI–MS Profile of the Tryptic Phosphopeptides of RIP140 [4]

Residues	Sequence	Unmodified			Modified		
		m/z (z)	M^{+2}	RT^1	m/z (z)	M^{+2}	RT^1
100–111	RLSD**S**IVNLNVK	453.26(3) 679.37(2)	1356.76	40.24	719.36(2)	1436.72	41.08
101–111	LSD**S**IVNNLNVK	601.33(2)	1200.66	41.74	641.31(3)	1280.63	42.87
101–112	LSD**S**IVNLNVKK	443.92(3) 665.36(2)	1328.75	37.86	705.35(2)	1408.72	39.45
199–212	SGPTLPDV**T**PNLIR	740.40(2)	1478.80	47.06	780.38(2)	1558.76	47.36
343–360	NNAATFQSPMGVVPS**S**PK	916.45(2)	1830.90	41.44	956.43(2)	1910.86	42.03
375–387	QAANN**S**LLLHLLK	478.94(3) 717.90(2)	1433.82	49.17 49.05	505.60(3) 757.90(2)	1513.80	53.03 52.90
476–492	IPGVDIKEDQDT**S**TNSK	923.94(2) 616.30(3)	1845.87	34.61	963.94(2)	1925.88	35.20
517–537	NA**S**PQDIHSDGTKFF**S**PQNYTR	591.53(4)	2362.12	35.31	815.02(2) 841.68(2)	2442.03 2522.02	35.91 36.55
538–548	TSVIE**S**PSTNR	595.80(2)	1189.59	30.83	635.78(2)	1269.56	30.75
670–678	LN**S**PLLSNK	493.28(2)	984.55	35.92	533.25(2)	1064.52	33.89
1001–09	TF**S**YPGMVK	515.24(2)	1028.49	39.21	555.24(2)	1108.46	41.23

[1] RT, retention time in minute.
[2] M^+, precursor mass.

430

TABLE 4.4-2. LC–ESI–MS Profile of the ACTEYLATED Tryptic Peptides of RIP140 [5]

Tryptic digests of RIP140 protein was subjected to LC–ESI–MS/MS. Three independent full-scan ion chromatograms from m/z 400–550, 550–750, and 750–1200 were recorded in an information-dependent acquisition (IDA) mode to acquire MS/MS data. The IDA data were searched online at MASCOT (http://www.matrixscience.com) MS/MS data search at the NCBI data bank. The MS/MS data were analyzed manually to confirm the sequence of the modified and the unmodified forms of the same peptide identified by the data bank search. The full scan chromatograms were analyzed to assign the charged state, retention time, and intensities of the peptides.

Residue	Sequence[1]	Unmodified $[m/z$ (z)], M^{+2}, RT[3] (min)	Modified m/z (z), M$^+$, RT(min)
101–112	LSDSINNLNV**K**K	[443.92 (3)], 1328.75, 38.09	[686.88 (2)], 11371.75, 41.37
155–170	QSL**K**EQGYALSHESLK	[606.65 (3)], 1816.93, 35.80	[620.98 (3)], 1859.91, 40.67
283–298	EHAL**K**TQNAHQVASER	Not detected	[621.31 (3)], 1860.91, 23.80
305–320	LQENGQ**K**DVGSSQLSK	[859.92 (2)], 1717.85, 29.34	[880.93(2)], 1759.84, 32.56
476–492	IPGVDI**K**EDQDTSTNSK	[923.95 (2)], 1845.90, 34.61	[945.43 (2)], 1888.88, 36.96
517–537	NASPQDIHSDGT**K**FSPQNYTR	[788.36 (3)], 2362.09, 35.35	[802.69 (3)], 2405.11, 37.65
606–630	G**K**ESQAEKPAPSEGAQNSATFSK	[836.40 (3)], 2506.17, 30.82	[850.74 (3)], 2459.19, 32.38
930–938	ES**K**SFNVLK	[526.27 (2)], 1050.58, 36.99	[547.79 (2)], 1093.56, 38.22

[1] The 43-unit mass difference between modified and unmodified peptides appeared because of acetylation of lysine along with deamidation of either asparagine (N) or glutamine (Q) residues present in the peptides.
[2] RT, retention time in minutes.
[3] M$^+$, precursor ion.

parameters were evaluated carefully while MS/MS data were analyzed to sequence the modified peptides. In addition, the MS/MS spectra of the modified peptide were always compared with the unmodified peptide (data not shown) for finger-printing purposes.

We provide an example of how an acetylation site can be identified by manual analysis of MS/MS spectrum. The acetylated peptide-spanning residues 101–112 displayed a doubly charged ion at 686.88 m/z (precursor 1371.75 m/z) in total ion chromatogram (TIC). The precursor ion showed a 43-unit mass shift caused by acetylation of a lysine and deamidation from an asparagine residue. The MS/MS spectrum of the modified peptide showed $y1$ ion at 147.1 m/z and $y1°$ ion at 129.1 m/z caused by loss of H$_2$O, which were identical to the unmodified peptide (Figure 4.4-3). The $y2$ ion appeared at 317.2 m/z instead of 275.2 m/z, resulting in

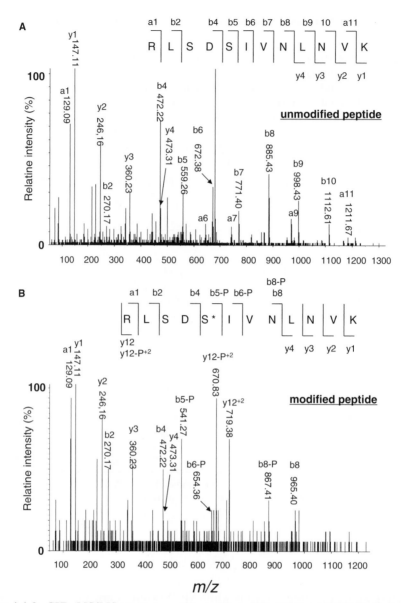

Figure 4.4-2. CID–MS/MS spectra peptide residues 100–111 form RIP140. Both spectra (unmodified, top and phosphorylated, bottom) showed identical b4 ions at m/z 472.22, but the b8 and b5 ions of the phosphopeptide showed 80 and −98 units mass shift, respectively, suggesting Ser-104 phosphorylation. (Courtesy: *Proteomics*, 5:2157–2166.)

a net mass shift of 42 Da, which confirmed the modification of Lys-111 by acetylation. The presence of a marker ion at 126 m/z specific for lysine acetylation confirmed this modification originated from acetylation rather than trimethylation [35, 36], which was further substantiated by the presence of consecutive y ions from y2 to y8 due to modified lysine residue. The y ions starting from y4 showed a 43-amu shift instead of 42, suggesting deamidation of Asn-109.

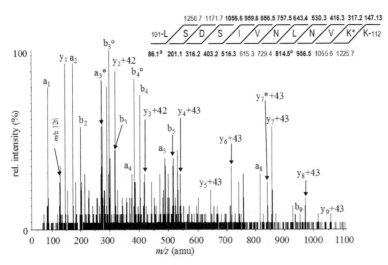

Figure 4.4-3. CID–MS/MS spectra of acetylated peptides spanning residues 101–112 from RIP140. The y1 ion at m/z 147.1 was identical to the native peptide, but the y2 and y3 ions showed a 42-unit mass shift, suggesting Lys-111 acetylation. (Courtesy: *Molecular & Cellular Proteomics*, 4:975–983.)

4.4.8   THE ROLE OF PTM IN CELLULAR FUNCTION AND CELL SIGNALING

Phosphorylation. Phosphorlation is the most well-studied PTM, and it occurs on a large number of proteins. We examined PTM of RIP140, which primarily functions as a lignad-dependent co-repressor for many nuclear receptors by recruiting HDACs in the transcription complex [34]. RIP140-null mice exhibited reproductive defects as well as abnormal energy homeostasis. We identified 11 phosphorylated sites including Ser^{104}, Thr^{202}, Thr^{207}, Ser^{358}, Ser^{380}, Ser^{488}, Ser^{519}, Ser^{531}, Ser^{543}, Ser^{672}, and Ser^{1003} [4, 34]. The MAP kinase-mediated phosphorylation of Thr^{202}, Thr^{207}, and Ser^{358} enhances the repressive potential of RIP140 by facilitating the recruitment of HDAC3. We generated constitutive dephosphorylated mutant by replacing the phosphorylated Ser/Thr with Ala and a constitutive positive mutant by replacing phospho-Ser/Thr with glutamic acid. The Ala-null mutant was unable to exert the repressive activity, whereas the constitutive positive mutant became more repressive [34], which demonstrates that phosphorylation of RIP140 critically modulates its biological activity.

In another study, we asked whether and how the orphan nuclear receptor 2 (TR2) can activate its target genes by phosphorylation without ligands [18]. We hypothesized that PTM of TR2 can modulate TR2 function without a putative ligand. We showed that, indeed, PKC-mediated phosphorylation of TR2 enhanced its stability and activation potential for its target RARβ2. We identified PKC-mediated phosphorylation on the ligand-binding domain of TR2 at Ser-461 and Ser-568, which enhanced its protein stability. We further identified phosphorylation of the DNA-binding domain (DBD) of TR2 at Ser-170 and Ser-185, which facilitated its DNA-binding ability and recruitment of coactivator p300/CBP-associated factor (P/CAF) [39]. Ser-185 is required for DNA binding, whereas both Ser-170 and Ser-185 are necessary for receptor interaction with P/CAF [18]. For a comprehensive review of phosphorylation of nuclear receptors, see the review by Rochette-Egly [14].

Acetylation. Acetylation is another widely known PTM. Many nuclear receptors (NRs) and transcription factors are known be modified by acetylation. We have also shown that RIP140 can be acetylated on eight lysine residues, including Lys^{111}, Lys^{158}, Lys^{287}, Lys^{311}, Lys^{482}, Lys^{529}, Lys^{607}, and Lys^{932}. The amino-terminal region [amino acids (aa) 1–495] was more repressive and accumulated more in the nuclei under a hyperacetylated condition, whereas hyperacetylation reduced the repressive activity and nuclear translocation of the central region (aa 336–1006). Hyperacetylation also enhanced the repressive activity of the full-length protein and triggered its export into the cytosol. This study revealed differential effects of PTM on various domains of RIP140. Recent reviews described nuclear receptors and transcription factors that can be modified by acetylation [22, 40].

Tyrosine Nitration. Tyrosine nitration is one of the protein modifications induced by reactive oxygen species. NO is an important factor that induces posttranslational modifications of proteins by cellular reduction and oxidation mechanism: cysteinyl-nitrosylation or Tyr nitration. Nuclear factor (NF)-kappaB activity can be rapidly suppressed by sodium nitroprusside, an NO donor. This effect was effectively reversed by peroxynitrite scavenger deferoxamine, suggesting a Tyr nitration-mediated mechanism. Tyr nitration of p65 induced its dissociation from p50, its association with IκBα, and subsequent sequestration of p65 in the cytoplasm by IkBa-mediated export. LCMS revealed specific nitration on Tyr-66 and Tyr-152 residues of p65, which were confirmed by mutation studies. These residues are important for the direct effects of NO on p65, which resulted in more p65 export and inactivation of NF-κB activity. This study identified a novel and efficient means in which NO rapidly inactivated NF-κB activity by inducing Tyr nitration on p65 [12].

4.4.9   PROTEIN POSTTRANSLATIONAL MODIFICATION AND DISEASES

The establishment of species- and tissue-specific protein databases provides a foundation for proteomics studies of diseases. Continual development will lead to functional proteomics studies, in which identification of protein modification in conjunction with functional data from established biochemical and physiological methods enables the examination of interplay between changes in a proteome and the progression of diseases. Recently, many investigations provided direct evidence for PTM in the pathophysiological progression of many diseases like diabetes, Alzheimer's diseases (AD), atherosclerosis, and oncogenesis.

Cardiovascular Diseases. In a comprehensive cardiovascular proteome analysis, various PTMs were documented in dilated cardiomyopathy [41]. A 2-DE analysis of dilated cardiomyopathy-diseased human myocardial tissue revealed more than 50 HSP27 proteins by iummunoblotting, illustrating a large number of PTMs potentially occur on a single protein [42].

Neurodegenerative Diseases. Mircotubule-associasted protein *tau* is documented to undergo several PTMs and aggregates into paired helical filaments (PHF) in AD and other taupathies [43]. PTMs of *tau* include hyperphosphorylation, glycosylation, ubiquitination, glycation, polyamination, nitration, and proteolysis. Hyperphosphorylation and glycosylation are crucial to the molecular pathogenesis of neurofibrillary degeneration of AD [43]. The others appeared to represent a failed mechanism for neurons to remove damaged, misfolded, and aggregated proteins. Therefore, it was proposed that modified *tau* can serve as a biomarker for the diagnosis of AD.

Diabetes. Very recent investigation showed methylglyoxal modification of mSIN3 protein and its linkage to diabetes retinopathy [44]. The report showed that in diabetes, because of impaired metabolism of glucose, the level of glyoxal concentration increased. This glyoxal can modify proteins and finally can modulate gene transcription.

Aging and Age-related Diseases. Oxidatively modified proteins increase as a function of age. Studies revealed an age-related increase in the level of protein carbonyl content, oxidized methionine, protein hydrophobicity, and cross-linked and glycated proteins [9]. Factors reducing protein oxidation increase the life span of experimental animals and vice versa. Furthermore, a number of age-related diseases are shown to associate with an elevated level of oxidized proteins. The accumulation of oxidatively modified protein can be attributed to a multitude factors that govern (1) the rate of formation of various kinds of reactive oxygen species, (2) the level of antioxidant defense that guards proteins against oxidative modification of proteins, (3) the sensitivity of proteins to oxidative attack, and (4) the repair and elimination of damaged proteins. The ROS are formed by ionizing radiation, activation of neutrophils and macrophages, oxidase catalyzed reaction, lipid peroxidation, and glycation/glycoxidation reactions. It is well established that the accumulation of oxidized protein is associated with a number of diseases. Elevated levels of protein carbonyls have been found in AD diseases, amyotrophic lateral scelerosis, cataractogenesis, systemic amyloidosis, muscular dystrophy, Parkinson's disease, progeria, Warner's syndrome, rheumatoid arthritis, and respiratory distress syndrome. Elevated levels of proteins modified by lipid peroxidation products are associated with Parkinson's diseases, cardiovascular diseases, iron-induced renal carcinogenesis, and experimental pancreatitis and atherosclerosis. Elevated levels of protein glycation/glycoxidation endproducts (AGEs) are associated with diabetes mellitus, ADs, atherosclerosis, Parkinson's diseases, and Down's syndrome. Elevated levels of protein nitrotyrosine damaged are associated with atherosclerosis, AD, lung injury, multiple sclerosis, and endotoxemia.

Cataractogenesis. Cataractogenesis has been noted in the human lens, due to PTM of lens protein αβ-crystallin by carbamylation and acetylation at Lys-92 [45].

Autoimmune Diseases. PTM can dictate an antigen in eliciting autoimmune diseases. For example, isoaspartyl posttranslational modification (conversion of aspartic acid to isoaspartic acid) has been shown to trigger autoimmune response to self-proteins. The presence of isoaspartyl proteins has been observed as a major component of the amyloid containing brain plaques of patients with AD [46].

Alcoholism. Many investigations showed that alcoholism modulates the level of PTM-like phosphorylation [47]. Alcohol consumption was shown to decrease the sialic acid conjugation to transferrin, an important carrier protein secreted from the liver to blood and other glycoproteins. This observation led to the development of a laboratory test for chronic alcohol use. Similar studies showed that direct production of alcohol metabolism (alpha-hydroxyethyl radicals, acetaldehyde and lipid peroxides) causes PTM that correlates with alcohol consumption in animal models and human subjects [48].

4.4.10   PTM AS A TARGET FOR THERAPEUTIC DEVELOPMENT

Histone deacetylase (HDAC) inhibitors are the new class of agents that modulate gene expression by altering chromatin structure and gene transcription. Several classes of HDAC inhibitors are known as therapeutics for tumors. The depsipeptides (FR901228 or FK228) are under clinical trial [49]. Three other classes including short-chain fatty acids (phenylbutyrate and valproic acid), benzamides (CI-994 and MS-27-275), and hydroxamic acid (suberoylanilide hydroxamic acid) are being developed. An attractive model is that the increase in histone acetylation leads to transcriptional activation of a few genes that can inhibit tumor growth. Ten structurally related HDACs have been described and fall into two classes [50, 51]. Class I HDACs consist of HDAC1, 2, 3, and 8; whereas class II HDACs consist of HDAC4, 5, 6, 7, 9, and 10. Members of a third class of HDACs (class III) are structurally unrelated to the human class I and class II HDACs, and they consist of homologues of the yeast Sir2 proteins [52]. The activity of class I and class II HDACs is inhibited by short-chain fatty acids and hydroxamic acids, but class III HDACs are not inhibited by these agents. Therefore, one major challenge is to identify specific HDACs inhibitors.

Another report showed curcumin (a component of many tropical spices) as a novel histone acetyl transferase (HAT) inhibitor [53]. The report showed that curcumin could block acetylation of histone and p53 *in vivo*, which led to apoptosis of Hela cells. Therefore, potential in the future exists to develop HAT inhibitors for managing cancers.

Similarly, histone arginine/lysine methylation can be the target for the development of therapeutics. The status of histone arginine methylation is intimately involved in gene transcription. Recently, several compounds were found to inhibit protein arginine methyltransferases (PRMTs) [54]. It can also be interesting to explore how this small molecule could be exploited for developing therapeutics.

Protein farnesylation is a lipid-conjugate-type posttranslational modification required for the cancer-causing activity of the GTPase Ras [55]. Although farnesyltransferase inhibitors (FTIs) are in clinical trials, their mechanism of action and the role of protein farnesylation in normal physiology are poorly understood. Protein farnesylation was found to be essential for early embryogenesis, dispensable for adult homeostasis, and critical for progression, but not initiation, of tumorigenesis. Preclinical work has revealed FTIs' ability to effectively inhibit tumor growth in vitro and in animal models across a wide range of malignant phenotypes. Acute myeloid leukemias (AMLs) are appropriate disease targets in that they express relevant biologic targets such as Ras, MEK, AKT, and others that may depend on farnesyl protein transferase activity to promote cell proliferation and survival. Phase I trials in AML and myelodysplasia have demonstrated biologic and clinical activities as determined by target enzyme inhibition, low toxicity, and both complete and partial responses. As a result, phase II trials have been initiated to further validate the clinical efficacy and to identify downstream signal transduction targets that may be modified by these agents [56].

We have recently reported rapid inactivation of NF-kappa B by tyrosine nitration [12]. As NF-kappaB is involved in many cellular functions in response to inflammatory responses [57], it can be a very important target in antioxidation therapy.

Nuclear receptors (NRs) orchestrate the transcription of specific gene networks in response to binding of their cognate ligands. They also act as mediators in a variety of signaling pathways by integrating diverse PTMs. NR phosphorylation concerns exist on all three major domains. Often, phosphorylation of NRs by kinases that are associated with general transcription factors (e.g., cdk7 within TFIIH) or activated in response to a variety of signals (MAPKs, Akt, PKA, PKC) facilitates the recruitment of coactivators and, therefore, cooperates with the ligand to enhance transcription activation. But phosphorylation can also contribute to the termination of the ligand response through inducing DNA dissociation, triggering NR degradation or decreasing ligand affinity. These different modes of regulation reveal an unexpected complexity of the dynamics of NR-mediated transcription. Therefore, small molecules that can modulate the phosphorylation status of NR can also be developed as therapeutics [14, 19]. For comprehensive reviews on PTM as targets in therapeutic development, the readers are referred to review Refs. 57–59.

FURTHER READING

Proteomic Resources

The readers are referred to the following websites, journals, and suppliers to acquire comprehensive knowledge about the proteomics resources.

World Wide Web (WWW) Resources (short listed)

http://www.matrixscience.com/: Theoretical digestion, database search option (intact protein) for PMF by MALDI–TOF MS data, LC–ESI–MS/MS data search for PMF and PTM identification. Other information, like MS/MS fragmentation and PTM, is also available.

http://proteinprospector.uscf.edu: Theoretical digestion using MS-Digest. One of the unique options in using MS–Digest is that one can generate theoretical MS of the modified peptide selecting a wide variety of PTMs. Database search for PMF can be done using MS-Fit and also mass peak list information from the common contaminants, for example, keratin and trypsin autolysis peaks in MS are available.

http://ca.expasy.org/: Protein database, calculation of protein MS and pI values, theoretical digestion. Prediction of varieties of PTMs (glucosylation, phosphorylation, sumoylation, etc.) can be done from the protein sequence.

http://www.ionsource.com/: Very helpful tutorials about PMF using MALDI–TOF MS and ESI–MS/MS, *de novo* protein sequencing and related information. This website provides very helpful tips and information for beginners in proteomics. One can also get many other website links for proteomics resources from this website.

Literature and Journals:

(1) Proteomics, (2) Journal of Proteome Research, (3) Molecular & Cellular Proteomics, (4) Electrophoresis, and (5) BBA protein-proteomics.

Vendors and Suppliers:

A comprehensive lists of vendors and suppliers related to proteomics resources are available on Ionsoure website (http://www.ionsource.com/).

4.4.11 CONCLUSION

The understanding of PTM will advance as more new PTMs are explored as bio-markers and more powerful techniques and tools are developed. It can be predicted that, within a few years, PTM analysis can become a routine procedure like HPLC in chemistry or PCR in biology laboratories. Currently, research of PTM is focusing on the basic principle and the studies of biological problems. Predictably, it will gradually be applied to the discovery of molecular targets for disease intervention and the development of therapeutics.

ACKNOWLEDGMENT

This work was supported by National Institutes of Health Grants DA11190, DA11806, DK54733, DK60521, and K02-DA13926 to L.-N. W.

REFERENCES

1. Gooley A A, Packer N H (1997). The importance of protein co- and post-translational modifications in proteome projects. In M.R. Wilkins; K.L. Williams; R.D. Appel; et al. (eds.), Springer–Verlag, *Proteome Research: New Frontiers in Functional Genomics: Berlin*. pp. 65–91.
2. Pandey A, Mann M (2000). Proteomics to study genes and genomes. *Nature*. 405: 837–846.
3. Liebler D C (2001). *Introduction to Proteomics: Tools for the New Biology*. Humana Press, Totowa, NJ.
4. Huq M D, Khan S A, Park S W, et al. (2005). Mapping of phosphorylation sites of nuclear corepressor receptor interacting protein 140 by liquid chromatography-tandem mass spectrometry. *Proteomics*. 5:2157–2166.
5. Huq M D, Wei L N (2005). Post-translational modification of nuclear corepressor receptor-interacting protein 140 by acetylation. *Mol. Cell. Proteomics*. 4:975–983.
6. Barlev N A, Liu L, Chehab N H, et al. (2001). Acetylation of p53 activates transcription through recruitment of coactivators/histone acetyltransferases. *Mol. Cell*. 8:1243–1254.
7. Bedford M T, Richard S (2005). Arginine methylation an emerging regulator of protein function. *Mol. Cell*. 18:263–272.

8. Wells L, Vosseller K, Hart G W (2001). Glycosylation of nucleocytoplasmic proteins: signal transduction and O-GlcNAc. *Science*. 291:2376–2378.

9. Stadman E R (2001). Protein oxidation in aging and age-related diseases. *Ann. N.Y. Acad. Sci*. 928:22–38.

10. Johnson E S (2004). Protein modification by SUMO. *Ann. Rev. Biochem*. 73:355–382.

11. Lee D Y, Teyssier C, Strahl B D, et al. (2005). Role of protein methylation in regulation of transcription. *Endocr. Rev*. 26:147–170.

12. Park S W, Huq M D, Hu, X, et al. (2005). Tyrosine nitration on p65: A novel mechanism to rapidly inactivate nuclear factor-kappaB. *Mol. Cell. Proteomics*. 4:300–309.

13. Teyssier C, Ma, H, Emter R, et al. (2005). Activation of nuclear receptor coactivator PGC-1alpha by arginine methylation. *Genes Dev*. 19:1466–1473.

14. Rochette-Egly C (2003). Nuclear receptors: integration of multiple signalling pathways through phosphorylation. *Cell. Signal*. 15:355–366.

15. Cote J, Boisvert F M, Boulanger M C, et al. (2003). Sam68 RNA binding protein is an in vivo substrate for protein arginine N-methyltransferase 1. *Mol. Biol. Cell*. 14:274–287.

16. Cheng X, Hart G W (2001). Alternative O-glycosylation/O-phosphorylation of serine-16 in murine estrogen receptor beta: post-translational regulation of turnover and transactivation activity. *J. Biol. Chem*. 276:10570–10575.

17. McCann J P, Mayes J S, Hendricks G R, et al. (2001). Subcellular distribution and glycosylation pattern of androgen receptor from sheep omental adipose tissue. *J. Endocrinol*. 169:587–593.

18. Khan S A, Park S W, Huq M, et al. (2005). Protein kinase C-mediated phosphorylation of orphan nuclear receptor TR2: effects on receptor stability and activity. *Proteomics*. 5:3885–3894.

19. Gronemeyer H, Gustafsson J A, Laudet V (2004). Principles for modulation of the nuclear receptor superfamily. *Nature Rev. Drug Disc*. 3:950–964.

20. Goody R S, Durek T, Waldmann H, et al. (2005). Application of protein semisynthesis for the construction of functionalized posttranslationally modified rab GTPases. *Methods Enzymol*. 403:29–42.

21. Acconcia F, Ascenzi P, Bocedi A, et al. (2004). Palmitoylation-dependent estrogen receptor alpha membrane localization: regulation by 17beta-estradiol. *Mol. Biol. Cell*. 16:231–237.

22. Xu W, Cho H, Evans R M (2003). Acetylation and methylation in nuclear receptor gene activation. *Methods Enzymol*. 364:205–223.

23. Fu M, Wang C, Zhang X, et al. (2004). Acetylation of nuclear receptors in cellular growth and apoptosis. *Biochem. Pharmacol*. 68:1199–1208.

24. Chen L F, Williams S A, Mu, Y, et al. (2005). NF-kappaB RelA phosphorylation regulates RelA acetylation. *Mol. Cell. Biol*. 25:7966–7975.

25. Chuikov S, Kurash J K, Wilson J R, et al. (2004). Regulation of p53 activity through lysine methylation. *Nature*. 432:353–360.

26. Chevillard-Briet M, Trouche D, Vandel L (2002). Control of CBP co-activating activity by arginine methylation. *EMBO J*. 21:5457–5466.

27. Rezai-Zadeh N, Zhang X, Namour F, et al. (2003). Targeted recruitment of a histone H4-specific methyltransferase by the transcription factor YY1. *Genes Dev*. 17:1019–1029.

28. Chan W M, Mak M C, Fung T K, et al. (2006). Ubiquitination of p53 at multiple sites in the DNA-binding domain. *Mol. Cancer Res*. 4:15–25.

29. Wurgler-Murphy S M, King D M, Kennelly P J (2004). The Phosphorylation Site Database: A guide to the serine-, threonine-, and/or tyrosine-phosphorylated proteins in prokaryotic organisms. *Proteomics.* 4:1562–1570.

30. Goy M F, Springer M S, Adler J (1977). Sensory transduction in Escherichia coli: role of a protein methylation reaction in sensory adaptation. *Proc. Nat. Acad. Sci. USA.* 74:4964–4968.

31. Eichler J, Adams M W (2005). Posttranslational protein modification in Archaea. *Microbiol. Mol. Biol. Rev.* 69:393–425.

32. Wu R C, Qin J, Yi P, et al. (2004). Selective phosphorylations of the SRC-3/AIB1 coactivator integrate genomic reponses to multiple cellular signaling pathways. *Mol. Cell.* 15:937–949.

33. Vo N, Fjeld C, Goodman R H (2001). Acetylation of nuclear hormone receptor-interacting protein RIP140 regulates binding of the transcriptional corepressor CtBP. *Mol. Cell. Biol.* 21:6181–6188.

34. Gupta P, Huq M D, Khan S A, et al. (2005). Regulation of co-repressive activity of and HDAC recruitment to RIP140 by site-specific phosphorylation. *Mol. Cell. Proteomics.* 4:1776–1784.

35. Zhang K, Yau, P M, Chandrashekhar B, et al. (2004). Differentiation between peptides containing acetylated or tri-methylated lysines by mass spectrometry: An application for determining lysine 9 acetylation and methylation of histone H3. *Proteomics.* 4:1–10.

36. Hirota J, Satomi Y, Yoshikawa K, et al. (2003). ε-*N,N,N*-trimethyllysine-specific ions in matrix-assisted laser desorption/ionization-tandem mass spectrometry. *Rapid Commun. Mass Spectrom.* 17:371–376.

37. Johnstone R W (2004). Deamidation of Bcl-X(L): a new twist in a genotoxic murder mystery. *Mol. Cell.* 10:695–697.

38. Ramasamy R, Yan S F, Schmidt A M (2006). Methylglyoxal comes of AGE. *Cell.* 124:258–260.

39. Khan S A, Park S W, Huq M D, et al. (2006). Ligand-independent orphan receptor TR2 activation by phosphorylation at the DNA-binding domain. *Proteomics.* 6:123–130.

40. Sterner D E, Berger S L (2000). Acetylation of histones and transcription-related factors. *Microbiol. Mol. Biol. Rev.* 64:435–459.

41. Arrell D K, Neverova I, Van Eyk J E (2001). Cardiovascular proteomics: evolution and potential. *Circ. Res.* 88:763–773.

42. Scheler C, Muller E C, Stahl J, et al. (1997). Identification and characterization of heat shock protein 27 protein species in human myocardial two-dimensional electrophoresis patterns. *Electrophoresis.* 18:2823–2831.

43. Gong C X, Liu F, Grundke-Iqbal I, et al. (2005). Post-translational modifications of tau protein in Alzheimer's disease. *J. Neural Transmis. (Vienna, Austria),* 112:813–838.

44. Yao D, Taguchi T, Matsumura T, et al. (2006). Methylglyoxal modification of mSin3A links glycolysis to angiopoietin-2 transcription. *Cell.* 124:275–286.

45. Lapko V N, Smith D L, Smith J B (2001). In vivo carbamylation and acetylation of water-soluble human lens alphaB-crystallin lysine 92. *Protein Sci.* 10:1130–1136.

46. Mamula M J, Gee R J, Elliott J I, et al. (1999). Isoaspartyl post-translational modification triggers autoimmune responses to self-proteins. *J. Biol. Chem.* 274:22321–22327.

47. Anni H, Israel Y (2002). Proteomics in alcohol research. *Alcohol Res. Health.* 26:219–232.

48. Anni H, Pristatsky P, Israel Y (2003). Binding of acetaldehyde to a glutathione metabolite: mass spectrometric characterization of an acetaldehyde-cysteinylglycine conjugate. *Alcoholism, Clin. Exper. Res.* 27:1613–1621.

49. Richon V M, O'Brien J P (2002). Histone deacetylase inhibitors: a new class of potential therapeutic agents for cancer treatment. *Clin. Cancer Res.* 8:662–664.

50. Marks P A, Rifkind R A, Richon V M, et al. (2001). Histone deacetylases and cancer: causes and therapies. *Nat. Rev. Cancer.* 1:194–202.

51. Kao H Y, Lee C H, Komarov A, et al. (2002). Isolation and characterization of mammalian HDAC10, a novel histone deacetylase. *J. Biol. Chem.* 277:187–193.

52. Landry J, Sutton A, Tafrov S T, et al. (2000). The silencing protein SIR2 and its homologs are NAD-dependent protein deacetylases. *Proc. Nat. Acad. Sci. USA.* 97:5807–5811.

53. Balasubramanyam K, Varier R A, Altaf M (2004). Curcumin, a novel p300/CREB-binding protein-specific inhibitor of acetyltransferase, represses the acetylation of histone/nonhistone proteins and histone acetyltransferase-dependent chromatin transcription. *J. Biol. Chem.* 279:51163–51171.

54. Cheng D, Yadav N, King R W, et al. (2004). Small molecule regulators of protein arginine methyltransferases. *J. Biol. Chem.* 279:23892–23899.

55. Sebti S M (2005). Protein farnesylation: implications for normal physiology, malignant transformation, and cancer therapy. *Cancer Cell.* 7:297–300.

56. Thomas X, Elhamri M (2005). Farnesyltransferase inhibitors: preliminary results in acute myeloid leukemia. *Bulletin du Cancer.* 92:227–238.

57. Garg A, Aggarwal B B (2002). Nuclear transcription factor-kappaB as a target for cancer drug development. *Leukemia.* 16:1053–1068.

58. Rohlff C (2000). Proteomics in molecular medicine: applications in central nervous systems disorders. *Electrophoresis.* 21:1227–1234.

59. Valaskovic G A, Kelleher N L (2002). Miniaturized formats for efficient mass spectrometry-based proteomics and therapeutic development. *Curr. Topics Med. Chem.* 2:1–12.

4.5

PEGYLATION: CAMOUFLAGE OF PROTEINS, CELLS, AND NANOPARTICLES AGAINST RECOGNITION BY THE BODY'S DEFENSE MECHANISM

Yong Woo Cho,[1] Jae Hyung Park,[2] Ji Sun Park,[1] and Kinam Park[3]

[1] Asan Institute for Life Sciences, University of Ulsan College of Medicine, Seoul, Korea
[2] College of Environment and Applied Chemistry, Kyung Hee University, Gyeonggi-do, Korea
[3] Purdue University, West Lafayette, Indiana

Chapter Contents

4.5.1   Introduction   444
4.5.2   PEGylated Proteins   444
    4.5.2.1   Chemical Modification   444
    4.5.2.2   Immunogenicity   448
    4.5.2.3   Clinical PEGylated Proteins   450
4.5.3   PEGylated Cells   451
    4.5.3.1   Immunocamouflage of Cells   451
    4.5.3.2   Red Blood Cells (RBCs)   452
    4.5.3.3   White Blood Cells   453
    4.5.3.4   Pancreatic Islets   454
4.5.4   Colloidal Nanoparticles with PEG-Exposed Surfaces   454
    4.5.4.1   Stealth Liposomes   455
    4.5.4.2   Polymeric Nanoparticles   456
4.5.5   Conclusions   456
    References   457

Handbook of Pharmaceutical Biotechnology, Edited by Shayne Cox Gad.
Copyright © 2007 John Wiley & Sons, Inc.

4.5.1 INTRODUCTION

Poly(ethylene glycol) (PEG) is a synthetic neutral polymer that is soluble in water as well as in many organic solvents, such as methylene chloride, ethanol, acetone, and chloroform. In aqueous solutions, PEG is highly hydrated due to the presence of the ether oxygen linkage as a hydrogen bond acceptor. The hydrated PEG has a high degree of segmental flexibility and large excluded volume. These properties enable PEG to behave as if 5–10 times larger than proteins of comparable molecular weights and to effectively exclude other polymers (both natural and synthetic) from PEG-grafted surfaces [1, 2]. As PEG is demonstrated to be nontoxic, the U.S. Food and Drug Administration (FDA) has approved it for use as a vehicle or base in foods, cosmetics, and pharmaceuticals, including injectable, topical, rectal, and nasal formulations. When injected into the bloodstream, PEG is rapidly removed from the body with clearance rates inversely proportional to polymer molecular weights [3]. PEG tends to accumulate in muscle, skin, bone, and liver to a higher extent than other organs irrespective of the molecular weights, and its elimination occurs either in the kidneys (for PEGs < 30 kDa) or in the feces (for PEGs > 20 kDa) [3, 4]. There have been no reports on the generation of antibodies to PEG under routine clinical administration of PEG–protein conjugates. A few reports have indicated production of antibodies to PEG under extreme experimental conditions using animals [5, 6]. It is important to note, however, that the minute quantities of PEG in therapeutic conjugates are generally considered to provide no chance of eliciting unwanted host responses.

PEGylation is the process of attaching PEG to various objects, such as peptides, proteins, cells, and other biologically active materials. As PEG has a unique set of properties in a biological environment, PEGylation of the target molecules has provided a wide range of biomedical applications. It has been demonstrated that PEGylated proteins have less toxicity [7, 8] and less immunogenicity [9, 10]. Increase in the molecular weights of proteins by PEGylation slows down kidney ultra-filtration, resulting in prolonged circulation in the bloodstream. The shielding effect of PEG reduces protein degradation by the enzymatic reaction and minimizes recognition of the immune system. In addition to proteins, recent efforts have made significant progress in designing PEGylated cells that show low immunogenic recognition, thus allowing transplantation. PEGylation has also been used to modify surfaces of drug and gene carriers to improve the blood circulation time and targeting efficiency.

4.5.2 PEGYLATED PROTEINS

4.5.2.1 Chemical Modification

Although therapeutic proteins are highly potent endogenous substances, they are generally susceptible to proteolytic degradation, aggregation, polymerization, and adsorption onto foreign surfaces. In addition, most proteins have short biological half-lives and often induce immunogenic responses when administrated into the body. In an attempt to stabilize proteins toward proteolysis and to increase blood circulation time in the body, various water-soluble polymers, such as polysaccharides [11, 12] and albumin [13], have been used for protein conjugation. Of the numerous natural and synthetic polymers, PEG is becoming a standard polymer for chemical modification of proteins and peptides to enhance their pharmaceutical properties. The pioneering first steps in PEGylation of proteins were taken in the late 1970s by Professor Frank Davis and his colleagues at Rutgers University [4].

PEG is a linear or branched polyether terminated with hydroxyl groups that is synthesized by anionic ring opening polymerization of ethylene oxide. The molecular weights of linear PEGs available for biomedical applications usually range between a few hundred to several tens of thousands of daltons. For PEGylation of proteins, the reactive functional groups at the chain end of PEG need to be introduced. The functional group is selected based on the type of available reactive groups on proteins, originating from lysine, cysteine, histidine, arginine, aspartic acid, glutamic acid, serine, threonine, tyrosine, N-terminal amino group, and C-terminal carboxylic acid.

PEGs for Carboxyl Conjugation. The nucleophilic PEG derivatives, such as PEG-amine and PEG-*p*-aminobenzylether [14], have been attached by the reaction with carbodiimide-activated carboxylic groups of proteins. For PEG-amine, this approach has been limited because the reactivity of PEG-amine is similar to that of the amino groups of proteins, resulting in poor selectivity. On the other hand, PEG-*p*-aminobenzylether is useful for selective reaction at a slightly acidic solution (pH 4.8–6.0) because it has the lower pKa value than the primary amino groups of the protein. Also, PEG-hydrazide (pKa ~ 3) can specifically react with the carbodiimide-activated carboxyl groups of proteins under acidic conditions (pH 4.5–5) at which primary amino groups of proteins are unreactive due to the protonation. It should be considered for protein modification, however, that the nucleophilic PEG derivatives often involve side reactions, including modification of tyrosyl and cysteinyl side groups and formation of *N*-acylurea derivatives [15, 16].

PEGs for Amine Conjugation. The most common reactive sites on proteins for PEGylation are the alpha or epsilon groups of lysine or the *N*-terminal amino groups of other amino acids [17]. To react with amino groups of proteins, PEG has been activated by a variety of chemical modifications. Figure 4.5-1 shows various chemical structures of activated PEGs.

Activated PEGs can be prepared by reacting the primary alcohol group of PEG with the chlorine group of trichloro-s-triazine (cyanuric chloride), by which two remaining chlorine groups are available for protein modification [18]. The resulting PEG-dichlorotriazine (Figure 4.5-1a) can react with nucleophilic functional groups on the surface of proteins such as lysine, serine, tyrosine, cysteine, and histidine [19]. One chloride is displaced during the reaction, whereas the other remains less reactive and may provide the possibility of the side reaction causing chemical cross-linking of protein molecules. PEG-tresylates (Figure 4.5-1b) were used to nonspecifically modify amino groups to form secondary amine linkages that are stable and does not change the total charge of the conjugate, compared with the native protein [20]. PEG-tresylates have more specific reactivity to amino groups than PEG-dichlorotriazine, whereas they may involve the side reactions such as formation of a degradable sulfamate linkage [21].

PEG-aldehydes (Figure 4.5-1c) are a useful electrophilic reagent capable of specifically reacting with amino groups of proteins without interference from other nucleophiles belonging to proteins [22]. The reaction of PEG-aldehyde with protein proceeds through a reversible Schiff base intermediate that is reduced in the

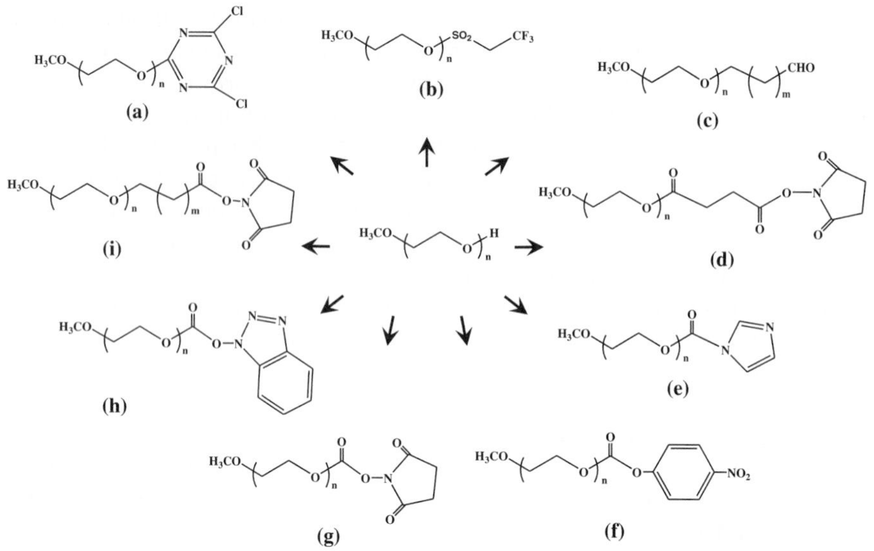

Figure 4.5-1. Chemical structures of activated PEGs for amine conjugation: (a) PEG-dichlorotriazine; (b) PEG-tresylate; (c) PEG-aldehydes; (d) PEG-succinimidyl succinate; (e) PEG-imidazole carbonate; (f) PEG-phenyl carbonate; (g) PEG-succinimidyl carbonate; (h) PEG-benzotriazole carbonate; and (i) PEG-active ester.

presence of sodium cyanoborohydride or sodium borohydride. Also, PEG-aldehyde has the ability to achieve selective modification between N-terminal α-amines and lysine ε-amines. When the reaction is performed under mild acidic conditions (pH 5–6), the α-amino groups are preferentially reacted with PEG-aldehyde. PEG-succinimidyl succinate (Figure 4.5-1d) is prepared by reaction of PEG with succinic anhydride, followed by carbodiimide-mediated condensation with N-hydroxy succinimide (NHS) [23]. This activated PEG reacts with proteins under mild conditions (pH < 7.8, room temperature), and the conjugates show comparable biological activities with native proteins [23, 24]. As the conjugates contain the ester linkage between PEG and protein, exposure to the physiological condition may induce hydrolysis, leading to loss of the benefits of PEG grafting. Katre et al. [25] demonstrated that the hydrolysis rate can be slowed by substituting the succinate group with a glutarate. PEG-imidazole carbonates (Figure 4.5-1e) and PEG-phenyl carbonates (Figure 4.5-1f) are prepared by reacting carbonylimidazole [17, 26] or chloroformate [27] with the terminal hydroxyl group of PEG, respectively. These derivatives have been used to produce conjugates via the formation of a carbamate bond between PEG and proteins. In general, they show much lower reactivity than PEG-succinimidyl succinate. The reaction time takes more than 24 h. The slower reaction rate is beneficial for selective modification of amino groups in proteins, thus producing conjugates with preserved biological activity.

PEG-succinimidyl carbonate (Figure 4.5-1g) is another activated PEG that specifically reacts with amino groups of proteins, thus forming a carbamate linkage [28]. It is found to be more selective than PEG-succinimidyl succinate for modification of the amino groups of proteins. In particular, PEG-succinimidyl carbonate is highly useful to prepare a minimally toxic conjugate because the liberated NHS produced by the reaction is known to be nontoxic. PEG-benzotriazole carbonate (Figure 4.5-1h) is similar to PEG-succinimidyl carbonate. PEG-benzotriazole carbonate forms the carbonate bonds via modification of proteins and is more selective than PEG-succinimidyl succinate in reacting with amino groups of proteins [29]. Furthermore, PEG-benzotriazole carbonate can react under mild conditions within a short reaction time (e.g., pH 8.5 at 25°C for 30 min). Despite their high selectivity, both PEG-succinimidyl carbonate and PEG-benzotriazole carbonate have been recently reported to react with histidine and tyrosine residues of proteins, which forms hydrolytically unstable linkages such as imidazolecarbamate [30].

In recent years, the most popular PEG derivatives for protein modification have been active esters of PEG carboxylic acids (Figure 4.5-1i) [17]. As the active esters react with primary amines to form the amide linkage under physiological conditions, they are useful for preparing stable conjugates. PEG carboxylic acids are readily activated by reacting them with NHS and carbodiimide. Of various PEG-active esters, the succinimidyl ester of carboxymethylated PEG was the first derivative developed for protein modification [31]. This derivative, however, is extremely reactive toward hydrolysis that limits the extended applications. The reactivity of PEG-active esters was found to be controlled by changing the distance between the active ester and the PEG backbone [32]. For example, the hydrolysis half-life of carboxymethylated PEG is 0.75 min at pH 8 and 25°C, whereas those of PEG-active esters bearing one and two additional methylene groups between PEG and active esters are 16.5 and 23 min, respectively, under the same condition.

PEGs for Cysteine Conjugation. Several PEG derivatives have been developed to specifically react with the thiol group of cysteines, including PEG-maleimide [33], PEG-orthopyridyl-disulfide [34], PEG-iodoacetamide [33], and PEG-vinyl sulfone [35]. These cysteine-targeting PEG derivatives are primarily designed for site-specific PEGylation of proteins because cysteine residues are rarely found on the surface of proteins, compared with lysine residues. If necessary, however, it is possible to add cysteine residues on the protein surfaces by genetic engineering, which may allow site-specific modification at areas that do not affect the biological activity of protein. Each derivative has its own characteristics. For example, the reaction of PEG-maleimide with the free thiol group is known to form a hydrolytically stable thioether linkage throughout the Michael-addition reaction. PEG-orthopyridyl-disulfide can react with protein sulfhydryl groups to form a disulfide bond capable of being cleaved to regenerate the native protein under reducing conditions. PEG-iodoacetamide reacts slowly with protein thiols by nucleophilic substitution via formation of a stable thioether linkage. For this polymer, the reaction should be carried out in a dark condition to prevent the generation of free iodine that may react with other amino acids such as tyrosine. PEG-vinyl sulfone is another activated PEG reacting with protein thiols to form a stable thioether bond.

PEGs for Selective Modification. In an attempt to produce PEGylated proteins with preserved biological activity, selective modifications are often required. In most cases, the successful protein modification depends on the balance between reactivity and selectivity. It is generally accepted that the selective modification is achieved by using less-reactive PEG derivatives, in which the reaction is performed at the low temperature for the long time period (>24h).

Based on the reversible protection of interfering nucleophilic groups, selected amino groups in proteins have been modified with PEG. The protected intermediates of proteins are first prepared by chemically attaching the protective reagent to the amino groups that play an important role in biological activity. PEG is then conjugated to the preferred sites on the surfaces of proteins. The subsequent deprotection of PEGylated intermediates affords the final products that may show comparable activities with native proteins. This approach allowed selective modifications with PEG for insulin [36, 37], vapreotide [38], and tumor necrosis factor-alpha (TNF-α) [39].

The site-specific modifications of proteins with PEG have also been achieved by the enzyme-mediated coupling reaction [40]. One representative enzyme for this approach is a genetically engineered glutamine aminotransferase that can be used to selectively conjugate PEG to glutamine residues in a recognized substrate sequence. Interleukin-2, site-specifically PEGylated by this method, showed good *in vitro* activities and increased half-lives, compared with the native protein [40].

Thiol-specific PEG derivatives are useful for selective modification of proteins because a few cysteine sulfhydryl groups are generally available on the surface of proteins. The addition of cysteine residues at specific sites in a protein sequence is possible by genetic engineering, allowing site-specific PEGylation of proteins. Chapman et al. [41] have developed PEGylated antibody fragments by site-specific modification using cysteine-selective PEG derivatives. The conjugates showed long *in vivo* half-lives and full retention of antigen-binding properties. A cysteine residue was also inserted into anti-HIV protein cyanovirin-N by site-directed mutagenesis, followed by chemical reaction with thiol-specific PEG derivatives [42]. The resulting conjugates were highly active, whereas nonspecific modification with PEG resulted in a total loss of antiviral activity.

PEG-aldehyde derivatives have been used to selectively modify N-terminal amino groups of proteins [2, 17, 43]. For example, PEGylated granulocyte colony stimulating factor (GCSF) was prepared using PEG-aldehyde derivatives that were attached to the N-terminal methionine residue [44, 45]. The reaction mixture contained more than 80% of the intended conjugates with preserved biological activity. PEGylated GCSF is currently approved by the FDA for clinical use.

4.5.2.2 Immunogenicity

Although recent advances in the recombinant DNA technology have allowed production of several proteins in large quantities, their applications are frequently limited due to immunogenic responses by the body. For example, recombinant human interleukin-2 (rhIL-2), expressed and purified from *Escherichia coli*, has received attention as an anticancer therapeutic agent. However, some patients undergoing clinical trials with rIL-2 have reported to develop antibodies to the protein [9]. It is also known that repeated administrations of human insulin, produced by genetic engineering, resulted in generation of antibody [46]. Immunogenicity of recombinant human proteins can lead to serious clinical consequences. Antibody binding to a protein may change its pharmacokinetics in the body [47]. Antibody formation can inhibit the therapeutic effect of proteins and/or neutralize the essential endogenous proteins. Long-term administration of recombinant human erythropoietin to patients induced formation of antibody capable of neutralizing endogenous erythropoietin [48, 49]. PEGylation has shown to suppress such adverse effects of recombinant proteins [18, 50]. In their pioneering studies on PEGylation, Abuchowski et al. demonstrated that administration of native albumin to rabbits resulted in rapid removal of protein by immuno-complex formation, whereas PEGylated albumin showed long circulation in the bloodstream [18]. Although a few references indicate generation of antibodies to PEG [5, 51], it is generally accepted that generation of anti-PEG antibodies is practically negligible, mainly due to its weak immunogenicity and the low amounts of PEGylated conjugates in clinical trials.

Suppression of immunogenicity by PEGylation is primarily based on shielding of the epitope (immunoreactive site) of a protein by flexible PEG chains. Improved immunological properties alter the pharmacokinetic profile of the modified protein. For example, when IgG was intravenously administered to monkeys, it was cleared rapidly because of an antiprotein immune response [52]. On the other hand, PEGylated IgG did not elicit an immune response, and their dose levels were 125% as compared with the native antibody. This indicated that PEGylation of a protein increases the circulation time in the bloodstream by reducing clearance by the immune system.

The extent of suppression of immunogenicity is known to depend on the PEG molecular weight, number of grafted PEG chains, conjugated sites, and PEG architecture (linear or branched) [46, 53]. Protein immunogenicity is generally reduced as the number of PEG, and its molecular weight increases. Conjugation of branched PEG is reported to be more effective than linear PEG in improving immunological and pharmacokinetic properties of the modified proteins. Asparaginase modified with 10-kDa PEG reduced the antigenic character by 10-fold, as compared with that of the same protein modified with 5-kDa PEG [46]. Branched PEG could reduce immunogenicity of uricase more efficiently than linear PEG [54]. The author suggested that the "umbrella-like" structure of branched PEG efficiently prevented the approach of antiprotein antibodies and immuno-competent cells (Figure 4.5-2). It

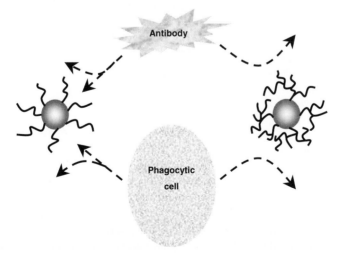

Figure 4.5-2. Molecular architectures of PEG grafted on proteins. The branched PEG protects more effectively from antibodies, phagocytic cells, and enzymes.

would be preferred to attach the PEG chains around the antigenic epitopes, if this does not deteriorate the biological activity of the protein [50].

4.5.2.3 Clinical PEGylated Proteins

In the early 1990s, two PEGylated proteins, PEG-adenosine deaminase (Pegademase) and PEG-L-asparaginase (Pegaspargase), were approved by the FDA as therapeutic agents replacing the parent compounds. Recently approved PEGylated protein drugs include Pegfilgrastim, Peginterferon alfa-2a, and Peginterferon alfa-2b. More than 20 PEGylated proteins are on the market or in clinical trials, as shown in Figure 4.5-3 [4, 17].

Pegademase was developed for the treatment of severe combined immunodeficiency (SCI), which is associated with an inherited deficiency of adenosine deaminase [55]. SCI patients are conventionally treated by transfusing red blood cells containing adenosine deaminase. However, the patients often suffer from iron overload and transfusion-associated viral infections [56, 57]. Use of Pegademas does not need to consider such adverse effects and allows minimizing the dose of therapeutic agents. L-Asparaginase has been used for treatment of acute lymphocytic leukemia, acute lymphoblastic leukemia, and chronic myelogenous leukemia. L-Asparaginase, however, requires frequent intramuscular injections and often induces allergic reactions [58]. Compared with unpegylated L-asparaginase,

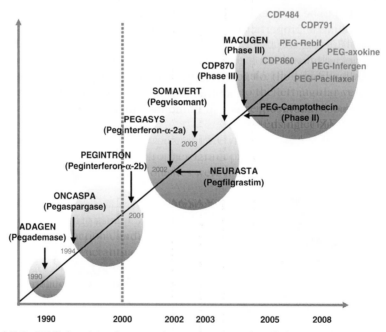

Figure 4.5-3. PEGylated products on the market or under clinical development. Since the first PEGylated protein was introduced, Pegademase (Adagen) entered the market in 1990 followed by a growing stream of PEGylated pharmaceuticals, as symbolized by circles in the figure. In 2004, a total of 26 PEGylated pharmaceuticals were on the market or under clinical trials.

Pegaspargase showed a substantially prolonged half-life. The half-lives of L-aspara-ginase and Pegaspargase were 20 and 357h, respectively [59]. PEGylation of L-asparaginase also reduced adverse immune responses [58]. Although the actual cost of Pegaspargase is higher than that of multiple injections of unpegylated L-asparaginase, en-hanced therapeutic effects and minimal complications may make overall the cost considerably less than that of conventional therapy.

Interferon-based therapeutic drugs have been approved for treatment of various diseases, such as chronic hepatitis B and C, renal cell carcinoma, chronic myeloge-nous leukemia, and malignant melanoma. However, the clinical use of interferon has been limited because of rapid absorption from the subcutaneous injection site, large volume of distribution, rapid clearance via the kidney, short serum half-life (~6h), and significant side effects such as depression and flu-like symptoms [60]. Interferons were initially PEGylated using a small, linear PEG (MW = 5kDa) at multiple sites via the urea linkage [61]. These conjugates were not suitable for further clinical applications because of pharmacokinetics needed to be improved. Interferon alfa-2b was then modified with longer PEG (MW = 12kDa) by forming a degradable linkage to improve the therapeutic effect [62]. The result was Pegin-terferon alfa-2b, which showed a long half-life with clinical efficacy, and is now marketed for treatment of chronic hepatitis C. The next generation of PEGylated interferon was interferon alfa-2a modified with a branched 40-kDa PEG via forma-tion of an amide linkage [63]. The resulting Peginterferon alfa-2a allowed sustained delivery of the protein and showed the ability to be detected in the serum for longer than 1 week. It is interesting to note that the different chemical linkages of the two PEGylated interferons led to different formulations for the market. Peginterferon alfa-2b is available as a lyophilized powder, thus requiring reconstruction before use. On the other hand, Peginterferon alfa-2a is provided as a ready-to-use solution because the conjugates are linked by a stable amide bond.

Several PEGylated proteins are currently undergoing clinical trials. PEGylated human growth hormone antagonist (Pegvisomant) is being tested for treatment of acromegaly. Pegvisomant has been approved in Europe and is waiting for FDA approval in the United States. The PEGylated inhibitor of tumor necrosis factor-alpha is being tested for the treatment of rheumatoid arthritis and Crohn's disease [64]. Patients treated with the PEGylated protein experienced improvement in physical function, pain, vitality, and mental health [65].

4.5.3 PEGYLATED CELLS

4.5.3.1 Immunocamouflage of Cells

Immunological recognition of allogeneic tissues is a primary concern in transfusion and transplantation medicine. Graft-versus-host-disease (GVHD) is induced by the transfer of allogeneic lymphocytes into immunocompromised individuals [66]. GVHD can cause significant morbidity and mortality among patients with tissue transplantation. Several pharmacological agents that inhibit the T-cell signaling and activation have been developed such as cyclosporine, azathioprine, and metho-trexate [67]. However, these drugs are less than optimal due to their substantial toxicity.

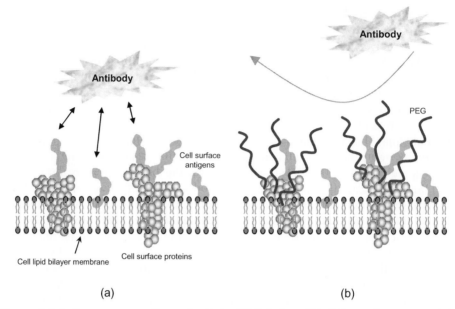

Figure 4.5-4. Immunocamouflage of cells by PEGylation: (A) Cell membrane containing surface antigens causes immune reactions, which destroy the transplanted cells. (B) PEGs grafted on the cell surface cytoprotect against antibodies.

Scott et al. [68–74] suggested a new method that can significantly diminish rejection episodes and may further enhance the induction of tolerance. They have demonstrated that the surface engineering of cells by PEGylation yielded antigenically and immunogically inert cells. Figure 4.5-4 schematically illustrates immunocamouflage of cells by surface PEGylation. The mechanisms underlying immunocamouflage by PEGylation are the loss of antigen recognition, impaired cell–cell interaction, and an inability of endogenous antibodies to effectively recognize and bind foreign epitopes. As a result of the high molecular flexibility and intense hydration, the grafted PEG can sterically block off a large three-dimensional volume thereby giving immunoprotective effects to membrane proteins and carbohydrates.

4.5.3.2 Red Blood Cells (RBCs)

Blood transfusion is a widely used therapy in medicine, particularly under surgery and trauma conditions. Surface engineering of red blood cells (RBCs) by PEGylation has been studied to camouflage the blood group antigens from their antibodies, thereby to generate universal RBCs. Several studies demonstrated that the surface engineering of RBCs with PEG significantly attenuated immunologic recognitions of surface antigens [68–71, 73–75]. PEGylation significantly decreased binding of human RBC blood group antibody and phagocytic destruction by heterologous phagocytes [68], PEGylation also dramatically diminished the immunogenicity of allogeneic and even xenogeneic RBCs in mouse models [71]. Importantly, this protective immunologic effect can be accomplished without adversely affecting the structure, function, or viability of the modified RBCs. PEGylation was found to have no significant detrimental effects on the RBC structure or function. PEGylation did not affect RBC lysis, morphology, and hemoglobin oxidation state. The functional aspects of the PEGylated RBCs were also maintained, as evidenced by normal oxygen binding and cellular deformability [71].

The immunocamouflage efficacy is largely dependent on the PEG grafting density, PEG length, and tomography of the membrane surface. Differently functionalized PEGs have been used to chemically graft PEG onto the membrane of cells. Typical functionalized PEGs include PEG-dichlorotriazine [68], PEG-benzotriazole carbonate [73], and PEG-succinimidyl succinate [73]. They all preferentially target the ε-amino group of exposed lysine residues on membrane proteins. PEG-dichlorotriazine is the most common reagent. PEGylation of RBCs with PEG-dichlorotriazine has been shown to mask Rh antigens to a reasonable extent, but masking of antigen A or B was partial. Recently, Nacharaju et al. suggested a new approach for PEGylation of RBC proteins using thiolation-mediated maleimide chemistry [75]. In the study, masking of blood group antigens was not directly correlated to the PEG-grafting density or PEG chain length. For optimizing the masking of A, B, D, and CE antigens, a combination of different size PEG chains was used. PEG-5000 (molecular weight of PEG 5 kDa) alone masked the most important antigens of the Rh system (C, c, E, e, and D) from their antibodies. However, the masking of the A and B antigens needed a combination of PEG-20000 and PEG-5000 to inhibit agglutination of RBCs completely with anti-A or anti-B [75]. Recently, another strategy for RBC PEGylation was suggested by Chung et al. [76]. RBC PEGylation was carried out by incorporation of lipid-PEG into RBC membranes. The incorporation was rapid and spontaneous at room temperature. No change was observed in the membrane shape of the lipid-PEG incorporated RBC. The incorporation of lipid-PEG dramatically attenuated the antigenic recognition of RBC. As many as 35% of individuals with sickle cell disease or thalassemias exhibit clinically significant alloimmunization to non-ABO antigens [77, 78]. Immunocamouflage of RBCs by PEGylation may significantly attenuate the risk of alloimmunization for patients requiring chronic RBC transfusions such as patients with sickle cell anemia and thalassemias.

4.5.3.3 White Blood Cells

T-lymphocytes play a central role in immune responses. Transfusion or transplantation of T lymphocytes into an allogeneic recipient can evoke potent immune responses. Extensive attention has been paid to inactivation of T cells as well as other leukocytes (B lymphocytes, mononuclear cells, and granulocytes) from blood products. PEGylation dramatically blocked allorecognition and proliferation in human peripheral blood mononuclear cells (PBMCs) and isolated murine splenocytes [69, 72, 79–81]. Loss of cellular proliferation was not due to cytotoxicity of grafted PEG but to the global camouflage of the cell surface, which inhibited cell–cell interactions. T-cell and antigen-presenting cell adhesion molecules (CD2, CD11a), signaling molecules (CD3ε, T-cell receptor), and costimulatory molecules (CD28, CD80) were efficiently camouflaged by PEGylation [81]. Immunocamouflage was clearly observed in murine models of transfusion-associated graft versus host disease (TA-GVHD) [81]. Mice receiving saline or the PEGylated splenocytes exhibited an identical survival curve, whereas mice receiving control splenocytes (unmodified) showed significant TA-GVHD. This improved survival was directly associated with the donor cell proliferation. Unmodified control splenocytes showed significant proliferation due to allorecognition of the host mice. In contrast, PEGylated cells failed to proliferate due to immunocamouflage and lack of allorecognition, indicating that PEGylation of donor splenocytes significantly decreased donor T-cell proliferation, thereby suppressing GVHD.

4.5.3.4 Pancreatic Islets

Transplantation of pancreatic islets has been one of the most challenging examples of tissue transplantation. Insulin-dependent diabetes mellitus (IDDM) is a disease from the autoimmune destruction of β cells of the islets of Langerhans in the pancreas. Tranplantation of islets into diabetic patients can be the most attractive mode of treatment, but the islets must be protected from the host's immune system to prevent graft rejection. Transplanted islets are recognized by host as antigens that activate immune cells, such as macrophages, granulocytes, and lymphocytes. Several studies demonstrated that camouflage of pancreatic islets by PEGylation could be an attractive approach to prevent activation of immune cells and secretion of cytokines, thereby improving transplantation of pancreatic islets [82–87].

PEG chains were covalently conjugated to the amino groups of collagen capsules of islet at physiological conditions through a stable amide bond [84]. PEG was not grafted onto the inner islet cell membrane but directly onto the collagen matrix of islet. Therefore, PEGylation did not damage the inner islet cells. Cell viability and functional activity remained intact after PEGylation. When free islets were cultured with lymphocytes, they completely lost the integrity of the collagen capsule [86]. The co-cultured lymphocytes were activated by free islets, thereby secreting a large amount of IL-2 and TNF-α cytokines. In contrast, the PEGylated islets were not damaged by co-culture with lymphocytes. The co-cultured lymphocytes were not activated by the PEGylated islets and rarely secreted IL-2 and TNF-α cytokines, suggesting that the grafted PEG effectively prevented cell–cell interaction necessary for the allogeneic rejection. The PEGylated islets were not affected by macrophoages, either. However, when co-cultured with lipopolysaccharide (LPS)-stimulated macrophages (activated macrophages), the PEGylated islets lost the integrity of the collagen capsule and were completely destroyed. Once macrophages are activated, they secrete several kinds of cytotoxic molecules, such as IL-1β, TNF-α, and nitric oxide (NO). Complete destruction of PEGylated islets indicates that the grafted PEG could not prevent infiltration of the cytotoxic molecules into the islets [84, 86].

4.5.4 COLLOIDAL NANOPARTICLES WITH PEG-EXPOSED SURFACES

At the interface of nanotechnology and biotechnology, nanosized colloidal particles (nanoparticles) have emerged as a new entity with great potential for target-specific therapy and imaging [88–90]. The fundamental advantages of colloidal nanoparticles for drug delivery result from their two main basic properties. First,

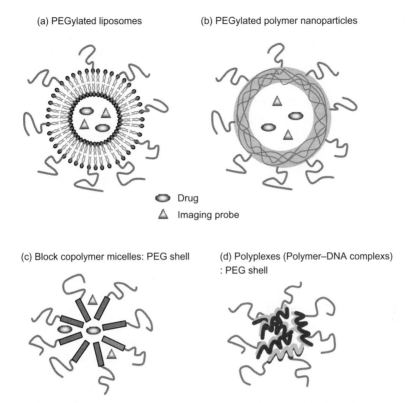

(a) PEGylated liposomes

(b) PEGylated polymer nanoparticles

● Drug

△ Imaging probe

(c) Block copolymer micelles: PEG shell

(d) Polyplexes (Polymer–DNA complexs) : PEG shell

Figure 4.5-5. Colloidal nanoparticles with PEG-exposed surfaces.

nanoparticles due to their small size can penetrate through capillaries and are taken up by cells, and this allows efficient drug accumulation at the target sites. Second, nanoparticles allow sustained drug release within the target site over a period of days or even weeks. Thus, nanoparticles could be effective delivery systems for therapeutic and imaging payloads. However, intravenously injected nanoparticles are also treated as foreign by the host, just like exogenously delivered proteins and cells. The nanoparticles should avoid recognition by the cells of the reticuloendothelial system (RES), namely macrophages of the liver, spleen, lungs, bone marrow, and lymphnodes. Adsorption of plasma proteins onto the surface of nanoparticles is also highly problematic for efficient delivery of drugs and diagnostic agents. As a result, there has been a growing interest in the surface engineering of colloidal nanoparticles to avoid rapid recognition by RES and to reduce adsorption of plasma proteins. These systems are often called "Stealth" nanoparticles, which are "invisible" to macrophages and have a prolonged half-life in the blood circulation. PEGylation has been increasingly exploited to camouflage colloidal nanoparticles against recognition by the body's defenses. Figure 4.5-5 schematically represents different colloidal nanopaticles with the PEG-exposed surface.

4.5.4.1 Stealth Liposomes

Development of PEGylated liposomes, also known as "stealth" liposomes, greatly broadened the application of liposomes. For incorporation of PEG into the liposomal bilayer, various functionalized lipids have been developed that contain a primary amino group, an epoxy group, or a diacylglycerol moiety [91, 92]. Alternatively, activated PEGs were anchored to reactive phospholipids groups of preformed liposomes [93]. In most cases, PEGylated liposomes with prolonged circulation time are in the size range of 70 to 200 nm and contain 3–7 mol% of PEG in addition to various amounts of phospholipids and cholesterol. The circulation half-lives of such stealth liposomes are 15–24 h in rodents and as high as 45 h in humans. In addition to PEG, various hydrophilic polymers have been grafted onto liposomes for prolonging the circulation time. They include polyvinylpyrrolidone, poly(vinyl alcohol), poly(2-methyl-2-oxazoline), poly(2-ethyl-2-oxazoline), polyacrylamide, poly(acryloyl morphine), and N-(2-hydroxylpropyl)methacrylamide copolymers [92, 94–98].

4.5.4.2 Polymeric Nanoparticles

Surface engineering of polymeric nanoparticles with PEG has been performed by physical adsorption, by incorporation during production of nanoparticles, or by covalent attachment to the surface. Typical examples of PEG-conjugated polymeric nanoparticles are PEG-poly(L-lactic acid) (PLA), PEG-poly(l-lactic-co-glycolic acid) (PLGA), PEG-poly(ε-caprolactone) copolymers, and PEG-polyphosphazene [92]. Although *in vitro* physico-chemical characterization studies, such as measurement of hydrodynamic volume and protein adsorption, have confirmed the steric barrier effects of PEG, the circulation half-lives of these nanoparticles were not as long as expected, generally less than a few hours. Many polymeric nanoparticles with a promising outlook based on *in vitro* results have faced the same problems when tested *in vivo*. Furthermore, little information is known with regard to the colloidal stability of nanoparticulate systems *in vivo* and their extent of interaction with blood and cellular components. This is an area that needs more understanding for effective use of nanoparticles.

4.5.5 CONCLUSIONS

PEGylation offers significant pharmacological advantages to proteins, cells, and colloidal nanoparticles. PEGylation improves protein solubility and molecular stability, and it reduces protein immunogenicity. By preventing rapid renal clearance, PEGylation extends the circulating life of proteins, thus allowing longer duration of action. PEGylation greatly enhances the clinical value of therapeutic proteins. Indeed, many PEGylated proteins have entered the market or are under clinical trials. PEGylation provides a unique strategy in avoiding immunological recognition of allogeneic cells and tissues. The global camouflage of the cell surface by PEGylation effectively blocks adhesion, recognition, and stimulatory pathways involved in the immune response. More importantly, unlike most therapeutics, PEGylation is nontoxic to modified cells as well as to the animal as a whole. PEGylation provides colloidal nanoparticles with a way to avoid rapid recognition by RES. PEGylation reduces deposition of plasma proteins to the surface of nanoparticles and thereby prolongs the circulating time in the bloodstream. Indeed, development of the PEGylation technology has been a cornerstone of the advances in protein therapeutics. Not only has this technology brought numerous clinically useful products, but also triggered explosive interest in polymers in the field of pharmaceutical biotechnology.

REFERENCES

1. Harris J M (1992). *Poly(ethylene glycol) chemistry: Biotechnical and biomedical applications*, Plenum Press, New York.
2. Harris J M, Zalipsky S (1997). *Polyethylene Glycol Chemistry and Biological Applications*, American Chemical Society, Washington, D.C.
3. Yamaoka T, Tabata Y, Ikada Y (1994). Distribution and tissue uptake of poly(ethylene glycol) with different molecular weights after intravenous administration to mice. *J. Pharm. Sci.* 83:601–606.
4. Harris J M, Chess R B (2003). Effect of pegylation on pharmaceuticals. *Nat. Rev. Drug Discov.* 2:214–221.
5. Richter A W, Akerblom E (1983). Antibodies against polyethylene glycol produced in animals by immunization with monomethoxy polyethylene glycol modified proteins. *Int. Arch. Allergy Appl. Immunol.* 70:124–131.
6. Cheng T L, Wu P Y, Wu M F (1999). Accelerated clearance of polyethylene glycol-modified proteins by anti-polyethylene glycol IgM. *Bioconjug. Chem.* 10:520–528.
7. Talpaz M (2001). Interferon-alfa-based treatment of chronic myeloid leukemia and implications of signal transduction inhibition. *Semin. Hematol.* 38:22–27.
8. Santhanam S, Decatris M, O'Byrne K (2002). Potential of interferon-alpha in solid tumours: Part 2. *BioDrugs.* 16:349–372.
9. Katre N V (1990). Immunogenicity of recombinant IL-2 modified by covalent attachment of polyethylene glycol. *J. Immunol.* 144:209–213.
10. Yang Z, Wang J, Lu Q, et al. (2004). PEGylation confers greatly extended half-life and attenuated immunogenicity to recombinant methioninase in primates. *Cancer Res.* 64:6673–6678.
11. Melton R G, Wiblin C N, Foster R L (1987). Covalent linkage of carboxypeptidase G2 to soluble dextrans–I. Properties of conjugates and effects on plasma persistence in mice. *Biochem. Pharmacol.* 36:105–112.
12. Melton R G, Wiblin C N, Baskerville A, et al. (1987). Covalent linkage of carboxypeptidase G2 to soluble dextrans–II. In vivo distribution and fate of conjugates. *Biochem. Pharmacol.* 36:113–121.
13. Wong K, Cleland L G, Poznansky M J (1980). Enhanced anti-inflammatory effect and reduced immunogenicity of bovine liver superoxide dismutase by conjugation with homologous albumin. *Agents Actions.* 10:231–239.
14. Pollak A, Whitesides G M (1976). Organic synthesis using enzymes in two-phase aqueous ternary polymer systems. *J. Am. Chem. Soc.* 98:289–291.
15. Zalipsky S, Gilon C, Zilkha A (1983). Attachment of drugs to polyethylene glycols. *Eur. Polym. J.* 19:1177–1183.
16. Zalipsky S (1995). Functionalized poly(ethylene glycol) for preparation of biologically relevant conjugates. *Bioconjug. Chem.* 6:150–165.
17. Roberts M J, Bentley M D, Harris J M (2002). Chemistry for peptide and protein PEGylation. *Adv. Drug Deliv. Rev.* 54:459–476.
18. Abuchowski A, van Es T, Palczuk N C, et al. (1977). Alteration of immunological properties of bovine serum albumin by covalent attachment of polyethylene glycol. *J. Biol. Chem.* 252:3578–3581.

19. Zalipsky S, Lee C (1992). Use of functionalized poly(ethylene glycol)s for modification of polypeptides. In J M Harris, S Zalipsky, (eds.), *Polyethylene Glycol Chemistry: Biotechnical and Biomedical Applications*: Plenum, New York, pp. 347–370.

20. Nilsson K, Mosbach K (1984). Immobilization of ligands with organic sulfonyl chlorides. *Methods Enzymol.* 104:56–69.

21. Gais H J, Ruppert S (1995). Modification and immobilization of proteins with polyethylene glycol tresylates and polysaccharide tresylates: Evidence suggesting a revision of the coupling mechanism and the structure of the polymer-polymer linkage. *Tetrahedron. Lett.* 36:3837–3838.

22. Harris J M, Herati R M (1993). Preparation and use of polyethylene glycol propionaldehyde. U.S. Patent 5,252,714.

23. Abuchowski A, Kazo G M, Verhoest, Jr. C R, et al. (1984). Cancer therapy with chemically modified enzymes. I. Antitumor properties of polyethylene glycol-asparaginase conjugates. *Cancer Biochem. Biophys.* 7:175–186.

24. Pasta P, Riva S, Carrea G (1988). Circular dichroism and fluorescence of polyethylene glycol-subtilisin in organic solvents. *FEBS Lett.* 236:329–332.

25. Katre N V, Knauf M J, Laird W J (1987). Chemical modification of recombinant interleukin 2 by polyethylene glycol increases its potency in the murine Meth A sarcoma model. *Proc. Natl. Acad. Sci. U.S.A.* 84:1487–1491.

26. Tondelli L, Laus M, Ferruti A S A (1985). Poly(ethylene glycol) imidazolyl formates as oligomeric drug-binding matrices. *J. Control Release.* 1:251–257.

27. Veronese F M, Largajolli R, Boccu E, et al. (1985). Surface modification of proteins. Activation of monomethoxy-polyethylene glycols by phenylchloroformates and modification of ribonuclease and superoxide dismutase. *Appl. Biochem. Biotechnol.* 11:141–152.

28. Miron T, Wilchek M (1993). A simplified method for the preparation of succinimidyl carbonate polyethylene glycol for coupling to proteins. *Bioconjug. Chem.* 4:568–569.

29. Dolence E K, Hu C, Tsang R, et al. (1997). Electrophilic polyethylene oxides for the modification of polysaccharides, polypeptides (proteins) and surfaces. U.S. Patent 5,985,263.

30. Wang Y S, Youngster S, Bausch J, et al. (2000). Identification of the major positional isomer of pegylated interferon alpha-2b, *Biochemistry* 39:10634–10640.

31. Zalipsky S, Barany G (1990). Facile synthesis of alpha-hydroxy-omega-carboxymethylpolyethylene oxide. *J. Bioact. Compat. Polym.* 227–231.

32. Harris J M, Kozlowski A (1997). Polyethylene glycol and related polymers monosubstituted with propionic or butanoic acids and fuctional derivatives thereof for biotechnical applications. U.S. Patent 5,672,662.

33. Kogan T P (1992). The synthesis of substituted methoxy-poly(ethylene glycol) derivatives suitable for selective protein modification. *Synth. Commun.* 22:2417–2424.

34. Woghiren C, Sharma B, Stein S (1993). Protected thiol-polyethylene glycol: a new activated polymer for reversible protein modification. *Bioconjug. Chem.* 4:314–318.

35. Morpurgo M, Veronese F M, Kachensky D (1996). Preparation of characterization of poly(ethylene glycol) vinyl sulfone. *Bioconjug. Chem.* 7:363–368.

36. Uchio T, Baudys M, Liu F, et al. (1999). Site-specific insulin conjugates with enhanced stability and extended action profile. *Adv. Drug Deliv. Rev.* 35:289–306.

37. Hinds K D, Kim S W (2002). Effects of PEG conjugation on insulin properties. *Adv. Drug Deliv. Rev.* 54:505–530.

38. Morpurgo M, Monfardini C, Hofland L J, et al. (2002). Selective alkylation and acylation of alpha and epsilon amino groups with PEG in a somatostatin analogue: tailored chemistry for optimized bioconjugates. *Bioconjug. Chem.* 13:1238–1243.

39. Tsunoda S, Ishikawa T, Yamamoto Y (1999). Enhanced antitumor potency of polyethylene glycolylated tumor necrosis factor-alpha: A novel polymerconjugation technique with a reversible amino-protective reagent. *J. Pharmacol. Exp. Ther.* 290: 368–372.

40. Sato H (2002). Enzymatic procedure for site-specific pegylation of proteins. *Adv. Drug Deliv. Rev.* 54:487–504.

41. Chapman A P, Antoniw P, Spitali M, et al. (1999). Therapeutic antibody fragments with prolonged in vivo half-lives. *Nat. Biotechnol.* 17:780–783.

42. Hinds K D (2005). Protein conjugation, cross-linking, and PEGylation. In R I Mahato, *Biomaterials for Delivery and Targeting of Proteins and Nucleic Acids.* CRC Press, Boca Raton, F. L., pp. 120–185.

43. Pepinsky R B, LePage D J, Gill A, et al. (2001). Improved pharmacokinetic properties of a polyethylene glycol-modified form of interferon-beta-1a with preserved in vitro bioactivity. *J. Pharmacol. Exp. Ther.* 297:1059–1066.

44. Kinstler O B, Brems D N, Lauren S L, et al. (1996). Characterization and stability of N-terminally PEGylated rhG-CSF. *Pharm. Res.* 13:996–1002.

45. Kinstler O B, Gabriel N E, Farrar C E, et al. (1999). N-terminally chemically modified protein compositions and methods. U.S. Patent 5,985,265.

46. Caliceti P, Veronese F M (2003). Pharmacokinetic and biodistribution properties of poly(ethylene glycol)-protein conjugates. *Adv. Drug Deliv. Rev.* 55:1261–1277.

47. Schernthaner G (1993). Immunogenicity and allergenic potential of animal and human insulins. *Diabetes Care.* 16(suppl 3):155–165.

48. Casadevall N, Nataf J, Viron B, et al. (2002). Pure red-cell aplasia and antierythropoietin antibodies in patients treated with recombinant erythropoietin. *N. Engl. J. Med.* 346:469–475.

49. Gershon S K, Luksenburg H, Cote T R, et al. (2002). Pure red-cell aplasia and recombinant erythropoietin. *N. Engl. J. Med.* 346:1584–1586.

50. Hermeling S, Crommelin D J, Schellekens H, et al. (2004). Structure-immunogenicity relationships of therapeutic proteins. *Pharm. Res.* 21:897–903.

51. Richter A W, Akerblom E (1984). Polyethylene glycol reactive antibodies in man: titer distribution in allergic patients treated with monomethoxy polyethylene glycol modified allergens or placebo, and in healthy blood donors. *Int. Arch. Allergy Appl. Immunol.* 74:36–39.

52. Chapman A P (2002). PEGylated antibodies and antibody fragments for improved therapy: A review. *Adv. Drug Deliv. Rev.* 54:531–545.

53. Veronese F M (2001). Peptide and protein PEGylation: a review of problems and solutions. *Biomaterials.* 22:405–417.

54. Caliceti P, Schiavon O, Veronese F M (2001). Immunological properties of uricase conjugated to neutral soluble polymers. *Bioconjug. Chem.* 12:515–522.

55. Hershfield M S (1995). PEG-ADA replacement therapy for adenosine deaminase deficiency: an update after 8.5 years. *Clin. Immunol. Immunopathol.* 76:S228–232.

56. Burnham N L (1994). Polymers for delivering peptides and proteins. *Am. J. Hosp. Pharm.* 51:210–218; 228–229.

57. Hilman B C, Sorensen R U (1994). Management options: SCIDS with adenosine deaminase deficiency. *Ann. Allergy.* 72:395–403; quiz 403–394, 407.

58. Graham M L (2003). Pegaspargase: A review of clinical studies. *Adv. Drug Deliv. Rev.* 55:1293–1302.

59. Holle L M (1997). Pegaspargase: An alternative? *Ann. Pharmacother.* 31:616–624.

60. Wills R J (1990). Clinical pharmacokinetics of interferons. *Clin. Pharmacokinet.* 19:390–399.

61. Monkarsh S P, Ma Y, Aglione A, et al. (1997). Positional isomers of monopegylated interferon alpha-2a: isolation, characterization, and biological activity. *Anal. Biochem.* 247:434–440.

62. Kozlowski A, Harris J M (2001). Improvements in protein PEGylation: Pegylated interferons for treatment of hepatitis C. *J. Control Rel.* 72:217–224.

63. Bailon P, Palleroni A, Schaffer C A, et al. (2001). Rational design of a potent, long-lasting form of interferon: a 40 kDa branched polyethylene glycol-conjugated interferon alpha-2a for the treatment of hepatitis C. *Bioconj. Chem.* 12:195–202.

64. Edwards C K 3rd (1999). PEGylated recombinant human soluble tumour necrosis factor receptor type I (r-Hu-sTNF-RI): Novel high affinity TNF receptor designed for chronic inflammatory diseases. *Ann. Rheum. Dis.* 58(suppl 1):I73–81.

65. Moreland L W, McCabe D P, Caldwell J R, et al. (2000). Phase I/II trial of recombinant methionyl human tumor necrosis factor binding protein PEGylated dimer in patients with active refractory rheumatoid arthritis. *J. Rheumatol.* 27:601–609.

66. Brubaker D B (1993). Immunopathogenic mechanisms of posttransfusion graft-vs-host disease. *Proc. Soc. Exp. Biol. Med.* 202:122–147.

67. Blazer B R, Korngold R, Vallera D A (1997). Recent advances in graft-versus-host disease (GVHD) prevention. *Immunol. Rev.* 157:79–109.

68. Scott M D, Murad K L, Koumpouras F, et al. (1997). Chemical camouflage of antigenic determinants: Stealth erythrocytes. *Proc. Natl. Acad. Sci. U.S.A.* 94:7566–7571.

69. Scott M D, Murad K L (1998). Cellular camouflage: Fooling the immune system with polymers. *Curr. Pharmeceut. Des.* 4:423–438.

70. Bradley A J, Test T T, Murad K L, et al. (2001). Complement interactions with methoxypoly(ethylene glycol)-modified human erythrocytes. *Transfusion* 41:1225–1233.

71. Murad K L, Mahany K L, Kuypers F A, et al. (1999). Structural and functional consequences of antigenic modulation of red blood cells with methoxypoly(ethylene glycol). *Blood* 93:2121–2127.

72. Murad K L, Gosselin E J, Eaton J W, et al. (1999). Stealth cells: Prevention of MHC class II mediated T cell activation by cell surface modification. *Blood.* 94:2135–2141.

73. Bradley A J, Murad K L, Regan K L, et al. (2002). Biophysical consequences of linker chemistry and polymer size on stealth erythrocytes: Size dose matter. *Biochim. Biophys. Acta.* 1561:147–158.

74. Scott M D, Chen A M (2004). Beyond the red cell: pegylation of other blood cells and tissues. *Transfus. Clin. Biol.* 11:40–46.

75. Nacharaju P, Boctor F N, Manjula B N, et al. (2005). Surface decoration of red blood cells with maleimidophenyl-polyethylene glycol facilitated by thiolation with iminothiolane: an approach to mask A, B, and D antigens to generate universal red blood cells. *Transfusion.* 45:374–383.

76. Chung H A, Kato K, Itoh C, et al. (2004). Casual cell surface remodeling using biocompatible lipid-poly(ethylene glycol) (n): Development of steal cells and monitoring of cell membrane behavior in serum-supplemented conditions. *J. Biomed. Mater. Res.* 70A:179–185.

77. Sirchia G, Zanella A, Parravicini A, et al. (1985). Red cell alloantibodies in thalassemia major. Results of an Italian cooperative study. *Transfusion.* 25:110–112.

78. Vichinsky E P, Earles A, Johnson R A, et al. (1990). Alloimmunization in sickle cell anemia and transfusion of racially unmatched blood. *N. Engl. J. Med.* 322.

79. Scott M D, Bradley A J, Murad K L (2000). Camouflaged red cells: low technology bioengineering for tranfusion medicine? *Transfus. Med. Rev.* 14:53–63.

80. Chen A M, Scott M D (2001). Current and future applications of immunologic attenuation via pegylation of cells and tissues. *BioDrugs.* 15:833–847.

81. Chen A M, Scott M D (2003). Immunocamouflage: Prevention of transfusion-induced graft-versus-host disease via polymer grafting of donor cells. *J. Biomed. Mater. Res.* 67A:626–636.

82. Panza J L, Wagner W R, Rilo H L R, et al. (2000). Treatment of rat pancreatic islets with reactive PEG. *Biomaterials.* 21:1155–1164.

83. Lee D Y, Yang K, Lee S, et al. (2002). Optimization of monomethoxy-polyethylene glycol grafting on the pancreatic islet capsules. *J. Biomed. Mater. Res.* 62:372–377.

84. Lee D Y, Nam J H, Byun Y (2004). Effect of polyethylene glycol grafted onto islet capsules on prevention of splenocyte and cytokine attacks. *J. Biomater. Sci. Polym. Ed.* 15:753–766.

85. Contreras J L, Xie D, Mays J, et al. (2004). A novel approach to xenotransplantation combining surface engineering and genetic modification of isloated adult porcine islets. *Surgery.* 136:537–547.

86. Jang J Y, Lee D Y, Park S J, et al. (2004). Immune reactions of lymphocytes and macrophages against PEG-grafted pancreatic islets. *Biomaterials.* 25:3663–3669.

87. Xie D, Smyth C A, Eckstein C, et al. (2005). Cytoprotection of PEG-modified adult porcine pancreatic islets for improved xenotransplantation. *Biomaterials.* 26:403–412.

88. Sahoo S K, Labhasetwar V (2003). Nanotech approaches to drug delivery and imaging. *Drug Disc. Today.* 8:1112–1120.

89. Barrat G M (2000). Therapeutic applications of colloidal drug carriers. *Pharm. Sci. Technol. Today.* 3:163–171.

90. Ferrari M (2005). Cancer nanotechnology: Opportunities and challenges. *Nature Rev. Cancer.* 5:161–171.

91. Drummond D C, Meyer O, Hong K, et al. (1999). Optimizing liposomes for delivery of chemotherapeutic agents to solid tumors. *Pharmacol. Rev.* 51:691–743.

92. Moghimi S M, Hunter A C, Murray J C (2001). Long-circulating and target-specific nanoparticles: Theory to practice. *Pharmacol. Rev.* 53:283–318.

93. Senior J, Delgado C, Fisher D, et al. (1991). Influence of surface hydrophilicity of liposomes on their interaction with plasma protein and clearance from the circulation: studies with poly(ethylene glycol)-coated vesicles. *Biochim. Biophys. Acta.* 1062:77–82.

94. Torchilin V P, Shtilman M I, Trubetskoy V S, et al. (1994). Amphiphilic vinyl polymers effectively prolong liposome circulation time in vivo. *Biochim. Biophys. Acta.* 1195:181–184.

95. Takeuchi H, Toyoda T, Toyobuku H, et al. (1996). Improved stability of doxorubicin-loaded liposomes by polymer coating. *Proc. Int. Symp. Control Rel. Bioact. Mater.* 23:409–410.

96. Torchilin V P, Trubetskoy V S, Whiteman K R, et al. (1995). New synthetic amphiphilic polymers for steric protection of liposomes in vivo. *J. Pharm. Sci.* 84:1049–1053.

97. Woodle M C, Engbers C M, Zalipsky S (1994). New amphiphathic polymer-lipid conjugates forming long-circulating reticuloendothelial system-evading liposomes. *Bioconjug. Chem.* 5:493–496.

98. Zalipsky S, Hansen C B, Oaks J M, et al. (1996). Evaluation of blood clearance rates and biodistribution of poly(2-oxazoline)-grafted liposomes. *J. Pharm. Sci.* 85:133–137.

4.6

UNEXPECTED BENEFITS OF A FORMULATION: CASE STUDY WITH ERYTHROPOETIN

DAVID KEAST

Parkwood Hospital, St. Joseph's Health Care, London, Ontario, Canada; University of Western Ontario, London, Ontario, Canada

Chapter Contents

4.6.1 Introduction   463
4.6.2 Erythropoietin as a Potential Growth Factor in Wound Healing   464
4.6.3 Clinical Experience with Recombinant Human Erythropoietin and Patients with Pressure Ulcers   466
4.6.4 Conclusions and Future Research   467
     References   467

4.6.1 INTRODUCTION

Recombinant human erythropoietin (rHuEPO) is a purified glycoprotein produced by recombinant DNA technology. It is indistinguishable from human urinary erythropoietin in its biologic and immunologic activity. It is commercially available and has absolute indications for the treatment of the anemia associated with chronic renal failure, prematurity, and platinum-based chemotherapy. Other uses have included the anemia associated with multiple myeloma, cancer, myelodysplasia, HIV infection, and other forms of chemotherapy. It has been used to potentiate preoperative autologous blood donation. Less common uses have been in the treatment of the anemia of chronic disease, for perisurgical augmentation and after allogenic bone marrow transplantation [1].

Endogenous erythropoietin (EPO) is produced primarily by the peritubular capillary endothelium of the kidney in response to tissue hypoxia or anemia. EPO is a

Handbook of Pharmaceutical Biotechnology, Edited by Shayne Cox Gad.
Copyright © 2007 John Wiley & Sons, Inc.

hormone that plays a crucial role in the regulation of hematopoiesis and induces the proliferation, maturation, and differentiation of erythroid (red blood cell) precursors. In renal failure, the nonfunctioning kidneys fail to produce adequate amounts of endogenous EPO to support erythropoiesis with resultant anemia. Before the introduction of exogenous erythropoietin in the 1980s, patients in renal failure required repeated blood transfusions.

After 1998 there was a significant increase in the number of cases of pure red cell aplasia (PRCA) associated with subcutaneous injections of rHuEPO in the renal dialysis population. PRCA is a relatively rare condition. It presents as a sudden onset of severe isolated anemia characterized by near-complete absence of erythroid precursors in the bone marrow [2]. Most of the initial reports in the literature before 1998 involved isolated cases, of which 50% had no known cause. After 1998, the PRCA was shown to be related to neutralizing antibodies to erythropoietin that cross-reacted with endogenous EPO [3]. Most of these reported antibody-mediated cases were for patients using Eprex (Ortho Biotech, a division of Janssen-Cilag, Bridgewater, NJ). In total between January 1, 1989 and June 30, 2004, the company reported a total of 206 cases, of which 155 occurred in the peak years of 2001–2003 [4]. The mean time from initial exposure to onset of PRCA was 9.1 months [5].

The rise in cases of PRCA corresponded with the introduction of polysorbate 80 stabilized Eprex in prefilled syringes with uncoated rubber stoppers, which was administered subcutaneously. Previously human serum albumin (HSA) was used to stabilize the product, but this practice was discontinued in several countries due to fears of disease transmission by HSA. Epidemiologic and immunologic data reported by Boven et al. [4] demonstrated that the rise in PRCA was most likely related to leachates from the uncoated stoppers in the presence of the polysorbate 80 stabilizer. These leachates may have potentiated the immunogenicity of the product when administered subcutaneously.

No cases for intravenous administration were reported. Since the introduction of coated stoppers in the prefilled syringes in April 2003 and a concomitant switch to intravenous administration in many dialysis units, the number of reported cases fell from 71 in 2003 to 2 reported cases in the first 4 months of 2004. Even at its peak, the incidence rate for PRCA was relatively low at 4.61/10,000 patient years of exposure to the product.

4.6.2   ERYTHROPOIETIN AS A POTENTIAL GROWTH FACTOR IN WOUND HEALING

In addition to its ability to stimulate erythropoiesis, EPO enhances cell phagocytosis and reduces macrophage activation, therefore modulating the inflammatory process. These anti-inflammatory effects of EPO may be able to reverse the chronic inflammatory condition that is believed to underlie chronic skin ulcers. By interfering with the chronic inflammatory process EPO may help reduce inflammatory cytokines and degradative enzymes that interfere with many wound healing processes and limit new tissue growth. Recent literature suggests that many chronic wounds persist because of inflammatory processes that alter the wound environment. Analysis of wound fluid taken from healing and chronic wounds revealed that chronic wounds have elevated levels of inflammatory mediators and degradative enzymes that interfere with the healing process [6]. Healing wounds are characterized by high mitogenic potential, rapid cellular migration, balanced inflammatory cytokines, low proteases, and good cellular response to growth factors. Chronic wounds, however, are characterized by poor mitogenic potential and cell migration, high proteases and inflammatory cytokines, and senescent cells unresponsive to growth factors [7]. EPO's effects on the wound healing process not only include restoring the normal wound environment, but it also has recently been found to interact with vascular endothelial growth factor (VEGF). Together EPO and VEGF stimulate endothelial cell mitosis and motility important in new vessel growth and wound healing [8]. Galeano et al. [9] demonstrated in an artificial wound model in genetically diabetic mice that erythropoietin injections increased VEGF mRNA expression, wound protein content, wound healing, and wound breaking strength. The additional benefits of EPO are its neurotrophic and neuro-protective functions that may ameliorate neurological damage after spinal cord and brain injury [10]. A specific EPO/EPO receptor system has been found in the central nervous system and in the cerebrospinal fluid, which is independent of the heatopoietic system [11].

Chronic skin ulcers experienced by patients who have low concentrations of circulating hemoglobin ($\leq 100\,g/L$) may be difficult to heal because of impairment in tissue oxygenation. There are various types of anemias with differing underlying causes, including nutrient deficiencies, such as iron deficiency anemia, and anemia as the result of chronic disease or chronic inflammatory processes. Iron deficiency anemia results from inadequate intake, absorption, or utilization of iron and/or acute or chronic blood loss [12, 13]. It is characterized by low hemoglobin and other hematological changes combined with microcytic, hypochromic red blood cells. There are low levels of stored iron as indicated by low serum ferritin values. Anemia of chronic disease (ACD) is characterized by low hemoglobin concentrations and other hematological changes with normocytic, normochromic red blood cells. In anemia of chronic disease, red blood cell production is impaired and there may be a shortened red blood cell life span. ACD is also characterized by a normal or elevated serum ferritin level. ACD is generally not an irondeficiency anemia, and it is refractory to an iron-enhanced diet, iron replacement therapy, either oral or intravenous, and transfusion. In fact, iron supplementation and transfusion are generally contraindicated in ACD because of the risk of iatrogenic hemochromatosis. ACD occurs in individuals with a chronic inflammatory process such as arthritis, critical illness, or chronic skin ulcers. It is thought to be the result of impaired responsiveness of erythroid progenitor cells due to a persistent elevated level of circulating inflammatory cytokines that are known to occur in patients with chronic inflammatory conditions [14]. Ferrucci et al. [15] demonstrated in a study of inflammatory serum markers in a sample of the residents of Chianti Italy that the anemia of inflammation evolved from a pre-anemic state in which normal hemoglobin was maintained in persons with high levels of inflammatory markers by increasing levels of erythropoietin to a clinical anemia in which erythropoietin levels were suppressed possibly through the inhibitory effect of inflammation on erythropoietin production.

However, despite the numerous documented effects of EPO on cellular and physiological events known to be important in the healing process and the management of ACD, few clinical studies are investigating the use of EPO to promote the healing of chronic wounds. In 1992 Turba et al. reported the successful treatment of ACD related to stage IV pressure ulcers [16]. In another hematological disorder, treatment with EPO resulted in the rapid and complete healing of chronic leg ulcer, further improvement in hematological parameters, and relief from chronic pain [17].

4.6.3   CLINICAL EXPERIENCE WITH RECOMBINANT HUMAN ERYTHROPOIETIN AND PATIENTS WITH PRESSURE ULCERS

Spinal cord-injured patients are at high risk of developing pressure ulcers, many of which often become associated with anemia of chronic disease. These ulcers are challenging to heal. Pressure redistribution, nutritional support, management of incontinence, and good local wound care are key components for healing these ulcers. Despite best-practice care, patients with anemia of chronic disease remain difficult to heal.

In an attempt to reverse the anemia of chronic disease, several patients in the Spinal Cord Injury rehabilitation unit at Parkwood Hospital, St. Joseph's Health Care, London, Canada, were treated with subcutaneous erythropoietin. A retrospective chart audit was conducted to review the effectiveness of 6 weeks of subcutaneous erythropoietin 75 IU/kg subcutaneously 3 times weekly in resolving refractory anemia of chronic disease and healing stage IV pressure ulcers [18]. The mean age of the patients was 59, and all had a stage IV pressure ulcer.

All patients received pressure off-loading, nutritional assessment and supplementation, and best-practice local wound care. Comorbid conditions such as diabetes were medically managed. All patients received rHuEPO 75 IU/kg subcutaneously three times weekly for 6 weeks. Iron supple-mentation with oral ferrous gluconate as indicated by ferritin levels was administered.

Mean initial hemoglobin was 88 g/L. After 6 weeks of rHuEPO injections, the mean hemoglobin for the four patients rose to 110 g/L. The individual results are shown in Figure 4.6-1. The mean number of ulcers decreased from 3 to 2.3, and

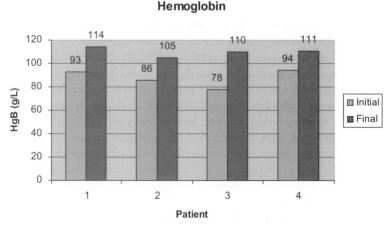

Figure 4.6-1. Comparison of initial and final hemoglobin after 6 weeks rHuEPO.

the mean surface area of the largest ulcer decreased from $42.3\,cm^2$ to $38.4\,cm^2$. Ulcer depth was decreased by half. The extent of undermining improved in all ulcers. Some patients showed an increased ability to fight intercurrent infections. All patients felt more energetic and were better able to participate in their rehabilitation activities. No adverse effects were observed.

The results were promising enough to suggest that a prospective study is warranted. Such a study would carefully collect all hematologic factors as well as use validated tools for determining ulcer size and appearance. In addition, collecting wound fluid before, during, and after treatment to determine the effect on chronic inflammatory mediators would be useful. Research into effects at the molecular level may be useful.

4.6.4   CONCLUSIONS AND FUTURE RESEARCH

The introduction of human recombinant erythropoietin significantly improved the quality of life for patients on real dialysis by managing the concomitant anemia, which results from end-stage renal failure. It has also been successfully employed in several other conditions in which erythropoiesis is suppressed either through disease processes or as a result of treatments.

The emergence of pure red cell aplasia as a complication of treatment with subcutaneous injection in the late 1990s was probably the result of the use of prefilled syringes with uncoated rubber stoppers in which the product was stabilized with polysorbate 80. Leachates from the stoppers most likely potentiated the immunogenicity of the product leading to the production of neutralizing antibodies. Since the reformulation of the product, the incidence of PRCA has been steadily dropping.

Human recombinant erythropoietin shows promise in resolving the refractory anemia of chronic disease associated with stage IV pressure ulcers, and further study is suggested. The results may also suggest that rHuEPO acts as a growth factor either alone or in conjunction with intrinsic factors in the wound. Studies of its role at the molecular level are indicated. Further studies of the neurotrophic and neuroprotective properties observed in rat models hold promise in the area of spinal cord injury research.

REFERENCES

1. Hoffman R (2005). Erythropetic growth factors. In R Hoffman, et al. (eds.), *Hematology Basic Principles and Practice*, 4th ed.: Churchill Livingstone, New York.
2. Dessypris E N (2005). Pure red cell aphasia. In R Hoffman, et al. (eds.), *Hematology Basic Principles and Practice*, 4th ed.: Churchill Livingstone, New York.
3. Casadevall N (2002). Antibodies against rHuRPO: Native and recombinant. *Nephrol. Dial. Transplant.* 17(suppl 5):42–47.
4. Boven K, Stryker S, Knight J, et al. (2005). The increased incidence of pure red call aphasia with Eprex formulation in uncoated stopper syringes. *Kidney Internat.* 67:2346–2353.

5. Bennett C, Luminari S, Nissenson A, et al. (2004). Pure red-cell aphasia and recombinant epoetin therapy—a follow-up report from the research on adverse drug events and reports project. *N. Engl. J. Med.* 351:1403–1408.

6. Mast B, Schultz G (1996). Interactions of cytokines, growth factors, and proteases in acute and chronic wounds. *Wound Rep. Reg.* 4:411–420.

7. Schultz G S, Sibbald R G, Falanga V, et al. (2003). Wound bed preparation: A systematic approach to wound management. *Wound Rep. Reg.* 11(suppl 2):1–28.

8. Buemi M, Aloisi C, Cavallaro E, et al. (2002). Recombinant human erythropoietin (rHuEPO): More than just the correction of uremic anemia. *J. Nephrol.* 15(2): 97–103.

9. Galeano M, Altavilla D, Cucinotta D, et al. (2004). Recombinant human erythropoietin stimulates angiogenesis and wound healing in the genetically diabetic mouse. *Diabetes.* 53(9):2509–2517.

10. Gorio A, Gokmen N, Erbayraktar S, et al. (2002). Recombinant human erythropoietin conteracts secondary injury and markedly enhances neurological recovery from experiment spinal cord injury. *Proc. Natl. Acad. Sci. U.S.A.* 99(14):9450–9455.

11. Buemi M, Cavallaro E, Floccari F, et al. (2002). Erythropoietin and the brain: from neurodevelopment to neuroprotection. *Clin. Sci. (London)* 103(3):275–282.

12. Woloschuk D M M (1997). A practical guide to anemias. *Pharm. Practice.* Aug 1–8.

13. Simmons-Holcomb S (2001). Anemia—pointing the way to a deeper problem. *Nursing* 31(7):36–42.

14. Spivak J L (2002). Iron and anemia of chronic disease. *Oncology.* 9(suppl 10):25–33.

15. Ferricci L, Guaralnik J M, Woodman R C, et al. (2005). Proinflammatory state and circulating erythropoietin in persons without anemia. *Amer. J. Med.* 118(11):1288e11–1288e19.

16. Turba R M, Lewis V L, Green D (1992). Pressure sore anemia: Response to erythropoietin. *Arch. Phys. Rehab. Rehabil.* 73(5):498–500.

17. Al-Momen A K (1991). Recombinant erythropoietin induced rapid healing of a chronic leg ulcer in a patient with sickle cell disease. *Acta. Haematol.* 86(1):46–48.

18. Keast D H, Fraser C (2004). The treatment of Chronic Skin Ulcers in individuals with anemia of chronic disease using recombinant human erythropoietin (EPO): A review of four cases. *Ostomy/Wound Managem.* 50(10):38–47.

5.1

CAPILLARY SEPARATION TECHNIQUES

DONG HEE NA[1] AND KANG CHOON LEE[2]

[1] *College of Pharmacy, Kyungsung University, Nam-ku, Busan, Korea*
[2] *Drug Targeting Laboratory, College of Pharmacy, SungKyunKwan University, Jangan-ku, Suwon, Korea*

Chapter Contents

5.1.1   Introduction   469
5.1.2   Principles of Capillary Electrophoresis   470
    5.1.2.1   Instrumentation   470
    5.1.2.2   Theory   472
    5.1.2.3   Separation Modes   474
    5.1.2.4   Detection   480
5.1.3   Applications   481
    5.1.3.1   Identity Determination   482
    5.1.3.2   Purity Control   484
    5.1.3.3   Heterogeneity Characterization of Glycoproteins   488
    5.1.3.4   Stability Study   495
    5.1.3.5   Characterization of PEGylated Peptides and Proteins   497
    5.1.3.6   Proteome Analysis   501
5.1.4   Conclusions   502
    References   502

5.1.1   INTRODUCTION

Capillary electrophoresis (CE) is a modern analytical method that is being extensively applied to the characterization of biotechnology-derived products like peptides and proteins [1]. Due to its ease of automation and facilitating the development of reproducible routine analysis, CE seems to be well suited for the quality control of biotechnological products, including process monitoring, purity assessments,

Handbook of Pharmaceutical Biotechnology, Edited by Shayne Cox Gad.
Copyright © 2007 John Wiley & Sons, Inc.

and stability studies [2]. Moreover, the high-resolution capacity of CE offers great potential for the analysis of heterogeneous protein products such as glycoproteins and polymer-conjugated proteins [3, 4]. Recently, the importance of CE in protein analysis is enhanced, with efforts being made to learn more about the compositions and functions of proteins [5]. As the field of proteomics becomes more important, the thousands of new proteins and peptides will be discovered, which will lead to the developments of many new pharmaceutical drugs.

Recently, CE has emerged as a powerful tool because it has many advantages over the conventional protein separation techniques such as polyacrylamide gel electrophoresis (PAGE) and high-performance liquid chromatography (HPLC). Compared with PAGE, CE is faster, easier, simpler, quantitative, and has automation capability. Like HPLC, CE has various different separation modes and is applicable to a wide range of analytes. CE requires a very small sample amount (nanoliters) and limited quantities of reagents (microliters of buffer), compared with HPLC that requires microliters of sample and milliliters of solvent. The main advantage of CE over PAGE and HPLC is the ability to produce a higher number of theoretical plates. The efficient separations result from the application of high electric fields (100–500 V/cm) and flat flow generated by the electro-osmotic flow (HPLC generates the pressure-driven parabolic flow). The high electrical resistance of a capillary and its favorable surface area-to-volume ratio, permitting efficient heat dissipation, enables the application of high electric fields without causing detrimental heat generation. Furthermore, the mild separation conditions of CE, which use mostly aqueous buffer solutions, present the compatibility to protein studies in native state, which can be important especially in the characterization of pharmaceutical proteins. The main disadvantage of CE is the low concentration limits of detection due to the short path length and the limited introduction of sample volume. The enhancements in this issue have been achieved by the advent of several preconcentration methods [6] and highly sensitive detection methods such as laser-induced fluorescence and mass spectrometry [7].

This chapter introduces the fundamental principles of CE and its application to the separation and characterization of peptides and proteins. One major characteristic of CE is the availability of various separation modes with different separation mechanisms based on the differences in charge-to-mass ratio (capillary zone electrophoresis), molecular size (capillary gel electrophoresis), isoelectric point (capillary isoelectric focusing), and hydrophobicity (micellar electrokinetic capillary chromatography and capillary electrochromatography). In the principles of CE, the instrumentation, theory, separation modes, and detection methods are described focusing on the analysis of peptides and proteins. In applications, CE methods for identification, purity assessment, heterogeneity characterization, and stability of peptide and protein products are highlighted with the characterization of Poly(ethylene glycol) (PEG)ylated biomolecules. Finally, the role of CE in proteomics research is discussed with the recent approaches.

5.1.2   PRINCIPLES OF CAPILLARY ELECTROPHORESIS

5.1.2.1   Instrumentation

The main components of a CE system are a high-voltage power supply, a capillary, inlet and outlet buffer vials, a detector, and a data output and handling device such as an integrator or computer (Figure 5.1-1A). A variety of commercial instruments are available with different capabilities, including injection methods, detectors, capillary cooling systems, and software [8]. CE is performed by filling inlet and outlet vials and the capillary with an electrolyte, usually an aqueous buffer solution. Electrodes connected to a high-voltage power supply are then immersed in the vials. Very small amounts of sample are introduced into one end of the capillary, and these are driven electrophoretically down the lumen of the capillary toward the opposite electrode. The species that migrate through the capillary are detected by an online optical detector near the capillary outlet, and these data are displayed as an electropherogram, in which separated compounds appear as peaks with different migration times.

Fused silica capillaries of length of 20–100 cm and inner diameters of 50–100 μm are typically used. The outer surface of capillary is coated with polyimide, which is strongly ultraviolet (UV) absorbent. A detection window can be made by simply burning or scraping off a small section of the polyimide outer capillary coating. This section of the capillary is then placed in the light path of the detector, and thus solutes are detected while in the capillary. Most commercial instruments have a capillary cartridge, which retains the capillary, provides mechanical support, and allows the capillary to be consistently aligned at the optical center of the detector. Cartridges are used in conjunction with cooling systems, which maintain capillary temperatures using coolant or cooled air.

Only a few microliters of sample are required for CE because usually only 1–50 nanoliters of sample are injected into a capillary. For example, a 50-μm inner diameter capillary 50 cm long has a volume of only ca. 1 μL [9]. Samples are injected

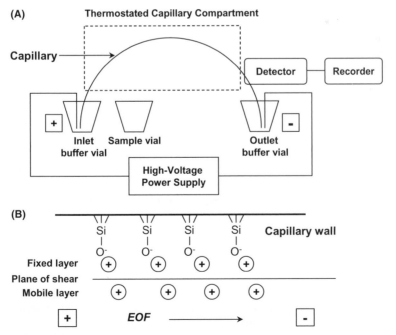

Figure 5.1-1. Schematic diagram of a capillary electrophoresis system (A) and electroosmotic flow (EOF) (B).

into capillaries using different techniques, such as hydrodynamically or electroki-netically. Hydrodynamic injection can be performed by pressure or siphoning. Pressure injection is performed by either pressurizing the sample vial or by apply-ing a vacuum to the outlet vial. Siphon injection is performed by raising the sample vial, which causes the sample to be siphoned into the capillary. In electrokinetic injection, an electric field is applied to the sample vial, causing the sample compo-nents to migrate into the capillary. Most commercially available instruments have autosamplers into which several sample vials can be loaded. The samples are then automatically injected into capillaries using one or more of the techniques above.

A variety of detectors have been used for CE, including UV absorbance, fluo-rescence, laser-induced fluorescence, and mass spectrometry. The most widely used is the UV absorbance detector. In some cases, two or more detectors are connected in series. A more detailed the description of the detection system is given in Section 5.1.2.4.

The power supply provides an electric field across the capillary with voltages up to 30kV, currents up to 300μA, and power up to 6W. Most instruments can be operated in either constant voltage, constant current, or constant power mode and have a reverse applied polarity facility. The constant voltage mode is most com-monly used. It is necessary to have a stable voltage, as any variations in voltage will cause changes in migration times.

Most CE systems are controlled by an external computer that controls all instru-mental functions. The operating parameters for each analysis are programmed by computer. The electropherograms obtained are plots of detector response versus time; thus, they resemble familiar HPLC or gas chromatography (GC) chromato-grams, which means that familiar data handling systems can be used. In addition, electronic integrators or computers are used for qualitative identification and quan-titation, as in HPLC or GC.

5.2.2.2   Theory

Electrophoresis is the phenomenon whereby ionic species in a conductive aqueous medium moves under the influence of an electric field. Thus, ionic molecules are separated due to differences in their electrophoretic mobilities. The electrophoretic mobility (μ) of a spherical ion is given by

$$\mu = q/6\pi\eta r$$

where q is the charge of the ion, η is the viscosity of the solution, and r is the hydrodynamic radius of the ion. This equation demonstrates the positive relation between charge-to-mass ratio (q/r) and electrophoretic mobility. The electropho-retic migration velocity (v) depends on the electrophoretic mobility (μ) and the applied electric field (E):

$$v = \mu E$$

Besides the electrophoretic migration, a fundamental electrophoretic pheno-menon occurring in CE is the electro-osmotic flow (EOF), which essentially is an electrical field-driven bulk solution flow from the anode to the cathode (Figure 5.1-1B). This flow occurs because acidic silanol groups on the inside of fused-silica capillary are ionized when in contact with buffer solution. At pH above 3, these silanol groups are deprotonated and form an electric double layer. When an electric field is applied, the net positively charged solution in a capillary moves toward the cathode. EOF is highly dependent on buffer pH; i.e., it increases on raising pH and plateaus at about pH 8, but it is not significant below pH 4. At neutral and higher pH, the EOF is sufficiently strong to allow all molecules, regardless of charge, to move toward the outlet and pass the detector. EOF can be effectively controlled by changing several experimental conditions, including separation buffer pH, ionic strength, addition of organic solvents, and buffer additives. The apparent migration velocity of the analyte depends on their electrophoretic mobility (μ_e), electro-osmotic mobility (μ_{eo}), the applied voltage (V), and capillary length (L):

$$v = (\mu_e + \mu_{eo})V/L$$

The electrophoretic mobility of peptides and proteins was first described by Offord [10], who proposed the following equation, which relates the mobility (μ) with valence (Z) and molecular mass (M):

$$\mu = kZM^{-2/3}$$

where k is an empirical constant. This equation states that frictional forces opposing electrophoretic migration are proportional to the surface areas of the species concerned. This implies that the electrophoretic mobility would be proportional to $1/r^2$ ($1/M^{2/3}$), rather than $1/r$ ($1/M^{1/3}$), which is suggested by the Stoke's model. This relationship was later confirmed and modified by several researchers in the CE field [11, 12]. Janini et al. presented the electrophoretic mobilities of 58 peptides that varied in size from 2 to 39 amino acids and in charge from 0.65 to 7.82 [13]. Figure 5.1-2 shows CE separation of several peptides in a 50-mM phosphate buffer

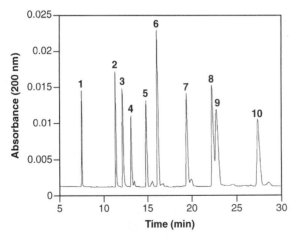

Figure 5.1-2. Electropherogram of a mixture of bioactive peptides separated at pH 2.5. Peptides: (1) reference; (2) bradykinin; (3) bradykinin fragment 1–5; (4)substance P; (5) [arg]–vasopressin; (6) luteinizing hormone releasing hormone; (7) bombesin; (8) leucine enkephalin; (9) methionine enkephalin; (10) oxytocin. (Reprinted from Ref. 13, with permission.)

at pH 2.5. The electrophoretic data were used to test existing theoretical models that correlate electrophoretic mobility with physical parameters. The best fit of the experimental data was obtained with the Offord model that correlates electrophoretic mobility with the charge-to-size parameter $q/M^{2/3}$.

5.1.2.3 Separation Modes

As HPLC has different separation modes of chromatography, including adsorption, partition (normal and reversed phase), ion-exchange, size-exclusion, and affinity, CE also has various separation modes, e.g., capillary zone electrophoresis (CZE), capillary gel electrophoresis (CGE), capillary isoelectric focusing (CIEF), micellar electrokinetic capillary chromatography (MEKC), and capillary electrochromatography (CEC) (Table 5.1-1). Moreover, these different CE separation modes can be used to complement each other, and thus, they greatly enhance the versatility of the technique. In many cases, it seems that no single method can separate peptides and proteins because these biomolecules are diverse and complex in terms of structure and composition. Therefore, different optimal strategies involving different CE separation modes are often used to solve the separation problems.

Capillary Zone Electrophoresis (CZE). CZE is a widely used CE technique and separates peptides and proteins based on differences in their charge-to-mass ratios. Separations occur in a capillary filled with a buffer of constant composition. For CZE, the run buffer choice is extremely important because it determines the charge on the analyte molecule and its migration rate. Thus, the type of buffer, its ionic strength, and its pH are optimized for particular separation problems. Buffers based on sodium phosphate, citrate, acetate, or combinations thereof with concentrations ranging from 10 to 200mM are frequently used [14].

TABLE 5.1-1. Separation Modes of CE for the Characterization of Peptides and Proteins

Technique	Separation Mechanism	Applications	Conventional Technique
CZE	Charge-to-mass ratios	Peptide mapping, purity and stability of peptides and proteins, and glycoprotein heterogeneity	—
SDS–CGE	Molecular size	Estimation of relative molecular mass of protein (10–200kDa) and purity and stability of peptides and proteins	SDS–PAGE
CIEF	Isoelectric point (p*I*)	p*I* determination of proteins, purity of proteins, and glycoprotein heterogeneity	Gel IEF
MEKC	Hydrophobicity	Purity of peptides and glycoprotein heterogeneity	RP–HPLC
CEC	Hydrophobicity	Peptide mapping and purity of peptides	RP–HPLC

Proteins contain both positively and negatively charged functional groups. The positively charged moieties in proteins, such as the guanidinium group of arginine residues, the amino groups of lysine residues and N-termini, and histidine residues, interact with negatively charged silanol groups on capillary walls at pH values above 3. These interactions between proteins and capillary walls (protein adsorption) lead to sample loss, peak broadening, poor resolution, and longer migration times.

Many strategies devised to overcome protein–wall interactions have focused either on selecting separation buffer conditions that reduce protein binding sites or on treating capillary surfaces to reduce interaction sites [15]. By using separation buffers with extreme pH values (e.g., below pH 2~3 or above pH 10), protein adsorption can sometimes be reduced. At pH values below around 2, the silanol groups of the capillary are fully protonated and surface charge approaches zero, whereas at pH values above 10, capillary surfaces are completely deprotonated and highly anionic. In these situations, protein adsorption is significantly reduced because of electrostatic repulsion. However, operations at extreme pH conditions can be problematic in finding suitable separation conditions due to reduction of differences of analytes in effective electrophoretic mobility. At low pH, EOF is negligible, making separations slow, and at high pH, EOF is generally high and resolutions are reduced. Additional problems encountered under extreme pH conditions include the effect on protein conformational changes. Unfolding or aggregation of proteins may occur, resulting in irreproducible data, often reduced efficiency, and a loss of biological activity. Therefore, the use of this approach is limited to a few selected applications.

To optimize separation conditions at moderate pH conditions, surface modifications of capillary can be helpful [16]. Such modifications can decrease protein–surface interactions by reducing the surface charge of the silica material. Several approaches have been proposed for modifying the inner walls of capillaries, such as chemical coating with bonded/cross-linked polymers, physical adsorption coating with cationic polymers, and dynamic coating with adsorbed surfactants. The most popular coating method is to covalently bind polymers such as polyacrylamide or poly(vinyl alcohol) to capillary walls [17]. Chemically coated capillaries are widely used and are available commercially. The preparations of this capillary type often involve several chemical reactions that may cause large variances between capillaries. In addition, these coatings are usually stable in restricted pH ranges and have limited lifetimes. Noncovalent capillary coatings are simply formed by treating the capillary with one or more charged polymers that are physically adsorbed onto internal walls. Initially, positively charged monolayer coatings of polybrene [18], polyethyleneimine [19], or poly(diallyldimethylammonium chloride) [20] were used. More recently, noncovalently bilayer-coated capillaries prepared by add anionic polymer to cationic polymer-coated capillary have been reported to be long-lived and to have an EOF that is stable over wide pH ranges (2–11) [21]. Katayama et al. reported on the use of a polymeric bilayer of polybrene and dextran sulfate for the separation of some model proteins [22]. Catai et al. demonstrated the use of capillaries coated with a bilayer of polybrene and poly(vinyl sulfonate) for the fast, highly reproducible, and efficient analysis of peptides and proteins [23, 24]. Another way of coating capillaries is to add the coating agents (usually low-molecular-weight compounds such as amines and surfactants) to the separation buffer for the dynamic coating of the inner capillary surface [25].

Sodium Dodecyl Sulfate-Capillary Gel Electrophoresis (SDS–CGE). SDS–CGE is a capillary-based version of SDS–polyacrylamide gel electrophoresis (SDS–PAGE) in the slab gel format, with advantages of shorter analysis times, ease of automation, and online detection and quantitation [26, 27]. In SDS–CGE, replaceable sieving polymers, such as linear polyacrylamide, poly(ethylene oxide), dextran, or pullulan, are used to achieve reproducible separations. These polymers permit the replacement of a separation matrix for each sample, thereby eliminating cross-contamination between samples and improving reproducibility. Best results are often obtained using chemically or dynamically coated capillaries.

For SDS–CGE, proteins are denatured and complexed with SDS before analysis. These SDS–protein complexes are then separated based on their sizes. The molecular mass of an unknown protein can be estimated by running protein standards as well. This assumes that migration is dependent on relative mass, on the basis of the following two assumptions. First, all SDS–protein complexes have the same charge-to-mass ratio in the presence of excess SDS (~1.4 g of SDS/1 g of protein). Second, the SDS–protein complexes have similar shapes, and thus, their sizes are linearly related to their molecular masses. However, these assumptions are not valid for some classes of proteins, e.g., basic proteins, hydrophobic membrane proteins, and glycoproteins, because deviations from predicted charge-to-mass ratio are often observed for their SDS–protein complexes. For example, basic proteins have lower charge-to-mass ratios because of the presence of positively charged amino acids, whereas hydrophobic membrane proteins have larger charge-to-mass ratios. Moreover, in the case of glycoproteins, the carbohydrate moieties do not bind with SDS, and this lowers the charge-to-mass ratio, leading to a decreased migration and overestimation of molecular mass [3, 28]. Figure 5.1-3 shows the SDS–CGE result for ricin toxin purified from the seeds of the castor bean (*Ricinus communis*) [29], in which two peaks of ricin toxin were partially separated, while only a single band appeared in the SDS–PAGE gel. Two peaks were identified to be ricin glycoform isomers with identical protein sequences but with different carbohydrate contents. When the molecular masses of two peaks were determined with a calibration curve made by protein standards (MW 14–200 kDa), they were significantly higher than those measured by matrix-assisted laser desorption/ionization time-of-flight mass spectrometry (MALDI–TOF MS). This shows the slower electrophoretic migration of glycoproteins in SDS-based gel electrophoresis compared with that of standard proteins. However, the presence of carbohydrate may improve the resolution between glycoforms in SDS–CGE [29].

In addition to the estimation of protein molecular mass, SDS–CGE is also the common choice for checking the presence of other proteins, protein aggregates, or protein degradation products during characterization, purity, or stability studies.

Capillary Isoelectric Focusing (CIEF). CIEF separates peptides and proteins according to isoelectric point (pI) differences. Proteins that differ by 0.005 pI units or even less have successfully been separated by CIEF. The sample is normally mixed with ampholytes (zwitterionic compounds), which have slightly different pI values spanning the desired pH range and act as a strong buffer at their pI values and then the capillary is filled with this mixture. The capillary inlet is placed in a vial containing acidic solution (anolyte) and the capillary outlet in a basic solution (catholyte), before applying an electric field. A pH gradient is quickly formed by

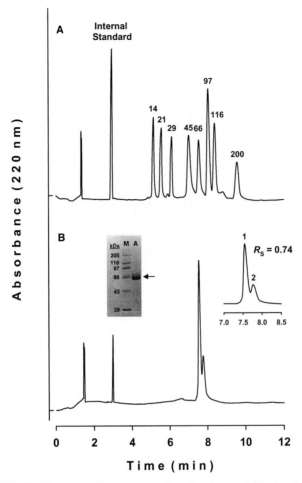

Figure 5.1-3. SDS–CGE electropherograms of molecular weight standard marker (14–200 kDa) (A) and ricin toxin glycoprotein under nonreducing conditions (B). The right insert is an enlargement of ricin peaks in SDS–CGE, presenting the partially separated two peaks ($R_s = 0.74$) compared with only a single band shown in the SDS–PAGE gel. (Reprinted from Ref. 29, with permission.)

the ampholytes ranging from low to high pH along the entire capillary length. According to their electrophoretic mobilities, proteins migrate to their pIs in the capillary where they are focused into very sharp zones. The overall focusing process can be monitored by observing current changes, which approach zero when ion movement inside a capillary stops. The CIEF process can be performed in one or two steps. One-step CIEF is performed in an uncoated capillary and usually employs polymers such as hydroxypropyl methyl cellulose or hydroxyl ethyl cellulose in order to reduce but not eliminate EOF [30]. At this point, either migration past a detection window or whole-column imaging is needed to visualize the separation [31]. In two-step CIEF, proteins are first focused in a coated capillary to eliminate EOF and then mobilized by adding a salt (sodium chloride) to the catholyte or by applying pressure to the capillary inlet (anolyte) [32, 33]. In general, the two-step

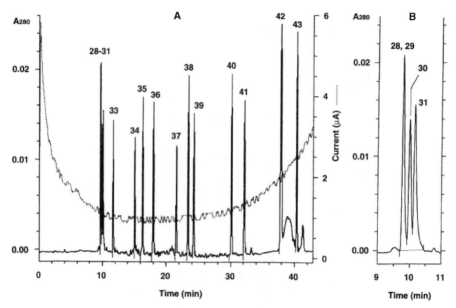

Figure 5.1-4. CIEF separation of the peptide p*I* markers ranging from pH 3.38 (peptide #43) to 10.17 (peptide #28). (B) Enlargement of the portion from 9 to 11 min of (A). (Reprinted from Ref. 35, with permission.)

method is preferred because it provides a much higher reproducibility [34]. As CIEF is usually run from positive to negative polarity, the most basic proteins pass the detector first. Absorbance must be monitored at UV 280 nm to avoid the strong absorption of ampholytes at lower wavelengths. The use of a coated capillary is required for high resolution because EOF interferes with maintenance of focused zones.

The estimation of p*I* in CIEF needs appropriate p*I* markers. Synthetic peptide and oligopeptide p*I* markers for CIEF with UV absorption detection have been developed [34]. Shimura et al. proposed a set of 16 synthetic oligopeptides as p*I* markers for CIEF that are fully compatible with UV detection [35]. Figure 5.1-4A shows the CIEF separation of 16 p*I* markers ranging from pH 3.38 (peptide number 43) to 10.17 (peptide number 28). Figure 5.1-4B is an enlargement of the region between 9 and 11 min, showing partial resolution of high p*I* markers and the overlap between peptide numbers 28 and 29 due to a lack of extension of the pH gradient toward more alkaline pH values. CIEF produces the best resolution for proteins and peptides among the CE separation modes, and it is a powerful tool for resolving modified proteins, for characterizing microheterogeneity, and for identifying glycoforms [36]. CIEF has also been found to be a powerful tool in proteomics studies [37].

Micellar Electrokinetic Capillary Chromatography (MEKC). MEKC separation is based on the partition of analytes between micelles and the surrounding aqueous phase [38]. This technique can be considered a type of chromatography where the stationary phase is essentially mobile and the mobile phase is electro-osmotically pumped. The micellar phase is composed of a surfactant added to the buffer above its critical micellar concentration (CMC). The most commonly used surfactants are SDS, bile salts, and hydrophobic chain quaternary ammonium salts. In general, neutral or alkaline buffer solutions are used to create a strong EOF that moves even anionic micelles in the capillary toward the cathode.

MEKC has been applied with great success to the analysis of a variety of small molecules. However, relatively few protein applications have been found for MEKC [39, 40], which may be because most proteins (>MW 5000) are too large to partition into the hydrophobic core of micelles. However, proteins can associate with micelles through hydrophobic, hydrophilic, and electrostatic mechanisms, and these interactions have been exploited to manipulate protein separation using MEKC [41]. By manipulating these protein–micelle interactions using the variables, such as surfactant concentration, pH, ionic strength, and the addition of an organic modifier, MEKC has been demonstrated to resolve proteins with minor structural variations and to allow the quantitative analysis of proteins present in complex matrices [3].

Capillary Electrochromatography (CEC). Capillary electrochromatography (CEC) is a hybrid technique of CE and HPLC that is generally carried out using packed capillary columns by the electro-osmostically driven mobile phase at high electric field strength. CEC has been rapidly developed in recent years as it has combined advantages of both techniques, i.e., the high selectivity of HPLC and the high efficiency of CE, with minimum consumption of both reagents and samples, and good compatibility with mass spectrometry [42]. The separation mechanism involved is based on chromatographic retention, and for charged analytes, on a combination of this and electrophoretic mobility. Initially, CEC was mainly used to separate neutral compounds such as polyaromatic hydrocarbons. Recently, the successful development of CEC column technology and improvements in instrumentation have encouraged research on CEC for peptide and protein separations [43]. Various capillary columns have been used for peptide and protein analysis, including packed capillaries, open-tubular capillaries, and monolithic columns.

In addition to the use of octadecylsilane (ODS) as a stationary phase, sulfonated poly(styrene-divinylbenzene), poly(2-sulfoethylaspartamide)-silica, open tubular silica, porous styrenic sorbents and octadecylsilica, pentofluorophenylsilica, trycontysilica, octadecylsilica, acrylate-based porous monoliths, and wide-pore stationary phases have been recently investigated [44, 45]. CEC separations using monolithic columns are being actively investigated because the stationary phases are easily prepared and supporting frits are unnecessary. In monolithic columns, the stationary phase is covalently bound to inner capillary walls. CEC separations of peptides and proteins in monolithic columns are usually performed in counter-directional mode; i.e., peptides and proteins migrate electrophoretically in a direction opposite to the EOF [46, 47].

Recently, a great deal of interest has been shown in the combination of CEC and mass spectrometry (MS) because CEC is considered to overcome many of the limitations of other CE techniques [48–50]. When combined with electrospray ionization (ESI)–MS, CZE is limited to a small number of buffer systems and suffers from low sample loading capacity. MEKC has relatively poor selectivity and difficult compatibility with MS because it requires the use of high surfactant concentrations. In contrast, CEC has good compatibility with MS, selectivity, sample loading capacity, and general applicability.

5.1.2.4   Detection

In the CE system, peptides and proteins are typically detected by UV absorbance, laser-induced fluorescence (LIF), or MS. UV absorbance is the most widely used detection method, but when higher sensitivity is required, LIF or MS are preferred. Recently, MS is being increasingly used to detect peptides and proteins separated by CE with the improvement of proteomics research.

UV Absorbance. UV detection of peptides and proteins is mostly performed at 200–220 nm, where the absorption is proportional to the number of peptide bonds, but sometimes around 254 or 280 nm, where the detection is based on the absorbance of the aromatic residues such as tryptophan, tyrosine, and phenylalanine. This detection method is most commonly used, but it has low sensitivity, which is a major disadvantage for the detection of analytes present at low concentrations. In CZE, UV detection is limited to micromolar or submicromolar concentrations for peptides and proteins. Several capillaries have been designed to improve the sensitivity in CE: (1) a rectangular capillary extended in the direction of the light path [51], (2) a Z-shaped capillary [52], and (3) a bubble cell capillary with an locally enlarged diameter in the detection region [53]. These capillaries help somewhat to improve sensitivity, but practical difficulties of availability and implementation into existing instruments remain.

Laser-induced Fluorescence (LIF). Fluorescence is inherently a sensitive detection method with lower detection limits than UV absorbance detection. In addition, fluorescence detectors are selective because only fluorescing molecules are detected. Typical detection limits of fluorescence detection lie in the range 10^{-7}–10^{-9} M. The light sources used are usually deuterium, tungsten, or xenon lamps. Lasers are also good sources of high-intensity radiation and are used for LIF detection. LIF is the most sensitive detection method in CE. Detection limits of LIF have been reported to fall in the range 10^{-18}–10^{-21} M [54–56]. A variety of lasers have been employed for the detection of peptides and proteins, which include Nd:YAG, argon ion, helium–cadmium, and helium–neon lasers. As proteins containing aromatic amino acid residues (tryptophan, tyrosine, and phenylalanine) have native fluorescence around 310 nm, the intrinsic fluorescence has been directly measured using Nd: YAG at 266 nm or helium-cadmium at 320 nm as a laser source [57, 58]. However, these systems are expensive and the helium–cadmium lasers have a problem of short lifetimes [59]. The most popular lasers for the detection of peptides and proteins are argon lasers at 488 and 514 nm and helium–neon lasers at 544, 593, and 633 nm that are relatively inexpensive, stable, and compact. However, they are often not suitable for exciting most proteins to induce native fluorescence. To overcome this disadvantage, the labeling of the molecules with a suitable fluorescent dye is required via pre-column, on-column, or post-column. Several fluorescence-labeling reagents are commercially available and conveniently used for peptides and proteins. Rhodamine, solvatochromic dyes (SYPRO Red, Nile Red), and

Albumin Blue 580 have been used to form stable and highly fluorescent complexes with proteins [59].

Mass Spectrometry (MS). MS is an important and powerful detection tool for the characterization of peptides and proteins by CE, as evidenced by recent developments of CE–MS methodology and its applications [60, 61]. MS detection considerably enhances the utility of CE by providing information about the identity of the separated molecules. The availability of CE–MS is also highly desirable for purity and stability studies of peptide and protein drugs. Especially, in proteomics, peptidomics, and peptide mapping, the importance of both on- and offline coupling of MS with CE separations of peptides and proteins is greatly growing [62].

The introduction of soft ionization techniques such as ESI and MALDI brought tremendous progress in on- and offline characterization of electrophoretically separated peptides and proteins by MS [50]. Combination of CE with MS techniques allows not only high-accuracy molecular mass determination of peptides and proteins separated by CE, but also it provides important structural data on amino acid sequence, the sites of posttranslational modifications, peptide mapping, and the noncovalent interactions of peptides and proteins.

ESI forms gaseous ions by applying a strong electric field to a fine spray of their liquid solutions. It allows the generation of multiply charged ions, which enables the detection of very large molecules even with instruments having a low mass range. The multiply charged peaks can be transformed into a singly charged peak by mathematical deconvolution, which enables the determination of the molecular weight of the original species. ESI is the preferred mode for online coupling CE with MS because molecules can be transferred directly from the capillary to the mass spectrometer via an interface. Three types of interfaces, i.e., sheathless, liquid–junction, and coaxial liquid sheath-flow, have been constructed for CE–ESI–MS coupling [63].

In MALDI, the analyte is cocrystalized with the matrix (small organic compounds having strong absorbance in the laser wavelength) and a laser beam is directed onto this crystal, causing vaporization of the matrix and desorption of the ions. The main advantage of MALDI is the generation of predominantly single-charged molecular ions of even macromolecules with a molecular mass up to 300 kDa with the exact determination of molecular mass with accuracy of ±0.1%. Therefore, the MALDI–MS is a more attractive option for the characterization of heterogeneous samples [7]. The MALDI–MS is combined with CE separations of peptides and proteins more in an offline mode than with an online liquid sample delivery connection [64]. In offline mode, fractions separated from the CE are collected and deposited on a MALDI target in the form of spots for MS analysis. Recently, approaches for the online coupling of CE with MALDI–MS have been developed to minimize sample handling and potential losses [65, 66].

5.1.3 APPLICATIONS

In pharmaceutical biotechnology, the development of analytical procedures to validate identity, strength, quality, and purity of biopharmaceutical products is one of the most important issues for the production of highly specialized biotechnology products, as described in the U.S. Food and Drug Administration's (FDA's) current Good Manufacturing Procedures (cGMP) requirements for drugs (Title 21 of the Code of Federal Regulations, Parts 210 and 211). In this chapter, we discuss the use of CE to assess the identity, purity, heterogeneity, and stability of peptides and proteins, and its applications of CE in proteomics research.

5.1.3.1   Identity Determination

The identity of peptides and proteins can be determined by specific activity assays, determination of amino acid composition and sequence, and assessment of such physico-chemical parameters as molecular mass and pI. Several CE techniques have been used for the identity of peptides and proteins, which include peptide mapping by CZE or CEC, CIEF for the determination of protein's pI, SDS–CGE for the determination of relative molecular masses of proteins, and CE–MS for direct molecular mass assignment of peaks separated by CE.

Traditionally, IEF and SDS–PAGE in slab gel systems have been routinely used to confirm the identity of the proteins. CE formats, CIEF and SDS–CGE, offer fast and reproducible separations and direct online detection without the need of staining and destaining procedures used in slab gel techniques. As a typical example, Hunt et al. showed CIEF and SDS–CGE methods for the qualitative analysis of recombinant humanized monoclonal antibody HER2 (rhuMAbHER2) [67]. The CIEF separated five charged isoforms of rhuMAbHER2 with estimated pI values in the range of 8.6–9.1. These results agreed well with the pI values determined on slab gel IEF. In SDS–CGE, the expected molecular masses of intact rhuMAbHER2 under nonreduced conditions and its heavy-chain and light-chain fragments under reduced conditions were identified with good correlation with SDS–PAGE analysis. This study demonstrated the feasibility of replacing the slab gel techniques with CE methods in a quality control environment.

CE–MS. In SDS–PAGE, the molecular mass of proteins is estimated based on comparison with reference proteins with relatively poor accuracy of ±5–10% [68]. When proteins are extracted from slab gels and subsequently transferred to ESI–MS or MALDI–MS, the mass accuracy somewhat improves, but the methodology still suffers from the interferences originated from gel [50]. Frequently, accurate molecular mass determination is necessary especially for the identification of unambiguous peptides and proteins. Since its introduction in 1987 by Olivares et al. [69], CE coupled to MS has gained increasing attention in peptide and protein research because of the CE ability to separate analyte from complex mixtures with high efficiency and minimal sample consumption. In CZE–ESI–MS, mass accuracies down to 0.0008% have been realized for model mixtures of proteins [70].

Tsuji et al. reported CZE–ESI–MS for the analysis of recombinant bovine and porcine somatotropins (rbSt and rpSt) [71]. The average molecular masses of rbSt and rpSt were determined to be 21,812.6 and 21,798.3, which were nearly identical to the theoretical values of 21,812.0 and 21,797.9, respectively. Yeung et al. reported on the CE–ESI–MS method for the analysis of high-mannose glycoproteins, ribo-nuclease B, and recombinant human bone morphogenetic protein-2 [72]. CE separations were performed with 50-mM beta-alanine buffer (pH 3.5) in a polyacrylamide-coated capillary and a coaxial sheath–liquid interface was used for CE–MS coupling. The identities of glycoform peaks separated by CE were determined by the combination of UV and MS detection data without the need of oligosaccharide release or derivatization treatments.

Na et al. reported on an offline combination of CZE and MALDI–TOF MS for the identification of salmon calcitonin (sCT) acylation products formed in the degrading poly(lactic-co-glycolic acid) (PLGA) microsphere formulations [64]. As shown in Figure 5.1-5, the peptides extracted from sCT microspheres incubated in drug release medium were analyzed by both CZE with UV detection (200 nm) and MALDI–TOF MS. After the incubation, the additional peak was observed in CE (Figure 5.1-5c) and mass peaks of m/z 3491.15 and 3549.16 except the intact sCT were presented in MALDI–TOF MS spectrum (Figure 5.1-5d). The identities of two peaks separated in CE (Figure 5.1-5c) were confirmed by reanalysis of each peak fraction using MALDI–TOF MS, which was also used for the quantitation of peptide loaded into microsphere formulations [73].

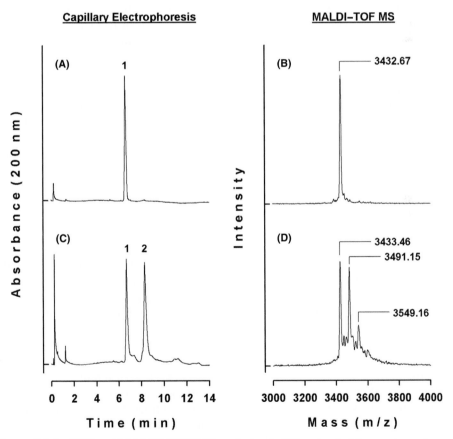

Figure 5.1-5. CZE and MALDI–TOF MS analysis of sCTs extracted from microsphere at 0 day (A and B) and 21 days (C and D) of incubation in release medium at 37°C. In the CE electropherogram, peak 1 is native sCT and 2 represents the degradation products. (Reprinted from Ref. 64, with permission.)

Recently, a variety of peptides and proteins were analyzed by CE–MS techniques, which include CZE–MALDI–TOF–MS, CZE–ESI–MS, CIEF–ESI–MS, and CEC–ESI–MS. Their applications are well summarized in a recent review paper [50].

Peptide Mapping. Peptide mapping is an important tool for protein identif-ication, primary structure determination, the detection of posttranslational modifications, the identification of genetic variants, and the determination of glycosylation and/or disulfide sites. For these reasons, peptide mapping is widely used for quality control and for the characterization of recombinant DNA-derived products. Moreover, the high resolution of CE makes it a powerful peptide mapping technique.

Rush et al. (1993) reported on the CE-based peptide mapping of recombinant human erythropoietin (EPO) derived from Chinese hamster ovary (CHO) cells. Using 100-mM heptanesulfonic acid in 40-mM sodium phosphate buffer (pH 2.5) as a separation buffer, this technique showed a baseline separation of 16 tryptic-digested peptides with one partial separation of two peptides [74]. This peptide map demonstrated the structural differences between the glycosylated form expressed in CHO cells and the non-glycosylated form expressed in *E. coli*, and it provided information on the heterogeneity in glycoforms of EPO. Zhou et al. presented the results of the CE–ESI–MS analysis of tryptic digests of EPO, and they found that the technique is complementary to HPLC–ESI–MS [75]. Boss et al. reported on an evaluation of CE–MS for the peptide mapping of EPO [76]. In this study, tryptic-digested peptides were first separated by reversed-phase (RP)–HPLC and collected fractions were analyzed by offline and online CZE–MS experiments employing a capillary coated with Polybrene in the presence of polyethylene glycol and 0.67-M formic acid as a separation buffer.

Despite the impressive ability of CE to separate peptides, a complete separation of all digested peptides is unlikely to be achieved because of the highly complex natures of the peptide maps of large proteins. Recently, multidimensional separations, such as two-dimensional (2D) electrophoresis, HPLC–CZE, HPLC–MS, CZE–MS, and HPLC–CZE–MS, have been applied for complete resolution of highly complex peptide mixtures [77]. A sequential offline combination of RP–HPLC and CZE was used for the peptide mapping of pepsin isoenzymes, recombinant human tissue plasminogen activator, cytochrome c, and myoglobulin [78–80]. Kang et al. presented unique "fingerprint" peptide maps of two closely related proteins, β-lactoglobulins A and B, by combining CZE separations performed in four different channels and MEKC separations performed in two different channels in a 96-capillary array [81]. He et al. developed a novel multiplex CE system for the high-throughput comprehensive peptide mapping of proteins [82]. By using multiple separation conditions using six CZE buffers and two MEKC buffers in a 20-capillary array, the peptide fragments of proteins digested by three different enzymes were readily resolved and showed unique fingerprints. These 20 capillaries were monitored simultaneously at 214 nm using a single photodiode array detector, and the overall analysis time from reaction to detection was about 40 min.

5.1.3.2 Purity Control

Purity control is necessary on peptide and protein products for process and quality control purposes [83–85]. Minor amounts of impurities and degradation products must be determined in the presence of much larger quantities of primary components. These components may be structurally very similar to the main component. CE can be applied as a sensitive control method for such determinations because it provides rapid and accurate qualitative and quantitative data on peptide and protein preparations.

Peptide Purity. Peptide purity checks have been routinely performed by RP–HPLC. Recently, CE has increasingly been used as a versatile technique for the analysis of peptide drugs in the pharmaceutical industry [84]. Because of the different separation mechanisms, CZE is recognized as an excellent complementary tool to RP–HPLC. CZE separates the peptide based on differences in mass-to-charge ratios, whereas RP–HPLC separation is per-formed based on hydrophobicity differences. Peptides with similar hydro-phobicities, which are difficult to be separated by RP–HPLC, can often be resolved based on charge-to-mass ratio differences. Therefore, peaks that appear pure by RP–HPLC are often resolved in multiple peaks by CZE. Moreover, CZE requires only nanoliter quantities of sample so that almost all of the sample can be used for subsequent sequence analysis.

Ridge and Hettiarachchi reported on CZE separations of bradykinin and its impurities using phosphate buffers at various pH values (pH 2.5, 3.5, and 4.5), which were not fully resolved by HPLC [86]. When buffer pH was increased from 2.5 to 4.5, several impurities were well separated from the major peak of bradykinin. In purity checks of peptide, the similar superiority of CZE over RP–HPLC is also found in publications by Chen et al. [87], Hettiarachchi et al. [88], and Moumakwa et al. [89].

Protein Purity. During the production of recombinant proteins, process monitoring is required to assure purity levels required in every step. The recovery and purification of product from fermentation broth typically involve various procedures, such as filtration, centrifugation, and chromatography. After each purification and final step, constituent levels must be determined to ensure that the desired levels of purity have been achieved. In addition to its control function, this purity information is also frequently used to further optimize purification processes. CZE and SDS–CGE can be mostly used for the purity checks of protein products.

Several CZE methods were reported for the purity determinations of recombinant human insulin. Human insulin consists of two peptide chains, A (21 amino acids) and B (30 amino acids), which are connected by two disulfide bonds. To determine the purity of insulin, two main degradation products, acidic and neutral desamido-insulin (desamidated at positions A-21 or A-21 and B-3, respectively), should be separated from the main compound [90]. According to U.S. Pharmacopoeia (USP), the relative amount of desamido insulin must not exceed 3% of the total amount of insulin and desamido insulin [91].

A CZE method for the separation of insulin and its deamidation products with untreated fused-silica capillaries using a run buffer containing 2-(*N*-cyclohexylamino) ethanesulfonic acid (CHES), triethylamine, and 10% acetonitrile has been reported [90]. This system separated acidic and neutral desamido-insulin in formulated human insulin with the relative standard deviation (RSD) of the migration times ≤1%, whereas RP–HPLC coeluted the neutral desamido-insulin with insulin. Sergeev et al. described an analytical scheme for monitoring recombinant human insulin during the various steps of its production [92]. The CZE method has been included, together with narrow-bore HPLC and MALDI–TOF MS, in the analytical scheme for monitoring the production of recombinant human insulin. Combinations of these complementary techniques allowed us to obtain unambiguous information about the purity and primary structure of all intermediates of recombinant human insulin production, and they enabled the optimization of some process parameters. Figure 5.1-6 shows the result of a CZE analysis performed to check the purity of insulin purified by ion-exchange chromatography. This CZE analysis showed the presence of several impurities from the preparation (Arg-insulin) and degradation (unidentified desamido-insulins) in the isolated insulin.

SDS–CGE provides an overall composition of a protein sample based on the molecular mass, even if it is less suited for the analysis of subtle changes in a protein. Thus, SDS–CGE is mainly used to check for the presence of other proteins, protein aggregates, or protein degradation fragments in the purity checks.

Hunt and Nashabeh showed the SDS–CGE method with LIF detection for the analysis of humanized recombinant IgG_1-monoclonal antibody (MAb) [93]. For LIF detection, the MAb was first derivatized with a 5-carboxytetramethylrhoda-

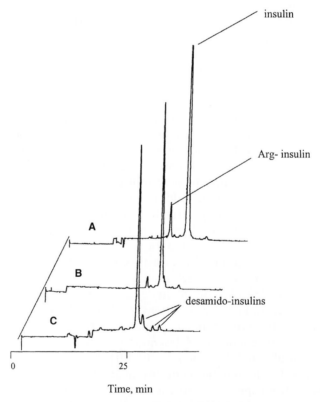

Figure 5.1-6. CZE analysis for insulin purity checks in three fractions obtained by ion-exchange chromatography. This CZE analysis demonstrated the presence of several impurities (Arg-insulin and desamido-insulins) in the isolated fractions. (Reprinted from Ref. 92, with permission.)

mine succinimidyl ester. The derivatized sample was then incubated with SDS, and the SDS–MAb complexes were separated by SDS–CGE using a hydrophilic polymer as a sieving matrix. The capabilities of SDS–CGE using UV detection or LIF detection after derivatization were compared. Figure 5.1-7 shows the results of the SDS–CGE analysis by UV detection at 220 nm and LIF detection using argon

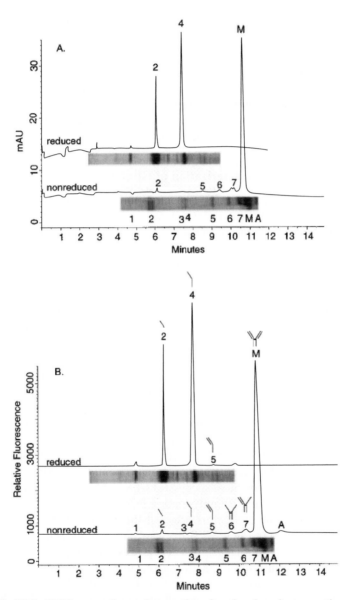

Figure 5.1-7. SDS–CGE separations of nonreduced and reduced preparations of a therapeutic recombinant MAb. Insets show silver-stained SDS–PAGE traces of the same sample preparations (M: monomer, A: aggregate). (A) UV detection of unlabeled samples at 220 nm. (B) LIF detection of labeled samples at 488-nm excitation and 560-nm emission. (Reprinted from Ref. 93, with permission.)

laser with excitation wavelength of 488 nm and emission wavelength of 560 nm. Under reducing conditions, the light and heavy chains originating from the original MAb were well separated, and the results of SDS–CGE under both reducing and nonreducing conditions generally showed good agreement with band profiles obtained by silver-stained SDS–PAGE. SDS–CGE using LIF detection allowed the detection of MAbs at a low-nanomolar concentration (~9 ng/mL), which was comparable with silver-stained SDS–PAGE and was a 140-fold increase over SDS–CGE using UV detection. This improved sensitivity also allowed the detection of low-level impurity peaks, which were not detected by UV detection. Repeated analysis showed that RSD values for the migration time were below 1%, and the RSD of the area of the main peak was lower than 0.6%. These applications demonstrate the usefulness of SDS–CGE–LIF for detecting manufacturing inconsistencies in recombinant protein production.

Several validated CZE and SDS–CGE methods for the purity analysis of peptides and proteins are provided in Table 5.1-2.

5.1.3.3   Heterogeneity Characterization of Glycoproteins

Glycoproteins produced from mammalian expression systems consist of a population of glycosylated variants (glycoforms). Glycosylation is one type of posttranslational modification that requires monitoring because variation in the carbohydrate composition may significantly alter the properties of a protein such as biological activity, pharmacokinetic properties, solubility, and stability. Several CE methods have been used for analyzing the intact glycoproteins and monitoring their production [3, 101]. Because of simplicity and high resolving capacity based on the charge-to-mass ratio, CZE has been routinely used for characterizing the heterogeneity of glycoproteins. In addition, CIEF has been increasingly used for separation of the glycoforms.

Erythropoietin (EPO). EPO is a glycoprotein produced primarily by the kidney and the main regulator of red blood cell production [102]. Human EPO consists of 165 amino acids that are heavily glycosylated (one O-linked and three N-linked carbohydrate chains corresponding to 40% of the molecular mass) [103]. Several glycoforms of human EPO exist with different degrees of glycosylation and sialic acid residue numbers [104]. Several reports are available on the application of CE for the separation of EPO glycoforms, involving different CE modes [105]. Among them, CZE and CIEF are the most successful CE modes in this respect. Table 5.1-3 provides an overview of the CE analyses of EPO glycoforms.

A CZE method for the characterization of the glycoform patterns of pharmaceutical EPO is included in the 2002 European Pharmacopoeia, as a substitute for the conventional isoelectric focusing test [113]. As shown in Figure 5.1-8, EPO was resolved with high resolution into its glycoforms using a separation buffer containing putrescine (1,4-diaminobutane) and urea [110]. Individual peaks correspond to multiple glycoforms with similar overall mass-to-charge ratios [3, 109, 114]. At pH 5.5, sialic acid residues are negatively charged and migrate against the EOF, and thus the glycoforms are considered to elute in the order of increasing number of sialic acid residues [109, 114]. The addition of putrescine to the separation buffer leads to a reduction in EOF and solute interaction with the capillary wall. Urea is

TABLE 5.1-2. CE Applications for the Purity Checks of Peptides and Proteins

Peptides and Proteins	CE Modes	Separation Buffers	Capillary	Detection	Ref.
Bradykinin	CZE	50-mM phosphate buffer, pH 2.5, 3.5, 4.5	Polyacrylamide-coated	UV (215 nm)	[86]
Biphalin	CZE	50-mM phosphate buffer, pH 2.5	Polyacrylamide-coated	UV (215 nm)	[88]
Enkephalin and dalagin analogs	CZE	(1) 100-mM phosphoric acid, 50-mM Tris, pH 2.25 (2) 100-mM iminodiacetic acid, pH 2.30 (3) 40-mM Tris, 40-mM tricine, pH 8.10	Uncoated	UV (206 nm)	[94]
Placental alkaline phosphatase	CZE	50-mM boric acid, 2-mM putrescine (pH 8.5)	Uncoated	UV (200 nm)	[95]
Placental alkaline phosphatase	SDS–CGE	CE–SDS protein kit (Bio-Rad)	Uncoated	UV (220 nm)	[95]
Protegrin IB-367	CZE	100-mM phosphate buffer, pH 2.6	Uncoated	UV (200 nm)	[87]
Recombinant acidic fibroblast growth factor	CZE	50-mM phosphate buffer (pH 2.5) with 0.25% hydroxypropyl-methylcellulose (HPMC)	Uncoated	UV (214 nm)	[96]
Recombinant hirudin	CZE	50-mM sodium phosphate, 40-mM aminoethanesulfonic acid, 25-mM hexanesulfonic acid, 10-mM sodium borate, 2-mM tetramethylammonium sulfate, 1-mM diaminopentane (pH 7)	Uncoated	UV (200 nm)	[97]

TABLE 5.1-2. *Continued*

Peptides and Proteins	CE Modes	Separation Buffers	Capillary	Detection	Ref.
Recombinant human insulin	CZE	50-mM acetate, 850-mM CHES, 10% acetonitrile (pH 7.8)	Uncoated	UV (214 nm)	[90]
Recombinant human insulin	CZE	100-mM sodium borate (pH 9.3)	Uncoated	UV (214 nm)	[98]
Recombinant human insulin	CZE	0.1-M tricine, 0.02-M NaCl, 1-M urea, 0.45-M ethylene glycol (pH 8.1)	Uncoated	UV (214 nm)	[91]
Recombinant human insulin	CZE	50-mM sodium phosphate (pH 10.6); 100-mM sodium phosphate-30% MeOH (pH 9)	Uncoated	UV (210 nm)	[92]
Recombinant human interleukin-6	ITP–CZE	20-mM ammonium acetate (pH 4.2)	Uncoated	UV (200 nm) ESI–MS	[99]
Recombinant monoclonal antibodies	SDS–CGE	CE–SDS protein kit (Bio-Rad)	Uncoated	UV (220 nm) LIF (Ex 488 nm, Em 560 nm)	[93]
Recombinant NADP$^+$-dependent formate dehydrogenase	SDS–CGE	eCAP SDS 14-200 kit (Beckman)	Uncoated	UV (214 nm)	[100]
Synthetic porcine secretin	CZE	50-mM phosphate buffer (pH 7.4 or 3.5)	Uncoated	UV (210 nm)	[89]

TABLE 5.1-3. CE Applications for the Heterogeneity Characterization of Recombinant Erythropoietin

Objective	CE Modes	Separation Buffers	Capillary	Detection	Ref.
Separation in drug formulations	CZE	200-mM sodium phosphate, 1-mM nickel chloride (pH 4.0)	eCAP amine (Beckman)	UV (200 nm)	[106]
Separation of glycoforms	CZE	100-mM acetate-phosphate (pH 4.0)	Uncoated	UV (214 nm)	[75]
Separation of glycoforms	CZE	10-mM sodium acetate, 0.5% HPMC (pH 5.7)	DB-1	UV (214 nm)	[107]
Discrimination of recombinant and urinary origin	CZE	10-mM tricine, 10-mM NaCl, 10-mM sodium acetate, 7-M urea, 3.9-mM putrescine (pH 5.5)	Uncoated	UV (214 nm)	[108]
Separation of glycoforms	CZE	300-mM acetic acid-ammonium acetate (pH 4.0–5.5)	Polybrene-coated	UV (214 nm)	[109]
Separation of NESP (novel EPO analog)	CZE	10-mM tricine, 10-mM NaCl, 10-mM sodium acetate, 7-M urea, 2.5-mM putrescine (pH 4.5–5.5)	Uncoated	UV (214 nm)	[110]
Separation of glycoforms	CIEF	Catholyte: 20-mM NaOH Anolyte: 91-mM phosphoric acid Ampholyte: Beckman ampholyte 3–10, Pharmalyte 2.5–5	polyacrylamide-coated	UV (280 nm)	[111]
Separation of glycoforms	CIEF	Catholyte: 20-mM NaOH Anolyte: 91-mM phosphoric acid Ampholyte: Beckman ampholyte 3–10, Pharmalyte 2.5–5	eCAP neutral (Beckman)	UV (280 nm)	[112]

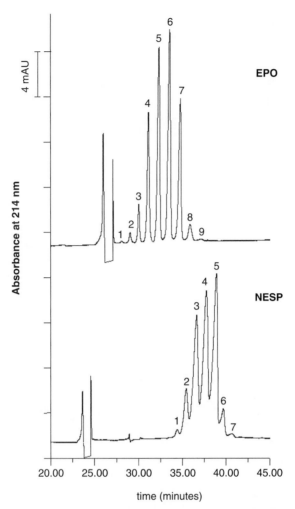

Figure 5.1-8. CZE separation of EPO and NESP glycoforms using a separation buffer containing 0.01-M Tricine, 0.01-M NaCl, 0.01-M NaAc, 7-M urea, and 2.5-mM putrescine (pH 5.5 adjusted with 2-M acetic acid) with bare fused-silica capillary (68.5 cm × 50 μm). (Reprinted from Ref. 110, with permission.)

used to inhibit protein aggregation by preventing the formation of intermolecular hydrogen bonds and disrupting hydrophobic and noncovalent interactions [105]. Sanz-Nebot et al. described the CZE separation of a novel erythropoiesis-stimulating protein (NESP), which is a recently approved hyperglycosylated analog of human EPO with a long-lasting effect [110]. As NESP, due to its two extra N-linked oligosaccharide chains, has a higher sialic acid content than EPO, its glycoforms are expected to carry a higher overall negative charge at pH 5.5. As shown in Figure 5.1-8, NESP glycoforms migrated slower than EPO glycoforms at pH 5.5 and the peaks corresponding to NESP glycoforms were only partially resolved. Figure 5.1-9 shows the influence of separation buffer pH on the separation of NESP glycoforms. The peak resolution was improved at lower pH condition, which may be due to the

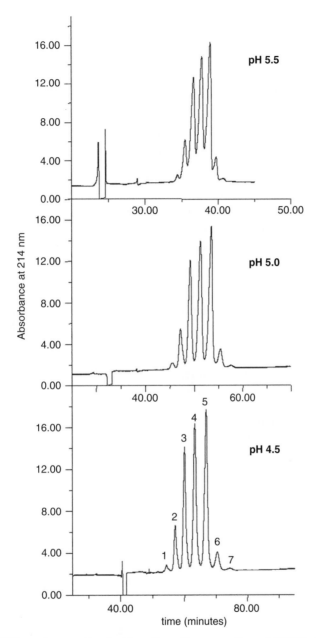

Figure 5.1-9. Effect of separation buffer pH on the separation of NESP glycoforms. The compositions of separation buffers are the same as Fig. 5.1-8, and the pH value was adjusted with 2-M acetic acid. (Reprinted from Ref. 110, with permission.)

lower p*I* of NESP compared with EPO. At pH 4.5, the seven peaks of NESP glycoforms were resolved to baseline.

CIEF is an important technique for analyzing the charged variants of EPO glycoforms on the basis of sialic acid content differences. Cifuentes et al. showed that EPO glycoforms could be resolved by using a mixture of broad and narrow pH-range ampholytes and by adding urea to the EPO sample [111]. The separation and quantification of seven EPO glycoforms with apparent pI values of 3.78–4.69 was accomplished using an optimized hydrodynamic mobilization method. Compared with conventional gle-IEF, better resolution was obtained in a shorter time. Lopez-Soto-Yarritu et al. described an improved CIEF method for the separation of EPO glycoforms with neutral-coated capillaries using ampholytes in the pH range 2–10 and bovine β-lactoglobulin-A as an internal standard [112]. CIEF analysis of EPO glycoforms was achieved with a migration time reproducibility of 0.05% and peak area reproducibility better than 3.4%.

Monoclonal Antibodies. Monoclonal antibodies (MAbs) are being in-creasingly used for a variety of diagnostic and therapeutic applications. As is the case for most recombinant proteins, preparations of MAbs often show considerable heterogeneity due to posttranslational modifications involving glycosylation. CIEF has been widely used for the separation of charged isoforms of several MAbs [67, 115–117].

Hunt et al. reported validation of CIEF method for recombinant MAb C2B8 [115]. As shown in Figure 5.1-10, CIEF separated four isoforms of MAb C2B8. Among them, three peaks were identified as charge isoforms present due to incomplete C-terminal lysine processing and represent MAb C2B8 with 2 (peak 1), 1 (peak 2), and 0 (peak 3) C-terminal lysines, respectively. The CIEF method was validated in accordance with International Conference on Harmonization (ICH) guidelines, and the result demonstrated that it is accurate, precise, linear, and highly specific for the determination of identity and charge distribution of MAb C2B8.

Tang et al. performed a systematic study on the routine analysis of a recombinant immunoglobulin G (IgG) by CIEF [116]. The method used a dimethyl siloxane-coated capillary (DB-1) and a separation matrix of 2% ampholytes in 0.4%

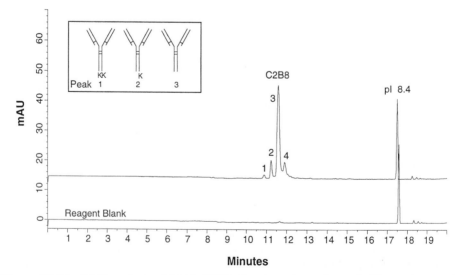

Figure 5.1-10. CIEF analysis of MAb C2B8. Box inset is a schematic illustration on the three C-terminal lysine variants. (Reprinted from Ref. 115, with permission.)

methylcellulose. The composition of various IgG samples with respect to isoform content was analyzed and compared quantitatively with gel-IEF. The reproducibility of the method was examined with the quantitative analysis of IgG sample in replicate over 3 days. The RSD of peak areas was below 2% intraday and 8% interday. The RSD for the mobilization times was below 1% intraday and 3% interday. Separation variability was examined over 150 runs. During this period, reproducible migration times and good resolution without peak shape deterioration were observed. This type of robustness demonstrates the potential of the CIEF method as a useful routine analysis tool.

Kats et al. employed the MEKC method to separate isoforms of chimeric MAb BR96 [118]. In the range from pH 2 to 12, MAb BR96 was separated into one to five isoforms by MEKC using a 12-mM borate buffer (pH 9.4) containing 25-mM SDS in uncoated capillary. Kats et al. further applied MEKC to separate structurally similar isoforms of a single-chain immunotoxin fusion protein BR96-sFV-PE40 using a 12-mM borate buffer (pH 9.6) containing 16-mM cholic acid [119].

Table 5.1-4 summarizes CE applications for the characterization of charge-based heterogeneity of several MAbs.

5.1.3.4 Stability Study

Peptides and proteins are complex molecules containing many functional groups, which can undergo a variety of degradation reactions, such as oxidation, reduction, deamidation, hydrolysis, arginine conversion, β-elimination, and racemization of amino acids [120]. Proteins further undergo physical changes in the secondary, tertiary, and quaternary structures [121]. These degradation reactions can alter the conformation, size, charge, and hydrophobicity of the peptides and proteins, thereby affecting the biological activity or even leading to formation of toxic compounds. As is recognized by the ICH guidelines for stability testing of biotechnological products, the monitoring of the stability of pharmaceutical peptides and proteins is of the utmost importance in the aspects of quality control and safety. CE has been used as a powerful tool for monitoring of stability and degradation of peptides and proteins in various conditions [122].

The high resolving power of CZE could be used for monitoring the stability and degradation reactions of peptide and protein products. Hoitink et al. applied CZE–MS for a stability study of goserelin, a luteinizing hormone-releasing hormone analog [123]. Using a 10% acetic acid as running buffer, high-resolution separation was obtained. In a stability study of goserelin at pH 5 and 9, the degradation of the C-terminal semi-carbazide group was observed. Lai et al. used CZE for the separation of asparagines-containing hexapeptide and its deamidation products and applied this method to a peptide stability study in the presence of polymers [124]. The offline CZE/MALDI–TOF MS combination has been used to monitor the stability of sCT, human parathyroid hormone 1-34 (PTH), and leuprolide in biodegradable PLGA microsphere formulations [64]. Figure 5.1-11 shows CZE monitoring of the sCT stability in the PLGA microspheres incubated in a 10-mM phosphate buffer saline (pH 7.4) containing 0.02% Tween 80 and 0.02% sodium azide at 37°C. CZE separations were performed using a 100-mM phosphate buffer (pH 2.5) in polyacylamide-coated capillary with UV detection at 200nm. During the *in vitro* drug release study, intact sCT peak rapidly decreased, whereas the

TABLE 5.1-4. CE Applications for the Heterogeneity Characterization of Monoclonal Antibodies

MAbs	CE Modes	Separation Buffers	Capillary	Detection	Ref.
MAb HER2	CIEF	Catholyte: 40-mM NaOH Anolyte: 20-mM phosphoric acid Ampholyte: Pharmalyte 8–10.5, Biolyte 7–9, Biolyte 3–10 (8:1:1) containing 0.5% TEMED, 0.2% HPMC	BioCAP LPA (Bio-Rad)	UV (280 nm)	[67]
MAb C2B8	CIEF	Catholyte: 40-mM NaOH Anolyte: 20-mM phosphoric acid Ampholyte: Pharmalyte 8–10.5, Biolyte 7–9, Biolyte 3–10 (8:1:1) containing 0.5% TEMED, 0.2% HPMC	BioCAP LPA (Bio-Rad)	UV (280 nm)	[115]
Recombinant IgGs	CIEF	Catholyte: 20-mM NaOH in 0.4% MC Anolyte: 120-mM phosphoric acid in 0.4% MC Ampholyte: 2% Pharmalyte 3–10 in 0.4% MC	DB-1	UV (280 nm)	[116]
Anti-TNF MAb D2E7	CIEF	Catholyte: 40-mM NaOH Anolyte: 20-mM phosphoric acid Ampholyte: Pharmalyte 8–10.5, Biolyte 7–9, Biolyte 3–10 (8:1:1) containing 0.5% TEMED, 0.2% HPMC	Neutral-coated (Beckman)	UV (280 nm)	[117]
MAb BR96	MEKC	12-mM borate buffer (pH 9.4) containing 25-mM SDS	Uncoated	UV (214 nm)	[118]
BR96 sFV-PE40 immunotoxin	MEKC	12-mM borate buffer (pH 9.6) containing 16-mM cholic acid	Uncoated	UV (214 nm)	[119]

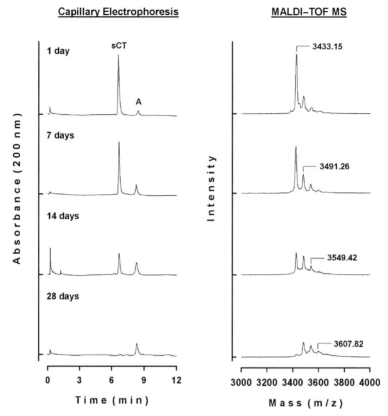

Figure 5.1-11. Monitoring of sCT stability in degrading PLGA microsphere formulations by CZE and MALDI–TOF MS. In CE, "A" represents the acylation products of sCT. (Reprinted from Ref. 64, with permission.)

additional peaks predominantly appeared after 28 days of incubation. MALDI–TOF MS data supported CE results with good correlation.

SDS–CGE can be also a useful stability-indicating method because it detects changes in fragmentation of proteins. Hunt et al. demonstrated that SDS–CGE could detect peaks resulting from the fragmention and aggregation of MAb HER2 when incubated at 37°C for 27 days [67]. Hunt and Nashabeh showed the potential of SDS-CGE for the separation and detection of proteolytic and deglycosylated fragments generated from MAbs [93].

5.1.3.5   Characterization of PEGylated Peptides and Proteins

The covalent attachment of PEG, PEGylation, is a well-established technique of overcoming several problems associated with the therapeutic uses of proteins. This technique has demonstrated reduced immunogenicity, extended circulating half-life, and improved stability for these therapeutic agents [125–127].

In general, the PEGylation process results in molecular heterogeneity with respect to the distribution in terms of number and positions of attached PEG molecules as well as the inherent polydispersity of PEG itself [128]. Therefore,

PEGylated molecules are among the most challenging products in analytical bio-chemistry, and their characterizations are becoming more important because these heterogeneities may confer different biological properties [129, 130]. When developing PEGylated biomolecules as therapeutic agents, various points should be considered, including the characterization of starting materials (i.e., proteins and PEG molecules), the determination of PEGylation sites, and the stoichiometry of PEG attachment [131]. The consistency of each preparation should also be established. For well-characterized PEGylated biomolecules, it is important to develop and validate suitable analytical techniques that enable consistent manufacturing processes to be established. Table 5.1-5 provides an overview of the applications of CE methods for the characterization of PEGylated peptides and proteins.

Cunico et al. described a charge-reversed CZE method for the analysis of PEGylated proteins [132]. A removable coating using ethylene glycol was applied to allow the negative capillary surface charge to be made positive, to prevent the protein adsorption to the capillary wall. Using this coated capillary, six different PEGylated molecules were characterized.

Bullock et al. characterized PEGylated superoxide dismutase (SOD) by CZE using a low-pH separation buffer (pH 2.05) [133]. At low pH, protein molecules are completely protonated and the charge differences between individual PEG-mers (i.e., mono-, di-, or tri-PEGmers) are minimized due to neutralization of lysine residues modified by PEG attachment. This results in a predominantly size-based separation of PEGylated proteins. Under low pH conditions, PEG–SOD conjugates were size-dependently separated and the separation pattern was very similar to the MALDI–TOF MS spectrum.

Li et al. used a semi-aqueous phosphate buffer (pH 2.5) of acetonitrile-water (1:1, v/v) to improve the resolution of PEGylated proteins [134]. They found that acetonitrile was mainly responsible for the good resolution observed, and this was attributed to reduced protein adsorption to the inner wall of the silica capillary used.

Na et al. determined PEGylation sites in three positional isomers of mono-PEGylated salmon calcitonins (mono-PEG-sCTs) using CZE-based peptide mapping analysis [135]. The resistance of PEGylation sites to proteolytic degradation resulted in different CE electropherogram patterns for tryptic digested mono-PEG–sCT isomers, and PEGylation sites were assigned accordingly and confirmed by MALDI–TOF MS.

Na and Lee reported on the characterization of PEGylated human parathyroid hormone (PEG–PTH) using CZE [136]. As shown in Figure 5.1-12, the CZE was used to optimize reaction conditions by monitoring the effects of reaction pH and the molar ratios of reactants on the PEGylation of PTH. The CZE method also allowed for determination of the extent of positional isomers in mono-PEG–PTH as well as for the identification of PEGylation sites.

Na et al. applied SDS–CGE to characterize PEGylated interferon alpha (PEG-IFN) [4]. The method well resolved the PEG–IFN species as well as the native IFN. As shown in Figure 5.1-13, four PEGylated IFNs (i.e., mono-, di-, tri-, and tetra-PEGylated IFNs) were detected and completely separated by SDS–CGE. The distribution of PEGylation reaction mixture in the SDS–CGE was found to be almost identical with that obtained by SDS–PAGE with Coomassie blue staining, although no band corresponding to tetra-PEGylated IFN was visualized.

TABLE 5.1-5. CE Applications for the Characterization of PEGylated Peptides and Proteins

Objective	CE Modes	Separation Buffers	Capillary	Detection	Ref.
Separation using charge-reversed mechanism	CZE	10-mM phosphate, 10% ethylene glycol (pH 2.0)	Uncoated	UV (215 nm)	[132]
Heterogeneity of PEG–SOD	CZE	15-mM boric acid, 15-mM 1,3-diaminopropane, 10-mM NaCl (pH 10.1)	Uncoated	UV (200 nm)	[137]
Separation of PEG–SOD and comparison with MALDI–TOF MS	CZE	40-mM phosphoric acid, 0.1-mg/mL PEG-114000, 4-mg/mL Jeffamine ED-600 (pH 2.05)	Uncoated	UV (200 nm)	[133]
Separation of PEG-lysozyme and peptide mapping	CZE	25-mM phosphate, 0.1-mg/mL PEG 600K (pH 2.7)	Uncoated	UV (200 nm)	[138]
Separation of semi-aqueous CE method	CZE	25-mM sodium phosphate (pH 2.5), acetonitrile-water (1:1, v/v)	Uncoated	UV (200 nm)	[134]
Identification of PEGylation sites in mono-PEG-salmon calcitonin	CZE	100-mM sodium phosphate (pH 2.5)	Polyacrylamide-coated	UV (200 nm)	[135]
Identification of PEGylation sites in mono-PEG-human parathyroid hormone (1–34)	CZE	100-mM sodium phosphate (pH 2.5)	Polyacrylamide-coated	UV (200 nm)	[136]
Separation of PEG-interferon alpha	SDS–CGE	CE–SDS protein kit (Bio-Rad)	Uncoated	UV (220 nm)	[4]

499

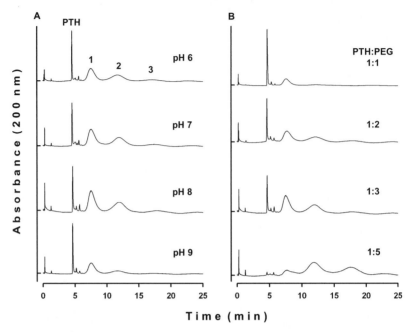

Figure 5.1-12. CZE monitoring of the PEGylation reaction of PTH through control of reaction pH (A) and molar ratios of reactants (B). The numbers above peaks represent the number of PEG molecules conjugated to PTH (i.e., 1: mono-, 2: di-, and 3: tri-PEGylated PTHs). (Reprinted from Ref. 136, with permission.)

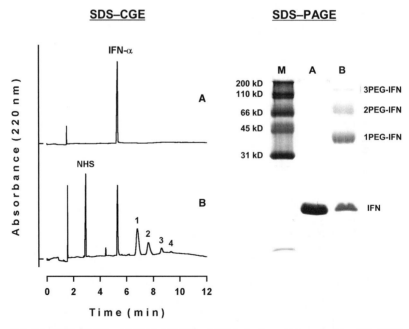

Figure 5.1-13. SDS–CGE and SDS–PAGE of IFNs before (A) and after (B) PEGylation reaction. In the SDS–CGE electropherogram, the numbers above the peaks represent the number of PEG molecules conjugated to IFN (i.e., 1: mono-, 2: di-, 3: tri-, and 4: tetra-PEGylated IFNs). (Reprinted from Ref. 4, with permission.) (This figure is available in full color at ftp://ftp.wiley.com/public/sci_ tech_med/pharmaceutical_biotech/.)

SDS–CGE was also useful for monitoring PEGylation reaction to optimize the reaction conditions such as the reaction molar ratio. This study demonstrated the potential of SDS–CGE for the characterization of PEGylated proteins with the advantages of speed, minimal sample consumption, and high resolution.

5.1.3.6 Proteome Analysis

Analysis of the proteome through the detection and identification of proteins from biological samples is gaining importance in the postgenomic era. The most traditional way to analyze proteome consists of protein separations by 2D gel electrophoresis (2-DE) and identifications by MS employing ESI or MALDI ionization [139]. However, the 2-DE–MS approach remains difficult for the detection of proteins expressed at extremely low levels or having extreme pI values or molecular masses and sensitivity [140]. In view of the limitations of 2-DE-based techniques, a considerable effort has been focused on the development of liquid phase-based analytical techniques, such as HPLC and CE, enabling the rapid, broad, and sensitive analysis of complex proteomic samples through the combination with MS or tandem MS analysis [5, 62, 141].

Various CE-based techniques have been widely used for proteome analysis, as has been described in numerous recent review articles [5, 62, 142, 143]. In addition to the complexity of protein samples, the large variation of the protein relative is one of the great challenges to the proteome analysis. For the broad proteome analysis, including the identification of lower abundance proteins, developments in capillary separations capable of providing extremely high resolving power and selective analyte enrichment are being highlighted [5]. As the resolving power of one-dimensional separation systems is generally not adequate for highly complex mixtures, multidimensional separations are becoming increasingly important for the comprehensive proteome analysis [7, 77].

Chen et al. developed an online combination system of CIEF with capillary RP–HPLC (CRPLC) using a microinjector as the interface for 2D separations of complex protein mixtures [144]. The resolving power of a combined CIEF–CRPLC system was demonstrated with tryptic digests of proteins from *Drosophila* salivary glands. The overall peak capacity was estimated to be around approximately 1800 over a run time of less than 8 hours.

Mohan and Lee developed a 2D CE separation system for proteomics by combining CIEF with transient capillary isotachophoresis (CITP)–CZE [145]. 2D separation of proteolytic peptides was applied to analyze tryptic digests of model proteins, including cytochrome c, ribonuclease A, and carbonic anhydrase II. Maximum peak capacity was estimated to be around 1600.

Ramsey et al. have combined MEKC with CZE, using switching of applied voltages to control fraction transport between the separation dimensions [146]. In analysis times of less than 15 min, 2D separation of tryptic digests of bovine serum albumin produced a peak capacity of 4200 (110 in the first dimension and 38 in the second dimension). The system was used to identify peptides from a tryptic digest of ovalbumin using standard addition and to distinguish between tryptic digests of human and bovine hemoglobin.

5.1.4 CONCLUSIONS

In biopharmaceutical analysis, CE is becoming the method of choice for the separation and characterization of peptide and protein pharmaceuticals. As shown in this chapter, the application of CE in this area has significantly increased during the past decade. The high efficient separation capability of CE provides an alternative and complementary technique to conventional analytical methods that are currently used to characterize biopharmaceuticals. Diverse separation modes of CE have been applied for the identity determination, purity control, heterogeneity characterization, stability study, and process consistency of biopharmaceuticals. In particular, CE has been recognized as a highly valuable tool for the efficient separation and characterization of the heterogeneous and complex protein products such as glycoproteins and PEGylated proteins. Recently, as the field of proteomics is becoming increasingly important, the role of CE in the strategy of protein analysis increases, with an emphasis on multidimensional separation and the combination of mass spectrometric techniques.

REFERENCES

1. Quigley W W, Dovichi N J (2004). Capillary electrophoresis for the analysis of biopolymers. *Analyt. Chem.* 76:4645–4658.

2. Patrick J S, Lagu A L (2001). Review applications of capillary electrophoresis to the analysis of biotechnology-derived therapeutic proteins. *Electrophoresis.* 22:4179–4196.

3. Pantazaki A, Taverna M, Vidal-Madjar C (1999). Recent advances in the capillary electrophoresis of recombinant glycoproteins. *Analyt. Chim. Acta.* 383:137–156.

4. Na D H, Park E J, Youn Y S, et al. (2004). Sodium dodecyl sulfate-capillary gel electrophoresis of polyethylene glycolylated interferon alpha. *Electrophoresis.* 25:476–479.

5. Cooper J W, Wang Y, Lee C S (2004). Recent advances in capillary separations for proteomics. *Electrophoresis.* 25:3913–3926.

6. Guzman N A, Majors R E (2001). New directions for concentration sensitivity enhancement in CE and microchip technology. *LC-GC.* 19:14–30.

7. Monton M R N, Terabe S (2005). Recent development in capillary electrophoresis-mass spectrometry of proteins and peptides. *Analyt. Sci.* 21:5–13.

8. Oda R P, Spelsberg T C, Landers J P (1994). Commercial capillary electrophoresis instrumentation. *LC-GC.* 12:50–51.

9. Altria K D (1996). *Capillary Electrophoresis Guidebook*, Humana Press Inc., Totowa, N.J., pp. 15–19.

10. Offord R E (1966). Electrophoretic mobilities of peptides on paper and their use in the determination of amide groups. *Nature.* 211:591–593.

11. Compton B J (1991). Electrophoretic mobility modeling of proteins in free zone capillary electrophoresis and its application to monoclonal antibody microheterogeneity analysis. *J. Chromatog.* 559:357–366.

12. Cifuentes A, Poppe H (1994). Simulation and optimization of peptide separation by capillary electrophoresis. *J. Chromatog. A.* 680:321–340.

13. Janini G M, Metral C J, Issaq H J, et al. (1999). Peptide mobility and peptide mapping in capillary zone electrophoresis. Experimental determination and theoretical simulation. *J. Chromatog. A.* 848:417–433.

14. Righetti P G, Gelfi C, Bossi A, et al. (2000). Capillary electrophoresis of peptides and proteins in isoelectric buffers: An update. *Electrophoresis.* 21:4046–4053.

15. Janini G M, Issaq H J (2001). Selection of buffer in capillary zone electrophoresis: Application to peptide and protein analysis. *Chromatographia.* 53:S18–S26.

16. Horvath J, Dolnik V (2001). Polymer wall coatings for capillary electrophoresis. *Electrophoresis.* 22:644–655.

17. Belder D, Deege A, Husmann H, et al. (2001). Cross-linked poly(vinyl alcohol) as permanent hydrophilic column coating for capillary electrophoresis. *Electrophoresis.* 22:3813–3818.

18. Li M X, Liu L, Wu J T, et al. (1997). Use of a polybrene capillary coating in capillary electrophoresis for rapid analysis of hemoglobin variants with on-line detection via an ion trap storage/reflectron time-of-flight mass spectrometer. *Analyt. Chem.* 69:2451–2456.

19. Erim F B, Cifuentes A, Poppe H, et al. (1995). Performance of a physically adsorbed high-molecular-mass polyethyleneimine layer as coating for the separation of basic proteins and peptides by capillary electrophoresis. *J. Chromatog. A.* 708:356–361.

20. Wang Y, Dubin P L (1999). Capillary modification by noncovalent polycation adsorption: Effects of polymer molecular weight and adsorption ionic strength. *Analyt. Chem.* 71:3463–3468.

21. Bendahl L, Hansen S H, Gammelgaard B (2001). Capillaries modified by noncovalent anionic polymer adsorption for capillary zone electrophoresis, micellar electrokinetic capillary chromatography and capillary electrophoresis mass spectrometry. *Electrophoresis.* 22:2565–2573.

22. Katayama H, Ishihama Y, Asakawa N (1998). Stable capillary coating with successive multiple ionic polymer layers. *Analyt. Chem.* 70:2254–2260.

23. Catai J R, Somsen G W, de Jong G J (2004). Efficient and reproducible analysis of peptides by capillary electrophoresis using noncovalently bilayer-coated capillaries. *Electrophoresis.* 25:817–824.

24. Catai J R, Tervahauta H A, de Jong G J, et al. (2005). Noncovalently bilayer-coated capillaries for efficient and reproducible analysis of proteins by capillary electrophoresis. *J. Chromatog. A.* 1083:185–192.

25. Righetti P G, Gelfi C, Verzola B, et al. (2001). The state of the art of dynamic coatings. *Electrophoresis.* 22:603–611.

26. Guttman A (1996). Capillary sodium dodecyl sulfate-gel electrophoresis of proteins. *Electrophoresis.* 17:1333–1341.

27. Takagi T (1997). Capillary electrophoresis in presence of sodium dodecyl sulfate and a sieving medium. *Electrophoresis.* 18:2239–2242.

28. Wu D, Regnier F E (1992). Sodium dodecyl sulfate-capillary gel electrophoresis of proteins using non-cross-linked polyacrylamide. *J. Chromatog.* 608:349–356.

29. Na D H, Cho C K, Youn Y S, et al. (2004). Capillary electrophoresis to characterize ricin and its subunits with matrix-assisted laser desorption/ionization time-of-flight mass spectrometry. *Toxicon.* 43:329–335.

30. Mazzeo J R, Krull I S (1991). Capillary isoelectric focusing of proteins in uncoated fused silica capillaries using polymeric additives. *Analyt. Chem.* 63:2852–2857.

31. Fang X, Tragas C, Wu J, et al. (1998). Recent developments in capillary isoelectric focusing with whole-column imaging detection. *Electrophoresis.* 19:2290–2295.

32. Hjerten S, Liao J L, Yao K Q (1987). Theoretical and experimental study of high-performance electrophoretic mobilization of isoelectrically focused protein zones. *J. Chromatog.* 387:127–138.

33. Schwer C (1995). Capillary isoelectric focusing: a routine method for protein analysis? *Electrophoresis.* 16:2121–2126.

34. Righetti P G (2004). Determination of the isoelectric point of proteins by capillary isoelectric focusing. *J. Chromatog. A.* 1037:491–499.

35. Shimura K, Wang Z, Matsumoto H, et al. (2000). Synthetic oligopeptides as isoelectric point markers for capillary isoelectric focusing with ultraviolet absorption detection. *Electrophoresis.* 21:603–610.

36. Kilar F (2003). Recent applications of capillary isoelectric focusing. *Electrophoresis.* 24:3908–3916.

37. Shimura K (2002). Recent advances in capillary isoelectric focusing: 1997–2001. *Electrophoresis.* 23:3847–3857.

38. Terabe S (2004). Micellar electrokinetic chromatography. *Analyt. Chem.* 76:241A–246A.

39. Kats M, Richberg P C, Hughes D E (1995). Conformational diversity and conformational transitions of a monoclonal antibody monitored by circular dichroism and capillary electrophoresis. *Analyt. Chem.* 67:2943–2948.

40. Jing P, Kaneta T, Imasaka T (2002). Micellar electrokinetic chromatography with diode laser-induced fluorescence detection as a tool for investigating the fluorescence labeling of proteins. *Electrophoresis.* 23:550–555.

41. Strege M A, Lagu A L (1997). Capillary electrophoresis of biotechnology-derived proteins. *Electrophoresis.* 18:2343–2352.

42. Fu H, Huang X, Jin W, et al. (2003). The separation of biomolecules using capillary electrochromatrography. *Curr. Opin. Biotechnol.* 14:96–100.

43. Krull I, Sebag A, Stevenson R (2000). Specific applications of capillary electrochromatography to biopolymers, including proteins, nucleic acids, peptide mapping, antibodies, and so forth. *J. Chromatog. A.* 887:137–163.

44. Bandilla D, Skinner C D (2004). Capillary electrochromatography of peptides and proteins. *J. Chromatog. A.* 1044:113–129.

45. Dearie H S, Smith N W, Moffatt F, et al. (2002). Effect of ionic strength on perfusive flow in capillary electrochromatography columns packed with wide-pore stationary phases. *J. Chromatog. A.* 945:231–238.

46. Ericson C, Hjerten S (1999). Reversed-phase electrochromatography of proteins on modified continuous beds using normal-flow and counterflow gradients. Theoretical and practical considerations. *Analyt. Chem.* 71:1621–1627.

47. Zhang S, Huang X, Zhang J, et al. (2000). Capillary electrochromatography of proteins and peptides with a cationic acrylic monolith. *J. Chromatog. A.* 887:465–477.

48. Choudhary G, Apffel A, Yin H, et al. (2000). Use of on-line mass spectrometric detection in capillary electrochromatography. *J. Chromatog. A.* 887:85–101.

49. Klampfl C W (2004). Review coupling of capillary electrochromatography to mass spectrometry. *J. Chromatog. A.* 1044:131–144.

50. Stutz H (2005). Advances in the analysis of proteins and peptides by capillary electrophoresis with matrix-assisted laser desorption/ionization and electrospray-mass spectrometry detection. *Electrophoresis.* 26:1254–1290.

51. Tsuda T, Sweedler J V, Zare R N (1990). Rectangular capillaries for capillary zone electrophoresis. *Analyt. Chem.* 62:2149–2152.

52. Moring S E, Reel R T, van Soest R E J (1993). Optical improvements of a Z-shaped cell for high-sensitivity UV absorbance detection in capillary electrophoresis. *Analyt. Chem.* 65:3454–3459.

53. Heiger D N, Kaltenbach P, Sievert H J (1994). Diode array detection in capillary electrophoresis. *Electrophoresis.* 15:1234–1247.

54. Drossman H, Luckey J A, Kostichka A J, et al. (1990). High-speed separations of DNA sequencing reactions by capillary electrophoresis. *Analyt. Chem.* 62:900–903.

55. Liu J, Shirota O, Novotny M V (1992). Sensitive, laser-assisted determination of complex oligosaccharide mixtures separated by capillary gel electrophoresis at high resolution. *Analyt. Chem.* 64:973–975.

56. Swerdlow H, Zhang J Z, Chen D Y, et al. (1991). Three DNA sequencing methods using capillary gel electrophoresis and laser-induced fluorescence. *Analyt. Chem.* 63:2835–2841.

57. Tseng W L, Chang H T (2000). On-line concentration and separation of proteins by capillary electrophoresis using polymer solutions. *Analyt. Chem.* 72:4805–4811.

58. Yeung E S (1999). Study of single cells by using capillary electrophoresis and native fluorescence detection. *J. Chromatog. A.* 830:243–262.

59. Lin Y W, Chiu T C, Chang H T (2003). Laser-induced fluorescence technique for DNA and proteins separated by capillary electrophoresis. *J. Chromatog. B.* 793:37–48.

60. Schmitt-Kopplin P, Frommberger M (2003). Capillary electrophoresis-mass spectrometry: 15 years of developments and applications. *Electrophoresis.* 24:3837–3867.

61. Schmitt-Kopplin P, Englmann M (2005). Capillary electrophoresis-mass spectrometry: survey on developments and applications 2003–2004. *Electrophoresis.* 26:1209–1220.

62. Simpson D C, Smith R D (2005). Combining capillary electrophoresis with mass spectrometry for applications in proteomics. *Electrophoresis.* 26:1291–1305.

63. von Brocke A, Nicholson G, Bayer E (2001). Recent advances in capillary electrophoresis/electrospray-mass spectrometry. *Electrophoresis.* 22:1251–1266.

64. Na D H, Youn Y S, Lee S D, et al. (2003). Monitoring of peptide acylation inside degrading PLGA microspheres by capillary electrophoresis and MALDI-TOF mass spectrometry. *J. Control. Rel.* 92:291–299.

65. Preisler J, Hu P, Rejtar T, et al. (2000). Capillary electrophoresis-matrix-assisted laser desorption/ionization time-of-flight mass spectrometry using a vacuum deposition interface. *Analyt. Chem.* 72:4785–4795.

66. Tegeler T J, Mechref Y, Boraas K, et al. (2004). Microdeposition device interfacing capillary electrochromatography and microcolumn liquid chromatography with matrix-assisted laser desorption/ionization mass spectrometry. *Analyt. Chem.* 76:6698–6706.

67. Hunt G, Moorhouse K G, Chen A B (1996). Capillary isoelectric focusing and sodium dodecyl sulfate-capillary gel electrophoresis of recombinant humanized monoclonal antibody HER2. *J. Chromatog. A.* 744:295–301.

68. Michalski W P, Shiell B J (1999). Strategies for analysis of electrophoretically separated proteins and peptides. *Analyt. Chim. Acta.* 383:27–46.

69. Olivares J A, Nguyen N T, Yonker C R, et al. (1987). On-line mass spectrometric detection for capillary zone electrophoresis. *Analyt. Chem.* 59:1230–1232.

70. Stutz H, Bordin G, Rodriguez A R (2004). Capillary zone electrophoresis of metal-binding proteins in formic acid with UV- and mass spectrometric detection using cationic transient capillary isotachophoresis for preconcentration. *Electrophoresis.* 25:1071–1089.

71. Tsuji K, Baczynskyj L, Bronson G E (1992). Capillary electrophoresis-electrospray mass spectrometry for the analysis of recombinant bovine and porcine somatotropins. *Analyt. Chem.* 64:1864–1870.

72. Yeung B, Porter T J, Vath J E (1997). Direct isoform analysis of high-mannose-containing glycoproteins by on-line capillary electrophoresis electrospray mass spectrometry. *Analyt. Chem.* 69:2510–2516.

73. Na D H, DeLuca P P, Lee K C (2004). Direct determination of the peptide content in microspheres by matrix-assisted laser desorption/ionization time-of-flight mass spectrometry. *Analyt. Chem.* 76:2669–2673.

74. Rush R S, Derby P L, Strickland T W, et al. (1993). Peptide mapping and evaluation of glycopeptide microheterogeneity derived from endoproteinase digestion of erythropoietin by affinity high-performance capillary electrophoresis. *Analyt. Chem.* 65:1834–1842.

75. Zhou G H, Luo G A, Zhou Y, et al. (1998). Application of capillary electrophoresis, liquid chromatography, electrospray-mass spectrometry and matrix-assisted laser desorption/ionization-time of flight-mass spectrometry to the characterization of recombinant human erythropoietin. *Electrophoresis.* 19:2348–2355.

76. Boss H J, Watson D B, Rush R S (1998). Peptide capillary zone electrophoresis mass spectrometry of recombinant human erythropoietin: an evaluation of the analytical method. *Electrophoresis.* 19:2654–2664.

77. Issaq H J, Chan K C, Janini G M, et al. (2005). Multidimensional separation of peptides for effective proteomic analysis. *J. Chromatog. B.* 817:35–47.

78. Hynek R, Kasicka V, Kucerova Z, et al. (1996). Application of reversed-phase high-performance liquid chromatography and capillary zone electrophoresis to the peptide mapping of pepsin isoenzymes. *J. Chromatog. B.* 681:37–45.

79. Wu S L (1997). The use of sequential high-performance liquid chromatography and capillary zone electrophoresis to separate the glycosylated peptides from recombinant tissue plasminogen activator to a detailed level of microheterogeneity. *Analyt. Biochem.* 253:85–97.

80. Issaq H J, Chan K C, Janini G M, et al. (1999). A simple two-dimensional high performance liquid chromatography/high performance capillary electrophoresis set-up for the separation of complex mixtures. *Electrophoresis.* 20:1533–1537.

81. Kang S H, Gong X, Yeung E S (2000). High-throughput comprehensive peptide mapping of proteins by multiplexed capillary electrophoresis. *Analyt. Chem.* 72:3014–3021.

82. He Y, Zhong W W, Yeung E S (2002). Multiplexed on-column protein digestion and capillary electrophoresis for high-throughput comprehensive peptide mapping. *J. Chromatog. B.* 782:331–341.

83. Catai J R, De Jong G J, Somsen G W (2005). *Methods for Structural Analysis of Protein Pharmaceuticals*, AAPS Press, Arlington, V.A. pp. 331–377.

84. Kasicka V (2003). Recent advances in capillary electrophoresis and capillary electro-chromatography of peptides. *Electrophoresis.* 24:4013–4046.

85. Hutterer K, Dolnik V (2003). Capillary electrophoresis of proteins 2001–2003. *Electrophoresis.* 24:3998–4012.

86. Ridge S, Hettiarachchi K (1998). Peptide purity and counter ion determination of bradykinin by high-performance liquid chromatography and capillary electrophoresis. *J. Chromatog. A.* 817:215–222.

87. Chen J, Fausnaugh-Pollitt J, Gu L (1999). Development and validation of a capillary electrophoresis method for the characterization of protegrin IB-367. *J. Chromatog. A.* 853:197–206.

88. Hettiarachchi K, Ridge S, Thomas D W, et al. (2001). Characterization and analysis of biphalin: an opioid peptide with a palindromic sequence. *J. Peptide Res.* 57:151–161.

89. Moumakwa B A, Crawley C D, Purich E, et al. (2005). Use of capillary electrophoresis in drug quality assessment of synthetic porcine secretin. *Biomed. Chromatog.* 19:68–79.

90. Mandrup G (1992). Rugged method for the determination of deamidation products in insulin solutions by free zone capillary electrophoresis using an untreated fused-silica capillary. *J. Chromatog.* 604:267–281.

91. Yomota C, Matsumoto Y, Okada S, et al. (1997). Discrimination limit for purity test of human insulin by capillary electrophoresis. *J. Chromatog. B.* 703:139–145.

92. Sergeev N V, Gloukhova N S, Nazimov I V, et al. (2001). Monitoring of recombinant human insulin production by narrow-bore reversed-phase high-performance liquid chromatography, high-performance capillary electrophoresis and matrix-assisted laser desorption ionization time-of-flight mass spectrometry. *J. Chromatog. A.* 907:131–144.

93. Hunt G, Nashabeh W (1999). Capillary electrophoresis sodium dodecyl sulfate nongel sieving analysis of a therapeutic recombinant monoclonal antibody: a biotechnology perspective. *Analyt. Chem.* 71:2390–2397.

94. Solinova V, Kasicka V, Barth T, et al. (2005). Analysis and separation of enkephalin and dalargin analogues and fragments by capillary zone electrophoresis. *J. Chromatog. A.* 1081:9–18.

95. Eriksson H J, Somsen G W, Hinrichs W L, et al. (2001). Characterization of human placental alkaline phosphatase by activity and protein assays, capillary electrophoresis and matrix-assisted laser desorption/ionization time-of-flight mass spectrometry. *J. Chromatog. B.* 755:311–319.

96. Roddy T P, Molnar T E, McKean R E, et al. (1997). Method of analysis of recombinant acidic fibroblast growth factor by capillary electrophoresis. *J. Chromatog. B.* 695:49–58.

97. Donges R, Brazel D (2002). Determination of recombinant hirudin structural deviants by capillary zone electrophoresis augmented with buffer additives. *J. Chromatog. A.* 979:217–226.

98. Klyushnichenko V E, Koulich D M, Yakimov S A, et al. (1994). Recombinant human insulin: III. High-performance liquid chromatography and high-performance capillary electrophoresis control in the analysis of step-by-step production of recombinant human insulin. *J. Chromatog. A.* 661:83–92.

99. Gysler J, Mazereeuw M, Helk B, et al. (1999). Utility of isotachophoresis-capillary zone electrophoresis, mass spectrometry and high-performance size-exclusion chromatography for monitoring of interleukin-6 dimer formation. *J. Chromatog. B.* 841:63–73.

100. Klyushnichenko V, Tishkov V, Kula M R (1997). Rapid SDS-Gel capillary electrophoresis for the analysis of recombinant NADP(+)-dependent formate dehydrogenase during expression in Escherichia coli cells and its purification. *J. Biotechnol.* 58:187–195.

101. Kakehi K, Honda S (1996). Analysis of glycoproteins, glycopeptides and glycoprotein-derived oligosaccharides by high-performance capillary electrophoresis. *J. Chromatog. A.* 720:377–393.

102. Krantz S B (1991). Erythropoietin. *Blood.* 77:419–434.

103. Lai P H, Everett R, Wang F F, et al. (1986). Structural characterization of human erythropoietin. *J. Biolog. Chem.* 261:3116–3121.

104. Davis J M, Arakawa T, Strickland T W, et al. (1987). Characterization of recombinant human erythropoietin produced in Chinese hamster ovary cells. *Biochem.* 26:2633–2638.

105. Yu B, Cong H, Liu H, et al. (2005). Separation and detection of erythropoietin by CE and CE–MS. *Trends Analyt. Chem.* 24:350–357.

106. Bietlot H P, Girard M (1997). Analysis of recombinant human erythropoietin in drug formulations by high-performance capillary electrophoresis. *J. Chromatog. A.* 759:177–184.

107. Kinoshita M, Murakami E, Oda Y, et al. (2000). Comparative studies on the analysis of glycosylation heterogeneity of sialic acid-containing glycoproteins using capillary electrophoresis. *J. Chromatog. A.* 866:261–271.

108. De Frutos M, Cifuentes A, Diez-Masa J C (2003). Differences in capillary electrophoresis profiles of urinary and recombinant erythropoietin. *Electrophoresis.* 24:678–680.

109. Sanz-Nebot V, Benavente F, Vallverdu A, et al. (2003). Separation of recombinant human erythropoietin glycoforms by capillary electrophoresis using volatile electrolytes. Assessment of mass spectrometry for the characterization of erythropoietin glycoforms. *Analyt. Chem.* 75:5220–5229.

110. Sanz-Nebot V, Benavente F, Gimenez E, et al. (2005). Capillary electrophoresis and matrix-assisted laser desorption/ionization-time of flight-mass spectrometry for analysis of the novel erythropoiesis-stimulating protein (NESP). *Electrophoresis.* 26:1451–1456.

111. Cifuentes A, Moreno-Arribas M V, de Frutos M, et al. (1999). Capillary isoelectric focusing of erythropoietin glycoforms and its comparison with flat-bed isoelectric focusing and capillary zone electrophoresis. *J. Chromatog. A.* 830:453–463.

112. Lopez-Soto-Yarritu P, Diez-Masa J C, Cifuentes A, et al. (2002). Improved capillary isoelectric focusing method for recombinant erythropoietin analysis. *J. Chromatog. A.* 968:221–228.

113. European Pharmacopoeia (2002). 4:1123–1128, 1316.

114. Watson E, Yao F (1993). Capillary electrophoretic separation of human recombinant erythropoietin (r-HuEPO) glycoforms. *Analyt. Biochem.* 210:389–393.

115. Hunt G, Hotaling T, Chen A B (1998). Validation of a capillary isoelectric focusing method for the recombinant monoclonal antibody C2B8. *J. Chromatog. A.* 800:355–367.

116. Tang S, Nesta D P, Maneri L R, et al. (1999). A method for routine analysis of recombinant immunoglobulins (rIgGs) by capillary isoelectric focusing (cIEF). *J. Pharmac. Biomed. Anal.* 19:569–583.

117. Santora L C, Krull I S, Grant K (1999). Characterization of recombinant human monoclonal tissue necrosis factor-α antibody using cation-exchange HPLC and capillary isoelectric focusing. *Analyt. Biochem.* 275:98–108.

118. Kats M, Richberg P C, Hughes D E (1997). pH-dependent isoform transitions of a monoclonal antibody monitored by micellar electrokinetic capillary chromatography. *Analyt. Chem.* 69:338–343.

119. Kats M, Richberg P C, Hughes D E (1997). Conformational transformations of the fusion protein BR96 sFv-PE40 as monitored by micellar electrokinetic capillary chromatography and circular dichroism. *J. Chromatog. A.* 766:205–213.

120. Reubsaet J L, Beijnen J H, Bult A, et al. (1998). Analytical techniques used to study the degradation of proteins and peptides: chemical instability. *J. Pharmac. Biomed. Anal.* 17:955–978.

121. Reubsaet J L, Beijnen J H, Bult A, et al. (1998). Analytical techniques used to study the degradation of proteins and peptides: physical instability. *J. Pharmac. Biomed. Anal.* 17:979–984.

122. International conferences on harmonization of technical requirements for registration of pharmaceuticals for human use. In Q5C, ICH Harmonised Tripartite Guideline, Quality of Biotechnological Products: Stability Testing Biotechnological/Biological Products. Available at: http://www.ich.org.

123. Hoitink M A, Hop E, Beijnen J H, et al. (1997). Capillary zone electrophoresis–mass spectrometry as a tool in the stability research of the luteinising hormone-releasing hormone analogue goserelin. *J. Chromatog. A.* 776:319–327.

124. Lai M, Skanchy D, Stobaugh J, et al. (1998). Capillary electrophoresis separation of an asparagine containing hexapeptide and its deamidation products. *J. Pharmac. Biomed. Anal.* 18:421–427.

125. Harris J M, Chess R B (2003). Effect of pegylation on pharmaceuticals. *Nature Rev. Drug Disc.* 2:214–221.

126. Na D H, Youn Y S, Park E J, et al. (2004). Stability of PEGylated salmon calcitonin in nasal mucosa. *J. Pharmac. Sci.* 93:256–261.

127. Na D H, DeLuca P P (2005). PEGylation of octreotide: I. Separation of positional isomers and stability against acylation by poly(D,L-lactide-co-glycolide). *Pharmac. Res.* 22:736–742.

128. Lee K C, Moon S C, Park M O, et al. (1999). Isolation, characterization, and stability of positional isomers of mono-PEGylated salmon calcitonins. *Pharmac. Res.* 16:813–818.

129. Na D H, Youn Y S, Lee K C (2003). Optimization of the PEGylation process of a peptide by monitoring with matrix-assisted laser desorption/ionization time-of-flight mass spectrometry. *Rapid Comm. Mass Spectrom.* 17:2241–2244.

130. Na D H, Lee K C, DeLuca P P (2005). PEGylation of octreotide: II. Effect of N-terminal mono-PEGylation on biological activity and pharmacokinetics. *Pharmac. Res.* 22:743–749.

131. Bailon P, Berthold W (1998). Polyethylene glycol-conjugated pharmaceutical proteins. *Pharmac. Sci. Technol. Today.* 1:352–356.

132. Cunico R L, Gruhn V, Kresin L, et al. (1991). Characterization of polyethylene glycol modified proteins using charge-reversed capillary electrophoresis. *J. Chromatog.* 559:467–477.

133. Bullock J, Chowdhury S, Johnston D (1996). Characterization of poly(ethylene glycol)-modified superoxide dismutase: Comparison of capillary electrophoresis and matrix-assisted laser desorption/ionization mass spectrometry. *Analyt. Chem.* 68:3258–3264.

134. Li W, Zhong Y, Lin B, Su Z (2001). Characterization of polyethylene glycol-modified proteins by semi-aqueous capillary electrophoresis. *J. Chromatog. A.* 905:299–307.

135. Na D H, Park M O, Choi S Y, et al. (2001). Identification of the modifying sites of mono-PEGylated salmon calcitonins by capillary electrophoresis and MALDI-TOF mass spectrometry. *J. Chromatog. B.* 754:259–263.

136. Na D H, Lee K C (2004). Capillary electrophoretic characterization of PEGylated human parathyroid hormone with matrix-assisted laser desorption/ionization time-of-flight mass spectrometry. *Analyt. Biochem.* 331:322–328.

137. Snider J, Neville C, Yuan L C, et al. (1992). Characterization of the heterogeneity of polyethylene glycol-modified superoxide dismutase by chromatographic and electrophoretic techniques. *J. Chromatog.* 599:141–155.

138. Roberts M J, Harris J M (1998). Attachment of degradable poly(ethylene glycol) to proteins has the potential to increase therapeutic efficacy. *J. Pharmac. Sci.* 87:1440–1445.

139. Aebersold R, Mann M (2003). Mass spectrometry-based proteomics. *Nature.* 422:198–207.

140. Gygi S P, Corthals G L, Zhang Y, et al. (2000). Evaluation of two-dimensional gel electrophoresis-based proteome analysis technology. *Proc. of the National Academy of Sciences of the United States of America.* 97:9390–9395.

141. Washburn M P, Wolters D, Yates III J R (2001). Large-scale analysis of the yeast proteome by multidimensional protein identification technology. *Nature Biotechnol.* 19:242–247.

142. Issaq H J (2001). The role of separation science in proteomics research. *Electrophoresis.* 22:3629–3638.

143. Shen Y, Smith R D (2002). Proteomics based on high-efficiency capillary separations. *Electrophoresis.* 23:3106–3124.

144. Chen J, Lee C S, Shen Y, et al. (2002). Integration of capillary isoelectric focusing with capillary reversed-phase liquid chromatography for two-dimensional proteomics separation. *Electrophoresis.* 23:3143–3148.

145. Mohan D, Lee C S (2002). On-line coupling of capillary isoelectric focusing with transient isotachophoresis-zone electrophoresis: A two-dimensional separation system for proteomics. *Electrophoresis.* 23:3160–3167.

146. Ramsey J D, Jacobson S C, Culbertson C T, et al. (2003). High-efficiency, two-dimensional separations of protein digests on microfluidic devices. *Analyt. Chem.* 75:3758–3764.

5.2

PHARMACEUTICAL BIOASSAY

Jun Li, Ming Zhao, and Shiqi Peng

School of Chemical Biology and Pharmaceutical Sciences, Capital Medical University, Beijing, China

Chapter Contents

5.2.1	Bioassay for Detecting Toxicity of Toxins from a Microorganism	513
	5.2.1.1 Colorimetric Yeast Assay for Detecting Trichothecene Mycotoxins	513
	5.2.1.2 Ciliate *T. thermophila* Assay for Detecting Trichothecene Mycotoxins	514
	5.2.1.3 MTT Assay for Detecting Fusarium Mycotoxins	515
	5.2.1.4 Mortality and Frass Production Assay for Detecting Toxicity of Bacterial Strains	516
	5.2.1.5 *L. sativum* Assay for Detecting Microcystin Toxicity	516
5.2.2	Toxicity Bioassay for Chemicals	517
	5.2.2.1 Lux-Fluoro Assay for Detecting Combined Genotoxicity and Cytotoxicity of Chemicals	518
	5.2.2.2 ALIC-Based Assay for Detecting Toxicity of Chemicals in SPM	518
	5.2.2.3 PWM-Induced IgM Assay for Detecting Toxicity of Chemicals	519
	5.2.2.4 GFP-Based Cell Assay for Detecting Toxicity of TCDD and Related Chemicals	520
	5.2.2.5 RACB Protocol for Detecting Ovarian Toxicity of Xenobiotics	521
	5.2.2.6 Uroepithelial Cell Assay for Detecting Urotoxicity of Chemicals	522
	5.2.2.7 Mating Efficiency Assay for Detecting Toxicity of Endocrine Disruptors	523
	5.2.2.8 Pnar-gfp Assay for Detecting Carcinogenic Toxicity of Nitrate	524
5.2.3	Antiviral and Anticancer Assay	524
	5.2.3.1 Human Breast Cancer Cell Assay for Detecting Anticancer Activity	525
	5.2.3.2 Ishikawa Cell and Rat Assay for Detecting Antiestrogens	525
	5.2.3.3 RT-PCR and Swine Assay for Detecting Anti-HEV Antibody	526
	5.2.3.4 Antitubercular and Cytotoxic Assay	527
	5.2.3.5 DNA Polymerase β Lyase Assay	527
	5.2.3.6 HIV-1 Protease and Reverse Transcriptase Kinetic Assay	528

	5.2.3.7	P-Glycoprotein Pump Assay	529
	5.2.3.8	KIRA Assay	529
5.2.4	Hepatoxicity and Hepatoprotective Assay		530
	5.2.4.1	GST-P Enzyme-Altered Foci Assay	530
	5.2.4.2	Partial Hepatectomy Assay	531
	5.2.4.3	Hepatoprotective Assay	531
	5.2.4.4	Immunological Assay	532
	5.2.4.5	Body Weight-Based Liver Tumor Incidence Assay	533
	5.2.4.6	Hepatocyte Primary Culture Assay	534
	5.2.4.7	CCL-64 Cell Growth Assay	536
5.2.5	Anti-Inflammatory Assay		536
	5.2.5.1	Adhesion Formation Assay	536
	5.2.5.2	PBMC Proliferation Assay	538
	5.2.5.3	COX-1, COX-2, and 5-LOX Assay	539
	5.2.5.4	CCR5 Receptor Binding Assay	539
	5.2.5.5	Tissue Binding Affinity Assay	539
	5.2.5.6	G93A-SOD1 Transgenic Mice Assay	540
	5.2.5.7	MAPK p44 (ERK1) and p42 (ERK2) Assay	541
	5.2.5.8	ELA4.NOB-1/CTLL Cell Assay	542
	5.2.5.9	FKBP51mRNA Assay	543
	5.2.5.10	Xylene-Induced Ear Edema Assay	544
5.2.6	Thrombus-Related Assay		544
	5.2.6.1	Antiplatelet Aggregation Assay	544
	5.2.6.2	Ferric Chloride-Induced Thrombosis Assay	544
	5.2.6.3	Electrical Stimulation-Induced Arterial Thrombosis Assay	545
	5.2.6.4	Thread-Induced Thrombosis Assay	545
	5.2.6.5	Euglobulin Clot Lysis Time (ECLT) Assay	545
	5.2.6.6	Fibrinolytic Area Assay	546
	5.2.6.7	Thrombolytic Assay	546
5.2.7	Immunomodulating Assay		546
	5.2.7.1	Rat Mast Cell and Rabbit Aortic Assay	546
	5.2.7.2	Dendritic Cell (DC) Assay	547
	5.2.7.3	Lymphoid Organ Assay	548
	5.2.7.4	IFN-γ Assay	549
	5.2.7.5	Anti-rHuEPO NAb Assay	550
	5.2.7.6	Radial Assay of Chemotaxis	550
5.2.8	Estrogen Assay		551
	5.2.8.1	Yeast Oestrogen Assay	551
	5.2.8.2	HELN α and HELN β Transfected Cell Assay	552
	5.2.8.3	Luciferase Assay	553
	5.2.8.4	Green Fluorescent Protein Expression Assay	554
5.2.9	Antimalarial Assay		555
	5.2.9.1	Plasmodium Falciparum and Murine P388 Leukemia Cell Assay	555
	5.2.9.2	Histidine-Rich Protein II Assay	556
	5.2.9.3	Antimalarial Activity Assay	556
	5.2.9.4	Survival of *Anopheles gambiae* Assay	556
	5.2.9.5	Chloroquine Assay	557
	5.2.9.6	Plasmodium Falciparum Clone and Dihydroartemisinin (DHA) Assay	557

5.2.10   Blood Pressure-Related Assay   558

      5.2.10.1   Human Plasma Assay   558

      5.2.10.2   Pulmonary Hypertension Assay   559

      5.2.10.3   Coronary Arteries (CA) Constriction Assay   560

      5.2.10.4   Middle Cerebral Artery Occlusion (MCAO) Assay and Vascular Heparan Sulfate Proteoglycans (HSPG) Perlecan Assay   561

      5.2.10.5   Temperature Assay in Awake Subjects   562

      References   563

5.2.1 BIOASSAY FOR DETECTING TOXICITY OF TOXINS FROM A MICROORGANISM

Some *Fusarium* mycotoxins and the metabolites of filamentous fungi and molds occur naturally in a variety of feeds, foodstuffs, crops, vegetables, fruits, water, and plants. Ingestion of the contaminative feeds, foodstuffs, crops, vegetables, fruits, water, and plants can produce health problems, such as reduced growth, anorexia, induced emesis, enhanced proliferation of estrogen-responsive tumor cells, human cervical cancer, and premature initial breast development in humans and animals. Among the common toxins, verrucarin A, T2-toxin, roridin A, deoxynivalenol (DON), zearalenone (ZEN), fumonisin B1 (FB1), moniliformin (MON), microcystin-LR (MCLR), and the toxins from *Bacillus thuringiensis* provide a worldwide threat to humans and animals. Based on these facts that the lactose-utilizing yeast *Kluyveromyces marxianus* exhibits well-characterized β-galactosidase activity and has chromogenic substrates, the toxicants interfering with any cellular function required for induction and expression of β-galactosidase gene can suppress β-galactosidase activity, DON and fumonisin B can inhibit cell growth, the dehydrogenase ensurveyed enzymes involved in mitochondrial functions can be inhibited by the toxicants, in axenic culture the melanin precursor overproducer mutant strain of the ciliate *Tetrahymena thermophila* can be grown to high concentration, in protecting plants from the hazardous effects of xenobiotics glutathione S-transferase (GST) and glutathione peroxidase (GPX) can play essential roles, colorimetric yeast assay [1], ciliate *T. thermophila* assay [2], MTT assay [3], mortality and frass production assay [4], and *Lepidium sativum* assay [5] are established.

5.2.1.1 Colorimetric Yeast Assay for Detecting Trichothecene Mycotoxins

To 50 mL of liquid medium, a single-cell colony of *K. marxianus* GK1005 (maintained routinely and grown on 1% (w/v) yeast extract, 1% (w/v) bacteriological peptone, and 2% (w/v) glucose (YPD), solidified when required with 2% (w/v) agar) on agar plate is transfered to prepare cultures that are incubated for 16 h at 35°C and 200 rpm in a rotary incubator. At 560 nm, the absorbance is measured to determine cell density, which is calibrated by direct hemocytometer counts. Mycotoxins containing sample is dissolved in spectroscopic grade methanol to make stock solutions typically at 0.1 mg/mL. Cetyl trimethyl ammonium bromide, polymyxin B sulphate, and polymyxin B nonapeptide are dissolved in water, filter-sterilized, and kept no more than a day as stock solutions. 5-Bromo-4-chloro-3-indolyl-β-D-galactopyranoside (X-gal) is dissolved in dimethylformamide (DMF) at 100 mg/mL and stored at −20°C in the dark, and before each assay, it is immediately diluted in aqueous DMF (2 parts water : 3 parts DMF) to prepare a working solution of 20-mg/mL X-gal assay. Before use o-nitrophenyl-β-D-galactopyranoside (ONPGal) is immediately dissolved in water at 4 mg/mL and subsequently discarded. Before use, MTT is dissolved in phosphate-buffered solution (PBS) and filter-sterilized to prepare a 5-mg/mL stock solution.

In the wells of a microtiter plate (sterile, flat-bottomed), 136 μL of growth medium consisted of 1% (w/v) yeast extract, 1% (w/v) bacteriological peptone, and 50-mM glucose—"YPD-50" are mixed with polymyxin B sulphate to give a final assay concentration of 15 μg/mL, to which 8 μL of stock solution of mycotoxin containing sample or methanol (controls) and 16 μL of yeast inoculum are successively added to give an initial assay cell density of approximately 2×10^8 cells/mL. Blank wells contain 152 μL of medium and 8 μL of methanol. The plate is mixed, sealed with a plate sealer, and incubated at 35°C for the duration of the assay, during which cell density is regularly monitored at 560 nm. After about 10 h, the control (mycotoxin-free) cultures reach the stationary phase; the activity of β-galactosidase or mitochondrial (MTT-cleavage) in the cultures is assayed.

To each well of the microtiter plate, 5 μL of sodium dodecyl sulfate (SDS) (0.1%, w/v) and 3 μL of chloroform are added to permeabilize the cells. For in vivo experiments and experiments examining the effects of different carbon sources, 1 μL of 100-mg/mL X-gal in DMF are added to each well. For examining the effects of altering glucose concentration and inoculum cell density, 5 μL of 20-mg/mL X-gal in DMF are added to each well. For methanol and ethanol toxicity experiments and the standardized bioassay, 8 μL of 20-mg/mL X-gal in aqueous DMF are added. After mixing the contents in the wells, the plates are incubated at 35°C for a maximum of 30 min. Using a test filter at 666 nm and a reference filter at 560 nm, the plates are read. β-Galactosidase activity is expressed as product formation (A_{666}–A_{560}), as a function of cell density (A_{560}) as well.

After the addition of 16 μL of MTT in PBS to each well and statically incubated at 35°C for 4 h, the medium is displaced by 200 μL of dimethyl sulfate (DMSO). Thorough mixing and dissolution the MTT-cleavage product is pipeted repeatedly. Using a test filter at 560 nm and a reference filter at 666 nm, the plates are read.

Into 50 mL of the solution consisting of 1% yeast extract, 1% bacteriological peptone, and 50-mM lactose, a single colony of *K. marxianus* GK1005 is inoculated and then incubated for 16 h at 35°C and 200 rpm in an orbital shaker. Into each of two universal bottles 10 mL of the samples of the culture are transferred, and then 0.2 mL of chloroform and 0.1 mL of SDS (0.1%, w/v) are added. To permeabilize the cells, to one bottle 20 μL of X-gal (100 mg/mL in DMF) are added, and to the other bottle 20 μL of DMF are added. In an orbital shaker, both bottles are incubated at 35°C and 200 rpm until in the bottle containing X-gal visible indigo precipitate is formed, of which the absorption spectrum is determined with DMF control sample as a reference.

5.2.1.2 Ciliate *T. thermophila* Assay for Detecting Trichothecene Mycotoxins

The phenotype characteristics of *T. thermophila* strain BI3840 are amicronucleated (amc), pigment producing (pig), and resistant to 25 μg/mL of cycloheximide (cy-r)

and mating type IV [6]. The PP210 medium consists of aqueous solution of pro-
teose peptone (2%, w/v), supplemented with 10 μM of $FeCl_3$, 250 μg/mL of streptomycin sulphate, and penicillin G. Mycotoxins containing samples are dissolved
in propyleneglycol or acetonitrile to make stock solutions at 5 mg/mL. *T. thermophila* strain BI3840 are grown axenically in PP210 medium for 24 h and maintained constantly at 28 ± 1°C. Then the exponential phase cellular suspensions are
distributed in microtiter plates (100 μL/well). According to the required final concentration, different dilutions of the stock solutions of mycotoxins containing
samples in PP210 medium are added into the cultures (100 μL/well). To each
100 μl/well of *T. thermophila* strain BI3840 culture, 100 μl/well of PP210 medium
without a mycotoxin containing sample is added to obtain one type of control, and
100 μl/well of PP210 medium plus the highest concentration of the corresponding
solvent added to obtain another type of control. In all cases, the microtiter plates
are incubated at 28 ± 1°C for 48 h and the wells with a dark-brown pigmentation
are considered as the positive growth wells.

5.2.1.3 MTT Assay for Detecting Fusarium Mycotoxins

Caco-2, C5-O, V79 cells (in Dulbecco's Modified Eagle Medium, DMEM), and
Chinese hamster ovary (CHO)-K1 cells (in DMEM/F-12) at passage numbers
between 30 and 50, and HepG2 cells (in Minimum Essential Medium) at passage
numbers between 80 and 100, are grown as monolayers in 80-cm^2 culture flasks,
which are harvested when they reach 80% confluence to maintain exponential
growth.

Using trypsin ethylenediaminetetraacetic acid (EDTA) (0.25% trypsin and
1-mM EDTA·4Na), the cell monolayers in exponential growth are harvested
and after repeated pipetting the single-cell suspensions are obtained. To 96
well plates, the single cell suspensions (1×10^2 to 5×10^4 cells per 200 μL medium/
well) are added by serial dilution. The medium are supplemented with 1.5-g/L
sodium bicarbonate, 0.11-g/L sodium pyruvate, 1% nonessential amino acid, 25-
mM HEPES, 100 U of penicillin/mL, 100 μg of streptomycin/mL, 25 ng of amphotericin B/mL, and 10% fetal bovine serum). Caco-2, C5-O, HepG2 (1×10^4
cells/100-μL medium), CHO-K1, and V79 (5×10^3 cells/100-μL medium) are
seeded to each well of the 96 well plates and incubated in a humidified atmosphere
of 5% CO_2 at 37°C for 24 h. A mycotoxin containing sample in 100 μL of medium
is added from high to low concentration to the wells and adjusted to final concentration needed. MTT is dissolved in PBS (5-mg/mL final concentration), filtered
(0.22-μm filter), and stored in the amber vials at 4°C for a month. After 48-h and
72-h incubation, to each well, 25 μL of MTT solution is added and incubated in a
humidified atmosphere of 5% CO_2 at 37°C for 4 h. At the end of the incubation
period, the media are discarded using a suction pump. The extraction buffer of
20% (w/v) SDS in a solution of 50% DMF in demineralized water (50:50, v/v) is
prepared at pH 4.7 and filtered (0.22-μm filter). The buffer in 100 μL of 20% (w/v)
of SDS [7] is extracted and added into each well to solubilize formazan crystals,
and the plates are shaked at 37°C overnight. The absorbance is measured at 570 nm
with 690 nm as a reference wavelength. The positive control contains an adjusted
seeding cell number in the log phase, of which the culture medium contains 0.1%
ethanol.

5.2.1.4   Mortality and Frass Production Assay for Detecting Toxicity of Bacterial Strains

B. thuringiensis subsp. *tenebrionis* and the six unidentified strains of *B. thuringiensis,* labeled A30, A299, A311, A409, A410, and A429, are grown on nutrient agar buffered with an equimolar concentration of KH_2PO_4 (50 mM, pH 7.0) at 30°C for 5 days. The bacterial culture consists of vegetative cells, sporangia, spores, and crystals. These stages are lyophilized and stored at −20°C. The spore-crystal suspensions of each strain with 1:1 in the spore:crystal ratio are used in the toxicity assays [8]. The concentration of extractable proteins (the final protein solution is referred to as extractable protein) is determined and adjusted to a given concentration of protein in the crystal-spore mixture. Spore-crystal mixtures are suspended in Triton X-100 solution (0.01%, v/v) and washed three times by centrifugation at 22°C and 11750 g for 3 min. To extract the proteins, the final pellet in a putatively selective crystal solubilizing buffer (40.5-mM Na_2CO_3, 0.5-mM phenylmethylsulfonylfluoride, 0.1-mM dithiothreitol, pH 10.0) is solubilized at 42°C by incubating for 2 h and vortexing every 30 min [9]. The suspension is centrifuged (22°C, 11750 g, 3 min), and the supernatant is assayed for total protein with bovine serum albumin as the standard [10]. The laboratory cultures of plants are periodically supplemented with pest insects to maintain hybrid vigor. The colony is reared in an incubator at 24°C under the light for 16 h and at 16°C in the darkness for 8 h. Abundant food is placed in plastic petri dishes (100 mm in diameter) with moist filter paper that contains 10 adults, the plastic petri dishes are incubated, and the number of frass pellets on filter paper is counted daily for 3 days.

Adult mortality and modified frass production assays are used to measure the toxicity of the extractable proteins of the bacterial strains against plants [11]. The modified frass production assay is regarded as a rapid method to determine the specificity of numerous bacterial strains toxins against pest insect. The food cylinders (6 mm in length, 36 mm in diameter, 10 cylinders/plate) of optimum medium for insect feeding are cut from foliage homogenate supplemented with agar. The cylinders are dipped into distilled water containing different concentration of the extractable proteins, agitated for 5 s, air-dried, and then coated with test spore-crystal suspensions. The concentration of extractable proteins is 75-, 150-, 225-, and 300-μg protein/mL with distilled water as a control. Three to eight days after eclosion, starved adults are collected and added to the diet in petri dishes (15 × 100 mm diameter) containing moistened filter paper (15 adult insects per plate per dose). In the mortality assay, insects are fed at 25°C in darkness for 8 days. Mortality is monitored daily, and the LC_{50} for each test strain is calculated by probit analysis. In the frass assay, insects are fed for 6 days. The total number of frass pellets is calculated.

5.2.1.5   *L. sativum* Assay for Detecting Microcystin Toxicity

Toxic *Microcystis aeruginosa* PCC 7806 is cultured, and sufficient aeration is provided for culture mixing and air supply. From the freeze-dried cells, toxin is extracted by 70% aqueous methanol and determined using the protein phosphatase inhibition assay from the difference in the change in absorbance at 410 nm [12, 13]. With 5% H_2O_2 for 5 min, the seeds of *L. sativum* are surface sterilized, washed three times with sterile water for 10 min, left in fresh water overnight to germinate, and then transferred to nutrient medium containing 6.5% N, 2.7% P, 13% K, 7% Ca, 2.2% Mg, 7.5% S, 0.15% Fe, 0.024% Mn, 0.0024% B, 0.005% Zn, 0.002% Cu, and 0.001% Mo in 0.8% agar. The container (12 seeds/container) is cultured at 27°C under continuous light (20 photons/m²/s). Before pouring, either 1- or 10-µg/L microcystin-LR (MCLR) toxin extracts are added to the medium. The fresh weights, root, and leaf lengths are measured daily in 2 days. GST and GPX activities are determined in 3 days.

The stems and roots of 10 plants from each of three containers from each group are prepared for 30% homogenate of plant material in buffer (0.01-M Tris-HCl pH 7.8, 7.5-mM PMSF, 2.5-mM EDTA, 325-µM bestatin, 3.5-µM E-64, 2.5-µM leupeptin, and 0.75-µM aprotinin). The homogenate is centrifuged at 4°C and 12,000 g in a benchtop centrifuge for 2 h, and the supernatant is stored at −80°C for the GST and GPX assay.

In the GST assay, to the supernatant fluid, a cocktail containing 0.1-M potassium phosphate (pH 6.5), 30-mM 1-chloro-2,4-dinitrobenzene and 20-mM glutathione are added and measured immediately at 340 nm for 4 min [14]. In the GPX assay, to the supernatant fluid, a cocktail of 0.1-mM potassium phosphate buffer (pH 7), glutathione reductase (4 ng/µL final), 10-mM glutathione, 30-mM EDTA, and 1-mM NADPH is added. The reaction is initated by adding cumene hydroperoxide and measured immediately at 340 nm for 3 min.

5.2.2 TOXICITY BIOASSAY FOR CHEMICALS

Environmental chemical contaminants as toxic agents may be suspended in air, absorbed in suspended particulate matter (SPM) and foods, and dissolved in water, and thus, they may result in several toxic effects such as DNA damage, decline in lung function, imbalance of immune function, activation of aryl hydrocarbon receptor (AhR), abnormal reproduction, and urinary bladder injuries. Lux-Fluoro assay is developed on specific lesions for cells induced by DNA damaging chemicals may lead to cell death or induce an error-prone repair pathway leading to mutagenesis and cancer induction [15]. Based on the air-liquid interface culture (ALIC) of the human alveolar type II cell line (A459) and culture of spleen cells obtained from BALB/c mice producing IgM after pokeweed mitogen (PWM) stimulation, ALIC-based assay and PWM-induced IgM assay are established [16, 17]. Exposure to extremely low concentrations of 2,3,7,8-tetrachlorodibenzo-p-dioxin (TCDD) results in activation of AhR and the induction of expression of a battery of genes, green fluorescent protein (GFP)-based cell assay is established [18]. In Reproductive Assessment by Continuous Breeding (RACB) protocol, differential follicle counts can provide a sensitive means of estimating the extent of ovarian toxicity in females exposed to xenobiotics [19]. Based on these facts that urinary bladder injuries induced by repeated oral administration of pharmaceuticals may be epithelial ulceration with edema and hemorrhage in the lamina propria and muscle layer of the urinary bladder, environmental chemical contaminants disrupt mammalian peptide signal transduction pathway via the interaction with the endocrine system in humans and various wildlife species and the cellular production of GFP is a function of nitrate concentration, uroepithelial cell assay [20], mating efficiency assay [21], and Pnar-gfp assay [22] are established and can be used in bioassays.

5.2.2.1   Lux-Fluoro Assay for Detecting Combined Genotoxicity and Cytotoxicity of Chemicals

The plasmid pPLS-1 (DSM10333), the genotoxicity sensing reporter component of the test panel, which carries the luxCDABFE genes downstream of a strong SOS-dependent promoter is constructed. The plasmid pGFPuv that carries the optimized "cycle 3" variant of GFP in frame with the lacZ initiation codon is the cytotoxicity sensing reporter component of the test panel. According to the modified "Hanahan" procedure, pPLS-1 and pGFPuv are used to transform the strain *S. typhimurium* TA1535 (ATCC: *S. choleraesuis* subsp. *choleraesuis* strain TA1535) [23]. For selecting plasmid carrying cells, the bacteria are cultured at 37°C in NB-medium supplemented with 50-µg/mL ampicillin. The DNA damaging agents containing sample are dissolved in distilled water at high concentration and diluted for the test. Each well of White LB96P-CMP Mikro Lumat Plates with a transparent bottom contains 10 µL of the solvent or different concentration of the test sample. The agar plate containing a single colony of bacteria, 10 mL of NB-medium with 50-µg/mL ampicillin, is shaken on a rotary shaker at 37°C for 16 h. Then the dilution (1:50) of bacteria in fresh warm NB-medium containing 50-µg/mL ampicillin are incubated at 37°C until the absorbance at 600 nm (A_{600}) reaches 0.2–0.3. Aliquots of 90 µL of this culture are added to each well, the microplate is covered with a gas-permeable self-adhesive seal and placed into the temperature-controlled microplate reader. The absorbance, luminescence, and fluorescence of the culture is successively and repeatedly determined by the programmed measurement cycle at 30°C for up to 8 h of continuous incubation with a duration of about 10 min per measurement cycle (total of 50 cycles). The measurement cycle of each well includes 120-s shaking, 0.2-s luminescence measurement without filter, 0.1-s absorbance measurement (490 nm, 20 nm bandwidth), and 0.1-s fluorescence measurement (excitation at 405 nm, emission at 510 nm).

5.2.2.2   ALIC-Based Assay for Detecting Toxicity of Chemicals in SPM

A549 cells and Hep G2 cells cultured in DMEM supplemented with 10% fetal bovine serum, 20-mM hydroxyethylpiperazine-N′-2-ethanesulfonic acid, penicillin (100 U/mL), streptomycin (100 µg/mL), amphotericin B (0.25 µg/mL), and 1.0% non-essential amino acid solution (only for Hep G2 cells) are subcultivated using 0.25% trypsin and 0.02% EDTA in PBS. After collection by trypsinization A549 cells are seeded at 1.0×10^5 cells/cm^2 onto the membrane culture insert precoated with a 0.03% type collagen solution. Initially, to both the apical (Ap) and the BL sides, the medium is added. Using a phase contrast microscope, the formation of the cell sheet is assessed. The evaluation of the development of tight junctions in the cell layers with a Millicell-ERS is performed based on the measurement of trans-epithelium electrical resistance (TEER). Until the TEER shows constant and saturated values (45–50 Ωcm^2 without the blank value of the membrane itself), the culture medium of A549 cells in the Ap side is removed and the ALIC is started. The chemical-free SPM samples are sterilized by ethylene oxide gas (EOG) performed with a commercially available sterilization system. After being measured carefully with a digital weighting machine, the EOG-sterilized SPM samples are loaded directly to the Ap side of an A549 cell layer in ALIC. In 12 well plates, A549 and Hep G2 cells are seeded at 1.0×10^5 cells/cm^2, cultured, and used after they reach confluence. In the preparation of the extracts of the SPM samples, the suspension of 24 mg of each sample and 200 μL of DMSO is sonicated for 15 min and centrifuged for 10 min at 15,000 rpm, and the supernatant is diluted with culture medium to adjust the final concentration of DMSO in the culture medium 0.5% and the highest concentration of SPM in the culture medium 600-μg-SPM/ mL-culture medium, which corresponds to the cell-surface-area-based load of the 158-μg-SPM/cm^2-cell surface.

The suspension of an SPM sample in 100 mL of CH_2Cl_2 is sonicated for 15 min and filtered with a membrane that has a 0.1-μm pore size, the filtrate is evaporated, and the residue is resolved in acetonitrile [24]. The concentration of the chemicals in the SPM extracts is determined using a multicolumn high-performance liquid chromatography (HPLC)/spectrofluorometer/computer system. According to EQ = S([PAH]$_1$EF$_1$ + ... + [PAH]$_n$EF$_n$), the chemically derived induction equivalent (EQ) is calculated to standardize the biological effect of the chemical on the EROD capacity of the Hep G2 cells [25].

After 48 h of exposure of SPM, each culture is rinsed twice with PBS, and to the Ap and BL sides, 500 μL and 1.5 mL of the substrate solution of acid phosphatase is added, respectively. For the monolayer culture, 1 mL of AP solution is added, the 12 well plates are incubated at 37°C for 2 h, the absorbance is measured at 405 nm with the value of cell-free well as control, and the cell survivability is calculated by referencing a predetermined calibration curve.

A549 and Hep G2 cells are cultured in 12 well plates, washed twice with PBS, and loaded with 10-μM 7-ethoxyresorufin in the presence of 10-μM dicumarol in culture medium. After 1 h of incubation, the fluorescence intensity is measured (530-nm excitation wavelength, 585-nm emission wavelength) to detect the formed resorufin. The intensity is calibrated to the resorufin concentration using a standard curve.

5.2.2.3 PWM-Induced IgM Assay for Detecting the Toxicity of Chemicals

The toxicants containing sample are dissolved in Ca, Mg-free PBS or DMSO, or the vehicle solvent (10 μL each). Spleen cells are aseptically removed from BALB/ cAnN mice (8–9 weeks old) [26], washed in Hanks' balanced salt solution supplemented with 10-mM HEPES buffer (pH 7.4), and diluted by RPMI 1640 medium containing 10% heat-inactivated fetal bovine serum, 2-mM glutamine, 0.4-M sodium pyruvate, 20-mM mercaptoethanol, non-essential amino acids, and antibiotics to prepare suspensions of 10^6 cells/mL. Cells are cultivated in a 24-well culture plate under 5% CO_2-air and the presence of PWM at 37°C for 2 to 6 days, supplemented with chemical toxicants containing a sample at the beginning of the culture. A total of 700 μL of culture media and residual cells in 300 μL of media are used for total IgM assay and cell proliferation assay, respectively.

The 96-well microtiter immunoplates are precoated overnight at 4°C with 100 μL of rabbit anti-mouse IgM antibody (IgG) in PBS (pH 7.4), washed with PBS containing 0.05% Tween20, and blocked overnight at 4°C with 150 μL of PBS containing 1% BSA. Each well is washed with PBS; into each well, 100-μL aliquots of culture media mentioned above (usually 11-fold diluted media) or reference mouse serum or standard mouse IgM diluted with 1% BSA-PBS are added, left at room temperature for 60 min, washed, and incubated; and 100-μL aliquots of horseraddish peroxidase-conjugated goat antimouse IgG in PBS are added. Sixty minutes later, to each well, 150-μL aliquots of substrate solutions containing 0.4% o-phenylenediamine and 0.01% hydrogen peroxide in 0.1-M citrate–0.2-M phosphate buffer are added and incubated. Fifteen minutes later, 50 μl of H_2SO_4 (0.5 M) is added to terminate the enzyme reaction. Using a microplate reader the optical density at 490 nm of each well is read.

After the addition of 30 μL of TetraColar One reagent, 300 μL of media mentioned above are incubated under 5% CO_2-air at 37°C for 2 h. The optical density at 450 nm of the media containing the water-soluble tetrazolium salt produced by the enzyme reaction is read, and the cell number in cultures is measured by the modification (WST-8 method) of the MTT assay for mitochondrial dehydrogenase activity in viable cells [27].

5.2.2.4 GFP-Based Cell Assay for Detecting the Toxicity of TCDD and Related Chemicals

By excising the 1846 base-pair (bp) Hind III fragment from the plasmid pGudLuc1.1, pGreen1.1 is created [28]. pGreen1.1 contains the 480-bp dioxin-responsive domain from the mouse CYP1A1 gene inserted upstream of the mouse mammary tumor virus (MMTV) promoter, confers dioxin responsiveness upon the MMTV promoter and adjacent reporter gene, and is inserted into the Hind III site immediately upstream of the enhanced GFP reporter gene in the plasmid pEGFP-1.

Plates of the mouse hepatoma (Hepa1c1c7, approximately 80% confluent) cells maintained in αMEM containing 10% fetal bovine serum are transfected with the construct pGreen1.1 (20 μg) using polybrene, grown for 24 h, split 1 to 10, and replated into selective media. After growth for 4 weeks, resistant clones are isolated and screened for the EGFP assay. H1G1.1c3 cells and DMSO (1% maximum final concentration) or TCDD (1 nM in DMSO) are cultured in 6-well culture plates at 37°C for 24 h, harvested by scraping into lysis buffer (50-mM NaH_2PO_4, 10-mM Tris-HCl pH 8, 200-mM NaCl), and lysed by repeated passage through a 27-gauge needle. The medium is centrifuged, and the fluorescence of an aliquot of the supernatant is determined (460-nm excitation wavelength, 510-nm emission wavelength). In the microtiter plate analysis of EGFP, the intact cells are plated into 96-well tissue culture dishes at 7.5×10^4 cells/well and allowed to attach for 24 h. The selective media are then changed to 100 μL of nonselective media containing the test chemical or DMSO (1% final solvent concentration). EGFP levels of the intact cells in the nonselective media are measured (at the indicated time points, 485-nm excitation wavelength, 515-nm emission wavelength). To normalize results between experiments, the instrument fluorescence gain setting is adjusted so that the level of EGFP induction by 1-nM TCDD produces a relative fluorescence of 9000 relative fluorescence units (RFUs). To photograph the cells, after growth on 25-mm round cover slips for 24 h and treated with DMSO or a 1-nM TCDD containing sample for 48 h, H1G1.1c3 cells are replaced into PBS. Cell fluorescence is visualized at a 490-nm excitation filter and 535-nm emission filter.

Recombinant mouse hepatoma (H1L1.1c2) cells grown in 24-well microplates are incubated with DMSO (10 μL/mL) and a TCDD containing sample in DMSO or its related chemical containing sample in DMSO at 37°C for 4 h; luciferase activity of cells in each well is determined and normalized to sample protein concentration using fluorescamine with bovine serum albumin as the standard [29, 30].

The complementary pair of 5′-GAT-CTG-GCTCTTCTCACGCAACTCCG-3′ and 5′-GATCCGGAGTTGCGTGAGAAGAGCCA-3′ (corr-esponding to the AhR binding site of DRE3 and designated as the DRE oligonucleotide) is radio-labeled with $[\gamma^{12}P]$ATP (6000 Ci/mmol). Gel retardation analysis of cytosolic AhR complexes transformed *in vitro* with a TCDD containing sample (20 nM) or related chemical containing sample is performed and protein–DNA complexes are visualized. The amount of $[^{32}P]$-labeled DRE in the induced protein–DNA complex is determined.

5.2.2.5 RACB Protocol for Detecting Ovarian Toxicity of Xenobiotics

RACB protocol consists of Task 1 (initial dose-setting study), Task 2 (continuous breeding phase), Task 3 (crossover mating trial), and Task 4 (F_1 offspring dose). In Task 2, control (n = 40 animals/sex) and up to three treatment groups (n = 20 animals/sex) are implicated. In Task 2, F_0 (parental) rodents are exposed to a chemical containing sample during a 7-day premating period, randomly assigned to a mating pair, and treated with the same chemical containing sample and dose throughout a 98-day period of continual cohabitation (during which multiple litters are born). To encourage immediate remating, in early pregnancies, neonates are removed from the dam within 12 h. The principal toxicity in Task 2 is aberrant reproductive performance in F_0 rodents as indicated by alterations in the number of litters per breeding pair or neonatal body weight and sex ratio. If a positive toxicity is observed in Task 2, Task 3 should be performed to determine whether males or females are more sensitive. For this phase, both male and female high-dose F_0 mice are paired to control F_0 mice of the opposite sex, and reproductive performance is compared to determine the affected sex (es). If a negative toxicity is observed in Task 2, Task 3 is omitted. After 98 days, breeding pairs are separated and continuously treated until the F_1 generation is delivered and weaned. In Task 4 the F_1 offspring of Task 2 parents are dosed until 74 ± 10 days of age, at which time male and female animals from the same treatment group but different litters are mated (n = 20 /sex/group) to generate F_2 litters. The F_1 offspring have been exposed to the chemical as gametes and as young adults during prenatal and postnatal development.

Ovaries from Task 1 (approximately 50 days old), F_0 parents from negative Task 2 (approximately 215 days old) or from Task 3 (approximately 240 days old), or F_1 offspring from Task 4 (approximately 120 days old) are removed at necropsy, trimmed of fat, fixed by immersion in Bouin's solution for 12 to 24 h, and transferred to 70% ethanol for storage and transport. An intact ovary from each animal is dehydrated in graded alcohols and xylene, embedded in a longitudinal orientation in separate paraffin blocks, and sectioned serially at 6 μm (approximately 400 sections per ovary). To each slide (approximately 40 slides per ovary), two rows of five sections retained in sequence are applied and stained with hematoxylin and eosin. Counts of ovaries from 10 mice per group are gathered. Ovaries are available from only 5 to 6 treated animals in F_1 offspring from two Task 4 EGME assays and from 9 treated animals from Task 2 oxalic acid study. To produce an equal number of control tissues for analysis from untreated animals of the same studies, the ovaries are also chosen randomly. Beginning with the first section of the first slide, sections from each ovary are examined. At least from the third or fourth section (a distance of 18 to 24 μm into the ovary), follicles are encountered and then categorized and enumerated. Differential counts from every tenth section or approximately 40 sections per ovary are made.

The ovarian follicles are categorized to small, growing, and antral ones by (1) the relative cross-sectional diameter of the follicle as measured from the outer margins of the granulosa cell layers, (2) the number of granulosa cell layers, and (3) the nature of the antral space. Small follicles, approximately 20 μm in mean diameters, consist of an isolated oocyte or an oocyte surrounded by a partial or unbroken, single layer of granulosa cells. Growing follicles, 20 to 70 μm, have an oocyte surrounded by a multilayered, solid mantle of granulosa cells. Antral follicles, more than 70 μm, are characterized by a central oocyte and fluid-filled space bordered by hundreds of layered granulosa cells [31]. Each counting session is limited to no more than 3 h to limit fatigue.

5.2.2.6   Uroepithelial Cell Assay for Detecting Urotoxicity of Chemicals

Male beagle dogs (31 months old) are individually housed at $23 \pm 2°C$ and a relative humidity of $60 \pm 20\%$ with a 12-h light/dark cycle. The dogs from a control group are anesthetized by sodium pentobarbital (25 mg/kg, i.v.) and killed by exsanguination. The bladder is aseptically excised, then lengthwise incised, and washed three times with the Krebs solution containing 110-mM NaCl, 5.8-mM KCl, 25-mM $NaHCO_3$, 1.2-mM KH_2PO_4, 2.0-mM $CaCl_2$, 1.2-mM $MgSO_4$, and 11.1-mM glucose (pH 7.4) at 4°C. After removal of excess fatty tissues, the bladder is transferred, mucosal side down, to a metal rack with 10 sharp metal pins along each edge [32], placed in 4°C Krebs solution, and the smooth muscle layer is carefully removed. The tissue is stretched, mucosal side up, across the metal pins on a $10 \times 10/cm^2$ plate, incubated in the minimum essential medium (MEM) containing 1% (v/v) penicillin/streptomycin/fungizon (PSF), 2.5-mg/mL dispase, and 20-mM MEM/PSF/dispase solution (pH 7.4) at 4°C for 24 h, the MEM/PSF/dispase solution is aspirated, the stripped mucosa is transferred to a sterile 150-mm culture dish, and uroepithelial cells are scraped from the connective tissues with cell scrapers, which are suspended in 20 mL of trypsin-EDTA (0.25% trypsin and 1-mM EDTA·4Na) and incubated at 37°C for 30 min. Later, the single cell suspension is diluted with MEM containing 1% PSF, 5% FBS, and 20-mM MEM/PSF/FBS solution (pH 7.4) to 50 mL, spun down at 4°C, and centrifuged (1000 g, 5 min). The supernatant is aspirated to prepare the suspension of the cells in 50 mL of MEM/PSF/FBS solution. The cells are then rewashed in 50 mL of the keratinocyte-SFM medium and resuspended at $6.0–7.0 \times 10^5$ cells/mL.

By mixing 5 mg of type IV collagen, 100 μL of glacial acetic acid, and 50 mL of distilled water, the collagen solution is prepared, kept overnight at 4°C, sterilized with a 0.22-μm bottle top filter, and stored at 4°C. The keratinocyte medium is added to both chambers and incubated at 37°C for 2 h. Before use, the collagen solution is diluted 1:9 with 10-mM Na_2CO_3-HCl (pH 9.0), of which 500 μL is added to each apical chamber after aspirating the keratinocyte medium and incubated at 37°C for 1 h. Before plating, the collagen solution is aspirated, 0.5 mL of the cell suspension is added to the apical chamber, and 1.5 mL of keratinocyte medium is added to the basal chamber. After incubation at 37°C for 3 days, when the transepithelial electric resistance (TER) reaches levels of approximately $1000 \Omega cm^2$ or higher, the apical and basal media are aspirated and replaced with 0.5 and 1.5 mL of the keratinocyte medium containing 1-mM $CaCl_2$ (KM/Ca solution), respectively, and TER is measured. Before use, by immersing in 70% ethanol, the electrodes are sterilized and washed with sterile PBS, and Ωcm^2 are calculated.

Cultured cells on a 12-mm transwell filter are fixed in 2% glutaraldehyde of 0.1-M phosphate buffer (pH 7.4), rinsed with 8.2% sucrose, postfixed in 1% OsO_4 of the buffer, dehydrated with alcohol, and embedded in epoxy resin. For light microscopic examination, the semi-thin sections are stained with 1% toluidine blue. To observe cells with a transmission electron microscope, ultrathin sections are stained with uranyl acetate and lead citrate. For immunofluorescence staining, the cultured cells on a 12-mm transwell filter are successively fixed with HISTO-CHOICE at room temperature for 1 min and acidic methanol (95% methanol and 5% glacial acetic acid) at $-20°C$ for 15 min, then incubated successively with anti-ZO-1 rabbit polyclonal antibody diluted 1:50 with PBS for 1 h and FITC-conjugated donkey antirabbit IgG diluted 1:40 with PBS for 30 min, or incubated successively with anti-E-cadherin mouse monoclonal antibody 1:100 with PBS and FITC-conjugated goat antimouse IgG diluted 1:10 with PBS. To observe cells with a confocal laser scanning microscope, the cells are washed three times for 5 min with PBS, the cell-grown transwell filters are cut, transferred to a slide glass, and mounted with the mounting medium.

After plating on a 12-mm transwell filter, TERs are monitored for 20 days, during which the medium is replaced every 3 days. When TER reaches $10,000 \Omega cm^2$ or more, the cultured cells are observed and their ZO-1 and E-cadherin are checked. Validating this culture system in the examination of the effects of cytochalasin-B on TER and ZO-1, cytochalasin-B is first dissolved in DMSO, diluted with the KM/Ca solution to make concentration of 1.6 µM (final DMSO concentration, 0.08%), 4 µM (final DMSO concentration, 0.2%), and 10 µM (final DMSO concentration, 0.5%), and sterilized with a 0.22-µm membrane filter. TER is measured 0, 1, 2, 4, 8, 24, and 48 h after exposure, and immunofluorescence for ZO-1 is observed 48 h after exposure, with DMSO diluted 1:200 with KM/Ca solution serving as a negative control solution.

5.2.2.7   Mating Efficiency Assay for Detecting Toxicity of Endocrine Disruptors

In the pheromone signaling pathway experiment, yeast *Saccharomyces* cerevisiae strains W303A (MATa; ade2, his3, trp1, ura3) with pSL307 plasmid fused to the FUS1-lacZ are cultured in SD medium (0.67% yeast nitrogen base and 2% glucose/L) supplemented with adenine, l-histidine, and l-tryptophan at 30°C overnight. To fresh YPD medium (10 g of Bacto-yeast extract, 20 g of Bacto-peptone/L, and 2% glucose) or SD medium one tenth of the overnight culture is transferred, incubated at 30°C for 3 h, and the cultures that are adjusted at the logarithmic phase (OD = 0.8–1.0) of growth are harvested and subsequently used at 30°C [33].

W303A and W303A with fused plasmid pSL307 are incubated in YPD or SD medium for 60 min with or without test compound, harvested, washed three times with distilled water, and incubated in fresh YPD or SD medium with 5-mM α factor for 120 min to observe shmoo formation. Using a β-galactosidase activity assay, the pheromone response pathway in the W303A strain with pSL307 fused to the

FUS1-lacZ reporter gene is quantified [34]. Yeast cells are centrifuged at 3000 g, pellets are washed twice in distilled water, resuspended in 1.0 mL of Z buffer (21.5-g $Na_2HPO_4 \cdot 12H_2O$, 6.2-g $NaH_2PO_4 \cdot 2H_2O$, 0.75-g KCl, 0.246-g $MgSO_4 \cdot 7H_2O$, 2.7-mL β-mercaptoethanol/l, pH 7.0) with glass beads, with which three drops of chloroform and two drops of 0.1% SDS are mixed by ortex mixing, and incubated at 28°C for 5 min. After the addition of 0.2 mL of 4-mg/mL p-nitrophenyl-β-D-galactopyranoside at 28°C, the reaction is started, and by the addition of 0.5 mL of 1-M Na_2CO_3, the reaction is stopped. On removal of the cells by centrifugation, the superant is measured at 420 nm. In the mating efficiency assay, the diluted W303A strain is treated with the solution of test chemicals in YPD medium for 60 min and spread on both SD and YPD media. The colony-forming units (CFUs) of W303A strain on YPD medium are incubated for 2–3 days to give the cardinal number of mating efficiency. Simultaneously, the 144-3A strain (MATa; ura3, leu2, his4) is incubated on YPD medium for 60 min and streaked with diluted W303A strain on SD medium. Conjugated diploid 144-3A strain/W303A strain on SD medium complements the auxotrophy of each haploid cell.

5.2.2.8 Pnar-gfp Assay for Detecting Carcinogenic Toxicity of Nitrate

Escherichia coli DH5α (Φ80dlacZΔM15 recA endA gyrA thi hsdR supE relA deoRΔ [lacZYA-argF]), pUC18 (for routine cloning procedures), pCRII (for TA cloning of PCR products), pGreen-TIR containing the gfp gene, and pMV4 containing the narG promoter are used [35, 36]. Minimal medium (pH 7.0) contains 3.9-mM KH_2PO_4, 6.1-mM K_2HPO_4, 1.5-mM $(NH_4)_2SO_4$, 0.2-mM $MgSO_4 \cdot 7H_2O$, 22.4-μM $MnSO_4 \cdot 4H_2O$, 0.9-μM $FeSO_4 \cdot 4H_2O$, 4.4-μM $CaCl_2 \cdot 2H_2O$, 400-mg/L casamino acids, 14.8-μM thiamine hydrochloride, and 22.2-mM glucose. LB medium (pH 7.5) contains 10-g/L bacto-tryptone, 5-g/L yeast extract, and 10-g/L NaCl. *E. coli* DH5α (pPNARGFP) cultures are supplemented with 100-μg/mL ampicillin. Primers J1 (5′-CATCGAATTCTCCTGTGGGAGCCT-3′) and J2 (5′-CTGGCATGCATTCACTTGCCGCCTT-3′) are designated for the amplification of the nar promoter [37, 38]. J1 contains an artificial GAATTC (EcoRI site) and anneals to nucleotides +3 to −23 in the nar region, where +1 is the first nucleotide of the start codon. J2 contains an artificial GCATGC (SphI site) and anneals to nucleotides −438 to −414. After purification using Qiagen Plasmid Midi Kits (Qiagen, USA) or Wizard Plus SV miniprep kits (Promega, USA), the DNA sequence is determined by MWG Biotech (Milton Keynes, UK). The whole bacterial cells are measured with an excitation wavelength of 480 nm (10-nm bandwidth) and an emission wavelength of 510 nm (5-nm bandwidth) on a fluorescence spectrometer. The nitrate-induced fluorescence image of *E. coli* DH5α (pPNARGFP) is obtained with a BP465-495 excitation filter and a Ba 520 emission filter on a fluorescence microscope equipped with epifluorescence optics. The measurements of optical density are performed on a spectrophotometer at 600 nm.

5.2.3 ANTIVIRAL AND ANTICANCER ASSAY

The elucidation of the mechanisms of special enzymes provides the bases for highly specific and sensitive bioassays focused on the target enzymes and cells. The alkaline phosphatase (AP) enzyme in the human endometrial adenocarcinoma cell line is sensitive to estrogen stimulation. Selective inhibition of the lyase activity of DNA polymerase sensitizes cancer cells to DNA-damaging agents. The human immunodeficiency virus type-1 reverse transcriptase (HIV-1 RT) and protease play key roles in HIV replication. The overexpression of the P-glycoprotein can produce cancer cell multidrug resistance (MDR). The receptor phosphorylation may induce ligand-induced receptor tyrosine kinase activation. Based on these facts, the human breast cancer cell assay [39], the Ishikawa cell and rat assay [40], the RT–PCR and swine assay [41], the antitubercular and cytotoxic assay [42], the DNA polymerase β lyase assay [43], the HIV-1 protease and reverse transcriptase kinetic assay [44], the P-glycoprotein pump assay [45], and the kinase receptor activation (KIRA) assay [46] are established and can be used in bioassays.

5.2.3.1   Human Breast Cancer Cell Assay for Detecting Anticancer Activity

Toxicity against brine shrimp *Artemia salina* nauplii is carried out in 96-well microplates using emetine hydrochloride as a positive control [47]. The selection of human breast cancer cell lines (MCF-7, MCF-7/ADR, MT-1, and MDA-MB-435) provides cells with a variety of receptor types; for instance, MCF-7 cells express high levels of estrogen receptors, MCF-7/ADR cells are resistant to doxorubicin, MT-1 cells have low estrogen receptor levels [48], the MDA-MB435 cell line has no estrogen receptors. In the determination of inhibition of cell growth in 96-well microtitre plates, the MTT assay is used [49]. The samples of test compounds in DMSO are diluted in medium so that the concentration of DMSO is more than 1% and does not affect cell growth. Overall, 180-µL aliquots of a cell suspension (1–2 $\times 10^4$ cells/mL) are plated into microplate wells and incubated overnight at 37°C in air containing 5% CO_2, and 150 µL of the culture medium is replaced with 150 µL of fresh medium, to which 20 µL of the solution of the test sample in medium is added to give final concentrations in the plate of 10, 1, 0.1, 0.01, and 0.001 µg/mL. Control wells with medium only and a positive control with 5-fluorouracil are used in each test. After 96 h of the incubation, 150 µL of medium is replaced with 150 µL of fresh medium, to which 20 µL of the 5-mg/mL MTT solution is added. After incubation for 4 h, 180 µL of medium is replaced with 180 µL of DMSO and carefully mixed and then determined at 540 nm using an enzyme-linked immunosorbent assay (ELISA) multiwell spectrophotometer.

5.2.3.2   Ishikawa Cell and Rat Assay for Detecting Antiestrogens

The cells are grown in 96-well plates in estrogen-free medium (phenol red free, with charcoal-stripped calf serum) and contain test compounds and antiestrogens at concentrations that are varied over several log orders. For the antiestrogen assay, to the cells, a range of concentrations of samples are added concurrently with 1-nM antiestrogens. After grown for 3 days, to determine AP activity, the cells are frozen, defrosted, and incubated with p-nitrophenylphosphate at room temperature, and the hydrolysis product p-nitrophenol is measured kinetically at 405 nm [50].

The specificity of the antiestrogenic activity is determined by using ERα and ERβ in ER element (ERE)-transfected human choriocarcinoma JAR cells, which are transfected with plasmids containing a consensus ERE fused to a firefly luciferase reporter gene and separately with the expression vectors for either human ERα or human ERβ. JAR cells are routinely cultured in RPMI 1640 supplemented with 10% fetal bovine serum, 0.5% non-essential amino acids, and 1% PEST (100-U penicillin/mL and 100-μg streptomycin/mL). After seeded in six-well plates for 24 h, cells are transfected using the Mirus Trans IT reagent in a serum- and antibiotic-free mixture of phenol-red free OptiMEM with 0.1–0.4-μg pC × N2 human ERα or pC × N2 h-ERβ, and a 0.75-μg 3 × ERE-TATA-Luc reporter constructed by introducing an HpaI/BglII fragment containing 3 × ERE-TATA into SmaI/BglII of the pGL3-Luc basic vector [51–53]. Medium is replaced with a phenol red-free RPMI containing 10% dextran-coated charcoal-treated calf serum and 0.5% non-essential amino acids. Twenty-four hours later, antiestrogens are added. Cells are incubated at 37°C in 5% CO_2 for 12 h, harvested in 10-mM Tris-HCl/10-mM EDTA/150-mM NaCl, and centrifuged at 4000 g for 4 min. The supernatant is removed, the cell pellets are lysed in Lysis Buffer 2, and luciferase activity is measured.

After the AP assay and washing three times with PBS, the Ishikawa cells are lysed using 1% Nonidet P-40 and 0.1% sodium dodecyl sulfate in the presence of protease inhibitors. Before the antibody incubation, proteins (25 μg/well) separated by SDS–PAGE on ice using 10% polyacrylamide gel are transferred to nitrocellulose membranes stained with Ponceau Red to ensure proper transfer. After blocking the membranes with 5% powdered milk in water, immunoblotting is performed. The blots are incubated with the ERα monoclonal antibody clone 6F11 overnight at 4°C. Using peroxidase-labeled horse antimouse secondary antibody and Chemiluminescence Reagent Plus, ERα is detected by Western blotting. Using a digital imaging analysis system, the intensity of the signal is analyzed. To normalize the amount of protein loaded in the gels, β-actin is used as an internal control.

The uterotrophic assay is performed with stimulation in immature rats [54]. Female SD rats (22 days old) are injected s.c. daily for 3 days with antiestrogens, control animals receive the vehicle (0.1-mL sesame oil), then animals are killed, and uteri are removed, dissected, blotted, and weighed. To determine whether antiestrogens have tissue-selective effects in cholesterol, uteri, and bone, ovariectomized female SD rats (250 g) are injected with antiestrogens s.c. for 35 days, and then killed by exsanguination under ether anesthesia. The total cholesterol concentration of the serum is determined by a commercial chromogenic assay. The uteri are dissected, weighed, fixed in formalin, and imbedded in paraffin to prepare for 5-μm sections. Using the Openlab image analysis system, endometrial luminal epithelium and glandular cell height are measured. The tibia free of extraneous tissue are histomorphometrically analyzed. The bones fixed in 70% ethanol are dehydrated in graded ethanol and cleared in toluene. Then the specimens are infiltrated with increasing concentrations of methymethacrylate (MMA) and embedded in MMA [55]. After polymerization, MMA blocks are cut to size, sanded, and polished to the appropriate level to prepare 4–5-μm sections. The sections are mounted on gelatin-coated slides, stained with toluidine blue, and measured [56].

5.2.3.3   RT–PCR and Swine Assay for Detecting Anti-HEV Antibody

To choose seronegative pigs for inoculation, 75 SPF pigs (2 weeks old) are tested by ELISA for swine Hepatitis E virus (HEV) IgG antibodies. Before inoculation, the pigs are allowed to acclimate to the research facilities for 1 week. Tissues (liver, heart, pancreas, or skeletal muscle) and feces of HEV-infected SPF pigs are collected at 3–7, 14–20, and 27–55 days postinoculation (DPI) and are pooled and stored at −70°C for inoculation [57]. Each inoculum is prepared as a tissue homogenate or fecal suspension (10%, w/v) in PBS and tested by a semiquantitative RT–PCR for swine HEV RNA. The positive control inoculum is a standard infectious pool of swine feces with a $10^{4.5}$ 50% pig infectious doses (PID_{50}) of swine HEV/mL.

Total RNA is extracted from 100 μL of each sample by TriZol reagent and tested by a nested PCR with primers located in the putative capsid gene (ORF2) [58, 59]. In the first round PCR, the fragment of 404 base pairs (bps) with the forward primer F1 (5′-AGCTCCTGTACCTGATGTTGACTC-3′) and the reverse primer R1 (5′-CTACAGAGCGCCAGCCTTGATTGC-3′) is formed. In the second round PCR, the fragment of 266 bps with the forward primer F2 (5′-GCTCACGTCATCT GTCGCTGCTGG-3′) and the reverse primer R2 (5′-GGGCTGAAC-CAAAA TCCTGACATC-3′) is formed. With R1 reverse primer and SuperScript II reverse transcriptase (GIBCO/BRL), total RNA is reverse transcribed at 42°C for 1 h and the resulting cDNA is amplified by PCR with ampliTaq Gold DNA polymerase. The PCR reaction is carried out for 39 cycles of denaturation at 94°C for 1 min, annealing at 52°C for 1 min, extension at 72°C for 1.5 min, and incubation at 72°C for 7 min. Overall, 10 μL of each round of PCR are mixed as the template. By gel electrophoresis, the amplified PCR products are examined. The virus titer of inocula is calculated and expressed as a genome equivalent (GE)/mL. After 7 to 21 days, serum samples of inoculated pigs are collected and tested by RT–PCR for swine HEV RNA.

Anti-HEV IgG antibodies in swine sera are detected using a standardized ELISA. A purified 55-kDa truncated recombinant putative capsid protein of human HEV strain Sar-55 is used as the antigen that cross-reacts well with the swine HEV. Peroxidase-labeled goat antiswine IgG is used as the secondary antibody. Duplicates per serum sample are used.

5.2.3.4 Antitubercular and Cytotoxic Assay

Using the microtiter plate alamarblue technique, the antitubercular activity is assessed against a nonvirulent strain of *Mycobacterium tuberculosis* ($H_{37}Ra$). As positive controls, isoniazid and kanamycin sulfate exhibit MICs in the ranges of 0.29–0.66 and 3.5–8.5 μM, respectively. According to the sulforhodamine B procedure, cytotoxicity is determined. MCF-7 (human breast adenocarcinoma), HeLa (human cervical carcinoma), KB (human oral epidermoid carcinoma), and HT-29 (colorectal carcinoma) are used as the target cell lines. After exposure to test samples, cell viability is determined colorimetrically at 515 nm. Dose-response evaluations yield a concentration mediating a 50% cytotoxic response (IC_{50}).

5.2.3.5 DNA Polymerase β Lyase Assay

Using terminal deoxynucleotidyltransferase + [α-^{32}P]ddATP, a 36-nucleotide oligodeoxyribonucleotide containing uridine at position 21 is labeled at its 3′-end and the product is subjected to 20% denaturing polyacrylamide gel electrophoresis for purification. Using autoradiography, the interesting band is visualized and excised. By heating to 70°C for 3 min and then slow cooling to 25°C, the DNA substrate is annealed to its complementary strand. To 200 μL of 354-nM [α-^{32}P]-labeled double-stranded oligodeoxynucleotide with uridine at position 21, 10-mM Hepes-KOH (pH 7.4), 5-mM MgCl$_2$, 50-mM KCl, 10-mg/mL bovine serum albumin, 2.4 U of uracil-DNA glycosylase, and 3 U of AP endonuclease are added, incubated at 37°C for 20 min, and an apurinic (AP) site is created in the [α-^{32}P]-labeled double-stranded oligodeoxynucleotide. To 5 μL of the above reaction mixture, the test samples and 0.172 U of rat DNA polymerase β are added. After incubation at room temperature for 30 min, the reaction is terminated by adding 0.5-M NaBH$_4$ (50-mM final concentration) and incubated at room temperature for 10 min. The reaction products are incubated at 70–80°C for another 20 min, separated on a 20% denaturing polyacrylamide gel, and visualized by autoradiography.

A total of 200 μL of culture samples containing approximately 1.0×10^4 A549 cells (maintained in Kaighn's modification of Ham's F12 medium (F12K) with 2-mM L-glutamine supplemented with 1.5-g/L sodium bicarbonate and 10% fetal bovine serum at 37 °C in air containing 5% CO$_2$) are placed in each well of 96-well culture plates, treated with test samples, and incubated at 37°C for 48 h in air containing 5% CO$_2$. In the determination of cytotoxicity, the culture medium is replaced with 15 μL of 5-mg/mL MTT per well, the samples are incubated at 37°C for 4 h in air containing 5% CO$_2$, 200 μL of DMSO is added, and the OD$_{570}$ value is obtained using a microplate reader.

5.2.3.6 HIV-1 Protease and Reverse Transcriptase Kinetic Assay

To the wells of a streptavidinecoated microtiter plate containing 20 μL of test sample solution and 4 ng of the HIV-1 RT in 20 μL of lysis buffer, 20 μL of the reaction mixture containing a homogenous template/primer hybrid ((rA)$_n$(dT)$_{15}$, 750-mA$_{260nm}$/mL final concentration) and a triphosphate substrate (dUTP/dTTP, 10-μM final concentration) are added. After the reaction is carried out at 37°C for 1 h, 200 μL of anti-digoxigenin-peroxide solution and ABTS [2,2–azino-bis-(3-ethyl benzothiazoline-6-sulfonic acid) diammonium salt] substrate are added for the coloring reaction. The absorbance of each well is measured at 405 nm with the reference wavelength at 490 nm and nevirapine as the reference compound. Using the same procedure mentioned above, except for the concentration of the enzyme solution (1 ng of the HIV-1 RT), the incubation time (30, 52, 80, 105, 130 min), and various concentrations of either the template/primer (1500, 750, 187.5 mA$_{260nm}$/mL) or the triphosphate substrate (20, 15, 10, 5, 2.5 mM) in the presence of the inhibitor, the enzyme kinetic assay is also performed. With a slight modified procedure to that mentioned above, the HIV-1 protease assay is performed. To the mixture of 1 μL of the solution of test compound in DMSO and 10.5 μL of the substr-ate solution (His-Lys-Ala-Arg-Val-Leu-(p-NO$_2$)-Phe-Glu-Ala-Nle-Ser-NH$_2$, 0.1 mg/mL in HIV-1 protease assay buffer), 0.5 μL of the recombinant protease solution (0.3 mg/mL) is added and incubated at 37°C for 15 min, the reaction is stopped by the addition of 1.2 μL of 10% trifluoroacetic acid, and the reaction mixture is diluted with 20 μL of water. The hydrolysate and the remaining substrate are quantitatively analyzed by HPLC [column, Inertsil ODS-3 (4.6 × 150 mm), a linear gradient of CH$_3$CN (15% to 40%) in 0.1% TFA, an injection volume of 20 μL, and a flow rate of 1.0 mL/min, detection at 280 nm]. The hydrolysate and substrate are eluted at 8.61 and 10.84 min, respectively.

5.2.3.7 P-Glycoprotein Pump Assay

MCF-7R cells (human breast adenocarcinoma cell line, resistant to adriamycin) maintained at 37°C in humidified atmospheric air containing 5% CO_2 in nutrient mixture (F10/HAM, contains L-glutamine, with 10% fetal calf serum, 1% non-essential amino acids, 60-μg/mL tylosin, and 1% antibiotic/antimycotic solution) are seeded in the wells of 96-well tissue microtiter culture plates at 3×10^4 cells in 200 μL of medium per well, incubated at 37°C for 24 h in a humidified atmospheric air containing 5% CO_2, and the medium is replaced with fresh medium containing 0.3-μM rhodamine 6 G together with samples. To thorough mixing of 50 μL of each of these prepared mixtures, 450 μL of fresh medium containing 0.3-μM rhodamine 6 G is added. After incubation of the test compound, reserpine (positive control) or rhodamine alone (negative control) is added and the plates are incubated at 37°C for another 3 h. Then the cells are washed twice with 200 μL of ice-cold PBS and trypsinized with 100 μL of phenol-red free trypsin solution for 15 min, which is then transferred onto empty wells. Overall, 100 μL of 4% SDS in PBS is added and the plates are shaken for 2 h to solubilize the cells and release rhodamine 6 G, which is determined by measuring the fluorescence of the dye at excitation and emission wavelengths of 530/25 and 590/35 nm, respectively.

On Lab-Tek chamber slides, MCF-7R cells are seeded at 5×10^4 cells per chamber and incubated at 37°C for 24 h in a humidified atmospheric air containing 5% CO_2, and the medium is replaced with fresh medium containing 0.3-μM rhodamine 6 G with or without 50-μM reserpine, with 100-μg/mL samples in separate chambers of the slide. The cells are incubated at 37°C for another 3 h and imaged using fluorescence microscopy with a 510–560-nm band-pass excitation filter and a 590-nm long-pass emission filter set.

5.2.3.8 KIRA Assay

There are two different approaches using the KIRA format. The first approach, IGF-I KIRA, uses adherent MCF-7 cells derived from a human breast adenocarcinoma expressing an endogenous IGF-I receptor to measure the bioactivity of IGF-I [60]. MCF-7 cells are seeded in each well of a flat-bottomed 96-well culture plate and incubated overnight, the supernatants are decanted, and the plates are lightly blotted on paper towels. To each well, the medium containing either experimental samples or the recombinant IGF-I standards are added. The cells are exposed to ligand for 15 min, the supernatants are decanted, and the plates are blotted. After addition of lysis buffer containing Triton X-100, sodium orthovanadate, and a cocktail of protease inhibitors, the crude lysates are generated and then transferred to an ELISA plate coated overnight with a 3B7 antibody (5.0 mg/mL) and blocked with 0.5% BSA. After removal of unbound material, the degree of receptor tyrosine phosphorylation is quantified with biotinylated anti-phosphotyrosine monoclonal antibody followed by HRP-conjugated dextranstreptavidin, and visualized with the development of a tetramethyl benzidine (TMB) substrate. The absorbance is determined at 450 nm with a 650-nm reference wavelength.

The second approach, N-terminal polypeptide D (gD)·trkA KIRA using CHO cells stably transfected with a recombinant human trkA receptor with an gD flag, is developed to measure the bioactivity of NGF. A polypeptide flag is cloned onto the N-terminus or C-terminus of the full-length recombinant human receptor stably transfected into CHO cells. A 26-amino-acid polypeptide derived from HSV gD as a capture reagent in the ELISA phase of the KIRA and a mAb 3C8 antibody as the capture antibody are used.

5.2.4   HEPATOXICITY AND HEPATOPROTECTIVE ASSAY

Mitochondrial proliferation of the chemicals related to hepatocellular carcinomas and their hepatocarcinogenecity are important factors for health. The renal adeno-carcinomagenecity, hepatic adenoma, and carcinomagenecity induced by chemicals can promote the formation of preneoplastic foci in rats or mice. The clonal growth of glutathione-S-transferase (GST-P) enzyme-altered foci and the histopathological change in the liver of rats or mice may relate to liver carcinogenesis of chemicals. In the liver injury induced by chemicals, such as CCl_4, Bacillus Calmette Guerin (BCG), and lipopolysaccharide (LPS), the activities of protein, glucose 6-phosphatase, amidopyrine *N*-demethylase, and aniline hydroxylase; the levels of hepatic triglycerides and lipid peroxidation; the concentrations of nitric oxide (NO), content of malondialdehyde (MDA), and superoxide dismutate (SOD); and the viability of thymocytes can be changed significantly. In the hepatocarcinogenesis induced by chemicals such as chloral hydrate, the body weight of B6C3F1 mice can be decreased significantly. Proliferation factor (PF) and hepatocyte growth factor (HGF) are involved in liver regeneration cascade and TGF-β-induced growth inhibition of CCL-64 cells; thus, PFs present in the partially hepatectomized rat serum can serve as an index of liver regeneration cascade and two cytokines (HGF and HGF) may result in an additive effect on proliferation. Based on these facts, the GST-P enzyme-altered foci assay [61], partial hepatectomy assay [62], hepatoprotective assay [63], immunological assay [64], body-weight-based liver tumor incidence assay [65], hepatocyte primary culture assay [66], and CCL-64 cell growth assay [67] are established.

5.2.4.1   GST-P Enzyme-Altered Foci Assay

According to the required levels (0%, 0.03%, 0.1%, and 0.3%), the test chemical is incorporated into an irradiated (6.0 kGy) powder diet MF. The stability of 0.1%, 0.03%, 0.5%, and 5.0% chemical in the prepared diets is previously confirmed for 6 weeks at room temperature. Male F344/DuCrj rats (5 week old) are given an approximately 1-week quarantine/acclimation period to monitor health conditions and body weights. The normal rats (6 week old) are randomly divided into 8 groups (18 rats each for groups 1–5, 9 rats each for groups 6–8). The rats of groups 1–5 receive an injection of the initiator N-nitrosodiethylamine (200-mg/kg body weight, i.p.) [68]. The rats of groups 6–8 receive an injection of the vehicle. Two weeks later, the rats receive an injection of the test chemical at the desirable dose for a suitable treating period. Three weeks after beginning the experiment, all rats are given two-thirds partial hepatectomy [69]. At week 8, all surviving rats are killed under ether anesthesia and their organs in the thoracic and abdominal cavities are examined macroscopically. Their livers are immediately excised and weighed to calculate the liver-to-body-weight ratio. For the 4–5-mm thick sections, the cranial and caudal parts of the right lateral lobe and the caudal part of the caudate lobe of all surviving rats are fixed in 10% buffered formalin solution, embedded in paraffin wax, sectioned, and stained immunohistochemically for glutathione S-transferase analysis (GST-P, ABC method) [70]. All GST-P positive hepatocytic foci larger than 0.2 mm in diameter (the lowest limit for reliable evaluation) are measured using a color image processor, and the numbers and areas (foci/cm^2) of the liver section are calculated. By microscopic analysis, BrdU-positive labeling indices (LIs) are quantified by randomly counting the number of positive nuclei per 1000 hepatocytes or number per unit area (mm^2) in sections stained immuno-histochemically for BrdU [71].

5.2.4.2 Partial Hepatectomy Assay

Male F344 rats (30 days old) acclimated for 4 weeks before the experiment are randomly divided into 3 groups. At week 0, the rats receive a single i.p. injection of the solution of N-nitrosodiethylamine (200 mg/kg) in 0.9% saline. Two weeks later, the rats receive daily gavage administration of corn oil or 0.1-mmol/kg test chemical in a corn oil vehicle through the remainder of the 8-week study. At week 3, all rats receive a partial hepatectomy. The rats are given food and water, and lighting is set on a 12-h light/dark cycle. On days 23, 26, 28, 47, and 56, at least five rats from each group are sacrificed by aortic exsanguination. Whole livers are removed, and the tissues are fixed in 10% neutral-buffered formalin, embedded in paraffin, serially sectioned at 5 μm, and mounted on microscope slides. Formalin-fixed sections are stained with hematoxylin and eosin to perform histopathological examination.

5.2.4.3 Hepatoprotective Assay

To induce liver injury, CF rats (150–200 g) and Swiss albino mice (20–30 g) of either sex are administered orally CCl$_4$ diluted with liquid paraffin. The animals of the vehicle control group are orally administered an equal volume of liquid paraffin. In hepatoprotective studies, four suitable doses of test compound and two standard doses of silymarin (25 and 50 mg/kg, p.o.) are fed to a respective group of rats 48 h, 24 h, and 2 h before and 6 h after CCl$_4$ (0.5 μL/kg, p.o.) intoxication. From the orbital sinus of all animals, blood is collected 18 h after CCl$_4$ intoxication for glutamic-pyruvic transaminase (GPT), glutamic oxaloacetic transaminase (GOT), and bilirubin analysis. In the posttreatment studies, the same doses of test compound and silymarin as mentioned for hepatoprotective studies are fed to rats 6 h, 24 h, and 48 h after CCl$_4$ (0.5 mL/kg, p.o.) intoxication. From all animals blood is collected 2 h after the last dose of test compound administration for GOT, GPT, and bilirubin analysis. The test compound and silymarin (50 mg/kg, p.o.) are fed to two different groups of rats at 48 h, 24 h, and 2 h before and 6 h after CCl$_4$ (100 μl/kg, p.o.) intoxi-cation, with the remaining two groups served as CCl$_4$ and vehicle control. Eighteen hours after CCl$_4$ intoxication, all animals of the four groups are fasted overnight and killed by decapitation. Livers are immediately excised and divided into two parts for preparing homogenate (10%, w/v), among which one part is homogenized in isotonic sucrose (0.25 M) for determining protein and glucose 6-phosphatase activity, and for preparing microsomes by calcium precipitation to determine amidopyrine N-demethylase and aniline hydroxylase activities with spectrophotometric methods, in which NADPH is used instead of the NADPH generating system, and another part is homogenized in isotonic PBS (0.01 M, pH 7.2, 0.15-M NaCl) for determining hepatic triglycerides and lipid peroxidation. The hepatoprotective activity is expressed as hepatoprotective percentage H and calculated by $H = [1 - (T - V)/(C - V)] \times 100$, wherein T is mean value of drug and CCl_4, C is mean value of CCl_4 alone, and V is the mean value of control-treated animals. In the determination of acute toxicity, different groups of mice (each ten) are fed with different doses of test compound, with one group with the same number of mice served as control. The animals are observed continuously for 1 h and then hourly for 4 h for any gross behavioral changes and further up to 72 h for any mortality.

5.2.4.4 Immunological Assay

A suspension of 2.5 mg of BCG (viable bacilli) in 0.2 mL of saline is injected via the tail vein into each mouse, and 10 days later a solution of 7.5 μg of LPS in 0.2 mL of saline is injected. The mice are anesthetized with ether, sacrificed by cervical dislocation 16 h after LPS injection, and trunk blood is collected into heparinized tubes (50 U/mL) and centrifuged at 4°C and 1500 rpm for 10 min. Serum is aspirated and stored at −70°C until assayed as described below. The liver is also removed and stored at −70°C until required.

For the *in vivo* experiment, the mice are equally divided into 5 groups randomly, including normal, model control, and test compound groups (3 different doses). Mice in test compound groups receive suitable doses using an 18-gauge stainless steel animal feeding needle for 10 days before LPS injection. Mice in the normal and model control groups are fed the same volume of vehicle only. For the *in vitro* experiment, the Kupffer cells isolated from normal and BCG priming rats are divided into 7 groups randomly, including control cells, cells added with LPS (5 μg/mL) alone, and cells added with LPS (5 μg/mL) and the test compound.

The liver of normal rat is initially perfused through the portal vein with D-Hank's until blood free and finally by recirculation with Hank's containing 0.5-g/L collagenase IV until the vessels are digested (up to 20 min). The liver is scraped using a cell scraper, filtered by a 100-μm filter, and stirred in Hank's containing 2.5-g/L pronase and 0.05-g/L DNase at 37°C for 20 min. After three times of centrifugation and washing at 4°C and 300 g for 10 min in Gey's balanced salt solution (GBSS), the cells are centrifuged in an 180-g/L Nycodenz gradient at 2500 g for 20 min. Kupffer cells are carefully sucked by cusp-straws at the pearl layer inderphase. The purified Kupffer cell fractions are finally collected by centrifugal elutriation. The Kupffer cells are washed with Hanks' and resuspended in RPMI 1640 medium containing antibiotics (100-U/mL penicillin, 100-mg/mL streptomycin), 2-mM glutamine, and 10% fetal calf serum. Overall, 1-mL aliquots containing 1×10^6 cells are added to 24-well culture plates. The cells are incubated for 60 min in a humidified atmosphere containing 5% CO_2 at 37°C. Nonadherent cells are removed, and adherent cells are washed twice with PBS. To observe the direct effect, the cells at a density of 1×10^6/mL are incubated with different concentrations of test compound. The cells

(1×10^6/well) are cultured for 48 h with 5-μg/mL LPS, the supernatants are collected, and the concentrations of TNF-α and NO are measured.

According to the *in vitro* liver injury model, BCG-induced Kupffer cells are isolated from the livers of the rats injected via the tail vein with 3 mg of BCG 10 days before, and hepatocytes are isolated from the normal rats. The hepatocytes (1×10^9/mL), different concentrations of test compound, and BCG-induced Kupffer cells (1×10^6/well) are cocultured for 48 h with 5-μg/mL LPS and the supernatants are collected. In a 96-well plate, 100 μL of cell culture supernatant and 100 μL of Griess reagent (10-g/L sulfanilamide and 1-g/L N-1-naphthylethylenediamine dihydrochloride in 2.5% phosphoric acid) are incubated at room temperature for 10 min. Absorbance at 540 nm is measured. The nitrite concentration is calculated by comparing samples with standard solutions of sodium nitrite produced in the culture medium.

Livers are thawed, weighed, homogenized with Tris-HCl (5 mM containing 2-mM EDTA, pH 7.4), and centrifuged (1000 g, 10 min, 4°C), and MDA and SOD in the supernatant are immediately analyzed. MDA in liver tissue is determined by the thiobarbituric acid method [72]. The assay for total SOD is based on its ability to inhibit the oxidation of oxyamine by the xanthine–xanthine oxidase system. The absorbance of the red nitrite produced by the oxidation of oxyamine is determined at 550 nm.

Thymocytes (2×10^6/well) from mice are cultured for 48 h in 96-well plates containing RPMI 1640 medium supplemented with 5-μg/mL concanavalin A and 0.1 mL of collected supernatant. Three hours before the termination of the culture, cells are pulsed with MTT stock (in sterile PBS, 5 mg/mL, stored in the dark at 4°C for up to 1 week, before use immediately filtered, 0.22 μm, to remove any formazan precipitate, 20 μL/well), returned to 37°C, and incubated for another 3 h. The plates are centrifuged at 1000 g for 10 min to form cell pellets and MTT formazan products. The supernatant is carefully aspirated without disturbing the pellets, and formazan is solubilized by adding isopropanol (100 μL of isopropanol : 200 μL of supernatant). Insoluble material is then removed by centrifugation at 1000 g for 10 min. The solubilized formazan in isopropanol is collected and distributed into 12-well flat-bottom ELASA plates at a final volume of 100 μL/well. Plates are read at 570 nm within 1 h of adding isopropanol.

5.2.4.5 Body Weight-Based Liver Tumor Incidence Assay

Addition to extra groups for high and low outliers, the B6C3F1 mice (at the 9-week age point) are assigned to 1 of 17 consecutive weight groups ranging from 20 to 57.5 g in 2.5-g intervals. Each weight group of mice developing a liver tumor is sorted for the calculation of the relative tumor risk. The mice with hepatocellular adenoma, carcinoma, or hepatoblastoma are designated as positive or otherwise designated as negative. The process is repeated for each age point to 68 weeks.

Approximately 4-, 5-, and 6-week-old male B6C3F1/Nctr BR mice are used at receipt, the initiation of controlled feeding, and on the first day of dosing, respectively. During the studies, the health of the mice is monitored. The dose for groups of 120 male mice receiving the solution of chloral hydrate in distilled water is 0, 25, 50, or 100 mg/kg (all at 5 mL/kg), 5 days per week for 104 to 105 weeks, each of which is divided into two dietary groups of 60 mice, and the vehicle controls received distilled water only. The *ad libitum* fed mice are fed available *ad libitum*, and the dietary controlled mice are fed in measured daily amounts. All mice have water available *ad libitum*. At the 71-week age point, 12 mice of each diet/dose group are euthanized for pathological and biochemical evaluation. At the 110-week age point, the remaining 48 mice of each diet/dose group are fasted overnight before necropsy. Among them the mice with liver tumors refer to mice bearing single or multiple hepatocellular adenomas, hepatocellular carcinomas, and/or hepatoblastomas.

From week 21, the weight dataset of the first experimental group is imported with the preceding week's dataset as a reference. According to the predefined sort criteria (including whether the weight value for the previous week is outside either the 5% or the 12% confidence limits of the idealized weight curve or whether the mouse gains or loses weight in the preceding week), each mouse weight from the dataset is sorted into its appropriate weight group and assigned either the corresponding *ad libitum* or the weight-reduced tumor risk value from the tumor risk tables. The resulting tumor risk dataset in the appropriate week's column of a new table is stored and proceeded to the next week's dataset. According to all weekly datasets from 21 to 68 weeks, the mean tumor risk for each mouse in the dataset is calculated, the survival time for each mouse is imported, and the Poly-3 weighting time at risk factor (a_{ij}) for each mouse is calculated. The mean (cumulative) tumor risk dataset is sorted into 2% incremental percentage tumor risk groups, and the corresponding a_{ij} value for each mouse is assigned into the appropriate tumor risk group. This procedure can be adapted to sort individual data for each week rather than the means of all 48 weeks. The resulting tumor risk group table is used to calculate the mean tumor risk estimate and standard deviation of the experimental group to predict the survival-adjusted liver tumor rate, over all liver tumor rates and the number of animals bearing tumors.

5.2.4.6 Hepatocyte Primary Culture Assay

On anesthetic male SD rats (225–250 g) with sodium pentobarbital (65 mg/kg i.p), tracheotomy is performed. A catheter is flushed with heparin (200 U/mL) to prevent blood clotting of PE 240 polyethylene tubing that is inserted into the trachea to facilitate the respiration. By inserting PE 50 polyethylene tubing, the right femoral artery and vein of the rats are cannulated, and the rats are infused with 0.5-mL saline containing 1-mg/mL sodium pentobarbital/100-g body weight/h through the vein by a Syringe Infusion Pump to supplement fluid loss and maintain the anesthetic level. On the abdomen posteriorly from the xiphoid process of the sternum, a 3-cm median-line incision, which is sutured with two layers of the muscle and skin, is made. Using surgical silk (size 0), left lateral and median liver lobes are ligated and given the partial hepatectomy (2/3 PHX). At various time points after PHX, the diaphragm is cut, and within 2 min from the right ventricle of the heart, a blood sample (5–10 mL per rat) is drawn and centrifuged (2500 g, 20 min) for collecting plasma. In the preparation of serum, the blood is allowed to clot on ice for 10 min, then spun down, with the serum of non-PHX rats as 0 h control. For sham-operated rats, the liver is manipulated without PHX.

In the liver perfusion, the male SD rat (300 g) is anesthetized with sodium pentobarbital (65 mg/kg, i.p.). By surgical silk an 18 G, 1.25-in (32 mm) i.v. catheter is used as the portal vein catheter tie to the vein. Through a catheter inserted into the inferior vena cava via the right atrium of the heart, the perfusion buffer is drained. The liver is perfused with 400 mL of nonrecycled oxygenated Ca^{2+}-Mg^{2+} free Dulbecco's PBS (DPBS, 2.68-mM KCl, 1.47-mM KH_2PO_4, 136.9-mM NaCl, 8.1-mM Na_2HPO_4, pH 7.4) containing 0.49-mM EDTA to reduce Ca^{2+} in the liver tissue, with 100 mL of DPBS without EDTA to wash out the EDTA in the system and with 100 mL of recycled oxygenated collagenase/Swim's 77 (0.25 mg/mL) with 5-mM Ca^{2+} for 10–15 min. Upon finishing perfusion, the liver is rinsed thoroughly with 10–20 mL of sterile DMEM. The liver is transferred into a sterile petri dish containing fresh medium. Using two forceps, the liver capsule membrane is slit, which is gently shaken and the cells are released into the medium. Two Spectra/Mesh N filters (70 µm, 40 µm, Spectrum) are used to filter the isolated liver cells to remove tissue chunks and cell debris. The cell suspension is filtered by low-speed differential centrifugation (300 g, 3 min, 4°C) for further purification. By aspiration the supernatant is discarded, the cell pellet is gently mixed with fresh medium, and the centrifugation procedure is repeated three times. Using the trypan blue exclusion method, viable cell concentration and percentage of nonhepatocytes is determined. In the purification of hepatocyte, 20 µL of trypan blue (0.4%) are added into 20 µL of cell suspension (dilution factor = 2), which is mixed and loaded onto the hemocytometer. All corner squares (64 squares/chamber) of the two chambers are counted (128 squares total; each square is 1 nL, 128 nL × 7.813 = 1 µL). Viable cells/µL = viable cells per 128 squares × 7.813 × 2. The final concentration is the average value of three separate countings. Viability (%) = viable cells/(viable + dead cells) × 100%. Nonhepatocyte (%) = nonhepatocyte/total cells × 100%.

Each well of six-well plates is coated with 20 µL of the solution of rat tail collagen (0.8-mg/mL double distilled H_2O with 0.1% acetic acid) and allowed to dry. To the well approximately 2 mL of DMEM/F-12 (supplemented with 25-mM sodium bicarbonate, 10-mM HEPES, 100-U/mL penicillin G, and 0.1-mg/mL streptomycin) is added. To each well the suspension of viable cells (final cell concentration, 100,000 cells/mL) are seeded and constantly mixed by agitation. The plates are incubated at 37°C overnight. When the attachment reaches the end, the medium is changed, mildly shaken, and aspirated. The wells are rinsed with 1 mL of medium and aspirated. From three randomly selected wells in the culture, the counts of the viable tarting-cell-number are calculated. Serum or plasma samples in 2 mL of fresh medium (10%) are added to wells, to which heparin (35-U/mL final) is added to prevent the medium from clotting during the culture. Cells grown in serum-free medium and in serum-free medium plus 100-ng/mL EGF and 20-mU/mL insulin are used as the negative and positive control, respectively. The medium is changed by aspiration at 24 h with no rinse. To each well, fresh medium with serum or plasma and other additives are added. At the end of 48 h, culture-attached hepatocytes are harvested and counted. The medium in each well is aspirated, serum or plasma-containing wells are rinsed once with saline and aspirated to wash off various plasma proteins. Overall, 400 µL of trypsin-EDTA is added to each well. The plate is covered and in the incubator for approximately 1 min and checked. The trypsin digestion is assessed. The plate is immediately placed on ice, and the serum of SD rats (40 µL, final concentration 10%) is added into each well to inhibit further trypsin digestion of the cells (440 µL total volume per well). On ice and using a pasteur pipet, the cells are gently blown from the bottom of the well. After all cells are lifted, the cell suspension is transferred into a 1.5-mL Eppendorf tube and kept on ice for an immediate counting with no centrifugation or trypsin removal. In a plate (dilution factor = 1.25) 40 μL of cells and 10 μL of 0.4% trypan blue are mixed and loaded onto the hemocytometer.

5.2.4.7 CCL-64 Cell Growth Assay

Serial dilutions of test samples are prepared on 96-well flat-bottomed microplates with 100 μL of CM per well. Before the assay, to activate latent TGF-B, the samples are transiently acidified by first adding HCl to pH 2 and subsequently neutralizing with NaOH. CCL-644 cells (Mv-1-Lu) grown in RPMI 1640 supplemented with 10% FCS, 2-mM L-glutamine, and 40-μg/mL gentamicin (Complete medium, CM) are seeded at 1×10^4 cells/well and grown for 24 h in a total volume of 0.2-mL CM. After 20 h, the cells are pulsed for 4 h with 1 μCi/well of [methyl-^{3}H]thymidine and harvested with a Micromate 196 cell harvester. The concentration of TGF-β in the sample is determined by the growth inhibition caused by the sample, compared with a standard curve obtained by testing serial dilutions of porcine TGF-β. When the assay is used to measure HGF, the concentration is determined by this cytokine's ability to reverse the growth inhibitory effect of 350 pg/mL of TGF-β.

5.2.5 ANTI-INFLAMMATORY ASSAY

In the development of serious diseases, inflammatory is frequently involved. The development of inflammatory events may be induced by chemicals such as glucocorticoid; are regulated by endogenous factors such as tumor necrosis factor alpha (TNFα), enzymes and proteins such as copper and zinc-superoxide dismutase (SOD1), proinflammatory peptide substance P (SP), RGD peptides, interleukin-4 (IL-4), IL-10, interferon-γ (IFN-γ), cyclooxygenase-1 (COX-1), cyclooxygenase-2 (COX-2), 5-lipoxygenase (5-LOX), macrophage inflammatory protein (MIP)-1R, glucocorticoid regulated protein CD163, FK506 binding protein 51 (FKBP51), and monocyte chemoattractant protein-1 (MCP-1); implicated by cytokine such as interleukin-1 (IL-1); and mediated by adhesions such as fibrous adhesions and cell adhesions resulting from the receptor, such as $α_{II}β_{III}$, human glucocorticoid receptor and chemokine receptors (CCR5), and banding. Based on these facts, the adhesion formation assay [73]; human peripheral blood mononuclear cell (PBMC) proliferation assay [74]; COX-1, COX-2, and 5-LOX assay [75]; CCR5 re-ceptor binding assay [76]; tissue binding affinity assay [77]; G93A-SOD1 transgenic mice assay [78]; mitogen-activated protein kinases (MAPK) p44 (ERK1) and p42 (ERK2) assay [79]; ELA4.NOB-1/cytotoxic T lymphocyte line (CTLL) cell assay [80]; and FKBP51 mRNA assay [81]; and Xylene-induced ear edema assay [82] are established.

5.2.5.1 Adhesion Formation Assay

For assessment of the neurokinin 1 receptor antagonist (NK-1RA) on peritoneal adhesion formation, a laparotomy is performed through a midline incision, and four ischemic buttons, spaced 1 cm apart, are created on both sides of the parietal peritoneum by grasping 5 mm of peritoneum with a hemostat and ligating the base of the segment with a 4-0 silk suture. To assess the effects of NK-1RA on adhesion formation, peritoneal adhesions are induced in 42 male Wistar rats (200–250 g) that are randomized to experimental groups receiving the specific non-peptide NK-1RA, the test compound, or vehicle. In the initial study, the rats in the experimental group receive a 0.2-mL i.p. injection of 25-mg/kg NK-1RA twice a day for 2 days. At the time of surgery, 1 ml of a 0.75-mg/mL test compound is given as a peritoneal lavage, and the rats then received i.p. injection for 7 days. Control rats are similarly injected/lavaged with sterile vehicle. This experiment is repeated with 10-mg/kg NK-1RA per day. At day 7, all the rats are killed and the adhesions are quantified in a blinded fashion. Each rat receives a percent adhesion score based on the number of ischemic buttons with attached adhesions.

For assessment of NK-1RA on peritoneal tissue plasminogen activator (tPA) and PAI-1 expression and activity, the temporal expression pattern of tPA and PAI-l mRNA in peritoneal tissue collected from a 0.5-cm radius of the ischemic buttons are determined by RT–PCR analysis at days 0, 1, 3, and 7 after surgery. Based on the results, the effects of NK-1RA administration on tPA and PAI-1 mRNA and protein levels are determined at postoperative day 1 in pedtoneal tissue and fluid by RT–PCR analysis and bioassay, respectively. The rats receive 5.0-mg/kg NK-1RA or vehicle per day. Control samples are collected from 6 nonoperated rats. All samples are immediately frozen in liquid nitrogen and stored at −80°C until used.

Total RNA is isolated from 50 mg of peritoneal tissue with the SV Total RNA Isolation System, and RT–PCR is conducted with the Gene-Amp RNA PCR System. To amplify tPA and PAI-1, the 28 cycles of 95°C, 60°C, and 72°C for 30 s each are used. In the amplification of tPA, the primer sets of 5′-TCTGACTTCGTCTGCCAGTG-3′ (sense) and 5′-GAG-GCCTTGGATGTGG TAAA-3′ (antisense) are used. In the amplification of PAI-1, the primer sets of 5′-ATCAACGACTGGGTGGAGAG-3′ (sense) and 5′-AGCCTGGTCATGTT GCTCTT-3′ (antisense) are used. Overall, 15 μL of PCR products are subjected to electrophoresis on 2% agarose gels containing 0.03-μg/mL ethidium bromide, and the quantitative level of the transcript is determined by scanned photographs of gels. Levels of mRNA expression are normalized to GAPDH, a constitutively expressed gene that does not vary among treatment groups.

Using the corresponding kits, the total levels of tPA and PAI-1 in peri-toneal fluid samples are measured. From 12 vehicle-administered, 12 NK-1RA-administered (10.0 mg/kg per day), and 12 nonoperated control rats, peritoneal fluid is collected in 5-mM citrate and 0.1-M acetate for assessment of fibrinolytic activity caused by tPA activation of plasminogen. Overall, 1.0 μL of peritoneal fluid samples are run on 10% SDS polyaerylamide gels containing 0.1% gelatin and 0.002% plasminogen. After electrophoresis the gels are washed twice in 2%Triton X-100 and incubated overnight at 37°C in 0.1-M glycine (pH 8.3). The gels are stained with a 0.25% Coomassie blue solution, and PA activity is visualized as clear bands produced by plasmin lysis of gelatin. Determining the contribution of serine pro-teases, a 10-mM serine protease inhibitor, PMSF, is added to the developing buffer. In the identification of the zones of lysis corresponding to tPA activity, tPA and/or uPA are immunoprecipitated from peritoneal fluid samples. To the mixture of 10 μL of peritoneal fluid and 10 μL of buffer containing 40-mM phosphate, 1-M

NaCl, 0.2% SDS, 2% Igepal CA-630, and 1% deoxycholate (pH 7.5), 1 μg of tPA and/or uPA antibodies are added and incubated at 4°C overnight. To each sample, 10 μL of a 50% UltraLink protein A/G slurry is added and incubated overnight at 4°C. Samples are centrifuged at 4°C and 16,000 g for 1 min, and the supernatant (50%) is analyzed by zymography for comparison with human recombinant tPA and uPA standards.

5.2.5.2   PBMC Proliferation Assay

On removal of plasma the heparinized human peripheral blood from 60 mL of healthy donors is centrifuged (4°C, 2000 g, 10 min), and the blood cells are diluted with PBS and centrifuged (1500 g, 30 min) [83]. After removal of red blood cells, the PBMC cell layers are collected, washed with cold distilled water and 10 × Hanks' buffer saline solution, and resuspended (2×10^6 cells/mL) in RPMI-1640 medium supplemented with 2% fetal calf serum, 100-U/mL penicillin, and 100-μg/mL streptomycin. In the determination of the lymphoproliferation, 100 μL of the PBMC suspension is deposited into a 96-well flat-bottomed plate with or without 5-μg/mL phytohemagglutinin (PHA, with cyclosporin as a positive control) [84], into which solutions of various concentrations of test compounds are added, the plates are incubated at 37°C for 3 days in humidified atmospheric air containing 5% CO_2, tritiated thymidine is added, incubated for another 16 h, and the cells are harvested on glass fiber filters using an automatic harvester and measured with a scintillation counting. In the determination of cell viability, approximately 2×10^5 T cells with or without PHA are cultured with 0.1% DMSO, solutions of various concentrations of test compounds are added, the plates are incubated for 3 days, the viable cell numbers are counted using a microscope with a hemocytometer following staining by trypan blue, and the percentage of viable cells is calculated. To analyze the cell cycle, 1 mL of the PBMC suspension is added into a 6-well flat-bottomed plate with or without 5-μg/mL PHA [85], 25 μg/mL of each test compound is added, the plates are incubated at 37°C for 3 days in humidified atmospheric air containing 5% CO_2, the medium is centrifuged, and the cells are harvested, washed with PBS, and fixed in 70% ethanol at −20°C for 30 min. DNA is then stained with 4-μg/mL propidium iodide containing 100-μg/mL ribonuclease A. Flow cytometric analysis is conducted.

After incubation with PHA alone or in combination with solutions of varying concentrations of test compounds for 3 days, the medium is centrifuged, and PBMC (2×10^5 cells/well) supernatants are collected to quantify the concentrations of IL-2, IL-4, IL-10, and IFN-γ by means of enzyme immunoassays.

After stimulation with or without PHA and coculture with 25-μg/mL, each compound for 18 h, the collected PBMC cells (5×10^6) are subjected to lysis with RNA-Beek and centrifuged, the supernatants are extracted with a phenol-chloroform mixture, and the extracted RNA is precipitated with isopropanol, pelleted by centrifugation, and redissolved in diethyl pyrocarbonate (DEPC)-treated H_2O. The concentration of the extracted RNA is measured using the optical density at 260 nm.

In the synthesis of the first-strand cDNA, 1-μg aliquots of RNA are reverse-transcribed. The mixture of 1-μg RNA in 12.5 μL of DEPC-treated H_2O and 20-μM oligodeoxythymidine (oligo dT) 18 is heated at 72°C for 2 min, and then quick-chilled on ice. To this mixture, 5.5 µL of concentrated synthesis buffer (50-mM Tris-HCl, pH 8.3, 75-mM KCl, 3-mM $MgCl_2$, 0.5-mM deoxynucleotides triphosphates (dNTPs), and 1 U ribonuclease inhibitor), and 10 U of moloney murine leukemia virus reverse transcriptase are added. The reaction mixture is incubated at 42°C for 1 h and then at 94°C for 5 min. To the reaction mixture, 80 µL of DEPC-treated H_2O is added and the mixture is stored at −20°C for use in the PCR. To 5 µL of the first-strand cDNA, 0.6-µM primers, 1.25-U Taq polymerase, 10-µL reaction buffer (2-mM Tri-HCl, pH 8.0, 0.01-mM EDTA, 0.1-mM dithiothreitol, 0.1% Triton X-100, 5% glycerol, and 1.5-mM $MgCl_2$) and 15.75 µL of water are added and the total volume is 25 µL. According to the air thermocycler, a denaturing temperature of 94°C for 45 s, annealing temperature of 58°C for 45 s, and elongation temperature of 72°C for 1 min for the first 25 cycles, and finally 72°C for 10 min, PCR is carried out and the amplified product is run on 1.8% agarose gels.

5.2.5.3 COX-1, COX-2, and 5-LOX Assay

For the COX-1 and COX-2 assay, the diluted solution of the test compound is incubated with COX-1 or COX-2 according to the standard method and then the enzyme reaction is initiated by the addition of [1-^{14}C]-arachidonic acid. From the incubation mediate, prostaglandin E_2 and prostaglandin D_2 (PED_2) are extracted to measure radioactivity. Indomethacin is used as a positive control. For the 5-LOX assay, the inhibitory activity of the test compound is evaluated. The test compound is incubated with 5-LOX (46 µg of protein) at 24°C for 10 min, and then the enzyme reaction is initiated by the addition of [1-^{14}C]-arachidonic acid (50 nCi). After 5 min, 4-M formic acid is added to terminate the enzyme reaction. Indomethacin is used as a positive control.

5.2.5.4 CCR5 Receptor Binding Assay

In the determination of chemokine receptor (CCR5) binding activity, a 96-well scintillation proximity assay (SPA) format and CHO cells overexpress-ing the human CCR5 receptor are used. From the CHO cells, MIP-1α and membranes are obtained, and after ^{125}I labeling, the former is converted into [^{125}I]-human MIP-1α. The solution of the test compound, 12.5% aqueous DMSO, 12 µg of membranes, 0.17-nM [^{125}I]-MIP-1α, 0.25 mg of Wheat Germ Agglutinin-SPA beads, and assay buffer (50-mM HEPES, 1-mM $CaCl_2$, 1-mM $MgCl_2$, 1% BSA, and a protease inhibitor cocktail), is incubated at room temperature for 5 h with shaking. The beads are settled for 2 h, and the total binding is measured via the radioactivity test. In the presence of 1-µM recombinant human MIP-1α, the nonspecific binding is defined. A IC_{50} of 2.7 nM of human MIP-1α is used as a reference.

5.2.5.5 Tissue Binding Affinity Assay

In hydrophobic teflon bags, the test compound and blood monocytes isolated from pooled buffycoats at a density of 2×10^8-cells/mL McCoy's medium supplemented with 20% fetal calf serum are cultured for 2 days. The monocytes are washed with cold PBS (pH 7.4), incubated with BSA (1%) at 4°C for 30 min, washed with PBS, incubated with monoclonal antibody anti-CD163 (5–10 µg/mL) at 4°C for 45 min, washed with PBS, and incubated with FITC-labeled secondary antibody goat-antimouse IgG_1 in 1% BSA at 4°C for 30 min. At the last 3 min of the incubation, Propidium iodide is added to determine cell viability and to exclude dead cells using activated cell sorter (FACS) analysis (488 nm, 250 mW, logarithmic amplification).

Human lung cancer-free tissue from patients with bronchial carcinomas and receiving no glucocorticoids for the last 4 weeks before surgery is immediately washed with Krebs-Ringer-HEPES buffer (118-mM NaCl, 4.84-mM KCl, 1.2-mM KH_2PO_4, 2.43-mM $MgSO_4$, 2.44-mM $CaCl_2 \cdot 2H_2O$, and 10-mM HEPES, pH 7.4) and sliced into pieces of $1 \, mm^3$ or deeply frozen immediately in liquid nitrogen and stored at –70°C.

The frozen human lung tissue is pulverized and homogenized in three aliquots buffer solution A (10-mM TRIS, 10-mM $NaMoO_4$, 30-mM NaCl, 10% glycerol, 4-mM DTT, 5-mM dichlorvos, and 1-mM Complete) with an Ultra Turrax mixer in an ice bath and then centrifuged at 4°C and 105,000 g for 1 h to prepare glucocorticoid receptors (30–60 fmol/mg cytosol [86]). The receptor binding experiments are performed according to the standard method [87]. The radiochemical purity of the labeled glucocorticoid is determined by HPLC, TLC, and scintillation counting.

5.2.5.6   G93A-SOD1 Transgenic Mice Assay

EOC-20 cells are grown in 75-cm² cell culture flasks with DMEM supplemented with 10% fetal calf serum and 20% L292 fibroblast-conditioned medium, transferred into 24-well cell culture plates, treated with the solution of test compound in DMSO or DMSO vehicle alone (1% final volume) for 30 min, and challenged with TNFα for 24 h. The culture medium is collected and assayed for determining the IC_{50} of the test compound that suppresses TNFα-induced NO_2^- by 50% via the determination of NO_2 [88]. The medium is removed, and DMEM lacking the phenol red indicator is added. To each well, MTS reagent [3-(4,5-dimethylthiazol-2-yl)-5-(3-carboxymethonyphenol)-2-(4-sulfophenyl)-2H-tetrazolium, inner salt] is added and incubated at 37°C for 30–45 min. The aliquots of media are collected and evaluated spectrophotometrically at 540 nm. The MTS solution prepared for the blank is incubated in the absence of cells. Viability is calculated as the ratio of OD_{540} in test compound-treated wells, relative to the same variable measured in wells treated with approximately 25-µM (typically 20 ng/mL) TNFα alone, after subtraction of the blank.

Transgenic mice expressing high copy numbers of human mutant G93A-SOD1 are maintained in the hemizygous state by mating G93A males with B6SJL-TGN females. The mice are fed *ad libitum* standard AIN93G diets or the same diets formulated with nordihydroguaiaretic acid at 2500 ppm; at 90 days of age, drug administration is started to demonstrate motor weakness and fine limb tremors [89]. At 90 days, 100 days of age, and subsequent 5-day intervals, mice are placed on a horizontal rod rotating at 1 rpm every 10 s until they fall from the rod; the mice that are no longer able to right themselves within 10 s of being placed on their sides are killed.

Using TRI reagent, total RNA is collected from the spinal cords of non-transgenic control and G93A+ transgenic mice. In the presence of avian myeloblastosis virus (AMV) RT using oligo(dT)$_{15}$ to prime the reaction, 5 μg of RNA samples are reverse transcribed. On completion, each reaction mixture is diluted with a TE buffer (10-mM Tris, 1-mM EDTA, pH 8.0) to a final volume of 50 μL. PCR amplification of a 309-bp 5LOX gene product is accomplished with Taq DNA polymerase, using the buffer and final concentrations of 1.5-mM MgCl$_2$, 0.2-mM dNTP, and 0.3-μM primer (5′-GGCACCGACGACTACATCTAC-3′, forward and 5′-CAATTTTG-CACGTCCATCCC-3′, reverse), of which the final volumes are 50 μL. A 353-bp PCR product of the β-actin primer (5′-CGGCCAGGTCATCACTATTG-3′, forward and ACT-CCTGCTTGCTGATCCAC-3′, reverse) is used as the normalization control. The optimal cycling conditions of PCR amplification of a 309-bp 5LOX gene product are at 94°C for 2 min, one cycle; at 94°C for 1 min, at 56°C for 1 min, and at 72°C for 1 min for 27 cycles; and at 72°C for 7 min, one cycle. The conditions for β-actin primers are the same as mentioned above, except that the annealing temperature is 54°C and performing 24 cycles. From each reaction, 25 μL of samples are collected, electrophoresed in 2% agarose/TBE [tris/borate/EDTA buffer (0.09-M Tris, 0.09-M borate, 0.002-M EDTA)] gels for 1.5 h, stained with ethidium bromide, and photographed with a NucleoVision imaging system. SL-29 fibroblast lysate is used as the positive control for 5LOX Western blots. On 4–20% gradient polyacrylamide gels, the electrophoresis is performed and the bands are visualized with chemiluminescence detection reagents. In the solution of 50-mM Tris-HCl (pH 6.8), 0.3-M NaCl, 1% β-mercaptoethanol, 1-mM phenylmethylsulfonyl fluoride, and 5-μM leupeptin, the spinal cord samples are homogenized and centrifuged (4°C, 30,000 g, 5 min). The supernanants are boiled for 10 min and centrifuged at 30,000 g and 4°C for 30 min, and the supernatants are dialyzed overnight against 50-mM Tris-HCl. From the heat-soluble fraction of total spinal cord lysate, microtubule-associated tau protein (C-tau) is cleaved and measured by sandwich ELISA, using affinity-purified monoclonal antibody 12B2 [90–92]. In immunohistochemical experiments, the terminally anesthetized mice are successively perfused transcardially with PBS (pH 7.4) and 4% paraformaldehyde in 0.1-M phosphate buffer. The spinal cord is collected, and the lumbar L5 region is processed for paraffin embedding. Serial cross sections of 5 μM thickness of the L5 spinal cord region are prepared for immunostain with antibodies to glial GFAP. In the absence of primary antibody, negative immunohistochemical controls are treated in the same way. GFAP-labeled sections are counterstained with hematoxylin.

5.2.5.7 MAPK p44 (ERK1) and p42 (ERK2) Assay

By RT–PCR from PMA-treated THP-1 cells, cDNA encoding CCR2B is isolated and subcloned into the expression vector pXMT3-neo [93], which contains an adenovirus late promoter and neo- and DHFR-selectable markers. After transfection with pXMT3-neo-CCR2B, selection with G418, amplification with methotrexate, and clone of CHO DUK-X cells, the resultant CHO-CCR2B cell line is cultured with modified MEMα (without ribonucleosides and deoxyribonucleosides) containing Glutamax-I, 100-U/mL penicillin, 100-μg/mL streptomycin, 10% (v/v) dialyzed FBS, and 80-nM methotrexate. Parental CHO cells are cultured with original MEMα containing 100-U/mL penicillin, 100-μg/mL streptomycin, and 10% (v/v)

FBS, split twice per week by incubation for 5 min with an enzyme-free cell dissociation buffer, and reseeded at a density of $1–2 \times 10^4$ cells/cm^2. Before an assay, the cells are seeded in methotrexate at 2×10^4 cells/well in flat-bottomed 96-well tissue culture plates, allowed to attach for 6 h, washed with 250-μL/well PBS, and incubated overnight in methotrexate containing endotoxin-free BSA (0.1%, w/v).

After preincubation of 60 μL of 20-nM MCP-1 with 60 μL of 0.4–1.8-μg/mL mAb at 37°C for 30 min, 50 μL of the MCP-1-mAb solution is added. The cells are preincubated at 37°C for 45 min with the solution of test compound in 50 μL of serum-free medium. After the addition of 50 μL of 20-nM MCP-1, the cells are incubated at 37°C for 5 min, the medium is replaced by 100-μL/well methanol equilibrated at –20°C, incubated at –20°C for another 20 min, and then washed three times with PBS containing Triton X-100 (0.1%, w/v). The cells are incubated with H_2O_2 (0.6%, v/v) in washing buffer for 20 min and washed as above, by which the activity of endogenous peroxidase is quenched. By adding a 250 μL/well assay buffer (washing buffer containing fraction V BSA, 5%, w/v) and incubating at room temperature for 1 h, the nonspecific binding sites are blocked. Then, the solution is replaced by a 100-μL/well monoclonal anti-phospho-ERK antibody (1 μg/mL in assay buffer). After incubating at 37°C for 1 h and washing three times with washing buffer, to the plate, a 100-μL/well goat-antimouse-IgG coupled with horseradish peroxidase (0.5 μg/mL in assay buffer) is added. After incubation at room temperature for 1 h and the plate is washed six times with washing buffer, 100 μL/well of tetramethylbenzidine is added. After color development for 10–30 min, the reaction is stopped by adding 50 μL of 2-M H_2SO_4 and the plates are read at 450 nm.

5.2.5.8 ELA4.NOB-1/CTLL Cell Assay

The ELA4.NOB-1 cells and CTLL cells maintained in RPMI 1640 containing 2-mM glutamine, 40-μg/mL gentamicin, and 100-U/mL penicillin, are transferred into a tissue culture medium (TCM) containing 5% heatinactivated fetal calf serum (FCS) and 10% FCS and recombinant human L-2 (100 U/mL), respectively. Cell lines are centrifuged (room temperature, 400 g, 1 min) and washed twice in fresh TCM supplemented with 10% FCS by resuspension and centrifugation. By enumeration using trypan blue (0.4%, w/v) in saline, ELA4.NOB-1 and CTLL cells are adjusted to 1×10^6 and 4×10^4 viable cells/mL, respectively. To 96-well round-bottom microtiter plates containing 100 μL of cytokine (control) or test compound, 50 μL aliquots of each cell suspension are placed. After incubation in air containing 5% CO_2 at 37°C for 24 h, the cells are treated with 50 μL of 2-μCi/mL tritiated thymidine, incubated for another 24 h, harvested onto printed filtermats that are dried at 30°C for 1 h and heat sealed into sample bags containing 4.5 mL of β-scintillant, and the radioactivity (cpm) is measured.

To each well containing 100-μL aliquots of the standard dilutions, 50 μL of ELA4.NOB-1 and 50 μL of CTLL cell suspensions are added. Using human recombinant IL-2 (0.025–1.6 U/mL) or test compound, the integrity of CTLL responses in the absence of ELA4.NOB-1 cells is measured. Human recombinant IL-1ra (0.1–100 ng/mL), ultrapure natural human TGF-β1 (0.01–10 ng/mL), and test compound are tested on ELA4.NOB-1 and CTLL cocultures in a similar way, generating standard dose response curves for these cytokines. In the tests of the effects of

IL-1ra and TGF-β1 on IL-1 induced activity in the coculture bioassay and on IL-2 induced CTLL proliferation, serial dilutions of IL-1β (0.78–50 pg/mL), IL-2 (0.025–1.6 U/mL), IL-1ra (1–100 ng/mL), TGF-β1 (0.01–10 ng/mL), and test compound in TCM supplemented with 10% FCS are used. Overall, 50-μL aliquots of each dilution of IL-1ra and IL-1β are dispensed and pipetted into 96-well microtiter plates, respectively, and 50 μL of the suspension of ELA4.NOB-1 and CTLL cells are aliquoted into each well, incubated, and treated with ^{3}H-thymidine as described above; similar assays are set up for IL-1ra and IL-2, for TGF-β1 and IL-1β, and for TGF-β1 and IL-2.

5.2.5.9   FKBP51 mRNA Assay

Within a 2-week period between 9 and 10 h by venapuncture, the blood of nine healthy controls without a history of GC medication (37 years old) and one GC hyposensitive patient harboring one nonfunctional GC receptor allele and resulting in a net functional GC receptor expression of approximately 50% is collected in sodium heparin tubes [94], and centrifuged, and the isolated PBMC are washed twice with RPMI 1640 and resuspended in assay medium consisting of RPMI 1640 supplemented with 4-mM L-glutamine, 100-U/mL penicillin, 100-μg/mL streptomycin, and 10% dextran-coated charcoal steroid-stripped fetal calf serum at 0.5 $\times$ 10^6 cells/mL. Overall, 1-mL aliquots of the PBMC suspension are added onto 24-well plates and incubated overnight at 37°C in humidified atmospheric air containing 5% CO_2. To the medium, dexamethasone (DXM) or the test compound is added and incubated for another 24 h to isolate RNA. For activation of PBMC, 0.1-mL aliquots of 0.5 $\times$ 10^5 cells are added onto a 96-well round-bottom plate and incubated with 1.5-μg/mL etanus toxoid for 9 h. Seventy two hours later, DXM is added. Then 96 h later, RNA is isolated. In a simultaneous control experiment, 78 h later, 1 μCi 3H-thymidine is added, the cells are harvested, and incorporated radioactivity is determined.

For purification of PBMC subsets, cells are isolated, washed twice, resuspended in PBS at 1 $\times$ 10^7 cells/mL, incubated at room temperature for 30 min with phycoerythrin labeled primary antibodies (mouse-antihuman CD20, CD4, CD8, or CD14), and washed twice with PBS. The PBMC is resuspended in PBS at 1 $\times$ 10^8 cells/mL, incubated at 4°C for 30 min and labeled with goat-anti-mouse IgG, and washed and resuspended in 0.5 mL of PBS. The labeled and unlabeled PBMC are separated, resuspended in assay medium, divided into 0.1-mL aliquots onto a 96-well round-bottom plate, and incubated overnight. To the wells, DXM is added and incubated for another 24 h to isolate RNA.

For real-time PCR performed on the LightCycler, after the isolation by use of the TriPure reagent and reverse transcription by use of MMLV-RT RNase H Minus of RNA, the synthesized cDNA is diluted 20–40 times in ddH_2O [95]. To calculate the relative expression level of the genes of interest, the relative expression level of the housekeeping gene β_2-microglobulin (β_2m) in each sample is determined. Real-time PCR is performed in 10-μL medium containing 5.0 μL of diluted cDNA (0.5 pmol/μL each primer), 10% either DNA master SYBR-green I solution (for the FKBP51 and β_2m PCR) or DNA hotstart mixture (for the FKBP12, FKBP13, FKBP22, FKBP25, FKBP52, and Cyp40 PCR), and an optimal concentration of MgCl$_2$ (4-mM β_2m PCR, 3-mM FKBP51 PCR). The mixture is denatured at 95°C for 30s and 10min for DNA Master I kit and Fast Start kit, respectively, and subjected to up to 40 amplification cycles: denaturing 15s at 95°C, annealing 10s at 56°C for β2m or at 67°C for FKBP51, and elongation 30s at 72°C. A single measurement of fluorescence is taken after each elongation step at 82°C.

5.2.5.10   Xylene-Induced Ear Edema Assay

Male Kunming mice (about 25g) are randomly divided into three groups of 12 mice, namely the test group, vehicle control group, and positive control group. The mice in the vehicle control group are administrated orally a suspension of Aspirin in CMC at a dosage of 20mg/kg, and a concentration of 0.3mg/mL, whereas the mice in the test group are administrated orally a suspension of test compound in CMC at a dosage of 20mg/kg, 4.0mg/kg, and 0.8mg/kg, and a concentration of 2.0mg/mL, 0.4mg/mL, and 0.08mg/mL. Thirty minutes later, 0.03mL of xylene is applied to both the anterior and the posterior surfaces of the right ear. The left ear is considered a control. Two hours after xylene application, the mice are killed and both ears are removed. Using a cork borer with a diameter of 7mm, several circular sections are taken and weighed. The increase in weight caused by the irritant is measured through subtracting the weight of the untreated left ear section from that of the treated right ear section.

5.2.6   THROMBUS-RELATED ASSAY

A series of events such as the platelet aggregation, fibrinogenesis, fibrin adhesion, fibrin aggregation, and vascular inner wall injury are implicated in the thrombosis. Thus the thrombosis, antithrombosis, and thrombolysis may relate to antiplatelet aggregation [96], chemical- and electrical-induced blood vessel injury [97, 98], thread-induced fibrin or platelet adhesion [99], euglobulin clot lysis time [100, 101], fibrinolytic area [102], and reduction of thrombus mass[103, 104].

5.2.6.1   Antiplatelet Aggregation Assay

Platelet-rich plasma is prepared by centrifugation of normal rabbit blood anticoagulation with sodium citrate at a final concentration of 3.8%. The platelet counts are adjusted to $2 \times 10^5/\mu L$ by the addition of autologous plasma. Platelet aggregation tests are conducted in an aggregometer using the standard turbidimetric technique. The agonists used may be either the usual platelet-activating factor (PAF, final concentration 10^{-5}–10^{-7}M) and adenosine diphosphate (ADP, final concentration 10^{-5}–10^{-7}M) or thrombin and collagen. The effects of the tested compound on PAF or ADP-induced platelet aggregation are observed. The maximal rate of the platelet aggregation (Am%) is represented by the peak height of the aggregation curve.

5.2.6.2   Ferric Chloride-Induced Thrombosis Assay

After overnight fasting, male SD rats (320–380g) are administered orally water (blank control), test compound (15, 30, 60mg/kg), and aspirin (positive control, 30mg/kg), and they are anesthetized with urethane (1.5g/kg i.p.) at 90min after the oral administration. The experiments are carried out according to the modified method described by Kurz et al. [97]. The left common carotid artery is isolated carefully, and a plastic sheet is placed under the vessel to separate it from the surrounding tissue. The surface of carotid artery is covered with a 4×0.5-cm cotton sheet saturated with $300\,\mu L$ of $FeCl_3$ solution (25%, w/v) for 15 min. Then, the injured carotid artery segment (0.6 cm) is cut off, from which the formed thrombus is taken out. After drying for 24 h at room temperature in a dehumidifier, the dried weight of the thrombus is measured.

5.2.6.3   Electrical Stimulation-Induced Arterial Thrombosis Assay

After overnight fasting, male SD rats (250–350 g) are anesthetized with urethane (1.5 g/kg, i.p.). The left carotid artery is isolated carefully. A plastic sheet is placed under the vessel to separate it from the surrounding tissue. Thrombus formation is induced with the modified Hladovec method [98]. The holder incorporates two electrodes, and the temperature sensor is fixed under the exposed carotid artery. Then, the rats are administered intravenously by NS (blank control), test compound (5, 10, 20 mg/kg), and aspirin (4 mg/kg), respectively. After 5 min, a current of 3 mA is delivered for 3 min. With thrombus formation, the temperature of carotid blood is abruptly decreased. The occlusion time (OT) is measured through the temperature sensor and timer on the electric thrombosis stimulator. The rate of thrombosis inhibition is calculated according to as follows: thrombosis inhibition (%) = (A_1 − A)/A × 100%, where A is the OT of the control group and A_1 is the OT of agent groups.

5.2.6.4   Thread-Induced Thrombosis Assay

Male SD rats (320–380 g) are treated with water (blank control), test compound (15, 30, 60, 120 mg/kg), and aspirin (positive control, 30 mg/kg, b.i.d. for 2.5 d i.g.) and anesthetized by intraperitoneal injection of urethane (1.5 g/kg) at 75 min after the last dose. The arteriovenous shunt operation is carried out as the method described by Umetsu and Sanai [99]. The left jugular vein and the right carotid artery are annulated with a 4-cm-long polyethylene tube (o.d. 1 mm) with heparin (50 U/kg) injection intravenously for anticoagulation. These catheters are connected to the ends of a 15-cm-long polyethylene tube (o.d. 2 mm) containing a 5-cm-long suture silk thread (no. 4) measured wet weight. At 120 min after the last treatment, the blood flowing through the shunt is confirmed for 15 min. The silk thread with thrombus is gently removed and measured immediately for the wet gross weight. The thrombus wet weight is determined by subtracting the premeasured silk wet weight from the gross weight. The rate of thrombosis inhibition is calculated as follows: thrombosis inhibition (%) = (A − A_1)/A × 100%, where A is the thrombus wet weight of the control (water) and A_1 is the weight after treatment with the agents.

5.2.6.5   Euglobulin Clot Lysis Time (ECLT) Assay

The rabbit euglobulin clots are prepared according to a published method [100, 101]. Plasma diluted at 1:20 in distilled water is precipitated at pH 5.7 with acetic acid (0.25%). After 30 min at 4°C, the suspension is centrifuged at 2000 g for 15 min and the precipitate is resuspended to the initial plasma volume with 50-mM sodium barbiturate buffer (pH 7.8, containing 1.66-mM $CaCl_2$, 0.68-mM $MgCl_2$, and 93.96-mM NaCl). To the rabbit euglobulin clots, NS (blank control), UK (positive control), or test compound is added and the ECLT or time to clot lysis is determined in a 96-well microtiter plate.

5.2.6.6   Fibrinolytic Area Assay

The fibrinogen–agarose mixture is prepared and coagulated with thrombin in plastic dishes according to a published procedure [102]. The fibrinogen–agarose mixture is prepared by mixing equal volumes of 0.3% of rabbit fibrinogen and 0.95% of agarose solutions, both dissolved in 50-mM sodium barbiturate buffer (pH 7.8). The fibrinogen–agarose mixture is coagulated with 100 mL of thrombin (100 U/mL) in plastic dishes (90 mm diameter × 1 mm depth). After 30 min at 4°C, an adequate number of wells, 5 mm in diameter, are perforated. To determine fibrinolytic activity, 30 μL of NS (blank control), UK (positive control), or test compound is added to the corresponding well. The plate is incubated, and areas of lysis are quantified by lysis area.

5.2.6.7   Thrombolytic Assay

Male Wistar rats (200–300 g) are anesthetized with pentobarbital sodium (80.0 mg/kg, i.p.). The right carotid artery and left jugular vein of the animals are separated. To the glass tube containing 1.0 mL of blood obtained from the right carotid artery of the rat, a stainless steel filament helix (15 circles; 15 mm × 1.0 mm) is added immediately. Fifteen minutes later, the helix with thrombus is carefully taken out and weighed. It is then put into a polyethylene tube that is filled with heparin sodium (50-U/mL NS), and one end is inserted into the left jugular vein [103, 104]. Heparin sodium is injected via the other end of the polyethylene tube as the anti-coagulant, after which the test compound is injected. The blood is circulated through the polyethylene tube for 90 min, after which the helix is taken out and weighed. The reduction of thrombus mass is recorded.

5.2.7   IMMUNOMODULATING ASSAY

Immunomodulation not only relates to a series of physiologic and pathologic phenomena, but also it can be estimated by related reaction cells, such as mast cells, DCs and NK cells, and level of cell factors such as interleukin, interferon, tumor necrosis factor, and transforming growth factor. Based on those facts, the rat mast cell and rabbit aortic assay [105], dendritic cell assay [106], lymphoid organ assay [107], IFN-γ assay [108], anti-rHuEPO NAb assay [109], and chemotaxic assay [110] are established.

5.2.7.1   Rat Mast Cell and Rabbit Aortic Assay

The ice-cold PBS (10 mL, 137-mM NaCl, 2.68-mM KCl, 0.91-mM $CaCl_2$, 8.1-mM Na_2PO_4, 1.47-mM KH_2PO_4, 0.91-mM $MgCl_2$, 5.6-mM glucose, and 20.0-mM

HEPES) is injected (i.p.) into male Wistar rats (200–250 g). Approximately 90–120 s later, PBS is collected and peritoneal is subsequently washed by 5 mL and 10 mL of PBS. The washing PBS are also collected and combined with first PBS, followed by centrifugation (200 g, 5 min, 4°C). After washing twice with PBS, the pellet is resuspended into 10 mL of PBS. From the rat, peritoneal lavage mast cells are isolated.

Preincubation of 1.8 mL of mast cell suspensions are carried out at 37°C for 10 min, and 0.1 mL of test compound is added to stimulate the mast cells, which are cooled in ice to terminate histamine release. After centrifugation (100 g, 10 min, 4°C), to the supernatant and the pellet, an equal volume of 0.8-M HClO$_4$ and twice volume of 0.4-M HClO$_4$ are added, respectively. A mixture of 125 μL of 5-M NaOH, 0.4 g of NaCl, and 2.5 mL of n-butanol is added to 1-mL solution of test compound, the mixture is centrifuged (200 g, 1 min, room temperature) and separated. The upper organic phase is mixed with 2 mL of 0.1-M NaOH saturated with NaCl and centrifuged (200 g, 1 min, room temperature), and the procedure is repeated, of which the upper organic phase is mixed with 2 mL of 0.1-M HCl and 7.6 mL of n-heptane. The lower aqueous phase (1 mL) is mixed with 0.1 mL of 10-M NaOH. By incubation with 0.1 mL of o-phthalaldehyde (10-mg/mL methanol) at room temperature for 4 min to perform the histamine-o-phthalaldehyde conjugation, and by addition of 0.6 mL of 3-M HCl to terminate the conjugation. The fluorescence of histamine-o-phthalaldehyde conjugate is assessed at 450 nm with emission excited at 360 nm.

The thoracic aorta of male Japanese White rabbits (3–3.5 kg) is cut into helical strips (approximately 4 mm wide and 20 mm long), and the endothelium is removed by gently rubbing the endothelial surface with cotton pellets. In 1 mL of organ bath containing the modified Krebs–Ringer–bicarbonate solution (120-mM NaCl, 4.8-mM KCl, 1.2-mM CaCl$_2$, 1.3-mM MgSO$_4$, 25.2-mM NaHCO$_3$, 1.2-mM KH$_2$PO$_4$, 5.8-mM glucose), the strips are mounted and suspended. With a force-displacement transducer connected to a polygraph muscle, the tensions of the strips are recorded isometrically. A passive tension of 1 g is initially applied and the strips are equilibrated for 60 min, after which, 60-mM NaCl in the modified Krebs–Ringer–bicarbonate solution is replaced by equimolar KCl for precontraction of the strips. Reaching a steady level of the response, the experiment is started. Contractile response to histamine is normalized with that of high K$^+$.

5.2.7.2 Dendritic Cell (DC) Assay

Before being used either for DC culture or for preparing T cells, human peripheral blood mononuclear cells (PBMCs) isolated from freshly leukapheresed blood are centrifuged and cryopreserved in 90% autologous serum and 10% DMSO. By negative depletion and using anti-HLA-DR monoclonal antibody-conjugated paramagnetic beads from allogeneic PBMCs, enriched T cells are prepared. The cells possessing potential costimulatory function are removed from the PBMCs, as suspension each batch of which is tested for the presence of any remaining B cells, monocytes, and 80–90% T cells. Using cell-specific paramagnetic bead preparations and by biomagnetic separation of the stimulators, B cells and T cells are purified and the purity of each cell type is greater than 90%, as determined by flow cytometry.

The cryopreserved PBMCs are thawed in warm AIM-V medium, washed with PBS, and resuspended in Opti-MEM medium supplemented with 1% heat-inactivated autologous plasma to prepare $(5–10) \times 10^6$ cells/mL suspension. The suspension of 1×10^9 cells is transferred into T-75 culture flasks and cultured for 1 h, the nonadherent cells are resuspended, aspirated out, and stringently washed with cold PBS to remove loosely adherent cells. To each flask, 1.5×10^9 Opti-MEM medium containing 5% heat-inactivated autologous plasma, 500-U/mL rhGM-CSF, and 500-U/mL rhIL-4 are added and the adherent cells are incubated for 6 days. These DCs are then treated either with BCG alone or BCG plus IFN-γ for 24 h.

In the COSTIM bioassay, allogeneic T cells are thawed in warm AIM-V culture media, and washed with and resuspended in PBS at 1×10^5 DCs or 1×10^6 T cells/mL. To each triplicate well of a U-bottom 96-well plate, 100 μL of 1×10^4 DCs and 100 μL of 1×10^5 allogeneic T cells are successively added, and with or without 0.005-μg/mL anti-CD3 monoclonal antibody, which is incubated at 37°C for 44 h in humidified atmospheric air containing 5% CO_2. A total of 0.5 μCi Tritiated (^{3}H)-thymidine in 50 μL of AIM-V is added to each well, cultured for the last 18 h, the cells are harvested, and the incorporated radioactivity is quantified. Before 1 h of adding T cells and anti-CD3, the sterile, azide-free, and low-endotoxin IgG1 monoclonal antibodies specific for CD54, CD80, CD86, and an isotype control (BD Pharmingen) are added to the DCs at 1 μg/well to observe the costimulatory molecule block. In the mixed lymphocyte reaction (MLR), the cryopreserved DCs (stimulators) and T cells (responders) are thawed in warm AIM-V media, washed with PBS, and resuspended in PBS at 1×10^5 DCs or 1×10^6 T cells/mL. To each well of a U-bottom 96-well plate, 100 μL of 1×10^4 DCs and 100 μL of 1×10^5 allogeneic T cells are successively added and incubated at 37°C for 6 days in humidified atmospheric air containing 5% CO_2. A total of 0.5 μCi Tritiated (^{3}H)-thymidine in 50 μL of AIM-V is added to each well, cultured for the last 18 h, the cells are harvested, and the incorporated radioactivity is quantified.

5.2.7.3 Lymphoid Organ Assay

After a 4-week acclimation period, male Wistar rats (4 weeks old, 225–250 g, housed in polypropylene cages in a environment-controlled room maintained at 22°C, 55% relative humidity, and a 12:12 h light/dark cycle) are allocated to six groups (n = 15–20 rats each). The untreated group is used as control (maintained on basal diet and sacrificed at weeks 4 and 30). During weeks 1 and 2, the DMBDD, DMBDD/PB, and DMBDD/2-AAF groups are sequentially treated with initiators DEN (N-nitrosodiethylamine, 100 mg/kg, i.p.), MNU (N-methyl-N-nitrosourea, 20 mg/kg, i.p., four times, two doses per week), and BBN (0.05% in drinking water during 2 weeks); during weeks 3 and 4, DHPN and DMH groups are sequentially treated with DHPN (0.1% in drinking water during 2 weeks) and DMH (1,2-dimethylhydrazine, 40 mg/kg, s.c., four times, two doses per week). At the end of the week 4, some animals of the DMBDD group are killed and the remainder is maintained on a basal diet until week 30. After the initiation, the DMBDD/PB and DMBDD/2-AAF groups are supplied with phenobarbital (PB, 0.05%) and 2-acetylaminofluorene (2-AAF, 0.01%) in the diet for 25 weeks, respectively. From the sixth week until week 30, two noninitiated groups receive PB or 2-AAF in the diet. At the week 4 or 30, all animals are killed under pentobarbital (45 mg/kg) anesthesia. The liver, kidneys, spleen, thymus, mesenteric lymph nodes, and bone marrow removed from all animals are fixed in buffered formalin for 48 h for tissue processing and histological analysis. Only at week 30 are the lung, small and large intestine, and Zymbal's gland examined. After removal of the liver, the kidneys, spleen, and thymus are weighed immediately. The spleen is cut in two halves for evaluation of cytokines and histological analysis, respectively. All removed organs are embedded in paraffin and stained with hematoxylin and eosin for histological analysis [111]. The suspensions of spleen cells dispersed in a Petri dish containing RPMI-1640 culture medium are centrifuged; resuspended in RPMI-1640 culture medium supplemented with 20-mg/mL gentamycin, 2-mM glutamine, and 10% inactivated fetal calf serum; and washed twice by centrifugation at 1500 g for 10 min. To aliquots of 2×10^6 cells/mL (500 µL/well) in 24-well flat-bottom microtiter plates, RPMI-1640 (500 µL/well) or Concanavalin A (CON A-2.5 µAg/mL, 500 µL/well) or Staphylococcus aureus Cowan's strain 1 (SAC-1:5000, 500 µL/well) is added and the plates are incubated at 37°C for 72 h in humidified atmospheric air containing 5% CO_2. After incubation, the collected supernatants are stocked at −70°C for quantification of cytokines. CONA in vitro stimulated samples are used to quantify IL-2, IFN-γ, IL-10, and TGF-β1. SAC in vitro stimulated samples are used to quantify TNF-α and IL-12. Using ELISA kits, cytokine production, IL-2, IL-12, TNF-α, IFN-γ, and IL-10 levels are measured. Samples are acidified by 1-M HCl and measured by Quantikine antihuman TGF-h1 kit for detection of the TGF-h1 immunoreactive form.

5.2.7.4   IFN-γ Assay

Human myelomonocytic KG-1 cells (ATCCCCL246) are incubated with RPMI-1640 medium supplemented with 10% FCS, 100-µg/mL penicillin, and 100-µg/mL streptomycin at 37°C in air containing 5% CO_2. By expression of the corresponding cDNA in E. coli HuIL-18 and MuIL-18 are prepared and purified to homogeneity. Using Pfu DNA polymerase (at 95°C for 45 s; at 72°C for 3.5 min; 10 cycles and at 95°C for 45 s; at 68°C for 3.5 min; 35 cycles) MuIL-18R cDNA (1.7 kbp) are amplified from murine liver RNA by RT–PCR and cloned into pCRScript Cam SK (+) to synthesize 5'-AGAGGAACCACCCACAACGATCCT-3' and 5'-TGAATAGGCACACGCAT-XGACCTCT-3' [112]. With the EF-1 promoter of pEF-BOS vector, the dihydrofolate reductase unit of pSV2dhfr (ATCC 37146) and the backbone of pRc/CMV vector pREF-XN is constructed. IL-18R cDNA is ligated into XhoI/NotI sites of the vector to form MuIL-18R expression vector pRcEFM18R. KG-1 cells (1×10^7) are washed twice with RPMI-1640 medium and transfected with 50 mg of pRcEFM18R by electroporation. In the presence of 400-µg/mL G-418, the transformed cells are selected and cloned. The suspension of 2×10^6 cells, on which the receptor binding of ^{125}I-labeled MuIL-18 or HuIL-18 has been examined, in RPMI-1640 containing 0.1% NaN_3 and 100-mM HEPES (pH 7.2) is incubated at 4°C for 1 h with approximately 4 ng of ^{125}I-labeled MuIL-18 or HuIL-18. After the separation of unbound IL-18, the cell-bound ^{125}I count is determined. Subtracting the nonspecific binding measured from 3 µg of unlabeled cognate ligand, the specific binding of IL-18 is obtained. To prepare mice serum containing endogenous MuIL-18, C57BL/6 mice are treated with 500 µg of heat-killed Propionibacterium acnes for 1 week and challenged with 1 µg of lipopolysaccharide for 2 h to induce endotoxic shock. From the heart, under proper anesthetization, blood samples are taken to prepare sera. The MuIL-18R-expressing KG-1 cells are washed and resuspended at 5×10^5 cells/mL with RPMI-1640 medium for 2 days. The cells are adjusted to 1×10^6 cells/mL with RPMI-1640 containing 10% FCS, to which the indicated amounts of MuIL-18, HuIL-18, or serum sample are added. One day later, the culture supernatants are recovered and the quantitative analysis of the produced IFN-γ is performed by ELISA. The culture supernatant is incubated with mAb-IFN-γ-15. The bound IFN-γ is further incubated with mAb-IFN-γ-6 and detected with hydrogen peroxide and o-phenylenediamine.

5.2.7.5   Anti-rHuEPO NAb Assay

32D-EPOR cells are incubated at 37°C in humidified atmospheric air containing 5% CO_2 with RPMI 1640 supplemented with 15% heat-inactivated FBS, 2-mM l-glutamine, and penicillin/streptomycin (1%, v/v) mixture. The EPO-dependent cells are incubated in RPMI 1640 supplemented with 10 U/mL of rHuEPO. Via two to three subcultures a week, cell densities are maintained between 3×10^4 and 1×10^6 cells/mL. Up to 30 days after thawing, the old cells in the cryopreserved cells are discarded and a vial of frozen cells is thawed and expanded in culture [113].

In the cell proliferation assay, 32D-EPOR cells are incubated overnight, harvested, and washed twice with RPMI 1640 lacking rHuEPO by centrifugation (200–300 g). The supernatant is discarded, the cell pellet is resuspended in RPMI 1640, and the suspention is recentrifuged. The formed cell pellet is resuspended in RPMI 1640 and adjusted to 5×10^5 cells/mL. Cells are incubated at 37°C (humidified atmospheric air containing 5% CO_2, without rHuEPO) for 16–24 h, centrifuged, resuspended in fresh RPMI 1640, and counted. Overall, 100 μL of 2×10^4 staged 32D-EPOR cells and 100 μL of prepared testing sample are incubated at 37°C for 44 h in humidified atmospheric air containing 5% CO_2. The solution of 2-μCi [methyl-3H] thymidine diluted in 50 μL of RPMI 1640 is added to each well and incubated for 4 h. The contents of the plate are harvested, 25 μL of scintillation fluid are added, and the cells are counted.

All assay controls are prepared in a mixture of 5% human serum and 15% pooled rat serum. In anti-rHuEPO NAb assay, the background control (N) consisting of cells only is prepared by mixing 40 μL of pooled human serum with 240 μL of RPMI 1640 and 120 μL of pooled rat serum. The maximum growth control (M) consisting of cells and 1-ng/mL rHuEPO is prepared by mixing 40 μL of pooled human serum with 120 μL of rHuEPO at 6.67-ng/ml, 120 μL of RPMI 1640 and 120 μL of pooled rat serum. The neutralizing antibody positive control (P) consisting of cells, 1-ng/mL rHuEPO and 500-ng/mL positive control antibody is prepared by mixing 120 μL of RPMI 1640 with 120 μL of pooled rat serum. Samples are prepared by mixing 40 μL of individual donor serum with 120 μL of rHuEPO at 6.67 ng/mL, 120 μL of RPMI 1640, and 120 μL of pooled rat serum. Before addition to the cells, all controls and samples are preincubated at room temperature for at least 30 min.

5.2.7.6   Radial Assay of Chemotaxis

From stock sorocarp cultures of *D. discoideum* grown on agar plates, spores from strain v12 are harvested and heat-shocked at 45°C for 30 min. The suspension of the spores is mixed with a full loop of *E. coli* B/r, the 200-μL aliquots in new SM-agar plates are cultured at 22°C for 24 h, the cells are harvested and maintained vegetatively by shanking (200 rpm in Lpp medium, approximately 10^6 cells/mL), or the cells are starved by washing four times by centrifugation (3000 g, 30 s). The cells are shaken in 15-mM Tris-HCl (pH 7.0) for 2 h or 4 h; the latter is for routine chemotaxis assays. Chemotaxis of *D. discoideum* amoebae is performed on thin agar plates. Overall, 1 mL of 1.0% agarose in 15-mM Tris-HCl (pH 7.0) is added to each Petri plate and agitated to allow even spreading. The chemoattractants cAMP or folic acid are added to the agar with or without agonists before pouring plates. Chemotactically competent cells are centrifuged to form a viscous suspension and spotted on the agar plates. Each plate containing six aliquots of cells is covered, incubated at 22°C for 3 h, uncovered and placed on a heater to fix cells and dessicate the agar. As a measure of chemotactic efficiency, the initial spot diameters are subtracted from the diameters of visible rings or "halos" formed by outwardly migrating cells. During a period of 60 min with a playback time of 3 min, the individual halos on the agar plates are monitored. Proteins (20 μg/lane) are separated from whole cell lysates by polyacrylamide gel electrophoresis in 12% sodium dodecyl-sulphate polyacrylamide gels and transferred to nitrocellulose membranes in a mini-trans-blot cell using a buffer containing 25-mM Tris, 192-mM glycine (pH 8.3), and 20% methanol. Transfer is with 100 V for 40–60 min employing a frozen cooling unit. Using standard curves produced by running prestained Rainbow markers in parallel lanes, relative molecular weights are estimated. Nitrocellulose blots are blocked in 4% BSA at 4°C for 16 h, incubated for 1 h in 1/500 monoclonal anti-phosphotyrosine PT-66, and diluted by a 1/1000 dilution of peroxidase conjugated goat-antimouse IgG at room temperature for 1 h to detect the proteins containing phosphorylated tyrosine residues. Between each incubation, blots are washed three times for 5 min and immunoreactive bands are visualized using the chromogenic substrate diaminobenzidine HCl and 0.025% peroxide.

5.2.8   ESTROGEN ASSAY

After transfection, some cells become special fused cells such as the GAIA–DNA domain binding estrogen receptor yeast [114], the HELNα and HELNβ transfected cells [115], the yeast stably expresses human estrogen receptor β (hERβ) [116], the yeast stably expresses human estrogen receptor α (hERα), and enhanced green fluorescent protein (yEGFP) in response to estrogens α (hERα) [117]. With the special characteristics and high response to estrogen stimulation, the mentioned cells and related substances are used for bioassays.

5.2.8.1   Yeast Oestrogen Assay

The yeast, of which the steroid-binding domain is fused by GAL4-VP16, is incubated with SC-medium without histidine at 30°C and 130 rpm overnight. By adding DMSO up to a final concentration of 15% (v/v), stock cultures are prepared from exponentially growing cultures and stored in 0.5-mL aliquots at −80°C. Exponentially growing overnight, cultures are diluted with SC-medium to an OD_{600nm} of 0.75. To 10-mL aliquots 100 μL of DMSO (negative controls), 100 μL of 17β-estradiol in DMSO (positive controls), or 100 μL of test compound in DMSO are added, incubated at 30°C and 130 rpm for 2 h, diluted to five-fold volume and determined to get OD_{600nm}. To 200 µL of the test culture, 600 µL of Z-buffer (60-mM $Na_2HPO_4 \cdot 7H_2O$, 40-mM $NaH_2PO_4 \cdot H_2O$, 10-mM KCI, 1-mM $MgSO_4 \cdot 7H_2O$, 35-mM β-mercaptoethanol), 20-µL SDS solution (3.5 mM), and 50 µL of chloroform are added, carefully mixed, and pre-incubated at 28°C for 5 min, and then 200 µL of o-nitrophenyl-3-D-galactopyranoside in Z-buffer (13.3 mM) are added to initiate the enzyme reaction. The cultures are incubated at 28°C until a significant yellow color develops. For 17β-estradiol-induced positive controls, the yellow color occurs within 20 min. For test chemical-induced assays, the yellow color occurs after 120 min. To the cultures, 500 µL of Na_2CO_3 (1 M) is added to stop the reaction, the cell debris is pelleted by centrifugation (25,500 g, 15 min), and the Ex_{420nm} of the supernatants is determined. The β-galactosidase activity of the test cultures is calculated according to $u[\mu mol/min] = Cs/t \cdot V \cdot ODs$, wherein t = incubation time (min) of the enzyme reaction, V = volume (0.2 cm³) of the used test culture aliquot, $ODs = OD_{600nm}$ of test culture, Cs = concentration of o-nitrophenyl-β-D-galactopyranoside (µM) in the reaction supermatant calculated according to $Cs(\mu M) = 10^6 \cdot [Exs-Ex_B]/\epsilon_N \cdot d$, wherein $Exs = Ex_{420nm}$ of the enzyme reaction supernatant of test compound, $ExB = Ex_{420nm}$ of the enzyme reaction supernatant of the blank control, $\epsilon_N = \epsilon$ for o-nitrophenyl-β-D-galactopyranoside in the enzyme assay reaction mixture (4666 × 10³ cm²/mole), and d = diameter of the cuvette (1 cm). EC_{50} are calculated from dose response curves obtained by fitting the data by $Y = D + [A − D]/[1 + (C/X)^B]$, wherein X = estrogen concentration in the test, Y = β-galactosidase activity, B = relative slope of the middle region of the curve as estimated from a linear/log regression of the linear part of the dose response curve, C = estrogen concentration at half maximal response, and D = minimum β-galactosidase activity. The relative β-galactosidase induction factor defined as the ratio of β-galactosidase activity in a chemical sample and in the corresponding negative control is used for test chemical sample screening results.

5.2.8.2   HELN α and HELN β Transfected Cell Assay

After incubation with 30 mg of C18 Oasis HLB batch for 10 min, 800 mL of every patient's serum are desteroided and centrifuged (13,000 g, 10 min) to separate the supernatant. To this stripped serum, a known amount of estradiol is added to obtain a set of estradiol standards with concentrations from 10^{-9} to 10^{-4} mM, and incubated for 4h at 37°C to permit an equilibrium bet-ween free estrogens and estrogens bound to sex hormone-binding globulin. HeLa cells transfected with ERE-β Glob-Luc-SVNeo and pSG5-Puro-hERα or β, so-called HELNα and HELNβ, are selected by geneticin and puromycin at 1 mg/mL and 0.5 µg/mL, respectively. Luminescent and inducible clones are identified using photon-counting cameras. HELN Erα is cultured in 150-cm² plastic flasks in DMEM without phenol red, supplemented with 5% dextran-coated and charcoal-treated FCS, 0.25-µg/mL puromycin, and 0.5-mg/mL geneticin. To the culture medium, the aromatase inhibitor aminogluthetimide (AG) is added at a concentration of 50 µM. The cells are seeded in 96-well plates (3 × 10⁴ cells per well) in the presence of DMEM without phenol red, supplemented with 3% dextran-coated and charcoal-treated FCS, 50-µM AG, and incubated for 8h at 37°C. Culture medium is replaced by 100 µL of DMEM without phenol red, supplemented with 50-µM AG, to which 20 µL of human serum in triplicate is added.

5.2.8.3   Luciferase Assay

The p403- and p405-GPD yeast expression vectors and the p406-CYC1 yeast expression vector are used to express the human estrogen receptor α and β and construct the reporter plasmid, respectively. On the isolated mRNA of T47D human breast cancer cells and intestinal Caco-2 cells, cDNA is synthesized. Using the T47D cDNA, marathon uterus cDNA, and the human intestine cDNA by PCRs, full-length human estrogen receptor β (ERβ) cDNA is obtained. To perform the first PCR, 34.2 μL of ultra-pure water, 5 μL of 25-mM $MgCl_2$, 5 μL of Expand HF 10 × concentrated buffer (without $MgCl_2$), 0.8 μL of 25-mM dNTP mix, 1 μL of the enzyme mixture, 2 μL of the different cDNAs, and 2 μL of a primer mixture containing 10 μM of each primer are pipetted into a thin-walled PCR tube and (1) denature template 3 min at 95°C, (2) denature template 30 s at 94°C, (3) anneal primers 1 min at 60°C, (4) elongation 2 min at 72°C, (5) go to step (2) and repeat 35 times, (6) elongation 7 min at 72°C, and (7) for over 10°C. The second PCR is performed with the same conditions, but 2 μL of the first PCR mixture is used instead of 2 μL of the different cDNAs. The 5′- and 3′-primer, 5′-CGTCTAGAGCT GTTATCTCAAGACATGGATATAA-3′ and 5′-TAGGATCCGTCACTGAGA CTGTGGGTTCTG-3′, contains a restriction site for *Xba*I just before the ATG start codon and for *Bam*HI just after the TGA stop codon, respectively. The full-length ERβ PCR product is isolated and ligated into a pGEM-T Easy Vector. The uterus/intestine human ERβ cDNA clone that fully corresponded to the ERβ sequence is cut out of the pGEM-T Easy plasmid with *Xba*I and *Bam*HI and cloned into the corresponding *Xba*I–*Bam*HI site of the p405-GPD-ERβ vector, which is used to transform Epicurian Coli XL-2 Blue Cells.

Yeast K20 is transfected with the p406-ERE2s2-CYC1-yEGFP reporter vector and integrated at the chromosomal location of the Uracil gene via homologous recombination to construct yeast hERα and hERβ cytosensors. The transformants are grown on MM/LH plates using PCR and Southern blot hybridization to select clones, in which the integration has occurred at the desired URA3 site with only a single copy of p406-ERE2s2-CYC1-yEGFP. This strain is transformed with the p403-GPD-Erα and the p405-GPD-ERβ, or both expression vectors and transformants are grown on MM/L, MM/H, or MM plates, respectively.

The plate with selective MM/L, MM/H, or MM medium is inoculated with −80°C stock yeast ERα, ERβ, or ERα/β cytosensor (20% glycerol, v/v), respectively; incubated at 30°C for 24–48 h; and stored at 4°C. The day before the assay, a single colony of the yeast cytosensor is inoculated with 10 mL of the corresponding selective medium overnight at 30°C with vigorous orbital shaking at 225 g. At the late log phase, the yeast ERα, ERβ, and ERα/β cytosensor culture is diluted (1 : 10) in MM/L, (1 : 20) in MM/H, and (1 : 20) in MM, respectively. This minimal medium consists of a yeast nitrogen base without amino acids or ammonium sulphate (1.7 g/L), dextrose (20 g/L), and ammonium sulphate (5 g/L). The MM/L and MM/H medium are supplemented with L-leucine (60 mg/L) or L-histidine (2 mg/ L), respectively.

A total of 200-μL aliquots of the yeast culture are pipetted into each well of 96 well plates, and then 1 μL of ethanol or DMSO stock solution of test chemical is added to result in 0.5% final concentration. The plates are included for 4 h and 24 h, fluorescence at 485 nm and emission at 530 nm are measured directly, and the OD of the yeast culture at 630 nm is determined to check the toxicity of test chemical for yeast.

5.2.8.4 Green Fluorescent Protein Expression Assay

Aliquots of 2 mL of blank calf urine and 17β-estradiol (E2β, 1 ng/mL), diethylstilbestrol (DES, 1 ng/mL) and 17α-ethynylestradiol (EE2, 1 ng/mL), α-zearalanol (30 or 50 ng/mL) and mestranol (10 ng/mL), and spiked calf urine samples are adjusted to pH 4.8 and 20 μL of β-glucuronidase/arylsulfatase (3 U/mL) are added. The samples are deconjugated by enzyme at 37°C overnight, treated with 2 mL of 0.25-M sodium acetate buffer (pH 4.8) and subjected to solid phase extraction (SPE) on a C18 column conditioned previously with 2.5 mL of methanol and 2.5 mL of sodium acetate buffer. The column is successively washed with 1.5 mL of 10% (w/v) sodium carbonate solution, 3.0 mL of water, 1.5 mL of sodium acetate buffer (pH 4.8), 3.0 mL of water, and 2 mL of methanol/water (50/50 v/v). The column is air-dried and eluted with 4 mL of acetonitrile. The eluate is applied to an NH2-column conditioned previously with 3.0 mL of acetonitrile. The acetonitrile eluate is evaporated to 2 mL by nitrogen gas stream, of which a 100-μL part (equivalent to 100-μL urine) is transferred to a 96-well plate and mixed with 50 μL of water and 2 μL of DMSO. The plate is dried overnight in a fume cupboard to remove the acetonitrile and screened on estrogenic activities with the yeast estrogen bioassay. In the same way, a reagent blank is prepared, using 2 mL of the 0.25-M sodium acetate buffer pH 4.8 instead of urine.

The yeast cytosensor (20% glycerol v/v) expressing the human estrogen receptor α (hERα) and yEGFP in response to estrogens is incubated in an agar plate containing the selective MM/L medium at 30°C for 24–48 h and then stored at 4°C. The day before the assay, a single colony of the yeast cytosensor is inoculated in 10 mL of selective MM/L medium overnight at 30°C with vigorous orbital shaking at 225 rpm. At the late log phase, the yeast ERα cytosensor is diluted in MM/L, giving an OD at 604 nm in the range of 0.07–0.13. For exposure in 96-well plates, aliquots of 200 μL of this diluted yeast culture are pipetted into each well, already containing the extracts of the urine samples. A E2β dose-response curve is included in each exposure experiment. Aliquots of 200 μL of the diluted yeast culture are pipetted into each well of a 96-well plate and exposed to different doses of E2βs performed through the addition of 2 μL of E2β stock solutions in DMSO. Each urine sample extract and each E2β stock is assayed. Exposure is performed for 0 h and 24 h. Fluorescence at these time intervals is measured directly using excitation at 485 nm and measuring emission at 530 nm. The densities of the yeast culture at these time intervals are also determined by measuring the OD at 630 nm to check whether a urine sample is toxic for yeast.

In the determination of the decision limit (CCα) and detection capability (CCβ) of the yeast estrogen bioassay, extracts of 20 blank calf urine, 20 spiked calf urine samples (E2β, DES and EE2 at 1 ng/mL, α-zearalanol at 30 or 50 ng/mL, and mestranol at 10 ng/mL), and a reagent blank are analyzed. After 24 h of exposure, the obtained fluorescence signals of the 20 blank urine samples, the 20 spiked urine samples, and a reagent blank are corrected for the signals obtained at 0 h (t_{24}–t_0). On three different days, the extracts of the blank urines and their corresponding spikes are prepared and analyzed in the yeast estrogen bioassay in three separate exposures. Within a time period of 10 days, these three sample treatments and exposures are performed. In another experiment, extracts of 20 blank urines and spikes of 50-ng zearalanol per mL urine are prepared in one day and are analyzed in the yeast estrogen bioassay in one exposure. In the context of EC Decision 2002/657, the mean signal of 20 blank calf urine samples plus 2.33 times the corresponding standard deviation is defined as the decision limit $CC\alpha$ ($\alpha = 1\%$), and the decision limit $CC\alpha$ plus 1.64 times the standard deviation of the signal of the spiked calf urine sample is defined as the detection capability $CC\beta$ ($\beta = 5\%$).

In the determination of the specificity of the yeast estrogen bioassay, three blank calf urine samples are spiked with a high dose of testosterone or progesterone (1000 ng/mL) and extracts are analyzed. In the determination of interference, these blank calf urine samples are spiked with a high dose of either testosterone or progesterone in combination with a low dose of estrogens (E2β, DES, and EE2 at 1 ng/mL, α-zearalanol at 50 ng/mL, and mestranol at 10 ng/mL).

5.2.9   ANTIMALARIAL ASSAY

The tremendous progress in the biology and the biochemistry of malaria parasites has led to the identification of drug targets that are both parasite specific and essential for parasite growth and survival, and some of them are now exploited in the mechanisms-based assays of various screening programs and determination of antimalarials' concentration such as plasmodium falciparum and murine P388 leukemia cell assay [118], histidine-rich protein II assay [119], anti-plasmodium activity assay [120], survival of Anopheles gambiae assay [121], chloroquine plus doxycycline assay[122], and plasmodium falciparum clone and dihydroartemisinin assay [123].

5.2.9.1   Plasmodium Falciparum and Murine P388 Leukemia Cell Assay

The required quantity of *P. falciparum* parasites maintained in a complete medium (RPMI-1640, 25-mM HEPES, 25-mM NaHCO$_3$, and 10% pooled human serum, with uninfected human red blood cells at 2.5% haematocrit) is introduced into flat-bottomed 96-well plates. The cell suspension (1% parasitaemia) is added to the plates (0.2 mL/well), which contains test compound in triplicate alongside untreated controls. The plates are shaken vigorously using a microculture plate shaker, and they are incubated at 37°C for 18h under microaerophilic conditions [124]. To each well, tritiating hypoxanthine with a specific activity of 14.1 Ci/mmol is added (0.5 µCi/well) and incubated at 37°C for another 24h. The contents of the well are frozen at −30°C, unfrozen at 50°C, filtrated, and washed several times with water. The disks are dried and added to toluene scintillator in vials, and the radioactivity incorporated into parasites is estimated.

Test compounds inhibiting 75% or more of the parasite growth are systematically submitted to cytotoxicity tests. Murine P388 leukemia cells grown in RPMI 1640 medium containing 0.01-nM β-mercaptoethanol, 10-mM L-glutamine, 100-U/mL G-penicillin, 100-µg/mL streptomycin, 50-µg/mL gentamycin, and 50-µg/mL nystatine, supplemented with 10% fetal calf serum, are incubated in humidified atmospheric air containing 5% CO$_2$ at 37°C. The inoculum seed at 10^4

cells/mL (0.1 mL/well) is transferred into flat-bottomed 96-well plates containing serial concentrations of test compound, and the plates are incubated in the required atmosphere at 37°C for 72 h. Thereafter, cells are incubated at 37°C with a 0.02% solution of neutral red in 1/9 methanol/water (0.1 mL/well) for 1 h, washed with 1N PBS, and lyzed with 1% SDS. After a brief agitation, the plates are read at 540 nm to measure the absorbance of the extracted dye, and cell viability is expressed as the percentage of cells incorporating dye relative to the untreated controls and IC_{50} values are determined by the linear regression method.

5.2.9.2 Histidine-Rich Protein II Assay

The serial dilutions of complement-inactivated plasma samples, which are from uncomplicated falciparum malaria patients and a healthy volunteer treated with oral sodium artesunate (100 mg followed by 50 mg orally every 12 h for 5 days), are applied in two columns to 96-well microculture plates at 50 μL/well. The serial dilutions of spiked plasma are applied in two columns and added to each plate as controls. On each 96-well plate, five unknown samples plus the controls are tested; in addition, one plate with six serial dilutions of known drug concentrations covering the whole test range is also tested. According to the literature, the culture and ELISA procedures are performed [125]. A total of 175-μL aliquots of the dilutions (0.1% parasite density, 1.7% hematocrit) of synchronized parasitized (clone W_2) blood samples from continuous culture are added to the plates (resulting in a total of 225 μL/well), with the frozen mixture of 1.05 mL of the remaining cell medium and 0.3 mL of drug-free plasma per plate as negative controls. Before being frozen-thawed, the plates are incubated at 37°C for 72 h in a candle jar or gas mixture (5% CO_2, 5% O_2, 90% N_2).

5.2.9.3 Antimalarial Activity Assay

The chloroquine resistant strain FcB1 of *P. falciparum* (50% inhibitory concentration [IC_{50}] of chloroquine = 62 ng/mL) maintained continuously in culture on human erythrocytes is used for the evaluation of antimalarial activities of test compounds [126]. The semiautomated microdilution technique [127] is modified to determine the *in vitro* antiplasmodial activity of test compounds. The serial medium dilutions of the stock solutions of test compounds in DMSO 10 mg/mL, of which the concentration never exceeds 0.1% and does not inhibit the parasite growth, are incubated with asynchronous parasite cultures (0.5% parasitemia and 1% final hematocrit) on 96-well plates at 37°C for 24 h and treated with [^{3}H] hypoxanthine (0.5 μCi/well) for 24 h. By comparison of the radioactivity incorporated into the treated culture with that in the control culture maintained on the same plate, the growth inhibition concentration for each test compound is defined. From the drug concentration-response curve, the IC_{50} is obtained and expressed as the mean determined from three independent experiments.

5.2.9.4 Survival of *Anopheles Gambiae* Assay

Adults of laboratory-reared *An. gambiae* mosquitoes (from a colony established from wild gravid females, all mosquito life stages are maintained under semi-field conditions) are given 6% glucose solution on filter-paper wicks, water, and routine human bloodmeals three times per week. Three days after each bloodmeal inside the cages, which are kept in screen houses under ambient conditions, oviposition cups are used to collect eggs on the following day. Under semi-field conditions, the larvae for both the colony and the experiments are maintained on fish food and the plastic pans (25 × 20 × 14 cm) are filled with fresh water [128].

In 30-cm cubic metal frame cages covered with mesh netting, approximately 100 pupae/cage and the test compound are for direct observation. *An. gambiae* mosquitoes are allowed to emerge, and evening observations of test compound feeding are conducted from 20:00 to 22:00 h, for which at intervals of 30 min, a flashlight is used. On a second night of observation, a mosquito net is placed over *Ricinus communis* in a screenhouse, under which a cup of water containing approximately 500 pupae is placed and emerging adults are allowed to feed on the test compound for 24 h. From 20:00 to 22:00 h in intervals of 30 min, on the following night, mosquitoes are observed until they are observed probing or feeding on the test compound.

For survival assay, mosquitoes are divided into regime groups. From the main mosquito colony, approximately 100 pupae per regime are harvested, transferred to plastic cups, and placed in cages and held in a screenhouse under ambient conditions. Each cage of mosquitoes has access to test compound sources; one group of mosquitoes has access to a cotton pad moistened with distilled water, representing negative controls; and another group of mosquitoes with neither test compound source nor water is used as an additional negative control. Dead mosquitoes are removed at 4-h intervals from 7:00 to 23:00 h until all die.

5.2.9.5 Chloroquine Assay

The blood samples (10 mL) from healthy volunteers who are on chloroquine (2 × 150-mg base weekly) and doxycycline (50 mg or 100 mg daily) malaria prophylaxis are centrifuged (1200 g, 10 min), and the plasma is separated and stored at −20°C. The chloroquine-sensitive FC27 strain of *P. falciparum* is used for the determination of antimalarial activity of each plasma sample [129]. On microculture plates, the plasma samples are diluted twofold with drug-free serum and inoculated with a suspension of parasitized erthrocytes in culture medium. The drug susceptibility and minimum concentration of spiked drug that inhibits parasites, relative to control, from developing to schizonts is recorded. The maximum inhibitory dilution (MID) of the volunteers' samples containing a choroquine concentration and the minimum inhibitory concentration (MIC) observed in the sensitivity test are recorded. The chloroquine equivalent concentration in the prediluted volunteers' specimen is estimated by multiplying the MID by the MIC.

5.2.9.6 Plasmodium Falciparum Clone and Dihydroartemisinin (DHA) Assay

By centrifugation for 5 s, 50 μL of Affigel protein A (binding capacity of 20 mg of purified human immunoglobulin G[IgG]/mg of gel) is washed twice with PBS (pH 7.2, 50 μL), and the RPMI 1640 medium (50 μL) that is supplemented with 5.94-g/L HEPES and 2.1-g/L sodium bicarbonate, centrifuged to discard the supernatant, and Affigel protein A gel is incubated with the plasma and serum (250 μL) spiked with dihydroartemisinin (DHA) or plasma from patients at room temperature for

30 min [130]. The stock solution of DHA in 70% ethanol (1 mg/mL) is kept at –10°C for up to 1 month before use, which is further diluted in plasma or serum to give a working concentration (1.25 to 100 ng/mL).

Affigel protein A-treated plasma (100 μL) and heat-inactivated serum (50 μL/well) are transferred to row A and B of a flat-bottom plate, respectively. Using the heat-inactivated serum added through G into row B as the diluent for plasma or serum containing DHA, twofold serial dilutions are prepared. To row H, 50 μL of heat-inactivated control plasma or serum is added for parasitizing and nonparasitizing erythrocyte controls. To all wells for rows A to G and eight wells for row H, the suspension of 175 μL of malaria parasite-infected erythrocytes (W2 clone; 0.5% parasitemia with 80% young rings at a 1.7% hematocrit) is added. To the remaining four wells of row H for nonparasitized controls, the similar suspension of uninfected erythrocytes is added. The microtiter plates are incubated at 37°C for 24 h and treated with 1 Ci/mmol [³H] hypoxanthine by the addition of 25 μL of 0.5 μCi isotope solution to each well. The microtiter plates are incubated for another 18 to 20 h. Harvesting the contents of the plates, the particulate material of the water-lysed cell suspension is collected on glass-fiber filter paper by filtration, and the filter paper is dried in an oven and placed in a plastic bag, to which 10 mL of scintillation fluid is added and the plastic bag is then sealed. By counting in a liquid scintillation counter, the level of incorporation of [³H] hypoxanthine by the malaria parasites is determined.

Fresh whole blood is collected to heparinized tubes, and the plasma and buffy coat are removed, and the erythrocytes are lysed and serially diluted (2- to 64-fold) with sterile water, of which 50-μL aliquots are diluted in 250 μL of plasma or medium spiked with DHA to give a final concentration of 50 ng/mL and a range of hemoglobin concentrations from 1.5 to 0.05 g%. The effect of hemolysate in both plasma and culture medium (RPMI 1640 medium with 15% heat-inactivated human serum) on the detection of DHA activity is examined. The final concentration of plasma and serum in each well is 30% and 15%, respectively, which is similar to that in the normal drug susceptibility test. In addition to testing hemolysates of normal erythrocytes, the plasma samples collected from patients with severe falciparum malaria before treatment with any antimalarial drugs can also be spiked with 50 ng of DHA/mL and subjected to the same bioassay.

5.2.10   BLOOD PRESSURE-RELATED ASSAY

Blood pressure can be regulated by receptor, enzyme, endogerous substance, and cell, such as bradykinin B2-receptor, ACE, 5-Hydroxytryptamine, and endothelia cell. Based on these facts, the human plasma assay [131], pulmonary hypertension assay [132], coronary arteries (CA) constriction assay [133], MCAO and HSPG assay [134], and temperature assay in awake subjects [135] are established.

5.2.10.1   Human Plasma Assay

The solution of test peptide is injected (i.v.) into adult male Wistar rats (housed in pairs, under a 12:12 h light/dark cycle, 0.4% NaCl rat chow, and water available *ad libitum*, after 3–5 days of adaptation to the facility, about 300 g) at a proper dose.

The inactin (100 mg/kg, i.p.; sodium ethyl-(1-methyl-propyl)-malonylthio-urea) anesthetized rats are treated with atropine (2.4 mg/kg, s.c.), and their ganglions are blockaded with pentolinium (19.2 mg/kg, s.c.). Overall, 10-mg/kg captopril is administered i.v. acutely to the rats in the captopril group, and it inhibits fully the pressor effect of 60-ng/kg angiotensin I.

To assist ventilation, the trachea is intubated and both vagi are severed to block reflex bradycardia. Into the right femoral vein and carotid artery, catheters are inserted for the injection of test compound and other agonists and antagonists, and for monitoring arterial blood pressure and heart rate and for blood sampling, respectively [136, 137], and blood pressure is monitored.

Responses, in terms of the increases in systolic (SBP) and diastolic (DBP) blood pressures, to test peptide and bradykinin are determined. After ganglion blockade with pentolinium (group GB) without or with added captopril (group GB+Cap), heart rate and plasma adrenaline and noradrenaline concentrations are determined. In some cases, the rats are not subjected to ganglion blockade and served as controls for animals in the GB group (control group).

To evaluate the contribution of bradykinin B2-receptor-mediated mechanisms, the selective bradykinin B2-receptor antagonist, HOE-140, is given to ganglion-blocked, the rats are captopril treated, and the increases in SBP, DBP, heart rate, and plasma adrenaline/noradrenaline induced by test peptide and bradykinin (group GB+Cap+HOE) are observed. To evaluate the contribution of AT1-receptor-mediated mechanisms, the selective AT1 receptor antagonist, losartan, is given to group GB rats before and after treatment with captopril (1 mg/kg, i.v.) and the increases in SBP, DBP, heart rate, and plasma adrenaline/noradrenaline induced by test peptide are observed.

For determining plasma adrenaline and noradrenaline arterially approximately 1 mL of blood is withdrawn via the carotid cannula at baseline (before injection of test peptide and bradykinin), at the peak of the SBP response, and after recovery from the response [138, 139]. Before each withdrawal, the rat is given a transfusion of approximately 1 mL of blood from a similarly prepared donor rat. The SBP peak is observed at 2–3 min after the injection of test peptide and at 1–2 min after the injection of bradykinin; blood sampling coincides with these intervals. Using HPLC and fluorimetric detection, plasma catecholamines are determined [140].

5.2.10.2   Pulmonary Hypertension Assay

The chest cavity of a mouse (10–20 weeks old, 22–30 g) anesthetized by pentobarbital sodium (80 mg/kg, i.p.) is opened, and the lungs are removed rapidly and placed in Krebs–Ringer bicarbonate solution (KRBS) containing 118.3-mM NaCl, 4.7-mM KCl, 1.2-mM $MgSO_4$, 1.2-mM KH_2PO_4, 2.5-mM $CaCl_2$, 25.0-mM Na_2CO_3, and 10.0-mM glucose, bubbled with 21% O_2. From the intrapulmonary artery (third to fourth generation, PA), rings (50–100 µm internal diameter, 2–3 mm long) are isolated using a dissection microscope. PA rings are mounted as ring preparations by threading two steel wires into the lumen and securing the wires to two supports, and they are placed in a small vessel wire myograph chamber. The support is attached to a micrometer for the control of ring circumference and to a force transducer for measurement of isometric tension, respectively. After removal of endothelial cells by gently rubbing the intraluminal surface with a steel wire, some vessels are successively perfused with 2-mL air bubbles and 2-mL of KRBS (perfusion pressure < 5 mmHg), before being mounted in the chamber. In the chamber filled with KRBS (pH 7.35–7.45), bubbled with 21% O_2–5% CO_2–balance N_2, the whole preparation is kept at 37°C. To control oxygen tension over the superfusate, Plexiglas is used as a cover over the chamber. The temperature and PA tension are recorded. At initial tension of 0 mN (1 g = 4.905 mN), isolated murine PA rings in the chamber are allowed to equilibrate for 10–15 min, which are increased to 5 mN in 2.5-mN steps at 4- to 5-min intervals and held constant thereafter. Through preliminary experiments, the resting tension for maximal constrictor response is optimized, in which the resting tension levels of 2.5, 5, and 7.5 mN are compared. In the evaluation of vascular viability, after the treatment of PA with 60-mM KCl, the tension is determined. The PA is washed extensively with KRBS, successively exposed to U-46619 (0.01 μM, thromboxane A_2 agonist) and 1-μM Ach, the resulting tension is recorded and stabilized for 5–10 min, and the agonists are washed out of the myograph chamber with KRBS.

Murine PA is isolated, placed in a confocal pressure myograph chamber, cannulated at both ends with glass micropipettes, and secured with a 12-0 nylon monofilament suture. To control transmural pressure, both cannulas are connected to a reservoir that can be raised or loared. In the chamber, PA is superfused at 37°C, constantly with KRBS, and gassed with 21%O_2–5%CO_2–balance N_2. To control oxygen tension over the superfusate, Plexiglas is used as a cover over the chamber. To the chamber, 0.1-mM dihydroethidum (DHE) is added, in which murine PA is incubated at 37°C for 45 min and then washed for 30 min. The reaction of ntracellular DHE and ROS produces a fluorescent oxidized product [141, 142] and PA images are scanned (480-nm-line argon laser, fluorescence 620 nm). Superoxide anion levels in isolated murine PA are measured by a lucigenin-enhanced chemiluminescence technique [143, 144] and scintillation counter, using a solution of 5-μM lucigenin in Krebs–HEPES buffer (10.0-mM HEPES acid, 135.3-mM NaCl, 4.7-mM KCl, 1.2-mM $MgSO_4$, 1.2-mM KH_2PO_4, 1.8-mM $CaCl_2$, 0.026-mM Na-EDTA, and 11.1-mM glucose), 1-mL total volume, after background chemiluminescence activity has stabilized for 5 min.

5.2.10.3   Coronary Arteries (CA) Constriction Assay

The left main coronary arteries (CA, 70–90 μm in diameter, 1-mm length) of a mouse are placed into a microvascular chamber, cannulated at both ends with glass micropipettes, and pressurized. By gently rubbing the intraluminal surface with a steel wire, the endothelial cells of some vessels are removed. The vascular intraluminal pressure (Ptm) is measured by using a pressure transducer positioned at the level of vessel lumen. The vessels in the chamber are superfused constantly with recirculating Krebs–Ringer bicarbonate solution containing 118.3-mM NaCl, 4.7-mM KCl, 1.2-mM $MgSO_4$, 1.2-mM KH_2PO_4, 2.5-mM $CaCl_2$, 25.0-mM $NaHCO_3$, and 11.1-mM glucose, which is gassed with 16% O_2, 5% CO_2, and balance N_2 (pH 7.35–7.45), and maintained at 37°C. To control oxygen tension over the superfusate the chamber is covered by Plexiglas, through which port an oxygen electrode is passed into the superfusate and positioned near the vessel to provide continuous measurement of oxygen tension. The vascular intraluminal diameter (ID) is measured continuously. The oxygen tension, vascular ID, and Ptm are recorded. The isolated CA in the chamber is allowed to equilibrate 30 min at a Ptm of 10 mm Hg, which is increased to 60 mm Hg in 10-mm Hg steps at 5–7-min intervals and held constant thereafter. When Ptm is increased to 60 mm Hg (time 0), the ID is recorded (appointed to ID60, namely ID at Ptm = 60 mm Hg) and the measurement is continued throughout the experiment.

The superoxide anions produced in isolated murine CA are measured using a lucigenin (bis-N-methyacidinium nitrate)-enhanced chemiluminescence technique. After background has stabilized chemiluminescence activity for 5 min, CA is placed in the chemiluminometer containing 1 mL of 5-μM lucigenin buffer solution, photon emission is recorded continuously, and the chemiluminescence signal is recorded as relative light units per second (RLU/s).

5.2.10.4 Middle Cerebral Artery Occlusion (MCAO) Assay and Vascular Heparan Sulfate Proteoglycans (HSPG) Perlecan Assay

Male baboons are randomly divided into treatment and control groups. In the treatment groups the animals undergo MCAO for 1 h or 2 h, or 3 h MCAO with subsequent reperfusion for 1 h or 4 h, or 1.5 h MCAO with 24 h reperfusion. In the control groups, the animals do not undergo any preparation procedure or suffer from MCAO at surgical implantation and has sustained hemiparesis for 7 days. The animals are transcardially perfused with isosmotic heparinized perfusate, and their brain tissues are removed under thiopental Na and prepared for frozen and paraffin sections [145, 146]. The samples of normal and 50 μL of ischemic brain tissue or 100 μL of purified reagents are added to the recipient tissues consisting of unfixed 10-μm-thick normal basal ganglia sections mounted on microscope slides.

When proteases are added to the normal recipient tissues, PBS (pH 7.0) is the incubation buffer for type-7 bacterial collagenase. The solution of matrix metalloproteinases (MMPs) and urokinase/plasminogen in a buffer containing 90-mM NaCl, 5-mM KCl, 1.5-mM MgCl$_2$, 23-mM Na gluconate, 27-mM NaAc, 10-mM CaCl$_2$, and 40-mM ZnCl$_2$ (pH 7.4) is prepared. The solution of cathepsin B and L in 200-mM NaAc containing 8-mM dithiothreitol, 1-mM EDTA, and 2.7-mM L-cysteine (pH 6.0) is prepared. The mixture of 5-U/mL plasminogen and 5-U/mL uPA is incubated at 37°C for 1 h; 1-μg/mL Pro-MMP-2 or 1-μg/mL pro-MMP-9 is added and incubated for 1 h. The recipient sections are incubated with 100 μL of each mixture or buffer alone at 37°C for 5 h (collagenase) or 18 h (MMPs and cathepsins) and washed with PBS and fixed.

From approximately 1.0-cm × 1.0-cm × 10-μm frozen sections, donor samples are derived. By centrifugation (300 g, 30 s) and repeated gentle pipetting, the mixture of 10 consecutive 10-μm sections of ischemic or normal basal ganglia is prepared. Overall, 50 μL of sample or PBS is added to the recipient sections, the mixture is incubated at 37°C for 18 h, and the recipient sections are washed with PBS and fixed in acetone or paraformaldehyde (PFA).

In the development of the specific antigens, the immunoperoxidase methods are used [147]. At 4°C, the overnight incubation of acetone-fixed frozen sections in the presence of the primary antibody is followed by developing immunoperoxidase with 3,3′-diaminobenzidine tetrahydrochloride. At 4°C, the overnight deparaffination of acetone-fixed frozen sections in the presence of the primary antibody is also followed by developing immunoperoxidase with 3,3′-diaminobenzidine tetrahydrochloride. Incorporating dUTP into nuclear DNA is used as the evidence of nuclear DNA scission/repair, and at 2 h, MCAO defines the ischemic core and peripheral regions of cellular neuronal injury. PFA-fixed cryosections are subject to the DNA polymerase I method [148]. All ischemic samples include ischemic core and ischemic peripheral regions. In the measurement of MMP-related activities, Gelatin zymography is performed with a modification to increase sensitivity [149]. Protease activities are identified by incubation of the gels in buffer containing GM6001 or APMSF.

5.2.10.5 Temperature Assay in Awake Subjects

The male SD rats (250–325 g) are caused reversible ischemia with the intraluminal filament occlusion of the middle cerebral artery (MCA) and maintained anesthesia with 2% halothane and mixed gas of nitrous oxide and oxygen (60:40) by face mask [150]. To expose the left carotid artery, a midline neck incision is made. The external carotid and pterygopalatine arteries are ligated with 5-0 silk. In the arterial wall, an incision is made. From the bifurcation of external and internal carotid arteries, a 4-0 heat blunted nylon suture of 18 mm is advanced. This distance reliably produces blockage of the origin of the MCA. The duration of occlusion for individual rats is varied to generate a range of ischemia durations for each experimental group. During surgery, the temporalis and brain temperature are monitored and maintained.

During the period (4–5 h) of post-anesthetic recovery, the awake rats are kept on a pad maintained at 37.5°C. To verify the effect of this procedure on brain temperature, an indwelling radio telemetry thermister is implanted into the rat of the nonischemic group. After induction of anesthesia, the rat is placed in a stereotaxic head frame. A thermister probe is placed in the left parietal cortex (3 mm deep to dura and 1 mm anterior, 3 mm lateral to bregma). To secure the probe to the skull, a plastic cap is fitted using methylmethacrylate [151]. After full recovery from the anesthesia, telemetered brain temperature is measured every 15 min. Temperatures are measured every 5 min for 24 h during a baseline period and 6 h after drug administration. After a drug treatment, some animals are placed on a heating pad for 2 h and their temperature is measured every 15 min. At 48 h and 72 h, the behavior of the rats is evaluated and classified to normal/abnormal according to a modified rodent examination scale [152]. All obtundation/reduced exploration, forepaw retraction on tail lifting, asymmetric forepaw grasp, axial twist, forced circling, or death are defined as abnormal.

To correctly position the filament, 0.2 mL of 2% Evan's Blue in saline is injected into each rat via tail vein and circulated for 30 to 60 min, after which the rat is perfused transcardially with 100-mL normal saline, and the brain is removed and immediately examined. Each day, two rats of the control group and four or five rats of the drug treatment group are injected test compound dissolved in saline via tail vein at 30, 60, 75, 120, 240, or 360 min after initiating occlusion to assess therapeutic efficacy.

To determine the effects of the test compound on blood gases, heart rate, and arterial blood pressure in separate groups, physiological determinations are performed. To observe the interaction between test compound and anesthesia, the determinations in awake or in anesthetized rats are performed. The ventral tail artery is cannulated with PE50 tubing for recording the arterial pressures and sampling arterial blood. Test compound is administered after the rat is stable for 30 min with no further adjustments needed to maintain homeostasis under 75–90 mmHg steady-state mean arterial pressure (MAP) and minimal halothane. To obtain similar data from awake subject, tail artery catheters are implanted under halothane inhaled anesthesia. After implantation, the rat is allowed to recover and is lightly restrained. One hour after stopping the anesthesia, the cannula is attached to the transducer and vital signs are recorded for a 30-min baseline period. After the subject is shown to be stable for 30 min, the test compound is given. Blood pressures and blood gases are recorded every 10 min for 1 h and then hourly.

REFERENCES

1. Engler K H, Coker R, Evans I H (1999). A novel colorimetric yeast bioassay for detecting trichothecene mycotoxins. *J. Microbiol. Meth.* 35:207–218.

2. Gonzaulez A M, Beniutez L, Soto T, et al. (1997). A rapid bioassay to detect mycotoxins using a melanin precursor overproducer mutant of the ciliate *Tetrahymena thermophila. Cell Biol. Int.* 21:213–216.

3. Cetin Y, Bullerman L B (2005). Cytotoxicity of *Fusarium* mycotoxins to mammalian cell cultures as determined by the MTT bioassay. *Food Chem. Toxicol.* 43:755–764.

4. Saadé F E, Dunphy G B, Bernier R L (1996). Response of the carrot weevil, *Listronotus oregonensis* (Coleoptera: Curculionidae), to strains of *Bacillus thuringiensis. Biol. Control.* 7:293–298.

5. Gehringer M M, Vijayne K, Nadya C (2003). The use of *Lepidium sativum* in a plant bioassay system for the detection of microcystin-LR. *Toxicon.* 41:871–876.

6. Kaney A R, Speare V J (1983). An amicronuleate mutant of *Tetrahymena thermophila. Exp. Cell Res.* 143:461–467.

7. Hansen M B, Nielsen S E, Berg K (1989). Re-examination and further development of precise and rapid dye method for measuring cell growth/cell kill. *J. Immunol. Methods.* 119:203–210.

8. Saadé F E (1993). Evaluation of strains of *Bacillus thuringiensis* as biological control agents of the adult stages of the carrot weevil, *Listronotus oregonensis* (Coleoptera; Curculionidae). M.Sc. Thesis. National Library, Ottawa, Canada.

9. Gringorten J L, Witt D P, Milne R E, et al. (1990). An *in vitro* system for testing *Bacillus thuringiensis* toxins: The lawn assay. *J. Invertebr. Pathol.* 56:237–242.

10. Bradford M (1976). A rapid and sensitive method for the quantification of microgram quantities of protein utilizing the principle of protein-dye binding. *Anal. Biochem.* 72:248–254.

11. van Frankenhuyzen K, Gringorten J L (1991). Frass failure and pupation failure as quantal measurements of *Bacillus thuringiensis* toxicity to Lepidoptera. *J. Invertebr. Pathol.* 58:465–467.

12. Gehringer M M, Downs K S, Downing T G, et al. (2003). An investigation into the effect of selenium supplementation on microcystin hepatotoxicity. *Toxicon.* 41:451–458.

13. Ward C J, Beattie K A, Lee E Y C, et al. (1997). Colorimetric protein phosphatase inhibition assay of laboratory strains and natural blooms of cyanobacteria: comparisons with high performance liquid chromatographic analysis for microcystins. *FEMS Microbiol. Lett.* 153:465–473.

14. Habig W H, Pabst M J, Jakoby W B (1974). Glutathione Stransferases: the first step in mercapturic acid formation. *J. Biol. Chem.* 249:7130–7139.

15. Baumstark-Khan C, Rode A, Rettberg P, et al. (2001). Application of the *Lux-Fluoro* test as bioassay for combined genotoxicity and cytotoxicity measurements by means of recombinant *Salmonella typhimurium* TA1535 cells. *Anal. Chim. Acta.* 437:23–30.

16. Shimizu K, Endo O, Goto S, et al. (2001). Bioassay-based evaluation of toxicity of suspended particulate matter in humans: integrated uses of alveolar cells (A549) in air liquid interface culture and hepatocarcinoma cells (Hep G2). *Biochem. Eng. J.* 22:1–9.

17. Takahashi K, Ohsawa M, Utsumib H (2002). A simple bioassay for evaluating immunotoxic properties of chemicals by use of *in vitro* antibody production system. *J. Health Sci.* 48:161–167.

18. Nagy S R, Sanborn J R, Hammock B D (2002). Development of a green fluorescent protein-based cell bioassay for the rapid and inexpensive detection and characterization of Ah receptor agonists. *Toxicol. Sci.* 65:200–210.

19. Bolon B, Bucci T J, Warbritton A R, et al. (1997). Differential follicle counts as a screen for chemically induced ovarian toxicity in mice: results from continuous breeding bioassays. *Fundam. Appl. Toxicol.* 39:1–10.

20. Yoshikazu I, Toshimasa J, Kazuhisa F (2003). Effect of nefiracetam, a neurotransmission enhancer, on primary uroepithelial cells of the canine urinary bladder. *Toxicol. Sci.* 72:164–170.

21. Fujita K, Nagaoka M, Komatsu Y, et al. (2003). Yeast pheromone signaling pathway as a bioassay to assess the effect of chemicals on mammalian peptide hormones. *Ecotox. Environ. Safe.* 56:358–366.

22. Taylor C J, Bain L A, Richardson D J, et al. (2004). Construction of a whole-cell gene reporter for the fluorescent bioassay of nitrate. *Anal. Biochem.* 328:60–66.

23. Sambrook J, Fritsch E F, Maniatis T (1989). *Molecular Cloning, A Laboratory Manual.* Cold Spring Harbor Laboratory Press, New York.

24. Koyano M, Endo O, Goto S, et al. (1998). Carcinogenic polynuclear aromatic hydrocarbons in the atmosphere in Chiang Mai, Thailand. *Jpn. J. Toxicol. Environ. Health.* 44.

25. Willett K L, Gradinali R P, Sericano J L, et al. (1996). Characterization of the H4IIE rat hepatoma cell bioassay for evaluation of environmental samples containing polynuclear aromatic hydrocarbons (PAHs). *Arch. Environ. Contam. Toxicol.* 32:442–448.

26. Ohsawa M, Masuko-Sato K, Takahashi K, et al. (1986). Strain differences in cadmium-mediated suppression of lymphocyte proliferation in mice. *Toxicol. Appl. Pharmacol.* 84:379–388.

27. Mosmann T (1983). Rapid colorimetric assay for cellular growth and survival: Application to proliferation and cytotoxicity assays. *J. Immunol. Meth.* 65:55–63.

28. Garrison P M, Tullis K, Aarts J M, et al. (1996). Species-specific recombinant cell lines as bioassay systems for the detection of 2,3,7,8-tetrachlorodibenzo-*p*-dioxin-like chemicals. *Fundam. Appl. Toxicol.* 30:194–203.

29. Denison M S, Phelan D, Winter G M, et al. (1998). Carbaryl, a Carbamate Insecticide, is a ligand for the Hepatic Ah (Dioxin) receptor. *Toxicol. Appl. Pharmacol.* 152: 406–414.

30. Ziccardi M H, Gardner I A, Denison M S (2000). Development and modification of a recombinant cell bioassay to directly detect halogenated and polycyclic aromatic hydrocarbons in serum. *Toxicol. Sci.* 54:183–193.

31. Pedersen T, Peters H (1968). Proposal for a classification of oocytes and follicles in the mouse ovary. *J. Reprod. Fertil.* 17:555–557.

32. Lewis S A, Hanrahan J W (1990). Physiological approaches for studying mammalian urinary bladder epithelium. In S. Fleischer and B. Fleischer, (eds.), *Methods in Enzymology* 192:632–650. Academic Press.

33. De Virgilio C, Burckert N, Bell W, et al. (1993). Disruption of TPS2, the gene encoding the 100-kDa subunit of the trehalose-6-phosphate synthase/phosphatase complex in *Saccharomyces cerevisiae*, causes accumulation of trehalose-6-phosphate and loss of trehalose-6-phosphate phosphatase activity. *Eur. J. Biochem.* 212:315–323.

34. Miller J H (1972). *Experiments in Molecular Genetics.* Cold Spring Harbor Laboratory, Cold Spring Harbor, NY, pp. 352–355.

35. Miller W G, Lindow S E (1997). An improved GFP cloning cassette designed for prokaryotic transcriptional fusions. *Gene.* 191:149–153.

36. Sodergren E J, Hsu P Y, DeMoss J A (1998). Roles of the narJ and narI gene products in the expression of nitrate reductase in *Escherichia coli. J. Biol. Chem.* 263:16156–16162.

37. Sambrook J, Fritsch E F, Maniatis T (1989). *Molecular Cloning: A Laboratory Manual, 2nd* ed.: Cold Spring Harbor Laboratory Press, Cold Spring Harbor, NY.

38. Li S F, DeMoss J A (1987). Promoter region of the nar operon of *Escherichia coli*: Nucleotide sequence and transcription initiation signals. *J. Bacteriol.* 169:4614–4620.

39. Khamis S, Bibby M C, Brown J E, et al. (2004). Phytochemistry and preliminary biological evaluation of *Cyathostemma argenteum*, a malaysian plant used traditionally for the treatment of breast cancer. *Phytother. Res.* 18:507–510.

40. Zhang J X, Labaree D C, Mor G, et al. (2004). Estrogen to antiestrogen with a single methylene group resulting in an unusual steroidal selective estrogen receptor modulator. *J. Clin. Endocrinol. Metab.* 89:3527–3535.

41. Kasorndorkbua C, Halbur P G, Thomas P J, et al. (2002). Use of a swine bioassay and a RT-PCR assay to assess the risk of transmission of swine hepatitis E virus in pigs. *J. Virol. Meth.* 101:71–78.

42. Wonganuchitmeta S N, Yuenyongsawad S, Keawpradub N, et al. (2004). Antitubercular sesterterpenes from the Thai sponge *Brachiaster* sp. *J. Nat. Prod.* 67:1767–1770.

43. Feng X, Gao Z, Li S, et al. (2004). DNA polymerase β lyase inhibitors from *Maytenus putterlickoides. J. Nat. Prod.* 67:1744–1747.

44. Ahn M J, Yoon K D, Min S Y, et al. (2004). Inhibition of HIV-1 reverse transcriptase and protease by phlorotannins from the brown alga *Ecklonia cava. Biol. Pharm. Bull.* 27:544–547.

45. Owusu M K, Kamuhabwa A R, Nshimo C (2004). Investigation of fractions present in the stem bark of *Annickia kummeriae* on their P-glycoprotein inhibitory pump activity. *Phytother. Res.* 18:652–657.

46. Sadick M D, Intintoli A, Quarmby V, et al. (1999). Kinase receptor activation (KIRA): A rapid and accurate alternative to end-point bioassays. *J. Pharm. Biomed. Anal.* 19:883–891.

47. Solis P N, Wright C W, Anderson M M, et al. (1993). A microwell cytotoxicity assay using *Artemia salina* (brine shrimp). *Planta Med.* 59:250–252.

48. Hambly R J, Double J A, Thompson M J, et al. (1997). Establishment and characterisation of new cell lines from human breast tumours initially established as tumour xenografts in NMRI nude mice. *Breast Cancer Res. Tr.* 43:247–258.

49. Mosmann T (1983). Rapid colorimetric assay for cellular growth and survival: Application to proliferation and cytotoxicity assays. *J. Immunol. Meth.* 65:55–63.

50. Littlefield B A, Gurpide E, Markiewicz L, et al. (1990). A simple and sensitive micro-titer plate estrogen bioassay based on stimulation of alkaline phosphatase in Ishikawa cells: Estrogenic action of Δ^5 adrenal steroids. *Endocrinol.* 127:2757–2762.

51. Kuiper G G, Lemmen J G, Carlsson B, et al. (1998). Interaction of estrogenic chemicals and phytoestrogens with estrogen receptor. *Endocrinol.* 139:4252–4263.

52. Mor G, Sapi E, Abrahams V M, et al. (2003). Interaction of the estrogen receptors with the Fas ligand promoter in human monocytes. *J. Immunol.* 170:114–122.

53. Niwa H, Yamamura K, Miyazaki J (1991). Efficient selection for high-expression transfectants with a novel eukaryotic vector. *Gene.* 108:193–199.

54. Emmens C W (1962). Estrogens. In R I, Dorfman (ed.), *Methods in Hormone Research.* Academic Press, New York, pp. 59–111.

55. Baron R, Vignery A, Neff L, et al. (1983). Processing of undecalcified bone specimens for bone histomorphometry. In RR, Recker (ed.), *Bone Histomorphometry: Techniques and Interpretation*, CRC Press. Boca Raton, FL.

56. Parfitt A. M, Drezner M. K, Glorieux F. H, et al. (1987). Bone histomorphometry: standardization of nomenclature, symbols, and units. Report of the ASBMR Histomorphometry Nomenclature Committee. *J. Bone Miner. Res.* 2:595–610.

57. Halbur P G, Kasorndorkbua C, Gilbert C, et al. (2001). Comparative pathogenesis of infection of pigs with hepatitis E virus recovered from a pig and human. *J. Clin. Microbiol.* 39:918–923.

58. Meng X J, Halbur P G, Haynes J S, et al. (1998). Experimental infection of pigs with the newly identified swine hepatitis E virus (swine HEV), but not with human strains of HEV. *Arch. Virol.* 143:1405–1415.

59. Meng X J, Halbur P G, Shapiro M S, et al. (1999). Genetic and experimental evidence for crossspecies infection by swine hepatitis E virus. *J. Virol.* 72:9714–9721.

60. Sadick M D, Sliwkowski M X, Nuijens A, et al. (1996). Analysis of heregulin-Induced ErbB2 phosphorylation with a high-throughput kinase receptor activation enzyme-linked immunosorbant assay. *Anal. Biochem.* 235:207–214.

61. Hagiwara A, Imai N, Doi Y, et al. (2003). Absence of liver tumor promoting effects of annatto extract (norbixin), a natural carotenoid food color, in a medium-term liver carcinogenesis bioassay using male F344 rats. *Cancer Lett.* 199:9–17.

62. Daniel L, Gustafson A L, Coulson L F, et al. (1998). Use of a medium-term liver focus bioassay to assess the hepatocarcinogenicity of 1,2,4,5-tetrachlorobenzene and 1,4-dichlorobenzene. *Cancer Lett.* 129:39–44.

63. Anand K K, Singh B, Saxena A K, et al. (1997). 3,4,5-trihydroxy benzoic acid (gallic acid), the hepatoprotective principle in the fruits of *Terminalia belerica*-bioassay guided activity. *Pharmacol. Res.* 36:315–321.

64. Wang H, Wei W, Shen Y X, et al. (2004). Protective effect of melatonin against liver injury in mice induced by *Bacillus Calmette-Guerin* plus lipopolysaccharide. *World J. Gastroentero.* 10:2690–2696.

65. Leakey J E A, Seng J E, Allaben W T (2003). Body weight considerations in the B6C3F$_1$ mouse and the use of dietary control to standardize background tumor incidence in chronic bioassays. *Toxicol. Appl. Pharmacol.* 193:237–265.

66. Wang H H, Lautt W W (1997). Hepatocyte primary culture bioassay: a simple tool to assess the initiation of the liver regeneration cascade. *J. Pharmacol. Toxicol. Meth.* 38:141–150.

67. Borset M, Waage A, Sundan A (1996). Hepatocyte growth factor reverses the TDF-β-induced growth inhibition of CCL-64 cells: a novel bioassay for HGF and implication for the TDF-β bioassay. *J. Immunol. Meth.* 189:59–64.

68. Ito N, Tsuda H, Tatematsu M, et al. (1988). Enhancing effect of various hepato-carcinogens on induction of preneoplastic glutathione S-transferase placental form positive foci in rats—an approach for a new medium-term bioassay system. *Carcinogen.* 9:387–394.

69. Hasegawa R, Tsuda H, Shirai T, et al. (1986). Effect of timing of partial hepatectomy on the induction of preneoplastic liver foci in rats given hepatocarcinogens. *Cancer Lett.* 32:15–23.

70. Hasegawa R, Ito N (1992). Liver medium-term bioassay in rats for screening of carcinogens and modifying factors in hepatocarcinogenesis. *Food Chem. Toxicol.* 30:979–992.

71. Ward J M, Hagiwara A, Anderson L M, et al. (1988). The chronic hepatic and renal toxicity of di(2-ethylhexyl)phthalate, acetaminophen, sodium barbital, and phenobarbital in male B6C3F1 mice: Autoradiographic, immunohistochemical, and biochemical evidence for levels of DNA synthesis not associated with carcinogenesis or tumor promotion. *Toxicol. Appl. Pharmacol.* 96:494–506.

72. Gavino V C, Miller J S, Ikharebha S O, et al. (1981). Effect of polyunsaturated fatty acids and antioxidants on lipid peroxidation in tissue cultures. *J. Lipid Res.* 22:763–769.

73. Reed K L, Fruin A B, Gower A C, et al. (2004). A neurokinin 1 receptor antagonist decreases postoperative peritoneal adhesion formation and increases peritoneal fibrinolytic activity. *P. Natl. Acad. Sci. USA.* 101:9115–9120.

74. Liu C P, Tsai W J, Lin Y L, et al. (2004). The extracts from *Nelumbo nucifera* suppress cell cycle progression, cytokine genes expression, and cell proliferation in human peripheral blood mononuclear cells. *Life Sci.* 75:699–716.

75. Li R W, Leach D N, Myers S P, et al. (2004). A new anti-inflammatory glucoside from *Ficus racemosa* L. *Planta Med.* 70:421–426.

76. Yoganathan K, Rossant C, Glover R P, et al. (2004). Inhibition of the human chemokine receptor CCR5 by variecolin and variecolol and isolation of four new variecolin analogues, emericolins A–D, from *Emericella aurantiobrunnea*. *J. Nat. Prod.* 67:1681–1684.

77. Valotis A, Neukam K, Elert O, et al. (2004). Human receptor kinetics, tissue binding affinity, and stability of mometasone furoate. *J. Pharm. Sci.* 93:1337–1350.

78. West M, Mhatre M, Ceballos A, et al. (2004). The arachidonic acid 5-lipoxygenase inhibitor nordihydroguaiaretic acid inhibits tumor necrosis factor α activation of microglia and extends survival of G93A-SOD1 transgenic mice. *J. Neurochem.* 91:133–143.

79. Terra J H, Montaño I, Schilb A, et al. (2004). Development of a microplate bioassay for monocyte chemoattractant protein-1 based on activation of p44/42 mitogen-activated protein kinase. *Anal. Biochem.* 327:119–125.

80. Ashwood P, Harvey R, Verjee T, et al. (2004). Competition between IL-1, IL-1ra and TGF-β1 modulates the response of the ELA4.NOB-1/CTLL bioassay: Implications for clinical investigations. *Inflamm. Res.* 53:60–65.

81. Vermeer H, Hendriks-Stegemana B I, van Suylekoma D, et al. (2004). An *in vitro* bioassay to determine individual sensitivity to glucocorticoids: induction of FKBP51 mRNA in peripheral blood mononuclear cells. *Mol. Cell. Endocrinol.* 218:49–55.

82. Bi L, Zhang Y, Zhao M, et al. (2005). Novel synthesis and anti-inflammatory activities of 2,5-disubstituted-dioxacycloalkanes. *Bioorg. Med. Chem.* 13:5640–5646.

83. Kuo Y C, Huang Y L, Chen C C, et al. (2002). Cell cycle progression and cytokine genes expression of human blood mononuclear cells modulated by *Agaricus blazei*. *J. Lab. Clin. Med.* 140:176–187.

84. Kuo Y C, Meng H C, Tsai W J (2001). Regulation of cell proliferation, inflammatory cytokines production and calcium mobilization in primary human T lymphocytes by emodin from *Polygonum hypoleucum Ohwi. Inflamm. Res.* 50:73–82.

85. Kuo Y C, Yang N S, Chou C J, et al. (2000). Regulation of cell proliferation, gene expression, production of cytokines, and cell cycle progression in primary human T lymphocytes by piperlactam S isolated from *Piper kadsura. Mol. Pharmacol.* 58:1057–1066.

86. Lowry O H, Rosebrough N J, Farr A L, et al. (1951). Protein measurement with the Folin phenol reagent. *J. Biol. Chem.* 193:265.

87. Högger P, Rohdewald P (1994). Binding kinetics of fluticasone propionate to the human glucocorticoid receptor. *Steroids.* 59:597–602.

88. Marzinzig M, Nussler A K, Stadler J, et al. (1997). Improved methods to measure end products of nitric oxide in biological fluids: nitrite, nitrate and S-nitrosothiols. *Nitric Oxide Biol. Chem.* 1:177–189.

89. Guégan C, Przedborski S (2003). Programmed cell death in amyotrophic lateral sclerosis. *J. Clin. Invest.* 111:153–161.

90. Zemlan F P (1999). Antibodies to CSF Tau: Characterization of antibodies recognizing cerebrospinal fluid structural proteins. *J. Neurochem.* 73:437–438.

91. Zemlan F P, Jauch E C (1999). Serum and CSF biomarker of neuronal damage in traumatic brain injury. *J. Neurotrauma.* 16:690.

92. Zemlan F P, Rosenberg W S, Luebbe P A, et al. (1999). Quantification of axonal damage in traumatic brain injury: affinity purification and characterization of cerebrospinal fluid tau proteins. *J. Neurochem.* 72:741–750.

93. Binkert C, Landwehr J, Mary J L, et al. (1989). Cloning, sequence analysis and expression of a cDNA encoding a novel insulin-like growth factor binding protein (IGFBP-2). *EMBO J.* 8:2497–2502.

94. Blackhurst G, McElroy P K, et al. (2001). Seasonal variation in glucocorticoid receptor binding characteristics in human mononuclear leukocytes. *Clin. Endocrinol.* (Oxf.) 55:683–686.

95. Vermeer H, Hendriks-Stegeman B I, van der Burg B, et al. (2003). Glucocorticoid-induced increase in lymphocytic FKBP51 mRNA expression: A potential marker for glucocorticoid sensitivity, potency and bioavailability. *J. Clin. Endocrinol. Metab.* 88:277–284.

96. Kluft C (1979). Studies on the fibrinolytic system in human plasma: quantitative determination of plasminogen activators and proactivators. *Thromb. Haemost.* 41:365–383.

97. Kurz K D, Main B W, Sandusky G E (1990). Rat model of arterial thrombosis induced by ferric chloride. *Thromb. Res.* 60:269–280.

98. Hladovec J (1974). Experimental arterial thrombosis in rats with continuous registration. *Thromb. Diath. Haemorrh.* 26:407–410.

99. Umetsu T, Sanai K (1978). Effect of 1-methyl-2-mercapto-5-(3-pyridyl)-imidazol(KC-6141), an antiaggregation compound, on experimental thrombosis in rats. *Thromb. Haemost.* 39:74–83.

100. Urano T, Sakakibara K, Rydzewski A (1990). Relationships between euglobulin clot lysis time and the plasma levels of tissue plasminogen activator and plasminogen activator inhibitor 1. *Thromb. Haemost.* 63:82–86.

101. Wardlaw J M, Warlow C P, Counsell C (1997). Systematic review of evidence on thrombolytic therapy for acute ischaemic stroke. *Lancet.* 350:607–614.

102. Kluft C (1979). Studies on the fibrinolytic system in human plasma: Quantitative determination of plasminogen activators and proactivators. *Thromb. Haemost.* 41:365–383.

103. Zhao M, Lin N, Wang C, et al. (2003). Synthesis and thrombolytic activity of fibrinogen fragment related cyclopeptides. *Bioorg. Med. Chem. Lett.* 9:961–964.

104. Zhao M, Liu J, Wang C, et al. (2005). Synthesis and biological activity of nitronyl nitroxide containing peptides. *J. Med. Chem.* 48:4285–4292.

105. Saito S, Tanaka M, Matsunaga K, et al. (2004). The combination of rat mast cell and rabbit aortic smooth muscle is the simple bioassay for the screening of antiallergic ingredient from methanolic extract of corydalis tuber. *Biol. Pharm. Bull.* 27:1270–1274.

106. Shankar Gopi, Fourrier M S, Grevenkamp M A, et al. (2004). Validation of the COSTIM bioassay for dendritic cell potency. *J. Pharm. Biomed. Anal.* 36:285–294.

107. Spinardi-Barbisan A L T, Kaneno R, Barbisan L F, et al. (2004). Chemically induced immunotoxicity in a medium-term multiorgan bioassay for carcinogenesis with Wistar rats. *Toxicol. Appl. Pharmacol.* 194:132–140.

108. Taniguchi M, Nagaoka K, Ushio S, et al. (1998). Establishment of the cells useful for murine interleukin-18 bioassay by introducing murine interleukin-18 receptor cDNA into human myelomonocytic KG-1 cells. *J. Immunol. Methods.* 217:97–102.

109. Wei X, Swanson S J, Gupta S (2004). Development and validation of a cell-based bioassay for the detection of neutralizing antibodies against recombinant human erythropoietin in clinical studies. *J. Immunol. Methods.* 293:115–126.

110. Browning D D, The T, O'Day D H (1995). Comparative analysis of chemotaxis in dictyostelium using a radial bioassay method: Protein tyrosine kinase activity is required for chemotaxix to folate but not to camp. *Cell. Signal.* 7:481–489.

111. Kuper C F, Harlerman J H, Richter-Reichelm H B, et al. (2000). Histopathologic approaches to detect changes indicative of immunotoxicity. *Toxicol. Pathol.* 28:454–466.

112. Parnet P, Garka K E, Bonner T P, et al. (1996). IL-1 Rrp is a novel receptor-like molecule similar to the type I interleukin-1 receptor and its homologues T1 r ST2 and IL-1R Acp. *J. Biol. Chem.* 271:3967.

113. Mason S, La S, Mytych D, et al. (2003). Validation of the BIACORE 3000 platform for detection of antibodies against erythropoietic agents in human serum samples. *Curr. Med. Res. Opin.* 19:651.

114. Rehmann K, Schramm K W, Kettrup A A (1999). Applicability of a yeast oestrogen screen for the detection of oestrogen-linke activity in environmental samples. *Chemosphere.* 38:3303–3312.

115. Paris F, Servant N, Térouanne B, et al. (2002). A new recombinant cell bioassay for ultrasensitive determination of serum estrogenic bioactivity in children. *J. Clin. Endocr. Metab.* 87:791–797.

116. Bovee T F H, Helsdingen R J R, Rietjens I M C M, et al. (2004). Rapid yeast estrogen bioassays stably expressing human estrogen receptors α and β, and green fluorescent protein: a comparison of different compounds with both receptor types. *J. Steroid Biochem. Mol. Biol.* 91:99–109.

117. Bovee T F H, Heskamp H H, Hamers A R M, et al. (2005). Validation of a rapid yeast estrogen bioassay, based on the expression of green fluorescent protein, for the screening of estrogenic activity in calf urine. *Anal. Chim. Acta.* 529:57–64.

118. Rasoanaivo P, Ramanitrahasimbola D, Rafatro H (2004). Screening extracts of Madagascan plants in search of antiplasmodial compounds. *Phytother Res.* 18:742–747.

119. Noedl H, Teja-Isavadharm P, Miller R S (2004). Nonisotopic, semiautomated plasmodium falciparum bioassay for measurement of antimalarial drug levels in serum or plasma. *Antimicrob. Agents Chemother.* 48:4485–4487.

120. Krief S, Martin M T, Grellier P, et al. (2004). Novel antimalarial compounds isolated in a survey of self-medicative behavior of wild chimpanzees in Uganda. *Antimicrob. Agents Chemother.* 48:3196–3199.

121. Impoinvil D E, Kongere J O, Foster W A, et al. (2004). Feeding and survival of the malaria vector *Anopheles gambiae* on plants growing in Kenya. *Med. Vet. Entomol.* 18:108–115.

122. Kotecka B, Edstein M D, Rieckmann K H (1996). Chloroquine bioassay of plasma specimens obtained from soldiers on cholo quine plus doxycycline for malaria priphylaxis. *Int. J. Parasitol.* 26:1325–1329.

123. Isavadharm T, Peggins J O, Brewer T G, et al. (2004). Plasmodium falciparum-based bioassay for measurement of artemisinin derivatives in plasma or serum paktiya. *Antimicrob Agents Chemother.* 48:954–960.

124. Trager W, Jensen J B (1977). Plasmodium falciparum in culture: use of outdated erythrocytes and description of the candle jar method. *J. Parasitol.* 63:883–887.

125. Noedl H, Wernsdorfer W H, Miller R S, et al. (2002). Histidine-rich protein II, a novel approach to antimalarial drug susceptibility testing. Antimicrob. *Agents Chemother.* 46:1658–1664.

126. Trager W, Jensen J B (1976). Human malarial parasites in continuous culture. *Science.* 193:673–675.

127. Desjardin R E, Canfield C J, Haynes J D, et al. (1979). Quantitative assessment of antimalarial activity in vitro by a semiautomated microdilution technique. *Antimicrob. Agents Chemother.* 16:710–718.

128. Knols B G J, Njiru B N, Mathenge E M, et al. (2002). MalariaSphere: a greenhouse-enclosed simulation of a natural *Anopheles gambiae* (Diptera: Culicidae) ecosystemin western Kenya. *Malaria J.* 1:19.

129. Kotecka B, Rieckmann K H (1993). Chloroquine bioassay using malaria microcultures. *Am. J. Trop. Med. Hyg.* 49:460–464.

130. Webster HK, Boundreau E F, Pavanand K, et al. (1985). Antimalarial drug susceptiblity testing of *Plasmodium falciparum* in Thailand using a microdilution radio-isotope method. *Am J. Trop. Med. Hyg.* 34:228–235.

131. Amfilochiadis A A, Papageorgiou P C, Kogan N, et al. (2004). Role of bradykinin B2-receptor in the sympathoadrenal effects of "new pressor protein" related to human blood coagulation factor XII fragment. *J. Hypertens.* 22:1173–1181.

132. Liu J Q, Folz R J (2004). Extracellular superoxide enhances 5-HT-induced murine pulmonary artery vasoconstriction. *Am. J. Physiol. Lung. Cell. Mol. Physiol.* 287:L111–118.

133. Liu J Q, Zelko I N, Folz R J (2004). Reoxygenation-induced constriction in murine coronary arteries: the role of endothelial NADPH oxidase (gp91phox) and intracellular superoxide. *J. Biol. Chem.* 279:24493–24497.

134. Fukuda S, Fini C A, Mabuchi T, et al. (2004). Focal cerebral ischemia induces active proteases that degrade microvascular matrix. *Stroke.* 35:998–1004.

135. Lyden P D, Jackson-Friedman C, Shin C, et al. (2000). Synergistic combinatorial stroke therapy: A quantal bioassay of a GABA agonist and a glutamate antagonist. *Exp. Neurol.* 163:477–489.

136. Mavrogiannis L, Trambakoulos D M, Boomsma F, et al. (2002). The sympathoadrenal system mediates the blood pressure and cardiac effects of human coagulation factor XII-related "new pressor protein." *Can. J. Cardiol.* 18:1077–1086.

137. Mavrogiannis L, Kariyaisam K P, Osmond D H (1997). Potent blood pressure raising effects of activated coagulation factor XII. *Can. J. Physiol. Pharmacol.* 75:1398–1403.

138. Papageorgiou P C, Pourdjabbar A, Amfilochiadis A A, et al. (2004). Are the cardiovascular and sympathoadrenal effects of human "new pressor protein" preparations attributable to human coagulation β-FXIIa? *Am. J. Physiol. Heart. Circ. Physiol.* 286: H837–H846.

139. Simos D, Boomsma F, Osmond D H (2002). Human coagulation factor XII-related "new pressor protein": role of PACAP in its cardiovascular and sympathoadrenal effects. *Can. J. Cardiol.* 18:1093–1103.

140. van der Hoorn F A, Boomsma F, Man in 't Veld AJ, et al. (1989). Determination of catecholamines in human plasma by high-performance liquid chromatography: Comparison between a new method with fluorescence detection and an established method with electrochemical detection. *J. Chromatogr.* 487:17–28.

141. Athar M, Elmets C A, Bickers D R, et al. (1989). A novel mechanism for the generation of superoxide anions in hematoporphyrin derivativemediated cutaneous photosensitization. Activation of the xanthine oxidase pathway. *J. Clin. Invest.* 83:1137–1143.

142. Becker L B, vanden Hoek T L, Shao Z H, et al. (1999). Generation of superoxide in cardiomyocytes during ischemia before reperfusion. *Am. J. Physiol. Heart Circ. Physiol.* 277:H2240-H2246.

143. Brandes R P, Barton M, Philippens K M, et al. (1997). Endothelial-derived superoxide anions in pig coronary arteries: Evidence from lucigenin chemiluminescence and histochemical techniques. *J. Physiol.* 500:331–342.

144. Guzik T J, Mussa S, Gastaldi D, et al. (2002). Mechanisms of increased vascular superoxide production in human diabetes mellitus: Role of NAD(P)H oxidase and endothelial nitric oxide synthase. *Circulation.* 105:1656–1662.

145. Hamann G F, Okada Y, Fitridge R, et al. (1995). Microvascular basal lamina antigens disappear during cerebral ischemia and reperfusion. *Stroke.* 26:2120–2126.

146. del Zoppo G J, Copeland B R, Harker L A, et al. (1986). Experimental acute thrombotic stroke in baboons. *Stroke.* 17:1254–1265.

147. Tagaya M, Haring H P, Stuiver I, et al. (2001). Rapid loss of microvascular integrin expression during focal brain ischemia reflects neuron injury. *J. Cereb. Blood Flow. Metab.* 21:835–846.

148. Tagaya M, Liu K F, Copeland B, et al. (1997). DNA scission after focal brain ischemia: temporal differences in two species. *Stroke.* 28:1245–1254.

149. Heo J H, Lucero J, Abumiya T, et al. (1999). Matrix metalloproteinases increase very early during experimental focal cerebral ischemia. *J. Cereb. Blood Flow Metab.* 19:624–633.

150. Longa E Z, Weinstein P R, Carlson S, et al. (1989). Reversible middle cerebral artery occlusion without craniectomy in rats. *Stroke.* 20:84–91.

151. Colbourne F, Sutherland G R, Auer R N (1996). An automated system for regulating brain temperature in awake and freely moving rodents. *J. Neurosci. Methods.* 67: 185–190.

152. Bederson J B. Pitts L H, Tsuji M, et al. (1986). Rat middle cerebral artery occlusion: Evaluation of the model and development of a neurologic examination. *Stroke.* 17:472–476.

5.3

ANALYTICAL CONSIDERATIONS FOR IMMUNOASSAYS FOR MACROMOLECULES

MARIAN KELLEY[1] AND BINODH DESILVA[2]

[1] *Centocor R&D, Inc. Randor, Pennsylvania*
[2] *Amgen, Inc. Thousand Oaks, California*

Chapter Contents

5.3.1	Introduction	574
5.3.2	Assessment Parameters	574
	5.3.2.1 Reference Material	574
	5.3.2.2 Assay Format/Reagent Selection	575
	5.3.2.3 Specificity and Selectivity	576
	5.3.2.4 Sample Collection Process/Matrix/Sample Preparation/Minimum Required Dilution	576
	5.3.2.5 Standard Calibrators and Standard Curves	577
	5.3.2.6 Precision/Accuracy: Role of QCs	578
	5.3.2.7 Range of Quantification	579
	5.3.2.8 Sample Stability (Bench Top, 4°C, −70°C)	579
	5.3.2.9 Dilutional Linearity and Parallelism	579
	5.3.2.10 Robustness and Ruggedness	580
	5.3.2.11 Run Acceptance Criteria	580
5.3.3	Partial Validation, Method Transfer, and Cross-Validation	580
	5.3.3.1 Partial Validation	580
	5.3.3.2 Method Transfer	581
	5.3.3.3 Cross-Validation	582
5.3.4	Total Error Versus 4–6–20 Rule	582
	5.3.4.1 Total Error	582
5.3.5	Documentation	582
5.3.6	Conclusion	583
	References	584

5.3.1   INTRODUCTION

Current guidance documents for the assessment of analytes and their metabolites in biological fluids are focused primarily on the quantification of small molecules using chromatographic and mass spectrometry platforms [1–7]. Specific recommendations coming out of these guidances fall short in the immunoassay realm. The antibody binding of analytes directly from biological fluids confers a practicality and specificity unique to these assays. These assays are typically nonlinear and experience specific issues associated with the protein matrix, e.g., the presence of endogenous materials, to name just one. Until the authorities propose a guideline targeted at best practices in developing and validating ligand binding assays (antibody-based immunoassays) to support pharmacokinetics/toxicokinetics (PK/TK), the following recommendations, based on the paper generated by a 10-member panel from Biotech, Pharma, and CROs [8], and supported by several previous publications on the topic [9, 10], are proposed.

Assay development and implementation are dynamic processes. The life-cycle steps of development, validation, and in-study assay monitoring can be reassessed throughout the process as the analyst benefits from the experience of extended use, adds stability data, or updates current precision and accuracy data.

During early immunoassay development, it is important to assess the "intended use" of the assay. When the intended use is to support pharmacokinetics in either nonclinical PK/TK or clinical bioequivalence (BE) studies, developing a validatable assay is of utmost importance. A well-thought-out validation plan and fully documented assay methods and results that support the validation are crucial. To withstand the scrutiny of the authorities, and to guarantee the optimal method is applied to your drug development process, the following parameters should be assessed critically, validated to specific criteria, and fully documented.

5.3.2   ASSESSMENT PARAMETERS

The goal of the development phase for immunoassays is to establish a method that can consistently produce a reliable result culminating in a validation plan with established target acceptance criteria for accuracy and precision. The following recommended parameters are assessed critically during early development, evaluated in late development, and finally, confirmed during validation.

5.3.2.1   Reference Material

Inherent to macromolecules is the associated variability (i.e., glycosylation and deamidation) of the material from different preparations. Therefore, it is not usually feasible to obtain a "reference standard" for a macromolecule drug for use in bioanalysis. Typically, what is obtained is a well-characterized product. Therefore, it is important to clearly state the source of the material and to refer to any documentation describing characteristics of that material. Whenever possible, the standards, the validation samples, and quality control (QC) samples should be prepared from separate vials of the same source material. In the case of lyophilized material, it is preferred to reconstitute two separate vials to prepare standards and QCs. In circumstances of limited availability of reference material (e.g., drug interaction studies where the pure standard is not available except from a limited source of commercial kits), standards and QC samples can be prepared from single aliquots after checking the comparability between lots or other commercial sources. In any circumstance, the lot numbers, batch numbers, and supporting documentation should be carefully monitored. This is a critical parameter that should be evaluated during method transfers and cross-validation studies.

5.3.2.2 Assay Format/Reagent Selection

Assay formats more recently have been defined as much by the platform (ELISA, RIA, Bioveris, Luminex) as by the type of immunoassay (direct or indirect binding, sandwich, competition or inhibition). Considering several different formats before selecting one is optimal, if time allows, because there are pros and cons to each format. Solid phase formats include glass, treated plastics, and coated plates (e.g., streptavidin). As antibody binding to the coated surface is highly dependent on pH, buffer content, salt concentration, detergents, nonspecific carrier protein, temperature, and shaking, any relevant conditions should be tested during the development of an immunoassay. When solid phase formats prove difficult, solution phase binding, for instance using coated beads, can result in a practical and sensitive assay.

The antibody or pair of antibodies selected for the assay format is the primary critical reagent for an immunoassay because it will ultimately define the specificity and sensitivity of the assay. Therefore, whenever possible, as many antibodies or pairs should be tested hand-in-hand with the selection of the assay format before moving into further development. This may include whole or fragmented monoclonals, polyclonals, and combinations as the assay format demands. The time it takes to develop and purify unique reagent antibodies demands that this start as early in the drug development process as possible.

Other critical reagents are those that impact the sensitivity, ruggedness, and consistency of the assay and especially include the detection system. It is a recommendation that a list of critical reagents and criteria for their acceptance be identified before initiating sample analysis. These must also be sourced, characterized, documented, and stored under conditions defined by stability investigation. When in-house reagents make up any part of the immunoassay, it is important to define and document the source, preparation, expiration, storage conditions, and acceptance criteria for their use to ensure a reliable result.

Once the final conditions are established, it is important to understand the assay's accuracy and precision capabilities because target criteria should be set for both before entering into validation. It is also good practice during validation to "stress the system" in a manner representing a true sample analysis run. Therefore, if the "batch" size is expected to be unusually large, a comparable batch, run during validation, should demonstrate that it could achieve the same acceptance criteria. It is particularly useful to evaluate the variability of QCs on every plate to predict the outcome expected during sample analysis.

The established/validated format should be used during sample analysis. Changes to the format must be revalidated to the extent that a consistent format is confirmed. The acceptability of changes to critical reagents should be documented before placing them into the sample analysis assays.

5.3.2.3   Specificity and Selectivity

One challenge of interpreting results derived from immunoassays is understanding exactly what the assay is detecting. Fundamental to this is knowing to what the antibody binds, that is, determining the antibody's specificity. A related parameter, selectivity, refers to the assay's ability to solely measure the target analyte in the presence of other protein constituents in the sample, some in very high concentrations.

Typically, experiments designed to determine the antibody's specificity involve assessing their cross-reactivity with similar compounds. When feasible, it is suggested that the sample matrix is spiked with physio-chemically similar compounds or variants of the target analyte.

Selectivity is the easier of the two parameters to assess experimentally. After the assay format is relatively fixed, multiple lots (at least 10 normals and 10 of the intended patient population in the case of clinical samples) of the sample matrix should be spiked near the lower limit of quantification and the recovery of the analyte should be calculated. The percent recovery of spike (expressed as relative error or %RE) is calculated as: 100× (observed mean conc − nominal concentration)/nominal concentration. This will aid in finally selecting a lower limit of quantification, before initiating assay validation.

An additional experiment to understand selectivity involves spiking both the diluent and the matrix with standard curve calibrators to detect matrix effects at all concentrations across the range of calibration.

Interfering substances, especially in patient populations, that require some evaluation may include prescription and over-the-counter medication, herbal medications, dietary substances, and even the presence of antidrug antibodies and associated immune complexes. During sample analysis, new matrix conditions may develop that would require further investigation and could include hemolysis, lipemia, rheumatoid factor, and so on.

Both specificity and selectivity experiments, if performed during the late development phase, after the assay format and critical reagents are fixed, may be cited in the validation report. Otherwise the experiments can be performed during validation. In both cases, the *a priori* acceptance criteria should be applied and acceptable recovery should be realized in at least 80% of the samples. Additionally, as lots of critical reagents are replenished, these types of assessments must reconfirm their acceptability.

5.3.2.4   Sample Collection Process/Matrix/Sample Preparation/Minimum Required Dilution

As mentioned, the ability of the antibody to bind to its target is the basis for assay sensitivity; therefore, maintaining the integrity of the analyte is vital. Frequently overlooked, the selection of sample matrix type and process for collection are crucial elements in bioanalysis. Rigorous investigation should go into the stability of the sample as it is collected, the anti-coagulant requirements, protease inhibitors, and other possible additives, and it should include the postprocessing storage. Furthermore, a process must be defined that provides for sample tracking "from cradle to grave" documenting the chain of custody, the sample ID, and the time spent at each location and temperature through analysis. After analysis, the archival and destruction process should be identified.

For practical reasons, the matrix of choice for immunoassays is typically serum. The plate washers or automated systems used during sample analysis easily clog when plasma is used, which then has the potential to impact the validity of the result. However, protein analytes, including cytokines that are identified as labile and prone to degradation in serum, would preferentially be collected in plasma. When collecting citrated plasma, it is wise to clearly identify the salt form of citrate (Na, K, Li) so that it is transparent what the comparable matrix should be for QC preparation.

Whenever possible, the matrix expected in the samples should be used in the preparation of the standard calibrators and QCs. Individual donors (found acceptable when assayed as blank and spiked) are pooled to produce the standard curve and QC matrix; then aliquots are spiked to produce the standard curve and QCs. Some matrices are difficult to obtain. To accommodate this, as long as the QCs prepared in the intended matrix demonstrate the accuracy and precision expected, non-matrix standard curve calibrators are acceptable.

QC accuracy assessments can be challenging when the therapeutic drug shares an endogenous protein. In this case, the endogenous protein concentration can either be added to or subtracted from the final value. Before establishing the calculations, the necessary experiments must be performed, and the consistency of the endogenous value contribution must be documented.

In cases where the endogenous interferents do not generate a linear signal, then a minimum required dilution (MRD) may be applied. A MRD is the dilution of a matrix that finally permits an accurate and precise measurement of analyte. Although the diluting sample reduces the sensitivity of the assay, it may be required to develop and validate an accurate and precise method.

5.3.2.5 Standard Calibrators and Standard Curves

As described in the Guidance for the Industry, May 2001 [7], the "calibration curve is the relationship between instrument response and known concentrations of the analyte" in the matrix of interest. In immunoassays, the affinity and avidity of the antibody(ies) used to develop the standard curve are fundamental to the range because they define the sensitivity of the assay. The antibody (or pairs) once selected should be used throughout validation and sample analysis. During the development phase, it is a good idea to use many more calibrators and replicates evenly across the range to thoroughly describe the concentration-response relationship and to investigate the "best fit." Thereafter, select at least six to eight calibrators, over the entire range of the curve before the start of validation. Once validation begins, the standard curve calibrators should not be adjusted. Spiked standard curve points may be used outside the quantification range of the curve to help fit the curve, but they are not used in the assessment.

"Best-fit" can be evaluated in a straightforward manner by fitting several regression methods to the data generated during the development phase. The relative error of the back-calculated standard calibrators should be calculated. As expected, the regression that consistently generates the most accurate and precise data is selected for confirmation in the validation phase. Usually a four- or five-parameter logistic with/without weighting describes the s-shaped immunoassay curve optimally; however, the smoothed spline has also been used with success with some platforms. It is important to critically assess whether the curves fall consistently above or below the expected concentration. This "trending" can invalidate a regression model.

5.3.2.6 Precision/Accuracy: Role of QCs

One function of the QC sample (or validation samples as they are sometimes called when used during the validation phase) is to mimic the study sample. Therefore, the therapeutic drug is spiked into a neat sample matrix, and not into a diluted or otherwise modified matrix. The QCs should be stored preferably along with the samples in the same freezer or at least in a comparable freezer. Documentation of freezer conditions will be critical to provide evidence that the conditions of samples and QCs are similar.

A second function of the QC sample is to measure the degree to which the assay is under control as determined by the assessment of accuracy and precision in multiple runs. Accuracy is described as the % relative error (%RE) or the amount the assayed value deviates from the nominal value. The precision is described as the % coefficient of variation (%CV) by dividing the standard deviation by the mean value recovered value.

During development, preliminary assessments are conducted to predict not only the QC concentration values to be used but also the target acceptance criteria. In this case, it is prudent to include more QC levels than will finally be used during validation. For instance, three levels prepared at the LLOQ will likely provide a clear indication of which QC is acceptable and where the assay cannot support the target acceptance. The same can be performed for the ULOQ. At least three levels between the two asymptotes are necessary because these will be used during sample analysis. For the validation process, five QC levels should be selected for testing, based on the accurate and precise readout: the final LLOQ to be confirmed; the LQC; less than three times the LLOQ; the MQC, placed about the middle of the curve; and the HQC, placed no more than 75% below the ULOQ and the ULOQ.

During validation, the five QCs should be tested in a minimum of six runs, at least in duplicate. It is not acceptable to discard any runs during this testing. The inter- and intrabatch precision and accuracy from all runs performed should be reported. It is generally expected that the precision and accuracy acceptance be at least 20–25% of the target range with a Total Error not exceeding 30%. (Total Error acceptance criterion is discussed later in this report.)

During sample analysis, only the LQC, MQC, and HQC defined during validation need to be run in duplicate with every assay. The precision and accuracy should conform to that established during validation. Typically, run acceptance criteria also require that at least one QC at each concentration level is acceptable, and that four out of the six QCs assayed must be acceptable. It is recommended that method acceptance criteria that are consistent with the validation be used during sample analysis [11, 12].

5.3.2.7   Range of Quantification

It is important to understand that the LLOQ and ULOQ have been selected because they define the true, accepted limits of the standard curve. Standard curve points below or above the LLOQ and ULOQ, respectively, are considered extrapolated and cannot be used to report sample concentrations. The LLOQ and the ULOQ validation samples and not the standard curve samples spiked into neat matrix define the standard curve range of quantification.

With certain toxicokinetic studies or therapeutic drugs dosed in high concentrations, the expected concentration in the sample may be well outside the quantification range as defined by the ULOQ. These samples must be diluted into the accepted range of the standard curve to determine an initial concentration and then multiplied by the dilution factor to determine the final concentration of the analyte in the sample. Although it is preferable to dilute the sample in matrix, in many cases, the process is performed in a buffer diluent.

5.3.2.8   Sample Stability (Bench Top, 4°C, –70°C)

The underlying reason that samples are tracked from "cradle to grave" is to document the integrity of the sample. Stability testing under all conditions to which a set of samples in a study is exposed supports the statement of integrity. Therefore, such testing of QC samples in matrix comprises evaluating stability during sample collection and processing: –20°C or –70°C long-term storage to cover the time from when the sample is processed to its final analysis, bench top under conditions similar to analysis, resident time in laboratory refrigerators and freezers, and freeze-thaw to cover the expected or possible cycles. Another consideration is to test the stability of the analyte in intermediate stock solutions if stored for later analysis.

The samples should be thawed under the same conditions used to thaw study samples. Similarly, refreeze the samples for a length of time that would ensure complete freezing, usually 12–24 hours. Samples should be assayed using a freshly prepared standard curve, and acceptable QCs should be used to monitor assay performance. The stability samples should meet the criteria for acceptance as defined for in the validation plan.

5.3.2.9   Dilutional Linearity and Parallelism

Dilutional linearity is the ability of a QC sample spiked at a high concentration (at least as high as the expected C_{max} in a study) to be acceptably diluted into the range of the standard curve so that the initial concentration determined off the regression and multiplied by the dilution factor produces a final concentration with an accuracy and precision expected for that assay. Usually, serial dilutions are employed to ensure that at least two to three fall on the acceptable range of the curve so that linearity can be demonstrated. When a study design produces samples that require a dilution greater than that established, it is advisable to employ an equivalent dilutional QC that is run in the assay.

To evaluate the "hook effect," dilute a very high QC (100- to 1000-fold greater than the ULOQ) so that the expected concentration read off the regression reads above the ULOQ.

Parallelism is determined in a manner similar to dilutional linearity but with incurred samples. This is usually not possible until after the initiation of a study. When incurred samples are available, one strategy is to pool the C_{max} samples to create parallelism QCs. Serially dilute the sample so that at least two to three fall on the acceptable range of the curve to demonstrate linearity.

5.3.2.10   Robustness and Ruggedness

Robustness and ruggedness experiments are used to demonstrate how reproducible a method is when conditions vary. An assay method protocol defines the exact steps to be followed to ensure an accurate and precise result. Unfortunately, slight-to-large deviations occur in the everyday process, which may or may not impact the result. Assay conditions like incubation temperature or exposure to light define the robustness of the assay, whereas changes to the routine, for instance, multiple analysts or different instruments, define the ruggedness. The actual batch size for a routine sample analysis run is frequently overlooked but is often an impactful ruggedness measurement that should be assessed.

5.3.2.11   Run Acceptance Criteria

Because validation QCs are used to assess the overall accuracy and precision of the method during the validation phase, no runs initiated for validation may be eliminated due to QC failure. Therefore, the acceptance criteria for the standard curves are used to accept each assay. At the conclusion of the validation, run acceptance criteria for QCs to be used in-study are summarized and run acceptance criteria are established based on this data.

During sample analysis, the standard curve should be evaluated for acceptance before evaluating the QCs. Once the curve has been accepted, then the performance of the QCs should be evaluated. The acceptance criteria established during the precision and accuracy (4–6–20 rule) should be followed as described.

5.3.3   PARTIAL VALIDATION, METHOD TRANSFER, AND CROSS-VALIDATION

Method validations can be classified into three broad categories, full, partial, and cross. A full validation is done for any new methods and involves method development, prestudy, and in-study validation. It is essential that a full validation be conducted for changes in species (e.g., rat to mouse), change in matrix within a species (e.g., rat serum to rat urine). However, a partial validation is sufficient when a minor change is made to an already fully validated method as described below.

5.3.3.1   Partial Validation

A partial validation is conducted when minor changes to a method are made. These may include method transfer, changes to anticoagulant (e.g., EDTA, heparin, citrate), changes in the reagents used in the method (especially pivotal reagents such as the primary antibody or secondary antibody), sample processing changes (e.g., how fast a clot needs to be spun, collection vessels, and storage condition), changes to sample volume (e.g., if the volume of sample is changed from 100 to 200 μL per well), extension of the concentration range, selectivity issues (e.g., different disease population and concomitant medication), conversion of a manual to an automated method, qualification of an analyst, and so on. Partial validations can range from a single intra-assay accuracy and precision run to a nearly full validation. Although sample-processing changes may only require one run, several runs would typically be expected for changes to lots of reagents. Transferring an analytical method may require substantially more experimentation. Typically, for partial validations, three accuracy and precision runs are conducted and the inter-assay accuracy and precision criteria are compared against the original validated assay. If these criteria are met, the change to the method will be accepted and the proper documentation should be made to the method or method standard operating procedure (SOP). Other validation parameters have to be compared depending on the situation where these experiments are conducted to scientifically justify the change; e.g., if the method is to be used with a diseased population different to the original intended use of this method, then additional selectivity experiments must be conducted to evaluate the matrix effect.

5.3.3.2 Method Transfer

Method transfer is the situation in which the method is fully validated in one laboratory (sending laboratory) and transferred to another laboratory (receiving laboratory) and requires at least a partial validation. The parameters to evaluate during this partial validation are accuracy and precision of the method, selectivity, benchtop stability, and so on.

In addition to the required documentation (e.g., method description, validation report, and certificate of analysis), the sending laboratory should provide information on those factors that may affect the ruggedness of the assay (e.g., identifying pivotal reagents and material). The method transfer process can be conducted in a multiphase manner, a transfer phase in which the sending laboratory participates in the transfer of the method at the receiving laboratory; the feasibility evaluation phase, in which the receiving laboratory runs the assay independently; the validation phase, which could be the phase where the partial validation is conducted; and finally a qualification phase in which both the sending and the receiving laboratory assays blinded samples. The method transfers require a plan or protocol that defines the process (e.g., experiments to be conducted) and the acceptance criteria. It is not uncommon for the sending laboratory to send personnel to physically demonstrate and train the personnel at the receiving laboratory.

Once the transferred method is validated at the receiving laboratory, an ideal scenario is to have both the sending and the receiving laboratories analyze blinded, 30 spiked samples covering the standard curve range, and 30 pooled incurred samples. The two sets of data may be compared by using a predefined statistical equivalence test [12, 13]. Alternatively, the differences between the two sets of data may be compared using an agreed to range of acceptability. Any acceptance criteria must be set *a priori* and documented in a plan or protocol before executing this phase of the method transfer process.

5.3.3.3 Cross-Validation

Cross-validation is conducted when two validated bioanalytical methods are used within the same study or submission, for example, ELISA assay to Biacore and ELISA to a liquid chromatography/mass spectrometry. It is recommended that test samples (spiked and/or pooled incurred samples) be used to cross-validate the bioanalytical methods. Data should be evaluated using an appropriate predefined acceptance criteria or statistical method [12, 13]. It should be cautioned that many times the methods that are being cross-validated may not have the same range of quantification. In these situations, it is necessary to prepare spiked samples within the range that are common to both methods for comparison.

5.3.4 TOTAL ERROR VERSUS 4–6–20 RULE

5.3.4.1 Total Error

The total error of a measurement takes into account both the systematic error (bias) and the random error components. Any measurement that is made during an experiment consists of both of these error parameters, and it is not possible to separate these two. Therefore, it is scientifically correct to use the total error criteria to assess the acceptability of a quality control result during a run.

As described in this chapter, prestudy validation runs are accepted based on the standard-curve acceptance criteria. No run acceptance criteria are applicable for prestudy validation sample assessments; i.e., no run can be rejected due to poor validation sample performance during accuracy and precision evaluation, and all data from the prestudy validation runs are reported without exceptions. In some cases, there may be assignable cause (e.g., technical issues) for removal of a validation sample data point before the calculation of the cumulative mean. Exclusion is applied at the end of the validation study period and must be documented as described in the documentation section.

For each in-study run, the standard curve must satisfy criteria described in the standard-curve section; however, run acceptance is based primarily on the performance of the QC samples. When using total error for ligand binding assays of macromolecules, the run acceptance criteria recommended in the precision and accuracy section requires that at least four of six (67%) QC results must be within 30% of their nominal values, with at least 50% of the values for each QC level satisfying the 30% limit. The recommended 4–6–30 rule imposes limits simultaneously on the allowable random error (imprecision) and systematic error (mean bias). If the application of an assay requires a QC target acceptance limit different than the 30% deviation from the nominal value, then prestudy acceptance criteria for precision and accuracy should be adjusted so that the limit for the sum of the interbatch imprecision and absolute mean RE is equal to the revised QC acceptance limit.

5.3.5 DOCUMENTATION

It is critical to document the information generated during the development of a method in a laboratory notebook or other acceptable format. Information on the following assessments should be generated during assay development: critical assay reagent selection and stability, assay format selection (antibody[ies], diluents, plates, detection system, etc.), standard-curve model selection, matrix selection, specificity of the reagents, sample preparation, preliminary stability, and a preliminary assessment of assay robustness. At the end of the development portion of the assay life cycle, a draft method or an assay worksheet should be generated to be referred to during the prestudy validation.

A validation plan should be written before the initiation of the prestudy validation experiments. Alternatively, reference to an appropriate SOP can be made to ensure that a documented outline exists for the experiments required for prestudy validation. This plan can be a stand-alone document or can be contained in a laboratory notebook or some comparable format. The documentation should include a description of the intended use of the method under consideration and a summary of the performance parameters to be validated that should include, but may not be limited to, standard curve, precision and accuracy, range of quantification, specificity and selectivity, stability, dilutional linearity, robustness, batch size, and run acceptance criteria. The plan should include a summary of the proposed experiments and the target acceptance criteria for each performance parameter studied.

After completion of the validation experiments, a comprehensive report should be written. The format of the report may be dictated by internal policies of the laboratory; however, such reports should summarize the assay performance data and deviations from the method SOP or validation plan and any other relevant information related to the conditions under which the assay can be used without infringing the acceptance criteria.

Cumulative standard curve and QC data tables containing appropriate statistical parameters should be generated and included with the study sample values in the final study report. Unlike the prestudy validation, failed runs are not included in these tables. Additional information to be included in the final report is a description of deviations that occurred during the study, a table of the samples that underwent repeat analysis and the reasons, and a table with details on all failed runs.

5.3.6 CONCLUSION

The primary focus in the analytical considerations of immunoassays for macromolecules should be in the development phase. By thoroughly defining each component of the immunoassay in the first phase of the assay life cycle, the resulting validation plan should be concise and the validation process should be straightforward. Several facets of the assay defined in the late development stage, such as specificity and dilutional linearity, can appropriately be included in the final validation report.

A typical validation will include at least six precision and accuracy assays to define the consistency of the assay. Within those assays, several parameters can be defined, including early stability, specificity, selectivity, and range of quantification. No assay run should be eliminated except for a true and documented analyst's error.

The validation QC samples define the range of the assay, and no values below the LLOQ or above the ULOQ may be reported. Within the six validation assay runs, the validation samples are used to define the cumulative precision and accuracy. During validation, no validation sample may be eliminated to show the true profile of the assay.

During sample analysis before assessing the QC samples for acceptance, the standard curve must be deemed appropriate by predetermined criteria. Only after the curve is accepted may the assessment of QC samples continue. QC sample results determine whether the assay run is valid. Acceptance criteria can be based on 4–6–20 rule or on Total Error and should be predicated on the criteria used in both the development and the prestudy validation phase. Overall, the immunoassay is a highly sensitive assay that can be used to quantify protein and peptide drugs in a biological matrix, often routinely in the pg/mL range.

REFERENCES

1. Shah V P, Midha K K, Dighe S V, et al. (1992). Analytical methods validation: Bioavailability, bioequivalence, and pharmacokinetic studies. *J. Pharm. Sci.* 81:309–312.
2. Shah V P, Midha K K, Dighe S, et al. (1992). Analytical methods validation: Bioavailability, bioequivalence, and pharmacokinetic studies. *Pharm. Res.* 9:588–592.
3. Riley C M, Rosanke T W (1996). *Development of validation of analytical methods: progress in pharmaceutical and biomedical analysis*, vol 3. Ed. Elsevier, Pergamon, NY.
4. Shah V P, Midha K K, Dighe S, et al. (1991). Analytical methods validation: Bioavailability, bioequivalence and pharmacokinetics studies. Conference Report. *Eur. J. Drug Metabol. Pharmacokinet.* 16:249–255.
5. Guideline on validation of analytical procedures: Definitions and terminology. *International Conference of Harmonization (ICH) of Technical Requirements for the Registration of Pharmaceuticals for Human Use.* Geneva, 1995. Switzerland, (1996).
6. Shah V P, Midha K K, Findlay J W A, et al. (2000). Bioanalytical method validation. A revisit with a decade of progress. *Pharm. Res.* 17:1551–1557.
7. Guidance for the Industry: Bioanalytical Method Validation (2001). US department of Health and Human Services FDA (CDER) and (CVM).
8. DeSilva B, Smith W, Weiner R, et al. (2003). Recommendations for the Bioanalytical Method Validation of Ligand-binding Assays to Support Pharmacokinetic Assessments of Macromolecules. *Pharmac. Res.* 20(11):1885–1900.
9. Findlay J W A, Smith W C, Lee J W, et al. (2000). Validation of Immunoassays for bioanalysis: A pharmaceutical industry perspective. *J. Pharm. Biomed. Anal.* 21:1249–1273.
10. Searle S R, Casella G, McCulloch C E, (1992). *Variance Components, Chapter 3.* Wiley, New York.
11. Mee R W (1984). β-expectation and β-content tolerance limits for balanced one-way ANOVA random model. *Technometrics.* 26:251–254.
12. Bland J M, Altman D G (1999). Measuring agreement in method comparison studies. *Stat. Meth. Med. Res.* 8:135–160.
13. Hartmann C, Smeyers-Verbeke J, Penninckx W, et al. (1995). Reappraisal of hypothesis testing for method validation: Detection of systematic error by comparing the means of two methods or two laboratories. *Analyt. Chem.* 67:4491–4499.

5.4

CHROMATOGRAPHY-BASED SEPARATION OF PROTEINS, PEPTIDES, AND AMINO ACIDS

JAN MOEBIUS,[1] REMCO SWART,[2] AND ALBERT SICKMANN[1]

[1]*Rudolf-Virchow-Center for Experimental Biomedicine, Würzburg, Germany*
[2]*LC-Packings—A Dionex Company, Amsterdam, The Netherlands*

Chapter Contents

5.4.1	Introduction	586
5.4.2	Basics of Chromatography	586
	5.4.2.1 Hardware Requirements	586
	5.4.2.2 HPLC Columns	588
	5.4.2.3 Miniaturization	589
5.4.3	Separation of Peptides	591
	5.4.3.1 Short Introduction	591
	5.4.3.2 Reversed Phase Separation of Peptides	591
	5.4.3.3 Column Media and Resins	593
	5.4.3.4 Multidimensional Separation of Peptides	594
5.4.4	Separation of Proteins	598
	5.4.4.1 Short Introduction	598
	5.4.4.2 Reversed Phase–Liquid Chromatography (RP–HPLC) Separation of Proteins	598
	5.4.4.3 Ion-exchange Separation of Proteins	602
	5.4.4.4 Multidimensional Liquid Chromatography of Proteins	603
5.4.5	Separation of Amino Acids—Amino Acid Analysis	605
	References	606

5.4.1 INTRODUCTION

Since the first commercial introduction of high-performance liquid chromatography (HPLC) in the 1970s, this versatile method has gained wide popularity, resulting in a phenomenal increase of successful applications described in literature. HPLC is a highly efficacious method in peptide and protein analysis especially for the preparation of samples for further analysis by electrospray mass spectrometry or quantification using amino acid analysis. This story of success is related with the continuous development and miniaturization of flow rates and column diameters and the high level of automation combined with the reliability of the instruments. Today HPLC can separate minute amounts of peptides and proteins and find those areas of biochemistry and biotechnology where only small amounts of sample are accessible. The following chapter is focused on the analytical aspects of liquid chromatography of biomolecules from amino acids to proteins.

5.4.2 BASICS OF CHROMATOGRAPHY

5.4.2.1 Hardware Requirements

The chromatography-based separation of amino acids, peptides, or proteins has to be divided into the different applications of preparative and analytical chromatography. For analytical HPLC, the identification and quantification of single components from a low complex mixture is the most important task. Therefore, the analytes of interest have to be separated as much as possible.

The preparative biochromatography is focused on the isolation and enrichment of a single protein species. Typically, the HPLC separation is performed with bigger columns providing higher loading capacities but relative short column lengths in order to keep separation time short. However, in this chapter, we will focus on the analytical aspects of HPLC separation.

Generally, solvent pumps with high pressure capability are required for HPLC due to the backpressure generated by the dense package of the column resin. The typical specifications are pressure limits of about 400 bar and 5800 psi, respectively, although this limit is normally not required. Recently, new column materials with small particle sizes of about 1.3 μm have been introduced that require HPLC pumps and tubings that resist pressures of up to 1000 bar (approximately 14,000 psi) [1–3]. For the separation of biomolecules, two different HPLC-pump types have been established: piston pumps and syringe pumps.

The basic principle of a syringe pump principle is simple: A motor activates a piston that pressurizes the solvent within a syringe. Although this type of a pump was widely used in the past, the syringe pump has some limitations; e.g., it usually cannot deliver the same high-pressure limits as stainless steel piston pumps. In addition, the amount of solvent for a single chromatographic separation is restricted to the syringe volume. Moreover, these pump systems are most often designed for binary gradients because for each solvent, an syringe pump is required. Recently, syringes have experienced a revival due to their scalability to smaller dimensions, which enables access to native nanoscale flow rates [4, 5]. Nevertheless, piston pumps are the pump systems that are provided by all major vendor companies because of several imminent advantages in flexibility and performance. A small piston pressurizes the solvent continually within the flow system. Check-valves prevent the backflush of the solvent. The refilling of the piston causes pulsation, which can be eliminated by using a second pump unit that performs the pump stroke while the first piston refills. The gradient is formed by a mixing device that can be located at the high- or low-pressure side, i.e., before or after the pump unit. Syringe pumps normally use a mixing-tee located directly after the two syringes. The gradient is formed by different movement speeds of the syringes.

Piston pumps can be designed for low-pressure mixing as well as for high-pressure gradient formation. The low-pressure mixing is realized by solenoids that open the flow path of a solvent for a discrete period of time. Subsequently, the solvents are mixed and pressurized. This setup applies only to a single pump unit, whereas high-pressure mixing in piston-pump systems requires a single pump unit for each solvent. The major advantage of this setup is the enhanced response time to changes in the gradient, which is required for steep and short gradients. Another aspect of HPLC pumps for biochromatography is the availability of bio-inert versions, which can be of importance for several applications covering extreme pH-values as well as several modified peptides that are retarded by stainless steel [6]. In contrast to stainless steel, all components of the flow-path are composed of inert materials like PEEK, Titanium, or Saphire (for the pistons).

Besides the pump as a solvent delivery device, the HPLC valves are important components for switching solvent lines (e.g., injection loops and parallel column setups). They are designed to resist the applied pressure and feature a small dead volume to prevent peak dispersion. HPLC valves are available in different switching modes and port numbers. Additionally, valves are also fashioned as bio-inert versions. For several applications, the HPLC column should be used in a column-oven for ensuring constant temperature during separation.

To introduce samples into the flow path, the easiest way is the usage of a sample loop that is connected to an HPLC valve. After loading and switching, the loop is integrated into the flow path and the sample is delivered to the separation column. Today, this task is done by autosamplers that can temper samples and provide 24 hours of fully automated and reproducible sample injection.

A further important component of each HPLC system is the detection device. Often, Ultraviolet (UV)-detectors are used for this purpose because many analytes absorb light in the range of 200–350 nm. UV detectors provide reasonable sensitivity and dynamic range as well as a low noise level.

Recently, multi-wavelength detectors that can switch between different wavelengths have been used. Diode array detectors (DADs) use a light source that is dispersed by a holographic grating after transmission through the flow cell. The light of the single wavelengths is reflected onto a diode array with several hundreds of diodes acquiring in real time a full-spectrum chromatogram. This leads to a better characterization of each eluted component.

Electrochemical detectors are applied for analytes that are capable of redox-reactions. The application field of biochromatography for PADs covers only a small spectrum of amino acids, whereas a broad range of applications is established for smaller molecules.

One of the most important detectors for proteins and peptides is a mass spectrometer that is necessary for structural analysis of peptides. The high sensitivity and capability of structural information are the two major reasons for the success of HPLC coupled to mass spectrometry in proteomics.

Often, a pressure is applied to generate the mobile phase flow though the column. Alternatively, in capillary electrochromatography (CEC), an electric field is applied along the column to drive the mobile phase through packed capillary columns with internal diameters (IDs) of 50–100 μm. This principle was proposed by Pretorius et al. [7] and further developed by other groups [8, 9]. In these electrically driven systems, the mobile phase flow is characterized by a flat velocity profile, which results in higher separation efficiencies. However, ineffective heat dissipation restricts the use of high electric field strengths for large ID columns and thereby limits the applicability of the technique.

5.4.2.2   HPLC Columns

The efficiency of an HPLC column is determined by the type of stationary phase particle and the quality of the column packing. The internal diameter of the column is not affecting the performance characteristics [10], although it has been deliberated that columns with low aspect ratio, i.e., ratio between internal diameter and particle diameter, have higher separation efficiency. A recent and extensive experimental study did not show convincing evidence for higher column efficiency for columns with an aspect ratio from 7.5 to 30 [11].

A more effective way to increase the efficiency in HPLC is by reducing the stationary phase particle size, as predicted by the well-known van Deemter equation [12], which describes the plate height in HPLC as the sum of different contributions. In this respect, submicron particles have been applied and efficiencies up to 730,000 plates/m have been reported for small molecules [13, 14].

However, as a result of the increased flow resistance of these submicron particle packed columns, extremely high pressures of 200–300 MPa are necessary to drive the mobile phase. Working at these high pressures, several phenomena should be taken into account: (1) increased viscosity of the mobile phase affecting the efficiency and pressure drop over the column, (2) Joule heating effects that may cause radial temperature gradients in analytical and narrow-bore columns, and (3) linear velocity surge caused by mobile-phase compression responsible for band broadening [15].

Recently ultra-performance liquid chromatography (UPLC) has been introduced commercially as a new direction in liquid chromatography. In UPLC columns packed with 1.7-μm, particles are operated at pressures up to 100 MPa. The main advantage is obviously the reduction in analysis time that has been demonstrated for pharmaceutical analysis [16] and metabolomics [17].

The use of monolithic columns in LC has advanced rapidly since their first introduction in the 1990s [18–21]. In contrast to capillary columns packed with particulate stationary phases, monolithic columns consist of a single continuous support. Monolithic stationary phases can be subdivided in two classes, i.e., polymer- and silica-based materials.

Polymer monoliths are prepared by *in situ* polymerization of monomers such as styrene, acrylate, methacrylate, acrylamide, or cyclic compounds and have been reviewed thoroughly [19]. The selectivity of the monolithic column can be changed by incorporation of functional monomers during the polymerization. A disadvantage of this procedure is that for each polymerization mixture, the conditions have to be optimized. Alternatively the functionality can be modified at the surface of the monolithic structure by (photo)grafting [22].

Silica monoliths are prepared by sol–gel technology starting from alkoxide precursors, such as tetraethoxy-silane or tetramethoxy-silane [23]. The structure of a silica monolith is characterized by an interconnected silica skeleton and bimodal distribution of 1–3-μm macropores (flow through pores) and 10–20 nm mesopores.

The main advantages of monolithic columns are the superior separation performance and low flow resistance. In addition due to their continuous nature, frits are not required to retain the stationary phase. The production process of monolithic columns is more flexible than that of packed columns; e.g., photo-polymerization can be applied to prepare monolithic structures or add selectivity locally. Both polymer- and silica-based monolithic capillary columns have been used for highly efficient separations in LC–mass spectrometry (MS) applications for proteomic research [24, 25].

5.4.2.3 Miniaturization of HPLC

Miniaturization in HPLC was started more than 25 years ago and was mainly driven by the need for analyzing minute samples [26–29]. The developments were accelerated by the widespread utilization of electrospray ionization (ESI) for interfacing LC and MS.

Reducing the ID of the LC column offers several advantages, among which are most importantly: (1) increased mass sensitivity for concentration sensitive detectors due to reduced chromatographic dilution, (2) easier coupling to MS interfaces [ESI and matrix-assisted laser desorption/ionization (MALDI)] due to the lower flow rates, and (3) a lower stationary phase and solvent consumption.

The increased mass sensitivity for concentration-sensitive detectors is the direct result of reduced dilution of the sample in small ID columns and is inversely proportional with the square of the reduction in column diameter. As a result of miniaturization in LC, a variety of new column formats were developed, which can be subdivided according to the ID. In Table 5.4-1, the ID, typical flow rate, and sample loading are listed for various HPLC techniques.

Analytical and narrow-bore HPLC columns are most commonly used for routine analysis, which require the highest possible robustness. Capillary- and nano-LC columns find application in various fields of biotechnology research, mainly for sample limited analysis of peptides and proteins [30–32].

TABLE 5.4-1. Typical Dimensions and Flow Rates for Analytical RP Chromatography

	Inner Diameter	Flow Rate	Relative Gain in Sensitivity
Conventional LC	4.6 mm	1000 μl/min	Baseline = 1
Narrowbore LC	2.1 mm	200 μL/min	5
Microbore LC	0.5–1.0 mm	50 μL/min	35
Capillary LC	180–300 μm	2–5 μL/min	200
Nano LC	50–100 μm	200 nL/min	3500

A significant reduction of the column ID put stringent demands on the instrumentation for micro-LC as all volumetric extra-column dispersion contributions must be down scaled accordingly. Initial developments in this area were achieved on modified standard LC instruments. For approximately 10 years, dedicated instrumentation for micro-, capillary-, and nano-LC has become commercially available and was recently discussed [33].

The reduction or downscaling factor in going from an analytical to a capillary column equals the ratio of the square of the column diameters. Downscaling from a 4.6-mm ID standard HPLC column to a 300-μm ID packed capillary column equals a theoretical factor of 200 in sensitivity (see Table 5.4-1). This factor applies to all components of the LC system, like flow rate, injection and detection volume, and connecting capillaries [34]. Likewise, for a 300-μm ID capillary column, the flow cell volume must be reduced to ~50 nL to prevent extra-column peak broadening. For injection of the sample, the downscaling factor implies that an injection volume of 20 μL on a 4.6-mm ID column equals an injection volume of ~90 nL on a 300-μm ID capillary column. Consequently autosamplers used for capillary LC must be able to inject nanoliter volumes in a reproducible manner. However, in gradient elution, the preferred mode of operation for separation of peptides and proteins, sample preconcentration can be applied. Relatively large volumes, often exceeding the column volume several times, are injected while the compounds are concentrated at the head of the column or trap column. In nano-LC with typical flow rates of 100–300 nL/min, column switching is a widespread technique, e.g., for peptide mapping when the sample amount is limited. A trap column is mounted onto a switching valve (see Figure 5.4-1) to reduce the sample volume from several microliters present in the sample loop to a few nanoliters (basically the volume of the trap column). After sample loading, the trap column is switched in line with the separation column before starting the HPLC gradient. An additional advantage of column switching is the ability of online sample desalting. A common setup for

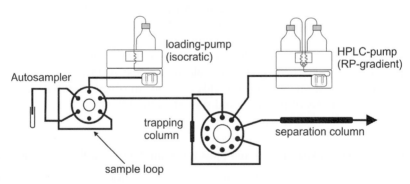

Figure 5.4-1. Schematic overview of typical reversed phase-HPLC with precolumn and column loading for peptide separation. The central switching valve has two positions: In position one, the sample is loaded onto the precolumn with the loading pump. After trapping and desalting of the sample, the valve is switched and the precolumn is integrated into the flow of the gradient pump. By increasing the amount of solvent B, the single peptides are separated on the separation column and afterward detected with a UV-detector and/or by mass spectrometry.

column switching in nano-LC is shown in Figure 5.4-1, although other configurations have been described as well [35, 36].

Typical flow rates applied in nano- and capillary-LC are in the range of 50 nL–10 µL/min according to Table 5.4-1. HPLC systems that can generate mobile phase gradients at these flow rates are available and are based on either flow splitting or dedicated micro- or nano-pumps. The generation of gradient profiles should be characterized by small delay volumes, efficient mixing, high accuracy, and acceptable reproducibility. Significant improvements in this respect have been achieved during the last couple of years, partly due to the incorporation of liquid flow sensors that can measure flow rates down to several nanoliters/minute. With current state-of-the art instrumentation, it is now possible to generate solvent gradients with a reproducibility better than 0.3% RSD at flow rates suitable for nano-LC.

5.4.3   SEPARATION OF PEPTIDES

5.4.3.1   Short Introduction

Reversed phase (RP) chromatography has been established as the superior separation technique for peptides. The chromatographic resolution of this method is higher compared with other methods like ion-exchange (IEX) or size exclusion chromatography (SEC).

5.4.3.2   Reversed Phase Separation of Peptides

RP–HPLC is well suited for direct online coupling with electrospray ionization–mass spectrometry (ESI–MS) due to solvent compatibility with the electrospray ionization technique.

The common gradient for peptide separation consists of a binary solvent system containing in solvent A: 0.1% formic acid and in solvent B: 0.1% formic acid and 84% acetonitrile. If no high-sensitive MS detection is required, formic acid can be replaced by TFA. For solvent B, the amount of TFA has to be reduced to 0.08% to prevent baseline shift during the gradient run time.

Gradients normally begin at 5% B with a linear increase to 50% B within 45 minutes. For rinsing the column, the percentage of B is increased to 95% and held for 5 minutes. Afterward the column is re-equilibrated to 5% B. Therefore, the complete gradient is finished within 1 hour. Extension to longer and thus shallower gradients is suggested if complex samples have to be separated.

The formic acid acts as an ion pair reagent that masks the peptide-charge and increases the hydrophobicity for better retention on reversed phase supports. Trifluoroacetic acid (TFA) has better characteristics for HPLC separation because of less UV absorption at 215 nm and superior properties as an ion pair reagent [37, 38]. Therefore, it is preferably used in solvents for loading on precolumns or separation without the need for MS detection. In this context, the usage of formic acid as an ion pair reagent instead of trifluoroacetic acid, which has less absorption in the UV range, is preferable because of its better ionization efficiency in MS ion sources. In contrast, TFA offers less UV absorption and superior separation results at the cost of inferior ionization rates.

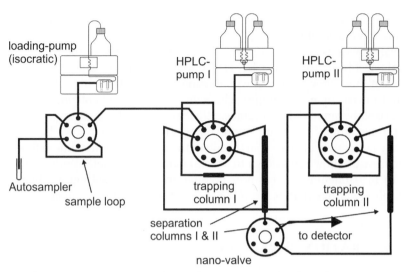

Figure 5.4-2. Set up of a dual-gradient HPLC system. This HPLC-system consist of a normal reversed phase separation system with an additional gradient pump and HPLC-valve. Two complete sets of pre- and separation columns are required. This setup enables the equilibration and sample preconcentration on the one column pair, whereas on the second column pair, the separation is performed. This reduces the dwell time of a connected mass spectrometer and thereby increases the sample throughput per day significantly. The final nano-valve directs the flow of the active separation column to the connected detector.

MS represents the most expensive part of a nano-LC–MS systems. Therefore, dwell times are expensive and should be reduced to a minimum. Parallel LC-systems consist of a more complex setup with two pairs of pre- and separation columns (see Figure 5.4-2).

CARLA Test. Dead volume is a serious obstacle for high-resolution RP separation using nano-HPLC. A dead volume acts as a passive mixing chamber in the flow path that levels the preformed gradient, thereby leading to peak broadening and reduced sensitivity. The definition of a dead volume is a relative subject because a 0.5-μL dead volume in a preparative HPLC with a 1-mL/min flow rate and a step-gradient is not recognized as such. But the chromatographic separation would fail in the case of a nano-HPLC with typical flow rates of 200 nL/min. For this reason, the detection of dead volumes that cannot be visually located is a very important task for the highest performance and sensitivity. The CARLA Test provides a very simple approach for online detection of dead volumes in HPLC systems [39]. For the test, a UV detector, a precolumn system with a loading pump using 0.1% TFA and the nano-HPLC-system using 0.1% FA, is required.

While the sample is loaded onto the trapping column and desalted by flushing with 0.1% TFA at an increased flow rate, the nano-HPLC branch is equilibrated with 0.1% FA. The UV baseline is set on the FA containing solution. By switching the HPLC valve, the trapping-column loop is then integrated into the nano-flow. Now, the nano-flow pushes the small volume of TFA solution through the

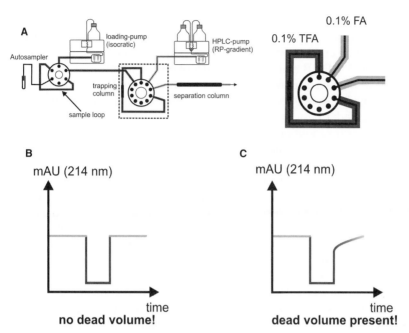

Figure 5.4-3. Carla test. (A) The Carla test requires the detection of the UV-trace at 214 nm and a precolumn setup with loading pump that delivers 0.1% TFA for peptide trapping and desalting and a nano-flow system working with 0.1% FA for peptide separation. The close-up view of the HPLC valve depicts the localization of TFA and formic acid. (B) Schematic UV-chromatogram after the switching event. The baseline is set to UV-absorption level of formic acid. The elution of TFA from the precolumn loop decreases the UV-signal to a significant level. If no dead volume is present, the signal has a rectangular shape. (C) If a dead volume acts as a small mixing chamber, the discrete volumes of TFA and FA are partially mixed, leading to long transitions back to baseline. In this case, eluted peptide peaks are broader and sensitivity level of such a HPLC-system is decreased.

trapping-column loop and the separation column to the UV detector that is reported by a signal drop. If no mixing occurs, the pattern should be shown as a sharp rectangular signal (see Figure 5.4-3B). However, if a dead volume is present, the two different solutions are mixed during passage through the flow path (see Figure 5.4-3C). This results in a smooth transition back to the FA baseline after an increased amount of time.

5.4.3.3   Column Media and Resins

The performance of HPLC separations depends on the quality of the applied column. Column dimensions as well as the used resins are the main features that have to be regarded. Although column systems with beads as chromatographic material have been extensively used, recently available monolithic columns represent a completely new technology.

Bead materials are normally made of silica spheres or synthetics like polystyrenedivinylbenzene (PS-DVB). The beads are then functionalized with the appropriate chemical group for interaction with analytes. In the case of peptides, C18- or

C8-phases are coupled to the silanol groups on the bead-surface with chlorotri-alkylsilane reagents. Often, not all available silanol groups are modified and are often blocked by end-capping to avoid ion exchange functionality interfering with the RP mode. PS-DVB materials are hydrophobic in nature and need only surface modification if other separation properties are required. The silica-bead packed columns provide a very good resolution for HPLC, but they are becoming unstable above pH 8. Under basic conditions, polymeric materials are therefore preferable. However, several vendors have developed chemical modifications on silica beads that slightly increase the stability of the material up to pH 9–10.

The pore diameter of the beads has a decisive influence on the binding capacity of the column, which means that small pore sizes increase the effective surface of the material and thereby the binding capacity. Nevertheless, if the pores are too small, the peptides cannot diffuse and therefore the interaction of the analyte and the stationary phase is suppressed. A pore diameter of 100 Å has proven to be very suitable for tryptic peptides, whereas high-molecular-weight peptides should be separated using 300-Å pore material. The bead size and the bead size distribution have a direct influence on the separation efficiency, which is expressed in the van Deemter-equation by the Height Equivalent to the Theoretical Plate (HETP). The smaller the bead size, the smaller the plate height and the higher the resolution. But with smaller bead sizes, the required flow rate has to be increased and therefore the backpressure rises significantly. Typically, analytical RP columns for peptides use bead sizes between 3 and 5 μm. The column diameter and the column length depend mainly on the required resolution and the available sample amount, respectively.

5.4.3.4 Multidimensional Separation of Peptides

Proteins differ in molecular weight, hydrophobicity, charge, and structure over a wide range, which renders it impossible to apply a single chromatographic separation for complex protein mixtures. Therefore, proteins are digested to peptides to reduce the huge differences in their physico-chemical properties that enables the application of the peptide mixture to chromatographic separation approaches. However, the digest increases the number of analyte species from the protein to the peptide level significantly. For instance, the expected number of proteins in a cell differs from 5 to 20,000 species. After digestion the expected number of peptides ranges between 50,000 and 200,000 species. Such complex samples can only be handled by multiple separation according different separation parameters. Generally, the concept of this approach is the subsequent separation and pre-fractionation of peptides with different chromatographic media. In most cases, the first dimension represents a strong IEX Chromatography where the peptides are fractionated during fixed time intervals. Afterward, each ion-exchange fraction is applied to an RP column as the final high-resolution separation dimension that is usually online-coupled to an MS representing an additional gas phase separation dimension with concurrent peptide analysis. Several different applications have been developed so far: offline or online MDLC, MudPIT, and the COFRADIC [40] technology.

The classic MDLC technique separates the peptides using a strong cation exchange (SCX) column for the first and C-18 RP material for the second dimension. Offline and online separation, respectively, differs in peptide transfer from the first to the second HPLC-separation dimension. For the online mode, eluted fractions are directly transferred to the next dimension, whereas during offline MDLC, the collected fractions are prepared for injection onto the RP column. The usage of fraction collectors and liquid handlers enables a semi-auto-mated sample preparation before reinjection for RP separation and subsequent analysis. The advantages of the offline mode are the application of continuous gradients and more flexibility in solvent selection, thereby leading to a more effec-tive elution and better resolved separation (see Figure 5.4-4). Additionally, both separation dimensions can be performed subsequently on a single HPLC system with less hardware requirements in contrast to the online system.

Online MDLC works in full automation mode and integrates first- and second-dimension separation into a single system, providing high sample throughput capa-bilities and short dwell times. The elution of the SCX column is performed with salt-plugs (see Figure 5.4-5) or step-gradients (see Figure 5.4-6) with increasing salt amounts. Subsequently, the eluted fraction is trapped onto an RP-precolumn where the peptides are desalted and concentrated. Then the HPLC valve is switched, and the precolumn is introduced into the flow path of the RP gradient. Peptides are separated and eluted similar to common RP separations. After the complete RP run, the precolumn is switched back and traps the next fraction eluting from the IEX column. In more complex MDLC systems, two or more precolumns are alter-nately used for trapping and elution to reduce the complete run time as well as to enable the application of continuous IEX gradients. A general disadvantage of the online system is the restriction to non-organic solvents for IEX elution. It has been shown that little amounts of organic modifiers increase the elution efficiency and prevent memory effects during the first dimension [41]. In online systems, this

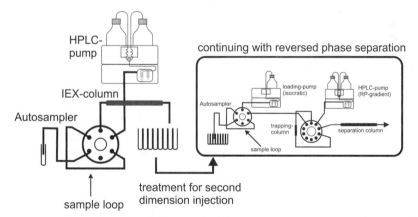

Figure 5.4-4. Scheme of an offline MDLC setup. The separation with this two-dimensional HPLC separation is divided into two steps. At the first step, the analyte is separated by an ion-exchange separation (IEX) and the eluted fractions are collected. Before the second step, the samples are concentrated and desalted and organic modifiers are removed if present. The second dimension is performed by a normal reversed phase-HPLC separation (see Figure 5.4-1). If the required capacities are not available, this type of MDLC run enables the separation of both dimensions on the same HPLC after a setup change of column and solvents.

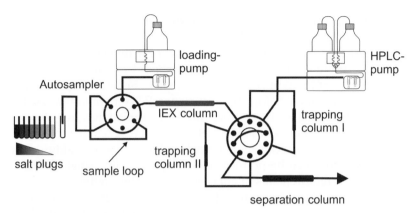

Figure 5.4-5. Online MDLC using salt-plugs. This HPLC separation mode uses salt-plugs with increasing salt concentrations that are injected by the autosampler onto the IEX-column to elute stepwise stronger bound peptides.

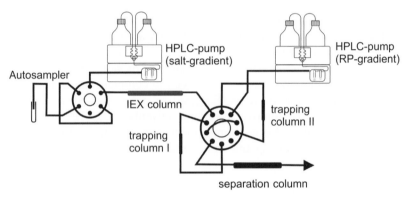

Figure 5.4-6. Online MDLC with gradient elution. This type of MDLC system is similar to the setup with salt-plug elution, but in this case, the elution is performed with a step- or linear-gradient that requires a second gradient capable HPLC-pump.

would result in inferior peptide recovery because the SCX-peptide fraction would not be trapped quantitatively on the precolumn due to the organic solvent.

This type of elution has the further disadvantage that peptides are present in different fractions due to memory effects. Additionally, the online systems consist of multiple HPLC pumps and valves, several connections, and a complex control program that gives rise to many potential sources of error as well as to difficulties in obtaining acceptable sensitivity. The multidimensional protein identification technology (MuDPIT) was designed to provide a high-resolution peptide separation technique for electrospray ionization mass spectrometry (ESI–MS) with a very good sensitivity [42] (see Figure 5.4-7). In contrast to the classic MDLC, the two separation phases are located in a single biphasic column. Thereby, the eluted peptides from the SCX phase are directly trapped on the RP resin without any dead volume. Additionally, the system is simplified due to the location of the two phases in the ESI-tip emitter. The sample is bound on the SCX phase, and fractions are eluted by salt-plug injection. The fractions are trapped on the RP phase and

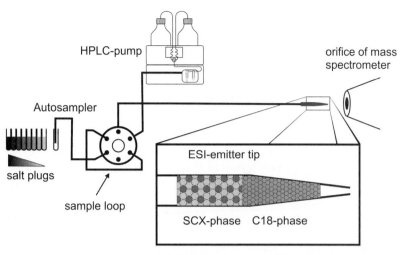

Figure 5.4-7. Schematic workflow of MuDPIT. The MuDPIT technique represents a relative simple setup compared with the other MDLC approaches. For the complete system, a single nano-gradient pump and an optional autosampler is required. The two-dimensional separation takes place in the biphasic ESI-Tip emitter, which results in a near dead-volume-free solution. Peptides are eluted from the strong cation exchange phase (SCX) with salt-plugs that are injected by the autosampler. Afterward the trapped peptides are separated with a linear gradient and directly sprayed into the mass spectrometer.

then separated by common RP gradients that are delivered by a single nano-HPLC pump. Therefore, the technical requirements are reduced in comparison with other online MDLC systems, and the complete separation is performed fully automated. The major disadvantage of this approach is the design of the biphasic separation column, because the two phases are in contact with a high salt concentration and high concentration of organic modifiers, respectively. This may lead to poor sample recovery and imminent loss of peptide species.

One major disadvantage of all presented approaches is the limited resolution in a two-dimensional separation. The discrete separation with tens of thousands of peptide species is still not possible. The combined fractional diagonal chromatography (COFRADIC) circumvents this problem by discarding most peptides [43]. The protein identification by MS requires only a single peptide that is specific for this particular protein. Therefore, the increase of sample complexity after enzymatic digest produces a lot of redundancy that is again reduced by the COFRADIC technique. The proteins are blocked at the cysteine and free amino-residues before proteolytic digestion. The resulting peptides are separated by RP chromatography and collected in minute fractions. Each fraction is then treated with TNBS to modify the newly generated free N-terminal amino groups and increase the hydrophobicity of these peptide species. Afterward, each fraction is rerun under the same conditions. All peptides that elute at the same retention time as before are original N-terminal peptides because they were blocked prior to TNBS-modification and are therefore not susceptible to retention time increase. These fractions can be collected for later MS analysis. The COFRADIC approach has also been extended to the separation of cysteine or methionine containing peptides [44]. Other targets like phosphorylations or other moieties that can be chemical modified may be targeted as well. Therefore, the COFRADIC approach is very suitable for directed global analysis of specific peptide subsets in complex peptide mixtures.

5.4.4   SEPARATION OF PROTEINS

5.4.4.1   Short Introduction

RP–HPLC separates compounds based on the reversible interaction with the hydrophobic stationary phase. RP–HPLC has become the most extensively practiced type of chromatography for proteins due to the unparalleled resolution with applications in purification, desalting, and analysis.

RP–HPLC separation of proteins has been the topic of several reviews and book chapters, e.g., Refs. 45 and 46.

RP chromatography of proteins is almost exclusively dominated by gradient elution. Besides the ability of gradient elution to separate a complex mixture of compounds with a broad range of hydrophobicity, there is another reason for the almost exclusive use of gradient elution.

In contrast to small molecules, proteins are amphiphilic and interact with the stationary phase only via a small part of the molecule, whereas the hydrophilic parts of the protein are in contact with the mobile phase. The mobile phase at the start of the gradient is typically an aqueous solution. During development of the gradient, a protein will almost completely desorb from the stationary phase at a specific organic solvent concentration. Before this concentration, the protein has an almost infinite retention on the stationary phase and will therefore not migrate through the column. This principle, often referred to as "on/off" retention mechanism, is also indicated by the steep adsorption isotherms of proteins. As a result, closely related proteins can be separated by small adjustments in the mobile phase composition. Hence, isocratic elution of proteins is not an attractive technique.

5.4.4.2   Reversed Phase–Liquid Chromatography (RP–HPLC) Separation of Proteins

Liquid chromatography is an essential analytical tool for the separation of proteins in biotechnology and pharmaceutical product development. Applications are found throughout the development process of biological products, from drug discovery to quality control. Molecular properties of proteins, like molecular mass, net charge, hydrophobicity, and structure give ample opportunities to use HPLC for their separation.

A variety of chromatography techniques is available for separation and analysis of proteins, i.e., RP, IEX, size exclusion, hydrophobic interaction, and affinity chromatography. Selection of the type of chromatography, scale of operation, and applied detection technique will be based on the sample availability and complexity and, on the other hand, the required information.

Stationary Phases.   The first step in the optimization of a chromatographic method for proteins is the selection of the stationary phase. Many stationary phases are available and are characterized by the material, particle shape and size, pore diameter, and functional group. For analytical purposes, regular-shaped, silica-based stationary phase particles are mostly employed. These particles are made of alternating silicon and oxygen atoms that form an incompressible three-dimensional network. The surface of the silica particle is covered with hydroxyl groups, called silanol groups, which are modified with a reactive alkyl silane to yield a siloxane bond. The length of the alkyl chain, which can vary from methyl (C1) to octadecyl (C18), determines the hydrophobicity of the RP sorbent and the binding strength of the protein. Unreacted, residual silanol groups are present after functionalization of the silica bead and have acidic properties that impair the RP separation. Ionic interaction between these silanol groups and the protein result in low protein recovery and peak tailing. To overcome this effect, end-capping of the stationary phase can be performed, i.e., the reaction of residual silanol groups with a small molecule.

Another limitation of silica-based stationary phases is the limited pH stability, which extends from around 3.0 to 7.0. Acidic solutions hydrolyze the siloxane bond between the bonded alkyl silane and the silica particle; at high pH, the silica substrate is readily dissolved. The hydrolytic stability of RP silica particles can be improved by shielding the siloxane bond with bulky alkyl groups or by surface coverage of the silica with a highly cross-linked aromatic layer [47]. Organic polymer-based supports have been developed as an alternative for the inorganic, silica-based stationary phases and have a high pH stability.

Resolution of the protein separation is affected by several stationary phase properties. Reduction of the particle diameter is obviously the most straightforward way to improve the efficiency. For analytical purposes, 3- to 10-μm particles are a good compromise between chromatographic performance and the required pressure. The pore size of the particles should be a minimum of 300 Å to provide accessibility for large proteins.

The large size of the proteins implies that diffusion coefficients are low and that mass transfer between the mobile and the stationary phase is slow. Several stationary phases have been designed that prevent excessive band broadening as a result of the low diffusivity of proteins. Perfusion beads are highly porous particles made from polystyrene divinylbenzene with pore sizes in the range of 100–8000 Å. The structure allows the mobile phase to perfuse through the particles and a convective transport of the molecules, not limited by diffusion. The analysis time in perfusion chromatography can be reduced significantly compared with conventional stationary phases, due to the low flow resistance and independence of flow rate on the resolution [48].

On the contrary, particles without pores enable fast, efficient separation of proteins due to the absence of intraparticle diffusion resistance. The absence of pores reduces the available surface area and thereby the loading capacity. Silica- [49] and polymer-based [50] nonporous particles have been applied for biomolecule separations. Also hybrid particles have been described that have a non/porous core and a 0.25-μm porous layer, composed of colloidal silica particles [51].

The application of monolithic columns for efficient RP–HPLC separation of proteins is rapidly developing. The advantage of monolithic columns for protein separations is the high mass transfer efficiency due to convective flow in the macroporous structure. As a result, very fast separations can be achieved.

Several polymer-based monolithic columns have become available commercially in different formats, e.g., disks, membranes, and LC columns with diameters in the range of 0.1–100 mm ID [52]. The preparation and application of organic polymer-based monoliths have been reviewed recently [20].

Extremely rapid RP separations of proteins have been demonstrated on poly(styrene-co-divinylbenzene) monolithic columns. A fast gradient and a flow rate of 10 mL/min were used to separate five model proteins in less than 20 s on a 50 × 4.6-mm ID column [53].The same type of monolithic structure has also been prepared in capillary columns of 200 μm ID and showed excellent performance in the identification of proteins by LC–MS through peptide mass fingerprinting and accurate intact molecular mass determination [24].

Acrylate monomers are used for the preparation of monolithic columns and are available in a broad variety of chemistries. For the RP type of monolithic columns, acrylates with an alkyl chain are employed.

Poly(butyl methacrylate-co-ethylene dimethacrylate monolithic columns have been prepared by photoinitiated polymerization and used for RP–HPLC of proteins [54]. The flow resistance of these monolithic columns was low, which enabled the use of high flow velocities of 87 mm/s. A mixture of four model proteins ribonuclease A, cytochrome c, myoglobin, and ovalbumin was baseline separated in around one minute and using flow rates of 100 μL/min. In Figure 5.4-8, chromatograms are shown that were obtained with three different gradients.

No effect of the pore size in the range of 0.7–2.2 μm was measured on the retention of proteins.

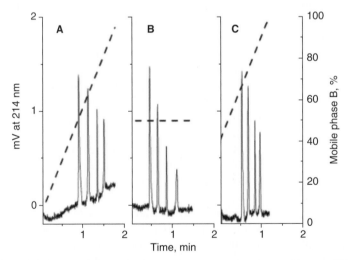

Figure 5.4-8. Effect of gradient on separation of a protein mixture using monolithic column I at 100 μL/min. Conditions: column size 100 mm × 0.2 mm. Mobile phase A, 0.1% TFA in water–acetonitrile (90:10, v/v); mobile phase B, 0.1% TFA in water–acetonitrile (10:90, v/v). Gradient in (a) 5–95% B from 0 to 1.8 min; gradient in (b) 5–50–50% B from 0 to 0.01 to 1.5 min, gradient in (c) 5–40–100% B from 0 to 0.01 to 1.21 min. Peaks: ribonuclease A, cytochrome *c*, myoglobin, and ovalbumin (elution order). (Figure and caption reprinted from Ref. 54, with permission from Elsevier.)

Mobile Phases and Gradients. Ion-pairing RP–HPLC is an ideal technique for the separation of proteins. A charged ion-pairing reagent with an opposite charge to that of the protein is added to the mobile phase to form a noncharged complex. The ion-pairing reagent increases the hydrophobicity and consequently the retention of the protein. TFA is the most commonly used ion-pairing reagent for proteins. Added in a concentration of 0.05% (v/v) to the mobile phase, the pH will be around 2.5. At this pH, the carboxylic groups of proteins are protonated and the trifluoroacetic anion binds to the protonated, positively charged amino groups of proteins.

Several ion-pairing reagents, including acids, bases, and salts, have been applied in RP liquid chromatography of proteins. TFA is preferred as it generally yields the highest separation performance. However, a disadvantage of TFA is that it can significantly lower the detection sensitivity in EIS–MS detection. The strong ion pairs are not easily broken apart under conditions used for EIS, thereby preventing ionization of the analyte. This effect is not easy predictable and has been found dependent on the ion-pairing reagent concentration, instrumental conditions, and the analyte. The influence of ion-pairing reagents on the sensitivity of detection in the coupling of RP–HPLC and ESI–MS for protein analysis has been reviewed [55]. In practice, a compromise between detection sensitivity and chromatographic resolution has to be found in the use of ion-pairing reagents for HPLC–ESI–MS applications. A more than 100-fold increase in detection sensitivity of proteins was reported when using 0.1% formic acid instead of 0.1% TFA at the cost of larger peak widths [56].

Several solvents can be chosen for the development of the gradient in RP chromatography. Acetonitril has a high volatility, low viscosity, and high UV transparency and is often preferred in LC/MS applications. For the analysis of large, hydrophobic proteins, alcohols with higher elution strength such as isopropanol can be employed. The gradient must be optimized for each application. Steep gradients are useful for protein desalting, whereas a shallow gradient should be used for complex samples.

Applications in Biotechnology. Biopharmaceutical products must be well characterized by means of analytical techniques. RP–HPLC with UV detection is suitable for the content assay and impurity profiling because of its high resolution. However, the bioactivity of protein-based pharmaceuticals not only relies on the primary structure but also on the higher order structure, i.e., the molecular confirmation, which is not revealed by UV absorbance. Circular dichroism and intrinsic fluorescence detection have been combined with UV detection in RP–HPLC to monitor the conformational properties of active compounds and impurities in biopharmaceutical products [57].

Characterization of (recombinant) proteins at the molecular level requires the use of MS detection.

RP–HPLC–MS can be applied to separate and identify proteins modifications such as deamidation, isomerization [58], and protein folding [59]. Even intact monoclonal IgG1 antibodies have been characterized by RP–HPLC–MS using a C8 stationary phase, elevated temperature of 65–70°C, and a combination of trifluoroacetic acid and heptafluorobutyric acid as ion-pairing agents.

Characterization of C-terminal lysine variants, glycosylation heterogeneity and degradation products of recombinant antibodies demonstrates the capability of this method [60].

The development of MS instrumentation has been accelerated during the last decade by proteomics research, the large-scale identification of the proteins in biological samples or tissues. Although nowadays the precise mass determination of even large proteins is possible on standard MS instruments, identification of proteins is still done at the peptide level after enzymatic digestion of the protein. An interesting development is internal gas phase protein fragmentation in the MS enabling accurate protein identification. In combination with a chromatographic separation, this development holds great promise for top-down proteomics [61].

5.4.4.3 Ion-exchange Separation of Proteins

In ion-exchange chromatography (IEC), retention is based on the interaction of charged amino acid residues of the protein with the counter-charge on the stationary phase media. In addition, protein modifications such as acetylation, phosphorylation, and glycosylation alter the charge of the protein and therefore the retention on the IEX column. IEX stationary phases carry either a positive charge (anion exchanger) or a negative charge (cation exchanger) and can be subdivided into weak and strong ion-exchanger resins. Strong ion-exchangers are ionized over a much wider pH range compared with their weak ion-exchanger and can be used over a broad pH range (~pH 2–12) Weak IEX columns lose their charge at pH > 9 (anion) and <6 (cation). The terms *strong* and *weak* do not refer to the strength of interaction; the interaction of the protein with the column is determined by the stationary phase, the mobile phase composition, and the protein. Elution of the protein is accomplished by competition with small counter-ions for the adsorption sites on the ion exchanger. Most often, linear salt gradients are used. Other means for protein elution are a pH gradient or a combined salt/pH gradient.

Method development is started with the selection of the type of ion-exchange resin and a buffer with appropriate pH to assure retention of the proteins to be separated. IEC separations are usually started with a low buffer and salt concentration with sufficient buffer capacity.

An advantage of IEC over RP–HPLC is that chromatographic conditions are very mild, thereby preserving the protein structure and bioactivity. Protein purification by means of IEC generally gives high yields and is an excellent technique for large-scale purification of pharmaceutical proteins in biotechnology.

Monolithic Columns for Ion-Exchange Chromatography of Proteins. Ionizable groups can be incorporated into monolithic structures to make them suitable for IEC in different ways. Starting with functional monomers has the advantage that the monolithic structure can be prepared in a single step. However, the process requires optimization for each new set of functional monomers and cross-linkers to obtain the required properties. Alternatively, the ionizable groups are introduced after the monolithic structure has been created in an additional step through reaction with the epoxide groups [62].

The epoxide groups of a polyglycidyl methacrylate-co-ethylene dimethacrylate monolith readily react with diethylamine to create a weak anion exchange column. This monolithic column is characterized by fast mass transport kinetics leading to excellent resolution. The columns seem to be stable, and the specific capacity is 40-mg/g medium for ovalbumin [60, 62]. More recently, graft polymerization within the large pores of the monoliths has been used to attach chains of reactive polymer. The advantage of the grafting procedure is the high density of surface groups and consequently high binding capacities that can be achieved, which are favorable for the separation of biopolymers [20].

Application in Biotechnology. A high-performance anion-exchange chromatography method was used for the determination of erythropoietin (EPO) in pharmaceutical products [63]. EPO is a glycoprotein with a molecular mass of 30–34 kDa. Besides being the main factor regulating red blood cell production, it is known for extensive misuse as a performance enhancer in endurance sports. The content of EPO in pharmaceutical preparations is currently measured by complex *in vivo* potency assays.

It was shown that the chromatography method was suitable to separate recombinant EPO from amounts of human serum albumin commonly present as a stabilizer in various pharmaceutical preparations. In addition, it was possible to obtain different elution profiles for EPO products with variations in the glycoforms. Fluorescence detection was applied for quantification and showed linear signals over the range of 10–200-μg/mL EPO.

5.4.4.4 Multidimensional Liquid Chromatography of Proteins

In proteomics the sample complexity is often high and requires high-resolution analytical separation techniques. Two-dimensional polyacrylamide gel electrophoresis (2D PAGE) fulfills this requirement and has been established as a core technique for complex protein mixtures. However, the small dynamic range and the inability to resolve very small and large proteins and proteins with extreme pI values have been serious limitations.

Multidimensional liquid chromatography is becoming an attractive alternative to gel-based separations and has been reviewed recently [64]. The power of multidimensional techniques is the ability to increase the peak capacity significantly by adding selectivity to the system. From a theoretical point of view, the peak capacity of a truly orthogonal 2D-HPLC system, i.e., the nature of the separation principle of both dimensions are independent, is given by multiplication of the peak capacity of each dimension [65]. If the peak capacity of both columns is 50, the total peak capacity for the 2D-LC system will be 2500. However in practice, the resolving power of multidimensional LC techniques is impaired by limited orthogonality and the slowness of the second dimension separation.

A wide variety of separation mechanisms have been combined in 2D-HPLC to separate complex protein mixtures. Size exclusion [66], affinity [67], IEX [68–70], and chromatofocusing [71] have been used as a first-dimension separation in combination with RP as a second-dimension separation. IEX–RP, the most extensively studied combination, has been coupled to EIS time-of-flight MS for separation and identification of ribosomal proteins from yeast [72]. Approximately 70% of the potential ribosomal subunits isoforms could be identified with an average mass error of ~50 ppm. In the experimental setup, two parallel RP columns were alternately switched in series with a strong anion-exchange column to reduce the analysis time.

Distribution of proteins over multiple second-dimension fractions is unwanted and influenced by several aspects. Peak splitting is unavoidable and depends on the elution profile and fraction size. In general, peak splitting can be worse for proteins with a molecular heterogeneity as a result of different modification forms or structural conformations. Also carryover or peak "ghosting" results in dispersion into multiple fractions and is generally more profound for high abundant and hydrophobic proteins.

The first dimension separation can be run with linear and step gradients.

An extensive comparison between these two gradient modes in IEX–RP separation of cytosolic proteins from *Escherichia coli* revealed a more effective fractionation for a linear salt gradient [70]. Fewer major proteins were distributed to multiple second-dimension fractions.

In the second RP dimension, gradient elution is required to elute proteins with a wide range in hydrophobicity. As gradient elution is typically a slow process, several approaches have been proposed to speed up the second-dimension separation, such as the use of short columns, high flow rates, and a parallel column setup [68].

Polystyrene–divinylbenzene monolithic columns have shown excellent separation performances for intact proteins [24]. A capillary PS–DVB monolithic column with 500 µm ID has been used in combination with IEX columns for offline 2D-LC of complex proteins samples. The power of this technique is illustrated with chromatograms of the separation of soluble proteins from *Salmonella typhimurium* in Figure 5.4-9. The ion-exchange separation has been developed on a strong cation- and anion-exchange column, coupled in series. The combination of these two IEX columns allow retention of both acidic and basic proteins. The resolution of the multidimensional separation is high. The PS–DVB monolithic column and IEX column have peak capacities of, respectively, 80 and 60. Giving the fact that 30 fractions were taken in the first dimension, the total peak capacity can be estimated as 2400 (80 × 30 fractions). The offline column configuration makes the method flexible; i.e., both dimensions can be optimized independently with respect to mobile phases, column dimensions, and sample loading.

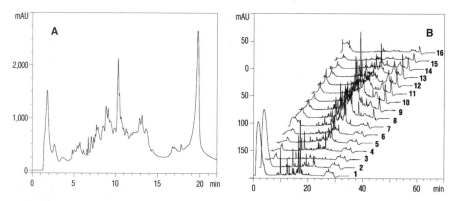

Figure 5.4-9. Offline 2D-LC of intact *S. typhimurium* proteins. (A) Separation of *S. typhimurium* protein extract on a ion-exchange ProPac SAX column. (B) Overview of the RP separations of Salmonella proteins on PS–DVB monolithic column 500 µm i.d. × 5 cm.

SEC. Another liquid chromatographic technique applied in biopharmaceutical characterization is SEC. In this technique, compounds are separated based on size and shape. In contrast to RP–HPLC, SEC has a low resolving power that has limited the application of the technique. The main application of SEC in pharmaceutical analysis is the determination of the native size of a protein and the quantification of protein aggregates. Typically protein-friendly conditions are employed in SEC that are expected not to affect the higher order structure of the protein.

5.4.5 SEPARATION OF AMINO ACIDS—AMINO ACID ANALYSIS

Amino acid analysis has become common in biotechnology, biomedical, and food analysis laboratories. In proteomics laboratories, this technique becomes now more important because quantitative proteomics requires knowledge about the exact sample amounts used in various kinds of experiments. Amino acid analysis is an exact, fast, and quantitative method for determining free and/or protein bound amino acids. Especially compared with Bradfort or Biuret, assays amino acid analysis is a more reliable technique with about an order-of-magnitude lower standard deviation. It is used to control protein yields from various protein preparations and to determine sample amounts prior to 2D-PAGE and 1D- and 2D-HPLC. Furthermore, it is necessary to measure exact protein amounts before stable isotope labeling with reagents like iTRAQ, ICPL, or ICAT [73–75]. If the ratio between chemical reagent and reactive side chain is chosen in the wrong way, a quantitative incorporation cannot be obtained and will lead to false results. Using relative quantification, the same protein amounts should be used. In the case of comparing sample A and B, the standard deviation will increase depending on the ratio of A to B and additional computation and data analysis are required to do normalization on the results.

Using the AQUA technology for absolute quantification of peptides from complex mixtures, it is mandatory to quantify the used stable isotope-labeled peptides before analysis, because peptides in solution undergo unspecific absorption to surfaces and will therefore alter their concentration [76]. However, even dried stable isotope-labeled peptides usually contain an amount of TFA salts from purification.

The method of choice to overcome these problems with the chemical quantification of proteins, peptides, or free amino acids are the classic amino acid analysis methods.

Amino acid analysis is an old method in protein chemistry (the first experiments were already described in 1820 by Braconnot). In 1972, Moore and Stein were awarded the Nobel Prize for establishing automated amino acid analysis. Roughly only two major methods for amino acid analysis exist: precolum derivatization and postcolumn derivatization. Today the sensitivity of amino acid analysis is at the femtomole level for fluorescence detection and at the picomole level for UV light detection methods. An overview about commonly used methods and reagents for amino acid analysis is shown in Table 5.4-2.

TABLE 5.4-2. Commonly Used Reagents for Amino Acid Analysis

Reagent	Method	Detection
Ninhydrin	Postcolumn derivatization	$\lambda = 570\,nm$ (prim. AA) $\lambda = 440\,nm$ (sec. AA)
Phenylisothiocyanate	Precolumn derivatization	$\lambda = 245\,nm$
Orthophthaldialdehyde	Precolumn derivatization	$\lambda = 338\,nm$
	Postcolumn derivatization	$\lambda ex = 230\,nm,\ 335\,nm$ $\lambda em = 455\,nm$
Fluorescamine	Postcolumn derivatization	$\lambda ex = 390\,nm$ $\lambda em = 475\,nm$
Fluorenylmethylchloroformate (FMOC)	Precolumn derivatization	$\lambda = 260\,nm$ $\lambda ex = 266\,nm$ $\lambda em = 305\,nm$
6-Aminoquinolyl-N-Hydroxy succinimidyl Carbamate (AQC)	Precolumn derivatization	$\lambda ex = 248\,nm$ $\lambda em = 395\,nm$
Dabsylchloride	Precolumn derivatization	$\lambda = 436\,nm$
Dansylchloride	Precolumn derivatization	$\lambda ex = 310\,nm$ $\lambda em = 540\,nm$

Abbreviations: λex = exitation wavelength; λem = emission wavelength.

REFERENCES

1. Churchwell M I, Twaddle N C, Meeker L R, et al. (2005). Improving LC-MS sensitivity through increases in chromatographic performance: Comparisons of UPLC-ES/MS/MS to HPLC-ES/MS/MS. *J. Chromatogr. B Analyt. Technol. Biomed. Life Sci.* 825(2):134–143.

2. Sherma J (2005). UPLC: Ultra-performance liquid chromatography. *J. AOAC Int.* 88(3):63A–67A.

3. Wren S A (2005). Peak capacity in gradient ultra performance liquid chromatography (UPLC). *J. Pharm. Biomed. Anal.* 38(2):337–343.

4. Exsigent Available: www.eksigent.com.

5. Proxeon Available: www.proxeon.com/proxeon/scienceEasyLC.do.

6. Tuytten R, Lemiere F, Witters E, Van Dongen W, Slegers H, Newton R P, Van Onckelen H, Esmans E L (2006). Stainless steel electrospray probe: A dead end for phosphorylated organic compounds? *J. Chromatography A.* 1104:209–221.

7. Pretorius V, Hopkins B J, Schieke, J D (1974). *J. Chromatogr.* 99:23.

8. Jorgenson J W, Lukacs K D (1981). *J. Chromatogr.* 218:209.

9. Eeltink S, Rozing G P, Kok W T (2003). Recent applications in capillary electrochromatography. *Electrophoresis.* 24(22–23):3935–3961.

10. Knox J H (1980). *J. Separation Sci.* 18:453.

11. Eeltink S, Rozing G P, Schoenmakers P J, et al. (2004). Study of the influence of the aspect ratio on efficiency, flow resistance and retention factors of packed capillary columns in pressure- and electrically-driven liquid chromatography. *J. Chromatogr. A.* 1044(1–2):311–316.

12. Meyer V R (2004). *Practical High-Performance Liquid Chromatography,* 4th ed. Wiley-VCH, Weinheim, Germany, p. 374.

13. Patel K D, Jerkovich A D, Link J C, et al. (2004). In-depth characterization of slurry packed capillary columns with 1.0-microm nonporous particles using reversed-phase isocratic ultrahigh-pressure liquid chromatography. *Anal. Chem.* 76(19):5777–5786.

14. Mellors J S, Jorgenson J W (2004). Use of 1.5-microm porous ethyl-bridged hybrid particles as a stationary-phase support for reversed-phase ultrahigh-pressure liquid chromatography. *Anal. Chem.* 76(18):5441–5450.

15. Jerkovich A D, Mellors J S, Thompson J W, et al. (2005). Linear velocity surge caused by mobile-phase compression as a source of band broadening in isocratic ultrahigh-pressure liquid chromatography. *Anal. Chem.* 77(19):6292–6299.

16. Novakova L, Matysova L, Solich P (2005). Advantages of application of UPLC in pharmaceutical analysis. *Talanta.* 68(3):908–918.

17. Plumb R, Castro-Perez J, Granger J, et al. (2004). Ultra-performance liquid chromatography coupled to quadrupole-orthogonal time-of-flight mass spectrometry. *Rapid Commun. Mass Spectrom.* 18(19):2331–2337.

18. Svec F, Frechet J M J (1992). Continuous Rods of Macroporous Polymer as High-Performance Liquid Chromatography Separation Media. *Anal. Chem.* 64: 820–822.

19. Ikegami T, Tanaka N (2004). Monolithic columns for high-efficiency HPLC separations. *Curr. Opin. Chem. Biol.* 8(5):527–533.

20. Svec F (2004). Preparation and HPLC applications of rigid macroporous organic polymer monoliths. *J. Sep. Sci.* 27(10–11):747–766.

21. Hjerten S, Liao J L, Zhang R (1989). *J. Chromatogr.* 473:273.

22. Rohr T, Hilder E F, Donovan J J, et al. (2003). Photografting and the control of surface chemistry in three-dimensional porous polymer monoliths. *Macromol.* 36:1677–1684.

23. Minakuchi H, Nakanishi K, Soga N, et al. (1996). Octadecylsilylated porous silica rods as separation media for reversed-phase liquid chromatography. *Anal. Chem.* 68: 3498–3501.

24. Walcher W, Oberacher H, Troiani S, et al. (2002). Monolithic capillary columns for liquid chromatography-electrospray ionization mass spectrometry in proteomic and genomic research. *J. Chromatogr. B Analyt. Technol. Biomed. Life Sci.* 782(1–2): 111–125.

25. Wienkoop S, Glinski M, Tanaka N, et al. (2004). Linking protein fractionation with multidimensional monolithic reversed-phase peptide chromatography/mass spectrometry enhances protein identification from complex mixtures even in the presence of abundant proteins. *Rapid Commun. Mass Spectrom.* 18(6):643–650.

26. Ishii D (1988). *Introduction to Mircoscale High-performance Liquid Chromatography.* VCH, Weinheim, Germany.

27. Scott R P W, Kucera P (1976). *J. Chromatogr.* 125:251.

28. Novotny M (1981). Microcolumns in liquid chromatography. *Anal. Chem.* 53(12): 1294A–1308A.

29. Ishii D, Asai K, Hibi K, et al. (1977). *J. Chromatogr.* 144:157.

30. Henzel W J, Bourell J H, Stults, J T (1990). Analysis of protein digests by capillary high-performance liquid chromatography and on-line fast atom bombardment mass spectrometry. *Anal. Biochem.* 187(2):228–233.

31. Moritz R L, Simpson R J (1992). Application of capillary reversed-phase high-performance liquid chromatography to high-sensitivity protein sequence analysis. *J. Chromatogr.* 599(1–2):119–130.

32. Griffin P R, Coffman J A, Hood L E, et al. (1991). Structural analysis of proteins by capillary HPLC electrospray tandem mass spectrometry. *Int. J. Mass Spectrom.* 111:131–149.

33. Rapp E, Tallarek U (2003). Liquid flow in capillary (electro)chromatography: Generation and control of micro- and nanoliter volumes. *J. Sep. Sci.* 26:453–470.

34. Chervet J P, Ursem M (1996). Instrumental requirements for nanoscale liquid chromatography. *Anal. Chem.* 68:1507–1512.

35. Meiring H D, van der Heeft E, ten Hove G J, et al. (2002). Nanoscale LC–MS(n): technical design and applications to peptide and protein analysis. *J. Sep. Sci.* 25: 557–568.

36. Licklider L J, Thoreen C C, Peng J, et al. (2002). Automation of nanoscale microcapillary liquid chromatography-tandem mass spectrometry with a vented column. *Anal. Chem.* 74(13):3076–3083.

37. Hearn M T W (1985). Ion-pair chromatography of amino acids, peptides, and proteins. In MTW Hearn (ed.), Ion-pair chromatography. *Theory and Biological and Pharmaceutical Applications.* Marcel Dekker, New York, p. 207.

38. Winkler G, Briza P, Kunz C (1986). Spectral properties of some ion-pairing reagents commonly used in reversed-phase high-performance liquid chromatography of proteins and peptides in acetonitrile gradient systems. *J. Chromatogr. A.* 361:191–198.

39. Mitulovic G, Smoluch M, Chervet J P, et al. (2003). An improved method for tracking and reducing the void volume in nano HPLC-MS with micro trapping columns. *Anal. Bioanal. Chem.* 376(7):946–951.

40. Gevaert K, Van Damme J, Goethals M, et al. (2002). Chromatographic isolation of methionine-containing peptides for gel-free proteome analysis: Identification of more than 800 Escherichia coli proteins. *Mol. Cell Proteomics.* 1(11):896–903.

41. Burke T W, Mant C T, Black J A, et al. (1989). Strong cation-exchange high-performance liquid chromatography of peptides. Effect of non-specific hydrophobic interactions and linearization of peptide retention behaviour. *J. Chromatogr.* 476:377–389.

42. Wolters D A, Washburn M P, Yates J R, III (2001). An automated multidimensional protein identification technology for shotgun proteomics. *Anal. Chem.* 73(23): 5683–5690.

43. Gevaert K, Goethals M, Martens L, et al. (2003). Exploring proteomes and analyzing protein processing by mass spectrometric identification of sorted N-terminal peptides. *Nat. Biotechnol.* 21(5):566–569.

44. Martens L, Van Damme P, Van Damme J, et al. (2005). The human platelet proteome mapped by peptide-centric proteomics: a functional protein profile. *Proteomics.* 5(12):3193–5204.

45. Kastner M (2000). *Protein liquid chromatography*, Vol. 61. Elsevier, Berlin, Germany, p. 976.

46. Aguilar M-I (2004). *Hplc of Peptides and Proteins: Methods and Protocols*, Vol. 251. Humana Press, Totowa, NJ.

47. Yang X, Ma L, Carr P W (2005). High temperature fast chromatography of proteins using a silica-based stationary phase with greatly enhanced low pH stability. *J. Chromatogr. A.* 24(1–2):213–220.

48. Afeyan N B, Gordon N F, Mazsaroff I, et al. (1990). Flow-through particles for the high-performance liquid chromatographic separation of biomolecules: Perfusion chromatography. *J. Chromatogr.* 519(1):1–29.

49. Janzen R, Unger K K, Giesche H, et al. (1987). Evaluation of advanced silica packings for the separation of biopolymers by high-performance liquid chromatography, IV. Mobile phase and surface-mediated effects on recovery of native proteins in gradient

elution on non-porous, monodisperse 1.5-microns reversed-phase silicas. *J. Chromatogr.* 397:81–89.

50. Maa Y F, Horvath C (1988). Rapid analysis of proteins and peptides by reversed-phase chromatography with polymeric micropellicular sorbents. *J. Chromatogr.* 445(1):71–86.

51. Kirkland J J, Truszkowski F A, Dilks C H Jr, et al. (2000). Superficially porous silica microspheres for fast high-performance liquid chromatography of macromolecules. *J. Chromatogr. A.* 890(1):3–13.

52. Jungbauer A, Hahn R (2004). Monoliths for fast bioseparation and bioconversion and their applications in biotechnology. *J. Sep. Sci.* 27(10–11):767–778.

53. Xie S, Allington R W, Svec F, et al. (1999). Rapid reversed-phase separation of proteins and peptides using optimized "moulded" monolithic poly(styrene-co-divinylbenzene) columns. *J. Chromatogr. A.* 865(1–2):169–174.

54. Lee D, Svec F, Frechet J M (2004). Photopolymerized monolithic capillary columns for rapid micro high-performance liquid chromatographic separation of proteins. *J. Chromatogr. A.* 1051(1–2):53–60.

55. Garcia M C (2005). The effect of the mobile phase additives on sensitivity in the analysis of peptides and proteins by high-performance liquid chromatography-electrospray mass spectrometry. *J. Chromatogr. B. Analyt. Technol. Biomed. Life Sci.* 825(2):111–123.

56. Huber C G, Premstaller A (1999). Evaluation of volatile eluents and electrolytes for high-performance liquid chromatography-electrospray ionization mass spectrometry and capillary electrophoresis-electrospray ionization mass spectrometry of proteins. I. Liquid chromatography. *J. Chromatogr. A.* 849(1):161–173.

57. Luykx D M, Goerdayal S S, Dingemanse P J, et al. (2005). HPLC and tandem detection to monitor conformational properties of biopharmaceuticals. *J. Chromatogr. B. Analyt. Technol. Biomed. Life Sci.* 821(1):45–52.

58. Zhang W, Czupryn J M, Boyle P T, Jr, et al. (2002). Characterization of asparagine deamidation and aspartate isomerization in recombinant human interleukin-11. *Pharm. Res.* 19(8):1223–1231.

59. Happersberger H P, Cowgill C, Glocker M O (2002). Structural characterization of monomeric folding intermediates of recombinant human macrophage-colony stimulating factor b (rhM-CSFb) by chemical trapping, chromatographic separation and mass spectrometric peptide mapping. *J. Chromatogr. B. Analyt. Technol. Biomed. Life. Sci.* 782(1–2):393–404.

60. Dillon T M, Bondarenko P V, Speed Ricci M (2004). Development of an analytical reversed-phase high-performance liquid chromatography-electrospray ionization mass spectrometry method for characterization of recombinant antibodies. *J. Chromatogr. A.* 1053(1–2):299–305.

61. Zhai H, Han X, Breuker K, et al. (2005). Consecutive ion activation for top down mass spectrometry: improved protein sequencing by nozzle-skimmer dissociation. *Anal. Chem.* 77(18):5777–5784.

62. Svec F, Frechet J M (1995). Modified poly(glycidyl methacrylate-co-ethylene dimethacrylate) continuous rod columns for preparative-scale ion-exchange chromatography of proteins. *J. Chromatogr. A.* 702(1–2):89–95.

63. Luykx D M, Dingemanse P J, Goerdayal S S, et al. (2005). High-performance anion-exchange chromatography combined with intrinsic fluorescence detection to determine erythropoietin in pharmaceutical products. *J. Chromatogr. A.* 1078(1–2):113–119.

64. Wang H, Hanash S (2003). Multi-dimensional liquid phase based separations in proteomics. *J. Chromatogr. B. Analyt. Technol. Biomed. Life Sci.* 787(1):11–18.

65. Giddings J C (1990). *The use of multiple dimensions in analytical separations.* In H J Cortes (ed.), *Multidimensional Chromatography: Techniques and Applications,* Marcel Dekker, New York.

66. Opiteck G J, Ramirez S M, Jorgenson J W, et al. (1998). Comprehensive two-dimensional high-performance liquid chromatography for the isolation of overexpressed proteins and proteome mapping. *Anal. Biochem.* 258(2):349–361.

67. le Coutre J, Whitelegge J P, Gross A, et al. (2000). Proteomics on full-length membrane proteins using mass spectrometry. *Biochem.* 39(15):4237–4342.

68. Wagner K, Racaityte K, Unger K K, et al. (2000). Protein mapping by two-dimensional high performance liquid chromatography. *J. Chromatogr. A.* 893(2):293–305.

69. Opiteck G J, Lewis K C, Jorgenson J W, et al. (1997). Comprehensive on-line LC/LC/MS of proteins. *Anal. Chem.* 69(8):1518–1524.

70. Millea K M, Kass I J, Cohen S A, et al. (2005). Evaluation of multidimensional (ion-exchange/reversed-phase) protein separations using linear and step gradients in the first dimension. *J. Chromatogr. A.* 1079(1–2):287–298.

71. Wall D B, Kachman M T, Gong S, et al. (2000). Isoelectric focusing nonporous RP HPLC: a two-dimensional liquid-phase separation method for mapping of cellular proteins with identification using MALDI-TOF mass spectrometry. *Anal. Chem.* 72(6):1099–1111.

72. Liu H, Berger S J, Chakraborty A B, et al. (2002). Multidimensional chromatography coupled to electrospray ionization time-of-flight mass spectrometry as an alternative to two-dimensional gels for the identification and analysis of complex mixtures of intact proteins. *J. Chromatogr. B Analyt. Technol. Biomed. Life Sci.* 782(1–2):267–289.

73. Ross P L, Huang Y N, Marchese J N, et al. (2004). Multiplexed protein quantitation in saccharomyces cerevisiae using amine-reactive isobaric tagging reagents. *Mol. Cell Proteomics.* 3(12):1154–1169.

74. Schmidt A, Kellermann J, Lottspeich F (2004). A novel strategy for quantitative pro-teomics using isotope-coded protein labels. *Proteomics.* 5(1):4–15.

75. Gygi S P, Rist B, Gerber S A, et al. (1999). Quantitative analysis of complex protein mixtures using isotope-coded affinity tags. *Nat. Biotechnol.* 17(10):994–999.

76. Gerber S A, Rush J, Stemman O, et al. (2003). Absolute quantification of proteins and phosphoproteins from cell lysates by tandem MS. *Proc. Natl. Acad. Sci. USA.* 100(12): 6940–6945.

5.5

BIOANALYTICAL METHOD VALIDATION FOR MACROMOLECULES

GERHARD K. E. SCRIBA

Friedrich-Schiller-University Jena, School of Pharmacy, Jena, Germany

Chapter Contents

5.5.1   Introduction   611
5.5.2   Method Validation   613
    5.5.2.1   Ligand Binding Assays   613
    5.5.2.2   Validation of Cell-based Bioassays   627
5.5.3   Conclusions and Future Considerations   628
5.5.4   Glossary of Terms   629
    References   630

5.5.1   INTRODUCTION

Before implementation of analytical methods for routine use, careful validation is required to demonstrate that the method is suitable for the intended purpose. Analytical methods employed for the quantitative determination of drug substances and their metabolites in biological media play a significant role in the evaluation and interpretation of pharmacokinetic data. To define the requirements and procedures for the validation of bioanalytical methods, a conference was held in 1990 in Crystal City in the Washington, DC, area, which was co-sponsored by the U.S. Food and Drug Administration (FDA), The Canadian Health Protection Branch, the American Association of Pharmaceutical Scientists (AAPS), and the Association of Official Analytical Chemists (AOAC), bringing together scientists from

Handbook of Pharmaceutical Biotechnology, Edited by Shayne Cox Gad.
Copyright © 2007 John Wiley & Sons, Inc.

regulatory authorities, industry, and academia [1]. Upon evaluation of the results of the first meeting, another conference was held 10 years later [2] that ultimately led to the publication of the FDA guideline "Guidance for Industry: Bioanalytical Method Validation" [3] complementing the general guidelines on method validation such as guidelines Q2A and Q2B by the International Conference on Harmonization (ICH) [4, 5] or pharmacopeial regulations [6]. From the beginning, it was realized that each analytical technique has its own characteristics, which will vary from analyte to analyte. Despite the fact that some similarities exist, a general difference between chemical methods, such as chromatography, and biological assays, such as immunoassays and microbiological assays, was acknowledged [1, 2].

Small, "conventional" drug molecules are preferentially analyzed by chromatographic techniques, specifically by liquid chromatography-mass spectrometry (LC-MS) and liquid chromatography-tandem mass spectrometry (LC-MS/MS), and, to date, most emphasis has been on the validation of bioanalytical methods for such molecules, which is also reflected in the FDA guideline [3]. However, because of the progress in recombinant DNA technology, the number of protein pharmaceuticals has increased dramatically. From 1996 through December 2005, the FDA has approved 253 so-called "biologics" for 384 indications [7]. Most of these products are proteins. As a result of their high potency and subsequent low applied doses resulting in extremely low concentrations in biological media, chromatographic techniques are not sensitive enough for bioanalysis. Thus, immunoassays are primarily used for the bioanalysis of protein drugs.

This divergence in analytical techniques for small molecules and macromolecules has triggered workshops and conferences on bioanalytical method validation of macromolecules focusing on issues such as quantitative immunoassays for therapeutic proteins, biomarkers, and drug neutralizing antibodies as well as bioassays. The results are documented in several publications of meeting reports [8–15], and these publications currently serve as quasi-guidance as no complete official document by the regulatory authorities of the United States or Europe on bioanalysis method validation of macromolecules exist to date. The FDA guidance on bioanalysis [3] acknowledges differences between the assay formats but does not cover all necessary topics for bioassays. A guideline by the European Agency for the Evaluation of Medical Products (EMEA) on pharmacokinetics of pharmaceutical proteins [16] also states specifics of immunoassays and lists points that should be addressed during method validation, but it provides no general guideline. Chromatographic assays are commonly applied to the analysis of protein drugs, but almost exclusively during product characterization release. Method validation of chromatographic techniques is basically identical for small molecules and macromolecules and is addressed in the ICH guidelines [4, 5], FDA guidelines [17], as well as in publications [18] and books [19, 20]. This chapter focuses on validation issues for macromolecule bioanalysis summarizing the current opinion according to the meeting reports [8–15] and further publications on method validation for macromolecules [21–25]. If possible, all terms related to assay validation are used in the sense of the ICH guideline Q2A [4]. A glossary of the most important terms is provided at the end of the chapter. A general guideline for method development of all different kinds of biological assays including validation aspects can be found on the Internet [26].

5.5.2 METHOD VALIDATION

Relevant macromolecular analytes in biological media can be classified into three categories: (1) pharmaceutical proteins administered as therapeutic agents; (2) biomarkers (i.e., endogenous substances that reflect physiological or pathophysiological processes or pharmacological responses to a therapeutic intervention); and (3) drug neutralizing antibodies that are generated as the response of the human organism to the application of a therapeutic protein. The primary assay formats for these molecules are ligand-binding assays (i.e., immunoassays and receptor-binding assays). Method validation of immunoassays will be the focus of the following discussion whereas cell-based assays will be only briefly addressed. Immunoassays can be roughly divided into competitive assays and sandwich assays, details are discussed in Chapter 5.3 of this volume. Validation will be discussed for immunoassays for the determination of pharmaceutical proteins in detail and, in subsequent sections, differences with regard to biomarkers and anti-drug antibodies will be mentioned. Not specifically addressed here, but evident in a good laboratory practice (GLP) environment, is the fact that proper documentation and standard operation procedures (SOPs) have to be written.

5.5.2.1 Ligand-Binding Assays

The term "ligand-binding assay" refers to methods that depend on the specific binding of an analyte to a biomolecule. Generally acknowledged inherent differences exist between bioanalytical chromatographic techniques and ligand-binding assays. The most relevant differences are summarized in Table 5.5-1. Whereas chromatography is based on physico-chemical principles, ligand-binding assays are based on a biological response because of the interaction of a ligand with an antibody or a receptor. Consequently, the reagents are derived from living organisms with the attendant variation typical for such reagents (i.e., poor batch-to-batch reproducibility). High-purity, well-characterized reference standards are most often not commercially available. Although specificity in chromatography is

TABLE 5.5-1. Comparison of the Characteristics of Chromatographic Assays and Immunoassays for Bioanalysis

Parameter	Chromatographic Assay	Immunoassay
Principle of measurement	physico-chemical properties of the analytes	antigen-antibody reaction
Detection	direct	indirect
Reagents, analytical standard	well characterized, high purity, widely available	not always completely characterized, unique, not widely available
Analytes	small molecules	small molecules and macromolecules
Sample pretreatment	typically pretreatment	usually no pretreatment
Calibration model	linear	nonlinear
Intermediate precision	high (<10%)	moderate (<20%)
Range	broad (several orders of magnitude)	limited

obtained by isolation of the typically small molecules from the matrix combined with analytical separation from other sample components and probably even detection by mass spectrometry, the isolation of macromolecules from biological media is impractical in most cases because of the low concentrations and the structural or physico-chemical similarity between the analytes and endogenous compounds. Thus, detection of the macromolecule analytes occurs in a complex physiological milieu and, therefore, highly depends on the specificity of the reagents and detection systems. Finally, whereas chromatographic assays display linear relationships between analyte concentration and detector response over a broad range covering 2–3 orders of magnitude, calibration curves of ligand-binding assays are typically nonlinear covering a rather limited range requiring dilution of very concentrated samples. On the time scale, the development of an immunoassay will be longer compared with high-performance liquid chromatography (HPLC) methods because of the need of the generation of antibodies. Once these antibodies have been obtained, the time frame for method validation between the assays is comparable. Sample throughput of immunoassays is excellent.

Standard Immunoassays. Validation is a continuing process through the whole life cycle of an assay. After selection of the assay format, preliminary data are obtained during method development, which are confirmed during prestudy validation and consequently applied during in-study validation (Table 5.5-2). This scenario is considered a "full validation." Partial validation is conducted when changes of a fully validated method occur that are considered minor, such as changes in the anticoagulant or changes in the used reagents. Partial validation can range from a single intra-assay accuracy and precision determination to nearly full validation. Method transfer from the developing laboratory to another laboratory or a production site requires at least partial validation. Cross-validation is conducted when two or more validated bioanalytical methods are used within the same study. Test samples (spiked samples or incurred test samples) should be used and the data should be evaluated using appropriate predefined acceptance criteria. Some, but not all, validation issues have been addressed in the FDA guideline on bioanalytical method validation [3].

Assay Format. Assay format selection is the first step in method development. Assay formats include competition, sandwich, direct and indirect binding, inhibition, solid-phase, and solution phase assays. Reagents, first of all, include the antibodies; but diluents and additives such as detergents also have to be considered. For solid-phase assays, selection of the solid support as well as the chemistry used for the immobilization of the antibody may be critical. Consideration should also be given to the selection of the assay detection system to provide good signal-to-noise ratio. Detection may be improved by switching from colorimetric detection to fluorimetric or chemiliminescesence or, seldom, to radiometric detection. All these variables are evaluated during method development and confirmed during prestudy validation. In addition to the individual components, the manner in which an array is set up and run should already be considered during method development. The assay configuration (i.e., the number and placement of standards) of quality control (QC) samples, and study samples on a plate in an attempt to mimic the anticipated size of the run batches during routine use should be established as

TABLE 5.5-2. Validation Parameters

Parameter	Method Development	Prestudy Validation	In-Study Validation
Assay format	establish	confirm	apply
Reagent selection, stability	identify and establish	confirm	monitor, lot change requires confirmation of performance
Specificity and selectivity	establish	confirm	may not apply, confirm
Matrix selection	establish	confirm	monitor, changes in matrix require demonstration of comparability
Minimum required dilution	establish	confirm	monitor
Standard curve, calibrators	select model, establish	confirm	monitor
Assay range	evaluate	establish	apply
Dilutional linearity	evaluate	establish	monitor, establish for dilutions not covered in prestudy validation
Parallelism	not applicable	investigate when possible	establish with incurred samples
Precision and accuracy	evaluate imprecision and bias	establish imprecision and bias	total error (4–6–30 rule)
Sample stability	initiate	establish	continuing assessment
Robustness and ruggedness	evaluate	establish	monitor
Run acceptance criteria	not applicable	based on standard curve acceptance criteria	standard curve and QC acceptance criteria, apply 4–6–30 rule

early as possible and confirmed during prestudy validation. It is recommended that at least 5% of the total samples of a given batch consist of QC samples.

Reagent Selection and Stability. Probably the most critical components of the assay are the antibodies used. These antibodies are produced by living organisms with the inherent variability of such reagents. They must be acquired in adequate amounts and sufficiently characterized. As antibodies are prone to lot-to-lot variations, ideally, different lots are evaluated during method development and prestudy validation. When an antibody (reagent) lot has to be changed during routine application, in-study validation must demonstrate comparable performance of the lots.

Assay performance and sensitivity will deteriorate upon degradation of the antibodies and other reagents. Therefore, it is important to investigate storage conditions to ensure the integrity of key reagents for the estimated period of time that they will be used. Stability testing is not addressed in the regulatory guidance documents [3–5], but it is an important aspect of method validation. If manufacturers provide expiration dates of reagents, they may be used instead of in-house stability determination. Reagent and antibody stability does not only concern storage at low temperatures (refrigerated or frozen), but standing times in laboratory equipment have to be considered. Thus, storage and handling conditions usually include bench-top stability, short-term and long-term storage, and stability to multiple freeze–thaw cycles. As it may also be desirable to store batches of assay microtiter plates for later use, the performance of a stored plate batch should be evaluated testing positive and negative samples. Stability is typically assessed during method development and confirmed during prestudy validation. It should be monitored during routine use.

Reference Material. In contrast to small molecules where reference standards are well characterized and certified standards are often commercially available from sources such as the U.S. Pharmacopeia, the European Pharmacopoeia, or the World Health Organization (WHO), proteins are often not as rigorously character- ized and their purity may vary from supplier to supplier. Variation in posttransla- tional modifications such as glycosylation and deamidation may be present. Thus, the proteins can vary in their potency and immunoreactivity. As the reference compounds are used as standard calibrators, validation sample and QC sample variation of the reference will have a profound impact on the assay performance. Therefore, it is important to clearly document the source of the material and to characterize the proteins as thoroughly as possible. Comparability between lots or sources should be evaluated if possible. If the analyte is not a new drug entity, the innovator company is typically the most reliable source of authentic material. As stated for the reagents, stability of the reference compounds is an issue that has to be ensured.

Specificity and Selectivity. With regard to antibodies, specificity is the ability to specifically bind to the protein of interest in the presence of related endogenous and exogenous components (i.e., without cross reactivity). Selectivity, a related concept, describes the ability to determine an analyte in the presence of other constituents in a sample. Chromatographic methods are selective because they separate and detect analytes in a complex sample, typically after sample prepara- tion steps for compound isolation. In contrast, ligand-binding assays measure ana- lytes in biological matrices without prior isolation. Thus, high specificity may be desirable. With regard to interferences with other sample and matrix components and nonspecific binding of an antibody to such components, the terms "specific nonspecificity" and "nonspecific nonspecificity" are also used [9] sometimes as synonyms for the ICH terms. Specific nonspecificity is the interference caused by compounds with structural similarity to the analyte of interest. Such compounds may be metabolites, degradation products of the analyte, isoforms and variants with posttranslational modifications, as well as endogenous substances. Causes for nonspecific nonspecificity may be unrelated matrix components, (patho) physiological factors interfering with ligand binding such as serum proteins and lipids, hemolysis, or anti-IgG antibodies. In addition, nonspecific adsorption to the microtiter plate may occur.

As macromolecular analytes often have structural elements in common with endogenous compounds, specificity may be difficult to achieve. Variant forms of proteins may not be available at the time of method development. In such cases, retrospective assessment may be acquired for the assay as more data become available over time. Rather than investigating specificity during method development, it is recommended to focus on reliable quantification of the analyte against a background of interfering matrix components [11].

Assay selectivity (nonspecific nonspecificity) is evaluated during method development by assaying spiked samples. Multiple lots of matrix should be evaluated, at least six [9] to 10 lots [11] are recommended. As selectivity problems occur, mostly at low concentrations, spiking should be performed at or near the lower limit of quantification (LLOQ). However, it may also be advisable to investigate higher concentrations. In case of interferences, it may be necessary to adjust the LLOQ before validation.

During prestudy validation, specificity and selectivity are confirmed. Selectivity may be expressed as acceptable recovery, applying the same principles as for the assessment of accuracy (see below). Acceptance criteria should be predefined. Acceptance criteria for selectivity and specificity typically do not exist for in-study validation. If potential interference may become a problem caused by the matrix from persons with the disease, specificity and selectivity should be confirmed by repeating suitable experiments once those disease-state matrices become available.

Matrix Selection. Matrix selection occurs in the development stage. Typically, biological fluids such as urine, plasma, serum, or cerebrospinal fluid as well as tissue samples may be collected. Additives such as anticoagulants, protease inhibitors, antioxidants, and so on may be present. It is necessary to document collection, processing, and storage conditions. The assay format may influence the choice of the matrix. For example, automated pipetting systems may be clogged by fibrin clots from plasma. In the absence of an endogenous signal, simple spike recovery experiments will determine the suitability of the matrix. In contrast, when the therapeutic protein is a recombinant version of an endogenous constituent, quantifiable amounts of the endogenous protein will be present in the matrix. Various strategies may be employed to limit or eliminate such interferences [9]. If the endogenous concentration is very low, and subsequently the percentage of the area under the curve obtained in pharmacokinetic studies caused by the endogenous compound is below 5%, the endogenous protein will introduce only a small bias and can be neglected. Alternatively, the endogenous level is determined from blank samples (no spike added) of a number of subjects and subsequently subtracted from the spiked samples. As already stated, 6–10 lots of the matrix should be evaluated. Further strategies include "stripping" of the matrix from the endogenous analyte either by nonspecific adsorption on charcoal or specific removal by affinity chromatography, the use of surrogate matrix from other species, or the use of protein-containing buffers. However, one has to keep in mind that the QC samples have to be prepared in the "original," unprocessed matrix. The presence or absence of matrix effects should be demonstrated by analyzing spiked samples at least at one concentration level. Differences between matrices obtained from healthy and diseased individuals may have to be considered.

In prestudy validation, the matrix selected during method development will be used to construct the calibration curves and validation samples. Once the effect of the matrix has been determined in method development, no further validation in later stages is required unless changes in the matrix occur.

Minimum Required Dilution. The minimum required dilution for an assay is the minimum magnitude of dilution of a sample with a defined diluent to optimize accuracy and precision in an assay. In many cases, dilution may not be necessary when analyzing plasma, serum, or other body fluids. Calibrator and validation samples are directly prepared in the matrix. For example, dilution may be required for samples with a background signal that is not caused by the endogenous version of the analyte. In the interest of a high signal-to-noise ratio (i.e., good accuracy and precision), sample dilution should be kept at a minimum. Once established during method development, this parameter should not be changed in later stages.

Assay Range. The assay range is defined as the validated concentration range between the LLOQ and the upper limit of quantitation (ULOQ), for which the results have an (predefined) acceptable level of precision and accuracy. The LLOQ and ULOQ are determined by the lowest and highest validation samples that show precision and accuracy of at least 25% expressed as relative error. The range is estimated during method development and validated by samples in the area of the anticipated LLOQ and ULOQ during prestudy validation. During routine use, samples that are above the ULOQ must be diluted (see "minimum required dilution" above). Samples that are below the LLOQ must be reported as "below LLOQ." The LLOQ during a batch run must be revised to higher concentrations if the QC samples at the lowest concentration fail to meet the 25% precision criteria.

Calibrators and Standard Curve. Standard calibrators are prepared by spiking known amounts of the (ideally, well-characterized) reference material into the matrix to obtain a standard curve from which the sample concentrations will be calculated. One of the major differences between chromatographic methods and immunoassays is that immunoassays display nonlinear relationships between the concentration and the measured response, which makes selection of the mathematical calibration curve model more complicated. Selection of the optimum calibration function is important to define the correct quantification range, to maximize accuracy and precision, and to achieve the quality control criteria. The mathematical model most widely used for immunoassay calibration curves is the four-parameter logistic function. If the curve is asymmetric, inclusion of a fifth parameter (i.e., using a five-parameter logistic function) may improve the fit to the data [25]. Algorithms that linearize the function such as logit-log may be used if goodness of fit is demonstrated, but such models are not recommended. Proper weighting of the data points in a calibration curve is also important to minimize bias and imprecision of the interpolated values near the LLOQ and ULOQ. Replicates with smaller variances are given greater weight compared with those with larger variances. The latter are normally found at the asymptotic ends of the curve [25].

As the standard concentrations should not be changed once validation has started, detailed investigation during method development is recommended using a greater number of data points and replicates compared with later validation stages [9, 11]. In the method development phase, the calibration curve should be constructed from a minimum of 10 non-zero standard points in duplicate spanning the anticipated concentration range about equally spaced on a logarithmic scale. Curve fit is achieved by a four-to-five-parameter logistic function. Whether to weight or not weight the responses should be supported by an evaluation of the relationship between the standard deviations of replicate values and the mean values at different concentration levels [25]. A minimum of three independent runs should be used to establish the calibration model with duplicate curves included in each run to estimated intrabatch standard curve repeatability. The absolute relative error for back-calculated standard point should be ≤20% ("intracurve"). Acceptability of a model is verified by evaluating the relative error (relative bias) between back-calculated and nominal concentrations of the calibration samples. A model is considered acceptable if the relative error for all back-calculated standard points does not exceed 10% ("intercurve") and the precision (coefficient of variation) for each calibrator is ≤15% [11]. Lack of fit may be caused by the use of an inappropriate mathematical function such as applying a four-parameter logistic function to asymmetric curves or inappropriate weighting of calibrators.

During prestudy validation, a minimum of six non-zero standards in duplicate are spaced evenly on the logarithmic concentration scale to fit the four-to-five-parameter logistic function. At this point, additional calibrators outside the range of quantification (so-called "anchor points") may be included to facilitate curve fitting. This approach is in agreement with the FDA guidance on bioanalytical method validation [3]. The regression model should be confirmed using at least six independent runs. Typically, the same runs are used to determine accuracy and precision (see below). For acceptable curves, the back-calculated values for at least 75% of the standard points not including the anchor points should be within 20% of the theoretical value (except at the LLOQ, where 25% are acceptable), and upon completion of the validation, the cumulative relative error and coefficient of variation for each calibrator should be ≤15% and ≤20% at the LLOQ.

The standard curve should be monitored during in-study validation with at least one set of calibrators per patch run. As for prestudy validation, the curve should be constructed from six concentrations in duplicate. Anchor points may be used. The final number of points used for curve fit must be either 75% of the total number or a minimum of six calibrator samples not including the anchor points. The relative error of the back-calculated samples should be ≤20% (≤25% at the LLOQ). If either the high or low calibrator standards have to be deleted, the range for this particular run must be limited to the next standard point. Samples out of range must be repeated.

Precision and Accuracy. Precision and accuracy are assay performance characteristics that describe the random (statistical) errors and systematic errors (bias) associated with repeated measurements of the same sample under specified conditions [3–5]. Precision is typically estimated by the percent coefficient of variation (% CV, also referred to as relative standard deviation or RSD) but may certainly also be reported as standard deviations. Method accuracy is expressed as the percent relative error (% RE) and is determined by the percent deviation of the weighted samples mean from samples with nominal reference values. A collection of validation sample statistics can be found in References 9, 11, and 25.

Spiked samples are analyzed over multiple runs with replicate determinations during method development and prestudy validation. QC samples are used during in-study validation to monitor the performance of the assay. Limits for minimum acceptable precision and accuracy should be established before or during method development and used throughout the life cycle of the assay.

It is recommended to determine accuracy and precision during method development with at least three batch runs using a minimum of eight sample concentrations analyzed in duplicate. The concentrations should span the whole validation concentration range of the assay. Recommended target limits for intrabatch and interbatch precision (% CV) are ≤20%, bias (% RE) should not exceed ±20%. At the LLOQ, a maximum of 25% for % CV and % RE is acceptable. These values are more lenient compared with the typical target values of chromatographic assays (% CV; 15% and 20% at LLOQ [3–5]) because immunoassays are inherently less precise than chromatographic assays [27]. In prestudy validation, at least six batch runs with a minimum of five different concentrations, one at the LLOQ, one at the ULOQ, and three concentrations in the lower, medium, and high range, analyzed in duplicate should be used for accuracy and precision determination. For each validation sample, the repeated measurements from all runs have to be statistically analyzed together. The target values for intrabatch and interbatch precision as well as accuracy (bias) are the same as the values applied in method development (20%, except for LLOQ, where 25% is acceptable). The total error of the method (i.e., the sum of % RE and % CV) should not exceed 30% (40% at the LLOQ). Further, for each in-study run, precision and accuracy are monitored by QC samples. Recommended run acceptance criteria are based on the total error, specifically on the deviation of the measured values from the nominal values, and not on statistical calculations such as the means or standard deviations. As QC samples, one set of at least three concentrations, one each in the low, medium, and high concentration range, are analyzed in duplicate in each batch run. As for small molecules, at least two thirds of the measured values of the QC samples must fall within a certain percentage of the corresponding nominal values and at least 50% of the results for each concentration of the QC samples must be within the specified limit. Thus, not all samples of a specific concentration are allowed outside the specifics. For chromatographic assays for small molecules, 15% has been adopted as limit (i.e., the "4–6–15 rule") [2]. At the LLOQ, 20% are acceptable. As a result of the lower precision of immunoassays, the error margin has been widened, a 6–4–25 rule is recommended by Findley et al. [9], whereas a 4–6–30 rule has been proposed at the macromolecule bioanalysis workshop [8]. The 30% margin is identical with the maximal acceptable total error of the method (sum of % RE and % CV) of the prestudy validation. Other statistical methods as acceptance criteria have been published [9, 28, 29] and can also be applied.

Dilutional Linearity. As the range of an immunoassay is usually limited, it may be necessary to dilute concentrated samples. Therefore, it has to be demonstrated that the analyte can still be reliably quantified upon dilution of high concentration samples so that they fall within the validated range of the assay. Moreover, a so-called prozone or "hook effect" can be identified. A hook effect is present when high concentration samples above the ULOQ display a lower response than ULOQ samples because of signal suppression caused by the high concentration of the analyte [30]. Dilutional linearity should not be confused with parallelism (see below).

Dilutional linearity is evaluated during method development, typically with spiking 100-fold to 1000-fold greater concentrations into the sample matrix followed by dilution with the assay matrix. Dilutions should be made so that several dilutions fall within the standard curve in the lower, middle, and upper parts of the curve. Dilution samples above the ULOQ can be included to evaluate a hook effect. The dilutions are further confirmed during prestudy validation. The back-calculated concentration for each diluted sample should be within 20% of the nominal or expected value. The precision of the cumulative back-calculated concentration should be ≤20%. If a sample has to be diluted during routine use of the method to a higher extend than assessed during prestudy validation, dilutional linearity should either be repeated or a dilutional QC sample should be included in the assay.

Parallelism. Parallelism is a characteristic that is typically assessed during in-study validation. It is conceptually similar to dilutional linearity with the difference that it is determined by dilutions of actual study samples (incurred samples). In some cases, samples from a preclinical pilot study may be available during prestudy validation so that assessment of parallelism may be performed at that time. Moreover, when an assay is validated with the aim to replace another assay, incurred samples from a previous study may be available for evaluation of this performance characteristic.

Parallelism is assessed using C_{max} samples from an actual study. Commonly, samples from several individuals are pooled to create a suitable validation sample. Using pooled samples eliminates the need for generation of multiple values for individual study samples. It is recommended that the relative standard deviation between samples of a dilutional series should be ≤30%.

Sample Stability. Experiments demonstrating sufficient stability of the analyte in the sample matrix must be included in prestudy validation. Such experiments should mimic as closely as possible the conditions under which study samples will be collected, stored, and processed. Assessment should include bench-top stability, refrigerator stability, whole-blood stability, freeze–thaw stability, and long-term freezer stability. Stability samples can be prepared by spiking the analyte reference at high and low concentrations into the sample matrix.

Bench-top stability refers to the conditions under which the samples are handled during processing of the samples at the analytical site and should be examined at room temperature for at least 2 h and at 2–8°C (refrigerator temperature) for a minimum of 24 h [31]. The stability in whole blood can be determined by spiking the analyte into freshly collected whole blood followed by incubation for up to 2 h and processing to obtain plasma or serum samples at certain intervals. The samples are subsequently processed and analyzed [31]. Typically, freeze–thaw stability evaluation includes three freeze–thaw cycles with at least 12 h between the thaws. The rate of freezing and thawing should mimic the manner in which samples will be handled as they are thawed before the assay. Long-term stability should demonstrate that the samples are stable throughout the lifetime of the study. The necessity to conduct studies on samples stored at −20°C and −70°C to −80°C may depend on the duration of the study.

Assessment of the stability is typically performed during prestudy validation and continued during in-study validation. If changes in sample handling or storage occur, additional stability evaluations must be carried out to reflect the altered conditions. The acceptance criteria for the stability evaluations, with the exception of the whole-blood stability, will be the same acceptance criteria applied for accuracy and precision of QC samples. If the measured value is within the acceptance criteria for accuracy, the sample is considered stable.

Robustness and Ruggedness. Robustness is the ability of an assay to withstand small but deliberate changes that may affect the assay. Such factors may, for example, include incubation temperature and duration, number of washes, light exposure, and lot-to-lot differences in key assay reagents or microtiter plates. Changes that may have an impact on the assay have to be clearly identified in the method description (SOP). The term ruggedness is not mentioned in the IHC guidelines [4, 5], but it is included in the USP monograph on validation [6]. The term describes the consistent performance of an assay under routine changes (i.e., different analysts, instruments, batch size). Such changes should have no significant impact on the consistency of an assay.

The extent of the assessment of robustness and ruggedness depends on the anticipated application of the method, the current status of the assay's life cycle, industry standards, and, last but not least, common sense. The majority of robustness testing will be conducted during method development to facilitate the early identification of factors that affect assay performance. Prestudy robustness and ruggedness validation may be limited to the conditions demonstrating acceptable performance under the anticipated in-study conditions, such as incubation temperature and time tolerances, batch sizes, and so on, but more formal evaluation can also be applied. Acceptable robustness and ruggedness during routine use are assumed when monitoring in-study QC samples demonstrating acceptable intra-assay and interassay precision.

Run Acceptance Criteria. Run acceptance criteria are used to accept or reject a run because of its performance. As a consequence, no defined run acceptance criteria are applicable during method development. Prestudy validation runs are accepted based on the standard curve acceptance criteria. No run can be rejected because of poor performance of a sample during precision and accuracy evaluation; all data from prestudy validation runs are reported unless there has been a clearly recognizable error during sample preparation or measurement. Despite the fact that the standard curve must satisfy the criteria described for standard curves above, in-study runs are accepted based primarily on the performance of the QC samples. As stated above in the paragraph on precision and accuracy, the 4–6–30 rule is recommended (i.e., at least four of six QC samples must be within 30% of their theoretical values and at least 50% of the values for each level must satisfy the 30% limit.

Biomarker Assays. A biomarker is defined as an endogenous substance that reflects physiological or pathophysiological processes or pharmacological responses to a therapeutic intervention [14], with a few exceptions such as viral load biomarkers are endogenous substances. It is a diverse class ranging from electrolytes to small molecules and macromolecules and a wide variety of analytical methodologies is used to quantify such substances. Although many assay characteristics apply to other analytical techniques and analytes as well, the following discussion will only touch on ligand-binding assays for the determination of macromolecules. As for the bioanalysis of therapeutic proteins, no official guidance documents are currently available for biomarker analysis. Moreover, differences exist between validation approaches according to GLP regulations, which are the basis for documents of the FDA and other regulatory agencies and the National Committee for Clinical Laboratory Standards (NCCLS) guidelines for diagnostic assays. Details can be found in References 13–15; the present discussion will only include the parameters used in a GLP regulatory environment.

Besides, in clinical diagnostics especially, novel biomarkers play an increasing role in drug development for the investigation of the pharmacologic response to drug treatment or as surrogate markers for clinical endpoints. Clinical and drug development decisions will depend on the quality of biomarker data. Thus, the utility and value of such data is ultimately determined by the validity of the assay and requires demonstration and documentation of performance characteristics, as mentioned in the previous subsection, for immunoassays such as accuracy, precision, specificity, range, stability, and so on. However, in contrast to bioassays for drug compounds, where quantitative results are achieved by calibration typically using well-defined reference standards, biomarker assays may differ considerably depending on the type of analytical measurement, the type of the analytical data that develop from the assay, or the intended use of the assay. Subsequently, different assay types and validation levels may apply. Rigorous method validation for a novel biomarker is not necessary for drug discovery-phase work. However, the design of the validation must change when the assay is transferred from screening to quantitative determination in later phases of drug development.

Biomarker assays (as other bioassays) may be classified into "definitive quantitative assays," "relative quantitative assays," "quasi-quantitative assays," and "qualitative assays" with varying degrees of validation requirements (Table 5.5-3) [14, 15]. For definitive quantitative assays, a well-defined or characterized standard of the biomarker is available. In the case of relative quantitative assays, calibration is performed with a standard that is not well characterized, not available in pure form, or not representative of the endogenous biomarker. Results from these assays are

TABLE 5.5-3. Assay Categories and Validation Characteristics

Parameter	Definitive Quantitative Assay	Relative Quantitative Assay	Quasi-Quantitative Assay	Qualitative Assay
Accuracy	✓	✓	—	—
Precision	✓	✓	✓	—
Sensitivity	✓ (LLOQ)	✓ (LLOQ)	✓	✓
Specificity	✓	✓	✓	✓
Range	✓ (LLOQ–ULOQ)	✓ (LLOQ–ULOQ)	✓	—
Dilutional linearity	✓	✓	—	—
Parallelism	✓	✓	—	—
Standard and reagent stability	✓	✓	—	—
Matrix stability	✓	✓	✓	✓

expressed as numerical values. Currently, most biomarker assays fall into the relative quantitative category. In quasi-quantitative assays, no reference material is available for the construction of a calibration curve. Nevertheless, the analytical result is expressed in numerical units. Examples are the measurement of enzymatic activity (expressed as units per volume) and vaccine or anti-drug antibodies where the response is reported as percent binding or titer. Qualitative assays use no standard either, and the results are reported non-numerically (i.e., "low, medium, and high" or "+, ++, and +++").

Although with respect to validation criteria a lot of similarities exist between ligand-binding assays for pharmaceutical proteins and biomarker assays, significant differences have to be noted, especially for novel biomarkers, in which a suitable reference standard is not present. If the intended reference is a recombinant protein, one should keep in mind that such proteins often have glycosylation patterns different from the endogenous equivalents. The glycosylation pattern of endogenous proteins is often heterogeneous so that it is virtually impossible to prepare glycoprotein standards that are identical to the natural circulating proteins. In the ideal case, a purified endogenous protein from the target species is used as a reference. This standard should be characterized in terms of analytical purity as thoroughly as possible. If a well-characterized standard is not available, the assay results will provide rather "relative" than "true" numerical values. As a result of the presence of endogenous biomarkers, no analyte-free matrix exists for the preparation of calibrator standards, which makes the establishment of the LLOQ particularly challenging. Standard curves may be prepared using pooled matrix from individuals with low baseline concentrations of the compound. Alternatively, (affinity) stripped matrix, a protein-containing buffer or matrix from an alternate species with negligible concentrations of the analyte, may be considered. Further strategies for minimization of interference from endogenous biomarkers can be used [9]. The preparation of calibrator standards not in the actual sample matrix is one major difference of biomarker assays from assays of protein drugs. When using such approaches, it has to be ensured that the use of the surrogate matrices does not introduce a bias in the assay. In contrast to calibration samples, QC samples must be prepared in an authentic matrix. In this case, matrix samples containing low basal levels can be selected and the middle and upper QC concentrations can be prepared by the addition of known biomarker amounts. Moreover, the disease state can have an impact not only on endogenous biomarker levels but also on the composition of the matrix itself. High concentrations of the marker produced during disease can cause a hook effect. A disease state may also alter the heterogeneity of the biomarker with altered cross reactivity to the antibody relative to the standard. The modified matrix composition can result in increased nonspecific binding (i.e., nonspecific nonspecificity).

The stages of validation of biomarker assays include establishment of the biomarker (development), so-called prevalidation, prestudy validation, and in-study validation [13–15]. The following short discussion will focus on the "GLP-like" definitive and relative quantitative assays. As the development and validation of an assay for novel biomarkers is quite diverse, the application of strict validation procedures appears problematic. Therefore, upon establishment of the prototype assay in the development phase, a formalized validation plan should be developed that defines the scope and purpose of the assay. Further activities during method development include the establishment of the reference standard, selection of antibodies and the assay format, as well as evaluation of key reagents. As mentioned above, selection of the matrix for the preparation of the calibrator concentrations may be challenging because of the presence of endogenous biomarker. Subsequently, the calibration curve model will be established. As for other immunoassays, nonlinear calibration using the four-to-five-parameter logistic model is the commonly acknowledged model for data fitting.

The prevalidation phase may be regarded as a method optimization phase where the calibration model is confirmed; range, LLOQ, and ULOQ are defined; and matrix interference is evaluated. The use of anchor points for the calibration curve may be feasible. As mentioned above, determination of the LLOQ may be difficult because of the presence of the endogenous analyte. Dilutional linearity may be evaluated as well. It is also considered useful to assess the biomarker in healthy and diseased individuals. For this purpose, at least 25 individuals should be tested [9] to account for intrasubject variability caused by circadian and seasonal fluctuations and intersubject variability.

Prestudy validation will additionally evaluate standard and reagent stability as well as matrix stability during collection processing and storage of the samples, will determine dilutional linearity and parallelism, and will confirm assay range and the calibration model. These criteria can be determined as described for immunoassays for pharmaceutical proteins above. In addition, accuracy and precision have to be determined. It is recommended to use validation samples at five different concentration levels analyzed at least in duplicate and in a minimum of six runs. One concentration should be at the anticipated LLOQ, one about 2–4 times the LLOQ, one in midrange on the log scale, one about 70–80% of the anticipated ULOQ, and one at the anticipated ULOQ. The following method acceptance criteria have been recommended [14]: Both accuracy (% RE) and precision (% CV) should be within 25%, except at the LLOQ where 30% is acceptable. Even more lenient criteria may be required in some cases, depending on the analyte or the type of assay and its limitations. It is important to note that such QC samples must be prepared in the actual matrix. The use of buffer, surrogate, or stripped matrix is not feasible, except for rare matrices such as cerebrospinal fluid or tears, where a surrogate matrix may be the only practical option. Spike recovery experiments should be performed on multiple individual lots of matrix to assess the accuracy, matrix effects, and interference. It may also be favorable to include a pilot study into prestudy validation runs.

As for immunoassays for pharmaceutical proteins, in-study validation of biomarker assays should include one set of calibrators to monitor the standard curve as well as a set of QC samples at three concentrations analyzed in duplicate for the decision to accept or reject a specific run. Recommended acceptance criterion is the 6–4–30 rule, but even more lenient acceptance criteria may be justified based on statistical rationale developed from experimental data [14].

Anti-drug Antibody Assays. Basically, the protein sequence of biophar-maceutical therapeutics can be nonhuman, chimeric, humanized, or fully human. Most such therapeutics elicit some level of antibody response against the product leading to potentially serious side effects or loss of drug efficacy. Thus, the immunogenicity of therapeutic proteins is a concern for clinicians, manufacturers, and regulatory agencies. For the detection of anti-drug antibodies, a number of assay methods including enzyme-linked immuno-sorbert assays (ELISA), immunoblotting, surface plasmon resonance, and bioassays are available, each technique having its own advantages and disadvantages [24]. Whereas binding assays identify the antibodies, immunoblotting provides information on the specificity. Surface plasmon resonance can show the antigen-antibody interaction in real time, whereas bioassays demonstrate the neutralizing potential of the antibodies. To date, microtiter plate-based ELISAs are still the most widely used format for testing for anti-drug antibodies because of their simplicity, sensitivity, and high-throughput capability. Generally, the validation parameters outlined for standard immunoassays as required by the regulatory authorities apply to immunoassay-based anti-drug antibody assays as well. These parameters include specificity, selectivity, accuracy, sensitivity, precision, robustness, ruggedness, and stability of reagents, analyte, and matrix. However, some differences exist because of the nature of the antibodies. These differences will be briefly addressed below. Further details can be found in References 22–24.

As a result of the lack of reference material, anti-drug antibody assays are generally quasi-quantitative assays (Table 5.5-3). In addition, the target analyte is generally polyclonal, consisting of antibodies of various isotype classes, specificities, and affinities (i.e., the analyte is not a defined molecule). Thus, no single positive control exists that accurately represents the target analyte. During the early validation phase, no clinical studies have typically been performed, so it may be a challenge to obtain a representative positive sample for method development and validation. When establishing specificity, accuracy, and sensitivity, several control analytes from different individuals or sources representing the test population should be investigated. Specificity of analyte binding can be assessed using immunodepleted samples.

Selectivity is a critical parameter determining the reliability of an antidrug antibody assay. In this context, one has to keep in mind that selectivity can vary between test samples because of the heterogenous nature of the antibodies. For the evaluation of selectivity, immunoglobulins and other potential interfering substances can be spiked into positive and negative samples at high but physiologically relevant concentrations. No substantial interference can be concluded if the recovery is 80–125%. The influence of different sample matrices, typically serum and plasma, should also be evaluated if both matrices may be analyzed with the same assay. The type of the matrix should not change the outcome of the assay. Furthermore, a comparison of specificity between normal- and disease-state matrices should be conducted to detect interfering substances that may be present in certain populations or disease states. Another unique property of anti-drug antibody assays is that the drug itself can act as an interfering substance, which can be mimicked with addition of the drug in varying concentrations to positive controls.

As a result of the quasi-quantitative nature of anti-drug antibody immunoassays and the lack of a reference standard a threshold value, the so-called "cutoff" or "cutpoint" is used to identify positive samples from nonspecific background noise. The assay cutoff is preferably determined by analyzing samples from healthy individuals and those affected by the disease. The data are subsequently used to calculate the cutoff value yielding 5% of false positives [23]. Moreover, one should consider that a low optical readout increases imprecision so that the cutoff level should not be too low for optical assays.

The sensitivity of an immunoassay is typically defined by its LLOQ, which can only be determined when a reference standard is available. Alternatively, a detection limit of qualitative assays is used where a distinction is essentially made between positive and negative results only. Typically, the antibody data are reported as "titer," the titer being the reciprocal of the highest dilution of a sample in which the instrument response is greater than the cutoff response [22]. At least two positive control analytes should be used. If a positive control antibody is available, a pseudo-calibration curve can be generated by a series of dilutions. However, one should keep in mind that "true" quantitation is impossible because of the lack of a true reference compound. Determination of the dilutional linearity is not so important when the result is reported as a titer. However, if the determination of positive samples is based on the interpolation from a reference standard curve, it is essential to demonstrate that the QC samples fall within the (limited) linear range of the calibration curve and not on a plateau or a region that may include a hook effect.

Precision should be assayed using positive controls, negative controls, and a diluent sample. Positive controls should be prepared at a high and a low concentration to demonstrate precision within this assay range. Typically, precision (% CV) of ≤30% is considered acceptable [22]. In-study monitoring of a batch run using QC samples consisting of at least one positive control, a matrix negative control, and a diluent negative control should be used to estimate assay performance.

5.5.2.2 Validation of Cell-based Bioassays

Bioassays use living systems that measure the biological activity of a therapeutic agent. Such assays may be used to study the effects of hormones or growth factors, but such systems can also address drug toxicity and side effects. Moreover, bioassays may also be applied to bioanalysis of biopharmaceutical proteins. Only *in vitro* assays using cell culture systems measuring a discrete response such as cell proliferation, differentiation, or survival will be briefly addressed. Bioassays may generate quantitative or quasi-quantitative data. Only a few considerations for bioassay method validation will be mentioned, as this type of assay is not frequently applied. Most validation procedures for quantitative assays, such as calibration and reagent and matrix stability, accuracy, and precision, are essentially identical to the procedures described for standard immunoassays above.

For a validated assay, an established immortal cell line is typically used. Thus, during method development, not only must a suitable reference standard be established, but also documentation of the cell line with respect to characteristics such as origin of the cell line, culture and passage history, morphology, surface markers, and receptors is required. It is advisable to study the effect of cell age (number of passages) on the measured response. Specificity may be another issue as cell lines proliferate, differentiate, or senesce and die in response to a large number of biomolecules that may be present in the samples obtained for bioanalytical studies. In addition, macromolecules can be metabolized leading to metabolites that may also be biologically active, which is not relevant when a pure compound is applied to the cell culture to study its effect on the cells. However, bioanalytical methods based on bioassays may not be specific for the analyte of interest. In this case, extensive study of interference because of nonspecificity should be conducted during prestudy validation using samples from a number of representative individuals.

5.5.3   CONCLUSIONS AND FUTURE CONSIDERATIONS

Despite the wide availability of chromatographic techniques hyphenated to mass spectrometry, immunoassays remain the most important methods for bioanalytical applications for monitoring macromolecules such as therapeutic proteins, biomarkers, or drug-induced antibodies. Current guidelines of regulatory authorities focus on chromatographic techniques with no or little reference to the specifics of ligand-binding assays. As some differences with respect to assay validation exist between chromatographic and ligand-binding assays, a number of issues require special attention. Reference material of the target analyte(s) is not always available in pure form, which is especially true for biomarkers and anti-drug antibodies. In addition, key reagents such as the antibodies for the immunoassay are frequently not commercially available. Stability issues of those key reagents, the reference material, and the biological matrix have to be considered.

The calibration curves of immunoassays are nonlinear, so that special attention should be paid to the selection of the correct calibration model. Anchoring points out of the validated range may optimize the curve fit. Selectivity of an immunoassay depends on the specificity of the antibody directed toward the analyte. Nonspecific interferences from the matrix as well as specific interferences (cross reactivity) from related compounds have to be considered. Special challenges occur if the analyte is an endogenous compound. In this case, analyte-free matrix may not exist, so that alternative strategies for the preparation of validation samples have to be explored. However, QC samples should be prepared in the original matrix if possible. Pathological states may alter the composition of the matrix or, in the case of biomarkers, the respective concentration, which has to be considered for an appropriate selection of the calibration standards and QC samples. Assays for anti-drug antibodies are quasi-quantitative so that complete GMP-like validation is normally not possible. Finally, as ligand-binding assays are inherently less precise than chromatographic assays, more lenient acceptance criteria for accuracy and precision as well as for run acceptance should be applied. The current opinion according to conference reports recommends as target values for precision (expressed as % CV) and accuracy (expressed as % RER) a maximum of 20% (25% at the LLOQ). Despite known disadvantages, application the 6–4–30 rule as run acceptance criteria during in-study validation has been adopted.

The current gap between the need for validated immunoassays according to GLP compliance and the lack of official guidance documents will certainly be closed in the near future because ongoing efforts at conferences between scientists from regulatory authorities, pharmaceutical companies, and scientific organizations will ultimately result in such guidance documents. However, technological advances in instrument automation and new technologies will continue to create new issues that have to be considered when developing and validating methods for bioanalysis. As each technique has and will have unique features, the challenge is the implementation of a dynamic, yet standardized and systematic approach for analytical method validation.

5.5.4 GLOSSARY OF VALIDATION TERMS

The terms are used according to the ICH guidelines [4, 5], where indicated, or according to the FDA guideline [3].

Acceptance criteria: Numerical limits, ranges, or other suitable measures for the acceptance of the results of analytical procedures.

Accuracy **(ICH)**: The closeness of agreement between a measured value and the theoretical true value. In statistics, accuracy is typically reported as % relative error (% RE).

Batch: A set of standard curve calibrators, validation samples or QC samples, or study samples that are analyzed in a single group of measurements; it is synonymous with run.

Bias: Systematic difference between measured test results and the theoretical true value. Bias is expressed as relative error (% RE) or as a ratio (% recovery).

Calibration curve: The functional relationship between the analyte concentration in standards (calibrators) and the measured response; it is synonymous with standard curve.

Calibrator: A solution of a biological matrix spiked with the analyte of interest for the construction of calibration curves; it is synonymous with standard.

Dilutional linearity: A condition where a spiked sample is serially diluted to result in a set of samples containing analyte concentrations that fall within the quantitative range of the assay and the whole set of dilutions can be measured with acceptable accuracy. Dilution of the sample does not result in biased measurement of the analyte concentration.

Intermediate precision **(ICH)**: Precision within a laboratory: different days, analysts, equipment, and so on.

Lower limit of quantitation (LLOQ): The lowest concentration of an analyte in a test sample that can be determined quantitatively with suitable accuracy and precision (ICH term: limit of quantitation).

Parallelism: A condition where spiked (calibrator) samples are serially diluted to result in a set of samples having analyte concentrations that fall within the quantitative range of the assay and do not exhibit an apparent trend toward increasing or decreasing estimates of the analyte over the range of dilutions.

Precision **(ICH)**: The closeness of agreement (degree of scatter) between a series of measurements obtained from multiple sampling of the same homogenous sample under defined conditions. Precision is considered at three levels: repeatability, intermediate precision, and reproducibility. In statistics, precision is typically reported as % coefficient of variation (% CV), also referred to as relative standard deviation (RSD).

Quality control (QC) sample: Prestudy or in-study samples with a known (nominal) concentration that are treated as unknowns in the assay. During in-study runs, QC samples are used as the basis for run acceptance or rejection.

Range **(ICH)**: The interval between the lower and upper concentrations of an analyte in samples for which it has been demonstrated that the analytical procedure has an acceptable level of accuracy and precision.

Repeatability **(ICH)**: The precision of repeated measurement under the same operating conditions over a short interval of time (within one run), also termed intrabatch or intrarun precision.

Reproducibility **(ICH)**: Precision of repeated measurements between analytical laboratories; also termed intralaboratory precision.

Robustness **(ICH)**: A measure of the capacity of analytical methods to remain unaffected by small but deliberate variations of the experimental parameters; indicates the reliability of the method during routine use.

Run: A set of standard curve calibrators, validation samples or QC samples, or study samples that are analyzed in a single group of measurements; it is synonymous with batch.

Selectivity **(ICH)**: Ability of a bioanalytical method to measure particular analyte(s) in a complex medium without the interference from other components of the mixture.

Specificity **(ICH)**: The ability to unequivocally measure the analyte in the presence of other components including related compounds or matrix components.

Upper limit of quantitation (ULOQ): The highest concentration of an analyte in a test sample that can be determined quantitatively with suitable accuracy.

Validation: Demonstration that a method is suitable for the intended use; it is typically demonstrated by establishing validation parameters that are applied to sample analysis.

Validation samples: Biological matrix samples spiked with the analyte at predetermined concentrations; it is used during prestudy validation to assess accuracy and precision.

REFERENCES

1. Shah V P, Midha K K, Dighe S, et al. (1992). Analytical methods validation: Bioavailability, bioequivalence and pharmacokinetic studies. *Pharm. Res.* 9:588–592.

2. Shah V P, Midha K K, Findlay J W A, et al. (2000). Bioanalytical method validation—a revisit with a decade of progress. *Pharm. Res.* 17:1551–1557.

3. Food and Drug Administration (2001). *Guidance for Industry: Bioanalytical Method Validation.* Available: http://www.fda.gov/cder/guidance/4252fnl.pdf.

4. International Conference on Harmonization (1994). *Guideline Q2A, Text on Validation of Analytical Procedures.* CPMP/ICH/381/95. Available: http://www.ich.org/LOB/media/MEDIA417.pdf.

5. International Conference on Harmonization (1996). *Guideline Q2B: Validation of Analytical Procedures : Methodology.* CPMP/ICH281/95. Available: http://www.ich.org/LOB/media/MEDIA418.pdf.

6. United States Pharmacopoeia 29 (2006). Section 1125: Validation of Compendial Methods, United States Pharmacopeia Convention, Rockville, MD.

7. Approved Biotechnology Drugs. Available: www.bio.org/speeches/pubs/er/approveddrugs.asp.

8. Findley J W A, Das I (1998). Validation of Immunoassay for Macromeolcules from Biotechnology. *J. Clin. Ligand Assay.* 21:249–253.

9. Findley J W A, Smith W C, Lee J W, et al. (2000). Validation of immunoassays for bio-analysis: A pharmaceutical industry perspective. *J. Pharm. Biomed. Anal.* 21:1249–1273.

10. Miller K J, Bowsher R R, Celniker A, et al. (2001). Workshop on Bioanalytical Methods Validation for Macromolecules: Summary Report. *Pharm. Res.* 18:1373–1383.

11. DeSilva B, Smith W, Weiner R, et al. (2003). Recommendations for the Bioanalytical Method Validation of Ligand-Binding Assays to Support Pharmacokinetic Assessments of Macromolecules. *Pharm. Res.* 20:1885–1900.

12. Smolec J, DeSilva B, Smith W, et al. (2005). Bioanalytical method validation for macromolecules in support of pharmacokinetic studies. *Pharm. Res.* 22:1425–1431.

13. Lee J W, Devanarayan V, Barrett Y C, et al. (2006). Fit-for-purpose method development and validation for successful biomarker validation. *Pharm. Res.* 23:312–328.

14. Lee J W, Weiner R S, Sailstad J M, et al. (2005). Method validation and measurement of biomarkers in nonclinical and clinical samples in drug development: A conference report. *Pharm. Res.* 22:499–511.

15. Lee J W, Nordbloom G D, Smith W C, et al. (2003). Validation of bioanalytical assays for novel biomarkers: Practical recommendations for clinical investigation of new drug entities. In J. Bloom (ed.), *Biomarkers in Clinical Drug Development*, Marcel Dekker, New York, pp. 119–149.

16. Dantus M M, Wells M L (2004). Regulatory Issues in Chromatographic Analysis in the Pharmaceutical Industry. *J. Liquid Chromatogr. Rel. Technol.* 27:1413–1442.

17. European Medicines Agency (2005). *Draft Guideline on the Clinical Investigation of the Pharmacokinetics of Therapeutic Proteins.* EMEA/CHMP/89249/2004. Available: http://www.emea.eu.int/pdfs/human/ewp/8924904en.pdf.

18. Food and Drug Administration (2001).*Guidance for Industry: IND Meetings for Human Drugs and Biologics. Chemistry, Manufacturing, and Controls Information.* Available: http://www.fda.gov/cder/guidance/3683fnl.pdf.

19. Deyl Z (ed.) (1997). *Quality Control in Pharmaceutical Analysis*, Elsevier, Amsterdam, the Netherlands.

20. Ermer J, Miller J H M (eds.) (2005). *Method Validation in Pharmaceutical Analysis.* Wiley-VCH, Weinheim, Germany.

21. Colburn W A, Lee J W (2003). Biomarkers, Validation and Pharmacokinetic-Pharmacodynamic Modeling. *Clin. Pharmacokin.* 42:997–1022.

22. Geng D, Shankar G, Schantz A, et al. (2005). Validation of immunoassays used to assess immunogenicity to therapeutic monoclonal antibodies. *J. Pharm. Biomed. Anal.* 39:364–375.

23. Mire-Sluis A R, Barrett Y C, Devanarayan V, et al. (2004). Recommendations for the design and optimization of immunoassays used in the detection of host antibodies against biotechnology products. *J. Immunol. Meth.* 289:1–16.

24. Wadhwa M, Bird C, Dilger P, et al. (2003). Strategies for detection, measurement and characterization of unwanted antibodies induced by therapeutic biologicals.*J. Immunol. Meth.* 278:1–17.

25. Smith W C, Sittampalam G S (1998). Conceptual and Statistical issues in the validation of analytic dilution assays for pharmaceutical applications. *J. Biopharm. Stat.* 8:509–532.

26. Assay Guidance Manual, version 4.1. Available: www.ncgc.nih.gov/guidance.

27. Braggio S, Barnaby R J, Grossi P, et al. (1996). A strategy for validation of bioanalytical methods. *J. Pharm. Biomed. Anal.* 14:375–388.

28. Hubert P, Chiap P, Crommen J, et al. (1999). The SFSTP guide on the validation of chromatographic methods for drug bioanalysis: From the Washington conference to the laboratory. *Anal. Chim. Acta.* 391:135–148.

29. Kringle R O (1994). An assessment of the 4–6-20 rule for acceptance of analytical runs in bioavailability, bioequivalence, and pharmacokinetic studies. *Pharm. Res.* 11:556–560.

30. Rodbard D, Feldman Y, Jaffe M L, et al. (1978). Kinetics of Two-Site Immunoradiometric (Sandwich) Assays. *Immunochemistry.* 15:77–82.

31. Kringle D, Hoffman D (2001). Stability methods for assessing stability of compounds in whole blood for clinical bioanalysis. *Drug Inf. J.* 35:1261–1270.

5.6

MICROARRAYS IN DRUG DISCOVERY AND DEVELOPMENT

Krishnanand D. Kumble[1] and Sarita Kumble[2]

[1] *Genesis Research, Parnell, Auckland, New Zealand*
[2] *Pictor Limited, Glendowie, Auckland, New Zealand*

Chapter Contents

5.6.1	Introduction	634
5.6.2	Array Formats	634
	5.6.2.1 Planar Arrays	634
	5.6.2.2 Solution Arrays	635
5.6.3	Factors for Consideration	635
	5.6.3.1 Surfaces	635
	5.6.3.2 Probes	639
	5.6.3.3 Sample Processing	640
	5.6.3.4 Signal Detection	642
	5.6.3.5 Data Analysis	644
5.6.4	Types of Arrays	646
	5.6.4.1 DNA Arrays	646
	5.6.4.2 Protein Arrays	647
	5.6.4.3 Tissue Arrays	649
	5.6.4.4 Other Arrays	649
5.6.5	Implications of Array Technology	650
	5.6.5.1 Experimental Design	650
	5.6.5.2 Molecular Network Analysis	651
	5.6.5.3 Drug Discovery and Development	652
	References	653

Handbook of Pharmaceutical Biotechnology, Edited by Shayne Cox Gad.
Copyright © 2007 John Wiley & Sons, Inc.

5.6.1 INTRODUCTION

Microarrays are grids of biomolecules (DNA, proteins, carbohydrates, small molecules) in which each element of the grid performs a specific assay such as identification of a specific binding molecule or measuring a particular enzymatic activity. A single sample is analyzed on each microarray providing simultaneous measurements for multiple parameters. Typically a complete experiment involves a large number of microarrays resulting in huge datasets. This deluge in data has led to approaches for integration of diverse datasets to enhance our understanding of both an individual molecular function and elaborate biological processes.

This potential of microarrays to facilitate global analysis of genes and gene products has led to its adoption in nearly every area of biological science—from basic research to clinical diagnostics. They are being used to tackle one of the major challenges in the postgenomic era—to understand the regulation, expression, and function of entire sets of genes, transcripts, and proteins. This information will help in characterizing complex biological processes at the molecular level in various cell types, both in normal and in disease states.

Since their conception in the mid-1990s, microarrays have seen widespread use in almost all aspects of biological research and their range of applications continues to expand. These include (1) biomarker discovery, to find genes that can be used for measuring and following disease progression; (2) target selectivity, to identify genes that can distinguish one disease from another as well as subclasses of disease; (3) pharmacology and toxicogenomics, to help in identification of poor compounds and to optimize the selection of promising leads; and (4) drug screening, identification of protein activity modulators. As the cost of microarrays continues to drop, it is clear that microarrays are becoming a more integral part of the drug discovery process. In short, microarrays are redefining the drug discovery and development process by providing greater knowledge at each step and by helping to understand the complex workings of biological systems.

This chapter provides a technical overview of microarrays, including information on array formats, production and use of microarrays for interrogating various molecular species—DNA, RNA, and proteins—and analysis of data. The challenges in experimental design to maximize microarray data quality, and the use of microarrays in understanding intracellular molecular networks as well as in drug discovery and development, are also discussed.

5.6.2 ARRAY FORMATS

5.6.2.1 Planar Arrays

High-Density Microarrays. High-density arrays are ideally suited for analyzing a small number of samples against thousands of genes, proteins, or small molecules. Early transcript profiling studies used high-density arrays in the context of target discovery with the objective of obtaining a short list of high-priority candidates that showed interesting expression patterns in disease states [1]. Arrays of oligonucleotides representing the entire human genome have been used for discovering polymorphic loci as well as the presence of thousands of alternative alleles [2, 3]. Other uses of high-density DNA arrays include determination of methylation patterns [4] and identification of transcription factor binding sequences [5].

Manufacture of high-density protein arrays presents a greater challenge due to the inherent heterogeneity in the physico-chemical properties and stability of proteins. However, arrays of recombinant proteins representing all yeast open reading frames (ORFs) have been manufactured and used for identification of novel binding activities [6]. *Escherichia coli* transformed with cDNA libraries to express mammalian proteins have been arrayed to identify autoantibodies in serum from the mouse model of systemic lupus erythematosus [7].

Low-density Microarrays. Low-density arrays can be used for simultaneously performing less complex analysis on many different samples [8]. The ability to rapidly screen thousands of biological samples for multiple parameters in the same assay makes the process cost-effective and easy to perform. This format is based on either using slides with hydrophobic barriers to create wells [9] or the 96-well microtiter plate formats, in which each well contains replicate mini-arrays of spots [10].

Low-density arrays are becoming increasingly popular for cytokine measurements due to the relative ease with which preexisting conventional enzyme-linked immunosorbent assay (ELISA) assays can be adapted to a miniaturized format [11]. In addition, low-density DNA arrays are being used for measuring polymorphisms to determine drug toxicity [12], and expression signatures obtained from high-density arrays are being used to guide the design of arrays with small sets of genes for use in cancer subclassification and prognosis [13].

Over the last few years, several different types of microarrays of various probe densities and content—DNA or protein—have become commercially available (Table 5.6-1). These provide microarray tools for analyzing samples derived not only from human tissues, but also from a wide spectrum of other species ranging from bacterial pathogens to Arabidopsis to mouse.

5.6.2.2   Solution Arrays

An alternative approach to planar arrays is three-dimensional arrays in solution. These have the advantage of parallel analysis and high throughput, in addition to better kinetics of binding in solution compared with planar arrays [14]. Positional information in these arrays is retained by a variety of methods such as fluorescence-encoded beads (two dyes with varying ratios incorporated into the bead set) and bar-coded nanoparticles (self-encoded with submicron metal stripes). The independence of each element offers the flexibility to multiplex either a few or thousands of measurements without the need to customize each assay. The high degree of reliability and reproducibility has resulted in the development of fluorescence-encoded beads for multiplex diagnostic assays [15].

This article will focus on technical issues, applications, and future directions in the use of planar arrays.

5.6.3   FACTORS FOR CONSIDERATION

5.6.3.1   Surfaces

Solid Supports. A wide variety of support materials have been used for manufacturing microarrays. Glass is the most commonly used support due to its

TABLE 5.6-1. A Partial List of Commercially Available Ready-to-Use Microarrays

Array Type	Company	URL	Brief Product Description
High-density oligonucleotide	Affymetrix	www.affymetrix.com	*In situ* synthesized oligonucleotides with multiple perfect match and mismatch probes per target gene.
High-density oligonucleotide	Agilent	www.agilent.com	60-mer probes in predesigned printed arrays; custom microarray design service is also available.
High-density oligonucleotide	Applied Biosystems	www.appliedbiosystems.com	60-mer probes designed to be within 1500 bases from the 3′ end of the transcript spotted on each chip.
Low density oligonucleotide	Mergen	www.mergen.com	60-mer probes printed in 16 replicate arrays per slide with 1000 probes in each array.
Low-density DNA	Superarray	www.superarray.com	Pathway-specific probes (up to 500 target genes) arrayed as 400-bp cDNA or 60-mer oligonucleotides.
Multiple content	Pamgene	www.pamgene.com	Printed oligonucleotides, peptides, or proteins as 4 or 96 replicate arrays on chips or plates, respectively.
Peptide	JPT Peptide Technologies	www.jpt.com	Printed peptide (12 to 15-mer) arrays for profiling of protease, kinase and phosphatase substrates.
Peptide	Pepscan systems	www.pepscan.nl	Arrays printed with 1200 protein kinase substrate peptides.
Protein	Invitrogen	www.invitrogen.com	Slides printed with recombinant proteins (yeast or human).
Antibody	Pierce	www.piercenet.com	Four to sixteen antibodies for ELISA assays printed in replicate arrays in 96-well format.
Antibody	BD Biosciences	www.clontech.com	Over 500 antibodies spotted for comparative profiling of protein levels from two samples.
Tissue	Biochain	www.biochain.com	Arrays or frozen tissue from 15 different species for determination of species cross-reactivity of antibodies.
Tissue	Chemicon	www.chemicon.com	Normal and cancer tissues arrayed in duplicate from paraffin-embedded tissue blocks.

low background fluorescence and amenability to automation [14]. Polystyrene with fluorescence and binding properties similar to glass is beginning to be increasingly adapted for microarray use [16]. These surfaces are usually modified by coating with one-, two-, or three-dimensional structures that provide either covalent attachment chemistry or enable binding of molecules through adsorption. Other microarray surfaces include membranes such as nitrocellulose and nylon and gold or silicon films [17–19]. Synthesis of oligonucleotides has been performed *in situ* on fused silica wafers [20], and peptide arrays have been synthesized on cellulose and polypropylene membranes [21].

Chemistry. Microarray surfaces must maintain the stability and activity of attached biomolecules, while remaining surface-bound through all processing steps. Spot morphology and background noise, either due to nonspecific sample binding or from the detection system used, are important issues to be considered in choosing the right immobilization chemistry. A variety of surfaces have been derivatized to expose various active groups; these determine the type of attachment—ionic or covalent—of the biomolecule to the surface [22].

Polylysine, which results in an amine surface carrying a positive charge, was among the early microarray surfaces used for binding to negatively charged biomolecules [23]. Membranes such as nitrocellulose and nylon that have traditionally been used for a variety of blotting applications have been modified for microarrays by application onto glass backing [17]. Table 5.6-2 provides commercial sources of commonly used surfaces for printing nucleic acids and proteins.

Oligonulceotides modified to carry an amino group or proteins through their lysine residues can be covalently attached to aldehyde- and epoxy-derivatized surfaces. Other reactive surfaces used to covalently link both DNA and proteins include N-hydroxy succinimide [20] and maleimide [24]. Proteins expressed with polyhistidine and biotin tags have been attached to surfaces coated with nickel chelate and streptavidin, respectively [25, 26]. This strategy is likely to result in the proper orientation of the displayed molecules on the surface, thereby improving their functionality and stability.

Heterobifunctional cross-linkers have been used to attach activated biomolecules on surfaces functionalized with either aminosilane [27] or mercaptopropylsilane [28]. These strategies have resulted in up to fourfold higher signal-to-noise ratios compared with the polylysine surface for arrayed antibodies [29]. Specific immobilization via interaction of surfaces functionalized with salicylhydroxamic acid derivatives and phenyldiboronic acid-labeled nucleic acids and proteins has also been demonstrated [30].

A new generation of chemistries has introduced poly(ethylene glycol)-functionalized surfaces that prevent nonspecific protein binding, thereby removing the need for a blocking step during processing [31]. This commercially available chemistry also offers a wide variety of additional functional groups. A large spacer between the support and the biomolecule results in higher analyte binding by avoiding steric hindrance [29]. Three-dimensional surfaces used for arraying biomolecules were shown to increase binding capacity and reduce denaturation of immobilized molecules, but they slowed down reaction kinetics due to reduction in diffusion rates [32]. These same chemistries are applied for biomolecule attachment to beads in the manufacture of suspension arrays.

TABLE 5.6-2. A Partial List of Commercially Available Microarray Surfaces

Company	Surface Chemistry	Application	URL
Telechem	Primary amine	Oligonucleotides and PCR products	www.arrayit.com
	Primary aldehyde	Amino-modified oligonucleotides and PCR products, or proteins	
Corning	Gamma aminopropyl silane	Long (>50-mer) oligonucleotides and cDNA	www.corning.com
Erie Scientific	Aluminized Aminopropyl silane	PCR products	www.eriemicroarray.com
	aldehyde silane or epoxide	Amino-modified oligonucleotides and PCR products, or proteins	
	Poly-L-lysine	PCR products or proteins	
Schleicher and Schuell	Nitrocellulose polymer	Proteins	www.schleicher-schuell.com
Pall Corporation	Modified nylon membrane	mRNA, PCR products and proteins	www.pall.com
Perkin Elmer	Specially formulated acrylamide-based polymer	Proteins	http://las.perkinelmer.com
NUNC	Aminopropylsilane,	Oligonucleotides and PCR products	www.nuncbrand.com
	Aldehyde and Epoxy	Amino-modified oligonucleotides and PCR products, or proteins	
	Maxisorp	Proteins	
Sigma	Poly-L-lysine	PCR products or proteins	www.sigma-aldrich.com
	Streptavidin	Biotin-tagged oligonucleotides or proteins	
	Maleimide	Amino-modified oligonucleotides and PCR products, or proteins	
Greiner	Aminosilane	Oligonucleotides and PCR products	www.greinerbiooneinc.com
	Aldehyde and Epoxy	Amino-modified oligonucleotides and PCR products, or proteins	
	Nickel	Histidine-tagged proteins	

5.6.3.2 Probes

Printing. Printing refers to spotting arrays of presynthesized biomolecules directly onto a solid surface. This versatile approach can be applied to generate microarrays of almost any biomolecule in conjunction with a proper im-mobilization method. Printing can generate many copies of the same array more efficiently than *in situ* synthesis because the immobilized molecules need to be synthesized only once. Several kinds of microarray printing technologies have been developed (Table 5.6-3), but the most commonly used method uses contact printing. In this method, pins are used to transfer samples from a source to the solid support by direct surface contact. Noncontact piezoelectric printers, which are also often used, can print more spots per unit time than contact printers by employing an electric current to accurately and rapidly dispense samples onto the solid support. Although the reproducibility of spot volumes delivered by contact printers typically has a

TABLE 5.6-3. A Partial List of Commercially Available Microarray Printers

Company/Arrayer	Features of the Printing Technology	URL
Affymetrix 417 Arrayer 427 Arrayer	Microcontact printing using a pin and ring system. Spot size of 150 to 200μ with 50-pL delivery volume.	www.affymetrix.com
Cartesian Dispensing SynQuad Hummingbird	High-speed microsolenoid valve with a high-resolution syringe pump dispense system. Delivery volumes from 20nL to μl.	www.cartesiantech.com
Genomic Solutions MicroGrid OmniGrid	Microcontact printing with solid or quill pins. Spot size and dispense volume dependent upon pin type.	www.genomicsolutions.com
Telechem SpotBot	Microcontact printing with solid or quill pins. Spot size and dispense volume dependent upon pin type. Low-cost arrayer with a footprint for bench-top use.	www.arrayit.com
LabNext Xact	Microcontact printing with solid or quill pins. Spot size and dispense volume dependent upon pin type. Lowest cost arrayer on the market.	www.labnext.com
Packard Bioscience SpotArray BioChip Arrayer	Non-contact printing with a piezo-electric printhead. Spot size of 75 to 200μ dependent upon the pin type.	www.packardbioscience.com

coefficient of variation (*cv*) below 20%, piezoelectric printers have greater control over the volume dispensed resulting in *cv*s of under 5%.

"Dropouts" or missing spots can be a problem, especially during high-throughput manufacture, and are primarily due to clogging of the dispensing components in the arrayer. To identify dropouts, microarrayers can be fitted with optical devices that can monitor dispensing and provide data on dropouts that can then be used to "fill in" missing spots. This system has vastly improved the quality of microarrays, reducing the need to print replicate spots of the same probe.

Other less-common deposition methods include electrospraying in a stable cone-jet mode to generate highly reproducible spots of as little as 50-pL volumes [33] and a laser transfer technique that allows the accurate deposition of picoliter volumes of proteins onto the solid surface [34].

In Situ Synthesis. Two typical *in situ* synthesis approaches are light-directed parallel synthesis and peptide synthesis on membrane (SPOT) synthesis. Although the former approach was initially developed for peptide synthesis [35], it has been adapted for the synthesis of DNA microarrays. This method uses a combination of photo-lithographic masks and combinatorial chemistry to synthesize oligonucleotides on fused silica wafers [36]. High-density arrays have also been manufactured by maskless methods using either digitally controlled aluminium mirrors to fabricate oligonucleotides by photodeposition [37] or by phos-phoramidite chemistry in microfluidic chips with three-dimensional nano-chambers [38].

The maskless method is also being used to create peptide arrays in microfluidic chips by parallel synthesis using digital photolithography and photogenerated acid during the deprotection step [39]. Peptide arrays have also been manufactured by SPOT synthesis using a combination of novel polymeric surfaces, linker and cleavage chemistries, as well as robotic liquid-handling systems [40]. These methods involve alternate synthesis and washing steps, increasing the possibility of contamination by reagents from adjacent spots, and thereby limiting array density [41].

5.6.3.3   Sample Processing

DNA Analysis. Detection of single nucleotide polymorphisms (SNPs) was one of the early applications of high-density oligonucleotide arrays [42, 43]. More recently these arrays have also been used to assess DNA copy numbers [44]. Each of these measurements requires the initial extraction of genomic DNA, which is then processed using several different methods each designed to incorporate labels for visualization of binding to specific probes on the array. Fragments from restriction digested genomic DNA were ligated with adaptors, amplified, and enzymatically end-labeled with biotin followed by hybridization to oligonucleotide arrays [42, 44]. Alternatively, labeling was done by incorporation of a biotin-tag during primer extension after hybridization of unlabeled amplified DNA fragments [43]. The biotin was then detected using Streptavidin conjugated to a fluorescent dye.

Genome-wide DNA–protein interactions have been mapped using intergenic oligonucleotide microarrays [45]. In this approach, called chromatin immunoprecipitation (ChIP), epitope-tagged proteins of interest are allowed to bind specific regions in genomic DNA and then cross-linked. The DNA-bound epitope-tagged proteins are immunoprecipitated using a tag-specific antibody. The DNA is then delinked from the protein and, after fluorescent labeling, hybridized to arrays for identification of regions that bind to the protein of interest. A similar approach has been used for identification of methylation sites in genomic DNA. In this method, oligonucleotide primer adaptors are attached to restriction-digested DNA, followed by a secondary digestion with a methylation-sensitive enzyme, labeling with a fluorescent dye, and hybridization to arrays [46]. These methods are providing a wealth of information on the factors that regulate gene expression and helping to understand these processes at a global level.

Transcript Analysis. Microarrays permit the rapid analysis of complex gene expression changes in cells and tissues during development, both normal and disease. These changes give distinct patterns that can be used for discovering new disease-specific therapeutic targets, and molecular diagnostics, as well as for following treatment efficacy and disease prognosis.

Typically for transcript profiling, a labeled nucleotide is incorporated during reverse transcription of the total cellular mRNA pools. Not only does this approach require a large amount of RNA (50 to 200 µg of total RNA) for hybridization, but RNA purity is also a critical factor in array performance due to potential nonspecific binding by other labeled macromolecules [47]. Arrays have been widely used for obtaining relative mRNA concentrations between two samples, test and reference, in which each sample is labeled with a different dye (typically Cy3 and Cy5), followed by mixing of the samples before hybridization [48]. An alternative method avoids incorporation of bulky dyes during reverse transcription by incorporating amino allyl labels to which N-hydroxyl succinimidyl dyes are chemically coupled in a later step [49].

The use of a common reference RNA allows for the comparison of data between various array experiments. Typically a pool of RNA derived from a variety of tissues is used as a reference sample. Efforts are being made to implement common standards (MIAME, minimal information about a microarray experiment) for transcript profiling through the MGED Society to enable data sharing between different groups [50].

Methods in which the amount of mRNA has been amplified to produce labeled cRNA, by incorporating a T7 RNA polymerase promoter site into one end of the cDNA followed by *in vitro* transcription are also widely used [51]. In this method, quantitative estimates of the amount of each transcript in a given sample can be calculated. In single-sample labeling experiments, a reference RNA may be spiked in during sample labeling and hybridization to facilitate quality control of the process as well as for comparison of data between different arrays.

Despite its current widespread use and the drive toward ensuring high-quality microarray data by the implementation of MIAME guidelines, very little is known about the extent to which application of different technologies influences the outcome of transcriptional profiles and differential expression. However, microarray users should be aware of studies that have attempted to present a comprehensive evaluation encompassing different probe types (oligonucleotides and cDNAs), labeling techniques and hybridization protocols [52].

Protein Analysis. The early adaptation of protein microarrays is the reformatting of already available ELISAs. In these assays, which were first described by Roger

Ekins [53], a protein antigen is identified and its quantity is measured using two antigen-specific antibodies—one surface-immobilized to capture the antigen, and the other chemically labeled to bind the captured antigen in the solution phase. The second antibody produces a detectable signal through the label. This method, in which the protein sample does not have to be labeled, has been widely used for measuring cytokine levels in a variety of samples such as serum and tissue culture supernatants [54, 55]. Protein expression profiles have also been measured by capturing dye-labeled protein lysates derived from cells and tissues on arrays of antibodies. These measurements have largely been ratiometric in which two samples each labeled with a different dye using protocols similar to those used for transcript analysis are mixed and incubated with arrays of antibodies [56, 57]. Single-sample analyses could be performed by spiking known amounts of control proteins into the samples before labeling.

5.6.3.4   Signal Detection

Detection Chemistries.   The most prevalent method for detecting binding of target to immobilized probes on microarrays are fluorescent dyes, of which Cy3 and Cy5 are the most widely used especially for differential transcript and protein profiling [58], whereas single-target hybridizations are commonly performed with fluorescein isothiocyanate [51]. Although Alexa dyes allow use of more than two colors in a single experiment, they are less commonly used [59]. Enzyme-labeled fluorescence (ELF), a phosphatase substrate that results in a precipitable product, has been used for both DNA and protein arrays [60].

Chemiluminescent detection is also used, mainly to detect antigen capture on antibody arrays by sequential incubations with biotinylated secondary antibodies and Streptavidin-conjugated horseradish peroxidase (HRP) [61]. Phosphorylation of arrayed kinase substrates in the presence of radiolabeled adenosine triphosphate (ATP) has been detected by autoradiography [62].

To increase sensitivity of detection, several signal amplification methods have been applied to the various types of molecular targets. Tyramide signal amplification (TSA) uses biotinyl-tyramide, an HRP substrate, to accumulate biotin at the reaction site. The "amplified" biotin is then detected using Streptavidin-HRP in conjunction with substrates that result in products detectable either by fluorescence, chemiluminescence, or colorimetry [63]. Other signal amplification methods include rolling circle amplification (RCA) in which an oligonucleotide-conjugated antibody binds to biotin on the target, followed by hybridization of a circular DNA molecule to the oligonucleotide. The circular DNA is then replicated using a DNA polymerase in the presence of a fluorescently labeled nucleotide [64]. Preformed branched DNA (dendrimers), each of which is attached to over 200 fluorescent dye molecules, have also been used for labeling array-bound targets [65].

Imaging Systems.   Assay format and cost are the primary determinants of the kind of devices used for imaging arrays. Table 5.6-4 lists some commercially available imaging systems used for microarray normalization. Fluorescence-based scanners that use lasers to excite fluorescent dyes attached to the target molecule (for transcript profiling) or to a specific secondary detection reagent (for ELISA assays) are the most widely used for signal detection and imaging. These instruments

TABLE 5.6-4. A Partial List of Commercially Available Scanners

Company/ Scanner	Scanner features	URL
Alpha Innotech NovaRay	CCD camera with a broadband excitation source with up to eight excitation and emission filters for imaging slides or microtiter plates.	www.alphainnotech.com
Kodak ImageStation 2000MM	CCD camera with selectable multi-wavelength illumination source for imaging microtiter plates; can also be used for radiography and whole animal imaging.	www.kodak.com
Tecan LS Reloaded	Scanning laser imaging system can be used for imaging slides or microtiter plates with a sensitivity of less than 0.1 fluor/μm^2.	www.tecan.com
Axon Instruments GenePix 4000B	Scanning laser imaging system with multiple filter options can be used for imaging slides with a sensitivity of less than 0.1 fluor/μm^2.	www.axon.com
Affymetrix 418 Array Scanner	Laser scanning microscope with two excitation filters can read Affymetrix' GeneChips with a sensitivity of less than 1 fluor/μm^2.	www.affymetrix.com
ChromaVision ACIS	Digital microscope with software for tissue scoring and quantitative analysis suited for tissue microarrays.	www.chromavision.com
Molecular Devices ImageXpress Micro	Inverted epifluorescent microscope with CCD camera and optional laser designed for scanning slides and microtiter plates.	www.moleculardevices.com

enable user-defined scanning resolutions and photomultiplier tube settings to adjust detection sensitivity. In addition, the availability of a wide range of fluorescent dyes makes this the most flexi-ble system for use in microarray image detection. Chemiluminescence-based detection systems using a charge-coupled device (CCD) together with conventional camera optics are also commonly used. Although the latter systems are flexible for applications in a wide range of assay formats, fluorescence scanners provide a higher range of sensitivity and dynamic range [66].

Label-independent methods such as surface plasmon resonance (SPR) can overcome variations caused by inconsistencies in labeling chemistries that are often seen in label-dependent detection systems. SPR has been used to measure affinities in binding reactions, especially between antigens and their cognate antibodies. These highly sensitive systems are beginning to be adapted for microarray-based measurements [67]. Another recently described method relies on the change in fluorescence decay times of tryptophan and tyrosine residues in proteins to identify protein–protein interactions. In this method, a change in binding behavior of proteins in solution toward protein partners arrayed on a solid support is detected with regard to binding specificity and protein amount [68].

5.6.3.5 Data Analysis

Most commercial microarray imagers supply data extraction software that can accommodate the unique parameters of scanned images. In addition, printing precision has simplified the process of detecting spot boundaries and measurement of inter-spot distances. A useful consideration is the format for storage of primary scanned images (usually as tiff files) so as to be able to take advantage of future developments in image analysis software [69]. Storage of raw image files retains maximum information, allowing the use of different normalization, image extraction, and quality metrics.

Several commercially available software packages are available that interpret and transform an array image into a dataset (Table 5.6-5). Software programs grid the elements of the array as a first step in processing the image. Background-corrected intensity values for each spots are obtained using one of several options: local (area around individual spots) or global (average signal in area outside of the grid) background corrections, or user-defined values such as those from negative control data points contained within the array. After background-corrected intensity values have been calculated for each spot, the data are normalized with respect to sources of systematic and biological variation [70]. The choice of the normalization method is critical because it impacts precision of data comparison between arrays.

Several methods for statistical analysis of microarray data are available depending on the experimental setup and the kind of biological question that needs to be addressed. Initial approaches to analyzing microarray data focused on

TABLE 5.6-5. A Partial List of Microarray Data Analysis Software

Name	Software Features	URL
Spotfinder	Software tool designed for microarray image processing using the TIFF image files generated by most microarray scanners.	www.tigr.org/software/
MIDAS	Tool for microarray data quality filtering and normalization that allows raw experimental data to be processed through various data normalizations, filters, and transformations via a user-designed analysis pipeline.	

TABLE 5.6-5. *Continued*

Name	Software Features	URL
ArrayViewer	Software tool designed to facilitate the presentation and analysis of microarray expression data, leading to the identification of genes that are differentially expressed.	
GeneTraffic	The application allows users to import, visualize, analyze, interpret, and store microarray data in a complete, secure data management environment. Software allows easy development of custom interface to relational database.	www.iobion.com
Genespring	The platform is designed to identify genes/pathways that are relevant to specific biological questions by comparing analysis results from expression, genotyping, protein, metabolite, and other data types.	www.agilent.com/chem/
BASE	Comprehensive database server manages biomaterial information, raw data, and images, and provides integrated and "plug-in"-able normalization, data viewing, and analysis tools. The system also has array production LIMS features that can be integrated with the data analysis.	http://base.thep.lu.se/
Image Pro Plus	Software features automated microscope control, advanced fluorescence acquisition, image deconvolution, and three-dimensional reconstruction and measurement.	www.mediacy.com
Rosetta Resolver	Comprehensive customizable gene expression analysis package that incorporates analysis tools with a scalable database. Software has capabilities to associate expression profiles to molecular pathways.	www.rosettabio.com
Pathiam	Hardware-independent, Web-enabled software for viewing and analyzing immunohistochemically stained slides. Pathologists can automatically generate reports containing a complete quantitative analysis of each sample.	www.bioimagene.com

unsupervised hierarchical clustering because these are simple ways in which data can be organized [71]. It is still the most commonly used analysis tool especially if the experiment has been designed to obtain a bird's eye view of transcript or protein profiles during a particular process. Machine learning techniques such as neural networks [72] and support vector machines [73] should be used for more advanced analyses when preliminary information can be used to guide data interpretation. Most software packages have a user-friendly graphical user interface (GUI) that enables performing a number of simple analysis, including data normalization, various kinds of clustering, and principal component analysis.

5.6.4   TYPES OF ARRAYS

5.6.4.1   DNA Arrays

Transcript Profiling. Transcript profiling was one of the earliest applications of microarray technology. In this application, fluorescently labeled cellular transcripts are hybridized either to arrays of spotted cDNA [19] or *in situ* synthesized oligonucleotides [36]. The former have generally been used for hybridization of a mixture of two transcript pools, each of which are derived from a different biological sample (such as diseased and normal tissue), and labeled with a different fluorescent dye (Cy3 and Cy5) before mixing [19]. A ratiometric analysis of fluorescence then helps to determine the relative expression levels of each transcript in the two samples. cDNA arrays have received wide acceptance within the academic community due to their low cost and ease of manufacture, as well as the ability to customize rapidly as new genomic sequence information becomes available. Customization of array content also easily accommodates the research interests of individual laboratories.

Oligonucleotide arrays, used for measuring transcript amounts from single samples, have the advantage of displaying a much larger number of very small spots (5μ), enabling the interrogation of multiple probes for each transcript, including mismatch probes for determination of hybridization specificity [36, 74]. Currently, oligonucleotide arrays that display probes covering all predicted genes in the human genome are commercially available from several vendors (Table 5.6-1).

The ability to obtain global transcript profiles has accelerated the process of discovery in all areas of biology ranging from basic discovery to drug development and diagnostics. Correlations between gene-expression patterns and disease states have led to the selection of the best drug targets for pharmaceutical development as well as for monitoring therapeutic outcomes [75]. One area in which DNA arrays are making a critical impact is cancer, where expression profiles from hundreds of cancer tissue samples have allowed subclassification of cancers based on the identification of specific transcript expression signatures. These are likely to guide therapy and improve prognosis for cancer patients in the near future [76–79].

Physical Characterization of Genes. Oligonucleotide arrays have found use in the physical characterization of the genome by helping to map transcription factor binding sites and methylation sites [80]. These arrays, with their ability to package hundreds of thousands of spots on each chip, have been widely used for SNP discovery [81]. Clinically significant SNPs have then been rearrayed in a low-density format for analyzing large numbers of samples to determine clinical validity. A U.S. Food and Drug Administration (FDA)-approved SNP genotyping array for CYP450 has recently been used to identify patients with a decreased ability to metabolize risperidone [10]. More recently, bacterial artificial chromosome (BAC) arrays have provided high-resolution maps of genetic changes, including gains and losses, as well as amplifications in chromosomal DNA by comparative genomic hybridization [82]. Other applications of oligonucleotide arrays have been to analyze splice variants by hybridization of labeled transcripts with arrays that combine exon and junction-derived probes, which are either specific or nonspecific to a splice event [83].

Cell Arrays. Immobilized arrays of plasmid DNA used to transfect cells in the presence of a transfection reagent are called cell arrays or reverse transfection arrays. These arrays are overlayed with cultured cells in medium and incubated so that cells superimposed on the DNA spot are transfected. The desired phenotypic change is visualized by staining the cells after a brief incubation period [84]. This format allows a variety of readouts, including cytoskeletal changes, apoptosis, and DNA replication, which can be either visualized with a laser scanner or a fluorescence microscope. Similar approaches have been used for increasing the throughput of loss-of-function studies using RNAi to monitor effects on cell phenotype [85, 86]. These studies enable functional whole genome screens without the need for expensive screening facilities.

5.6.4.2 Protein Arrays

Protein Profiling

Antibody Arrays. Antibodies are the most commonly arrayed protein class due to their structural similarity and stability. Many monoclonal antibodies generated over the years for binding to various epitopes on proteins offer a diverse source of capture and detection antibodies. These are being used in antibody arrays, especially for determination of growth factor and cytokine levels from a variety of biological samples [87]. Arrays have been multiplexed with antibodies that were developed and used for conventional ELISAs to measure levels of more than 50 cytokines with a high degree of sensitivity (pg/mL) from low sample volumes (less than 50 µL) [64]. However, each set of arrayed antibodies needs to be characterized for specificity of binding to the target molecule to ensure data quality. Recent studies have shown significant cross-reactivity of arrayed antibodies to proteins other than their intended targets [88].

Identification of disease biomarkers is a rapidly growing area of proteomic research. These biomarkers can enable better predictive capabilities in disease diagnosis and prognosis, as well as in drug development by identifying potential drug toxicities and side-effects. Although it has been suggested that antibody arrays can be used to profile serum and cell lysates, these measurements are complicated by the fact that protein concentrations in these samples cover 10 to 14 orders of magnitude [87], requiring systems that can simultaneously detect both low- and high-abundance proteins within a single array. The sample complexity issue may be alleviated by reduction in sample complexity before profiling. This can be accomplished using either liquid phase protein separation systems or by profiling the protein complement of various cell organelles separately [89]. Samples can be labeled with either fluorescent tags followed by direct detection after capture or with other haptens such as biotin followed by detection with Streptavidin conjugated to a reporter molecule [90].

As well-characterized antibodies are available to only a small subset of total cellular proteins, profiling complex protein samples could be performed using arrays of *in vitro* synthesized antibody libraries [91]. This format has the advantage of being able to directly array bacterial cells rather than purified recombinant antibodies [92]. Once disease-relevant proteins have been identified through screening for global protein changes, arrays containing the "diagnostic" subset of antibodies that recognize disease-specific proteins can be used for high-throughput, large-sample analysis [66].

Antigen Arrays. Several studies have been performed to demonstrate disease identification by analysis of serum antibodies to clinically relevant proteins [93–96]. One study used arrays of autoantigen diagnostic markers to measure autoantibody titers in serum of patients with autoimmune disease [95]. In an extension of this study, autoimmune patient sera screened for binding to arrayed antigens identified specificity of autoantibody responses defining autoantigens relevant in human disease [97]. More recently, the diversity of B-cell response as a function of disease severity was measured using an array of over 225 distinct epitopes derived from several myelin-associated proteins. These arrays could identify distinct sets of epitopes and could demonstrate a correlation between reduced epitope spreading and improved disease outcome. This information is now being used to guide the development of a tolerizing DNA vaccine, which could potentially limit epitope spreading and improve disease outcome [98]. These antigen arrays have been shown to have a much higher sensitivity than conventional ELISAs with a three-log linear dynamic range [97].

Protein Function Arrays. The development of high-throughput expression and purification methods have made a large number of proteins available for arraying. A small amount of material (pg to ng) is sufficient for printing numerous arrays that can then be used to perform rapid, functional screens. The first genome-wide protein display, an array of all yeast ORFs, was used to screen for binders to calmodulin and phospholipids [6]. In this study, 5800 different yeast proteins with hexahistidine tags were arrayed on nickel-coated glass slides. The immobilized proteins were then probed with various labeled phospholipids resulting in the identification of more than 150 proteins that were shown to bind phospholipids for the first time. Other protein binding measurements include identification of protein–protein and protein–DNA interactions [99].

In addition to binding assays, functional properties are also being measured on arrays. Protein kinase activity has been determined either by immobilizing fluorogenic substrates followed by the addition of specific kinases [100] or on kinase arrays incubated with peptide substrates in the presence of radiolabeled ATP [101]. Protein kinase arrays have also been used for identification of inhibitors by incubating small-molecule binders in the presence of substrates demonstrating the ability of arrays to screen for kinase inhibitors [102].

G-protein-coupled receptors (GPCRs) currently make up the largest class of therapeutic drug targets and are an obvious choice for microarray-based drug screening. An early study demonstrating the capability of microarrayed GPCRs for agonist and antagonist screening showed retention of activity through several activation/ deactivation cycles after ligand binding on arrayed rhodopsin [103]. Competitive binding assays in which mixtures of fluorescently labeled ligands and unlabeled inhibitors were used to demonstrate binding selectivity to different receptor subtypes have also been described [104].

Other Types of Protein Arrays. Peptides, arrayed either by deposition on the surface [105] or by *in situ* synthesis [106], have been used for epitope-mapping [107], measuring enzyme activity [108], and cell capture [105]. They have also been used for identification of peptide antagonists to proteins that are likely drug targets [109]. In addition, protease specificities were identified using peptidyl coumarin substrate arrays created by immobilizing the fluorescent coumarin leaving group via the C-terminus to the solid support, enabling synthesis of a high-diversity peptide library at the N-terminal end [110].

Arrays of peptide–major histocompatibility complex (MHC) complexes have been created on acrylamide gel-coated glass surfaces and used for ligand-specific capture of CD4(+) and CD8(+) lymphocytes. This approach should be useful to characterize multiple epitope-specific T-cell populations during immune responses associated with infection, cancer, autoimmunity, and vaccination [111].

Covalent mRNA–protein fusions were displayed on single-stranded DNA arrays through nucleic acid hybridization of the mRNA in the fusion molecule [112]. Similarly, capture agents such as nucleic acid aptamers have also been proposed for use in protein binding [113, 114].

5.6.4.3 Tissue Arrays

Tissue microarrays (TMAs) are displays of several tens to hundreds of tissue specimens on a single slide for parallel analysis [115]. Paraffin embedded tissue are generally used for arraying [116]; however, arrays have also been constructed from frozen tissue [117]. The ability to array archival paraffin embedded tissue opens up vast archives of patient samples for medical research. Arrayed tissue sections processed either for cytological staining or *in situ* hybridization, combined with automated image analysis systems, provides a powerful molecular profiling tool. TMAs are commonly used to confirm results obtained from expression microarrays, as well as in the development of diagnostic and prognostic markers for clinical applications.

5.6.4.4 Other Arrays

Although DNA and protein arrays continue to be widely used, other types of molecules are being arrayed and used for a variety of applications. These include small-molecule arrays in which individual members of a chemical library are deposited on a surface and used for performing binding assays with target proteins [102, 118, 119]. Several strategies are being applied for displaying small molecules on surfaces. The simplest strategy is to spot molecules on glass after mixing with glycerol to maintain "wetness' after deposition. This strategy was used to create and screen arrays by spraying the target protein, human caspase 6, along with its fluorigenic substrate and screening for dark spots of enzyme inhibitors against noninhibitor fluorescent spots [120]. An elegant method that takes advantage of

DNA hybridization specificities to provide an address for arrays of protein nucleic acid (PNA)-tagged small molecules has been described [121]. More sophisticated libraries of molecules that selectively modulate activities of mutated kinases have been used to identify chemical inhibitors that switch off individual kinases [122].

Carbohydrate arrays have been developed and used as novel high-throughput analytical tools for monitoring carbohydrate–protein interactions such as profiling protein binding to various sugars and to measure enzymatic activity on sugar substrates [123]. These microarrays have also been used to discover anti-adhesion therapeutics by identifying carbohydrate binding specificities of pathogenic bacteria. The display of carbohydrate ligands on a surface in a manner that mimics interactions at the cell–cell interface is ideal for whole-cell applications [124]. Similarly, monosaccharide-based arrays of N-acetyl galactosamine and N-acetylneuraminic acid derivatives were used to identify binders to cholera and tetanus toxins [125].

5.6.5   IMPLICATIONS OF ARRAY TECHNOLOGY

5.6.5.1   Experimental Design

Microarrays are observed as tools for "descriptive" research and not for "hypothesis-driven" research, which is primarily driven to answer specific questions related to a known set of molecules. However, most microarray-based studies do have clear objectives and are designed to answer well-defined questions. The plan for specimen selection should follow from the objectives of the microarray study. Studies may either be exploratory, whose results should be confirmed, or designed to test various models; in which case, the experiments have to be designed in a focused manner [126].

Appropriate design of microarray studies is critical to draw valid and useful conclusions from the large amount of data that are invariably obtained from each experiment. Criteria for design will vary depending on the type of array used—DNA, protein, and small molecule—as well as the experimental objectives, and sources of variability within the system. Issues such as selection of samples, number of replicates needed, allocation of samples to dyes for ratiometric profiling, and sample size are important considerations in these studies [127].

Three sources of variation should be considered in the design of a microarray experiment: (1) biological, at the level of setting up of the cells, cell lines or animals, and sample treatment; (2) technical, which is introduced during sample processing; and (3) analytical, which highlights signal bias and readout [128]. To minimize each of these variations, the experiment should be set up such that the biological material to be used for analysis should be handled separately from the initiation of the experiment providing biological replicates. If this results in an unwieldy experimental setup, pooling the samples at some stage of processing may still allow a reduction in biological variance. This approach is risky in the event of the presence of an extreme outlier in the pool, because this can unduly influence the expression values obtained from the pool. Samples from every stage of the experiment with a likelihood of introduction of variation should be treated separately. Replicate measurements from the same biological sample helps to reduce technical variation.

TABLE 5.6-6. Experimental Design for Two Color Microarray-Based Comparisons of Transcript Profiles Between Disease and Healthy Tissue Samples

Design	Array 1		Array 2		Array 3	
	Dye 1 (e.g., Cy3)	Dye 2 (e.g., Cy5)	Dye 1	Dye 2	Dye 1	Dye 2
Reference	Disease Sample 1	Healthy sample	Disease Sample 2	Healthy sample	Disease Sample 3	Healthy sample
Loop	Disease Sample 1	Healthy Sample 1	Healthy Sample 1	Disease Sample 2	Disease Sample 2	Healthy Sample 2
Balanced Block	Disease Sample 1	Healthy Sample 1	Healthy Sample 2	Disease Sample 2		

Ratiometric analysis of signal by comparing signal intensities between two samples from the same spot reduces errors due to printing differences. This has led to the use of a reference sample in transcript profiling experiments, and it is an integral part of differential profiling, making the choice of the reference sample critical. The most important considerations in choosing the appropriate standard are that it is readily available, homogeneous, and stable over a period of time. The reference sample does not need to have any biological relevance, because it merely serves to compare expression profiles between different experiments performed in different laboratories. Among the standards that have been used are mRNA populations obtained from a wide variety of cells or tissues with the aim of lighting up every element on the array. Alternatively, a reference sample is prepared from the samples that will be assayed in the experiment to ensure that every transcript present in the test samples will be represented in the reference standard [126]. Care has to be taken, however, that no transcript in the reference sample saturates the signal in the detection system.

If a direct comparison needs to be made between two samples, for example, samples from pooled healthy tissue with samples from disease tissue, it would be necessary to perform forward and reverse labeling to remove any dye bias that may alter the results [129]. Swapping dyes between samples reduces systematic bias in the data that can be corrected during the normalization step [130]. Alternative design strategies that have been suggested are the balanced block design [131] and the loop design [132]. The same considerations can be applied to two-color antibody array-based protein profiling [133]. These experimental design strategies have been briefly illustrated in Table 5.6-6.

5.6.5.2   Molecular Network Analysis

Computational analysis is essential to transform the large amount of data generated by microarrays into a mechanistic understanding of various modules connecting at hubs and nodes to form molecular networks. Methods that identify the regulatory mechanisms underlying these modules and processes by an integrative analyses in the context of other data sources are often capable of extracting deeper biological insight from the data. Such integrative computational approaches include meta-analysis, functional enrichment analysis, interactome analysis, transcriptional network analysis, and integrative model system analysis. In addition, comparative analysis, combining human data with model organisms, can lead to more robust findings [134, 135].

Such methods that analyze biological processes in terms of higher level modules can identify robust signatures of disease mechanisms. Application of these methods to understand human cancers have delineated molecular subtypes of cancer associated with disease progression and treatment response [76–78, 136, 137].

A closely integrated data warehouse has been constructed to link different kinds and numbers of biological networks to experimental results such as those coming from microarrays. The basic idea is to consider and store biological networks as graphs in which the nodes represent promoters, proteins, genes, and transcripts, and the edges specify the relationship between the various nodes. Direct links to underlying sequences (exons, introns, promoters, amino acid sequences) in a systematic way enable close interoperability to sequence analysis methods. This approach allows us to store, query, and update a wide variety of biological information in a way that is semantically compact without requiring changes at the database schema level when new kinds of biological information are added. Such a system can be set up using software available from the public domain [138].

5.6.5.3   Drug Discovery and Development

Target Discovery.   Microarrays are increasingly being used in drug discovery for a wide range of applications, including target discovery and selection, pharmacology, and toxicogenomics [139]. The objective of using transcript profiling in target discovery is to identify a shortlist of candidate genes with distinct expression patterns during disease. Typical selection criteria include disease-specific changes in expression levels and tissue or cell-type selectivity [140]. A secondary screen is then typically used to examine the role(s) of short-listed genes by analyzing additional samples, and identifying correlations with disease progression. Validation experiments are performed to follow up the candidate genes using animal disease models or mice in which the target gene has been knocked out.

Network analysis is especially relevant to antimicrobial drug discovery. The relatively small size of microbial genomes enables data from profiling experiments to be used for modeling gene networks because these would be several orders of magnitude lower in complexity compared with human cellular networks. In addition, the availability of whole-genome nucleotide sequence data from a growing list of microbial genomes allows designing of probes for DNA arrays. Such microarrays are already being used to obtain a global perspective of host–pathogen interactions. This is beginning to make an impact on our understanding of pathogenesis and the strategies taken to combat infectious diseases, including the identification of novel drug targets [141].

Target Validation.   Several approaches, including those that involve the use of arrays, can be taken to validate short-listed therapeutic candidates. Genes whose transcript profiles suggest additional analysis can be profiled using TMAs in tens to hundreds of tissues by immunohistochemistry [142]. TMAs can be used to identify heterogeneities between primary tumors and their metastases, as demonstrated in the analyses of erbB2 in primary and metastatic breast cancers

[143]. The expression of a target gene in normal tissue from various vital organs can also be assessed using TMA panels. It is well known that mRNA levels do not reflect protein quantities and function, so proteomic analysis may help in making better decisions on targets to be developed for therapeutics [144].

Cell-based gene knockouts can closely mimic the actions of potential drugs to identify phenotypic changes and potential side effects. The successful adaptation of RNA interference in a microarray format to facilitate efficient suppression of gene expression has enabled high-throughput target validation for a wide variety of cell types [85, 86].

Pharmacology. A study that monitored the expression patterns of the entire complement of yeast genes under a variety of experimental conditions and genetic backgrounds was used to categorize drugs into various classes based on the expression pattern changes they induced. This database is being used to stratify novel drugs based on their mechanism of action by monitoring the changes in expression induced by them [1].

Drug safety in animals is tested by monitoring a diverse spectrum of events, most commonly liver and kidney toxicity, and fatty liver. Transcriptional activation of drug metabolizing enzymes in livers of drug-treated animals or primary human hepatocytes has been widely used to identify drug metabolizing pathways. This helps to identify drugs early in development that might have toxicity issues due to inadequate metabolism [145–147]. In human populations, certain polymorphisms in the cytochrome P450 genes result in slow drug metabolism leading to toxicity. Additionally, the identification of SNPs in the cytochrome P450 genes of patients during clinical trials can help to stratify patients in whom the treatment is likely to be effective [148].

REFERENCES

1. Hughes T R, Marton M J, Jones A R, et al. (2000). Functional discovery via a compendium of expression profiles. *Cell.* 102:109–126.
2. Lipshutz R J, Fodor S P, Gingeras T R, et al. (1999). High density synthetic oligonucleotide arrays. *Nat. Genet.* 21:20–24.
3. Kidgell C, Winzeler E A (2005) Elucidating genetic diversity with oligonucleotide arrays. *Chromos. Res.* 13:225–235.
4. Gitan R S, Shi H, Chen C M, et al. (2002). Methylation-specific oligonucleotide microarray: a new potential for high-throughput methylation analysis. *Genome Res.* 12:158–164.
5. Lieb J D, Liu X, Botstein D, et al. (2001). Promoter-specific binding of Rap1 revealed by genome-wide maps of protein-DNA association. *Nature Genet.* 28:327–334.
6. Zhu H, Bilgin M, Bangham R (2001). Global analysis of protein activities using proteome chips. *Science.* 293:2101–2105.
7. Gutjahr C, Murphy D, Lueking A, et al. (2005). Mouse protein arrays from a TH1 cell cDNA library for antibody screening and serum profiling. *Genomics.* 85:285–296.
8. Huang J X, Mehrens D, Wiese R, et al. (2001). High-throughput Genomic and Proteomic Analysis Using Microarray Technology. *Clini. Chem.* 4:1912–1916.
9. Haab B B, Geierstanger B H, Michailidis G, et al. (2005). Immunoassay and antibody microarray analysis of the HUPO Plasma Proteome Project reference specimens:

Systematic variation between sample types and calibration of mass spectrometry data. *Proteomics.* 5:3278–3291.

10. Wiese R, Belosludtsev Y, Powdrill T, et al. (2001). Simultaneous multianalyte ELISA performed on a microarray platform. *Clin. Chem.* 47:1451–1457.

11. Kumble K D (2003). Protein microarrays: new tools for pharmaceutical development. *Analytical and Bioanalytical Chemistry.* 377:812–819.

12. de Leon J, Susce M T, Pan R M, et al. (2005). The CYP2D6 poor metabolizer phenotype may be associated with risperidone adverse drug reactions and discontinuation. *J. Clin. Psychia.* 66:15–27.

13. Perez E A, Pusztai L, Van de Vijver M (2004). Improving patient care through molecular diagnostics. *Semin Oncol.* 31:14–20.

14. Zhou H, Roy S, Schulman H, et al. (2001). Solution and chip arrays in protein profiling. *Trends Biotechnol.* 19:S34–S39.

15. Martins T B, Burlingame R, von Muhlen C A, et al. (2004). Evaluation of multiplexed fluorescent microsphere immunoassay for detection of autoantibodies to nuclear antigens. *Clin. Diagn. Lab. Immunol.* 11:1054–1059.

16. Available: NUNC Brand Products, Denmark. http://www.nuncbrand.com

17. Stillman B A, Tonkinson J L (2000). FAST slides: A novel surface for microarrays. *Biotechniques.* 29:630–635.

18. Kanda V, Kariuki J K, Harrison D J, et al. (2004). Label-free reading of microarray-based immunoassays with surface plasmon resonance imaging. *Analyt. Chem.* 76:7257–7262.

19. Manning M, Redmond G (2005). Formation and characterization of DNA microarrays at silicon nitride substrates. *Langmuir.* 21:395–402.

20. Venkatasubbarao S (2004). Microarrays-status and Prospects. *Trends Biotechnol.* 22:630–637.

21. Zander N (2004). New planar substrates for the *in situ* synthesis of peptide arrays. *Molec. Divers.* 8:189–195.

22. Angenendt P, Glokler J (2004). Evaluation of antibodies and microarray coatings as a prerequisite for the generation of optimized antibody microarrays. *Methods Molec. Bio.* 264:123–134.

23. Schena M, Shalon D, Davis R W, et al. (1995). Quantitative monitoring of gene expression patterns with a complementary DNA microarray. *Science.* 270:467–470.

24. Vaidya A A, Norton M L (2004). DNA attachment chemistry at the flexible silicone elastomer surface: toward disposable microarrays. *Langmuir.* 20:11100–11107.

25. Meredith G D, Wu H Y, Allbritton N L (2004). Targeted protein functionalization using His-tags. *Bioconjug. Chem.* 15:969–982.

26. Wacker R, Schroder H, Niemeyer C M (2004). Performance of antibody microarrays fabricated by either DNA-directed immobilization, direct spotting, or streptavidin-biotin attachment: A comparative study. *Analyt. Biochem.* 330:281–287.

27. Chrisey L A, Lee G U, O'Ferrall C E (1996). Covalent attachment of synthetic DNA to self-assembled monolayer films. *Nucleic Acids Res.* 24:3031–3039.

28. Rogers Y H, Jiang-Baucom P, Huang Z J, et al. (1999). Immobilization of oligonucleotides onto a glass support via disulfide bonds: A method for preparation of DNA microarrays. *Analyt. Biochem.* 266:23–30.

29. Kusnezow W, Hoheisel J D (2003) Solid supports for microarray immunoassays. *J. Biomolec. Recog.* 16:165–176.

30. Springer A L, Gall A S, Hughes K A, et al. (2003). Salicylhydroxamic acid functionalized affinity membranes for specific immobilization of proteins and oligonucleotides. *J. Biomolec. Technol.* 14:183–190.

31. Angenendt P, Glokler J, Sobek J, et al. (2003). Next generation of protein microarray support materials: evaluation for protein and antibody microarray applications. *J. Chromatog. A.* 1009:97–104.

32. Guschin D, Yershov G, Zaslavsky A, et al. (1997). Manual manufacturing of oligonucleotide, DNA, and protein microchips. *Analyt. Biochem.* 250:203–211.

33. Moerman R, Frank J, Marijnissen J C M, et al. (2001). Miniaturized electrospraying as a technique for the production of microarrays of reproducible micrometer-sized protein spots. *Analyt. Chem.* 73:2183–2189.

34. Ringeisen B R, Wu P K, Kim H, et al. (2002). Picoliter-scale protein microarrays by laser direct write. *Biotechnol. Prog.* 18:1126–1129.

35. Fodor S P, Read J L, Pirrung M C, et al. (1991). Light-directed, spatially addressable parallel chemical synthesis. *Science.* 251:767–773.

36. Lockhart D J, Dong H, Byrne M C, et al. (1996). Expression monitoring by hybridization to high-density oligonucleotide arrays. *Nature Biotechnol.* 14:1675–1680.

37. Singh-Gasson S, Green R D, Yue Y, et al. (1999). Maskless fabrication of light-directed oligonucleotide microarrays using a digital micromirror array. *Nature Biotechnol.* 17:974–980.

38. Gao X, LeProust E, Zhang H, et al. (2001). A flexible light-directed DNA chip synthesis gated by deprotection using solution photogenerated acids. *Nucleic Acids Res.* 29:4744–4750.

39. Pellois J P, Zhou X, Srivannavit O, et al. (2002) Individually addressable parallel peptide synthesis on microchips. *Nature Biotechnol.* 20:922–926.

40. Reimer U, Reineke U, Schneider-Mergener J (2002). Peptide arrays: from macro to micro. *Cur. Opin. Biotechnol.* 13:315–320.

41. Xu Q, Lam K S (2003) Protein and Chemical Microarrays—Powerful Tools for Proteomics. *J. Biomed. Biotechnol.* 3:257–266.

42. Matsuzaki H, Loi H, Dong S, et al. (2004). Parallel genotyping of over 10,000 SNPs using a one-primer assay on a high-density oligonucleotide array. *Genome Res.* 14:414–425.

43. Gunderson K L, Steemers F J, Lee G, et al. (2005). A genome-wide scalable SNP genotyping assay using microarray technology. *Nature Gen.* 37:549–554.

44. Bignell G R, Huang J, Greshock J, et al. (2004). High-resolution analysis of DNA copy number using oligonucleotide microarrays. *Genome Res.* 14:287–295.

45. Iyer V R, Horak C E, Scafe C S, et al. (2001). Genomic binding sites of the yeast cell-cycle transcription factors SBF and MBF. *Nature.* 409:533–538.

46. Huang T H, Perry M R, Laux D E (1999). Methylation profiling of CpG islands in human breast cancer cells. *Human Molec. Gen.* 8:459–470.

47. Duggan D J, Bittner M, Chen Y, et al. (1999). Expression profiling using cDNA microarrays. *Nature Gen.* 21:10–14.

48. Grunenfelder B, Winzeler E A (2002). Treasures and traps in genome-wide data sets: case examples from yeast. *Nature Rev. Gen.* 3:653–661.

49. Manduchi E, Scearce L M, Brestelli J E, et al. (2002). Comparison of different labeling methods for two-channel high-density microarray experiments. *Physiol. Genom.* 10:169–179.

50. Brazma A, Hingamp P, Quackenbush J, et al. (2001). Minimum information about a microarray experiment (MIAME)-toward standards for microarray data. *Nat Genet.* 29:365–371.

51. Lockhart D J, Dong H, Byrne M C, et al. (1996). Expression monitoring by hybridization to high-density oligonucleotide arrays. *Nature Biotechnol.* 14:1675–1680.
52. Yauk C L, Berndt M L, Williams A, et al. (2004). Comprehensive comparison of six microarray technologies. *Nucleic Acids Res.* 32:e124.
53. Ekins R P. (1998). Ligand assays: from electrophoresis to miniaturized microarrays. *Clin. Chem.* 44:2015–2030.
54. Tam S W, Wiese R, Lee S, et al. (2002). Simultaneous analysis of eight human Th1/Th2 cytokines using microarrays. *J. Immunolog. Meth.* 261:157–165.
55. Kastenbauer S, Angele B, Sporer B, (2005). Patterns of protein expression in infectious meningitis: A cerebrospinal fluid protein array analysis. *J. Neuroimmunol.* 164:134–139.
56. Nielsen U B, Cardone M H, Sinskey A J, et al. (2003). Profiling receptor tyrosine kinase activation by using Ab microarrays. *Proc. of the National Academy of Sciences U S A.* 100:9330–9335.
57. Sreekumar A, Nyati M K, Varambally S, et al. (2001). Profiling of cancer cells using protein microarrays: discovery of novel radiation-regulated proteins. *Cancer Res.* 61:7585–7593.
58. Holloway A J, van Laar R K, Tothill R W, et al. (2002). Options available—from start to finish—for obtaining data from DNA microarrays II. *Nature Gen.* 32:481–489.
59. Forster T, Costa Y, Roy D, et al. (2004). Triple-target microarray experiments: a novel experimental strategy. *BMC Genomics.* 5:13.
60. Mendoza L G, McQuary P, Mongan A, et al. (1999). High-throughput microarray-based enzyme-linked immunosorbent assay (ELISA). *Biotechniques.* 27:778–780.
61. Paweletz C P, Charboneau L, Bichsel V E, et al. (2001). Reverse phase protein microarrays which capture disease progression show activation of pro-survival pathways at the cancer invasion front. *Oncogene.* 20:1981–1989.
62. Falsey J R, Renil M, Park S, et al. (2001). Peptide and small molecule microarray for high throughput cell adhesion and functional assays. *Bioconjug. Chem.* 12:346–353.
63. Bobrow M N, Shaughnessy K J, Litt G J (1991). Catalyzed reporter deposition, a novel method of signal amplification. II. Application to membrane immunoassays. *J. Immunolog. Meth.* 137:103–112.
64. Schweitzer B, Roberts S, Grimwade B, et al. (2002). Multiplexed protein profiling on microarrays by rolling-circle amplification. *Nature Biotechnol.* 20:359–365.
65. Stears R L, Getts R C, Gullans S R (2000). A novel, sensitive detection system for high-density microarrays using dendrimer technology. *Physiolog. Genom.* 3:93–99.
66. Ault-Riche D, Atkinson B, Kumble K D (2005). *Methods for Structural analysis of Protein Pharmaceuticals.* AAPS Press, Arlington, V.A. pp: 545–571.
67 Kanda V, Kariuki J K, Harrison D J, et al. (2004). Label-free reading of microarray-based immunoassays with surface plasmon resonance imaging. *Analyt. Chem.* 76:7257–7262.
68. Striebel H M, Schellenberg P, Grigaravicius P, et al. (2004). Readout of protein microarrays using intrinsic time resolved UV fluorescence for label-free detection. *Proteomics.* 4:1703–1711.
69. Geschwind D H (2001). Sharing gene expression data: An array of options. *Nature Rev. Neurosci.* 2:435–438.
70. Bilban M, Buehler L K, Head S, et al. (2002). Defining signal thresholds in DNA microarrays: exemplary application for invasive cancer. *BMC Genomics.* 3:19.
71. Slonim D K. (2002). From patterns to pathways: gene expression data analysis comes of age. *Nature Genetics.* 32:502–508.

72. Xu Y, Selaru F M, Yin J, et al. (2002). Artificial neural networks and gene filtering distinguish between global gene expression profiles of Barrett's esophagus and esophageal cancer. *Cancer Res.* 62:3493–3497.

73. Brown M P, Grundy W N, Lin D, et al. (2002). Knowledge-based analysis of microarray gene expression data by using support vector machines. *Proc. of the National Academy of Sciences U S A.* 97:262–267.

74. Harbig J, Sprinkle R, Enkemann S A (2005). A sequence-based identification of the genes detected by probesets on the Affymetrix U133 plus 2.0 array. *Nucleic Acids Res.* 33:e31.

75. Petricoin E F III, Hackett J L, Lesko L J, et al. (2002). Medical applications of microarray technologies: a regulatory science perspective. *Nature Gen.* 32:474–479.

76. Golub T R, Slonim D K, Tamayo P (1999). Molecular classification of cancer: Class discovery and class prediction by gene expression monitoring. *Science.* 286:531–537.

77. Alizadeh A A, Eisen M B, Davis R E, et al. (2000). Distinct types of diffuse large B-cell lymphoma identified by gene expression profiling. *Nature.* 403:503–511.

78. van de Vijver M J, He Y D, van't Veer L J, et al. (2002). A gene-expression signature as a predictor of survival in breast cancer. *New Eng. J. Med.* 347:1999–2009.

79. van't Veer L J, Paik S, Hayes D F (2005). Gene expression profiling of breast cancer: a new tumor marker. *J. Clin. Oncol.* 23:1631–1635.

80. Pollack J R, Iyer V R (2002). Characterizing the physical genome. *Nature Genetics.* 32:515–521.

81. Shi M M (2001). Enabling large-scale pharmacogenetic studies by high-throughput mutation detection and genotyping technologies. *Clin. Chem.* 47:164–172.

82. Cowell J K (2005). High throughput determination of gains and losses of genetic material using high resolution BAC arrays and comparative genomic hybridization. *Combin. Chem. High Throughput Screen.* 7:587–596.

83. Fehlbaum P, Guihal C, Bracco L, et al. (2005). A microarray configuration to quantify expression levels and relative abundance of splice variants. *Nucl. Acids Res.* 33:e47.

84. Ziauddin J, Sabatini D M (2001). Microarrays of cells expressing defined cDNAs. *Nature.* 411:107–110.

85. Wheeler D B, Carpenter A E, Sabatini D M (2005). Cell microarrays and RNA interference chip away at gene function. *Nature Gen.* 37:S25–S30.

86. Vanhecke D, Janitz M (2004). High-throughput gene silencing using cell arrays. *Oncogene.* 23:8353–8358.

87. Patterson S D, Aebersold R H (2003). Proteomics: the first decade and beyond. *Nature Gen.* 33:311–323.

88. Michaud G A, Salcius M, Zhou F, et al. (2003). Analyzing antibody specificity with whole proteome microarrays. *Nature Biotechnol.* 21:1509–1512.

89. Madoz-Gurpide J, Wang H, Misek D E, et al. (2001). Protein based microarrays: a tool for probing the proteome of cancer cells and tissues. *Proteom.* 1:1279–1287.

90. Haab B B (2005). Antibody arrays in cancer research. *Molec. Cell. Proteom.* 4:377–383.

91. Wingren C, Ingvarsson J, Lindstedt M, et al. (2003). Recombinant antibody microarrays—a viable option? *Nature Biotechnol.* 21:223.

92. de Wildt R M, Mundy C R, Gorick B D, et al. (2000). Antibody arrays for high-throughput screening of antibody-antigen interactions. *Nature Biotechnol.* 18:989–994.

93. Bacarese-Hamilton T, Mezzasoma L, Ingham C, et al. (2002). Detection of allergen-specific IgE on microarrays by use of signal amplification techniques. *Clin. Chem.* 48:1367–1370.

94. Mezzasoma L, Bacarese-Hamilton T, Di Cristina M (2002). Antigen microarrays for serodiagnosis of infectious diseases. *Clin. Chem.* 48:121–130.

95. Joos T O, Schrenk M, Hopfl P (2000). A microarray enzyme-linked immunosorbent assay for autoimmune diagnostics. *Electrophor.* 21:2641–2650.

96. Jain K K (2004). Applications of biochips: from diagnostics to personalized medicine. *Curr. Opin. Drug Disc. Devel.* 7:285–289.

97. Robinson W H, DiGennaro C, Hueber W (2002). Autoantigen microarrays for multi-plex characterization of autoantibody responses. *Nature Med.* 8:295–301.

98. Robinson W H, Fontoura P, Lee B J (2003). Protein microarrays guide tolerizing DNA vaccine treatment of autoimmune encephalomyelitis. *Nature Biotechnol.* 21:1033–1039.

99. Ge H (2000). UPA, a universal protein array system for quantitative detection of protein-protein, protein-DNA, protein-RNA and protein-ligand interactions. *Nucl. Acids Res.* 28:e3.

100. Houseman B T, Huh J H, Kron S J, et al. (2002). Peptide chips for the quantitative evaluation of protein kinase activity. *Nature Biotechnol.* 20:270–274.

101. Zhu H, Klemic J F, Chang S (2000). Analysis of yeast protein kinases using protein chips. *Nature Gen.* 26:283–289.

102. MacBeath G, Schreiber S L (2000). Printing proteins as microarrays for high-throughput function determination. *Science.* 289:1760–1763.

103. Bieri C, Ernst O P, Heyse S, et al. (1999). Micropatterned immobilization of a G protein-coupled receptor and direct detection of G protein activation. *Nature Biotechnol.* 17:1105–1108.

104. Fang Y, Webb B, Hong Y (2004). Fabrication and application of G protein-coupled receptor microarrays. *Methods Molec. Biol.* 264:233–243.

105. Falsey J R, Renil M, Park S, et al. (2001). Peptide and small molecule microarray for high throughput cell adhesion and functional assays. *Bioconjug. Chem.* 12:346–353.

106. Toepert F, Knaute T, Guffler S (2003). Combining SPOT synthesis and native peptide ligation to create large arrays of WW protein domains. *Angewandte Chemie Internat. Ed.* 42:1136–1140.

107. Reineke U, Kramer A, Schneider-Mergener J (1999). Antigen sequence- and library-based mapping of linear and discontinuous protein-protein-interaction sites by spot synthesis. *Curr. Topics Microbiol. Immunol.* 243:23–36.

108. Houseman B T, Huh J H, Kron S J, et al. (2002). Peptide chips for the quantitative evaluation of protein kinase activity. *Nature Biotechnol.* 20:270–274.

109. Reineke U, Sabat R, Kramer A, et al. (1996). Mapping protein-protein contact sites using cellulose-bound peptide scans. *Molec. Divers.* 1:141–148.

110. Salisbury C M, Maly D J, Ellman J A (2002). Peptide microarrays for the determina-tion of protease substrate specificity. *J. Amer. Chem. Soc.* 124:14868–14870.

111. Soen Y, Chen D S, Kraft D L, et al. (2003). Detection and characterization of cellular immune responses using peptide-MHC microarrays. *PLoS Biol.* 1:E65.

112. Weng S, Gu K, Hammond P W, et al. (2002). Generating addressable protein microar-rays with PROfusion covalent mRNA-protein fusion technology. *Proteomics.* 2:48–57.

113. Brody E N, Willis M C, Smith J D, et al. (1999). The use of aptamers in large arrays for molecular diagnostics. *Molec. Diagnos.* 4:381–388.

114. Collett J R, Cho E J, Lee J F (2005). Functional RNA microarrays for high-throughput screening of antiprotein aptamers. *Analyt. Biochem.* 338:113–123.

115. Braunschweig T, Chung J Y, Hewitt S M (2004). Perspectives in tissue microarrays. *Combinat. Chem. High Throughput Screen.* 7:575–585.

116. Callagy G, Cattaneo E, Daigo Y (2003). Molecular classification of breast carcinomas using tissue microarrays. *Diagnos. Molec. Pathol.* 12:27–34.

117. Miyaji T, Hewitt S M, Liotta L A, et al. (2002). Frozen protein arrays: a new method for arraying and detecting recombinant and native tissue proteins. *Proteomics.* 2:1489–1493.

118. Barnes-Seeman D, Park S B, Koehler A N, et al. (2003). Expanding the functional group compatibility of small-molecule microarrays: discovery of novel calmodulin ligands. *Angewandte Chemie Internat. Ed.* 42:2376–2379.

119. Koehler A N, Shamji A F, Schreiber S L (2003). Discovery of an inhibitor of a transcription factor using small molecule microarrays and diversity-oriented synthesis. *J. Amer. Chem. Soc.* 125:8420–8421.

120. Gosalia D N, Diamond S L (2003). Printing chemical libraries on microarrays for fluid phase nanoliter reactions. *Proc. of the National Academy of Sciences U S A.* 100:8721–8726.

121. Winssinger N, Ficarro S, Schultz P G, et al. (2002). Profiling protein function with small molecule microarrays. *Proc. of the National Academy of Sciences U S A.* 99:11139–11144.

122. Bishop A C, Ubersax J A, Petsch D T, et al. (2000). A chemical switch for inhibitor-sensitive alleles of any protein kinase. *Nature.* 407:395–401.

123. Shin I, Cho J W, Boo D W (2004). Carbohydrate arrays for functional studies of carbohydrates. *Combinat. Chem. High Throughput Screen.* 7:565–574.

124. Disney M D, Seeberger P H (2004). The use of carbohydrate microarrays to study carbohydrate-cell interactions and to detect pathogens. *Chem. Biol.* 11:1701–1707.

125. Ngundi M M, Taitt C R, McMurry S A (2005). Detection of bacterial toxins with monosaccharide arrays. *Biosens. Bioelectr.* 21:1195–1201.

126. Simon R, Radmacher M D, Dobbin K (2002). Design of studies using DNA microarrays. *Genet. Epidemiol.* 23:21–36.

127. Dobbin K, Simon R (2005). Sample size determination in microarray experiments for class comparison and prognostic classification. *Biostatist.* 6:27–38.

128. Churchill G A (2002). Fundamentals of experimental design for cDNA microarrays. *Nature Genet.* 32:490–495.

129. Yang Y H, Speed T (2002). Design issues for cDNA microarray experiments. *Nature Rev. Genet.* 3:579–588.

130. Kerr M K, Churchill G A (2001). Statistical design and the analysis of gene expression microarray data. *Genet. Res.* 77:123–128.

131. Dobbin K, Shih J H, Simon R (2003). Questions and answers on design of dual-label microarrays for identifying differentially expressed genes. *J. Nation. Cancer Instit.* 95:1362–1369.

132. Kerr M K, Churchill G A (2001). Experimental design for gene expression microarrays. *Biostatist.* 2:183–201.

133. Eckel-Passow J E, Hoering A, Therneau T M, et al. (2005). Experimental design and analysis of antibody microarrays: Applying methods from cDNA arrays. *Cancer Res.* 65:2985–2989.

134. Rhodes D R, Chinnaiyan A M (2005). Integrative analysis of the cancer transcriptome. *Nature Genet.* 37:S31–S37.

135. Segal E, Friedman N, Kaminski N (2005). From signatures to models: understanding cancer using microarrays. *Nature Genet.* 37:S38–S45.

136. Sorlie T (2004). Molecular portraits of breast cancer: tumour subtypes as distinct disease entities. *Euro. J. Cancer.* 40:2667–2675.

137. Greer B T, Khan J (2004). Diagnostic classification of cancer using DNA microarrays and artificial intelligence. *Annals N. Y. Acad. Sci.* 1020:49–66.

138. Koehler J, Rawlings C, Verrier P (2005). Linking experimental results, biological networks and sequence analysis methods using Ontologies and Generalised Data Structures. *In Silico Biol.* 5:33–44.

139. Butte A (2002). The use and analysis of microarray data. *Nature Rev. Drug Disc.* 1:951–960.

140. Gerhold D L, Jensen R V, Gullans S R (2002). Better therapeutics through microarrays. *Nature Genet.* 32:547–551.

141. Shaw K J, Morrow B J (2003). Transcriptional profiling and drug discovery. *Curr. Opin. Pharmacol.* 3:508–512.

142. Sauter G, Simon R, Hillan K (2003). Tissue microarrays in drug discovery. *Nature Rev. Drug Disc.* 2:962–972.

143. Simon R, Nocito A, Hubscher T (2001). Patterns of her-2/neu amplification and over-expression in primary and metastatic breast cancer. *J. Nat. Cancer Inst.* 93:1141–1146.

144. Griffin T J, Gygi S P, Ideker T (2002). Complementary profiling of gene expression at the transcriptome and proteome levels in Saccharomyces cerevisiae. *Molec. Cell. Proteom.* 1:323–333.

145. Kato N, Shibutani M, Takagi H (2004). Gene expression profile in the livers of rats orally administered ethinylestradiol for 28 days using a microarray technique. *Toxicol.* 200:179–192.

146. Meneses-Lorente G, de Longueville F, Dos Santos-Mendes S (2003). An evaluation of a low-density DNA microarray using cytochrome P450 inducers. *Chem. Res. Toxicol.* 16:1070–1077.

147. Harris A J, Dial S L, Casciano D A (2004). Comparison of basal gene expression profiles and effects of hepatocarcinogens on gene expression in cultured primary human hepatocytes and HepG2 cells. *Mutat. Res.* 549:79–99.

148. Rodrigues A D, Rushmore T H (2002). Cytochrome P450 pharmacogenetics in drug development: in vitro studies and clinical consequences. *Curr. Drug Metab.* 3:289–309.

5.7

GENETIC MARKERS AND GENOTYPING ANALYSES FOR GENETIC DISEASE STUDIES

ANNE E. KWITEK AND MICHAEL OLIVIER

Medical College of Wisconsin, Milwaukee, Wisconsin

Chapter Contents

5.7.1   Introduction   661
5.7.2   Simple Sequence Length Polymorphisms (SSLPs)   663
    5.7.2.1   General Description   663
    5.7.2.2   Discovery   663
    5.7.2.3   Methods for Genotyping   665
    5.7.2.4   Tools and Resources   669
5.7.3   Single Nucleotide Polymorphisms (SNPs)   670
    5.7.3.1   General Description   670
    5.7.3.2   Discovery   671
    5.7.3.3   Methods for Genotyping   673
    5.7.3.4   Linkage Disequilibrium and Haplotypes   682
5.7.4   Application for Genetic Analyses   683
    5.7.4.1   Linkage Analysis   683
    5.7.4.2   Association Analysis   684
    Internet Resources   685
    References   686

5.7.1   INTRODUCTION

Over the past decade, major research efforts have resulted in a wealth of sequence information for many genomes. In addition to the human genome, several other mammalian genomes have also been completely sequenced, providing a tremendous resource for biomedical researchers to examine gene sequences, their

Handbook of Pharmaceutical Biotechnology, Edited by Shayne Cox Gad.
Copyright © 2007 John Wiley & Sons, Inc.

regulation, and their role in health and disease. However, although we commonly talk about "the genome" when we refer to the genome sequence, every individual of a species has a slightly different genome sequence. It is this variation in the genome sequence that accounts for phenotypic differences, and for different susceptibilities to a large number of diseases. In humans, if we knew which sequence differences predisposed individuals to common human disorders such as hypertension or Alzheimer's disease, we could develop preventive treatments that could prevent disease, despite the genetic risk. It is this vision that has driven researchers involved in the Human Genome Project over the past years. Likewise, if we understood the genetic variation in model organisms, such as rats or mice, responsible for disease susceptibility or variability in drug metabolism and efficacy, we could use this information to test novel drugs more efficiently and target specialized drugs toward individuals with specific genetic predispositions.

Although there has been significant progress in biomedical research to suggest that this vision may one day become reality, we do not yet have a good understanding of the genetic differences in the genome sequence, and how they influence the development of disease. One reason why progress has been slow is the fact that there are so many differences. The sequence of any two human genomes differs in at least three million bases. Most differences in mammalian genomes are single base pair differences in the DNA sequence, but others include length variation in repeated sequences and chromosomal rearrangements. This presents a tremendous challenge to geneticists and biomedical researchers: How do you distinguish between DNA sequence variation that has no functional consequences and the differences that affect gene function and possibly the well-being of the organism? After all, only about 5% of any mammalian genome encodes for genes, and only small portions of the remaining 95%, have discernible functions. As a consequence, it is likely that most DNA sequence variation has no effect on the organism because it is located in these regions without (known) function.

To further identify the determinants of genetic disorders, different approaches have been used to zoom in on regions of the genome that contain the mutation(s) that are responsible for the genetic defect. Both commonly used approaches, family-based linkage studies and association analyses, require genetic markers, i.e., sequences that are variable between individuals and can be used to distinguish the DNA from different individuals. Even before the completion of the various genome sequencing projects, linkage analyses identified numerous loci in the genomes of both humans and model organisms that harbor genes involved in the development of a variety of common disorders. These studies used genetic polymorphisms, i.e., differences in the DNA sequence, to distinguish individuals with and without the disease and to identify regions shared by affected individuals. It is this initial use of DNA sequence variation as markers for genetic studies that this chapter will focus on. We will describe two types of genetic variation commonly used as markers in genetic studies, describe how they were discovered and can be interrogated by genotyping in large numbers of individuals, and will end by briefly describing the approaches of linkage and association studies that are aided by and rely on these genetic markers. Numerous review articles have summarized different aspects of this chapter, and we will focus primarily on the methodologies underlying the different genotyping approaches commonly used today in genetic laboratories.

5.7.2 SIMPLE SEQUENCE LENGTH POLYMORPHISMS (SSLPs)

5.7.2.1 General Description

Eukaryotic genomes contain interspersed repetitive sequence elements, many of which have no known function. Incredibly, these "nonfunctional" sequences comprise about 20% of the human genome [1]. It is thought that these repetitive motifs were generated by transposons or retrotransposons that replicate and then "jump" to other locations to propagate their genome, in the form of either DNA (in the case of transposons) or RNA (in the case of retrotransposons). Some repetitive sequence elements are long, about 1000–5000 base pairs (bp) in length, and they are known as LINES (long interspersed elements), whereas others are shorter, about 200–500 bp, and are known as SINES (short interspersed elements). The Alu element, approximately 300 bp in length, is the most abundant SINE in the human genome [2] and comprises about 3% of the human genome.

Another type of repetitive element consists of very short sequences (2 to 5 bp), arrayed in tandem. These sequences, called microsatellites, simple sequence length polymorphisms (SSLPs), simple sequence repeats (SSRs), or short tandem repeat polymorphisms (STRPs), tend to vary in unit size (from 5 to 30 units). The number of these tandem elements varies, and as a result, so does their length, making these sequences *polymorphic* (having different forms). The most abundant class of microsatellite repeat, is the dinucleotide repeat of cytosine and adenine, or (CA) · (GT) repeats [3, 4]. In the human genome, these dinucleotide repeat microsatellites occur every 30 to 60 kbp, corresponding to approximately 50,000 to 100,000 microsatellites in the 3 billion base pair human genome [5]. Other common microsatellites include trinucleotide and tetranucleotide repeats (3 and 4 bp repeats, respectively) [6].

The length of the microsatellite repeat is thought to be due to errors in DNA replication [7]. During DNA replication, the polymerase machinery is thought to "slip," either eliminating or adding repeat units. The deletion or addition of a single repeat unit occurs most frequently (91%) [8]. The frequency of this type of mutation, on average, is 1.2×10^{-3} per microsatellite, translating to roughly 1 in 1000 meioses (as determined by examining how often a parent passes an expanded or contracted microsatellite repeat to his/her offspring) [8]. The range of mutation frequency for any given microsatellite is 0 to 8×10^{-3} and seems to occur more frequently in the male vs. the female germline.

With a few exceptions, microsatellites do not have functional consequences on clinical outcomes. They are most commonly found in intergenic or intronic portions of a eukaryotic genome. However, it is important to keep in mind that, although the microsatellite is repeated throughout the genome, there is a unique sequence flanking it, corresponding to the location within the genome in which it integrated. Therefore, this class of polymorphic repeat is ideal for an assay that can tag a specific genome location for use in genetic studies, which will be discussed in detail in this chapter.

5.7.2.2 Discovery

The microsatellite was first discovered in 1981 by Miesfeld et al. [9] by screening a human gene clone library with a mouse 6-kbp (6 kilobase pair) ribosomal gene nontranscribed spacer probe (rDNA NTS). They found this sequence was interspersed not only throughout the human and mouse genomes, but also in every other eukaryotic organism they studied—pigeon, frog, slime mold, and yeast. However, it was not identified in *Escherichia coli*. Therefore, the rDNA NTS has been conserved throughout all eukaryotes, but not down to prokaryotes. Miesfeld et al. then narrowed down the conserved element, which they called ECl (evolutionarily conserved), to a 251-bp fragment of the rDNA NTS that contained a tandem array of the (TG) · (CA) dinucleotide. However, the reason for its dramatic evolutionary conservation remained a mystery.

In 1989, two back-to-back manuscripts were published, by Weber and May [4] and Litt and Luty [3], reporting the use of microsatellites as genetic markers; i.e., they could be used to identify inheritance patterns from parent to offspring. By cloning and sequencing several microsatellites, it was found that each tandem repeat was flanked by a unique genomic sequence, thereby providing the ability to identify an individual's "genotype" at a specific DNA location (locus) containing a microsatellite. By sequencing 59 of these microsatellites, Weber and May found that more than half of them contained repeats of 13 or greater dinucleotides, with the longest being 30 units in length [4]. They went on to test 10 microsatellites for Mendelian inheritance. All microsatellites tested displayed expected Mendelian inheritance in three-generation families.

Figure 5.7-1 displays a cartoon example of the inheritance of a single SSLP marker. Each band represents the length of the microsatellite fragment, separated according to its length by gel electrophoresis. Each parent, being diploid, has two alleles for the microsatellite, each having a different number of dinucleotide repeats. The mother and father each contribute one of these alleles to their offspring, with a 50% probability of a child inheriting either allele. The inheritance of each allele from their parents is shown in the children, in the expected Mendelian ratios.

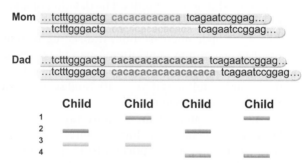

Figure 5.7-1. Schematic of SSLP genotypes. The top two bars represent a mother's chromosomal fragments containing two different alleles of a $(CA)_n$ repeat, the first having six tandem repeats (red allele) and the second having seven tandem repeats (green allele). The bottom two bars represent a father's alleles (eight repeats = blue; nine repeats = purple) for the same SSLP. The bottom represents genotypes from four children of the above parents, as they might appear on an electrophoretic gel. The four columns indicate four different children's genotypes for the SSLP, with the numbers indicating the four possible alleles. The colored bars correspond to the alleles inherited from the parents, representing all possible allele combinations according to Mendelian inheritance. (This figure is available in full color at ftp://ftp.wiley.com/public/sci_ tech_med/pharmaceutical_biotech/.)

The discovery of the microsatellite as a highly polymorphic genetic marker, and its ability to generate genotypes in a high-throughput manner, revolutionized genetic analyses of disease and greatly facilitated the mapping and sequencing of not only the human genome but also several model organisms, as discussed in the following sections.

5.7.2.3   Methods for Genotyping

The major advantage of genotyping using SSLPs over previously used genetic markers, e.g., restriction fragment length polymorphisms (RFLPs), is twofold. First, SSLPs are highly polymorphic, with alleles ranging anywhere from just a few to over 30 nucleotide repeat units. This greatly increases the likelihood that an individual will have alleles of differing sizes for any assayed SSLP, i.e., the marker is informative, and can be used across multiple families for genetic analyses [10]. In fact, many tens of thousands of SSLP assays have been developed for organisms ranging from human to mouse, rat, dog, cat, cow, chicken, pig, fish, frog, fruit fly, corn, rice, poplar, and yeast.

Second, the relatively short stretch of DNA containing a microsatellite allows it to be readily assayed using the polymerase chain reaction (PCR). PCR is a chemical reaction that can exponentially amplify DNA fragments [11]. In the case of SSLP assays, short oligonucleotide primers (~18–25 nucleotides in length) are generated that are complimentary to sequences flanking either side of the tandem repeat. The double-stranded genomic DNA is denatured (unraveled) by heating to 94°C, after which the reaction is cooled to approximately 55°C to 60°C. This temperature allows the oligonucleotide primers to anneal to the denatured genomic DNA. Finally the reaction is heated to 72°C, which allows a temperature stable polymerase, generally *Taq* polymerase isolated from a thermostable bacterium (*Thermus aquaticus*), to extend the oligonucleotide chain, using the complimentary genomic DNA strand as its template and the four dNTPs (dATP, dTTP, dCTP, and dGTP) as building blocks to extend the DNA fragment. This procedure is repeated multiple times, doubling the copy of the sequence contained between the primers with each cycle. Therefore, at the end of 30 cycles of consecutive denaturing, annealing, and extension, there are 2^{30} copies of the desired fragment, which is a sufficient amount of product for detection using radioactive isotopes or fluorescent dyes, as described below. Because PCR can be carried out in high-density microtiter plates (up to 1536 wells/plate) on commercial thermocyclers (machines that heat and cool the reaction samples required for PCR), genotyping of thousands of individuals can be performed in a single day.

Radioactive Detection. Initially, PCR amplified product (amplicon) containing an SSLP, was most commonly detected using gel electrophoresis and a radioactively labeled nucleotide. This was done by either directly incorporating a radioactive tag in the PCR amplicon or by labeling the oligonucleotide primer. For direct incorporation, a radioactively tagged nucleotide, either ^{32}P or ^{35}S attached to the α position of one of the dNTPs (i.e., $\alpha^{32}P$-dCTP or $\alpha^{35}S$-dATP), is added to the PCR reaction (for detailed protocol, see Ref. 12). During each extension step of the reaction, labeled nucleotides are directly incorporated into the amplicon. An alternative approach used to detect the PCR amplicon is to label one primer used in PCR. In this approach, ^{32}P attached to the γ position of a dNTP (generally γ^{32}P-dATP or γ^{32}P-dCTP) is added to the 3′ end of one PCR primer, using T4 DNA kinase (for detailed protocol, see Ref. 12) before the PCR reaction. During PCR, this labeled primer is annealed to the target sequence and extended, thereby labeling the PCR amplicon. The advantage of direct incorporation is that only a single step amplifies and labels the desired PCR fragment; however, this method can also lead to labeling of nonspecific PCR product, making results more difficult to interpret. End-labeling of the PCR primer can eliminate the detection of nonspecific product, but also it requires high levels of radioactivity and an extra experimental step. However, both approaches have been used extensively and successfully, and which approach is used depends on individual preference.

To detect the radio-labeled SSLP amplicon, one must be able to separate the labeled fragments, based on their length. With radioactively labeled amplicons, this is done using denaturing gel electrophoresis. Reaction samples are mixed with a loading buffer containing blue tracking dye and formamide, and then heated to 94°C to denature the product. The formamide keeps the product denatured. Samples are then loaded onto a standard denaturing polyacrylamide sequencing gel. As the fragments migrate through the gel, the smaller sized fragments (containing fewer tandem repeats) migrate more quickly, whereas the larger fragments move more slowly, separating the two alleles. The blue dye allows tracking of the migration so that the product does not travel off the end of the gel. The polyacrylamide gel is then transferred to a piece of paper, dried, and exposed to autoradiographic film for detection and genotype determination. Figure 5.7-2 shows an example of an autoradiograph result of the radioactive genotyping method.

Fluorescence Detection. The use of radioactivity has been largely supplanted by fluorescent dyes for genotyping, due to the obvious reduction of health hazards and a greater flexibility (as discussed later). As with radioactive labeling, the PCR amplification of the specific SSLP is largely the same, except that one dNTP is labeled with a fluorescent dye (fluorophore). Also, rather than detecting the electrophoresed product by autoradiography, the fluorescently labeled amplicon is

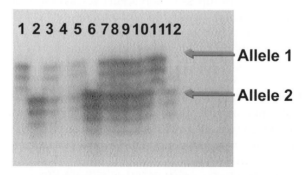

Figure 5.7-2. X-ray autoradiographic film from radioactive genotyping. As in Figure 5.7-1 each column (labeled lanes 1–12) is the genotype of a single SSLP marker in 12 individuals. The two possible alleles are indicated by the red arrows. (This figure is available in full color at ftp://ftp.wiley.com/public/sci_ tech_med/pharmaceutical_biotech/.)

detected as it passes a laser detector, either on a gel-based (e.g., an ABI 377 automated slab gel sequencer) or capillary-based DNA sequencer (e.g., an ABI 3700 or 3730 capillary sequencer). A major advantage of using fluorescent detection lies in the automation of genotype determination. As the labeled product passes by the laser detector, the signal is transferred to a computer algorithm that automatically converts the signal detection to a genotype. This greatly reduces manual genotype interpretation, reducing technical time and error as well as increasing overall genotyping throughput. Another important advantage of fluorescent geno-typing lies in the ability to label different SSLP amplicons with different fluorophores. Up to three different fluorophores are commonly used, 6-FAM, HEX, and NED, each of which has a distinct excitation and emission profile. A fourth dye, ROX, is used to label a common internal size standard, allowing accurate determination of product size [13]. The ability to distinguish multiple SSLP amplicons, each labeled with a different fluorophore, allows for simultaneous detection of multiple SSLPs in a single assay, increasing genotyping throughput. This approach is also amenable to high-throughput robotic sample preparation, as it eliminates the precautions necessary for handling radioactivity.

Again, as with radioactive detection, the fluorescent tags can either be directly incorporated into the amplicon or added to the PCR primer. A disadvantage of the labeled primer is that each SSLP must be synthesized with the fluorescent tag attached, which greatly increases the cost of the primer. Oetting et al. [14] devised a method to overcome this cost issue by combining unlabeled template-specific primers and a labeled, common M13 primer. By this method, one template-specific primer for each SSLP is synthesized with an 18-bp tail specific to the M13 phage on its 5′ end. This primer is combined with its template-specific mate, as well as a common M13 primer labeled with a particular fluorophore, and subject to PCR amplification. Figure 5.7-3 shows the general strategy for this method. The amplification begins with the tailed primer incorporating into the template sequence in the initial cycles. In subsequent reactions, the incorporated sequence becomes the docking site for the M13 primer-dye conjugate. The single M13 tailed primer-dye conjugate is common for all SSLPs, greatly reducing the cost of genotyping.

Multiplex Analysis. As discussed, using fluorescent detection of SSLPs greatly improves genotyping throughput because of the availability of multiple fluorophores and the ability to multiplex (determine genotypes of multiple SSLPs simultaneously). The amplicon size of SSLPs varies not only within a single SSLP but also from one SSLP to another, depending on the distance from the repeat the flanking primers are chosen. This allows for the ability to detect multiple SSLPs within a single sample on a gel or capillary, as again each SSLP migrates according to size. For example, the allele size range for one SSLP amplicon may be 120–140 bp, whereas another may range from 200 to 250 bp, and a third may range from 325 to 341 bp. Therefore, all three amplicons could be combined in a single reaction, as the laser can detect the size differences and their ranges are sufficiently different to differentiate the genotypes of the three SSLPs. Furthermore, because the fluorophores have distinct excitation/emission profiles, multiple fluorophores can also be combined in a single sample. Therefore, by combining amplicons of distinct sizes labeled with one fluorophore with those labeled with other fluorophores, one

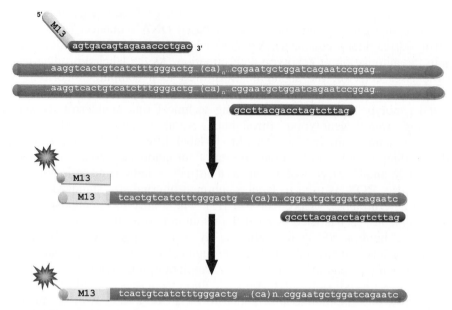

Figure 5.7-3. Schematic of fluorescent genotyping using an M13-tailed unique PCR primer (green with pink 5′ tail), an untagged unique PCR primer (green), and a common M13-dye conjugated primer (yellow). In the first PCR cycles (top), the unique primers hybridize to the DNA flanking either side of the (CA)$_n$ repeat. The M13 tag is incorporated into the PCR product (middle), creating a site for the fluorescently tagged complementary M13 primer to anneal to and label the PCR product with a particular fluorescent tag (FAM; dark blue star). This product is then detected by either a gel-based or capillary-based DNA sequencer. (This figure is available in full color at ftp://ftp.wiley.com/public/sci_tech_med/pharmaceutical_biotech/.)

can genotype up to nine SSLPs in a single sample. Figure 5.7-4 displays the results of multiplexing six SSLPs of different sizes and colors (indicating different fluorophores) on a single electrophoretic gel.

Two means of multiplexing are possible, either at the PCR or the electrophoresis steps. It is possible to combine PCR primers specific for multiple SSLPs in a single PCR reaction, whereby each SSLP is amplified simultaneously. The advantage of this approach is the reduction in PCR costs, as the overall reaction size is not modified. However, this approach requires up-front knowledge of the allele-size range of each SSLP in the samples being genotyped. Furthermore, some level of up-front optimization of the PCR reaction may be required so that each SSLP amplicon is generated with the same efficiency within the reaction. Once a set of two or three SSLPs are optimized for PCR multiplexing, these combinations are generally not modified. This approach is most efficient when the same set of SSLPs is used time and time again with varying DNA samples. The other multiplexing approach involves independent PCR amplification of each SSLP, followed by pooling of the amplicons before electrophoresis (gel or capillary). The advantage of this approach lies in the freedom to mix and match different marker sets. Marker sets may be slightly different depending on the population being genotyped. In some cases, the entire genome may be screened for initial linkage results. In other cases, an

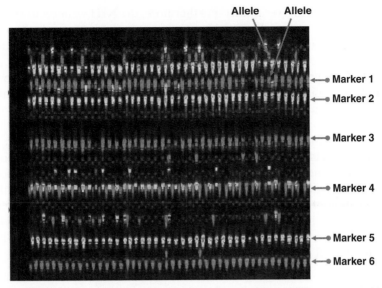

Figure 5.7-4. Multiplex genotyping results from an ABI 377 DNA sequencer. PCR product from six different SSLPs in the same DNA sample were pooled and electrophoresed on the ABI 377 slab gel sequencer. Three SSLPs of independent size ranges were labeled with FAM (blue), and three SSLPs of independent size ranges were labeled with NED (yellow) according to the method outlined in Figure 5.7-3. The red product is a ROX-labeled internal size standard that allows automated determination of allele sizes and conversion to individual genotypes. The laser detection can distinguish between the overlapping FAM-labeled and the NED-labeled SSLP products. (This figure is available in full color at ftp://ftp.wiley.com/public/sci_ tech_med/pharmaceutical_biotech/.)

individual locus may be the focus of higher density mapping, for instance, to narrow a disease gene interval by identifying critical recombinant individuals. This assay allows the individual to mix and match markers from a characterized set to do these specific scans. By this approach, multiple pooled amplicons, regardless of primer sequence, have a high success rate.

The use of fluorescent detection, combined with robotics and automated data analysis pipelines, facilitate highly automated, high-throughput genotyping platforms, resulting in thousands of genotypes per day at a relatively low cost.

5.7.2.4   Tools and Resources

To increase the efficiency of genotyping, major efforts have been made to generate a well-characterized and highly efficient marker set for genetic linkage analysis. One such effort was the Cooperative Human Linkage Center (CHLC). This National Institutes of Health (NIH)-funded project characterized thousands of di-, tri-, and tetranucleotide repeat polymorphisms to determine their informativeness (i.e., the number and frequency of SSLP alleles) [6, 10] and to array them according to their genetic position in the human genome (i.e., to develop high-quality genetic maps) [15]. Dr James Weber, a member of this Center, and his colleagues developed several genome-wide marker sets (http://research.marshfieldclinic.org/genetics/

Genotyping_Service/mgsver2.htm). Furthermore, the NIH went on to fund a centralized genotyping center, allowing investigators to submit research proposals for genome-wide linkage scans of their study populations (http://www.nhlbi.nih.gov/resources/medres/genotype.htm). A European group, Genethon, also led a major effort to develop human SSLPs and generate high-density genetic maps for use in genetic linkage studies [16]. These efforts have also reached the commercial arena, where companies such as DeCode will generate genotypes on a genome-wide or region-specific basis on a fee-for-service basis (http://www.decode.com/).

The discovery and use of microsatellites have revolutionized the ability to perform genome-wide genetic linkage studies, resulting in the positional cloning of many important genetic diseases ranging from mental retardation, glaucoma, heart defects, to neurological disease. They are particularly powerful for the localization and identification of single-gene disorders. Furthermore, the high-density genetic maps generated using microsatellite markers were instrumental to laying the groundwork for the eventual sequencing of not only the human genome, but also several other model organisms.

5.7.3   SINGLE NUCLEOTIDE POLYMORPHISMS (SNPs)

5.7.3.1   General Description

DNA is constantly modifying and mutating. During DNA replication, the duplication of the genetic material during cell divisions and inaccurate copying of the existing DNA leads to minor sequence changes in the copied version of the genome. Likewise, environmental influences such as mutagenic chemicals, ultraviolet (UV), or other radiation can also alter the DNA. On the basis of these constantly occurring changes in the DNA sequence of any organism, it becomes obvious that all individuals of a species are slightly different based on their DNA sequence. Most of these differences are single base changes in the DNA. Essentially one nucleotide in the sequence of one individual is replaced by a different nucleotide in another individual. These differences, illustrated in Figure 5.7-5, are called *single nucleotide*

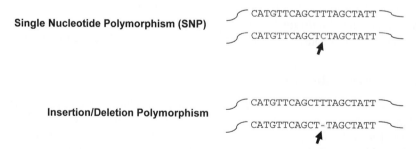

Figure 5.7-5. Single nucleotide polymorphisms and insertion/deletion polymorphisms. Two sections of a DNA strand are shown. The top of the figure shows a SNP, a single nucleotide difference between the two DNA strands (highlighted by a red nucleotide and the arrow). In the bottom diagram, one base is missing in the lower DNA strand (deletion polymorphism), leading to a frame shift in the DNA sequence. (This figure is available in full color at ftp://ftp.wiley.com/public/sci_ tech_med/pharmaceutical_biotech/.)

polymorphisms (SNPs). SNPs are the most common genome sequence variation. For the human genome, a comparison of two individuals reveals on average one SNP every 800 bases [17]. As a consequence, two humans differ in their DNA sequence at over 3 million positions, or approximately 0.1%. Similarly, other mammalian species show significant sequence variation as well. In mice, one SNP can be found every 250 bases across the genome when 13 inbred (i.e., genetically identical) and two wild strains are compared [18]. Inbred rat strains show a similar rate of SNPs [19]. These sequence differences are responsible for all genetically caused differences between individuals, such as eye color, hair or coat color, or disease susceptibility. If we could unravel the DNA sequence of every human individual, we would be able to decipher all genetic mutations that cause the most common diseases such as diabetes, stroke, or Alzheimer's disease. Unfortunately, though, the cost of sequencing every patient far exceeds the available budgets for these studies. Therefore, scientists have focused on discovering the most important and most common SNPs in the human genome, and now use this subset of SNPs in the human genome to investigate the genetic causes of these disorders [20]. Similar efforts are under way in other mammalian (and nonmammalian) species that serve as model organisms for specific diseases or traits in humans. Below we will describe and discuss the methods commonly used to discover SNPs, and the methodologies used to interrogate these SNPs in large numbers of individuals simultaneously by genotyping.

5.7.3.2 Discovery

Often, SNPs are discovered by sequencing the same genomic sequence in a number of unrelated samples. For this approach, a region of genomic DNA is amplified by PCR from different DNA samples. The resulting amplicons are resequenced using common sequencing technology. Approaches using gel-based resequencing and hybridization-based oligonucleotide array sequencing have successfully been used for large-scale SNP discovery efforts in the human and mouse genome as well as in other species. Analysis of the resulting sequences reveals base pair differences, or SNPs. Software tools have been developed to automate the SNP discovery from sequencing data [21], and these tools are routinely used today in a variety of organisms for SNP discovery. An example of an SNP as it would appear in sequencing data is shown in Figure 5.7-6.

Although sequencing is commonly used for SNP discovery today, other indirect methods exist to uncover genomic sequences containing SNPs. Given the cost of sequencing, it may be desirable to preselect these DNA fragments for subsequent sequencing, and to eliminate DNA sequences that do not contain polymorphisms.

One of the first reported methodologies for SNP discovery was the PCR-based single-stranded conformational polymorphism method [22]. In this approach, a region of DNA is amplified using PCR. The resulting amplicons are denatured, and they are allowed to re-anneal under conditions that favor the formation of single-stranded secondary structures. As the formation of secondary structures is sequence-dependent, any difference in the sequence of the PCR amplicon would lead to different folding. These differently folded products can be separated through non-denaturing polyacrylamide gel electrophoresis. Samples with different

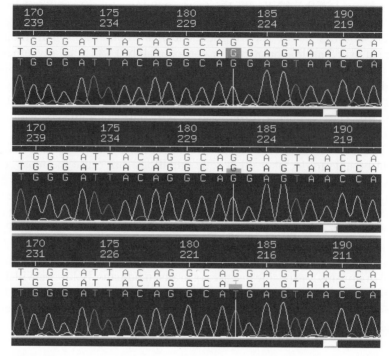

Figure 5.7-6. SNP discovery using the consed and polyphred software tools. The figure depicts three DNA traces from different individuals. Sequencing was performed using BigDye Terminator Chemistry. The white vertical bar highlights a variant nucleotide in the sequence. The top two traces have a "G" at this position, and the bottom trace shows a "T". (This figure is available in full color at ftp://ftp.wiley.com/public/sci_tech_med/pharmaceutical_biotech/.)

sequences will fold differently and, as a result, migrate differently in the gel, allowing for the detection of SNPs. This method has been used extensively for targeted screening of genes for potential disease-causing mutations in a variety of species. Recently, it has been adapted for capillary electrophoresis to facilitate large-scale screening (for a recent review, see Ref. 23).

Another approach to identify PCR amplicons that contain sequence variants uses denaturing high-performance liquid chromatography (DHPLC) [24]. In this approach, PCR amplicons are first denatured and then reannealed. In contrast to SSCP analysis, products are then allowed to form double-stranded duplexes. If the PCR product contains an SNP, this re-annealing process will generate homoduplexes (matching sequences reanneal) and heteroduplexes (strands containing the different two alleles of an SNP will anneal and form a single base pair mismatch). When these duplexes are separated on a liquid chromatography column under partially denaturing conditions (separation at temperatures close to the melting temperature of the duplex), the elution times of homo- and heteroduplexes will differ because the melting temperature of the heteroduplex will be slightly lower than the homoduplex due to the base pair mismatch in the sequence. The method and instrumentation has been automated, and it is commercially available. DHPLC has been used successfully for the identification of mutations in a large number of disease genes, and it has been adapted for the analysis of cancer samples and chromosomal abnormalities [25–32]. Other modifications of the heteroduplex analysis exist, and they have been used by laboratories for a variety of studies. These include mobility analysis [33] or enzymatic cleavage of heteroduplexes [34].

One additional methodology for SNP discovery should be mentioned because it is based on an entirely different experimental approach. In 1995, Faham and Cox [35] described a method called mismatch repair detection (MRD) that used the bacterial mismatch repair system in *E. coli* to enrich for DNA fragments containing SNPs. The template for mismatch repair in *E. coli* is hemi-methylated double-stranded DNA (only one strand of the DNA is methylated), which is formed by mixing and annealing different single-stranded DNA samples grown in methylation-competent and -incompetent *E. coli* strains. The method has been developed into a high-throughput screening platform that can simultaneously analyze pooled PCR amplicons from large numbers of individuals [36, 37]. The method has been shown to enhance the discovery of rare SNPs that would be missed in pooled sequencing (or would require large numbers of individual samples to be sequenced).

In summary, several methods exist to identify SNPs in genomic DNA. Although some of them may be useful to enrich for SNP-containing PCR amplicons before traditional DNA sequencing, the commonly used approach of discovering SNPs from aligned sequence data from different individuals remains the "gold standard" and the basis of the majority of SNPs publicly available in databases.

5.7.3.3   Methods for Genotyping

Numerous methods have been developed and are being marketed for SNP genotyping, i.e., the specific determination of the nucleotide present at a polymorphic site in a genome. Theoretically, this information could be obtained by resequencing each and every sample using the same approach as for SNP discovery. However, this process is labor- and cost-intensive and provides significantly more information than is needed; i.e., it delineates the complete flanking sequence of the SNP, which, in most cases, will not differ between samples, and therefore is redundant. Instead, genotyping methods provide a cost-efficient alternative that can quickly generate genotype information on large numbers of samples. No attention is paid to the remainder of the genome sequence and its potential variation; the only information obtained is the nucleotide present in the sequence at the specific site of a known SNP. To develop methodologies for this genotyping procedure, the SNP in question, its possible nucleotides (alleles) that can be present at the site, and the sequence adjacent to the SNP need to be known. All methods discussed below use this information to obtain genotyping information for a large number of DNA samples. As outlined below, the different methods allow the analysis (genotyping) of individual SNPs in large numbers of samples, or varying numbers of SNPs for one sample at a time. All methods have been shown to result in reliable genotyping results with low error rates. The choice of the appropriate methodology primarily depends on the number of SNPs and the number of samples that need to be interrogated.

Essentially all genotyping approaches used to date fall into one of four categories: allele-specific hybridization, primer extension, oligonucleotide ligation, or invasive cleavage. Figure 5.7-7 illustrates these different approaches. We will first

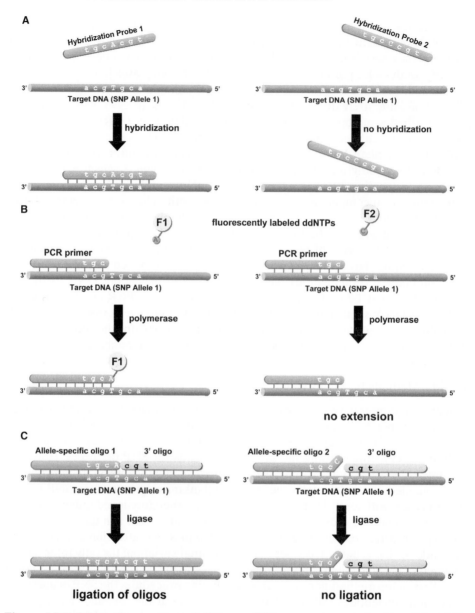

Figure 5.7-7. Schematic overview of different SNP genotyping approaches. (**A**) Allelic hybridization. (**B**) Primer extension. (**C**) Oligonucleotide ligation. (**D**) Invasive cleavage. (This figure is available in full color at ftp://ftp.wiley.com/public/sci_tech_med/pharmaceutical_biotech/.)

briefly describe the principles of each approach and then outline commercially available genotyping platforms based on these approaches for different study applications.

Allele-specific Hybridization. Allele-specific hybridization distinguishes the allele present at an SNP using two different oligonucleotide probes. Each probe

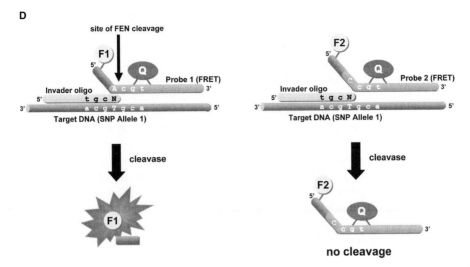

Figure 5.7-7. *Continued*

contains the SNP position in the middle and is complementary to one allele of the SNP. For hybridization to genomic DNA or PCR amplicons, conditions are chosen that allow the hybridization of the matching probe but not the mismatch probe. The successful hybrids between the target DNA and the oligonucleotide probes can be detected using a variety of approaches. Methods include the use of fluorescence resonance energy transfer (FRET) probes (e.g., in the 5′ exonuclease TaqMan assay, described below), molecular beacon probes, or oligonucleotide arrays (Affymetrix GeneChip system, see below).

Primer Extension. Primer extension uses the standard principle of PCR. Here, an oligonucleotide primer extends up to the SNP site. The assay then determines what nucleotide is incorporated into the extended PCR product by the polymerase. Often, the assay is performed in two steps. First, the genomic interval is amplified by PCR. The amplified product is subsequently mixed with another primer that anneals adjacent to the SNP site. Then individual dideoxynucleotides are added to the reaction. Upon incorporation of the matching nucleotide at the SNP site, the resulting extension product can no longer be extended further because it lacks a 3′–OH group required for nucleotide incorporation. Thus, the reaction essentially stops once the nucleotide has been incorporated.

Dideoxynucleotides are commercially available with different fluorescent tags; therefore, the resulting extension products can be examined using laser scanners to see which fluorophore has been incorporated. This method directly reveals the genotype of the template DNA. The method can be multiplexed and is commercially available (SNaPshot, see below). Alternatively to the use of dideoxynucleotides, regular deoxynucleotides can be used sequentially in the extension reaction. Here, the primer is essentially extended as in a regular PCR reaction. The only difference is the sequential addition of one nucleotide at a time. This method is commercially available (see pyrosequencing, described below).

Oligonucleotide Ligation. Similar to DNA polymerase, DNA ligase is a highly specific enzyme that repairs nicks in the DNA. Nicks are missing phosphodiester bonds in one of the two strands of the DNA molecule. DNA ligase detects and repairs these nicks, provided the nucleotides on both sides of the missing bond are complementary to the nucleotides in the intact DNA strand. Landergren et al. [38] first described a novel approach for SNP genotyping that uses specific oligonucleotides that will hybridize to single-stranded DNA templates. One oligonucleotide will align with the sequence upstream of an SNP, including the actual nucleotide at the SNP site. A second oligonucleotide is complementary to the downstream template sequence, starting with the first nucleotide adjacent to the SNP. Although the latter oligonucleotide can be used to interrogate both alleles of the SNP, the first oligonucleotide is specific for one allele of the SNP, because it will end with the specific SNP nucleotide. Only the oligonucleotides perfectly matching the template sequence will be ligated by the ligase, whereas a mismatch will prevent the ligation of the alternate oligonucleotide. The method has been incorporated in a variety of commercial platforms such as the SNPlex platform or the Illumina GoldenGate technology, which is described below in more detail.

Invasive Cleavage. The last SNP genotyping method we describe is invasive cleavage, which is commercially available as the Invader assay. This method depends on the ability of flap endonucleases to recognize specific three-dimensional structures that are formed when two overlapping oligonucleotides hybridize perfectly to a target DNA molecule [39]. The overlap occurs at the SNP site, and only the oligonucleotide combination that is complementary to the SNP allele will be cleaved by the enzyme. The reaction is highly specific, and a mismatch to the target DNA will prohibit the enzymatic reaction. This methodology has been adapted for SNP genotyping assays, as described in more detail below.

Analysis of Individual SNPs. Historically, investigators have used a variety of PCR-based methods to genotype individual SNPs. For all of these methods, the genomic sequence around the SNP was known and amplified using specific PCR primers that selectively amplified the genomic interval, including the SNP. SNPs that alter a restriction enzyme recognition site could then be genotyped by performing a restriction digest of the PCR product and by separating the resulting DNA segments by agarose gel electrophoresis. PCR products that contain the restriction enzyme recognition site will be cleaved into two products and will result in two bands of different size on the agarose gel. In contrast, PCR products from samples that contain an SNP in the recognition site will no longer be cleaved by the enzyme, and will result in a single band of larger molecular weight upon gel electrophoresis. Finally, PCR products from samples heterozygous for the SNP would result in a total of three bands, one band for the uncleaved product and two additional bands from the portion of the PCR product that can be cleaved. A schematic of the procedure is shown in Figure 5.7-8. The method enjoys widespread use to this day because it is robust and uses basic molecular biology technology available to most laboratories. With the ever increasing number of commercially available restriction enzymes uncovered from various microbes, the repertoire of SNPs that can be genotyped by this approach has been steadily increasing.

Unfortunately, many SNPs do not lead to an alteration of a restriction enzyme recognition site in the genome sequence. For these SNPs, alternative approaches have been developed that use PCR and gel electrophoresis. In one approach, one

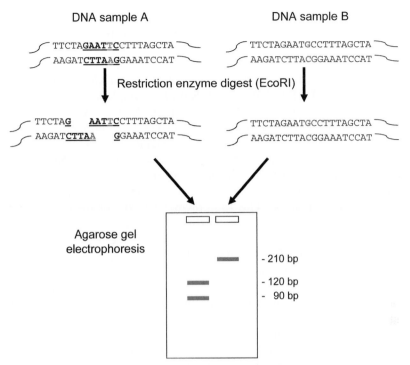

Figure 5.7-8. Schematic overview of PCR–RFLP. Double-stranded DNA molecules are shown for DNA sample A and B. These two samples are PCR amplicons from genomic DNA. The two sequences differ for the base pair highlighted in red. Sample A contains a restriction enzyme recognition site for *Eco*RI (underlined), and sample B does not. After restriction enzyme digestion, agarose gel electrophoresis results in different DNA fragments for the two samples. (This figure is available in full color at ftp://ftp.wiley.com/public/ sci_ tech_med/pharmaceutical_biotech/.)

initial PCR primer is designed so that it includes the polymorphic SNP site at its 3′ end. Two separate primers are used in two PCR reactions. One reaction includes the forward primer ending with one nucleotide present at the SNP, and the other reaction includes a primer ending with the alternative nucleotide. As the thermo-stable DNA polymerase used in PCR requires a perfect match of the primer at the 3′ end to attach additional nucleotides and amplify the DNA segment, the PCR reaction will only yield a PCR product when the forward primer perfectly matches the DNA template. Accordingly, the presence or absence of PCR product depending on the primer used indicates the genotype for the template DNA. Although this method works well for established assays, each assay needs to be optimized individually to ensure that both alleles of the SNP are successfully amplified when present. Furthermore, the design of the PCR primers may be difficult for some SNP sequences.

As an alternative to the allele-specific PCR and the restriction digest of PCR products (often called PCR–RFLP), several commercial platforms are available to genotype individual SNPs in large numbers of samples. In our discussion here, we will focus on three commonly used platforms: the 5′-exonuclease TaqMan assay, based on allele-specific hybridization; pyrosequencing, based on the primer extension principle; and the Invader assay, based on the invasive cleavage reaction of a flap endonuclease.

TaqMan Assay. The TaqMan assay uses the 5′ exonuclease activity of Taq polymerase [40]. Two FRET oligonucleotide probes complementary to the sequence around an SNP are hybridized to the target DNA. Each contains a different fluorescent dye that is quenched in the FRET probe. Each probe is specific for one of the two SNP alleles; i.e., it contains the complementary base to the SNP allele in the middle of the probe. The hybridization temperature is chosen such that only the matching probe hybridizes efficiently. After hybridization, a normal PCR extension reaction begins starting from an oligonucleotide primer annealed upstream of the FRET probe. The Taq polymerase extends the primer and, upon reaching the hybridized FRET probe, begins to remove it from the template DNA by cleaving one nucleotide at a time. The reaction is mediated by the specific 5′ exonuclease activity of the *Taq* polymerase. Once the entire oligonucleotide FRET probe has been cleaved, the fluorophore and the quencher molecule have been separated, and the energy transfer can no longer take place because the two molecules are no longer in close physical proximity. As a consequence, the fluorophore will now emit light upon excitation. As the two FRET probes contain different fluorophores depending on the allele, the resulting fluorescent color unambiguously determines the allele of the SNP in the target DNA.

TaqMan assays are commercially available for many SNPs and can be designed on demand by Applied Biosystems. The method requires a real-time PCR instrument or a fluorescent plate reader to measure the fluorescence of the two dyes at the end of the PCR reaction.

Pyrosequencing. Pyrosequencing [41] has been described as a sequencing-by-synthesis method. In this approach, four different enzymes and specific substrates are used in a sequential reaction to produce light whenever a nucleotide complementary to the template DNA strand is incorporated. If the added nucleotide is not complementary to the base in the template DNA, no light is generated. The light-generating reaction is depended on the pyrophosphate that is released during the DNA polymerase reaction. If the added nucleotide forms a base pair, the DNA polymerase incorporates the nucleotide and pyrophosphate will be released. The released pyrophosphate will be converted to ATP by another enzyme (ATP sulfurylase). Luciferase then uses the ATP to generate a detectable light signal. The light intensity is proportional to the number of the ATP molecules, which in turn is proportional to the number of the incorporated nucleotides. After the completion of the reaction, the excess of each nucleotide is degraded by apyrase.

The reaction is repeated in the same order, with a different nucleotide added. If the added nucleotide does not form an incorporated base pair with DNA template, no light will be produced. The four nucleotides are added to the mixture in a defined order (e.g., ACGT), and thus, the sequence of the template DNA can be deduced by matching the nucleotide with the light signal generation.

Pyrosequencing provides rapid real-time determination of 20–30 nucleotides of target DNA sequence. With this technology, not only the SNP alleles but also the adjacent nucleotides are determined, providing greater confidence in the correct SNP detection. The reaction is automated and requires a dedicated instrument.

Invader Assay. The Invader assay [39] uses two oligonucleotide probes that are hybridized to target DNA containing an SNP site. The two oligonucleotides hybridize to the single-stranded target and form an overlapping invader structure at the site of the SNP, as illustrated in Figure 5.7-7. One oligonucleo-tide, the Invader oligo, is complementary to the target sequence 3′ of the polymorphic site, and it ends with a nonmatching base overlapping the SNP nucleotide. The second oligonucleotide, the allele-specific probe, contains the complementary base of the SNP allele, and extends to the sequence 5′ of the polymorphic site. This probe can also extend on its 5′ site with additional noncomplementary nucleotides. Once the two oligonucleotides anneal to the target DNA, they form a three-dimensional invader structure over the SNP site that can be recognized by a specific enzyme, cleavase, or flap endonuclease. The enzyme cleaves the probe 3′ of the base complementary to the polymorphic site (i.e., 3′ of the overlapping invader structure). If the probe is designed as a fluorescence resonance energy transfer (FRET) molecule containing a fluorophore at the 5′ end and an internal quencher molecule, the cleavage reaction will separate the fluorophore from the quencher and generate a fluorescent signal. If, in contrast, the probe oligonucleotide does not match the SNP allele present in the target DNA (i.e., the probe is complementary to the alternate SNP allele), then no overlapping invader structure is formed, and the probe is not cleaved. This distinction is highly specific, with only minimal unspecific cleavage of the mismatch probe.

Often, the Invader oligonucleotide is designed to permanently anneal to the target DNA at the assay temperature. In contrast, the probe oligonucleotide is designed to have a melting temperature close to the assay temperature. As a consequence, the probe constantly anneals and detaches. During the annealing, cleavase can cleave the oligonucleotide, the remnant detaches, and a new uncleaved oligonucleotide probe can reanneal to the same site. This design ensures the cleavage of a large number of probes, resulting in a strong signal even from small amounts of template DNA. Therefore, SNPs can be genotyped directly from genomic DNA using Invader assays, eliminating the need for additional PCR amplification. In addition to the FRET probes, other methods of detection (polarized light, mass spectrometry) have been used [42–44]. However, the benefit of the fluorescence detection is the ability to read the assay after isothermic incubation in a standard fluorescence plate reader.

Multiplex Analysis. Although the methods described above work well and reliably, they are not suited for genotyping multiple SNPs in large sets of samples. As each SNP has to be assayed individually, all SNPs have to be interrogated sequentially rather than in parallel. However, numerous other platforms exist that allow the genotyping of several SNPs at the same time. Here, we will discuss methods that allow up to 10 SNPs to be genotyped in one reaction. In addition, we will discuss high-throughput approaches in the next section.

Both methods described here are based on the primer extension reaction. In both platforms, oligonucleotide primers are designed that extend up to the nucleotide adjacent to the SNP site. The following primer extension reaction will use ddNTPs to incorporate the matching base for the SNP allele present in the target DNA. The resulting products are then separated and detected by different methods.

The first detection method, as implemented in the MassEXTEND platform from Sequenom, uses mass spectrometry to determine the accurate molecular weight of the resulting extension reaction product, using matrix-assisted laser desorption/ionization time-of-flight (MALDI–TOF) mass spectrometry. For this approach, the reaction products are mixed with a matrix and deposited on the surface of a metal plate. The matrix and the reaction products are hit with a pulse from a laser beam, resulting in the vaporization of a small number of DNA molecules from the reaction product. The vaporized molecules are transferred into a vacuum flight tube and charged with an electrical field pulse. This accelerates the resulting ions toward the detector at the end of the tube. The time between the application of the electrical field pulse and the collision of the ions with the detector is referred to as the time of flight. As the time it takes for the ion to pass through the flight tube to the detector is directly proportional to the mass of the ion (with larger molecules flying slower), this is a very precise measure of the molecular weight of the DNA products. The detectable mass differences are small, and the difference between an incorporated ddA and ddT of 9 mass units can be clearly distinguished. The assay allows further multiplexing even for reactions resulting in very similar mass for the products by adding noncomplementary tail sequences to the 5′ end of the primer, thus increasing the mass of the extension product. A 12-plex has been described by Ross et al. [45] and even a 20-plex by Kim et al. [46].

Alternatively, the single base extension products can be analyzed using the SNaPShot technology from Applied Biosystems. In this approach, fluorescently labeled ddNTPs are used in the primer extension reaction, and the resulting extension products are separated and detected using commercially available capillary sequencers. Each ddNTP is labeled with a different fluorescent dye; thus, the different alleles of a SNP can be distinguished by the different color. Similar to the MassEXTEND approach, multiplexing can be improved by adding noncomplementary tail sequences to the primer. According to the manufacturer, 10-plex reactions can be performed with high accuracy.

High-Throughput Approaches. Currently, three high-throughput platforms for SNP genotyping are commercially available. All platforms allow the genotyping of large numbers of SNPs (1000–500,000) in a single experiment for one DNA sample. Although this significantly reduces the genotyping cost per SNP, it results in a tremendous cost for a project when 1000–2000 DNA samples need to be genotyped for thousands of SNPs. In addition, all approaches require specialized equipment and therefore are only of interest to laboratories that intend to perform large numbers of genotyping projects.

The first assay platform, developed by Illumina, combines allele-specific ligation with an extension reaction. The method, called the GoldenGate assay, uses the BeadArray technology, a method using fiberoptic substrates with randomly assembled arrays of beads. Detailed information and flow charts of the methodology can be found at the Illumina website (http://www.illumina.com/products/prod_snp.ilmn).

In the assay, three oligonucleotides are designed for each SNP locus. Two oligos are specific to each allele of the SNP site, called the Allele-Specific Oligos (ASOs).

A third oligo that hybridizes downstream from the SNP site is the Locus-Specific Oligo (LSO). All three oligonucleotide sequences are complementary to the target sequence and contain universal PCR primer sites. During the primer hybridization process, the assay oligonucleotides hybridize to the genomic DNA sample bound to paramagnetic particles. Hybridization occurs before any PCR amplification step; therefore, no amplification bias is introduced into the assay. Extension of the appropriate ASO and ligation of the extended product to the LSO joins information about the genotype present at the SNP site to a unique address sequence on the LSO. These joined, full-length products provide a template for PCR using universal PCR primers, two of which are fluorescently labeled. As a last step, the single-stranded, dye-labeled amplification products are hybridized to the complement bead type through their unique address sequences, and the labeled beads can be analyzed using arrayed optical fibers. Up to 1536 SNPs may be interrogated simultaneously in this manner using this Golden-Gate technology.

Similar degrees of multiplexing can be achieved using oligonucleotide array technology from Affymetrix. These arrays contain 25-mers that are synthesized directly on chip surfaces. In recent iterations, Affymetrix has developed arrays that allow the interrogation of 10,000, 100,000, or even 500,000 SNPs per reaction. The 500K array methodology uses an approach called whole-genome sampling analysis to selectively amplify regions of the genome containing SNPs, and subsequently interrogate the alleles present in the amplification products by hybridization to allele-specific oligonucleotides.

In the first step of the analysis, a single genomic DNA sample is digested with a restriction enzyme. After complete digestion, adaptors are ligated to the digestion products that allow the amplification of a subset of the fragments using a universal primer pair complementary to the adaptor sequence. Any SNP located in close proximity to the restriction site will be amplified by PCR and can be interrogated on the chip. The resolution is primarily driven by the restriction enzyme used. The more it cuts the genomic DNA, the more small amplification products can be obtained and hybridized to the chip. Accordingly, the different chips, (10K, 100K, 500K) use different enzymes in the initial step of the reaction, but subsequent steps are identical.

Although the assay allows the interrogation of a large number of SNPs simultaneously, the user has no influence on the composition of the chip, i.e., the SNPs that will be interrogated. The complete panel of 500,000 SNPs has been preselected during the design of the chip, primarily driven by the technical requirements of the restriction digest and amplification described above. However, a novel methodology developed by Parallele Biosciences (now Affymetrix) uses the chip-based hybridization approach for the ultimate analysis, but it uses a different approach for the SNP interrogation. A more detailed description and illustrations of the method can be found at http://www.affymetrix.com/technology/mip_technology. affx. The approach, termed molecular inversion probe (MIP) technology, uses a long oligonucleotide that will hybridize with its end sequences in inverted fashion to the flanking sequences of an SNP [47, 48]. Using one dNTP in four different reactions, the SNP site can be filled in to form a circular padlock probe around the SNP site. The probe will only circularize when the dNTP added is complementary to the SNP allele present. The circular probe is resistant to DNAse digestion. Once the nonfilled probes have been digested using DNAse, the padlock probe is opened, using a cleavage site within the MIP to form a linear template for a PCR reaction using universal primers. The amplification product contains a unique tag sequence for each MIP that can be detected by hybridization to an Affymetrix TrueTag array. Assays can be designed and genotyped in parallel for up to 10,000 SNPs. This platform offers a new tool to design custom panels of SNPs for a specific study, rather than depend on a preselected set of SNPs on other Affymetrix chips.

Overall, all three approaches allow the efficient genotyping of large numbers of SNPs. All require specialized equipment, and although the cost of genotyping per SNP is low, the total cost of any genotyping project is high due to the large number of SNPs. These technologies are only suitable for large-scale studies, and they are best performed in collaboration with or support from institutional core facilities.

5.7.3.4 Linkage Disequilibrium and Haplotypes

Most SNPs in a genome probably arose from individual single mutation events at an early time during the history of the species hundreds of generations ago. From the time of each mutation event, the new allele (i.e., the new "mutant" nucleotide) is located on an individual chromosome that has specific alleles of other SNPs that arose previously on the same chromosome. This physical arrangement of SNP alleles along a chromosome is called a *haplotype*. Over multiple successive generations, recombination and novel mutation events will lead to a rearrangement or modification of this ancestral haplotype around the new SNP allele. As a consequence, the new allele only remains on the same portion of the ancestral haplotype with other SNP alleles that are on a short distance away; i.e., recombination events are less likely to separate the two alleles. This nonrandom arrangement of SNPs along the chromosome, i.e., the maintenance of a small segment of the ancestral haplotype, is called *linkage disequilibrium* (LD) or *allelic association*. This arrangement is used in disease association analyses that are based on the idea that it should be possible to identify the effect of any common disease-causing SNP allele by determining the ancient haplotype segment on which it is located. By using a sufficient number of SNP markers in a genetic study, any common variant, even if it was not assayed directly, should display significant LD with a neighboring marker SNP. This approach has been used successfully in human genetic studies to identify genes responsible for Mendelian disorders such as Hirschprung's Disease or cystic fibrosis [49, 50].

The LD structure has been primarily studied in the human genome [51–53], although initial analyses have been performed for the mouse, rat, and dog genome as well. In the human genome, it has become clear over the past several years that the extent of LD and haplotypes in the human genome is not simply a function of distance between SNPs. Rather, the size of regions of significant LD is highly variable in different regions of the genome and reflects the complex history and interplay of recombination and mutation on different regions of the genome. If SNPs are not in LD, the alleles of the SNPs occur in seemingly random combination on individual chromosomes. As a consequence, the alleles of neighboring SNPs, in the absence of LD between them, can form a large number of different haplotypes (2^n for n SNPs). In contrast, regions where neighboring SNPs are in significant LD, only a small number of resulting haplotypes is observed. These haplotypes are representative of the ancient haplotypes that have not been broken up by recombination. Therefore, an analysis of haplotype patterns would not only identify regions of significant LD between SNPs, but it would also identify those common (presumably ancient) haplotype patterns that represent the majority of chromosomes in that particular genomic interval. Knowledge of these common haplotypes would then permit the identification of "tag SNPs," individual representative nonredundant SNPs that would unambiguously differentiate all major haplotypes without analyzing all SNPs in that particular region.

To facilitate the selection of these tag SNPs, the International HapMap Consortium, a publicly funded effort by the National Institutes of Health, has analyzed the genome-wide LD structure comprehensively in four different human populations: A cohort of North Americans of Northern European descent, an African population, a cohort of Han Chinese, and Japanese individuals [20]. In total, over 5.8 million SNPs have been genotyped in each of these populations, and the data of this effort are publicly available. It is the hope that selecting an informative subset of SNPs based on LD information from this study will provide a genome-wide coverage in whole genome disease association studies [54–57]. Here, SNPs would be selected so that all other SNPs from the HapMap study not included in the representative set would be in LD with one or more of the selected SNPs. It has been estimated that this comprehensive coverage of the human genome will require around 500,000 SNPs across all chromosomes. Undoubtedly, commercial providers of genotyping platforms will offer high-throughput SNP panels based on the HapMap information in the near future.

5.7.4 APPLICATION FOR GENETIC ANALYSES

As mentioned in Section 5.7.1, genotyping of SSLPs and SNPs is essential for the analysis of the genetic basis of common disorders. In the following section, we will briefly discuss the use of these polymorphisms in linkage and association analysis.

5.7.4.1 Linkage Analysis

A powerful means of identifying the location of genes causing disease is by genetic linkage mapping. During meiosis, when germ cells are being generated, homologous chromosomes can recombine, or cross over. As a result, offspring will have different combinations of parental alleles. We can use the percentage of recombinant offspring as a measure of physical distance, making a genetic linkage map. Linkage maps are based on probability. For instance, if two genes or loci are on different chromosomes, the probability of cosegregation of these loci is 50%. However, the closer loci are on the genome, the lower the probability of recombination between them. The measure of genetic distance is approximately the recombination frequency between loci. Although linkage is defined as <50% recombination, 47% recombination, for example, is not a reassuring or reliable measure of distance. Therefore, a genetic map needs many genetic markers to be reliable. Because of their relatively high abundance, their high informativeness, and their representation across all eukaryotes, SSLPs are the backbone of nearly every eukaryotic genetic map. To generate a genetic linkage map, large families, such as the three-generation families from The Foundation Jean Dausset-Centre d'Etude du Polymorphisme Humain (CEPH), are typically genotyped with hundreds or thousands of SSLPs. Analytical computer programs determine the frequency of recombination between parents and offspring, thus determining the distance between markers, and arrange the SSLPs according to their order and distance on each chromosome.

High-density genetic maps spanning the entire human genome have been generated by several groups and are commonly used for mapping disease genes and quantitative trait loci (QTL) [15, 16]. Just as recombination determines the genetic distance between SSLPs, it also determines the distance between a disease gene and an SSLP. If a particular gene determines a specific phenotype or disease, then one can genotype families with SSLPs spanning the genome and identify cosegregation of an SSLP allele and a clinical phenotype. If recombination is significantly lower than expected between an SSLP allele and a clinical outcome, e.g., hemophilia, then the gene causing the disease is *linked* to that marker. Because the genome location of that marker is known, the location of that disease gene is also known. The first disease linked by the use of microsatellite genotyping was facioscapulohumeral muscular dystrophy in 1990 [58]. Since that time, thousands of loci have been mapped using this approach. Genetic linkage studies are most powerful when studying large, multigenerational families affected by a single gene (Mendelian) disease. However, linkage has proven more of a challenge with common complex diseases. Because multiple genes with relatively small effects are thought to be involved in common diseases such as hypertension or diabetes, combined with the fact that these diseases often affect older patients, it is often impossible to find large families with multiple generations. To address these issues, association studies offer the advantages of being able to study cases and controls or small families, which are much more readily ascertained. Furthermore, the large number of SNPs and high-throughput SNP genotyping technologies now makes genome-wide association studies feasible, as discussed below.

5.7.4.2   Association Analysis

When it proves difficult to recruit a sufficient number of families for a genetic study, a popular alternative is an association study to elucidate the genetic basis of common diseases. In contrast to linkage studies where SSLP markers are used to identify a region of the genome that is inherited by affected offspring from affected parents, association studies aim to identify alleles that are found more often in individuals affected by a disease than in a control group. In the perfect association study, one would test the causal mutation for association with disease. However, in almost all cases, the causal mutation is unknown. As a consequence, association studies resort to indirect approaches relying on linkage disequilibrium (LD) between the causal mutation and another SNP that is being genotyped. If the SNP was associated with the disease, one would expect to find a higher frequency of the associated allele of the SNP in individuals affected by the disease (cases) when compared with a healthy control group.

Association studies can be performed in family-based cohorts. Numerous statistical approaches have been developed to test for association of SNP alleles with disease in small trios, nuclear, or even extended families. However, most association studies use unrelated patients in their analysis. In these cohorts, it is assumed that individuals are affected because they share a similar genetic susceptibility. However, as it is possible that some individuals may be affected for reasons other than genetic susceptibility (most common human disorders are also significantly influenced by environmental factors, e.g., lifestyle), it is imperative that a large number of patients is recruited for the study so that most individuals share the same genetic susceptibility genes and mutations. This cohort of affected individuals is then matched with a cohort of unaffected, healthy controls. Here, it is important that these control individuals match the patient cohort in the distribution of age, sex, ethnicity, and potential environmental factors that may influence the development of the disease.

Once the case and control cohorts have been collected, the DNA samples can be used for SNP genotyping. Depending on the scope of the study, samples are either genotyped for SNPs in candidate genes that were preselected for the study or a genome-wide association analysis using a comprehensive SNP set across the entire genome will be performed. Published data to date primarily include candidate gene association analyses. A recent review by Newton-Cheh and Hirschhorn [59] discusses in detail the important considerations for the design of these association studies and the potential problems that can be encountered. The number of SNPs to be analyzed would determine the best genotyping platform to be used in the study. Clearly, candidate gene SNPs can be genotyped with common approaches for individual SNP genotyping or moderate multiplexing platforms. However, genome-wide association studies will require high-throughput approaches.

Although the recent efforts to elucidate the LD structure of the human genome (HapMap Consortium) promise to facilitate the selection of informative subsets of SNPs for these types of association studies, it remains to be seen whether whole genome association studies will help uncover the genetic determinants for such common diseases as diabetes, asthma, or cardiovascular disorders. Clearly, candidate gene studies have yielded several well-replicated associations, but they account only for a small portion of the overall genetic susceptibility. Hopefully, comprehensive association studies of these disorders will help uncover additional genes responsible for the disease susceptibility. These discoveries would significantly advance our ability to predict, diagnose, and hopefully treat these disorders in affected individuals.

INTERNET RESOURCES

SSLP Genotyping Resources

DeCode: fee-for-service Genotyping—http://www.decode.com/

Marshfield Medical Research Foundation: Genetic markers and maps—http://research. marshfieldclinic.org/genetics/Genotyping_Service/mgsver2.htm

Fondation Jean Dausset—CEPH: DNA from multigenerational families, Human Diversity Panel, Genotype database, genetic maps—http://www.cephb.fr/

CIDR (Center for Inherited Disease Research): NIH-funded genotyping (Multi-instutional)—http://www.cidr.jhmi.edu/

RSnG (Resequencing and Genotyping Service): NIH-funded resequencing and genotyping (NHLBI)—http://rsng.nhlbi.nih.gov/scripts/index.cfm

SNP Genotyping Resources

dbSNP: NCBI-maintained central database and repository for SNPs and genotyping information—http://www.ncbi.nlm.nih.gov/SNP/

HapMap homepage: http://www.hapmap.org/

Affymetrix: Information on GeneChip arrays—http://www.affymetrix.com/index.affx

Applied Biosystems: Information on SNaPshot, TaqMan, fluorescently labeled ddNTPs—http://myscience.appliedbiosystems.com/navigation/mysciApplications.jsp?tabName
Attribute=applSnp

Illumina: Information on GoldenGate SNP genotyping technology—http://www.illumina.com/products/prod_snp.ilmn

Parallele: Information on MIP genotyping platform—http://www.affymetrix.com/technology/mip_technology.affx

Pyrosequencing: http://www.pyrosequencing.com/

Sequenom: Information on MassEXTEND MALDI-TOF-based SNP genotyping approaches, fiberoptic bead arrays—http://www.sequenom.com/

Third Wave Technologies: Information on Invader assay technology—http://www.twt.com/

REFERENCES

1. Griffiths A J F (2000). *An Introduction to Genetic Analysis*, 7th ed. W.H. Freeman, New York, p. 860.
2. Houck C M, Rinehart F P, Schmid C W (1979). A ubiquitous family of repeated DNA sequences in the human genome. *J. Mol. Biol.* 132(3):289–306.
3. Litt M, Luty J A (1989). A hypervariable microsatellite revealed by in vitro amplification of a dinucleotide repeat within the cardiac muscle actin gene. *Am. J. Hum. Genet.* 44(3):397–401.
4. Weber J L, May P E (1989). Abundant class of human DNA polymorphisms which can be typed using the polymerase chain reaction. *Am. J. Hum. Genet.* 44(3):388–396.
5. Hamada H, et al. (1984). Characterization of genomic poly(dT-dG).poly(dC-dA) sequences: Structure, organization, and conformation. *Mol. Cell Biol.* 4(12):2610–2621.
6. Sheffield V C, et al. (1995). A collection of tri- and tetranucleotide repeat markers used to generate high quality, high resolution human genome-wide linkage maps. *Hum. Mol. Genet.* 4(10):1837–1844.
7. Jeffreys A J, Wilson V, Thein S L (1985). Hypervariable "minisatellite" regions in human DNA. *Nature.* 314(6006):67–73.
8. Weber J L, Wong C (1993). Mutation of human short tandem repeats. *Hum. Mol. Genet.* 2(8):1123–1128.
9. Miesfeld R, Krystal M, Arnheim N (1981). A member of a new repeated sequence family which is conserved throughout eucaryotic evolution is found between the human delta and beta globin genes. *Nucleic Acids Res.* 9(22):5931–5947.
10. Weber J L (1990). Informativeness of human (dC-dA)n.(dG-dT)n polymorphisms. *Genomics.* 7(4):524–530.
11. Mullis K B, Faloona F A (1987). Specific synthesis of DNA in vitro via a polymerase-catalyzed chain reaction. *Methods Enzymol.* 155:335–350.
12. Ausubel F M (1992). *Short protocols in molecular biology : A compendium of methods from Current protocols in molecular biology*, 2nd ed. Greene Pub. Associates, New York.

13. Moreno C, et al. (2004). Genome-Wide Scanning With SSLPs in the Rat. *Methods. Mol. Med.* 108:131–138.

14. Oetting W S, et al. (1995). Linkage analysis with multiplexed short tandem repeat polymorphisms using infrared fluorescence and M13 tailed primers. *Genomics.* 30(3):450–458.

15. Murray J C, et al. (1994). A comprehensive human linkage map with centi-morgan density. Cooperative Human Linkage Center (CHLC). *Science.* 265(5181):2049–2054.

16. Dib C, et al. (1996). A comprehensive genetic map of the human genome based on 5264 microsatellites. *Nature.* 380(6570):152–154.

17. Sachidanandam R, et al. (2001). A map of human genome sequence variation containing 1.42 million single nucleotide polymorphisms. *Nature.* 409(6822):928–933.

18. Frazer K A, et al. (2004). Segmental phylogenetic relationships of inbred mouse strains revealed by fine-scale analysis of sequence variation across 4.6 mb of mouse genome. *Genome Res.* 14(8):1493–1500.

19. Smits B M, et al. (2004). Genetic variation in coding regions between and within commonly used inbred rat strains. *Genome Res.* 14(7):1285–1290.

20. The international HapMap project. *Nature.* 2003. 426(6968):789–796.

21. Nickerson D A, Tobe V O, Taylor S L (1997). PolyPhred: Automating the detection and genotyping of single nucleotide substitutions using fluorescence-based resequencing. *Nucleic Acids Res.* 25(14):2745–2751.

22. Hayashi, K (1991). PCR-SSCP: A simple and sensitive method for detection of mutations in the genomic DNA. *PCR Methods Appl.* 1(1):34–38.

23. Andersen P S, et al. (2003). Capillary electrophoresis-based single strand DNA conformation analysis in high-throughput mutation screening. *Hum. Mutat.* 21(5):455–465.

24. Xiao W, Oefner P J (2001). Denaturing high-performance liquid chromatography: A review. *Hum. Mutat.* 17(6):439–474.

25. Yu B, et al. (2005). Denaturing high performance liquid chromatography: High throughput mutation screening in familial hypertrophic cardiomyopathy and SNP genotyping in motor neurone disease. *J. Clin. Pathol.* 58(5):479–485.

26. Han W, et al. (2004). Using denaturing HPLC for SNP discovery and genotyping, and establishing the linkage disequilibrium pattern for the all-trans-retinol dehydrogenase (RDH8) gene. *J. Hum. Genet.* 49(1):16–23.

27. Mellai M, et al. (2003). Prolactin and prolactin receptor gene polymorphisms in multiple sclerosis and systemic lupus erythematosus. *Hum. Immunol.* 64(2):274–284.

28. Huang H, et al. (2002). Screening of single nucleotide polymorphisms in nasopharyngeal carcinoma associated genes by denaturing high-performance liquid chromatography. *Di Yi Jun Yi Da Xue Xue Bao.* 22(7):602–604.

29. Shi J, et al. (2002). Sequence variations in the mu-opioid receptor gene (OPRM1) associated with human addiction to heroin. *Hum. Mutat.* 19(4):459–460.

30. Ribas G, Neville M J, Campbell R D (2001). Single-nucleotide polymorphism detection by denaturing high-performance liquid chromatography and direct sequencing in genes in the MHC class III region encoding novel cell surface molecules. *Immunogenet.* 53(5):369–381.

31. Hecker K H, et al. (2000). Mutation detection in the human HSP7OB' gene by denaturing high-performance liquid chromatography. *Cell Stress Chaper.* 5(5):415–424.

32. Giordano M, et al. (1999). Identification by denaturing high-performance liquid chromatography of numerous polymorphisms in a candidate region for multiple sclerosis susceptibility. *Genomics.* 56(3):247–253.

33. Upchurch D A, Shankarappa R, Mullins J I (2000). Position and degree of mismatches and the mobility of DNA heteroduplexes. *Nucleic Acids Res.* 28(12):E69.

34. Youil R, Kemper B W, Cotton R G (1995). Screening for mutations by enzyme mismatch cleavage with T4 endonuclease VII. *Proc. Natl. Acad. Sci. U.S.A.* 92(1):87–91.

35. Faham M, Cox D R (1995). A novel in vivo method to detect DNA sequence variation. *Genome Res.* 5(5):474–482.

36. Faham M, et al. (2005). Multiplexed variation scanning for 1000 amplicons in hundreds of patients using mismatch repair detection (MRD) on tag arrays. *Proc. Natl. Acad. Sci. U.S.A.* 102(41):14717–14722.

37. Faham M, et al. (2001). Mismatch repair detection (MRD): High-throughput scanning for DNA variations. *Hum. Mol. Genet.* 10(16):1657–1664.

38. Landegren U, et al. (1988). A ligase-mediated gene detection technique. *Science.* 241(4869):1077–1080.

39. Lyamichev V, et al. (1999). Polymorphism identification and quantitative detection of genomic DNA by invasive cleavage of oligonucleotide probes. *Nat. Biotechnol.* 17(3):292–296.

40. Livak K J (1999). Allelic discrimination using fluorogenic probes and the 5' nuclease assay. *Genet. Anal.* 14(5–6):143–149.

41. Fakhrai-Rad H, Pourmand N, Ronaghi M (2002). Pyrosequencing: An accurate detection platform for single nucleotide polymorphisms. *Hum. Mutat.* 19(5):479–485.

42. Kwok P Y (2002). SNP genotyping with fluorescence polarization detection. *Hum. Mutat.* 19(4):315–323.

43. Hsu T M, et al. (2001). Genotyping single-nucleotide polymorphisms by the invader assay with dual-color fluorescence polarization detection. *Clin. Chem.* 47(8):1373–1377.

44. Berggren W T, et al. (2002). Multiplexed gene expression analysis using the invader RNA assay with MALDI-TOF mass spectrometry detection. *Anal. Chem.* 74(8):1745–1750.

45. Ross P, et al. (1998). High level multiplex genotyping by MALDI-TOF mass spectrometry. *Nat. Biotechnol.* 16(13):1347–1351.

46. Kim S, et al. (2003). Multiplex genotyping of the human beta2-adrenergic receptor gene using solid-phase capturable dideoxynucleotides and mass spectrometry. *Anal. Biochem.* 316(2):251–258.

47. Hardenbol P, et al. (2005). Highly multiplexed molecular inversion probe genotyping: Over 10,000 targeted SNPs genotyped in a single tube assay. *Genome Res.* 15(2):269–275.

48. Hardenbol P, et al. (2003). Multiplexed genotyping with sequence-tagged molecular inversion probes. *Nat. Biotechnol.* 21(6):673–678.

49. Kerem B, et al. (1989). Identification of the cystic fibrosis gene: Genetic analysis. *Science.* 245(4922):1073–1080.

50. Puffenberger E G, et al. (1994). A missense mutation of the endothelin-B receptor gene in multigenic Hirschsprung's disease. *Cell.* 79(7):1257–1266.

51. Reich D E, et al. (2001). Linkage disequilibrium in the human genome. *Nature.* 411(6834):199–204.

52. Stephens J C, et al. (2001). Haplotype variation and linkage disequilibrium in 313 human genes. *Science.* 293(5529):489–493.

53. Daly M J, et al. (2001). High-resolution haplotype structure in the human genome. *Nat. Genet.* 29(2):229–232.

54. Halldorsson B V, Istrail S, De La Vega F M (2004). Optimal selection of SNP markers for disease association studies. *Hum. Hered.* 58(3–4):190–202.

55. Stram D O (2004). Tag SNP selection for association studies. *Genet. Epidemiol.* 27(4):365–374.

56. Sebastiani P, et al. (2003). Minimal haplotype tagging. *Proc. Natl. Acad. Sci. U.S.A.* 100(17):9900–9905.

57. Ke X, Cardon L R (2003). Efficient selective screening of haplotype tag SNPs. *Bioinform.* 19(2):287–288.

58. Wijmenga C, et al. (1990). Facioscapulohumeral muscular dystrophy gene in Dutch families is not linked to markers for familial adenomatous polyposis on the long arm of chromosome 5. *J. Neurol. Sci.* 95(2):225–229.

59. Newton-Cheh C, Hirschhorn J N (2005). Genetic association studies of complex traits: Design and analysis issues. *Mutat. Res.* 573(1–2):54–69.

6.1

PROTEINS: HORMONES, ENZYMES, AND MONOCLONAL ANTIBODIES—BACKGROUND

WAYNE MATERI, ZOLTAN GOMBOS, AND DAVID S. WISHART

National Institute for Nanotechnology, University of Alberta, Edmonton, Alberta, Canada

Chapter Contents

6.1.1	Introduction	691
6.1.2	Protein Pharmaceuticals	695
6.1.3	Peptide and Protein Hormone Biopharmaceuticals	698
	6.1.3.1 Examples of Hormone Biopharmaceuticals	699
	6.1.3.2 The Future of Hormone Biopharmaceuticals	712
6.1.4	Antibody Biopharmaceuticals	713
	6.1.4.1 Examples of Antibody Biopharmaceuticals	721
	6.1.4.3 The Future of Antibody Biopharmaceuticals	722
6.1.5	Enzyme Biopharmaceuticals	724
	6.1.5.1 Examples of Enzyme Biopharmaceuticals	728
	6.1.5.2 The Future of Enzyme Biopharmaceuticals	730
6.1.6	Conclusions	731
	References	732

6.1.1 INTRODUCTION

The human body is a veritable pharmacopia of useful protein drugs. After all, it is the collection of our own, endogenously produced enzymes, hormones, and antibodies that are responsible for maintaining homeostasis, stabilizing wounds, fighting infections, neutralizing toxins, keeping cancerous cells in check, and generally keeping us alive. It is only when we lose the function of certain enzymes, hormones, or antibodies—or when our bodies are overwhelmed with some kind of trauma (blood loss, stroke, heart attack, massive infection, or heavy tumor burden)—that we need some supplementation to our natural protein drugs. The fact that our bodies or body fluids contain some kind of miraculous healing or "vital" substances was recognized even in ancient times by the Chinese, Moche Indians, Maasai, ancient Scythians, Gypsies, and early Christians when would-be warriors or the sick and sinful were encouraged to drink such body fluids as blood or urine to imbue health, healing, or strength. However, the general failure of these primitive protein cocktails was probably evident to most patients and many physicians especially by the Middle ages. Given the abysmal failure of oral protein delivery, Italian physicians, in the late 1400s began experimenting with a different approach, namely intravenous delivery via blood transfusions. The first recorded instance of a therapeutic blood transfusion was for the treatment of Pope Innocent VIII who fell into a stroke-induced coma in 1492 [1]. Although this effort ultimately failed, efforts to refine or improve blood transfusions from animals to humans or humans to humans as a way to treat wounds and other illnesses continued for another 300 years. The first therapeutically successful blood transfusion, and perhaps the first example of the successful use of a protein drug (albeit impure albumin), was performed in England by James Blundell in 1825 to treat a woman suffering from postpartum hemorrhaging [2]. By 1844 Blundell and his colleague Samual Armstrong Lane became the first to use blood (i.e., factor VIII) to treat a genetic disease—hemophilia [3]. Throughout the 1800s and early 1900s, blood transfusions became more refined (through blood typing and the use of aseptic conditions) and blood became a therapeutic protein product to treat wounds (albumin), infections (antibodies, antiglobulins), and other genetic disorders.

Obviously blood is not the only source of protein drugs. As far back as 1796 other sources of protein pharmaceuticals were beginning to be recognized. In particular, the work of Edward Jenner in developing an effective small-pox vaccine [4] could possibly be described as the first example of using protein drugs for prophylactic purposes. Keeping with the ancient theme of using "vital" body fluids for medical applications, Jenner used the fluid from cow-pox pustules as the source of his protein pharmaceutical (cow-pox viral coat proteins). In the 1870s Louis Pasteur extended Jenner's ideas by developing vaccines for cholera, anthrax, and rabies using deactivated bacterial cells or dried nerve tissues. Pasteur formulated the dried bacterial or viral protein components into injectable solutions. In this regard Pasteur was the first to introduce the concept of "synthetic" or *ex vivo* biological products as potential drugs. However the idea of working with a purified, active ingredient, especially with biological material, was still many years away. The principle of prophylactic vaccines, developed by Jenner and refined by Pasteur, was also extended to the development of "therapeutic vaccines" or antivenoms in the late 1800s. Antivenoms are generated using a method developed by Albert Calmette in 1895, which he devised to treat victims of cobra bites. Antivenoms are created by injecting a small amount of the targeted venom into animals such as horses, sheep, or rabbits. This leads to the production of different antibodies against the toxin. The mixture of antibodies is then harvested from the animal's blood and serves as the antivenom for a specified source venom.

Up until the mid-1800s the true source of the therapeutic powers of blood or the prophylactic powers of pustules or bacterial extracts was not known. Then in 1838 the concept of proteins was introduced by Jons Jakob Berzelius [5]. Berzelius argued that proteins (he is credited for creating the name) found in blood and bacteria were special organic substances that behaved or functioned as chemical compounds. This led to the realization that many different proteins existed in blood and other biological tissues and that they could potentially be isolated and treated as pure substances, just like all other chemical or small-molecule drug entities. However, it was not until 1922 that the technology to isolate therapeutically useful proteins was put to good use. Thanks largely to the work of four Canadian scientists (Banting, Best, Collip, and MacLeod), insulin became the first "pure" protein therapeutic to be used in the successful treatment of a human disease—diabetes. Before 1922, type I diabetes was a near-certain death sentence characterized by an agonizingly long and painful wasting process. The fact that it frequently afflicted adolescent children made it particularly devastating. Although the cause of diabetes was partially known by 1910, early efforts to use unpurified pancreatic extracts met with repeated failure and severe allergic reactions. The use of purified insulin, on the other hand, had absolutely stunning results. It is hard not to overstate the kind of spectacular recoveries observed among diabetic patients nor the effect that insulin, as a drug, had on the public as well as on the medical and pharmaceutical community. Indeed, the insulin "effect" not only saved the lives of millions of diabetics, but it essentially launched the modern concept of "pure" protein pharmaceuticals and laid the foundation to today's modern biotech industry.

Most of today's protein pharmaceuticals fall into 5 general classes: (1) hormones, (2) vaccines, (3) antibiotics, (4) antibodies, and (5) enzymes. Three of these classes, hormones, antibodies, and enzymes, are primarily used to treat noninfectious or endogenous diseases (i.e., genetic diseases or diseases of aging), whereas the remaining two classes, vaccines and antibiotics, are used to treat or prevent infectious or exogenous diseases (i.e., bacterial or viral infections). Interestingly all five drug classes were essentially identified or defined only in the past 80 years. Insulin, a hormone, was the first purified protein drug to be marketed (1922). Shortly after, in 1923, purified diptheria toxoid (a 58-kDa protein) was introduced, making it the first purified vaccine product [6]. The first purified peptide antibiotic, gramicidin S, was introduced in Russia in 1943 as a topical wound-healing agent. The first purified antivenoms or antitoxin antibodies were introduced in the 1950s and the first monoclonal antibody—OKT3 (muronomab or Orthoclone)—was approved for human use in 1986 [7]. The first purified enzyme to be approved for therapeutic use, Activase, was marketed in 1987. Interestingly, the first human recombinant protein used in disease management was also insulin (Humulin) [8], which was introduced in 1982 by Eli Lilly.

Up until 1982 all peptide and protein pharmaceuticals were isolated from "natural" sources, meaning they were purified from animal (or human) biofluids and tissues. Working with natural sources is exceedingly difficult, expensive, and can lead to the transmittance of undetected infectious materials (viruses, bacteria, prions) to patients. Because most proteins of pharmaceutical interest are relatively scarce, it was often necessary to process tens or hundreds of kilograms of tissue or fluid to produce a few milligrams of the desired product in pure form. For instance, up until the 1980s the preparation of sufficient human growth hormone (somatotropin) for a single treatment to combat dwarfism would require the extraction and homogenization of several adult human (cadaver) brains [6].

For the more abundant proteins (such as albumin) it is possible to find them in concentrations of about 35 g/L in selected tissues or biofluids [9]. This makes their isolation and purification relatively simple. However, many important protein pharmaceuticals are only found in concentrations of 1 or 2 pg/L. At such low concentrations, it is almost impossible to isolate and purify these compounds from the thousands of other proteins that are in the body. Furthermore, even if one wanted to produce enough material for commercial purposes, it would probably require the processing or homogenization of the entire world supply of cattle, pigs, or humans. Today most protein pharmaceuticals are produced through recombinant methods, including insulin, growth hormone, and most monoclonal antibodies. This allows large quantities of even the rarest protein to be purified and prepared under tightly controlled manufacturing conditions without the concerns over contamination with host viruses (HIV, HCV), toxins, or prions (that cause diseases such as CJD or BSE). Although mast protein drugs are biosynthetically produced (via cell culture), a small number of peptide pharmaceuticals (<20 residues) are produced chemically via automated peptide synthesis. These include Leuprolide (Eligard), Oxytocin (Oxytocin), Calcitonin (Miacalcin), and Abarelix (Plenaxis), for example. Peptides and proteins that continue to be isolated from natural sources include Menotropin (Repronex) and Hyaluronidase (Vitrase), for example.

Since the introduction of insulin in 1922, more than 110 distinct protein or peptide drugs have been approved for human use by the U.S. Food and Drug Administration (FDA) [10], with 50 being hormones, 2 being vaccines, 2 being peptide antibiotics, 26 being antibodies or antibody mixtures, and 17 being enzymes (see Tables 6.1-1–6.1-3). These peptide/protein pharmaceuticals range in size from less than 10 amino acids to over 1000 and may contain a variety of covalent modifications such as carbohydrate chains, D-amino acids, or amino acids with unusual side chains. More details about the structure and chemical composition of most protein pharmaceuticals can be found in subsequent chapters in this book as well as in the DrugBank database under its collection of biotech drugs [10]. According to the Biotechnology Industry Organization (www.bio.org), there are now more than 300 peptide and protein pharmaceuticals currently in clinical trials or under review. Just as with the FDA-approved biotech drugs, most of these newer products or potential products fall into the three major classes of endogenous or human-derived proteins: (1) hormones, (2) antibodies, and (3) enzymes. It is these three classes of pharmaceutically important proteins that are the focus of this chapter.

Beyond providing a historical perspective to proteins and protein pharmaceuticals, this chapter is intended to provide the reader with an overview of the general features and characteristics of pharmaceutically important hormones, antibodies, and enzymes. Together these three classes of protein therapies account for nearly three quarters of all approved biotech drugs and more than 90% of all biotech drug sales. In addition to providing a general outline about what hormones, enzymes, and antibodies are, this chapter will also provide the reader with specific examples and detailed descriptions of a few of the more important or historically interesting hormones, enzymes, and antibodies used today. The chapter is divided into five sections, an introduction to the history of protein pharmaceuticals, a general discussion about the advantages and disadvantages of protein pharmaceuticals, a detailed discussion on hormones, a detailed description of enzymes, and a detailed discussion on antibodies. Each section on hormones, enzymes, and antibodies provides a working description or definition of class of these protein/peptide products; a general discussion on certain unique aspects of their formulation or delivery; several specific examples of important FDA-approved hormones, enzymes, or antibodies; and a brief discussion of some of the new enzymes, hormones, and antibodies that are under development or in clinical trials. A complete discussion concerning all aspects of hormones, enzymes, and antibodies is far beyond the scope of this chapter. Readers who are interested in learning more might want to consider reading several excellent books [11–13] or referring to the DrugBank database (http://redpoll.pharmacy.ualberta.ca/drugbank/) for more extensive biological, pharmaceutical, and biochemical data.

6.1.2   PROTEIN PHARMACEUTICALS

For a peptide or protein to be a useful drug, it must generally be water soluble, relatively stable, nonimmunogenic (i.e., identical or near identical to a human homologue), mostly monomeric, and causal or preventative to some disease process. Not all polypeptides match these requirements, and therefore, not all polypeptides are potentially suitable as drugs. Typically those that are not (yet) useful drugs are highly polymeric structural proteins such as tubulin, actin, or myosin. Equally impractical or improbable drugs are membrane-bound receptor proteins (G-protein-coupled receptors, laminins, etc.). These classes of proteins, which may account for as much as 60% of the human proteome, are usually most suitable as drug targets—not drugs. Certain other proteins, such as polymerases, histones, and ribosomal proteins, perform such vital functions that they are "essential" to life and so cannot generally serve as useful protein drugs either, although DNaseI (Pulmozyme) is a notable exception. Using these relatively simple criteria about what constitutes a viable drug versus a viable drug targets, it is possible to scan the human proteome and identify likely classes of potential protein drugs. When this is done one is typically left with three major groups of peptides or proteins: (1) hormones, (2) enzymes, and (3) immunological molecules (antibodies).

Peptide and protein pharmaceuticals are different from small-molecule drugs. For one, polypeptide drugs are generally much larger, ranging in size from 1000 daltons to more than 200,000 daltons (compared with <600 daltons for most small-molecule drugs). For another, the noncovalent structure or three-dimensional fold (i.e., the tertiary structure) of most polypeptides is absolutely critical to their function. This is in distinct contrast to small-molecule drugs, wherein their covalent or "primary" structure defines their function. Smaller polypeptides have relatively little tertiary structure or exhibit substantial tertiary structure only when bound to a target receptor. Larger (>40 residues) polypeptides generally have a stable, well-defined tertiary structure. As a general rule, polypeptides with less than 40 amino acids are called peptides, whereas those with more than 40 residues are called proteins. Typically most hormones are peptides (<40 residues), although some hormones such as somatotropin (human growth hormone) can be up to 200 residues in length. Many peptide hormones are actually fragments of larger precursor proteins or pro-proteins. Although most hormones are relatively small polypeptides, enzymes and antibodies are generally much larger. As a rule, most enzymes are between 200 and 800 residues in length, whereas most antibodies are typically

1200 residues in length, although smaller antibody fragments (Fabs, single-chain antibodies) can also be used as protein drugs.

Perhaps the most impressive and appealing quality about peptide and protein pharmaceuticals is their exquisite selectivity and specificity. Bioactive proteins target their receptors or bind their small-molecule ligands with a precision that matches a cruise missile. Almost no cross-reactivity to other targets or other receptor molecules is observed with most known hormones, enzymes, or antibodies. Additionally most protein drugs also exhibit very appealing pharmacokinetics and excellent bioavailability. These favorable pharmaceutical properties are not entirely unexpected. Indeed, the process of natural selection over hundreds of millions of years has evolved bioactive proteins such as hormones, enzymes, and antibodies to perform their functions in a fashion that is optimal to the health and viability of each organism. In contrast to so-called large-molecule (i.e., peptide and protein) drugs, almost no known small-molecule drug exhibits comparable selectivity or sensitivity. Perhaps the only small molecules that do are those drugs that are identical to the naturally occurring chemicals in the body (i.e., estrogen, testosterone, thyroxine, and epinephrine). Indeed, these small molecules likely coevolved with their target proteins.

The fact that polypeptides are large molecules means that far more functional and operational information can be chemically encoded into them than small molecules. In other words, there is much more room on a protein scaffold to add, remove, or substitute chemical moieties to change a given protein's specificity, bioavailability, kinetics, or function. This chemical flexibility can allow a protein chemist to engineer a potential or promising new protein drug into a more useful product. Indeed, many peptide and protein pharmaceuticals on the market today are engineered, via site-directed mutagenesis, or chemically modified to enhance their stability, selectivity, or efficacy. For example, the cytokine hormone Infergen is a "designed" interferon derived from a consensus alignment of bioactive mammalian interferon alpha sequences. This designed molecule was found to give the protein greater bioactivity. Another example is pegademase, a PEGylated version of the enzyme adenosine deaminase [14]. This protein is decorated with poly(ethylene glycol) (PEG) subunits covalently attached to exposed lysine side chains. This covalent chemical modification extends the *in vivo* half-life of the protein by a factor of 10 or more. Yet another example of an engineered or rationally designed protein is the hormone LisPro Insulin, a derivative of insulin in which Lysine 28 and Proline 29 have been interchanged [15]. This chemical change was found to enhance the activity and extend the lifetime of this form of insulin relative to natural insulin. Beyond these specific examples of engineered hormones and enzymes is a much larger collection of chimeric murine-human antibodies such as Infliximab, Rituximab, and Abciximab that have been designed to have murine variable domains and human constant domains. These chimeric features reduce the antigenicity of monoclonal murine antibodies and thereby greatly enhance the tolerance and half-life of these molecules.

Selectivity, specificity, and "programmability" are what make protein pharmaceuticals so attractive to today's pharmaceutical and biotech industry. As a result, enormous efforts are now going into the discovery, production, and FDA approval of dozens of new protein drugs. In 2005, the market capitalization of the biotechnology industry in the United States was estimated to be $311 billion. In 2004, 40 new protein pharmaceuticals (drug types, indications, or formulations) were approved by the (FDA)—more than in any prior year. With the completion of the Human Genome Project and the launch of many large-scale proteomic efforts [16, 17], the potential to find even more pharmaceutically useful hormones, enzymes, and antibodies will likely grow even further. How much further can it grow? The "universe" of endogenous protein drugs can be estimated by considering the size of the human proteome and the current diversity of functions or roles that have been identified to date. In particular, the approximately 25,000 genes identified in the human body probably code for around 40,000 different proteins (isozymes or splice variants) and up to 1,000,000 different posttranslationally modified variants, including chemically modified or cleaved proteins [18]. Current annotations suggest that approximately 3000 different human enzymes exist, along with over 150 different hormones. Given the number of enzymes and hormones available (which may be used for diseases requiring replacement therapy), this suggests that there could be at least 3500 antibody targets (which may be needed for antagonizing the unwanted effects of certain enzymes or hormones). If we include the number of known receptors (estimated to be 4000) for which antibodies may be targeted, then the number of potential, pharmaceutically important antibodies may be closer to 8000 molecules. Given that there are less than 400 protein drugs that are either approved or in the drug development pipeline, this rough calculation suggests we still have a long way to go before we exhaust the supply of potential endogenous protein drugs.

Although there is much to be optimistic about the future of protein pharmaceuticals, there are still many unique problems with their development, production, and delivery. Among the more obvious problems with protein drugs is the fact that they are much more delicate than small-molecule drugs. Proteins such as hormones, antibodies, and enzymes cannot normally be "compounded" or pressed into dry pills or emulsified or concentrated into tinctures. This type of conventional pharmaceutical manufacturing and formulation would destroy the activity of most protein pharmaceuticals. Similarly most peptide hormones, antibodies, and enzymes cannot be stored indefinitely at room temperatures in nonsterile containers; instead they must be kept in a cool, dark, aqueous, sterile environment for no more than a few weeks. These limitations to protein preparation and formulation have created a significant challenge to pharmaceutical chemists. Potential solutions to these problems are discussed in Chapter 4 of this book.

Yet another challenge in working with or producing peptide and protein drugs is their cost. Relative to small-molecule drugs, manufacturing costs for peptide and protein pharmaceuticals are enormous, with many protein drugs being priced at 10 to 1000 times the cost of a small-molecule drug (on a per-milligram basis). This price difference exists because most polypeptide drugs cannot be synthesized using classicl synthetic organic methods—they are simply too large and too complex for today's technology. Currently the only cost-effective method for producing most polypeptide hormones, antibodies, and enzymes is through recombinant gene expression in bacterial or mammalian cell culture. This is a time-consuming, low-yield (100mg—10g per liter) process that requires the use of large fermentor systems, carefully monitored growth conditions, and multiple purification steps.

Beyond the obvious problems of storage and production of protein pharmaceuticals lies one of the more discouraging facts about protein drugs: The body is not a particularly friendly place for foreign proteins. Indeed, our bodies have developed a huge arsenal of tricks to quickly and efficiently dispose of proteins that are swallowed, ingested, or injected. For instance, proteins that are swallowed are immediately attacked by enzymes (amylases [19]) in the mouth, which are designed to remove the protective sugars that cover many proteins. Moving down the throat, these "naked" proteins are adsorbed and denatured by surface interactions with the mucosal cells. As the surviving proteins enter the stomach and small intestine, they are exposed to extremely acidic conditions (which instantly denature most proteins) and a host of acid-stable proteases that digest these denatured molecules into short, inactive peptides. Similarly, if a peptide or protein pharmaceutical is injected into the bloodstream, it may be attacked by antibodies or swallowed by T cells and cut into fragments. Alternatively, it may be adsorbed by albumins or lipoproteins and permanently removed from the circulation. Even if an injected protein survives these insults, a foreign protein may still be attacked by plasma proteases and demolished into tiny fragments. Foreign proteins are "ill treated" due to the design of our bodies to use proteins as a source of food (if ingested) or to identify pathogens (such as bacteria, viruses, and parasites) that have invaded our circulatory system. This makes most proteins "enemies" of the body, something to be digested or eliminated at every opportunity. Even for its own "friendly" proteins (i.e., the proteins needed for cellular functions), the body has established several mechanisms to turn over or eliminate these proteins every few days. This regular turnover prevents old proteins from malfunctioning and keeps tight control over regulatory proteins that need only be around for minutes or hours at a time. These finely tuned mechanisms, which have evolved over the past two billion years, make it very difficult for pharmaceutical companies to use protein drugs as injectables or oral consumables.

6.1.3 PEPTIDE AND PROTEIN HORMONE BIOPHARMACEUTICALS

A hormone is classically defined as a chemical messenger, either a small molecule or a large macromolecule, which is synthesized in an organ or tissue and then secreted into the circulatory system where it subsequently affects separate target organs whose cells bear an appropriate receptor. Hormone actions include the stimulation or inhibition of growth, the induction or suppression of programmed cell death (apoptosis), the modulation of the immune system, the regulation of metabolism, and the preparation for a new activity (e.g., fighting, fleeing, and mating) or phase of life (e.g., puberty, childbirth, and menopause). In many cases, one hormone may regulate the production and release of other hormones. Many hormones can be described as messengers serving to regulate the metabolic activity of an organ or tissue. Hormones also control the reproductive cycle of virtually all multicellular organisms.

Animal hormones essentially fall into four different chemical classes: (1) amine-derived hormones, (2) peptide/protein hormones, (3) steroid hormones, and (4) lipid or phospholipid hormones. Amine-derived hormones such as catecholamines and thyroxine are derivatives of the amino acids tyrosine and tryptophan. Peptide hormones including insulin, growth hormone, and vasopressin consist of polypeptides ranging in length from 5 to 200 residues. Steroid hormones such as estrogen and testosterone are derived from cholesterol. The adrenal cortex and the gonads are primary sources for these hormones. Lipid and phospholipid hormones are derived from lipids such as linoleic acid and phospholipids such as arachidonic acid. For this chapter we are primarily concerned with peptide hormones.

The functional diversity of peptide/protein hormones is enormous. There are as many potential classes of these hormones as there are classes of hormonally regulated biological functions. If we also include potential antagonists to natural hormonal function (i.e., peptides that may bind competitively or noncompetitively to hormone receptors), we could double the number of classes. However we can generally group subcategories so that pharmaceutically important peptide hormones might be classified into the following broad conceptual categories:

1. Homeostatic regulators control basic physiological conditions such as blood glucose levels (insulin), water retention (vasopressin, desmopressin), or uterine muscle contractions (oxytocin).
2. Fertility regulators can be used to either inhibit or promote fertility. Although luteinizing hormone (LH) and follicle stimulating hormone (FSH) are regulated indirectly through the steroid hormone (estrogen and progesterone) levels to suppress fertility, direct introduction of the peptide hormones themselves is used to induce development and release of multiple ova.
3. Growth/division regulators may either include hormones regulating growth of the entire body (e.g., human growth hormone) or of a specific tissue type (e.g., erythropoeitin for hematopoeisis, palifermin for mucositis, or aldesleukin for lymphocytes). Some fertility hormones (e.g., leuprolide) may act indirectly as general or specific growth inhibitors.
4. Immunomodulatory hormones modulate the normal immune or inflammatory responses. This class can include hormones that upregulate the immune response (e.g., interferon) or that suppress it (e.g., cyclosporine).

If the definition of a hormone were expanded to include localized secreted or cell-surface signaling molecules, many other kinds of regulators would join this list, including molecules with roles in developmental fate specification and differentiation (e.g., netrins, semaphorins, and morphogens). Because the effects of such signals are highly localized, their use as effective therapeutic agents would require greater precision in administration than current systemic delivery methods. Table 6.1-1 provides a complete list of the peptide/protein hormones that are currently approved by the FDA. More details about some of these are provided below.

6.1.3.1 Examples of Hormone Biopharmaceuticals

Homeostatic Hormones. Insulin is produced by beta cells found in the islets of Langerhans that are dispersed throughout the pancreas. The islet cells constitute only 1% of total pancreatic tissue. Each islet contains about 3000 endocrine cells with a core of beta cells surrounded by other endocrine cells such as the glucagon-secreting alpha cells, the somatostatin-secreting delta cells, and the pancreatic polypeptide-secreting PP cells. The beta cells first synthesize a 109-amino-acid preproinsulin that is subsequently processed to an 86-amino-acid proinsulin. The

TABLE 6.1-1. FDA-Approved Protein/Peptide Hormone Pharmaceuticals

Generic Name	Chemical Name	Brand Name(s)	Description	Indication
Abarelix	acetyl-D-β-naphthylAla-D-4-chloroPhe-D-3-pyridylAla-L-Ser-L-N-methylTyr-D-Asp-L-Leu-L-N(e)-isopropyl-Lys-L-Pro-D-Ala	Plenaxis	Anti-Testosterone, Antineoplastic. Synthetic decapeptide antagonist to gonadotropin releasing hormone (GnRH)	Palliative treatment of advanced prostate cancer
Aldesleukin	Human interleukin 2 (modified)	Proleukin	Antineoplastic. Produced in recombinant *E. coli.* Nonglycosylated. N-terminal Ala deleted. Cys125Ser substitution.	adults with metastatic renal cell carcinoma
Anakinra	Human interleukin-1 receptor antagonist protein	Kineret	Antirheumatic, Immunomodulatory. Recombinant human interleukin-1 receptor antagonist. Produced in *E. coli.* 153 residues.	adult rheumatoid arthritis
Becaplermin	Human platelet-derived growth factor	Regranex	Topical anti-ulcer. Recombinant protein produced in the yeast, *Saccharomyces cerevisiae.* Homodimer.	topical treatment of skin ulcers (from diabetes)
Calcitonin, salmon	Salmon calcitonin	Miacalcin	Antihypocalcemic, Antiosteporotic. Synthetic peptide, 32 residues long formulated as a nasal spray.	osteoporitis
Cetrorelix	acetyl-D-3-(2'-naphtyl)-ala-D-4-chlorophe-D-3-(3'-pyridyl)-ala-L-ser-L-tyr-D-citruline-L-leu-L-arg-L-pro-D-ala-amide	Cetrorelix, Cetrorelixum, Cetrotide	Infertility Agent. Synthetic hormone blocks premature ovulation by antogonizing GnRH-stimulated LH release.	inhibition of premature LH surges in women undergoing controlled ovarian stimulation

Name	Description	Trade Names	Properties	Indication
Choriogonadotropin alfa	Human chorionic gonadotropin	Novarel, Ovidrel; Pregnyl, Profasi	Fertility Agent. Has 2 subunits, alpha = 92 residues, beta = 145 residues, each with N-and O-linked carbohydrates.	female infertility
Darbepoetin alfa	Human erythropoietin	Aranesp	Antianemic. Human erythropoietin with 2aa substitutions to enhance glycosylation. Produced in recombinant CHO cells.	anemia (from renal transplants or certain HIV treatment)
Denileukin diftitox	Diptheria toxin-Interleukin-2 fusion protein	Ontak	Antineoplastic. Recombinant diphtheria toxin fragments A and B (1–387) fused with interleukin-2 (1–133). Produced in *E. coli.*	cutaneous T-cell lymphoma
Desmopressin	1-desamino-8-D-arginine vasopressin	Concentraid; DDAVP, Stimate	Renal Agent. Synthetic modification of vasopressin. Deaminated and Arg8 is D not L.	primary nocturnal enuresis, diabetes insipidus, temporary polyuria and polydipsia
Desmopressin acetate	Desmopressin acetate	Stimate	Antidiuretic. Synthetic analogue of the natural pituitary hormone 8-arginine vasopressin (ADH). 3-mercaptopropionic acid replaces N-terminal Cys and D-Arg replaces Arg.	diabetes insipidus, prevention of polydipsia, polyuria and dehydration
Epoetin alfa	Human erythropoietin	Epogen, Epogin, Epomax, Eprex, NeoRecormon	Antianemic. Produced in recombinant CHO cells.	anemia (from renal transplants or certain HIV Treatment)

TABLE 6.1-1. *Continued*

Generic Name	Chemical Name	Brand Name(s)	Description	Indication
Eptifibatide	N 6-(aminoiminomethyl)- N 2-(3-mercapto-1-oxo Pro-L-Lys-L-Gly-L- Asp-L-Try-L-Pro-L-Cys	Integrilin	Anticoagulant, Antiplatelet. Synthetic cyclic hexapeptide binds to platelet receptor glycoprotein and inhibits platelet aggregation.	myocardial infarction, acute coronary syndrome
Felypressin	2-(L-Phenylalanine)-8-L- lysine-vasopressin	Felipresina, Felipressina, Felypressin, Felypressine, Felypressinum	Renal Agent, Vasoconstrictor. Hemostatic. Synthetic analog of Lypressin (Lys-vasopressin) with more vasoconstrictor than antidiuretic action.	localizing agent for anesthetics
Filgrastim	Human granulocyte colony stimulating factor	Neupogen	Antineutropenic, Antiinfective, Immunomodulatory. 175 residues, expressed in *E. coli.*	Increases leukocyte production. Treatment of non-myeloid cancer, neutropenia, and bone marrow transplant.
Follitropin alfa/beta	Human follicle stimulating hormone	Gonal-F	Fertility Agent. Fertility Agent. Recombinant.Two non- covalently linked, non-identical glycoproteins with 92 and 111 amino acids.	female infertility
Follitropin beta	Human follicle stimulating hormone	Follistim	Fertility Agent. Recombinant. Two non-covalently linked, non-identical glycoproteins with 92 and 111 amino acids.	female infertility
Glucagon recombinant	Human Glucagon	GlucaGen, Glucagon	Antihypoglycemic Agent. 29 residue peptide hormone. Produced in *E. coli.*	severe hypoglycemia, also used in gastrointestinal imaging

Drug	Composition	Brand	Description	Indication
Goserelin	pyro-Glu-His-Trp-Ser-Tyr-D-Ser(Bu^t)-Leu-Arg-Pro-Azgly-NH$_2$ acetate	Zoladex	Antineoplastic Agent. Synthetic hormone. Stops production of testosterone in men. Decreases production of estradiol in women.	breast cancer, prostate cancer, endometriosis
Insulin Glargine recombinant	Recombinant human insulin	Lantus	Hypoglycemic agent. Produced in *E. coli*. IDiffers from human insulin in that Asn21 is replaced by Gly and two Arg are added to the C-terminus of B chain.	diabetes (type I and II)
Insulin Lyspro recombinant	Human Insulin	Humalog, Insulin Lispro	Hypoglycemic Agent. Monomeric form of regular insulin hexamer. Absorbed more rapidly than regular insulin. Lsy28 and Pro29 swapped in B chain	diabetes (type I and II)
Insulin recombinant	Human insulin	Novolin R	Hypoglycemic Agent. 51 residues. Produced as 2 chains in recombinant *E. coli*. Chains linked chemically *in vitro*.	diabetes (type I and II)
Insulin, porcine	Porcine insulin	Iletin II	Hypoglycemic agent, Isolated from pro-insulin of pig pancreas. Forms hexameric structure.	diabetes (type I and type II)
Interferon alfa-2a	Human interferon alpha 2a	Roferon A	Antineoplastic, Antiviral. 165 residues. Produced in recombinant *E. coli*.	chronic hepatitis C, hairy cell leukemia, AIDS-related Kaposi's carcoma, chronic myelogenous leukemia

TABLE 6.1-1. *Continued*

Generic Name	Chemical Name	Brand Name(s)	Description	Indication
Interferon alfacon-1	Consensus sequence interferon alpha	Infergen	Antineoplastic, Antiviral, Immunomodulatory. Consensus sequence derived Interferon alpha, differs from human interferon alpha by 20 residues (88% identical to IFN alpha 2a)	hairy cell leukemia, malignant melanoma, and AIDS-related Kaposi's sarcoma
Interferon alfa-n1	Human interferon alpha	Wellferon	Antiviral, Immunomodulatory. Purified, natural glycosylated human interferon alpha. 166 residues.	venereal or genital warts caused by the Human Papiloma Virus
Interferon alfa-n3	Human interferon alpha (2a, 2b and 2c)	Alferon	Antineoplastic, Antiviral, Immunomodulatory. Purified, natural human interferon alpha. proteins (includes 2a, 2b and 2c)	hairy cell leukemia, malignant melanoma, AIDS-related Kaposi's sarcoma, laryngeal papillomatosis in children, some types of hepatitis
Interferon beta-1a	Human interferon beta	Avonex	Antiviral, Immunomodulatory. 166 residues, glycosylated. Produced in recombinant CHO cells. Primary sequence identical to natural human interferon beta.	relapsing/remitting multiple sclerosis. Also condyloma acuminatum.
Interferon beta-1b	Human interferon beta	Betaseron	Antiviral, Immunomodulatory. 165 residues. Cysteine 17 substituted with serine. Produced in *E. coli*, nonglycosylated.	relapsing/remitting multiple sclerosis
Interferon gamma-1b	Human interferon gamma-1b	Actimmune	Antiviral, Immunomodulatory. Produced from recombinant *E coli*. Purification by column chromatography.	chronic granulomatous disease, osteopetrosis

Generic name	Chemical name	Trade name	Description	Indication
Lepirudin	Hirudo medicinalis hirudin (variant 1)	Refludan	Anticoagulants, Antithrombotic. Identical to natural hirudin except for substitution of leu for ile at N-terminal and absence of a sulfate group on tyr63. Produced in yeast cells.	heparin-induced thrombocytopenia
Leuprolide	5-oxo-pro-his-tryp-ser-tyr-leu-leu-arg-N-ethyl-pro	Eligard	Antineoplastic, analog of gonadotropin releasing hormone.	prostate cancer, endometriosis, uterine fibroids, premature puberty
Lutropin alfa	Human luteinizing hormone	Luveris	Fertility Agent. Recombinant human LH produced in yeast with 2 subunits, alpha = 92 residues, beta = 121 residues.	female infertility
menotropins	mixture of LH and FSH	Gengraf, Neoral, Restasis, Sandimmune	Immunomodulatory. Mixture of LH and FSH extracted from urine of postmeonopausal women.	transplant rejection, rheumatoid arthritis, severe psoriasis
Octreotide acetate	Octreotide acetate	Sandostatin LAR	Anabolic Agent, Hormone replacement. Acetate salt of cyclic octapeptide. Pharmacologic properties mimick somatostatin.	acromegaly and side effects from cancer chemotherapy
Oprelvekin	Human interleukin 11	Neumega	Coagulant, Thrombotic.	Increases reduced platelet levels due to chemotherapy
Oral interferon alfa	Human interferon alpha	Veldona	Antiinfective; Antineoplastic. Interferon alpha, 165 amino acids.	oral warts arising from HIV infection
Oxytocin	Oxytocin	Oxytocin, Pitocin, Syntocinon	Antiocolytic. Synthetic 9 residue cyclic peptide. Prepared synthetically to avoid possible contamination with vasopressin.	assist in labor, labor induction, uterine contraction induction

TABLE 6.1-1. *Continued*

Generic Name	Chemical Name	Brand Name(s)	Description	Indication
Palifermin	Human keratinocyte growth factor	Kepivance	Antimucositis. 140 residues long, expressed in *E. coli.* First 23 residues removed to improve stability.	mucositis (mouth sores)
Pegfilgrastim	PEGylated human granulocyte colony stimulating factor	Neulasta	Anti-infective, Antineutropenic, Immunomodulatory. 175 residues. Produced in recombinant *E. coli.*	non-myeloid cancer, neutropenia, bone marrow transplant
Peginterferon alfa-2a	PEGylated human interferon alpha 2a	Pegasys	Antineoplastic, Antiviral, Immunomodulatory. Covalent conjugate of recombinant alfa-2a interferon with single polyethylene glycol (PEG) chain. Produced in *E. coli.* 165 aa.	hairy cell leukemia, malignant melanoma, and AIDS-related Kaposi's sarcoma
Peginterferon alfa-2b	PEGylated human interferon alpha 2b	PEG-Intron	Antineoplastic, Antiviral, Immunomodulatory. 165 amino acids. Has a single amino acid substitution (Lys23Arg).	hairy cell leukemia, malignant melanoma, and AIDS-related Kaposi's sarcoma
Pegvisomant	PEGylated human growth hormone	Somavert	Anabolic Agent, Hormone replacement. Selectively binds to GH receptors on cell surfaces and blocks binding of endogenous GH.	acromegaly
Sargramostim	Human granulocyte macrophage colony stimulating factor	Leucomax, Leucine	Antineoplastic, anti-infective. Recombinant GM-CSF is expressed in yeast. 127 residue glycoprotein.	cancer and bone marrow transplant
Secretin, synthetic	Human secretin	Secremax, SecreFlo	Diagnostic Agent. Produced in the S cells of the duodenum in response to low local pH. Stimulates secretion of bicarbonate from liver and pancreas.	diagnosis of pancreatic exocrine dysfunction and gastrinoma

Sermorelin	Human growth hormone-releasing hormone fragment	Geref	Anabolic Hormone replacement. Acetate salt of an amidated synthetic 29-amino acid peptide from amino-terminal segment of the naturally occurring growth hormone-releasing hormone.	dwarfism, HIV-induced weight loss
Somatropin recombinant	Human growth hormone	BioTropin, Genotropin, Humatrope, Norditropin, Nutropin, NutropinAQ, Protropin, Saizen, Serostim, Tev-Tropin	Anabolic Agent, Hormone replacement. 191 residues. Produced in recombinant *E. coli*.	dwarfism, acromegaly, and prevention on HIV-induced weight loss
Thyrotropin Alfa	Human alpha thyrotropin	Thyrogen	Diagnostic. Heterodimeric glycoprotein with two noncovalently linked subunits identical to human pituitary TSH.	detection of residueal or recurrent thyroid cancer
Urofollitropin	Human follicle stimulating hormone	Fertinex, Metrodin	Fertility Agent. Purified from urine of postmenopausal women.	female infertility
Vasopressin	Cys-Tyr-Phe-Gln-Asn-Cys-Pro-Arg-Gly-NH2 (Cys linked by- S-S- bridge)	Pitressin; Pressyn	Antidiuretic, Vasoconstrictor. Nine amino acid peptide secreted from posterior pituitary. Binds receptors in collecting tubules of the kidney and promotes reabsorbtion of water.	enuresis, polyuria, diabetes insipidus, polydipsia, oesophageal varices with bleeding

A-chain and B-chain regions are connected by the C-peptide residues and by two disulfide bridges. Proinsulin is further cleaved by the Golgi apparatus before secretion, releasing the C-peptide (35 residues) from the two insulin chains (51 residues total) joined by disulfide bonds. Insulin is secreted into the blood by the beta cells and carried to its major target tissues in the liver, adipocytes, and muscles. Insulin binds to the insulin receptor tyrosine kinase localized in the plasma membrane of target cells leading to receptor autophosphorylation and a signal transduction cascade that is known to regulate more than 100 downstream genes.

Banting, Best, Collip, and MacLeod first isolated insulin in 1921 and demonstrated its therapeutic potential in treating type I diabetes (diabetes mellitus) in 1922 [20]. Before the development of recombinant forms of the hormone, it was isolated from bovine or porcine pancreata. Recombinant human insulin is produced in either *Escherichia coli* bacteria or *Saccharonlyces cerivisiae* yeast strains that have been genetically engineered. Active insulin consists of two peptide chains (called A and B) joined by disulfide linkages (Figure 6.1-1). The A chain is 21 amino acids long, whereas the B chain is 30 amino acids long. *E. coli* lack the protease necessary to cleave proinsulin, so A and B chains are either produced in separate cells, purified, and oxidized chemically to form the disulfide linkages, or full-length proinsulin is produced and cleaved in a separate *in vitro* proteolytic reaction (reviewed in Ref. 21). Insulin produced in *S. cerivisiae* is genetically modified so that A and B chains are linked directly or by a short synthetic sequence. The final product is converted into human insulin by a trypsin-mediated transpeptidation reaction carried out in an aqueous–organic solvent media, in the presence of excess threonine ester. Human insulin was the first animal protein to have been made in bacteria in such a way that its structure is absolutely identical to that of the natural molecule. Modifications to the original recombinant insulin have included reversing selected amino acid positions (insulin Lispro) and replacing and

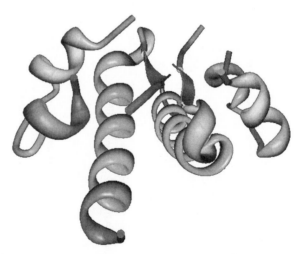

Figure 6.1-1. The x-ray crystal structure of a human insulin homodimer (A/B/C/D) at 1.5-Å resolution (PDB ID: *1ZEH*). Each component of the dimer is composed of a smaller chain (A/C of 21 residues) linked by disulfide bridges to a larger chain (B/D of 30 residues).

adding selected amino acids (insulin Glargine and Aspart). The purpose of these modifications is to improve the kinetics or duration of drug action.

Insulin is generally administered parenterally (subcutaneous injection), although many researchers and companies are developing alternative administration methods, i.e., oral, buccal, (Oral-lyn) or pulmonary (Exubera) [10]. These methods require greatly increased dosage in the case of buccal or pulmonary administration or additives to improve the bioavailability of orally administered insulin (including encapsulation, permeabilization, or chemical stabilization additives). No intestinally absorbed (oral) version of insulin has yet been proven effective and received FDA approval.

Another group of important homeostatic regulatory hormones are oxytocin and vasopressin. These two hormones are structurally related 9-mer oligopeptides produced in the hypothalamus and then transported along axons to the posterior pituitary where they are stored until they are secreted. Oxytocin and vassopressin are identical in sequence and structure except for residues three and eight; yet they have very different physiological effects. Oxytocin stimulates mammary milk secretion and contraction of uterine smooth muscle and likely plays a role in labor initiation. Vasopressin (also known as antidiuretic hormone—ADH) is a vasoconstrictor and reduces the water content of urine by increasing the reclamation of urinary water by the renal collecting duct. However, due to their structural similarity, there is a slight overlap of physiological response to these hormones; oxytocin has about 1% the antidiuretic potency of vasopressin, whereas vasopressin has about 15% the mammary secretion potency of oxytocin [11]. Both proteins are cyclic peptides with cysteines at positions one and six being joined by disulfide bonds, forming a six-residue ring structure with a flexible three-residue amidated carboxyl-terminal tail. Both hormones act through binding to specific receptors that are coupled to G-protein signal transduction pathways. G-protein activation leads to activation of inositol triphosphate and diacylglycerol, resulting in internal Ca^{2+} release and a MAP kinase cascade. The end result is phosphorylation of the myosin light chain leading to smooth muscle contraction. Oxytocin receptors are found in mammary myoepithelial and uterine smooth muscle cells, whereas vasopressin receptors are found predominantly in the cortical glomerular mesangial cells and medullary vascular smooth muscle cells of the kidney. In addition to stimulating contraction of vascular cells, vasopressin also causes translocation of aquaporin water channels to the apical surface of cells in the renal collecting duct. Traumatic or metastatic damage to the hypothalamus as well as rare hereditary disease can result in a lack of vasopressin, leading to the development of diabetes insipidus [11]. Those inflicted with this disease produce large amounts of dilute urine and must drink large amounts of fluid to replace what is lost.

Vasopressin is administered parenterally, but its synthetic analog, desmopressin acetate, can be administered by rhinal tube or orally, by tablet, for the treatment of diabetes insipidus [10]. Oxytocin and Vasopressin are manufactured through solid-phase synthesis [22, 23]. Antidiuretics are in fairly widespread use, having been prescribed in about 15% of all physician visits in the United Sates in 2002–2003 up from about 10% only 7 years before [24]. Oxytocin is used to induce labor by parenteral administration, either through subcutaneous or intravenous injection, or can be administered nasally to increase milk production in nursing mothers. Oxytocin induction of labor occurred in approximately 20% of births in the United States in 2003 [24].

Fertility Hormones. LH and FSH are heavily N-glycosylated glycoprotein dimers containing a common alpha chain of 92 amino acids and a unique 121 (LH) or 111 (FSH) residue beta chain. These hormones regulate human sex steroid production (LH) and gametogenesis (LH and FSH) in both males and females. They are expressed in a cyclic fashion by the pituitary gonadotrope cells. In males, for example, there is one LH pulse of consistent amplitude approximately every 90 minutes, but in females the amplitude and frequency of LH pulses varies throughout the ovarian cycle [11]. LH activates a seven-transmembrane G-protein-coupled receptor that is found primarily in the Leydig cells of the testes of males and in granulosa and theca cells in the female ovarian follicle. The FSH seven-transmembrane G-protein-coupled receptors are found only in Sertoli cells in the male testes or in granulosa cells in the female ovarian follicle. In males, LH induces the formation of testosterone in its target cells, whereas FSH induces production of androgen. In females, the complex cooperative activation of LH and FSH receptors regulates the levels of the primary sex steroids, progesterone, and estradiol. LH and FSH can be isolated from the urine of postmenopausal women, or they can be made in recombinant form by expression of transgenes in Chinese hamster (CHO) cells. LH and FSH are usually administered together to treat female infertility by stimulating ovulation and follicular maturation of the ovulated follicle into a corpus luteum. LH and FS (Lutropin—LH—or Repronex—combined LH and FSH) are administered parenterally. Of the approximately 6.7 million women with fertility problems in the United States in 1995, 35% were treated with some form of ovulation-inducing drug [25].

Growth Regulatory Hormones. Growth hormone (also known as somatotropin or somatropin) was first isolated in 1944 [11]. It is an anabolic hormone responsible for inducing cell division and controlling an individual's height. Growth hormone also increases calcium retention, strengthens and increases the mineralization of bone, increases muscle mass, induces protein synthesis, and stimulates the immune system. Mature growth hormone is a single-chain, nonglycosylated 191 residue polypeptide formed in the anterior pituitary. It is derived from a larger prohormone that includes a cleaved secretory signal peptide. In addition to the pituitary, growth hormone is also expressed at a much lower level in other tissues, including the central nervous system, mammary glands, gonads of both sexes, and hematopoietic and immune system cells. The growth hormone receptor was the first identified cytokine receptor and is an approximately 620-residue single-pass transmembrane protein that is highly glycosylated and ubiquitinated. A proteolytic fragment of the receptor, growth hormone binding protein, circulates throughout the bloodstream and increases the half-life of growth hormone by reducing its rate of degradation and clearing. The growth hormone receptor has been observed throughout the body in tissues as diverse as the gastrointestinal tract; the musculoskeletal system; the cardiorespiratory system; hematopoietic and immune systems; the central nervous system; integument, renal, and urinary systems; and the endocrine system. It is also found on cells of all major lineages in the developing fetus. However, in adults, it achieves its highest level of expression in the liver. A single growth hormone ligand binds to two receptors causing dimerization. This trimer, lacking its own kinase function, then recruits a member of the Janus tyrosine kinase family (JAK2) leading to activation of STAT-mediated gene transcription along with activation of the Ras-Raf-MEK, phosphatidylinositol 3-kinase (PI 3-kinase) and protein kinase C pathways [26, 27]. As might be expected, this has diverse and widespread effects on physiology and morphology. The main effect is to stimulate both chondrocyte and osteoblast proliferation leading to longitudinal bone growth. In addition, muscle mass increase is favored as an increase in lipolytic activity in adipose tissue leads to a reduction in total fat. Growth hormone also affects differentiation within the central nervous system with consequent cognitive effects such as enhancement of long-term memory, learning, and rapid eye movement sleep, although only when administered at supraphysiological levels.

A reduction or loss of growth hormone production, an autosomal recessive disorder, leads to obvious signs of dwarfism as early as 6 months of age. Laron syndrome is a defect in the growth hormone receptor also leading to sharply reduced stature but is resistant to exogenous treatment with growth hormone [29]. By contrast, chronic overproduction of growth hormone can lead to acromegaly, a sometimes fatal condition due to resulting cardiovascular, cerebrovascular, respiratory, and metabolic diseases.

Recombinant human growth hormone for injection is produced in *E. coli*. The mature product is identical to the natural human hormone, although it is initially translated with a signal peptide that is cleaved by the cell as the peptide is synthesized and exported into the periplasm. Growth hormone production can also be increased through administration of semorelin acetate, the acetate salt of an amidated synthetic 29-amino acid peptide corresponding to the amino-terminal residues of human growth hormone releasing hormone [11]. By contrast overproduction of growth hormone leading to acromegaly can be treated by a PEGylated version of growth hormone, known as Pegvisomant [30]. This chemically altered version of growth hormone competitively binds to the receptors without activating them, thus interfering with the action of the elevated endogenous hormone.

Another example of a growth regulatory hormone is erythropoietin. Erythropoeitin (EPO) is a 166-residue cytokine, produced primarily by interstitial cells near the proximal tubular cells of the kidney. EPO stimulates erythropoiesis, the formation of red blood cells. The hormone was initially partially purified by Goldwasser and colleagues from the urine of anemic patients in 1977, and the cDNA was cloned by Lin et al. and Jacobs et al. in 1985 [31, 32]. The protein is highly N- and O-glycosylated, with up to 40% of its weight being carbohydrate. Each glycosylation chain is terminated with a sialic acid, which may also be incorporated into the chain, and the total sialylation affects both receptor-binding affinity and serum half-life (the latter is of greater importance in total EPO biological activity). Thus, the kidney sets the hematocrit (the percent of whole blood that is composed of red blood cells) at a normal value of 45% by regulating red cell mass through EPO production and plasma volume through excretion of salt and water [11, 33]. EPO released into the blood travels to the bone marrow and binds to its cell-surface receptor on erythroid-specific precursors to stimulate red blood cell production. Like the related cytokine, human growth hormone, the binding of erythropoietin to its cognate receptor leads to dimerization and subsequent recruitment of the JAK2 tyrosine kinase signal transduction pathways to stimulate erythrocyte proliferation and differentiation. EPO has been approved for the treatment of anemia associated with chronic renal failure, with chemotherapy in cancer patients, and anemia associated with surgery. EPO is also beneficial for treatment of anemia associated with Zidovudine (e.g., AZT) therapy for patients infected with HIV. Because the functional protein must be correctly glycosylated posttranslationally, it is prepared from recombinant CHO cells expressing the erythropoietin precursor, including the 27-amino-acid signal peptide. A variant form (darbepoietin alpha) includes 2 amino acid substitutes that add N-glycosylation sites. The drug is currently only available for parenteral administration.

Immunomodulatory Hormones. Interferons constitute a family of species-specific vertebrate proteins (type I: IFN-α, IFN-β, IFN-ε, IFN-κ, and IFN-ω and type II: IFN-γ in humans) that confer general resistance to a broad range of viral infections, affect cell proliferation, and modulate immune responses [34]. Interferons were originally characterized in 1957 by Isaacs and Lindenmann as soluble proteins that induce antiviral activity in chicken cells. IFN-β and IFN-ω have only one functional copy each, but IFN-α is found in 13 different alleles producing 12 functional protein isoforms. The various IFN-α members each exhibit a distinct profile of antiviral, antiproliferative, and immunostimulatory effects, but variants of the IFNA2 gene (IFN-α2a, IFN-α2b, and IFN-α2c) provide the basis for most interferon therapeutics. Although the type I interferons invoke different cellular responses, they share a common dimeric receptor, IFNAR, which is found in low levels on the plasma membrane of all cell types. Each IFNAR subunit is composed of a single-pass transmembrane protein associated with a member of the JAK tyrosine kinase family (TYK2 for IFNAR-1 and JAK1 for IFNAR-2c). Upon binding of the ligand, the receptor dimerizes and cross-autophosphorylates the cytoplasmic receptor tails along with TYK2 and JAK1. This induces the phosphorylation and subsequent nuclear translocation of a STAT signaling complex and activation of gene-specific transcription [36]. It is not yet known how the many different species of IFN-α can lead to different cellular responses while acting through the same receptor, although data have indicated different ligands may have slightly different binding sites and may involve other receptors in the activated complex.

Parenteral administration of natural and recombinant interferons for the treatment of hairy cell leukemia, malignant melanoma, AIDS-related Kaposi's sarcoma, and a variety of viral diseases has been approved by the FDA. The type I IFNs exhibit a wide breadth of biological activities, including antiviral and antiproliferative effects as well as stimulation of MHC I antigen presentation and NK-cell activation. This makes them ideal for the treatment of many serious illnesses but can also result in undesirable side effects, especially at the high dosages normally employed. Oral administration of low-dose interferons has been studied in clinical trials but has not yet received approval for general use. Interferon-alpha was approved by the FDA on February 25, 1991 as a treatment for hepatitis C [37, 38]. In January 2001, the FDA approved PEGylated interferon-alpha. The PEGylated form is injected once weekly, rather than three times per week for conventional interferon-alpha.

6.1.3.2   The Future of Hormone Biopharmaceuticals

Research into new peptide hormone pharmaceuticals continues along several lines. Probably the most active area of research is into alternative delivery methods such as oral, nasal, buccal, transdermal, or pulmonary. All of these alternatives share issues of barrier penetration (whether mucosal or epidermal), protein stability, rate of clearance, and bioavailability.

The epithelial mucosa lining intestinal, buccal, nasal, and pulmonary tracts present significant barriers to protein or peptide hormone access to the circulatory system. Increasing the mucosal adhesiveness of therapeutic proteins delivered through the mucosa leads to an increase in the duration of residence at absorption sites. This is typically accomplished through encapsulation of the protein in micro- or nanoparticles made of mucoadhesive polymers [39, 40]. Increased duration in the mucosa also exposes the protein drug to increased extracellular proteolytic activity. Therefore therapeutic proteins and peptides can be chemically altered to reduce their susceptibility to proteases (e.g., in the intestine). Oral desmopressin acetate, for example, is partly protected against proteolytic degradation through amidation of the carboxyl-terminus, replacement of L-Arg with D-Arg and replacement of the amino-terminal residue with mercaptopropionic acid [41].

In addition to the mucosa, the epithelial cells also present a formidable barrier to the uptake of therapeutic proteins. A variety of strategies have been used to overcome this barrier, most notably through changing the hydrophobicity of the protein by the covalent attachment of lipophilic moieties. For example, the oral hexyl insulin monoconjugate 2 (HIM2) includes seven to nine units of PEG and a hexyl alkyl chain covalently attached to Lys29 and has proven safe and somewhat effective in controlling postprandial hyperglycemia in patients with type 1 diabetes mellitus during phase I/II clinical trials [42]. More recently, epithelial permeation enhancers, such as cationic or polymeric derivatives of the hydrophilic polysaccharide, chitosan, have proven effective in enhancing the effectiveness of oral [43], nasal [44], or pulmonary [39] administration of insulin or calcitonin in experimental settings. The precise mechanism by which these various permeant enhancers work is not yet understood, although, at least in the intestine, the modification of the tight junctions between cells in the intestinal epithelia is thought to be crucial [40].

In addition to novel delivery methods, several new peptide hormones with therapeutic potential have been recently identified. Ghrelin is a 28-residue, posttranslationally modified peptide that is secreted primarily from cells in the stomach and that has been implicated in the regulation of feeding behavior [11]. There is evidence that its orexigenic (appetite stimulating) activity is mediated, in part, by the hypothalamic hormone neuropeptide Y. Ghrelin is also the endogenous ligand for the growth hormone secretagogue receptor; it stimulates release of growth hormone from the pituitary independently of hypothalamic growth hormone releasing hormone. Repeated administration of Ghrelin has recently been shown to improve muscle mass and functional capacity in cachectic patients with heart failure or chronic obstructive pulmonary disease [45] and to alleviate chemotherapy-induced dyspepsia in rodents [46]. Leptin is an adipocyte-derived cytokine with an appetite-suppressing effect, likely mediated, in part, through neuropepetide Y. Along with adiponectin, another adipocyte-derived hormone, leptin may increase fat oxidation and promote insulin sensitivity via the AMP-activated protein kinase, thus reducing circulating fatty acids and triacylglycerol [47].

6.1.4   ANTIBODY BIOPHARMACEUTICALS

Antibodies or immunoglobulins are a class of disease-fighting proteins produced by B cells found in the lymphatic system or bone marrow. Antibodies are produced to bind a specific antigen that has stimulated the immune system. Their selectivity and tunability to recognize and bind an almost infinite range of large- and small-molecule substrates is what makes antibodies so important to the immune system. Once bound, the antigen can be ingested and destroyed by other cells of the immune system, primarily macrophages. Each antibody consists of four polypeptides—two heavy chains and two light chains joined by disulfide linkages in a hinge region to form a "Y"-shaped molecule (Figure 6.1-2a). The arms of the Y are known as the antigen-binding fragments (Fab). The region of the heavy chain associated with the light chains in the Fab contains one variable domain and one constant domain. The light chains each contain one constant domain and one variable domain. The variable domains of both the light and the heavy chain are known collectively as the variable fragment (Fv), and each contains hypervariable regions (known as complementarity-determining regions–CDRs) that are primarily responsible for determining antigen recognition. The lower portion of the "Y," consisting primarily of the heavy chains, is called the constant region. The constant region determines the mechanism used to destroy antigen. There are five classes of immunoglobulins (antibodies) in humans: IgM, IgG, IgE, IgA, and IgD. Most biopharmaceuticals are based on IgG.

Antibodies not only recognize and bind to antigens, but they also direct the process by which the antigens are eliminated or killed. These are called the effector function. The base of the heavy chain (called the constant fragment, Fc) is species-

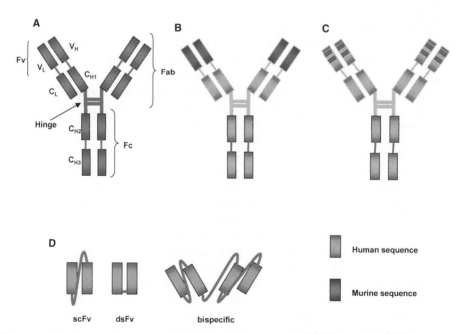

Figure 6.1-2. IgG antibodies and antibody fragments. (A) Murine IgG antibody showing variable (Fv), antigen-binding (Fab), and constant (Fc) fragments. Heavy and light chains are indicated by suffix H and L, respectively. (B) Chimeric IgG with murine Fv regions fused to human constant regions (C_L, C_{H1}, and Fc). (C) Humanized IgG with fused murine CDRs in Fv regions. (D) Antibody fragments. scFv—single-chain Fv; dsFc—disulfide-stabilized Fv; bispecific—two linked scFvs with different antigen recognition.

specific and mediates effector functions, including antibody-dependent cellular cytotoxicity (ADCC) and complement-dependent cytotoxicity (CDC). In ADCC, antibodies bind to Fc receptors on the surface of natural killer (NK) cells or macrophages, and trigger phagocytosis or lysis of the targeted cells. In CDC, antibodies kill the targeted cells by triggering the complement cascade at the cell surface. IgG is most efficient in inducing both ADCC and CDC and, therefore, is most suitable for use as a therapeutic molecule.

The first monoclonal antibody (mAb) tested in 1986 as a therapeutic in humans was a murine antibody, Muromonab (OKT3) [6]. Despite the high expectations of mAb therapy, OKT3 failed as a good treatment for transplantation rejection, primarily as a result of severe human anti-murine antibody response in patients. The second mAb marketed for therapeutic use in humans, Abciximab (ReoPro), required nearly 9 years of development to produce a chimeric Ab with mouse Fab and human Fc regions [48]. Since that time, more than 27 recombinant antibody therapeutics have been approved by the FDA, accounting for 24% of all approved protein/peptide therapeutics (Table 6.1-2) The current scheme for classifying mAbs is based on variation in their structure (Figure 6.1-2b–d).

1) Standard mAbs are produced by fusing murine cultured melanoma cells with isolated immune-reactive spleen cells to produce hybridoma fusions. These hybridomas are assayed *in vitro* for reactivity to the selected antigen and clones of single reactive cells are grown by intraperitoneal injection into naïve mice. Thus, all reactive cells from a single mouse result from a single reactive clone, and hence, the resulting antibodies are called monoclonal.

2) Fc-humanized mAbs are the result of genetic engineering to replace the murine Fc with human Fc in the melanoma cell lines [49].

3) An alternative approach to producing humanized Abs is to graft murine CDRs and a few structural residues from the heavy chain into a human IgG heavy chain [50]. In combination with the normal mouse light chain produced by the hybridoma, this CDR-grafting leads to the development of Daclizumab (Zenapax) in 1997. Instead of generating CDRs in whole animals by antigen immunization, they can also be selected by panning combinatorial phage libraries with recombinant Fabs [51]. Adalimumab (Humira) was the first therapeutic antibody developed with this technique in 2002. Yet another strategy for producing humanized Abs immunizes transgenic mice engineered to express human IgG against the selected antigen [52, 53]. Several therapeutic mAbs derived from such transgenic "humanized" mice are in late-phase clinical trial, including Panitumumab, an anti-EGFR Ab for the treatment of advanced metastatic colorectal cancer.

4) Recombinant bacterial techniques, coupled with the recognition that the entire antibody is not always required, have led to the formation of many therapeutic antibody fragments (Figure 6.1-3d). Whole IgG molecules are about 150 kDa, and subsequently, they diffuse into tissue slowly and are cleared from the body slowly as well. These characteristics make them poor candidates in diagnostic imaging or radiotherapy applications. The Fv region is the smallest portion of the Ab that maintains its antigen specificity, and it is small enough to be easily expressed in *E. coli*. Single-chain Fvs (scFvs) are fusion proteins containing both heavy and light variable regions connected

TABLE 6.1-2. FDA-Approved Antibody Pharmaceuticals

Generic Name	Chemical Name	Brand Name(s)	Drug Category	Indication
Abciximab	Chimeric human-murine (anti CD41) Ab	ReoPro	Anticoagulant, Antiplatelet. Fab fragment of chimeric human-murine monoclonal antibody binds to the glycoprotein IIb/IIIa receptor of human platelets and inhibits aggregation.	myocardial infarction, adjunct to percutaneous coronory intervention, unstable angina
Adalimumab	Humanized anti-TNF antibody	Humira	Antirheumatic, Immunomodulatory. Created using phage display technology. Has human-derived heavy- and light-chain variable regions and human constant regions. It is produced by recombinant DNA technology in a mammalian cell expression system.	for treatment of rheumatoid arthritis
Alefacept	Human LFA-3/IgG1 fusion protein	Amevive	Immunomodulatory, Immunosuppresant. Dimeric fusion of extracellular CD2-binding portion of leukocyte function antigen-3 linked to the Fc (hinge, CH2 and CH3 domains) portion of human IgG1. Produced in recombinant CHO cells.	moderate-to-severe chronic plaque psoraisis
Alemtuzumab	Humanized anti-CD52 antibody	Campath	Antineoplastic. Humanized mAb specific to lymphocyte antigen (cell surface glycoprotein,CD52). IgG1 with human variable framework and constant regions and murine CDRs. Produced in CHO cells.	B-cell chronic lymphocytic leukemia

Name	Type	Trade name	Description	Use
Antithymocyte globulin	Rabbit anti-thymocyte globulin	Atgam, Thymoglobulin	Immunomodulatory. Polyclonal antibody depletes T cells responsible for acute organ rejection in transplant patients.	renal transplant rejection
Arcitumomab	AntiCEA antibody	CEA-Scan	Diagnostic, Imaging Agent. Fab fragment of murine IgG1 mAb IMMU-4 (also called NP-4 with specificity for carcinoembryonic antigen (CEA) covalently labeled with Technitium 99.	imaging colorectal tumors
Basiliximab	Chimeric murine/human anti-CD25 antibody	Simulect	Immunomodulatory, Immunosuppressant. Blocks IL-2R alpha on surface of activated T lymphocytes. Produced by mouse myeloma cell line. Contains human heavy- and light-chain constant regions and murine heavy- and light-chain variable regions.	kidney transplant rejection
Bevacizumab	Humanized anti-VEGF antibody	Avastin	Antiangiogenesis, Antineoplastic. Contains human framework regions and the murine anti-VEGF CDRs. Produced in recombinant CHO cells.	metastatic colorectal cancer
Capromab	Murine anti-7E11 antibody	ProstaScint	Imaging Agent. Murine IgG1 mAb conjugated to linker chelator peptide group for indium attachment.	diagnosis of prostate cancer and detection of intra-pelvic
Cetuximab	Humanized anti-EGFR antibody	Erbitux	Antineoplastic. Epidermal growth factor receptor binding Fab. Has variable regions of murine EGFR mAb for the N-terminal portion of human EGFR with human IgG1 heavy- and kappa light-chain constant (framework) regions.	metastases metastatic colorectal cancer

717

TABLE 6.1-2. *Continued*

718

Generic Name	Chemical Name	Brand Name(s)	Drug Category	Indication
Daclizumab	Humanized anti-CD25 antibody	Zenapax	Immunomodulatory, Immunosupressant. Humanized IgG1 Mab that binds to the human IL-2R (anti-CD25). Composite of 90% human (IgG1 constant plus variable framework regions of the Eu myeloma Ab) and 10% murine Ab (CDR) sequences.	renal transplant rejection
Digibind	Anti-digoxin antibody fragment	Digibind	Sheep antibody Fab fragment specific to Digoxin.	digitoxin overdose or digitalis glycoside toxicity
Efalizumab	Humanized anti-CD11a antibody	Raptiva	Immunomodulatory, Immunosupressant. Humanized IgG1 anti-CD11a mAb. Produced in recombinant CHO cells.	psoriasis
Etanercept	Human tumor necrosis factor receptor fusion protein	Enbrel	Antirheumatic, Immunomodulatory. Dimeric fusion protein consisting of extracellular ligand-binding portion of human p75 TNFR linked to IgG1 Fc portion lacking CH1 domain. Produced by recombinant DNA technology in CHO cells.	severe adult and juvenile rheumatoid arthritis
Gemtuzumab ozogamicin	Humanized chimeric anti-CD33 antibody	Mylotarg	Antineoplastic. Recombinant humanized IgG4 conjugated with cytotoxic antitumor antibiotic, calicheamicin. Binds to cell-surface CD33 antigen. Produced in mammalian cell culture.	acute myeloid leukemia
Ibritumomab	Murine anti-CD20 antibody	Zevalin	Antineoplastic. Indium conjugated murine IgG1 anti-CD20 mAb. CD20 antigen is found on the surface of normal and malignant B lymphocytes. Produced in recombinant CHO cells.	non-Hodgkin's lymphoma

Name	Antibody type	Trade name	Description	Indication
Immune globulin	Human antibodies	Flebogamma, Gamunex	Anti-infective, Immunomodulatory. Mixture of IgG1 and other Abs derived from human plasma via Cohn fractionation. 70.3% IgG1, 24.7% IgG2, 3.1% IgG3, and 1.9% IgG4.	immunodeficiencies, thrombocytopenic purpura, Kawasaki disease, gammablobulinemia, leukemia, bone transplant
Infliximab	Chimeric mouse/human anti-TNF-alpha antibody	Remicade	Immunomodulatory, Immunosuppresants. TNF-alpha binding antibody. Chimeric Ab composed of human constant and murine variable regions. Produced by a recombinant cell line cultured by continuous perfusion	Crohn's disease, psoriasis, rheumatoid arthitis and ankylosing spondylitis
Muromonab	Murine anti-CD3 antibody	Orthoclone OKT3	Immunomodulatory, Immunosuppressant. Purified murine (mouse) mAb, directed against CD3 (T3) receptor on the surface of T lymphocytes. Cultured from murine ascites. 93% monomeric IgG2a.	prevention of organ rejection
Natalizumab	Humanized anti a4 integrin antibody	Tysabri, Antegren	Immunomodulatory. Humanized IgG4 mAb produced in murine myeloma cells. Contains human framework regions and murine anti-a4 itegrin CDR.	multiple sclerosis
Omalizumab	Humanized anti-IgE antibody	Xolair	Anti-Asthmatic, Immunomodulatory. Recombinant humanized IgG1 mAb binds to human IgE. Produced by recombinant CHO cells.	allergy-induced asthma

TABLE 6.1-2. *Continued*

Generic Name	Chemical Name	Brand Name(s)	Drug Category	Indication
Palivizumab	Humanized anti-RSV antibody	Synagis	Antiviral. Humanized mAb (5% murine) directed to an epitope in the A antigenic site of the F protein of respiratory syncytial virus (RSV). Produced by a stable murine myeloma cell line.	respiratory diseases caused by respiratory syncytial virus
Rituximab	Chimeric murine/human monoclonal anti-CD20 Ab	Rituxan	Antineoplastic. Chimeric murine/human monoclonal anti-CD20 Ab. Contains murine light- and heavy-chain variable region sequences and human constant region sequences. C20 is found on the surface of normal and malignant B lymphocytes.	B-cell non-Hodgkins lymphoma (CD20 positive)
Satumomab Pendetide	Murine anti-TAG72 antibody conjugated with Indium 111	OncoScint	Imaging Agent. Tumor associated glycoprotein mAb conjugated with Indium 111 for radioimaging colon tumors.	diagnosis of extrahepatic malignant cancers
Tositumomab	Murine anti-CD20 antibody	Bexxar	Antineoplastic. Murine IgG2 anti-CD20 mAb. Produced in cultured mammalian cells. Can be covalently linked to radioactive Iodine 131.	non-Hodgkin's lymphoma
Trastuzumab	Humanized anti-HER2 antibody	Herceptin	Antineoplastic. A recombinant humanized mAbbinds to the extracellular domain of human EGFR. Produced in CHO cell culture.	HER2-positive metastatic breast cancer

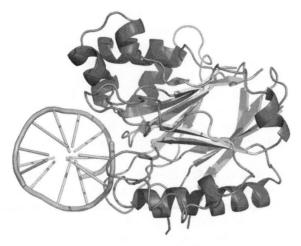

Figure 6.1-3. The x-ray crystal structure of DNase I in the presence of the palindromic d(GCGATCGC) DNA fragment at 2.0-Å resolution (PDB ID: *2DNJ*).

through a short peptide linker. Disulfide-stabilized Fvs (dsFvs) connect the two variable regions through engineered cysteine disulfide linkages. Diabodies are homodimers of scFvs that are covalently linked by a short peptide linker so that the V domains are forced to make intermolecular complexes. Bispecific Fvs are fusions of two complete Fv regions with a different antigen specificity. Variations on these themes incorporating part of the constant region are also possible.

6.1.4.1   Examples of Antibody Biopharmaceuticals

Muromonab (Orthoclone OKT3) is a purified murine monoclonal antibody specific to the CD3 receptor on the surface of human T cells (T lymphocytes). Used in organ transplant prophylaxis, Muromonab is administered parenterally. It has been effective in reversing corticosteroid-resistant acute rejection in renal, liver, and cardiac transplant recipients. Muromonab binds to the epsilon subunit of the surface glycoprotein CD3 that is associated with the T-cell receptor. It seems to kill CD3 positive cells by inducing Fc mediated apoptosis, antibody mediated cytotoxicity, and complement-dependent cytotoxicity. Immediately after administration CD3-positive T lymphocytes are abruptly removed from circulation, most likely by NK-mediated cytoxicity directed toward the T cells themselves [54] or by inducing T-cell apoptosis [55]. The OKT3-dependent reduction in alloreactive T lymphocytes severely downmodulates transplant rejection. However, because OKT3 contains the mouse Fc regions, its application led to several unfortunate side effects, including acute cytokine overstimulation, flu-like symptoms, and, in rare cases, pulmonary edema, central nervous system disorders, or an anaphylactic reaction to the murine Fc (reviewed in Ref. 56.)

Abciximab (ReoPro) was the first partially humanized antibody approved for therapeutic use. It was produced from transgenic murine melanoma cells containing human heavy-chain genes so that the resulting Fab fragment is a chimera of human and mouse Ab fragments. Abciximab binds to the glycoprotein IIb/IIIa receptor (a member of the integrin family of adhesion receptors and the major platelet surface receptor involved in platelet aggregation) and competitively inhibits fibrinogen-mediated platelet aggregation and clot formation [57]. Thus it is used in the treatment of myocardial infarction, unstable angina, and as an adjunct to percutaneous coronary intervention. Because the constant regions of the Fab fragment are derived from a human gene, side effects such as those observed with Muromonab are dramatically reduced. Nevertheless, in rare cases, acute profound thrombocytopenia can occur after Abciximab administration leading to a recommendation to monitor platelet count in patients within 11 to 21 hours after treatment [58].

Daclizumab is a humanized IgG mAb that binds to the human IL-2 receptor (anti-CD25). Daclizumab is a composite of human (90%) and murine (10%) antibody sequences. The human sequences were derived from the constant domains of human IgG1 and the variable framework regions of the Eu myeloma antibody [59]. The antibody is produced by expression in a mammalian cell line. The murine sequences were derived from the complementarity-determining regions of a murine anti-CD25 antibody [50]. Daclizumab competitively inhibits the cytokine IL-2 from binding to its CD25 receptor and phosphorylating the IL-2R beta and gamma-chains. This leads to an inhibition of downstream STAT activator genes and a subsequent immunosuppressive effect [60]. Early evaluations of clinical effectiveness of Daclizumab have been promising with reduced side effects relative to OKT3.

Adalimumab (Humira) is an anti-TNF antibody used in the treatment of rheumatoid arthritis. Elevated levels of TNF are found in the synovial fluid of rheumatoid arthritis patients and play an important role in both the pathologic inflammation and the joint destruction that are hallmarks of the disease. Adalimumab binds specifically to TNF-alpha and blocks its general cytokine effects through its interaction with the p55 and p75, thereby reducing TNF-induced inflammation and halting joint destruction. Adalimumab also lyses surface TNF expressing cells *in vitro* in the presence of complement. It was the first FDA-approved antibody therapeutic created using phage display technology [48] resulting in an antibody with human-derived heavy- and light-chain variable regions and human IgG constant regions. Adalimumab is produced by recombinant DNA technology in a mammalian cell expression system from which it is purified and administered by subcutaneous injection. Like other drugs that block TNF, use of Adalimumab has been associated with reactivation of previous serious infections such as tuberculosis, sepsis, and fungal infections. Patients with active infections should not be treated with Adalimumab.

6.1.4.3   The Future of Antibody Biopharmaceuticals

The development of future therapeutic antibodies revolves around several central issues: (1) further humanization to reduce unwanted Fc-based immune responses, (2) generation of novel CDRs against selected targets, and (3) selection of novel therapeutic Ab targets. Humanization of mAbs has been extensively reviewed elsewhere [61, 62]. A leading focus of recent activity is the use of transgenic mice that have been engineered by a combination of modified traditional transgenic techniques, yeast artificial chromosome (YAC) technology, and microcell-mediated chromosome transfer. The use of these techniques is required due to the large size of the immunoglobulin gene clusters; the human heavy-, λ-light-, and κ-light-chain loci (located on chromosomes 14, 22, and 2, respectively) span over a megabase each. In addition, murine native light- and heavy-chain regions are deleted in these heavily engineered mice to reduce the production of murine antibodies. Several mAbs produced from these mice are currently under development, including Zanolimumab (anti-CD4 for treatment of cutaneous and peripheral T-cell lymphoma [63]), Panitumumab (anti-epidermal growth factor receptor (EGFR) for treatment of advanced metastatic colorectal and other cancers [64]), and Denosumab (anti-receptor activator of nuclear factor kappa B ligand (RANKL)—a key mediator of the resorptive phase of bone remodeling—for treatment of osteoporosis, rheumatoid arthritis, and cancer with skeletal metastasis [65]).

We have discussed techniques using combinatorial phage libraries to pan for CDRs responding to therapeutic targets. In addition, libraries of novel CDRs can be displayed on the surface of *E. coli* or *S. cerevisiae* or produced *in vitro*, using ribosome display or semi- or fully synthetic antibody libraries (reviewed in Ref. 48). Libraries can be based on CDRs of inoculated or naïve individuals and extract variable domains from IgG or IgM regions by polymerase chain reaction (PCR) and subsequent cloning into expression vectors for display (e.g., in phage or *E. coli*). Although naïve libraries (those that have been derived from nonimmunized individuals) have greater complexity than libraries from inoculated individuals, they also tend to have reduced expression and increased toxicity in *E. coli*. *In vitro* methods circumvent these issues and permit harsher selection conditions that may lead to the production of higher affinity antibodies. Phage display is the only technology that has been developed sufficiently that it can be used commercially, and several therapeutic antibodies produced using this approach are in various stages of development, including Mapatumumab (an agonistic mAB that activates the TNF receptor apoptosis induced ligand (TRAIL-R1) to induce apoptosis in multiple types of cancer cells), Metelimumab (anti-TGF-β1 for treatment of sclerodoma), and AbThrax (anti-*Bacillus anthracis* Protective Antigen prevents anthrax from inducing toxin-mediated cell death).

Although novel therapeutic targets continue to be sought routinely, antibody development has mainly been focused on traditional cell-surface and extracellular antigens. Development of novel potential therapeutic targets is dependent on discoveries from fields as diverse as cell biology, cancer biology, microbiology, and developmental genetics. Recently, attention has been focused on the development of antibody derivatives, called intrabodies, that are directed toward intracellular targets [66]. The reducing environment found within the cells makes it difficult for most antibodies to form the disulfide bonds necessary for molecular stability. Intrabodies are formed from a subset of naturally occurring immunoglobulin frameworks that are capable of supporting a variety of antigen-binding structures inside this reducing environment. Alternative methods for discovering intrabodies involve screening scFvs, yeast surface display, phage display, intracellular antibody capture, and directed modifications to the Fc region to improve molecular stability in the cell. Experimental intrabodies that bind H-RAS and block oncogenic transformation of NIH 3T3 cells have recently been developed from scFvs [67] as have anti-angiogenic adenoviral-delivered scFvs [68].

6.1.5 ENZYME BIOPHARMACEUTICALS

Enzymes are produced by living organisms and function as biological catalysts in specific biochemical reactions of substrates. Enzymes accelerate reactions usually through stabilization of a reactive intermediate at their active site, the place where the reaction occurs. The enzymes, themselves, may be chemically altered during the reaction but are restored to their initial state by its completion. Enzymes are critical to the functioning of most cellular systems as most essential metabolic or catabolic reactions would occur too slowly, or would lead to different products without enzymes. Because of their importance, the mutation, overproduction, underproduction, or deletion of a single critical enzyme can lead to severe disease. For example, mucopolysaccharidosis type VI (or Maroteaux–Lamy syndrome), is a lethal syndrome that develops when patients are deficient in an enzyme called N-acetylgalactosamine 4-sulfatase (galsulfase) and, thus, cannot degrade glycosaminoglycans sufficiently. These compounds accumulate in various tissues leading to permanent, progressive cellular damage affecting an individual's appearance, physical abilities, organ and system functioning, and, in most cases, mental development [69].

More than 3000 human enzymes have been identified and named. Typically the suffix-*ase* is added to the name of the substrate (e.g., lactase is the enzyme that catalyzes the breakdown of lactose) or the type of reaction (e.g., DNA polymerase catalyzes the assembly of DNA). However, this is not always the case, especially when enzymes modify multiple substrates. For this reason, enzymes are formally named and classified based on the recommendations of the Nomenclature Committee of the International Union of Biochemistry, which assigns an Enzyme Commission (EC) number and a formal name. There are six main classes of enzymes: oxidoreductases (EC 1), transferases (EC 2), hydrolases (EC 3), lyases (EC 4), isomerases (EC 5), and ligases (EC 6). Most enzyme pharmaceuticals in current use belong to the hydrolase class such as galsulfase and agalsidase. These enzymes are involved in the hydrolysis (i.e., cleaving) of various bonds. However, there are also examples of pharmaceutical enzymes that belong to oxidoreductases, for example, rasburicase. Some enzymes are highly specific to certain substrates, such as adenosine deamidase, making them ideal drug candidates. On the other hand, some enzymes have wide specificities, such as vitamin A esterase, making these less desirable for drug development.

The potential utility of enzymes as pharmaceuticals was noted many decades ago, and since then, nearly two dozen enzymes have been developed to treat a variety of diseases. Almost all enzyme therapies developed to date are used to deal with a loss of function defect (a mutation that diminishes activity, a low level of production, a deletion). Hence, most enzyme drugs are used as enzyme replacement therapies (ERT) for relatively rare, inborn errors of metabolism (IEMs). As a result, many enzyme therapeutics fall under the FDA's Orphan Drug Designation. However, a few enzyme therapies can also be used to treat much more common conditions such as cancer, heart attacks, and stroke. In the United States, the first enzyme to receive FDA approval was a tissue plasminogen activator called alteplase. This protein, which is now commonly used to treat strokes, was introduced in 1987 as Activase. Since then at least 16 other enzyme drugs have been introduced into the marketplace. Some of these are described in more detail below and in Table 6.1-3.

TABLE 6.1-3. FDA-Approved Enzyme Pharmaceuticals

Generic Name	Brand Name(s)	Indication	Description
Agalsidase beta	Fabrazyme	Fabry's disease	Recombinant human α-galactosidase A. The mature protein is composed of 2 subunits of 398 aa. Protein is glycosylated and produced by CHO cells.
Alglucerase Imiglucerase	Ceredase Cerezyme	Gaucher's disease	Human β-glucocerebrosidase or β-D-glucosyl-N-acylsphingosine glucohydrolase. 497 aa protein (59.3 kDa) with N-linked carbohydrates. Alglucerase is prepared by modification of the oligosaccharide chains of human β-glucocerebrosidase. The modification alters the sugar residues at the nonreducing ends of the oligosaccharide chains of the glycoprotein so that they are predominantly terminated with mannose residues.
Alteplase	Activase	Acute myocardial infarction, acute ischemic stroke, and acute pulmonary emboli	Human tissue plasminogen activator, purified, glycosylated, 527 aa purified from CHO cells.
Anistreplase	Eminase	Acute pulmonary emboli, intracoronary emboli, and myocardial infarction	Human tissue plasminogen activator, purified, glycosylated, 527 residues purified from CHO cells. Eminase is a lyophilized formulation of anistreplase, the p-anisoyl derivative of the primary Lys-plasminogen-streptokinase activator complex (a complex of Lys-plasminogen and streptokinase). A p-anisoyl group is chemically conjugated to a complex of bacterial-derived streptokinase and human plasma-derived Lys-plasminogen proteins.
Asparaginase	Elspar	Acute lymphocytic leukemia & non-Hodgkin's lymphoma	L-asparagine amidohydrolase from *Escherichia coli.*
Collagenase	Santyl	Chronic dermal ulcers & severe skin burns	The enzyme collagenase is derived from fermentation of *Clostridium histolyticum.*

TABLE 6.1-3. *Continued*

Generic Name	Brand Name(s)	Indication	Description
Dornase alfa DNase	Dilor Lufyllin Neothylline Pulmozyme	Cystic fibrosis	Recombinant human deoxyribonuclease I. Delivered by aerosol mist. The protein is produced by genetically engineered CHO cells containing DNA encoding for the native human protein, DNase I. 260aa.
Hyaluronidase	Vitrase	For increase of absorption and distribution of other injected drugs, for rehydration and diabetic retinopathy	Highly purified sheep hyaluronidase for injection into the vitreous cavity of the eye.
Laronidase	Aldurazyme	Mucopolysaccharidosis	Human recombinant α-L-iduronidase, 628aa (mature form), produced by recombinant DNA technology in CHO cells. Laronidase is an 83kDa glycoprotein. The nucleotide sequence is identical to a polymorphic form of human α-L-iduronidase. It contains 6 N-linked oligosaccharide modification sites.
Pancrelipase	Cotazym Pancrease Ultrase Zymase	Cystic fibrosis, chronic pancreatitis, and pancreatic duct blockage	Protein mixture isolated from porcine or bovine pancreas, sometimes called pancreatin. Contains 3 enzymes: amylase, lipase, and protease (chymotrypsin).
Pegademase	Adagen	Severe combined immunodeficiency disease	Bovine adenosine deaminase derived from bovine intestine that has been extensively PEGylated for extended serum half-life.
Pegaspargase	Oncaspar	Acute lymphoblastic leukemia	PEGylated L-asparagine amidohydrolase from *Escherichia coli*. PEGylation substantially (factor of 4) extends the protein half-life.
Rasburicase	Elitek	Hyperuricemia	Rasburicase (34.1kDa) reduces elevated plasma uric acid levels due to tumor lysis (from chemotherapy). It catalyzes the oxidation of uric acid into an inactive and soluble metabolite (allantoin).

Reteplase	Retavase	Acute pulmonary emboli, intracoronary emboli, and myocardial infarction	Human tissue plasminogen activator, purified from CHO cells, glycosylated, 355 aa. Retavase is considered a "third-generation" thrombolytic agent, genetically engineered to retain and delete certain portions of human tPA. Retavase is a mutant of human tPA formed by deleting various aa's present in endogenous human tPA. Retavase contains 355 of the 527 aa of native human tPA (aa 1–3 and 176–527), and retains the activity-related kringle-2 and serine protease domains of human tPA. Three domains are deleted from retavase—kringle-1, finger, and EGF.
Streptokinase	Streptase	Acute evolving transmural myocardial infarction, pulmonary embolism, deep vein thrombosis, arterial thrombosis or embolism, and occlusion of arteriovenous cannulae	A purified preparation of a bacterial protein from group C (β)-hemolytic *Streptococcus*.
Tenecteplase	TNKase	Myocardial infarction and intracoronary emboli	Tenecteplase is a 527 aa glycoprotein developed by introducing the following modifications to the cDNA for natural human tPA: Thr_{103} to Asp_{103}, Asp_{117} to Gln_{117} (both within the kringle 1 domain), and the tetra-Ala substitution at aa 296–299 in the protease domain.
Urokinase	Abbokinase	Pulmonary embolism, coronary artery thrombosis, and *IV* catheter clearance	Low-molecular-weight form of human urokinase that consists of an A chain of 2 kDa linked by a SH bond to a B chain of 30.4 KDa. Recombinant urokinase plasminogen activator.

6.1.5.1 Examples of Enzyme Biopharmaceuticals

Dornase alfa (DNase I; EC 3.1.21.1) was introduced over a decade ago as an optional drug to treat cystic fibrosis (CF). Cystic fibrosis is an autosomal, recessive, hereditary disease that affects the lungs, sweat glands, and the digestive system, causing chronic respiratory and digestive problems. It is caused by mutations in the cystic fibrosis transmembrane conductance regulator (CFTR) protein and is the most common fatal autosomal recessive disease in Caucasians. Approximately 1 in 25 individuals of European descent is a carrier of a CF mutation. Dornase breaks down the built-up sputum in the lungs of affected individuals through the digestion of the accumulated extracellular DNA from the breakdown of neutrophils that collect due to the local inflammation of the lungs [70]. As the drug has a short duration of action, it is administered continuously through daily inhalation by a nebulizer or a compressor system. This treatment benefits about 40% of patients. The drug is usually well tolerated, although some side effects such as chest pain and skin rashes, have been reported [71]. The three-dimensional structure of DNase I in the presence of a short DNA segment is illustrated in Figure 6.1-3.

Galsulfase (Aryplase) or N-acetylgalactosamine 4-sulfatase (ARSB; EC 3.1.6.12) is a hydrolase drug used in the treatment of mucopolysaccharidosis (MPS) type IV (or Maroteaux–Lamy syndrome), a lysosomal storage disease. Patients suffering from this lethal syndrome are deficient in Galsulfase expression leading to a block of the degradation of glycosaminoglycans, which causes an accumulation of these compounds in various tissues [69]. Galsulfase treatment is well tolerated, although some patients develop antibodies against the enzyme. However, these enzyme-specific antibodies generally decrease significantly in these patients after about a year and a half of treatment [72]. Another drug has been available for the treatment of MPS type I since 2003. This disease is caused by a deficiency in α-L-iduronidase (laronidase; Aldurazyme; EC 3.2.1.76) [73]. Treatment with laronidase for the milder forms of MPS type I has been shown to be safe, well tolerated, and effective. No significant side effects are associated with the administration of this enzyme-drug, although most patients develop antibodies against the enzyme; however, with time the antibody titers subside [74]. Injection of laronidase must be done on a weekly basis [73].

Agalsidase (Fabrazyme) is used in the treatment of Fabry disease, a lethal inborn error of metabolism disorder in which patients lack lysosomal hydrolase α-galactosidase A (αGalA; EC 3.2.1.22). This enzyme is involved in the breakdown of glycosphingolipids leading to the accumulation of α-galactose in tissues and plasma [75]. This accumulation can lead to skin lesions, febrile episodes, and burning sensations in the extremities. Death in early adulthood is usually due to renal failure because of proteinuria induced hypertension. The bimonthly injection of agalsidase-α (Replagal) or agalsidase-β (Fabrazyme) alleviates the severe pain associated with this condition. The cost of these drugs can exceed $170,000 per year.

Pegademase (adenosine deaminase; EC 3.5.4.4) is a PEGylated enzyme used for treating patients with adenosine deaminase (ADA) deficiency [76] and suffering from a form of severe combined immunodeficiency syndrome (SCID), a disease that occurs in about 1/50,000 births. SCID patients have a severely crippled immune system and cannot clear or recover from the mildest microbial or viral infections.

SCID patients, if untreated, usually die within 1 year due to severe, recurrent infections such as chronic diarrhea, ear infections, recurrent pneumonia, and profuse oral candidiasis. The ADA deficiency leads to an accumulation of purine metabolites, in particular dGTP, which are cytotoxic to lymphoid stem cells. Restoration of normal ADA levels removes the dGTP and allows the lymphoid stem cells to propagate. Pegademase is administered intravenously, stored at 4°C, and should never be frozen [77].

Imiglucerase (alglucerase) is used in the treatment of type 1 Gaucher disease (GD), a lysosomal storage disorder due to a deficiency of glucocerebrosidase (EC 3.2.1.45). This enzyme deficiency leads to the accumulation of glucosylceramide in cells [78]. Imiglucerase has been used to treat Gaucher disease for the past decade and is given intravenously. Its application leads to a marked improvement in the clinical manifestation of the disease [79]. However, this drug cannot reverse all effects of Gaucher disease, such as fibrotic and necrotic tissue damage [80]. The drug is well tolerated and has few side effects. The most common side effect is antibody generation to imiglucerase [81]. However, the excellent safety profile of imiglucerase enables patient self-administration and prevents the morbidity associated with the disease.

Although most enzyme therapies are used to treat relatively rare metabolic or genetic disorders, some enzymes are used to treat much more common afflictions such as heart attacks and stroke. Tissue plasminogen activators (tPA; EC 3.4.21.68) are intravenously administered fibrinolytic agents that are used in reperfusion for the treatment of acute ischemic stroke [82]. It has been shown that the use of tPAs such as reteplase (Retavase), tenecteplase (TNKase), alteplase (Activase), and anistreplase (Eminase) within 3 hours post-stroke-onset affords at least an order of magnitude more benefit to patients than aspirin, the only other beneficial drug for this condition. About one third of patients treated with intravenous fibrinolytic therapy exhibit an improvement poststroke [83]. Furthermore, tPAs are also used in the treatment of acute myocardial infarction [84]. Tenecteplase is known to have a long half-life; therefore, a single bolus injection is sufficient for treatment [85]. The half-life of reteplase is even longer allowing its administration as a double-bolus injection. However, this increased half-life seems to be only useful in reducing the number of injections, as it provides no additional health benefits to the patient [86]. Streptokinase (Streptase; EC 3.4.99.0) is another thrombolytic agent that, in some instances, has been shown to be more effective than tPAs [87]. Lastly, urokinase (Abbokinase; EC 3.4.21.73) is another type of thrombolytic agent that activates plasminogen through another mechanism and is used in the treatment of pulmonary embolisms.

Several enzyme therapies are also used in the treatment of cancer. For instance, rasburicase (uricase; Elitek; EC 1.7.3.3) is used in the treatment of hyperuricemia due to tumor lysis syndrome (TLS). This is a potentially life-threatening condition that is associated with rapidly developing tumors, such as those found in lymphoma and leukemia, in patients undergoing chemotherapy. The incidence of hyperuricemia in these patients is close to 20% [88]. Uricase is an interesting enzyme, because it is found in most mammals, with the exception of humans, where the gene contains a nonsense mutation [89]. Rasburicase is well tolerated, has a very fast onset of action, and is administered intravenously once a day. However, the enzyme cannot be used on patients with glucose-6-phosphate dehydrogenase deficiency [90]. The

PEGylated formulation of uricase is less immunogenic and has a longer half-life than its "naked" form [89].

Asparaginase (EC 3.5.1.1) is an example of another enzyme-based anticancer therapy. Asparaginase, derived from *E. coli*, is frequently used in the treatment of acute lymphoblastic leukemia [91]. Asparaginase works by depriving leukemic cells of L-asparagine. Asparaginase hydrolyzes the amino acid L-asparagine to L-aspartic acid and ammonia. Most human tissues do not require L-asparagine because it can be synthesized from L-glutamine by the action of asparagine synthase (AS). However, acute lymphoblastic leukemia cells generally have very low levels of AS and cannot upregulate the AS gene during conditions of L-asparagine depletion. Asparaginase is an effective way of depleting L-asparagine from the body, and this depletion results in selective cell death of those cells that cannot make L-asparagine. Asparaginase has a relatively short half-life (10–20 hours). However, PEGylated asparaginase has a half-life that is 10–15 times longer. PEGylated-asparaginase (Pegaspargase) was developed in the 1970s and 1980s. The drug has undergone extensive testing and seems to retain its antileukemic effectiveness while allowing less-frequent administration than the native compound. Although the actual cost to patients for Pegaspargase is greater than that of multiple injections of other Asparaginases, the reduced need for physician visits and treatment of complications of therapy may make overall treatment costs considerably less than that of the conventional Asparaginases.

Also some drug formulations contain more than one enzyme. For example, pancrelipase or Zymase contains amylase (for processing starch), lipase (for breaking down fats), and chymotrypsin (a protease) in various combinations. This protein mixture is isolated from porcine or bovine pancreatic tissues and is sometimes called pancreatin. This enzyme cocktail is sold to help digestion in people who do not produce enough of their own digestive enzymes. Such a condition can arise due to pancreatic disfunction caused by pancreatitis, cystic fibrosis, or pancreatic cancer.

6.1.5.2   The Future of Enzyme Biopharmaceuticals

The field of enzyme therapy is not static. New enzymes are being developed and tested to treat a variety of diseases or conditions all the time. Some of the most active work has been in the area of lysosomal storage diseases. Indeed there are now several enzyme pharmaceuticals that are entering phase II and phase III trials to treat several fatal or debilitating syndromes associated with sugar metabolism or sugar catabolism. One example is acid maltase, which is associated with a rare, fatal condition called Pompe's disease. Pompe's disease, also known as infantile glycogen storage disease type 2 (GSD-II), or acid maltase deficiency, results from a genetic deficiency of the acid maltase enzyme, a protein that normally breaks down glycogen (sugar stored in cells). In the absence of acid maltase, glycogen accumulates to toxic levels in cardiac and skeletal muscles, causing the muscles to waste away. Early experiments with acid maltase on several newborns with Pompe's disease indicated good success, particularly with those with residual acid maltase activity [92]. Several other lysosomal storage diseases are also being investigated with enzyme replacement therapy, including Hurler's disease and Maroteaux–Lamy syndrome. Hurler's disease is characterized by a deficiency of α-iduronidase activity that causes accumulation of dermatan sulfate and heparin sulfate in patients. This leads to stiff joints, reduced range of motion, restrictive pulmonary disease, sleep apnea, and other problems with the eyes, liver, joints, and skeleton. Recent clinical trials with α-iduronidase ERT have shown significant improvement in some patients, especially with their range of motion, flexibility, and reduced sleep apnea (reviewed in Ref. 93). Maroteaux–Lamy syndrome is caused by the accumulation of excessive quantities of dermatan sulfate because of a deficiency of the enzyme N-acetylgalactosamine-4-sulphatase (arylsulfatase B). An ERT trial has recently been conducted with recombinant arylsulfatase B given once every week. In this trial, urinary glycosaminoglycans were reduced, walking ability improved, the range of shoulder motion increased, and joint pain lessened [72]. However, several adverse reactions were also noted and all patients developed antibodies to arylsulfatase.

Not all enzyme therapies are likely to be successful, at least not until better methods for formulating and delivering enzymes are developed. One central challenge to almost all ERTs is the delivery of the proteins to the right tissues. For example, among the lysosomal storage diseases, new and improved formulation and delivery technologies need to be developed to deliver therapeutic enzymes effectively to cardiac muscle and kidney in Fabry's disease, skeletal muscle in patients with Pompe's disease, and to joint tissues and structures in patients with Hurler's disease and Maroteaux–Lamy syndrome. Improved tissue targeting would no doubt help many other enzyme therapeutics used to treat cancer, heart attacks, and strokes.

Another challenge with ERT is the enormous cost and the sometimes questionable benefits or small improvements to patient quality of life. For some IEMs, the rarity of the disease means that production costs will always remain high. For Gaucher's disease, in which ERT has been the most effective, the annual cost of ERT ranges from \$40,000 to 320,000 per patient. The cost of ERT for Fabry's disease is about \$160,000 per year. Even for more common disorders such as cancer, heart attacks, and stroke, the costs of enzyme-based drugs are sometimes observed as too high. In fact, similar cost–benefit debates have swirled around the issue of tPA versus aspirin and heparin treatments for stroke for more than 15 years [94]. In some cases, the enzyme drug is determined to be very cost-effective. In other cases it is not. Certainly as the market size grows for some enzyme therapies, the costs will inevitably drop. Likewise as formulations are improved and delivery mechanisms are enhanced, it is likely that the cost–benefit balance will change markedly.

6.1.6 CONCLUSIONS

Pharmaceuticals based on natural and recombinant proteins have proven their therapeutic value. Whether for hormone or enzyme replacement therapies or to block or inhibit other protein signals, pharmaceutical proteins exhibit exquisite selectivity and specificity in their targets. Over 100 proteins have been approved by the U.S. Food and Drug Administration since their inception and many more are currently in various phases of development. Despite their potency, many issues hinder the widespread use of protein pharmaceuticals. Proteins require special (expensive) preparation techniques, usually involving recombinant engineering and expression in cell culture followed by lengthy and costly purifications. They must be stored and shipped under special conditions as they are susceptible to denaturation and degradation by proteolysis. The barriers to easy oral administration and bioavailability are formidable, as the body has many natural defenses to exogenous proteins. Nevertheless, although there are many hurdles to be overcome, protein biopharmaceuticals are certain to see continued rapid development and growth because their potential for providing a more specific and potent therapeutic approach is unparalleled.

REFERENCES

1. Lindeboom G A (1954). The story of a blood transfusion to a Pope. *J. Hist. Med. Allied Sci.* 9(4):455–459.

2. Baskett T F (2002). James Blundell: The first transfusion of human blood. *Resuscita.* 52(3):229–233.

3. Farr A D (1981). Treatment of haemophilia by transfusion: the first recorded case. *J. Royal Soc. Med.* 74(4):301–305.

4. Jenner E (1801). *An inquiry into the causes and effects of the Variolae Vaccinae, A Disease Discovered in Some of the Western Counties of England, Particularly Gloucestershire, and Known by the Name of the Cow-pox,* 3rd ed.: D.N. Shury, London.

5. Dunsch L (1986). *Jèons Jacob Berzelius,* 1. Aufl. ed, Biographien hervorragender Naturwissenschaftler, Techniker und Mediziner., Bd. 85. BSB B.G. Teubner Leipzig.

6. Raben M S (1957). Preparation of growth hormone from pituitaries of man and monkey. *Science.* 125(3253):883–884.

7. du Toit D F, Heydenrych J J (1986). The application, mechanism of action and side-effects of immunosuppressive agents in clinical transplantation. *S. Afr. Med. J.* 70(11):687–691.

8. Federlin K, Laube H, Velcovsky H G (1981). Biologic and immunologic *in vivo* and *in vitro* studies with biosynthetic human insulin. *Diabet. Care.* 4(2):170–174.

9. Don B R, Kaysen G (2004). Serum albumin: relationship to inflammation and nutrition. *Semin. Dial.* 17(6):432–437.

10. Wishart D S, et al (2006). DrugBank: a comprehensive resource for *in silico* drug discovery and exploration. *Nucleic Acids Res.* 34(Database iss.):D668–D672.

11. Henry H L, Norman A W (2003). *Encyclopedia of Hormones.* Academic Press, Amsterdam, the Netherlands.

12. Bugg T (2004). *Introduction to Enzyme and Coenzyme Chemistry.* Blackwell Publishers, Malden, MA.

13. Lo B K C (2004). *Antibody engineering: Methods and Protocols, Methods in molecular biology,* vol. 248. Humana Press, Clifton, N.J.

14. Hershfield M S, et al (1987). Treatment of adenosine deaminase deficiency with polyethylene glycol-modified adenosine deaminase. *New Eng. J. Med.* 316(10):589–596.

15. Howey D C, et al (1994). [Lys(B28), Pro(B29)]-human insulin. A rapidly absorbed analogue of human insulin. *Diabetes.* 43(3):396–402.

16. Venter J C, et al (2001). The sequence of the human genome. *Science.* 291(5507): 1304–1351.

17. Lander E S, et al (2001). Initial sequencing and analysis of the human genome. *Nature.* 409(6822):860–921.

18. Christopher S (2004). Has the yo-yo stopped? An assessment of human protein-coding gene number. *PROTEOMICS*. 4(6):1712–1726.

19. Svensson B, et al (2004). Proteinaceous α-amylase inhibitors. *Biochim. Biophy. Acta.* 1696(2):145–156.

20. Rosenfeld L (2002). Insulin: discovery and controversy. *Clin. Chem.* 48(12):2270–2288.

21. Walsh G (2005). Therapeutic insulins and their large-scale manufacture. *Appl. Microbiol. Biotechnol.* 67(2):151–159.

22. Meienhofer J, Sano Y (1968). A solid-phase synthesis of [lysine]-vasopressin through a crystalline protected nonapeptide intermediate. *J. Amer. Chem. Soc.* 90(11):2996–2997.

23. Bayer E, Hagenmaier H (1968). Solid phase synthesis of oxytocin. *Tetrahed. Lett.,* 172037–172039.

24. Martin J, et al (2005). *Births: Final Data for 2003. National Vital Statistics Reports,* vol. 54. National Center for Health Statistics. Hyattsville, MD.

25. Stephen E H, Chandra A (2000). Use of infertility services in the United States: 1995. *Family Plan. Perspect.* 32(3):132–137.

26. Vanderkuur J A, et al (1997). Signaling molecules involved in coupling growth hormone receptor to mitogen-activated protein kinase activation. *Endocrinol.* 138(10):4301–4307.

27. Frank S J (2002). Receptor dimerization in GH and erythropoietin action—it takes two to tango, but how? *Endocrinol.* 143(1):2–10.

28. Schneider H J, Pagotto U, Stalla G K (2003). Central effects of the somatotropic system. *Euro. J. Endocrinol.* 149(5):377–392.

29. Hull K L, Harvey S (1999). Growth hormone resistance: Clinical states and animal models. *J. Endocrinol.* 163(2):165–172.

30. Kopchick J J (2003). Discovery and mechanism of action of pegvisomant. *Euro. J. Endocrinol.* 148(Suppl 2):S21–S25.

31. Lin F-K, et al (1985). *Cloning and Expression of the Human Erythropoietin Gene,* Proc. National Academies of Sciences, vol. 82. pp. 7580–7584.

32. Jacobs K, et al (1985). Isolation and characterization of genomic and cDNA clones of human erythropoietin. *Nature.* 313(6005):806–810.

33. Binley K, et al (2002). Long-term reversal of chronic anemia using a hypoxia-regulated erythropoietin gene therapy. *Blood.* 100(7):2406–2413.

34. Pestka S, Krause C D, Walter M R (2004). Interferons, interferon-like cytokines, and their receptors. *Immunol. Rev.* 202(1):8–32.

35. Isaacs A, Lindenmann J (1957). Virus interference. I. The interferon. *Proc. of the Royal Society of London, Series B.* 147(927):258–267.

36. Stark G R, et al (1998). How cells respond to interferons. *Ann. Rev. Biochem.* 67227–67264.

37. Thomson CenterWatch, *CenterWatch Clinical Trials Listing Service,* 2006.

38. Wikipedia contributors. *Interferon.* Wikipedia, The Free Encyclopedia 2006, Available: http://en.wikipedia.org/w/index.php?title=Interferon&oldid=40341679.

39. Takeuchi H, Yamamoto H, Kawashima Y (2001). Mucoadhesive nanoparticulate systems for peptide drug delivery. *Adv. Drug Del. Rev.* 47(1):39–54.

40. Salamat-Miller N, Johnston T P (2005). Current strategies used to enhance the paracellular transport of therapeutic polypeptides across the intestinal epithelium. *Internat. J. Pharm.,* 294(1–2):201–216.

41. Lundin S, et al (1989). Degradation of [mercaptopropionic acid1, D-arginine8]-vasopressin (dDAVP) in pancreatic juice and intestinal mucosa homogenate. *Pharmacol. Toxicol.* 65(2):92–95.

42. Clement S, et al (2004). Oral modified insulin (HIM2) in patients with type 1 diabetes mellitus: results from a phase I/II clinical trial. *Metabolism.* 53(1):54–58.

43. Prego C, et al (2005). Transmucosal macromolecular drug delivery. *J. Control. Rel.* 101(1–3):151–162.

44. Varshosaz J, Sadrai H, Heidari A (2006). Nasal delivery of insulin using bioadhesive chitosan gels. *Drug Del.* 13(1):31–38.

45. Nagaya N, Kojima M, Kangawa K (2006). Ghrelin, a novel growth hormone-releasing peptide, in the treatment of cardiopulmonary-associated cachexia. *Inter. Med.* 45(3): 127–134.

46. Liu Y L, et al (2006). Ghrelin alleviates cancer chemotherapy-associated dyspepsia in rodents. *Cancer Chemother. Pharmacol.* 58(3):326–333.

47. Greenberg A S, Obin M S (2006). Obesity and the role of adipose tissue in inflammation and metabolism. *Amer. J. Clin. Nutrit.* 83(2):461S–465S.

48. Kim S J, Park Y, Hong H J (2005). Antibody engineering for the development of therapeutic antibodies. *Molec. Cells.* 20(1):17–29.

49. Morrison S L, et al (1984). Chimeric human antibody molecules: mouse antigen-binding domains with human constant region domains. *Proc. Natl. Acad. Sci. USA.* 81(21): 6851–6855.

50. Jones P T, et al (1986). Replacing the complementarity-determining regions in a human antibody with those from a mouse. *Nature.* 321(6069):522–525.

51. Zebedee S L, et al (1992). Human combinatorial antibody libraries to hepatitis B surface antigen. *Proc. Nat. Acad. Sci. USA.* 89(8):3175–3179.

52. Green L L, et al (1994). Antigen-specific human monoclonal antibodies from mice engineered with human Ig heavy and light chain YACs. *Nature Genetics.* 7(1):13–21.

53. Fishwild D M, et al (1996). High-avidity human IgG kappa monoclonal antibodies from a novel strain of minilocus transgenic mice. *Nature Biotechnol.* 14(7):845–851.

54. Nizet Y, et al (2000). The experimental (in vitro) and clinical (in vivo) immunosuppressive effects of a rat IgG2b anti-human CD2 mAb, LO-CD2a/BTI-322. *Transplantat.* 69(7):1420–1428.

55. Popma S H, Griswold D E, Li L (2005). Anti-CD3 antibodies OKT3 and hOKT3gamma1(Ala-Ala) induce proliferation of T cells but impair expansion of alloreactive T cells; aspecifc T cell proliferation induced by anti-CD3 antibodies correlates with impaired expansion of alloreactive T cells. *Internat. Immunopharmacol.* 5(1): 155–162.

56. Sgro C (1995). Side-effects of a monoclonal antibody, muromonab CD3/orthoclone OKT3: Bibliographic review. *Toxicol.* 105(1):23–29.

57. Coller B S (1997). Platelet GPIIb/IIIa antagonists: the first anti-integrin receptor therapeutics. *J. Clin. Investigat.* 100(suppl 11):S57–S60.

58. Berkowitz S D, et al (1997). Acute profound thrombocytopenia after C7E3 Fab (abciximab) therapy. *Circulation.* 95(4):809–813.

59. Morgan E L, Hugli T E, Weigle W O (1982). Isolation and identification of a biologically active peptide derived from the CH3 domain of human IgG1. *Proc. Nat. Acad. Sci. USA.* 79(17):5388–5391.

60. Goebel J, et al (2000). Daclizumab (Zenapax) inhibits early interleukin-2 receptor signal transduction events. *Transpl. Immunol.* 8(3):153–159.

61. Weiner L M (2006). Fully human therapeutic monoclonal antibodies. *J. Immunother.* 29(1):1–9.

62. Lonberg N (2005). Human antibodies from transgenic animals. *Nature Biotechnol.* 23(9):1117–1125.

63. Bayes M, Rabasseda X, Prous J R (2005). Gateways to clinical trials. *Meth. Findings Experim. Clin. Pharmacol.* 27(5):331–372.

64. Bayes M, Rabasseda X, Prous J R (2005). Gateways to clinical trials. *Meth. Findings Experim. Clin. Pharmacol.* 27(8):569–612.

65. Bayes M, Rabasseda X, Prous J R (2005). Gateways to clinical trials. *Meth. Findings Experim. Clin. Pharmacol.* 27(10):711–738.

66. Stocks M (2005). Intrabodies as drug discovery tools and therapeutics. *Curr. Opin. Chem. Biol.* 9(4):359–365.

67. Tanaka T, Rabbitts T H (2003). Intrabodies based on intracellular capture frameworks that bind the RAS protein with high affinity and impair oncogenic transformation. *Embo J.* 22(5):1025–1035.

68. Popkov M, et al (2005). Targeting tumor angiogenesis with adenovirus-delivered anti-Tie-2 intrabody. *Cancer Res.* 65(3):972–981.

69. Litjens T, Hopwood J J (2001). Mucopolysaccharidosis type VI: Structural and clinical implications of mutations in N-acetylgalactosamine-4-sulfatase. *Human Mutat.* 18(4): 282–295.

70. Shah P L, et al (1996). *In vivo* effects of recombinant human DNase I on sputum in patients with cystic fibrosis. *Thorax.* 51(2):119–125.

71. Davies J, et al (1997). Retrospective review of the effects of rhDNase in children with cystic fibrosis. *Pediat. Pulmonol.* 23(4):243–248.

72. Harmatz P, et al (2005). Direct comparison of measures of endurance, mobility, and joint function during enzyme-replacement therapy of mucopolysaccharidosis VI (Maroteaux-Lamy syndrome): Results after 48 weeks in a phase 2 open-label clinical study of recombinant human N-acetylgalactosamine 4-sulfatase. *Pediatr.* 115(6):e681–e689.

73. Miebach E (2005). Enzyme replacement therapy in mucopolysaccharidosis type I. *Acta Paediatr. Supplem.* 94(447):58–60.

74. Kakavanos R, et al (2003). Immune tolerance after long-term enzyme-replacement therapy among patients who have mucopolysaccharidosis I. *Lancet.* 361(9369): 1608–1613.

75. Desnick R J, et al (2003). Fabry disease, an under-recognized multisystemic disorder: Expert recommendations for diagnosis, management, and enzyme replacement therapy. *Ann. Intern. Med.* 138(4):338–346.

76. Blackburn M R, Kellems R E (2005). Adenosine deaminase deficiency: Metabolic basis of immune deficiency and pulmonary inflammation. *Adv. Immunol.* 86:1–41.

77. Hershfield M S (1995). PEG-ADA: An alternative to haploidentical bone marrow transplantation and an adjunct to gene therapy for adenosine deaminase deficiency. *Human Mutat.* 5(2):107–112.

78. Beutler E (1995). Gaucher disease. *Adv. Genet.* 3217–3249.

79. Pastores G M, et al (2004). Therapeutic goals in the treatment of Gaucher disease. *Semin. Hematol.* 41(4 Suppl 5):4–14.

80. Weinreb N J, et al (2005). Guidance on the use of miglustat for treating patients with type 1 Gaucher disease. *Amer. J. Hematol.* 80(3):223–229.

81. Rosenberg M, et al (1999). Immunosurveillance of alglucerase enzyme therapy for Gaucher patients: Induction of humoral tolerance in seroconverted patients after repeat administration. *Blood.* 93(6):2081–2088.

82. Hacke W, et al (2004). Association of outcome with early stroke treatment: pooled analysis of ATLANTIS, ECASS, and NINDS rt-PA stroke trials. *Lancet.* 363(9411): 768–774.

83. Saver J L (2004). Number needed to treat estimates incorporating effects over the entire range of clinical outcomes: Novel derivation method and application to thrombolytic therapy for acute stroke. *Archives of Neurol.* 61(7):1066–1070.

84. Menon V, et al (2004). Thrombolysis and adjunctive therapy in acute myocardial infarction: the Seventh ACCP Conference on Antithrombotic and Thrombolytic Therapy. *Chest.* 126(suppl 3):549S–575S.

85. Davydov L, Cheng J W (2001). Tenecteplase: A review. *Clin. Therapeu.* 23(7):982–997.

86. Noble S, McTavish D (1995). Levocabastine. An update of its pharmacology, clinical efficacy and tolerability in the topical treatment of allergic rhinitis and conjunctivitis. *Drugs.* 50(6):1032–1049.

87. Capstick T, Henry M T (2005). Efficacy of thrombolytic agents in the treatment of pulmonary embolism. *Euro. Respirat. J.* 26(5):864–874.

88. Annemans L, et al (2003). Incidence, medical resource utilisation and costs of hyperuricemia and tumour lysis syndrome in patients with acute leukaemia and non-Hodgkin's lymphoma in four European countries. *Leukemia and Lymphoma.* 44(1):77–83.

89. Bomalaski J S, et al (2002). Uricase formulated with polyethylene glycol (uricase-PEG 20): biochemical rationale and preclinical studies. *J. Rheumatol.* 29(9):1942–1949.

90. Pui C H (2002). Rasburicase: a potent uricolytic agent. *Expert Opin. Pharmacother.* 3(4):433–442.

91. Pui C H, Evans W E (2006). Treatment of acute lymphoblastic leukemia. *New Engl. J. Med.* 354(2):166–178.

92. Amato A A (2000). Acid maltase deficiency and related myopathies. *Neurol. Clin.* 18(1):151–165.

93. Wraith J E (2005). The first 5 years of clinical experience with laronidase enzyme replacement therapy for mucopolysaccharidosis I. *Expert Opin. Pharmacother.* 6(3):489–506.

94. Schneck M J, Biller J (2005). New treatments in acute ischemic stroke. *Curr. Treat. Opt. Neurol.* 7(6):499–511.

6.2

FORMULATION AND DELIVERY ISSUES OF THERAPEUTIC PROTEINS

SATHY V. BALU-IYER,[1,2] RAZVAN D. MICLEA,[2] AND VIVEK S. PUROHIT[3]

[1] *University at Buffalo, State University of New York, Amherst, New York*
[2] *Roswell Park Cancer Institute, Buffalo, New York*
[3] *R&D, Eurand, Inc., Vandalia, Ohio*

Chapter Contents

6.2.1 Introduction 737
    6.2.1.1 Recombinant DNA Technology 738
    6.2.1.2 Specificity and Potency 738
    6.2.1.3 Molecular Medicine 738
    6.2.1.4 Therapeutic, Subunit Vaccines 738
6.2.2 Stability of Therapeutic Proteins 739
    6.2.2.1 Chemical Instability 739
    6.2.2.2 Physical Instability 740
    6.2.2.3 Formulation and Delivery Approaches to Overcome Instability 741
6.2.3 Immunological Properties of Protein Therapeutics 745
    6.2.3.1 Immnogenicity of Proteins 745
    6.2.3.2 Formulation and Delivery Considerations to Reduce Immunogenicity 747
6.2.4 Pharmacokinetics and Pharmacodynamics 748
    6.2.4.1 Formulation and Delivery Approaches to Overcome PK Issues 749
    References 751

6.2.1 INTRODUCTION

The protein pharmaceuticals such as human growth hormone and insulin have been known for years. However, in the past decade, an increased number of protein

Handbook of Pharmaceutical Biotechnology, Edited by Shayne Cox Gad.
Copyright © 2007 John Wiley & Sons, Inc.

products have come on the market and into the development pipeline as a result of the following reasons.

6.2.1.1 Recombinant DNA Technology

Advances in recombinant technology alleviated the problem of protein production and its purity. Through the process of recombinant DNA technology and use of nonhuman cell lines, human proteins can be manufactured free of viral contamination. This process enables production of large quantities of proteins previously difficult to obtain from human sources. Further isolation of protein from human sources is associated with a high risk of viral contamination. One such example, plasma derived clotting factor VIII isolated from human blood, resulted in transmission of viral diseases such as hepatitis and AIDS [1, 2].

6.2.1.2 Specificity and Potency

Protein products are specific as they are identified based on their mechanism of action. Further, protein products are attractive because of their potency, as only μg quantities or a few thousand IU are required to exert the pharmacological effect.

6.2.1.3 Molecular Medicine

The term molecular medicine is defined as understanding disease at the molecular level. For example, the deficiency of factor VIII causes a bleeding disorder, Hemophilia A [3]. Given the accomplishments made in the area of human genome and the emergence of proteomics, functional proteins that have important therapeutic value have been developed, and more will be identified in the future.

6.2.1.4 Therapeutic, Subunit Vaccines

Protein and peptide antigen-based vaccines are safe. The search for safe, single-shot vaccines against infectious and other diseases is a major focus of current research. Several peptide-and protein-based antigens are at different stages of development.

However, the protein therapeutics have unique handling and development requirements relative to the small organic molecule-based drugs and pose exclusive challenges for the formulation scientist. The formulation and delivery issues of therapeutic proteins can contribute to difficulties in pharmaceutical development such as

- Stability (in vivo and in vitro)
- Immune toxicity
- Pharmacokinetics / pharmacodynamics-related problems

Formulation strategies, such as addition of specific functional excipients, freeze-drying, protein engineering, and delivery strategies, have been used to improve the stability and circulation half-life and reduce immunogenicity, which in turn improve therapeutic efficacy of protein drugs. This chapter will focus on formulation and delivery strategies of therapeutic proteins.

6.2.2 STABILITY OF THERAPEUTIC PROTEINS

As a result of the complex structure of the proteins, formulation of protein therapeutics pose unique difficulties as it is susceptible to physical and chemical instabilities. The complexity develops from the hierarchical nature of its structure: primary, secondary, tertiary, and quaternary structures. Primary structure is the amino acid sequence of the polypeptide chains; secondary structure refers to local-ordered conformation; tertiary structure deals with the spatial arrangement of secondary structural elements (often referred as global fold); and the quaternary structure is the spatial arrangement of subunits. In general, chemical instability is related to primary structure of the protein, whereas physical instability is associated with the global fold or 3D structure of the molecule. The common problems encountered for protein products are listed in Table 6.2-1.

6.2.2.1 Chemical Instability

Chemical instability refers to covalent modifications of the molecule that result in a new chemical entity. Some amino acids in the protein are chemically labile and undergo degradation to form a new chemical entity. The chemical reactions that affect the proteins are as follows:

Deamidation. The side chain of amide linkage in a Gln or Asn residue is hydrolysed to form a free carboxylic acid that gives mixtures of peptides in which the polypeptide backbone is attached via an alpha-carboxyl linkage (Asp/Glu) or is attached via beta carboxyl linkage (iso-Asp/Glu) [4]. The key step in the reaction is the formation of a five- or six-membered cyclic imide structure [5], and this degradative reaction most often occurs at a higher rate at the sequence Asn-Gly.

Oxidation. Oxidation is an important chemical degradation pathway for several proteins and can occur during all steps of its pharmaceutical development. The amino acids that are labile to oxidation are Met, His, Cys, Trp, Phe, Tyr, and Pro. Met has been identified as one of the most easily oxidized amino acids in proteins

TABLE 6.2-1.

Chemical Instability	Physical Instability
Hydrolysis	Denaturation
Deamidation (Gln and Asn)	Adsorption to surfaces
Oxidation (His, Met, Cys, Trp, and Tyr)	Aggregation
Proteolysis	Precipitation
Incorrect disulfide formation (Cysteine)	
Racemization	
Beta elimination	

[6, 7]. Free radicals generated by metal ion catalysis and photo or thermal degradation of buffer or other excipients can accelerate the oxidative degradation reactions.

Beta Elimination. The beta elimination can occur in amino acids that have beta side chains. The amino acids that are labile for beta elimination are Cys, Ser, Thr, Phe, and Lys [6]. The rate of this degradative reaction is influenced by pH, temperature, and presence of metal ions. The beta elimination of Cys leads to incorrect disulfide formation, as this reaction produces thiols [6].

Incorrect Disulfide Formation. This reaction occurs in neutral and alkaline pH conditions. Under such conditions, deamidation also occurs. The thiols (formed by beta elimination) can promote the disufide bond formation [8]. In an acidic environment, sulfenium ion promotes the incorrect disulfide bonds [9]. Proteins with scrambled disulfide bridges have been shown to yield native structure by incubating the protein with small amounts of mercaptoethanol or Cys.

Racemization. All amino acids except Gly are chiral and are subjected to base catalyzed racemization. This reaction leads to the formation of nonmetabolizable D-enantiomers and is often associated with loss of activity.

The chemical instability studies are often performed in simulated pharmaceutical stress conditions such as pH shifts or thermal stress conditions in the presence and in the absence of metal ions or buffer salts. The degradative products are analyzed by sequencing techniques such as mass spectrometry, HPLC techniques, and by employing specific enzymatic assays.

6.2.2.2 Physical Instability

Physical stability refers to changes in the higher-order structure (secondary and above) and is manifested as aggregation, precipitation, and surface adsorption [10]. Proteins, because of their complex structure and global fold, undergo physical instability [11]. In the global fold, the exposure of the hydrophobic groups is minimized, and this conformation is required for optimal biological activity of the protein. Denaturation is a molecular process that disrupts the global fold of the molecule.

Denaturation. Denaturation refers to an alteration in the global fold of the molecule (i.e, a disruption in tertiary and frequently secondary). The global fold of the molecule can be altered by environmental conditions that are encountered in pharmaceutical development of proteins, including changes in temperature, pH, pressure, and concentration of denaturing agents such as surfactants. This molecular process is generally associated with inactivation of the protein. Denaturation of the protein may also be followed by additional phenomena of physical degradation such as surface adsorption, aggregation, or precipitation.

Aggregation. The existence of I (folding intermediate) states in several proteins leads to the inactivation of the protein by aggregation. Moderate amounts of stress can generate I states, which retains secondary structure but tertiary structural features such as intramolecular contacts are lost [12]. These intermediates associate to form large aggregates.

Precipitation. Precipitation is the macroscopic manifestation of aggregation. It has been observed that the formation of finely divided precipitate of insulin occurs on the walls of the container [13]. The precipitation of proteins is also observed during the manufacturing stages. Upon expression in recombinant systems, many proteins form aggregates called inclusion bodies [14].

Surface Adsorption.   Adhesion of proteins to surfaces is a well-known phenomenon. The denatured or I states of the protein expose hydrophobic domains. This exposure promotes the binding of the protein to the walls of the container and other surfaces it comes in contact with. It has been well documented that insulin binds to surfaces of delivery pumps, to glass and plastic containers, and to the inside of the intravenous bags [15].

6.2.2.3   Formulation and Delivery Approaches to Overcome Instability

The risk of chemical instability can be assessed from the primary sequence of the protein. The sequence containing labile amino acids such as Asn-Gly and Met would be indicative of potential instability issues. The rate of chemical reactions that alter the primary sequence of the protein is higher in solution conditions and can limit the shelf-life of protein therapeutics. As the mobility of reactants is minimized in the solid state, freeze-drying is often attempted to improve the stability [16]. In such instances, physical instability is a major issue to be dealt with. Freeze-drying, also termed as lyophilization, is a dessication process in which the solvent (usually water) is first frozen and then is removed by sublimation in a vacuum [17]. In other words, the protein in solution is frozen, producing discrete ice and solute crystals. The solid ice is sublimed. Controlled heating desorbs any of the tightly bound water.

Three steps constitute the freeze-drying process [17].

- Freezing
- Primary drying (sublimation) to remove bulk, free water
- Secondary drying (desorption) to remove structured, more tightly bound water

To prevent denaturation/unfolding of the protein during the harsh freeze-drying process, excipients are added, including bulking agents, suitable buffers for pH and osmotic adjustment, cryoprotectants, protein structure stabilizers, and phase-state modifiers (glass transition temperature modifiers) [17, 18]. As a result of the potency of therapeutic proteins, a low concentration of the protein is commonly used, which may lead to loss of therapeutic ingredients during freeze-drying [17]. To prevent protein loss, bulking agents such as mannitol, glycine, lactose, and sucrose are added to increase the bulk mass. These bulking agents can also be used as cryoprotectants and to increase the collapse temperature, thereby improving product stability [18]. Cryoprotectants stabilize the proteins during the freezing and drying/storage process and improve the collapse temperature. The collapse temperature is closely linked to product temperature, which in turn, determines the storage temperature [16]. It is generally desired to store the product at room temperature, or 4–8°C, and, thus, excipients combined with a bulking agent that increases the glass (product) temperature are desired. Many informative review articles on freeze-drying are available [16–18].

The screening of protein structure stabilizers for long-term stability is generally achieved by performing folding/unfolding studies in solution conditions. As a first step to perform solution studies, a reasonable quantity of the protein is required. As protein folding is a complex but interesting area of research, this section of formulation development involves basic research. It is essential that suitable analytical, spectroscopic, and thermal methods are developed to investigate the protein folding [19, 20]. In addition, solubility, buffer and pH conditions, and biochemical properties such as size, charge, and PI are determined. Acceptable buffers can be obtained from FDA guidelines and generally regarded as safe (GRAS) list. Commonly used buffers include phosphates, acetates and carbonates, lactates, ascorbates, and citrates. Addition of sodium chloride and other salts should also be investigated as these buffers and salts can increase the stability of a protein.

It is a common practice to employ thermal stress and pH changes to unfold the protein [21, 22], as these are often encountered during the pharmaceutical development of protein drugs. A pH range of 5 to 7 is desired for formulation development [23]. Temperature-dependent studies yield thermodynamic information such as Tm and ΔG and the stability of the protein. To determine the Tm and ΔG, a simple two-state model for unfolding is assumed [24, 25].

$$N \Leftrightarrow D$$

The equilibrium constant for the unfolding is given by

$$K = [D]/[N]$$

The Gibbs free energy changes can be deduced from the equilibrium constant using the formula $\Delta G = (G_D - G_N) = -RT\ln K$, where R is the gas constant and T is the absolute temperature. The Tm and ΔG are determined by spectroscopic methods [circular dichroism (CD), fluorescence, second derivative ultraviolet absorption spectroscopy] or by thermal methods (Differential Scanning Calorimetry, DSC) [20]. The aggregation of the protein is monitored using gel filtration, by light-scattering measurements, analytical ultracentrifugation, and field flow fractionation techniques.

The spectroscopic methods allow the estimation of [D] and [N] at any temperature. At the melting temperature Tm, the ΔG is zero and ΔS is equal to ΔH/Tm where [D] and [N] are equal. The measured free energy is typically 5–20 Kcal/mol. Thus, the unfolding of the protein decreases the stability. A strategy to screen for excipients involves compounds that increase the Tm and prevent aggregation and surface adsorption. The above rational strategy has been used to identity excipients for acidic fibroblast growth factor [21].

The existence of intermediate states complicates this simple approach. The reversible denaturation or unfolding step can be followed by the formation of irreversible inactivated forms of the protein (I). The inactivated forms are chemical reaction (intermolecular disulfide bridges, intermolecular beta strands), aggregation, dissociation of oligomers to monomers, adsorption to surfaces, and incorrect refolding.

$$N \Leftrightarrow D \rightarrow I$$

K is the equilibrium constant that defines $N \Leftrightarrow D$ transition and thermodynamic stability. k is the rate constant for the irreversible transition from $D \rightarrow I$ and provides information regarding the long-term stability. Although the equation shows the inactivation after denaturation, the inactivation can proceed from the native state through the formation of intermediate states [26].

It has been observed for several proteins that the intermediate structures are formed as the protein unfolds from N state to D state [26]. As the protein unfolds, protein loses tertiary structure and, frequently, secondary structure. In some instances, the secondary structure remains intact while the tertiary structure is lost [12], which is clear from spectral studies that measure loss of secondary and tertiary structural changes. One spectroscopic technique that is sensitive to tertiary structure (e.g., fluorescence) would detect changes, whereas other techniques that are sensitive to secondary structures (e.g., far UV CD) do not show any spectral changes. This molecular property is defined as molten globule or structured intermediate [12]. These intermediates expose hydrophobic domains, and thus promote aggregation or surface adsorption.

The mechanisms of protein aggregation and stability are further complicated by the existence of aggregation-prone states in the unfolding of the protein that mimics the native state. Proteins subjected to thermal stress can aggregate from the fully unfolded state, but partially unfolded states have more frequently been implicated in aggregation processes; an example of the latter is the transition to molten globule states, in which proteins retain the majority of their secondary structural features while losing most of their tertiary structure. Such partially unfolded states have more exposed apolar regions and are more prone to aggregation than native or unfolded conformations. For proteins such as interferon gamma (IFN-γ) [27] and granulocyte colony stimulating factor (GCSF) [28], it has been clearly demonstrated that extensive unfolding or formation of molten globule states is not a prerequisite for aggregation and can occur even under solution conditions that are considered to be physiological. Rather, aggregation seems to occur from an expanded state that more resembles the native protein than substantially disrupted structures. In many such cases, the rate-limiting step is not the generation of extensively unfolded states. In the case of rhIFN-γ, aggregation has been shown to occur through the formation of a transiently expanded conformational species, whose surface area is greater than the native state by only 9% [27]. Similarly, native rhGCSF has been shown to aggregate through the formation of a monomeric transition state in which the surface area increase was only 15% [28]. The role of such small conformational fluctuations and their effect on long-term stability has received considerable attention recently.

The large therapeutic proteins, such as factor VIII, are often organized into distinct domains that play a critical role in their structure and function [29, 30]. The folding/unfolding behavior of these individual domains, and the interactions between them, may coordinate various functions in multidomain proteins.

Furthermore, the modular assembly of multidomain proteins often results in complex unfolding characteristics and the existence of intermediate species, which may lead to physical instability including a tendency to aggregate. In the case of rFVIII, neither complete unfolding of the protein nor the generation of partially unfolded states of the protein appears to be a prerequisite for the protein to aggregate. Rather, aggregation appears to involve subtle conformational changes in the C2 domain encompassing the lipid-binding region [31]. The excipients that bind to this domain have been developed as a formulation excipient that promotes the stability [31, 32].

Another complication that can impact the protein folding and stability is the irreversibility and "kinetic" aspects of protein folding. If kinetic components exist, they can interfere with the excipient screening process. It has been observed that the unfolding of rFVIII after thermal perturbation of the protein leads to irreversible aggregation [32, 33]. Based on thermal denaturation studies of multidomain proteins like phosphoglycerate kinase (PGK) and thermolysin, Sanchez-Ruiz et al. demonstrated that irreversible denaturation defies analysis by standard equilibrium methods and suggested that the unfolding may be a kinetically controlled process [34, 35]. It is a general practice to acquire the unfolding profile at different heating rates to determine the Tm. If Tm is found to be a function of the heating rate, it is important to develop excipients to minimize this kinetic contribution to protein instability. Such heating rate dependency on unfolding transitions has also been observed for other proteins including IFN-γ, Immunoglobulin G (IgG), and streptokinase [30, 36, 37].

In general, the heating rate dependence of protein thermal structural perturbations can be understood in terms of the Lumry–Eyring model [38], as shown below. According to this framework, a reversible unfolding step is followed by an irreversible event:

$$N \underset{k2}{\overset{k1}{\rightleftharpoons}} U \overset{k3}{\longrightarrow} A \qquad\qquad \text{(scheme 6.2-1)}$$

where A is the final (aggregated) state of the native protein N, irreversibly arrived at from the reversible altered form U. In this scheme, the first-order kinetic constants for each reaction j are represented by k_j, which varies with temperature according to the Arrhenius equation. If $k_3 \gg k_2$, then all of U will be converted into A, and the process can be represented by the simpler kinetic scheme as discussed in detail by Sanchez-Ruiz et al. [34]:

$$N \overset{k}{\longrightarrow} A \qquad\qquad \text{(scheme 6.2-2)}$$

This scheme represents the limiting case of the Lumry–Eyring model consisting of only two populated states, the native (N) and the final aggregated state (A). This scheme predicts that k is a first-order kinetic constant that varies with temperature, as given by the Arrhenius equation. Based on this simple kinetic model, T_m should vary with heating rate (v) according to the equation:

$$\ln(v/T_m^2) = \ln(AR/E_a) - E_a/RT_m$$

where A is the frequency factor, E_a is the activation energy of the unfolding step, and R is the gas constant. Therefore, a linear plot of $\ln(v/T_m^2)$ versus $1/T_m$ should indicate a first-order reaction with slope of $-E_a/R$. The activation energy (E_a) for factor VIII was found to be ~535 KJ/Mole (~128 Kcal/Mole) and compares well with the E_a associated with the two transitions observed for the multidomain protein IgG [E_a for the two transitions were reported to be 456 KJ/mole (~109 Kcal/mole) and 692 KJ/mole (~165 Kcal/mole)] [30]. Thus, analysis of the data in terms of kinetic scheme 6.2-2 suggests that, under the experimental conditions used in this study, the rate-controlling step in the aggregation of rFVIII may be a unimolecular reaction involving protein conformational changes. Based on our antibody binding assays, we believe that the conformational changes involve at least the lipid-binding region, 2303–2332 of the C2 domain [31]. O-phospho-L-serine (OPLS) is known to bind to the lipid-binding region in the C2 domain [31]. The addition of ligand capable of binding to this region inhibited conformational changes and interfered with the aggregation process.

It is important to study the protein folding at different concentrations of the protein once the selection of excipients is narrowed [39], which is critical for scale up and for proteins such as monoclonal antibody products where higher doses are desired. The kinetic aspects of protein folding and concentration-dependent aggregation could potentially lead to instability. Overall, equilibrium unfolding is often used to screen excipients for the development of protein formulations. However, minor conformational changes that resemble the native state can lead to aggregation interfering with the equilibrium unfolding measurements. In addition, use of higher protein concentrations can accelerate the kinetically controlled aggregation process [39]. Thus, formulation development requires the characterization of minor conformational changes or aggregation prone states and their kinetics. Previous findings on the molecular details of the aggregation prone state of rFVIII and their kinetics [32] have contributed toward the rational design of a rFVIII-phosphoserine (PS) complex with improved physical characteristics [31].

6.2.3 IMMUNOLOGICAL PROPERTIES OF PROTEIN THERAPEUTICS

6.2.3.1 Immunogenicity of Proteins

Proteins and peptides are more immunogenic than conventional xenobiotics, and they are more prone to evoke immunotoxicity reactions (Table 6.2-2). The immunotoxicity is usually manifested in the form of a humoral immune response involving the formation of antibodies [40–42]. The antibodies can abrogate the efficacy of the administered protein by neutralizing its activity [41–43] or by altering its pharmacokinetics, thus complicating the dose-response relationships. The risk of generating an immune response is dictated by a host of product-specific factors such as product origin, molecular structure, impurities/degradation products, and formulation [44]. Proteins of animal or microbial origin can invariably be expected to elicit an immune response as they will be recognized as foreign proteins (neoantigen) triggering a classic immune response involving the B and T cells [45]. Human proteins, expressed in microbes such as *Escherichia Coli* can be immunotoxic because of lack of glycosylation [46], which can expose immunogenic neoepitopes with the possibility of epitope spreading leading to immune responses against conserved regions in the long term. Even proteins of human origin, when

TABLE 6.2-2.

Therapeutic Protein	Detected Incidence of Anti-product Antibody
OKT3	~80%
Remicade	10–57%
Roferon	20–50%
Proleukin (interleukin-2)	47–74%
Betaferon	44%
Advate	15–30%
Bone morphogenic protein-7	13–38%
Enbrel (TNF receptor / anti-TNF IgG)	16%
Thrombopoietin	≤10%
PEGylated-MGDF	0.5–1.6%

administered to humans, genetically deficient of the respective protein, will trigger a classic immune response as the patients are not immunotolerant to the protein. The above is illustrated by the observations in patients having severe forms (negligible endogenous protein production) of Fabry's disease [47], an X-linked deficiency of α-galactosidase A enzyme or Hemophilia A, where almost 30% of the patients develop antibodies upon therapy with the respective protein [48, 49]. Proteins used as therapeutic agents to supplement endogenous proteins (e.g., erythropoeitin, IL-2) have the potential of immunotoxicity because of the breaking of immune tolerance to self proteins. The breaking of immunotolerance to self proteins results more because of product-related factors (impurities, degradation products, and formulation excipients) [50] and dosing regimens employed for therapy [51].

A major challenge faced by a protein formulation scientist is the inability to predict or asses the risk of immunotoxicity of the final product. The prediction of immune response requires a comprehensive consideration of protein structure, putative genetic background of the patient population, the formulation excipients, as well as frequency and route of administration [44]. The ability to asses the potential of immunotoxicity of a protein therapeutic is mainly hampered by the lack of predictive *in vitro* or *in vivo* models. Animal models, widely used in preclinical drug testing, cannot predict immune responses in humans because of issues of immunotolerance and basic differences in the structure and function of the immune system that may not be readily translated across species. Mouse models have been the foundation for the majority of the work done toward understanding immune responses. However, when it comes to prediction of immunotoxicity of therapeutic proteins, significant discrepancies in adaptive and innate immunity preclude any extrapolation to humans. Nevertheless, animal models can still be used to conduct preclinical immunotoxicity studies, provided one appreciates all the caveats associated with use of such animal models. Evidence suggests that nonhuman primates and transgenic/knockout mice models may be useful for predicting the relative immunogenicity of protein therapeutics and assessing the risk of immunotoxicity [52, 53].

As mentioned above, several product-related factors can influence the immunogenicity of protein therapeutics. The impact of trace impurities and excipients used in the formulation on the immunogenicity of the protein relative to the pure native protein should be evaluated in relevant animal models (nonhuman primates or transgenic/knockout mice) very early during the development cycle. Close attention should be paid to the impact of excipients and impurities on the structure of the protein. Structural changes can promote formation of neo-epitopes, which can initiate immune response in tolerant individuals followed by epitope spreading and thus resulting in immunotoxicity. The methods used to detect immune responses should be extremely sensitive to detect even small amounts of antibodies in the animal models, as even small amounts of inhibitory antibodies observed in animal models may indicate the potential for significant immunotoxicity in humans. Further, cellular basis for generation of the immune response should be thoroughly investigated in the animal models with the emphasis on mechanism. The above information can greatly enhance our ability to anticipate and devise strategies to prevent or minimize immunotoxicity issues in humans early on during the product development cycle. Aggregates are one of the major degradation products formed in protein therapeutics. Both chemical and physical degradation can give rise to the formation of aggregates. It is generally believed that the aggregates can enhance the immunogenicity of therapeutic protein preparations [44]. Although the presence of aggregates in protein preparations can influence its immunotoxicity, the generalization of the above statement may not be appropriate. Several studies with aggregates of various proteins have shown that aggregates do not always enhance the development of antibodies [54]. Aggregation can be associated with structural changes of the protein. The mechanism (physical or chemical) causing the aggregation of the protein dictates the structural features of the aggregated protein. These structural changes can significantly alter the epitope repertoire of the aggregated protein relative to the monomeric protein, thereby altering its properties as an immunogen [55]. Additionally, the altered structure can also change the processing of the immunogen by the endosomal proteases in the antigen-presenting cells, which is essential for presentation by the major histocompatibility complex [55]. Hence, to understand the risk of immunotoxicity associated with the administration of aggregates, one needs to have the explicit understanding of the correlation between structure and immunogenicity. Aggregates that retain the structure of the native protein are more likely to cause immunotoxicity. However, aggregates that do not retain the structure of the protein should not be considered safe as the exact immunotoxicological consequences are not known and should be investigated.

Individuals having a genetic predisposition for generating an immune response against a therapeutic agent are unlikely to benefit from the prior knowledge generated during the preclinical immunotoxicity testing. However, methods and techniques identifying their genetic predisposition will greatly benefit this subset of patients as alternatives can be explored early or before initiating therapy. *In vitro* systems using dendritic cells and B-cells isolated from patients need to be developed and hold promise as an individualized immunotoxicity screening method before initiating therapy.

6.2.3.2 Formulation and Delivery Considerations to Reduce Immunogenicity

Although the mechanism is not clear, immunodominant epitopes in the protein sequence and aggregates or impurities in the protein product can facilitate the immunogenicity of the protein. Immunodominant epitopes (linear or conformational) are subunits of the protein that are most easily recognized by the immune system and can evoke a specific immune response [56, 57]. Proteins having these immunodominant epitopes when administered in humans do run the risk of breaking tolerance. Additionally, in humans who are intolerant to a given protein, the immunodominant epitopes are most likely to evoke the immune response. A strategy used to reduce immunogenicity of a protein is to identify the immunodominant regions in the protein and replace them with innocuous sequences while retaining the original structure and function of the protein. An example of the above approach is the development of hybrid porcine/human FVIII molecule in which the immunodominant regions in the human molecule were replaced with less immunogenic regions of the porcine FVIII [58]. The above strategy has the potential for development of protein molecules that are less immunogenic by virtue of their design.

Another strategy used frequently to reduce immunogenicity of protein molecules by tampering with its basic chemistry is the use of "pegnology." The term signifies the science of attaching polyethylene glycol polymer chains to specific residues in the protein molecule, thereby influencing its properties of immunogenicity and circulation half-lives of the parent protein molecule. The mechanism by which the pegylation of the protein molecule can reduce immunogenicity is the shielding of specific immunodominant epitopes. The shielding phenomenon prevents the recognition of specific epitopes by B-cells, thereby preventing initiation of an immune response. Adagen (pegylated form of adenosine deaminase) and Oncaspar (Pegylated asparaginase) are examples of pegylated proteins currently in clinical use where immunogenicity has been reduced relative to the parent protein by pegylation. Another desirable property of pegylation is that it usually reduces the clearance of the protein molecule, which serves a dual purpose as it increases the circulation half-life of the protein molecule, thereby reducing the frequency of administration. Hence, the overall affect of pegylation is an improved therapeutic protein.

6.2.4   PHARMACOKINETICS AND PHARMACODYNAMICS

It is generally accepted that the successful development of new therapeutic entities is strictly dependent on their pharmacokinetic/pharmacodynamic properties. The classic PK concepts can be successfully applied to large molecules, leading to optimization of the dosing regimen to achieve the desired pharmacological effect. However, the pharmacokinetic scientists usually encounter challenging difficulties in interpreting the concentration-time profiles for the protein drugs, partly because of the limitations of the detection systems in terms of lower accuracy and precision of the bioassays. Moreover, the presence of endogenous proteins in some instances and the low bioavailability and the rapid elimination from the systemic circulation in others will further decrease the ability to determine the true levels of therapeutic proteins. The recent advances in immunoassays and mass spectrometry analysis improved our ability for quantification. However, caution should be taken when analyzing data from immuno-/radio-labeled assays or LC/MS, because of the inability of those methods to discern between active and inactive protein molecules (native state vs. proteolytic fragments/denatured states). Extensive review articles on the subject of validation of bioanalytical methods are available in the literature [59].

6.2.4.1 Formulation and Delivery Approaches to Overcome PK Issues

The lack of protein stability within the gastrointestinal tract leads [60] to the administration of therapeutic biomolecules through the parenteral routes. Among those, the i.v. route grants complete bioavailability, a rapid distribution within the blood with achievement of high drug concentrations [61]. On the other hand, the s.c. and i.m. administration of therapeutics protein is very appealing based on the possibility of an increase in residence time because of slow absorption from the injection site [44]. Hence, efforts are being made to develop new strategies for alternative parenteral routes. The extent of absorption from the interstitial space varies from case to case and is dependent on the extent of proteolytic degradation at the injection site. In many instances, the proteolytic degradation is a saturable process and an increase in the therapeutic dose usually translates to increased bioavailability [62]. However, a higher dose might break the tolerance level and lead to undesired immunological properties, as discussed in the previous section. Hence, to exploit the advantages offered by parenteral routes other than i.v., a need exists to design improved, novel methodology or delivery vehicles that would limit the proteolytic degradation and prevent the rapid elimination from the systemic circulation.

PEGylation. Poly (ethylene glycol) (PEG) is a nontoxic, highly soluble polymer [63] that can be linked to a protein via several chemical modifications. This approach includes attaching PEG to ε-amino (lysine), thiol (cysteine), and hydroxyl (tyrosine, serine) groups or carbohydrate moieties (glycoproteins) as well as at the N- or the C-terminal of peptides and proteins [64]. The foundation of this technology resides in the steric hindrance factor induced by the bulky covalently attached polymer. For instance, the covalent linkage of PEG to protein surfaces leads to shielding of putative immunodominant epitopes that will directly affect the antibody recognition process [65]. Apart from the improved immunological properties, the shielding factor exerted by the PEG alters the accessibility of proteolytic enzymes [66]. It is also responsible for a decrease in receptor-mediated endocyotsis with a subsequent decrease in clearance. In addition, the net increase in hydrodynamic radius (two or three water molecules—tightly bound per ethylene glycol monomer) will lead to a decrease in renal clearance, further improving the pharmacokinetic properties of the therapeutic proteins and peptides [67].

The prolonged circulation time, in turn, reduces the frequency of administration that can further reduce the adverse immune response and can maximize the pharmacological effects [44].

PEGylation of asparaginase (Oncaspar) resulted in greater than 96% reduction in the incidence of detectable immunogenicity [68]. In a phase III study involving 2000 patients, a native interferon induced antibodies in 14.5% of recipients in contrast to only 1.5% of patients receiving PEGlyated derivative. In addition, the renal filtering was reduced by ~100-fold for PEG-interferon relative to the unpegylated one. Thus, PEGylation seems to be a promising strategy for reducing immunogenecity and for reducing the frequency of dosing.

Historically, the first generation of PEG derivatives targeted the labile, solvent-accessible amino groups present either at the N-terminal or on the side chain of positively charged amino acid residues [68]. Lysine is one of the amino acids commonly found on protein surfaces. Hence, it was relatively easy to covalently couple PEG or other molecules at this position. However, the high occurrence of lysine in proteins was also responsible for a high degree of heterogeneity in the first generation of PEGylated proteins. Therefore, both Oncaspar [68] and Adagen [69] (PEG adenosine deamidase) suffered from this design flaw. However, the use of different PEGylated isomers received FDA approval as long as data was provided to support batch-to-batch reproducibility.

The second generation of PEG derivatives were merely designed to overcome the heterogeneity concerns present in the first generation of PEGylated proteins. To accomplish PEGylation at specific positions, the novel PEG reagents took advantage of the subtle differences in the pKa of α and ε amino groups present on the side chains of amino acid residues [70]. Another successful approach was to target the thiol group of cysteine residues that are rarely occurring on proteins surfaces and, hence, should lead to a more homogeneous population of PEG derivatives [71].

However, despite those advances in PEG technology, several issues still plague the formation of PEGylated derivatives. The PEGylation approaches are limited by substantial loss of specific activity of the protein, leading to inconsistent therapeutic effects. It is worthwhile to mention that new strategies are being developed to address the loss of biological activity and function. In this line, Peleg-Shulman et al. [72] adopted a reversible PEGylation approach in the case of interferon $\alpha2$, in which the PEG derivative undergoes limited hydrolysis under physiological conditions. The advantage of the reversible PEG-interferon lies in the fact that the released interferon can penetrate the tissue effortlessly and therefore exert its biological function. On the other hand, the PEG derivative retains its PEGylation advantages in terms of PK parameters such as the long circulation half-life.

To maintain the protein bioactivity, other valuable technologies incorporate a blocking step of the active site before PEGylation through the use of ligands/enzyme inhibitors. Those inhibitors can be free in solution or bound to a solid phase, facilitating the subsequent removal procedure of free PEG.

Altogether, the PEGylation approaches presented herein have been successful to small proteins that, overall, have a limited number of active sites on the surface. Such approaches might not be applicable to all proteins, particularly for the larger ones. It is too early to speculate whether those limitations might be responsible for the low anti-PEG antibody response evoked following administration in humans of PEG-uricase [73].

Lipid Structures as Drug Delivery Vehicles. Liposomes and lipid-based nanoparticles can act as a sustained release system for protein drugs. The formulation of Interferon Gamma in negatively charged liposomes [74] increases the residence time of the protein by approximately 19 times upon s.c. administration. Further, liposomal formulation of factor VIII improves physical stability, reduces immunogenicity, and prolongs the circulation time of factor VIII (unpublished results). Liposomes can be PEGylated to further extend their blood circulation time. Compared with classic liposomes, PEGylated vesicles show increased half-life, decreased plasma clearance [75], and a shift in distribution in favor of diseased tissues. PEG is incorporated into the lipid bilayer of the liposome, forming a hydrated shell that protects it from destruction by proteins. Pegylated liposomes are also less extensively taken up by the reticuloendothelial system [76] and are less likely to leak drug while in circulation. This approach of PEGylation avoids the covalent modification of the protein and the complication developing from these modifications.

In addition, liposomes can act as protein stabilizers that reduce the protease degradation and improve the physical stability (unpublished data, [62]). The aggregation promoted by the existence of intermediate states is a common problem. As these intermediate structures expose hydrophobic domains, the addition of liposomes at critical stages of the protein unfolding stabilize the protein against aggregation and help to refold the protein to native conformation. This property is referred to as chaperones, and the use of chaperones as a formulation excipient has not received considerable attention. Another mechanism, lipidic particles, can stabilize the protein by competing with the aggregation kinetics. For example, protein inactivation after the unfolding step is controlled by a kinetic process. Lipidic particles can interfere with this aggregation step. We have shown in our laboratory that the aggregation of factor VIII is kinetically controlled and undergoes irreversible denaturation [32]. However, in the presence of lipidic particles, the recovery of factor VIII upon refolding is much higher compared to free factor VIII.

PLGA Microspheres. Polymers such as PLGA microspheres have been used as delivery vehicles for peptides. In this delivery system, the peptide is embedded in a polymer matrix [77]. Two biodegradable polymeric systems are currently approved by FDA for use, Lupron [78] and Zoladex [79]. Several peptide drugs have been formulated in the biodegradable polymers. These PLGA polymers have been shown to release the peptide over the period of 1 to 3 months, which alleviates the pain caused by daily injections and improves the patient's compliance [78]. The monomers lactic and glycolic connected by ester bonds undergo ester hydrolysis that releases the peptide [77]. The peptide encapsulation is performed in methylene chloride. Such harsh conditions denature protein molecules but are acceptable for peptides. The microsphere encapsulation provides sustained release of IL-12 at the tumor site and avoids systemic toxicity associated with the administration of IL-12 [80]. However, there is substantial loss of activity.

REFERENCES

1. Suiter T M (2002). First and next generation native rFVIII in the treatment of hemophilia A. What has been achieved? Can patients be switched safely? Semin. Thromb. Hemost. 28(3):277–284.

2. Rosendaal F R, Smit C, Briet E (1991). Hemophilia treatment in historical perspective: A review of medical and social developments. Ann. Hematol. 62(1):5–15.

3. Larner A J (1987). The molecular pathology of haemophilia. *Q. J. Med.* 63(242):473–491.

4. Xie M, Schowen R L (1999). Secondary structure and protein deamidation. *J. Pharm. Sci.* 88(1): 8–13.

5. Meinwald Y C, Stimson E R, Scheraga H A (1986). Deamidation of the asparaginylglycyl sequence. *Int. J. Pept. Protein Res.* 28(1):79–84.

6. Manning M C, Patel K, Borchardt R T (1989). Stability of protein pharmaceuticals. *Pharm. Res.* 6(11):903–918.

7. Fransson J, et al. (1996). Oxidation of human insulin-like growth factor I in formulation studies: Kinetics of methionine oxidation in aqueous solution and in solid state. *Pharm. Res.* 13(8):1252–1257.

8. Costantino H R, Langer R, Klibanov A M (1995). Aggregation of a lyophilized pharmaceutical protein, recombinant human albumin: Effect of moisture and stabilization by excipients. *Biotechnol.* 13(5):493–496.

9. Robert Liu W, R.L.A.M.K. (1991). Moisture-induced aggregation of lyophilized proteins in the solid state. *Biotechnol. Bioengine.* 37(2):177–184.

10. Wang W (1999). Instability, stabilization, and formulation of liquid protein pharmaceuticals. *Int. J. Pharm.* 185(2):129–188.

11. Chi E Y, et al. (2003). Physical stability of proteins in aqueous solution: Mechanism and driving forces in nonnative protein aggregation. *Pharm. Res.* 20(9):1325–1336.

12. Kuwajima K (1996). The molten globule state of alpha-lactalbumin. *Faseb. J.* 10(1): 102–109.

13. Benson E A, et al. (1988). Flocculation and loss of potency of human NPH insulin. *Diabetes Care.* 11(7):563–596.

14. Baneyx F, Mujacic M (2004). Recombinant protein folding and misfolding in Escherichia coli. *Nat. Biotechnol.* 22(11):1399–1408.

15. Sefton M V, Antonacci G M (1984). Adsorption isotherms of insulin onto various materials. *Diabetes.* 33(7):674–680.

16. Carpenter J F, et al. (1997). Rational design of stable lyophilized protein formulations: Some practical advice. *Pharm. Res.* 14(8):969–975.

17. Tang X, Pikal M J (2004). Design of freeze-drying processes for pharmaceuticals: Practical advice. *Pharm. Res.* 21(2):191–200.

18. Chang L L, et al. (2005). Mechanism of protein stabilization by sugars during freeze-drying and storage: Native structure preservation, specific interaction, and/or immobilization in a glassy matrix? *J. Pharm. Sci.* 94(7):1427–1444.

19. Pelton J T, McLean L R (2000). Spectroscopic methods for analysis of protein secondary structure. *Anal. Biochem.* 277(2):167–176.

20. Bruylants G, Wouters J, Michaux C (2005). Differential scanning calorimetry in life science: Thermodynamics, stability, molecular recognition and application in drug design. *Curr. Med. Chem.* 12(17):2011–2020.

21. Tsai P K, et al. (1993). Formulation design of acidic fibroblast growth factor. *Pharm. Res.* 10(5):649–659.

22. Remmele R L, Jr., et al. (1998). Interleukin-1 receptor (IL-1R) liquid formulation development using differential scanning calorimetry. *Pharm. Res.* 15(2):200–208.

23. Osterberg T, Fatouros A, Mikaelsson M (1997). Development of freeze-dried albumin-free formulation of recombinant factor VIII SQ. *Pharm. Res.* 14(7):892–898.

24. Baldwin R L, Rose G D (1999). Is protein folding hierarchic? I. Local structure and peptide folding. *Trends. Biochem. Sci.* 24(1):26–33.

25. Privalov P L (1979). Stability of proteins: Small globular proteins. *Adv. Protein Chem.* 33:167–241.

26. Fink A L (1998). Protein aggregation: Folding aggregates, inclusion bodies and amyloid. *Fold. Des.* 3(1):R9–23.

27. Kendrick B S, et al. (1998). A transient expansion of the native state precedes aggregation of recombinant human interferon-gamma. *Proc. Natl. Acad. Sci. USA.* 95(24): 14142–14146.

28. Krishnan S, et al. (2002). Aggregation of granulocyte colony stimulating factor under physiological conditions: Characterization and thermodynamic inhibition. *Biochemi.* 41(20):6422–6431.

29. Jaenicke R (1999). Stability and folding of domain proteins. *Prog. Biophys. Mol. Biol.* 71(2):155–241.

30. Vermeer A W, Norde W (2000). The thermal stability of immunoglobulin: Unfolding and aggregation of a multi-domain protein. *Biophys. J.* 78(1):394–404.

31. Ramani K, et al. (2005). Lipid binding region (2303–2332) is involved in aggregation of recombinant human FVIII (rFVIII). *J. Pharm. Sci.* 94(6):1288–1299.

32. Ramani K, et al. (2005). Aggregation kinetics of recombinant human FVIII (rFVIII). *J. Pharm. Sci.* 94(9):2023–2029.

33. Grillo A O, et al. (2001). Conformational origin of the aggregation of recombinant human factor VIII. *Biochem.* 40(2):586–595.

34. Sanchez-Ruiz J M, et al. (1988). Differential scanning calorimetry of the irreversible thermal denaturation of thermolysin. *Biochem.* 27(5):1648–1652.

35. Galisteo M L, Mateo P L, Sanchez-Ruiz J M (1991). Kinetic study on the irreversible thermal denaturation of yeast phosphoglycerate kinase. *Biochem.* 30(8):2061–2066.

36. Kendrick B S, et al. (1998). Aggregation of recombinant human interferon gamma: Kinetics and structural transitions. *J. Pharm. Sci.* 87(9):1069–1076.

37. Azuaga A I, et al. (2002). Unfolding and aggregation during the thermal denaturation of streptokinase. *Eur. J. Biochem.* 269(16):4121–4133.

38. Lumry R, Eyring H (1954). Conformation changes of proteins. *J. Phys. Chem.* 58:110–120.

39. Kim Y S, et al. (2003). Congo red populates partially unfolded states of an amyloidogenic protein to enhance aggregation and amyloid fibril formation. *J. Biol. Chem.* 278(12):10842–10850.

40. Kulkarni R, et al. (2001). Therapeutic choices for patients with hemophilia and high-titer inhibitors. *Am. J. Hematol.* 67(4):240–246.

41. Grauer A, Ziegler R, Raue F (1995). Clinical significance of antibodies against calcitonin. *Exp. Clin. Endocrinol. Diabetes.* 103(6):345–351.

42. Casadevall N, et al. (2002). Pure red-cell aplasia and antierythropoietin antibodies in patients treated with recombinant erythropoietin. *N. Engl. J. Med.* 346(7):469–475.

43. Scandella D H, et al. (2001). In hemophilia A and autoantibody inhibitor patients: The factor VIII A2 domain and light chain are most immunogenic. *Thromb. Res.* 101(5):377–385.

44. Braun A, et al. (1997). Protein aggregates seem to play a key role among the parameters influencing the antigenicity of interferon alpha (IFN-alpha) in normal and transgenic mice. *Pharm. Res.* 14(10):1472–1478.

45. Chance R E, Root M A, Galloway J A (1976). The immunogenicity of insulin preparations. *Acta. Endocrinol. Suppl.* 205:185–198.

46. Gribben J G, et al. (1990). Development of antibodies to unprotected glycosylation sites on recombinant human GM-CSF. *Lancet.* 335(8687):434–437.

47. Linthorst G E, et al. (2004). Enzyme therapy for Fabry disease: Neutralizing antibodies toward agalsidase alpha and beta. *Kidney Int.* 66(4):1589–1595.

48. Jacquemin M G, Saint-Remy J M (1998). Factor VIII immunogenicity. *Haemophil.* 4(4):552–557.

49. Lollar P, et al. (2001). Factor VIII inhibitors. *Adv. Exp. Med. Biol.* 489:65–73.

50. Schellekens H (2002). Bioequivalence and the immunogenicity of biopharmaceuticals. *Nat. Rev. Drug. Discov.* 1(6):457–462.

51. Ross C, et al. (2000). Immunogenicity of interferon-beta in multiple sclerosis patients: Influence of preparation, dosage, dose frequency, and route of administration. Danish Multiple Sclerosis Study Group. *Ann. Neurol.* 48(5):706–712.

52. Palleroni A V, et al. (1997). Interferon immunogenicity: Preclinical evaluation of interferon-alpha 2a. *J. Interferon. Cytokine Res.* 17 (Suppl 1):S23–27.

53. Wierda D, Smith H W, Zwickl C M (2001). Immunogenicity of biopharmaceuticals in laboratory animals. *Toxicol.* 158(1–2):71–74.

54. Koch C, et al. (1996). A comparison of the immunogenicity of the native and denatured forms of a protein. *Apmis.* 104(2):115–125.

55. Purohit V S, Middaugh C R, Balasubramanian S V (2006). Influence of aggregation on immunogenicity of recombinant human Factor VIII in hemophilia A mice. *J. Pharm. Sci.* 95(2):358–371.

56. Reding M T, et al. (2003). Human CD4+ T-cell epitope repertoire on the C2 domain of coagulation factor VIII. *J. Thromb. Haemost.* 1(8):1777–1784.

57. Reding M T, et al. (2004). Epitope repertoire of human CD4(+) T cells on the A3 domain of coagulation factor VIII. *J. Thromb. Haemost.* 2(8):1385–1394.

58. Barrow R T, et al. (2000). Reduction of the antigenicity of factor VIII toward complex inhibitory antibody plasmas using multiply-substituted hybrid human/porcine factor VIII molecules. *Blood.* 95(2):564–568.

59. Miller K J, et al. (2001). Workshop on bioanalytical methods validation for macromolecules: Summary report. *Pharm. Res.* 18(9):1373–1383.

60. Blum P M, et al. (1981). Survival of oral human immune serum globulin in the gastrointestinal tract of low birth weight infants. *Pediatr. Res.* 15(9):1256–1260.

61. Tang L, et al. (2004). Pharmacokinetic aspects of biotechnology products. *J. Pharm. Sci.* 93(9):2184–2204.

62. Fatouros A, Liden Y, Sjostrom B (2000). Recombinant factor VIII SQ–stability of VIII: C in homogenates from porcine, monkey and human subcutaneous tissue. *J. Pharm. Pharmacol.* 52(7):797–805.

63. Yamaoka T, Tabata Y, Ikada Y (1994). Distribution and tissue uptake of poly(ethylene glycol) with different molecular weights after intravenous administration to mice. *J. Pharm. Sci.* 83(4):601–606.

64. Veronese F M, Pasut G, (2005). PEGylation, successful approach to drug delivery. *Drug Discov. Today.* 10(21):1451–1458.

65. Hershfield M S, et al. (1991). Use of site-directed mutagenesis to enhance the epitope-shielding effect of covalent modification of proteins with polyethylene glycol. *Proc. Natl. Acad. Sci. U S A.* 88(16):7185–7189.

66. Delgado C, Francis G E, Fisher D (1992). The uses and properties of PEG-linked proteins. *Crit. Rev. Ther. Drug Carrier Syst.* 9(3–4):249–304.

67. Molineux G (2004). The design and development of pegfilgrastim (PEG-rmetHuG-CSF, Neulasta). *Curr. Pharm. Des.* 10(11):1235–1244.

68. Graham M L (2003). Pegaspargase: A review of clinical studies. *Adv. Drug Deliv. Rev.* 55(10):1293–1302.

69. Levy Y, et al. (1988). Adenosine deaminase deficiency with late onset of recurrent infections: Response to treatment with polyethylene glycol-modified adenosine deaminase. *J. Pediatr.* 113(2):312–317.

70. Kinstler O B, et al. (1996). Characterization and stability of N-terminally PEGylated rhG-CSF. *Pharm. Res.* 13(7):996–1002.

71. Goodson R J, Katre N V (1990). Site-directed pegylation of recombinant interleukin-2 at its glycosylation site. *Biotechno.* 8(4):343–346.

72. Peleg-Shulman T, et al. (2004). Reversible PEGylation: A novel technology to release native interferon alpha2 over a prolonged time period. *J. Med. Chem.* 47(20):4897–4904.

73. Ganson N J, et al. (2005). Control of hyperuricemia in subjects with refractory gout, and induction of antibody against poly(ethylene) glycol (PEG), in a phase I trial of subcutaneous PEGylated urate oxidase. *Arthritis. Res. Ther.* 8(1):R12.

74. Van Slooten M L, et al. (2001). Liposomes as sustained release system for human interferon-gamma: Biopharmaceutical aspects. *Biochim. Biophys. Acta.* 1530(2–3):134–145.

75. Allen C, et al. (2002). Controlling the physical behavior and biological performance of liposome formulations through use of surface grafted poly(ethylene glycol). *Biosci. Rep.* 22(2):225–250.

76. Allen T M, et al. (1991). Liposomes containing synthetic lipid derivatives of poly(ethylene glycol) show prolonged circulation half-lives in vivo. *Biochim. Biophys. Acta.* 1066(1):29–36.

77. Cohen S, et al. (1991). Controlled delivery systems for proteins based on poly(lactic/glycolic acid) microspheres. *Pharm. Res.* 8(6):713–720.

78. Cox M C, Scripture C D, Figg W D (2005). Leuprolide acetate given by a subcutaneous extended-release injection: Less of a pain? *Expert Rev. Anticancer. Ther.* 5(4):605–611.

79. Perry C M, Brogden R N (1996). Goserelin. A review of its pharmacodynamic and pharmacokinetic properties, and therapeutic use in benign gynaecological disorders. *Drugs.* 51(2):319–346.

80. Hill H C, et al. (2002). Cancer immunotherapy with interleukin 12 and granulocyte-macrophage colony-stimulating factor-encapsulated microspheres: Coinduction of innate and adaptive antitumor immunity and cure of disseminated disease. *Cancer Res.* 62(24):7254–7263.

6.3

PHARMACOKINETICS

NOBUHITO SHIBATA,[1] YUKAKO ITO,[2] AND KANJI TAKADA[2]

[1]*Doshisha Women's College of Liberal Arts, Kyoto, Japan*
[2]*Kyoto Pharmaceutical University, Kyoto, Japan*

Chapter Contents

6.3.1   Introduction   758
6.3.2   Insulin   759
    6.3.2.1   Background   759
    6.3.2.2   Pharmacokinetics of Human Insulin after Intravenous
          Administration   760
    6.3.2.3   Pharmacokinetics of Human Insulin after Subcutaneous
          Administration   761
    6.3.2.4   Pharmacokinetics of Rapid-acting Insulin Analog after
          Subcutaneous Administration   762
    6.3.2.5   Other Formulations of Human Insulin   763
6.3.3   Erythropoietin   765
    6.3.3.1   Background   765
    6.3.3.2   Pharmacokinetics of Erythropoietin in Healthy Subjects   766
    6.3.3.3   Pharmacokinetics of Erythropoietin in Patients   768
    6.3.3.4   Erythropoietin Formulation   770
6.3.4   Granulocyte-colony Stimulating Factor (G-CSF)   771
    6.3.4.1   Background   771
    6.3.4.2   Pharmacokinetics of G-CSF in Animals   772
    6.3.4.3   Pharmacokinetics of G-CSF in Humans   773
    6.3.4.4   G-CSF Formulation   775
6.3.5   Interferon   778
    6.3.5.1   Background   778
    6.3.5.2   Pharmacokinetics of Interferon in Animals   779
    6.3.5.3   Pharmacokinetics of Interferon in Humans   780
    6.3.5.4   Interferon Formulation   781

Handbook of Pharmaceutical Biotechnology, Edited by Shayne Cox Gad.
Copyright © 2007 John Wiley & Sons, Inc.

6.3.6   Growth Hormone   783
    6.3.6.1   Background   783
    6.3.6.2   Pharmacokinetics of Growth Hormone in Animals   784
    6.3.6.3   Pharmacokinetics of Growth Hormone in Humans   785
    6.3.6.4   Growth Hormone Formulation   785
6.3.7   Leuprolide   788
    6.3.7.1   Background   788
    6.3.7.2   *In vitro* Permeation Study of Leuprolide   789
    6.3.7.3   Clinical Pharmacokinetics of Depot Leuprolide   789
    6.3.7.4   Transdermal Iontophoretic Delivery of Leuprolide 790
6.3.8   Desmopressin   791
    6.3.8.1   Background   791
    6.3.8.2   Pharmacokinetics of Desmopressin in Humans   791
    6.3.8.3   Renal and Biliary Excretion, and Sex Differences in
                Desmopressin Pharmacokinetics   792
    6.3.8.4   Desmopressin Formulation   793
6.3.9   Antibodies   794
    6.3.9.1   Background   794
    6.3.9.2   Pharmacodynamics and Pharmacokinetics of mAbs   795
    6.3.9.3   Muromonab CD-3 Pharmacokinetics   797
    6.3.9.4   Basiliximab Pharmacokinetics   797
    6.3.9.5   Trastuzumab Pharmacokinetics   799
    6.3.9.6   Infliximab Pharmacokinetics   799
    6.3.9.7   Rituximab Pharmacokinetics   800
    6.3.9.8   Palivizumab Pharmacokinetics   800
6.3.10 Recommendations for Future Study   801
    References   801

6.3.1   INTRODUCTION

Recombinant DNA technology produced many protein drugs in large scale and many important protein drugs like erythropoietin (EPO) for anemia-associated diseases, granulocyte-colony stimulating factor (G-CSF) for leucopoenia, and insulin for diabetes are now clinically available [1–3]. In addition, peptide hormones like leuprolide [an agonist of luteinizing hormone-releasing hormone (LHRH)] are also clinically used. However, these peptide/protein drugs are unstable because of hydrolytic degradation in the body [4]. Peptide/protein drugs also have a barrier for their administration because of low absorption efficiency. Therefore, their is parenteral preparation, *main dosage form* in which case intravenous (i.v.) or subcutaneous (s.c.) injection is used. After i.v. injection of these drugs, the administered drug molecules are completely available to the systemic circulation (i.e., bioavailability (BA) is 100%). On the other hand, BA becomes lower when these drugs are injected s.c. or i.m. In some protein drugs like interferon gamma, BAs after s.c. and i.m. injection are greater than 100% (i.e., 200% and 300%, respectively [5–7]. Furthermore, it is recognized that protein drugs whose molecular weight is larger than about 20 kDa are absorbed into the systemic circulation via lymphatic route. As a result, a large T_{max} value is obtained, when peak serum or plasma drug level appears after administration; for example, T_{max} is approximately 8 h for s.c.-injected EPO. After the serum drug level reached its maximum level, C_{max}, serum drug concentration decreases. In the case of organic compound drugs, the elimination process is composed of hepatic metabolism and renal excretion pathways. However, for peptide/protein drugs, we must consider the hydrolytic degradation process because of poor tolerability of these drugs to endogenous hydrolytic enzymes. Therefore, the elimination half-life of peptide/protein drugs is generally shorter than that of organic compound drugs. To increase the stability of peptide/protein drugs, several approaches have been performed and second-generation peptide/protein drugs have been developed and launched on the market [4], including PEGylated protein drugs (i.e., pegylation) and attachment of polyethylene glycol (PEG) to protein molecules. In the case of PEG-EPO, the administration frequency has decreased to once per week as compared with three times per week for the conventional EPO preparations; although almost the same serum EPO levels are obtained in the two types of EPO preparations. Most of the current peptide/protein drugs are injectable, and there is great commercial potential for delivery systems that use alternative administration routes like the pulmonary route and the nasal route, as suggested from insulin new dosage forms. In these cases, the pharmacokinetic profile of the peptide/protein drugs shows a completely different pattern as compared with that obtained after i.v. injection.

In this section, the pharmacokinetics of clinically important peptide/protein drugs, such as insulin, EPO, G-CSF, interferon, growth hormone, leuprolide, desmopressin, and antibodies, are described in relation to their administration routes and formulations (i.e., dosage forms).

6.3.2 INSULIN

6.3.2.1 Background

The finding of insulin in 1922 by Ranting and Best is one of the greatest discoveries in the history of medicine [8], and now it has been already proven that adequate control of blood glucose delays or prevents the progression of diabetic complications. To achieve the suggested targets for glycemic control necessary to reduce the incidence of diabetic complications, it has been established that a more intensive insulin regimen requiring multiple insulin injections is required for patients with type 1 diabetes mellitus. From a patient's point of view, however, noninvasive insulin delivery systems have been required for a long time. Indeed, a major limitation for advancing to intensive insulin therapy is that the only viable way to administer insulin is through injection. Several other methods of insulin delivery are now available or in development, including: continuous subcutaneous insulin infusion by a wearable infusion pump; total or segmented transplantation of a pancreas; transplantation of isolated islet cells; implantation of a programmable insulin pump; oral, nasal, rectal, and transdermal mechanisms of insulin delivery; insulin analogs; implantation of polymeric capsules that give continuous or time-pulsed release of insulin; and implantation of a biohybrid artificial pancreas, which uses encapsulated islets [9]. Many of these methods of insulin delivery are aimed at achieving a more physiological means of delivery of the insulin, at improving glycemic control, and, hopefully, at minimizing the secondary complications of diabetes.

On the other hand, during the nearly 80 years of insulin therapy, insulin preparations have undergone dramatic changes in their purification, production, and formulation. Despite these changes, and despite the development of different injection regimens and delivery systems, normalizing glycemic control still has not been adequately achieved. Recently, specific amino acid substitutions of the β-chain of the insulin molecule by recombinant DNA techniques have resulted in insulin analogs that have different pharmacokinetic properties than currently available insulins [10]. Some of these insulin analogs appear to have less variation in absorption than conventional insulins [11], and analogs have been developed with both very rapid [12] and delayed absorption [11]. Those analogs with a very rapid absorption may be useful in better mimicking the physiological insulin profiles seen postprandially than currently available native insulins [11]. All of these advances have been accomplished through protein engineering to produce insulin analogs with a decreased tendency for self-association (monomeric insulin), which in turn increases the subcutaneous absorption of these insulins [13]. These monomeric insulin analogs have been found to have a faster onset of action and are better able to reduce postprandial glucose excursions than conventional human insulin preparations, which subcutaneously become hexamers and slow absorption [14]. These insulin analogs, especially the monomeric types, appear to be less immunogenic and antigenic [13] where the pharmacokinetic aspects of human insulin and an analog insulin of aspart and lispro are described. In addition, other delivery options for insulin besides subcutaneous administration, such as transdermal and nasal approaches, are described.

6.3.2.2   Pharmacokinetics of Human Insulin after Intravenous Administration

In the 1970s, pork insulin had been widely used for the treatment of diabetics. In the 1980s, biosynthetic and recombinant DNA technology came to be introduced. In 1981, Halban et al. [15] compared the biologic potency of pancreatic human insulin with that of biosynthetic human insulin, which was manufactured by recombinant DNA techniques in bacteria, after intravenous injection in rats. The metabolic clearance rates of A14-mono-^{125}I-insulin (pork) were found to be similar to those observed for semi-synthetic [^{3}H]insulin (pork), being 20.8 ± 0.8 mL/min/kg and 23.6 ± 1.0 mL/min/kg, respectively. On the other hand, the metabolic clearance rates for A14-mono-^{125}I-biosynthetic human insulin was 24.6 ± 2.2 mL/min/kg, which was significantly higher than the value for A14-mono-^{125}I-pork insulin. In addition, the specific activity of the pancreatic human insulin was 32 U/mg, whereas that for biosynthetic human insulin was 27 U/mg. These values were not significantly different but were higher than the value for pancreatic pork insulin (25 U/mg). Biosynthetic human insulin is thus no less active than pork insulin in rats, and possibly somewhat more active. Moreover, in 1987, Brogden and Heel [16] reviewed that human insulin, whether produced by recombinant DNA techniques (biosynthetic, insulin crb) or enzymatic modification of porcine insulin (semi-synthetic, insulin emp) is equivalent in biological activity to porcine insulin after intravenous administration. There were no differences in the pharmacokinetic profiles among them. However, slight differences between human and porcine insulin in hypogly-cemic activity after subcutaneous injection appear to be related to differences in absorption and are unlikely to be of major clinical importance.

Turnheim and Waldhausl [17] characterized pharmacokinetic aspects of recom-binant human insurin in human. After intravenous injection, the plasma concentra-tion of insulin declines with at least two exponentials. Inslin distributes rapidly from the intravascular space to tissue compartments with a distribution half-life of 2.4 min, whereas insulin disappears slowly with an elimination half-life of 50–130 min, which is reflecting the elimination from the interstitial fluid and the tissues that use insulin. Total body clearance of insulin, which is the result of metabolic degradation, ranges between 700 and 800 mL/min. The sites of degradation are primarily the liver (hepatic insulin clearance: 320–400 mL/min) and the kidneys (renal insulin clearance: 190–270 mL/min); hence, insulin disposal depends on the function of these organs. The apparent volume of distribution for insulin is approxi-mately equal to the extracellular space. Insulin kinetics appears not to be altered in diabetes mellitus, except in cases with insulin antibodies or in insulin-resistant patients, in which insulin removal may be retarded.

6.3.2.3 Pharmacokinetics of Human Insulin after Subcutaneous Administration

There is a continuing search for improved insulin formulations to imitate the physi-ological pattern of insulin secretion as closely as possible, and thereby to minimize the complications of diabetes mellitus. The major advances achieved to date are in the area of human insulin analog synthesis resulting from the introduction of recom-binant DNA techniques and in improved delivery systems that use noninvasive or minimally invasive modes of administration. Usually, insulin is administered sub-cutaneously to control blood sugar levels in type 1 and type 2 diabetics. There are plenty of insulin pharmaceutics, and insulin's physico-chemical differences provide different profiles in the insulin pharmacokinetics after subcutaneous injection. Kang et al. [18], investigated absorption mechanisms of insulins with different physico-chemical properties after subcutaneous administration in humans. Based on a single-blind randomized comparison study of seven healthy male volunteers 22–43 years of age, equimolar dosages of ^{125}I-labeled forms of soluble hexameric 2 Zn^{2+} human insulin and human insulin analogs with differing association states at pharmaceutical concentrations (AspB10, dimeric; AspB28, mixture of monomers and dimers; AspB9, GluB27, monomeric) were administred. The initial fractional disappearance rates for the four insulin preparations were 20.7 ± 1.9 (hexameric soluble human insulin), 44.4 ± 2.5 (dimeric analog AspB10), 50.6 ± 3.9 (analog AspB28), and $67.4 \pm 7.4\%/h$ (monomeric analog AspB9, GluB27). The absorption of the dimeric analog was significantly faster than that of hexameric human insulin, and the absorption of monomeric insulin analog AspB9, GluB27 was significantly faster than that of dimeric analog AspB10. There was a negative linear correlation between association state and the initial fractional disappearance rates ($r = -0.98$, $p < 0.02$). A log-linear scale analysis for the elimination of insulin showed that only the monomeric analog had a monoexponential elimination. On the other hand, two phases in the rates of absorption were identified for the dimer and three for hexa-meric human insulin. The fractional disappearance rates calculated by log-linear regression analysis were monomer $73.3 \pm 6.8\%$/h; dimer $44.4 \pm 2.5\%$/h from 0 to 2 h and $68.9 \pm 3.5\%$/h from 2.5 h onward; and hexameric insulin $20.7 \pm 1.9\%$/h from 0 to 2 h, $45.6 \pm 5.0\%$/h from 2.5 to 5 h, and $70.6 \pm 6.3\%$/h from 5 h onward. The author concluded that the lag phase and the subsequent increasing rate of subcutaneous soluble insulin absorption can be explained by the associated state of native insulin in pharmaceutical formulation and its progressive dissociation into smaller units during the absorption process.

6.3.2.4 Pharmacokinetics of Rapid-acting Insulin Analog after Subcutaneous Administration

By the end of the 1980s, to accommodate postprandial hyperglycemia, monomeric insulin formulations, insulin lispro (the Lys-Pro analog) and insulin aspart (the Asp-Pro analog) had been developed for clinical use [19, 20] (Figure 6.3-1).

Formulations including insulin lispro have a rapid rate of absorption after subcutaneous injection and, therefore, have to be administered at meal time. Their residence time is also about two fold shorter than regular human insulin, minimizing the risk of the excessive hypoglycemic effect that characterizes regular human insulin formulations. Becker et al. compared the pharmacokinetics and pharmacodynamics of insulin lispro with regular human insulin after subcutaneous injection in obese subjects in a single-dose, randomized, double-blind, cross-over euglycemic clamp study [21]. Insulin lispro had more rapid-acting profiles than regular human insulin. Fractional glucose infusion rate (GIR)-area under the curve (AUC) of the GIR curve and maximum GIR were greater for insulin lispro compared with regular human insulin. Time to 20% (early glucose disposal) and 80% (bulk of activity) of total GIR-AUC were shorter for insulin lispro than regular human

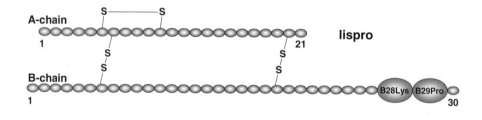

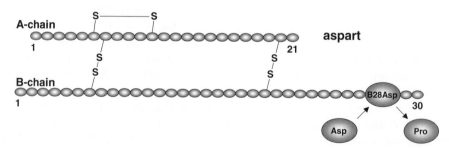

Figure 6.3-1. Schematic presentation of primary structure of human insulin lispro and aspart.

insulin, which was corroborated by more rapid and shorter residing pharmacokinetic profiles of insulin lispro than regular human insulin, evidenced by shorter times to 20% of total AUC of insulin, C_{max} of insulin, and mean residence time. On the other hand, Osterberg et al. compared the pharmacokinetics and pharmacodynamics of insulin aspart with regular human insulin in healthy subjects in a cross-over trial. There were statistically significant differences between most of the human insulin and insulin aspart pharmacokinetic parameters, including the sigmoidicity of the time-course of absorption (1.5 for human insulin vs. 2.1 for insulin aspart), elimination rate constant ($0.010\,min^{-1}$ vs. $0.016\,min^{-1}$, respectively). In addition, the pharmacodynamic model parameters were mostly not different, except for the rate of insulin action ($0.012\,min^{-1}$ vs. $0.017\,min^{-1}$, respectively) [20]. Moreover, Homko et al. compared insulin levels and actions in patients with type 1 diabetes after subcutaneous injection of insulin analogs, aspart and lispro [22]. Both insulin analogs produced similar serum insulin levels (250–300 pmol/l) at approximately 30 min and disappeared from serum after approximately 4 h. Insulin aspart and lispro had similar effects on glucose and fat metabolism. Effects on carbohydrate metabolism (glucose uptake, glucose oxidation, and endogenous glucose production) peaked after approximately 2–3 h and disappeared after approximately 5–6 h. Effects on lipid metabolism (plasma free fatty acid, ketone body levels, and free fatty acid oxidation) appeared to peak earlier (at approximately 2h) and disappeared earlier (after approximately 4h) than the effects on carbohydrate metabolism. The author concluded that both insulin aspart and lispro are indistinguishable from each other with respect to blood levels and that they are equally effective in correcting abnormalities in carbohydrate and fat metabolism in patients with type 1 diabetes.

6.3.2.5   Other Formulations of Human Insulin

Inhaled Insulin. Subcutaneous insulin has been used to treat diabetes since the 1920s; however, despite a number of different formulations, intensive insulin therapy with multiple daily injections has not gained widespread clinical acceptance. Great efforts have been made in searching for noninvasive administration modes of insulin that will avoid the need for parenteral administration of subcutaneous injection. Attempts to find effective, well-tolerated, nonenteral routes for delivering insulin began in the 1920s, and, over the years, have included ocular [23], buccal [24], rectal [25], vaginal [26], oral [27], nasal [28] and uterine [29] delivery systems, where various enhancers or formulations have been tested to increase the bioavailability of insulin from each route. Until recently, it was believed that insulin delivered noninvasively was associated with a bioavailability that was too low to offer a realistic clinical approach. However, progress in the pharmaceutical approach of insulin suggests that inhaled insulin is an effective, well-tolerated, noninvasive alternative to subcutaneous regular insulin.

Oral inhalation of insulin potentially offers noninvasive treatment and better glycemic control in diabetes by virtue of its apparently faster absorption into the systemic circulation compared with subcutaneous injection. However, the pharmacokinetics of inhaled insulin in the human lung has yet to be fully clarified because of the complexity of insulin-glucose physiology and the difficulty in approximating the inhaled dose. In 2004, Sakagami conducted a meta-analysis of insulin pharmacokinetics in the lung after oral inhalation of insulin in humans [30], where, based on the data from well-controlled clinical studies of inhaled insulin published in the literature, a physiologically realistic insulin-glucose kinetic model was derived. The model defined the first-order absorption ($k_{a,L}$) and parallel non-absorptive loss ($k_{mm,L}$), the latter primarily occurring via metabolism and mucociliary clearance in the lung with two systemic compartments. The significant estimate for $k_{a,L}$ was found to be from 0.020 to $0.032\,h^{-1}$, effectively unchanged across doses (from 0.3 to 1.8 IU/kg), formulations (powder vs. liquid), and subjects (healthy vs. diabetic), suggesting passive diffusive absorption of insulin from the lung. In contrast, the values for $k_{mm,L}$ were much larger (from 0.5 to $1.6\,h^{-1}$) and decreased with increasing inhaled dose. From this perspective, the author considered that a dose-dependent saturable lung metabolism regulates the value of $k_{mm,L}$, alongside mucociliary clearance. As a result, the absolute bioavailability of insulin through the oral inhalation ranged from 1.5% to 4.8%.

Henry et al. [31] evaluated the AERx insulin Diabetes Management System (AERx iDMS) to the deep lung for systemic absorption, which was an aerosol of liquid human insulin produced by Aradigm (Hayward, CA). This study examined the effects on pulmonary function, pharmacokinetics, and pharmacodynamics of inhaled insulin in asthmatic and nonasthmatic subjects without diabetes, where a total of 28 healthy and 17 asthmatic subjects were enrolled in a two-part, open-label trial. The area under the time versus concentration curve (AUC) of insulin after inhalation showed significantly greater value for healthy subjects than for asthmatic subjects, whereas no difference was observed for maximum concentration in the two groups. In addition, a greater reduction of serum glucose level was observed in healthy subjects. Moreover, asthmatic subjects had greater intra-individual variations in insulin AUC, maximum concentration values, and glucose levels than healthy subjects. In this report, the author concluded that, after inhaling insulin using the AERx iDMS, asthmatic subjects absorbed less insulin than healthy subjects, resulting in less reduction of serum glucose.

The rationale behind developing a pulmonary drug delivery system is to ensure that insulin powder is delivered deep into the lungs, where it is easily absorbed into the bloodstream, in a handheld inhalation device. More recently, Nektar Therapeutics (formerly Inhale Therapeutic Systems) has developed a pulmonary drug delivery system for insulin [32]. This device converts the insulin powder particles into an aerosol cloud for the patient to inhale. The inhaler requires no power source, and the clear chamber ensures that the patient knows immediately when all the insulin has been inhaled. In addition, Nektar Therapeutics is using its Advanced PEGylation technology to develop a dry powder-inhaled polyethylene glycol (PEG) formulation for delivering peptides efficiently across the lungs and to promote prolonged serum concentration of the peptide. PEGylation is designed to increase the size of the active molecule and ultimately improve drug performance by optimizing pharmacokinetics, increasing bioavailability, and decreasing immunogenicity and dosing frequency. The investigation has begun with inhaled, long-acting (PEGylated) insulin, and is funded by Pfizer.

Intranasal Insulin.   Insulin administered nasally has considerable potential for the treatment of both type 1 and type 2 diabetes. Using enhancers or novel delivery systems such as adequate bioadhesive microspheres, it is believed that the low bioavailability of simple formulations of insulin can be greatly improved [33].

Therefore, several efforts on developing the nasal delivery system of insulin with nontoxic and nonirritant properties have been done.

In 1992, Drejer et al. [31] investigated the pharmacokinetics of intranasal insulin containing a medium-chain phospholipid (didecanoyl-L-alpha-phosphatidylcho-line) as absorption enhancer in 11 normal volunteers. Intranasal insulin was absorbed in a dose-dependent manner with a mean plasma insulin peak 23 ± 7 min after administration. Mean plasma glucose nadir was seen after 44 ± 6 min, 20 min after intravenous injection. Moreover, intranasal administration of insulin resulted in a faster time-course of absorption than subcutaneous injection, and the bioavailability for the nasal formulation was 8.3% relative to an intravenous bolus injection when plasma insulin was corrected for endogenous insulin production estimated by C-peptide.

On the other hand, there are negative opinions on the nasal delivery of insulin. In 1995, Hilsted et al. evaluated metabolic control and safety parameters (hypoglycemia frequency and nasal mucosa physiology) in 31 insulin-dependent diabetic patients treated with intranasal insulin in an open, cross-over randomized trial [35]. Serum insulin concentrations increased more rapidly and decreased more quickly during intranasal as compared with subcutaneous insulin administration. Metabolic control deteriorated, as assessed by hemoglobin A1c concentrations, slightly but significantly after intranasal as compared with subcutaneous insulin therapy. The bioavailability of intranasally applied insulin was low; therefore, intranasal insulin doses were approximately 20 times higher than subcutaneous doses. The frequency of hypoglycemia was similar during intranasal and subcutaneous insulin therapy, and nasal mucosa physiology was unaffected after intranasal insulin. The author concluded that because of low bioavailability and a high rate of therapeutic failure, intranasal insulin treatment was not a realistic alternative to subcutaneous insulin injections.

In the early twenty-first century, nanoparticles based on nanotechnology have altered the direction of investigation in the drug delivery system. Nanoparticles, an evolvement of nanotechnology, are increasingly considered as a potential candidate to carry therapeutic agents safely into a targeted compartment in an organ, particularly a tissue or a cell [36]. In 2002, Dyer et al. investigated effects of chitosan-formulated nanoparticles on the insulin pharmacokinetics and pharmacodynamics as a nasal drug delivery system [37]. The nasal absorption of insulin after administration in chitosan nanoparticle formulations and in chitosan solution and powder formulations was evaluated in rats and in conscious sheep. Insulin-chitosan nanoparticle formulations produced a pharmacological response in the two animal models, although in both cases the response in terms of lowering the blood glucose levels was less than that of the nasal insulin-chitosan solution formulation. The insulin-chitosan solution formulation was found to be significantly more effective than the complex and nanoparticle formulations. The authors concluded that chitosan nanoparticles did not improve the nasal absorption.

6.3.3 ERYTHROPOIETIN

6.3.3.1 Background

Erythropoietin (EPO), a glycoprotein hormone with a molecular mass of 30 KDa, is mainly synthesized by the kidney in response to hypoxia. Serum EPO concentrations in humans normally range from 6 to 32 U/L [38]. The plasma half-life of EPO is reported to range from 2 to 13 hours, with a volume of distribution close to plasma volume. As expected for a large sialoglycoprotein, less than 10% of EPO is excreted in the urine [39]. EPO is known to stimulate the proliferation and differentiation of erythrocytic progenitors in the bone marrow, leading to reticulocytosis and increased erythrocyte numbers in the blood [38].

Recombinant human EPO (rHuEPO) has proven beneficial for treating renal anemia as well as anemia of other chronic disorders, including autoimmune diseases, malignancies, and AIDS. Evidence from clinical trials indicates that anemia associated with cancer chemotherapy generally responds well to a standard rHuEPO regimen of 150 IU/kg subcutaneous (s.c.) three times per week [40, 41]. Several dosage recommendations have been made for surgery patients depending on the type of surgery and practicality of an autologous blood donation program. In patients scheduled for major orthopedic surgery, three weekly 600 IU/kg s.c. doses of rHuEPO, with the last one on the day of the surgery, have been suggested [42]. Goldberg et al. present here pharmacokinetic/pharmacodynamic (PK/PD) modeling from four comparable, placebo-controlled, parallel-group studies in healthy subjects who were administered single rHuEPO doses of 300 to 2400 IU/kg as well as multiple doses of 600 IU/kg/wk and 150 IU/kg/tiw for 4 weeks.

Clinical trials of recombinant product began in Seatle and London in 1986, and its efficacy was established at an early stage [43–46]. The early studies confirmed that intravenously administered EPO was highly effective in correcting the anemia of end-stage renal failure in hemodialysis. The intravenous route was, however, clearly impractical for patients maintained on continuous ambulatory peritoneal dialysis (CAPD) who had no ready vascular access, which necessitated an examination of the intraperitoneal and subcutaneous routes of administration, and several pharmacokinetic studies have compared these in CAPD patients [47, 48]. These three routes of erythropoietin administration are the only ones to have been investigated either from a pharmacokinetic or a therapeutic point of view. Although Storring et al. [49] and Charles et al. [50] compared the two forms of rHuEPO (i.e., EPO-alpha and EPO-beta), they reported that the volume of distribution after intravenous EPO-beta compared with EPO-alpha and that delayed subcutaneous absorption of EPO-beta compared with EPO-alpha. They believed the pharmacokinetic differences were caused by their glycosylation.

6.3.3.2 Pharmacokinetics of Erythropoietin in Healthy Subjects

Intravenous administration in the dialysis center has been practical and effective for patients undergoing hemodialysis [51, 52], but impractical for patients without vascular access who are receiving peritoneal dialysis. Studies of subcutaneous administration showed lower but more sustained peak plasma EPO concentrations (C_{max}) than occurred using the intravenous route. The efficacy of subcutaneous administration supports the concept of a sustained minimal effective EPO concentration. High C_{max} values after intravenous administration appear unnecessary; the dose response relationship seen with intravenous administration most likely correlates with serum EPO concentrations maintained above a minimal effective level.

Flaharty et al. presented data on the pharmacokinetics of EPO in healthy normal subjects [53]. To assess phamacokinetics to human recombinant erythropoietin

(EPO-beta), single intravenous doses (10, 50, 150, and 500 IU/kg) were administered at monthly intervals to 16 healthy subjects in a two-panel, placebo-controlled, double-blind ascending-dose trial. A 1000-IU/kg dose was subsequently administered in an open manner. EPO concentrations were detemined in serum by radioimmunoassay. Mean EPO apparent half-lives ranged fiom 4.42 to 11.02 hours. The apparent volume of distribution was between 40 and 90 mL/kg, consistent with plasma water, and the apparent clearance values ranged from 4 to 15 mL/kg/hr, with both parameters having the highest values at the 10-IU/kg dose level. Clearance tended to decrease as a function of dose. The use of EPO in healthy individuals will have clinical relevance if EPO is found to be effective for use in autologous transfusion programs. Although EPO was well tolerated in this study, clinical trials in patients with end-stage renal disease have shown adverse effects, including hypertension, seizures, and thrombocytosis, related to the increase in erythropoiesis. These effects are not expected in normal subjects, but careful monitoring should be performed if EPO is used in repeated doses in healthy subjects.

Salmonson et al. [54] also described the pharmacokinetics of rHuEPO after intravenous and subcutaneous administration of 50 U/kg to six healthy male volunteers. The calculated mean values for volume of distribution at steady state and clearance after an i.v. dose were 76 ± 33 mL/kg and 12.0 ± 3.0 mL/h/kg, respectively. Serum concentrations of rHuEPO peaked at 13.0 ± 6.0 h after the s.c. dose and the bioavailability over 72 h was $36 \pm 23\%$. The mean residence time and half-life of rHuEPO were 6.2 ± 1.0 and 4.5 ± 0.9 h after i.v. and 46 ± 18 and 25 ± 12 h after s.c. administration. They found that the serum concentration time profiles after i.v. adminisitaration followed a mono- or bi-exponential decline, and the elimination after s.c. dose was described by a mono-exponential decline.

McMahon et al. [55] reported a double-blind, placebo-controlled study of the pharmacokinetics and safety of multiple doses of rHuEPO 150 or 300 U/kg, either by i.v. bolus or s.c. in normal male subjects, which demonstrated that rHuEPO had a dose-related effect on the hematocrit independent of the route of administration, and that multiple doses of rHuEPO had no direct pressor effects. When rHuEPO was injected by i.v. route, a mono-exponential decrease in serum EPO level was observed for 18 to 24 hours postdose. Absorption of s.c.-injected rHuEPO occurred more slowly with relatively low serum EPO levels being maintained for 48 hours. As a result of the prolonged elevation in serum EPO concentrations with s.c. injection, with injections every 48 hours, s.c. resulted in higher preinjection serum EPO concentrations than with i.v. injection of the same dose.

Moreover, Ramakrishnan et al. [56] described the pharmacokinetic and pharmacodynamic modeling of rHuEPO after single and multiple doses in healthy volunteers. A one-compartment model with limited distribution and nonlinear elimination characterizes the data from the i.v. dosing. Most of the rHuEPO after s.c. dosing was rapidly absorbed within 2 to 3 days by zero-order rate constant. A dual-absorption model (fast zero-order and slow first-order inputs) with nonlinear disposition characterized the pharmacokinetics of s.c. rHuEPO. A similar pharmacokinetic model describes the time-course of plasma concentrations of rHuEPO on monkeys [57]. On the other hand, Cheung et al. [58] demonstrated the pharmacokinetics of rHuEPO after single s.c. doses of 300, 450, 600, 900, 1200, 1350, 1800, and 2400 IU/kg and in multiple s.c. dose regimens: 150 IU/kg three times a week for four weeks and 600 IU/kg once per week for four weeks in open-label, randomized placebo-controlled studies in healthy volunteers. The absorption rate of rHuEPO after s.c. administration was independent of dose, whereas clearance was dose-dependent, in which it decreased with increasing dose. In an advanced investigation, Cheung et al. [59] further described the comparison of the pharmacokinetics administered s.c. once weekly (qw), 150 IU/kg, and three times weekly (tiw), 40,000 IU. The C_{max} values for serum rHuEPO qw were six times and AUC values three times that of the tiw regimen. Subsequent study reported by Krzyzanski et al. [60] showed that a phamacokinetic and pharmacodynamic (PK/PD) model for rHuEPO in healthy subjects was developed to describe the time profiles of changes in serum rHuEPO and the pharmacological responses of percent reticulocytes, total red blood cell counts, and hemoglobin after single and multiple subcutaneous administration of rHuEPO. Data used is the same data as Wing Cheung et al. [59]. A dual-absorption model (fast zero-order and slow first-order inputs) with linear disposition kinetics was used to characterize the pharmacokinetics of erythropoietin after subcutaneous administration. Flip-flop kinetics was apparent in the pharmacokinetics as the absorption rate was slower (ka = 0.7 day^{-1}) than the elimination rate (CL/Vd = 1.2 – 9.2 day^{-1}).

6.3.3.3   Pharmacokinetics of Erythropoietin in Patients

rHuEPO is established as an effective therapeutic agent for anemia in patients undergoing long-term hemodialysis. After an intravenous dose of EPO 24 to 240 U/kg, serum clearance was 0.66 ± 0.42 L/h; terminal elimination half-life was 6 to 9 h and did not change after repeated doses given for more than two months [61]. In hemodialysis patients, the pharmacokinetics of rHuEPO was studied by Brockmöller et al. [62]. They investigated in 12 patients under chronic hemodialysis on a thrice weekly intravenous rHuEPO treatment. The pharmacokinetics of rHuEPO was also assessed after a subcutaneous injection during the initial period and during maintenance treatment. After the first i.v. dose, plasma EPO concentrations were best described by a mono-exponential disposition function with a mean (±s.d.) elimination half-life of 5.4 ± 1.7 h. The volume of distribution was 70 ± 5.2 mL/kg, and the clearance was administered intravenously as 10.1 ± 3.5 mL/h/kg. After three months of continuous therapy, the plasma half-life of rHuEPO decreased by 15% (mean half-life during steady state: 4.6 ± 2.8 h), while mean clearance and volume of distribution remained constant. On the other hand, the absorption of rHuEPO after s.c. injection was prolonged and flip-flop kinetics applied. The extent of systemic availability of the s.c. dose ranged from 28% to 100% (mean 44%).

Gladziwa et al. [63] described the pharmacokinetics of rHuEPO after long-term therapy in patients undergoing hemodialysis and hemofiltration. In 17 patients with end-stage renal disease (ESRD), studies were performed in three groups to assess pharmacokinetics during the intertreatment interval and during hemofiltration and dialysis treatment. After an intravenous bolus injection of EPO 150 U/kg, the half-life was 7.7 h, steady-state volume of distribution was 0.066 L/kg, and total plasma clearance was 5.4 mL/min. The mean steady-state serum concentration during multiple-dose administration was 656 U/L. The drug was not eliminated by hemofiltration or dialysis. Therefore, long-term treatment of ESRD patients with EPO does not significantly alter the pharmacokinetic profile of the drug. EPO dosage adjustment or substitution after hemofiltration and dialysis is not necessary. On the other hand, peritoneal dialysis (PD) has gained recognition worldwide as an alternative to hemodialysis in the management of patients with end-stage renal disease. The most common is continuous ambulatory peritoneal dialysis (CAPD) in which most patients perform four to five exchanges daily. As the peritoneal catheter provides direct access to the peritoneum, intraperitoneal drug administration has become widely used for the administration of certain drugs. The instillation and drainage of PD fluids contribute to the total body clearance of drugs, in which peritoneal clearance is low. Taylor III et al. [67] reported the pharmacokinetic studies after rHuEPO administration to patients receiving PD. Intraperitoneal (i.p.) drug administration, via a PD catheter, produces a concentration gradient between the peritoneal and vascular compartments, favoring drug absorption. Absorption is enhanced if the plasma concentration of drug is low relative to the concentration of drug in the peritoneal fluid. Lymphatics draining the diaphragm contribute to the absorption of albumin and dextrans from the peritoneum and may be important in the absorption of protein drugs or other large-molecular-weight compounds like rHuEPO. The pharmacokinetic parameters observed in patients undergoing CAPD following i.v. and s.c. administration are similar to those determined for hemodialysis patients [65, 66]. s.c. adminisitration was found efficacious in several studies of PD patients [67–71], and allows reduction of the dose of EPO to 50–60% of the i.v. dose. i.p. administration is convenient and circumvents any pain associated with s.c. administration. In several studies, EPO was administered i.p. 1 L to 2 L of dialysate and allowed to dwell for varying time periods [72–75]. Low absorption of EPO from the peritoneal cavity is seen if dialysate is present. Hematological response similar to those observed with either s.c. or i.v. administration were seen following i.p. administration using this technique, but substantially higher doses were required [76–78]. Other investigators have reported poor efficacy, although this may result partly from the low doses administered [79, 80]. Based on these observations, the i.p. administration of EPO in large volumes of dialysate appears prohibitively costly to achieve the desired results. An improved bioavailability was observed in humans after administration of undiluted EPO 400 U/kg followed by an 8-hour dry dwell [81]. A greater than nine fold increase in bioavailability was observed with dry administration, compared with diluting the dose in 2-L dialysate followed by an 8-hour dwell. The dry-dose administration area under the concentration-time curve (AUC) approached that seen with s.c. dosing. Ateshkadi et al. [82] administered EPO 100 U/kg IP to eight patients undergoing CAPD using a 4-hour dry dwell followed by an infusion of 2-L dialysate that was allowed to dwell for 6 hours. The i.p. EPO serum-concentration time profiles were similar to those observed with s.c. administration, with a mean bioavailability of 11.4% compared with 22.8% for s.c. administration. Ongoing studies are examining the pharmacokinetics and efficacy of administration using an 8-hour dry dwell. Single-dose pharmacokinetics of EPO were studied in pediatric dialysis patients [83, 84]. EPO appears to be better absorbed and more rapidly cleared in children than it is in adults. Differences in the volume of distribution of EPO have also been noted between children and adults. These differences may be result of methodological differences between studies or true physiological differences relating to the plasma volume or tissue distribution of the drug. A recent study of i.p. administration in five children on continuous cycling peritoneal dialysis (CCPD) suggests that this route is safe and efficacious [85].

6.3.3.4   Erythropoietin Formulation

Humoral factors and cytokines are glycosylated with variable numbers of carbohydrate chains, each of which has variable carbohydrate structures. Although the biological function is typically determined by the protein component, carbohydrate can play a role in molecular stability, solubility, *in vivo* activity, serum half-life, and immunogenicity. The sialic component of carbohydrate, in particular, can extend the serum half-life of protein therapeutics. One aspect of glycoengineering is to introduce N-linked glycosylation consensus sequences into desirable positions in the peptide backbone to generate proteins with increased sialic acid-containing carbohydrate, thereby increasing *in vivo* activity because of a longer serum half-life [86].

DarbEPO alfa (DA) is a hyperglycosylated analog of recombinant human erythropoietin (rHuEPO) generated by glycoengineering. This rHuEPO is used to treat the anemia associated with chronic kidney disease and cancer. The discovery that the sialic acid-containing carbohydrate content of rHuEPO was directly proportional to the serum half-life and *in vivo* bioactivity [87] led to the hypothesis that additional carbohydrate on rHuEPO might further enhance *in vivo* activity. DA contains two additional N-linked glycosylation sites and has increased *in vivo* activity and prolonged duration of action because serum half-life is increased three fold above rHuEPO [88, 89], which allows for less frequent dosing with subsequent increased convenience for the patient and caregiver and improved patient compliance.

Allon et al. [90] investigated the pharmacokinetics of darbEPO alfa and rHuEPO after repeated intravenous dosing in patients with chronic kidney disease receiving hemodialysis. Overall, 47 patients were randomized to receive darbEPO alfa administered once weekly or three times weekly or EPO administered three times weekly for up to 52 weeks. Pharmacokinetic profiles were measured during weeks 1 and 12. The terminal half-life of darbepoietin alfa was 17.8 hours, which was approximately two to three times longer than that of EPO (6.3 hours). The clearance of darbepoietin alfa (2.00 mL/h/kg) was approximately four times slower than those of EPO (8.58 mL/h/kg). They conclude that darbepoietin alfa can be administered less frequently than EPO in patients with chronic kidney disease receiving hemodaialysis.

Lerner et al. [91] also studied the pharmacokinetics of darbepoietin alfa in pedisatric patients with chronic kidney disease (CKD). Twelve patients 3–16 years of age with CKD were randomized and received a single 0.5 μg/kg dose of darbepoietin alfa administered i.v. or s.c. After a 14- to 16- day washout period, patients received an identical dose of darbepoietin alfa by the alternate route. After i.v. administration, the mean clearance of darbepoietin alfa was 2.3 mL/h/kg with a mean terminal half-life of 22.1 h. After s.c. administration, absorption was rate limiting, with a mean terminal half-life of 42.8 h and a mean bioavailability of 54%. Comparison of these results with those from a previous study of darbepoietin alfa in adult patients [89] indicated that the disposition of darbepoietin alfa administered i.v. or s.c. is similar in adult and pediatric patients, although absorption may be slightly more rapid in pediatric patients after s.c. dosing.

6.3.4   GRANULOCYTE-COLONY STIMULATING FACTOR (G-CSF)

6.3.4.1   Background

Colony stimulating factors (CSFs) as well as hematopoietic growth factors are a group of glycoproteins regulating the survival, proliferation, and differentiation of hematopoietic progenitor cells as well as the function and activation of the mature cells. The CSFs are produced by a variety of cells and range in molecular mass from 14 to 90 kDa. Recently, these factors have been purified, cloned, and produced through recombinant DNA techniques. The recombinant factors have been shown to have biologic properties and actions that are similar to the naturally occurring factors. The availability of quantities of the recombinant factors has resulted in their introduction into clinical trials and into the market.

The pharmacokinetics of CSFs has been evaluated widely in animals and in man. Early studies, however, used impure or poorly characterized fractions and the interpretation of these results is difficult. With the availability of recombinant proteins, these limitations were overcome and definitive studies have been completed. In general, the biologic effects of these factors can be quantitated by measuring the effects on the target hematopoietic cell population, which has provided a basis for relating the pharmacokinetic profile of the factor to the pharmacodynamic response. As clinical trials and the therapeutic applications of these factors are expanded, the knowledge and understanding of the pharmacokinetics of the drug will be critical to guide the clinician in the choice of dosing routes and schedule.

The generation of granulocytes from immature hematopoietic progenitor cells depends on the presence of several hormone-like glycoproteins. CSFs are acidic glycoproteins required for the survival, proliferation, and differentiation of hematopoietic progenitor cells [92–95]. Granulocyte-colony stimulating factor (G-CSF) was identified because it induced differentiation in a murine myelomonocytic leukemic cell line and stimulated granulocyte colony formation by normal progenitor cells [96]. The hematopoietic actions of this regulator are restricted exclusively to cells of neutrophilic granulocyte lineage [97–99]. Murine and human G-CSFs have been purified from various sources [96–98] and the genes encoding human and murine G-CSF have been isolated [100–102]. This factor supports the growth and proliferation of relatively late granulocyte progenitor cells already committed to the neutrophil lineage and enhances the function and activation of mature neutrophils [103, 104]. This factor has also been cloned and expressed in both bacterial cells and mammalian cells [104, 105].

Characterization of the gene encoding G-CSF has led to the production of this protein by recombinant DNA techniques. The recombinant form of human G-CSF, rhG-CSF, is also capable of supporting the formation of granulocytic colonies from committed precursor cells. The availability of large quantities of molecularly homogeneous and biologically active human G-CSF by recombinant DNA technology has made it possible to explore the use of human G-CSF as a therapeutic agent. rhG-CSF shows broad species cross-reactivity with respect to its characteristic biologic response, a rapid and dose-dependent neutrophilia. The potential clinical applications for rhG-CSF include myelorestoration after bone marrow transplantation, mitigating chemotherapy-induced neutropenia, boosting host defense against infections, and augmenting effecter cell purpose [106].

6.3.4.2 Pharmacokinetics of G-CSF in Animals

Cohen et al. [107] also reported biexponential clearance of *E. coli*-derived rhG-CSF administrated intravenously to hamsters; the α and β serum half-lives were 0.5 and 3.8 hr, respectively (all data were normalized to the amount recovered at 5 min after injection). Tanaka and Kaneo [108] reported the pharmacokinetics of rhG-CSF (KRN8601) in male Sprague-Dawley rats at a dose of 5 μg/kg as a pre-clinical investigation. The serum concentrations of rhG-CSF were monitored using a specific sandwich enzyme immunoassay. For i.v. administration, the serum concentration-time curve showed a rapid disappearance of rhG-CSF from the systemic blood with a mean residence time (MRT) of 1.34 hr. For s.c., i.m., and i.p. administration, lower peak serum levels were observed; but after 2 to 3 hr, rhG-CSF levels were higher than those for i.v. administration with the MRTs of 3.92, 2.90, and 3.54 hr, respectively. Moreover, Tanaka et al. [109, 110] examined the influence of dose level and route of administration on the pharmacokinetics of *E. coli*-derived G-CSF after i.v. administration to Sprague-Dawley rats at doses of 10 or 100 μg/kg. Biexponential plasma clearance was noted after an i.v. dose of 100 μg/kg; the half-lives at distribution phase (α) and elimination phase (β) were 25 and 102 min, respectively. In addition, the total body clearance of rhG-CSF in the rat after i.v. administration showed an approximate 30% decrease from 80 to 56 mL/h/kg as the dose was increased from 1 to 10 μg/kg; no further decreases in the total body clearance were noted up to 100 μg/kg. However, the volume of distribution calculated was equivalent at each dose. For the serum concentration profiles of rhG-CSF after extravascular administration, an i.m. injection gave slightly faster absorption kinetics of rhG-CSF from the injection site into systemic blood than did s.c. and i.p. injections [108]. After s.c. administration, biological effects were noted at 2 h and were sustained for 36 h, suggesting rapid absorption. Peak serum concentrations were seen at approximately 2–3 h postdose. The calculated bioavailability of the subcutaneous dose was approximately 78%. There were nonproportional increases in the AUC value; a 100-fold increase in the dose resulted in a 164-fold increase in the AUC. Bioavailability after subcutaneous dosing ranged from approximately 50% to 70%. The results of this study suggested that two mechanisms may be involved in the clearance of G-CSF. One pathway is saturated at low concentrations and represents high-affinity, low-capacity receptor-mediated uptake by target cells (neutrophils). The other mechanism represents clearance by nonspecific pathways such as renal excretion.

In multiple-dose studies, rhG-CSF was injected into animals by i.v. and s.c. at 5 μg/kg/day for 7 days [108]. When the serum concentration-time profiles on day 7 after multiple dosing of rhG-CSF were compared with single dosing, the AUC after i.v. multiple dosing decreased by 17.4%. However, the half-lives and the volume of distribution were not significantly different between single and multiple dosing groups. In addition, the AUC after s.c. multiple dosing decreased by 25.6% without significant differences in the bioavailability and observed maximum serum concentration of rhG-CSF. These results showed that the clearance of rhG-CSF increased after multiple dosing, although the mechanism of increased clearance was not apparent.

Moreover, Tanaka et al. reported the effect of sex differences and hepatic or renal failure on the pharmacokinetics of rhG-CSF in the rats. After i.v. and s.c. administration of rhG-CSF to male and female Sprague-Dawley rats at a dose of 5 and 100 µg/kg, area under the concentration versus time curve and elimination half-lives of rhG-CSF in female rats were smaller than those for male rats. The volume of distribution of rhG-CSF in female rats was not significantly different from that in male rats. After s.c. administration, the values of AUC, MRT, and half-lives of elimination phase in female rats were smaller than those for male rats. In the *in vitro* biological activities of rhG-CSF using [^{3}H]thymidine uptake assay in cultures of bone marrow cells obtained from male and female rat femur, female rat bone marrow cells showed a similar dose-response profile to rhG-CSF to that of male rat bone marrow cells. The effect of rhG-CSF administration in rats was a specific activity on the neutrophil lineage with an increase of neutrophils in peripheral blood. On the other hand, the *in vivo* effects of rhG-CSF after i.v. and s.c. adrninistration to male and female rats at 5 and 100 µg/kg doses, namely the neutrophil count in female rats, was similar to that in male rats in the early period; however, the neutrophil count in female rats was lower than that in male rats 24 h after administration. These results indicate the sex differences in pharmacokinetics and the *in vivo* effects of rhG-CSF as well as the close relationship between pharmacokinetics and the *in vivo* effects of rhG-CSF [111]. In rats with experimental renal and hepatic failure, the total body clearance of rhG-CSF after i.v. administration at a dose of 10 µg/kg was 44.48 mL/h/kg in sham-operated rats compared with 9.429 mL/h/kg in bilaterally nephrectomized rats. In sham-operated rats, the elimination half-life of rhG-CSF at β phase was 1.51 h, and it increased to 5.33 h after nephrectomy. In contrast, the volumes of distribution were identical in both rats. In rats with acute renal failure induced by uranyl nitrate, the total body clearance and volume of distribution were identical to those of control rats, but the elimination half-life at β phase was slightly shorter. In 70% hepatectomized rats, the clearance of rhG-CSF decreased from 42.08 mL/h/kg to 31.93 mL/h/kg, and similar half-lives were observed in rats in both the sham-operated and hepatectomized groups. However, the volume of distribution decreased after hepatectomy. In rats with hepatic failure induced by CCl$_4$, the pharmacokinetic changes were similar to those observed in hepatectomized rats. These results suggest that renal clearance makes a major contribution to total body clearance compared with hepatic clearance [112].

6.3.4.3 Pharmacokinetics of G-CSF in Humans

In the early period, the pharmacokinetics of *E. coli*-derived rhG-CSF as part of phase I and II clinical trial was evaluated by Layton et al. [113]. The protein was administered by i.v. or s.c. routes at dose levels of 3 to 30 µg/kg. The elimination half-life after i.v. dosing was 1.4 h at a dose of 1 µg/kg, and increased to 4.2 h at a dose of 60 µg/kg. These results indicated saturation mechanisms in the total body clearance at doses greater than 10 µg/kg. When a continuous s.c. infusion was performed for 5 days, steady state was not achieved and there was a rapid reduction in serum G-CSF levels during the last 2 days of administration. Neutrophil levels increased during this period, and these results were interpreted as neutrophil receptor-mediated regulation of G-CSF uptake and degradation. The peak levels of G-CSF after initiation of s.c. dosing were achieved at 6 h, and these levels were maintained for up to 16 h. When high doses of melphalan were coadministered with G-CSF, the neutrophil levels did not increase and serum G-CSF levels remained constant.

For the pharmacokinetics of rhG-CSF in humans, it has been reported that the absorption and clearance of rhG-CSF follow first-order kinetics without any apparent concentration dependence [114]. When rhG-CSF was administered by 24-h constant i.v. infusion at a dose level of 20 µg/kg, the mean serum concentration achieved 48 ng/mL. Constant i.v. infusion for 11 to 20 days produced steady-state serum concentrations over the infusion period. Subcutaneous administration of rhG-CSF at doses of 3.45 and 11.5 µg/kg resulted in peak serum concentrations of 4 and 49 ng/mL, respectively. The mean value of volume of distribution was 150 mL/kg. The elimination half-life was 3.5 h after either i.v. routes or s.c. routes, with a clearance rate of 0.5–0.7 mL/min/kg. The administration of a daily dose for 14 consecutive days did not affect the half-life.

More recently, Hayashi et al. reported the results of pharmacokinetic and pharmacodynamic (PK/PD) analysis for rhG-CSF (lenograstim) after s.c. administration [115]. A total of 72 adult healthy volunteers were administered 1 µg/kg of rhG-CSF. There was no correlation between the maximal plasma concentration (C_{max}) and an increase in peripheral neutrophil count, whereas a negative correlation between AUC and this increase was observed. It is considered that the mechanism of this phenomenon is probably based on the correlation between the elimination rate constant (ke) and neutrophil increase, because the ke probably has a close relationship with uptake by neutrophil and its progenitor the G-CSF receptor. Indeed, a volunteer with higher ke showed a greater increase in neutrophil count. Therefore, AUC is proportional to the rhG-CSF remainder, that is, the proportion that is not consumed in the course of increasing the neutrophil count. From this perspective, the bioavailability calculated from the AUC is unlikely to indicate the absorbed amount. A two-compartment pharmocokinetic model with zero-order absorption and first-order elimination provided a good curve fit, suggesting that the absorption process is not a first-order process but is a zero-order process.

As the clearance of rhG-CSF is known to decrease with a rise in dose and is known to be saturable, the average clearance after i.v. administration will be lower than that after s.c. administration. Therefore, the apparent absolute bioavailability with subcutaneous administration calculated from the AUC ratio is expected to be a conservative estimate. In a second study by Hayashi et al., the estimation of rhG-CSF absorption kinetics after s.c. administration with a nonlinear elimination pharmacokinetic model using a modified Wagner–Nelson method was studied [116], and the results of the bioequivalency study between two rhG-CSF formulations with a dose of 2 µg/kg were described. The apparent absolute bioavailability for s.c. administration was 56.9% and 67.5% for each formulation, and the ratio between them was approximately 120%. The true absolute bioavailability was, however, estimated to be 89.8% and 96.9%, respectively, and the ratio was approximately 108%. The absorption pattern was applied to other doses, and the values of predicted clearance for s.c. and i.v. administrations were then similar to the values for several doses reported in the literature. Using this pharmacokinetic model, the underestimation of bioavailability was around 30%, and the amplification of difference was 2.5 times, from 8% to 20%, because of the nonlinear pharmacokinetics. The neutrophil increases for each formulation were identical, despite the different bioavailabilities. Therefore, it is considered that there is an upper limit to the transfer rate of rhG-CSF from subcutaneous tissue to blood, and that the amount eliminated through the saturable process, which indicates the amount consumed by the G-CSF receptor, was identical for each formulation.

Wang et al. also reported the PK/PD relationship of the granulopoietic effects of r-metHuG-CSF (filgrastim) in healthy volunteers through a population pharmacokinetic approach [117]. Healthy male volunteers were enrolled into a four-way crossover clinical trial. Subjects received four single doses of r-metHuG-CSF (375 and 750 μg for i.v. and s.c., respectively) with an intervening washout period of 7 days, and an absolute neutrophil count (ANC) was determined. Data analysis was performed using mixed-effects modeling as implemented in the NONMEM software package. The final PK/PD model incorporates a two-compartment PK model with bisegmental absorption from the s.c. site, first-order and saturable elimination pathways, and an indirect PD model. A sigmoidal E_{max} model for the stimulation of ANC input rate (k_{in}) was superior to the conventional E_{max} model ($E_{max} = 12.7 \pm 1.7$; $EC_{50} = 4.72 \pm 0.72$ ng/mL; Hill = 1.34 ± 0.19). In addition, a time-variant scaling factor for ANC observations was introduced to account for the early transient depression of ANC after r-metHuG-CSF administration. The absolute bioavailability after s.c. administration of r-metHuG-CSF was estimated to be 0.619 ± 0.058 and 0.717 ± 0.028 for 375 μg and 750 μg s.c. doses, respectively. The time profiles of concentration and ANC, as well as the concentration approximate ANC relationship of r-metHuG-CSF in healthy volunteers, were well described by the developed population PK/PD model.

On the other hand, Hernandez-Bernal et al. evaluated the equivalence of the pharmacokinetic, pharmacodynamic, and safety properties of two commercially available recombinant G-CSF formulations [Hebervital (Heber Biotec, Havana, formulation A) and Neupogen (Hofman-La Roche S.A, formulation B)] in 24 healthy male volunteers using a standard two-way randomized crossover doubleblind study, with a 3-week washout period [118]. A single 300-μg G-CSF dose was administered subcutaneously. Absolute neutrophils (ANC), white blood cells (WBC), and $CD3^{4+}$ cell counts were the pharmacodynamic variables measured up to 120 h. Other clinical and laboratory determinations were used as safety criteria. The pharmacokinetic parameters for formulation A and B were very close to each other (i.e., AUC, 235.9 vs. 270.0 ngh/mL; C_{max}, 29.2 vs. 33.4 ng/mL; T_{max}, 4.2 vs. 4.7 h; half-life, 3.2 vs. 2.8 h; CL, 260.9 vs. 277.2 mL/h; Vd, 1.2 vs. 1.1 L; and MRT, 7.58 vs. 7.38 h). The pharmacodynamics showed high similarity because ANC and WBC had the same profiles for both products and no differences were detected for the estimated parameters. The $CD3^{4+}$ cell count increments were evident for both formulations in a similar way as well. According to the overall results, these formulations could be considered as clinically comparable.

6.3.4.4 G-CSF Formulation

Generally, peptide/protein drugs undergo extensive hydrolysis in the gastrointestinal tract and have short circulating half-lives in the blood. For optimal clinical efficacy, therefore, they must be given by daily injections. Recently, it has been found that attaching a polyethylene glycol (PEG) moiety (PEGylation) improves the pharmacokinetic and pharmacodynamic profiles of peptide/protein drugs [119]. Conjugating peptide/protein drugs with PEG, a process known as pegylation, is now an established method for increasing the circulating half-life of protein and liposomal pharmaceuticals. PEGs are nontoxic water-soluble polymers that, owing to their large hydrodynamic volume, create a shield around the pegylated drug, thus protecting it from renal clearance, enzymatic degradation, and recognition by cells of the immune system [120].

For the formulation of G-CSF, PEGylated filgrastim (pegfilgrastim) has a longer half-life than unmodified filgrastim and, when administered as a single dose, has been shown to be at least as efficacious as daily filgrastim. As a result of its reduced renal clearance, the elimination of pegfilgrastim is predominantly neutrophil-mediated, so its clearance is self-regulated. In addition, pegfilgrastim can be administered at a fixed dose instead of in weight-based doses. For these reasons, pegfilgrastim can potentially increase patient adherence and acceptance of treatment, thus having a beneficial effect on their quality of life. Furthermore, treatment is likely to cost less because of the reduced need for medical interventions. Similarly, pegylated liposomal doxorubicin has a longer half-life than unmodified doxorubicin and has been shown, because of its reduced reticuloendothelial system clearance, to produce higher concentrations of doxorubicin in tumors and to have greater clinical efficacy than doxorubicin in the treatment of some solid tumors. Pegylated liposomal doxorubicin is also associated with less myelosuppression and febrile neutropenia.

For the pegylated formulation for subcutaneous route, van Der Auwera et al. produced a pegylated G-CSF formulation, Ro25-8315, which is produced by conjugation of rhG-CSF mutant with PEG [121], where the pharmacodynamics and pharmacokinetics of Ro25-8315 was evaluated in comparison with rhG-CSF (filgrastim). Subjects received single subcutaneous doses of Ro25-8315 ranging from 10 to 150μg under a double-blind, randomized, placebo-controlled design. Filgrastim was administered as a single dose (5 or 10μg/kg) and, following a 14-day washout period, daily for 7 days. Ro25-8315 increased absolute neutrophil count (ANC) by six- to eight-fold and CD3^{4+} cell count more than 30-fold at the highest doses tested. Single doses (60–150μg) of Ro25-8315 and multiple doses of filgrastim had similar effects on ANC and CD34^{+}. The phamacokinetic profile of Ro25-8315 was dose-dependent, with peak concentrations and the AUC increasing 100-fold over the range of doses studied. Time to reach peak concentration (T_{max}) and half-life of Ro25-8315 averaged 20–30h at all doses, approximately three times longer than with filgrastim. Adverse events were not serious and occurred with similar frequency with both products. Pegylation of rhG-CSF mutant results in more desirable pharmacokinetic properties and a longer duration of action with effective increases in ANC and measures of peripheral blood progenitor cell mobilization for at least 1 week.

More recently, Maeda et al. developed a new sustained-release formulation of rhG-CSF [122]. The formulation is a double-layer minipellet (DL-MP) that is designed to maintain a sustained release of rhG-CSF without an initial burst. The DL-MP is composed of a core of collagen matrix and a coating layer with collagen. This formulation of rhG-CSF was then prepared, and its characteristics were determined, in normal rats. After subcutaneous administration, it was found that blood rhG-CSF concentration was maintained for about 1 week. Moreover, after administration of the DL-MP rhG-CSF formulation with additional condroitin sulfate, a persistent increase in white cell count was found. Therefore, the DL-MP formulation system was useful as a long-acting formulation of rhG-CSF characterized by excellent long-acting properties without an initial burst.

In general, an oral route is most popular form of administration; however, peptide/protein drugs without some pharmaceutical modifications should not be applicable because of poor absorption based mainly on extensive hydrolysis in the gastrointestinal tract. Taking this background into consideration, Eiamtrakarn et al. designed a new formulation of G-CSF for oral administration [123], where a new gastrointestinal mucoadhesive patch system (GI-MAPS) has been designed for the oral delivery of G-CSF in beagle dogs. The system consists of four layered films, $3.0 \times 3.0\,mm^2$, contained in an enteric capsule. The 40-μm backing layer is made of a water-insoluble polymer, ethyl cellulose (EC). The surface layer is made of an enteric pH-sensitive polymer such as hydroxypropylmethylcellulose phthalate (HP-55), Eudragit L100 or S100, and was coated with an adhesive layer. The middle layer, a drug-containing layer, made of cellulose membrane is attached to the EC backing layer by a heating press method. Both drug and pharmaceutical additives including an organic acid, citric acid, and a non-ionic surfactant, polyoxyethylated castor oil derivative (HCO-60), were formulated in the middle layer. The surface layer was attached to the middle layer by an adhesive layer made of carboxyvinyl polymer (Hiviswako103). The same three kinds of GI-MAPSs containing 125 μg of rhG-CSF were prepared and orally administered to dogs, and the increase in total white blood cell (WBC) counts were measured as the pharmacological index for G-CSF. Comparison with the total increase of WBCs after i.v. injection of the same amount of G-CSF (125 μg) indicated the pharmacological availabilities (PA) of G-CSF were 23%, 5.5%, and 6.0% for Eudragit L100, HP-55, and Eudragit S100 systems. By decreasing the amount of HCO-60 and citric acid, the PA of G-CSF decreased. These results suggest the usefulness of GI-MAPS for the oral administration of protein.

For pulmonary drug delivery, the opportunity for systemic administration of peptides and proteins that are, at present, usually administered parenterally has been available [124]. As a result of the large surface area, good vascularization, immense capacity for solute exchange, and ultra-thinness of the alveolar epithelium are unique features of the lung, which can facilitate systemic delivery via pulmonary administration of peptide/protein drugs. Physical and biochemical barriers, lack of optimal dosage forms, and delivery devices limit the systemic delivery of biotherapeutic agents by inhalation. Current efforts exist to overcome these difficulties to deliver metabolic hormones. Systemic delivery of rhG-CSF can readily induce an increase in circulating levels of natural GCSF to approximately three to five times greater than baseline. Niven et al. [125] showed that rhG-CSF induced systemic response after delivery by aerosol in hamsters. The absorption from the lung was rapid, with a concomitant increase in white blood cells to four times baseline. The bioavailability was 45.9% of the administered dose and 62.0% of the dose reached the lung lobes. In a study that compared pulmonary administration of rhG-CSF powder with solution [126], a normal systemic response was obtained, indicating that rhG-CSF retains its activity in the solid state after formulation. Dissolution and absorption of rhG-CSF from powders were not rate limiting because the plasma concentration versus time profiles peaked at similar times in both powder and solution administration.

More recently, Miyamoto et al. developed the *in vivo* nasal absorption system of the FITC-dextrans with a mean molecular weight ranging from 4.3 to 167 kDa and rhG-CSF [127], in which the effect of poly-L-arginine (poly-L-Arg) on the *in vivo* nasal absorption in rats were studied. When FITC-dextrans were coadministered intranasally with 1.0 w/v% poly-L-Args of different molecular weight [MW, ca. 45.5 and 92 kDa, poly-L-Arg (50) and poly-L-Arg (100)], the bioavailability increased markedly compared with that after administration of FITC-dextran alone. However, the bioavailability decreased exponentially with the increasing molecular weight of FITC-dextrans. There was no significant difference between the enhanced nasal absorption of FITC-dextrans achieved by the coadministration of poly-L-Arg (50) and poly-L-Arg (100). Moreover, the relationship between the bioavailability and the molecular weight of FITC-dextrans indicated that the molecular weight of protein drugs, which exhibited efficient absorption with poly-L-Arg, was about 20 kDa, whereas the lower limit of bioavailability for developing a potent transnasal delivery system was assumed to be about 10%. Indeed, the nasal absorption of rhG-CSF, which has a molecular weight of 18.8 kDa, was also increased after coadministration of 1.0 w/v% poly-L-Arg (50) and the bioavailability was about 11%. It seems likely that poly-L-Arg can be used to provide adequate nasal absorption of various protein drugs, which have a molecular weight of about 20 kDa, thereby allowing the successful development of a variety of transnasal drug delivery systems.

6.3.5 INTERFERON

6.3.5.1 Background

Interferons (IFNs) are classified into three major groups that reflect antigenic and structural differentiation. IFN-α (leukocytic interferon) is produced by B lymphocytes, null lymphocytes, and macrophages; its production can be induced by foreign, virus-infected, tumor or bacterial cells [128]. Human IFN-α is a family of more than 15 subtypes, with molecular masses of 17.5–23 kDa. These proteins are composed of 165–166 amino acid residues and show approximately 80% sequence homology. The secondary structure of human IFN-α consists of 55–70% α-helix and less than 16% β-sheet. In general, human IFN-α does not contain N-linked glycosylation sites. There are 13 species of subtype IFN-α, and each IFN-α subtype has a different affinity for IFN receptor or a different antiviral activity [129]. These IFNs tend to be acidic, with isoelectric points of 5.7–7.0. In commercial settings, native IFN-α (Ly-IFN-Ly) and recombinant IFNs (IFN-α 2a, IFN-α 2b, and IFN-α con) are available [130]. IFN-β (fibroblast interferon) is a product of fibroblasts, epithelial cells, and macrophages that can be induced by viral and other nucleic acids. Human IFN-β is a single species with the molecular size of the natural glycoprotein 23 kDa with 166 amino acid residues. IFN-β shows 29% structural homology to IFN-α. The secondary structure consists of 36% α-helix and 33%-β sheet. The isoelectric point falls between 6.8 and 7.8. Now, native IFN-β and recombinant IFN-β, IFN-β 1b are available [131]. IFN-γ (immune interferon) is a product of activated T lymphocytes stimulated by foreign antigens and is composed of 143 amino acids. Human IFN-γ is heterogeneous, the molecular size depending on glycosylation and possibly oligomerization. Higher molecular sizes of 50–70 kDa have been reported, suggesting oligomerization. It shows no statistically significant sequence homology to IFN-α or IFN-β. IFN-γ is acid-labile and highly basic, with an isoelectric point of 8.6–8.7 [132].

A lot of works involving IFNs has been done with natural or cloned human IFN preparations. In the 1970s, human testing of IFNs was restricted by their limited availability, purity, and specific activity; however, recombinant DNA technology has provided high-purity (>99%) and high-specific-activity preparations for large-scale human testing [133]. Human leukocyte interferon was the first preparation available. Early phase I studies, intending to establish the safety, tolerance, pharmacokinetics, and efficacy of interferons, targeted a host of viral and oncologic diseases using animal species [134–136]. In clinical use, the results were often disappointing. However, the importance of the route and rate of administration as treatment variables had been identified. On the other hand, the pharmacokinetics of conventional drugs commonly demonstrates species specificity. In case of IFNs, however, the catabolism and excretion of IFNs are similar across most species. Therefore, animals may serve as suitable models for the pharmacokinetics of IFNs in humans.

6.3.5.2 Pharmacokinetics of Interferon in Animals

Absorption of IFN-α from intramuscular [137, 138], subcutaneous [139, 140], intraperitoneal [141], intradermal [142], duodenal, and rectal [143] sites has been reported. In general, IFN-α absorption from these sites is prolonged, and maximum serum concentrations occur 1–6h post injection, followed by measurable concentrations through 8–24h post injection. The concentration-time profile appears to be independent of the purity or source of IFN-α, namely, partially purified, natural, or recombinant. In addition, the disposition profiles were similar across the tested species, including mice, rabbits, dogs, and monkeys. Several of these studies have determined the absolute bioavailability from the intramuscular site to be 42% in dogs [139] and 93% [144] and 56% [136] in monkeys.

The absorption of IFN-β from intramuscular, subcutaneous, and intraperitneal sites is similar to that observed for IFN-α. The concentration-time curves are also similar to those seen with IFN-α, namely, prolonged absorption with concentrations persisting from 9 to 24h after the injection. Estimates of absolute bioavailability from the intramuscular site can be obtained from published data: 60% in rats [141], 33% [145] and 60% in rabbits, and 43% in monkeys [145]. Recently, a bioavailability of 2.2% was reported for IFN-β after intranasal administration to rabbits [146]. However, absorption appears to be independent of species and interferon source. Contraly to this appearance, the absorption of IFN-γ has not been studied as extensively as that of IFN-α or -β. According to the results of Weck et al., serum concentrations of IFN-γ were very low or nondetectable after intramuscular injection of nonglycosylated IFN-γ in monkeys [147].

After bolus injection, the initial decline in serum concentrations following intravenous administration is rapid for IFN-α, -β, and -γ for all species tested with the half-lives of initial distribution on the order of minutes. Serum concentrations initially decrease rapidly, then decline more slowly, with terminal elimination half-lives ranging from minutes (30–40min for IFN-α in mice) [148] to hours (7–12h for IFN-β in rabbits) [145] depending on the species selected and the type of interferon.

Both IFN-α and IFN-β have a volume of distribution after intravenous administration that ranges from 20% to 100% of body weight in mice [148], rats [149], rabbits [145], dogs [139], and monkeys [138], suggesting distribution into a volume that approximates total body water. Similar data for IFN-γ are not available. The tissue content of the interferons generally parallels that found in serum or plasma. Measurable concentrations or titers of interferons have been demonstrated in the brain, spleen, lung, liver, and kidney. For IFN-β and -γ, the amount of interferon determined in specific organs or tissues reflected the amount found in the serum or plasma, suggesting no uptake of these interferons into the sample organs or tissues. For IFN-α, the uptake results on tissue distribution studies have been mixed. Heremans et al. [141] reported the uptake of IFN-α by liver, lung, and spleen after intraperitoneal administration of homologous mouse IFN-α to mice. In contrast, Bohoslawec et al. [148] showed no appreciable uptake of murine IFN-α into brain, liver, lung, or spleen after intravenous administration to mice. However, the kidney showed an appreciable uptake of murine IFN-α. Two human IFN-α subtypes, A and D, and one human hybrid, A/D, displayed a tissue distribution profile similar to that of biologically active mouse IFN-α [148].

The catabolism of interferons has been the most widely researched area of preclinical IFN pharmacokinetics. Undoubtedly, the gastrointestinal tract is the most effective and efficient system for catabolizing proteins. The liver has also been shown to play a role in the catabolism of dycosylated proteins, and the kidney is the major catabolic organ for circulating low-molecular-mass (<60 kDa) proteins. In addition, other organs and tissues, such as the lungs and muscles, are thought to play minor roles in protein catabolism. For the catabolism of interferons, IFN-α has been the most extensively studied of the three types of interferons. In general, IFN-α is filtered through the glomeruli of the kidney followed by luminal endocytosis and proximal tubular reabsorption [150]. During reabsorption, IFN-α undergoes proteolytic degradation by lysosomal enzymes; therefore, negligible amounts of intact IFN-α are excreted in the urine. In nephrectomized animals (rats and rabbits), therefore, the clearance was significantly decreased [151, 152]. In addition to the above findings, Bocci et al. demonstrated that the liver has been shown to play a small role in the catabolism of IFN-α [153]. On the other hand, both IFN-β and IFN-γ undergo renal catabolism [153], but to a much smaller extent than IFN-α. Natural IFN-β and -γ are glycoproteins that contain sialic acid groups; they are therefore subject to the liver catabolism characteristic of this class of proteins. For recombinant IFN-α, there is an evidence that it reduces hepatic drug metabolism activity in mice [154]. Moreover, it has been demonstrated that the greater the antiviral activity of a given IFN in murine cells, the greater its ability to depress hepatic microsomal cytochrome P450 content in mice [154].

6.3.5.3 Pharmacokinetics of Interferon in Humans

The pharmacokinetics of interferons has been evaluated using a variety of dosing regimens varying from intermittent to continuous administration. Although definitive pharmacokinetics has not been reported in all cases, enough information is available to define their basic disposition characteristics. Serum interferon concentrations after intravenous administration decline rapidly in a biexponential manner for IFN-α and IFN-β [155, 156]. On the other hand, monoexponential decline has been observed for IFN-γ [157]. IFN concentrations fall several orders of magnitude over the measurable serum concentration-time course. Terminal half-lives range from 4 to 16 h for IFN-α, 1 to 2 h for IFN-β, and 25 to 35 min for IFN-γ. At the doses tested, serum concentrations are generally measurable for between 8 and 24 h after dosing of IFN-α and up to 4 h after injection for IFN-β and -γ. The total body clearance of IFN-α has been reported to range from 4.9 to 21 liters/h or 24 liters/h per/m^2. For IFN-β, the total body clearance is much higher, 19–38 liters/h per/m^2, whereas that of IFN-γ falls between the two at 12.66–32.4 liters/h.

Oral absorption of intact proteins has met with limited success because of the natural proteolytic capabilities of the gastrointestinal tract. The systemic absorption of interferons from sites other than the gastrointestinal tract has been remarkably good, considering the size of these molecules. Intramuscularly and subcutaneously administered IFN-α [158, 159] and nonglycosylated IFN-γ [160] are well absorbed, >80% for IFN-α and 30–70% for IFN-γ. These routes exhibit protracted absorption, which results in maximum serum or plasma concentrations occurring after 1 to 8 h. Concentrations are measurable for 4 to 24 h after injection for both IFN-α and IFN-γ. Maximum serum concentrations following these routes of administration are at least an order of magnitude less than the highest concentration observed after intravenous administration of an equal dose. The absorption of IFN-β from muscle or skin has not been sufficient to produce much greater serum concentrations because of the limits of assay detection.

The volume of distribution is similar for both IFN-α and IFN-γ, ranging from 12 to 40 liters [158, 161]. Although this volume is not physiological, it is approximately 20% to 60% of body weight. Information regarding IFN-β has not been established. On the other hand, the penetration of interferons across the human blood-brain barrier has been actively studied. Partially purined [162], natural [163], and recombinant [162] interferons do not readily cross the blood-brain barrier intact after intravenous, intramuscular, or subcutaneous administration. These data suggest that direct effect of parent interferon for the nervous system is not responsible for the occurrence of neurotoxicity, a common side effect of interferon therapy.

6.3.5.4   Interferon Formulation

Other routes of administration have been evaluated, such as inhalation, intralesional, intranasal, intraperitoneal, intraventricular, and intraocular rates, in clinical studies. For the most part, these alternative routes were attempts to improve the delivery of interferon to sites not easily accessible via the systemic circulation. These dosing strategies have provided adequate concentrations of interferon in cerebrospinal fluid, lymph, nasal mucosa, and peritoneal fluid, but have not led to clinical success, undoubtedly reflecting the lack of understanding of the inherent mechanism of interferon action [164].

The recommended twice-weekly administration of IFN-α does not maintain sustained plasma concentrations of the drug, thereby adversely affecting the potential biological response [165]. During the last few years, "pegylation technology" has been used to develop long-acting forms of IFN-α that help to avoid most of these problems [166]. Polyethylene glycols are nontoxic water-soluble polymers that, owing to their large hydrodynamic volume, create a shield around the pegylated drug, thus protecting it from renal clearance, enzymatic degradation, and recognition by cells of the immune system. Two separate compounds, peginterferon α-2a (PEGASYS) and peginterferon α-2b (PEG-Intron), are now both approved for use alone or in combination with ribavirin for the treatment of chronic hepatitis C [167]. Interferon therapy for chronic hepatitis C is not a cure, but it is able to decrease the viral load and may decrease the risk of complications (e.g., cirrhosis, liver failure, and liver cancer) [168]. Pegylation of the interferon increases the amount of time the interferon remains in the body by increasing the size of the interferon molecule. Increasing molecule size slows the absorption, prolongs the half-life, and decreases the rate of interferon clearance [169]. Thus the duration of biological activity is increased with pegylated INF over nonpegylated INF. The pegINF-α products offer an advantage over nonpegylated INF-α products because of less frequent administration. Tolerability of the pegylated INF is comparable with the nonpegylated formulations. Monotherapy with these agents produces a better response in some patients than monotherapy with the nonpegylated formulation [170]. Combination therapy with ribavirin is more effective than monotherapy. In 1995, Nalin et al. demonstrated that conjugation of IFN-α 2a to a linear polyethylene glycol (PEG) (M, 5000) via a urea linkage increased the half-life, and plasma residence time parameters of IFN-α 2a were one and one-half fold and two fold greater than those corresponding to the native IFN-α 2a in rats [171], with reduced immunogenicity in mice [172]. In humans, a twice-weekly dosing of PEG_{5K}-IFN-α 2a is required to achieve antiviral levels similar to those attained with unconjugated IFN, as shown in a phase II trial involving chronic hepatitis C patients [173]. Ramon et al. reported that the conjugation of INF-α 2b (IFN-α 2b) to a branched-chain (40,000) polyethylene glycol ($PEG_{2,40K}$-IFN-α 2b) markedly improved the resistance to tryptic degradation and the thermal stability of IFN-α 2b. The serum half-life of $PEG_{2,40K}$-IFN-α 2b showed 330-fold longer, whereas plasma residence time was increased 708 times compared with native IFN [174]. Pegylation alters the pharmacokinetic properties of IFN-α and allows for once-weekly administration. Pegylated IFNs contain either linear PEG chains of small molecular weight, as in pegylated IFN-α-2b (12 kD), or larger branched moieties, as in pegylated IFN-α-2a (40 kD). There are pharmacokinetic and pharmacodynamic differences between these two IFNs. The much-increased sustained virological response rates observed with pegylated IFN-α-2a (40 kD) and pegylated IFN-α-2b (12 kD) support the rationale for pegylation of IFN [175]. The absorption half-life of standard IFN-α is 2.3 h, whereas absorption half-lives for pegylated IFN-α-2a and α-2b are 50 h and 4.6 h, respectively. The volume of distribution for pegylated IFN-α-2a is considerably restricted, whereas the volume of distribution for pegylated IFN-α-2b is only approximately 30% lower than that for conventional IFN. As a result of its large size, the 40-kD pegylated IFN-α-2a has a more than 100-fold reduction in renal clearance compared with conventional IFN-α. Clearance of pegylated IFN-α-2b is about one tenth that of unmodified IFN-α. Although data are limited, both drugs appear to show differences in the initial viral decay pattern in patients with chronic hepatitis C. However, it remains unknown whether these differences in the initial viral decline predict differences in the primary clinical endpoint, sustained virological response.

The development of new IFN-α delivery strategies is a key issue to simplify its administration and improve its therapeutic effects while reducing its dose-related side effects. One of the most attractive approaches toward this purpose is the encapsulation of IFN-α into microspheres by a biodegradable polymer, poly(lactic-glycolic acid) (PLGA). Sanchez et al. developed the method of encapsulation of IFN-α into biodegradable microparticles and nanoparticles using PLGA and poloxamer 188, which is a stabilizing agent, and encapsulated IFN-α within PLGA/poloxamer blend microspheres or nanospheres [176] where their preparation techniques led to the efficient encapsulation of IFN-α and the modulation of their particle size ranging from 280nm to 40µm. The antiproliferative activity of the IFN-α varied depending on the formulation. Moreover, PLGA/poloxamer blend microspheres were able to provide significant amounts of active IFN-α for up to 96 days.

In 2004, Peleg-Shulman et al. developed a new type of pegylated INF-α-2 conjugate, peg$_{40}$-FMS-IFN-α-2, which is capable of regenerating native INF-α-2 at a slow rate under physiological conditions [177]. A 2-sulfo-9-fluorenylmethoxycarbonyl (FMS)-containing bifunctional reagent, MAL-FMS-NHS, has been synthesized, enabling the linkage of a 40-kDa peg-SH to IFN-α-2 through a slowly hydrolyzable bond. The *in vitro* rate of regeneration of native interferon from this preparation was estimated to have a half-life of 65h. Following subcutaneous administration to rats and monitoring of circulating antiviral activity, active IFN-α-2 levels peaked at 50h, with substantial levels still being detected 200h after administration. This value contrasts with a half-life of about 1h measured for unmodified interferon. The concentration of active IFN-α-2 scaled linearly with the quantity injected. Comparing subcutaneous to intravenous administration of peg40-FMS-IFN-α-2, the long circulatory lifetime of IFN-α-2 was affected both by the slow rate of absorption of the pegylated protein from the subcutaneous volume and by the slow rate of discharge from the peg in circulation.

More recently, Ito et al. demonstrated the usefulness of the oral patch preparations for INF-α [178]. The patch preparations were composed of three layers; a water-insoluble backing layer, a drug-containing layer with an absorption enhancer, and a surface layer containing pH-dependent polymer were prepared. As absorption enhancer, three surfactants, Gelucire44/14 (lauroyl macrogol-32 glycerides), Labrasol (caprylocaproyl macrogol-8 glycerides), and HCO-60 (polyoxyethylated hydrogenerated castor oil) were used in preparing the IFN-α patch preparations, an in vitro release study showed that the time when half of the formulated IFN-α is released from the patches were 3.4 ± 0.1 min for HCO-60, 7.8 ± 0.1 min for Gelucire44/14, and 11.4 ± 0.1 min for Labrasol preparations. An in vivo absorption study showed that Gelucire44/14 preparation showed a higher C_{max}, 7.66 ± 0.82 IU/mL, and AUC, 12.85 ± 1.49 IU h/mL, than Labrasol (6.51 ± 0.89 and 8.30 ± 1.34 IU h/mL) and HCO-60 (6.02 ± 1.14, 7.53 ± 1.84 IU h/mL) preparations, respectively. By comparing the AUC obtained after s.c. injection of the same dose of IFN-α with rats, bioavailability was estimated to be 7.8% in the Gelucire44/14 preparation.

6.3.6   GROWTH HORMONE

6.3.6.1   Background

Growth hormone (GH) is an endocrine hormone produced and stored in the anterior pituitary gland, which is secreted into the blood circulation in a pulsation pattern. Deficiency of GH results in abnormally poor growth. This disease state, namely hypopituitary dwarfism, has been recognized for several years. By the end of 1950s, it was found that a patient with hypopituitary dwarfism given injections of extracts of human pituitary tissue grew remarkably well. However, the supply of human GH (hGH) was extremely limited, because the only source of hGH was from cadaver pituitaries. Specific criteria were established for identifying patients who were clearly GH-deficient and who stood the best chance of benefiting from therapy. If identified, a child would receive intramuscular injections of these pituitary extracts three times per week. The dosing levels were limited and the continuing therapy was always threatened by shortages of material. In addition to being in thin supply, the problem of purifying the material provided a further challenge. As with any therapeutic drug that is derived from human tissue, contaminants of other protein hormones as well as those of viral origin had to be removed. However, more recently, the availability of synthesized hGH, produced through recombinant DNA technology, greatly reduced these concerns and provided a plentiful supply of hormone. As part of the process for proving the safety and efficacy of the use of hGH in humans, pharmacokinetic studies have been conducted in a number of species including mice, rats, monkeys, and humans.

6.3.6.2 Pharmacokinetics of Growth Hormone in Animals

In 1982, Sigel et al. studied that the disappearance rates of hGH and a 20-kDa variant over 10 min after intravenous injection of iodinated material in male and female mice [179]. The half-lives of hGH and the 20-kDa variant in males were 4.1 and 3.9 min, respectively, whereas the half-lives in females were 4.7 and 4.0 min. There were no significant differences in the protein half-lives between male and female mice. In addition, there was no apparent relationship between the half-lives, the dissimilar binding characteristics for the target sites, and growth-promoting activities. In 1988, the pharmacokinetics of biosynthetic human growth hormone (B-hGH) and pit-hGH were compared in rats [180]. Normal and hypophysectomized male and female rats were injected subcutaneously and intramuscularly with 0.1 mg/kg B-hGH and pit-hGH and intravenously with 0.06 mg/kg B-hGH and pit-hGH. A smaller distribution volume and a slower metabolic clearance rate were found for B-hGH compared with pit-hGH. Moreover, a smaller distribution volume and a slower metabolic clearance rate were found for hypophysectomized rats compared with normal rats for both proteins. The plasma half-lives was estimated to be about 3–7 min in the initial elimination phase and 29 min in the terminal elimination phase. The plasma levels of hGH were higher after subcutaneous compared with intramuscular administration for both proteins. From these results, it was suggested that an extensive local degradation took place at the subcutaneous and intramuscular injection sites. In addition, comparative tissue distribution studies with radioiodinated B-hGH and pit-hGH showed no differences between the growth hormones. The females accumulated relatively more label in the liver than the males, whereas the males accumulated relatively more label in the kidneys.

On the other hand, the total body clearance (CL) of synthetic hGH was studied in eight adult rhesus monkeys [181], where four monkeys were lean (<20% body fat) and the other four were obese (>20% body fat). The monkeys were given a single i.v. bolus injection of hGH (2.5 mg/kg), followed by a constant infusion of hGH (250 mg/hr) for 2.5 h. Venous blood samples were collected before the infusion and every 10 min during the infusion. In both groups, steady state was reached 70 min after the start of the infusion. The mean CL of synthetic hGH was 12.7 ± 1.7 liters/24 h in the lean group and 19.5 ± 2.9 liters/24 h in the obese group (p < 0.007). However, the CL values that standardized by body weight were the same in both groups. These observations suggest that the CL of hGH is directly proportional to body weight, and the lower plasma GH levels in obesity may be caused by an increase in its CL not compensated for by an appropriate increase in the rate of GH secretion.

As for the metabolism, the metabolic fate of GH in peripheral tissues has not been clearly defined as with other protein hormones. The clearance of GH from the systemic circulation is thought to occur by receptor-mediated uptake at the liver and filtration at the kidney [182]. After intravenous injection of [^{125}I]hGH to female rats, hGH was speciacally taken up by the liver with radioactivity ultimately becoming associated with the lysosomal subfraction. The specific uptake of [^{125}I]hGH into rat liver was associated with accumulation in the Golgi apparatus [183, 184]. Interestingly, in each of these studies, radioactivity in the liver was characterized by TCA solubility and membrane binding as an intact hormone. These findings suggest that receptor-mediated uptake of GH in the liver is not directly linked to hydrolytic inactivation of the hormone.

6.3.6.3   Pharmacokinetics of Growth Hormone in Humans

In the early phases of product development, data from laboratory animals can sometimes be used to predict the phamacokinetic profile in humans [183]. Dosing levels for initial clinical trials can thus be selected. In the later stages of development, when human data were available, allometeric analysis is useful for evaluating the clinical relevance of pharmacological data collected in laboratory animals. Actually, the phamacokinetic parameters in humans can often be predicted by using pharmacokinetic data collected in laboratory animals by extrapolation by following equations based on body weight. Application of this method assumes that the pharmacokinetics are dose-proportional over the dosage range under condition. The allometric approach uses the following equation: $Y = aW^b$, where Y represents a pharmacokinetic parameter and W represents body weight in kg; constants, a and b represent the allometric constants and the allometric exponent. Using the allometric equations, that pharmacokinetic parameters (total body clearance, initial volume of distribution, and volume of distribution at steady state for rhGH in mice, rats, monkeys, and humans) show a strong association among these pharmacokinetic parameters as a function of body weight. This good interspecies agreement supports the use of pharmacokinetic data collected in laboratory animals to predict the distribution of rhGH and other proteins in humans [183].

6.3.6.4   Growth Hormone Formulation

GH is a pituitary hormone responsible for postnatal growth. Like all polypeptide hormones, GH was long thought to be circulating exclusively in the free form rather than being bound to plasma proteins, which was in contrast to some other hormones of hydrophobic nature (e.g., steroids), which required protein binding for solubilization in the aqueous plasma environment. It is a 20–22-kDa polypeptide that exists in several molecular forms. GH acts by binding to GH receptors in multiple tissues; it initiates a cascade of biochemical events, among which the generation of insulin-like growth factor-I (IGF-I) is important for the growth process at the local tissue level. Other important actions of GH are those involved in anabolism, such as intracellular transport of amino acids and other cellular constituents, nitrogen, and protein synthesis. The biochemical events immediately following GH binding to receptors (intracellular signaling) are only incompletely understood. In 1987, the hepatic GH receptor was cloned and structurally characterized; it is a single-chain glycoprotein with a 620-amino acid backbone, of which 246 are extracellular, 24 form a hydrophobic transmembrane domain, and 350 are intracellular [185]. Like all polypeptide hormones, GH was long thought to be circulating exclusively in the free form rather than being bound to plasma proteins. At the end of 1980s, however, it was found that GH also circulates by binding to specific GH-binding proteins in plasma [186]. At least two GH-binding proteins (GH-BP) have been identified in human plasma; one binds GH with high affinity, the other with lower affinity.

hGH is currently administered by injection daily or three times a week. To improve patient compliance, there have been a lot of research efforts for the development of long-acting formulations of protein drugs. The parenteral delivery route often requires frequent dosing because of the short half-lives of these drugs. A few clinical studies have shown that a continuous infusion of hGH via a pump resulted in growth velocity and insulin-like growth factor-I (IGF-I) levels comparable with those of daily injections. The results indicate that pulsatile hGH release may not be required for clinical efficacy, and the sustained-release formulation may improve patient compliance and efficacy with a suitable pharmacokinetics release profile [187]. For a development of sustained-release formulation of protein drugs, one of the key requirements may be the integrity of the protein drugs within microparticles during and after formulation processes contrary to small-molecular-weight drugs; proteins have three-dimensional structures to retain their bioactivity. Some stresses (e.g., temperature, pH, or organic solvents) associated with the preparation of sustained-release formulation can cause deamidation, oxidation, denaturation, or aggregation of the protein, which can turn the protein biologically inactive and even immunogenic.

From the end of the 1990s to the early 2000s, rapid development in molecular biology and recent advancement in recombinant technology increased identification and commercialization of potential protein drugs. It is well known that traditional forms of administration for the peptide and protein drugs often rely on their parenteral injection, as the bioavailability of these therapeutic agents is poor when administered nonparenterally. Tremendous efforts by numerous investigators across the world have occured to improve protein formulations and, as a result, a few successful formulations have been developed including sustained-release hGH. At the end of 1990s, a sustained-release system was developed for rhGH using biodegradable microsphere formulations composed of polymers of lactic and glycolic acid (PLGA). Lee et al. reported that long-acting formulations of rhGH were prepared by stabilizing and encapsulating the protein into three different injectable, biodegradable microsphere formulations composed of PLGA [188]. In this study, the microsphere formulations were compared in juvenile rhesus monkeys by measuring the serum levels of rhGH and two proteins induced by hGH, insulin-like growth factor-I (IGF-I) and IGF binding protein-3 (IGFBP-3) after single s.c. administration. The results showed that PLGA formulations induced a higher level of IGFBP-3 than was induced by injections of the same amount of rhGH in solution. Moreover, Brodbeck et al. evaluated the effects of altering dynamics of PLGA solution depot on the sustained-release delivery of hGH by subcutaneous injection in rats, and obtained about 28 days of long-term sustained-release profiles from the PLGA solution depot of hGH in normal rats at serum levels of 10–200 ng/mL [189].

In the early 2000s, several other formulations of hGH have been developed. In 2001, Maeda et al. developed a sustained-release collagen film of hGH and evaluated its effect on wound healing in db/db mice [190]. The release profiles of hGH from the collagen films varied with composition and preparation conditions; however, the film prepared by air-drying of the mixture of hGH and collagen solution released hGH continuously over 3 days both in vitro and in vivo. By application of collagen film containing 3 mg of hGH twice at an interval of 6 days to wounds, the area of wounds on day 21 was significantly reduced compared with that of nontreated wounds in mice. In 2002, Kim et al. reported the usefulness of the synthesized HA/Pluronic composite hydrogels [temperature-sensitive hyaluronic acid (HA) hydrogels, which were synthesized by photopolymerization of vinyl group modified HA in combination with acrylate group end-capped poly(ethylene glycol)-poly(propylene glycol)-poly(ethylene glycol) tri-block copolymer (Pluronic F127)] [191]. This HA/Pluronic composite hydrogels gradually collapsed with increasing temperature over the range of 5 to 40°C suggesting that the Pluronic component formed self-associating micelles in the hydrogel structure. The mass erosion occurred much faster at 37°C than at 13°C, indicating that at the higher temperature, the ester linkage between the Pluronic and acrylate group might be more exposed to an aqueous environment and thus be more readily hydrolyzed because of Pluronic micellization. This HA/Pluronic composite hydrogels is applicable for a formulation of rhGH in sustained-release profiles that followed a mass erosion pattern.

More recently, aiming at once-a-week injection, a novel sustained-release formulation of rhGH using sodium hyaluronate (SR-rhGH) was developed for the treatment of children who have growth failure caused by lack of adequate secretion of endogenous GH [192, 193]. SR-rhGH was produced in the form of solid microparticles by spray-drying technology. A single administration of a prototype formulation of SR-rhGH with a ratio of hGH:HA-1:1 to cymomolgus monkeys through a fine 26-gauge needle induced continuous elevation of serum IGF-I level for 6 days, demonstrating the bioactivity of hGH released from the prototype formulation. When the ratio of hGH to HA changed from 1:1 to 1:3, hGH released more slowly in vitro from SR-rhGH with almost complete release of hGH loaded. According to phamacokinetic and pharmacodynamic studies in beagle dogs, sustained release of hGH from the optimized formulation of SR-rhGH continued for a more extended period longer than 72 h with a lower C_{max} than those of prototype formulations. The single administration resulted in an elevation of serum insulin-like growth factor-I (IGF-I) level for 6 days, with a maximum value that was higher than the baseline level by 350 ng/mL, which supported the possibility of SR-rhGH as a once-a-week injection formulation of hGH. The bioavailability of both formulations was comparable with that of hGH daily injection formulation. Finally, toxicity studies revealed no evidence of adverse effect in both cymomolgus monkeys and beagle dogs [193].

As for a formulation using another administration route, Leitner et al. developed a nasal delivery system of hGH [194] based on the thiomer polycarbophil-cysteine (PCP-Cys) in combination with the permeation mediator glutathione (GSH). Microparticles were prepared by dissolving PCP-Cys/GSH/hGH (7.5:1:1.5), PCP/hGH (8.5:1.5), and mannitol/hGH (8.5:1.5) in demineralized water, followed by lyophilization and micronization. PCP-Cys/GSH/hGH and PCP/hGH microparticles showed a comparable size distribution (80% in the range of 4.8 to 23 μm) and swelled to almost four fold size in phosphate-buffered saline. Both formulations exhibited almost identical sustained drug release profiles. The intranasal administration of the PCP-Cys/GSH/hGH microparticulate formulation resulted in a relative bioavailability of 8.11%, which represents a three fold and a 3.3-fold improvement compared with that of PCP/hGH microparticles and mannitol/hGH powder, respectively. The nasal microparticulate formulation based on PCP-Cys/GSH/hGH might represent a promising novel tool for the systemic delivery of hGH.

6.3.7   LEUPROLIDE

6.3.7.1   Background

Leuprolide acetate is a synthetic and superpotent agonist of luteinizing hormone-releasing hormone (LHRH) with a biphasic effect on the pituitary (Figure 6.3-2).

It is a decapeptide hormone that is effective in the treatment of hormone-dependent diseases such as prostate and mammary tumors and endometriosis [195, 196]. It initially stimulates the secretion of both luteinizing hormone (LH) and follicle stimulating hormone (FSH), and on long-term continuous administration it inhibits secretion of LH and FSH, and the concentration of testosterone in men and estrogen in women drop sharply. It is administered pulsatively for adjuvant hormonal therapy and continuously for hormone-related anti-neoplastic therapy comprising prostatic and breast cancer and endometriosis [197]. As leuprolide is a peptide, it is orally inactive and generally given subcutaneously or intramuscularly. Recently, however, several attempts are considered to improve bioavailability and effectiveness of leuprolide as well as a patient's convenience for taking the medicine.

LH-RH

5-OXO-Pro His Trp Ser Tyr Gly Leu Arg Pro Gly-NH$_2$

leuprolide acetate

5-OXO-Pro His Trp Ser Tyr D-Leu Leu Arg Pro-NH-CH$_2$-CH$_3$-COOH

Figure 6.3-2. Schematic presentation of primary structure of luteinizing hormone-releasing hormone (LH-RH) and leuprolide acetate.

6.3.7.2 *In vitro* Permeation Study of Leuprolide

For most polypeptides, oral administration is often limited by enzyme instability in the gastrointestinal tract. Furthermore, because peptides are highly ionized, and thus are mostly hydrophilic, their potential for permeating gastrointestinal absorption barriers is poor. Attempts to understand the degradation and permeation behavior in the intestine are beneficial to improve oral efficacy. Intramuscular and intranasal routes are the common routes of administration. Undoubtedly, the oral route is the most convenient. However, plasma levels of leuprolide after oral administration of an aqueous solution to rats and humans are mostly below the limit of detection (0.2 ng/mL) [198]. Therefore, it is important to understand and elucidate the absorption and degradation mechanisms of leuprolide in the intestinal tract. In 2004, Guo et al. reported the transport mechanism of leuprolide across rat intestine, rabbit intestine, and Caco-2 cell monolayer [199]. Using techniques based on everted gut sac and Caco-2 cell monolayer, it was found that flux of leuprolide increased with increasing concentration of drug, showing a passive diffusion pathway. At a low concentration of leuprolide, trypsin inhibitor had a strong enhancement effect on the permeability of drug by protecting enough drug for permeation. Chitosan had no affect on the activity of α-chymotrypsin. In addition, the authors found that the increase in permeation of leuprolide was caused by opening of the tight junctions and interaction with cells. Therefore, the authors concluded that both inhibition of proteolytic enzymes and opening the tight junctions to allow for paracellular transport of leuprolide improved the intestinal absorption.

6.3.7.3 Clinical Pharmacokinetics of Depot Leuprolide

In 1984, for cancer treatment, a sustained-release intramascular formulation based on biodegradable microspheres had been developed by Saunders et al. and thereafter marketed (Lupron@ Depot, TAP Pharmaceuticals) [200]. This injectable microsphereiscapable of sustaining required drug levels over a period of 1, 3, and 4 months after a single administration. These systems, in addition to improving the patient convenience and compliance, reduce the needed dose to one-quarter to one-eighth of the injected aqueous solution by sustaining therapeutic drug levels at target receptor sites [201]. More recently, sustained-release depot formulations, in which the hydrophilic leuprolide is entrapped in biodegradable highly lipophilic synthetic polymer microspheres based on polylactic/glycolic acid (PLGA) or polylactic acid (PLA), have been developed to avoid daily injections [202], where leuprolide is released from these depot formulations at a functionally constant daily rate for 1, 3, or 4 months, depending on the polymer type (PLGA for a 1-month depot and PLA for depot of >2 months), with doses ranging between 3.75 and 30 mg. The mean peak plasma leuprolide concentrations of 13.1, 20.8–21.8, 47.4, 54.5, and 53 μg/L occur within 1–3 h of depot subcutaneous administration of 3.75, 7.5, 11.25, 15, and 30 mg, respectively, compared with 32–35 μg/L at 36–60 min after a subcutaneous injection of 1 mg of a nondepot formulation. Sustained drug release from the PLGA microspheres maintains plasma concentrations between 0.4 and 1.4 mg/L over 28 days after single 3.75-, 7.5-, or 15-mg depot injections. The mean areas under the concentration-time curve are similar for subcutaneous or intravenous injection of short-acting leuprolide 1mg; a significant dose-related increase in the AUC from 0 to 35 days is noted after depot injection of leuprolide 3.75, 7.5, and 15mg. The mean volume of distribution of leuprolide is 37L after a single subcutaneous injection of 1 mg, and 36, 33, and 27L after depot administration of 3.75, 7.5, and 15mg, respectively. The total body clearance is 9.1 L/h and the elimination half-life is 3.6h after a subcutaneous 1-mg injection; corresponding values after intravenous injection are 8.3L/h and 2.9h. A 3-month depot PLA formulation of leuprolide 11.25mg ensures a mean plasma peak concentration of around 20μg/L at 3h after subcutaneous injection, and continuous drug concentrations of 0.43 to 0.19μg/L from day 7 until before the next injection. In addition, an implant administration of these sustained-release systems of leuprolide provided a lasting delivery of leuprolide for 1 year. Serum leuprolide concentrations remained at a steady mean of 0.93mg/L until week 52, suggesting zero-order drug release from the implant.

6.3.7.4   Transdermal Iontophoretic Delivery of Leuprolide

As for another route of administration of leuprolide, Kochhar et al. demonstrated an application of transdermal delivery of leuprolide using heat-separated human epidermal membrane from cadaver skin of multiple donors. The author used an iontophoresis technique to deliver the drug and examined the effect of constant voltage or constant current application [203, 204]. When the heat-separated human epidermal membrane was subjected to constant voltage within the range of 250 to 1000mV during the iontophoretic phase, iontophoretic enhancement of leuprolide permeation at pH 7.2 was greater than at 4.5. A developed model to account for iontophoretic enhancement yielded that, first, the porosity increased with the applied voltage to as much as three times the original at 1000mV. Second, the lipid pathway scontributed approximately 20% to the total permeation during the passive phase. Third, the electro-osmotic flow contributed significantly to the enhancement and its direction was from anode to cathode at pH 7.2 and the opposite at pH 4.5. The magnitude of the electro-osmotic flow was at pH 4.5 somewhat lower than at pH 7.2. Repeated iontophoretic applications of 250mV on the same skin specimen resulted in the same enhancement every time and did not cause any barrier alterations when applied for 1h every 24h, which was not the case if the duration between the two iontophoretic applications was only 3h. In contrast, keeping the current densities at 0.5 to 2.3μA/cm², the permeation rate of leuprolide increased linearly with the current density for the universal buffer and at pH 7.2 was almost double that at pH 4.5 despite the greater ionic valence of the drug at pH 4.5 compared with pH 7.2. The drug transference number at both pH values was approximately 0.5%. Replacement of the universal buffer with polymaleic acid yielded higher drug permeation rates and increased its transference number at a comparable pH. Transference number of the drug with polymaleic acid appeared to increase with current density. From these in vitro investigations, the authors concluded that the fluxes obtained for both electrolyte systems with the present experimental arrangement could be extrapolated to deliver therapeutically relevant doses of the drug.

6.3.8 DESMOPRESSIN

6.3.8.1 Background

Desmopressin (the vasopressin analog 1-deamino-8-D-arginine vasopressin, dDAVP), a potent synthetic peptide hormone of the natural pituitary hormone vasopressin, is used chiefly for treatment of enuretic disorders, such as central idiopathic or secondary diabetes insipidus [205, 206] and nocturia and nocturnal enuresis in young children [207]. Desmopressin has also proven to be useful in the treatment of bleeding diseases by activating and mobilizing factor VIII and von Willebrand's factor reserves in hemophilia A and von Willebrand–Jurgens syndrome [208]. The 1100-Da molecule, which is typically taken in doses of 1–20 μg, shows variable and low oral and nasal bioavailability (0.1% and 3.4%, respectively) [1]. Although injectable formulations demonstrate better bioavailability, they are poorly suited for routine use in young children. For the treatment of these disorders, desmopressin (Desmopressin®) in usually administered intranasally by use of sprays or drops. This route of desmopressin administration was shown to be more efficacious than the oral route, because bypassing the gastrointestinal tract increases the absolute bioavailability from less than 1% to approximately 5% [209]. A more acceptable route of administration with potentially good bioavailability could be offered by transdermal delivery. It is well documented that the stratum comeum, the outermost layer of the skin, constitutes an impermeable barrier to hydrophilic or high-molecular-weight drugs. These molecules can only be delivered into or through the skin if the barrier function of the stratum corneum is disrupted by any of a number of available methods. In this field, ultrasound and iontophoresis, which are known to enhance transdermal delivery of small molecules, have also been proposed for use with peptides and macromolecules.

6.3.8.2 Pharmacokinetics of Desmopressin in Humans

As a basic study, Rembratt et al. investigated the pharmacokinetic and pharmacodynamic profiles of desmopressin after oral and intravenous ad-mini-stration in men with a high incidence of nocturia [210]. Based on an open-randomized and four-way crossover design, desmopressin was administered orally (0.2 mg) and intravenously (2 μg) at daytime and nighttime. The concentration-time curve after 2 μg intravenous desmopressin was best fitted by a biexponential equation. The mean (95% CI) AUC at night was 302 (272–335) pg h/mL and in the day was 281 (253–312) pg h/mL. No statistically significant differences were detected between night and day except for terminal half-life, which was 3.1 h at night and 2.8 h in the daytime (p = 0.02). Peak plasma concentration (C_{max}) was 6.2 (5.1–7.5) pg/mL at night and 6.6 (5.5–7.9) pg/mL in the daytime, respectively. Median time to reach C_{max} (t_{max}) was 1.5 (range 1.0–4.1) h at night and 1.5 (range 0.5–0.3) h in the day, and the oral bioavailability was 0.08%. The pharmacodynamic effects of oral and intravenous desmopressin given in the daytime were similar during the first 6 h after dosing. The nighttime dosing and daytime intravenous dose resulted in antidiuresis throughout the measuring period, whereas the effect of the daytime peroral dose receded after 6 h. The authors concluded that the terminal half-life was longer at night than in the daytime, but the difference is considered too small to be of clinical importance. Moreover, they also concluded that, despite low bioavailability, the pharmacodynamic effects of oral desmopressin were similar in magnitude to those after intravenous dose at night and during the first 6 h after daytime administration.

More precisely, d'Agay-Abensour et al. investigated the absolute bioavailability of an aqueous solution of desmopressin from different regions of the gastrointestinal tract (stomach, duodenum, jejunum, ileum, colon, rectum) in 6 healthy male volunteers aged 24 to 35 years [211]. For intravenous administration, the subjects received 4 μg of desmopressin. For intestinal administration, 400 μg of desmopressin was directly applied to six distinct sites in the GI tract via two- or four-channel tubes with or without a distal occlusive balloon. After intravenous administration, the elimination half-life, total body clearance, amount excreted in urine, and renal clearance of desmopressin were 60.0 min, 1.7 mL/min/kg, 2.0 μg, and 0.8 mL/min/kg, respectively. The mean bioavailability of desmopressin after gastric, duodenal, jejunal, distal ileal, proximal colonic, and rectal application was 0.19%, 0.24%, 0.19%, 0.03%, 0.04%, and 0.04%, respectively. Thus, absorption was significantly higher in the three upper GI regions in comparison with the three lower regions.

6.3.8.3 Renal and Biliary Excretion and Sex Differences in Desmopressin Pharmacokinetics

Agerso et al. investigated the differences in pharmacokinetic profiles of desmopressin between healthy subjects and renally impaired patients after intravenous administration [212]. Overall, 18 renal impairment and 6 healthy volunteers were enrolled in the study. Each subject received a single intravenous dose of 2-μg desmopressin, and blood and urine samples were collected for 24 h, where plasma concentrations and the amounts of desmopressin excreted in the urine were analyzed simultaneously by means of a technique of mixed effects modeling. Their results showed that both renal and nonrenal clearances of desmopressin were found to vary with the creatinine clearance (CrCL). A decrease of 1.67% in the CrCL was found to cause a 1.74% decrease in the renal clearance and a 0.95% decrease in the nonrenal clearance. The fall in renal failure caused the amount of desmopressin excreted in urine to decrease from 47% in healthy subjects to 21% in the patients with severe renal impairment. The mean total body clearance of desmopressin in healthy subjects and patients with severe renal impairment was 10 L/h and 2.9 L/h, respectively. Correspondingly, the mean terminal half-life was 5.7 h in healthy subjects and 10 h in patients with severe renal impairment. The authors concluded that great caution should be taken when deciding on an efficient dosage regimen if patients with moderately or severely impaired renal function are to be treated with desmopressin at all, despite the drug appearing to be safe and well tolerated by patients with impaired renal function.

The pharmacokinetic profiles of desmopressin in the liver were also investigated. Lundin et al. demonsted the biliary excretion of drug after intrajugular venous, intraportal venous, and intraduodenal venous administration in conscious pigs [213]. In all cases of administration routes, the biliary excretion was less than 1% of the administered dose with the plasma/bile concentration ratio of less than 1.0. A significant first-pass effect was found when the liver was exposed to a high intraportal dose of desmopressin. The uptake study of [^{3}H]desmopressin by incubating with liver tissue slices showed that desmopressin was rapidly removed from the incubation medium, indicating desmopressin was markedly metabolized in the liver.

As another parameter to alter the pharmacodynamic profile of desmopressin, a sex difference is most important. In 2004, Odeberg et al. demonstrated that there was a sex difference in a human pharmacokinetic (PK) and pharmacodynamic (PD) study [214], where desmopressin was administered intravenously as a single dose (for the PK study, a 2-μg dose; for the PD study, a 0.2-μg dose), and parameters for urine flow and urine osmolality were estimated. The pharmacokinetics of desmopressin after a fixed bolus injection were influenced neither by piroxicam nor by sex of subjects. However, the pharmacodynamics of desmopressin showed a sex difference where females exhibited a more pronounced antidiuretic effect than males, which was statistically significant when the effects were submaximal (>4.5 h after dose). The sex differences were diminished after pretreatment with an NSAID, piroxicam, indicating a prostaglandin PGE_2-mediated mechanism.

6.3.8.4   Desmopressin Formulation

Usually, desmopressin is administered intranasally by use of sprays or drops. This administration route of desmopressin was considered to be more efficacious than the oral route, because bypassing the gastrointestinal tract increases the absolute bioavailability from less than 1% to approximately 5%. However, nasal application of desmopressin is accompanied by high intersubject and intrasubject variability in plasma pharmacokinetics [215]. Therefore, there have been several pharmaceutical research efforts to improve nasal delivery of desmopressin.

In 1987, Lethagen et al. demonstrated a comparison study, which was made of intranasal administration of 300-μg desmopressin by spray, with intravenous administration of 0.2-, 0.3-, and 0.4-μg/kg desmopressin in 10 healthy volunteers [216], where the effect of desmopressin was measured on the antihemophilia factor, factor VIII (F VIII), and on plasminogen activator release. Plasma levels of desmopressin showed a clear dose response with the maximum levels at 0.4 μg/kg intravenous dose. The affect of the spray approximated the 0.2-μg/kg response. However, the maximum response in both F VIII complex and plasminogen activator release was obtained after a 0.3-μg/kg i.v. dose. The response at a 0.4-μg/kg i.v. dose was not significantly different from the response at a 0.3-μg/kg dose, indicating that maximum stimulation was reached with 0.3 μg/kg. There was no correlation between plasma levels of F VIII and desmopressin, indicating that the biological response to the drug is subject to saturation kinetics. The reproducibility of the effect of the spray dose on F VIII was 21% (c.v.) and 27% for the intraindividual and interindividual variation, respectively, and compared favorably with intravenous administration. Therefore, they concluded that intranasal desmopressin (300 μg) is as effective as 0.2 μg/kg intravenously and provides an accurate, reproducible, and convenient alternative to parenteral administration. Moreover, in 1988, Harris et al. reported the effects of concentration and dose volume on the nasal bioavailability and biological response to desmopressin in humans [217]. A nasal formulation of 300 μg of desmopressin was administered using a premetered spray device in doses of either 1 × 50-, 2 × 50-, or 1 × 100-μL actuations to both nostrils.

Intravenous administration of 0.2 µg/kg was also given as a reference for bioavailability calculations. The biological response was determined by measuring circulating levels of F VIII. Peak plasma levels of desmopressin were greatest after the 2 × 50-µL dose, followed by the 1 × 50- and 1 × 100-µL doses. The bioavailability of desmopressin from the 2 × 50-µL dose was 20%, which was significantly greater than the 11% after the 1 × 50-µL ($p < 0.01$) and 9% after the 1 × 100-µL ($p < 0.001$) doses. The biological response was clearly enhanced after the 2 × 50-µL dose compared with the 1 × 50- and 1 × 100-µL doses. The interindividual response in F VIII levels to nasal desmopressin ranged from 20% (CV) to 30%, which compared favorably with the 36% variation after intravenous administration. These results show that by optimizing concentration, volume, and splay apparatus, a significant enhancement can be obtained in bioavailability and clinical efficacy of desmopressin.

More recently, as a new apparatus for desmopressin delivery, Cormir et al. reported the usefulness of a coated microneedle array patch system (Macronux technology) to deliver desmopressin through the skin [218]. Although it is available in injectable, intranasal, and oral formulations, intranasal and oral administration result in low and variable bioavailability. However, administering desmopressin transdermally using Macronux technology, which uses a microneedle array to overcome the skin barrier, is markedly useful to improve the bioavailability of desmopressin. In their reports, the tips of microneedles in 2-cm^2 arrays were covered with a solid coating of various amounts of desmopressin and applied to the skin of hairless guinea pigs for 5 or 15 min. Pharmacologically relevant amounts of desmopressin were delivered after 5 min. Bioavailability was as high as 85% and showed acceptable variability (30%). Immunoreactive serum desmopressin reached peak levels after a T_{max} of 60 min. Elimination kinetics for serum desmopressin was similar after other transdermal and intravenous delivery, suggesting the absence of a skin depot. These results suggest that transdermal delivery of desmopressin by Macroflux technology is a safe and efficient alternative to currently available routes of administration.

6.3.9   ANTIBODIES

6.3.9.1   Background

Antibodies serve two important functions: They bind and modulate antigens and they bind complement and immune effector cells, such as natural killer cells and monocytes. Recently, targeted therapies using monoclonal anti-bodies (mAbs) have gained increased therapeutic application. As shown in Table 6.3-1, overall, 19 mAbs are currently approved by FDA to treat various disease states, such as oncology, inflammation, infectious disease, and cardiovascular disease [219]. All approved mAbs are of the IgG class. Each IgG molecule contains two identical heavy chains and two identical light chains, 13 of which are intact mAbs, three are conjugated, and one is an mAb fragment (Fab). A feature of mAb therapeutics is the high specificity conferred by the antibody interaction with a specific region on the targeted antigen. The antigen-binding site is formed by the intertwining of the light-chain variable domain (V_L) and the heavy-chain variable domain (V_H). Each V domain contains three short stretches of peptide known as the complementarily

TABLE 6.3-1. Therapeutic Antibodies Approved by FDA

Approved	Generic name	Trade mame	Disease indication
1986	muromanab–CD3	0KT3	Allograft regction
1994	abciximab	ReoPro	Adjunct to PTCA
1995	edrecolomab	Panorex	Colorectal cancer
1997	rituximab	Rituxan	Non-Hodgkins lymphoma
1997	daclizumab	Zenapax	Prevention of kidney Transplantr rejection
1998	trastuzumab	Herceptin	Metastatic breast cancer
1998	palivizumab	Synagis	RSV propylaxis
1998	basiliximab	Simulect	Prevention of kidney Transplantr rejection
1998	infliximab	Remicade	Rheumatoid arthrists, Crohn's disease
2000	gemtuzumab ozogamicin	Mylotarg	CD33-acute myeloid leukemia
2001	alemtuzumab	Campath	B-cell chronic lymphocyte leukemia
2002	ibritumomab tiuxetan	Zevalin	Non-Hodgkings lymphoma
2002	adalimumab	Humira	Rheumatoid arthrists, Crohn's disease
2003	omalizumab	Xolair	Asthma
2003	efalizumab	Raptiva	Prorisis
2003	tosirtumomab	Bexxar	Non-Hodgkings lymphoma
2004	cetuximab	Erbitux	Colorctal cancer
2004	bevacizumab	Avastin	NSCL cancer

determining regions (CDRs); the CDRs are the prominent determinants of antigen-binding affinity and specificity. The light chain contains one constant domain, C_L. The heavy chain contains three constant domains, C_H1, C_H2, and C_H3. The C_H2 and C_H3 domains allow interactions of the IgG molecule with various components of the immune system by either binding C1q, which activates the complement cascade and elicits complement-dependent cytotoxicity, or by binding to Fcγ receptors on immune effector cells, which elicits antibody-dependent cellular cytotoxicity (Figure 6.3-3).

These same variable and constant domains of the molecule also affect IgG catabolism and elimination [220]. In addition to their contributions to mAb pharmacological activities, the FcγRs could also regulate elimination and pharmacokinetics of mAbs. Other factors, such as mAb structure and engineering, host factors, concurrent medications, and immunogenicity, can alter the pharmacokinetic and pharmacodynamic profiles. Hence, understanding the factors that affect the pharmacokinetics of mAb is of high importance for effective therapeutic application.

6.3.9.2 Pharmacodynamics and Pharmacokinetics of mAbs

mAbs can exert their pharmacological effects via several different mechanisms, such as neutralizing antigen function, activating receptors by mimicking

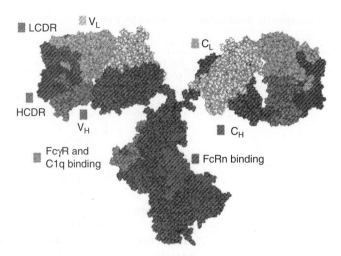

Figure 6.3-3. Representation of the space-filling model of an IgG molecule.

endogenous receptor ligands, delivering toxins to specific cells (targeted delivery), and eliciting effector functions in conjunction with antigen modulation [221]. Under some circumstances, where antigen expression is high in crucial organs (e.g., heart, lung, and vasculature), the effector function might not be desirable and could be deleterious; however, in other instances, such as applications in hematological malignancies, effector functions might be a significant part of the mechanism of action and maximizing the antibody effector functions might be highly desirable. Clearance of antibodies through the cells of the reticuloendothelial system (RES) can be regulated through the interaction with various Fc receptors. FcRns are expressed on phagocytic cells of the RES and protect IgG from rapid clearance. Similarly, cells of the RES express various types of Fcγ receptors. Interactions of IgG antibodies with this subset of receptors could potentially impact antibody clearance. As discussed, polymorphism in Fcγ receptors impacted the therapeutic response to rituximab [222]. Although the favorable clinical outcome was attributed to a more efficient Fcγ RIIIa-dependent cytotoxicity, it is not clear if polymorphism in Fcγ receptors could have any potential impact on antibody clearance. Polymorphism in Fcγ RIIIa was shown to impact in vivo clearance of red blood cells (RBC) coated with an anti-D IgG3 antibody in humans [223]. The faster clearance was attributed to a more efficient elimination of opsonized RBC by phagocytic cells in the spleen of the subjects homozygous for Fcγ RIIIa-F/F158 [223]. These results are consistent with previous data demonstrating the impaired clearance of IgG-sensitized RBC in the presence of an anti-Fcγ RIIIa antibody [224]. In addition, downregulation of Fcγ receptor on cells of monocyte lineage by immunomodulatory drugs, such as methotrexate (MTX), might potentially impact the clearance of mAbs such as adalimumab.

As the site on IgG that is responsible for binding to FcRn has been mapped and well characterized, the trend in the field of antibody engineering is to mutate the FcRn-binding site. Therefore, an increased FcRn binding provides altering the pharmacokinetics of mAbs. In 2002, Dall'Acqua et al. found that major improvement in FcRn binding occurred when mutations were introduced at positions 252,

254, 256, 433, 434, and 436, which are at the interface of the Fc-FcRn-binding region [225]. More recently, Hinton et al. discovered two mutations, T250Q and M428L, that caused, respectively, a three fold and seven fold increase in FcRn binding and, when combined together, caused a 28-fold increase in FcRn binding [226]. After intravenous administration of the M428L or T250Q/M428L mutant, about a two fold increase in plasma half-life was observed in rhesus monkeys compared with the parental molecule. As these positions are conserved among different human IgG isotypes, the authors proposed that the beneficial effect of these mutations should hold good for other isotypes as well.

As another way to alter the pharmacokinetics of mAbs, there is a PEGylation technology (the process of conjugating polyethylene glycol chains to the antibody fragments) of mAbs. PEGylation of proteins and liposomes has been a time-tested and successful technique that offered the advantage of reducing immunogenicity, increasing the plasma half-life, increasing solubility, and reducing protease sensitivity [227]. Therefore, the science of antibody PEGylation has two primary aims, which are (1) to preserve the antigen-binding activity completely and (2) to link the PEG molecule to the antibody in a stable manner. These aims are achieved by performing site-specific PEGylation using maleimide chemistry. Site-specific PEGylation was done by introducing a free cysteine to the end of the hinge region in a Fab [228] or by incorporating the hinge region on the C-terminus of a Fab and scFv [229]. The increase in half-life observed with PEGylated antibody fragments is usually caused by a prolongation of the distribution phase that represents the redistribution of a molecule in the extravascular environment. It, therefore, appears that PEGylation slows the redistribution of the mAbs from the systemic circulation to the interstitial compartment. In the following section, representative mAbs will be described.

6.3.9.3 Muromonab CD-3 Pharmacokinetics

In 1986, Muromonab CD-3 (OKT-3), which is a monoclonal antibody was first introduced to clinical practice, and the agent has been effective in reversing corticosteroid-resistant acute rejection in renal, liver, and cardiac transplant recipients. OKT-3 is administered only by intravenous injection and has a harmonic half-life of approximately 18h. It binds specifically to the CD-3 complex, which is involved in antigen recognition and cell stimulation, on the surface of T lymphocytes. Immediately after administration, CD-3-positive T lymphocytes are abruptly removed from the circulation. OKT-3 may be removed by opsonization by the reticuloendothelial system when bound to T lymphocytes, or by human antimurine antibody production [230].

6.3.9.4 Basiliximab Pharmacokinetics

Basiliximab (Simulect) is a chimeric monoclonal antibody for immunoprophylaxis against acute rejection in renal transplantation. Basiliximab is a glycoprotein produced by recombinant technology. It is used to prevent white blood cells from acute renal transplantation rejection. It specifically binds to and blocks the alpha chain of interleukin-2 receptors (IL-2R alpha), also known ascD25 antigen, on the surface of activated T lymphocytes. In 1997, to identify a single-dose regimen providing IL-2R-saturating serum concentrations in the critical first posttransplant month,

Kovarki et al. examined the pharmacokinetic aspects of basiliximab in a multi-center open-label, prospective dose-escalation study [231]. Thirty-two recipients (20 men and 12 women, 47 ± 11 years old, 65 ± 12 kg body weight) of primary mis-matched cadaver kidneys were enrolled. The immunosuppression regimen consisted of steroids and azathioprine from day 0 and cyclosporine from day 10. Basiliximab was infused over 30 min as a single dose (40 mg or 60 mg). Peak concentration and area under the concentration versus time curves increased proportionally with dose. Postinfusion concentrations declined in a biphasic manner with a terminal half-life of 6.5 ± 2.1 days. Widely dispersed correlations were noted between body weight and distribution volume (r = 0.29) or between it and total body clearance (r = 0.45), suggesting no clinical relevance for weight-adjusted dosing. Moreover, there were no apparent gender-related differences in basiliximab disposition. As serum concentrations in excess of 0.2 μg/mL are sufficient to saturate IL-2R epitopes on circulating T lymphocytes, the authors concluded that 40 mg and 60 mg of infusion dose was sufficient to keep this threshold level for 26 ± 8 days and for 32 ± 11 days, respectively.

Moreover, Kovarki et al. reported the results of a randomized, open-label prospective study with recipients of primary cadaveric liver allografts to characterize the disposition and immunodynamics of basiliximab for immunoprophylaxis of acute rejection [232]. Patients received a total intravenous dose of 4-mg basiliximab in addition to baseline dual immunosuppression consisting of cyclosporine and steroids. The central volume of distribution was 5.6 ± 1.7 L, with a steady-state volume of 7.5 ± 2.5 L. The total body clearance and elimination half-life were 75 ± 24 mL/h and 4.1 ± 2.1 days, respectively. Basiliximab was measurable in drained ascites fluid, and clearance by this route was an average of 20% of total clearance. Total body clearance correlated positively with the volume of postoperative blood loss (r = 0.5253, p = 0.0101), suggesting that bleeding may represent an additional route of drug removal. More recently, they conducted a population pharmacokinetic screen to identify demographic-clinical covariates of basiliximab in kidney and liver transplantation [233, 234]. In kidney transplantation (169 basiliximab-treated patients), basiliximab clearance was 36.7 ± 15.2 mL/h, distribution volume 8.0 ± 2.4 L, and half life 7.4 ± 3.0 days. Body weight (range, 44–131 kg) and age (range, 20–69 years) each contributed to the variability in clearance and volume. However, gender, ethnic group, and the presence of proteinuria had no clinically relevant influences on basiliximab disposition. Receptor-saturating basiliximab concentrations were maintained for 36 ± 14 days (range, 12–91) when patients received 4-mg basiliximab [233]. On the other hand, in liver transplantation (184 basiliximab-treated patients), basiliximab clearance was 55 ± 26 mL/h, the distribution volume was 9.7 ± 4.2 L, and the half-life was 8.7 ± 6.7 days. Patient weight, age, sex, ethnicity, history of alcoholism, hepatitis C seropositivity, and notable postoperative bleeding had no clinically relevant influences on basiliximab disposition; however, the cumulative volume of drained ascites fluid in the first week was positively correlated with the clearance. Receptor-saturating basiliximab concentrations (≧0.1 μg/mL) were maintained for 38 ± 16 days, which was negatively correlated with the cumulative volume of drained ascites fluid in a week [234]. However, there was no apparent relationship between the incidence or day of onset of acute rejection episodes during CD25 saturation and basiliximab concentration in kidney and liver transplants.

6.3.9.5 Trastuzumab Pharmacokinetics

Trastuzumab (Herceptin) is a monoclonal antibody for human epidermal growth factor receptor 2 (HER2)-positive metastatic breast cancer (MBC). A phase II study by Baselga et al. demonstrated the safety, efficacy, and pharmacokinetics of trastuzumab monotherapy given as first-line treatment once every 3 weeks in women with HER2-positive MBC [235]. Patients received a loading dose of trastuzumab, 8 mg/kg intravenously and then 6 mg/kg at 3-week intervals until disease progression or patient withdrawal. In total, 105 patients received a median of five cycles of therapy. The overall response rate was 19% and the clinical benefit rate was 33%. Median time to progression was 3.4 months. The most common treatment-related adverse events were rigors, pyrexia, headache, nausea, and fatigue. Median baseline left ventricular ejection fraction was 63%. The mean trough levels of trastuzumab in plasma were lower and the peak levels were higher with 3-weekly trastuzumab compared with weekly treatments. The author concluded that administering higher doses on a three-times-per-week schedule did not compromise the efficacy and safety of trastuzumab in women with HER2-positive MBC, and average exposure was similar to that observed with weekly therapy. In addition, a population pharmacokinetic analysis was performed by Bruno et al. to investigate precise aspects of trastuzumab pharmacokinetics [236]. A nonlinear mixed-effect model was applied for the pharmacokinetic data from phase I, II, and III studies of 476 patients. The phase I study enrolled patients with advanced solid tumors. The phase II and III studies enrolled patients with HER2-positive MBC. Patients in the pivotal phase II and III studies were treated with a 4-mg/kg loading dose of trastuzumab followed by 2 mg/kg weekly for up to 840 days. A two-compartment linear pharmacokinetic model best described the data and accounted for the long-term accumulation observed after weekly administration of trastuzumab. Population estimates for total body clearance, volume of distribution of the central compartment, and terminal half-life of trastuzumab were 0.225 L/day, 2.95 L, and 28.5 days, respectively. Interpatient variabilities in the total body clearance and volume of distribution were 43% and 29%, respectively. The number of metastatic sites, plasma level of extracellular domain of the HER2 receptor, and patient weight were significant baseline covariates for total body clearance, volume of distribution, or both. However, these covariate effects on trastuzumab exposure were modest and not clinically important in comparison with the large interpatient variability of CL. Concomitant chemotherapy (anthracycline plus cyclophosphamide, or paclitaxel) did not appear to influence clearance.

6.3.9.6 Infliximab Pharmacokinetics

Infliximab (Remicade) is a chimeric monoclonal antibody against tumor necrosis factor (TNF)-alpha that has shown efficacy in Crohn's disease and rheumatoid arthritis with a disease-modifying activity and rapid onset of action [237]. It is administered intravenously, generally in a schedule with initial infusions at 0, 2, and 6 weeks, followed by administration once every 8 weeks. Infliximab is effective in the treatment of patients with moderately to severely active Crohn's disease with an inadequate response to other treatment options or those with fistulizing disease. In combination with methotrexate, infliximab reduced signs and symptoms and

delayed disease progression in patients with active, methotrexate-refractory rheumatoid arthritis and in those with early disease. However, direct pharmacokinetic
trials of infliximab in healthy volunteers or patients are not yet available. Infliximab
is predominantly distributed to the vascular compartment and eliminated with a
t1/2beta between 10 and 14 days. No accumulation was observed when it was
administered at intervals of 4 or 8 weeks Methotrexate may reduce the clearance
of infliximab from serum [238].

6.3.9.7   Rituximab Pharmacokinetics

Rituximab (Rituxan) is a B-cell-depleting anti-CD20 chimeric IgG κ monoclonal
antibody being investigated for the treatment of rheumatoid arthritis and is approved
for the treatment of B-cell lymphoma [239]. Kim et al. investigated pharmacokinetic profiles of rituximab in rabbits following intravitreal administration [240].
After intravitreal injections of 1-mg rituximab, the elimination half-life of rituximab, estimated based on the two compartments, was 4.7 days. Fitting the data to
a two-compartment model yielded a clearance from the aqueous humor of 1.2 µL/
min. The clearance was less than the reported rate of aqueous humor outflow,
indicating that elimination by this route could have been sufficient to account for
the disappearance of the drug from the eye. The duration of time over which sustained levels of rituximab were achieved suggests that intravitreal administration
warrants further investigation as an approach to treating vitreous and anterior
chamber infiltrates in patients with primary intraocular lymphoma. In addition, a
population pharmacokinetic analysis of rituximab using a total of 102 patients by
Ng et al. [239] demonstrated that body surface area and gender were the most significant covariates for both a total body clearance and a volume of distribution in
central compartment. Body surface area alone only explained about 19.7% of the
total interindividual variability of the total body clearance. In a simulation study,
body surface area-based dosing normalized drug exposure over a wide range of
body surface area but did not seem to improve the predictability of rituximab AUC
in rheumatoid arthritis patients. Therefore, no rationale for body surface area-
based dosing for rituximab in rheumatoid arthritis patients was found.

6.3.9.8   Palivizumab Pharmacokinetics

Respiratory syncytial virus (RSV) infection represents a major cause of pediatric
respiratory hospitalizations. RSV is the most common cause of lower respiratory
tract infection in infants. Palivizumab (Synagis) is a humanized monoclonal antibody to the fusion protein of RSV, and is active in animal models for prevention
of pulmonary RSV replication [241]. Based on a phase I/II multicenter, randomized, double-blind, placebo-controlled, dose escalation trial by Subramanian et al.
in premature infants or infants with bronchopulmonary dysplasia, it was found that
the mean half-life of 20 days was comparable with that of other immunoglobulin
G preparations. Mean trough serum concentrations 30 days after infusion were 6.8,
36.1, and 60.6 µg/mL for the 3-, 10-, and 15-mg/kg dose groups, respectively. Mean
serum concentrations of palivizumab that have been shown to produce a 2-log
reduction in pulmonary RSV titer in cotton rats were maintained when 10- or
15-mg/kg palivizumab was given every 30 days to pediatric patients at high risk

for serious RSV disease. In addition, based on another phase I/II multicenter, randomized, double-blind, placebo-controlled, dose escalation trial in 99 healthy children hospitalized with acute RSV infection by Saez-Llornes et al., it was found that mean serum concentrations of palivizumab in the 5- and 15-mg/kg groups, respectively, were 61.2 and 303.4 μg/mL at 60 min and 11.2 and 38.4 μg/mL after 30 days. There were no significant differences in clinical outcomes between placebo and palivizumab groups for either dose. A single 15-mg/kg dose achieved serum palivizumab concentrations above the 25–30-μg/mL concentration associated with 2-log reduction of pulmonary RSV titer in the cotton rat model [242].

6.3.10 RECOMMENDATIONS FOR FUTURE STUDY

Recombinant DNA technology has produced many proteins and peptides drugs in large scale, and they are becoming increasingly important as therapeutic agents. The successful development of recombinant proteins depends on the proper characterization of their pharmacokinetics and an understanding of the relationship between drug exposure or dose and the pharmacological response. Although pharmacokinetic characterization of proteins is often complicated because of the absence of a specific assay, difficulties exist in identifying metabolites, endogenous circulating concentrations, and the presence of binding proteins. However, nanobiotechnologically derived formulations for peptide/protein drugs also show some new challenges. Numerous methods have been explored for increasing the residence time of proteins in the systemic circulation or oral bioavailability and enhancing the pharmacological effect. Some of the techniques have been employed to include site-specific mutagenesis, polymer modifications, and fusion with immunoglobulins, targeted through the use of liposomes or conjugation to toxins, radionuclides, and other proteins. In the future, these methods should provide adequate formulations of peptide drugs to increase the successful therapeutic efficacy when formulations are administered through the alternative administration routes.

REFERENCES

1. Walsh G (1998). *Biopharmaceuticals: Biochemistry and Biotechnology*, John Wiley & Sons, Ltd, England, 1998, pp. 1, 41.
2. Reichert J M, Paquette C (2003). Clinical development of therapeutic recombinant proteins. *Biotechn.* 35:176–185.
3. Reichert J M, Paquette C (2003). Therapeutic recombinant proteins: Trends in US approvals 1982 to 2002. *Curr. Opin. Molec. Therapeu.* 5:139–147.
4. Fletcher K (1998). Drug delivery: Strategies and technologies. *Pharmaceut. Sci. Technol.* 1:49–51.
5. Kurzrock R, Rosenblum M G, Sherwin S A, et al. (1985). Pharmacokinetics, single-dose tolerance, and biological activity of recombinant gamma-interferon in cancer patients. *Cancer Res.* 45:2866–2872.
6. Radwanski E, Perentesis G, Jacobs S, et al. (1987). Pharmacokinetics of interferon α-2b in healthy volunteers. *J. Clin. Pharmacol.* 27:432–435.

7. McLennan D N, Christopher J H, Porter E, et al. (2004). Lymphatic absorption is the primary contributor to the systemic availability of epoetin alfa following subcutaneous administration to sheep. *J. Pharmacol. Experim. Therapeut.* 313:345–351.

8. Pillai O, Panchagnula R (2001). Insulin therapies-past, present and future. *Drug Deliv. Ther.* 6:1056–1061.

9. Kennedy F K (1991). Recent developments in insulin delivery techniquescurrent status and future potential. *Drugs.* 42:213–227.

10. Brange J, Ribel U, Hansen J F, et al. (1988). Monomeric insulins obtained by protein engineering and their medical omplications. *Nature.* 333:679–682.

11. Jorgensen S, Vaag A, Langkjaer L, et al. (1989). NovoSol basal: Pharmaco-kinetics of a novel soluble long-acting insulin analogue. *Br. Med. J.* 299:415–419.

12. Vola J P, Owens D R, Dolben J, et al. (1988). Recombinant DNAderived monomeric insulin analogue: Comparison with soluble human insulin in normal subjects. *Br. Med. J.* 297:1236–1239.

13. Brange J, Owens D R, Kany S, et al. (1990). Monomeric insulins and their experimental and clinical implications. *Diabetescare.* 13:923–954.

14. Kang S, Owens D R, Vola J P, et al. (1990). Comparison of insulin analogue B9Asp B27Glu and soluble human insulin in insulin treated diabetes. *Lancet.* 335:303–306.

15. Halban P A, Berger M, Gjinovci A, et al. (1981). Biologic activity and pharmacokinetics of biosynthetic human insulin in the rat. *Diabetescare.* 4:238–243.

16. Brogden R N, Heel R C (1987). Human insulin: A review of its biological activity, pharmacokinetics and therapeutic use. *Drugs.* 34:350–371.

17. Turnheim K, Waldhausl W K (1988). Essentials of insulin pharmacokinetics. *Wien. Klin. Wochenschr.* 100:65–72.

18. Kang S, Brange J, Burch A, et al. (1991). Subcutaneous insulin absorption explained by insulin's pharmacochemical properties. Evidence from absorption studies of soluble human insulin and insulin analogues in humans. *Diabetescare.* 14:942–948.

19. Hoffman A, Ziv E (1997). Pharmacokinetic considerations of new insulin formulations and routes of administration. *Clin. Pharmacokinet.* 33:258–301.

20. Osterberg O, Erichsen L, Ingwersen S H, et al. (2003). Pharmacokinetic and pharmacodynamic properties of insulin aspart and human insulin. *J. Pharmacokinet. Pharmacodyn.* 30:221–235.

21. Becker R H, Frick A D, Burger F, et al. (2005). Insulin gulisine, a new rapid-acting insulin analogue, displays a rapid time-action profile in obese nondiabetic subjects. *Exp. Clin. Endocrinol. Diabetes.* 113:435–443.

22. Homko C, Deluzio A, Jimenz C, et al. (2003). Comparison of insulin aspart and lispro: Pharmacokinetic and metabolic effects. *Diabetescare.* 26:2027–2031.

23. Chiou G C Y, Chuang C Y, Chang M S (1989). Systemic delivery of insulin through eyes to lower the glucose concentration. *J. Ocular. Pharmacol.* 5:81–91.

24. Starokadomskyy P L, Dubey I Y (2006). New absorption promorter for the buccal delively: preparation and characterization of lysalbinic acid. *Int. J. Pharm.* 308:149–154.

25. Aungst B J, Roger N J, Shefter T S (1988). Comparison of nasal, rectal buccal, sublingual, and intramuscular insulin efficacy and the effect of a bile salt absorption promoter. *J. Pharmacol. Exp. Ther.* 244:23–27.

26. Degim Z, Degim T, Acarturk F, et al. (2005). Rectal and vaginal administration of insulin-chitosan formulations: An experimental study in rabbits. *J. Drug. Target.* 13:563–572.

27. Spangler R S (1990). Insulin administration via liposomes. *Diabetescare.* 13:911–922.

28. Kennedy F K (1991). Recent developments in insulin delivery thechniquescurrent states and future. *Drugs.* 42:213–227.

29. Chetty D J, Chien Y W (1998). Novel methods of insulin delively: An update. *Crit. Rev. Ther. Drug. Carrier. Syst.* 15:629–670.

30. Sakagami M (2004). Insulin disposition in the lung following oral inhalation in humans. A meta-analysis of its pharmacokineticsclin. *Pharmacokinet.* 43:539–552.

31. Henry R R, Mudaliar S R, Howland W C III, et al. (2003). Inhaled insulin using the AERx Insulin diabetes management system in healthy and asthmatic subjects. *Diabetescare.* 26:764–769.

32. (2004). Insulin inhalation—Pfizer/Nektar therapeutic: HMR 4006, inhaled PEG-insulin—Nektar, PEGylated insurin—Nektar. *Drugs. R.D.* 5:166–170.

33. Illum L, Davis S S (1992). Intranasal insulin. Clinical pharmacokineticsclin. *Pharmacokinet.* 23:30–41.

34. Drejer K, Vagg A, Bech K, et al. (1992) Intranasal administration of insulin with phospholipid as absorption enhancer: Pharmacokinetics in normal subjects. *Diabet. Med.* 9:335–340.

35. Hilsted J, Madsbad S, Hvidberg A, et al. (1995) Intranasal insulin therapy: the clinical realities. *Diabetologia.* 38:680–684.

36. Yih T C, Al-Fandi M (2006). Engineered nanoparticles as precice drug delivery system. *J. Cell. Biochem.* 11:107–112.

37. Dyer A M, Hinchcliffe M, Watts P, et al. (2002). Nasal delivery of insulin using novel chitosan based formulations: A comparative study in two animal models between simple chitosan formulations and chitosan nanoparticles. *Pharm. Res.* 19:998–1008.

38. Jelkmann W (1992). Erythropoietin: structure, control of production, and function. *Physiol. Rev.* 72:449–489.

39. Lappin T R, Rich I N (1996). Erythropoietin—the first 90 years. *Clin. Lab. Haematol.* 18:137–145.

40. Markham A, Bayson H M (1995). Epoetin alfa: A review of its pharmacodynamic and pharmacokinetic properties and therapeutic use in nonrenal applications. *Drugs.* 49:232–254.

41. Thatcher N (1998). Management of chemotherapy-induced anemia in solid tumors. *Semi. Oncol.* 25:23–26.

42. Goldberg M A, McCutchen J W, Jove W, et al. (1996). A safty and efficacy comparison study of two dosing regimens of epoetin alfa in patients undergoing major orthopedic surgery. *Am. J. Orthop.* 25:544–552.

43. Bommer J, Alexiou U, Muller-Buhl E, et al. (1987). Recombinant human erythropoietin therapy in haemodialysis patients-dose determination and clinical experience. *Nephrol. Dial. Transpl.* 2:238–242.

44. Casati S, Passerini P, Campise M R, et al. (1987). Benefits and risks of protracted treatment with human recombinant erythropoietin in patients having haemodialysis. *Br. Med. J.* 295:1017–1020.

45. Eshbach J W, Egrie J C, Downing M R, et al. (1987). Correction of the anemia of end-stage renal disease with recombinant human erythropoietin. *N. Engl. J. Med.* 316:73–78.

46. Winearls C G, Oliver D O, Pippard M J, et al. (1986). Effect of human erythropoietin derived from recombinant DNA on the anaemia of patients maintained by chronic haemodialysis. *Lancet.* 2:1175–1178.

47. Boelaert J R, Schurgers M L, Matthys E G, et al. (1989). Comparative pharmacokinetics of recombinant erythropoietin administered by the intravenous, subcutaneous, and intraperitoneal routs in continuous ambulatory peritoneal dialysis patients. *Perit. Dial. Int.* 9:95–98.

48. Hughes R T, Cotes P M, Oliver D O, et al. (1989). Correction of the anaemia of chronic renal failure with erythropoietin: pharmacokinetic studies in patients on haemodialysis and CAPD. *Contrib. Nephrol.* 76:122–130.

49. Storring P L, Tiplady R J, Gaines Das R E, et al. (1998). Epoetin alfa and beta differ in their erythropoietin isoform compositions and biological properties. *Br. J. Haematol.* 100:79–89.

50. Halstenson C E, Marcres M, Katz S A, et al. (1991). Comparative pharmacokinetics and pharmacodynamics of epoetin alfa and epoetin beta. *Clin. Pharmacol. Ther.* 50:702–712.

51. Casati S, Passerini P, Campise M R, et al. (1987). Benefits and risks of protracted treatment with human recombinant erythropoietin in patients having haemodialysis. *Br. Med. J.* 295:1017–1020.

52. Eschbach J W, Egrie J C, Downung M R, et al. (1987). Correction of the anemia of end-stage renal disease with recombinant human erythropoietin: Results of a combined phase I and Iiclinical trial. *N. Engl. J. Med.* 316:73–78.

53. Flaharty K K, Caro J, Erslev A, et al. (1990). Pharmacoldnetics and erythropoietic response to human recombinant erythropoietin in healthy men. *Clin. Pharmacol. Ther.* 47:557–564.

54. Salmonson T, Danielson B G, Wikström B (1990). The pharmacokinetics of recombinant human erythropoietin after intravenous and subcutaneous administration to healthy subjects. *Br. J. Clin. Pharmacol.* 29:709–713.

55. McMahon F G, Vargas R, Ryan M, et al. (1990). Pharmacokinetics and effects of recombinant human erythropoietin after intravenous and subcutaneous injections in healthy volunteers. *Blood.* 76:1718–1722.

56. Ramakrishnan R, Cheung W K, Wacholtz M C (2004). Pharmacokinetic and pharmacodynamic modeling of recombinant human erythropoietin after single and multiple doses in healthy volunteers. *J. Clin. Pharmacol.* 44:991–1002.

57. Ramakrishnan R, Cheung W K, Farrell F, et al. (2003). Pharmacokinetic and pharmacodynamic modeling of recombinant humanerythropoietin after intravenous and subcutaneous dose administration in cynomolgus monkeys. *J. Pharmacol. Exp. Ther.* 306:324–331.

58. Cheung W K, Goon B L, Guilfoyle M C, et al. (1998). Pharmacokinetics and pharmacodynamics of recombinant human erythropoietin after single and multiple subcutaneous doses to healthy subjectsclin. *Pharmacol. Ther.* 64:412–423.

59. Cheung W, Minton N, Gunawardena K (2001). Pharmacokinetics and pharmacodynamics of epoetin alfa once weekly and three times weekly. *Eur. J. Clin. Pharmacol.* 57:411–418.

60. Krzyanski W, Jusko W J, Wacholtz M C, et al. (2005). Pharmacokinetic and pharmacodynamic modeling of recombinant human erythropoietin after multiple subcutaneous doses in healthy subjects. *Eur. J. Pharm. Sci.* 26:295–306.

61. Kndler J, Eckardt K-U, Ehmer B, et al. (1989). Single-dose pharmacokinetics of recombinant human erythropoietin in patients with various degrees of renal failure. *Nephrol. Dial. Transplant.* 4:345–349.

62. Brockmöllerm J, Köchling J, Weber W, et al. (1992). The pharmacokinetics and pharmacodynamics of recombinant human erythropoietin in haemodialysis patients. *Br. J. Clin. Pharmacol.* 34:499–508.

63. Gladziwa U, Klotz U, Bäumer K, et al. (1993). Pharmacokinetics of epoetin (recombinant human erythropoietin) after long term therapy in patients undergoing haemodialysis and haemofiltration. *Clin. Pharmacokinet.* 25:145–153.

64. Taylor III C A, Abdel-Rahman E, Zimmerman S W, et al. (1996). Clinical pharmacokinetics during continuous ambulatory peritoneal dialysis. *Clin. Pharmacokinet.* 31:293–308.

65. Macdougall I C, Roberts D E, Coles G A, et al. (1991). Clinical pharmacokinetics of epoetin (recombinant human erythropoietin). *Clin. Pharmacokinet.* 20:99–113.

66. Korbet S M (1993). Anemia and erythropoietin in hemodialysis and continuous ambulatory peritoneal dialysis. *Kidney. Int.* 43:S111–S119.

67. Bommer J, Barth H P, Schwöbel B (1990). RhEPO treatment of anemia in uremic patientscontrib. *Nephrol.* 87:59–67.

68. Lubrich-Birkner I, Schollmeyer P, Steinhauer H B (1990). One year experience with subcutaneous human erythropoietin in CAPD: Correction of renal anemia and increased ultrafiltration. *Adv. Perit. Dial.* 6:302–307.

69. Miranda B, Selgasc–Riñon R, et al. (1990). Treatment of the anemia with human recomnbinant erythropoietin in CAPD patients. *Adv. Perit. Dial.* 6:296–301.

70. Piraino B, Johnston J R (1990). The use of subcutaneous erythropoietin in CAPD patientsclin. *Nephrol.* 33:200–202.

71. Cheng I K P, Cy C, Chan M K, et al. (1991). Correction of anemia on patients on continuous ambulatory peritoneal dialysis with subcutaneous recombinant erythropoietin twice a week: A long-term study. *Clin. Nephrol.* 35:207–212.

72. Boelaert J R, Schurgers M L, Matthys E G, et al. (1989). Comparative pharmacokinetics of recombinant erythropoietin administered by the intravenous, subcutaneous, and intraperitoneal routes in continuous ambulatory peritoneal dialyasis patients. *Perit. Dial. Int.* 9:95–98.

73. Kampf D, Kahl A, Passlick J, et al. (1989). Single-dose kinetics of recombinant human erythropoietin after intravenous, subcutaneous and intraperitoneal administration. *Contrib. Nephrol.* 76:106–111.

74. Macdougall I C, Roberts D E, Nuebert P, et al. (1989). Pharmac kinetics of recombinant human erythropoietin in patients on continuous ambulatory peritoneal dialysis. *Lancet. I.* 8635:425–427.

75. Stockenhuber F, Loibl U, Gottsauner-Wolf M, et al. (1991). Pharmckinetics and dose response after intravenous and subcutaneous administration of recombinant erythropoietin in patients on regular haemodialysis treatment or continuous ambulatory peritoneal dialysis. *Nephron.* 59:399–402.

76. Icardi A, Paoletti E, Molinelli G (1990). Efficacy of recombinant erythropoietin after subcutaneous or intraperitoneal administration to patients on CAPD. *Adv. Peril. Dial.* 6:292–295.

77. Frenken L A M, Struijk D G, Coppens P J W, et al. (1992). Intraperitoneal administration of recombinant human erythropoietin. *Perit. Dial. Int.* 12:378–383.

78. Nasu T, Mitui H, Shinohara Y, et al. (1992). Effect of erythropoietin in continuous ambulatory peritoneal dialysis patientscomparison between intravenous and intraperitoneal administration. *Perit. Dial. Int.* 12:373–377.

79. Lui S F, Chung W W M, Leung C B, et al. (1990). Pharmacokinetics and pharmacodynamics of subcutaneous and intraperitoneal administration of recombinant human erythropoietin in patients on continuous ambulatory peritoneal dialysisclin. *Nephrol.* 33(1):47–51.

80. Huang T P, Lin C Y (1995). Intraperitoneal recombinant human erythropoietin therapy: Influence of the duration of continuous ambulatory peritoneal dialysis treatment and peritonitis. *Am. J. Nephrol.* 15:312–317.

81. Bargman J M, Jones J E, Petro J M (1992). The pharmacokinetics of intraperitoneal erythropoietin administered undiluted or diluted in dialysate. *Perit. Dial. Int.* 12:369–372.

82. Ateshkadi A, Johnson C A, Oxton L L, et al. (1993). Pharmacokinetics of intraperitoneal, intravenous, and subcutaneous recombinant human erythtopoietin in patients on continuous ambulatory peritoneal dialysis. *Am. J. Kidney. Dis.* 21:635–642.

83. Evans J H C, Brocklebank J T, Bowmer C J, et al. (1991). Pharmacokinetics of recombinant human erythropoietin in children with renal failure. *Nephrol. Dial. Transpl.* 6:709–714.

84. Montini G, Zacchello G, Perfurno F, et al. (1993). Pharmacokinetics and hematologic response to subcutaneous administration of recombinant human erythropoietin in children undergoing long-term peritoneal dialysis: A multicenter study. *J. Pediatr.* 122:297–302.

85. Valentini R P, Sedman A B, Smoyer W E, et al. (1995). The safety and efficacy of intraperitoneal (IP) erythropoiethl (EPO) in children on CCPD. *J. Am. Soc. Nephrol.* 6:514.

86. Elliott S, Lorenzini T, Asher S, et al. (2003). Enhancement of therapeutic protein in vivo activities through glycoengineering. *Nat. Biotehnol.* 21:414–421.

87. Egrie J C, Brownw J K (2001). Development and characterization of novel erythropoiesis stimulating protein (NESP). *Br. J. Cancer.* 84:3–10.

88. Egrie J C, Dwyer E, Brown J K, et al. (2003). Darbepoetin alfa has a longer circulating half-life and greater in vivo potency than recombinant human erythropoietin. *Exp. Hematol.* 31:290–299.

89. Macdougall I C, Gray S J, Elsto O, et al. (1999). Pharmacokinetics of novel erythropoiesis stimulating protein compared with epoetin alfa in dialysis patients. *J. Am. Soc. Nephrol.* 10:2392–2395.

90. Allon M, Kleinman K, Walczyk M, et al. (2002). Pharmacokinetics and pharmacodynamics of darbepoetin alfa and epoetin in patients undergoing dialysisclin. *Pharmacol. Ther.* 72:546–555.

91. Lerner G, Kale A S, Wardy B A, et al. (2002). Pharmacokinetics of darbepoetin alfa in pediatric patients with chronic kidney disease. *Pediatr. Nephrol.* 17:933–937.

92. Burgess A W, Metcalf D (1980). The nature and action of granulocytemacrophage colony stimulating factors. *Blood.* 56:947–958.

93. Metcalf D (1984). *The Hemopoietic Colony Stimulationg Factors.* Elsevier, Amsterdam, The Netherlands.

94. Begley C G, Metcalf D, Lopez A F, et al. (1985). Fractionated populations of normal human marrow cells respond to both human coloney-stimulating factors with granulocyte-macrophage activity. *Exp. Hematol.* 13:956–962.

95. Metcalf D (1986). The molecular biology and functions of the granulocytemacrophage coloney-stimulating factors. *Blood.* 67:257–267.

96. Nicola N A, Metcalf D, Matsumoto M, et al. (1983). Purification of a factor indicating differentiation in murine myelomomocyte leukemia cells: Identification as granulocyte colony stimulating factor (G-CSF). *J. Biol. Chem.* 258:9017–9023.

97. Nomura H, Imazeki I, Oheda M, et al. (1986). Purification and characterization of human granulocyte colony-stimulating factor (G-CSF). *EMBO J.* 5:871–876.

98. Zsebo K, Cohen A M, Murdock D C, et al. (1986). Reconbinant human granulocyte colony syimulating factor: molecular and biological characterization. *Immunobiol.* 172:175–184.

99. Welte K, Platzer E, Lu L, et al. (1985). Prification and biochemical characterization of human pluripotent hematopoietic coloney-stimulating factor. *Proc. Natl. Acad. Sci. U.S.A.* 82:1526–1530.

100. Souza L M, Boone T C, Gabrilove J, et al. (1986). Recombinant human granulocyte colony-stimulating factor: effects on normal and leukemic myeloid cells. *Science.* 232:61–65.

101. Tsutiya M, Asano S, Kaziro Y, et al. (1986). Isolation and characterization of the cDNA for murine granulocyte coloney-stimulating factor. *Proc. Natl. Acad. Sci. U.S.A.* 83:7633–7637.

102. Nagata S, Tsuchiya M, Asano S, et al. (1986). Molecular cloning and expression of cDNA for human granulocyte colony-stimulating factor. *Nature.* 319:415–418.

103. Burgess A W, Mrtcalf D (1977). Serum half-life and organ distribution of radiolabeled colony stimulating factor in mice. *Exp. Hematol.* 5:456–464.

104. Nicola N A, Metcalf D, Matsumoto M, et al. (1983). Purification of a factor inducing differentiation in murine myelomonocytic leukemia cells: Identification as granulocyte colony-stimulating factor (G-CSF). *J. Biol. Chem.* 258:9017–9023.

105. Souza L M, Boone T C, Gabrilove J, et al. (1986). Recombinant human granulocyte colony-stimulating factor: Effects of normal and leukemic myeloid cells. *Science.* 232:61–65.

106. Morstyn G, Lieschke G J, Sheridan W, et al. (1989). Clinical experience with recombinant human granulocyte colony-stimulating factor and granulocyte macrophage coloney-stimulating factor. *Semin. Hematol.* 26:9–13.

107. Choen A M, Zsebo K M, Inoue H, et al. (1987). In vivo stimulation of granulopoiesis by recombinant human granulocyte colony-stimulating factor. *Proc. Natl. Acad. Sci. U.S.A.* 84:2484–2488.

108. Tanaka H, Kaneko T (1990). Pharmacokinetics of recombinant human granulocyte colony-stimulating factor in the rat. *Drug. Metabol. Dispos.* 19:200–203.

109. Tanaka H, Tokiwa T (1989). Pharmacokinetics of recombinant human granulocyte colony stimulating factor studied in the rat by a sandwich enzyme-linked immunosorbent assay. *J. Pharmacol. Exp. Ther.* 255:724–729.

110. Tanaka H, Okada Y, Mayumi K, et al. (1989). Pharmacokinetics and pharmacodynamics of recombinant human granulocyte-coloney stimulating factor after intravenous and subcutaneous administration in the rat. *J. Pharmacol. Exp. Ther.* 251:1199–1989.

111. Tanaka H, Kaneo T (1991). Sex differences in the pharmacokinetics of recombinant human granulocyte colony-stimulating factor in the rat. *Drug. Metabol. Dispos.* 19:1034–1039.

112. Tanaka H, Tokiwa T (1990). Influence of renal and hepatic failure on the pharmacokinetics of recombinant human granulocyte colony-stimulating factor (KRN8601) in the rat. *Cancer Res.* 50:6615–6619.

113. Layton J E, Hockman H, Sheridan W P, et al. (1989). Evidence for a novel in vivo control mechanism of granulopoiesis: Maturee cell-related control of a regulatory growth factor. *Blood.* 4:1303–1307.

114. Ferraiolo B L, Mohler M A, Gloff C A (1992). *Protein Pharmacokinetics and Metabolism.* Plenum, New York.

115. Hayashi N, Kinoshita H, Yukawa E, et al. (1999). Pharmacokinetic and pharmacodynamic analysis of subcutaneous recombinant human granulocyte coloney stimulating factor (lenograstim) administration. *J. Clin. Pharmacol.* 39:583–592.

116. Hayashi N, Aso H, Higashida M, et al. (2001). Estimation of rhG-CSF absorption kinetics after subcutaneous administration using a modified Wagner-Nelson method with a nonlinear elimination model. *Eur. J. Pharmaceu. Sci.* 13:151–158.

117. Wang B, Ludden T M, Cheung E N, et al. (2001). Population pharmacokinetic-pharmacodynamic modeling of filgrastim (r-metHuG-CSF) in healthy volunteers. *J. Pharmacokinet. Pharmacodyn.* 28:321–342.

118. Hernandez-Bernal F, Garacia-Garcia I, Gonzalez-Delgado C A, et al. (2005). Bioequivalence of two recombinant granulocyte colony-stimulating factor: Formulations in healthy male volunteers. *Biopharmac. Drug. Dispos.* 26:151–159.

119. Yowell S L, Blackwell S (2002). Novel effects with polyethylene glycol modified pharmaceuticals. *Cancer Treat. Rev.* 28(suppl A):2–6.

120. Molineux G (2002). Pegylation: Engineering improved pharmaceutical for enhanced therapy. *Cancer Treat. Rev.* 28(suppl A):13–16.

121. van Der Auwera P, Platzer E, Xu Z X, et al. (2001). Pharmacodynamics and pharmacokinetics of single doses of subcutaneous pegylated human G-CSF mutant (Ro 25-8315) in healthy volunteerscomparison with single and multiple daily doses of filgrastim. *Am. J. Hematol.* 66:245–251.

122. Maeda H, Nakagawa T, Adachi N, et al. (2003). Design of long-acting formulation of protein drugs with a double-layer structure and its application to rhG-CSF. *J. Control. Rel.* 91:281–297.

123. Eiamtrakarn S, Itoh Y, Kishimoto J, et al. (2002). Gastrointestinal mucoadhesive patch system (GI-MAPS) for oral administration of G-CSF, a model protein. *Biomat.* 23:145–152.

124. Agu R U, Ugwoke M I, Armand M, et al. (2001). The lung as a route for systemic delivery of the therapeutic proteins. *Respir. Res.* 2:198–209.

125. Niven R W, Loft F D, Cribbs J M (1993). Pulmonary absorption of recombinant methionyl human granulocyte coloney stimulating factor (r-huG-CSF) after intratracheal instillation to the hamster. *Pharm. Res.* 10:1604–1610.

126. Niven R W, Loft F D, Ip A Y, et al. (1994). Pulmonary delivery of powders and solutionscontaining granulocyte coloney stimulating factor (r-huG-CSF) to the rabbit. *Pharm. Res.* 11:1101–1109.

127. Miyamoto M, Natsume H, Satoh I, et al. (2001). Effect of poly-L-arginine on the nasal absorption of FITC-dextran of different molecular weights and recombinant human granulocyte coloney-stimulating factor (rhG-CSF) in rats. *Int. J. Pharmaceu.* 226:127–138.

128. Cantell K, Pyhala L (1976). Pharmacokinetics of human leukocyte interferon. *J. Infect. Dis.* 133(suppl A):A6–A12.

129. Blatt L M, Davis J M, Klein S B, et al. (1996). The biologic activity and molecular characterization of a novel synthetic interferon-alpha speciesconsensus interferon. *J. Interferon Cytokine Res.* 16:489–499.

130. Kohase M (2003). New IFN products for clinical use. *Sogo. Rinsyo.* 52:2540–2545.

131. Higuchi Y, Hashida M (2003). Pharmacokinetics of interferon. *Sougo. Rinsyo.* 52:2449–2505.

132. Anderson P M, Hanson D C, Hasz D E, et al. (1994). Cytokins in liposomes: Preliminary studies with IL-1, IL2, IL6, GM-CSF, and interferon-γ. *Cytokine.* 6:92–101.

133. Goddel D V, Yelverton E, Ullrick A, et al. (1980). Human leukocyte interferon produced by E. coli is biologically active. *Nature.* 287:411–416.

134. Gutterman J U, Fine S, Quesada J, et al. (1984). Recombinant leukocyte A interferon: Pharmacokinetics, single-dose tolerance, and biologic effects in cancer patients. *Cancer Res.* 44:4164–4171.

135. Hawkins M J, Borden E C, Merritt J A, et al. (1984). Comparison of the biological effects of two recombinant human interferon alpha (rA and rD) in humans. *J. Clin. Oncol.* 2:221–226.

136. Krown S E, Real F X, Cunningham-Rundles S, et al. (1983). Preliminary observations on the effect of recombinant leukocyte A interferon in homosexual men with Kaposi's sarcoma. *N. Engl. J. Med.* 308:1071–1076.

137. Cantell K, Fiers W, Hirvones S, et al. (1984). Circulating interferon in rabbits after simultaneous intramuscular administration of human alpha and gamma interferon. *J. Interferon. Res.* 4:291–292.

138. Collins J M, Riccardi R, Trown P, et al. (1985). Plasma and cerebrospinal fluid pharmacokinetics of recombinant interferon alpha A in monkeyscomparison of intravenous, intermuscular and intraventricular delivery. *Cancer Drug. Del.* 2:247–253.

139. Gibson D M, Cotler S, Speigel H E, et al. (1985). Pharmacokinetics of recombinant leukocyte A interferon following various routes and modes of administration to the dog. *J. Interferon Res.* 5:403–408.

140. Sarkar F H (1982). Pharmacokinetic comparison of leukocyte and Escherrichia coli-derived human interferon type alpha. *Antiviral Res.* 2:103–106.

141. Heremans H, Billiau A, DeSorner P (1980). Interferon in experimental viral interferon in mice: tissue interferon levels resulting from virus infection and from exogenous interferon therapy. *Infect. Immun.* 30:513–522.

142. Bocci V, Muscettola M, Naldini A (1986). The lymphatic route IV: Pharmacokinetics of human recombinant interferon alpha and natural interferon beta administered intradermally in rabbits. *Gen. Pharmacol.* 17:93–96.

143. Naito S, Tanaka S, Mizuno M, et al. (1984). Concentrations of human interferons alpha and beta in rabbit body fluids. *Int. J. Pharm.* 18:117–125.

144. Wills R J, Spiegel H E, Soike K F (1984). Pharmacokinetics of recombinant leukocyte A interferon following iv infusion and bolus, im, and po administration to African green monkeys. *J. Interferon Res.* 4:399–409.

145. Gomi K, Morimoto M, Inoue A, et al. (1984). Pharmacokinetics human recombinant interferon-beta in monkeys and rabbits. *Gann.* 75:292–300.

146. Maitani Y, Igawa T, Machida Y, et al. (1986). Intranasal administration of beta-interferon in rabbits. *Drug. Design Del.* 1:65–70.

147. Weck P K, Shalaby M R, Apperson S, et al. (1982). Comparative biological properties of human alpha, beta and gamma INF's derived from bacteria. *Third International Congress for Interferon Research*, Miami, FL.

148. Bohoslawec O, Trown P W, Wills R J (1986). Pharmacokinetics and tissue distribution of recombinant human A,D,A/D(Bgl) and I interferons and mouse alpha-interferon in mice. *J. Interferon Res.* 6:207–213.

149. Abreu S L (1983). Pharmacokinetics of rat fibroblast interferon. *J. Pharmacol. Exp. Ther.* 226:197–220.

150. Bino T, Edery H, Gertler A, et al. (1982). Involvement of the kidney in catabolism of human leukocyte interferon. *J. Gen. Virol.* 59:39–45.

151. Bocci V, Pacini A, Muscettola M, et al. (1981). Renal metabolism of rabbit serum interferon. *J. Gen. Virol.* 55:297–304.

152. Tokazewski-Chen S A, Marafino B J Jr, Stebbing N (1983). Effects of nephrectomy on the pharmacokinetics of variousclosed human interferons in rats. *J. Pharmacol. Exp. Ther.* 227:9–15.

153. Bocci V, Pacini A, Babdinelli L, et al. (1982). The role of liver in the catabolism of human alpha-and beta-interferon. *Gen. Virol.* 60:397–400.

154. Perkinsdon A, Lasker J, Kramer M J, et al. (1982). Effects of three recombinant human leukocyte interferons on drug metabolism in mice. *Drug. Metab. Dispos.* 10:579–585.

155. Wells R J, Weck P K, Baehner R L, et al. (1988). Interferon in children with recurrent acute lymphocytic leukemia: A phase I study of pharmacokinetics and tolerance. *J. Interferon Res.* 8:309–318.

156. Liberati A M, Biscottini B, Fizzotti M, et al. (1989). A phase I study of human natural interferon-beta in cancer patients. *J. Interferon Res.* 9:339–348.

157. Vadhan-Raj S, Nathan C F, Sherwin S A, et al. (1986). Phase I trail of recombinant interferon gamma by 1-hr iv infusion. *Cancer Treat Rep.* 70:609–614.

158. Bornemann L D, Speigel H E, Dziewanowska Z E, et al. (1985). Intravenous and intramuscular pharmacokinetics of recombinant leukocyte A interferon. *Eur. J. Clin. Pharmacol.* 28:469–471.

159. Hawkins M J, Borden E C, Merritt J A, et al. (1984). Comparison of the biological effects of two recombinant human interferon alpha (rA and rD) in humans. *J. Clin. Oncol.* 2:221–226.

160. Thompson J A, Cox W W, Lindgren C G, et al. (1987). Subcutaneous recombinant gamma interferon in cancer patients: toxicity, pharmacokinetics and immunomodulatory effects. *Cancer Immunol. Immunother.* 25:47–53.

161. Gutterman J U, Fine S, Quesada J, et al. (1982). Recombinat leukocyte A interferon: Pharmakokinetics, single-dose tolerance, and biologic effects in cancer patients. *Ann. Intern. Med.* 96:549–556.

162. Smith R A, Kingsubry D, Alksne J, et al. (1985). Distribution of alpha interferon in serum and cerebrospinal fluid after systemic administration. *Clin. Pharmacol. Ther.* 37:85–88.

163. Priestman T J, Johnston M, Whiteman P D (1982). Peameability observations on the pharmacokinetics of human lymphoblastoid interferon given by intramuscular injection. *Clin. Oncol.* 8:265–269.

164. Ferraiolo B L, Mohler M A, Gloff C A (1992). Protein pharmacokinetics and metabolism, in Pharmaceutical Biotechnology, vol. 1. Plenum Press, New York, p. 136.

165. Nakamura H (2005). Pegylated interferon alfa-2b formulation. *Farmacia.* 41:769–761.

166. Reddy K R (2004). Development and pharmacokinetics and pharmacodynamics of pegylated interferon alfa-2a (40 kDa). *Semin. Liver Dis.* 24:33–38.

167. Pedder S C (2003). Pegylation of interferon alpha: Structural and pharmacokinetic properties. *Semin. Liver Dis.* 23:19–22.

168. Molineux G (2002). Pegylation: Engineering improved pharmaceuticals for enhanced therapy. *Cancer Treat Rev.* 28:13–16.

169. Baker D E (2001). Pegylated interferons. *Rev. Gastroenterol. Disord.* 1:87–99.

170. Crawford J (2002). Clinical uses of pegylated pharmaceuticals in oncology. 28:7–11.

171. Nalin C, Rosen P (1995). U.S. Patent 5, 382,657.

172. Bailon P, Palleroni A, Schaffer C A, et al. (2001). Rational design of a potent, long-lasting form of interferon: A 40 kDa balanced polyethylene glycol-conjugated interferon alpha-2a for the treatment of hepatitisc. *Bioconjug. Chem.* 12:195–202.

173. Caliceti P (2004). Pharmacokinetics of pegilated interferons: What is misleading? *Dig. Liver Dis.* 36:S334–S339.

174. Ramon J, Saez V, Baez R, et al. (2005). PEGylated interferon-α2b: A branced 40 K polyethylene glycol derivative. 22:1374–1386.

175. Zeuzem S, Welsch C, Herrmann E (2003). Pharmacokinetics of peginterferons. *Semin. Liver Dis.* 23:23–28.

176. Sanchez A, Tobio M, Gonzalez L, et al. (2003). Biodegradable micro- and nanoparticles as long termdelively vehicles for interferon-alpha. *Eur. J. Pharmaceu. Sci.* 18: 221–229.

177. Peleg-Shulman T, Tsubery H, Mironchik M, et al. (2004). Reversible PEGylation: A novel technology to release native interferon α2 over a prolonged time period. *J. Med. Chem.* 47:4897–4904.

178. Ito Y, Tosh B, Togashi Y, et al. (2005). Absorption of interferon alpha from patches in rats. *J. Drug. Target.* 13:383–390.

179. Sigel M B, Vanderlaan W P, Kobrin M S, et al. (1982). The biological half-life of human grouth hormone and a biologically active 20,000-dalton variant in mouse blood. *Endocrin. Rescommun.* 9:67–77.

180. Jorgensen K D, Monrad J D, Brondum L, et al. (1988). Pharmacokinetics of biosynthetic and pituitary human growth hormones in rats. *Pharmacol. Toxicol.* 63:129–134.

181. Duby A K, Hanukoglu A, Hansen B C, et al. (1988). Metabolic clearance rates of synthetic human growth hormone in lean and obese male rhesus monkeys. *J. Clin. Endoc. Metab.* 67:1064–1067.

182. Postel-Vinay M C, Kayseer C, Desbuquios B (1982). Fate of injected human growth hormone in the female rat in vivo. *Endocrinol.* 111:244–251.

183. Mordenti J, Chen S A, Moore J A, et al. (1991). Interspecies scaling og clearance and volume of distribution and for five therapeutic proteins. *Pharm. Res.* 8:1351–1359.

184. Josefsberg Z, Posner B I, Patel B, et al. (1979). The uptake of prolactin into female rat liver. Concentration of intact hormone in the Golgi apparatus. *J. Biol. Chem.* 254:209–214.

185. Leung D W, Spencer S A, Cachianes G, et al. (1987). Grouth hormone receptor and serum binding protein: Purification, cloning and expression. *Nature.* 330:537–543.

186. Baumann G, Shaw M A (1990). Plasma transport of the 20,000 dalton variant of human growth hormone (20K): Evidence for a 20K-specific binding site. *J. Clin. Endocrinal. Metab.* 71:1339–1343.

187. Cook D M, Biller B M K, Vance M L, et al. (2002). The pharmacokinetic and pharmacodynamic characteristics of a long-acting growth horomone (GH) preparation (Nutropin Depot) in GH-deficient adults. *J. Clin. Endocrinol. Metabol.* 87:4508–4514.

188. Lee H J, Rriley G, Johnson O, et al. (1997). In vivo characterization of sustained-release formulations of Human Growth Hormone. *J. Pharmacol. Exp. Ther.* 281:1431–1439.

189. Brodbeck K J, Pushpala S, McHugh A J (1999). Sustained release of human growth hormone from PLGA solution. *Depos. Pharmaceu. Res.* 16:1825–1829.

190. Maeda M, Kadota K, Kajihara M, et al. (2001). Sustained release of human growth hormone (hGH) from collagen film and evaluation of effect on wound healing in db/db mice. *J. Control. Rel.* 77:261–272.

191. Kim M R, Park T G (2002). Temperature-responsive and degradable hyaluronic acid/pluroic composite hydrogels for controlled release of human growth hormone. *J. Control Rel.* 80:69–77.

192. Hahn S K, Kim S J, Kim M J, et al. (2004). Characterization and in vivo study of sustained-release formation of human growth hormone using sodium hyaluronate. *Pharmaceu. Res.* 21:1374–1391.

193. Kim S J, Hahn S K, Kim M J, et al. (2005). Development of a novel sustained release formulation of recombinant human growth hormone using sodium hyaluronate microparticles. *J. Controlled Rel.* 104:323–335.

194. Leitner V M, Guggi D, Krauland A H, et al. (2004). Nasal delivery of human growth hormone: In vitro and in vivo evaluation of a thiomer/glutathione microparticulate delivery system. *J. Controlled Rel.* 100:87–95.

195. Adjei A, Love S, Jhonson E, et al. (1993). Effect of formation adjuvants on gastrointestinal adsorption of leuprolide acetate. *J. Drug. Target.* 1:251–258.

196. Fu lu M, Reiland T (1994). Compositions and method for the sublingual or buccal administration of therapeutic agents. U.S. Patent 5, 248,657.

197. Kuret J A, Murad F (1992). Adenohypophyseal hormones and related sub-stances. In A Goodman, T W Gilman, A S Rall, et al. (eds.), *The Pharmacological Basis of Therapeutics*: McGraw-Hill, Singapore, pp. 1335–1360.

198. Zheng Y, Lu F, Qui Y, et al. (1999). Enzymatic degradation of leuprolide in rat intestinal mucosal homogenates. *Pharm. Dev. Technol.* 4:539–544.

199. Guo J, Ping Q, Jiang G, et al. (2004). Transport of leuprolide across rat intestine, rabbit intestine, and Caco-2 cell monolayer. *Int. J. Pharmeceu.* 278:415–422.

200. Sanders L M, Kent J K, McRae G I, et al. (1984) Controlled release of a leutenising hormone releasing hormone analogue from (D,L-lactide, glycolide) microspheres. *J. Pharm. Sci.* 73:1294–1297.

201. Okada H, Toguchi H (1995). Biodegradable microspheres in drug delivery. *Clin. Rev. Ther. Drug. Syst.* 12:1–99.

202. Periti P, Mazzei T, Mini E (2002). Clinical pharmacokinetics of depot leuprorein. *Clin. Pharmacokinet.* 41:458–504.

203. Kochhar C, Imanidis G (2002). In vitro transdermal iontophoretic deliveru of leuprolide-mechanism under constant voltage application. *J. Pharm. Sci.* 92:84–96.

204. Kochhar C, Imanidis G (2004). In vitro transdermal iontophoretic delivery of leuprolide-mechanism under constant current application. *J. Control. Rel.* 98:25–35.

205. Andersson K E, Arrner B (1972). Effects of DDAVP, a tynthetic analogue of vasopressin, in patients with cranial diabetes in sipidus. *Acta. Med. Scand.* 192:21–27.

206. Becker D J, Foley T P (1978). 1-Deamino-8-arginine vasopressin in the treatment of central diabetes insipidus in childhood. *J. Pediatr.* 92:1011–1015.

207. Dimson S B (1977). Desmopressin as a treatment for enuresis. *Lancet.* 1:1260–1261.

208. Mannuicci P M, Ruggeri Z M, Pareti F I, et al. (1977). D.D.A.V.P. in haemophillia. *Lancet.* 2:1171–1172.

209. Fjellestad-Paulsen A, Lundin H P, Paulsen O (1993). Pharmacokinetics of 1-Deamino-8-arginine vasopressin after various routes of administration in healthy volunteers. *Clin. Endocortinol.* 38:177–182.

210. Rembratt A, Graugaard-Jensen C, Senderovitz T, et al. (2004). Pharmacokinetics and pharmacodynamics of desmopressin administred orally versus intravenously at daytime versus night-time in healthy men aged 55–70 years. *Eur. J. Clin. Pharmacol.* 60:397–402.

211. Agay-Abensour L, Ajellestad-Paulsen A, Hoglund P, et al. (1993). Absolute bioavailability of an aqueous solution of 1-Deamino-8-arginine vasopressin from different regions of gastrointestinal tract in man. *Eur. J. Clin. Pharmacol.* 44:473–476.

212. Agerso H, Laesen L S, Riis A, et al. (2004). Pharmacokinetics and renal excretion of desmopressin after intravenous administration to healthy subjects and renally impaired patients. *Br. J. Clin. Pharmacol.* 58:352–358.

213. Lundin S, Pierzynovski S G, Westrom B R, et al. (1991). Biliary excretion of the vasopressin analogue DDAVP after intraduodenal, intrajugular and intraportal administration in the conscious pig. *Pharmacol. Toxicol.* 68:177–180.

214. Odeberg J M, Callreus T, Lundin S, et al. (2004). A pharmacokinetics and pharmacodynamics study of desmopressin: Evaluationg sex differences and the effect of pretreatment with piroxicam, and further validation of an indirect response model. *J. Pharm. Pharmacol.* 56:1389–1398.

215. Lethagen S, Harris A S, Sjorin E, et al. (1987). Intranasal and intravenous administration of desmopressin: Effect on F VIII/vW, pharmacokinetics and reproducibility. *Thromb. Haematost.* 58:1033–1036.

216. Harris A S, Ohlin M, Lethagen S, et al. (1988). Effects of concentration and volume on nasal baioavailability and biological response to desmopressin. *J. Pharm. Sci.* 77:337–339.

217. Joukhader C, Schenk B, Kaehler S T, et al. (2003). A replicate study design for testing bioequivglence: A case study on two desmopressin nasal spray preparations. *Pharmacokinet. Dispos.* 59:631–636.

218. Cormier M, Bjohnson M, Ameri K, et al. (2004). Transdermal delivery of desmopressin using a coated microneedle array patch system. *J. Controlled Rel.* 97:503–511.

219. Kim S J, Park Y, Hong H J (2005). Autibody engineering for the development of therapeutic antibodies. *Mol. Cells.* 20:17–29.

220. Tabrizi M A, Tseng C M, Roskos L K (2006) Elimination mechanisms of therapeutic monoclinal antibodies. *Drug Discov. Today.* 11:81–88.

221. Houghton A N, Scheinberg D A (2000). Monoclonal antibody therapies—a constant treatment to cancer. *Nat. Med.* 6:373–374.

222. Cartron G, Dacheux L, Salles G, et al. (2002). Therapeutic activity of humanized anti-CD20 monoclonal antibody and polymorphism in IgG Fc receptor FcgammaRIIIa gene. *Blood.* 99:754–758.

223. Kumpel B M, De Haas M, Koene H R, et al. (2003). Clearance of red cell by monoclonal IgG3 anti-D in vivo is affected by the VF polymorphism of Fcgamma RIIIa(CD16). *Clin. Exp. Immunol.* 132:81–86.

224. Clarkson S B, et al. (1986). Treatment of refractory immune thrombocytopenic purpura with anti-Fc gamma-receptor antibody. *N. Engl. J. Med.* 314:1236–1239.

225. Dall'Acqua W F, Woods R M, Ward E S, et al. (2002). Increasing the affinity of a human IG1 for the neonatal Fc receptor: Biological consequences. *J. Immunol.* 169:5171–5180.

226. Hinton P R, Johlfs M G, Xiong J M, et al. (2003). Engineered human IgG antibodies with longer serum half-lives in primates. *J. Biol. Chem.* 279:6213–6216.

227. Chapman A P (2002). PEGylated antibodies and antibody fragments for improved therapy: A review. *Adv. Drug. Deliv. Rev.* 54:531–545.

228. Chapman A P, Antoniw A, Spitali M, et al. (1999). Therapeutic antibody fragments with prolonged in vivo half-lives. *Nat. Biotechnol.* 17:780–783.

229. King D J, Turner A, Farnsworth A P, et al. (1994). Improved tumor targeting with chemically cross-linked recombinant antibody fragments. *Cancer Res.* 54:6176–6185.

230. Hooksc M A, Wade S, Millikan W J Jr (1991). Muromonab CD-3: a review of its pharmacology, pharmacokinetics, and clinical use in transplantation. *Pharmacother.* 11:26–37.

231. Kovarik J, Wolf P, Isterne J M, et al. (1997). Disposition of Basiliximab, an interleukin-2 receptor monoclonal antibody, in recipients of mismatched cadaver renal allografts. *Transplant.* 64:1701–1705.

232. Kovarik J, Breidenbach T, Gerbeau C, et al. (1998). Disposition and immunodynamics of basiliximab in liver allograft recipients. *Clin. Pharmacol. Ther.* 64:66–72.

233. Kovarki J M, Kahan B D, Rajagopalan P R, et al. (1999). Population pharmacokinetics and exposure-response relationships for basiliximab in kidney transplantation. *Transplant.* 68:1288–1294.

234. Kovarki J M, Nashan B, Neuhaus P, et al. (2001). A population pharmacokinetic screen to identify demographic-clinical covariates of basiliximab in liver transplantation. *Clin. Pharmacol. Ther.* 69:201–209.

235. Baselga J, Carbonell X, Castaneda-Soto N J, et al. (2005). PhaseII study of efficacy, safety, and pharmacokinetics of trastuzumab monotherapy administered on a 3-weekly schedule. *J. Clin. Oncol.* 23:2162–2171.

236. Bruno R, Washington C B, Lu J F, et al. (2005). Population pharmacokinetics of trastuzumab in patients with HER2+mesenteric breast cancer. *Cancer Chemother Pharmacol.* 56:361–369.

237. Siddiqui M A, Scott L J (2005). Infliximab: A review of its use in Crohn's disease and rheumatoid arthritis. *Drugs.* 65:2179–2208.

238. Schwab M, Klotz U (2001). Pharmacokinetic considerations in the treatment of inflammatory bowel disease. *Clin. Pharmacokinet.* 40:723–751.

239. Ng C M, Bruno R, Combs D, et al. (2005). Population pharmacokinetics of rituximab (anti-CD20 monoclonal antibody) in rheumatoid arthritis patients during a phase II clinical trial. *Clin. Pharmacol.* 45:792–801.

240. Kim H, Csaky K G, Chan C C, et al. (2006). The pharmacokinetics of rituximab following an intraviteral injection. *Exp. Eye Res.* 82:760–766.

241. Subramanian K N, Weisman L E, Rhodes T, et al. (1999). Safety, tolerance and pharmacokinetics of a humanized monoclonal antibody to respiratory syncytail virus in premature, infants and infants with bronchopulmonary dysplasia. MEDI-493 study group. *Pediatr. Infect. Dis. J.* 17:110–115.

242. Saez-Llorens X, Moreno M T, Ramilo O, et al. (2004). MEDI-493 study group: Safety and pharmacokinetics of palivizumab therapy in children hospitalized with respiratory syncytail virus infection. *Pediatr. Infect. Dis. J.* 23:707–712.

6.4

IMMUNOGENICITY OF THERAPEUTIC PROTEINS

Suzanne Hermeling,[1,2] Daan J.A. Crommelin,[1]
Huub Schellekens,[1,2] and Wim Jiskoot[1,3]

[1] *Utrecht Institute for Pharmaceutical Sciences (UIPS), Utrecht University, The Netherlands.*
[2] *Central Laboratory Animal Institute, Utrecht University, The Netherlands.*
[3] *Leiden/Amsterdam Center for Drug Research (LACDR), Leiden University, Leiden, The Netherlands.*

Chapter Contents

6.4.1 Introduction 816
6.4.2 Immune Mechanisms 817
    6.4.2.1 Classic Immune Response 817
    6.4.2.2 Breaking of Immune Tolerance 818
6.4.3 Factors Influencing Immunogenicity 818
    6.4.3.1 Product Characteristics 818
    6.4.3.2 Patient Characteristics 820
    6.4.3.3 Treatment Characteristics 820
    6.4.3.4 Other Factors 820
6.4.4 Detection 821
    6.4.4.1 Assays 821
    6.4.4.2 The Design of an Assay Strategy 823
    6.4.4.3 Predictive Models 823
    6.4.4.4 Prediction on Basis of Structure and Sequence 824
6.4.5 Biological and Clinical Consequences 825
6.4.6 A Risk-based Approach of Immunogenicity 825
6.4.7 Examples 825
    6.4.7.1 Epoetin: How a Formulation Change Increased the Immunogenicity 825
    6.4.7.2 RhIFNβ: Immunogenicity of Glycosylated and Nonglycosylated Variants 827

Handbook of Pharmaceutical Biotechnology, Edited by Shayne Cox Gad.
Copyright © 2007 John Wiley & Sons, Inc.

6.4.7.3   RhIFNα2: Reduction of the Immunogenicity by Optimizing
Production and Formulation   828
6.4.8   Conclusions   829
References   829

6.4.1   INTRODUCTION

Since the introduction of recombinant DNA techniques, it is possible to produce proteins on a large scale. These include proteins that are identical or nearly identical to endogenous human proteins, but also foreign proteins like streptokinase and asparaginase. Not only foreign proteins, but also recombinant human protein therapeutics are potentially immunogenic, and indeed most of these products induce antibodies in some patients after a certain period of treatment (see Table 6.4-1 for examples). Patients will first develop binding antibodies (BAbs), and this can be followed by the formation of neutralizing antibodies (NAbs). BAbs usually do not cause major complications, but NAbs bind to the active site of the protein and may interfere with the therapeutic effect. NAbs sometimes also cross-react with the endogenous protein, which can lead to serious complications [3, 4].

This chapter gives an overview of the current knowledge about the immunogenicity of therapeutic proteins. First the mechanisms by which antibodies are formed will be discussed, followed by the factors that can influence the immune response. An assay strategy for the detection of antibodies in human sera is discussed. The

TABLE 6.4-1.   Examples of Recombinant Therapeutic Proteins with Reported Immunogenicity[1]

Type of Protein	Protein
Hormones	Insulin
	Growth hormone
Cytokines	Interferon alpha
	Interferon beta
	Interleukin 2, 3 and 12
Enzymes	Factor VIII
	DNase
	Tissue plasminogen activator
Antibodies	Anti-CD3 (murine antibody)
	Anti-Her2 (humanized antibody)
	Anti-IgE (humanized antibody)
	Anti-respiratory syncytial virus (humanized antibody)
	Anti-IL-2 receptor (humanized antibody)
Growth factors	G–CSF
	GM–CSF
	Erythropoietin
	Thrombopoietin

[1]   Adapted from Refs. 1 and 2.

biological and clinical consequences are summarized, and the U.S. Food and Drug Administration (FDA) procedure to handle immunogenicity concerns of novel products is explained. Next, we give some examples of therapeutic proteins with immunogenicity problems, followed by the final conclusions.

6.4.2 IMMUNE MECHANISMS

Depending on the type of therapeutic protein, antibodies can be formed via two pathways (Table 6.4-2): a classic immune response to foreign proteins and breaking of immune tolerance to self-proteins [1].

6.4.2.1 Classic Immune Response

A classic immune response occurs after administration of foreign proteins. Antigen-presenting cells (APCs) will take up the protein, digest it, and present peptides on major histocompatibility complex (MHC)-molecules on their surface. T cells will recognize these peptides in combination with the MHC-molecules and will activate B cells to produce antibodies against the foreign protein. This type of reaction is observed when proteins of animal, microbial, or plant origin are administered to patients. The antibody formation is usually fast, within days to weeks, and often occurs after a single injection. The antibodies are mostly NAbs and persist for a long time.

A classic immune response can also occur on administration of recombinant human proteins to patients with an innate deficiency, e.g., children lacking growth hormone. These children lack the immune tolerance that normally exists for self-proteins.

TABLE 6.4-2. Types of Immune Reaction Against Therapeutic Proteins[1]

	Classic Immune Response	Breaking of Immune Tolerance
Properties of Product	Microbial or Plant Origin	Human Homologue
Characteristics of antibody formation	Fast Often after a single injection High incidence Neutralizing antibodies Long duration	Slow After prolonged treatment Low incidence Mainly binding antibodies Disappear after stopping of treatment and sometimes during treatment
Cause of immunogenicity	Presence of non-self-antigens	Impurities and presence of aggregates Mechanism of antibody formation is unknown
Consequences	Loss of efficacy in most cases	In most patients, no consequences

[1] Taken from Ref. 5.

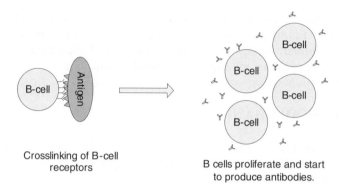

Crosslinking of B-cell
receptors

B cells proliferate and start
to produce antibodies.

Figure 6.4-1. Schematic representation of a possible mechanism of breaking B-cell tolerance. The antigen contains multiple epitopes that cross-link B-cell receptors. The cross-linking of the receptors is observed as a danger signal because of which the B cells start to proliferate to plasma cells and produce antibodies.

6.4.2.2   Breaking of Immune Tolerance

Most recombinant human proteins are homologous to their endogenous counterparts. The immune system of the patients will therefore recognize the protein as a self-protein, for which the patient is tolerant. Antibody formation against self-proteins can occur through breaking of this immune tolerance. The antibody formation via this process is slow and often becomes apparent only after months of chronic administration of the protein to the patient. Usually the antibodies disappear when treatment is stopped. The exact mechanism of breaking of immune tolerance is not known. Multimeric antigen presentation with a narrow (~5–10nm) spacing, however, is known to break immune tolerance, possibly by cross-linking of B-cell receptors (Figure 6.4-1). The only natural proteins showing this closely spaced multimeric antigen presentation are microbial antigens, and apparently during evolution, there was a strong selective pressure to react vigorously to this type of antigen presentation [6–8].

6.4.3   FACTORS INFLUENCING IMMUNOGENICITY

Many factors influence the immune response induced by therapeutic proteins. These can be divided into product, patient, and treatment characteristics, and several factors are still unknown.

6.4.3.1   Product Characteristics

Proteins are complex structures consisting of primary, secondary, tertiary, and sometimes quaternary structures. Changes in one of these structural levels might influence the immune response. How a protein is formulated influences the chemical and physical stability of the protein. Therefore the formulation can also influence the immunogenicity. Moreover, contaminants and impurities play an important role.

Primary Sequence. Based on amino acid sequence, no definite predictions about the immunogenicity of a product can be made. A single amino acid change in insulin was enough to elicit a strong immune response [9]. Consensus interferon alpha (co-IFNα), on the other hand, has several amino acid changes as compared with the naturally occurring human IFNαs (10–23 amino acid differences and on average 89% homology with naturally occurring IFNαs), but it has no increased immunogenicity [10]. Foreign proteins, such as streptokinase, staphylokinase, bovine adenosine deaminase, and salmon calcitonin will elicit a classic immune response in patients.

Chemical changes of the primary structure, such as oxidation or deamidation of amino acids, can lead to novel epitopes, which can induce a classic immune response [11]. Moreover, the chemical change can induce aggregation leading to multimeric antigen presentation and, thus, breaking of immune tolerance.

Glycosylation. Glycosylation is a common posttranslational modification and is cell and species specific. The glycans can differ in chain length, sequence, and linkage position to the peptide backbone [12]. Endogenous proteins often consist of several glycosylated isoforms [13]. The carbohydrates can play a role in molecular stability, *in vivo* activity, serum half-life, and immunogenicity. Carbohydrates can decrease the immunogenicity of therapeutic proteins, by shielding immunogenic epitopes [12, 14] or by increasing the solubility of the protein and thereby preventing aggregation [15].

Production of therapeutic proteins in plants is being developed, but concerns are raised about the immunogenicity of plant glycans. Glycosylation of proteins in plants and humans differ in fine detail [16]. Especially the plant-specific α(1,3)-fucose and β(1,2)-xylose groups are considered immunogenic in humans [12]. Metabolic engineering of the plant N-glycan biosynthesis pathway is being pursued to prevent the insertion of these sugars in the glycan chains [16].

PEGylation. The most successful approach to increase the mean plasma half-life of proteins is by chemically coupling poly(ethylene glycol) (PEG) moieties to the protein. The attachment of PEG (PEGylation) decreases the overall rate of clearance of the protein, shields the protein from proteolytic enzymes, and masks immunogenic sites [17, 18]. The PEG molecules can differ in conjugation type, molecular weight, and conformation (i.e., linear, branched, or multiple-branched) [19]. PEGylation can decrease the immunogenicity of a therapeutic protein by blocking antibody binding sites, promoting solubility, and permitting less frequent dosing [20]. Branched PEG is more effective than linear PEG in improving the immunological properties of the protein [21]. Although in most cases PEGylation decreases the immunogenicity of a therapeutic protein, two examples of therapeutic proteins are known in which the PEGylated protein was more immunogenic than the non-PEGylated variant: PEGylated recombinant human megakaryocyte growth and development factor (PEG-rhMGDF) [4] and recombinant methionyl human tumor necrosis factor binding protein PEGylated dimer [22]. The reason for this increased immunogenicity is not known, but it might be related to the increased plasma half-lives, which leads to prolonged exposure to the immune system.

Impurities and Contaminants.   Recombinant proteins can be produced in different expression systems, each having advantages and disadvantages [16]. Impurities from the expression systems can be of influence on the immunogenicity of the final product. They can act as adjuvants or be immunogenic themselves [11]. Patients receiving *Escherichia coli*-derived granulocyte-macrophage colony-stimulating factor (GM–CSF) were shown to develop antibodies to GM–CSF as well as *E. coli* proteins [23].

Secondary Structure.   Many studies have shown the importance of aggregation in inducing and/or increasing an immune response [24–27]. Aggregation usually occurs after partial unfolding of the protein due to, e.g., shear/shaking or high temperature [28]. Aggregation may expose new epitopes on the surface of the protein for which the immune system is not tolerant. This will lead to a classic immune response. Aggregation can also lead to multimeric antigen presentation, which is known to break B-cell tolerance. Therefore, assessing the presence of aggregates in protein formulations is considered to be very important, although not all aggregates will induce an immune response as was shown in our laboratory [29]. The analytical tools for characterizing the aggregates in a formulation are described elsewhere in this book.

6.4.3.2   Patient Characteristics

Patient characteristics, such as genetic background and the disease status, are known to influence rate and type of immune reactions [11]. Hemophilia patients with severe genetic defects are more prone to antibody formation than patients with minor genetic defects [30, 31]. Eprex (Ortho Biotech, Tilburg, The Netherlands) associated pure red cell aplasia (PRCA) was only observed in patients with renal failure and not in cancer patients [32].

6.4.3.3   Treatment Characteristics

Also the treatment characteristics can have an influence on the immunogenicity of therapeutic proteins. Usually antibodies are only induced after prolonged treatment of the protein. The intramuscular (i.m.) route of administration is less immunogenic than subcutaneous (s.c.) administration. Intravenous (i.v.) administration usually is the least immunogenic route of administration.

6.4.3.4   Other Factors

When evaluating the immunogenicity of therapeutic proteins, one also has to consider the immunomodulatory effects of the protein. For instance, IFNα may increase the antibody production because of its innate immunomodulatory effects [33]. The timing of blood sampling may influence the amount of antibody present in the sera. If a blood sample is taken too early after the administration of the protein, antibodies can still be complexed to the antigen and the sample may prove to be false-negative. If the sample is taken too long after the last injection, antibodies may already have disappeared [34].

6.4.4   DETECTION

Detection of antibodies can be done by several methods (Table 6.4-3). As there are no standardized assays for antibodies to most products and laboratories mostly use their own in-house methods, comparison of results from different studies is in principle impossible. Ideally, a combination of standardized assays should be performed to ensure that all possible types (low affinity and high affinity, classes and isotypes, binding and neutralizing) of antibodies are detected.

6.4.4.1   Assays

Antibody assays can be divided into two main categories: binding assays and bioassays. The binding assays detect in principle any antibodies with sufficient affinity

TABLE 6.4-3. Advantages and Disadvantages of the Most Commonly Used Assays to Detect Binding Anti-Drug Antibodies in Patient Sera[1]

Type of Assay/ Method	Advantages	Disadvantages
ELISA (bridging format; coating with drug and detecting with labeled drug)	• Detection of all isotype responses • Species independent • High throughput • Easy to use • Good sensitivity	• Low-affinity antibodies may be missed • Epitopes may be masked by immobilization • Protein conformation may becompromised by immobilization • Antibody classes/isotypes are not discriminated
ELISA (direct format; coating with drug and detecting with labeled anti-Ig)	• High throughput • Easy to use • Good sensitivity • Isotype detection possible (by using isotype specific conjugates)	• Immobilization • Protein conformation may be compromised by immobilization • Low-affinity antibodies may be missed
ELISA (indirect format; coating with a specific Mab or biotin and then drug)	• Coating plate with specific MAb keeps drug in oriented position • Consistent coating and epitope exposure	• MAb should not alter epitopes' accessibility • MAb epitope should be clinically irrelevant • Isotype detection determined by conjugate • Species specificity determined by conjugate • Low-affinity antibodies may be missed
Radioimmuno (precipitation) assay (RIP/RIA)	• Solution phase, which retains the conformation of the antigen • Moderate-to-high throughput • Generally good sensitivity	• Uses radioactivity • Conjugation chemistry might degrade or alter molecule • May trap labeled drug nonspecifically in precipitate

TABLE 6.4-3. *Continued*

Type of Assay/ Method	Advantages	Disadvantages
Immunoblotting	• Dissects the specificity of the antibodies • Profile of reactivity against subcomponents of the product • Sensitivity is good	• Nonquantitative • Low throughput • Conformational epitopes are missed
Surface plasmon resonance	• Solution phase, which mostly retains the conformation of the antigen • Automated • No detection conjugate required • Not species-specific • Detection of all isotypes • Determination of association and dissociation constants • Enables detection of low-affinity antibodies	• Linking chemistry may affect molecule or mask epitopes • Regeneration step may degrade molecule • Low throughput • Requires dedicated equipment • Sensitivity usually lower than ELISA or RIA (except for low-affinity antibodies)

[1] Adapted from Refs. 35 and 36.

for the therapeutic protein. A bioassay tests whether the antibodies detected by the binding assay can neutralize the biological effects of the therapeutic protein. In most cases, BAbs have no clinical effect, but the binding assay is used as a screening tool to identify samples with possible neutralizing activity.

Binding Assays. Several binding assays are available that can detect antibodies against therapeutic proteins, with solid phase binding assays being the most commonly used for antibody detection. Table 6.4-3 summarizes the advantages and disadvantages of the most commonly used techniques. The advantage of surface plasmon resonance over the other methods is the possibility to measure the binding of the antibodies in real time, which gives information about the affinity of the antibodies. Figure 6.4-2 shows a sensorgram of sera containing antibodies with different association and dissociation profiles. A Western blot gives information about the specificity of the antibodies in the serum and may show antibodies to impurities as was shown in patients receiving GM–CSF [23]. For a more detailed description of these methods, we refer to specialized literature [35–37].

Neutralizing Assays. The detection of neutralizing antibodies is performed in assays that can show inhibition of the biological activity of the therapeutic protein. In general this will be a bioassay. The therapeutic protein is incubated with the serum to be tested, and the mixture is tested in the bioassay. If the antibodies in the serum have neutralizing capacities, the biological activity of the mixture is reduced, as compared with the pure therapeutic protein.

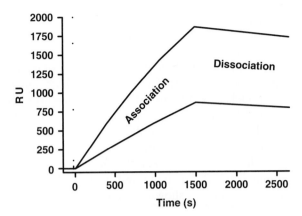

Figure 6.4-2. Sensorgram obtained with a BIAcore 3000 using anti-rhIFNα2b sera with different association and dissociation profiles.

6.4.4.2   The Design of an Assay Strategy

Long-term studies are needed to fully assess the immunogenicity of a product. Also, the timing of sampling is important. A pretreatment blood sample should be taken, and several time points should be considered because antibodies may be induced only after prolonged treatment. If a sample is taken too early after administration of the therapeutic protein, the circulating protein may interfere with the antibody assay [9]. Conversely, if the sampling is too long after the last administration of the protein, antibody levels might already have dropped below the detection limit of the assays. Both situations will lead to an underestimation of the incidence of antibody positive patients.

To fully assess the immunogenicity of a therapeutic protein, a combination of assays is necessary. In general, sera are screened first by an enzyme-linked immunosorbent assay (ELISA) or radioimmunoassay (RIA) type of binding assay, which have a high through-put and are easy to perform. A screening assay is optimized for sensitivity and therefore suffers from a relative high number of false-positive results. So, all initial positive sera should be confirmed, e.g., by showing a reduced binding after adding the product to the serum. The confirmed positive sera are then tested for neutralizing activity. Further characterization of the antibody response may follow to determine isotype, affinity, and so on. Whatever approach is used, it is important to validate the assays to show reproducibility, precision, robustness, and so on [35, 38]. International reference preparations are only available for a limited number of assays, but they are crucial for making comparison between different laboratories possible [39].

6.4.4.3   Predictive Models

Most of the data about immunogenicity of therapeutic proteins come from clinical trials or postmarketing surveillance. It would of course be better if the immunogenicity of a product could be predicted before the clinical phase of development.

Conventional Animals. Conventional animal models may be useful to predict the immunogenicity of microbial products as staphylokinase and streptokinase, which are foreign proteins for humans as well as animals [40]. Conventional animals can also be used if the therapeutic protein shows a high degree of intraspecies sequence homology. An example is human thrombopoietin, which induced neutralizing antibodies in animal models as well as in patients [41]. However, most recombinant human therapeutic proteins are foreign proteins for animals that will elicit an immune response in all cases. So their predictive value for the immunogenicity of human proteins in humans is low. Nonhuman primates that share a higher level of sequence homology with humans have been shown to be excellent models for some products such as human growth hormone and lys-pro insulin [42], but for other proteins, their predictive value has shown to be limited.

Immune Tolerant Animals. In most cases, the induction of antibodies to human therapeutic proteins in patients is based on breaking B-cell tolerance. This process makes animals tolerant for the protein the best model for predicting immunogenicity in humans. Mice are the species of choice to induce tolerance, which can be achieved by chronic administration of the protein in large quantities [43] or by making the animals transgenic for the gene expressing the protein [44]. Obvious drawbacks of the first method are the large quantities of protein required and the need for testing the tolerance of each animal. Mouse lines expressing the transgene only need to be evaluated for immune tolerance once and only need testing by polymerase chain reaction (PCR) to identify transgene positivity after breeding. Transgenic animals, immune tolerant for insulin [45], tissue plasminogen activator (tPA) [46], IFNα2 [26], and IFNβ [47] are available. The insulin and tPA models were used to evaluate whether amino acid substitutions in the protein introduced new epitopes [43, 44]. The interferon models were used to study the effect of aggregates and other degradation products on immunogenicity [26, 29, 47].

Some therapeutic proteins might have immune modulatory effects that influence the induction of antibodies. If the human protein lacks this effect in mice, their predictive value may suffer. This deficiency can be prevented by challenging both the human as well as the murine homologue. Braun et al. showed that mixtures of murine and human IFNα2a caused an increase in immune response in wild-type mice to the human protein [26]. Another point of consideration when using transgenic mice as predictors for immunogenicity is the difference between mouse and human MHC and their presentation of T-cell epitopes [39, 41]. Also, important individual patient characteristics, such as disease burden and concomitant therapy influencing the antibody response, are difficult to reproduce in the transgenic immune-tolerant animal model. Although these models will never become predictors of the individual patient's antibody response, they will be very useful in the development phase of a new protein therapy or after a formulation or production change. The transgenic immune-tolerant mice can also be used to study the mechanism of antibody induction, because they share tolerance for the protein with patients.

6.4.4.4   Prediction on Basis of Structure and Sequence

Other strategies, based on structural analysis, have been described to predict the immunogenicity of therapeutic proteins. Immunogenic epitopes can be predicted based on amino acid sequence analysis and MHC-binding. Modifications in an immuno-dominant T-cell epitope by a single amino acid was reported to reduce the immunogenicity of rhIFNβ-1b in BALB/cByJ mice [48]. Analysis of B-cell epitopes is more difficult, but it is also ongoing [49]. These types of analyses may help to reduce the immunogenicity, in case the product is a foreign protein inducing a classic immune response. As discussed, in most of cases, the induction of antibodies in patients is based on breaking B-cell tolerance. Although the exact mechanism is still unknown, we know that the product quality, independent of the amino acid sequence, plays an important role.

6.4.5 BIOLOGICAL AND CLINICAL CONSEQUENCES

The clinical effects of antibody formation vary from no effect to severe life-threatening situations [50]. Binding antibodies can alter the pharmacokinetic properties of the therapeutic protein [51, 52]. Neutralizing antibodies cannot only affect the pharmacokinetics but also block the pharmacological effect of the therapeutic protein [51–53]. Severe side effects occur if NAbs cross-react with an endogenous protein, as was observed for PEG–rhMGDF and recombinant human erythropoietin (epoetin). Patients receiving PEG–rhMGDF developed NAbs cross-reacting with endogenous thrombopoietin, which resulted in a severe thrombocytopenia [4]. Patients receiving Eprex (an epoetin formulation marketed in Europe, Canada, and Australia) developed NAbs that cross-reacted with endogenous erythropoietin, resulting in PRCA [3]. The number of patients with antibody-mediated PRCA rapidly increased after a formulation change of Eprex, as will be discussed below. Besides the formation of antibodies, allergic reactions can occur.

6.4.6 A RISK-BASED APPROACH OF IMMUNOGENICITY

The FDA evaluates immunogenicity concerns of novel products in development and for major changes in manufacturing or clinical uses via a risk-based approach. A schematic representation of this approach is given in Figure 6.4-3. Three elements are considered important for the risk assessment strategy: the severity of consequences of the immune response to a therapeutic protein, host-specific factors that impact the immunogenicity positively or negatively, and product-specific factors that impact the immunogenicity positively or negatively. Based on their expertise and literature data from marketed products, the FDA decides whether immunogenicity testing should be performed before or during clinical trials.

6.4.7 EXAMPLES

6.4.7.1 Epoetin: How a Formulation Change Increased the Immunogenicity

Erythropoietin is a protein produced by the kidneys that stimulates the production of erythrocytes. Epoetin is mainly used in patients suffering from anemia associated with chronic renal failure or cancer. Three forms of epoetin are commercially

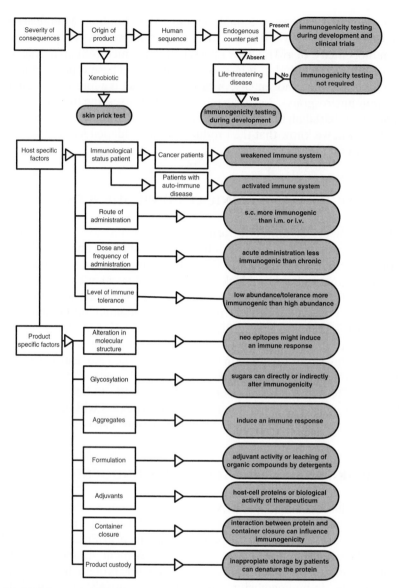

Figure 6.4-3. Schematic representation of the risk-based approach that the FDA uses for novel products in development and for major changes in manufacture and clinical use of licensed products. Adapted from Refs. 54 and 55.

available, epoetin alfa, epoetin beta, and darbepoetin alfa, which is a hyperglycosylated analog [32]. In 1997 the first patient with antibody-mediated PRCA was reported [56]. Antibody-mediated PRCA is caused by antibodies induced by exogenous epoetin that cross-react with endogenous erythropoietin. In 2002, 13 patients were reported that had developed antibody-mediated PRCA, after administration of Eprex [3]. These cases occurred after 1998. Since then, more patients with antibody-mediated PRCA were identified. In 1998 the formulation of Eprex was changed: Human serum albumin (HSA) was replaced by glycine and polysorbate 80. Several hypotheses have been postulated to explain the increase in PRCA cases. Hermeling et al. hypothesized that some epoetin molecules were solubilized in micelles, which might have led to multimeric antigen presentation [57]. Others claim that the polysorbate 80 extracted leachates out of the uncoated rubber stoppers, which had an adjuvant effect, leading to an increase in antibodies [58]. An adjuvant, however, can only increase an existing immune response and cannot break B-cell tolerance by itself. It will be very diffcult to pinpoint the exact cause of the increased number of PRCA cases. In 2002, actions were taken to decrease the number of PRCA cases: emphasizing strict adherence to storage and handling conditions, the introduction of a contraindication for s.c. administration of Eprex in chronic renal failure in many countries and the introduction of coated rubber stoppers. These actions decreased the number of patients with antibody-mediated PRCA, but the question remains of what exactly caused the increased immunogenicity.

6.4.7.2 RhIFNβ: Immunogenicity of Glycosylated and Nonglycosylated Variants

Interferon beta is a cytokine with anti-inflammatory, antitumor, antiviral, and cell-growth-regulatory effects. It is mainly produced by macrophages and epithelial and fibroblast cells [59, 60]. Administration of rhIFNβ has been established as a treatment for relapsing-remitting multiple sclerosis [60]. Natural hIFNβ is glycosylated, contains 166 amino acids, and has a molecular weight of approximately 25 kDa [38]. Three forms of rhIFNβ are available on the market (Table 6.4-4). RhIFNβ-1a (Rebif and Avonex) has an amino acid sequence similar to that of endogenous hIFNβ, is produced in Chinese hamster cells, and is glycosylated. RhIFNβ-1b (Betaseron) is produced in *E. coli* and thus not glycosylated. Moreover, Cys-17 has been mutated to Ser-17 and the N-terminal methionine is lacking.

All three formulations induce neutralizing antibodies of the IgG-type, reducing the efficacy of the therapy [24, 61–63]. The antibodies are usually detectable after the first 6 to 12 months of the treatment [61, 63], but the clinical effects do not appear until after 18–24 months after the start of the treatment [64]. RhIFNβ-1b induces antibodies in more patients than does rhIFNβ-1a [59, 65]. This increased incidence of immunogenicity of rhIFNβ-1b as compared with rhIFNβ-1a is probably due to the lack of glycosylation, which makes the protein more prone to aggregation. Size-exclusion chromatography showed that 60% of rhIFNβ-1b in Betaseron is heavily aggregated [15]. It was shown that Betaseron can break the tolerance of transgenic immune tolerant mice [47], which suggests that these animals may be predictive for the immunogenicity observed in patients.

TABLE 6.4-4. Currently Marketed rhIFNβ Formulations

rhIFNβ-1a	rhIFNβ-1b[1]
Avonex (Biogen-Idec, Cambridge, MA)	Betaseron (Berlex Laboratories, Montville, NJ)
Rebif (Serono, Rockland, MA)	Betaferon (Schering, Berlin, Germany)

[1] Betaseron and Betaferon are the same formulations, marketed by two companies.

6.4.7.3 RhIFNα2: Reduction of the Immunogenicity by Optimizing Production and Formulation

RhIFNα2 is used for the treatment of a variety of malignancies and viral diseases [66]. It inhibits viral replication, increases class I MHC expression, stimulates Th1 cells, and inhibits proliferation of many cell types [2]. Two regular and two PEGylated forms of rhIFNα2 are on the market (Table 6.4-5). The PEGylated forms are less immunogenic than the non-PEGylated forms. RhIFNα2a and rhIFNα2b only differ in 1 amino acid. RhIFNα2b is produced in *E. coli* cells, which implies that the protein is nonglycosylated. Natural hIFNα2 is O-glycosylated [67]. Patients receiving rhIFNα2a initially produced higher levels of antibodies and at higher incidences than rhIFNα2b. This difference was shown not to be related to the difference in amino acid sequence. The antibodies induced by rhIFNα2a fully cross-react with rhIFNα2b and vice versa [68]. Co-IFNα is a non-naturally occurring synthetic recombinant type I interferon. Its amino acid sequence (166 amino acids) is created by taking for each position the amino acid most commonly observed in 13 IFNα subtypes. Despite the nonsimilarity of co-IFNα with human IFNα species (10–23 amino acid differences and on average 89% homology with naturally occurring IFNαs) [69], the number of patients producing antibodies against co-IFNα is similar to the number of patients producing antibodies after rhIFNα2b treatment [70].

Table 6.4-6 shows that optimization of the production and formulation process decreases the immunogenicity [34]. Storing the formulation at room temperature

TABLE 6.4-5. Currently Marketed rhIFNα2 Formulations

rhIFNα2a	rhIFNα2b
Roferon (Roche, Woerden, The Netherlands)	Intron A (Schering-Plough, Kenilworth, NJ)
Pegasys (PEGylated; Roche, Woerden, The Netherlands)	PEG-Intron A (PEGylated; Schering-Plough, Kenilworth, NJ)

TABLE 6.4-6. Immunogenicity of rhIFNα2a Formulations After Production and Formulation Optimization[1]

	Patients with NAbs	Peak Titers (INUs[2])
A[3] HSA-containing, lyophilized formulation stored at 25°C	24% (n = 190)	435–113,100
B HSA-containing, lyophilized formulation stored at 4°C	29% (n = 86)	400–19,200
C HSA-containing, liquid formulation stored at 4°C	13% (n = 110)	400–19,200
D HSA-free liquid formulation stored at 4°C	11% (n = 81)	588–3191
E HSA-containing lyophilized formulation stored at 4°C	12% (n = 74)	672–2441

[1] Adapted from Ref. 34.
[2] INUs: IFN neutralizing units.
[3] Formulations A–C were made before the production process was fully optimized and therefore also contained oxidized rhIFNα2a [72]. Formulations D and E contained ultra-pure material.

leads to oxidation and increases the immunogenicity of a rhIFNα2a formulation (formulation A vs. formulation B). Better purification methods further reduce the immunogenicity (formulation C vs. formulations D and E). Formulations A–C were shown to contain aggregates responsible for the increased immune response [26].

Recently it has been shown that not all aggregates, but only aggregates of rhIFNα2b with a native-like structure, can elicit an immune response in transgenic mice immune tolerant for hIFNα2 [71]. It was also shown that not the oxidation, but the aggregation, accompanying the oxidation was the reason for the immunogenicity.

6.4.8 CONCLUSIONS

Many factors influence the immune response against therapeutic proteins. Unfortunately, it is still impossible to fully predict the immunogenicity of a therapeutic protein before going into clinical trials. The presence of (native-like) aggregates in a formulation is one main factor known to increase the immune response. Formulation changes might have an effect on the immunogenicity as was observed for epoetin and rhIFNα2a. The effect of a formulation change is also difficult to predict, but it can be evaluated in immune-tolerant transgenic mice. Although these tests will not fully predict the immunogenicity of a product, it can give information about the immunogenicity of a new formulation/product as compared with previous formulations/products. As most therapeutic proteins only induce antibodies in a small number of patients, postmarketing surveillance is important. Also basic studies to link physical–chemical properties with immunogenicity are important, and these studies may avoid animal testing in the future.

REFERENCES

1. Porter S (2001). Human immune response to recombinant human proteins. *J. Pharm. Sci.* 90:1–11.

2. Frost H (2005). Antibody-mediated side effects of recombinant proteins. *Toxicol.* 209:155–160.

3. Casadevall N, Nataf J, Viron B, et al. (2002). Pure red-cell aplasia and antierythropoietin antibodies in patients treated with recombinant erythropoietin. *N. Engl. J. Med.* 346:469–475.

4. Li J, Yang C, Xia Y, et al. (2001). Thrombocytopenia caused by the development of antibodies to thrombopoietin. *Blood.* 98:3241–3248.

5. Schellekens H (2005). Factors influencing the immunogenicity of therapeutic proteins. *Nephrol. Dial. Transplant* 20(Suppl 6):vi3–vi9.

6. Chackerian B, Lowy D R, Schiller J T (2001). Conjugation of a self-antigen to papillomavirus-like particles allows for efficient induction of protective autoantibodies. *J. Clin. Invest.* 108:415–423.

7. Fehr T, Bachmann M F, Bucher E, et al. (1997). Role of repetitive antigen patterns for induction of antibodies against antibodies. *J. Exp. Med.* 185:1785–1792.

8. Bachmann M F, Rohrer U H, Kundig T M, et al. (1993). The influence of antigen organization on B cell responsiveness. *Science.* 262:1448–1451.

9. Hermeling S, Crommelin D J A, Schellekens H, et al. (2004). Structure-immunogenicity relationships of therapeutic proteins. *Pharm. Res.* 21:897–903.

10. Schellekens H (2002). Bioequivalence and the immunogenicity of biopharmaceuticals. *Nat. Rev. Drug. Discov.* 1:457–462.

11. Chirino A J, Mire-Sluis A (2004). Characterizing biological products and assessing comparability following manufacturing changes. *Nat. Biotechnol.* 22:1383–1391.

12. Lis H, Sharon N (1993). Protein glycosylation. Structural and functional aspects. *Eur. J. Biochem.* 218:1–27.

13. Skibeli V, Nissen-Lie G, Torjesen P (2001). Sugar profiling proves that human serum erythropoietin differs from recombinant human erythropoietin. *Blood.* 98:3626–3634.

14. Sinclair A M, Elliott S (2005). Glycoengineering: The effect of glycosylation on the properties of therapeutic proteins. *J. Pharm. Sci.* 94:1626–1635.

15. Runkel L, Meier W, Pepinsky R B, et al. (1998). Structural and functional differences between glycosylated and non-glycosylated forms of human interferon-beta (IFN-beta). *Pharm. Res.* 15:641–649.

16. Chen M, Liu X, Wang Z, et al. (2005). Modification of plant N-glycans processing: the future of producing therapeutic protein by transgenic plants. *Med. Res. Rev.* 25:343–360.

17. Chaffee S, Mary A, Stiehm E R, et al. (1992). IgG antibody response to polyethylene glycol-modified adenosine deaminase in patients with adenosine deaminase deficiency. *J. Clin. Invest.* 89:1643–1651.

18. Hinds K D, Kim S W (2002). Effects of PEG conjugation on insulin properties. *Adv. Drug Deliv. Rev.* 54:505–530.

19. Chapman A P (2002). PEGylated antibodies and antibody fragments for improved therapy: A review. *Adv. Drug Deliv. Rev.* 54:531–545.

20. Chirino A J, Ary M L, Marshall S A (2004). Minimizing the immunogenicity of protein therapeutics. *Drug Discov. Today* 9:82–90.

21. Caliceti P, Veronese F M (2003). Pharmacokinetic and biodistribution properties of poly(ethylene glycol)-protein conjugates. *Adv. Drug Deliv. Rev.* 55:1261–1277.

22. Moreland L W, McCabe D P, Caldwell J R, et al. (2000). Phase I/II trial of recombinant methionyl human tumor necrosis factor binding protein PEGylated dimer in patients with active refractory rheumatoid arthritis. *J. Rheumatol.* 27:601–609.

23. Wadhwa M, Skog A L, Bird C, et al. (1999). Immunogenicity of granulocyte-macrophage colony-stimulating factor (GM-CSF) products in patients undergoing combination therapy with GM-CSF. *Clin. Cancer Res.* 5:1353–1361.

24. Bertolotto A, Malucchi S, Sala A, et al. (2002). Differential effects of three interferon betas on neutralising antibodies in patients with multiple sclerosis: a follow up study in an independent laboratory. *J. Neurol. Neurosurg. Psychiatry* 73:148–153.

25. Bertolotto A, Malucchi S, Milano E, et al. (2000). Interferon beta neutralizing antibodies in multiple sclerosis: neutralizing activity and cross-reactivity with three different preparations. *Immunopharmacol.* 48:95–100.

26. Braun A, Kwee L, Labow M A, et al. (1997). Protein aggregates seem to play a key role among the parameters influencing the antigenicity of interferon alpha (IFN-alpha) in normal and transgenic mice. *Pharm. Res.* 14:1472–1478.

27. Moore W V, Leppert P (1980). Role of aggregated human growth hormone (hGH) in development of antibodies to hGH. *J. Clin. Endocrinol. Metab.* 51:691–697.

28. Wang W (2005). Protein aggregation and its inhibition in biopharmaceutics. *Int. J. Pharm.* 289:1–30.

29. Hermeling S (2005). Structural Aspects of the Immunogenicity of Therapeutic Proteins; Transgenic Animals as Predictors for Breaking Immune Tolerance. Utrecht University, Utrecht, The Netherlands, pp. 208.

30. Prescott R, Nakai H, Saenko E L, et al. (1997). The inhibitor antibody response is more complex in hemophilia A patients than in most nonhemophiliacs with factor VIII autoantibodies. Recombinate and Kogenate Study Groups. *Blood*. 89:3663–3671.

31. Gilles J G, Jacquemin M G, Saint-Remy J M (1997). Factor VIII inhibitors. *Thromb. Haemost*. 78:641–646.

32. Schellekens H (2005). Immunologic mechanisms of EPO-associated pure red cell aplasia. *Best Pract. Res. Clin. Haematol*. 18:473–480.

33. Palleroni A V, Aglione A, Labow M, et al. (1997). Interferon immunogenicity: Preclinical evaluation of interferon-alpha 2a. *J. Interferon Cytokine Res*. 17(Suppl 1): S23–27.

34. Ryff J C (1997). Clinical investigation of the immunogenicity of interferon-alpha 2a. *J. Interferon Cytokine Res*. 17(Suppl 1):S29–33.

35. Mire-Sluis A R, Barrett Y C, Devanarayan V, et al. (2004). Recommendations for the design and optimization of immunoassays used in the detection of host antibodies against biotechnology products. *J. Immunol. Methods*. 289:1–16.

36. Wadhwa M, Bird C, Dilger P, et al. (2003). Strategies for detection, measurement and characterization of unwanted antibodies induced by therapeutic biologicals. *J. Immunol. Methods*. 278:1–17.

37. Kersten G F, Westdijk J (2005). Immunoassays. In W Jiskoot and D J A Crommelin (eds.), *Methods for Structural Analysis of Protein Pharmaeuticals*, Vol. III, Biotechnology: Pharmaceutical Aspects, AAPS Press, Arlington, V.A., pp. 501–526.

38. Brickelmaier M, Hochman P S, Baciu R, et al. (1999). ELISA methods for the analysis of antibody responses induced in multiple sclerosis patients treated with recombinant interferon-beta. *J. Immunol. Methods*. 227:121–135.

39. Schellekens H (2002). Immunogenicity of therapeutic proteins: Clinical implications and future prospects. *Clin. Ther*. 24:1720–1740.

40. Bugelski P J, Treacy G (2004). Predictive power of preclinical studies in animals for the immunogenicity of recombinant therapeutic proteins in humans. *Curr. Opin. Mol. Ther*. 6:10–16.

41. Rosenberg A S, Worobec A (2005). A risk-based approach to immunogenicity concerns of therapeutic protein products, Part 3: Effects of manufacturing changes in imuunogenicity and the utility of animal immunogenicity studies. *Biopharm. Int*. Jan. 1, 2005:1–3.

42. Wierda D, Smith H W, Zwickl C M (2001). Immunogenicity of biopharmaceuticals in laboratory animals. *Toxicol*. 158:71–74.

43. Simioni P U, Fernandes L G, Gabriel D L, et al. (2004). Induction of systemic tolerance in normal but not in transgenic mice through continuous feeding of ovalbumin. *Scand. J. Immunol*. 60:257–266.

44. Zhang J, Xu L, Haskins M E, et al. (2004). Neonatal gene transfer with a retroviral vector results in tolerance to human factor IX in mice and dogs. *Blood*. 103:143–151.

45. Ottesen J L, Nilsson P, Jami J, et al. (1994). The potential immunogenicity of human insulin and insulin analogues evaluated in a transgenic mouse model. *Diabetologia*. 37:1178–1185.

46. Stewart T A, Hollingshead P G, Pitts S L, et al. (1989). Transgenic mice as a model to test the immunogenicity of proteins altered by site-specific mutagenesis. *Mol. Biol. Med*. 6:275–281.

47. Hermeling S, Jiskoot W, Crommelin D J A, et al. (2005). Development of a transgenic mouse model immune tolerant for human interferon beta. *Pharm. Res.* 22:847–851.

48. Yeung V P, Chang J, Miller J, et al. (2004). Elimination of an immunodominant CD4+ T cell epitope in human IFN-beta does not result in an in vivo response directed at the subdominant epitope. *J. Immunol.* 172:6658–6665.

49. Saha S, Bhasin M, Raghava G P (2005). Bcipep: A database of B-cell epitopes. *BMC Genomics.* 6:79.

50. Koren E, Zuckerman L A, Mire-Sluis A R (2002). Immune responses to therapeutic proteins in humans-clinical significance, assessment and prediction. *Curr. Pharm. Biotechnol.* 3:349–360.

51. Antonelli G (1997). In vivo development of antibody to interferons: an update to 1996. *J. Interferon Cytokine Res.* 17(Suppl 1):S39–46.

52. Thorpe R, Swanson S J (2005). Current methods for detecting antibodies against erythropoietin and other recombinant proteins. *Clin. Diagn. Lab Immunol.* 12:28–39.

53. Quesada J R, Rios A, Swanson D, et al. (1985). Antitumor activity of recombinant-derived interferon alpha in metastatic renal cell carcinoma. *J. Clin. Oncol.* 3:1522–1528.

54. Rosenberg A S, Worobec A (2004). A risk-based approach to immunogenicity concerns of therapeutic protein products, Part 1: Considering consequences of the immune response to a protein. *Biopharm. Int.* Nov. 1, 2004:1–4.

55. Rosenberg A S, Worobec A (2004). A risk-based approach to immunogenicity concerns of therapeutic protein products, Part 2: Considering host-specific and product-specific factors impacting immunogenicity. *Biopharm. Int.* Dec. 1, 2004:1–7.

56. Prabhakar S S, Muhlfelder T (1997). Antibodies to recombinant human erythropoietin causing pure red cell aplasia. *Clin. Nephrol.* 47:331–335.

57. Hermeling S, Schellekens H, Crommelin D J A, et al. (2003). Micelle-associated protein in epoetin formulations: A risk factor for immunogenicity? *Pharm. Res.* 20:1903–1907.

58. Sharma B, Bader F, Templeman T, et al. (2004). Technical investigations into the cause of the increased incidence of antibody-mediated pure-red cell aplasia associated with Eprex®. *Eur. J. Hosp. Pharm.* 5:86–91.

59. Antonelli G, Dianzani F (1999). Development of antibodies to interferon beta in patients: technical and biological aspects. *Eur. Cytokine Netw.* 10:413–422.

60. Mayorga C, Luque G, Romero F, et al. (1999). Antibodies to commercially available interferon-beta molecules in multiple sclerosis patients treated with natural interferon-beta. *Int. Arch. Allergy Immunol.* 118:368–371.

61. Malucchi S, Sala A, Gilli F, et al. (2004). Neutralizing antibodies reduce the efficacy of betaIFN during treatment of multiple sclerosis. *Neurol.* 62:2031–2037.

62. Sorensen P S, Ross C, Clemmesen K M, et al. (2003). Clinical importance of neutralising antibodies against interferon beta in patients with relapsing-remitting multiple sclerosis. *Lancet.* 362:1184–1191.

63. Pachner A R (2003). Anti-IFNbeta antibodies in IFNbeta treated MS patients. *Neurol.* 61:S1–S5.

64. Bertolotto A (2004). Neutralizing antibodies to interferon beta: implications for the management of multiple sclerosis. *Curr. Opin. Neurol.* 17:241–246.

65. Cook S D, Quinless J R, Jotkowitz A, et al. (2001). Serum IFN neutralizing antibodies and neopterin levels in a cross-section of MS patients. *Neurol.* 57:1080–1084.

66. Pfeffer L M, Dinarello C A, Herberman R B, et al. (1998). Biological properties of recombinant alpha-interferons: 40th anniversary of the discovery of interferons. *Cancer Res.* 58:2489–2499.

67. Adolf G R, Kalsner I, Ahorn H, et al. (1991). Natural human interferon-alpha 2 is O-glycosylated. *Biochem. J.* 276 (Pt 2):511–518.

68. Antonelli G, Currenti M, Turriziani O, et al. (1991). Neutralizing antibodies to interferon-alpha: relative frequency in patients treated with different interferon preparations. *J. Infect. Dis.* 163:882–885.

69. Intermune. *Infergen (Interferon alfacon-1)*. Available: http://www.infergen.com/pdf/infergen_pi.pdf. Accessed July 26, 2005.

70. Tong M J, Reddy K R, Lee W M, et al. (1997). Treatment of chronic hepatitis C with consensus interferon: A multicenter, randomized, controlled trial. Consensus Interferon Study Group. *Hepatol.* 26:747–754.

71. Hermeling S, Aranha L, Damen J M A, et al. (2005). Structural characterization and immunogenicity in wildtype and immune tolerant mice of degraded recombinant human interferon alpha2b. *Pharm. Res.* In Press.

72. Hochuli E (1997). Interferon immunogenicity: technical evaluation of interferon-alpha 2a. *J. Interferon Cytokine Res.* 17(Suppl 1):S15–21.

6.5

DEVELOPMENT AND CHARACTERIZATION OF HIGH-AFFINITY ANTI-TOPOTECAN IgG AND FAB FRAGMENTS

JIN CHEN AND JOSEPH P. BALTHASAR

University at Buffalo, The State University of New York, Buffalo, New York

Chapter Contents

6.5.1 Introduction 835

6.5.2 Materials and Methods 836

    6.5.2.1 Development of Hybridomas Secreting Anti-TPT Antibodies 836

    6.5.2.2 Characterization of Anti-Topotecan IgG (ATI) 837

    6.5.2.3 Production and Purification of ATI and ATF 838

    6.5.2.4 Development of ELISA for Quantification and ATF in Rat Plasma 839

    6.5.2.5 Pharmacokinetic Studies of ATI and ATF in Rats 839

6.5.3 Results 840

    6.5.3.1 Development of Hybridomas Secreting Anti-TPT Antibodies 840

    6.5.3.2 Characterization of ATI 840

    6.5.3.3 Production and Purification of ATI and ATF 842

    6.5.3.4 ELISA Validation 845

    6.5.3.5 Pharmacokinetics of ATI and ATF 846

6.5.4 Discussion 847

    References 849

6.5.1 INTRODUCTION

This laboratory is investigating a drug targeting strategy that employs anti-drug antibodies to increase drug binding in blood, minimize systemic unbound drug exposure, and reduce the systemic drug toxicity after intraperitoneal (i.p.) chemotherapy [1, 2]. The approach uses i.p. administration of an anticancer drug with

Handbook of Pharmaceutical Biotechnology, Edited by Shayne Cox Gad.

simultaneous intravenous (i.v.) administration of anti-drug antibodies, where the antibodies are employed to impart regio-specific alterations in drug disposition.

To test the feasibility of this approach, earlier studies in this laboratory used methotrexate (MTX) as a model drug [2, 3]. Pharmacokinetic investigations demonstrated that systemic administration of polyclonal anti-MTX Fab with i.p. administration of MTX led to a decrease in the systemic exposure of unbound MTX, while not altering peritoneal exposure [2]. To facilitate further evaluation of anti-MTX antibodies on MTX induced toxicity, monoclonal anti-MTX IgG and Fab were produced and purified [4]. Subsequent studies demonstrated that after this combination therapy, anti-MTX Fab could allow a fivefold increase of the maximum tolerated dose of i.p. MTX and enhance the median survival time of mice bearing peritoneal tumors [5].

Based on these promising results, we have proposed to apply this targeting strategy to enhance the selectivity of i.p. topotecan (TPT) chemotherapy, as TPT has been shown to be the most active second-line therapy for refractory, metastatic ovarian cancer [6–9]. To facilitate testing hypotheses related to i.p. TPT chemotherapy, it was necessary to develop anti-TPT antibodies. Relative to polyclonal antibodies, monoclonal antibodies would demonstrate more reproducible properties (i.e., affinity and specificity) and may be more feasibly produced in large quantities. In this chapter, we describe the development of hybridomas secreting monoclonal antibodies against TPT, and the production, purification, characterization, and pharmacokinetics of an anti-TPT monoclonal antibody and of monoclonal anti-TPT Fab fragments.

6.5.2   MATERIALS AND METHODS

6.5.2.1   Development of Hybridomas Secreting Anti-TPT Antibodies

Preparation and Characterization of KLH-TPT Immunogen.   Topotecan was provided by Drug Synthesis & Chemistry Branch, Developmental Therapeutics Program, Division of Cancer Treatment and Diagnosis, National Cancer Institute, National Institutes of Health (Bethesda, MD). TPT was conjugated to keyhole limpet hemocyanin (KLH) via the Mannich reaction using Pierce PharmLink Immunogen Kit (Pierce, Rockford, IL). Briefly, 0.9-mg TPT was linked with 2-mg KLH by condensation with formaldehyde. The reaction was incubated at 50°C overnight, followed by dialysis against PBS (pH 7.4) for 2 h, and then at 4°C overnight. The synthesized immunogen was further diluted by PBS to a final concentration of 1 mg/mL. The degree of conjugation was estimated by ultraviolet (UV) spectroscopy at 411 nm for topotecan and at 280 nm for KLH.

Development of Hybridomas that Secrete Anti-TPT Antibodies.   Approximately 100-μg KLH–TPT was dissolved in PBS and emulsified with an equal volume of Freund's incomplete adjuvant (Sigma, St. Louis, MO). Female Balb/c mice (Harlan, Indianapolis, IN) were immunized with 200-μL emulsion by i.p. injection every 3 weeks. Animals were bled via the saphenous vein 7 days after each dose of KLH–TPT. Whole blood was collected and centrifuged. The resulting plasma was used to assess anti-TPT activity via a previously validated HPLC assay [10]. Briefly, plasma was diluted with PBS and incubated with 250-ng/mL TPT for 2 h. Free TPT was then separated from bound TPT through ultra-filtration using Centrifree tubes (10-KD molecular-weight cutoff; Bellerica, MA). The high-performance liquid chromatography (HPLC) assay employed an isocratic mobile phase of a mixture of 10-mM KH_2PO_4 water–methanol–triethylamine (72:26:2, v/v/v, pH 3.5), with fluorescence detection at an excitation wavelength of 361 nm and an emission wavelength of 527 nm. Blank mouse plasma was used as the control. The mouse with the highest activity (i.e., the lowest ratio of free TPT to that of control) was sacrificed, and the spleen was aseptically removed.

Spleen cells were obtained by teasing the spleen apart with forceps and by passing tissue fragments through a stainless steel screen. The spleen cells were washed in RPMI media (Invitrogen, Grand Island, NY) twice by centrifugation at 500 g in tabletop centrifuge (International Equipment Co., Needham, MS). Myeloma cells, SP2/0-Ag 14 (ATCC, Manassas, VA), were washed once in RPMI. Spleen cells were mixed with myeloma cells and centrifuged. The cell pellets were carefully suspended in 1-mL 50% poly(ethylene glycol) (PEG, Sigma) for 1 min by gently swirling in 37°C water. Cells were diluted and washed again in RPMI media by centrifugation and then resuspended in RPMI HAT (hypoxanthine, aminopterin, and thymidine) medium (Invitrogen) with 10% fetal bovine serum (Invitrogen). Cells suspended in HAT medium were dispensed into the wells of several 96-well plates, which were then incubated at 37°C. Clones, which became visible in 2–3 weeks, were then transferred into the wells of 24-well plates.

Screening of Anti-Topotecan Activity by HPLC Assay. Anti-TPT activities within cell culture supernatants were assessed via HPLC, using similar methods as described above. Initial screening was conducted by pooling six hybridoma supernatants together, and the pooled culture supernatants (195 μL) were incubated with 5 μL of TPT solution (in saline) to a final concentration of 25 ng/mL for 2 h. Unbound TPT was separated from bound TPT through ultra-filtration and then quantified by HPLC. The lower limit of quantitation (LOQ) for TPT was 0.02 ng on column (corresponding to 1 ng/mL in 20-μL samples). Blank RPMI HAT media spiked with TPT was used as a control. After demonstration of binding from pooled supernatants, the individual wells from that pool were further tested as described above.

6.5.2.2   Characterization of Anti-Topotecan IgG (ATI)

Isotype and Binding Affinity/Capacity of ATI. The isotypes of secreted antibodies were determined through the use of a mouse monoclonal antibody isotyping kit (Cell Sciences, Norwood, MA). An anti-TPT IgG1 antibody (ATI), secreted by hybridoma 8C2, was selected for further characterization. The 8C2 supernatant (with a constant ATI con-centration) was incubated with increasing concentrations of TPT (100, 125, 150, 175, 200, 225, and 250 ng/mL) for 2 h to allow binding between TPT and ATI. Free TPT was then separated from antibody-bound TPT by ultra-filtration and assayed via HPLC.

The concentration of bound TPT was calculated by subtracting the measured unbound TPT concentration from the known total TPT concentration. TPT–ATI binding was characterized through the construction of Woolf and Rosenthal binding plots, according to the relationships:

$$\frac{F}{B} = \frac{1}{K_A \times nPt} + \frac{F}{nPt} \text{ and } \frac{B}{F} = nPt \times K_A - K_B \times B$$

respectively. In each equation, B represents the molar concentration of TPT bound to ATI, F is the molar concentration of unbound TPT, K_A is the equilibrium association constant (i.e., a measure of binding affinity), and nPt is the total antibody binding capacity. Initial estimates for K_A were determined as the inverse of the intercept times the slope of the Woolf plot or as the negative slope of the Rosenthal plot. The TPT binding capacity (nPt) was calculated as the inverse of the slope of the Woolf plot or as the negative ratio of intercept over slope of the Rosenthal plot. Final parameter estimates were determined by fitting the parameters of the equation:

$$B = \frac{nPt \times F}{1/K_A + F}$$

to the unbound and bound TPT concentration data using WinNonLin v.2.1 software (Pharsight, CA).

6.5.2.3   Production and Purification of ATI and ATF

Production of ATI. The hybridoma cells secreting ATI were grown in serum-free medium supplemented with 0.5% gentamicin (Hybridoma SFM, Invitrogen). Large quantities of ATI were produced using a Unisyn Cell Pharm hollow fiber bioreactor ($10\,ft^2$, 10-KDa molecular-weight cutoff, BioVest International, Minneapolis, MN). Briefly, seed hybridoma cells were first grown in culture within 1-L spinner flasks kept in a CO_2 incubator (Model 2100, VWR, West Chester, PA). Approximately 1 $\times 10^9$ hybridoma cells in 60 mL of serum-free medium were loaded aseptically into the extracapillary space of the bioreactor cartridge. The temperature was maintained at 37°C, and RPMI 1640 medium (Fisher, Pittsburgh, PA), supplemented with 5-g/L glucose and 5-mg/L gentamicin, was fed into the intracapillary space. Medium containing the anti-TPT monoclonal antibody was harvested from the extracapillary space at a rate of 30–75 mL/day and was replaced with fresh serum-free medium. The collected media were centrifuged for 10 min at 7000 rpm and filtered with a sterile 0.22-μm cellulose acetate bottle-top filter (Corning) before purification.

Purification of ATI. ATI was purified from culture medium via protein-G affinity chromatography (HiTrap Protein-G, Pharmacia, Piscataway, NJ) using an automated BioLogic medium-pressure chromatography system (Bio-Rad, Hercules, CA). Briefly, culture medium was loaded onto the column, which was then washed with 20-mM Na_2HPO_4 (pH 7.0), and antibody was eluted using 100-mM glycine buffer (pH 2.8). The ATI concentrations were assessed by UV absorbance at 280 nm, with the assumption that 1-mg/mL ATI corresponds to 1.35 AU [11].

Production and Purification of Anti-TPT Fab (ATF). ATF was prepared by papain digestion of ATI. Briefly, ATI at a concentration of 5 mg/mL in sodium acetate buffer (pH 5.5) was incubated in 37°C water in the presence of papain, cysteine, and ethylene diamine tetraacetic acid (EDTA) for 6 h. The reaction was stopped by the addition of incubation of 305-mg iodoacetamide, and subsequent incubation with the reaction mixture for 30 min. The digested mixture was dialyzed against 5-mM KH_2PO_4 (pH 6.0) overnight. ATF fragments were then purified from the dialysate by hydroxyapatite chromatography (BioRad) with 5-mM KH_2PO_4 (pH 6.0) as loading buffer and 400-mM K_2HPO_4 (pH 6.0) as elution buffer. The concentration of ATF was also assessed by UV absorbance as described above.

Assessment of Antibody Purity by SDS-PAGE. The purity of ATI and ATF was assessed through sodium dodecyl sulphate–polyacrylamide gel electrophoresis (SDS–PAGE) using a 10% polyacrylamide gel with a 4% stacking gel. Purified ATI (2 μg), purified ATF (2 μg), and the ATI digest mixture were prepared in a nonreducing sample buffer and boiled for 5 min. Samples were then loaded on the gel, and electrophoresed at 140 V for 15 min and 200 V for 35 min. The gel was stained with Coomasie blue R-250 (Bio-Rad) and destained with 40% methanol/10% acetic acid overnight, and the migration distances of ATI and ATF were compared with the migration distances of molecular-weight standards (MW-SDS-200, Sigma).

6.5.2.4   Development of ELISA for Quantification of ATI and ATF in Rat Plasma

Using a similar approach as described by Hansen and Balthasar [12], a species-specific ELISA was developed to quantify ATI and ATF in rat plasma. Briefly, goat anti-mouse IgG (250 μL, 1:500 dilution with 20-mM Na_2HPO_4 buffer, no pH adjustment, Sigma) was immobilized on Nunc Maxisorp 96 well microplates (Roskilde, Denmark) overnight at 4°C. On the next day, wells were washed with phosphate buffer containing 0.05% Tween (Sigma) to remove unbound protein, 250 μL of standards or quality control (QC) samples were then added, and the plates were incubated for 2 h. The wells were washed again and incubated with 250 μL of 1:500 diluted Fab-specific goat anti-mouse antibody-alkaline phosphatase conjugate (in 20-mM Na_2HPO_4 buffer, no pH adjustment) for 1 h. After washing, 200 μL of p-nitro phenyl phosphate (Pierce), 4 mg/mL in diethanolamine buffer (pH 9.8), was added and the change of absorbance at 405 nm was monitored with a micro-plate reader (Spectra Max 250; Molecular Devices, Sunnyvale, CA).

Standard curves were constructed by plotting the change in absorbance with time (dA/dt) versus concentration of ATI or ATF. Standards for the assay were prepared by dilution of stock solutions of ATI or ATF to the appropriate concentrations (0, 25, 50, 100, 175, and 250 ng/mL for ATI; 0, 5, 10, 20, 50, and 100 ng/mL for ATF) with PBS (pH 7.4), with the addition of 1% (v/v) blank rat plasma. Assays were validated with respect to precision and accuracy, by analysis of QC samples at 25, 100, 250 ng/ml for ATI and 5, 20, 100 ng/ml for ATF, respectively. Intra-assay and inter-assay variability were determined through the analysis of QC samples.

6.5.2.5   Pharmacokinetic Studies of ATI and ATF in Rats

Female Sprague–Dawley rats (Harlan, Indianapolis, IN), 180–200 g, were instrumented with jugular vein cannulas under ketamine/xylazine (90/10 mg/kg) anesthesia. Phar-macokinetic studies for ATI were conducted 2–3 days after surgery by i.v. administration of 1-, 15-, or 100-mg/kg ATI (n = 4/group) via jugular vein cannula. Blood samples (200 μL) were collected through cannula at 1, 3, 6, 12, 24, 48, 96, and 168 h. For ATF studies, the antibody fragments were administered i.v.

at doses of 1, 15, and 60 mg/kg (n = 4/group) via the jugular vein cannula, and blood samples were collected at 0.25, 0.5, 1, 3, 6, 12, and 24 h. After the collection of blood, plasma was immediately isolated by centrifugation at 10,000 g for 2 min. Plasma samples were stored at −20°C before to analysis by ELISA.

Pharmacokinetic Analysis. Standard noncompartmental analyses were conducted to assess ATI and ATF pharmacokinetics using WinNonlin software (v. 2.1) (Pharsight, Mountain View, CA). The areas under the plasma concentration versus time curve from time zero to infinity (AUC_{inf}) were determined via the log-linear trapezoidal method. The terminal half-life was determined from the relationship of $t_{1/2} = \ln 2/\lambda$, where λ is the negative slope of the terminal phase of the lnC versus time plot. Systemic clearance (CL) was estimated by dividing the administered dose by AUC_{inf}. The volume of distribution at steady state (V_{ss}) was determined by the product of clearance and the mean residence time.

Statistics. One-way analysis of variance and Bonferroni's post hoc tests were used to compare values of half-life and systemic clearance for ATI and ATF. Differences were considered to be significance when $p < 0.05$.

6.5.3   RESULTS

6.5.3.1   Development of Hybridomas Secreting Anti-TPT Antibodies

TPT is a small hapten with an MW of 421 D, and to stimulate an immune response, TPT was coupled to keyhole limpet hemocyanin via the Mannich reaction. The resultant immunoconjugate preparation was found to contain 90 ± 42 molecules of TPT per molecule of KLH.

After the immunization of mice with KLH–TPT, each mouse was found to show high TPT binding activity in plasma. Fusion of spleen cells with myeloma cells led to the generation of >500 hybridoma cells. Hybridoma 8C2 was identified as showing significant anti-TPT activity via HPLC. As shown in Figure 6.5-1, the ultra-filtrate of control media (blank RPMI HAT media) that was spiked with TPT demonstrated a TPT peak area of 2.5×10^6 units. The response from culture supernatant pool #17, which consisted of media pooled from six individual hybridomas, showed a >50% decrease in the TPT peak area. Further analysis of each hybridoma contributing to this pooled mixture revealed that the media in the well of 8C2 contained antibody against TPT, as the unbound TPT peak area decreased to 5.2% of control (Figure 6.5-1C).

6.5.3.2   Characterization of ATI

The antibody secreted by 8C2 was identified as an IgG1 (data not shown). Good linearity was obtained in the Woolf plot ($R^2 = 0.9999$, Figure 6.5-2) and in the Rosenthal plot ($R^2 = 0.9997$, Figure 6.5-3). Additionally, both plots provided very close estimates of the apparent binding affinity ($K_A = 4.28 \times 10^8 M^{-1}$ from Woolf plot and $5.01 \times 10^8 M^{-1}$ from Rosenthal plot) and identical values for binding capacity ($nPt = 3.2 \times 10^{-7} M$ from both plots). The curves were consistent with the existence of a single "type" of binding site, as expected for a monoclonal antibody. Final

A

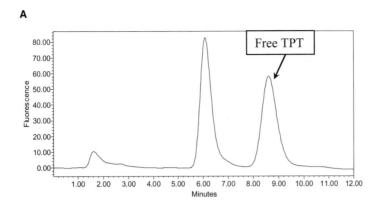

B

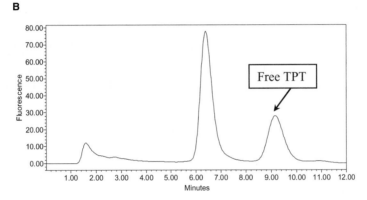

C

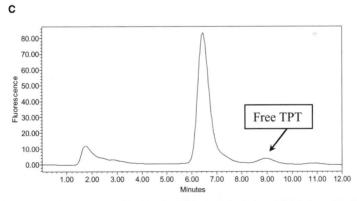

Figure 6.5-1. HPLC screening of positive hybridomas secreting ATAb (see detailed descriptions in text). (A) Negative control: blank HAT media spiked with TPT. (B) Response from the plate # 17 column 1 (well A1, B1, C1, D1, E1, and F1). (C) Response from 17A1 (8C2).

estimates of binding parameters obtained via nonlinear least-squares fitting were $K_A = 4.8 \times 10^8 \, M^{-1}$ (CV = 5.2%) and $nPt = 3.2 \times 10^{-7} \, M$ (CV = 0.46%, Figure 6.5-4). We could not detect unbound TPT concentrations below the dissociation constant K_D value ($K_D = 1/K_A = 1.01 \, ng/L$) due to limitations in the sensitivity of the HPLC assay; nonetheless, the accuracy of the estimated value of K_A is supported by the

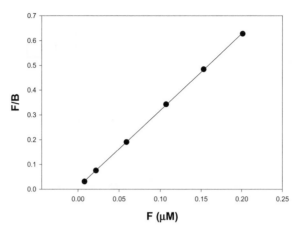

Figure 6.5-2. The Woolf plot was constructed according to the relationship: $\frac{F}{B} = \frac{1}{K_A \times Pt} + \frac{F}{nPt}$, where B is the molar concentration of TPT bound to ATI and F is the molar concentration of free TPT. K_A is the binding affinity of the anti-TPT antibody, and nPt is the total antibody capacity within the media. The slope is $3.13\,\mu M^{-1}$, and the intercept is 0.0073.

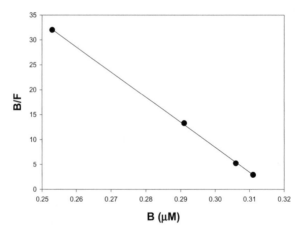

Figure 6.5-3. The Rosenthal plot was constructed according to the relationship: $\frac{B}{F} = nPt \times K_A - K_A \times B$, where B is the molar concentration of TPT bound to ATI and F is the molar concentration of free TPT. K_A is the binding affinity, of the anti-TPT antibody and nPt is the total antibody, capacity within the media. The slope is $-501.42\,\mu M^{-1}$, and the intercept is 158.85.

excellent linearity obtained in both Woolf and Rosenthal plots and the low coefficient of variance associated with the fitted value.

6.5.3.3 Production and Purification of ATI and ATF

Antibody production from 8C2 cells growing within the hollow fiber bioreactor reached 240 mg/wk by the third week. ATI was purified from the media through protein G affinity chromatography (Figure 6.5-5). Final ATI concentrations in the

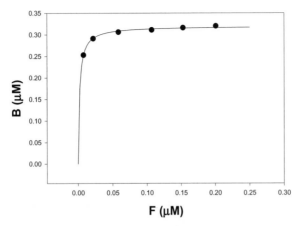

Figure 6.5-4. Bound vs. free TPT plot for 8C2 by fitting the free and bound TPT concentration data to the following equation: $B = \dfrac{nPt \times F}{1/K_A + F}$, where B is the molar concentration of TPT bound to ATI, F is the molar concentration of free TPT, K_A is the binding affinity of the anti-TPT antibody, and nPt is the total antibody capacity within the media.

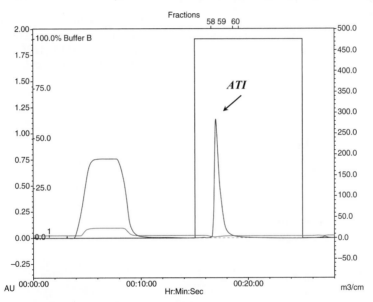

Figure 6.5-5. Purification of anti-topotecan IgG by a protein G column chromatograph. ATI containing media was loaded onto the column using 20-mM Na_2HPO_4 (pH 7.0) and eluted using 100-mM glycine buffer (pH 2.8) at a flow rate of 4 mg/mL (the line represents the percentage of elute buffer onto the column, indicated by buffer B on the y-axis).

elution buffer were typically 0.5–1.2 mg/mL. ATF was produced from the digestion of ATI with papain, and ATF was then purified from undigested ATI, Fc fragments, and secondary digestion products via hydroxyapatite chromatography (Figure 6.5-6). The purity of purified ATI and ATF were assessed by SDS–PAGE, and both ATI and ATF migrated as clear, single bands, apparently free from high- and low-molecular-weight contaminants (Figure 6.5-7). Estimated molecular weights were consistent with expectations for IgG (~150 KD) and Fab (~50 KD).

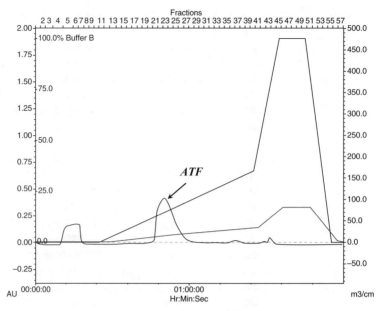

Figure 6.5-6. Purification of anti-topotecan Fab (ATF) by a hydroxyapatite chromatograph. ATF fragments were then purified from the dialysate by hydroxyapatite chromatograph with 5-mM KH_2PO_4 (pH 6.0) as loading buffer and 400-mM K_2HPO_4 (pH 6.0) as elution buffer (the line represents the percentage of elute buffer onto the column, indicated by buffer B on the y-axis).

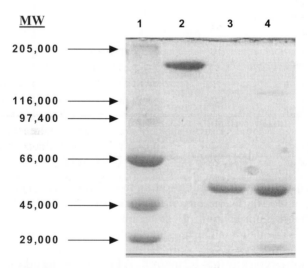

Figure 6.5-7. SDS–PAGE analysis of the purity of ATI and ATF. Lane 1 shows the molecular weight standards; lane 2 was the purified ATI (2 μg) collected from a protein G chromatograph of 8C2 serum-free media; lane 3 was the purified ATF (2 μg) collected from a hydroxyapatite chromatograph from a digested ATI mixed solution; lane 4 was an ATI digested mixed solution.

6.5.3.4   ELISA Validation

Representative standard curves are shown in Figure 6.5-8 (ATI) and Figure 6.5-9 (ATF). The curves were linear over the range of 0–250 ng/mL for ATI and 0–100 ng/mL for ATF. Intra-assay recovery for ATI ranged from 88.3% to 103%, with CVs less than 12%. The interassay recovery for ATI ranged from 94.4% to 102%, with CVs less than 8% (Table 6.5-1A). Similarly, the assay for ATF demonstrated an intra-assay recovery of 94.2% to 108%, with CVs less than 11%, and interassay recoveries ranged from 95.1% to 95.8%, with CVs less than 13% (Table 6.5-1B).

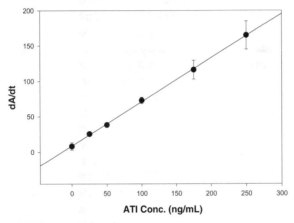

Figure 6.5-8. Representative ELISA standard curve for ATI over the range of 0–250 ng/mL ($r^2 = 0.999$). Error bars represent the standard deviation of the mean of the three replicates.

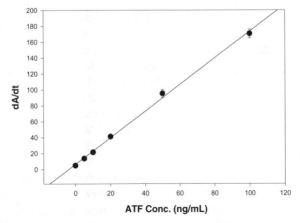

Figure 6.5-9. Representative ELISA standard curve for ATF over the range of 0–100 ng/mL. Analysis of linear regression shows an $r^2 = 0.998$. Error bars represent the standard deviation of the mean of the three replicates.

TABLE 6.5-1. The Accuracy and Precision (CV%) Of ELISA Associated with Assayed Concentrations of Quality Control Samples for (A) ATI and (B) ATF in Rat Plasma Samples (n = 3)

(A) ATI

	QC Conc. (ng/mL)	Assayed Conc. (ng/mL)	CV (%)	Recovery (%)
Intra-assay	25	22.1	11.9	88.3
	100	103	3.0	103
	250	252	5.0	101
Interassay	25	23.9	7.1	95.7
	100	102	4.0	102
	250	236	8.0	94.4

(B) ATF

	QC Conc. (ng/mL)	Assayed Conc. (ng/mL)	CV (%)	Recovery (%)
Intra-assay	5	5.4	9.7	108
	20	18.9	10.6	94.4
	100	94.2	4.5	94.2
Interassay	5	4.8	12.5	95.8
	20	19.0	9.9	95.1
	100	95.1	3.2	95.1

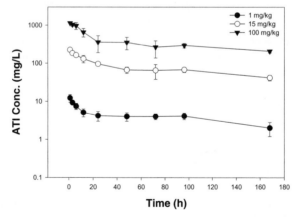

Figure 6.5-10. Plasma concentration-time profile after i.v. bolus of ATI at 1, 15, and 100 mg/kg to rats (n = 4/group). Blood samples were collected at 1, 3, 6, 12, 24, 48, 96, and 168 h. Plasma concentrations of ATI were determined by ELISA.

6.5.3.5 Pharmacokinetics of ATI and ATF

ATI plasma concentration versus time profiles after i.v. administration of 1, 15, and 100 mg/kg are shown in Figure 6.5-10, whereas Figure 6.5-11 shows ATF plasma concentration versus time profiles after i.v. bolus administration of 1, 15, and 60 mg/kg to rats. Noncompartmental analyses of the parameters for ATI and ATF are summarized in Table 6.5-2A and in Table 6.5-2B, respectively. ATI systemic clearance and terminal half-life seem to be dose independent, as no statistical differences were found over the dose range of 1 to 100 mg/kg ($p > 0.05$). No differences

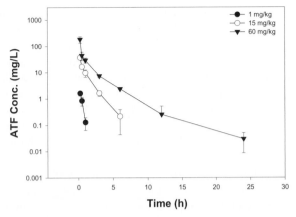

Figure 6.5-11. Plasma concentration-time profile after i.v. administration of ATF at 1, 15, and 60 mg/kg in rats (n = 4/group). Blood samples were collected at 0.25, 0.5, 1, 3, 6, 12, and 24 h. Plasma concentrations for ATF were determined by ELISA.

TABLE 6.5-2. Pharmacokinetic Parameters Estimated by Noncompartmental Analysis for (A) ATI and (B) ATF

(A) ATI

Parameter	Unit	1 mg/kg	15 mg/kg	100 mg/kg
AUC_{inf}	mg/L*h	689 ± 82.8	$(1.46 \pm 0.227) \times 10^4$	$(1.04 \pm 0.292) \times 10^5$
$t_{1/2}$	h^{-1}	105 ± 92.9	118 ± 18.1	154 ± 53.5
V_{ss}	L/kg	0.120 ± 0.030	0.110 ± 0.020	$0.200 \pm 0.030*$
CL	mL/h/kg	1.08 ± 0.330	0.750 ± 0.100	1.00 ± 0.250

(B) ATF

Parameter	Unit	1 mg/kg	15 mg/kg	60 mg/kg
AUC_{inf}	mg/L*h	1.17 ± 0.280	38.8 ± 6.80	189 ± 39.7
$t_{1/2}$	h^{-1}	0.360 ± 0.320	0.850 ± 0.24	1.97 ± 1.32
V_{ss}	L/kg	0.330 ± 0.060	0.340 ± 0.030	0.310 ± 0.160
CL	L/h/kg	$0.890 \pm 0.220*$	0.400 ± 0.070	0.330 ± 0.060

$*p < 0.01$

were found for the estimated steady-state volume of distribution of ATI after dosing with 1 and 15 mg/kg, but the value of V_{ss} estimated for the 100-mg/kg group was significantly higher ($p < 0.01$). ATF terminal half-life and steady-state volume of distribution were not dependent on dose ($p > 0.05$ in the dose range from 1 to 60 mg/kg). No differences were found for the systemic clearance of ATF between 15 and 60 mg/kg, but ATF clearance estimated from 1 mg/kg was found to be significantly higher than that observed for the other doses ($p < 0.01$).

6.5.4 DISCUSSION

This laboratory is investigating the use of anti-drug antibodies within an inverse targeting strategy that attempts to increase the pharmacokinetic and therapeutic selectivity of intraperitoneal chemotherapy. The approach combines i.p. chemotherapy with systemic administration of anti-drug antibodies or antibody fragments (collectively named ADAb) [1, 2]. It is anticipated that ADAb in the blood may rapidly bind drug that diffuses out of the peritoneum, effectively reducing peak plasma-free (i.e., unbound) drug concentrations, the cumulative systemic exposure to unbound drug, and the extravascular distribution of the anticancer drug. In addition, it is anticipated that ADAb will likely show very slow entry into the peritoneal cavity, leading to little change in the cumulative peritoneal exposure to unbound drug. As such, the proposed targeting strategy may produce desirable regio-specific alterations in drug disposition, which may enhance to therapeutic selectivity of i.p. chemotherapy.

These hypotheses have been supported by a series of preclinical studies that employed MTX and anti-MTX Fab as model drugs [2, 4, 5]. However, MTX is not an ideal drug for future clinical applications of the inverse targeting strategy, due to its limited ability to penetrate solid tumors [13] and due to its relatively low potency. TPT may be an ideal drug for use in the proposed approach; this agent is a preferred therapy for metastatic, refractory ovarian cancer, and TPT is very potent, with a maximum tolerated dose of ~5 mg in humans (via i.p. instillation) [14].

To allow evaluation of TPT within the inverse targeting strategy, we have developed anti-TPT monoclonal antibodies. We applied traditional fusion techniques [11], which led to the generatation of >500 clones per fusion (in our hands). In our initial attempts to screen for hybridomas secreting anti-TPT antibodies, we first developed an antibody capture ELISA [11]. However, the antibody capture assay produced high background responses to a variety of proteins (e.g., bovine serum albumin and ovalbumin), even under stringent blocking and washing conditions (e.g., with various blocking agents such as 3% albumin and 20% milk). Displacement ELISAs also showed high background values when incubated with saline relative to TPT (data not shown). Due to the lack of selectivity associated with these immunoassays, we elected to screen for anti-TPT activity via HPLC.

For screening purposes, HPLC is not an ideal method due to its relatively low sensitivity and due to the low throughput of HPLC (i.e., relative to ELISA). In most cases, the identification of antibody binding via HPLC would require that the assay limit of quantification (LOQ) be substantially lower than the concentration of antibody in the supernatant (nPt), and in most cases, identification of antibody by HPLC would require that the product of LOQ and K_A be approximately equal to or greater than 1. Given that the LOQ of our HPLC assay for TPT was ~2 × 10^{-9} M, we anticipated that we would only identify those hybridomas with high antibody production and those that secrete high-affinity anti-TPT antibodies (K_A ~ 5 × 10^8 M^{-1}). From our studies with ~1000 hybridomas generated, two hybridomas were found to secrete antibodies in sufficient concentration and with sufficient affinity to allow detection by HPLC. One hybridoma was found to secrete an IgA anti-TPT antibody, and the other hybridoma, 8C2, was found to secrete an IgG1 anti-TPT antibody. Hybridoma 8C2 was selected for further use as, relative to IgA, it is generally easier to produce, purify, and digest IgG into Fab fragments. Consistent with our expectations, nonlinear least-squares regression revealed that 8C2 secreted an anti-TPT IgG with a high binding affinity (K_A = 4.8 × 10^8 M^{-1}).

To facilitate pharmacokinetic studies for ATI and ATF, a species-specific ELISA was developed and validated to quantify murine ATI and ATF in rat plasma, based on the method of Hansen and Balthasar [12]. The current assay demonstrated similar CV% values for inter- and intra-assay variability for IgG (Hansen and Balthasar reported less than 8.5% of CV% value for intra-assay variability and less than 25.1% for inter assay variability) [12].

Noncompartmental analysis of ATI plasma data indicated that ATI distributed into a steady-state volume of distribution of 110–200 mL/kg, and that ATI was cleared at a rate of 0.75–1.08 mL/h/kg, with a terminal half-life of 105–154 h. These values are similar to those reported by Bazin-Redureau et al. after single-dose (0.7 mg/kg) i.v. administration of a monoclonal murine IgG1 to rats [15]. In their study, they reported $V_{ss} = 125 \pm 4.0$ mL/kg, systemic clearance = 0.48 ± 0.05 mL/h/kg, and a terminal half-life = 194 ± 19 h.

After i.v. administration, ATF distributed into a steady-state volume of distribution of 0.31–0.33 L/kg, similar to the value (0.38 ± 0.03 L/kg) reported by Bazin-Redureau et al. [15] and similar to the value that reported by Pentel (0.43 ± 0.06 L/kg) in a study of high-dose (7.5 g/kg) human Fab administered to rats [16]. The terminal half-life of ATF was determined to be 0.36–1.97 h, which is comparable with a reported value of 1.71 ± 0.25 h, after administration of radio-iodinated Fab to rats [17], and to the half-life of an anti-MTX Fab (1.36 h) [18], but considerably shorter than the values reported by Bazin-Redureau et al. (9.84 ± 0.74 h) [15] and by Pentel et al. (16.3 ± 2.4 h) [16]. The systemic clearance of ATF was 0.33–0.89 L/h/kg, which is higher than the value reported by McClurkan et al. [19] (0.16 ± 0.05 L/h/kg, after i.v. administration of 120-mg/kg Fab in rat) and markedly higher than the value found by Pentel et al. (for high dose of Fab to rats, 0.03 ± 0.004 L/h/kg) [16] and the value reported by Bazin-Redureau et al. (0.03 ± 0.0004 L/h/kg) [15]. Some differences in IgG or Fab pharmacokinetic parameters may be due to the differences in the species of origin of IgG or Fab (human [16] or rat [17] versus mouse in our study), differences in the types of antibodies preparation (polyclonal versus monoclonal), differences in the dose level studied (possible dose-dependent kinetics), or differences in the sensitivity of the analytical techniques used for quantifying Fab concentrations (radiolabeled antibodies quantification versus ELISA). Additional experiments are underway to investigate the pharmacokinetics of anti-TPT Fab more completely (i.e., over a wider dose range, and using more extensive sampling) in animals bearing peritoneal tumors.

In summary, we describe the production, purification, characterization, and pharmacokinetics of a monoclonal anti-TPT IgG and Fab fragments. The anti-TPT antibodies have demonstrated high affinity and sufficient purity for use in evaluating their use in "inverse targeting" strategies to optimize TPT i.p. chemotherapy.

REFERENCES

1. Balthasar J, Fung H L (1994). Utilization of antidrug antibody fragments for the optimization of intraperitoneal drug therapy: studies using digoxin as a model drug. *J. Pharmacol. Experim. Therapeu.* 268:734–739.

2. Balthasar J P, Fung H L (1996). Inverse targeting of peritoneal tumors: Selective alteration of the disposition of methotrexate through the use of anti-methotrexate antibodies and antibody fragments. *J. Pharm. Sci.* 85:1035–1043.

3. Balthasar J P, Fung H L (1995). High-affinity rabbit antibodies directed against metho-trexate: production, purification, characterization, and pharmacokinetics in the rat. *J. Pharm. Sci.* 84:2–6.

4. Lobo E D, Soda D M, Balthasar J P (2003). Application of pharmacokinetic-pharma-codynamic modeling to predict the kinetic and dynamic effects of anti-methotrexate antibodies in mice. *J. Pharm. Sci.* 92:1665–1676.

5. Lobo E D, Balthasar J P (2005). Application of anti-methotrexate Fab fragments for the optimization of intraperitoneal methotrexate therapy in a murine model of perito-neal cancer. *J. Pharm. Sci.* 94:1957–1964.

6. Markman M (1997). Topotecan: An important new drug in the management of ovarian cancer. *Semin. Oncol.* 24(1 Suppl 5):S5–1.

7. Herben V M, ten Bokkel Huinink W W, Beijnen J H (1996). Clinical pharmacokinetics of topotecan. *Clin. Pharmacokinet.* 31:85–102.

8. Hsiang Y H, Liu L F (1988). Identification of mammalian DNA topoisomerase I as an intracellular target of the anticancer drug camptothecin. *Cancer Res.* 48:1722–1726.

9. Hsiang Y H et al. (1985). Camptothecin induces protein-linked DNA breaks via mam-malian DNA topoisomerase I. *J. Biol. Chem.* 260:14873–14878.

10. Chen J, Balthasar J P (2005). High-performance liquid chromatographic assay for the determination of total and free topotecan in the presence and absence of anti-topotecan antibodies in mouse plasma. *J. Chromatogr. B. Analyt. Technol. Biomed. Life Sci.* 816:183–192.

11. Harlow E, Lane D (1988). *Antibodies, A Laboratory Manual.* Cold Spring Harbor Laboratory, Cold Spring Harbor, M.I., p. 276.

12. Hansen R J, Balthasar J P (1999). An ELISA for quantification of murine IgG in rat plasma: An application to the pharmacokinetic characterization of AP-3, a murine anti-glycoprotein IIIa monoclonal antibody, in the rat. *J. Pharmaceu. Biomed. Anal.* 21:1011–1016.

13. West G W, Weichselbaum R, Little J B (1980). Limited penetration of methotrexate into human osteosarcoma spheroids as a proposed model for solid tumor resistance to adjuvant chemotherapy. *Cancer Res.* 40:3665–3668.

14. Plaxe S C et al. (1998). Phase I and pharmacokinetic study of intraperitoneal topotecan. *Invest. New Drugs.* 16:147–153.

15. Bazin-Redureau M et al. (1998). Interspecies scaling of clearance and volume of dis-tribution for horse antivenom F(ab′)2. *Toxicol. Appl. Pharmacol.* 150:295–300.

16. Pentel P R et al. (1988). Pharmacokinetics and toxicity of high doeses of antibody Fab fragments in rats. *Drug Metabol. Dispos.* 16:141–145.

17. Arend W P, Silverblatt F J (1975). Serum disappearance and catabolism of homologous immunoglobulin fragments in rats. *Clin. Exp. Immunol.* 22:502–513.

18. Tayab Z R, Balthasar J P (2004). Development and validation of enzyme-linked immu-nosorbent assays for quantification of anti-methotrexate IgG and Fab in mouse and rat plasma. *J. Immunoassay Immunochem.* 25:335–344.

19. McClurkan M B et al. (1993). Disposition of a monoclonal anti-phencyclidine Fab frag-ment of immunoglobulin G in rats. *J. Pharmacol. Experim. Therapeu.* 266:1439–1445.

6.6

RECOMBINANT ANTIBODIES FOR PATHOGEN DETECTION AND IMMUNOTHERAPY

NICHOLAS J. POKORNY, JEANINE I. BOULTER-BITZER, J. CHRIS HALL, JACK T. TREVORS, AND HUNG LEE

University of Guelph, Guelph, Ontario, Canada

Chapter Contents

6.6.1   Introduction   851
6.6.2   General Properties of Antibodies   852
 6.6.2.1   Binding Properties   852
 6.6.2.2   Monoclonal Antibodies Produced from Hybridoma Cell Lines   853
 6.6.2.3   Recombinant Antibody (rAb) Technology   853
 6.6.2.4   Antibody Libraries   856
 6.6.2.5   Soluble Antibody Production   861
6.6.3   Immunodetection of Pathogens   861
 6.6.3.1   Formats of Detection   862
6.6.4   Immunotherapy   865
 6.6.4.1   Passive Immunotherapy   866
 6.6.4.2   Vaccine/Active Immunotherapy   870
6.6.5   Conclusions   872
 Acknowledgments   872
 References   872

6.6.1   INTRODUCTION

Antibody technology has existed since the establishment of the hybridoma monoclonal system [1]. In recent years, there has been increasing interest in developing recombinant antibodies (rAb) for immunodiagnostics and immunotherapeutical

Handbook of Pharmaceutical Biotechnology, Edited by Shayne Cox Gad.
Copyright © 2007 John Wiley & Sons, Inc.

applications, with the expectation that recombinant antibody fragments will capture a significant share of the US\$6 billion per year diagnostic market [2]. Recombinant antibodies have reduced conventional monoclonal antibodies into smaller capture molecules, while holding the promise of improving the sensitivity and specificity of many diagnostic assays. This chapter summarizes some new developments in rAb with a particular emphasis in pathogen detection and immunotherapy.

6.6.2   GENERAL PROPERTIES OF ANTIBODIES

Within the mammalian immune system, there are five general classes of immunoglobulin molecules, IgA, IgD, IgE, IgG, and IgM, differentiated by the structure of their constant regions and their immune function [3]. Immunoglobulins are antigen binding glycoproteins present on the B-lymphocyte (or B-cell) membrane and secreted by plasma cells. IgG is the most abundant class of immunoglobulins in sera. As most antigens offer multiple epitopes, a variety of B-cell clones are induced, each derived from a B-cell that recognizes a particular epitope. The resulting polyclonal antibody serum comprises a heterogeneous mixture of antibodies, each specific to one epitope. The antibodies produced from a single B-cell clone are monoclonal in that they are all specific to a single epitope [3].

The general immunoglobulin structure is a 4-chain "Y"-shaped molecule of two heavy-chain and two light-chain polypeptides joined together by disulfide bridges [3]. The greatest variability of immunoglobulins occurs at the N-terminal, or the tips, of the "Y" molecule. This is where the antibody interacts with the antigen. This variable region includes the ends of both the heavy (V_H) and the light (V_L) chains. Within the variable regions are three discontinuous regions of the greatest variability, known as the hypervariable (HV) or complementarity-determining region (CDR), which directly contacts a portion of the antigen's surface [2]. Separating the HV regions are four highly conserved framework regions that serves as a scaffold to hold the HV regions together and that forms a unique β-barrel antigen-recognition site. In general, more amino acid residues of the heavy-chain CDRs interact with the antigen than in the light-chain CDRs. In many cases, binding of the antigen induces a conformational change in the antibody, antigen, or both [4].

Antibodies have been used as research tools for cell, antigen, or pathogen identification; pathogenesis studies; ligands for column chromatography and molecule purification; diagnostic reagents; therapeutic antibody preparations; and the identification of protective antigens/epitopes in vaccine development.

It is important that antibodies being used for immunodetection of pathogens or for the immunotherapeutic purposes be able to bind strongly and specifically to the pathogen of interest. The most desired antibodies should possess high affinity and specificity for a given pathogen, with little-to-no cross-reactivity to other antigenic targets.

6.6.2.1   Binding Properties

The binding properties observed during antibody interactions with their target antigens are generally described in terms of affinity, avidity, and specificity.

Affinity. Affinity is the strength of interaction between an antibody and an antigen at a single antigenic site. Within each antigenic site, the variable region interacts through weak noncovalent bonds with antigen at numerous sites; the more interactions, the stronger the affinity. Antibodies with more available antigen-binding domains will possess a higher intrinsic antigen affinity.

Avidity. Avidity is a measure of the overall stability or strength of the antibody–antigen complex. It is controlled by three major factors: antibody-epitope affinity; the valence of both the antigen and the antibody; and the structural arrangement of the interacting ligands.

Specificity. Specificity is based on the combination of both affinity and avidity to determine the likelihood that the particular antibody is binding to a precise antigen epitope. The matrix on which the antigen is tested may influence specificity and sensitivity by interfering with the antigen–antibody complex. The more specific an antibody, the less chance there is for cross-reactivity with a nontarget antigen and subsequent false-positive results in immunodiagnostic assays. In immunotherapeutic applications, the desired antibody should only target the harmful pathogen and not any indigenous microorganisms.

Cross-Reactivity. Cross-reactivity refers to an antibody or population of antibodies binding to epitopes on other antigens. This can be caused either by low avidity or specificity of the antibody or by multiple distinct antigens having identical or very similar epitopes.

6.6.2.2 Monoclonal Antibodies Produced from Hybridoma Cell Lines

The fusion of an antibody-producing B-cell with a myeloma cell results in a hybrid cell, called a hybridoma. These cells possess the immortal growth properties of the myeloma cell and secrete the antibody produced by the B-cell. All antibodies produced by the individual hybridoma cell lines have the same antigenic specificity. Monoclonal antibodies (mAbs) require a great investment of resources, as their production is labor intensive and costs much more to develop, produce, validate, and stockpile than polyclonal antibodies [5]. However, once a hybridoma cell line is developed, mAbs can be easily made in large quantities without the lot-to-lot variation observed in polyclonal antibody preparations [6]. Lack of reliable myeloma partners has prevented the hybridoma fusion technique from widespread use in non-rodent species [6]. Monoclonal antibodies are generally of the IgG class. Because of their bivalency, their functional affinity and avidity are greatly increased as compared with monovalent recombinant antibodies [2]. Monoclonal antibodies have been traditionally used in immunoassays for the detection of pathogenic microorganisms.

6.6.2.3 Recombinant Antibody (rAb) Technology

With the advent of recombinant DNA technology, antibody genes can be amplified and selected through phage display, cell-surface display, or cell-free display systems (i.e., ribosome display). A major advantage shared by these systems is the direct link between the genotype and the phenotype of displayed antibodies during selection. This allows for simultaneous coselection of the desired antibodies and their encoding genes as a result of the binding characteristics of the displayed antibodies. The gene encoding the desired antibody can be further manipulated for improvements in affinity and/or specificity, increased expression, posttranslation modification, or fusion of a secondary protein [7].

Recombinant DNA technology and antibody engineering have increased the available avenues of antibody use, particularly in the application of immunotherapies. Recombinant monoclonal antibodies from various animals can now be produced without a myeloma cell line as a fusion partner. Animals can be immunized by exposure to a desired antigen, and the antibody genes from these animals are amplified by polymerase chain reaction (PCR) and expressed in different expression systems in various formats, such as the Fab, single-chain variable fragment (scFv), and single-domain antibodies (V_H, V_{HH}, and V_{NAR}). Figure 6.6-1 illustrates some different recombinant antibody fragments and the parent antibodies from which they were derived. Recombinant Fab [8], scFv [9], V_H [10], V_{HH} [11], and

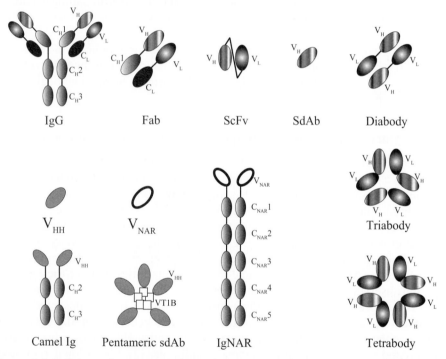

Figure 6.6-1. Schematic representations of common antibody formats and recombinant antibodies. The structure of the immunoglobulin IgG molecule is shown as the "gold-standard" for antibody structure. Homodimeric Camelid Ig and shark IgNAR are indicated along with their representative single-domain antibodies, V_{HH} and V_{NAR}. A variety of common antibody fragments are depicted, as well as some of their multimeric formats. V_H, variable domain of heavy chain; C_H, constant domain of heavy chain; V_L, variable domain of light chain; C_L, constant domain of light chain; V_{HH}; variable domain of Camelid immunoglobulin; V_{NAR}, variable domain of IgNAR; C_{NAR}, constant domain of IgNAR; VT1B, B subunit of Verotoxin 1.

V_{NAR} [12] have been synthesized and found to possess binding characteristics comparable with their respective parental monoclonal antibodies.

The major advantage of these recombinant antibody fragments is that they have increased the available avenues of antibody use, particularly in immunotherapies. They are smaller in size and therefore easier to manipulate genetically and can be expressed in bacterial systems in large quantities with little lot-to-lot variation. The rAbs can be expressed in various surface-display systems (i.e., phage display or cell-surface display) while retaining their functionality during the panning process. As the development of rAbs does not depend on species-specific cell fusion partners, rAbs can be developed from any animal for which one can design an appropriate set of primers for immunoglobulin amplification. Expression and purification of rAbs by bacterial fermentation is less expensive, easier to perform, and less time consuming than production of either polyclonal or monoclonal antibodies [13].

Recombinant antibodies have three key disadvantages when compared with conventional monoclonal antibodies: (1) they are less stable *in vivo* than natural antibodies; (2) they cannot cross-link antigens; and (3) they may lack critical domains necessary for certain biological functions [14].

Numerous differing types of rAbs exist that vary depending on their size, conformation, and source. The following describes the general features of some rAb types.

Fab. Fab fragments, as the classic monovalent antibody fragment, can be produced from an IgG by treatment with papain. The Fab fragment is composed of the entire light antibody chain and part of the heavy chain (Figure 6.6-1). This configuration retains the antigen-binding domain. The interchain disulfide bond remains intact to provide proper tertiary protein conformation. Fab fragments have been shown to have an *in vivo* half-life on the order of hours [15], making them attractive for use in immunotherapeutic applications, as compared with some other rAbs.

Scfv. This antibody fragment is composed of the variable segments of both the heavy and the light antibody chains. The two fragments are typically joined by a flexible peptide linker, most commonly the 15 amino acids of $(Gly_4Ser)_3$, to improve scFv folding and stability. However, the linker between V_H and V_L can interfere with folding in some cases, and this renders the scFv susceptible to proteolysis [16]. Due to rapid *in vivo* blood clearance of scFv, the half-life is in the range of minutes to tens of minutes [17]. The smaller size of scFv antibodies allows them to more easily penetrate tissues.

V_H. These single-domain antibodies are composed entirely of a single-variable antibody domain. Despite early excitement surrounding the functional activity of these, in those that have been studied, they rarely retained the affinity of the parent antibody and were poorly soluble and often prone to aggregation. Exposure of the hydrophobic surface of the V_H to the solvent, which normally interacts with the V_L, causes a sticky behavior of the isolated V_Hs [18]. Because of their small size, it is possible for V_H antibodies to potentially target some epitopes that are inaccessible to larger antibody fragments [10].

V_{HH}. Members of the *Camelidae* family have evolved high-affinity single V-like domains that are mounted on an Fc-equivalent constant domain framework. Their binding domains consist only of the heavy-chain variable domains, referred to as V_{HH}, to distinguish it from the conventional V_H antibody fragment. The variable domain V_{HH} is immediately followed by the hinge region consisting of the C_H2 and C_H3 domains. Affinities of V_{HH}s are in the nanomolar range and are comparable with those of Fab and scFv fragments, and they can be stable under high temperatures, with specific binding being retained at temperatures up to 90°C [11, 18].

V_{NAR}. The immunoglobulin isotype novel antigen receptor (IgNAR) is a homodimeric heavy-chain complex found in the serum of the nurse shark (*Ginglymostoma cirratum*). IgNAR do not contain light chains [19]. Each molecule consists of a single-variable domain (V_{NAR}) and five constant domains (C_{NAR}). Similarities between the V_{NAR} and the V_{HH} include the presence of disulfide bonds and binding affinities in the nanomolar range [20].

Antibody Variants. Fab and scFv fragments have been engineered into dimeric, trimeric, or tetrameric conjugates using either chemical or genetic cross-links. By reducing the scFv linker length to between zero and five residues, self-assembly is directed into multimers [2]. Bivalent diabodies are the results of joining the V_H of one scFv to the V_L of a second scFv molecule, and vice versa, with stabilization introduced by disulfide bridges [21]. Single-domain antibodies can be fused to the B-subunit of *Escherichia coli* verotoxin, or shiga-like toxin, which self-assembles to form a homopentamer, resulting in simultaneous antibody pentamerization and introduction of increased avidity [22]. Fusion antibodies have also been produced in which secondary proteins, such as green fluorescent protein, are expressed along with the rAb. This is done as a means to facilitate monitoring of the antibodies, making antibody-based pathogen detection easier [23].

6.6.2.4 Antibody Libraries

Three general types of libraries are constructed for selection of rAbs: (1) immune, (2) nonimmune or naïve, and (3) or synthetic/semisynthetic. Immune libraries are generated by cloning antibody genes from B-cells of immunized animals. This approach takes advantage of the fact that antibodies directed against the antigen are enriched in such donors, and that they have been subjected to an *in vivo* affinity maturation by the host immune system. These libraries have the added advantage that many isolated antibodies are already antigen specific [24]. Hybridoma cell lines can also be used for the formation of immune libraries. As each hybridoma cell line theoretically produces only a single type of functional antibody, only one set of immunoglobulin genes is available. However, depending on the myeloma cell line used in the fusion process, rearranged but nonfunctional heavy- and light-chain genes in the hybridoma may be present and preferentially amplified and cloned [25]. To select for functional immunoglobulin genes from hybridoma cell lines, intensive screening of cloned genes through panning of phage-display anti-bodies is necessary [25].

Nonimmune or naïve libraries are constructed in the same manner as an immune library, but B-cells from nonimmunized donors are used as sources for antibody genes. The diversity of isolated antibodies increases with the library size, and increasing the library size improves the chance of obtaining high-affinity antibodies [24]. Theoretically, a single universal naïve library can be constructed and used to provide rAbs with any desired specificity. In practice, however, such antibodies have affinities that may be suboptimal [26].

Synthetic/semisynthetic libraries are artificially constructed by using PCR to randomly assemble the genes encoding the HV regions from a naïve B-cell. By employing this process, a unique set of V_H and V_L genes are recombined. These are sometimes referred to as single-pot libraries, because each such library can be source of antibodies directed against multiple antigens. This technology can also be used to display the immunoglobulin genes of conventional hybridomas for the purpose of rescuing unstable clones or improving their binding specificities and/or affinities through genetic manipulations such as *in vitro* mutagenesis or heavy/light-chain shuffling [26].

The use of antibody libraries has numerous advantages over hybridoma antibody technology: (1) They allow for affinity-selection of binding clones and not just to screen for binding activity; (2) recombinant antibody libraries can be rescreened against many different antigens; (3) negative subtraction screening can be done against irrevelent nontarget background proteins, thus reducing the number of nonspecific binders; (4) rAbs can be developed from any species for which primers exist; (5) rAb libraries can be applied to high-throughput screening technologies; and (6) they are well suited to many downstream molecular technologies.

Panning Antibody Libraries. Panning is the methodology by which desirable rAbs are selected from an antibody library. During panning, recombinant antibody libraries are incubated with the target antigen to bind clones bearing specificities of interest. Unbound or weak clones are washed away, and the rAbs that remain bound to the antigen are eluted and amplified for further propagation. After two to four rounds of panning, all binders should be specific to the antigen of interest. Antibody-displaying complexes bearing expressed rAb can be used directly in diagnostic assays or can be genetically modified to express soluble antibody molecules [7, 27].

Because the screening for positive clones takes places in one step, millions of clones are evaluated during each round of selection. Thus, this process requires only a few hours instead of days to months required for screening conventional hybridoma cells for monoclonal antibodies. The clones are indefinitely stable and capable of self-replication when supplies are low. They can be stored as bacterial stocks, phage stocks, or plasmid DNA.

Phage Display. Phage display is often used to screen for recombinant antibodies. It is based on the functional expression of peptides and proteins on the surfaces of bacteriophages. The filamentous bacteriophages are a group of viruses that contain a circular single-stranded DNA genome encased in a long cylinder protein capsule. The infection process is facilitated by attachment to the bacterial pilus. The Ff class of filamentous phages has been most extensively studied. They use the F pilus specific to *E. coli* cells containing the F plasmid [27]. These bacteriophages do not kill their host during reproduction. Rather, the bacterial host is used by the phage for DNA replication as well as for synthesizing phage proteins. There are 11 genes associated with the f1 bacteriophage genome [27]. DNA replication is performed by the cooperation between three phage genes and the host system. Three phage gene products are used in the assembly of phage particles. There is a single phage gene encoding a major capsid protein and four genes encoding minor capsid proteins. Expression of phage proteins is performed by the bacterial host system. The assembly of the filamentous phage is a membrane-associated event requiring the five capsid proteins, the three assembly proteins, adenosine triphosphate (ATP), a proton motive force, and at least one bacterial protein, thioredoxin [28].

Specific peptides and proteins can be displayed on the phage particles as a result of the fusion of specific proteins to the periplasmic portion of the capsid proteins before bacteriophage assembly. Most phage display libraries of peptides and antibodies have been made using the pIII minor capsid protein as the fusion vector, resulting in a chimeric pIII fusion protein. It has the disadvantage that a maximum of five expressed antibody molecules can be displayed per phage host [29]. The main advantage is that pIII fusion proteins containing large inserts can be packaged into the phage. The gene encoding the displayed peptide is inserted between the signal sequence and the beginning of the first domain of the pIII capsid protein. The insertion of the molecule here places the displayed protein at a single polar end of the phage particle. The pVIII major coat protein can also be used for peptide expression, but it is limited in that only peptides of six or eight amino acids in length can be displayed. The advantage is that many copies of the peptide (up to 2500 copies) are displayed along the entire surface of the bacteriophage [30].

Peptides can also be displayed on smaller filamentous particles referred to as phagemids. The phagemid genome contains the phage origin of replication and an independent gene encoding antibiotic resistance. Chimeric genes encoding peptide-phage protein fusions can be placed under the control of a specific promoter in these phagemid genomes. Superinfection of the bacterial host with a filamentous helper phage activates the phage origin of replication. Subsequent transcription and translation provide all phage-derived proteins and enzymes required for phagemid replication. Using a helper phage containing a defective packaging signal allows most phage particles produced to contain the phagemid ss-DNA. When compared with phage vectors, higher yields of phagemid vectors are obtained by plasmid preparation, and larger inserts are more readily maintained [27]. Bacterial transformation efficiency is higher with phagemid than phage vector, and this may result in greater library diversity [31, 32]. Generally, one to three copies of the pIII fusion protein will be expressed by each phage, whereas up to 200 copies of the pVIII fusion peptides can be expressed on the phage capsid [29]. The main disadvantage of using a phagemid vector is the high background during selection due to the presence of wild-type phages, which have been shown to exhibit nonspecific binding during affinity selection. As a result, the conventional M13K07 helper phage has been modified. The modified helper phage either requires that the host bacterial strain provides the pIII proteins [33] or possesses a trypsin cleavage site within the pIII gene that is absent in the fusion protein [34]. Trypsin treatment of phage after selection can eliminate most phages that do not carry the recombinant coat protein displaying the peptide of interest. Because of the need for a helper phage to rescue phagemid vectors, they are less efficient than phage vectors, which allow higher display levels and make panning more efficient, with more binders being isolated in fewer rounds [35]. The lower valency of phagemid vectors allows the selection of binders with higher affinity. Recombinant antibodies from phagemid vectors generally have affinities that are 5–10 times higher than those from phage vectors, on average [35]. The higher avidity associated with phage display may result in the selection of lower affinity clones.

Bacterial Display. The display of recombinant peptides on the surface of bacteria such as *E. coli* was first described more than a decade ago. Several peptides have been successfully displayed on bacterial surfaces [36], including antigenic determinants, heterologous enzymes, antibody fragments, and peptide libraries. Numerous reviews detail the various classes of proteins that have been displayed on both Gram-positive and Gram-negative bacteria [24, 29, 37].

The major obstacle for export of foreign peptides to the surface of a Gram-negative bacterial cell is the presence of an extensive network of macromolecules found within the bilayered cell envelope. It is necessary to select a leader sequence that translocates the peptide through the membrane, as well as an anchor protein that displays the rAb without disrupting the membrane integrity. Initially, short gene fragments were inserted into genes for the *E. coli* outer membrane proteins: the maltoporin LamB [38]; outer membrane protein OmpA [39]; and the phosphate-inducible porin PhoE [40], with the gene products being displayed on the surface of the recombinant bacteria. Lipoproteins [41, 42], fimbriae [43] and flagellar proteins [44], as well as dedicated systems with coupled translocation and surface anchoring have also been used for surface display of peptides [45]. Most current studies of bacterial display of antibodies involve the lipoprotein–porin (Lpp–OmpA') fusion system, with expression of recombinant proteins fused to the sorting signal that directs their incorporation on the cell surface [46]. The expressed antibodies retain their binding ability and approximately 60,000 scFv molecules can be displayed per cell [47]. Selection of good binders can be done by fluorescence-assisted cell sorting (FACS) flow cytometry, using fluorescently labeled antigens, thus avoiding the immobilization of the antigen, elution of bound phages, and reinfection of bacteria with eluted phages.

Expression of scFv on the surfaces of Gram-positive bacteria has been investigated. ScFv antibodies have been displayed on the surfaces of *Staphylococcus xylosus* and *Staphylococcus carnosus* through the fusion to the C-terminus of protein A [48]. Recently, spores of *Bacillus subtilis*, a dormant form of living cells known for their resistance to harsh environments, have been used for display of recombinant antibodies. ScFv antibodies have been fused to CotB, a surface-exposed coat component of the *B. subtilis* spore coat [49]. Spores represent a unique opportunity for surface display because of the potential of providing a durable supporting matrix similar to that of chemical polymer beads. There are three advantages of using Gram-positive over Gram-negative bacteria for surface display. First, larger molecules may be displayed on Gram-positive than Gram-negative bacteria. Second, the single-layered cell wall of Gram-positive bacteria may have better efficiency of fusion protein secretion and folding than the double-layered Gram-negative bacterial cell surface. Third, the rigid Gram-positive cell wall may allow the cells to withstand harsh conditions during affinity selection [7].

Yeast Display. The eukaryotic system of yeast offers an advantage of posttranslational modification and processing of mammalian proteins and is, therefore, better suited for expression and secretion of human-derived antibody fragments than a host such as *E. coli*. Yeasts displaying antibodies share many of the same advantages as bacterial display. Yeast have a rigid, thick cell wall that enables the stable maintenance of surface-displayed proteins [50]. As in bacterial display, the large size of yeast cells allows for efficient selection using FACS flow cytometry [51]. Several systems can be used for expression of rAb on yeast cell surfaces. Yeast display primarily uses the α-agglutinin yeast adhesion receptor to display recombinant proteins on the surface of *S. cerevisiae* [52]. The α-agglutinin receptor acts as an adhesion molecule to stabilize cell–cell interactions and facilitate fusion between yeast cells of opposite mating types. The receptor consists of two proteins, Aga1 and Aga2. The Aga1 protein is secreted from the cell and becomes covalently attached to the extracellular matrix of the yeast cell wall. The rAb–Aga2 fusion proteins bind to Aga1 after secretion and remain attached to the cell via Aga1, displaying the recombinant antibody on the yeast cell surface [53, 54]. The Aga2–fusion rAb can be displayed on the cell surface at 10,000 to 100,000 copies/cell [53].

Glycosylated-inositol (GPI)-anchored proteins have also been used to anchor foreign peptides for cell-surface display in *S. cerevisiae* [55] and *Hansenula polymorpha* [56]. Four different *gpi* genes and their encoded proteins were characterized for the purpose of displaying fusion protein on the cell surface. The *FLO1* gene encodes the Flo1p protein, a lectin-like cell-wall protein thought to form stem-like structures that cover the distance of the cell wall. This protein has also been used to express peptides on the yeast surface [57].

In bacterial or yeast surface display, the phenotype and genotype are linked through the display of the protein on the surface of a cell, in which the DNA can also be propagated and expressed. No subcloning step is required after each selection. Yeast display is particularly beneficial as yeast cells are generally more robust and can express large quantities of heterologous proteins with appropriate post-translational modifications [52]. However, the display of multiple antibody fragments on bacterial or yeast surfaces contributes to an avidity effect, potentially contributing to the selection of rAbs with lower binding affinities [58].

Ribosome Display. This cell-free display system uses the same principle as the phage-display systems in bacteria. The phenotype and genotype of a peptide are linked together and, therefore, can be simultaneously selected based on the function of the peptide. However, the linkage is a physical one in that the genetic material is covalently linked to its encoded product in the formation of an antibody–ribosome–mRNA (ARM) complex through *in vitro* transcription/translation [59].

Ribosome display is an *in vitro* method, with all steps being nucleic acid-based. This allows the assembly of a large-size library with up to 10^{15} members. Mutations can be incorporated in each cycle of selection using PCR-based mutagenesis, thereby allowing for the introduction of diversification during selection. Antibodies with equilibrium dissociation constants improved from the nanomolar to picomolar range using ribosome display have been reported [60, 61].

The key advantage of ribosome display over phage display systems is the freedom from any bacterial transformation step, which typically limits the size of antibody libraries. The library size in phage, bacterial, or yeast display is determined by the transformation efficiency. Generally, 10^{10} to 10^{11} members represent the upper possible limit of the size of an *E. coli* library. Due to the generally lower transformation efficiency in yeasts, large yeast-display libraries are composed of up to 10^9 clones [53]. Another advantage of ribosome display systems is that the expression of the target peptides is not limited by the host. With *in vivo* display systems, proteins that are toxic or detrimental to cell growth will inhibit or kill the host cell, and this can potentially limit the library size for screening. Furthermore, bacteria are prone to introduce natural mutations into foreign DNA to evade selection pressure. Mutations and/or recombinations within the plasmid are frequently found that may provide benefits to the cells. Undesirable selection pressure on individuals can easily be avoided in a cell-free system. Ribosome display is especially useful for construction of a naïve library, which provides a better library complexity and diversity than its phage display counterpart [7]. The use of a ribosome display cannot only select and enrich high-affinity binders from a library, but also it provides *in vitro* hypermutation for the maturation of selected clones to improve binding characteristics [61, 62].

6.6.2.5   Soluble Antibody Production

Numerous systems are available for the expression of soluble recombinant antibodies. Bacteria, yeasts, plants [63], insects and mammalian cell lines [64], and cloned transgenic animals have all been used for rAb expression [65]. Bacteria, in particular *E. coli*, are favored for expression of small, nonglycosylated Fab, scFv, and V_H fragments and diabodies. ScFv fragments are generally expressed to a higher level over Fab in bacteria, although yields may vary for each rAb fragment [2]. Bacterial expression of rAbs is typically achieved by fusion to N-terminal signal peptides, which targets the protein to the periplasmic space of *E. coli*. This periplasmic expression may render a large fraction of the produced rAb insoluble. Soluble bacterial antibody production can also be directed into the culture medium. The risk is that the production of inclusion bodies often results in insoluble protein aggregates. This necessitates laborious renaturation steps and subsequent *in vitro* folding, resulting in a low final yield [18]. Secretion of rAbs can be associated with cell lysis and subsequent product loss. Also, if the rAb being produced is toxic to the bacteria, production levels will be drastically deceased. Because bacteria cannot carry out eukaryotic posttranslational modification, their use is not suitable when glycosylation of the rAb is required.

Purification of Recombinant Antibodies. Purification of bacterial expressed rAb is assisted by the presence of fusion polypeptides such as c-Myc, poly-His, and the FLAG epitope, which have been added to allow for affinity purification. Subsequent cleavage of the fusion protein yields the rAb of interest.

6.6.3   IMMUNODETECTION OF PATHOGENS

Diagnostic tests for viruses, bacteria, fungi, and their spores are often based on culture-dependent methods. While these methods are very sensitive and generally simple to perform, they can be time consuming, requiring several days, and possibly weeks, to complete. If microorganisms are nonculturable, microscopy can be used for identification, as is the case for numerous protozoan parasites. Microscopy requires specific training and may be subject to user bias and error. Also, it may be hard to differentiate between microorganisms at a species level. Epifluorescent microscopy, with fluorochrome-conjugated antibodies specific to the target organism, can make this method much easier to perform. Molecular-based methods such as PCR can be used for quick identification of specific target pathogens, but can be expensive and, like microscopy, requires specific training for the operator. However, PCR cannot distinguish between those detected organisms that are viable and infective from those that are dead. Isolation of DNA from some matrices can be problematic because of the presence of inhibitors that can adversely affect PCR [66].

An antibody to be used in immunodiagnostics should ideally have high binding specificity and affinity to the target antigen of interest. As such, many commercially available immunodiagnostic antibodies are produced as monoclonal antibodies. As the production of rAbs becomes more routine, their availability is sure to increase. Table 6.6-1 lists some developed rAbs that could be used in immunodiagnostics. One major advantage of recombinant antibodies is that they are more easily selected for when the target pathogen may be harmful to an animal being immunized for the production of monoclonal antibodies.

6.6.3.1 Formats of Detection

Successful immunodiagnostic applications are based on the availability of antibodies with high affinity and specificity. Initially, polyclonal antibodies from hyperimmunized animals or monoclonal antibodies from hybridoma cell lines were used. More recently, rAbs are being constructed that possess affinities rivaling those of monoclonal antibodies. Therefore, it is anticipated that rAbs will begin to be deployed in greater frequency in immunodiagnostics in the short to medium term.

The efficiency of immunoassays relies on four components: (1) the target antigen to be detected; (2) the antibody used for detection; (3) the method to separate bound antigen and antibody complexes form unbound reactants (if a heterogenous mixture is used); and (4) the detection method. At the most fundamental level, the efficacy of any given immunoassay is dependent on two major factors: the efficiency of antigen–antibody complex formation and the ability to detect these complexes. It is desirable that the antibodies used have high affinity for the target antigen with low cross-reactivity to minimize false positives. Greater sensitivity of immunoassays is generally associated with higher affinity antibodies.

The use of polyclonal antibodies in sera has inherent disadvantages. Laboratory animals are required, and results can vary drastically in their ability to respond to different antigens, resulting in significant variation among different lots of antibodies. The antibody response to a given antigen tends to be broad. This covers both specific and cross-specific epitopes, causing problems such as high-background or unwanted reactivity for contaminants in the antigen preparation. Polyclonal immunoassays can be highly sensitive but may not be very specific [6, 100].

There is an inherent disadvantage when using highly specific antibodies. Antigenic drift or shift can make using monoclonal antibodies difficult. The evolution

TABLE 6.6-1. A List of Some Pathogens Detected by Recombinant Antibodies

Pathogen	rAb Type	Source Library	Reference
Bacterial			
Bacillus subtilis	scFv	Human Naïve	[67]
Burkholderia pseudomallei exotoxin	scFv	Mouse Immune	[68, 69]
Chlamydia psittaci	scFv	Human Naïve	[70]
Chlamydia trachomatis	scFv	Human Naïve	[70]
Cyanobacterial Hepatoxin Microcystins	scFv	Human Semi-Synthetic	[71]
Clostridium difficile Toxin B	scFv	Mouse Immune	[72]
Escherichia coli O157:H7	scFv	Human Naïve	[73]
Moraxella catarrhalis	scFv	Human Semi-Synthetic	[74]
Streptococcus mutans	scFv	V_H of Mouse Immune, V_L of Human Naïve	[75]
Streptococcus suis	scFv	Human Semi-Synthetic	[76]
Fungal			
Candida albicans	scFv	Human Naïve	[77]
Fusarium Mycotoxin Deoxynivalenol	scFv	Mouse Immune	[78]
Phytophthora infestans	scFv	Naïve	[79]
Protozoan			
Cryptosporidium parvum	Fab	Mouse Immune	[80]
Entamoeba histolytica	Fab	Human Immune	[81]
Plasmodium chabaudi	scFv	Mouse Immune	[82]
Plasmodium falciparum	scFv	Mouse Immune	[81]
Viral			
Beet Necrotic Yellow Vein Virus	scFv	Unknown	[83]
Bluetongue Virus	scFv	Mouse Immune	[84]
Cucumber Mosaic Virus	scFv	Mouse Immune	[85]
Ebola Virus	Fab	Human Immune	[86]
Foot and Mouth	scFv	Mouse Immune	[87]
Grapevine Virus B	scFv	Human Semi-synthetic	[88]
Herpes Simplex Virus 1	Fab	Human Immune	[89]
Herpes Simplex Virus 2	Fab	Human Immune	[90]
Human cytomegalovirus	Fab	Human Immune	[91]
Infectious Bursal Disease Virus	scFv	Chicken Immune	[92]
Influenza A	Fab	Human Naïve	[93]
Severe Acute Respiratory Syndrome Coronavirus (SARS-CoA)	scFv	Mouse Semi-synthetic	[94]
Vaccinia Virus	Fab	Human Immune	[95]
Venezuelan Equine Encephalomyelitis Virus	scFv	Mouse Immune	[96, 97]
Other			
Prion – PrP^{Sc}	scFv	Mouse Immune	[98]
Androctonus australis hector (Scorpion) Venom Toxin Aah1	scFv	Mouse Immune	[99]

Abbreviation: PrP^{Sc} = The modified form of PrP^c (a normal cellular protein thought to be involved in synaptic function), which may cause disease; i.e., the prion is known as PrP^{Sc} (for scrapie).

of an antigen by either genetic modification or active selection can produce targets capable of evading detection systems. Also monoclonal antibodies rarely possess affinities that match the avidities of good polyclonal preparations [100].

Immunoassays. A variety of methods are available to detect antibody–antigen interactions. Some are well established, whereas others have only been developed recently.

A common way to detect antibody–antigen binding in an immunoassay is the enzyme-linked immunosorbant assay (ELISA). ELISA is based on the target antigen being captured onto a plastic multiwell or tube by a capture antibody previously bound to the solid matrix. Bound antigen is then detected using a secondary antibody. The detector antibody can be directly labeled with a signal-generating molecule, or it can be detected with another antibody that is labeled with an enzyme. These enzymes catalyze the chemical reaction with a substrate that results in a colorimetric change. The color intensity can be measured by a spectrophotometer. The ELISA immunoassay is commonly used by clinical laboratories, can be adapted for high-throughput screening, and is relatively inexpensive. Electrochemiluminescence (ECL) is similar to ELISA except that the detector antibody is directly labeled with a chemiluminescence label. Time-resolved fluorescence (TRF) immunoassay is a sandwich-type assay similar to those used for ELISA except that the detector antibodies are directly labeled with lanthanide chelates. TRF exploits the differential fluorescence life span of lanthanide chelate labels compared with background fluorescence [100, 101].

Lateral flow immunoassays are commonly recognized as dipstick or handheld immunochromatographic assays. Detection of target antigens depends on formation of an antigen–antibody complex that migrates along the matrix by capillary action. Assessment of result is qualitative and subject to operator bias. The sensitivity is generally 1-log worse than that of ELISA [101]. Novel rapid immunoassays have been recently developed for the detection of HIV-1 antibodies in clinical samples [102].

Flow cytometry is based on the use of a fluorescent probe being bound by the target antigen, which is streamed past a laser beam where the probe is excited. A detector analyzes the fluorescent properties of the antigen–antibody complex as it passes through the laser beam. This immunoassay can be used in multiplex formats with little or no loss of sensitivity. However, flow cytometry lacks the sensitivity of ELISA and associated immunoassays, and the system requires training and expertise to operate [100]. *Cyrptosporidium parvum* oocysts have been detected in various matrices by flow cytometry and, recently, in soil samples by using antibodies conjugated to fluorescein isothiocyanate [103].

One of the more popular immunassays for pathogen detection involved the use of antibodies in combination with epifluorescent microscopy, usually referred to as an immunofluorescence assay (IFA). Monoclonal and polyclonal antibodies are conjugated to fluorochromes, so that the pathogen can be more easily detected. However, a lack of specificity sometimes results in these IFA assays, indicating the presence of more than one type of pathogenic organism, especially in species that are closely related [104]. Antibodies developed against *Eimeria acervulina* have been found to cross-react with several different *Eimeria* spp., *Neospora caninum*, *Toxoplasma gindii*, and *C. parvum* [104].

6.6.4   IMMUNOTHERAPY

Immunotherapy is a form of medical treatment intended to stimulate or restore the ability of the immune system to fight infection and disease. It can be either nonspecific or specific. Cytokines, growth factors, and leukocytes can enhance host defencse against a variety of pathogens and are examples of nonspecific immunotherapies. In contrast, the administration of antibodies is a form of specific immunotherapy that, similar to antimicrobial therapy, is directed specifically at the pathogen or toxin [105]. Immunotherapy can also be either replacement or reconstitution therapy, which is intended to correct the underlying immunological defects that predispose people to infection, or augmentative therapy, which is intended to enhance the body's immune function against the pathogen [105].

The first use of antibodies for treatment of ailments was performed in 1890 by Behring and Kitasato. They demonstrated that immunity could be transferred to naïve animals by serum of animals that had been challenged with nonlethal doses of a crude toxin preparation. As all original antibody preparations were derived from the serum of immunized animals or immune human donors, the treatment was called serum therapy. Even though the therapy was effective, the administration of large amounts of foreign animal proteins was often associated with side effects that ranged from immediate hypersensitivity reactions to serum sickness [106]. The use of serum therapy was abandoned due to several drawbacks, such as the occurrence of serum sickness, the risk of disease transmission, and lot-to-lot variations of different serum preparations. In the mid-1930s, the introduction of antibiotics replaced the use of antibodies for therapy against most infectious pathogens. Antibodies retained a niche as a treatment for venoms, toxins, and certain viral infections [106]. With the establishment of hybridoma monoclonal antibody production, there was hope for the rapid development of many therapeutic applications, especially in field of oncology [106].

A renewed interest in the production of antibodies for therapeutic use has occurred as a result of numerous driving forces: (1) the increased incidence of pathogen resistance to antibiotics; (2) the lack of effective chemotherapeutic agents; (3) the toxicity of available drugs; (4) the emergence of new microorganisms and viral diseases; (5) the epidemiological evidence of an important role for the humoral immune system in host control of infection; (6) the increased incidence of chronic disease in individuals with various antibody deficiencies; and (7) the use of antibodies is currently the only means to provide immediate immunological protection against biological agents [105, 107–110].

The action of antibodies in immunotherapy is directed against pathogen stages that are susceptible to antibodies. By blocking the pathogen's interaction with the host cell, antibodies can prevent pathogen attachment and/or invasion [109]. Direct lysis of pathogens can occur, by interfering with microbial gene expression and growth, and by enhancing and activating immune cells or a combination thereof [111]. In bacterial disease, antibodies neutralize toxins, facilitate opsonization, and with complement, promote bacterial lysis [112]. In viral disease, antibodies are the best option for therapy and prophylaxis [113]. Antibodies block viral entry into uninfected cells, promote antibody-directed cell-mediated cytotoxicity by natural killer cells, and neutralize virus alone or with the participant of complement [112]. That a microbe inside a host cell is separated from serum antibody has led to the belief that serum antibody cannot be effective against intracellular pathogens. However, at some point in the infectious cycle, most intracellular pathogens reside in the extracellular space, where they are susceptible to antibody action [114]. A hallmark of protozoan parasites is their ability to evade the host's immune system. In many cases, this is accomplished by stage-specific expression of unique antigens, causing immune responses to stages that are subsequently lost during development [115, 116].

From the point of view of antibody production for immunotherapy, careful target selection is of the utmost importance. Although it is important that the antibodies are pathogen specific, monospecific antibodies, such as monoclonal antibodies, may not be a good choice because they are too specific. Narrow specificity is advantageous because it prevents development of resistance among nontargeted microorganisms and avoids disturbance of the normal microflora. However, narrow specificity becomes a disadvantage in cases of mixed infections that cannot be treated with a single antibody preparation. It also decreases the potential market for the drug [113]. Moreover, the high specificity is at a disadvantage with the emergence of pathogen variants that lack the determinant that the antibody recognizes, such as viral escape mutants [106]. The use of cocktails of antibodies that are specific for several antigens could obviate this concern. However, this approach would also have the drawback of increasing the cost of production and the complexity of regulatory issues involving efficacy and safety [106]. Therapeutically relevant targets can also be difficult to identifiy.

Most antibodies currently being used for immunotherapy are monoclonal immunoglobulins (i.e., IgG and IgM). Their multivalency and high specificity and affinity make them attractive for therapies. Brekke and Løset [117] and Casadevall [118] have reviewed the various monoclonal antibodies currently available for immunotherapies as well as those undergoing clinical trials. Most antibodies slated for therapeutic and clinical applications are being developed for use in cancer treatment [118]. However, it has been demonstrated that in cancer immunotherapy, the fraction of injected monoclonal antibody that effectively reaches the tumor is usually less than 1% of the injected dose, which is caused by poor tumor penetration and limited target specificity. By reducing the antibody size and increasing its affinity for the target, it should be easier to obtain better targeting molecules [119]. Recombinant antibodies are smaller and can be matured for increased affinity. Therefore, it is not surprising that their use for therapy against viruses and intracellular pathogens is becoming more attractive. Currently, relatively few rAbs are being used for immunotherapy, as compared with hybridoma monoclonals. Table 6.6-2 lists some rAbs currently being studied for immunotherapeutic activities. Because of the need for therapeutic antibodies to be specific, it is feasible that any antibody used for therapy could also be used for immunodiagnostic purposes.

6.6.4.1 Passive Immunotherapy

Passive immunotherapy involves administering preformed antibodies from external sources to the recipient patient, as opposed to therapies like vaccines, which encourage the body to make its own antibodies. Numerous rAbs are specific to various pathogens (see Table 6.6-2). Pathogens that affect humans, plants, and commercial/domestic animals are all being looked at for their susceptibility to antibodies. For

TABLE 6.6-2. A List of Some Recombinant Antibodies that May Be Theoretically Effective in Immunotherapy

Antibody Target	rAb Type	Therapeutic Status	Reference
Bacterial			
Bacillus anthracis Toxin	Fab	Rat; *In Vivo*	[120]
Neisseria meningitidis	scFv	Mouse; *In Vivo*	[121]
Staphlococcus aureus – Methicillin Resistant	Aurograb	Human; Clincal Trial	NeuTec Pharma plc
Fungal			
Candida albicans	scFv	Rat; *In Vivo*	[122, 123]
Candida spp.	Mycograb	Human; Clincal Trial	NeuTec Pharma plc; [124]
Cryptococcus neoformas	scFv	Mouse; *In Vivo*	[108]
Fusarium oxysporum	scFv	In Planta	[125]
Protozoan			
Plasmdodium falciparum	Recombinant Peptide	Cell Culture; *In Vitro*	[126]
Plasmodium yoelii	scFv	Mouse; *In Vivo*	[127]
Plasmodium berghei	scFv	Mosquito; *In Vivo*	[128]
Viral			
Cucumber Mosaic Virus	scFv	In Planta	[129]
Foot and Mouth Disease	Fab	PRNT; *In Vitro*	[130]
Herpes Simplex Virus 1,2	Fab	Cell Culture; *In Vivo*	[131]
HIV-1 Reverse Transcriptase	scFv	Cell Culture; *In Vitro*	[132]
HIV-1 Vif Protein	scFv	Cell Culture; *In Vitro*	[133]
Japanese Encephalitis Virus	Fab	PRNT; *In Vitro*	[134]
Measles Virus	Fab	PRNT; *In Vitro*	[135]
Rabies Virus	scFv-Fc	Mouse; *In Vivo*	[136]
Rotavirus	Fab	Cell Culture; *In Vitro*	[137]
Varicella-zoster Virus	scFv	Cell Culture; *In Vitro*	[138]
Other			
Prion PcP^{Sc}	scFv	Cell Culture; *In Vitro*	[139]
Prion PcP^{Sc}	Fab	Cell Culture; *In Vitro*	[140]

Abbreviation: PRNT = Plaque Reduction Neutralization Test. PcP^{Sc} = The modified form of PrP^c (a normal cellular protein thought to be involved in synaptic function), which may cause disease; i.e., the prion is known as PrP^{Sc} (for scrapie). Aurograb = Commercially available recombinant antibody from NeuTec Pharma plc. Mycograb = Commercially available recombinant antibody from NeuTec Pharma plc.

passive immunotherapy to be effective, antibodies must be present at the site of infection long enough to interact with the appropriate infective stage of the organism's life cycle, bind to the appropriate antigen(s), and activate the appropriate effector mechanisms (if any) [109]. For intestinal pathogens, considerations for the effective therapy include dosage schedule, gastrointestinal transit time and residence of the antibody in the small intestine (the primary site of infection), gastrointestinal degradation of the antibody and ways of protecting the antibody activity (formulation), as well as the antigenic specificity and antigenic relatedness of strains or isolates of the pathogen [109]. Passive antibody therapy for infectious diseases is aided by the large antigenic differences between the microorganism and the host and can be optimized by identifying and targeting molecules essential for infectivity and by understanding mechanisms of antibody-mediated neutralization [106, 141]. In some gastrointestinal diseases caused by microorganisms, prophylactic uses of specific antibodies have proven to be much more effective than conventional supportive treatment, such as in human studies of enterotoxigenic *E. coli*-induced "travelers diarrhea" [142] and *Shigella* infections [143]. Unfortunately, rAbs have been found to possess a weaker neutralization effect on pathogens when compared with the corresponding monoclonal antibody [8].

Numerous variables can lead to a negative result from passive immunotherapy. Too little antibody can result in a lack of protective efficiency. Too much antibody can produce a prozone-like effect, whereby more antibody provides less protection than less antibody. The antibodies used in therapy can also be affected by affinity/avidity, specificity, isotype, idiotype, preparation, and route of administration. The genetic background and immune competency of the host can also affect treatment as can route of pathogen infection, genetic background, and inoculum size of the pathogen [114]. A dramatic example of the limitations of passive antibody transfer experiments is provided by the observation that transfer of either too little or too much antibody can result in no protection [114].

Passive antibody therapy and prophylaxis have a natural place in animal health because mammalian species such as pigs, horses, sheep, and cows do not transmit maternal immunity prenatally (antibodies do not cross the barrier of the placenta, as is the case in humans) but postnatally through colostral antibodies [113].

Humanized Antibodies. For human immunotherapeutic applications, the optimal antibody should be of human origin. Immune libraries are perhaps the best way to produce neutralizing antibody responses to protective epitopes on infectious pathogens [26]. For therapeutic purposes, the efficacy of nonhuman-derived monoclonal antibodies is limited by the induction of anti-mouse immune response. Using hybridoma methods to immortalize human B-cells for the construction of human monoclonal antibodies would avoid these issues. However, the absence of a suitable fusion partner and other technical issues has made methods that rely on human B-cell immortalization problematic, forcing reliance on nonhuman organisms [6].

The development of human(ized) antibody molecules is mostly aimed at reduction of unwanted immunological properties in medical applications. Useful antibodies have been isolated from mice and murine-based antibody libraries. However, these antibodies are immunogenic in humans. To use these molecules as successful therapeutic agents, they must be redesigned to reduce their immunogenicity, thereby reducing the human–anti-mouse (HAMA) immune response [144]. The antigenicity of therapeutic antibodies can be reduced by chimerization, in which nonhuman domains are replaced by human domains, though fusion of unmodified variable domains to the constant regions of the light and heavy chains of a human antibody [144]. Humanization involves not only fusion of the (usually murine) variable domains to human constant regions, but also the alteration of selected murine variable domain framework residues to more closely match the most related available human variable template sequence [18]. CDR grafting involves the substitution of nonhuman CDR from an antibody into the most closely related human antibody sequence available, so that only the CDR domains are nonhuman in origin [144]. Veneering involves changing only exposed residues so that they reflect the human consensus sequences, whereas those residues that are buried are left unmodified [144]. These strategies may often preserve the binding affinity of the murine antibody while significantly reducing immunogenicity [145].

Antibody-based therapies that use human or humanized antibodies have low toxicities and high specificities. The high specificity of antibodies is both an advantage and a disadvantage. The advantage of high specificity is that antibody-based therapies target only the pathogenic organism that causes disease and, therefore, should not alter the host flora or select for resistance among nontargeted microorganisms. However, high specificity also means more than one antibody preparation might be required to target microorganisms with high antigenic variation [106].

Human naïve or semisynthetic libraries, such as the Tomlinson or Griffin libraries, can be screened for pathogen specific rAbs [146]. This eliminates the need for antibody humanization, making them readily available for immunotherapeutic applications in humans.

Delivery of Antibodies. Antibodies can be administered as human or animal plasma or serum, as pooled human immunoglobulin for intravenous (IVIG) or intramuscular (IG) use, as high-titer human IVIG or IG from immunized or convalescing donors, and as monoclonal antibodies [112]. Oral administration of large doses of antibodies resulting in high local concentrations can be given without the serious safety concerns related to systemic administration of foreign proteins (as would be the case for xenogeneic systemic administration). However, oral administration of antibodies is only feasible for treatment of certain types of mucosal infectious diseases, such as *C. parvum*-associated infections. There is a restriction to oral antibody delivery, specifically the possibility of gastrointestinal destruction of the antibody given the low gastric pH and the presence of gastric and intestinal proteases. As a result, duration of the antibody availability due to peristalsis is limited to the order of 1–2 hours [109]. These problems could be remedied by coadministration of antacids, enteric coatings (pH dependant release), bioerodable coating (delayed release, pH independent), or liposomal formulation (enzyme induced release). In addition to protective coatings, formulations must be able to disperse rapidly when the coating is shed [109]. Otherwise, antibodies intended for therapeutic applications need to be administered systemically [113] to provide a much longer duration of activity, on the order of days rather than hours [109]. An important potential advantage of antibody therapies is that they can be synergistic or additive when combined with conventional antimicrobial chemotherapy against bacterial and viral diseases [106].

As antibodies are natural products, they must be produced in cell lines or other live expression systems. This raises the theoretical concern that there could be contamination of antibody preparations by infectious agents such as prions or viruses. Although tight regulation and regulatory vigilance and surveillance can reduce this concern, the need for ongoing monitoring and testing for contamination contributes to the high cost of developing and administering antibody therapies [106].

6.6.4.2 Vaccine/Active Immunotherapy

The advent of recombinant DNA technologies has increased the available avenues for vaccine development. Although not related to rAbs, active immunotherapy is important for the prevention of illness due to infection by pathogenic microorganisms. Active immunity is the response generated by the immune system after exposure to a vaccine or microorganism. Both antibodies and cell-mediated immunity can be generated by active immunization. Because antibody therapy can be protective against an infectious disease, a vaccine that elicits similar antibodies should be protective against the relevant pathogen. Ideal vaccines should be safe, elicit the appropriate immune response, and provide long-term protection at low doses [147]. Vaccines against infectious organisms have been traditionally developed by administration of whole live attenuated or inactivated microorganisms [148]. Although live attenuated viral vaccines can elicit responses from cytotoxic T lymphocytes (CTLs) and from the humoral system, inactivated viruses and recombinant proteins are generally not potent inducers of CTLs [147]. However, subunit and recombinant vaccines hold promise for agents of bioterror because they deliver defined antigens, perhaps for multiple agents, and have the potential to reduce the immunogenicity of whole cell preparations [110]. Table 6.6-3 indicates a few infectious pathogens whose disease can be prevented, or minimized, by the use of recombinant antigens, recombinant viruses, or DNA vaccines.

Recombinant Antigens. Recombinant antigens are synthesized to represent the most immunogenic antigens from a particular pathogen. However, when purified

TABLE 6.6-3. Representative List of Pathogens Being Treated by Active Immunotherapy

Antibody Target	Type of Vaccine	Therapeutic Status	Reference
Chlamydia pneumoniae	DNA/Recomb Virus	Mouse; *In Vivo*	[149]
Cryptosporidium parvum	Recomb Antigen	Bovine; *In Vivo*	[150]
Entamoeba histolytica	Recomb Antigen	Mouse; *In Vivo*	[151]
HIV-1 Envelop Protein	Recomb Virus	Mouse: *In Vivo*	[152]
Listeria monocytogenes	DNA Vaccine	Mouse; *In Vivo*	[153]
Mycobacterium avium	Recomb Antigen	Lamb; *In Vivo*	[154]
Plasmdodium falciparum	Recomb Virus	Human; *In Vivo*	[155]
Vibrio anguillarum	Recomb Peptide	Piscine; *In Vivo*	[156]
Virulent infectious Bursal Disease Virus	DNA Vaccine	Chicken; *In Vivo*	[157]

recombinant antigen is administered using traditional immunization protocols, the levels of cellular immunity induced are generally low and not capable of eliciting complete protection against diseases caused by intracellular microbes [148]. Most injected proteins are not efficiently presented by major histocompatibility complex (MHC) class I, making this a major limitation for diseases in which cellular immune responses play a key role [147]. Recombinant antigens that have been used as vaccines are considered safe, but they do have limitations in their use. For example, a recombinant protein can have posttranslational modifications from the pro-tein present in the natural infection and may not elicit optimal responses. Immunization with recombinant antigens requires that parasite epitopes involved in invasion and attachment are known and available as well as amenable to re-combinant DNA technology, and they can be produced in a pure form. Molecular techniques for producing recombinant antigens that elicit neutralizing antibodies have been developed that may serve as a means to provide vaccination against pathogens.

Recombinant Viruses. This type of vaccination is based on the use of recombinant viruses encoding specific antigens as immunization tools. Recombinant viruses can express the foreign antigens directly inside cells of the host organism, as would happen in natural infection. Antigens so expressed are made available to the intracellular antigen-processing machinery. Displayed antigens can be bound by MHC molecules, thus favoring their presentation to T lymphocytes [148]. The poxviruses, adenoviruses, and influenza viruses are three of the most attractive and efficient vectors that can be used for vaccination purposes [148]. However, re-exposure of the immune system to the same recombinant antigens expressed by the viral vectors may not recall the memory T-cell responses induced during the primary exposure, as subsequent administrations of the same virus were cleared faster from the organism due to immune responses induced against the virus itself [158]. In cases where immunization with one viral vector does not elicit sufficient levels of immunity, researchers have successfully elicited immunization by the sequential administration of a different recombinant virus expressing the same antigen [158].

DNA Vaccines. DNA vaccines were introduced less than a decade ago, but they have already been applied to a wide range of infectious and malignant diseases, including those that are parasitic and difficult to control [159]. DNA vaccines involve the use of plasmids containing viral, parasitic, or bacterial genes that are expressing the protective antigen in mammalian cells after their introduction [160]. Uptake and expression of the DNA has been demonstrated in several species, for several parasitic and bacterial infections (reviewed in Ref. 147). DNA vaccination may also be useful to define key genes that play a role as antigens for diseases that are not well understood. Vaccination with DNA have been administered via (1) direct injection of naked DNA, (2) administration of DNA complexed with liposome formulations or other compounds that target specific receptors, or (3) compact DNA associated with gold particle bombardment [147].

The main safety issues for DNA vaccines include integration of the plasmid into the host genomic DNA, production of pathogenic anti-DNA antibodies to the plasmid that could induce or exacerbate diseases such as systemic lupus erythema-tosus, or immunologic tolerance to the antigen being expressed [147]. As a result, to produce vaccines safely and efficiently, the use of plasmids that do not replicate in mammalian cells is desirable for gene expression. The risk of chromosomal integration and subsequent activation of an oncogene or disruption of a tumor suppressor gene or other regulatory gene is another safety issue [147].

6.6.5 CONCLUSIONS

With increased concern about the pathogen status in food and water, the application of antibodies in the detection of these pathogens also continues to increase. Design of new rAbs that are more specific and have a higher affinity for their target is becoming more commonplace. These new rAbs are making immunoassays more sensitive. When combined with advances in our understanding of the pathogenic nature of infectious diseases, rAbs are making immunotherapy more relevant in the prevention and treatment of diseases caused by pathogens. With the increased number of available antibody libraries and ease with which rAbs can be obtained, a rapid process exists for generation of specific, high-affinity rAbs against virtually any target. As the parameters for improving *in vivo* efficacy of rAbs becomes more understood, and bacterial fermentation provides a cost effective route to scale up production of many formats of engineered rAbs, their presence in pathogen diagnostic and therapies will be enhanced.

ACKNOWLEDGMENTS

We would like to thank the Canadian Water Network (CWN), the Natural Sciences and Engineering Research Council (NSERC) of Canada, Ontario Ministry of Agriculture and Food (OMAF), and Ontario Ministry of Environment for their support of our research on recombinant antibodies and *Cryptosporidium parvum*. We would also like to acknowledge GAP EnviroMicrobial Services and the NIH AIDS Research & Reference Reagent Program for their cooperation in our studies.

REFERENCES

1. Kohler G, Milstein C (1975). Continuous cultures of fused cells secreting antibody of predefined specificity. *Nature.* 256:495–497.
2. Holliger P, Hudson P J (2005). Engineered antibody fragments and the rise of single domains. *Nature Biotechnol.* 23:1126–1136.
3. Goldsby R A, Kindt T J, Osborne B A, et al. (2003). *Immunology.* W.H. Freeman, New York.
4. Stanfield R L, Takimoto-Kamimura M, Rini J M, et al. (1993). Major antigen-induced domain rearrangements in an antibody. *Structure.* 1:83–93.
5. Hoogenboom H R, Winter G (1992). By-passing immunisation. Human antibodies from synthetic repertoires of germline VH gene segments rearranged in vitro. *J. Mol. Biol.* 227:381–388.
6. Berry J D (2005). Rational monoclonal antibody development to emerging pathogens, biothreat agents and agents of foreign animal disease: The antigen scale. *Vet. J.* 170:193–211.

7. Yau K Y, Lee H, Hall J C (2003). Emerging trends in the synthesis and improvement of hapten-specific recombinant antibodies. *Biotechnol. Adv.* 21:599–637.

8. Thullier P, Lafaye P, Megret F, et al. (1999). A recombinant Fab neutralizes dengue virus in vitro. *J. Biotechnol.* 69:183–190.

9. Park K J, Park D W, Kim C H, et al. (2005). Development and characterization of a recombinant chicken single-chain Fv antibody detecting *Eimeria acervulina* sporozoite antigen. *Biotechnol. Lett.* 27:289–295.

10. Ward E S, Gussow D, Griffiths A D, et al. (1989). Binding activities of a repertoire of single immunoglobulin variable domains secreted from *Escherichia coli. Nature.* 341:544–546.

11. van der Linden R H, Frenken L G, de Geus B, et al. (1999). Comparison of physical chemical properties of llama VHH antibody fragments and mouse monoclonal antibodies. *Biochim. Biophys. Acta.* 1431:37–46.

12. Nuttall S D, Humberstone K S, Krishnan U V, et al. (2004). Selection and affinity maturation of IgNAR variable domains targeting *Plasmodium falciparum* AMA1. *Proteins.* 55:187–197.

13. Emanuel P A, Dang J, Gebhardt J S, et al. (2000). Recombinant antibodies: a new reagent for biological agent detection. *Biosens. Bioelectron.* 14:751–759.

14. Crowe J E, Jr, Cheung P Y, Wallace E F, et al. (1994). Isolation and characterization of a chimpanzee monoclonal antibody to the G glycoprotein of human respiratory syncytial virus. *Clin. Diagn. Lab. Immunol.* 1:701–706.

15. Lamarre A, Talbot P J (1995). Protection from lethal coronavirus infection by immunoglobulin fragments. *J. Immunol.* 154:3975–3984.

16. Whitlow M, Bell B A, Feng S L, et al. (1993). An improved linker for single-chain Fv with reduced aggregation and enhanced proteolytic stability. *Protein Eng.* 6: 989–995.

17. Lamarre A, Yu M W, Chagnon F, et al. (1997). A recombinant single chain antibody neutralizes coronavirus infectivity but only slightly delays lethal infection of mice. *Eur. J. Immunol.* 27:3447–3455.

18. Joosten V, Lokman C, Van Den Hondel C A, et al. (2003). The production of antibody fragments and antibody fusion proteins by yeasts and filamentous fungi. *Microb. Cell Fact.* 2:1–15.

19. Greenberg A S, Avila D, Hughes M, et al. (1995). A new antigen receptor gene family that undergoes rearrangement and extensive somatic diversification in sharks. *Nature.* 374:168–173.

20. Nuttall S D, Krishnan U V, Doughty L, et al. (2003). Isolation and characterization of an IgNAR variable domain specific for the human mitochondrial translocase receptor Tom70. *Eur. J. Biochem.* 270:3543–3554.

21. Hudson P J, Kortt A A (1999). High avidity scFv multimers; diabodies and triabodies. *J. Immunol. Meth.* 231:177–189.

22. Zhang J, Tanha J, Hirama T, et al. (2004). Pentamerization of single-domain antibodies from phage libraries: a novel strategy for the rapid generation of high-avidity antibody reagents. *J. Mol. Biol.* 335:49–56.

23. Kim I S, Shim J H, Suh Y T, et al. (2002). Green fluorescent protein-labeled recombinant fluobody for detecting the picloram herbicide. *Biosci. Biotechnol. Biochem.* 66:1148–1151.

24. Benhar I (2001). Biotechnological applications of phage and cell display. *Biotechnol. Adv.* 19:1–33.

25. Krebber A, Bornhauser S, Burmester J, et al. (1997). Reliable cloning of functional antibody variable domains from hybridomas and spleen cell repertoires employing a reengineered phage display system. *J. Immunol. Meth.* 201:35–55.

26. Siegel D L (2002). Recombinant monoclonal antibody technology. *Transfus. Clin. Biol.* 9:15–22.

27. Barbos C F III, Burton D R, Scott J K, et al. (2001). Phage Display: A Laboratory Manual. Cold Sprng Harbor Laboratory, Cold Spring Harbor, N.Y.

28. Feng J N, Russel M, Model P (1997). A permeabilized cell system that assembles filamentous bacteriophage. *Proc. Natl. Acad. Sci. (USA).* 94:4068–4073.

29. Fernandez L A (2004). Prokaryotic expression of antibodies and affibodies. *Curr. Opin. Biotechnol.* 15:364–373.

30. Iannolo G, Minenkova O, Petruzzelli R, et al. (1995). Modifying filamentous phage capsid: limits in the size of the major capsid protein. *J. Mol. Biol.* 248:835–844.

31. MacKenzie R, To R (1998). The role of valency in the selection of anti-carbohydrate single-chain Fvs from phage display libraries. *J. Immunol. Meth.* 220:39–49.

32. Tout N L, Yau K Y, Trevors J T, et al. (2001). Synthesis of ligand-specific phage-display ScFv against the herbicide picloram by direct cloning from hyperimmunized mouse. *J. Agric. Food Chem.* 49:3628–3637.

33. Rondot S, Koch J, Breitling F, et al. (2001). A helper phage to improve single-chain antibody presentation in phage display. *Nature Biotechnol.* 19:75–78.

34. Goletz S, Christensen P A, Kristensen P, et al. (2002). Selection of large diversities of antiidiotypic antibody fragments by phage display. *J. Mol. Biol.* 315:1087–1097.

35. O'Connell D, Becerril B, Roy-Burman A, et al. (2002). Phage versus phagemid libraries for generation of human monoclonal antibodies. *J. Mol. Biol.* 321:49–56.

36. Georgiou G, Poetschke H L, Stathopoulos C, et al. (1993). Practical applications of engineering gram-negative bacterial cell surfaces. *Trends Biotechnol.* 11:6–10.

37. Samuelson P, Gunneriusson E, Nygren P A, et al. (2002). Display of proteins on bacteria. *J. Biotechnol.* 96:129–154.

38. Charbit A, Wang J, Michel V, et al. (1998). A cluster of charged and aromatic residues in the C-terminal portion of maltoporin participates in sugar binding and uptake. *Mol. Gen. Genet.* 260:185–192.

39. Mejare M, Ljung S, Bulow L (1998). Selection of cadmium specific hexapeptides and their expression as OmpA fusion proteins in *Escherichia coli. Protein Eng.* 11:489–494.

40. Janssen R, Tommassen J (1994). PhoE protein as a carrier for foreign epitopes. *Int. Rev. Immunol.* 11:113–121.

41. Fuchs P, Breitling F, Dubel S, et al. (1991). Targeting recombinant antibodies to the surface of *Escherichia coli*: fusion to a peptidoglycan associated lipoprotein. *BioTechnol.* 9:1369–1372.

42. Fuchs P, Weichel W, Dubel S, et al. (1996). Separation of E. coli expressing functional cell-wall bound antibody fragments by FACS. *Immunotechnol.* 2:97–102.

43. Hedegaard L, Klemm P (1989). Type 1 fimbriae of *Escherichia coli* as carriers of heterologous antigenic sequences. *Gene.* 85:115–124.

44. Lu Z, Murray K S, Van Cleave V, et al. (1995). Expression of thioredoxin random peptide libraries on the *Escherichia coli* cell surface as functional fusions to flagellin: A system designed for exploring protein-protein interactions. *BioTechnol.* 13: 366–372.

45. Kornacker M G, Pugsley A P (1990). The normally periplasmic enzyme beta-lactamase is specifically and efficiently translocated through the *Escherichia coli* outer

membrane when it is fused to the cell-surface enzyme pullulanase. *Mol. Microbiol.* 4:1101–1109.

46. Francisco J A, Stathopoulos C, Warren R A, et al. (1993). Specific adhesion and hydrolysis of cellulose by intact *Escherichia coli* expressing surface anchored cellulase or cellulose binding domains. *BioTechnol.* 11:491–495.

47. Chen G, Cloud J, Georgiou G, et al. (1996). A quantitative immunoassay utilizing *Escherichia coli* cells possessing surface-expressed single chain Fv molecules. *Biotechnol. Prog.* 12:572–574.

48. Gunneriusson E, Samuelson P, Uhlen M, et al. (1996). Surface display of a functional single-chain Fv antibody on staphylococci. *J. Bacteriol.* 178:1341–1346.

49. Du C, Chan W C, McKeithan T W, et al. (2005). Surface display of recombinant proteins on *Bacillus thuringiensis* spores. *Appl. Environ. Microbiol.* 71:3337–3341.

50. Georgiou G, Stathopoulos C, Daugherty P S, et al. (1997). Display of heterologous proteins on the surface of microorganisms: from the screening of combinatorial libraries to live recombinant vaccines. *Nature Biotechnol.* 15:29–34.

51. Colby D W, Kellogg B A, Graff C P, et al. (2004). Engineering antibody affinity by yeast surface display. *Methods Enzymol.* 388:348–358.

52. Boder E T, Wittrup K D (2000). Yeast surface display for directed evolution of protein expression, affinity, and stability. *Meth. Enzymol.* 328:430–444.

53. Feldhaus M J, Siegel R W (2004). Yeast display of antibody fragments: a discovery and characterization platform. *J. Immunol. Meth.* 290:69–80.

54. Kieke M C, Cho B K, Boder E T, et al. (1997). Isolation of anti-T cell receptor scFv mutants by yeast surface display. *Protein Eng.* 10:1303–1310.

55. Washida M, Takahashi S, Ueda M, et al. (2001). Spacer-mediated display of active lipase on the yeast cell surface. *Appl. Microbiol. Biotechnol.* 56:681–686.

56. Kim S Y, Sohn J H, Pyun Y R, et al. (2002). A cell surface display system using novel GPI-anchored proteins in *Hansenula polymorpha. Yeast.* 19:1153–1163.

57. Matsumoto T, Fukuda H, Ueda M, et al. (2002). Construction of yeast strains with high cell surface lipase activity by using novel display systems based on the Flo1p flocculation functional domain. *Appl. Environ. Microbiol.* 68:4517–4522.

58. VanAntwerp J J, Wittrup K D (2000). Fine affinity discrimination by yeast surface display and flow cytometry. *Biotechnol. Prog.* 16:31–37.

59. He M, Taussig M J (1997). Antibody-ribosome-mRNA (ARM) complexes as efficient selection particles for in vitro display and evolution of antibody combining sites. *Nucl. Acids. Res.* 25:5132–5134.

60. Hanes J, Jermutus L, Weber-Bornhauser S, et al. (1998). Ribosome display efficiently selects and evolves high-affinity antibodies in vitro from immune libraries. *Proc. Natl. Acad. Sci. (USA).* 95:14130–14135.

61. Hanes J, Schaffitzel C, Knappik A, et al. (2000). Picomolar affinity antibodies from a fully synthetic naive library selected and evolved by ribosome display. *Nature Biotechnol.* 18:1287–1292.

62. Jermutus L, Honegger A, Schwesinger F, et al. (2001). Tailoring in vitro evolution for protein affinity or stability. *Proc. Natl. Acad. Sci. (USA).* 98:75–80.

63. Olea-Popelka F, McLean M D, Horsman J, et al. (2005). Increasing expression of an anti-picloram single-chain variable fragment (ScFv) antibody and resistance to picloram in transgenic tobacco (*Nicotiana tabacum*). *J. Agric. Food Chem.* 53:6683–6690.

64. Peipp M, Saul D, Barbin K, et al. (2004). Efficient eukaryotic expression of fluorescent scFv fusion proteins directed against CD antigens for FACS applications. *J. Immunol. Meth.* 285:265–280.

65. Kamihira M, Ono K, Esaka K, et al. (2005). High-level expression of single-chain Fv-Fc fusion protein in serum and egg white of genetically manipulated chickens by using a retroviral vector. *J. Virol.* 79:10864–10874.

66. Miller D N, Bryant J E, Madsen E L, et al. (1999). Evaluation and optimization of DNA extraction and purification procedures for soil and sediment samples. *Appl. Environ. Microbiol.* 65:4715–4724.

67. Zhou B, Wirsching P, Janda K D (2002). Human antibodies against spores of the genus *Bacillus*: a model study for detection of and protection against anthrax and the bioterrorist threat. *Proc. Natl. Acad. Sci. (USA).* 99:5241–5246.

68. Lim K P, Li H, Nathan S (2004). Expression and purification of a recombinant scFv towards the exotoxin of the pathogen, *Burkholderia pseudomallei. J. Microbiol.* 42:126–132.

69. Nathan S, Li H, Mohamed R, Embi N (2002). Phage display of recombinant antibodies toward *Burkholderia pseudomallei* exotoxin. *J. Biochem. Mol. Biol. Biophys.* 6:45–53.

70. Lindquist E A, Marks J D, Kleba B J, et al. (2002). Phage-display antibody detection of *Chlamydia trachomatis*-associated antigens. *Microbiol.* 148:443–451.

71. McElhiney J, Drever M, Lawton L A, et al. (2002). Rapid isolation of a single-chain antibody against the cyanobacterial toxin microcystin-LR by phage display and its use in the immunoaffinity concentration of microcystins from water. *Appl. Environ. Microbiol.* 68:5288–5295.

72. Deng X K, Nesbit L A, Morrow K J, Jr (2003). Recombinant single-chain variable fragment antibodies directed against *Clostridium difficile* toxin B produced by use of an optimized phage display system. *Clin. Diagn. Lab. Immunol.* 10:587–595.

73. Kanitpun R, Wagner G G, Waghela S D (2004). Characterization of recombinant antibodies developed for capturing enterohemorrhagic *Escherichia coli* O157:H7. *Southeast Asian J. Trop. Med. Public Health.* 35:902–912.

74. Boel E, Bootsma H, de Kruif J, et al. (1998). Phage antibodies obtained by competitive selection on complement-resistant *Moraxella (Branhamella) catarrhalis* recognize the high-molecular-weight outer membrane protein. *Infect. Immun.* 66:83–88.

75. Kuepper M B, Huhn M, Spiegel H, et al. (2005). Generation of human antibody fragments against *Streptococcus mutans* using a phage display chain shuffling approach. *BMC Biotechnol.* 5:1–12.

76. de Greeff A, van Alphen L, Smith H E (2000). Selection of recombinant antibodies specific for pathogenic *Streptococcus suis* by subtractive phage display. *Infect. Immun.* 68:3949–3955.

77. Haidaris C G, Malone J, Sherrill L A, et al. (2001). Recombinant human antibody single chain variable fragments reactive with *Candida albicans* surface antigens. *J. Immunol. Meth.* 257:185–202.

78. Choi G H, Lee D H, Min W K, et al. (2004). Cloning, expression, and characterization of single-chain variable fragment antibody against mycotoxin deoxynivalenol in recombinant *Escherichia coli. Protein Expr. Purif.* 35:84–92.

79. Gough K C, Li Y, Vaughan T J, et al. (1999). Selection of phage antibodies to surface epitopes of *Phytophthora infestans. J. Immunol. Methods.* 228:97–108.

80. Chen L, Williams B R, Yang C Y, et al. (2003). Polyclonal Fab phage display libraries with a high percentage of diverse clones to *Cryptosporidium parvum* glycoproteins. *Int. J. Parasitol.* 33:281–291.

81. Tachibana H, Cheng X J, Watanabe K, et al. (1999). Preparation of recombinant human monoclonal antibody Fab fragments specific for *Entamoeba histolytica. Clin. Diagn. Lab. Immunol.* 6:383–387.

82. Fu Y, Shearing L N, Haynes S, et al. (1997). Isolation from phage display libraries of single chain variable fragment antibodies that recognize conformational epitopes in the malaria vaccine candidate, apical membrane antigen-1. *J. Biol. Chem.* 272: 25678–25684.

83. Kerschbaumer R J, Hirschl S, Kaufmann A, et al. (1997). Single-chain Fv fusion proteins suitable as coating and detecting reagents in a double antibody sandwich enzyme-linked immunosorbent assay. *Anal. Biochem.* 249:219–227.

84. Nagesha H S, Wang L F, Shiell B, et al. (2001). A single chain Fv antibody displayed on phage surface recognises conformational group-specific epitope of bluetongue virus. *J. Virol. Meth.* 91:203–207.

85. Chae J S, Choi J K, Lim H T, et al. (2001). Generation of a murine single chain Fv (scFv) antibody specific for cucumber mosaic virus (CMV) using a phage display library. *Mol. Cells.* 11:7–12.

86. Maruyama T, Parren P W, Sanchez A, et al. (1999). Recombinant human monoclonal antibodies to Ebola virus. *J. Infect. Dis.* 179(suppl 1):S235–239.

87. ShengFeng C, Ping L, Tao S, et al. (2003). Construction, expression, purification, refold and activity assay of a specific scFv fragment against foot and mouth disease virus. *Vet. Res. Commun.* 27:243–256.

88. Saldarelli P, Keller H, Dell'Orco M, et al. (2005). Isolation of recombinant antibodies (scFvs) to grapevine virus B. *J. Virol. Methods.* 124:191–195.

89. Cattani P, Rossolini G M, Cresti S, et al. (1997). Detection and typing of herpes simplex viruses by using recombinant immunoglobulin fragments produced in bacteria. *J. Clin. Microbiol.* 35:1504–1509.

90. Bugli F, Manzara S, Torelli R, et al. (2004) Human monoclonal antibody fragment specific for glycoprotein G in herpes simplex virus type 2 with applications for serotype-specific diagnosis. *J. Clin. Microbiol.* 42:1250–1253.

91. Williamson R A, Lazzarotto T, Sanna P P, et al. (1997). Use of recombinant human antibody fragments for detection of cytomegalovirus antigenemia. *J. Clin. Microbiol.* 35:2047–2050.

92. Sapats S I, Heine H G, Trinidad L, et al. (2003). Generation of chicken single chain antibody variable fragments (scFv) that differentiate and neutralize infectious bursal disease virus (IBDV). *Arch. Virol.* 148:497–515.

93. Desogus A, Burioni R, Ingianni A, et al. (2003). Production and characterization of a human recombinant monoclonal Fab fragment specific for influenza A viruses. *Clin. Diagn. Lab. Immunol.* 10:680–685.

94. Liu H, Ding Y L, Han W, Liu M Y, et al. (2004). Recombinant scFv antibodies against E protein and N protein of severe acute respiratory syndrome virus. *Acta Biochim. Biophys. Sin.* 36:541–547.

95. Schmaljohn C, Cui Y, Kerby S, et al. (1999). Production and characterization of human monoclonal antibody Fab fragments to vaccinia virus from a phage-display combinatorial library. *Virol.* 258:189–200.

96. Duggan J M, Coates D M, Ulaeto D O (2001). Isolation of single-chain antibody fragments against Venezuelan equine encephalomyelitis virus from two different immune sources. *Viral Immunol.* 14:263–273.

97. Hu W G, Alvi A Z, Chau D, et al. (2003). Development and characterization of a novel fusion protein composed of a human IgG1 heavy chain constant region and a

single-chain fragment variable antibody against Venezuelan equine encephalitis virus. *J. Biochem.* 133:59–66.

98. Cardinale A, Filesi I, Vetrugno V, et al. (2005). Trapping prion protein in the endoplasmic reticulum impairs PrPC maturation and prevents PrPSc accumulation. *J. Biol. Chem.* 280:685–694.

99. Devaux C, Moreau E, Goyffon M, et al. (2001). Construction and functional evaluation of a single-chain antibody fragment that neutralizes toxin AahI from the venom of the scorpion *Androctonus australis hector*. *Eur. J. Biochem.* 268:694–702.

100. Andreotti P E, Ludwig G V, Peruski A H, et al. (2003). Immunoassay of infectious agents. *Biotechn.* 35:850–859.

101. Payne W J, Jr, Marshall D L, Shockley R K, et al. (1988). Clinical laboratory applications of monoclonal antibodies. *Clin. Microbiol. Rev.* 1:313–329.

102. Arens M Q, Mundy L M, Amsterdam D, et al. (2005). Preclinical and clinical performance of the Efoora test, a rapid test for detection of human immunodeficiency virus-specific antibodies. *J. Clin. Microbiol.* 43:2399–2406.

103. Walker M, Redelman D (2004). Detection of *Cryptosporidium parvum* in soil extracts. *Appl. Environ. Microbiol.* 70:1827–1829.

104. Matsubayashi M, Kimata I, Iseki M, et al. (2005). Cross-reactivities with *Cryptosporidium* spp. by chicken monoclonal antibodies that recognize avian *Eimeria* spp. *Vet. Parasitol.* 10:47–57.

105. Casadevall A, Pirofski L A (2001). Adjunctive immune therapy for fungal infections. *Clin. Infect. Dis.* 33:1048–1056.

106. Casadevall A, Dadachova E, Pirofski L A (2004). Passive antibody therapy for infectious diseases. *Nature Rev. Microbiol.* 2:695–703.

107. Casadevall A, Pirofski L A (2004). New concepts in antibody-mediated immunity. *Infect. Immun.* 72:6191–6196.

108. Cenci E, Bistoni F, Mencacci A, et al. (2004). A synthetic peptide as a novel anticryptococcal agent. *Cell Microbiol.* 6:953–961.

109. Crabb J H (1998). Antibody-based immunotherapy of cryptosporidiosis. *Adv. Parasitol.* 40:121–149.

110. Hassani M, Patel M C, Pirofski L A (2004). Vaccines for the prevention of diseases caused by potential bioweapons. *Clin. Immunol.* 111:1–15.

111. Hengel H, Masihi K N (2003). Combinatorial immunotherapies for infectious diseases. *Int. Immunopharmacol.* 3:1159–1167.

112. Keller M A, Stiehm E R (2000). Passive immunity in prevention and treatment of infectious diseases. *Clin. Microbiol. Rev.* 13:602–614.

113. Berghman L R, Abi-Ghanem D, Waghela S D, et al. (2005). Antibodies: an alternative for antibiotics? *Poult. Sci.* 84:660–666.

114. Casadevall A (2003). Antibody-mediated immunity against intracellular pathogens: two-dimensional thinking comes full circle. *Infect. Immun.* 71:4225–4228.

115. Joiner K A, Dubremetz JF (1993). *Toxoplasma gondii*: a protozoan for the nineties. *Infect. Immun.* 61:1169–1172.

116. Nardin E H, Nussenzweig R S (1993). T cell responses to pre-erythrocytic stages of malaria: role in protection and vaccine development against pre-erythrocytic stages. *Annu. Rev. Immunol.* 11:687–727.

117. Brekke O H, Løset G A (2003). New technologies in therapeutic antibody development. *Curr. Opin. Pharmacol.* 3:544–550.

118. Casadevall A (2004). The methodology for determining the efficacy of antibody-mediated immunity. *J. Immunol. Meth.* 291:1–10.

119. Chester K A, Hawkins R E (1995). Clinical issues in antibody design. *Trends Biotechnol.* 13:294–300.

120. Wild M A, Xin H, Maruyama T, et al. (2003). Human antibodies from immunized donors are protective against anthrax toxin in vivo. *Nature Biotechnol.* 21:1305–1306.

121. Beninati C, Arseni S, Mancuso G, et al. (2004). Protective immunization against group B meningococci using anti-idiotypic mimics of the capsular polysaccharide. *J. Immunol.* 172:2461–2468.

122. Magliani W, Conti S, Salati A, et al. (2003). Biotechnological approaches to the production of idiotypic vaccines and antiidiotypic antibiotics. *Curr. Pharm. Biotechnol.* 4:91–97.

123. Polonelli L, Magliani W, Conti S, et al. (2003). Therapeutic activity of an engineered synthetic killer antiidiotypic antibody fragment against experimental mucosal and systemic candidiasis. *Infect. Immun.* 71:6205–6212.

124. Matthews R C, Rigg G, Hodgetts S, et al. (2003). Preclinical assessment of the efficacy of mycograb, a human recombinant antibody against fungal HSP90. *Antimicrob. Agents Chemother.* 47:2208–2216.

125. Peschen D, Li H P, Fischer R, et al. (2004). Fusion proteins comprising a *Fusarium*-specific antibody linked to antifungal peptides protect plants against a fungal pathogen. *Nature Biotechnol.* 22:732–738.

126. Li F, Dluzewski A, Coley A M, et al. (2002). Phage-displayed peptides bind to the malarial protein apical membrane antigen-1 and inhibit the merozoite invasion of host erythrocytes. *J. Biol. Chem.* 277:50303–50310.

127. Vukovic P, Chen K, Qin Liu X, et al. (2002). Single-chain antibodies produced by phage display against the C-terminal 19 kDa region of merozoite surface protein-1 of *Plasmodium yoelii* reduce parasite growth following challenge. *Vaccine.* 20:2826–2835.

128. Yoshida S, Matsuoka H, Luo E, et al. (1999). A single-chain antibody fragment specific for the *Plasmodium berghei* ookinete protein Pbs21 confers transmission blockade in the mosquito midgut. *Mol. Biochem. Parasitol.* 104:195–204.

129. Villani M E, Roggero P, Bitti O, et al. (2005). Immunomodulation of cucumber mosaic virus infection by intrabodies selected in vitro from a stable single-framework phage display library. *Plant Mol. Biol.* 58:305–316.

130. Kim Y J, Lebreton F, Kaiser C, et al. (2004). Isolation of foot-and-mouth disease virus specific bovine antibody fragments from phage display libraries. *J. Immunol. Meth.* 286:155–166.

131. De Logu A, Williamson R A, Rozenshteyn R, et al. (1998). Characterization of a type-common human recombinant monoclonal antibody to herpes simplex virus with high therapeutic potential. *J. Clin. Microbiol.* 36:3198–3204.

132. Shaheen F, Duan L, Zhu M, et al. (1996). Targeting human immunodeficiency virus type 1 reverse transcriptase by intracellular expression of single-chain variable fragments to inhibit early stages of the viral life cycle. *J. Virol.* 70:3392–3400.

133. Goncalves J, Silva F, Freitas-Vieira A, et al. (2002). Functional neutralization of HIV-1 Vif protein by intracellular immunization inhibits reverse transcription and viral replication. *J. Biol. Chem.* 277:32036–32045.

134. Wu S C, Lin Y J, Chou J W, et al. (2004). Construction and characterization of a Fab recombinant protein for Japanese encephalitis virus neutralization. *Vaccine.* 23:163–171.

135. de Carvalho Nicacio C, Williamson R A, Parren P W, et al. (2002). Neutralizing human Fab fragments against measles virus recovered by phage display. *J. Virol.* 76:251–258.

136. Ray K, Embleton M J, Jailkhani B L, et al. (2001). Selection of single chain variable fragments (scFv) against the glycoprotein antigen of the rabies virus from a human synthetic scFv phage display library and their fusion with the Fc region of human IgG1. *Clin. Exp. Immunol.* 125:94–101.

137. Higo-Moriguchi K, Akahori Y, Iba Y, et al. (2004). Isolation of human monoclonal antibodies that neutralize human rotavirus. *J. Virol.* 78:3325–3332.

138. Kausmally L, Waalen K, Lobersli I, et al. (2004). Neutralizing human antibodies to varicella-zoster virus (VZV) derived from a VZV patient recombinant antibody library. *J. Gen. Virol.* 85:3493–3500.

139. Donofrio G, Heppner F L, Polymenidou M, et al. (2005). Paracrine inhibition of prion propagation by anti-PrP single-chain Fv miniantibodies. *J. Virol.* 79:8330–8338.

140. Peretz D, Williamson R A, Kaneko K, et al. (2001). Antibodies inhibit prion propagation and clear cell cultures of prion infectivity. *Nature.* 412:739–743.

141. Riggs M W, Stone A L, Yount P A, et al. (1997). Protective monoclonal antibody defines a circumsporozoite-like glycoprotein exoantigen of *Cryptosporidium parvum* sporozoites and merozoites. *J. Immunol.* 158:1787–1795.

142. Freedman D J, Tacket C O, Delehanty A, et al. (1998). Milk immunoglobulin with specific activity against purified colonization factor antigens can protect against oral challenge with enterotoxigenic *Escherichia coli*. *J. Infect. Dis.* 177:662–667.

143. Tacket C O, Binion S B, Bostwick E, et al. (1992). Efficacy of bovine milk immunoglobulin concentrate in preventing illness after *Shigella flexneri* challenge. *Am. J. Trop. Med. Hyg.* 47:276–283.

144. Hayden M S, Gilliland L K, Ledbetter J A (1997). Antibody engineering. *Curr. Opin. Immunol.* 9:201–212.

145. Nabel GJ (2004). Genetic, cellular and immune approaches to disease therapy: past and future. *Nature Med.* 10:135–141.

146. Griffiths A D, Williams S C, Hartley O, et al. (1994). Isolation of high affinity human antibodies directly from large synthetic repertoires. *Embo. J.* 13:3245–3260.

147. Montgomery D L, Ulmer J B, Donnelly J J, et al. (1997). DNA vaccines. *Pharmacol. Ther.* 74:195–205.

148. Rocha C D, Caetano B C, Machado A V, et al. (2004). Recombinant viruses as tools to induce protective cellular immunity against infectious diseases. *Int. Microbiol.* 7:83–94.

149. Penttila T, Tammiruusu A, Liljestrom P, et al. (2004). DNA immunization followed by a viral vector booster in a *Chlamydia pneumoniae* mouse model. *Vaccine.* 22:3386–3394.

150. Hong-Xuan H, Lei C, Cheng-Min W, et al. (2005). Expression of the recombinant fusion protein CP15-23 of *Cryptosporidium parvum* and its protective test. *J. Nanosci. Nanotechnol.* 5:1292–1296.

151. Melzer H, Baier K, Felici F, et al. (2003). Humoral immune response against proteophosphoglycan surface antigens of *Entamoeba histolytica* elicited by immunization with synthetic mimotope peptides. *FEMS Immunol. Med. Microbiol.* 37:179–183.

152. Gonzalo R M, Rodriguez D, Garcia-Sastre A, et al. (1999). Enhanced CD8+ T cell response to HIV-1 env by combined immunization with influenza and vaccinia virus recombinants. *Vaccine.* 17:887–892.

153. Simon B E, Cornell K A, Clark T R, et al. (2003). DNA vaccination protects mice against challenge with *Listeria monocytogenes* expressing the hepatitis C virus NS3 protein. *Infect. Immun.* 71:6372–6380.

154. Sechi L A, Mara L, Cappai P, et al. (2006). Immunization with DNA vaccines encoding different mycobacterial antigens elicits a Th1 type immune response in lambs and

protects against *Mycobacterium avium* subspecies paratuberculosis infection. *Vaccine.* 3:229–235.

155. Moorthy V S, Pinder M, Reece W H, et al. (2003). Safety and immunogenicity of DNA/modified vaccinia virus ankara malaria vaccination in African adults. *J. Infect. Dis.* 188:1239–1244.

156. Xia Y J, Wen W H, Huang W Q, et al. (2005). Development of a phage displayed disulfide-stabilized Fv fragment vaccine against *Vibrio anguillarum*. *Vaccine.* 23:3174–3180.

157. Sun J H, Yan Y X, Jiang J, et al. (2005). DNA immunization against very virulent infectious bursal disease virus with VP2-4-3 gene and chicken IL-6 gene. *J. Vet. Med. B Infect. Dis. Vet. Public. Health.* 52:1–7.

158. Murata K, Garcia-Sastre A, Tsuji M, et al. (1996). Characterization of in vivo primary and secondary CD8+ T cell responses induced by recombinant influenza and vaccinia viruses. *Cell Immunol.* 173:96–107.

159. He H, Zhao B, Liu L, et al. (2004). The humoral and cellular immune responses in mice induced by DNA vaccine expressing the sporozoite surface protein of *Cryptosporidium parvum*. *DNA Cell Biol.* 23:335–339.

160. Ivory C, Chadee K (2004). DNA vaccines: designing strategies against parasitic infections. *Genet. Vaccines Ther.* 2:17.

6.7

THE RADIOPHARMACEUTICAL SCIENCE OF MONOCLONAL ANTIBODIES AND PEPTIDES FOR IMAGING AND TARGETED *IN SITU* RADIOTHERAPY OF MALIGNANCIES

RAYMOND M. REILLY

University of Toronto, Toronto, Ontario, Canada

Chapter Contents

6.7.1 Introduction   884

6.7.2 Tumor and Normal Tissue Targeting Properties of mAbs and Peptides   885

6.7.3 Selection of a Radionuclide for Tumor Imaging   886

6.7.4 Selection of a Radionuclide for Targeted *in situ* Radiotherapy   890

6.7.5 Labeling mAbs and Peptides with Radiohalogens   892

     6.7.5.1 Iodine Radionuclides   892

     6.7.5.2 Bromine Radionuclides   896

     6.7.5.3 Fluorine Radionuclides   898

     6.7.5.4 Astatine Radionuclides   900

6.7.6 Labeling mAbs and Peptides with Radiometals   902

     6.7.6.1 Technetium Radionuclides   902

     6.7.6.2 Rhenium Radionuclides   906

     6.7.6.3 Indium Radionuclides   908

     6.7.6.4 Yttrium Radionuclides   913

     6.7.6.5 Gallium Radionuclides   915

     6.7.6.6 Copper Radionuclides   916

     6.7.6.7 Lutetium Radionuclides   918

     6.7.6.8 Lead, Bismuth, and Actinium Radionuclides   918

6.7.7 Characterization of Radiolabeled mAbs and Peptides   919

     6.7.7.1 Evaluation of the Homogeneity of Radiopharmaceutical Bioconjugates   919

     6.7.7.2 Measurement of Radiochemical Purity (RCP)   921

Handbook of Pharmaceutical Biotechnology, Edited by Shayne Cox Gad.
Copyright © 2007 John Wiley & Sons, Inc.

6.7.7.3   Measurement of Immunoreactivity or Receptor-Binding Properties   921
6.7.7.4   Evaluation of *In Vitro* and *In Vivo* Stability   923
6.7.7.5   Preclinical Biodistribution, Tumor Imaging, and Dosimetry Studies   924
6.7.7.6   Preclinical Studies to Evaluate AntiTumor Effects and Normal
           Tissue Toxicity   925
6.7.7.7   Kit Formulation and Associated Pharmaceutical Testing   926
6.7.8   Summary   926
         Acknowledgment   927
         References   927

6.7.1   INTRODUCTION

Cancer is a major health problem worldwide. It is estimated that more than 10 million new cases of cancer are diagnosed annually around the globe and that more than 4 million individuals die each year from the disease [1]. Intense research over the past few decades has led to important discoveries by cancer biologists that are just now stimulating the development of potentially more effective and safer biologically targeted therapies for the disease. One promising strategy for treating malignancies that exploits their biological phenotype is targeted *in situ* radiotherapy [2]. In this approach, monoclonal antibodies (mAbs) that recognize tumor-associated antigens or peptide ligands that specifically bind to cell-surface growth factor receptors are used as targeting vehicles to selectively deliver radionuclides to cancer cells for *in situ* radiation therapy. These two approaches are known as radioimmunotherapy (RIT) or peptide-directed radiotherapy (PDRT), respectively. An extension of RIT or PDRT is to employ radiolabeled mAbs or peptides for imaging metastatic deposits for detection or to noninvasively characterize their phenotype *in situ*, known as molecular imaging [3]. Phenotypic characterization of tumors will be critical in the future to be able to appropriately select patients for treatment with new biologically targeted anticancer agents, including targeted radiotherapeutics. For example, it is known that breast cancers that exhibit high levels of amplification of the human epidermal growth factor receptor-2 (HER2/neu) gene respond best to treatment with the humanized anti-HER2/neu mAb, trastuzumab (Herceptin; Roche Pharmaceuticals Ltd., Nutley, NJ) [4]. HER2/neu gene amplification is currently evaluated in a primary breast cancer biopsy by immunohistochemical staining for the HER2/neu protein or by fluorescence *in situ* hybridization (FISH) for the gene copy number. A recent report, however, suggests that molecular imaging of HER2/neu expression in breast tumors using indium-111 (^{111}In)-labeled trastuzumab may reliably predict which patients respond to Herceptin as well as identify those who may be at risk for toxicity from the drug [5]. Moreover, imaging studies using radiolabeled forms of biopharmaceuticals could allow a noninvasive *in situ* assessment of their pharmacokinetic properties and, in particular, their tumor and normal tissue uptake and elimination. This may provide insight into their effectiveness as antitumor agents as well as their potential sites of normal organ toxicity. There have been many comprehensive reviews on the topics of imaging and targeted *in situ* radiotherapy of malignancies with radiolabeled mAbs and peptides [2, 3, 6]. In this chapter, the radiopharmaceutical science that provides the foundation for these biomolecules for targeting radionuclides to tumors is discussed.

6.7.2   TUMOR AND NORMAL TISSUE TARGETING PROPERTIES OF MABS AND PEPTIDES

The tumor and normal tissue targeting properties of mAbs and their fragments and peptides must be considered when designing a radiopharmaceutical for molecular imaging or targeted *in situ* radiotherapy of cancer. Intact IgG mAbs (M_r = 150 kDa; Figure 6.7-1) are macromolecules that are cleared slowly from the blood with an elimination half-life of 2–3 days for murine forms and 4 days for chimeric and humanized mAbs [7]. This slow elimination provides multiple passes through a tumor at which extravasation and interaction with tumor cells may occur. Thus, the tumor accumulation [percent injected dose/g (% i.d./g)] of radiolabeled intact IgG mAbs is much greater than that of smaller and more rapidly cleared antibody fragments (e.g., Fab or scFv; Figure 6.7-1) or that of small peptides. This property renders intact IgG mAbs more attractive for RIT, because it maximizes the amount of radioactivity that is delivered to the tumor and, thus, the radiation absorbed dose delivered. However, the macromolecular properties of intact IgG mAbs severely restricts their tumor penetration. It has been shown that radiolabeled mAbs remain close to tumor blood vessels, whereas Fab and single-chain Fv (scFv) fragments penetrate more deeply into tumors [8]. This differential depth of penetration for intact IgG mAbs compared with their fragments may be related in part to antigen-binding affinity and avidity, because it has been hypothesized that a "binding-site barrier" restricts the penetration of antibodies into tumors as a consequence of interaction with antigens on cancer cells proximal to the blood vessels [9, 10]. Antibody fragments (Fab and scFv) are monovalent, which diminishes their avidity and affinity compared with divalent IgG mAbs. The poor tumor penetration

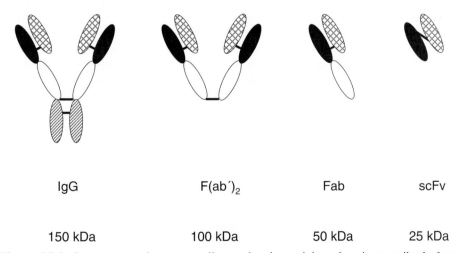

Figure 6.7-1. Structures and corresponding molecular weights of various antibody forms used for imaging or targeted *in situ* radiotherapy of cancer.

properties of intact IgG mAbs produces heterogeneous distribution of radioactivity within a tumor. Moreover, the Fc-domain of intact mAbs is recognized by asialo-glycoprotein receptors on hepatocytes, which promotes liver accumulation of radioactivity [7]. Retention of radioactivity in the liver interferes with detection of hepatic metastases, especially when radiometal-labeled mAbs and peptides are used for imaging. This is less problematic for radioiodinated biomolecules due to their intracellular catabolism and release of radioiodine from hepatocytes.

For tumor imaging, antibody fragments [F(ab')$_2$, Fab, or scFv] or peptides are the most useful because they are rapidly eliminated from the blood and most normal tissues (except the kidneys), which minimizes circulating background radioactivity and yields high tumor/blood (T/B) and tumor/normal tissue (T/NT) ratios that occur at early times after injection. The major challenge with radiolabeled mAb fragments and peptides is their high accumulation in the kidneys. Kidney uptake is thought to be due to glomerular filtration of the fragments or peptides followed by charge interactions between cationic amino acid residues (e.g., lysine and arginine) in the proteins and the negatively charged renal tubular cell membrane [11]. High renal uptake of the somatostatin octapeptide analog, DOTATOC labeled with the β-emitter, ^{90}Y is associated with serious kidney toxicity in patients when used for PDRT of neuroendocrine malignancies [2]. However, renal toxicity can be avoided by co-administering intravenous solutions of lysine or arginine that competitively inhibit the interaction between renal tubular cells and radiolabeled mAb fragments or peptides [11, 12]. Furthermore, renal toxicity is not associated with the use of these radiopharmaceuticals for tumor imaging, because they are administered at much lower doses and labeled with low linear energy transfer (LET) γ-emitting radionuclides.

6.7.3   SELECTION OF A RADIONUCLIDE FOR TUMOR IMAGING

Radionuclides suitable for labeling mAbs or peptides for tumor imaging may be single γ-photon emitters (Table 6.7-1) or positron-emitters (Table 6.7-2). Images using single γ-photons are usually acquired in three-dimensional mode, known as single-photon emission computerized tomography (SPECT; Figure 6.7-2). The single γ-photons are detected by thallium-doped sodium iodide [NaI(Tl)] scintillation crystals that are housed in two opposing heads of a γ-camera. The heads are rotated 180° around the subject to obtain a series of images. Images of the body, organ, or tissue of interest are reconstructed by back-extrapolation of the lines of detection acquired by each camera head. This technique allows the reconstructed images to be "sliced" to visualize a single image plane separate from any overlying or underlying interfering planes. Optimal γ-energies for SPECT imaging are 100–300 keV. The corresponding imaging technique for positron-emitters is called positron-emission tomography (PET). Positrons are β$^+$ particles that are emitted when a proton is converted to a neutron in the decay scheme of certain radionuclides to stable elements. The positrons travel a distance (0.7 to 5 mm) that is directly proportional to their energy before they are annihilated through interaction with an electron in tissues. Positron annihilation creates two 511-keV γ-photons that travel out at approximately 180° from one another at the site of annihilation. PET relies on the simultaneous detection of these two γ-photons within a narrow

TABLE 6.7-1. Radionuclides Suitable for Labeling Biomolecules for SPECT Imaging of Tumors

Radionuclide	Production Method	Eγ (abundence)	T$_{1/2}$phys	Labeling Methods
^{99m}Tc	^{99}Mo/^{99m}Tc generator	140.5 keV (98.9%)	6.0 hours	Binding to thiols; chelation by tetradentate complexes; HYNIC; interaction of carbonyl complex with histidine residues
^{111}In	^{112}Cd(p,2n)^{111}In	171.3 keV (90.2%), 245.4 keV (94.0%)	2.8 days	Chelation by DTPA, SCN-Bz-DTPA or DOTA
^{67}Ga	^{68}Zn(p,2n)^{67}Ga	93.3 keV (35.7%), 184.6 keV (19.7%), 300.2 keV (16.0%)	3.3 days	Chelation by desferrioxamine (DFO) or DOTA
^{123}I	^{124}Xe(p,2n)^{123}Cs→^{123}Xe →^{123}I; ^{124}Xe(p,pn)^{123}Xe→^{123}I	159 keV (83.4%)	13.2 hours	Direct radio-iodination with chloramine-T or Iodogen; indirect conjugation using ATE, SIPC, SGMIB; indirect conjugation using carbohydrate adducts such as TCB or dextran
^{131}I	^{130}Te(n,γ)^{131}Te→^{131}I	364 keV (81.2%)	8.0 days	Same as for ^{123}I

TABLE 6.7-2. Radionuclides Suitable for Labeling Biomolecules for PET Imaging of Tumors

Radionuclide	Production Method	β⁺ Energy (Abundence)	Intrinsic Spatial Resolution	T₁/₂phys	Labeling Methods
^{124}I	^{124}Te(p,n)^{124}I or ^{124}Te(d,2n)^{124}I	0.8–2.1 MeV (23%)	2.3 mm	4.2 days	Direct radio-iodination with chloramine-T or Iodogen
^{76}Br	^{76}Se(p,n)^{76}Br	3.4 MeV (54%)	5.0 mm	16.2 hours	Direct radiobromination with chloramine-T; indirect conjugation using SPBrB, HPEM, brom-3-pyridine-carboxylate, *closo*-dodecaborate, *nido*-undecaborate
^{18}F	^{18}O(p,n)^{18}F	0.6 MeV (100%)	0.7 mm	110 minutes	Indirect conjugation using SFBS, SFB, NPFP, FB-CHO, AFP
^{94m}Tc	^{94}Mo(p,n)^{94m}Tc	2.5 MeV (72%)	3.3 mm	52 minutes	N₄-tetradentate chelator
^{110m}In	^{110}Cd(p,n)^{110m}In	2.3 MeV (62%)	3.0 mm	69 minutes	Chelation by DTPA
^{86}Y	^{86}Sr(p,n)^{86}Y	1.2–1.5 MeV (33%)	1.8 mm	14.7 hours	Chelation by DOTA
^{68}Ga	^{68}Ge/^{68}Ga generator	1.9 MeV (89%)	2.4 mm	68 minutes	Chelation by DOTA
^{64}Cu	^{64}Ni(p,n)^{64}Cu	0.7 MeV (19%)	0.7 mm	12.7 hours	Chelation by DOTA, BAT, or TETA

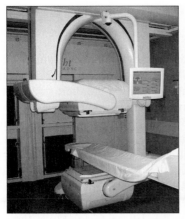
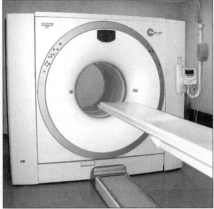

SPECT   PET

Figure 6.7-2. Clinical imaging systems used in nuclear medicine to visualize the distribution of radiopharmaceuticals labeled with single photon-emitting radionuclides or with positron-emitting radionuclides. *Left Panel*: A gamma camera used for single photon-emission computed tomography (SPECT). *Right Panel*: A positron-emission tomograph used for PET scanning. (Photographs generously provided by Deborah Scollard.) (This figure is available in full color at ftp://ftp.wiley.com/public/sci_ tech_med/pharmaceutical_ biotech/.)

time window ("time of flight") by a ring of lutetium silicate (LSO) or bismuth subgerminate (BGO) scintillation crystals surrounding the subject (Figure 6.7-2). Back-extrapolation of the lines of detection allows accurate identification of the site of positron annihilation and yields high spatial resolution images. However, it is important to recognize that the spatial resolution of PET is impacted by the finite distance that the positron travels before its annihilation. This distance ranges from 0.7 mm for the low-energy, 0.6-MeV β^+ particles of ^{18}F to 5.3 mm for the higher energy 1.7-MeV β^+-particles emitted by ^{76}Br (Table 6.7-2) [13].

It is important to consider the pharmacokinetic properties of the vehicle in selecting an appropriate radionuclide for tumor imaging. Intact IgG mAbs require several days to reach maximum tumor uptake and to be cleared from the blood and normal tissues. Therefore, single γ-photon-emitters such as ^{111}In or ^{131}I with a half-life of 2.8 or 8 days, respectively (Table 6.7-1), or longer lived positron-emitters such as ^{124}I (half-life of 4 days; Table 6.7-2) are most suitable for labeling these vehicles for tumor imaging. Antibody fragments such as Fab or scFv are cleared more rapidly from the blood and most normal tissues (except the kidneys), yielding high T/B and T/NT ratios within 24 hours. Therefore, these vehicles may be labeled with short-to-intermediate half-life single γ-photon emitters such as ^{99m}Tc, ^{123}I, or ^{67}Ga for SPECT (Table 6.7-1) or with ^{64}Cu or ^{86}Y for PET (Table 6.7-2). The shorter half-life positron-emitters such as ^{18}F and ^{68}Ga are reserved for labeling peptides and scFv fragments of mAbs that exhibit very rapid tumor uptake and elimination from the blood and normal tissues. Ultra-short-lived positron-emitters such as ^{15}O, ^{13}N, or ^{11}C or (half-lives of 2, 10 or 20 minutes, respectively) that have been used for labeling small molecules for PET, are not feasible for imaging using mAbs or peptides.

6.7.4 SELECTION OF A RADIONUCLIDE FOR TARGETED *IN SITU* RADIOTHERAPY

Radionuclides suitable for targeted *in situ* radiotherapy of tumors (Table 6.7-3) emit either α-particles, β-particles, or Auger and conversion electrons [2]. The important differences between these different forms of radiation are their range in tissues and their LET. α-Particles consist of two protons and two neutrons and carry a 2^+ charge. These particles have the highest LET ($100\,keV/\mu m$), are densely ionizing, and travel $50–100\,\mu m$ (5–10 cell diameters) in tissues. α-emitters such as ^{211}At, ^{212}Bi, or ^{225}Ac are most useful for eradicating small clusters of cancer cells or micrometastases. β-particles are high-energy electrons that have a 1^- charge and travel 2–12 mm in tissues (200–1200 cell diameters). β-particles deposit most of their energy at the end of their track length in tissues. However, for comparison with α-particles, the average LET of the β-particles emitted by ^{131}I ($E\beta = 0.6\,MeV$) over their 2-mm track length is $0.3\,keV/\mu m$. The average LET of the β-particles emitted by ^{90}Y ($E\beta = 2.3\,MeV$) over their 10–12-mm track length is $0.2\,keV/\mu m$. Due to the long range of the β-particles emitted by ^{131}I, ^{90}Y, $^{186}Re/^{188}Re$, or ^{175}Lu conjugated to mAbs or peptides that are targeted to tumor cells, it is possible to irradiate and kill more distant nontargeted tumor cells ("cross-fire" effect). This is advantageous for larger lesions (i.e., 2–10 mm in diameter) in which there is likely to be incomplete targeting of tumor cells. Inadequate targeting could allow some cells to survive. However, the "cross-fire" effect of the β-particles also contributes to dose-limiting bone marrow toxicity in RIT, due to nonspecific irradiation of hematopoietic stem cells by circulating radioactivity perfusing the bone marrow [2].

Auger electrons are very low-energy electrons emitted by radionuclides that decay by electron capture (EC). In EC, a proton in the nucleus captures an electron from an inner orbital shell, creating a vacancy in the shell. This vacancy is filled by the decay of an electron from a higher shell. The excess energy released is transferred to an outer orbital electron that is then ejected from the atom, creating a 2^+ atomic species. Because of their very low energy ($<30\,keV$), Auger electrons travel only a few nanometers to at most a few micrometers in tissues (less than one cell diameter). Their LET approaches that of α-emitters ($100\,keV/\mu m$). However, an antibody or peptide carrying an Auger electron-emitting radionuclide such as ^{125}I, ^{123}I, ^{111}In, or ^{67}Ga must be internalized and, ideally, translocated to the nucleus for the electrons to be the most damaging to DNA. Biomolecules that recognize peptide growth factor receptors [e.g., epidermal growth factor receptors (EGFRs) or somatostatin receptors (SMSRs)] are frequently internalized into cells and in some cases translocated to the nucleus, which allows them to be employed for targeted Auger electron radiotherapy of malignancies. The advantage of Auger electron-emitting radionuclides is that it is possible to restrict killing to cells that specifically bind and internalize radiolabeled mAbs or peptides. There is no "cross-fire" effect from Auger electron-emitting radionuclides, which should obviate any major nonspecific radiotoxicity to bone marrow stem cells, particularly if these cells do not express the target epitopes/receptors. On the other hand, in contrast to the more energetic β-emitters, the lack of a "cross-fire" effect from Auger electron-emitters does not permit killing of nontargeted cancer cells, although a "bystander effect" has been reported [14]. Auger electron-emitting radionuclides are therefore most useful for treating small tumor deposits or micrometastases for which delivery of radiolabeled biomolecules is more homogeneous.

TABLE 6.7-3. Radionuclides Suitable for Labeling Biomolecules for Targeted *in situ* Radiotherapy of Tumors

Radionuclide	Production Method	Particulate Emissions (Energy)	Maximum Range in Tissues	$T_{1/2phys}$	Labeling Methods
^{125}I	^{124}Xe$(n,\gamma)^{125}$Xe $\rightarrow^{125}$I	Auger electrons (<30 keV)	<10 μm	59.4 days	Direct radio-iodination with chloramine-T or Iodogen; indirect conjugation using ATE, SIPC, SGMIB
^{123}I	^{124}Xe$(p,2n)^{123}$Cs$\rightarrow^{123}$Xe $\rightarrow^{123}$I	Auger electrons	<10 μm	13.2 hours	Same as for ^{125}I
^{131}I	Neutron irradiation of ^{130}Te	β-particles (0.6 MeV)	2 mm	8.0 days	Same as for ^{125}I
^{211}At	^{209}Bi$(\alpha,2n)^{211}$At	α-particles (5.9–7.4 MeV)	50–100 μm	7.2 hours	Indirect conjugation of ATE, SAB, SAPC
^{186}Re	^{185}Re$(Re,\gamma)^{186}$Re	β-particles (1.1 MeV)	3 mm	3.7 days	Binding to thiols; chelation by tetradentate complexes; HYNIC; interaction of carbonyl complex with histidine residues; trisuccin
^{188}Re	^{188}W$/^{188}$Re generator	β-particles (2.1 MeV)	8 mm	17 hours	Same as for ^{186}Re
^{90}Y	^{90}Sr$/^{90}$Y generator	β-particles (2.3 MeV)	12 mm	2.7 days	Chelation by DOTA
^{67}Ga	^{68}Zn$(p,2n)^{67}$Ga	Auger electrons (<30 keV)	<10 μm	3.3 days	Chelation by DFO and DOTA
^{64}Cu	^{68}Zn$(p,\alpha n)^{64}$Cu	β-particles (0.6 MeV)	2 mm	12.7 hours	Chelation by DOTA, BAT, or TETA
^{67}Cu	^{nat}Zn$(p,2p)^{67}$Cu or ^{68}Zn$(p,2p)^{67}$Cu	β-particles (0.4–0.6 MeV)	2 mm	2.6 days	Same as for ^{64}Cu
^{177}Lu	^{176}Yb$(n,\gamma)\rightarrow^{177}$Yb $\rightarrow^{177}$Lu	β-particles (0.5 MeV)	2 mm	6.6 days	Chelation by DOTA or CHX-DTPA
^{213}Bi	^{225}Ac$/^{213}$Bi generator	α-particles (8 MeV)	50–100 μm	46 minutes	Chelation by DOTA or CHX-DTPA
^{225}Ac	^{233}U$\rightarrow^{225}$Ac	α-particles (several duaghter radionuclides with different energies)	50–100 μm	10 days	Chelation by DOTA

6.7.5 LABELING MABS AND PEPTIDES WITH RADIOHALOGENS

6.7.5.1 Iodine Radionuclides

Radioiodination is one of the simplest ways to radiolabel a biomolecule. Several radionuclides of iodine are available for SPECT or PET (^{123}I, [124]I, ^{131}I; Tables 6.7-1 and 6.7-2) or for targeted *in situ* radiotherapy of cancer (^{125}I and ^{131}I; Table 6.7-3). Iodine is present in a valence state of 1⁻ in the alkaline solution in which it is supplied. It requires oxidation to a 1⁺ valence state for electrophilic substitution into tyrosine amino acids in antibodies and peptides [15]. Reaction of radioiodine with histidine, cysteine, methionine, phenylalanine, and tryptophan residues is possible but less likely. The most commonly used oxidizing agents are chloramine-T (N-chloro-4-methylbenzene sulfonamide) [16] and Iodogen (1,3,4,6-tetrachloro-3α, 6α-diphenylglycouril; Pierce Chemical Co. Rockford, IL) [17]. Chloramine-T provides higher radiolabeling yields than Iodogen (70–90% vs. 40–60%) but because it is a stronger oxidizing agent, it may damage mAbs, especially if the conditions are not carefully controlled. A typical chloramine-T-mediated radio-iodination involves diluting the antibody or peptide in a slightly alkaline buffer and incubating it with radio-iodine and 10–20 μg of chloramine-T in a glass tube for 30–60 seconds at room temperature [18]. The radio-iodination reaction is stopped by adding 20–40 μg of the reducing agent, sodium metabisulfite. Radio-iodinated antibodies and peptides can be purified from free radioiodide by size-exclusion chromatography (SEC); alternatively, peptides may be purified by reversed-phase chromatography. Iodogen is a water-insoluble oxidizing agent that is dissolved in chloroform; 10–20 μg are then aliquoted into a glass tube and the chloroform is evaporated using a gentle stream of nitrogen to leave a coating of Iodogen on the inside surface of the tube. The biomolecule contained in a suitable buffer and radioiodide are then added to the tube and incubated for 1–2 minutes at room temperature. The reaction is stopped simply by transferring the radio-iodination mixture to a chromatography column for purification, leaving the water-insoluble Iodogen remaining in the tube. Iodogen is a more gentle oxidizing agent than chloramine-T and is preferred for radio-iodinating mAbs and their fragments.

The main challenge for radio-iodination of mAbs and peptides is their instability *in vivo* to proteolysis, deiodination, and loss of radioactivity from tumor cells. It is widely recognized that radio-iodinated biomolecules are proteolytically degraded in cells to radio-iodotyrosine, which is efficiently exported from the cells by membrane amino acid transporters. Released radio-iodotyrosine is deiodinated by deiodinases found in tissues, and the free radio-iodine redistributes and accumulates in organs with sodium iodide symporter expression, particularly the thyroid, stomach, and salivary glands. For tumor imaging applications, these catabolic processes diminish the tumor signal and increase normal tissue uptake of radioactivity. Moreover, in radiotherapeutic applications, these processes diminish the radiation absorbed dose to the tumor and increase the dose to normal tissues, thus narrowing the therapeutic index. To address this issue, several new radio-iodination methods have been developed that retain the radioactive catabolites within cells; these methods are known as "residualizing" techniques.

Zalutsky and Narula [19] first reported a method for residualizing radio-iodine in cells in 1987. This group synthesized the N-succinimidyl ester of 3-(tri-n-butylstannyl) benzoate (ATE), a precursor that could be radio-iodinated in anhydrous chloroform by iodo-destannylation using anhydrous t-butylhydroperoxide (TBHP) as an oxidant. Radio-iodinated ATE, termed N-succinimidyl-3-iodobenzoate (SIB; Figure 6.7-3), was purified on a silica gel Sep-Pak (Waters) and conjugated to the N-terminus or ε-amino groups of lysine residues on IgGs through its reactive N-succinimidyl ester side chain. In this method, radio-iodine is situated at a position on the aromatic ring of ATE that is not ortho- to a phenolic hydroxyl group (as in the case of radio-iodotyrosine); this renders the molecule resistant to *in vivo* de-iodination. In mice, thyroid uptake of radioactivity was substantially decreased compared with IgGs labeled with radio-iodine using Iodogen. A paired-label experiment comparing the tumor and normal tissue uptake of an F(ab')$_2$ fragment of the OC125 mAb labeled with ^{125}I using SIB or ^{131}I using Iodogen in mice implanted with OVCAR-3 ovarian carcinoma xenografts, revealed that thyroid uptake of radio-iodine was reduced 100-fold using ATE [20]. There was also more rapid elimination of radioactivity from normal tissues, and at 96 hours post-injection (p.i.), T/NT ratios were fourfold higher. Similar promising results were observed in mice implanted with s.c. D-54 MG glioblastoma xenografts administered anti-tenascin mAb 81C6 labeled with ^{125}I using ATE compared with labeling with ^{131}I using Iodogen [21]. Zalutsky et al. [22] extended this residualizing strategy with an alternative radio-iodination agent, N-succinimidyl 5-[^{131}I]iodo-3-pyridinecarboxylate (SIPC; Figure 6.7-3). SIPC was used to radio-iodinate 81C6 mAb IgG and the

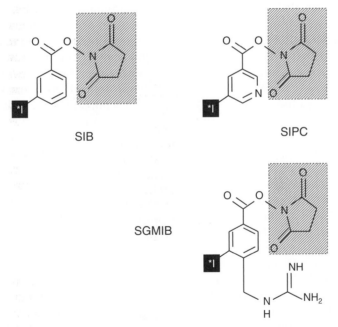

Figure 6.7-3. Precursors used for residualizing radio-iodination of monoclonal antibodies and peptides. Chemically reactive groups on the precursors are shaded, and site for radio-iodination of the precursor is indicated in black and with an asterisk. SIB: N-succinimidyl-3-iodobenzoate; SIPC: N-succinimidyl 5-iodo-3-pyridinecarboxylate; SGMIB: N-succinimidyl 4-guanidinomethyl-3-iodobenzoate.

F(ab')$_2$ fragment of anti-melanoma mAb Mel-14. It was hypothesized that SIPC, an iodopyridine, would be more dissimilar than SIB to iodotyrosine, and thus more resistant *in vivo* to de-iodination. The normal tissue distribution in mice of the 81C6 mAb and Mel-14 F(ab')$_2$ fragment radio-iodinated using SIPC or ATE were similar and there was very low thyroid uptake of radioactivity [<0.2–0.3% injected dose (% i.d.) at 7 days p.i.] using either reaagent. The tumor-specific anti-EGFRvIII mAb L8A4, which internalizes into receptor-positive cells and is highly susceptible to de-iodination, was radio-iodinated using SIPC resulting in improved tumor cell retention of radioactivity *in vitro* and increased accumulation in EGFRvIII positive tumor xenografts in mice providing enhanced T/NT ratios in comparison with L8A4 antibodies radio-iodinated using Iodogen [23, 24]. Radio-iodination using SIPC has also been applied to two 13-mer peptides: α-melanocyte-stimulating hormone (α-MSH) and its analog, [Nle^4, D-Phe^7]-α-MSH, resulting in preservation of receptor-binding properties *in vitro* and stability against de-iodination *in vivo* in mice [25]. However, conjugation of the peptides with SIPC increased their hydrophobicity. The catabolite of mAbs and peptides radio-iodinated with SIPC was found to be lysine-iodobenzoic acid (Lys-IBA).

One strategy that has been investigated to further enhance the retention of radio-iodine in cells is to use labeling techniques that generate a charged catabolite after intracellular proteolysis, which cannot traverse the lysosomal membrane and is thus resistant to exocytosis. Vaidyanathan et al. [26] synthesized N-succinimidyl 4-guanidinomethyl-3-[^{131}I]iodobenzoate (SGMIB; Figure 6.7-3), a radio-iodinating reagent that was expected to generate a positively charged catabolite at the acidic pH in lysosomes. They used this reagent to radio-iodinate anti-EGFRvIII mAb L8A4. There was three to fourfold greater retention of radioactivity in receptor-positive U87MG glioblastoma cells when these antibodies were radio-iodinated using SGMIB than when Iodogen was used. Analysis confirmed that the final catabolites were cationic. SGMIB has advantages over SIPC as a residualizing radio-iodination reagent, in that a twofold improvement in tumor retention of radioactivity was observed for L8A4 mAbs in mice implanted s.c. with D-256 glioblastoma xenografts expressing EGFRvIII. Again, similar to the SIPC reagent, thyroid radioactivity was very low (<0.35% i.d.) [27]. An analogous approach was reported by Shankar et al. [28] who used N-succinimidyl 3-[^{131}I]iodo-4-phosphonomethylbenzoate ([^{131}I]SIPMB) to radio-iodinate mAb L8A4; this reagent gen-erates an anionic catabolite that is retained within cells.

A different strategy for residualizing radio-iodine in cells uses radio-iodinated diethylenetriaminepentaacetic acid (DTPA)-containing peptides composed of one or more D-amino acids including D-tyrosine that are conjugated through a maleimide functional group to chemically reduced IgG mAbs [29]. Two such peptides: R–Gly–D–Tyr–D–Lys[1-(p-thiocarbonylaminobenzyl)DTPA], termed IMP-R1, and [R–D–Ala–D–Tyr–D–Tyr–D–Lys]$_2$(CA-DTPA), termed IMP-R2, were described by Govindan et al. [29]. The BOC-protected peptides were radio-iodinated at the D-tyrosine amino acid using chloramine-T and derivatized at their N-terminus with sulfo-SMCC to introduce maleimide groups. After deprotection, the maleimide-derivatized and radio-iodinated peptides were conjugated to thiol groups generated by dithiothreitol (DTT) reduction of some disulfide linkages on the anti-CD20 mAb LL2 or the anti-EGP-1 mAb RS7. The premise of including DTPA in the peptides was that as it had been shown that mAbs labeled with ^{111}In through introduction of a DTPA metal chelator (see section 6.7.3) were catabolized to ^{111}In–DTPA–lysine, which was retained within the cells [30], it would be expected that radio-iodinated peptides containing this group would be trapped. The DTPA groups in the peptides are not used for radiolabeling. D-amino acids were included because these are more resistant to proteolysis than L-amino acids, and in particular, D-iodotyrosine is more resistant to *in vivo* de-iodination than L-tyrosine [31]. The immunoreactivity of the RS7 antibody was maintained using this labeling strategy (radio-iodinated LL2 was not tested for immunoreactivity), and both antibodies exhibited greater retention of radioactivity in tumor cells *in vitro* than antibodies radio-iodinated using chloramine-T. A series of radio-iodinated DTPA–D-amino acid peptides (IMP-R1 to IMP-R8) that differed in their hydrophobicity and charge was synthesized and conjugated to a DTT-reduced RS7 mAb [32]. These peptides varied in their DTPA content and extent of maleimide derivatization. Increasing the maleimide substitution in the peptides from one to two functional groups per molecule increased the conjugation efficiency with the R27 mAb from 30% to >80%. In mice implanted s.c. with Calu-3 lung carcinoma xenografts, tumor radioactivity was enhanced for mAb RS7 radio-iodinated using any of the DTPA–D-amino acid peptides compared with chloramine-T radioiodinated mAbs. However, normal tissue uptake was highest for the most hydrophobic peptides. IMP-R4: MCC–Lys(MCC)–Lys(1-((p-CSNH)benzyl)DTPA)–D-Tyr–D-Lys(3-((p-CSNH)benzyl)DTPA)–OH, where MCC represents the maleimide groups, provided the greatest retention of radio-iodine in tumors and the lowest normal tissue accumulation. This approach has been optimized using a one-vial kit labeling procedure under GMP conditions to yield at least 100 mCi of highly pure (>95%) ^{131}I-labeled humanized anti-CEA mAb hMN-14 that exhibits preserved immunoreactivity (>95%) for targeted *in situ* radiotherapy of malignancies [33]. A similar strategy was recently described by Foulon et al. [34]. This group used a polycationic peptide composed of D-amino acids: D–Lys–D–Arg–D–Tyr–D–Arg–D–Arg to radio-iodinate anti-EGFRvIII mAb L8A4. These peptides were first radio-iodinated using Iodogen and then conjugated in 60% overall yield via a maleimide group to mAb L8A4, which was thiolated using 2-iminothiolane. Paired label experiments in mice implanted s.c. with U87 glioblastoma xenografts expressing EGFRvIII and administered mAb L8A4 labeled with ^{125}I using this approach showed up to a fivefold higher tumor accumulation compared with L8A4 labeled directly with ^{131}I using Iodogen. However, use of the peptides for radio-iodinating L8A4 increased accumulation in the kidneys, perhaps because of binding of the positively charged catabolites to renal tubular cells [34].

Radio-iodine may also be residualized in cells by substitution onto an aromatic residue that is linked to a carbohydrate moiety attached to the biomolecule [35]. A sevenfold increase in tumor retention of radioactivity was observed in mice implanted with Calu-3 lung carcinoma xenografts at 7 days p.i. of radio-iodinated mAb RS7 using a dilactitol-tyramine (DLT) carbohydrate adduct compared with chloramine-T radio-iodination (38.0% vs. 5.5% i.d./g, respectively) [36]. Similar results were achieved using tyramine cellobiose (TCB)-radio-iodinated anti-EGFRvIII mAb L8A4 in mice bearing receptor-positive tumor xenografts [37]. However, a limitation of this approach seems to be the higher liver and spleen uptake of radioactivity of mAbs radio-iodinated using TCB [38]. Epidermal growth factor (EGF), a 53-amino acid peptide ligand for the EGFR present on many epithelial malignancies, has been labeled using an analogous approach by radio-iodinating a tyrosine-modified dextran and then conjugating this dextran molecule to EGF [39]. Glioblastoma cells expressing EGFR showed significantly greater retention of radioactivity when incubated with dextran-EGF in which the radio-iodine was present on the dextran moiety compared with EGF or dextran-EGF where the radio-iodine was substituted into tyrosine amino acids on the EGF molecule itself.

Since the advent of high-resolution PET/computed tomography (CT) [40] and small animal PET tomographs [41] and building on the success of ^{18}F-fluorodeoxyglucose (^{18}F-FDG) for PET of malignancies [42], there has been a growing interest in labeling mAbs and peptides with positron-emitters. ^{124}I is a positron-emitter with a sufficiently long half-life of 4.2 days that is feasible for labeling mAbs and peptides for PET. Nevertheless, ^{124}I has limitations compared with ^{18}F. These limitations include its lower abundance of positron decay compared with ^{18}F (23% versus 100%); the higher β^+ energies for ^{124}I compared with ^{18}F (0.8–2.1 MeV vs. 0.6 MeV), which result in poorer spatial resolution (2.3 mm vs. 0.7 mm); and the emission of high-energy γ-photons by ^{124}I that degrade the PET image and contribute along with the higher positron energy and longer half-life to higher radiation absorbed doses [13]. Nevertheless, intact IgG mAbs [43–46], genetically engineered antibody fragments (e.g., minibodies and diabodies) [47, 48], and peptides [49] have been labeled with ^{124}I, in most cases using chloramine-T or Iodogen. It is important to add a small amount of ascorbic acid to the formulation after labeling biomolecules with ^{124}I in order to protect them from radiolytic decomposition caused by the high-energy positrons [46]. ^{124}I-labeled antibodies have been successfully used for PET imaging of receptor/antigen-positive tumors in mouse xenograft models [43, 45–48] and in one study in a child with neuroblastoma [44]. One advantage of PET using ^{124}I-labeled antibodies is that it allows accurate quantification of radioactivity uptake in tumors and normal tissues. This could be useful for predicting the radiation absorbed doses for subsequent RIT using the corresponding ^{131}I-labeled antibodies [47, 50].

6.7.5.2 Bromine Radionuclides

Although there has been interest in labeling mAbs and peptides with ^{124}I for PET, another attractive positron-emitter is bromine-76 (^{76}Br). ^{76}Br decays with a half-life of 16.2 hours, emitting positrons in 54% abundance with energy of 3.4 MeV. The almost sixfold higher positron energy of ^{76}Br compared with that of ^{18}F (0.6 MeV) provides poorer spatial resolution (>5 mm vs. 0.7 mm, respectively) and a higher radiation absorbed dose [13]. Nevertheless, ^{76}Br is more practical for labeling biomolecules for PET than ^{124}I due to its greater abundance of positron emissions, and because it can be produced using low-energy biomedical cyclotrons by the ^{76}Se(p,n)^{76}Br reaction [51]. Chloramine-T was initially studied for radiobromination of mAbs and peptides using ^{76}Br. The weaker oxidizing agent, Iodogen, does not seem to be useful for ^{76}Br labeling because of the greater resistance to oxidation compared with radio-iodine [52]. The colon cancer mAbs A33, 3S193, and 38S1 as well as EGF have been labeled with ^{76}Br using chloramine-T in yields of 63–77% and with good preservation of antigen/receptor-binding characteristics [53]. Nevertheless, in one study, the immunoreactivity of ^{76}Br-mAb 38S1 was diminished to

45% and only radiobromination using bromoperoxidase was found to generate radiolabeled antibodies with preserved immunoreactivity [52]. Human colon cancer xenografts implanted into nude rats have been visualized by PET using ^{76}Br-mAb 38S1 [54, 55].

Despite the ability to directly radiobrominate some antibodies using chloramine-T, many investigators obtained inconsistent and low yields and as mentioned, in some cases, decreased immunoreactivity using this approach. Therefore, newer strategies were explored for labeling mAbs with ^{76}Br using N-succinimidyl para-[^{76}Br]bromobenzoate (^{76}Br-SPBrB; Figure 6.7-4) [56]. ^{76}Br-SPBrB was generated by bromodestannylation of N-succinimidyl-para-tri-n-butylstannylbenzoate (SPMB) using chloramine-T [56, 57]. The resulting ^{76}Br-SPBrB was purified by high-performance liquid chromatography (HPLC) and conjugated through the activated N-succinimidyl moiety to ε-amino groups of lysines on the antibodies. Using this approach, mAb 38S1 was labeled with ^{76}Br in 49% yield and with good preservation of immunoreactivity (69–76%) [56]. Building on the strategy described by Vaidya-nathan et al. [26] to radio-iodinate antibodies using reagents (e.g., SGMIB) that yield positively charged catabolites that are retained in cells, Mume et al. [58] synthesized N-succinimidyl 5-[^{76}Br]bromo-3-pyridinecarboxylate (Figure 6.7-4) by bromodestannylation of N-succinimidyl-5-(tributylstannyl)3-pyridinecarboxylate and conjugated it to the HER-2/neu mAb trastuzumab (Herceptin). The labeling yield was 45%, but after purification, the immunoreactivity with HER2/neu positive SKOV-3 ovarian carcinoma cells was >75%. However, this method of radiobromination did not improve cellular retention of radioactivity compared with trastuzumab labeled with ^{76}Br–SPBrB. A site-specific radiobromination technique was recently reported for labeling the affibody molecule $(Z_{HER2-4})_2$–Cys with ^{76}Br at a cysteine residue [59]. $(Z_{HER2-4})_2$–Cys affibodies are small recombinant proteins [$M_r = 7$ kDa (monomer) or 15 kDa (dimer)] that bind with high specificity to HER2/neu receptors, similar to antibody fragments but produced by phage display techniques [60]. A bifunctional reagent, ((4-hydroxy-phenyl)ethyl)maleimide (HPEM; Figure 6.7-4), reactive with thiols was synthesized and radiobrominated with ^{76}Br using chloramine-T. The radiobrominated HPEM was conjugated through its maleimide group to the free thiol on the cysteine residue of the $(Z_{HER2-4})_2$–Cys affibodies. ^{76}Br–$(Z_{HER2-4})_2$–Cys exhibited preserved binding to HER2/neu-positive SKOV-3 cells *in vitro* and achieved high tumor uptake (5% i.d./g) and T/NT ratios (up to 31:1) at 4 hours p.i. *in vivo* in mice implanted with SKOV-3 tumor xenografts.

Finally, a totally new approach to labeling antibodies with ^{76}Br employed derivatives of polyhedral boron clusters such as *closo*-dodecaborate (2$^-$) or *nido*-undecaborate (1$^-$) anions (Figure 6.7-4) [61, 62]. These structures form strong boron–bromine bonds and were labeled with ^{76}Br using chloramine-T. Once labeled with ^{76}Br, they were conjugated through their benzylisothiocyanato side chain to ε-amino groups of lysine residues on mAbs. The idea was that, as ^{76}Br is attached through a bromine–boron bond to a charged molecular structure that is completely foreign to the body, enzymatic debromination and exocytosis of radioactivity from cells would be substantially reduced. Bruskin et al. [61] labeled trastuzumab (Herceptin) with ^{76}Br in >80% yield using this approach, resulting in a preparation that was immunoreactive with HER2/neu-positive SKBR-3 human breast cancer cells. Trastuzumab has also been labeled with ^{76}Br using the related *nido*-undecaborate (1$^-$) anion [62].

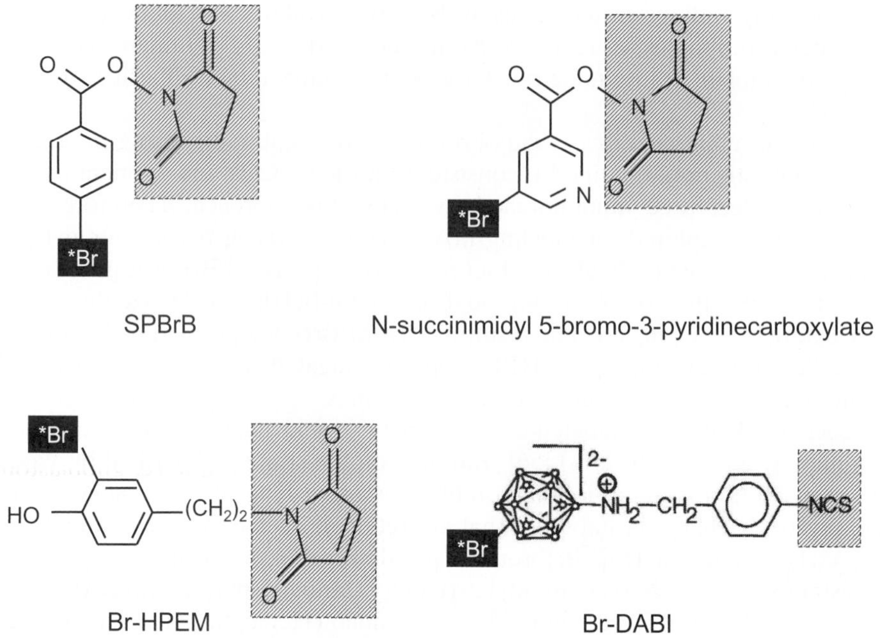

SPBrB   N-succinimidyl 5-bromo-3-pyridinecarboxylate

Br-HPEM   Br-DABI

Figure 6.7-4. Precursors used for residualizing radiobromination of monoclonal anti-bodies and peptides. Chemically reactive groups on the precursors are shaded, and site for radiobromination of the precursor is indicated in black and with an asterisk. SPBrB: N-succinimidyl para-bromobenzoate; Br-HPEM: bromo-((4-hydroxy-phenyl)ethyl) maleimide; Br-DABI: bromo-(4 isothiocyanatobenzyl-ammonio)-bromo-decahydro-closo-dodecaborate.

6.7.5.3   Fluorine Radionuclides

Fluorine-18 (^{18}F) is the most widely used radionuclide for PET of tumors, usually in the form of ^{18}F-FDG [42]. Due to its relatively low positron energy (0.6 MeV), ^{18}F provides excellent spatial resolution (0.7 mm); it also is associated with lower radiation absorbed doses compared with ^{124}I or ^{76}Br [13]. There has been consider-able interest in labeling mAb fragments and peptides with ^{18}F [63]. The pharma-cokinetics of intact IgG mAbs, in particular their slow elimination from the blood, do not lend themselves to the use of ^{18}F, due to its short half-life of 110 minutes. In contrast, antibody fragments (e.g., F(ab')$_2$, Fab, and scFv) and peptides are accu-mulated rapidly in tumors and eliminated quickly from the blood and most normal tissues. Antibody fragments and peptides labeled with ^{18}F could therefore be useful for PET. The principal challenge in labeling biomolecules with ^{18}F is that nucleo-philic substitution reactions are required; direct electrophilic radiofluorination of biomolecules requires carrier fluoride that yields low specific activity that is unsuit-able for imaging epitopes/receptors on tumor cells [63]. To solve this problem, biomolecules are labeled with ^{18}F by first substituting ^{18}F onto a prosthetic group and then chemically linking this group through a reactive side chain to the biomol-ecules [63]. One primary consideration in designing methods for labeling antibody fragments and peptides with ^{18}F is the amount of time required to complete the labeling and conjugation procedures due to the short 110-minute half-life of ^{18}F. It is desirable that the labeling procedure, including quality control testing, be completed in 1–1.5 hours, to minimize losses in yield, simply due to the physical decay of ^{18}F.

Garg et al. [64] described an approach for labeling F(ab')$_2$ and Fab fragments of the antimyosin antibody R11D10 using 4-[^{18}F]fluorobenzylamine succinimidyl ester ([^{18}F]SFBS; Figure 6.7-5). Although this method was rapid with a total labeling time of 1.5 hours, synthesis of [^{18}F]SFBS required three separate reactions followed by HPLC purification of the reagent; the yield of [^{18}F]SFBS was 25–40%. Conjugation of [^{18}F]SFBS to antibody fragments through reaction of the succinimidyl ester with ε-amino groups on lysines was performed in a fourth step, which required purification of the ^{18}F-labeled fragments from excess [^{18}F]SFBS by SEC. The immunoreactivity of the ^{18}F-labeled F(ab')$_2$ and Fab fragments of R11D10 were relatively preserved (89% and 75%, respectively). Using [^{18}F]SFBS as a prosthetic group, Garg et al. [65] labeled F(ab')$_2$ fragments of the anti-glioma mAb Mel-14 and obtained high T/NT ratios (up to 14:1 at 4 hours p.i.) in a s.c. glioblastoma mouse tumor xenograft model. They also labeled Fab fragments of the TP-3 antibody with [^{18}F]SFBS, which permitted PET of osteosarcoma tumors in dogs [66]. An analogous approach for labeling biomolecules with ^{18}F uses the prosthetic group, [^{18}F]-N-succinimidyl 4-(fluoromethyl)benzoate ([^{18}F]SFB; Figure 6.7-5)

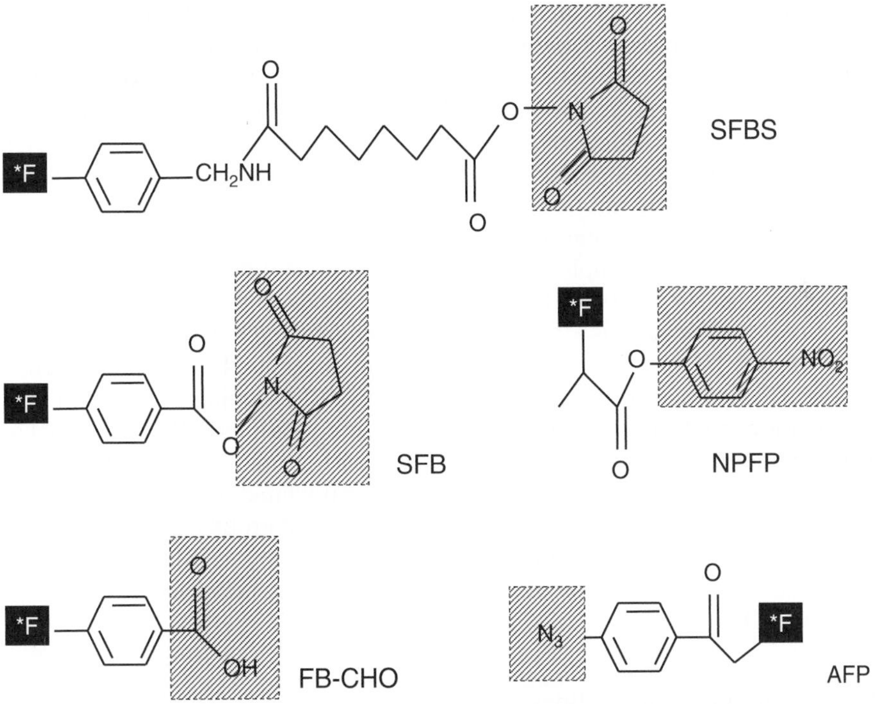

Figure 6.7-5. Prosthetic precursors used for radiofluorination of monoclonal antibody fragments and peptides. Chemically reactive groups on the precursors are shaded, and site for radiofluorination of the precursor is indicated in black and with an asterisk. SFBS: 4-fluorobenzylamine succinimidyl ester; SFB: N-succinimidyl 4-(fluoromethyl)benzoate; NPFP: 4-nitrophenyl 2-fluoropropionate; FB-CHO: 4-fluorobenzaldehyde; AFP: 4-azidofluorophenacyl.

[67]. The synthesis of [^{18}F]SFB was optimized [68], and the reagent was used for labeling F(ab')$_2$ fragments of the anti-glioma mAb Mel-14. The immunoreactivity of ^{18}F-labeled Mel-14 F(ab')$_2$ fragments using [^{18}F]SFBS or [^{18}F]SFB was indistinguishable (65% vs. 64%, respectively), and the radiolabeled fragments demonstrated similar tissue distribution and pharmacokinetics in dogs [69]. The 13-amino acid α-MSH peptide analog, N-acetyl-Ser-Tyr-Ser-NorLeu-Glu-His-D-Phe-Arg-Trp-Gly-Lys-Pro-Val-NH$_2$, has been labeled with [^{18}F]SFB with good preservation of receptor-binding affinity [70].

Another reagent that has been used for labeling biomolecules with ^{18}F is 4-nitrophenyl 2-[^{18}F]fluoropropionate ([^{18}F]NPFP; Figure 6.7-5) [71]. [^{18}F]NPFP is a smaller prosthetic group that may have less impact on the receptor-binding properties and/or physico-chemical properties of peptides than [^{18}F]SFBS or [^{18}F]SFB. Accordingly, [^{18}F]NPFP was used for labeling the octreotide peptide analog, SDZ 223-228, with ^{18}F resulting in full retention of biological activity [72]. It was necessary to Boc-protect the ε-amino group on lysine-5 in SDZ 223-228 during labeling with [^{18}F]NPFP and then deprotect afterward, because this residue is critical for SMSR-binding. Nevertheless, a radiofluorination technique using 4-[^{18}F]fluorobenzaldehyde ([^{18}F]FB–CHO; Figure 6.7-5) that allows site-specific labeling of aminoxyacetic acid functionalized octreotide analogs without the need to protect/deprotect the ε-amino group on lysine-5 has been recently reported [73]. This method allowed PET imaging of s.c. transplantable AR42J rat pancreatic tumors in athymic mice at 1 hour p.i. of the ^{18}F-labeled octreotide derivatives. In another study, D–Phe^1–octreotide labeled with ^{18}F using [^{18}F]NPFP exhibited higher liver accumulation and hepatobiliary clearance of radioactivity than the corresponding radio-iodinated derivative in Lewis rats bearing pancreatic islet cell tumors, suggesting that [^{18}F]NPFP may increase the hydrophobicity of peptides [74]. Nevertheless, T/B ratios were 5.2:1 at 1 hour p.i. and 4.2:1 at 2 hours p.i., due to the rapid blood clearance of the ^{18}F-D-Phe^1-octreotide. [^{18}F]NPFP and [^{18}F]SFB have also been employed for ^{18}F labeling of peptides that contain the arginine-glycine-aspartic acid (RGD) sequence that binds to the αvβ3 integrin implicated in tumor angiogenesis [75, 76]. In a study comparing labeling of proteins with [^{18}F]NPFP, [^{18}F]SFB, and another fluorinating reagent, 4-azidofluorophenacyl-[^{18}F]AFP (Figure 6.7-5), it was found that [^{18}F]NPFP-conjugated human serum albumin was partially unstable under slightly basic conditions [77]. It was concluded that [^{18}F]SFB may be the most suitable radiofluorinating agent. Finally, [^{18}F]FB-CHO (Figure 6.7-5) has been used to site-specifically label human serum albumin with ^{18}F at hydrazinenicotinamide (HYNIC) functional groups introduced into the protein using HYNIC N-hydroxysuccinimide ester [78]. This method has similarities with the use of HYNIC for labeling biomolecules with ^{99m}Tc (section 6.7.6.1).

6.7.5.4 Astatine Radionuclides

Astatine-211 (^{211}At; Table 6.7-3) is an α-particle emitter with a half-life of 7.2 hours that is useful for labeling mAbs and peptides for targeted *in situ* radiotherapy of malignancies [2, 79]. ^{211}At is produced in a cyclotron using the ^{209}Bi(α,2n)^{211}At reaction. However, the astatine–carbon bond is unstable [80] and direct electrophilic astatination of biomolecules at tyrosine residues using oxidizing agents such as chloramine-T or hydrogen peroxide results in low yield [81] and susceptibility *in vivo* to de-astatination [82]. Zalutsky and Narula [83] addressed this problem by labeling N-succinimidyl 3-(tri-n-butylstannyl) benzoate (ATE) with [211]At. The N-succinimidyl 3-[[211]At]astatobenzoate (SAB; Figure 6.7-6) was conjugated to ε-amino groups on lysine residues in antibodies, which is similar to the strategy used for radio-iodination. Using SAB, intact mAb 81C6 IgG and F(ab')$_2$ fragments of mAb Mel-14 directed against glioblastoma were labeled with [211]At with preservation of immunoreactivity *in vitro* and excellent targeting *in vivo* to s.c. D-54 MG human glioblastoma xenografts in mice [84]. Due to its short half-life (7.2 hours) and the short range of the α-particles (50–100μm), F(ab')$_2$ fragments are more suitable for labeling with [211]At than intact mAb IgG's because they penetrate deeper into tumors and are cleared more quickly from the blood [7]. Despite their more rapid blood clearance than F(ab')$_2$, Fab fragments are not appropriate for labeling with [211]At because they accumulate to high levels in the kidneys, thus posing a potential radiotoxicity hazard. Paired label experiments in normal mice of the anti-CEA mAb C110 or its F(ab')$_2$ fragment labeled with [211]At or ^{131}I using the ATE method revealed that there was greater tissue retention of [211]At than for [131]I [85]. In another study, mice implanted s.c. with TK-82 human renal cell carcinoma xenografts showed similar tumor and normal tissue uptake after administration of [211]At or ^{125}I-labeled A6H F(ab')$_2$ [86]. Tumor uptake in this mouse xenograft model was 30% i.d./g for [211]At-A6H F(ab')$_2$ versus 19% i.d./g for ^{125}I-A6H F(ab')$_2$ at 19 hours p.i. T/B ratios were 3:1 for both [211]At and [125]I. [211]At-labeled anti-tenascin mAb 81C6 has been produced using the SAB reagent in sufficient quantities (2–10mCi) under GMP conditions for use in a phase I clinical trial in glioblastoma patients [87]. Radiolytic decomposition of SAB due to the α-particles emitted by [211]At and subsequent low radiolabeling yields is nevertheless a major challenge in producing clinical quality [211]At-labeled mAbs [88, 89]. It seems that radiolytic decomposition is especially problematic when chloroform is used as a solvent for synthesizing SAB; benzene and methanol are alternatives that are less susceptible to the radiolytic effects of [211]At.

Reist et al. [90] extended their previous residualizing radio-iodination approach using N-succinimidyl 5-[^{131}I]iodo-3-pyridinecarboxylate (SIPC) to [211]At. Anti-EGFRvIII mAb L8A4 was labeled by conjugation with N-succinimidyl 5-[[211]At]astato-3-pyridinecarboxylate (SAPC; Figure 6.7-6). Again, the premise is

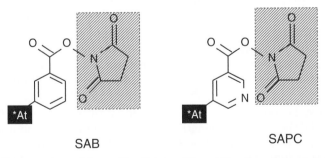

SAB

SAPC

Figure 6.7-6. Precursors used for residualizing radio-astatination of monoclonal antibodies and peptides. Chemically reactive groups on the precursors are shaded, and site for radio-astatination of the precursor is indicated in black and with an asterisk. SAB: N-succinimidyl 3-astatobenzoate; SAPC: N-succinimidyl 5-astato-3-pyridinecarboxylate.

that proteolysis of the [211]At-labeled antibodies would yield a positively charged [211]At catabolite that would be unable to traverse the lysosomal membrane or be exocytosed. Immunoreactivity was maintained, and [211]At- and ^{131}I-labeled L18A4 showed similar tumor and normal tissue accumulation in mice implanted s.c. with EGFRvIII-positive U87MG glioblastoma xenografts. Other biomolecules that have been labeled with [211]At include octreotide [91] and the anti-CD20 mAb rituximab (Rituxan; Roche Pharmaceuticals Ltd.) directed against non-Hodgkin's B-cell lymphomas [92].

6.7.6 LABELING MABS AND PEPTIDES WITH RADIOMETALS

6.7.6.1 Technetium Radionuclides

Technetium-99m (^{99m}Tc; Table 6.7-1) is the most widely available, least expensive, and most commonly used radionuclide in nuclear medicine and is thus an attractive candidate for labeling mAbs, their fragments, as well as peptides for tumor imaging. ^{99m}Tc is a metastable form of ^{99}Tc, which decays by internal conversion to its ground state (^{99}Tc) with a half-life of 6 hours, emitting a γ-photon of 140 keV that is easily imaged by gamma cameras available in all nuclear medicine facilities. ^{99m}Tc is produced from the decay of molybdenum-99 (^{99}Mo) using a commercially available ^{99}Mo/^{99m}Tc generator at very low cost (less than $0.50 per mCi). Its relatively short half-life minimizes the radiation exposure to patients undergoing imaging procedures, but it necessitates rapid labeling procedures for biomolecules. There are several approaches to labeling antibodies and peptides with ^{99m}Tc [93, 94]. These approaches include (1) direct methods that rely on the binding of ^{99m}Tc to endogenous thiols generated by reduction of disulfide linkages in mAbs or introduced chemically by reaction with thiolating agents such as 2-iminothiolane, and (2) indirect methods that involve binding of ^{99m}Tc to a chelating agent that is then conjugated to the mAb or peptide ("pre-formed chelator" approach) or to a chelator already incorporated into the biomolecule. There are advantages and disadvantages of each of these strategies with respect to ease of use, *in vivo* stability, and amenability to kit formulation.

Intact mAb IgGs can be labeled simply and directly with ^{99m}Tc by reduction of a small proportion (<5%) of the inter- or intrachain disulfide bonds to free thiols by treatment with a 2000-fold molar excess of 2-mercaptoethanol (2-ME) for 30 minutes (Figure 6.7-7) [94]. The reduced IgGs are purified from excess 2-ME by SEC and labeled to high efficiency (>90%) by transchelation of ^{99m}Tc from ^{99m}Tc-glucoheptonate or ^{99m}Tc-methylene diphosphonate (^{99m}Tc–MDP). ^{99m}Tc-glucoheptonate or ^{99m}Tc–MDP are prepared from commercial kits and ^{99m}Tc pertechnetate (^{99m}TcO$_4^-$) eluted from a ^{99}Mo/^{99m}Tc generator. This method of labeling mAbs with ^{99m}Tc was introduced by Schwartz and Steinstrasser in 1987 [95] and later optimized by Mather and Ellison in 1990 [96]. A recent modification of the Schwartz method employed exposure of the anti-CD20 mAb rituximab (MabThera; Roche) to ultraviolet (UV) light of wavelength 320 nm for 20 minutes to photoreduce some disulfide linkages to thiols, which were then labeled wth ^{99m}Tc by transchelation from ^{99m}Tc–MDP [97]. This photoreduction method was originally described by Stalteri and Mather in 1996 [98]; it does not require removal of any chemical reducing agents and is somewhat simpler and potentially more controllable than chemical

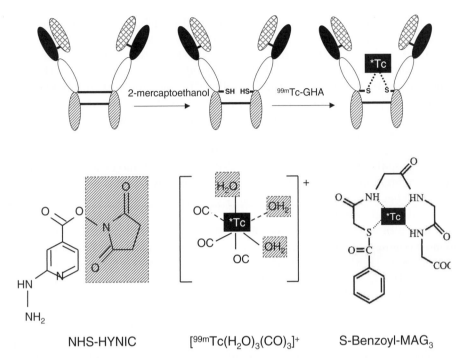

Figure 6.7-7. *Top Panel*: Method for direct labeling of intact monoclonal antibodies with ^{99m}Tc. Some disulfide linkages between the antibody chains are chemically reduced with 2-mercaptoethanol. The free thiols produced are sites for binding ^{99m}Tc-glucoheptonate (^{99m}Tc-GHA). *Bottom Panel*: Metal chelators that can be conjugated to monoclonal antibodies or peptides for labeling with ^{99m}Tc. Chemically reactive groups on the precursors are shaded, and complexation of ^{99m}Tc is indicated in black and with an asterisk. NHS-HYNIC: N-hydroxysuccinimide ester of hydrazinenicotinamide; $[^{99m}Tc(H_2O)_3(CO)_3]^+$: ^{99m}Tc(I)-carbonyl complex; S-benzoyl-MAG$_3$: S-benzoyl-mercaptoacetyl-glycyl-glycyl-glycine.

reduction with 2-ME. The advantage of the Schwartz technique is that purified reduced IgGs can be dispensed into vials and stably maintained either frozen or lyophilized until required for labeling with ^{99m}Tc, thus generating a kit formulation [99]. The major disadvantage of the method is that if too great an amount of reducing agent is used, too many disulfide linkages are reduced, increasing the risk for disrupting key linkages needed to maintain protein folding, integrity, and immunoreactivity. This is especially a problem for mAb fragments such as F(ab')$_2$, Fab, and scFv. Likewise the method cannot be used for labeling small peptides with ^{99m}Tc that harbor disulfide bonds required for maintaining biological activity (e.g., somatostatin analogs). Careful control and optimization of reduction conditions through measurement of the number of free thiols generated using Ellman's reagent [100, 101] is needed to successfully apply this method for labeling biomolecules with ^{99m}Tc. Another disadvantage is that IgGs labeled with ^{99m}Tc through direct binding to thiols are subject to loss of the radiolabel *in vivo* over time by exchange with endogenous thiol containing molecules such as cysteine and glutathione [94]. Nevertheless, the Schwartz method has been used for labeling several mAbs with ^{99m}Tc providing preserved immunoreactivity and the ability to target and image human cancers in mouse xenograft models [102, 103] and in patients [104, 105].

Due to the limitations of the Schwartz method for labeling mAb fragments (e. g., F(ab')$_2$ or Fab) and peptides with ^{99m}Tc, alternative strategies have been investigated. One method uses hydrazinenicotinamide (HYNIC) and coligands such as ethylenediaminodiacetic acid (EDDA), ethlenediaminetetraacetic acid (EDTA), tricine, or glucoheptonate to form a stable ^{99m}Tc complex with the biomolecule (Figure 6.7-7) [106]. This method was described by Abrams et al. in 1990 [107]. HYNIC is conjugated to the mAb (or its fragments) or peptides using an excess of HYNIC N-hydroxysuccinimide ester (or other chemically reactive form), which forms an amide linkage with the N-terminus or ε-amino groups in lysines. After chromatographic purification, the HYNIC-derivatized biomolecule is labeled with ^{99m}Tc by incubation with ^{99m}TcO$_4^-$ in the presence of SnCl$_2$ and coligand. Coligands are needed because HYNIC occupies only one or two of the six coordination sites of ^{99m}Tc [106]. SnCl$_2$ is included to reduce ^{99m}Tc from its 7$^+$ valence state in ^{99m}TcO$_4^-$ to a valence of 4$^+$ or 5$^+$ for complexation with HYNIC-coligands. Our group has labeled Fab fragments of the HER2/neu mAb trastuzumab (Herceptin) with ^{99m}Tc using HYNIC [108]. ^{99m}Tc-HYNIC-trastuzumab Fab showed preserved binding affinity for HER2/neu receptors on SKBR-3 human breast cancer cells *in vitro* (K$_d$ = 1.6 × 10^{-8} M^{-1}) and demonstrated good tumor targeting *in vivo* in mice implanted s.c. with HER2/neu-positive BT-474 breast cancer xenografts (T/B ratio = 3:1 at 24 hours p.i.). BT-474 tumors were visualized by imaging. Others have labeled mAb Fab fragments with ^{99m}Tc using HYNIC achieving preserved immunoreactivity and good tumor targeting [109, 110]. As mentioned, the Schwartz method is not amenable to labeling peptides that harbor key intramolecular disulfide linkages. However, the somatostatin analog, D-Phe^1-Tyr^3-octreotide (TOC), which contains a disulfide bond essential for somatostatin receptor (SMSR)-binding, has been labeled with ^{99m}Tc by conjugation to HYNIC using a variety of coligands [111–113]. Similarly, the 53-amino-acid peptide, EGF, which harbors three disufide bonds required for maintenance of its tertiary structure and receptor-binding properties has been labeled with ^{99m}Tc using HYNIC and tricine as a coligand [114]. In each case, receptor-binding was preserved *in vitro*, allowing tumor imaging in mouse xenograft models [113, 114] or in patients [111, 112].

An interesting finding is that the choice of coligand for HYNIC drastically affects the plasma protein binding, elimination from the blood, and tissue distribution of the ^{99m}Tc-labeled biomolecules. This effect was first reported by Babich and Fischman in 1995 [115] who noted differences in radioactivity accumulation in the lungs, kidneys, liver, spleen, and gastrointestinal tract of rats for the chemotactic peptide (N–For–Met–Leu–Phe–Lys) conjugated to HYNIC and labeled with ^{99m}Tc using glucarate, glucoheptonate, mannitol, or glucamine as coligands. Use of EDDA as a coligand instead of tricine for ^{99m}Tc labeling of the somastatin analogs HYNIC–TOC or RC160 produced a more hydrophilic complex that cleared more quickly from the blood in rats and favored renal over hepatobiliary clearance [116]. However, substitution of the backbone of EDDA with dimethyl, diethyl, or dibenzyl moieties increased the hydrophobicity of ^{99m}Tc–HYNIC–RC160, resulting in a fivefold to sixfold increased plasma protein binding *in vitro* and substantially increased liver uptake and hepatobiliary elimination in rats [117]. Tricine as a coligand for ^{99m}Tc–HYNIC–RC160 produced liver and intestinal radioactivity uptake intermediate between that of the EDDA and substituted EDDA derivatives. HYNIC–TOC labeled with ^{99m}Tc using EDDA as a coligand showed excellent tumor localization in a study of 10 patients with SMSR-positive tumors allowing imaging of lesions at 4 hours p.i. [112]. Similarly, in a study of 13 patients, 12 of whom had SMSR-positive malignancies, administered ^{99m}Tc–HYNIC–TOC labeled using tricine as a coligand, tumors were imaged as early as 10 minutes p.i. [111]. A comparison between these two studies revealed that the circulating blood background radioactivity was higher and that urinary excretion of radioactivity was lower in patients receiving ^{99m}Tc–tricine–HYNIC–TOC than those administered ^{99m}Tc–EDDA–HYNIC–TOC, likely due to the increased plasma protein binding of ^{99m}Tc–tricine–HYNIC–TOC [118]. In almost all cases, the ^{99m}Tc–HYNIC-labeled analogs detected the same lesions as ^{111}In–DTPA–D–Phe^{1}–octreotide, the "gold standard" for SMSR tumor imaging, but the images were clearer and lesions were detected earlier using the ^{99m}Tc analogs. ^{99m}Tc–tricine–HYNIC–TOC missed one liver lesion in a patient, which was detected with ^{111}In–DTPA–D–Phe^{1}–octreotide, probably due to its higher liver uptake [111]. The effect of a coligand on plasma protein binding and blood clearance does not seem to be restricted to small peptides such as ^{99m}Tc–HYNIC–TOC. Ono et al. [119] found that 3-benzoylpyridine (BP) ternary complexes of ^{99m}Tc–HYNIC-conjugated Fab fragments [^{99m}Tc–HYNIC–Fab(tricine)(BP)] exhibited lower plasma protein binding *in vitro* than the corresponding binary complexes [^{99m}Tc–HYNIC–Fab(tricine)$_2$] and were cleared more quickly from the blood in mice (0.35% vs. 0.98% i.d./g at 24 hours p.i., respectively). ^{99m}Tc–HYNIC–Fab(tricine)(BP) also exhibited lower liver uptake than ^{99m}Tc–HYNIC–Fab(tricine)$_2$. They concluded that the ^{99m}Tc–tricine coligand exchanges *in vivo* with plasma proteins as well as lysosomal proteins in tissues, causing slow blood clearance and high liver retention of radioactivity. This loss of radioactivity was decreased using the more stable ternary ^{99m}HYNIC–Fab(tricine)(BP) complexes. Our group similarly found a slower blood clearance in mice for trastuzumab Fab labeled with ^{99m}Tc through HYNIC using glucoheptonate as a coligand than for ^{111}In-labeled trastuzumab Fab (3.2% vs. 1.4% i.d./g at 24 hours p.i., respectively) [108, 120]. There was also higher liver uptake for ^{99m}Tc–HYNIC–trastuzumab Fab than for ^{111}In–trastuzumab Fab at 24 hours p.i. (3.8% vs. 2.4% i.d./g, respectively). As ^{99m}Tc labeling efficiencies for antibodies and peptides are high (>90%) using HYNIC, kit formulation is possible, and indeed kits have been created for labeling HYNIC–TOC with ^{99m}Tc [111].

A particularly useful method for labeling biomolecules with ^{99m}Tc, especially those produced by protein engineering techniques and that contain polyhistidine affinity tags, employs a ^{99m}Tc(I)–carbonyl complex (Figure 6.7-7) [121, 122]. This method, described by Waibel et al. in 1999 [122], involves a simple one-vial synthesis of the organometallic aqua–ion complex [^{99m}Tc(H$_2$O)$_3$(CO)$_3$]$^+$, which then efficiently complexes through release of its water molecules with imidazole nitrogens in histidine-containing biomolecules. The [^{99m}Tc(H$_2$O)$_3$(CO)$_3$]$^+$ complex is formed by heating ^{99m}TcO$_4^-$, Na$_2$CO$_3$, and NaBH$_4$ flushed with carbon monoxide (CO) at 75°C for 30 minutes. The yield of [^{99m}Tc(H$_2$O)$_3$(CO)$_3$]$^+$ is >95%. A commercial kit for producing the [^{99m}Tc(H$_2$O)$_3$(CO)$_3$]$^+$ complex (Iso-Link; Mallinckrodt, Petten, Belgium) is available. Labeling with ^{99m}Tc is achieved simply by mixing the [^{99m}Tc(H$_2$O)$_3$(CO)$_3$]$^+$ complex with the histidine-containing protein or peptide and heating for 20–30 minutes at 37°C. An scFv containing a polyhistidine tag was labeled with the ^{99m}Tc(I)–carbonyl complex resulting in a stable complex *in vitro* in human serum and retaining 87% of its radioactivity over 24 hours at 37°C [122]. It was also stable to challenge *in vitro* with a 5000-fold molar excess of histidine, retaining 94% of its radioactivity [122]. This technique is attractive because recombinant antibodies and peptides often incorporate polyhistidine affinity tags for their purification, thus allowing them to be directly and easily labeled using the ^{99m}Tc(I)–carbonyl complex. Moreover, Re(I)–carbonyl complexes have also been prepared for this method, allowing extension to radionuclides of rhenium (e.g., ^{186}Re or ^{188}Re; Table 6.7-3) for targeted *in situ* radiotherapy of malignancies [123].

Finally, tetradentate chelators such as N$_3$S (triamidothiols), N$_2$S$_2$ (diamidodithiols), or N$_2$S$_4$ (diaminotetrathiols) have been used for labeling mAbs and peptides with ^{99m}Tc (Figure 6.7-7) [106]. The advantage of these chelators is that they form stable, well-defined ^{99m}Tc complexes, but they require introduction into biomolecules through reaction of a chemically reactive ester in a side chain with ε-amine groups on lysines. This conjugation step is nonspecific and has the potential to target a key lysine required for maintenance of antigen or receptor-binding. This problem can be solved in the case of small peptides by incorporating the tetradentate chelator at a specific position into the biomolecule during its solid phase synthesis using sequences of amino acids such as cysteine-glycine-glycine-glycine (N$_3$S) [124]. Moreover, there is the possibility that ^{99m}Tc will bind to low-affinity endogenous metal-binding sites in antibodies in addition to the high-affinity sites introduced with the tetradentate chelator. Therefore, to avoid this possibility, a "preformed chelator" approach was reported by Fritzberg et al. [125]. In this preformed chelator approach, a stable ^{99m}Tc–N$_2$S$_2$ complex is synthesized first and then conjugated to the mAbs through a reactive tetrafluorophenyl ester side chain. However, this strategy is complex and time consuming; the conjugation step is inefficient; and it does not lend itself easily to kit formulation. Despite these limitations, tetradentate chelators have been employed for labeling various mAbs and peptides with ^{99m}Tc [94, 106]. An interesting application of a tetradentate chelator was described for ^{94m}Tc labeling of the somatostatin analog, demotate [126]. ^{94m}Tc is a positron-emitter (β$^+$ 72%; Eβ$^+$ max 2.5 MeV; Table 6.7-2) with a half-life of 52 minutes. Demotate incorporating the 1,4,8,11-tetraazaundecane (N$_4$) tetradentate chelator was labeled with ^{94m}Tc in the presence of SnCl$_2$. ^{94m}Tc-demotate exhibited preserved receptor-binding to A-427 non-small cell lung cancer cells infected with the AdHASSTR2 adenovirus encoding the SMSR subtype 2. A-427 xenografts infected with AdHASSTR2 implanted s.c. into athymic mice were visualized by PET at 1 hour p.i. of ^{94m}Tc-demotate.

6.7.6.2   Rhenium Radionuclides

Methods for labeling antibodies and peptides with radionuclides of rhenium have been developed from techniques used for labeling with ^{99m}Tc, due to the similarity in the chemistry of technetium and rhenium [127]. Two radionuclides of rhenium (^{186}Re and ^{188}Re) are useful for targeted *in situ* radiotherapy of cancer (Table 6.7-3). ^{186}Re decays with a half-life of 3.7 days emitting a β-particle with maximum energy (E$_{max}$) of 1.07 MeV as well as a low abundance (9%) γ-photon of 137 keV, which is useful for imaging. ^{188}Re decays with a 17-hour half-life emitting a β-particle with E$_{max}$ of 2.12 MeV and a low abundance (15%) γ-photon of 155 keV that can be imaged. The twofold higher β-particle energy of [188]Re compared with [186]Re provides a threefold longer range in tissues (8 vs. 3 mm, respectively), making it more useful for treating larger tumors (i.e., >1 cm in diameter) or tumors in which there is incomplete targeting of radiolabeled mAbs or peptides to cancer cells. On the other hand, the relatively short half-life of [188]Re (17 hours) may limit its feasibility for labeling intact IgG mAbs for radiotherapeutic applications due to their slow kinetics of tumor uptake and clearance from the blood and normal tissues. Antibody fragments and peptides may be more suitable for labeling with [188]Re. A major advantage of [188]Re is that it can be produced carrier-free and in high purity using a [188]W/[188]Re generator system [128], which would make the radionuclide available at low cost in any nuclear medicine facility.

Both direct and indirect methods have been used for labeling antibodies with [186]Re/[188]Re [127]. Direct methods rely on the binding of reduced [186]Re/[188]Re to free thiols on the antibodies generated by reduction with 2-ME as described for [99m]Tc (Section 6.7.6.1). However, the chemistries of technetium and rhenium are not identical. In particular, rhenium is more difficult to reduce from its 7+ valence state to its lower valence states of 4+ or 5+ for labeling biomolecules, and it is more easily re-oxidized [129, 130]. Thus, a greater amount of $SnCl_2$ reducing agent is required. The optimal pH for labeling with [186]Re/[188]Re is also lower than that for ^{99m}Tc-labeling (pH 4.5–5.0 vs. pH 7.0–7.5, respectively) and pH values >5.0 can result in re-oxidation of rhenium [129]. Moreover, radiolysis of antibodies and peptides caused by the high-energy β-particles emitted by [186]Re/[188]Re can be a problem, and thus, radioprotectants such as ascorbic acid, gentisic acid or human serum albumin are often incorporated into the formulations. In one study, ascorbic acid was employed as both a radioprotectant and a reducing agent for the disulfide bonds for labeling IgG with [186]Re [130]. Much longer incubation times (17–24 hours) have been used for direct labeling of mAbs with [186]Re/[188]Re [130, 131] than for ^{99m}Tc labeling (30 minutes) [132]. However, in one study [133], a kit formulation was developed for labeling the anti-CD20 mAb ritiximab (Rituxan; Roche) with [188]Re. A labeling efficiency >97% was obtained in only 1–1.5 hours after the addition of [188]Re perrhenate eluted from a [188]W/[188]Re generator to the kit.

Tetradentate chelators (e.g., N_3S; Figure 6.7-7) have also been used for labeling mAbs and their F(ab')$_2$ fragments [134, 135] as well as the somatostatin peptide RC-160 [136] with [186]Re/[188]Re. In the case of antibodies, the "preformed chelate" approach was used, whereas for RC-160, the N_3S chelator was introduced during solid phase peptide synthesis. One method that has been used for direct labeling of antibodies using [188]Re employs the hydroxamic acid, trisuccin. Trisuccin differs from most tetradentate ligands (e.g., N_3S structures) in that it does not contain a free thiol. Therefore, it can be conjugated directly to a mAb and used for labeling with [186]Re/[188]Re without the risk of the free thiol reacting with disulfides on the antibodies. Safavy et al. [137] labeled two different humanized forms of the tumor-associated glycoprotein-72 (TAG-72) mAb CC49 with [188]Re using trisuccin conjugated through a 6-oxoheptanoic acid linker molecule. The labeling efficiency ranged from 80% to 98%, and the immunoreactivity ranged from 69% to 77%. High tumor uptake of radioactivity (18%–23% injected dose/g) was found in mice implanted with s.c. TAG-72-positive LS174T human colon cancer xenografts. Finally, bombesin, a 14-amino acid analog of gastrin-releasing peptide (GRP) that binds to GRP receptors on prostate, breast, lung, and pancreatic cancers, has been labeled using a ^{188}Re(H$_2$O)(CO)$_3$–carbonyl complex similar to that described for labeling biomolecules with ^{99m}Tc (Section 6.7.6.1) [138]. Targeting of PC-3 prostate cancer xenografts in mice was achieved with ^{188}Re-labeled bombesin.

6.7.6.3   Indium Radionuclides

Indium-111 (^{111}In; Table 6.7-1) is a γ-emitting radionuclide (Eγ = 172 and 245 keV) with a half-life of 67 hours that is routinely used for labeling mAbs and peptides for SPECT imaging of tumors. The positron-emitter, ^{110m}In (Table 6.7-2), has a half-life of 69 minutes and is useful for labeling peptides for PET. The β$^+$ energy of ^{110m}In (2.26 MeV) provides a spatial resolution of 3.0 mm compared with 0.7 mm for ^{18}F [13]. Octreotide was labeled with ^{110m}In and used for PET of a patient with a SMSR-positive intestinal carcinoma metastasis [139]. ^{114m}In and its daughter product, ^{114}In, are long-lived radionuclide impurities in ^{111}In, but they have potential for targeted *in situ* radiotherapy of malignancies. ^{114m}In decays to ^{114}In with a half-life of 49.5 days emitting an imageable γ-photon of 190 keV, as well as Auger and conversion electrons that can kill cancer cells. ^{114}In decays to ^{114}Cd (0.5%) or ^{114}Sn (99.5%) with a half-life of 72 seconds emitting β$^-$ particles (E$_{max}$ = 1.98 MeV) that are also useful for treatment of tumors. Octreotide has been labeled with ^{114m}In [140]. Antibodies and peptides are labeled with ^{111}In by introducing the chelator, diethylenetriaminepentaacetic acid (DTPA), by reaction of ε-amino groups on lysine residues or the N-terminal amine of the biomolecules with reactive forms of DTPA (Figure 6.7-8), such as DTPA dianhydride (cDTPAA) [141], DTPA mixed anhydride [142], or DTPA p-benzylisothiocyanate (SCN-Bz-DTPA) [143, 144]. Conjugation with cDTPAA involves suspending cDTPAA in anhydrous chloroform, dispensing an

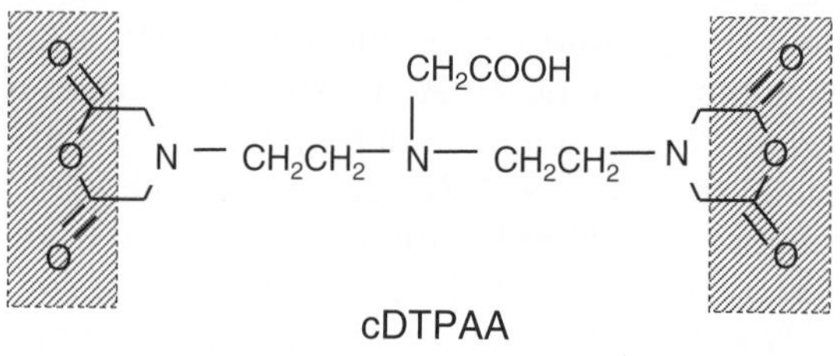

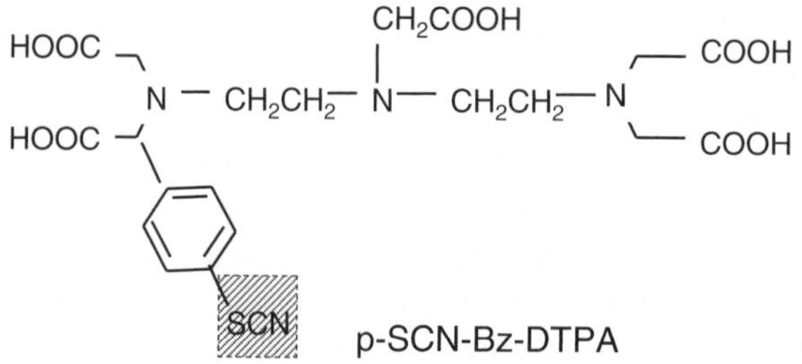

Figure 6.7-8. Metal chelators that can be conjugated to monoclonal antibodies or peptides for labeling with ^{111}In. cDTPAA: Bicyclic anhydride of diethylenetriaminepentaacetic acid; p-SCN-Bz-DTPA: p-isothiocyanatobenzyl-diethylenetriaminepentaacetic acid. Chemically reactive groups for conjugation to the antibodies or peptides are shaded.

aliquot of the suspension containing a known amount of cDTPAA into a clean, dry glass tube, and evaporating the chloroform to dryness using a gentle stream of nitrogen to leave a film of cDTPAA on the inside surface of the tube. The antibody or peptide dissolved in 50-mM sodium bicarbonate buffer, pH 7.5, is then added to the tube. Reaction with cDTPAA occurs rapidly within 1–2 minutes, although 15–30 minutes are often allowed. DTPA conjugation efficiency is dependent on protein concentration, pH, and the molar ratio of cDTPAA: antibody/peptide. The DTPA-derivatized biomolecule is separated from excess DTPA by SEC and/or by ultrafiltration. Pure DTPA-conjugated antibodies or peptides can be dispensed into unit-dose vials to produce kits that can be labeled to high radiochemical purity (RCP > 90%) simply by adding ^{111}In to the vial and incubating at room temperature for 15–30 minutes [145]. Radiolabeling is achieved by transchelation of ^{111}In from acetate or citrate complexes to DTPA. The acetate or citrate counterions are used to maintain the solubility of ^{111}In at pH 5–7.5 used for labeling. ^{111}In–acetate or – citrate complexes are formed by mixing ^{111}InCl$_3$ with 0.5–1-M sodium acetate or citrate buffer, pH 5.0. One critical parameter that must be considered for labeling biomolecules with ^{111}In is the presence of trace amounts of divalent or trivalent metal ions (e.g., Fe, Al, Cd, and Zn) in the labeling reaction. These trace metals may exist at even higher levels than ^{111}In, which is in commercial ^{111}InCl$_3$ solutions [146], and can interfere with labeling by occupying the small number of DTPA metal-binding sites present in biomolecules. Trace metal contamination is minimized by acid-washing of glassware, by using trace-metal-free plasticware, through storing DTPA-conjugated proteins frozen (to avoid leaching of metals into the solution from the glass container), and by purification of conjugation buffers on a cation-exchange column (e.g., Chelex-100) [147]. A method has been reported for ultra-purification of ^{111}In from trace metals in ^{111}InCl$_3$ solution by selective extraction of ^{111}In as an iodide complex into anhydrous diethyl ether [146].

Hnatowich et al. [141] found that the reaction between cDTPAA and IgG is dependent on pH (optimum between pH 7.5 and 8.5) and protein concentration (optimum >15 mg/mL). There was an inverse relationship between the molar ratio of cDTPAA : IgG and the conjugation efficiency with the greatest efficiency (>70%) found at a 1:1 ratio using a protein concentration of 15 mg/mL. However, it may not always be possible to achieve these high protein concentrations, and thus, greater molar ratios of cDTPAA : antibody/peptide (e.g., 5:1 to 20:1) are used to compensate for the lower conjugation efficiency at lower concentrations (e.g., 2–5 mg/mL). One limitation of the cDTPAA method is that the reagent contains two anhydride moieties that can react with ε-amino groups on lysines or the N-terminal amine of biomolecules. Reaction of cDTPAA with lysines on two biomolecules causes intermolecular cross-linking, leading to the formation of dimers and higher molecular weight polymers. Reaction of cDTPAA with two lysines on the same biomolecule generates intramolecularly cross-linked species. This latter possibility is especially troublesome because in contrast to intermolecularly linked species, intramolecularly cross-linked biomolecules are not easily detected or measured by chromatographic techniques but may significantly diminish immunoreactivity or receptor-binding properties due to protein misfolding [148]. The proportion of intermolecularly linked molecules is directly proportional to the molar ratio of cDTPAA : IgG and protein concentration in the conjugation reaction, and the resulting substitution level (moles DTPA/mole biomolecule) of the conjugate.

Hnatowich et al. [141] reported that the proportion of polymers increased from 0.3% when IgG was conjugated with cDTPAA at a 1:1 molar ratio (cDTPAA:IgG) to as much as 40% at a molar ratio of 10:1 when the IgG concentration was 15 mg/mL. Our group found less than 11% IgG dimers when the HER-2/neu mAb trastuzumab (Herceptin) was modified with a fourfold molar excess of cDTPAA using a protein concentration of 5 mg/mL. In this instance, there were two DTPA molecules/trastuzumab IgG [149]. Another limitation of the cDTPAA method is that one of the five carboxylic acid groups of DTPA is used to form an amide linkage with the antibody or peptide. This converts DTPA to diethylentriaminetetraacetic acid (DTTA), which forms a much less stable heptadentate complex with [111]In. This instability is manifested by a moderate rate of transchelation *in vivo* (7–10% per day) of ^{111}In from DTPA-conjugated antibodies to transferrin, which then causes deposition of radioactivity in the liver and bone marrow [150, 151]. In contrast, the transchelation rate from an octadentate ^{111}In–DTPA complex to transferrin is 1–2% per day [151, 152]. Despite these limitations of cDTPAA as a bifunctional chelator, its has been used for ^{111}In labeling of intact IgG mAbs [18, 120, 147, 151], single-chain Fv fragments [153], and peptides [18, 154, 155]. In the case of peptides that harbor critical lysine residues necessary for receptor-binding, these amino acids must be Boc-protected during conjugation of the N-terminal amine with DTPA, and then deprotected afterward [155].

To address the limitations of cDTPAA, other forms of DTPA have been synthesized that position a reactive group on the methylene carbon of one carboxymethyl arm instead of using one carboxylic acid group to link with an antibody or peptide [143, 144]. One of these DTPA bifunctional chelators is the p-benzylisothiocyanate derivative of DTPA (p–SCN–Bz–DTPA; Figure 6.7-8). The p–SCN–Bz–DTPA chelator reacts with ε-amino groups or the N-terminus of biomolecules to form a thiourea linkage. As there is only one reactive moiety on the p–SCN–Bz–DTPA chelator, there is no possibility of inter- or intramolecular cross-linking, which helps to preserve immunoreactivity or receptor-binding affinity [143, 144]. Our group found that the immunoreactive fraction of mAb 2G3 reactive with a 330-kDa glycoprotein on breast and ovarian cancer was 0.52 when conjugated with 0.5–1.5 moles of DTPA using cDTPAA, but it was 0.77 when conjugated with p–SCN–Bz–DTPA [151]. Importantly, the retention of all five carboxylic acid groups in the DTPA molecule for binding ^{111}In preserves the highly stable octadentate ^{111}In–DTPA complex, which diminishes the loss of ^{111}In to transferrin in plasma. The B-cell lymphoma antibody Lym-1 conjugated with p–SCN–Bz–EDTA (a chelator similar to p–SCN–Bz–DTPA) and labeled with ^{111}In showed a lower rate of loss of ^{111}In in serum than Lym-1 conjugated with DTPA using cDTPAA (<1% vs. 14% over 5 days, respectively) [156]. Similar results were obtained *in vitro* in serum and *in vivo* in plasma in mice for other ^{111}In–SCN–Bz–EDTA or SCN–Bz–DTPA immunoconjugates [157]. However, our group did not observe increased stability of ^{111}In-labeled mAb 2G3 conjugated to SCN–Bz–DTPA *in vitro* in serum compared with ^{111}In–DTPA–mAb 2G3 (both had transchelation rates of 7% per day), but we did find a lower rate of loss of ^{111}In from ^{111}In–SCB–Bz–DTPA–2G3 to ascites fluid from ovarian cancer patients (5% vs. 11% per day) [151]. Avoidance of polymerization and increased serum stability of p–SCN–Bz–DTPA conjugated biomolecules decreases liver uptake and retention of radioactivity, a common problem with ^{111}In-labeled mAbs and peptides [158].

Methods have been developed to site-specifically conjugate DTPA to the Fc domain of antibodies for labeling with ^{111}In in order to better preserve their immunoreactivity [159]. Site-specific conjugation of DTPA was achieved by reaction of the N-terminal amine of the tripeptide, glycine–tyrosine–lysine–DTPA (GTK–DTPA) with aldehydes generated in the Fc-domain by sodium periodate oxidation [160]. Stabilization of the resulting Schiff base linkage between GTK–DTPA and the antibodies was achieved by sodium borohydride reduction. This approach positions the DTPA chelator at a position that is remote from the Fab antigen-binding region. Nevertheless, the sodium periodate oxidation step can diminish antibody immunoreactivity if not controlled [161]. The number of aldehydes generated by sodium periodate oxidation can be measured by a spectrophotometric assay that relies on the reaction of the aldehyde groups with dinitrophenylhydrazine (DNPH), generating a derivative that absorbs at 360 nm [162]. Our group recently reported a novel strategy for site-specific labeling of recombinant biomolecules with ^{111}In that takes advantage of its high affinity for transferrin ($K_a = 10^{28}$ L/mole). We fused the gene for the n-lobe of human transferrin (hn-Tf) through a DNA sequence that encoded a flexible polypeptide linker $[(GGGGS)_3]$ to the gene for vascular endothelial growth factor ($VEGF_{165}$) and expressed the recombinant fusion protein in *Pichia pastoris* [163]. The hnTf–VEGF protein bound ^{111}In directly through its hnTf moiety and retained its binding affinity for VEGF receptors on human umbilical vascular endothelial cells (HUVECs). The protein did not bind to transferrin receptors on cells, because such binding requires both the n- and the c-lobes of transferrin. ^{111}In–hnTf–VEGF localized specifically in angiogenic U87MG glioblastoma xenografts implanted s.c. in athymic mice permitting tumor imaging at 72 hours p.i. However, ^{111}In–hnTf–VEGF exhibited relatively high liver uptake and there was a moderately rapid loss of [111]In *in vitro* from the protein to transferrin in plasma (21% per day). This transchelation was not likely due to a lower affinity of the hnTf moiety for ^{111}In, but it was caused by competition with the higher concentrations of transferrin present in plasma.

Limiting the metal chelator substitution of a biomolecule to one to two DTPA or EDTA groups per molecule in order to preserve its immunoreactivity or receptor-binding affinity restricts the specific activity that can be achieved for labeling with [111]In. For example, monosubstitution of EGF with DTPA restricts the maximum specific activity that can be practically achieved with ^{111}In to 40 MBq/μg (2.4×10^5 MBq/μmole). At this low specific activity, only one in eight EGF molecules carries an ^{111}In atom and almost 90% of EGFRs in a tumor would be targeted by non-radiolabeled EGF. To address this issue, Remy et al. [164] conjugated maleimide-derivatized EGF with a thiol-containing multibranched peptide containing four EDTA-like metal chelators for [111]In. However, the MCP–4–EDTA–S–MB–EGF conjugate showed a 40-fold decrease in receptor-binding affinity in a competition assay with MDA–MB–468 human breast cancer cells compared with unmodified EGF. In contrast, our group found that derivatization of EGF directly with one to two DTPA metal chelators and labeling with ^{111}In yielded a radiopharmaceutical with receptor-binding affinity identical to that of ^{125}I-labeled EGF ($K_a = 7.3 \times 10^8$ L/mole) [18]. Greater success was achieved by our group in maximizing the specific activity of ^{111}In–EGF by conjugating maleimide-derivatized EGF to thiolated human serum albumin (HSA), which presents 60 lysine residues for DTPA conjugation [165]. Conjugation of EGF to HSA diminished its receptor-binding affinity 15-fold ($K_a = 5.1 \times 10^7$ L/mole), but there were no further decreases in affinity when up to 23 DTPA chelators were preferentially substituted into the HSA moiety. The specific activity of ^{111}In–DTPA–HSA–EGF was increased 10-fold compared with ^{111}In–DTPA–hEGF. ^{111}In–DTPA–HSA–EGF retained its receptor-mediated internalization and nuclear translocation properties in MDA–MB–468 cells and was fourfold more growth-inhibitory toward the cells. An analogous strategy was reported by Manabe et al. [166] that used a polylysine peptide carrier to conjugate as many as 42 DTPA molecules to the anti-HLA mAb H-1 with > 90% retention of immunoreactivity. Similarly, starburst dendrimers that display 64 amine groups on their surface were derivatized with up to 43 1B4M DTPA-like chelators, and then were conjugated through a maleimido bond to mAb OST7 [167]. The maximum specific activity achieved for labeling these starburst dendrimer immunoconjugates with [111]In (8.4 MBq/μg; 1.3×10^6 MBq/μmole) was 48-fold higher than directly conjugated 1B4M-OST7. In another study, poly(ethylene glycol) (PEG) containing an amine functional group was used to conjugate DTPA chelators to the anti-EGFR mAb C225 with good preservation of immunoreactivity [168]. This approach provides a method of labeling with ^{111}In while minimizing nonspecific retention of radioactivity in the liver due to conjugation with PEG [169].

High retention of radioactivity in the liver with ^{111}In–DTPA-conjugated mAbs is thought to be due to intracellular trapping of ^{111}In-catabolites. ^{111}In–DTPA-labeled IgG antibodies interact with glycoprotein receptors on hepatocytes through their Fc-domain and are internalized into endosomes. The antibodies are routed to lysosomes for proteolytic degradation. The ultimate catabolite of proteolysis has been identified as ^{111}In–DTPA–ε–Lys [30, 170]. This catabolite cannot easily cross the lysosomal membrane due to its positive charges or be exocytosed due to its poor recognition by cell membrane amino acid transporters and, thus, becomes trapped. An analogous mechanism has been proposed for retention of ^{111}In-labeled mAb fragments and peptides by renal tubular cells. ^{111}In-labeled mAb fragments or peptides are filtered by the glomerulus and reabsorbed by renal tubular cells. Proteolytic catabolism within renal tubular cells of ^{111}In–DTPA–F(ab')$_2$ fragments yields ^{111}In–DTPA–ε–Lys, which is trapped within the cells [171]. Proteolysis of ^{111}In–DTPA–D–Phe[1]–octreotide or ^{111}In–DTPA–L–Phe[1]–octreotide similarly results in retention of ^{111}In–DTPA–D–Phe[1]–OH or ^{111}In–DTPA–L–Phe[1]–OH catabolites, respectively, in renal tubular cells [172, 173].

The macrocyclic chelator, 1,4,7,10-tetraazacyclododecane N,N′,N″,N‴-tetraacetic acid (DOTA) (Section 6.7.6.4) has been less commonly used for labeling mAbs and peptides with ^{111}In because the kinetics of binding ^{111}In by DOTA are much slower than those of DTPA. Elevated temperatures of 37–43°C for mAbs [174] and heating at 100°C for peptides [175] combined with longer incubation times (30–45 minutes) are required to obtain complete incorporation of ^{111}In into DOTA-conjugated bio-molecules. Nevertheless, intact IgG mAbs [174, 176], Fab fragments [177] and diabodies [178], as well as octreotide [175] have been labeled with ^{111}In using DOTA. The stability of ^{111}In–DOTA-conjugated mAbs in serum is greater than that of ^{111}In–DTPA-conjugated antibodies. Lewis et al. [174] found that there was less than 0.8% loss of ^{111}In in serum from ^{111}In–DOTA–cT84.66 mAb over 10 days compared with 13.0% for ^{111}In–DTPA–cT84.66. The advantage of conjugating mAbs or peptides with DOTA for labeling with ^{111}In is that it provides an analog for the corresponding yttrium-90 (^{90}Y)–DOTA–biomolecule used for targeted *in situ* radiotherapy. Imaging with the ^{111}In–DOTA–biomolecule can be used to predict radiation dosimetry estimates to tissues for the ^{90}Y–DOTA–biomolecule [179, 180]. Use of an ^{111}In–DTPA–conjugated mAb or peptide for this purpose may yield inaccurate dosimetry estimates because of major differences in stability *in vivo* between ^{111}In– and ^{90}Y–DTPA complexes.

6.7.6.4   Yttrium Radionuclides

Two radionuclides of yttrium, ^{86}Y and ^{90}Y, are available for labeling biomolecules for imaging or radiotherapeutic purposes, respectively. ^{86}Y (Table 6.7-2) is a positron-emitter with a half-life of 14.7 hours that is produced in a cyclotron using the ^{86}Sr(p,n)^{86}Y reaction [181]. The spatial resolution for detection of the positron annihilation from ^{86}Y is 1.8 mm compared with 0.7 mm for ^{18}F [13]. ^{90}Y (Table 6.7-3) is a pure β-emitter with a half-life of 2.7 days that is produced by a strontium-90 (^{90}Sr)/^{90}Y generator [182]. ^{90}Y is an attractive radionuclide for *in situ* radiotherapy of cancer because it does not emit γ-radiation, which minimizes the radiation exposure to health-care personnel and family members from a patient-administered ^{90}Y-labeled mAb or peptide. For example, patients with non-Hodgkin's B-cell lymphoma treated with ^{90}Y-ibritumomab tiuxetan (Zevalin; Biogen Idec, Cambridge, MA) do not need to be isolated for radiation safety reasons and can be treated as outpatients, whereas patients receiving ^{131}I-labeled tositumomab (Bexxar; Glaxo-SmithKline, Uxbridge, Middlesex, UK) require special radiation safety precautions to be taken [183]. However, as ^{90}Y does not emit γ-radiation, the tissue distribution of ^{90}Y-labeled mAbs and peptides cannot be easily imaged (only poor-resolution Bremstrahlung images can be acquired). Therefore, the radiation absorbed doses to tissues in patients from ^{90}Y-labeled biomolecules are often estimated by imaging studies with ^{111}In-labeled analogs [179, 180]. However, as mentioned (see Section 6.7.3), differences in the *in vivo* stability of ^{111}In- and ^{90}Y-labeled biomolecules can produce discrepancies in their pharmacokinetics and normal organ distribution, yielding inaccuracies in the dosimetry estimates. PET imaging using ^{86}Y-labeled analogs has been suggested as a means of more accurately estimating the radiation absorbed doses from ^{90}Y-labeled biomolecules [184].

The macrocyclic chelator, DOTA (Figure 6.7-9), is the chelating agent of choice for labeling mAbs and peptides with yttrium-90 (^{90}Y) due to the high stability of ^{90}Y-DOTA complexes ($K_a = 10^{24}$ L/mole) [185]. Initially, DTPA was used as a chelator for ^{90}Y-labeling [186], but it was discovered that ^{90}Y–DTPA complexes were unstable *in vivo*, which caused bone accumulation of free ^{90}Y released from the immunoconjugates. The stability of ^{90}Y-labeled biomolecules is of paramount importance because even a small amount of free ^{90}Y sequestered in the bone can contribute significantly to bone marrow toxicity due to the long range (10–12 mm) of the β-particles ("cross-fire" effect). This is one reason that the maximum tolerated dose of ^{90}Y-labeled mAbs (e.g., Zevalin) is 32 mCi, whereas up to 150 mCi of ^{131}I-labeled mAbs can be safely administered [2]. Bz–SCN–DTPA (Figure 6.7-8) yields more stable ^{90}Y complexes than those formed with DTPA, but the complexes are not as stable as ^{90}Y–DOTA complexes. In one study [187], the bone uptake of radioactivity in mice implanted s.c. with SK–RC–52 renal cell carcinoma xenografts at 7 days p.i. of ^{90}Y–DOTA-conjugated chimeric G250 mAbs was only 0.4% i.d./g,

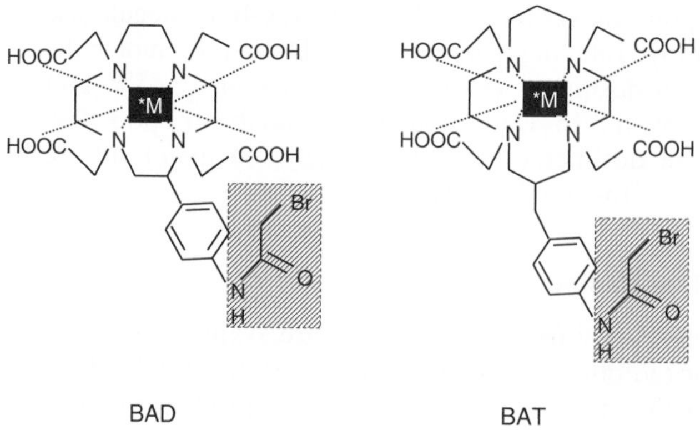

Figure 6.7-9. Chemically reactive forms of two macrocyclic chelators that can be conjugated to monoclonal antibodies or peptides for labeling with radiometals. Groups for conjugation to the antibodies or peptides are shaded. BAD: p-bromoacetamidobenzyl-DOTA complexes ^{90}Y, ^{67}Ga, ^{177}Lu, ^{212}Pb, ^{212}Bi or ^{225}Ac. BAT: bromoacetamidobenzyl-TETA complexes ^{64}Cu or ^{67}Cu.

compared with 1.2% i.d./g for Bz–SCN–DTPA- and 10.7% i.d./g for DTPA-conjugated antibodies. Similarly, a threefold lower bone uptake of radioactivity at 10 days p.i. in mice was observed for ^{88}Y–DOTA–hLL2 anti-CD22 mAbs compared with ^{88}Y–Mx–DTPA–hLL2 mAbs [188].

DOTA, which was first described by McCall et al. in 1990 [189], can be conjugated to mAbs by conversion to its reactive form, p-bromoacetamidobenzyl–DOTA (BAD; Figure 6.7-9) followed by reaction of BAD with free thiols introduced into the antibodies by reaction with 2-iminothiolane (2-IT) [190]. 2-IT also creates a spacer between the mAbs and the DOTA chelator that allows more efficient labeling with ^{90}Y. An alternative approach involves reaction of an N-hydroxysuccinimidyl ester of DOTA with ε-amino groups on lysine residues or the N-terminus of biomolecules [191]. Labeling of DOTA-conjugated mAbs is achieved by incubation with ^{90}Y chloride (^{90}YCl$_3$) mixed with 0.5-M ammonium acetate buffer, pH 7.0–7.5, for 30 minutes at 37°C [192]. Similar conditions have been employed for labeling DOTA-conjugated peptides with ^{90}Y, except that a temperature of 80–100°C was used to accelerate the incorporation of ^{90}Y [193]. Labeling of biomolecules with ^{90}Y using DOTA is highly susceptible to the effects of divalent trace metal ion contamination, especially Ca^{2+}, Fe^{2+}, and Zn^{2+}. In fact, the affinity constant (K$_a$) for binding of Fe^{2+} by DOTA is 100,000 times higher than that for binding ^{90}Y [185]. It is critical to exclude trace-metals as much as possible from the ^{90}Y labeling reaction. One recent Letter to the Editor even suggests that minor trace-metal contamination of pipette tips can seriously diminish the labeling efficiency of DOTA-conjugated peptides with ^{90}Y [194]. Another important issue in labeling biomolecules with ^{90}Y is the potential for radiolysis. Interaction of the moderate energy (E$_{max}$ 2.2 MeV) β-particles emitted by ^{90}Y with water molecules in the buffers used to formulate the radiopharmaceuticals generates highly reactive free radicals that degrade the metal chelator as well as the biomolecules themselves. The radiolysis effect is dependent on specific activity and on radioactivity concentration. For example, ^{90}Y–BAD–2–IT–Lym–1 antibodies formulated at a specific activity of 1–2 mCi/mg remained pure and relatively immunoreactive (>75%) in storage over a 3-day period [195]. However, at a specific activity of 4 mCi/mg, the radiochemical purity (RCP) of ^{90}Y–BAD–2–IT–Lym–1 dropped to 65% and the immunoreactivity decreased to 28%. At a specific activity of 9.4 mCi/mg, the RCP of ^{90}Y–BAD–2–IT–Lym–1 decreased to 21% and the immunoreactivity was virtually abolished (3%). Radiolysis can be minimized by inclusion of radioprotectants such as ascorbic acid, gentisic acid, or human serum albumin in the formulation and by freezing the radiolabeled biomolecules to minimize diffusion of free radicals in the aqueous solutions [195, 196]. In one study, it was shown that ascorbic acid may also act as a suitable buffering agent, thereby removing the need for the ammonium acetate buffer [197].

A controversial issue with respect to the use of DOTA as a metal chelator for ^{90}Y and other radiometals is its potential immunogenicity in humans. In a phase I/II clinical trial in which six ovarian cancer patients received ^{90}Y–DOTA–HMFG1 murine mAbs intraperitoneally, all patients developed anti-DOTA antibodies and three patients developed serum sickness [198]. Four of eight patients who received ^{111}In–DOTA-conjugated mAbs intravenously developed anti-DOTA antibodies. The immune response seems to be directed against the DOTA ring structure and not the benzyl-containing side chain. The immune response to DOTA is also dependent on the antigenicity of the biomolecule to which it is conjugated, because rabbit IgG conjugated to DOTA did not induce anti-DOTA antibodies in rabbits, whereas mouse IgG–DOTA immunoconjugates administered to rabbits stimulated an immune response toward the chelators [199]. Moreover, an immune response to DOTA has not been found with all mAbs. Anti-DOTA antibodies were not detected in the serum of 18 lymphoma patients administered multiple doses of ^{111}In–BAD–2–IT–Lym–1 mAbs, although these patients typically do not mount a strong immune response to radioimmunoconjugates [200]. Similarly, ^{90}Y–DOTATOC, an octapeptide somatostatin analog, does not seem to be immunogenic, again suggesting that the biomolecule carrier must be considered in assessing the potential immunogenicity of DOTA [201].

6.7.6.5 Gallium Radionuclides

Gallium-67 (^{67}Ga; Table 6.7-1) is a cyclotron-produced radionuclide with a half-life of 78.2 hours that has been used for many years in nuclear medicine as ^{67}Ga citrate for tumor imaging [202]. Despite its history of use, the properties of ^{67}Ga are not ideal for imaging, however, due to the high energy of two of its γ-photons (Eγ = 300 and 393 keV), which make it difficult to collimate. Nevertheless, ^{67}Ga has recently received attention as a radiolabel for mAbs for *in situ* radiotherapy of cancer, exploiting its abundant Auger and conversion electron emissions [203]. Most of the recent interest in gallium radionuclides has been focused on labeling peptides with ^{68}Ga for PET (Table 6.7-2) [204]. Antibodies have not been labeled with ^{68}Ga for PET, because their kinetics of tumor uptake and elimination from the blood and normal tissues is too slow. ^{68}Ga is conveniently produced using a germanium-68 (^{68}Ge)/^{68}Ga generator system that could allow production of the radionuclide for up to one year in a nuclear medicine facility using a single generator, due to the long half-life of ^{68}Ge (270 days) [205]. ^{68}Ga decays with a half-life of 68 minutes to ^{68}Zn. Its positron energy is 1.92 MeV, which provides a spatial resolution of 2.4 mm versus 0.7 mm for [18]F [13]. [66]Ga is a longer lived positron-emitting radionuclide (half-life of 9.5 hours), which has also been studied for labeling peptides for PET [206].

Initially, desferrioxamine (DFO) was used as a chelator for labeling peptides with [68]Ga [74, 207], but more recently DOTA has been used (Figure 6.7-9) [204, 208]. Most PET tumor imaging studies with ^{68}Ga-labeled peptides have focused on DOTATOC, a synthetic octapeptide analog of somatostatin. Labeling DOTATOC involves adjusting the pH of [68]GaCl$_3$, which is eluted in 0.5-M HCl from the [68]Ge/^{68}Ga generator to pH 4.8 with 50-mM sodium acetate buffer. The required amount of ^{68}Ga acetate complex is then mixed with DOTATOC (10–20 nmols) and heated at 95°C for 15 minutes in a heating block. Early studies in which DOTATOC was labeled with ^{68}Ga using this technique yielded labeling efficiencies that were about 50%, thus requiring postlabeling purification on a C-18 Sep-Pak cartridge [208, 209]. The final radiochemical purity was >95%. The large volumes of ^{68}Ga eluates eluted from the [68]Ge/^{68}Ga generator system as well as contamination with trace metals were believed to be responsible for the low labeling efficiencies for DOTATOC with [68]Ga. Improvements in the concentration of ^{68}Ga eluates using an ion-exchange cartridge combined with more homogeneous microwave heating of the labeling reaction for 10–20 minutes have recently yielded almost complete incorporation of ^{68}Ga in to extremely small quantities (<1 nmole) of DOTATOC, thus increasing the specific activity by almost 100-fold [210].

Interestingly, DOTATOC labeled with ^{68}Ga has a fivefold higher binding affinity for SMSR subtype 2 (IC$_{50}$ = 2.5 nmol/L) compared with ^{90}Y–DOTATOC (IC$_{50}$ = 11 nmol/L) and a ninefold higher affinity than that of ^{111}In–DTPA–octreotide (IC$_{50}$ 22 nmol/L) [211]. Combined with the high spatial resolution and sensitivity of PET, this provided a greater sensitivity for imaging small meningioma lesions (100% vs. 85%) in patients compared with SPECT imaging with ^{111}In–DTPA–octreotide [209, 212]. The somatostatin analog DOTANOC, which binds to SMSR subtypes 2 and 5, has been labeled with ^{68}Ga and used for PET of a patient with metastases from a neuroendocrine tumor [213]. Preclinical studies have been performed in mouse tumor xenograft models using ^{68}Ga–DOTA–α–melanocyte-stimulating hormone (α-MSH) for imaging melanoma [214] or ^{68}Ga–DOTA–EGF to image gliomas or epidermoid carcinoma xenografts [215]. Due to the convenient availability of a generator system for ^{68}Ga, it is not necessary for a nuclear medicine facility to have access to a cyclotron, and therefore, it is likely that more studies ultimately leading to clinical application of many different ^{68}Ga-labeled peptides for PET imaging of tumors will be performed in the near future.

6.7.6.6 Copper Radionuclides

Several radionuclides of copper are suitable for labeling mAbs, their fragments, or peptides for PET (e.g., [60]Cu, [61]Cu, or ^{64}Cu; Table 6.7-2) [216]. [64]Cu decays with a half-life of 12.7 hours by three different pathways: (1) electron capture (41%) emitting Auger and conversion electrons, (2) positron (β$^+$) emission (19%), and (3) β$^-$ emission (40%). The β$^-$ emissions make the radionuclide useful not only for PET but also for radiotherapy of tumors. Indeed, in one study, ^{64}Cu-labeled octreotide inhibited the growth of CA20948 rat pancreatic tumors in Lewis rats [217]. In another study, ^{64}Cu-labeled 1A3 mAbs were used to treat hamsters implanted s.c.

with GW39 human colon carcinoma xenografts [218]. ^{64}Cu is most commonly produced in a biomedical cyclotron by the ^{64}Ni(p,n)^{64}Cu reaction using a ^{64}Ni-enriched target [219, 220]. The positron energy of ^{64}Cu is almost identical to that of ^{18}F (0.657 MeV vs. 0.635 MeV, respectively), and the intrinsic spatial resolution is indistinguishable (0.73 vs. 0.70 mm, respectively) [13, 221]. However, due to the lower abundance of positron emission with ^{64}Cu compared with ^{18}F (19% vs. 97%), the sensitivity for PET with ^{64}Cu is about fivefold lower than that with ^{18}F [221]. The short-lived positron-emitter, ^{61}Cu (half-life 3.32 h), has a greater abundance of positron emission than ^{64}Cu (60% vs. 19%) and has been used for labeling octreotide for PET of CA20948 tumors in rats [222]. ^{67}Cu (Table 6.7-3) can be used for labeling mAbs and peptides for targeted *in situ* radiotherapy. ^{67}Cu decays with a half-life of 2.6 days emitting β$^-$ particles with energies of 0.395 (51%), 0.484 (28%), and 0.577 (20%) MeV as well as several γ-photons with energies ranging from 91 to 300 keV that are imageable [223]. ^{67}Cu is produced in a biomedical cyclotron using the ^{nat}Zn(p,2p)^{67}Cu or ^{68}Zn(p,2p)^{67}Cu reactions or in a nuclear reactor using the ^{nat}Zn(n,p)^{67}Cu reaction [223].

DTPA (Figure 6.7-8) and benzyl-EDTA were initially examined for chelating ^{67}Cu, but it was soon found that these complexes formed were unstable, releasing 70–95% of their radiolabel *in vitro* in serum over 5 days [224]. In contrast, the macrocyclic chelator, 1,4,8,11-tetraazacyclotetradecane-N′,N″,N‴-tetraacetic acid (TETA; Figure 6.7-9), provided a more stable ^{67}Cu complex that lost only 2–6% of the radiolabel in serum when conjugated to Lym-1 antibodies [224]. This greater stability of ^{67}Cu–TETA vs. ^{67}Cu–DTPA or Bz–EDTA complexes is not due to a higher thermodynamic stability, but it is believed to be due to the structural rigidity of the ^{67}Cu–TETA complex, which shields the radionuclide from attack by endogenous copper binding proteins. Copper is bound in plasma by albumin and transcuprein, which transport the metal to hepatocytes [216]. Copper internalized by hepatocytes is used to synthesize metalloenzymes such as superoxide dismutase (SOD), is stored bound to metallothionein (MT), or is secreted back into the plasma complexed to ceruloplasmin. DOTA (Figure 6.7-9) and its analogs form stable complexes with copper radionuclides, but TETA is the most widely used chelator [216].

BAT (Figure 6.7-9) is a modified form of TETA, which contains a bromo-acetamidobenzyl side chain that is reactive with antibodies modified with 2-minothiolane (2-IT) to present a free thiol. The inclusion of the 2-IT spacer improves the labeling efficiency of the antibodies with ^{67}Cu and ^{64}Cu [225]. DOTA and TETA chelators can be introduced into peptides during solid phase synthesis for labeling with ^{64}Cu/^{67}Cu [226, 227]. Labeling of TETA-conjugated mAbs or peptides is achieved by incubation with ^{64}Cu/^{67}Cu in a 100-mM ammonium citrate buffer, pH 5.5, at room temperature for 30–60 minutes [226, 228]. Disodium EDTA is added to chelate any unbound radiometal, and the ^{64}Cu/^{67}Cu-labeled biomolecules are purified by SEC. ^{64}Cu/^{67}Cu labeled peptides are usually purified on a C-18 Sep-Pak (Waters Associates Inc., Milford, MA). Using TETA, mAb1A3 IgG and its F(ab′)$_2$ fragments were labeled with ^{64}Cu for PET in a phase I/II trial of 36 patients with colorectal carcinoma [229] and Lym-1 antibodies were labeled with ^{67}Cu for RIT of 12 patients with non-Hodgkin's lymphoma [228]. Several different peptides have been labeled with ^{64}Cu using DOTA or TETA for PET of tumors, including octreotide analogs [226, 227, 230, 231], RGD peptides that recognize $\alpha_v\beta_3$

integrins [232], vasoactive intestinal polypeptide (VIP) analogs [233], and bombesin derivatives [234].

$^{64}Cu/^{67}Cu$ labeled intact IgG antibodies exhibit high accumulation in the liver, whereas $^{64}Cu/^{67}Cu$-labeled F(ab')$_2$ fragments and peptides are retained by the kidneys. In rats administered 1A3 mAbs labeled with ^{67}Cu using four different chelators, including 1-[(1,4,8,11-tetraazacyclotetradec-1-yl)methyl]benzoic acid (CPTA), a macrocycle related to TETA, it was found that there was a slow rate of transchelation of ^{67}Cu to SOD in the liver [235]. In contrast, ^{67}Cu-labeled F(ab')$_2$ fragments of mAb 1A3 were rapidly catabolized to ^{67}Cu–CPTA–ε–lysine. Similarly, transchelation of radiometal to SOD in the liver was observed for ^{64}Cu–TETA–octreotide in rats [236]. However, in a clinical study, in which 10 patients received ^{67}Cu–2IT–BAT–Lym-1 antibodies [237], only a small fraction (0.8–7.8%) of the injected dose of ^{67}Cu was recycled by the liver and, in this case, mostly to ceruloplasmin. There was no transchelation to SOD detectable. The higher concentration of SOD in the liver of rats compared with humans may account for this species-dependent catabolic route for $^{64}Cu/^{67}Cu$-labeled biomolecules [236].

6.7.6.7 Lutetium Radionuclides

Lutetium-177 (^{177}Lu; Table 6.7-3) is a β-emitting radionuclide (Eβ = 0.495 MeV) with a half-life of 6.7 days that is useful for labeling mAbs and peptides for targeted *in situ* radiotherapy of tumors. In addition, ^{177}Lu emits γ-photons of energies 113 and 208 keV that can be imaged. ^{177}Lu is produced in a nuclear reactor by neutron irradiation of $^{176}Lu_2O_3$. Several chelators have been employed for labeling mAbs and peptides with ^{177}Lu including derivatives of DOTA (e.g., PA–DOTA and CA–DOTA) and analogs of DTPA (e.g., CHX–A–DTPA and SCN–Bz–DTPA) [238]. The difference between the two DOTA chelators is that in PA–DOTA, the benzylisothiocyanate side chain used for conjugation is attached to one nitrogen in the macrocycle, whereas in CA–DOTA, it is attached to a methylene carbon. PA–DOTA forms a less stable complex with ^{177}Lu than CA–DOTA or CHX–DTPA [238]. The instability of the ^{177}Lu–PA–DOTA complex may account for the prolonged retention of radioactivity in the reticuloendothelial system (RES), including the bone marrow of patients administered ^{177}Lu–PA–DOTA–CC49 mAbs for RIT of TAG-72 positive malignancies and, consequently, the observed dose-limiting hematopoietic toxicity [239]. ^{177}Lu released from biomolecules is expected to be sequestered in the skeleton because lutetium competes with calcium for bone deposition [240]. It has been shown that by adding DTPA to ^{177}Lu–DOTA^0–Tyr^3]octreotate used for PDRT of SMSR-positive tumors, the small amount (<2–5%) of free ^{177}Lu impurity in the radiopharmaceutical can be rapidly eliminated from the body by renal excretion as ^{177}Lu–DTPA, thus minimizing bone uptake of radioactivity [241]. In addition to labeling intact IgG CC49 mAbs with ^{177}Lu [239, 242], dimeric scFv fragments of this antibody have been labeled with ^{177}Lu for radiotherapeutic purposes [243].

6.7.6.8 Lead, Bismuth, and Actinium Radionuclides

DOTA (Figure 6.7-9) is a also useful bifunctional chelator for labeling antibodies with radionuclides of lead, bismuth, and actinium for targeted *in situ* radiotherapy

[2]. Lead-212 (^{212}Pb) decays to the α-emitter, bismuth-212 (^{212}Bi). The α-emitter, actinium-225 (^{225}Ac), decays to a series of daughter radionuclides that are α-emitters (^{221}Fr, ^{217}At, ^{213}Bi) or β-emitters (^{213}Po, ^{209}Tl, ^{209}Pb, ^{209}Bi). The use of ^{225}Ac as a radiolabel for mAbs has been termed an "atomic nanogenerator" because the decay of ^{225}Ac generates the daughter radionuclides locally in tumors that then emit several different forms of radiation that kill cancer cells [244]. The main concern for employing radionuclides such as ^{212}Pb or ^{225}Ac that do not decay directly to stable elements is that the decay process may decrease the stability of the metal chelate complex due to the transformation of one element into another (i.e., ^{212}Pb is converted into ^{212}Bi). In particular, bismuth has an affinity for the kidneys, and it was shown that even low doses of ^{225}Ac-labeled HuM195 antibodies caused severe renal toxicity and anemia in monkeys, likely due to release of ^{213}Bi from the antibodies and redistribution of the radionuclide to the kidneys [245]. Bismuth radionuclides are also stably bound by CHX–A–DTPA and CHX–B–DTPA, two analogs of DTPA [246, 247]. HER2/neu mAbs AE1 and trastuzumab (Herceptin) have been labeled with ^{212}Pb or ^{225}Ac, respectively, using DOTA as the metal chelator [248, 249].

6.7.7   CHARACTERIZATION OF RADIOLABELED MABS AND PEPTIDES

It is important to fully characterize the properties of radiolabeled mAbs and peptides intended for tumor imaging or targeted *in situ* radiotherapy of malignancies. Characterization includes (1) analytical tests that evaluate homogeneity and radiochemical purity; (2) radioligand binding assays that assess the ability to specifically bind target antigens/receptors *in vitro*; (3) stability studies that assess the loss of radiolabel *in vitro* in biologically relevant media; and (4) biodistribution, pharmacokinetic, dosimetry, and imaging studies that reveal tumor and normal tissue uptake *in vivo* in an animal model and predict radiation absorbed doses in humans.

6.7.7.1   Evaluation of the Homogeneity of Radiopharmaceutical Bioconjugates

Evaluation of homogeneity is especially important for mAbs or peptides conjugated to chelators for labeling with radiometals (Section 6.7.6). Substitution of these biomolecules with too many chelators may significantly decrease their immunoreactivity or receptor-binding affinity [159]. The substitution level (moles chelators/ mole of biomolecule) may be measured by spectrophotometric, fluorescence, or radiochemical assays. For example, HYNIC substitution in biomolecules intended to be labeled with ^{99m}Tc or ^{186}Re/^{188}Re (see sections 6.7.1 and 6.7.2) can be measured by a colorimetric assay that relies on reaction of the HYNIC groups with p-nitrobenzaldehyde to form a complex that absorbs at 385 nm [162]. DTPA substitution for labeling biomolecules with ^{111}In (see Section 6.7.6.3) can be measured by several different techniques: (1) radiochemical assays that rely on the binding of ^{111}In or ^{57}Co by DTPA [141, 250]; (2) spectrophotometric assays that generate a colored arsenazo III DTPA complex [251]; (3) fluorescence assays that measure the binding of europium(III) by DTPA [252]; or (4) immunoelectrofocusing (IEF) that reveals changes in pI associated with DTPA substitution [253]. The most common method is simply to radiolabel an aliquot of the impure DTPA conjugation mixture with a trace amount of ^{111}In and separate the resulting ^{111}In–DTPA–biomolecule from free ^{111}In–DTPA by silica gel instant thin-layer chromatography (ITLC–SG) developed in 100-mM sodium citrate, pH 5.0. The conjugation efficiency is then calculated from the ITLC–SG results and multiplied by the molar ratio of cDTPAA : antibody/peptide used in the reaction to estimate the average number of moles of DTPA per mole of biomolecule. Typically, a substitution level of one to two moles DTPA per mole of biomolecule is desirable to minimize polymerization and/or interference in immunoreactivity or receptor-binding affinity [159]. The number of DOTA chelators introduced into antibodies or peptides for labeling with ^{90}Y, ^{68}Ga, ^{177}Lu, ^{212}Bi, or ^{225}Ac (see Sections 6.7.4, 6.7.5, 6.7.7, and 6.7.8) can be measured by a spectrophotometric assay, which measures the decrease in absorbance at 656 nm for a Pb(II)–arsenazo III complex upon transchelation of Pb^{2+} to DOTA [254].

The polymerization of biomolecules as a consequence of metal chelator substitution may be assessed by size-exclusion HPLC with UV detection (Figure 6.7-10) or by sodium dodecylsulfonate–polyacrylamide gel electrophoresis (SDS–PAGE; Figure 6.7-10). Polymerization is most commonly due to cross-linking of mAbs or

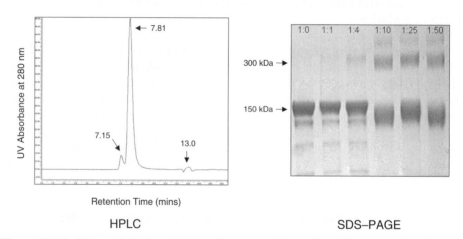

HPLC   SDS–PAGE

Figure 6.7-10. Two analytical methods used to assess the purity and homogeneity of metal-chelated monoclonal antibodies or peptides. *Left Panel*: Size-exclusion HPLC analysis of trastuzumab (Herceptin) IgG conjugated with the bicyclic anhydride of DTPA (cDTPAA) on a BioSep-SEC-S2000 column eluted with 150-mM sodium chloride/10-mM sodium phosphate buffer, pH 6.8, at a flow rate of 0.6 mL/min with UV detection at 280 nm. The molar ratio of cDTPAA : IgG was 4:1. Peak with a retention time (t_R) of 7.81 minutes represents a DTPA-conjugated trastuzumab monomer. Peak with t_R of 7.15 minutes represents a DTPA–trastuzumab dimer caused by cross-linking of the antibody molecules through the two chemically reactive groups on cDTPAA. Peak with t_R of 13.0 minutes represents a small amount of unconjugated DTPA (detected through a change in refractive index). Right Panel: SDS–PAGE analysis of trastuzumab (Herceptin) IgG conjugated with cDTPAA on a 4–10% Tris HCl gradient mini-gel stained with Coomassie Brilliant Blue. There is an increase in the proportion of dimerized (300 kDa) and polymerized (>300 kDa) IgG species as the molar ratio of cDTPAA : IgG is increased from 1:1 to 50:1. (Data generously provided by Kristin McLarty.)

peptides by bifunctional chelators such as cDTPAA that contain two reactive groups that interact with ε-amino groups on lysines on two separate biomolecules (intermolecular linkage) or within one biomolecule (intramolecular linkage) [141, 151]. Such polymerization can be eliminated by using a bifunctional chelator that contains only a single reactive group (e.g., SCN–Bz–DTPA). The level of polymerization that is tolerable depends on the biomolecule, but less than 10% of higher molecular weight species is desirable to avoid diminishing immunoreactivity or receptor-bindng affinity, as well as to minimize sequestration by the reticuloendothelial system (RES) [158].

6.7.7.2   Measurement of Radiochemical Purity (RCP)

In our laboratory, we determine the radiochemical purity (RCP) of radiolabeled mAbs or peptides by size-exclusion HPLC on a BioSep SEC-S 2000 column (Phenomenex, Canada) eluted with 100-mM sodium phosphate buffer, pH 7.4 at a flow rate of 1.0 mL/minute, and interfaced with a PerkinElmer diode array detector set at 280 nm and an FSA radioactivity detector (Figure 6.7-11). The RCP of radiolabeled peptides could also be measured by reversed-phase HPLC on a C-18 column eluted with trifluoroacetic acid/methanol/water combinations. Through overlaying the UV and radioactivity traces, these HPLC chromatograms confirm that the biomolecule is radiolabeled. Furthermore, they identify and quantify any radiochemical impurities, such as free radiolabeled chelators or radionuclides. However, for rapid measurement of RCP, paper chromatography or ITLC systems are more commonly used. The level of RCP purity required for radiolabeled mAbs and peptides depends on their intended clinical application and the toxicity associated with the impurities. An RCP > 90% would be acceptable for mAbs or peptides labeled with a single γ-photon-emitter (e.g., ^{99m}Tc or ^{111}In) or a short-lived positron-emitter (e.g., ^{18}F or ^{68}Ga) and intended for tumor imaging. On the other hand, biomolecules labeled with α-emitters (e.g., ^{211}At) or β-emitters (e.g., ^{131}I or ^{90}Y) for targeted *in situ* radiotherapy of malignancies require a higher RCP (>95%) because of the larger amounts of radioactivity administered and the inherent toxicity of the radiochemical impurities, particularly if they concentrate in a sensitive organ (e.g., bone or thyroid). For example, free ^{90}Y is sequestered *in vivo* by bone, increasing the risk for bone marrow toxicity [255].

6.7.7.3   Measurement of Immunoreactivity or Receptor-Binding Properties

The immunoreactivity or receptor-binding characteristics of mAbs or peptides conjugated to chelators for radiometal labeling must be assessed because these effect tumor uptake *in vivo* [159]. Similarly, these properties should be evaluated for the final radiolabeled biomolecules. Two parameters are measured: (1) the antigen/receptor-binding affinity and (2) the immunoreactive fraction (IRF) or receptor-binding fraction (RBF). The affinity constant (K_a) or dissociation constant (K_d) are measured in direct or indirect competition antigen/receptor-binding assays and compared with those of the unmodified mAbs or peptides. Direct binding assays (previously known as Scatchard assays) are performed by incubating increasing concentrations of radiolabeled mAbs or peptides with tumor cells that express the target epitopes/receptors. These assays may also be performed by incubating the

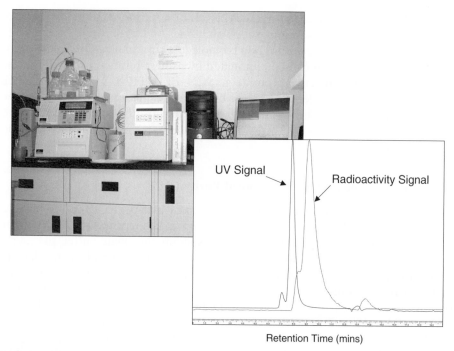

Retention Time (mins)

Figure 6.7-11. *Left Panel*: A size-exclusion HPLC system with a diode array UV detector and FSA flow-through radioactivity detector for evaluating the radiochemical purity and homogeneity of radiolabeled monoclonal antibodies and peptides. *Right Panel*: Typical HPLC chromatograms obtained for analysis of ^{111}In–DTPA-conjugated trastuzumab (Herceptin). The radioactivity signal is offset from the UV signal because of the distance of tubing between the two detectors, which are in sequence. The radioactivity signal has less resolution than the UV signal due to its larger flow cell. ^{111}In–DTPA–trastuzumab IgG in this example has less than 5% free ^{111}In–DTPA impurity (indicated by small peak on the radioactivity trace at $t_R = 13.8$ minutes). The major peak with t_R of 7.7 minutes in the UV trace and 9.5 minutes in the radioactivity trace represents a ^{111}In–DTPA–trastuzumab IgG monomer. The smaller peak with t_R of 7.1 minutes in the UV trace and 8.25 minutes in the radioactivity trace represents a ^{111}In–DTPA–trastuzumab IgG dimer. (Data generously provided by Kristin McLarty.) (This figure is available in full color at ftp://ftp.wiley.com/public/sci_ tech_med/pharmaceutical_biotech/.)

radiolabeled biomolecules with purified antigens/receptors coated onto wells in a microELISA plate. The binding of the radiolabeled biomolecules to the cells is measured in the absence [total binding (TB)] or presence [nonspecific binding (NSB)] of an excess (e.g., 100 nM) of non-radiolabeled biomolecules to saturate the epitopes/receptors on the cells. Specific binding (SB) is obtained by subtracting NSB from TB and is plotted versus the concentration of radiolabeled mAbs or peptides (Figure 6.7-12). The curve is fitted to a direct antigen/receptor-binding model using nonlinear fitting software (e.g., Prism; Graphpad Software, San Diego, CA) and the K_a and maximum number of binding sites/cell (B_{max}) estimated. Competition radioligand binding assays can be performed by measuring the binding of radiolabeled mAbs or peptides to cells displaying the target epitopes or receptors in the presence of increasing concentrations of unmodified biomolecules. The K_d is

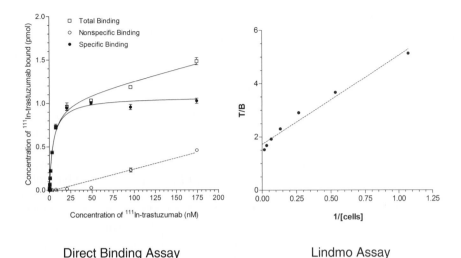

Direct Binding Assay   Lindmo Assay

Figure 6.7-12. *Left Panel*: Results for a direct binding assay for a ^{111}In–DTPA–trastuzumab (Herceptin) IgG incubated with SK–BR-3 human breast cancer cells overexpressing HER2/neu. The K_a and B_{max} in this example were 1.2×10^8 L/mole and 1.0×10^6 receptors/cell, respectively. *Right Panel*: Results for measurement of the immunoreactive fraction (IRF) of a ^{111}In–DTPA–trastuzumab IgG incubated with SK–BR-3 cells using the Lindmo method [256]. In this example, the intercept on the ordinate (1/IRF) was 1.7 and the IRF was therefore 0.59. (Data generously provided by Kristin McLarty.)

then estimated from the concentration of unmodified antibodies or peptides required to displace 50% of the initial binding of the radiolabeled biomolecules to the cells. Finally, the IRF or RBF of radiolabeled mAbs or peptides may be measured by the Lindmo assay [256]. In this assay, a small amount of radiolabeled biomolecules (e. g., 5–10 ng) is incubated with increasing concentrations of cancer cells that display the target epitopes/receptors. The total radioactivity counts added divided by the bound counts (T/B) is plotted versus the inverse of the cell concentration (1/[cells]). The intercept on the ordinate of this plot is 1/IRF or 1/RBF (Figure 6.7-12, right panel). The IRF or RBF provides information that is different than that of K_a or K_d, in that these parameters describe the fraction of radiolabeled biomolecules that can bind their target epitopes/receptors at "infinite antigen excess," irrespective of their binding affinity. Ideally, measurement of K_a or K_d as well as determination of IRF or RBF should be performed to assess immunoreactivity or receptor-binding characteristics.

6.7.7.4   Evaluation of *In Vitro* and *In Vivo* Stability

Studies to evaluate the *in vitro* and *in vivo* stability of radiolabeled mAbs and peptides are an integral part of their development. Kits used to prepare radiolabeled biomolecules (Section 6.7.7.7) must meet specifications established for labeling efficiency, immunoreactivity/receptor binding, homogeneity, sterility and apyrogenicity, and other quality control tests over the expected storage period [145]. Similarly, the radiolabeled biomolecule must maintain the specified level of RCP over the interval from the time of labeling until it is administered. The results of these stability studies will define the expiry times for the kits and/or radiolabeled mAbs or peptides [145]. The stability of the radiolabeled biomolecules *in vitro* in biologically relevant media such as plasma or serum must also be determined. The anticipated mechanism of loss of radiolabel *in vivo* should be taken into account in designing these studies. For example, studies examining the transchelation of radiometal to transferrin in serum/plasma are necessary for [111]In- or ^{67}Ga-labeled biomolecules [151], whereas studies that measure transchelation to albumin or other copper-binding proteins are required for ^{64}Cu/[67]Cu-labeled biomolecules [224]. Cysteine-challenge studies are required for ^{99m}Tc-labeled biomolecules, especially those labeled using the Schwartz technique [94, 257]. Frequently, stability *in vitro* does not predict stability *in vivo*, because the radiolabeled biomolecules may be catabolized in tissues, resulting in release and redistribution of the radionuclide. For example, radioiodinated proteins are stable *in vitro* in serum/plasma but *in vivo* may be internalized by tumor and normal cells, in which they are rapidly catabolized by lysosomal proteases and deiodinases, resulting in release and redistribution of free radioiodine. This instability *in vivo* can be addressed by employing residualizing radio-iodination techniques (Section 6.7.5.1). Nevertheless, due to the limitations of *in vitro* stability studies, it is important to evaluate the stability of radiolabeled mAbs and peptides *in vivo* by chromatographic analysis of plasma/serum samples to identify any circulating radioactive catabolites, as well as by monitoring changes in tissue distribution of radioactivity over time by imaging or biodistribution studies (see Section 6.7.7.5). In these studies, it is critical to sample tissues that are known to sequester the radionuclide, should it be released from the biomolecule, i.e., bone for ^{90}Y or ^{177}Lu and thyroid and stomach for iodine radionuclides. Noninvasive SPECT or PET imaging studies in humans allow these stability studies to be continued into clinical trials of the radiolabeled biomolecules.

6.7.7.5   Preclinical Biodistribution, Tumor Imaging, and Dosimetry Studies

It is important to evaluate the tumor and normal tissue distribution of radiolabeled biomolecules in an animal model. For radiolabeled mAbs and peptides intended to target solid tumors, these studies are performed in athymic mice implanted subcutaneously (s.c.) with human tumor xenografts [18, 108, 120, 132, 258]. In the case of hematological malignancies (e.g., B-cell lymphoma or leukemias), athymic or non-obese diabetic severe combined immunodeficient (NOD-scid) mice may be engrafted s.c. or inoculated intravenously (i.v.) with malignant cells to establish a tumor model [259–261]. Tumor-bearing mice are injected i.v. with 25–100 μCi (0.9–3.7 MBq) of radiolabeled mAbs or peptides. Control groups may consist of (1) mice bearing the same tumors but injected with radiolabeled nonspecific biomolecules that do not recognize the target epitopes/receptors, (2) mice bearing these tumors but injected with the radiolabeled mAbs or peptides mixed with a large excess of unlabeled biomolecules to saturate the epitopes/receptors, or (3) mice implanted with a tumor that does not express the target epitopes/receptors and receiving the radiolabeled mAbs or peptides. Goups of five to six mice are sacrificed at selected times postinjection (p.i.) depending on the anticipated pharmacokinetics. The tumor and samples of blood and normal tissues are obtained and weighed, and the radioactivity in each is measured by γ-scintillation counting. For pure β-emitters (e.g., ^{90}Y), liquid scintillation counting may be used to measure the radioactivity

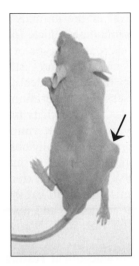

Figure 6.7-13. Imaging of an athymic mouse bearing a subcutaneous HER2/neu-positive BT-474 human breast cancer xenograft (arrow) at 72 hours postintravenous (tail vein) injection of ^{111}In-DTPA-trastuzumab (Herceptin). The tumor is clearly visualized using the radiopharmaceutical. (Image generously provided by Kristin McLarty and Deborah Scollard.) (This figure is available in full color at ftp://ftp.wiley.com/public/sci_ tech_ med/pharmaceutical_biotech/.)

in tissues, or alternatively, the Bremstrahlung radiation caused by interaction of the β-particles with the tissues can be measured in a γ-counter. The tumor and normal tissue uptake is expressed as the percent injected dose per gram of tissue (% i.d./g). The tumor/blood (T/B) and T/NT ratios are calculated. Tumor uptake varies from 1% to 2% i.d./g for radiolabeled peptides to as high as 10% to 20% i.d./g for intact IgG mAbs [18]. Tumor uptake should be significantly greater than that in control groups of mice. The T/B ratio should be >2:1 and, ideally, as high as 5:1 to 10:1. In addition to biodistribution studies, imaging studies of γ-emitting or positron-emitting biomolecules may be performed using dedicated high-resolution (1–2 mm) microSPECT or microPET small animal imaging devices (Figure 6.7-13) [262, 263]. These images can be quantified by region-of-interest (ROI) analysis and are useful to follow any changes in the biodistribution of radioactivity in an individual animal over time. In addition, preclinical small animal imaging studies provide "proof-of-principle" for radiolabeled biomolecules intended for tumor imaging applications in humans. The results of biodistribution and imaging studies can be used to predict the radiation absorbed doses to organs in humans for subsequent clinical trials using computer software such as OLINDA [264, 265].

6.7.7.6 Preclinical Studies to Evaluate Anti-Tumor Effects and Normal Tissue Toxicity

Radiolabeled mAbs and peptides intended for RIT or PDRT of malignancies need to be first evaluated preclinically in mouse tumor xenograft models. These studies involve administration of increasing doses of the radiolabeled biomolecules to mice and monitoring the tumor growth or survival of the animals over time [258]. Control groups of mice receive either the formulation vehicle (e.g., saline) or the nonspecific radiolabeled mAbs of the same class and isotype or peptides that do not recognize the target epitopes/receptors. Generalized and gastrointestinal toxicity is monitored by weighing the animals every few days to identify any weight loss (>10%), whereas bone marrow toxicity is assessed by determining leukocyte, red blood cell, and platelet counts in blood samples. Liver toxicity is evaluated by following serum transaminases, whereas kidney toxicity is determined by measuring serum creatinine. Hematopoietic toxicity, if present, is usually observed within 2–3 weeks after injection of the radiolabeled biomolecules and normalizes within 6–8 weeks, whereas liver and kidney toxicity may require an extended observation period (4–8 weeks). An ideal radiotherapeutic agent will exhibit clearly evident and specific antitumor effects with minimal–moderate and manageable normal tissue toxicity [258].

6.7.7.7   Kit Formulation and Associated Pharmaceutical Testing

Labeling methods for mAbs and peptides that involve the chelation of radiometals (e.g., ^{99m}Tc, ^{111}In, ^{68}Ga, ^{90}Y, ^{177}Lu, and ^{64}Cu) allow formulation of radiopharmaceutical kits [99, 145]. These kits consist of a unit dose of a solution of the metal chelator–biomolecule conjugates dispensed into a glass vial sealed with a rubber septum and aluminum crimp. The kits may be stored at 2–8°C, kept frozen at −10°C, or lyophilized, depending on the particular mAb or peptide. Radiolabeling is performed simply by adding a buffered solution of the radionuclide to the vial and incubating for a predetermined period of time and temperature. The kits are designed such that the labeling efficiency is >90–95%, thus requiring no post-labeling purification and only minimal quality control testing (e.g., assay for total radioactivity, pH measurement, and RCP testing). Other quality control testing procedures are performed on the kit formulation before its use for labeling biomolecules. These tests include protein/peptide purity and homogeneity, evaluation of immunoreactivity or receptor-binding properties, estimation of the level of chelator substitution, measurement of protein concentration and pH, and stability studies. Key pharmaceutical tests such as the USP Sterility Test and USP Bacterial Endotoxins Test also need to be performed on the kits. The radiolabeled biomolecules may be tested retrospectively for sterility and apyrogenicity, provided that the kits have been validated using trial batches to produce a sterile, apyrogenic product.

6.7.8   SUMMARY

Radiolabeled mAbs and peptides offer the opportunity for molecular imaging of tumors in order to probe their phenotypic properties. Various strategies have been developed for labeling these biomolecules with single γ-photon-emitters or positron-emitters for SPECT and PET imaging, respectively. These approaches can be extended for labeling biomolecules with α- or β-emitters or radionuclides that emit Auger and conversion electrons for targeted *in situ* radiotherapy of malignancies. Establishment of specifications for radiolabeled mAbs and peptides supported by comprehensive characterization testing procedures assures the quality of these novel biopharmaceuticals for human evaluation in clinical trials.

ACKNOWLEDGMENT

The author would like to acknowledge the financial support for research originating in his laboratory at the University of Toronto described in this chapter. Funding was received from the U.S. Department of Defense Breast Cancer Research Program, the Canadian Institutes of Health Research, the Canadian Breast Cancer Research Alliance, the Canadian Breast Cancer Foundation, the Cancer Research Society, Inc., the Susan G. Komen Breast Cancer Foundation, the Natural Sciences and Engineering Research Council of Canada, the James Birrell Neuroblastoma Research Fund, and the Ontario Cancer Research Network through funding provided by the Province of Ontario. The author would also like to acknowledge the outstanding contributions by the graduate students, post-doctoral fellows, and research technicians who have been members of his research group over the years.

REFERENCES

1. Ferlay J, Bray F, Pisani P, et al. (2000). Cancer incidence, prevalence and mortality worldwide. *IARC CancerBase, 5th ed*: IARC Press, Lyon, France.

2. Reilly R M (2005). Biomolecules as targeting vehicles for in situ radiotherapy of malignancies. In J Knaeblein, R Mueller, (eds.), *Modern Biopharmaceuticals: Design, Development and Optimization*. Wiley-VCH, Weinheim, Germany, pp. 497–526.

3. Goldenberg D M (1989). Future role of radiolabeled monoclonal antibodies in oncological diagnosis and therapy. *Semin. Nucl. Med.* 19(4):332–339.

4. Baselga J (2001). Clinical trials of Herceptin (trastuzumab). *Eur. J. Cancer.* 37: S18–S24.

5. Behr T M, Béhé M, Wörmann B (2001). Trastuzumab and breast cancer. *N. Engl. J. Med.* 345:995–996.

6. Krenning E P, Kwekkeboom D J, Bakker W H, et al. (1993). Somatostatin receptor scintigraphy with [^{111}In-DTPA-D-Phe^{1}]- and [^{123}I-Tyr^{3}]-octreotide: The Rotterdam experience with more than 1000 patients. *Eur. J. Nucl. Med.* 20:716–731.

7. Reilly R M, Sandhu J, varez-Diez T M, et al. (1995). Problems of delivery of monoclonal antibodies. Pharmaceutical and pharmacokinetic solutions. *Clin. Pharmacokinet.* 28(2):126–142.

8. Yokota T, Milenic D E, Whitlow M, et al. (1992). Rapid tumor penetration of a single-chain Fv and comparison with other immunoglobulin forms. *Cancer Res.* 52(12): 3402–3408.

9. Saga T, Neumann R D, Heya T, et al. (1995). Targeting cancer micrometastases with monoclonal antibodies: a binding-site barrier. *Proc. Natl. Acad. Sci. USA.* 92(19):8999–9003.

10. van Osdol W, Fujimori K, Weinstein J N (1991). An analysis of monoclonal antibody distribution in microscopic tumor nodules: consequences of a "binding-site barrier". *Cancer Res.* 51:4776–4784.

11. Behr T M, Goldenberg D M, Becker W (1998). Reducing the renal uptake of radiolabeled antibody fragments and peptides for diagnosis and therapy: Present status, future prospects and limitations. *Eur. J. Nucl. Med.* 25:201–212.

12. Jamar F, Barone R, Mathieu I, et al. (2003). ^{86}Y-DOTA^{0}-D-Phe^{1}-Tyr3-octreotide (SMT487)- a phase 1 clinical study: pharmacokinetics, biodistribution and renal pro-

tective effect of different regimens of amino acid co-infusion. *Eur. J. Nucl. Med.* 30:510–518.

13. Pagani M, Stone-Elander S, Larsson S A (1997). Alternative positron emission tomography with non-conventional positron emitters: effects of their physical properties on image quality and potential clinical applications. *Eur. J. Nucl. Med.* 24:1301–1327.

14. Xue L Y, Butler N J, Makrigiorgos G M, Adelstein S J, Kassis A I (2002). Bystander effect produced by radiolabeled tumor cells in vivo. *J. Nucl. Med.* 43(Suppl): 276P–277P.

15. Finn R, Cheung N-K V, Divgi C, et al. (1991). Technical challenges associated with the radiolabeling of monoclonal antibodies utilizing short-lived, positron emitting radionuclides. *Nucl. Med. Biol.* 18:9–13.

16. Hunter W M, Greenwood F C (1962). Preparation of ^{131}I labeled human growth hormone of high specific activity. *Nature.* 194:495–496.

17. Fraker P J, Speck J C, Jr (1978). Protein and cell membrane iodinations with a sparingly soluble chloroamide, 1,3,4,6-tetrachloro-3a,6a-diphrenylglycoluril. *Biochem. Biophys. Res. Commun.* 80(4):849–857.

18. Reilly R M, Kiarash R, Sandhu J, et al. (2000). A comparison of EGF and MAb 528 labeled with 111In for imaging human breast cancer. *J. Nucl. Med.* 41(5):903–911.

19. Zalutsky M R, Narula A S (1987). A method for the radiohalogenation of proteins resulting in decreased thyroid uptake of radioiodine. *Appl. Radiat. Isot.* 38:1051–1055.

20. Zalutsky M R, Narula A S (1988). Radiohalogenation of a monoclonal antibody using an N-succinimidyl 3-(tri-n-butylstannyl) benzoate intermediate. *Cancer Res.* 48:1446–1450.

21. Zalutsky M R, Noska M A, Colapinto E V, et al. (1989). Enhanced tumor localization and in vivo stability of a monoclonal antibody radioiodinated using N-succinimidyl 3-(tri-n-butylstannyl) benzoate. *Cancer Res.* 49:5543–5549.

22. Garg S, Garg P K, Zalutsky M R (1991). N-succinimidyl 5-(trialkylstannyl)-3-pyridinecarboxylates: A new class of reagents for protein radioiodination. *Bioconjug. Chem.* 2:50–58.

23. Reist C J, Garg P K, Alston K L, et al. (1996). Radioiodination of internalizing monoclonal antibodies using N-succinimidyl 5-iodo-3-pyridinecarboxylate. *Cancer Res.* 56:4970–4977.

24. Reist C J, Archer G E, Wikstrand C J, et al. (1997). Improved targeting of an anti-epidermal growth factor receptor variant III monoclonal antibody in tumor xenografts after labeling using N-succinimidyl 5-iodo-3-pyridinecarboxylate. *Cancer Res.* 57: 1510–1515.

25. Garg P K, Alston K L, Welsh P C, et al. (1996). Enhanced binding and inertness to dehalogenation of α-melanotropic peptides labeled using N-succinimidyl 3-iodobenzoate. *Bioconjug. Chem.* 7:233-239.

26. Vaidyanathan G, Affleck D J, Li J, et al. (2001). A polar substituent-containing acylation agent for the radioiodination of internalizing monoclonal antibodies: N-succinimidyl 4-guanidinomethyl-3-[^{131}I]iodobenzoate ([^{131}I]SGMIB). *Biocon-jug. Chem.* 12:428–438.

27. Vaidyanathan G, Affleck D J, Bigner D D, et al. (2002). Improved xenograft targeting of tumor-specific anti-epidermal growth factor receptor variant III antibody labeled using N-succinimidyl 4-guanidinomethyl-3-iodobenzoate. *Nucl. Med. Biol.* 29:1–11.

28. Shankar S, Vaidyanathan G, Affleck D, et al. (2003). N-succinimidyl 3-[^{131}I]iodo-4-phosphonomethylbenzoate ([^{131}I]SIPMB), a negatively charged substituent-bearing

acylation agent for the radioiodination of peptides and mAbs. *Bioconjug. Chem.* 14:331–341.

29. Govindan S V, Mattes M J, Stein R, et al. (1999). Labeling of monoclonal antibodies with diethylenetriaminepentaacetic acid-appended radioiodinated peptides containing D-amino acids. *Bioconjug. Chem.* 10:231–240.

30. Franano F N, Edwards W B, Welch M J, et al. (1994). Metabolism of receptor targeted ^{111}In-DTPA-glycoproteins: Identification of ^{111}In-DTPA-ε-lysine as the primary metabolic and excretory product. *Nucl. Med. Biol.* 21:1023–1034.

31. Dumas P, Maziere B, Autissier N, et al. (1973). Specificite de L'iodotyrosine desiodase des microsomes thyroidiens et hepatiques. *Biochim. Biophys. Acta.* 293:36–47.

32. Stein R, Govindan S V, Mattes M J, et al. (2003). Improved iodine radiolabels for monoclonal antibody therapy. *Cancer Res.* 63:111–118.

33. Govindan S V, Griffiths G L, Stein R, et al. (2005). Clinical-scale radiolabeling of a humanized anticarcinoembryonic antigen monoclonal antibody, hMN-14, with residualizing ^{131}I for use in radioimmunotherapy. *J. Nucl. Med.* 46:153–159.

34. Foulon C F, Reist C J, Bigner D D, et al. (2000). Radioiodinatin via D-amino acid peptide enhances cellular retention and tumor xenograft targeting of an internalizing anti-epidermal growth factor receptor variant III monoclonal antibody. *Cancer Res.* 60:4453–4460.

35. Thorpe S E, Baynes J W, Chroneos Z C (1993). The design and application of residualizing labels for studies of protein catabolism. *FASEB.* 7:399–405.

36. Stein R, Goldenberg D M, Thorpe S R, et al. (1995). Effects of radiolabeling monoclonal antibodies with a residualizing iodine radiolabel in the accretion of radioisotope in tumors. *Cancer Res.* 55:3132–3139.

37. Reist C J, Archer G E, Kurpad S N, et al. (1995). Tumor-specific anti-epidermal growth factor receptor variant III monoclonal antibodies: use of the tyramine-cellobiose radioiodination method enhances cellular retention and uptake in tumor xenografts. *Cancer Res.* 55:4375–4382.

38. Zalutsky M R, Xu F J, Yu Y, et al. (1999). Radioiodinated antibody targeting of the HER-2/neu oncoprotein: Effects of labeling method on cellular processing and tissue distribution. *Nucl. Med. Biol.* 26:781–790.

39. Sundberg Ä L, Blomquist E, Carlsson J, et al. (2003). Cellular retention of radioactivity and increased radiation dose. Model experiments with EGF-dextran. *Nucl. Med. Biol.* 30:303–315.

40. Townsend D W, Carney J P J, Yap J T, et al. (2004). PET/CT today and tomorrow. *J. Nucl. Med.* 45(suppl.):4S–14S.

41. Chatziionnou A F (2002). Molecular imaging of small animals with dedicated PET tomographs. *Eur. J. Nucl. Med.* 29:98–114.

42. Gambhir S S, Czernin J, et al. (2001). A tabulated summary of the FDG PET literature. *J. Nucl. Med.* 42(suppl.):1S–93S.

43. Bakir M A, Eccles S A, Babich J W, et al. (1992). c-erbB2 Protein overexpression in breast cancer as a target for PET using iodine-124-labeled monoclonal antibodies. *J. Nucl. Med.* 33:2154–2160.

44. Larson S M, Pentlow K S, Volkow N D, et al. (1992). PET scanning of iodine-124-3F9 as an approach to tumor dosimetry during treatment planning for radioimmunotherapy in a child with neuroblastoma. *J. Nucl. Med.* 33:2020–2023.

45. Lee F T, Hall C, Rigopoulos A, et al. (2001). Immuno-PET of human colon xenograft-bearing BALB/c nude mice using ^{124}I-CDR-grafted humanized A33 monoclonal antibody. *J. Nucl. Med.* 42:764–769.

46. Verel I, Visser G W M, Vosjan M J W D, et al. (2004). High-quality [124]I-labelled monoclonal antibodies for use as PET scouting agents prior to ^{131}I-radioimmunotherapy. *Eur. J. Nucl. Med. Mol. Imaging.* 31:1645–1652.

47. Robinson M K, Doss M, Shaller C, et al. (2005). Quantitative immuno-positron emission tomography imaging of HER2-positive tumor xenografts with an iodine-124 labeled anti-HER2 diabody. *Cancer Res.* 65:1471–1478.

48. Sundaresan G, Yazaki P J, Shively J E, et al. (2003). [124]I-labeled engineered anti-CEA minibodies and diabodies allow high-contrast, antigen-specific small-animal PET imaging of xenografts in athymic mice. *J. Nucl. Med.* 44:1962–1969.

49. Iozzo P, Osman S, Glaser M, et al. (2002). *In vivo* imaging of insulin receptors by PET: preclinical evaluation of iodine-125 and iodine-124 labelled human insulin. *Nucl. Med. Biol.* 29:73–82.

50. Eary J F (2001). PET imaging for planning cancer therapy. *J. Nucl. Med.* 42:770–771.

51. Tolmachev V, Lovqvist A, Einarsson L, et al. (1998). Production of [76]Br by a low-energy cyclotron. *Appl. Radiat. Isot.* 49:1537–1540.

52. Lovqvist A, Sundin A, Ahlstrom H, et al. (1995). [76]Br-labeled monoclonal anti-CEA antibodies for radioimmuno positron emission tomography. *Nucl. Med. Biol.* 22:125–131.

53. Sundin J, Tolmachev V, Koziorowski J, et al. (1999). High yield direct [76]Br-bromination of monoclonal antibodies using chloramine-T. *Nucl. Med. Biol.* 26:923–929.

54. Lovqvist A, Sundin A, Roberto A, et al. (1997). Comparative PET imaging of experimental tumors with bromine-76-labeled antibodies, fluorine-18-fluorodeoxyglucose and carbon-11-methionine. *J. Nucl. Med.* 38:1029–1035.

55. Lovqvist A, Sundin A, Ahlstrom H, et al. (1997). Pharmacokinetics and experimental PET imaging of a bromine-76-labeled monoclonal anti-CEA antibody. *J. Nucl. Med.* 38:395–401.

56. Hoglund J, Tolmachev V, Orlova A, et al. (2000). Optimized indirect [76]Br-bromination of antibodies using N-succinimidyl para-[[76]Br]bromobenzoate for radioimmuno PET. *Nucl. Med. Biol.* 27:837–843.

57. Wilbur D S, Hylarides M D (1991). Radiolabeling of a monoclonal antibody with N-succinimidyl para-[[77]Br]bromobenzoate. *Nucl. Med. Biol.* 18:363–365.

58. Mume E, Orlova A, Malmström P-U, et al. (2005). Radiobromination of humanized anti-HER2 monoclonal antibody trastuzumab using N-succinimidyl 5-bromo-3-pyridinecarboxylate, a potential label for immunoPET. *Nucl. Med. Biol.* 32:613–622.

59. Mume E, Orlova A, Larsson B, et al. (2005). Evaluation of ((4-hydroxyphenyl)ethyl) maleimide for site-specific radiobromination of anti-HER2 affibody. *Bioconjug. Chem.* 16:1547–1555.

60. Steffen A-C, Wikman M, Tolmachev V, et al. (2005). In vitro characterization of a bivalent anti-HER-2 affibody with potential for radionuclide based diagnostics. *Cancer Biother. Radiopharm.* 20:239–248.

61. Bruskin A, Sivaev I, Persson M, et al. (2004). Radiobromination of monoclonal antibody using potassium [[76]Br] (4 isothiocyanatobenzyl-ammonio)-bromo-decahydro-closo-dodecaborate (bromo-DABI). *Nucl. Med. Biol.* 31:205–211.

62. Winberg K J, Persson M, Malmström P-U, et al. (2004). Radiobromination of anti-HER2/neu/ErbB-2 monoclonal antibody using the p-isothiocyanatobenzene derivative of the [[76]Br]undecahydro-bromo-7,8-dicarba-nido-undecaborate(1-) ion. *Nucl. Med. Biol.* 31:425–433.

63. Okarvi S M (2001). Recent progress in fluorine-18 labelled peptide radiopharmaceuticals. *Eur. J. Nucl. Med.* 28:929–938.

64. Garg P K, Garg S, Zalutsky M R (1991). Fluorine-18 labeling of monoclonal antibodies and fragments with preservation of immunoreactivity. *Bioconjug. Chem.* 2:44–49.

65. Garg P K, Garg S, Bigner D D, et al. (1992). Localization of fluorine-18-labeled Mel-14 monoclonal antibody F(ab′)$_2$ fragment in a subcutaneous xenograft model. *Cancer Res.* 52:5054–5060.

66. Page R L, Garg P K, Garg S, et al. (1994). PET imaging of osteosarcoma in dogs using a fluorine-18-labeled monoclonal antibody Fab fragment. *J. Nucl. Med.* 35:1506–1513.

67. Lang L, Eckelman W C (1994). One-step synthesis of ^{18}F labeled [^{18}F]-N-succinimidyl 4-(fluoromethyl) benzoate for protein labeling. *Appl. Radiat. Isot.* 45:1155–1163.

68. Vaidyanathan G, Zalutsky M R (1994). Improved synthesis of N-succinimidyl 4-[^{18}F]fluorobenzoate and its application to the labeling of a monoclonal antibody fragment. *Bioconjug. Chem.* 5:352–356.

69. Page R L, Garg P K, Vaidyanathan G, et al. (1994). Preclinical evaluation and PET imaging of 18F-labeled Mel-14 F(ab′)$_2$ fragment in normal dogs. *Nucl. Med. Biol.* 21:911–919.

70. Vaidyanathan G, Zalutsky M R (1997). Fluorine-18-labeled [Nle^4,D-Phe^7]-α-MSH, an a-melanocyte stimulating hormone analogue. *Nucl. Med. Biol.* 24:171–178.

71. Guhlke S, Coenen H H, Stocklin G (1994). Fluoroacylation agents based on small n.c.a. [^{18}F]fluorocarboxylic acids. *Appl. Radiat. Isot.* 45:715–727.

72. Guhlke S, Wester H-J, Bruns C, et al. (1994). (2-[^{18}F]fluoropropionyl-(D)phe^1)-octreotide, a potential radiopharmaceutical for quantitative somatostatin receptor imaging with PET: synthesis, radiolabeling, *in vitro* validation and biodistribution in mice. *Nucl. Med. Biol.* 21:819–825.

73. Poethko T, Schottelius M, Thumshirn G, et al. (2004). Two-step methodology for high-yield routine radiohalogenation of peptides: ^{18}F-labeled RGD and octreotide analogs. *J. Nucl. Med.* 45:892–902.

74. Wester H-J, Brockman J, Rosch F, et al. (1997). PET-pharmacokinetics of ^{18}F-octreotide: A comparison with ^{67}Ga-DFO- and ^{86}Y-DTPA-octreotide. *Nucl. Med. Biol.* 24:275–286.

75. Chen X, Park R, Shahinian A H, et al. (2004). ^{18}F-labeled RGD peptide: initial evaluation for imaging brain tumor angiogenesis. *Nucl. Med. Biol.* 31:179–189.

76. Haubner R, Kuhnast B, Mang C, et al. (2004). [^{18}F]galacto-RGD: Synthesis, radiolabeling, metabolic stability, and radiation dose estimates. *Bioconjug. Chem.* 15:61–69.

77. Wester H-J, Hamacher K, Stocklin G (1996). A comparative study of n.c.a. fluorine-18 labeling of proteins via acylation and photochemical conjugation. *Nucl. Med. Biol.* 23:365–372.

78. Chang Y S, Jeong J M, Lee Y-S, et al. (2005). Preparation of ^{18}F-human serum albumin: A simple and efficient protein labeling method with ^{18}F using a hydrazone-formation method. *Bioconjug. Chem.* 16:1329–1333.

79. Couturier O, Supiot S, Degraef-Mougin M, et al. (2005). Cancer radioimmunotherapy with alpha-emitting nuclides. *Eur. J. Nucl. Med. Mol. Imaging.* 32:601–614.

80. Visser G W M, Diemer E L, Kaspersen F M (1981). The nature of the astatine-protein bond. *Int. J. Appl. Radiat. Isot.* 32:905–912.

81. Aaij C, Tschroots W R J M, Lindner L, et al. (1975). The preparation of astatine labelled proteins. *Int. J. Appl. Radiat. Isot.* 26:25–30.

82. Vaughan A T M, Fremlin J H (1978). The preparation of astatine labelled proteins using an electrophilic reaction. *Int. J. Nucl. Med. Biol.* 5:229–230.

83. Zalutsky M R, Narula A S (1988). Astatination of proteins using an N-succinimidyl tri-n-butylstannyl benzoate intermediate. *Appl. Radiat. Isot.* 39:227–232.

84. Zalutsky M R, Garg P K, Friedman H S, et al. (1989). Labeling monoclonal antibodies and F(ab′)$_2$ fragments with the α-particle-emitting nuclide astatine-211: Preservation of immunoreactivity and *in vivo* localizing capacity. *Proc. Natl. Acad. Sci. USA.* 86:7149–7153.

85. Garg P K, Harrison C L, Zalutsky M R (1990). Comparative tissue distribution in mice of the alpha-emitter [211]At and ^{131}I as labels of a monoclonal antibody and F(ab′)$_2$ fragment. *Cancer Res.* 50:3514–3520.

86. Wilbur D S, Vessella R L, Stray J E, et al. (1993). Preparation and evaluation of *para*-[[211]At]astatobenzoyl labeled anti-renal cell carcinoma antibody A6H F(ab′)$_2$. *In vivo* distribution comparison with *para*-[^{125}I]iodobenzoyl labeled A6H F(ab′)$_2$. *Nucl. Med. Biol.* 20:917–927.

87. Zalutsky M R, Zhai X-C, Alston K L, et al. (2001). High-level production of α-particle-emitting [211]At and preparation of [211]At-labeled antibodies for clinical use. *J. Nucl. Med.* 42:1508–1515.

88. Pozzi O R, Zalutsky M R (2005). Radiopharmaceutical chemistry of targeted radio-therapeutics, Part I: Effects of solvent on the degradation of radiohalogenation precursors by [211]At α-particles. *J. Nucl. Med.* 46:700–706.

89. Pozzi O R, Zalutsky M R (2005). Radiopharmaceutical chemistry of targeted radio-therapeutics, Part 2: Radiolytic effects of [211]At α-particles influence N-succinimidyl 3-[211]At-astatobenzoate synthesis. *J. Nucl. Med.* 46:1393–1400.

90. Reist C J, Foulon C F, Alston K, et al. (1999). Astatine-211 labeling of internalizing anti-EGFRvIII monoclonal antibody using N-succinimidyl 5-[[211]At]astato-3pyridinecarboxylate. *Nucl. Med. Biol.* 26:405–411.

91. Vaidyanathan G, Affleck D, Welsh P, et al. (2000). Radioiodination and astatination of octreotide by conjugation labeling. *Nucl. Med. Biol.* 27:329–337.

92. Aurlien E, Larsen R H, Bruland O S (2000). Demonstration of highly specific toxicity of the α-emitting radioimmunoconjugate [211]At-rituximab against non-Hodgkin's lymphoma cells. *Br. J. Cancer.* 83:1375–1379.

93. Liu S, Edwards D S, Barrett J A (1997). [99m]Tc labeling of highly potent small peptides. *Bioconjug. Chem.* 8:621–636.

94. Reilly R M (1993). Immunoscintigraphy of tumors using [99]Tc[m]-labelled monoclonal antibodies: A review. *Nucl. Med. Commun.* 14(5):347–359.

95. Schwartz A, Steinstrasser A (1987). A novel approach to Tc-99m labeled monoclonal antibodies. *J. Nucl. Med.* 28:721.

96. Mather S J, Ellison D (1990). Reduction-mediated technetium-99m labeling of monoclonal antibodies. *J. Nucl. Med.* 31:692–697.

97. Stopar T G, Milinaric-Rascan I, Fettich J, et al. (2006). ^{99m}Tc-rituximab radiolabelled by photoactivation: a new non-Hodgkin's lymphoma imaging agent. *Eur. J. Nucl. Med. Mol. Imaging.* 33:53–59.

98. Stalteri M A, Mather S J (1996). Technetium-99m labelling of the anti-tumor antibody PR1A3 by photoactivation. *Eur. J. Nucl. Med.* 23:178–187.

99. Morales A A, Nunez-Gandolff G, Perez N P, et al. (1999). Freeze-dried formulation for direct ^{99m}Tc-labeling ior-egr/r3 mab: Additives, biodistribution, and stability. *Nucl. Med. Biol.* 26:717–723.

100. Ellman G L (1958). A colorimetric method for determining low concentrations of mercaptans. *Arch. Biochem. Biophys.* 74:443–450.

101. Iznaga-Escobar N, Morales A, Nunez G (1996). Micromethod for quantification of SH groups generated after reduction of monoclonal antibodies. *Nucl. Med. Biol.* 23:641–644.

102. Marks A, Ballinger J R, Reilly R M, et al. (1995). A novel anti-seminoma monoclonal antibody (M2A) labelled with technetium-99m: Potential application for radioimmunoscintigraphy. *Br. J. Urol.* 75(2):225–229.

103. Steffens M G, Oosterwijk E, Kranenborg M H G C, et al. (1999). In vivo and in vitro characterizations of three ^{99m}Tc-labeled monoclonal antibody G250 preparations. *J. Nucl. Med.* 40:829–836.

104. Vallis K A, Reilly R M, Chen P, et al. (2002). A phase I study of ^{99m}Tc-hR3 (DiaCIM), a humanized immunoconjugate directed towards the epidermal growth factor receptor. *Nucl. Med. Commun.* 23(12):1155–1164.

105. Torres L A, Perera A, Batista J F, et al. (2005). Phase I/II clinical trial of the humanized anti-EGF-r monoclonal antibody h-R3 labelled with ^{99m}Tc in patients with tumor of epithelial origin. *Nucl. Med. Commun.* 26:1049–1057.

106. Fichna J, Janecka A (2003). Synthesis of target-specific radiolabeled peptides for diagnostic imaging. *Bioconjug. Chem.* 14:3–17.

107. Abrams M J, Juweid M, tenKate C I, et al. (1990). Technetium-99m-human polyclonal IgG radiolabeled via the hydrazino nicotinamide derivative for imaging focal sites of infection in rats. *J. Nucl. Med.* 31(12):2022–2028.

108. Tang Y, Scollard D, Chen P, et al. (2005). Imaging of HER2/neu expression in BT-474 human breast cancer xenografts in athymic mice using [^{99m}Tc]-HYNIC-trastuzumab (Herceptin) Fab fragments. *Nucl. Med. Commun.* 26(5):427–432.

109. Ono M, Arano Y, Mukai T, et al. (2001). Plasma protein binding of ^{99m}Tc-labeled hydrazino nicotinamide derivatized polypeptides and peptides. *Nucl. Med. Biol.* 28(2):155–164.

110. Ultee M E, Bridger G J, Abrams M J, et al. (1997). Tumor imaging with technetium-99m-labeled hydrazinenicotinamide-Fab′ conjugates. *J. Nucl. Med.* 38: 133–138.

111. Bangard M, Behe M, Guhike S, et al. (2000). Detection of somatostatin receptor-positive tumors using the new ^{99m}Tc-tricine-HYNIC-D-Phe[1]-Tyr[3]-octreotide: First results in patients and comparison with ^{111}In-DTPA-D-Phe[1]-octreotide. *Eur. J. Nucl. Med.* 27:628–637.

112. Decristoforo C, Mather S J, Cholewinski W, et al. (2000). ^{99m}Tc-EDDA/HYNIC-TOC: A new ^{99m}Tc-labelled radiopharmaceutical for imaging somatostatin receptor-positive tumors: First clinical results and intra-patient comparison with ^{111}In-labelled octreotide derivatives. *Eur. J. Nucl. Med.* 27:1318–1325.

113. Decristoforo C, Melendez-Alafort L, Sosabowski J K, et al. (2000). ^{99m}Tc-HYNIC-[Tyr[3]]-octreotide for imaging somatostatin-receptor-positive tumors: Preclinical evaluation and comparison with ^{111}In-octreotide. *J. Nucl. Med.* 41:1114–1119.

114. Cornelissen B, Kersemans V, Burvenich I, et al. (2005). Synthesis, biodistribution and effects of farnesyltransferase inhibitor therapy on tumur uptake in mice of ^{99m}Tc labelled epidermal growth factor. *Nucl. Med. Commun.* 26:147–153.

115. Babich J W, Fischman A J (1995). Effect of "co-ligand" on the biodistribution of ^{99m}Tc-labeled hydrazino nicotinic acid derivatized chemotactic peptides. *Nucl. Med. Biol.* 22:25–30.

116. Decristoforo C, Mather S J (1999). Technetium-99m somatostatin analogues: Effect of labelling methods and peptide sequence. *Eur. J. Nucl. Med.* 26:869–876.

117. Decristoforo C, Mather S J (1999). 99m-Technetium-labelled peptide-HYNIC conjugates: Effects of lipophilicity and stability on biodistribution. *Nucl. Med. Biol.* 26:389–396.

118. Decristoforo C, Cholewinski W, Donnemiller E, et al. (2000). Detection of somatostatin receptor-positive tumors using the new ^{99m}Tc-tricine-HYNIC-D-Phe[1]-Tyr[3]-octreotide: first results in patients and comparison with ^{111}In-DTPA-D-Phe1-octreotide. *Eur. J. Nucl. Med.* 27:1580.

119. Ono M, Arano Y, Mukai T, et al. (2001). ^{99m}Tc-HYNIC-derivatized ternary ligand complexes for ^{99m}Tc-labeled polypeptides with low in vivo protein binding. *Nucl. Med. Biol.* 28(3):215–224.

120. Tang Y, Wang J, Scollard D A, et al. (2005). Imaging of HER2/neu-positive BT-474 human breast cancer xenografts in athymic mice using ^{111}In-trastuzumab (Herceptin) Fab fragments. *Nucl. Med. Biol.* 32(1):51–58.

121. Sattelberger A P, Atcher R W (1999). Nuclear medicine finds the right chemistry. *Nat. Biotechnol.* 17:849–850.

122. Waibel R, Alberto R, Willuda J, et al. (1999). Stable one-step technetium-99m labeling of His-tagged recombinant proteins with a novel Tc(I)-carbonyl complex. *Nat. Biotechnol.* 17:897–901.

123. Schibi R, Schwarzbach R, Alberto R, et al. (2002). Steps toward high specific activity labeling of biomolecules for therapeutic application: Preparation of precursor [Re-188(H_2O)(3)(CO)(3)](+) and synthesis of tailor-made bifunctional ligand systems. *Bioconjug. Chem.* 13:750–756.

124. Blok D, Feitsman H I J, Kooy Y M C, et al. (2004). New chelation strategy allows for quick and clean ^{99m}Tc-labeling of synthetic peptides. *Nucl. Med. Biol.* 31:815–820.

125. Fritzberg A R, Abrams P G, Beamier P L, et al. (1988). Specific and stable labeling of antibodies with technetium-99m with a diamide diothiolate chelating agent. *Proc. Natl. Acad. Sci. USA.* 85:4025–4029.

126. Rogers B E, Parry J J, Andrews R, et al. (2005). MicroPET imaging of gene transfer with a somatostatin receptor-based reporter gene and [94m]Tc-demotate. *J. Nucl. Med.* 46:1889–1897.

127. Griffiths G L, Goldenberg D M, Jones A L, et al. (1992). Radiolabeling of monoclonal antibodies and fragments with technetium and rhenium. *Bioconjug. Chem.* 3(91):99.

128. Griffiths G L, Goldenberg D M, Knapp Jr. F F, et al. (1991). Direct radiolabeling of monoclonal antibodies with generator-produced rhenium-188 for radioimmunotherapy: Labeling and animal biodistribution studies. *Cancer Res.* 51:4594–4602.

129. Iznaga-Escobar N (2001). Direct radiolabeling of monoclonal antibodies with rhenium-188 for radioimmunotherapy of solid tumors—a review of radiolabeling characteristics, quality control and in vitro stability studies. *Appl. Radiat. Isot.* 54:399–406.

130. John E, Thakur M L, DeFulvio J, et al. (1993). Rhenium-186-labeled monoclonal antibodies for radioimmunotherapy: preparation and evaluation. *J. Nucl. Med.* 34: 260–267.

131. Rhodes B A, Lambert C R, Marek M J, et al. (1996). Re-188 labelled antibodies. *Appl. Radiat. Isot.* 47:7–14.

132. Reilly R M, Ng K, Polihronis J, et al. (1994). Immunoscintigraphy of human colon cancer xenografts in nude mice using a second-generation TAG-72 monoclonal antibody labelled with [99]Tcm. *Nucl. Med. Commun.* 15(5):379–387.

133. Ferro Flores G, Torres-Garcia E, Garcia-Pedroza L, et al. (2005). An efficient, reproducible and fast preparation of [188]Re-anti-CD20 for the treatment of non-Hodgkin's lymphoma. *Nucl. Med. Commun.* 26:793–799.

134. Goldrosen M H, Biddle W C, Pancock J, et al. (1990). Biodistribution, pharmacokinetic, and imaging studies with [186]Re-labeled NR-LU-10 whole antibody in LS174T colonic tumor-bearing mice. *Cancer Res.* 24:7973–7978.

135. Visser G W, Gerretsen M, Herscheid J D, et al. (1993). Labeling of monoclonal antibodies with rhenium-186 using the MAG3 chelate for radioimmunotherapy of cancer: A technical protocol. *J. Nucl. Med.* 34:1953–1963.

136. Zamora P O, Gulhke S, Bender H, et al. (1996). Experimental radiotherapy of receptor-positive human prostate adenocarcinoma with 188Re-RC-160, a directly-radiolabeled somatostatin analogue. *Int. J. Cancer.* 65:214–220.

137. Safavy A, Khazaeli M B, Safavy K, et al. (1999). Biodistribution study of [188]Re-labeled trisuccin-HuCC49 and trisuccin-HuCC49ΔCH$_2$ conjugates in athymic nude mice bearing intraperitoneal colon cancer xenografts. *Clin. Cancer Res.* 5(suppl.): 2994S–3000S.

138. Smith C J, Sieckman G L, Owen N K, et al. (2003). Radiochemical investigations of [[188]Re(H$_2$O)(CO)$_3$-diaminopropionic acid-SSS-bombesin(7-14)NH$_2$]: Synthesis, radiolabeling and in vitro/in/vivo GRP receptor targeting studies. *Anticancer Res.* 23:63–70.

139. Lubberink M, Tolmachev V, Widström C, et al. (2002). [110m]In-DTPA-D-Phe[1]-octreotide for imaging of neuoendocrine tumors using PET. *J. Nucl. Med.* 43:1391–1397.

140. Tolmachev V, Bernhardt P, Forssell-Aronsson E, et al. (2000). [114m]In, a candidate for radionuclide therapy: low-energy cyclotron production and labeling of DTPA-D-Phe[1]-octreotide. *Nucl. Med. Biol.* 27:183–188.

141. Hnatowich D J, Childs R L, Lanteigne D, et al. (1983). The preparation of DTPA-coupled antibodies radiolabeled with metallic radionuclides: An improved method. *J. Immunol. Meth.* 65:147–157.

142. Krejcarek G E, Tucker K L (1977). Covalent attachment of chelating groups to macromolecules. *Biochem. Biophys. Res. Commun.* 77:581–585.

143. Brechbiel M W, Gansow O A, Atcher R W, et al. (1986). Synthesis of 1-(p-isothiocyanatobenzyl) derivatives of DTPA and EDTA. Antibody labeling and tumor-imaging studies. *Inorg. Chem.* 25:2772–2781.

144. Westerberg D A, Carney P L, Rogers P E, et al. (1988). Synthesis of novel bifunctional chelators and their use in preparing monoclonal antibody conjugates for tumor targeting. *J. Med. Chem.* 32:236–243.

145. Reilly R M, Scollard D A, Wang J, et al. (2004). A kit formulated under good manufacturing practices for labeling human epidermal growth factor with ^{111}In for radiotherapeutic applications. *J. Nucl. Med.* 45(4):701–708.

146. Zoghbi S S, Neumann R D, Gottschalk A (1986). The ultrapurification of indium-111 for radiotracer studies. *Invest. Radiol.* 21:710–713.

147. Reilly R, Sheldon K, Marks A (1989). Labelling of monoclonal antibodies 10B, 8C, and M2A with indium-111. *Int. J. Appl. Radiat. Isot.* 40(4):279–283.

148. Carney P L, Rogers P E, Johnson D K (1989). Dual isotope study of iodine-125 and indium-111-labeled antibody in athymic mice. *J. Nucl. Med.* 30:374–384.

149. McLarty K, Cornelissen B, Scollard D, et al. (2006). Imaging of HER-2/neu positive human breast cancer xenografts in CD1[nu/nu] mice using ^{111}In-labeled trastuzumab (Herceptin). *Canadian Breast Cancer Research Alliance: Reasons for Hope Conf.* Montreal, Quebec, Canada.

150. Hnatowich D J, Griffin T W, Kosciuczyk C, et al. (1985). Pharmacokinetics of an indium-111-labeled monoclonal antibody in cancer patients. *J. Nucl. Med.* 26:849–858.

151. Reilly R M, Marks A, Law J, et al. (1992). *In-vitro* stability of EDTA and DTPA immunoconjugates of monoclonal antibody 2G3 labelled with In-111. *Appl. Radiat. Isot.* 43:961–967.

152. Goodwin D A, Meares C F, McTigue M, et al. (1986). Metal decomposition rates of ^{111}In-DTPA and EDTA conjugates of monoclonal antibodies *in vivo. Nucl. Med. Commun.* 7:831–838.

153. Reilly R M, Maiti P K, Kiarash R, et al. (2001). Rapid imaging of human melanoma xenografts using an scFv fragment of the human monoclonal antibody H11 labelled with ^{111}In. *Nucl. Med. Commun.* 22(5):587–595.

154. Orlova A, Briskin A, Sjöström A, et al. (2000). Cellular processing of ^{125}I- and ^{111}In-labeled epidermal growth factor (EGF) bound to cultured A431 tumor cells. *Nucl. Med. Biol.* 27:827–835.

155. Bakker W H, Albert R, Bruns C, et al. (1991). [^{111}In-DTPA-D-Phe^1]-octreotide, a potential radiopharmaceutical for imaging of somatostatin receptor-positive tumors: Synthesis, radiolabeling and in-vitro validation. *Life. Sci.* 49:1583–1591.

156. Cole W C, DeNardo S J, Meares C F, et al. (1987). Comparative serum stability of radiochelates for antibody radiopharmaceuticals. *J. Nucl. Med.* 28:83–90.

157. Deshpande S V, Subramanian R, McCall M J, et al. (1990). Metabolism of indium chelates attached to monoclonal antibody: Minimal transchelation of indium from benzyl-EDTA chelate in vivo. *J. Nucl. Med.* 31:218–224.

158. Blend M J, Greager J A, Atcher R W, et al. (1988). Improved sarcoma imaging and reduced hepatic activity with indium-111-SCN-Bz-DTPA linked to MoAb 19–24. *J. Nucl. Med.* 29:1810–1816.

159. Reilly R (2005). The immunoreactivity of radiolabeled antibodies—its impact on tumor targeting and strategies for preservation. *Cancer Biother. Radiopharm.* 19:669–672.

160. Rodwell J D, Alvarez V L, Lee C, et al. (1986). Site-specific covalent modification of monoclonal antibodies: In vitro and in vivo evaluation. *Proc. Natl. Acad. Sci. USA.* 83:2632–2636.

161. Abraham R, Moller D, Gabel D, et al. (1991). The influence of periodate oxidation on monoclonal antibody avidity and immunoreactivity. *J. Immunol. Meth.* 144:77–86.

162. King T P, Zhao S W, Lam T (1986). Preparation of protein conjugates via intermolecular hydrazone linkage. *Biochem.* 25(19):5774–5779.

163. Chan C, Sandhu J, Guha A, et al. (2005). A human transferrin-vascular endothelial growth factor (hnTf-VEGF) fusion protein containing an intergrated binding site for ^{111}In for imaging tumor angiogenesis. *J. Nucl. Med.* 46:1745–1752.

164. Remy S, Reilly R M, Sheldon K, et al. (1995). A new radioligand for the epidermal growth-factor receptor—In-111 labeled human epidermal growth-factor derivatized with a bifunctional metal-chelating peptide. *Bioconj. Chem.* 6(6):683–690.

165. Wang J, Chen P, Su Z F, et al. (2001). Amplified delivery of indium-111 to EGFR-positive human breast cancer cells. *Nucl. Med. Biol.* 28(8):895–902.

166. Manabe Y, Longley C, Furnanski P (1986). High-level conjugation of chelating agents onto immunoglobulins: Use of an intermediary poly(L-lysine)-diethylenetriaminepentaacetic acid carrier. *Biochim. Biophys. Acta.* 883:460.

167. Kobayashi H, Sato N, Saga T, et al. (2000). Monoclonal antibody-dendrimer conjugates enable radiolabeling of antibody with markedly high specific activity with minimal loss of immunoreactivity. *Eur. J. Nucl. Med.* 27:1334–1339.

168. Wen X, Wu Q-P, Lu Y, et al. (2001). Poly(ethylene glycol)-conjugated anti-EGF receptor antibody C225 with radiometal chelator attached to the termini of polymer chains. *Bioconjug. Chem.* 12:545–553.

169. Wen X, Wu Q-P, Ke S, et al. (2001). Conjugation with ^{111}In-DTPA-poly(ethylene glycol) improves imaging of anti-EGF receptor antibody C225. *J. Nucl. Med.* 42:1530–1537.

170. Duncan J R, Welch M J (1993). Intracellular metabolism of indium-111-DTPA labeled receptor targeted proteins. *J. Nucl. Med.* 34:1728–1738.

171. Rogers B E, Franano F N, Duncan J R, et al. (1995). Identification of metabolites of ^{111}In-diethylenetriaminepentaacetic acid-monoclonal antibodies and antibody fragments in vivo. *Cancer Res.* 55:5714–5720S.

172. Akizawa H, Arano Y, Uezono T, et al. (1998). Renal metabolism of ^{111}In-DTPA-D-Phe[1]-octreotide in vivo. *Bioconjug. Chem.* 9:662–670.

173. Bass L A, Lanahan M V, Duncan J R, et al. (1998). Identification of the soluble in vivo metabolites of indium-111-diethylenetriaminepentaacetic acid-D-Phe[1]-octreotide. *Bioconjug. Chem.* 9:192–200.

174. Lewis M R, Raubitschek A, Shively J E (1994). A facile, water-soluble method for modification of proteins with DOTA. Use of elevated temperature and optimized pH to achieve high specific activity and high chelate stability in radiolabeled immunoconjugates. *Bioconjug. Chem.* 5:565–576.

175. de Jong M, Bakker W H, Krenning E P, et al. (1997). Yttrium-90 and indium-111 labelling, receptor binding and biodistribution of [DOTA[0],D-Phe[1],Tyr[3]]octreotide, a promising somatostatin analogue for radionuclide therapy. *Eur. J. Nucl. Med.* 24:368–371.

176. Williams L E, Lewis M R, Bebb G C, et al. (1998). Biodistribution of [111]In- and ^{90}Y-labeled DOTA and maleimidocysteinamido-DOTA ocnjugated to chimeric anticarcinoembryonic antigen antibody in xenograft-bearing nude mice: comparison of stable and chemically labile linker systems. *Bioconjug. Chem.* 9:87–93.

177. Tsai S W, Li L, Williams L E, Anderson A L, et al. (2001). Metabolism and renal clearance of ^{111}In-labeled DOTA-conjugated antibody fragments. *Bioconjug. Chem.* 12:264–270.

178. Li L, Olafsen T, Anderson A L, et al. (2002). Reduction of kidney uptake in radiometal labeled peptide linkers conjugated to recombinant antibody fragments. Site-specific conjugation of DOTA-peptides to a Cys-diabody. *Bioconjug. Chem.* 13:985–995.

179. Cremonesi M, Ferrari M, Zoboli S, et al. (1999). Biokinetics and dosimetry in patients administered with ^{111}In-DOTA-Tyr[3]-octreotide: Implications for internal radiotherapy with ^{90}Y-DOTATOC. *Eur. J. Nucl. Med.* 26:877–886.

180. Onthank D C, Liu S, Silva P J, et al. (2004). [90]Y and ^{111}In complexes of a DOTA-conjugated integrin $\alpha_v\beta_3$ receptor antagonist: different but biologically equivalent. *Bioconjug. Chem.* 15:235–241.

181. Yoo J, Tang L, Perkins T A, et al. (2005). Preparation of high specific activity [86]Y using a small biomedical cyclotron. *Nucl. Med. Biol.* 32:891–897.

182. Chinol M, Hnatowich D J (1987). Generator-produced yttrium-90 for radioimmunotherapy. *J. Nucl. Med.* 28:1465–1470.

183. Harwood S J, Gibbons L K, Goldner P J, et al. (2002). Outpatient radioimmunotherapy with Bexxar. *Cancer.* 94:1358–1362.

184. Garmestani K, Milenic D E, Plascjak P S, et al. (2002). A new and convenient method for purification of ^{86}Y using a Sr(II) selective resin and comparison of biodistribution of ^{86}Y and ^{111}In labeled Herceptin™. *Nucl. Med. Biol.* 29:599–606.

185. Liu S, Edwards D S (2001). Bifunctional chelators for therapeutic lanthanide radiopharmaceuticals. *Bioconjug. Chem.* 12:7–34.

186. Hnatowich D J, Virzi F, Doherty P W (1985). DTPA-coupled antibodies labeled with yttrium-90. *J. Nucl. Med.* 26:503–509.

187. Brouwers A H, van Eerd J E M, Frielink C, et al. (2004). Optimization of radioimmunotherapy of renal cell carcinoma: labeling of monoclonal antibody cG250 with ^{131}I, ^{90}Y, ^{177}Lu or ^{186}Re. *J. Nucl. Med.* 45:327–337.

188. Griffiths G L, Govindan S V, Sharkey R M, et al. (2003). ^{90}Y-DOTA-hLL2: An agent for radioimmunotherapy of non-Hodgkin's lymphoma. *J. Nucl. Med.* 44:77–84.

189. Deshpande S V, DeNardo S J, Kukis D L, et al. (1990). Yttrium-90-labeled monoclonal antibody for therapy: Labeling by a new macrocyclic bifunctional chelating agent. *J. Nucl. Med.* 31:473–479.

190. McCall M J, Diril H, Meares C F (1990). Simplified method for conjugating macrocyclic bifunctional chelating agents to antibodies via 2-iminothiolane. *Bioconjug. Chem.* 1:222–226.

191. Lewis M R, Kao J Y, Anderson A-L J, et al. (2001). An improved method for conjugating monoclonal antibodies with N-hydroxysulfosuccinimidyl DOTA. *Bioconjug. Chem.* 12:320–324.

192. Kukis D L, DeNardo S J, DeNardo G L, et al. (1998). Optimized conditions for chelation of yttrium-90-DOTA immunoconjugates. *J. Nucl. Med.* 39:2105–2110.

193. Breeman W A P, de Jong M, Visser T J, et al. (2003). Optimising conditions for radiolabelling of DOTA-peptides with ^{90}Y, ^{111}In and ^{177}Lu at high specific activities. *Eur. J. Nucl. Med. Mol. Imaging.* 30:917–920.

194. Hainsworth J E S, Mather S J (2005). Regressive DOTA labelling performance with indium-111 and yttrium-90 over a week of use. *Eur. J. Nucl. Med. Mol. Imaging.* 32:1348.

195. Salako Q A, O'Donnell R T, DeNardo S J (1998). Effects of radiolysis on yttrium-90-labeled Lym-1 antibody preparations. *J. Nucl. Med.* 39:667–670.

196. Liu S, Edwards D S (2001). Stabilization of ^{90}Y-labeled DOTA-biomolecule conjugates using gentisic acid and ascorbic acid. *Bioconjug. Chem.* 12:554–558.

197. Liu S, Ellars C E, Edwards D S (2003). Ascorbic acid: Useful as a buffer agent and radiolytic stabilizer for metalloradiopharmaceuticals. *Bioconjug. Chem.* 14:1052–1056.

198. Kosmas C, Snook D, Gooden C S, et al. (1992). Development of humoral immune responses against a macrocyclic chelating agent (DOTA) in cancer patients receiving radioimmunoconjugates for imaging and therapy. *Cancer Res.* 52:904–911.

199. Watanabe N, Goodwin D A, Meares C F, et al. (1994). Immmunogenicity in rabbits and mice of an antibody-chelate conjugate: Comparison of (S) and (R) macrocyclic enantiomers and an acyclic chelating agent. *Cancer Res.* 54:1049–1054.

200. DeNardo G L, Mirick G R, Kroger L A, et al. (1996). Antibody responses to macrocycles in lymphoma. *J. Nucl. Med.* 37:451–456.

201. Perico M E, Chinol M, Nacca A, et al. (2001). The humoral immune response to macrocyclic chelating agent DOTA depends on the carrier molecule. *J. Nucl. Med.* 42:1697–1703.

202. Saha G B (2004). Diagnostic uses of radiopharmaceuticals in nuclear medicine. In *Fundamentals of Nuclear Pharmacy*, 5th ed. Springer, New York, 247–329.

203. Michel R B, Brechbiel M W, Mattes M J (2003). A comparison of 4 radionuclides conjugated to antibodies for single-cell kill. *J. Nucl. Med.* 44:632–640.

204. Maecke H R, Hofmann M, Haberkorn U (2005). ^{68}Ga-labeled peptides in tumor imaging. *J. Nucl. Med.* 46:172S–178S.

205. Schumacher J, Maier-Borst W (1981). A new [68]Ge/^{68}Ga radioisotope generator system for production of 68Ga in dilute HCl. *Int. J. Appl. Radiat. Isot.* 32:31–36.

206. Ugur O, Kothari P J, Finn R D, et al. (2002). Ga-66 labeled somatostatin analogue DOTA-D-Phe1-Tyr3-octreotide as a potential agent for positron emission tomography imaging and receptor mediated internal radiotherapy of somatostatin receptor positive tumors. *Nucl. Med. Biol.* 29:147–157.

207. Smith-Jones P M, Stolz B, Bruns C, et al. (1994). Gallium-67/gallium-68-[DFO]-octreotide-a potential radiopharmaceutical for PET imaging of somatostatin receptor-positive tumors: Synthesis and radiolabeling in vitro and preliminary in vivo studies. *J. Nucl. Med.* 35:317–325.

208. Breeman W A P, de Jong M, de Blois E, et al. (2005). Radiolabelling DOTA-peptides with [68]Ga. *Eur. J. Nucl. Med. Mol. Imaging.* 32:478–485.

209. Henze M, Schuhmacher J, Hipp P, et al. (2001). PET imaging of somatostatin receptors using [^{68}Ga]DOTA-D-Phe[1]-Tyr[3]-octreotide: First results in patients with meningiomas. *J. Nucl. Med.* 42:1053–1056.

210. Velikyan I, Beyer G L, Langström B (2004). Microwave-supported preparation of ^{68}Ga bioconjugates with high specific activity. *Bioconjug. Chem.* 15:554–560.

211. Reubi J C, Schar J-C, Waser B, et al. (2000). Affinity profiles for human somatostatin receptor subtypes SST1-SST5 of somatostatin radiotracers selected for scintigraphic and radiotherapeutic use. *Eur. J. Nucl. Med.* 27:273–282.

212. Hofmann M, Maecke H, Börner A R, et al. (2001). Biokinetics and imaging with the somatostatin receptor PET radioligand ^{68}Ga-DOTATOC: preliminary data. *Eur. J. Nucl. Med.* 28:1751–1757.

213. Wild D, Macke H R, Waser B, et al. (2005). ^{68}Ga-DOTANOC: A first compound for PET imaging with high affinity for somatostatin receptor subtypes 2 and 5. *Eur. J. Nucl. Med. Mol. Imaging.* 32:724.

214. Froideveaux S, Calame-Christe M, Schumacher J, et al. (2004). A gallium-labeled DOTA-α-melanocyte-stimulating hormone analog for PET imaging of melanoma metastases. *J. Nucl. Med.* 45:116–123.

215. Velikyan I, Sundberg ÄL, Lindhe Ö, et al. (2005). Preparation and evaluation of ^{68}Ga-DOTA-hEGF for visualization of EGFR expression in malignant tumors. *J. Nucl. Med.* 46:1881–1888.

216. Blower P J, Lewis J S, Zweit J (1996). Copper radionuclides and radiopharmaceuticals in nuclear medicine. *Nucl. Med. Biol.* 23:957–980.

217. Anderson C J, Jones A L, Bass L A, et al. (1998). Radiotherapy, toxicity and dosimetry of copper-64-TETA-octreotide in tumor-bearing rats. *J. Nucl. Med.* 39:1944–1951.

218. Connett J M, Anderson C J, Guo L Y, et al. (1996). Radioimmunotherapy with a ^{64}Cu-labeled monoclonal antibody: a comparison with [67]Cu. *Proc. Natl. Acad. Sci. USA.* 93:6814–6818.

219. McCarthy D W, Shefer R E, et al. (1997). Efficient production of high specific activity ^{64}Cu using a biomedical cyclotron. *Nucl. Med. Biol.* 24:35–43.

220. Obata A, Kasamatsu S, McCarthy D W, et al. (2003). Production of therapeutic quantities of ^{64}Cu using a 12 MeV cyclotron. *Nucl. Med. Biol.* 30:535–539.

221. Williams H A, Robinson S, Julyan P, et al. (2005). A comparison of PET imaging characteristics of various copper radioisotopes. *Eur. J. Nucl. Med. Mol. Imaging.* 32:1473–1480.

222. McCarthy D W, Bass L A, Cutler P D, et al. (1999). High purity production and potential applications of copper-60 and copper-61. *Nucl. Med. Biol.* 26:351–358.

223. Novak-Hofer I, Schubiger P A (2002). Copper-67 as a therapeutic nuclide for radioimmunotherapy. *Eur. J. Nucl. Med.* 29:821–830.

224. Cole W C, DeNardo S J, Meares C F, et al. (1986). Serum stability of [67]Cu chelates: comparison with ^{111}In and [57]Co. *Nucl. Med. Biol.* 13:363–368.

225. Anderson C J, Schwarz S W, Connett J M, et al. (1995). Preparation, biodistribution and dosimetry of copper-64-labeled anti-colorectal carcinoma monoclonal antibody fragments 1A3-F(ab')$_2$. *J. Nucl. Med.* 36:850–858.

226. Lewis J S, Srinivasan A, Schmidt M A, et al. (1999). In vitro and in vivo evaluation of ^{64}Cu-TETA-Tyr^3-octreotate. A new somatostatin analog with improved target tissue uptake. *Nucl. Med. Biol.* 26:267–273.

227. Li W P, Lewis J S, Kim J, et al. (2002). DOTA-D-Tyr^1-octreotate: A somatostatin analogue for labeling with metal and halogen radionuclides for cancer imaging and therapy. *Bioconjug. Chem.* 13:721–728.

228. O'Donnell R T, DeNardo G L, Kukis D L, et al. (1999). A clinical trial of radioimmunotherapy with [67]Cu-2-IT-BAT-Lym-1 for non-Hodgkin's lymphoma. *J. Nucl. Med.* 40:2014–2020.

229. Philpott G W, Schwarz S W, Anderson C J, et al. (1995). RadioimmunoPET: detection of colorectal carcinoma with positron-emitting copper-64-labeled monoclonal antibody. *J. Nucl. Med.* 36:1818–1824.

230. Anderson C J, Dehdashti F, Cutler P D, et al. (2001). ^{64}Cu-TETA-octreotide as a PET imaging agent for patients with neuroendocrine tumors. *J. Nucl. Med.* 42:213–221.

231. Anderson C J, Pajeau T S, Edwards W B, et al. (1995). In vitro and in vivo evaluation of copper-64-octreotide conjugates. *J. Nucl. Med.* 36:2315–2325.

232. Chen X, Hou Y, Tohme M, et al. (2004). Pegylated Arg-Gly-Asp peptide: [64]Cu labeling and PET imaging of brain tumor $\alpha_v\beta_3$-integrin expression. *J. Nucl. Med.* 45:1776–1783.

233. Thakur M L, Aruva M R, Gariepy J, et al. (2004). PET imaging of oncogene overexpression using ^{64}Cu-vasoactve intestinal peptide (VIP) analog: comparison with ^{99m}Tc-VIP analog. *J. Nucl. Med.* 45:1381–1389.

234. Rogers B E, Bigott H M, McCarthy D W, et al. (2003). MicroPET imaging of a gastrin-releasing peptide receptor-positive tumor in a mouse model of human prostate cancer using a ^{64}Cu-labeled bombesin analogue. *Bioconjug. Chem.* 14:756–763.

235. Rogers B E, Anderson C J, Connett J M, et al. (1996). Comparison of four bifunctional chelates for radiolabeling monoclonal antibodies with copper radioisotopes: biodistribution and metabolism. *Bioconjug. Chem.* 7:511–522.

236. Bass L A, Wang M, Welch M J, et al. (2000). In vivo transchelation of copper-64 from TETA-octreotide to superoxide dismutase in rat liver. *Bioconjug. Chem.* 11:527–532.

237. Mirick G R, O'Donnell R T, DeNardo S J, et al. (1999). Transfer of copper from a chelated [67]Cu-antibody conjugate to ceruloplasmin in lymphoma patients. *Nucl. Med. Biol.* 26:841–845.

238. Milenic D E, Garmestani K, Chappell L L, et al. (2002). In vivo comparison of macrocyclic and acyclic ligands for radiolabeling of monoclonal antibodies with ^{177}Lu for radioimmunotherapeutic applications. *Nucl. Med. Biol.* 29:431–442.

239. Mulligan T, Carrasquillo J A, Chung Y, et al. (1995). Phase I study of intravenous ^{177}Lu-labeled CC49 murine monoclonal antibody in patients with advanced adenocarcinoma. *Clin. Cancer Res.* 1:1447–1454.

240. Müller W A, Linzner U, Schäffer E H (1978). Organ distribution studies of lutetium-177 in mouse. *Int. J. Nucl. Med. Biol.* 5:29–31.

241. Breeman W A P, van der Wansem K, Bernard B F, et al. (2003). The addition of DTPA to [^{177}Lu-DOTA^0,Tyr^3]octreotate prior to administration reduces rat skeleton uptake of radioactivity. *Eur. J. Nucl. Med.* 30(312):315.

242. Alvarez R D, Partridge E E, Khazaeli M B, et al. (1997). Intraperitoneal radioimmunotherapy of ovarian cancer with ^{177}Lu-CC49: A phase I/II study. *Gynecol. Oncol.* 65:94–101.

243. Chauhan S C, Jain M, Moore E D, et al. (2005). Pharmacokinetics and biodistributon of ^{177}Lu-labeled multivalent single-chain Fv construct of the pancarcinoma monoclonal antibody CC49. *Eur. J. Nucl. Med. Mol. Imaging.* 32:264–273.

244. McDevitt M R, Dangshe M, Lai L, et al. (2001). Tumor therapy with targeted atomic nanogenerators. *Science.* 294:1537–1540.

245. Miederer M, McDevitt M R, Sgouros G, et al. (2004). Pharmacokinetics, dosimetry, and toxicity of the targetable atomic generator, ^{225}Ac-HuM195, in nonhuman primates. *J. Nucl. Med.* 45:129–137.

246. Milenic D E, Roselli M, Mirzadeh S, et al. (2001). In vivo evaluation of bismuth-labeled monoclonal antibody comparing DTPA-derived bifunctional chelates. *Cancer Biother. Radiopharm.* 16:133–146.

247. Yao Z, Garmestani K, Wong K J, et al. (2001). Comparative cellular catabolism and retention of astatine-, bismuth-, and lead-radiolabeled internalizing monoclonal antibody. *J. Nucl. Med.* 42:1538–1544.

248. Ballangrud A M, Yang W-H, Palm S, et al. (2004). Alpha-particle emitting atomic generator (actinium-225)-labeled trastuzumab (Herceptin) targeting of breast cancer spheroids: Efficacy versus HER2/neu expression. *Clin. Cancer Res.* 10:4489–4497.

249. Horak E, Hartmann F, Garmestani K, et al. (1997). Radioimmunotherapy targeting of HER2/neu oncoprotein on ovarian tumor using lead-212-DOTA-AE1. *J. Nucl. Med.* 38:1944–1950.

250. Meares C F, McCall M J, Rearden D T, et al. (1984). Conjugation of antibodies with bifunctional chelating agents: isothiocyanate and bromoacetamide reagents: Methods of analysis and subsequent addition of metal ions. *Anal. Biochem.* 142:68.

251. Pippin C J, Parker T A, McMurry T J, et al. (1992). Spectrophotometric method for the determination of a bifunctional DTPA ligand in DTPA-monoclonal antibody conjugates. *Bioconjug. Chem.* 3:342–345.

252. Hartikka M, Vihko P, Södervall M, et al. (1989). Radiolabelling of monoclonal antibodies: Optimization of conjugation of DTPA to F(ab′)$_2$-fragments and a novel measurement of the degree of conjugation using Eu(III) labelling. *Eur. J. Nucl. Med.* 15:157–161.

253. Pham D T, Kaspersen F M, Bos E S (1995). Electrophoretic method for the quantitative determination of a benzyl-DTPA ligand in DTPA monoclonal antibody conjugates. *Bioconjug. Chem.* 6:313–315.

254. Dadachova E, Chappell L L, Brechbiel M W (1999). Spectrophotometric method for determination of bifunctional macrocyclic ligands in macrocyclic ligand-protein conjugates. *Nucl. Med. Biol.* 26:977–982.

255. Breeman W A P, de jong M T h M, de Jong M, et al. (2004). Reduction of skeletal accumulation of radioactivity by co-injection of DTPA in [^{90}Y-DOTA^0,Tyr^3]octreotide solutions containing free ^{90}Y^{3+}. *Nucl. Med. Biol.* 31:821–824.

256. Lindmo T, Boven E, Cuttitta F, et al. (1984). Determination of the immunoreactive fraction of radiolabelled monoclonal antibodies by linear extrapolation of binding to infinite antigen excess. *J. Immunol. Meth.* 72:77–89.

257. Mardirossian G, Wu C, Rusckowski M, et al. (1992). The stability of $^{99}Tc^m$ directly labelled to an Fab′ antibody via stannous ion and mercaptoethanol reduction. *Nucl. Med. Commun.* 13:503–512.

258. Chen P, Cameron R, Wang J, et al. (2003). Antitumor effects and normal tissue toxicity of 111In-labeled epidermal growth factor administered to athymic mice bearing epidermal growth factor receptor-positive human breast cancer xenografts. *J. Nucl. Med.* 44(9):1469–1478.

259. Sharkey R M, Karacay H, Chang C H, et al. (2005). Improved therapy of non-Hodgkin's lymphoma xenografts using radionuclides pretargeted with a new anti-CD20 bispecific antibody. *Leukemia.* 19:1064–1069.

260. Michel R B, Rosario A V, Brechbiel M W, et al. (2003). Experimental therapy of disseminated B-cell lymphoma xenografts with ^{213}Bi-labeled anti-CD74. *Nucl. Med. Biol.* 30:715–723.

261. Bonnet D, Bhatia M, Wang J Y C, et al. (1999). Cytokine treatment or accessory cells are required to initiate engraftment of purified primitive human hematopoietic cells transplanted at limiting doses into NOD/SCID mice. *Bone. Marrow. Transplant.* 23:203–209.

262. Iwata K, MacDonald L R, Hwang A B, et al. (2002). CT-SPECT for small animal imaging with A-SPECT. *J. Nucl. Med.* 43(suppl.):10P–Abst. No. 35.

263. Lewis J S, Achilefu S, Garbow J R, et al. (2002). Small animal imaging: Current technology and perspectives for oncological imaging. *Eur. J. Cancer.* 38:2173–2188.

264. Stabin M G, Sparks R B, Crowe E (2005). OLINDA/EXM: the second-generation personal computer software for internal dose assessment in nuclear medicine. *J. Nucl. Med.* 46:1023–1027.

265. Reilly R M, Chen P, Wang J, et al. (2006). Preclinical pharmacokinetic, biodistribution, toxicology and dosimetry studies of ^{111}In-DTPA-hEGF—an Auger electron-emitting radiotherapeutic agent for EGFR-positive breast cancer. *J. Nucl. Med.* In press.

7.1

GENE THERAPY—BASIC PRINCIPLES AND THE ROAD FROM BENCH TO BEDSIDE

Gabor M. Rubanyi

Cardium Therapeutics, Inc., San Diego, California

Chapter Contents

7.1.1   Introduction   943
7.1.2   Basic Principles   944
    7.1.2.1   From Nucleic Acid Delivery to Therapeutic Effects   944
    7.1.2.2   Essential Components and Selected Technical Features of a
        Gene Therapy Product   946
    7.1.2.3   Disease Targets for Gene Therapy   953
7.1.3   From Bench to Bedside: The Development of Ad5FGF-4 to Treat
    Patients with Recurrent Angina   956
    7.1.3.1   Preclinical Efficacy   956
    7.1.3.2   Toxicology and Biodistribution   958
    7.1.3.3   cGMP Manufacturing and Characterization of the Gene
        Therapy Vector   958
    7.1.3.4   Clinical Studies   963
    7.1.3.5   Lessons Learned from Past Failures   966
7.1.4   Conclusions   966
    References   967

7.1.1   INTRODUCTION

Human gene therapy (HGT) is defined as the transfer of nucleic acids (DNA or RNA) to somatic cells of a patient, which results in a therapeutic effect, by either (1) correcting genetic defects, (2) overexpressing proteins that are therapeutically useful, or (3) inhibiting the production of "harmful" proteins (e.g., by shRNA).

Handbook of Pharmaceutical Biotechnology, Edited by Shayne Cox Gad.
Copyright © 2007 John Wiley & Sons, Inc.

Although it has several theoretical advantages, so far HGT has not delivered the promised results: Convincing clinical efficacy in pivotal phase III clinical trials has not yet been demonstrated in the United States or Europe (although two gene therapy products for cancer treatment have recently been approved and marketed in China). HGT is a complex process, involving multiple steps in the human body (delivery to organs, tissue targeting, cellular trafficking, regulation of gene expression level and duration, biological activity of therapeutic protein, safety of the vector and gene product, to name just a few), most of which are not completely understood.

In most applications, gene therapy represents a new, innovative drug delivery system making use of the technical and scientific advances of the last two decades in microbiology, virology, organic chemistry, molecular biology, biochemistry, cell biology, genetics, genomics, and genetic engineering. It is more than "gene transfer," which is only a part of the complex, multiphase process of identification, manufacturing, preclinical testing, and clinical development of gene therapy products (see Section 7.1.3).

The prerequisites of successful HGT include therapeutically suitable genes (with a proven role in pathophysiology of the disease either by its lack, mutation, or overexpression), appropriate gene delivery systems (e.g., viral and nonviral vectors), proof of principle of efficacy and safety in appropriate preclinical models, and suitable manufacturing and analytical processes to provide well-defined HGT products for clinical investigations.

The principle of gene therapy has specific therapeutic advantages over existing therapeutic modalities (such as small molecules or biologics) in certain disease indications. These include (1) correction of the genetic cause of a disease, (2) selective treatment of affected (diseased) cells and tissues (the cells and tissues produce their own "remedy"), and (3) long-term treatment after a single application. It may be best suited for so-called "nondrugable" targets (i.e., the modulation of target molecule is not feasible with small-molecule or protein drugs). Based on these theoretical principles, at the time of its first introduction more than 17 years ago, gene therapy promised to be an effective and safe treatment modality, which will soon cure diseases and replace classic therapies.

Over the past few years, significant progress has been made in various enabling technologies and in the molecular understanding of diseases and the manufacturing of vectors. These advances, combined with the growing experience with gene transfer in humans, promises to contribute to the ultimate success of this new class of therapeutics.

7.1.2 BASIC PRINCIPLES

7.1.2.1 From Nucleic Acid Delivery to Therapeutic Effects

Gene therapy consists of multiple biological processes in the body, the exact nature of which is in most cases still unknown (Figure 7.1-1). Gene therapy is initiated with the introduction of an appropriate vector (viral or nonviral) either into the body locally (direct tissue injection), into body cavities (e.g., peritoneum or cerebrospinal fluid), or into the bloodstream (systemic delivery). The vector needs to "find"

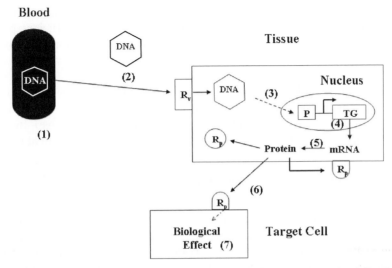

Figure 7.1-1. Multiple biological processes of in vivo gene transfer. The following main steps are required for successful gene transfer in the human body: (1) vector (viral, nonviral, cell-based) delivery (localized tissue delivery or systemic delivery via blood circulation) of a gene (DNA); (2) vector "recognition" by specific receptors (Rv) on cells in target tissue; (3) uptake of the vector by cells, trafficking to the nucleus and delivery of vector DNA in to the nucleus; (4) transcription (expression) of therapeutic (trans) gene in the nucleus; (5) translation of mRNA into therapeutic protein in the cytoplasm; (6) interaction of therapeutic protein with its receptors (Rp) within the "producing" cell (intracrine mechanism), on the surface of "producing" cell (autocrine mechanism) or on neighboring "target" cell (paracrine mechanism). For some applications, the therapeutic protein enters the circulation and acts distant from the target tissue (endocrine mechanism) (e.g., erythropoietin, coagulation factors VIII and IX, and growth hormone); and (7) after interaction with its receptor, the protein induces a biological effect, which results in therapeutic benefits. (Reproduced with permission from Pergamon Press.)

its target tissue, after which it enters the target cell membrane and traffics through the cytoplasm to reach and enter the nucleus. Once there, the therapeutic (trans)gene needs to be transcribed and the formed mRNA needs to be appropriately translated into the therapeutic protein. The protein then acts on its receptor(s) either on the cell that produced it (intracrine or autocrine mechanism), on neighboring cells (paracrine mechanism), or at distant sites after entering the blood circulation (endocrine mechanism, e.g., erythropoietin, coagulation factors, and growth hormone). Finally, after interacting with its receptor, the protein needs to induce an appropriate biological effect that results in therapeutic benefits. For gene correction or gene knock-down approaches, the steps are similar to those outlined above, except that the last step is modification of the genome ("gene correction") or blockade of mRNA transcription (siRNA/shRNA) of endogenous genes, respectively.

Although the factors needed for successful gene therapy are not different from any new therapeutic modality (which include technical [gene delivery and expression], clinical [therapeutic efficacy and safety], and socioeconomic factors), the specific technical success factors are unique for gene therapy approaches. They include the choice of appropriate therapeutic gene(s) (with a proven role in the pathomechanism of the disease, specifically targeted and of sufficient potency), gene delivery systems (of sufficient targeting ability, transfection efficiency, and safety), and gene expression regulation systems to control the level and timing of transgene expression.

The therapeutic and socioeconomic success of gene therapy products includes the requirement that the benefits of gene therapy should outweigh the risks and should offer advantages over conventional (usually less expensive) treatments, before this new approach will become accepted in the general medical practice.

7.1.2.2   Essential Components and Selected Technical Features of a Gene Therapy Product

Gene Delivery Vectors.   The goal in the discovery and development of most gene therapy vectors is to deliver and express genes at the appropriate site and at therapeutically meaningful levels in a controlled manner. The first-generation gene therapy vectors have used the ability of viral systems to deliver genetic information to human cells. Attempts are also made to develop nonviral synthetic vectors and hybrid synthetic-viral systems that are potentially safer alternatives for gene delivery. A third approach uses isolated human cells (stem cells, progenitor cells, or somatic cells) as a means to introduce the therapeutic genes into specific human cell populations where the therapeutic product is required.

Viral Vectors.   Viruses "acquired" numerous biological properties over millions of years of evolution that allow them to effectively recognize and enter cells, traffic within the cytosol to the nucleus, translocate into the nucleus, and express their genes in the host cell (Figure 7.1-2).

The most frequently used viral vectors in clinical trials so far are retroviruses and adenoviruses (Table 7.1-1). Several other viral vectors are in preclinical development or are under clinical evaluation, including adeno-associated virus (AAV), lentivirus, herpes simplex virus (HSV), and others.

Retroviruses can lead to a stable integration of the transfected gene into the host genome and therefore produce long-lasting gene transfer. Replication-deficient retroviruses are produced *in vitro* in specific packaging cells transfected previously with retroviral genes (G, P, E) that have been deleted from the genome of the therapeutic retroviruses. Major limitations of the retroviruses are their low titers, their inability to infect nondividing cells, and the potential risk of insertional mutagenesis. The development of new pseudotyped retroviruses has increased virus titers that will permit more efficient gene transfer.

There are more than 50 known serotypes of *adenoviruses*, but until recently, serotype 5 (Ad5) and to a lesser extent serotype 2 (Ad2) (both class C adenoviruses) have been used as recombinant, nonreplicating gene delivery vectors. *Recombinant Ad5* can be produced in high titer. Ad5 vectors do not lead to stable integration of the transgene into the host genome (only at very low frequency in cell culture [2]), and they usually remain extrachromosomal and cause only a transient transgene expression. Replication-deficient adenoviruses are produced *in vitro* in specific packaging cells that complement gene products (e.g., *E1* and *E3*) deleted from the genome of the therapeutic adenoviruses. They give effective tran-

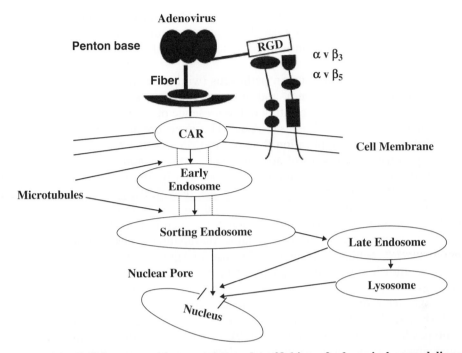

Figure 7.1-2. Cellular recognition, uptake, and trafficking of adenoviral gene delivery vectors. Receptor-mediated entry of adenovirus into cells depends on interaction of two of its coat proteins with two different cell-surface receptors. The viral fiber knob protein mediates attachment to the cell via the CAR. After attachment, an RGD tripeptide motif in the penton base protein binds to intergrins ($\alpha v\beta 3$ or $\alpha v\beta 5$), which mediates internalization. Knowledge of these domains and processes helped to design novel strategies to change the natural tropism of adenoviral vector ("re-targeting"; see text for details.) Once inside the cell, the adenovirus effectively and quickly (within less than 60 minutes) reaches the nucleus (via the endosome, microtubules, and the lysosome) and translocates into the nucleus via nuclear pores with the help of the nuclear localization signal (NLS) peptide in the viral coat protein. (Reproduced with permission from Pergamon Press.)

TABLE 7.1-1. Gene Delivery Systems (Vectors) Used in Gene Therapy Clinical Trials (1989–Present)[1]

Delivery System	Number of Trials
Retrovirus	254
Adenovirus	240
Naked DNA	132
Lipofection	85
Poxvirus	52
Vaccinia virus	30
Herpes simplex virus	26
Adeno-associated virus	19
RNA transfer	10

[1] From Ref. 1.

sient gene expression in proliferating and nonproliferating cells, but first-generation adenoviruses have the disadvantage of producing immunological and inflammatory reactions. These complications should be lessened with second-generation adeno-viral vectors [3].

Some adenoviruses are human pathogens (wild-type Ad5, for example, causes the "common cold" syndrome), and most patients have already been exposed to them during their lifetime, resulting in the presence of various levels of circulating neutralizing anti-viral antibodies (NABs). This exposure may hinder the effectiveness of their systemic application. Use of viral vectors that are not human pathogens (e.g., various serotypes of adenoviruses, AAV, or nonhuman adenoviruses) can avoid this problem. Immune response evoked by the first application of the viral vector (human or nonhuman) may interfere with their repeated application (although the immune response may depend on the route of delivery and dose used). First- and second-generation adenoviral vectors have limited cloning size (<10 kb), which prevents the use of large therapeutic genes, multiple genes, or complex gene regulatory elements. The use of "gutless" adenovirus avoids this problem, allowing a cloning capacity of up to ~30 kb, but filling the vector with "useless" DNA is a challenge, and so is the proper manufacturing of these modified vectors, which require the presence of helper viruses.

AAVs have been used for effective gene transfer to muscle [4]. Numerous AAV serotypes have been identified (AAV1–9) recently, with a different tissue tropism, which gives us hope for using them in a tissue-specific manner [5].

In addition to pseudotyping (retroviruses) or deletion of nonessential viral genes (adenoviruses), several other strategies are pursued to improve viral vectors for human therapeutic use. This is a very active field, the overview of which is beyond the scope of this chapter. Some notable activities include changing viral coat proteins (fibers, pentons, etc.) to redirect the tropism of the virus and incorporation of specific gene regulatory systems (gene-switches, tissue-specific promoters, etc.) to allow proper timing, duration, extent, and localization of therapeutic gene expression.

Nonviral Vectors. The existing synthetic vectors (naked DNA, cationic liposomes, etc.) are far from being perfect delivery systems. Although they are less pathogenic and may have reduced toxicity compared with some existing viral vectors, depending on the dose injected, liposomes may aggregate in the blood and can cause severe toxic reactions [6].

Plasmid and liposome complexes are easy to produce and are safer, but they have low gene-transfer efficiency. However, novel lipid formulations and synthetic cationic polymer carriers have clearly improved the effectiveness of plasmid-mediated gene transfer [7–9].

In the future, the "perfect" gene delivery vector may be synthetic, incorporating many advantages of viruses (which over their evolution acquired the perfection of gene delivery to host organisms, such as dense DNA "packaging," cell recognition, cellular uptake, cytosolic trafficking, efficient nuclear uptake, and gene expression in the host cell nucleus, Figure 7.1-3), but avoiding the unwanted properties of viruses (e.g., off-target tropism, pathogenicity, cell toxicity, and immune and inflammatory reactions).

Dense DNA packaging (a prerequisite for efficient gene transfer) can be achieved by using, for example, protamine sulfate (a "trick" borrowed from sperm cells,

Nonviral (Synthetic) Vectors

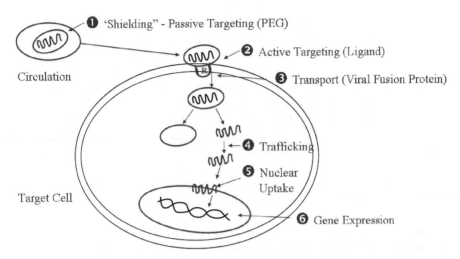

Figure 7.1-3. The "ideal" synthetic (nonviral) gene delivery vector. After dense DNA packaging is accomplished (e.g., by protamine sulfate), the surface of synthetic particles (which is usually positively charged) needs to be "shielded" (e.g., by poly (ethylene-glycol) [PEG]) so that they do not attach to blood elements or to each other and, therefore, have an extended circulating plasma half-life (1) (passive targeting to "leaky" vessels"). The surface of the particles will contain specific ligands for active targeting to selected cells/tissues (2). By engineering viral fusion proteins to the particle coat, cell entry is facilitated (3). Cellular trafficking will be enhanced and intracellular degradation of DNA prevented (4). Nuclear uptake will be facilitated by viral nuclear localization signal (NLS) peptides (5). Chromosomal localization will be augmented (e.g., by the *rep* gene, which allows targeting to chromosome 19) and gene expression regulated by specific transcriptional control elements (6). (Reproduced with permission from Pergamon Press.)

which similar to viruses package DNA efficiently). The particle should be "shielded" from binding to plasma proteins, blood cells, or to each other to allow longer circulating plasma half-life and more efficient uptake in target tissues (e.g., at the site of leaky vessels in tumors) ("passive" targeting). They will be engineered to contain cell recognition ligands (e.g., transferring for proper "active" targeting of cancer cells), cell membrane fusion proteins, and nuclear localization signals (all "borrowed" from viruses) for efficient cellular trafficking. Chromosomal localization and insertion for long-term expression of the therapeutic gene will be achieved by adding, for example, the *rep* gene, which allows targeting and insertion of the DNA to chromosome 19 [10]. These examples are just a few that are being tested today. However, even more sophisticated systems can be expected in the future, along with appropriate manufacturing and analytical processes, which will allow their introduction to human subjects.

Cell-Based Delivery of Therapeutic Genes. Although cell therapies have been used in medicine for several decades (e.g., blood transfusion), the use of cells manipulated *ex vivo* with therapeutic genes and then reintroduced into patients offers a new strategy by which to deliver therapeutic genes.

Human hematopoietic stem cells, mesenchymal stem cells, neuronal stem cells, embryonic stem cells, and peripheral T cells are the focus of ongoing research efforts [11, 12]. The panel of stem cells available for gene therapy purposes will increase as isolation and culturing procedures improve and appropriate factors are identified that can be used to drive their differentiation along distinct cell lineage pathways. Issues currently being addressed include the development of vectors (e.g., lentiviruses) for efficient stem cell gene transduction, expression and regulation of therapeutic genes during lineage progression from stem cells to differentiated cells, control of stem cell growth, expansion *ex vivo*, and engraftment and differentiation *in vivo*. Genetically engineered stem cells or less pluripotent progenitor cells are currently being tested for therapeutic angiogenesis (endothelial cell progenitors), Parkinson's disease (neuronal stem cells), bone marrow transplantation (hematopoiectic stem cells), and AIDS (e.g., hematopoietic stem cells transfected with the *RevM10* gene) [11–13].

"Customized" Gene Delivery Vectors. It is important to emphasize that there will not be a "universal" vector that is optimally useful for all indications. On the contrary, each disease target will have a specific set of technical requirements, and the "perfect" vector for a specific disease should be optimized according to these specific criteria. For example, some diseases will require local delivery (e.g., ischemia, retinitis pigmentosa, and Parkinson's disease), whereas others necessitate systemic delivery (e.g., cancer). For certain diseases, the gene of a secreted protein (e.g., coagulation factors VIII and IX for hemophilia A and B, respectively; growth hormone; and erythropoietin) can be expressed in almost any tissue of the body. Sometimes only transient, short-lived gene expression will be needed (e.g., therapeutic angiogenesis and cancer), whereas in other cases long-term (sometime lifelong) gene expression duration will be necessary (e.g., most monogenic diseases, such as familial hypercholesterolemia, hemophilia, and SCID). For certain disease targets, most, if not all, target cells need to be transfected (cancer), whereas in other cases, this will not be necessary (e.g., with most secreted therapeutic proteins).

In certain diseases, tight control of gene expression will not be important (e.g., coagulation factors), whereas in others very tight regulation of the gene expression will be essential (e.g., insulin). Some diseases will require specific targeting of the vector for efficient and safe delivery after systemic application (e.g., cancer). Other disease targets will require tissue-or disease-specific promoter elements (e.g., arteriosclerosis and cancer). In some instances, conditionally inducible gene expression regulation (gene-switch) will allow precise dosing and timing of gene expression, which will also be an essential element for safety reasons (i.e., ability to "turn off" the gene if serious side effects occur).

Most vectors optimized for a certain disease target will consist of multiple "components," each fulfilling some necessary technical, biological, or therapeutic need. As the different elements will probably be perfected (and patented) by different companies, the introduction of the optimal multicomponent vector may be difficult because of intellectual property rights and commercial obstacles.

Gene Delivery Targeting. The effectiveness of gene therapy is determined by a combination of the effects of gene delivery into the target tissue, the entry of the new genetic material into cells, and the expression of the transfected gene in the target tissue (Figure 7.1-1). When specific physical or biological targeting methods are available, they generally improve the expression of the transfected gene in the target organ and reduce gene expression in off-target tissues, which may lead to unwanted side effects.

Physical Targeting. A variety of physical gene delivery methods has been introduced to achieve better local tissue targeting of vectors. An example of the effective physical targeting is catheter-mediated gene transfer to various regions of the body (e.g., feed arteries of organs, such as leg muscles, heart, and liver, or retrograde injection via veins) [14–16]. Intramuscular injection of plasmid DNA or viral vectors encoding angiogenic growth factors has been used in ischaemic myocardium [17] and peripheral vascular disease [18, 19]. Another approach to local delivery to small arterioles and capillaries is injection of biodegradable microspheres coated with recombinant growth factors or plasmid DNA [20]. Ultrasonography [21], alone or in combination with microbubbles [22], can also potentially be used to improve the efficiency of gene transfer. For facilitation of intramuscular, intratumoral, or intradermal delivery of naked, plasmid, or cationic liposome-carried DNA, gene gun technology [23] and electroporation [24] can be used. Although these delivery methods offer certain advantages in specific disease targets, they are being progressively replaced by more specific biological targeting methods.

Biological Targeting. Biological vector targeting uses modification of viral coat proteins (for viral vectors) or surface properties of synthetic vectors (e.g., liposomes).

   Passive targeting makes use of alteration of the pharmacokinetics of vectors by "shielding" them from binding to blood cells, plasma proteins, immunoglobulins, unwanted tissues, or to each other, allowing them to circulate for longer periods of time in the blood and accumulate in specific tissues with "leaky" blood vessels (such as tumors).

   Active (vector) targeting aims at directing vector binding and uptake to specific cells in the body by selective "targeting" molecules attached to the surface of the vectors by a variety of technologies (e.g., chemical and genetic). Recent advances in the biological understanding of adenovirus structure and adenovirus receptor interactions have lead to the development of targeted adenovirus vectors. Receptor-mediated entry of adenovirus serotype 5 (Ad5) into cells has been found to depend on two of its coat proteins. The fiber knob protein mediates primary attachment to the cell via the coxsackie-adenovirus receptor (CAR) protein [25]. After cell attachment via fiber–CAR interaction, an RGD tripeptide motif in the penton base protein binds to integrins ($\alpha_v\beta_3$ and $\alpha_v\beta_5$) that mediate cellular internalization [26, 27] (Figure 7.1-2). Other adenovirus serotypes use different cell-surface receptors (e.g., the group B adenoviruses bind to CD46 [28]), most of which have not been identified yet.

   Two basic requirements are necessary to create a (re)targeted Ad5 vector: Interaction of adenovirus with its native receptors (e.g., CAR) must be first removed, and novel, tissue-specific ligands must be added to the virus [29–31]. Three general approaches have been used to achieve these basic requirements. In the "two-component" approach, a bispecific molecule is complexed with the adenovirus [29, 32]. The bispecific component simultaneously blocks native receptor (CAR) binding and redirects virus binding to a tissue-specific receptor (e.g., to integrins using the RGD motif). In the "one-step" approach, the adenovirus is genetically modified in the fiber protein domains to remove native receptor interactions and a novel ligand is genetically incorporated into one adenovirus coat protein [31–33]. The third approach is to use specific polymers to cover the entire virus and "shield" it from unwanted interactions. This method creates a truely "stealth" virus, preventing its interaction, not only with its primary (CAR, via fiber) and secondary cell-surface receptors (integrins, via penton RGD), but also with plasma proteins (including neutralizing antibodies), blood cells, and most tissue sites, providing an ideal situation for extended persistence in the circulation and *passive targeting* [34]. The polymer coat allows covalent attachment of specific "ligands" of choice to its surface, effectively retargeting the vector to a desired tissue (*active targeting*) [35].

Transcriptional Targeting. Receptor targeting technology can be combined with "transcriptional targeting" approaches (by the use of cell-, tissue- or disease-specific promoters).

A tissue- or disease-specific promoter will allow therapeutic gene expression only in cells that express transcription factor proteins binding to these specific promoter sites. Some tissue- or disease-specific promoters, identified and tested in animal models to date, include the promoter of the prostate-specific antigen (PSA) gene [36], osteocalcin gene [37], glucose "sensing" molecules for insulin gene delivery, and hypoxia response element (HRE) activated by hypoxia inducible factors (HIFs) in hypoxic/ischemic tissues (e.g., tumors) [38].

Gene Expression Control Systems. Regulating therapeutic gene expression will be necessary for both the clinical efficacy and the safety of most HGT applications. Gene therapy offers the promise of replacing frequent injections of an expensive therapeutic protein or antibody with an infrequent or even one-time administration of a gene delivery vector, which would then provide continuous therapeutic protein production at the desired site. An appropriate system of gene expression will control and allow titration of protein levels, dosing to be adjusted as the disease evolves, and therapy to be initiated repeatedly or terminated at will. Regulation can be achieved by a physiological/pathological signal (e.g., glucose and hypoxia) [39,40] or via dose-dependent ligand binding and activation of chimeric transcription factor proteins, which then interact with DNA elements incorporated into the vector and regulate therapeutic gene expression [41].

Much progress has been made on control systems, where gene expression is regulated pharmacologically by a small-molecule drug. Four major systems are known to date that have already been tested in animals: those regulated by the antiobiotic tetracycline (Tet) [42], the insect steroid ecdysone or its analogs [43], the anti-progestin mifepristone (RU486) [44], and chemical dimerizers such as the immunosuppressant rapamycin and its analogs [45–47]. They all involve small-molecule-dependent recruitment of a transcriptional activation domain to a basal

GeneSwitch Regulated Expression System - *Two Plasmid System*

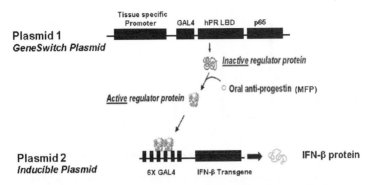

Figure 7.1-4. The two-plasmid GeneSwitch gene expression control system. Plasmid 1 contains the DNA of the mutated ligand binding domain (LBD) of the human progesterone receptor (hPR) driven by a tissue specific promoter (e.g., actin promoter for muscle applications). It also contains the GAL4 DNA-binding domain and the p65 activator domain (all combined called the "regulator protein"). After expression, the regulator protein remains inactive until an anti-progestin small molecule compound (e.g., mifepriston, MFP) binds to it, and initiates conformation change and dimerization. The active regulator protein binds to the promoter region of Plasmid 2, which induces the expression of the therapeutic gene of interest (in this example interferon β, IFNβ).

promoter driving the gene of interest, but they differ in the mechanism of recruitment. Such a pharmacological gene expression regulation system should meet the following criteria. Basal expression should be very low and inducible to high levels over a wide dose range. Induction should be a positive effect (adding rather than removing a drug) and use an orally active small molecule that has no untoward pharmacological effects in humans. The regulatory protein(s) should have no effects on endogenous gene expression and should be of human origin to minimize immunogenicity. An example of one of the most advanced systems (GeneSwitch), developed by Valentis, Inc. (Burlingame, CA) is shown in Figure 7.1-4. It has been successfully used for the timing (Figure 7.1-5) and dosing (Figure 7.1-6) of circulating plasma levels of therapeutic proteins (such as IFNβ) expressed in the hind limb muscle of mice after i.m. delivery of their DNA in the form of AAV-1 (Figure 7.1-5) or naked plasmid combined with electroporation (Figure 7.1-6).

7.1.2.3 Disease Targets for Gene Therapy

Ever since the first clinical attempt of gene transfer more than 17 years ago, numerous monogenic (Table 7.1-2) and complex (multigenic) diseases (Table 7.1-3) were targeted with a large variety of gene therapy strategies.

In 1989, Rosenberg et al. performed the first human gene therapy trial [49] when they used a retrovirus to introduce the gene coding for resistance to neomycin into human tumor-infiltrating lymphocytes before infusing them into five patients with advanced melanoma.

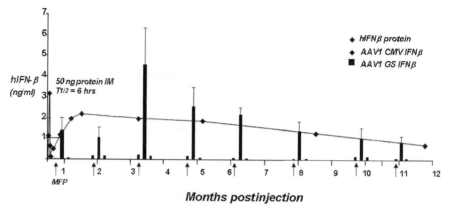

Months postinjection

Figure 7.1-5. Long-term regulated expression of human Interferon β (hIFNβ) in mice.
Three groups of male C57B16/J mice (n = 10 each) were injected with hIFNβ protein (50 ng,
intramuscular, i.m.; diamonds), adeno-associated virus serotype 1 (AAV1) carrying the
hIFNβ gene driven by the constitutive CMV promoter (10^{10} vp/animal; i.m.; dotted line),
or AAV1 carrying the hIFNβ gene regulated by the GeneSwitch (GS) inducible promoter
(columns). Circulating plasma hIFNβ level (ng/mL) was measured by ELISA. I.m. injection
of IFNβ protein resulted in transient increase in plasma IFNβ levels with very rapid clearance (T1/2: 6 hours). In contrast, single i.m. injection of AAV1 CMV IFNβ resulted in
long-term (at least up to 12 months) elevation of plasma hIFNβ protein levels. IFNβ protein
expression could be repeatedly induced by intraperitoneal (i.p.) administration of MFP
(once every 6 weeks up to 12 months; arrows) after single i.m. administration of AAV1GS
IFNβ. The induced plasma hIFNβ levels from the GS construct were similar to the sustained plasma levels produced by the constitutive (CMV) promoter.

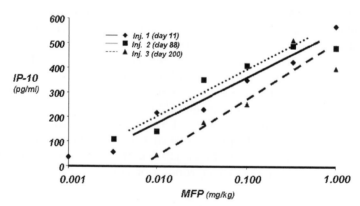

Figure 7.1-6. Reproducible dose-dependent induction of IFNβ by Mifepristone (MFP).
C57B16/J mice (n = 35) were injected intramuscularly (followed by electroporation) with a
plasmid (250 μg) containing the murine (m) IFNβ gene regulated by GeneSwitch upon
induction by MFP. Groups of mice (n = 5 each) were given different doses (0 to 1 mg/kg)
of MFP intraperitonally (i.p.). Plasmid injection and MFP dose-response were repeated
three times, every other month, for 6 months. The level of mIFNβ protein in plasma was
estimated by the IFNβ-induced marker protein, IP-10. Note the linear dose-response to
MFP over a two log dose-range (0.01–1.0 mg/kg), which was reproducible upon readministration of the plasmid.

TABLE 7.1-2. Disease Targets for Gene Therapy: Monogenic Diseases

Disease	Gene(s)	Number of Clinical Protocols/Trials (1990–1999)[1]
Cystic fibrosis	CFTR, α-1-anti-trypsin	24
Severe combined immunodeficiency (SCID)	ADA	3
Gaucher disease	Glucocerebrosidase	3
Canavan disease	Aspartoacylase	2
Hemophilia A	Factor VIII	2
Hemophilia B	Factor IX	2
Chronic granulomatous disease	Gp91 phox	2
ALS	CNTF	1
Familial hypercholesterolemia	LDL-R	1
Hunter disease	ldurinate-2-sulfatase	1
Leukocyte adherence deficiency	CD 18	1
Muscular dystrophy	Sarcoglycan, dystrophin, utrophin	1
Fanconi anemia	Group A gene	1
Purine nucleoside phosphorylase deficiency	PNP	1
Ornithin transcarbamylase deficiency	OTC	1

[1] From Ref. 48.

TABLE 7.1-3. Disease Targets for Gene Therapy: Multigenic Diseases[1]

Cancer
Gynecological: breast, ovary, cervix
Nervous system: glioblastoma, leptomeningeal
    carcinomatosis, glioma, astrocytoma,
    neuroblastoma
Gastrointestinal: colon, colorectal, liver
    mestastases, post-hepatitis liver cancer
Genito urinary: prostate, renal
Skin: melanoma
Head and neck
Lung: adenocarcinoma, small cell, non-small cell
Mesothelioma
Hematological: leukemia, lymphoma, multiple
Myeloma
Sarcoma
Germ cell tumors

Infectious
HIV/AIDS
Tetanus
CMV infection
Adenovirus infection

Vascular disease
Peripheral arterial disease
Coronary heart disease
Venous ulcers
Vascular complications of
    diabetes

Other diseases
Inflammatory bowel disease
Rheumatoid arthritis
Chronic renal disease
Carpal tunnel syndrome
Alzheimer's disease
Fractures
Diabetic neuropathy
Parkinson's disease
Erectile dysfunction
Superficial corneal opacity
Retinitis pigmentosa
Glaucoma

[1] From Ref. 1.

The first trials have been designed to establish feasibility and safety, to demonstrate the expression of therapeutic protein(s) *in vivo* by the genes transferred, and in some cases, to show therapeutic benefit.

As of January 31, 2004, there were 918 trials in 24 countries on record [1]. The U.S. accounts for two thirds of these trials. Cancer is by far the most common disease indication, followed by inherited monogenic diseases and cardiovascular diseases. Viral vectors have been the most frequently used vehicles for transferring genes into human cells, with retroviruses and adenoviruses representing the vast majority. Over 100 distinct genes have been transferred.

7.1.3 FROM BENCH TO BEDSIDE: THE DEVELOPMENT OF AD5FGF-4 TO TREAT PATIENTS WITH RECURRENT ANGINA

A brief summary of the preclinical and early clinical development and manufacturing of the Ad5FGF-4 gene therapy product for the facilitation of angiogenesis in patients with chronic myocardial ischemia (recurrent angina) is provided below for the purpose of illustrating the complex processes involved in the many stages of preclinical and clinical development of an HGT product.

7.1.3.1 Preclinical Efficacy

Specific criteria need to be fulfilled for angiogenic gene therapy to be successful. The gene selected should code for a protein with proven angiogenic activity, the vector should provide high gene-transfer efficiency, the delivery technique should target the ischemic tissue, and the procedure should be safe, in both the short and the long term [50]. In addition, appropriate current Good Manufacturing Practices (cGMP) need to be established (for details, see Section 7.1.3.3). Preclinical and initial clinical studies indicate that angiogenic gene transfer with Ad5FGF-4, delivered by a single intracoronary injection, may meet these criteria [16, 50–55].

Proof-of-Concept Study. Preclinical investigation of Ad5FGF-4 involved demonstration of myocardial angiogenesis in a porcine model of stress-induced myocardial ischemia [51]. In this experimental model, an ameroid constrictor is placed around the proximal left circumflex coronary artery, leading to gradual closure of the artery over the following weeks. The subsequent development of collateral vessels allows normal myocardial function and blood perfusion at rest, but blood flow is insufficient to prevent ischemia during periods of increased oxygen demand. In the proof-of-concept studies, stress-induced (atrial pacing) left ventricular function and blood flow changes were evaluated by two-dimensional echocardiography approximately 35 days after ameroid placement. Ad5FGF-4 (10^{11} viral particle [v.p.] per animal) (n = 6) was then administered by intracoronary injection. Ventricular function and blood flow were reassessed 14 days later, and the animals were killed to quantify the presence of adenoviral DNA and FGF-4 mRNA and protein expression in the myocardium.

Polymerase chain reaction (PCR) and reverse-transcription–PCR (RT–PCR) analysis demonstrated the presence of Ad5 DNA and FGF-4 mRNA, respectively, in the injected pig hearts [51]. Immunoblotting showed that pigs treated with

Ad5FGF-4 expressed FGF-4 protein in heart tissue but not in other organs, such as the liver and eye. Contractile function, assessed as the degree of wall thickening of the ischemic region during atrial pacing, had improved significantly 2 weeks after FGF-4 gene transfer compared with preinjection (Figure 7.1-7A). This functional

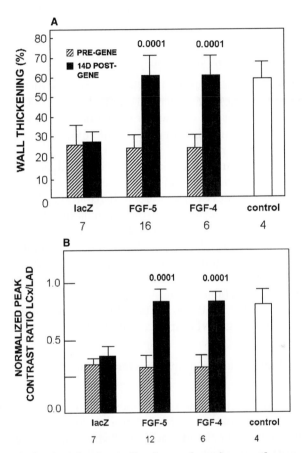

Figure 7.1-7. Proof-of-principle pig studies for angiogenic gene therapy after intracoronary delivery of Ad5FGF-4. (A) Regional contractile function. Basal wall thickening in the ischemic region of the ameroid-equipped pig heart was normal, but atrial pacing (200 bpm) was associated with reduced wall thickening. At 2 weeks after intracoronary Ad5FGF-4 gene transfer (10^{12} v.p.), there was a 2.3-fold increase in wall thickening in the ischemic region during pacing ($p < 0.0001$); similar to results observed after Ad5FGF-5 gene transfer. Pigs receiving Ad5LacZ showed a similar degree of pacing-induced deficit in the ischemic region before and 2 weeks after gene transfer. (B) Regional myocardial blood flow. Before gene transfer, animals showed a deficit in blood flow in the ischemic region during pacing. At 2 weeks after Ad5FGF-4 gene transfer, the animals showed improved flow in the ischemic region ($p < 0.0001$); this was similar to results observed after Ad5FGF-5 gene transfer. Pigs receiving Ad5LacZ showed a similar degree of pacing-induced deficit in the ischemic region before and 2 weeks after gene transfer. Columns represent mean values for number of experiments shown below the columns; error bars denote 1 SD. (Reproduced with permission from Mary Ann Liebert, Inc.)

improvement was associated with a normalization of regional blood flow during atrial pacing (Figure 7.1-7B) [51]. Thus, this study provided experimental evidence that a single intracoronary injection of Ad5FGF-4 ameliorates deficits in myocardial blood flow and function in the setting of chronic myocardial ischemia in pigs.

Follow-up studies in the same pig model showed that the effect of single Ad5FGF-4 injection lasted at least for 3 months, and that the effect was dose dependent [51].

7.1.3.2   Toxicology and Biodistribution

A Good Laboratory Practices (GLP) toxicology and biodistribution study was performed to determine the potential adverse effects of Ad5FGF-4 in healthy pigs after either intracoronary or left ventricular (systemic) administration (10^{12} v.p. left ventricular, 10^{10}–10^{12} v.p. intracoronary). Systemic biodistribution of the product was also assessed.

This study found no significant test article-related toxicologic effects [51]. PCR analysis for adenoviral DNA and RT–PCR analysis for the transgene (FGF-4) mRNA were performed to assess vector biodistribution. Adenoviral DNA was detected in 27 of 110 organs examined from animals injected intracoronary with 10^{12} v.p. Ad5FGF-4. Adenoviral DNA was detected in the lung, liver, spleen, and testis. Detectable adenoviral DNA typically decreased over time in most organs. The presence of viral DNA in the testis was observed at 5 days but not at 28 or 84 days. In no case in which adenoviral DNA was present in extracardiac sites was transgene expression detectable at the mRNA level after RT–PCR analysis [51]. These results (along with cGMP; see Section 7.1.3.3) supported the submission of an investigational new drug (IND) application in 1998.

7.1.3.3   cGMP Manufacturing and Characterization of the Gene Therapy Vector

This section summarizes the various activities required in the course of cGMP manufacturing and testing of the Ad5FGF-4 product for the initiation and advancement of its clinical development and reaching a "well characterized product" status with the Regulatory Authorities.

Vector Construction. Construction of the recombinant replication incompetent (E1/2 deleted) adenovirus serotype 5 (Ad5) expressing the human FGF-4 cDNA required three components: the FGF-4 transgene, a plasmid shuttle vector to carry the transgene as well as 5′ Ad5 sequences, and a second plasmid carrying the bulk of the Ad5 genome.

The full-length cDNA for human FGF-4 (the "transgene") was isolated from a cDNA library, which was constructed from mRNA of Kaposi's sarcoma DNA transformed NIH3T3 cells [56]. The cDNA that encodes the FGF-4 peptide is approximately 1.2 kB in size, and the FGF-4 protein has 206 amino acids, including a 33-amino-acid signal peptide at the N-terminus.

Plasmid pACCMVplpASR(-) (the "shuttle vector") contains the heterologous CMV promoter, a polylinker, and SV40 polyadenylation sequences flanked by partial human Ad5 sequences [57]. The FGF-4 cDNA was subcloned, as an Eco R1 fragment, into this adenovirus shuttle vector at its single Eco R1 site.

Plasmid pJM17 (the "adenovirus plasmid") contains the required Ad5 sequences except that the E1 region is disrupted by the insertion of pBR322 sequences [58]. A unique feature of this plasmid is the presence of some nonviral sequences in the E3 region [59] that become relevant for the viral vector identity testing.

To generate the Ad5FGF-4 viral vector, plasmids pACSR/FGF-4 and pJM17 were co-transfected into HEK 293 cells using a calcium phosphate method. The cells were overlaid with nutrient agarose. Homologous recombination between the vectors created E1-deleted, FGF-4 gene-containing vector genomes capable of replication in HEK 293 cells.

Master Virus Bank. Six clones were isolated from one round of virus plaque purification and were screened for protein expression. One clone was selected, amplified on HEK 293 cells, purified by cesium chloride (CsCl) ultracentrifugation, and used for preparation of a Master Virus Bank (MVB). A purified Ad5FGF-4 virus seed was propagated through three consecutive rounds of plaque purification. A final plaque was then expanded and purified by anion exchange chromatography, sterilized by filtration, and stored at −70°C. This virus stock was checked for sterility and absence of measurable replication competent adenovirus (RCA). An MVB was then created by one additional cycle of propagation in serum-free suspension culture of HEK 293 cells, and this virus was purified, aliquoted, and frozen. The MVB was tested and confirmed to be free of RCA as well as adventitious agents. The MVB was stored at −70°C.

Master Cell Bank and Manufacturer's Working Cell Bank. A scalable manufacturing process to meet clinical, as well as commercial, needs for recombinant adenovirus vector would only be attractive if it could be performed without using bovine serum in the culture medium. For these reasons, attachment and serum-dependent HEK 293 cells were first adapted to suspension culture by direct transfer into shake flasks in a modified William's essential medium containing 2% fetal bovine serum (Hyclone, Gamma irradiated). The cultures were passaged continually for over 6 weeks in the same environment until the cells exhibited a consistent growth pattern of less than 48 hours of doubling time and greater than 90% viability. These suspension-adapted cells were subsequently adapted to serum-free medium (modified IS293, Irvine Scientific) by gradual weaning. Cells from this bank were thawed and expanded and then used to generate a Master Cell Bank (MCB). Culture identification as human cells, growth on soft agarose to assess potential tumorigenicity, and screening for adventitious agents were also performed (according to current cell-line testing guidelines). Starting from the MCB, a Manufacturer's Working Cell Bank (MWCB) was prepared and then similarly tested before routine use.

Manufacture of the Bulk Drug Substance. The Ad5FGF-4 vector was manufactured under cGMP in a state-of-the-art, validated facility, following Biosafety Level 2 (BL2) practices. For the preparation of the Ad5FGF-4 drug substance, cells from the MWCB and virus from the MVB were employed. The virus was propagated in suspension cultures of HEK 293 cells using serum-free medium and the virus progeny was purified using a combination of anion exchange chromatography (AIEX) and ultra-filtration (UF). The purified bulk drug substance was stored at −70°C.

Drug Substance (Virus) Purification. Medium components and extracellular (nonviral) contaminants were largely removed from adenoviral infected HEK 293 cells using three successive physiological buffer solution (PBS) washes and centrifugation of the cells at the time of harvest. The harvested cells were then frozen in a solution of PBS with 2% sucrose and stored at −70°C, until purification. The virus purification process was initiated with two additional cycles of freeze and thaw steps. After the final thaw, the ruptured cell suspension was centrifuged to remove cell debris. Virus purification was established with a protocol based on traditional protein purification techniques: column chromatography and ultra-filtration. Separation from nucleic acid contaminants was achieved by combining strategic peak collection and utilization of a tangential flow ultra-filtration step with a 100-kDa molecular-weight cutoff pore size membrane. Both of these purification procedures are readily and linearly scaled up and thus can support the large-scale production of Ad5FGF-4 for both clinical testings and commercial use.

Manufacture Scale-Up. The adenovirus harvest process consisted of three general steps: concentration, lysis, and clarification. In the small-scale process, the intact adenovirus-infected cells in culture medium were aliquoted and batch centrifuged to achieve a 30-fold concentration, and the supernatant was manually removed and replaced with a smaller volume of the freeze buffer. Although sufficing for harvest volumes of no more than 10 L, such a manual process would prove to be inadequate on a larger manufacturing scale.

A successful harvest method was identified using continuous centrifugation in a Westfalia CSC-6 disk-stack device with a hydroheratic feed system (HHFS). Scale-up of the adenovirus harvest process using a continuous centrifuge with HHFS has been shown to efficiently separate and concentrate an infected cell pellet while allowing the relatively easy exchange of growth medium for freezing buffer without open, manual manipulation. This patented process step combined separation of the cell pellet from culture medium, medium/buffer exchange, cell concentration, and then cell lysis in a single-unit operation step. The resulting lysate could be clarified and the virus purified.

Analytical Techniques and Release Specifications of Bulk Virus Lots. The crude viral harvest was tested for adventitious agents as needed to comply with current regulatory guidelines. Purified virus was tested to confirm the absence of RCAs, residual host cell proteins and DNA, and endotoxin, and to quantify total and infectious vector particles and, thus, the infectivity ratio. Bulk virus lots were released after meeting specifications, which included <1 endotoxin unit/mL and infectivity ratios of >2% (infectious titer/total viral particles).

Chromatographic Assays for the Quantitation of Intact Viral Particles. One main challenge in the development of the adenovirus purification process is the quantitation of intact viral particles in crude samples. Nonaggregated, purified viral samples were easily quantified by the reverse phase (RP)–high-performance liquid chromatography (HPLC) method [60]. The RP–HPLC assay quantitates the individual structural proteins with UV absorbance at 214 nm as the basis for the measurement of the adenovirus concentration. Unfortunately this assay could not be used to calculate the yields of the purification steps because it could not reliably quantitate crude samples. To solve this problem, an anion exchange (AIEX)–HPLC procedure was developed using a trimethyl anion exchange (TMAE) resin. The TMAE–HPLC was suitable for the analysis of all in-process samples ranging from clarified cell harvest to the final formulated, purified recombinant adenovirus product.

Endpoint Dilution (EPD) Assay for Infectivity. To determine infectivity, an endpoint dilution assay was developed [61]. The precision of this assay has a standard deviation of 0.2 when viral titers are expressed in $\log_{10}$.

Replication Competent Adenoviruses (RCA). Recombination between the viral E1 gene carried by HEK 293 cells and the E1 deleted recombinant adenovirus can and does occur at low frequency to generate RCAs, essentially wild-type Ad5 lacking the FGF-4 transgene. A highly sensitive assay involving viral amplification and cytopathic effect (CPE) readout has been described [62]. The limit of detection of this assay was one RCA in 3.2×10^{12} virus particles. Virus banks and bulk Ad5FGF-4 virus batches testing positive for RCA were not used in clinical studies.

Host Cell Protein Detection by ELISA. Typically biotechnology products have been produced in cell lines not of human origin. In such cases, residual host cell proteins (HCPs) are perceived as a safety concern to the patient. In the case of Ad5FGF-4 produced in HEK 293 cells, it is expected that the patient may respond to viral proteins with an immune response, but it is less clear that human HCPs will be immunogenic or pose a serious safety concern. One approach to detection of such HCP is by enzyme-linked immunosorbent assay (ELISA) using a polyclonal antibody reagent such as the HEK 293 HCP detection kit offered by Cygnus Technologies (Renchesler, MD). This ELISA was used to show a consistent, high HCP titer in crude cell lysates and to show that the diethyl-aminoethyl (DEAE) chromatography step is sufficient and reproducible to reduce HCPs to below the level of quantitation (20 ng/mL).

Residual DNA. Two methods were used to quantitate residual host cell (HEK 293) DNA in Ad5FGF-4 preparations. The first, using membrane hybridization of an oligonucleotide probe followed by chemiluminescent detection, was based on a procedure described by Walsh et al. [63]. Extracted samples were blotted onto a positively charged nylon membrane using a slot blot apparatus, the blot was then hybridized with a biotinylated oligonucleotide probe specific to the primate alpha-satellite sequence D17Z1, and the bound probe was detected with a streptavidin–alkaline phosphatase conjugate, followed by a chemiluminescent alkaline phosphatase substrate. The second method used TaqMan PCR quantitation of *Alu* repeat sequences in virus samples extracted using the Qiagen QIAAmp Viral RNA Mini Kit. In-process sample testing showed that the two methods gave similar results, both demonstrating a greater than 3 log reduction in the level of host cell DNA by the purification process, with final purified material containing less than 0.5-ng/mL DNA.

Product Identity Tests by PCR. The adenoviral backbone of the Ad5FGF-4 genome is derived from *dl*309, an Ad5 mutant, commonly used for the construction of E1-deleted adenoviral vectors. This mutant has a short stretch of foreign DNA inserted in place of a portion of the E3 region [59]. PCR confirmation of the type of adenovirus backbone (*dl*309 or wild-type Ad5 E3 region) served as a test for product identity and purity. It was possible to confirm transgene orientation as well as differentiate the product digest pattern from those obtained from either wt Ad5, RCA, or vectors containing different transgenes.

Product Potency Tests. Two assays were developed that measure the potency of the FGF-4 protein. In the first case, a "one-step" growth promotion assay is conducted on normal, human retinal pigment epithelial cells (ARPE-19). The assay measures metabolic activity (Alamar blue dye metabolism) after infection of ARPE-19 cells with a serial dilution of the drug substance (Ad5FGF-4). The increase in metabolic activity correlated with FGF-4 protein production (determined by an FGF-4 ELISA), increased *de novo* DNA synthesis (measured by BrdU incorporation), and an increase in cell number.

The second assay measured the production of FGF-4 protein produced in and secreted by A549 cells after infection with different doses of Ad5FGF-4. An FGF-4 specific ELISA (R&D Systems) allowed quantitation of FGF-4 protein present in the cell culture medium. The FGF-4 produced was confirmed as biologically active by stimulating the growth of ARPE-19 cells (see above).

DNA Sequencing of the Entire Viral Vector. The complete DNA sequence of the Ad5FGF-4 genome was determined using double-stranded DNA sequencing of Ad5FGF-4 DNA extracted from an MVB expansion. A total of 166 oligonucleotide primers were used to obtain overlapping readings from both strands of DNA. The observed sequence was compared with that predicted for Ad5FGF-4 by insertion of the FGF-4 transgene sequence into the wild-type Ad5 sequence [64], (Genbank Accession Number M73260, originally deposited as NC001406), into which the *dl*309 specific mutations had been incorporated [59]. The observed sequence matched the predicted 5′ 4.9 kB of vector sequence including the FGF-4 transgene. The sequence data distal to the transgene (3' end) confirmed that the Ad5FGF-4 backbone was derived from the Ad5 mutant *dl*/309.

Genome Analysis. As a further safety precaution, and as a specific regulatory request, the DNA sequence of the FGF-4 transgene was analyzed for other possible open reading frames (ORFs) that might encode unexpected foreign proteins upon infection of target cells with the recombinant adenovirus. This analysis also included any reading frames that crossed the transgene/virus junctions and might therefore involve viral sequences. The genome analysis confirmed that only FGF-4 protein is made from the transgene carried in Ad5FGF-4.

The cGMP manufacturing and analytical process (outlined above) and the facility where it took place (Berlex Biosciences, Richmond, CA) have been approved by both the U.S. Food and Drug Administration (FDA) and the European Regulatory Agencies. Based on the test data, the FDA qualified Ad5FGF-4 as a "well characterized product."

7.1.3.4   Clinical Studies

There are several goals of Ad5FGF-4 angiogenic gene therapy for patients with chronic myocardial ischemia. It should promote new collateral vessel formation in the heart. In turn, this should increase perfusion of ischemic regions, leading to improved myocardial oxygen delivery and left ventricular function. From the patient's perspective, the ultimate goal of treatment should be to reduce or ameliorate symptoms of angina, increase exercise capacity, improve quality of life, and decrease the long-term risk of acute coronary events (such as acute myocardial infarction).

The clinical development program was initiated in 1998 to determine whether the improvement in cardiac perfusion and function (without any product-related adverse effects) detected in pigs translates into a clinical therapeutic benefit in patients with chronic stable angina. Four clinical studies [Angiogenic Gene Therapy (AGENT) phase 1/2 trial, AGENT 2 phase 2 trial, and AGENT 3 and 4 phase 2B/3 trials] have been initiated, and some (AGENT and AGENT-2) have been completed to date.

The Phase 1/2 AGENT Trial. The Angiogenic Gene Therapy Trial (AGENT) was the first ever randomized, double-blind, placebo-controlled (12 sites) U.S. clinical trial of angiogenic gene therapy for myocardial ischemia. The objectives of the AGENT trial were to evaluate the safety and anti-ischemic effects of five ascending doses of Ad5FGF-4 (3.2×10^8 to 3.2×10^{10} v.p.), randomizing in a ratio of 1:2 (placebo:active) in patients with chronic stable exertional angina, who had recurring angina despite optimal medical treatment and were not in need of immediate surgical revascularization.

A total of 79 patients with chronic stable angina were enrolled. Patients could exercise for ≥3 minutes in an exercise treadmill test (ETT) using the modified Balke protocol. The adenovirus vector was infused over 90 seconds through subselective catheters into all major patent coronary arteries and grafts, 40% into the right coronary distribution and 60% into the left coronary distribution. Repeat treadmill exercise tests were performed at 4 and 12 weeks posttreatment.

The increase from baseline in treadmill exercise duration at weeks 4 and 12 was greater among patients receiving 10^{10} v.p. Ad5FGF-4 than among those receiving placebo (the difference at week 4 was statistically significant) (Figure 7.1-8) [52]. Analysis of the subgroup of patients with a baseline ETT time of <10 minutes showed the gene therapy produced a statistically significant increase in exercise capacity compared with placebo at 12 weeks (27% vs. 12%; $p = 0.01$).

The Phase 2 AGENT-2 Trial. AGENT 2 was designed to assess whether Ad5FGF-4 improved myocardial perfusion in patients with stable angina. It was also designed to further evaluate safety. Based on the results of the AGENT trial, a dose of 10^{10} v.p. was selected. The primary endpoint was the change in stress-related (adenosine-induced) reversible perfusion defect size (RPDS) as assessed by single-photon emission computed tomography (SPECT), 8 weeks after treatment [52].

A total of 52 patients underwent double-blind randomization (35 to Ad5FGF-4, 17 to placebo). Total perfusion defect size (PDS) at baseline was 32% and RPDS was 20%.

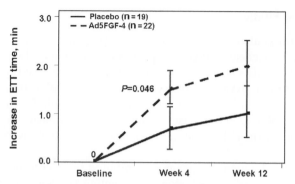

Figure 7.1-8. AGENT Trial ETT Results. Mean (± SEM) change in total exercise treadmill test (ETT) time for patients at 4 and 12 weeks after intracoronary infusion of 10^{10} v.p. of Ad5FGF4 (n = 22) or placebo (n = 19).

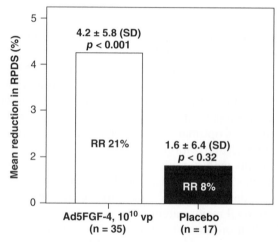

Figure 7.1-9. AGENT-2 Trial Perfusion Results. Mean change in reversible perfusion defect size (RPDS) compared with baseline for all active (n = 35) and placebo (n = 17) patients at 4 and 8 weeks after intracoronary infusion of Ad5FGF-4 (10^{10} v.p.) or placebo. RR = relative reduction in RPDS compared with baseline. (Reproduced with permission from Elsevier.)

The mean reduction in RPDS from baseline at 8 weeks posttreatment in the Ad5FGF-4 group was 4.2 ± 5.6% ($p < 0.001$), a 21% decrease from baseline, versus a reduction in the placebo group of only 1.6 ± 5.4% ($p = 0.32$), corresponding to a decrease of 8% from baseline (Figure 7.1-9). Similar results were observed for the change in total PDS from baseline at 8 weeks posttreatment: Ad5FGF-4 decreased total PDS by a mean of 4.6 ± 5.6% ($p < 0.001$) compared with a reduction of 2.4 ± 6.5% with placebo ($p = $ NS).

More of the patients who received the active product than those who received placebo reported complete resolution of angina (30% vs. 13%) and no nitroglycerin use (43% vs. 17%) at 8 weeks. In addition, the incidence of worsening/unstable angina and revascularation by coronary artery bypass grafting or angioplasty at 12 months was considerably lower in the Ad5FGF-4 group (6% and 6%, respectively) compared with those in the placebo group (24% and 16%, respectively).

Safety in AGENT and AGENT 2 Trials. In the AGENT and AGENT-2 trials combined, a total of 131 patients (95 on Ad5FGF-4 and 36 on placebo) were treated and followed for 12 months. Overall, Ad5FGF-4 was well tolerated. There was no rise in cardiac enzymes, electrocardiographic change, or clinical evidence of myocarditis associated with the treatment [52, 53].

Vector distribution after intracoronary administration was examined in the AGENT trial [52]. Adenovirus could be detected by endpoint dilution infectivity assay in the pulmonary artery during intracoronary infusion and in peripheral venous blood 1 hour later. The frequency of positive samples increased with increasing doses of Ad5FGF-4 (Figure 7.1-10). Virus was not detected in the urine collected over 6 hours after treatment. Neutralizing antibody titer to Ad5 increased in most patients, but no FGF-4 protein was detected in the circulation at any time. Semen samples (n = 8) were tested by PCR (8 weeks posttreatment) and found to be negative for Ad5FGF-4 DNA. Ad5FGF-4-related adverse effects included dose-related transient fever (n = 8[8%]) and transient increase in liver enzymes (n = 3 [3%]).

Overall, the safety profile was reassuring and consistent with safety data in other cardiovascular adenoviral gene therapy trials [65, 66]. There was no evidence of myocarditis, retinal neovascularization, or angioma formation.

The positive trend toward efficacy and the safety observed in these two trials were in good agreement with the preclinical safety and efficacy data obtained in pigs, and they provided the basis to progress the clinical development to larger, multinational phase 2b/3 trials (AGENT-3 and AGENT-4), the results of which will be reported in the future.

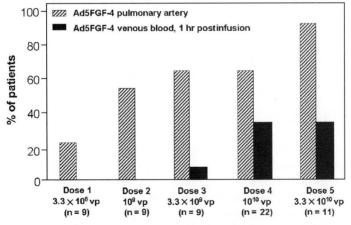

Figure 7.1-10. AGENT Trial Virus Distribution. Detection of adenovirus in the pulmonary artery during intrcoronary infusion and in venous blood at 1 hour after infusion in patients treated with increasing doses of Ad5FGF-4 in the AGENT trial (n = 60). (Reproduced with permission from Lippincott, Williams & Wilkins.)

7.1.3.5   Lessons Learned from Past Failures

The remarkable clinical safety observed in most cardiovascular gene therapy trials to date [52, 53, 65, 66] is of no coincidence. Lessons learned from past failures at the manufacturing, preclinical efficacy and safety, and clinical development levels were all taken into account in the development of Ad5FGF-4.

Failures were due primarily to two factors: safety issues and lack of clinical efficacy.

The tragic death of Jesse Gelsinger at the University of Pennsylvania treated with a very high dose (>10 e13 vp) of a first-generation adenoviral vector raised the issue of dose and innate immune response. We learned that side effects of Ad5 are dose dependent, which follow a so-called "elbow" shape (i.e., no effect up to a certain dose and severe effects at the next dose). Therefore we have tested the dose-response relationship of efficacy, biodistribution, side effects, and toxic effects very thoroughly in our preclinical models, and we chose the doses of 10e9 and 10e10 vp (more than three orders of magnitude lower than the dose used in the "Gelsinger" case).

Another gene therapy clinical trial that led to serious side effects was the SCID trial in France, where several of the children treated developed leukemia. These later were attributed to retrovirus integration at a site in the genome, which led to significant clonal expansion of the treated lymphocytes. Therefore, the choice of vector could be of great importance. In our product candidate, we intentionally avoided the use of integrating vectors (e.g., retrovirus, lentivirus, and AAV). As relatively short-term and transient gene expression is sufficient for therapeutic angiogenesis, we did not need to use these integrating vectors.

Both the "Gelsinger" case and the "SCID Trial" intensified regulatory scrutiny, which served the field very well. Small (mostly physician sponsored) trials using poorly characterized material are not allowed anymore. Both the FDA and the European Regulatory Agencies require well-controlled cGMP Manufacturing to be in place before clinical testing can start.

The recent failure of Hemofilia B trials due to lack of efficacy raised other important issues. All animal tests using the AAV.fIX product showed positive results from mouse to subhuman primates. The lack of efficacy in humans is now attributed (at least in part) to a memory T-cell mediated immune response against the AAV vector, which is absent in animals. However, no human-specific immune response to Ad5 vectors has been identified to date.

7.1.4   CONCLUSIONS

The progress of gene therapy in the past 17 years has been slower than was expected, which is due to several factors. Gene therapy is a pioneering new therapeutic modality based on complex biological systems occurring at the leading edge of biomedical knowledge. Incomplete knowledge of the genes involved in the pathomechanism of diseases constitutes a limit to generate clinically effective gene therapies, especially in complex diseases with multiple interacting genetic and environmental factors. Stringent and time-consuming safety studies are needed as is the establishment of new regulatory frameworks essential to control the applications of gene therapy and ensure safety to the patient and the population at large.

High costs are involved in the R&D of gene therapy and complex issues of intellectual property and commercial rights need to be resolved.

Despite the early high exceptions and the subsequent setbacks, one has to recognize that this new therapeutic modality is still in its infancy. It will neither deliver medical "miracles" (as its early proponents predicted) nor will it "disappear" because of a few disappointing setbacks (as some of its recent antagonists predict). As with all new technologies, gene therapy has to run its course in its current "development phase" before it reaches "maturation," when its full potential will be exploited.

REFERENCES

1. Edelstein M L, Abedi M R, Wixon J, Edelstien R M (2004). Gene therapy clinical trials worldwide: 1989–2004—an overview. *J. Gene Med.* 6:597–602.

2. Harui A, Suzuki S, Kochanek S, et al. (1999). Frequency and stability of chromosomal integration of adenovirus vectors. *J. Virol.* 73:6141–6146.

3. Wilson J M (1996). Adenoviruses as gene delivery vehicles. *New Eng. J. Med.* 334: 1185–1187.

4. Svensan E C, Marshall D J, Woodard K, et al. (1999). Efficient and stable transduction of cardiomyocytes after intramocardial injection or intracoronary perfusion with recombinant adeno-associated virus vector. *Circulation.* 99:201–205.

5. Gao G P, Alvira M R, Wang L, et al. (2002). Novel adeno-associated viruses from rhesus monkeys as vectors for human gene therapy. *Proc. Nat. Acad. Sci. USA.* 99(18):11854–11859.

6. Li S, Huang L. (2000). Nonviral gene therapy: Promises and challenges. *Gene Ther.* 7:31–34.

7. Plank C, Mechtler K, Szoka Jr F C, et al. (1996). Activation of the complement system by synthetic DNA complexes: A potential barrier for intravenous gene delivery. *Human Gene Ther.* 7:1437–1446.

8. Stephan D J, Yang Z-Y, San H, et al. (1996). A new cationic liposome DNA complex enhances the efficiency of arterial gene transfer in vivo. *Human Gene Ther.* 7:1803–1812.

9. Turunun M P, Hiltunen M O, Ruponen M, et al. (1999). Efficient adventitial gene delivery to rabbit carotid artery with cationic polymer-plasmid complexes. *Gene Ther.* 6:6–11.

10. Young Jr S M, McCarty D M, Degtyareva N, et al. (2000). Roles of adeno-associated virus Rep protein and human chromosome 19 in site-specific recombination. *J. Virol.* 74:3953–3966.

11. Asahara T, Kalka C, Isner J M (2000). Stem cell therapy and gene transfer for regeneration. *Gene Ther.* 7:451–457.

12. Gage F H (1998). Cell therapy. *Nature.* 392(suppl.):18–24.

13. Su L, Lee R, Bonyhadi M, et al. (1997). Hematopoietic stem-cell based gene therapy for acquired immuno-deficiency syndrome: Efficient transduction and expression of RevM10 in myeloid cells in vivo and in vitro. *Blood.* 89:2283–2290.

14. Boekstegers P, von Degenfeld G, Giehrl W, et al. (2000). Myocardial gene transfer by selective pressure regulated retroinfusion of coronary veins. *Gene Ther.* 7:232–240.

15. Takeshita S, Zheng L P, Brogi E, et al. (1994). Therapeutic angiogenesis: A single intraarterial bolus of vascular endothelial growth factor augments revascularization in a rabbit ischemic hind limb model. *J. Clini. Invest.* 93:662–670.

16. Giordano F J, Ping P, McKirnan M D, et al. (1996). Intracoronary gene-transfer of fibroblast growth factor-5 increases blood flow and contractile function in an ischemic region of the heart. *Nature Med.* 2:534–539.

17. Mack C A, Aptel S R, Schwarz E A, et al. (1998). Biologic bypass with the use of adenovirus-mediated gene transfer of the complementary deoxyribonucleic acid for vascular endothelial growth factor 121 improves myocardial perfusion and function in the ischemic porcine heart. *J. Thorac. Cardiovasc. Surg.* 115:168–177.

18. Shyu K G, Manor O, Magner M, Yancopoulos G D, Isner J M (1998). Direct intramuscular injection of plasmid DNA encoding angiopoietin-1 but not angiopoietin-2 augments revascularization in the rabbit ischemic hindlimbs. *Circulation.* 98:2081–2087.

19. Tsurumi Y, Takeshita S, Chen D, et al. (1996). Direct intramuscular gene transfer of naked DNA encoding vascular endothelial growth factor augments collateral development and tissue perfusion. *Circulation.* 94:3281–3290.

20. Banai S, Jaklitsch M T, Shou M, et al. (1994). Angiogenic-induced enhancement of collateral blood flow to ischemic myocardium by vascular endothelial growth factor in dogs. *Circulation.* 89:2183–2189.

21. Lawrie A, Brisken A F, Francis S E, et al. (1999). Ultrasound enhances reporter gene expression after transfection of vascular cells in vitro. *Circulation.* 99:2617–2620.

22. Villanueva F S, Jankowski R J, Klibanov S, et al. (1998). Microbubbles targeted to intercellular adhesion molecule-1 bind to activated coronary artery endothelial cells. *Circulation.* 98:1–5.

23. Yang N-S, Sun W H (1995). Gene gun and other non-viral approaches for cancer gene therapy. *Nature Med.* 1:481–483.

24. Mir L M, Burean M F, Gehl J, et al. (1999). High-efficiency gene transfer into skeletal muscle mediated by electric pulses. *Proc. Nat. Acad. Sci. USA.* 96:4262–4267.

25. Bergelson J M, Cunningham G D, Kurt-Jones E A, et al. (1997). Isolation of a common receptor for coxsacki B viruses and adenoviruses 2 and 5. *Science.* 275:1320–1323.

26. Nemerow G R, Stewart P L (1999). Role of αv integrins in adenovirus cell entry and gene delivery. *Microbiol. Molec. Biol. Rev.* 63:725–734.

27. Wickham T J, Mathias P, Cheresh D A, et al. (1993). Integrins αvβ3 and αvβ5 promote adenovirus internalization but not virus attachment. *Cell.* 73:309–319.

28. Gaggar A, Shayakhmetov D M, Lieber A (2003). CD46 is a cellular receptor for group B adenoviruses. *Nature Med.* 9(11):1408–1412.

29. Douglas J T, Rogers B E, Rosenfeld M E, et al. (1996). Targeted gene delivery by tropism-modified adenoviral vectors. *Nature Biotechnol.* 14:1574–1578.

30. Wickham T J, Roelvink P W, Brough D E, et al. (1996). Adenovirus targeted to heparin-containing receptors increases its gene delivery efficiency to multiple cell types. *Nature Biotechnol.* 14:1570–1573.

31. Wickham T J, Tzeng E, Shears L L, et al. (1997). Increased in vitro and *in* vivo gene transfer by adenovirus vectors containing chimeric fiber proteins. *J. Virol.* 71:8221–8229.

32. Krasnykh V N, Douglas J T, van Beusechem V W (2000). Genetic targeting of adenoviral vectors. *Molec. Ther.* 1:391–405.

33. Wickham T J (2000). Targeting adenovirus. Gene *Ther.* 7:110–114.

34. Green N K, Herbert C W, Hale S J, et al. (2004). Extended plasma circulation time and decreased toxicity of polymer-coated adenovirus. *Gene Ther.* 11:1256–1263.

35. Fisher K D, Stallwood Y, Green N K, et al. (2001). Polymer-coated adenovirus permits efficient retargeting and evades neutralizing antibodies. *Gene Ther.* 8(5):341–348.

36. Pang S, Dannull J, Kaboo R, et al. (1997). Identification of a positive regulatory element responsible for tissue-specific expression of prostate-specific antigen. *Cancer Res.* 57:495–499.

37. Ko S C, Cheon J, Kao C, et al. (1996). Osteocalcin promoter-based toxic gene therapy for the treatment of osteosarcoma in experimental models. *Cancer Res.* 56:4614–4619.

38. Shibata T, Giaccia A J, Brown J M (2000). Development of a hypoxia-responsive vector for tumor-specific gene therapy. *Gene Ther.* 7:493–498.

39. Thule P M, Liu J, Phillips L S (2000). Glucose regulated production of human insulin in rat hepatocytes. *Gene Ther.* 7:205–214.

40. Varley A W, Munford R S (1998). Physiologically responsive gene therapy. *Molecular Med. Today.* 4:445–451.

41. Clackson T (2000). Regulated gene expression systems. *Gene Ther.* 7:120–125.

42. Gossen M, Bujard H (1992). Tight control of gene expression in mammalian cells by tetracycline-responsive promoters. *Proc. Nat. Acad. Sci. USA.* 89:5547–5551.

43. No D, Yao T P, Evans R M (1996). Ecdysone-inducible gene expression in mammalian cells and transgenic mice. *Proc. Nat. Acad. Sci. USA.* 93:3346–3351.

44. Wang Y, O'Malley Jr B W, Tsai S Y, et al. (1994). A regulatory system for use in gene transfer. *Proc. Nat. Acad. Sci. USA.* 91:8180–8184.

45. Ho S N, Biggar S R, Spencer D M, et al. (1996). Dimeric ligands define a role for transcriptional activation domains in reinitiation. *Nature.* 382:822–826.

46. Magari S R, Rivera V M, Iuliucci J D, et al. (1997). Pharmacologic control of humanized gene therapy system implanted into nude mice. *J. Clin. Invest.* 100:2865–2872.

47. Rivera V M, Clackson T, Natesan S, et al. (1996). A humanized system for pharmacologic control of gene expression. *Nature Med.* 2:1028–1032.

48. Rosenberg S A, Blaese R M, Brenner M K, et al. (2000). Human gene marker/therapy clinical protocols. *Hum Gene Ther.* 11:919–979.

49. Rosenberg S A, Aebersold P, Cornetta K, et al. (1990). Gene transfer into humans—immunotherapy of patients with advanced melanoma, using tumor-infiltrating lymphocytes modified by retroviral gene transduction. *New Eng. J. Med.* 323:570–578.

50. Rubanyi G M (2001). The future of human gene therapy. *Molec. Aspects Med.* 22:113–142.

51. Gao M H, Lai N C, McKirnan M D, et al. (2004). Increased regional function and perfusion after intracoronary delivery of adenovirus encoding fibroblast growth factor-4: Report of preclinical data. *Human Gene Ther.* 15:574–587.

52. Grines C L, Watkins M, Mahmarian J, et al. (2003). A randomized double blind placebo-controlled trial of Ad5FGF-4 gene therapy and its effect on myocardial perfusion in patients with stable angina. *J. Amer. Coll. Cardiol.* 42:1339–1347.

53. Grines C L, Watkins M W, Helmer G, et al. (2002). Angiogenic Gene Therapy (AGENT) trial in patients with stable angina pectoris. *Circulation.* 105(11):1291–1297.

54. Rubanyi G M (2004). The design and preclinical testing of Ad5FGF-4 to treat chronic myocardial ischemia. *Euro. Heart J.* 6(suppl. E):E12–E17.

55. Watkins M W, Rubanyi G M (2003). Gene therapy for coronary artery disease: Preclinical and initial clinical results with intracoronary administration of Ad5FGF-4. In G. M., Rubanyi, S., Ylä-Herttuala, (eds.), *Human Gene Therapy: Current Opportunities and Future Trends*, Springer, Berlin, Germany, pp. 61–80.

56. Delli Bovi P, Curatola A M, Kern F G, et al. (1987). An oncogene isolated by transfection of Kaposi's sarcoma DNA encodes a growth factor that is a member of the FGF family. *Cell.* 50:729–737.

57. Gomez-Foix A M, Coats W S, Baque S, et al. (1992). Adenovirus-mediated transfer of the muscle glycogen phosphorylase gene into hepatocytes confers altered regulation of glycogen metabolism. *J. Biol. Chem.* 267:25129–25134.

58. McGrory W J, Bautista D S, Graham F L (1988). A simple technique for the rescue of early region 1 mutations into infectious human Adenovirus Type 5. *Virol.* 163:614–617.

59. Bett A J, Krougliak V, Graham F L (1995). DNA sequence of the deletion/insertion in early region 3 of Ad5 dl309. *Virus Res.* 39:75–82.

60. Lehmberg E, Traina J A, Chakel J A, et al. (1999). Reversed-phase high performance liquid chromatographic assay for the adenovirus type 5 proteome. *J. Chromatog. B.* 732:411–423.

61. Lehmberg E, McCaman M, Traina J, et al. (2004). Analytical assays to characterize adenoviral vectors and theoretic applications. In G. Subramanian, (ed.), *Manufacturing of Gene Therapeutics*, Kluwer Academic/Plenum Press, Dordrecht, the Netherlands, pp. 210–225.

62. Murakami P, Pungor E, Files J, et al. (2002). A single short stretch of homology between adenoviral vector and packaging cell line can give rise to cytopathic effect-inducing, helper-dependent E1-positive particles. *Human Gene Ther.* 13:909–920.

63. Walsh P S, Varlaro J, Reynolds R (1992). A rapid chemiluminescent method for quantitation of human DNA. *Nucleic Acids Res.* 20:5061–5065.

64. Chroboczek J, Bieber F, Jacrot B (1992). The sequence of the genome of adenovirus type 5 and its comparison with the genome of adenovirus type 2. *Virology.* 186(1):280–285.

65. Isner J M, Vale P R, Symes J F, et al. (2001). Assessment of risks associated with cardiovascular gene therapy in human subjects. *Circulation Res.* 89:389–400.

66. Ylä-Herttuala S, Martin J F (2000). Cardiovascular gene therapy. *Lancet.* 355:213–222.

FURTHER READING

Isner J M (1997). Angiogenesis and collateral formation. In K.L. March, (ed.), *Gene Transfer in the Cardiovascular System: Experimental Approaches and Therapeutic Implications.* Kluwer Academic Publishers, Boston, M.A., pp. 307–330.

7.2

PLASMID DNA AND MESSENGER RNA FOR THERAPY

STEVE PASCOLO

University of Tübingen, Institute for Cell Biology, Tübingen, Germany

Chapter Contents

7.2.1 Nonviral Coding Nucleic Acids: Definition and Characteristics in Relation to Potential Therapeutic Use   971
    7.2.1.1 Historical Overview   971
    7.2.1.2 DNA: The Stable Genetic Information Store   974
    7.2.1.3 RNA: The MultiFunction Molecule   975
    7.2.1.4 Recognition of Nucleic Acids by Immune Cells   978
    7.2.1.5 Production of Plasmid DNA and Messenger RNA   979
    7.2.1.6 Packaging of Nucleic Acids for Therapy   980
7.2.2 Plasmid DNA and Messenger RNA for Therapy   982
    7.2.2.1 Design and Optimization of pDNA and mRNA Vectors   983
    7.2.2.2 Vaccination   986
    7.2.2.3 Gene Therapy   998
    7.2.2.4 Production and Regulations   1002
7.2.3 Conclusions and Perspectives   1004
    References   1006

7.2.1 NONVIRAL CODING NUCLEIC ACIDS: DEFINITION AND CHARACTERISTICS IN RELATION TO POTENTIAL THERAPEUTIC USE

7.2.1.1 Historical Overview

Extracted from the nucleus of human cells in the kitchen of Tübingen's castle (Germany), Nuclein is an heterogeneous product that was introduced in 1868 by Friedrich Miescher (1844–1895). Miescher's pupil, Richard Altmann (1852–1900),

Handbook of Pharmaceutical Biotechnology, Edited by Shayne Cox Gad.

further fractionated Nuclein and isolated in 1889 the substance that he named "nucleic acids." These molecules were characterized by Albrecht Kossel (1853–1927, winner of the Nobel Prize for Medicine in 1910). Kossel used chemical hydrolysis of nucleic acids to identify their basic components: the nucleotides. Five nucleotides are found in the Nuclein: two purines [adenylic acid (A) and guanylic acid (G)] and three pyrimidines [cytidylic acid (C), uridylic acid (U), and thymidylic acid (T)]. In 1910, Kossel could also distinguish between two families of nucleic acids present in Nuclein: one is degraded at high pH and contains A, C, G, and U; one is resistant and contains A, C, G, and T (Figure 7.2-1). The precise chemical structure of the nucleotides was deciphered by a student of Kossel: Phoebus Levene (1869–1940) together with Walter Abraham Jacob (1883–1967) in 1929. These researchers found that nucleotides consist of a nitrogen base (Adenine, Guanine, Cytosine, Thymine, or Uracile), a sugar (pentose), and a phosphate group. They also found that nucleotides are linked linearly in the order of phosphate-sugar thanks to phosphodiester bonds. This way, they generate a polymer with a phosphodiester backbone. The pentose in the high-pH-sensitive nucleic acid is a ribose, whereas in the high-pH-resistant nucleic acid, it is a 2'-deoxyribose. Based on this chemical difference are the names of the two nucleic acids: RNA for

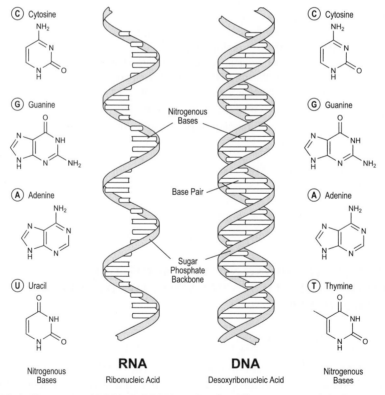

Figure 7.2-1. Structure of DNA and RNA molecules. The structure of the four nucleotides that compose RNA (on the left) or DNA (on the right) is shown. The representation of linear RNA and DNA molecules is shown in the middle. Actually, RNA makes a secondary structure through intra-strand base pairing (not shown).

the ribo-nucleic acid and DNA for the deoxyribo-nucleic acid. Although the Nuclein contains RNA and DNA, the cytosole of cells was shown by Tornbjörn Caspersson (1911–1998) and Jean Brachet (1909–1988) to contain only RNA. In 1953, Jim Watson (1928–), Francis Crick (1916–2004), and Maurice Wilkins (1917–), who received the Nobel Prize in 1962 for their resolution of the structure of DNA, demonstrated that, thanks to specific hydrogen bonds between their base moieties, nucleotides are complementary: A pairs with T or U using two hydrogen bonds and G pairs with C using three hydrogen bonds. Helped by the x-ray photograph of DNA (called photograph 51) obtained by Rosalind Franklin (1920–1958), they proposed that DNA is a molecule with two antiparallel strands shaped in a regular helix of a 2 nm diameter and 3.4 nm step (10 nucleotides for one turn) with its phosphodiester backbone on the outside and the bases on the inside of the helix (Figure 7.2-1). One year earlier, in 1952, Alfred Hershey (1908–1997, Nobel Prize winner in 1969) and Martha Chase (1927–2003) demonstrated using T2 phages that DNA is the genetic material in bacteria. There results confirmed the conclusions of Oswald Theodore Avery (1877–1955) who showed in 1943 that the "transforming principle" that can modify *Pneumococcus*'s phenotype from "smooth" to "rough" is DNA. By deduction and because of the location and behavior of DNA in eukaryotic cells, it was admitted, before being experimentally demonstrated, that DNA is the genetic material in higher organisms as well.

Meanwhile, the characterization of the cytosolic RNA-rich granules called ribosomes and identified by Albert Claude (1898–1983, Nobel Prize in 1974) indicated that RNA is involved in protein synthesis [1, 2]. Such a function of RNA was first hypothesized by Torbjoern Caspersson (1910–1997) and Jean Brachet (1909–1998) at the end of the 1930s and was then based on the relation between the RNA content of cells and their activity (proliferation). Ribosomes contain ribosomic RNA (rRNA) and are the site of protein production. The key intermediate between nucleic acids and proteins is the hybrid aminoacyl-RNA molecule called transfer RNA (tRNA) identified in 1958 [3]. It consists in a family of at least 20 members in any organism. A tRNA is a short (between 75 and 85 nucleotides) RNA molecule linked at its 3′ end to one amino-acid [4, 5]. Finally, the identification of an unstable RNA called messenger (mRNA) [6, 7] which is a copy of genes to be translated into proteins completed the basic description of the fundamental protein machinery. It is summarized in Crick's central dogma of molecular biology (Figure 7.2-2): DNA contained in the nucleus is transcribed into (messenger) RNA, which is translated into proteins [8]. Since the formulation of this central dogma, the molecular mechanisms involved in these processes have been characterized in detail. Of interest, we now know that the transient nature of all RNA molecules compared with DNA is not due to the chemical structure difference (in cells, pH are neutral or acidic: DNA and RNA are practically equally stable at the chemical level) but due to the abundance inside and outside the cells of potent and stable ribonucleases (RNases). We have also learned that RNA has more functions than initially postulated: It can replace DNA as genome in viruses and eventually (i.e., in retroviruses) be reverse transcribed into DNA. Moreover, several types of RNA molecules with various biological activities (structure, enzyme activities, containment of gene information) are involved in each step of gene expression (see Section 7.2.1.3). Nowadays, thanks to the molecular biology methods, we can also introduce designed foreign DNA and RNA molecules into cells (Figure 7.2-2). As proposed by Crick

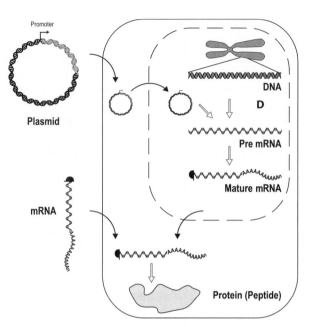

Figure 7.2-2. Central Dogma. The genes located on the chromosomes are transcribed into mRNA, which is exported from the nucleus to the cytosole where it is translated. Thanks to molecular biology techniques, it is possible to bring exogenous genetic information. Foreign DNA molecules must cross two membranes: the cytoplasmic and the nuclear envelopes, whereas mRNA must cross only the cytoplamic membrane to be active.

et al. in 1961, through a genetic code based on three letters (three consecutive bases: a codon), the nucleic acid information copied from the DNA genome into mRNA is translated in proteins. Each possible codon ($4^3 = 64$ possibilities) corresponds to one precise amino acid. This genetic code is universal: It is conserved between all living organisms. As there are only 20 amino acids, most of them are encoded by more than one codon. Some codons are favored (more frequently found in highly expressed genes) in a species-specific manner. For example, alanine, which is encoded by four codons would be preferably encoded by GCA in *Bacillus subtilis*, by GCC in *Homo sapiens*, by GCG in *Escherichia coli*, and GCT in *Saccharomyces cerevisiae*. The correspondence between codons and amino acids, i.e., the fundamental translation process, is performed by specialized enzymes called aminoacyl tRNA synthetases. There are at least 20 different aminoacyl tRNA synthetases in any organism. Each is responsible for loading one amino acid on the adequate tRNA(s). By transferring a specific amino acid on a specific tRNA, which has the adequate anticodon (complementary to the codon in the mRNA), aminoacyl tRNA synthetase convert genetic information (nucleotides) into protein information (amino acids).

7.2.1.2   DNA: The Stable Genetic Information Store

In eukaryotic cells, DNA molecules are located in the nucleus. During the interphase (between two cell divisions) of the cell cycle, DNA molecules are packed by histones into an heterogenic structure called the chromatin. Through intense condensation of the chromatin, each DNA molecule can be visualised as a chromosome during mitosis or meiosis. A human cell contains 23 pairs of DNA molecules that consist altogether ca. 3.15×10^9 bases. These molecules stretched would represent altogether a ca. 2-m-long chain. They have a stable nucleotide sequence that is eventually modified only at the ends (the telomeres). At this location, an inducible ribo–nucleo–protein complex that contains a reverse transcriptase (telomerase) uses a matrix RNA molecule to elongate the DNA molecules with a repetitive sequence motif. In proliferating cells not expressing the telomerase, the telomeres shorten at each division due to incomplete replication of DNA molecules at their extremities. Below a certain size of the telomeres (the Hayflick limit), the cell undergoes senescence and starts a self-destruction program (apoptosis). Thus the shortening of telomeres at each replication cycle is a method to "date" a cell: to know how many times it divided. For cells that need to proliferate (germ and stem cells, for example), the telomerase complex is induced and maintains or extends the telomeres' length, keeping the cell "young." Aside from the telomeres, the sequence of the chromosomic DNA molecules is unmodified in healthy cells. However, genomic DNA molecules have a dynamic structure in constant reorganization and modification: Methylations of residues, condensation into a closed chromatin configuration, opening into an accessible chromatin configuration, or changes of the double-helix structure (A, B or Z) can affect the availability of the DNA sequences for gene expression. Several important conserved sequences define a functional DNA stretch. Many of them have a role that is not yet characterized. Only ca. 2% of the human genome is functionally identified as being genes. These entities are characterized by a promoter that is the starting point of mRNA transcription. Promoters are recognized by the RNA polymerase II (the RNA polymerase responsible for the production of mRNA in eukaryotes). They have a minimal core sequence of a few nucleotides [usually a "TATA(A/T)A (A/T)" sequence called a TATA box or Hogness box], 19 to 27 bases upstream of the initiation site of the mRNA. The TATA box is usually surrounded by G- and C-rich sequences. Its efficacy largely depends on upstream proximal control sequences. The so-called "CAAT Box" (consensus GG(C/T)CAATCT) and "GC box" (consensus GGGCGG) enhance the activity of the promoter, i.e., its utilization by the RNA polymerase II. Distal upstream and downstream sequences called "Enhancers" also impact the activity of the promoter. An efficacious promoter in eukaryotic cells requires ca. 100 bases upstream from the TATA box. Highly specialized promoters such as "Locus Control Regions" that also affect the local chromatin structure and keep it in an opened configuration can increase the efficacy of transcription. Downstream of the promoter is the transcribed sequence that is segmented between exons (sequences that will be present in the cytosolic mRNA) and introns (sequences that will be spliced-out from the neo-transcribed mRNA in the nucleus). Transcription of mRNA stops at a transcription stop signal, which is usually a U-repeat. Before it leaves the nucleus, the immature mRNA is further processed as described bellow.

7.2.1.3 RNA: The MultiFunction Molecule

As opposed to DNA that has a well-determined unique function that is to stably store the genetic information, RNA is a pluri- (or even omni-)potent molecule.

RNA molecules can have all functions of leaving: storage of genetic information, enzymatic activity, and construction of defined three-dimensional structures. RNA is the only biomolecule that has all of these activities. Thus, it was proposed as the primordial macromolecule that, on its own, started the concept of life 4.2 billion years ago [9].

Structure-related Functions (rRNA, tRNA, aptamers). Because of its single-stranded nature, RNA refolds according to the most thermodynamically favored base-pairing. This way, it generates stem loops which position each other into complex three-dimensional structures. It is still hard to predict the dominant structures that will adopt a precise RNA sequence, but several algorithms can help in predicting them (see www.bioinfo.rpi.edu/applications/mfold/rna/form1.cgi).

The sequence-dictated three-dimensional conformation of an RNA molecule is essential in the functionality of several RNA types: The ribosomic RNAs fold into structures that scaffold the ribosomes, and the tRNAs fold into defined structures that allow their recognition by both amino-acyl tRNA synthetase and ribosomes. In both cases, some nucleotide sequences that remain single stranded are involved in the enzymatic processes performed by the molecules. Thus, the sequence of an RNA dictates its structure and allows it to perform its function. Both are strictly connected. One single mutation in a rRNA or a tRNA may completely change the favored three-dimensional structure of the molecule and totally impair the RNA's function(s).

The scaffolding characteristic of RNA molecules is used in drugs under the generic name of aptamers. Selected for their capacity to interact precisely with a target structure similarly to the antibody–antigen interaction, aptamers can specifically bind to any molecule or complex of interest. The first pharmaceutical aptamer is Macugen™ (Eyetech Pharmaceuticals, Inc., and Pfizer, Inc.). It recognizes and neutralizes specifically the vascular endothelial growth factor, thus inhibiting pathologic blood vessel growth. It is used as treatment against age-related acute macular degeneration.

Regulation of Protein Expression-related Functions (antisense, siRNA, miRNA). Small RNA molecules of ca. 20 nucleotides in length can remain linear with all their bases available to pair with a complementary sequence present on another RNA molecule. Such "antisense" RNA oligonucleotides can this way impair the function (block translation) of mRNA. Another type of short RNA molecules of ca. 20 nucleotides in length is double stranded and has two bases overhangs. It is called small inhibitory RNA (siRNA) when naturally produced in the cytosole out of foreign double-stranded RNA molecules or micro-RNA (miRNA) when encoded in the DNA genome. Through its recognition by a proteic complex called RISK, one strand is discarded (the sense strand) and the other strand (the antisense) is used as a guide to direct RISK to a complementary sequence in a mRNA and induce a cleavage on this target molecule. Thus, siRNA or miRNA are a catalytic version of antisense RNA: one molecule can inhibit (induce the degradation of) many mRNA. Some miRNAs can also modify the DNA sequence in the promoter region by inducing methylations. This way, they dictate the level of packaging into the chromatin and the expression of a gene. For more information on small RNA molecules as regulators of gene expression and their use as therapeutics, refer to Chapter 7.6.

Enzymatic Functions (ribozymes). Through a fixed three-dimensional structure and the availability of certain nonpaired residues, RNA molecules can fold into enzymes called ribozymes. The evidence of such an enzymatic capacity of RNA was discovered by Altmann [10] and Cech [11]. It was acknowledged with a Nobel Prize (Chemistry, 1989). Initially, this enzymatic activity was restricted to the capacity of ribozymes to recognize and cleave or religate in an autonomous way a target RNA (mRNA) sequence. Later, other enzymatic capacities of ribozymes were evidenced: Selected ribozymes can catalyse the synthesis of nucleotides [12] or the creation of amide bonds [13].

Coding Functions (mRNA). The pre-messenger RNA produced by transcription of the genomic DNA is processed before leaving the nucleus: Splicing of introns is made by spliceosomes that are ribo–proteic complexes. Within these structures, the RNA provides the sequence specificity of the spliceosome's activity. An exon does not have any sequence recognized by the spliceosome. It is the intron, and its borders that are recognized and excised. Evolutionarily, introns may derive from movable genetic elements called transposons. Thus, theoretically, introns can be introduced naturally (through evolution) or experimentally (thanks to molecular biology methods) within any coding sequence. They present the advantage that they enhance the efficacy of gene expression, probably through the facilitation of mRNA export to the cytosole. Moreover, thanks to regulated alternative splicing mechanisms, some specific exons may be spliced out with flanking introns. Such alternatively spliced mRNAs encode proteins with missing domain(s) and thus with different function(s). Meanwhile, the deregulation of the splicing mechanisms in dysfunctional cells, such as tumor cells, may allow introns to remain in the mature mRNA. Then, a protein with an extra domain and eventually truncated (if the intron contains a stop codon in the frame of translation) is generated. Aside from being spliced, the immature mRNA is also capped at its 5′ end (the addition of a methylated guanine followed by three phosphates bonds) and poly-adenylated at its 3′ end (the addition of several hundreds of adenine residues). The spliced, capped, poly-adenylated mRNA is exported to the cytoplasm through nuclear pores. In the cytosole, several elongation initiation factors recognize the Cap structure, whereas the poly-A binding protein binds to the 3′ poly A tail. These proteins allow the ribosomes to bind to the mRNA and start to scan from the 5′ end the nucleotide sequence until it finds a start codon (ATG) in a "Kozak" surrounding (A/GNNNAUGG). At this location, the first amino acid, always a Methionine, is brought by a tRNA that has the 5′ CAU 3′ anticodon. Then, the ribosome translocates toward the 3′ end to the next codon where an adequate tRNA will bring the second amino acid encoded by the gene. The process continues toward the 3′ end of the mRNA and stops when the ribosome meets a UGA, UAA, or UAG "termination" codon. The neosynthesized protein and the ribosomes are then released. Several ribosomes are sitting on one mRNA. According to its stability and efficacy of translation, one mRNA can produce many proteins. Ultimately, the mRNA is degraded by intracellular RNases. This catabolic process is highly regulated: Several sequences in the mRNA, and especially at its ends, before and after the coding sequence (UnTranslated Regions: UTR), have primary and secondary structures that are recognized by specific proteins in the cytosole. Those proteins may induce or, on the contrary, prevent the detection and degradation of the mRNA by the intracellular RNases, which are in general exonucleases [10].

This means that they degrade the mRNA starting by the 5′ or 3′ ends. Both ends are protected in the mature mRNA: the Cap on the 5′ and the poly-A tail covered by the poly-A binding proteins on the 3′. Specific and regulated mechanisms of decapping and deadenylation will render the mRNA accessible to the exonucleases. Accelerated degradation (shortening mRNA half-life) can be induced by the presence of AU-rich sequences called AUREs in the 3′ UTR, or by the presence of a specific restriction site. This latter possibility is illustrated by the regulation of iron metabolism: An iron-induced endoribonuclease recognizes a sequence present in the mRNA coding for the transferrin receptor (TfR) that is responsible for iron-loaded transferin uptake. The presence of high iron stocks in the cell induces the endonuclease, which cleaves the TfR-encoding mRNA. Thus, it prevents TfR expression and further iron uptake. Inhibition of mRNA degradation (increasing mRNA half-life) involves other untranslated sequences as illustrated with the very stable globin mRNAs: They contain in their 3′ UTR, pyrimidine-rich sequences that are recognized by a ubiquitous protein complex called the alpha-complex. It stabilizes the poly-A binding proteins associated with the poly-A tail and secures the integrity of the 3′ end of the mRNA, thus increasing the half-life of the whole RNA molecule. Depending on the presence of stabilization and destabilization sequences in its UTRs, its length, structure, and efficacy of translation, the intracellular mRNA can have a half-life varying from minutes to weeks [11]. Outside the cells, naked mRNA are degraded within seconds because of the abundant, ubiquitous, stable, and processive extracellular RNases [14]. The presence of such an efficacious machinery to degrade extracellular RNA may have been developed as a protection against viruses, especially those with RNA genomes. It may also control the secretion and recapture of RNA by neighboring cells, which is hypothesized to represent a way for cells to communicate [15].

7.2.1.4 Recognition of Nucleic Acids by Immune Cells

In eukaryotic cells, DNA is strictly in the nucleus and in energy-producing sub-compartments (mitochondria and chloroplast), whereas RNA is strictly in those same compartments plus the cytosole. In healthy situations there is no nucleic acid in the endoplasmic reticulum, golgi apparatus, endosomes or lysosomes, and associated vesicles. The immune system developed the tools to scrutinize for the presence of nucleic acids in those compartments because it would be associated with a pathogenic situation such as the presence of an infectious agent. A family of receptors similar to immunity-related receptors from flies called Tolls receptors is dedicated to the detection of such "danger signals." This family of Toll-like receptors (TLRs) contains 11 known members. Four of them were shown to recognize nucleic acids: TLR3 is specific for double-stranded RNA (dsRNA), TLR7 and TLR8 recognize single-stranded RNA (ssRNA), and TLR9 recognizes an unmethylated DNA pattern ("CpG motif") that is found specifically in the genome of microorganisms such as bacteria (Figure 7.2-3). These four receptors are located in the lumen of cytoplasmic vesicles (mainly endosomes) in immune cells like B cells, NK cells, granulocytes, macrophages, or dendritic cells (DCs). TLR3, 7, and 8 detect the infection by viruses such as influenza (dsRNA genome) or HIV (ssRNA genome), whereas TLR9 signals the presence of bacteria in the endosomes. Upon engagement of TLRs, immune cells are activated. They trigger an adaptive (involving T

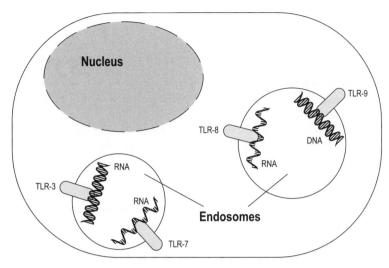

Figure 7.2-3. Recognition of nucleic acids by TLR. TLR 3, −7, −8, and −9 are located in the endosomes and recognize dsRNA (TLR3), stabilized ssRNA (TLR7 and TLR8), and bacteria DNA (TLR9).

and B lymhocytes) immune response specific for the antigens associated with or encoded by the foreign nucleic acid(s). This natural immune detection of exogenous nucleic acids is an advantage for the utilization of these molecules in vaccines but a disadvantage for the utilization of nucleic acids in gene therapy.

7.2.1.5 Production of Plasmid DNA and Messenger RNA

Plasmid DNA. Plasmid DNA (pDNA) are small (from 2 to 15 kb) circular DNA molecules that contain a bacterial origin of replication. They can proliferate within a bacteria and represent up to 1% of the microorganims's mass. They are manipulated to code for a protein that neutralizes an antibiotic. Moreover, they can contain any desired sequence, such as for example a promoter, a gene from bacterial or non-bacterial origin, and a poly-adenylation signal. Plasmids are purified from bacteria grown in the presence of the antibiotic that they can neutralize. This guarantees that the bacteria will conserve and amplify their plasmids during fermentation. Frequently, the antibiotic is ampicillin, which is deactivated by the protein β-lactamase encoded by many standard plasmid vectors. The bacteria are pelleted, usually by centrifugation, and cells are lysed, usually thanks to sodium dodecyl sulfate (SDS) and sodium hydroxide. After neutralization with potassium acetate, most proteins and the genomic DNA precipitate while the plasmids remain in the aqueous phase. Several methods exist to purify pDNA from these solutions, but the most commonly used is anionic exchange chromatography (see, for example, the protocole by Qiagen at www.qiagen.com). Protocols are available to eliminate most lipopolysaccharides coming from the bacteria and that may affect the functionality of the pDNA for therapeutic utilizations. However, standard purification methods will provide the pDNA with three topologies: Most is supercoiled circular DNA, but some is in the form of a relaxed circle (one strand has at least one nick) or linear (both strands were broken at the same place). The supercoiled plasmid is the form that should be used for research and therapy because, based on *in vivo* transcription studies, it is the most active form [16, 17]. Although the nicked pDNA molecules are most often not strongly affecting the functionality of the plasmid *in vitro* or *in vivo*, it is custom to use pDNA with a low amount of these topologies for preclinical and clinical utilizations. At laboratory scale, the purification of the supercoiled pDNA can be obtained using caesium chloride gradients. For larger scale and clinical trials, specialized companies offer the production of high-quality pDNA for research and therapy [Good Manufacturing Practice (GMP) certification; see, for example, plasmid factory at www.plasmidfactory.com]. One liter of bacteria culture can yield up to 30 mg of pDNA.

Messenger RNA. Messenger RNA are produced from purified pDNA using bacterial RNA polymerases in a reaction called "*in vitro* transcription." The pDNA used for *in vitro* transcription must contain a specific promoter recognized by the bacteriophage RNA polymerases T3, T7, or SP6 in front of the coding sequence of interest. It is linearized using a unique restriction site that must be at the end of the sequence to be transcribed. Then, the linearized plasmid is mixed with the RNA polymerase and the four nucleotides plus, in standard reactions, an excess of Cap analog (m7G(5′)ppp(5′)G) compared with GTP. Usually the Cap analog is at ca. 6 mM in the reaction; ATP, UTP, and CTP are at 4 mM; and GTP is at 1.5 mM in the final reaction. The RNA polymerase synthesizes more than 100 mRNA molecules from one pDNA template. After the synthesis of the mRNA, 1 hour at 37°C, the DNA is destroyed by a DNase and the mRNA is recovered by precipitation, usually with lithium chloride. Afterward, the mRNA is resuspended in the adequate solution.

7.2.1.6 Packaging of Nucleic Acids for Therapy

Although some cells in skin, muscle, brain, and so on can spontaneously take up exogenous injected naked pDNA or mRNA [18], this process is relatively inefficient and can consequently be used only when a local expression of a protein is desired, i.e., for vaccination. Even if physical methods exist to enhance the uptake of naked nuclei acids (electroporation or hydrodynamic delivery; see Section 7.2.2.2), increasing the efficacy of transfection *in vivo* is best reached using encapsulation of the nucleic acids. Two methods are used: cationic polymers or liposomes.

Polymeric Particles. Cationic polymers can spontaneously associate with the anionic nucleic acids. Such complexes can keep a positively charged surface that allows an interaction with the negatively charged cell surfaces. Commonly used cationic polymers for the production of micro- or nanoparticles include chitosan, protamine, poly-L-lysine (PLL), poly-ethylenimine (PEI), and polyamidoamine dendrimers [19]. Usually, a simple mixing of the cationic polymers and the nucleic acid, eventually with high-speed homogenization, is enough to obtain particles. Thus, production is very easy and upscalable. However, some of these cationic polymers, such as PEI or PLL, are toxic or have secondary effects (for example, chitosan induces hypocholesterolemia). Moreover, particles generated by simple mixing of nucleic acids and cationic polymers may be very heterogenous in size and physical properties, making them hard to qualify for pharmaceutical use. A more controlled method to produce particles is based on poly(lactide co-glycolide) (PLG). It is a biodegradable and biocompatible molecule used for several biomedical products including sutures. As it is a negatively charged molecule, it needs to be mixed with cationic polymers such as those listed above or chemical products [cetyltrimethylammonium bromide (CTAB), dimethyl dioctadecyl ammonium bromide (DDA), or 1,2-dioleoyl-1,3-trimethylammoniopropane (DOTAP), for example]. Optimally, a mixture of several reagents allows the production of homogenous particles with a defined size (from ca. 100 nm up to μm) and surface charge (zeta potential). When using PLG, protocols based on the solvent evaporation process are used [20]. Cationic particles can be manufactured and then coated with the nucleic acid or produced in the presence of the nucleic acid. In the first case, the genetic material is on the outside of the particles, and in the second in case, it is inside. Because the production of the PLG particles involves high shear, organic/aqueous interface, high temperature, and freeze-drying, the nucleic acid may be denatured or degraded during the process. Thus, for a production that is reliable, reproducible, and easy to upscale in GMP conditions, the coating of nucleic acids on the surface of premade cationic PLG particles is preferred. In both cases, the nucleic acid associated to the particles is protected from the activity of nucleases and can be applied by injection (the particles are phagocytosed by cells) or, for vaccination, by gene-gun [21]. The release of the nucleic acid is largely determined by the nature of the polymers used for the preparation of the particles. *In vitro*, the particles may readily release quickly a significant amount of nucleic acid in the injection solution. For example, PLG/CTAB microparticles released ca. 35% of their loaded pDNA within 1 day [20]. For this reason, particles should be stored lyophilized and injected quickly after reconstitution in the injection's buffer. *In vivo*, the biodegradability of the polymers and the site/method of delivery will dictate the release of the genetic information. For some particles, the release may be performed slowly, over several days [20]. Polymers with specific degradation patterns (for example, in endosomes: at low pH) can be engineered to increase the efficacy of delivery and regulate the time frame in which nucleic acids are available for expression within the cell. Some layer-per-layer production processes allow the production of particles with several layers of nucleic acids separated by layers of polymers (see www.capsulution.de). The choice of the polymer(s) will dictate the degradation profile of the particles and thus the release of the several layers of nucleic acids. Theoretically, such a strategy allows the slow release of nucleic acids within cells over a long time.

Liposomes. Liposomes are vesicles that consist of an aqueous compartment enclosed in a phospholipids bilayer. As they lack proteinaceous components, liposomes are not immunogenic. A broad variety of cationic liposomes containing eventually modified lipids have been used to deliver pDNA and mRNA for therapeutic utilizations. These formulations are usually based on a cationic lipid such as 1,2-dioleol-3-trimethylammonium propane (DOTAP) and a fusogenic lipid such as 1,2-dioleoyl-sn-glycero-3-phosphoethanolamine (DOPE) [22, 23]. Despite their ease of production and efficacy *in vitro* for transfecting cultured cells, cationic liposomes face toxicity issues *in vivo* [24, 25]. Meanwhile, their heterogeneous size and instability are drawbacks for a pharmaceutical production and qualification. Moreover, cationic liposomes are often inactivated in the presence of serum. As possible improvements of cationic liposomes, immunoliposomes, which are liposomes coated with an antibody, are developed. Thanks to their targeting, they are more efficacious than classic liposomes and can consequently be administered at a lower dosage. This way, the immunoliposomes are less toxic. Another variation of liposomes are called stealth liposomes. They consist in poly(ethylene glycol) (PEG)-coated liposomes. PEG prevents adsorption of the liposomes by endothelial cells after intravenous delivery. Thus, stealth liposomes are remaining in the circulation and are distributed more homogenously through the body than classic liposomes.

7.2.2 PLASMID DNA AND MESSENGER RNA FOR THERAPY

Plasmid DNA (pDNA) and messenger RNA (mRNA) are attractive tools for therapy because they are simple molecules that are easy to produce, store, detect, and manipulate using molecular biology techniques. They can be used to quickly turn into a drug any genetic information. Alternatives to pDNA and mRNA are virus- or bacteria-derived vehicles that can contain the genetic information to be expressed as a drug. The drawbacks of these methods are as follows:

1. Safety issues: The modified infectious agents may revert to a wild-type phenotype or may recombine with wild-type viruses to produce new sorts of pathogens.
2. Specificity: Not only the gene of interest is injected into a patient but also the proteins and eventually parts of the genetic information of the pathogen. The function of these proteins may interfere with the function of the transgene (in vaccination, for example, some pathogen-specific mechanisms may interfere with the priming of the immune response against the transgene), and moreover, an immune response against the vehicle (virus or bacteria) may be dominant. Not only could this hide the response against the relevant antigen for vaccination, but it could also prevent the repeated or chronic delivery of the drug for gene therapy.

Plasmid DNA and mRNA, which are both nonviral coding nucleic acid vehicles, are minimal genetic vectors: in their human host, they express only the therapeutic protein of interest. The two main utilizations of pDNA and mRNA for therapy are as follows:

1. Vaccination using simply the gene sequences of pathogens (viruses, bacteria, parasites, or tumors).
2. Genetic complementation after the gene responsible for a genetic disease is characterized.

Vaccination is by far the most exploited feature of nonviral coding nucleic acids. This is due to

1. The fact that a local, transient expression of a foreign protein is enough to trigger an immune response.
2. The natural capacity of pDNA and mRNA to stimulate the immune system through TLR.

As the production of nucleic acids is barely affected by their sequence, generic production facilities and methods were developed to produce GMP-certified pDNA or mRNA. Several specialized bioindustries throughout the world can supply within a few weeks any nucleic acid to be used in therapy. This feature is unique because all other polymeric biomolecules used in therapy (synthetic peptides, recombinant proteins, or viruses) face more or less severe production and storage issues that are changing depending on their sequence. Thus, pDNA and mRNA seem to be the most versatile and simplest biomolecules to be used as active pharmaceutical ingredients in new drugs. Both nucleic acids were used in clinical trials and are known to be safe. Efficacy was reported, but pDNA- and mRNA-based drugs should be optimized before such treatments are available on the market.

7.2.2.1   Design and Optimization of pDNA and mRNA Vectors

Coding Sequence. As mentioned in Section 7.2.1, most amino acids are encoded by several codons, some of them being more favorable because they correspond to more abundant tRNA. Unfavorable codons or "rare" codons (as opposed to frequent, i.e., favorable codons) force the ribosome to pause during translation, and this may result in the premature ending of translation that generates abortive proteins. In any case, unfavorable codons decrease the efficacy of translation of full-length proteins. This phenomenon is particularly pronounced when the vaccine nucleic acid is a copy of a gene from a non-mammalian organism: Genes from bacteria or parasites have a codon usage very different from mammals. Moreover, some human virus genes and some endogenous human genes have an unfavorable codon usage that may be part of the mechanisms of regulation of their specific expression. For example, the HIV GAG gene has a "bad" codon usage that will limit the efficacy of gag production. This is actually important in the regulation of the virus replication's cycle. Thus, for human genes and more importantly for nonhuman genes, a codon optimization should preferably be performed in the coding part of therapeutic nucleic acids. The design of the gene is performed starting from the protein sequence to be expressed. Through algorithms that are based on the precise human codon usage a coding sequence will be proposed that should be ideally translated by the human cell machinery. The corresponding gene is synthesized by assembly of synthetic oligonucleotides. This process has another advantage, which is to generate a totally synthetic gene that can be very well documented and consequently easily introduced in a GMP manufacturing process (see Section 7.2.2.4).

Noncoding Sequences. The nucleic acid vector can be optimized in the noncoding (untranslated) parts of the mRNA. Mainly, as mentioned in Section 7.2.1, the AURES destabilizing sequences should be deleted because they would destabilize the mRNA and limit the expression of the antigen. In general, only the coding sequence of the gene of interest should be used for constructing the therapeutic vector; any sequence upstream of the ATG and downstream of the stop codon should be avoided. Meanwhile, stabilization thanks to 3′ UTR [26] in the form, for example, of the ca. 180 bases or ca. 80 bases UTR from the β- or α-globin genes, respectively, can be added after the coding sequence of interest (after the stop codon). Both of these sequences are recognized by the ubiquitous α-complex, which stabilizes the mRNA and enhances its translation rate. They are routinely used in the pDNA and mRNA vectors used for therapy [27–30].

Specific Optimization of pDNA Vectors. The promoter used for the expression (transcription) of the transgene offers the possibility to control the level, site, and eventually time of expression of the transgenic protein. Concerning the level of expression, strong ubiquitous promoters helped by enhancers will be adequate when a high and broad expression of the transgene is needed. However, if an expression of the transgene is expected in specific cell types, investigators can use the large panel of restricted promoters that are active thanks to cell-type-specific transcription factors. This way, the expression of the transgene can be restricted to the liver (albumin promoter, for example), lymphocytes (immunoglobulin promoter or CD2 locus control region, for example), antigen presenting cells (fascin promoter for example), and so on. The drawback of these promoters is that they may lead to a relatively weak transcription activity compared with ubiquitous promoters. The control of time of expression is a critical parameter for all pDNA-based approaches because a long-term expression may have serious negative effects on the efficacy of the treatment (immune tolerance instead of immune activation, for example) or unaccept-able side effects (overexpression of growth factors or other proteins with physiological activity that are used in gene therapy). A way to finely tune the expression of a transgene is to use an inducible promoter such as the Tet-On or Tet-Off promoters [30a]. Their activity depends on the Tet-On or Tet-Off transcription factors that must be constitutively encoded by the therapeutic pDNA construct. The Tet-On and Tet-Off proteins should be expressed thanks to ubiquitous promoters. The presence of tetracyclin or its analog Doxicyclin, which can just be added in the drinking water, will activate (Tet-On) or deactivate (Tet-Off) the transcription capacity of the Tet protein. Accordingly, it will turn ON (Tet-On) or OFF (Tet-Off) the transcription of the transgene of interest. The drawback of these systems is that they are "leaky," thus always giving a basal expression of the transgene even in conditions where it should be silenced (in the presence of Doxicyclin for Tet-off and in its absence for Tet-On).

The other functional parts of the plasmid such as the transcription termination signal, the bacterial origin of replication, and the antibiotic used for production of the plasmid do not impact the efficacy of the pDNA as a therapeutic vector. However, the antibiotic-resistance gene expressed by the plasmid when it is in a bacteria (the gene is expressed thanks to bacteria-specific controlling sequences) is relevant for the regulatory aspects (see Section 7.2.2.4). Moreover, the overall sequence of the plasmid is active at the level of immune stimulation: The more the plasmid contains unmethylated CpG immnostimulating motifs, the more it may induce signaling through TLR9 in human immune cells, and consequently, the stronger may be the priming of the immune response. Thus, for vaccination approaches, pDNA vectors should be enriched in CpG motifs, whereas for gene therapy approaches, pDNA should be depleted of such sequences.

Specific Optimization of mRNA Vectors. As mentioned, a 3′ globin (α or β) UTR at the end of the mRNA provides higher intracellular stability to the mRNA, which is of advantage whether the mRNA is the vaccine vector or a product of the vaccine vector (with pDNA vaccines as mentioned in the above paragraph). Aside from this, specific modifications of the mRNA vector are the poly-A tail and the 5′ Cap structure. Concerning the poly A tail, it should be of a minimum of 30 residues and preferably exceed 100 residues [31]. To generate the poly-A tail, there are two methods: Either it is encoded in the plasmid DNA that is used as a matrix for *in vitro* transcription or it is added by a specific polymerase (Poly(A) polymerase) at the end of the mRNA transcripts. In the first case, a A-tail can be of a maximum of ca. 120 As because more residues would titrate the adenosine triphosphate in the transcription reaction and limit its output. The other method is to submit the *in vitro* transcribed mRNA to the activity of the Poly(A) polymerase, which is a commercially available enzyme that polymerizes A residues at the 3′ end of a ribonucleic acid, in a matrix-independent manner. The enzyme can add several hundred A residues at the 3′ end of the mRNA molecules. This method is preferred to produce efficacious and stable mRNA for research purposes. However, the production of such poly-adenylated mRNA in GMP conditions is limited by the fact that the poly-adenylation process is not totally controlled and may generate heterogeneous molecules with more or less long A tails. This could make endproduct controlling, characterization of a batch, and batch-to-batch reproducibility more difficult than when the poly-A tail is encoded in the plasmid and is consequently clearly defined in length.

As mentioned in Section 7.2.1, at the 5′ end of mRNA molecules to be translated by eukaryotic cells, there is the characteristic Cap structure. It consists of a 7-methylated guanine residue linked through three phosphates to the first residue of the mRNA. A noncapped mRNA is not translated in eukaryotic cells. Not only is this structure essential for recognition of the mRNA by the translation machinery (the cap is recognized by the initiation factor 4: eIF4), but it also protects the 5′ end of the mRNA against exonucleases. Similarly to what was described for the poly-A tail, either the Cap structure is incorporated during the transcription process or it is added afterward on the *in vitro* transcribed mRNA.

In the first case, a fourfold excess of synthetic Cap analog (m7G(5′)ppp(5′)G) versus GTP in the transcription mixture guarantees that ca. 80% of the produced molecules will start with a Cap instead of a canonical G residue (the T7, T3, or SP6 promoters are made in a way that they will start transcription with a G residue). The drawback of this method is that the GTP content in the transcription reaction must be lowered compared with the other residues and thus becomes the limiting factor for the transcription. It significantly reduces the amount of recovered mRNA compared with an *in vitro* transcription reaction without Cap and with an identical concentration of all four residues. However, this method to produce capped mRNA is efficacious, fast, and reliable. Some modifications of the Cap analog can improve the efficacy of the mRNA vector: The standard synthetic Cap is a di-guanine molecule with one of its G methylated and three phosphates in-between the nucleotides. It can be incorporated at the start of the mRNA in two orientations: the methylated G being at the 5′ end or the non-methylted G at the 5′ end. Only the former situation will give a functional mRNA. Thus, statistically half of the capped mRNA molecules are nonfunctional. Strategies to avoid this phenomenon include the anti-reverse Cap structure (ARCA) [32]. It consists in a Cap analog in which the sugar carrying the methylated guanine is modified in its 3′ position (OCH₃). Thus, the dinucleotide can be used as the start nucleotide by the RNA polymerase only in the functional orientation where the 3′OH of the non-methylated based is used to extend the mRNA. ARCA Cap enhances the functionality of *in vitro* transcribed mRNA vectors. However, ARCA Cap is relatively difficult to synthesize and consequently more costly than the standard Cap analog.

The second method for capping mRNA takes advantage of the activity of the vaccinia virus capping enzyme, also known as guanylyltransferase. This enzyme is commercially available. In the presence of GTP and S-adenosyl methionine (SAM), it can add a natural Cap structure (7-methylguanosine) to the 5′ triphosphate of a RNA molecule. As it is an enzymatic reaction, it can bring a correct Cap to all mRNA molecules, and thus, it is optimal compared with *in vitro* transcription in the presence of standard Cap but similar theoretically to *in vitro* transcription in the presence of ARCA Cap.

Because of the lack of a reliable quantitative assay to control for the presence of the Cap structure, aside from a functional assay of the mRNA expression (transfection of the mRNA in cells and detection of the translated protein), the enzymatic capping of mRNA is rarely used to produce mRNA for research or therapy. The utilization of the synthetic Cap (nonmodified analog or ARCA) in the *in vitro* transcription reaction is the standard method to produce capped mRNA.

Aside from the 5′ and 3′ ends, some internal modifications of the mRNA such as a phosphorothiate backbone or 2′ modifications of residues can be brought into the mRNA as long as the modified residues are substrate for the RNA polymerases routinely used for *in vitro* transcription (T7, T3, or SP6 RNA polymerases) and do not interfere with the translation process. It seems that only a few modifications such as phosphorothioate nucleotides and 2′ amino residues can be used [14]. However, they lower the efficacy of transcription and translation while they do not noticeably increase the stability of the mRNA toward RNases. Some more work may allow the identifications of modifications that may allow the efficacious production of mRNA, which resists extracellular RNases and remains well translatable by the ribosome. Such mRNA would be optimized compared with native mRNA for therapeutic uses.

7.2.2.2   Vaccination

Introduction. Upon entry of a pathogen (virus, bacteria, or parasite) or upon pathologic genetic modifications of a cell of the body (tumors), pathogen associated molecular patterns (PAMPs) or "danger signals" such as dsRNA or stabilized ssRNA (virus), CpG DNA or flagelin (bacteria), uric acid, or heat shock proteins (tumors), for example, are recognized by immune receptors such as TLR receptors or scavenger receptors. These activating receptors are expressed by cells of the innate immunity such as macrophages and DCs that are professional antigen presenting cells (APCs). Some PAMPs also stimulate B cells, NK cells, or NKT cells. Professional APCs loaded with a pathogen's derived material and stimulated by the associated PAMP migrate from the site of detection of the danger signal to the draining secondary lymphoid organs such as lymph nodes and activate antigen-specific T and B lymphocytes through their clonotypic receptors: T-cell receptor

(TCR) for the former and surface immunoglobulin for the latter. There are approximately 5×10^9 T cells (ca. 70% are CD4 positive helper cells and ca. 30% are CD8 positive cytotoxic cells called CTLs) and 2×10^9 B cells in an adult human body. Generally, only a few hundred naïve lymphocytes specific for a precise antigen are available (frequency ca. 10^{-7}) [33]. Upon an encounter with the antigen-loaded activated APC, specific lymphocytes are triggered and proliferate. A lymphocyte clone may expand so much that it reaches a frequency of 10^{-2} [34]. Activated lymphocytes become armed effector cells. Armed B lymphocytes secrete the clonotypic immunoglobulin that sprays through the body, recognize its antigen, and inactivate it or target it to destruction by effector mechanisms of the innate immunity. Armed CD4-positive T lymphocytes secrete cytokines (mainly interferon-γ for Th1 cells and Il-4 for Th2 cells) that are necessary for immune cells' activity and they stimulate APCs. Armed CD8-positive T lymphocytes migrate to the site of antigen expression and kill cells expressing it. Upon disappearance of the antigen (complete elimination of the pathogen), armed effector cells disappear (contraction of the immune response) and some memory cells remain for the individual's whole life. Not only does this memory cell allow a higher number of specific lymphocytes to be present in the body at an eventual second appearance of the antigen than at the time of the primary encounter, but also memory lymphocytes have the peculiarity to react very quickly and efficiently (higher affinity of the antibodies for example). Thus, in the case of reappearance of the antigen, the neutralization of the pathogen (infectious agent or tumor cells) will be much faster than at the time of first priming. Prophylactic vaccination mimics a first contact between the immune system and a pathogen. It guaranties that, should the pathogen be present in the body, its recognition and elimination will be efficient and fast, preventing acute or chronic (persistent) disease. Therapeutic vaccination or immunotherapy aims at boosting the more or less preexisting immune response (triggered naturally by the already present pathogen), sustaining its strength (stimulating the proliferation of specific lymphocytes) and broadness (stimulating a larger spectrum of lymphocytes specific for the pathogen) to enhance its efficacy. As shown for several pathogens such as for example, Hepatitis B Virus (HBV), Human Immunodeficiency Virus (HIV), or tumor, a strong immune response against the pathogen is associated with a better control of the disease. High morbidity and mortality is often due to a weak immune response against those pathogens. Thus, immunotherapy aims at enhancing the natural capacity of the immune system to help control the pathogens. Another type of vaccination is used to shift or tolerize a preexisting pathologic immune response: in allergies or autoimmune diseases. Those immunotherapeutic regimen allow the manipulation of a preexisting immune response to render it armless.

Nucleic Acids for Vaccination. In all cases where the antigen of interest is a protein, nucleic acids can be used as a method to vaccinate. An adequate pDNA or mRNA vector can be produced *in vitro* (see above Section 7.2.2.1), formulated for the delivery (see Section 7.2.1.6) and eventually co-injected with an adjuvant in the form of a danger signal or cytokine. The local expression of the foreign nucleic acid-encoded protein associated with the local activation of innate immunity will trigger a T- and B-lymphocyte-mediated antigen-specific immune response that can protect against a challenge with the pathogen (infectious agent or tumor cell) that expresses this protein. It can also modify an existing pathologic immune response (autoimmunity, allergy) to render it armless. Both pDNA- and mRNA-based vaccines were reported to be capable of inducing humoral and cellular immunity (review by Liu and Ulmer for pDNA [35] and Pascolo for mRNA [36]). The particularly advantageous feature of this vaccination technique is that because antigens are made by the cells of the body, their epitopes are presented on major histocompatibility complex (MHC) I molecules and a CTL response is generated. Other vaccine methods based on proteins or inactivated infectious agents do not usually efficiently induce CTL. These cells are particularly relevant for antivirus, anti-intracellular pathogen and antitumor immunity. Thus, nucleic-acid based vaccines are theoretically very attractive modalities in the context of these pathogens.

It is usually accepted, although not firmly demonstrated, that the stronger is the transgene's expression (the highest is the amount of produced antigen), the better is the triggering of the immune response. Thus, enhancing the injected nucleic acid's translation (see 7.2.2.1) is a common method for improving the efficacy of nucleic acid-based vaccination. Improving the efficacy of the processing and presentation of the coded antigen was also reported to strongly impact the vaccine's potential (see below). However, of foremost importance is probably the site of delivery, the formulation, and the used adjuvant, which have a very strong and documented impact on the efficacy of the pDNA- or mRNA-based vaccines.

Optimization of the Antigen Processing and T-cell Priming. When the antigen encoded by the nucleic acid is to be recognized by T lymphocytes, its derived peptide epitopes must be efficaciously presented by MHC class I or class II molecules. Although any cytosolic, membrane or secreted protein can be used to produce MHC I and MHC II associated epitopes, several methods to enhance this presentation were described.

For augmenting the presentation by MHC I molecules, a sequence coding for ubiquitin can be inserted at the beginning of the gene. The produced protein will be cleaved after the ubiquitin, thus exposing a non-methionine N-terminal residue [37]. Should this residue be a "destabilizing" one, the protein is quickly degraded by the proteasome and its derived epitope efficiently presented on MHC I proteins. Another method to enhance MHC I epitope presentation is to insert a sequence coding for Herpes Simplex Virus (HSV) VP22 protein or heat schock protein 60 or calreticulin at the beginning of the gene [38–40]. HSV VP22, HSP60, and calreticulin moieties have the particularity to enhance MHCI-antigen presentation of the conjugated protein.

For augmenting the presentation by MHC II molecules, the engineering of genes encoding Invariant chain (Ii) [41] or lysosome-associated membrane protein-1 (LAMP-1) [42] fused to the antigen is recommended. Both Ii and LAMP-1 can direct the foreign nucleic acid-encoded protein to the cellular compartments where the degradation of antigens and the loading of MHC II molecules take place.

Shipping the antigen to the extracelullar compartment and at the same time targeting it to APC is another method to enhance T-cell priming: Whatever cell expresses the foreign nucleic acid, the antigen will be released and captured by APCs. To this end, the antigen can be fused for example to CD40L, Flt-3L, or CTLA4 [43–45] because these molecules bind to APC-specific surface proteins.

These strategies enhanced the efficacy of pDNA-based vaccines but may have the theoretical drawback of triggering an immune response against the self-protein used as fusion. Such an immune response would lead to autoimmune disorders targeting the immune system.

Finally, for enhancing the priming of the T cell, especially when the immune system may be tolerant (self-proteins used in anticancer vaccination), the addition of a "foreign" sequence to the antigen such as for example the Pan DR Epitope (PADRE: AKFVAAWTLKAAA) [46] or the utilization of a xenogenic antigen (the sequence of the antigen in species other than human) that has a ca. 90% homology to the nominal antigen are shown to increase the vaccine potential of nucleic acids [47, 48].

Specific Functional Optimization of pDNA Vectors for Vaccination. For a good expression of the vaccine transgene, a strong promoter must be used. Investigators can chose between (1) ubiquitous promoters such as the commonly used promoter of the early genes of CMV, eventually enhanced using additional regulatory elements (Operators [49]); and (2) cell-type specific promoters such as those that can promote transcription only in APCs (promoter exclusively used by dermal dendritic cells such as the fascin gene [50], for example) or locus control regions [12]. In the first case, independently of the site of injection and of the cell type that take up the pDNA, there will be a local expression of the protein, and through direct presentation (when the plasmid was taken by APC) or cross-presentation (when the plasmid was taken up by a somatic cell and the produced protein phagocytosed by neighboring APCs), specific lymphocytes are activated. In the latter case, expression of the antigen is limited to APCs. This has several advantages: As APCs are terminally differentiated cells, they cannot persist or proliferate; thus, long-term uncontrolled expression of the transgene and risks of transformation (when the transgene is an oncogene to be used in antitumor vaccination) are excluded. The induction of immune tolerance due to long-term expression of the antigen by somatic cells is also excluded. Moreover, the delivery of vaccine plasmids that have an expression restricted to APCs was reported to enhance the efficacy of vaccination compared with ubiquitously expressed plasmids [13].

Should the plasmid contain no CpG motifs, the introduction of some of these sequences outside of the functional sequences (i.e., origin of replication, antibiotic resistance, and relevant gene expression cassette) would be needed to render the plasmid functional for vaccination purposes.

Delivery of Nonviral Coding Nucleic Acids for Vaccination

Injection of Naked Nucleic Acids. Although *in vitro* cultured cells do not usually take up naked nucleic acids (pDNA or mRNA) spontaneously, skin cells and muscle cells do in vivo. As published in the milestone article by Wolff et al. [18] and presented in Figure 7.2-4, the direct injection of pDNA or mRNA resuspended in an isotonic buffer results in the local uptake of the nucleic acid that can be visualized by the expression of the encoded protein (luciferase, EGFP, CAT, or β-galactosidase are easily detectable markers) or by the induction of an immune response directed against the encoded protein. This is true whether the injection is made intramuscular, intradermal, or subcutaneous (review by Wolff and Budker

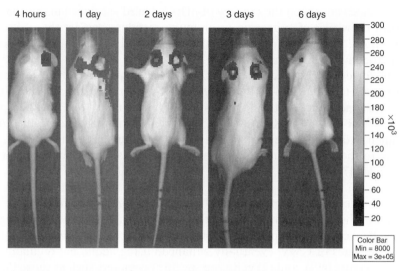

Figure 7.2-4. Expression of luciferase *in vivo* after injection of nonviral coding nucleic acids. Mice were injected in the ear pina with 10 μg of naked mRNA (right ear "R") or 10 μg of naked pDNA (left ear "D") coding luciferase. At the indicated time points, mice received a systemic injection of luciferin and were monitored by an *in vivo* imaging system for luciferase-mediated light emission. The injection of mRNA quickly results in strong luciferase expression at the site of application and totally vanishes after a few days. Plasmid DNA gives an expression at the site of injection that peaks later than the one obtained with mRNA injections, and that remains for a long time (up to weeks, depending on the injections).

[50]). To get an immune response, standard injections in mice consist of 50 to 200 μg of nucleic acids. A great optimization is obtained by direct intra-lymph-node injections, which proved in preclinical models to be much more efficacious than other routes, i.e., giving similar immunity than a 100-fold higher dose injected intramuscularly or intradermally [51]. Such a method was evaluated in clinical trials and found to be a feasible and safe approach [52].

The mechanisms and molecules necessary for the spontaneous uptake of exogenous nucleic acids are not fully characterized. However, it is probable that both pDNA and mRNA are phagocytosed, eventually in a receptor-mediated way (the reporter gene expression, after pDNA or mRNA injection, is a saturating process that can be competed with irrelevant nucleic acids), and travel through endosomes before they can reach, in an unknown way, the cytosole and, for pDNA, subsquently the nucleus (Figure 7.2-2). As pDNA and mRNA vectors contain intrinsic danger signals (nonmethylated CpG sequences recognized by TLR9 for the former and any nonmodified ribonucleic acids recognized by TLR7 and TLR8 for the latter) that are detected in the endosomes of cells that have phagocytosed the exogenous nucleic acid (Figure 7.2-3), the expression of the encoded protein parallels a local activation of the immune system. This results in the development of a B-cell- and T-cell-mediated immune response specific for the protein encoded by the injected nucleic acid. After the antigen is not expressed any longer, the triggered immune response becomes a memory response that provides an advantage to the treated individual, should he be re-exposed to the nominal antigen expressed by a pathogen.

An *in vivo* electroporation method was developed and tested in order to enhance the penetration of injected naked nuclei acids in cells at the site of delivery (for review, see Prud'homme et al. [53]). After the injection of the nucleic acid, a short electric pulse is given through the injected tissue thanks to electrodes connected to a programmed generator. This method allows a strong enhancement of the uptake of the injected pDNA. So far, however, it has not been successful in enhancing the uptake of injected mRNA (unpublished personal observation). For pDNA vaccination, *in vivo* electroporation is becoming a standard method to enhance the efficacy of vaccination in mice and it is being tested in a phase I trial in humans (see www.clinicaltrials.com).

Delivery of Vaccine Nucleic Acids Using Particles or Cells. One problem when using injections of naked nucleic acids for vaccination is that a big amount of vector is needed to vaccinate mice and *a fortiori* humans. This is partly due to the quick degradation of the extracellular vector (half-life within seconds for the mRNA [14] and eventually minutes for pDNA [54]) and its relatively poor uptake by cells. Only a few thousand plasmids are expressed at the site of injection [55]. For vaccination, three strategies were developed that can protect the nucleic acid and/or deliver it efficiently to its functional site (cytosol for mRNA and nucleus for pDNA): Gene Gun, *in vitro* transfection of professional APCs before adoptive transfer, and finally encapsulation in particles or liposomes.

GENE-GUN. The gene-gun technology is based on a ballistic delivery. The nucleic acid is precipitated on micrometric gold particles that are resuspended in a buffer and quickly injected into an horizontal tubing of 4 mm in diameter and 75 cm in length. The particles are left there to sediment. Thereafter, the buffer is slowly discarded and the tube is turned 180° along its axis in a way that the initial bead's bed is on the "roof" and disperses slowly through gravity on the inside of the tubing, covering it homogeneously. After it is dried in the inside, the tube is chopped into the size of cartridge (ca. 1.25-cm-long tubes). The cartridges are placed in the cartridge holder of a gun that is connected to pressured helium. Upon firing, the gas goes through the cartridge and propels the beads with high speed. Should the gun be a few centimeters from the skin, the beads penetrate the *stratum corneum* and stop in the dermis. They release the nucleic acid in the hit cells or cells' nuclei. Within these cells are dermis DCs and Langerhans cells. They are activated probably through signaling by TLR9 (pDNA vaccines [56]) or TLR7 and TLR8 (mRNA vaccines [57]) and migrate to the draining lymph node where they can activate antigen-specific lymphocytes.

ADOPTIVE TRANSFER OF *IN VITRO* TRANSFECTED APCs. As the key event in vaccination is the presentation of the antigen and its derived epitopes by professional APCs, researchers investigated the possibility of generating a pure APC population out of blood cells and of modifying these cells with exogenous nucleic acids before re-implanting them in the body. Because the transfection of pDNA in APCs such as DCs is inefficient and, when successful, induces apoptosis, the methods of genetic modification of APCs were principally developed using mRNA vectors. As APC, the monocyte-derived dendritic cells are most often used. They are easily produced, even in GMP conditions, starting from blood monocytes that are extracted from total blood either using their characteristic to stick quickly on plastic surfaces or by sorting (magnetic beads or FACS) with monoclonal antibodies specific for monocytes such as CD14. The monocytes are cultivated 1 week with GM–CSF and interleukine-4 to generate immature DCs. These cells can be easily transfected with mRNA using simple co-incubation or, for a more efficient, more controlled uptake, electroporation. Eventually, the cells are rendered mature by the addition of one or several danger signals and further cultivated 1 or 2 days. Then the cells are re-injected into the patient. Although the method was shown to work in animals and in humans (see Section 7.2.2.1), the optimal danger signal(s) and site of injection are still to be precisely defined. For the maturation, a cocktail containing IL-1β, IL-6, TNF-α, and PGE2 may be optimal to generate mature DCs that can migrate to the lymph nodes draining the site of adoptive transfer without dying too early from overstimulation [58]. Using this type of cocktail, DCs injected subcutaneous or intradermally are expected to go to the secondary lymphoid organs and prime antigen-specific lymphocytes. Alternatively, the direct intra-lymph-node injection of mature transfected DCs may be the best method to guarantee that most *in vitro* generated APCs are at their site of activity after adoptive re-implantation. The utilization of this method in many human clinical trials is ongoing throughout the world, and optimized protocols will soon be deducted from these studies.

PARTICLES. Both pDNA and mRNA vectors entrapped in particles or coated on particles were shown to be potent vaccines. In all cases, the particle strategy allowed a drastic reduction of the amount of the nucleic acid needed to get a significant immunity. For mRNA vectors, Martinon et al. [59] were the first to report that influenza nucleoprotein coding mRNA entrapped in liposomes can induce an immune response against influenza virus in mice. It was then shown that minimal amounts of mRNA (down to 1 μg of encapsulated mRNA injected intradermal or intravenous) are enough to prime an antigen-specific immune response [60]. This technology has not yet been transferred to human clinical trial. Encapsulated pDNA, however, were shown earlier to be potent vaccine formulations. Plasmid DNA entrapped [61] or coated [20] on cationic microparticles or enclosed in liposomes [62] are more potent than naked pDNA as vaccines. Using either intramuscular or intravenous routes of delivery, doses as low as 1 μg of encapsulated pDNA were shown to prime humoral and cellular immune responses in mice. Using these formulations, oral [61, 63, 64], intranasal [65], or other mucosal delivery routes (intrarectal [66], for example) can be used to stimulate mucosal and systemic immunity in animals (mice or macaques).

Adjuvants. Although nucleic acids have the intrinsic capacity to signal "danger" through TLR7, 8, or 9, extra-immunostimulation can enhance or polarize the immune response. It can be achieved by injection of cytokines or nucleic-acid-coding cytokines together with the vaccine. The cytokine GM–CSF is one of the most studied adjuvants in the context of nucleic acid vaccinations. It was shown to enhance and to polarize toward Th1 immune responses after pDNA [67] or mRNA [68] applications. The optimal timing of application of such an adjuvant is not clearly established. Theoretically, the local delivery of GM–CSF should attract immature DCs that would render the site more reactive to a consecutive nucleic acid injection. However, practically, in mice it looks like the injection of GM–CSF should at best follow the injection of pDNA or mRNA. The optimal timing of GM–CSF injections in humans would need to be deducted from clinical trials especially designed to decipher this matter. Inflammatory cytokines such as IL-12 are also foreseen as promising adjuvants to be used in combination with nucleic acid vaccines.

Preclinical Results and Clinical Applications. Both pDNA- and mRNA-based vaccinations were demonstrated to be efficacious in animal models as prophylactic or therapeutic immunotherapies against tumors, infectious diseases, and allergy. Two pDNA-based vaccines are commercialized for veterinary use: an anti-equine fever and an anti-infectious hematopoietic necrosis virus (IHNV) for farm-raised salmons. In humans, several formulations of nucleic acid vaccines are tested in clinical trials (see the actualized list of trials at www.clinicaltrials.gov). Although pDNA-based vaccine trials were reported in the context of antitumor, antivirus (HIV, influenza virus, HBV) and antiparasite (*Plasmodium falciparum*) approaches, mRNA-based vaccines were up to now tested only as immunotherapies against cancer (review by Liu and Ulmer for pDNA [35] and Pascolo for mRNA [36]).

Anti-infectious Diseases Vaccines (HIV, Plasmodium falciparum, influenza virus, HBV, Mycobacterium tuberculosis). Theoretically, any infectious agent can be recognized by the immune system, and thus, vaccination is of interest to prevent acute infection or reduce chronic replication. Preclinical results have shown at least in mice that for all economically or socially relevant pathogens, pDNA-based vaccination is a robust method to prevent or treat the infections. Several trials in the human population are ongoing (see www.clinicaltrials.com) and address the safety of pDNA vaccines designed to trigger immunity against HIV, *Plasmodium falciparum*, influenza virus, HBV, *Mycobacterium tuberculosis*, severe acute respiratory syndrome (SARS) virus, West Nile Virus, or Ebola virus. Most are phase I trials and thus address only the safety and dose. The following paragraphs describe the results obtained for several of these test vaccines.

AIDS. The Acquired Immunodeficiency Syndrome (AIDS) is the fourth cause of death in the world (an estimated 3.5 million people die from AIDS every year). An estimated 40.3 million persons are living with HIV, and 5 million people per year become infected. An affordable and efficacious anti-AIDS vaccine is urgently needed (review by Girard et al. [69]). Due to the potency of pDNA vaccines to induce CTL in mice and the known efficacy of these cells to control HIV replication in humans, the first trials using direct injection of nucleic acids were performed in HIV-infected individuals in the mid-1990s. Follow-up trials were performed in uninfected volunteers. These studies showed that anti-HIV pDNA vaccines are feasible and safe [70]. The induced immune responses were less strong than expected based on preclinical studies. However, CTL responses could be recorded after injection of naked pDNA [70]. Interestingly, in infected patients, the pDNA vaccines could induce new anti-HIV immune specificities, not triggered by the infection itself. Thus, in the context of immunotherapy, pDNA vaccination has indeed the potential to boost the preexisting immunity but also to broaden it by inducing new virus-specific CTL and antibodies. Up to now, no clinical efficacy of pDNA vaccines against HIV could be demonstrated. However, strategies using pDNA as a prime followed by recombinant viruses or proteins as boost are being evaluated as vaccine protocols in high-risk populations [71, 72].

MALARIA. *Plasmodium falciparum* (*P. falciparum*) is the most pathogenic parasite causative of malaria. Yearly, 300–400 millions clinical cases of *P. falciparum* infections are reported, among which 2–3 million are deadly. Moreover, an increase of the incidence of the disease has been observed due to the emergence of drug-resistant parasites and of insecticide-resistant mosquitos. For these reasons, a protective and cost-effective vaccine against *P. falciparum* is needed. *P. falciparum*, being an intracellular parasite, its control requests the recognition by CTL. Thus again, nucleic acid-based immunotherapies are theoretically adapted for the development of a preventive or therapeutic vaccine against malaria. The same year as the report of anti-HIV pDNA-based vaccines was published so was the result of the first anti-malaria pDNA-based vaccine [73]. In this and follow-up studies, the induction by the pDNA vaccine of CTL against *P. falciparum* epitopes was documented. However, no antibody response was induced. The codon usage of *P. falciparum* is very inadequate for mammal cells. Thus, the pDNA vaccines based on native *P. falciparum* genes may have been relatively inefficiently translated *in vivo*. This may account for the generation of CTL, which are recognizing fragments of the protein but fail to generate antibodies that require the production of the full-length foreign protein. To conclude, although the documented induction of anti-malaria T cells by pDNA applications [74, 75] is an encouraging progress toward the development of an antimalaria vaccine, a randomized placebo-controlled phase III trial is required to see whether pDNA-induced anti-*P. falciparum* immunity results in the reduction of the burden of infection, morbidity, and mortality within populations living in endemic areas.

HEPATITIS B. The hepatitis B virus (HBV), which may have infected more than one third of the population, is responsible for approximately 1 to 2 million deaths every year. After infection, most people clear the virus through an efficient immune response. However, the virus persists in ca. 10% of the people and eventually leads to liver diseases such as hepatocellular cancer. Although the anti-HBV vaccine based on the recombinant surface antigen is efficient and safe, 5% to 10% of healthy immunocompetent subjects do not respond to it [75]. For those people and the ca. 350 million carriers, an alternative vaccine that would trigger a B-cell but also a T-cell response is of great interest. Indeed, a potent T-cell response of the Th1 type is believed to lead to the control of HBV. Several prophylactic and therapeutic trials are ongoing. One of the most successful methods to trigger immunity against HBV using pDNA is based on the utilization of the gene-gun: A trial showed that ballistic delivery of pDNA coding HBS could induce CD4 and CD8 T cells as well as protective levels of antibodies in all treated hepatitis-naïve volunteers [76]. Moreover, no or low responders to the standard vaccine were found to respond to the pDNA applied with the gene-gun (12 out of 16 patients responded [77]). Each dose of the vaccine contained no more than 4 µg of plasmid. Subjects received between one and three doses, with a two-month interval in-between applications. Thus pDNA-based vaccination is a safe, feasible, cost-effective, and promising prophylactic and therapeutic approach against HBV. The results of phase II and III trials are needed to prove clinical efficacy and allow commercialization of this vaccine.

TUBERCULOSIS. *Mycobacterium tuberculosis* is the etiologic agent of tuberculosis. The disease is due to the immune response against the pathogen more than to the pathogen itself. Immunotherapy using the whole inactivated pathogen or BCG may enhance rather than decrease the disease. Nucleic acid vaccination offers the possibility to perform immunotherapy with the immunogenic parts of the pathogen that are associated with a protective immune response rather than with a pathologic immune response. Heat Shock Protein (HSP) 65 is a dominant mycobacterial antigen. Vaccination with HSP65-encoding pDNA in mice was superior to infection in priming protective T cells [78]. In this model, the same method was found to be efficacious in a therapeutic setting [79]. However, these results were not confirmed by another laboratory, where vaccination of mice with pDNA encoding mycobacterium HSP65 enhanced the pathology. These contradictive results may be due to the ratio of Th1 versus Th2 cells primed by the vaccine, which may be linked with the purity of the nucleic acid (the presence of endotoxin, which orientate the response toward Th2), the site of injection (intradermal delivery orientates the response toward Th2), or the level of immunological naivety of mice that depends on the clean conditions in which they are kept. As an improvement of the approach, a combination between efficacious anti-mycobacterium chemotherapy and pDNA vaccination may allow a sustained immune control of the pathogen [80]. This regimen may be used in infected patients suffering from tuberculosis. However, the risk of increasing rather than decreasing the pathologies associated with *Mycobacterium tuberculosis* using pDNA vaccination is a drawback that limits the testing of this immunotherapy.

Antitumor Vaccines. Many, if not all, tumor cells express characterized specific antigens (self-proteins or virus-encoded proteins) that make them recognizable and controllable by the immune system, especially by T cells. The capacity of immune-stimulation approaches, among them, nucleic acid-based vaccines, to control tumors was clearly proven in many animal models. For this reason, nucleic acid vaccines coding tumor antigens such as virus-encoded oncogenes (for example, HPV 16 E7), self-proteins (prostate antigens for prostate cancer, clonotypic antibodies for B-cell leukemia, melanocyte-specific proteins for melanoma, or shared tumor antigens such as testis-antigens for many tumor types, for example) or mutated oncogenes (p53 or cRas, for example) were tested as immunotherapy in cancer patients [81]. Prophylactic vaccination against cancer is validated in mice [82] but not yet transferred to healthy, tumor-prone individuals identified thanks to family history and detection through genotyping using known tumor-predisposition markers.

PLASMID DNA-BASED ANTI-TUMOR VACCINES. One of the first reports of an antitumor pDNA-vaccine phase I trial used a dual expression plasmid coding HBS and carcino embryonic antigen (CEA), a protein overexpressed on many epithelial cancers. The plasmid was repetitively injected intramuscularly. Per injection, up to 2 mg was applied. Some patients developed a good immune response toward HBS, but there was only a very limited response toward CEA. No objective clinical response was recorded. Similar results were obtained in another phase I trial using a pDNA coding gp100 (an enzyme involved in melanine synthesis and expressed in most melanoma [83]) whether the DNA was injected intramuscularly or intradermally. As a superior vaccination strategy that may allow breaking tolerance, the intranodal delivery of pDNA coding tyrosinase (like gp100, it is a protein expressed in melanocytes and melanoma) was tested in melanoma patients [52]. Patients received up to 800 µg of pDNA every 2 weeks for four cycles. The treatment was feasible and safe. Nearly half of the patients (11 of 26) showed the triggering of an immune response against tyrosinase. Overall survival in this small cohort was found to be higher than expected. Meanwhile the utilization of an adjuvant in the form of recombinant cytokines (IL-2 and GM-CSF) applied subctaneously was tested in a phase I trial involving prostate tumor patients [84]. The pDNA (up to 900 µg per injection, five cycles) was given intramuscularly and intradermally. Cellular and humoral immune responses targeted to PSA could be detected in two of three patients injected with the highest dose. Biochemical regression was recorded in these two immunologically responding patients. Finally, as an immunotherapy strategy targeting the idiotype of B-cell lymphoma, pDNA expressing patient-specific clonotype were delivered with a needleless device (Biojector). Patients received monthly injections of up to 1800 µg of pDNA [85]. Here again, an immune response (B or T cell) could be triggered against the tumor-specific antigen (the antibody's idiotype) in most patients (7 out of 12).

Viral oncogenes are a privileged target for antitumor immunotherapies: Being foreign molecules, they are usually efficaciously recognized by the immune system and can be used as prophylactic antitumor vaccination targets. Human Papilloma Virus-16 (HPV-16) encodes two oncogens: E6 and E7. PLG-encapsulated pDNA coding for a few HLA-A2 epitopes of HPV-16 E7 administered intramuscularly could induce an immune response against HPV-16 in most treated patients suffering from HPV-16 associated anal dysplasia [86] or cervical intraepithelial neoplasia [87]. Clinical responses (histological responses) were observed in some patients (3 out of 12 in the study by Klencke et al.; 5 out of 15 in the study by Sheets et al.). Thus, pDNA-based vaccinations against HPV can be envisioned as a possible prophylactic and therapeutic antitumor immunotherapy regimen. This technology may become a standard anti-HPV therapy in the relatively near future.

MESSENGER RNA-BASED ANTITUMOR VACCINES. The only published results where mRNA vectors were used to vaccinate cancer patients are based on the utilization of mRNA-transfected DCs as described above. This popular method first described in 1996 [88] was validated in several preclinical mouse models and used to vaccinate prostate or renal cancer patients with different mRNA preparations (review by Gilboa [89]). In most patients, an increased T-cell immune response against the antigen(s) encoded by the mRNA(s) could be observed. Thus, mRNA-based vaccines using *in vitro* transfection of autologous DCs is a feasible and efficacious approach to stimulate a T-cell immune response against tumor antigen. Phase II and III trials are needed to show the clinical efficacy of mRNA-transfected DCs as anticancer immunotherapy. However, the need for a large amount of blood to prepare a few vaccine doses, the costly and variegating production of DC in GMP-conditions, as well as the logistic issues restrict the utilization of this technology. As an alternative, direct injections of naked or protamine-encapsulated mRNA was shown to be efficacious for priming cellular and humoral immune responses in mice [27, 90]. It is being evaluated in human as anticancer immunotherapy (ongoing phase I/II studies).

CONCLUSION ON ANTITUMOR VACCINES BASED ON PDNA OR MRNA. Most published results show limited efficacy of nucleic acid-based vaccines as anticancer immunotherapies, highlighting the need for improving these strategies. However, the possibility of triggering an immune response against self-tumor antigens with pDNA or mRNA vaccines sustains the hope that once the correct antigens and optimized immunization methods will be identified, antitumor vaccinations would be a way to get regression or at least stabilization of tumor diseases in many patients. The fundamental knowledge of immune-tolerance and control of autoimmunity will help to find the protocols that will allow nucleic-acid based vaccines to become potent antitumor therapies. Of particular and new interest is the role of regulatory T cells (Treg) that can suppress immunity and are usually increased in number in the blood of tumor patients. Antitumor vaccination may enhance the activity of suppressive cells rather than the one of effector antitumor cells, thus leading to a failure of the therapeutic approach [91]. Controlling, i.e., reducing, the activity of Treg and of other immunomodulatory cells or molecules may be crucial for the success of anticancer immunotherapies. Methods such as chemotherapies, which are lymphoablative or specific inhibition of gene expression using siRNA (anti-TGF-β or IL-10, for example), may allow a transient reduction of natural immunosuppression that could be used for intensive vaccination protocols. Such combinations are being evaluated in clinical trials.

Anti-allergy Vaccines. Allergic diseases affect approximately one third of the population, and for unclear reasons, prevalence of these diseases is increasing. Through the triggering of Th2 type of T helper cells, environmental antigens induce the release of Il-4, Il-5, and Il-13 activate basophil, eosinophil, mast cells, and IgE, which eventually induce the development of inflammatory allergic diseases such as asthma, eczema, hay fever, and rhinitis. The desensitization of a patient can be obtained through repeated injections of increasing doses (starting at very low doses) of the nominal allergen over the course of several years. However, this protocol has been reported to have inconsistent efficacy and in rare cases very strong toxicity [92]. As vaccination with pDNA was reported to induce a Th1 response and inhibit the Th2 response, it was anticipated that pDNA vaccines could have a potential for the prevention and cure of allergic diseases. Indeed, it was demonstrated in animal models that the injection of pDNA coding for an allergen (β-galactsidase in experimental models or relevant allergen such as house dust mite Derp5, latex Hevb5, peanut Arah2, honeybee venom phospholipase A2, pollen allergen Cryj1, or Betv) can be a prophylactic or therapeutic treatment against allergies [93]. However, a clear control of the Th2 response, eventually using the co-injection of the vaccine with nucleic acids coding Il-12 or IL-15 as enhancers of the Th1 response, is needed before pDNA desensitization protocols can be planned for testing in the human population.

Conclusion on Vaccination Using Nonviral Coding Nucleic Acids. Both pDNA and mRNA vectors can induce antigen-specific humoral and cellular immunity of the Th1 and/or Th2 types. When using direct intramuscular injections of naked plasmid, a minimum of 800 μg of pDNA seems to be necessary to get a detectable immune response. Optimization of the vectors as described in Section 7.2.2.1 is a standard way to increase the efficacy of the nucleic acid-based vaccines and should be used for any clinical trials. The choice of the relevant pathogen-derived antigen, especially in the case of antitumor immunotherapies, is critical: This antigen should be well immunogenic and recognized by T and/or antibodies on pathogenic cells (infected and malignant) or by antibodies on infectious agents. The delivery method should be cost-efficient, praticable, and result in the induction of a strong immune response. Although the efficient intra-lymph-node delivery method may be adequate for therapy of a limited number of patients with advanced nontreatable diseases (tumor patients, for example), the gene-gun delivery sounds like the most favorable delivery method for easy, painless, efficacious, and cost-effective nucleic-acid-based vaccination of large populations. The utilization of a nontoxic adjuvant is always relevant, and at the moment, GM–CSF, which is the most frequently used cytokine in the context of nucleic acid vaccination trials, looks like a fairly good candidate. Meanwhile the control of Tregs is fundamental to trigger efficient immunity, especially in anticancer immunotherapies. The choice of the vehicle (pDNA or mRNA) needs to be addressed in clinical trials where both vectors are compared. Although mRNA looks safer due to its transient nature, pDNA may be more efficacious and require less injections for reaching protective immunity (high antibody titers and high T-cell frequencies). On another hand, long-term persistence of antigen expression when using pDNA may interfere with the development of memory T cells [94] and thus render a prophylactic vaccine inefficient. The induction of memory cells by mRNA- and pDNA-based vaccines needs to be evaluated and compared in human populations, especially when prophylactic vaccination is envisioned. Thus, it could be stated that the prevention of diseases or the treatment of chronic non-life-threatening diseases (i.e, allergy) may be most adequately addressed by mRNA vectors, whereas the treatment of late-stage patients (AIDS or tumor patients, for example) may be at the moment best addressed with pDNA vectors.

7.2.2.3 Gene Therapy

Introduction. Known mutated genes that are the cause of recessive genetic diseases are, for example, adenosine deaminase in severe combined immunodeficiency disease (SCID), cystic fibrosis transmembrane conductance regulator gene (CFTR) in cystic fibrosis, hypoxanthine-guanine phosphoribosyl transferase in Lesch-Nyhan disease, glucocerebrosidase enzyme in Gacher disease, dystrophin in Duchenne muscular dystrophy, and factors XIII and XI for some hemophilia. In these cases, using foreign recombinant nucleic acids, it is possible to bring the protein that is missing or deficient. This technology is called gene complementation. The challenge of nucleic-acid-based gene therapies is to bring in all defined target cells an amount of nucleic acid that would provide a functional quantity of the encoded protein without over- or under-expression by some cells and without stimulating immunity or inducing apoptosis. General optimizations of gene expression as described for vaccination with pDNA and mRNA vectors in Section 7.2.2.1 should be performed to obtain the highest possible expression, thus efficacy, from the nuclei-acid-based drugs. At the same time, the vectors should be designed to have low immunogenicity, for example, by eliminating CpG sequences in pDNA vehicles. Using pDNA vectors, the utilization of promoters that specifically drive gene expression in the adequate tissue or cell type is a method to lower side effects such as expression of the exogenous vector in irrelevant organs or cells. Although delivery of naked nucleic acid is evidenced *in vivo* and can be used for some approaches, the efficacious systemic delivery of nucleic acids needed for gene therapy is best obtained thanks to encapsulation [19]. Both methods are presented below. None of them has been used in human clinical trials. Up to now, gene therapy approaches tested in humans are based on virus-derived constructs that can efficiently penetrate cells and integrate in the nucleus. However, because of safety issues, pDNA and eventually mRNA (not used in preclinical models up to now) may appear again as favorable vehicles for gene complementation.

Naked pDNA for Gene Therapy. As published by Wolff et al. in 1990, the injection of naked pDNA in the mouse skeletal muscle results in local transgene expression [18]. Since then, other species such as rats, cats, or monkeys and other sites such as skin, liver, brain, urological organs (assisted by *in vivo* electroporation), thyroid, and tumors were shown to be permissive to the local uptake of injected naked pDNA [95]. However, the local transgene expression obtained by direct injection of naked pDNA into an organ is usually too low to be used in the context of genetic disease, even when the missing function is contained in a soluble protein secreted in the bloodstream (for example, coagulation factors in hemophilia). It was estimated that after injection of naked pDNA, only a few thousand plasmids are functionally retained within cells of the injected tissue [56]. To treat a genetic disease, a physiological expression of the therapeutic molecule by all cells that need the missing function or a functional level of the protein in the body fluids is required. For this reason, gene therapy approaches based on naked pDNA use a systemic delivery method such as intravenous or, at best when organs other than lungs are targeted, intra-arterial injections. Indeed, the intravascular injection of naked pDNA surprisingly results in the expression of the encoded protein, especially in the liver [50]. As it is the case for intramuscular and intraskin injections, the intravascular injection of naked pDNA was initially a (negative) control of experiments designed to test the potency of transfection reagents. To reach a good expression level in mouse hepatocytes, the pDNA should be injected in a relatively large volume and short delivery time ("hydrodynamic" delivery) through the tail vein, the portal vein, the hepatic vein, or the bile duct. One injection can transfect about 1% of the hepatocytes throughout the entire mouse liver. This method can be used to produce *in situ* liver-specific proteins and soluble proteins such as coagulation factors XIII and XI that are missing in hemophilic patients. The production of critical proteins such as growth hormones with this technology is theoretically possible but would be hazardous because the amount of protein produced and the duration of the expression are not controlled.

Although naked siRNA can be delivered using the same procedure, naked mRNA would not be suitable because it is degraded within seconds in the serum [14].

Using hydrodynamic injection in a muscle's artery or vein, the expression of naked pDNA can also be obtained in the muscle. However, for a high expression, the blood flow through the targeted vessel must be stopped surgically during the injection [96, 97]. Expression of the naked pDNA can also be obtained in the cardiac myocytes after injection in the left ventricular wall.

These preclinical observations made in mice, rats, and monkeys allow us to foresee the utilization of naked pDNA as a treatment of muscle diseases such as Duchenne muscular dystrophy, peripheral limb ischemia, and cardiac ischemia. However, the methods described above for animals (hydrodynamic injection, transient stop of blood flow through a muscle) are difficult to transpose to the human situation. Instead of an hydrodynamic delivery, human may benefit from the utilization of a vasodilator before pDNA injection. This treatment could increase the permeability of discontinuous endothelium such as those in liver, spleen, and bone marrow and induce permeability through continuous endothelium such as those in the brain. This way, plasmid DNA may go through the vascular wall and reach the target cells of the organs. As mentioned, the utilization of pDNA for gene therapy in humans has not yet been evaluated in clinical trials. Virus-based gene complementation is still observed as the most feasible method although safety issues are raised. Encapsulated pDNA is possibly an improved version of naked pDNA for gene therapy and has better safety features than virus-based delivery systems.

Encapsulated pDNA for Gene Therapy. Both cationic liposomes and polymeric particles were used to deliver efficiently pDNA in animal models for gene therapy approaches. Although in these preclinical assays the expression of the transgene vanished after few days, the uncontrolled long-term expression as well as a possible overexpression after repeated delivery of encapsulated pDNA are safety issues that must be addressed before these methods are evaluated in clinical trials. However, thanks to the recent knowledge on gene regulation (utilization of inducible promoters such as Tet On and Tet Off promoters), immune activation (CpG motifs), immune modulation (Tregs for immunosuppression, for example), we can anticipate new generations of safe (controllable and nonimmunogenic) encapsulated plasmid constructs that will be the basis of future drugs designed to treat genetic diseases.

Preclinical Results Obtained with Liposome Formulations. One of the most frequent genetic diseases with high morbidity and mortality for which the deficient gene has been identified is cystic fibrosis. It is due to mutations in the CFTR, which is a transporter of chloride anions. This deficiency is recessive: Bringing the correct gene in the cells can correct it. As cystic fibrosis affects the lung, local gene therapy approaches where the genetic vehicle is delivered through airways is foreseen as an efficacious and safe method. As early as 1992, liposomes encapsulating pDNA coding for CFTR were found to be capable of bringing this protein function to airway epithelium of the lung after intratrachea instillation in mice [98]. Optimization of the delivery can be obtained by modifying the liposome formulation [99]. This promising method was not evaluated in patients.

Meanwhile, the intravenous delivery of lipid-entrapped pDNA was shown to efficaciously lead to gene expression, especially in liver, lung, and kidneys [100]. The addition of targeting molecules can efficiently target the vesicles to specific cells. For example, the utilization of galactosyl-cholesterol or mannosylated-cholesterol in the liposome formulation allows efficient delivery of the pDNA to the liver in mice [101, 102].

To conclude, cationic liposome formulations were shown in animal models to be efficient, safe, and versatile delivery vehicles for pDNA molecules. The engineering of liposomes coated with targeting molecules is one method to enhance the efficacy and specificity of these vehicles. Clinical trials in humans are needed to demonstrate the feasibility, safety, and efficacy of these approaches.

Preclinical Results Obtained with Particles. Poly-L-lysine and poly-L-Ornithine were used to generate particles that can entrap pDNA. The cationic polymers can be coupled to a targeting molecule such as galactose or mannose for enhanced delivery to the liver after intravenous or intraportal injections in mice [102–104]. Expression in other organs than the liver (for example, kidneys) is observed; however, expression in the liver using these particles is dominant. Such particles are found to be more efficient and to persist for a longer time in the circulation than liposomes. Hepatic expression of the pDNA delivered by poly-L-Lysine particles can be found several months after delivery [105]. These promising formulations were not evaluated in humans.

Conclusion on Gene Therapy Using Nonviral Coding Nucleic Acids. As opposed to vaccination strategies, the local delivery of naked pDNA in an organ is not foreseen as a possible gene therapy approach because the protein expression is too low. However, the systemic delivery through the blood circulation of naked or entrapped pDNA was found to result in strong expression of the transgene, mainly in the liver. The utilization of hepatocyte-specific promoters in the pDNA would allow us to turn these methods into liver-specific expression systems. They may be used to produce soluble proteins such as coagulation factors XIII and XI or erythropoietin in patients suffering from genetic deficiencies for these molecules. Encapsulated pDNA expressed by hepatocytes could also theoretically be used for the sustained production of therapeutic monoclonal antibodies (the anti-Her-2/neu antibody: Herceptin, for example) by the body's own cells in tumor patients. Aside from that, some methods (hydrodynamic delivery, occlusion of veins, particles) can allow delivery of pDNA in the muscle as a possible treatment for myopathies. Finally, inoculation through the airways can result in gene expression in the lung and could be used as a treatment against cystic fibrosis. Thus, although they are not yet tested in humans, the gene therapy approaches based on simple pDNA coding for a therapeutic protein are very exciting and promising approaches that offer a broad range of applications. However, several issues have to be addressed:

1. The natural immunogenicity of the pDNA. Even without any known CpG sequence in the plasmid, repeated applications may still induce the activation of the immune system. This immune response could neutralize the produced transgenic protein and result in the specific killing of cells expressing it. As a remedy, the utilization of immunosuppressive drugs concomitantly with the application of pDNA-based gene therapies should probably be envisioned.

2. The uncontrolled persistence and expression of the transgene prevents its utilization for the production of active molecules such as growth factors. Repeated applications may result in some patients in the accumulation of

pDNA and overexpression of the transgene. This would have negative effects on the producing organs (necrosis) and pathogenic systemic effect due to the protein's own activity. The utilization of inducible promoters and a careful monitoring of the patients before each application of the pDNA would be needed to perform safe clinical trials.

Cationic liposomes that were frequently used and validated in several laboratories and animal models may be the first approaches used for gene therapy using pDNA in humans. Meanwhile, the utilization of encapsulated mRNA appears theoretically as a method to circumvent all limitations of pDNA-based approaches for gene therapy. Encapsulated mRNA were not yet reported as gene therapy tools but because they are transient molecules, readily expressed in the cytosole, and if nonstabilized, barely activate immunity (TLR7 and TLR8), they can be foreseen as safe (transient), efficient (expressed in the cytosole), and nonimmunogenic (not stabilized) genetic information vehicles to be used in gene therapy.

7.2.2.4 Pharmaceutical Production and Regulations

7.2.2.4.1 Plasmid DNA. Several companies in the United States and Europe offer the contract manufacture of documented pDNA for preclinical use and of GMP-certified pDNA for clinical utilisations (see, for example, the Plasmid Factory at www.plasmidfactory.com). It is regulatory practice that ampicilin is not used in GMP production in order to avoid cross-contaminations and concerns with penicillin-allergies. Instead, plasmids containing a kanamycin-resistance gene are frequently used. At best, the starting material is a pDNA containing a synthetic gene constructed by assembly of oligonucleotides. This way, the history of the pDNA can be clearly reported in the production's file. The whole manufacturing process is free of bovine-derived products. The plasmid of interest is tested in different *E. coli* strains and several culture media to find out the best conditions for producing the highest amount and best quality plasmid. The final product should meet regulatory guidelines such as those of the FDA: http://www.fda.gov/cberg/gdlns/plasdnavac.htm. It should appear as a clear and colorless solution by visual inspection. The final controls must meet the following criteria:

1. Identity: Sequencing should show 100% identity with the expected sequence. Electrophoresis or chromatography studies should be used to show that the pDNA is predominantly supercoiled. The restriction enzyme digest pattern analyzed by gel electrophoresis can be included. Susceptibility to DNase can also be used as a proof of molecular identity.
2. Content: Quantification should be performed by absorbance at 260 nm. Osmolarity and pH should also be measured using standard methods.
3. Purity: Residual proteins, chromosomal bacteria DNA, RNA, and endotoxin must be below specified limits. Sterility must be controlled by standard microbiological assays.

Moreover, some tests can be performed to check counter-ions (should be NaCl, thanks to precipitation of the pDNA with alcohol plus NaCl), residual solvents (if they are used during the production), and potency (functional assay using, for example, transfection of cells and verification of the expression of the protein of interest or testing of the immune response after injection in an animal model).

Production of pDNA using up to 100-m^3 scale fermentation is performed by specialized industries [106]. Batches of more than 1 kg of pDNA can be produced. As functional doses for vaccination in humans were of ca. 800 µg (intramuscular injection) or 4 µg (gene-gun), 1 kg of pDNA is enough for 1,250,000 or 250,000,000 doses, respectively. This is largely enough for pDNA-based therapies to be commercialized worldwide as treatments against infectious diseases, cancer, or genetic diseases.

Preclinical toxicology studies should address the standard systemic and local reactogenicity, histopathology, and toxicity (acute and chronic dosage) in rodent and nonrodent animals. Also, as pDNA-based therapies are classified as gene therapy approaches, additional strict safety issues must be addressed before starting a clinical trial (for a review, see Ref. 35). The potency of the pDNA to integrate in the genome of the injected animal must be tested using assays that can detect one integrated pDNA in more than 150,000 nuclei. The potential of the pDNA preparation to induce anti-DNA antibodies (the etiologic agents of systemic lupus erythematosus) must be evaluated. The biodistribution and persistence of the pDNA must also be measured: Tissues and body fluids are analyzed at different time points for the presence of the pDNA using PCR methods. Of particular relevance is the study of pDNA in germ cells because such an event could theoretically generate transgenic descendants.

These toxicology studies were performed many times using many different pDNA batches. Close to no integration of pDNA in genomes, no induction of pathogenic anti-DNA antibodies, and a clearance of injected pDNA within days were reported. However, long-term (months) antigen expression in mice using intramuscular injections of a pDNA that may replicate in mammalian cells was found [107]. In this case, the cellular response triggered by the vaccine was weak and may in part account for the persistence of the foreign nucleic acid. The antigen coded by the vaccine in this preclinical study (Hepatitis Surface Antigen) associated with specific antibodies formed circulating immune complexes that caused pathologic lesions in liver and kidney. Thus, special care should be devoted to the design of the plasmid to guarantee that it cannot replicate in mammalian cells. Meanwhile, the regulatory authorities acknowledged the fact that since the first report of trials in humans using injections of up to 800 µg of nonviral coding pDNA (1998), no adverse events have ever been registered. Thus, especially for phase I trials, the toxicology studies can now be facilitated in agreement with the relevant authorities. For vaccination, gene-gun offers the extra-safety advantages that very low doses are used and that the pDNA is delivered exclusively in the dermis, avoiding systemic distribution and potential side effects while guaranteeing the elimination of most of the injected product through natural skin regeneration.

7.2.2.4.2 Messenger RNA.

Two companies offer contract manufacture of mRNA at laboratory and clinical grade (GMP-certified): Asuragen, a spin-off of Ambion, in the United States and CureVac in Europe. Here again, the whole manufacturing process is free of bovine-derived products. There are no official guidelines for the production of mRNA; however, it can be proposed that the final product should be checked by the standard quality controls. It should appear as a clear and colorless solution by visual inspection. The following criteria would have to be met by the results of the final controls:

1. Identity: Sequencing of the plasmid used for *in vitro* transcription should show 100% identity with the expected sequence. At best, reverse transcription and sequencing of the final mRNA could be performed. It should also show 100% identity with the expected sequence. Susceptibility to RNase could additionally be used as a proof of molecular identity.
2. Content: Quantification should be performed by absorbance at 260 nm. Osmolarity and pH should also be measured using standard methods.
3. Purity: Residual proteins, chromosomal bacteria DNA, plasmid DNA, bacteria RNA, aberrant mRNA transcripts (smaller or larger by-products of the transcription) and endotoxin shoud be below specified limits. Sterility must be controlled by standard microbiological assays.

As is the case for pDNA, some physical tests can be performed to check counterions, residual solvents (if they are used during the production), and potency (functional assay using, for example, transfection of cells and verification of the expression of the protein of interest or testing of the immune response after injection in an animal model).

Preclinical toxicology studies should address the standard systemic and local reactogenicity, histopathology, and toxicity (acute and chronic dosage) in rodent and nonrodent animals. Since the FDA and some European authorities decided to classify mRNA-based therapies as no-gene therapy (for nonreplicative mRNA as depicted in this chapter), the implementation of clinical trials does not require additional specific toxicology testing.

7.2.3 CONCLUSIONS AND PERSPECTIVES

Since the initial publication by Wolff et al. showing that the injection of pDNA or mRNA simply coding for a protein can result in gene expression *in vivo*, many studies have been performed in animal models and humans showing that this method is safe and can elicit the expected results, which is the expression of the protein. Because pDNA and stabilized mRNA have the natural capacity to activate the immune system through the triggering of TLR, pDNA- and mRNA-based therapeutic utilizations are mostly tested in the field of vaccination. However, since gene therapies based on viruses have shown severe toxicity features in recent human trials, some pDNA or mRNA formulations may now be intensely developed as safer alternatives. Of special interest is the yet unexplored use of (encapsulated) mRNA for gene therapy. As this transient nucleic acid cannot persist or accumulate and would not generate unwanted long-term side effects even after repeated injections, it may seem as the optimal active pharmaceutical ingredient for gene therapies. Meanwhile, vaccination strategies based on nonviral coding nucleic acids are very advanced in terms of time to commercialization. The injection of naked pDNA and the adoptive transfert of APCs transfected with mRNA were demonstrated in many human clinical trials to be feasible and safe and to result in the development of the expected antigen-specific immunity. Optimal formulation, dosage, frequency of application, and adjuvant still need to be clearly defined.

Improvements of the methods include the optimization of the gene sequence, the optimization of the antigen processing, the enhancement of uptake using for example electrical methods, or the use of particles for formulating the vaccines. Concerning the last point, three types of particles were evaluated: liposomes, cationic degradable polymers, and coated gold particles delivered by gene-gun. This latter technology seems to be at the moment the most efficacious, versatile, cost-efficient method for vaccination with nucleic acids. Moreover, it is painless for the patient and very reproducible because the operator cannot influence the delivery (as opposed to injections, especially intradermal injections). The company Powdermed (www.powdermed.com) is developing and implementing the utilization of gene-gun-based therapies to deliver pDNA as vaccines against infectious diseases and cancer. Meanwhile, for autologuous vaccination, which may be needed for example with the mutating and patient-specific HIV virus, vaccination with adoptively transferred mRNA-transfected DCs as developed by Argos pharmaceuticals (www.argospharmaceuticals.com) may be of high potency. The choice between pDNA and mRNA for clinical utilizations is a matter of cost and efficacy. Although there was an intense development and optimization of pDNA-based vaccines, accompanied by the implementation of industrial production facilities, mRNA was until recently, relatively unexplored. However, as summarized in Table 7.2-1, mRNA offers many advantages over pDNA in the context of therapeutic utilizations. Being active in the cytosole, mRNA does not need to cross the very selective nuclear envelope to be expressed. Being transient, mRNA can allow the controlled production in terms of quantity and time frame of the protein of interest. Moreover, as it is a single-stranded molecule, mRNA does not have the discrete topologies that have pDNA and that could affect its clinical efficacy. This feature allows mRNA to have theoretically a higher functional batch-to-batch reproducibility. At the level of efficacy for vaccination, pDNA is assumed currently to be superior to mRNA; however, the direct comparison of optimal formulations of pDNA and mRNA using gene-gun delivery, for example, as a consistent and reliable method, still needs to be performed. Meanwhile the unique possibility that offers mRNA to transfect, without inducing death, DCs *in vitro*, is a guarantee of vaccine efficacy using a minimum amount of nucleic acid. Thus, although pDNA-based vaccination approaches are very advanced and may soon be commercialized, it can be

TABLE 7.2-1.

Parameter	pDNA	mRNA
Delivery to the functional site	Hard (Nucleus)	Easy (Cytosole)
Control of level of expression	Hard	Easy
Control of duration of expression	Hard (Days to months)	Easy (Hours to few days)
Large-scale GMP production	Easy	Easy
Conservation	Easy	Easy
Batch-to-batch reproducibility	Low (Three topologies)	High
Regulatory issues	Severe (Gene therapy)	Easy
Immunogenicity (vaccines)	High	Low
Usage for gene therapy	Efficacious	Not tested

envisioned that due to their safety features, mRNA-based approaches will be the second generation of nucleic-acid-based vaccination strategies.

Several companies active in the field of nucleic-acid-based vaccines will probably manage to get their product(s) and methods brought to the market for disease-specific utilizations. Prevention and treatments of hepatitis B, AIDS, influenza, and some cancer are the most advanced. Thus, in the near future, drugs containing pDNA or mRNA as an active pharmaceutical ingredient should be available in pharmacies and help in treating or preventing pandemic infections (HIV, influenza virus, *plasmodium falciparum*, etc.) as well diseases with unmet medical needs such as certain types of cancers. The continuous optimization of these methods ensures that the future pDNA- or mRNA-based drugs will get more and more efficacious and will be used for a large area of therapeutic applications.

REFERENCES

1. Littlefield J W, Keller E B, Gross J, et al. (1955). Studies on cytoplasmic ribonucleo-protein particles from the liver of the rat. *J. Biol. Chem.* 217:111–123.
2. Palade G E (1955). A small particulate component of the cytoplasm. *J. Biophys. Biochem. Cytol.* 1:59–68.
3. Hoagland M B, Stephenson M L, Scott J F, et al. (1958). A soluble ribonucleic acid intermediate in protein synthesis. *J. Biol. Chem.* 231:241–257.
4. Chapeville F, Lipmann F, Von Ehrenstein G, et al. (1962). On the role of soluble ribonucleic acid in coding for amino acids. *Proc. Natl. Acad. Sci. U.S.A.* 48:1086–1092.
5. Feldmann H, Zachau H G (1964). Chemical evidence for the 3′-linkage of amino acids to s-RNA+. *Biochem. Biophys. Res. Commun.* 15:13–17.
6. Brenner S J F M M (1961). An unstable intermediiate carrying information from genes to ribosome for protein synthesis. *Nature.* 190:576–581.
7. Gros F, Gilbert W, Hiatt H H, et al. (1961). Molecular and biological characterization of messenger RNA. *Cold Spring Harb. Symp. Quant. Biol.* 26:111–132.
8. Crick F (1970). Central dogma of molecular biology. *Nature.* 227:561–563.
9. Joyce G F (2002). The antiquity of RNA-based evolution. *Nature.* 418:214–221.
10. Tourriere H, Chebli K, Tazi J (2002). mRNA degradation machines in eukaryotic cells. *Biochimie.* 84:821–837.
11. Ross J (1995). mRNA stability in mammalian cells. *Microbiol. Rev.* 59:423–450.
12. Festenstein R, Kioussis D (2000). Locus control regions and epigenetic chromatin modifiers. *Curr. Opin. Genet. Dev.* 10:199–203.
13. Sudowe S, Ludwig-Portugall I, Montermann E, et al. (2006). Prophylactic and thera-peutic intervention in IgE responses by biolistic DNA vaccination primarily targeting dendritic cells. *J. Allergy Clin. Immunol.* 117:196–203.
14. Probst J, Brechtel S, Scheel B, et al. (2006). Characterization of the ribonuclease activity on the skin surface. *Genet. Vaccines. Ther.* 4:4.
15. Benner S A (1988). Extracellular 'communicator RNA'. *FEBS Lett.* 233:225–228.
16. Kano Y, Miyashita T, Nakamura H, et al. (1981). In vivo correlation between DNA supercoiling and transcription. *Gene.* 13:173–184.
17. Sekiguchi J M, Swank R A, Kmiec E B (1989). Changes in DNA topology can modu-late in vitro transcription of certain RNA polymerase III genes. *Mol. Cell Biochem.* 85:123–133.

18. Wolff J A, Malone R W, Williams P, et al. (1990). Direct gene transfer into mouse muscle in vivo. *Science*. 247:1465–1468.

19. Patil S D, Rhodes D G, Burgess D J (2005). DNA-based therapeutics and DNA delivery systems: A comprehensive review. *AAPS. J.* 7:E61–E77.

20. Singh M, Briones M, Ott G, et al. (2000). Cationic microparticles: A potent delivery system for DNA vaccines. *Proc. Natl. Acad. Sci. U.S.A.* 97:811–816.

21. Uchida M, Natsume H, Kishino T, et al. (2004). Immunization by particle bombardment of antigen-loaded poly-(DL-lactide-co-glycolide) microspheres in mice. *Vaccine.* 24:2120–2130.

22. Felgner J H, Kumar R, Sridhar C N, et al. (1994). Enhanced gene delivery and mechanism studies with a novel series of cationic lipid formulations. *J. Biol. Chem.* 269:2550–2561.

23. Hofland H E, Shephard L, Sullivan S M (1996). Formation of stable cationic lipid/DNA complexes for gene transfer. *Proc. Natl. Acad. Sci. U.S.A.* 93:7305–7309.

24. Dokka S, Toledo D, Shi X, et al. (2000). Oxygen radical-mediated pulmonary toxicity induced by some cationic liposomes. *Pharm. Res.* 17:521–525.

25. Filion M C, Phillips N C (1997). Toxicity and immunomodulatory activity of liposomal vectors formulated with cationic lipids toward immune effector cells. *Biochim. Biophys. Acta.* 1329:345–356.

26. Holcik M, Liebhaber S A (1997). Four highly stable eukaryotic mRNAs assemble 3′ untranslated region RNA-protein complexes sharing cis and trans components. *Proc. Natl. Acad. Sci. U.S.A.* 94:2410–2414.

27. Hoerr I, Obst R, Rammensee H G, et al. (2000). In vivo application of RNA leads to induction of specific cytotoxic T lymphocytes and antibodies. *Eur. J. Immunol.* 30:1–7.

28. Klencke B, Matijevic M, Urban R G, et al. (2002). Encapsulated plasmid DNA treatment for human papillomavirus 16-associated anal dysplasia: a Phase I study of ZYC101. *Clin. Cancer Res.* 8:1028–1037.

29. Conry R M, LoBuglio A F, Wright M, et al. (1995). Characterization of a messenger RNA polynucleotide vaccine vector. *Cancer Res.* 55:1397–1400.

30. Malone R W, Felgner P L, Verma I M (1989). Cationic liposome-mediated RNA transfection. *Proc. Natl. Acad. Sci. U.S.A.* 86:6077–6081.

30a. Gossen M, Bonin AL, Freundlieb S, Bujard H (1994). Inducible gene expression for higher eukaryotic cells. *Curr. Opin. Biotechnol. Oct.* 5(5):516–520 (Review).

31. Mockey M, Goncalves C, Dupuy F P, et al. (2006). mRNA transfection of dendritic cells: synergistic effect of ARCA mRNA capping with Poly(A) chains in cis and in trans for a high protein expression level. *Biochem. Biophys. Res. Commun.* 340: 1062–1068.

32. Stepinski J, Waddell C, Stolarski R, et al. (2001). Synthesis and properties of mRNAs containing the novel "anti-reverse" cap analogs 7-methyl(3′-O-methyl)GpppG and 7-methyl (3′-deoxy)GpppG. *RNA.* 7:1486–1495.

33. Boon T, Coulie P G, Van Den Eynde B J, et al. (2006). Human T cell responses against melanoma. *Annu. Rev. Immunol.* 24:175–208.

34. Selin L K, Vergilis K, Welsh R M, et al. (1996). Reduction of otherwise remarkably stable virus-specific cytotoxic T lymphocyte memory by heterologous viral infections. *J. Exp. Med.* 183:2489–2499.

35. Liu M A, Ulmer J B (2005). Human clinical trials of plasmid DNA vaccines. *Adv. Genet.* 55:25–40.

36. Pascolo S (2004). Messenger RNA-based vaccines. *Expert. Opin. Biol. Ther.* 4: 1285–1294.

37. Chau V, Tobias J W, Bachmair A, et al. (1989). A multiubiquitin chain is confined to specific lysine in a targeted short-lived protein. *Science.* 243:1576–1583.

38. Anthony L S, Wu H, Sweet H, et al. (1999). Priming of CD8+ CTL effector cells in mice by immunization with a stress protein-influenza virus nucleoprotein fusion molecule. *Vaccine.* 17:373–383.

39. Chhabra A, Mehrotra S, Chakraborty N G, et al. (2004). Cross-presentation of a human tumor antigen delivered to dendritic cells by HSV VP22-mediated protein translocation. *Eur. J. Immunol.* 34:2824–2833.

40. Peng S, Trimble C, He L, et al. (2006). Characterization of HLA-A2-restricted HPV-16 E7-specific CD8(+) T-cell immune responses induced by DNA vaccines in HLA-A2 transgenic mice. *Gene Ther.* 13:67–77.

41. Momburg F, Fuchs S, Drexler J, et al. (1993). Epitope-specific enhancement of antigen presentation by invariant chain. *J. Exp. Med.* 178:1453–1458.

42. Ruff A L, Guarnieri F G, Staveley-O'Carroll K, et al. (1997). The enhanced immune response to the HIV gp160/LAMP chimeric gene product targeted to the lysosome membrane protein trafficking pathway. *J. Biol. Chem.* 272:8671–8678.

43. Boyle J S, Brady J L, Lew A M (1998). Enhanced responses to a DNA vaccine encoding a fusion antigen that is directed to sites of immune induction. *Nature.* 392:408–411.

44. Hung C F, Hsu K F, Cheng W F, et al. (2001). Enhancement of DNA vaccine potency by linkage of antigen gene to a gene encoding the extracellular domain of Fms-like tyrosine kinase 3-ligand. *Cancer Res.* 61:1080–1088.

45. Xiang R, Primus F J, Ruehlmann J M, et al. (2001). A dual-function DNA vaccine encoding carcinoembryonic antigen and CD40 ligand trimer induces T cell-mediated protective immunity against colon cancer in carcinoembryonic antigen-transgenic mice. *J. Immunol.* 167:4560–4565.

46. Alexander J, Sidney J, Southwood S, et al. (1994). Development of high potency universal DR-restricted helper epitopes by modification of high affinity DR-blocking peptides. *Immunity.* 1:751–761.

47. Gregor P D, Wolchok J D, Turaga V, et al. (2005). Induction of autoantibodies to syngeneic prostate-specific membrane antigen by xenogeneic vaccination. *Int. J. Cancer.* 116:415–421.

48. Pupa S M, Iezzi M, Di Carlo E, et al. (2005). Inhibition of mammary carcinoma development in HER-2/neu transgenic mice through induction of autoimmunity by xenogeneic DNA vaccination. *Cancer Res.* 65:1071–1078.

49. Barouch D H, Yang Z Y, Kong W P, et al. (2005). A human T-cell leukemia virus type 1 regulatory element enhances the immunogenicity of human immunodeficiency virus type 1 DNA vaccines in mice and nonhuman primates. *J. Virol.* 79:8828–8834.

50. Wolff J A, Budker V (2005). The mechanism of naked DNA uptake and expression. *Adv. Genet.* 54:3–20.

51. Maloy K J, Erdmann I, Basch V, et al. (2001). Intralymphatic immunization enhances DNA vaccination. *Proc. Natl. Acad. Sci. U.S.A.* 98:3299–3303.

52. Tagawa S T, Lee P, Snively J, et al. (2003). Phase I study of intranodal delivery of a plasmid DNA vaccine for patients with Stage IV melanoma. *Cancer.* 98:144–154.

53. Prud'homme G J, Glinka Y, Khan A S, et al. (2006). Electroporation-enhanced non-viral gene transfer for the prevention or treatment of immunological, endocrine and neoplastic diseases. *Curr. Gene. Ther.* 6:243–273.

54. Robertson J S, Griffiths E (2001). Assuring the quality, safety, and efficacy of DNA vaccines. *Mol. Biotechnol.* 17:143–149.

55. Ledwith B J, Manam S, Troilo P J, et al. (2000). Plasmid DNA Vaccines: Investigation of Integration into host cellular DNA. *Intervirology.* 43:258–272.

56. Tang D C, DeVit M, Johnston S A (1992). Genetic immunization is a simple method for eliciting an immune response. *Nature.* 356:152–154.

57. Qiu P, Ziegelhoffer P, Sun J, et al. (1996). Gene gun delivery of mRNA in situ results in efficient transgene expression and genetic immunization. *Gene Ther.* 3:262–268.

58. Jonuleit H, Kuhn U, Muller G, et al. (1997). Pro-inflammatory cytokines and prostaglandins induce maturation of potent immunostimulatory dendritic cells under fetal calf serum-free conditions. *Eur. J. Immunol.* 27:3135–3142.

59. Martinon F, Krishnan S, Lenzen G, et al. (1993). Induction of virus-specific cytotoxic T lymphocytes in vivo by liposome-entrapped mRNA. *Eur. J. Immunol.* 23:1719–1722.

60. Hess P R, Boczkowski D, Nair S K, et al. (2006). Vaccination with mRNAs encoding tumor-associated antigens and granulocyte-macrophage colony-stimulating factor efficiently primes CTL responses, but is insufficient to overcome tolerance to a model tumor/self antigen. *Cancer Immunol. Immunother.* 55:672–683.

61. Jones D H, Corris S, McDonald S, et al. (1997). Poly(DL-lactide-co-glycolide)-encapsulated plasmid DNA elicits systemic and mucosal antibody responses to encoded protein after oral administration. *Vaccine.* 15:814–817.

62. Perrie Y, Frederik P M, Gregoriadis G (2001). Liposome-mediated DNA vaccination: the effect of vesicle composition. *Vaccine.* 19:3301–3310.

63. Chen S C, Jones D H, Fynan E F, et al. (1998). Protective immunity induced by oral immunization with a rotavirus DNA vaccine encapsulated in microparticles. *J. Virol.* 72:5757–5761.

64. Herrmann J E, Chen S C, Jones D H, et al. (1999). Immune responses and protection obtained by oral immunization with rotavirus VP4 and VP7 DNA vaccines encapsulated in microparticles. *Virol.* 259:148–153.

65. Singh M, Vajdy M, Gardner J, et al. (2001). Mucosal immunization with HIV-1 gag DNA on cationic microparticles prolongs gene expression and enhances local and systemic immunity. *Vaccine.* 20:594–602.

66. Sharpe S, Hanke T, Tinsley-Bown A, et al. (2003). Mucosal immunization with PLGA-microencapsulated DNA primes a SIV-specific CTL response revealed by boosting with cognate recombinant modified vaccinia virus Ankara. *Virol.* 313:13–21.

67. Kusakabe K-I, Xin K-Q, Katoh H, et al. (2000). The timing of GM-CSF expression plasmid administration influences the Th1/Th2 response induced by an HIV-1-specific DNA vaccine. *J. Immunol.* 164:3102–3111.

68. Carralot J-P, Probst J, Hoerr I, et al. (2004). Polarization of immunity induced by direct injection of naked sequence-stabilized mRNA vaccines. *Cell. Molec. Life Sci.* 61(18):2418–2424.

69. Girard M P, Osmanov S K, Kieny M P (2006). A review of vaccine research and development: the human immunodeficiency virus (HIV). *Vaccine.* 24:4062–4081.

70. Calarota S, Bratt G, Nordlund S, et al. (1998). Cellular cytotoxic response induced by DNA vaccination in HIV-1-infected patients. *Lancet.* 351:1320–1325.

71. Calarota S A, Weiner D B (2004). Approaches for the design and evaluation of HIV-1 DNA vaccines. *Expert. Rev. Vacc.* 3:S135–S149.

72. Giri M, Ugen K E, Weiner D B (2004). DNA vaccines against human immunodeficiency virus type 1 in the past decade. *Clin. Microbiol. Rev.* 17:370–389.

73. Wang R, Doolan D L, Le T P, et al. (1998). Induction of antigen-specific cytotoxic T lymphocytes in humans by a malaria DNA vaccine. *Science*. 16:476–480.

74. Wang R, Richie T L, Baraceros M F, et al. (2005). Boosting of DNA vaccine-elicited gamma interferon responses in humans by exposure to malaria parasites. *Infect. Immun.* 73:2863–2872.

75. Zuckerman A J (2006). HIV variants and hepatitis B surface antigen mutants. *J. Med. Virol.* 78(suppl 1):S1–S2.

76. Roy M J, Wu M S, Barr L J, et al. (2000). Induction of antigen-specific CD8+ T cells, T helper cells, and protective levels of antibody in humans by particle-mediated administration of a hepatitis B virus DNA vaccine. *Vaccine*. 19:764–768.

77. Rottinhaus S T, Poland G A, Jacobson R M, et al. (2003). Hepatitis B DNA vaccine induces protective antibody responses in human non-responders to conventional vaccination. *Vaccine*. 21:4604–4608.

78. Lowrie D B, Silva C L, Colston M J, et al. (1997). Protection against tuberculosis by a plasmid DNA vaccine. *Vaccine*. 15:834–838.

79. Lowrie D B, Tascon R E, Bonato V L, et al. (1999). Therapy of tuberculosis in mice by DNA vaccination. *Nature*. 400:269–271.

80. Ha S-J, Jeon B-Y, Kim S-C, et al. (2003). Therapeutic effect of DNA vaccines combined with chemotherapy in a latent infection model after aerosol infection of mice with Mycobacterium tuberculosis. *Gene Ther.* 10(18):1592–1599.

81. Shaw D R, Strong T V (2006). DNA vaccines for cancer. *Frontiers Biosci.* 11: 1189–1198.

82. Lollini P L, De Giovanni C, Pannellini T, et al. (2005). Cancer immunoprevention. *Future Oncol.* 1:57–66.

83. Rosenberg S A, Yang J C, Sherry R M, et al. (2003). Inability to immunize patients with metastatic melanoma using plasmid DNA encoding the gp 100 melanoma-melanocyte antigen. *Hum. Gene Ther.* 14(8):709–714.

84. Pavlenko M, Roos A K, Lundqvist A, et al. (2004). A phase I trial of DNA vaccination with a plasmid expressing prostate-specific antigen in patients with hormone-refractory prostate cancer. *Br. J. Cancer.* 91(4):688–694.

85. Timmerman J M, Singh G, Hermanson G, et al. (2002). Immunogenicity of a plasmid DNA vaccine encoding chimeric idiotype in patients with B-Cell lymphoma. *Cancer Res.* 62:5845–5852.

86. Klencke B, Matijevic M, Urban R G, et al. (2002). Encapsulated plasmid DNA treatment for human papillomavirus 16-associated anal dysplasia: a Phase I study of ZYC101. *Clin. Cancer Res.* 8:1028–1037.

87. Sheets E E, Urban R G, Crum C P, et al. (2003). Immunotherapy of human cervical high-grade cervical intraepithelial neoplasia with microparticle-delivered human papillomavirus 16 E7 plasmid DNA. *Am. J. Obstet. Gynecol.* 188:916–926.

88. Boczkowski D, Nair S K, Snyder D, et al. (1996). Dendritic cells pulsed with RNA are potent antigen-presenting cells in vitro and in vivo. *J. Exp. Med.* 184:465–472.

89. Gilboa E, Vieweg J (2004). Cancer immunotherapy with mRNA-transfected dendritic cells. *Immunol. Rev.* 199:251–263.

90. Granstein R D, Ding W, Ozawa H (2000). Induction of anti-tumor immunity with epidermal cells pulsed with tumor-derived RNA or intradermal administration of RNA. *J. Invest. Dermatol.* 114:632–636.

91. Nishikawa H, Kato T, Tanida K, et al. (2003). CD4$^+$ CD25$^+$ T cells responding to serologically defined autoantigens suppress antitumor immune responses. *Proc. Natl. Acad. Sci. U.S.A.* 100(19):10902–10906.

92. Wolf B L, Hamilton R G (2002). Near-fatal anaphylaxis after Hymenoptera venom immunotherapy. *J. Aller. Clin. Immunol.* 102(3):527–528.

93. Chua K Y, Huangfu T, Liew L N (2006). DNA vaccines and allergic diseases. *Clin. Exper. Pharmacol. Physiol.* 33(5–6):546–550.

94. Wherry E J, Barber D L, Kaech S M, et al. (2004). Antigen-independent memory CD8 T cells do not develop during chronic viral infection. *Proc. Natl. Acad. Sci. U.S.A.* 101:16004–16009.

95. Nishikawa M, Hashida M (2002). Nonviral approaches satisfying various requirements for effective in vivo gene therapy. *Biol. Pharm. Bull.* 25:275–283.

96. Budker V, Zhang G, Danko I, et al. (1998). The efficient expression of intravascularly delivered DNA in rat muscle. *Gene Ther.* 5(2):272–276.

97. Zhang G, Budker V, Williams P, et al. (2001). Efficient expression of naked dna delivered intraarterially to limb muscles of nonhuman primates. *Hum. Gene Ther.* 12(4): 427–438.

98. Yoshimura K, Rosenfeld M A, Nakamura H, et al. (1992). Expression of the human cystic fibrosis transmembrane conductance regulator gene in the mouse lung after *in vivo* intratracheal plasmid-mediated gene transfer. *Nucleic Acids Res.* 20:3233–3240.

99. McCluskie et al. (1998). *Antisense Nucl. Acid Drug Devel.* 8:401–414.

100. Liu Y, Mounkes L C, Liggitt H D, et al. (1997). Factors influencing the efficiency of cationic liposome-mediated intravenous gene delivery. *Nature Biotechnol.* 15:167–173.

101. Kawakami S, Fumoto S, Nishikawa M (2000). In vivo gene delivery to the liver using novel galactosylated cationic liposomes. *Pharm. Res.* 17(3):306–313.

102. Kawakami S, Sato A, Nishikawa M, et al. (2000). Mannose receptor-mediated gene transfer into macrophages using novel mannosylated cationic liposomes. *Gene Ther.* 7:292–299.

103. Wu G Y, Wu C H (1988). Receptor-mediated gene delivery and expression in-vivo. *J. Biol. Chem.* 263:14621–14624.

104. Nishikawa M, Takemura S, Yamashita F (2000). Pharmacokinetics and in vivo gene transfer of plasmid DNA complexed with mannosylated poly(L-lysine) in mice. *J. Drug Target.* 8:29–38.

105. Perales J C, Ferkol T, Beegen H, et al. (1994). Gene Transfer in vivo: Sustained expression and regulation of genes introduced into the liver by receptor-targeted uptake. *Proc. Natl. Acad. Sci. U.S.A.* 91:4086–4090.

106. Hoare M, Levy M, Bracewell D G, et al. (2005). Biopocess engineering issues that would be faced in producing a DNA vaccine at up to 100 m^3 fermentation scale for an influenza pandemic. *Biotechnol. Prog.* 21:1577–1592.

107. Zi, Yao, Zhu, et al. (2006). *Euro. J. Immuno.* 36(4):875–886.

7.3

FORMULATIONS AND DELIVERY LIMITATIONS OF NUCLEIC-ACID-BASED THERAPIES

TATIANA SEGURA

University of California, Los Angeles, Los Angeles, California

Chapter Contents

7.3.1 Introduction — 1013
7.3.2 Systemic and Direct Injection Delivery — 1015
    7.3.2.1 Formulations — 1016
7.3.3 Matrix-Based Delivery — 1031
    7.3.3.1 Direct Encapsulation and Release — 1033
    7.3.3.2 Matrix-Tethered Delivery — 1037
7.3.4 Delivery Limitations and Current Solutions — 1038
    7.3.4.1 Extracellular Limitations — 1038
    7.3.4.2 Intracellular Limitations — 1040
7.3.5 Outlook — 1044
    References — 1045

7.3.1 INTRODUCTION

The ability to control the gene expression of a desired cellular population via the delivery of nucleic acids has inspired applications in clinical and basic science. In the past, nonviral gene delivery referred to the delivery of plasmid DNA (pDNA) whereby the expression of a target protein, which could restore normal biological function or aid during wound healing, would be increased. More recently, however, the term *gene delivery* has expanded to include the delivery of small interfering RNA (siRNA) and oligonucleotides (ON), which can be used to decrease or

Handbook of Pharmaceutical Biotechnology, Edited by Shayne Cox Gad.
Copyright © 2007 John Wiley & Sons, Inc.

downregulate the expression of a target protein. Ongoing clinical trials using gene delivery as a therapeutic agent include cancer [1], cystic fibrosis [2–4], adenosine deaminase deficiency [5], and infectious diseases such as acquired immunodeficiency syndrome (AIDS) [6].

Gene delivery employs both viral and nonviral vectors to deliver nucleic acids. Although viral delivery strategies are in general more efficient than their nonviral counterparts, an increasing concern about immune responses to viral vectors *in vivo*, which could pose additional risks for patients undergoing gene therapy [7], has motivated the development of nonviral delivery vectors that aim to synthetically mimic biologically produced viruses. Nonviral delivery vectors have to be engineered to be able to package plasmid DNA to form particles that can (1) travel through biological fluids without losing activity, (2) target appropriate cells, (3) be internalized by the targeted cell population, and (4) finally trafficked to the intracellular site of action (e.g., nucleus and/or cytosol; Figure 7.3-1). Furthermore, the ideal delivery system would achieve the above-mentioned points while being nonimmunogenic and nontoxic. Although there have been numerous approaches to the engineering of delivery vectors, this chapter will focus mainly on nonviral delivery vectors that show promising approaches to overcome the delivery limitations mentioned above or that have been tested *in vivo*.

Strategies to administer nonviral delivery vectors *in vivo* can be divided into two main approaches: systemic delivery or direct injection of DNA nanoparticles and the implantation of a matrix that releases "naked" DNA or DNA nanoparticles. Systemic delivery aims at engineering delivery vectors that can be delivered noninvasively through a bodily fluid such as intravenously, transdermally, and orally, while targeting an appropriate tissue (e.g., tumor) or organ (e.g., lung). Local delivery, on the other hand, aims at bypassing the tissue-targeting step by delivering the nonviral vectors directly (or nearby) to the site of action, typically with the use of a polymeric matrix. The polymeric matrix can be designed to slowly release the

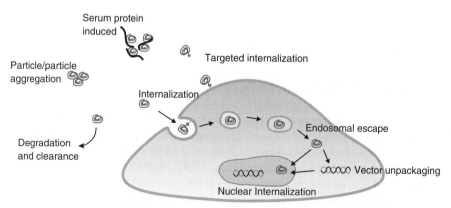

Figure 7.3-1. Limitations of nonviral gene delivery formulations are shown. Limitations can be divided into extracellular limitations, such as (1) interaction with serum components, (2) particle–particle interaction, (3) clearance from target site, and (4) targeting, and intracellular limitations, such as (1) internalization, (2) endosomal escape, (3) vector unpackaging, and (4) nuclear internalization.

DNA nanoparticles to the local environment where cells can internalize them or can be designed so that it can be infiltrated with local cells that can take up the DNA nanoparticles residing inside the matrix.

Design parameters for targeted cellular internalization and intracellular trafficking are similar for systemic and local delivery approaches; however, the design of the vector physical properties such as surface coating and particle size must be tailored to the method of application. This chapter will be divided into sections with Section 7.3.2 discussing systemic and direct injection formulations (Tables 7.3-1–7.3-4), Section 7.3.3 discussing matrix-based delivery formulations (Tables 7.3-2–7.3-5), and Section 7.3.4 discussing limitations with nonviral nucleic acid delivery and current solutions (Tables 7.3-4 and 7.3-7).

7.3.2 SYSTEMIC AND DIRECT INJECTION DELIVERY

In vivo production and secretion of therapeutic proteins by DNA delivery can be achieved through either systemic or local administration, each providing a unique opportunity for gene therapy. Systemic delivery allows noninvasive access to many target cells and tissue that are not accessible by direct administration. The most common approach involves the complexation of naked DNA with cationic lipids (lipoplexes) or cationic polymers (polyplexes) that package the DNA into nano-sized aggregates. The major factors limiting the *in vivo* effectiveness for systemically delivered lipoplexes and polyplexes is colloidal instability and ineffective targeting. For intravenous administration, complexes are rapidly eliminated from the blood stream (<30 minutes) and found mostly in the liver and to a lesser extent in organs with fine capillary beds like skin, muscle, and intestine [8]. The interaction of biomolecules from the physiological fluid with the complexes induces aggregation and dissociation of the complexes, which limits bioavailability and internalization [9, 10]. The association of serum components (albumin, heparin, lipoprotein, or specific opsonins) with polyplexes and lipoplexes, which can target the complex for clearance by macrophages and is affected by the surface charge density and surface morphology of complexes [11]. Cationic formulations (i.e., lipoplexes and polyplexes) can provide an enhanced stability against degradation and may allow for a more efficient interaction with the negatively charged endothelial cell; however, noncationic or less cationic formulations may be required to avoid accumulation in the reticuloendothelial system. Variations in the charge ratio may be used as part of a passive targeting strategy. For lipoplexes delivered intravenously, different formulations (e.g., charge ratio) can be used to target the lung, liver, blood cells, or tumors [12, 13].

To avoid the difficulties of crossing the endothelium and nonspecific uptake by the liver, DNA can be delivered directly to the target tissue by injection of polyplexes, lipoplexes, or microspheres with encapsulated DNA. Injection of lipoplexes and polyplexes into the desired tissue can be used to achieve *in vivo* gene transfer. However, transfection depends on the physico-chemical properties of the complexes and the rate of clearance from the tissue, which varies from tissue to tissue and depends on factors such as the lymph supply. Nevertheless, polyplexes have been found to be sufficiently small and stable so as to diffuse throughout the brain ventricular spaces after a local injection injection [14].

7.3.2.1 Formulations

Unprotected plasmid DNA (naked DNA) is unstable under *in vivo* conditions, due to rapid degradation by serum nucleases [10, 15, 16]. Therefore, carriers or vectors are necessary to protect DNA or RNA from degradation to facilitate uptake into specific cells, and to transfer the DNA or RNA their target location (e.g., nucleus or cytosol). Delivery vectors for systemic delivery are designed to increase gene transfer function by enhancing the stability of DNA, the efficiency of cellular uptake and intracellular trafficking, and the biodistribution of DNA. In this section, the formulations that have been developed to deliver nucleic acids such as plasmid DNA (pDNA), oligonucleotide (ON), or small interfering RNA (siRNA) will be reviewed. Tables 7.3-1 and 7.3-2 summarize the most widely used cationic polymer and cationic lipid formulations.

Cationic Polymers. Cationic polymers have been used since the late 1980s [20] as formulations for nucleic acid delivery. Cationic polymers contain high densities of primary, secondary, or tertiary amines, some of which are protonated at neutral pH. This high density of positive charges allows the cationic polymers to form stable condensed structures with pDNA, termed polyplexes, which are capable of entering the cell. Furthermore, polyplexes protect the DNA from nucleases found in serum and other extracellular environments. Table 7.3-1 shows the structure of frequently used cationic polymers, which have been used *in vivo* to deliver therapeutic genes, although many novel polymers are being developed, including polyallylamine [29], peptoids [30], poly(dimethyl aminoethyl methacrylate) [31, 32], poly(trime-thyl aminoethyl methacrylate) [33], poly(β-amino ester) [34,35], and poly(phosphoester) [36–38]. Cationic polymers vary widely in molecular architecture, ranging from linear to highly branched molecules, which influences their complexation with nucleic acids as well as their transfection efficiency. In addition to providing positive charges for DNA complexation, the primary amines also serve as functional groups to chemically modify the polymers with ligands and peptides that can enhance one or more of the steps in the transfection process (Section 7.3.4, Tables 7.3-3, 7.3-4). Furthermore, tertiary and some secondary amines are typically neutral at physiological pH and have been found to aid in intracellular trafficking by facilitating the release of the polyplexes from the endosome though endosomal buffering (Section 7.3.4.2). The use of cationic polymers and lipids to deliver therapeutically relevant genes and their therapeutic outcome is summarized in Tables 7.3-5, 7.3-6, and 7.3-7 for formulations delivered systemically or via a matrix-based delivery approach.

Interaction of Cationic Polymers with DNA. Cationic polymers form aggregates with DNA, termed polyplexes, via electrostatic interactions, which protect the nucleic acid from degradation [121]. During the complexation of DNA with cationic polymers, the extended structure of DNA is changed to a more condensed configuration forming aggregates in the nanometer size range. Polyplexes typically give rise to particles with spherical, globular, or rod-like structures, which are 20 to 200 nm in diameter [122, 123] but can reach up to 1000 nm. Evidence that electrostatic interactions mediate the complexation of cationic polymers include Fourier transform infrared resonance (FTIR) data showing a reduction of the asymmetric

TABLE 7.3-1. Structure of Commonly Used Cationic Polymer Formulations for Nonviral Gene Delivery

PEI, Poly(ethylene imine) [17]	PAMAM, Poly(amidoamine) dendrimer [18]	Cyclodextrin containing polymers [19]

Poly(amino acids) [20]	Chitosan [21]	Dextran Spermine [22]

Table 7.3-2. Structure of Commonly Used Lipid Formulations for Nonviral Gene Delivery

Cationic lipids	Neutral lipids
DOTMA [2,3-bis(oleoyl)propyl]trimethyl ammonium chloride [23]	DOPE dioleoylphosphatidyl ethanolamine [24]
DMRIE 1,2-dimyristyloxypropyl-3-dimethyl-hydroxyethyl ammonium bromide [25]	DOPC dioleoylphosphatidyl choline, Chol: cholesterol [26]
DODAB dioctadecyldimethyl ammonium bromide [27]	Chol cholesterol [28]
DOTAP 1,2-diacyl-3-trimethylammonium propane [29a]	DC-Chol 3[N-(N′,N′-dimethylaminoethane)-carbamoyl] cholesterol [28]

TABLE 7.3-3. Extracellular limitations to nonviral gene delivery

Vector Stabilization	Mechanism	Refs.
PEG	Provides stability against serum components	[39]
Cyclodextrin	Reduces toxicity and inflammation, provides stability against serum components	[40] (Rev. [41])
Chitosan	Provides stability against serum components	[42]
Poly(glucaramidoamine)	Provides stability against serum components	[43]
CHKKKKKKHC	Covalent stabilization	[44]

Targeting	Cell Surface Target	Target Cell/Tissue	Refs.
Folate	Folate receptor	Cancer cells/ tumor	[45] (Rev. [46])
Transferrin	Transferring receptor	Cancer cells/ tumor	[47] (Rev. [48])
Mannose	Mannose receptor	Dendritic cells	[49, 50]
Galactose	Gal/GalNAc receptor	Hepatocytes/dendritic cells	[51]
Anti-OV-TL16 antibody fragment	OA3 receptor	Ovarian carcinoma cells	[52]
Anti-HER2 antibody	Human epidermal growth factor receptor-2	Breast, ovarian cancer cells	[53]
Anti-PECAM antibody	Platelet endothelial cell adhesion molecule	Lung endothelia	[54]
RGD peptide	Integrin	Broad range of cells	[55]
Anti-CD3	CD3 receptor	Lymphocytes	[56]
EGF (epidermal growth factor)	EGF receptor	Cancer cells	[57]

phosphate stretching vibration of plasmid DNA after complexation with polyethyleneimine (PEI) [124]. Furthermore, microcalorimetric measurements have shown that complex formation between the graft copolymer poly(ethylene oxide)-PEI and a poly(dTA) result from the formation of ion pairs between ionized amino groups of PEI segments of the copolymer and the phosphate groups of DNA [125]. An increase in salt concentration generally results in a decrease binding affinity between the cationic polymer and the DNA probably due to a charge shielding effect at the higher salt concentrations [125–127].

The complexation and condensation behavior depend on the polymer's physical properties, including molecular weight, density of charges, whether it is linear or branched, and the ratio of polymer to DNA [10, 15, 127]. The molecular weight of the cationic polymer influences both the condensation behavior as well as the complex size. In general, for most cationic polymers, an increase in molecular weight results in a decrease of complex size until a condensation limit is achieved. For example, for PEI, it has been found that molecular weights higher than 25 kDa showed no further increase of complex size, whereas molecular weights of 2 kDa or lower result in a decreased ability to complex DNA to form small complexes

TABLE 7.3-4. Intracellular Limitations to Nonviral Gene Delivery

Toxicity	Mechanism	Refs.
PEG	Charge shielding	[47, 58, 59]
Cyclodextrin	Charge shielding	[40] (Rev. [41])
Poly(glucaramidoamine)	Charge shielding	[43]
P(EI-co-LSA)	Degradability	[60]
PEI-PEG-PEL	Degradability	[61]

Endosomal Escape	Mechanism	Sequence	Refs.
Secondary and tertiary amines	Endosomal buffering	Not applicable	[17, 18]
INF (influenza virus-derived sequence)	Endosomolytic at pH 5.0 by adopting an amphipathic α-helical conformation	GLFEAIEGFIENGWEGMIDGWYG	[62]
KALA	pH-dependent endosomolytic activity by adopting an α-helical conformation	WEAKLAKALAKALAKHLAKALA KALKACEA	[63, 64]
H5WYG	pH-dependent endosomolytic between pH 7 and 6.	GLFHAIAHFIHGGWHGLIHGWYG	[65]
GALA	pH-dependent membrane fusion	WEAALAEALAEALAEHLAEALA EALEALAA	[66, 67]
poly(α-alkyl acrylic acids)	pH-dependent hydrophilic-to-hydrophobic conversion causes membrane disruption	Poly(propyl acrylic acid)	[68, 69]

Vector Unpackaging	Mechanism	Refs.
Polylysine	Partial carbohydrate substitutions to reduce electrostatic interactions	[70]
Polylysine	Partial polyhydroxyalcanoyl (gluconoyl) substitution to reduce electrostatic interactions	[71]
Polylysine	Short polylysine facilitates unpacking	[72]
PEI	Partial acetylation PEI decreases the physiological buffering capacity	[70, 71, 73]
Chitosan	The chitosan is degraded by the predelivered chitosanase	[74]
Chitosan-NIPAAm	Thermoresponsive enhances decomplexation below critical solubility temperature	[75]
CHKKKKKHC	Environmentally sensitive disulfide bonds	[44]

Nuclear Localization	Mechanism	Sequence	Refs.
SV40	Importin-α	PKKKRKVEDPY	[76]
M9	Transportin	GNQSSNFGPMKGGNFGGRSSGP YGGGGQYFAKPRNQGGY	[77]
Importin-β	NPC	Importin-β (1-643aa)	[78]
Mu	Unknown	MRRAHHRRRRASHRRMRGG	[79]

TABLE 7.3-5. *In vivo* Systemic and Direct Injection Delivery Studies Using Polyplexes Vectors to Deliver Therapeutically Relevant Genes

Formulation	Therapeutic Target	Nucleic Acid	Gene	Species/Delivery	Ref.
PEI	Viral infection	siRNA	Influenza virus	Mice/retro-orbitally	[80]
PEI	Cancer	siRNA	HER-2	Mice/i.p. injection	[81]
PEI	Cancer	DNA	sst2 receptor	Mice, hamster/local	[82]
PEI	Cancer	DNA	IL-12	Mice/inhalation	[83]
PEI	Cancer	DNA	p53	Mice/inhalation	[83]
PEG-PEI-Chol	Cancer	DNA	pmIL-12	Mice/Local	[84]
PEI/Pluronic F127	Vascular diseases	DNA	Hu-uPA	Rats/Local	[85]
Chitosan	Viral infection	DNA	RSV antigens	Mouse/Intranasal	[86]
Cyclodextrin	Cancer	siRNA	EWS-FLI1	Mice/Tail vein injection	[87]
Cyclodextrin-PEG	Cancer	DNA	DNAzyme	Mice/Intraveneous and intraperitoneal	[88]

TABLE 7.3-6. *In vivo* Systemic and Direct Injection Studies Using Lipoplex Formulations to Deliver Therapeutically Relevant Nucleic Acids

Formulation	Therapeutic Target	Nucleic Acid	Gene	Species/Delivery	Ref.
DOTAP:Chol	Cancer	DNA	Mda-7/IL-24	Mice/intravenous, local	[89]
DC-Chol:DOPE	Cancer	DNA	HLA-A2	Human/direct injection	[90]
DC-Chol:DOPE	Cancer	DNA	HLA-B13	Human/direct injection	[90]
DC-Chol:DOPE	Cancer	DNA	H-2K	Human/direct injection	[90]
DC-Chol:DOPE	Cancer	DNA	HLA-B7	Human/intratumorally and intraveneous	[91]
DC-Chol	Cancer	DNA	HER-2/neu	Mice/intraperitoneal injection	[92]
DC-Chol	Cancer	DNA	HER-2/neu	Human/intracavitary injection	[93]
LSFV	Cancer	DNA	IL-12	Human/intraveneously	[94]
DMRIE:DOPE	Cancer	DNA	HLA-B7	Human/intratumorally	[95]
TMAG:DLPC:DOPE	Cancer	DNA	IFN-β	Human/direct injection	[96]
DMRIE:DOPE	Cancer	DNA	HLA-B7 and $\beta 2$-microglobulin	Human/intratumorally	[97]
DC-Chol:DOPE	Cystic fibrosis	DNA	CFTR	Human/direct instillation	[98]
DC-Chol:DOPE	Cystic fibrosis	DNA	CFTR	Human/topical application	[99]
67A (GL-67:DOPE:DMPE:PEG)	Cystic fibrosis	DNA	CFTR	Human/aerosolization	[100]
DC-Chol:DOPE	Cystic fibrosis	DNA	CFTR	Human/direct instillation	[101]
DOTAP	Cystic fibrosis	DNA	CFTR	Human/aerosolization	[102]
DOTMA:DOPE	Anti-inflammatory	siRNA	α_1-Antitrypsin	Human/direct application	[103]
Oligofectamine	Viral infection	siRNA	Nucleoprotein acidic polymerase	Mice/intranasal	[104]

1023

TABLE 7.3-7. Matrix Delivery *in vivo* Studies Using Nonviral Vectors to Deliver Therapeutically Relevant Genes

Material/Matrix	Nucleic Acid	Result	Gene	Species/Delivery	Ref.
Collagen	DNA	Bone formation	BMP-4/hPTH1-34	Rat/femur	[105]
Collagen	DNA	Bone formation	hPTH1-34	Canine/bone	[106]
Atelocollagen	DNA	Increase platelet number	FGF-4	Mice/inttramuscular	[107]
Atelocollagen	siRNA	Tumor growth suppression	FGF-4	Mice/intratumorally	[108]
Atelocollagen	siRNA	Tumor growth suppression	VEGF	Mice/intratumorally	[109]
Collagen	DNA	Granulation tissue and ephithelization	PDGF	Rabbit ear	[110]
Collagen	DNA/PLL	Survival of axotomized RGCs	FGF2, BDNF, NT-3	Rat optic nerve	[111]
PLG	DNA	Small blood vessel formation	Endothelial locus-1	Mouse subdermal	[112]
PLG	DNA	Enhanced matrix deposition and vascularization	PDGF	Rat/subdermal	[113]
PLGA	DNA	Bone formation	BMP-4	Rat cranium	[114]
PLG	DNA	Increase blood vessel density	VEGF	Mice/subcutaneous	[115]
Gelatin	DNA	Decreased blood urea nitrogen	MMP-1	Mouse/renal subcapsule	[116]
Gelatin	DNA	Angiogenesis	FGF-4	Rabbit/hindlimb	[117]
EVAc	DNA	Induction of specific IgA	LDH-C4	Mice/intravaginal	[118]
Fibrin	DNA/PLL	Mature blood vessel formation	pSG5-VEGF$_{165}$	Rabbit/hindlimb schemia model	[119]
Fibrin	DNA/PLL	Mature blood vessel formation	HIF-1αODD	Mice/back wound	[120]

[128]. The stability and size of the polyplexes formed between cationic polymers and DNA have also been correlated to primary amine content [10, 129]. Low branched PEI (low density of primary amines) requires higher N/P rations to complete condensation compared with highly branched derivatives (high density of primary amines) [130]. Last, complex size tends to decrease with an increasing polymer-to-DNA ratio [10, 121, 131].

Polyplexes as Cellular Transfection Agents. In addition to reducing the size of DNA and protecting the DNA from degradation by serum nucleases, binding of polycations to DNA typically results in polyplexes having a net positive charge. This positive charge has been found to be essential for efficient transfection [14, 17, 130]. It is thought that the positively charged polyplexes electrostatically interact with the negatively charged proteoglycans of the cell membrane [132], resulting in nonspecific adsorptive endocytosis (Figure 7.3-1). The polyplexes, which are now trapped inside the endosomes, need to escape before endosome–lysosome fusion occurs, which would result in polyplex degradation inside the lysosomes (Section 7.3.4.2). Cationic polymers such as PEI have the advantage over other cationic polymers [e.g., poly(amino acids)] in that they do not require an endosomal lysotropic agent to be added to achieve efficient transfection [17, 18]. Once released from the endosome the polyplexes need to transport the DNA to the nucleus where it can be transcribed or siRNA or ON to the cytosol, where they can mediate gene downregulation. The transition from the cytosol to the nucleus is not well understood, and it depends on the physical properties of the polymer and the cell cycle and degree of mitotic activity of the cells. To aid the transition from the cytosol to the nucleus cationic polymers have been modified with nuclear localization sequences typically derived from viruses (Section 7.3.4.2). Full unpackaging or decomplexation of the siRNA and ON in the cytosol is critical for their action. Strategies to enhance decomplexation involve the introduction of environmentally sensitive bonds [44], reducing the affinity of the cationic polymer for the DNA [70–73] and the delivery of enzymes that can degrade the cationic polymer [47] (Table 7.3-4 in Section 7.4.3).

Although positively charged polyplexes promote efficient transfection *in vitro*, they lead to aggregation and low transfection *in vivo* with most polyplexes accumulated in the liver after systemic administration [133]. The aggregation of polyplexes has been attributed to the presence of negatively charged proteins in serum, which can mediate bridging of multiple polyplexes together [8]. Strategies to generate neutral complexes, which do not aggregate or interact with serum components, have been developed and are discussed in Section 7.3.4.1.

Poly(ethylenimine). In recent years, poly(ethylenimine) (Table 7.3-1, PEI) has emerged as a widely used cationic polymer to mediate gene transfer *in vivo* and *in vitro* (for review, see Ref. 10), resulting in two commercially available transfection products ExGene (Fermentas, Inc., Hanover, MD) and jetPEI (Polyplus-transfection, San Marcos, CA). PEI has a protonatable hydrogen every two carbons in its structure, making it the polycation with the highest density of protonable amines [17]. Branched PEI is synthesized using the acid catalyzed ring opening polymerization of aziridine, resulting in a theoretical 1:2:1 ratio of primary to secondary to tertiary amines. However, C-13 nuclear resonance spectroscopy measurements have shown that the degree of branching is closer to 1:1:1 for most commercially available PEIs [134], which indicates a higher degree of branching. Linear PEI, on the other hand, is synthesized by ring opening polymerization of 2-ethyl-2-oxazoline followed by acid hydrolysis [135]. Linear and branched PEI can be purchased in a wide variety of molecular weights ranging from less than 1000 Da to 1.6×10^3 kDa; however, the range most typically used for gene delivery is 5 to 25 kDa, mostly due to toxicity caused by higher molecular weight PEIs (Section 7.3.4.2). PEI forms complexes with DNA ranging from 77 to 300 nm in size at N/P ratios ranging from 6 to 10 [10]. High-molecular-weight PEI forms smaller, more stable particles and achieves higher transfection efficiencies than low-molecular-weight PEI; however, it is more toxic [128]. Similarly, highly branched PEI forms smaller particles that are more stable than low-branched PEI and achieves higher transfection efficiencies, but again, it is more toxic than low-branched PEI.

PEI has be used extensively *in vivo* to deliver DNA and siRNA locally to the brain [14, 136] cornea [137], tumors [82, 84], and vasculature [85] and systemically to the lung [80, 83] and tumors [81]. The charge of the DNA/PEI complexes delivered *in vivo* have a profound effect on their organ targeting. Delivery of positively charged complexes in the tail vein in mice resulted in gene transfer primarily in the lungs, whereas surface shielding transferring-PEI–DNA complexes resulted in preferential gene delivery to distantly growing tumors [138]. Tables 7.3-5, 7.3-6 (systemic), and 7.3-7 (matrix) summarize *in vivo* studies involving nonviral vectors and their therapeutic outcome.

Poly(amino acids). Cationic poly(amino acids) (peptide, Table 7.3-1) are some of the most commonly used cationic polymers for gene delivery. Cationic peptides such as polylysine (Table 7.3-1, PLL) are commercially available in a variety of molecular weights. However, commercially available PLLs are highly polydisperse. To synthesize monodisperse peptides, solid phase synthesis is used that involves the immobilization of the growing peptide on a solid support and a series of protecting/deprotecting synthetic steps (e.g., Fmoc chemistry). The sequential addition of each amino acid provides peptides an absolute level of control over the sequence of the growing peptide, which allows for the specific attachment of targeting ligands or other desired modifications anywhere along the molecule.

PLL and its derivatives are the most commonly used cationic peptides for gene delivery, typically used at charge ratios (+/−) ranging from 3:1 to 6:1. As increasing amounts of PLL are added to DNA, the structure changes from circular to thick, flattened to compact, and finally to toroids and rods at a charge ratio of 6:1 [122]. The diameter and cross section of the toroids are approximately 140 nm and 44 nm, respectively [122]. As with other cationic polymers, the ideal length of the PLL represents a balance between two competing effects: effective condensation and cytotoxicity. Relative to low molecular weight, the high-molecular-weight PLL forms tighter, smaller condensates that are more resistant to the effects of salt concentration and sonication [139]. Cell transfection by PLL is typically lower than other cationic polymers that possess a buffering capacity such as poly(ethylene imine) and to cationic lipids. Extensive modifications aiming at improving the transfection efficiency of PLL have been performed, including the partial modification of the PLL side chains with histidine and imidazole groups, to enhance endosomal buffering and escape (Section 7.3.4.2) [65, 140], the addition of PEG to enhance polyplex stability, the addition of targeting ligands to enhance internalization and targeting, and the addition of covalent bonds within the polyplex to enhance its stability and polyplex unpackaging (Section 7.3.4). Furthermore, peptides containing multiple lysines have been synthesized such as Cys–His–(Lys)$_6$–His–Cys, which enhances endosomal escape, while providing covalent stabilization of the polyplexes via disulfide bonds [44].

PLL and its derivatives have not been as extensively used *in vivo* as other cationic polymers to deliver therapeutically relevant genes; however, they have found some success in systemic delivery to the liver [141] and matrix-based delivery to enhance angiogenesis [120]. Tables 7.3-5, 7.3-6 (systemic), and 7.3-7 (matrix) summarize the *in vivo* studies involving nonviral vectors and their therapeutic outcome.

Cationic Dendrimers. Dendrimers are polymers that branch out from a multifunctional core in a symmetric fashion. After each subsequent monomer addition (a generation), the number of attachment points increases symmetrically generating more branches. Cationic dendrimers have an attractive architecture for gene transfer because their well-defined structure and robust chemistry enables the synthesis of many generations of protonatable amines (for review of dendrimers as gene delivery vectors, see 142). Dendrimers such as poly(amidoamine) dendrimers (PAMAM, Table 7.3-1) and poly(propylenimine) (PPI, Table 7.3-1) have been shown to be capable of mediating gene delivery *in vitro* and *in vivo* [18, 131, 143, 144]. PAMAM is commercially available and is synthesized from either an ammonia or ethylenediamine core by successive addition of methyl acrylate and ethylenediamine [142]. PPI is also commercially available and is synthesized from a butylenediamine core by successive addition of acrylonitrile to a primary amino group followed by hydrogenation of nitrile groups to primary amino groups. The surface charge and diameter of the dendrimers is determined by the number of synthetic steps (i.e., number of generations) [142]. The generation of PAMAM used to complex DNA determines particle size and transfection efficiency. Complexation of DNA with fifth-generation (G5) PAMAM dendrimers produces monodisperse condensates with a radius below 200 nm [18]. At charge ratios (+/−) above or equal to one, no free DNA is observed [18]. Although PAMAM dendrimers of generations 3 (G3) to 10 (G10) can form stable complexes with DNA, the higher generations of dendrimers (G5 to G10) can transfect cells at higher efficiencies [18, 145]. Transfection efficiency of PAMAM is enhanced by partial degradation of the dendrimer structure, forming a less homogeneous structure with a higher density of primary amines [144]. The partially degraded PAMAM dendrimer has resulted in the transfection reagent Superfect. Cationic dendrimers have been used *in vivo* to deliver nucleic acids locally to the cornea [146], tumors [147], heart [148], and lung [145] and systemically to the lung [149], liver [16, 131], spleen [149], and tumors [16]. Tables 7.3-5, 7.3-6 (systemic), and 7.3-7 (matrix) summarize the *in vivo* studies involving nonviral vectors and their therapeutic outcome.

Polysaccharide Containing Polymers. Chitosan (Table 7.3-1) is composed of 2-amino-2-deoxy β-D-glucan and is prepared from naturally occurring chitin via alkaline deacetylation. Unlike other cationic polymers chitosan is nontoxic and biodegradable, making it an ideal candidate for therapeutic applications. Although the density of positive charges of chitosan is lower than for other cationic polymers, it has been shown to form particles with DNA, which protect the DNA from degradation [21, 150, 151]. Chitosan/DNA complexes have been found to be in the order of 100 nm with a spherical shape [21, 151]. Like other cationic polymers, chitosan has been chemically modified with a variety of ligands or other functionalities to enhance its transfection efficiency [21]. Chitosan has been recently used to deliver DNA mainly intranasally to the lungs for genetic immunization applications. After intranasal administration, DNA/chitosan complexes encoding for RSV antigen could attenuate pulmonary inflammation and reduce viral titer and viral antigen load [86].

Recently dextran grafted with spermine (Table 7.3-1, D-SPM) has been investigated to mediate DNA delivery. D-SPM are synthesized via reductive amination and can form aggregates with plasmid DNA and mediate efficient gene transfer *in vitro* and *in vivo* [22, 152]. Interestingly, the addition of hydrophobic domains, N-oleyl (ODS), to D-SPM (D-SPM–ODS) enhanced the stability of complexes in serum containing media. Furthermore, the addition of ODS groups achieved high transfection in the presence of serum similar to that achieved with linear PEI [153]. These results suggest that the hydrophobic nature of the complexes may be another strategy to improve transfection in the presence of serum. However, the addition of the ODS groups also increased the toxicity of the D-SPM–ODS when compared with D-SPM, but the toxicity was similar to linear PEI and branched PEI [153].

Cyclodextrins (Table 7.3-1, CDs) are cyclic, cup-shaped molecules with a hydrophilic exterior and a hydrophobic interior, which allows them to form inclusion complexes with hydrophobic molecules. Inclusion complexes are bimolecular complexes in which the "host" forms a cavity into which the "guest" molecule binds through noncovalent interactions. CDs are composed of six, seven, or eight glucose units, termed α, β, and γ, respectively; are water soluble; and are U.S. Food and Drug Administration (FDA) approved as solublizing agents for hydrophobic drugs. Davis et al. have investigated cationic molecules containing CDs as gene delivery vectors since 1999 [19]. Cationic CDs could form stable complexes with DNA with sizes in the 100-nm range. More recently CD–PAMAM [149, 154, 155] and CD–PEI [156, 157] have been investigated and were found to efficiently transfect cells and be less toxic than their parent molecules. Furthermore, both CD–PAMAM and CD–PEI have been successfully used *in vivo* [155, 156]. Furthermore, cyclodextrin containing polymers have been recently used to deliver siRNAs to inhibit tumor growth [87]. As mentioned for other cationic polymers, modifications to include targeting functionality are often desirable of a gene delivery formulation. A key advantage of cyclodextrin containing gene delivery formulations is that functionality can be added through inclusion complexes without covalent modifications. Therefore, targeting molecules, which contain a hydrophobic molecule bound to them, can be "tethered" to the polyplex via cyclodextrin. This approach has been used to introduced PEG [158], galactose [158], transferring [88, 159], and insulin [157] to cationic cyclodextrin formulations.

Cationic Lipids. One of the most investigated approaches for condensing nonviral DNA for efficient gene transfer is the use of cationic lipids. Felgner et al. [23] and Bennett et al. for ON [160] used cationic lipids for the first time to deliver DNA and ONs. More recently, siRNA has been delivered to mammalian cells using commercially available lipids [161–163]. The structures of several frequently used cationic lipids are shown in Table 7.3-2. Although some cationic lipids are used individually to deliver nonviral DNA (e.g., DOTAP), many formulations of cationic lipids also contain a zwitterionic or neutral colipid, such as DOPE (Table 7.3-2) or cholesterol, to enhance transfection. Formulations of cationic lipids have been widely applied for *in vitro* nucleic acid transfection, and more than 30 products are commercially available for this purpose, including Lipofectin (a 1:1 mixture of DOTMA and DOPE), Transfectam, Lipofectase, Lipofect-AMINE, and LipoTaxi [164].

The main components of a cationic lipid are a hydrophilic lipid anchor, a linker group, and a positively charged head group. The lipid anchor is typically either a fatty chain (e.g., derived from oleic or myristic acid) or a cholesterol group, which determines the physical properties of the lipid bilayer, such as flexibility and the rate of lipid exchange [24]. The linker group is an important determinant of the chemical stability, biodegradability, and transfection efficiency of the cationic lipid. Biodegradable lipids are being developed, which can be metabolized by various enzymes (e.g., esterases and peptidases) to minimize toxicity [165, 166]. The linker can also provide sites for the introduction of novel side chains to enhance targeting, uptake, and trafficking. The positively charged head group on the cationic lipid self-assembles with the negatively charged DNA and is a critical determinant of the transfection and cytotoxic properties of liposome formulations. The head groups differ markedly in structure and may be single- or multiple-charged as primary, secondary, tertiary, and/or quaternary amines. The synthesis of novel cationic lipids from libraries of building blocks has provided insight into some structure-activity relationships [167–169]. The hydrophobicity of the lipid moiety has a crucial effect on *in vitro* gene transfer. Multivalent head groups, such as spermine, in a "T-shape" configuration tend to be more effective than their mon-ovalent counterparts at facilitating gene transfer [24, 168, 170]. Generally, increases in the linker length correspond to increases in the gene delivery activity [169]. Continued progress toward a comprehensive relationship among lipid structure, complexation with DNA, and subsequent interaction with the biological environ-ment (e.g., cell membrane and extracellular membrane components) is necessary to facilitate the design of cationic lipids with optimal properties.

Mixing of DNA and cationic lipid results in the collapse of DNA to form a condensed structure (lipoplex), in which nucleic acids are buried within the lipid. The thermodynamic driving force for association of the DNA and lipid is the entropy increase from the release of counter-ions and bound water associated with DNA and the lipid surface [29, 171]. Liposome association with DNA has resulted in tube-like bilayers [28, 172], multilameller complexes [27, 29, 173, 174], as well as structures containing non-bilayer elements [26]. Multilamellar organization, in which DNA is intercalated between cationic lipid bilayers, has been reported for lipolexes resulting from the interaction of DNA with unilameller vesicles of a cat-ionic lipid, such as DOTAP [29] and DODAB [27], respectively, combined with DOPE and cholesterol. Both of these studies used x-ray scattering to show that a periodicity of 6.5 nm occurs in the lamellar structure. The cationic lipid bilayer thickness was 3.9 nm, and the thickness of the water layer was 2.6 nm, which is suf-ficient to include a hydrated DNA double helix with a total diameter of 2.5 nm. A second periodicity was observed, which was attributed to a DNA–DNA correla-tion, which ranged from 2.45 nm to 5.71 nm as the concentration of helper lipid was increased. These lamellar complexes can be converted into hexagonal complexes with DNA confined in inverted lipid micelles by changing the membrane spontaneous curvature or the membrane flexibility [175]. This hexagonal arrangement has demonstrated increased transfection, which is attributed to the relative instability of the complexes due to rapid fusion with anionic vesicles and subsequent DNA release. X-ray diffraction studies and microscopy have illustrated the high degree of variability in lipoplex structure and the need to understand and control the parameters that govern the organization of these structures.

The colloidal properties (e.g., size and stability) of the lipoplexes are principally determined by the cationic lipid/DNA charge ratio and not the composition of the lipid or the helper lipid [176]. The charge ratio (+/−) is typically defined as the number of amines on the cationic lipid relative to the number of phosphate groups on the DNA. A neutral charge ratio (1:1 charge ratio for lipid/DNA) is typically avoided because it results in the formation of large aggregates (>1 μm) [176]. Lipoplexes prepared at a positive charge ratio and a negative charge ratio likely represent structures with different lipid and DNA packaging [176]. At a positive charge ratio, large multilamellar vesicles (LMVs, diameter 300–700 nm) transfected cells more efficiently than the small unilameller vesicles (SUVs, diameter 50–200 nm) [24, 177, 178]. These observations were consistent whether the structures were formed as LMVs or as SUVs that aggregated to form LMVs [177]. The order in which DNA and lipid are mixed is critical and significantly affects the lipid and DNA packing [171, 176]. For the addition of DNA to lipid, a gradual increase in size was observed. When adding lipid to DNA, the particle size remains roughly constant until the amount of lipid positive charge exceeds the nucleic acid negative charge, whereupon the particles grow rapidly in size [171].

The net charge on the lipoplex affects its interactions with other components present *in vivo* and *in vitro* (e.g., media, serum, extracellular matrix glycoproteins, and mucosal secretions), which can limit the transfection efficiency. A positive charge ratio, which facilitates interactions with the cell membrane, is frequently used for *in vitro* studies (3:1), whereas *in vivo* studies may require the charge ratio to be altered because of interactions with components of the physiological environment [11]. The charge ratio of the complex determines the zeta potential, which ranges from −55 mV to +55 mV as the charge ratio is increased [179]. Multivalent anions present in the serum or media can facilitate fusion of the lipids causing an increase in the size of the particle. Polyanions with adequate anionic charge density (e.g., heparin) release DNA from the complex by binding the cationic lipid [179]. Serum can be a complicating factor for positively charged complexes, possibly causing premature release of the DNA from the complex and enhancing degradation by nucleases. For ON:lipid complexes, the various components of serum (e.g., BSA, lipoproteins, macroglobulin) interact with the complexes and alter the complex diameter, zeta potential, and interfere with cellular uptake and nuclear trafficking [180].

Aggregation of lipoplexes in polyelectrolyte solutions occurs rapidly, which can result in loss of activity in less than 24 hours; thus, strategies are being developed to stabilize the particles and prolong their shelf life. To improve lipoplex stability, PEG–PE has been incorporated into the cationic liposome. PEG containing liposomes are prevented from aggregating and interacting with serum components, which increases their stability [181–183]. Alternatively, new preparation methods are being developed in which a detergent is present in solution with the cationic lipid and DNA [184]. Removal of the detergent by dialysis allows the formation of uniform complexes. This process yielded a lipid/DNA suspension that was able to transfect tissue culture cells up to 90 days after formation with no loss in activity. Lyophilization, which is a common approach used for many pharmaceuticals, is also being applied to lipoplexes to increase their shelf life. Cryoprotectants (e.g., sucrose and trehalose) are typically added to prevent aggregation and fusion of plasmid/lipid complexes during lyophilization. Complexes lyophilized in the presence of 0.5-M sucrose or trehalose maintained transfection rates and the sizes of rehydrated complexes as compared with nonlyophhilized controls [185].

Cationic polymers such as poly(amino acids) have also been used in combination with liposomes. DNA is initially complexed with polylysine (PLL) at low charge ratios and cationic lipids are subsequently added to completely condense the DNA. Alternatively, the PLL condensed DNA containing a net positive charge can subsequently be complexed with an anionic lipid [168]. Precondensation with polylysine has been shown to reduce serum inhibition and to enhance the transfection efficiency [186, 187].

Catinic polymer formulations have been extensively used to deliver nucleic acids *in vivo* for the treatment of cancer [89–97] and cystic fibrosis [98–102] as an anti-inflammatory agent [103] and to prevent or reduce viral infections [104]. Table 7.3-6 summarizes the use of cationic polymers to deliver nucleic acids *in vivo* and for a therapeutic purpose.

7.3.3 MATRIX-BASED DELIVERY

Matrix-based delivery has been proposed to enhance gene transfer *in vivo* by delaying clearance from the desired tissue, protecting the nucleic acid from degradation, and extending opportunities for internalization. Furthermore, matrix-based delivery can provide sustained delivery to maintain the vector at effective levels within the target tissue. Many of these properties have been observed for controlled release systems that deliver proteins [188]. Gene transfer from tissue engineering matrices may increase the number of cells expressing the transgene along with the extent of transgene expression, while minimizing the quantity of vector used. Additionally, matrix-based delivery may reduce the number of dosages or the required cumulative dose [188]. Matrix-based delivery of nucleic acids can be divided into two delivery approaches: (1) direct encapsulation and release, where naked DNA, polyplexes, or lipoplexes encapsulated into the matrix for later release; and (2) matrix-tethered delivery (also called substrate-mediated or solid phase delivery), where polyplexes or lipoplexes are immobilized to the matrix for release after matrix or tether degradation (Figure 7.3-2). To date most matrix-based delivery approaches involve the encapsulation of naked DNA and its subsequent release to transfect surrounding cells. Polymer encapsulation strategies can shield the vector against degradation, clearance, and an immune response. Drug release from the matrix into the tissue can be designed to occur rapidly, as in a bolus delivery, or over an extended period of time, which may affect the local concentration and cellular internalization. For rapid release, levels would be expected to quickly rise and decline as the DNA is cleared or degraded. For sustained delivery, the

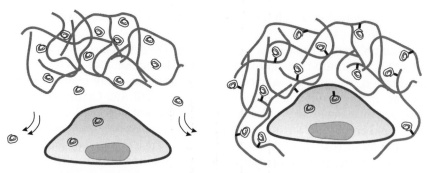

Figure 7.3-2. Matrix-based DNA delivery can be divided into encapsulation and release approaches, where the nucleic acid is encapsulated for later release, and matrix-tethered delivery, where nucleic acid polyplexes or lipoplexes are immobilized directly to a matrix that also supports cell adhesion. These approaches are typically used for applications in tissue engineering where the delivery of nucleic acids is used to augment tissue formation.

concentration may be maintained within an appropriate range by adjusting the release rate (e.g., through the polymer choice). Release of DNA from tissue engineering scaffolds has been observed for times ranging from hours to several weeks [113, 115, 189–191]. Variations in the polymer composition and physical form, which includes properties such as porosity, mass, and size, affects the diffusion of DNA from the vehicle and hence the release rate.

Tethered delivery approaches aim to significantly slow down release kinetics by immobilizing polyplexes or lipoplexes directly to the matrix. Because the polyplexes are immobilized to the matrix, the matrix must also support cell adhesion and infiltration by the surrounding cells rather than function as a nucleic acid reservoir. Internalization of the immobilized polyplexes results after the tether between the polyplex and the matrix is degraded, the matrix is degraded or the affinity of the polyplex for the matrix is reduced due to environmental changes. The release kinetics of the polyplexes in this case depends on the chemistry used for polyplex immobilization and the chemistry of the matrix. Biotynilated polyplexes immobilized to a NeutrAvidin coated substrate was the first example of matrix-tethered delivery [192]. This matrix-tethered delivery strategies resulted in a 100-fold increase of transgene expression compared with bolus delivery, likely due to an increase of the polyplex concentration at the cellular microenvironment. Furthermore, immobilization of polyplexes to the material surface has shown to enhace the stability of the polyplexes in polyelectrolyte solutions, preventing their aggregation [193]. The strength of the tether bond or the number of tethers affects both the density of immobilized polyplexes as well as the transfection efficiency. A high density of tethers results in high surface densities of DNA with low ability to transfect, probably due to the polyplex being too tightly bound to the matrix [192]. More recently, matrix-tethered strategies rely on electrostatic interactions between poly or lipoplexes and the matrix to immobilize the complexes [194, 195].

Matrix-based delivery has been investigated mostly for regenerative medicine applications. Regenerative medicine aims to regenerate tissue by implanting biocompatible and biodegradable scaffolds at sites of injury or disease. The implanted scaffold must provide the mechanical support for the growing tissue and the bioactive signals needed for proper tissue formation while allowing for cell attachment and growth (for reviews on tissue engineering, see Refs. 196 and 197. Delivery of nucleic acids from tissue engineering scaffolds could be used as the bioactive signals needed to recreate the environments needed for tissue formation allowing for direct manipulation of cellular gene expression patterns. In the following sections, matrix-based nucleic acid delivery will be overviewed with Section 7.3.3.1 focusing on encapsulation and release approaches and Section 7.3.3.2 on matrix-tethered delivery approaches. Table 7.3-7 summarizes *in vivo* studies using matrix-based nucleic acid delivery.

7.3.3.1   Direct Encapsulation and Release

Direct encapsulation and release strategies have been proposed to enhance gene transfer *in vivo* by delaying clearance from the desired tissue, protecting the DNA from degradation and extending the opportunities for internalization (Figure 7.3-2). Furthermore, matrix-based delivery can provide sustained delivery to maintain the vector at effective levels within the target tissue.

Synthetic Polymers. Polymers composed of lactide and glycolide (PLGA) are perhaps the most widely used and recognized biodegradable synthetic polymers because these polymers are FDA approved and are generally considered to be biocompatible. Hydrolytic degradation of the polymer produces lactic acid and glycolic acid, two substances naturally involved in metabolic pathways of the body. The degradation rate can be controlled through the composition of the polymer and molecular weight of the chains. These polymers exhibit relatively even polymer chain scission throughout their bulk after being placed *in vivo*. These properties have led to the use of these materials for drug delivery and as scaffolds for tissue engineering [198]. Drug release from these systems typically occurs by a combination of polymer degradation and drug diffusion from the polymer. Gene delivery from polymeric matrices has been applied to tissue repair and wound healing [106, 113]. These matrices are typically implanted at a specific anatomic location where they serve multiple roles as described previously [139]. Initially, the matrix functions to create and maintain a space *in vivo*. However, the matrix also acts as a scaffold to support cell migration, proliferation, and differentiation of healthy cells from the surrounding tissue. As cells invade the matrix, they encounter DNA that is either released from or entrapped within the matrix.

A gas foaming/particulate leaching process can be employed to fabricate interconnected open pore structures of PLGA for controlled release of DNA [113, 199]. This process employs carbon dioxide to process a mixture of polymer and porogen, in order to fuse adjacent polymer particles into an interconnected structure. The DNA can be lyophilized with the microspheres [113] or encapsulated within the microspheres [200, 201]. Lyophilization of DNA with the microspheres can provide large quantities of incorporated DNA, with relatively rapid release kinetics. Incorporation of DNA into the microspheres provides for a more sustained release relative to the lyophilization method [200], with the release kinetics dependent on the polymer molecular weight and microsphere size [201]. DNA can be incorporated into polymer microspheres using several approaches [201–205]. Subcutaneous implantation of scaffolds results in transfected cells observed within the scaffold and the tissue immediately adjacent to the scaffold, with protein production sufficient to promote physiological responses [113, 115]. An alternative approach to fabrication of PLG scaffolds for DNA delivery is electrospinning [206]. Electrospinning creates nonwoven, nanofibered membranous structures that release DNA, with maximal release occurring at approximately 2 hours.

Microspheres can also be employed to deliver DNA to tissues. Microspheres loaded with nonviral DNA can be fabricated from nondegradable and degradable polymers in sizes ranging from 0.1 to 100μm and have been used for applications such as DNA vaccines and systemic protein delivery [198]. One of the main advantages for delivery vehicles of this size is that they can be administered in a minimally invasive manner (e.g., direct injection and oral delivery). The use of biodegradable polymers provides the additional advantage of not having to retrieve the implant after DNA release. Unfortunately, these two qualities (size and degradability) make removal of the devices difficult, should the therapy need to be terminated prematurely. The loading of DNA into the polymer ranges from 0.1- to 10-μg DNA per milligram of polymer using various techniques based on either an emulsion or phase inversion process [204, 207, 208]. A double emulsion process has been used to incorporate aqueous solutions of nonviral DNA, both supercoiled and complexed with poly-L-lysine, into polymer. Incorporation efficiencies ranging from 20% to 80% have been obtained, although the typical incorporation efficiency is approximately 30% [106]. When supercoiled DNA is incorporated using this procedure, the incorporated DNA is structurally intact with 39% of the DNA in the supercoiled conformation. Protecting the DNA by either condensation [207] or using a cryogenic approach [204] can be used to increase the DNA present in the supercoiled form to more than 80%.

DNA polyplexes have been encapsulated and released from polymer microspheres, which may enable these microspheres to be fabricated into matrices using an approach such as the gas-foaming procedure. PLL/DNA complexes have been incorporated into PLG microspheres using a double emulsion process. DNA is incorporated with efficiencies ranging from 30% to 45%, is released over approximately 35 days, and retains its integrity [207, 208]. Alternatively ONs complexed with PEI have been incorporated and released from PLG microspheres. The release profile of the ON/PEI complexes depended on the size, loading, and pore structure of the microspheres. The sustained release of ON/PEI complexes resulted in improved intracellular penetration of the delivered vector as compared with uncomplexed DNA [209, 210]. PLG/DNA scaffolds have successfully been used *in vivo* to enhance matrix deposition [113], angiogenesis [112, 113, 115], and bone formation [114].

Synthetic PEG hydrogels have been extensively investigated as cell transplantation vehicles and as tissue engineering scaffolds. Synthetic hydrogels allow for the complete engineering of the extracellular environment given that they lack direct cellular interactions and have limited protein absorption. Thus, every aspect of the extracellular matrix environment must be engineered, including integrin binding sites, growth factors, and other bioactive signals. DNA delivery from synthetic hydrogels based on PEG have shown that the release rate can be modulated depending on the degree of cross-linking and degradation rate of the hydrogel [211, 212]. Naked DNA encapsulated in photo-cross-linked PEG hydrogels could modulate the release profile of the DNA ranging from linear to a delayed release profile [211].

Naked DNA encoding for transforming growth factor beta-1 encapsulated in ther-mosensitive hydrogel scaffolds resulted in reepitelization of wounds in diabetic mice if applied at early stages during wound healing.

Natural Polymers. Collagens are the major component of mammalian connective tissue and thus have been widely investigated as a biomaterial for cell growth and drug delivery. At least 14 types of collagen are distributed throughout the body. The most abundant is type I collagen. Type I collagen is found in high concentrations in tendon, skin, bone, and fascia, and thus, these tissues are sources for its isolation. After isolation, typically from bovine or porcine sources, collagen is preserved using techniques such as fixation with aldehydes or other chemical preservatives, gamma-irradiation, or lyophilization [213, 214]. Collagen can be lyophilized to form a spongy product or can be formed into a hydrogel by using its natural ability to self-assemble at neutral pH and body temperature (37°C) or by covalently cross-linking it using aldehydes or carbodiimides [105–107, 215].

Collagen has been used as a matrix for numerous tissue engineering applica-tions, including bone, skin, nerve, and cartilage [216–218]. The collagen serves as a scaffold for the migration of repair cells into the matrix and serves to either retain the DNA within the scaffold [106] or provide gradual release [191]. Collagen-based release may limit vector degradation and can induce transgene expression for up to 40 days [191]. Naked DNA delivery from collagen matrices has been employed to promote tissue formation by transfecting invading fibroblasts. Transfected fibro-blasts within DNA-loaded collagen scaffolds, also termed gene-activated matrices (GAMs), subsequently act as bioreactors for localized production of tissue induc-tive factors. Matrices were prepared by lyophilization of type I collagen and sub-sequent immersion in a DNA solution. For applications in tissue engineering, collagen/DNA constructs have been implanted into an adult rat femur [105] and a canine bone defect model [106]. Matrices loaded with 1 mg of DNA were capable of transfecting cells *in vivo*, which resulted in protein production for up to 3 weeks postimplant. For the canine model, however, regeneration required 100 mg of plasmid delivered from the matrix for regeneration [106]. Atelocollagen/DNA constructs have also been used for therapeutic applications, including implanted constructs in the muscle to increase platelet number [191], and in the rabbit ear to aid in would healing [110]. Furthermore, atelocollagen/siRNA constructs have been delivered intratumorally to suppress tumor growth [108, 109].

DNA complexed with cationic lipids or cationic polymers can also be incorpo-rated and released from collagen-based matrices, while maintaining their activity. Naked DNA delivery produces substantial transgene expres-sion *in vivo*, but it results in low levels of transgene expression *in vitro*. DNA complexes can transfect cells *in vitro*; thus, the ability to release DNA complexes can extend the applicabil-ity of nonviral DNA release matrices to the engineering of tissues *in vitro*. Note, however, that release of DNA complexes may differ significantly from that for naked DNA, due to the different physical properties of complexes relative to naked DNA. Collagen matrices loaded with DNA by pipetting solutions of naked DNA, PEI/DNA complexes, and lipoplexes onto collagen showed different release kinet-ics [189]. *In vitro* release studies demonstrated that naked DNA was rapidly released, PEI/DNA and lipid complexes were slowly released, and that PEI/DNA complexes with a protective copolymer had intermediate release kinetics. The

PEI/DNA complexes with the protective copolymer gave the highest transfection *in vitro* and *in vivo*, with the highest *in vivo* expression occurring at 4 days and measurable quantities observed at 7 days [189]. PLL/DNA complexes encapsulated in a collagen sponge have been implanted into severed rat optic nerves as a means to promote neuron survival and promote regeneration [111]. The PLL was modified with bFGF to facilitate the internalization and intracellular trafficking. DNA was detected in the retina for up to 3 months. Nerve terminals were observed extending into the collagen, and they seemed appeared capable of transporting the DNA by retrograde transport.

Hydrogels based on agarose, fibrin, hyaluronic acid (HA), and chitosan have been employed independently as biomaterials for fabrication of tissue engineering matrices [219–221] or as materials to regulate DNA delivery. Fibrin matrixes have been extensively used as tissue engineering scaffolds, sealants, and drug delivery matrices. Fibrin sealants have been employed for the delivery of plasmids to promote angiogenesis, with fibrin-based delivery providing similar responses to delivery in PBS solution [119].

Agarose gels have been used to encapsulate PLL/DNA polyplexes and mediate sustained release to transfect smooth muscle cells *in vitro* with an efficiency less than that obtained by freshly formed complexes, but greater than that obtained with naked DNA [222]. Hydrogels based on fibrin and agarose primarily function to limit the release of DNA; however, hydrogels employing chitosan offer the potential to condense the DNA. Chitosan is a positively charged, naturally occurring polysaccharide that can form complexes with DNA. Chitosan/DNA complexes exhibit minimal cytotoxicity, destabilize the lipid bilayer to facilitate internalization [223], and produce high levels of transfection *in vitro* and *in vivo* [42, 224]. Tissue engineering matrices based on these hydrogels, or combinations of these materials, provide a variety of approaches to regulate DNA delivery *in vitro* and *in vivo*. HA-based matrices have not been used extensively for gene delivery, although their potential as tissue engineering scaffold has long been recognized. DNA loaded, cross-linked HA-based delivery systems have been formed in the form of matrices and microspheres. A solution of HA and plasmid DNA was lyophilized to form a spongy material before cross-linking with adipic acid dihydrazide to form a stable three-dimensional matrix. This HA matrix demonstrated the capacity for sustained release of naked DNA, with release likely occurring after degradation of the matrix [225]. The release rate of naked DNA from the matrices, some of which may be associated with HA fragments, could be modulated by the extent of cross-linking in the hydrogel. DNA-loaded hyaluronic acid microspheres have been formed and found to be able to deliver genes both *in vivo* and *in vitro* [226]. The microspheres were formed using a water-in-oil emulsion and covalent cross-linking with adipic acid dihydrazide to stabilize the microspheres. These microspheres could then be implanted to mediate *in vivo* gene delivery or delivered to plated cells. In both cases, the released DNA mediates transgene expression.

HA has also been cross-linked with collagen and used as gene delivery matrices. The procedure used can be similar to that used for collagen gene delivery systems or that used to form HA hydrogels. HA is mixed with collagen and lyophilized to form a spongy material. This material is then cross-linked using carbodiimide chemistry, dehydrothermal treatment, or exposure to ultraviolet light. After cross-linking, the matrices are soaked in a DNA solution [227] to introduce the DNA to the porous scaffolds. These matrices have been shown to mediate transgene expression *in vivo* and *in vitro* [227].

7.3.3.2   Matrix-Tethered Delivery

Material-tethered delivery combines cationic polymer and cationic lipid formulations with matrix delivery to generate a more efficient delivery strategy by immobilizing the polyplexes or lipoplexes directly to a matrix that also supports cell adhesion and cellular infiltration (Figure 7.3-2). This approach places the nucleic acid directly at the cellular microenvironment, removing the need for diffusion of the complexes to the cell membrane Matrix tethered has not been as extensively used *in vivo* as the direct encapsulation methods described above; however, it has found success mediating DNA delivery from PLGA scaffolds and hyaluronic acid and fibrin hydrogels.

Synthetic Polymers. Matrix-tethered delivery approaches in synthetic polymeric matrices have focused on nonspecific interactions (adsorption) to immobilize the complexes to the matrix. PLGA scaffolds have been coated with a variety of polyplexes such as PAMAM/DNA and PEI/DNA. PAMAM/DNA complexes were dried on a porous PLG scaffold and shown to retain their activity, being able to mediate gene transfer both *in vitro* and *in vivo*, and were found to be a function of DNA concentration and charge ratio [228]. PEI/DNA complexes were absorbed to PLG scaffolds without allowing them to dry on the surface, in contrast to the method described above. It was found that delivery of DNA via matrix-based delivery resulted in similar levels of expression compared with bolus delivery; however, a lower dose of DNA was used. The amount of DNA immobilized was dependent on the time of incubation of the polyplexes with the scaffold, the N/P ratio used, and the initial amount of DNA used [195]. Interestingly, higher amounts of immobilized DNA did not result in higher transgene expression, with 45 ng resulting in more efficient transfection than 600 ng [195].

Natural Polymers. Hyaluronic acid and fibrin were employed independently as biomaterials for fabrication of tissue engineering matrices [221, 229–231] or as materials to mediate gene delivery [232, 233]. HA has been cross-linked with a wide variety of chemistries to form stable hydrogels. DNA/PEI complexes were immobilized to hyaluronic acid hydrogels via biotin–avidin bonds. PEI was modified with biotin groups and subsequently used to complex DNA to form biotynilated polyplexes. The polyplexes were then immobilized to a HA hydrogel, which had been chemically modified to introduce neutravidin, a biotin binding protein. Fibroblasts, which were directly in contact with the polyplex/HA hydrogel, were transfected; however, those directly adjacent to the hydrogel were not, indicating that direct contact of the cell with the tethered polyplexes is essential for transfection [233]. The size of the tethered polyplexes resulted in different transgene expression levels and percent of cells expressing the transgene. Small complexes (~150 nm) resulted in lower levels of transgene expression compared with large complexes (~1000 nm); however, small complexes resulted in ~50% of the cells transfected, which is double that of large complexes. Futhermore, the HA hydrogels used in these studies were topographically patterned with groves and ridges, which resulted in spatially controlled gene expression along the ridges of the hydrogel surface [232].

Unlike HA, fibrinogen forms a stable hydrogel matrix upon thrombin activation to form a fibrin matrix. Fibrin matrices can be further stabilized by the addition of factor XIIIa, which forms covalent bonds between glutamine and lysine residues. Schense and Hubbell have used factor XIIIa chemistry extensively to bind biologically active peptides [234] and growth factors [235] to fibrin matrices. More recently, Hubbell et al. used the same strategy to immobilize polyplexes to fibrin hydrogels [120, 232]. Two different 21 amino acid peptides were designed to have a DNA binding sequences (Cys–His–(Lys)6–His–Cys) in combination with a transglutaminase substrate site (Asn–Gln–Glu–Gln–Val–Ser–Pro–Leu) or a nuclear localization site (from SV40; see Table 7.3-4). The polyplexes were formed with mixtures of the two poly(amino acids) and mixed with fibrinogen, thrombin, and factor XIIIa, resulting in the covalent immobilization of the polyplexes within the fibrin matrix. Transfection of COS-7 cells in a two-dimensional (2D) sandwich assay, in which cells are plated on top of a polyplex/fibrin hydrogel and then a second gel is cast on top, resulted in ~25% of the cells transfected [232]. These hydrogels have been used *in vivo* to deliver a mutant HIF-1α plasmid, which has the transcription factor constantly active, and resulted in enhanced wound healing in a skin wound model [120]. Furthermore, the delivery of a HIF-1α encoding plasmid was able to result in the formation of more mature blood vessels when compared with the delivery of a vascular endothelial growth factor (VEGF) encoding plasmid.

7.3.4 DELIVERY LIMITATIONS AND CURRENT SOLUTIONS

The ultimate applicability of the various biomaterials for gene delivery rests on their ability to effectively deliver the nucleic acids to their target intracellular location (Figure 7.3-1). An effective gene delivery system will protect the nucleic acid from degradation, target the appropriate cell population, be efficiently internalized by the cell, avoid degradative pathways, and ultimately localize to the nucleus (DNA) or cytosol (siRNA or ON, Figure 7.3-1). The materials under development can specifically enhance one or more of these steps in the transfection process. The following sections describe how the materials interact with biological systems and how they can be designed to overcome the obstacles to effective gene transfer. Table 7.3-3 also summarizes this data.

7.3.4.1 Extracellular Limitations

Vector Stabilization and Fast Clearance from Body. Interaction of nonviral formulations with serum components results in the deactivation of the lipoplexes or polyplexes due to aggregation [9, 10]. Furthermore, aggregated lipoplexes and polyplexes can lead to rapid clearance of the polyplexes by phagocytic cells and the reticuloendotheial sytem [8]. Hydrophilic polymers such as poly(ethylene glycol) (PEG) [39, 47, 236], N-(2-hydroxypropyl)methacrylamide) (HPMA) [236–239], and oligosaccharides [40, 41, 43, 236, 240] have been found to stabilize complexes against salt and protein aggregation, which results in longer circulation times. The increased stability probably results from steric effects that (1) prevent interaction with serum proteins, cells, and tissue; (2) increase solubility of the complexes in aqueous milieu; and (3) prevent particle–particle interactions [241]. Traditional cationic polymers for gene delivery such as PEI, PAMAM, and PLL have been made more biocompatible using this approach [18, 241–245].

Targeted Internalization. The plasma membrane of the cell provides a protective shell that serves to limit the transport of undesired molecules, such as DNA. However, cells must communicate with their environment and have thus developed mechanisms to transport high- and low-molecular-weight macromolecules across the membrane. Nonclathrin-coated pit internalization can occur through smooth invaginations of 150–300 nm or via potocytosis, which involves the invagination of caveolae-rich 50–100-nm-diameter vesicles from the cell surface. DNA complexes are thought to enter cells primarily through clathrin-coated pits. These pits, with diameters of approximately 150 nm, are internalized from the plasma membrane to form coated vesicles. Interestingly, recent studies using a series of inhibitors for different endosomal pathways have elucidated different mechanisms of internalization for lipoplexes and polyplexes [246]. Lipoplex internalization and lipoplex-mediated nucleic acid delivery was strongly inhibited by potassium depletion or using chlorpromazine, indicating that clathrin endocytosis is the primary internalization pathway of lipoplexes [246]. On the contrary, polyplex internalization was inhibited by 25% when the caveolae pathway was inhibited using filipin or genistein and by 20% when clathrin-mediated endocytosis was inhibited [246], indicating that polyplexes use two internalization pathways. Gene transfer mediated by polyplexes was completely abolished when the caveolae pathway was inhibited but not when the clathrin-mediated pathway was inhibited [246], suggesting that internalization via caveolae-mediated endosomes leads to efficient transfection. Particle size also affects the endosomal internalization pathway used. Particles with a diameter <200 nm were found to be internalized via clathrin-mediated endocytosis and were ultimately delivered to the lysosome. In contrast, particles that were >500 nm entered the cells via caveolae and never reached the lysosomal compartment [247]. Furthermore, multiple studies have shown that large PEI/DNA complexes >500 nm were more efficient at mediating high transgene expression than small complexes <200 nm [233, 248], suggesting that it is due to mode of internalization and intracellular trafficking.

Receptor-mediated gene delivery offers a promising approach to create a specific interaction with the cell surface and to target a particular internalization pathway. Synthetic materials termed molecular conjugates are being developed and are composed of two domains: a DNA binding domain and a receptor binding domain. The DNA binding domain is in charge of linking the DNA to the rest of the molecular conjugate molecule. The DNA binding domain is typically composed of a cationic polymer or a cationic lipid, which self-assembles with DNA. The receptor binding domain is frequently attached to functional groups on the cationic polymer or lipid before complexation with DNA [169]; however, complexes can be initially formed and subsequently coupled to a ligand [182, 249]. The function of the receptor binding domain is to direct the DNA/molecular conjugate complex to a receptor and guide the complex through the internalization pathway that leads to endosomal entrapment. The intracellular fate of DNA complexes can depend on the type of endocytic process involved in its internalization [246]. The diversity and abundance of membrane bound receptors and ligands and the high efficiency of internalization and turnover ratio makes receptor-mediated endocytosis a powerful tool for controlling uptake. Targets for which the attached ligand binds include the receptors for asialoglycoprotein [20, 250, 251], transferrin [47, 252–254] (review [48]), folate [45, 255, 256] (review [46]), manose [49, 50], galactose [51], lectins [257–259], integrins [55, 179] antibodies [52–54, 56] and EGF [57, 260]. The attachment of a targeting ligand to the lipoplex has been shown to increase transfection by 1000-fold as compared with lipid with no targeting ligand [261]. In principle, any monoclonal antibody, Fab fragment of a monoclonal antibody, peptide, peptide mimetic, proteins, or peptide fragment can be added to the molecular conjugate using either a covalent or an ionic attachment.

Maximal gene delivery through the receptor-mediated endocytosis pathway requires that the design parameters be optimized, which includes properties such as the ligand binding affinity, the length of the linker between the ligand and the complex, the complex charge ratio and structure, and the number of ligands per complex [260, 262]. Ligand–receptor interactions can be very specific and may be negatively affected when covalently bound to a molecular conjugate. Knowing whether the receptor binding domain is interacting to its receptor with the same efficiency as the free ligand is important to achieve high gene delivery levels. In a model using an EGF–polylysine conjugate, it was found that a longer spacer arm, between the two domains, resulted in more native-like EGF–receptor binding and higher transfection efficiency than shorter spacer arms [260]. The presence of the ligand can alter the structure of the nucleic acid complex. Incorporation of asialoorosomucoid into PLL/DNA complexes was more efficient at condensing DNA than PLL alone, probably due to an aggregation of many PLLs around the negatively charged protein [122]. At low charge ratios, specific (ligand mediated) and nonspecific endocytosis had insignificant levels of transfection. As the charge ratio was increased to 4, the transfection efficiency by specific binding enhanced transfection by a factor of 3 over nonspecific endocytosis [260]. Finally, the number of ligands per complex can affect the binding of the complexes to cells. An excessive number of ligands may inhibit the binding of complexes [260]. Optimization of the parameters that affect the surface binding of the molecular conjugates improves the specificity and gene transfer efficiency for gene delivery.

7.3.4.2 Intracellular Limitations

The steps after internalization of naked-DNA, polyplexes or lipoplexes, endosomal escape, and nuclear localization are thought to be rate limiting for the transfection of many cell types. Internalization of the plasmid does not necessarily correlate to transfection [263]. Much of the DNA that is internalized into the endosome is either retained or degraded within the endosome. The DNA that does escape the endosome and enters the cytoplasm must subsequently avoid degradation and be transported to the nucleus for successful gene transfer [264]. *In vitro* studies have demonstrated that although greater than 95% of the cells were positive for plasmid (>100,000 copies per cells), less than 50% of the cells were expressing the transgene [25, 265]. The following sections describe how the materials interact with intracellular biological systems and how they can be designed to overcome the obstacles for effective gene transfer. Table 7.3-4 also summarizes this data.

Toxicity. One common drawback of efficient nonviral gene delivery strategies is that they are generally toxic. This finding is not surprising given that efficient nucleic acid delivery formulations are also efficient at getting into the cell and moving through the intracellular space most likely in an unspecific manner. For example, PEI has been found to colocalize with the nucleous and with the delivered DNA, whereas PLL has not [266]. As reviewed, PEI is more efficient at gene transfer than PLL and PEI is more toxic than PLL. Nuclear PEI may nonspecifically interact with genomic DNA, preventing its normal transcription, thus causing toxicity. Furthermore, the efficient endosomal buffering of PEI and PAMAM dendrimers, while improving their effectiveness to deliver DNA, may also contribute toxicity by preventing the natural acidification of endosomal vesicles and degradation of unwanted endosomal cargo. Thus, can we have an efficient nucleic acid delivery strategy without some toxicity? Is it possible to have an effective therapeutic approach with absolutely no side effects? The answer is that it needs to be attempted although the blockbuster of nonviral gene delivery formulations may not be completely nontoxic. Some general characteristics of cationic polymers and lipids make them toxic such as molecular weight, density of positive charges, and lack of degradability. This section briefly outlines the most common features associated with toxicity and their current solutions.

Nonviral formulations are generally more efficient and more toxic as the polymer-to-nucleic acid ratio is increased (higher N/P ratios), suggesting a role of free uncomplexed polymer in the solution in efficient delivery and toxicity. A recent report has shown that polyplexes of DNA and PEI contain an average of 3.5 plasmids (5800 base pairs) and 30 PEI (25 kDa) molecules when prepared at N/P ratios of 6 and 10, assuming that the DNA is completely complexed [267]. Based on these calculations, there is 86% of free PEI in the complex mixture [267, 268]. Purification of the PEI/DNA complexes by dialysis has shown a reduction of toxicity; however, it also reduced transfection efficiency [269]. Efficient gene transfer was restored when free PEI was added to the mixture [269].

High-molecular-weight cationic polymers are generally more toxic than low-molecular-weight polymers [124], most likely due to aggregation of free polymer with the cell membrane [130] or other cellular components. Together these findings suggest that free high-molecular-weight polymer leads to efficient but toxic gene transfer. Strategies to prevent such toxicity have focused on reducing the density of positive charges (number of primary amines) without affecting the buffering capacity of the polymer and, thus, reducing its interaction with the cell membrane and other cellular components.

Another structural aspect of cationic polymers that contributes to their efficiency as gene transfer agents and their toxicity is the degree of branching and backbone flexibility. Highly branched cationic polymers form smaller particles and mediate more efficient gene transfer than low-branched polymers but are more toxic compared with low-branched or linear polymers [130]. The high toxicity of branched polymers has been associated with polymer backbone stiffness; more flexible, hyperbranched PEI derivatives with additional secondary and tertiary groups show lower toxicity *in vitro* than commercially available branched PEI [270].

One approach that could reduce the toxicity associated with a high molecular weight and a high degree of branching is making the delivery formulation degradable. Degradable PEIs are of particular interest because the basic structure of PEI is not biologically or chemically degradable. Efforts to make PEI degradable typically involve the cross-linking of low-molecular-weight PEI (600–1800 kDa) with disulfide linkages [61] or oligo(L-lactic acid cosuccinic acid) [60] to form the higher molecular weight PEIs required to achieve efficient transfection. Furthermore, low-molecular-weight PEI has been cross-linked with PEI through a degradable ester bond [242]. Cytotoxicity experiments showed decreased toxicity or complexes formed with degradable PEG–PEI/DNA complexes when compared with complexes formed with 25-kDa PEI [242]. However, transfection studies showed that the degradable PEI was less efficient at mediating gene transfer than 25-kDa PEI [242]. Although PAMAM dendrimers have been found to be three orders of magnitude less toxic than PEI (600–1000 kDa) and PLL (36.6 kDa), they still suffer from toxic effects [124]. The toxicity of PAMAM dendrimers seems to have a different mechanism from that of toxicity caused by linear cationic polymers in that it does not cause membrane hemolysis at a similar concentration of PEI and PLL [271]. It has been reported that primary amines are more toxic than secondary or tertiary amines for PEI-type polymers [10]. However, for PAMAM dendrimers, this relationship was not true. Although cytotoxicity increased with generation (higher molecular weight), it was surprisingly independent of surface charge [272, 273]. Strategies to reduce toxicity of PAMAM dendrimers include quaternization [131], reducing or reversal of surface charge [274], and sterically shielding the surface groups by binding C12 lauroyl groups or PEG2000 [272] Cytoxicity of PLL, although low relative to many nondegradable delivery agents, is not negligible. Dendritic PLL has been found to have cytoxicity that is much lower than that of PLL and is nearly the same as naked DNA, although the PLL itself leads to cell viability of about 80% of naked DNA as determined by 3-(4,5-dimethylthiazol-2-yl)-2,5-diphenyltetrazolium bromide (MTT) assay [275].

Endosomal Escape. After internalization, the coated vesicle transforms into an early endosome, which is accompanied by acidification of the vesicular lumen that continues into the late endosomal and lysosomal compartments, reaching a final pH in the perinuclear lysosome of approximately 4.5. For lipoplex-mediated delivery, the interaction of the lipids with the endosomal membrane is thought to facilitate escape of the DNA to the cytoplasm before its degradation in the lysosome. Although it is not well understood, some lipoplexes preferentially fuse with the early endosome, whereas others fuse with the late endosome [24]. pH-sensitive liposomes can take advantage of the acidification process to facilitate the release of plasmids into the cytoplasm before lysosomal degradation [276]. Release of the DNA from the complex may occur at the wall of the endosome [179, 277]. Xu and Szoka proposed a model in which destabilization of the endosomal membrane causes a flip-flop of anionic lipids from the cytoplasmic facing monolayer, which diffuse laterally into the complex and form a charge neutral ion pair with the cationic lipid [179]. This pairing results in displacement of the DNA from the cationic lipid and release of the DNA into the cytoplasm. Displacement of the cationic lipid from the DNA before it enters the nucleus is critical for the ultimate expression of the gene.

Materials containing secondary and tertiary amines that are protonated at acidic pH, and not at neutral pH like primary amines, can act as proton sponges that buffer the decrease in pH and ultimately cause DNA release into the cytoplasm. For example, PEI shows a level of protonation of 20% at pH = 7.4 compared with about 45% at pH = 5 [278]. This proton sponge effect is thought to result from the protonation of amino moieties on the polymer as the pH decreases inside the endocytic vesicle. The influx of counter Cl⁻ ions, which occurs to maintain electro-neutrality, induces osmotic swelling and rupture of the vesicle membrane [29]. Evidence for the proton sponge hypothesis include decelerated acidification of endosomal vesicles, as well as elevated chloride accumulation and a 140% increase in the relative volume in PEI-containing endosomes [279]. Modifications investigated for PLL to introduce endosomal buffering include the partial modification of the PLL side chains with imidazoles [140, 280] or histidine [65]. Both approaches have been found to enhance the release of DNA complexed with PLL from the endosome.

Endosomal disrupting peptides, also termed fusogenic peptides, can be covalently incorporated into DNA/lipid and DNA/cationic polymer complexes to enhance escape from the endosome [281, 282]. Viruses and bacteria have evolved sophisticated endosomal release pathways [283]. Based on the viral and bacterial pathways, peptides for gene delivery have been identified from the viral fusion proteins and have been used successfully to enhance endosomal escape. An extensive list of fusogenic peptides can be found in Ref. 284. The peptides range in length from 15 to 30 amino acids and form stable amphipathic α-helices primarily due to alternating hydrophilic and hydrophobic amino acids. Most peptides are amphipathic pH sensitive being only active at low pH [283]. There are two mechanisms by which fusion peptides disrupt endosomal membranes. First, the peptide causes the rearrangement of the lipid packaging, thereby changing the membrane integrity and causing release of the endosomal contents. Second, the peptide causes pore formation within the endosomal membrane, without affecting the membrane integrity [283]. Most peptides induce endosomal release by membrane disruption rather than by membrane fusion [283]. For example, GALA, which was one of the first synthetic peptides to be designed and synthesized, undergoes a conformational change to form an alpha-helix at acidic pHs [66, 67, 284]. The α-helix formed induces endosomal leakage by disrupting the endosomal membrane. A similar peptide, KALA, also functions to disrupt endosomal membranes but is positively charged and is capable of condensing DNA and mediating gene delivery [63, 64].

More recently, a family of acid responsive polymers based on α-alkyl acrylic acids such as methacrylic acid (MAA), ethylacrylic acid (EAA), propylacrylic acid, and butylacrylic acid (BAA) and their copolymers with alkyl acrylates or methacrylates have been explored for their ability to disrupt membranes and their potential as endosomal disruption polymers. The key feature of these poly(α-alkyl acrylic acids) is that they switch from a hydrophilic to a hydrophobic character as they become protonated (pH sensitive), and the switch to a hydrophobic character has been shown to disrupt membranes [68, 69]. The polymer, PPAA, has been shown to be 15 times more effective than PEAA at membrane disruption and to have maximum hemolytic activity at pH ≤ 6, which is in the range of endosomal pH [68, 285]. PPAA has been successfully been used to enhance lipolyplex-mediated gene transfer [286] *in vitro* and wound healing by altering extracellular matrix organization and greater vascularization *in vivo* [286].

Nuclear Localization. The transport of DNA from the cytoplasm to the nucleus may be the most significant limitation to successful gene transfer. Plasmids injected far from the nuclei (60–90 μm) had less protein expression than plasmids injected near the nuclei [287]. In addition to cytoplasmic transport limitations, the size of DNA is problematic for crossing into the nucleus. The nuclear pores allow free diffusion entry of only small particles (less than approximately 70 kDa) [288]. Entry of the DNA can be facilitated by the breakdown of the nuclear membrane, which occurs during cell divi-sion [289]. However, for delivery to nondividing cells, nuclear localization sequences (NLSs) can be incorporated into the DNA complexes to direct the plasmid into the nucleus.

NLSs are short peptide sequences (5–25 amino acids) that are necessary and sufficient for nuclear localization of their respective proteins [288]. These sequences can be incorporated into complexes with cationic lipids [290] and cationic polymers [291, 292], or they can be directly linked to the plasmid [293, 294]. Some nuclear localization signals have stretches of positive charge, which has led to speculation that cationic polymers function as nuclear localization signals [186]. The cationic polymer PEI rapidly accumulates in the nucleus of cells and to a greater extent than PLL [266]. Nevertheless, the attachment of NLS does significantly enhance nuclear accumulation [291]. For the transfection of nondividing endothelial cells, lipoplex transfection resulted in 5% of the cells testing positive for transfection, whereas the incorporation of an NLS resulted in more than 80% of the cells testing positive [290]. Examples of nuclear localization sequences include PKKKRKVEDPY (SV40) [76], GNQSSNFGPMKGGNFGGRSSGPYGGGGQYFAKPRNQGGY (M9) [77], importin-β (1–643aa) [78], and MRRAHHRRRRASHRRMRGG (mu) [79].

Vector Decomplexation. Recently, vector decomplexation has been viewed as another limiting step to efficient gene transfer [72]. Vector decomplexation involves the events that need to take place for the dissociation of the delivery formulation from the nucleic acid. This step is critical for efficient gene transfer to take place given that without it DNA cannot be transcribed and siRNA/ON cannot efficiently mediate gene downregulation. Strategies to enhance decomplexation involve the introduction of environmentally sensitive bonds (mostly disulfide bonds) [244, 295] and reducing the affinity of the cationic polymer for the DNA [70, 71, 73]. More recently, a thermosensitive copolymer, N-isopropylacrylamide-co-vinyl laurate, was covalently coupled to chitosan and employed to enhance unpacking. At temperatures below the polymer's critical solution temperature, the polymer extended and was soluble, causing the enhanced unpackaging of the chitosan/DNA polyplexes [75]. Furthermore, enzymatically degradable formulations (chitosan) have been used to enhance unpackaging and gene transfer by encouraging intracellular enzymatic degradation (via chitosinase) of the polyplexes [74].

7.3.5 OUTLOOK

The design and construction of ideal formulations for nonviral gene delivery will continue to focus on generating materials that are "smarter," taking advantage of the cell's natural environments and processes. One approach to the design and synthesis of smarter nonviral gene delivery formulations is to better understand the intracellular and extracellular limitations at the molecular level and to understand the role (positive or negative) of the intracellular and extracellular molecules and environments that the lipoplexes and polyplexes come in contact with during the transfection process. The cell is a complex structure that contains a myriad of molecules and events occurring simultaneously, which result in specific cell phenotypes. How the addition of lipoplexes and polyplexes affect this harmonious environment is essential for the design and generation of nonviral formulations that are less toxic and more effective, and this design is likely to be different for different cell types. For example, recent studies have shown that the rigidity of the matrix where the cells are attached affects polyplex uptake and transgene expression, with stiffer materials resulting in enhanced uptake, unpackaging, and gene expression [296]. Although the molecular mechanism for this enhancement was not elucidated in this study, it is clear that the cellular environment (extracellular and intracellular) and cell "state" affects the effectiveness of nonviral formulations. During the past decade, nonviral gene delivery has identified key limiting steps to polyplex and lipoplex formulations and has addressed these limitations with clever solutions such as the attachment of specific cell binding ligands, fusogenic peptides, nuclear localization sequences, and degradability. The next decade will undoubtedly come with more advances for nonviral formulations that can more elegantly bypass the different barriers encountered during gene transfer.

REFERENCES

1. Merdan T, Kopecek J, Kissel T (2002). Prospects for cationic polymers in gene and oligonucleotide therapy against cancer. *Adv. Drug Deliv. Rev.* 54:715–758.

2. Ziady A G, et al. (2003). Transfection of airway epithelium by stable PEGylated poly-L-lysine DNA nanoparticles in vivo. *Mol. Ther.* 8:936–947.

3. Ferrari S, Geddes D M, Alton E W (2002). Barriers to and new approaches for gene therapy and gene delivery in cystic fibrosis. *Adv. Drug Deliv. Rev.* 54:1373–1393.

4. Griesenbach U, et al. (2002). Gene therapy progress and prospects: Cystic fibrosis. *Gene Ther.* 9:1344–1350.

5. Onodera M, Sakiyama Y (2000). Adenosine deaminase deficiency as the first target disorder in gene therapy. *Expert Opin. Investig. Drugs.* 9:543–549.

6. Buchschacher G L, Jr, Wong-Staal F (2001). Approaches to gene therapy for human immunodeficiency virus infection. *Hum. Gene Ther.* 12:1013–1019.

7. Gansbacher B (2003). Report of a second serious adverse event in a clinical trial of gene therapy for X-linked severe combined immune deficiency (X-SCID). Position of the European Society of Gene Therapy (ESGT). *J. Gene Med.* 5:261–262.

8. Dash P R, et al. (1999). Factors affecting blood clearance and in vivo distribution of polyelectrolyte complexes for gene delivery. *Gene Ther.* 6:643–650.

9. Ledley F D (1996). Pharmaceutical approach to somatic gene therapy. *Pharm. Res.* 13:1595–1614.

10. Neu M, Fischer D, Kissel T (2005). Recent advances in rational gene transfer vector design based on poly(ethylene imine) and its derivatives. *J. Gene Med.* 7:992–1009.

11. Mahato R I, Smith L C, Rolland A (1999). Pharmaceutical perspectives of nonviral gene therapy. *Adv. Genet.* 41:95–156.

12. Cohen-Sacks H, et al. (2002). Novel PDGFbetaR antisense encapsulated in polymeric nanospheres for the treatment of restenosis. *Gene Ther.* 9:1607–1616.

13. Takakura Y, Mahato R I, Hashida M (1998). Extravasation of macromolecules. *Adv. Drug Deliv. Rev.* 34:93–108.

14. Goula D, et al. (1998). Size, diffusibility and transfection performance of linear PEI/DNA complexes in the mouse central nervous system. *Gene Ther.* 5:712–717.

15. Mahato R I (1999). Non-viral peptide-based approaches to gene delivery. *J. Drug Target.* 7:249–268.

16. Dufes C, et al. (2005). Synthetic anticancer gene medicine exploits intrinsic antitumor activity of cationic vector to cure established tumors. *Cancer Res.* 65:8079–8084.

17. Boussif O, et al. (1995). A versatile vector for gene and oligonucleotide transfer into cells in culture and in vivo: Polyethylenimine. *Proc. Natl. Acad. Sci. USA.* 92:7297–7301.

18. Haensler J, Szoka F C, Jr. (1993). Polyamidoamine cascade polymers mediate efficient transfection of cells in culture. *Bioconjug. Chem.* 4:372–379.

19. Gonzalez H, Hwang S J, Davis M E (1999). New class of polymers for the delivery of macromolecular therapeutics. *Bioconjug. Chem.* 10:1068–1074.

20. Wu G Y, Wu C H (1987). Receptor-mediated in vitro gene transformation by a soluble DNA carrier system. *J. Biol. Chem.* 262:4429–4432.

21. Mansouri S, et al. (2004). Chitosan-DNA nanoparticles as non-viral vectors in gene therapy: Strategies to improve transfection efficacy. *Eur. J. Pharm. Biopharm.* 57:1–8.

22. Hosseinkhani H, et al. (2004). Dextran-spermine polycation: An efficient nonviral vector for in vitro and in vivo gene transfection. *Gene Ther.* 11:194–203.

23. Felgner P L, et al. (1987). Lipofection: A highly efficient, lipid-mediated DNA-transfection procedure. *Proc. Natl. Acad. Sci. USA.* 84:7413–7417.

24. Felgner J H, et al. (1994). Enhanced gene delivery and mechanism studies with a novel series of cationic lipid formulations. *J. Biol. Chem.* 269:2550–2561.

25. Zabner J, et al. (1995). Cellular and molecular barriers to gene transfer by a cationic lipid. *J. Biol. Chem.* 270:18997–19007.

26. Mok K W, Cullis P R (1997). Structural and fusogenic properties of cationic liposomes in the presence of plasmid DNA. *Biophys. J.* 73:2534–2545.

27. Lasic D D, et al. (1997). The structure of DNA-liposome complexes. *J. Amer. Chem. Soc.* 119:832–833.

28. Sternberg B, Sorgi F L, Huang L (1994). New structures in complex formation between DNA and cationic liposomes visualized by freeze-fracture electron microscopy. *FEBS Lett.* 356:361–366.

29. Boussif O, et al. (1999). Synthesis of polyallylamine derivatives and their use as gene transfer vectors in vitro. *Bioconjug. Chem.* 10:877–883.

29a. Radler J O, et al. (1997). Structure of DNA-cationic liposome complexes: DNA intercalation in multilamellar membranes in distinct interhelical packing regimes. *Science.* 275:810–814.

30. Murphy J E, et al. (1998). A combinatorial approach to the discovery of efficient cationic peptoid reagents for gene delivery. *Proc. Natl. Acad. Sci. USA.* 95:1517–1522.

31. van de Wetering P, et al. (1999). Structure-activity relationships of water-soluble cationic methacrylate/methacrylamide polymers for nonviral gene delivery. *Bioconjug. Chem.* 10:589–597.

32. van de Wetering P, et al. (1998). 2-(Dimethylamino)ethyl methacrylate based (co)polymers as gene transfer agents. *J. Control. Rel.* 53:145–153.

33. Wolfert M A, et al. (1996). Characterization of vectors for gene therapy formed by self-assembly of DNA with synthetic block co-polymers. *Hum. Gene Ther.* 7:2123–2133.

34. Jon S, Anderson D G, Langer R (2003). Degradable poly(amino alcohol esters) as potential DNA vectors with low cytotoxicity. *Biomacromol.* 4:1759–1762.

35. Akinc A, et al. (2003). Synthesis of poly(beta-amino ester)s optimized for highly effective gene delivery. *Bioconjug. Chem.* 14:979–988.

36. Wen J, et al. (2004). Biodegradable polyphosphoester micelles for gene delivery. *J. Pharm. Sci.* 93:2142–2157.

37. Zhao Z, et al. (2003). Polyphosphoesters in drug and gene delivery. *Adv. Drug Deliv. Rev.* 55:483–499.

38. Wang J, et al. (2002). Enhanced gene expression in mouse muscle by sustained release of plasmid DNA using PPE-EA as a carrier. *Gene Ther.* 9:1254–1261.

39. Kim J K, et al. (2003). Enhancement of polyethylene glycol (PEG)-modified cationic liposome-mediated gene deliveries: Effects on serum stability and transfection efficiency. *J. Pharm. Pharmacol.* 55:453–460.

40. Popielarski S R, Mishra S, Davis M E (2003). Structural effects of carbohydrate-containing polycations on gene delivery. 3. Cyclodextrin type and functionalization. *Bioconjug. Chem.* 14:672–678.

41. Davis M E, et al. (2004). Self-assembling nucleic acid delivery vehicles via linear, water-soluble, cyclodextrin-containing polymers. *Curr. Med. Chem.* 11:179–197.

42. Leong K W, et al. (1998). DNA-polycation nanospheres as non-viral gene delivery vehicles. *J. Control. Rel.* 53:183–193.

43. Liu Y, et al. (2004). New poly(d-glucaramidoamine)s induce DNA nanoparticle formation and efficient gene delivery into mammalian cells. *J. Am. Chem. Soc.* 126:7422–7423.

44. McKenzie D L, et al. (2000). Low molecular weight disulfide cross-linking peptides as nonviral gene delivery carriers. *Bioconjug. Chem.* 11:901–909.

45. Gottschalk S, et al. (1994). Folate receptor mediated DNA delivery into tumor cells: potosomal disruption results in enhanced gene expression. *Gene Ther.* 1:185–191.

46. Ward C M (2000). Folate-targeted non-viral DNA vectors for cancer gene therapy. *Curr. Opin. Mol. Ther.* 2:182–187.

47. Ogris M, et al. (1999). PEGylated DNA/transferrin-PEI complexes: Reduced interaction with blood components, extended circulation in blood and potential for systemic gene delivery. *Gene Ther.* 6:595–605.

48. Li H, Qian Z M (2002). Transferrin/transferrin receptor-mediated drug delivery. *Med. Res. Rev.* 22:225–250.

49. Diebold S S, et al. (1999). Mannose polyethylenimine conjugates for targeted DNA delivery into dendritic cells. *J. Biol. Chem.* 274:19087–19094.

50. Sato A, et al. (2001). Enhanced gene transfection in macrophages using mannosylated cationic liposome-polyethylenimine-plasmid DNA complexes. *J. Drug Target.* 9:201–207.

51. Sagara K, Kim S W (2002). A new synthesis of galactose-poly(ethylene glycol)-polyethylenimine for gene delivery to hepatocytes. *J. Control. Rel.* 79:271–281.

52. Merdan T, et al. (2003). Pegylated polyethylenimine-Fab' antibody fragment conjugates for targeted gene delivery to human ovarian carcinoma cells. *Bioconjug. Chem.* 14:989–996.

53. Chiu S J, Ueno N T, Lee R J (2004). Tumor-targeted gene delivery via anti-HER2 antibody (trastuzumab, Herceptin) conjugated polyethylenimine. *J. Control. Rel.* 97:357–369.

54. Li S, et al. (2000). Targeted gene delivery to pulmonary endothelium by anti-PECAM antibody. *Am. J. Physiol. Lung Cell Mol. Physiol.* 278:L504–511.

55. Kunath K, et al. (2003). Integrin targeting using RGD-PEI conjugates for in vitro gene transfer. *J. Gene Med.* 5:588–599.

56. Buschle M, et al. (1995). Receptor-mediated gene transfer into human T lymphocytes via binding of DNA/CD3 antibody particles to the CD3 T cell receptor complex. *Hum. Gene Ther.* 6:753–761.

57. Blessing T, et al. (2001). Different strategies for formation of pegylated EGF-conjugated PEI/DNA complexes for targeted gene delivery. *Bioconjug. Chem.* 12:529–537.

58. Erbacher P, et al. (1999). Transfection and physical properties of various saccharide, poly(ethylene glycol), and antibody-derivatized polyethylenimines (PEI). *J. Gene Med.* 1:210–222.

59. Kwoh D Y, et al. (1999). Stabilization of poly-L-lysine/DNA polyplexes for in vivo gene delivery to the liver. *Biochim. Biophys. Acta.* 1444:171–190.

60. Petersen H, et al. (2002). Poly(ethylenimine-co-L-lactamide-co-succinamide): A biodegradable polyethylenimine derivative with an advantageous pH-dependent hydrolytic degradation for gene delivery. *Bioconjug. Chem.* 13:812–821.

61. Gosselin M A, Guo W, Lee R J (2001). Efficient gene transfer using reversibly cross-linked low molecular weight polyethylenimine. *Bioconjug. Chem.* 12:989–994.

62. Kichler A, et al. (1997). Influence of membrane-active peptides on lipospermine/DNA complex mediated gene transfer. *Bioconjug. Chem.* 8:213–221.

63. Wyman T B, et al. (1997). Design, synthesis, and characterization of a cationic peptide that binds to nucleic acids and permeabilizes bilayers. *Biochem.* 36:3008–3017.

64. Baru M, et al. (1998). Lysosome-disrupting peptide increases the efficiency of in-vivo gene transfer by liposome-encapsulated DNA. *J. Drug Target.* 6:191–199.

65. Pichon C, Goncalves C, Midoux P (2001). Histidine-rich peptides and polymers for nucleic acids delivery. *Adv. Drug Deliv. Rev.* 53:75–94.

66. Li W, Nicol F, Szoka F C, Jr. (2004). GALA: A designed synthetic pH-responsive amphipathic peptide with applications in drug and gene delivery. *Adv. Drug Deliv. Rev.* 56:967–985.

67. Subbarao N K, et al. (1987). pH-dependent bilayer destabilization by an amphipathic peptide. *Biochem.* 26:2964–2972.

68. Murthy N, et al. (1999). The design and synthesis of polymers for eukaryotic membrane disruption. *J. Control. Rel.* 61:137–143.

69. Thomas J L, Barton S W, Tirrell D A (1994). Membrane solubilization by a hydrophobic polyelectrolyte: surface activity and membrane binding. *Biophys. J.* 67:1101–1106.

70. Erbacher P, et al. (1995). Glycosylated polylysine/DNA complexes: gene transfer efficiency in relation with the size and the sugar substitution level of glycosylated polylysines and with the plasmid size. *Bioconjug. Chem.* 6:401–410.

71. Erbacher P, et al. (1997). The reduction of the positive charges of polylysine by partial gluconoylation increases the transfection efficiency of polylysine/DNA complexes. *Biochim. Biophys. Acta.* 1324:27–36.

72. Schaffer D V, et al. (2000). Vector unpacking as a potential barrier for receptor-mediated polyplex gene delivery. *Biotechnol. Bioeng.* 67:598–606.

73. Forrest M L, et al. (2004). Partial acetylation of polyethylenimine enhances in vitro gene delivery. *Pharm. Res.* 21:365–371.

74. Liang D C, et al. (2006). Pre-deliver chitosanase to cells: A novel strategy to improve gene expression by endocellular degradation-induced vector unpacking. *Int. J. Pharm.* 314:63–71.

75. Sun S, et al. (2005). A thermoresponsive chitosan-NIPAAm/vinyl laurate copolymer vector for gene transfection. *Bioconjug. Chem.* 16:972–980.

76. Ritter W, et al. (2003). A novel transfecting peptide comprising a tetrameric nuclear localization sequence. *J. Mol. Med.* 81:708–717.

77. Bremner K H, et al. (2004). Factors influencing the ability of nuclear localization sequence peptides to enhance nonviral gene delivery. *Bioconjug. Chem.* 15:152–161.

78. Nagasaki T, et al. (2005). Enhanced nuclear import and transfection efficiency of plasmid DNA using streptavidin-fused importin-beta. *J. Control. Rel.* 103:199–207.

79. Akita H, et al. (2006). Evaluation of the nuclear delivery and intra-nuclear transcription of plasmid DNA condensed with micro (mu) and NLS-micro by cytoplasmic and nuclear microinjection: A comparative study with poly-L-lysine. *J. Gene Med.* 8:198–206.

80. Thomas M, et al. (2005). Full deacylation of polyethylenimine dramatically boosts its gene delivery efficiency and specificity to mouse lung. *Proc. Natl. Acad. Sci. USA.* 102:5679–5684.

81. Urban-Klein B, et al. (2005). RNAi-mediated gene-targeting through systemic application of polyethylenimine (PEI)-complexed siRNA in vivo. *Gene Ther.* 12:461–466.

82. Carrere N, et al. (2005). Characterization of the bystander effect of somatostatin receptor sst2 after in vivo gene transfer into human pancreatic cancer cells. *Hum. Gene Ther.* 16:1175–1193.

83. Densmore C L (2003). Polyethyleneimine-based gene therapy by inhalation. *Expert Opin. Biol. Ther.* 3:1083–1092.

84. Fewell J G, et al. (2005). Synthesis and application of a non-viral gene delivery system for immunogene therapy of cancer. *J. Control. Rel.* 109:288–298.

85. Zaitsev S, et al. (2004). Polyelectrolyte nanoparticles mediate vascular gene delivery. *Pharm. Res.* 21:1656–1661.

86. Mohapatra S S (2003). Mucosal gene expression vaccine: a novel vaccine strategy for respiratory syncytial virus. *Pediatr. Infect. Dis. J.* 22:S100–103; discussion S103–104.

87. Hu-Lieskovan S, et al. (2005). Sequence-specific knockdown of EWS-FLI1 by targeted, nonviral delivery of small interfering RNA inhibits tumor growth in a murine model of metastatic Ewing's sarcoma. *Cancer Res.* 65:8984–8992.

88. Pun S H, et al. (2004). Targeted delivery of RNA-cleaving DNA enzyme (DNAzyme) to tumor tissue by transferrin-modified, cyclodextrin-based particles. *Cancer Biol. Ther.* 3:641–650.

89. Ramesh R, et al. (2004). Local and systemic inhibition of lung tumor growth after nanoparticle-mediated mda-7/IL-24 gene delivery. *DNA Cell Biol.* 23:850–857.

90. Hui K M, et al. (1997). Phase I study of immunotherapy of cutaneous metastases of human carcinoma using allogeneic and xenogeneic MHC DNA-liposome complexes. *Gene Ther.* 4:783–790.

91. Nabel G J, Felgner P L (1993). Direct gene transfer for immunotherapy and immunization. *Trends Biotechnol.* 11:211–215.

92. Xing X, Yujiao Chang J, Hung M (1998). Preclinical and clinical study of HER-2/neu-targeting cancer gene therapy. *Adv. Drug Deliv. Rev.* 30:219–227.

93. Hortobagyi G N, Hung M C, Lopez-Berestein G (1998). A Phase I multicenter study of E1A gene therapy for patients with metastatic breast cancer and epithelial ovarian

cancer that overexpresses HER-2/neu or epithelial ovarian cancer. *Hum. Gene Ther.* 9:1775–1798.

94. Ren H, et al. (2003). Immunogene therapy of recurrent glioblastoma multiforme with a liposomally encapsulated replication-incompetent Semliki forest virus vector carrying the human interleukin-12 gene—a phase I/II clinical protocol. *J. Neurooncol.* 64:147–154.

95. Nabel G J, et al. (1996). Immune response in human melanoma after transfer of an allogeneic class I major histocompatibility complex gene with DNA-liposome complexes. *Proc. Natl. Acad. Sci. USA.* 93:15388–15393.

96. Yoshida J, et al. (2004). Human gene therapy for malignant gliomas (glioblastoma multiforme and anaplastic astrocytoma) by in vivo transduction with human interferon beta gene using cationic liposomes. *Hum. Gene Ther.* 15:77–86.

97. Stopeck A T, et al. (2001). Phase II study of direct intralesional gene transfer of allovectin-7, an HLA-B7/beta2-microglobulin DNA-liposome complex, in patients with metastatic melanoma. *Clin. Cancer Res.* 7:2285–2291.

98. Gill D R, et al. (1997). A placebo-controlled study of liposome-mediated gene transfer to the nasal epithelium of patients with cystic fibrosis. *Gene Ther.* 4:199–209.

99. Caplen N J, et al. (1995). Liposome-mediated CFTR gene transfer to the nasal epithelium of patients with cystic fibrosis. *Nat. Med.* 1:39–46.

100. Ruiz F E, et al. (2001). A clinical inflammatory syndrome attributable to aerosolized lipid-DNA administration in cystic fibrosis. *Hum. Gene Ther.* 12:751–761.

101. Hyde S C, et al. (2000). Repeat administration of DNA/liposomes to the nasal epithelium of patients with cystic fibrosis. *Gene Ther.* 7:1156–1165.

102. Porteous D J, et al. (1997). Evidence for safety and efficacy of DOTAP cationic liposome mediated CFTR gene transfer to the nasal epithelium of patients with cystic fibrosis. *Gene Ther.* 4:210–218.

103. Brigham K L, et al. (2000). Transfection of nasal mucosa with a normal alpha1-antitrypsin gene in alpha1-antitrypsin-deficient subjects: comparison with protein therapy. *Hum. Gene Ther.* 11:1023–1032.

104. Tompkins S M, et al. (2004). Protection against lethal influenza virus challenge by RNA interference in vivo. *Proc. Natl. Acad. Sci. USA.* 101:8682–8686.

105. Fang J, et al. (1996). Stimulation of new bone formation by direct transfer of osteogenic plasmid genes. *Proc. Natl. Acad. Sci. USA.* 93:5753–5758.

106. Bonadio J, et al. (1999). Localized, direct plasmid gene delivery in vivo: prolonged therapy results in reproducible tissue regeneration. *Nat. Med.* 5:753–759.

107. Ochiya T, et al. (1999). New delivery system for plasmid DNA in vivo using atelocollagen as a carrier material: The Minipellet. *Nat. Med.* 5:707–710.

108. Minakuchi Y, et al. (2004). Atelocollagen-mediated synthetic small interfering RNA delivery for effective gene silencing in vitro and in vivo. *Nucleic Acids Res.* 32:e109.

109. Takei Y, et al. (2004). A small interfering RNA targeting vascular endothelial growth factor as cancer therapeutics. *Cancer Res.* 64:3365–3370.

110. Tyrone J W, et al. (2000). Collagen-embedded platelet-derived growth factor DNA plasmid promotes wound healing in a dermal ulcer model. *J. Surg. Res.* 93:230–236.

111. Berry M, et al. (2001). Sustained effects of gene-activated matrices after CNS injury. *Mol. Cell Neurosci.* 17:706–716.

112. Eliaz R E, Szoka F C, Jr. (2002). Robust and prolonged gene expression from injectable polymeric implants. *Gene Ther.* 9:1230–1237.

113. Shea L D, et al. (1999). DNA delivery from polymer matrices for tissue engineering. *Nat. Biotechnol.* 17:551–554.

114. Huang Y C, et al. (2005). Bone regeneration in a rat cranial defect with delivery of PEI-condensed plasmid DNA encoding for bone morphogenetic protein-4 (BMP-4). *Gene Ther.* 12:418–426.

115. Jang J H, Rives C B, Shea L D (2005). Plasmid delivery in vivo from porous tissue-engineering scaffolds: Transgene expression and cellular transfection. *Mol. Ther.* 12:475–483.

116. Aoyama T, et al. (2003). Local delivery of matrix metalloproteinase gene prevents the onset of renal sclerosis in streptozotocin-induced diabetic mice. *Tissue Eng.* 9:1289–1299.

117. Kasahara H, et al. (2003). Biodegradable gelatin hydrogel potentiates the angiogenic effect of fibroblast growth factor 4 plasmid in rabbit hindlimb ischemia. *J. Am. Coll. Cardiol.* 41:1056–1062.

118. Shen H, Goldberg E, Saltzman W M (2003). Gene expression and mucosal immune responses after vaginal DNA immunization in mice using a controlled delivery matrix. *J. Control. Rel.* 86:339–348.

119. Jozkowicz A, et al. (2003). Delivery of high dose VEGF plasmid using fibrin carrier does not influence its angiogenic potency. *Int. J. Artif. Organs.* 26:161–169.

120. Trentin D, et al. (2006). Peptide-matrix-mediated gene transfer of an oxygen-insensitive hypoxia-inducible factor-1{alpha} variant for local induction of angiogenesis. *Proc. Natl. Acad. Sci. USA.* 103:2506–2511.

121. Abdelhady H G, et al. (2003). Direct real-time molecular scale visualisation of the degradation of condensed DNA complexes exposed to DNase I. *Nucleic Acids Res.* 31:4001–4005.

122. Golan R, et al. (1999). DNA toroids: stages in condensation. *Biochem.* 38:14069–14076.

123. Dunlap D D, et al. (1997). Nanoscopic structure of DNA condensed for gene delivery. *Nucleic Acids Res.* 25:3095–3101.

124. Fischer D, et al. (2003). In vitro cytotoxicity testing of polycations: influence of polymer structure on cell viability and hemolysis. *Biomat.* 24:1121–1131.

125. Bronich T, Kabanov A V, Marky L A (2001). A thermodynamic characterization of the interaction of a cationic copolymer with DNA. *J. Phys. Chem. B.* 105:6042–6050.

126. Kabanov V A, Kabanov A V (1998). Interpolyelectrolyte and block ionomer complexes for gene delivery: physico-chemical aspects. *Adv. Drug Deliv. Rev.* 30:49–60.

127. Chen W, Turro N J, Tomalia D A (2000). Using ethidium bromide to probe the interactions between DNA and dendrimers. *Langmuir.* 16:15–19.

128. Kunath K, et al. (2003). Low-molecular-weight polyethylenimine as a non-viral vector for DNA delivery: Comparison of physicochemical properties, transfection efficiency and in vivo distribution with high-molecular-weight polyethylenimine. *J. Control. Rel.* 89:113–125.

129. Tang M X, Szoka F C (1997). The influence of polymer structure on the interactions of cationic polymers with DNA and morphology of the resulting complexes. *Gene Ther.* 4:823–832.

130. Fischer D, et al. (1999). A novel non-viral vector for DNA delivery based on low molecular weight, branched polyethylenimine: Effect of molecular weight on transfection efficiency and cytotoxicity. *Pharm. Res.* 16:1273–1279.

131. Schatzlein A G, et al. (2005). Preferential liver gene expression with polypropylenimine dendrimers. *J. Control. Rel.* 101:247–258.

132. Erbacher P, Remy J S, Behr J P (1999). Gene transfer with synthetic virus-like particles via the integrin-mediated endocytosis pathway. *Gene Ther.* 6:138–145.

133. Thomas M, Klibanov A M (2003). Non-viral gene therapy: polycation-mediated DNA delivery. *Appl. Microbiol. Biotechnol.* 62:27–34.

134. von Harpe A, et al. (2000). Characterization of commercially available and synthesized polyethylenimines for gene delivery. *J. Control. Rel.* 69:309–322.

135. Brissault B, et al. (2003). Synthesis of linear polyethylenimine derivatives for DNA transfection. *Bioconjug. Chem.* 14:581–587.

136. Abdallah B, et al. (1996). A powerful nonviral vector for in vivo gene transfer into the adult mammalian brain: polyethylenimine. *Hum. Gene Ther.* 7:1947–1954.

137. Kuo C N, et al. (2005). Dehydrated form of plasmid expressing basic fibroblast growth factor-polyethylenimine complex is a novel and accurate method for gene transfer to the cornea. *Curr. Eye Res.* 30:1015–1024.

138. Kircheis R, et al. (2001). Tumor targeting with surface-shielded ligand–polycation DNA complexes. *J. Control. Rel.* 72:165–170.

139. Adami R C, et al. (1998). Stability of peptide-condensed plasmid DNA formulations. *J. Pharm. Sci.* 87:678–683.

140. Pack D W, Putnam D, Langer R (2000). Design of imidazole-containing endosomolytic biopolymers for gene delivery. *Biotechnol. Bioeng.* 67:217–223.

141. Harada-Shiba M, et al. (2002). Polyion complex micelles as vectors in gene therapy–pharmacokinetics and in vivo gene transfer. *Gene Ther.* 9:407–414.

142. Dufes C, Uchegbu I F, Schatzlein A G (2005). Dendrimers in gene delivery. *Adv. Drug Deliv. Rev.* 57:2177–2202.

143. Qin L, et al. (1998). Efficient transfer of genes into murine cardiac grafts by Starburst polyamidoamine dendrimers. *Hum. Gene Ther.* 9:553–560.

144. Bielinska A U, Kukowska-Latallo J F, Baker J R, Jr. (1997). The interaction of plasmid DNA with polyamidoamine dendrimers: mechanism of complex formation and analysis of alterations induced in nuclease sensitivity and transcriptional activity of the complexed DNA. *Biochim. Biophys. Acta.* 1353:180–190.

145. Kukowska-Latallo J F, et al. (1996). Efficient transfer of genetic material into mammalian cells using Starburst polyamidoamine dendrimers. *Proc. Natl. Acad. Sci. USA.* 93:4897–4902.

146. Hudde T, et al. (1999). Activated polyamidoamine dendrimers, a non-viral vector for gene transfer to the corneal endothelium. *Gene Ther.* 6:939–943.

147. Maruyama-Tabata H, et al. (2000). Effective suicide gene therapy in vivo by EBV-based plasmid vector coupled with polyamidoamine dendrimer. *Gene Ther.* 7:53–60.

148. Wang Y, et al. (2001). Combination of electroporation and DNA/dendrimer complexes enhances gene transfer into murine cardiac transplants. *Am. J. Transplant.* 1:334–338.

149. Kihara F, et al. (2003). In vitro and in vivo gene transfer by an optimized alpha-cyclodextrin conjugate with polyamidoamine dendrimer. *Bioconjug. Chem.* 14:342–350.

150. Cui Z, Mumper R J (2001). Chitosan-based nanoparticles for topical genetic immunization. *J. Control. Rel.* 75:409–419.

151. Illum L, et al. (2001). Chitosan as a novel nasal delivery system for vaccines. *Adv. Drug Deliv. Rev.* 51:81–96.

152. Eliyahu H, et al. (2006). Dextran-spermine-based polyplexes–evaluation of transgene expression and of local and systemic toxicity in mice. *Biomat.* 27:1636–1645.

153. Eliyahu H, et al. (2005). Novel dextran-spermine conjugates as transfecting agents: comparing water-soluble and micellar polymers. *Gene Ther.* 12:494–503.

154. Arima H, et al. (2001). Enhancement of gene expression by polyamidoamine dendrimer conjugates with alpha-, beta-, and gamma-cyclodextrins. *Bioconjug. Chem.* 12:476–484.

155. Kihara F, et al. (2002). Effects of structure of polyamidoamine dendrimer on gene transfer efficiency of the dendrimer conjugate with alpha-cyclodextrin. *Bioconjug. Chem.* 13:1211–1219.

156. Pun S H, et al. (2004). Cyclodextrin-modified polyethylenimine polymers for gene delivery. *Bioconjug. Chem.* 15:831–840.

157. Forrest M L, Gabrielson N, Pack D W (2005). Cyclodextrin-polyethylenimine conjugates for targeted in vitro gene delivery. *Biotechnol. Bioeng.* 89:416–423.

158. Pun S H, Davis M E (2002). Development of a nonviral gene delivery vehicle for systemic application. *Bioconjug. Chem.* 13:630–639.

159. Bellocq N C, et al. (2003). Transferrin-containing, cyclodextrin polymer-based particles for tumor-targeted gene delivery. *Bioconjug. Chem.* 14:1122–1132.

160. Bennett C F, et al. (1998). Structural requirements for cationic lipid mediated phosphorothioate oligonucleotides delivery to cells in culture. *J. Drug Target.* 5:149–162.

161. Xia H, et al. (2002). siRNA-mediated gene silencing in vitro and in vivo. *Nat. Biotechnol.* 20:1006–1010.

162. Bertrand J R, et al. (2002). Comparison of antisense oligonucleotides and siRNAs in cell culture and in vivo. *Biochem. Biophys. Res. Commun.* 296:1000–1004.

163. Perelson A S, et al. (1984). Mechanism of cell-mediated cytotoxicity at the single cell level. VIII. Kinetics of lysis of target cells bound by more than one cytotoxic T lymphocyte. *J. Immunol.* 132:2190–2198.

164. Ferrari M E, et al. (1998). Analytical methods for the characterization of cationic lipid-nucleic acid complexes. *Hum. Gene Ther.* 9:341–351.

165. MacDonald R C, et al. (1999). O-ethylphosphatidylcholine: A metabolizable cationic phospholipid which is a serum-compatible DNA transfection agent. *J. Pharm. Sci.* 88:896–904.

166. Wang J, et al. (1998). Synthesis and characterization of long chain alkyl acyl carnitine esters. Potentially biodegradable cationic lipids for use in gene delivery. *J. Med. Chem.* 41:2207–2215.

167. Byk G, et al. (1998). Novel non-viral vectors for gene delivery: synthesis of a second-generation library of mono-functionalized poly-(guanidinium)amines and their introduction into cationic lipids. *Biotechnol. Bioeng.* 61:81–87.

168. Lee E R, et al. (1996). Detailed analysis of structures and formulations of cationic lipids for efficient gene transfer to the lung. *Hum. Gene Ther.* 7:1701–1717.

169. Byk G, et al. (1998). Synthesis, activity, and structure–activity relationship studies of novel cationic lipids for DNA transfer. *J. Med. Chem.* 41:229–235.

170. Falk K, et al. (2001). Reduction of experimental adhesion formation by inhibition of plasminogen activator inhibitor type 1. *Br. J. Surg.* 88:286–289.

171. Kennedy M T, et al. (2000). Factors governing the assembly of cationic phospholipid-DNA complexes. *Biophys. J.* 78:1620–1633.

172. Gershon H, et al. (1993). Mode of formation and structural features of DNA-cationic liposome complexes used for transfection. *Biochem.* 32:7143–7151.

173. Templeton N S, et al. (1997). Improved DNA: liposome complexes for increased systemic delivery and gene expression. *Nat. Biotechnol.* 15:647–652.

174. Gustafsson J, et al. (1995). Complexes between cationic liposomes and DNA visualized by cryo-TEM. *Biochim. Biophys. Acta.* 1235:305–312.

175. Koltover I, et al. (1998). An inverted hexagonal phase of cationic liposome-DNA complexes related to DNA release and delivery. *Science.* 281:78–81.

176. Xu Y, et al. (1999). Physicochemical characterization and purification of cationic lipoplexes. *Biophys. J.* 77:341–353.

177. Turek J, et al. (2000). Formulations which increase the size of lipoplexes prevent serum-associated inhibition of transfection. *J. Gene Med.* 2:32–40.

178. Ross P C, Hui S W (1999). Lipoplex size is a major determinant of in vitro lipofection efficiency. *Gene Ther.* 6:651–659.

179. Xu Y, Szoka F C, Jr. (1996). Mechanism of DNA release from cationic liposome/DNA complexes used in cell transfection. *Biochem.* 35:5616–5623.

180. Zelphati O, et al. (1998). Effect of serum components on the physico-chemical properties of cationic lipid/oligonucleotide complexes and on their interactions with cells. *Biochim. Biophys. Acta.* 1390:119–133.

181. Mok K W, Lam A M, Cullis P R (1999). Stabilized plasmid-lipid particles: Factors influencing plasmid entrapment and transfection properties. *Biochim. Biophys. Acta.* 1419:137–150.

182. Meyer O, et al. (1998). Cationic liposomes coated with polyethylene glycol as carriers for oligonucleotides. *J. Biol. Chem.* 273:15621–15627.

183. Hong K, et al. (1997). Stabilization of cationic liposome-plasmid DNA complexes by polyamines and poly(ethylene glycol)-phospholipid conjugates for efficient in vivo gene delivery. *FEBS Lett.* 400:233–237.

184. Hofland H E, Shephard L, Sullivan S M (1996). Formation of stable cationic lipid/DNA complexes for gene transfer. *Proc. Natl. Acad. Sci. USA.* 93:7305–7309.

185. Anchordoquy T J, Carpenter J F, Kroll D J (1997). Maintenance of transfection rates and physical characterization of lipid/DNA complexes after freeze-drying and rehydration. *Arch. Biochem. Biophys.* 348:199–206.

186. Vitiello L, et al. (1996). Condensation of plasmid DNA with polylysine improves liposome-mediated gene transfer into established and primary muscle cells. *Gene Ther.* 3:396–404.

187. Gao X, Huang L (1996). Potentiation of cationic liposome-mediated gene delivery by polycations. *Biochem.* 35:1027–1036.

188. Langer R (1998). Drug delivery and targeting. *Nature.* 392:5–10.

189. Scherer F, et al. (2002). Nonviral vector loaded collagen sponges for sustained gene delivery in vitro and in vivo. *J. Gene Med.* 4:634–643.

190. Cleek R L, et al. (1997). Inhibition of smooth muscle cell growth in vitro by an antisense oligodeoxynucleotide released from poly(DL-lactic-co-glycolic acid) microparticles. *J. Biomed. Mater. Res.* 35:525–530.

191. Ochiya T, et al. (2001). Biomaterials for gene delivery: Atelocollagen-mediated controlled release of molecular medicines. *Curr. Gene Ther.* 1:31–52.

192. Segura T, Shea L D (2002). Surface-tethered DNA complexes for enhanced gene delivery. *Bioconjug. Chem.* 13:621–629.

193. Segura T, Volk M J, Shea L D (2003). Substrate-mediated DNA delivery: Role of the cationic polymer structure and extent of modification. *J. Control. Rel.* 93:69–84.

194. Bengali Z, et al. (2005). Gene delivery through cell culture substrate adsorbed DNA complexes. *Biotechnol. Bioeng.* 90:290–302.

195. Jang J H, et al. (2006). Surface adsorption of DNA to tissue engineering scaffolds for efficient gene delivery. *J. Biomed. Mater. Res. A.* 77:50–58.

196. Langer R, Vacanti J P (1993). Tissue engineering. *Science.* 260:920–926.

197. Lutolf M P, Hubbell J A (2005). Synthetic biomaterials as instructive extracellular microenvironments for morphogenesis in tissue engineering. *Nat. Biotechnol.* 23:47–55.

198. Saltzman M W, Baldwin S P (1998). Materials for protein delivery in tissue engineering. *Adv. Drug Deliv. Rev.* 33:71–86.

199. Harris L D, Kim B S, Mooney D J (1998). Open pore biodegradable matrices formed with gas foaming. *J. Biomed. Mater. Res.* 42:396–402.

200. Nof M, Shea L D (2002). Drug-releasing scaffolds fabricated from drug-loaded microspheres. *J. Biomed. Mater. Res.* 59:349–356.

201. Jang J H, Shea L D (2003). Controllable delivery of non-viral DNA from porous scaffolds. *J. Control. Rel.* 86:157–168.

202. Luo D, et al. (1999). Controlled DNA delivery systems. *Pharm. Res.* 16:1300–1308.

203. Hsu Y Y, Hao T, Hedley M L (1999). Comparison of process parameters for micro-encapsulation of plasmid DNA in poly(D,L-lactic-co-glycolic) acid microspheres. *J. Drug Target.* 7:313–323.

204. Ando S, et al. (1999). PLGA microspheres containing plasmid DNA: Preservation of supercoiled DNA via cryopreparation and carbohydrate stabilization. *J. Pharm. Sci.* 88:126–130.

205. Hedley M L, Curley J, Urban R (1998). Microspheres containing plasmid-encoded antigens elicit cytotoxic T-cell responses. *Nat. Med.* 4:365–368.

206. Luu Y K, et al. (2003). Development of a nanostructured DNA delivery scaffold via electrospinning of PLGA and PLA-PEG block copolymers. *J. Control. Rel.* 89:341–353.

207. Capan Y, et al. (1999). Influence of formulation parameters on the characteristics of poly(D, L-lactide-co-glycolide) microspheres containing poly(L-lysine) complexed plasmid DNA. *J. Control. Rel.* 60:279–286.

208. Capan Y, et al. (1999). Preparation and characterization of poly (D,L-lactide-co-glycolide) microspheres for controlled release of poly(L-lysine) complexed plasmid DNA. *Pharm. Res.* 16:509–513.

209. De Rosa G, et al. (2002). Biodegradable microparticles for the controlled delivery of oligonucleotides. *Int. J. Pharm.* 242:225–228.

210. De Rosa G, et al. (2003). Long-term release and improved intracellular penetration of oligonucleotide-polyethylenimine complexes entrapped in biodegradable microspheres. *Biomacromol.* 4:529–536.

211. Quick D J, Anseth K S (2004). DNA delivery from photocrosslinked PEG hydrogels: Encapsulation efficiency, release profiles, and DNA quality. *J. Control. Rel.* 96:341–351.

212. Lee P Y, Li Z, Huang L (2003). Thermosensitive hydrogel as a Tgf-beta1 gene delivery vehicle enhances diabetic wound healing. *Pharm. Res.* 20:1995–2000.

213. Pachence J M (1996). Collagen-based devices for soft tissue repair. *J. Biomed. Mater. Res.* 33:35–40.

214. Schoen F J, Levy R J (1999). Tissue heart valves: Current challenges and future research perspectives. *J. Biomed. Mater. Res.* 47:439–465.

215. Pieper J S, et al. (1999). Preparation and characterization of porous crosslinked collagenous matrices containing bioavailable chondroitin sulphate. *Biomat.* 20:847–858.

216. Pieper J S, et al. (2002). Crosslinked type II collagen matrices: Preparation, characterization, and potential for cartilage engineering. *Biomat.* 23:3183–3192.

217. Verdu E, et al. (2002). Alignment of collagen and laminin-containing gels improve nerve regeneration within silicone tubes. *Restor. Neurol. Neurosci.* 20:169–179.

218. Wisser D, Steffes J (2003). Skin replacement with a collagen based dermal substitute, autologous keratinocytes and fibroblasts in burn trauma. *Burns.* 29:375–380.

219. Bellamkonda R, Ranieri J P, Aebischer P (1995). Laminin oligopeptide derivatized agarose gels allow three-dimensional neurite extension in vitro. *J. Neurosci Res.* 41:501–509.

220. Madihally S V, Matthew H W (1999). Porous chitosan scaffolds for tissue engineering. *Biomat.* 20:1133–1142.

221. Sakiyama-Elbert S E, Panitch A, Hubbell J A (2001). Development of growth factor fusion proteins for cell-triggered drug delivery. *Faseb. J.* 15:1300–1302.

222. Meilander N J, et al. (2003). Sustained release of plasmid DNA using lipid microtubules and agarose hydrogel. *J. Control. Rel.* 88:321–331.

223. Fang N, et al. (2001). Interactions of phospholipid bilayer with chitosan: effect of molecular weight and pH. *Biomacromol.* 2:1161–1168.

224. Mao H Q, et al. (2001). Chitosan-DNA nanoparticles as gene carriers: Synthesis, characterization and transfection efficiency. *J. Control. Rel.* 70:399–421.

225. Kim A, et al. (2003). Characterization of DNA-hyaluronan matrix for sustained gene transfer. *J. Control. Rel.* 90:81–95.

226. Yun Y H, et al. (2004). Hyaluronan microspheres for sustained gene delivery and site-specific targeting. *Biomat.* 25:147–157.

227. Samuel R E, et al. (2002). Delivery of plasmid DNA to articular chondrocytes via novel collagen-glycosaminoglycan matrices. *Hum. Gene Ther.* 13:791–802.

228. Bielinska A U, et al. (2000). Application of membrane-based dendrimer/DNA complexes for solid phase transfection in vitro and in vivo. *Biomat.* 21:877–887.

229. Campoccia D, et al. (1998). Semisynthetic resorbable materials from hyaluronan esterification. *Biomat.* 19:2101–2127.

230. Bulpitt P, Aeschlimann D (1999). New strategy for chemical modification of hyaluronic acid: preparation of functionalized derivatives and their use in the formation of novel biocompatible hydrogels. *J. Biomed. Mater. Res.* 47:152–169.

231. Duranti F, et al. (1998). Injectable hyaluronic acid gel for soft tissue augmentation. A clinical and histological study. *Dermatol. Surg.* 24:1317–1325.

232. Trentin D, Hubbell J, Hall H (2005). Non-viral gene delivery for local and controlled DNA release. *J. Control. Rel.* 102:263–275.

233. Segura T, Chung P H, Shea L D (2005). DNA delivery from hyaluronic acid-collagen hydrogels via a substrate-mediated approach. *Biomat.* 26:1575–1584.

234. Schense J C, Hubbell J A (1999). Cross-linking exogenous bifunctional peptides into fibrin gels with factor XIIIa. *Bioconjug. Chem.* 10:75–81.

235. Zisch A H, et al. (2001). Covalently conjugated VEGF—fibrin matrices for endothelialization. *J. Control. Rel.* 72:101–113.

236. Toncheva V, et al. (1998). Novel vectors for gene delivery formed by self-assembly of DNA with poly(L-lysine) grafted with hydrophilic polymers. *Biochim. Biophys. Acta.* 1380:354–368.

237. Howard K A, et al. (2000). Influence of hydrophilicity of cationic polymers on the biophysical properties of polyelectrolyte complexes formed by self-assembly with DNA. *Biochim. Biophys. Acta.* 1475:245–255.

238. Dash P R, et al. (2000). Decreased binding to proteins and cells of polymeric gene delivery vectors surface modified with a multivalent hydrophilic polymer and retargeting through attachment of transferrin. *J. Biol. Chem.* 275:3793–3802.

239. Oupicky D, et al. (2000). Steric stabilization of poly-L-Lysine/DNA complexes by the covalent attachment of semitelechelic poly[N-(2-hydroxypropyl)methacrylamide]. *Bioconjug. Chem.* 11:492–501.

240. Wang W, Tetley L, Uchegbu I F (2001). The level of hydrophobic substitution and the molecular weight of amphiphilic poly-L-lysine-based polymers strongly affects their assembly into polymeric bilayer vesicles. *J. Colloid Interface Sci.* 237:200–207.

241. Katayose S, Kataoka K (1997). Water-soluble polyion complex associates of DNA and poly(ethylene glycol)-poly(L-lysine) block copolymer. *Bioconjug. Chem.* 8:702–707.

242. Ahn C H, et al. (2002). Biodegradable poly(ethylenimine) for plasmid DNA delivery. *J. Control. Rel.* 80:273–282.

243. Choi Y H, et al. (1999). Characterization of a targeted gene carrier, lactose-polyethylene glycol-grafted poly-L-lysine and its complex with plasmid DNA. *Hum. Gene Ther.* 10:2657–2665.

244. Carlisle R C, et al. (2004). Polymer-coated polyethylenimine/DNA complexes designed for triggered activation by intracellular reduction. *J. Gene Med.* 6:337–344.

245. Shuai X T, et al. (2003). Novel biodegradable ternary copolymers hy-PEI-g-PCL-b-PEG: Synthesis, characterization, and potential as efficient nonviral gene delivery vectors. *Macromol.* 36:5751–5759.

246. Rejman J, Bragonzi A, Conese M (2005). Role of clathrin- and caveolae-mediated endocytosis in gene transfer mediated by lipo- and polyplexes. *Mol. Ther.* 12:468–474.

247. Rejman J, et al. (2004). Size-dependent internalization of particles via the pathways of clathrin- and caveolae-mediated endocytosis. *Biochem. J.* 377:159–169.

248. Ogris M, et al. (2001). DNA/polyethylenimine transfection particles: influence of ligands, polymer size, and PEGylation on internalization and gene expression. *AAPS PharmSci.* 3:e21.

249. Templeton N S, Lasic D D (1999). New directions in liposome gene delivery. *Mol. Biotechnol.* 11:175–180.

250. Baatz J E, et al. (1994). Utilization of modified surfactant-associated protein B for delivery of DNA to airway cells in culture. *Proc. Natl. Acad. Sci. USA.* 91:2547–2551.

251. Plank C, et al. (1992). Gene transfer into hepatocytes using asialoglycoprotein receptor mediated endocytosis of DNA complexed with an artificial tetraantennary galactose ligand. *Bioconjug. Chem.* 3:533–539.

252. Zenke M, et al. (1990). Receptor-mediated endocytosis of transferrin-polycation conjugates: an efficient way to introduce DNA into hematopoietic cells. *Proc. Natl. Acad. Sci. USA.* 87:3655–3659.

253. Wagner E, et al. (1990). Transferrin-polycation conjugates as carriers for DNA uptake into cells. *Proc. Natl. Acad. Sci. USA.* 87:3410–3414.

254. Cotten M, et al. (1990). Transferrin-polycation-mediated introduction of DNA into human leukemic cells: stimulation by agents that affect the survival of transfected DNA or modulate transferrin receptor levels. *Proc. Natl. Acad. Sci. USA.* 87:4033–4037.

255. Weitman S D, et al. (1992). Cellular localization of the folate receptor: potential role in drug toxicity and folate homeostasis. *Cancer Res.* 52:6708–6711.

256. Leamon C P, Low P S (1991). Delivery of macromolecules into living cells: a method that exploits folate receptor endocytosis. *Proc. Natl. Acad. Sci. USA.* 88:5572–5576.

257. Stewart A J, et al. (1996). Enhanced biological activity of antisense oligonucleotides complexed with glycosylated poly-L-lysine. *Mol. Pharmacol.* 50:1487–1494.

258. Erbacher P, et al. (1996). Gene transfer by DNA/glycosylated polylysine complexes into human blood monocyte-derived macrophages. *Hum. Gene Ther.* 7:721–729.

259. Midoux P, et al. (1993). Specific gene transfer mediated by lactosylated poly-L-lysine into hepatoma cells. *Nucleic Acids Res.* 21:871–878.

260. Schaffer D V, Lauffenburger D A (1998). Optimization of cell surface binding enhances efficiency and specificity of molecular conjugate gene delivery. *J. Biol. Chem.* 273:28004–28009.

261. Remy J S, et al. (1995). Targeted gene transfer into hepatoma cells with lipopolyamine-condensed DNA particles presenting galactose ligands: A stage toward artificial viruses. *Proc. Natl. Acad. Sci. USA.* 92:1744–1748.

262. Ziady A G, et al. (1999). Chain length of the polylysine in receptor-targeted gene transfer complexes affects duration of reporter gene expression both in vitro and in vivo. *J. Biol. Chem.* 274:4908–4916.

263. Reimer D L, Kong S, Bally M B (1997). Analysis of cationic liposome-mediated inter-actions of plasmid DNA with murine and human melanoma cells in vitro. *J. Biol. Chem.* 272:19480–19487.

264. Lechardeur D, et al. (1999). Metabolic instability of plasmid DNA in the cytosol: a potential barrier to gene transfer. *Gene Ther.* 6:482–497.

265. Tseng W C, Haselton F R, Giorgio T D (1997). Transfection by cationic liposomes using simultaneous single cell measurements of plasmid delivery and transgene expres-sion. *J. Biol. Chem.* 272:25641–25647.

266. Godbey W T, Wu K K, Mikos A G (1999). Tracking the intracellular path of poly(ethylenimine)/DNA complexes for gene delivery. *Proc. Natl. Acad. Sci. USA.* 96:5177–5181.

267. Clamme J P, Krishnamoorthy G, Mely Y (2003). Intracellular dynamics of the gene delivery vehicle polyethylenimine during transfection: Investigation by two-photon fluorescence correlation spectroscopy. *Biochim. Biophys. Acta.* 1617:52–61.

268. Clamme J P, Azoulay J, Mely Y (2003). Monitoring of the formation and dissociation of polyethylenimine/DNA complexes by two photon fluorescence correlation spectros-copy. *Biophys. J.* 84:1960–1968.

269. Boeckle S, et al. (2004). Purification of polyethylenimine polyplexes highlights the role of free polycations in gene transfer. *J. Gene Med.* 6:1102–1111.

270. Banerjee P, et al. (2004). Novel hyperbranched dendron for gene transfer in vitro and in vivo. *Bioconjug. Chem.* 15:960–968.

271. Ruponen M, Yla-Herttuala S, Urtti A (1999). Interactions of polymeric and liposomal gene delivery systems with extracellular glycosaminoglycans: physicochemical and transfection studies. *Biochim. Biophys. Acta.* 1415:331–341.

272. Malik N, et al. (2000). Dendrimers: relationship between structure and biocompatibil-ity in vitro, and preliminary studies on the biodistribution of 125I-labelled polyami-doamine dendrimers in vivo. *J. Control. Rel.* 65:133–148.

273. Zinselmeyer B H, et al. (2002). The lower-generation polypropylenimine dendrimers are effective gene-transfer agents. *Pharm. Res.* 19:960–967.

274. Jevprasesphant R, et al. (2003). The influence of surface modification on the cytotoxic-ity of PAMAM dendrimers. *Int. J. Pharm.* 252:263–266.

275. Ohsaki M, et al. (2002). In vitro gene transfection using dendritic poly(L-lysine). *Bioconjug. Chem.* 13:510–517.

276. Budker V, et al. (1996). pH-sensitive, cationic liposomes: A new synthetic virus-like vector. *Nat. Biotechnol.* 14:760–764.

277. Zelphati O, Szoka F C, Jr. (1996). Mechanism of oligonucleotide release from cationic liposomes. *Proc. Natl. Acad. Sci. USA.* 93:11493–11498.

278. SUH J, Paik H J, Hwang B K (1994). Ionization of Poly(Ethylenimine) and Poly(Allylamine) at Various Phs. *Bioorg. Chem.* 22:318–327.

279. Sonawane N D, Szoka F C, Jr, Verkman A S (2003). Chloride accumulation and swelling in endosomes enhances DNA transfer by polyamine-DNA polyplexes. *J. Biol. Chem.* 278:44826–44831.

280. Midoux P, Monsigny M (1999). Efficient gene transfer by histidylated polylysine/pDNA complexes. *Bioconjug. Chem.* 10:406–411.

281. Cristiano R J, et al. (1993). Hepatic gene therapy: efficient gene delivery and expression in primary hepatocytes utilizing a conjugated adenovirus-DNA complex. *Proc. Natl. Acad. Sci. USA.* 90:11548–11552.

282. Curiel D T, et al. (1991). Adenovirus enhancement of transferrin-polylysine-mediated gene delivery. *Proc. Natl. Acad. Sci. USA.* 88:8850–8854.

283. Plank C, Zauner W, Wagner E (1998). Application of membrane-active peptides for drug and gene delivery across cellular membranes. *Adv. Drug Deliv. Rev.* 34:21–35.

284. Kakudo T, et al. (2004). Transferrin-modified liposomes equipped with a pH-sensitive fusogenic peptide: An artificial viral-like delivery system. *Biochem.* 43:5618–5628.

285. Cheung C Y, et al. (2001). A pH-sensitive polymer that enhances cationic lipid-mediated gene transfer. *Bioconjug. Chem.* 12:906–910.

286. Kyriakides T R, et al. (2002). pH-sensitive polymers that enhance intracellular drug delivery in vivo. *J. Control. Rel.* 78:295–303.

287. Dowty M E, et al. (1995). Plasmid DNA entry into postmitotic nuclei of primary rat myotubes. *Proc. Natl. Acad. Sci. USA.* 92:4572–4576.

288. Jans D A, Chan C K, Huebner S (1998). Signals mediating nuclear targeting and their regulation: application in drug delivery. *Med. Res. Rev.* 18:189–223.

289. Tseng W C, Haselton F R, Giorgio T D (1999). Mitosis enhances transgene expression of plasmid delivered by cationic liposomes. *Biochim. Biophys. Acta.* 1445:53–64.

290. Subramanian A, Ranganathan P, Diamond S L (1999). Nuclear targeting peptide scaffolds for lipofection of nondividing mammalian cells. *Nat. Biotechnol.* 17:873–877.

291. Chan C K, Jans D A (1999). Enhancement of polylysine-mediated transferrinfection by nuclear localization sequences: Polylysine does not function as a nuclear localization sequence. *Hum. Gene Ther.* 10:1695–1702.

292. Singh D, et al. (1999). Peptide-based intracellular shuttle able to facilitate gene transfer in mammalian cells. *Bioconjug. Chem.* 10:745–754.

293. Zanta M A, Belguise-Valladier P, Behr J P (1999). Gene delivery: A single nuclear localization signal peptide is sufficient to carry DNA to the cell nucleus. *Proc. Natl. Acad. Sci. USA.* 96:91–96.

294. Sebestyen M G, et al. (1998). DNA vector chemistry: The covalent attachment of signal peptides to plasmid DNA. *Nat. Biotechnol.* 16:80–85.

295. Wetzer B, et al. (2001). Reducible cationic lipids for gene transfer. *Biochem. J.* 356:747–756.

296. Kong H J, et al. (2005). Non-viral gene delivery regulated by stiffness of cell adhesion substrates. *Nat. Mater.* 4:460–464.

7.4

PHARMACOKINETICS OF NUCLEIC-ACID-BASED THERAPEUTICS

JOHN C. SCHMITZ,[1] ALEKSANDRA PANDYRA,[2,3] JAMES KOROPATNICK,[2,3,4] AND RANDAL W. BERG[2,3,4]

[1] *VACT Healthcare System, VA Cancer Center, and Yale Cancer Center, Yale University School of Medicine, West Haven, Connecticut*
[2] *Cancer Research Laboratory Program, London Regional Cancer Program, London, Ontario, Canada*
[3] *The University of Western Ontario, London, Ontario, Canada*
[4] *Lawson Health Research Institute, London Health Sciences Centre, London, Ontario, Canada*

Chapter Contents

7.4.1 Introduction: Nucleic-Acid-Based Therapeutics — 1062
7.4.2 Antisense Chemistries—Modifications to DNA/RNA Structure — 1064
7.4.3 Additional Synthetic Chemistries — 1065
7.4.4 Methods for Detection of ASOs in Tissues and Fluids — 1065
    7.4.4.1 Labeled ASOs — 1065
    7.4.4.2 Nondenaturing Polyacrylamide and Capillary Gel Electrophoresis — 1066
    7.4.4.3 Hybridization-Based Approaches — 1066
7.4.5 Routes of Administration — 1066
7.4.6 PK (ADME) of ASOs Administered by Injection — 1067
    7.4.6.1 Absorption and Distribution — 1067
    7.4.6.2 Metabolism and Elimination — 1068
    7.4.6.3 PK Properties of 2′-MOE, PS, and PD ASOs — 1068
    7.4.6.4 PK Analyses from Selected Clinical Trials — 1068
    7.4.6.5 Target Downregulation in Tissues of Interest — 1069
7.4.7 RNA Interference — 1069
    7.4.7.1 siRNA Specificity — 1070
    7.4.7.2 Cell Culture — 1071
    7.4.7.3 *In vivo* Studies with "Naked" siRNAs — 1071
    7.4.7.4 Hydrodynamic High-Pressure "Naked" siRNA Injection — 1072
    7.4.7.5 Chemically Modified siRNAs — 1072

Handbook of Pharmaceutical Biotechnology, Edited by Shayne Cox Gad.
Copyright © 2007 John Wiley & Sons, Inc.

7.4.7.6   Ligand-Mediated siRNA Delivery   1073
7.4.7.7   Intranasal siRNA Injection   1073
7.4.7.8   *In vivo* Studies with Encapsulated siRNAs   1074
7.4.7.9   Linear Polyethylenimine   1074
7.4.7.10   Cationic Lipids   1075
7.4.7.11   siRNAs in Clinical Trials   1076
7.4.8   Conclusions   1076
7.4.8.1   Nucleic-Acid-Based Therapeutics   1076
7.4.8.2   PK (ADME) and Toxicology Differences   1077
7.4.8.3   Relative Clinical Utility of the Different Approaches   1077
References   1077

7.4.1   INTRODUCTION: NUCLEIC-ACID-BASED THERAPEUTICS

Nucleic-acid-based therapies include various forms of gene and antisense therapeutics. Gene therapy encompasses the use of plasmid vectors, retroviral and lentiviral vectors, adeno-associated virus (AAV) and other methods of introducing a gene expression or gene silencing nucleic acid sequence into a target tissue or cell. Gene silencing has been achieved with ribozymes, small interfering double-stranded RNAs (siRNAs), and synthetic antisense oligonucleotides (ASOs), whether these are unmodified or chemically modified DNA or RNA structures (Figure 7.4-1). Additional chemically synthesized, but not natural, oligonucleotides include DNA-like peptide nucleic acids (PNAs) and phosphorodiamidate morpholino oligonucleotides (PMOs) (Figure 7.4-1). Some nucleic-acid-based therapeutics are designed to influence the splicing of pre-mRNA to modify or alter the protein produced, rather than to silence the gene expression entirely [1, 2, 3]. Many gene therapy and antisense approaches have unique applications based on local delivery, which limits the distribution of the therapeutic in the body and precludes considerations of their pharmacokinetics (PK). We have included key citations describing PK for some gene-based and oligonucleotide-based approaches in Table 7.4-1, but the remainder of the chapter is limited to ASOs and siRNAs for brevity and clarity.

Most PK studies of nucleic-acid-based therapies are, in fact, devoted to description of the PK properties of ASOs. Many academic and industry researchers have chosen to focus on the use and development of synthetic ASOs as therapeutics. RNA- and DNA-based ASOs designed against a large number of targets have progressed through preclinical stages and into clinical trials. As a consequence, there is a large body of literature describing the PK properties of these ASOs in animals and humans. We will first discuss the various chemical structures of ASOs being developed as therapeutics, and then we will focus on the published animal and human PK data regarding ASOs of two chemistries, exemplified by Genasense (Oblimersen, G3139, targeting the mRNA for the anti-apoptotic protein Bcl-2) and OGX-011 (targeting the mRNA encoding the chaperone protein clusterin).

The second major (and more recent) focus in the literature is on the PK properties of various embodiments of siRNA therapeutics. Excitement has grown recently with our increased understanding of the mechanisms responsible for the endogenous pathways of posttranscriptional gene silencing, also known as RNA inter-

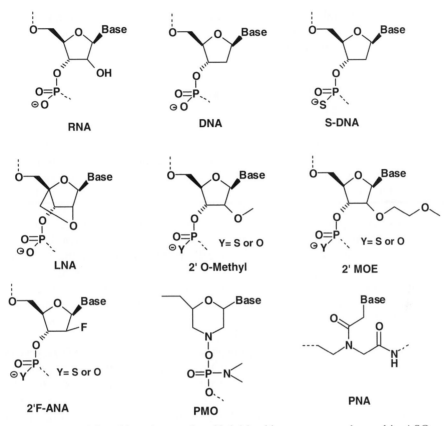

Figure 7.4-1. Nucleic acid analogs and artificial backbones commonly used in ASOs and siRNAs.

TABLE 7.4-1. Selected PK Studies of Nucleic-Acid-Based Therapies

Category	Comments	Citations
Plasmid and viral	Reviews of PK and clinical trials	[4–6]
Ribozymes	Targeting HEP C virus in mice	[7]
	Review of anti-VEGF-R1	[8]
	PK in monkeys	[9]
	Trial in healthy volunteers	[10]
	Trials in cancer patients	[11, 12]
PMO	AVI-4126, targeting c-myc	[13–15]
LNA		[16–18]
PNA	PK in rats	[19]
	PK in mice	[20]

ference (RNAi). This understanding has led to design and synthesis of expression vectors, short hairpin RNA (shRNA), and small interfering RNA (siRNA) molecules to reduce or silence expression of target genes. Although use of RNAi technology is widespread in laboratories, therapeutic applications are not yet as far advanced as for ASOs. Studies describing PK properties of siRNA in animal models are limited, and only three siRNA molecules are currently in human clinical trials. The relevant preclinical PK studies and the ongoing clinical trials are discussed in the latter portion of the chapter.

7.4.2 ANTISENSE CHEMISTRIES—MODIFICATIONS TO DNA/RNA STRUCTURE

ASOs are designed to reduce expression of a target protein based on double-stranded Watson–Crick base pairing with the mRNA coding for the protein. RNA-based ASOs act by inhibiting translation of the target mRNA, and subsequent normal protein turnover (degradation) results in reduced target protein levels in cells [21, 22]. DNA-based ASOs, also known as oligodeoxynucleotides (ODNs), mediate mRNA degradation by activation of endogenous ribonuclease H and subsequent exonuclease cleavage of the target mRNA [23–26]. Again, routine protein turnover reduces target protein levels in cells in the absence of the mRNA to provide template for new protein synthesis. Advantages of ASOs over other nucleic-acid-based strategies include ease of delivery into cells and tissues, improved specificity for individual mRNA target molecules, flexibility in target choice and reagent design, and relative ease of chemical synthesis that facilitates large-scale production of pharmaceutical-grade material [22, 27–29].

Early studies with unmodified phosphodiester (PD) DNA ASOs revealed various problems, including delivery into target cells and nuclease sensitivity in body fluids and in cells. Since then, numerous chemical modifications to the DNA structure have been incorporated into ASOs (see Figure 7.4-1). One of the first improvements was inclusion of phosphorothioate internucleotide linkages (PS DNA), where the nonbridging oxygen is replaced with a sulfur atom. This improved stability against endo- and exonucleases inside cells and in body fluids [22, 30]. "Second generation" ASOs were designed to include 2′-methoxyethoxylation (2′-MOE), 2′-O-methyl, or other modifications of the ribose sugar group [31–33]. These modifications at the 2′ position enhance uptake and stability of antisense ODNs [22, 32], but they preclude RNase H activity. ASOs containing 2′ modifications at the 5 or 6 nucleotides on each end, with DNA bases in the central region (called "gapmers"), retain the ability to provide a suitable target for ribonuclease H [25, 32, 34, 35]. Several PS DNA ASOs and gapmer ASOs have progressed through preclinical development stages, including PK and toxicological studies, and many are currently in human clinical trials (see Table 7.4-2).

In terms of PK properties, ASOs of the same chemistry but with different nucleotide base sequences seem to behave identically. In contrast, differences in PK profiles depend critically on the ASO chemistry and modifications. That is, ASOs with the same base sequence but having PS versus PD, or DNA versus RNA versus 2′-MOE modifications will have slightly or vastly different PK profiles [32, 34]. Furthermore, separately from PK properties, differences in the nucleotide sequence incorporated into ASOs with identical chemistry do affect physiological responses, including toxicities and off-target effects. For example, ASOs containing CpG dinucleotide motifs often activate particular toxicities and immune responses [62–64]. Toxicological changes are also dependent on backbone and sugar modifications, such as activation of the complement cascade by ASOs with PS inter-

TABLE 7.4-2. Selected Clinical Trials of DNA-Based ASOs

ASO	Target	Chemistry	Citations
G3139	Bcl-2	PS DNA	[36–40]
GTI-2040	RNR	PS DNA	[41]
OL(1)p53	p53	PS DNA	[42, 43]
MG98	DNMT1	PS DNA	[44]
GEM91	HIV gag	PS DNA	[45, 46]
ISIS 5132	c-raf-1	PS DNA	[47–49]
ISIS 2503	H-ras	PS DNA	[50, 51]
ISIS 2302	ICAM-1	PS DNA	[52–54]
ISIS 3521	PKC-α	PS DNA	[47, 55, 56]
LY900003	PKC-α	(= ISIS 3521)	[57]
GEM231	PKA	2′-Me Gapmer	[58, 59]
ISIS 104838	TNF-α	2′-MOE Gapmer	[60]
OGX-011	Clusterin	2′-MOE Gapmer	[61]

nucleotide linkages but not by those with PD backbones [45, 65, 66]. These toxicological issues are not discussed in detail here, but they have been reviewed recently elsewhere [62, 65, 67].

7.4.3 ADDITIONAL SYNTHETIC CHEMISTRIES

Several groups of researchers have been pursuing alternative synthetic chemistries to achieve the desirable properties of an ideal ASO. Some of these designs that are in more-advanced stages of development as potential drug candidates are shown in Figure 7.4-1. These include locked nucleic acids (LNAs) [16, 17, 68–71], 2′-fluoroarabinonucleic acids (FANAs) [72–74], PNAs [75–79], and PMOs [13, 14, 80–83]. Only a limited number of ASOs with these chemistries have progressed to PK studies in rodent models, primate models, or human clinical trials. Relevant PK studies regarding this group of nucleic-acid-based ASOs are given in Table 7.4-1. For brevity and due to the lack of human PK data, we will limit our discussion of these other chemistries.

7.4.4 METHODS FOR DETECTION OF ASOs IN TISSUES AND FLUIDS

7.4.4.1 Labeled ASOs

In our work [84], we have used T4 polynucleotide kinase to end-label ASOs using radioactive [^{32}P]-γ-adenosine triphosphate (ATP). Methods to end-label or synthesize ASOs with biotin, fluorescein, digoxigenin, and so on, are also available. ASOs are amenable to internal labeling with radioactive halogens [85]. Radioactively labeled ASOs are tracked using phospho-imaging, gel electrophoresis, and other techniques. Fluorescence microscopy is used to monitor fluorescently labeled ASOs, whereas detection of biotin-labeled ASOs is done with anti-biotin antibodies or avidin-conjugated enzymes and appropriate substrates.

7.4.4.2   Nondenaturing Polyacrylamide and Capillary Gel Electrophoresis

Acrylamide gels (15% to 20%) can be used effectively to separate ASOs from longer intact nucleic acids and shorter degradation products [72]. Radioactively labeled ASOs are detected using standard autoradiography or phospho-imaging techniques [35]. Using acrylamide gels cast in capillary tubes, ASOs can also be separated, identified, and quantified after electrophoresis. Absorbance at 260 nm, mass spectrometry, and so on, can be used to detect ASOs as they leave the capillary gels [35, 86–88].

7.4.4.3   Hybridization-Based Approaches

Methods have been developed to detect ASOs in plasma and other biological samples, based on hybridization with labeled complementary probes [89]. The probes are tagged at one end with biotin and the other end with digoxigenin. After hybridization and binding to neutravidin-coated 96-well plates, nuclease S1 is added to degrade unhybridized probe. Anti-digoxigenin antibodies and enzyme-linked secondary antibodies are then added sequentially, followed by enzyme substrate (for example, AttoPhos, which fluoresces after enzymatic cleavage) for detection and quantitation [60, 61].

We have developed a hybridization approach to monitor the intracellular uptake and distribution of ASOs in cell culture experiments [R. Berg et al., unpublished]. Cells treated with ASOs were fixed in paraformaldehyde and then biotin-labeled complementary oligonucleotides were added in hybridization buffer. Detection was with an avidin–horseradish peroxidase conjugate, which cleaves diaminobenzidine to produce a colored precipitate within the cells. An alternative detection approach uses fluorescein-labeled streptavidin, which is visualized by fluorescence and/or confocal microscopy [S. Fard et al., manuscript submitted]. We are currently adapting this protocol for use on tissue and tumor sections taken from mice treated systemically with ASOs.

A hybridization-based approach has also been developed to quantitate levels of ribozymes in serum using paired complementary oligonucleotide probes, one labeled with biotin and the other with digoxigenin [90]. The annealed "triplex" is first collected on streptavidin-coated 96-well plates, then an anti-digoxigenin antibody conjugated with alkaline phosphatase is added, followed by addition of the enzyme substrate p-nitrophenyl phosphate, which is cleaved into a soluble colored product that is quantitated by absorbance at 405 nm.

7.4.5   ROUTES OF ADMINISTRATION

As with any drug, the PK profile of nucleic-acid-based therapeutics varies according to the route of administration. Before focusing on the PK properties of ASOs administered intravenously, we briefly discuss the published PK analysis of topical, inhaled, and oral formulations of ASOs.

Topical administration of ASOs has been investigated for a small number of indications, and it has been the subject of several recent reviews [91–93]. To date, topical administration has not progressed past preclinical animal models, although these include human skin transplanted (grafted) onto immunocompromised mice [81, 94]. The distribution of ASOs applied to the skin is generally limited to epiderimis and dermis, and it may be limited in intact skin but more advantageous in conditions that include inflammation or other damage, especially psoriasis [93, 95]. Distinct chemistries of ASOs, such as second-generation versus PMOs, may behave differently when applied topically. Furthermore, the formulation of the application compound [96–98] and the use of delivery methods such as sonophoresis (ultrasound) [99, 100] may also affect the delivery and PK properties of ASOs applied topically.

Respirable, or inhaled, ASOs have been developed as potential therapeutics for treatment of asthma, allergic rhinitis, and other pulmonary diseases [101–104]. In animal models including rodents and rabbits, inhaled ASOs are detectable in all cell types in the lungs but only in small amounts (<5–10% of the administered dose) in extra-pulmonary organs, and they exhibit minimal systemic bioavailability of intact, full-length ASO [103, 105, 106]. Less than 5% of the delivered dose was detected in the lung 72 hours after treatment (the elimination half-life is approximately 30 hours). At 72 hours, nearly 70% of the administered dose was excreted in the urine, the major pathway of elimination. ASOs designed and formulated for delivery by inhalation may have particular strengths as potential therapeutics for pulmonary diseases. One ASO, EPI-2010 (targeting the adenosine signaling pathway in asthma), completed phase I and II human trials, but further development was not pursued due to reduced efficacy in important patient markets [103].

Stability of unmodified DNA ASOs is limited after oral administration, due to degradation in the stomach and small intestine. Although backbone and ribose modifications substantially improved the stability of ASOs administered orally [107], intestinal permeability of ASOs is limited due to hydrophilicity, high negative charge, and high molecular mass [108, 109]. Formulation of ASOs in sodium caprate further improves stability and intestinal permeability of orally administered ASOs [108, 110]. Oral formulations have progressed to animal studies in rodents, pigs, and dogs [107–110], but not yet to human trials.

Lipid delivery vehicles such as stealth liposomes [111] and cationic liposome formulations [112–114], and additional vehicles such as polymers of chitosan [115], are being pursued as potential means to improve uptake and distribution of ASOs. Although these novel formulations and alternative routes of administration might ultimately prove to be more successful, the intravenous route has been the most successful thus far in the development of therapeutic ASOs.

7.4.6   PK (ADME) OF ASOs ADMINISTERED BY INJECTION

The PK properties of PS ASOs in animal models have been extensively studied and reviewed [33, 62, 116–118]. Several routes of administration, including intraperitoneal (i.p.), intravenous (i.v.), intradermal (i.d.), or subcutaneous (s.c.) injection, have been compared in animal models. PK profiles of ASOs in animals correlate well with those observed in human clinical trials, where ASOs are usually administered by i.v. or s.c. injection.

7.4.6.1   Absorption and Distribution

ASOs are highly bound to albumin and α2-macroglobulin in the blood, with binding affinities in the micro- to millimolar range [32, 65, 119]. In mice, rats, monkeys, and humans, more than 96% of administered PS ASO was bound to plasma proteins [32, 119]. This high protein binding affinity limits renal clearance, but PS ASOs are rapidly cleared from plasma to tissues. Plasma clearance is dose-dependent, with plasma half-life ranging from 30 to 60 minutes [32, 119]. Widespread systemic distribution of ASOs is observed after i.d. and s.c. injections [65], but s.c. injection resulted in increased bioavailability and steady-state ASO levels [120]. Minimal differences in distribution were observed when comparing i.p and i.v. injection [116, 118]. In animal models, ASOs accumulated to highest levels in the liver and kidney, followed by spleen and lung, with only minor distribution to the brain and testes [65, 120, 121]. Distribution was independent of ASO length and nucleotide sequence [65]. Subcutaneous tumor xenografts accumulated levels of ASOs similar to spleen and lung [84]. Recent clinical trials of G3139 also measured similar disposition of intact and partially shortened ASO in humans, as discussed below [122, 363].

7.4.6.2 Metabolism and Elimination

Metabolism of ASOs occurs primarily through degradation by exonucleases and endonucleases. Degradation products have lower protein binding affinity and higher solubility, facilitating their excretion in urine and exhaled air [65, 121, 123–125].

7.4.6.3 PK Properties of 2′-MOE, PS, and PD ASOs

In a recent study the PK profiles of three distinctly modified ASOs were compared after i.v. injection into monkeys [34]. They compared a PD ASO with 2′-MOE modifications on all nucleotides, a PS ASO with 2′-MOE modifications only on the eight 3′-end residues, and a PS ASO with 2′-MOE modifications on the six nucleotides on each end (i.e., a gapmer). The plasma half-lives of the two PS ASOs were similar (~45 minutes), whereas that of the PD ASO was markedly shorter (~14 minutes). The two PS ASOs also had very similar tissue distribution, with the highest concentrations in kidney, liver, lymph nodes, and spleen. However, the gapmer exhibited a significantly longer tissue half-life, compared with the ASO with only the 3′ MOE modifications (~22 days vs. ~5 days). This suggests that the presence of both 5′-end and 3′-end MOE modifications on the gapmer protects the ASO from degradation in tissues and that the position of the modifications could significantly affect biological lifetime and efficacy in a clinical setting. This study confirmed that the chemistry of the ASO contributes to PK behavior and nuclease sensitivity, as suggested by earlier studies [119]. Although the extent of the 2′-MOE modification did not alter the distribution properties, the presence of this modification is important in the tissue distribution patterns and in determining nuclease resistance within tissues.

7.4.6.4 PK Analyses from Selected Clinical Trials

Several clinical trials using Genasense (G3139, a PS DNA ASO targeting the anti-apoptotic protein Bcl-2) have included detailed PK analyses. One of the earliest trials of G3139 used a 14-day continuous subcutaneous infusion at doses ranging from 4.6 to 195.8 mg/m^2/day [126]. The mean plasma half-life of G3139 was 7.46 hours. More recent trials measured PK profiles of G3139 administered at 7 mg/kd/day by continuous i.v. infusion for 8 days [36], 10 days [37], or 5 days [122]. All showed similar PK results for plasma clearance, metabolism, and urinary excretion.

The first report of administration into humans of a 2'-MOE gapmer (ISIS 104838, targeting tumor necrosis factor α) included extensive PK analysis of blood and urine samples [60]. This trial also compared i.v. and s.c. dosing regimens. The plasma half-life of the ASO was 1 to 1.8 hours at doses greater than 1 mg/kg. Urinary excretion of full-length ASO during the initial 24 hours postadministration was less than 10%, whereas excretion of 8 to 12 nucleotide metabolites (i.e., degradation products) increased over time thereafter.

Similarly, a recent clinical trial of the 2'-MOE gapmer OGX-011 (targeting the chaperone protein clusterin) in prostate cancer patients indicated dose-dependent PK and pharmacodynamic properties [61]. Plasma distribution half-lives ranged from 0.476 to 3.83 hours, and other plasma PK parameters were similar to the PS ASO G3139 [126]. Estimates of ASO concentrations in the target tissue ranged from 1.67 to 4.82 μg/g of prostate tissue, depending on the dose of ASO delivered. Metabolism and elimination were not measured in the study.

7.4.6.5 Target Downregulation in Tissues of Interest

Drug efficacy for ASOs can be measured in terms of reductions in target mRNA and protein. The outcomes of drug treatment are not formally PK issues, but they are within the realms of pharmacodynamics and toxicology. It is important to consider, however, how the PK profile of a systemically administered ASO affects the pharmacodynamic and toxicological properties of the drug. Many clinical trials of ASOs include some measures of target mRNA and protein in target or surrogate marker tissues. In one of the early trials of the PS DNA ASO G3139 in patients with non-Hodgkin's lymphoma, Bcl-2 mRNA and protein were demonstrably reduced in isolated peripheral blood lymphocytes, but only in 7 of 16 patients [126]. Recent clinical trials have shown similar interpatient variability in target downregulation [36, 37]. In the case of ASOs targeting clusterin, *in vitro* experiments revealed the improved efficacy of 2'-MOE modified gapmers compared with PS DNA ASOs [35]. Superior target downregulation and increased antitumor activity were also demonstrated in animal models of prostate cancer tumor growth [35]. Dose-dependent reductions in clusterin mRNA and protein in prostate tumor samples were also shown in the phase I clinical trial of OGX-011 [61], indicating that the intended biological activity was achieved in the intended target tissue. This represents a tremendous success for antisense therapeutics and should provide encouragement for others developing ASOs as potential clinically useful drugs.

7.4.7 RNA INTERFERENCE

RNA interference (RNAi) is a cellular regulatory mechanism in which double-stranded RNA (dsRNA) induces the specific degradation of its homologous target mRNA. This process was discovered by Guo et al. who found that injection of both antisense and sense RNAs into *Caenorhabditis elegans* resulted in a decrease of their target mRNA [127]. Several years later, Fire et al. demonstrated that this observation was due to the single-stranded RNA populations containing small amounts of double-stranded RNAs. Using highly purified RNA preparations, they showed that dsRNA injected into *C. elegans* resulted in specific and potent interference of cellular gene expression [128]. This inhibition was much greater than that observed with injection of either the antisense or the sense strands by themselves. The molecular mechanism of RNAi involves cleavage of dsRNAs by an RNase III-like endonuclease termed Dicer into 21–25 nucleotide (nt) dsRNAs. These small interfering dsRNAs (siRNAs) are bound in a multiprotein RNA-induced silencing complex (RISC), which unwinds the siRNA, and one strand is then used as a guide to target complementary mRNAs for degradation [129–131].

RNAi gained rapid acceptance as a tool for functional genomics in *C. elegans* and *Drosophila* model systems. It was initially unclear as to whether RNAi existed in mammalian cells, because treatment of mammalian cells with dsRNA results in global degradation of RNAs and repression of protein synthesis. As researchers toiled diligently to understand the detailed molecular mechanisms involved in the RNAi pathway, Elbashir et al. identified 21- to 22-nt dsRNAs as the effector molecules that induce RNAi in *Drosophila* embryos [132]. Additionally, they demonstrated that chemically synthesized siRNAs could be transfected into mammalian cells and induce RNAi without inducing global RNA degradation [133]. Since this discovery, the field of siRNA and related methodologies has grown at an exponential rate. Although siRNAs have competed with ASOs as tools for functional genomic analysis and target validation, one primary interest is to develop siRNAs as therapeutic agents.

7.4.7.1 siRNA Specificity

One of the most important issues regarding development of new pharmaceuticals is their specificity of effect. The overall intent is to design molecules that inhibit their intended target without undesirable side effects. As with ASOs and other nucleic-acid-based therapies, demonstrating that the observed cellular affects are due to inhibition of the intended mRNA target remains a challenge. Initial studies suggested that one or two mismatches could alleviate the silencing ability of an siRNA [132]. However, recent studies have identified siRNAs with effects on unintended targets with as little as seven complementary nucleotides [134]. Moreover, in addition to the G:U wobble base pair, it has been shown that mismatches formed between adenine (A) and cytosine (C) are well tolerated [135]. Thus, the ability of siRNAs to specifically downregulate predicted targets is being called into question. Many investigators have turned to microarray analysis to validate the specificity of their siRNAs. This approach may nonetheless provide researchers with a false sense of specificity. Saxena et al. showed that siRNAs with multiple mismatches located in the center of the molecule can function as micro-RNAs that inhibit expression of the target protein through translational arrest and not mRNA degradation [136]. Thus, the microarray patterns from two different siRNAs targeting the same mRNA may be similar, but the siRNAs could still inhibit protein expression through translational inhibition. It remains unclear to what extent each of these potential nonspecific effects may impact their potential therapeutic use. With a more thorough understanding of the RNAi mechanism, new selection guidelines will identify optimal target sites that account for these sequence-based nonspecific effects. Identification of the most active siRNA with limited nonspecific effects will be critical for the success of siRNA molecules as therapeutic agents.

7.4.7.2 Cell Culture

Since their discovery only a few years ago, many researchers in both academic and industrial settings have used siRNAs to knock down mRNA targets within cultured cells. Unlike ASOs, which can enter cells by themselves when used at high concentrations, siRNAs cannot cross the cellular membrane by themselves due to their two negatively charged phosphate backbones. Cationic lipids such as Lipofectamine 2000 (L2K) and Oligofectamine are commonly used to complex the siRNA and allow uptake into cultured cells. However, a recent publication has demonstrated that cationic lipids can have unintended effects. Fedorov et al. transfected siRNAs into HeLa cells using either L2K or electroporation and performed microarray studies [137]. Although both transfection methods resulted in a similar knockdown of the target mRNA, the number of genes affected by each method differed significantly. Lipid transfection resulted in increased expression of 65 genes, whereas electroporation increased only 11 genes. Over 50% of the upregulated genes could be directly attributed to L2K. Because L2K is also commonly used for transfecting DNA plasmids and ASOs, investigators should thoroughly examine the effect that the transfection reagent has on gene expression patterns. In addition to these lipids, other cationic molecules such as cell-penetrating peptides and polyethyleneimine have also been used to transport siRNAs across the cell membrane [138, 139]. Further studies are needed to determine what affects these transport molecules have on global gene expression.

7.4.7.3 *In vivo* Studies with "Naked" siRNAs

Although most siRNA researchers will readily acknowledge that the siRNAs are more potent than the ASOs and result in a more potent and reproducible target suppression, others have shown siRNAs to be equivalent to ASOs [140]. The real difference between these two approaches becomes clear once the nucleic acids are injected into animals. ASOs have therapeutic activity even when they are injected intravenously without any cationic carrier molecule, as discussed in the preceding section. Although most studies show that naked siRNA injections have no effect, some have demonstrated that uncomplexed, unmodified siRNAs can be injected into animals and inhibit the target protein in tissues. Verma et al. intravenously injected approximately 3 µg of a β-catenin-targeted siRNA into the tail vein of tumor-bearing nude mice. After 72 hours, tumors were harvested and significant decreases in β-catenin protein levels could be observed by Western blot analysis [141]. Similarly, Filleur et al. injected intraperitonealy 3 µg of siRNAs targeting either luciferase or vascular endothelcal growth factor (VEGF) and could demonstrate a 50% reduction in luciferase activity and a 70% reduction in VEGF expression. VEGF siRNA injections also delayed the onset of tumors [142]. Furthermore, Duxbury et al. delivered similar amounts of an siRNA targeting ribonucleotide reductase mRNA. After 6 weeks of biweekly tail vein siRNA injections, the levels of target protein in the tumor were suppressed and a small decrease in the tumor growth rate was observed. More importantly, the combination of the siRNA and gemcitabine resulted in a synergistic suppression of tumor growth. The control siRNA did not enhance the antitumor affect of gemcitabine [143]. All of these results using naked siRNAs are surprising because siRNAs are rapidly degraded in plasma with a half-life of about 3 minutes [144]. Given that many investigators have shown that unmodified, uncomplexed siRNAs have little to no activity (as discussed below), it remains unclear as to whether systemic injection of unprotected siRNAs is a viable therapeutic option.

7.4.7.4   Hydrodynamic High-Pressure "Naked" siRNA Injection

To overcome the problem of siRNA delivery into tissues, several investigators have employed the high-pressure hydrodynamic method. This method involves rapid injection of siRNA in a large volume (1–2 mL) of saline or phosphate-buffered solution (PBS). Surprisingly, the mice can tolerate this harsh administration quite well [145]. Song et al. was able to reduce Fas mRNA and protein levels in mouse liver for up to 10 days after hydrodynamic injection (50 μg siRNA in 1-mL PBS). They also demonstrated that Fas siRNA in-jection could protect mice from fulminant hepatitis [146]. Another group targeted caspase 8 with siRNAs and showed decreased liver expression of caspase 8 and prevented Fas (CD95)-mediated apoptosis after hydrodynamic injection (120-μg siRNA in 2-mL PBS). As with the previous Fas-targeted approach, caspase 8 siRNAs improved the survival of two animal models of acute liver failure [147]. Layzer et al. used quantitative whole-body imaging to demonstrate that hydrodynamic coinjections of luciferase reporter plasmids and luciferase-targeted siRNAs could reduce luciferase expression by 85%. Injection of a nuclease-resistant 2′-F siRNA resulted in a similar reduction in luciferase expression. However, if siRNAs were injected under low pressure, no decrease in luciferase activity was observed [148]. Although the use of hydrodynamic injection in animal models is relatively simple and effective, it is not feasible in the clinical setting. An additional limitation of this methodology is that the liver is the primary organ for siRNA uptake with only limited uptake in other tissues.

7.4.7.5   Chemically Modified siRNAs

Contrary to the studies demonstrating activity with unmodified siRNAs, the major limitation for the use of nucleic acids as therapeutic agents is their susceptibility to attack from serum and intracellular nucleases. The antisense field has identified a host of ribose ring and phosphate backbone modifications that enhance the stability of ASOs [Figure 7.4-1]. Extensive analysis has been performed to demonstrate which chemical modifications enhance siRNA stability and still allow for efficient silencing [149–151]. Altering the phosphate backbone of an siRNA to phosphorothioate reduces its ability to induce RNAi by 50% and increases nonspecific cellular toxicity. Designing siRNAs entirely with 2′-O-methyl moieties improves their nuclease stability dramatically and eliminates their silencing ability. One nucleotide modification that seems to be the most promising is the 2′-fluorine modification on the cytidines and uridine residues. 2′-F-modified siRNAs have plasma half-lives of days instead of minutes and can inhibit their targets with similar potency as the unmodified siRNAs. Researchers at Sirna Therapeutics have developed siRNAs in which all 2'-OH groups have been substituted with deoxy-, 2'-O-methyl-, or 2'-fluoro-nucleotides, which are termed "short interfering nucleic acid" (siNA). In human serum, siNAs have a half-life on the order of 3 days [144]. Using a hepatitis B viral mouse model, they injected HBV-targeted siNAs by standard low-pressure tail vein injections (60–600-μg siRNA) thrice daily for 2 days. After 24 hours, the levels of serum HBV DNA were significantly reduced in a dose-dependent manner. The inverted control siNA had little to no effect on serum levels of HBV DNA. No significant antiviral activity was observed with unmodified siRNAs at any dose. Although this study demonstrated *in vivo* activity of a fully modified siRNA, it is uncertain whether these siNAs are effectively taken up into tissues by standard intravenous injections. One assumes that lower serum titers are the result of lower viral production from the tissues, but no data were presented regarding decreased HBV tissue expression. Given the high doses required for activity, more studies are needed to validate the therapeutic effectiveness of siNAs.

7.4.7.6 Ligand-Mediated siRNA Delivery

Another approach to improve tissue delivery of siRNAs is to attach small molecules on the end of the siRNA. Soutschek et al. conjugated a cholesterol molecule on the 3'-end of the sense strand of a partial chemically modified siRNA to generate "chol-siRNA" [152]. A chol-siRNA targeting luciferase could reduce luciferase expression in HeLa cells with an IC50 of about 200 nM without any transfection reagents. Upon intravenous injection into animal models, the addition of cholesterol improved the siRNAs pharmacokinetic properties by increasing the elimination half-life and reducing plasma clearance in rodents. Chol-siRNAs, as detected by RNase protection assay, were broadly distributed throughout the tissues of the mouse confirming the improved pharmacological properties. No detectable amounts of unconjugated siRNAs were observed in any tissues. Intravenous injections of a chol-siRNA targeting apolipoptrotein B (1 mg) daily for 3 days resulted in inhibition of apoB mRNA in liver and intestine, decreased serum levels of apoB protein, and a reduction of total plasma cholesterol. Hence, this study demonstrated the importance of conjugating a small molecule to the siRNA that will enhance tissue uptake. However, further optimization is required to identify modifications that will permit more clinically acceptable doses and schedules.

7.4.7.7 Intranasal siRNA Injection

As an alternative approach to systemic siRNA injections, several groups have demonstrated that intranasal administration of siRNA can be effective for certain disease applications [153–155]. X. Zhang et al. [154, 155] were the first to demonstrate that unmodified siRNA could be administered intranasally and silence the intended target mRNA without the need for transfection agents. Bitko et al. [153] designed siRNAs targeting the respiratory syncytial virus and parainfluenza virus. A single intranasal siRNA administration (70 μg) provided significant protection against respiratory infection from viral challenge. Two days after intranasal injection, siRNAs could still be detected in the mouse lungs. There are several advantages for this route of administration. It is relatively painless and noninvasive. The risk of potential side effects is also reduced because no cationic carrier molecules are required for cellular uptake. It remains unclear as to the mechanism of cellular uptake after intranasal siRNA administration.

7.4.7.8 *In vivo* Studies with Encapsulated siRNAs

An alternative approach for *in vivo* delivery of siRNAs is to encapsulate or complex the siRNA with cationic molecules. These positively charged molecules serve several functions. After interacting with the negatively charged phosphate backbone of the nucleic acid, they condense the nucleic acid into a more compact structure, protect against degradation by plasma nucleases, and enhance tissue uptake. The use of these cationic synthetic molecules is not limited to siRNA delivery. ASOs and DNA plasmid vectors can also be delivered with these molecules. As well, it remains unclear as to whether chemical modifications to the siRNA are necessary if one is using cationic molecules for delivery. In cell culture, lipid-complex siRNAs are protected from degradation by serum nucleases. One benefit of using natural, unmodified siRNAs is that when they actually do degrade, the nucleotides should not have any nonspecific effects of cellular metabolism.

7.4.7.9 Linear Polyethylenimine

The cationic molecule polyethylenimine (PEI) has been shown to effectively transport nucleic acids into cells in both the *in vitro* and the *in vivo* setting. PEI condenses nucleic acids into 0.2–0.4-nm positively charged particles, is taken up into cells by endocytosis, and then releases the nucleic acid contents into the cytoplasm efficiently via a process called the "proton sponge" [156]. Urban-Klein et al. demonstrated that PEI complexation of an unmodified siRNA protects the siRNA from nuclease degradation. In cultured ovarian carcinoma cells, they showed that linear low-molecular-weight PEI could transport their HER2-targeted siRNA into cells, inhibit HER2 protein expression by 70%, and reduce colony formation by 50% with a single PEI/siRNA transfection. In a tumor-bearing nude mouse model, intraperitoneal injection of PEI/siRNA resulted in a rapid uptake of the P32-labeled siRNA into the tumor, muscle, liver, and kidney with little uptake into the lungs. No siRNA was observed in the tissues after injection of uncomplexed siRNAs. As they observed no histological changes in the lungs and the absence of visible side effects, intraperitoneal injections of PEI/siRNA seem to avoid systemic PEI-associated toxicity previously reported [157]. Furthermore, they showed that intraperitoneal injections of PEI/HER-2 siRNA (10μg) two to three times a week significantly reduced growth of tumor xenografts. HER2 mRNA levels were inhibited by 50% by this treatment regimen. As important controls for specificity, injection of either a nonspecific siRNA complexed with PEI or uncomplexed HER-2 siRNA had no effect on tumor growth. However, a recent report demonstrated that tail vein injections of PEI along with other cationic molecules can have antitumor activity by themselves [158]. Although this effect may be cell-line dependent, caution should be used when interpreting the therapeutic effects of siRNAs when cationic molecules are used for delivery.

One main concern with using PEI as a nucleic acid transporter is the lack of tissue specificity. To improve the ability of PEI to target tumors, Schiffelers et al.

designed ligand-targeted, sterically stabilized nanoparticles (RPPs) for complexation with siRNAs [159]. These particles comprise branched PEI that is PEGylated with an RGD peptide on the end of the polyethylene glycol (PEG). Once the siRNA is complexed in the nanoparticles, the RGD peptide should preferentially target the tumor neovasculature. One hour after intravenous injection of a fluorescently labeled siRNA complexed with RPP, significant siRNA uptake was observed in the tumor with poor liver and lung accumulation. Conversely, siRNA complexed with only branched PEI did not accumulate in the tumor to any appreciable extent but did accumulate in the liver and lung. To verify whether the siRNA distributed into the tumor was active, they evaluated the expression of luciferase in the tumor and its knockdown by a siRNA specific for luciferase. Thus, luciferase was complexed with either PEI or RPP particles and injected into tumor-bearing mice. Although luciferase levels in the tumor were equivalent for both PEI and RPP particles, only RPP could selectively target the tumor. Two hours after plasmid injections, luciferase siRNA–RPP nanoparticles were injected into the same mice. After 24 hours, the luciferase activity had decreased by 90% demonstrating sequence-specific inhibition. To evaluate whether these nanoparticles have therapeutic potential, siRNAs targeting the VEGF receptor were complexed with RPP and injected into tumor-bearing mice every 3 days at a dose of 40μg administered in the tail vein. They observed a strong inhibition of tumor growth, a reduction in blood vessel formation around the tumor, and a decrease in the expression of tumor VEGF R2 protein levels. The control siRNA had no effect, demonstrating specificity.

7.4.7.10 Cationic Lipids

In addition to PEI-related molecules, various cationic lipids have been used for nucleic acid transport *in vivo*. Lipids such as the commercially available DOTAP have been effective in delivering siRNAs into tissues such as kidney and spleen [160]. Others have developed their own delivery lipids. Yano et al. designed a less-toxic cationic liposome (LIC-101) and demonstrated its *in vitro* ability to deliver a Bcl-2-targeted siRNA into four different cell lines and inhibit Bcl-2 protein expression [161]. Furthermore, tail vein injection of LIC-101/Bcl-2 siRNA inhibited tumor growth in two different mouse models. Chien et al. developed a novel cationic liposome based on a synthetic cationic cardiolipin analog (CCLA) [162]. They showed their lipid had significantly less-toxic side effects than DOTAP and could suppress tumor growth by >70% in a severe combined immunodeficiency (SCID) mouse xenograft model. Injection of a mismatched siRNA complexed with CCLA had no effect on tumor growth. Morrissey et al., in an attempt to improve delivery of their HBV-targeted siRNA, incorporated their siRNA into a specialized liposome to form a stable nucleic-acid–lipid particle (SNALP) [163]. In a mouse model, an unmodified siRNA had a plasma elimination half-life of approximately 2 minutes. Injection of a nuclease-stabilized siRNA improved substantially the elimination half-life to 49 minutes. However, siRNA encapsulation in SNALP improved the plasma pharmacokinetics most dramatically with an elimination half-life of 6.5 hours. They demonstrated that SNALP–siRNA could significantly suppress serum levels of HBV in HBV-challenged mice with much lower doses and reduced dosing frequency than previously shown with uncomplexed siRNA [144].

7.4.7.11 siRNAs in Clinical Trials

Although some uncertainly remains for the feasibility and effectiveness of siRNA therapeutics, several companies are rapidly moving ahead with siRNA molecules in clinical trials. In the Fall of 2004, Acuity Pharmaceuticals initiated the first human siRNA clinical trial for treatment of wet age-related macular degeneration (AMD) with their lead compound Cand5, which silences VEGF. Phase I studies showed that intravitreal injections of an unmodified siRNA, Cand5, was safe and well tolerated [164]. Pharmacokinetic analysis revealed that the siRNA was not detected in the plasma of any patient. Lack of systemic exposure due to localized siRNA injection likely prevented the adverse side effects of anti-VEGF therapy. However, given that unmodified siRNAs are very unstable in plasma, it is unclear whether the siRNA could not escape the vitreal cavity or whether it was degraded so rapidly after entry into the bloodstream that it could not be detected in plasma. Acuity Pharmaceuticals has already initiated phase II studies with its lead siRNA candidate. The second siRNA to enter clinical trials was Sirna-027 from Sirna Therapeutics. Although both Cand5 and Sirna-027 are being investigated for treatment of AMD, Sirna-027 targets the VEGF receptor as opposed to VEGF, and is chemically modified to enhance stability. Interim phase I results were recently reported and demonstrated that Sirna-027 seems to be safe and well tolerated [165]. All patients experienced improvement or stabilization of visual acuity, and 23% of the patients showed a clinically significant improvement in visual acuity. A third company (Alnylam Pharmaceuticals) has recently initiated a phase I study with its siRNA (ALN-RSV01), which targets respiratory syncytial virus. Using their Direct RNAi approach, the siRNA will be administered directly to the lungs and silence a key RSV gene, which hopefully will prevent spreading of the viral infection.

Other pharmaceutical companies are also seeking to initiate their own siRNA clinical trials. An Australian company (Benitec) in collaboration with the City of Hope is planning to enter their RNA-based HIV drug into clinical trials sometime in 2006. Unlike Cand5 and Sirna-027, which are siRNA molecules, Benitec specializes in RNAi plasmid and vector expression systems. Patients who have become resistant to HAART therapy would receive a transplant of their own stem cells containing an RNAi expression system targeting HIV to generate HIV resistance. Nucleonics, Inc. also uses RNAi expression systems for gene silencing and expects to file its first IND for hepatitis B and hepatitis C in 2006. CytRx Corporation is rapidly advancing their preclinical siRNA studies for the treatment of obesity, type 2 diabetes, and Lou Gehrig's disease. Thus, many pharmaceutical companies have decided to develop siRNAs as therapeutic agents. However, as with ASOs, the fate of therapeutic siRNAs will be determined from the initial siRNA clinical trials.

7.4.8 CONCLUSIONS

7.4.8.1 Nucleic Acid-Based Therapeutics

Plasmids and viral vectors have been developed to deliver gene therapy in animal models and human clinical trials. These approaches aim to express a gene of interest in target tissue. Many of these approaches use local delivery, and thus, PK analysis is limited to the short-range disposition, metabolism, and excretion of the nucleic acid therapeutic. ASOs and siRNA molecules are being developed as gene therapeutics to reduce the expression of target genes, through double-stranded nucleic acid interactions with target mRNA. Development of orally bioavailable formulations and lipid-encapsulated antisense molecules is being pursued, but most preclinical studies and clinical trials employ intravenous injection to deliver these therapeutics.

7.4.8.2 PK (ADME) and Toxicology Differences

For ASOs, the differences in PK and toxicology properties depend largely on the chemistry of the backbone and the modifications to the sugars. Thus, the ADME characteristics are independent of nucleotide base sequence and, therefore, do not change significantly according to the gene or protein being targeted. The toxicological properties do depend to a certain extent on the nucleotide sequence, for example, the effects of CpG dinucleotides and G-quartets. Whether these principles will hold for siRNA therapeutics will likely been seen in the next few years, as the development of siRNA with novel chemical modifications progresses.

7.4.8.3 Relative Clinical Utility of the Different Approaches

The relative effectiveness of ASOs and siRNAs, in terms of target downregulation, have been studied in cell culture and animal experiments [29, 140, 166, 167]. Although it seems that siRNA may be more effective than ASOs in most experimental circumstances, the development of modified ASOs with improved PK properties, including stability to nucleases, uptake into target tissues, and tissue half-life, has progressed much further than for siRNA. Indeed, although many ASOs of various chemistries have progressed into phase II and III clinical trials and thus the PK studies are widely reported, siRNAs are just now entering human clinical trials and their PK properties are not yet fully described or understood. It seems likely that both types of antisense therapeutics will demonstrate clinical utility in a variety of indications in the coming years.

REFERENCES

1. Wilton S D, Fletcher S (2005). RNA splicing manipulation: strategies to modify gene expression for a variety of therapeutic outcomes. *Curr. Gene Ther.* 5:467–483.
2. Kole R, Vacek M, Williams T (2004). Modification of alternative splicing by antisense therapeutics. *Oligonucleotides.* 14:65–74.
3. Sazani P, Kole R (2003). Therapeutic potential of antisense oligonucleotides as modulators of alternative splicing. *J. Clin. Invest.* 112:481–486.
4. Edelstein M L, Abedi M R, Wixon J, Edelstein R M (2004). Gene therapy clinical trials worldwide 1989–2004-an overview. *J. Gene Med.* 6:597–602.
5. Shah P B, Losordo D W (2005). Non-viral vectors for gene therapy: Clinical trials in cardiovascular disease. *Adv. Genet.* 54:339–361.
6. Nishikawa M, Takakura Y, Hashida M (2005). Pharmacokinetics of plasmid DNA-based non-viral gene medicine. *Adv. Genet.* 53PA:47–68.

7. Lee P A, Blatt L M, et al. (2000). Pharmacokinetics and tissue distribution of a ribozyme directed against hepatitis C virus RNA following subcutaneous or intravenous administration in mice. *Hepatol.* 32:640–646.

8. Schubert S, Kurreck J (2004). Ribozyme- and deoxyribozyme-strategies for medical applications. *Curr. Drug Targets.* 5:667–681.

9. Parry T J, Bouhana K S, Blanchard K S, et al. (2000). Ribozyme pharmacokinetic screening for predicting pharmacodynamic dosing regimens. *Curr. Issues Mol. Biol.* 2:113–118.

10. Sandberg J A, Parker V P, et al. (2000). Pharmacokinetics and tolerability of an anti-angiogenic ribozyme (ANGIOZYME) in healthy volunteers. *J. Clin. Pharmacol.* 40:1462–1469.

11. Kobayashi H, Eckhardt S G, et al. (2005). Safety and pharmacokinetic study of RPI.4610 (ANGIOZYME), an anti-VEGFR-1 ribozyme, in combination with carboplatin and paclitaxel in patients with advanced solid tumors. *Cancer Chemother. Pharmacol.* 56:329–336.

12. Weng D E, Masci P A, et al. (2005). A phase I clinical trial of a ribozyme-based angiogenesis inhibitor targeting vascular endothelial growth factor receptor-1 for patients with refractory solid tumors. *Mol. Cancer Ther.* 4:948–955.

13. Amantana A, Iversen P L (2005). Pharmacokinetics and biodistribution of phosphorodiamidate morpholino antisense oligomers. *Curr. Opin. Pharmacol.* 5:550–555.

14. Devi G R, Beer T M, et al. (2005). In vivo bioavailability and pharmacokinetics of a c-MYC antisense phosphorodiamidate morpholino oligomer, AVI-4126, in solid tumors. *Clin. Cancer Res.* 11:3930–3938.

15. Arora V, Devi G R, Iversen P L (2004). Neutrally charged phosphorodiamidate morpholino antisense oligomers: uptake, efficacy and pharmacokinetics. *Curr. Pharm. Biotechnol.* 5:431–439.

16. Jepsen J S, Sorensen M D, Wengel J (2004). Locked nucleic acid: a potent nucleic acid analog in therapeutics and biotechnology. *Oligonucleotides.* 14:130–146.

17. Fluiter K, ten Asbroek A L, et al. (2003). In vivo tumor growth inhibition and biodistribution studies of locked nucleic acid (LNA) antisense oligonucleotides. *Nucleic Acids Res.* 31:953–962.

18. Wahlestedt C, Salmi P, et al. (2000). Potent and nontoxic antisense oligonucleotides containing locked nucleic acids. *Proc. Natl. Acad. Sci. U.S.A.* 97:5633–5638.

19. McMahon B M, Mays D, et al. (2002). Pharmacokinetics and tissue distribution of a peptide nucleic acid after intravenous administration. *Antisense Nucleic Acid Drug Dev.* 12:65–70.

20. Kristensen E (2002). In vitro and in vivo studies on the pharmacokinetics and metabolism of PNA constructs in rodents. *Methods Mol. Biol.* 208:259–269.

21. Crooke S T (2004). Antisense strategies. *Curr. Mol. Med.* 4:465–487.

22. Kurreck J (2003). Antisense technologies. Improvement through novel chemical modifications. *Eur. J. Biochem.* 270:1628–1644.

23. Giles R V, Spiller D G, Tidd D M (1993). Chimeric oligodeoxynucleotide analogues: enhanced cell uptake of structures which direct ribonuclease H with high specificity. *Anticancer Drug Des.* 8:33–51.

24. ten Asbroek A L, van Groenigen M, Nooij M, Baas F (2002). The involvement of human ribonucleases H1 and H2 in the variation of response of cells to antisense phosphorothioate oligonucleotides. *Eur. J. Biochem.* 269:583–592.

25. Zamaratski E, Pradeepkumar P I, Chattopadhyaya J (2001). A critical survey of the structure-function of the antisense oligo/RNA heteroduplex as substrate for RNase H. *J. Biochem. Biophys. Meth.* 48:189–208.

26. Stein C A, Cohen J S (1988). Oligodeoxynucleotides as inhibitors of gene expression: A review. *Cancer Res.* 48:2659–2668.

27. Achenbach T V, Brunner B, Heermeier K (2003). Oligonucleotide-based knockdown technologies: Antisense versus RNA interference. *Chembiochem.* 4:928–935.

28. Scanlon K J (2004). Anti-genes: siRNA, ribozymes and antisense. *Curr. Pharm. Biotechnol.* 5:415–420.

29. Grunweller A, Wyszko E, et al. (2003). Comparison of different antisense strategies in mammalian cells using locked nucleic acids, 2′-O-methyl RNA, phosphorothioates and small interfering RNA. *Nucleic Acids Res.* 31:3185–3193.

30. Micklefield J (2001). Backbone modification of nucleic acids: synthesis, structure and therapeutic applications. *Curr. Med. Chem.* 8:1157–1179.

31. Geary R S, Watanabe T A, et al. (2001). Pharmacokinetic properties of 2′-O-(2-methoxyethyl)-modified oligonucleotide analogs in rats. *J. Pharmacol. Exp. Ther.* 296:890–897.

32. Geary R S, Yu R Z, Levin A A (2001). Pharmacokinetics of phosphorothioate antisense oligodeoxynucleotides. *Curr. Opin. Investig. Drugs.* 2:562–573.

33. Agrawal S (1999). Importance of nucleotide sequence and chemical modifications of antisense oligonucleotides. *Biochim. Biophys. Acta.* 1489:53–68.

34. Yu R Z, Geary R S, et al. (2004). Tissue disposition of 2′-O-(2-methoxy) ethyl modified antisense oligonucleotides in monkeys. *J. Pharm. Sci.* 93:48–59.

35. Zellweger T, Miyake H, et al. (2001). Antitumor activity of antisense clusterin oligonucleotides is improved in vitro and in vivo by incorporation of 2′-O-(2-methoxy)ethyl chemistry. *J. Pharmacol. Exp. Ther.* 298:934–940.

36. Tolcher A W, Chi K, et al. (2005). A phase II, pharmacokinetic, and biological correlative study of oblimersen sodium and docetaxel in patients with hormone-refractory prostate cancer. *Clin. Cancer Res.* 11:3854–3861.

37. Marcucci G, Stock W, et al. (2005). Phase I study of oblimersen sodium, an antisense to Bcl-2, in untreated older patients with acute myeloid leukemia: Pharmacokinetics, pharmacodynamics, and clinical activity. *J. Clin. Oncol.* 23:3404–3411.

38. Tolcher A W, Kuhn J, et al. (2004). A Phase I pharmacokinetic and biological correlative study of oblimersen sodium (genasense, g3139), an antisense oligonucleotide to the bcl-2 mRNA, and of docetaxel in patients with hormone-refractory prostate cancer. *Clin. Cancer Res.* 10:5048–5057.

39. Rudin C M, Kozloff M, et al. (2004). Phase I study of G3139, a bcl-2 antisense oligonucleotide, combined with carboplatin and etoposide in patients with small-cell lung cancer. *J. Clin. Oncol.* 22:1110–1117.

40. Morris M J, Tong W P, et al. (2002). Phase I trial of BCL-2 antisense oligonucleotide (G3139) administered by continuous intravenous infusion in patients with advanced cancer. *Clin. Cancer Res.* 8:679–683.

41. Desai A A, Schilsky R L, et al. (2005). A phase I study of antisense oligonucleotide GTI-2040 given by continuous intravenous infusion in patients with advanced solid tumors. *Ann. Oncol.* 16:958–965.

42. Bishop M R, Iversen P L, et al. (1996). Phase I trial of an antisense oligonucleotide OL(1)p53 in hematologic malignancies. *J. Clin. Oncol.* 14:1320–1326.

43. Bayever E, Iversen P L, et al. (1993). Systemic administration of a phosphorothioate oligonucleotide with a sequence complementary to p53 for acute myelogenous leukemia and myelodysplastic syndrome: initial results of a phase I trial. *Antisense Res. Dev.* 3:383–390.

44. Stewart D J, Donehower R C, et al. (2003). A phase I pharmacokinetic and pharma-codynamic study of the DNA methyltransferase 1 inhibitor MG98 administered twice weekly. *Ann. Oncol.* 14:766–774.

45. Sereni D, Tubiana R, et al. (1999). Pharmacokinetics and tolerability of intraven-ous trecovirsen (GEM 91), an antisense phosphorothioate oligonucleotide, in HIV-positive subjects. *J. Clin. Pharmacol.* 39:47–54.

46. Zhang R, Yan J, et al. (1995). Pharmacokinetics of an anti-human immunodeficiency virus antisense oligodeoxynucleotide phosphorothioate (GEM 91) in HIV-infected subjects. *Clin. Pharmacol. Ther.* 58:44–53.

47. Tolcher A W, Reyno L, et al. (2002). A randomized phase II and pharmacokinetic study of the antisense oligonucleotides ISIS 3521 and ISIS 5132 in patients with hormone-refractory prostate cancer. *Clin. Cancer Res.* 8:2530–2535.

48. Rudin C M, Holmlund J, et al. (2001). Phase I trial of ISIS 5132, an antisense oligo-nucleotide inhibitor of c-raf-1, administered by 24-hour weekly infusion to patients with advanced cancer. *Clin. Cancer Res.* 7:1214–1220.

49. Stevenson J P, Yao K S, et al. (1999). Phase I clinical/pharmacokinetic and pharma-codynamic trial of the c-raf-1 antisense oligonucleotide ISIS 5132 (CGP 69846A). *J. Clin. Oncol.* 17:2227–2236.

50. Adjei A A, Dy G K, et al. (2003). A phase I trial of ISIS 2503, an antisense inhibitor of H-ras, in combination with gemcitabine in patients with advanced cancer. *Clin. Cancer Res.* 9:115–123.

51. Cunningham C C, Holmlund J T, et al. (2001). A Phase I trial of H-ras antisense oli-gonucleotide ISIS 2503 administered as a continuous intravenous infusion in patients with advanced carcinoma. *Cancer.* 92:1265–1271.

52. Kahan B D, Stepkowski S, et al. (2004). Phase I and phase II safety and efficacy trial of intercellular adhesion molecule-1 antisense oligodeoxynucleotide (ISIS 2302) for the prevention of acute allograft rejection. *Transplan.* 78:858–863.

53. Yacyshyn B R, Barish C, et al. (2002). Dose ranging pharmacokinetic trial of high-dose alicaforsen (intercellular adhesion molecule-1 antisense oligodeoxynucleotide) (ISIS 2302) in active Crohn's disease. *Aliment. Pharmacol. Ther.* 16:1761–1770.

54. Glover J M, Leeds J M, et al. (1997). Phase I safety and pharmacokinetic profile of an intercellular adhesion molecule-1 antisense oligodeoxynucleotide (ISIS 2302). *J. Pharmacol. Exp. Ther.* 282:1173–1180.

55. Marshall J L, Eisenberg S G, et al. (2004). A phase II trial of ISIS 3521 in patients with metastatic colorectal cancer. *Clin. Colorectal Cancer.* 4:268–274.

56. Mani S, Rudin C M, et al. (2002). Phase I clinical and pharmacokinetic study of protein kinase C-alpha antisense oligonucleotide ISIS 3521 administered in combina-tion with 5-fluorouracil and leucovorin in patients with advanced cancer. *Clin. Cancer Res.* 8:1042–1048.

57. Villalona-Calero M A, Ritch P, et al. (2004). A phase I/II study of LY900003, an antisense inhibitor of protein kinase C-alpha, in combination with cisplatin and gem-citabine in patients with advanced non-small cell lung cancer. *Clin. Cancer Res.* 10:6086–6093.

58. Goel S, Desai K, et al. (2003). A safety study of a mixed-backbone oligonucleotide (GEM231) targeting the type I regulatory subunit alpha of protein kinase A using a continuous infusion schedule in patients with refractory solid tumors. *Clin. Cancer Res.* 9:4069–4076.

59. Chen H X, Marshall J L, et al. (2000). A safety and pharmacokinetic study of a mixed-backbone oligonucleotide (GEM231) targeting the type I protein kinase A by two-

hour infusions in patients with refractory solid tumors. *Clin. Cancer Res.* 6:1259–1266.

60. Sewell K L, Geary R S, et al. (2002). Phase I trial of ISIS 104838, a 2'-methoxyethyl modified antisense oligonucleotide targeting tumor necrosis factor-alpha. *J. Pharmacol. Exp. Ther.* 303:1334–1343.

61. Chi K N, Eisenhauer E, et al. (2005). A phase I pharmacokinetic and pharmacodynamic study of OGX-011, a 2'-methoxyethyl antisense oligonucleotide to clusterin, in patients with localized prostate cancer. *J. Natl. Cancer Inst.* 97:1287–1296.

62. Jason T L, Koropatnick J, Berg R W (2004). Toxicology of antisense therapeutics. *Toxicol. Appl. Pharmacol.* 201:66–83.

63. Krieg A M (1999). Mechanisms and applications of immune stimulatory CpG oligodeoxynucleotides. *Biochim. Biophys. Acta.* 1489:107–116.

64. Ioannou X P, Gomis S M, et al. (2002). CpG-containing oligodeoxynucleotides, in combination with conventional adjuvants, enhance the magnitude and change the bias of the immune responses to a herpesvirus glycoprotein. *Vaccine.* 21:127–137.

65. Levin A A (1999). A review of the issues in the pharmacokinetics and toxicology of phosphorothioate antisense oligonucleotides. *Biochim. Biophys. Acta.* 1489:69–84.

66. Shaw D R, Rustagi P K, et al. (1997). Effects of synthetic oligonucleotides on human complement and coagulation. *Biochem. Pharmacol.* 53:1123–1132.

67. Gleave M E, Monia B P (2005). Antisense therapy for cancer. *Nat. Rev. Cancer.* 5:468–479.

68. Fluiter K, Frieden M, et al. (2005). On the in vitro and in vivo properties of four locked nucleic acid nucleotides incorporated into an anti-H-Ras antisense oligonucleotide. *Chembiochem.* 6:1104–1109.

69. Jepsen J S, Wengel J (2004). LNA-antisense rivals siRNA for gene silencing. *Curr. Opin. Drug Discov. Devel.* 7:188–194.

70. Vester B, Wengel J (2004). LNA (locked nucleic acid): high-affinity targeting of complementary RNA and DNA. *Biochem.* 43:13233–13241.

71. Frieden M, Christensen S M, et al. (2003). Expanding the design horizon of antisense oligonucleotides with alpha-L-LNA. *Nucleic Acids Res.* 31:6365–6372.

72. Lok C N, Viazovkina E, et al. (2002). Potent gene-specific inhibitory properties of mixed-backbone antisense oligonucleotides comprised of 2'-deoxy-2'-fluoro-D-arabinose and 2'-deoxyribose nucleotides. *Biochem.* 41:3457–3467.

73. Mangos M M, Min K L, et al. (2003). Efficient RNase H-directed cleavage of RNA promoted by antisense DNA or 2'F-ANA constructs containing acyclic nucleotide inserts. *J. Am. Chem. Soc.* 125:654–661.

74. Wilds C J, Damha M J (2000). 2'-Deoxy-2'-fluoro-beta-D-arabinonucleosides and oligonucleotides (2'F-ANA): synthesis and physicochemical studies. *Nucleic Acids Res.* 28:3625–3635.

75. Kaihatsu K, Huffman K E, Corey D R (2004). Intracellular uptake and inhibition of gene expression by PNAs and PNA-peptide conjugates. *Biochem.* 43:14340–14347.

76. Nielsen P E (2005). Gene targeting using peptide nucleic acid. *Methods Mol. Biol.* 288:343–358.

77. Marin V L, Roy S, Armitage B A (2004). Recent advances in the development of peptide nucleic acid as a gene-targeted drug. *Expert. Opin. Biol. Ther.* 4:337–348.

78. Foubister V (2003). New PNA tool makes great antisense. *Drug Discov. Today.* 8:520–521.

79. Koppelhus U, Nielsen P E (2003). Cellular delivery of peptide nucleic acid (PNA). *Adv. Drug Deliv. Rev.* 55:267–280.

80. Stephens A C (2004). Technology evaluation: AVI-4126, AVI BioPharma. *Curr. Opin. Mol. Ther.* 6:551–558.

81. Pannier A K, Arora V, Iversen P L, et al. (2004). Transdermal delivery of phosphorodiamidate Morpholino oligomers across hairless mouse skin. *Int. J. Pharm.* 275:217–226.

82. Iversen P L (2001). Phosphorodiamidate morpholino oligomers: favorable properties for sequence-specific gene inactivation. *Curr. Opin. Mol. Ther.* 3:235–238.

83. Iversen P L, Arora V, Acker A J, et al. (2003). Efficacy of antisense morpholino oligomer targeted to c-myc in prostate cancer xenograft murine model and a Phase I safety study in humans. *Clin. Cancer Res.* 9:2510–2519.

84. Berg R W, Werner M, et al. (2001). Tumor growth inhibition in vivo and G2/M cell cycle arrest induced by antisense oligodeoxynucleotide targeting thymidylate synthase. *J. Pharmacol. Exp. Ther.* 298:477–484.

85. Kuhnast B, Dolle F, et al. (2000). General method to label antisense oligonu-cleotides with radioactive halogens for pharmacological and imaging studies. *Bioconjug. Chem.* 11:627–636.

86. Yu R Z, Geary R S, Levin A A (2004). Application of novel quantitative bioanalytical methods for pharmacokinetic and pharmacokinetic/pharmacodynamic assessments of antisense oligonucleotides. *Curr. Opin. Drug Discov. Devel.* 7:195–203.

87. Gilar M, Fountain K J, et al. (2003). Characterization of therapeutic oligonucleotides using liquid chromatography with on-line mass spectrometry detection. *Oligonucleotides.* 13:229–243.

88. Freudemann T, von Brocke A, Bayer E (2001). On-line coupling of capillary gel electrophoresis with electrospray mass spectrometry for oligonucleotide analysis. *Anal. Chem.* 73:2587–2593.

89. de Serres M, McNulty M J, Christensen L, et al. (1996). Development of a novel scintillation proximity competitive hybridization assay for the determination of phosphorothioate antisense oligonucleotide plasma concentrations in a toxicokinetic study. *Anal. Biochem.* 233:228–233.

90. Brown-Augsburger P, Yue X M, et al. (2004). Development and validation of a sensitive, specific, and rapid hybridization-ELISA assay for determination of concentrations of a ribozyme in biological matrices. *J. Pharm. Biomed. Anal.* 34:129–139.

91. Brand R M (2001). Topical and transdermal delivery of antisense oligonucleotides. *Curr. Opin. Mol. Ther.* 3:244–248.

92. Brand R M, Iversen P L (2005). Transdermal delivery of antisense oligonucleotides. *Methods Mol. Med.* 106:255–269.

93. White P J, Atley L M, Wraight C J (2004). Antisense oligonucleotide treatments for psoriasis. *Expert. Opin. Biol. Ther.* 4:75–81.

94. Dokka S, Cooper S R, Kelly S, et al. (2005). Dermal delivery of topically applied oligonucleotides via follicular transport in mouse skin. *J. Invest Dermatol.* 124:971–975.

95. White P J, Gray A C, et al. (2002). C-5 propyne-modified oligonucleotides penetrate the epidermis in psoriatic and not normal human skin after topical application. *J. Invest Dermatol.* 118:1003–1007.

96. Ramakumar S, Phull H, et al. (2005). Novel delivery of oligonucleotides using a topical hydrogel tissue sealant in a murine partial nephrectomy model. *J. Urol.* 174:1133–1136.

97. Bochot A, Couvreur P, Fattal E (2000). Intravitreal administration of antisense oligonucleotides: potential of liposomal delivery. *Prog. Retin. Eye Res.* 19:131–147.

98. Klimuk S K, Semple S C, et al. (2000). Enhanced anti-inflammatory activity of a liposomal intercellular adhesion molecule-1 antisense oligodeoxynucleotide in an acute model of contact hypersensitivity. *J. Pharmacol. Exp. Ther.* 292:480–488.

99. Tezel A, Dokka S, Kelly S, Hardee G E, Mitragotri S (2004). Topical delivery of anti-sense oligonucleotides using low-frequency sonophoresis. *Pharm. Res.* 21:2219–2225.

100. Regnier V, Preat V (1998). Localization of a FITC-labeled phosphorothioate oligodeoxynucleotide in the skin after topical delivery by iontophoresis and electroporation. *Pharm. Res.* 15:1596–1602.

101. Sandrasagra A, Leonard S A, et al. (2002). Discovery and development of respirable antisense therapeutics for asthma. *Antisense Nucleic Acid Drug Dev.* 12:177–181.

102. Nyce J (2002). Respirable antisense oligonucleotides: a new, third drug class targeting respiratory disease. *Curr. Opin. Allergy Clin. Immunol.* 2:533–536.

103. Ball H A, Van Scott M R, Robinson C B (2004). Sense and antisense: therapeutic potential of oligonucleotides and interference RNA in asthma and allergic disorders. *Clin. Rev. Allergy Immunol.* 27:207–217.

104. Ball H A, Sandrasagra A, et al. (2003). Clinical potential of respirable antisense oligonucleotides (RASONs) in asthma. *Am. J. Pharmacogenom.* 3:97–106.

105. Ali S, Leonard S A, et al. (2001). Absorption, distribution, metabolism, and excretion of a respirable antisense oligonucleotide for asthma. *Am. J. Respir. Crit. Care Med.* 163:989–993.

106. Templin M V, Levin A A, et al. (2000). Pharmacokinetic and toxicity profile of a phosphorothioate oligonucleotide following inhalation delivery to lung in mice. *Antisense Nucleic Acid Drug Dev.* 10:359–368.

107. Geary R S, Khatsenko O, et al. (2001). Absolute bioavailability of 2′-O-(2-methoxyethyl)-modified antisense oligonucleotides following intraduodenal instillation in rats. *J. Pharmacol. Exp. Ther.* 296:898–904.

108. Raoof A A, Chiu P, et al. (2004). Oral bioavailability and multiple dose tolerability of an antisense oligonucleotide tablet formulated with sodium caprate. *J. Pharm. Sci.* 93:1431–1439.

109. Agrawal S, Zhang X, et al. (1995). Absorption, tissue distribution and in vivo stability in rats of a hybrid antisense oligonucleotide following oral administration. *Biochem. Pharmacol.* 50:571–576.

110. Raoof A A, Ramtoola Z, et al. (2002). Effect of sodium caprate on the intestinal absorption of two modified antisense oligonucleotides in pigs. *Eur. J. Pharm. Sci.* 17:131–138.

111. Yu R Z, Geary R S, et al. (1999). Pharmacokinetics and tissue disposition in monkeys of an antisense oligonucleotide inhibitor of Ha-ras encapsulated in stealth liposomes. *Pharm. Res.* 16:1309–1315.

112. Miyano-Kurosaki N, Barnor J S, et al. (2004). In vitro and in vivo transport and delivery of phosphorothioate oligonucleotides with cationic liposomes. *Antivir. Chem. Chemother.* 15:93–100.

113. Yoshida J, Mizuno M (2003). Clinical gene therapy for brain tumors. Liposomal delivery of anticancer molecule to glioma. *J. Neurooncol.* 65:261–267.

114. Rudin C M, Marshall J L, et al. (2004). Delivery of a liposomal c-raf-1 antisense oligonucleotide by weekly bolus dosing in patients with advanced solid tumors: A phase I study. *Clin. Cancer Res.* 10:7244–7251.

115. Springate C M, Jackson J K, Gleave M E, et al. (2005). Efficacy of an intratumoral controlled release formulation of clusterin antisense oligonucleotide complexed with

chitosan containing paclitaxel or docetaxel in prostate cancer xenograft models. *Cancer Chemother. Pharmacol.* 56:239–247.

116. Agrawal S, Temsamani J, Tang J Y (1991). Pharmacokinetics, biodistribution, and stability of oligodeoxynucleotide phosphorothioates in mice. *Proc. Natl. Acad. Sci. U.S.A.* 88:7595–7599.

117. Dvorchik B H (2000). The disposition (ADME) of antisense oligonucleotides. *Curr. Opin. Mol. Ther.* 2:253–257.

118. Saijo Y, Perlaky L, Wang H, et al. (1994). Pharmacokinetics, tissue distribution, and stability of antisense oligodeoxynucleotide phosphorothioate ISIS 3466 in mice. *Oncol. Res.* 6:243–249.

119. Geary R S, Henry S P, Grillone L R (2002). Fomivirsen: clinical pharmacology and potential drug interactions. *Clin. Pharmacokinet.* 41:255–260.

120. Cotter F E, Waters J, Cunningham D (1999). Human Bcl-2 antisense therapy for lymphomas. *Biochim. Biophys. Acta.* 1489:97–106.

121. Raynaud F I, Orr R M, et al. (1997). Pharmacokinetics of G3139, a phosphorothioate oligodeoxynucleotide antisense to bcl-2, after intravenous adminis-tration or continu-ous subcutaneous infusion to mice. *J. Pharmacol. Exp. Ther.* 281:420–427.

122. O'Brien S M, Cunningham C C, et al. (2005). Phase I to II multicenter study of obli-mersen sodium, a Bcl-2 antisense oligonucleotide, in patients with advanced chronic lymphocytic leukemia. *J. Clin. Oncol.* 23:7697–7702.

123. Lopes de Menezes D E, Mayer L D (2002). Pharmacokinetics of Bcl-2 antisense oli-gonucleotide (G3139) combined with doxorubicin in SCID mice bearing human breast cancer solid tumor xenografts. *Cancer Chemother. Pharmacol.* 49:57–68.

124. Cossum P A, Truong L, et al. (1994). Pharmacokinetics of a 14C-labeled phosphoro-thioate oligonucleotide, ISIS 2105, after intradermal administration to rats. *J. Phar-macol. Exp. Ther.* 269:89–94.

125. Cossum P A, Sasmor H, et al. (1993). Disposition of the 14C-labeled phosphorothioate oligonucleotide ISIS 2105 after intravenous administration to rats. *J. Pharmacol. Exp. Ther.* 267:1181–1190.

126. Waters J S, Webb A, et al. (2000). Phase I clinical and pharmacokinetic study of bcl-2 antisense oligonucleotide therapy in patients with non-Hodgkin's lymphoma. *J. Clin. Oncol.* 18:1812–1823.

127. Guo S, Kemphues K J (1995). par-1, a gene required for establishing polarity in C. elegans embryos, encodes a putative Ser/Thr kinase that is asymmetrically distributed. *Cell.* 81:611–620.

128. Fire A, Xu S, et al. (1998). Potent and specific genetic interference by double-stranded RNA in Caenorhabditis elegans. *Nature.* 391:806–811.

129. Zamore P D (2001). RNA interference: Listening to the sound of silence. *Nat. Struct. Biol.* 8:746–750.

130. Sharp P A (2001). RNA interference–2001. *Genes Dev.* 15:485–490.

131. Bernstein E, Denli A M, Hannon G J (2001). The rest is silence. *RNA.* 7:1509–1521.

132. Elbashir S M, Martinez J, Patkaniowska A, et al. (2001). Functional anatomy of siRNAs for mediating efficient RNAi in Drosophila melanogaster embryo lysate. *EMBO J.* 20:6877–6888.

133. Elbashir S M, Harborth J, et al. (2001). Duplexes of 21-nucleotide RNAs mediate RNA interference in cultured mammalian cells. *Nature.* 411:494–498.

134. Lin X, Ruan X, et al. (2005). siRNA-mediated off-target gene silencing triggered by a 7 nt complementation. *Nucleic Acids Res.* 33:4527–4535.

135. Du Q, Thonberg H, Wang J, et al. (2005). A systematic analysis of the silencing effects of an active siRNA at all single-nucleotide mismatched target sites. *Nucleic Acids Res.* 33:1671–1677.

136. Saxena S, Jonsson Z O, Dutta A (2003). Small RNAs with imperfect match to endogenous mRNA repress translation. Implications for off-target activity of small inhibitory RNA in mammalian cells. *J. Biol. Chem.* 278:44312–44319.

137. Fedorov Y, King A, et al. (2005). Different delivery methods-different expression profiles. *Nat. Methods.* 2:241.

138. Muratovska A, Eccles M R (2004). Conjugate for efficient delivery of short interfering RNA (siRNA) into mammalian cells. *FEBS Lett.* 558:63–68.

139. Urban-Klein B, Werth S, Abuharbeid S, et al. (2005). RNAi-mediated gene-targeting through systemic application of polyethylenimine (PEI)-complexed siRNA in vivo. *Gene Ther.* 12:461–466.

140. Vickers T A, Koo S, et al. (2003). Efficient reduction of target RNAs by small interfering RNA and RNase H-dependent antisense agents. A comparative analysis. *J. Biol. Chem.* 278:7108–7118.

141. Verma U N, Surabhi R M, Schmaltieg A, Becerra C, Gaynor R B (2003). Small interfering RNAs directed against beta-catenin inhibit the in vitro and in vivo growth of colon cancer cells. *Clin. Cancer Res.* 9:1291–1300.

142. Filleur S, Courtin A, et al. (2003). SiRNA-mediated inhibition of vascular endothelial growth factor severely limits tumor resistance to antiangiogenic thrombospondin-1 and slows tumor vascularization and growth. *Cancer Res.* 63:3919–3922.

143. Duxbury M S, Ito H, Zinner M J, et al. (2004). RNA interference targeting the M2 subunit of ribonucleotide reductase enhances pancreatic adenocarcinoma chemosensitivity to gemcitabine. *Oncogene.* 23:1539–1548.

144. Morrissey D V, Blanchard K, et al. (2005). Activity of stabilized short inter-fering RNA in a mouse model of hepatitis B virus replication. *Hepatol.* 41:1349–1356.

145. Lewis D L, Hagstrom J E, Loomis A G, et al. (2002). Efficient delivery of siRNA for inhibition of gene expression in postnatal mice. *Nat. Genet.* 32:107–108.

146. Song E, Lee S K, et al. (2003). RNA interference targeting Fas protects mice from fulminant hepatitis. *Nat. Med.* 9:347–351.

147. Zender L, Hutker S, et al. (2003). Caspase 8 small interfering RNA prevents acute liver failure in mice. *Proc. Natl. Acad. Sci. U.S.A.* 100:7797–7802.

148. Layzer J M, McCaffrey A P, et al. (2004). In vivo activity of nuclease-resistant siRNAs. *RNA.* 10:766–771.

149. Chiu Y L, Rana T M (2003). siRNA function in RNAi: A chemical modification analysis. *RNA.* 9:1034–1048.

150. Czauderna F, Fechtner M, et al. (2003). Structural variations and stabilising modifications of synthetic siRNAs in mammalian cells. *Nucleic Acids Res.* 31:2705–2716.

151. Harborth J, Elbashir S M, et al. (2003). Sequence, chemical, and structural variation of small interfering RNAs and short hairpin RNAs and the effect on mammalian gene silencing. *Antisense Nucleic Acid Drug Dev.* 13:83–105.

152. Soutschek J, Akinc A, et al. (2004). Therapeutic silencing of an endogenous gene by systemic administration of modified siRNAs. *Nature.* 432:173–178.

153. Bitko V, Musiyenko A, Shulyayeva O, Barik S (2005). Inhibition of respiratory viruses by nasally administered siRNA. *Nat. Med.* 11:50–55.

154. Zhang W, Yang H, et al. (2005). Inhibition of respiratory syncytial virus infection with intranasal siRNA nanoparticles targeting the viral NS1 gene. *Nat. Med.* 11:56–62.

155. Zhang X, Shan P, et al. (2004). Small interfering RNA targeting heme oxygenase-1 enhances ischemia-reperfusion-induced lung apoptosis. *J. Biol. Chem.* 279:10677–10684.

156. Boussif O, Lezoualc'h F, et al. (1995). A versatile vector for gene and oligonucleotide transfer into cells in culture and in vivo: polyethylenimine. *Proc. Natl. Acad. Sci. U.S.A.* 92:7297–7301.

157. Chollet P, Favrot M C, Hurbin A, et al. (2002). Side-effects of a systemic injection of linear polyethylenimine-DNA complexes. *J. Gene Med.* 4:84–91.

158. Dufes C, Keith W N, et al. (2005). Synthetic anticancer gene medicine exploits intrinsic antitumor activity of cationic vector to cure established tumors. *Cancer Res.* 65:8079–8084.

159. Schiffelers R M, Ansari A, et al. (2004). Cancer siRNA therapy by tumor selective delivery with ligand-targeted sterically stabilized nanoparticle. *Nucleic Acids Res.* 32: e149.

160. Sioud M, Sorensen D R (2003). Cationic liposome-mediated delivery of siRNAs in adult mice. *Biochem. Biophys. Res. Commun.* 312:1220–1225.

161. Yano J, Hirabayashi K, et al. (2004). Antitumor activity of small interfering RNA/cationic liposome complex in mouse models of cancer. *Clin. Cancer Res.* 10:7721–7726.

162. Chien P Y, Wang J, et al. (2005). Novel cationic cardiolipin analogue-based liposome for efficient DNA and small interfering RNA delivery in vitro and in vivo. *Cancer Gene Ther.* 12:321–328.

163. Morrissey D V, Lockridge J A, et al. (2005). Potent and persistent in vivo anti-HBV activity of chemically modified siRNAs. *Nat. Biotechnol.* 23:1002–1007.

164. Prenner J L, Thompson J T, Miller D G, et al. (2005). An Open Label Study for the Evaluation of Tolerability of Five Dose Levels of Intravitreous VEGF siRNA (Cand5) in Patients With Wet Age-Related Macular Degeneration. *American Academy of Ophthalmology Annual Meeting,* poster 463.

165. Quinlan E, Nguyen Q, Kaiser P, et al. (2005). A Phase I, open-label, dose escalation trial of intravitreal injection of small interfering RNA molecule in subjects with neovascular AMD. *American Academy of Ophthalmology Annual Meeting.* Available: http://www.sirna.com/media/pdfs/AAOSIRNA.pdf.

166. Miyagishi M, Hayashi M, Taira K (2003). Comparison of the suppressive effects of antisense oligonucleotides and siRNAs directed against the same targets in mammalian cells. *Antisense Nucleic Acid Drug Dev.* 13:1–7.

167. Bertrand J R, Pottier M, et al. (2002). Comparison of antisense oligonucleotides and siRNAs in cell culture and in vivo. *Biochem. Biophys. Res. Commun.* 296:1000–1004.

7.5

CASE STUDIES—DEVELOPMENT OF OLIGONUCLEOTIDES

EZHARUL HOQUE CHOWDHURY AND TOSHIHIRO AKAIKE

Tokyo Institute of Technology, Yokohama, Japan

Chapter Contents

7.5.1	Introduction	1087
7.5.2	Oligonucleotides in Functional Genomics for Drug Target Discovery	1088
	7.5.2.1 Northern Blotting	1089
	7.5.2.2 Polymerase Chain Reaction (PCR)	1089
	7.5.2.3 High-Density Oligonucleotide Arrays	1094
7.5.3	Oligonucleotides in Drug Target Validation and Potential Therapeutics	1096
	7.5.3.1 Diversity in Antisense Technology	1096
7.5.4	Oligonucleotides in Genotyping Polymorphisms	1099
	7.5.4.1 Gel-Based Genotyping for Detecting Known Polymorphisms	1099
	7.5.4.2 Non-Gel-Based High-Throughput Genotyping Technologies	1100
7.5.5	Conclusion	1103
	References	1104

7.5.1 INTRODUCTION

The human genome is estimated to consist of 30,000 genes, up to 20% of which are considered to be expressed in a cell at a certain time. The human genome has now been completely sequenced, and huge amounts of data are being presented in public databases [1, 2]. However, these data do not provide information about the function of the genes or the significance of their expression in the specific state of

Handbook of Pharmaceutical Biotechnology, Edited by Shayne Cox Gad.
Copyright © 2007 John Wiley & Sons, Inc.

a disease. Thus, to fully exploit the massive DNA sequence information, smart technologies are necessary for rigorous analysis of gene expression. The collection of genes that are transcribed from genomic DNA, referred to as the "transcriptome," is a major determinant of cellular phenotype and function. The transcription of genomic DNA to produce mRNA is the first step for protein synthesis, and differences in the process are responsible for both morphological and phenotypic differences as well as indicative of cellular responses to environmental stimuli and perturbations. Unlike the genome, the transcriptome is highly dynamic and changes rapidly and dramatically in response to perturbations or during normal cellular events [3, 4]. Understanding the function of genes with respect to the time, location, and degree of expression is central to elucidating the activity and biological roles of its encoded protein. In addition, changes in the multigene expression patterns can provide clues about regulatory mechanisms and broader cellular functions and biochemical pathways. Thus, the knowledge gained from these types of measurements through implementation of proper and efficient technology can help determine the causes and consequences of disease, how drugs and drug candidates work in cells and organisms, and what gene products might have therapeutic uses or may be appropriate targets for therapeutic intervention.

Oligonucleotides have emerged as essential tools in pharmaceutical biotechnology by virtue of their small size, low-cost chemical synthesis, and precise base pairing capability with other complementary sequences either in the transcriptome or in the genome, according to the Watson–Crick model. Functional genomics-based screening, identification, and validation of drug targets, development of target protein-specific new generation therapeutics, and genotyping of many polymorphisms to develop tailor-made drugs for individuals are all being accomplished by many smart technological appro-aches using synthetic oligonucleotides of variable sizes and sequences.

7.5.2   OLIGONUCLEOTIDES IN FUNCTIONAL GENOMICS FOR DRUG TARGET DISCOVERY

The classic sciences of chemistry, pharmacology, microbiology, and biochemistry have shaped the course of drug discovery. Additionally, molecular biology has exerted a profound influence on drug discovery, allowing the concept of genetic information to be dealt with in very concrete chemical and biochemical terms. Although the earlier influence of molecular biology was restricted to cloning and expressing genes that encode therapeutically useful protein drugs (mostly recombinant proteins and monoclonal antibodies), recently the main promise of molecular biology for drug discovery lies in the potential to understand disease processes at the molecular genetic level and determine the optimal targets of drug intervention. The next step is to design novel pharmaceutical compounds either by high-throughput screening (HTS), a process in which batches of compounds are tested for binding activity or biological activity against target molecules, or simulation of drug–target interactions in order to design a drug to fit a target structure precisely.

However, the molecular biology techniques that are most commonly used for drug target discovery are fully or partially based on synthetically designed oligonucleotides for detection and quantitation of specific mRNAs as representatives of target proteins.

7.5.2.1 Northern Blotting

Northern analysis remains a standard method for detection and quantitation of mRNA levels despite the advent of powerful techniques, such as reverse transcriptase–polymerase chain reaction (RT–PCR), real-time PCR, and microarray. It is used to quantitate a specific mRNA, determine transcript size, detect alternative splice variants of a gene, and identify closely related species. RNA samples are first separated by size via electrophoresis in an agarose gel under denaturing conditions, transferred to a membrane, cross-linked, and hybridized with a labeled probe, such as oligonucleotide (cDNA or RNA can also be applicable). Sensitivity can be improved with oligo deoxythymidine (dT) selection for enrichment of mRNAs because the dT oligonucleotide anchor prime [oligo(dT)] in a column hybridizes to the 3′-poly(A) tail of the mRNA. The radiolabeled or nonisotopically labeled probe has a sequence complementary to that of target mRNA facilitating efficient hybridization between them. Nonisotopic labeling of oligonucleotides can be broadly classified into two categories: indirect and direct. The indirect method involved the incorporation of a hapten into the probe molecule that is subsequently detected by affinity recognition with an antibody or a binding moiety, such as streptavidin. The binding moiety is, in turn, coupled to a signaling enzyme, such as horseradish peroxidase. Direct methods, on the other hand, covalently attach the signaling enzymes to the probe molecules. Factors that influence the hybridization efficiency include temperature, ionic strength, destabilizing agents, mismatched base pairs, duplex length, viscosity, and base composition. High salt favors hybridization reactions (i.e., less specificity and greater background), whereas decreased salt and/or increased detergent or temperature increases hybridization specificity and reduces background.

7.5.2.2 Polymerase Chain Reaction (PCR)

The PCR is a test-tube system for DNA replication that allows a "target" DNA sequence to be selectively amplified several million-fold in just a few hours. During PCR, high temperature is used to separate the DNA molecules into single strands, and synthetic sequences of single-stranded DNA (20–30 nucleotides) serve as primers to enable DNA polymerase to synthesize the complementary strand. Two different primer sequences are used to bracket the target region to be amplified. One primer is complementary to one DNA strand at the beginning of the target region; a second primer is complementary to a sequence on the opposite DNA strand at the end of the target region. Thus, PCR technology consists of three steps: denaturation of double-stranded target DNA at 94–96°C, annealing (hybridization) of primers to their complementary sequences on either side of the target sequence at 50–65°C and finally extending a complementary DNA strand from each primer with the help of a heat-stable DNA polymerase at 72°C. As amplification proceeds, the DNA sequence between the primers doubles after each cycle. After 30 such cycles, a theoretical amplification factor of one billion is attained.

For the discovery of drug target proteins based on functional genomics, some advanced PCR techniques are now playing the vital roles.

Reverse Transcriptase (RT)–PCR. RT–PCR is the most sensitive technique currently available for mRNA detection and quantitation. Compared with the two other commonly used techniques (Northern blot analysis and RNase protection assay) for quantifying mRNA levels, RT–PCR can be used to quantify mRNA levels from much smaller samples. In fact, this technique is sensitive enough to enable quantitation of RNA from a single cell. RT–PCR is a technique in which an RNA strand is "reverse" transcribed into its DNA complement, followed by amplification of the resulting DNA using PCR. Transcribing an RNA strand into its DNA complement is termed reverse transcription (RT), and it is accomplished through the use of an RNA-dependent DNA polymerase (reverse transcriptase). Afterward, a second strand of DNA is synthesized through the use of a deoxyoligonucleotide primer and a DNA-dependent DNA polymerase. The complementary DNA and its antisense counterpart are then exponentially amplified via PCR. The original RNA template is degraded by RNase H leaving behind DNA only.

Real-Time PCR.   Over the last several years, the development of novel chemistries and instrumentation platforms enabling detection of PCR products on a real-time basis has led to widespread adoption of real-time RT–PCR as the method of choice for quantitating changes in gene expression. It enables quantitation of the initial amount of the template most specifically, sensitively, and reproducibly by monitoring the fluorescence emitted during the reaction as an indicator of amplicon production during each PCR cycle (i.e., in real time), and it is a preferable alternative to other forms of quantitative RT–PCR that detect the amount of final amplified product at the endpoint [5, 6]. Furthermore, real-time RT–PCR has become the preferred method for validating results obtained from array analyses and other techniques that evaluate gene expression changes on a global scale.

The real-time PCR system is based on the detection and quantitation of a fluorescent reporter [7, 8]. This signal increases in direct proportion to the amount of PCR product in a reaction. By recording the amount of fluorescence emission at each cycle, it is possible to monitor the PCR reaction during the exponential phase where the first significant increase in the amount of PCR product correlates to the initial amount of target template. The higher the starting copy number of the nucleic acid target, the sooner a significant increase in fluorescence is observed. A significant increase in fluorescence above the baseline value measured during the 3–15 cycles indicates the detection of accumulated PCR product.

Real-time quantitative PCR analysis can be performed with oligonucleotide-based hydrolysis probes or hybridization probes. Hydrolysis probes include TaqMan probes [9], molecular beacons [10, 11], and scorpions [12]. They use the fluorogenic 5′ exonuclease activity of Taq polymerase to measure the amount of target sequences in cDNA samples (Figure 7.5-1). TaqMan probes, Molecular Beacons, and Scorpions depend on fluorescence resonance energy transfer (FRET) to generate the fluorescence signal via the coupling of a fluorogenic dye molecule and a quencher moiety to the same or different oligonucleotide substrates.

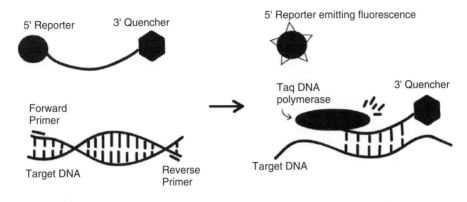

Figure 7.5-1. A model for dual-labeled fluorogenic probes for real-time PCR. The probe is labeled with two different dyes: A reporter dye is located on the 5' end, and a quencher dye is located on the 3' end. The quencher dye inhibits the natural fluorescence emission of the reporter dye by FRET. During the elongation step, Taq DNA polymerase 5'–3' exonuclease activity hydrolyzes the probe bound to the specific DNA template, releasing the reporter dye from the target/oligonucleotide-quencher hybrid and causing an increase of fluorescence in proportion to the amount of target DNA.

Hydrolysis Probes

TAQMAN PROBES. These are oligonucleotide probes longer than the primers (20–30 bases long with a Tm value of 10°C higher) that contain a fluorescent dye usually on the 5' base and a quenching dye (usually TAMRA) typically on the 3' base. When irradiated, the excited fluorescent dye transfers energy to the nearby quenching dye molecule in the FRET mechanism [13]. Thus, the close proximity of the reporter and quencher prevents emission of any fluorescence while the probe is intact. TaqMan probes are designed to anneal to an internal region of a PCR product. When the polymerase replicates a template on which a TaqMan probe is bound, its 5' exonuclease activity cleaves the probe [14]. This ends the activity of quencher (no FRET), and the reporter dye starts to emit fluorescence, which increases in each cycle proportional to the rate of probe cleavage. Accumulation of PCR products is detected by monitoring the increase in fluorescence of the reporter dye. Because the cleavage occurs only if the probe hybridizes to the target, the origin of the detected fluorescence is specific amplification.

MOLECULAR BEACONS. Like TaqMan probes, molecular beacons also use FRET to detect and quantitate the synthesized PCR product via a fluorescent dye coupled to the 5' end and a quencher attached to the 3' end of an oligonucleotide substrate. Unlike TaqMan probes, molecular beacons are designed to remain intact during the amplification reaction, and they must rebind to the target in every cycle for signal measurement. Molecular beacons form a stem-loop structure when free in solution. Thus, the close proximity of the fluorescent and quencher dyes prevents the probe from fluorescing. When a molecular beacon hybridizes to a target, the fluorescent and quencher dyes are separated, FRET does not occur, and the fluorescent dye emits light upon irradiation.

SCORPIONS. With scorpion probes, sequence-specific priming and PCR product detection is achieved using a single oligonucleotide. The scorpion probe maintains a stem-loop configuration in the unhybridized state. The fluorophore is attached to the 5′ end and is quenched by a moiety coupled to the 3′ end. The 3′ portion of the stem also contains a sequence that is complementary to the extension product of the primer. This sequence is linked to the 5′ end of a specific primer via a nonamplifiable monomer. After extension of the Scorpion primer, the specific probe sequence can bind to its complement within the extended amplicon, thus opening the hairpin loop. This prevents the fluorescence from being quenched and a signal is observed.

Hybridization Probes

LIGHTCYCLER. A LightCycler probe (Roche Applied Science, Indianapolis, IN) is a pair of single-stranded fluorescent-labeled oligonucleotides. Oligo Probe I is labeled at its 3′ end with a donor fluorophore dye, and Oligo Probe 2 is labeled at its 5′ end with one of two available acceptor fluorophore dyes. The free 3′ hydroxyl group of Probe 2 must be blocked with a phosphate group to prevent Taq DNA polymerase extension. To avoid any steric problems between the donor and the acceptor fluorophores on both sides, there should be a spacer of one to five nucleotides to separate the two probes from each other. During the annealing step, the PCR primers and the LightCycler probes hybridize to their specific target regions, which causes the donor dye to come into close proximity to the acceptor dye. When the donor dye is excited by light from the LightCycler instrument, energy is transferred by FRET from the donor to the acceptor dye, which causes the acceptor dye to emit light at a longer wavelength than the light emitted by the instrument. The acceptor fluorophore's emission wavelength is detected by the LightCycler instrument's optical unit. The significant increase of measured fluorescence signal is directly proportional to the amount of target DNA.

Differential Display (DD)–PCR. DD–PCR is a PCR-based technique that can detect differentially expressed genes. It provides a picture of the transcript pool of cells or tissues by displaying subsets of mRNAs. Subsets obtained from different cell types or tissues can be compared and used for isolation of the genes of interest. The general strategy for DD–PCR consists of the combination of (1) reverse transcription using a dT oligonucleotide anchor primer (oligo(dT)$_n$) to produce cDNA; (2) performing PCR using the cDNA as a template with the anchor primer and an arbitrary primer; and (3) separation of the PCR product by electrophoresis and visualization. The differential display method is thus far unique in its potential to visualize all expressed genes in a eukaryotic cell in a systematic and sequence-dependent manner by using multiple primer combinations.

Serial Analysis of Gene Expression (SAGE). SAGE analysis is a method derived to provide a readout, via sequencing, of the spectrum of genes being expressed in a cell or tissue in both a qualitative and a quantitative manner [15, 16]. Three principles underlie the SAGE methodology:

1. A short sequence tags (10–15 bp) contains sufficient information to uniquely identify a transcript provided that the tag is obtained from a unique position within each transcript.

2. Sequence tags can be linked together to form long serial molecules that can be cloned and sequenced.

3. Quantification of the number of times a particular tag is observed provides the expression level of the corresponding transcript.

A detailed description of SAGE is provided in Figure 7.5-2.

Important advantages of SAGE over DD–PCR are that SAGE data are quantitative and cumulative. Accurate, quantitative transcript profiles describing the abundance of all genes expressed in a cell or tissue are generated by SAGE, provided sufficient sequencing is completed. The resulting data may then be compared with all existing and future SAGE databases constructed in a similar manner.

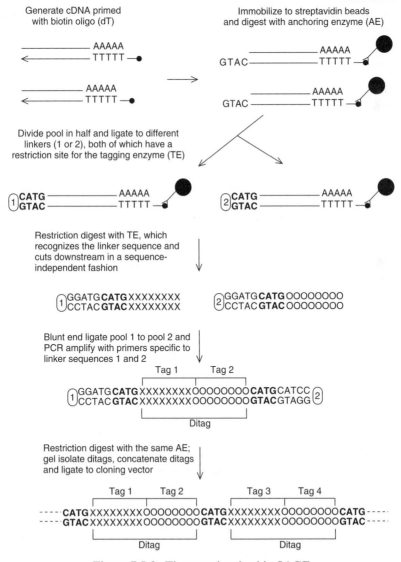

Figure 7.5-2. The steps involved in SAGE.

7.5.2.3 High-Density Oligonucleotide Arrays

High-density arrays of oligonucleotides are the most powerful and versatile tools for functional genomics. They work by hybridization of labeled RNA or DNA in solution to DNA molecules attached at specific locations on a surface. So DNA microarrays may be defined as miniaturized, ordered arrangements of nucleic acid fragments derived from individual genes located at defined positions on a solid support, which enables the analysis of thousands of genes in parallel by specific hybridization. DNA microarrays provide information on how several genes are abnormally regulated in a disease. For example, a microarray of 100 genes that have a role in inflammation was used to examine rheumatoid tissue. The analysis revealed the upregulation of genes encoding interleukin 6 and several matrix metalloproteinases [17]. The microarray technology will continue to contribute to the understanding of responses to drug treatments [18]. Based on the fabrication technique used, planar microarrays can be broadly classified into microarrays prepared by *in situ* synthesis of oligonucleotides and direct spotting of oligonucleotides on glass, membrane, or other derivatized surfaces by microspotting or inkjet printing methods.

Microarrays Fabrication by in-situ Synthesis. A combination of photolithography and combinatorial chemistry is used to develop very high-density DNA chips by Affymetrix (http://www.affymetrix.com, Santa Clara, CA). Fused silica wafers are hydroxylated and then silanized. The silanes contain a linker molecule and a protective group that can be activated by light. The silane film provides uniform hydroxyl density to initiate the microarray fabrication process. Photolithographic masks are then used to illuminate specific locations on the silanized wafer. Ultraviolet (UV) exposure causes the deprotection of groups on the silane, after which the activated groups couple nucleotides onto the silica wafer. Additional capping steps prevent unattached molecules from becoming probes. The side chains on the nucleotides are protected to prevent branched-chain formation. These steps of deprotection, coupling, and capping are continued until probes of full length are fabricated (Figure 7.5-3).

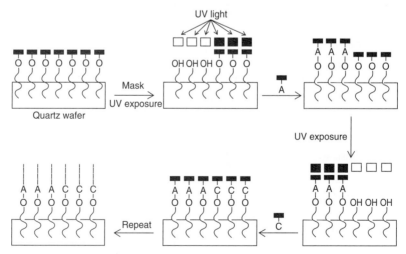

Figure 7.5-3. Fabrication of high-density oligonucleotide array by photoactivation and deprotection of nucleic acids. Photomasks are used to pattern UV light at localized regions to selectively synthesize patterned array.

In an alternative approach developed by NimbleGen (http://www.nimblegen. com/, Madison, WI), arrays are built using photodeposition chemistry based on a maskless array synthesizer (MAS). The MAS system is a high-density DNA fabrication instrument that uses a maskless light projector as a "virtual mask" instead of the physical chromium masks used by Affymetrix. The virtual mask is an array of hundreds of thousands of individually addressable aluminum mirrors on a computer chip. These mirrors function as virtual masks that reflect the desired pattern of UV light and are controlled by the computer.

In situ synthesized oligonucleotide probes generally do not exceed 25 bases in length because of reaction efficiency limits. As a result, mismatches and spurious target-probe binding can take place due to the limited specificity and binding affinity for a 25-residue oligonucleotide. To overcome this problem, a series of oligonucleotides that differ by a one-base mismatch from the gene-specific probe is also included on the array and can be used to determine the amount of mismatch hybridization, which can then be substracted from the signal [19, 20]. The Affymetrix arrays contain between 40,000 and 60,000 probes (including multiple mismatch controls for each gene) and provide the highest density of probes of any array.

Microarrays Fabrication by Direct Spotting. Microarrays can be fabricated by directly spotting oligonucleotides on microscope glass slides, membranes, or other surfaces. Various types of spotters and derivatized substrate surfaces have been developed to fabricate microarrays [21–23]. The surface groups on the derivatized surface determine how the oligonucleotides will be immobilized on the surface. DNA molecules contain phosphate groups, which are commonly used for immobilization on positively charged surfaces. Amine surfaces carry a positive charge at neutral pH. Therefore, negatively charged DNA molecules form ionic bonds with the positively charged amine-derivatized surface. Often, UV or thermal treatment is used to covalently link the DNA molecules to amine-derivatized slides that are most suitable for immobilizing long oligonucleotides. Sometimes, to improve the binding efficiency, a longer chain spacer is attached to the aminated surface to provide sufficient separation between the amino groups and substrate surface. Poly-1-lysine-coated slides containing a dense layer of amino groups are similarly used to attach DNA molecules via ionic interactions. Additionally, amino-modified can be covalently immobilized on aldehyde- and epoxy-derivatized glass slides.

Microarray is now an essential tool in functional genomics for drug target discovery. However, in addition to the obvious use of functional genomics in basic research and target discovery, there are many other specific uses, including biomarker determination to find genes that correlate with and presage disease progression; pharmacology to determine differences in gene expression in tissues exposed to various doses of compounds; toxico-genomics to find gene expression patterns in a model tissue or organism exposed to a compound and their use as early predictors of adverse events in humans; target selectivity to define a compound by the gene expression patterns; prognostic tests to find a set of genes that accurately distinguishes one disease from another; and disease subclass determination to find multiple subcategories of tumors in a single clinical diagnosis [24].

7.5.3 OLIGONUCLEOTIDES IN DRUG TARGET VALIDATION AND POTENTIAL THERAPEUTICS

Because their sequence is complementary to that of mRNA, antisense oligonucle-otides (ON) form a very powerful weapon for studying gene function (functional genomics) and for discovering new and more specific treatments of human diseases (antisense therapeutics). The binding or hybridization of antisense nucleic acid sequences to a specific mRNA target will interrupt normal cellular processing of the genetic message of a gene through at least three mechanisms: (1) the oligonucle-otide: RNA duplex may form substrate for endogenous RNase H, leading to mRNA cleavage; (2) the duplex may prevent the productive assembly of the ribosomal complex preventing translation; and (3) the duplex may arrest a ribosomal complex already engaged in translation leading to a truncated protein [25]. However, the successful therapeutic application of antisense ONs has been delayed because of delivery problems. Natural ONs are not membrane-permeable, and even after delivery into cells with the help of amphipathic cations or liposomes, antisense ONs can still be hydrolyzed by cellular nucleases before reaching their intended target mRNAs. To improve their bioavailability, various types of chemically modified antisense ONs have been produced [26–29]. However, chemical modification has also led to undesirable complications such as decreased sequence-specificity and general toxicity. On the other hand, the discovery of RNA interference (RNAi) started a new era in antisense technology [30–32]. Double-stranded short interfer-ing RNAs (siRNA) could silence their target mRNA almost 100 times more effec-tively than single-stranded ONs [33]. However, delivery of siRNA is still a problem as successful administration often requires the use of plasmids or viral vectors, leading to the risk of random integration into chromosomal DNA, which is a serious threat to therapeutic safety [34].

7.5.3.1 Diversity in Antisense Technology

Many types of chemically modified antisense ONs have been manufactured with the changes in the overall electronic charge as well as the incorporation of non-phosphate oligonucleotide backbones. They are commonly grouped into three "generations" based on the type of modifications (Figure 7.5-4).

First-Generation Antisense ONs. These ONs were manufactured through the substitution of one nonbridging oxygen atom in the phosphate group of ONs with either a sulphur or a methyl group. Those DNA analoges with a sulfur group are known as phosphorothioate ONs, which is the most widely used antisense ON to date. The main advantages of phosphorothioate ONs are resistance against nucleases, ability to recruit RNase H to cleave target mRNA, ease of synthesis, and attractive pharmacokinetic properties. The first U.S. Food and Drug Administration (FDA)-approved antisense drug, Vitravene from Isis (Carlsbad, CA), which treats a condition called cytomegalovirus (CMV) retinitis in people with AIDS, and the majority of antisense compounds in clinical trials to date are based on this chemical design.

Second Generation Antisense ONs. These contain nucleotides with alkyl modification at the 2′ position of the ribose. 2′-O-methyl and 2′-O-methoxyethyl

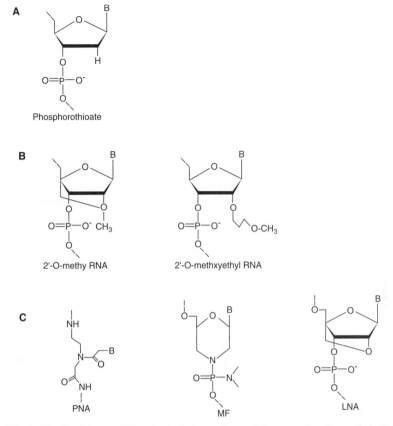

Figure 7.5-4. Nucleotide modifications for use in antisense technology. (A) The first-generation phosphorothioate backbone modification. (B) The second ribonucleotides modified at the 2′ hydroxyl by adding a methyl or methoxy–ethyl group. (C) The third-generation modifications involving the complete replacement of phosphodiester linkages as in PNA and MF or the conformational lock as in LNA.

RNAs are the most important representatives. The significant improvements are a reduction in general toxicity, increased hybrid stability, and increased nuclease resistance. However, the only disadvantage is that 2′-O-methyl RNA cannot induce the RNase H cleavage of the target mRNA. To induce RNase H cleavage, mixed backbone ONs (MBOs) were developed by surrounding a phosphorothioate-modified deoxyribose core that retains RNase H activity with nuclease-resistant arms, such as 2′-O-methyl ribonucleosides [35, 36]. The antisense activity, pharmacokinetics, *in vivo* degradation, and safety profile of an MBO can be modulated by combining appropriate oligonucleotide segments and backbone modifications at defined sites. GEM231, which targets the RI(alpha) subunit of protein kinase A (PKA), is built on an MBO platform and is being developed by Hybridon (Cambridge, MA) as a cancer therapeutic [37]. It is an 18-mer oligonucleotide with four 2′-O-methyl ribonucleosides at both the 3′- and the 5′-ends surrounding the remaining deoxynucleosides, all with phosphorothioate internucleotide linkages.

Third-Generation Antisense ONs. These are DNA and RNA analoges with modified phosphate linkages or ribose as well as nucleotides with a completely different chemical moiety replacing furanose ring. Peptide nucleic acids (PNAs), morpholino phosphoroamidates (MFs), and locked nucleic acids (LNAs) are three interesting RNA/DNA analoges in this class. These compounds are essentially nuclease resistant while maintaining good hybridization affinity with their complementary mRNA.

PNA and MF possess the structures where the phosphodiester linkages are completely replaced, respectively, with a polyamide (peptide) and phosphoroamidate backbone (Figure 7.5-4) [38, 39]. They both form tight bonds with their RNA targets and probably exert their effects by blocking translation, as neither molecule effectively activates RNase H. Noncharged backbones prevent PNA and MF from binding to proteins that normally recognize polyanions in a nonspecific manner. However, due to the electrostatically neutral property, solubility and cellular uptake are serious problems for both PNA and MF. The ability of PNA to recognize duplex DNA makes PNA more promising candidates for modulating gene expression or inducing mutations by strand invasion of chromosomal duplex DNA [40] because the PNA:DNA duplex is more stable than a DNA:DNA duplex [41]. Effective gene knockdown has also been shown by antisense MF, and one such MF targets the *c-myc* oncogene [42] and is being developed by AVI BioPharma (Portland, OR).

LNAs [also known as bridged nucleic acids (BNAs)] are another type of novel high-affinity molecules that provide major improvements in several key properties. They contain a methylene bridge connecting the 2′-oxygen of the ribose with the 4′-carbon (Figure 7.5-4), leading to improved binding to complementary DNA and RNA sequences [43]. Although the 2′–4′ linkage reduces or eliminates activation of mRNA cleavage by RNase H, it is straightforward to synthesize chimeric "gapmers," in which a central DNA portion is flanked by LNA to enhance the stability of binding and enable recruitment of RNase H. LNAs have been successfully used to suppress the expression of an RNA polymerase, resulting in the inhibition of tumor growth in a xenograft model [44].

siRNAs and Traditional Antisense Technology. siRNA are the effector molecules of the RNAi pathway in which endogenously produced double-stranded RNA is cleaved by the Dicer enzyme within the cells into 21- to 28-nucleotide siRNA for targeted degradation of complementary mRNA through the formation of RNA-induced silencing complexes (RISCs). siRNA can be synthetically produced or expressed from vectors transcribing short double-stranded hairpin-like RNAs that are processed into siRNAs inside the cell. Unlike ONs, siRNA cannot effectively target pre-mRNA for degradation in mammalian cells [45]. However, siRNA is now a reliable gene-silencing tool in functional genomics for the validation of potential drug targets identified by the techniques described above. The low concentration of siRNA required to elicit effective gene silencing and the fact that siRNAs are specifically and rapidly incorporated into RISC diminish the potential for the nonspecific binding with proteins as happened in phosphorothioate-modified ONs, which makes them toxic at a high concentration. However, it is important to identify effective siRNAs so that the lowest possible concentration of siRNA can be allowed for gene silencing [46–48]. If siRNAs are to be used for therapeutic purposes, methods must be developed that protect the possible side effects of siRNAs by either modification of the siRNA backbone [49–51] or avoidance of immunostimulatory sequence motifs [51, 52].

7.5.4 OLIGONUCLEOTIDES IN GENOTYPING POLYMORPHISMS

Single nucleotide polymorphisms (SNPs), the most frequent DNA sequence variations found in the human genome (about 1 per 350-bp frequency), are being uncovered and assembled into large SNP databases that promise to enable the dissection of the genetic essence of disease and drug response [53–57]. The identification of a complex set of genes that cause a disease will require both linkage and association analyses of thousands of polymorphisms across the human genome in thousands of individuals. In addition, certain genetic polymorphisms cause significantly different responses among individuals on exposure to a particular drug [58]. Understanding the role of genetic polymorphisms in drug responses will help to increase drug efficacy and decrease adverse effects by tailoring medications to patients' genetic makeup. Advances in the field, which is known as "pharmacogenetics," have important clinical implications and practical value for the design of dosing regimens. There are two broad areas in pharmacogenetics, specifically pharmacokinetics (PK) describing drug adsorption, distribution, metabolism and elimination, and pharmacodynamics (PD) elucidating the pharmacological effects of a drug on the body (either desired or undesired) and differences in either PK or PD can lead to variable drug efficacy or toxicity risk [59]. The Human Genome Project has provided a windfall of sequence polymorphism data, and because of the collaborative SNP discovery initiatives such as the SNP Consortium (TSC), millions of human SNPs have been catalogued, many of which are publicly available in the TSC and NCBI dbSNP repositories (http://snp.cshl.org; http://www.ncbi.nlm.nih.gov/SNP/). The high-density SNP map will allow researchers to expand their capabilities for identification of critical diseases and drug-response genes. Genotyping SNPs in large-scale pharmacogenetic studies is an integrated part of the drug discovery and development process. Advanced technologies to identify genetic polymorphisms rapidly, accurately, and economically are becoming a priority in the implementation of pharmacogenetics to drug development, clinical trials, and clinical monitoring for drug efficacy and toxicity [60]. SNP genotyping is now performed by several advanced technologies, most of which are based on synthetic oligonucleotide primers or probes.

7.5.4.1 Gel-based Genotyping for Detecting Known Polymorphisms

PCR–Restriction Fragment Length Polymorphism (RFLP) Analysis. Commonly used methods include gel-electrophoresis-based techniques such as PCR coupled with RFLP analysis. Specific regions of DNA sequences can be PCR amplified by using specific primers. The PCR products are then digested with appropriate restriction enzymes and visualized by staining the gel after electrophoresis. If the genetic polymorphism produces a gain or loss of the restriction site, a different restriction digestion pattern will be obtained [60]. A major limitation is the requirement that the polymorphisms alter a restriction enzyme cutting site [61].

Oligonucleotide Ligation Assay (OLA) Genotyping. OLA relies on hybridization with specific oligonucleotide probes that can effectively dis-criminate between the wild-type and variant sequences. Three oligonucleotide probes are used in OLA: two allele-specific probes (one specific for the wild-type allele and the other specific for the mutant allele) and a fluorescent common probe. The 3′ ends of the allele-specific probes are immediately adjacent to the 5′ end of the common probe. The gene fragment containing the polymorphic site is amplified by PCR and incubated with the probes. In the presence of thermally stable DNA ligase, ligation of the fluorescent-labeled probe to the allele-specific probe(s) occurs only when there is a perfect match between the variant or the wild-type probe and the PCR product template. These ligation products are then separated by electrophoresis, thus enabling the identification of the wild-type genotypes, the variants, the heterozygotes, and the variants (Figure 7.5-1). By varying the combinations of color dyes and probe lengths, multiple mutations can be detected in a single reaction [62].

7.5.4.2 Non-Gel-Based High-Throughput Genotyping Technologies

TaqMan Genotyping. The assay is based on TaqMan real-time PCR technology as described above. However, SNP genotyping requires two TaqMan probes that differ at the polymorphic site, with one probe comple-mentary to the wild-type allele and the other to the variant allele. A 5′ reporter dye and a 3′ quencher dye are covalently linked to both wild-type and variant allele probes. During the PCR annealing step, the TaqMan probes hybridize to the targeted polymorphic site. During the PCR extension phase, only the perfectly hybridized probe will be cleaved by the 5′ nuclease activity of the *Taq* polymerase, leading to an increase in the characteristic fluorescence of the reporter dye. A mismatched probe will not be recognized by the *Taq* polymerase, resulting in the quenching of fluorescence due to the physical proximity of the reporter and quencher dyes.

Molecular Beacons. As described, the ends of molecular beacons are complementary to each other, forming a stem structure, whereas the intervening loop is com-plementary to a sequence within the PCR amplified product. When hybridized to the right target sequence, the two fluorescent dyes (donor and acceptor) at opposite ends of the probes are separated and the fluorescence is increased dramatically [63]. Mismatched probetarget hybrids dissociate at a substantially lower temperature than exactly complementary hybrids. This thermal instability of mismatched hybrids increases the specificity of molecular beacons. For SNP genotyping, two molecular beacons with exact sequence matches to the wild-type and variant alleles are used in the same PCR reaction. These two molecular beacons are labeled with different fluorophores that emit fluorescent light at distinct optical wavelengths. The use of two differentially labeled molecular beacons in the same PCR reaction allows simultaneous determination of three possible allelic combinations.

Invader Assay. The Invader assay developed by Third Wave Technologies, Inc. (Madison, WI) is an attractive FRET-based technique to genotype SNPs without PCR amplification (Figure 7.5-5). Two oligonucleotides, a wild-type or variant signal probe plus an upstream Invader probe, are used in each reaction. These

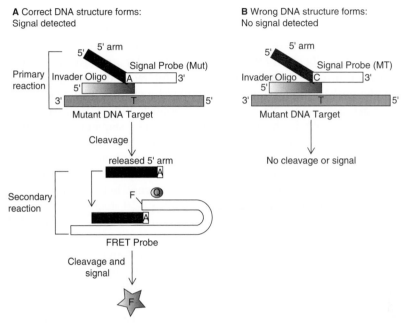

Figure 7.5-5. Invader SNP assay.

probes hybridize in tandem to a specific region of genomic DNA. When the 3′ end of an upstream oligonucleotide (Invader probe) overlaps the hybridization site of the 5′ end of a downstream oligonucleotide (signal probe) by at least one base pair, the structure will be recognized and cleaved by Cleavase [64, 65]. A single-nucleotide mismatch immediately upstream of the cleavage site renders the conformation unrecognized by Cleavage. Each cleavage product then serves as an invader oligonucleotide in a secondary reaction where it directs the cleavage of a combined labeled FRET probe-template construct. The secondary oligonucleotide probe is 5′ end-labeled with the donor flurophore, which is quenched by an internal acceptor dye. When the DNA is cleaved, the donor and acceptor dyes are no longer in close proximity, the quenching is abolished and fluorescence is generated. The whole process repeats multiple times, producing linear amplification of the fluorescent signal. Assays are read with a fluorescence plate reader and genotypes are assigned after determination of the net wild-type/variant signal ratio for each sample [64, 66].

FRET-based PCR–OLA. The three dye-labeled ligation probes for each SNP are designed to have low melting temperatures. A 5′ donor dye-labeled common probe terminates one base immediately upstream from the polymorphic site. Two allele-specific 5′-phosphorylated, 3′-acceptor dye-labeled probes have polymorphic nucleotides at the 5′ end. A thermostable DNA polymerase with no 5′ nuclease activity and a thermostable DNA ligase are used. The first stage of PCR reaction is kept at high temperature, and the ligation probes cannot to anneal. After sufficient PCR products are generated, the second stage of the reaction, with a low annealing temperature, allows ligation to occur. By analyzing the fluorescence signals of all dyes, individual genotypes can be determined directly after one reaction using real-time PCR or by endpoint signal analysis using a fluorescent plate reader [67].

Rolling Circle Amplification (RCA).   Like the Invader assay, RCA is also sensitive enough to work directly from genomic DNA [68, 69]. For each SNP, two linear, allele-specific probes are designed that circularize when they anneal to the target sequence. Each probe consists of a single oligonucleotide, 80–90 bases in length. The 5'-end of the probe is phosphorylated and bears a sequence of 20 nucleotides that will hybridize to the region immediately 5' of the SNP. The 3'-end of the probe contains 10–20 nucleotides complementary to the region immediately 3' of the SNP. Both allele-specific probes are identical with the exception of the 3'-base, which is varied to complement the polymorphic site. Sandwiched between the allele-specific probe arms is a generic backbone sequence of 40–50 nucleotides that provides binding sites for two RCA amplification primers to be used in the second stage of the assay.

The first stage of the assay is the allele discrimination step. Target genomic DNA is denatured to make it single stranded and then hybridized with a pair of single-allele-specific, open-circle oligonucleotide probes resulting in circularization of the probe. Allele discrimination is performed by a thermophilic DNA ligase. Ligation of the probe will preferentially occur when the 3'-base exactly complements the target DNA. If there is a mismatch between the 3'-allele-specific base and the target, the ligation product will be produced at a much-reduced rate. In a homozygous sample, the ligation of one allele-specific probe predominates over the other, and in a heterozygous sample, both probes are ligated with equal efficiency.

Subsequent detection of the circularized oligonucleotide probes is by RCA, using the two non-cross-reacting primers. The first primer is complementary to 20 bases of the circularized oligonucleotide probe in the backbone region and is extended by an exonuclease-deficient DNA polymerase. The extending probe eventually displaces itself at its 5'-end once one revolution of the circularized probe is completed. Continued polymerization and displacement result in the generation of a single-stranded product of complementary concatamers. The second primer is the same sense as the probe. It binds to each tandem repeat of the first strand product at sites corresponding to a second region of the backbone, where it primes DNA synthesis at each site. As these multiple binding events elongate, strand displacing activity of the DNA polymerase causes these nascent strands to be displaced and a multitude of new recognition sites for the first primer are exposed in a process known as branching. Eventually, strand displacement results in the release of double-stranded DNA fragments from the parent molecule, thus generating up to 10^9 copies of each circular oligonucleotide probe under isothermal conditions. A FRET-based RCA assay called SNIPER has been developed by Amersham Pharmacia Biotech. [70], in which the second amplification primer is an Amplifluor labeled oliogonucleotide with a 5'-hairpin loop containing FRET dyes. As the complementary strands to the hairpin primers are synthesized, the polymerase reads through the stem of the hairpin loop and opens out the 5'-end. In this linearized conformation, the fluorophore is no longer quenched and fluorescence can be detected. Probes that fail to hybridize do not give fluorescence signal in solution, eliminating the need for purification or separation steps, which makes automation of the RCA assay easier.

DNA Microarray Genotyping. A high-density oligonucleotide microarray offers simultaneous analysis of many polymorphisms. The DNA sample of interest is PCR amplified to incorporate fluorescently labeled nucleotides and then hybridized to hundreds of thousands of oligonucleotides attached to a solid silicon surface in an ordered array. Each oligonucleotide in the high-density array acts as an allele-specific probe. Perfectly matched sequences hybridize more efficiently to their corresponding oligomers on the array and, therefore, give stronger fluorescent signals over mismatched probe-target combinations [71, 72]. The hybridization signals are quantified by high-resolution fluorescent scanning and analyzed by computer software.

Sequencing by hybridization can be an efficient method to monitor many SNPs. It is possible to array a set of short oligonucleotides covering the entire DNA fragment on a DNA chip. Because the precise sequence of the oligonucleotide at each location on the chip is known, the pattern of hybridization can be determined using fluorescently labeled DNA probes. The advantage of this method is that a large DNA fragment or a large collection of small PCR products can be scanned in one hybridization. More than 100 such arrays have been used for high-throughput screening of SNPs covering 2.3 Mb of genomic DNA [73] and thousands of SNPs have been screened rapidly by use of chip-based resequencing [74].

As mentioned, almost all currently available SNP genotyping starts with a locus-specific amplification step, followed by an allele discrimination step [59, 73, 75]. At capacities of 1000 SNPs or less, locus-specific amplification is economically feasible in terms of oligonucleotide synthesis and other reagent costs. At capacities of 10,000 SNPs and greater, the costs of designing, synthesizing, and managing such an enormous number of oligonucleotides become prohibitive. Additionally, large amounts of starting sample DNA are required to genotype tens of thousands of SNPs in a locus-specific manner. Thus, considering the high complexity of human genome, the most common strategy is now to reduce the complexity by amplifying portions of the genomic DNA by PCR. This can be done in a random or semi-random fashion, for example, by using restriction enzyme-based adapter ligation PCR [76]. This representational approach, which is used to detect changes in genomic copy number, has also been applied to genotyping of 10^4–10^5 or more SNPs from a single sample preparation [77, 78]. Genomic complexity reduction assay involves five primary steps, starting with restriction digestion, ligation of adaptor, amplification, fragmentation, and labeling, before hybridization to the oligonucleotide array [78]. Highly multiplexed PCR-based approaches have recently been developed to genotype specific loci of interest, enabling targeted genotyping of 10^3–10^4 or more SNPs from a single sample [79]. These targeted genotyping approaches include a pre-PCR enzymatic SNP scoring step, the addition of universal priming sites by ligation, and subsequent universal primer PCR. Today, almost all high-throughput genotyping is done using one of these methods [80].

7.5.5 CONCLUSION

The automated oligonucleotide synthesizers now enable large-quantity, low-cost synthesis of precisely designed different sequences of oligonucleotide based in a parallel fashion. On the other hand, depending on the specificity of sequences for precise base pairing to each other, oligonucleotides have the enormous potential to facilitate manufacturing of innovative numerous molecular machinery with the help of molecular biology enzymes, for being used as genomics tools as well as therapeutic tools. It should be emphasized that the new high-throughput approaches, such as microarrays, do not replace the conventional standard methods. The standard methods, such as northern blots or RT–PCR, are simply used in a more targeted fashion to complement the broader measurements and to follow up on the genes, pathways, and mechanisms implicated by array results.

REFERENCES

1. Lander E S, et al. (2001). Initial sequencing and analysis of the human genome. *Nature.* 409:860–921.
2. Venter J C, et al. (2001). The sequence of the human genome. *Science.* 291:1304–1351.
3. Spellman P T, et al. (1998). Comprehensive identification of cell cycle-regulated genes of the yeast *Saccharomyces cerevisiae* by microarray hybridization. *Mol. Biol. Cell.* 9:3273–3297.
4. Lockhart D J, Winzeler E A (2000). Genomics, gene expression and DNA arrays. *Nature.* 405:827–836.
5. Freeman W M, Walker S J, Vrana K E (1999). Quantitative RT-PCR: Pitfalls and potential. *Biotechniques.* 26:112–122, 124–125.
6. Raeymaekers L (2000). Basic principles of quantitative PCR. *Mol. Biotechnol.* 15:115–122.
7. Lee L G, Connell C R, Bloch W (1993). Allelic discrimination by nick-translation PCR with fluorogenic probes. *Nucleic Acids Res.* 21:3761–3766.
8. Livak K J, Flood S J, Marmaro J, et al. (1995). Oligonucleotides with fluorescent dyes at opposite ends provide a quenched probe system useful for detecting PCR product and nucleic acid hybridization. *PCR Methods Appl.* 4:357–362.
9. Heid C A, Stevens J, Livak K J, et al. (1996). Real time quantitative PCR. *Genome Res.* 6:986–994.
10. Tan W, Wang K, Drake T J (2004). Molecular beacons. *Curr. Opin. Chem. Biol.* 8:547–553.
11. Vet J A M, Marras S A E (2004). Design and optimization of molecular beacon real-time polymerase chain reaction assays. *Oligonucleotide Syn. Meth. Appl.* 288: 273–290.
12. Solinas A, et al. (2001). Duplex Scorpion primers in SNP analysis and FRET applications. *Nucleic Acids Res.* 29:e96.
13. Chen X, Zehnbauer B, Gnirke A, et al. (1997). Fluorescence energy transfer detection as a homogeneous DNA diagnostic method. *Proc. Natl. Acad. Sci. U.S.A.* 94:10756–10761.
14. Holland P M, Abramson R D, Watson R, et al. (1991). Detection of specific polymerase chain reaction product by utilizing the 5′–3′ exonuclease activity of Thermus aquaticus DNA polymerase. *Proc. Natl. Acad. Sci. U.S.A.* 88:7276–7280.
15. Velculescu V E, Zhang L, Vogelstein B, et al. (1995). Serial analysis of gene expression. *Science.* 270:484–487.
16. Tuteja R, Tuteja N (2004). Serial analysis of gene expression: Applications in Human studies. *J. Biomed. Biotechnol.* 2:113–120.

17. Heller R A, et al. (1997). Discovery and analysis of inflammatory disease-related genes using cDNA microarrays. *Proc. Natl. Acad. Sci. U. S. A*. 94:2150–2155.

18. Lennon G L (2000). High-throughput gene expression analysis for drug discovery. *Drug Discov. Today*. 5:59–66.

19. Lockhart D J, et al. (1996). Expression monitoring by hybridization to high-density oligonucleotide arrays. *Nat. Biotechnol*. 14:1675–1680.

20. Wodicka L, Dong M, Mittmann M H, et al. (1997). Genome-wide expression monitoring in Saccharomyces cerevisiae. *Nat. Biotechnol*. 15:1359–1366.

21. Bowtell D D (1999). Options available–from start to finish–for obtaining expression data by microarray. *Nat. Genet*. 21(suppl 1):25–32.

22. Jain K K (2000). Applications of biochip and microarray systems in pharmacogenomics. *Pharmacogenom*. 1:289–307.

23. Okamoto T, Suzuki T, Yamamoto N (2000). Microarray fabrication with covalent attachment of DNA using bubble jet technology. *Nat. Biotechnol*. 18:384–385.

24. Butte A (2002). The use and analysis of microarray data. *Nat. Rev. Drug. Discov*. 1:951–960.

25. Dagle J M, Weeks D L (2001). Oligonucleotide-based strategies to reduce gene expression. *Differentiation*. 69:75–82.

26. Kurreck J (2003). Antisense technologies. Improvement through novel chemical modifications. *Eur. J. Biochem*. 270:1628–1644.

27. Cho-Chung Y S, et al. (1997). Antisense-protein kinase A: A single-gene-based therapeutic approach. Antisense *Nucleic Acid Drug Dev*. 7:217–223.

28. Srivastava R K, et al. (1999). Growth arrest and induction of apoptosis in breast cancer cells by antisense depletion of protein kinase A-R1 alpha subunit: p53-independent mechanism of action. *Mol. Cell. Biochem*. 195:25–36.

29. Wang H, et al. (1999). Antitumor activity and pharmacokinetics of a mixed-backbone antisense oligonucleotide targeted to the R1alpha subunit of protein kinase A after oral administration. *Proc. Natl. Acad. Sci. U.S.A*. 96:13989–13994.

30. Elbashir S M, et al. (2001). Duplexes of 21-nucleotide RNAs mediate RNA interference in cultured mammalian cells. *Nature*. 411:494–498.

31. Fire A, et al. (1998). Potent and specific genetic interference by double-stranded RNA in Caenorhabditis elegans. *Nature*. 391:806–811.

32. Hamilton A J, Baulcombe D C (1999). A species of small antisense RNA in posttranscriptional gene silencing in plants. *Science*. 286:950–952.

33. Miyagishi M, et al. (2003). Comparison of the suppressive effects of antisense oligonucleotides and siRNAs directed against the same targets in mammalian cells. Antisense *Nucleic Acid Drug Dev*. 13:1–7.

34. Stevenson M (2004). Therapeutic potential of RNA interference. *N. Engl. J. Med*. 351:1772–1777.

35. Agrawal S, Zhao Q (1998). Mixed backbone oligonucleotides: Improvement in oligonucleitide-induced toxicity in vivo. Antisense *Nucleic Acid Drug Dev*. 8:135–139.

36. Shen L X, et al. (1998). Impact of mixed-backbone oligonucleotides on target binding affinity and target cleaving speficity and selectivity by *Escherichia coli* RNase H. *Bioorg. Med. Chem*. 6:1695–1705.

37. Goel S, et al. (2003). A safety study of a mixed-backbone oligonucleotide (GEM231) targeting the type 1 regulatory subunit alpha of protein kinase A using a continuous schedule in patients with refractory solid tumors. *Clin. Cancer Res*. 9:4069–4076.

38. Summerton J (1999). Morpholino antisense oligomers: the case for an RNase H-dependent structure type. *Biochim. Biophys. Acta*. 1489:141–154.

39. Nielsen P E, Egholm M (1999). An introduction to peptide nucleic acid. *Curr. Issues Mol. Biol.* 1:89–104.

40. Braasch D A, Corey D R (2002). Novel antisense and peptide nucleic acid strategies for controlling gene expression. *Biochem.* 41:4503–4510.

41. Egholm M, et al. (1993). PNA hybridizes to complementary oligonucleotides obeying the Watson-Crick hydrogen-bonding rules. *Nature.* 365:566–568.

42. Kurreck J (2003). Antisense technologies. Improvement through novel chemical modifications. *Eur. J. Biochem.* 270:1628–1644.

43. Braasch D A, Corey D R (2001). Locked nucleic acid (LNA): Fine-tuning the recognition of DNA and RNA. *Chem. Biol.* 8:1–7.

44. Fluiter K, et al. (2003). *In vivo* tumor growth inhibition and biodistribution studies of locked nucleic acid (LNA) antisense oligonucleotides. *Nucleic Acids Res.* 31:953–962.

45. Zeng Y, Cullen B R (2002). RNA interference in human cells is restricted to the cytoplasm. *RNA.* 8:855–860.

46. Persengiev S P, Zhu X, Green M R (2004). Nonspecific, concentration-dependent stimulation and repression of mammalian gene expression by small interfering RNAs (siRNAs). *RNA.* 10:12–18.

47. Sledz C A, Holko M, de Veer M J, et al. (2003). Activation of the interferon system by short-interfering RNAs. *Nature Cell Biol.* 5:834–839.

48. Bridge A J, Pebernard S, Ducraux A, et al. (2003). Induction of an interferon response by RNAi vectors in mammalian cells. *Nature Genet.* 34:263–264.

49. Dorsett Y, Tuschl T (2004). siRNAs: Application in functional genomics and potential as therapeutics. *Nature Rev. Drug Disc.* 3:318–329.

50. Hornung V, et al. (2005). Sequence-specific potent induction of IFN-alpha by short interfering RNA in plasmacytoid dendritic cells through TLR7. *Nat. Med.* 11:263–270.

51. Marques J, Williams R G (2005). Activation of the mammalian immune system by siRNAs. *Nat. Biotechnol.* 23:1399–1405.

52. Judge A D, et al. Sequence-dependent stimulation of the mammalian innate immune response by synthetic siRNA. *Nat. Biotechnol.* 23:457–462.

53. Collins F S, Guyer M S, Chakravarti A (1997). Variations on a theme: Cataloging human DNA sequence variation. *Science.* 278:1580–1581.

54. Wang D G, et al. (1998). Large-scale identification, mapping and genotyping of single-nucleotide polymorphisms in the human genome. *Science.* 280:1077–1082.

55. Collins F S, Patrinos A, Jordan E, et al. (1998). New goals for the U.S. Human Genome Project 1998–2003. *Science.* 282:682–689.

56. McCarthy J J, Hilfiker R (2000). The use of single-nucleotide polymorphism maps in pharmacogenomics. *Nat. Biotechnol.* 18:505–508.

57. Marshall E (2000). Drug firms to create public database of genetic mutations. *Genomics.* 284:406–407.

58. Kleyn P W, Vesell E S (1998). Genetic variation as a guide to drug development. *Science.* 281:1820–1821.

59. Johnson J A (2003). Pharmacogenetics: Potential for individualized drug therapy through genetics. *Trends Gene.* 19:660–666.

60. Shi M M (2001). Enabling large-scale pharmacogenic studies by high-throughput mutation detection and genotyping technologies. *Clin. Chem.* 47:164–172.

61. Shi M M, Bleavins M R, de la Iglesia F A (1999). Technologies for detecting genetic polymorphisms in pharmacogenomics. *Mol. Diagn.* 4:343–351.

62. Baron H, Fung S, Aydin A, et al. (1996). Oligonucleotide ligation assay (OLA) for diagnosis of familial hypercholesterolemia. *Nat. Biotechnol.* 14:1279–1282.

63. Tyagi S, Bratu D P, Kramer F R (1998). Multicolor molecular beacons for allele discrimination. *Nat. Biotechnol.* 16:49–53.

64. Lyamichev V, et al. (1999). Polymorphism identification and quantitative detection of genomic DNA by invasive cleavage of oligonucleotide probes. *Nat. Biotechnol.* 17:292–296.

65. Kaiser M W, et al. (1999). A comparison of eubacterial and archaeal structure-specific 5'-exonucleases. *J. Biol. Chem.* 274:21387–21394.

66. Kwiatkowski R W, Lyamichev V, de Arruda M, et al. (1999). Clinical, genetic and pharmacogenic applications of the Invader assay. *Mol. Diagn.* 4:353–364.

67. Chen X, Livak K J, Kwok P-Y (1998). A homogeneous, ligase-mediated DNA diagnostic test. *Genome Res.* 8:549–556.

68. Lizardi P, Huang X, Zhu Z, et al. (1998). Mutation detection and single-molecule counting using isothermal rolling-circle amplification. *Nat. Genet.* 19:225–232.

69. Baner J, Nilsson M, Mendel-Hartvig M, et al. (1998). Signal amplification of padlock probes by rolling circle replication. *Nucleic Acids Res.* 26:5073–5078.

70. Faruqi F A, et al. (2001). High-throughput genotyping of single nucleotide polymorphisms with rolling circle amplification. *BMC Genom.* 2:1–10.

71. Chee M, et al. (1996). Accessing genetic information with high-density DNA arrays. *Science.* 274:610–614.

72. Lipshutz R J, Fodor S P A, Gingeras T R, et al. (1999). High density synthetic oligonucleotide arrays. *Nat. Genet.* 21(suppl 1):20–24.

73. Wang D G, et al. (1998). Large-scale identification, mapping and genotyping of single nucleotide polymorphisms in the human genome. *Science.* 280:1077–1082.

74. Hacia J G, et al. (1999). Determination of ancestral alleles for human single-nucleotide polymorphisms using high-density oligonucleotide arrays. *Nat. Genet.* 22:164–167.

75. Kwok P Y (2001). Methods for genotyping single nucleotide polymorphisms. *Annu. Tev. Genomics Hum. Genet.* 2:235–258.

76. Lucito R, et al. (1998). Genetic analysis using genomic representations. *Proc. Natl. Acad. Sci. U.S.A.* 95:4487–4492.

77. Kennedy G C, et al. (2003). Large-scale genotyping of complex DNA. *Nat. Biotechnol.* 21:1233–1237.

78. Matsuzaki H, et al. (2004). Parallel genotyping over 10,000 SNPs using a one-primer assay on a high-density oligonucleotide array. *Genome Res.* 14:414–425.

79. Hardenbol P, et al. (2003). Multiplexed genotyping with sequence tagged molecular inversion probes. *Nat. Biotechnol.* 21:673–678.

80. Gunderson K L, Steemers F J, Lee G, et al. (2005). A genome-wide scalable SNP genotyping assay using microarray technology. *Nat. Genetics.* 37:549–554.

7.6

RNA INTERFERENCE: THE NEXT GENE-TARGETED MEDICINE

ANDREW V. OLEINIKOV[1] AND MATTHEW D. GRAY[2]

[1] Seattle Biomedical Research Group, Seattle, Washington
[2] MDG Associates, Inc., Seattle, Washington

Chapter Contents

7.6.1	Introduction	1110
7.6.2	Basic Principles and Molecular Mechanisms of RNAi	1111
7.6.3	RNAi as a Tool for Identifying and Validating Drug Targets	1114
	7.6.3.1 Low-Throughput Screening for Drug Targets	1116
	7.6.3.2 High-Throughput Screening for Drug Targets	1117
7.6.4	RNAi as an Instrument for Therapeutic Use	1119
	7.6.4.1 Cancer	1122
	7.6.4.2 AIDS and Other Viral Infections	1123
	7.6.4.3 Parasite Infections	1124
	7.6.4.4 Neurological Disorders	1125
7.6.5	Basic Theoretical and Practical Concerns for Application of RNAi in Therapy	1125
	7.6.5.1 Variable Efficiency of Different siRNAs	1126
	7.6.5.2 Nonspecific Systemic and Off-Target Effects, and the Potential for Pleiotropic Effects	1126
7.6.6	Technical Challenges of siRNA Therapy: *In vivo* Delivery and Stability	1128
	7.6.6.1 Delivery of RNAi Compounds	1129
	7.6.6.2 Stability of RNAi Compounds	1130
7.6.7	RNAi Drugs under Development	1130
	7.6.7.1 Cancer	1131
	7.6.7.2 Viral and Respiratory Diseases	1132
	7.6.7.3 Arthritis	1133
	7.6.7.4 Autoimmune Diseases	1133
	7.6.7.5 Neurodegenerative Disorders	1133
	7.6.7.6 Ocular Disorders	1134

Handbook of Pharmaceutical Biotechnology, Edited by Shayne Cox Gad.
Copyright © 2007 John Wiley & Sons, Inc.

7.6.8   Comparison of siRNAs with other Pharmaceuticals (Targeted and
Nontargeted Type), Including Closely Related Gene Specific
Antisense-Based Drugs   1134
7.6.9   Biotech and Drug Companies Currently Invested in RNAi Drug
Discovery and Development   1136
7.6.10   Concluding Remarks   1137
         Acknowledgment   1137
         References   1138

7.6.1   INTRODUCTION

Advances in medical technology have depended on cellular and molecular research to gain an in-depth understanding of the basic mechanisms of homeostasis. Medical investigators have used their knowledge of cellular processes to search for small chemical compounds that can alter specific biochemical reactions in cells and tissues. Many of these compounds function in a general fashion through an alteration in biochemical pathways or via stimulation of the immune system. Drugs that can modulate gene expression in a targeted fashion, however, offer the potential of a much higher level of specificity and control over cellular processes. The newest laboratory method for modulating gene expression, RNA interference (RNAi), has gained rapid interest throughout the world of molecular research, in general, and is being actively pursued as the next "gene-targeted medicine."

RNAi is a process in which double-stranded RNA (dsRNA) directs the specific degradation of a corresponding target mRNA. The mediators of this process are small dsRNAs, of 21 bp in length, called short interfering RNAs (siRNAs). The phenomenon of RNAi was first reported in 1998, by Fire et al. [1], who observed that dsRNA could inhibit gene expression in *Caenorhabditis elegans*. Although the underlying mechanism of this effect was poorly understood at the time, the generality of the phenomenon and its utility in the laboratory as a method of posttranscriptional gene silencing were quickly confirmed by the work of several groups [2–8]. The process whereby dsRNA is cleaved into siRNAs was initially described in plants [9] and then in *Drosophila* [10, 11]. The cellular protein factors associated with this conversion in the cell were identified soon afterward ([12] and recently reviewed in Refs. 13 and 14).

The rapid growth and popularity of RNAi as a research topic of interest and as a useful tool in the laboratory is evidenced by the tremendous yearly increase in publications on the subject since its initial description less than a decade ago. A current search of the scientific literature, using the terms "RNAi" and "siRNA" as keywords, reveals that over 10,000 articles have been published on the subject within the last 5 years, since the beginning of this century (Figure 7.6-1). This type of response to a technical discovery is wholly unprecedented and could only perhaps be reasonably compared with the "explosion" of significant and widespread advances that the invention of the polymerase chain reaction (PCR) [15–17] contributed to biological research and current biotechnology toward the end of the last century.

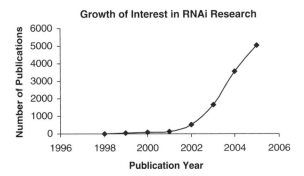

Figure 7.6-1. Growth of interest in RNAi research.

Our aim in this chapter is not to give an extensive review of the rapidly expanding literature in this field but, rather, to describe the main ideas, approaches, and technological directions that may guide the medical research-er interested in using RNAi methodologies to identify and/or develop new drug candidates. We have included several examples and discussions of the concepts involved to provide a clear picture of the potential advantages and limitations of specific RNAi-based gene-targeted medicine.

7.6.2   BASIC PRINCIPLES AND MOLECULAR MECHANISMS OF RNAi

For many years scientists involved in biomedical research have searched for simple methods for targeting the regulation of a gene's activity. These methods would be used for characterizing gene functions as well as for developing targeted medicines to treat and correct a variety of diseases and disorders. In 1990, when Napoli et al. reported on the specific repression of the chalcone synthase (CS) gene in petunia flowers by introduction of a chimeric CS gene and demonstrated that it was the result of a reduction in the level of CS mRNA [18], it would have been almost impossible to suggest that in just a few years time there would be a revolution in our ability to specifically regulate the expression of virtually any gene in any cell and in many organisms.

Work starting from that of Andrew Fire et al. [1] describing the phenomenon of RNAi in *C. elegans* in 1998, up until now, has shown that RNAi is a conserved and universal mechanism of defense [19] (against viruses and mobile genetic elements, retrotransposons) and gene regulation that is widely spread among different biological kingdoms [20]. For the efficient silencing of any target gene, a corresponding homologous dsRNA can be delivered into the cells and tissues of organisms by several different methods: direct injection, feeding, soaking, electroporation, transfection, and so on. RNAi occurs by two parallel and, in many respects, similar pathways involving many common cellular proteins: the so-called short interfering RNA (siRNA) and microRNA (miRNA) pathways (Figures 7.6-2 and 7.6-3). Both pathways lead to the specific silencing of genes, either through degradation or a translational block of the corresponding mRNA.

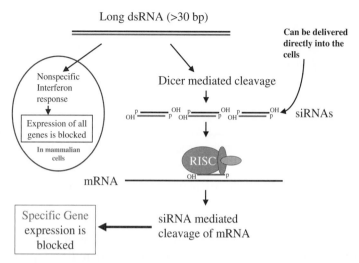

Figure 7.6-2. Mechanism for inhibition of gene expression via the siRNA branch of the RNAi pathway. Short interfering RNAs can be delivered directly into the target cells to avoid nonspecific interferon (IFN) response.

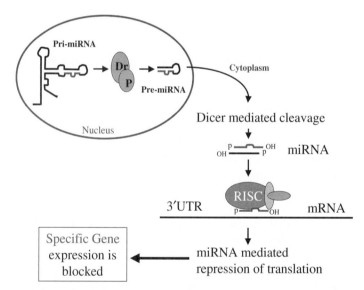

Figure 7.6-3. Mechanism for inhibition of gene expression via the miRNA branch of the RNAi Pathway. Proteins involved in Microprocessor: Dr—Drosha, P—Pasha, UTR—untranslated region of mRNA.

The first step in the siRNA pathway is degradation of the long dsRNA into short double-stranded fragments, later termed siRNAs (Figure 7.6-2). siRNAs, which have two overhanging 3′-nucleotides and 5′-phosphate groups [9–11, 21], are converted from longer dsRNAs by a cellular enzyme with RNase III activity called dicer ([22] and reviewed in Ref. 23). At the next step, siRNAs bind to the cellular complex called the RNA-induced silencing complex (RISC) [22, 24], of which the main component is a protein called argonaute [19, 24]. Only one strand of the siRNA molecule specifically interacts with, and hence "enters," RISC [25, 26] and determines the specificity of target RNA cleavage.

With the success of RNAi-dependent targeted inhibition of gene expression in different lower organisms, immediate efforts were made to apply this technology in mammalian cells and organisms by the intracellular delivery of long dsRNAs cognate to the gene assigned to downregulation. However, in mammalian cells, there is a well-known nonspecific antiviral defense mechanism based on activation of the interferon (IFN) pathway in response to intracellular invasion by viral dsRNA [27]. In this pathway, several IFN-induced genes (including the RNA-dependent protein kinase (PKR), the 2′,5′-oligoadenylate synthetase (OAS) and RNase L, and the Mx protein GTPase) orchestrate a complex cascade of events leading to the degradation of the foreign dsRNA, a general arrest of the translational machinery of the host cell, and other downstream effects important for host cell defenses over viral infections. Moreover, PKR can be differentially modulated by different species of dsRNA. This effect may depend on the sequence and secondary structure of a dsRNA, and PKR can be stimulated with as few as 11 bp of dsRNA [28]. In any case, this complex mixture of different nonspecific responses can completely abolish the sequence-specific gene-targeted effects of dsRNAs.

Nonetheless, an understanding of the molecular mechanisms of RNAi proceeding through the endogenous generation of siRNAs [10, 21] allowed Elbashir et al. [29] to design a smart way to avoid most of these nonspecific IFN-related effects. They designed and delivered into the cells gene-specific dsRNA as short dsRNAs (siRNAs, 21–22 bp), as they occur in their mature form during the natural RNAi mechanism. Using this modification, most of the IFN response, which is induced by longer dsRNAs (>30 bp), is avoided. Their experiments demonstrated the use of siRNA compounds as efficient and specific downregulators of gene expression. This breakthrough began a new era for the use of RNAi technology in biomedical research for pathway profiling, drug candidate identification, and targeted therapy using mammalian cells and organisms. Nevertheless, some types of nonspecific effects still present a concern for the use of RNAi technology for therapeutic applications, as will be discussed later on in this chapter.

A detailed understanding of the molecular mechanism underlying RISC activity has led to improvements in the design of siRNA compounds. The current principles for siRNA design are (1) the strand with the more loosely base-paired 5′ end in the siRNA duplex enters RISC, (2) the 5′ end of this strand has to be phosphorylated, and (3) the siRNA duplex needs to have 3′ overhanging ends of 2-nucleotides in length. Despite this knowledge, only about 20% of siRNAs designed are active in RNAi, and hence, all require experimental validation of their activity. Furthermore, the efficiency of repression of the target gene by siRNA may not be absolute, as the degree of inhibition observed depends on several factors. These factors include the chemical stability and specific activity of the siRNA compound; the "effective" concentration delivered to the cells, which depends on the number and duration of treatments; the physiological state of the target cell; the level of target mRNA; and, especially, the level and turnover rate of the target protein in the cell or tissue. A number of these factors will be discussed in detail below.

With intense research efforts in the field of RNAi, it became clear that another cellular pathway of gene regulation, discovered earlier in *C. elegans* and based on the transcription of small untranslated RNAs [30, 31], is closely related to siRNA-dependent gene regulation [32]. This second branch of RNAi, which has also remained extremely conserved during evolution, was termed the microRNA, or miRNA, pathway (Figure 7.6-3). MicroRNAs have been identified across many different biological kingdoms (from plants to nematodes to humans), and they function to regulate the expression of different genes. Both pathways, siRNA and miRNA, share common elements of their core machinery [33], such as dicer and RISC; however, there are some obvious differences and specialized components. The miRNA process begins in the nucleus, where long (up to several thousand bp [34]) capped pri-miRNA molecules are transcribed by RNA-polymerase II from corresponding miRNA genes. These complex RNA molecules are processed into hairpin-like pre-miRNA molecules (about 65–70 bp) by a multiprotein complex, called microprocessor, containing at least an RNaseIII enzyme Drosha and its partner Pasha [35]. Pre-miRNA molecules are then transported to the cytoplasm where they are converted to miRNAs by dicer (Figure 7.6-3) [23, 32]. The main difference between the two pathways is that miRNA molecules are not 100% homologous with the target mRNA and may contain up to several mismatches. In addition, miRNA-dependent inhibition of expression, which most likely takes place in specialized subcellular compartments called P-bodies [36], occurs through the interaction of RISC-bound miRNAs with the 3′-untranslated region of the target mRNA and results in an arrest of translation rather than degradation of the message, as in the siRNA pathway.

Currently, it is not clearly understood what targets a short dsRNA down the miRNA pathway. It is possible that the presence of some nonhomologous regions between an siRNA molecule and some unintended target RNA may direct this complex into the miRNA pathway and produce some off-target effects. In any case, a more detailed understanding of the miRNA pathway will clearly provide additional insights for the further development of powerful RNAi-based tools for gene regulation and therapy, and for the evasion of nonspecific and off-target effects.

Since its discovery less than a decade ago, the RNAi field has progressed at an incredible pace, permitting manipulations of gene expression that were previously impossible. As a result, this technology immediately acquired an important role in pharmaceutical science for the screening, identification, and validation of drug targets, a process that, before the discovery of RNAi, usually expended several years of intense effort. Moreover, the specific features of RNAi technology, such as high target specificity, efficiency, and to some extent, the ease of compound design, have implicated this technology as a potential instrument for therapeutic use. In the following sections we discuss a variety of biological and technical aspects of applying RNAi technologies to the discovery and development of customized, state-of-the-art, gene-targeted medicines of the future.

7.6.3 RNAi AS A TOOL FOR IDENTIFYING AND VALIDATING DRUG TARGETS

The traditional drug discovery process implemented by many pharmaceutical companies includes drug target identification, characterization of the target, high-throughput screening (HTS) of small-molecule libraries, and medicinal chemistry optimization. The bottleneck of this scheme has traditionally been the identification of targets due to the absence of appropriate high-throughput screening approaches. The recent discovery and development of RNAi technology, however, has provided a new approach that will accelerate the rate and efficiency of the initial and overall screening process.

With the emergence of siRNA technology, its first useful application was obvious: to switch off (or downregulate) a certain gene of interest to determine what phenotypic changes might occur within the cell (or organism). These phenotypic changes can be monitored at different levels by measuring (1) the relative expression of selected genes of interest or the global changes in expression of all (or most) genes within the cell (organism), and (2) the inhibition or stimulation of other physiological parameters of the cell or organism, such as its growth rate, shape, biochemical markers, and specific functions. If downregulation/switching off a particular gene leads to the desirable phenotypic effect, this gene or gene product is then selected as a target for the development of a drug, which will have a similar effect on the inhibition of activity of the corresponding protein. By screening multiple genes (one by one, or in combinations) for downregulation of their expression and estimating the presence and magnitude of the desirable phenotypic outcome, novel genes can be identified as targets for the future development of therapeutic compounds. In this respect, systematic genome-wide screens using RNAi technology and libraries of siRNA compounds seem to offer the most powerful approach for the identification of multiple drug target genes [37].

At the same time, RNAi technology may also make it possible to identify any undesirable effects due to the downregulation of certain genes of interest, which could have detrimental effects upon the pathway, cell, or organism. Therefore, this approach may provide an effective means for the early elimination of such genes from the list of drug target candidates; which is likely to save time and money in the subsequent drug development process. Below, we will review the methods currently being employed using siRNA technology for drug target identification and validation, as well as the specific target genes and respective diseases that are currently under investigation using this approach.

The most common way of exploiting the siRNA approach for target validation is to perform experiments in cultured cells. There are several methods of introducing siRNAs into mammalian cells. Chemically synthesized siRNAs or siRNAs prepared by *in vitro* transcription can be delivered into cells through a process similar to cell transfection with DNA vectors, using several different agents to facilitate such delivery [38]. Recently, several companies have developed transfection reagents specifically designed to improve the delivery of siRNA oligonucleotides in cell culture (Ambion, Austin, TX; Invitrogen, Carlsbad, CA; QIAGEN, The Netherlands).

The second popular method for induction of RNAi is to transfect cells with vectors containing siRNA expression constructs. This is similar to the expression of exogenous genes using vector-mediated transfection except that the result of this process is not the translation of the transcribed mRNA but the assembly of a double-stranded siRNA or hairpin loop siRNA, called shRNA, within the cell. Both forms can efficiently inhibit the expression of target genes [39–43]. A major advantage of "DNA-directed" RNAi is its potential for a more sustained effect, because the siRNA-expression vector may persist in the cells for several generations providing a constant supply of siRNA molecules. Also, the recombinant expression of the siRNA can be controlled by specifically selected regulatory elements incorporated into the vector, providing more precise control over the timing, dose, and cell specificity of target gene inhibition. It is also possible to introduce siRNA vectors into cells using DNA electroporation [44]. This method is not as convenient as transfection when working with cell cultures *in vitro*, but it may be especially well suited for the localized delivery of siRNAs when working *in vivo*.

7.6.3.1   Low-Throughput Screening for Drug Targets

The first direct approach to identifying plausible drug targets would be the application of siRNA technology to well-characterized biochemical pathways involved in some disease of interest. When proteins involved in a particular pathway are known as well as their functions, it is straightforward to select a few target mRNAs against which siRNAs can be prepared and optimized for their effect. Also, with fewer initial targets, the outcome of RNAi screening can be studied in greater detail. Below, we describe several examples where siRNA technology has been used to identify critical checkpoints in a known pathway and to test whether desirable phenotypic effects could be obtained.

Cancer. Grillo et al. [45] used siRNA to validate the cyclin D1 and CDK4 protein complex as a drug target for cancer drug discovery. The inhibition of endogenous cyclin D1 or CDK4 expression by RNAi resulted in hypophosphorylation of the retinoblastoma gene product and accumulation of MCF-7 breast cancer cells in G1. These results support the prevailing view that pharmacological inhibition of cyclin D1/CDK4 complexes is an effective strategy for inhibiting the growth of tumors.

Another recent example demonstrates that inhibition of an enzyme in the same pathway in cells of different species may have different, and even opposite, effects on phenotypic outcome. This emphasizes that careful selection of model cells and characterization of molecular detail of the biochemical pathway, including tissue- or cell-specific expression of different proteins, is critical to selecting a suitable drug target. Hill et al. [46] demonstrated that rat (NMU and 13762), but not human (MCF-7 and T47D), breast cancer cell lines express Heme oxygenase-1 (HO-1), whereas indoleamine 2,3 dioxygenase (IDO) is expressed by both. IDO inhibition with siRNA resulted in diminished proliferation of rat cells, whereas HO-1-negative human cell lines increased proliferation upon IDO inhibition. As IDO inhibits the antiproliferative effects of HO-1, these authors concluded that IDO has the opposite effect on proliferation, dependent upon the co-expression of HO-1. This example clearly shows that results obtained on cells from other organisms are not automatically transferable to human cells, even if they seem to have highly homologous proteins and very conserved pathways. It also shows that a drug specifically targeting the expression or activity of a particular gene might have the opposite effect in cell types that differ in their expression profiles. These potential problems need to be considered when screening for therapies directed at a particular target protein.

Respiratory Diseases. An interesting example of hypothesis-driven identification of a potential drug target *in vivo* using siRNA is described in the work of

Lomas-Neira et al. [47]. In their study of acute lung injury (ALI), they hypothesized that local silencing of keratinocyte-derived chemokine (KC) or macrophage-inflammatory protein-2 (MIP-2), via the local administration of siRNA against KC or MIP-2, after traumatic shock/hemorrhage would suppress signaling for the influx of polymorphonuclear neutrophils (PMNs) to the lung, thereby reducing ALI associated with a secondary septic challenge (cecal ligation and puncture). This hypothesis was tested using intratracheal instillation in an *in vivo* model (mice). They found that 24h after this treatment intratracheal MIP-2 siRNA significantly reduced tissue and plasma interleukin-6, tissue MIP-2, as well as neutrophil influx. In contrast, KC siRNA treatment reduced plasma KC, tissue KC, and IL-6 but produced no significant reduction in plasma IL-6 or neutrophil influx, thus identifying MIP-2, but not KC, as a potent drug target. In addition, this study demonstrates the potential use of siRNA as a therapeutic.

Inflammation. The exact roles of many cytokine genes that influence rheumatoid arthritis are still unknown. Inhibition of some of these genes can strongly reduce joint inflammation [48]. In an *in vivo* arthritic mouse model, Schiffelers et al. [49] used electroporation [44] to deliver a tumor-necrosis factor alpha (TNFα) siRNA-coding vector into joint tissue. They demonstrated not only that TNFα is an effective target for reducing joint inflammation, but also a new strategy for therapeutic intervention for rheumatoid arthritis. This approach, hence, may clearly serve as a model for the study of arthritis disease pathways through gene-specific loss-of-function phenotypes.

Parkinson's Disease. Animal disease models play an extremely important role in identification of drug lead candidates. Some animal models can be created simply by making gene knockouts. However, if the targeted gene is an essential gene for survival of the organism, these animals cannot be easily created. For example, mice in which the gene encoding tyrosine hydroxylase (TH) has been knocked out die shortly after birth, making it impossible to study the neurological effects of this alteration in the fully developed animal. RNAi technology can help to overcome this obstacle by creating an animal disease model through the local and/or temporal knocking out of essential genes. Hommel et al. [50] demonstrated that injection of an adeno-associated virus encoding a TH-specific shRNA into the midbrain of 9-week-old mice generated an attenuated locomotor response to amphetamine, an effect that is also characteristic for toxin-induced rat models of Parkinson's disease. These authors, hence, suggest that similar knockout strategies may be effectively used in other disease model systems. In addition, as we mentioned above, because the RNAi effect obtained by treatment with siRNA compounds may not be absolute, there may also be a certain advantage in a knock-down vs. knockout phenotype; in that, a knock down might be more appropriate for modeling some complex human diseases [51].

7.6.3.2 High-Throughput Screening for Drug Targets

The second valuable approach to using RNAi for drug target identification is a high-throughput strategy (HTS), where a library of siRNAs directed against many genes is screened simultaneously or sequentially to identify target genes and their products for further evaluation as drug leads. As described above, this approach might be used against a restricted set of genes that are members of a certain pathway, for example, a "pathway approach," or against most of the genes in the genome, a "global approach." For these approaches, traditional HTS equipment, such as robotic systems, can be implemented, facilitating the processing of thousands of cell-based screening protocols a day [52]. Both chemically synthesized siRNAs as well as vectors coding for functional si/shRNAs can be delivered into the cells using this high-throughput format. Retroviral [53] and lentiviral [54] vectors have been shown to be particularly useful vehicles for high-throughput screening of siRNA and shRNA libraries due to their robust infection and expression in a wide range of mammalian cell types. To simplify the transfection procedure on multiwell plates or slides, Chang et al. [55] suggested the pretreatment of surfaces with cationic polymers. This provides for the simple addition of naked DNA vectors or siRNAs directly to the cells in these pretreated wells in order to achieve efficient transfection. Another method of high-throughput screening using vector-encoded or chemically synthesized siRNAs is the so-called "reverse-transfection," or solid phase, method [56–58], originally designed for the expression of cDNA libraries [59]. In this method, a number (up to several thousands) of individual siRNAs or siRNA-encoding vectors are spotted in a predetermined array onto a microscope glass slide, which is then treated with transfection reagent and covered with recipient cells. Individual cells seeded directly over each spot of material take up only the siRNAs or vectors localized at that spot. That, hence, results in a virtual array of siRNA-treated cell colonies. After 24–48 hours, the array of treated cells is evaluated for the desired phenotype to identify effective siRNAs and, more importantly, the corresponding target gene. It has been estimated that, by this method, between 100 and 500 reverse-transfections could be performed using the same amount of reagents required for a single transfection in a well of a 96-well plate [58].

Cancer. Using a human cell line and an shRNA library (containing 23,742 distinct shRNAs, 3 shRNAs per gene) cloned into a retroviral vector and directed against of 7914 genes, Berns et al. [60] identified genes involved in the p53 pathway. This protein is a known tumor-supressor and induces cell cycle arrest in cells that have undergone DNA damage. By inhibiting an array of genes using their shRNA library, the authors identified five new genes, and p53 itself, whose expression are required for proliferation arrest of cells. After infection of cells with the pooled shRNA library, the relevant genes associated with this specifically desired phenotype were rapidly identified by unique "molecular bar-codes," which were introduced into each siRNA construct.

The HTS–RNAi approach is extremely promising in providing rapid identification of proteins, which control or are associated with disease-relevant phenotypes. A lentiviral library of shRNA targeting 9610 human and 5563 mouse genes (human homologues) was efficiently used to identify proteins that compromise the 26S proteasome function [61]. The 26S proteasome, a well-studied biochemical machinery involved in the critical degradation of many unwanted proteins including some oncoproteins (e.g., c-Myc), has been implicated in certain diseases. About 30 proteins, mostly proteasome subunits, including five out of five proteasome ATPases and the two largest non-ATPases, were identified in their screen. This work also exploited a "mole-cular bar-coding" system, which facilitated the identification of effective shRNAs.

A single, well-constructed library can be used to screen for gene products involved in completely different disease-relevant pathways. The most important step in applying this approach is the design of a functional assay that will clearly discriminate between the genes whose inhibition results in the desired phenotypes to be obtained by screen. In each case, the specialized development of an appropriate cell-based assay is required. Principles and strategies for developing such assays for the identification of drug-targets for different diseases, including rheumatoid arthritis, osteoarthritis, asthma, osteoporosis, Alzheimer's disease, and cancer, are discussed in the recent review by van Es and Arts [62].

The combination of a high-throughput microarray analysis of differential gene expression followed by RNAi inhibition of selected genes also represents a powerful methodology for the identification of new drug targets [63]. Using this approach, Williams et al. [64] analyzed the differential expression of genes in normal tissue versus colon tumors and polyps isolated from 20 patients. In this study, they identified 574 genes that were upregulated in tumor patients. Out of this large set, they selected a single gene, survivin, a potent inhibitor of apoptosis, and downregulated its expression by RNAi. Inhibition of survivin expression via RNAi resulted in severely reduced tumor growth *in vitro* as well as in an *in vivo* mouse xenograft model. Future analyses using RNAi inhibition of other upregulated genes will undoubtedly make it possible to identify additional precise targets for drug development.

Using this combined approach plus the construction of gene knockout mice, Gunton et al. [65] identified the transcription factor ARNT as an important factor involved in impaired glucose-stimulated insulin release and in causing changes in gene expression similar to those in human type 2 islets. Hence, RNAi was used here to identify and validate a new mouse model for type 2 diabetes.

The reverse application of these two technologies, where cells are first subjected to RNAi treatment and the resulting changes in global gene expression are then assessed by microarray analysis, allows for the dissection of cellular pathways of interest [66]. Such analyses are extremely important: (1) for understanding the global changes that occur in the cellular response to the downregulation of the selected target, (2) for elucidating the role of the target gene in intracellular pathways, (3) for identification of unwanted or detrimental effects, and (4) for selecting the best checkpoints for therapeutic intervention.

These examples demonstrate that RNAi technology offers scientific promise for the identification of effectual drug targets and an excellent perspective for pursuing the development of corresponding small molecules to modulate these targets. Nevertheless, some caution should be reserved in its application because, although RNAi technology may provide a clear understanding of a particular protein's role in a pathway and the phenotypic effects of its downregulation, a small drug directed at the same cellular checkpoint or mechanism might produce different effects [67].

7.6.4 RNAi AS AN INSTRUMENT FOR THERAPEUTIC USE

With the apparent great success of using RNAi technology for the inhibition of specific genes in cultured cells, and in some animal models, there has been a corresponding great expectation of efforts to directly transfer this technology for therapeutic use in humans. Several companies are currently developing RNAi-based drugs to target different diseases (see Table 7.6-1, and discussion below). A variety of serious challenges for the development of such drugs will be discussed in the next two sections. Herein we will review the basic principles and methods for therapeutic treatment, as well as a few prospective directions for the application of RNAi in medicinal therapies.

The most important aspect of using siRNAs as therapeutic agents is their specificity to the target RNA. An optimally designed siRNA for therapeutic use should cause the destruction or repression of only its specific target RNA. This principle puts particular restrictions upon the choice of potential target sequences. For example, if several variants of a protein exist, among which only some forms are disease-relevant and, hence, represent a subset of targets for repression, the drug must be designed based on sequential differences between the intended target and the nontarget forms of the protein. Such variants can be the products of different alleles of the same gene, the results of alternative splicing of a single gene, or the products of a family of related genes. In this respect, dominant mutations that define specific disorders are particularly good candidates for RNAi-based treatment due to the inherent potential ability of an siRNA to discriminate between the normal gene product and the "mutant target" RNAs containing just a few, or even a single, nucleotide mutation(s) [68–70].

As RNAi acts at the RNA level, either through the degradation of the target mRNA by an siRNA-directed mechanism or through the inhibition of the target mRNA's translation by the miRNA pathway, its final phenotypic effect will depend significantly on the amounts, stabilities, and turnover rates of the target RNA and its corresponding protein. In the general case, when the ultimate goal of treatment

TABLE 7.6-1. Companies Developing RNAi Drugs

Company	Targeted Disease or Agent and Stage
Sirna	AMD—phase I
	HC, Asthma, DB, Cancer—preclinical
	HD—proof of principle
Alnylam Pharmaceuticals	RSV—phase I
	AMD, DR, DME, SCI—preclinical
	INFL, CF, PD—research
Acuity Pharmaceuticals	AMD, DR—phase I
Benitech	HC, HIV—preclinical (scheduled for phase I in 2006)
Genesis R&D Corp. Ltd.	Allergy—preclinical
ToleroTech Inc.	Transplant rejection—preclinical
Intradigm Corp.	Cancer (anti-angiogenesis)—preclinical, IND in 12 months
CytRx	Obesity, Type 2 DB, ALS, CMV—research

Data obtained from company websites.
Abbreviations: AMD, age-related macular degeneration; HC, Hepatitis C; DB, Diabetes; HD, Huntington's Disease; RSV, Respiratory Syncytial Virus; DR, diabetic retinopathy; DME, diabetic macular edema; SCI, Spinal Cord Injury, INFL, Influenza; CF, Cystic Fibrosis; PD, Parkinson's Disease; HIV, Human Immunodeficiency Virus; ALS, amyotrophic lateral sclerosis; CMV, Cytomegalovirus.

is to completely abolish a target protein, a successful siRNA compound should be targeted to a gene encoding a relatively short-lived protein with a high turnover rate, preferably one with a low total amount in the cell or organism. Following these criteria, the inhibition of the production of the targeted gene product will most effectively result in a significant decrease in the concentration of this protein in the cell and, hence, will optimally affect the downstream biochemical events in order to treat the disease condition.

A promising class of gene target candidates for siRNA intervention is genes with temporal profiles of expression. If an siRNA drug is delivered to cells before or at the beginning of expression of these genes, there are better chances that the production of their products will be inhibited sufficiently to produce the desirable therapeutic effect; even better if these proteins are normally produced in small quantities. A good example of such genes is viral genes that start to express early after viral infection. Certain specific viral transcription factors, generally expressed in small quantities, should be particularly well suited as targets for siRNA treatment. Several examples have demonstrated that siRNA directed against viral transcription factors can efficiently inhibit viral replication. Lee et al. successfully inhibited the replication of human immunodeficiency virus type 1 (HIV-1) by Tat-specific siRNA [71]. Simultaneous inhibition of viral transcription factor and surface antigen by a vector expressing two siRNAs specific for HBx and HBs significantly reduced hepatitis B virus replication in cell culture [72].

Another suitable group of targets for siRNA treatment in the case of viral infection is host proteins involved in infection and propagation of the virus. The key consideration here is that downregulation of these molecules must not be detrimental to the host. In the study mentioned above, by Lee et al. [71], inhibition of HIV-1 was achieved not only by Tat-specific siRNA but also by siRNAs specific for the CCR-5 cellular coreceptor. Using RNAi, Chiu et al. [73] inhibited expression of human positive transcription elongation factor P-TEFb. P-TEFb is involved in transcriptional regulation of cellular genes as well as HIV-1 genes. Targeting of endogenous P-TEFb resulted in the inhibition of Tat transactivation and HIV-1 replication in host cells. Interestingly, P-TEFb knockdown was not lethal to the host cells, which showed normal P-TEFb kinase activity. These authors suggested that there may be a different critical threshold of P-TEFb activity required for cell viability than for HIV replication. This example shows that RNAi-mediated gene inhibition may provide an opportunity to specifically inhibit a disease-causing agent while minimizing the effect on normal cellular function. Nevertheless, the inhibition of any cellular gene for the purpose of disease treatment or prevention should be considered very carefully, not only at cellular level but at the level of the whole organism as well.

As mentioned in Section 7.6.2, using siRNA it is rare to achieve an effect of 100%, where the target RNA is completely abolished or "knocked out." More often the result is a "knock-down" effect. In some cases, such a partial reduction of gene expression is sufficient for obtaining the desired phenotypic or therapeutic effect. It is also prudent to be able to switch "on" and "off" the expression of siRNA at certain times or in specific tissues. Fritsch et al. designed a CRE-lox-based strategy that allows one to repress gene activity through shRNA expression in a time-dependent manner in cells, and in a time- or tissue-dependent manner in animals [74, 75]. Other vectors with chemically inducible promoters have been designed for the controlled expression of shRNAs using tetracycline [76, 77], ecdisone [78], macrolide [77], and butyrolactone [79]. Still, in some cases, an incomplete shut-off of the target gene may not produce the desired therapeutic effect. In these cases, RNAi therapy could be used effectively in combination with other drugs directed to the same, or a different, checkpoint in order to achieve a more satisfactory therapeutic effect. Such parallel treatments, with drugs having different modes of action, also make it possible to overcome potential antagonistic factors due to mutations or alternative pathways leading to a disease.

A prolonged treatment with any drug may encounter the emergence of mutations causing resistance to the treatment. RNAi technology is inherently susceptible to this problem. However, expression vectors able to express two shRNAs simultaneously have been designed and used to help overcome potential loss of RNAi-drug efficacy [80]. Using two siRNAs directed against the 3D-RNA-dependent RNA polymerase of coxsackievirus B3, Shubert et al. obtained efficient inhibition of virus propagation in HeLa cells and reduced virus titers by up to 90%. Introducing point mutations into either target site can abrogate the inhibitory effect of that siRNA, when used separately. These authors constructed an siRNA double-expression vector (SiDEx) to achieve simultaneous expression of both siRNAs from a single plasmid. Their "double-drug" construct successfully downregulated the point-mutated target gene. In addition, these vectors could be used for the simultaneous knockdown of two different targeted genes for functional studies or therapeutic use.

In Section 7.6.3, we described the application of RNAi for drug target identification and validation in different diseases. In examples where inhibition of expression of specific genes led to the desirable phenotypic effects, those targets may consequently be considered for RNAi therapeutic use. Below we provide additional examples of how the principles and methods described in this section can be applied to the development of therapies for different diseases.

7.6.4.1   Cancer

A significant proportion of all cancers result from mutations in cellular proto-oncogenes. Nearly half of these cancers contain point mutations in the tumor-suppressor gene, p53. The ability of an siRNA compound to discriminate between wild-type (WT) and mutated forms of p53 mRNA allowed for the specific knockdown of a dominant-negative mutation, resulting in an inhibition of tumor growth and restoration of WT protein function [68].

Fusion proteins, which result from chromosomal translocations, are commonly associated with the growth transformation that occurs in some cancers. Leukemic fusion gene AML1/MTG8 is associated with up to 15% of all *de novo* cases of acute myeloid leukemia. Heidenreich et al. [81] used a human leukemic cell line, Kasumi-1, to demonstrate that efficient suppression of the disease-associated fusion protein AML1/MTG8, but not WT AML1, can be achieved by siRNAs directed against the fusion site of the AML1/MTG8 mRNA. This treatment, in combination with TGF beta(1)/vitamin D(3), significantly reduced the clonogenicity of Kasumi-1 cells.

One subtype of acute lymphoblastoid leukemia is characterized by a t(1;19) chromosomal translocation resulting in the expression of a novel fusion protein

E2A-Pbx1. This mutant protein activates the expression of a secreted Wnt16 gly-coprotein [82], which is widely involved in cell proliferation, differentiation, and oncogenesis, and promotes transformation and leukemogenesis. Mazieres et al. [83] demonstrated that only one of two Wnt isoforms (Wnt16b) is overexpressed in t(1;19)-containing cell lines. Using isoform-specific siRNAs and an anti-Wnt16 antibody, they showed that targeted inhibition of the overexpressed Wnt16b leads to apoptotic cell death.

A common problem in modern cancer therapy is the development of cellular resistance to conventional chemotherapeutic compounds. In many cases, this resistance is due to the overexpression of the multidrug resistance gene, MDR1. Targeting of this gene using RNAi technology may significantly decrease or even completely reverse the multidrug-resistance phenotype [84–86]. Thus, again, suggesting the potential utility of using RNAi in combination with traditional strategies.

Inhibition of oncogene expression, or of other components of molecular path-ways involved in cell proliferation, may slow the growth of cancer cells in a tumor without leading to apoptotic cell death; and hence, it may not always eliminate these cells from the organism. In addition, because RNAi rarely leads to the com-plete knockout of target gene expression, it might be advantageous to use a com-bined regimen of RNAi and a small drug and/or radiation treatments for more efficient combat against cancer. RNAi compounds can be directed against genes responsible for anti-apoptotic effects [87–89], against tumor-suppressor genes [90], against dsDNA-brake repair proteins [91, 92], or against other genes in cellular survival pathways [93]. In this scheme, RNAi-induced suppression of normally protective, repair-activated, or damage-induced genes in tumor cells may result in even greater increases in proliferation of tumor cells and, hence, an increased sen-sitivity to the combined chemotherapeutics or radiation.

7.6.4.2   AIDS and Other Viral Infections

Interestingly, to date, almost all HIV-1 genes have been targeted by RNAi, at least in cell culture experiments. It has proven possible to block the replication of HIV-1 in cultured cells by siRNAs targeted to viral sequences [94, 95]. The main obstacle to therapy, though, is the specific delivery of siRNAs to cells infected or susceptible to HIV-1 infection *in vivo*. We shall discuss this problem of cell-specific targeting in detail later. However, it has been shown recently that HIV-1 can suppress the function of the host cellular RNAi machinery [96]. This suppression of RNA silenc-ing may potentially affect the efficiency of RNAi drugs directed against HIV-1 gene targets, and against other target viruses that may have evolved similar anti-host mechanisms. In addition, HIV infections are particularly difficult to treat, in part, due to the viruses' extreme ability to transform as a result of high mutation rates, especially of surface receptors, which gives rise to the high sequence diversity of HIV genomes among infected subjects. One measure against this tendency for viral sequence diversity might be the simultaneous use of several siRNAs directed against several viral genes. On the other hand, nonvariable human cell receptors for HIV infection represent a great target for siRNA treatment. Martinez et al. [97] demonstrated that RNAi directed against CD4 coreceptors CXCR4 and CCR5 efficiently blocked acute infection by X4 (NL4-3) and R5 (BaL) HIV-1 strains but had no effect on CD4 cells or control green fluorescence protein expression. In another example, Nair et al. [98] presented evidence that siRNA directed against the CD4 independent attachment receptor (DC–SIGN) significantly inhibits HIV infection of dendritic cells. Nevertheless, as mentioned above, any targeting of host genes should be evaluated for unwanted or detrimental effects, especially at the organism level. Some negative effects could significantly restrict the usage of endogenous gene targets.

Many other viral infections, including the ones most threatening to humans, like hepatitis C, B, and A viruses (HCV, HBV, HAV), human papillomavirus viruses (HPVs), West Nile virus, and human herpes viruses (i.e., HSV, CMV, EBV, and HHV-8) are also the targets of intense research for the development of effective RNAi treatments. It has been demonstrated that RNAi can disrupt the replication of many viruses, including HCV, HBV, and HAV in cell culture [99, 100] and, as later demonstrated in mouse models, *in vivo* [101, 102]. The major limitation, again, is the delivery and stability of RNAi compounds for application as human therapies in clinically relevant systems, which will be discussed further below. These examples demonstrate, however, that RNAi technology has great potential for the treatment of viral disease, yet clearly requires further development for clinical applications.

7.6.4.3   Parasite Infections

Despite the biotechnological and drug development booms of the last several decades, parasite infections (malaria, African sleeping sickness, leishmaniasis, Chagas disease, etc.) remain the foremost health burden and source of mortality and morbidity on our planet, especially in the tropics. Malaria, alone, contributes to about 500 million cases of disease per year that result in more than 2–2.5 million deaths, mostly in sub-Saharan Africa [103]. Not only are newer drugs still expensive and unavailable to most of the population in these regions, but parasites also develop drug resistance rapidly, especially to the drugs extensively used, like chloroquine [104]. Due to many reasons, vaccine development has not yet been successful against these diseases. Therefore, the appearance of RNAi technology was perceived with great interest and was immediately explored in studies aimed at defining treatments for parasite infections.

It was quickly discovered that RNAi works in *Trypanosome brucei* as well as it does in *C. elegans* [105]. Since that time, RNAi mechanisms have been extensively studied in *T. brucei* [106]. The technology is now being used extensively to dissect different biochemical pathways in *T. brucei* [107] as well as to study potential therapeutic applications [109–110].

Unfortunately, it is still not clear whether RNAi functions in malaria parasites. Very few publications have described RNAi in malaria [111–114]. Genomic data mining does not clearly identify any members of the RNAi machinery in these parasites by homology searches. Given the enormous increase in the number of publications on RNAi in other systems, and the likely numerous efforts made in malaria research to use RNAi, there seems to be a surprising lack of published evidence demonstrating RNAi in malaria. It is possible that in this parasite the RNAi pathway was lost or has diverged during evolution to perform other functions in the organism, which is incompatible with the classic utilization of the RNAi mechanisms being explored in other organisms [106]. It is also possible that the effects described in the papers published on malaria actually reflect other processes and pathways, such as the abundant presence of antisense RNAs [106] as recently described in malaria [115, 116].

7.6.4.4 Neurological Disorders

Many neurological disorders, including Alzheimer's disease, Parkinson's disease, fragile-X syndrome, and amyotrophic lateral sclerosis (ALS), result from dominant mutations in a single allele. Therefore, as discussed above, this class of "disease genes" is perfectly suited as candidates for treatment by sequence-specific RNAi therapy. In the example of ALS, given below, stepwise improvements led to progress in the development and use of RNAi technology for therapeutic treatment of the disease in an animal model.

ALS is a familial neurodegenerative disease characterized by motor neuron degeneration, paralysis, and death. One cause of this disease (2–3% of cases) is mutations in the Cu,Zn superoxide dismutase (SOD1) gene. Ding et al. [117] identified a specific shRNA that could specifically degrade mutant SOD1 mRNA. Later, Maxwell et al. [118] investigated the functional effects of RNAi-mediated silencing of mutant SOD1 in cultured murine neuroblastoma cells and found that silencing of mutant SOD1 protects these cells against cyclosporin A-induced cell death. Recently, Raoul et al. [119] demonstrated that in SOD1 (G93A) transgenic mice, a model for familial ALS, intraspinal injection of a lentiviral vector that produces RNAi-mediated silencing of SOD1 substantially retards both the onset and the progression rate of the disease. Another group [120] injected lentiviral vector expressing SOD1 RNAi compound into various muscle groups of SOD1 (G93A) mice. This treatment also resulted in an efficient and specific reduction of SOD1 expression and improved survival of vulnerable motor neurons in the brain stem and spinal cord. They observed a considerable delay in the onset of ALS symptoms by more than 100% and an extension in survival by nearly 80% of their normal life span. Despite this success in mice, there is an additional complexity with this disease in humans. There are more than 100 different known mutations in the SOD1 allele. Whereas it is difficult even to propose to design efficient RNAi compounds to address each of these mutations in individual patients, Xia et al. [121] suggested another elegant approach based on RNAi technology to overcome this problem. They designed a replacement RNAi strategy, where all mutant and wild-type forms of the gene are inhibited by RNAi. The wild-type SOD1 function is then replaced by a custom designed wild-type SOD1 gene that is resistant to the RNAi. These ALS studies clearly demonstrate how therapeutic strategies based on RNAi technology can be improved over a short period of time with the accumulated incorporation of different molecular aspects of a certain disease into improvements in the application of the technology itself.

7.6.5 BASIC THEORETICAL AND PRACTICAL CONCERNS FOR APPLICATION OF RNAi IN THERAPY

Despite the great potential of RNAi technology for therapeutic applications, there are a number of intrinsic features of this process, which present a considerable challenge to its practical use.

7.6.5.1   Variable Efficiency of Different siRNAs

As discussed in the second section, different dsRNA fragments cognate to the target RNA have different efficiencies in inducing its degradation. Even more uncertainties exist about the miRNA-dependent mechanism of inhibiting mRNA translation. Because the relative efficiencies of different siRNAs in RNAi may differ significantly, currently there is no way to predict their individual potentials. Screening for the most efficient siRNAs using cell cultures can be a laborious and expensive procedure. In addition, natural nucleotide polymorphism and mutations in a gene target site may affect the efficacy of RNAi [122–124]. Therefore, efforts have been made to overcome this problem of variable efficiency of different siRNAs.

Several methods based on the use of siRNA mixtures, which may contain a particular efficient siRNA (or several), have been developed. These include the preparation of siRNA mixtures using RNaseIII [125] or dicer [126] enzymes to digest longer dsRNAs. The short RNAs produced as a result of these digestions have been found to be efficient in RNAi. The disadvantage of these digestive approaches, however, is in their potential for nonspecific effects. If a randomly chopped long dsRNA contains areas of homology to other genes, or splice variants of the same gene, the expression of these genes might also be inhibited by cognate siRNAs (see below). In addition, obtaining long dsRNAs also adds to the potential total cost of, and time required for, these approaches.

It would be more advantageous to use defined mixtures of siRNAs, complex enough to inhibit the desired gene with high probability, but devoid of any siRNAs with unwanted homologies. Besides creating higher target efficiency, the use of such mixtures would also alleviate the need to screen for individually active compounds. We developed a simple, fast, and inexpensive method for the preparation of defined-siRNA mixtures using custom DNA oligonucleotide microarrays as a starting material [127]. In this study, we demonstrated that siRNA pools, prepared using array-derived DNA oligonucleotide mixtures, are an efficient tool for the inhibition of exogenous as well as endogenous genes in cell cultures [128]. In this approach, many (up to 30) different siRNAs are designed against a single gene. As the sequence of the human genome is known, any unwanted homologous siRNA sequences could be excluded from the designed pool. Many different siRNA pools can be synthesized using a single DNA oligonucleotide microarray, which makes this approach extremely affordable relative to the construction of defined complex mixtures via the synthesis and/or cloning of many different individual siRNA species.

Although the approaches using siRNA mixtures prepared as described above are efficient for use in cultured cells for pathway validation and drug target identification experiments, their use as therapeutic agents might, nevertheless, be problematic due to manufacturing issues as well as issues related to the specific delivery of siRNAs to the target cell or tissue (see below). In any case, the identification of an efficient single or a mixture of RNAi compounds requires a careful selection process and experimental testing to maximize gene silencing and minimize potential off-target and nonspecific effects.

7.6.5.2   Nonspecific Systemic and Off-Target Effects, and the Potential for Pleiotropic Effects

We described, in the second section, how long dsRNAs may induce nonspecific effects in mammalian cells due to the activation of interferon and the induction of the cellular dsRNA-dependent protein kinase response. Therefore, to avoid these nonspecific effects, the use of siRNA (<~30 bp) was proposed [29] and proved to be an efficient technology for posttranscriptional gene silencing. Nevertheless, it was later noted that even the short dsRNAs used to generate RNAi effects, and designed in the form of either siRNA or shRNA, can still induce some mammalian genes involved in the interferon response [129], although results from different laboratories were conflicting [130]. To circumvent these effects, different measures can be implemented, including different administration routes, dose of vectors, selection of promoters, and use of additional immunosuppression [131].

Interferon-independent nonspecific effects also exist. Scacheri et al. found that siRNAs could induce dramatic and significant changes in the levels of p53 and p21 proteins that were unrelated to silencing of the target gene [132]. These authors speculated that partial complementary sequence matches to off-target genes may have resulted in a microRNA-like inhibition of translation. In this respect, it is, actually, difficult to distinguish between nonspecific and sequence-specific off-target effects (see below). It has also been reported that si/shRNAs induce immune activation in macrophages and dendritic cells through toll-like receptor 3 [133, 134]. Moreover, Persengiev et al. [135] found that 21-bp siRNAs nonspecifically stimulated or repressed more than 1000 genes involved in diverse cellular functions. The effects on gene expression were dependent on siRNA concentration and were stable throughout the course of siRNA treatment. Again, here too, it is difficult to distinguish the nonspecific systemic effects from off-target effects. In addition, there is a potential for pleiotropic effects due to a genuine physiological response of downstream genes due to the downregulation of the target genes. If off-target genes become downregulated, these may also produce a cascade of unintended physiological responses. Such a cascade of multiple events could result in the global dysregulation of many unrelated genes. Therefore, potential widespread, nonspecific effects on mammalian gene expression must be carefully considered in the design of therapeutic RNAi applications.

Besides nonspecific cellular effects due to the RNAi compounds themselves, there are additional sources for potential nonspecific effects; these include the vehicles for, and methods of, delivery of RNAi compounds to cells and tissues. For example, viral vectors, as well as other components such as transfection reagents [136], chemical modifications to RNAi compounds, or chemical substances used to facilitate the delivery of RNAi compounds, might produce unwanted immune and other cellular responses [137]. These responses should be considered as additional possible sources for nonspecific effects in therapeutic applications of RNAi.

In addition to the aforementioned nonspecific effects, one additional problem has recently become evident: that the cellular RNAi machinery, normally involved in the regulation of some endogenous genes' expression and in maintaining a stable chromosome structure, might be indirectly affected by large amounts of exogenous siRNAs introduced by transfection or vector-directed transcription, for example. This may result in an imbalance or saturation of the cellular resources of endogenous molecules in the pathway.

The possibility of saturating the RNAi machinery was raised in early studies [122], and it has recently been supported by several observations [138–140]. Moreover, it has been suggested that perturbation of the cellular RNAi machinery might be involved in different disorders and diseases [141]. Therefore, exploiting the RNAi machinery for therapeutic interventions could present additional nonspecific complications for cellular and organismal homeostases that should be carefully evaluated in the design and use of RNAi compounds, including the consequences of the dosage and timing of treatment.

Off-target effects are another serious concern in the application of RNAi for therapy. Such effects might be due to the partial homology of an RNAi compound to an irrelevant, or related but different, gene, which is not a target for downregulation. These effects will clearly be different for different sequences, and the parameters and degree of homology, which confer the extent of off-target effects caused by a particular RNAi compound, are complex and not fully understood. It has been shown that, depending on the sequence, most RNAi compounds can tolerate one [142], or even several [143], mismatches, though, typically with reduced efficiency. Moreover, it was shown recently that in some cases mismatches can even enhance the activity of RNAi compounds, possibly through the miRNA-regulatory pathway [144]. In fact, as few as 11 base pairs of homology with single-stranded RNA can be sufficient to produce an inhibitory effect [145]. Therefore, careful elimination of possible antagonistic homologies is an important consideration in the design of RNAi compounds. This is feasible with the great wealth of sequence data and bioinformatics currently available from human and pathogen genomes. In this respect, treatment with complex mixtures of defined siRNA molecules [128] may also offer a unique advantage over current methods. Because each efficient siRNA has an optimum concentration when its effect is maximal (the practical effective concentration range in cell culture is 10 to 100nM), if such an siRNA molecule, homologous to an off-target gene, were present in a defined mixture, it is likely that its individual concentration would be well below the effective range. Thus, because a complex mixture targeted at a single gene may include 30–50 different siRNAs at an active total concentration of 10–100nM, this may, theoretically, lessen the off-target effects of each siRNA molecule contained within the mixture. Nevertheless, despite the incorporation of even the most conscientious theoretical design elements to minimize nonspecific and off-target effects, the empirical testing of each compound is requisite for its evaluation and development for therapeutic application.

7.6.6   TECHNICAL CHALLENGES OF SIRNA THERAPY: *IN VIVO* DELIVERY AND STABILITY

The use of any drug for clinical application involves two important issues: the delivery of the drug to the affected cells or tissues, and the stability of the drug while in the organism. These factors directly influence the determination and choice of effective dosage, delivery time, and routes. RNAi compounds are not unique in this respect, and their physical–chemical and biological characteristics determine the limits and specifics of RNAi therapy. The physical stability of any compound might depend significantly on its particular route of delivery, and therefore, these parameters are potentially restrictive to each other and may, hence, limit different options for use.

7.6.6.1 Delivery of RNAi Compounds

RNAi compounds for therapeutic treatment can be generated in two ways: chemically synthesized si/shRNA molecules and vector-encoded molecules, which are synthesized by the cells after intracellular delivery of the corresponding expression vectors. The choice between these two approaches will determine the specifics of treatment: different targets and routes of delivery, different stability of compounds and efficiency of treatment, different possible side effects, etc. In many respects, vector-encoded delivery seems to offer advantages over the more traditional "pharmaceutical" approach of drug delivery. For example, the duration of effect delivered by chemically synthesized si/shRNA molecules might be shorter than the duration expected from vector-encoded molecules due to their continuous supply by intracellular transcription. In addition, these "internally synthesized" RNAi compounds are not subject to the potentially destructive environments of the organism, such as chemically synthesized and systemically or locally delivered siRNAs might be, for example, through gastrointestinal (GI), intravenous, or intramuscular delivery routes. It is usually sufficient to deliver only a small number of siRNA-expressing vector molecules per cell to obtain efficient silencing. Therefore, the delivery of RNAi compounds by these two approaches should have completely different dosage and timing strategies. In turn, the vectors for delivering RNAi compounds can be constructed using viral or DNA plasmid systems, which also might provide for or necessitate different delivery routes. It is also probable that siRNA-expression vectors will have to be customized for the delivery of targeted compounds to specific tissues (i.e., employing tissue-specific promoters, or viruses with tissue-type limited host ranges). There are currently three kinds of viral vectors in use for RNAi compound delivery: adenovirus, adeno-associated virus, and lentivirus [146]. In general, plasmid and viral RNAi-vectors, as methods of gene therapy, could have some advantages over siRNA oligonucleotides in delivery, stability, and efficacy. However, vector-associated gene therapies have their own long history of safety concerns and potential for nonspecific effects [147].

There are two fundamental means of drug delivery: local delivery and systemic delivery; each with its inherent advantages and disadvantages. Local delivery is not always practical or feasible, although it might be conceptually preferred for its potential efficiency in targeting the specific tissue or even a pool of target cells. Systemic delivery suffers from low efficiency of the relevant cells being targeted. Morrissey et al. [148] calculated that only 1% of intravenously administered (30 mg/kg) siRNAs reached the targeted organ (liver). Combined with the aforementioned differences in the forms of RNAi compounds, all these issues make the delivery problem quite complex. In these respects, nasal delivery represents one of the most convenient and powerful methods of delivery for both local and systemic delivery [149]. The substantial previous experience gained by advances made in the delivery of antisense drugs (one of which, Vitravene, was approved by the FDA in 1998) may contribute significantly to the related RNAi-based therapies [150].

Recently, substantial success in the targeted delivery of chemically synthesized siRNAs was achieved using siRNAs in the form of a supramolecular complex with a fusion protein consisting of a protamine-antibody directed at a specific cell-surface receptor [151, 152]. These complexes were found to be efficient in silencing target genes when delivered either locally (by intratumoral injection) or systemically (by intravenous injection). The use of viral vectors might also prove efficient for systemic delivery and, as discussed above, are likely to provide long-term efficacy. Moreover, specially designed viral vectors, targeting specific tissues or even specific cells within the target tissue, might significantly improve delivery of RNAi compounds to the intended target cells. Gou et al. [153] recently demonstrated, both *in vitro* and *in vivo*, that the use of a cell-specific promoter introduced into an adenoviral vector could direct the specific expression of shRNA in one type of cell without interfering with other cell systems present. This approach seems extremely useful for the delivery of RNAi drugs. For example, as these authors suggested, to target the expression of an oncogene in cancer cells, but not in normal cells, a telomerase reverse transcriptase promoter (active in cancer cells, but inactive in normal somatic cells) [154] could be used to drive shRNA expression specifically in cancer cells.

7.6.6.2 Stability of RNAi Compounds

As with any drug delivered into an organism, an RNAi compound is subject to inactivation, degradation, and elimination. The good news, though, is that because RNAi drugs share the same basic chemical composition independent of the targeted disease, any findings on stabilization of compound(s) in one model should be applicable to other models and compounds, if the compounds are delivered through a similar route.

si/shRNAs are subject to enzymatic degradation in the body, as are any RNA molecules. Due to their small size and double-stranded nature, though, these structures seem to be more stable than many types of RNA and even more stable than the single-stranded DNA oligonucleotides used as antisense drugs [155]. Nevertheless, because degradation is a common problem, several different approaches have been undertaken to create compounds that may resist destruction in the cell. One way to stabilize a nucleic acid compound in the biological environment is to chemically modify it so that it will become resistant to enzymatic degradation while maintaining its intended biological function. Several such modifications to siRNAs, and the influence of these modifications on their pharmacokinetic properties, have been described [148, 156–158]. Another way of stabilizing RNAi compounds is the formation of supramolecular complexes, such as nanoparticles, which have been extensively explored for the delivery of RNAi compounds [159–163], or of liposome-based complexes, which also can be used for plasmid vector-based siRNA delivery [164–166]. Despite the method(s) chosen for delivery of chemically prepared si/shRNAs, the achievement of intracellular delivery at therapeutically effective concentrations in the target cells still represents a major challenge to the practicable use of RNAi technology for therapy, especially via systemic delivery routes.

7.6.7 RNAi DRUGS UNDER DEVELOPMENT

No RNAi drugs are currently on the market for use in humans. Only a few compounds have entered phase I clinical trials, and some will be starting soon (see below). In the last few years, a tremendous investment has been made in the development and application of RNAi compounds as drugs. Several recent reviews describe these extensive efforts and the progress made in several small animal models [51, 167–169]. The effective use of RNAi–*in vivo* animal models may not only lead to considerable savings for the drug industry in the expensive development of small drugs, but it will also likely indicate whether particular RNAi compounds have significant potential for further development into drugs. Our aim in this section is to describe several interesting recent examples of RNAi drugs under development for different diseases in animal models. These studies illustrate the application of many principles and concerns discussed above.

7.6.7.1 Cancer

Hu-Lieskovan et al. [170] described systemic RNAi therapy for metastatic cancer in a murine model of metastatic Ewing's sarcoma. They prepared siRNA against the EWS-FLI1 gene product and delivered it using a targeted, nonviral delivery system. SiRNA was packaged into cyclodextrin-containing polycation complexes, which contained not only siRNA but also transferrin as a targeting ligand for specific delivery to transferrin receptor-expressing tumor cells. This delivery system transiently reduced expression of the target gene in metastatic cells and slowed tumor growth. Using long-term twice-weekly treatments, they demonstrated absence of any tumor growth in 80% of mice. Control experiments revealed that only the transferrin-targeted system was efficient in antitumor effects. Moreover, they found that low-pressure, low-volume tail-vein administrations were efficient in the delivery of RNAi compounds and resulted in significant improvements over the high-pressure, high-volume intravenous injections previously used in many studies of RNAi therapy in mouse models [171–173], but that are untenable and unacceptable in humans in routine clinical settings. It was also very important to the success of this study that no abnormalities in interleukin-12 and IFN-alpha, liver and kidney function tests, complete blood counts, or pathology of major organs, i.e., no off-target side effects, were observed from the treatments described. This delivery system was found to be ineffective for brain tumors, however, because the siRNA complexes used did not cross the blood-brain barrier. Consequently, different RNAi approaches will have to be developed to specifically address the treatment of brain tumors.

Yano et al. [174] used mouse models for liver metastasis and prostate cancer to test a human bcl-2 oncogene-specific siRNA complexed with a novel cationic liposome, LIC-101. These complexes were delivered by bolus intravenous injections into the livers of disease-model mice and completely suppressed the growth of liver tumors. The injection volume used was 0.1 mL per 10 g of body weight, which is normally acceptable for human therapy. In addition, in their prostate cancer model, siRNA complexes were administered subcutaneously (0.1 mg in 0.1 mL per mouse) near the tumor five times a week for 2 weeks in mice bearing PC-3 xenografts. The average tumor volume on day 36 of the treatment was $1300 \, mm^3$ in the control group, compared with $487 \, mm^3$ in the siRNA compound-treated group. The authors of this study also indicated that other cationic liposomes caused much greater toxicity than LIC-101, as judged by hemolysis activity on human erythrocytes, although these data were not presented.

KITENIN, a member of the Tetraspanin family, is expressed by many different tumors. Its expression is correlated with the acquisition of metastatic phenotypes

[175]. Lee et al. [176] treated mice with established colon tumors with KITENIN-targeted siRNAs cloned into pSUPER vectors. Weekly, or semiweekly, injections of these siRNA-encoding vectors, delivered in suspension with FuGENE 6 transfection reagent (Roche) for 1 month into tail veins, resulted in a marked regression of tumor size and inhibited metastases. Delivery of this siRNA also significantly prolonged the survival of mice compared with the vector alone (77 ± 12.2 versus 25 ± 9.9 days, $p < 0.01$). In this study, it was demonstrated that the antitumor effects of KITENIN siRNA derived from both the generation of a tumor-specific immune response *in vivo* and the suppression of tumor invasion.

7.6.7.2   Viral and Respiratory Diseases

Hepatitis B virus (HBV) infections cause acute and chronic hepatitis and hepatocellular carcinomas. Uprichard et al. [177] demonstrated, in a mouse model of established Hepatitis B infection, that infection with an adenovirus vector carrying shRNA directed at HBV-specific viral RNA could suppress preexisting HBV gene expression and replication to almost undetectable levels for at least 26 days. Morrissey et al. [148] obtained similar results in a study using hydrodynamic tail vein injections of HBV RNA-specific siRNA stabilized by means of chemical modifications.

Influenza A virus infects about 15–20% of the population each year and is a major source of morbidity and mortality worldwide. Antiviral vaccines only afford protection against a limited range of strains, due to substantial mutation rates in these viruses. Ge et al. [178] demonstrated that siRNAs specific for conserved regions of influenza virus genes can be used to treat and prevent influenza virus infections in mice. In this study, slow i.v. administration of small volumes containing siRNAs complexed with a polycationic carrier was used. Virus titers in lungs were reduced from 10- to 1000-fold by single doses of 60–120 µg of siRNA. Similar effects were also observed when mice were given DNA vectors by i.v. or intranasally, from which siRNA precursors were transcribed. In another study using the same siRNAs, but different carriers, doses, strains of mice, and virus strains, Tompkins et al. [179] observed a significant increase in the survival of mice after lethal challenge with viruses. In both studies, the effects were siRNA-specific and not due to nonspecific antiviral IFN responses. These experiments provide a good basis for studies of prophylaxis and therapy of influenza virus infections in human populations using siRNA compounds.

Respiratory syncytial virus (RSV) and parainfluenza virus (PIV) are leading causes of respiratory disease in infants, young children, immuno-compromized patients, and the elderly. Bitko et al. [140] used viral-specific siRNAs for the treatment of these respiratory syndromes in an animal model. In this study, either, or both, viral infections could be specifically prevented and inhibited when corresponding siRNAs were infused intranasally in mice, with or without transfection reagents. These investigators noted an excellent correlation between the activities observed in their siRNAs in *in vitro* cell cultures and those observed in their animal model. This type of approach presents a fast screen for efficient siRNA sequences before more expensive and demanding applications are pursued in animals. In addition, nasal administration of siRNA compounds seems to provide a convenient and potent route for delivery of these drugs directed against, at least, respiratory viral diseases in humans, and perhaps against many others.

7.6.7.3 Arthritis

We described, in Section 7.6.3, the use of electroporation for the local delivery of αTNF-specific siRNAs to inhibit joint inflammation in a mouse model of collagen-induced arthritis [50]. In another example, Inoue et al. [180] used electro-transfer of an αTNF-specific siRNA-polyamine complex to significantly ameliorate collagen-induced arthritis in rats. As αTNF plays an important role in many processes in different organs, local delivery was used, in this case, to restrict the effects to the target organ, the joint, and minimize the potential effects on other αTNF-dependent processes elsewhere in the organism.

7.6.7.4 Autoimmune Diseases

Autoreactive T cells play an important role in many autoimmune diseases, such as multiple sclerosis (MS), type 1 diabetes, systemic lupus erythematosus (SLE), psoriasis, asthma, and experimental autoimmune encephalomyelitis (EAE). Lovett-Racke et al. [181] studied inhibition of T-bet, a member of the T-box family of transcription factors, which is expressed in Th1 cells, but not Th2 cells, in EAE, an inflammatory and demyelinating disease of the central nervous system and model for the human disease MS. A single injection (20 or 50 μg) into the tail vein of T-bet-specific naked siRNA at the time of immunization (to induce EAE) reduced the incidence of EAE disease by as much as 75% depending on the siRNA dose. It is possible that modification of the delivery method to one more appropriate for human routine clinical settings, such as those that employ cell-specific targeting as described above in the subsection on *Cancer*, might provide a better basis for clinical trials of RNAi compounds in autoimmune disorders. In addition, the use of a targeted cell delivery system, for example, using an epitope from the T-cell receptor, similar to the transferrin-directed complexes used in [170], might provide more specific elimination of disease-relevant T cells.

7.6.7.5 Neurodegenerative Disorders

Alzheimer disease is a neurodegenerative disorder characterized by an accumulation of β-amyloid (Aβ), a proteolytic product of the amyloid precursor protein (APP) [182]. Increase in β-secretase (BACE1) activity has been directly associated with the production and accumulation of Aβ [183–185]. Singer et al. [186] used lentiviral vectors expressing BACE1-targeted siRNAs delivered locally by intracranial injection into the hippocampi of APP transgenic mice (a model of Alzheimer disease). This treatment specifically reduced both the cleavage of APP into Aβ and neurodegeneration in this *in vivo* model. RNAi-mediated reduction of Aβ-reactive plaques was observed only in the hippocampus and not in other parts of the brain. Behavioral deficits were, nevertheless, also improved by this treatment. Another important issue in this study was that siRNA-targeted inhibition downregulated, rather than completely knocked out, the target gene; this may have been advantageous, in this case, because BACE1 has several potential substrates and the complete switching off of its activity could produce undesired or deleterious effects. In the effective case of RNAi-mediated downregulation of a target gene, therefore, disease improvements might be observed by lowering the level of production of a relevant protein below some critical threshold level, without significant effect on normal functions.

Huntington's disease (HD) is a dominantly inherited neurodegenerative disorder associated with polyglutamine expansion in the huntingtin (htt) protein [187], for which, currently, there is no effective treatment. Wang et al. [188] used an siRNA compound directed against human mutant-htt, containing expanded polyglutamine tracts, delivered locally into the brains of newborn HD transgenic model mice by intraventricular injections. This treatment significantly increased the longevity of HD mice, resulted in improved motor function, and slowed HD-associated loss of body weight. Despite these successes, this method is currently problematic for human therapeutic applications; not only because of the delivery route used, but also due to the challenge of designing an siRNA sequence that is highly selective for the mutant form of htt, because reduced expression of the wild-type human gene could be deleterious.

7.6.7.6 Ocular Disorders

In age-related macular degeneration, vascular endothelial growth factor (VEGF) plays a crucial role in destructive vascularization. Downregulation of VEGF by antibody [189], or by siRNA [190], has been shown to reduce vascular invasion. Recently, Tolentino et al. [191] used intravitreal injection of VEGF-specific siRNA in a non-human primate model of laser-induced choroidal neovascularization (CNV). They demonstrated that VEGF-specific siRNA treatment is capable of inhibiting vascular growth and permeability in a dose-dependent manner. This study served as a preclinical "proof-of-principle," and phase I clinical studies of VEGF siRNA in patients with exudative age-related macular degeneration are currently in progress [192].

7.6.8 COMPARISON OF SIRNAS WITH OTHER PHARMACEUTICALS (TARGETED AND NONTARGETED TYPE), INCLUDING CLOSELY RELATED GENE SPECIFIC ANTISENSE-BASED DRUGS

Most drugs on the market today are small-molecule compounds, which can be systemically delivered in the form of pills. This is certainly the most convenient method. However, it might not be the most efficient method of drug delivery, due to the different specific physical-chemical nature of the drugs. Nevertheless, the consumer market is heavily affected by a demand for consumer convenience, and pharmacological companies have generally followed this trend with much success. There is a substantial number of drugs, however, which are not pills and could not be pills due to their physical-chemical, biological, and pharmacological properties. RNAi compounds will most likely belong to this class of drugs, at least in the near future, unless substantial efforts will be invested in the specific and robust protection of the extremely sensitive and reactive RNA molecules from the enzymes of the digestive tract. It is impractical, perhaps, to expect such efforts until the efficacy and safety of these drugs are proven in clinical trials, which will most likely be decided over the next 5–7 years.

Efficient delivery of a drug to its target organ(s), tissue(s), cell(s), and molecule(s) is an important requirement for the success of any therapeutic treatment. Any single improvement in targeted delivery, even of existing compounds, may significantly reduce the exposure of irrelevant cells and organs, which would result in less potential side effects and improved drug efficacy.

Many common drugs target proteins. The interaction of small molecules with proteins often treats or lessens the disease by correcting some biochemical aspect of a pathological pathway. RNAi drugs are completely different from these common drugs in that their immediate target is an RNA molecule, and not a protein. The effect on the "target protein" is a secondary one, by downregulation of the abundance or translation of the mRNA coding for this protein. Therefore, there may be a substantial delay in the desired effects of RNAi if the turnover rate of the target protein is low, or if the protein is localized in a different tissue than where it is expressed so that this protein and its mRNA are spatially separated. For example, a small drug directed at a protein present in the blood can be delivered directly into the bloodstream by a variety of methods. To achieve a similar effect, an RNAi compound must be delivered specifically to the organs and cells that produce this protein. If a particular serum protein has a very slow turnover rate, it might not even be feasible to obtain the desired effect in a reasonable period of time using RNAi. Another example of a potentially deceptive RNAi target is anucleated cells, such as erythrocytes, which are deficient in active mRNA and protein syntheses. In this case, to make a correction in these cells, RNAi compounds should be directed at their progenitors. Yet another difference is that RNAi drugs act by reducing the *amount* of the target protein, and not by inhibiting its activity as many small-molecule drugs do. Reducing the amount of a target protein might produce effects different from the inhibition of its activity, because the mere concentration of a particular protein in a cell or tissue could determine the regulation of other processes or biochemical pathways in the organism. All of these potentially problematic aspects that derive from differences between protein and RNA targeting should, therefore, be carefully considered when conceiving the targeted design of an RNAi drug against a specific protein.

A clear advantage of RNAi drugs is their common physical-chemical nature, unlike the disparate nature of the kingdom of small molecules. Therefore, advances made in any aspect of development of a specific RNAi compound (e.g., delivery, stability, and modification) will likely translate to improvements of other RNAi compounds directed against different targets. Other obvious advantages to RNAi drugs are ease of design and synthesis, high specificity, ease of target validation, and fast initial development and redevelopment [193]. It is much easier do design and test RNAi compounds against several targets or against several sites of the same target, a task not feasible for the common small drug. This property of RNAi drugs should significantly improve efficacy and resistance against mutations and natural diversity or polymorphism.

Most similar to RNAi compounds in their application and chemical nature are antisense compounds; one of which, Vitravene, reached the market as an FDA-approved drug in 1998 [194, 195]. Antisense oligonucleotides have been studied for their ability to knock down gene expression since 1978 [196, 197]. Later studies support that the mechanism of antisense inhibition acts either through the inhibition of translation or through RNase H-dependent degradation of mRNA [198].

Both technologies, RNAi and antisense, can successfully inhibit target gene expression; and both share many practical problems, such as delivery, stability, and potential off-target effects. Nevertheless, a significant difference between these two technologies is their relative efficiency of compound design. An efficient antisense oligonucleotide must be able to bind its target mRNA under the physiological conditions that exist inside a cell. However, mRNAs in cells have folded structures, with more than 90% of their sequence unavailable for hybridization with short oligonucleotides [199]. Therefore, not all possible antisense oligonucleotides would be able to hybridize to a target mRNA. Although the RNAi mechanism involves a specialized apparatus for RNA targeting and efficient hybridization of siRNA, antisense oligonucleotides do not have this help. Hence, the ability to efficiently hybridize to a specific mRNA directly determines the potential success or failure of an antisense experiment. As a result, where one out of every five randomly chosen double-stranded short RNA molecules is likely to be effective in RNAi, only about one out of every hundred antisense oligonucleotides would be expected to have some activity. As physical forces playing a role in the hybridization process are complex, a more reliable means to screen for efficient antisense oligonucleotides would be an empirical approach [200], which can be an expensive, labor, and time-consuming process. In addition, the concentrations of RNAi compounds required for inhibition are lower than those required for antisense oligonucleotides, which might result in less nonspecific toxicity. For an extensive description and detailed comparison of both technologies, see the comprehensive review by Achenbach et al. [201]. In general, and by many criteria, RNAi technology seems to provide a more convenient and powerful technology for therapeutic treatment. Nevertheless, therapeutic application of RNAi technology is still in its infancy; although therapeutic applications of antisense technology have been under development for decades and have achieved some substantial success, with one drug on the market and several promising drug candidates in advanced stages of clinical trials [150, 194]. In fact, many advances in antisense technology (e.g., modification and stabilization of compounds, delivery routes, aversion of side effects) have been, and will continue to be, applicable to the related field of RNAi technology. Hence, it is, perhaps, currently difficult to judge which technology may be superior, but the potential of both for delivering effective therapeutic applications will definitely continue to be explored in the future.

7.6.9   BIOTECH AND DRUG COMPANIES CURRENTLY INVESTED IN RNAI DRUG DISCOVERY AND DEVELOPMENT

Naturally, recognizing its great utility and success in multiple research applications and its high potential for therapy, numerous industrial companies and research institutes have taken an interest and joined the RNAi-therapy game. Over a dozen biotech companies are currently competing in the development of RNAi pharmaceuticals. There seem to be even more businesses exploiting the RNAi boom by providing different kinds of RNAi-related products and services to desirous research teams, such as the synthesis of RNAi compounds and libraries; the design, construction, and distribution of RNAi vectors; and customized transfection reagents. (e.g., Invitrogen, Ambion, QIAGEN, Dharmacon). Changes are occurring fast on this playing field, as some companies have quickly been "taken out of the game" by others (e.g., Atugen was recently acquired by SR Pharma plc), and additional companies join in by changing their focus to RNAi technology (e.g., Genta). Several biotech companies have filed for FDA approval of clinical trials with RNAi compounds (Table 7.6-1).

Collectively, the companies invested in the RNAi drug business have over 40 RNAi compounds engaged in therapeutic projects, which cover a wide range of diseases, including cancers (10 compounds), infectious diseases (mostly viral, 11 compounds), neurological disorders (5 compounds), allergy and inflammation (2 compounds), diabetes, obesity, and others [202]. Some major players have been Sirna Therapeutics, Alnylam Pharmaceuticals, and Acuity Pharmaceuticals, which all compete in the development of RNAi drugs directed against age-related macular degeneration (AMD). This common choice of target disease does not seem to be coincidental, as (1) no cure is currently available for AMD, (2) the target gene product VEGF has been clearly implicated in the development of the disease, and (3) VEGF-specific siRNA compounds were found in prior studies to inhibit its expression and the growth and vascular permeability associated with the disease [189–191] (also, as discussed in Section 7.6.7). Most importantly, local delivery through topical application is feasible in this disease of the eye, assuring appropriate therapeutic concentrations of the compound and minimizing potential side effects. If there are no complications due to safety issues, in a few years we may see who will win the race to establish this new class of gene targeted medicines [203]. In the near future, the market for RNAi technology and related products will surely continue its rapid expansion [204], and more exciting news from biotech companies on advances in RNAi drug development will come into play.

7.6.10 CONCLUDING REMARKS

Extensive research and development efforts in RNAi technology during the last few years have resulted in enormous success in our ability to downregulate the expression of almost any gene in cell culture and in many different animal models. This success has created great enthusiasm and hope that the exploitation of this powerful technology will soon result in highly effective and safe new drugs, able to cure diseases, which were incurable or difficult to cure before. A significant concentration of research efforts in four key areas: target selection, targeted delivery, compound stability, and minimization of off-target effects should result in superior RNAi drugs. We shall see, probably within the next 5–7 years, whether expectations can be realized and deliver the first RNAi-based drug for human use on the market.

ACKNOWLEDGMENT

We thank Susan Francis and Vlad Malkov for their critical review of the manuscript and valuable discussion.

REFERENCES

1. Fire A, et al. (1998). Potent and specific genetic interference by double-stranded RNA in Caenorhabditis elegans. *Nature.* 391(6669):806–811.

2. Powers J, et al. (1998). A nematode kinesin required for cleavage furrow advancement. *Curr. Biol.* 8(20):1133–1136.

3. Kennerdell J R, Carthew R W (1998). Use of dsRNA-mediated genetic interference to demonstrate that frizzled and frizzled 2 act in the wingless pathway. *Cell.* 95(7):1017–1026.

4. Kennerdell J R, Carthew R W (2000). Heritable gene silencing in Drosophila using double-stranded RNA. *Nat. Biotechnol.* 18(8):896–898.

5. Li Y X, et al. (2000). Double-stranded RNA injection produces null phenotypes in zebrafish. *Dev. Biol.* 217(2):394–405.

6. Zhou Y, et al. (2002). Post-transcriptional suppression of gene expression in Xenopus embryos by small interfering RNA. *Nucleic Acids Res.* 30(7):1664–1669.

7. Svoboda P, et al. (2000). Selective reduction of dormant maternal mRNAs in mouse oocytes by RNA interference. *Development.* 127(19):4147–4156.

8. Svoboda P, Stein P, Schultz R M (2001). RNAi in mouse oocytes and preimplantation embryos: Effectiveness of hairpin dsRNA. *Biochem. Biophys. Res. Commun.* 287(5):1099–1104.

9. Hamilton A J, Baulcombe D C (1999). A species of small antisense RNA in post-transcriptional gene silencing in plants. *Science.* 286(5441):950–952.

10. Zamore P D, et al. (2000). RNAi: Double-stranded RNA directs the ATP-dependent cleavage of mRNA at 21 to 23 nucleotide intervals. *Cell.* 101(1):25–33.

11. Hammond S M, et al. (2000). An RNA-directed nuclease mediates post-transcriptional gene silencing in Drosophila cells. *Nature.* 404(6775):293–296.

12. Moss E G (2001). RNA interference: It's a small RNA world. *Curr. Biol.* 11(19): R772–R775.

13. Filipowicz W, et al. (2005). Post-transcriptional gene silencing by siRNAs and miRNAs. *Curr. Opin. Struct. Biol.* 15(3):331–341.

14. Tang G (2005). siRNA and miRNA: An insight into RISCs. *Trends Biochem. Sci.* 30(2):106–114.

15. Mullis K, et al. (1986). Specific enzymatic amplification of DNA in vitro: The polymerase chain reaction. *Cold Spring Harb. Symp. Quant. Biol.* 51(Pt 1):263–273.

16. Scharf S J, Horn G T, Erlich H A (1986). Direct cloning and sequence analysis of enzymatically amplified genomic sequences. *Science.* 233(4768):1076–1078.

17. Saiki R K, et al. (1986). Analysis of enzymatically amplified beta-globin and HLA-DQ alpha DNA with allele-specific oligonucleotide probes. *Nature.* 324(6093):163–166.

18. Napoli C, Lemieux C, Jorgensen R (1990). Introduction of a chimeric chalcone synthase gene into petunia results in reversible co-suppression of homologous genes in trans. *Plant Cell.* 2(4):279–289.

19. Tabara H, et al. (1999). The rde-1 gene, RNA interference, and transposon silencing in C. elegans. *Cell.* 99(2):123–132.

20. Hammond S M, Caudy A A, Hannon G J (2001). Post-transcriptional gene silencing by double-stranded RNA. *Nat. Rev. Genet.* 2(2):110–119.

21. Elbashir S M, Lendeckel W, Tuschl T (2001). RNA interference is mediated by 21- and 22-nucleotide RNAs. *Genes Dev.* 15(2):188–200.

22. Bernstein E, et al. (2001). Role for a bidentate ribonuclease in the initiation step of RNA interference. *Nature.* 409(6818):363–366.

23. Hammond S M (2005). Dicing and slicing: The core machinery of the RNA interference pathway. *FEBS Lett.* 579(26):5822–5829.

24. Hammond S M, et al. (2001). Argonaute2, a link between genetic and biochemical analyses of RNAi. *Science.* 293(5532):1146–1150.

25. Schwarz D S, et al. (2003). Asymmetry in the assembly of the RNAi enzyme complex. *Cell.* 115(2):199–208.

26. Khvorova A, Reynolds A, Jayasena S D (2003). Functional siRNAs and miRNAs exhibit strand bias. *Cell.* 115(2):209–216.

27. Samuel C E (2001). Antiviral actions of interferons. *Clin. Microbiol. Rev.* 14(4):778–809.

28. Karpala A J, Doran T J, Bean A G (2005). Immune responses to dsRNA: Implications for gene silencing technologies. *Immunol. Cell Biol.* 83(3):211–216.

29. Elbashir S M, et al. (2001). Duplexes of 21-nucleotide RNAs mediate RNA interference in cultured mammalian cells. *Nature.* 411(6836):494–498.

30. Lee R C, Feinbaum R L, Ambros V (1993). The C. elegans heterochronic gene lin-4 encodes small RNAs with antisense complementarity to lin-14. *Cell.* 75(5):843–854.

31. Wightman B, Ha I, Ruvkun G (1993). Posttranscriptional regulation of the heterochronic gene lin-14 by lin-4 mediates temporal pattern formation in C. elegans. *Cell.* 75(5):855–862.

32. Ruvkun G, Wightman B, Ha I (2004). The 20 years it took to recognize the importance of tiny RNAs. *Cell.* 116(suppl 2):S93–96, 2 p following S96.

33. Tijsterman M, Plasterk R H (2004). Dicers at RISC; the mechanism of RNAi. *Cell.* 117(1):1–3.

34. Cai X, Hagedorn C H, Cullen B R (2004). Human microRNAs are processed from capped, polyadenylated transcripts that can also function as mRNAs. *Rna.* 10(12): 1957–1966.

35. Denli A M, et al. (2004). Processing of primary microRNAs by the Microprocessor complex. *Nature.* 432(7014):231–235.

36. Liu J, et al. (2005). A role for the P-body component GW182 in microRNA function. *Nat. Cell Biol.* 7(12):1161–1166.

37. Carpenter A E, Sabatini D M (2004). Systematic genome-wide screens of gene function. *Nat. Rev. Genet.* 5(1):11–22.

38. Keown W A, Campbell C R, Kucherlapati R S (1990). Methods for introducing DNA into mammalian cells. *Methods Enzymol.* 185:527–537.

39. Brummelkamp T R, Bernards R, Agami R (2002). A system for stable expression of short interfering RNAs in mammalian cells. *Science.* 296(5567):550–553.

40. Sui G, et al. (2002). A DNA vector-based RNAi technology to suppress gene expression in mammalian cells. *Proc. Natl. Acad. Sci. USA.* 99(8):5515–5520.

41. Yu J Y, DeRuiter S L, Turner D L (2002). RNA interference by expression of short-interfering RNAs and hairpin RNAs in mammalian cells. *Proc. Natl. Acad. Sci. USA.* 99(9):6047–6052.

42. Miyagishi M, Taira K (2002). U6 promoter-driven siRNAs with four uridine 3′ overhangs efficiently suppress targeted gene expression in mammalian cells. *Nat. Biotechnol.* 20(5):497–500.

43. Paddison P J, et al. (2002). Short hairpin RNAs (shRNAs) induce sequence-specific silencing in mammalian cells. *Genes Dev.* 16(8):948–958.

44. Katahira T, Nakamura H (2003). Gene silencing in chick embryos with a vector-based small interfering RNA system. *Dev. Growth Differ.* 45(4):361–367.

45. Grillo M, et al. (2005). Validation of cyclin D1/CDK4 as an anticancer drug target in MCF-7 breast cancer cells: Effect of regulated overexpression of cyclin D1 and siRNA-mediated inhibition of endogenous cyclin D1 and CDK4 expression. *Breast Cancer Res. Treat.* 95:185–194.

46. Hill M, et al. (2005). Heme oxygenase-1 inhibits rat and human breast cancer cell proliferation: Mutual cross inhibition with indoleamine 2,3-dioxygenase. *Faseb J.* 19(14):1957–1968.

47. Lomas-Neira J L, et al. (2005). In vivo gene silencing (with siRNA) of pulmonary expression of MIP-2 versus KC results in divergent effects on hemorrhage-induced, neutrophil-mediated septic acute lung injury. *J. Leukoc. Biol.* 77(6):846–853.

48. Arend W P (2001). Physiology of cytokine pathways in rheumatoid arthritis. *Arthritis Rheum.* 45(1):101–106.

49. Schiffelers R M, et al. (2005). Effects of treatment with small interfering RNA on joint inflammation in mice with collagen-induced arthritis. *Arthritis Rheum.* 52(4): 1314–1318.

50. Hommel J D, et al. (2003). Local gene knockdown in the brain using viralmediated RNA interference. *Nat. Med.* 9(12):1539–1544.

51. Leung R K, Whittaker P A (2005). RNA interference: From gene silencing to gene-specific therapeutics. *Pharmacol. Ther.* 107(2):222–239.

52. Xin H, et al. (2004). High-throughput siRNA-based functional target validation. *J. Biomol. Screen.* 9(4):286–293.

53. Devroe E, Silver P A (2002). Retrovirus-delivered siRNA. *BMC Biotechnol.* 2:15.

54. Abbas-Terki T, et al. (2002). Lentiviral-mediated RNA interference. *Hum. Gene Ther.* 13(18):2197–2201.

55. Chang F H, et al. (2004). Surfection: A new platform for transfected cell arrays. *Nucleic Acids Res.* 32(3):e33.

56. Kumar R, Conklin D S, Mittal V (2003). High-throughput selection of effective RNAi probes for gene silencing. *Genome Res.* 13(10):2333–2340.

57. Mousses S, et al. (2003). RNAi microarray analysis in cultured mammalian cells. *Genome Res.* 13(10):2341–2347.

58. Silva J M, et al. (2004). RNA interference microarrays: High-throughput loss-of-function genetics in mammalian cells. *Proc. Natl. Acad. Sci. USA.* 101(17):6548–6552.

59. Ziauddin J, Sabatini D M (2001). Microarrays of cells expressing defined cDNAs. *Nature.* 411(6833):107–110.

60. Berns K, et al. (2004). A large-scale RNAi screen in human cells identifies new components of the p53 pathway. *Nature.* 428(6981):431–437.

61. Paddison P J, et al. (2004). A resource for large-scale RNA-interference-based screens in mammals. *Nature.* 428(6981):427–431.

62. van Es H H, Arts G J (2005). Biology calls the targets: Combining RNAi and disease biology. *Drug Discov. Today.* 10(20):1385–1391.

63. Zhou A, et al. (2003). Identification of NF-kappa B-regulated genes induced by TNFalpha utilizing expression profiling and RNA interference. *Oncogene.* 22(13): 2054–2064.

64. Williams N S, et al. (2003). Identification and validation of genes involved in the pathogenesis of colorectal cancer using cDNA microarrays and RNA interference. *Clin. Cancer Res.* 9(3):931–946.

65. Gunton J E, et al. (2005). Loss of ARNT/HIF1beta mediates altered gene expression and pancreatic-islet dysfunction in human type 2 diabetes. *Cell.* 122(3):337–349.

66. Semizarov D, Kroeger P, Fesik S (2004). siRNA-mediated gene silencing: A global genome view. *Nucleic Acids Res.* 32(13):3836–3845.

67. Fitzgerald K (2005). RNAi versus small molecules: Different mechanisms and specificities can lead to different outcomes. *Curr. Opin. Drug Discov. Devel.* 8(5):557–566.

68. Martinez L A, et al. (2002). Synthetic small inhibiting RNAs: Efficient tools to inactivate oncogenic mutations and restore p53 pathways. *Proc. Natl. Acad. Sci. USA.* 99(23):14849–14854.

69. Fluiter K, et al. (2003). Killing cancer by targeting genes that cancer cells have lost: Allele-specific inhibition, a novel approach to the treatment of genetic disorders. *Cell Mol. Life Sci.* 60(5):834–843.

70. Miller V M, et al. (2003). Allele-specific silencing of dominant disease genes. *Proc. Natl. Acad. Sci. USA.* 100(12):7195–7200.

71. Lee M T, et al. (2003). Inhibition of human immunodeficiency virus type 1 replication in primary macrophages by using Tat- or CCR5-specific small interfering RNAs expressed from a lentivirus vector. *J. Virol.* 77(22):11964–11972.

72. Wu K L, et al. (2005). Inhibition of Hepatitis B virus gene expression by single and dual small interfering RNA treatment. *Virus Res.* 112(1–2):100–107.

73. Chiu Y L, et al. (2004). Inhibition of human immunodeficiency virus type 1 replication by RNA interference directed against human transcription elongation factor P-TEFb (CDK9/CyclinT1). *J. Virol.* 78(5):2517–2529.

74. Fritsch L, et al. (2004). Conditional gene knock-down by CRE-dependent short interfering RNAs. *EMBO Rep.* 5(2):178–182.

75. Ventura A, et al. (2004). Cre-lox-regulated conditional RNA interference from transgenes. *Proc. Natl. Acad. Sci. USA.* 101(28):10380–10385.

76. Matsukura S, Jones P A, Takai D (2003). Establishment of conditional vectors for hairpin siRNA knockdowns. *Nucleic Acids Res.* 31(15):e77.

77. Malphettes L, Fussenegger M (2004). Macrolide- and tetracycline-adjustable siRNA-mediated gene silencing in mammalian cells using polymerase II-dependent promoter derivatives. *Biotechnol. Bioeng.* 88(4):417–425.

78. Gupta S, et al. (2004). Inducible, reversible, and stable RNA interference in mammalian cells. *Proc. Natl. Acad. Sci. USA.* 101(7):1927–1932.

79. Weber W, et al. (2005). Engineered Streptomyces quorum-sensing components enable inducible siRNA-mediated translation control in mammalian cells and adjustable transcription control in mice. *J. Gene Med.* 7(4):518–525.

80. Schubert S, et al. (2005). Maintaining inhibition: siRNA double expression vectors against coxsackieviral RNAs. *J. Mol. Biol.* 346(2):457–465.

81. Heidenreich O, et al. (2003). AML1/MTG8 oncogene suppression by small interfering RNAs supports myeloid differentiation of t(8;21)-positive leukemic cells. *Blood.* 101(8):3157–3163.

82. McWhirter J R, et al. (1999). Oncogenic homeodomain transcription factor E2A-Pbx1 activates a novel WNT gene in pre-B acute lymphoblastoid leukemia. *Proc. Natl. Acad. Sci. USA.* 96(20):11464–11469.

83. Mazieres J, et al. (2005). Inhibition of Wnt16 in human acute lymphoblastoid leukemia cells containing the t(1;19) translocation induces apoptosis. *Oncogene.* 24(34):5396–5400.

84. Nieth C, et al. (2003). Modulation of the classical multidrug resistance (MDR) phenotype by RNA interference (RNAi). *FEBS Lett.* 545(2–3):144–150.

85. Peng Z, et al. (2004). Reversal of P-glycoprotein-mediated multidrug resistance with small interference RNA (siRNA) in leukemia cells. *Cancer Gene Ther.* 11(11):707–712.

86. Stege A, et al. (2004). Stable and complete overcoming of MDR1/P-glycoprotein-mediated multidrug resistance in human gastric carcinoma cells by RNA interference. *Cancer Gene Ther.* 11(11):699–706.

87. Futami T, et al. (2002). Induction of apoptosis in HeLa cells with siRNA expression vector targeted against bcl-2. *Nucleic Acids Symposium Ser.* 2(1):251–252.

88. Monks N R, Biswas D K, Pardee A B (2004). Blocking anti-apoptosis as a strategy for cancer chemotherapy: NF-kappaB as a target. *J. Cell Biochem.* 92(4):646–650.

89. Tarnawski A, et al. (2005). Rebamipide inhibits gastric cancer growth by targeting survivin and Aurora-B. *Biochem. Biophys. Res. Commun.* 334(1):207–212.

90. Mack P C, et al. (2004). Enhancement of radiation cytotoxicity by UCN-01 in non-small cell lung carcinoma cells. *Radiat. Res.* 162(6):623–634.

91. Peng Y, et al. (2002). Silencing expression of the catalytic subunit of DNA-dependent protein kinase by small interfering RNA sensitizes human cells for radiation-induced chromosome damage, cell killing, and mutation. *Cancer Res.* 62(22):6400–6404.

92. Collis S J, et al. (2005). The life and death of DNA-PK. *Oncogene.* 24(6):949–961.

93. Tagliaferri P, et al. (2005). Antitumor therapeutic strategies based on the targeting of epidermal growth factor-induced survival pathways. *Curr. Drug Targets.* 6(3):289–300.

94. Bagasra O (2005). RNAi as an antiviral therapy. *Expert Opin. Biol. Ther.* 5(11):1463–1474.

95. Cullen B R (2005). Does RNA interference have a future as a treatment for HIV-1 induced disease? *AIDS Rev.* 7(1):22–25.

96. Bennasser Y, et al. (2005). Evidence that HIV-1 encodes an siRNA and a suppressor of RNA silencing. *Immunity.* 22(5):607–619.

97. Martinez M A, et al. (2002). Suppression of chemokine receptor expression by RNA interference allows for inhibition of HIV-1 replication. *Aids.* 16(18):2385–2390.

98. Nair M P, et al. (2005). RNAi-directed inhibition of DC-SIGN by dendritic cells: Prospects for HIV-1 therapy. *Aaps J.* 7(3):E572–578.

99. Gitlin L, Andino R (2003). Nucleic acid-based immune system: The antiviral potential of mammalian RNA silencing. *J. Virol.* 77(13):7159–7165.

100. Randall G, Rice C M (2004). Interfering with hepatitis C virus RNA replication. *Virus Res.* 102(1):19–25.

101. Giladi H, et al. (2003). Small interfering RNA inhibits hepatitis B virus replication in mice. *Mol. Ther.* 8(5):769–776.

102. Klein C, et al. (2003). Inhibition of hepatitis B virus replication in vivo by nucleoside analogues and siRNA. *Gastroenterol.* 125(1):9–18.

103. Nahlen B L, et al. (2005). Malaria risk: Estimating clinical episodes of malaria. *Nature.* 437(7056):E3; discussion E4–5.

104. Plowe C V (2005). Antimalarial drug resistance in Africa: Strategies for monitoring and deterrence. *Curr. Top. Microbiol. Immunol.* 295:55–79.

105. Ngo H, et al. (1998). Double-stranded RNA induces mRNA degradation in Trypanosoma brucei. *Proc. Natl. Acad. Sci. USA.* 95(25):14687–14692.

106. Ullu E, Tschudi C, Chakraborty T (2004). RNA interference in protozoan parasites. *Cell Microbiol.* 6(6):509–519.

107. Djikeng A, et al. (2004). Analysis of gene function in Trypanosoma brucei using RNA interference. *Meth. Mol. Biol.* 265:73–83.

108. Montalvetti A, et al. (2003). Farnesyl pyrophosphate synthase is an essential enzyme in Trypanosoma brucei. In vitro RNA interference and in vivo inhibition studies. *J. Biol. Chem.* 278(19):17075–17083.

109. Witola W H, et al. (2004). RNA-interference silencing of the adenosine transporter-1 gene in Trypanosoma evansi confers resistance to diminazene aceturate. *Exp. Parasitol.* 107(1–2):47–57.

110. Sheader K, et al. (2005). Variant surface glycoprotein RNA interference triggers a precytokinesis cell cycle arrest in African trypanosomes. *Proc. Natl. Acad. Sci. USA.* 102(24):8716–8721.

111. McRobert L, McConkey G A (2002). RNA interference (RNAi) inhibits growth of Plasmodium falciparum. *Mol. Biochem. Parasitol.* 119(2):273–278.

112. Kumar R, et al. (2002). Characterisation and expression of a PP1 serine/threonine protein phosphatase (PfPP1) from the malaria parasite, Plasmodium falciparum: Demonstration of its essential role using RNA interference. *Malar. J.* 1:5.

113. Malhotra P, et al. (2002). Double-stranded RNA-mediated gene silencing of cysteine proteases (falcipain-1 and -2) of Plasmodium falciparum. *Mol. Microbiol.* 45(5):1245–1254.

114. Mohmmed A, et al. (2003). In vivo gene silencing in Plasmodium berghei–a mouse malaria model. *Biochem. Biophys. Res. Commun.* 309(3):506–511.

115. Patankar S, et al. (2001). Serial analysis of gene expression in Plasmodium falciparum reveals the global expression profile of erythrocytic stages and the presence of antisense transcripts in the malarial parasite. *Mol. Biol. Cell.* 12(10):3114–3125.

116. Militello K T, et al. (2005). RNA polymerase II synthesizes antisense RNA in Plasmodium falciparum. *RNA.* 11(4):365–370.

117. Ding H, et al. (2003). Selective silencing by RNAi of a dominant allele that causes amyotrophic lateral sclerosis. *Aging Cell.* 2(4):209–217.

118. Maxwell M M, et al. (2004). RNA interference-mediated silencing of mutant superoxide dismutase rescues cyclosporin A-induced death in cultured neuroblastoma cells. *Proc. Natl. Acad. Sci. USA.* 101(9):3178–3183.

119. Raoul C, et al. (2005). Lentiviral-mediated silencing of SOD1 through RNA interference retards disease onset and progression in a mouse model of ALS. *Nat. Med.* 11(4):423–428.

120. Ralph G S, et al. (2005). Silencing mutant SOD1 using RNAi protects against neurodegeneration and extends survival in an ALS model. *Nat. Med.* 11(4):429–433.

121. Xia X G, et al. (2005). An RNAi strategy for treatment of amyotrophic lateral sclerosis caused by mutant Cu, Zn superoxide dismutase. *J. Neurochem.* 92(2):362–367.

122. Parrish S, et al. (2000). Functional anatomy of a dsRNA trigger: Differential requirement for the two trigger strands in RNA interference. *Mol. Cell.* 6(5):1077–1087.

123. Sharp P A (2001). RNA interference–2001. *Genes Dev.* 15(5):485–490.

124. Gitlin L, Karelsky S, Andino R (2002). Short interfering RNA confers intracellular antiviral immunity in human cells. *Nature.* 418(6896):430–434.

125. Yang D, et al. (2002). Short RNA duplexes produced by hydrolysis with Escherichia coli RNase III mediate effective RNA interference in mammalian cells. *Proc. Natl. Acad. Sci. USA.* 99(15):9942–9947.

126. Myers J W, et al. (2003). Recombinant Dicer efficiently converts large dsRNAs into siRNAs suitable for gene silencing. *Nat. Biotechnol.* 21(3):324–328.

127. Oleinikov A V, et al. (2003). Self-assembling protein arrays using electronic semiconductor microchips and in vitro translation. *J. Proteome Res.* 2:313–319.

128. Oleinikov A V, Zhao J, Gray M D (2005). RNA interference by mixtures of siRNAs prepared using custom oligonucleotide arrays. *Nucl. Acids Res.* 33(10):e92.

129. Sledz C A, et al. (2003). Activation of the interferon system by short-interfering RNAs. *Nat. Cell Biol.* 5(9):834–839.

130. Robbins M A, Rossi J J (2005). Sensing the danger in RNA. *Nat. Med.* 11(3):250–251.

131. Zhou H S, Liu D P, Liang C C (2004). Challenges and strategies: The immune responses in gene therapy. *Med. Res. Rev.* 24(6):748–761.

132. Scacheri P C, et al. (2004). Short interfering RNAs can induce unexpected and divergent changes in the levels of untargeted proteins in mammalian cells. *Proc. Natl. Acad. Sci. USA.* 101(7):1892–1897.

133. Kariko K, et al. (2004). Small interfering RNAs mediate sequence-independent gene suppression and induce immune activation by signaling through toll-like receptor 3. *J. Immunol.* 172(11):6545–6549.

134. Kariko K, et al. (2004). Exogenous siRNA mediates sequence-independent gene suppression by signaling through toll-like receptor 3. *Cells Tissues Organs.* 177(3):132–138.

135. Persengiev S P, Zhu X, Green M R (2004). Nonspecific, concentration-dependent stimulation and repression of mammalian gene expression by small interfering RNAs (siRNAs). *RNA.* 10(1):12–18.

136. Ma Z, et al. (2005). Cationic lipids enhance siRNA-mediated interferon response in mice. *Biochem. Biophys. Res. Commun.* 330(3):755–759.

137. Tomanin R, Scarpa M (2004). Why do we need new gene therapy viral vectors? Characteristics, limitations and future perspectives of viral vector transduction. *Curr. Gene Ther.* 4(4):357–372.

138. McManus M T, et al. (2002). Small interfering RNA-mediated gene silencing in T lymphocytes. *J. Immunol.* 169(10):5754–5760.

139. Holen T, et al. (2002). Positional effects of short interfering RNAs targeting the human coagulation trigger Tissue Factor. *Nucl. Acids Res.* 30(8):1757–1766.

140. Bitko V, et al. (2005). Inhibition of respiratory viruses by nasally administered siRNA. *Nat. Med.* 11(1):50–55.

141. Gong H, et al. (2005). The role of small RNAs in human diseases: Potential troublemaker and therapeutic tools. *Med. Res. Rev.* 25(3):361–381.

142. Pusch O, et al. (2003). Nucleotide sequence homology requirements of HIV-1-specific short hairpin RNA. *Nucl. Acids Res.* 31(22):6444–6449.

143. Saxena S, Jonsson Z O, Dutta A (2003). Small RNAs with imperfect match to endogenous mRNA repress translation. Implications for off-target activity of small inhibitory RNA in mammalian cells. *J. Biol. Chem.* 278(45):44312–44319.

144. Holen T, et al. (2005). Tolerated wobble mutations in siRNAs decrease specificity, but can enhance activity in vivo. *Nucl. Acids Res.* 33(15):4704–4710.

145. Jackson A L, et al. (2003). Expression profiling reveals off-target gene regulation by RNAi. *Nat. Biotechnol.* 21(6):635–637.

146. Devroe E, Silver P A (2004). Therapeutic potential of retroviral RNAi vectors. *Expert Opin. Biol. Ther.* 4(3):319–327.

147. Yi Y, Hahm S H, Lee K H (2005). Retroviral gene therapy: Safety issues and possible solutions. *Curr. Gene Ther.* 5(1):25–35.

148. Morrissey D V, et al. (2005). Activity of stabilized short interfering RNA in a mouse model of hepatitis B virus replication. *Hepatol.* 41(6):1349–1356.

149. Johnson P H, Quay S C (2005). Advances in nasal drug delivery through tight junction technology. *Expert Opin. Drug. Deliv.* 2(2):281–298.

150. Patil S D, Rhodes D G, Burgess D J (2005). DNA-based therapeutics and DNA delivery systems: A comprehensive review. *Aaps J.* 7(1):E61–77.

151. Song E, et al. (2005). Antibody mediated in vivo delivery of small interfering RNAs via cell-surface receptors. *Nat. Biotechnol.* 23(6):709–717.

152. Williams B R (2005). Targeting specific cell types with silencing RNA. *N. Engl. J. Med.* 353(13):1410–1411.

153. Gou D, et al. (2004). Gene silencing in alveolar type II cells using cell-specific promoter in vitro and in vivo. *Nucl. Acids Res.* 32(17):e134.

154. Gu J, et al. (2002). A novel single tetracycline-regulative adenoviral vector for tumor-specific Bax gene expression and cell killing in vitro and in vivo. *Oncogene.* 21(31): 4757–4764.

155. Bertrand J R, et al. (2002). Comparison of antisense oligonucleotides and siRNAs in cell culture and in vivo. *Biochem. Biophys. Res. Commun.* 296(4):1000–1004.

156. Soutschek J, et al. (2004). Therapeutic silencing of an endogenous gene by systemic administration of modified siRNAs. *Nature.* 432(7014):173–178.

157. Chen X, et al. (2005). Chemical modification of gene silencing oligonucleotides for drug discovery and development. *Drug Discov. Today.* 10(8):587–593.

158. de Fougerolles A, et al. (2005). RNA interference in vivo: Toward synthetic small inhibitory RNA-based therapeutics. *Meth. Enzymol.* 392:278–296.

159. Guo S, et al. (2005). Specific delivery of therapeutic RNAs to cancer cells via the dimerization mechanism of phi29 motor pRNA. *Hum. Gene Ther.* 16(9):1097–1109.

160. Kakizawa Y, Furukawa S, Kataoka K (2004). Block copolymer-coated calcium phosphate nanoparticles sensing intracellular environment for oligodeoxynucleotide and siRNA delivery. *J. Control. Rel.* 97(2):345–356.

161. Schiffelers R M, et al. (2004). Cancer siRNA therapy by tumor selective delivery with ligand-targeted sterically stabilized nanoparticle. *Nucl. Acids Res.* 32(19):e149.

162. Lochmann D, et al. (2005). Albumin-protamine-oligonucleotide nanoparticles as a new antisense delivery system. Part 1: Physicochemical characterization. *Eur. J. Pharm. Biopharm.* 59(3):419–429.

163. Khaled A, et al. (2005). Controllable self-assembly of nanoparticles for specific delivery of multiple therapeutic molecules to cancer cells using RNA nanotechnology. *Nano. Lett.* 5(9):1797–1808.

164. Safinya C R (2001). Structures of lipid-DNA complexes: Supramolecular assembly and gene delivery. *Curr. Opin. Struct. Biol.* 11(4):440–448.

165. Paul C P, et al. (2002). Effective expression of small interfering RNA in human cells. *Nat. Biotechnol.* 20(5):505–508.

166. Sioud M (2005). siRNA delivery in vivo. *Meth. Mol. Biol.* 309:237–249.

167. Uprichard S L (2005). The therapeutic potential of RNA interference. *FEBS Lett.* 579(26):5996–6007.

168. Pai S I, et al. (2005). Prospects of RNA interference therapy for cancer. *Gene Ther.* 8:8.

169. Spankuch B, Strebhardt K (2005). RNA interference-based gene silencing in mice: The development of a novel therapeutical strategy. *Curr. Pharm. Des.* 11(26):3405–3419.

170. Hu-Lieskovan S, et al. (2005). Sequence-specific knockdown of EWS-FLI1 by targeted, nonviral delivery of small interfering RNA inhibits tumor growth in a murine model of metastatic Ewing's sarcoma. *Cancer Res.* 65(19):8984–8992.

171. Lewis D L, et al. (2002). Efficient delivery of siRNA for inhibition of gene expression in postnatal mice. *Nat. Genet.* 32(1):107–108.

172. McCaffrey A P, et al. (2002). RNA interference in adult mice. *Nature.* 418(6893): 38–39.

173. Song E, et al. (2003). RNA interference targeting Fas protects mice from fulminant hepatitis. *Nat. Med.* 9(3):347–351.

174. Yano J, et al. (2004). Antitumor activity of small interfering RNA/cationic liposome complex in mouse models of cancer. *Clin. Cancer Res.* 10(22):7721–7726.

175. Lee J H, et al. (2004). KAI1 COOH-terminal interacting tetraspanin (KITENIN), a member of the tetraspanin family, interacts with KAI1, a tumor metastasis suppressor, and enhances metastasis of cancer. *Cancer Res.* 64(12):4235–4243.

176. Lee J H, et al. (2005). Suppression of progression and metastasis of established colon tumors in mice by intravenous delivery of short interfering RNA targeting KITENIN, a metastasis-enhancing protein. *Cancer Res.* 65(19):8993–9003.

177. Uprichard S L, et al. (2005). Clearance of hepatitis B virus from the liver of transgenic mice by short hairpin RNAs. *Proc. Natl. Acad. Sci. USA.* 102(3):773–778.

178. Ge Q, et al. (2004). Inhibition of influenza virus production in virus-infected mice by RNA interference. *Proc. Natl. Acad. Sci. USA.* 101(23):8676–8681.

179. Tompkins S M, et al. (2004). Protection against lethal influenza virus challenge by RNA interference in vivo. *Proc. Natl. Acad. Sci. USA.* 101(23):8682–8686.

180. Inoue A, et al. (2005). Electro-transfer of small interfering RNA ameliorated arthritis in rats. *Biochem. Biophys. Res. Commun.* 336(3):903–908.

181. Lovett-Racke A E, et al. (2004). Silencing T-bet defines a critical role in the differentiation of autoreactive T lymphocytes. *Immunity.* 21(5):719–731.

182. Selkoe D J, (2004). Alzheimer disease: Mechanistic understanding predicts novel therapies. *Ann. Intern. Med.* 140(8):627–638.

183. Holsinger R M, et al. (2002). Increased expression of the amyloid precursor beta-secretase in Alzheimer's disease. *Ann. Neurol.* 51(6):783–786.

184. Fukumoto H, et al. (2002). Beta-secretase protein and activity are increased in the neocortex in Alzheimer disease. *Arch. Neurol.* 59(9):1381–1389.

185. Li R, et al. (2004). Amyloid beta peptide load is correlated with increased beta-secretase activity in sporadic Alzheimer's disease patients. *Proc. Natl. Acad. Sci. USA.* 101(10):3632–3637.

186. Singer O, et al. (2005). Targeting BACE1 with siRNAs ameliorates Alzheimer disease neuropathology in a transgenic model. *Nat. Neurosci.* 8(10):1343–1349.

187. Gusella J F, MacDonald M E (2000). Molecular genetics: Unmasking polyglutamine triggers in neurodegenerative disease. *Nat. Rev. Neurosci.* 1(2):109–115.

188. Wang Y L, et al. (2005). Clinico-pathological rescue of a model mouse of Huntington's disease by siRNA. *Neurosci. Res.* 53(3):241–249.

189. Krzystolik M G, et al. (2002). Prevention of experimental choroidal neovascularization with intravitreal anti-vascular endothelial growth factor antibody fragment. *Arch. Ophthalmol.* 120(3):338–346.

190. Reich S J, et al. (2003). Small interfering RNA (siRNA) targeting VEGF effectively inhibits ocular neovascularization in a mouse model. *Mol. Vis.* 9:210–216.

191. Tolentino M J, et al. (2004). Intravitreal injection of vascular endothelial growth factor small interfering RNA inhibits growth and leakage in a nonhuman primate, laser-induced model of choroidal neovascularization. *Retina.* 24(1):132–138.

192. Check E (2005). A crucial test. *Nat. Med.* 11(3):243–244.

193. Barik S (2004). Development of gene-specific double-stranded RNA drugs. *Ann. Med.* 36(7):540–551.

194. Graham M J (1999). Oligonucleotide Therapeutics–IBC Sixth International Conference. *La Jolla, CA. IDrugs.* 2(7):653–655.

195. Orr R M (2001). Technology evaluation: Fomivirsen, Isis Pharmaceuticals Inc/CIBA vision. *Curr. Opin. Mol. Ther.* 3(3):288–294.

196. Zamecnik P C, Stephenson M L (1978). Inhibition of Rous sarcoma virus replication and cell transformation by a specific oligodeoxynucleotide. *Proc. Natl. Acad. Sci. USA.* 75(1):280–284.

197. Stephenson M L, Zamecnik P C (1978). Inhibition of Rous sarcoma viral RNA translation by a specific oligodeoxyribonucleotide. *Proc. Natl. Acad. Sci. USA.* 75(1): 285–288.

198. Wagner R W (1994). Gene inhibition using antisense oligodeoxynucleotides. *Nature.* 372(6504):333–335.

199. Sohail M (2001). Antisense technology: Inaccessibility and non-specificity. *Drug. Discov. Today.* 6(24):1260–1261.

200. Sohail M, Southern E M (2002). Oligonucleotide scanning arrays: Application to high-throughput screening for effective antisense reagents and the study of nucleic acid interactions. *Adv. Biochem. Engineer./Biotechno.* 77:43–56.

201. Achenbach T V, Brunner B, Heermeier K (2003). Oligonucleotide-based knockdown technologies: antisense versus RNA interference. *Chembiochem.* 4(10):928–935.

202. Beal J (2005). Silence is golden: Can RNA interference therapeutics deliver? *Drug Discov. Today.* 10(3):169–172.

203. Fishman M C, Porter J A (2005). Pharmaceuticals: a new grammar for drug discovery. *Nature.* 437(7058):491–493.

204. Howard K (2003). Unlocking the money-making potential of RNAi. *Nat. Biotechnol.* 21(12):1441–1446.

7.7

DELIVERY SYSTEMS FOR PEPTIDES/ OLIGONUCLEOTIDES AND LIPOPHILIC NUCLEOSIDE ANALOGS

R.A. SCHWENDENER[1] AND HERBERT SCHOTT[2]

[1] *Institute of Molecular Cancer Research, University of Zurich, Zurich, Switzerland*
[2] *Institute of Organic Chemistry, University of Tübingen, Tübingen, Germany*

Chapter Contents

7.7.1	State of the Art of Nanosized Delivery Systems	1150
7.7.2	Comparison of Viral vs. NonViral Gene Delivery Systems	1150
7.7.3	Viral Systems	1152
	7.7.3.1 Adenoviruses	1152
	7.7.3.2 Poxviruses	1152
	7.7.3.3 Herpex Simplex Virus	1153
	7.7.3.4 Lentiviruses	1153
	7.7.3.5 Adeno-Associated Virus	1153
7.7.4	NonViral Systems	1154
	7.7.4.1 Cationic Polymers, Lipoplexes	1155
	7.7.4.2 NonLipidic Polycation Gene Delivery Systems	1155
	7.7.4.3 Other Methods	1156
	7.7.4.4 Virosomes	1157
7.7.5	Liposomes and Cationic Liposome–DNA Complexes (Lipoplexes)	1158
7.7.6	Peptide and Oligonucleotide Liposome Vaccine Formulations	1162
7.7.7	Liposomes as Carriers of Lipophilic and Amphiphilic Nucleoside Analogs	1163
7.7.8	Outlook and Future Directions	1167
	References	1167

7.7.1   STATE OF THE ART OF NANOSIZED DELIVERY SYSTEMS

The earliest developments of drug delivery systems (DDS) date back to the 1950s where the first microencapsulated drugs were introduced. In the 1960s polymer-based slow release systems appeared, and as the first nanosized DDS with spherical shape properties, the liposomes were recognized as models to study membranes and as carriers of both hydrophilic and lipophilic molecules. Liposomes, small unilamellar phospholipid bilayer vesicles, undoubtedly represent the most extensively and advanced drug delivery vehicles. After a long period of research and development efforts, liposome-formulated drugs have now entered the clinics to treat cancer and systemic fungal infections, mainly because they are biologically inert and biocompatible and do not cause unwanted toxic or antigenic reactions [1–3]. For the delivery of genetic material (DNA, ribozymes, DNAzymes, aptamers, (antisense-) oligonucleotides, small interfering RNAs), the liposomes, in particular lipid–DNA complexes termed lipoplexes, compete with viral gene transfection systems, as will be outlined in the following sections. Nanoparticles, nanospheres, polymersomes, nanogels, micelles, dendrimers, and virosomes are other main types of nanocarrier systems used for drug delivery [4–7]. As shown in Figure 7.7-1, these drug delivery systems vary in their compositions, shapes, sizes, drug loading capacity, as well as their pharmacokinetic and organ or tissue targeting properties [8].

DDS are developed for drugs with nonideal properties that include (1) poor solubility, where a conventional pharmaceutic formulation is difficult to prepare as poorly water soluble drugs may precipitate in aqueous media. (2) Tissue damage caused by inadvertent extravasation of drugs, e.g., tissue necrosis caused by cytotoxic drugs. (3) Loss of drug activity after administration, e.g., enzymatic and fast metabolic degradation. (4) Unfavorable pharmacokinetic properties and poor biodistribution. (5) Lack of selectivity for target organs or tissues. Systemic drug distribution may cause toxic side effects, and low concentrations in target tissues may cause suboptimal therapeutic effects.

The formulation of pharmacologically active drug molecules in DDS can improve or abolish these unfavorable properties. However, there are also drawbacks in DDS development, such as system complexity, unwanted biologic effects, stability, costs of development and scale-up, as well as intellectual property issues. In the limited format of this review, it is not possible to cover all methods and references in the field. Hence, we concentrate this review on DDS for the delivery of peptides, DNA, plasmids, oligodeoxynucleotides, siRNA, and lipophilic nucleoside derivatives. Vaccine delivery systems will also be mentioned, and examples will be provided to demonstrate the general development trends.

7.7.2   COMPARISON OF VIRAL VS. NONVIRAL GENE DELIVERY SYSTEMS

Viral gene delivery has become an important therapeutic strategy for the development of novel treatment approaches. A deeper understanding of vector biology and the molecular mechanisms of disease together with remarkable advances in molecular biology and vector technology have considerably advanced the field of human

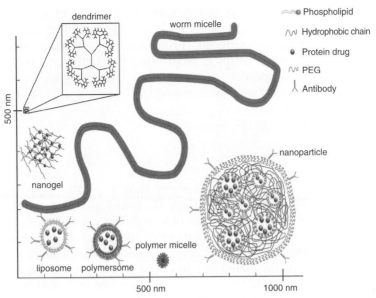

Figure 7.7-1. Nanocarriers for vascular drug delivery. Schematic representation of several main types of nanocarriers depicted in relative size scale with limited and oversimplified structural features. All nanocarriers can be surface-conjugated with targeting antibodies (or alternative affinity moieties) and PEG polymer providing stealth features. Dendrimers are the smallest of nanocarriers, in the maximum range of tens of nanometers. They possess multiple end groups suitable for a high extent of coupling targeting or active agents. Liposomes, composed of biologically derived phospholipids, are the most common form of nanocarriers with large aqueous loading potential; yet they are also the least stable of carriers. Polymersomes, one of three self-assembled polymeric nanocarriers, are the synthetic polymer analog of liposomes, possessing enhanced stability, dense PEG coating, and therefore prolonged circulation *in vivo*. Worm micelles are long, flexible cylindrical polymer micelles that possess one of the longest circulation times recorded *in vivo* with a yet unknown potential for therapeutic applications. Polymeric micelles are the smallest of the self-assembling polymer aggregate carriers and are the least stable self-assembler. Nanogels are composed of cross-linked polymers with drug entrapped into the ensuing matrix. Their circulation and potential therapeutic uses remain to be studied. Nanoparticles are solid polymer structures formed through processing rather than through self-assembly methods. They represent the largest of the carriers, have the greatest active protein loading capacity measured to date, and can protect encapsulated therapeutic enzymes from external proteolysis. (With permission from Ref. 8.)

gene therapy development. However, most viral gene delivery systems used to date have demonstrated limitations in practicality and safety, mainly due to low levels and short duration of recombinant transgene expression, induction of host immunogenicity to vector constituents, and suboptimal transgene expression to tissues or cells. A recent, additional cause for concern over using viral vectors is the phenomenon known as insertional mutagenesis, in which the chromosomal integration of viral gene material either interrupts the expression of a tumor suppressor gene or activates an oncogene, leading to the malignant transformation of cells.

Thus, safer, nonviral delivery approaches are needed and the latest advances indicate that efficient, long-term gene expression can be achieved by nonviral means. In particular, integration of DNA can be targeted to specific genomic sites without harmful consequences, and it is possible to maintain transgenes as small episomal plasmids or artificial chromosomes. The application of these approaches to human gene therapy is progressively becoming a reality. Here, we briefly compare use, properties, advantages, and disadvantages of viral and nonviral gene delivery, the two main types of DDS that are used in gene therapy and vaccine approaches.

7.7.3 VIRAL SYSTEMS

The lack of efficient nontoxic gene delivery systems is still the major impediment to the successful application of gene therapy. Having evolved to deliver their genes to target cells, viruses are effective means of gene delivery and they can be manipulated to express therapeutic genes or to replicate specifically in certain cells.

The first viral vector systems were developed more than 25 years ago [9], and since then viral gene therapy strategies has been progressively developed [10]. A variety of virus vectors has been employed and modified to deliver genes to cells to provide either transient, such as adenovirus [11–13], poxviruses (vaccinia) [14], or herpes virus [15], or permanent, such as retroviruses (lentivirus) [16–18] and adeno-associated virus [19], transgene expression; each approach has its characteristic advantages and disadvantages.

7.7.3.1 Adenoviruses

Replication-defective adenoviruses are vectors of choice for delivering corrective genes into human cells. Major efforts are directed to design new generations of adenoviral vectors that feature reduced immunogenicity and improved targeting ability. Various targeting strategies have been attempted aiming at obtaining optimized and specific cellular transduction, including that of genetic manipulation of the viral capsid. Modification of the tropism-determining fiber protein and other capsid proteins has yielded vectors that are superior to the first-generation adenoviruses employed for gene therapy [11]. Adenoviral-based vectors are susceptible both to cytotoxic T-lymphocyte and humoral immune responses. In addition, leaky adenoviral genes also render transduced cells susceptible to host immune responses. These are the main reasons why adenoviral-based vectors are not suitable to correct genetic disorders, which require long-term expression of the transgene [20]. The production of adenoviral vectors for gene therapy applications still faces several challenges that limit the availability of high-quality material for clinical applications [21, 22].

7.7.3.2 Poxviruses

Poxviruses represent a heterogenous group of DNA viruses that have been used to express a multitude of foreign genes. Vaccinia virus is the prototypical recombinant poxvirus that can generate potent antibody and T-cell responses. Recombinant vaccinia viruses (rVVs) as nonreplicating viral vectors have been demonstrated for their great potential as vaccines, diminished cytopathic effects, high levels of protein expression, and strong immunogenicity, and they are relatively safe in animals and in human patients. These properties have led to the use of rVVs as vaccines against HIV and cancer [14, 23].

7.7.3.3   Herpes Simplex Virus

Attenuated genetically engineered herpes simplex virus (HSV) vectors are potential vectors for several human therapy applications. These include delivery and expression of human genes to cells of the nervous systems, selective destruction of cancer cells, prophylaxis against infection with HSV or other infectious diseases, and targeted infection to specific tissues or organs [15].

7.7.3.4   Lentiviruses

Lentiviruses, members of the retroviral family, have the ability to infect cells at both mitotic and postmitotic stages of the cell cycle, thus opening the possibility to target nondividing cells and tissues. Human Immunodeficiency Virus (HIV)-based vectors have been used *in vitro* and *in vivo* in several situations; however, safety concerns still exist. Therefore, the development of vector systems based on primate as well as nonprimate lentiviruses is ongoing. Recent developments in the modification of the virus coat allow more targeted approaches and open new possibilities for the systemic delivery of therapeutic viruses [24]. However, the specific mechanisms used by different retroviruses to efficiently deliver their genes into cell nuclei remain largely unclear. Understanding these molecular mechanisms may reveal features to improve the efficacy of current retroviral vectors [25].

7.7.3.5   Adeno-Associated Virus

Vectors based on the adeno-associated virus (AAV) have attracted much attention as potent gene-delivery vehicles, mainly because of the persistence of this non-pathogenic virus in the host cell and its sustainable therapeutic gene expression. The principal historical limitation of this vector system, efficiency of recombinant AAV-mediated (rAAV) transduction, has recently observed a dramatic increase as the titer, purity, and production capacity of rAAV preparations have improved [26]. AAV vectors have been used in phase I clinical trials for the treatment of neurological disorders, such as Parkinson's and Canavan's diseases. Indeed, AAV-mediated gene transfer is a promising tool for the delivery of the therapeutic gene into the central and peripheral nervous systems. AAV-mediated gene transfer was also applied in phase I and II clinical trials for the treatment of cystic fibrosis and in phase I trials for the treatment of hemophilia B. In the context of cancer, the ability of attenuated viruses to replicate specifically in tumor cells has already yielded some impressive results in clinical trials, allowing the design of new therapeutic approaches, particularly when combined with other approved anticancer therapies. Despite the remark-able progress that has been reported, the design of further optimized vectors is still required. As it stands, AAV-mediated gene transfer has a limited capacity in accommodating foreign genes. In addition, some pre-clinical studies have shown that AAV-derived vectors can cause tumors in animals due to mutagenic random vector integration into the genome [27]. To circumvent this problem, a novel approach to AAV-mediated gene therapy based on gene targeting through homologous recombination has been introduced that allows efficient, high-fidelity, nonmutagenic gene repair in a host cell [28].

7.7.4 NONVIRAL SYSTEMS

Currently, the most severe limitations of the nonviral gene therapy systems are low transfection efficiency of gene material into the target cells, physico-chemical instability, and cytotoxicity. The major obstacles encountered in the transfer of foreign genetic material into the body include interactions with blood components and vascular endothelial cells and uptake by the mononuclear phagocyte system (MPS). The degradation of DNA by serum nucleases is another major obstruction for functional delivery to the target. In addition to targeting a specific cell type, an ideal nonviral vector (liposomes, lipoplexes, virosomes), once taken up by a target cell, must manifest an efficient endosomal escape, provide sufficient protection of DNA in the cytosol and help facilitate an easy passage of cytosolic DNA to the nucleus (Figure 7.7-2).

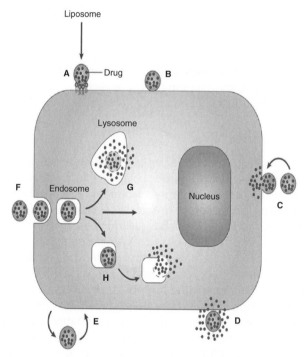

Figure 7.7-2. Liposome–cell interactions. Drug-loaded liposomes can specifically (**A**) or nonspecifically (**B**) adsorb onto the cell surface. Liposomes can also fuse with the cell membrane (**C**) and release their contents into the cell cytoplasm, or they can be destabilized by certain cell membrane components when adsorbed on the surface (**D**) so that the released drug can enter the cell via micro-pinocytosis. Liposome can undergo the direct or transfer-protein mediated exchange of lipid components with the cell membrane (**E**) or be taken up by specific or nonspecific endocytosis (**F**). In the case of endocytosis, a liposome can be delivered by the endosome into the lysosome (**G**), or en route to the lysosome, the liposome can provoke endosome destabilization (**H**), which results in drug liberation into the cytoplasm. (With permission from Ref. 29.)

7.7.4.1 Cationic Polymers, Lipoplexes

Cationic polymers have a great potential for DNA complexation and have shown to be useful as nonviral vectors for gene therapy applications. For the past 15–20 years liposomes composed of cationic lipids, termed lipoplexes, have routinely been used for the delivery of nucleic acids such as plasmids, oligodeoxynucleotides, and siRNA to cells in culture and *in vivo*. Many of these reagents are commercially available or can be formulated in the laboratory [30–32]. Most cationic lipid–DNA complexes form a multilayered structure with DNA sandwiched between the cationic lipids. Much more rarely, an inverted hexagonal structure with single DNA strands encapsulated in lipid tubules is observed [33]. Among other advantages, lipoplexes have the ability to transfer very large genes into cells. However, as the understanding of their mechanisms of action is still incomplete, their transfection efficiencies remain low compared with those of viruses. Particularly in cultured cells, toxicity remains a significant problem. In addition, these complexes are immunostimulatory, a fact that may either be harmful or beneficial. The development of cationic lipids that are safe to use, especially for *in vivo* applications, and possess enhanced transfection capabilities is an ongoing process. The lipoplexes are described in detail in Section 7.7.5.

7.7.4.2 NonLipidic Polycation Gene Delivery Systems

Of the many nonlipidic polycation gene delivery systems developed in the past decades, poly(L-lysine) (PLL) was the first polycation used for nonviral gene delivery [34]. Among a vast number of other positively charged polymers, polyethylenimine (PEI) has been widely used for nonviral transfection *in vitro* and *in vivo* and has an advantage over other polycations in that it combines strong DNA condensation capacity with an intrinsic endosomolytic activity [35–38].

Other synthetic and natural polycations developed as nonviral vectors are polyamidoamine dendrimers, the synthetic cationic polymer poly(2-dimethylamino)ethyl methacrylate (PDMAEMA) [39], and chitosan. Polyamidoamine (PAMAM) dendrimers represent a novel class of polycationic synthetic polymers that can be used for gene transfer [40, 41]. The three-dimensional spherical structure of dendrimers offers synthesis control of the molecule in terms of degree and generation of branching. The control of branching of the dendrimers during synthesis allows the production of polymer particles with a very high degree of monodispersity, which is a significant advantage over other polymers such as polylysine that generate highly polydisperse particles (see Figure 7.7-1). Low polydispersity can lead to reproducible gene delivery and a clinically reliable formulation. The cationic amino acid residues in the polymeric structure of PAMAM dendrimers can help in DNA condensation and endosome release. Dendrimers that protect oligonucleotides from serum nucleases have been used to enhance oligonucleotide delivery [42].

Chitosan, a natural-based polymer obtained by alkaline deacetylation of chitin, is nontoxic, biocompatible, and biodegradable. These properties make chitosan a promising candidate for conventional and novel drug delivery systems. Because of the high affinity of chitosan for cell membranes, it has been used as a coating agent for liposome formulations [43–45].

7.7.4.3 Other Methods

Various other methods, such as DNA electrotransfer, electroporation-based gene transfer, calcium phosphate nanoparticles, peptide nucleic acids, and cell penetrating peptides, round off the currently used methods in the field of nonviral gene delivery techniques.

DNA electrotransfer, the use of electric pulses to transfect various types of cells, is well known and regularly used *in vitro* for bacteria and eukaryotic cell transformation. Electric pulses can also be delivered *in vivo* either through the skin or with electrodes in direct contact with the target tissues. After injection of naked DNA in a tissue, appropriate local electric pulses can result in a very high expression of the transferred genes [46].

Electroporation has been applied in preclinical autoimmune and/or inflammatory diseases to deliver either cytokines and anti-inflammatory agents or immunoregulatory molecules. The method is also effective for the intratumoral delivery of therapeutic vectors, and it strongly boosts DNA vaccination against infectious agents or tumor antigens. Electroporation gene therapy has become a widely used method for nonviral gene delivery, including applications for intramuscular and intratumoral electro-gene transfer and for the transfection of dendritic and stem cells [47, 48]. The current challenges faced by both *in vitro* and *in vivo* applications comprise the enhancement of transfection efficiency, extention of the duration of gene expression, and increase of the survival rate for *in vitro* cell transfections.

Virus- or liposome-like-sized calcium phosphate nanoparticles of 20–200 nm mean diameter have been found to overcome many of the known limitations in delivering genes to the nucleus of specific cells. It has been demonstrated that calcium ions play an important role in endosomal escape, cytosolic stability, and enhanced nuclear uptake of DNA through nuclear pore complexes. The role of exogenous calcium ions to overcome the major obstacles encountered in the practical accomplishment of gene delivery suggests that calcium phosphate nanoparticles can be designated as a new generation of nonviral vectors [49].

Peptide nucleic acid (PNA) is a powerful new biomolecular tool with a wide range of important applications. PNA mimics the behavior of DNA and binds complementary nucleic acid strands and RNA sequences with high affinity and selectivity. The unique chemical, physical, and biological properties of PNA are exploited to produce powerful biomolecular tools, antisense and antigene agents, molecular probes, and biosensors [50–53].

Cell penetrating peptides (CPPs) have proven to be an efficient intracellular delivery system overcoming the lipophilic barrier of cell membranes [54]. CPPs can deliver a wide range of large cargo molecules such as proteins, peptides, oligonucleotides, and even small particles as liposomes to a variety of cell types and to different cellular compartments. The CPPs are basic, lysine-, or arginine- rich amphipathic peptides. The peptides originate from different sources, either as naturally occurring peptide sequences, virally derived (TAT, VP22), from transcription factors (pAntp), as chimeric (transportan) or as synthetic peptides (polyarginines) and others [55].

As depicted in Figure 7.7-3, CPPs can either form complexes with peptides, proteins, plasmids, oligonucleotides, or siRNA molecules or they can be covalently linked to these cargo molecules [54]. Larger structures such as liposomes have also

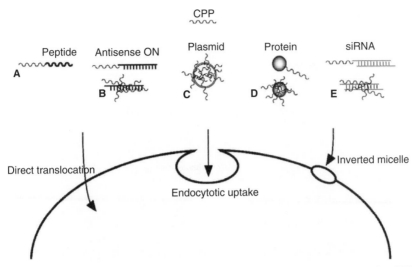

Figure 7.7-3. Cell penetrating peptides, CPPs. Suggested uptake mechanisms for CPPs and examples of delivered cargoes. (A) CPP and peptide in single amino acid chain. (B) Oligo-deoxynucleotides either in complex or covalently linked. (C) Plasmid in complex by electrostatic interaction. (D) Protein either as fusion protein or in complex with CPP. (E) siRNA, covalently linked or as complex. (With permission from Ref. 54.)

been decorated with the TAT [56] or pAntp CPPs [57], demonstrating higher cell uptake rates *in vitro*.

Unfortunately, their usefulness as drug delivery systems is hampered by their ability to penetrate virtually any cell type both *in vitro* and *in vivo*. This feature makes CPP applications as target specific drug delivery systems complicated and their application as therapeutic drug carrier systems seems unlikely, unless their target cell specificity can be significantly improved.

7.7.4.4 Virosomes

Virosomes were developed from liposomes by combining liposomes with fusogenic viral envelope proteins. Almeida et al. [58] were the first to report on the generation of lipid vesicles containing viral spike proteins derived from influenza virus. Using preformed liposomes and hemagglutinin (HA) and neuraminidase (NA), purified from influenza virus, they succeeded to generate membrane vesicles with spike proteins protruding from the vesicle surface. Visualization of these vesicles by electron microscopy revealed that they very much resembled native influenza virus. Consequently, they were named virosomes. Reconstituted viral envelopes (virosomes, artificial viral envelopes) appear to be ideally suited as vaccine formulations for the delivery of protein antigens to the cytosol of antigen presenting cells (APCs), and thus for the introduction of antigenic peptides into the MHC class I presentation pathway. Cytotoxic T-cell activity can be induced by immunization of mice with an antigenic peptide or an entire protein encapsulated in virosomes. As shown schematically in Figure 7.7-4, the action of influenza virosomes is likely to involve both delivery of the enclosed antigen to the cytosol of antigen presenting

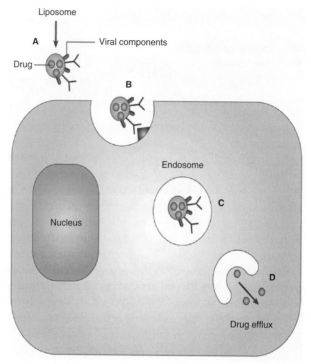

Figure 7.7-4. Virosome–cell interactions. Liposome modified with specific viral components (**A**) and loaded with a drug can specifically interact with cells (**B**), provoke endocytosis, and via the interaction of viral components with the inner membrane of the endosome (**C**), allow for drug efflux into the cell cytoplasm (**D**). (With permission from Ref. 29.)

cells and the powerful helper activity of the virosomal hemagglutinin. Although the immune responses elicited by DNA-virosomes are moderate, they are promising and warrant further research to ultimately develop effective DNA-based virosomal vaccines [59–61].

7.7.5 LIPOSOMES AND CATIONIC LIPOSOME-DNA COMPLEXES (LIPOPLEXES)

Liposomes have become known as one of the most versatile tools for the delivery of DNA, DNA-related, and many other therapeutic molecules [6, 7, 29]. Liposomes are spherical vesicles that consist of an aqueous compartment enclosed in a phospholipid bilayer. If multiple bilayers of lipids are formed around the primary core, the structures that are generated are known as multilamellar vesicles (MLVs). MLVs are formed spontaneously by reconstitution of lipid films in aqueous media. Small unilamellar vesicles (SUVs) of specific size (100–500 nm) are produced by high-pressure extrusion of MLVs through polycarbonate membranes. SUVs (25–90 nm) are also obtained by sonication of MLVs or larger SUVs, by detergent dialysis [62] and by many other, less important methods. Both hydrophilic and hydrophobic drugs can be entrapped in the liposomes, and the choice of the lipid composition as well as the surface modification of the liposomes provides them with a high versatility such as long circulation half-life and sustained and targeted delivery (Figure 7.7-5) [29, 62].

Liposomes can be used as DNA drug delivery systems either by entrapping the DNA-based therapeutics inside the aqueous core or by complexing them to positively charged lipids (lipoplexes, see below). Liposomes offer significant advantages over viral delivery systems; for example, liposomes are generally nonimmunogenic because of the absence of proteinaceous components. As the phospholipid composition in the liposome bilayers can be varied, liposomal delivery systems are of high versatility and customized formulations can be easily engineered to obtain desired sizes, surface charge, composition, and morphology. Liposome encapsulated DNA molecules are protected from nuclease activity and thus improve their biological stability.

Long circulating ("stealth") liposomes are sterically stabilized liposomal formulations that include poly(ethylene glycol) (PEG)-conjugated lipids or other hydrophilic coating molecules. PEGylation prevents the opsonization and recognition of the liposomal vesicles by the MPS. PEGlyation has also been used in conjunction with other polymeric delivery systems such as poly(L-lysine) to achieve longer circulation half-lives.

Immunoliposomes are complex drug or gene delivery systems that can be used for cell targeting by the incorporation of functionalized antibodies attached to lipid bilayers. The "state-of-the-art" immunoliposomes are long circulating PEG-liposomes to which receptor specific molecules are attached, preferably at the distal tips of the PEG chains [62–65]. Immunoliposomes target specific receptors and facilitate receptor-mediated endocytosis for cell uptake (see Fig. 2). Immunoliposomes decorated with single-chain antibody fragments against the ED-B isoform of fibronectin were successfully used in targeted delivery of cytotoxic drugs into tumors *in vivo* [66]. Tissue-specific gene delivery using immunoliposomes has also been achieved by antitransferrin receptor immunoliposomes [67].

To release encapsulated material into the cytoplasm, pH-sensitive liposomes can be generated by the inclusion of dioleyl-phosphatidylethanolamine (DOPE) into liposomes composed of acidic lipids such as cholesterylhemisuccinate or oleic acid. At the neutral cellular pH 7, these lipids have the typical bilayer structure; however, upon endosomal compartmentalization, they undergo protonation and collapse into nonbilayer structures, thereby leading to the disruption and destabilization of the endosomal bilayer, which in turn helps in the rapid release of encapsulated molecules into the cytoplasm [7].

There is an important difference between liposome vesicles and cationic lipid–DNA complexes, the lipoplexes. Lipoplexes are cationic lipid–DNA complexes that are formed spontaneously in aqueous media upon mixing DNA and preformed liposome vesicles composed of cationic and neutral lipids [68]. From a physical point of view, lipoplexes are ordered, self-assembled, composite aggregates whose spatial geometry and phase behavior are controlled by the electrostatic interactions between the positively charged lipids and the negatively charged DNA molecules. Numerous studies have demonstrated the use of cationic liposome formulations for the delivery of different plasmid constructs in a wide range of cells, both *in vivo* and *in vitro*. Despite their cytotoxicity, their nonimmunogenic nature and the

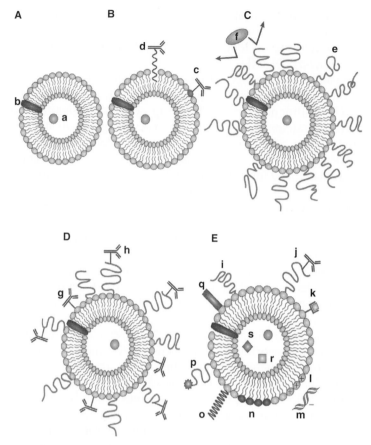

Figure 7.7-5. Evolution of liposomes. (**A**) Early traditional "plain" liposomes with water-soluble drug (**a**) entrapped into the aqueous liposome interior, and lipophilic drug (**b**) incorporated into the liposomal membrane. (**B**) Antibody-targeted immunoliposome with antibody covalently coupled (**c**) to the reactive phospholipids in the membrane, or hydrophobically anchored (**d**) into the liposomal membrane after preliminary modification with a hydrophobic moiety. (**C**) Long circulating liposome grafted with a protective polymer (**e**) such as PEG, which shields the liposome surface from the interaction with opsonizing proteins (**f**). (**D**) Long circulating immunoliposome simultaneously bearing both protective polymer and antibody, which can be attached to the liposome surface (**g**) or, preferably, to the distal end of the grafted polymeric chain (**h**). (**E**) New-generation liposome, the surface of which can be modified (separately or simultaneously) by different ways. Among these modifications are the attachment of protective polymer (**i**) or protective polymer and targeting ligand, such as antibody (**j**); the attachment/incorporation of a diagnostic label (**k**); the incorporation of positively charged lipids (**l**) allowing for the complexation with DNA yielding lipoplex structures (**m**); the incorporation of stimuli-sensitive lipids (**n**); the attachment of a stimuli-sensitive polymer (**o**); the attachment of a cell-penetrating peptide (**p**); and the incorporation of viral components (**q**). In addition to a drug, liposomes can be loaded with magnetic particles (**r**) for magnetic targeting and/or with colloidal gold, silver particles, or fluorescent molecules (**s**) for microscopic analysis. (Adapted with permission from Ref. 29.)

simplicity of production of these systems make them attractive tools for gene transfer. Currently, many gene therapy trials in progress employ nonviral liposomal vectors for transgene delivery.

Lipoplex formulations generally consist of mixtures of cationic and neutral (zwitterionic) lipids. Commonly used cationic lipids are 1,2-dioleoyl-3-trimethylammonium propane (DOTAP), N-[1-(2,3-dioleyloxy)propyl]-N,N,N-trimethylammonium chloride (DOTMA), 2,3-dioleoyloxy-N-[2-(sperminecarboxamido)ethyl]-N,N-dimethyl-1-propanaminium (DOSPA), dioctadecyl amido glycil spermine (DOGS), and 3,[N-(N1,N-dimethylethylenediamine)-carbamoyl] cholesterol (DC-chol) [69–71]. Commonly used neutral molecules, also known as helper or colipids, are DOPE and cholesterol. The cationic lipids in the liposomal formulation serve as a DNA complexation and DNA condensation agent during the formation of the lipoplex. The positive charge also helps in cellular association. The colipids facilitate membrane perturbation and fusion. Many proprietary reagents of cationic lipids such as Lipofectamine (Invitrogen, Carlsbad, CA), Effectene (Qiagen, Valencia, CA), and Tranfectam (Promega, Madison, WI) are commercially available. However, most of these transfection reagents can only be used for *in vitro* gene transfection applications. Despite the appreciable success of cationic lipids in gene transfer, toxicity is a main issue for both *in vitro* and *in vivo* applications. Inflammatory toxicity represents a typical toxicity associated with systemic administration of lipoplexes. Results obtained from *in vivo* studies indicate that lipoplex gene delivery systems mediate uptake of plasmid DNA by the liver, mainly by Kupffer cells, in which a large amount of cytokines is produced [72, 73]. Upon administration via the airways, cytokine-mediated pulmonary toxicity and TNF-α induction by cationic lipids in lung tissue have been reported [74]. Reduction in toxicity was observed by a modification of DOPE polyplexes with cetylated PEI, resulting in remarkable transgene efficiency with low cytotoxicity [72]. In another study, a sterically stabilized immunolipoplex composed of a p53 DNA–lipid complex to which PEG molecules and an antitransferrin receptor single-chain antibody fragment were attached resulted in improved delivery of the complex to tumor cells *in vivo* [75]. The negative factors of lipoplex-mediated gene transfer are low transfection efficiencies, which have been attributed to the heterogeneity and instability of the lipoplex formulations. Lipoplex size heterogeneity also adversely affects their quality control, scale-up, and long-term shelf stability, which are important issues for their pharmaceutical development. Compared with viral vectors, the transfection efficiencies of cationic liposomal vectors are significantly lower. Another drawback in the use of cationic lipids is the rapid inactivation of their cargo in the presence of serum [76].

As an alternative to cationic lipids, the potential of anionic lipids for DNA delivery has been investigated. However, in recent years, only a few studies using anionic liposomal DNA delivery vectors have been reported [77, 78]. These vectors have limited applications, mainly because of inefficient entrapment of DNA molecules within anionic liposomes and lack of toxicity data. Lacking progress of these systems may be attributed to the poor association between DNA molecules and anionic lipids, which is caused by electrostatic repulsion between these negatively charged species.

7.7.6 PEPTIDE AND OLIGONUCLEOTIDE LIPOSOME VACCINE FORMULATIONS

With the increasing availability of defined antigens such as highly purified proteins or synthetic peptides, more effective and safer vaccines are developed. As many antigens are often poorly immunogenic when administered alone, the development of suitable adjuvants, possessing the ability to potentiate the immunogenicity of a given antigen, preferably with little or no side effects, is required. Based on their principal mechanisms of action, adjuvants can be divided into two groups: (1) vaccine delivery systems and (2) immunostimulatory adjuvants. As described in this chapter, vaccine delivery systems are generally particulate delivery systems of size dimensions comparable with pathogens as bacteria and viruses (e.g., liposomes, microemulsions, immunostimulatory complexes, and other nano- or microparticle systems in the size range of 20–500 nm). Their function is mainly to target antigens to antigen-presenting cells (APCs; dendritic cells, macrophages) and to act as adjuvants. The rationale for the development of an optimal vaccine is to ensure that both antigen and adjuvant are delivered into the same population of APCs, thereby reducing systemic distribution and minimizing the potential to induce adverse reactions. Small unilamellar liposomes have an important potential as delivery systems for the coadministration of antigens and of immunostimulatory adjuvants, including synthetic oligodeoxynucleotides containing immunostimulatory deoxycytidylyl-deoxyguanosine dinucleotides (CpGs) or DNA encoding antigens [79–81]. Additionally, the efficiency of the liposomes can be improved by targeting them more effectively and specifically to the APCs by exploiting various scavengers and other receptors as their targets or by enhancing their cell uptake by modification with cell penetrating peptides as recently shown by us [57].

Using the lymphocytic choriomeningitis virus (LCMV) model system, we successfully prepared efficient peptide vaccines with liposomes as the carrier [82]. Liposome-encapsulated antigenic peptides were highly immunogenic when administered intradermally, and they elicited protective antiviral immunity. An optimized formulation contained immunostimulatory oligonucleotides leading to activation of dendritic cells and antitumor immunity in LCMV peptide transfected EL4 thymomas. In a follow-up study we could confirm the efficacy of such peptide–liposome vaccines in a hepatitis C virus model in HLA-A2.1 transgenic mice [83]. These findings clearly indicate that liposomal antigen delivery *in vivo* is a promising approach to induce efficient antiviral and antitumor immune responses with relevance for human applications.

A cautionary note to the potential dangers of all viral gene products, transgenes, viral proteins and peptides, and CpG DNA sequences in siRNA or plasmids formulated in DDS has to be given. Immune responses induced by these molecules may lead to problems such as transient gene expression, nonefficient readministration of the same vectors and to severe side effects in clinical trials [84]. Due to their particulate nature, the DDS are recognized as foreign in an organism that reacts with an immune response. However, the immunomodulating activities of the DDS depend largely on their composition, size, and homogeneity. Synthetic polymers can exhibit significant immunomodulating activity, whereas liposomes prepared with natural phospholipids and cholesterol are known to be less immunogenic [85].

7.7.7 LIPOSOMES AS CARRIERS OF LIPOPHILIC AND AMPHIPHILIC NUCLEOSIDE ANALOGS

Most applications of liposomes as therapeutic drug carrier systems are based on the encapsulation of water-soluble molecules within the trapped aqueous volume of the liposomes. Long circulating PEG-modified liposomes with cytotoxic anti-tumor drugs doxorubicine, paclitaxel, vincristine, and methotrexate are examples of clinically applied chemotherapeutic liposome formulations [86–88]. In contrast to the extensive exploitation of the trapped aqueous volume of the liposomes that serves as a nanocontainer for the water-soluble molecules, the phospholipid bilayer has not been given the same attention for its use as the carrier matrix for lipophilic drugs. Hence, the development of liposomal drug formulations with lipophilic drugs is less popular. This difference may have several reasons, the main probably consisting in the chemistry required to transform water-soluble molecules into lipophilic compounds that allow incorporation into the lipid bilayer core. The most favorable chemical modifications to obtain a molecule that intercalates in a stable fashion into the lipophilic moiety of a lipid bilayer consist in the attachment of long-chain fatty acyl or alkyl residues, for example saturated or unsaturated fatty acids, preferably palmitic or stearic acid and alkylamines, preferably hexadecyl- or octadecylamine to a suitable functional group of the hydrophilic part of the molecule. Some recent examples of lipophilic modifications of antitumor drugs and their formulation in liposomes are gemcitabine, 5-iodo-2′-deoxyuridine, methotrexate, paclitaxel, cytosine arabinoside, and a lipophilic topoisomerase inhibitor, DB67 [89–96].

Drugs that are highly lipophilic by their own nature, e.g., taxanes, epothilones, and cyclosporins, can only be used therapeutically by the addition of possibly toxic solubilizing agents (e.g., Cremophor EL) in complex pharmaceutical formulations [97–99]. One of several feasible means of obtaining nontoxic parenterally applicable formulations of such drugs is their incorporation into the bilayer matrix of phospholipid liposomes.

Nucleoside analogs are a major class of chemotherapeutic agents for the treatment of cancer and viral diseases. Natural endogenous nucleosides must be phosphorylated to corresponding 5′-triphosphates to be incorporated into the DNA or RNA synthesized within the cell. The first phosphorylation step, leading to the formation of nucleoside-5′-monophosphate, is commonly performed by a nucleoside kinase encoded by either the host cell or the virus infecting the host cell. Hence, cellular and virally encoded kinases play a vital role in the metabolism and replication of cells and viruses. Nucleoside analogs used for chemotherapy of cancer and viral infections are in essence prodrugs because they must be phosphorylated in the cytoplasm like the natural nucleosides to triphosphates before they can exert their activities. Thus, administration of phosphorylated nucleoside analogs would circumvent the enzymatic phosphorylation step, which transforms inactive nucleosides into active drugs. Phosphate groups possess an anionic charge at nearly all physiological pH values, making them very polar. This high polarity can be responsible for many deficiencies in terms of efficient drug delivery. Nucleotides are too hydrophilic to penetrate the lipid-rich cell membrane. In addition, blood and cell-surface phosphohydrolases rapidly metabolize the nucleotides to the corresponding nucleosides. Due to their polarity, nucleotides often exhibit a low volume of distribution and are therefore efficiently cleared by renal elimination. In an attempt to overcome these shortcomings, various prodrugs or "pronucleotide" approaches have been devised and investigated for the *in vivo* delivery of pharmacologically active nucleotides. The general aim of these approaches has been to promote passive diffusion through cell membranes and to increase the bioavailability of phosphorylated nucleoside analogs [100, 101]. A simple solution for the delivery of nucleoside monophosphate analogs into cells consists in the chemical neutralization of the ionizable phosphate group via chemical derivatization for example by esterification, resulting in nucleotide phosphodi- and phosphotriester derivatives or by the synthesis of nucleotide phophorodiamidates, cyclic phosphoramidates, and phosphoramidate mono- and diesters. These compounds have a neutral charge and should be capable of entering cells via passive diffusion. To retain the advantage of nucleoside phosphotriesters with regard to their improved cellular uptake, a variety of biolabile moieties have been evaluated as potential protecting groups for nucleoside phosphotriesters.

We chose the approach of the chemical transformation of water-soluble nucleosides of known cytotoxic and antiviral properties into lipophilic drugs or prodrugs. Thus, we reversed the paradigm of transforming lipophilic molecules into hydrophilic derivatives [102]. The first cytotoxic nucleoside that we transformed into lipophilic derivatives was 1-β-D-arabino-furanosyl cytosine (ara-C) because its major clinical disadvantages are a very short plasma half-life and rapid degradation by deamination to the inactive metabolite 1-β-D-arabino-furanosyluracil (ara-U), a shortcoming that also impedes the oral application of ara-C. To reduce these limitations, a large number of 5'- and N^4-substituted ara-C derivatives have been synthesized and characterized in the past (reviewed in Ref. 103). We synthesized a new class of N^4-alkyl-ara-C derivatives with alkyl chain lengths ranging between 6 and 22 C-atoms, demonstrating a typical structure-activity correlation between the length of the alkyl side chains and their antitumor activity profile [104]. The most effective derivative, N^4-octadecyl-ara-C (NOAC), which is highly lipophilic and extremely resistant toward deamination exerted excellent antitumor activity after oral and parenteral therapy in several mouse tumor models and showed to have distinct pharmacological properties compared with ara-C [105].

Consequently, we further modified NOAC by the synthesis of a new generation of lipophilic/amphiphilic heterodinucleoside phosphate derivatives, termed "duplex drugs" that combine the clinically used antitumor drugs ara-C and 5-fluorodeoxyuridine (5-FdU) with NOAC yielding the heterodinucleoside phosphates arabinocytidylyl-N^4-octadecyl-1-β-D-arabino-furanosylcytosine (ara-C-NOAC) and 2'-deoxy-5-fluorouridylyl-N^4-octadecyl-1-β-D-arabinofuranosy-lcytosine (5-FdU-NOAC) (Figure 7.7-6) [106, 107]. Ethynylcytidine (1-(3-C-ethynyl-β-D-ribopenta-furanosyl)-cytosine, ETC) is a novel nucleoside that was found to be highly cytotoxic [108–110]. Thus, by combination of ETC with NOAC, we obtained the lipophilic duplex drug ETC-NOAC (3'-C-ethynylcytidylyl-(5' → 5')-N^4-octadecyl-1-β-D-arabinofuranosylcytosine).

Due to the combination of the effects of both active molecules that can be released into the cytoplasm as monomers or as the corresponding monophosphates (MPs), the cytotoxic activity of the duplex drugs is expected to be more pronounced as compared with the monomeric drugs. Furthermore, it can be anticipated that the monophosphorylated nucleosides ara-CMP, 5-FdU-MP, and ETC-MP, respectively, are directly released into the cell after enzymatic cleavage of the duplex drugs. Thus, monophosphorylated molecules would not have to pass the first phosphorylation step, which is known to be rate limiting [100, 101].

The lipophilic side chains allow a stable incorporation of these duplex drugs into liposomes, allowing the exploitation of the advantages liposome formulations are offering. Due to their high polarity, the nonderivatized heterodinucleoside phosphodiesters are less suited for liposomal formulations. In comparison with the nonpolar heterodinucleoside phosphotriesters that have good properties to be taken up by cells, but whose capacity to be cleaved enzymatically is limited, the cleavage of the natural phosphodiester bond of the duplex drugs is not constrained. A delayed intracellular release of nucleoside and nucleotide analogs from the duplex molecules provides a depot effect that may be therapeutically of advantage.

A structure-related disadvantage of the duplex drugs is that upon enzymatic cleavage of the phosphodiester bonds, a 1-to-1 ratio of nucleoside to nucleotide is obtained (see Figure 7.7-6). Thus, the desired 5′-phosphorylated nucleotide is only formed at maximally 50%. Additionally, one of the two nucleosides has to be transformed into a lipophilic derivative without loss of antitumor or antiviral activity. On the other hand, it was shown that the lipophilic derivatization of nucleosides can result in enhanced activity and modulation of cell specificity [106].

The disadvantage of obtaining only a 50% yield of 5′-phosphorylated nucleoside analogs, which results from the enzymatic cleavage of the duplex drugs, can be avoided by the synthesis of glycerol–lipid–heteronucleotides, the so-called "multiplex drugs." Figure 7.7-7 shows an example of such a compound where the two cytotoxic nucleosides ara-C and 5-FdU are linked at their 5′-hydroxyl groups via a phosphodiester to the 1,3-hydroxy groups of glycerol. To introduce amphiphilic/

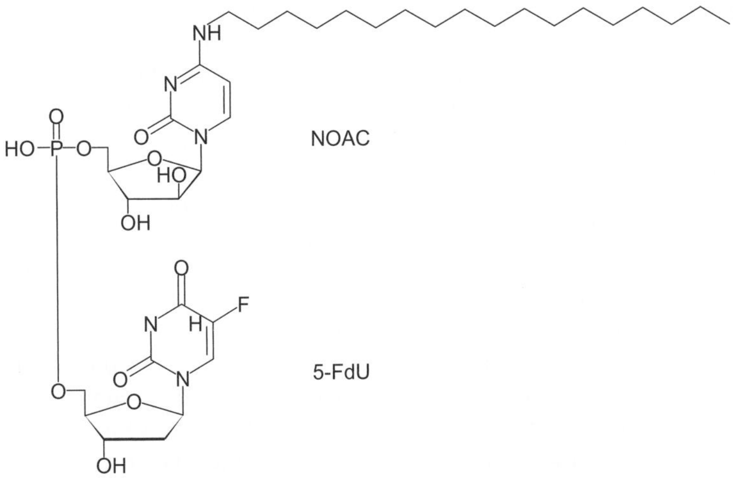

Figure 7.7-6. Duplex drugs. Structure of the anticancer amphiphilic heterodinucleosidyl phosphodiester (Duplex-drug) combining the hydrophilic 5-fluoro-deoxyuridine (5-FdU) with the lipophilic N^4-octadecylarabino-furanosylcytosine (NOAC). Upon enzymatic hydrolysis of the duplex drugs, different anticancer nucleotides and nucleosides are formed with additive or synergistic activities.

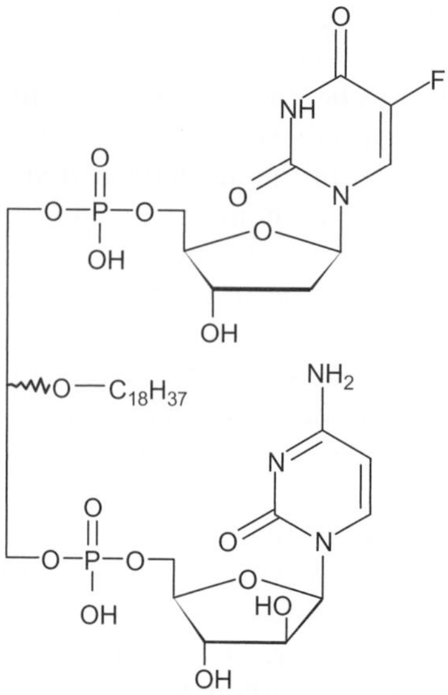

Figure 7.7-7. Multiplex drugs. Example of an amphiphilic glyceryllipid-heterodinucleotide (Multiplex drug). The two hydrophilic nucleoside-5′-monophosphates p5-FdU and paraC are esterified with the terminal hydroxyl residues of the lipophilic 2-octadecylglycerol. By the enzymatic cleavage, several different anticancer metabolites with additive or synergistic activities are obtained.

lipophilic properties, an octadecyl chain is coupled to the 2-hydroxy position of glycerol.

After metabolic degradation by phosphodiesterase, both nucleosides are released as 5′-nucleotides. In case of different hydrolysis kinetics, other active intermediate products may be formed. The structure of the glycerol–lipid–heteronucleotides provides a programmed release of differently active drugs that can penetrate a cell membrane when the cleavage takes place outside of a cell. These compounds may also be formulated in DDS, such as liposomes or micellar systems.

In vitro tests of the glycerol–lipid–heteronucleotides shown in Figure 7.7-7 revealed that these multiplex drugs inhibited colony formation of 5-FU sensitive and resistant human colon tumor cell lines and induced dose-dependent apoptosis in colon tumor cells as well as in mouse leukemia cells. No significant difference in the cytotoxicity could be observed between 5-FU sensitive and resistant cells, indicating that the multiplex drugs might be useful for the treatment of 5-FU resistant tumors [111, 112]. The effectiveness of the postulated mechanisms and advantages of the multiplex drugs will have to be elucidated in *in vivo* experiments.

We conclude that the chemical modification of water-soluble molecules by attachment of long alkyl chains and their stable incorporation into the bilayer membranes of small unilamellar liposomes represent a very promising example of taking advantage of the high loading capacity lipid bilayers offer for lipophilic drugs. The combination of chemical modifications of water-soluble drugs with their pharmaceutical formulation in liposomes is a valuable method for the development of novel pharmaceutical preparations not only for the treatment of tumors or infectious diseases, but also for many other disorders.

7.7.8 OUTLOOK AND FUTURE DIRECTIONS

The development of DDS is a challenging venture that combines research efforts of experts in various areas, including bioengineering, nanotechnology, biomaterials, pharmaceutics, biochemistry, and cell and molecular biology. Specific characteristics of pathological processes and cell or tissue types that are the subject of therapeutic interventions govern the path from target selection to the development of specific DDS formulations. The identification of novel cellular targets, for example, easily accessible vascular endothelial cells, in contrast to tumor cells or other less reachable tissues, will lead to optimized pharmaceutical drug delivery formulations and preparation technologies. Refinement of DDS to overcome unwanted properties such as toxicity, unspecific tissue distribution, and uncontrolled release of entrapped active molecules will be the major challenges in the field. Future DDS will mostly be based on DNA therapeutics such as DNA, ribozymes, DNAzymes, aptamers, (antisense-) oligonucleotides, small interfering RNAs, and their next-generation analogs and derivatives. The use of high-throughput systems for lead identification will yield many new therapeutic targets. Validation and optimization of these disease targets will provide a tremendous stimulus in developing newer potent molecules and their corresponding delivery systems. Further advances in the study of gene function and identification of single-nucleotide polymorphisms will not only help in fine-tuning DNA-based therapeutics but will also fulfill the ultimate goal of providing individualized therapies and medicines.

REFERENCES

1. Bangham A (1992). Liposomes: Realizing their promise. *Hospital Prac.* 27:51–62.
2. Kim S (1993). Liposomes as carriers of cancer chemotherapy. *Drugs.* 46:618–638.
3. Drummond D C, Meyer O, Hong K, et al. (1999). Optimizing liposomes for delivery of chemotherapeutic agents to solid tumors. *Pharmacol. Rev.* 51:691–743.
4. Haag R, Kratz F (2006). Polymer therapeutics: Concepts and applications. *Angewandte Chemie Internat. Edi. Eng.* 45:1198–1215.
5. Tiera M J, Winnik F O, Fernandes J C (2006). Synthetic and natural polycations for gene therapy: state of the art and new perspectives. *Curr. Gene Ther.* 6:59–71.
6. Patil S D, Rhodes D G, Burgess D J (2005). DNA-based therapeutics and DNA delivery systems: A comprehensive review. *AAPS J.* 7:E61–E77.
7. Torchilin V P (2006). Recent approaches to intracellular delivery of drugs and DNA and organelle targeting. *Annual Rev. Biomedi. Engin.* 8:1.1–1.31.
8. Ding B-S, Dziubla T, Shuvaev V V, et al. (2006). Advanced drug delivery systems that target the vascular endothelium. *Molec. Intervent.* 6:98–112.
9. Wei C M, Gibson M, Spear P G, et al. (1981). Construction and isolation of a transmissible retrovirus containing the src gene of Harvey murine sarcoma virus and the thymidine kinase gene of herpes simplex virus type 1. *J. Virol.* 39:935–944.
10. Young L S, Searle P F, Onion D, et al. (2006). Viral gene therapy strategies: from basic science to clinical application. *J. Pathol.* 208:299–318.
11. Noureddini S C, Curiel D T (2005). Genetic targeting strategies for adenovirus. *Molec. Pharmaceuti.* 2:341–347.

12. Barouch D H (2006). Rational design of gene-based vaccines. *J. Pathol.* 208:283–289.

13. Nadeau I, Kamen A (2003). Production of adenovirus vector for gene therapy. *Biotechnol. Adv.* 20:475–489.

14. Moroziewicz D, Kaufman H L (2005). Gene therapy with poxvirus vectors. *Curr. Opin. Molec. Therapeuti.* 7:317–325.

15. Argnani R, Lufino M, Manservigi M, et al. (2005). Replication-competent herpes simplex vectors: design and applications. *Gene Ther.* 12:*S1*, S170–S177.

16. Wiznerowicz M, Trono D (2005). Harnessing HIV for therapy, basic research and biotechnology. *TRENDS Biotechnol.* 23:42–47.

17. Sinn P L, Sauter S L, McCray Jr P B (2005). Gene therapy progress and prospects: Development of improved lentiviral and retroviral vectors—design, biosafety, and production. *Gene Ther.* 12:1089–1098.

18. Trono D (2000). Lentiviral vectors: turning a deadly foe into a therapeutic agent. *Gene Ther.* 7:20–23.

19. Vasileva A, Jessberger R (2005) Precise hit: adeno-associated virus in gene targeting. *Nat. Rev. Microbiol.* 3:837–847.

20. Romano G (2006). The controversial role of adenoviral-derived vectors in gene therapy programs: Where do we stand? *Drug News Perspecti.* 19:99–106.

21. Kaplan J M (2005). Adenovirus-based cancer gene therapy. *Curr. Gene Ther.* 5:595–605.

22. Kanerva A, Hemminki A (2005). Adenoviruses for treatment of cancer. *Ann. Med.* 37:33–43.

23. Guo Z S, Bartlett D L (2004). Vaccinia as a vector for gene delivery. *Expert Opin. Biolog. Ther.* 4:901–917.

24. Bartosch B, Cosset F L (2004). Strategies for retargeted gene delivery using vectors derived from lentiviruses. *Curr. Gene Ther.* 4:427–443.

25. Anderson J L, Hope T J (2005). Intracellular trafficking of retroviral vectors: Obstacles and advances. *Gene Ther.* 12:1667–1678.

26. Kapturczak M H, Flotte T, Atkinson M A (2001). Adeno-associated virus (AAV) as a vehicle for therapeutic gene delivery: improvements in vector design and viral production enhance potential to prolong graft survival in pancreatic islet cell transplantation for the reversal of type 1 diabetes. *Curr. Molec. Med.* 1:245–258.

27. Romano G (2005). Current development of adeno-associated viral vectors. *Drug News Perspecti.* 18:311–316.

28. Vasileva A, Jessberger R (2005). Precise hit: Adeno-associated virus in gene targeting. *Nat. Rev. Microbiol.* 3:837–847.

29. Torchilin V P (2005). Recent advances with liposomes as pharmaceutical carriers. *Nat. Rev. Drug Disc.* 4:145–160.

30. Ewert K, Evans H M, Ahmad A, et al. (2005). Lipoplex structures and their distinct cellular pathways. *Adv. Genet.* 53:119–155.

31. Spagnou S, Miller A D, Keller M (2004). Lipidic carriers of siRNA: differences in the formulation, cellular uptake, and delivery with plasmid DNA. *Biochem.* 43:13348–13356.

32. Shuey D J, McCallus D E, Giordano T (2002). RNAi: gene-silencing in therapeutic intervention. *Drug Disc. Today.* 7:1040–1046.

33. Safinya C R (2001). Structures of lipid-DNA complexes: supramolecular assembly and gene delivery. *Curr. Opin. Struct. Biol.* 11:440–448.

34. Wu G Y, Wu C H (1988). Evidence for targeted gene delivery to Hep G2 hepatoma cells *in vitro. Biochem.* 27:887–892.

35. Lungwitz U, Breunig M, Blunk T, et al. (2005). Polyethylenimine-based non-viral gene delivery systems. *Euro. J. Pharm. Biopharm.* 60:247–266.

36. Neu M, Fischer D, Kissel T (2005). Recent advances in rational gene transfer vector design based on poly(ethylene imine) and its derivatives. *J. Gene Med.* 7: 992–1009.

37. Cho Y W, Kim J D, Park K (2003). Polycation gene delivery systems: escape from endosomes to cytosol. *J. Pharm. Pharmacol.* 55:721–734.

38. Demeneix B, Behr J P (2005). Polyethylenimine (PEI). *Adv. Genet.* 53:217–230.

39. Haensler J, Szoka F C Jr (1993). Polyamidoamine cascade polymers mediate efficient transfection of cells in culture. *Bioconj. Chem.* 4:372–379.

40. Eichman J D, Bielinska A U, Kukowska-Latallo J F, et al. (2000). The use of PAMAM dendrimers in the efficient transfer of genetic material into cells. *Pharmaceutical Sci. Technol. Today.* 3:232–245.

41. Zinselmeyer B H, Mackay S P, Schatzlein A G, et al. (2002). The lower-generation polypropylenimine dendrimers are effective gene-transfer agents. *Pharmaceut. Res.* 19:960–967.

42. Yoo H, Juliano R L (2000). Enhanced delivery of antisense oligonucleotides with fluorophore-conjugated PAMAM dendrimers. *Nucl. Acids Res.* 28:4225–4231.

43. Lee K Y, Kwon I C, Kim Y H, et al. (1998). Preparation of chitosan self-aggregates as a gene delivery system. *J. Control. Rel.* 51:213–220.

44. Prabaharan M, Mano J F (2005). Chitosan-based particles as controlled drug delivery systems. *Drug Del.* 2:41–57.

45. Shi C, Zhu Y, Ran X, et al. (2006). Therapeutic potential of chitosan and its derivatives in regenerative medicine. *J. Surg. Res.* 133:185–192.

46. Andre F, Mir L M (2004). DNA electrotransfer: Its principles and an updated review of its therapeutic applications. *Gene Ther. S1,* S33–42.

47. Li S (2004). Electroporation gene therapy: New developments *in vivo* and *in vitro. Curr. Gene Ther.* 4:309–316.

48. Prud'homme G J, Glinka Y, Khan A S, et al. (2006). Electroporation-enhanced non-viral gene transfer for the prevention or treatment of immunological, endocrine and neoplastic diseases. *Curr. Gene Ther.* 6:243–273.

49. Maitra A (2005). Calcium phosphate nanoparticles: Second-generation nonviral vectors in gene therapy. *Exp. Rev. Molec. Diagnos.* 5:893–905.

50. Nielsen P E, Egholm M (1999). An introduction to peptide nucleic acid. *Curr. Iss. Molec. Biol.* 1:89–104.

51. Marin V L, Roy S, Armitage B A (2004). Recent advances in the development of peptide nucleic acid as a gene-targeted drug. *Exp. Opin. Biolog. Ther.* 4:337–348.

52. Simonson O E, Svahn M G, Tornquist E, et al. (2005). Bioplex technology: Novel synthetic gene delivery pharmaceutical based on peptides anchored to nucleic acids. *Curr. Pharmac. Design.* 11:3671–3680.

53. Nielsen P E, Egholm M, Buchardt O (1994). Peptide nucleic acid (PNA). A DNA mimic with a peptide backbone. *Bioconj. Chem.* 5:3–7.

54. Järver P, Langel U (2006). Cell-penerating peptides—A brief introduction. *Biochim. Biophys. Acta.* 1758:260–263.

55. Wagstaff K M, Jans D A (2006). Protein transduction: cell penetrating peptides and their therapeutic applications. *Curr. Med. Chem.* 13:1371–1387.

56. Torchilin V P, Rammohan R, Weissig V, et al. (2001). TAT peptide on the surface of liposomes affords their efficient intracellular delivery even at low temperature and in the presence of metabolic inhibitors. *Proc. Nati. Acad. Sci. USA.* 98:8786–8791.

57. Marty C, Meylan C, Schott H, et al. (2004). Enhanced heparan sulfate proteoglycan-mediated uptake of cell-penetrating peptide-modified liposomes. *Cell. Molec. Life Sci.* 61:1785–1794.

58. Almeida J D, Edwards D C, Brand C M, et al. (1975). Formation of virosomes from influenza subunits and liposomes. *Lancet.* 2:899–901.

59. Kaneda Y (2000). Virosomes: Evolution of the liposome as a targeted drug delivery system. *Adv. Drug Del. Rev.* 43:197–205.

60. Felnerova D, Viret J F, Gluck R, et al. (2004). Liposomes and virosomes as delivery systems for antigens, nucleic acids and drugs. *Curr. Opin. Biotechnol.* 15:518–529.

61. Daemen T, de Mare A, Bungener L, et al. (2005). Virosomes for antigen and DNA delivery. *Adv. Drug Del. Rev.* 57:451–463.

62. Allen T M (2002). Ligand-targeted therapeutics in anticancer therapy. *Nat. Rev. Cancer.* 2:750–763.

63. Fenske D B, Cullis P R (2005). Entrapment of small molecules and nucleic acid-based drugs in liposomes. *Meth. Enzymol.* 391:7–40.

64. Sapra P, Tyagi P, Allen T M (2005). Ligand-targeted liposomes for cancer treatment. *Curr. Drug Del.* 2:369–381.

65. Park J W, Benz C C, Martin F J (2004). Future directions of liposome- and immuno-liposome-based cancer therapeutics. *Seminars Oncol.* 6(suppl 13):196–205.

66. Marty C, Odermatt B, Schott H, et al. (2002). Cytotoxic targeting of F9 teratocarci-noma tumours with anti-ED-B fibronectin scFv antibody modified liposomes. *British J. Cancer.* 87:106–112.

67. Xu L, Huang C C, Huang W, et al. (2002). Systemic tumor-targeted gene delivery by anti-transferrin receptor scFv-immunoliposomes. *Molec. Cancer Ther.* 1:337–346.

68. May S, Ben-Shaul A (2004). Modeling of cationic lipid-DNA complexes. *Curr. Med. Chem.* 11:151–167.

69. Ferrari M E, Rusalov D, Enas J, et al. (2002). Synergy between cationic lipid and co-lipid determines the macroscopic structure and transfection activity of lipoplexes. *Nucleic Acids Res.* 30:1808–1816.

70. Tranchant I, Thompson B, Nicolazzi C, et al. (2004). Physicochemical optimisation of plasmid delivery by cationic lipids. *J. Gene Med.* 6:S24–S35.

71. Zhdanov R I, Podobed O V, Vlassov V V (2002). Cationic lipid-DNA complexes-lipoplexes-for gene transfer and therapy. *Bioelectrochem.* 58:53–64.

72. Matsuura M, Yamazaki Y, Sugiyama M, et al. (2003). Polycation liposome-mediated gene transfer *in vivo. Biochim. Biophys. Acta.* 1612:136–143.

73. Zhang J S, Liu F, Huang L (2005). Implications of pharmacokinetic behavior of lipo-plex for its inflammatory toxicity. *Adv. Drug Del. Rev.* 57:689–698.

74. Freimark B D, Blezinger H P, Florack V J, et al. (1998) Cationic lipids enhance cyto-kine and cell influx levels in the lung following administration of plasmid: cationic lipid complexes. *J. Immunol.* 160:4580–4586.

75. Yu W, Pirollo K F, Rait A, et al. (2004). A sterically stabilized immunolipoplex for systemic administration of a therapeutic gene. *Gene Ther.* 11:1434–1440.

76. Audouy S, Molema G, De Leij L, et al. (2000). Serum as a modulator of lipoplex-mediated gene transfection: dependence of amphiphile, cell type and complex stabil-ity. *J. Gene Med.* 2:465–476.

77. Son K K, Tkach D, Hall K J (2000). Efficient *in vivo* gene delivery by the negatively charged complexes of cationic liposomes and plasmid DNA. *Biochim. Biophys. Acta.* 1468:6–10.

78. Lakkaraju A, Dubinsky J M, Low W C, et al. (2001). Neurons are protected from excitotoxic death by p53 antisense oligonucleotides delivered in anionic liposomes. *J. Biol. Chem.* 276:32000–32007.

79. Rothenfusser S, Tuma E, Wagner M, et al. (2003). Recent advances in immunostimulatory CpG oligonucleotides. *Curr. Opin. Molec. Ther.* 5:98–106.

80. Wang H, Rayburn E, Zhang R (2005). Synthetic oligodeoxynucleotides containing deoxycytidyl-deoxyguanosine dinucleotides (CpG ODNs) and modified analogs as novel anticancer therapeutics. *Curr. Pharmaceu. Des.* 11:2889–2907.

81. Chen W C, Huang L (2005). Non-viral vector as vaccine carrier. *Adv. Gene.* 54: 315–337.

82. Ludewig B, Barchiesi F, Pericin M, et al. (2001). *In vivo* antigen loading and activation of dendritic cells via a liposomal peptide vaccine mediates protective antiviral and anti-tumour immunity. *Vaccine.* 19:23–32.

83. Engler O B, Schwendener R A, Dai Wen J, et al. (2004). A liposomal peptide vaccine inducing CD8+ T cells in HLA-A2.1 transgenic mice, which recognise human cells encoding hepatitis C virus (HCV) proteins. *Vaccine.* 23:58–68.

84. Zhou H, Liu D, Liang C (2004). Challenges and strategies: The immune responses in gene therapy. *Med. Res. Rev.* 24:748–761.

85. Storni T, Kündig T M, Senti G, et al. (2005). Immunity in response to particulate antigen-delivery systems. *Adv. Drug Del. Rev.* 57:333–355.

86. Rose P G (2005). Pegylated liposomal doxorubicin: optimizing the dosing schedule in ovarian cancer. *Oncol.* 10:205–214.

87. Cattel L, Ceruti M, Dosio F (2004). From conventional to stealth liposomes: A new Frontier in cancer chemotherapy. *J. Chemother.* S4:94–97.

88. Straubinger R M, Arnold R D, Zhou R, et al. (2004). Antivascular and antitumor activities of liposome-associated drugs. *Anticancer Res.* 24(2A):397–404.R

89. Immordino M L, Brusa P, Rocco F, et al. (2004). Preparation, characterization, cytotoxicity and pharmacokinetics of liposomes containing lipophilic gemcitabine prodrugs. *J. Control. Rel.* 100:331–346.

90. Bergman A M, Kuiper C M, Noordhuis P, et al. (2004). Antiproliferative activity and mechanism of action of fatty acid derivatives of gemcitabine in leukemia and solid tumor cell lines and in human xenografts. *Nucleosides Nucleot. Nucleic Acids.* 23:1329–1333.

91. Harrington K J, Syrigos K N, Uster P S, et al. (2004). Targeted radiosensiti-sation by pegylated liposome-encapsulated 3′, 5′-O-dipalmitoyl 5-iodo-2′-deoxyuridine in a head and neck cancer xenograft model. *British J. Cancer.* 91:366–373.

92. Pignatello R, Puleo A, Puglisi G, et al. (2003). Effect of liposomal delivery on in vitro antitumor activity of lipophilic conjugates of methotrexate with lipoamino acids. *Drug Del.* 10:95–100.

93. Stevens P J, Sekido M, Lee R J (2004). A folate receptor-targeted lipid nanoparticle formulation for a lipophilic paclitaxel prodrug. *Pharmaceu. Res.* 21:2153–2157.

94. Lundberg B B, Risovic V, Ramaswamy M, et al. (2003). A lipophilic paclitaxel derivative incorporated in a lipid emulsion for parenteral administration. *J. Control. Rel.* 86:93–100.

95. Bergman A M, Kuiper C M, Myhren F, et al. (2004). Antiproliferative activity and mechanism of action of fatty acid derivatives of arabinosylcytosine (ara-C) in leukemia and solid tumor cell lines. *Nucleosides Nucleot. Nucleic Acids.* 23:1523–1526.

96. Lopez-Barcons L A, Zhang J, Siriwitayawan G, et al. (2004). The novel highly lipophilic topoisomerase I inhibitor DB67 is effective in the treatment of liver metastases of murine CT-26 colon carcinoma. *Neoplasia.* 6:457–467.

97. Fahr A, van Hoogevest P, May S, et al. (2005). Transfer of lipophilic drugs between liposomal membranes and biological interfaces: consequences for drug delivery. *Euro. J. Pharmaceu. Sci.* 26:251–265.

98. Strickley R G (2004). Solubilizing excipients in oral and injectable formulations. *Pharmaceu. Res.* 21:201–230.

99. ten Tije A J, Verweij J, Loos W J, et al. (2003). Pharmacological effects of formulation vehicles: Implications for cancer chemotherapy. *Clin. Pharmacokin.* 42:665–685.

100. Krise J P, Stella V J (1996) Prodrugs of phosphates, phosphonates, and phosphinates. *Adv. Drug Del. Rev.* 19:287–310.

101. Wagner C R, Iyer V V, McIntee E J (2000). Pronucleotides: Toward the in vivo delivery of antiviral and anticancer nucleotides. *Med. Res. Rev.* 20:417–451.

102. Schwendener R A, Schott H (2005). Lipophilic arabinofuranosyl cytosine derivatives in liposomes. *Meth. Enzymol.* 391:58–70.

103. Hamada A, Kawaguchi T, Nakano M (2002). Clinical pharmacokinetics of cytarabine formulations. *Clinical Pharmakinet.* 41:705–718.

104. Schwendener R A, Schott H (1996). Lipophilic 1-β-D-arabino-furanosylcytosine derivatives in liposomal formulations for oral and parenteral antileukemic therapy in the murine L1210 leukemia model. *J. Cancer Res. Clini. Oncol.* 122:723–726.

105. Schwendener R A, Friedl K, Depenbrock H, et al. (2001). In vitro activity of liposomal N⁴octadecyl-1-β-D-arabino-furanosylcytosine (NOAC), a new lipophilic derivative of 1-β-D-arabino-furanocylcytosine on biopsized clonogenic human tumor cells and hematopoietic precursor cells. *Investigat. New Drugs.* 19:203–210.

106. Cattaneo-Pangrazzi R M C, Schott H, Wunderli-Allenspach H, et al. (2000). The novel heterodinucleoside dimer 5-FdU-NOAC is a potent cytotoxic drug and a p53-independent inducer of apoptosis in the androgen-independent human prostate cancer cell lines PC-3 and DU-145. *Prostate.* 45:8–18.

107. Cattaneo-Pangrazzi R M C, Schott H, Wunderli-Allenspach H, et al. (2000). New amphiphilic heterodinucleoside phosphate dimers of 5-fluorodeoxyuridine (5FdUrd): Cell cycle dependent cytotoxicity and induction of apoptosis in PC-3 prostate tumor cells. *Biochem. Pharmacol.* 60:1887–1896.

108. Takatori S, Kanda H, Takenaka K, et al. (1999). Antitumor mechanisms and metabolism of the novel antitumor nucleoside analogues, 1-(3-C-ethynyl-β-D-ribopento-furanosyl)cytosine and 1-(3-C-ethynyl-β-D-ribopento-furanosyl)-uracil. *Cancer Chemother. Pharmacol.* 44:97–104.

109. Tabata S, Tanaka M, Endo Y, et al. (1997). Anti-tumor mechanisms of 3′-ethynyluridine and 3′-ethynylcytidine as RNA synthesis inhibitors: Development and characterization of 3′-etyhynyluridine-resistant cells. *Cancer Lett.* 116:225–231.

110. Yokogawa T, Naito T, Kanda H, et al. (2005). Inhibitory mechanisms of 1-(3-C-ethynyl-β-D-ribopentofuranosyl)uracil (EUrd) on RNA synthesis. *Nucleosides Nucleot. Nucleic Acids.* 24:227–232.

111. Ludwig P S, Schwendener R A, Schott H (2005). Synthesis and anticancer activities of amphiphilic 5-fluoro-2′-deoxyuridylic acid prodrugs. *Euro. J. Med. Chem.* 40:494–504.

112. Saiko P, Horvarth Z, Bauer W, et al. (2004). *In vitro* and *in vivo* antitumor activity of novel amphiphilic dimers consiting 5-fluorodeoxyuridine and arabinofuranosylcytosine. *Internat. J. Oncol.* 25:357–364.

8.1

GROWTH FACTORS AND CYTOKINES

MANDEEP K. MANN AND BRIAN E. CAIRNS

University of British Columbia, Vancouver, British Columbia, Canada

Chapter Contents

8.1.1 Background   1173
    8.1.1.1 Neuropathic and Nociceptive Pain Mechanisms   1173
    8.1.1.2 Introduction to Nerve Growth Factor   1175
    8.1.1.3 Introduction to Tumor Necrosis Factor α 1177
    8.1.1.4 Clinical Significance   1179
    8.1.1.5 NGF and Experimental Muscle Pain   1181
    8.1.1.6 TNFα and Experimental Joint Pain   1182
8.1.2 Case Studies   1184
    8.1.2.1 Effect of IM injection of NGF in humans   1184
    8.1.2.2 NGF effects on masticatory muscle nociceptors   1186
    References   1188

8.1.1 BACKGROUND

8.1.1.1 Neuropathic and Nociceptive Pain Mechanisms

Role of Growth Factors and Cytokines in These Pain Mechanisms (Peripheral). Pain is generally classified into two main categories: neuropathic and nociceptive (inflammatory) pain. Nociceptive pain is caused by the activation of primary afferent fibers known as nociceptors [1]. Primary nociceptive afferents (Aδ and C fibers) can become sensitized as a result of various factors such as tissue trauma, inflammation, and ischemia [2]. Peripheral injury (tissue damage) results in the synthesis and release of various inflammatory mediators that induce inflammation and edema as part of the healing process. These mediators include bradykinin, substance P, histamine, serotonin (5-HT), glutamate, acetylcholine (Ach), ATP, ions (H^+, K^+), cholecystokinin, nerve growth factor (NGF) [3], cytokines such as tumor necrosis factor α (TNFα), interleukin-1β (IL-1β) and interleukin-6 (IL-6), and eicosanoids (such as prostaglandin E_2, prostaglandin I_2, and leukotriene B_4), which excite and sensitize nociceptors as well as recruit additional nociceptors to enhance pain perception [1, 4]. Peptide neurotransmitters such as substance P (SP) and calcitonin gene-related peptide (CGRP) can be released from the nerve terminals of nociceptors into damaged tissues, a process that serves to augment the effects of inflammatory agents by causing vasodilation and increased capillary permeability. Bradykinin (released from blood vessels) and other chemical agents such as 5-HT released from platelets, histamine from mast cells, and eicosanoids from various cellular elements further contribute to sensitization of nociceptors, by either opening ion channels or activating second messenger systems [5].

The process of peripheral sensitization is manifested by a decrease in the activation threshold of nociceptors (allodynia) and an increase in the response of nociceptors to noxious stimulation (hyperalgesia) [6, 7]. In addition, peripheral sensitization may be enhanced due to the recruitment of "silent" nociceptors [6, 8], which are normally unresponsive to thermal or mechanical stimuli in healthy tissue. The release of various chemical mediators in the periphery also causes repetitive activity in primary afferent fibers. This barrage of input results in the central release of neurotransmitters and neuropeptides that act on brain cells or neurons in the central nervous system to increase their response to tissue stimulation. This phenomenon is known as central sensitization and results in the amplification and persistence of nociceptive signals that are eventually perceived as pain. It is theorized that central sensitization is the mechanism underlying pain referral and clinical signs of secondary hyperalgesia, which refers to the sensitization of noninflamed tissue surrounding the site of inflammation in the periphery [9–12].

Normally, as the injured tissue heals and inflammation subsides, the phenomena of peripheral and central sensitization dissipate [1]. However, under certain conditions, pain does not resolve with tissue healing and becomes chronic [1, 4]. It is thought that the transition from acute to chronic pain under these conditions may reflect damage or dysfunction of the peripheral and/or central nervous system, which results in abnormal processing of sensory input, and thus, this type of pain is called neuropathic pain [1, 13, 14]. The pathophysiology of neuropathic pain syndromes (such as postherpetic neuralgia, diabetic neuropathy, posttraumatic neuralgia etc.) is not completely understood, but it is considered to be complex and multifactorial [1, 13–15]. Neuropathic pain syndromes and peripheral nerve injury are generally accompanied by a cascade of inflammatory events ("neuroinflammatory cascade") mediated by a group of cytokines and other molecules, as mentioned above [4]. Cytokines such as TNFα and IL-1β, released at the site of inflammation, bind to their receptors on primary afferent fibers and activate and/or sensitize nociceptors, thus inducing and maintaining pain as well as a state of hyperalgesia [4]. Studies have shown that the expression of the receptors of these cytokines increases after peripheral nerve injury [16–18] and likely contributes to increased neuronal excitability. Also, use of cytokine antagonists has been shown to reduce nociceptive behavior in rodents [19–21], suggesting a role for TNFα and IL-1β in mediating neuropathic-like pain. Similarly, neurotrophic factors such as NGF have also been shown to be involved in nociceptive and neuropathic pain conditions such as diabetic peripheral neuropathies.

This chapter will focus on NGF and TNFα and their role in the mechanisms of neuropathic and nociceptive pain.

8.1.1.2   Introduction to Nerve Growth Factor

Discovery of NGF. Rita Levi-Montalcini, an Italian scientist, discovered NGF in the early 1950s. Levi-Montalcini's research on the central nervous system and its peripheral targets led her to hypothesize that "the failure of neurons to thrive in the absence of peripheral target tissue was because of a degenerative process" [22]. Through her research collaboration with Victor Hamburger, a neuroembryologist, Levi-Montalcini gained further evidence that confirmed her hypothesis. Their work on chick embryos led them to hypothesize that "developing nerve cells depend on feedback signals that are in limited supply, and the neuronal targets provide a specific signal that is required for neuronal survival" [22, 23]. This "specific signal" was later identified to be "a diffusible agent that stimulated the growth and differentiation of nerve cells" and thus named the "nerve growth factor." The biological characterization (isolation and purification) of NGF was achieved through experiments performed by Levi-Montalcini in collaboration with Stanley Cohen, a biochemist who discovered the epidermal growth factor [22]. In 1986, both Rita Levi-Montalcini and Stanley Cohen received a Nobel Prize in medicine for their discovery of NGF [23].

NGF was the first neurotrophic factor to be characterized [24]. The structure of NGF comprises 2α, 2β, and 2γ subunits and one or two zinc ions [25]. The β subunit of the complex is released as a result of autocatalytic cleavage initiated by the loss of zinc ions. The biologically active form of NGF is NGF-β, which is a dimeric molecule (2β subunits) weighing 26.8 kDa, with each monomer consisting of 118 amino acids and three disulfide bridges [25, 26].

Role/Function of NGF. NGF is a neurotrophic protein molecule essential for the growth and survival of sympathetic and small-diameter afferent neurons (nociceptors) [27, 28] and is involved in neuronal function and plasticity [24]. NGF belongs to the neurotrophin family, which also includes the brain-derived neurotrophic factor (BDNF), glial-derived neurotrophic factor (GDNF), neurotrophin-3 (NT-3), and neurotrophin 4/5 (NT-4/5) [29, 30].

Sources of NGF. Neurons as well as non-neuronal cells such as mast cells, fibroblasts, eosinophils, T and B lymphocytes, and epithelial cells synthesize NGF [31, 32]. NGF and other neurotrophins can also be synthesized and secreted by sympathetic and sensory target organs [24]. High levels of NGF are present in the central nervous system (CNS) where it is known to play a crucial role in growth and plasticity [29]. During development, NGF is secreted by target tissues of sensory axons/neurons [24]. Also, during peripheral nerve injury, NGF synthesis is initiated in Schwann cells and fibroblasts within the injured nerve by cytokines released in response to the injury (for survival and regeneration of the injured nerve) [24].

NGF Receptors. NGF exerts its biological effects by acting on two different receptors found on target cells: tyrosine kinase receptor A (TrkA; tropomyosin related kinase A) and p75 receptor [29]. The TrkA receptor is a tyrosine kinase consisting of intracellular, transmembrane, and extracellular domains, and NGF binding results in phosphorylation of key transduction-related proteins or ion channels causing downstream receptor modulation [29, 32]. The TrkA receptor has a very slow dissociation rate for NGF and is therefore called a high-affinity NGF receptor [33]. In the periphery, TrkA receptors are expressed on small-diameter sensory neurons, inflammatory cells, and sympathetic neurons [30, 34]. The p75 receptor is a member of the superfamily of TNF (tumor necrosis factor) receptor-related molecules that has extracellular and transmembrane domains. The p75 receptor can bind all neurotrophins [29]. The p75 receptor is known as the low-affinity NGF receptor because it has fast association and dissociation rates for NGF binding [33]. The binding of NGF to TrkA and p75 receptors results in the activation of downstream signal transduction pathways that mediate the physiological effects of NGF [35].

Downstream Signaling Pathways of NGF. NGF binding to the TrkA receptors causes receptor activation and dimerization, which results in transautophosphory-lation and activation of intracellular signaling cascades like extracellular signal-regulated kinase (ERK), phosphatidylinositol 3-kinase (PI3K), and phospholipase Cγ (PLC-γ) pathways [24, 35]. The Src homologous and collagen-like (Shc) adaptor protein binds to the activated Trk receptor, and its phosphorylation results in an increase in the activities of PI3K and Akt (protein kinase B), which are involved in neuronal survival [35]. The Shc binding to TrkA receptors also increases the activities of Ras and ERK, which can influence transcriptional events such as induction of cyclic AMP-response element binding (CREB) transcription factor that is involved in neurite outgrowth, cell cycle, and synaptic plasticity [35]. It is thought that sustained activation of the ERK pathway requires the internalization of the NGF–TrkA receptor complex into membrane vesicles [24, 36]. The PLC-γ signaling cascade is initiated by direct PLC-γ binding to the activated Trk receptor and results in protein kinase C activation and the release of inositol phosphates and calcium [24, 35]. It has been suggested that activation of these signaling pathways may result in the modulation of ion channels such as the capsaicin receptor channel known as the TRPV1 or VR1 channel [35, 37]. The p75 receptor also activates signaling components such as Jun N-terminal kinase (JNK), nuclear factor-κB (NF-κB), and ceramide to mediate processes such as apoptosis or cell death [35].

NGF Tissue Levels. As compared with neurotransmitters, neurotrophic factors such as NGF exist in very low concentrations; therefore, their detection is very difficult [38]. The maximum peak level of NGF in the interstitial fluid of the brain of patients with severe head injury was found to be 1100 pg/mL [39]. A study using an enzyme-linked immunosorbent assay (ELISA) (measuring unbound β-NGF concentration) has shown that men have significantly higher serum NGF concentration (40.8 ± 10.8 pg/mL) than women (8.2 ± 1.4 pg/mL in the follicular phase and 14.4 ± 2.9 pg/mL in the luteal phase) [40]. As differences in the NGF concentrations in both the follicular and the luteal phase of the menstrual cycle were reported to be statistically significant, it has been suggested that female sex hormones (and androgens) influence circulating NGF levels [40].

8.1.1.3   Introduction to Tumor Necrosis Factor α

Discovery of TNFα. In 1975, Carswell et al. published a study that showed that endotoxin treated serum of bacillus Calmette-Guerin (BCG)-infected mice induces the release of a substance that is selectively toxic for tumor/malignant cells [41]. They reported that this substance was as effective at causing necrosis of tumors as endotoxin itself and named it "the tumor necrosis factor" (TNF) [41]. TNF is now considered a part of a large and diverse group of polypeptides called cytokines, whose major role is to mediate interactions between the inflammatory and the immune system [42, 43]. There are two forms of TNF: TNFα and TNFβ. TNFα is synthesized as a prohormone, and its soluble form (17 kD, nonglycosylated protein) is released from the 26-kD transmembrane protein via specific proteolytic cleavage [44, 45]. TNFβ or lymphotoxin shares 28% sequence homology with TNFα [46].

Role/Function of TNFα. TNFα belongs to a family of cytokines that also includes interleukins, interferons, and transforming growth factors [42]. TNFα is involved in various processes such as inflammation, cytotoxicity, (apoptotic) cell death, metabolism (insulin resistance), thrombosis, and fibrinolysis [44, 47]. TNFα is also involved in exerting endocrine, paracrine, and autocrine control of inflammatory responses. It can activate macrophages [46], initiate a cascade of other cytokines (IL-1β, IL-6, and IL-8) and growth factors (NGF), and can recruit circulating inflammatory cells to the local site of inflammation to induce edema [47, 48]. TNFα-induced inflammatory response also involves the release of arachidonic acid, which results in the production of pro-inflammatory mediators such as prostaglandins and leukotrienes [49]. TNFα plays a major role in the pathogenesis of septic shock syndrome that results from the body's reaction to a bacterial, viral, or parasitic infection [49]. In the CNS, TNFα has both neurotoxic and neuroprotective effects [50, 51]. Recent research has suggested that TNFα (and other cytokines) link the immune and nervous system and may be involved in the generation of pain and hyperalgesia [52]. It has also been suggested that cytokines like TNFα can induce the production of NGF in both the peripheral and the central nervous system [53].

Sources of TNFα. Various stimuli such as endotoxins, superantigens, osmotic stress, injury, inflammation, and radiation can cause the release of TNFα [48]. TNFα is produced mainly by monocytes and/or macrophages and T cells [45, 48]. However, small amounts of TNFα can be released by various other cell types [44, 52]. For instance, Schwann cells can produce TNFα after injury. Like several other cytokines, TNFα can induce its own production as well [52].

TNFα Receptors. TNFα exerts its effect through two high-affinity, cell-surface receptors: p55 and p75 receptors [42, 52]. Both TNF receptors are expressed in the

CNS [42]. Both receptors are glycoproteins and have a single-membrane spanning hydrophobic segment [49]. At least one of these two TNF receptors is present on virtually all cells, explaining the pleiotropic nature of TNFα [44, 45]. It has been shown that tissues including kidney, liver, adipose, and muscle tissue express the p55 receptor [46]. Most of the effects of TNFα are mediating via its binding and activation of the p55 receptor. Some processes such as cytotoxicity and thymocyte proliferation are also associated with the p75 receptor [49]. The binding of TNFα to its receptors results in the activation of a multitude of complex signaling pathways. TNFβ also binds to the same receptors as TNFα and mediates similar biological actions [49].

Downstream Signaling Pathways of TNFα. The binding of TNFα to its receptors results in the activation of multiple signal transduction pathways, kinases, and transcription factors that activate several cellular genes [45]. These signaling pathways include those that activate transcription factors (such as NF-κB and AP1), protein kinases (such as mitogen activated protein kinase or MAPK, JNK, p38), and proteases [44]. The binding of TNFα to the p55 receptor results in the activation of phospholipase C (PLC) and production of diacylglycerol (DAG), which then activates a calcium-independent protein kinase C (PKC) isotype [49]. Through an unknown pathway, PKC then activates Jun and Fos proteins that are components of AP-1 transcription factor [49, 54] and therefore results in the induction of AP-1 responsive genes [54]. DAG can also activate a C type phospholipase known as SMase, which hydrolyzes sphingomyelin (SM) to ceramide, a second messenger for processes such as apoptosis. In addition, p55 receptor activation can result in the phosphorylation of phospholipase A_2 (PLA$_2$), which triggers the release of arachidonic acid (AA) and production of eicosanoids involved in the process of inflammation [49]. PLA$_2$ activation is also thought to be involved in TNFα-induced cytotoxicity. TNFα can also activate transcription factors such as NF-κB and JNK. Several adaptor proteins that initiate these pathways and link TNF receptors to their downstream targets have been identified. One such protein is called the death domain protein, which is involved in the process of apoptosis [44, 51]. Another adaptor protein is part of the TNF receptor associated factors (TRAFs) family, which consists of six distinct proteins, but only TRAF2, TRAF5, and TRAF6 are thought to mediate NF-κB and JNK activation [44]. Another molecule required for TNFα mediated NF-κB activation is known as the receptor interacting protein (RIP) [44]. NF-κB activation results in the synthesis of new proteins, some of which are involved in cell death, whereas others result in cell proliferation. Therefore, depending on the cell cycle, TNFα can either initiate apoptosis (programmed cell death) or be protective [51].

TNFα Tissue Levels. Cytokines are produced on demand, and they only travel over short distances. It is thought that *in vivo* concentrations of cytokines are in the range of a few picograms to nanograms per milliliter [52]. In a study conducted on baboons, infusion of a lethal dose of *Escherichia coli* increased plasma TNFα concentration to a peak level of $20,500 \pm 9890 \, pg/mL$ within 90 minutes [55]. Similarly, it has been reported that in human volunteers, a bolus injection of endotoxin results in a burst increase in TNFα levels (peak level: $358 \pm 166 \, pg/mL$) followed by a rapid decline to undetectable levels [45, 56].

8.1.1.4   Clinical Significance

Currently available surgical techniques and powerful analgesics such as opioids are relatively ineffective for the treatment of painful neuropathies or, in the case of opioid analgesics, have other limitations such as adverse side effects [4]. Therefore, researchers have targeted other cellular and molecular mechanisms and mediators that are thought to be involved in neuropathic and inflammatory pain conditions in an effort to develop new drugs and therapeutic strategies. One such molecule is the nerve growth factor.

NGF.   NGF levels change during some neurodegenerative disease states and other pathological conditions [57, 58]. NGF levels have been increased exogenously in experiments investigating diabetic peripheral neuropathies. There is a loss of dorsal root ganglion neurons and degeneration of Schwann cells and small-diameter sensory neuronal fibers during diabetic peripheral neuropathies [59]. Experimentally, it has been shown that induction of diabetes with streptozotocin causes a decrease in NGF levels in the sensory neurons and target tissues of rats [60, 61] and that human diabetic patients have low NGF levels in the skin [62]. Also, studies have shown that NGF uptake and retrograde transport mechanisms are impaired in animal models of diabetes [63, 64]. Therefore, it was thought that NGF replacement therapy might have a beneficial effect on the treatment of this condition due to its involvement in the promotion of neuronal growth and survival as well as its ability to prevent neuronal damage to small-diameter sensory and autonomic neurons [65].

Initial animal studies showed that increasing NGF levels in experimental models of peripheral neuropathy had beneficial effects (prevented the decrease in SP and CGRP levels in dorsal root ganglion neurons and prevented the loss of sensitivity to thermal noxious stimuli, i.e., prevented the onset of symptoms associated with polyneuropathy) [65–67]. This led to phase I (0.03–1-µg/kg human NGF) and II (0.1- or 0.3-µg/kg human NGF three times a week for 6 months) clinical trials that involved subcutaneous administration of human NGF into healthy human volunteers and diabetic patients suffering from polyneuropathy [68, 69]. Although symptoms of polyneuropathy were decreased by NGF treatment, these trials also revealed that human NGF administration results in serious dose-limiting, pain-related systemic side effects such as hyperalgesia at the site of injection, diffuse myalgias or arthralgias, and leg cramps (more frequent and severe with higher doses) [68, 69]. The positive results from phase II clinical trials resulted in a phase III clinical trial (0.1-µg/kg human NGF three times a week for 48 weeks) that also reported pain-related side effects in response to human NGF administration but failed to show the efficacy of human NGF in treating diabetic polyneuropathy [70]. Although initially NGF seemed to be an ideal therapeutic agent for the treatment of diabetic neuropathy, the failure of these subsequent clinical trials has served to highlight the potentially important role altered NGF levels may play in the development of chronic joint and muscle pain.

NGF is also considered to be a novel therapeutic target for the treatment of other types of neuropathic pain. Studies conducted on animal models of neuropathic pain suggest that there might be a relationship between neuropathic pain and NGF [71–74]. Procedures used to induce neuropathic pain such as sciatic nerve

constriction [73] or spinal nerve ligation [75] result in pain behaviors such as allodynia and mechanical and thermal hyperalgesia [71, 73, 74] and lead to increases
in the levels of NGF [75, 76]. Administration of NGF antiserum has been shown
to inhibit mechanical and thermal hyperalgesia caused by elevated endogenous
NGF levels [72, 77], which suggests that NGF is a mediator of mechanical sensitivity. Also, neuropathic patients with chronic hyperalgesia and allodynia have
increased levels of NGF in the area of peripheral neuropathy [78]. However, some
studies conducted on animal models of nerve constriction injury have reported
contradictory results. For example, exogenous NGF administration (chronic infusion of $10\,\mu L$ of 0.5-mg/mL NGF) has been shown to abolish behavioral hyperalgesia (decrease in mechanical threshold) caused by chronic constriction injury of
the rat sciatic nerve [79]. Also, a study has shown that intrathecal rat NGF infusion
restores the antiallodynic and antihyperalgesic effects of morphine in a rat model
of sciatic nerve constriction injury [80]. Overall, these studies suggest that there
might be a correlation between NGF levels in the periphery and pain behaviors
such as mechanical allodynia and thermal hyperalgesia that are hallmarks of
neuropathic pain.

Tissue inflammation is known to result in an increase in the production and
release of NGF [32, 81–83]. Studies have shown that inflammatory mediators such
as interleukin-1 (IL-1), IL-4, IL-5, tumor necrosis factor-α, and interferon-γ induce
NGF release. In turn, NGF can augment neurogenic inflammation by promoting
the release of inflammatory mediators from basophils, mast cells, macrophages,
and T and B lymphocytes [32, 34]. Various inflammatory diseases such as rheumatoid arthritis [84], multiple sclerosis [31], and systemic lupus erythematosus [31, 81]
lead to upregulation of NGF synthesis. Anti-NGF antibodies have been shown to
significantly reduce inflammatory hypersensitivity caused by high levels of NGF
[85]. Therefore, NGF is considered to be a key mediator in the production of
inflammatory pain. It has been suggested that NGF influences inflammatory
responses by affecting immune cell function, altering neuropeptide (SP and CGRP)
levels in sensory fibers [1, 31, 32, 86], or sensitizing nerve terminals through receptor phosphorylation [34]. Therefore, elevated endogenous levels of NGF seem to
be associated with neuropathic and inflammatory pain. However, the use of anti-
NGF antibodies for the treatment of these conditions is still under research, as
some studies have produced contradictory results.

TNFα. Another novel therapeutic target that has been identified is TNFα. TNFα
(and IL-1) plays a major role in the pathogenesis of inflammatory diseases such as
arthritis [47]. It has been shown that there is an increased expression of TNFα
producing cells in the synovial tissue samples of osteoarthritis patients [87] and
that TNFα levels have also been detected in the synovial fluid and synovium of
rheumatoid arthritis patients [47, 88]. It is thought that TNFα (acting synergistically
with IL-1) increases the synthesis of matrix degrading proteins like interstitial
collagenase and stromelysin [49, 89] and causes tissue (cartilage) destruction
observed in rheumatoid arthritis, osteoarthritis, and other joint diseases [46, 51,
88]. Injections of TNFα have been shown to accelerate the onset or increase the
severity of collagen-induced arthritis in rats and mice [90, 91]. The use of anti-
TNFα antibodies or other TNFα inhibitors has been shown to treat collagen-
induced arthritis in mice [92, 93]. Also, a study has shown that anti-TNFα antibodies

can reduce inflammation (foot swelling) and joint destruction even when used after disease onset [94].

The positive results of these animal studies led to clinical trials with monoclonal antibodies of TNFα [95–97]. An open label trial was conducted with chimeric (75% human and 25% murine) monoclonal anti-TNFα antibody cA2 (infliximab) to test its safety and efficacy in the treatment of rheumatoid arthritis [97, 98]. Infliximab was infused two to four times over 2 weeks at a total dose of 20-mg/kg body weight in 20 patients with rheumatoid arthritis unresponsive to disease-modifying anti-rheumatic drugs (DMARD) therapy. It was found that this treatment led to a rapid improvement in patient symptoms; i.e., patients had decreased number of swollen/ tender joints, decreased degree of pain, and stiffness in the joints. Also, there was a decrease in the laboratory measurements of inflammatory activity [erythrocyte sedimentation rate and C-reactive proteins (CRP)]. The maximum clinical response (change from baseline) was achieved within the first month of treatment and lasted for 8–22 weeks. The infusions were reported to be safe as they were well tolerated and did not cause any adverse effects. It was reported that TNFα monoclonal antibody treatment might be useful for long-term therapy (for, e.g., for controlling chronic diseases) as seven patients that were retreated (given repeated infusions) after a relapse also showed significant improvement [96].

Another TNFα monoclonal antibody known as etanercept (a fusion protein of human IgG and two p75 TNF receptors) [98] that is a competitive inhibitor of TNF's cell-surface receptor has been developed. Clinical studies have shown that etanercept therapy results in significant clinical benefits and is well tolerated and safe for long-term use [99–101]. Etanercept is given subcutaneously (25 mg) twice weekly for the treatment of rheumatoid arthritis [98]. It is also used in the treatment of psoriatic arthritis, and a recent clinical trial has reported that in addition to decreasing joint pain and fatigue, etanercept also decreases symptoms of depression associated with this chronic disease [102].

Recently, a fully human monoclonal TNFα antibody called adalimumab (a humanized IgG₁) [98] has also been developed for the treatment of rheumatoid arthritis. Double-blind, placebo-controlled clinical trials have shown that the addition of adalimumab to methotrexate therapy resulted in significant, long-term improvement (slowed progression of structural joint damage and a reduction in signs and symptoms) in rheumatoid arthritis patients [103, 104]. Therefore, adalimumab has been approved for treating rheumatoid arthritis as a monotherapy and in combination with methotrexate [98]. Adalimumab is administered subcutaneously every 2 weeks (45 mg) [98]. Currently, all three of these TNFα inhibitors are being used in the treatment of various inflammatory conditions, such as rheumatoid arthritis, ankylosing spondylitis, psoriatic arthritis, and Crohn's disease.

8.1.1.5   NGF and Experimental Muscle Pain

Animal Research. Studies in adult rats [83, 105] have shown that NGF administration leads to hyperalgesia and sensitization of cutaneous nociceptive neurons [6, 82, 106–108]. Local or systemically administered mouse or human NGF resulted in two phases of hyperalgesia in response to thermal and mechanical stimuli in adult male and female rats: an early phase starting 30 minutes after NGF administration and a late, but longer lasting phase starting several hours after NGF administration and lasting for several days [30, 105–107, 109]. Thermal hyperalgesia occurred during both phases but mechanical hyperalgesia only during the late phase. Local, but not systemic, administration of NGF resulted in hyperalgesia that was limited to the site of injection. Interestingly, injection of human NGF has not been reported to result in any overt pain behavior in these animals [30]. It was subsequently demonstrated that human NGF applied directly to the receptive fields of cutaneous afferent fibers decreased the thermal or heat threshold without changing the mechanical or cold threshold [30, 107]. These results suggest that the first phase of thermal hyperalgesia is due to the peripheral effects of NGF, whereas the second phase of thermal and mechanical hyperalgesia are likely a result of central mechanisms.

Human Research. In healthy men and women, single intravenous or subcutaneous doses of human NGF (0.03 to 1 µg/kg) produced mild-to-moderate muscle pain, especially pain in the masseter muscle that increased with chewing, beginning 1–1.5 hours postinjection [69]. Beginning at 60–90 minutes after human NGF administration, subjects reported muscle pain that increased at 4 to 6 hours and then slowly decreased over 2 to 8 days. Female subjects reported greater muscle pain as compared with male subjects, and the effect was found to be dose-dependent, which suggests that human NGF may have sex-dependent effects [69]. Subcutaneous injection of human NGF resulted in hyperalgesia to touch and heat at the site of the injection (in addition to mild, diffuse myalgias), which lasted for 7 weeks and was found to be dose-dependent [69]. This study demonstrated that human NGF administration results in diffuse muscle pain (including pain in the orofacial region) that peaks hours after injection and increases with muscle use. Another study conducted by Dyck et al. [110] had reported that healthy human subjects injected with intradermal injection of human NGF in the forearm (1 or 3 µg) developed pressure allodynia and heat-pain hyperalgesia within 3 hours on the NGF-injected side. The authors suggested that local tissue mechanisms might be responsible for the rapid onset of these processes [110].

8.1.1.6   TNFα and Experimental Joint Pain

Animal Research. In 1992, Cunha et al. conducted a study in rats to determine the hyperalgesic effects of various cytokines including TNFα [111]. They found that TNFα played a crucial role in the development of inflammatory hyperalgesia and blockade of TNFα activity completely abolished the inflammatory response. They reported that TNFα-induced hyperalgesia occurs through both prostaglandin and sympathetically mediated pathways [111]. Another study conducted on male rats has shown that repeated intraplantar injections of TNFα into the rat paw (2.5 pg daily for 30 days) caused persistent mechanical hypersensitivity in nociceptors that lasted for at least 30 days after the treatment was stopped [112]. The authors reported that hypersensitivity was not induced when indomethacin and atenolol were injected with TNFα, suggesting that TNFα-induced mechanical hypersensitivity was mediated via the release of eicosanoids and sympathomimetic mediators, as suggested by the Cunha et al. study [112]. Various other studies have also shown that intraplantar injections of TNFα can induce mechanical allodynia and hyperalgesic effects in rats [52, 111, 113, 114].

Researchers have also used an *in vitro* rat skin model to determine the effects of TNFα on heat-induced release of CGRP (neuropeptide) from nociceptors [115]. They found that, in contrast to the excitatory effects of TNFα *in vivo*, it did not induce any CGRP release at 32°C; however, TNFα produced a transient (short-term) and dose-dependent heat sensitization. Thus, the authors speculated that, in addition to their role in mediating long-term sensitization in various chronic inflammatory conditions [52, 116], cytokines such as TNFα might also play a role in the acute phase of inflammation, i.e., cause heat sensitization through receptor-associated mechanisms (phosphorylation of ion channels) [115] rather than through signaling pathways. A recent study has also suggested that TNFα-induced rapid thermal and mechanical sensitization processes (after peripheral administration) are probably due to a direct effect on afferent fibers [117].

TNFα is also thought to be one of the mediators involved in formalin-induced orofacial nociception in behavioral rat models [118] and carrageenan (CG)-induced inflammation in mice [119, 120]. A study has shown that intraarticular injections of TNFα could cause an intense incapacitation response in carrageenan-primed rat knee joints that lasts for more than 8 hours [121]. The treatment of knee joints with anti-TNFα antibodies (intraarticular injection) significantly inhibited CG-induced incapacitation response, which suggests that TNFα is a mediator of CG-induced inflammatory incapacitation and is involved in maintaining nociceptive responses over the long term [121]. Another study has shown that induction of arthritis in mice (streptococcal cell wall arthritis model) caused the release of TNFα, which resulted in significant knee joint swelling [122]. The use of anti-TNFα treatment markedly reduced the joint swelling, which suggests a role for TNFα in this arthritis model [122]. The contribution of TNFα to inflammatory pain has been assessed in another experimentally induced model of arthritis (complete Freund's adjuvant or CFA) [123]. It was found that CFA-induced joint inflammation in rats was associated with increased expression of both TNFα receptors in the dorsal root ganglion cells. Also, the number of macrophages was correlated with the development of hyperalgesia, which suggests that TNFα induces inflammatory hyperalgesia through its direct actions on neuronal receptors as well as by increasing the expression of macrophages [123].

The effects of subcutaneously injected TNFα on the mechanical sensitivity and activity of C fibers have also been tested in anesthetized rats [124]. It was found that the 5-ng dose of TNFα decreased the mechanical thresholds of ~67% of the C fibers, and this sensitization began within 30 minutes of the injection and lasted for at least 2 hours. Also, the authors reported that TNFα caused significant plasma protein extravasation that was dose-dependent [124]. Therefore, these studies suggest that TNFα released after tissue injury or disease is involved in the generation of increased pain sensitivity and inflammation, which is observed in patients suffering from inflammatory diseases. Further evidence for TNFα's role in pain conditions comes from studies that have shown strong correlations among tissue levels of TNFα, pain, and hyperalgesia [52, 118, 125–127].

Human Research. Animal research has provided strong evidence for the role of TNFα in mediating inflammatory pain sensitivity and tissue destruction associated with chronic inflammatory conditions, and the successful use of anti-TNFα antibodies in animals has led to their use in human subjects. Various clinical trials have shown positive results (as discussed) and led to the approval of these agents for use in the treatment of conditions like rheumatoid arthritis.

TNFα has also been thought to be involved in the pathogenesis of other conditions such as migraine headaches, multiple sclerosis, and depression. Recently, a study was conducted to determine the role of TNFα in a migraine. It was found that migraine patients had significantly higher plasma TNFα levels during an attack as compared with headache-free periods, which suggests a potential role for TNFα [178]. Similarly, serum TNFα levels have been reported to be significantly higher in patients suffering from depression and multiple sclerosis than in healthy controls [129]. Also, it has been reported that TNFα may be important in the pathogenesis of temporomandibular disorders (TMDs) and the measurement of TNFα levels in the synovial aspirate may be useful for the diagnosis of temporomandibular joint (TMJ)-related pain conditions [130–133].

8.1.2   CASE STUDIES

8.1.2.1   Case Study I: Effect of Intramuscular Injection of NGF in Humans

To investigate the effects of intramuscular injection of NGF on muscle pain, a study was conducted on 12 male volunteers with a mean age of 24.8 years. A baseline measurement of pressure pain threshold (PPT) and pressure tolerance threshold (PTOL) was conducted with a pressure algometer for both the masseter and the temporalis muscles. PPT is a measure of mechanical allodynia and refers to the lowest pressure that the subjects perceived to be painful. PTOL is considered to be a measure of mechanical hyperalgesia and refers to the maximum pressure that the subjects could tolerate. Each subject received two injections (buffered isotonic saline control and ~0.1-μg/kg recombinant human NGF; injection volume = 0.2 mL), one in each masseter muscle. The side and sequence of injections was randomized and blinded. After the first injection, the subjects were asked to rate the pain experienced on an electronic visual analog scale (VAS) of 1 (no pain) to 10 (worst imaginable pain). The pain scores were used to determine the pain intensity felt by the subjects. Fifteen minutes after the first injection, the second injection was made in the contralateral masseter muscle and the pain score measurement was repeated. Subjects were also asked to fill out a pain questionnaire and rate their pain intensity on a 0–10 numerical rating scale (NRS) during activities involving jaw movements such as chewing, yawning, and talking. Both PPT and PTOL measurements were repeated 1, 7, 14, 21, and 28 days after the injections [134].

The results showed that intramuscular injection of NGF did not evoke any significant pain in the masseter muscle, as the VAS scores after the NGF injection were not significantly different from the low VAS scores reported after the control saline injection. There was no significant difference in the baseline PPT and PTOL values in either temporalis or masseter muscles. The NGF injection caused a decrease in the PPT values (as compared with the baseline) in the masseter muscle, with significantly lower PPTs at 1 and 7 days after injection (Figure 8.1-1). The saline injection also caused a significant decrease in PPTs 1 day after injection; however, the PPT values increased significantly above baseline after 14 and 21 days. At 1 day after injection, the mean PPT value in the NGF-injected muscles was significantly lower than that in the saline-injected muscles (Figure 8.1-1). The PPT

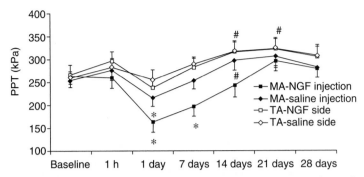

Figure 8.1-1. PPT measured in four test muscles for 28 days. NGF or isotonic saline was injected into the masseter (MA) muscle. Temporalis (TA) muscle PPTs were also assessed bilaterally on the side of NGF and saline injection. Each point indicates mean ± SEM (n = 12). The symbol (*) indicates significantly lower and (#) significantly higher compared with baseline (Dunnett: $p < 0.05$). (Reproduced from Ref. 134, with the permission of the International Association for the Study of Pain.)

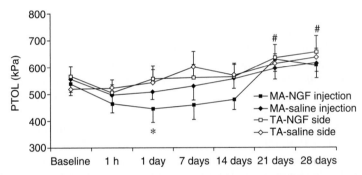

Figure 8.1-2. Pressure pain tolerances (PTOL) measured on four test muscles over 28 days. NGF or isotonic saline was injected into the masseter (MA) muscle. Temporalis (TA) muscle PPTs were also assessed bilaterally on the side of NGF and saline injection. Each point indicates mean ± SEM (n = 12). The symbol (*) indicates significantly lower and (#) significantly higher compared with baseline ($p < 0.05$). (Reproduced from Ref. 134, with the permission of the International Association for the Study of Pain.)

values in both temporalis muscles increased after (control saline or NGF) injections in the masseter muscles. The NGF injection also caused a decrease in the PTOL values, with significantly lower PTOL values at 1 day (compared with baseline) (Figure 8.1-2). However, at 21 days, the mean PTOL value in NGF-injected muscles increased significantly above the baseline value. The control saline injection did not cause any significant change in the PTOL values. The PPT values of the temporalis muscle ipsilateral to saline-injected masseter muscle were significantly higher at 21 and 28 days after injection. However, there was no significant change in the PPTs of the temporalis muscle ipsilateral to the NGF-injected masseter muscle. Subjects reported pain upon chewing and yawning in the NGF-injected masseter muscle, whereas activities such as talking, swallowing, and smiling did not evoke pain in the NGF-injected masseter muscle (Figure 8.1-3) [134].

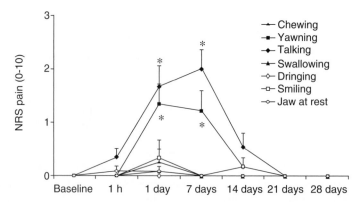

Figure 8.1-3. Perceived pain intensity measured on a 0–10 NRS in response to various jaw functions before and after injection of NGF into the masseter muscle. Each point indicates mean ± SEM ($n = 12$). The symbol (*) indicates significantly higher NRS pain scores compared with baseline (Dunnett: $p < 0.05$). (Reproduced from Ref. 134, with the permission of the International Association for the Study of Pain.)

The results of this study suggested that intramuscular injection of NGF did not evoke any significant pain but caused mechanical sensitization at the site of injection in the masseter muscle that lasted for at least 7 days and resulted in TMD-like symptoms. A significant reduction in the PPT and PTOL values in the NGF-injected muscles indicated that NGF can cause mechanical allodynia and hyperalgesia. Also, a significant number of subjects reported pain during oral functions such as chewing and yawning, which is also a characteristic symptom of TMD. The authors suggested that intramuscular administration of NGF into the masseter muscle could be used to model TMD-like pain and sensitization and to further study the pathophysiology of TMD. In addition, the authors commented that it was not clear whether NGF was causing mechanical sensitization through a peripheral or central mechanism [134]. Therefore, the mechanisms responsible for NGF-induced sensitization need to be further investigated.

8.1.2.2   Case Study II: NGF Effects on Masticatory Muscle Nociceptors

An animal study was undertaken to determine whether NGF significantly alters the masseter muscle afferent fiber mechanical threshold as part of the mechanism whereby it decreases PPTs in human subjects. In this study, *in vivo* single afferent neuron recordings were conducted on anesthetized, adult female Sprague–Dawley rats ($n = 17$). A recording electrode was lowered into the trigeminal ganglion to record orthodromic action potentials of afferent fibers activated by applying mechanical force to the masseter muscle. A blunt mechanical search stimulus was used to find afferent fibers in the rat masseter muscle. An antidromic collision technique [135] was used to confirm the projection of a fiber to the central nervous system (Figure 8.1-4). The mechanical threshold (MT) of each masseter muscle afferent fiber was assessed with an electronic von Frey (VF) hair (Figure 8.1-5). For each fiber, a baseline MT recording was conducted every minute for 10 minutes. Each primary afferent fiber was assigned to one of the following two groups: vehicle

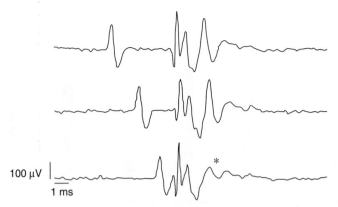

100 μV

1 ms

Figure 8.1-4. An example of a collision between an orthodromic (activated by mechanical stimulation of the masseter muscle) and an antidromic (activated by electrical stimulation of the caudal brain stem) is illustrated. *Collision resulted in the disappearance of the antidromic spike. This method was used to confirm the projection of the masseter afferent fiber from the masseter muscle to the caudal brain stem, a region believed to be responsible for the integration of muscle pain input.

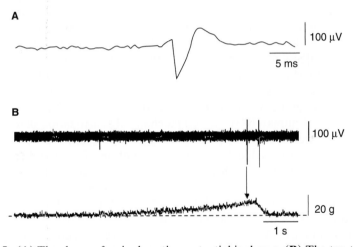

Figure 8.1-5. (**A**) The shape of a single action potential is shown. (**B**) The top trace shows the action potentials recorded from the trigeminal ganglion. The bottom trace shows an increase in the force applied with an electronic von Frey (VF) hair on the masseter muscle until the fiber started firing action potentials. The threshold of the fiber (7.71 g) was determined by subtracting the baseline from the minimum force required to activate the fiber.

control (10 μL of phosphate-buffered isotonic saline and albumin, $n = 10$) or 25-μg/mL human NGF (1 μg/kg, respectively, 10 μL, $n = 7$) dissolved in phosphate-buffered isotonic saline and albumin. After the baseline recording, an injection was made into the masseter muscle and the evoked response was recorded for 10 minutes. Sixty minutes after the injection, 10 consecutive MT recordings were conducted at 1-minute intervals. These recordings were repeated every hour for a total of 3 hours. Also, the effect of NGF on plasma protein extravasation in the masseter muscle was determined.

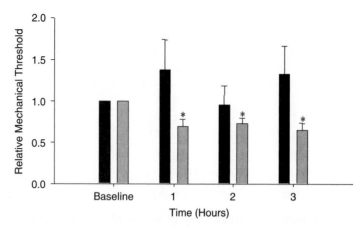

Figure 8.1-6. The vertical bar graph illustrates the mean relative mechanical threshold of afferent fibers at various time points after a 10-μL injection of phosphate-buffered saline control (black bars, $n = 10$) or 25-μg/mL human NGF (gray bars, $n = 7$) into the masseter muscle. The error bars represent the standard error of the mean. *$p < 0.05$: ANOVA and post hoc Holm–Sidak method compared with the baseline relative mechanical threshold.

Intramuscular injection of human NGF into the rat masseter muscle failed to evoke afferent discharge; however, it decreased the MT of masseter afferent fibers compared with the baseline (Figure 8.1-6). There was a significant decrease in the MT 1 hour postinjection (as compared with the baseline), which lasted for 3 hours. The postinjection MT of saline-injected fibers was not significantly different from the baseline. There was no significant difference between the level of plasma protein extravasation between control saline and human NGF groups.

The injection of human NGF caused mechanical sensitization that lasted for at least 3 hours; however, there was no indication that human NGF produced these effects indirectly through local muscle inflammation. The peripheral effects of NGF are likely mediated through a receptor mechanism, for example, TrkA receptor mediated pathways [136]. It is possible that nonspecific mechanisms, such as foreign protein reactions, decreased degradation of human NGF by rat proteases, slowed clearance from the muscle, modulation of intracellular calcium concentration (calcium uptake) [137, 138], actions on bradykinin receptors [139], capsaicin receptors [37, 140], or sodium channels [141, 142] might contribute to the effect of human NGF in rats.

The results of this study suggested that intramuscular injection of human NGF at a concentration equivalent to that which resulted in mechanical sensitization in humans [134] caused a significant decrease in the mechanical threshold of rat masseter muscle afferent fibers. Therefore, the method of injecting human NGF into the rat masseter muscle could be used to study peripheral mechanisms associated with the development of NGF-induced masseter muscle sensitization in humans.

REFERENCES

1. Pasero C (2004). Pathophysiology of neuropathic pain. *Pain Manage. Nursing.* 5:3–8.

2. Mense S (1993). Nociception from skeletal muscle in relation to clinical muscle pain. *Pain.* 54:241–289.

3. DeLeo J A, Winkelstein B A (2002). Physiology of chronic spinal pain syndromes: From animal models to biomechanics. *Spine.* 27:2526–2537.

4. White F A, Bhangoo S K, Miller R J (2005). Chemokines: integrators of pain and inflammation. *Nat. Rev. Drug Disc.* 4:834–844.

5. Willis W D (1998). The somatosensory system. In R M Berne, M N Levy, B M Koeppen, et al. (eds.), *Physiology*: Mosby, Inc., St. Louis, MD, pp. 109–128.

6. Dmitrieva N, McMahon S B (1996). Sensitisation of visceral afferents by nerve growth factor in the adult rat. *Pain.* 66:87–97.

7. Meyer R A, Campbell J N, Raja S N (1994). Peripheral neural mechanisms of nociception. In P D Wall, R Melzack (eds.), *Textbook of Pain*: Churchill Livingstone, Edinburgh, Scotland, pp. 13–44.

8. Almeida T F, Roizenblatt S, Tufik S (2004). Afferent pain pathways: A neuroanatomical review. *Brain Res.* 1000:40–56.

9. McMahon S B, Lewin G R, Wall P D (1993). Central hyperexcitability triggered by noxious inputs. *Curr. Opin. Neurobiol.* 3:602–610.

10. Ren K, Dubner R (1999). Central nervous system plasticity and persistent pain. *J. Orofacial Pain.* 13:155–163.

11. Sessle B J (2000). Acute and chronic craniofacial pain: Brainstem mechanisms of nociceptive transmission and neuroplasticity, and their clinical correlates. *Crit. Rev. Oral Biol. Med.* 11:57–91.

12. Woolf C J (1991). Generation of acute pain: central mechanisms. *British Med. Bull.* 47:523–533.

13. Dworkin R H, Backonja M, Rowbotham M C, et al. (2003). Advances in neuro-pathic pain: diagnosis, mechanisms, and treatment recommendations. *Archives Neurol.* 60:1524–1534.

14. Harden R N (2005). Chronic neuropathic pain. Mechanisms, diagnosis, and treatment. *Neurolog.* 11:111–122.

15. Hansson P T, Dickenson A H (2005). Pharmacological treatment of peripheral neuropathic pain conditions based on shared commonalities despite multiple etiologies. *Pain.* 113:251–254.

16. Ohtori S, Takahashi K, Moriya H, et al. (2004). TNF-alpha and TNF-alpha receptor type 1 upregulation in glia and neurons after peripheral nerve injury: studies in murine DRG and spinal cord. *Spine.* 29:1082–1088.

17. Schafers M, Geis C, Svensson C I, et al. (2003). Selective increase of tumour necrosis factor-alpha in injured and spared myelinated primary afferents after chronic constrictive injury of rat sciatic nerve. *Euro. J. Neurosci.* 17:791–804.

18. Shubayev V I, Myers R R (2000). Upregulation and interaction of TNFalpha and gelatinases A and B in painful peripheral nerve injury. *Brain Res.* 855:83–89.

19. Lindenlaub T, Teuteberg P, Hartung T, et al. (2000). Effects of neutralizing antibodies to TNF-alpha on pain-related behavior and nerve regeneration in mice with chronic constriction injury. *Brain Res.* 866:15–22.

20. Raghavendra V, Tanga F, DeLeo J A (2003). Inhibition of microglial activation attenuates the development but not existing hypersensitivity in a rat model of neuropathy. *J. Pharmacol. Exper. Therapeut.* 306:624–630.

21. Sommer C, Lindenlaub T, Teuteberg P, et al. (2001). Anti-TNF-neutralizing antibodies reduce pain-related behavior in two different mouse models of painful mononeuropathy. *Brain Res.* 913:86–89.

22. Aloe L (2004). Rita Levi-Montalcini: the discovery of nerve growth factor and modern neurobiology. *Trends Cell Biol.* 14:395–399.

23. Levi-Montalcini R (1987). The nerve growth factor 35 years later. *Sci.* 237:1154–1162.

24. Huang E J, Reichardt L F (2001). Neurotrophins: roles in neuronal development and function. *Ann. Rev. Neurosci.* 24:677–736.

25. Holland D R, Cousens L S, Meng W, et al. (1994). Nerve growth factor in different crystal forms displays structural flexibility and reveals zinc binding sites. *J. Molec. Biol.* 239:385–400.

26. McDonald N Q, Lapatto R, Murray-Rust J, et al. (1991). New protein fold revealed by a 2.3-A resolution crystal structure of nerve growth factor. *Nature.* 354:411–414.

27. Aloe L, Bracci-Laudiero L, Bonini S, et al. (1997). The expanding role of nerve growth factor: From neurotrophic activity to immunologic diseases. *Allergy.* 52:883–894.

28. Owolabi J B, Rizkalla G, Tehim A, et al. (1999). Characterization of Antiallodynic Actions of ALE-0540, a Novel Nerve Growth Factor Receptor Antagonist, in the Rat. *J. Pharmacol. Exper. Therapeut.* 289:1271–1276.

29. Aloe L, Allevab E, Fiore M (2002). Stress and nerve growth factor: Findings in animal models and humans. *Pharmacol. Biochem. Behavior.* 73:159–166.

30. Andreev N Y, Dimitrieva N, Koltzenburg M, et al. (1995). Peripheral administration of nerve growth factor in the adult rat produces a thermal hyperalgesia that requires the presence of sympathetic post-ganglionic neurons. *Pain.* 63:109–115.

31. Bracci-Laudiero L, Aloe L, Buanne P, et al. (2002). NGF modulates CGRP synthesis in human B-lymphocytes: a possible anti-inflammatory action of NGF? *J. Neuroimmunol.* 123:58–65.

32. de VRIES A, Dessing M C, Engels F, et al. (1999). Nerve Growth Factor Induces a Neurokinin-1 Receptor-Mediated Airway Hyperresponsiveness in Guinea Pigs. *Amer. J. Respir. Criti. Care Med.* 159:1541–1544.

33. Chao M V, Hempstead B L (1995). p75 and Trk: A two-receptor system. *Trends Neurosci.* 18:321–326.

34. Woolf C J, Ma Q P, Allchorne A, et al. (1996). Peripheral cell types contributing to the hyperalgesic action of nerve growth factor in inflammation. *J. Neurosci.* 16:2716–2723.

35. Chao M V (2003). Neurotrophins and their receptors: A convergence point for many signaling pathways. *Neurosci.* 4:299–309.

36. York R D, Molliver D C, Grewal S S, et al. (2000). Role of phosphoinositide 3-kinase and endocytosis in nerve growth factor-induced extracellular signal-regulated kinase activation via Ras and Rap1. *Molec. Cell. Biol.* 20:8069–8083.

37. Chuang H, Prescott E D, Kong, H, et al. (2001). Bradykinin and nerve growth factor release the capsaicin receptor from PtdIns(4,5)P2-mediated inhibition. *Nat. Rev.* 411:957–962.

38. Humpel C, Ebendal T, Olson L (1996). Microdialysis: a way to study in vivo release of neurotrophic bioactivity: a critical summary. *J. Molec. Med.* 74:523–526.

39. Winter C D, Iannotti F, Pringle A K, et al. (2002). A microdialysis method for the recovery of IL-1beta, IL-6 and nerve growth factor from human brain in vivo. *J. Neurosci. Meth.* 119:45–50.

40. Martocchia A, Sigala S, Proietti A, et al. (2002). Sex-related variations in serum nerve growth factor concentration in humans. *Neuropept.* 36:391–395.

41. Carswell E A, Old L J, Kassel R L, et al. (1975). An endotoxin-induced serum factor that causes necrosis of tumors. *Proc. Nat. Acad. Sci. USA.* 72:3666–3670.

42. Allan S M, Rothwell N J (2001). Cytokines and acute neurodegeneration. *Nat. Rev. Neurosci.* 2:734–744.

43. Rothwell N J, Luheshi G, Toulmond S (1996). Cytokines and their receptors in the central nervous system: physiology, pharmacology, and pathology. *Pharmacol. Therapeut.* 69:85–95.

44. Darnay B G, Aggarwal B B (1999). Signal transduction by tumour necrosis factor and tumour necrosis factor related ligands and their receptors. *Annals Rheumat. Dis.* 58(suppl 1):I2–I13.

45. Vilcek J, Lee T H (1991). Tumor necrosis factor. New insights into the molecular mechanisms of its multiple actions. *J. Biol. Chem.* 266:7313–7316.

46. Sherry B, Cerami A (1988). Cachectin/tumor necrosis factor exerts endocrine, paracrine, and autocrine control of inflammatory responses. *J. Cell Biol.* 107:1269–1277.

47. Burger D, Dayer J M (2002). Cytokines, acute-phase proteins, and hormones: IL-1 and TNF-alpha production in contact-mediated activation of monocytes by T lymphocytes. *Annals NY Acad. Sci.* 966:464–473.

48. Van Deventer S J (1997). Tumour necrosis factor and Crohn's disease. *Gut.* 40:443–448.

49. Heller R A, Kronke M (1994). Tumor necrosis factor receptor-mediated signaling pathways. *J. Cell Bio.* 126:5–9.

50. Allan S M (2000). The role of pro- and antiinflammatory cytokines in neurodegeneration. *Annals NY Acad. Sci.* 917:84–93.

51. Dinarello C A (2000). Proinflammatory cytokines. *Chest.* 118:503–508.

52. Sommer C, Kress M (2004). Recent findings on how proinflammatory cytokines cause pain: peripheral mechanisms in inflammatory and neuropathic hyperalgesia. *Neurosci. Lett.*, 361:184–187.

53. Steiner P, Pfeilschifter J, Boeckh C, et al. (1991). Interleukin-1 beta and tumor necrosis factor-alpha synergistically stimulate nerve growth factor synthesis in rat mesangial cells. *Amer. J. Physiol.* 261:F792–798.

54. Brenner D A, O'Hara M, Angel P, et al. (1989). Prolonged activation of jun and collagenase genes by tumour necrosis factor-alpha. *Nature.* 337:661–663.

55. Tracey K J, Fong Y, Hesse D G, et al. (1987). Anti-cachectin/TNF monoclonal antibodies prevent septic shock during lethal bacteraemia. *Nature.* 330:662–664.

56. Hesse D G, Tracey K J, Fong Y, et al. (1988). Cytokine appearance in human endotoxemia and primate bacteremia. *Surgery Gynecol. Obstet.* 166:147–153.

57. Chaldakov G N, Fiore M, Hristova M G, et al. (2003). Metabotrophic potential of neurotrophins: implication in obesity and related diseases? *Med. Sci. Monitor.* 9: HY19–21.

58. Hellweg R, Hartung H D (1990). Endogenous levels of nerve growth factor (NGF) are altered in experimental diabetes mellitus: A possible role for NGF in the pathogenesis of diabetic neuropathy. *J. Neurosci. Res.* 26:258–267.

59. Leinninger G M, Vincent A M, Feldman E L (2004). The role of growth factors in diabetic peripheral neuropathy. *J. Periph. Nerv. Syst.* 9:26–53.

60. Brewster W J, Fernyhough P, Diemel L T, et al. (1994). Diabetic neuropathy, nerve growth factor and other neurotrophic factors. *Trends Neurosci.* 17:321–325.

61. Fernyhough P, Diemel L T, Brewster W J, et al. (1994). Deficits in sciatic nerve neuropeptide content coincide with a reduction in target tissue nerve growth factor messenger RNA in streptozotocin-diabetic rats: effects of insulin treatment. *Neurosci.* 62:337–344.

62. Anand P, Terenghi G, Warner G, et al. (1996). The role of endogenous nerve growth factor in human diabetic neuropathy. *Nature Med.* 2:703–707.

63. Hellweg R, Raivich G, Hartung H D, et al. (1994). Axonal transport of endogenous nerve growth factor (NGF) and NGF receptor in experimental diabetic neuropathy. *Exper. Neurol.* 130:24–30.

64. Jakobsen J, Brimijoin S, Skau K, et al. (1981). Retrograde axonal transport of transmitter enzymes, fucose-labeled protein, and nerve growth factor in streptozotocin-diabetic rats. *Diabetes.* 30:797–803.

65. Apfel S C (2002). Nerve growth factor for the treatment of diabetic neuropathy: what went wrong, what went right, and what does the future hold? *Internat. Rev. Neurobio.* 50:393–413.

66. Apfel S C, Arezzo J C, Brownlee M, et al. (1994). Nerve growth factor administration protects against experimental diabetic sensory neuropathy. *Brain Res.* 634:7–12.

67. Fernyhough P, Diemel L T, Hardy J, et al. (1995). Human recombinant nerve growth factor replaces deficient neurotrophic support in the diabetic rat. *Euro. J. Neurosci.* 7:1107–1110.

68. Apfel S C, Kessler J A, Adornato B T, et al. (1998). Recombinant human nerve growth factor in the treatment of diabetic polyneuropathy. NGF Study Group. *Neurol.* 51:695–702.

69. Petty B G, Cornblath D R, Adornato B T, et al. (1994). The effect of systemically administered recombinant human nerve growth factor in healthy human subjects. *Ann. Neurol.* 36:244–246.

70. Apfel S C, Schwartz S, Adornato B T, et al. (2000). Efficacy and safety of recombinant human nerve growth factor in patients with diabetic polyneuropathy: A randomized controlled trial. rhNGF Clinical Investigator Group. *J. Amer. Med. Associ.* 284:2215–2221.

71. Deng Y S, Zhong J H, Zhou X F (2000). Effects of endogenous neurotrophins on sympathetic sprouting in the dorsal root ganglia and allodynia following spinal nerve injury. *Exper. Neurol.* 164:344–350.

72. Li L, Xian C J, Zhong J H, et al. (2003). Lumbar 5 ventral root transection-induced upregulation of nerve growth factor in sensory neurons and their target tissues: a mechanism in neuropathic pain. *Molec. Cell. Neurosci.*, 23:232–250.

73. Ramer M S, French G D, Bisby M A (1997). Wallerian degeneration is required for both neuropathic pain and sympathetic sprouting into the DRG. *Pain.*, 72:71–78.

74. Ruiz G, Ceballosc D, Banos J-E (2004). Behavioral and histological effects of endoneurial administration of nerve growth factor: Possible implications in neuropathic pain. *Brain Res.* 1011:1–6.

75. Oh E J, Yoon Y W, Lee S E, et al. (2000). Changes in nerve growth factor levels in dorsal root ganglia and spinal nerves in a rat neuropathic pain model. *Exper. Brain Res.* 130:93–99.

76. Heumann R, Korsching S, Bandtlow C, et al. (1987). Changes of nerve growth factor synthesis in nonneuronal cells in response to sciatic nerve transection. *J. Cell Biol.* 104:1623–1631.

77. Theodosiou M, Rush R A, Zhou X F, et al. (1999). Hyperalgesia due to nerve damage: role of nerve growth factor. *Pain.* 81:245–255.

78. Anand P (2004). Neurotrophic factors and their receptors in human sensory neuropathies. *Prog. Brain Res.* 146:477–492.

79. Ren K, Thomas D A, Dubner R (1995). Nerve growth factor alleviates a painful peripheral neuropathy in rats. *Brain Res.* 699:286–292.

80. Cahill C M, Dray A, Coderre T J (2003). Intrathecal nerve growth factor restores opioid effectiveness in an animal model of neuropathic pain. *Clin. Neuropharmacol.* 45:543–552.

81. Bielefeldt K, Ozaki N, Gebhart G F (2003). Role of nerve growth factor in modulation of gastric afferent neurons in the rat. *Amer. J. Phys. Gastrointest. Liver Physio.* 284: G499–G507.

82. Djouhri L, Dawbarn D, Robertson A, et al. (2001). Time course and nerve growth factor dependence of inflammation-induced alterations in electrophysiological membrane properties in nociceptive primary afferent neurons. *J. Neurosci.* 21:8722–8733.

83. Woolf C J, Safieh-Garabedian B, Ma Q P, et al. (1994). Nerve growth factor contributes to the generation of inflammatory sensory hypersensitivity. *Neurosci.* 62:327–331.

84. Aloe L, Tuveri M A, Carcassi U, et al. (1992). Nerve growth factor in the synovial fluid of patients with chronic arthritis. *Arthritis Rheumat.* 35:351–355.

85. Woolf C J (1996). Phenotypic modification of primary sensory neurons: the role of nerve growth factor in the production of persistent pain. *Philosoph. Trans. Royal Soc. London. Series B, Biolog. Sci.* 351:441–448.

86. Lindsay R M, Harmar A J (1989). Nerve growth factor regulates expression of neuropeptide genes in adult sensory neurons. *Nature Rev.* 337:362–364.

87. Benito M J, Veale D J, FitzGerald O, et al. (2005). Synovial tissue inflammation in early and late osteoarthritis. *Annals Rheumatic Dis.* 64:1263–1267.

88. Feldmann M (1996). The cytokine network in rheumatoid arthritis: definition of TNF alpha as a therapeutic target. *J. Royal Coll. Physicians of London.* 30:560–570.

89. Fiers W (1991). Tumor necrosis factor. Characterization at the molecular, cellular and in vivo level. *FEBS Lett.* 285:199–212.

90. Brahn E, Peacock D J, Banquerigo M L, et al. (1992). Effects of tumor necrosis factor alpha (TNF-alpha) on collagen arthritis. *Lymphokine Cytokine Res.* 11:253–256.

91. Cooper W O, Fava R A, Gates C A, et al. (1992). Acceleration of onset of collagen-induced arthritis by intra-articular injection of tumour necrosis factor or transforming growth factor-beta. *Clin. Experim. Immunol.* 89:244–250.

92. Piguet P F, Grau G E, Vesin C, et al. (1992). Evolution of collagen arthritis in mice is arrested by treatment with anti-tumour necrosis factor (TNF) antibody or a recombinant soluble TNF receptor. *Immunol.* 77:510–514.

93. Thorbecke G J, Shah R, Leu C H, et al. (1992). Involvement of endogenous tumor necrosis factor alpha and transforming growth factor beta during induction of collagen type II arthritis in mice. *Proc. Nat. Acad. Sci. USA.* 89:7375–7379.

94. Williams R O, Feldmann M, Maini R N (1992). Anti-tumor necrosis factor ameliorates joint disease in murine collagen-induced arthritis. *Proc. Nat. Acad. Sci. USA.* 89:9784–9788.

95. Elliott M J, Maini R N, Feldmann M, et al. (1994). Randomised double-blind comparison of chimeric monoclonal antibody to tumour necrosis factor alpha (cA2) versus placebo in rheumatoid arthritis. *Lancet.* 344:1105–1110.

96. Elliott M J, Maini R N, Feldmann M, et al. (1994). Repeated therapy with monoclonal antibody to tumour necrosis factor alpha (cA2) in patients with rheumatoid arthritis. *Lancet.* 344:1125–1127.

97. Elliott M J, Maini R N, Feldmann M, et al. (1993). Treatment of rheumatoid arthritis with chimeric monoclonal antibodies to tumor necrosis factor alpha. *Arthrit. Rheumat.* 36:1681–1690.

98. Nash P T, Florin T H (2005). Tumour necrosis factor inhibitors. *Med. J. Australia.* 183:205–208.

99. Flagg S D, Meador R, Hsia E, et al. (2005). Decreased pain and synovial inflammation after etanercept therapy in patients with reactive and undifferentiated arthritis: An open-label trial. *Arthrit. Rheumatism.* 53:613–617.

100. Moreland L W, Schiff M H, Baumgartner S W, et al. (1999). Etanercept therapy in rheumatoid arthritis. A randomized, controlled trial. *Ann. Intern. Med.* 130:478–486.

101. Weinblatt M E, Kremer J M, Bankhurst A D, et al. (1999). A trial of etanercept, a recombinant tumor necrosis factor receptor:Fc fusion protein, in patients with rheumatoid arthritis receiving methotrexate. *New Eng. J. Med.* 340:253–259.

102. Tyring S, Gottlieb A, Papp K, et al. (2006). Etanercept and clinical outcomes, fatigue, and depression in psoriasis: double-blind placebo-controlled randomised phase III trial. *Lancet.* 367:29–35.

103. Keystone E C, Kavanaugh A F, Sharp J T, et al. (2004). Radiographic, clinical, and functional outcomes of treatment with adalimumab (a human anti-tumor necrosis factor monoclonal antibody) in patients with active rheumatoid arthritis receiving concomitant methotrexate therapy: a randomized, placebo-controlled, 52-week trial. *Arthrit. Rheumatism.* 50:1400–1411.

104. Weisman M H, Moreland L W, Furst D E, et al. (2003). Efficacy, pharmacokinetic, and safety assessment of adalimumab, a fully human anti-tumor necrosis factor-alpha monoclonal antibody, in adults with rheumatoid arthritis receiving concomitant methotrexate: A pilot study. *Clin. Therapeut.* 25:1700–1721.

105. Lewin G R, Ritter A M, Mendell L M (1993). Nerve growth factor-induced hyperalgesia in the neonatal and adult rat. *J. Neurosci.* 13:2136–2148.

106. Rueff A, Dawson A J, Mendell L M (1996). Characteristics of nerve growth factor induced hyperalgesia in adult rats: Dependence on enhanced bradykinin-1 receptor activity but not neurokinin-1 receptor activation. *Pain.* 66:359–372.

107. Rueff A, Mendell L M (1996). Nerve Growth Factor and NT-5 induce increased thermal sensitivity of cutaneous nociceptors in vitro. *J. Neurophysiol.* 76:3596–3596.

108. Thompson S W N, Dray A, McCarson K E, et al. (1995). Nerve growth factor induces mechanical allodynia associated with novel A fibre-evoked spinal reflex activity and enhanced neurokinin-1 receptor activation in the rat. *Pain.* 62:219–231.

109. Lewin G R, Rueff A, Mendell L M (1994). Peripheral and central mechanisms of NGF-induced hyperalgesia. *Euro. J. Neurosci.* 6:1903–1912.

110. Dyck P J, Peroutka S, Rask C, et al. (1997). Intradermal recombinant human nerve growth factor induces pressure allodynia and lowered heat-pain threshold in humans. *Neurol.* 48:501–505.

111. Cunha F Q, Poole S, Lorenzetti B B, et al. (1992). The pivotal role of tumour necrosis factor alpha in the development of inflammatory hyperalgesia. *British J. Pharmacol.* 107:660–664.

112. Sachs D, Cunha F Q, Poole S, et al. (2002). Tumour necrosis factor-alpha, interleukin-1beta and interleukin-8 induce persistent mechanical nociceptor hypersensitivity. *Pain.* 96:89–97.

113. Perkins M N, Kelly D, Davis A J (1995). Bradykinin B1 and B2 receptor mechanisms and cytokine-induced hyperalgesia in the rat. *Canadian J. Physiol. Pharmacol.* 73:832–836.

114. Safieh-Garabedian B, Poole S, Allchorne A, et al. (1995). Contribution of interleukin-1 beta to the inflammation-induced increase in nerve growth factor levels and inflammatory hyperalgesia. *British J. Pharmacol.* 115:1265–1275.

115. Opree A, Kress M (2000). Involvement of the proinflammatory cytokines tumor necrosis factor-alpha, IL-1 beta, and IL-6 but not IL-8 in the development of heat hyperalgesia: effects on heat-evoked calcitonin gene-related peptide release from rat skin. *J. Neurosci.* 20:6289–6293.

116. Sommer C, Schmidt C, George A (1998). Hyperalgesia in experimental neuropathy is dependent on the TNF receptor 1. *Exper. Neurol.* 151:138–142.

117. Jin X, Gereau R W T (2006). Acute p38-mediated modulation of tetrodotoxin-resistant sodium channels in mouse sensory neurons by tumor necrosis factor-alpha. *J. Neurosc.* 26:246–255.

118. Tak P P, Smeets T J, Daha M R, et al. (1997). Analysis of the synovial cell infiltrate in early rheumatoid synovial tissue in relation to local disease activity. *Arthrit. Rheumatism.* 40:217–225.

119. Cunha T M, Verri W A, Jr., Silva J S, et al. (2005). A cascade of cytokines mediates mechanical inflammatory hypernociception in mice. *Proc. Nat. Acad. Sci. USA.* 102:1755–1760.

120. Frode T S, Souza G E, Calixto J B (2001). The modulatory role played by TNF-alpha and IL-1 beta in the inflammatory responses induced by carrageenan in the mouse model of pleurisy. *Cytokine.* 13:162–168.

121. Tonussi C R, Ferreira S H (1999). Tumour necrosis factor-alpha mediates carrageenin-induced knee-joint incapacitation and also triggers overt nociception in previously inflamed rat knee-joints. *Pain.* 82:81–87.

122. Kuiper S, Joosten L A, Bendele A M, et al. (1998). Different roles of tumour necrosis factor alpha and interleukin 1 in murine streptococcal cell wall arthritis. *Cytokine.* 10:690–702.

123. Inglis J J, Nissim A, Lees D M, et al. (2005). The differential contribution of tumour necrosis factor to thermal and mechanical hyperalgesia during chronic inflammation. *Arthrit. Res. Ther.* 7:R807–816.

124. Junger H, Sorkin L S (2000). Nociceptive and inflammatory effects of subcutaneous TNFalpha. *Pain.* 85:145–151.

125. Barnes P F, Chatterjee D, Brennan P J, et al. (1992). Tumor necrosis factor production in patients with leprosy. *Infec. Immunity.* 60:1441–1446.

126. Lindenlaub T, Sommer C (2003). Cytokines in sural nerve biopsies from inflammatory and non-inflammatory neuropathies. *Acta Neuropathol.* 105:593–602.

127. Shafer D M, Assael L, White L B, et al. (1994). Tumor necrosis factor-alpha as a biochemical marker of pain and outcome in temporomandibular joints with internal derangements. *J. Oral Maxillofac Surg.* 52:786–791.

128. Perini F, D'Andrea G, Galloni E, et al. (2005). Plasma cytokine levels in migraineurs and controls. *Headache.* 45:926–931.

129. Mikova O, Yakimova R, Bosmans E, et al. (2001). Increased serum tumor necrosis factor alpha concentrations in major depression and multiple sclerosis. *Euro. Neuropsychopharmacol.* 11:203–208.

130. Emshoff R, Puffer P, Rudisch A, et al. (2000). Temporomandibular joint pain: relationship to internal derangement type, osteoarthrosis, and synovial fluid mediator level of tumor necrosis factor-alpha. *Oral Surg., Oral Med., Oral Pathol., Oral Radiol., Endodon.* 90:442–449.

131. Emshoff R, Puffer P, Strobl H, et al. (2000). Effect of temporomandibular joint arthrocentesis on synovial fluid mediator level of tumor necrosis factor-alpha: implications for treatment outcome. *Internat. J. Oral Maxillofacial Surg.* 29:176–182.

132. Kaneyama K, Segami N, Sun W, et al. (2005). Analysis of tumor necrosis factor-alpha, interleukin-6, interleukin-1beta, soluble tumor necrosis factor receptors I and II,

interleukin-6 soluble receptor, interleukin-1 soluble receptor type II, interleukin-1 receptor antagonist, and protein in the synovial fluid of patients with temporomandibular joint disorders. *Oral Surg., Oral Med., Oral Pathol., Oral Radiol., Endodon.*, 99:276–284.

133. Kaneyama K, Segami N, Sun W, et al. (2005). Levels of soluble cytokine factors in temporomandibular joint effusions seen on magnetic resonance images. *Oral Surg., Oral Med., Oral Pathol., Oral Radiol., Endodont.* 99:411–418.

134. Svensson P, Cairns B E, Wang K, et al. (2003). Injection of nerve growth factor into human masseter muscle evokes long lasting mechanical allodynia and hyperalgesia. *Pain.* 104:241–247.

135. Cairns B E, Gambarota G, Svensson P, et al. (2002). Glutamate-induced sensitization of rat masseter muscle fibers. *Neurosci.* 109:389–399.

136. Malik-Hall M, Dina O A, Levine J D (2005). Primary afferent nociceptor mechanisms mediating NGF-induced mechanical hyperalgesia. *Euro. J. Neurosci.* 21:3387–3394.

137. Jiang H, Guroff G (1997). Actions of the neurotrophins on calcium uptake. *J. Neurosc. Res.* 50:355–360.

138. Jiang H, Takeda K, Lazarovici P, et al. (1999). Nerve growth factor (NGF)-induced calcium influx and intracellular calcium mobilization in 3T3 cells expressing NGF receptors. *J. Biolog. Chem.* 274:26209–26216.

139. Dray A, Perkins M (1993). Bradykinin and inflammatory pain. *Trends Neurosci.* 16:99–104.

140. Zhu W, Galoyan S M, Petruska J C, et al. (2004). A developmental switch in acute sensitization of small dorsal root ganglion (DRG) neurons to capsaicin or noxious heating by NGF. *J. Neurophysiol.* 92:3148–3152.

141. Fjell J, Cummins T R, Dib-Hajj S D, et al. (1999). Differential role of GDNF and NGF in the maintenance of two TTX-resistant sodium channels in adult DRG neurons. *Brain Res. Molec. Brain Res.* 67:267–282.

142. Fjell J, Cummins T R, Fried K, et al. (1999). In vivo NGF deprivation reduces SNS expression and TTX-R sodium currents in IB4-negative DRG neurons. *J. Neurophysiol.* 81:803–810.

8.2

GROWTH FACTORS, CYTOKINES, AND CHEMOKINES: FORMULATION, DELIVERY, AND PHARMACOKINETICS

HEPING CAO AND RUI LIN

Peace Technology Development, North Potomac, Maryland

Chapter Contents

8.2.1 Types of Growth Factors, Cytokines, and Chemokines 1197

8.2.2 Functions of Growth Factors, Cytokines, and Chemokines 1198

8.2.3 Some Drugs Involving Growth Factors, Cytokines, and Chemokines 1211

8.2.4 Formulations of Growth Factors, Cytokines, and Chemokines 1212

8.2.5 Delivery of Growth Factors, Cytokines, and Chemokines 1213

8.2.6 Pharmacokinetics of Growth Factors, Cytokines, and Chemokines 1213

8.2.7 Potential Applications of Tristetraprolin Family Proteins in Inflammation-Related Diseases by Causing Cytokine mRNA Instability 1214

References 1221

8.2.1 TYPES OF GROWTH FACTORS, CYTOKINES, AND CHEMOKINES

Growth factors refer to a large class of proteins capable of stimulating cellular proliferation and differentiation. Examples of growth factors include cytokines and hormones that bind to specific receptors on the surface of their target cells.

Cytokines are a subclass of growth factors. Cytokines are primarily secreted from leukocytes. Cytokines that are secreted from lymphocytes are called lymphokines and those secreted by monocytes or macrophages are called monokines. Growth factors are often used interchangeably with cytokines. Traditionally,

Handbook of Pharmaceutical Biotechnology, Edited by Shayne Cox Gad.
Copyright © 2007 John Wiley & Sons, Inc.

cytokines are associated with hematopoietic cells and the immune system, including lymphocytes, spleen, thymus, and lymph nodes. However, some of the same signaling proteins are also being used by other cells and tissues. Growth factors imply positive effects on cell division, but cytokines are neutral in terms of their functions. Some cytokines promote cell growth or proliferation, such as granulocyte colony stimulating factor (GCSF) and granulocyte macrophage colony stimulating factor (GM-CSF). Other cytokines exhibit inhibitory effects on these processes, such as Fas ligand as "death" signal causing target cells to undergo programmed cell death or apoptosis.

Chemokines are a family of pro-inflammatory activation-inducible cytokines or small protein signals secreted by cells. Chemokines induce directed chemotaxis in nearby responsive cells and therefore the name *chemo*tactic cyto*kines*. Four types of chemokines exist: (1) C chemokine: the first and the third conserved cysteine residues in the mature protein are missing; (2) CC chemokine: the first two conserved cysteine residues are adjacent in the mature protein; (3) CXC chemokine: one amino acid residue separates the first two conserved cysteine residues in the mature protein; and (4) CX3C chemokine: three amino acid residues separate the first two conserved cysteine residues in the mature protein.

Table 8.2-1 lists examples of the major types of growth factors, cytokines, and chemokines. This list is not intended to be complete. New members of growth factors, cytokines, and chemokines are discovered continuously.

8.2.2 FUNCTIONS OF GROWTH FACTORS, CYTOKINES, AND CHEMOKINES

Growth factors bind to receptors on the cell surface and subsequently activate cellular proliferation or differentiation. Many growth factors stimulate cellular division in numerous cell type, whereas others are specific to particular cell types. They often promote cell differentiation and maturation, whereas varies between growth factors. For example, bone morphogenic proteins stimulate bone cell differentiation, whereas vascular endothelial growth factors stimulate blood vessel differentiation.

Cytokines stimulate the humoral and cellular immune responses and activate phagocytic cells. Each cytokine binds to a specific cell-surface receptor, initiates intracellular signaling cascades, and alters cellular functions. The functional consequences may include upregulation or downregulation of gene expression, resulting in the production of other cytokines; increases in the number of surface receptors for other molecules; or the suppression of their own effects by feedback inhibition. The major functions of cytokines could be classified into three categories: (1) autocrine: cytokine acts on the cells that secrete it; (2) paracrine: cytokine action is restricted to the immediate vicinities of cytokine's secretion; and (3) endocrine: cytokine diffuses to distant regions of the body to affect different tissues. However, cytokines exhibit considerable "redundancy" because many cytokines share similar functions. Cytokines are also pleiotropic factors because they act on many different cell types. The reasons for these effects of cytokines are that a certain type of cell may express receptors for more than one cytokine and different cells may express receptors for the same cytokine.

TABLE 8.2-1. Types of Growth Factors, Cytokines, and Chemokines

Type	Representative growth factors, cytokines, and chemokines
Adipose-derived cytokine	Leptin
Angiogenic cytokine	VEGFa, VEGFb
C chemokine	XCL1 (Lymphotactin), XCL2 (SCM-1β)
CC chemokine	CCL1 (I-309), CCL2 (MCP-1), CCL3 (MIP-1α), CCL4 (MIP-1β), CCL5 (RANTES), CCL6 (C10), CCL7 (MIP-3), CCL8 (MCP-2), CCL9 (MIP-1γ), CCL10 (MRP-2), CCL11 (Eotaxin), CCL12 (MCP-5), CCL13 (MCP-4), CCL14 (HCC-1), CCL15 (MIP-5), CCL16 (HCC-4), CCL17 (TARC), CCL18 (MIP-4), CCL19 (MIP-3β), CCL20 (MIP-3α), CCL21 (Exodus-2), CCL22 (MDC), CCL23 (MIP-3), CCL24 (Eotaxin-2), CCL25 (TECK), CCL26 (Eotaxin-3), CCL27 (CTACK), CCL28 (MEC)
CXC chemokine	CXCL1 (GRO-α), CXCL2 (GRO-β/MIP-2), CXCL3 (GRO-γ), CXCL4 (PF-4), CXCL5 (ENA-78/LIX), CXCL6 (GCP-2), CXCL7 (NAP-2), CXCL8 (IL-8), CXCL9 (MIG), CXCL10 (IP-10), CXCL11 (I-TAC), CXCL12 (SDF-1α/1β), CXCL13 (BCA-1), CXCL14 (BRAK), CXCL15 (Lungkine), CXCL16 (small inducible cytokine B6)
CX3C chemokine	CX3CL1 (Fractalkine)
Hematopoietic growth factor	IL-20, M-CSF, SCF, SCGF-α, SCGF-β
Immunoregulatory lymphokine	IL-2, IL-9, IL-10, IL-13, IL-15
Interferon	INF-α, INF-β, INF-γ, INF-λ
Lectin	GALECTIN-1, GLECTIN-3
Neurotrophic factor	ART, BDNF, CNTF, GDNF, MK, NTN/NRTN, β-NGF, NNT-1, NT-3, NT-4, PSP, PTN
Osteoinductive cytokine	BMP-2
Other growth factor	EGF, EG-VEGF, Flt3 ligand, G-CSF, GM-CSF, HGF, IGF-1, IGF-II, IL-3, IL-5, IL-7, IL-25 (SF20), KGF/FGF-7, OSM, PDGF, TGFA, TPO
Pleiotropic cytokine	CT-1, IL-4, IL-6
Pro-inflammatory cytokine	IL-1α, IL-1β, IL-8, IL-17, IL-19, TNF-α, TNF-β, MIP-1α, MIP-1β, MIP-1γ, MIP-2, MIP-3, MIP-3α, MIP-3β, MIP-4, MIP-5
TGF superfamily	TGF-α, TGF-β1 to TGF-β4, GDF-1 to GDF-15, BMP-1 to BMP-15
TNF superfamily	TNFSF1 (TNF-β), TNFSF2 (TNF-α), TNFSF3 (TNFC/p33), TNFSF4 (TXGP1/gp34), TNFSF5 (CD40L/gp39), TNFSF6 (FasL), TNFSF7 (CD70), TNFSF8 (CD30L), TNFSF9 (CD137L), TNFSF10 (TRAIL/Apo2L), TNFSF11 (RANKL), TNFSF12 (TWEAK/Apo3L), TNFSF13 (APRIL/TALL2), TNFSF13B (THANK/TNFSF20), TNFSF14 (LIGHT), TNFSF15 (TL1/VEG1), TNFSF18 (AITRL/TL6)

Chemokines induce directed chemotaxis in nearby responsive cells. They are released from various cells in response to bacteria and viruses infection and in response to agents that cause physical damage such as silica or urate crystals. The main functions of chemokines are chemoattractants for leukocytes. They help to recruit monocytes, neutrophils, and other effector cells from the blood to the sites of infection or damage. They serve to guide cells involved in innate immunity and in the adaptive immune system. Some chemokines have other roles in the development of lymphocytes, migration, and the growth of new blood vessels.

Table 8.2-2 lists examples of the major functions of growth factors, cytokines, and chemokines. This list is not intended to be complete. New functions of growth factors, cytokines, and chemokines are discovered rapidly.

TABLE 8.2-2. Major Functions of Growth Factors, Cytokines, and Chemokines

No.	Abbreviation	Full Name	Major Function
1	AITRL	Activation-inducible TNF receptor ligand	Regulates T-cell proliferation and survival; promotes the interaction between T lymphocytes and endothelial cells
2	Apo-SAA	Apo-serum amyloid A protein	Circulates with high-density lipoproteins; increased by inflammatory stimulus
3	APRIL	A proliferating-inducing ligand	Stimulates proliferation of various tumor cell lines and B and T cells
4	ART	Artemin	Supports the survival of peripheral ganglia
5	4-1BBL		Provides a co-stimulatory signal for T cell activation and expansion
6	BAFF	B cell activating factor	Stimulates B and T cell function
7	BCA-1	B cell attracting chemokine	Behaves as a potent chemoattractant for B lymphocytes; induces a weak chemotactic response in T cells and macrophages
8	BD-1, -2, -3	β-Defensins	Provides antimicrobial activity; behaves as chemoattractant for immature dendritic cells and memory cells
9	BDNF	Brain-derived neurotrophic factor	Supports neuron proliferation and survival
10	BMP	Bone morphogenetic protein	Modulates cell proliferation, differentiation, matrix synthesis, apoptosis; initiates, promotes, and regulates the development, growth, and remodeling of bone and cartilage
11	BMP-2	Bone morphogenetic protein-2	Induces bone and cartilage formation; plays an important role in cardiac morphogenesis

TABLE 8.2-2. *Continued*

No.	Abbreviation	Full Name	Major Function
12	BMP-13	Bone morphogenetic protein-13	Regulates articular cartilage; repairs oteochondral defects
13	BMP-14	Bone morphogenetic protein- 14	Plays a role in long bones during embryonic development and postnatally in articular cartilage
14	BRAK	Breast and kidney-expressed chemokine	Behaves as a highly selective monocyte chemoattractant
15	C10	CCL6	Behaves as a chemoattractant B cells, CD4+ T cells, monocytes, and NK cells; exhibits suppressive activity on colony formation
16	CT-1	Cardiotrophin-1	Induces myocyte hypertrophy; enhances the survival of cardiomyocyte and different neuronal populations
17	CD40L	CD40-Ligand	Plays an important role in B cell proliferation and differentiation, and immunoglobulin class switching
18	CNTF	Ciliary neurotrophic factor	Provides a vital role in the survival of neural cells
19	CTACK	Cutaneous T cell-attracting chemokine	Attracts CLA+ T cells and directs them into the skin
20	CXCL16		Attracts lymphocyte subsets during inflammation; facilitates certain immune response
21	EGF	Epidermal growth factor	Stimulates proliferation of epidermal and epithelial cells; inhibits gastric secretion; involves in wound healing
22	EG-VEGF	Endocrine-gland-derived vascular endothelial growth factor	Induces proliferation and migration fenestration in capillary endothelial cells
23	EMAP-II	Endothelial-monocyte activating polypeptide II	Inhibits endothelial cell proliferation, vasculogenesis, neovessel formation; inhibits angiogenesis of vascular beds; suppresses tumor growth; induces apoptosis and myeloperoxidase activity from neutrophils
24	ENA-78	Epithelial neutrophil activating peptide 78	Binds to CXCR2 receptor
25	Eotaxin	CCL11	Binds to CCR3 receptor; chemoattracts eosinophils; plays a key role in the regulation of eosinophil recruitment in asthmatic lung and allergic reactions
26	Eotaxin-2	CCL24	Binds to CCR3 receptor; chemoattracts cells expressing CCR3 including eosinophils,

TABLE 8.2-2. *Continued*

No.	Abbreviation	Full Name	Major Function
			basophils, Th2 T cells, mast cells; inhibits the proliferation of multipotential hematopoietic progenitor cells
27	Eotaxin-3	CCL26	Similar to Eotaxin and Eotaxin-2
28	Exodus-2	CCL21	Binds to CCR7 receptor; chemoattracts T and B lymphocytes; inhibits hematopoiesis
29	FasL	Fas ligand	Binds to Fas receptor; induces apoptosis of Fas-containing cells; kills T cells; activates B cells leading to downregulation of immune response
30	FGF family	Fibroblast growth factor family	Promotes cell proliferation and differentiation
31	Flt3L	Flt3-ligand	Binds to cells expressing tyrosine kinase receptor Flt3; regulates proliferation of early hematopoietic cells
32	FS	Follistatin	Binds to TGF-β ligands; inhibits their access to their signaling receptors
33	Fractalkine	CX3CL1	chemoattracts monocytes, microglia cells, and NK cells
34	Galectin-1	S-LAC lectin-1	Binds to β-galactoside moieties; interacts with CD3, CD4, and CD45; induces apoptosis of activated T cells and T leukemia cells; inhibits the protein phosphatase activity of CD45
35	Galectin-3	Galactose-specific lectin-3	Binds to β-galactoside moieties; regulates embryogenesis, inflammatory responses, cell progression, and metastasis
36	gAcrp30	Globular domain of adipocyte complement-related protein of 30 kDa	Binds to AdipoR1 and AdipoR2 receptors; decreases hyperglycemia and reverses insulin resistance; promotes fat loss
37	GCP-2	Granulocyte chemotactic protein-2	Attracts neutrophils; exhibits anti-angiogenic activity
38	G-CSF	Granulocyte colony-stimulating factor	Stimulates the development of committed progenitor cells to neutrophils; enhances the functional activities of the mature end-cell; facilitates hematopoietic recovery after bone marrow transplantation or chemotherapy
39	GDF-11	Growth/differentiation factor-11	Suppresses neurogenesis through a myostatin-like pathway

TABLE 8.2-2. *Continued*

No.	Abbreviation	Full Name	Major Function
40	GDNF	Glial cell line-derived neurotrophic factor	Promotes dopamine uptake, survival and differentiation of midbrain neurons; improves bradykinesia, rigidity, and postural instability
41	GM-CSF	Granulocyte/macrophage colony-stimulating factor	Stimulates the development of neutrophils and macrophages; promotes proliferation and development of early erythroid megakaryocytic and eosinophilic progenitor cells; inhibits neutrophil migration; enhances the functional activities of the mature end-cells; increases the recovery rate of hematopoietic cells after bone marrow transplantation
42	GRO	Growth regulated protein/ melanoma growth stimulatory activity family (MGSA)	Chemoattracts and activates neutrophils and basophils
43	HCC-1	Hemafiltrate CC chemokine-1	Chemoattracts blood monocytes
44	HGF	Hepatocyte growth factor	Induces cell proliferation, motility, morphogenesis; inhibits cell growth; enhances neuron survival; helps liver regeneration
45	I-309	CCL1	Chemoattracts monocytes and Th2-differentiated T cells
46	IFN-α	Interferon alpha	Induces nonspecific resistance against viral infections; regulates expression of MHC class I antigens; affects cell proliferation; modulates immune responses; treats viral infection and neoplasia
47	IFN-γ	Interferon gamma	Increases the surface expression of class I MHC antigens to T-helper (CD4+) cells; regulates the antigen-specific phases of the immune responses; stimulates lymphoid cell functions
48	IFN-λs	Interferon-lambdas	Induces antiviral defense
49	IGF-I/II	Insulin-like growth factor I and II	Stimulates the proliferation and survival of various cell types similar to insulin but with higher growth-promoting activity
50	IL-1α	Interleukin-1 alpha	Stimulates thymocyte proliferation by inducing IL-2 release; promotes B cell maturation and proliferation; exhibits mitogenic FGF-like activity; stimulates the release of prostaglandin and collagenase from synovial calls

TABLE 8.2-2. *Continued*

No.	Abbreviation	Full Name	Major Function
51	IL-1β	Interleukin-1 beta	Similar to IL-1α. IL-α is a mainly cell-associated cytokine, but IL-β is a secreted cytokine
52	IL-2	Interleukin-2	Stimulates growth and differentiation of B cells, NK cells, lymphokine-activated killer cells, monocytes, macrophages, and oligodendrocytes; enhances the immune system of patients for the treatment of cancer and infectious diseases
53	IL-3	Interleukin-3	Promotes the survival, differentiation, and proliferation of committed progenitor cells of the megakaryocyte, granulocyte-macrophage, erythroid, eosinophil, basophil, and mast cell lineages; enhances thrombopoiesis, phagocytes, and antibody-mediated cellular cytotoxicity; activates monocytes
54	IL-4	Interleukin-4	Regulates diverse T and B cell responses, including cell proliferation, survival, and gene expression; regulates the differentiation of native CD4+ T cells into helper Th2 cells; regulates immunoglobulin class switching to the IgG1 and IgE isotypes
55	IL-5	Interleukin-5	Stimulates the proliferation and activation of eosinophils; induces cell-mediated immunity against parasitic infections and some tumors
56	IL-6	Interleukin-6	Regulates immunes and inflammatory responses; stimulates B cell differentiation and antibody production; induces expression of hepatic acute-phase proteins; regulates bone metabolism
57	IL-7	Interleukin-7	Affects early B and T cells mainly; costimulates mature T cells with other factors
58	IL-8	Interleukin-8	Chemoattracts and activates neutrophils
59	IL-9	Interleukin-9	Enhances the proliferation of T lymphocytes, mass cells, erythroid precursor cells, and megakaryoblastic leukemia cells

TABLE 8.2-2. *Continued*

No.	Abbreviation	Full Name	Major Function
60	IL-10	Interleukin-10	Inhibits the expression of pro-inflammatory cytokines such as IL-1 and TNF-α; enhances humoral immune responses and attenuates cell-mediated immune responses
61	IL-11	Interleukin-11	Regulates hematopoiesis by stimulating growth of myeloid, erythroid, and megakaryocyte progenitor cells; regulates bone metabolism; inhibits production of pro-inflammatory cytokines; protects against gastromucosal injury
62	IL-12	Interleukin-12	Induces IFN-γ expression by NK and T cells; promotes the growth and activity of activated NK, CD4+, and CD8+ cells; induces the development of INF-γ-producing Th1 cells
63	IL-13	Interleukin-13	Plays a role in the expulsion of gastrointestinal parasites; exhibits anti-inflammatory effects on monocytes and macrophages; inhibits pro-inflammatory cytokine expression; enhances B cell proliferation; increases IgE production by inducing isotype switching
64	IL-15	Interleukin-15	Stimulates the proliferation of T lymphocytes; relates to rheumatoid arthritis, inflammatory bowel diseases
65	IL-16	Interleukin-16	Induces chemotaxis of CD4+ T cells and monocytes and eosinophils; induces expression of IL-2R and MHC class II proteins on CD4+ T cells
66	IL-17A	Interleukin-17A	Induces production of pro-inflammatory and hematopoietic molecules
67	IL-17B	Interleukin-17B	Stimulates the release of TNF-α and IL-1β from cells of monocyte lineage
68	IL-17D	Interleukin-17D	Stimulates the production of IL-6, IL-8, and GM-CSF; inhibits hemopoiesis of myeloid progenitor cells
69	IL-17E	Interleukin-17E	Stimulates the secretion of IL-8; induces activation of NFκB in cells expressing IL-17BR receptor

TABLE 8.2-2. *Continued*

No.	Abbreviation	Full Name	Major Function
70	IL-17F	Interleukin-17F	Stimulates proliferation and activation of T cells and PBMCs; regulates cartilage matrix turnover; inhibits angiogenesis
71	IL-19	Interleukin-19	Upregulates IL-6 and TNF-α; induces apoptosis through TNF-α
72	IL-20	Interleukin-20	Stimulates colony formation by CD34+ multipotential progenitors; induces inflammatory skin diseases
73	IL-22	Interleukin-22	Inhibits IL-4 production by Th2 cells; induces acute-phase reactants in the liver and pancreas
74	IP-10	γ-interferon inducible protein 10	Chemoattracts Th1 lymphocytes and monocytes; inhibits cytokine-stimulated hematopoietic progenitor cell proliferation; behaves angiostatic and mitogenic for vascular smooth muscle cells
75	I-TAC	Interferon inducible T-cell alpha chemokine	Chemoattracts IL-2-activated T cells
76	JF	CCL2/MCP-1	Chemoattracts and activates monocytes, activated T cells, basophils, NK cells, and immune dendritic cells
77	KC	CXCL1/GROα	Chemoattracts and activates neutrophils and basophils
78	KGF	Keratinocyte growth factor	Regulates keratinocyte growth
79	LAG-1	Lymphocyte activation gene-1 protein	Chemoattracts monocytes; exhibits activity in HIV suppressive factor
80	LD78β	CCL3L1	Exhibits activity in HIV suppression assays
81	LEC	Liver-expressed chemokine (CCL16)	Chemoattracts monocytes and lymphocytes
82	Leptin	Obesity protein	Suppresses appetite; increases thermogenesis; reduces body weight, food consumption, and plasma glucose levels
83	LIGHT	TNFSF14	Activates MFlB; costimulates the activation of lymphocytes; induces apoptosis in some tumors
84	Lymphotactin	XCL-1	Chemoattracts lymphocytes
85	M-CSF	Macrophage colony-stimulating factor	Regulates cell proliferation, differentiation, and survival of blood monocytes, tissue macrophages, and their

TABLE 8.2-2. *Continued*

No.	Abbreviation	Full Name	Major Function
			progenitor cells; modulates dermal thickness and male and female fertility; treats infection, malignancies, and atherosclerosis; facilitates hematopoietic recovery after bone marrow transplantation
86	MCPs	Monocyte chemoattractant proteins	Chemoattracts and activates monocytes, activated T cells, basophils, NK cells, and immune dendritic cells
87	MDC	Macrophage-derived chemokine (CCL22)	Chemoattracts monocytes, NK cells, and dendritic cells; exhibits HIV suppressive activity
88	MEC	Mucosae-associated epithelial chemokine (CCL28)	Chemoattracts resting CD4 and CD8 T cells and eosinophils
89	MK	Midkine	Chemoattracts embryonic neurons, neutrophils, and macrophages; exhibits angiogenic, growth, and survival activities during tumorgenesis
90	MIG	Monokine-induced by interferon-γ (CXCL9)	Chemoattracts Th1 lymphocytes; inhibits tumor growth, angiogenesis, and colony formation of hematopoietic progenitors
91	MIP-1α/1β	Macrophage inflammatory protein-1 alpha/-1 beta	Regulates the trafficking and activation state of inflammatory cells such as macrophages, lymphocytes, and NK cells; chemoattracts B cells, eosinophils, and dendritic cells
92	MIP-1γ	Macrophage inflammatory protein-1 gamma	Chemoattracts neutrophils; inhibits colony formation of bone marrow myeloid immature progenitors
93	MIP-2	Macrophage inflammatory protein-2 (CXCL2/GROβ)	Chemoattracts and activates neutrophils and basophils
94	vMIP-2	Viral macrophage inflammatory protein-2	Chemokine analog of human herpes virus; exhibits antagonistic activity toward several chemokine receptors
95	MIP-3	Macrophage inflammatory protein-3 (CCL23)	Chemoattracts monocytes and resting T lymphocytes; inhibits colony formation of bone marrow myeloid immature progenitors

TABLE 8.2-2. *Continued*

No.	Abbreviation	Full Name	Major Function
96	MIP-3α	Macrophage inflammatory protein-3 alpha (CCL20)	Chemoattracts lymphocytes and dendritic cells; promotes the adhesion of memory CCD4+ T cells; inhibits colony formation of bone marrow myeloid immature progenitors
97	MIP-3β	Macrophage inflammatory protein-3 beta (CCL19)	Chemoattracts T and B lymphocytes and myeloid progenitor cells
98	MIP-4	Macrophage inflammatory protein-4 (CCL18)	Chemoattracts lymphocytes; exhibits activity on CD4+ and CD8+ T cells
99	MIP-5	Macrophage inflammatory protein-5 (CCL15)	Chemoattracts T cells and monocytes
100	Myostatin	GDF-8	Inhibits skeletal muscle growth; interacts with Activin type I and II receptors; suppresses myoblast proliferation by arresting cell cycle in G1 phase
101	Myostatin propeptide		Binds to and inhibits myostatin; increases skeletal muscle growth
102	NAP-2	Neutrophil activating protein-2 (CXCL7)	Chemoattracts and activates neutrophils
103	NTN	Neurturin	Promotes neuron development and survival
104	β-NGF	Beta-nerve growth factor	Promotes neuron development and preservation
105	NNT-1	Novel neurotrophin-1/ B-cell stimulating factor-3 (BSF-3)	Binds to and activates glycoprotein 130 and leukemia inhibitory factor receptor member β; induces tyrosine phosphorylation of these receptors
106	Noggin		Binds to ligands of TGF-β family and regulates their activity by interfering their access to signaling receptors
107	NP-1	Neutrophil peptide-1	Exhibits antimicrobial activity and chemotactic activity on dendritic cells
108	NT-3	Neurotrophin-3	Promotes the growth and survival of nerve and glial cells
109	NT-4	Neurotrophin-4	Promotes the survival of peripheral sensory sympathetic neurons
110	OSM	Oncostatin M	Regulates neurogenesis, osteogenesis, and hematopoiesis; stimulates the proliferation of fibroblast, smooth muscle cells, and Kaposi's sarcoma cells; inhibits some tumor cells; promotes cytokine release from

TABLE 8.2-2. *Continued*

No.	Abbreviation	Full Name	Major Function
			endothelial cells; enhances the expression of low-density lipoprotein receptor in hematoma cells
111	OPG	Osteoprotegerin	Binds to sRANKL; inhibits osteoclastogenesis by interrupting the signaling between stromal cells and osteoclastic progenitor cells and leads to excess accumulation of bone and cartilage
112	PDGF	Platelet-derived growth factor (AA, BB, AB forms)	Acts as mitogens for cells like smooth muscle cells, connective tissue cells, bone and cartilage cells, and blood cells; relates to hyperplasia, chemotaxis, embryonic neuron development, and respiratory tubule epithelial cell development
113	PSP	Persephin	Promotes the survival of ventral mid-brain dopaminergic neurons and motor neurons after sciatic nerve oxotomy; promotes ureteric bud branching
114	PF-4	Platelet factor 4 (CXCL4)	Chemoattracts neutrophils and monocytes; inhibits angiogenesis
115	PTN	Pleiotrophin	Functions in neurogenesis, cell migration, secondary organogenetic induction, and mesoderm-epithelial interaction
116	PTHrP	Parathyroid hormone-related protein	Regulates extracellular concentrations of calcium and phosphorous
117	RANKL	Receptor activator of NFκB ligand	Regulates specific immunity and bone turnover; binds to RANK receptor; stimulates naïve T cell proliferation; promotes survival of RANK +T cells; regulates T cell-dependent immune response; binds to OPG; inhibits osteoclastogenesis and leads to excess accumulation of bone and cartilage
118	RANK	Receptor-activator of NF- kappa B	Binds to RANKL
119	RANTES	Regulated upon activation normal T cell express sequence	Chemoattracts monocytes, memory T cells (CD4+/ CD45R+), basophils, and eosinophils; inhibits some strains of HIV-1, HIV-2, and simian immunodeficiency virus (SIV)

TABLE 8.2-2. *Continued*

No.	Abbreviation	Full Name	Major Function
120	RELMα	Resistin-like molecule alpha	Tissue-specific cytokine; physiological roles unknown
121	RELMβ	Resistin-like molecule beta	Tissue-specific cytokine; physiological roles unknown
122	Resistin	Adipose tissue-specific secretory factor	Tissue-specific cytokine; physiological roles unknown
123	SCF	Stem cell factor	Essential for the survival, proliferation, and differentiation of hematopoietic cells committed to the melanocyte and germ cell lineages
124	SCGFα/β	Stem cell growth factor alpha/beta	Supports the growth of primitive hematopoietic cells; promotes proliferation of erythroid or myeloid progenitor cells
125	SDF-1α/β	Stromal cell derived factor-1 alpha/beta (CXCL12)	Chemoattracts T and B cells; induces migration of CD34+ stem cells; exhibits HIV suppressive activity in cells expressing CXCR4 receptor
126	SF20	IL-25	Binds to the surface of cells expressing TSA-1 (Thymic shared Ag-1) receptor; stimulates the proliferation of FDCP2 cells and lymphoid cells
127	Shh	Sonic Hedgehog	Regulates developmental processes
128	TACI	Transmembrane activator and CAML interactor (TNFRSF13B)	Binds to APRIL and BAFF; stimulates the activation of NFκB and AP-1; mediates calcineurin-dependent activation of NF-AT; stimulates B and T cell function
129	TARC	Thymus and activation regulated chemokine (CCL17)	Chemoattracts T cells
130	TECK	Thymus expressed-chemokine (CCL25)	Chemoattracts activated macrophages, thymocytes, and dendritic cells
131	TGF-α	Transforming growth factor-alpha	Stimulates the proliferation of epidermal and endothelial cells; induces transformation and anchorage independence in cultured cells
132	TGF-β1	Differentiation inhibiting factor and cartilage-inducing factor	Regulates cell proliferation, growth, differentiation, motility, synthesis and deposition of extracellular matrix, embryogenesis, tissue remodeling, and wound healing
133	TGF-β2	Glioblastoma-derived T cell suppressor factor	Regulates cell proliferation, growth, differentiation, motility, synthesis and deposition of extracellular matrix, embryogenesis, tissue remodeling, and wound healing

TABLE 8.2-2. *Continued*

No.	Abbreviation	Full Name	Major Function
134	TGF-β3		Regulates cell proliferation, growth, differentiation, motility, synthesis and deposition of extracellular matrix, embryogenesis, tissue remodeling, and wound healing
135	TNF-α	Tumor necrosis factor alpha	Inhibits tumor cells; mediates immune response against bacterial infection; induces septic shock, autoimmune diseases, rheumatoid arthritis, inflammation, and diabetes
136	TNF-β	Tumor necrosis factor beta	Mediator of inflammatory and immune response; similar to TNF-α
137	TPO	Thrombopoietin	Stimulates megakaryocyte proliferation and maturation
138	TRAIL	TNF-related apoptosis-inducing ligand (Apo2L)	Activates rapid apoptosis in tumor cells
139	TWEAK	TNF-related weak inducer of apoptosis	Induces NFκB activation and chemokine secretion; promotes endothelial cell proliferation and migration
140	VEGF	Vascular endothelial growth factor	Stimulates endothelial cell proliferation and migration; promotes angiogenesis and vascular permeability

8.2.3 SOME DRUGS INVOLVING GROWTH FACTORS, CYTOKINES, AND CHEMOKINES

It was estimated that the revenue potential involving growth factors, cytokines, and chemokines could reach at least tens of billions of U.S. dollars. The research based on these proteins is changing the treatment of various diseases such as inflammation and infectious and neoplastic diseases. Current product development is focused on autoimmunity, allergy, cancer, infection, transplantation, and wound healing. A number of the first-generation cytokine-based drugs, including IFN-1β and TNF-α, are being replaced by next-generation drugs developed from studies using more refined approaches, causing the knowledge of growth factors/cytokines/chemokine networks to become more clarified. The increased market potential of approved drugs targeting growth factors, cytokines, chemokines, and their receptors has encouraged major biotechnology and pharmaceutical companies to devote their pipelines to the development of new therapeutics.

Table 8.2-3 lists a few examples of the therapeutic targets currently under development targeting growth factors, cytokines, and chemokines.

TABLE 8.2-3. Representative Drugs Involving Growth Factors, Cytokines, and Chemokines

Type	Representative Drug	Company
TNF superfamily	HGS-ETR1 (agonistic human monoclonal antibody to TRAIL receptor 1)	Human Genome Sciences
	HGS-ETR2 (agonistic human monoclonal antibody to TRAIL receptor 2)	Human Genome Sciences
	HGS-TR2J (agonistic human monoclonal antibody to TRAIL receptor 2)	Human Genome Sciences
Growth factor	Neupogen (Filgrastim, r-metHuG-CSF)	Amgen
	Neulasta (Pegfilgrastim, PEG-r-metHuG-CSF)	Amgen
	Nutropin [somatropin (rDNA origin) for injection]	Amgen
	Growth hormone	GENENTECH
	Kepivance (a recombinant form of human keratinocyte growth factor)	Amgen
Pro-inflammatory cytokine	Enbrel (etanercept) (TNF blocker)	Amgen
	Kineret (a recombinant form of a naturally occurring protein that regulates interleukin-1)	Amgen
C chemokine	CCR5 mAb	Human Genome Sciences
Angiogenic cytokine	Avastin (bevacizumab) (anti-VEGF antibody)	GENENTECH
Interferon	INTRON A (recombinant interferon α-2b)	Schering-Plough Corp.
	Albuferon (albumin-interferon α)	Human Genome Sciences

8.2.4 FORMULATIONS OF GROWTH FACTORS, CYTOKINES, AND CHEMOKINES

The formulations for growth factors, cytokines, and chemokines include low concentrations of nontoxic detergents, such as Tween 80, and osmotic agents, such as sorbitol. Additional agents are also included for providing optimum pH for their activity. Some salts are usually included in the formulations. A number of formulations for growth factors, cytokines, and chemokines exist. These formulations are based on saline, liposome, conjugation, and coadministration. The goals of developing optimum formulations are designed to protect the activities of these proteins and to effectively deliver the biologics to the targets.

Table 8.2-4 lists a few examples of the formulations for growth factors, cytokines, and chemokines (please refer to Refs. 1–12 for more details).

TABLE 8.2-4. Formulation of Growth Factors, Cytokines, and Chemokines

Formulation	Growth factor and cytokine	Reference
Saline	G-CSF	[1]
	IL-2	[2]
Liposome	G-CSF/ProGelz interleukin-2	[1]
	GM-CSF and TNF-alpha	[2]
	VEGF/PLGA	[3]
Conjugation	PEGfilgrastim, MonoPEGylated INF-alpha2b	[4]
	bio-bFGF/OX26-SA	[5]
	Fc-EPO (NDS) fusion protein	[6]
	(99m)Tc-IL-8	[7]
	PEG-rHuMGDF	[8]
	PEGlylated INF	[9]
Coadministration	IL-2/INF-alpha/histamine	[10]
	TNF-alpha/Doxil	[11]
	TNF-alpha/MTX	[12]

8.2.5 DELIVERY OF GROWTH FACTORS, CYTOKINES, AND CHEMOKINES

A number of routes for the delivery of growth factors, cytokines, and chemokines have been studied in detail. Those routes include: (1) oral administration; (2) intravenous injection; (3) intraperitoneal injection; (4) intramuscular injection; and (5) subcutaneous injection. Most of the delivery methods for these compounds are accomplished through injection because of their protein nature.

The U.S. FDA recently approved the first inhaled version of insulin (Exubera) for adults with type 1 or type 2 diabetes. Pfizer and Aventis developed Exubera, in collaboration with Nektar Therapeutics (formerly Inhale Therapeutics, a company that specializes in finding delivery solutions for oral, injectable, and pulmonary drug administration). It weighs about 4 ounces and is about the size of an eyeglass case when closed. It is portable, but not necessarily discrete. Exubera is a short-acting powder form of insulin that can be taken before meals. Inhaled insulin enters the bloodstream more rapidly than that delivered by subcutaneous injection. However, most people with diabetes still need to get long-acting insulin via an injection. Some side effects are associated with the inhaled insulin, including coughing, shortness of breath, sore throat, and dry mouth. It is also possible that overdose for smokers may be developed because smokers get more of the inhaled insulin entering their bloodstream than nonsmokers. Much still needs to be done to effectively deliver growth factors, cytokines, and chemokines to the patients.

Table 8.2-5 lists a few examples of the delivery for growth factors, cytokines, and chemokines (please refer to Refs 2, 3, 5, 9, 10, 13–17 for more details).

8.2.6 PHARMACOKINETICS OF GROWTH FACTORS, CYTOKINES, AND CHEMOKINES

Numerous published papers exist dealing with the pharmacokinetics of growth factors, cytokines, and chemokines. Those studies were performed using cells,

TABLE 8.2-5. Routes of Administration of Growth Factors, Cytokines, and Chemokines

Route of delivery	Growth factor and cytokine	Reference
Oral administration	TLF	[13]
Intravenous injection	VEGF	[3]
	BFGF	[5]
	IL-2	[2]
	Apo2L/TRAIL	[14]
	IL-1	[15]
Intraperitoneal injection	INF-gamma	[16]
Intramuscular injection	INFs	[9]
Subcutaneous injection	VEGF	[3]
	IL-2/IFN-alpha	[10]
	G-CSF analog	[17]
	INFs	[9]

tissues, animals, and patients. The experiments tested the various formulations and delivery methods mentioned above.

Table 8.2-6 is a summary of the selected studies on the formulation, delivery, and pharmacokinetics for growth factors, cytokines, and chemokines (Please refer to Refs. 1–18 for more details).

8.2.7 POTENTIAL APPLICATIONS OF TRISTETRAPROLIN FAMILY PROTEINS IN INFLAMMATION-RELATED DISEASES BY CAUSING CYTOKINE mRNA INSTABILITY

Tristetraprolin (TTP) and related proteins are well-established factors that destabilize cytokine mRNAs [19, 20]. TTP is the best-understood member of a small family of tandem CCCH zinc finger proteins. Similar zinc finger sequences are found in numerous species in the GenBank database ranging from human to yeast and plants (including human, cow, mouse, rat, sheep, chimpanzee, dog, horse, pig, *Xenopus*, carp, zebrafish, *Drosophila*, *C. elegans*, baker's yeast, fission yeast, oyster, rice, and *Arabidopsis*.) In humans, three members of this family have been characterized: TTP (ZFP36, TIS11, G0S24, and NUP475), ZFP36L1 (TIS11B, cMG1, ERF1, BRF1, and Berg36), and ZFP36L2 (TIS11D, ERF2, and BRF2) [19]. They are encoded by different genes and their patterns of cell- and tissue-specific expression and agonist-stimulated expression are quite different. However, they share certain properties: All have highly conserved tandem zinc finger domains, in which each C8xC5xC3xH zinc finger is preceded by the sequence R/KYKTEL, and the two fingers are separated by 18 amino acids [19]; all are nuclear-cytoplasmic shuttling proteins [21, 22]; and all are capable of binding AREs within single-stranded RNA [23–29] and promoting the deadenylation and subsequent destruction of those transcripts, both in transfection studies and in cell-free experiments [26–28, 30].

TTP is an extraordinarily low-abundance, inducible, stable cytosolic, and hyperphosphorylated mRNA binding protein [24, 25, 31–33]. TTP mRNA and protein

TABLE 8.2-6. Summary of the Formulation, Delivery, and Pharmacokinetics for Growth Factors, Cytokines, and Chemokines

Drug	Experimental system	Formulation	Delivery	Pharmacokinetics	Reference
Apo2L/ TRAIL	Mice model of human colon carcinoma	Apo2L/ TRAIL	Intravenous injection	Intact [125I]-Apo2L/TRAIL was detectable in the solid tumor at all time points and was the only tissue in which radioactivity transiently increased over time. Apo2L/ TRAIL was stable in the circulation, localized to human solid xenograft tumors, and was primarily eliminated through the kidney	[14]
Epo	Mice myeloma cells	Fc-Epo(NDS) fusion protein		Fc-Epo(NDS) was secreted almost exclusively as a dimer, was relatively stable to the removal of N-linked oligosaccharides, had much improved pharmacokinetic properties, and had a significantly improved effect on RBC production	[6]
bFGF	Rat	bFGF, bio-bFGF, and bio-bFGF/OX26-SA conjugation to a blood-brain barrier peptide drug delivery vector	Intravenous injection	The brain uptake of the [125I]-bio-bFGF/OX26-SA was increased five-fold, although the uptake of the conjugate by peripheral tissues was decreased relative to the unconjugated bio-bFGF. Conjugation of bio-bFGF to a BBB drug delivery vector (1) caused only a minor decrease in affinity for the bFGF receptor,	[5]

TABLE 8.2-6. *Continued*

Drug	Experimental system	Formulation	Delivery	Pharmacokinetics	Reference
				(2) decreased the peripheral organ uptake of the bFGF, and (3) increased the brain uptake of the neurotrophin	
G-CSF	Mice	ProGelz (PG) (17% poloxamer-407 and 5% hydroxypropyl methylcellulose) vs. saline	One injection in PG vs. multiple injections in saline	PG increased bioavailability (>1.5-fold), the time required to achieve the maximum concentration (Tmax) (six fold), and prolonged elimination (Tbeta) half-life (>three fold). Mobilized HPC was transported more rapidly to the spleen than that to the blood (289-fold vs. eight-fold increase)	[1]
GM-CSF & TNF-α	Normal and tumor-bearing mice	Liposomal cytokine vs. soluble cytokines		Liposome-entrapped GM-CSF and TNF-α was more efficacious immunomodulators than the soluble cytokines: (1) the entrapped cytokine displayed high stability; (2) release of TNF-α, but not of GM-CSF, from the liposomes was required for their biological activity in vitro; (3) plasma half-lives and the area under the curve of the entrapped cytokines were 10–20 times greater than those of the soluble cytokines; (4) the toxicity of liposomal TNF-α was 1/3–1/7 that of soluble TNF-α; (5) liposomal GM-CSF led to a 2–4-fold increase	[18]

1216

Drug	Species	Comparison	Route	Findings	Ref.
G-CSF analog	Monkeys		Subcutaneous injection	vs. soluble GM-CSF in the number of peritoneal and spleen leukocytes and of GM colony-forming cells in the spleen; and (6) liposomal GM-CSF markedly enhanced the level of blood granulocytes. The analog enhanced stability and resulted in greater systemic exposure, probably because of improved absorption from the subcutaneous compartment	[17]
IFN-α2b	Rat	MonoPEGylated IFN-α2b vs. native IFN-α		PEGylation markedly enhanced both the resistance to tryptic degradation and the thermal stability of IFN-alpha2b. The serum half-life of 40K PEG-IFN was 330-fold longer, while plasma residence time was increased 708 times compared with native IFN. The PEG2,40K conjugate of IFN-α2b has higher stability than the native cytokine	[4]
IFN-γ	Mice		Intraperitoneal and intravenous injection	IFN-γ pretreatment for 24h resulted in retardation of plasma elimination of the drug with a concomitant increase of its tissue levels in liver, kidney, and intestine. A more pronounced effect might be elicited by IFN-γ for common substrates of P-gp and CYP3A	[16]
IFNs		Pegylated interferon	Intramuscular, muscular, intravenous, and subcutaneous injection	The decline in serum concentrations of IFNs was rapid after intravenous administration. The volume of distribution was approximately 20–60% of body weight. Terminal elimination half-lives ranged from	[9]

TABLE 8.2-6. *Continued*

Drug	Experimental system	Formulation	Delivery	Pharmacokinetics	Reference
				4 to 16 hrs, 1 to 2 hrs, and 25 to 35 min for alpha, beta, and gamma, respectively. Intramuscular and subcutaneous administration of interferons alpha and beta resulted in protracted but fairly good absorption >80% for interferon alpha to 70% for interferon gamma. Pegylated interferon safety and pharmacodynamic profiles were comparable. Pegylated interferon demonstrated delayed clearance compared with non-pegylated interferon. The severity of adverse effects was dose-dependent. Most patients treated had influenza-like syndrome within 2 to 8 hrs of drug administration. Other effects such as fatigue, lethargy, and anorexia were usually dose-limiting. Neuropsychiatric reactions might also become dose-limiting. Interferon induced the formation of serum neutralizing antibodies in approximately 10–20% of treated patients	
IL-1α	Phase I trial		Intravenous injection	IL-1 had measurable effects on tumor blood flow and caused a significant decrease in blood flow. This decrease was temporally associated	[15]

Cytokine	Subject	Treatment	Route	Result	Ref
				with a significant leukopenia and an increased expression of the adhesion integrin CD11b on the circulating cell surface. IL-1 caused decreased tumor blood flow in vivo in human cancer patients, an effect that was temporally related to cytokine-induced peripheral blood cellular changes	
IL-2, IFN-α	Patients with stage IV melanoma	IL-2 2.4MIU/m2, IFN-alpha 3MIU, histamine as an adjunct	Subcutaneous injections	Histamine, IL-2, and IFN-α treatment was safe, well-tolerated, and tumor responses were observed	[10]
IL-2	Mice	Liposomal IL-2 vs. soluble IL-2	Intravenous injection	Liposomal IL-2 exhibited a superior immunomodulatory activity and more efficiently stimulated spleen cell proliferation and lymphokine-activated killer (LAK) cell activation in vitro, fastened the circulation time, and increased in the leukocyte levels in blood, spleen, and peritoneal exudates	[2]
IL-8	Neutropenic rabbits	IL-8 was coupled with (99m)Tc		(99m)Tc-IL-8 was localized in the abscess, mainly bound to peripheral neutrophils. Accumulation in the abscess was a highly specific, neutrophil-driven process. Total fraction that accumulated in the inflamed tissue was extremely high	[7]
MGDF	Monkey bone marrow-derived mononuclear cell	Pegylated MGDF (PEG-rHuMGDF)	Infusion	The study confirmed predictions from PK/PD modeling of PEG-rHuMGDF (pegylated recombinant human megakaryocyte growth and development factor) that	[8]

TABLE 8.2-6. *Continued*

Drug	Experimental system	Formulation	Delivery	Pharmacokinetics	Reference
				thrombocytopenia is preventable after AuBMT	
TLF	Patients with advanced solid tumors		Oral administration	Significant levels of talactoferrin were undetectable in circulation, but a statistically significant increase in circulating IL-18 occured, a pharmacodynamic indicator of talactoferrin activity	[13]
TNF-α	Melanoma-bearing mice	Co-administered Doxil	Coinjection	Systemic application of clinically tolerable doses of TNF might improve drug distribution and tumor response	[11]
TNF-α	Patients with RA	Adalimumab (a human anti-TNF α monoclonal antibody) added to continuing MTX therapy	Coadministration of adalimumab and MTX	Adalimumab exhibited linear pharmacokinetics. Among patients with active RA who had not had an adequate response to MTX, addition of adalimumab to MTX achieved long-term improvement compared with placebo plus MTX	[12]
VEGF	Rat	Poly(lactic-co-glycolic) acid microspheres vs. protein solutions	Subcutaneous vinjection of microspheres vs. intravenous injection of solutions	VEGF in microspheres administered subcutaneously was detected with low plasma concentrations and high subcutaneous concentrations over a period of 7 weeks. VEGF in solution administered intravenously was rapidly cleared with high plasma concentrations as expressed by rapid absorption and elimination	[3]

are detected in a number of tissues including spleen, thymus, lymph node, lung, liver, and intestine, consistent with its proposed antiinflammatory role [*31, 34*]. The expression levels of TTP mRNA and protein in mammalian cells are increased in response to several kinds of stimuli, including insulin and other growth factors, and stimulators of innate immunity, such as the endotoxin lipopolysaccharide (LPS) [*31, 34*]. TTP is an mRNA-binding protein with high binding specificity for the so-called class II AREs within the 3′-untranslated mRNAs [*23–27, 29, 30, 35, 36*]. TTP and its related proteins can bind to the 3′-untranslated AREs of certain clinically important mRNAs, such as those coding for TNFα [*24–27*], GM-CSF [*30*], IL-3 [*36*], cyclooxygenase-2 [*37, 38*], and plasminogen activator inhibitor type 2 [*39*]. This specific binding of TTP to the AREs results in the destabilization of the mRNAs in transfected human embryonic kidney (HEK) 293 cells [*26, 27*]. In animals, TTP deficiency causes a profound inflammatory syndrome with cachexia, dermatitis, erosive arthritis, autoimmunity, and myeloid hyperplasia, apparently because of excessive production of TNFα and GM-CSF, both of whose mRNAs are direct targets of TTP and are stabilized in cells from the knockout (KO) mice [*27, 30, 40*]. For these reasons, TTP can be thought of as an anti-inflammatory or arthritis-suppressor protein [*25, 40, 41*]. Therefore, TTP and related proteins are potential therapeutic agents for inflammation-related diseases.

REFERENCES

1. Robinson S N, Chavez J M, Blonder J M, et al. (2005). Hematopoietic progenitor cell mobilization in mice by sustained delivery of granulocyte colony-stimulating factor. *J. Interferon Cytokine Res.* 25:490–500.

2. Kedar E, Gur H, Babai I, et al. (2000). Delivery of cytokines by liposomes: Hematopoietic and immunomodulatory activity of interleukin-2 encapsulated in conventional liposomes and in long-circulating liposomes. *J. Immunother.* 23:131–145.

3. Kim T K, Burgess D J (2002). Pharmacokinetic characterization of 14C-vascular endothelial growth factor controlled release microspheres using a rat model. *J. Pharm. Pharmacol.* 54:897–905.

4. Ramon J, Saez V, Baez R, et al. (2005). PEGylated interferon-alpha2b: A branched 40K polyethylene glycol derivative, *Pharm. Res.* 22:1374–1386.

5. Wu D, Song B W, Vinters H V, et al. (2002). Pharmacokinetics and brain uptake of biotinylated basic fibroblast growth factor conjugated to a blood-brain barrier drug delivery system. *J. Drug Target* 10:239–245.

6. Way J C, Lauder S, Brunkhorst B, et al. (2005). Improvement of Fc-erythropoietin structure and pharmacokinetics by modification at a disulfide bond. *Protein Eng. Des. Sel.* 18:111–118.

7. Rennen H J, Boerman O C, Oyen W J, et al. (2003). Kinetics of 99mTc-labeled interleukin-8 in experimental inflammation and infection. *J. Nucl. Med.* 44:1502–1509.

8. Farese A M, MacVittie T J, Roskos L, et al. (2003). Hematopoietic recovery following autologous bone marrow transplantation in a nonhuman primate: effect of variation in treatment schedule with PEG-rHuMGDF. *Stem Cells.* 21:79–89.

9. Arnaud P (2002). The interferons: Pharmacology, mechanism of action, tolerance and side effects, *Rev. Med. Inter.* 23 (suppl 4):449s–458s.

10. Lindner P, Rizell M, Mattsson J, et al. (2004). Combined treatment with histamine dihydrochloride, interleukin-2 and interferon-alpha in patients with metastatic melanoma. *Anticancer Res.* 24:1837–1842.

11. Brouckaert P, Takahashi N, van Tiel S T, et al. (2004). Tumor necrosis factor-alpha augmented tumor response in B16BL6 melanoma-bearing mice treated with stealth liposomal doxorubicin (Doxil) correlates with altered Doxil pharmacokinetics. *Int. J. Cancer.* 109:442–448.

12. Weisman M H, Moreland L W, Furst D E, et al. (2003). Efficacy, pharmacokinetic, and safety assessment of adalimumab, a fully human anti-tumor necrosis factor-alpha monoclonal antibody, in adults with rheumatoid arthritis receiving concomitant methotrexate: a pilot study. *Clin. Ther.* 25:1700–1721.

13. Hayes T G, Falchook G F, Varadhachary G R, et al. (2006). Phase I trial of oral talactoferrin alfa in refractory solid tumors. *Invest. New Drugs.* 24:233–240.

14. Xiang H, Nguyen C B, Kelley S K, et al. (2004). Tissue distribution, stability, and pharmacokinetics of Apo2 ligand/tumor necrosis factor-related apoptosis-inducing ligand in human colon carcinoma COLO205 tumor-bearing nude mice. *Drug Metab. Dispos.* 32:1230–1238.

15. Logan T F, Jadali F, Egorin M J, et al. (2002). Decreased tumor blood flow as measured by positron emission tomography in cancer patients treated with interleukin-1 and carboplatin on a phase I trial. *Cancer Chemother. Pharmacol.* 50:433–444.

16. Kawaguchi H, Matsui Y, Watanabe Y, et al. (2004). Effect of interferon-gamma on the pharmacokinetics of digoxin, a P-glycoprotein substrate, intravenously injected into the mouse. *J. Pharmacol. Exp. Ther.* 308:91–96.

17. Luo P, Hayes R J, Chan C, et al. (2002). Development of a cytokine analog with enhanced stability using computational ultrahigh throughput screening. *Protein Sci.* 11:1218–1226.

18. Kedar E, Palgi O, Golod G, et al. (1997). Delivery of cytokines by liposomes. III. Liposome-encapsulated GM-CSF and TNF-alpha show improved pharmacokinetics and biological activity and reduced toxicity in mice. *J. Immunother.* 20:180–193.

19. Blackshear P J (2002). Tristetraprolin and other CCCH tandem zinc-finger proteins in the regulation of mRNA turnover. *Biochem. Soc. Trans.* 30:945–952.

20. Carrick D M, Lai W S, Blackshear P J (2004). The tandem CCCH zinc finger protein tristetraprolin and its relevance to cytokine mRNA turnover and arthritis. *Arthritis Res. Ther.* 6:248–264.

21. Murata T, Yoshino Y, Morita N, et al. (2002). Identification of nuclear import and export signals within the structure of the zinc finger protein TIS11. *Biochem. Biophys. Res. Commun.* 293:1242–1247.

22. Phillips R S, Ramos S B, Blackshear P J (2002). Members of the tristetraprolin family of tandem CCCH zinc finger proteins exhibit CRM1-dependent nucleocytoplasmic shuttling. *J. Biol. Chem.* 277:11606–11613.

23. Blackshear P J, Lai W S, Kennington E A, et al. (2003). Characteristics of the interaction of a synthetic human tristetraprolin tandem zinc finger peptide with AU-rich element-containing RNA substrates. *J. Biol. Chem.* 278:19947–19955.

24. Cao H, Dzineku F, Blackshear P J (2003). Expression and purification of recombinant tristetraprolin that can bind to tumor necrosis factor-alpha mRNA and serve as a substrate for mitogen-activated protein kinases. *Arch. Biochem. Biophys.* 412:106–120.

25. Cao H (2004). Expression, purification, and biochemical characterization of the anti-inflammatory tristetraprolin: a zinc-dependent mRNA binding protein affected by posttranslational modifications. *Biochem.* 43:13724–13738.

26. Carballo E, Lai W S, Blackshear P J (1998). Feedback inhibition of macrophage tumor necrosis factor-alpha production by tristetraprolin. *Sci.* 281:1001–1005.

27. Lai W S, Carballo E, Strum J R, et al. (1999). Evidence that tristetraprolin binds to AU-rich elements and promotes the deadenylation and destabilization of tumor necrosis factor alpha mRNA. *Mol. Cell. Biol.* 19:4311–4323.

28. Lai W S, Carballo E, Thorn J M, et al. (2000). Interactions of CCCH zinc finger proteins with mRNA. Binding of tristetraprolin-related zinc finger proteins to Au-rich elements and destabilization of mRNA. *J. Biol. Chem.* 275:17827–17837.

29. Worthington M T, Pelo J W, Sachedina M A, et al. (2002). RNA binding properties of the AU-rich element-binding recombinant Nup475/TIS11/tristetraprolin protein. *J. Biol. Chem.* 277:48558–48564.

30. Carballo E, Lai W S, Blackshear P J (2000). Evidence that tristetraprolin is a physiological regulator of granulocyte-macrophage colony-stimulating factor messenger RNA deadenylation and stability. *Blood.* 95:1891–1899.

31. Cao H, Tuttle J S, Blackshear P J (2004). Immunological characterization of tristetraprolin as a low abundance, inducible, stable cytosolic protein. *J. Biol. Chem.* 279:21489–21499.

32. Cao H, Deterding L J, Venable J D, et al. (2006). Identification of the antiinflammatory protein tristetraprolin as a hyperphosphorylated protein by mass spectrometry and site-directed mutagenesis. *Biochem. J.* 394:285–297.

33. Carballo E, Cao H, Lai W S, et al. (2001). Decreased sensitivity of tristetraprolin-deficient cells to p38 inhibitors suggests the involvement of tristetraprolin in the p38 signaling pathway. *J. Biol. Chem.* 276:42580–42587.

34. Lai W S, Stumpo D J, Blackshear P J (1990). Rapid insulin-stimulated accumulation of an mRNA encoding a proline-rich protein. *J. Biol. Chem.* 265:16556–16563.

35. Bakheet T, Williams B R G, Khabar K S A (2003). ARED 2.0: An update of AU-rich element mRNA database. *Nucl. Acids Res.* 31:421–423.

36. Lai W S, Blackshear P J (2001). Interactions of CCCH zinc finger proteins with mRNA: Tristetraprolin-mediated AU-rich element-dependent mRNA degradation can occur in the absence of a poly(A) tail. *J. Biol. Chem.* 276:23144–23154.

37. Sawaoka H, Dixon D A, Oates J A, et al. (2003). Tristetraprolin binds to the 3′-untranslated region of cyclooxygenase-2 mRNA. a polyadenylation variant in a cancer cell line lacks the binding site. *J. Biol. Chem.* 278:13928–13935.

38. Sully G, Dean J L, Wait R, et al. (2004). Structural and functional dissection of a conserved destabilizing element of cyclo-oxygenase-2 mRNA: evidence against the involvement of AUF-1 [AU-rich element/poly(U)-binding/degradation factor-1], AUF-2, tristetraprolin, HuR (Hu antigen R) or FBP1 (far-upstream-sequence-element-binding protein 1). *Biochem. J.* 377:629–639.

39. Yu H, Stasinopoulos S, Leedman P, et al. (2003). Inherent instability of plasminogen activator inhibitor type 2 mRNA is regulated by tristetraprolin. *J. Biol. Chem.* 278:13912–13918.

40. Taylor G A, Carballo E, Lee D M, et al. (1996). A pathogenetic role for TNF alpha in the syndrome of cachexia, arthritis, and autoimmunity resulting from tristetraprolin (TTP) deficiency. *Immunity.* 4:445–454.

41. Phillips K, Kedersha N, Shen L, et al. (2004). Arthritis suppressor genes TIA-1 and TTP dampen the expression of tumor necrosis factor alpha, cyclooxygenase 2, and inflammatory arthritis. *Proc. Natl. Acad. Sci. U. S. A.* 101:2011–2016.

9

PROTEIN ENGINEERING WITH NONCODED AMINO ACIDS: APPLICATIONS TO HIRUDIN

VINCENZO DE FILIPPIS

University of Padova, Padova, Italy

Chapter Contents

9.1	Introduction	1226
9.2	Hirudin–Thrombin Interaction	1227
9.3	Selection of Amino Acid Replacements	1229
	9.3.1 Val1	1229
	9.3.2 Ser2	1230
	9.3.3 Tyr3	1230
9.4	Structure-Activity Relationships	1230
	9.4.1 Val1→X	1231
	9.4.2 Ser2→X	1232
	9.4.3 Tyr3→X: Effect of Side-Chain Volume and Hydrophobicity	1232
	9.4.4 Tyr3→X: Orientation Effects	1232
	9.4.5 Tyr3→X: Electronic Effects	1235
9.5	Cumulative Amino Acid Substitutions in Hirudin Yield a Highly Potent Thrombin Inhibitor	1238
9.6	Structural Mapping of Thrombin Recognition Sites in the Na^+-Bound and Na^+-Free Form	1239
9.7	Incorporation of Noncoded Amino Acids as Spectroscopic Probes in the Study of Protein Folding and Binding	1242
	9.7.1 7-Azatryptophan	1243
	9.7.2 3-Nitrotyrosine	1248
9.8	Concluding Remarks	1253
	Acknowledgment	1254
	References	1254

9.1   INTRODUCTION

During the past decade, the incorporation of noncoded amino acids into proteins has emerged as a novel and promising approach in protein science. In its infancy, the approach of protein engineering was essentially restricted to the possibility to chemically modify, in a rather unspecific fashion, particular amino acid side chains in proteins [1]. More recently, the advent of recombinant DNA technology allowed the site-specific alteration of a given polypeptide chain at a glance, thus, greatly expanding the tools available for studying the molecular mechanisms of protein folding, stability, and function [2]. Nevertheless, a quantitative description of the physical and chemical basis that makes a polypeptide chain efficiently fold into a stable and functionally active conformation is still elusive [3]. This mainly originates from the fact that nature combined, in a yet unknown manner, different properties (i.e., hydrophobicity, conformational propensity, polarizability, and hydrogen bonding capability) into the 20 standard protein amino acids, thus making it difficult, if not impossible, to univocally relate the change in protein stability or function to the variation of physico-chemical properties caused by amino acid exchange(s). In this view, incorporation of noncoded amino acids with tailored side chains, allowing investigators to finely tune the structure at a protein site, would facilitate to dissect the effects of a given mutation in terms of one or a few physico-chemical properties, thus, greatly expanding the scope of physical-organic chemistry in the study of proteins [4].

Incorporation of noncoded amino acids has been widely exploited in the study of bioactive peptides [5]. However, the results of these studies provide only a qualitative picture of the mechanism of the ligand–receptor interaction, and thus, they are of limited predictive power, as also documented by the explosion in the last few years of serendipity-based approaches in peptide design and drug discovery [6]. These difficulties stem primarily from the fact that introduction of single or multiple amino acid exchanges into a short peptide is expected not only to alter the interaction energy at the binding site(s) of the receptor, but also to affect (in a yet unpredictable way) fundamental properties of the free ligand, including electrostatic potential, hydration energy, and conformational entropy. Contrary to short peptides (5–20 amino acids), longer polypeptide chains (>50 amino acids) are generally characterized by a well-defined and stable three-dimensional (3D) structure, thus serving as macromolecular scaffolds, by keeping the global properties of the molecule rather constant and allowing the changes in binding free energy to be related in a more predictable way to the local variations of the physico-chemical properties at the mutation site(s). Hence, the incorporation of noncoded amino acids into proteins represents a further, (almost) obligatory extension of the studies aimed at elucidating the relationships existing among the structure, stability, and function in proteins (i.e., protein engineering).

In this view, several different strategies have been pursued to incorporate noncoded amino acids, including peptide synthesis [7] native chemical ligation [8, 9], enzyme-catalyzed semisynthesis [10, 11], biosynthetic incorporation *via* auxotrophic bacterial strain expression [12, 13], and nonsense suppression methodologies in cell-free [4] or whole-cell [14] expression systems. Both native chemical ligation and enzymatic semi-synthesis require careful tailoring of the experimental procedures, whereas biosynthetic incorporation methods are not site-specific and are restricted to those cases when the structure of the noncoded amino acid (e.g., p-fluorophenylalanine, selenomethionine, and 7-zatryptophan) is similar to that of the natural counterpart to be replaced. On the other hand, despite significant advancements during the last decade [15–17], genetic methods are still limited by exceedingly low amounts of the resulting mutant protein, usually micrograms, that impair a thorough structural and functional characterization. However, stepwise solid-phase chemical synthesis remains the easiest and fastest approach to site-specifically incorporate in high yields any noncoded amino acid into even long (50–80 amino acids) polypeptide chains approaching the size of real proteins [18].

Relevant applications from our laboratory, regarding protein engineering with noncoded amino acids as a tool in the study of protein folding and function, will be presented. In particular, we will discuss the use of noncoded amino acids in structure-activity relationship (SAR) studies of hirudin binding to thrombin, as well as the incorporation of noncoded analogs of tryptophan and tyrosine (i.e., 7-azatryptophan and 3-nitrotyrosine) as spectroscopic probes for studying the hirudin–thrombin interaction.

9.2   HIRUDIN–THROMBIN INTERACTION

Thrombin is a serine protease of the chymotrypsin family that plays a key role at the interface among coagulation, infammation, and cell growth [19], and it exerts either procoagulant or anticoagulant functions in hemostasis [20]. The procoagulant role entails conversion of fibrinogen into fibrin and platelets activation, whereas the anticoagulant role regards the activation of protein C [21]. The most effective modulator of thrombin function in solution is Na^+, which triggers the transition of the enzyme from an anticoagulant (*slow*) form to a procoagulant (*fast*) form [22–24]. The Na^+-bound (*fast*) form displays procoagulant properties, because it cleaves more specifically fibrinogen and the protease-activated receptor (PAR), whereas the Na^+-free (*slow*) form is anticoagulant because it retains the normal activity toward protein C but cannot promote acceptable hydrolysis of procoagulant substrates [23, 24]. The importance of Na^+ on thrombin function is outlined by the fact that under physiological conditions the two forms are almost equally populated and that natural thrombin variants with compromised Na^+ binding result in bleeding phenotypes [24]. In this view, molecules capable of affecting the equilibrium between these two forms can have great potential therapeutic impact on the treatment of coagulative disorders [25].

Among natural anticoagulants, hirudin is the most potent and specific inhibitor of thrombin, with a dissociation constant in the 20–200-fM range [26, 27]. Due to its important pharmacological implications, the hirudin–thrombin pair has been the object of thorough structural and biochemical investigations. With respect to this, structural studies conducted on hirudin in the free [28–30] and thrombin-bound state [31] indicate that this inhibitor is composed of a compact N-terminal region, encompassing residues 1–47 and cross-linked by three disulfide bridges, and a flexible negatively charged C-terminal tail that binds to the fibrinogen-recognition site (exosite I) on thrombin. The N-terminal domain covers the active site of thrombin and through its first three amino acids extensively penetrates into the specificity pockets of the enzyme (see Figure 9-1). Notably, the N-terminal tripeptide makes about half of the total contacts observed for the binding of the core domain 1–47 to thrombin (PDB code 4htc.pdb) [31] and accounts for ~30% of the total free energy of binding [33]. By limited proteolysis of full-length hirudin, we could produce the peptide fragment corresponding to the N-terminal domain 1–47 of hirudin HM2 [27]. Although far less active (at least 2×10^5-fold) than intact hirudin, this fragment (like the parent hirudin molecule) binds ~30-fold more tightly to the *fast* form of thrombin than to the *slow* form [34], which suggests that the structural determinants for this behavior are stored in the N-terminal domain. Hence, mutational studies on hirudin fragment 1–47 can be useful not only to investigate the physico-chemical determinants responsible for the extraordinary

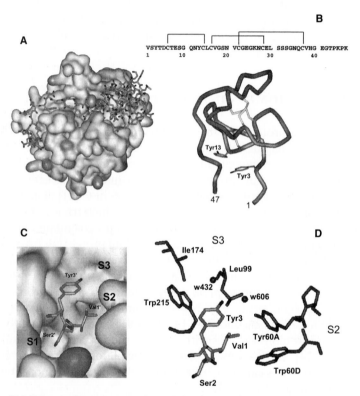

Figure 9-1. (**A**) Schematic representation of the interaction of full-length hirudin (stick) with thrombin (van der Waals surface), based on the crystal structure of the hirudin–thrombin complex [31]. (**B**) Amino acid sequence [32] and schematic representation of the solution structure [30] of the N-terminal fragment 1–47 of hirudin HM2 from *Hirudinaria manillensis*. Disulfide bonds are indicated by plain lines. (**C–D**) Structural details of the interaction of the N-terminal tripeptide of hirudin with thrombin. Val1' of hirudin contacts the S2 site of thrombin, shaped by Tyr60a and Trp60d; Ser2' covers, but does not penetrate, the S1 site, containing Asp189 at the bottom; Tyr3' fills the apolar S3 site, formed by Trp215, Leu99, and Ile174. Structural water molecules at the hirudin–thrombin interface in the S2–S3 sites are also indicated. The ribbon drawing was generated using the program WebLab ViewerPro vs. 4.0 (Molecular Simulations Inc., 2000).

affinity and specificity of hirudin for thrombin, but also to probe the structural properties of the enzyme in the *slow* or *fast* forms.

9.3   SELECTION OF AMINO ACID REPLACEMENTS

9.3.1   Val1

Nuclear magnetic resonance (NMR) studies indicate that Val1 is almost fully exposed to solvent and highly flexible in the free hirudin [29, 30, 35]. Conversely, in the thrombin-bound state, Val1 is completely buried into the active site of the enzyme and fixed into a single side-chain conformation [31]. In particular, the α-NH$_2$ group of Val1 forms a hydrogen bond with the Oγ of the catalytic Ser195, whereas its side chain makes numerous hydrophobic contacts with Tyr60A and Trp60D, shaping the S2 of the enzyme (Figure 9-1). This loop, absent in other homologous trypsin-like proteases, defines the S2 specificity site on thrombin and narrows the access to the active site such that only small-sized apolar residues are allowed.

Considering the structural properties of hirudin in the free and bound state and the steric requirements at the S2 site of thrombin, we shaved the Val1 side chain with Ala or replaced it with *tert*-butylglycine (*t*Bug) (Figure 9-2). In fact, both Val and *t*Bug have comparable side-chain volume [36] and strong β-forming propensities [37], with minimum energy points at φ/ψ = −90°, 100° for Val and φ/ψ = −130°,

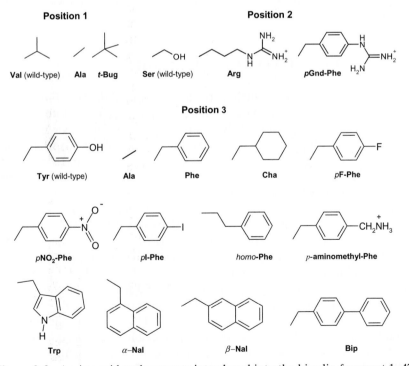

Figure 9-2. Amino acid replacements introduced into the hirudin fragment 1–47.

140° for *t*Bug [37]. On the other hand, the addition of a methyl group to the Cβ of valine is expected to restrict the backbone conformations available to *t*Bug to about one third of those allowed to Val [37].

9.3.2   Ser2

Analysis of the 3D structure of hirudin thrombin complex reveals that position 2 of hirudin is located at the entrance to, but does not enter, the primary specificity (S1) site [31]. Hence, the occupancy of the S1 site is not strictly required for binding. To evaluate the effects of perturbation of Asp189, located at the bottom of the S1 site, on the affinity of hirduin fragment 1–47 to thrombin, we have replaced Ser2 with Arg and its more rigid analog, *p*-guanidophenylalanine (*p*Gnd-Phe).

9.3.3   Tyr3

Contrary to the high conformational flexibility observed for the first two residues of hirudin, Tyr3 has a well-defined conformation in both the free and the bound state [29–31]. In the hirudin–thrombin complex, the side chain of Tyr3 projects into the apolar binding site of thrombin (the S3 site), which is formed by a large hydrophobic cavity comprising residues Trp215, Leu99, and Ile174 [31]. The importance of position 3 is confirmed by the fact that Tyr3 is highly conserved through the hirudin family [38]. To probe the S3 site of thrombin, we introduced at position 3 of hirudin 1–47 relatively large structural and chemical diversity by systematically replacing Tyr3 with natural and non-natural amino acids having different side-chain volume, hydrophobicity, electronic, and conformational properties (Figure 9-2).

9.4   STRUCTURE-ACTIVITY RELATIONSHIPS

Chemical synthesis of peptide analogs of the N-terminal domain 1–47 of hirudin was carried out by a combination of manual and automated solid-phase synthesis using the synthetic strategy previously described [34, 39]. The crude peptides with the Cys-residues in the reduced state were allowed to fold (2 mg/mL) under air-oxidizing conditions in bicarbonate buffer, pH 8.3, in the presence of 100-μM β-mercaptoethanol [39]. As an example, the reversed phase (RP)–high-performance liquid chromatography (HPLC) analyses of the crude synthetic analog Tyr3*homo*-Phe in the reduced and disulfide-oxidized state are reported in Figure 9-3. The chemical identity of the disulfide folded peptides was established by N-terminal sequence analysis and electrospray ionization (ESI)–time-of-flight (TOF) mass spectrometry, and for some synthetic peptides, the exact topology of disulfide bonds was established by enzymatic fingerprint analysis [39]. All peptides were purified by preparative RP–HPLC, lyophilized, and used for subsequent conformational and functional characterization. The results of conformational characterization, conducted by far- and near-Ultraviolet (UV) circular dichroism (CD) indicated that amino acid exchanges do not appreciably affect the conformation of hirudin fragment 1–47, allowing us to interpret the differences in affinity to thrombin exclusively on the basis of the variation of the physico-chemical properties at

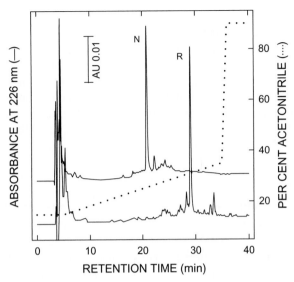

Figure 9-3. RP–HPLC analysis of the crude synthetic analog Tyr3*homo*-Phe in the reduced (*lower trace*) and disulfide oxidized state (*upper trace*). The chromatographic separations were conducted on a Vydac C18 analytical column (4.6 × 250 mm, 5 μm) eluted with a linear acetonitrile-0,05% TFA (. . . .) gradient, at a flow rate of 0.8 mL/min.

the mutation site. The inhibitory potency of the synthetic analogs toward the pro-coagulant *(fast)* and anticoagulant *(slow)* form of thrombin was determined by measuring at 405 nm the release of *p*-nitroaniline from the synthetic substrate D-Phe-Pro-Arg-*p*-NA, under temperature (25°C) and salt conditions in which the enzyme predominantly (>90%) exists in the *fast* (0.2-M NaCl) or in the *slow* (0.2-M choline chloride ChCl) form [22, 23] (Table 9-1).

9.4.1   Val1 → X

Shaving of Val1 with Ala (Val1Ala) reduces affinity for the *fast* and *slow* form of thrombin by 15- and 3-fold, respectively. On the other hand, replacement of Val1 with *tert*-butylglycine (*t*Bug) enhances the affinity of fragment 1–47 for the fast form of thrombin by about 3-fold, with a gain in the free energy of binding (ΔG_b) of 0.65 kcal/mol. This result is unprecedented, considering that almost all mutations at this position reported so far dramatically decrease binding [33, 41].

Model building studies indicate that *t*Bug fits snugly into the S2 subsite of the enzyme without steric hindrance and buries approximately the same amount of apolar surface as Val1 upon binding to thrombin. Likely, the enhanced binding of V1tBug is due to the higher symmetry of the *tert*-butyl group of *t*Bug over the isopropyl side chain of Val. In the case of *t*Bug, three energetically equivalent side-chain rotamers of *t*Bug can bind thrombin into the functionally active confor-mation, leading to a reduction of the entropy change of binding compared with Val ($\Delta S_{b,\text{Val}\to t\text{Bug}}$), which binds thrombin in only one totamer (i.e., the *trans* rotamer). An estimate of $\Delta S_{b,\text{Val}\to t\text{Bug}}$ can be obtained from the equation $\Delta S_{b,\text{Val}\to t\text{Bug}} = R \cdot \ln (\gamma_{b,t\text{Bug}}/\gamma_{b,\text{Val}}) = R \cdot \ln (3/1)$ [42], where $\gamma_{b,t\text{Bug}}$ and $\gamma_{b,\text{Val}}$ are the number of func-tionally active conformations allowed to *t*Bug and Val in the hirudin–thrombin complex, respectively. Thus, a relative stabilization of $\Delta G_{\text{Val} \to t\text{Bug}} = -T \cdot \Delta S_{\text{Val} \to t\text{Bug}} = -0.65\,\text{kcal/mol}$ at 298K is expected for the Val1tBug analog over the natural species, in excellent agreement with the experimental value ($\Delta G_{\text{Val} \to t\text{Bug}} = -0.63\,\text{kcal/mol}$) reported in Table 9-1. Our results suggest that proper introduction of symmetric groups into amino acid side chains can significantly improve binding by increasing the entropy of the ligand in the bound state, thus reducing the overall change in ΔS_b. With this respect, amino acid substitutions like valine $\to$ *tert*-butylglycine or leucine $\to$ *tert*-butylalanine represent safe mutations, enabling us to improve binding with minimal steric requirements.

9.4.2   Ser2 → X

Replacement of Ser2 with Arg or its more rigid analog pGnd-Phe induces a strong enhancement of the affinity of fragment 1–47 for thrombin (see Table 9-1), which is partly anticipated by modeling studies showing that the Arg side chain can be electrostatically coupled to Asp189, positioned at the bottom of the S1 site. Notably, perturbation of Asp189 enhances the affinity of hirudin preferentially for the *slow* form of the thrombin (see below).

9.4.3   Tyr3 → X: Effect of Side-Chain Volume and Hydrophobicity

The effects of coarse variations in the side-chain volume at position 3 on the binding to the *fast* or *slow* form were investigated by replacing Tyr3 with a smaller amino acid, like Ala, and with much larger residues, like Trp, α- and β-naphthylalanine (αNal and βNal), and biphenylalanine (Bip) (see Figure 9-2) [34, 43]. Thrombin dinding data, reported in Table 9-1, indicate that shaving of Tyr3 at Cβ reduces affinity almost exclusively for the *fast* form of thrombin, whereas enlargement of the side-chain volume at position 3 enhances binding in all cases, but preferentially to the procoagulant *fast* form of the enzyme. The lower affinity of Tyr3Ala can be reasonably explained either by the lower hydrophobicity of Ala or by the loss of numerous van der Waals contacts with the S3 site of the enzyme. Modeling of the Tyr3Ala–thrombin complex reveals that shaving of the Tyr3 side chain creates a cavity of about $150\,\text{Å}^3$ at the inhibitor–enzyme interface within the S3 region, which allows penetration of water at the binding interface, with a resulting destabilization of the hirudin–thrombin complex. Increasing the side-chain volume at position 3 also increases the hydrophobic effect, due to the burial of larger apolar surface upon binding. Strikingly, replacement of Ala with Bip enhances the affinity of hirudin 1–47 for thrombin by more than 1.3×10^4 times.

Clearly, thrombin binding data indicate that hydrophobicity is a major driving force for interaction. However, in the following we show that, beyond hydrophobicity, both side-chain orientation and electronic effects also play an important role in binding.

9.4.4   Tyr3 → X: Orientation Effects

As shown in Table 9-2, Trp is less hydrophobic than αNal. Nevertheless, the corresponding 1–47 analogs Tyr3Trp and Tyr3αNal display very similar inhibitory activity. Conversely, although αNal and βNal have the same hydrophobicity value,

TABLE 9-1. Thermodynamic Data for the Binding of Synthetic Analogs of Fragment 1–47 to the Procoagulant (*fast*) and Anticoagulant (*slow*) Form of Thrombin[1]

Side Chain	*Fast form*		*Slow form*		
	K_d (nM)	$\Delta\Delta G_b$ (kcal/mol)	K_d (nM)	$\Delta\Delta G_b$ (kcal/mol)[2]	ΔG_c (kcal/mol)[3]
Position 1					
Val (wild-type)	42 ± 0.5	—	1460 ± 20	—	–2.12
Ala	630 ± 7	1.61	4080 ± 40	0.61	–1.10
*t*Bug	14 ± 0.4	–0.63	1000 ± 80	–0.24	–2.5
Position 2					
Ser (wild-type)	42 ± 0.5	—	1460 ± 20	—	–2.12
Ser2Arg	1.7 ± 0.02	–1.90	12 ± 2	–2.84	–1.16
Ser2*p*GdnPhe	2.5 ± 0.1	–1.67	16 ± 2	–2.67	–1.10
Position 3					
Tyr (wild-type)	42 ± 0.5	—	1460 ± 20	—	–2.12
Ala	2750 ± 23	2.47	2600 ± 30	0.32	0.03
Phe	4 ± 0.1	–1.37	203 ± 3	–1.18	–2.31
Cha	49 ± 0.6	0.09	1740 ± 23	–0.09	–2.12
*p*F-Phe	18 ± 0.3	–0.48	583 ± 6	0.56	–2.04
*p*NO$_2$-Phe	185 ± 3.0	0.88	2035 ± 20	0.22	–1.46
*p*I-Phe	24 ± 0.2	–0.33	1020 ± 12	–0.23	–2.22
homo-Phe	157 ± 2.5	0.79	3290 ± 38	0.46	–1.79
*p*AM-Phe	140 ± 2.0	0.72	1590 ± 22	0.03	–1.44
Trp	6.6 ± 0.7	–1.08	550 ± 40	–0.60	–2.6
αNal	6.4 ± 0.4	–1.10	530 ± 30	–0.62	–2.6
βNal	1.1 ± 0.1	–2.14	94 ± 4	–1.64	–2.64
Bip	0.2 ± 0.01	–3.17	12 ± 1	–2.84	2.43
BugArgNal	$15.0 \pm 3\,\text{pM}$	–4.67	$220 \pm 50\,\text{pM}$	–5.23	–1.6
r-Hirudin HM2	$0.20 \pm 0.03\,\text{pM}$	–7.25	$5.5 \pm 0.6\,\text{pM}$	–7.36	–1.9

[1] All measurements were carried out at 25°C in 5-mM Tris, pH 8.0, containing 0.1% PEG in the presence of 200-mM NaCl for the *fast* form or 200-mM ChCl when the *slow* form was being studied.

[2] $\Delta\Delta G_b$ is the difference in the free energy change of binding to thrombin between the synthetic analog (ΔG_b^*) and the natural fragment (ΔG_b^{wt}): $\Delta\Delta G_b = \Delta G_b^* - \Delta G_b^{wt}$. A negative value of $\Delta\Delta G_b$ indicates that the mutated species binds more tightly to thrombin than the natural fragment. Errors are ±0.1 kcal/mol or less.

[3] ΔG_c is the free energy of coupling to thrombin measured as $\Delta G_c = \Delta G_{b,\text{fast}} - \Delta G_{b,\text{slow}}$ [40]. The value of ΔG_c is negative if the inhibitor binds to the *fast* form with higher affinity than to the *slow* form. Binding data for the triple mutated species BugArgNal and recombinant hirudin HM2 variant from *Hirudinaria manillensis* are also reported.

Tyr3βNal is sixfold more potent than Tyr3αNal. Model building studies and accessible surface area (ASA) calculations provide reasonable explanation for these results. In fact, both Trp and αNal have similar side-chain orientation and bury approximately the same amount of apolar surface area upon complex formation. In addition, the side-chain of the residue in position 3 points toward the S2 site of thrombin in both Tyr3Trp and Tyr3αNal. Conversely, the βNal isomer in the analog Tyr3βNal buries a larger amount of apolar surface area upon binding, and it favor-

TABLE 9-2. Physico-Chemical Properties of Amino Acid Side Chains at Position 3 of Hirudin Fragment 1–47

Side Chain at Position 3	Volume (Å^3)[1]	logP (π)[2]	F^3	R^3	μ^4 (Debye)
Tyr (wild-type)	138	1.97 (0.96)	0.29	−0.64	−1.57 2.02[5]
Ala	38	0.60 (0.31)			0.0
Phe	127	2.73 (1.79)	0.00	0.00	0.36
Cha	146	3.88 (2.72)			0.0
pF-Phe	133	2.87	0.43	0.34	1.79
pNO$_2$-Phe	157	2.42 (1.96)	0.67	0.16	4.38
pI-Phe	160	3.85	0.40	−0.19	1.72
homo-Phe	147	3.15 (2.10)			0.39
pAM-Phe	165	−0.80			n.a.
Trp	170	2.60 (2.25)			n.a.
αNal	180	3.87 (3.08)			n.a.
βNal	180	4.00 (3.15)			0.44
Bip	215	4.63			n.a.

[1] Side-chain molecular volumes are referred to the volume within van der Waals surface and are calculated for the corresponding organic compounds taken as suitable models for the amino acid side chains (e.g., p-cresol for Tyr, toluene for Phe).

[2] LogP values are those of the corresponding model compounds determined experimentally at 25°C and reported by Sangster [44]. When available, the hydrophobic substituent constants, π, of amino acid side chains are given in parenthesis [36].

[3] F and R are the field (F) and resonance (R) contribution to the Hammett electronic substituent constant, σ_p, according to Hansch et al. [45]. Electron-withdrawing groups possess positive F and R values, whereas electron-releasing groups have negative R values.

[4] μ is the electric dipole moment of organic model compounds of the amino acid side chains determined at 25°C in benzene [46].

[5] The μ value of Tyr was corrected for the absence of resonance effects. A positive sign of μ stands for a negative endpointing away from the phenyl ring, toward the para-X substituent.

Abbreviation: n.a., not available.

Note: A detailed description of the procedures employed to get estimates of side-chain physico-chemical properties is reported in Ref 43.

ably interacts with in the S3 site (see Figure 9-1D). Further support to the importance of orientation effects to the binding of hirudin to thrombin comes from the low affinity of the analog of hirudin 1–47 in which Tyr3 was replaced by *homo*-Phe. Although *homo*-Phe is much more hydrophobic than Tyr and Phe (Table 9-2), the analog Tyr3*homo*Phe is less potent than the wild-type species and Tyr3Phe analogs by 4- and 40-fold, respectively (Table 9-1). Due to steric clashes, the side chain of *homo*-Phe cannot favorably interact with the apolar S3 site, but more likely it points toward Tyr60A and Trp60D in the S2 site, leaving an uncompensated cavity at the S3 site.

An important aspect emerging from our work is that the intrinsic aversion of nonpolar groups for water is the dominant driving force for ligand binding only when the removal of these groups from the aqueous solvent leads to favorable specific interactions with the receptor binding-site(s), which suggest that ligand–receptor association is strongly influenced by both hydrophobic and shape-dependent packing effects.

9.4.5 Tyr3 → X: Electronic Effects

Aromatic–aromatic interactions have been identified as important factors for protein stability and binding [47]. Indeed, edge-to-face interaction between two aromatic side chains allows the δ^+ hydrogen atoms of the edge of one aromatic ring to approach the δ^- π-electron cloud of the other ring, favorably contributing to protein stability and binding by 0.6–1.3 kcal/mol [48]. Analysis of the 3D structure of the hirudin–thrombin complex reveals that the δ^+ hydrogens on the edge of the Tyr3 ring can favorably interact in a T-shaped conformation with the δ^- π-electron cloud of the aromatic ring of Trp215 at the S3 site on thrombin (see Figure 9-1D). To probe the importance of aromatic–aromatic interactions in the hirudin–thrombin system, we replaced Tyr3 with Phe and its saturated analog, cyclohexylalanine (Cha). Compared with Tyr or Phe, Cha is similar in size but significantly more hydrophobic (Table 9-2). Hence, if hydrophobicity were the dominant driving force for binding, Tyr3Cha would be expected to be a more potent inhibitor than the wild-type and Tyr3Phe derivatives. Conversely, the Tyr3 → Phe exchange improves affinity for thrombin by 10-fold, whereas saturation of the aromatic ring of Phe (by Phe → Cha substitution) reduces binding by a similar amount. These findings clearly indicate that the enhanced affinity of Tyr3Phe and wild-type species over Tyr3Cha is primarily due to the stabilizing interaction with Trp215, which is possible only for the aromatic side chain of Phe or Tyr and not for its saturated analog, Cha, thus demonstrating that, beyond hydrophobicity, electronic effects, and specifically aromatic–aromatic interactions, can markedly modulate binding.

With the aim to strengthen aromatic–aromatic interactions, we systematically replaced Tyr3 with noncoded amino acids retaining the aromatic nucleus of Tyr, as well as similar size and hydrophobicity, but possessing electron-withdrawing substituents on the aromatic ring (i.e., *p*-fluoro-, *p*-iodo-, *p*-nitro-Phe, *p*-aminomethyl-Phe; see Table 9-2). The presence of electron-attracting groups in the *para*-position of Phe3 was expected to enhance binding by making the hydrogens of the aromatic nucleus at position 3 more electron-deficient, thus reinforcing the aromatic electrostatic interaction with the π-electron cloud of Trp215. Contrary to expectations, electron-withdrawing substituents failed to improve thrombin binding. Conversely, their presence resulted in a significant reduction in the affinity by 4 (i.e., Tyr3*p*F-Phe) to 46-fold (i.e., Tyr3*p*NO$_2$-Phe) in respect to that of Tyr3Phe (Table 9-1). Rather surprisingly, the presence of *p*-aminomethyl-Phe (*p*AM-Phe), carrying a net positive charge at position 3 (*p*K$_a$ 9.36), reduced affinity for the *fast* form of thrombin by only threefold and did not affect the binding strength to the *slow* form. Notably, the charged amino-group of *p*AM-Phe is shielded from water and buried into the apolar S3 site. Both theoretical and experimental work estimated an energy cost of 9–10 kcal/mol to bury a charge in the protein interior [49, 50]. In our case, the difference of hydrophobicity (see Table 9-2) between p-aminomethyl-Phe and Tyr or Phe would yield an unfavourable increase of ΔG_b to thrombin by 3.8 and 4.8 kcal/mol, respectively. Therefore, the low energetic penalty of 0.72 and 0.03 kcal/mol experimentally derived for the binding of Tyr3pAM-Phe to the *fast* or *slow* form of the enzyme would lead us to reconsider the apolar properties of the S3 site or, more reasonably, to invoke specific interactions of the charged amino group within the S3 site (see below) that would compensate for its unfavorable desolvation properties.

Given the apolar character of the S3 site, traditionally referred to as the "aryl binding site" of thrombin [51], we first attempted to relate the affinity of the synthetic analogs of hirudin fragment 1–47 for the fast form of thrombin to the hydrophobicity value of the amino acid side chain at position 3. The contribution of hydrophobicity to the free energy change of binding (ΔG_b) was estimated using the approach of Eisenberg and McLachlan [52], which assumes that desolvation free energy change of binding, $\Delta G_{desolv} = G_{desolv}(\text{complex}) - [G_{desolv}(\text{thrombin}) + G_{desolv}(\text{hirudin})] = -\Sigma_i \Delta\sigma_i \times \Delta ASA_i$, is related to the amount of polar and apolar surface area of both ligand and receptor that becomes buried upon complex formation [53]. ΔASA_i is the change in the ASA for the atom-type i upon binding of hirudin to thrombin, and $\Delta\sigma_i$ (i.e., the atomic solvation parameter) is the solvation free energy change per unit area of atom type i that becomes buried upon binding. Desolvation free energy change of binding for a mutated hirudin analog (M) was calculated relatively to that of the wild-type fragment 1–47 (WT) as $\Delta\Delta G_{desolv} = \Delta G_{desolv}(\text{M}) - \Delta G_{desolv}(\text{WT})$.

Data reported in Figure 9-4A clearly indicate that the experimental free energy change of ΔG_b is linearly related to the variation of desolvation free energy (ΔG_{desolv}), which is calculated on theoretical grounds. Only two data points strongly deviate from the regression line, namely Tyr3pI-Phe and Tyr3pAM-Phe. In particular, these analogs were found to bind thrombin much more tightly than predicted by desolvation free energy calculations. Structural analysis of the corresponding complexes with thrombin reveals that both the iodine atom of pI-Phe and the charged nitrogen of pAM-Phe can productively interact with the aromatic nucleus of Trp215 at the S3 site, thus compensating (or overwhelming) their unfavorable desolvation free energy of binding. In the case of the Tyr3pI-Phe analog, the favorable interaction of iodine with Trp215 is predominantly driven by dispersive forces related to the high polarizability of iodine. In this regard, I_2 forms stable complexes with benzene [54]. On the other hand, the surprisingly high affinity of the Tyr3pAM-Phe analog can be rationalized by taking into account a specific charge–π interaction between the protonated amino group of p-aminomethyl-Phe and the π-electrons of Trp215. Since the seminal work of Burley and Petsko [47] on "weakly polar interactions," a great deal of experimental evidences have been accumulated over the years, leading to the conclusion that "nonconventional" hydrogen bonds of the type X–H . . . π and cation–π interactions can play a key role in protein structure and stability, as well as in ligand recognition [55]. With respect to this, theoretical calculations indicate that, among many other possible combinations of cations and aromatics, the interaction of NH_4^+ (or K^+) with indole provides the most favorable interaction energy [56]. These predictions have been confirmed by a comprehensive structural analysis of X–H . . . π hydrogen bonding in proteins, in which the indolyl

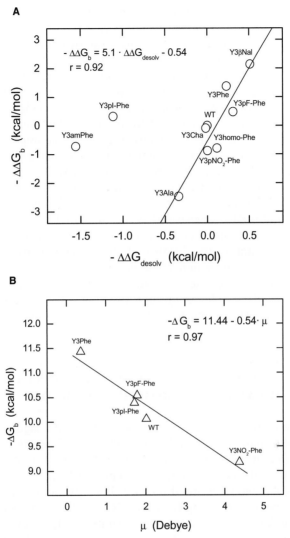

Figure 9-4. (**A**) Plot of the experimental free energy change of binding to the *fast* form of thrombin ($-\Delta\Delta G_b$) for the synthetic analogs of fragment 1–47 mutated at position 3 as a function of free energy change of desolvation ($\Delta\Delta G_{desolv}$) upon hirudin–thrombin interaction. (**B**) Plot of ΔG_b *versus* the electric dipole moment μ of the amino acid side chain at position 3. A positive sign of μ stands for a negative endpoint away from the phenyl ring toward the *para*-X substituent. For details, see Ref 43. **r** is the correlation coefficient of the linear fitting.

nucleus of Trp was found by far to be the most frequent π-acceptor of hydrogen and Lys-Trp to be the most represented donor–acceptor pair [57].

To better understand the physical nature of these forces driving the hirudin–thrombin interaction, we plotted the experimental ΔG_b of hirudin analogs *versus* the electric dipole moment, μ, of the *para*-X substituted (X = –H, –F, –I, –OH and –NO$_2$) amino acid side chain at position 3 (see Table 9-2). The data shown in Figure 9-4B indicate that the affinity of hirudin fragment 1–47 for thrombin is inversely related to the value of μ, which suggests that hirudin–thrombin interaction is desta-bilized by the presence of a partial negative charge at the *para*-position of the phenyl ring, as in the case of pNO2-Phe, where the two oxygen atoms are strongly electron-dense. A reasonable explanation for these results may be found in the specific electrostatic interactions with the electronic π-system of Trp215, which for p-aminomethyl-Phe are attractive, whereas for pNO2-Phe are repulsive.

Taken together, our results indicate that the relative change in the affinity of hirudin 1–47 analogs can be accounted for by the apolar character of the ligand quite well, but on the other hand, they also emphasize the importance of "hidden interactions," mainly of electrostatic nature, in strengthening hirudin–thrombin binding. If a reasonable estimate of the hydrophobic effect can be obtained by calculating the free energy change due to desolvation of apolar surfaces that become buried upon ligand–receptor association, electrostatic effects are much more diffi-cult to evaluate and predict, because they are mediated by weakly polar interac-tions, whose strength is strongly dependent on the electron density and polarizability, as well as on the orientation and distance of the interacting groups, such that even subtle perturbations in the ligand or receptor structure may dramatically alter binding. In this perspective, the possibility to introduce at a given protein site noncoded amino acids with larger structural and chemical diversity will improve our understanding of the mechanisms dictating molecular recognitions in proteins.

9.5 CUMULATIVE AMINO ACID SUBSTITUTIONS IN HIRUDIN YIELD A HIGHLY POTENT THROMBIN INHIBITOR

Hirudin offers numerous advantages over the existing anticoagulants heparins and cumarins [58, 59], and its use in clinical practice has been recently introduced. However, hirudin administration necessitates careful dose-titration, and bleeding effects are not rare [60]. These problems stem primarily from the intrinsic instabil-ity of the highly flexible C-terminal tail of full-length hirudin to degradation by endogenous proteases, generating truncated N-terminal fragments that, however, are dramatically less potent as thrombin inhibitors than the intact molecule [39, 61].

To minimize the hirudin sequence binding to thrombin and to improve its thera-peutic profile, several N-terminal fragments of hirudin have been prepared as potential anticoagulants. Their use would provide a more predictable effect *in vivo* compared with intact hirudin, due to their stability to denaturants and proteolytic attack and lower immunogenicity [61]. Moreover, N-terminal core fragments tar-geting solely the active site of thrombin are expected to have a safer therapeutic profile, in keeping with the notion that active-site reversible inhibitors of thrombin display a better antithrombotic/hemorrhagic balance than bivalent inhibitors [62–64]. However, a major limitation of these fragments resides in their poor affinity for thrombin (K_d = 30–400 nM), compared with hirudin (K_d = 0.2–1.0 pM) [27, 33, 61].

To possibly obtain a mini-hirudin retaining the highly potent antithrombin activ-ity of full length hirudin, but lacking the susceptibility to proteolysis, we selected the best-performing amino acid exchanges tested in our previous work (i.e., Val1*t*Bug,

Ser2Arg, and Tyr3βNal) [34] and combined them in the same analog, denoted as BugArgNal [65], which was synthesized in high yields by standard Fmoc chemistry. The introduction of only three mutations at the N-terminal end of hirudin fragment 1–47 yields a molecule that inhibits the *fast* or *slow* form by 2670- and 6820-fold more effectively than the natural counterpart and that binds exclusively at the active site of thrombin with an affinity ($K_{d,fast}$ = 15 pM, $K_{d,slow}$ = 220 pM) comparable with that of full-length hirudin ($K_{d,fast}$ = 0.2 pM, $K_{d,slow}$ = 5.5 pM). Strikingly, BugArgNal induced a change in the coagulative parameters (i.e., thrombin time, prothrombin time, and activated partial thromboplastin time) comparable with that evoked by intact hirudin (unpublished data). BugArgNal is also highly stable to low pH and resistant to the action of numerous proteases, including trypsin, chymotrypsin, thermolysin, and pepsin, and like full-lemgth hirudin, it displays almost absolute selectivity for thrombin over other closely related, physiologically important serine proteases, including plasmin, factor Xa, and tissue plasminogen activator, up to the highest concentration of inhibitor tested (10 μM). Only a slight inhibition was observed for factor Xa, with an estimated K_d value higher than 8 μM.

The issue of protease selectivity is crucial in the design of novel thrombin inhibitors, because inhibition of other physiologically relevant serine-proteases (e.g., tissue plasminogen activator and plasmin) can impair their clinical use [66–68]. The results reported in this study demonstrate that the presence of an Arg-residue at the N-terminal end of hirudin fragment 1–47 strongly improves binding, while retaining the extraordinary selectivity of the natural product. This is in contrast with the results obtained for low-molecular-weight inhibitors, where it is paradigmatic that the introduction of a positive charge at P1-position improves binding, but strongly reduces selectivity [66–68], because the positive charge of the inhibitor interacts in the S1 site of thrombin with Asp189, which is highly conserved among the endogenous enzymes prevalent in the vascular system. Likely, in the case of BugArgNal, many weak favorable contacts operate at the inhibitor–thrombin interface to cooperatively encode protease specificity.

As shown in Table 9-1, the effects of the amino acid replacements are additive in both the *fast* and the *slow* forms of thrombin, indicating that S1, S2, and S3 sites behave independently; that is, perturbation of a given site of the enzyme does not affect the binding properties of the other two sites [69]. This result is of particular relevance, because it would allow the properties of a multiple mutant to be inferred directly from those of the singly mutated species and to engineer incremental increases in binding strength and selectivity for either allosteric form of thrombin. For instance, substitution of Val1 with *t*Bug, Ser2 with Arg, and Tyr3 with Bip would yield a synthetic analog about 15,750 more potent than the wild-type species, with a predicted K_d value of about 2.5 pM.

9.6 STRUCTURAL MAPPING OF THROMBIN RECOGNITION SITES IN THE NA⁺-BOUND AND NA⁺-FREE FORM

The effect of Na⁺ binding on thrombin function is allosteric in nature [23, 70], and several crystal structures of the enzyme with and without Na⁺ bound have been recently reported [71–74]. The structures of the pseudo-wild-type thrombin mutant (Arg77aAla), reported by Di Cera et al. [71, 72], in the presence or absence of Na⁺, display only small changes in the side-chain orientation of Ser195 in the active site, Asp189 in the S1 site, Glu192 and Asp222 on the protein surface, and some rearrangement of the water molecules filling the S1 site. Conversely, the structures of the *fast* and *slow* form reported by Huntington et al. [73, 74] show significant differences at the level of the S2 and S3 sites, which in the *slow* form protrude onto the protein surface and limit the access to the catalytic pocket. In particular, the apolar cavity of the S3 site is restricted by protrusion of Trp215, which is possibly caused by reorientation of the underlying Phe227 and 168–182 disulfide bond. Partial unfolding of the Na$^+$ site is observed, with a collapse of the 148-loop onto the groove leading to the catalytic pocket. Furthermore, all putative structures of the *fast* and slow forms reported so far display numerous contacts between thrombin monomers in the crystal lattice [71–74], and therefore, the packing effects can also influence the thrombin structure in the crystal.

Hence, we decided to use the N-terminal hirudin domain 1–47 as a molecular probe of the solution conformation of thrombin recognition sites in the *fast* and *slow* form [75]. The guiding idea is that structural information on thrombin sites in the two allosteric forms can be gained from the physico-chemical properties of the mutated residue at position 1–3 of hirudin and from the effects of the perturbations introduced on the value of coupling free energy ΔG_c, which is the difference in standard free energy of binding of the inhibitor to the *fast* and *slow* form of thrombin, $\Delta G_c = \Delta G^f - \Delta G^s$. Thrombin binding data reported in Table 9-1 can be summarized as follows: First, Ala-shaving at either position 1 or position 3 reduces affinity almost exclusively for the fast form; second, side-chain enlargement at position 3 with bulky and hydrophobic amino acids (i.e., βNal and Bip) strongly enhances affinity for both forms, but preferentially for the fast form; third, electrostatic perturbation of the primary specificity site S1, by Ser2 → Arg or Ser2 → *p*Gnd-Phe exchange, enhances affinity preferentially for the slow form. These data were analyzed within the theoretical framework of site-specific thermodynamics [75–77] and used to extract structural information on the specificity sites of thrombin in the two allosteric forms:

$$
\begin{array}{ccc}
 & \Delta G^s & \\
 & S \Leftrightarrow SI & \\
\Delta G^0 \updownarrow & & \updownarrow \Delta G^1 \\
 & F \Leftrightarrow FI & \\
 & \Delta G^f &
\end{array}
$$

The *slow* (S) and *fast* (F) forms bind the inhibitor (I) with a standard free energy change ΔG^s and ΔG^f, whereas ΔG^0 and ΔG^1 represent the free energy changes for switching from the *slow* to the *fast* form in the absence or presence of the inhibitor. The coupling free energy (ΔG_c) for the cycle is given by the equation $\Delta G_c = \Delta G^f - \Delta G^s = \Delta G^1 - \Delta G^0$, where ΔG^f and ΔG^s can be determined experimentally. The preferential loss (or gain) in affinity of a mutated inhibitor for one of the two allosteric forms of the enzyme is a measure of the energetic contribution of the interactions being lost (or gained) upon mutation and provide strong, albeit indirect, means for identifying those regions on thrombin that have different structural features in the *slow* or *fast* form. On the other hand, mutations that affect both forms to the same extent reveal that the perturbations introduced in the inhibitor are important for binding to thrombin and that the site is not involved in the *slow* → *fast* transition or, otherwise, that the entity of the perturbation introduced is too small to elicit a different behavior at that site in the two allosteric forms. In addition, because Na^+ binds the hirudin–thrombin complex with ~20-fold higher affinity than the free enzyme ($\Delta G^\circ \neq \Delta G^1$) [40], then from the linkage principles mentioned above, it follows that the inhibitor must bind with different affinity to the *slow* and *fast* forms ($\Delta G^f \neq \Delta G^s$), ruling out other possibilities.

In the case of Tyr3 → Ala and Val1 → Ala exchanges, elimination of the interactions of Tyr and Val side chains beyond the Cβ strongly reduces affinity for the *fast* form, by 65- and 15-fold, respectively, whereas it is practically ineffective on the *slow* form, which suggests that the presence of Tyr and Val is crucial to enhance binding exclusively to the procoagulant *fast* form. On the other hand, the presence of the larger side chain of βNal at position 3 enhances affinity for the *fast* form by about 40-fold and by only 15-fold for the *slow* form. Taken together these results suggest that the S3 site of thrombin in the procoagulant (*fast*) form is in a more open and accessible conformation in respect to the less forgiving structure it acquires in the anticoagulant *slow* form. Also consistent with our model are the effects of the replacement of Ser2 with Arg or *p*Gnd-Phe, whose long, charged side chain is expected to facilitate penetration of the inhibitor into thrombin recognition sites, to a greater extent in the case of the more closed *slow* form than in the case of the *fast* form of the enzyme, which is already accessible for binding. From the effects of amino acid substitutions on the affinity of fragment 1–47 for the enzyme allosteric forms, we conclude that the specificity sites of thrombin in the Na^+-bound form are in a more open and permissible conformation, compared with the more closed structure they assume in the Na^+-free form [75].

The structural picture of thrombin allosteric forms proposed above is consistent with detailed molecular dynamics (MD) simulations carried out for 18 ns in full explicit water [43], showing two well-defined conformational minima on the energy landscape. After about 5 ns, a concerted conformational transition, involving the S2/S3 sites, the 148-loop, and the fibrinogen binding site, leading the thrombin molecule from a more compact and closed form, which we propose can be related to the anticoagulant (*slow*) form in the Na^+-free state, to a more open and accessible conformation, which we propose can be related to the procoagulant (*fast*) form in the Na^+-bound state (see Figure 9-5). Moreover, the results

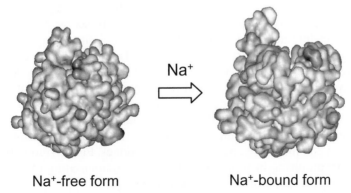

Na^+

Na⁺-free form   Na⁺-bound form

Figure 9-5. Surface representation of thrombin in the putative *slow* and *fast* form. The model was built on the basis of the coordinates derived form 18-ns MD simulations in explicit water solvent [43].

of MD analysis outline the high degree of correlation existing between the motions of all these regions of thrombin, all occurring after about 5 ns, and suggest that a structural network is present, capable of communicating the conformational changes, induced by Na^+ binding, between different structural domains of the enzyme.

Our model also provides reasonable explanation for the fact that those substrates that are related to the procoagulant activities of thrombin (e.g., fibrinogen, PAR-1, and factor XIII) orient a bulky side chain deep into the S3 site of the enzyme and, as expected, are cleaved by the *fast* form of thrombin 20–40-fold more efficiently than by the *slow* form [23]. In particular, fibrinogen interacts at the S3 site of the enzyme through the bulky Phe8 having a side-chain volume of 127 $Å^3$ [1bbr.pdb; 78], PAR-1 through Leu38 (Leu = 100 $Å^3$) [1nrs.pdb; 79], and factor XIII through two Val-residues (Val29 and Val34; Val = 79 $Å^3$) [1de7.pdb; 80]. On the other hand, protein C, which is related to the anticoagulant function of thrombin, does not seem to extensively interact with the S3 site of the enzyme [43], and as expected, it is cleaved with similar specificity by either the *slow* or the *fast* form ($\Delta G_c = 0.2$ kcal/mol) [23].

9.7 INCORPORATION OF NONCODED AMINO ACIDS AS SPECTROSCOPIC PROBES IN THE STUDY OF PROTEIN FOLDING AND BINDING

A major application of protein engineering with noncoded amino acids regards the introduction into proteins of biophysical probes possessing physico-chemical properties (e.g., side-chain volume, hydrophobicity) similar to those of the corresponding natural amino acids, but spectral features distinct from those of the natural counterparts and highly sensitive to the chemical environment in which the probe is located [81]. Hence, by the use of the so-called "spectrally enhanced proteins" [82], it should be possible to effectively monitor the local structure and dynamics of the mutated protein during key events, such as protein folding and denaturation or ligand binding, without significantly perturbing the kinetics and equilibrium properties of the process under investigation. With respect to this, in a recent study, Cohen et al. could site-specifically introduce 6-dimethylamino-2-acyl-naphthylalanine (Aladan) into the B1 domain of staphylococcal protein G to obtain estimates of the local dielectric constant of the protein at different sites [83].

More specifically, the development of new spectroscopic tools for studying protein–protein interactions is central to many disciplines, including structural biology, biotechnology, and drug discovery [84]. Traditionally, the change in tryptophan (Trp) fluorescence has been exploited to study ligand–protein interactions [85]. However, the fluorescence signal of many proteins is insensitive to ligand binding [86], because fluorescence changes are mostly restricted to those cases where Trp-residues are embedded in the ligand–protein interface or when the ligand binding induces conformational changes in the protein, remote from the binding region and involving one or more Trp-residues. Furthermore, the presence of multiple tryptophans in proteins may lead to compensating effects that often complicate interpretation of the fluorescence data [85].

To overcome these problems, several extrinsic spectroscopic probes, characterized by well-defined spectral properties, have been covalently bound to protein functional groups (i.e., Cys and Lys), to act as energy donors or acceptors in fluorescence resonance energy transfer (FRET) studies [84, 87–89]. This approach, however, is limited by possible labeling heterogeneity, nonquantitative modification, structural alteration of the proteins resulting from the labeling *per se*, and perturbation of the binding process, due to the large size of the fluorescent labels used [87, 88].

In the following, we show the utility of two noncoded analogs of tyrosine and tryptophan, namely 3-nitrotyrosine (NT) and 7-azatryptophan (AW), in the study of hirudin folding and binding to thrombin [90, 91].

9.7.1   7-Azatryptophan

Among the noncoded tryptophan analogs studied so far (i.e., 5-hydroxy- and 5-metoxy-Trp, benzo[b]thiophenylalanine and the spectrally silent fluorotryptophans), 7-azatryptophan (AW), an isostere of tryptophan (W), displays interesting absorption and fluorescence properties [92]. Unlike tryptophan, free AW displays single exponential fluorescence decay and the presence of a nitrogen-atom at position 7 in the indolyl-nucleus results in a red shift of 10 nm in the absorption and 46 nm in the emission of AW compared with Trp. Furthermore, the fluorescence λ_{max} and quantum yield of 7-azaindole (7AI) are strongly influenced by the polarity of the chemical environment. In particular, on going from cyclohexane to water the emission fluorescence of 7AI is shifted from 325 to 400 nm and the quantum yield is decreased by 10-fold. The quantum yield of AW increases from 0.01 in aqueous solution, pH 7, to 0.25 in acetonitrile [93]. Hence, it should be possible to selectively excite the fluorescence of AW at the red edge of its absorption (between 310 and 320 nm), where Tyr does not absorb and the contribution of Trp is negligible and, thus, investigate protein folding and binding processes through variation of the AW fluorescence signal.

Recently, we exploited the unique spectroscopic properties of AW to probe the disulfide-coupled folding of the hirudin N-terminal domain 1–47 and the binding to its the target enzyme, thrombin [90]. Before chemical synthesis, the resolution of the commercially available enantiomeric mixture of AW was carried out by treating the racemic mixture with acetic anhydride and subsequent enantioselective deacylation with immobilized *Asperigillus oryzae* acylase-I to yield L-AW (Scheme 9-1). Purified L-AW was then reacted with 9-fluorenylmethoxycarbonyl chloride

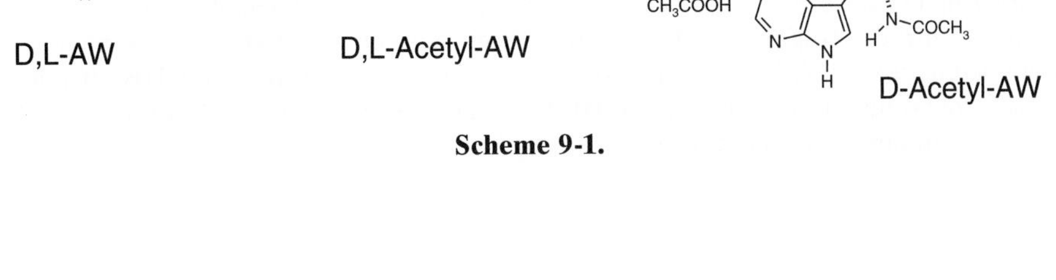

Scheme 9-1.

(Fmoc-Cl), to quantitatively obtain the Fmoc-derivative, which was subsequently used in the solid-phase synthesis of the analog of hirudin 1–47 in which Tyr3 was replaced by AW.

The replacement of tryptophan with the isosteric 7-azatryptophan leads to a significant reduction (~10-fold) in the affinity of fragment 1–47 for the *fast* form of thrombin, indicating that exchange of even a single atom ($C \rightarrow N$) can substantially affect binding to the enzyme. These results can be explained by the lower hydrophobicity of 7-azatryptophan compared with that of tryptophan, as given by the values of octanol→water partition coefficient (logP) of indole ($\log P_{indole} = 2.33$) and 7-azaindole ($\log P_{azaindole} = 1.72$), determined experimentally [90].

The results of spectroscopic characterization indicate that the λ_{max} values in the absorption and fluorescence spectra of Y3AW are red-shifted by ~10 and 40 nm, respectively, compared with those of Y3W (Figure 9-6). The fluorescence spectra of Y3AW with the six Cys-residues in the fully reduced or correctly folded state are shown in Figure 9-7 and compared with those of the corresponding Y3W analog. The emission spectrum of the reduced, unfolded Y3AW reveals the presence of two distinct, well-resolved bands at 305 and 397 nm, assigned to the contribution of Tyr13 and AW at position 3, respectively. In the folded state the Tyr-band disappears, whereas the fluorescence of AW is blue-shifted to 390 nm and enhanced by about 20%. Our results can be rationalized by considering that in the reduced state, hirudin fragment 1–47 is in a random coil conformation (unpublished results), with the donor (Tyr13) and acceptor (AW3) amino acids far apart in space. In the native state, the chain folding brings Tyr13 in close proximity to AW3 and allows the energy absorbed by Tyr to be efficiently transferred to AW. In the case of Y3W, a single band at 350 nm is observed in the native state, whereas in the reduced state the contribution of Tyr13 appears as a very weak shoulder at 303 nm, overwhelmed by the stronger emission of Trp3 at 355 nm (Figure 9-7C).

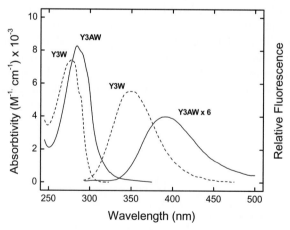

Figure 9-6. UV absorption and emission fluorescence spectra of Y3AW (—) and Y3W (--) analogs of hirudin fragment 1–47. The fluorescence spectrum of free Y3AW analog was multiplied by sixfold. All measurements were carried out at 25°C in 5-mM Tris-HCl buffer, pH 8.0, containing 0.2-M NaCl and 0.1% PEG-8000, at a protein concentration of 2 μM. Sample excitation was at 280 nm.

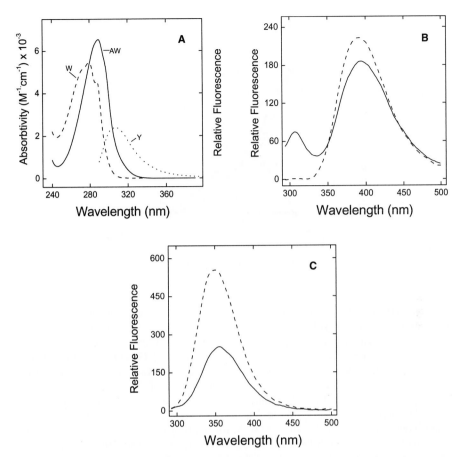

Figure 9-7. Disulfide oxidative folding of Y3AW analog monitored by fluorescence spectroscopy. (**A**) UV-absorption spectra of free tryptophan (W, --) and 7-azatryptophan (AW, —), acting as energy acceptors, and fluorescence spectrum of tyrosine (Y, ···), acting as an energy donor. Emission fluorescence spectra of Y3AW (**B**) and Y3W (**C**) in the reduced (—) and disulfide oxidized, native state (--). All spectra were taken at 25°C by exciting the samples (5 µM) at 280 nm in 0.1-M NaHCO$_3$, pH 8.3, except for that of the fully reduced form, which was recorded in 0.1-M morpholinoethane sulfonic acid buffer, pH 6.0. This pH value is sufficiently low to impair disulfide formation for at least 4 hours (not shown) and, concomitantly, avoid protonation of N^7-atom of the azaindole nucleus, which has a pK$_a$ value of 4.5 [94].

This actually makes it difficult (if not impossible) to follow the folding process of hirudin by Tyr → Trp energy transfer measurements.

The strong dependence of the fluorescence properties of AW on solvent polarity [95] was exploited to investigate the binding of Y3AW analog to thrombin (Figure 9-8). To minimize the contribution of the nine Trp-residues present in the thrombin sequence, we excited AW at the red edge of its absorption range (320 nm), where the contribution of Trp is expected to be negligible [93]. The spectra reported in Figure 9-8 clearly indicate that the fluorescence of AW is strongly quenched upon binding to thrombin. Given the known solvent-dependent emission of AW and the apolar character of the S2/S3 binding sites of thrombin, this result is surprising. In

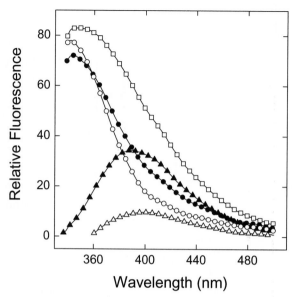

Figure 9-8. Hirudin–thrombin interaction probed by fluorescence spectroscopy. Fluorescence emission spectra of isolated thrombin (○–○) and Y3AW analog (▲–▲) and equimolar mixture of the enzyme and inhibitor (●–●). All spectra were recorded at 25°C in 5-mM Tris-HCl buffer, pH 8.0, containing 0.2-M NaCl and 0.1% PEG-8000, at a protein concentration of 2 µM, and by exciting the samples at 320 nm. Theoretical sum spectrum (□–□), obtained by adding the spectrum of free thrombin to that of the Y3AW analog. The contribution of Y3AW in the thrombin-bound form is estimated by the difference spectrum (△–△), obtained by subtracting the spectrum of isolated thrombin from that of enzyme-inhibitor complex. Given a K_d value of 60 nM for the thrombin–Y3AW complex and a concentration of enzyme and inhibitor of 2 µM, it is expected that about 20% of Y3AW remains in the free form. Notably, the intensity of the difference spectrum accounts for 25% of that of the free inhibitor, which suggests that the fluorescence of Y3AW in the complex with thrombin is almost totally quenched.

fact, we would have expected that binding of Y3AW to thrombin is accompanied by a blue-shifted emission of AW and a substantial increase in its fluorescence intensity. The fluorescence spectra of the model compound 7-azaindole (7AI) in different solvents (Figure 9-9) well document the extraordinary dependence of the fluorescence signal of 7AI on the solvent polarity, shifting from 325 nm in cyclohexane (not shown) to 345 nm in diethylether and 362 nm in acetonitrile. In particular, in water-restricted environments (e.g., water-saturated diethyl ether), the fluorescence signal is dramatically reduced, due to the formation of a 1:1 7AI-H$_2$O cyclic adduct that promotes formation in the excited state of a "tautomer" species that is poorly fluorescent [96–98].

In the light of these considerations, the strong quenching effect on the fluorescence of 7AW, observed upon binding of Y3AW to thrombin, can be explained on the basis of the model structure of Y3AW bound to thrombin. As shown in Figure 9-10, the –NH group of AW interacts with water molecule w432 (B-factor: 36 Å^2; occupancy: 1.0), whereas N^7 may be linked to Tyr60a through a water bridge involving the structural water molecule w606 (B-factor: 22 Å^2; occupancy: 0.52) [31].

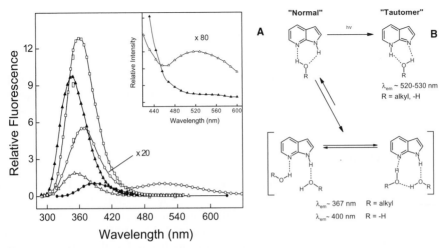

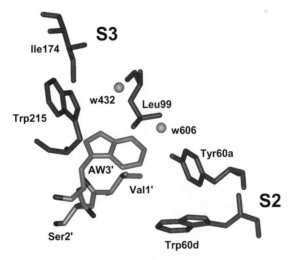

Figure 9-9. (**A**) Fluorescence emission spectra of 7-azaindole (7AI) in different solvents: acetonitrile (□–□), diethyl ether (▲–▲), *n*-propanol (○–○), water-saturated diethyl ether (△–△), and deionized water (●–●). The emission of 7AI in *n*-propanol is multiplied by 20-fold. *Inset*, fluorescence spectra of 7AI in diethyl ether (▲–▲) and water-saturated diethyl ether (△–△) expanded by a factor of 80 to show the growth of the second band at ~520 nm in organic-water solution. All spectra were recorded at 25°C and 2-μM 7AI, by exciting the samples at 280 nm. (**B**) Consensus scheme explaining the emission properties of 7AI in alcohol (R = alkyl) or water solution (R = –H). The polyhydrated species in water are more populated and emit radiation at 400 nm. In water-restricted environments (e.g., water-saturated diethyl ether), the concentration of the "normal" monohydrated species increases, and upon excitation, it is converted into the "tautomer" form, which is responsible for the longer wavelength emission at ~520 nm (*Inset* to panel **A**) [90].

Figure 9-10. Schematic representation of the interaction of the N-terminal tripeptide of hirudin analog Y3AW with thrombin. The model structure of Y3AW was obtained by keeping the position of all atoms unchanged and using the same dihedral angles that Tyr3 has in the x-ray structure of wild-type hirudin complexed to thrombin ($\chi^1 = -61°$, $\chi^2 = -56°$, [4htc.pdb; 31]. Water molecule w432 (B-factor 36 Å^2, occupancy 1.0) is in the same plane of the azaindole ring of AW at position 3, whereas N^7 occupies the same position of the –OH group in Tyr3, which is linked to Tyr60a through a water bridge involving w606 (B-factor 22 Å^2, occupancy 0.52).

Hence, the rigid, structural water molecules at the hirudin–thrombin interface can have a crucial role in quenching the fluorescence of 7AW, because they promote the nonradiative decay of AW in the excited state more effectively than the labile water molecules solvating the 7AW in the free Y3AW analog.

In conclusion, our data demonstrate that the incorporation of 7-azatryptophan into proteins can be of broad applicability in structure-activity relationship studies, where a Trp-isostere is required, or as a spectroscopic probe in the study of protein folding and binding [90].

9.7.2 3-Nitrotyrosine

3-Nitrotyrosine (NT) is produced *in vivo* by reaction of protein tyrosines with peroxynitrite [99]. The NT side chain is only $30\,\text{Å}^3$ larger than the unmodified Tyr, and the presence of the electron-withdrawing nitro-group makes the phenolic hydrogen of free NT about 10^3-fold more acidic (pK_a 6.8) [100]. At $pH < pK_a$, where the neutral form is predominant, NT is more hydrophobic than Tyr, whereas at higher pH, where NT exists in the ionized form, it is much more polar [101, 102]. NT can form an internal hydrogen bond, and its absorption properties are strongly pH-dependent (see Figure 9-11). In particular, at basic pH, the UV/Vis spectrum

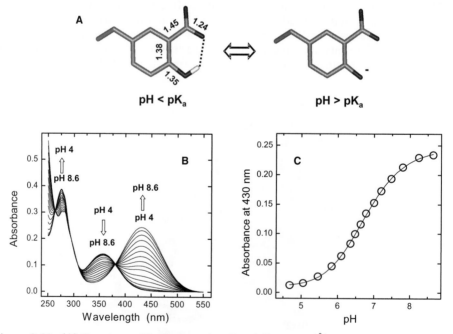

Figure 9-11. (**A**) Structure of 3-nitrotyrosine. Bond distances (Å) are taken from the crystallographic structure of free NT [103]. (**B–C**) Spectrophotometric titration of Y3NT by UV/Vis absorption spectroscopy. Absorption spectra (**B**) and plot of the absorbance values at 430 nm of Y3NT (2 mL, 48 μM) as a function of pH (**C**). Measurements were carried out in 2-mM citrate–borate–phosphate buffer, at the indicated pH. Fitting data points yields a pK_a value of 6.74 ± 0.02 for NT.

of free NT displays a major band at 422 nm, characteristic of the ionized form, whereas at acidic pH a prominent band appears at 355 nm, assigned to the contribution of the neutral form [100]. NT is essentially nonfluorescent and absorbs radiation in the wavelength range where both Tyr and Trp emit fluorescence, with a Trp-to-NT Förster's distance (i.e., the donor-acceptor distance at which the FRET efficiency is 50%) as large as 26 Å [104]. For these reasons, NT has great potential as an energy acceptor in FRET studies, and indeed, direct chemical nitration of Tyr was used to investigate the structural and folding properties of calmodulin [104] and apomyoglobin [105–107]. However, very little is known about the possibility of exploiting the unique spectral properties of NT to study molecular recognition [100, 108, 109].

Hence, we chose the hirudin–thrombin system as a suitable model for evaluating the potentialities of NT as a spectroscopic probe in the study of protein–protein interactions [91]. To this aim, we synthesized two analogs of the N-terminal domain (residues 1–47) of hirudin: Y3NT, in which Tyr3 was replaced by NT, and S2R/Y3NT, containing the cumulative substitutions Ser2→Arg and Tyr3→NT. In the presence of saturating concentrations of Y3NT or S2R/Y3NT, the fluorescence of thrombin is strongly quenched and approaches a similar value, under either *fast* (0.2-M NaCl) and *slow* (0.2-M ChCl) conditions (Figure 9-12A, B). We have demonstrated that quenching of fluorescence is mainly caused by FRET, occurring between (some of) the Trp-residues of thrombin (i.e., the donors) and the single 3-nitrotyrosine of the inhibitors (i.e., the acceptor).

FRET is a nonradiative decay process occurring between a donor and an acceptor, which interact *via* electromagnetic dipoles transferring the excitation energy of the donor to the acceptor. For a one-donor–one-acceptor system, the efficiency of energy transfer depends on the extent of spectral overlap of the emission spectrum of the donor with the absorption spectrum of the acceptor, on the donor quantum yield, on the inverse sixth power of the distance separating the donor and acceptor, and on their orientation [88]. In the case of hirudin–thrombin interaction, there is an extensive overlap of the emission spectrum of the enzyme (i.e., the donor) with the absorption spectrum of the inhibitor (i.e., the acceptor) (see Figure 9-12C). In addition, Trp-to-NT energy transfer is also favored by the relatively short distances separating Trp-residues and NT in the enzyme–inhibitor complex [31, 91]. To estimate the possible contribution of spectroscopic effects other than FRET (e.g, unspecific binding, dynamic or static quenching, inner filter effect), the fluorescence of thrombin was measured in the presence of increasing concentrations of free NT. The data reported in Figure 9-12D indicate that NT slightly (~14%) reduces the fluorescence of the enzyme, in keeping with the notion that nitrocompounds (e.g., nitromethane and nitrobenzene) quench the emission of polycyclic aromatic hydrocarbons mainly by a dynamic mechanism [110]. Moreover, we found that for concentrations of NT-containing analogs lower than 20 µM, inner filter effect can be neglected. These considerations allow us to conclude that quenching of thrombin fluorescence by Y3NT (or S2R/Y3NT) is mainly caused by Trp-to-NT energy transfer.

The quenching data reported above were used to obtain quantitative estimates (i.e., K_d values) of the binding of hirudin analogs to thrombin allosteric forms (see Figure 9-13 and Table 9-3). For both analogs, the excellent fit of the experimental data to the curve describing one-site binding mechanism is a stringent, albeit

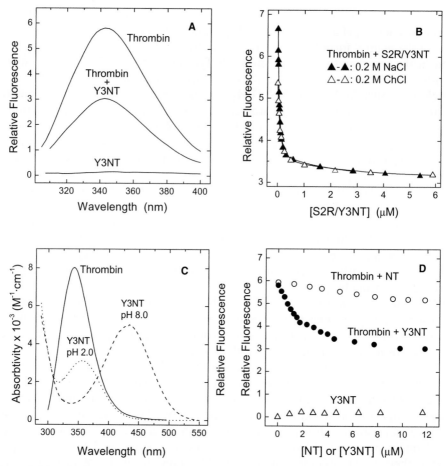

Figure 9-12. Binding of Y3NT and S2R/Y3NT to thrombin, monitored by Trp-to-NT fluorescence energy transfer. (**A**) Fluorescence spectra of thrombin alone (50 nM) and in the presence of Y3NT (10 μM). For comparison, the spectrum of the free inhibitor Y3NT (10 μM) is also reported. (**B**) Change in the fluorescence of thrombin as a function of S2R/Y3NT concentration, under fast (▲–▲, 0.2-M NaCl) and slow (△–△, 0.2-M ChCl) conditions. (**C**) Superimposition of the fluorescence spectrum of thrombin (continuous line) with the absorption spectra of Y3NT at pH 2 and 8.0 (dashed/dotted lines). (**D**) Change in thrombin fluorescence as a function of Y3NT concentration (●–●). As a control, the fluorescence intensity of thrombin in the presence of free NT (○–○) is reported. The signal of Y3NT alone (△–△) is also included. All measurements were carried out at 25°C by exciting the protein samples at 295 nm in 5-mM Tris-HCl buffer, pH 8.0, containing 0.1% (w/v) PEG 8000 and 0.2-M salt, as indicated, and recording the fluorescence signal at 342 nm.

indirect, proof of 1:1 binding stoichiometry. The replacement of Tyr3 with NT resulted in a drop in the affinity of Y3NT for thrombin, which was restored in the doubly substituted analog S2R/Y3NT by replacing Ser2 with Arg. The structural model of Y3NT bound to thrombin (see Figure 9-14), based on the crystallographic structure of the hirudin–thrombin complex [31], reveals that the –NO₂ group of NT might be easily accommodated into the S2 specificity site of the enzyme without

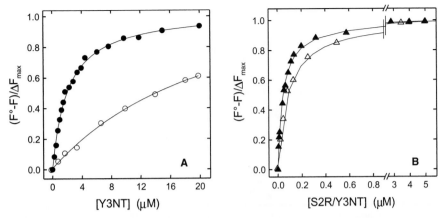

Figure 9-13. Determination of the dissociation constant (K_d) of the complexes formed by the synthetic analogs Y3NT (**A**) (●, ○) and S2R/Y3NT (**B**) (▲, △) with thrombin, under fast (filled symbols) and slow (empty symbols) conditions. Continuous lines represents the best fit of the data points to the equation describing tight binding [91], which allowed us to obtain the K_d values reported in Table 9-3.

TABLE 9-3. Thrombin Binding Data of the Synthetic Hirudin Analogs, as Obtained by Fluorescence Energy Transfer[1] and Enzyme Inhibition Assays[2]

1–47 Analogs	Fast Form		Slow Form		
	K_d (nM)	$\Delta\Delta G_b$ (kcal/mol)	K_d (nM)	$\Delta\Delta G_b$ (kcal/mol)	ΔG_c (kcal/mol)
WT[2] (Tyr3)	42 ± 0.5	—	1,460 ± 20	—	−2.10
Y3NT[1]	1,325 ± 80	2.06	17,000 ± 500	1.45	−1.51
S2R[2]	1.7 ± 0.02	−1.90	12 ± 2	−2.84	−1.16
S2R/Y3NT[1]	45 ± 2.2	0.06	91 ± 4.2	−1.64	−0.41

[1] The K_d values of Y3NT and S2R/Y3NT were obtained by fitting FRET data to the equation describing tight binding [91].
[2] The inhibitory potency of the wild-type (WT) and S2R analog was determined at 25°C by measuring at 405 nm the release of *p*-nitroanilide, *p*NA, from the synthetic substrate D-Phe-Pro-Arg-*p*NA. Additivity of mutational effects was calculated by the equation $\Delta G_I = \Delta\Delta G_b(\text{S2R/Y3NT}) - [\Delta\Delta G_b(\text{Y3NT}) + \Delta\Delta G_b(\text{S2R})]$, where ΔG_I is the free energy term that accounts for the energetic interaction between the mutated sites [9].

requiring steric distortion. Likely, the lower affinity of Y3NT reflects the lower hydrophobicity of NT at pH 8.0, where it exists by ~95% in the ionized form. With respect to this, the logP value of 2-nitrophenol, taken as a suitable model of the NT side chain, is −1.47 at pH 8.0 [101], whereas that of phenol, taken as a model of Tyr, is +1.50 [102]. Besides hydrophobicity, the presence of the nitro-group introduces a net (i.e., at pH 8.0) negative charge at position 3 of hirudin, which can oppose binding through unfavorable electrostatic interaction with the strong negative potential of the thrombin active site [111], in agreement with our previous structure-activity relationship studies [43]. The reliability of these data was verified by comparing the K_d values of Y3NT and S2R/Y3NT, obtained by FRET

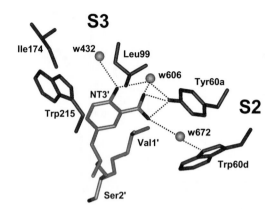

Figure 9-14. Schematic representation of the interaction of the N-terminal tripeptide of Y3NT with thrombin. The structure of Y3NT was modeled on the structure of the hirudin–thrombin complex [4htc.pdb; 31]. Relevant NT-thrombin distances, in the 2.5–3.5-Å range, are indicated by dashed lines. Of note, w606, which in the wild-type hirudin–thrombin structure connects Tyr3′ of the inhibitor to Tyr60a of the enzyme, is well suited as a hydrogen bond donor to stabilize the NT ring system [91].

measurements, with those determined by classic enzyme inhibition experiments, in which the rate of thrombin-mediated substrate hydrolysis was measured as a function of inhibitor concentration [91]. Strikingly, the K_d values for the binding of Y3NT and S2R/Y3NT to thrombin *fast* form were determined as $1.4 \pm 0.1\,\mu M$ and $41 \pm 2\,nM$, respectively, in agreement (8–10%) with those obtained by FRET (Table 9-3).

The absorption spectra of S2R/Y3NT, recorded in the absence and presence of thrombin, are reported in Figure 9-15. Under *fast* conditions, the 430-nm band of S2R/Y3NT, assigned to the contribution of the ionized form of NT at pH 8.0, is reduced by 54%, whereas an additional band of similar intensity appears at about 362 nm, characteristic of NT in the neutral form. Of note, the binding of Y3NT to thrombin *fast* form yields very similar results [91]. These observations can be explained on the basis of the modeled structure of Y3NT bound to thrombin and assuming that NT interacts with the enzyme in the neutral form.

As shown in Figure 9-14, three structural water molecules at the enzyme–inhibitor interface (i.e., w432, w606, and w672), characterized by low thermal factors and high occupancy values, can variably interact with NT. In particular, w606, which in the structure of the wild-type hirudin–thrombin complex connects Tyr3′ of the inhibitor to Tyr60a of the enzyme [31], is suitably positioned as a hydrogen bond donor to stabilize the six-membered ring system of NT (see Figure 9-11A and Figure 9-14). As a result, the contribution of the protonated NT in the bound form appears as a distinct band at about 362 nm in the absorption spectrum of the thrombin–S2R/Y3NT complex (Figure 9-15). The residual intensity of the 430-nm band is contributed by the ionized form of NT in the free inhibitor that exists in equilibrium with the thrombin-bound form. Hence, we conclude that the phenate moiety of NT in the free state becomes protonated to phenol upon binding to thrombin and that a water molecule at the hirudin–thrombin interface, likely w606, functions as a

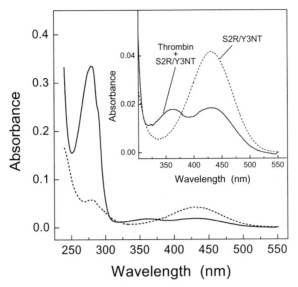

Figure 9-15. Binding of the synthetic analog S2R/Y3NT to thrombin, monitored by UV/Vis absorption spectroscopy. Spectra of the inhibitor (8.6 μM) were taken at 25° in 5-mM Tris-HCl buffer, pH 8.0, containing 0.1% PEG 8000 and 0.2 M NaCl, in the absence (--) and presence (--) of thrombin (4.1 μM). For clarity, the *Inset* shows the spectra in the wavelength range 300–550 nm.

hydrogen donor. Notably, w432 and w606 are conserved in the structure of thrombin bound to hirugen (i.e., the 53–64 peptide of hirudin) [1hah.pdb; 112], where the specificity sites of the enzyme are unoccupied, thus suggesting that these water molecules represent constant spots in the solvation shell of thrombin and, perhaps, key elements for molecular recognition, in keeping with the key role that protein–water interactions play in ligand binding [113].

Our results demonstrate that NT is a suitable spectroscopic probe for investigating ligand–protein interactions and suggest that its incorporation into proteins may have vast applications for identifying ligand–protein interaction.

9.8 CONCLUDING REMARKS

It is widely accepted that even the knowledge of the structure of the ligand–receptor complex provides only partial information for predicting ligand binding energetics [114]. In this view, we have demonstrated here that it is possible to transfer the quantitative structure-activity relationship (QSAR) approach, traditionally applied to low-molecular-weight bioactive compounds, to the study of recognition in macromolecular systems, such as the hirudin–thrombin complex. Recent advances in chemical and genetic methodologies will, hopefully, allow the researchers to extend this approach to other protein systems of relevant pharmacological application. Moreover, we have shown that incorporation of noncoded amino acids, possessing peculiar spectroscopic properties, can have great potentialities in biotechnology and pharmacological screening.

ACKNOWLEDGMENT

This work was supported by grants PRIN-2003 and PRIN-2005 from the Italian Ministry of University and Scientific Research. I would like to thank Prof. Angelo Fontana for stimulating discussions. The support of Dr. Olmetta Iadicicco is also gratefully acknowledged.

REFERENCES

1. Freedman R B (1971). Applications of the chemical reactions of proteins in studies of their structure and function. *Q. Rev. Chem. Soc.* 25:431–462.
2. Fersht A, Winter G (1992). Protein engineering. *Trends Biochem. Sci.* 17:292–294.
3. Smock R G, Gierasch L M (2005). Finding the fittest fold: using the evolutionary record to design new proteins. *Cell.* 122:832–834.
4. Cornish V W, Mendel, D, Schultz P G (1995). Probing protein structure and function with an expanded genetic code. *Angew. Chem.* 34:621–633.
5. Giannis A, Rübsam F (1997). Peptidomimetics in drug design. *Adv. Drug Res.* 29:1–78.
6. Hogan J G Jr (1997). Combinatorial chemistry in drug discovery. *Nat. Biotech.* 15:328–330.
7. Kent S B (1988). Chemical synthesis of peptides and proteins. *Annu. Rev. Biochem.* 57:957–989.
8. Dawson P, Kent S B (2000). Synthesis of native proteins by chemical ligation. *Annu. Rev. Biochem.* 69:923–960.
9. Nilsson B L, Soellner M B, Raines R T (2005). Chemical synthesis of proteins. *Annu. Rev. Biophys. Biomol. Struct.* 34:91–118.
10. Wallace C J (1993). Understanding cytochrome c function: Engineering protein structure by semisynthesis. *FASEB J.* 7:505–515.
11. De Filippis V, De Antoni F, Frigo M, et al. (1998). Protein stabilization by Ala→Aib replacements. *Biochem.* 37:1686–1696.
12. Wong C-Y, Eftink M R (1997). Biosynthetic incorporation of tryptophan analogues into staphylococcal nuclease: Effect of 5-hydroxytryptophan and 7-azatryptophan on structure and stability. *Protein Sci.* 6:689–697.
13. Budisa N, Minks C, Mediano F, et al. (1998). Residue-specific bioincorporation of non-natural, biologically active amino acid into proteins as possible drug carriers: Structure and stability of the *per*-thiaproline mutant of annexin V. *Proc. Natl. Acad. Sci.* 95:455–459.
14. Dougherty D A (2000). Unnatural amino acids as probes of protein structure and function. *Curr. Opin. Chem. Biol.* 4:645–652.
15. Cropp A T, Schultz P G (2004). An expanding genetic code. *Trends Genet.* 20:625–630.
16. England P M (2004). Unnatural amino acid mutagenesis: A precise tool for probing protein structure and function. *Biochem.* 43:11623–11629.
17. Hendrickson T L, de Crécy-Lagard V, Schimmel P (2004). Incorporation of nonnatural amino acids into proteins. *Annu. Rev. Biochem.* 73:147–176.
18. Albericio F (2004). Developments in peptides and amide synthesis. *Curr. Opin. Chem. Biol.* 8:211–221.

19. Minami T, Sugiyama A, Wu S O, et al. (2004). Thrombin and phenotypic modulation of the endothelium. *Arterioscler. Thromb. Vasc. Biol.* 24:41–53.

20. Griffin J H (1995). The thrombin paradox. *Nature.* 378:337–338.

21. Esmon C T (2003). The Protein C pathway. *Chest.* 124:26S–32S.

22. Dang Q D, Vindigni A, Di Cera E (1995). An allosteric switch controls the procoagulant and anticoagulant activities of thrombin. *Proc. Natl. Acad. Sci. USA* 92:5977–5981.

23. Di Cera E, Dang Q D, Ayala Y M (1997). Molecular mechanisms of thrombin function. *Cell Mol. Life Sci.* 53:701–730.

24. Di Cera E (2003). Thrombin interactions. *Chest.* 124:11S–17S.

25. Berg D T, Wiley M R, Grinnell B W (1996). Enhanced Protein C activation and ihibition of fibrinogen cleavage by a thrombin modulator. *Science.* 273:1389–1391.

26. Braun P J, Dennis S, Hofsteenge J, et al. (1988). Use of site-directed mutagenesis to investigate the basis for the specificity of hirudin. *Biochem.* 27:6517–6522.

27. Vindigni A, De Filippis V, Zanotti G, et al. (1994). Probing the structure of hirudin from *Hirudinaria manillensis* by limited proteolysis: Isolation, characterization and thrombin-inhibitory properties of N-terminal fragments. *Eur. J. Biochem.* 226:323–333.

28. Folkers P J M, Clore G M, Driscoll P C, et al. (1989). Solution structure of recombinant hirudin and the Lys-47→Glu mutant: A nuclear magnetic resonance and hybrid distance geometry-dynamical simulated annealing study. *Biochem.* 28:2601–2617.

29. Szyperski T, Güntert P, Stone S R, et al. (1992). Nuclear magnetic resonance solution structure of hirudin (1–51) and comparison with corresponding three-dimensional structures determined using the complete 65-residue hirudin polypeptide chain. *J. Mol. Biol.* 228:1193–1205.

30. Nicastro G, Baumer L, Bolis G, et al. (1997). NMR solution structure of a novel hirudin variant HM2, N-terminal 1–47 and N64 → V+G mutant. *Biopolym.* 41:731–749.

31. Rydel T J, Tulinsky A, Bode W, et al. (1991). Refined structure of the hirudin-thrombin complex. *J. Mol. Biol.* 221:583–601.

32. Scacheri E, Nitti G, Valsasina B, et al. (1993). Novel hirudin variants from the leech *Hirudinaria manillensis.* Amino acid sequence, cDNA cloning and genomic organization. *Eur. J. Biochem.* 214:295–304.

33. Betz A, Hofsteenge J, Stone S R (1992). Interaction of the N-terminal region of hirudin with the active-site cleft of thrombin. *Biochem.* 31:4557–4562.

34. De Filippis V, Quarzago D, Vindigni A, et al. (1998). Synthesis and characterization of more potent analogues of hirudin fragment 1–47 containing non-natural amino acids. *Biochem.* 37:13507–13515.

35. Szyperski T, Güntert P, Stone S R, et al. (1992). Impact of protein-protein contacts on the conformation of thrombin-bound hirudin studied by comparison with the nuclear magnetic resonance solution structure of hirudin (1–51). *J. Mol. Biol.* 228:1206–1211.

36. Fauchère J-L, Charton M, Kier L B, et al. (1988). Amino acid side-chain parameters for correlation studies in biology and pharmacology. *Int. J. Pept. Prot. Res.* 32:269–278.

37. Paterson Y, Leach S J (1978). The effect of side-chain branching on the theoretically predicted conformational space available to amino acid residues. *Macromol.* 11:409–415.

38. Steiner V, Knecht R, Börsen K O, et al. (1992). Primary structure and function of novel *O*-glycosylated hirudins from the leech *Hirudinaria manillensis*. *Biochem.* 31:2294–2298.

39. De Filippis V, Vindigni A, Altichieri L, et al. (1995). Core domain of hirudin from leech *Hirudinaria manillensis*: Chemical synthesis, purification and characterization of a Trp3 analog of fragment 1–47. *Biochem.* 34:9552–9564.

40. Ayala Y, Di Cera E (1994). Molecular Recognition by thrombin. Role of the slow → fast transition, site-specific ion binding energetics and thermodynamic mapping of structural components. *J. Mol. Biol.* 235:733–746.

41. Wallace A, Dennis S, Hofsteenge J, et al. (1989). Contribution of the N-terminal region of hirudin to its interaction with thrombin. *Biochem.* 28:10079–10084.

42. Wang J, Szewczuk Z, Yue S-Y, et al. (1995). Calculation of relative binding energies and configurational entropies: A structural and thermodynamic analysis of the nature of non-polar binding of thrombin inhibitors based on hirudin. *J. Mol. Biol.* 253:473–492.

43. De Filippis V, Colombo G, Russo I, et al. (2002). Probing hirudin-thrombin interaction by incorporation of noncoded amino acids and molecular dynamics simulation. *Biochem.* 43:1537–1550.

44. Sangster J (1989). Octanol-water partition coefficients of simple organic compounds. *J. Phys. Chem.* 18:1111–1229.

45. Hansch C, Leo A, Unger S H, et al. (1973). "Aromatic" substituent constants for structure-activity correlations. *J. Med. Chem.* 16:1207.

46. Lien E J, Guo Z-R, Li R-L, et al. (1982). Use of dipole moment as a parameter in drug-receptor interaction and quantitative structure-activity relationship studies. *J. Pharm. Sci.* 71:641–655.

47. Burley S K, Petsko G A (1988). Weakly polar interactions in proteins. *Adv. Protein Chem.* 39:125–189.

48. Serrano L, Bycroft M, Fersht A R (1991). Aromatic-aromatic interactions and protein stability. *J. Mol. Biol.* 218:465–475.

49. Gilson M K, Rashin A, Fine R, et al. (1985). On the calculation of electrostatic interactions in proteins. *J. Mol. Biol.* 184:503–516.

50. Dao-Pin S, Anderson D E, Baase W A, et al. (1991). Structural and thermodynamic consequences of burying a charged residue within the hydrophobic core of T4 lysozyme. *Biochem.* 30:115121–11529.

51. Berliner L J, Shen Y Y L (1977). Physical evidence for an apolar binding site near the catalytic centre of human α-thrombin. *Biochem.* 16:4622–4626.

52. Eisenberg D, McLachlan A D (1986). Solvation energy in protein folding and binding. *Nature.* 319:199–203.

53. Horton N, Lewis M (1992). Calculation of the free energy of association for protein complexes. *Protein Sci.* 1:169–181.

54. Legon A C (1999). Prereactive complexes of dihalogens XY with Lewis bases B in the gas phase: A systematic case for the halogen analogue B–XY of the hydrogen bond B–HX. *Angew. Chem., Int. Ed.* 38:2686–2698.

55. Kim K S, Tarakeshwar P, Lee J Y (2000). Molecular clusters of pi-systems: Theoretical studies of structures, spectra, and origin of interaction energies. *Chem. Rev.* 100:4145–4185.

56. Mecozzi S, West A P, Dougherty D A (1996). Cation-pi interactions in aromatics of biological and medicinal interest: Electrostatic potential surfaces as a useful qualitative guide. *Proc. Natl. Acad. Sci. USA.* 93:10566–10571.

57. Steiner T, Koellner G (2001). Hydrogen bonds with pi-acceptors in proteins: Frequencies and role in stabilizing local 3D structures. *J. Mol. Biol.* 305:535–557.

58. Markwardt F (1994). The development of hirudin as an antithrombotic drug. *Thromb. Res.* 74:1–23.

59. Pineo G F, Hull R D (1995). Hirudin and hirudin analogues as new anticoagulant agents. *Curr. Opin. Hematol.* 2:380–385.

60. Fenton J W Jr, Ofosu F A, Brezniak D V, et al. (1998). Thrombin and antithrombotics. *Semin. Thromb. Hemostasis.* 24:87–91.

61. Chang J-Y (1990). Production, properties, and thrombin inhibitory mechanism of hirudin amino-terminal core fragments. *J. Biol. Chem.* 265:22159–22166.

62. Callas D, Fareed J (1995). Comparative pharmacology of site directed antithrombin agents. Implication in drug development. *Thromb. Haemostasis.* 74:473–781.

63. Hursting M J, Alford K L, Becker J P, et al. (1997). Novastan®: A small-molecule, direct thrombin inhibitor. *Sem. Thromb. Hemostasis.* 23:503–516.

64. Verstraete M (1997). Direct thrombin inhibitors: Appraisal of the antithrombotic/hemorrhagic balance. *Thromb. Haemostasis.* 78:357–363.

65. De Filippis V, Russo I, Vindigni A, et al. (1999). Incorporation of noncoded amino acids into the N-terminal domain 1–47 of hirudin yields a highly potent and selective thrombin inhibitor. *Protein Sci.* 8:2213–2217.

66. Tapparelli C, Metternich R, Ehrhardt C, et al. (1993). Synthetic low-molecular weight inhibitors: Molecular design and pharmacological profile. *Trends Pharmacol. Sci.* 14:366–376.

67. Das J, Kimball D S (1995). Thrombin active site inhibitors. *Bioorg. Med. Chem.* 3:999–1007.

68. Sanderson P E J, Naylor-Olsen A M (1998). Thrombin inhibitor design. *Curr. Med. Chem.* 5:289–304.

69. Wells J A (1990). Additivity of mutational effects in proteins. *Biochem.* 29:8509–8517.

70. Di Cera E, Guinto E R, Vindigni A, et al. (1995). The Na^+ binding site of thrombin. *J. Biol. Chem.* 270:22089–22092.

71. Pineda A O, Savvides S N, Waksman G, et al. (2002). Crystal structure of the anticoagulant *slow* form. *J. Biol. Chem.* 277:40177–40180.

72. Pineda A O, Carrell C J, Bush L A, et al. (2004). Molecular dissection of Na^+ binding to thrombin. *J. Biol. Chem.* 279:31842–31853.

73. Huntington J A, Esmon C (2003). The molecular basis of thrombin allostery revealed by a 1.8-Å structure of the "slow" form. *Structure.* 11:469–479.

74. Carter W J, Myles T, Gibbs C S, et al. (2004). Crystal structure of anticoagulant thrombin variant E217K provides insights into thrombin allostery. *J. Biol. Chem.* 279:26387–26394.

75. De Filippis V, De Dea E, Lucatello F, et al. (2005). Effect of Na^+ binding on the conformation, stability, and molecular recognition properties of thrombin. *Biochem. J.* 390:485–492.

76. Wyman J (1968). Regulation in macromolecules as illustrated by haemoglobin. *Q. Rev. Biophys.* 1:35–80.

77. Di Cera E (1998). Site-specific analysis of mutational effects in proteins. *Adv. Protein Chem.* 51:59–119.

78. Stubbs M T, Oschkinat H, Mayr I, et al. (1992). The interaction of thrombin with fibrinogen: A structural basis for its specificity. *Eur. J. Biochem.* 206:187–195.

79. Mathews I I, Padmanabhan K P, Ganesh V, et al. (1994). Crystallographic structures of thrombin complexed with thrombin receptor peptides: Existence of expected and novel binding modes. *Biochem.* 33:3266–3279.

80. Sadasivan C, Yee V C (2000). Interaction of the factor XIII activation peptide with α-thrombin: Crystal structure of its enzyme-substrate analog complex. *J. Biol. Chem.* 275:36942–36948.

81. Cornish V W, Benson D R, Altenbach C A, et al. (1994). Site-specific incorporation of biophysical probes into proteins. *Proc. Natl. Acad. Sci.* 91:2910–2924.

82. Wong C-Y, Eftink M R (1998). Incorporation of tryptophan analogues into staphylococcal nuclease: Stability toward thermal and guanidine-HCl induced unfolding. *Biochem.* 37:8947–8953.

83. Cohen B E, McAnaney T B, Park E S, et al. (2002). Probing protein electrostatics with a synthetic fluorescent amino acid. *Science.* 296:1700–1703.

84. Hovius R, Vallotton P, Wohland T, et al. (2000). Fluorescence techniques: Shedding light on ligand-receptor interactions. *Trends Pharmacol. Sci.* 21:266–273.

85. Eftink M R (1997). Fluorescence methods for studying equilibrium macromolecule-ligand interactions. *Methods Enzymol.* 278:221–257.

86. Jameson D M, Croney J C, Moens P D (2003). Fluorescence: Basic concepts, practical aspects, and some anecdotes. *Methods Enzymol.* 360:1–43.

87. Wu P, Brand L (1994). Resonance energy transfer: Methods and applications. *Anal. Biochem.* 18:1–13.

88. Selvin P R (1995). Fluorescence resonance energy transfer. *Methods Enzymol.* 246:300–334.

89. Yan Y, Marriott G (2003). Analysis of protein interactions using fluorescence technologies. *Curr. Opin. Chem. Biol.* 7:635–640.

90. De Filippis V, De Boni S, De Dea E, et al. (2004). Incorporation of the fluorescent amino acid 7-azatryptophan into the core domain 1–47 of hirudin as a probe of hirudin folding and thrombin recognition. *Protein Sci.* 13:1489–1502.

91. De Filippis V, Frasson R, Fontana A (2006). 3-Nitrotyrosine as a spectroscopic probe for investigating protein-protein interactions. *Protein Sci.* 15:976–986.

92. Twine S M, Szabo A G (2003). Fluorescent amino acid analogs. *Methods Enzymol.* 360:104–127.

93. Ross A, Szabo A G, Hogue C W V (1997). Enhancement of protein spectra with tryptophan analogs: Fluorescence spectroscopy of protein-protein and protein-nucleic acid interactions. *Methods Enzymol.* 278:151–190.

94. Negrerie M, Gai F, Bellefeuille S M, et al. (1991). Photophysics of a novel optical probe: 7-Azaindole. *J. Phys. Chem.* 95:8663–8670.

95. Chapman C F, Maroncelli M (1992). Excited-state tautomerization of 7-azaindole in water. *J. Phys. Chem.* 96:8430–8441.

96. Chou P-T, Martinez M L, Cooper W C, et al. (1992) Monohydrate catalysis of excited-state double-proton transfer in 7-azaindole. *J. Phys. Chem.* 96:5203–5205.

97. Smirnov A V, English D S, Rich R L, et al. (1997). Photophysics and biological applications of 7-azaindole and its analogs. *J. Phys. Chem. B.* 101:2758–2769.

98. Mente S, Maroncelli M (1998). Solvation and the excited-state tautomerization of 7-azaindole and 1-azacarbazole: Computer simulations in water and alcohol solvents. *J. Phys. Chem.* 102:3860–3876.

99. Halliwell B (1997). What nitrates tyrosine? Is nitrotyrosine specific as a biomarker of peroxynitrite formation *in vivo*? *FEBS Lett.* 411:157–160.

100. Riordan J F, Sokolovsky M, Vallee B L (1967). Environmentally sensitive tyrosyl residues. Nitration with tetranitromethane. *Biochem.* 6:358–361.

101. Csizmadia F, Tsantili-Kkoulidou A, Panderi I, et al. (1997). Prediction of distribution coefficient from structure. 1. Estimation method. *J. Pharm. Sci.* 7:865–871.

102. Abraham M H, Du C M, Platts J A (2000). Lipophilicity of the nitrophenols. *J. Org. Chem.* 65:7114–7118.

103. Mostad A, Natarajan S (1990). Crystal and molecular structure of 3-nitro-4-hydroxy-phenylalanine nitrate. *Z. Kristall.* 193:127–136.

104. Steiner R F, Albaugh S, Kilhoffer M-C (1991). Distribution separations between groups in an engineered calmodulin. *J. Fluoresc.* 1:15–22.

105. Rischel C, Poulsen F M (1995). Modification of a specific tyrosine enables tracing of the end-to-end distance during apomyoglobin folding. *FEBS Lett.* 374:105–109.

106. Rischel C, Thyberg P, Rigler R, et al. (1996). Time-resolved fluorescence studies of the molten globule state of apomyoglobin. *J. Mol. Biol.* 257:877–885.

107. Tcherkasskaya O, Ptitsyn O B (1999). Direct energy transfer to study the 3D structure of non-native proteins: AGH complex in the molten globule state of apomyoglobin. *Protein Eng.* 12:485–490.

108. Juminaga D, Albaugh S A, Steiner R F (1994). The interaction of calmodulin with regulatory peptides of phosphorylase kinase. *J. Biol. Chem.* 269:1660–1667.

109. Mezo A R, Cheng R P, Imperiali B (2001). Oligomerization of uniquely folded mini-protein motifs: Development of a homotrimeric ββα peptide. *J. Am. Chem. Soc.* 123:3885–3891.

110. Dreeskamp H, Koch E, Zander M (1975). On the fluorescence quenching of polycyclic aromatic hydrocarbons by nitromethane. *Z. Naturforsch.* 30a:1311–1314.

111. Karshikov A, Bode W, Tulinsky A, et al. (1992). Electrostatic interactions in the association of proteins: An analysis of the thrombin-hirudin complex. *Protein Sci.* 1:727–735.

112. Vijayalakshmi J, Padmanabhan K P, Mann K G, et al. (1994). The isomorphous structures of prethrombin-2, hirugen-, and PPACK-thrombin: Changes accompanying activation and exosite binding to thrombin. *Protein Sci.* 3:2254–2271.

113. Mattos C (2002). Protein-water interactions in a dynamic world. *Trends Biochem. Sci.* 27:203–208.

114. Engh R A, Brandstetter H, Sucher G, et al. (1996). Enzyme flexibility, solvent and weak interactions characterize thrombin-ligand interactions: Implications for drug design. *Structure.* 4:1353–1362.

10.1

PRODUCTION AND PURIFICATION OF ADENOVIRUS VECTORS FOR GENE THERAPY

D. M. F. PRAZERES AND J. A. L. SANTOS

Instituto Superior Técnico, Lisbon, Portugal

Chapter Contents

10.1.1	Gene Therapy	1262
	10.1.1.1 Milestones	1262
	10.1.1.2 Vectors	1263
10.1.2	Adenovirus in Gene Therapy	1264
	10.1.2.1 Introduction	1264
	10.1.2.2 Biology and Properties	1265
	10.1.2.3 Recombinant Adenovirus Vectors	1267
	10.1.2.4 Targeting Strategies	1269
	10.1.2.5 Applications	1270
	10.1.2.6 Safety Aspects	1273
10.1.3	Adenovirus Manufacturing	1273
	10.1.3.1 GMPs and Validation	1274
	10.1.3.2 Product Specifications and Quality Control	1274
	10.1.3.3 Environmental and Safety Issues	1275
	10.1.3.4 Process Considerations	1276
	10.1.3.5 Production	1277
	10.1.3.6 Downstream Processing	1279
10.1.4	Analysis and Evaluation of an Adenovirus Production Process	1281
	10.1.4.1 Process Description	1282
	10.1.4.2 Inventory Analysis	1285
	10.1.4.3 Cost Analysis and Economic Assessment	1285
10.1.5	Concluding Remarks and Outlook	1288
	References	1289

10.1.1   GENE THERAPY

10.1.1.1   Milestones

Gene therapy is a conceptually simple and attractive process, which consists of the introduction of one or more functional genes in a human/non-human receptor (*in vivo*, *in situ*, or *ex vivo*). It constitutes a promising alternative for the treatment, diagnosis, or cure of genetic defects such as cystic fibrosis [1] or acquired diseases like cancer [2] and AIDS [3]. DNA vaccines can also be developed on the basis of genes [4] to provide immunity against infectious agents (e.g., malaria [5]) or treat noninfectious diseases such as tumors and allergies [6].

Although gene therapy is a recent endeavor of the human mind, it has nevertheless come a long way since the early, nonauthorized administration of Shope papilloma virus to argininemia-suffering patients by Stanfield Rogers et al. [7, 8] (Table 10.1-1). Despite the flawed design and consequent failure of this clinical trial, Rogers was one of the first scientists to anticipate the therapeutic potential of viruses as carriers of genetic information [8]. The first federally approved gene therapy clinical trials took place in 1990 when an adenosine deaminase (ADA)-deficient patient was given her own T cells engineered with a retroviral vector carrying a normal ADA gene [9]. This experiment paved the way for further clinical trials. In 1993 an adenovirus vector (AdV) was first used in a clinical trial designed to evaluate the potential of direct transfer of cystic fibrosis transmembrane conductance regulator (CFTR) cDNA in the treatment of cystic fibrosis (CF) [10].

TABLE 10.1-1. Gene Therapy Milestones

Year	Vector	Target Disease	Comments	Reference
1970	Shope virus	Argininemia	First human trial, unauthorized	[7, 8]
1990	Retrovirus	ADA-SCID	First federally approved human trial	[9]
1993	Adenovirus	Cystic fibrosis	First clinical trial with an adenovirus vector	[10]
1999	Adenovirus	OTC deficiency	Death of patient due to an adverse reaction to the adenovirus vector	[11, 12]
2000	Retrovirus	SCID-X1 syndrome	Apparent clinical cure of two recipients	[13]
2003	Retrovirus	SCID-X1 syndrome	Development of leukemia-like syndrome in recipients due to retrovirus integration	[14]
2004	Adenovirus	Head and neck carcinoma	The first human GT product, Gendicine, received approval from the Chinese FDA	[15]

Abbreviations: ADA = adenosine deaminase; OTC = ornythine transcarbamylase; SCID = severe combined immunodeficiency.

The first setback faced by gene therapy came in 1999 when a patient suffering from ornithine transcarbamylase (OTC) deficiency died after administration of an adenovirus vector encoding OTC [11, 12]. The year 2000 saw gene therapy's first major success: A gene therapy protocol could correct the phenotype of an X-linked severe combined immunodeficiency (SCID-X1) syndrome in two patients who had been re-infused with autologous CD34 bone marrow cells transduced *ex vivo* with a retrovirus vector encoding the γC receptor gene [13]. These successes were later shadowed by the development of leukemia-like syndrome in recipients of the treatment as a consequence of retrovirus integration in proximity to the LMO2 proto-oncogene promoter [14]. In October 16, 2003, a recombinant adenovirus vector expressing the tumor-suppressor gene p53 was approved by the State Food and Drug Administration (SFDA) of China for the treatment of head and neck squamous cell carcinoma [15–17]. Developed and manufactured by Shenzen SiBiono GeneTech (China) and trademarked under the name Gendicine, it became the first human gene therapy product to reach the market in April 2004 [15]. As of September 2005, 2600 patients had been treated with Gendicine, with projections estimating 50,000 patients to receive the product by 2006 [18].

10.1.1.2 Vectors

The transport of the therapeutic transgenes toward the nuclei of the target cells can be carried out both by viral [19] and nonviral vectors such as plasmid DNA [20]. Plasmid DNA molecules are extra-chromosomal carriers of genetic information that have the ability to replicate autonomously. These vectors constitute an attractive gene transfer system because they are safer and easier to produce when compared with viral vectors [20–22]. However, plasmid DNA vectors are less effective in transfecting cells when compared with viral vectors, which have a natural ability to deliver and express their genes in a wide variety of cell types and tissues [11, 23].

The major drawback of viral vectors is undoubtedly related to safety aspects. When confronted with viral vector particles, the human immune system, which has evolved to tackle wild-type infections, will likewise generate an immune response [19]. The potency and severity of this response can ultimately lead to death, as occurred in the 1999 OTC gene therapy clinical trial (Table 10.1-1) [12]. Thus, a safe and effective use of viral vectors in gene therapy requires the modification of natural viruses to impair replication and expression of viral proteins and thus minimize toxicity and immunogenicity [19]. The infection pathway of these recombinant viral vectors should of course remain unaltered. Further goals of virus modification are to provide room in the viral genome for transgenes, to modulate transgene expression, and to improve the selectivity of infection.

Most recombinant viral vectors used in gene therapy clinical trials belong to one of the following categories (Figure 10.1-1): (1) adenovirus, (2) retrovirus/lentivirus, (3) pox/vaccinia virus, (4) adeno-associated virus (AAV), and (5) herpes simplex-1 virus (HSV-1). According to the data provided in the *Journal of Gene Medicine Database* (Wiley Database, http://www.wiley.co.uk/genmed/clinical/) and updated in July 2005, 25% of the total 1076 gene therapy clinical trials were using AdVs (Figure 10.1-1). The popularity and high expectations generated toward AdVs is well expressed by the fact that both the first gene therapy tragedy and success used adenovirus to deliver transgenes (Table 10.1-1).

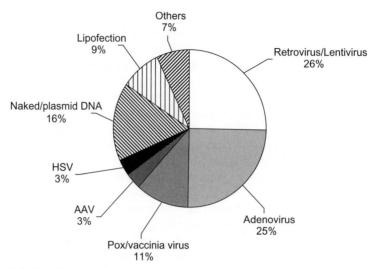

Figure 10.1-1. Breakdown of gene therapy clinical trials in terms of vector type ($N = 1076$) (source http://www.wiley.co.uk/genmed/clinical/, July 2005). Abbreviations: AAV—adeno-associated virus; HSV—herpes simplex virus.

TABLE 10.1-2. Characteristics of Adenoviruses Relevant for Human Gene Therapy Applications

Efficient transduction of a large variety of quiescent and actively dividing cells
Relatively high capacity for transgene insertion
Convenient and simple methods of vector construction
Ability to replicate at high titers in complementing cell lines
High stability, which allows efficient purification and long-term storage
Biology has been extensively studied and is well understood
Transient transgene expression due to episomal location of viral genome
Immunogenicity
Insert-size limit of 7.5 kb

10.1.2   ADENOVIRUS IN GENE THERAPY

10.1.2.1   Introduction

Adenoviruses were discovered in cultures of human adenoids in 1953 by Rowe et al. [24] and have since then been implicated in several respiratory, ocular, and gastrointestinal human diseases [25, 26]. Nonhuman adenoviruses have also been found in many other mammalian (dogs, horses, sheeps, chimpanzees, etc.) and nonmammalian (ducks, fowl, geese, etc.) species [26]. Overall, more than 100 members have been included in the *Adenoviridae* family. This section will only give a brief description of the major characteristics of the adenovirus and its vectors because this topic has been covered in depth in several recent reviews [26–31].

The extensive studies that followed the discovery of adenoviruses unveiled many features that make them a popular gene delivery vector (Table 10.1-2) [11, 26, 32]. For instance, adenoviruses are extremely efficient in the transduction of both quiescent and actively dividing cells in most tissues [19, 26] and their genome can be manipulated easily to generate recombinant adenoviruses with improved properties. It is also possible to propagate them to high titers, and thus, it becomes relatively easy to generate sufficient amounts for research purposes and small-scale clinical trials [26]. On the down side, adenoviruses can generate potent immunogenic reactions in human recipients [28], although in some instances this immunogenicity may enhance antitumor effects [19] and vaccination efficiency [33]. Furthermore, and because adenoviruses persist in the nucleus of transduced cells as extra-chromosomal episomes, the expression of transgenes is transient [19, 26].

10.1.2.2 Biology and Properties

The 51 distinct serotypes of human adenovirus (Ad1 to Ad51) are classified into six groups (A to F) on the basis of sequence homology and hemagglutination properties [25]. Most AdVs used in gene therapy are derived from serotype 2 (Ad2) and 5 (Ad5) of the subgroup C [26]. The efficacy of Ad2 and Ad5 derived vectors, however, may be limited by the preexistence of humoral and/or cellular immunity to these serotypes in most human populations. Thus, other serotypes such as Ad11 and Ad35 to which most humans do not have neutralizing antibodies may be clinically useful [34, 35]. The antivector immunity may also be circumvented by the physical shielding of the adenovirus coat and the use of nonhuman adenovirus serotypes [36].

Adenoviruses are nonenveloped viruses with 26- to 45-kbp-long linear genomes of double-stranded DNA (the genome of human Ad2 comprises 35,937 base pairs) [37]. The genome is encapsidated in an icosahedral protein coat (12 vertices, 20 surfaces) made essentially of 240 nonvertex hexons and 12 vertex pentons, each with one or two protruding fibers (Figures 10.1-2 and 10.1-3). Each hexon protein is a trimer of the identical polypeptide II (pII), and each penton protein is formed by the interaction of five polypeptides (pIII). These pentons are tightly associated with one or two fibers made of three polypeptides (pIVs) each [37]. Overall, the fiber protein is composed of an N-terminal tail for penton binding, a rigid shaft, and a distal globular knob domain responsible for interaction with host cell receptors [38, 39]. The net charge of each hexon monomer in the Ad5 serotype is −23.8. This makes adenoviral capsids highly negative with an overall surface charge of over −17,000 [40, 41]. Other minor protein components can be found in the capsid, including protein IIIa, VI, VIII, and IX [37]. The size of the adenovirus particle is around 70–110 nm, and its molecular weight falls within the range $150–180 \times 10^6$ [39]. The buoyant density of an adenoviral particle in CsCl is $1.32–1.35\,g\,cm^{-3}$ [39]. Assuming a 170×10^6 MW and a 35,937-bp-long genome [42], the protein and DNA content of one Ad2 viral particle (VP) can be estimated to be 24×10^{-5} (86% of dry weight) and 4×10^{-5} (14% of dry weight) pg/VP, respectively. The corresponding water content, if a 100-nm viral particle is assumed, is approximately 60%.

The core of an adenoviral particle consists of the DNA genome complexed with four polypeptides (pV, pVII, mu, TP). Figure 10.1-4 schematically shows the structure of an adenovirus genome. The extremities contain inverted terminal repeat (ITR) sequences (100–140 bp), covalently linked to terminal proteins (TPs), which function as replication origins [23]. The nearby ψ sequence at the left end of the

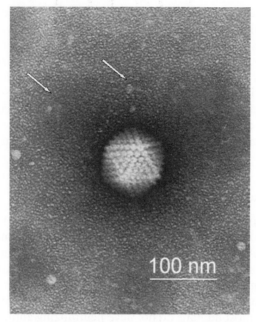

Figure 10.1-2. Human adenovirus C. The bar represents 100 nm. The arrows, upper left, point to fiber knobs. (Micrograph reprinted with authorization from Dr. H.-W. Ackermann, Department of Microbiology Medical Faculty Laval University, Quebec, Canada.)

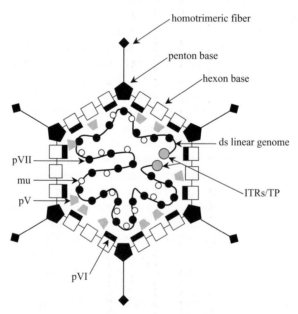

Figure 10.1-3. Schematic structure of an adenovirus particle. Major capsid (hexon, penton, fiber) and core (pV, pVII, Mu) proteins are shown. Abbreviation: ITR—inverted terminal repeat; TP—terminal protein. (Adapted from Ref. 27.)

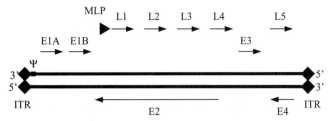

Figure 10.1-4. The adenovirus genome. The early (E1–E4) and late (L1–L5) transcription units are identified by the arrows. The inverted terminal repeats (ITRs), packaging sequence (ψ), and major late promoter (MLP) are also shown. (Adapted from Ref. 26.)

genome consists of a series of seven repeats and is required for efficient packaging [26]. The genes in both strands are grouped into early and late transcriptional units. The basis of this classification is the two-phase infectious cycle (30–40h), which is characteristic of adenoviruses. In the "early" phase, the virus particles enter the host cell through binding of the homotrimeric protruding fibers to the coxsackievirus B and adenovirus receptor (CAR) [26]. After endosomal uptake, release, and capsid dismantling, the viral genome is transported and delivered to the nucleus [38]. The early genes E1 to E5 are then selectively transcribed and translated to modulate functions of the host cell and thus create an optimal environment for virus replication [26]. Specifically, this involves driving the host cell into S-phase (E1A gene), suppressing the host's cell apoptotic machinery (E1B gene), modifying the host's immunological environment (E3 gene), encoding proteins for viral DNA replication (E2 gene), and blocking cellular protein synthesis [43].

The "late" phase of the infectious cycle involves the synthesis of structural proteins of the virus, capsid assembly, genome encapsidation, and maturation of fully infectious viral particles. These events are associated with the transcription and translation of the "late" genes L1 to L5 [26, 27]. A key player in the transcription of the late genes is the major late promoter (MLP), which is activated early on via the E2 gene [27, 29].

10.1.2.3   Recombinant Adenovirus Vectors

A safe and effective use of AdVs requires the engineering of natural viruses into useful recombinant adenovirus vectors by using molecular biology tools [26]. These modifications are essentially designed to: (1) impair replication and expression of viral proteins and thus minimize toxicity and immunogenicity [19], (2) provide room in the viral genome for the therapeutic transgenes, and (3) modulate transgene expression. Once these recombinant vectors are available, transgenes can be inserted in their genome. In many cases, tissue- and cell-specific heterologous promoters are inserted along side to provide better expression [27, 29].

First-Generation Vectors. Most first-generation AdVs have been constructed by deleting the viral early gene E1 (Figure 10.1-5). The goal underlying this deletion is to impair viral replication and production of capsid proteins and thus prevent the vectors from causing disease in the gene therapy recipient [27]. Simultaneously, room is made for transgenes up to 6.5kb, which are usually expressed under the

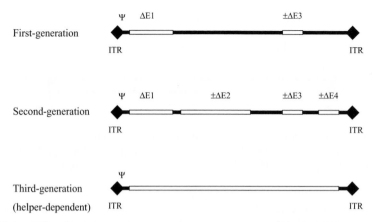

Figure 10.1-5. The three generations of adenovirus vectors. (A) First-generation vectors lack the E1 and sometimes the nonessential E3 region. (B) Second-generation vectors contain additional deletions of the E2 or E4 region. (C) Third-generation or "gutless" vectors are constructed by removing all or most viral genes and maintaining the *cis*-acting elements ITR and ψ. (Adapted from Ref. 26.)

control of an heterologous promoter [27]. The propagation of these E1 replication defective vectors can only be accomplished in complementing cells lines such as the human embryonic kidney 293 (HEK-293) [44] or the human retinoblast (PER. C6)[45]. These cells are transformed with the E1 gene and therefore transcomplement its deficiency in the recombinant vectors.

First-generation AdVs are characterized by several problems. For instance, recombination between the replication-defective virus and the E1 sequences in the complementing HEK-293 cell line can occur during propagation, giving rise to replication-competent adenoviruses (RCA) that will contaminate the recombinant virus being produced [26]. This problem has been circumvented in the PER.C6 cell line in which a rational design of the E1 transgene prevents generation of RCAs [45]. Another problem with first-generation vectors is that despite the deletion of the E1 gene, viral protein expression is not completely shut off. The concomitant low level of viral replication taking place in the transduced host cells, thus, can generate a cytotoxic-T lymphocyte (CTL) immune response that in the end will destroy the transgene-expressing cells and contribute to gene silencing [26]. Despite this residual immunogenicity, first-generation vectors may find application in cases such as cancer therapy and vaccination where a short-term transgene expression is desirable [26]. It is worth mentioning that the transgene product itself, depending on its own intrinsic immunogenicity, can also trigger a cellular immune response capable of destroying the transduced cells [29]. However, the magnitude and nature of this immune response seems to vary considerably depending on the specific transgene product being expressed [46].

Second-Generation Vectors. Second-generation AdVs have been constructed by additionally deleting the viral regulatory E2 and/or E4 coding sequences (Figure 10.1-5), on the expectation that progressive deletions should reduce viral antigen expression, increase *in vivo* persistence, and reduce antiviral immune response

[47]. The E2 gene encodes proteins that are required for initiation and elongation of viral DNA synthesis and the activation of the MLP, whereas E4 encodes proteins involved in the accumulation, splicing, and transport of early and late viral mRNA and in DNA replication and virus particle assembly [47]. As in the first generation, complementing cell lines able to express the missing functions must be generated and isolated to ensure propagation of the recombinant viruses. The HEK-293 cell line can be conveniently adapted to this end [48]. A first advantage of these doubly deleted vectors is the reduced probability of the emergence of RCAs, because this would require simultaneous reversions in the E1 and E2 or E4 regions [47]. Several studies with E1/E2 and E1/E4 deleted vectors have shown reduced synthesis of viral proteins, extended transgene expression, and reduced toxicity [49–51]. Other studies, however, have challenged these results [47].

Third-Generation Vectors. A third generation of AdVs has been further developed by deleting most or virtually all viral genes in the genome (Figure 10.1-5) [52–55]. These gutless or helper-dependent vectors can accommodate inserts up to 30 kbp and have shown reduced immunogenicity and prolonged transgene expression in mice [55]. The reduced immunogenicity still observed should be attributed either to the injected viral proteins or to the transgene product [29]. Growth of helper-dependent vectors depends on coinfection of an E1-complementing cell line such as HEK-293 by an E1-deleted helper virus (hence, the name helper-dependent) that provides missing replication and assembly functions *in trans* [26, 52]. The simultaneous propagation of the helper virus can be limited by resorting to a Cre-lox mechanism that allows the packaging sequence to be deleted [53]. In short, the helper genome is provided with loxP-sites that flank the ψ signal, whereas complementing 293 cells are engineered to stably express the enzyme Cre-recombinase. During infection, Cre-recombinase excises the loxP-flanked signal rendering the helper genome unpackable [29]. Nevertheless, the two types of viral particles (helper-dependent and helper) must be separated after propagation, a task that is difficult to accomplish [52].

Oncolytic Adenoviruses. A class of AdVs has been developed to specifically kill cancer cells, an approach that has been termed virotherapy [56]. Contrary to the approach pursued with first-, second-, and third-generation adenovirus vectors, these oncolytic or conditionally replicating adenoviruses (CRAds) are engineered in such a way that they retain the ability to replicate. This replication however, is tumor selective—the virus will only infect and proli-ferate in malignant cells with mutations in specific tumor suppressor genes [2, 19, 56]. Thus, a transgene is actually not required for the therapeutic effect to take place.

10.1.2.4 Targeting Strategies

Several therapeutically relevant human cells (e.g., skeletal and smooth muscle cells, endothelial cells, hematopoietic cells, and some tumor cells) are not easily trans-duced by adenovirus vectors due to low or inexistent levels of expression of the CAR receptor. On the other hand, the broad tropism of adenoviruses may lead to the undesirable expression of the transgene in nontarget cells. Thus, an improve-ment in the transduction selectivity of existing adenovirus vectors is crucial to circumvent these limitations and improve efficacy in many clinical applications. The following strategies have been devised to meet this end: (1) structural retargeting, (2) tropism ablation, and (3) transcriptional targeting [38].

Structural targeting relies on the structural modification of the adenoviral capsid by genetic incorporation of peptides [57], IgG-binding domains [58], and fiber proteins from other serotypes [59], metabolic biotinylation [60], and PEGylation [61]. These ligands are selected or designed to direct vector attachment to alternative cell receptors in CAR-deficient cells [38]. The goal of tropism ablation, on the other hand, is to reduce transduction of nontarget, CAR-expressing cells. This "de-targeting" strategy can be accomplished by constructing "knobless" vectors or by introducing point mutations in the fiber protein knob [30, 38]. The capsid of these mutant vectors is then modified with ligands (e.g., peptides) adequate for transduction of the target cells (retargeting) [62–64].

In transcriptional targeting, cell-type-specific promoters are used to restrict transgene expression to specific tissues. The adenoviral particles may infect different cells, but the transgene is expressed only in those cells that actively express transcription factors required to drive expression from the cell-specific promoter [19, 33]. For instance, this approach has been used in the context of adenoviral-mediated treatment of gastrointestinal cancer [65]. This application has been limited by the undesirable expression of the transgene (which codes for HSVtk-herpes simplex virus thymidine kinase) in the liver due to the vector hepatotropism. Thus, the transgene was placed under the control of the cyclooxygenase-2 (cox-2) promoter, which is inactive in liver cells but active in many gastrointestinal cancers. Experiments showed that the cox-2 promoter could confine the cytocidal effect of HSVtk specifically to cyclooxygenase-2-positive gastrointestinal cancer, while mitigating the otherwise fatal hepatotoxicity [65].

10.1.2.5 Applications

Most adenovirus vector applications and clinical trials described and reported in the literature have targeted genetic diseases and cancer. This section will briefly mention the most significant applications. The reader is directed to several reviews for a more comprehensive description of applications [2, 19, 27, 33, 56, 66].

Genetic Diseases. In the case of genetic diseases, the transgene encodes a protein that is missing or is defective in the host organism. A typical and well-studied example is cystic fibrosis, the first human disease targeted in an adenovirus gene therapy clinical trial [10]. Cystic fibrosis is an inherited, recessive disease caused by a variety of mutations in the gene encoding the CFTR protein [1, 67, 68]. Although multiple organs are affected, the lung is the life-threatening organ [25]. The virus-vector mediated shuttling of a normal copy of the CFTR gene toward the affected lung cells could potentially prevent the onset, or halt the progression, of the disease. The choice of adenoviruses as CFTR gene delivery vectors is logical given their natural ability to infect lung cells. Nevertheless, barriers such as the lack of CARs in airway epithelial cells and alveolar macrophages, and the specific pulmonary-associated T-helper cell response, have prevented a successful transgene expression [27]. Significantly, all clinical trials currently under way involving AdVs and directed toward genetic diseases target cystic fibrosis (*Journal of Gene Medicine Database*, http://www.wiley.co.uk/genmed/clinical/).

Adenoviral gene therapy has also targeted muscular dystrophy, a genetic disease characterized by progressive muscle weakness [69]. Muscular dystrophy is caused by mutations in the X-linked dystrophin gene (DMD), which lead to prematurely aborted dystrophin synthesis. The lack of this important structural protein in muscle cells causes fiber damage and membrane leakage. In this context, gene therapy attempts to deliver a dystrophin expression vector to the nuclei of striated muscle cells. Given the huge size of the DMD gene (2.6 Mb), mini-gene cassettes (14 kb) have been generated that are capable of expressing therapeutic levels of a functional dystrophin protein [70]. Nevertheless, vectors with a capacity large enough to accommodate the 14-kb dystrophin cDNA are still required. Unlike first- and second-generation AdVs that are limited to 7–8-kb transgenes, third-generation adenovirus vectors adequately meet this requirement. The use of such vectors has resulted in a prolonged expression of the transgene in mice muscle cells [55, 69, 70]. Despite these promising results, barriers such as the lack of adenovirus receptors in the target muscle cells and potent immune response have to be overcome before a therapeutic use is developed [27, 69]. Further examples of the application of AdVs to treat genetic diseases include OTC deficiency, factor VIII deficiency, Tay-Sachs disease, and glycogen storage disease II [27].

Cancer. Different strategies have been pursued in an attempt to treat cancer via gene therapy, by taking advantage of molecular differences between normal and tumor cells [2, 27, 33]. Once transferred to the target cells, expression of the transgene delivers the antitumor effect. Clinical data suggest excellent safety when AdVs are injected locally. Synergistic effects with treatment options such as radiotherapy and chemotherapy have further improved the efficacy of cancer gene therapy [33]. Most efforts have relied on one of the following approaches: (1) tumor suppression, (2) suicide therapy, (3) cancer vaccination, and (4) virotherapy.

The tumor suppression approach attempts to induce apoptosis by delivering a tumor suppressor gene that is missing or defective in the tumor cells [2]. On the contrary, normal cells infected by the tumor suppressor delivery vector will not be detrimentally affected. The approach is perfectly illustrated with p53, a gene whose mutations have been associated with several tumors. The delivery of wild-type p53 gene efficiently induces apoptosis in cells of different tumors, as demonstrated in several phase I and II clinical trials [43, 71]. Additionally, this toxic effect can extend to neighbor, uninfected tumor cells, a phenomenon known as the "bystander effect". This effect has been attributed to the ability of p53 to block angiogenesis [2]. Combination of p53 with immunomodulatory genes, cytotoxic drugs, or radiotherapy may further improve efficacy [2, 27]. Recombinant adenoviruses encoding the p53 gene have been used to treat more than 20 kinds of cancer indications, including head and neck squamous cell carcinoma, lung cancer, breast cancer, and liver cancer [15]. Studies have focused both on the effect of adenoviral gene therapy alone or in combination with conventional therapies such as chemotherapy, radiotherapy, and surgery [15]. The efficiency of the adenovirus/p53 strategy is well illustrated with the results of a phase II/III clinical trial using the commercial product Gendicine. In this study, 135 patients with head and neck squamous cell carcinoma were divided into two groups. The first group (63) received Gendicine in combination with radiotherapy, whereas the second group (72) received radiotherapy alone. Significant difference in terms of complete or partial tumor regression was shown between the two groups, with 93% of the patients responding in the Gendicine/radiotherapy group *versus* 79% in the radiotherapy group [15].

The second approach used to kill tumor cells is suicide gene therapy, which combines the delivery of suicide transgenes with the separate administration of a harmless prodrug. Once reaching the target cancer cells, the suicide gene expresses an enzyme that metabolizes the prodrug into a cytotoxic agent that kills cells. The diffusion of the cytotoxic agent into neighbor cells further generates a "bystander" effect that increases the efficacy of the strategy [2, 27]. An example of such a transgene/prodrug combination that has been tested in the clinic is herpes simplex virus thymidine kinase (HSVtk)/ganciclovir [2, 27, 33]. The suicide HSVtk transgene is first delivered to cancer cells, for example, via AdVs. The administered ganciclovir is then metabolized by the expressed HSVtk into ganciclovir triphosphate, a nucleotide analoge that blocks DNA synthesis [72–75]. When a DNA strand incorporates this analoge, chain termination results and cells die upon induction of apoptosis [76]. This effect is, of course, more pronounced in tumor cells, which divide much more actively when compared with native cells. Another feature that contributes to the attractiveness of the approach is the use of a drug such as Ganciclovir that is widely used clinically [76]. Another example of suicide cancer therapy is the combined use of *Escherichia coli* cytosine deaminase (CD) with the prodrug 5-fluorocytosine. In this case, 5-fluorocytosine is metabolized by the expressed CD into 5-fluorouracil, a pyrimidine antagonist that blocks DNA and RNA synthesis [2]. The subsequent direct and bystander inhibitory effect can be enhanced further by combination with radiotherapy [77, 78].

Cancer can also be treated by the adenovirus-mediated delivery of transgenes to tumor cells with the goal of boosting antitumor immunity [27]. These adenoviral "cancer vaccines" can harbor either immunomodulatory genes (e.g., IL2 or IL12) and/or tumor antigens (e.g., MART 1 or gp 100 melanoma antigens), which once expressed should induce tumor regression. Human dendritic cells isolated from patients have also been transduced with AdVs designed to express tumor antigens. The therapeutic effect is achieved after reinfusion of the transformed cells [27].

Virotherapy constitutes a fourth approach that has been attempted clinically to treat cancer. The oncolytic vector Onyx-015 (previously d-1042) is a characteristic example of virotherapy. The mode of action can be explained as follows. Both adenoviruses and tumor cells need to block the p53 function to replicate. In normal cells the intact p53 function blocks replication of the Onyx-015 vector [2, 19, 56]. In tumor cells, however, and because the p53 gene is mutated, infection with the adenovirus vector Onyx-015 is followed by replication and cell destruction. Furthermore, because after cell destruction an increased number of viral particles is released, the infection/propagation/cell death process can continue in neighbor cells [43]. This approach has been tested in several phase I and phase II clinical trials as described in Ref. 43. In one phase II study, the intratumoural injection of Onyx-015 combined with chemotherapy resulted in an 83% tumor response in head and neck cancer patients [79]. The efficacy of oncolytic vectors can be further improved by resorting to transductional targeting, such has the genetic insertion of an integrin binding RGD-4C motif in the fiber proteins of the vector. This modification resulted in enhanced infectivity in ovarian cancer cells, which suggests improvements in clinical efficacy [33].

10.1.2.6 Safety Aspects

Adenovirus infection triggers both cellular and humoral responses. Viral proteins (either synthesized *de novo* or not) are processed by antigen-presenting cells and presented to CD8+ T cells (by means of MHC class I molecules) and CD4+ T-helper cells (by means of MHC class II molecules). This induces proliferation of cytotoxic T lymphocytes that specifically destroy the infected cells. CD4+ T-helper cells are also involved in the production of adenovirus-specific neutralizing antibodies directed toward the virus capsid [29]. The risks associated with this immunogenicity have been vividly demonstrated by the death of a patient in the 1999 OTC clinical trial. After the direct administration of a high dose of the adenoviral vector (3.8×10^{13} viral particles) to the liver, wide dissemination into the circulation triggered a massive activation of innate immunity followed by systemic inflammation that led to fever, intravascular coagulation, and multiorgan failure [19]. Responsibility for eliciting the immune response was attributed to viral capsid proteins rather than the transgene [19]. In subsequent years, a wealth of evidence accumulated confirming that AdVs may induce harmful immune and inflammatory responses, especially when large doses are administered systemically [19]. Adverse effects to adenovirus administration, however, may vary from individual to individual and depend on predisposing and underlying conditions [19].

The improvement of safety through minimization of toxicity and immunogenicity has been one of the major drivers in the development of recombinant AdVs. Nevertheless, even when third-generation "gutless" vectors are used, reduced immunogenicity due to the injected viral capsid proteins or to the transgene product is still observed [29, 55]. Thus, a need exists for the development of safe and effective methods capable of downregulating the host immune response against both adenoviral capsid proteins and transgene products. For instance, conventional immunosuppressive agents such as cyclosporine and FK506 may partially reduce immune responses and thus increase transduction and long-term transgene expression [46]. However, low doses of these agents must be used to prevent substantial organ toxicity. Improved results have been obtained by combining the use of immunomodulatory immunoglobulins and immunosuppressive agents. With this strategy, the immune response against adenovirus proteins and the transgene product dystrophin has been abrogated to a degree not achievable with the use of either agent alone [46].

Another problem associated with immunity to AdVs is the generation of memory cells. Upon subsequent administration of the same vector, the immune response is boosted by these cells, effectively reducing the efficacy of the repeated dosing [11].

10.1.3 ADENOVIRUS MANUFACTURING

The increasing number of adenoviral gene therapy applications that are moving from the laboratory to the clinic is creating a need for large amounts of highly purified recombinant AdVs. This demand is expected to increase as the first products reach the market. Thus, there is a clear need for a parallel development of efficient, scalable, and reproducible adenovirus manufacturing processes capable of delivering high amounts of infective AdV particles [80].

10.1.3.1   GMPs and Validation

Recombinant AdVs, such as all products that are to be administered to humans or animals, must be manufactured in accordance with a set of regulations issued by regulatory authorities such as the U.S. Food and Drug Administration (FDA) or the European Medicines Agency (EMEA) in the European Union. These regulations are known as current Good Manufacturing Practices (cGMPs) and cover all aspects of the production, from choosing and testing raw materials, to utilities, packaging, shipping, and transferring of final products to the clinic [32, 81]. If these items are not in conformity with GMPs, the product is deemed to be adulterated and cannot be legally approved [82]. In the specific case of AdVs, the facilities and processes used in manufacturing should be designed carefully to guarantee maximum protection to the product, to personnel, and to the environment [32, 83]. For instance, dedicated facilities should be used to prevent cross-contamination from previous or parallel batches of other bioproducts [32]. Confinement of production operations to class C clean rooms is also advisable to avoid dissemination of recombinant adenoviruses into the air [32].

One cornerstone provision of cGMPs is validation, a concept introduced to assure product consistency [84]. The validation of downstream processing operations aims to prove that they are capable of consistently removing impurities (e.g., host cell components, process-related materials, adventitious agents) to acceptable levels. Additionally, acceptance limits and operating ranges for each step must be determined [84]. Validation studies usually lead to optimized processes with reduced variability and, as a consequence, to a decrease in the number of failed batches. Thus, the development of adenovirus manufacturing processes (and associated facility) should be undertaken with validation in mind, not only to improve quality assurance and accelerate approval, but also to reduce costs [85, 86].

10.1.3.2   Product Specifications and Quality Control

One core concept hovering GMPs is quality assurance (QA). Among other attributes, QA is a means of guaranteeing the excellence, security, and dependability of the manufacturing process or its product [81]. QA is a key issue in process development, validation, and product approval, as well as on the assessment of the endproduct quality in comparison with product specifications. The characterization of recombinant AdVs thus constitutes a crucial aspect in all steps of product development, from basic research to clinical trials [87]. As it is the case for other biologics, adenovirus products have an inherent variability in their composition, stability, and potency. They are also subject to the variability inherent to the biological nature of some of the methods used to test them. An AdV should be well characterized to demonstrate that it is consistent in composition, exhibits long-term physico-chemical and biological stability, and is free of adventitious agents (microorganisms, adeno-associated viruses), contaminants, and impurities [88]. This requires the development and setup of a range of "validatable" analytical methodologies capable of fully characterizing the product during processing and in its final formulation, while ensuring the production of a consistent product [87]. The molecular structure of AdVs, hence their biological activity (potency), may be sensitive to factors such as temperature (labile for $T > 40°C$), pH (labile for pH < 6), and freeze/thawing cycles [89, 90]. Thus, a correct evaluation of product stability requires the use of convenient analytical techniques (e.g., circular dichroism and light scattering [89]). Analytical techniques also play an important role in the analysis of source materials, in the assessment of the impact of changes in manufacturing processes, and in the validation of processes and cleaning.

The exact final specifications (identity, efficacy, safety, potency, and purity) for an adenovirus vector product will usually depend on the intended therapeutic use, and thus, they are defined during clinical trials. However, regulatory agencies such as the FDA, the EMEA, or the SFDA provide guidelines and quality standards that are helpful during product and process development [91–93]. Table 10.1-3 exemplifies some of the assays, methods, and specifications used in the quality control of Gendicine (adapted from Refs. 15 and 93).

10.1.3.3   Environmental and Safety Issues

An assessment of the environmental impact of a process is an essential part of the design [94, 95]. The costs associated with the treatment and disposal of the waste generated by a particular process solution should be estimated beforehand. This information is one key element in the final decision on which process to select. Environmentally unfriendly operations such as those that generate large amounts

TABLE 10.1-3. Specifications and Recommended Assays for Assessing AdV Purity, Safety, and Potency (adapted and abbreviated from Refs. 15 and 93).

Parameter	Analytical Method	Specification
	Adenovirus-transgene	
Appearance	Opalescence	Visual inspection
pH	pH meter	8.0–8.5
Identity		
adenovirus genome	Restriction mapping	According to sequence
p53 gene	PCR	396 bp
Viral titer		
viral particles (VP)	Absorbance at 260 nm	$\geq 6.7 \times 10^{11}$ VP/mL
infectious units (IU)	TCID$_{50}$	$2\text{--}4 \times 10^{10}$ IU/mL
IU/VP	—	$\geq 3.3\%$
Purity	A260/A280 ratio	1.2–1.3
Potency		
expression	Western blot	Positive
bioactivity	Saos-2 cell bioassay	Positive
	Impurities	
RCA	A549 cell bioassay	≤ 1 RCA/3×10^{10} VP/mL
AAV	PCR	Negative
Proteins	ELISA	≤ 66 ng/mL
Genomic DNA	Southern blot	≤ 6.6 ng/mL
Endotoxins	*Limulus* ameobocyte lysate (LAL)	≤ 6.6 EU/mL

Abbreviations: TCID$_{50}$ = tissue culture infective dose; RCA = replication-competent adenovirus; AAV = adeno-associated virus.

of hazardous solvents and materials, which are usually costly to dispose of, should definitely be avoided. Additionally, adequate systems should be in place to provide an efficient decontamination of solid (e.g., autoclaving and incineration) and liquid wastes (e.g., heat or caustic inactivation) before release from the manufacturing area [32, 96, 97]. Once inactivated, these wastes should pose limited risk to the environment. Nevertheless, process optimization should always be geared toward minimization of waste generation [94, 95]. The formation of aerosols during processing should be minimized because adenovirus can spread via airborne transmission and contribute to dissemination within the manufacturing rooms and facility. Unit operations that need special safety precautions (e.g., explosion-proof tanks, blow-out walls, emission containment, personnel protection, etc.) to operate should also be carefully considered, because these requirements can dramatically increase the cost of equipment, building design, and construction. Overall, environmental and safety issues require an intimate knowledge of the process technology solutions available.

10.1.3.4   Process Considerations

A process for the manufacture of an AdV consists of several activities aimed at the production of a certain amount (measured as number of viral particle [VP] units or infectious viral particle [IVP] units) of the target product at an acceptable cost and quality (Figure 10.1-6). The generation of fully qualified cell banks and virus seeds is at the forefront of these activities [32, 83]. These banks should be extensively tested for contaminants, including RCAs [83]. A rigorous screening of raw materials, especially those of biological origin, is also mandatory to guarantee that extraneous organisms and contaminants are not introduced into the process [83]. Upstream and downstream processing unit operations are selected, arranged, designed, and operated to manufacture a bulk product. After filling and finishing, the product can be distributed and shipped. These activities should be GMP-compliant and thoroughly scrutinized under a quality control program, as discussed in the previous sections (Figure 10.1-6).

Flowsheets for the recovery and purification of biologicals are usually established with several rules of thumb and on the basis of accumulated experience with the target product [95]. Simulation tools can also be used to rapidly evaluate bioprocess alternatives and speed up development [95, 98–100]. The specifics of adenovirus purification were first addressed by researchers that have developed a range of

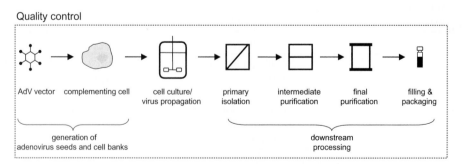

Figure 10.1-6. Outline of the activities involved in a typical AdV production process.

efficient lab-scale protocols, most of them based on CsCl density-gradient ultra-centrifugation [101]. Unfortunately, many of these protocols use reagents (e.g., CsCl is toxic) that are not acceptable for the manufacturing of a biological pharmaceutical. Furthermore, ultra-centrifugation is not amenable to scale-up due to the limited capacity of commercial ultra-centrifuges [41]. Nevertheless, many published adenovirus production processes include modifications or adaptations of specific steps used in these laboratory procedures [40, 101–107].

Cell culture is usually optimized to obtain high cell densities (cells/mL) and specific titers of infectious viral particles (IVPs/cell). The final goal is of course to maximize overall mass production (specific titer × cell density = IVP/mL) [80]. Next, a sequence of unit operations must be set up to recover the viral particles and eliminate host cell impurities (genomic DNA, RNA, proteins, RCAs, etc.) until the desired level of purity is met (Table 10.1-3). For a product with an intended use in humans, this removal of impurities is mandatory to avoid side effects upon administration to patients. The downstream processing unit operations can be grouped into three different stages: primary isolation, intermediate purification, and final purification (Figures 10.1-6 and 10.1-7). Ideally, the overall process should have a limited number of high-recovery steps, so that processing costs are reduced and acceptable yields are obtained [108]. The process should also use Generally Regarded As Safe (GRAS) reagents. Furthermore, lengthy operations and processes should be avoided to cut costs from overhead, amortization of equipment, and direct labor charges [109]. Processes with overall adenoviral particle yields of 32% [105], 50% [107], 60% [106], and 71% [104] have been reported in the literature.

10.1.3.5 Production

Most E1-deficient recombinant adenoviruses are propagated in the HEK-293 and PER.C6 cell lines (and derivatives thereof) as discussed in Section 10.1.2.3. The generation of RCAs is minimal in PER.C6 but remains a concern for HEK-293. Both cell lines have been documented for GMP manufacturing [110]. The production of these cells can be accomplished by a variety of methods, which depend on whether adherent or suspension cell lines are being used.

An adenoviral production process with HEK-293 and PER.C6 is a two-phase process. In the first phase, cultures are started at an appropriate seeding density (e.g., 0.3×10^6 cells/mL) and allowed to grow until a cell density around $0.5–1 \times 10^6$ cells/mL is reached. The second phase starts at this point with the infection of the culture at a ratio of virus titer to cell density (known as multiplicity of infection—MOI) adequately chosen (e.g., MOI = 10). An exchange of medium is usual, but not mandatory, before infection [111–113]. The cell density at the time of infection is a crucial parameter, with specific viral particle productivity (IVP/cell) usually dropping for cell densities higher than 0.5×10^6 cells/mL [80]. After infection, viral particles propagate and accumulate in the cells with maximum titers ($\sim 10^{10}–10^{11}$ VP/mL) typically obtained 48 hours postinfection (hpi). Maximum cell densities of 1×10^7 cells/mL have been reported [111]. Loss of viability and cell death ensue thereafter [80].

Adherent Cell Culture. Adherent or anchorage-dependent cell lines require attachment to a surface for their survival and replication. These cultures are labor

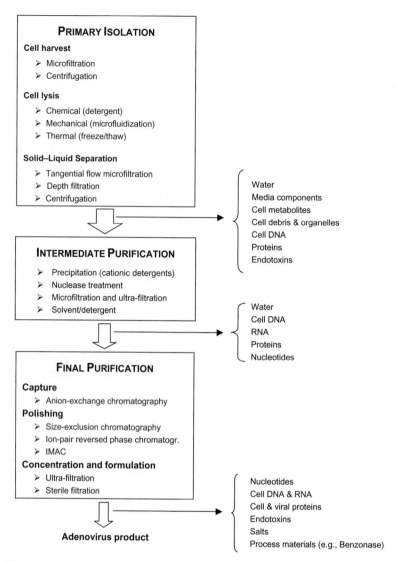

Figure 10.1-7. Generic block of downstream processing of AdV vectors showing unit operation options in each section.

intensive, require both a large amount of space and specific equipment handling, and have a limited potential for scale-up. Thus, they are more adequate during the early stages of product and process development [110, 114, 115].

Adherent cells can be grown in stationary or microcarrier cultures [116]. Stationary cultures use devices such as flasks, roller bottles, and cell factories that provide wall surface area for cell attachment [117]. Microcarrier cultures, on the other hand, can be implemented in agitated bioreactors [114, 116]. These microcarriers are small porous particles (0.2 mm) that provide large surface areas for cell attachment and can be suspended in the culture medium by gentle agitation. The culture environment is thus easily controlled and the scale-up is done by increasing the bioreactor volume. In comparative studies, stationary cultures usually yield higher adenovirus titers (up to 1.1×10^4 VP/cell) than microcarrier cultures (up to 8.5×10^3 VP/cell) [114, 116].

Suspension Cell Culture. Suspension cell lines are more convenient for large-scale manufacturing [111, 112]. A further advantage is that they can be grown in serum-free media, without the addition of bovine-derived components [110]. This advantage reduces costs and batch-to-batch variability and facilitates downstream processing and validation [111]. Conveniently, both the HEK-293 [80, 111–114] and the PER.C6 [118, 119] cell lines can be adapted to grow on suspension cultures. When grown in serum-free media, these cell lines allow high growth rates and cell densities as well as high yields of adenoviral particles [111]. The tendency of cell derivatives to aggregate, however, is usually a recurrent concern with suspension cultures, especially at high cell densities [111, 120].

The cultivation of cells and adenovirus propagation in suspension bioreactors has been performed in batch, fed-batch, and perfusion modes. In batch operation, no nutrients are added during the course of growth or infection. Although batch cultures are simple to perform and contamination risks are low, essential nutrients are rapidly depleted from the medium and cell viability decreases [110, 121].

The fed-batch mode is used to maintain cell viability after infection by supplementing limiting nutrients (e.g., glucose, glutamine, and amino acids) or reducing the accumulation of toxic metabolites (e.g., ammonia and lactate) [48, 80, 112, 113]. It also improves viral titers by increasing the cell density at which cells can be infected up to 2×10^6 cells/mL, without reducing the per-cell yield of product [121]. Both the medium composition as well as the feeding strategy are important parameters [112]. Nutrient depletion and the buildup of toxic metabolites cannot be completely alleviated by fed-batch cultures [121]. Operation under perfusion mode can overcome this limitation by ensuring a constant renewal of medium in the bioreactor. Briefly, cells are retained at high concentrations inside the bioreactor by using devices such as hollow fibers and membrane units, whereas fresh medium is continuously supplemented [121]. Consequently, the cell density at which cells can be infected can be increased. With this strategy, HEK-293 cell densities of 8×10^6 cells/mL could be achieved at the time of infection, and viral titers of 7.8×10^9 IVP/mL were obtained at 60 hpi [121].

10.1.3.6   Downstream Processing

Primary Isolation and Intermediate Purification. The suspension generated during cell culture constitutes the starting point for downstream processing. In the primary isolation stage, cells are harvested and viral particles are released from the cells together with other impurities. A significant reduction in volume occurs, and the most plentiful impurities such as extracellular liquid, proteins, genomic DNA, and cell debris are removed. Cells are typically harvested by centrifugation or microfiltration. The resulting cell slurry is then resuspended in an adequate volume (1/10 to 1/20 of the culture volume) of a suitable buffer or of the culture medium itself. The subsequent cell lysis operation is typically performed using freeze/thaw cycles [40, 106, 118, 119], osmotic shock [80], sonication [101, 122], microfluidization [102], or by the addition of detergents such as Triton X-100 or

Tween-20 [104, 107]. Unlike lysis by freeze/thawing and sonication, cell shearing by microfluidization is more rapid, reproducible, and easy to scale-up [102].

After lysis, cleared lysates are obtained by removing cell debris and larger unruptured organelles with centrifugation [40, 118, 119], depth filtration [107], or tangential flow microfiltration [102, 106] operations. Treatment with nucleases (e. g., Benzonase [Merck, Gibbstown, NJ], Pulmozyme [Roche, Basel, Switzerland], DNase, RNase T1, and RNase I) is usually performed before or after clarification, to reduce the cellular DNA and RNA load [40, 80, 104]. This step not only improves purity, hence the safety of the viral product [40], but also it reduces agglomeration of viral particles, which is usually induced by adhesion of nucleic acids [104]. In some cases, nuclease treatment can be carried out at the end of the process, after chromatographic purification [106].

As an alternative to nuclease digestion, cationic detergents can be added during lysis to selectively precipitate cellular DNA. Recent experiments with the detergent domiphen bromide have shown that it is possible to obtain three logs of DNA clearance without losses in the infectivity of purified viral particles [107]. Thus, the use of DNA removal operations such as nuclease treatment and anion exchange chromatography may be eliminated or reduced.

The viral particle-containing solutions resulting from nuclease treatment and cationic detergent precipitation are typically filtered, concentrated, and conditioned before chromatographic purification [40, 80, 101, 106, 107]. The inclusion of a solvent/detergent step at this stage may be included to inactivate potential enveloped viruses that could have been coamplified [32].

Final Purification. A pharmaceutical product such as an AdV requires a high degree of purity. This level of purity is usually achieved with a combination of chromatography and filtration operations. The goal is to separate the viral particles from the most recalcitrant impurities that persist in the streams. The chromatographic operations described in the literature for virus purification explore properties such as size, charge, hydrophobicity, and metal affinity [41, 80, 102–107].

Anion-exchange (AEX) chromatography is widely used as a viral particle capture step. It explores the interaction between the negatively charged hexons in viral capsids and the stationary phases bearing positively charged ligands such as quaternary amines [41]. Impurities (media components, low-molecular-weight DNA, penton, hexon, and fiber proteins are eluted at low salt (<0.25-M NaCl), whereas bound capsids are displaced with a salt gradient [41, 102, 103, 106]. A recent study shows that the NaCl concentration required to elute viral capsids from the anion exchange column is a function of the serotype being purified (ranging from 0.27 to 0.45 M), and it correlates very well to the electrostatic properties of the hexon protein in each serotype [41]. Tentacular (e.g., DEAE-Fractogel; Merck, Gibbstown, NJ), perfusion (e.g., POROS.RTM.50D; PerSeptive Biosystems, Framingham, MA), and soft gel in rigid shell (e.g., Ceramic HyperD.TM.F; BioSepra, Villeneuve-La-Garenne, France) materials have all been used for AEX purification of adenovirus particles [102, 106]. The reported AEX viral particle yields range from 63% [104] to 80% [80].

AEX can also be performed in expanded-bed adsorption (EBA) mode [102]. Under this mode of operation, cell lysates can be applied directly to the column from below. Large debris and unlysed cells that can move freely around the anion-exchanger beads eventually leave through the top of the column, whereas adenoviral particles bind to the anion-exchange matrix. Extensive washing from below limits nonspecific interactions between the particulates and the resin. Finally, the flow is reversed, the anion-exchanger beads are allowed to pack, and the viral particles are eluted under a salt gradient [102]. Due to its early use in the downstream processing, EBA can be considered as an intermediate purification unit operation.

Adenovirus particles have also been purified by affinity chromatography by taking advantage of an ingenious targeting approach. Briefly, the fiber capsid protein of adenovirus 5 vectors was genetically fused to a biotin acceptor peptide (BAP). These BAP-modified fibers were then metabolically biotinylated during virus propagation in the HEK-293 cell line. The resulting covalently biotinylated viral particles could then be purified from crude lysates by avidin-affinity chromatography [122].

AEX alone is unlikely to be sufficient to produce an adenovirus product with the required purity [106]. The inclusion of a chromatographic polishing step may thus be necessary as a means to remove traces of DNA and protein impurities. Size-exclusion [80, 103], ion-pair reversed phase [106], and metal affinity [105] chromatography have all been used toward this end. In the case of size-exclusion, if an adequate matrix is selected (e.g., Superdex 200; GE Healthcare, Piscataway, NJ), virus elute in the flowthrough because of its large size, whereas low-molecular-weight impurities are retarded. The column loading volume can be increased up to 20% of the bed volume, because a group separation is achieved [102]. An extra advantage of size exclusion as a polishing step is that it can also be used to exchange buffer [103]. This enables formulation to be carried out simultaneously with polishing. In the case of ion-pair reverse phase chromatography, elution conditions are carefully selected to retain impurities in a PolyFlo matrix (Puresyn, Inc., Malvern, PA) and to allow the viral particles not to bind. Under this negative chromatography operation mode, viral particles are recovered in the flowthrough with an 87% recovery yield [106]. This leads to some dilution of the adenovirus stream, as is always the case when products are recovered in the flowthrough.

Processes for adenovirus purification typically end with concentration, formulation, and sterile filtration operations [40, 80, 106]. Concentration and formulation are usually carried out in ultra-filtration units equipped with 100–300-kDA membranes [40, 106]. The exact composition of the formulation buffer will depend on the intended application, mode of administration (injectable, aerosol), and required short-term and shelf stability [104, 123]. A typical liquid formulation may include an aqueous buffer supplemented with cryoprotectants (e.g., sucrose) and stabilizers such as the nonionic-surfactant polysorbate-80, the chelating agent EDTA, and the oxidation inhibitors ethanol and histidine [123]. Filtration under sterile conditions is typically performed with 0.22-μm membranes [103, 106].

10.1.4 ANALYSIS AND EVALUATION OF AN ADENOVIRUS PRODUCTION PROCESS

In this section, a pilot-scale process that has been developed specifically for the production and purification of AdVs for gene therapy [80] is analyzed and evaluated for large-scale manufacturing with the use of the process simulator software SuperPro Designer (Intelligen, Inc., Scotch Plains, NJ). The major objective of the analysis presented here is to estimate the cost of the production of an adenovirus therapeutic product. Results further provide insights into process weaknesses and strengths, effectively directing bioprocess engineers toward better processes.

10.1.4.1 Process Description

As a design basis we have assumed a plant capacity of around 1×10^{18} viral particles of purified recombinant adenovirus produced per year. This amount of product is sufficient to treat 125,000 patients per year on the basis of 1×10^{12} VP doses given weekly for a total of 8 weeks, as described in a Gendicine clinical trial [15]. The plant is designed to operate 330 days a year, with a new batch initiated every 11 days—this corresponds to 30 batches at 33.3×10^{15} VP/batch. The total batch time is around 17 days. Guidelines and quality standards issued by regulatory agencies have been used to set up product specifications in terms of final purity (Table 10.1-3). Finally, we have assumed that, at the end of the process, the bulk adenoviral product will be distributed in vials, each containing a 1×10^{12} VP dose of 0.6 mL of sterile formulation buffer. This corresponds to a viral titer of 167×10^{11} VP/mL, which is higher that the specified value of 6.7×10^{11} VP/mL (Table 10.1-3).

The entire flowsheet for the production of adenovirus is divided into cell culture and virus propagation (Figure 10.1-8) and downstream processing (Figure 10.1-9)

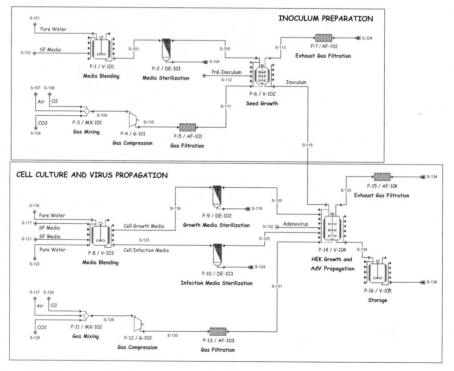

Figure 10.1-8. Adenoviral vector production flowsheet showing the cell culture and virus propagation section (implemented with the process simulator software SuperPro Designer [Intelligen, Inc., Scotch Plains, NJ]).

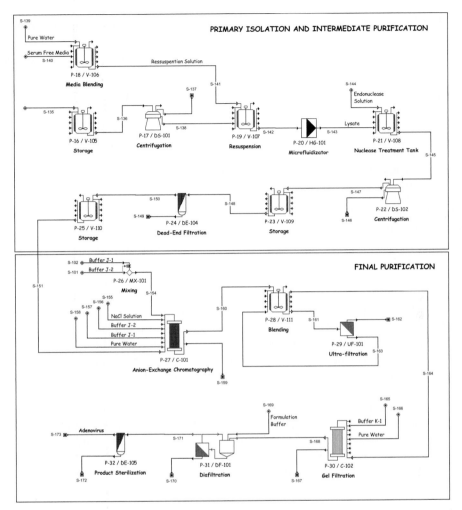

Figure 10.1-9. Adenoviral vector production flowsheet showing downstream processing section (implemented with the process simulator software SuperPro Designer [Intelligen, Inc., Scotch Plains, NJ]).

sections. The overall adenovirus particle recovery yield per batch is around 67% (33.3×10^{15} VP are recovered out of the 5×10^{16} VP that are present in the whole cell lysate). This figure is within yields reported in the literature for similar processes [104]. Input data used in the simulation software SuperPro Designer were taken from the reference publication [80] and supplemented with information from other published processes [40, 106, 107]. Educated guesses were made to provide for missing data. Table 10.1-4 summarizes the most important data—process volumes, step yields, and viral particle titers.

Cell Culture Section. The activities in this section include media sterilization, inoculum preparation, cell growth, and virus propagation (Figure 10.1-8). Serum-free (SF) medium is sterilized by 0.2-μm filtration and used for both inoculum preparation and cell growth/virus propagation. The culture is assumed to take

TABLE 10.1-4. Summary of Input Data Used in the Bioprocess Simulator SuperPro Designer

Step	Volume (L)	AdV Titer ($\times 10^{10}$ VP/mL)	Step Yield (%)
Cell culture	1000	5.0	—
Harvest + resuspension	109	46	100
Cell Lysis	109	45	98
Benzonase digestion	109	45	100
Centrifugation + dead-end filtration	106	46	99
Anion exchange chromatography	118	36.5	89
Ultrafiltration	11.8	350.0	96
Size exclusion chromatography	59.0	66.5	95
Ultrafiltration	19.7	169.3	85
Sterile filtration	19.7	169.3	100

Notes: An overall yield of 67% is obtained. The final viral titer is higher than the specified value (67 $\times 10^{10}$ VP/mL).

place in fed-batch mode (with a single nutrient addition), in a bioreactor with a working volume of 1000 L. This size is a reasonable assumption, given that cultures with volumes as high as 10,000 L are currently being developed for production of HIV adenoviral vaccines [118]. The inoculum is added to 500 L of culture medium at a seeding density of 3×10^5 cell/mL, and cells are allowed to grow for 6 days. Then, 500 L of fresh SF medium are added, and cells at a concentration of 0.5×10^6 cell/mL are infected at a MOI of 10. Cell growth and virus propagation are conducted at 37°C and pH 7.2, with constant addition of a gaseous mixture (80% air, 10% CO_2, 10% O_2) at 0.05 vvm [80]. Harvest is performed 48 hpi, with typical values assumed for the final cell concentration (1×10^6 cells/mL), adenovirus titer (5×10^4 VP/cell), and ratio of infectious viral particle to total viral particles (1:10) [80].

Downstream Processing Section. The activities in this section include cell harvest and cell lysis, nuclease digestion, and chromatographic purifications (Figure 10.1-9). Cells from the suspension culture (1 m^3) are harvested by continuous centrifugation, resuspended in SF medium (1/10 of the original volume), and lysed by microfluidization (e.g., 2000 psi and 1 pass) [80]. Cell DNA is then drastically reduced ($\approx$4.5 logs) by adding Benzonase up to a final concentration of 150 IU/mL [107]. After removal of debris by centrifugation, the supernatant is clarified in a 0.2-μm dead-end filtration unit [80]. Adenovirus capture and purification is performed by AEX chromatography column (Fractogel EMD-DEAE, 0.4-m bed height, 0.6-m bed diameter) according to the instructions presented by Tang et al. [103]. Briefly, each cycle comprises six distinct operations: (1) equilibration with 22 bed volumes (BVs) of buffer J1 (50-mM sodium phosphate pH 7.5, 265-mM NaCl, 2-mM $MgCl_2$, 2% (w/v) sucrose) at 4 cm/min, (2) loading of 106 L of feed at 1 cm/min, (3) washing with 4 BVs of buffer J1 at 2 cm/min, (4) washing with 8 BVs of 94% buffer J1 and 6% buffer J2 (50-mM sodium phosphate pH 7.5, 600-mM NaCl, 2-mM $MgCl_2$, 2% (w/v) sucrose), (5) elution of bound adenovirus particles with 10 BVs of linear gradient from 6% to 100% buffer J2 at 2 cm/min, and (6) column cleaning with 4 BVs of 1-M NaCl at 4 cm/min. The total cycle time is 12.26 h. The bound viral particles are eluted with an overall yield of 89% [103]. The adenovirus pool is then concentrated 10 times by ultra-filtration using a 100-kDa membrane [80]. A step yield of 96% is assumed based on data published for ultra-filtration of different viral particles [124]. A size exclusion chromatography column (Superdex 200, 0.8-m bed height, 0.4-m bed diameter) is included as a polishing step (95% step yield) [103]. The column is loaded with the totality of the viral solution (12% of the bed volume), and upon elution with an adequate buffer (e.g., 20-mM sodium phosphate pH 8.0, 100-mM NaCl, 2-mM $MgCl_2$, 2% (w/v) at 0.43 cm/min), a pool of adenoviral particles is obtained that is five times diluted relatively to the feed [103]. The total cycle time is 18.3 h. Concentration and formulation into an adequate buffer (2.5% glycerol, 25-mM NaCl, 20-mM Tris, pH 8 [80]) is subsequently carried out in an ultra-filtration unit equipped with a 100-kDA membrane. Finally, the adenovirus product is sterile filtered into glass vials (1×10^{12} VP in 0.6 mL).

Process Scheduling. The scheduling and equipment utilization for one production batch is shown in Figure 10.1-10. The plant batch time is approximately 405 hours, with a new batch (i.e., inoculum preparation) initiated every 264 hours. This batch start time roughly corresponds to the fourth day of the cell culture in the previous batch. The inoculum preparation, cell culture, and virus propagation procedures, with a duration of approximately 367 hours, are clearly identified in the chart as the time bottleneck. Comparatively, the downstream processing is complete within a mere 38 hours. The time bottleneck in this section is the size-exclusion chromatography polishing step, which takes about 18 hours.

10.1.4.2 Inventory Analysis

The overall material balance per year is summarized in Table 10.1-5. Remarkably, the annual mass production of purified adenovirus vector constitutes a minor fraction (~0.3 ppm) of the total amount of materials required for its production. Apart from the adenovirus vector, cell debris, and gases, all output materials end up in liquid waste streams, which are disposed of after adequate treatment (e.g., heat or caustic inactivation) to minimize environmental impacts. Water is the major raw material used (~97%), most of it for equipment cleaning. This is typical in the production of biopharmaceuticals as can be seen by checking IgG [95], recombinant β-glucuronidase [125], and plasmid DNA [100] production examples. Serum-free medium (~0.2%) and gases (~1%) are used in the cell culture and virus propagation step. Large amounts of sodium chloride, sodium phosphate, and sucrose (~0.8%) are required as buffer components in the chromatographic operations. A substantial amount of sodium hydroxide (~0.5%) is also used for equipment cleaning.

10.1.4.3 Cost Analysis and Economic Assessment

Table 10.1-6 shows the key economic evaluation results for this project. Economic evaluations were based on the following assumptions: (1) the entire direct fixed capital is depreciated linearly over a period of 10 years assuming a 10% salvage

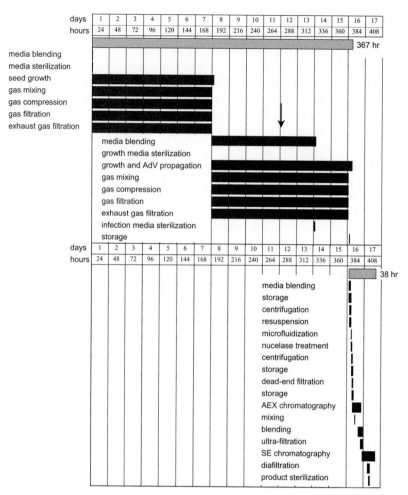

Figure 10.1-10. Gantt chart for production scheduling of one batch of adenoviral vector. The top and bottom charts show the duration of each operation in the cell culture and downstream processing sections, respectively. The total process time is 405 hours. The arrow indicates the start of a new batch at 264 hours.

value for the entire plant, (2) the project lifetime is 15 years, and (3) 1×10^{18} VP of adenoviral product will be produced per year. For a plant of this capacity, the total capital investment is around $17.8 million. The unit production cost is ~$7.2/dose ($10^{12}$ VP) or $24,715/g. This figure is considerably higher when compared with production costs of other biopharmaceuticals such as β-glucuronidase ($43.0/g [125]), insulin ($42.2/g [95]), IgG ($908/g [95]), and plasmid DNA ($375/g, [100]).

The total equipment purchase cost was estimated to be around $2.8 million. The most expensive piece of equipment is the bioreactor used for cell culture and virus propagation, priced at $506,000. The cost of unlisted equipment (including the equipment used in the inoculum preparation section) was assumed to represent 20% of the total equipment cost.

TABLE 10.1-5. Overall Material Balances for the Annual Production of 1×10^{18} VP (~0.285 kg) of AdV (kg/year). This Amount Corresponds to 67% of the Total AdV Propagated in the Culture Section.

Component	Total Inlet	Total Outlet
Adenovirus	0	0.425
Carbon dioxide	1,050	1,077
Debris	0	17
DNase	0.095	0.095
Glycerol	91	91
Magnesium chloride	30	30
Nitrogen	6,444	6,444
Nucleic acids	0	0
Nucleotides	0	9
Oxygen	3,006	2,995
Proteins	0	81
SF media	1,628	1,498
Sodium chloride	3,582	3,582
Sodium hydroxide	5,013	5,013
Sodium phosphate	1,066	1,066
Sucrose	3,171	3,171
Tris base	11	11
Process water	734,446	734,446
Water for injection	210,860	210,860
Total	970,398	970,392

TABLE 10.1-6. Key Economic Evaluation Results for Adenovirus Production

Direct fixed capital	$16,593,000
Total capital investment	$17,820,000
Plant throughput	0.29 kg (10^{18} VP) AdV/year
Operating cost	$7,149,000/year
Unit production cost	$7.17/dose

The breakdown of the annual operating cost (AOC) is shown in Figure 10.1-11. Facility-dependent cost (44% of the AOC) is the principal operating cost in this process, as is typical for high-value products that are produced in small quantities [95]. Labor-related costs come next, accounting for 34% of the AOC. The annual cost of raw materials is around $1.0 million, 92% of which are associated with the culture SF medium (priced at $28/L). The cost of consumables is around $193,000, representing approximately 3% of the AOC. The chromatographic resins in the AEX and size-exclusion chromatographic steps represent 84% of this value. The cost of utilities (electricity, steam, and cooling agents) is minimal, representing less than 0.1% of the AOC.

Figure 10.1-12 shows the return on investment (ROI) and payback time for different product selling prices ($/dose). The ROI increases around 3.5% for every $1 increase in the selling price of each AdV dose, whereas the payback time declines for every $1 increase. If one assumes that a dose of product is sold at $30,

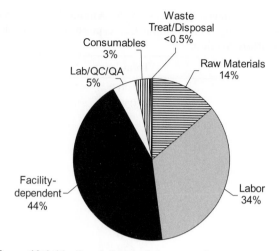

Figure 10.1-11. Breakdown of the annual operating cost.

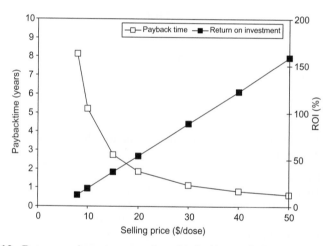

Figure 10.1-12. Return on investment and payback time at different AdV selling prices.

the annual revenue will amount to $30,662,000 for an annual production of 1×10^{18} VP. This corresponds to a gross profit (i.e., revenue-operating cost) of $23,513,000. At this selling price, the payback time is 1.14 years with an ROI of 88%. This economic picture could change slightly by the inclusion of costs that were not accounted for in this case study (filling & packaging section, R&D, process validation).

10.1.5   CONCLUDING REMARKS AND OUTLOOK

Adenoviral gene therapy has matured to the point where several products should be hitting the market in the wake of Gendicine. Many of the initial drawbacks and barriers (e.g., immunogenicity, lack of specificity, and transient expression of transgenes) have been tackled with innovative solutions and strategies (targeting, gutless vectors) that have proved valuable in some cases. To provide sufficient material for the ongoing gene therapy clinical trials, production and purification processes must be developed and optimized in parallel to deliver the required selectivity, yield, efficiency, and productivity. The design of these processes under cGMPs should be based on a profound knowledge of the target molecule and associated impurities and on the performance of the available unit operations. Although several well-established processes have been documented in the literature, new developments are likely to surface within the coming years because of increased investments from academia and industry in the area. The analysis presented in this chapter indicates that the production and purification of a therapeutic adenovirus product is economically viable, even with a suboptimized process. Process improvements will certainly reduce costs. As an educated guess, we may estimate selling prices to be in the range $20–$50/10^{12} VP.

REFERENCES

1. Davies J C, Geddes D M, Alton E W (2001). Gene therapy for cystic fibrosis. *J. Gene Med.* 3:409–417.

2. McCormick F (2001). Cancer gene therapy: Fringe or cutting edge? *Nat. Rev. Cancer.* 1:130–141.

3. Strayer D S, Akkina R, Bunnell B A, et al. (2005). Current status of gene therapy strategies to treat HIV/AIDS. *Mol. Ther.* 11:823–842.

4. Donnelly J J, Wahren B, Liu M A (2005). DNA vaccines: Progress and challenges. *J. Immunol.* 175:633–639.

5. Tuteja R (2002). DNA vaccine against malaria: A long way to go. *Crit. Rev. Biochem. Molec. Biol.* 37:29–54.

6. Leitner W W, Thalhamer J (2003). DNA vaccines for non-infectious diseases: New treatments for tumour and allergy. *Expert Opin. Biol. Ther.* 3:627–638.

7. Anderson W (1997). Human gene therapy: The initial concepts. In K Brigham (ed.), *Gene Therapy for Diseases of the Lung*: Marcel Dekker Inc, New York, pp. 3–16.

8. Friedmann T (2001). Stanfield Rogers: Insights into virus vectors and failure of an early gene therapy model. *Molec. Ther.* 4:285–288.

9. Blaese R M, Culver K W, Miller A D, et al. (1995). T lymphocyte-directed gene therapy for ADA-SCID: Initial trial results after 4 years. *Science.* 270:475–480.

10. Zabner J, Couture L A, Gregory R J, et al. (1993). Adenovirus-mediated gene transfer transiently corrects the chloride transport defect in nasal epithelia of patients with cystic fibrosis. *Cell.* 75:207–216.

11. Somia N, Verma I M (2000). Gene therapy: Trials and tribulations. *Nature Rev. Genet.* 1:91–99.

12. Teichler Zallen D (2000). US gene therapy in crisis. *Trends. Genet.* 16:272–275.

13. Cavazzana-Calvo M, Hacein-Bey S, de Saint Basile G, et al. (2000). Gene therapy of human severe combined immunodeficiency (SCID)-X1 disease. *Science.* 288:669–672.

14. Hacein-Bey-Abina S, Von Kalle C, Schmidt M, et al. (2003). LMO2-associated clonal T cell proliferation in two patients after gene therapy for SCID-X1. *Science.* 302:415–419.

15. Peng Z (2005). Current status of gendicine in China: Recombinant human Ad-p53 agent for treatment of cancers. *Human Gene Ther.* 16:1016–1027.

16. Surendran A (2004). News in brief: China approves world's first gene therapy drug. *Nat. Med.* 10:9.

17. Rolland A (2005). Gene medicines: The end of the beginning? *Adv. Drug Del. Rev.* 57:669–673.

18. Wilson J M (2005). Gendicine: The first commercial gene therapy product. *Human Gene Ther.* 16:1014–1015.

19. Thomas C E, Ehrhardt A, Kay M A (2003). Progress and problems with the use of viral vectors for gene therapy. *Nat. Rev. Genet.* 4:346–358.

20. Li S, Huang L (2000). Nonviral gene therapy: Promises and challenges. *Gene Ther.* 7:31–34.

21. Ledley F D (1995). Nonviral gene therapy: The promise of genes as pharmaceutical products. *Human Gene Ther.* 6:1129–1144.

22. Prazeres, D M F, Ferreira, G N M, Monteiro G A, et al. (1999). Large-scale production of pharmaceutical-grade plasmid DNA for gene therapy: Problems and bottlenecks. *Trends Biotechnol.* 17:169–174.

23. Robbins P D, Tahara H, Ghivizzani S C (1998). Viral vectors for gene therapy. *Trends Biotechnol.* 16:35–40.

24. Rowe W P, Huebner R J, Gilmore L K, et al. (1953). Isolation of a cytopathogenic agent from human adenoids undergoing spontaneous degeneration in tissue culture. *Proc. Soc. Exper. Biol. Med.* 84:570–573.

25. Wolff G (2004). Adenovirus for Gene Therapy. In T Blankenstein (ed.), *Gene Therapy: Principles and Applications*: Birkhäuser Verlag, Basel, Switzerland, pp. 29–45.

26. McConnell M J, Imperiale M J (2004). Biology of adenovirus and its use as a vector for gene therapy. *Human Gene Ther.* 15:1022–1033.

27. Russell W C (2000). Update on adenovirus and its vectors. *J. Gen. Virol.* 81:2573–2604.

28. Muruve D A (2004). The innate immune response to adenovirus vectors. *Human Gene Ther.* 15:1157–1166.

29. Schagen F H, Ossevoort M, Toes R E, et al. (2004). Immune responses against adenoviral vectors and their transgene products: A review of strategies for evasion. *Crit. Rev. Oncol. Hematol.* 50:51–70.

30. Nicklin S A, Wu E, Nemerow G R, et al. (2005). The influence of adenovirus fiber structure and function on vector development for gene therapy. *Molec. Ther.* 12:384–393.

31. Fabry C M S, Rosa-Calatrava M, Conway J F, et al. (2005). A quasi-atomic model of human adenovirus type 5 capsid. *EMBO J.* 24:1645–1654.

32. Lusky M (2005). Good manufacturing practice production of adenoviral vectors for clinical trials. *Human Gene Ther.* 16:281–291.

33. Bauerschmitz G J, Barker S D, Hemminki A (2002). Adenoviral gene therapy for cancer: from vectors to targeted and replication competent agents (review). *Internat. J. Oncol.* 21:1161–1174.

34. Seshidhar Reddy P, Ganesh S, Limbach M P, et al. (2003). Development of adenovirus serotype 35 as a gene transfer vector. *Virol.* 311:384–393.

35. Lemckert A A, Sumida S M, Holterman L, et al. (2005). Immunogenicity of heterologous prime-boost regimens involving recombinant adenovirus serotype 11 (Ad11) and Ad35 vaccine vectors in the presence of anti-ad5 immunity. *J. Virol.* 79:9694–9701.

36. Holterman L, Vogels R, van der Vlugt R, et al. (2004). Novel replication-incompetent vector derived from adenovirus type 11 (Ad11) for vaccination and gene therapy: Low seroprevalence and non-cross-reactivity with Ad5. *J. Virol.* 78:13207–13215.

37. Regenmortel M H V V, Fauquet C M, Bishop D H L (eds.) (2001). *Virus Taxonomy: Seventh Report of the International Committee on Taxonomy of Viruses*, Academic Press Inc, London.

38. Volpers C, Kochanek S (2004). Adenoviral vectors for gene transfer and therapy. *J. Gene Med.* 6(suppl 1):S164–171.

39. ICTVdB Management C (2006). 00.001 Adenoviridae. In C Büchen-Osmond (ed.), *ICTVdB—The Universal Virus Database*, version 3, Columbia University, New York.

40. Konz J O, Lee A L, Lewis J A, et al. (2005). Development of a purification process for adenovirus: controlling virus aggregation to improve the clearance of host cell DNA. *Biotechnol. Progress.* 21:466–472.

41. Konz J O, Livingood R C, Bett A J, et al. (2005). Serotype specificity of adenovirus purification using anion-exchange chromatography. *Human Gene Ther.* 16:1–8.

42. van Oostrum J, Burnett R M (1985). Molecular composition of the adenovirus type 2 virion. *J. Virol.* 56:439–448.

43. Vecil G G, Lang F F (2003). Clinical trials of adenoviruses in brain tumors: A review of Ad-p53 and oncolytic adenoviruses. *J. Neurooncol.* 65:237–246.

44. Graham F L, Smiley J, Russell W C, et al. (1977). Characteristics of a human cell line transformed with by DNA from human adenovirus type 5. *J. Gen. Virol.* 36:59–74.

45. Fallaux F J, Bout A, van der Velde I, et al. (1998). New helper cells and matched early region 1-deleted adenovirus vectors prevent generation of replication-competent adenoviruses. *Human Gene Ther.* 9:1909–1917.

46. Guibinga G H, Lochmuller H, Massie B, et al. (1998). Combinatorial blockade of calcineurin and CD28 signaling facilitates primary and secondary therapeutic gene transfer by adenovirus vectors in dystrophic (mdx) mouse muscles. *J. Virol.* 72: 4601–4609.

47. Lusky M, Christ M, Rittner K, et al. (1998). In vitro and in vivo biology of recombinant adenovirus vectors with E1, E1/E2A, or E1/E4 deleted. *J. Virol.* 72:2022–2032.

48. Nadeau I, Sabatie J, Koehl M, et al. (2000). Human 293 cell metabolism in low glutamine-supplied culture: Interpretation of metabolic changes through metabolic flux analysis. *Metabol. Engineer.* 2:277–292.

49. Dedieu J F, Vigne E, Torrent C, et al. (1997). Long-term gene delivery into the livers of immunocompetent mice with E1/E4-defective adenoviruses. *J. Virol.* 71:4626–4637.

50. Engelhardt J F, Ye X, Doranz B, et al. (1994). Ablation of E2A in recombinant adenoviruses improves transgene persistence and decreases inflammatory response in mouse liver. *Proc. Nat. Acad. Sci. U S A.* 91:6196–6200.

51. Gao G P, Yang Y, Wilson J M (1996). Biology of adenovirus vectors with E1 and E4 deletions for liver-directed gene therapy. *J. Virol.* 70:8934–8943.

52. Sandig V, Youil R, Bett A J, et al. (2000). Optimization of the helper-dependent adenovirus system for production and potency in vivo. *Proc. Nat. Acad. Sci. U S A.* 97:1002–1007.

53. Parks R J, Chen L, Anton M, et al. (1996). A helper-dependent adenovirus vector system: Removal of helper virus by Cre-mediated excision of the viral packaging signal. *Proc. Nat. Acad. Sci. U S A.* 93:13565–13570.

54. Parks R J, Graham F L (1997). A helper-dependent system for adenovirus vector production helps define a lower limit for efficient DNA packaging. *J. Virol.* 71:3293–3298.

55. Chen H H, Mack L M, Kelly R, et al. (1997). Persistence in muscle of an adenoviral vector that lacks all viral genes. *Proc. Nat. Acad. Sci. U S A.* 94:1645–1650.

56. Kirn D, Martuza R L, Zwiebel J (2001). Replication-selective virotherapy for cancer: Biological principles, risk management and future directions. *Nat. Med.* 7:781–787.

57. Ghosh D, Barry M A (2005). Selection of muscle-binding peptides from context-specific peptide-presenting phage libraries for adenoviral vector targeting. *J. Virol.* 79:13667–13672.

58. Volpers C, Thirion C, Biermann V, et al. (2003). Antibody-mediated targeting of an adenovirus vector modified to contain a synthetic immunoglobulin g-binding domain in the capsid. *J. Virol.* 77:2093–2104.

59. Stevenson S C, Rollence M, Marshall-Neff J, et al. (1997). Selective targeting of human cells by a chimeric adenovirus vector containing a modified fiber protein. *J. Virol.* 71:4782–4790.

60. Campos S K, Parrott M B, Barry M A (2004). Avidin-based targeting and purification of a protein IX-modified, metabolically biotinylated adenoviral vector. *Molec. Ther.* 9:942–954.

61. Eto Y, Gao J Q, Sekiguchi F, et al. (2005). PEGylated adenovirus vectors containing RGD peptides on the tip of PEG show high transduction efficiency and antibody evasion ability. *J. Gene Med.* 7:604–612.

62. Henning P, Andersson K M, Frykholm K, et al. (2005). Tumor cell targeted gene delivery by adenovirus 5 vectors carrying knobless fibers with antibody-binding domains. *Gene Ther.* 12:211–224.

63. Nicklin S A, Von Seggern D J, Work L M, et al. (2001). Ablating adenovirus type 5 fiber-CAR binding and HI loop insertion of the SIGYPLP peptide generate an endothelial cell-selective adenovirus. *Molec. Ther.* 4:534–542.

64. van Beusechem V W, van Rijswijk A L, van Es H H, et al. (2000). Recombinant adenovirus vectors with knobless fibers for targeted gene transfer. *Gene Ther.* 7:1940–1946.

65. Yamamoto M, Alemany R, Adachi Y, et al. (2001). Characterization of the cyclooxygenase-2 promoter in an adenoviral vector and its application for the mitigation of toxicity in suicide gene therapy of gastrointestinal cancers. *Molec. Ther.* 3:385–394.

66. St George J A (2003). Gene therapy progress and prospects: Adenoviral vectors. *Gene Ther.* 10:1135–1141.

67. Matsuse T, Teramoto S (2000). Progress in adenovirus-mediated gene therapy for cystic fibrosis lung disease. *Curr. Therapeu. Res.* 61:422–434.

68. Harvey B G, Leopold P L, Hackett N R, et al. (1999). Airway epithelial CFTR mRNA expression in cystic fibrosis patients after repetitive administration of a recombinant adenovirus. *J. Clin. Invest.* 104:1245–1255.

69. van Deutekom J C, van Ommen G J (2003). Advances in Duchenne muscular dystrophy gene therapy. *Nat. Rev. Genet.* 4:774–783.

70. Chamberlain J S (2002). Gene therapy of muscular dystrophy. *Human Molec. Genet.* 11:2355–2362.

71. Lang F F, Bruner J M, Fuller G N, et al. (2003). Phase I trial of adenovirus-mediated p53 gene therapy for recurrent glioma: biological and clinical results. *J. Clin. Oncol.* 21:2508–2518.

72. Moolten F L (1986). Tumor chemosensitivity conferred by inserted herpes thymidine kinase genes: Paradigm for a prospective cancer control strategy. *Cancer Res.* 46:5276–5281.

73. Tsuchiyama T, Kaneko S, Nakamoto Y, et al. (2003). Enhanced antitumor effects of a bicistronic adenovirus vector expressing both herpes simplex virus thymidine kinase

and monocyte chemoattractant protein-1 against hepatocellular carcinoma. *Cancer Gene Ther.* 10:260–269.

74. Lanuti M, Gao G P, Force S D, et al. (1999). Evaluation of an E1E4-deleted adenovirus expressing the herpes simplex thymidine kinase suicide gene in cancer gene therapy. *Human Gene Ther.* 10:463–475.

75. Mizuguchi H, Hayakawa T (2002). Enhanced antitumor effect and reduced vector dissemination with fiber-modified adenovirus vectors expressing herpes simplex virus thymidine kinase. *Cancer Gene Ther.* 9:236–242.

76. Vile R G (2004). Thymidne kinases. In T Blankenstein (ed.), *Gene Therapy: Principles and Applications*. Birkhäuser Verlag, Basel, Switzerland, pp. 247–266.

77. Ueno M, Koyama F, Yamada Y, et al. (2001). Tumor-specific chemo-radio-gene therapy for colorectal cancer cells using adenovirus vector expressing the cytosine deaminase gene. *Anticancer Res.* 21:2601–2608.

78. Anello R, Cohen S, Atkinson G, et al. (2000). Adenovirus mediated cytosine deaminase gene transduction and 5-fluorocytosine therapy sensitizes mouse prostate cancer cells to irradiation. *J. Urol.* 164:2173–2177.

79. Khuri F R, Nemunaitis J, Ganly I, et al. (2000). A controlled trial of intratumoral ONYX-015, a selectively-replicating adenovirus, in combination with cisplatin and 5-fluorouracil in patients with recurrent head and neck cancer. *Nat. Med.* 6:879–885.

80. Kamen A, Henry O (2004). Development and optimization of an adenovirus production process. *J. Gene Med.* 6(suppl 1):S184–192.

81. Kanarek A D (2001). A guide to Good Manufacturing Practice. In S C DiClemente (ed.), *D&MD How-to Guides*: D&MD, Westborough, MA.

82. Doblhoff-Dier O, Bliem R (1999). Quality control and assurance from the development to the production of biopharmaceuticals. *Trends Biotechnol.* 17:266–270.

83. Boyd J E (1999). Facilities for large-scale production of vectors under GMP conditions. In A Meager (ed.), *Gene Therapy Technologies, Applications and Regulations—from Laboratory to Clinic*: John Wiley & Sons LTD, Chichester, UK, pp. 383–400.

84. Sofer G, Hagel L (1997). Handbook of process chromatography: A guide to optimization, scale-up and validation, 1st ed. In G Sofer, L Hagel (eds.), Academic Press; San Diego, CA.

85. Akers J, McEntire J, Sofer G (1994). Biotechnology product validation, part 2: A logical plan. *Pharmaceut. Technol. Euro.* 6:230–234.

86. Tolbert W R, Merchant B, Taylor J A, et al. (1996). Designing an initial gene therapy manufacturing facility. *BioPharm Internat.* Nov.: 32–40.

87. Roitsch C, Achstetter T, Benchaibi M, et al. (2001). Characterization and quality control of recombinant adenovirus vectors for gene therapy. *J. Chromatog. B.* 752:263–280.

88. Meager A, Vocke T, Zimmermann, G. (1999). The development of the regulatory process in Europe for biological medicines: How it affects gene therapy products. In A Meager (ed.), *Gene Therapy Technologies, Applications and Regulations—from Laboratory to Clinic.* John Wiley & Sons LTD, Chichester, UK, pp. 319–346.

89. Rexroad J, Wiethoff C M, Green A P, et al. (2003). Structural stability of adenovirus type 5. *J. Pharmaceut. Sci.* 92:665–678.

90. McGann L E, Yang H Y, Walterson M (1988). Manifestations of cell damage after freezing and thawing. *Cryobiol.* 25:178–185.

91. CBER (1996). Addendum to the points to consider in Human somatic cell and gene therapy (1991). US FDA, Center for Biologics Evaluation and Research, Rockville, MD.

92. EMEA (1999). Note for guidance on the quality, preclinical and clinical aspects of gene transfer medicinal products. The European Agency for the Evaluation of Medicinal Products, London.

93. Peng Z (2004). Points to consider for human gene therapy and product quality control State Food and Drug Administration of China. *BioPharm Internat.* May: 6–9.

94. Prazeres, D M F, Ferreira G N M (2004). Design of flowsheets for the recovery and purification of plasmids for gene therapy and DNA vaccination. *Chem. Engin. Proc.* 43:615–630.

95. Petrides D (2003). Bioprocess Design. In R G Harrison, P W Todd, S R Rudge, et al. (eds.), *Bioseparations Science and Engineering.* Oxford University Press, Oxford, UK, pp. 319–372.

96. Gregoriades N, Luzardo M, Lucquet B, et al. (2003). Heat inactivation of mammalian cell cultures for biowaste kill system design. *Biotechnol. Prog.* 19:14–20.

97. Jannat R, Hsu D, Maheshwari G (2005). Inactivation of adenovirus type 5 by caustics. *Biotechnol. Prog.* 21:446–450.

98. Petrides D P, Koulouris A, Lagonikos P T (2002). The role of process simulation in pharmaceutical processes development and product commercialization. *Pharmaceut. Engin.* 22:1–8.

99. Steffens M A, Fraga E S, Bogle I D (2000). Synthesis of bioprocesses using physical properties data. *Biotechnol. Bioengin.* 68:218–230.

100. Freitas S S, Santos J A L, Prazeres D M F (2006) Plasmid DNA production. In E Heinzle, A Biwer, C Cooney (eds.), *Development of Sustainable Bioprocesses: Modeling and Assessment.* Wiley, New York, pp. 271–285.

101. Ugai H, Yamasaki T, Hirose M, et al. (2005). Purification of infectious adenovirus in two hours by ultra-centrifugation and tangential flow filtration. *Biochem. Biophys. Res. Comm.* 331:1053–1060.

102. Carrion M E, Menger M, Kovesdi I (2003). Efficient purification of adenovirus. US Patent appl. 6586226.

103. Tang J C-T, Vellekamp G, Bondoc L L (2001). Methods for purifiying viruses. US Patent appl. 20010036657.

104. Zhang S, Thwin C, Wu Z, Cho T (2001). Method for the production and purification of adenoviral vectors. US Patent appl. 6194191 B1.

105. Huyghe B G, Liu X, Sutjipto S, et al. (1995). Purification of a type 5 recombinant adenovirus encoding human p53 by column chromatography. *Human Gene Ther.* 6:1403–1416.

106. Green A P, Huang J J, Scott M O, et al. (2002). A new scalable method for the purification of recombinant adenovirus vectors. *Human Gene Ther.* 13:1921–1934.

107. Goerke A R, To B C, Lee A L, et al. (2005). Development of a novel adenovirus purification process utilizing selective precipitation of cellular DNA. *Biotechnol. Bioengin.* 91:12–21.

108. Knight P (1989). Downstream processing. *Bio/technol.* 7:777–782.

109. Wheelwright S M (1987). Designing downstream processes for large-scale protein purification. *Bio/technol.* 5:789–793.

110. Nadeau I, Kamen A (2003). Production of adenovirus vector for gene therapy. *Biotechnol. Adv.* 20:475–489.

111. Cote J, Garnier A, Massie B, et al. (1998). Serum-free production of recombinant proteins and adenoviral vectors by 293SF-3F6 cells. *Biotechnol. Bioengin.* 59:567–575.

112. Nadeau I, Gilbert P A, Jacob D, et al. (2002). Low-protein medium affects the 293SF central metabolism during growth and infection with adenovirus. *Biotechnol. Bioengin.* 77:91–104.

113. Ferreira T B, Ferreira A L, Carrondo, M J T, et al. (2005). Effect of refeed strategies and non-ammoniagenic medium on adenovirus production at high cell densities. *J. Biotechnol.* 119:272–280.

114. Iyer P, Ostrove J M, Vacante D (1999). Comparison of manufacturing techniques for adenovirus poduction. *Cytotechnol.* 30:169–172.

115. Braas G, Searle P F, Slater N K, et al. (1996). Strategies for the isolation and purification of retroviral vectors for gene therapy. *Biosepar.* 6:211–228.

116. Wu S C, Huang G Y, Liu J H (2002). Production of retrovirus and adenovirus vectors for gene therapy: A comparative study using microcarrier and stationary cell culture. *Biotechnol. Prog.* 18:617–622.

117. Okada T, Nomoto T, Yoshioka T, et al. (2005). Large-scale production of recombinant viruses by use of a large culture vessel with active gassing. *Human Gene Ther.* 16:1212–1218.

118. Xie L, Metallo C, Warren J, et al. (2003). Large-scale propagation of a replication-defective adenovirus vector in stirred-tank bioreactor PER.C6 cell culture under sparging conditions. *Biotechnol. Bioengin.* 83:45–52.

119. Xie L, Pilbrough W, Metallo C, et al. (2002). Serum-free suspension cultivation of PER.C6(R) cells and recombinant adenovirus production under different pH conditions. *Biotechnol. Bioengin.* 80:569–579.

120. Tsao, Y-S, Condon R, Schaefer E, et al. (2001). Development and improvement of a serum-free suspension process for the production of recombinant adenoviral vectors using HEK293 cells. *Cytotechnol.* 37:189–198.

121. Cortin V, Thibault J, Jacob D, et al. (2004). High-titer adenovirus vector production in 293S cell perfusion culture. *Biotechnol. Prog.* 20:858–863.

122. Parrott M B, Adams K E, Mercier G T, et al. (2003). Metabolically biotinylated adenovirus for cell targeting, ligand screening, and vector purification. *Molec. Ther.* 8:688–700.

123. Evans R K, Nawrocki D K, Isopi L A, et al. (2004). Development of stable liquid formulations for adenovirus-based vaccines. *J. Pharmaceu. Sci.* 93:2458–2475.

124. Azari M, Boose J A, Burhop K E, et al. (2000). Evaluation and validation of virus removal by ultrafiltration during the production of diaspirin crosslinked haemoglobin (DCLHb). *Biologicals.* 28:81–94.

125. Evangelista R L, Kusnadi A R, Howard J A, et al. (1998). Process and economic evaluation of the extraction and purification of recombinant beta-glucuronidase from transgenic corn. *Biotechnol. Prog.* 14:607–614.

10.2

ASSESSING GENE THERAPY BY MOLECULAR IMAGING

Pascal Delépine and Claude Férec

EFS Bretagne—Site de Brest, INSERM U613, Brest, France

Chapter Contents

10.2.1   Technological Advances Applied to Imaging of Gene Transfer   1297

10.2.2   *In Vitro* Gene Imaging Studies   1299

     10.2.2.1   Characterization of Gene Transfer Agents   1299

     10.2.2.2   Gene Transfer Agents Pathway: Localisation and Persistence of the Transgene at its Target   1301

     10.2.2.3   Evaluation of Gene Transfer Efficiency   1303

10.2.3   *In Vivo* Imaging Experiments   1303

     10.2.3.1   Imaging Biodistribution of Gene Transfer Agents   1303

     10.2.3.2   Imaging Gene Transfer Efficacy   1304

     10.2.3.3   Advantages and Drawbacks of Imaging Approaches   1309

10.2.4   Conclusion   1310

     References   1311

10.2.1 TECHNOLOGICAL ADVANCES APPLIED TO IMAGING OF GENE TRANSFER

At the dawn of the twenty-first century, cancer and cardiovascular disease are the first causes of mortality and the focus of considerable attention and mobilization of financial and human resources. Increased understanding of pathogenesis and its underlying molecular processes has led to proven or investigational therapies, such as gene therapy. The concept of gene therapy (and gene transfer) first appeared in the 1960s when the structure of DNA was defined, and much effort has since been devoted to developing suitable strategies. Although gene therapy was initially considered primarily in terms of treatment of genetic disease, the concept has evolved,

Handbook of Pharmaceutical Biotechnology, Edited by Shayne Cox Gad.
Copyright © 2007 John Wiley & Sons, Inc.

and now the two main fields of gene therapy are cancer and cardiovascular disease. Gene therapy consists of overexpressing a functional gene, replacing a mutated gene, or expressing an exogenous gene with the aim of correcting a malfunction, inducing an immune response, or stimulating angiogenesis. Currently, stem cell research is generating new hope in the field of cell therapy, with which gene transfer is increasingly associated.

The first challenge was to transfer DNA into cells, and *in vitro* and *in vivo* gene transfer techniques were developed: mechanical (gene gun), physical (calcium phosphate precipitation), chemical (synthetic gene transfer agents), and biological (viruses). Some of these techniques are commonly used, at least *in vitro*, and have been evaluated in clinical trials, whereas others are currently under investigation. Clinical testing is contingent on technical quality and reproducibility and finding answers to a variety of questions:

- Does the vector or gene transfer procedure work as expected? Optimally?
- Is the preparation optimal?
- What about the route of transfer?
- What is the fate of the nucleic acid and of the vector (if any)?
- When do the vector and nucleic acid dissociate?
- Does the nucleic acid reach its target (and only its target)?
- Does the procedure have any deleterious effect?
- How long is it before expression occurs?
- How long does the transgene persist in the targeted cell or organ?
- How long is transgene expression effective?

Before addressing these questions, it is first essential to ensure that the nucleic acid is transferred into the cell (or organ) and is active. Transfer can be observed *in vitro* by means of microscopy or electron microscopy. Using nucleic acid probes tagged with gold particles, it is possible to observe the presence of a DNA sequence [1]. Direct labeling is also possible using radionuclides, such as phosphorus 32, and is mostly applied *in vivo* in animal models because of its low resolution. Direct labeling is sometimes impractical, so indirect labeling has been developed. Staining techniques have been improved in parallel.

Although imaging approaches can be used to observe, understand, and document all phenomena involved in gene transfer, many are destructive or "invasive" and often require sacrifice of the animal, as in all histopathological, immunohisto-logical, and immunofluorescent techniques. These techniques yield information that is valuable but that only offers a snapshot of a particular moment in time: Researchers are usually interested in a longer-term image of induced phenomena. The use of several groups of the same animal species to obtain this longer-term image is not really satisfying, and so noninvasive techniques tend to replace invasive or destructive ones. These noninvasive techniques can track processes in a single group of animals and have the ethical advantage that they reduce the number of animals used. Also, invasive approaches can be applied to *in vitro* and *in vivo* research experiments, but not to clinical studies, which may explain why we lack information on the fate of gene transfer agents and transgenes in humans. This technological drawback may be partially responsible for the slowness with which gene therapy is being transferred from the lab bench to the bedside.

The most studied imaging methods are positron emission tomography (PET), single-photon emission (computed) tomography (SPE(C)T), bioluminescence, magnetic resonance imaging (MRI), and gamma imaging. Whereas TEP, SPECT, MRI, and gamma imaging are easily performed on animals, bioluminescence is largely limited because of tissue absorption, with a 90% decrease in signal for each centimeter of tissue. Given this limitation, it is clear that better reporter proteins and reporter genes are needed. Optical imaging is usually performed with proteins that emit green light [luciferase or green fluorescent protein (GFP)], and a shift toward red or near-infrared emission is considered to restrict this absorption and allow the signal to leave the animal's body. Bioluminescence usually yields just planar information, but never devices give 3D information. One technique combines laser scanning, which defines the animal's topography, and filters to select the emitted photons (different wavelengths). Software is used to reconstruct a 3D image of the animal. In another, tomographic, approach, a charge-coupled device (CCD) camera rotates around the animal giving whole-body information on photon emission.

Improvments in molecular biology also exists in the MRI field. The ferritin reporter gene was developed for use with MRI. Overexpression of the transgene in the modified cells (various methods of gene transfer are possible) leads to intracellular ferritin excess and iron entrapment. Ferritin associates with iron, which accumulates (aggregates) in the cell as nanomagnets that can be imaged by MRI [2]. Initially, the very common reporter gene beta-galactosidase was used for MRI. The paramagnetic ion is "blocked" using a galactopyrannoside screen and so is not accessible to water. When the reporter gene beta-galactosidase is expressed, its enzymatic activity alters the galactopyrannoside structure and frees the paramagnetic ion, which can then be imaged by MRI [3]. Another approach for MRI uses a reporter gene that encodes a membrane transferrin receptor. The transgene receptor is modified to be resistant to downregulation, so the receptor is largely overexpressed in comparison with untransfected cells. Internalization of superparamagnetic structures (such as MION) conjugated to transferrin is increased in modified cells [4].

As noted above, gene transfer is increasingly associated with cell therapy. Imaging can provide information on the fate of the reinfused modified cells and also on the expression of the transgene or the corrected gene. By loading the cells of interest with contrast agents, it is possible to use MRI to track their movements through the animal's body [5]. The exitation devoted to quantum dots also spreads to cell therapy. Indeed, recently the fate of cells reinjected to mice was studied after labeling those cells by means of self-illuminating quantum dots. It was then possible to image the cells migration by fluorescence imaging *in vivo* [6].

10.2.2 *IN VITRO* IMAGING STUDIES

10.2.2.1 Characterization of Gene Transfer Agents

Structure. Unless DNA is transferred by a physical method, it is necessary to generate a structure able to carry the nucleic acid. When virus capsid or synthetic molecules are involved, they must be characterized physically to allow rational studies on how they ensure vectorization. The structures of gene transfer agents have to be defined to understand how they interact or associate with other components. Visualization of the structures alone is currently possible only by microscopy. Classically, transmission electron microscopy (TEM) is performed after negative staining with uranyl acetate [7] (Figure 10.2-1). Vector characterization is frequently combined with characterization of the "nucleic acid-vector" association. TEM is also used, for example, to confirm that the DNA has intertwined with the polymer assembly [8]. Atomic force microscopy is a relatively new tool that enables the visualization of nanostructures. It will probably give much information on the

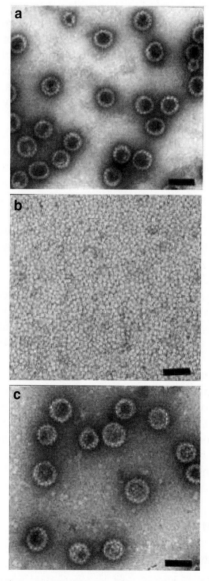

Figure 10.2-1. Electron microscopy to image gene transfer structures. In this example, the authors have imaged the gene transfer structure they developed (human papillomavirus-like particles) after uranyl acetate negative staining. Reprinted from Ref. 7, with permission from Oxford Journals.

structures obtained according to the preparation way and also on the association of the nucleic acid with its vector.

Nucleic Acid—Vector Interaction and Kinetics. One way to confirm that the vector interacts with or encapsulates the nucleic acid is to visualize the interaction. Scanning force microscopy has shown that polymers and cationic lipids interact with DNA in a similar way [9] (Figure 10.2-2). TEM has been used to observed this kind of interaction between DNA and synthetic gene transfer agents [8]. Although fluorescence resonance energy transfer (FRET) data are usually represented graphically [10], it is theoretically possible to obtain images of the fusion events using fluorescence microscopy, which shows how lipids used to prepare lipoplexes interact and mix and also gives information on the capacity of a structure to interact with a membrane model representing the cells.

10.2.2.2 Gene Transfer Agent Pathways: Localization and Persistence of the Transgene at its Target

To improve gene transfer agents, we need to know which hurdles they will have to overcome. Early studies of intracellular trafficking of the vectors were performed by means of electron microscopy.

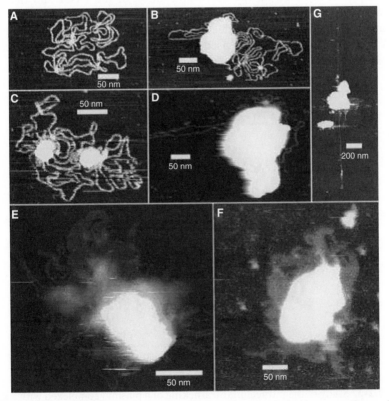

Figure 10.2-2. Scanning force microscopy to image the association between the nucleic acid and the vector. The authors have imaged DNA and its condensation by synthetic gene transfer agents. Reprinted from Ref. 9, with permission from Oxford Journals.

Transmission Electron Microscopy. To observe polyplexes, thin cell sections are negatively stained using uranyl acetate and observed by TEM [8], which can track gene transfer agents. Existing structures can be observed, but when the polymer dissociates, nothing is visible any more. To observe the nucleic acid either free or associated with its vectors, a labeling using gold-labeled probes can be used. The nucleic acid sequence is recognized by a probe that has been tagged with gold particles. The intracellular progression of the nucleic acid can thus be visualized [1] (Figure 10.2-3).

Fluorescence Microscopy. This microscopic approach is used to study the trafficking of either the vector or the nucleic acid. By a fluorescent labeling of the structure of interest (with fluorochrome such as FITC or TRITC), it is possible to visualize the internalization and trafficking processes using fluorescence microscopy or confocal microscopy. Using this approach, Midoux's group studied the internalization of polyplexes in HepG2 cells [11]. An interest of this last technique is that both the vector and the nucleic acid can be labeled. The complex evolution (and its disruption) can be observed.

Magnetic Resonance Imaging. MRI can be used to track complexes during transfection even at the cellular level [12]. MRI microscopy achieves a resolution of about $1\,\mu m$. It is based on the co-complexation of the DNA and the contrast agents with a cationic polymer (theoretically, it is also possible with all cationic gene transfer agents). By adding selective molecules, it is possible to target specific cells. A gadolinium derivative is used to follow the fate of the complex (DNA

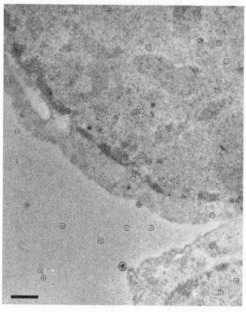

Figure 10.2-3. Study of the intracellular trafficking of a plasmid complexed with a synthetic phosphonolipid using electonic microscopy. Cell lines were transfected with lipoplexes. Then, DNA probes specific to the plasmid used and tagged with gold particles was incubated with the cells pellet sections. The DNA migration was also observed. Reprinted from Ref. 1, with permission from Elsevier.

polymer) within the cell. This real-time visualization reveals only the contrast agent aggregates. No information is given on the organization of the visualized structure, and it is not possible to say whether or when the complexes have been disrupted.

10.2.2.3 Evaluation of Gene Transfer Efficiency

Given how hard it is to demonstrate the functional activity of genes, a nucleic acid delivery system is usually developed using reporter genes encoding proteins that are easily detected through enzymatic reaction or their intrinsic fluorescence.

Fluorescence Microscopy. To ensure that a delivery system or a vector efficiently transfers a nucleic acid into a cell, the easiest approach is to use a reporter gene encoding a protein that can be visualized directly. Reporter genes such as GFP or its new derivatives RFP (red) or BFP (blue) are satisfying because gene expression can be easily visualized by fluorescence microscopy. Moreover, by using a fusion protein, this approach not only reveals expres-sion but also defines cellular localization, particularly when using confocal microscopy. Indirect visualization can also be performed using a nucleic acid that encodes a tagged protein. By incubating the cells with a fluorescent ligand that specifically forms a covalent bond with the tagged protein, the transfected cells appears fluorescent.

Bioluminescence. The bioluminescence approach is not frequently used for *in vitro* studies because of the scarcity of the equipment required. The classic reporter gene luciferase can be used to visualize changes in cells, which is of particular interest when working with a nucleic acid that encodes both luciferase and its substrate, luciferin.

Magnetic Resonance Imaging. When used with MRI, classic contrast agents can track complexes but give no information on gene expression, which is why new reporter genes were developed to modify and hence activate contrast agents, including the gene-encoding ferritin, transfer agents, and enzymes that enable "smart" MRI contrast agent activation [13]. These smart contrast agents are prepared in a weak relaxivity state. When they enter a cell genetically modified by gene transfer, they are transformed into a strong relaxivity state, which is the case of EGad, a gadolinium complex. When Egad interacts with β-galactosidase, a widely used reporter protein, the Egad chelate is disrupted and the relaxation properties of gadolinium can be imaged.

All these MRI approaches applied to cells and to animal models are potentially applicable to humans. MRI is particularly interesting because it gives real-time information potentially at high resolution.

10.2.3 *IN VIVO* IMAGING EXPERIMENTS

10.2.3.1 Imaging Biodistribution of Gene Transfer Agents

It is essential to study the biodistribution of components involved in gene transfer (virus, synthetic vector, or nucleic acid) to identify all *in vivo* barriers. This study indicates what proportion of the administered dose actually reaches the target, as well as any unexpected localizations (according to the quantity at each site) or blocking sites, which is of particular interest because it allows prediction of the dose that will produce a significant therapeutic effect. Biodistribution studies indicate the time when the nucleic acid is released from its vector. Using markers specific to each component (by labeling or conjugating), it is possible to follow both of them separately and also to observe their separation and fate (i.e., their elimination route and metabolism).

The most frequently used approach is direct, in which a radionuclide (usually a β-emitter) is used to label the different components. Indirect studies are also possible, such as the use of markers that are supposed to mimic the behavior of the studied molecules. Invasive imaging methods are used less frequently. In some, the compounds involved are radiolabeled and administered before sacrifice of the animal. The animal's body is thinly sectioned using a cryomicrotome and visualization is done either by classic autoradiography or phosphorus imaging. Gene vectors may also be labeled with fluorescent markers and visualized on microsections. As a result of whole-body dilution, this approach is essentially used to pinpoint a component in the cells of a defined organ, after localization by scintigraphy or autoradiography. Although invasive approaches have been widely used, live imaging is now largely preferred, mainly using the methods described below.

Radioisotopic Imaging. Radioisotopic imaging (using γ-emitters) efficiently tracks the distribution of viral [14] and nonviral vectors [15, 16]. In these two cases, the adenovirus and the GLB43 lipid vector (Figure 10.2-4) were labeled with ^{99m}Tc pertechnetate using a stannous tin procedure. Although most available devices give planar imaging, this approach is quantitative. Scintigraphic imaging can be applied to all kinds of animals and also to humans (depending on the radioactive dose). The method used to label the vector is not always suitable for the nucleic acid, as nucleic acid chelation may inhibit its expression. It is, therefore, necessary to choose between information on expression and on location.

Magnetic Resonance Imaging. MRI is a multivalent approach that also predicts gene transfer efficiency (i.e., whether the transgene actually reached the targeted site). The transgene pathway can be mimicked using a contrast agent, which is administered as the nucleic acid would have been (i.e., associated with a gene vector or directly injected into the targeted tissue). Thus, a gene transfer approach can also be validated by MRI. Moreover, as MRI visualizes tissues and organs, any deleterious effects after gene transfer can be monitored [17].

10.2.3.2   Imaging Gene Transfer Efficacy

The sites of gene transfer clearly must be defined whatever the animal model used or clinical application. For most applications, a specific target is considered. Different administration routes will result in particular expression levels and accuracy of targeting. Direct visualization on thin sections is a common approach in which immunofluorescent probes of the expressed protein are used to study transgene expression (Figure 10.2-5). For example, Bartoli et al. visualized the transgenic protein (α-sarcoglycan) by immunohistochemistry using a secondary antibody conjugated to the fluorochrome Alexa488 in animal models of muscular dystrophies [18].

Animal care and ethical issues are of increasing importance even in research, and so when suitable technology is available, it is preferable to use noninvasive methods

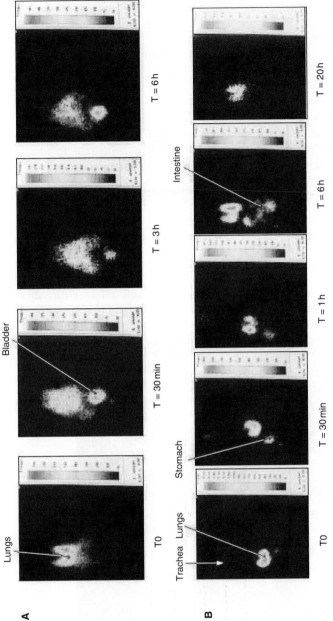

Figure 10.2-4. Scintigraphic biodistribution study. A plasmid encoding the reporter protein luciferase was labeled with ^{99m}Tc. The lipoplexes, resulting from the association of the phosphonolipid GLB43 with the radiolabeled plasmid, was administrated to mice either by intravenous injection or endotracheal nebulization. The animals were imaged using a gamma camera dedicated to small animals. Reprinted from Ref. 15, with permission from Nature Publication Group. (This figure is available in full color at ftp://ftp.wiley.com/public/sci_tech_med/pharmaceutical_biotech/.)

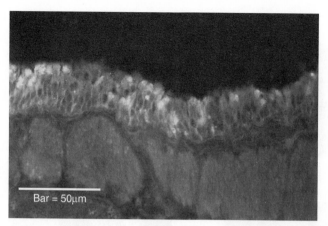

Figure 10.2-5. Immunohistofluorescence imaging to visualize the cells that express the transgene. After mice transfection with a reporter gene (luciferase) associated with the synthetic amphiphiphilic vector (KLN47), the transgene expression was visualized using a primary antibody directed against the luciferase protein. The secondary antibodies were FITC labeled. The expressing cells are thus identified in the targeted tissue (from Delépine et al., unpublished data). (This figure is available in full color at ftp://ftp.wiley.com/ public/sci_tech_med/pharmaceutical_biotech/.)

Figure 10.2-6. *In vivo* bioluminescence imaging. The luciferase transgene expression was observed using a CCD camera 24 hours after an intravenous injection of a lipoplexe (KLN20 phosphonolipid + pCMVLuc) [19]. (This figure is available in full color at ftp://ftp. wiley.com/public/sci_tech_med/pharmaceutical_biotech/.)

to study gene transfer. Tomography is useful in this regard as it gives information on gene expression in terms of time lag, duration, magnitude, and location.

Bioluminescence [19] *(Figure 10.2-6).* Since a pioneering 1998 study of bioluminescence in living mammals [20], bioluminescence technology has advanced greatly and is probably now the most accessible and common approach to *in vivo* imaging. It is, however, limited by the data acquisition, as the most frequently used devices equipped with a CCD camera do not give tomographic information (because of the cost of the tomographic apparatus) and the main part of the emitted light is absorbed by the tissues. Moreover, this absorption depends on tissue type and depth, so it is difficult to extrapolate the results from one animal to another. Another limitation is the diffusion of the substrate. In most cases, the luciferase substrate is injected intraperitoneally. If the transfected site is less accessible because of disease or another cause, the substrate concentration is reduced locally, as is the luminescent signal. This bias can be prevented by using a genetic construction that simultaneously expresses both luciferase and its substrate.

New tomographic devices based on bioluminescence are able to generate a 3D reconstruction, using software to calculate photon absorption for the various tissues. Two advantages accrue from this ability: First, this approach is quantitative and, second, it more precisely localizes the source(s) of luminescence induced by the gene transfer procedure. Although bioluminescence techniques are evolving, they are still only applicable to small animals. The hurdles inherent to this imaging approach and to the use of such reporter genes prevent its application to humans.

Fluorescence. Most bioluminescence imaging devices are also equipped for fluorescence imaging. Fluorescent markers are less sensitive than bioluminescent ones and so are used less. Most of the frequently used reporters emit at a wavelength similar to that of natural molecules, particularly GFP and DsRed. The background is also high with these probes, so few markers are wholly satisfying. However, whole-body fluorescence imaging has been used to visualize gene expression [21].

Positron Emission Tomography. PET can be used to study gene transfer expression in living animals. It has a satisfactory resolution (around 1 mm) for definition of organ targeting and, above all, gives a quantitative signal. The most commonly used positron-emitting radioisotopes are ^{18}F, ^{15}O, ^{13}N, and ^{11}C. PET imaging is used diagnostically in humans to monitor function or localize tumors. Its potential has been progressively enhanced and the increased resolution of microPET methods enables imaging in small laboratory animals [22] (Figure 10.2-7). The three main PET approaches to *in vivo* imaging involve an intracellular enzyme-encoding reporter gene, a membrane receptor-encoding reporter gene, or a membrane transporter-encoding reporter gene [23, 24].

The enzyme approach consists of a genetic modification using an enzyme-encoding reporter gene, usually encoding an HSV-tk mutant (HSV1-sr39tk) [25]. When the genetic modification is performed, the neo-synthesized enzyme is able to phosphorylate uracil derivatives such as $[^{124}I]$FIAU and acycloguanosine derivatives such as $[^{18}F]$FHBG. The phosphorylated substrate is unable to leave the cell and the intracellular accumulation of radiolabeled molecules generates a detectable signal. Using the reporter gene HSV-tk, a significant correlation has been demonstrated in rats (after adenovirus gene transfer) between the transgene expression and tissue accumulation of the reporter probe $[^{18}F]$FGCV [26].

The receptor approach consists of a genetic modification using a reporter gene encoding a receptor, usually the dopamine D2 receptor [27] or the SSTr2 receptor [28]. Using positron-emitting labeled probes that specifically target the receptor, it is possible to identify the sites where the genetic modification occurred.

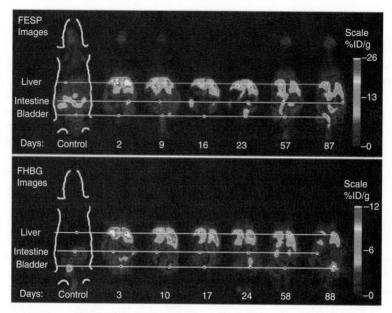

Figure 10.2-7. MicroPET imaging to visualize the kinetics of gene expression in mice. Mice were transduced with either a D2R PET or a HSV1-sr39TK PET reporter gene by intravenous injection of adDTm. Reprinted from Ref. 22, with permission from Elsevier. (This figure is available in full color at ftp://ftp.wiley.com/public/sci_tech_med/pharmaceutical_biotech/.)

The transporter PET strategy consists of modifying the cells of interest with a gene that encodes a transporter able to generate intracellular accumulation of a radiolabeled reporter probe. The sodium iodide symporter (NIS) is one of the most used transporters in this application [29].

Using sequences that enable simultaneous expression of two genes (such as IRES), it is theoretically possible to define expression of the gene of interest in terms of location, magnitude, and duration by studying reporter gene expression [30].

Magnetic Resonance Imaging. MRI can be used to assess gene transfer expression, which is visualized and localized in the animal's body through planar or spatial (2D or 3D) images. *In vivo* transgene expression can be observed in real time, thereby giving information on the time lag between gene transfer and gene expression. MRI has the great advantage of visualizing surrounding tissues. It is therefore possible not only to assess gene transfer efficiency but also any deleterious (inflammation, necrosis, and so on) or beneficial (such as tumor regression) effects [31]. Any effects of gene transfer are defined early and at high resolution and sensitivity, whatever the gene transfer process adopted. MRI can be used for whole-body imaging of gene expression. In mouse brain, gene expression can be monitored after stereotaxic injection of recombinant adenovirus that encodes ferritin [2] (Figure 10.2-8). The relevance of gene transfer to rat muscle by electrotransfer has also been documented by MRI. Expression of the reporter gene luciferase was well correlated with contrast agent entrapment (Gd-DOTA) when administered by the same procedure [17]. Coadministration of the contrast agent and the reporter gene

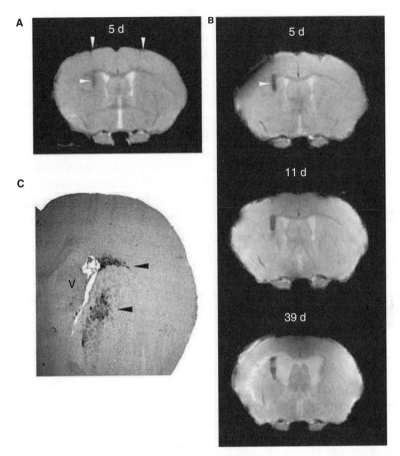

Figure 10.2-8. MRI imaging for gene expression study. In this example, Genove et al. [2] have documented the expression of a ferritin-encoding gene after stereotaxic injection of a recombinant adenovirus. In this study, the MRI imaging approach was validated by an X-Gal staining after injection of an AdV-LacZ. MRI allows a kinetic study of the gene expression. Reprinted from Ref. 2, with permission from Nature Publication Group. (This figure is available in full color at ftp://ftp.wiley.com/public/sci_tech_med/pharmaceutical_ biotech/.)

does not modify reporter gene expression. So, once the experimental parameters have been optimized, MRI can be used to indirectly evaluate gene transfer expression and location.

10.2.3.3 Advantages and Drawbacks of Imaging Approaches

Gene transfer approaches are not only used to study gene transfer agents. Some reporter genes introduce tags into *ex vivo* modified cells, thereby allowing imaging of their fate (migration, homing, and so on).

To complete the gene transfer studies, safety can be studied in animal models by using ultrasound or X-ray computed tomography to document any tissue or organ alteration. Planar imaging is valuable and easy to perform, but does not really reflect the true situation, as photon emission depends on tissue type and depth. Moreover, emission is multidirectional, whereas collection is planar, which leads to a large loss of information. PET is currently only used to characterize gene expression (time lag, level, and duration), which may seem quite limiting with regard to the questions initially posed.

Biodistribution studies using PET can be imagined in which a positron emitter is used to label vectors (either viral or nonviral) or the nucleic acid. Such an approach will be contingent on efficient labeling that does not alter the various components.

MRI of cells or small biological samples requires high magnetic fields (often up to 11.7 T), and the requisite equipment is limited in availability because of its prohibitive cost. Experimentally, MRI is very satisfying because the signal obtained not only yields an image but also provides information on sample content (MR spectroscopy). This twofold analysis will become increasingly powerful as higher magnetic field strengths are used. MRI will soon be able to give information simultaneously on changes in structure (by imaging) and in chemical composition (by spectroscopy) of the targeted site.

Bioluminescence devices are also able to give 3D images, but quantification is indirect and based on mathematical treatment that involves considerable extrapolation.

10.2.4   CONCLUSION

Advances in gene transfer have gone hand in hand with developments in the imaging technologies used to assess transfer efficiency. In gene expression studies, the most frequently used imaging methods are optical (particularly bioluminescence) and nuclear (PET). Whereas optical approaches are limited by the tissue absorption of the signal, the nuclear approach suffers from insufficient resolution. Magnetic resonance offers high resolution, but higher magnetic fields and more relevant and sensitive probes are needed.

No single imaging method can fulfill all research requirements. For large-scale screening or preliminary validation of a new approach, the least expensive method is to be preferred. More costly methods can be applied once the gene transfer approach has been shown to be feasible for gene therapy. Some molecular systems enable two imaging strategies. Using a fusion protein encoding both luciferase and an HSVtk enzyme, gene transfer has been documented using optical bioluminescence and PET [32]. Such fusion proteins are very promising and current improvements might be of great value in gene transfer imaging. An alternative to these fusion proteins is to use two reporter genes (delivered by the same route) to enable a combination of methods, such as PET and fluorescence imaging. Reporter genes may also be coupled with a therapeutic gene through an IRES sequence to monitor modified cells after subcutaneous injection in the mouse [33]. This approach visualizes where expression occurs and its therapeutic effects. But what of functional changes induced by the reporter gene? In the case of PET, three approaches can be envisioned, involving enzymes, receptors, or transporters. How might these proteins impact on cell function? Is therapeutic gene expression really representative of the effects in normal conditions? And how reliable a picture of reality is given by imaging techniques? Proposed solutions to such outstanding questions may be interesting in research terms but are not always appropriate in a clinical setting. It seems best to tackle these problems by comparing and combining data from various imaging approaches.

REFERENCES

1. Montier T, Cavalier A, Delépine P, et al. (2003). The use of in situ hybridization to study the transgene pathway following cellular transfection with cationic phosphonolipids. *Blood Cells Molec. Dis.* 30:112–123.
2. Genove G, DeMarco U, Xu H, et al. (2005). A new transgene reporter for in vivo magnetic resonance imaging. *Nat. Med.* 11:450–454.
3. Louie A Y, Huber M M, Ahrens E T, et al. (2000). In vivo visualization of gene expression using magnetic resonance imaging. *Nat. Biotechnol.* 18:321–325.
4. Weissleder R, Moore A, Mahmood U, et al. (2000). In vivo magnetic resonance imaging of transgene expression. *Nat. Med.* 6:351–355.
5. Dodd S J, Williams M, Suhan J P, et al. (1999). Detection of single mammalian cells by high-resolution magnetic resonance imaging. *Biophys. J.* 76:103–109.
6. So M K, Xu C, Loening A M, et al. (2006). Self-illuminating quantum dot conjugates for in vivo imaging. *Nat. Biotechnol.* 24:339–343.
7. Touze A, Courasaget P (1998). *In vitro* gene transfer using human paillomavirus-like particles. *Nucleic Acids Res.* 26:1317–1323.
8. Labat-Moleur F, Steffan A M, Brisson C, et al. (1996). An electron microscopy study into the mechanism of gene transfer with lipopolyamines. *Gene Ther.* 3:1010–1017.
9. Dunlap D D, Maggi A, Soria M R, et al. (1997). Nanoscopic structure of DNA condensed for gene delivery. *Nucleic Acids Res.* 25:3095–3101.
10. Kumar V V, Pichon C, Refregiers M, et al. (2003). Single histidine residue in headgroup region is sufficient to impart remarkable gene transfection properties to cationic lipids: evidence for histidine-mediated membrane fusion at acidic pH. *Gene Ther.* 10:1206–1215.
11. Goncalves C, Mennesson E, Fuchs R, et al. (2004). Macropinocytosis of polyplexes and recycling of plasmid via the clathrin-dependent pathway impair the transfection efficiency of human hepatocarcinoma cells. *Molec. Ther.* 10:373–385.
12. Kayyem J F, Kumar R M, Fraser S E, et al. (1995). Receptor-targeted co-transport of DNA and magnetic resonance contrast agents. *Chem. Biol.* 2:615–620.
13. Bell J D, Taylor-Robinson S D (2000). Assessing gene expression in vivo: Magnetic resonance imaging and spectroscopy. *Gene Ther.* 7:1259–1264.
14. Lerondel S, Le Pape A, Sene C, et al. (2001). Radioisotopic imaging allows optimization of adenovirus lung deposition for cystic fibrosis gene therapy. *Human Gene Ther.* 12:1–11.
15. Delépine P, Montier T, Guillaume C, et al. (2002). Visualization of the transgene distribution according to the administration route allows prediction of the transfection efficacy and validation of the results obtained. *Gene Ther.* 9:736–739.
16. Delépine P, Guillaume C, Montier T, et al. (2003). Biodistribution study of phosphonolipids: A class of non-viral vectors efficient in mice lung-directed gene transfer. *J. Gene Med.* 5:600–608.
17. Leroy-Willig A, Bureau M F, Scherman D, et al. (2005). In vivo NMR imaging evaluation of efficiency and toxicity of gene electrotransfer in rat muscle. *Gene Ther.* 12:1434–1443.

18. Bartoli M, Poupiot J, Goyenvalle A, et al. (2006). Noninvasive monitoring of therapeutic gene transfer in animal models of muscular dystrophies. *Gene Ther.* 13:20–28.

19. Delépine P, Montier T, Lerondel S, et al. (2002). Imaging of gene distribution and gene expression: A new approach for gene therapy studies on living little rodents. *10th meeting of the ESGT. Antibes-Juan les Pins, France.*

20. Contag P R, Olomu I N, Stevenson D K, et al. (1998). Bioluminescent indicators in living mammals. *Nature Med.* 4:245–247.

21. Yang M, Baranov E, Moossa A R, et al. (2000). Visualizing gene expression by whole-body fluorescence imaging. *Proc. Nat. Acad. Sci. USA.* 97:12278–12282.

22. Liang Q, Gotts J, Satyamurthy N, et al. (2002). Noninvasive, repetitive, quantitative measurement of gene expression from a bicistronic message by positron emission tomography, following gene transfer with adenovirus. *Molec. Ther.* 6(1):73–82.

23. Penuelas I, Boan J, Marti-Climent J M, et al. (2004). Positron emission tomography and gene therapy: Basic concepts and experimental approaches for *in vivo* gene expression imaging. *Molec. Imag. Biol.* 6:225–238.

24. Dharmarajan S, Schuster D P. (2005). Molecular imaging of pulmonary gene expression with positron emission tomography. *Proc. Amer. Thoracic Soc.* 2:549–552, 514–516.

25. Gambhir S S, Bauer E, Black M E, et al. (2000). A mutant herpes simplex virus type 1 thymidine kinase reporter gene shows improved sensitivity for imaging reporter gene expression with positron emission tomography. *Proc. Nat. Acad. Sci. U S A.* 97:2785–2790.

26. Gambhir S S, Barrio J R, Phelps M E, et al. (1999). Imaging adenoviral-directed reporter gene expression in living animals with positron emission tomography. *Proc. Nat. Acad. Sci. U S A.* 96:2333–2338.

27. MacLaren D C, Gambhir S S, Satyamurthy N, et al. (1999). Repetitive, non-invasive imaging of the dopamine D2 receptor as a reporter gene in living animals. *Gene Ther.* 6:785–791.

28. Zinn K R, Buchsbaum D J, Chaudhuri T R, et al. (2000). Noninvasive monitoring of gene transfer using a reporter receptor imaged with a high-affinity peptide radiolabeled with 99mTc or 188Re. *J. Nucl. Med.* 41:887–895.

29. Groot-Wassink T, Aboagye E O, Glaser M, et al. (2002). Adenovirus biodistribution and noninvasive imaging of gene expression in vivo by positron emission tomography using human sodium/iodide symporter as reporter gene. *Human Gene Ther.* 13:1723–1735.

30. Inubushi M, Tamaki N (2005). Positron emission tomography reporter gene imaging in the myocardium: for monitoring of angiogenic gene therapy in ischemic heart disease. *J. Cardiac Surg.* 20:S20–24.

31. Kuehn B M (2005). MRI reveals gene activity in vivo. *JAMA* 293:2584.

32. Ray P, Wu A M, Gambhir S S (2003). Optical bioluminescence and positron emission tomography imaging of a novel fusion reporter gene in tumor xenografts of living mice. *Cancer Res.* 63:1160–1165.

33. Luker G D, Sharma V, Pica C M, et al. (2002). Noninvasive imaging of protein-protein interactions in living animals. *Proc. Nat. Acad. Sci. USA.* 99:6961–6966.

11

OVERVIEW OF STEM AND ARTIFICIAL CELLS

ALEJANDRO SOTO-GUTIERREZ, NALU NAVARRO-ALVAREZ, JORGE DAVID RIVAS-CARRILLO, AND NAOYA KOBAYASHI

Okayama University Graduate School of Medicine and Dentistry, Okayama, Japan

Chapter Contents

11.1	Introductory Remarks	1314
11.2	Embryonic Stem Cells	1315
	11.2.1 Introduction	1315
	11.2.1.1 Properties of Embryonic Stem Cells	1315
	11.2.1.2 Use and Applications of ES Cells	1324
	11.2.2 Conclusions	1325
	References	1326
11.3	Adult Somatic Stem Cells or Postnatal Stem Cells	1331
	11.3.1 Introduction	1331
	11.3.1.1 General Concepts of Stem Cells	1332
	11.3.1.2 General Classification of Stem Cells	1333
	11.3.1.3 General Characteristics of Stem Cells	1334
	11.3.1.4 Hematopoietic Stem Cells (HSCs)	1334
	11.3.1.5 Mesenchymal Stem Cells (MSCs)	1339
	11.3.1.6 Multipotent Adult Progenitor Cells (MAPCs)	1341
	11.3.1.7 Side-Population Phenotype Cells (SPs)	1342
	11.3.1.8 Tissue-Specific Cell Progenitors (TSCPs)	1343
	11.3.1.9 Umbilical Cord Blood-Derived Stem Cells (UCBSCs)	1346
	11.3.1.10 Heterogenous Populations of Stem Cells in Bone Marrow	1347
	11.3.1.11 Uses and Applications of Stem Cells in Toxicology	1347
	References	1348
11.4	Artificial Cells	1360
	11.4.1 Introduction	1360
	11.4.1.1 Immortalization of Human Cells	1361
	11.4.1.2 Tightly Immortalized Human Cell Lines	1366
	11.4.1.3 Conditionally Immortalized Cell Lines	1366

Handbook of Pharmaceutical Biotechnology, Edited by Shayne Cox Gad.
Copyright © 2007 John Wiley & Sons, Inc.

	11.4.1.4	Ultra-Transform Immortalized Transgenic Cell Lines	1367
	11.4.1.5	Reversibly Immortalized Cell Lines	1367
	11.4.1.6	Applications	1369
11.4.2	Conclusions		1369
	References		1369

11.1 INTRODUCTORY REMARKS

Research on stem cells allows us to get the information about how an organism grows and develops from a single cell and how healthy normal cells replace damaged cells in adult organisms. This promising area of stem cell studies is fascinating scientists to investigate the possibility of cell-based therapies to treat a wide range of diseases, which is referred to as regenerative medicine. Stem cells have two important characteristics that distinguish them from other types of cells. First, they are capable of renewing themselves for long periods through cell division. Second, under some experimental conditions, the cells can be induced to be the functional cells, such as the beating cells of the heart muscle, the albumin-producing cells of the liver, or the insulin-secreting cells of the pancreas. Thus, it is now hypothesized by researchers that stem cells may, in the near future, play a basic role in treating diseases such as heart disease, liver disease, and diabetes. Toward that goal, it is important to understand stem cell biology and its therapeutics. Eventually, control of the growth and differentiation of stem cells will be a big tool in the fields of regenerative medicine, tissue engineering, drug discovery, and toxicity testing.

Although normal human cells are ideal to develop cell therapy, it is unlikely that human cells can be isolated on a scale sufficient to treat many patients. The use of animal cells results in the concerns related to the transmission of infectious pathogens and immunologic and physiologic incompatibilities between the donor and humans. Human embryonic stem cells and bone marrow multipotent adult progenitor cells have receive much attention as a possible source for such cell therapy and drug discovery. It is unlikely that perfect control of differentiation of these multipotent cells will be achieved in very near future. Another attractive cell source is human-derived cell lines. In particular, the use of tightly regulated clonal human cell lines are of value. Such cell lines grow economically in tissue culture and provide the advantage of uniformity, sterility, and freedom of pathogens. Reversible immortalization mediated by Cre/loxP site recombination seems to be the most reliable approach to construct human cells for the clinical setting. In Section 11.2, the authors review the nature of embryonic stem cells and their potential. In Section 11.3, somatic stem cells are discussed. In Section 11.4, the authors describe the general concepts of senescence, crisis, telomeres, telomerase, and immortalization of the cells. Finally, the authors introduce the construction of artificial human liver cells and pancreatic beta cells by the use of Cre/loxP-based reversible immortalization.

11.2   EMBRYONIC STEM CELLS

11.2.1   Introduction

Embryonic stem cells have become a very important source for basic research and possible clinical applications since more than 20 years ago when the establishment of mouse embryonic stem cells (ES cells) was achieved by Evans and Kaufman [1] and Martin [2]. They established the pluripotent cells from the inner cell mass of the blastocyst, mouse ES cells, and developed the culture conditions for the cells *in vitro*. The achievement evolved the studies on teratocarcinomas, tumors that arise in the gonads of several inbred strains and consist of an array of somatic tissues juxtaposed together in a disorganized fashion, and gave the origins of the concept of embryonal carcinoma (EC). As teratocarcinomas could also induced by grafting the blastocyst to ectopic sites, it was likely that pluripotent cell lines could be derived directly from the blastocyst. As expected, a stable diploid cell line that could differentiate into all embryonic cell types was established and formed functional germ cells after transplantation into chimeric mice [1–3]. Testicular teratocarcinomas occur spontaneously in humans, and pluripotent cell lines were derived from the teratocarcinomas [4]. In 1998, human ES cell lines were established from pre-implanted embryos by Thomson et al. [5]. Frozen human embryos that were produced by *in vitro* fertilization at an early stage were thawed and cultured to the blastocyst stage. Fourteen inner cell masses were isolated, and five ES cell lines were established. During the first 8 months of culture, no period of replicative crisis was observed in any cell lines. The principal characteristics of the established cells were that the cells expressed high levels of telomerase activity, which suggests that their life span exceeds that of somatic cells. These cells also expressed surface markers that are typical of human embryonic carcinoma cells, such stage-specific embryonic antigen (SSEA)-3, SSEA-4, TRA-1-60, TRA-1-81, and alkaline phosphatase. The cells also expressed the Transcription Factor Octamer binding protein 4 (Oct-4) to undergo somatic differentiation [6]. ES cells could be used for many different purposes, including early development research [7], toxicology and drug screening [8], and gene and protein screening. One of the most important uses is for regenerative medicine and cell therapy, which involves the transplantation of healthy, functional, and propagating cells to restore the viability or function of deficient tissues and organs [9]. The availability of reliable cells is essential for the toxicology and drug discovery process. The primary cells, immortalized cells, and genetically modified cells have been used for drug discovery; however, the inconsistent availability of the primary cells and the genetic abnormalities of transformed cells are the current problems with such application. ES cells could offer considerable advantages due to their plasticity, proliferative capacity, and the ability to undergo homologous recombination at relatively high frequency. In conclusion, ES cells may offer several important advantages in the field of basic biology, drug discovery, and the future cell therapies in various human diseases.

11.2.1.1   *Properties of Embryonic Stem Cells*

Pluripotency

MOUSE ES CELLS.   Culturing mouse ES cells from the inner cell mass of the preimplanted blastocyst were first reported more than 20 years ago [1, 2]. To date only three species of mammals have yielded long-term cultures of self-renewing ES cells: mice, monkeys, and humans [1, 5, 10]. Mouse ES cells lines have shown an unlimited capacity of proliferation and ability to contribute to all cell lineages, which are defined as pluripotency to all definitive tissues: ectoderm, mesoderm, and endoderm. Leukemia inhibitory factor (LIF), a cytokine belonging to the IL-6 family, has been identified as an exogenous signal to maintain ES self-renewal [11]. LIF was initially identified by its activity to induce differentiation of M1 leukemia cells [12], whereas it mediates the opposite cellular responses in ES cells. The signal transduction of LIF consists in the LIF-specific receptor subunit LIFRβ and the common signal transducer gp130, which is shared between the members of the IL-6 cytokine family; the gp130 signaling regulates several cell functions through signal transduction and activation of transcription factor STAT3 that may interact and affect the function of common target genes. Cytokines, including IL-6, IL-11, oncostatin M, ciliary neurotrophic factor, and cardiotrophin-1, show similar properties maintaining the pluripotency of mES cells [12]. A coculture of mouse ES cells on an inactivated mouse embryonic fibroblast layer is also required to maintain the undifferentiated stage. Thus, production of some critical factors of the fibroblast layer is required either to promote self-renewal of mouse ES cells or to suppress their differentiation. The activation of STAT3 is essential to the LIF signaling pathway, but it plays an accessory role to maintain ES cell identity [13]. Moreover, there are two major pathways of intracellular signal transduction downstream of gp130; the Jak-Stat pathway and the Shp2-Erk pathway. As Jak and Shp2 interact with separate subdomains of the intracellular domain of gp-130, it has been demonstrated that the activation of Jak but not Shp2 is sufficient to preserve an undifferentiated status of mouse ES cells. The finding means that the LIF signal is mainly transmitted to the nuclei by the Jak-STAT signal pathway [14] and the Shp2-Erk pathway does not contribute directly to stem cell renewal as demonstrated when adding Erk kinase inhibitor PD98059 in the medium resulting in self-renewal [14]. Although STAT3 acts as a transcription factor to activate target genes, there is only one gene whose specific function in pluripotent cell population is confirmed, that is, the POU-family transcription factor octamer-3/4 (Oct-3/4) encoded by Pou5f1 [12]. The essential role of Oct-3/4 in mouse development has been revealed by targeting gene deletion [15]. Oct-3/4-deficient embryos fail to initiate fetal development because the prospective founder cells of the ICM do not require pluripotency and become diverted into the trophoectoderm lineage, which indicates that Oct-3/4 is essential to establish a pluripotent cell population in preimplantation development [15]. ES cells require a critical level of Oct-3/4 to maintain stem cell renewal, and at least a twofold increase in the expression of Oct-3/4 causes differentiation of mouse ES cells into the endoderm and mesoderm, whereas reduction to less than 50% of the normal expression level of Oct-3/4 triggers de-differentiation of mouse ES cells into the trophectoderm [16]. Recently identification of the homeodomain protein Nanog as another key regulator of pluripotentiality has opened another door of the complicated system of pluripotency; the dosage of Nanog is a critical determinant of cytokine-independent colony formation, and the forced expression of this protein confers constitutive self-renewal in ES cells without gp-130 stimulation; Nanog may act to restrict the differentiation-inducing potential of Oct-3/4 [17]. Other investigators have reported that the induction of the expression of Inhibitor of differentiation (Id) by addition of TGF-β1/bone morphogenetic protein

(BMP) in combination with LIF sustains self-renewal via the Smad pathway [18]. Recent studies have implicated the importance of Wnt-signaling pathways in the maintenance of ES cell pluripotency [19]. Components of the β catenin Wnt-signaling pathway are expressed in ES cells, and it is likely that activation of the Wnt-signaling pathway using a glycogen synthase kinase (GSK)-3-specific inhibitor (BIO) that maintains the pluripotent state of human ES and mouse ES cells [19] (see Figure 11.2-1).

The pluripotency of ES cells depends on the balance of various signaling molecules, and thus, such imbalance causes differentiation of ES cells. Many other molecules, such as Genesis [20], Rex-1 [21], Sox2 [22], GBX2 [23], and UTF1 [24], have been identified with a potential role in defining pluripotency.

Human ES Cells. The murine models of isolation, derivation, culture, and characterization in ES cells have provided valuable information to the generation of human ES (hES) cell lines derived from the embryos at the preimplantation stage, which involves culturing embryos to the morula or blastocyst stage (see Figure 11.2-2). These embryos are donated for research to establish hES cell lines. Thomson et al. isolated from the ICM of human blastocysts, placed on inactivated

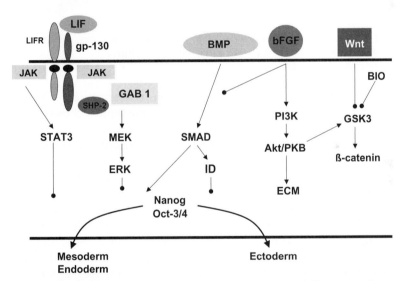

Figure 11.2-1. *Regulation pathways of self-renewal in the undifferentiated stage of ES cells.* The undifferentiation of ES cells is regulated by Nanog, Oct-3/4, and interactions between LIF-dependent JAK/STAT3 pathways principally in mouse ES cells. In human ES cells, the mechanism involved is the BMP-dependent activation of Id target genes; the role of bFGF is to activate the (PI3K)/Akt/PKB pathway, which subsequently down-regulates the expression of ECM molecules; and finally the Wnt pathway activation by specific pharmacological inhibitor BIO of GSK-3 maintains the undifferentiated phenotype in both mouse and human ES cells. The balanced expression level of Oct-3/4 determines the fate of ES cells. Adapted from Niwa.[30] Abbreviations: LIF (Leukemia Inhibitor factor), BMP (bone morphogenetic protein), bFGF (basic fibroblast growth factor), ECM (extracellular matrix), BIO (6-bromoinduribin-3′-oxime), GSK3 (glycogen synthase kinase-3). (This figure is available in full color at ftp://ftp.wiley.com/public/sci_tech_med/pharmaceutical_biotech/.)

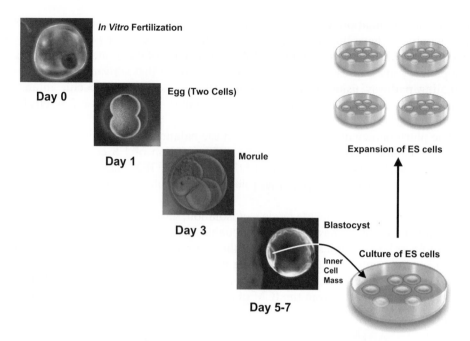

Figure 11.2-2. *Developmental origin of embryonic stem cell lines.* Illustration demonstrates the derivation of embryonic stem cells and their culture.

murine feeder cells, and successfully performed initial derivations of hES cell lines [5]. Since that time, many laboratories have applied the same techniques to derivate hES cell lines (see Figure 11.2-2). hES cells have been currently characterized by a set of markers and their differentiation capacity. These criteria include the expression of several surface markers and transcription factors that associated with an undifferentiated state. In addition, maintenance of extended proliferating capacity, pluripotency, and normal euploid karyotype without marked change in the epigenetic status of hES cells is a crucial issue for the future use of the cells in clinical trials. Several surface markers have been identified to characterize hES cells, in which glycolipids and glycoproteins, such as SSEA-4, TRA-1-60, and TRA-1-81, that are expressed in human embryocarcinoma cells are present in hES cells. Some surface antigens initially described in other stem cells are AC133, CD-9, CD-117, and CD-135, which are also expressed in hES cells [25]. The stability of the expression of these surface markers in hES cells after culture for prolonged periods of time has been maintained.

Several critical transcription factors that play a critical role in maintaining self-renewal of stem cells have now been identified, and these analyses are also useful for characterization of hES cells. One of these transcription factors is Oct-3/4, and several Oct-3/4-targeting genes have been identified in hES cells, which are Utf-1, Rex-1, PDGFαR, Otx-2, Lefty-1, and Nanog. However, their specific roles have not been identified, and their expression has been retained in hES cells for over a year in culture [15–19, 25]. Additionally, all hES cell lines express high levels of telomerase, an enzyme that helps to maintain telomeres that protect the end of the chromosomes. Telomerase activity and long telomeres are characteristics of proliferating cells in embryonic tissue and germ cells. As cells divide and differentiate throughout the life span of an organism or cell line, the telomeres become progressively shortened and lose the ability to maintain their length. The functions of telomeres and telomerase seem to be important in the cell division, normal development, and aging [26]. Mouse ES cells have shown to require activation of the gp-130/STAT3 pathway to maintain the cells in the undifferentiated stage. This activation is generally achieved by the addition of LIF to the culture medium. In contrast, it has been demonstrated that the addition of exogenous LIF to hES cell culture does not maintain the pluripotent capacity of hES cells [27]. However, common intracellular signaling pathways exist between mES cells and hES cells to regulate self-renewal and to maintain an undifferentiated state. These signaling pathways have not yet been clarified, although transcriptional profiling or gene expression technology has identified several genes, transcription factors, ligand/receptor pairs, and secreted inhibitors of signaling pathways. hES cells can be maintained in several different conditions by the use of growth factors belonging to the TGF-β1/BMP superfamily, fibroblast growth factor (FGF) family, and Wnt family.

Previous studies have demonstrated that prolonged propagation of undifferentiated hES cells requires culture of the cells on embryonic fibroblast feeder layers. The use of bFGF has become a common thread in the culture medium formulas, which have been recently developed by using feeder-free conditions [28]. hES cells can be maintained in serum replacement-containing medium that has been supplemented with 36–40 ng/mL of bFGF, in which culture condition the cells can retain the expression of surface markers, transcription factors, normal karyotype, telomerase activity, and pluripotency of hES cells conventionally cultured on embryonic fibroblast feeder layers. The mechanism of bFGF activity in hES cell culture remains unclear, but it has been demonstrated that treatment with 40–100 ng/mL of bFGF inhibits BMP signaling in hES cells. A further role of bFGF in hES cells to maintain pluripotentiality is activation of the phosphatidylinositol 3-kinase (P13K/Akt, PKB) pathway, which subsequently enhances the expression of extracellular matrix molecules (ECMs). Removal of bFGF in hES cell culture or the treatment of the cells with chemical inhibitors of the PKB pathway results in downregulation of the expression of hES cell markers as well as a decrease in ECM components [29] (see Figure 11.2-1).

Differentiation. Pluripotency is one of the defining features of ES cells. The most definitive test of pluripotency of the cells is the formation of chimeras in mice in which the mES cells are injected into the blastocyst. The approach cannot be applied to assess pluripotency of hES cells, therefore, and teratoma formation after injection of embryonic bodies (EBs) of hES cells *in vitro* into immunocompromised mice is currently used to validate the pluripotency of the established hES cell lines in culture. In the case of mES cells, once differentiation of ES cells has started, the cells representing the primary germ layers spontaneously develop *in vitro* in the absence of LIF. The culture conditions to form EBs include hanging drops [30], suspension mass culture [31], or the use of methylcellulose [32]. Initially, an outer layer of endoderm-like cells forms within ESs, followed by the development of an ectodermal layer and subsequent specification of mesodermal cells over a period of a few days [33]. The generation of specific functional cell types from hES cells has been demonstrated both *in vitro* and *in vivo*. In fact, with the rapid interest in gene targeting for the development of genetically modified mice, most of the efforts directed toward ES cells have been made in the maintenance of ES cells in an undifferentiated state. Although research on *in vitro* differentiation of ES cells has been limited, investigators have demonstrated differentiation protocols, which assess the differentiated cells via expression analysis of cell-specific markers. However, very few markers are specific for one cell type, and for that reason, panels of markers must be used in these experiments. Understanding of cellular differentiation during embryogenesis has led to methods for enriching populations of specific cell types: First, genetic manipulation of ES cells facilitates differentiation and serves as a framework for genetic engineering of ES cells by key transcription factors to regulate their cell fate [34]. Second, culture conditions supplemented with established growth factors can be used. Several samples of clinically and pharmacologically relevancy have been proposed, including almost any kind of cells, such as chondrocytes [35], atrial and ventricular cardiomyocytes [30], hepatocytes [36], pancreatic islet cells [37], and motor neurons [38]. Third, coculture conditions have been reported to facilitate and produce differentiated cells [39] (see Figure 11.2-3). Fourth, ectopic implantation of ES cells into syngenic or immune-compromised mice produces specific cell types [40]. Although such transplantation is a powerful approach, this is not a practical method for deriving cells to be evaluated in clinical experiment. Currently the desired cells cannot be developed on a stable large scale. Inductive differentiation protocols have generated many cell types to further enrich *in vitro* differentiated populations by the use of selective methods that are based on the expression of specific marker proteins. Such

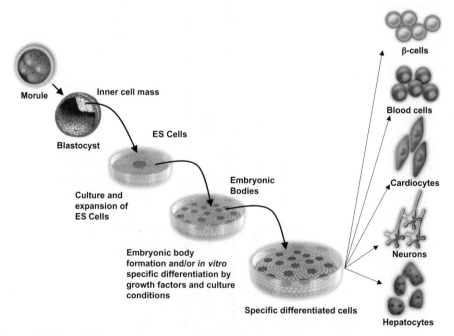

Figure 11.2-3. *Differentiation potential of embryonic stem cells.* Illustration shows the possible application of ES cells for cell therapy.

selective protocols can include promoter-based recovering strategies, based on GFP-dependent cell sorting [41], or antibiotic selection [42], and have allowed the enrichment of increasingly defined phenotypes from ES cells. Cell sorters have been applied to enrich the populations by using specific cell surface markers of ES cell derivatives.

Investigators have used the forced gene expression to influence *in vitro* differentiation of ES cells. For example, constitutive overexpression of murine *Pax 4* in ES cells combined with an inductive protocol has resulted in an enrichment of nestin-positive progenitors and insulin-producing cells among other cells found in pancreatic islets [43]. Finally, the next step in developing clinical trials and drug discovery is the use of internal bioimplants using bioactive materials that provide biological signals at the site of damaged tissues *in vivo*. This next generation of implantable bioartificial devices can permit the evaluation of the potential of ES cell derivatives *in vivo* without having direct contact with the blood circulation. Such an approach allows researchers to avoid a risk of tumor formation of ES cell derivatives and to examine the fate of such cells in an isolation manner from the body.

ENDODERMAL DIFFENTIATION. The pancreas and liver are the derivatives of the definitive endoderm; hepatic and pancreatic cells are of special therapeutic interest for the treatment of diabetes mellitus and hepatic failure. Theoretically, both cells can be generated from ES cells. Researchers have demonstrated different strategies *in vitro* to differentiated mES cells in hepatocyte-like cells; these cells have shown specific transcription factors and proteins of normal hepatocytes; they also have been shown the presence of the two lineages of the liver cells, bile duct epithelial cells and oval cells. Once the cells are transplanted, these cells can integrate into the hepatic parenchyma and function in the host liver. It has been previously demonstrated that ES cells can differentiate into hepatocyte-like cells [36, 40, 44] through *in vivo* differentiation of mES cells in the damaged host liver and subsequent recovery of mES cell-derived hepatocytes by the use of an albumin-promoter-derived GFP expression system [40]. Briefly, mES cells in which an albumin-promoter-derived GFP expression cassette was externally introduced were injected into the spleen of the hepatic-injured mice and tumors were developed after 3 weeks. Examination of the tumors showed the presence of hepatocytes derived from mES cells. These hepatocytes were positive for hepatic markers for at least 3 weeks and were capable of proliferating. The cells were highly characterized by revealing correspondence in 98% of gene expression profiles compared with primary mouse hepatocytes [45]. It is known that the formation of EBs itself can raise the cell population expressing hepatocyte phenotype [46], but such a population in EBs is estimated to be considerably small. Differentiation and isolation of hepatocyte-like cells from hES cells has been demonstrated by using hES cells that are stably transfected with the reporter gene of GFP fused to an albumin-promoter [47]. The generation of ES-derived insulin-producing cells may represent a critical cell source for the treatment of diabetes. In this field, ES cells hold a great hope as a source of β-cells, but unfortunately this has proven to be more complicated than expected. Both mouse and human ES cells can be manipulated to contain insulin and even to regulate insulin secretion [37, 48, 49]. Several methods have been used to obtain enriched populations by the stable transfection with a cDNA construct containing a neomycin-resistance gene under the control of the insulin or Nkx6.1 promoter [49], by the selection with nestin-positive cells [37, 50], or induction of critical transcription factors, such as pax4 and pdx-1 [43, 51]. Although these reports have encouraged that ES cells can generate cells containing insulin and pancreatic β-cell markers, a recent report has demonstrated that insulin-positivity of ES cell derivatives was considerable due to insulin uptake from culture media by apoptotic cells [52]. Measurement of C-peptide, which is excised from proinsulin in the process of maturation of insulin, has been strongly recommended when demonstrating differentiation of ES cells to pancreatic beta cells. Differentiation protocols have to clarify that ES cells are directed to definitive endoderm, not to visceral endoderm, which has similar markers of gene expression but different pathways from pancreatic beta cell lineage. According to the latest research, the differentiation ratio of insulin-positive cells from ES cell populations is about 2.7% compared with less than 1% in undifferentiated controls [53].

ECTODERMAL DIFFERENTIATION. Epithelial cell differentiation from ES cells has been identified by the presence of cytokeratines and specific keratinocyte markers [54]. Enrichment of keratinocytes *in vitro* from ES cells has been achieved by seeding the cells on various extracellular matrices in the presence of bone morphoprotein (BMP-4) and/or ascorbate; such protocols promote the formation of epidermal equivalents. It has been reported that the resulting tissue displays patterns similar to the embryonic skin. The cells express the late differentiation markers of fibroblasts, which suggests that ES cells have the capacity to reconstitute fully differentiated skin *in vitro* [55]. Researchers have reported a differentiation capacity of ES cells into neurons and glial cells [56, 57]. The neural differentiation ratios have significantly improved by the introduction of numerous strategies, including lineage selections (dopaminergic, serotonergic [57–63], or γ-aminobutyric acid (GABA)-ergic neurons [57]), astrocytes, oligodendrocytes [64], glutamatergic and cholinergic neurons [58], and growth factors. The possibility of producing dopaminergic neurons from ES cells has been demonstrated by the use of both fibroblast growth factor 8 (FGF8) and sonic hedgehog, which were implicated as tandem initiators of dopaminergic neurogenesis [58]. Later, the enrichment of dopaminergic neurons has been acheived by mimicking the oxygen tension of the developing midbrain [59]. When such cells were implanted into 6-OHDA-lesioned rats, where nigrostrial dopaminergics afferents are largely lost, the brain function restored normalcy to the dopamine-depleted animals [60]. Similar results were observed in animal experiments by the transplantation of monkey ES cell-derived dopaminergic neurons into 1-methyl-4-phenyl-1,2,3,6-tetrahydropyridine (MPTP)-lesioned adult cymolgus monkeys, in which treatment-associated behavioral improvement was noted by 10 weeks after transplantation [61]. In general, proliferation of neural precursor cells is induced by the addition of FGF and epidermal growth factor (EGF) into the culture medium. Thereafter, neural differentiation can be facilitated by the addition of neural differentiation factors, such as glial cell line-derived neurotrophic factors (GDNFs), neurturin (NT), TGF-β3, and IL-1β [62]. Although the ability of hES cells to generate derivatives of the neural ephitelium has been demonstrated, the selective derivation of neuron subtypes has been difficult to achieve until now. Therefore, long-term survival of the grafted cells has not been successful in experimental models, and a potential risk of teratoma formation by undifferentiated cell populations after implantation should be carefully studied before any therapeutic clinical trials can be considered.

MESODERMAL DIFFERENTIATION. The mesodermal germ layer has the capacity to differentiate into muscle, bone, cartilage, blood, and connective tissues. ES cells also have been successfully used to reconstitute mesodermal developmetal processes *in vitro* by generating several mesodermal cell types, such as adipogenic cells [63], chondrogenic cells [65], osteoblasts [66], and myogenic cells [67]. One of the mostly studied cell lineages in ES cells is cardiomyocytes. Similar to other differentiation protocols, the generation of cardiomyocytes from ES cells requires an initial aggregation step to form EBs. Within EBs, cardiomyocytes are located between an epithelial layer and a basal layer of mesenchymal cells [30]. Cardiomyocytes are easy to identify *in vitro*, because they spontaneously contract after 3–4 days of culture. The number of spontaneously beating cells can be increased by adding differentiation-inducing factors, such as dimethyl sulfoxide (DMSO), retinoic acid, Dynorphin B, and cardiogenol derivatives [30, 68–73]. Developmental changes of cardiomyocytes can be correlated with the length of time in culture and divided into three stages: early (pacemaker-like cells), intermediate, and terminal (atrail-, ventricular-, nodal-, His-, and purkinje-like cells) [30]. hES cell-derived cardiomyocytes have showed the expected molecular, structural, electrophysiologic, and contractile properties of nascent embryonic myocardium [68, 69]. hES cell-derived cardiomyocytes differ from those of mouse in an important and potencially exploitable capacity, that is, proliferation. In contrast to the limited proliferation of mouse ES cell-derived cardiomyocytes [70], human ES cell-derived cardiomyocytes show sustained cell cycle activity both *in vitro* [71] and *in vivo* (in the heart of the nude rats) [72]. *In vivo* studies with hES cell-derived cardiomyocytes have increased lately [72, 73] and suggested that hES cell-resulting cardiomyocytes engraft and integrate into the host myocardium in experimental immunosuppressed animals [73]. These data represent an exciting proof-of-concept evidence for the potential application of hES cell-derived cardiomyocytes in the formation of biological pacemakers. Several important challenges include long-term follow-up of the grafts to confirm whether such cells maintain their pace-making ability over time and development of efficient protocols for large-scale production of highly purified cells without any risks of teratoma formation of the cells *in vivo*.

Hematopoietic cells and blood vessels are believed to arise from common progenitor cells, hemangioblasts. Cystic EBs increase the generation of blood islands containing erythrocytes and macrophages [31], and differentiation of the cells on semi-solid medium is efficient for the formation of neutrophils, mast cells, macrophages, and erythrocytes [32]. Application of fetal calf serum (FCS), cytokines such as IL-1 and IL-3, or granulocyte–macrophage colony stimulating factor (GM–CSF) in ES cell culture generates early hematopoietic precursor cells expressing both embryonic z globin (βH1) and adult β major globin RNAs. Experiments to identify potential inducers of the hematopoietic lineage indicate that Wnt3 is one of the most important signaling molecules that play a significant role in enhancing hematopoietic commitment during *in vitro* differentiation of ES cells [74].

11.2.1.2 Use and Applications of ES Cells   One of the uses of mES cells is the development of genetically modified mice. Researchers have conducted genetic modifications in mES cells to generate genetically modified mice to evaluate specific functions of the targeted genes. These engineered mice have selectively inactivated specific genes to evaluate their target functions and toxicity [75]. This technology can be applied to *in vivo* assays to test for clinical conditions of the central nervous system, cardiovascular diseases, tumors, and metabolic changes in muscle, fat, and bone. The approach has become a crucial method for the utility of different targets in toxicology. Another significant use of ES cells is achieved by a homologous recombination that allows the preplanned replacement of mutations in endogenous murine alleles. Base changes allow an inactivation of specific domains or an alteration of binding sites of a specific protein. For example, single-base changes allow the expression of conserved human mutations. An allelic variant in presenilin 1 is associated with early onset of Alzheimer's disease [76].

Undifferentiated ES Cells. The use of undifferentiated ES cells has been rapidly facilitated to gain functional information and to generate and evaluate knockout mice. An example was the development and evaluation of a nuclear factor-κβ precursor p105 in ES cells [77]. The genetically modified ES cells had specific deletions in the C-terminal region of the p105 gene. These cells demonstrated that p105 was important in the control of NF-κβ-binding activity, detailing the role of NF-κβ in inflammatory responses.

Researchers have also used ES cells deficient in specific genes in the undifferentiated state to evaluate the gene function in the cell development *in vitro* [78].

Differentiated ES Cells. Researchers have demonstrated that cells derived from ES cells *in vitro* reflect the cellular physiology of relevant to primary cells. Gene inactivation has been applied to study events in signaling pathways that could be relevant in specific cell types. For example, there is the role of mitogen-activated protein kinase kinase-1 (MEKK1) in a cellular model of postischemic reperfusion injury using ES cell-derived cardiomyocytes [79]. This approach clearly allows time- and cost-saving evaluation of target genes. In addition, results from knockout cells derived from ES cells can provide clues as to the developmental cause of embryonic lethality.

Teratoma formation has been used to evaluate the role of target gene expression on the cell proliferation and differentiation of the resulting tumor formation. This kind of strategy was applied to demonstrate a reduced proliferative capacity of ES cells null for cyclooxygenase 2 compared with cyclooxygenase 1-deficient or wild-type ES cells [80]. The strategy can be used to evaluate the role of potential cancer-inducing targets on cell proliferation. Reproductive toxicity is a target where ES cells could be applied to *in vitro* teratology assays [81]. Such a procedure could reduce the use of animal procedures. It has been reported that 78% of correlation exists between the data of *in vivo* and *in vitro* experiments using ES cells as an alternative to animal testing in teratology and embryo-toxicity testing [82]. The aim is to overcome the limitation of availability of primary cells or the genetic abnormality of immortalized cell lines. Generation of specific cell and tissue types is necessary rather than heterogeneous populations that can result from spontaneous differentiation of ES cells. For this reason, ES cell lines expressing GFP selectively in the specific differentiation to cardiomyocytes [83] or hepatocytes [40] have been designed [84]. Finally, therapeutic concepts of ES cells must be well argued. Some difficulties can be identified in the application of adult stem cells or embryonic stem cells, in order to develop stem cell-replacement therapy. We need to meticulously select the best candidates of cells in terms of the facilities to propagate, manipulate, and select the most purified population of the desired cell type. We need to produce immune tolerance when allogenic cells are used for stem cell-based cell therapy. In addition, ES cells bring certain advantages against the rest of stem cells. Firstly, adult stem cells are difficult to isolate and hard to propagate in culture; in contrast, ES cells are derived easily once an embryo has been obtained and they can grow indefinitely in culture. ES cells can be manipulated genetically by homologous recombination to correct a genetic defect [85]. On the other hand, adult stem cells can be genetically manipulated only by the introduction of retroviral gene delivery, which overexpress the transduced genes and could happen in an insertional mutagenesis [86]. ES cells have the capacity to differentiate into any kind of tissue of the body, that is, pluripotent, but adult stem cells are multipotent and their differentiation capacity seems to be restricted. Protocols must be well developed in ES cells to generate an enrichment population of cells of a specific lineage or cell type. Furthermore, if therapeutic application is the final goal of ES cells, culturing techniques need to be scaled up for mass production of clinically relevant quantities of the specified cells. Finally, immunological barriers must be overcome. Regarding this issue some promising alternatives have been suggested lately. Nuclear transplantation denotes the introduction of a nucleus from an adult donor cell into an enucleated oocyte to generate a cloned embryo. When transferred to the uterus of a female, this embryo has the potential to become an infant that is a clone of the adult donor cell; this process is referred as to "reproductive cloning" (see Figure 11.2-4). When explanted in culture, this embryo can give rise to ES cells that have the potential to become almost any type of cells present in the adult body. The resulting ES cells by nuclear transfer are genetically identical to the donor and thus potentially useful for therapeutic applications; this process is called "therapeutic cloning" (see Figure 11.2-4). Therapeutic cloning may substantially improve the treatment for neurodegenerative diseases, blood disorders, or diabetes, because the therapy for such diseases is currently limited by the availability or immunocompatibility of tissue transplants. Indeed, experiments in animals have shown that nuclear cloning combined with gene and cell therapy represents a valid strategy for treating genetic disorders [85].

11.2.2   Conclusions

Considerable progress has been made toward the generation of more defined culture conditions of ES cells since the initial isolation and growth conditions were described. There have been many modifications of the procedures, including growth of ES cells, differentiation potential, and cell lines establishment. The availability of hES cells represents an extraordinary opportunity for cell transplantation that may be applicable to humans in the future. Despite exciting advantages and advances in the field of hES cell research, many challenges have to be addressed in the near future. The culture conditions have to improved and be humanized. The cells cultured in xenogeneic conditions are likely to be considered and

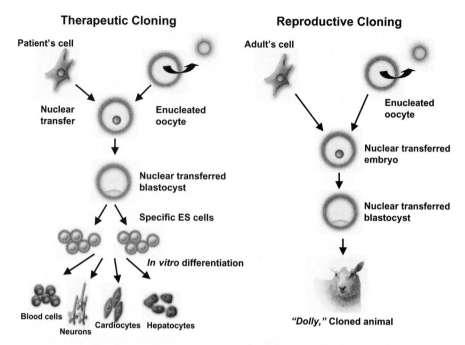

Figure 11.2-4. *Comparison of therapeutic cloning and reproductive cloning.* In therapeutic cloning (left panel), the diploid nucleus of an adult donor cell is introduced into an enucleated oocyte and it divides into cloned blastocysts that will provide lines of embryonic stem cells that can be differentiated *in vitro* into any types of cell for therapeutic purposes. In contrast, reproductive cloning (right panel) requires the transfer of the cloned blastocysts into surrogate mothers and a few of the cloned blastocysts will give rise to the cloned animals.

regulated. Concerns about the infection from nonhuman pathogens are the most discussed issues for the clinical application. Although ES cells offer great promise for regenerative medicine, the near-term applications could be in drug discovery. Researchers might have a store of stem cells lines derived from many human populations from different ethnicities and gene variations. These cells can be of extreme importance in the development of effective and safe therapeutics.

References

1. Evans M J, Kaufman M H (1981). Establishment in culture of pluripotential cells from mouse embryos. *Nature.* 292:154–156.

2. Martin G R (1981). Isolation of a pluripotent cell line from early mouse embryos cultured in medium conditioned by teratocarcinoma stem cells. *Proc. Natl. Acad. Sci. USA.* 78:7634–7638.

3. Bradley A, Evans M, Kaufman M H, et al. (1984). Formation of germ-line chimaeras from embryo-derived teratocarcinoma cell lines. *Nature.* 309:255–256.

4. Andrews P W, Damjanov I, Simon D, et al. (1984). Pluripotent embryonal carcinoma clones derived from the human teratocarcinoma cell line Tera-2. Differentiation in vivo and in vitro. *Lab. Invest.* 50:147–162.

5. Thomson J A, Itskovitz-Eldor J, Shapiro S S, et al. (1998). Embryonic stem cell lines derived from human blastocysts. *Science.* 282:1145–1147.

6. Reubinoff B E, Pera M F, Fong C Y, et al. (2000). Embryonic stem cell lines from human blastocysts: Somatic differentiation in vitro. *Nat. Biotechnol.* 18:399–404.

7. Pera M F, Trounson A O (2004). Human embryonic stem cells: Prospects for development. *Develop.* 131:5515–5525.

8. McNeish J (2004). Embryonic stem cells in drug discovery. *Nat. Rev. Drug Discov.* 3:70–80.

9. Strom T B, Field L J, Ruediger M (2002). Allogeneic stem cells, clinical transplantation and the origins of regenerative medicine. *Curr. Opin. Immunol.* 14:601–605.

10. Thomson J A, Kalishman J, Golos T G, et al. (1995). Isolation of a primate embryonic stem cell line. *Proc. Natl. Acad. Sci. USA.* 92:7844–7848.

11. Williams R L, Hilton D J, Pease S, et al. (1988). Myeloid leukaemia inhibitory factor maintains the developmental potential of embryonic stem cells. *Nature.* 336:684–687.

12. Niwa H (2001). Molecular mechanism to maintain stem cell renewal of ES cells. *Cell Struct. Funct.* 26:137–148.

13. Niwa H, Burdon T, Chambers I, et al. (1998). Self-renewal of pluripotent embryonic stem cells is mediated via activation of STAT3. *Genes Dev.* 12:2048–2060.

14. Burdon T, Stracey C, Chambers I, et al. (1999). Suppression of SHP-2 and ERK signalling promotes self-renewal of mouse embryonic stem cells. *Dev. Biol.* 210:30–43.

15. Nichols J, Zevnik B, Anastassiadis K, et al. (1998). Formation of pluripotent stem cells in the mammalian embryo depends on the POU transcription factor Oct4. *Cell.* 95:379–391.

16. Niwa H, Miyazaki J, Smith A G (2000). Quantitative expression of Oct-3/4 defines differentiation, dedifferentiation or self-renewal of ES cells. *Nat. Genet.* 24:372–376.

17. Chambers I, Colby D, Robertson M, et al. (2003). Functional expression cloning of Nanog, a pluripotency sustaining factor in embryonic stem cells. *Cell.* 113:643–655.

18. Ying Q L, Nichols J, Chambers I, et al. (2003). BMP induction of Id proteins suppresses differentiation and sustains embryonic stem cell self-renewal in collaboration with STAT3. *Cell.* 115:281–292.

19. Paling N R, Wheadon H, Bone H K, et al. (2004). Regulation of embryonic stem cell self-renewal by phosphoinositide 3-kinase-dependent signaling. *J. Biol. Chem.* 279:48063–48070.

20. Sutton J, Costa R, Klug M, et al. (1996). Genesis, a winged helix transcriptional repressor with expression restricted to embryonic stem cells. *J. Biol. Chem.* 271:23126–23133.

21. Hosler B A, Rogers M B, Kozak C A, et al. (1993). An octamer motif contributes to the expression of the retinoic acid-regulated zinc finger gene Rex-1 (Zfp-42) in F9 teratocarcinoma cells. *Mol. Cell Biol.* 13:2919–2928.

22. Avilion A A, Nicolis S K, Pevny L H, et al. (2003). Multipotent cell lineages in early mouse development depend on SOX2 function. *Genes. Dev.* 17:126–140.

23. Chapman G, Remiszewski J L, Webb G C, et al. (1997). The mouse homeobox gene, Gbx2: Genomic organization and expression in pluripotent cells in vitro and in vivo. *Genomics.* 46:223–233.

24. Okuda A, Fukushima A, Nishimoto M, et al. (1998). UTF1, a novel transcriptional coactivator expressed in pluripotent embryonic stem cells and extraembryonic cells. *Embo. J.* 17:2019–2032.

25. Carpenter M K, Rosler E S, Fisk G J, et al. (2004). Properties of four human embryonic stem cell lines maintained in a feeder-free culture system. *Dev. Dyn.* 229:243–258.

26. Amit M, Carpenter M K, Inokuma M S, et al. (2000). Clonally derived human embryonic stem cell lines maintain pluripotency and proliferative potential for prolonged periods of culture. *Dev. Biol.* 227:271–278.

27. Sato N, Sanjuan I M, Heke M, et al. (2003). Molecular signature of human embryonic stem cells and its comparison with the mouse. *Dev. Biol.* 260:404–413.

28. Xu C, Rosler E, Jiang J, et al. (2005). Basic fibroblast growth factor supports undifferentiated human embryonic stem cell growth without conditioned medium. *Stem Cells.* 23:315–323.

29. Kim S J, Cheon S H, Yoo S J, et al. (2005). Contribution of the PI3K/Akt/PKB signal pathway to maintenance of self-renewal in human embryonic stem cells. *FEBS Lett.* 579:534–540.

30. Boheler K R, Czyz J, Tweedie D, et al. (2002). Differentiation of pluripotent embryonic stem cells into cardiomyocytes. *Circ. Res.* 91:189–201.

31. Doetschman T C, Eistetter H, Katz M, et al. (1985). The in vitro development of blastocyst-derived embryonic stem cell lines: Formation of visceral yolk sac, blood islands and myocardium. *J. Embryol. Exp. Morphol.* 87:27–45.

32. Wiles M V, Keller G (1991). Multiple hematopoietic lineages develop from embryonic stem (ES) cells in culture. *Develop.* 111:259–267.

33. Wobus A M, Boheler K R (2005). Embryonic stem cells: prospects for developmental biology and cell therapy. *Physiol. Rev.* 85:635–678.

34. Chung S, Sonntag K C, Andersson T, et al. (2002). Genetic engineering of mouse embryonic stem cells by Nurr1 enhances differentiation and maturation into dopaminergic neurons. *Eur. J. Neurosci.* 16:1829–1838.

35. Hegert C, Kramer J, Hargus G, et al. (2002). Differentiation plasticity of chondrocytes derived from mouse embryonic stem cells. *J. Cell Sci.* 115:4617–4628.

36. Lavon N, Yanuka O, Benvenisty N (2004). Differentiation and isolation of hepatic-like cells from human embryonic stem cells. *Different.* 72:230–238.

37. Lumelsky N, Blondel O, Laeng P, et al. (2001). Differentiation of embryonic stem cells to insulin-secreting structures similar to pancreatic islets. *Science.* 292:1389–1394.

38. Wichterle H, Lieberam I, Porter J A, et al. (2002). Directed differentiation of embryonic stem cells into motor neurons. *Cell.* 110:385–397.

39. Yamane T, Kunisada T, Yamazaki H, et al. (1997). Development of osteoclasts from embryonic stem cells through a pathway that is c-fms but not c-kit dependent. *Blood.* 90:3516–3523.

40. Yamamoto H, Quinn G, Asari A, et al. (2003). Differentiation of embryonic stem cells into hepatocytes: Biological functions and therapeutic application. *Hepatol.* 37:983–993.

41. Kolossov E, Fleischmann B K, Liu Q, et al. (1998). Functional characteristics of ES cell-derived cardiac precursor cells identified by tissue-specific expression of the green fluorescent protein. *J. Cell Biol.* 143:2045–2056.

42. Li M, Pevny L, Lovell-Badge R, et al. (1998). Generation of purified neural precursors from embryonic stem cells by lineage selection. *Curr. Biol.* 8:971–974.

43. Blyszczuk P, Czyz J, Kania G, et al. (2003). Expression of Pax4 in embryonic stem cells promotes differentiation of nestin-positive progenitor and insulin-producing cells. *Proc. Natl. Acad. Sci. USA.* 100:998–1003.

44. Yamada T, Yoshikawa M, Kanda S, et al. (2002). In vitro differentiation of embryonic stem cells into hepatocyte-like cells identified by cellular uptake of indocyanine green. *Stem Cells.* 20:146–154.

45. Yamamoto Y, Teratani T, Yamamoto H, et al. (2005). Recapitulation of in vivo gene expression during hepatic differentiation from murine embryonic stem cells. *Hepatol.* 42:558–567.

46. Chinzei R, Tanaka Y, Shimizu-Saito K, et al. (2002). Embryoid-body cells derived from a mouse embryonic stem cell line show differentiation into functional hepatocytes. *Hepatol.* 36:22–29.

47. Heins N, Englund M C, Sjoblom C, et al. (2004). Derivation, characterization, and differentiation of human embryonic stem cells. *Stem Cells.* 22:367–376.

48. Soria B, Roche E, Berna G, et al. (2000). Insulin-secreting cells derived from embryonic stem cells normalize glycemia in streptozotocin-induced diabetic mice. *Diabetes.* 49:157–162.

49. Assady S, Maor G, Amit M, et al. (2001). Insulin production by human embryonic stem cells. *Diabetes.* 50:1691–1697.

50. Kahan B W, Jacobson L M, Hullett D A, et al. (2003). Pancreatic precursors and differentiated islet cell types from murine embryonic stem cells: an in vitro model to study islet differentiation. *Diabetes.* 52:2016–2024.

51. Miyazaki S, Yamato E, Miyazaki J (2004). Regulated expression of pdx-1 promotes in vitro differentiation of insulin-producing cells from embryonic stem cells. *Diabetes.* 53:1030–1037.

52. Rajagopal J, Anderson W J, Kume S, et al. (2003). Insulin staining of ES cell progeny from insulin uptake. *Science.* 299:363.

53. Ku H T, Zhang N, Kubo A, et al. (2004). Committing embryonic stem cells to early endocrine pancreas in vitro. *Stem Cells.* 22:1205–1217.

54. Bagutti C, Wobus A M, Fassler R, et al. (1996). Differentiation of embryonal stem cells into keratinocytes: Comparison of wild-type and beta 1 integrin-deficient cells. *Dev. Biol.* 179:184–196.

55. Coraux C, Hilmi C, Rouleau M, et al. (2003). Reconstituted skin from murine embryonic stem cells. *Curr. Biol.* 13:849–853.

56. Fraichard A, Chassande O, Bilbaut G, et al. (1995). In vitro differentiation of embryonic stem cells into glial cells and functional neurons. *J. Cell Sci.* 108(Pt 10):3181–3188.

57. Goridis C, Rohrer H (2002). Specification of catecholaminergic and serotonergic neurons. *Nat. Rev. Neurosci.* 3:531–541.

58. Ye W, Shimamura K, Rubenstein J L, et al. (1998). FGF and Shh signals control dopaminergic and serotonergic cell fate in the anterior neural plate. *Cell.* 93:755–766.

59. Studer L, Csete M, Lee S H, et al. (2000). Enhanced proliferation, survival, and dopaminergic differentiation of CNS precursors in lowered oxygen. *J. Neurosci.* 20:7377–7383.

60. Kim J H, Auerbach J M, Rodriguez-Gomez J A, et al. (2002). Dopamine neurons derived from embryonic stem cells function in an animal model of Parkinson's disease. *Nature.* 418:50–56.

61. Takagi Y, Takahashi J, Saiki H, et al. (2005). Dopaminergic neurons generated from monkey embryonic stem cells function in a Parkinson primate model. *J. Clin. Invest.* 115:102–109.

62. Rolletschek A, Chang H, Guan K, et al. (2001). Differentiation of embryonic stem cell-derived dopaminergic neurons is enhanced by survival-promoting factors. *Mech. Dev.* 105:93–104.

63. Bost F, Caron L, Marchetti I, et al. (2002). Retinoic acid activation of the ERK pathway is required for embryonic stem cell commitment into the adipocyte lineage. *Biochem. J.* 361:621–627.

64. Angelov D N, Arnhold S, Andressen C, et al. (1998). Temporospatial relationships between macroglia and microglia during in vitro differentiation of murine stem cells. *Dev. Neurosci.* 20:42–51.

65. Kramer J, Hegert C, Hargus G, et al. (2005). Mouse ES cell lines show a variable degree of chondrogenic differentiation in vitro. *Cell Biol. Int.* 29:139–146.

66. Bourne S, Polak J M, Hughes S P, et al. (2004). Osteogenic differentiation of mouse embryonic stem cells: differential gene expression analysis by cDNA microarray and purification of osteoblasts by cadherin-11 magnetically activated cell sorting. *Tissue Eng.* 10:796–806.

67. Rohwedel J, Maltsev V, Bober E, et al. (1994). Muscle cell differentiation of embryonic stem cells reflects myogenesis in vivo: developmentally regulated expression of myogenic determination genes and functional expression of ionic currents. *Dev. Biol.* 164:87–101.

68. Kehat I, Kenyagin-Karsenti D, Snir M, et al. (2001). Human embryonic stem cells can differentiate into myocytes with structural and functional properties of cardiomyocytes. *J. Clin. Invest.* 108:407–414.

69. He J Q, Ma Y, Lee Y, et al. (2003). Human embryonic stem cells develop into multiple types of cardiac myocytes: Action potential characterization. *Circ. Res.* 93:32–39.

70. Klug M G, Soonpaa M H, Field L J (1995). DNA synthesis and multinucleation in embryonic stem cell-derived cardiomyocytes. *Am. J. Physiol.* 269:H1913–1921.

71. Snir M, Kehat I, Gepstein A, et al. (2003). Assessment of the ultrastructural and proliferative properties of human embryonic stem cell-derived cardiomyocytes. *Am. J. Physiol. Heart Circ. Physiol.* 285:H2355–2363.

72. Kehat I, Khimovich L, Caspi O, et al. (2004). Electromechanical integration of cardiomyocytes derived from human embryonic stem cells. *Nat. Biotechnol.* 22:1282–1289.

73. Xue T, Cho H C, Akar F G, et al. (2005). Functional integration of electrically active cardiac derivatives from genetically engineered human embryonic stem cells with quiescent recipient ventricular cardiomyocytes: insights into the development of cell-based pacemakers. *Circul.* 111:11–20.

74. Lako M, Lindsay S, Lincoln J, et al. (2001). Characterisation of Wnt gene expression during the differentiation of murine embryonic stem cells in vitro: Role of Wnt3 in enhancing haematopoietic differentiation. *Mech. Dev.* 103:49–59.

75. Abuin A, Holt K H, Platt K A, et al. (2002). Full-speed mammalian genetics: In vivo target validation in the drug discovery process. *Trends Biotechnol.* 20:36–42.

76. Grilli M, Diodato E, Lozza G, et al. (2000). Presenilin-1 regulates the neuronal threshold to excitotoxicity both physiologically and pathologically. *Proc. Natl. Acad. Sci. USA.* 97:12822–12827.

77. Ishikawa H, Ryseck R P, Bravo R (1996). Characterization of ES cells deficient for the p105 precursor (NF-kappa B1): Role of p50 NLS. *Oncogene.* 13:255–263.

78. Gowen L C, Johnson B L, Latour A M, et al. (1996). Brca1 deficiency results in early embryonic lethality characterized by neuroepithelial abnormalities. *Nat. Genet.* 12:191–194.

79. Minamino T, Yujiri T, Papst P J, et al. (1999). MEKK1 suppresses oxidative stress-induced apoptosis of embryonic stem cell-derived cardiac myocytes. *Proc. Natl. Acad. Sci. USA.* 96:15127–15132.

80. Zhang X, Morham S G, Langenbach R, et al. (2000). Lack of cyclooxygenase-2 inhibits growth of teratocarcinomas in mice. *Exp. Cell Res.* 254:232–240.

81. Klemm M, Genschow E, Pohl I, et al. (2001). Permanent embryonic germ cell lines of BALB/cJ mice—an in vitro alternative for in vivo germ cell mutagenicity tests. *Toxicol. In Vitro.* 15:447–453.

82. Seiler A, Visan A, Buesen R, et al. (2004). Improvement of an in vitro stem cell assay for developmental toxicity: The use of molecular endpoints in the embryonic stem cell test. *Reprod. Toxicol.* 18:231–240.

83. Bremer S, Worth A P, Paparella M, et al. (2001). Establishment of an in vitro reporter gene assay for developmental cardiac toxicity. *Toxicol. In Vitro.* 15:215–223.

84. Paparella M, Kolossov E, Fleischmann B K, et al. (2002). The use of quantitative image analysis in the assessment of in vitro embryotoxicity endpoints based on a novel embryonic stem cell clone with endoderm-related GFP expression. *Toxicol. In Vitro.* 16:589–597.

85. Rideout W M, 3rd, Hochedlinger K, Kyba M, et al. (2002). Correction of a genetic defect by nuclear transplantation and combined cell and gene therapy. *Cell.* 109:17–27.

86. Check E (2003). Second cancer case halts gene-therapy trials. *Nature.* 421:305.

11.3   ADULT SOMATIC STEM CELLS OR POSTNATAL STEM CELLS

11.3.1   Introduction

In the past few decades, a true revolution has been occurring in a spectacular way in the field of medicine and biology. Regenerative medicine with the intention of tissue regeneration, and thus the curative treatment of diseases, has awoken maximum interest from scientific communities all over the world [1].

The likelihood that the human body contains the cells capable of regenerating and repairing the damaged or disease tissues has turned from an implausible subject to a virtual belief. It has been well known that stem cells with differentiating potential to replenish progeny are present in postnatal tissues of mammals. Recently many studies have demonstrated the abilities of stem cells to form multiple types of cells and the presence of such cells in an increasing number of tissues. We attempt here to provide an overview of adult stem cells in terms of possible mechanisms of their differentiation and their potential in therapeutic use.

The presence of adult somatic stem cell or cell progenitor has been clearly identified in several tissues. It has primarily been easily identified in tissues with high cell replication. These cells contribute to repair and regeneration throughout the life span in adult tissues, which also have the capacity of cell renewal [2, 3]. Recently, the concept of stem cell niches was proposed, in which stem cells exist in small numbers in silence under normal conditions but become activated after tissue injury or other pathological conditions [4]. Some stem cells possess enormous plasticity to differentiate into any cell types derived from the germinal layers (mesoderm, ectoderm, and endoderm). In contrast, other stem cells can only terminally differentiate into the same germinal layer from which they originated. Several factors, which are not clearly understood, are involved in the regulation of stem cells in terms of their potential of differentiation, tissue repair, and regeneration. On the other hand, the tragedy includes the development and progression of malignancy due to their longevity with respect to the body life span and their ability to give rise to multiple cell phenotypes [5]. Based on the nature of stem cells, there is a worldwide interest in stem cell research to understand the natural growth and senescence of the cells and to balance tissue repair and regeneration through the life span. Studies on the mechanism that controls the maintenance of stem cells and differentiation signals is useful to understand how external (toxins, radiation, etc.) and internal injuries (cytogenetic alterations, mutation, etc.) potentially modify the fate of stem cells and eventually lead them to malignant transformation [5–7]. Stem cells can withstand over stressful environmental events associated with tissue damage after surgical procedures [8], exposure to toxics agents [8, 9], and extreme cold [10] and then repopulate and repair adult tissues. Although such *in vivo* environments seem to be transient and hostile against host cells and mature cells are easily destroyed, stem cells can survive, differentiate, and regenerate the tissue [8]. Their differentiation capacities investigated under a variable controlled environment may allow us a window into the regenerative process of adult tissues. To facilitate stem cell research, *in vitro* culture should mimic native niches of stem cells *in vivo* by application of matrices, growth factors, and drug delivery systems [11–15].

11.3.1.1 General Concepts of Stem Cells The name *stem cells* is commonly used to refer the cells that are relatively undifferentiated while retaining the ability to divide and proliferate throughout postnatal life, providing progenitor cells that can differentiate into tissue-specific cells. The ability of stem or progenitor cells to give rise to different populations of terminally differentiated cells is referred to as plasticity. Thus, their potential is called totipotent, pluripotent, or multipotent. The term *totipotent* should be strictly reserved for the unique stem cells that can form embryonic and extra-embryonic membranes [6, 16], which are the most primitive about 4 days after fecundation [17]. The term *pluripotent* is used to refer to the stem cells' ability to form all cell types of the proper embryo, except for the extra-embryonic membranes and their tissues. In contrast, the term *multipotent* is used to name those that give rise to a subset of cell lineages. During the development process, the totipotent stem cells originate and pluripotent stem cells come out of the germ layers. Afterward multipotent stem cells are generated [16]. These cells, in turn, form the oligopotent progenitor cells in the developing organs. Moreover, intestinal progenitor cells are considered to be quadripotent; they can form progeny that become mucous, absorptive, neuroendocrine, or Paneth cells. In the case of bronchial lining cells, the progenitors are tripotent: Progeny turn into neuroendocrine, mucous, or ciliated cells. The oval cells of the liver are bipotent; they can derivate into duct cells and hepatocytes. Other type of cells, such as epidermal progenitors, are unipotent to produce only a single progeny [6, 16, 17].

Tissue Homeostasis and Stem Cell Renewal. Physiological tissue renewal is accomplished by a delicate balance among specialized cells, young progeny, and their tissue-specific stem cells. They perpetually renew, which is evident in many tissues, such as, blood, skin, gastrointestinal tract, respiratory tract, and testis. In contrast, other tissues, including cardiac and neural tissues, seem to have a limited response under regenerative circumstances. Thus, it was considered that cardiac and neural tissues had no regenerative capacities [16, 18]. Asynchronous division is a typical *in vivo* pattern of stem cells; it is defined by the division of one stem cell to give rise to one daughter cell that remains as a stem cell while another undergoes the process of differentiation. The transiently amplifying cells provide an expanded population that differentiates into more mature cells, which can no longer proliferate and eventually die [3, 6, 17]. Such phenomenon is observed in most tissues, however, others, such as the liver [19] and pancreas [20], are mainly driven by replication of mature cells rather than stem cell division and differentiation under normal conditions. Occasionally, in those tissues under toxic DNA-damage conditions when replication of mature cells is compromised, facultative stem cells accomplish the tissue regeneration [19].

Stem Cells and Their Niches. The microenvironment or the stem cell niche tightly regulates the behavior of the cells. A combination of cells and extracellular matrix components, soluble factors delivered to the tissue from the vasculature, and the growth factors produced by the cells govern all aspects of the behavior of stem cells. For example, in the intestinal mucosa, the pericryptal myofibroblasts that surround the crypts may serve as niche cells, whereas in the hair follicles, the region just below the sebaceous glands seems to be a stem cell niche [17, 18]. The niches themselves control many dynamic facets of the stem cells, intrinsically regulating the internal signaling, synthesis of structural and metabolic proteins, their mitotic activities, axes, and growth pattern. These events certainly play a pivotal role in leading intrinsic and extrinsic factors that will define the physiological function of stem cells or the first step in the pathological transformation of stem cells in carcinogenesis [17, 18].

11.3.1.2   General Classification of Stem Cells   Because adult somatic stem cells consist of a different population of the cells that share some common characteristics, it is difficult to rigorously divide the cells into some classifications. A new theory has currently proposed that stem cells are generated from a single cell source [21–24]. Thus, we review here adult stem cells according to the commonest classification as follows:

1. Hematopoietic stem cells (HSCs)
2. Mesenchymal stem cells (MSCs)
3. Multipotent adult progenitor cells (MAPCs) isolated by fluorescent-activated cell sorting (FACS) from the bone marrow
4. Side-population phenotype cells (SPs)
5. Tissue-specific cell progenitors (TSCPs)
6. Umbilical cord blood-derived stem cells (UCBDSs)

Bone Marrow Stem Cells. The bone marrow has been considered to provide an adequate microenvironment for stem cells to survive forever. Stem cells in the bone marrow can migrate into blood vessels, circulate around the body, and return home when needed [25]. A long time ago, only a few stem cells were recognized in humans and they were thought to be a restricted population with a limited potential to differentiate in a single organ system. Such reasoning was examined through the study of bone marrow, which contains a wide range of cell populations. The complex constitution of bone marrow also attracted the interest of scientists. Such investigations changed the original thought to a new one, which implies that stem cells also have the ability to give rise to the cells of other tissue types as well as to the cells of the original, referred to as the "plasticity of stem cells." Bone marrow has been described to contain at least two [26, 27] different types of stem cells. One type includes HSCs, which produce the entire progeny of blood cells of the body, and the other includes MSCs, which are a promising source for tissue repair [28]. It is also believed that another population exists, called MAPCs [27], which could be generated from the adult bone marrow of several species, and can differentiate into multiple cell types *in vitro* [29, 30] (Figure 11.3-1).

11.3.1.3 *General Characteristics of Stem Cells* Stem cells possess a wide range of different characteristics, which have made them an attractive cell source for cell biology, ontogeny, toxicological studies, and cell therapy [12–16, 18, 24, 31]. Several types of stem cells have been reported, and each possesses particular characteristics that determine their potential (Table 11.3-1).

11.3.1.4 *Hematopoietic Stem Cells (HSCs)*

Characteristics. For the past three decades, HSCs have been considered important cells. In 1963, the origin of hematopoietic cells was first reported [33]. One of the main characteristics of HSCs is the capacity to give rise to intermediate precursor

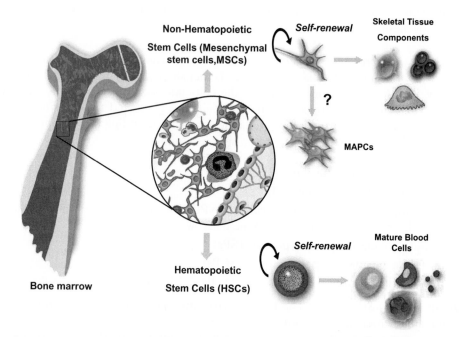

Figure 11.3-1. *Schematic model of bone marrow niche.* Bone marrow (BM) harbors a heterogeneous population of progenitor cells, which can generate a diversity of different cells. BM is composed of the osteoblastic zone and the vascular zone. In the vascular zone, HSCs exist that are capable of giving rise to all hematopoietic progeny, in direct contact with MSCs, which are known as progenitors for skeletal tissue components such as bone, cartilage, hematopoietic-suporting stroma, and adipose tissue. (This figure is available in full color at ftp://ftp.wiley.com/public/sci_tech_med/pharmaceutical_biotech/.)

TABLE 11.3-1. Stem Cell Types and Their General Characteristics

	Tissue-specific cell progenitors[1]	Pluripotent and multipotent stem cells[2]	Embryonic stem cells[3]
Cell commitment	Lineage-commitment*	Uncommitment	Uncommitment
Cell quiescence under neither serum supplemented medium nor analogous knockout replacement	Quiescence	Quiescence	Non-quiescence
Telomerase activity	Absent	Most present but variable activity	High activity
Cell life span	Hayflick's limited	Extended	Unlimited
Cell growth at confluence	Inhibited by cell contact	Inhibited by cell contact	Non-contact inhibition
Inducible plasticity	Limited*	Usually unlimited	Unlimited
In vitro preservation of undifferentiated stage	Possible but limited	Possible but limited	Certainly possible
Cell membrane markers or antigens	Often lineage-commitment*	Various according to their cell phenotypes	Alkaline phosphatase SSEA-1,-3,-4 Oct-3/4

[1] Tissue-specific cell progenitors are commonly quadri-, tri-, or bipotent; therefore, even though they give rise to progeny with different phenotypes, the fate of the progenitor cells is lineage-specific. Each progenitor cell has a unique profile on the cell surface markers, but the profile is similar to that of both the tissue and the progeny that they will originate [6, 16, 32].

[2] Pluripotent and multipotent stem cells are heterogeneous population. These cells are practically the entire populations of postnatal stem cells [16]. They include HSCs, MSCs, SPs, MAPCs, and UCBDSs. Although they have different phenotypes, some of their characteristics are frequently shared.

[3] Embryonic stem cells are the prototype of all stem cells. Characteristics are well-defined because ES cell lines were established.

* Even though these populations are localized in adult tissues having lineage-commitment and exhibiting limited plasticity and local cell markers, some of these cells are capable of de-differentiating and then becoming multipotent stem cells.

or progenitor cell populations that partially differentiate and commit to various types of blood cell lineages [34, 35]. HSCs need to possess the hallmark properties to equilibrate cell self-renewal, whereas the cells quickly generate progenitors as a workforce as well as additional stem cells without depleting the reserves. Therefore, HSCs need to be multipotent; that is, a single HSC can produce several different lineages of mature blood cells and proliferate to yield a broad number of mature progeny. HSCs in the bone marrow of the mice are a rare population with a frequency of 1 in 10,000 to 100,000 of total blood cells, and the cells may be even less in humans [36–38]. HSCs are thought to be relatively in-active in the adult hematopoietic system, with 1–3% in the progression of the cell cycle and approximately 90% in the cell cycle G0. The cells divide only a few or not at all until they are required to differentiate [39–42]. Abnormal clonal expansion of HSCs may result in chronic myelogenous leukemia, polycythemia vera, and myelodysplastic syndromes. HSCs are believed to have the ability to live for a long period of time, probably a lifetime in the recipient after bone marrow transplantation. In fact, HSCs require the bone marrow microenvironment, which regulates their migration, proliferation, and differentiation, to maintain active hematopoiesis throughout their lifetime [43, 44].

Isolation and Phenotypic Characteristics of HSCs. Recent experiments have demonstrated that HSCs in bone marrow from many different species can be purified by FACS as SP cells [40, 45, 46]. In 1994, multipotent cells committed to the hematopoietic lineage in mouse according to their different array of cell-surface markers were isolated [39]. This cells were described as KTLS c-kit$^+$ (K), Thy-1.1^{low} (a marker on stem cells) (T), Lin$^{-/low}$ (Lineage-marker) (L), and Sca-1$^+$ (Stem cell antigen-1) (S) [1]. The cells showed >80% for hematopoietic multilineage differentiation and represented only 1/2000 cells in the bone marrow. HSCs were proposed to be divided into three compartments based on both the expression of the surface markers and their self-renewal ability [39, 47].

1) Long-term HSCs (LT–HSCs): LT–HSCs habitually reside in the bone marrow and have for all intents and purposes six developmental alternatives: remain quiescent, differentiate, self-renew, migrate, enter senescence, or undergo programmed cell death. The cells represent only 0.007% of cells in the bone marrow. In young adult mice, approximately 8% of LT–HSCs arbitrarily enters into cell division per day [42], and half of their progeny are LT–HSCs to maintain the level of steady state [48]. LT–HSCs perform self-renewal perpetually without depleting the pool of stem cells, and the cells are in charge of producing proliferative short-term HSCs. LT–HSCs express the surface markers of Thy1.1^{low}Flk-2$^-$. The expression of Flk-2 is upregulated and the expression of Thy-1.1 is downregulated as self-renewal capacity of the cells diminishes [49].

2) Short-term HSCs (ST–HSCs): ST–HSCs engender the lineage-committed progenitors to produce the billions of differentiated hematopoietic cells in the peripheral blood daily. The self-renewal life span of ST–HSCs is 6–8 weeks, and afterward, the cells fade away from the bone marrow. ST–HSCs represent 0.01% of the cells in the bone marrow of young adult C57BL mice and have the phenotype of thy1.1^{low} Flk-2$^+$.

3) Multipotent progenitor cells (MPs): MPs have restricted self-renewal potential for less than 2 weeks. The expression of thy1.1$^-$Flk-2$^+$ is also identified in MPs [49]. The offspring of HSCs have been characterized and lineage restricted oligopotent progenitor cells for *lymphoid*, common lymphoid progenitor (CLP), and *myeloid*, common myeloid progenitor (CMP), granulocyte-monocyte progenitor (GMP), and megakaryocyte-erythrocyte progenitor (MEP) lineages (Figure 11.3-2).

Comparable Phenotypic Markers of HSCs in Human and Mouse. CD34 was the first marker found in human hematopoietic progenitors [50]. Most human HSCs express CD 34, which is also expressed in the committed progenitors [51] and

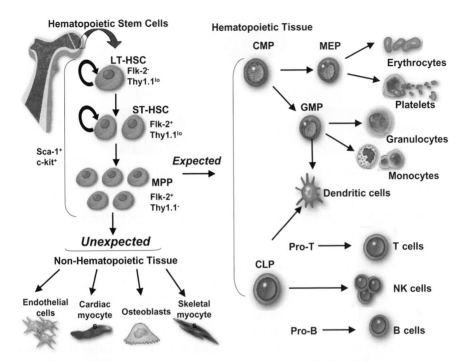

Figure 11.3-2. *Model of hematopoietic stem cell lineage.* Long-term hematopoietic stem cells (LT–HSCs) have the ability of unlimited self-renewal and of generating other precursors that give rise to population of the hematopoietic tissue cells (expected), including mature blood cells, CMPs (common myeloid progenitors), CLPs (common lymphoid progenitors), MEPs (megakaryocytes/erythrocyte progenitors), and GMPs (granulocytes/monocyte progenitosr). HSCs also generate a population of the non-hematopoietic tissue cells (unexpected), including ostoeblasts, skeletal myocytes, cardiac myocytes, and endothelial cells. (This figure is available in full color at ftp://ftp.wiley.com/public/sci_tech_med/pharmaceutical_biotech/.)

nonhematopoietic progenitors [52, 53]. The characteristic phenotype of human HSCs includes the lack of expression of lineage markers (Lin-) and expression of Thy-1, c-Kit, and Sca-1, CD45 [54] without CD38 expression [51, 55, 56]. Nonetheless, some human HSCs can be found in the fraction of CD34-negative population [57, 58]. CD34- negative HSCs are a precursor fraction of CD34-positive HSCs. Human HSCs also express Bcrp, known also as ABCG2 transporter, which outflows particular molecules as Hoechst-33342 staining [40, 59]. Detection of CD34 expression in HSC could be performed by the use of HCC-1 antibody in CD59 family members [60] (the sca-1 [61] antibody for mouse detects CD59 family members). Recently CD133 has been identified and could be used as another marker for human HSCs instead of CD34 [62]. It is important to remember that the CD34$^+$ bone marrow population denotes 1–6% of the cells in bone marrow, whereas the HSC compartment corresponds to only 0.05% [38]. Hence the CD34$^+$ population includes HSCs and a small fraction of the non-HSC cell population. Most quiescent LT–HSCs are CD34$^+$Thy$^+$Lin$^-$CD38 [63] (Figure 11.3-3).

Mechanism of Activation Cycle of HSCs. During the late phase of embryonic development, HSCs are located in the fetal liver, where they undergo massive

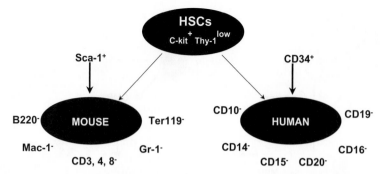

Figure 11.3-3. *Marker profiles of HSCs in human and mouse.* Human HSCs are positive for c-kit[+], Thy-1[low], and CD34[+], and mouse HSCs are positive for c-kit[+], Thy-1[low], and Sca-1[+]. (This figure is available in full color at ftp://ftp.wiley.com/public/sci_tech_med/pharmaceutical_biotech/.)

expansion before they enter into the bone marrow and express AA4.1 and Mac 1, which are usually absent in HSCs in a quiescent state [64]. It has been generally believed that the microenvironment affects the fate of the cells. When HSCs are unperturbed in their quiescence niche, the cells express receptors associated with metabolism and aging (IGF1R) and show the activity of tyrosine kinase Tie1, which allows the cells to respond to multiple mitogenic signals. HSCs also express high levels of transcription factors, such as c-fos and GATA-2, which enable quick activation of the cells. HSCs are ready to act in response to changes in their environment "state of readiness." Immediately encountering the stress, HSCs pause totally by remaining quiescent in their niche, while they prepare to proliferate. This phase is mediated by upregulation of TIMP and serine proteinase inhibitor A-3g and by antiproliferative genes, such as Tob1, p21, and Btg3. Interferon-induced genes are also upregulated, indicating that HSCs are responding to pro-inflammatory signals caused by the stress. The signals induce a proliferative status, which is divided into early and late proliferation phases. In the early phase, the expression of genes involved in the regulation of replication and repair of DNA is enhanced in HSCs [65]. In the late phase when most of HSCs are in cycle, other genes related to energy production are expressed, indicating an increase in the metabolic activity of HSCs. In this context, two proposed approaches are associated with the proliferation state, mobilization [66] and migration [67]. HSCs need to move out of their quiescence niches and enter into a proliferative zone. Downregulation of α-4 integrin is necessary for proliferation of HSCs, and downregulation of c-kit has been linked to mobilization of HSCs [68]. At the final phase, HSCs are required to return to their niches in order to stay at the initial state of bone marrow. This step usually starts the day when the damage is ameliorated, when several cells in cycle decrease, and is related with the expression of antiproliferative genes [69, 70].

Plasticity of HSCs. The hematopoietic tissues such as the bone marrow and peripheral blood have heterogeneous stem cell populations, including hematopoietic stem cells, mesenchymal stem cells, multipotent adult progenitor cells, and endothelial precursor cells. HSCs have been widely used to study the plasticity of adult stem cells [23]. Stem cell plasticity can be defined as a unique property of tissue-specific adult stem cells. Approximately more than 80% of studies reporting plasticity of adult stem cells have been performed using the cells derived from the bone marrow or mobilized peripheral stem cells [23, 71, 72]. Although a previous concept was that differentiation of HSCs was restricted only to a hematopoietic lineage, current studies have reported that HSCs derived from bone marrow can give rise to hematopoetic precursors and multiple "unexpected" cell types, such as neural cells [73–75], hepatic cells [54, 76–78], cardiac muscle cells [79–81], and skeletal muscle cells [82, 83]. Many attempts have been proposed to define the mechanism for how stem cell plasticity occurs. Despite these efforts, there is still no clear evidence regarding whether stem cell plasticity really exists. Some investigators have suggested "trans-differentiation" as a possible cause of stem cell plasticity [84, 85], and others have proposed the cell fusion effect [86]. Even though many theories about plasticity have been proposed, nobody knows for certain whether stem cell plasticity is common or infrequent, whether it might be a vestige of the potential expression during embryonic development, or whether it has a physiologic role in the repair and homeostasis of the tissues. The plasticity of adult HSCs, even though it is rare, has suggested that the environment can reprogram the fate of the cells.

Potential Implications of HSCs. Studies on bone marrow cells differentiating into nonhematopoietic lineages have proposed that the cells are a representa-tive source of cell "replacement" in the treatment of numerous diseases. To amplify and generalize this issue for clinical translation, we should accumulate the useful processes of regeneration of blood formation by HSCs [36], skin replacement by putative epidermal stem cells [87], and the benefits reported in treating patients with myocardial infarction with bone marrow cells, mobilized peripheral HSCs, or hematopoietic progenitors [88, 89].

11.3.1.5 Mesenchymal Stem Cells (MSCs)

Characteristics. MSC is a wide population that exists in the niche of the bone marrow and has attracted the attention of researchers by their potential for self-renewal and differentiation into functional cell types in the intrinsic tissue where they reside. These cells have also been isolated from the fat tissues [90]. MSCs can be expanded in culture for several passages without losing their differentiation potential and have been widely accepted as stem cells due to their usefulness in the clinical treatment of osteogenesis imperfecta [91], bone tissue regeneration [92], and hematopoietic recovery [93].

Self-Renewal and Multilineage Differentiation Potential of MSCs. Characterization of MSCs is subject of active investigation, because diversities of techniques to isolate the cells and different ways to analyze the self-renewal ability of the cells have been documented. Considering the lack of reliable specific markers and locating site of MSCs, further experiments are needed to address these issues. To identify MSCs, a combination of several monoclonal antibodies has been tested [94]. As of now, markers for MSCs include SH-2 [95], SH-3, SH-4 [96], SB-10 [97], CD29, CD44, CD90, and STRO-1 [98]. In addition, MSCs are negative to hematopoietic markers CD34 and CD45 and the cells are human leukocyte antigen (HLA) class I-positive and HLA class II-negative [99]. The first experiment supporting the presence of MSCs was conducted in the early 1960s [100, 101]. Aspirated bone marrow cells were cultured at low density. The resultant cells formed colonies of fibroblasts, and they were responsible for osteogenesis [102, 103]. Murine MSCs showed self-renewal activity, and the cells differentiated into many types of cell lineages, such as osteoblasts and chondroblasts [104–109], adipocytes [110], neuronal progenitors [111], and myocytes [83] with the stimulation of cytokines, growth factors, and chemicals *in vitro*. However, it is controversial whether such plasticity of MSCs is induced *in vivo* even after minimal injury. To address this issue, researchers have used animal models of injury [112, 113]. Such experiments have shown that MSCs contribute to the repair or regeneration of damaged bone, cartilage, and infracted myocardial tissues [114–116].

Isolation of MSCs. Since 1980 when the characterization of human bone marrow fibroblast colony-forming cells (CFU-Fs) was performed, which was the gold standard to identify MSCs, many researchers have attempted to develop different methods for isolation of MSCs [117]. Now MSCs can be identified by their ability to adhere to a static surface and their proliferat-ing potential. Approximately 30% of human bone marrow-aspirated cells adhering to plastic culture dishes are considered to be MSCs [110]. MSCs in the bone marrow are a heterogeneous population that contains not only putative cells but also tripotent (ability to differentiate into osteocyte, chondrocyte, and adipose lineage-osteo/chondro/adipo), bipotent (osteo/chondro), or unipotent cells (osteo) [118–121] (Figure 11.3-4).

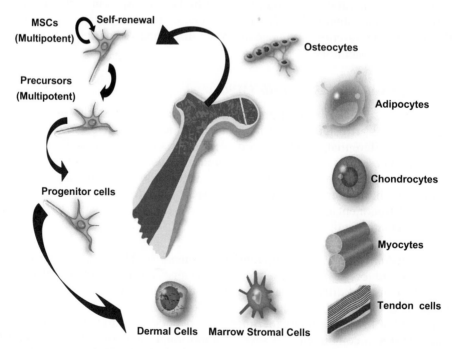

Figure 11.3-4. *Multilineage differentiation of mesenchymal stem cells.* Mesenchymal stem cells (MSCs) have the ability to differentiate into all types of cells of connective tissue. (This figure is available in full color at ftp://ftp.wiley.com/public/sci_tech_med/ pharmaceutical_biotech/.)

Future Prospectives of MSCs. Adult MSCs have shown great promise in cell therapy in humans due to their multipotentiality, capacity for extensive self-renewal, and lack of induction of immune response. MSCs may be a valuable source to introduce foreign genes and could be a useful tool for gene therapy for bone regeneration and tissue engineering [122]. Researchers have also proposed MSCs as a useful means to protect the brain tissue from ischemic damage [123, 124]. MSCs ameliorate functional deficits after stroke in rats probably by releasing protective cytokines by the cells. Another potential clinical application is the use of MSCs as a drug delivery vehicle for the treatment of invasive malignant tumors [125]. The behavior of MSCs has been exploited as a tumor-targeting gene therapy in gliomas. Many studies regarding genetically modified MSCs have revealed extraordinary antitumor effects of the cells in experimental models of gliomas [126–127]. MSCs have the advantage of easy propagation *in vitro*. Moreover, implantation of autologous MSCs into patients with malignant gliomas is ethically nonproblematic.

11.3.1.6 Multipotent Adult Progenitor Cells (MAPCs)

MAPCs are bone marrow-derived stem cells with an extensive *in vitro* expansion ability, more than 80 population doublings, as well as a capacity to differentiate *in vivo* and *in vitro* into tissue cells of all three germinal layers; ectoderm, mesoderm, and endoderm. These cells have been isolated from three different species, including mouse, rat, and human [27].

Isolation and Culture of MAPCs. MAPCs of different species require different isolation and culture procedures. Human MAPCs are isolated after seeding bone marrow mononuclear cells (BMMNCs) of CD45$^-$ GlyA$^-$ at low densities (1–$3 \times 10^3/cm^2$) onto fibronectin-coated plastic dishes, with <2% FCS (fetal calf serum), EGF (epidermal growth factor), and PDGF-BB (platelet-derived growth factor). During the culture, depletion of the mono-nuclear cells is observed. The remnant clones that emerge over time are then harvested, and approximately 20% of the total cells seeded. The cells should be plated again at a density of 0.5–1×10^3 cells/cm^2. Maintenance of the stem cell phenotype is critically dependent on such low-density inoculation. The culture conditions of mouse MAPCs are similar to those described in human MAPC culture, except for the addition of the leukemia inhibitor factor (LIF).

Phenotypic Characteristics and Proliferation Capacity of MAPCs. Human MAPCs are characterized by the lack of hematopoietic markers such as CD34 and CD45 and the presence of the low expression of VCAM, CD44, MHC class I, and endoglin [29, 128]. The cells can be cultured *in vitro* for more than 50–80 population doublings, while preserving their normal karyotype. Mouse MAPCs are positive for CD34, CD45, CD44, c-Kit, CD3, Gr-1, Mac-1, and CD19 and negative for major histocompatibility complex (MHC) class I and class II. The cells express a lower level of Flk1 and Sca1 and a higher level of CD13 and SSEA-1 [129]. Mouse MAPCs express pluripotent markers of Oct-4, rex-1, and nanog, which were observed only in ES cells [129]. The cells can be cultured for more than 80–150 population doublings, but the karyotype of the cells becomes unstable.

In Vitro Differentiation of MAPCs. The differentiation of MAPCs to several lineages is achieved when the cells are re-plated at high density ($1-2 \times 10^4$ cells/cm^2). The cells were cultured in the absence of the serum, but with lineage-specific cytokines required for the differentiation of the desired tissues.

1) **Mesoderm.** By the use of VEFG, the differentiation of MAPCs has been demonstrated into functional mature endothelial cells expressing endothelial markers such as CD31, flk-1, and vWF. Such MAPC-derived endothelial cells are capable of contributing to neoangionesis and wound healing *in vivo* [29, 130].

2) **Ectoderm.** Verfaille et al. have shown the potential characterization of MAPCs to differentiate into functional neuroectodermal cells, including astrocytes, oligodendrocytes, and neurons [27, 131, 132].

3) **Endoderm.** MAPCs can differentiate into hepatocyte-like cells by using FGF4 and HGF. Such cells expressed hepatic markers of CK19, AFP, CK18, albumin, HepPar-1, and CD26 and produced albumin, urea, and glycogen [30].

In vivo Differentiation of MAPCs. The potential of multipotency of MAPCs has also been examined *in vivo* [27] by intravenous transplantation of the cells into postnatal murine recipients treated either without radiation or with sublethal irradiation. The observation of chimaerism in many somatic tissues of the mice derived from blastocysts demonstrated the engraftment of MAPCs in various tissues, including hematopoietic, lung, gut, and liver. In these tissues, MAPCs have acquired phenotypic characteristics of the respective cells, indicating their multipotency.

Future Prospectives of MAPCs. MAPCs have been compared with ES cells, for their similar *in vitro* potential to give rise to tissue cells of all three germ layers. MAPCs have shown the same ability as ES cells *in vivo* experiments, although MAPCs do not develop tumors. Due to the nontumorigenecity, MAPCs can serve as a source for the production of a wide spectrum of transplantable cells with an enormous promise for the treatment of many degenerative or inherited diseases. Taking it into account that MAPCs can be recovered from the patients themselves, the clinical use of the cells would be advantageous, avoiding the need for immunosuppressive therapy. MAPCs provide a prospectively beneficial tool to study developmental biology of the stem cells as well as drug discovery.

11.3.1.7 Side-Population Phenotype Cells (SPs)

Purification of stem cells was reported based on an efflux of fluorescent dyes, such as Rhodamine 123 and Hoechst 33342 [133]. The purified cell population, referred to as SP, has been identified as a small fraction in the bone marrow in all species examined [45]. SP cells can yield about 0.05–0.1% from total bone marrow cells by the use of a flow cytometry cell sorting [40, 45, 59]. Despite being a small population, it possesses interesting capacities, including extensive proliferation and multipotential to generate the entire adult hematopoietic cell lineages [40]. Indeed, only a single transplantation of 200 SP cells can fully repopulate the bone marrow of lethally irradiated mice [40, 134]. It has been reported that a successful expression of dystrophin is achieved in a mouse model of Duchenne's muscular dystrophy after transplantation of SP cells [82]. SP cells have contributed to regenerate the infarcted myocardium by developing cardiocytes and vascular endothelium cells [81]. SP cells have been identified in others several tissues, such as Posterior, by identifying the expression of the ABC transporter Bcrp1/ABCG2 [59]. Because of Hoechst 33342-related toxicity [46], other strategies have been proposed to recover SP cells by using monoclonal antibodies. Even though SP cells seem to be a small source for basic research and possible cell therapy, the cells have shown greater potential. Since SP cells were identified in the bone marrow, the cells have held great promise due to their plasticity and replicating capacity *in vivo* without any tumorigenic transformation [40, 45, 46, 81, 82]. *In vivo* differentiation of SP cells has succeeded in only mesoderm tissues (skeletal muscle, cardiocytes, and vascular cells) [81, 82]. Studies have shown that hematopoietic stem cells in the bone marrow are capable of trans-differentiating and giving rise to neuronal progenitor cells [73, 75, 135], but that such trans-differentiation has not occurred in SP cells [136].

Tissue-Specific SP Phenotype Cells from Different Compartments. For a period of time, the SP phenotype cells were identified in many other different tissues, including the brain [137, 138], pituitary gland [139], skin [140–143], corneal [144] and limbal epithelial ocular tissue [145], mammary glands [146], skeletal muscle [138, 143, 147, 148], lung [134, 138, 149–151], heart [138, 152, 153], liver [138, 154], spleen [138], pancreas [155], small intestine [138], kidney [138, 156–158], testis [159, 160], peripheral blood [138, 161], and in umbilical cord blood [45].

Main Characteristics of SP Cells from Diverse Compartments. A variable number of SP cells have been distributed in the different tissues of the body. The SP cells are a heterogeneous population that possesses the ability to give rise to their lineage-specific progenitor cells [138]. Some of these cells have been analyzed and found to be multipotent [139]. *In vitro* cultures of SP cells have shown differences in their propagation capacities and quiescent state [140, 145, 161].

11.3.1.8 Tissue-Specific Cell Progenitors (TSCPs)

Neural Cell Progenitor. The cells involving the development of mature central nervous system (CNS) are precisely regulated by both temporal and local patterns of differentiation, which determine the appropriate cell function. The role of niches shows plasticity coordinated by both the intrinsic factors and the extrinsic soluble signals. Therefore, the specific neural differentiation is affected by this regional patterning. Neurogenesis involves the opposing soluble signals that determine dorsal and ventral patterning: Sonic hedgehog (Shh), bone morphogenetic protein (BMP) antagonists, chordin, and noggin are secreted from the floor plate, whereas other signals originating from the roof plate result in the creation of the gradient of signal concentrations [162, 163]. Precise concentration and ratio of each independent signal derive the development of the specific neuronal phenotypes at different points in accordance with these gradients. Such phenomena cause the different expression patterns of the cells and create the groups of neurons with different patterns of cell division and differentiation [163]. At the beginning, the CNS stem cells divide symmetrically to enrich their population pool, or asymmetrically to generate more differentiated progeny, which become mature cells of neuronal and glial lineages. The spinal cord can produce both oligodendrocytes and neurons from common precursors, and the final decision of the cell fate depends on extrinsic signals and specific patterns of transcriptional activation [164].

General Characteristics of Neural Progenitor Cells. In rodent models, neural progenitor cells have been identified in specific regions of adult CNS. The cells isolated from different sites of CNS are not identical and thus show different growth characteristics, trophic factor requirements, and specific patterns of differentiation [36, 165]. Growth factor requirements can define at least two major types of neural progenitor cells isolated from CNS. One cell type requires a high dose of EGF and can be expanded as floating neurospheres, while preserving their phenotype after many passages, and eventually they come to be FGF-responsive [166, 167]. In contrast, another cell type isolated from CNS is exclusively FGF-dependent. Such cells can be propagated, as adherent cultures with FGF-dependent and the cells do not respond to EGF. The cells do not express EGF receptor either *in vivo* or *in vitro*. As these cells have been isolated from multiple brain areas, it is likely that both of neural precursor cells coexist. Both cells can self-renew and differentiate into the three neural cell types *in vitro*. Differentiation of the cells in tissue culture can be induced by withdrawal of growth factors or by induction of specific signals [166–168]. It seems that distinct stem cell populations are programmed to differentiate into specific cell types during the CNS development based on their response to growth factors. Moreover, some of these phenotypes expressed in progenitor cells in CNS have been observed in progenitors isolated from the pancreas of adult mice, which can generate neural and pancreatic lineages [169].

Epidermal Progenitor Cells. Keratinocytic populations can be grown *in vitro*, while displaying clonal growth. Keratinocytes can generate normal epidermis when they are appropriately grafted [170, 171]. Furthermore, the proliferative heterogeneity of the keratinocytes reveals the presence of stem cells, that is, epidermal progenitor cells. β1 integrin seems to play a central role in the regulation of the progenitor cells. When β1 integrin binds its ligand of type IV collagen, the cells exhibit high proliferative capacity, while maintaining their undifferentiated state. Therefore, the number and patterning of progenitor cells *in vivo* are autoregulated by the stem cell niche and turnover of the cells provides regulatory feedback signals [172]. The differentiating potential of epidermal progenitor cells has totally denied the previous speculation, implying that the cells were not multipotent [173–175].

Progenitor Cells in the Respiratory Tract. The respiratory tract seems to have a local progenitor population, which gives rise to mature respiratory cells and preserves the native architecture along the tract. Consequently, tissue homeostasis is maintained by these cells, in which the trachea and bronchus segments, bronchiole, and alveolus are maintained by mucous secreting basal cells, by Clara cells, and by type II pneumocytes, respectively. These cells serve as structural and functional parenchyma and work for quick regeneration if any injury occurs in the respiratory tract [176]. In addition, the recent reports have demonstrated that progenitors existing in the populations of basal cells, Clara cells, and type II pneumocytes are not only able to provide tissue function, but also possess renewal capacities [177, 178]. These progenitor cell populations contain both SP cells and hematopoietic markers-expressing cells and are capable of repopulating the bone marrow as HSCs do. These findings have sustained the theory that bone marrow stem cells serve as the whole postnatal stem cells and their own progenies to provide a common interchangeable cell source [23, 177, 178].

Skeletal Muscle Progenitor Cell or Satellite Cells. The embryonic mesoderm generates skeletal musculature. Satellite cells are skeletal muscle progenitor cells that are responsible for postnatal growth and repair [179, 180]. Injury in skeletal musculature results in detriment in skeletal muscle mass, but injury itself triggers muscle regeneration of muscle tissue, which is accomplished by differentiation of native progenitor cells [180]. The tissue-specific cell progenitors are so-called satellite cells [179–181]. Despite the replenishment of the satellite cell pool during muscle growth, the number of satellite cells, which is highest in postnatal muscle, declines with aging [181]. This unique population of the cells exists between the muscle fiber sarcolemma and its covering basement membrane. Anatomically, the satellite cells can be clearly defined. The cells are quiescent lying outside the myofibers but beneath the basement membranes [179, 180]. Quiescent muscle satellite cells are characterized by the expression of surface markers such as M-cadherin, syndecan 3 and 4, and CD34 [180]. Even using a standard dissociation technique of the muscle tissue, an efficient isolation of myogenic progentior cells has not been achieved [179, 180]. Many growth factors are implicated in the regulation, chemotaxis, proliferation, and differentiation of satellite cells [179, 181]. Basic fibroblast growth factor (bFGF), platelet-derived growth factor (PDGF), transferring, and hepatocyte growth factor (HGF) have been identified as potent mitogens for satellite cells. HGF, bFGF, and IGF-1 and TGF-β can also promote chemotaxic activity of satellite cells in tissue culture [179, 182]. Based on their SP phenotype, Hoechst 33342 dye efflux-used FACS has become the easiest way to isolate skeletal muscle progenitor cells. This SP population can generate terminally differentiated skeletal muscle cells and completely repopulate the bone marrow after transplantation [179].

Hepatic Cell Progenitors. Hepatocytes possess the ability to rapidly respond to the parenchymal loss with turning on active mitosis. After massive loss of hepatocytes induced by two-thirds partial hepatectomy, the remaining hepatocytes make a cell-cycle progression at two to three times to restore preoperative cell volume. Such an enormous repopulating capacity is con-ducted by unknown potential and properties of hepatocytes [183]. These crucial properties that defines a stem cell are capable of generating a large progeny. This phenomenon can be observed at least in some hepatocytes [184–187]. Putative hepatic stem cells have been identified by the cell membrane markers, such as the lowered expression of the asialoglyco-protein receptor [188, 189]. Induction of a massive liver damage compromises the regeneration of hepatocytes, and then facultative hepatic progenitor cells distributed along the small branches of the intrahepatic biliary trees are activated to divide. Oval cells that express the same markers as HSCs (c-kit, flt-3, Thy-1, and CD34) transiently amplify the biliary population and then differentiate into hepatocytes

[19, 190, 191]. This particular phenomenon may provide information as to how the liver provides niches for facultative stem cells or offers an opportunity for cell fusion [19].

Progenitor Cells in the Gastrointestinal Tract. The presence of progenitor cells in the intestinal crypts has been observed [17]. Intestinal crypts and gastric glands are considered as specific niches that contain noncommitment and probably multipotent stem cells [17, 192–195]. The epithelial intestinal cells form the lining invaginates with numerous crypts and finger-shaped projections along the gastrointestinal tract and possess a rich population of neuroenteroendocrine cells disseminated throughout their epithelium. The plasticity of progenitor cells in the intestinal epithelium has been shown by differentiation of the cells to insulin-secreting cells using GLP-1 [196]. Intestinal crypts and gastric glands are enclosed by protective fenestrated intestinal subepithelial myofibroblasts and disperse in the lamina propria with merging blood vessels. The myofibroblasts secrete HGF, TGF-beta, and KGF to regulate the differentiation of the intestinal epithelial progenitor cells. Thus, myofibroblast cells play a central role in regulating the differentiation of the progenitor cells located in Lieberkühn crypts [197, 198].

Adult Pancreatic Progenitor Cells. The adult pancreatic tissue is composed of the exocrine and endocrine cells. During embryological development, the pancreatic tissue is generated from differentiation of the ductal epithelium. The exocrine tissue is first differentiated and then endocrine tissue is differentiated [199]. Islets of Langerhans under physiological conditions turn over continuously, but slowly. Terminally differentiated pancreatic beta cells have a life span of approximate 50 days [200, 201]. The cells have a balance of apoptosis and replacement, which is conducted by replication of differentiated beta cells [20]. The replicating capacity of the beta cells is limited [202]. The presence of pancreatic progenitor cells that can give rise to pancreatic and neuronal cell lineages has been demonstrated [169]. Ductal and acinar cells have been proven to have plasticity to differentiate into insulin-secreting cells or to trans-differentiate into hepatocytes [203–209]. The beta cells certainly replicate; however, the senescence of the cells still determines their potential [200, 202]. Otherwise, beta cells may be reprogrammed by cell fusion. *In vitro* culture human islets gradually lose their insulin expression, but they preserve the expression of PDX-1 [208, 210, 211]. Recent research has suggested that adult pancreatic islets contains progenitor cells [206]. Pancreatic progenitor cells may also exist in the non-endocrine tissue [208]. These progenitor cells are capable of generating terminally differentiated endocrine and exocrine progeny [155, 203–206, 212]. They are identified by the positive expression of nestin, referred as to nestin-positive progenitor cells (NIPs) [155, 206, 212], and are dispersed in the whole pancreatic tissue [213]. In addition, SP phenotype cells were observed in NIPs [155] and the cells can give rise to vascular and neuronal progenies [169, 214].

11.3.1.9   Umbilical Cord Blood-Derived Stem Cells (UCBSCs)   Stem cells were isolated from umbilical cord blood (UCB) and cultured *in vitro* [215]. Umbilical cord blood-derived stem cells (UCBSCs) had been proposed as an alternative source of regeneration of hemopoietic tissue by allogeneic transplantation

[216]. The successful hematopoietic reconstitution in a patient with Fanconi's anemia has proved the potential of UCBSCs [217, 218]. Usually small cell fractions are recovered from the cord blood, but the primitive hematopoietic population possesses a high proliferative capacity. UCB cells contain a larger hematopoietic population of $CD34^-$ as well as a subset of $CD34^+$ and $CD38^-$, of which population ratio is about fourfold higher than that in bone marrow cells [219]. Moreover, the most primitive phenotype ($CD34^+CD38^-CD45RA^{low}CD71^{low}Thy-1^+c-kit^{low}Rh^{low}$)- expressing cells present in UCB at 0.003% of the entire population of nucleated cells [220]. Their proliferative capacity reach to a 50-fold expansion of colony forming cells [221, 222]. Successful *in vitro* expansion of the cells depends on the cytokines contained in the culture. Such cytokines include SCF, IL-1, IL-3, IL-6, GM-CSF, G-CSF, M-CSF, erythropoietin, and thrombopoietin [223]. The cells also possess multipotent characteristics and outstanding plasticity [220]. Therefore, there have been several approaches of differentiation of non-hematopoietic tissues ongoing in the clinical tissue regeneration settings [224–236] and *in vitro* toxi- cological studies [237].

11.3.1.10 *Heterogeneous Populations of Stem Cells in Bone Marrow*   Bone marrow contains two main well-defined populations of stem cells, HSCs and MSCs. They renew by themselves and give rise to all of the terminally differentiated blood lineages. In addition, bone marrow possesses a small subset population of non- hematopoietic stem cells (NHSCs). This population of NHSCs is identified by the expression of $CXCR4^+$, $CD34^+$, $AC133^+$, lin^-, and $CD45^-$ [22, 23]. Moreover, there is evidence that tissue damage in the body itself is capable of triggering the bone marrow stem cells to get into the circulation [238–242]. Subsequently, the stem cells modify their phenotype in accordance with the tissue where they migrate. Specific activation of membrane surface receptors and subsequent activation of transcription factors determine a modulation of the phenotypes and migratory properties of the stem cells. Chemotactic or homing signaling is involved in the regulation of the migration. The bone marrow stem cells have an exceptional ability to adapt and survive in different microenvironments. The cells can respond to several cytokines and growth factor, such as HGF, VEGF, LIF, and bFGF. The axis $SDF-1-CXCR4^+$ seems to be the most crucial regulatory factor, which is involved in honing of the stem cells. Therefore, it may be considered that bone marrow stem cells can serve as the source of all adult somatic stem cells [23, 238–242]. UCBSCs are considered as HSCs because of their strong proliferative capacity. UCBSCs certainly display higher potentials compared with adult stem cells [217, 220] and have been proposed as the best candidate for regenerating the damaged tissues in the clinical setting among the stem cells [231, 232].

11.3.1.11 *Uses and Applications of Stem Cells in Toxicology*   Besides the further clinical application of stem cells, utilization of the cells in the pharmacological field is also of great interest. As mentioned, stem cells are specialized cells found within many tissues of the body, in which they have the role to maintain homeostasis and repair of the damage tissue. The advantages of unlimited proliferation and subsequent differentiation of the stem cells can be of great use for *in vitro* screening of the newly developed drugs, allowing us to predict possible adverse effects without

the need to test them in animals. For example, hepatocytes derived from stem cells can be routinely screened for new drugs and chemicals to evaluate their liver toxicity.

References

1. Lagasse E, Shizuru J A, Uchida N, et al. (2001). Toward regenerative medicine. *Immunity.* 14:425–436.

2. Punzel M, Ho A D (2001). Divisional history and pluripotency of human hematopoietic stem cells. *Ann. N Y Acad. Sci.* 938:72–81; discussion 81–82.

3. Punzel M, Liu D, Zhang T, et al. (2003). The symmetry of initial divisions of human hematopoietic progenitors is altered only by the cellular microenvironment. *Exp. Hematol.* 31:339–347.

4. Nagy P (1995). The facultative stem cell: A new star in liver pathology. *Pathol. Oncol. Res.* 1:23–26.

5. Almeida-Porada G, Porada C, Zanjani E D (2001). Adult stem cell plasticity and methods of detection. *Rev. Clin. Exp. Hematol.* 5:26–41.

6. Sell S (2004). Stem cell origin of cancer and differentiation therapy. *Crit. Rev. Oncol. Hematol.* 51:1–28.

7. Huntly B J, Gilliland D G (2005). Leukaemia stem cells and the evolution of cancer-stem-cell research. *Nat. Rev. Cancer.* 5:311–321.

8. Fausto N, Campbell J S (2003). The role of hepatocytes and oval cells in liver regeneration and repopulation. *Mech. Dev.* 120:117–130.

9. Allard E K, Boekelheide K (1996). Fate of Germ Cells in 2,5-Hexanedione-Induced Testicular Injury. *Toxicol. Appl. Pharmacol.* 137:149–156.

10. Miyagi K, Yamazaki T, Tsujino I (2001). Application of hypothermia to autologous stem cell purging. *Cryobiol.* 42:190–195.

11. Czyz J, Wobus A (2001). Embryonic stem cell differentiation: The role of extracellular factors. *Different.* 68:167–174.

12. Rohwedel J, Guan K, Hegert C, et al. (2001). Embryonic stem cells as an in vitro model for mutagenicity, cytotoxicity and embryotoxicity studies: present state and future prospects. *Toxicol. In Vitro.* 15:741–753.

13. Wobus A M, Guan K, Pich U (2001). In vitro differentiation of embryonic stem cells and analysis of cellular phenotypes. *Methods Mol. Biol.* 158:263–286.

14. Davila J C, Cezar G G, Thiede M, et al. (2004). Use and application of stem cells in toxicology. *Toxicol. Sci.* 79:214–223.

15. Wobus A M, Boheler K R (2005). Embryonic stem cells: Prospects for developmental biology and cell therapy. *Physiol. Rev.* 85:635–678.

16. Wagers A J, Weissman I L (2004). Plasticity of adult stem cells. *Cell.* 116:639–648.

17. Alison M R, Poulsom R, Forbes S, et al. (2002). An introduction to stem cells. *J. Pathol.* 197:419–423.

18. Blau H M, Brazelton T R, Weimann J M (2001). The evolving concept of a stem cell: entity or function? *Cell.* 105:829–841.

19. Grompe M (2005). The origin of hepatocytes. *Gastroenterol.* 128:2158–2160.

20. Dor Y, Brown J, Martinez O I, et al. (2004). Adult pancreatic beta-cells are formed by self-duplication rather than stem-cell differentiation. *Nature.* 429:41–46.

21. Kucia M, Ratajczak J, Ratajczak M Z (2005). Are bone marrow stem cells plastic or heterogenous—that is the question. *Exp. Hematol.* 33:613–623.

22. Kucia M, Ratajczak J, Ratajczak M Z (2005). Bone marrow as a source of circulating CXCR4+ tissue-committed stem cells. *Biol. Cell.* 97:133–146.

23. Kucia M, Reca R, Jala V R, et al. (2005). Bone marrow as a home of heterogenous populations of nonhematopoietic stem cells. *Leukemia.* 19:1118–1127.

24. Majka M, Kucia M, Ratajczak M Z (2005). Stem cell biology—a never ending quest for understanding. *Acta Biochim. Pol.* 52:353–358.

25. Lapidot T, Dar A, Kollet O (2005). How do stem cells find their way home? *Blood.* 106:1901–1910.

26. Mazurier F, Doedens M, Gan O I, et al. (2003). Rapid myeloerythroid repopulation after intrafemoral transplantation of NOD-SCID mice reveals a new class of human stem cells. *Nat. Med.* 9:959–963.

27. Jiang Y, Jahagirdar B N, Reinhardt R L, et al. (2002). Pluripotency of mesenchymal stem cells derived from adult marrow. *Nature.* 418:41–49.

28. Bianco P, Riminucci M, Gronthos S, et al. (2001). Bone marrow stromal stem cells: nature, biology, and potential applications. *Stem Cells.* 19:180–192.

29. Reyes M, Lund T, Lenvik T, et al. (2001). Purification and ex vivo expansion of post-natal human marrow mesodermal progenitor cells. *Blood.* 98:2615–2625.

30. Schwartz R E, Reyes M, Koodie L, et al. (2002). Multipotent adult progenitor cells from bone marrow differentiate into functional hepatocyte-like cells. *J. Clin. Invest.* 109:1291–1302.

31. Vats A, Bielby R C, Tolley N S, et al. (2005). Stem cells. *Lancet.* 366:592–602.

32. Young H E, Black A C, Jr (2004). Adult stem cells. *Anat. Rec. A Discov. Mol. Cell. Evol. Biol.* 276:75–102.

33. Becker A J, Mc C E, Till J E (1963). Cytological demonstration of the clonal nature of spleen colonies derived from transplanted mouse marrow cells. *Nature.* 197:452–454.

34. Weissman I L, Anderson D J, Gage F (2001). Stem and progenitor cells: origins, phenotypes, lineage commitments, and transdifferentiations. *Annu. Rev. Cell Dev. Biol.* 17:387–403.

35. Kondo M, Scherer D C, King A G, et al. (2001). Lymphocyte development from hematopoietic stem cells. *Curr. Opin. Genet. Dev.* 11:520–526.

36. Weissman I L (2000). Stem cells: Units of development, units of regeneration, and units in evolution. *Cell.* 100:157–168.

37. Bonnet D (2002). Haematopoietic stem cells. *J. Pathol.* 197:430–440.

38. Morrison S J, Uchida N, Weissman I L (1995). The biology of hematopoietic stem cells. *Annu. Rev. Cell Dev. Biol.* 11:35–71.

39. Morrison S J, Weissman I L (1994). The long-term repopulating subset of hemato-poietic stem cells is deterministic and isolatable by phenotype. *Immunity.* 1:661–673.

40. Goodell M A, Brose K, Paradis G, et al. (1996). Isolation and functional properties of murine hematopoietic stem cells that are replicating in vivo. *J. Exp. Med.* 183:1797–1806.

41. Bradford G B, Williams B, Rossi R, et al. (1997). Quiescence, cycling, and turnover in the primitive hematopoietic stem cell compartment. *Exp. Hematol.* 25:445–453.

42. Cheshier S H, Morrison S J, Liao X, et al. (1999). In vivo proliferation and cell cycle kinetics of long-term self-renewing hematopoietic stem cells. *Proc. Natl. Acad. Sci. USA.* 96:3120–3125.

43. Nibley W E, Spangrude G J (1998). Primitive stem cells alone mediate rapid marrow recovery and multilineage engraftment after transplantation. *Bone Marrow Trans.* 21:345–354.

44. Lanzkron S M, Collector M I, Sharkis S J (1999). Hematopoietic stem cell tracking in vivo: A comparison of short-term and long-term repopulating cells. *Blood*. 93:1916–1921.

45. Goodell M A, Rosenzweig M, Kim H, et al. (1997). Dye efflux studies suggest that hematopoietic stem cells expressing low or undetectable levels of CD34 antigen exist in multiple species. *Nat. Med.* 3:1337–1345.

46. Uchida N, Fujisaki T, Eaves A C, et al. (2001). Transplantable hematopoietic stem cells in human fetal liver have a CD34(+) side population (SP)phenotype. *J. Clin. Invest.* 108:1071–1077.

47. Morrison S J, Wandycz A M, Hemmati H D, et al. (1997). Identification of a lineage of multipotent hematopoietic progenitors. *Develop.* 124:1929–1939.

48. Morrison S J, Wandycz A M, Akashi K, et al. (1996). The aging of hematopoietic stem cells. *Nat. Med.* 2:1011–1016.

49. Christensen J L, Weissman I L (2001). Flk-2 is a marker in hematopoietic stem cell differentiation: A simple method to isolate long-term stem cells. *Proc. Natl. Acad. Sci. USA*. 98:14541–14546.

50. Civin C I, Strauss L C, Brovall C (1984). Antigenic analysis of hematopoiesis. III. A hematopoietic progenitor cell surface antigen defined by a monoclonal antibody raised against KG-1a cells. *J. Immunol.* 133:157–165.

51. Larochelle A, Vormoor J, Hanenberg H (1996). Identification of primitive human hematopoietic cells capable of repopulating NOD/SCID mouse bone marrow: Implications for gene therapy. *Nat. Med.* 2:1329–1337.

52. Asahara T, Murohara T, Sullivan A, et al. (1997). Isolation of putative progenitor endothelial cells for angiogenesis. *Science*. 275:964–967.

53. Peichev M, Naiyer A J, Pereira D, et al. (2000). Expression of VEGFR-2 and AC133 by circulating human CD34(+) cells identifies a population of functional endothelial precursors. *Blood*. 95:952–958.

54. Lagasse E, Connors H, Al-Dhalimy M, et al. (2000). Purified hematopoietic stem cells can differentiate into hepatocytes in vivo. *Nat. Med.* 6:1229–1234.

55. Sutherland H J, Eaves C J, Eaves A C, et al. (1989). Characterization and partial purification of human marrow cells capable of initiating long-term hematopoiesis in vitro. *Blood*. 74:1563–1570.

56. Kawashima I, Zanjani E D, Almaida-Porada G, et al. (1996). CD34+ human marrow cells that express low levels of Kit protein are enriched for long-term marrow-engrafting cells. *Blood*. 87:4136–4142.

57. Nakamura Y, Ando K, Chargui J, et al. (1999). Ex vivo generation of CD34(+) cells from CD34(−) hematopoietic cells. *Blood*. 94:4053–4059.

58. Verfaillie C M, Almeida-Porada G, Wissink S, et al. (2000). Kinetics of engraftment of CD34(−) and CD34(+) cells from mobilized blood differs from that of CD34(−) and CD34(+) cells from bone marrow. *Exp. Hematol.* 28:1071–1079.

59. Zhou S, Schuetz J D, Bunting K D, et al. (2001). The ABC transporter Bcrp1/ABCG2 is expressed in a wide variety of stem cells and is a molecular determinant of the side-population phenotype. *Nat. Med.* 7:1028–1034.

60. Hill B, Rozler E, Travis M, et al. (1996). High-level expression of a novel epitope of CD59 identifies a subset of CD34+ bone marrow cells highly enriched for pluripotent stem cells. *Exp. Hematol.* 24:936–943.

61. Bradfute S B, Graubert T A, Goodell M A (2005). Roles of Sca-1 in hematopoietic stem/progenitor cell function. *Exp. Hematol.* 33:836–843.

62. Yin A H, Miraglia S, Zanjani E D, et al. (1997). AC133, a novel marker for human hematopoietic stem and progenitor cells. *Blood*. 90:5002–5012.

63. Uchida N, Tsukamoto A, He D, et al. (1998). High doses of purified stem cells cause early hematopoietic recovery in syngeneic and allogeneic hosts. *J. Clin. Invest.* 101:961–966.

64. Jordan C T, McKearn J P, Lemischka I R (1990). Cellular and developmental properties of fetal hematopoietic stem cells. *Cell.* 61:953–963.

65. Randall T D, Weissman I L (1997). Phenotypic and functional changes induced at the clonal level in hematopoietic stem cells after 5-fluorouracil treatment. *Blood.* 89:3596–3606.

66. Wright D E, Cheshier S H, Wagers A J, et al. (2001). Cyclophosphamide/granulocyte colony-stimulating factor causes selective mobilization of bone marrow hematopoietic stem cells into the blood after M phase of the cell cycle. *Blood.* 97:2278–2285.

67. Venezia T A, Merchant A A, Ramos C A, et al. (2004). Molecular signatures of proliferation and quiescence in hematopoietic stem cells. *PLoS. Biol.* 2:e301.

68. Heissig B, Hattori K, Dias S, et al. (2002). Recruitment of stem and progenitor cells from the bone marrow niche requires MMP-9 mediated release of kit-ligand. *Cell.* 109:625–637.

69. Levy D E, Lee C K (2002). What does Stat3 do? *J. Clin. Invest.* 109:1143–1148.

70. Raz R, Lee C K, Cannizzaro L A, et al. (1999). Essential role of STAT3 for embryonic stem cell pluripotency. *Proc. Natl. Acad. Sci. USA.* 96:2846–2851.

71. Krause D, Cantley L G (2005). Bone marrow plasticity revisited: Protection or differentiation in the kidney tubule? *J. Clin. Invest.* 115:1705–1708.

72. Kashofer K, Bonnet D (2005). Gene therapy progress and prospects: Stem cell plasticity. *Gene. Ther.* 12:1229–1234.

73. Brazelton T R, Rossi F M, Keshet G I, et al. (2000). From marrow to brain: Expression of neuronal phenotypes in adult mice. *Science.* 290:1775–1779.

74. Eglitis M A, Mezey E (1997). Hematopoietic cells differentiate into both microglia and macroglia in the brains of adult mice. *Proc. Natl. Acad. Sci. USA.* 94:4080–4085.

75. Mezey E, Chandross K J, Harta G, et al. (2000). Turning blood into brain: Cells bearing neuronal antigens generated in vivo from bone marrow. *Science.* 290:1779–1782.

76. Petersen B E, Bowen W C, Patrene K D, et al. (1999). Bone marrow as a potential source of hepatic oval cells. *Science.* 284:1168–1170.

77. Theise N D, Badve S, Saxena R, et al. (2000). Derivation of hepatocytes from bone marrow cells in mice after radiation-induced myeloablation. *Hepatol.* 31:235–240.

78. Alison M R, Poulsom R, Jeffery R, et al. (2000). Hepatocytes from non-hepatic adult stem cells. *Nature.* 406:257.

79. Orlic D, Kajstura J, Chimenti S, et al. (2001). Bone marrow cells regenerate infarcted myocardium. *Nature.* 410:701–705.

80. Orlic D, Kajstura J, Chimenti S, et al. (2001). Mobilized bone marrow cells repair the infarcted heart, improving function and survival. *Proc. Natl. Acad. Sci. USA.* 98:10344–10349.

81. Jackson K A, Majka S M, Wang H, et al. (2001). Regeneration of ischemic cardiac muscle and vascular endothelium by adult stem cells. *J. Clin. Invest.* 107:1395–1402.

82. Gussoni E, Soneoka Y, Strickland C D, et al. (1999). Dystrophin expression in the mdx mouse restored by stem cell transplantation. *Nature.* 401:390–394.

83. Ferrari G, Cusella-De Angelis G, Coletta M, et al. (1998). Muscle regeneration by bone marrow-derived myogenic progenitors. *Science.* 279:1528–1530.

84. Krause D S (2005). Engraftment of bone marrow-derived epithelial cells. *Ann. NY Acad. Sci.* 1044:117–124.

85. Orlic D (2005). BM stem cells and cardiac repair: where do we stand in 2004? *Cytother.* 7:3–15.

86. Terada N, Hamazaki T, Oka M, et al. (2002). Bone marrow cells adopt the phenotype of other cells by spontaneous cell fusion. *Nature.* 416:542–545.

87. Pellegrini G, Ranno R, Stracuzzi G, et al. (1999). The control of epidermal stem cells (holoclones) in the treatment of massive full-thickness burns with autologous keratinocytes cultured on fibrin. *Transplant.* 68:868–879.

88. Stamm C, Westphal B, Kleine H D, et al. (2003). Autologous bone-marrow stem-cell transplantation for myocardial regeneration. *Lancet.* 361:45–46.

89. Strauer B E, Brehm M, Zeus T, et al. (2002). Repair of infarcted myocardium by autologous intracoronary mononuclear bone marrow cell transplantation in humans. *Circulat.* 106:1913–1918.

90. Zuk P A, Zhu M, Mizuno H, et al. (2001). Multilineage cells from human adipose tissue: implications for cell-based therapies. *Tissue Eng.* 7:211–228.

91. Horwitz E M, Gordon P L, Koo W K, et al. (2002). Isolated allogeneic bone marrow-derived mesenchymal cells engraft and stimulate growth in children with osteogenesis imperfecta: Implications for cell therapy of bone. *Proc. Natl. Acad. Sci. USA.* 99:8932–8939.

92. Petite H, Viateau V, Bensaid W, et al. (2000). Tissue-engineered bone regeneration. *Nat. Biotechnol.* 18:959–963.

93. Koc O N, Gerson S L, Cooper B W, et al. (2000). Rapid hematopoietic recovery after coinfusion of autologous-blood stem cells and culture-expanded marrow mesenchymal stem cells in advanced breast cancer patients receiving high-dose chemotherapy. *J. Clin. Oncol.* 18:307–316.

94. Haynesworth S E, Baber M A, Caplan A I (1992). Cell surface antigens on human marrow-derived mesenchymal cells are detected by monoclonal antibodies. *Bone.* 13:69–80.

95. Barry F P, Boynton R E, Haynesworth S, et al. (1999). The monoclonal antibody SH-2, raised against human mesenchymal stem cells, recognizes an epitope on endoglin (CD105). *Biochem. Biophys. Res. Commun.* 265:134–139.

96. Barry F, Boynton R, Murphy M, et al. (2001). The SH-3 and SH-4 antibodies recognize distinct epitopes on CD73 from human mesenchymal stem cells. *Biochem. Biophys. Res. Commun.* 289:519–524.

97. Bruder S P, Ricalton N S, Boynton R E, et al. (1998). Mesenchymal stem cell surface antigen SB-10 corresponds to activated leukocyte cell adhesion molecule and is involved in osteogenic differentiation. *J. Bone Miner. Res.* 13:655–663.

98. Simmons P J, Torok-Storb B (1991). Identification of stromal cell precursors in human bone marrow by a novel monoclonal antibody, STRO-1. *Blood.* 78:55–62.

99. Le Blanc K, Tammik C, Rosendahl K, et al. (2003). HLA expression and immunologic properties of differentiated and undifferentiated mesenchymal stem cells. *Exp. Hematol.* 31:890–896.

100. Friedenstein A J, Piatetzky S, II, Petrakova K V (1966). Osteogenesis in transplants of bone marrow cells. *J. Embryol. Exp. Morphol.* 16:381–390.

101. Friedenstein A J, Gorskaja J F, Kulagina N N (1976). Fibroblast precursors in normal and irradiated mouse hematopoietic organs. *Exp. Hematol.* 4:267–274.

102. Keating A, Singer J W, Killen P D, et al. (1982). Donor origin of the in vitro haematopoietic microenvironment after marrow transplantation in man. *Nature.* 298:280–283.

103. Owen M, Friedenstein A J (1988). Stromal stem cells: Marrow-derived osteogenic precursors. *Ciba Found Symp.* 136:42–60.

104. Beresford J N (1989). Osteogenic stem cells and the stromal system of bone and marrow. *Clin. Orthop. Relat. Res.* 240:270–280.

105. Caplan A I (1991). Mesenchymal stem cells. *J. Orthop. Res.* 9:641–650.

106. Bruder S P, Jaiswal N, Haynesworth S E (1997). Growth kinetics, self-renewal, and the osteogenic potential of purified human mesenchymal stem cells during extensive subcultivation and following cryopreservation. *J. Cell. Biochem.* 64:278–294.

107. Jaiswal N, Haynesworth S E, Caplan A I, et al. (1997). Osteogenic differentiation of purified, culture-expanded human mesenchymal stem cells in vitro. *J. Cell. Biochem.* 64:295–312.

108. Bruder S P, Kurth A A, Shea M, et al. (1998). Bone regeneration by implantation of purified, culture-expanded human mesenchymal stem cells. *J. Orthop. Res.* 16:155–162.

109. Kadiyala S, Young R G, Thiede M A, et al. (1997). Culture expanded canine mesenchymal stem cells possess osteochondrogenic potential in vivo and in vitro. *Cell Transplant.* 6:125–134.

110. Prockop D J (1997). Marrow stromal cells as stem cells for nonhematopoietic tissues. *Science.* 276:71–74.

111. Woodbury D, Schwarz E J, Prockop D J, et al. (2000). Adult rat and human bone marrow stromal cells differentiate into neurons. *J. Neurosci. Res.* 61:364–370.

112. Hofstetter C P, Schwarz E J, Hess D, et al. (2002). Marrow stromal cells form guiding strands in the injured spinal cord and promote recovery. *Proc. Natl. Acad. Sci. USA.* 99:2199–2204.

113. Kopen G C, Prockop D J, Phinney D G (1999). Marrow stromal cells migrate throughout forebrain and cerebellum, and they differentiate into astrocytes after injection into neonatal mouse brains. *Proc. Natl. Acad. Sci. USA.* 96:10711–10716.

114. Bianco P, Gehron Robey P (2000). Marrow stromal stem cells. *J. Clin. Invest.* 105:1663–1668.

115. Pereira R F, O'Hara M D, Laptev A V, et al. (1998). Marrow stromal cells as a source of progenitor cells for nonhematopoietic tissues in transgenic mice with a phenotype of osteogenesis imperfecta. *Proc. Natl. Acad. Sci. USA.* 95:1142–1147.

116. Toma C, Pittenger M F, Cahill K S, et al. (2002). Human mesenchymal stem cells differentiate to a cardiomyocyte phenotype in the adult murine heart. *Circulat.* 105:93–98.

117. Castro-Malaspina H, Gay R E, Resnick G (1980). Characterization of human bone marrow fibroblast colony-forming cells (CFU-F) and their progeny. *Blood.* 56:289–301.

118. Pittenger M F, Mackay A M, Beck S C, et al. (1999). Multilineage potential of adult human mesenchymal stem cells. *Science.* 284:143–147.

119. Muraglia A, Cancedda R, Quarto R (2000). Clonal mesenchymal progenitors from human bone marrow differentiate in vitro according to a hierarchical model. *J. Cell. Sci.* 113 (Pt 7):1161–1166.

120. Bianco P, Cossu G (1999). Uno, nessuno e centomila: Searching for the identity of mesodermal progenitors. *Exp. Cell. Res.* 251:257–263.

121. Bianco P, Riminucci M, Kuznetsov S, et al. (1999). Multipotential cells in the bone marrow stroma: Regulation in the context of organ physiology. *Crit. Rev. Eukaryot. Gene Expr.* 9:159–173.

122. Tsuda H, Wada T, Ito Y, et al. (2003). Efficient BMP2 gene transfer and bone formation of mesenchymal stem cells by a fiber-mutant adenoviral vector. *Mol. Ther.* 7:354–365.

123. Kurozumi K, Nakamura K, Tamiya T, et al. (2004). BDNF gene-modified mesenchymal stem cells promote functional recovery and reduce infarct size in the rat middle cerebral artery occlusion model. *Mol. Ther.* 9:189–197.

124. Kurozumi K, Nakamura K, Tamiya T, et al. (2005). Mesenchymal stem cells that produce neurotrophic factors reduce ischemic damage in the rat middle cerebral artery occlusion model. *Mol. Ther.* 11:96–104.

125. Nakamura K, Ito Y, Kawano Y, et al. (2004). Antitumor effect of genetically engineered mesenchymal stem cells in a rat glioma model. *Gene Ther.* 11:1155–1164.

126. Benedetti S, Pirola B, Pollo B (2000). Gene therapy of experimental brain tumors using neural progenitor cells. *Nat. Med.* 6:447–450.

127. Ehtesham M, Kabos P, Kabosova A (2002). The use of interleukin 12-secreting neural stem cells for the treatment of intracranial glioma. *Cancer Res.* 62:5657–5663.

128. Reyes M, Verfaillie C M (2001). Characterization of multipotent adult progenitor cells, a subpopulation of mesenchymal stem cells. *Ann. NY Acad. Sci.* 938:231–233, discussion 233–235.

129. Thomson J A, Itskovitz-Eldor J, Shapiro S S, et al. (1998). Embryonic stem cell lines derived from human blastocysts. *Science.* 282:1145–1147.

130. Reyes M, Dudek A, Jahagirdar B, et al. (2002). Origin of endothelial progenitors in human postnatal bone marrow. *J. Clin. Invest.* 109:337–346.

131. Keene C D, Ortiz-Gonzalez X R, Jiang Y (2003). Neural differentiation and incorporation of bone marrow-derived multipotent adult progenitor cells after single cell transplantation into blastocyst stage mouse embryos. *Cell Transplant.* 12:201–213.

132. Jiang Y, Henderson D, Blackstad M, et al. (2003). Neuroectodermal differentiation from mouse multipotent adult progenitor cells. *Proc. Natl. Acad. Sci. USA.* 100(suppl 1):11854–11860.

133. Wolf N S, Kone A, Priestley G V, et al. (1993). In vivo and in vitro characterization of long-term repopulating primitive hematopoietic cells isolated by sequential Hoechst 33342-rhodamine 123 FACS selection. *Exp. Hematol.* 21:614–622.

134. Summer R, Kotton D N, Sun X, et al. (2004). Translational physiology: Origin and phenotype of lung side population cells. *Am. J. Physiol. Lung. Cell Mol. Physiol.* 287: L477–483.

135. Mezey E, Chandross K J (2000). Bone marrow: A possible alternative source of cells in the adult nervous system. *Eur. J. Pharmacol.* 405:297–302.

136. Castro R F, Jackson K A, Goodell M A, et al. (2002). Failure of bone marrow cells to transdifferentiate into neural cells in vivo. *Science.* 297:1299.

137. Murayama A, Matsuzaki Y, Kawaguchi A, et al. (2002). Flow cytometric analysis of neural stem cells in the developing and adult mouse brain. *J. Neurosci. Res.* 69:837–847.

138. Asakura A, Rudnicki M A (2002). Side population cells from diverse adult tissues are capable of in vitro hematopoietic differentiation. *Exp. Hematol.* 30:1339–1345.

139. Chen J, Hersmus N, Van Duppen V, et al. (2005). The adult pituitary contains a cell population displaying stem/progenitor cell and early embryonic characteristics. *Endocrinol.* 146:3985–3998.

140. Terunuma A, Jackson K L, Kapoor V, et al. (2003). Side population keratinocytes resembling bone marrow side population stem cells are distinct from label-retaining keratinocyte stem cells. *J. Invest. Dermatol.* 121:1095–1103.

141. Yano S, Ito Y, Fujimoto M, et al. (2005). Characterization and localization of side population cells in mouse skin. *Stem Cells.* 23:834–841.

142. Montanaro F, Liadaki K, Volinski J, et al. (2003). Skeletal muscle engraftment potential of adult mouse skin side population cells. *Proc. Natl. Acad. Sci. USA.* 100:9336–9341.

143. Montanaro F, Liadaki K, Schienda J, et al. (2004). Demystifying SP cell purification: Viability, yield, and phenotype are defined by isolation parameters. *Exp. Cell Res.* 298:144–154.

144. Budak M T, Alpdogan O S, Zhou M (2005). Ocular surface epithelia contain ABCG2-dependent side population cells exhibiting features associated with stem cells. *J. Cell Sci.* 118:1715–1724.

145. Umemoto T, Yamato M, Nishida K, et al. (2006). Limbal epithelial side-population cells have stem cell-like properties, including quiescent state. *Stem Cells.* 24:86–94.

146. Alvi A J, Clayton H, Joshi C, et al. (2003). Functional and molecular characterisation of mammary side population cells. *Breast Cancer Res.* 5:R1–8.

147. Liadaki K, Kho A T, Sanoudou D, et al. (2005). Side population cells isolated from different tissues share transcriptome signatures and express tissue-specific markers. *Exp. Cell Res.* 303:360–374.

148. Asakura A, Seale P, Girgis-Gabardo A, et al. (2002). Myogenic specification of side population cells in skeletal muscle. *J. Cell. Biol.* 159:123–134.

149. Giangreco A, Shen H, Reynolds S D, et al. (2004). Molecular phenotype of airway side population cells. *Am. J. Physiol. Lung Cell. Mol. Physiol.* 286:L624–630.

150. Majka S M, Beutz M A, Hagen M, et al. (2005). Identification of novel resident pulmonary stem cells: Form and function of the lung side population. *Stem Cells.* 23:1073–1081.

151. Summer R, Kotton D N, Liang S, et al. (2005). Embryonic lung side population cells are hematopoietic and vascular precursors. *Am. J. Respir. Cell. Mol. Biol.* 33:32–40.

152. Pfister O, Mouquet F, Jain M, et al. (2005). CD31– but Not CD31+ cardiac side population cells exhibit functional cardiomyogenic differentiation. *Circ. Res.* 97:52–61.

153. Meissner K, Heydrich B, Jedlitschky G, et al. (2005). The ATP-binding cassette transporter ABCG2 (BCRP), a marker for side population stem cells, is expressed in human heart. *J. Histochem. Cytochem.* 54:215–221.

154. Shimano K, Satake M, Okaya A, et al. (2003). Hepatic oval cells have the side population phenotype defined by expression of ATP-binding cassette transporter ABCG2/BCRP1. *Am. J. Pathol.* 163:3–9.

155. Lechner A, Leech C A, Abraham E J, et al. (2002). Nestin-positive progenitor cells derived from adult human pancreatic islets of Langerhans contain side population (SP) cells defined by expression of the ABCG2 (BCRP1) ATP-binding cassette transporter. *Biochem. Biophys. Res. Commun.* 293:670–674.

156. Hishikawa K, Marumo T, Miura S, et al. (2005). Musculin/MyoR is expressed in kidney side population cells and can regulate their function. *J. Cell. Biol.* 169:921–928.

157. Hishikawa K, Marumo T, Miura S, et al. (2005). Leukemia inhibitory factor induces multi-lineage differentiation of adult stem-like cells in kidney via kidney-specific cadherin 16. *Biochem. Biophys. Res. Commun.* 328:288–291.

158. Iwatani H, Ito T, Imai E, et al. (2004). Hematopoietic and nonhematopoietic potentials of Hoechst(low)/side population cells isolated from adult rat kidney. *Kidney Int.* 65:1604–1614.

159. Falciatori I, Borsellino G, Haliassos N, et al. (2004). Identification and enrichment of spermatogonial stem cells displaying side-population phenotype in immature mouse testis. *Faseb. J.* 18:376–378.

160. Lassalle B, Bastos H, Louis J P (2004). "Side Population" cells in adult mouse testis express Bcrp1 gene and are enriched in spermatogonia and germinal stem cells. *Develop.* 131:479–487.

161. Preffer F I, Dombkowski D, Sykes M, et al. (2002). Lineage-negative side-population (SP) cells with restricted hematopoietic capacity circulate in normal human adult blood: Immunophenotypic and functional characterization. *Stem Cells.* 20:417–427.

162. Briscoe J, Chen Y, Jessell T M, et al. (2001). A hedgehog-insensitive form of patched provides evidence for direct long-range morphogen activity of sonic hedgehog in the neural tube. *Mol. Cell* 7:1279–1291.

163. Briscoe J, Ericson J (2001). Specification of neuronal fates in the ventral neural tube. *Curr. Opin. Neurobiol.* 11:43–49.

164. Jessell T M (2000). Neuronal specification in the spinal cord: Inductive signals and transcriptional codes. *Nat. Rev. Genet.* 1:20–29.

165. Morrison S J (2001). Neuronal potential and lineage determination by neural stem cells. *Curr. Opin. Cell Biol.* 13:666–672.

166. Ciccolini F, Svendsen C N (1998). Fibroblast growth factor 2 (FGF-2) promotes acquisition of epidermal growth factor (EGF) responsiveness in mouse striatal precursor cells: Identification of neural precursors responding to both EGF and FGF-2. *J. Neurosci.* 18:7869–7880.

167. Weiss S, Dunne C, Hewson J, et al. (1996). Multipotent CNS stem cells are present in the adult mammalian spinal cord and ventricular neuroaxis. *J. Neurosci.* 16:7599–7609.

168. Palmer T D, Takahashi J, Gage F H (1997). The adult rat hippocampus contains primordial neural stem cells. *Mol. Cell Neurosci.* 8:389–404.

169. Seaberg R M, Smukler S R, Kieffer T J (2004). Clonal identification of multipotent precursors from adult mouse pancreas that generate neural and pancreatic lineages. *Nat. Biotechnol.* 22:1115–1124.

170. Compton C C, Nadire K B, Regauer S, et al. (1998). Cultured human sole-derived keratinocyte grafts re-express site-specific differentiation after transplantation. *Different.* 64:45–53.

171. Jones P H, Watt F M (1993). Separation of human epidermal stem cells from transit amplifying cells on the basis of differences in integrin function and expression. *Cell.* 73:713–724.

172. Levy L, Broad S, Diekmann D (2000). beta1 integrins regulate keratinocyte adhesion and differentiation by distinct mechanisms. *Mol. Biol. Cell.* 11:453–466.

173. Fernandes K J, McKenzie I A, Mill P, et al. (2004). A dermal niche for multipotent adult skin-derived precursor cells. *Nat. Cell. Biol.* 6:1082–1093.

174. Liang L, Bickenbach J R (2002). Somatic epidermal stem cells can produce multiple cell lineages during development. *Stem Cells.* 20:21–31.

175. Toma J G, Akhavan M, Fernandes K J, et al. (2001). Isolation of multipotent adult stem cells from the dermis of mammalian skin. *Nat. Cell. Biol.* 3:778–784.

176. Borthwick D W, Shahbazian M, Krantz Q T, et al. (2001). Evidence for stem-cell niches in the tracheal epithelium. *Am. J. Respir. Cell. Mol. Biol.* 24:662–670.

177. Griffiths M J, Bonnet D, Janes S M (2005). Stem cells of the alveolar epithelium. *Lancet.* 366:249–260.

178. Kotton D N, Summer R, Fine A (2004). Lung stem cells: New paradigms. *Exp. Hematol.* 32:340–343.

179. Charge S B, Rudnicki M A (2004). Cellular and molecular regulation of muscle regeneration. *Physiol. Rev.* 84:209–238.

180. Montarras D, Morgan J, Collins C, et al. (2005). Direct isolation of satellite cells for skeletal muscle regeneration. *Science.* 309:2064–2067.

181. Morgan J E, Partridge T A (2003). Muscle satellite cells. *Int. J. Biochem. Cell. Biol.* 35:1151–1156.

182. Bischoff R (1997). Chemotaxis of skeletal muscle satellite cells. *Dev. Dyn.* 208: 505–515.

183. Forbes S, Vig P, Poulsom R, et al. (2002). Hepatic stem cells. *J. Pathol.* 197:510–518.

184. Overturf K, Al-Dhalimy M, Finegold M, et al. (1999). The repopulation potential of hepatocyte populations differing in size and prior mitotic expansion. *Am. J. Pathol.* 155:2135–2143.

185. Overturf K, al-Dhalimy M, Ou C N, et al. (1997). Serial transplantation reveals the stem-cell-like regenerative potential of adult mouse hepatocytes. *Am. J. Pathol.* 151:1273–1280.

186. Tateno C, Yoshizato K (1996). Growth and differentiation in culture of clonogenic hepatocytes that express both phenotypes of hepatocytes and biliary epithelial cells. *Am. J. Pathol.* 149:1593–1605.

187. Tateno C, Yoshizato K (1996). Long-term cultivation of adult rat hepatocytes that undergo multiple cell divisions and express normal parenchymal phenotypes. *Am. J. Pathol.* 148:383–392.

188. Hirose S, Ise H, Uchiyama M, et al. (2001). Regulation of asialoglycoprotein receptor expression in the proliferative state of hepatocytes. *Biochem. Biophys. Res. Commun.* 287:675–681.

189. Ise H, Sugihara N, Negishi N, et al. (2001). Low asialoglycoprotein receptor expression as markers for highly proliferative potential hepatocytes. *Biochem. Biophys. Res. Commun.* 285:172–182.

190. Paku S, Schnur J, Nagy P, et al. (2001). Origin and structural evolution of the early proliferating oval cells in rat liver. *Am. J. Pathol.* 158:1313–1323.

191. Petersen B E, Goff J P, Greenberger J S, et al. (1998). Hepatic oval cells express the hematopoietic stem cell marker Thy-1 in the rat. *Hepatol.* 27:433–445.

192. Cohn S M, Roth K A, Birkenmeier E H, et al. (1991). Temporal and spatial patterns of transgene expression in aging adult mice provide insights about the origins, organization, and differentiation of the intestinal epithelium. *Proc. Natl. Acad. Sci. USA.* 88:1034–1038.

193. Gordon J I, Schmidt G H, Roth K A (1992). Studies of intestinal stem cells using normal, chimeric, and transgenic mice. *Faseb. J.* 6:3039–3050.

194. Karam S M (1999). Lineage commitment and maturation of epithelial cells in the gut. *Front. Biosci.* 4:D286–298.

195. Roth K A, Hermiston M L, Gordon J I (1991). Use of transgenic mice to infer the biological properties of small intestinal stem cells and to examine the lineage relationships of their descendants. *Proc. Natl. Acad. Sci. USA.* 88:9407–9411.

196. Suzuki A, Nakauchi H, Taniguchi H (2003). Glucagon-like peptide 1 (1–37) converts intestinal epithelial cells into insulin-producing cells. *Proc. Natl. Acad. Sci. USA.* 100:5034–5039.

197. Powell D W, Mifflin R C, Valentich J D (1999). Myofibroblasts. II. Intestinal subepithelial myofibroblasts. *Am. J. Physiol.* 277:C183–201.

198. Powell D W, Mifflin R C, Valentich J D, et al. (1999). Myofibroblasts. I. Paracrine cells important in health and disease. *Am. J. Physiol.* 277:C1–9.

199. Gittes G K, Galante P E, Hanahan D, et al. (1996). Lineage-specific morphogenesis in the developing pancreas: Role of mesenchymal factors. *Develop.* 122:439–447.

200. Finegood D T, Scaglia L, Bonner-Weir S (1995). Dynamics of beta-cell mass in the growing rat pancreas. Estimation with a simple mathematical model. *Diabetes.* 44:249–256.

201. Hunziker E, Stein M (2000). Nestin-expressing cells in the pancreatic islets of Langerhans. *Biochem. Biophys. Res. Commun.* 271:116–119.

202. Bonner-Weir S, Toschi E, Inada A, et al. (2004). The pancreatic ductal epithelium serves as a potential pool of progenitor cells. *Pediatr. Diabetes.* 5(suppl 2):16–22.

203. Lechner A, Nolan A L, Blacken R A, et al. (2005). Redifferentiation of insulin-secreting cells after in vitro expansion of adult human pancreatic islet tissue. *Biochem. Biophys. Res. Commun.* 327:581–588.

204. Song K H, Ko S H, Ahn Y B, et al. (2004). In vitro transdifferentiation of adult pancreatic acinar cells into insulin-expressing cells. *Biochem. Biophys. Res. Commun.* 316:1094–1100.

205. Lardon J, De Breuck S, Rooman I, et al. (2004). Plasticity in the adult rat pancreas: transdifferentiation of exocrine to hepatocyte-like cells in primary culture. *Hepatol.* 39:1499–1507.

206. Zulewski H, Abraham E J, Gerlach M J, et al. (2001). Multipotential nestin-positive stem cells isolated from adult pancreatic islets differentiate ex vivo into pancreatic endocrine, exocrine, and hepatic phenotypes. *Diabetes.* 50:521–533.

207. Heller R S, Stoffers D A, Bock T, et al. (2001). Improved glucose tolerance and acinar dysmorphogenesis by targeted expression of transcription factor PDX-1 to the exocrine pancreas. *Diabetes.* 50:1553–1561.

208. Bonner-Weir S, Taneja M, Weir G C, et al. (2000). In vitro cultivation of human islets from expanded ductal tissue. *Proc. Natl. Acad. Sci. USA.* 97:7999–8004.

209. Bogdani M, Lefebvre V, Buelens N, et al. (2003). Formation of insulin-positive cells in implants of human pancreatic duct cell preparations from young donors. *Diabetol.* 46:830–838.

210. Beattie G M, Montgomery A M, Lopez A D, et al. (2002). A novel approach to increase human islet cell mass while preserving beta-cell function. *Diabetes.* 51:3435–3439.

211. Beattie G M, Itkin-Ansari P, Cirulli V, et al. (1999). Sustained proliferation of PDX-1+ cells derived from human islets. *Diabetes.* 48:1013–1019.

212. Kim S Y, Lee S H, Kim B M, et al. (2004). Activation of nestin-positive duct stem (NPDS) cells in pancreas upon neogenic motivation and possible cytodifferentiation into insulin-secreting cells from NPDS cells. *Dev. Dyn.* 230:1–11.

213. Ueno H, Yamada Y, Watanabe R, et al. (2005). Nestin-positive cells in adult pancreas express amylase and endocrine precursor Cells. *Pancreas.* 31:126–131.

214. Treutelaar M K, Skidmore J M, Dias-Leme C L, et al. (2003). Nestin-lineage cells contribute to the microvasculature but not endocrine cells of the islet. *Diabetes.* 52:2503–2512.

215. Knudtzon S (1974). In vitro growth of granulocytic colonies from circulating cells in human cord blood. *Blood.* 43:357–361.

216. Nakahata T, Ogawa M (1982). Hemopoietic colony-forming cells in umbilical cord blood with extensive capability to generate mono- and multipotential hemopoietic progenitors. *J. Clin. Invest.* 70:1324–1328.

217. Gluckman E, Broxmeyer H A, Auerbach A D, et al. (1989). Hematopoietic reconstitution in a patient with Fanconi's anemia by means of umbilical-cord blood from an HLA-identical sibling. *N. Engl. J. Med.* 321:1174–1178.

218. Cohen Y, Nagler A (2004). Umbilical cord blood transplantation–how, when and for whom? *Blood Rev.* 18:167–179.

219. de Wynter E A, Testa N G (2001). Interest of cord blood stem cells. *Biomed. Pharmacother.* 55:195–200.

220. Mayani H, Lansdorp P M (1998). Biology of human umbilical cord blood-derived hematopoietic stem/progenitor cells. *Stem Cells.* 16:153–165.

221. Donaldson C, Denning-Kendall P, Bradley B, et al. (2001). The CD34(+)CD38(neg) population is significantly increased in haemopoietic cell expansion cultures in serum-free compared to serum-replete conditions: dissociation of phenotype and function. *Bone Marrow Trans.* 27:365–371.

222. Hows J M (2001). Status of umbilical cord blood transplantation in the year 2001. *J. Clin. Pathol.* 54:428–434.

223. Moore M A, Hoskins I (1994). Ex vivo expansion of cord blood-derived stem cells and progenitors. *Blood Cells.* 20:468–479; discussion 479–481.

224. Buzanska L, Habich A, Jurga M, et al. (2005). Human cord blood-derived neural stem cell line-Possible implementation in studying neurotoxicity. *Toxicol In Vitro.* 19(7):991–999.

225. Bieback K, Kern S, Kluter H, et al. (2004). Critical parameters for the isolation of mesenchymal stem cells from umbilical cord blood. *Stem Cells.* 22:625–634.

226. Sanberg P R, Willing A E, Garbuzova-Davis S, et al. (2005). Umbilical cord blood-derived stem cells and brain repair. *Ann. N.Y. Acad. Sci.* 1049:67–83.

227. Chen N, Hudson J E, Walczak P, et al. (2005). Human umbilical cord blood progenitors: The potential of these hematopoietic cells to become neural. *Stem Cells.* 23:1560–1570.

228. Watt S M, Contreras M (2005). Stem cell medicine: Umbilical cord blood and its stem cell potential. *Semin. Fetal Neonat. Med.* 10:209–220.

229. Schmidt D, Breymann C, Weber A, et al. (2004). Umbilical cord blood derived endothelial progenitor cells for tissue engineering of vascular grafts. *Ann. Thorac. Surg.* 78:2094–2098.

230. Ma N, Stamm C, Kaminski A, et al. (2005). Human cord blood cells induce angiogenesis following myocardial infarction in NOD/scid-mice. *Cardiovasc. Res.* 66:45–54.

231. Leor J, Guetta E, Feinberg M S (2006). Human umbilical cord blood-derived CD133+ cells enhance function and repair of the infarcted myocardium. *Stem Cells.* 24:772–780.

232. Korbling M, Robinson S, Estrov Z, et al. (2005). Umbilical cord blood-derived cells for tissue repair. *Cytother.* 7:258–261.

233. Ishikawa F, Yasukawa M, Yoshida S, et al. (2004). Human cord blood- and bone marrow-derived CD34+ cells regenerate gastrointestinal epithelial cells. *Faseb J.* 18:1958–1960.

234. Pessina A, Eletti B, Croera C, et al. (2004). Pancreas developing markers expressed on human mononucleated umbilical cord blood cells. *Biochem. Biophys. Res. Commun.* 323:315–322.

235. Gang E J, Jeong J A, Hong S H, et al. (2004). Skeletal myogenic differentiation of mesenchymal stem cells isolated from human umbilical cord blood. *Stem Cells.* 22:617–624.

236. Gang E J, Hong S H, Jeong J A, et al. (2004). In vitro mesengenic potential of human umbilical cord blood-derived mesenchymal stem cells. *Biochem. Biophys. Res. Commun.* 321:102–108.

237. Jeong J A, Gang E J, Hong S H, et al. (2004). Rapid neural differentiation of human cord blood-derived mesenchymal stem cells. *Neuroreport.* 15:1731–1734.

238. Kucia M, Dawn B, Hunt G, et al. (2004). Cells expressing early cardiac markers reside in the bone marrow and are mobilized into the peripheral blood after myocardial infarction. *Circ. Res.* 95:1191–1199.

239. Kucia M, Jankowski K, Reca R, et al. (2004). CXCR4-SDF-1 signalling, locomotion, chemotaxis and adhesion. *J. Mol. Histol.* 35:233–245.

240. Kucia M, Ratajczak J, Reca R, et al. (2004). Tissue-specific muscle, neural and liver stem/progenitor cells reside in the bone marrow, respond to an SDF-1 gradient and are mobilized into peripheral blood during stress and tissue injury. *Blood Cells Mol. Dis.* 32:52–57.

241. Ratajczak M Z, Kucia M, Reca R, et al. (2004). Stem cell plasticity revisited: CXCR4-positive cells expressing mRNA for early muscle, liver and neural cells "hide out" in the bone marrow. *Leukemia.* 18:29–40.

242. Wojakowski W, Tendera M, Michalowska A, et al. (2004). Mobilization of CD34/CXCR4+, CD34/CD117+, c-met+ stem cells, and mononuclear cells expressing early cardiac, muscle, and endothelial markers into peripheral blood in patients with acute myocardial infarction. *Circulat.* 110:3213–3220.

11.4 ARTIFICIAL CELLS

11.4.1 Introduction

A major limitation to the clinical application of cell therapy and drug discovery is the current inability to isolate an adequate number of functional and transplantable cells [1]. Unfortunately, the number of human organs available for transplantation or cell isolation is severely limited. Considering the cost and difficulty in some cases of cell isolation and the need for immediate availability of consistent and functionally uniform cell preparations, human cells cannot be isolated on a scale sufficient to treat more than a fraction of the patients who need organ transplantation. The use of animal cells would result in additional concerns related to the transmission of infectious pathogens and immunologic and physiologic incompatibilities between donors and humans [2, 3]. Thus, other alternative cell sources, such as stem cells, have been explored. In particular, embryonic stem cells have unlimited proliferative capacity and theoretically can differentiate all kinds of cell types. In contrast, somatic stem cells have finite proliferation ability in the currently available culture system [4]. The cultivation of mammalian cells is an important experimental procedure that is a fundamental tool in biological studies [5]. Establishment of methods to control differentiation and proliferation in stem cells is not yet achieved. Thus, possible tumor development will be occurred after de-differentiated or trans-differentiated stem cells. To overcome this issue, great efforts have been made to construct immortalized human cell lines with an unlimited replicative potential.

However, despite the widespread use of the culture system, we have recently begun to understand the essential mechanisms that control cell growth and division. This section discusses recent advances in our understanding of the molecular mechanisms that regulate the life span of cell lineages *in vitro* and an approach to construct human cell lines with currently available genetic manipulation for cell therapy and the study of pharmacology and toxicology. The definition of cellular immortality is infinite survival with an unlimited proliferative potential [6]. In fact, primary human cells in tissue culture rarely undergo spontaneous immortalization process [7]. Thus, researchers tried to immortalize human cells by using x-ray [8] and chemical carcinogens [7], but it turned out to be very inefficient. Currently, human cells have been immortalized by an introduction of the transforming genes of DNA tumor viruses, such simian virus 40 large T antigen (SV40LT) [9], papillomaviruses [10], and the catalytic unit of the human telomerase reverse transcriptase [11].

11.4.1.1 *Immortalization of Human Cells*

Limited Proliferative Life Span or Senescence of Human Cells. Senescence is a terminally arrested growth state of the cells. It differs from the nonproliferative state of terminally differentiated cells. As senescent cells remain viable, senescence can be distinguished with cell death processes such as apoptosis or necrosis. Senescent cells are arrested at the G1/S phase of the cell cycle and are thus distinct from nondividing G0-arrested quiescent cells [5, 12]. Human cells are defined as senescent when they fail to respond to mitogens and the population of the cells does not divide in a certain period of time, for example, 30 days. Additional characteristics of senescent cells include (1) expression of β-galactosidase, (2) increased lysossomal biogenesis, (3) decreased rates of protein synthesis and degradation that are distinct from their pre-senescent ancestors, (4) increased cell size, (5) multinucleation, (6) cytoplasmic vacuolation, and (7) decreased membrane fluidity. The rate of protein, DNA, and RNA syntheses is reduced in senescent cells. Senescent cells accumulate altered macromolecules and exhibit DNA alterations such as shortened telomeres, decreased chromatin condensation, increased karyotypic abnormalities, and decreased methylation. Senescence does not lead directly to cell death, and senescent cells can persist in culture for up to 2 years if fed regularly [13].

Senescent cells may accumulate in aged tissue [14] and could compromise tissue function by both their altered pattern of gene expression and their nonproliferative state, resulting in aged phenotypes of the tissues [15]. A link between *in vitro* senescence and *in vivo* aging is suggested by the observation that the *in vitro* life span of cells from various mammalian species correlates with their *in vivo* life span and that the cells from individuals with premature ages suffering from disorders such as Werner's syndrome, Down's syndrome, and progeria have a shorter *in vitro* life span than that of the cells from normal individuals. However, evidence that the *in vitro* life span of normal human cells correlates with the donor age remains controversial [16].

Immortal or Extended Life Span of Human Cells. More than 30 years ago, it was demonstrated for the first time that SV40 [17], a DNA tumor virus of the papova virus family isolated by Sweet and Hillman [18], can morphologically transform human fetal and adult skin fibroblast. The transformation of mammalian cells by SV40 is known to require expression of only the early region of the viral genomes, which encodes two proteins of large T-antigen (94 kd) and small t-antigen (17 kd) [18]. Transfer and expression of specific oncogenes, such as simian virus 40 large T antigen (SV40 LT), in primary human cells can generate cell populations that propagate for extended periods of time *in vitro*, presumably because they bypass the first senescence crisis through the inactivation of p53 and Rb, and thus extend their life span by about 20 population doubling (PD) [19]. This extended life span ends in M2 crisis stage (Figures 11.4-1 and 11.4-2), which typically coincides with

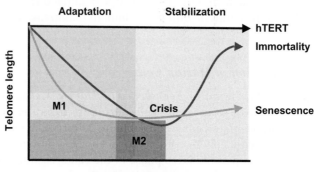

Figure 11.4-1. *Cellular senescence and immortalization.* Telomere length is maintained by telomerase, and most human somatic cells have lower levels of telomerase. The cells are telomerase-negative and experience telomere shortening with each cell division. Shortened telomeres may start the cells to enter senescence at the Hayflick limit, or M1. This proliferative checkpoint can be overcome by an inactivation of pRB/p16 or p53, for example, by the use of SV40 or human papilloma virus oncoproteins. Such cells continue to suffer telomere erosion and ultimately enter crisis, or M2, characterized by cell death. Quite a few surviving cells acquire stabilization of telomere length and unlimited proliferative potential, mostly due to activation of telomerase. (This figure is available in full color at ftp://ftp.wiley.com/public/sci_tech_med/pharmaceutical_biotech/.)

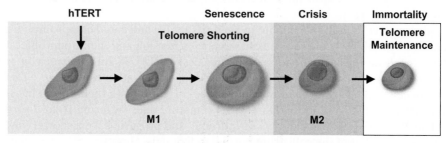

Figure 11.4-2. *Immortalization of cells with hTERT.* Ectopic expression of hTERT allows cells to bypass proliferation barriers and become immortal. Telomeres are very important in booth senescence (M1) and crisis (M2) as hTERT introduction either before or after M1 results in cell immortalization. If introduction of hTERT does not result in immortalization in the cells, another type of growth arrest called "premature senescence" or "cell culture shock" will be occurred, which is telomere-independent. (This figure is available in full color at ftp://ftp.wiley.com/public/sci_tech_med/pharmaceutical_biotech/.)

dangerously shortened telomeres, and every mitosis is accompanied by enormous chromosomal instability in the cells. Most of the cells fail to survive crisis, and the frequency rate of the cells that overcome the crisis is very low in humans. The *in vitro* spontaneous immortalization of human cells using SV40T is approximately 10^6–10^7 and is believed to result from activation of the endogenous telomerase [20]. Such cells are infrequently tumorogenic, and tumorigenicity occurs only after long periods of continuous culture of the cells [10]. These phenomena have suggested that establishment of permanent cell lines in culture involves two distinct processes: (1) an initial adaptation of cells to grow in the unnatural environment of culture dish, and (2) an acquired ability of these adapted cells to proliferate indefinitely in tissue culture (Figure 11.4-1). The life span of cells is measured in cell divisions [5]; thus, cells must possess a molecular "clock" that counts the number of times they have divided. In a telomere model of senescence, telomere length acts as this counting mechanism. Nucleoprotein structures that constitute the ends of linear chromosomes became the primary candidates to fulfill this role in human cells. It was for the first time described molecularly in 1981 [21]; telomeres serve to protect the ends of chromosomes from illegitimate fusions and other damage and shield the ends of chromosomes from recognition by the cellular DNA repair machinery [22]. Telomeric DNA is maintained at a constant length, and telomeres shorten at a constant rate with progressive cell divisions in human cells (Figure 11.4-3). Telomere shortening occurs for at least two reasons: (1) Telomeres shorten with successive replication because the polymerases involved in conventional DNA replication cannot fully copy telomeric DNA [23] and (2) telomerase, an enzyme responsible for telomere maintenance, is tightly repressed in most human somatic cells [24]. Most immortalized human cells express telomerase [24], and telomere length is stable. These observations have suggested that some elements of telomere structures monitor cell proliferation and order a signal of the onset of replicative senescence. The ectopic expression of telomerase in pre-senescent cells halts telomere shortening and permits cells to avoid replicative senescence. Telomerase is a ribonucleoprotein composed of an RNA subunit, *hTERC*, that is ubiquitously expressed [25] and a reverse transcriptase protein catalytic subunit, *hTERT*, whose expression correlates with telomerase activity in immortal cells [26].

hTERT is indeed undetectable in most types of normal human cells, but it is readily detectable in most immortal cell lines, cells derived from human cancers, and human tumor samples. Expression of *hTERT* in pre-senescent human cells confers telomerase activity [10], resulting in telomere lengthening or stabilization, and it allows the cells to bypass replicative senescence. Cells expressing *hTERT* maintain the normal karyotype and continue to respond in the same manner as pre-senescent, nonimmortalized cells [27]. Thus, activation of telomerase by expression of *hTERT* permits the cells to bypass senescence and become immortalized in a single step. Although maintenance of telomere plays an important role in replicative senescence, it has been proved that both pRB and p53 tumor suppressor pathways also play a prominent role in regulating the onset of replicative senescence (Figure 11.4-4). When cells expressing SV40LT continue to repress hTERT expression and lack a detectable telomerase activity, the telomere length of the cells gradually shortens with continued cell proliferation. Eventually such cells enter the second period of diminished growth, which referred is to as crisis or M2 [28]. Telomere shortening may be one signal that activates the pRB or p53 pathway and initiates replicative senescence. These observations have suggested that both pRB

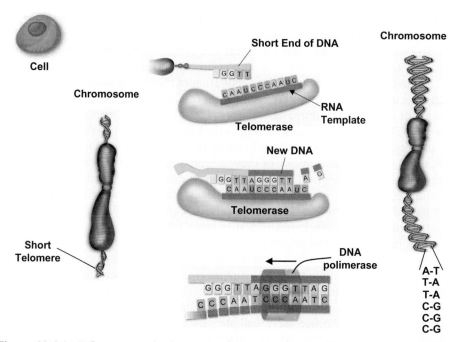

Figure 11.4-3. *Telomeres and telomerase.* Telomere is a repeating sequence of double-stranded DNA located at the ends of chromosomes. Long telomere length is associated with immortalized cell lines, cancer cells, and embryonic stem cells. Cells experience telomere shortening with each cell division. Telomerase is an enzyme that lengthens telomeres by adding on repeating sequences of DNA. Telomerase binds the ends of the telomere via an RNA template that is used for the attachment of a new strand of DNA. Telomerase adds several repeated DNA sequences, then releases to a second enzyme, DNA polymerase, and attaches the opposite or complementary strand of DNA, completing the double-stranded extension of the chromosome ends. The function of telomere and telomerase seem to be important in cell division, normal development, and cancer research. (This figure is available in full color at ftp://ftp.wiley.com/public/sci_tech_med/pharmaceutical_biotech/.)

and p53 pathways in conjunction with telo-mere shortening play significant roles in governing replicative senescence (Figure 11.4-1).

Crisis stage. Crisis (M2) is distinguished from senescence by cell death in the presence of ongoing cell division [29]. Although most cells that enter crisis die by apoptosis, a few cells overcome crisis and become immortal [30]. Immortal cells that have survived crisis typically exhibit aneuploidy and extensive nonreciprocal chromosomal translocations, which suggests that substantial genomic rearrangements accompany the selection of these rare surviving cells. These observations indicate that crisis is precipitated by critically shortened telomeres that have lost their ability to protect chromosomes in the setting of p53 loss. Immortal cells that emerge from crisis show the stabilized telomere lengths with extended passage, which suggests that only those cells that acquire the ability to maintain stable telomere lengths survive. It is important to note that, in general, immortal cells fail to form tumors in immunodeficient animals [31]. Experimentally, the introduction of *hTERT* in post-senescent, pre-crisis cells confers telomerase activity, stabilizes telomere

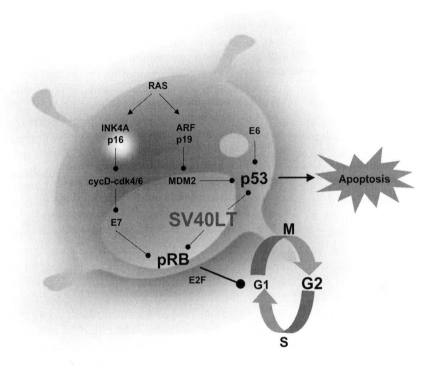

Figure 11.4-4. *pRB and p53 tumor suppressors pathways.* These pathways play critical roles in the cell physiology and mutations. Their affectation is found in most human cancers. The viral oncoprotein SV40 LT targets pRB and p53 pathways. (This figure is available in full color at ftp://ftp.wiley.com/public/sci_tech_med/pharmaceutical_biotech/.)

length, and permits these telomerase-expressing cells to become immortal [10, 27, 32–35]. Human cells immortalized in this manner often exhibit near-diploid karyotypes, which strongly suggests that critical telomere shortening in the absence of pRB and p53 functions initiates crisis. Two barriers limit proliferative lifespan in human cells: (1) replicative senescence and (2) crisis. Furthermore, it is obvious that telomerase, pRB, and p53 play critical roles in regulating an entry of human cells into these two states.

Immortality. Several groups have demonstrated that the introduction of pairs of cooperating oncogenes, such as myc and ras [36], or the adenoviral E1A protein and ras [37] into the primary rodent cells leads to direct transformation of the cells. However, several laboratories have now shown that telomere biology differs in important ways between human and murine cells [38]. Unlike most human cells in which telomerase is tightly repressed, most murine cells express detectable telomerase activity. In addition, telomeres are maintained at much longer lengths in cells derived from inbred mice than are observed in human cells [38]. These observations have suggested that the replicative life span of murine cells is not limited by telomere length. With extended cell division, telomere shortening activates either replicative senescence or crisis and telomeres play in protecting chromosomal integrity, which suggests that telomeres play an important role in repressing tumor formation. In contrast, the maintenance of telomere length confers cell immortality. These observations also indicate that telomeres serve as a factor that restricts and/or promotes malignant transformation. Under most circumstances, short telomeres limit the life span of the cells, depending on the status of the pRB and p53 pathways, to greatly reduce the pool of premalignant cells (Figure 11.4-2).

To address cellular immortality, researchers have made great efforts to establish an immortalized human cell lines. Such cell lines provide the advantage of uniformity and sterility, grow in unlimited quantity, and are far less costly than isolating the primary cells. One approach to construct clonal cell lines is transduction of the primary cells with genes from DNA tumor viruses, such as SV40, human papillomavirus, and Epstein–Barr virus [39]. Transformation of normal cells with early region genes of SV40, typically by transfection with expression plasmids, remains a common immortalization technique. For example, human fetal hepatopcytes were successfully transduced with pSV3neo DNA containing both large and small T antigens of the early region of SV40 and bacterial neomycin phosphotransferase gene (Figure 11.4-3). After G418 treatment, one line of the surviving clones, OUMS-29, grew well in the chemically defined serum-free medium without any crisis and showed liver-specific functions [40]. When transplanted into immunodeficient mice, OUMS-29 cells were not tumorigenic. The potential risk of malignant formation of OUMS-29 cells cannot be precluded in humans. Safeguards, including the introduction of suicide genes, should be considered in immortalized cells for human application.

In parallel, cells are recovered from surgically resected tumor specimens and cultured. Cell lines have been established for the study of different pathophysiologies. These cell lines grow indefinitely in tissue culture and have been widely used all over the world. However, genetic constitution and alterations of such cells are often observed, and several important specific functions, including several receptors and transporters, have been unfortunately deregulated [41].

11.4.1.2 Tightly Immortalized Human Cell Lines Transduction of immortalized cell lines with these suicide genes would provide a way to eliminate the cells after transplantation. Cells modified to express a herpes simplex virus–thymidine kinase (HSV–TK) gene become sensitive to ganciclovir (GCV) [42]. Thus, researchers have introduced the HSV–TK gene into cell lines. For example, OUMS-29/TK cells, immortalized human fetal hepatocytes expressing HSV–TK, were more than 100 times sensitive to GCV treatment than unmodified parental OUMS-29 cells. OUMS-29/TK cells stopped proliferation in the presence of 5-μM GCV [40].

11.4.1.3 Conditionally Immortalized Cell Lines An approach to create a conditionally immortalized cell line is the use of a transforming gene containing a temperature-sensitive mutation. Primary rat hepatocytes were successfully immortalized with a thermolabile mutant SV40T (encoded by the early region mutant tsA58) and functioned as well as primary hepatocytes after transplantation [43–45]. However, the continued presence of SV40T in the transplanted cells may increase the risk of malignant transformation of the cells after transplantation in the recipients.

11.4.1.4 Ultra-Transform Immortalized Transgenic Cell Lines This approach has proven to be successful because the resulting cell lines showed stability in culture and sensitivity to chemical exposure. The strategy implicates the development of a bi-transgenic hepatocyte cell line to evaluate the ability of various organic and inorganic chemicals to induce the expression of the HSP70-driven reporter gene. Development of two types of transgenic mice is necessary in advance. One is a transgenic mice (Hsp70/hGH) [46] secreting high levels of human growth hormone (hGH), and another is transgenic model (AT/cytoMet) [46] allowing the reproducible immortalization of untransformed hepatocytes retaining liver functions. Both strains are crossed, and the resulting transgenic strain permits a reproducible immortalization of untransformed hepatocytes. Several stable hepatic cell lines (MMH–GH) showing highly differentiated phenotypes have been generated from the double transgenic animals. This strategy is valuable in the field of toxicology and in the development of chemical and physical xenobiotics. The technology provides a simple biological system that reduces the need for animal experimentation and/or continuously isolating fresh hepatocytes [46].

11.4.1.5 Reversibly Immortalized Cell Lines To generate an immortalized hepatocyte cell line more suitable for clinical use, a more tightly regulated system for cell growth should be considered. An attractive system using site-specific recombination has been documented [47]. DNA sequences intervened by loxP recombination targets can be excised after expression of Cre recombinase [47]. A Cre/loxP system allows the construction of reversibly immortalized cell lines. To provide more stringent control over the expression of transforming genes, human hepatocytes were transduced with a retro-viral vector SSR#69 expressing SV40T and selectable positive (hygromycin resistance gene) and negative (HSV-TK) markers that were intervened by a pair of loxP recombination targets and subsequently excised by Cre/loxP recombination [31, 48]. One emerging clone after SSR#69-transduction, NKNT-3, was a highly differentiated hepatocyte cell line. NKNT-3 cells were sensitive to 5-µM GCV. Adenovirus-mediated Cre recombinase expression was efficiently performed to remove SV40T from NKNT-3 cells. Intrasplenic transplantation of such reverted NKNT-3 cells significantly improved the survival of 90% hepatectomized rats [31]. Cre/loxP-based reversible immortalization has been achieved in several types of human cells [31, 48, 49]. In our studies, SV40T-transduced normal human endothelial cells (ECs) acquired an extensive proliferation capacity to population doubling level (PDL) 65 to 80, but complete immortalization was not achieved [49]. This was explained by the absence of spontaneous activation of endogenous telomerase, which is known as one of the essential participants in cellular immortalization processes in SV40T-transduced ECs [50–52]. To achieve immortalization of normal human ECs, a retroviral vector SSR#197 expressing hTERT and GFP cDNAs flanked by a pair of loxB target sequences was constructed. Cotransduction of human ECs with retroviral vectors SSR#69 and SSR#197 facilitated establishment of a completely immortalized cell line TMNK-1 that expresses a differentiated endothelial phenotype. Feasibility of reversible immortalization in TMNK-1 cells was demonstrated by using TAT-derived HIV-mediated Cre/loxP recombination followed by GFP-negative cell sorting and drug selection [27]. Lately, the system has been applied to establish hepatic stellate cell lines [33], cholangiocyte cell lines [10], hepatic progenitor cell lines [53], bone marrow-derived human cells [54], and human pancreatic beta cell lines [32]. To establish immortalized human pancreatic beta cell lines, freshly isolated human pancreatic islet cells were transduced with SSR#69, followed by hygromycin selection. Then, the resultant cells were super-infected with SSR#197 for immortalization (Figure 11.4-5). At the first screening, tumorigenic assay of the resulting cell lines was performed in immunodeficient mice. Then, gene expression analysis was conducted in nontumorigenic clones. Based on these findings, NAKT-15 turned out to be highly differentiated. To obtain the reverted form of NAKT-15 cells, the cells were infected with a recombinant adevovirus virus vector (AxCANCre) expressing Cre recombinase tagged with a nuclear localization signal (NLS) expressing the recombinant adenovirus vector. After AxCANCre infection, GFP-

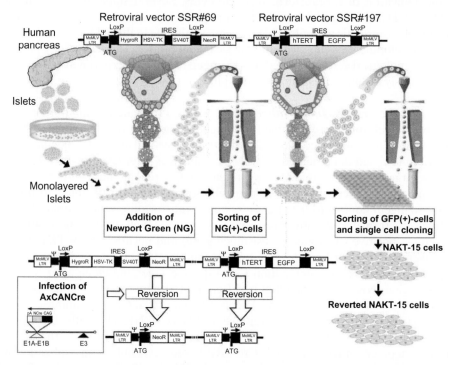

Figure 11.4-5. *Scheme for the establishment of reversibly immortalized human β-cell lines.* Freshly isolated human islets were cultured in monolayer and transduced with the retroviral vector SSR#69. After hygromycin selection of SSR#69-transduced cells, Newport Green dye was added and Newport Green-positive cells were sorted. After transduction with SSR#197, EGFP-positive cells were sorted and subjected to single-cell cloning, yielding the clone NAKT-15. Transient expression of Cre recombinase was induced by infecting cells with a recombinant adenovirus expressing Cre recombinase (AxCANCre), which removed the SV40T and TERT genes that were flanked by loxP sites. Neomycin (G418) selection and EGFP-negative cell sorting resulted in the collection of reverted NAKT-15 cells. CAG, promoter consisting of cytomegalovirus IE enhancer, chicken b-actin promoter, and rabbit b-globin polyadenylation signal; HSV-TK, herpes simplex virus thymidine kinase; HygroR, hygromycin-resistance gene; IRES, internal ribosomal entry site; MoMLV LTR, long terminal repeat of Moloney murine leukemia virus leader sequence; NCre, nuclear localizing signal-tagged Cre recombinase; pA, polyadenylation signal; j, packaging signal. (This figure is available in full color at ftp://ftp.wiley.com/public/sci_tech_med/ pharmaceutical_biotech/.)

negative cell populations were recovered by a MoFlo cell sorter and then cultured in the presence of neomycin analog G418 [32]. These procedures are well illustrated in Figure 11.4-5. This strategy can be applicable to various types of human cells.

11.4.1.6 Applications Numerous genomics-based technologies are now routinely applied to drug discovery. The central role for these technologies is validating the next generation of therapeutics from novel targets identified through genomics. The aim of such technologies is to accurately identify the next generation of targets that demonstrate therapeutic efficiency and safety. Immortal human cell lines will have unique attributes that can be used for drug discovery, and if cell therapy is the goal of such cells, the addition of redundant safeguards into the cells and accumulation of experimental data will be required before clinical trials. The ultimate goal of cell transplantation is an autologous setting in which the patient's own cells are genetically modified *ex vivo* with a reversible immortalization method, and then a reverted form of the cells can be transplanted back to the patient. Immortalized or reversibly immortalized cells offer continuous availability, uniformity, and sterility, and the cells can be functionally cryopreserved for the future use. Production of such cell lines may overcome the shortage of donors, and thus, the established human cell clones may be considered as a potential cell source for the treatment of diabetic patients or liver failure patients with transplantation. We anticipate that the production of artificial cells will be useful not only to provide carcinogenesis models, but also to develop drug discovery and possible cell therapy in the future. Reversible immortalization of human cells has important applications in biological research, biotechnology, and medicine.

11.4.2 Conclusions

The ethical issue of transplanting immortalized cells is still controversial, but clinical approval of a protocol involving transplantation of these cells depends on the balance of risks and benefits. If patients with high risk were to receive such transplants with considerable benefits, approval would be obtained after reliable accumulation of data regarding the safety and efficacy of immortalized cells. We here represent an important step in the development of a useful strategy for resolving the organ shortage that now limits the use of normal human cells for cell therapies. Such technology can be applicable to a variety of somatic cells and would potentially be used to treat a large number of patients with clinically significant pathologic conditions.

References

1. McNeish J (2004). Embryonic stem cells in drug discovery. *Nat. Rev. Drug Discov.* 3:70–80.
2. van der Laan L J, Lockey C, Griffeth B C, et al. (2000). Infection by porcine endogenous retrovirus after islet xenotransplantation in SCID mice. *Nature.* 407:90–94.
3. Butler D (2002). Xenotransplant experts express caution over knockout piglets. *Nature.* 415:103–104.

4. Hayflick L, Moorhead P S (1961). The serial cultivation of human diploid cell strains. *Exp. Cell Res.* 25:585–621.

5. Masters J R (2000). Human cancer cell lines: Fact and fantasy. *Nat. Rev. Mol. Cell Biol.* 1:233–236.

6. Cooper C S, Park M, Blair D G, et al. (1984). Molecular cloning of a new transforming gene from a chemically transformed human cell line. *Nature.* 311:29–33.

7. Rhim J S, Park J B, Jay G (1989). Neoplastic transformation of human keratinocytes by polybrene-induced DNA-mediated transfer of an activated oncogene. *Oncogene.* 4:1403–1409.

8. Borek C (1980). X-ray induced in vitro neoplastic transformation of human diploid cells. *Nature.* 283:776–778.

9. Rhim J S, Jay G, Arnstein P, et al. (1985). Neoplastic transformation of human epidermal keratinocytes by AD12-SV40 and Kirsten sarcoma viruses. *Science.* 227:1250–1252.

10. Munger K, Howley P M (2002). Human papillomavirus immortalization and transformation functions. *Virus Res.* 89:213–228.

11. Maruyama M, Kobayashi N, Westerman K A, et al. (2004). Establishment of a highly differentiated immortalized human cholangiocyte cell line with SV40T and hTERT. *Transplant.* 77:446–451.

12. Sherwood S W, Rush D, Ellsworth J L, et al. (1988). Defining cellular senescence in IMR-90 cells: a flow cytometric analysis. *Proc. Natl. Acad. Sci. USA.* 85:9086–9090.

13. Goldstein S (1990). Replicative senescence: the human fibroblast comes of age. *Science.* 249:1129–1133.

14. Dimri G P, Lee X, Basile G, et al. (1995). A biomarker that identifies senescent human cells in culture and in aging skin in vivo. *Proc. Natl. Acad. Sci. USA.* 92:9363–9367.

15. Kipling D, Cooke H J (1990). Hypervariable ultra-long telomeres in mice. *Nature.* 347:400–402.

16. Cristofalo V J, Allen R G, Pignolo R J, et al. (1998). Relationship between donor age and the replicative lifespan of human cells in culture: a reevaluation. *Proc. Natl. Acad. Sci. USA.* 95:10614–10619.

17. Shein H M, Enders J F (1962). Transformation induced by simian virus 40 in human renal cell cultures. I. Morphology and growth characteristics. *Proc. Natl. Acad. Sci. USA.* 48:1164–1172.

18. Sweet B H, Hilleman M R (1960). The vacuolating virus, S.V. 40. *Proc. Soc. Exp. Biol. Med.* 105:420–427.

19. Fanning E (1992). Simian virus 40 large T antigen: the puzzle, the pieces, and the emerging picture. *J. Virol.* 66:1289–1293.

20. Ray F A, Kraemer P M (1993). Iterative chromosome mutation and selection as a mechanism of complete transformation of human diploid fibroblasts by SV40 T antigen. *Carcinogen.* 14:1511–1156.

21. Blackburn E H, Chiou S S (1981). Non-nucleosomal packaging of a tandemly repeated DNA sequence at termini of extrachromosomal DNA coding for rRNA in Tetrahymena. *Proc. Natl. Acad. Sci. USA.* 78:2263–2267.

22. Blackburn E H (2001). Switching and signaling at the telomere. *Cell.* 106:661–673.

23. Meyerson M (2000). Role of telomerase in normal and cancer cells. *J. Clin. Oncol.* 18:2626–2634.

24. Kim N W, Piatyszek M A, Prowse K R, et al. (1994). Specific association of human telomerase activity with immortal cells and cancer. *Science.* 266:2011–2015.

25. Feng J, Funk W D, Wang S S, et al. (1995). The RNA component of human telomerase. *Science.* 269:1236–1241.

26. Nakamura T M, Morin G B, Chapman K B, et al. (1997). Telomerase catalytic subunit homologs from fission yeast and human. *Science.* 277:955–959.

27. Matsumura T, Takesue M, Westerman K A, et al. (2004). Establishment of an immortalized human-liver endothelial cell line with SV40T and hTERT. *Transplant.* 77:1357–1365.

28. Counter C M, Avilion A A, LeFeuvre C E, et al. (1992). Telomere shortening associated with chromosome instability is arrested in immortal cells which express telomerase activity. *Embo. J.* 11:1921–1929.

29. Wei W, Sedivy J M (1999). Differentiation between senescence (M1) and crisis (M2) in human fibroblast cultures. *Exp. Cell. Res.* 253:519–522.

30. Macera-Bloch L, Houghton J, Lenahan M, et al. (2002). Termination of lifespan of SV40-transformed human fibroblasts in crisis is due to apoptosis. *J. Cell Physiol.* 190:332–344.

31. Kobayashi N, Fujiwara T, Westerman K A, et al. (2000). Prevention of acute liver failure in rats with reversibly immortalized human hepatocytes. *Science.* 287:1258–1262.

32. Narushima M, Kobayashi N, Okitsu T, et al. (2005). A human beta-cell line for transplantation therapy to control type 1 diabetes. *Nat. Biotechnol.* 10:1274–1282.

33. Watanabe T, Shibata N, Westerman K A, et al. (2003). Establishment of immortalized human hepatic stellate scavenger cells to develop bioartificial livers. *Transplant.* 75:1873–1880.

34. Shibata N, Watanabe T, Okitsu T, et al. (2003). Establishment of an immortalized human hepatic stellate cell line to develop antifibrotic therapies. *Cell Transplant.* 12:499–507.

35. Okitsu T, Kobayashi N, Jun H S, et al. (2004). Transplantation of reversibly immortalized insulin-secreting human hepatocytes controls diabetes in pancreatectomized pigs. *Diabetes.* 53:105–112.

36. Land H, Parada L F, Weinberg R A (1983). Tumorigenic conversion of primary embryo fibroblasts requires at least two cooperating oncogenes. *Nature.* 304:596–602.

37. Ruley H E (1983). Adenovirus early region 1A enables viral and cellular transforming genes to transform primary cells in culture. *Nature.* 304:602–606.

38. Wright W E, Shay J W (2000). Telomere dynamics in cancer progression and prevention: fundamental differences in human and mouse telomere biology. *Nat. Med.* 6:849–851.

39. Katakura Y, Alam S, Shirahata S (1998). Immortalization by gene transfection. *Methods Cell Biol.* 57:69–91.

40. Kobayashi N, Miyazaki M, Fukaya K, et al. (2000). Transplantation of highly differentiated immortalized human hepatocytes to treat acute liver failure. *Transplant.* 69:202–207.

41. Blumrich M, Zeyen-Blumrich U, Pagels P, et al. (1994). Immortalization of rat hepatocytes by fusion with hepatoma cells. II. Studies on the transport and synthesis of bile acids in hepatocytoma (HPCT) cells. *Eur. J. Cell Biol.* 64:339–347.

42. Culver K W, Ram Z, Wallbridge S, et al. (1992). In vivo gene transfer with retroviral vector-producer cells for treatment of experimental brain tumors. *Science.* 256:1550–1552.

43. Fox I J, Chowdhury N R, Gupta S, et al. (1995). Conditional immortalization of Gunn rat hepatocytes: an ex vivo model for evaluating methods for bilirubin-UDP-glucuronosyltransferase gene transfer. *Hepatol.* 21:837–846.

44. Schumacher I K, Okamoto T, Kim B H, et al. (1996). Transplantation of conditionally immortalized hepatocytes to treat hepatic encephalopathy. *Hepatol.* 24:337–743.

45. Nakamura J, Okamoto T, Schumacher I K, et al. (1997). Treatment of surgically induced acute liver failure by transplantation of conditionally immortalized hepatocytes. *Transplant.* 63:1541–1547.

46. Sacco M G, Amicone L, Cato E M, et al. (2004). Cell-based assay for the detection of chemically induced cellular stress by immortalized untransformed transgenic hepatocytes. *BMC Biotechnol.* 4:5.

47. Sternberg N, Hamilton D, Austin S, et al. (1981). Site-specific recombination and its role in the life cycle of bacteriophage P1. *Cold Spring Harb. Symp. Quant. Biol.* 45(Pt 1):297–309.

48. Westerman K A, Leboulch P (1996). Reversible immortalization of mammalian cells mediated by retroviral transfer and site-specific recombination. *Proc. Natl. Acad. Sci. USA.* 93:8971–8976.

49. Noguchi H, Kobayashi N, Westerman K A, et al. (2002). Controlled expansion of human endothelial cell populations by Cre-loxP-based reversible immortalization. *Hum. Gene. Ther.* 13:321–334.

50. Bodnar A G, Ouellette M, Frolkis M, et al. (1998). Extension of life-span by introduction of telomerase into normal human cells. *Science.* 279:349–352.

51. Halvorsen T L, Leibowitz G, Levine F (1999). Telomerase activity is sufficient to allow transformed cells to escape from crisis. *Mol. Cell Biol.* 19:1864–1870.

52. Zhu J, Wang H, Bishop J M, et al. (1999). Telomerase extends the lifespan of virus-transformed human cells without net telomere lengthening. *Proc. Natl. Acad. Sci. USA.* 96:3723–3728.

53. Delgado J P, Parouchev A, Allain J E, et al. (2005). Long-term controlled immortalization of a primate hepatic progenitor cell line after Simian virus 40 T-Antigen gene transfer. *Oncogene.* 24:541–551.

54. Nishioka K, Fujimori Y, Hashimoto-Tamaoki T, et al. (2003). Immortalization of bone marrow-derived human mesenchymal stem cells by removable simian virus 40T antigen gene: analysis of the ability to support expansion of cord blood hematopoietic progenitor cells. *Int. J. Oncol.* 23:925–932.

12.1

REGULATION OF SMALL-MOLECULE DRUGS VERSUS BIOLOGICALS VERSUS BIOTECH PRODUCTS

María de los Angeles Cortés Castillo and José Luis Di Fabio

Pan American Health Organization, Washington, DC

Chapter Contents

12.1.1 Historical Perspective on Regulation 1373
12.1.2 Issues Related to Regulation for Biologicals (Vaccines) Compared with Regulation for Other Medicinal Products 1376
    12.1.2.1 Regulation of Small Molecules 1377
    12.1.2.2 Regulation of Classic Vaccines and Biologicals 1378
    12.1.2.3 Regulation of Biotechnology Products 1380
12.1.3 Main Regulatory Agencies 1381
    12.1.3.1 Food and Drug Administration (FDA) 1381
    12.1.3.2 European Medicines Agency (EMEA) 1383
    12.1.3.3 International Committee for Harmonization (ICH) 1385
12.1.4 Global Role of WHO in Drugs and Biologicals Regulation 1385
    12.1.4.1 WHO Procedure for Assessing Drug Regulatory Authorities 1386
12.1.5 Other Initiatives for Harmonized Regulation 1387
12.1.6 Future Perspective of Regulation 1387
    References 1389

12.1.1 HISTORICAL PERSPECTIVE ON REGULATION

Regulation is the logical consequence of the search from health institutions and governments to ensure public protection from unsafe products.

The regulation of drugs emerged first in developed countries. The first initiatives of regulation appeared in 1862 with the Bureau of Chemistry in the Department of Agriculture of the United States, lead by Harvey Wiley, trying to regulate food,

Handbook of Pharmaceutical Biotechnology, Edited by Shayne Cox Gad.
Copyright © 2007 John Wiley & Sons, Inc.

drug, and biological products. At that time, adulterated food and drugs as The Great Blood Purifier System (Peter's specific) and Mrs. Winslow's Soothing Syrup for teething and colicky babies (which was related with morphine, without a label indication) were the main targets of regulation [1].

The Bureau of Chemistry enforced a 1906 law until 1927, when it was reorganized to form the Food, Drug and Insecticide Administration, later renamed, in 1931, as the Food and Drug Administration [1]. The 1906 Act prohibited the sale of adulterated or misbranded drugs and prohibited interstate transport of disallowed food and drugs. It also mandated that a variety of information be stated on the drug package. Changes to labeling requirements and a provision to label certain dangerous ingredients led many patent medicine makers to reformulate their products because many of the products contained harmful chemicals [1].

After an event in 1937, when a marketed anti-infective called Elixir Sulfanimide caused over 100 deaths regulators were forced to change the law to require that drugs meet certain safety requirements before being marketed. A new 1938 Act called for evidence of drug safety, prohibited false therapeutic drug claims, and also included cosmetics and therapeutic devices under FDA regulation. The 1938 Act established a listing of active ingredients on the drug label and required that drugs be labeled with adequate directions for safe use [1].

Even after all these efforts to ensure the safety of drugs were implemented, many incidents (included the thalidomide tragedy) occurred and the U.S. Congress had to pass major amendments to the Act in 1962. In addition to safety issues, for the first time drug manufacturers were required to prove the effectiveness of their drug products before marketing.

In parallel with the development of drug regulations, biological regulations were historically developed in connection to safety concerns. The 1902 Biologics Control Act enacted by the U.S. Congress established the initial concepts used for the regulation of biologicals. These provisions were revised in 1944 and expanded in the 1950s. Attention to regulation of vaccines and toxins was enforced until a major tragedy occurred in the United States, and only then were actions taken by the federal government to ensure public protection from unsafe products. The incident occurred in 1901, in St. Louis, Missouri, when 20 children became ill and 14 died after administration of an equine-derived diphtheria antitoxin (prepared from horse serum) contaminated with tetanus toxin. The serum had been manufactured in local establishments designed without any of the current knowledge of Good Manufacturing Practices, particularly without procedures that guarantee the product's safety and potency. This event forced legislation to regulate the sale of biologicals. In July 1902, Congress passed the Biologics Control Act also known as the "Virus-Toxin" law [2], which is the first time that the United States' government took control over biologicals' production, consequently with the responsibility to ensure their safety for the public. In this act, the need to include the control of the manufacturing establishments and facilities was recognized, as assurance of purity could not be demonstrated if control was limited to inspection and testing of a final product. This act settled the essential concepts for ensuring the safety of biologicals. These ideas are used as the basis for ensuring safety and effectiveness of biologicals throughout the world [3].

The history of vaccine control and regulation in developed nations has been one of increasing complexity. By 1955, many biologicals, including blood products and vaccines, had been licensed and the need for appropriate regulation was evident. In the United States, the use of inactivated polio vaccines from different manufacturers, licensed in the country, provoked a number of children to develop poliomyelitis, and some of the children died. The reason was incompletely inactivated vaccines [3]. This incident led to the expansion of the biologicals control function to the establishment of the Division of Biological Standards (DBS) within the National Institutes of Health. Later, in 1972, the DBS was transferred to FDA and became the Bureau of Biologics. In 1982, the Bureau of Biologics was renamed the Office of Biologics Research and Review (OBRR) and combined with the Office of Drugs Research and Review (ODRR) formed the Center for Drugs and Biologics. In 1987, the OBRR was separated and renamed the Center for Biologics Evaluation and Research (CBER) [4].

The need to develop and institute an independent evaluation of medicinal products before they are allowed into the market for public consumption was reached at different times in different regions, generally associated with the level of technological development. In Japan, government regulations requiring all medicinal products to be registered for sale started in the 1950s. In many countries in Europe, the thalidomide tragedy of the 1960s was the trigger. This tragedy revealed that the new generation of synthetic drugs, which were revolutionizing medicine at the time, had the potential to harm as well as to cure [5].

Between 1960 and 1970, important developments transpired regarding laws, regulations, and guidelines for reporting and evaluation of data on safety, quality, and efficacy of new medicinal products. At the same time, the industry was becoming more international and seeking new global markets, but still the registration of medicines remained a national responsibility.

The urgent need to rationalize and harmonize regulation has been forced by the need of countries to have access to drugs and biologicals to cure, treat, and prevent diseases according to the development of important technological advances in science and responding to the effort of manufacturers to gain access to more and new markets.

For the purpose of this review, drugs will be classified as small molecules, biologicals, and biotech products:

- Small molecules are most medicines or drugs chemically synthesized with molecular weights of typically less than 500 Da applicable to prevention, treatment, or cure of diseases or injuries in man.
- Biologicals are any virus, therapeutic serum, toxin, antitoxin, vaccine, blood component or derivative, allergenic extract, or analogous product applicable to the prevention, treatment, or cure of diseases or injuries in man, generally with molecular weights larger than thousands of Daltons.
- Biotechnological products are any biological or pharmaceutical product made using recombinant DNA techniques and intended to prevent, treat, or cure diseases or injuries in man. This group includes mainly proteins of high molecular weight.

Drugs or small molecules of natural origin that are not produced through chemical synthesis are not considered in this review.

12.1.2 ISSUES RELATED TO REGULATION FOR BIOLOGICALS (VACCINES) COMPARED WITH REGULATION FOR OTHER MEDICINAL PRODUCTS

WHO has established that effective drug regulation promotes and protects public health by ensuring that [6]:

- medicines are of the required quality, safety, and efficacy;
- health professionals and patients have the necessary information to enable them to use the medicines rationally;
- all premises, persons, and practices engaged in the development, manufacture, importation, exportation, wholesale, supply, dispensing, and promotion of drugs comply with approved standards, norms, procedures, and requirements;
- product information is unbiased, accurate, and appropriate;
- illegal manufacturing and trade are detected and adequately sanctioned;
- promotion and advertising are fair, balanced, and aimed at rational drug use; and
- access to medicines is not hindered by unjustified regulatory work.

National drug budgets, as a proportion of total health budgets, currently range from 7% to 66% worldwide. The proportion is higher in developing countries (24–66%) than in developed countries (7–30%) [7]. People and governments willingly spend money on drugs because of the role they can play in saving lives, restoring health, preventing diseases, and stopping epidemics. However, to accomplish these goals, drugs must be safe, effective, and of good quality and used appropriately, which means, in turn, that their development, production, importation, exportation, and subsequent distribution must be regulated to ensure that they meet prescribed standards.

In the search of implementing regulations for drugs, biologicals, and all medicines to guarantee the protection and benefit of the users, governments have established the national drug regulatory authorities (DRAs). These institutions must be developed with a clear mission, solid legal basis, realistic objectives, appropriate organizational structure, adequate number of qualified staff, sustainable financing, access to technical literature and information, and with the capacity to exert effective market control. DRAs must be accountable to both the government and the public and their decision-making processes should be transparent. Monitoring and evaluating mechanisms should be built into the regulatory system to assess attainment of established objectives [6].

Regulatory authority is generally based on laws, which represent policy choices. This authority is assigned to designated organizations, generally part of the bureaucratic institutions, with the mission to carry out drug regulation. Several factors regarding the authority and the capacity for exercising such authority affect the operation and drug regulatory activities [7]. Among the most important are:

- availability of standards, procedures, and guidelines to be used as guidance in performing the authorized functions;

- consideration of strategies to overcome structural and resources constraints, including financial adequacy and sustainability;
- establishment of clear strategies for planning, monitoring, and evaluation;
- availability of appropriate human resources including number, qualifications, remuneration, and human resources development; and
- definition of the scope of regulatory authority, in relation to type and extent of the activities carried out to implement legal provisions, including sanctions for noncompliance.

The value of drug regulatory activities depends on whether they produce the intended outcomes in terms of the following [7]:

- the quality of pharmaceutical products marketed;
- the proportion of licensed pharmaceutical facilities meeting international standards in Good Manufacturing Practices, Good Laboratory Practices, Good Clinical Practices, and so on;
- the number of illegal products on the market; and
- the number of illegal facilities marketing pharmaceutical products.

In addition to effectiveness, policy makers must also address regulatory efficiency, transparency, and accountability when evaluating regulatory policies. The cost effectiveness of drug regulation; the cost for pharmaceutical businesses and consumers of regulatory delays; political and commercial influence over regulatory decisions, public access to regulatory procedures and decision-making criteria; communication among the regulatory authority, its clients, and the consumer; and the accountability for the results of regulatory actions are performance indicators that must be assessed in the evaluation of regulatory agencies.

12.1.2.1 Regulation of Small Molecules

Since the mid-1930s, many new pharmaceutical products have emerged, and trade in the pharmaceutical industry has taken on international dimensions. At the same time, however, the circulation of toxic, substandard, and counterfeit drugs on the markets has increased, mainly because of ineffective regulation of production and trade of pharmaceutical products in exporting and importing countries. Problems relating to drug safety and efficacy are generally caused by the use of drugs containing toxic substances or impurities, drugs whose claims have not been verified or that have unknown severe adverse reactions, as well as substandard preparations or counterfeits drugs. All these problems can be tackled effectively only by establishing an effective drug regulatory system.

The development process stages for a small molecule drug that are relevant to the regulatory scrutiny are as follows:

- drug discovery,
- drug purification,
- preclinical research, and
- clinical trials (phase I, II, and III).

When the dossier of a drug is submitted to the DRA, an exhaustive review of the medical/clinical, chemistry/manufacture, pharmacology/toxicology aspects, and their statistical validity is conducted. The main element of the evaluation is to certify its safety (clinical trials phase I or II) and, during this stage of the approval, the manufacturer can be required to submit additional information to ensure the safety of the product. When safety is acceptable, the DRA proceeds with complete reviews to ensure efficacy, stability, and all other specifications that guarantee the product complies with the required norms. At any stage of the trial, the DRA of the country where the trial is being conducted should have the authority to put it on hold if deficiencies or safety issues appear and until they are resolved. If results of clinical trials demonstrate that the drug is safe and efficacious over existing treatment drugs, an application is made in most of the regulatory agencies to seek approval for marketing the drug.

As part of the new drug approval process, the DRA conducts inspection for Good Manufacturing Practices (GMP inspection) of the facilities where the drug is being manufactured. GMP is a quality concept and consists of a set of policies and procedures for manufacturing processes to guarantee that drugs are pure, consistent, safe, and effective. These policies and procedures describe the facilities, equipment, methods, and controls for producing drugs with the intended quality. The guiding principle for GMP is that quality cannot be tested in a product, but must be designed and built into each batch of the drug product throughout all aspects of its manufacturing processes [8].

12.1.2.2 Regulation of Classic Vaccines and Biologicals

Well-managed programs of vaccination have brought about profound reductions in the impact of diseases in terms of morbidity and mortality in the majority of countries of the world. One major disease, smallpox, has been totally eradicated largely as a result of vaccination and another, poliomyelitis, is absent from large areas of the world and on target for eradication during the next few years, and the incidence of measles has been greatly reduced. Vaccines provide one of the most cost-effective public health interventions and are among the safest medicinal products [9] (Table 12.1-1).

These enormous achievements have been possible in part because internationally agreed principles and procedures are in place to secure high levels of safety, efficacy, and quality of vaccines. Vaccines differ from therapeutic medicines first because of the biological and thus inherently variable nature of the products themselves, the raw material used in their production, and the biological methods used to test them. Thus, special expertise and procedures are needed for their manufacture, control, and regulation. The use of appropriate standard materials and reference preparations, whenever they exist, is fundamental to the standardization and control of vaccines [9].

Vaccines are unique in the fact that they are usually administered to very large numbers of healthy people, mostly infants, in national immunization programs. Thus, safety and quality are of high relevance. Although vaccines have a key role in preventive medicine, recent history of their use has shown a general high level of safety compared with their benefit. In most cases, minor adverse reactions may occur, but these reactions do not challenge the risk-benefit advantage of vaccina-

TABLE 12.1-1. Classification and Characteristics of Vaccines

- **Classic vaccines:** products of a whole or part (subunit) of a microorganism. They can be attenuated or inactivated vaccines, subunits, or toxoids products of the microorganism's metabolism, for instance, Diphtheria-Tetanus-Pertussis (DTP), Measles–Mumps–Rubella (MMR), oral polio vaccine (OPV), inactivated polio vaccine (IPV), smallpox, rotavirus, and so on.
- **Recombinant vaccines:** obtained as products of gene modification of different organisms (yeast or bacteria) with posterior purification to get the immunogenic protein (example Hepatitis B vaccine).
- **Synthetic vaccines:** Part or the whole antigen is a product of chemical synthesis, proteic nature (peptides and oligopeptides), or carbohydrates (saccharides and oligosaccharides) (polyribosylribitol phosphate (PRP) synthetic antigen conjugated to tetanus toxoid).
- **Combined vaccines**: including different combination of classic, recombinant, or synthetic vaccines (DTP-HepB-Hib).
- New approach of vaccines: DNA vaccines and plant-derived vaccines, under development but still not licensed.

tion. A number of potential and theoretical risks are implicit in their use. They include, in particular, the presence of adventitious agents derived from the source materials or introduced during manufacture or, as is the case of live vaccines, the presence of virulent organisms caused by reversions of the vaccine virus or bacteria to its virulent form. In these cases, a possible risk to the community at large may exist in addition to the vaccinees [9].

A remarkable difference between vaccines and biologicals and other synthetic pharmaceutical drugs consists of the difficulty of evaluating the activity for those biological products, and thus the differences among them in their control and regulation. In general, for synthetic drugs and medicinal products, the active ingredient is totally defined and the activity can be evaluated by quantitative methods. They are synthesized in the laboratory in a precisely controlled way by a defined sequence of chemical reactions, and the nature of these molecules is such that, once they are made, their exact structure and composition can be determined by laboratory tests. Today, technology allows the modeling of small drug-receptor interactions that can be studied using computational chemistry (*in silico*) methods and extraction of valuable data and information from databases (bioinformatics). The modeled drug is then synthesized using combinatorial chemistry. The application of technologies such as X-ray crystallography and Nuclear Magnetic Resonance [8] are standard procedures for the 3D structural determinations and drug-receptor interactions.

For biologicals, the biological activity and its consequences must be evaluated in an integrated way before starting clinical trials. The complexity and challenge that regulators face with vaccines can be exemplified with the case of combined vaccines, where different epitopes are being introduced in the human body, inducing different responses to the regulatory mechanisms of the immune system, while not causing interference in the responses between them. New vaccines challenge regulators with the development of novel procedures for the proper evaluation of their efficacy and safety.

It is important that the standardization and control of vaccines by the national regulatory authority and the manufacturer are continually reviewed and modified so as to reflect the current state of science and technology incorporating an improved understanding of quality and safety issues. Regulatory authorities must thus be proactive and maintain an acute awareness of scientific developments in the vaccine field [9].

12.1.2.3   Regulation of Biotechnology Products

Recombinant insulin was the first biotechnology product, which emerged in 1982. Today, biopharmaceutical products have diverged to encompass not only recombinant forms of natural proteins and biologicals derived from natural sources, including recombinant plant-derived pharmaceutical proteins, but also monoclonal antibody (mAb)-based therapeutics. All these products require special consideration from a regulatory point of view.

In the past, microbial and animal cell culture was the most common method of producing recombinant proteins. Currently, the interest is focusing increasingly on transgenic animals because of their apparent ability to produce complex proteins at a high volume and a low cost [6]. Many of these products are now in different phases of clinical trials, although some of them without encouraging results [10].

During 2002 and 2003, regulators in North America and Europe approved a total of 64 biopharmaceuticals for human use. These approved drugs included hormones, blood factors, thrombolytics, vaccines, interferons, monoclonal antibodies, and therapeutic enzymes (not including new indications for products already approved) [11]. Data published by Walsh [11] suggest that during the past three years, over a quarter of all new drug approvals were biopharmaceuticals, taking into account that these numbers do not reflect duplication for products with different formulations and different commercial names for the same active ingredient. All of the new biopharmacuticals approved in United States in 2003 are protein-based drugs (rather than nucleic acid-based drugs). According to estimates at that time, around 500 biopharmaceuticals were undergoing clinical trials [11].

To date, no gene therapy drug has gained marketing approval as a direct consequence of the technical, manufacturing, and regulatory difficulty that still beset this class of therapeutic compounds. Besides, it is important to consider that a bad reputation on these products exists as a consequence of reported incidents of adverse reactions after their use [11]. Monoclonal antibody therapeutics have been hit particularly hard regarding warnings mainly related to hypersensitivity or anaphylactic reactions or cytokine storms [12].

The major target indications of biopharmaceuticals currently undergoing clinical trials include cancer, cardiovascular, and infectious diseases, the major killers within the developed world. Although in excess of 30 nucleic acid-based drugs are currently being evaluated for gene therapy, vaccines, and other applications, the majority remain in the early stages of clinical development (phase I/II) [11].

Another kind of biotechnological product that is acquiring interest as a therapeutic agent is the biopharmaceuticals derived from plants. As a result of the limited production capacity of many pharmaceutical products, the production in plants represents an innovative tool that can expand the yields of active principles because the large volume of biomass that can be developed at the fields. An extra advantage is that these products are free of human diseases and mammalian viral vectors. In products derived from bacteria, the bacterial endotoxin content of the final product has always been of concern, which is not an issue in products derived from plants.

To date, only proteins have been developed through genetically modified plant, including mammalian antibodies, blood components, coagulation factors, vaccines (edible vaccines), hormones (insulin, somatotropin, and erythropoietin), various interferons, and other therapeutics agents such as enzymes and interleukins [13].

Monoclonal antibodies for therapeutic and diagnostic use seem to be the best potential successful products produced in plants, but they currently have not reached the marketplace. Tobacco plants are the most commonly used plant in these developments. Hepatitis B surface protein and rabies virus glycoprotein are examples of advances in edible vaccines.

In the production of biopharmaceuticals derived from genetically modified plants, several steps are part of the process, and they commonly include:

- selection of host plant;
- selection of the gene expression system;
- location of gene expression within the plant (green matter, the seed, or other tissues);
- production of biomass; and
- implementation of purification systems for the protein.

Several of these steps will require new and special regulatory developments. The products obtained from genetically modified plants must be carefully studied by regulators focusing on some main issues such as the development of Good Manufacturing Practices appropriate for the use in the fields that can guarantee the consistency of production during the manufacturing process. Environmental issues also exist that must be taken into account. Potential effects on nontarget species such as butterflies, honeybees, and other wildlife at or near the growing sites are important to be considered [13]. The possible transgene escape through pollen or seed dispersal, the potential for recombinant molecules to enter the food chain [14], and the proper evaluation of the impact that the introduction of virus from plants into the human body can cause are issues that have to be resolved for the success of this modern biotechnology tool.

Table 12.1-2 summarizes some of the main differential issues regarding small molecules, biologicals, and biotechnological products.

12.1.3   MAIN REGULATORY AGENCIES

12.1.3.1   Food and Drug Administration (FDA)

FDA is recognized as the strongest regulatory agency in the world. Although the agency has limited its regulatory functions to products to be used in the United States, this agency has been the pioneer in the development of the regulation for drugs, including vaccines, blood products, and a lot of new biotechnology products emerging in the field that are used worldwide by other regulatory agencies.

Since 1962, the key elements of product safety and efficacy have not changed, although the agency's level of participation, collaboration, and transparency has

TABLE 12.1-2. Major Differences that Impact the Regulation of Small Products versus Biologicals versus Biotech Products

Activity or issue	Small molecules	Biologicals	Biotech products
Control of starting materials	Generally purified and well-characterized chemicals	Most of culture media for fermentation processes are not totally synthetic. Use of animal-derived additives (fetal calf serum) is requested	Requirements for: Baseline data on the host and the gene (vector) used for production; stability of the host-vector expression system under storage and recovery conditions
Control of the manufacturing process	In general, well-established procedures are available for synthesis and purification of drugs and medicines	Great diversity of production and purification procedures, depending on strain of origin. Procedures developed case by case	Great diversity of production processes depending on the host and gene used. Requirements for control at genetic, post-transcriptional, post-translational level or during production and purification
Quality control tests	Well-established qualitative and quantitative control tests are available to measure the activity of medicines and drugs. Mainly required for final product	Difficulties in the evaluation of biological activity. In many cases, "in process" quality control tests are requested to ensure the safety and effectiveness of the final product. It is very important that the correlation of biological activity with efficacy is shown in clinical trials. Need to show absence of potentially hazardous contaminants from animal origin, cell substrates, or additives	Similar to biologicals, "in process" quality control tests are requested to ensure the safety and effectiveness of the product. Need to show absence of potentially hazardous contaminants mainly oncogenic DNA

increased as demonstrated with review goal dates, meetings with industry, guidance documents, and public and private partnerships. Today, according to FDA estimates, 10,510 approved drugs and more than 100,000 over-the-counter drugs are on the market [15].

As discussed by Ray and Stein [16], "the current regulatory process does not have systematic provisions for obtaining important data needed to guide clinical practice. Planned data collection happens almost exclusively during pre-marketing testing. The FDA approves medications on the basis of studies of limited duration that include relatively small numbers of patients who are often healthier than the target populations for the new drug. Although many important effects of a new medication almost certainly will be unknown at the time of licensing, there are no systematic provisions for post-marketing studies."

This problem is common for regulatory agencies around the world. The lack of information presented by manufacturers regarding the safety during the use by populations at large, the identification of the best indicators for efficacy, and the differentiation that manufacturers try to obtain between products with the same active ingredient, frequently make the regulation function difficult to perform. In addition, in developing countries, the lack of expertise in the different areas of regulation, particularly on clinical trial evaluation and interpretation of statistical results, increases the high uncertainty of the regulatory agencies' results.

12.1.3.2 European Medicines Agency (EMEA)

The European Medicines Agency (EMEA) is a decentralized body of the European Union created with the objective to harmonize regulatory activities in Europe. EMEA began its activities in 1995, when the European system for authorizing medicinal products was introduced, providing for a centralized and a mutual recognition procedure. Its main responsibility is the protection and promotion of public and animal health through the evaluation and supervision of medicines for human and veterinary use.

Before 1995, the legal framework in Europe for drugs was 25 independent national regulatory authorities and thus 25 parallel national review processes, with the same number of independent marketing authorizations. The consequence was poor resource utilization, divergent scientific opinions, and confusion by physicians, nurses, and health workers in general when decisions had to be made. At the present time, EMEA coordinates the evaluation and supervision of medicinal products throughout the European Union. The Agency brings together the scientific expertise of the 25 European Union Member States in a network of 42 national competent authorities. It cooperates closely with the international partners, reinforcing the European Union contribution to global harmonization [17].

At the beginning, the EMEA set up two procedures for product authorization in Europe, the centralized and decentralized procedures.

The centralized procedure is as follows [17]:

- submission of the marketing authorization application to EMEA;
- a unique evaluation is carried out through the Committee for Medicinal Products for Human Use (CHMP);

- if the committee concludes that the product has proven its quality, safety, and efficacy, it adopts a positive opinion; and
- the positive opinion is sent to the commission to be transformed into a single market authorization valid for the whole of the European Union.

The decentralized procedure establishes a mutual recognition between Member States. Briefly, this procedure is based on:

- a marketing license application is submitted to one Member State;
- if the national regulatory authority concludes that the product has proven its quality, safety, and efficacy, it grants a marketing authorization to the product; and
- the marketing authorization is extended to one or more additional Member States.

Today, EMEA uses both procedures, but is primarily involved in the centralized procedure [17]. The regulation 2309/93 has defined a mandatory centralized procedure for medicinal products developed by means of one of the following biotechnological processes: recombinant DNA, controlled expression of genes coding, hybridoma, and monoclonal antibody methods. In addition, after 2005, the regulation establishes the mandatory centralized procedure for medicinal products containing new active substances for AIDS, cancer, diabetes, and neurodegenerative diseases, including orphan designated medicinal products. The regulation still allows the mutual recognition procedure for generic centralized products or for medicinal products with a significant therapeutic benefit, products of scientific or technical innovation, or products answering the interest of patients or animal health at the community level.

   The fact that more and more products are being included in the mandatory centralized procedure is indicative of a general reliance on this procedure more than on the mutual recognition procedure; it can also reflect that not all national regulatory authorities have developed the same regulatory capabilities to authorize the marketing of new products for the whole European Community.

   Although European countries have faced many challenges in the establishment of regulatory activities of the agency, mainly because the diverse conflicting priorities between the Member States, the centralized procedure has generated harmonization in the scientific opinion in Europe, getting one marketing authorization valid in the European Community, one common name, and, most important for the impact on health professionals and users (potentially 370 million), one common product information.

   After 10 years of been established, EMEA is a good example of harmonization, with both centralized and mutual recognition procedures, that can be extrapolated to other regions of the world, mainly in the developing world where the few countries that have developed strong regulatory systems can share technical knowledge and expertise in the different areas of regulations with those countries in the corresponding regions with limited resources to the benefit of the whole region.

12.1.3.3 International Committee for Harmonization (ICH)

The ICH is not a regulatory agency, it is a unique project that brings together the regulatory authorities of Europe, Japan, and the United States and experts from the pharmaceutical industry in the three regions to discuss scientific and technical aspects of product registration.

The purpose is to make recommendations on ways to achieve greater harmonization in the interpretation and application of technical guidelines and requirements for product registration to reduce or obviate the need to duplicate the testing carried out during the research and development of new medicines. The objective of such harmonization is a more economical use of human, animal, and material resources and the elimination of unnecessary delay in the global development and availability of new medicines while maintaining safeguards on quality, safety, and efficacy and meeting regulatory obligations to protect public health. The guidelines established by ICH are followed by many countries around the world, sometimes as the only guide available to establish harmonized standards on quality, safety, and efficacy for their products. However, its impact in classic vaccine regulation is limited, being more developed regarding biotechnological products.

Six founding members of ICH are directly involved in the decision-making process. They represent the regulatory bodies and the industry in the European Union, Japan, and the United States. They are the European Commission and the European Federation of Pharmaceutical Industries and Associations (EFPIA) in Europe; the Ministry of Health Labor and Welfare and the Japan Pharmaceutical Manufacturers Association (JPMA) in Japan; the FDA and the Pharmaceutical Research and Manufacturers of America (PhRMA); and a group of observers including WHO, The European Free Trade Area (EFTA represented at ICH by Swissmedic Switzerland), and Canada represented by Health Canada.

12.1.4 GLOBAL ROLE OF WHO IN DRUGS AND BIOLOGICALS REGULATION

Although WHO is not a regulatory agency, it provides guidelines for acceptability of products taking in consideration the specifications needed in regions where the regulatory institutions have not fully developed. With the responsibility to guarantee the quality, safety, and efficacy of products all around the world, WHO develops guidelines joining efforts from experts in the most diverse regions, getting a generalized opinion about the best way to establish technical guidelines. WHO provides relevant expertise ant technical assistance through such activities as guidelines development; workshops and training courses; coordination and promotion of anti-counterfeiting measures; prequalification of medicines for priority diseases; pharmacovigilance for global medicine safety, regulatory, and other information exchange; and review of narcotic and psychotropic substances for scheduling within the 1961 Convention on Narcotic Drugs and the 1971 Convention on Psychotropic Substances.

12.1.4.1   WHO Procedure for Assessing Drug Regulatory Authorities

As WHO establishes [9], the overall objective of a DRA is to ensure that medicinal products (pharmaceuticals, biologicals including vaccines, blood products, and other biologicals) are of acceptable quality, safety, and efficacy and that they are manufactured and distributed in ways that ensure their quality until they reach the patient/consumer and their commercial promotion is accurate.

According to WHO [9], the main functions of a DRA are as follows:

- registration (licensing) of products,
- inspection and licensing of manufacturers,
- inspection and licensing of distributors,
- post-marketing surveillance, and
- regulation of claims that can be made for commercial promotion of products as well as authorization of clinical trials.

A DRA can be effective only if it has the following:

- a legal basis for all its functions in legislation and regulations,
- sufficient human and financial resources,
- access to appropriate scientific expertise, and
- access to a quality control laboratory.

Particularly for vaccines and biologicals, in comparison with medicines, because the inherent variability of these products caused by the biological nature of their starting materials, their manufacturing processes, and their test methods, WHO has identified six essential regulatory functions to evaluate a regulatory system as effective [9]. These functions are as follows:

- a published set of clear requirements for licensing,
- surveillance of vaccine field performance (safety and efficacy),
- system of lot release,
- use of laboratory when needed,
- regular inspections of manufacturer for Good Manufacturing Practices compliance, and
- evaluation of clinical performance in clinical trials.

An extra function not included in the six established by WHO must be considered during the evaluation of regulatory authority for vaccines. This function must include the distribution and vaccine's shipment mechanisms; it must be able to guarantee the delivery of vaccines to all points of use with an acceptable level of quality.

Finally, research in the regulatory agencies, looking to have experts in each function updated in all relevant aspects involved in the regulation activities, is an aspect that must be considered during the implementation of regulatory systems in the countries.

The difference of the regulatory activities that WHO performs in comparison with the activities of other regulatory institutions is the impact that WHO guidelines, norms, and procedures have all over the world. The documents are developed looking for consensus with globally recognized experts. They are a reflection of the activities that are being developed in the world, and it facilitates their implementation.

12.1.5   OTHER INITIATIVES FOR HARMONIZED REGULATION

The Pan American Network for Drug Regulatory Harmonization (PANDRHA) is an initiative to support the process of drug regulatory harmonization throughout the Americas, with the mission to promote drug regulatory harmonization for all aspects of quality, safety, and efficacy of pharmaceutical products as a contribution to the quality of life and health care of the citizens of the member countries of the Americas. It includes the participation of all regulatory authorities in the region along with representatives of academia, industries, and other experts. The main forum for developing the PANDRHA activities is the Pan-American Conference on Drug Regulatory Harmonization, where there is also participation from other organizations such as the Caribbean Common Market (CARICOM), the Southern Countries Common Market (Mercado Comun del Sur, MERCOSUR), the North American Free Trade Agreement (NAFTA), and the Central American Integration System (Sistema de Integracion Centro Americano, SICA) and constitutes a way to disseminate the decision on drug regulatory harmonization of global initiatives such as the ICH. The network functions through working groups assigned to different areas including: good manufacturing practices, bioequivalence, good clinical practices, combat to drug counterfeiting, pharmacopoeia, drug classification, medicinal plants, pharmacovigilance, good laboratory practices, vaccines, and drug registration.

Other initiatives around the world exist that are incorporating, among many objectives, the harmonization of drug regulation. Although not all of these organizations have been developed exclusively for regulation, such initiatives can be found in the Association of South-East Asian Nations (ASEAN) [18], Andean Community of Nations (CAN), the Collaboration Agreement of Drug Regulatory Authorities in European Union Associated Countries (CADREAC), the European Union, the Gulf Central Committee for Drug Registration (GCC-DR), the Southern Common Market (MERCOSUR), and the Southern African Development Community (SADC).

12.1.6   FUTURE PERSPECTIVE OF REGULATION

Regulators must be prepared to face challenges resulting from the fast scientific, technological, and methodological development of drugs, biologicals, and biotechnological products. This effort will require that DRAs establish stronger alliances with the research community, universities, and academia in general so as to have available experts in all fields. Still, much remains to be done in the development of clinical trials with products derived from new technologies, mainly focusing on those that target the immune system's natural control mechanisms, including therapeutic monoclonal antibodies, interleukins, and some vaccine's adjuvant, to avoid the recent disastrous example in Europe with the use of a monoclonal antibody [10].

Regulatory authorities have been receiving important pressure from industry to decrease approval times. Some agencies are trying to reduce approval times by making changes to the review system such as: improving communication and consultation by applicants [19], which leads to better preparation of dossiers; eliminating unnecessary administrative time loss; interacting with similar regulatory agencies; and increasing the number of employed reviewers with appropriate expertise. The latter has impacted through the accumulation of experience, know-how, and proficiency within the regulatory agency. However, it must be clear that from the point of view of society and improving public health, approval times alone are not the only goal. It is more critical to make the best decision to secure safety, efficacy, and effectiveness of the new drugs.

At the same time, for product developers, the recommendation to integrate regulation in the very early stages of their research and development work exist. This strategy can contribute as a success factor instead of a bottleneck at the later stages of licensing and market authorization. Incorporating a regulatory perspective will have a positive effect on the cost and time frame of product development. Furthermore, having a good understanding of regulatory requirements will also assist the developer in identifying nonviable projects at an early stage, thus reducing the risk of the loss of investment.

Regulation of drugs must balance the extreme efforts to regulate and control the presence of adverse effects while trying to address health care needs of the population in general. New products are costly to develop and produce. To guarantee their safety, specifically for possible rare events associated to its use, the need of designing and conducting large clinical trials to show statistical significance exist. This cost will be included in the product's price. As a recent example, in the vaccine area, large clinical trials had to be conducted to demonstrate safety for two new rotavirus vaccines. Over 60,000 children were involved in trials for each vaccine. These regulatory requirements add not only onto the cost of development but also onto the timing for the products to access the markets.

Post-marketing surveillance activities follow programs based in passive collection of spontaneous reports for ensuring safety using voluntary reporting of adverse reactions from physicians and other health workers is a common practice in developed countries, in many cases without or with poor participation from the regulatory agencies. This procedure shows deficiencies such as underreporting, difficulty in calculating rates because incomplete numerator data along with unreliable denominators, and limited ability to establish cause and effect [20, 21].

Recent withdrawals from the market of high-profile drugs are more often being carried out by the manufacturer who proceeds to withdraw the product before the regulatory agency does so, in principle to protect the public from further harm, but, at the same time, the manufacturer avoids the drastic consequences of penalties that the law establishes if the recall is done by the regulatory agency. Although much has been done to strengthen quality and safety during the development and production processes, post-marketing surveillance involvement by the regulatory authorities is still a pending issue.

The strongest regulatory agencies are still looking for reforms on drug regulation [16, 22] focused to improve the current regulatory process. Modifications proposed for the regulation in developed countries generally are focused on solving the particular problems in terms of the country or region that is involved, and, in general, are not applicable to regulation in developing countries. To be able to address the enormous challenges faced by regulation of new products, regulatory agencies in developing countries will have to urgently strengthen their networks and establish new harmonized regulatory and methodological procedures.

REFERENCES

1. Chabra R, Kremzner M E, Killany B J (2005). FDA policy on unappproved drug products: Past, present and future. *Ann. Pharmacother.* 39:1260–1264.
2. Bren L (2006). The road to the biotech revolution: Highligths of 100 years of biologics regulation. *FDA Consumer Mag.* Centen. Edn. 1–7.
3. Offit PA (2005). The Cutter Incident. Yale University Press, New Haven, CT, pp. 83–131.
4. Baylor N W, Midthun K (2004). Regulation and testing of vaccines. In B A Plotkin, W A Orenstein (eds.), *Vaccines*, 4th ed.: Saunders, Philadelphia, PA, pp. 1539–1555.
5. Structure of ICH. Available: www.ich.org/cache/html/355-272-1.html.
6. WHO/Drug regulation. Available: www.who.int/medicines/areas/quality_safety/.
7. Ratanawijitrasin S, Wondemagegnehu E (2002). Effective drug regulation. A multi-country study. *WHO.* pp. 7–10.
8. Ng R (2004). *Drugs. From Discovery to Approval.* John Wiley and Sons, Inc., New York, pp. 51, 176–178, 212.
9. World Health Organization, Geneva (1999). Regulation of vaccines: Building on existing drug regulatory authorities. WHO/V&B/99. pp. 1–28.
10. Wadman M (2006). London's disastrous drug trial has serious side effects for research. *Nature.* 440:388–389.
11. Walsh G (2003). Biopharmaceutical benchmarks. *Nat. Biotechnol.* 21:865–870.
12. Hopkin M (2006). Can super-antibody drugs be tamed? *Nature.* 440:855–859.
13. Goldstein D A, Thomas J A (2004). Biopharmaceuticals derived from genetically modified plants. *QJ Med.* 97:705–716.
14. Ma J K, Barros E, Bock R, et al. (2005). Molecular farming for new drugs and vaccines. Current perspectives on the production of pharmaceuticals in transgenic plants. The European Union Framework 6 Pharma-Planta Consortium. *EMBO Reports.* 6(7):593–599.
15. Walsh G (2000). Biopharmaceuticals benchmarcks. *Nat. Biotechnol.* 18:831–833.
16. Wayne A, Ray W A, Stein M (2006). Reform on drug regulation—beyond and independiente drug safety board. *N. Engl. J. Med.* 354:194–200.
17. About the EMEA—Structure. Available: www.emea.eu.int.
18. Awang D C M Z B C (2003). ASEAN Initiatives toward pharmaceutical regulatory harmonization. *Drug Info. J.* 37:55–58.
19. Heidenreich K (2006). Dialogue between companies and the EMEA regarding the centralized procedure from an industry's point of view. *Drug Info. J.* 40:15–21.
20. Fontanarosa P B, Rennie D, De Angelis C D (2004). Post marketing surveillance—lack of vigilance, lack of trust. *JAMA.* 292(21):2647–2650.

21. Lexchin J (2006). Is there still a role for spontanous reporting of adverse drug reactions? *CMAJ. J.* 174:191–192.

22. Nutt D J (2006). Informed consent—a new approach to drug regulation? *J. Psychopharmacol.* 20:3–4.

12.2

INTELLECTUAL PROPERTY AND BIOTECHNOLOGY

Tania Bubela[1] and Karen Lynne Durell[2]

[1] University of Alberta, Edmonton, Alberta, Canada
[2] McGill University, Montreal, Quebec, Canada

Chapter Contents

12.2.1 Introduction   1392
12.2.2 Overview of Intellectual Property Rights for Biotechnology   1392
    12.2.2.1 Patents   1392
    12.2.2.2 Other Intellectual Property Rights   1394
12.2.3 Biotechnology Patents, Public Policy, Public Opinion, and Morality   1396
12.2.4 International Initiatives   1400
    12.2.4.1 The World Trade Organization (WTO)   1401
    12.2.4.2 World Intellectual Property Office (WIPO)   1403
    12.2.4.3 The World Health Organization (WHO)   1404
    12.2.4.4 Union for the Protection of New Varieties of Plants (UPOV) 1404
    12.2.4.5 Convention on Biological Diversity (CBD)   1405
    12.2.4.6 Food and Agricultural Organization (FAO)   1406
12.2.5 National Patent Laws   1406
    12.2.5.1 Substantive Patent Law   1407
    12.2.5.2 Procedural Patent Law   1412
    12.2.5.3 Legal Challenges: Infringement, Exemptions to Infringement, and Validity 1416
12.2.6 The Effect of the Regulatory Environment   1420
    12.2.6.1 Biosimilars   1421
12.2.7 Reform Proposals for Innovation Governance in Biotechnology   1422
    12.2.7.1 Strengthening Research Exemptions   1422
    12.2.7.2 Patent Pooling   1423
    12.2.7.3 Open Source Patents   1424
    12.2.7.4 Compulsory Licensing   1425
    References   1426

Handbook of Pharmaceutical Biotechnology, Edited by Shayne Cox Gad.
Copyright © 2007 John Wiley & Sons, Inc.

12.2.1   INTRODUCTION

Intellectual property laws and regulations govern the creation, use, and exploitation of intellectual activity in the industrial, scientific, literary, or artistic fields within national boundaries [1, 2]. They establish property protection over intangible items such as inventions, signs, and information. In theory, intellectual property protection provides incentives for the production and distribution of new knowledge that may benefit society by establishing a mechanism for the creator of that knowledge to obtain compensation for the use of protected works.

Intellectual property laws and regulations are designed to manage the competing interests of those who produce new knowledge, those who want to make use of new knowledge, and those who want simply to buy the products and services made through the use of new knowledge. It is becoming increasingly difficult to balance these competing desires in light of rapid advances in fields such as biotechnology and information technology. More broadly, designing sets of laws and regulations, along with the governmental institutions and practices that administer them in keeping with business practices and public interests, is becoming increasingly difficult, which is especially so given the lack of empirical data on how the entire intellectual property system works for innovation in emerging technological fields [3, 4]. The central issue is the appropriate balance to be struck between setting intellectual property rights that encourage and reward innovators while still permitting affordable and ready access to the public and other users. The latter concern is paramount in the context of health biotechnology.

This chapter provides an overview of intellectual property rights available for biotechnological innovation and the national and international laws and regulations governing those rights. The focus is on patent rights, the most widely sought intellectual property right in the field of biotechnology. We compare the national patent laws of the United States, Europe, Japan, and Canada.

Unfortunately, most patent laws and regulations were designed for inanimate matter. Significant difficulties develop in the application of arcane laws to technological innovation involving biological material, especially self-replicating life forms. Patent rights have become a flash point for many of the social and ethical concerns that derive in the context of biotechnology. Although the controversy centers on patenting of biotechnological innovation, the underlying concern is the increasing commercialization of biotechnological research. We conclude with widely raised policy reforms designed to address issues of balance, access, and public support for research and innovation in health biotechnology.

12.2.2   OVERVIEW OF INTELLECTUAL PROPERTY RIGHTS FOR BIOTECHNOLOGY

12.2.2.1   Patents

Patents, the focus of this chapter, are the most prevalent form of intellectual property right (IPR) sought for biotechnological innovation [5]. Patent rights are negative rights because they confer upon the holder the right to prevent others from selling, using, importing, and making the invention for a set period of time [6–8]. In effect, patent rights grant a holder exclusive control over an invention—a monopoly right. This broad exclusionary right is an important and valuable tool for industry as it can translate into the ability to limit entry by others into a market sector and to set the market price for an invention.

Patent rights may be granted over several aspects of an invention including the product itself and the process to manufacture the product. Several formal steps exist that must be completed to obtain a patent. The two most important are: (1) submission of a patent application, drafted to meet legislated format and content requirements; and (2) a successful review of the patent application by a national patent office based on the patent criteria. The most crucial aspect of the patent application is the claims; only those elements of the invention included in the claims will be granted patent protection. Nonclaimed elements will not be granted patent protection [9, 10]. The patent criteria, with some variation between national patent systems, are that the invention be new, useful, and involve an inventive step. An invention that fails to meet any of the criteria, or any other requirements set out in national patent laws, will not be granted a patent.

As patent rights are national in scope, inventors who want to secure patent rights in more than one nation must submit a patent application and comply with the patent legislation in each country individually. Patent rights granted by one country will not be respected in any other country. For example, if an inventor is granted a patent in the United States but fails to seek patent protection in Canada, the inventor's patent rights will end at the U.S./Canadian border and Canadians will be free to make use of the invention. Moreover, if an inventor is granted a first patent in the United States and a second patent in Canada for the same invention, the U.S. legislation will govern the use of the patent rights granted in the United States and the Canadian legislation will govern the use of the patent rights granted in Canada. Generally, no overlap of patent legislation exists between nations.

The modern version of patent rights has evolved and changed significantly from its historical role. For example, in Europe, patents were originally granted as letters patent, which were intended to reward inventors for their industrial activities. The original letters patent were different from today's patents in that they granted privileges rather than property rights. The sovereign had the right to grant a privilege to an inventor, and in so doing, encourage the development of new industries within a nation [11]. Letters patent were not necessarily granted for inventions that were absolutely new but for those that were newly introduced to the nation [11]. Copying a foreign invention and introducing it to a new country was an entirely permissible and desired outcome. For example, a patent was granted in France by the French crown "for glassware according to the manner of Venice" [11], although it was common knowledge that the glassware technique was not novel but was previously practiced in Venice. The fact that it was new to France was sufficient to garner a patent right.

The modern patent system is largely a creation of the nineteenth century and is popularly understood to respond to the argument that inventors have a natural, inherent property right in their inventions that the state has an obligation to protect [11]. This argument is directly related to the realization that innovation can represent significant commercial value. Accordingly, the state agrees to protect the inventor's natural property right in an innovation, but considers public benefit by requiring the inventor to disclose publicly the details of the invention. Disclosure allows for the generation and circulation of technical information, which is known as the "information function" of the patent system [2].

The second main justification for patents is that they provide an incentive for the production of new innovation and improvements on old innovations. The key incentive is the 20-year patent term. The establishment of 20 years as a firm and unbending time period during which inventions may be exploited solely by the inventor has many benefits. In particular, it offers certainty in the length of the right, a major factor in attracting investment for research and development.

Modern patent legislation, which emerged in the United States and the United Kingdom in the 1970s, formalized the information function of patent law. It requires an inventor to describe the invention in such a way that it can be put into practice by a person skilled in the art. Modern-day patent law also rationalizes the procedural component of patent examination by establishing set criteria that must be met for an invention to merit a grant of patent rights [11]. The outcome of these changes is that skilled examiners are now required to evaluate whether a patent application is complete and ultimately whether an invention is patentable.

12.2.2.2 Other Intellectual Property Rights

A number of other forms of IPRs, besides patent rights, may also apply to protect elements of biotechnological innovations. Here, we present a brief overview of these IPRs, which include trade secret, plant breeders' rights, copyright, database rights, and trade mark.

Trade Secret. Trade secret grants a right over information that is treated as confidential by a corporation or other organization. The right translates into the ability to sue for a breach of confidentiality if someone, for example, a key employee, discloses confidential information to a competitor. The wronged party is entitled to a remedy usually in the form of monetary compensation for the harm caused by the wrongful disclosure of the secret information.

Trade secret provides a necessary supplement to other IPRs because it can be used to protect technical information and know how during research, development, and testing stages of a biotechnology project. Unlike other IPRs, trade secret is rarely rooted in statute, but is usually established by the collective history of judgments from litigated cases known as jurisprudence [12, 13].

Trade secret protection is acknowledged when information is treated as confidential, which usually means those to whom it is properly disclosed realize that it is secret and agree to protect the confidentiality of the information. In many cases, companies establish secrecy through the use of confidentiality agreements with employees and other customers or partners. As the right relies on absolute secrecy, trade secret protection is fragile; the right can be lost immediately upon nonconfidential disclosure of the secret information. Companies therefore usually opt to seek patent protection once a product is introduced to the market and only rely on trade secret during the research and development stages. It is particularly true for innovations that can be easily reversed-engineered. Such inventions cannot be adequately protected by way of trade secret because they are subject to copying.

However, in some cases, trade secret protection may constitute a competitive advantage by sustaining marketplace confidentiality [14]. If the technical details of an innovation remain secret, it is often difficult for competitors to recreate a product, invent around it, or improve on it, especially if the innovation is complex. These latter types of activities are commonly known as competitor piggy-backing. If adequate confidentiality measures are put in place and enforced, trade secret can be an effective method of protecting information from the public and securing a market sector monopoly [15].

Plant Breeders' Rights. Generally, plant breeders' rights are granted to a person who breeds a new plant variety and for this reason are helpful tools for many biotechnology projects. Like patents, plant breeders' rights require a formal application and review process, which is undertaken nationally. The strength of plant breeders' rights varies greatly between nations. The rights granted may extend to the plant, as in the United States where plant rights are a subset of the patent system, or may only include the propagating material and not the plant, as set out by Canadian legislation. It is argued that strong plant breeders' rights negatively affect farmers because plant breeders may exercise their right to limit the subsequent uses of their plants. For example, a farmer who grows a crop from protected seeds may be prevented from saving the seeds from that crop for the purpose of cultivating a subsequent crop [16, 17]. Concerns over farmers' rights vary globally resulting in an international collection of plant breeders' rights statutes that are inconsistent in strength and scope.

Copyright. Copyright regulates the creation and use of a range of original works such as books, songs, films, and computer programs. Copyright protects the expression of an idea, not the idea itself. Thus, it protects the work that is produced but, unlike a patent, cannot prevent the making, use, or sale of an identical work by an independent originator [18]. Copyright differs from patent and trade mark in another important way—it arises automatically and confers upon the owner the right to prohibit copying or performance of a work. For example, a researcher has a copyright in the scientific papers they seek to publish, although he or she generally transfers or assigns those rights to the journal before publication. Copyright is granted by national statutes for the lifetime of the author plus anywhere from 50 to 70 years after the death of the author, depending on the country. However, because of international treaties, the national foundation of copyright law is less oppressive than for patent rights. Copyright can cross borders—a work created in the United States will be acknowledged as a copyright protected work in Canada as well [19]. Copyright is of limited use for the products and processes of biotechnology, even in the area of collections of biological data, gene sequences, and plants.

Database & Material Deposit Rights. One emerging area of biotechnology that is testing the limits of intellectual property regimes is the production of academic data collections, especially biological and genetic data linked to long-term health of the individuals who provided the data. Such databases require substantial intellectual investment. Traditional IPRs, however, such as patents and copyright, do not protect mere collections of data. Indeed, the prevailing thought in legal discourse over IPRs and data is that data *per se* should not be protected based on the public nature of many large data collection efforts [20]. However, there may be need for some form of property right to prevent misappropriation and misuse of data. One such model exists, namely the European Database Directive, which was introduced in 1996 by the European Union. The directive vests an exclusive right in the producer of a database to grant permission to extract and reuse the contents of the database [11]. The debate about the appropriateness of database rights in the context of biological data is ongoing [20].

It should be noted that some national patent laws require the deposit of biological and genetic materials to supplement the disclosure of an invention. Materials are submitted to authorized depositaries that eventually find themselves in possession of a collection of biological samples. Biological material depositaries are governed at the international level by the Budapest Treaty on the International Recognition of the Deposit of Microorganisms for the Purpose of Patent [21], as established in 1977 (discussed in more detail below). The treaty and national patent rules cooperatively establish authorized depositories and sets rules for the deposit as well as the availability to the public of the material.

Trade Mark. National trade mark legislation grants protection for certain words or symbols, usually trade names such as Microsoft or pictures such as the Nike swoosh, when the words or symbols are used in connection with specific marketplace goods or services [2]. Trade mark holders are granted the right to prevent others from using the protected words or symbols to represent the same or similar goods or services. Generally, trade marks must be registered at a national office to be effective, although some jurisdictions acknowledge that a right can be established through consistent long-term use.

12.2.3 BIOTECHNOLOGY PATENTS, PUBLIC POLICY, PUBLIC OPINION, AND MORALITY

The patent system is traditionally viewed as a tool for promoting economic goals [22]. It is understood to work by encouraging the advancement of new industries, research and development, or innovation. In contrast, noneconomic factors such as health, human rights, environment, or ethics have historically been viewed as extrinsic to the activities regulated by the patent system or as undesirable side effects that require mitigation. Accordingly, patent law has been largely unconcerned with moral arguments. A few notable exceptions exist including medical treatments and inventions designed for purely criminal applications.

With the advent of the patenting of the products and processes of biotechnology, moral arguments have entered the patent arena with more force than ever, especially in Europe [23]. Indeed, the commercialization and patenting of biotechnological innovation has become a focal point for opposition to biotechnology *per se*. The core objection is that some applications of biotechnology, such as the genetic manipulation of higher organisms or embryonic stem cell research, are in themselves wrong because they are against God, nature, or human dignity. The patenting of human genes also draws special criticism as establishing property rights over the common heritage of humankind.

Intuitively, biotechnological inventions do not fit neatly into the customary model of patentable subject matter because the original law was developed for mechanical inventions. Two reasons why the patent system is not necessarily a neat fit with biological matter are the ability of living organisms to self-replicate, which has significant implications for the exclusive rights patents grant over the use and manufacture of a product, and because many biological products, such as genes, may be seen more as discoveries than as new inventions. But, despite the ill fit, patents have been issued over life forms since the mid-1800s and biological inventions have been patented since the 1970s [5]. That said, some forms of biotechnology remain unpatentable; for example, genes and gene fragments are not patentable in their natural state, but only in isolated form.

In the field of biotechnology, policy makers must achieve a delicate balance between the advancement of research and biomedical innovation and the many social, legal, and ethical challenges raised by this field of science. The latter, coupled with the fast pace of advances in biomedical research, makes it essential that there be an ongoing public review accompanied by appropriate legislative and regulatory responses [23]. Unfortunately, legislative and regulatory responses generally lag behind research and corresponding moral and ethical dilemmas. It is therefore necessary to keep law making and regulatory processes flexible and responsive to scientific advances, including those in the realm of IPRs.

There is considerable debate about the appropriateness of using the patent system to regulate innovation on the basis of public morality [3]. Those opposed state that the patent system is ethically neutral, concerning itself only with the technical requirements for patentability spelled out in national laws and influenced by international standards. Historically, patents are concerned primarily with economics and innovation while other regulatory mechanisms are both more flexible and better at optimizing health, safety, and environmental outcomes. However, others argue that the balance has shifted too far toward granting property rights over biotechnological innovation at the expense of the beneficial social outcomes of information diffusion and access to new innovation [3]. Globally, opposition to the patenting of living organisms, genes, and gene fragments may be viewed as broad opposition to the commercialization or commodification of biological matter [24]. Public unease may be augmented by distorted media reporting of sensitive issues, or these reports may merely reflect public perceptions. Either way, such reports may lead to a general public mistrust of government regulators and politicians [25].

Although such concerns may seem unfounded, it is not wise simply to discard public sentiment that opposes biotechnology. It is overly simplistic to state that public opinion, particularly public concern, is based on a combination of ignorance and misinformation. Proponents of genetic technologies and products use this argument to delegitimize public opposition. Media sensationalism is "blamed for amplifying and exploiting that public ignorance" [26]. Indeed, public opinion does not necessarily show a correlation between a level of education and support for biotechnology but depends, instead, on the segment of the public and on the specific genetic technology or product [25].

On the whole, polling studies have shown that the public is largely supportive of biomedical research [27], especially when that research is perceived to be independent of industry and conducted at publicly funded research institutions [24, 28]. Public support for agricultural biotechnology, on the other hand, is lower, especially in Europe, partly because the public has not been convinced of the utility of the applications of agricultural biotechnology, which translates into a higher level of risk tolerance by the public for health biotechnology. That said, the scientific community must be willing to accept that some research avenues are simply not acceptable to the majority of the public, and should potentially be reconsidered. Rather than blaming the media and other information conduits that are generally presenting a pro-science position, especially in the biomedical research, scientists need to demonstrate the utility of their research along with a fair assessment of the risk [29, 30].

In a democratic society, it is important to maintain public trust and confidence because the lay public can exert substantial influence on their public representatives who in turn fund or regulate scientific research and the use of science-based technologies [27]. One substantial set of studies estimates that public opinion, as measured by polling, becomes functioning law about two-thirds of the time [31]. Public representatives may also ban or criminalize scientific research that is perceived as immoral, dangerous, or unjust. Indeed, such steps have been taken in Canada with the recent enactment of the *Assisted Human Reproduction Act*, which has criminalized a range of research endeavors such as human cloning, somatic cell nuclear transfer, and germ-line alteration [32].

One factor that may significantly influence public opinion and public support for health biotechnology is the increasing trend toward commercialization of publicly funded research. Many major public funding agencies and research institutions, including publicly funded universities, now set commercialization of research activities as an explicit goal. Indeed, the rise of modern biotechnology is often explained by two events: First, the United States Supreme Court allowed the first patent of a genetically modified bacterium in 1980 [33], giving industry an incentive to invest and engage in biotechnological research; and second, the United States Congress passed the Bayh–Dole Act in 1980, providing universities with the right to obtain patents in the results of federally funded research and encouraging universities to transfer technology to industry. The success of technology transfer between universities and industry is measured by the annual survey of the Association of University Technology Managers [34].

However, evidence exists that the public is becoming increasingly suspicious of the ties between industry and researchers, at a time when academic biomedical research receives unprecedented levels of funding from industry and research institutions are increasingly focused on policies aimed at commercialization [35–38]. Survey data show that a perceived connection with commercial forces has an adverse impact on the perceived credibility of researchers. If there is a perception that funding comes from less independent sources, credibility deteriorates. For example, one recent survey of the Canadian and U.S. public on biotechnology issues found that publicly funded university researchers are highly trusted and credible. Only the World Health Organization (WHO) and peer-reviewed scientific journals were rated higher. In contrast, scientists working for biotechnology companies and university researchers funded by industry were not considered credible [39]. A 2000 focus group study done on behalf of the Canadian government came to a similar conclusion, finding that "many people say university scientists are much more credible than other scientists because it is assumed they are free from funding pressures and therefore more 'independent'" [40].

Some of the most highly publicized controversies in science policy have been a direct result of the growing ties between biomedical science and commerce. For example, in the United States, the death of Jesse Gelsinger raised concerns about whether financial considerations had had an inappropriate impact on the running of the gene therapy research trial [41]. Other concerns about commercialization are directly related to the patenting process, the Myriad Genetics controversy being the most obvious. As described by Professor Caulfield, "the decision by Myriad, a Utah based company, to enforce its patents over the BRCA1/2 genes was, arguably, the most significant recent catalyst of national policy-making activity, media coverage, and public outrage on the topic of gene patents. In 2001, the company sent a 'cease-and-desist' letter to most Canadian provinces that used the test as part of the publicly funded health care system. The case was viewed, rightly or not, as a harbinger of the policy challenges created by gene patents" [42].

Other high-profile biotechnology patent stories exist, such as Harvard College's attempts to patent a mouse, genetically modified to be susceptible to cancer, known as the "onco-mouse." A patent was granted both for the method of making the mouse and the mouse itself by the United States Patent and Trademark Office (USPTO) in 1989, but the application for the mouse patent was initially rejected by the European Patent Office (EPO). Unlike the United States, Europe allows for the rejection of patents considered to be contrary to public morality or the *ordre public*. The patent was eventually granted, but the EPO developed a statement on the patenting of transgenic animals. The analysis developed by the EPO requires an evaluation of the risks, including moral risks, and benefits of a new technology, with the recognition that every new technology is accompanied by some risk. In the case of the onco-mouse, the benefits to humankind of the mouse model for cancer research outweighed the public morality concerns about protecting the environment and the suffering of the experimental animals. The risks were further allayed by the lack of research alternatives to mouse models and the additional efficiency of using the onco-mouse in reducing the number of mice required for experimentation. The primary importance of this analysis is that biotechnology is to be treated the same as any other new technology in evaluating risks and benefits, and genetically modified organisms are not excluded out of hand from patentability on the basis of immorality.

The Canadian courts have taken a different stance on the patentability of higher life forms in a number of high-profile cases. In *Harvard College v. Canada (Commissioner of Patents)*, the Supreme Court of Canada (SCC), in a slim majority, concluded that the onco-mouse was not patentable subject matter under Canadian patent law. The SCC held that the drafters of the Patent Act had not intended higher life forms to fit within the definition of patentable subject matter [43]. To confuse the issue further, the patentability of higher life forms was addressed again only two years later in the 2004 case of *Monsanto Canada Inc. v. Schmeiser* [44]. In that case, a farmer had grown a crop of Monsanto's genetically modified canola without authorization from Monsanto. In its patent, Monsanto had claimed all products and processes related to the genetically modified canola except for the canola plant itself, because higher life forms are not patentable in Canada. The SCC did not explicitly overturn its decision that higher life forms are not patentable, but ruled that if a nonclaimed element, in this case the canola plant, is determined to have an "important role in production," then patent protection will extend to that element as well as the claimed elements [44]. In other words, having a patent over a gene inserted into a higher life form grants patent protection over the higher life form as a whole, even though technically the higher life form cannot be claimed as an invention in Canada.

The effect of the decision was to grant patent protection to a nonpatentable plant (a higher life form) because its bioengineered cells are patentable [44]. In application, this doctrine means that patent scope can be understood to cover both the claimed invention as well as any broader structure that encapsulates or incorporates the invention [44]. These Canadian decisions have been referred to as "the most significant decisions in the area of biotech patent law anywhere in the world," and the esteem attributed to them makes their inconsistencies that much more disconcerting [45].

These examples in Canada and Europe have generated a high degree of public and policy-making interest. Indeed, in these regions, a high level of public discomfort exists with the patenting of higher life forms and human genes, which is augmented by the morally contentious nature of much biotechnology research, for example, the ongoing debates about embryonic stem cell research. In controversial areas, public trust is fragile. Professor Caulfield summarizes the dilemma as follows: "The public may be setting aside moral concerns based on the belief that the research is being done in the public's interest. If the public comes to believe that commercial interests are dominating the research process, trust could easily be lost. And, it should not be forgotten that many in the public are already suspicious of the motivations behind biotechnology innovation" [42].

In addition, numerous other policy challenges exist that stem back to the social benefit side of the patent bargain. Significant concerns exist that a commercialization agenda for biomedical research will skew the research agenda away from basic research and that new biotechnological innovations will drive up the cost of health care in publicly funded systems, such as Europe and Canada, and increase the disparity between patients or regions that can afford new technologies and those who cannot. These concerns may partially be addressed by a well-formulated and well-balanced patent system and government regulation.

12.2.4   INTERNATIONAL INITIATIVES

As a result of the national scope of many intellectual property laws, problems sometimes develop where inventions, such as biopharmaceuticals, are traded internationally. International trade can bring divergent national legislation into conflict. Countries that are net exporters of innovations protected by IPRs, such as the United States, the European Union, and Japan, advocate strongly in the international arena for the strengthening of national intellectual property laws and international harmonization of these laws [46]. Although international harmonization has not been achieved, intercountry trade has lead to a number of international initiatives that significantly impact the intersection between IPRs and biotechnological innovation.

Here, we review the main international treaties relating to either IPRs or biotechnology and the institutions that administer them. The treaties range from attempts to coordinate the filing of patents at an international level to guidelines for the sharing of benefits derived from biologic resources. Depending on the form of a biotechnological innovation, several of these treaties may be applicable to a single invention. Thus, simultaneous compliance with several treaties may be required. As the treaties have not been drafted to reflect and consider each other, but do in some instances overlap, compliance can be difficult in some areas of biotechnological research and commercialization.

12.2.4.1 The World Trade Organization (WTO)

The function of the WTO, to establish trade standards at an international level, is particularly important in today's era of globalization. The WTO engages in "analysis and debate about the relationship between international trade and investment, and its implications for economic growth and development" [47]. It promotes trade and eliminates inefficiencies [47] by imposing obligations on members to accept common rules and arbitration methods. The WTO addresses trade issues in a number of different categories, one of which is the protection of IPRs and trade.

The WTO membership includes countries at varying levels of development, and from divergent political and legal systems, which are differentially impacted by globalization. This breadth of input lends credibility to both the organization and the standards created by WTO membership [48]. Despite criticism, the membership of the WTO causes its principles to be adopted widely around the world.

Trade-Related Aspects of Intellectual Property Rights (TRIPs). TRIPs, the most influential multilateral agreement on IPRs, was created by the WTO at its 1986–1994 Uruguay Round of negotiations [49]. TRIPs regulates trade exchanges through uniform transnational rules and encompasses issues related to biotechnology within its framework. The TRIPs negotiations were instigated by the realization in the 1970s of the negative effect of counterfeiting of trade-marked products on international trade [50]. Although there was opposition to the WTO assuming a role in global intellectual property issues, a move that was thought to usurp the authority of the World Intellectual Property Organization (WIPO), a General Agreement on Tariffs and Trade decision of January 28, 1987 established guidelines for the negotiation process that eventually produced TRIPs [50].

The official goal of TRIPs is to "reduce distortions and impediments to international trade" posed by divergent national IPRs [49]. It presents a common minimum standard of IPRs for the adoption within national intellectual property laws of each WTO member country. The common standards are enforced by the WTO, which adjudicates disagreements between nations [50]. Each WTO member country retains the right to legislate more stringent IPR regimes [50].

TRIPs sets standards for many forms of IPRs—copyright, plant breeders' rights, and trade marks—but the provisions relevant to patent rights are particularly significant for biotechnological innovations. To be eligible for patent protection,

TRIPs requires an invention to be new, to have an industrial application, and to have an inventive step [49]. Inventions that meet these criteria are protected against use, offer for sale, and sale and importation by anyone who is not the rights-holder for a minimum term of 20 years [49]. Moreover, TRIPs requires nations to grant patent rights for all fields of technology without discrimination and for both inventive products and processes [49].

Despite the harmonizing nature of TRIPs, a number of controversies have developed over the implementation of provisions related to biotechnological innovation and higher life forms. Article 27.3(b) states that "plants and animals other than microorganisms, and essentially biological processes for the production of plants or animals other than non-biological and microbiological processes" may be excluded from patent protection. However, plant varieties must be protected within national legislation through patents, *sui generis* protection, or a combination of both. Patents are well understood and *sui generis* protection is commonly provided as plant variety protection or plant breeders' rights legislation, compliant with UPOV, discussed below. The correct blend of these methods of protection has been subject to varying interpretations, particularly in the field of biotechnological invention. The issue of integrating the protection of traditional knowledge about genetic resources and benefit sharing with traditional knowledge holders and plant variety protection is particularly controversial.

Another controversial aspect of TRIPs is the compulsory licensing provisions in Article 31, which allow a government to override the rights to exclusive use and manufacture of a product or process granted by a patent. National IPR regimes may allow for compulsory licensing under limited circumstances. Conditions for granting a compulsory license include a national emergency and public noncommercial use for public health, national defense, or environmental protection. The provisions of a compulsory license may likewise be limited by national IPR laws. For example, negotiations with the holder of patent rights must start before the granting of a license, the scope and duration of the license must be limited, the rights holder must be compensated, and the license must be nonexclusive. However, these strictures have not quelled the debate, which is fueled by the characterization of the compulsory license issue as a struggle between developed nations, who support very limited compulsory licensing, and developing nations, who are interested in the extension of broad compulsory licensing rights to combat public health crises such as the HIV/AIDS pandemic [50].

TRIPs does address some concerns specific to developing countries. Article 66 exempts developing countries from compliance with most provisions of TRIPs for a term of 10 years because of "their economic, financial, and administrative constraints, and their need for flexibility to create a viable technological base" [49]. A further extension may be granted on request. Moreover, TRIPs admonishes developed nations to engage in technology transfer initiatives with less-developed countries to facilitate the establishment of a sound and viable technological base in those countries [49].

TRIPs also establishes minimum standards and definitions for undisclosed information or trade secrets.

Sanitary and Phytosanitary Agreement (SPS). The SPS is a multilateral mechanism formulated to protect human, animal, and plant health in WTO member countries. The SPS details how governments can apply food safety and animal and plant health measures. These measures are based on standards articulated by a number of international organizations including the food safety standards formulated by Codex Alimentarius Commission, the animal health and disease control measures drafted by the International Office of Epizootics, and the steps for plant quarantine and phytosanitary certificate regulation articulated by the International Plant Protection Convention. SPS measures are based on a risk assessment—risks should be evaluated and the measures adopted should be commensurate with the risk. The SPS recognizes the sovereignty of nations and acknowledges that a country may maintain measures that are stricter than the SPS measures. Each nation has a right to determine its appropriate level of risk. Set standards will only be questioned if they result in discrimination or a disguised restriction on trade [51].

12.2.4.2 World Intellectual Property Office (WIPO)

WIPO is a specialized agency of the United Nations (UN) system of organizations, with a mandate to administer intellectual property matters recognized by the member states of the UN [52]. It administers 23 treaties and various programs that aim to harmonize national intellectual property legislation and procedures, provide services for international applications for industrial property rights, exchange intellectual property information, provide legal and technical assistance to developing and other countries, facilitate the resolution of private intellectual property disputes, and marshal information technology as a tool for storing, accessing, and using valuable intellectual property information.

Patent Cooperation Treaty (PCT). The Patent Cooperation Treaty (PCT) makes it possible to reserve the right to seek patent protection for an invention simultaneously in each of a large number of countries by filing an "international" patent application [53]. Such an application may be filed by anyone who is a national or resident of a contracting state. It may be filed in the applicant's national or regional patent office. Copies of a filed application are then forwarded to the International Bureau and an International Search Authority for processing and review.

The aim of the PCT system is to consolidate and streamline patenting procedures. A PCT application reserves the right to file patent applications in all of the member states at a future date. It gives the applicant more time to submit patent applications to national offices than is allotted under national patent systems. This additional period of time (up to 42 months depending on the residence of the applicant and whether a priority filing has been made) gives an applicant more time to decide where to continue with national patent applications. The PCT process also provides the applicant with valuable information about the potential patentability of the invention through an international search report and the optional international preliminary examination report.

The PCT system is expanding rapidly: the number of member states, including many European countries, Japan, and Canada, has more than doubled in the last eight years, to 125, and the number of international applications has grown from 2600 in 1979 to about 110,065 in 2003.

Budapest Treaty on the International Recognition of the Deposit of Microorganisms for the Purposes of Patent Procedure (Budapest Treaty) [21]. Patents require the disclosure of the invention, usually in writing, with sufficient detail that a person skilled in the art can reproduce the invention, which raises difficulties when the invention involves a microorganism or the use of a microorganism. The Budapest Treaty allows for disclosure to be made by depositing a sample of the microorganism with a specialized institution [54]. In practice, the term "microorganism" is interpreted in a broad sense, covering a range of biological material the deposit of which is necessary for the purposes of disclosure, in particular regarding inventions relating to the food and pharmaceutical fields.

An "international depositary authority" (IDA) is a scientific institution—typically a "culture collection"—that is capable of storing microorganisms. Such institutions have been formally nominated by the contracting state within which they reside and have agreed to abide by the treaty provisions. Under the treaty, the deposit with any IDA satisfies disclosure requirements of national patent offices of contracting states if those states allow or require deposits of microorganisms in their national patent legislation. The regional offices of The African Regional Intellectual Property Organization (ARIPO), the Eurasian Patent Organization (EAPO), and the European Patent Office (EPO) have recognized this method of disclosure, as have the United States, the United Kingdom, Japan, and Canada.

12.2.4.3 The World Health Organization (WHO)

The WHO is the UN specialized agency for health. Although it does not administer any intellectual property treaties, it is involved in intellectual property issues in the field of health. Indeed, the independent Commission on Intellectual Property Rights, Innovation and Public Health, established by the World Health Assembly in 2003 has recently completed a report entitled, *Public Health, Innovation and Intellectual Property Rights* [55]. This report considers "intellectual property rights, innovation, and public health, including the question of appropriate funding and incentive mechanisms for the creation of new medicines and other products against diseases that disproportionately affect developing countries . . ." [56].

12.2.4.4 Union for the Protection of New Varieties of Plants (UPOV)

UPOV is an intergovernmental organization with headquarters in Geneva, Switzerland [57]. It administers the UPOV System of Protection of Plant Varieties, established in Paris in 1961, with the adoption of the International Convention for the Protection of New Varieties of Plants [58]. The goal of the UPOV Act is to encourage the development of new plant varieties and to benefit society through the grant of a *sui generis* IPR to plant breeders.

Members of UPOV are required to grant and protect plant breeders' rights, including exclusive rights for the production, reproduction, and conditioning, for the purpose of propagation, offering for sale, selling or marketing, importing, exporting, and stocking, of plant varieties. The UPOV Act also sets criteria for plant variety protection, namely novelty, distinctness, uniformity, and stability [59].

The impetus behind plant breeders' rights varies globally, and, thus, the rights granted vary from nation to nation, despite the harmonizing role of UPOV. For example, countries that allow for plant patents offer broader, stronger enforcement options than countries with *sui generis* plant breeders' rights. The extent of rights granted to plant breeders reflects issues of national concern, such as the rights of farmers to save and reuse seeds, among others. The stronger rights for plants provided by patents are favored by multi-national agribusiness and countries such as the United States.

Recent changes to the UPOV Act have resulted in significant controversy about the appropriate scope of plant variety protection. Specifically, the act was extended to provide protection to an "essentially derived variety" of plant, which has left confusion about exactly which plants are now covered by UPOV. Moreover, debate rages about provisions that allow a breeder to impose royalty payments on farmers who save seeds for the purpose of reusing them in the future [17]. Royalty payments may be a significant hurdle for some farmers, particularly those in developing countries, and the provisions may augment food insecurity issues [60].

Another controversial modification to the act allows for *ordre public* considerations to supersede infringement if equitable remuneration is offered, which opens the door to a system in which *ordre public* may be invoked to avoid seeking the consent of a rights-holder to use a protected plant variety. Together, these changes to the UPOV Act are worrisome to those who perceive it to be unjust and those who consider the act to offer insufficient solutions for the problems facing farmers in developing countries [61].

12.2.4.5   Convention on Biological Diversity (CBD)

The CBD, established in 1992, provides a coordinated international framework for the commercialization and sustainable use of genetic resources [62]. It forms the basis for a fair and equitable means of sharing benefits derived from the commercial use of genetic resources. The CBD tries to balance the sovereign rights of states to draft legislation regarding the natural environment within their borders with obligations states owe to other nations. For example, Article 3 of the CBD acknowledges a state's right to exploit its own resources, but simultaneously asserts the responsibility of a state to refrain from harming areas lying outside its borders. Article 15 addresses genetic resources specifically. It permits states to regulate access to genetic resources. However, at the same time, that access must be subject to prior informed consent obtained from the party providing the resources. This provision is intended to ensure benefits flow to indigenous or other marginalized communities, which are often the custodians of traditional knowledge and rich biodiversity within their territories. The CBD is binding for its signatories, which presently include 188 parties. However, it does not include the United States.

The CBD is not formally recognized by the WTO and, consequently, is considered to lack significant legal weight. Developing countries in particular continue to advocate for the adoption of the CBD by the WTO. They argue that the CBD promotes fair and equitable use of genetic resources and, therefore, its integration with the WTO is necessary to formulate balanced international standards for genetic resources. The CBD standards may also complement and enhance the

TRIPs provisions, which promote technology transfer between developed and developing countries.

Parties attending the conference that adopted the CBD have also accepted a supplementary agreement known as the Cartagena Protocol on Biosafety as of January 29, 2000 [63]. The Protocol protects biological diversity from potential risks posed by genetically modified organisms (GMOs). One initiative is for an advance informed agreement procedure that will ensure all relevant information is provided to a country before the importation of GMOs. The protocol also establishes a biosafety clearinghouse, intended to facilitate an exchange of information on GMOs and aid the implementation of the protocol.

12.2.4.6   Food and Agricultural Organization (FAO)

The FAO administers the International Treaty on Plant Genetic Resources for Food and Agriculture (FAO Treaty), in force as of June 29, 2004, which encompasses all plant genetic resources that fall within the categories of food or agriculture [64]. It sets out access and benefit sharing rules for these resources. The provisions of the FAO Treaty accord with the CBD, although its subject matter is limited in comparison. The FAO Treaty sets up a multilateral system to facilitate access to plant genetic resources for food and agriculture, allowing countries that have ratified the treaty to work together to create equitable material transfer agreements, in compliance with the treaty. It also coordinates a pool of crop resources, collected from the members, which facilitates sharing of information among farmers and breeders. In general, the FAO Treaty aims to protect genetic resources relating to specific plants and is one of the first treaties to propose an operational system of plant genetic resource distribution at an international level.

12.2.5   NATIONAL PATENT LAWS

In this section, we compare patent rules and patent criteria for biotechnological innovation in the United States, Europe, Japan, and Canada. Although attempts have been made to harmonize patent laws globally, so far it has been impossible to achieve international consensus. TRIPs sets minimum standards that must be included in national patent legislation enforced by all WTO members, but it has not had the effect of creating consistent national patent laws.

Some attempts at regional harmonization of patent rules have been implemented. The European Patent Convention (EPC) guides the granting of patents in European countries by offering an alternative means of filing a patent in several European jurisdictions simultaneously [65]. A party seeking a patent in Europe has two options: file an application in national offices or file with the European Patent Office (EPO). An application filed with the EPO, upon grant of patent, is equivalent to having a patent in each member state of the EPC [not necessarily the same as the member states of the European Union (EU)]. All patent applications filed under the EPO system are required to submit to EPC rules. However, the jurisdiction and activities of the EPC are limited to the granting of patents. The EPC does not deal with issues developing after the grant of a patent such as infringement. Thus, the EPC governs the adjudication of the granting of patens, but national laws govern post-grant challenges.

Regional filing initiatives aside, patent laws are generally national in scope and apply only to the use of patented inventions within national borders. A patent application must, therefore, be filed in each country where patent protection is sought. The problem is that the patent law of each country is slightly different. TRIPs has created consistency in some aspects of patenting such as the patent term, which must be at least 20 years from the filing date. However, national variation may affect the subject matter that can be patented, the criteria used to judge whether a patent may be awarded, and the scope of the right granted.

It is tempting to formulate judgments about the relative strength of national patent laws based solely on patentable subject matter and the scope of the right to exclusivity. However, it is important to note that patent laws have many elements. Subject matter and scope are only two aspects of any bundle of patent rights. Patent subject matter, criteria, permissible invalidity and infringement challenges, claims construction tests, and drafting rules all play important roles in patent law as well.

12.2.5.1 Substantive Patent Law

Patentable Subject Matter. A threshold issue for patentability is whether the subject matter of the patent falls within the categories of patentable subject matter set by national patent law. In some jurisdictions, such as the United States, what is considered patentable subject matter is broadly interpreted. In the 1980 United States Supreme Court decision of *Diamond v. Chakrobarty*, which revolved around the patentability of a genetically modified bacterium capable of metabolizing oil, the Supreme Court noted that U.S. patent law embodies Thomas Jefferson's philosophy that "ingenuity should receive a liberal encouragement." [33]. The court made the famous observation that an inventor should be able to patent "anything under the sun that is made by man." Based on this justification, the United States has gone on to grant patent rights over innovations that are excluded from patentability in other countries. One such category is plant patents. The USPTO also grants patents over culture methods, differentiated cells derived from human embryonic stem cells, and the human embryonic stem cells themselves [66], subject matter that is considered not to be patentable in other jurisdiction because of moral concerns.

Adopting a narrower definition of patentable subject matter, other countries have drawn lists, of varying lengths, of inventions explicitly excluded from patentability. In Europe, the EU adopted Directive 98/44/EC of the European Parliament on the legal protection of biotechnological inventions as a standard to be adopted by all its member states and complied with by the EPO [67]. The provisions of the directive outline a restrictive approach to patenting biotechnological inventions, stating in Article 6 that inventions, including "(1) processes for cloning human beings; (2) processes for modifying the germ line genetic identity of human beings; (3) uses of embryos for industrial or commercial purposes; processes for modifying the genetic identity of animals which are likely to cause them suffering without any substantial medical benefit to man or animal, and also (4) animals resulting from such processes," and "where their commercial exploitation would be contrary to *ordre public* or morality" are not patentable. Recital 38 of the directive confirms that this list is not exhaustive and that any process whose application offends against human dignity is also excluded from patentability.

The EPC specifically limits patentable subject matter for patent applications filed with the EPO, excluding inventions that are contrary to *ordre public* or morality, as do many national European patent systems [65, 67]. The EPC further identifies discoveries, scientific theories, mathematical methods, aesthetic creations, games, business methods, computer programs, presentations of information, and methods for treatment of the human or animal body by surgery or therapy and diagnostic methods as unpatentable [65]. Furthermore, the human body, at the various stages of its formation and development, and the simple discovery of one of its elements, including the sequence or partial sequence of a gene, cannot constitute patentable inventions. However, an element isolated from the human body or otherwise produced by means of a technical process may constitute a patentable invention, even if the structure of that element is identical to a natural element. Plant and animal varieties and essential biological processes for production of plants and animals are also explicitly excluded from protection [65]. In keeping with the requirements of TRIPs, however, microbiological processes and their resultant products are patentable [65]. In compliance with Article 6 of the directive, the EPO has amended the EPC to include Rule 23d, which states that the list of exclusions included in Article 6 of the directive will likewise be unpatentable under the EPC.

In practice, genetically modified mice have been found to be patentable because they provide research platforms with significant benefits for research into human diseases [43]. Human embryonic stem cells, on the other hand, are not patentable on moral grounds because of the constraints imposed by Rule 23d(c) of the EPC, which prohibits the patenting of the "human embryo"[66]. The EPO, in two leading rulings now under appeal, has interpreted Rule 23(d) to exclude not only process patents for creating stem cell lines, thereby destroying the embryo, but also patents derived from already existing stem cell lines. Adult stem cells are not caught within the moral exclusion of subject matter from patentability.

The national patent laws of European nations have not been as well tested as the EPC provisions and the application of the directive within jurisdictions is open to interpretation and therefore can be adjudicated in variant manners. For example, the United Kingdom has interpreted the legal prohibition of patenting human embryos more narrowly than the EPO [66]. The United Kingdom Patent Office has issued a Practice Notice stating that although patents will not be granted for processes of obtaining stem cells from human embryos, nor for human totipotent cells because these have the potential to develop into a human being, patents will be granted for pluripotent stem cells, which develop from further division of totipotent cells and no longer have the ability to develop into a human being. Both the United Kingdom and Sweden have issued stem cell patents filed by the Wisconsin Alumni Research Foundation. The U.K. practice has not yet been challenged on moral grounds in the national courts.

Japanese patent laws are different still from those of either the United States, EPO, or individual European nations. Japan identifies a broad range of inventions as patentable. It does, however, impose two limits on patentability; those innovations that contravene public order or morality are deemed unpatentable [68].

The Canadian Patent Act identifies several innovations as nonpatentable subject matter, including games, medical treatments, scientific principles, and abstract theorems. As has been discussed earlier, officially higher life forms are not patentable in Canada [43], although recent case law suggests that the rights associated with patent protection are extended to genetically modified higher life forms, such as plants, that are made up of patented genes [44].

There is a lack of clarity on the patentability of biotechnological innovation, which can have a number of unwanted outcomes. The scope of the right over some biotechnology inventions, such as genetically modified plants, can be difficult to determine. Such uncertainty may lead to increased litigation for lucrative inventions. The ethical and moral boundaries of patent rights are unknown and are differentially applied, which can create problems for international technology transfer. The same invention can receive very different treatment as patentable subject matter depending on the region or nation where an application is filed.

Patent Criteria. Patent criteria are applied to determine whether a product or process described in a patent application is an invention. If the product or process fails to meet the criteria, the innovation is not patentable. The deliberation on whether an invention meets the patent criteria is distinct from whether the invention falls within the category of patentable subject matter. An invention may meet all the patent criteria, but its subject matter may fall into a category that *a priori* cannot receive patent protection.

Most countries have three criteria that must be met for patent protection to be granted. Among countries, the criteria are often similar in nature and intent, but not identical. These differences can have a significant effect for biotechnological innovations. In the United States and Canada, the criteria are utility, novelty, and non-obviousness. The EPC defines the criteria as industrial application, novelty, and inventive step. In Japan, the criteria are industrial application, novelty, and inventive step.

Utility or Industrial Application. Utility and industrial application are similar, but not identical, criteria. In the United States, the definition of utility is not mere functionality but, rather, whether the invention is useful for its described purposes. For inventions involving DNA sequences, utility is considered to be established if a substantial, credible, and specific use is demonstrated in the patent application. In Canada, utility is understood to mean whether the invention works [69]. It does not have to work well or for a prolonged period of time, nor does a model or prototype have to be presented. Simple evidence that the invention, as it is claimed in the patent application, will work is sufficient.

Industrial application, the standard applied by the EPO and several European countries, is a narrower criterion than utility because it requires an invention to be applicable for use in industry. However, the interpretation of what constitutes an industrial application varies among national patent laws. For example, in the United Kingdom, an invention must be capable of being made or used in some kind of industry, which is broadly defined as anything that is distinct from a purely intellectual or aesthetic activity. Industry does not mean commercial use *per se*, and it includes agriculture [70]. In France, industrial application is more narrowly defined than in the United Kingdom, as industry is interpreted to mean an endeavor that turns a profit. Sweden applies an interpretation similar to the United Kingdom: An invention having industrial application must be of a technical nature, have technical effect, and be reproducible [71]. In Germany, an invention has industrial application if it can be made or used in any kind of industry, including agriculture. The German definition is similarly worded to the provisions of the United Kingdom and France, but it is much more narrowly interpreted so that manufacture or use must occur in an industrial enterprise.

The European Biotechnology Directive, which must be adhered to by all EU member states and the EPO, sets out specific requirements for the industrial application of DNA patents. The directive states that an invention involving DNA must indicate the function of the DNA and "in order to comply with the industrial application criterion, it is necessary in cases where a sequence or partial sequence of a gene is used to produce a protein or part of a protein, to specify which protein or part of a protein is produced or what function it performs" [67].

The definition of industrial application included in the Japanese patent law is similar to that imposed by the EPO in the EPC [72]. However, the fact that Japanese patent applications are rarely refused for lack of industrial application suggests that Japan's application of this criterion may be less stringent than that of the EPO.

Novelty. The novelty criterion simply identifies whether the innovation is truly new, meaning it is not previously known or existing. To determine novelty, patent offices conduct a search for prior disclosure, including prior art materials that describe the invention and were available publicly before the date the patent application was filed. Patent offices in different countries conduct searches with varying degrees of diligence.

One factor that determines the breadth of the search undertaken is the definition of what constitutes disclosure as spelled out in national patent law. The United States considers use of the invention, printed publications, or patents from any country that pre-date the day when the applicant created the invention [73]. The EPO criterion includes use as well as both oral descriptions and written descriptions of an invention pre-dating the filing of the patent application [65]. Oral disclosure is a hot issue in the debate over the patentability of traditional knowledge, as traditional knowledge is often passed down orally rather than in written form, and, accordingly, patent laws that only recognize written disclosure do not consider oral information to represent a barrier to patentability. Japan encompasses public knowledge or public working of inventions and printed publications that precede the filing date of the patent application [68]. In Canada, written descriptions of the invention as well as public disclosure, which may include oral disclosure, pre-dating the filing of the patent application all constitute disclosure [74].

Some prior disclosure of an invention is permissible. In the United States and Canada, the inventor is granted a one-year grace period during which he can disclose the invention before filing a patent application without the disclosure barring the right to a patent [73, 74]. Without this grace period, as soon as an invention is disclosed, it can no longer be considered novel, and therefore is unpatentable. Japan and the EPO allow only a six-month grace period [65]. The conditions for the grace period under the EPC are very stringent; only disclosures because of inclusion in an officially recognized international exhibition or that is caused by an "evident abuse" (e.g., breach of a confidentiality agreement) are permissible. Some European national patent laws, such as those in force in the United Kingdom, do not permit a patent application to be filed after any public disclosure of an invention [75]. A prudent inventor will delay disclosure of an invention until he is certain of rules in the jurisdictions where patent rights will be sought or until after an application is filed.

Another factor affecting the scope of prior art searches is the amount of time that examiners have to devote to a search, which is directly related to the staffing levels and expertise of examiners working in national patent offices. For some examiners, searching time is short and prior art searches are consequently limited [76, 77], which has implications if a patent is later challenged on the ground of novelty. A search for prior art conducted by an examiner that is limited in scope because of time constraints may fail to discover prior art, which may later be discovered by persons wishing to oppose the patent. Any prior art can be used to challenge a patent and ultimately render the patent invalid for lack of novelty.

The novelty criterion further identifies a distinction that is particularly relevant to biotechnological innovation—the difference between an invention and a discovery. An invention is a product or process that did not exist previously, whereas a discovery refers to something that already exists and is a product of nature [78]. In practice, albeit somewhat counterintuitively, the discovery/invention distinction means that a gene sequence that exists *in situ* is not patentable, whereas an isolated gene sequence, with further restrictions in some countries, is patentable [79]. When a patent is sought for a gene, it is, in fact, sought for the complementary DNA (cDNA) sequence. That molecule is a manmade product derived from nature. Thus, although patentability of human body parts is controversial, isolated human genes, proteins, and cell lines are patentable.

Non-Obviousness or Inventive Step. Non-obviousness or inventive step is the third criterion applied to establish whether an innovation meets the definition of an invention set by patent legislation. Non-obviousness invokes the expertise of a person skilled in the art and asks whether such a person would have considered the invention to be obvious in light of the ordinary knowledge of the art [80]. The date of the relevant knowledge of the art differs between countries. The United States requires a consideration of knowledge existing when the invention was made, whereas Canada requires it as of the claim date of the patent application. Canadian jurisprudence distinguishes obviousness and novelty in the following manner: saying that an invention is obvious if a person skilled in the art sees it and says, "any fool could have done that," whereas novelty invokes a reaction of "your invention, though clever, was already known" [81].

Inventive step is a similar criterion to non-obviousness. The EPO actually defines inventive step as existing if an invention is non-obviousness in regard to the state of the art. Japan's test states that there is no inventive step if the invention could have been easily made before the filing date by a person skilled in the art on the basis of pre-existing inventions [68].

As is true of utility/industrial application and novelty as well, each of the provisions for non-obviousness and inventive step are similar. It is, therefore, the wording and interpretation of each nation's criterion that is critical because it ultimately determines if an invention is patentable within that jurisdiction. These variations in national patent criteria may cause an invention to be denied patent protection in one country while being granted such protection in another.

12.2.5.2 Procedural Patent Law

The Patent Application. A patent is a legal and technical document generally drafted by a patent agent. It at once discloses information about the invention and establishes the boundary around the rights given to the patent holder and is intended to give fair warning about infringing activities. To this end, the drafting of patents is a highly technical art, controlled by strict rules and procedural requirements [82]. For example, patents must be drafted clearly and completely so that the invention can be directly performed by a person skilled in the art. However, a patent rarely results in such clear and compre-hensible disclosure. Key details of the optimum manufacture or working of an invention are often withheld to preserve marketplace advantage because patent documents are available for public perusal.

Patents are generally made up of four parts: an abstract, a description of the invention, one or more claims, and any drawing or technical illustrations referred to in the description or claims. The exact nature of these components may vary from nation to nation. The abstract is a concise summary of the invention and generally identifies the field of art and the technical problem the invention addresses. In some jurisdictions, a word limit is imposed on the abstract [83]. The drawings provide a representation of embodiments of the invention. Generally, elements of the invention that are depicted are numbered for the ease of reference. The description, also known as the specification, provides background information about the field of art of the invention and prior art, addresses why the invention overcomes a technical problem not addressed by the prior art, and offers a complete description of the invention in its various embodiments. The description may reference the drawings for the purpose of better describing the invention. As has been discussed, if an invention involves biological material that cannot be adequately described, the applicant may deposit a sample of this biological material at a recognized institution or depository, depending on the rules of the jurisdiction where the patent is filed [21]. Special rules also exist in some jurisdictions for applications relating to nucleotide and amino acid sequences [65]. Above all else, the description is intended to support the claims. To fulfill this role, the description may be relied on to offer insight into the intention of the inventor for the purpose of any interpretation of the claims.

The claims are the most important element of a patent because only the elements of an invention that are claimed are granted patent protection. Claims are highly technical and challenging to draft because of the strict rules governing their formulation and organization.

Claims can describe various types of inventions or embodiments of the same invention. The most common types of claims are product claims and process claims. Product claims confer protection over any aspects of a product specified, regardless of how that product was derived. Process claims describe a particular means of making a product, and it is the described process that is granted patent protection. Also relevant for biotechnology patents are Markush claims, which are used where embodiments of the invention can include a class of compounds (e.g., chemicals or DNA sequences). By opting to use a Markush claim, the applicant need not spell out every possible combination individually, instead he can claim a range of products that may be used in an embodiment of the invention. This range is compiled on the basis of a limited number of representative samples.

Every patent application must include all of the sections prescribed by national patent law. A patent right may be refused on the basis of an application that is deficient. However, a well-drafted patent application can result in a grant of significant patent rights and great economic value for an owner.

Patent Filing

First-to-File versus First-to-Invent. The United States applies a "first-to-invent" patent system, which means that if two different patent applications describe the same invention, the patent right will be issued to the invention that can be proven to have been invented first [84]. By contrast, the patent law of Canada, the EPO, and Japan, impose a "first-to-file" system, meaning that if two patent applications for the same invention are filed, the patent right will be granted to the applicant who filed an application at the earliest date [85], which is true even if the two identical inventions were arrived at independently. For this reason, first-to-file systems are often referred to as creating a "high stakes race" [86].

When to File. The first step toward obtaining patent protection for an invention is to file a patent application. The date the patent application is filed is a key reference point for the patent examination and establishes the beginning of the patent term. In accordance with TRIPs, most countries now grant patent rights for 20 years beginning on the earliest filing date [49]. Considerations of the desired patent term and the possibility of being usurped by competitors should dictate when a patent application is filed. Patent term is important because an invention that is in early-stage development and will not reach the market for many years may benefit from a decision to delay filing. A later filing date will ensure that a greater period of the patent term will occur while the invention is marketable (e.g., if the invention takes 15 years post-filing to complete research and development, it will only be on the market during 5 years of the patent term, so waiting to file can extend the period of patent protection while the product is on the market). The position of competitors is important because a delay in filing a patent means that a competitor working on an identical or similar invention may file an application first, and thereby extinguish patent rights for others in the future.

The decision of when to file a patent application can be tricky for complex biotechnological inventions that involve a long research and development period. In the case of the NF-κB patent application, it took 16 years to complete the review at the USPTO before the patent was issued [87]. This significant review period was, at least in part, caused by the large number of claims that were required because of the complexity of the invention.

How to File. There are three possible ways to file a patent application, as a national patent, a regional patent, or a Patent Cooperation Treaty (PCT) filing. The decision of which type of patent filing to pursue will depend on commercial, strategic, and cost considerations. For example, the cost of filing a PCT patent is greater than filing in a national patent office. Strategic considerations may relate to the speed of the grant of a patent under each type of application, which is often related to the type of patent review employed. The PCT includes an extended review period of 30 months pre-national filing, the EPO requires a full examination of the patent, whereas some national patent offices require no examination (e.g., Belgium, the Netherlands, Switzerland, and Ireland).

National patent offices act within the boundaries of the country they exist within and apply that country's patent law. Patent applicants may file applications directly with national patent offices in each country where patent protection is required. The decision of where to file an application is generally based on the predicted potential markets for the invention. For example, a crop plant genetically modified to be resistant to frost may only be marketable in countries with colder climates and with markets open to genetically modified foods. Another consideration is the patentable subject matter allowed in each jurisdiction. For this product, an inventor can seek patent protection for a plant in the United States, but not Canada or Europe. The most cost-effective strategy for an invention that is only marketable in a small number of countries is to file patent applications directly with individual national patent offices.

A further consideration is the differences in disclosure rules set by different countries, as discussed above under "novelty." As has been discussed, if an invention is publicly disclosed before an application is filed, it may limit the countries in which patent protection may be sought. For example, once disclosure is made, an inventor may be barred from obtaining a patent in the United Kingdom. For this reason, disclosure should never occur before the decision about where patent rights will be sought, as patent applications themselves are a form of disclosure.

In some circumstances, patent filings can rely on each other for the purpose of establishing an early filing date. The Paris Convention allows for priority filings among signatory nations. A priority filing is one that relies on the filing date of a previously filed patent application. For example, if on April 19, 2006 an inventor files a patent application for protein A in the United States, a Paris Convention signatory, a second patent application for protein A can be filed in Canada, also a Paris Convention signatory, up to 12 months later. If the Canadian application is identified as a priority filing based on the U.S. application, then the priority date granted to the Canadian application will be the filing date of the U.S. filing—April 19, 2006—which takes off some of the pressure of deciding exactly what type of patent to file immediately while holding an early filing date for subsequent filings to rely on, which is crucial in a first-to-file patent system.

Inventors who wish to obtain patent rights in multiple countries represented by a region may file applications with a regional office, if such an office exists. The EPO is a regional patent office serving several European countries, and ARIPO accepts applications for multiple African nations. Regional patent offices allow for a filing process that reserves rights in all nations within the region simultaneously. For example, once an inventor files a patent application with the EPO, the right to national patent protection in each of the participating European nations is reserved, so that after a successful examination, the inventor is granted a patent right in each of the individual nations represented by the EPO.

Regional patent offices are distinct from national patent offices. National patent offices not only undertake the filing of patents, they also administer the national patent laws. Regional patent offices are limited to undertaking the grant of patents. Post-grant patent rights will be dealt with in accordance with the national patent laws of individual nations. For example, an infringement dispute erupting in France over a patent application that was filed with the EPO will be resolved in accordance with the national patent laws of France.

Regional offices, in developing nations in particular, play another key role besides granting patent applications. They allow for the integration of services and free flow of information. ARIPO was mainly established to pool the resources of its member countries in industrial property matters to avoid duplication of financial and human resources [88].

The PCT system facilitates the filing of a patent application on a even broader scale, globally [53]. Upon filing, a PCT patent application reserves for an inventor the right to seek patent protection in all countries that are signatories to the PCT at the time of filing. This option is generally chosen by an inventor for two reasons, first, because the inventor is unsure of the scope of the market for the invention, but believes it could be global, and second, because the inventor needs time to raise investor funds. The time advantage is achieved because the process of PCT examination of a patent application can add several months (at least 30 months) to the filing process, a significant period of time to be able to put off a final determination about where patent rights are required internationally for an invention, especially considering the pressure to file patent applications as soon as possible imposed by investors and competitors, and the expense that is incurred to seek patent protection in multiple countries. A PCT filing offers some breathing space to an inventor. Moreover, an additional time advantage can be achieved if a PCT filing is made as a priority filing based on a previous foreign filing. For example, a filing made in the United States on April 19, 2006 for a widget can be identified as the priority filing for a subsequent PCT application for the widget, so that the PCT filing would be granted the priority date of April 19, 2006. Through this process, the 30-month lapse during which an inventor can decide which countries to pursue patent rights in is increased to 42 months.

It is important to remember that the PCT is not a patent-granting system but only a patent-filing system [89]. Once a PCT patent application has been examined and deemed patentable, the inventor is then required to file national applications in each of the countries where patent rights are desired. The initial PCT filing merely reserves the right to file in each of these nations (so that applications will not be refused because they missed disclosure deadlines, etc.), it does not take the place of national filings—the PCT is a preliminary step to national filings. However, as the patent application has already been examined by PCT examiners, the expectation is that the national patent filing will be a simple and straightforward process, devoid of extensive discussion between the inventor and the examiner. The drawback of a PCT filing is that it is slightly more expensive than normal national filings; a fee is charged for the PCT filing in addition to the filing fees that must be paid in each country where patent applications are subsequently sought. However, for inventions with a truly global appeal, such as with some types of biotechnology, PCT filing may be worth the added cost.

The determination of how a patent application will be filed—nationally, regionally, or PCT—can greatly affect the ultimate scope of the freedom to operate (FTO) achieved for a commercial patent owner. Evidence exists that for biotech start-up companies, FTO considerations are surpassing the previous focus on exclusion of competition [90]. Issues such as freedom from third-party contracts and properly timed disclosure can also affect FTO; but, ultimately, securing the right to seek patent rights in key marketplaces will ensure significant FTO success.

12.2.5.3   Legal Challenges: Infringement, Exemptions to Infringement, and Validity

The Scope of Patent Protection. The scope of patent protection granted to an invention defines the breadth of the patent right, sometimes referred to as the fence around the invention [80, 91]. Only when armed with an understanding of patent scope can a party know which activities will constitute infringement of the patent. The interpretation of patent protection differs by jurisdiction. In some countries, patent protection is limited to commercial activities involving the invention whereas in other jurisdictions the range of activity is broader. For example, in the United States and Canada, patent legislation does not limit patent protection to the commercial working of the invention. In contrast, in Japan, the patent holder's right is limited to the commercial application of the patented invention. For patents granted by the EPO, the applicable patent protection applied in an infringement proceeding is prescribed by the national patent law in the jurisdiction in which the alleged infringement occurred; but, generally, most European states impose a level of industrial application as a boundary for patent protection. The infringement test and specific details of infringement may also differ from one jurisdiction to the next.

Infringement. The test for infringement varies by jurisdiction. A U.S. patent is infringed when an unauthorized person makes, uses, offers to sell, or sells any patented invention within the United States or imports a patented invention into the United States [73]. Those who actively induce infringement or who sell or import critical components of inventions, knowing the sale will aid others to infringe the patent, may also be found liable for contributory infringement.

Infringement proceedings occur before a national court in the United States. The test for patent infringement asks whether the patent has been either literally or substantively infringed. The test determines the scope of the invention claimed in the patent and decides whether the patent has been infringed as a related consideration. Literal infringement means that the product or process of the invention has been duplicated as it was claimed. Substantive infringement, on the other hand, also known as the doctrine of equivalents, considers the "spirit of the invention" [82]. The question here is whether the alleged infringing product or process copies aspects of the invention that, although not being expressly included in the text of the patent, should nevertheless be considered part of the invention as it was conceived of or intended by the inventor. Thus, the substantive test is not limited to the text of the patent, but can be read into the document.

A further aspect of the U.S. infringement test that is unique is the doctrine of prosecution history estoppel. The doctrine prevents a patent holder from alleging substantive infringement based on a claim, or portion of a claim, that formed part of the original patent application but was removed during the office action stage [92]. The office action stage is the examination process when the patent applicant and patent examiner enter into discussion about a filed patent application. Generally, during the office action stage, the patent examiner will refuse claims included in the patent application on the basis of prior art; the patent applicant will either agree or disagree with the challenge. As a result of this process, the claims of the patent application will be modified at the behest of the patent examiner. The result is that the claims included in the filed patent application are often entirely different from those included in the granted patent, as claims are often reduced in scope or removed altogether during this process. Prosecution history estoppel holds that any element of an invention removed during the office action stage cannot later be considered protected by the patent as a result of any interpretation of the claims.

The remedies for infringement available in the United States include an injunction (court order to stop the infringing activity or to destroy the infringing product) or damages (monetary or other compensation calculated to reimburse the patent holder for the losses suffered) [73]. Attorney fees may also be claimed.

As has been stated, infringement of patents granted by the EPO will be evaluated in accordance with the national patent law of the jurisdiction where infringement is alleged to occur. As an example, infringement occurs in the United Kingdom if there is disposal of, offer to dispose of, use, or importation of a patented product or a product created using a patented process or unauthorized use of a patented process in the United Kingdom [93]. Both substantive and contributory infringement are prohibited. Proceedings may occur before a court or comptroller, and possible infringement claims include an injunction, an order to deliver or destroy the patented product, damages, an accounting of profits (a calculation of the profits of the infringer derived from the infringement), and a declaration that a patent is valid and has been infringed. The test applied to infringement is the Catnic test [94], which first establishes patent scope by reducing the invention into essential and nonessential elements. Only the essential elements are then considered in the subsequent evaluation as to whether infringement occurred. The considerations of patent scope and infringement are distinct steps in the test, and scope is determined in a manner that takes no account of the infringement allegation.

Japan applies a literal/substantive test, similar to that used in the United States and has adopted the doctrine of equivalents [95]. The Japanese version of the test, however, also takes into account the weight or importance of the claimed elements of the invention in a manner similar to the United Kingdom [96, 97]. Infringement proceedings occur within the courts but rarely rely on oral arguments as the majority of the proceedings are presented in the form of written briefs. Damages may be claimed as a remedy for infringement, as can lost profits. Infringement awards were low for a long period in Japan, but evidence exists that these have been rising steadily since 2002 [98]. The burden of proof is considered to be higher in Japanese infringement proceedings than in U.S. cases (i.e., it is more difficult to prove infringement in Japan than in the United States).

In Canada, anyone who makes, constructs, uses, or sells a patented invention without authorization to do so can be found guilty of infringement. An infringement action may proceed in the higher courts in each province or in the more specialized federal court [74]. Remedies include damages, injunction, and accounting of profits. In Canada, as is true in most jurisdictions, an infringement allegation can only be raised after a patent is granted; however, Canada is unique in that the damages claimed can include a period of time between the date when the patent application was disclosed to the public and the patent was granted, in addition to the post-grant period [74].

The test applied to determine infringement in Canada is the purposive construction test [99, 100]. This test is based on the U.K. Catnic test, although the tests are not absolutely identical. The first step of the test is a determination of the scope of a patent, which involves the break down of the claims into essential and nonessential elements. As is true of the U.K. test, this step is undertaken without any consideration of the infringement allegation. An important distinction between the Canadian test and that of the United States is that it is limited to the "four corners" of the patent document, whereas the U.S. test allows for consideration of extrinsic documents, such as the prosecution history, otherwise known as the file wrapper information. Once patent scope is determined, the Canadian test compares the alleged infringing innovation to the essential elements of the patented invention. No infringement occurred if an essential element is different or omitted in the alleged infringing device; but infringement may have occurred if nonessential elements are substituted or omitted.

Infringement is a common challenge raised in the biotechnology patent world against competitors. Myriad Genetics, as described above, applied the threat of an infringement suit to great advantage. Ariad Pharmaceuticals, owner of the NF-κB patent, likewise has "gone to extraordinary lengths to ensure that anything that comes within so much as a whiff of NF-κB will be drawn into the '516' patent's black hole" [87].

Exemptions to Infringement: Exemptions for Research and Generic Drug Manufacturing. Certain activities are exempt from patent infringement challenges, and, again, these vary according to national patent laws. Exemptions may be raised to defend against allegations of infringement. The most pertinent exemption for biopharmaceuticals and biotechnology is the research exemption. The research or experimental use exemption curtails a patent holder's rights by permitting researchers and research institutions to make certain uses of a patented invention without compensating the patent holder. Many policy documents argue that the research exemption for free access to basic research materials should be strengthened (see discussion below) [101]. However, it is becoming increasingly difficult to justify a research exemption as the lines between private-sector and public-sector research blur and the commercialization activities of public-sector research institutions increase.

In the United States, no statutory research exemption exists; instead, a very narrow experimental use exemption has been carved out by judges and exists in the common law [102, 103]. Previously, the exemption did not apply if the research has "the slightest commercial implication" [104]. However, the decision of the United States Court of Appeals for the Federal Circuit in *Madey v. Duke University* [105] has further narrowed the exemption. Now, any conduct by a research institution that is in keeping with its legitimate business, regardless of the commercial implications of the research, cannot rely on the research exemption. In the case of *Madey*, Duke University was using patented equipment for teaching and research. The court characterized both of these activities as part of the legitimate business activities of the university and therefore decided that Duke University could not be exempted from patent infringement. The result is that the research exemption has been restricted only to the most exceptional circumstances [102].

Most importantly, for biopharmaceuticals, however, and the eventual entry of generic products of biotechnological research into the marketplace (see discussion below) is the Restoration Act or Hatch–Waxman Amendments to the Drug Price Competition and Patent Term Restoration Act of 1984 [106]. This act creates an exemption to the normal rules governing infringement for parties seeking Food and Drug Administration (FDA) approval for generic drugs. Section 271(e) of the U.S. Patent Act now states that "use of a patented invention solely for purposes reasonably related to gathering data to support an FDA application for generic versions of [patented] drugs" does not constitute infringement. However, the generic drug manufacturer, although having the ability to do research to show that their drug is equivalent to the patented original, must still wait until the patent expires before it can seek FDA approval.

In Europe, the issue of infringement is still largely determined by national patent law. Article 31 of the Community Patent Convention creates an exemption for acts "done for experimental purposes relating to the subject matter of the patented invention" [107]. All members of the EU except Austria have introduced a similar provision into their national patent laws [102]. However, the interpretation of the scope of the exemption still varies between countries. For example, in Germany, the scope of the experimental use exemption is broad and applies to activities that test the viability and potential for development of patented inventions. In the United Kingdom, however, the exemption is narrower. It does not apply for experimentation in the context of regulatory approval in contrast to Canada, the United States and other countries of the EU [102].

Infringement exceptions in Japan's patent law include some research uses [68]. Further exemptions include, in the United States patented inventions in use upon vessels, aircraft, or vehicles that are in the Unites States on a temporary basis (and will be used solely aboard the vessel, aircraft, or vehicle); and if a person other than the patent holder reduced the invention to practice at least one year before the filing date of the patent, that person can continue to use the invention [73].

The United Kingdom also exempts patented inventions used on an aircraft that is temporarily landing in or crossing over the United Kingdom [93]. Canada offers a similar exemption for ships, vessels, aircraft, or land vehicles within Canada, allowing temporary use of patented inventions solely for their needs and upon the craft; and they also acknowledge that any prior holder (meaning a person who had the patented product in their possession before the claim date of the patent) has a right to continue to use the product [74].

Challenging the Validity of a Patent. The validity of the claims made in a patent is likely to be challenged in one of two situations: directly in a post-grant opposition or validity challenge proceeding, depending on the jurisdiction; or alternatively, as a defense to an action for infringement. The latter use of a validity challenge may be formulated in a variety of ways, but will most likely argue that the invention is not patentable subject matter, that the patent criteria is not met (e.g., prior art exists that establishes that the invention is not new), or that claims are invalid (e.g., claims are not supported by the description or are affected by a technical deficiency). In some countries, pre-grant validity challenges may be raised as well; for example, in Canada, a prior art filing process can be invoked [74]. This challenge involves the delivery of prior art that proves that the invention was not novel at the date of filing to the patent office by a person wishing to thwart the grant of a patent. In Canada, such challenges are unpopular because the person filing the prior art has no right to involvement in the proceedings after delivery to the patent office and is bound by the decision. Canadian validity challenges are normally only raised after a patent has been granted, at which point the challenger can have full involvement in the proceedings and a right of appeal.

An opposition proceeding is a form of post-grant validity challenge available for EPO patents [65]. As the EPO is solely a patent-granting organization, the effect of the issuance of an EPO patent is effectively to create multiple patents from a single application, one patent in each of the European countries designated by the applicant during the patent prosecution stage. However, if an EPO opposition proceeding is launched within nine months of the granting of the EPO patent, it has the effect of pulling back each of the national patents, so that a single EPO patent exists [108]. It is the validity of this single EPO patent that is challenged. The benefit of the opposition proceeding is that only one challenge needs to be made; this option is more cost-effective than pursuing multiple proceedings in each individual European nation where the patent exists post-grant. An opposition proceeding may be brought by any person and is based on three grounds.

(1) that the subject matter is not patentable under European law;
(2) that the European patent does not disclose the invention so that it could be carried out by a person skilled in the art; and
(3) that the subject matter of the European patent extends beyond the content.

In the United States, post-grant oppositions are under discussion as a major feature of proposed patent reform [109–111]. Advocates of opposition proceedings hope that allowing an opposition to be filed in the USPTO within a limited time window, immediately after the patent issues, would help weed out invalid patents in a procedurally simple, relatively inexpensive forum. Presently, a re-examination process is set out in the U.S. patent law, which is a challenge to the validity of a patent based on the presentation of prior art [73]. However, the outcome of the re-examination is binding and no right to appeal is granted, and, like the Canadian filing of prior art process, the re-examination process is rarely used [108].

In Japan, the opposition procedure after an examiner's decision to grant a patent was abandoned in 2005. The alternative is a trial for the invalidation of the patent [112].

Canada does not offer an opposition procedure, but includes a reexamination procedure in its patent law. Re-examination may be requested for any claim or claims at any time during the term of the patent, but the evaluation is undertaken solely on the basis of prior art [74]. The Commissioner of Patents appoints a three-person re-examination board to undertake the review. Unlike the re-examination process in the United States, the board's decision is not the end, and an appeal can be launched within three months.

12.2.6   THE EFFECT OF THE REGULATORY ENVIRONMENT

Biotechnological innovations that are eligible for patent or other IPR protection can simultaneously be regulated by relevant IPR laws and other legislation or regulation. For example, an overlap between patent law and legislation directed specifically at biologic products may affect biopharmaceuticals. Biologics, as defined by the FDA, "in contrast to drugs that are chemically synthesized, are derived from living sources (such as humans, animals, and microorganisms). Most biologics are complex mixtures that are not easily identified or characterized, and many biologics are manufactured using biotechnology. Biological products often represent the cutting-edge of biomedical research and, in time, may offer the most effective means to treat a variety of medical illnesses and conditions that presently have no other treatments available" [113]. A biotechnological innovation that meets the definition of biologic and is granted patent rights will need to comply with food and drug legislation as well as patent law.

Simultaneous compliance with two sources of legislation has the potential to affect the scope of patent rights. For example, the parameters for the introduction of generic versions of the patented drug into the marketplace are set not only by patent law but also by food and drug legislation. In Canada, the patent regime has been tied to the food and drug legislation to address concerns that patent rights over drugs may negatively affect public health by limiting access to drugs. Cooperation between the food and drug regime and the patent regime also ensures that the term of a patented drug is respected. Specifically, legislation prohibits generic drug manufacturers from introducing an equivalent drug into the marketplace during the patent term, thereby introducing patent considerations into the regulatory approval system.

12.2.6.1 Biosimilars

An emerging issue in the field of biopharmaceuticals is the issue of biogenerics and how these may be introduced into the marketplace. Biogenerics are the generic version of the original, usually patented, biologic product; the same relationship is found between the original pharmaceutical and its generic counterpart. The primary distinguishing feature between biologics and traditional pharmaceuticals is the former uses recombinant DNA technology to produce macromolecules whereas the latter conforms to a "chemical synthesis/small molecule paradigm" [114]. Diverse biotechnological products include hormones, cellular growth factors, enzymes, clotting and anticlotting factors, monoclonal antibodies, and fusion proteins. The very diversity and complexity of biologics leads to problems in setting the regulatory environment.

To date, no specific regulations or universally applicable guidelines exist governing biogenerics. Each product is reviewed on a case-by-case basis according to existing regulations governing generic pharmaceuticals. Such regulations exist in most jurisdictions and provide a simplified approval process based on the proof of "essential similarity" or product equivalence of the generic pharmaceutical compared with the patented pharmaceutical already available in the marketplace. Equivalency may be in terms of chemical composition, pharmaceutical equivalence or dosage forms, and thera-peutic equivalence. Generic pharmaceuticals are almost identical to the original product or even contain an active ingredient for which the patent has expired.

The problem with biopharmaceuticals is that these are generally characterized by properties that may complicate the demonstration of equivalence because of the difficulty of characterizing the physico-chemical makeup [115]. The product properties, such as the impurity profile, glycosylation, and protein folding, may be highly dependent on the process by which they are manufactured, adding to the complexity of a determination of equivalency [115].

A further problem is the issue of the patent thicket that surrounds much biotechnological innovation. There may be multiple patents held by multiple parties for constituent ingredients, processes required for the development of biopharmaceuticals, and the products involved in those processes (e.g., gene and protein patents). Navigating the thicket to ensure that a biogeneric violates no extant patents may be quite challenging. Thickets greatly impede FTO.

These problems will become increasingly significant as the patents for top-selling biopharmaceuticals expire. Several of the top-15 selling biopharmaceuticals come off patent in the next five years and a number of large generic manufacturers have begun development of biogenerics [115]. Indeed, if a similar market penetration is assumed for biogenerics as for generic pharmaceuticals (10–15%), the world-wide market for biogenerics will be $2–3B U.S. [114, 115].

Some hope exists that the regulatory framework will become responsive to the emergence of biogenerics. The shift will require a move from essential similarity (or demonstration that the generic is almost identical to the origi-nal pharmaceutical) to product comparability [115] and clearer working definitions for different classes of biologics [114]. These issues are under consideration by the FDA, and the European Commission has passed legislation for the approval of "similar biological medicinal products" [116].

12.2.7 REFORM PROPOSALS FOR INNOVATION GOVERNANCE IN BIOTECHNOLOGY

The rapidly changing area of biotechnology innovation does not necessarily fit comfortably within the confines of traditional intellectual property laws. Some reforms have been made at national and international levels to adapt intellectual property law, particularly patent laws, to biotechnology. These reforms generally take the form of specific provisions and adjunct statutes, regulations, and practice guidelines within patent offices designed to address patent rights in biotechnology. Many of the reforms are driven by moral and ethical concerns that develop from the research *per se* and discomfort with the commodification of products and processes that are perceived to belong to nature or life. However, some more pragmatic concerns exist, developing from considerations of the appropriate balance to be struck between the rights of innovators in biotechnology and access to the innovation by the public, by government, and by researchers, among others. In this final section, we explore some of the reforms that have been proposed as solutions or compromises to moral and ethical dilemmas and to the issue of striking the right balance for patent protection of biotechnology innovation.

12.2.7.1 Strengthening Research Exemptions

Many jurisdictions are currently considering reform proposals for a broad research exemption from patent infringement [117, 118]. The concerns prompting the call for policy reform are set against a backdrop of rapid growth of DNA-related patents, particularly in the United States with growth of about 5% per annum [119]. It is estimated that over 3000 new DNA-related patents per year have been issued in the United States since 1998 and more than 40,000 such patents have been granted [120]. On the human front, at least 20% of the human genes are associated with at least one U.S. patent, and many have multiple patents [121].

The first concern, raised by Heller and Eisenberg, is the plethora of overlapping patents that may result, in the research context, in a "tragedy of the anti-commons" [122]. That is, the large number and diversity of ownership of gene patents may make it difficult to acquire all of the rights necessary for products and processes required for research (e.g., a patent thicket), resulting in the underuse of valuable technologies and the abandonment of promising avenues of research. Second, concern exists that patent holders will exercise their right of exclusivity in an overly restrictive manner that prevents others from developing or accessing a particular technology, for example, by refusing to license the technology for use by others [123]. The case of Myriad Genetics' zealous enforcement of its BRCA1/2 gene patents to prevent government-run diagnostic laboratories in Ontario from testing for genetic predisposition to breast cancer was such an example. Denial of access to patented genes and other biological products raises particular concerns because of the likely inability to invent around product patents because of the unique role these molecules play in biological processes. Third, the increasing focus on commercialization of public-sector research institutions will undermine norms of open scientific collaboration, especially the sharing of research results and materials [124, 125]. Increasing levels of industry funding will augment this effect and contribute to delays in the publication and dissemination of scientific research.

To date, the empirical evidence indicates that most of these concerns have not materialized [126]. However, real worries exist about the sharing of research materials, resulting in an increasingly competitive environment, which is likely tied to the commercialization ethos, industry funding, and the practical difficulties, such as time limitations, of transferring research materials [126].

12.2.7.2 Patent Pooling

A combination of the controversial nature of some areas of biotechnological innovation and a general distrust of commercially driven research are two factors that will influence public support and policy setting for both the private research sector and the public research sector. Patent pools may be a potential solution for maintaining public confidence in the research enterprise [42].

A patent pool is an arrangement in which "two or more patent owners agree to license certain of their patents to one another and/or third parties" [127, 128]. Patent pools bring together patent holders in a specific area of innovation, such as a viral genome, to facilitate the efficient use and development of a technology. The patents are "pooled" in the sense that the arrangement allows inventors in the pool to use all patented invention under favorable licensing terms. Any benefits that may materialize are then shared among the group. Patent pools have been used successfully in the motion picture industry, in aeronautics, and in the development of video and DVD technology [129].

As commercial interests are part of the problem, patent pools could be set up as independent, nonprofit corporations with an independent governance structure in which researchers are removed from direct involvement with industry and commercial forces [42, 130]. The governing body of the patent pool would make decisions about commercialization, access, and licensing. These decisions would be open to public scrutiny to ensure that research is in the public interest and not solely concerned with the profit motive. Such a structure is being considered for the Canadian Stem Cell Network and research related to the SARS genome, for example.

Of course, many challenges face the setting up of a patent pool. Patent pools may trigger anticompetition laws [127]. Researchers may not join in the patent pool because, although patent pools lower research transaction costs and spread risk, they also decrease the potential for the attainment of large profits from an invention. In the end, "the major incentive for all parties is economic benefit. In order for a patent pool to be an effective solution, the right balance has to be achieved between the cost of creating a pool and the prospect of adequate revenue generated by royalties on the endproduct" [131].

12.2.7.3 Open Source Patents

Another governance model designed to improve access to information and biotechnological research is open source patents. Open source patent systems share the goal of promoting free dissemination of biotechnological research. The aim is to foster an environment of sharing between inventors and the public rather than marketplace monopolies. Open source systems can be directed at endproducts or research tools used to develop products. Some consider open source to represent a grassroots movement to return to precommercialization sensibilities about scientific research and development, where there was a greater ethos of freely sharing scientific information among members of the scientific community [132].

The concept of open source rights as an alternative to IPRs originated with copyright. The rights granted by copyright arise automatically with the creation of a work. Copyright holders are then free to decide who may use the work. Open source systems, often found to protect software available for download on the World Wide Web, are created when copyright holders agree to license their works in a nonexclusive manner to anyone who agrees to abide by certain terms. The terms usually ensure that the copyrighted works themselves and any derivative works will be made available to future users in the same nonexclusive manner. As patent rights do not arise automatically, but must be actively applied for, open source patent systems have more options available for their formulation. Open source patent systems operate in a diverse manner, and some include inventions that are protected by patent rights whereas others apply to innovations that are not patent-protected. The similarity with all open source copyright is that open source patent initiatives build a free flow of information upon the foundation of innovation [133].

Several functioning examples of open source patent systems exist. The Public Patent Foundation (PPF) is one such initiative that facilitates the creation of a commons wherein patents may be pooled and made freely available to other participants [134]. The specific method applied by the PPF is to grant a nonexclusive and payment-free license to participants. The Biological Innovation of Open Society (BIOS) also involves technologies for which patent rights have been granted.

Its operation is focused on research tools rather than final products, and its goal is to "assemble groups of enabling technologies that together provide the pieces necessary for a particular form of research investigation" [134]. Both PPF and BIOS set licensing terms that achieve their specific goals [134].

The Tropical Disease Initiative (TDI) is an example of an open source patent system in which inventions need not be protected by patent rights [135]. Its aim is to create an accessible Web database to facilitate "searching for new targets, finding chemicals to attack known targets, and posting data from related chemistry and biological experiments" [135], which makes information readily accessible by researchers in developing and developed nations. Eventually, TDI hopes its open source system will aid virtual pharmas—organizations that are not engaged in in-house development, but that use a grouping of commercial and academic partnerships to create a portfolio of promising drug innovations [135].

12.2.7.4 Compulsory Licensing

The debate about compulsory licenses for patented products occurred in the context of access to HIV/AIDs drugs in Africa. At issue was the effect of patent rights on the accessibility and affordability of pharmaceutical products and vaccines in developing nations. Similar concerns around access and cost are likely to develop with biopharmaceuticals and other innovations resulting from research in health biotechnology.

One policy option for ensuring the accessibility and affordability of biotechnological innovation is for the state to implement compulsory licenses. Compulsory licenses override the exclusive patent right granted to a patent holder. A number of reasons why a state may take such a step exist, including public emergency and public health. Many national patent laws include provisions that permit the application of compulsory licenses. TRIPs also permits WTO members to include compulsory licensing within national patent legislation, in the instances of either "national emergency or extreme urgency," and sets out that the invention is to be applied to a "public noncommercial use" [49].

At its Cancun Ministerial Meeting in 2003, the WTO decided to implement changes to enable the least-developed countries to import cheaper generic drugs manufactured under compulsory licensing in other countries in the event that they were themselves unable to manufacture the pharmaceuticals. Calling it an "historic agreement for the WTO," Director General Supachai Panitchpakdi recognized that it was the first time that the WTO both validated and facilitated the full use of TRIPs' flexibility by developing countries [136].

Canada was the first country to introduce legislation to provide access to affordable medicines under the Cancun agreement. Canada's Patent Act has been amended to permit the award of compulsory licenses to export medications to developing countries in need. The amendments stipulate the conditions under which the government can issue a compulsory licence to a pharmaceutical manufacturer for the manufacture and export of a patented drug to the world's developing and least-developed countries. However, it is too early to tell whether the compulsory licensing regime will actually result in increased access to medicines [137, 138].

REFERENCES

1. World Intellectual Property Office (WIPO), World Intellectual Property Organization Convention (1967). Available: http://www.wipo.int/treaties/en/convention/trtdocs_wo029.html.

2. Bently L, Sherman B (2004). *Intellectual Property 2d*, Oxford University Press, New York. 1131p.

3. Gold R, et al. (2004). The unexamined assumptions of intellectual property: Adopting an evaluative approach to patenting biotechnological innovation. *Public Aff. Quart.* 18:299–344.

4. Centre for Intellectual Property Policy (CIPP). Available: http://www.cipp.mcgill.ca/en. Accessed 21 July 2006.

5. Westerlund L (2002). *Biotech Patents: Equivalency and Exclusions under European and US Patent Law*, Kluwer Law International, New York, pp. 351.

6. Carrier M (2004). Cabining intellectual property through a property paradigm. *Duke L.* 54:1–124.

7. Seymore S (2006). My patent, your patent, or our patent? Inventorship disputes within academicresearch groups. *Alb. L.J. Sci. & Tech.* 16:125–167.

8. Tomlinson N (2006). Tax abuse halting progress? An inside look at patent donations and their tax deductibility. *SW. U. L. Rev.* 35:183–206.

9. *Whirlpool Corp. v. Camco Inc.* (2000). 2. S.C.R. 1067.

10. *Mallinckrodt, Inc. v. Medipart, Inc.* (1992). 976 Fed. Cir. 700.

11. Walterscheid W (1994). The early evolution of the United States patent law: Antecedents. *J. Pat. & Trademark Off. Soc.* 76:697–715.

12. *Lac Minerals Ltd. v. International Corona Resources Ltd* (Canada). (1989). 2 S.C.R. 74.

13. *Coco v. Clark* (UK). (1969). *RPC* 41.

14. Isaacs R (2004). How not to tell all: Find out how to preserve the company's competitive edge by preventing proprietary information from being given away. *Security Mgmt.* 48(5):101–106.

15. The Week's Business: Coca Cola Caper, *AFX International Focus*. 7 July 2006.

16. Shand H (1999). Legal and technological measures to prevent farmers from saving seed and breeding their own plant varieties. In J Alexandria (ed.), *Perspectives on New Crops and New Uses*: ASHS Press, Alexandria, VA, pp. 124–126.

17. National Farmers Union (2004). Undermining farmers' rights to their seed. Available: http://nfu.ca/seedsavercampaign/NFU_Seeds_Fact_Sheet_2.pdf. Accessed 20 July 2006.

18. Vaver D (1997). *Intellectual Property Law*. Irwin Law, Concord, Ontario, p. 22.

19. *Berne Convention for the Protection of Literary and Artistic Works* (1971). Paris, France.

20. Bovenberg J A (2004). Inalienably yours? The new case for an inalienable property right in human biological material: Empowerment of sample donors or a recipe for a tragic Anti-Commons? *J. Law Technol.* 1(4):591–616.

21. World Intellectual Property Organization (1997). *Budapest Treaty on the International Recognition of the Deposit of Microorganisms for the Purpose of Patent Procedure.* Geneva, Switzerland, 1989.

22. Arai H (1999). *Intellectual Property Policies for the Twenty-First Century: The Japanese Experience in Wealth Creation. WIPO*, pp. 1–107.

23. Mills O (2005). *Biotechnology Inventions: Moral Restraints and Patent Law,* Ashgate Publishing, Ltd. Aldershot, UK, p. 195.

24. Einsiedel E F (2001). *Biotechnology and the Canadian Public: 1997 and 2000.* University of Calgary, Alberta, Canada.

25. Einsiedel E F (2000). Cloning and its discontents- a Canadian perspective. *Nature Biotech.* 18:943–944.

26. Priest S, Gillespie A (2000). Seeds of discontent: Expert opinion, mass media messages, and the public image of agricultural biotechnology. *Sci. Engin. Ethics.* 6:529–539.

27. Condit C (2001). What is "public opinion" about genetics? *Nature Rev. Genet.* 2:811–815.

28. Gaskell G, Bauer M, Durant J, et al. (1999). Worlds apart? The reception of genetically modified foods in europe and the U.S. *Science.* 285:384–387.

29. Bubela T, Caulfield T (2005). Media Representations of Genetic Research. In E F Einsiedel, F Timmermans (eds.), *Crossing Over. Genomics in the Public Arena,* University of Calgary Press, Alberta, Canada.

30. Holtzman N, Bernhardt B, Mountcastle-Shah E, et al. (2005). The quality of media reports on discoveries related to human genetic diseases. *Community Genet.* 8:133–144.

31. Page B, Shapiro R (1992). *The Rational Public: Fifty Years of Trends in Americans' Policy Preferences,* University of Chicago Press, Chicago, IL, pp. 67–116.

32. Parliament of Canada (2003). *Assisted Human Reproduction Act.* Bill C-13: LS-434E.

33. *Diamond, Commissioner of Patents & Trademarks v. Chakrabarty.* 1980. 447 U.S. 303.

34. Association of University Technology Managers (AUTM). Available: http://www.autm.net/index.cfm.

35. Caulfield T (2005). ASHG: Human stem cell biology: Principles, applications and ethical considerations—patent law, *American Society of Human Genetics, 55th Annual Meeting,* Salt Lake City, Utah.

36. Caulfield T (2006). Stem cell patents and social controversy: A speculative view from Canada. *Med. Law Internat.* 7:219–232.

37. Morin K, et al. (2002). Managing conflicts of interest in the conduct of clinical trials. *JAMA.* 287:78–84.

38. Lemmens T (2004). Leopards in the temple: Restoring scientific integrity to the commercialized research scene. *J. Law, Med. Ethics.* 32:641–657.

39. Canadian Biotechnology Secretariat Industry Canada (2005). International public opinion research on emerging technologies: Canada-U.S. survey results. Available: http://www.biostrategy.gc.ca/CMFiles/E-POR-ET_200549QZS-5202005-3081.pdf. Accessed 22 July 2006.

40. Pollara and Earnscliffe Research and Communications (2000). Third wave: Executive summary (Prepared for and presented to the Biotechnology Assistant Deputy Minister Coordinating Committee (BACC), Government of Canada). Available: http://www.biostrategy.gc.ca/english/view.asp?x=551&all=true. Accessed 22 July 2006.

41. Nelson D, Weiss R (2000). Penn Researchers Sued in Gene Therapy Death. *The Washington Post.* September 19, A:03.

42. Caulfield T (2004). Biotechnology Patents, Public Trust and Patent Pools: The Need for Governance? Health Biotechnology and Intellectual Property: A New Framework. Hosted at the European University Institute, Florence, Italy.

43. *Harvard College v. Canada (Commissioner of Patents).* (2002). 4 S.C.R. 45.

44. *Monsanto Canada Inc. v. Schmeiser.* (2004). 1 S.C.R. 902.

45. Morrow D, Ingram C (2005). Of transgenic mice and round-up ready canola: The decisions of the supreme court of canada in Harvard College v. Canada and Monsanto v. Schmeiser. *U.B.C. Law Rev.* 38:189–222.

46. McCalman P (1999). Reaping what You Sow: An Empirical Analysis of International Patent Harmonization. Available: http://www2.cid.harvard.edu/cidtrade/Issues/mccalman.pdf. Accessed 20 July 2006.

47. Hamilton C, Rochwerger P (2005). Trade and Investment: Foreign Direct Investment Through Bilateral and Multilateral Treaties. *N.Y. Internat. Law Rev.* 18:1–59.

48. Wolfe R (2005). Decision-making and Transparency in the 'Medieval' WTO: Does the Sutherland report have the right prescription? *J. Int. Econ. L.* 8:632–633.

49. World Trade Organization (1995). Trade-Related Aspects of Intellectual Property Rights (TRIPs).

50. Blakeney M (1996). *Trade Related Aspects of Intellectual Property Rights: A Concise Guide to the TRIPS Agreement*, Sweet & Maxwell, London, pp. 1–158.

51. World Trade Organization (1998). Agreement on the Application of Sanitary and Phytosanitary Measures. Available: http://www.wto.org/English/tratop_e/ sps_e/sps_e.htm.

52. World Intellectual Property Organization (WIPO). Available: http://www.wipo.int/portal/index.html.en.

53. World Intellectual Property Organization (WIPO) (1970). Patent Cooperation Treaty (PCT). Available: http://www.wipo.int/pct/en/texts/articles/atoc.htm.

54. World Intellectual Property Organization (WIPO) (1970). Summary of the Patent Cooperation Treaty (PCT). Available: http://www.wipo.int/treaties/en/registration/pct/summary_pct.html. Accessed 20 July 2006.

55. Commission on Intellectual Property Rights, Innovation and Public Health. April 2006. Public health, Innovation and Intellectual Property Rights.

56. World Health Organization (2003). *Public Health, Innovation and Intellectual Property Rights.* Available: http://www.who.int/intellectualproperty/documents/thereport/en/index.html. Accessed 20 July 2006.

57. Union for the Protection of New Varieties of Plants (UPOV). Available: http://www.upov.int.

58. World Intellectual Property Organization (1961). Contracting parties and signatories to the International Convention for the Protection of New Varieties of Plants (UPOV). Available: http://www.upov.int/en/publications/conventions/1991/pdf/act1991.pdf.

59. Halewood M (1999). Indigenous and local knowledge in international law: A preface to Sui Generis intellectual property protection. *McGill Law J.* 44:953–996.

60. Shiva V (2005). The Indian seed act and patent act: sowing the seeds of dictatorship. *ZNet,* Available: http://www.zmag.org/content/print_article.cfm. Accessed 20 July 2006.

61. Wendt J, Izquierdo J (2001). Biotechnology and development: A balance between IPR and protection and benefit-sharing. *Electr. J. Biotechnol.* 3, Available: http://www.agbios.com/docroot/articles/01-361-002pdf. Accessed 21 July 2006.

62. Convention on Biological Diversity (CBD). Available: http://www.biodiv.org/default.shtml.

63. Secretariat of the Convention on Biological Diversity (2000). Cartagena Protocol on Biosafety, Geneva.

64. Food and Agriculture Organization of the United Nations (2001). International Treaty on Plant Genetic Resources for Food and Agriculture. Available: http://www.fao.org/ag/cgrfa/itpgr.htm.

65. European Patent Convention (EPC) (1973). (rev. 2002). European Patent Office. Available: http://www.european-patent-office.org/legal/epc/e/ma1.html.

66. Porter G, Denning C, Plomer A, et al. (2006). The patentability of human embryonic stem cells in Europe. *Nat. Biotechnol.* 24:653–655.

67. Directive 98/44/EC of the European Parliament and of the Council of 6 July 1998 on the legal protection of biotechnological inventions. Available: http://europa.eu.int/smartapi/cgi/sga_doc?smartapi!celexapi!prod!CELEXnumdoc&lg=EN&numdoc=31998L0044&model=guichett.

68. *Japan Patent Law*, as amended (2003).

69. Canadian Intellectual Property Office (CIPO) (2006). A Guide to Patents: Utility. Available: http://strategis.ic.gc.ca/sc_mrksv/cipo/patents/pat_gd_protect-e.html#sec2. Accessed 20 July 2006.

70. UK Patent Office. What is a Patent? (2006). Available: http://www.patent.gov.uk/patent/ definition.htm. Accessed 20 July 2006.

71. Swedish Patent and Registration Office (2006). Which Inventions are Patent-able? Available: http://www.prv.se/english/patents/which.html. Accessed 20 July 2006.

72. Japan Patent Office (2006). Examination Guidelines for Patent and Utility Model in Japan: Part II Requirements for Patentability. Available: http://www.jpo.go.jp/tetuzuki_e/index.htm. Accessed 20 July 2006.

73. US Patent Act (2004). 35 U.S.C.

74. *Patent Act*, R. S. C. 1985, C. p-4, as amended.

75. UK Patent Office (2005). Confidentiality and Confidentiality Agreements, DDU85/01-05. Available: http://www.patent.gov.uk/patent/info/cda.pdf. Accessed July 21 2006.

76. Elder G (1985). A practical guide for proving fraud on the patent and trademark office: J.P. Stevens & Co. v. Lex Tex Ltd. *Am. U. L. Rev.* 34:729–746.

77. *Control Data Canada Ltd. v. Senstar Corp.* (1989). 23 C.P.R. 3d.

78. Patent Appeal Board (1982). Application of Abitibi Co. 62 C.P.R. 2d.

79. UK Patent Office (2005). Examination Guidelines for Patent Applications relating to Biotechnological Inventions in the UK Patent Office.

80. Gold R, Durell K (2005). Innovating the skilled reader: Tailoring patents to new technologies. *Intellectual Prop. J.* 19:189–226.

81. *Beloit v. Valmet* (1986). 8 C.P.R. 3d.

82. Hitchman C, MacOdrum D (1990). Don't fence me in: Infringement in substance in patent actions. *C.I.P.R.* 7:167–225.

83. Canada, Patent Rules (SOR/96-423), R. 79(6).

84. Mossinghoff G (2005). Small Entities and the "First to Invent" Patent System: An Empirical Analysis. *Washington Legal Foundation.* Available: http://www.wlf.org/upload/MossinghoffWP.pdf. Accessed 20 July 2006.

85. Canadian Intellectual Property Office (2006). A Guide to Patents: Glossary. Available: http://strategis.ic.gc.ca/sc_mrksv/cipo/patents/pat_gd_gloss-e.html. Accessed 20 July 2006.

86. Cameron D, Davis M (2001). Patents on Bay Street? *The Globe and Mail.* 24 August.

87. Marshall A, et al. (2006). A license to print money? *Nat. Biotechnol.* 24:593.

88. African Regional Industrial Property Organization (ARIPO) (2006). Background on the ARIPO. Available: http://www.aripo.wipo.net/background.html. Accessed 20 July 2006.

89. World Intellectual Property Organization (2006). Protecting your Inventions Abroad: Frequently Asked Questions about the Patent Cooperation Treaty. Available: http://www.wipo.int/pct/en/basic_facts/faqs_about_the_pct. Accessed 20 July 2006.

90. Anderegg M, et al. (2006). Trendspotting: A shift in intellectual property focus. *Nat. Biotechnol.* 24:609–611.

91. *Minerals Separation North American Corp. v. Noranda Mines, Ltd.* (1947). Ex. C.R. 306.

92. *Festo Corp. v. Shoketsu Kinzoku Kogyo Kabushiki Co.* (2002). 304 Fed. Cir. 3d 1289.

93. *Patent Act*, 1977 (UK), c. 37, as amended.

94. *Catnic Components Ltd. v. Hill & Smith Ltd.* (1982). R.P.C. 183 H.L.

95. Takenaka T (2000). Patent infringement damages in Japan and the United States: Will increased patent infringement damage awards revive the Japanese economy? *Re-Engin. Patent Law.* 2:301–309.

96. *THK Co Ltd v. Tsubakimoto Precision Products Co Ltd.* (1994). Tokyo High court.

97. *Genentech Inc. v. Sumitomo Pharmaceuticals Company* Ltd. (1996). Osaka High Court.

98. Ladas & Parry LLP (2002). Japan—Damage Awards in Patent Infringement Actions. August Newsletter.

99. *Free World Trust v. Électro Santé Inc.* (2000). 2 S.C.R. 1024.

100. *Whirlpool Corp. v. Camco Inc.* (2000). 2. S.C.R. 1067.

101. Kleiman M, Sontag L (2006). Patenting Pathways. *Nat. Biotechnol.* 24:616–617.

102. Center for Intellectual Property Policy and Health Law Institute (2004). The Research or Experimental Use Exemption: A Comparative Analysis. Report prepared for Health Canada. Available: http://www.cipp.mcgill.ca/data/publications/ Accessed 23 July 2006.

103. Gold E, Joly Y, Caulfield T (2005). Genetic research tools, the research exception and open science. *GenEdit.* 3:1–11.

104. *Embrex v. Service Engineering* (2000). 216 F.3d 1343.

105. *Madey v. Duke University* (2002). No. 01-1567 CAFC.

106. Drug Price Competition and Patent Term Restoration Act (1984). 35 U.S.C. §271(e).

107. Convention for the European Patent for the Common Market *Community Patent Convention.* 1975, 1976 O.J. L 17.

108. Grubert A (2006). Understanding European Patent Opposition In Light Of U.S. Patent Practice. *Intellect. Prop. Report.* 4:39. Available: http://www.imakenews.com/bakerbotts/e_article000274940.cfm. Accessed 21 July 2006.

109. Cochran D (2006). United States: Toward a Post-Grant Opposition Proceeding in the United States. *Jones Day.* 21 April 2006. Available: http://www.mondaq.com/article.asp?articleid=39280. Accessed 23 July 2006.

110. Kirk M (2004). Patent quality improvement: Post-grant opposition. *Statement before the Subcommittee on Courts, the Internet, and Intellectual Property Committee on the Judiciary U.S. House of Representatives.* Available: http://www.aipla.org/Content/ContentGroups/Legislative_Action/108th_Congress1/Testimony2/Testimony_on_Patent_Quality_Improvement_Post-Grant_Opposition.htm. Accessed 24 July 2006.

111. United States Congress (2005). Bill 2795 H.R.

112. Trilateral Statistical Report (2004). *Patent Activity at Trilateral Offices.* Trilateral Cooperation, Munich, Germany. Available: <http://www.trilateral.net/tsr/tsr_2004/tsr2004.pdf. Accessed 24 July 2006.

113. U.S. Food and Drug Administration, Centre for Biologics Evaluation and Research (2006). About Us. Available: http://www.fda.gov/cber/about.htm. Accessed 20 July 2006.

114. Dudzinski D (2005). Reflections on historical, scientific, and legal issues relevant to designing approval pathways for generic versions of recombinant protein-based therapeutics and monoclonal antibodies. *Food Drug Law J.* 60:143–259.

115. Polastro E (2001). Biogenerics: Myth or reality? *Innov. Pharm. Technol.* Dec:63–65.

116. Class S. Biogenerics: Waiting for the Green Light. *IMS Company Profiles.* Available: http://www.imshealth.com/web/content/0,3148,64576068_63872702_70261006_71226827,00.html. Accessed 22 July 2006.

117. Ontario Ministry of Health (2002). Genetics, Testing and Gene Patenting: Charting New Territory in Healthcare. Government of Ontario, Toronto, Canada.

118. National Academy of Sciences (2004). *A Patent System for the 21st Century*, The National Academies Press, Washington, D.C.

119. Williams-Jones B (2002). History of a gene patent: Tracing the development and application of commercial BRCA testing. *Health Law J.* 10:123–146.

120. National Academy of Sciences (2005). *Reaping the benefits of Genomic and Proteomic Research: Intellectual Property Rights, Innovation, and Public Health*; National Academies Press, Washington, D.C.

121. Jensen K, Murray F (2005). Enhanced: Intellectual property landscape of the human genome. *Science.* 310:239–240.

122. Heller M, Eisenberg R (1998). Can patents deter innovation? The anticommons in biomedical research. *Science.* 280:698–701.

123. Caulfield T (2005). Policy conflict: Gene patents and health care in Canada. *Community Genet.* 8:223–227.

124. Nelson R R (2004). The market economy and the scientific commons. *Res. Policy.* 33:455–471.

125. David P A (2004). Can 'Open Science' be protected form the evolving regime of IPR protection? *J. Theoret. Institut. Econ.* 160:1–26.

126. Walsh J P, Cho C, Cohen W M (2005). View from the bench: Patents and material transfers. *Science.* 309:2002–2003.

127. US Department of Justice and the Federal Trade Commission (1995). Antitrust Guidelines for the Licensing of Intellectual Property. 4 Trade Reg. Rep. CCH.

128. Goldstein J, et al. (2005). Patent pools as a solution to the licensing problem of diagnostic genetics. *Drug Dis. World.* Spring, 86–90.

129. Clarke J, et al. (2000). Patent Pools: A Solution to the Problem of Access in Biotechnology Patents?" United States Patent and Trademark Office. Available: http://www.uspto.gov/web/offices/pac/dapp/opla/patentpool.pdf.

130. Resnik D B (2002). The commercialization of human stem cells: Ethical and policy issues. *Health Care Anal.* 10:127–136.

131. Verbeure B, Van Zimmeren E, Matthijs G, et al. (2006). Patent pools and diagnostic testing. *Trends Biotechnol.* 24:114–120.

132. Opderbeck D (2004). The penguin's genome, or coase and open source biotechnology. *Harv. J. Law & Tec.* 18:167–227.

133. Burke D (2001). Open Source Genomics. *B.U. J. Sci. & Tech. Law*. 8:255–256.

134. Feldman R (2004). The open source biotechnology movement: Is it patent misuse? *Minn. J.L. Sci. & Tech*. 6:117–167.

135. Maurer S, et al. (2004). Finding cures for tropical diseases: Is open source and answer? *Minn. J.L. Sci. & Tech*. 6:169–175.

136. World Trade Organization (2003). Decision Removes Final Patent Obstacle to Cheap Drug Imports. Press/350/Rev. Available: http://www.wto.org/english/news_e/pres03_e/pr350_e.htm. Accessed August 2005.

137. Canadian HIV/AIDS Legal Network (2004). Global Access to Medicines: Will Canada Meet the Challenge? *Standing Committee on Industry, Science and Technology regarding Bill C-9, An Act to amend the Patent Act and the Food and Drugs Act*. Available: http://www.aidslaw.ca/Maincontent/issues/cts/patent-amend/SCIST%20Submission_Feb2604.PDF. Accessed August 2005.

138. Canadian HIV/AIDS Legal Network (2003). Amendment to Canada's Patent Act to Authorize Export of Generic Pharmaceuticals—Update. Available: http://www.aidslaw.ca/Maincontent/issues/cts/patent-amend/PatentActAmendment_Update.pdf. Accessed August 2005.

12.3

COMPARABILITY STUDIES FOR LATER-GENERATION PRODUCTS— PLANT-MADE PHARMACEUTICALS

PATRICK A. STEWART

Arkansas State University, State University, Arkansas

Chapter Contents

12.3.1 Introduction   1433
12.3.2 Benefits Offered by Plant-Made Pharmaceuticals (PMPs)   1435
12.3.3 Threats Posed by Plant-Made Pharmaceuticals (PMPs)   1438
     12.3.3.1 Starlink Bt Corn   1438
     12.3.3.2 Prodigene PMP Corn   1439
     12.3.3.3 Syngenta Bt Corn   1440
     12.3.3.4 Conclusions   1441
12.3.4 Public Response to Plant-Made Pharmaceuticals (PMPs)   1441
     12.3.4.1 Interested Public/Pressure Group Comments   1441
     12.3.4.2 Public Opinion Concerning PMPs   1442
     12.3.4.3 Conclusions   1445
12.3.5 Regulation of Plant-Made Pharmaceuticals (PMPs) in the Fields   1446
12.3.6 Conclusions   1449
     References   1450

12.3.1 INTRODUCTION

The future of pharmaceutical drugs in the United States is likely that of continuing economic growth, as there is increasing use of pharmaceutical drugs in the general population, with an estimated 43% of Americans using at least one pharmaceutical drug between 1999 and 2000 [1]. This fact, in conjunction with an aging population needing and demanding more and cheaper pharmaceutical drugs, with an average of 30 filled prescriptions for non-institutionalized Medicare enrollees aged 65 and older [2], suggests a need for alternative approaches to traditional "vat-based" mammalian cell-based production. Although such production approaches as animal-based "pharming" are promising, among the most promising of approaches is that of Plant-Made Pharmaceuticals (PMPs). These PMPs use genetically engineered plants, a technology used for over 20 years, to express substances that may be used to treat and cure such diseases as cancer, Alzheimer's, diabetes, HIV, heart disease, and others. By using naturally occurring processes in the plant, along with the ability to scale up production quickly, PMPs are a promising technology. However, as with any new technology, the potential pitfalls are plentiful, chief among which are the perceptions of the public, which is largely unaware of this technology, and whose perceptions place it on a threshold of acceptance or rejection, making it a "boundary technology".

PMPs may be considered a boundary technology for a variety of reasons. First, it is a subset of a relatively new technology, that of genetically engineered plants. As a result of its newness, risks from genetic engineering of plants are not well characterized. Although recent events suggest the potential risks, and the current inadequacy of the regulatory system to either prevent or respond to crises (see Section 12.3.3), from genetically engineered plants have not been tangibly experienced by the average American. Instead of actual events, characterized risks tend to exist as either extrapolations from laboratory experiments or as hypothetical events. Although laboratory experiments do provide a good deal of insight into the mechanisms by which events may occur, they do not replicate the reality of the fields in which PMPs may be grown with the field's dynamic, nonlinear, complex systems affected by multiple interacting factors such as weather, soil, cropping systems, chemical usage, and so on. Hypotheticals about PMPs tend to be extreme in nature and politically charged, with Cassandras of doom preaching worst-case scenarios and Candides of optimism trumpeting the best possible future. Neither perspective, however, appreciates the probabilistic nature of risk. Here, risk is defined as hazard multiplied by exposure. And, in the words of Howard and Donnelly, "(I)n order for the risk to be zero, either the exposure or the hazard must be zero" [3], which, of course, is unlikely if not impossible, especially considering increased sensitivity of detection methods and understanding of the effect of different substances on different individuals.

A second reason to consider PMPs as a boundary technology is that they promise to radically change public perceptions by drawing attention to the changed character of the plants themselves. It is here that hypothetical scenarios concerning PMPs make their mark, leading to acceptance or rejection of this technology. Although genetically engineered plants are currently a pervasive part of the agricultural landscape and consumed extensively, with the staple crops of corn and soybean grown in genetically engineered variants in the United States and comprising 45% and 85% of total acreage in 2004 [4], respectively, there is little public awareness of their presence in the food supply [5–7], which is mainly a result of them being genetically engineered to express agronomic qualities, such as herbicide tolerance and pest resistance, which are typically not brought to the attention of the public. Although plants have long been used for medicinal purposes and food can have healing properties, genetically engineered plants, especially dealing with food crops, will likely draw public attention by underlining the transgenic nature of the products that span species and may even incorporate human DNA. Some precedent in public response to genetic engineering may be seen in "second-generation" food products with enhanced product quality, which, although offering useful characteristics, have yet to be economically successful.

The final reason PMPs may be considered a boundary technology derives from the previous reason. PMPs may be recognized as both a food and a medicinal application of genetic engineering. Although it is highly unlikely that the same plant will be processed for both food and pharmaceutical purposes, it is quite likely the public will perceive them as a unified technology. In terms of public response to date, there have been quite different paths in the reception of genetically engineered pharmaceuticals and medical products and that of genetically engineered plants. There has been little controversy over genetically engineered pharmaceutical products, likely because people are willing to accept risk when it comes to recovering the health they have lost, whereas people are likely to be risk averse when it comes to changing their diet, especially when they perceive nothing wrong with it. By exhibiting both characteristics, PMPs are likely to arouse ambiguity in response, depending on the personal salience of particular pharmaceutical products being developed.

Considering these reasons, this chapter will explore PMPs and their potential with an eye toward public response and policy actions. In this chapter, we first consider potential benefits of new products moving toward the market, such as enhanced access and decreased price. We next look at the risks of this boundary technology, especially adventitious presence in which PMPs might accidentally enter the food supply, which is illustrated by recounting three recent events in which the food chain was either breached or threatened by unapproved food products. We then consider public opinion toward PMPs, both in response to these incidents, as seen in comments to Federal Register changes, and through a regional survey of public opinion. Finally, we consider the federal regulations currently in place and recent political developments before drawing conclusions.

12.3.2   BENEFITS OFFERED BY PLANT-MADE PHARMACEUTICALS (PMPs)

The benefits of PMPs have been well characterized and documented by proponents of this technology. Indeed, much of the research and development of this technology is driven by optimism that PMPs will have tremendous health benefits for an aging and growing population and the potential for economic gain, especially when compared with current production approaches. This outlook is bolstered by awareness of the sheer size of the pharmaceutical drug market. Of this market, an increasing proportion builds on genetic engineering.

According to Elbehri [8], three factors commend PMPs over current approaches to express pharmaceuticals, such as bacteria, yeast, and mammalian cells, and more recent technologies, such as insect cells and transgenic animals. These three factors are cost advantage, production scalability, and safety. The cost advantage of PMPs over the current use of mammalian cell cultures is great, with mammalian cell bioreactors ranging from $106–650 per gram of antibody, whereas PMPs would likely cost four to five times less for the same pharmaceutical.

The second factor is the capacity to more easily scale-up production than is currently the case with mammalian cell-based systems, which is based on the ability to increase production by growing more plants on added acreage in a short period of time, avoiding the need to build more production facilities with buildings, vats, and other associated capital costs [8]. However, this benefit is offset by production being driven by growing seasons and the ability of plants to reproduce seeds enough to scale-up production [9]. In other words, pharmaceuticals will be produced on the basis of an agricultural cycle, with the inherent difficulties of predicting weather, pest outbreaks, and other factors that influence natural systems. Although storage may not prove problematic, especially if the pharmaceutical is expressed in the seed (which might not be the case), predicting market demand is yet another difficult factor. Another difficulty with scaling up lies with enforcing regulatory controls. Expanding production acreage implies increased potential for inadvertent movement of the PMPs' genes, whether by cross-breeding with other plants or their being blown into neighboring fields, bringing with it legal liabilities for environmental and health damage if they occur. Therefore, expanded acreage implies expanded implementation of safeguards and increased surveillance.

The third and final factor is the belief that PMPs are safer than using proteins from recombinant microorganisms or cells, which is because "PMPs do not carry potentially harmful human or animal viruses into the drug" [8], as plants do not serve as hosts for such human pathogens as HIV, prions, hepatitis, and so on [10]. Furthermore, plants are typically constructed to express a portion of a pathogen or toxin, reducing possible infection or innate toxicity [11]. This, however, does not alleviate concerns over potential allergenicity, which will be present in most pharmaceuticals.

In spite of the benefits offered, PMP development is still in its infancy, with retardation of its progress caused by reasons discussed in the next section. However, analysis of U.S. Department of Agriculture (USDA) data concerning release permits for "pharmaceuticals, industrial, and value-added proteins" over the last two years, during which time stricter regulatory standards were enforced, suggests a still-fertile field of research. Overall, 35 permits were applied for during 2004 and 2005, with permit applications increasing from 15 in 2004 to 20 in 2005, of which nearly 70% were granted and 60% were released into the fields.

Institutions doing research in PMPs tend to be dominated by small research companies (see Figure 12.3-1). Although research universities, mainly Iowa State

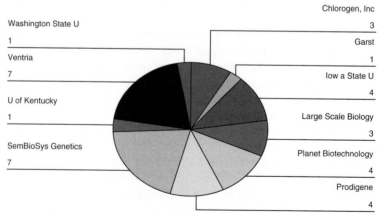

Figure 12.3-1. Institution.

University (four applications) and the University of Kentucky and Washington State University (one application apiece), are involved in research, companies such as SemBioSys Genetics and Ventria (seven applications apiece), Planet Biotechnology and Prodigene (four applications), Large Scale Biology and Chlorogen, Inc. (three applications), and Garst (one application) are the mainstays of research currently being sent to the fields.

The five plants serving as hosts for the experimental activities reflect research realities (see Figure 12.3-2). Tobacco is the preferred host, having nearly one third of the permit applications, reflecting its long-time use as a research plant in genetic engineering. Although corn has long history of experimentation and being used extensively in the field, and, as a result, is the second most popular plant for experimentation, the change in regulations in response to critiques of the regulatory system, and fears of it entering the food supply, has led to its reduced relevance in 2005, where only two university-based field trials were proposed, compared with its more extensive use (seven applications) in 2004. The remaining plants, safflower (seven applications), rice (six applications), and barley (two applications), might prove to be the preferred future host plants for PMPs and industrial products.

Examples of products being developed include Ventria BioScience's work on human recombinant lactoferrin and lysozyme, which may be used as an oral rehydration solution [12]; SemBioSys has been developing PMP-based human insulin (to meet the needs of an increasing population of diabetics) and human APO IP, which reduces and stabilizes arterial plaque associated with "acute coronary syndrome" such as heart attacks and angina, as well as strokes [13]; Planet Biotechnology's treatments of tooth decay by bacterial infection, hair loss through specific forms of chemotherapy, and treatment for the common cold by blocking infection by rhinovirus are likewise promising [14]. Prodigene has brought to market recombinant trypsin, normally produced by the pancreas, which may be used in the production of insulin, human and veterinary vaccines, cell cultures, and wound care [15]. It has also developed a vaccine targeting swine transmissible gastroenteritis [16]. Finally, Chlorogen is developing a "chloroplast transformation technology" that would introduce genes into plants, allowing large-scale production without potential genetic transfer through pollen [17]. These products and others promise

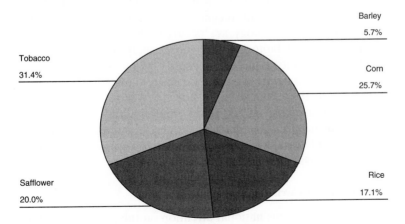

Figure 12.3-2. Host plant.

increased product availability and a range of new products, with health benefits and economic growth implied.

Although the future of PMPs may, like any technology, be difficult to predict, the experimental activity concerning these plants appears to be relatively robust considering recent regulatory changes and increased public skepticism. Although the amount of permit applications are relatively small, especially compared with the sheer numbers of experimentation involving genetically engineered plants, PMPs and associated technologies appear to be growing in importance and in the fields. And given the benefits they offer in the production of new pharmaceutical drugs, this growth can be expected.

12.3.3 THREATS POSED BY PLANT-MADE PHARMACEUTICALS (PMPs)

In spite of the benefits specifically offered by PMPs, and by genetically engineered plants generally, apprehension exists over their release into the environment. According to the National Research Council report "Bioconfinement of Genetically Engineered Organisms" [18], three concerns exist over the unintentional movement of genes placed into target plants. The first is that the inserted genes will lead to the target plant or relatives cross-breeding and becoming weeds by obtaining benefits that enable it to overrun or disrupt ecosystems. The second concern is that wild relatives, especially those with economic, social, or moral values, will be at risk of extinction by either being outcompeted or through hybridization with genetically engineered plants. The third concern, that the genetic material will spread to other domesticated varieties [18], is perhaps one of the highest profile concerns for PMPs, especially when considering that many PMPs are currently being expressed in food plants. Although these concerns are currently being researched and will likely continue to pose a threat as genetically engineered plants dominate the commodity agriculture landscape, no evidence of these concerns coming to fruition have come to light to date.

Another major concern, if not the major concern, is that PMPs will somehow find their way into the food supply. As a result, it is feared the public will respond with decreased institutional trust and will make market decisions that negatively affect food producers. As a result, recent changes to the regulatory structure concerning agricultural biotechnology are occurring because of concerns "that the expansion in agricultural biotechnology increasingly will put pressure on seed production and commodity handling systems" [19] to segregate and control its products. Specifically, concerns exist that the current regulatory scheme is not able to adequately address the amount and variety of new genetically engineered plants, with PMPs chief among them. Doubts over the agricultural biotechnology regulatory system have been raised by three recent events, calling into question the scientific basis for regulation, the effectiveness of regulatory enforcement, and the integrity of the food system.

12.3.3.1 StarLink Bt Corn

The first major event to focus national attention on the potential violability of the U.S. food chain occurred when StarLink corn made its way into the human food supply in 1999 and 2000. StarLink corn used the Cry9c variant of *Bacillus thuringiensis* (Bt), a protein that kills targeted insects by eating through their guts, leading to sepsis and the inability to digest food. StarLink corn was not deemed fit for either human consumption or export because of presumed possible human allergic reactions. As a result, it was to be used only for animal feed or nonfood industrial purposes.

In September and October of 2000, attention was drawn to the presence of StarLink corn in the food chain when the consumer group Genetically Engineered Foods Alert performed tests on taco shells and other corn-based products being sold in grocery stores and in fast food restaurants [20]. Estimates by the USDA's Economic Research Service suggest 123.8 million bushels were co-mingled in 2000, whereas Aventis', the company responsible for StarLink, estimates that over 430 million bushels were co-mingled in 1999 and 2000 [21] with an estimated 340,908 acres in 28 states planted with StarLink Bt corn [20]. Overall, a total of nearly 300 products were recalled, including 70 types of corn chips, 80 types of taco shells, and almost 100 restaurant food products.

In an attempt to minimize economic loss and public concern, Aventis and USDA bought back existing grain supplies and recalled food containing StarLink corn. However, significant damage was done. In addition to illustrating the failure of the U.S. regulatory system concerning genetically engineered plants, there was disruption of the downstream food manufacturing marketplace with corn exports being affected. Indeed, within a single year, of 110,000 grain tests by federal inspectors, StarLink corn showed up in one-tenth [22]. Although the complexity of market forces allows only estimation, a significant decrease in trade with Japan and South Korea occurred immediately after StarLink corn was found [21]. Since then, the European Union has required labeling and tracing requirements for genetically engineered goods that are seen as greatly hindering U.S. farm product competitiveness, especially with respect to corn and soybean products. Overall, estimates suggest the StarLink event and the recall in its aftermath have cost the food industry $1 billion [23].

12.3.3.2 Prodigene PMP Corn

The second event, that of Prodigene's PMP corn, is one immediately relevant for PMP production. In September and October of 2002, in Iowa and Nebraska respectively, APHIS found "volunteer" corn plants genetically engineered to produce a pharmaceutical to prevent "traveler's diarrhea" growing in soybean fields in violation of permit conditions. Specifically, Prodigene did not abide by the conditions of their field release permit to eradicate all traces of the experimental crop from the fields, as small quantities of this corn ended up in soybean that was to be processed and sold for human consumption.

Although none of this corn made it into the food supply, this near miss drew attention to the regulatory system's lack of readiness in dealing with the next generation of genetically engineered crops, those producing pharmaceutical and industrial products. Prodigene had to pay a civil penalty of $250,000; destroy 500,000 bushels, or $2.7 million dollars worth, of soybean in Nebraska; and incinerate 155 acres of corn in Iowa because of concern that cross-pollination occurred, as well as posting a $1 million bond and accede to higher compliance standards for future field tests [24]. Perhaps more important in terms of long-term political implications, the Grocery Manufacturers of America (GMA) and other food processing interest groups expressed concern over PMP field test regulations, with John R. Cady, CEO of the National Food Processors Association commenting "(it is) nothing short of alarming to know that at the earliest stages of development of crops for plant-made pharmaceuticals, the most basic preventative measures were not faithfully observed. This apparent violation of rules . . . very nearly placed the integrity of the food supply in jeopardy" [25].

12.3.3.3 Syngenta Bt Corn

The third and most recent event came to light in March 2005 when the seed company Syngenta revealed that it had accidentally sold approximately 146,000 tons of experimental Bt-10 corn that had been approved for experimental use only by USDA but was grown on 37,000 acres over a four-year period (2001–2004). The antibiotic marker gene is often used in the experimental phase of many genetically engineered plant studies to test for the presence or absence of altered genetics.

Although the fact that the Syngenta Bt-10 corn inadvertently entered the food chain raises questions about the effectiveness of the USDA regulatory system, especially after regulatory changes were put in place in response to the StarLink and Prodigene situations, the antibiotic marker gene entering the food supply has raised concern because of perceived problems with increased antibiotic resistance with human infections. Although many other factors contribute to antibiotic resistance, including their extensive use in farm animals, concern over the use of antibiotics in genetically engineered crops date back to the earliest days of genetically engineering when, in 1990, Calgene asked for an opinion concerning their use from FDA, and it was found that there was reasonable certainty no harm would result from its use. However, since that time, concerns are still being raised concerning their near ubiquitous use, whether in genetically engineered plants or in farm animals [26].

Although Syngenta officials argue that the release was relatively small, affecting 0.01% of all corn grown from 2001 to 2004 [23], and has not appeared to affect U.S. consumers, response from abroad suggests that U.S. producers will be affected in the export marketplace. Specifically, on April 15, 2006, the European Union placed a temporary ban on U.S. corn imports unless they are proven to be Syngenta Bt-10-free; on May 25, 2006, they enforced this ban by impounding a U.S. shipment of corn gluten upon its arrival in Ireland when it was found to be tainted with Bt-10. Likewise, in South Korea, imports of U.S. corn are being investigated with certification of future shipments as Bt-10-free required for entry, while most recently (June 1, 2006), a shipment of corn to Japan found to be Bt-10-contaminated was found and held [27].

As a result, Syngenta has been fined $375,000 and ordered to sponsor a compliance training conference by USDA [28–30]. However, the regulatory response appears to be subdued at best, especially in light of previous failures by biotechnology companies to implement federal regulations.

12.3.3.4   Conclusions

According to the 1984 Coordinated Framework for the Regulation of Biotechnology, USDA, EPA, and FDA are responsible for regulating agricultural biotechnology. Although this framework has proved to be serviceable for most of its existence, recent events suggest it is not meeting the challenges posed by the increased volume and variety of genetically engineered crops. The instances discussed above illustrate some of the problems facing U.S. regulators, especially the reduced trust in institutions. However, what is most important for producers is that an inefficient and ineffective regulatory system might push consumers elsewhere.

Although critics argue that the above cases show the system is working by identifying flaws in the regulatory system and changing practices of both government agencies and industry, in the end, consumer trust is what matters most. Although lost consumer trust in domestic markets is a threat, with StarLink-tainted products being recalled from grocer's shelves in 2000, long-lasting effects in the U.S. market have not materialized, especially when the recent Syngenta Bt-10 corn case is considered. International trade disruption, however, has occurred in both StarLink and Syngenta cases and has been costly in terms of trade with the European Union and with Asian trade partners South Korea and Japan. Perhaps most important, trust in the U.S. regulatory system by our trade partners is waning, with a recent editorial title concerning the Syngenta case in the respected science journal *Nature* underscoring diminishing respect: "Don't rely on Uncle Sam" [30].

12.3.4   PUBLIC RESPONSE TO PLANT-MADE PHARMACEUTICALS (PMPs)

Public response to PMPs suggests they are of two minds concerning PMPs. On the one hand, PMPs promise to offer tangible benefits that directly enhance the quality of life by producing less expensive pharmaceutical drugs that may treat a broader range of ailments. This perspective is bolstered in a recent PEW report that showed a strong relationship between perception of PMPs as being a good use of genetic engineering and the likelihood of it having a positive impact on the respondent's family [6].

On the other hand, the risks of PMPs entering the food supply or remaining in the environment as weeds gives pause to critics of the technology. The StarLink and Syngenta Bt-10 cases stand out as examples of the regulatory and food systems' failure to prevent the unwanted presence of genetically engineered crops in the food supply. Although the Prodigene case was a near miss, it also focused attention on the flaws in the regulatory system and may have had a chilling effect on the development of both PMPs and PMIPs. Although these three cases reflect "growing pains" in regulating the technology as more plant products with a greater variety of uses enter the fields, the question remains as to whether this problem will be endemic or a "mere blip" that has been dealt with and will not recur.

12.3.4.1   Interested Public/Pressure Group Comments

The amount of comments in response to the 2003 Federal Register notices dealing with changes to the regulation of PMPs and PMIPs, which in turn are arguably in response to the Starlink and Prodigene incidents discussed in the previous section, reflect the new-found importance of the field release of genetically engineered plants to the public. The March 10, 2003 Federal Register notice concerning PMP and PMIP field testing requirements attracted at least 847 comments (of which 77 were late). In comparison, changes to the APHIS regulations in 1993 garnered 84 comments, whereas even more wide-ranging regulatory changes in 1997 attracted only 50 comments [31].

Although the numbers of comments received in response to PMP regulatory changes are exponentially higher than those received in response to prior Federal Register notices, multiple factors are responsible. E-government with electronic dockets easing comments on Federal rule-making has expanded public participation, especially with salient issues. For example, proposed organic standards posted in the Federal Register received an unprecedented 275,000 comments. Here, although the usual policy actors interested in genetically engineered plants are still involved, a high percentage of comments were sent by individuals and organizations not normally associated with this debate [32].

As expected, support for the new technology came from biotechnology and biotechnology-related companies, agricultural organizations, state agricultural departments, universities, and agricultural sector pressure groups. Critiques were also received from those who appeared to have ties with the organic movement or environmental groups, as evidenced by the roughly 600 comments e-mailed as cut-and-paste forwards.

Unexpectedly, especially in comparison with previous regulatory changes, concerns were raised by other powerful groups. Namely, the GMA and affiliated groups expressed concern over uncontained field release of PMPs and PMIPs, especially in food and feed plants. Interestingly enough, concern by consumer groups and traditional biotechnology opponents was tempered, likely mitigated by the potential for medical benefits from this new technology.

Comments by the interested public in response to the changes to PMP regulations give insight into concerns the general public may have once their awareness of these new technologies are raised, especially if PMPs and PMIPs enter into widespread production. A few themes stand out when comments of those critiquing the proposed USDA regulations are analyzed. The first theme concerns the science used and its effectiveness in regulating the threats posed by PMPs and PMIPs to the food supply and the environment. The next two themes expand on institutional distrust suggesting that governmental institutions, namely USDA, are either negligent, showing disregard for American farmers and the public by allowing these "dangerous" plants to be grown, or have been captured by corporate interests and are doing their bidding despite the threat posed. The risk perceptions of those commenting on the regulations suggest this technology is unknown, unnatural, uncontrolled, and likely to end in disasters disproportionately affecting future generations [32]. As can be expected by the foregoing responses, PMPs and PMIPs engender a good deal of fear and anger, emotional responses we deal with more systematically in the next section.

12.3.4.2 Public Opinion Concerning PMPs

Although the general public's opinion on PMPs is best analyzed through surveys, little research on their opinion has been carried out in the United States. In

Canada, Einsiedel and Medlock [33] carried out a series of focus groups analyzing attitudes toward PMPs and industrial products after providing information on plant molecular farming generally and five specific examples of this technology. Findings by them suggest greater support for novel medical products benefiting human health or the environment. Risk perceptions were driven by concerns over contamination of the food supply or the environment with higher risk seen as deriving from these products being grown outdoors in food plants that are able to go to seed or that flower, thus dispersing their genetic material to food plants that might be consumed by humans. When risk-benefit analyses were made, Einsiedel and Medlock found "acceptability is predicated on the idea that there are stringent approval processes and long-term measures in place to ensure safety" [33].

One exception to the lack of systematic public opinion research on PMPs is a study carried out by Stewart and McLean [7] using a telephone survey examining consumers living in Arkansas, Louisiana, Texas, Oklahoma, and New Mexico. This study was conducted between May 9 and June 10, 2004 by the Arkansas State University Center for Social Research using computer-assisted telephone interviewing (CATI) technology and had a very good response rate (a total of 680 interviews completed for a response rate of 61% and a margin of error of +/−3.5%). Exact wording of the questions are contained in Appendix A.

Five questions were asked of participants and ascertained personal benefits, likelihood of PMPs entering the U.S. food supply, the likelihood of respondents eating food from PMPs, how worried the respondents would be if they ate food with PMPs in it, and how angry they would be if they ate it without their knowledge. PMPs were characterized as "plants genetically modified to produce pharmaceutical drugs. These are plants that are modified to produce compounds used in manufacturing vaccines for diarrhea, antibodies to fight cancer, and drugs to treat such illnesses as cystic fibrosis."

Findings suggest the respondents have mixed feelings about PMPs. Generally speaking, PMPs are seen as presenting personal benefits to respondents, with two-thirds seeing either "a great deal" or "some benefit" (Figure 12.3-3). Of these, a plurality (36.5%) see a great deal of benefit for themselves, whereas only 12% see themselves as receiving no benefit at all from PMP-based products. Finally, only 12% had no opinion about PMPs.

Although respondents see benefits from PMPs, a majority see adventitious presence as likely to occur, with nearly 60% perceiving PMPs as accidentally entering the U.S. food supply. Only one fourth of respondents see this event as either "somewhat" or "very unlikely" to occur (Figure 12.3-4). Whether this pattern of response is because of the Starlink or Prodigene controversies discussed earlier in this chapter, because of a lack of faith in governmental and food production institutions to prevent this event from occurring, or the result of some other unspecified factor is not known; it may be assumed, however, that a relatively high level of public skepticism exists.

Although respondents see themselves as less likely to eat food with PMP components than see the potential for adventitious presence, those seeing it as either "very" or "somewhat likely" make up a substantial proportion of respondents, with a combined 45% (Figure 12.3-5). At the same time, 38% perceive themselves as able to avoid such foods, responding to the question that they are either "somewhat" or "very unlikely" to eat food coming from PMP plants.

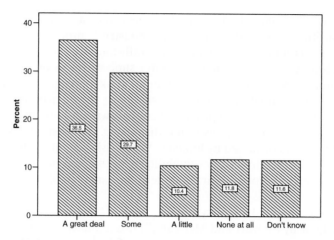

Figure 12.3-3. How much benefit would you get from PMP GM plants?

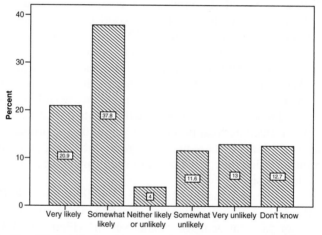

Figure 12.3-4. How likely is it that PMPs might accidentally enter the U.S. food supply?

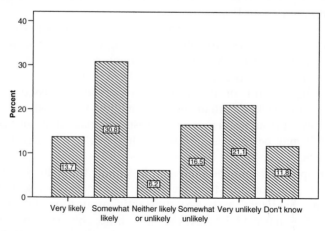

Figure 12.3-5. How likely is it that you would eat food coming from PMP GM plants?

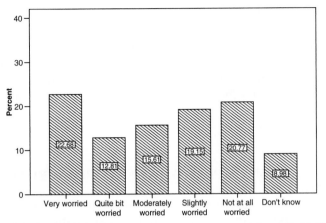

Figure 12.3-6. How worried would you be if you ate food from PMP GM plants?

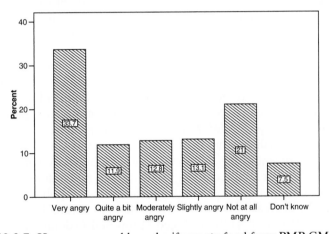

Figure 12.3-7. How angry would you be if you ate food from PMP GM plants?

Emotional response to eating food with PMPs in it suggests survey respondents are at least somewhat worried about the potential for it to occur (Figure 12.3-6). Overall, 23% are "very worried" and 13% are "quite a bit worried" about this event occurring. However, an increasing proportion of respondents show lesser levels of concern, with 16% "moderately worried," 19% "slightly worried," and 21% "not at all worried."

Although levels of concern are not as high as can be expected, great potential exists for upsetting the public. The pattern of response that can be seen in Figure 12.3-7 suggests polarized opinions. Specifically, one-third of respondents state they would be "very angry" if they ate food with PMPs in it without their knowledge, whereas only one-fifth of respondents would be "not at all angry."

12.3.4.3   Conclusions

Research available concerning public attitudes toward PMPs emphasizes it's "boundary technology" nature. Great promise and great risk are inherent in PMPs because of the personal benefits of increased access to a wider array of pharmaceutical drugs balanced by their potential for accidental release into the food supply or the environment. These ambivalent feelings are likely to persist until either the benefits of PMPs are felt by their products entering the marketplace or an environmental or health catastrophe involving these products occurs. In either case, the absence of awareness of PMPs (and even genetically engineered plants) is a driving issue.

12.3.5 REGULATION OF PLANT-MADE PHARMACEUTICALS (PMPs) IN THE FIELDS

The regulation of PMPs will establish a different, potentially more circuitous, path to the marketplace because of its "boundary technology" nature. Although, as a pharmaceutical, it will be regulated in a traditional manner by FDA, as a plant, it must adhere to standards set forth by USDA. As a genetically engineered plant, it triggers a set of regulations specific to this technology as a whole, and as a PMP specifically, as discussed below.

On the whole, the regulation of agricultural biotechnology has attracted relatively little public involvement until recently, which is most likely because of its highly technical nature and the downstream nature of most of its products, which have predominantly been concerned with replacing or reducing production inputs such as herbicides and pesticides. More recently, the promise of new products and dearth of experienced difficulties in the field experiments have led to relaxed regulations. However, recent events detailed above have publicized and politicized the regulation of agricultural biotechnology with stricter regulations and greater scrutiny by an expanding range of parties with economic and political clout [34].

To address the decreasing trust in the regulatory structure, the Office of Science and Technology Policy (OSTP) published "Proposed Federal Actions To Update Field Test Requirements for Biotechnology Derived Plants and To Establish Early Food Safety Assessments for New Proteins Produced by Such Plants" in August 2002. Specifically, the notice was published to provide guidance to USDA, EPA, and FDA to update field testing requirements for food and feed crop plants and establish early food safety assessments for new plant proteins, most specifically PMPs and PMIPs, in line with the 1986 Coordinated Framework. Three principles in the document are relied on in updating the Coordinated Framework: (1) The level of field test confinement should be consistent with the level of environmental, human, and animal health risk associated with the introduced proteins and trait(s); (2) if a trait or protein presents an unacceptable or undetermined risk, field test confinement requirements would be rigorous enough to restrict outcrossing or comingling of seed and, further, the occurrence of these gene or gene products from these field tests would be prohibited in commercial seed, commodities, and processed food and feed; and (3) even if these traits or proteins do not present a health or environmental risk, field test requirements should still minimize the occurrence of outcrossing and comingling of seed, although low levels of genes and gene products could be found acceptable based on meeting applicable regulatory standards [19].

In light of concerns raised by increased experimentation with PMPs and PMIPs, and addressed by OSTP in their notice [19], USDA's Animal and Plant Health Inspection Service (APHIS) changed rules concerning their field testing of PMPs in March 2003 [35]. Although the resulting regulations are expected to be modified further over the coming years, they currently incorporate significant changes in how PMPs and PMIPs are regulated [36]. Specifically, for all plants genetically engineered to produce pharmaceutical or industrial compounds and field tested under permit, APHIS established seven conditions that can be grouped into three categories. The first considers field test siting, the second the dedication of equipment and facilities to their production, and the third considers procedural matters.

Field test siting regulations proposed by APHIS provide two conditions to be met, with special consideration for pharmaceutical corn. First, the perimeter fallow zone will be increased from 25 feet to 50 feet to prevent inadvertent comingling with plants to be used for food or feed. Second, production of food and feed plants at the field test site and perimeter fallow zone will be restricted for the following season to prevent inadvertent harvesting. Furthermore, specific permit conditions for pharmaceutical corn have been instituted, likely a result of corn being the organism of choice until the Prodigene event drew public concern and regulatory changes were put in place in its wake.

A second theme concerns the dedication of farm equipment and facilities to the production of such crops. First, APHIS requires planters and harvesters to be dedicated to the test site for the duration of the tests, and although tractors and tillage attachments do not have to be dedicated, they have to be cleaned in accordance with APHIS protocols. The equipment and regulated articles must be stored in dedicated facilities for the field test's duration.

The final three requirements from the proposed rules concern procedural aspects of dealing with field tests of PMPs and plants producing industrial compounds. First, APHIS requires submission of cleaning procedures to minimize risk of seed movement. Second, procedures for seed cleaning and drying are required to be submitted and approved to confine plant material and minimize risk of seed loss or spillage. Finally, permittees will be required to implement an APHIS-approved training program to successfully comply with the stated permit conditions [35].

A key factor in any regulatory arrangement is the ability to ensure that those regulated are complying with the requirements set forth. As a result of the potentially contentious nature of PMPs and plant-produced industrial compounds, APHIS plans to increase the number of field site inspections "to correspond with critical times relevant to the confinement measures" [35]. Therefore, in addition to maintaining records of activities related to meeting permitting conditions, and increasing the likelihood of audit-ing them to verify required permit conditions were met, APHIS might inspect permitted field tests up to five times during the growing season: once at pre-planting to evaluate the site location, once at the planting stage to verify site coordinates and adequate cleaning of planting equipment, at midseason to verify reproduction isolation protocols and distances, at harvest to verify cleaning of equipment and their appropriate storage, and again at post-harvest to verify cleanup of the field site. In addition, two post-harvest inspections may occur to verify that the regulated articles do not persist in the environment. Finally, APHIS may inspect more frequently if deemed necessary [35].

As a result of the potential for both PMPs and PMIPs entering into the food supply, and using the PMP regulatory changes as a starting point, APHIS took immediate action to remove the notification track option, requiring complete permit track review in their recent (August 6, 2003) interim rule and request for comments [36].

The rationale given in the interim rule and request for comments was that although 14 field releases (9 notifications, 5 permits) had been carried out to date, the type of genetic engineering being carried out was to enhance such nutritional components as oil content. However, recent genetic modifications have been for "non-food traits with which APHIS has little regulatory experience or scientific familiarity" [36]. As such, the definition of PMIPs has three criteria: (1) the plants produce compounds new to the plant; (2) this compound has not normally been used in food or feed; and (3) the compound is being expressed for non-food/feed purposes [36].

An administrative reorganization of how USDA-APHIS regulates biotechnology created the Biotechnology Regulatory Services (BRS) to address concerns raised by PMPs and PMIPs specifically and genetically engineered organisms generally. According to USDA-APHIS, "Given the growing scope and complexity of biotechnology, now more than ever, APHIS recognizes the need for more safeguards and greater transparency of the regulatory process to ensure that all those involved in the field testing of GE crops understand and adhere to the regulations set forth by BRS."

Changes instituted by BRS include new training for APHIS inspectors in auditing and inspections of field trials, the use of new technologies such as global positioning systems, and analysis of historical trends to inform monitoring and inspection. According to APHIS, there are nine overarching goals that the changes will serve with key components being: (1) enhanced and increased inspections in which risk-based criteria, along with other factors, will be used to assess field test sites with higher-risk sites being inspected at least once a year, with other sites being randomly selected for yearly inspections; (2) auditing and verification of records of businesses and organizations to verify accuracy and implementation; (3) remedial measures to protect "agriculture, the food supply, and the environment in the event of compliance infraction" with the establishment of a "first-responder" group to deal with serious infractions; (4) standardized infraction resolution in which criteria will be established to determine the extent of an infraction and the response, whether further investigation, the issuance of a guidance letter, the issuance of a written warning, or referral to APHIS' Investigative and Enforcement Services (IES) unit for further action; (5) documentation, in which a database will be set up to track field test inspections and resulting compliance infractions, and transparency to keep stakeholders and the public informed on the regulatory decision-making process; (6) continuous process improvements where, as the science of biotechnology advances, so too will regulations and permit conditions to allow safe field testing; (7) an emergency response protocol, being developed with input from EPA and FDA, in which a quick response plan will be put in place "to counteract potential impacts on agriculture, the food supply, and the environment"; (8) training for field test inspectors in their dealings with PMP and PMIP field test sites, as well as the latest in auditing; and (9) certification concerning compliance with the highest level of auditing standards [37].

Although the reorganization can be seen as streamlining and focusing enforcement efforts, the potential for unduly high levels of workload stresses placed on this 26-member unit can be foreseen. First, BRS draws on APHIS inspectors to inspect field tests; however, more than 2600 of these agriculture quarantine inspectors have been transferred to the Department of Homeland Security [36]. The current agreement between USDA-APHIS and DHS allows for continued access by APHIS and BRS, although it can be expected that problems might occur as a result of split responsibilities and duties.

Implementation considerations are still an issue. Although thousands of field tests have been carried out in a variety of sites through the USDA-APHIS field permitting, notification, and petitioning program, these tend to be short-term tests, typically lasting for one growing season and collecting limited data that considers the ecological effect. Little work has been done in the way of long-term effects on the ecology of the regions in which they are sited.

Also, although genetically engineered plants have been in the fields for at least a decade, in the case of herbicide-tolerant and pest-resistant plants, their management has left something to be desired with large numbers of farmers not following federal regulatory standards [38, 39]. Although USDA states that of the 7402 field tests carried out between 1990 and 2001 and regulated by APHIS, only 115 resulted in compliance infractions and, for the most part, these were relatively minor infractions that did not raise public concern [40], concerns still exist. However, regarding PMPs, because of the regulations concerning them and the potential for controversy, it is likely that the regulations will be more stringently followed.

12.3.6 CONCLUSIONS

The "boundary technology" nature of PMPs places it firmly into limbo in terms of social acceptability and, ultimately, economic viability. In terms of benefits, the opportunities afforded by this subset of genetic engineering of plants are in terms of increased diversity of products, reduced production costs, scalability on relatively short notice, and reduced potential for health risks of toxicity and pathogenicity because of less-easily crossed species barriers. The risks of unintentional movement of genes into other plants or the entry of PMP products into the food supply, however unrealized to date, loom large. These threats are especially salient in light of likely scaling up of plantings of PMPs and greater diversity of PMP products. And when current regulatory difficulties are taken into account, it is an especially daunting challenge.

Although the potential benefits of PMPs are large and the risks may be perceived as negligible, even if such products enter the food supply as feared, public opinion looms large in whether this boundary technology is accepted. Given that most Americans are not aware of their current consumption of genetically engineered foods and the extent of their market presence, PMPs entering the food supply would likely be their first "real" exposure to the risks of genetically engineered plants. The resulting response—apathy, hysteria, or somewhere in between—would have an extremely large impact on the future of PMPs and their growth, which is not to discount the effect of other factors that may hasten or retard the development of PMPs. Economic climate, government policies and regulations, international trade, and a slew of other factors will impact the future of PMPs. What remains is the question of how far and how fast this technology will develop.

REFERENCES

1. NCHS (National Center for Health Statistics) (2005). FASTATS—Drug Use Health Facts. Available: http://www.cdc.gov/nchs/drugs/htm. Accessed 7 November 2005.

2. Federal Interagency Forum on Aging Related Statistics (2005). Federal Forum Reports Americans Aging Well, But Gaps Remain. Press release. Available: http://www.aging-stats.gov. Accessed 7 November 2005.

3. Howard J A, Donnelly K C (2004). Quantitative safety assessment model for transgenic protein products produced in agricultural crops. *J. Agricul. Environ. Ethics.* 17:545–558.

4. PEW (Pew Initiative on Food and Biotechnology) (2004). Factsheet: Genetically Modified Crops in the United States. Available: http://pewagbiotech.org/resources/factsheets/. Accessed 23 September 2005.

5. Hallman W K, Hebden W C, Cuite C L, et al. (2004). Americans and GM Food: Knowledge, Opinion and Interest in 2004. (Publication number RR-1104-007). Food Policy Institute, New Brunswick, NJ.

6. PEW (Pew Initiative on Food and Biotechnology) (2004). *Public Sentiment About Genetically Modified Food.* Mellman Group, Inc./Public Opinion Strategies for the Pew Initiative. Pew Initiative on Food and Biotechnology: Washington, D.C. Available: http://pewagbiotech.org/research/2004update/overview.pdf. Accessed 27 May 2005.

7. Stewart P A, McLean W P (2005). Public Opinion Toward the First, Second and Third Generations of Plant Biotechnology. *In Vitro Cell Developmen. Biol.—Plant* 41(6): 718–724.

8. Elbehri A (2005). Biopharming and the food system: Examining the potential benefits and risks. *AgBioForum.* 8(1):18–25.

9. Delaney D E (2002). Choice of crop species and development of transgenic product lines. In E E Hood, J A Howard (eds.), *Plants As Factories for Protein Production*: Kluwer Academic Publishers: Boston, MA. pp. 139–158.

10. Larrick J W, Yu L, Naftzger C, et al. (2002). Human pharmaceutical products in plants. In E E Hood, J A Howard (eds.), *Plants As Factories for Protein Production.* Kluwer Academic Publishers: Boston, MA. pp. 79–102.

11. Jilka J M, Streatfield S J (2002). Animal health. In E E Hood, J A Howard (eds.), *Plants As Factories for Protein Production.* Kluwer Academic Publishers: Boston, MA. pp. 103–118.

12. BIO PMP&AMP Update: Quarterly News (2005). Lactoferrin and Lysozyme may address health concerns in the developing world. *Biotechnology Industry Organization.* 1–2.

13. SemBioSys Website (2005). Pipeline: Pharmaceutical Products. Available: http://www.goodmedia.com/equicom/sembiosys/Main.aspx. Accessed 2 November 2005.

14. Planet Biotechnology—Products. Available: http://www.planetbiotechnology.com/products.html. Accessed 7 November 2005.

15. Prodigene. Prodigene Brings Plant-Produced Recombinant Trypsin to Market. Available: http://www.prodigene.com/news_releases/04-04-28_TrypZean(tm).html. Accessed 7 November 2005.

16. Prodigene. Prodigene Shows Lactogenic Immunity With an Oral Swine Vaccine. Available: http://www.prodigene.com/news_releases/04-06-23_TGEV.html. Accessed 7 November 2005.

17. Chlorogen News. Chlorogen, Sigma-Aldrich Sign Joint Development Agreement to Produce First Ever Products from Chloroplasts. Available: http://www.chlorogen.com/news.htm. Accessed 7 November 2005.

18. NRC (National Research Council) (2004). *Biological Confinement of Genetically Engineered Organisms*. The National Academies Press: Washington, D.C.

19. OSTP (Office of Science and Technology Policy) (2002). Proposed Federal Actions to Update Field Test Requirements for Biotechnology Derived Plants and To Establish Early Food Safety Assessments for New Proteins Produced by Such Plants; Notice." *67 Fed. Reg.* 149:50577–50580.

20. Harl N, Moline S (2003). The StarLink situation. Ames: Iowa State University. White Paper. Available: http://www.exnet.iastate.edu/Pages/grain/publications/starlink.html. Accessed 5 June 2005.

21. Lin W, Allen E (2001). Starlink: Impacts on the U.S. corn market and world trade. Feed Yearbook/April 2001. Economic Research Service, USA, pp. 40–48. Available: http://www.ers.usda.gov/Briefing/Biotechnology/starlinkarticle.pdf. Accessed 20 August 2004.

22. Ellstrand N C (2003). Going to "great lengths" to prevent the escape of genes that produce specialty chemicals. *Plant Physiol.* 132:1770–1774.

23. Macilwain C (2005). U.S. launches probe into sales of unapproved transgenic corn. *Nat.* 434:807.

24. USDA (U.S. Department of Agriculture) (2003). USDA Announces Actions Regarding Plant Protection Act Violations Involving ProdiGene, Inc. Press Release. December 6. Available: http://www.usda.gov/news/releases/2002/12/0498.htm. Accessed 7 November 2005.

25. Fox J L (2003). Puzzling industry response to prodigene fiasco. *Nat. Biotechnol.* 21(1):3–4.

26. Nestle M (2003). *Safe food: Bacteria, biotechnology, and bioterrorism*. University of California Press: Berkeley, CA.

27. Reuters News Service (2005). Chronology-recent events in Bt-10 corn controversy. Available: http://www.reuters.com/newsArticle.jhtml? type=topNews&storyID=8669 951. Accessed 4 June 2005.

28. Herrera S (2005). Syngenta's gaffembarrasses industry and White House. *Nat. Biotechnol.* 23:514.

29. USDA (U.S. Department of Agriculture) (2005). What are the result of BRS' 2005 compliance investigations? Available: http://www.aphis.usda.gov/brs/compliance12.html. Accessed 5 December 2005.

30. Nature (2005). Don't rely on Uncle Sam. *Nat.* 434:807.

31. NRC (National Research Council) (2002). *Environmental Effects of Transgenic Plants: The Scope and Adequacy of Regulation*. National Academies Press: Washington, D.C.

32. Stewart P A, McLean W P (2005). Fear and hope over the third generation of agricultural biotechnology. *AgBioForum: J. Agrobiotechnol. Manag. Econom.* 7(3):133–141.

33. Einsiedel E F, Medlock J (2005). A public consultation on plant molecular farming. *AgBioForum.* 8(1):26–32.

34. Stewart P A, Knight A J (2005). Trends affecting the next generation of U.S. agricultural biotechnology: Politics, policy and plant made pharmaceuticals. *Technolog. Forecas. Soc. Change.* 72(5):521–534.

35. USDA (U.S. Department of Agriculture) (2003). USDA Announces Actions Regarding Plant Protection Act Violations Involving ProdiGene, Inc. Press Release. December 6. Available: http://www.usda.gov/news/releases/2002/12/0498.htm. Accessed 5 December 2005.

36. USDA (U.S. Department of Agriculture) (2003). Field testing of plants engineered to produce pharmaceutical and industrial compounds. *68 Fed. Reg.* 46:11337–11340.

37. USDA (U.S. Department of Agriculture) (2003). United States Department of Agriculture Pre-Briefing for reporters on USDA's Federal Register Notice on Field Testing of Pharmaceutical-Producing Plants. March 6. Available: www.usda.gov/news/releases/2003/03/084.htm. Accessed 4 June 2005.

38. Jaffe G (2003). Planting Trouble: Are Farmers Squandering Bt. Corn Technology? An Analysis of USDA Data Showing Significant Non-Compliance with EPA's Refuge Requirements. Center for Science in the Public Interest. Available: www.cspinet.org. Accessed 27 May 2005.

39. Stewart P A, Harding D, Day E (2002). Regulating the New Agricultural Biotechnology by Managing Innovation Diffusion. *Amer. Rev. Public Admin.* 32(1):78–99.

40. USDA (U.S. Department of Agriculture) (2003). Biotechnology Regulatory Services: Compliance and Enforcement. Agricultural Biotechnology Website. Available: www.aphis.usda.gov/brs/compliance. Accessed 27 May 2005.

APPENDIX I: PLANT-MADE PHARMACEUTICALS QUESTIONS

I would next like to ask you a few questions about plants genetically modified to produce pharmaceutical drugs. These plants are modified to produce compounds used in manufacturing vaccines for diarrhea, antibodies to fight cancer, and drugs to treat such illnesses as cystic fibrosis.

How much benefit do you believe you would personally get from this type of plant? (1, a great deal; 2, some; 3, a little; 4, none at all; 8, DK; 9, refused)

How likely is it that these types of plants might accidentally enter the U.S. food supply? (1, very likely; 2, somewhat likely; 3, neither likely nor unlikely; 4, somewhat unlikely; 5, very unlikely; 8, DK; 9, refused)

How likely would you be to eat food coming from this type of plant? (1, very likely; 2, somewhat likely; 2, neither likely nor unlikely; 4, somewhat unlikely; 5, very unlikely; 8, DK; 9, refused)

How worried would you be if you ate food coming from this type of plant? (1, very worried; 2, quite a bit worried; 3, moderately worried; 4, slightly worried; 5, not at all worried; 8, DK; 9, refused)

How angry would you be if you ate food coming from this type of plant without your knowledge? (1, very angry; 2, quite a bit angry; 3, moderately angry; 4, slightly angry; 5, not at all angry; 8, DK; 9, refused)

12.4

BIOSIMILARS

H. SCHELLEKENS, W. JISKOOT, AND D.J.A. CROMMELIN
*Utrecht Institute for Pharmaceutical Sciences, Utrecht University, Utrecht,
The Netherlands*

When a patent of a drug expires, competitors are allowed to introduce copies of the drug to the market. In the case of classic drugs, which are in general produced by chemical synthesis, a submission for marketing approval needs to contain data showing the generic to be chemically identical to the innovative drug, e.g., with spectroscopic and chromatographic separation techniques, and pharmaceutically acceptable [1]. In addition, in the case of oral products, the bioavailability of the generic product must be shown to be equivalent to that of the reference product as assessed by the results of pharmacokinetic studies demonstrating an equivalent rate and extent of absorption. Bioavailability studies are typically conducted in healthy volunteers in whom the geometric means for pharmacokinetic parameters such as area under the drug plasma concentration-time curve (AUC) and maximum observed plasma concentration (C_{max}) fall within an acceptable range (usually no more than $\pm 20\%$ with a 90% degree of confidence). These bioequivalence requirements make the assumption that two chemically identical products that demonstrate an identical pharmacokinetic profile will also have identical clinical effects. Thus, the randomized clinical trials conducted to demonstrate safety and efficacy that are required for the approval of innovator products are not necessary for the approval of conventional generics as the data generated with the original product can be extrapolated to the generic. Long-term experience with chemical drugs and the availability of sophisticated methods for analysis ensure that the generic products are safe and effective. Because the company marketing a generic product does not need to make the large investments necessary to develop and launch an innovative drug and can take advantage of the extensive experience and possible extensions of the indications gained during the period the drug was patent protected, the price of a generic is relatively low. The introduction of a generic also leads to

Handbook of Pharmaceutical Biotechnology, Edited by Shayne Cox Gad.
Copyright © 2007 John Wiley & Sons, Inc.

a price drop of the original product. Because reducing health costs is a political priority, most countries promote the introduction and use of generics. The competition with generics is also an important incentive for the innovator to improve the original poduct. So generics are an essential part of the innovation cycle of drugs.

For several reasons the classic generic paradigm can only partly be extrapolated to therapeutic proteins [2–5]. The major reasons are listed in Box 12.4-1. Protein pharmaceuticals are mostly large complex molecules showing heterogeneity. This heterogeneity can be due to natural processes in the host cells such as variations in glycosylation or protein clipping. Modifications of the product can also be introduced during production, purification, formulation, and storage. Besides these product-related impurities, host cells and (biological) materials used during production and purification may introduce process-related impurities.

Therapeutic proteins are manufactured by complicated processes, and all steps in production, purification, and formulation may influence the biological and clinicial properties of the final product. The different steps are, therefore, carefully monitored by analytical methods using in-house standards. These analytical methods have, in part, been developed (or at least refined) for each specific product. For many products and especially the first for which the patents will expire shortly, the production process has undergone continuous improvement based on manufacturing and clinical experience.

The features of a particular biopharmaceutical are the result of the basic characteristics of the molecule such as amino acid sequence and three-dimensional structure as well as the specific production, purification, formulation, and storage conditions (Box 12.4-2). To produce a biopharmaceutical of constant required quality, a company also needs the experience and the in-house standards to apply the methods used to analyze the structure of a given product. There are various guidelines of the European Medicine Evaluation Agency, the Food and Drug Administration, the Japanese Ministry of Health and Welfare, and the ICH, which require manufacturers to show that they control the production process and are capable of reproducibly manufacturing batches that not only meet product specifications, but also conform to the definition of the product as established through full characterization. Modifications of the established process are only accepted if the manufacturer can show that the product of the new process is comparable with the initially manufactured product. Comparability studies include revalidation of

---

BOX 12.4-1. CHARACTERISTICS OF THERAPEUTIC PROTEINS

- Size
  - 100–500 times larger than classic drugs
  - Cannot be completely characterized by physico-chemical methods
- Immunogenicity
- Structural heterogeneity
- Relatively high biological activity
- Relatively unstable

BOX 12.4-2. FACTORS INFLUENCING THE ACTIVITY OF THERAPEUTIC PROTEINS

- Gene and promotor
- Host cell
- Culture conditions
- Purification
- Formulation
- Storage and handling
- Unknown factors

BOX 12.4-3. WHAT IS IN A NAME?

- Biogenerics
- Second entry biologicals
- Subsequent entry biologicals
- Off-patent biotech products
- Multisource products
- Follow-up biologics
- Biosimilars
- Similar biological medicinal products

---

the process, reevaluation of process and product-related impurities, recharacterization of the product in side-by-side analyses by all available state-of-the-art methods, and stability studies with the product from the new process in addition to release testing. When considered relevant, these comparability studies also may include clinical studies examining the pharmacokinetic, pharmacodynamic, and immunogenic properties of the new product and some efficacy and safety studies as well.

Little, if any, of the expertise, analytical methods and in-house standards, specifics of the production process, historical process, and validation data or full characterization data required for comparability assessment of therapeutic proteins are available in the public domain. As a rule, they are proprietary knowledge. It is inconceivable, in most cases, that another manufacturer, on the basis of the patent or published data, is able to manufacture a protein pharmaceutical that can be assumed similar enough to the original innovative product that only a limited documentation of physico-chemical characteristics would be sufficient to show equivalence. In most cases only limited data are available in pharmacopoeial monographs and scientific papers. Moreover, even the most sophisticated analytical tools are not sensitive enough to fully predict the biological and clinical characteristics of the product.

Because the generic approach is not applicable to protein drugs, the term *biogenerics* is considered misleading. Other terms have been used over the years, which are listed in Box 12.4-3. Similar biological medicinal products is the official

TABLE 12.4-1. Expiration of Biologics in the United States (US) and European Union (EU)

Pioneer Company	Product	Indication(s)	US Patent/ Market Exclusivity Expires	EU Patent/ Market Exclusivity Expires
Genentech	Nutropin™ (somatropin)	Growth disorders	Expired	Expired
Abbott	Abbokinase™ (eudurase urokinase)	Ischemic events	Expired	Expired
Eli Lilly	Humulin™ (recombinant insulin)	Diabetes	Expired	Expired
Genzyme	Ceredase™ (alglucerase); Cerezyme™ (imiglucerase)	Gaucher disease	Expired	Expired
AstraZeneca	Streptase™ (streptokinase)	Ischemic events	Expired	Expired
Biogen/Roche	Intron A™ (IFN-alfa-2b)	Hepatitis B and C	2002	2003 (France) 2007 (Italy)
Serono	Serostim™ (somatropin)	AIDS wasting	2003	NA
Eli Lilly	Humatrope™ (somatropin)	Growth disorders	ODE*** 2003	NA
Amgen	Epogen™, Procrit™, Eprex™ (erythropoietin)	Anemia	2013	2004
Roche	NeoRecormon™ (erythropoietin)	Anemia	NA	2005
Genentech	TNKase™ (tenecteplase TNK-tPA)	Acute myocardial infarction	2005	2005
InterMune	Actimmune™ (IFN-gamma-1b)	Chronic granulomatous disease (CGD), malignant osteopetrosis	2005, 2006, 2012	2002, 2004
Genentech	Activase™, Alteplase™ (tPA)	Acute myocardial infarction	2005, 2010	2005
Chiron	Proleukin™ (IL-2)	HIV	2006, 2012	2005
Amgen	Neupogen™ (filgrastim G-CSF)	Anemia, leukemia, neutropenia	2015	2006

terminology in the European Union (EU), but biosimilars has become the preferred terminology both in scientific and regulatory discussions and publications and will also be used in this chapter.

Table 12.4-1 shows the expiration dates of the patents of several recombinant DNA-derived therapeutic proteins. The patents of products as interferon alpha and epoetin alpha have already expired, and others will follow shortly. Table 12.4-1 also shows the patent and data protection to last longer in the United States than in Europe. This partly explains why the regulations regarding the conditions for marketing authorization in the EU are more developed than in the United States.

In the EU legislation, which became effective in November 2005, biosimilars have been set apart from the generic drugs. Clinical data have legally become mandatory for a marketing authorization submission for biosimilars to the EMEA, which like innovative biotechnology products will need to be evaluated through a centralized procedure and not by national regulatory approval. However, the European regulators do not require a full dossier; the idea of extrapolation of data as with classic generics is maintained, as it is considered unethical to ask for a complete set of clinical data. The CHMP, the scientific committee of the EMEA, has issued several guidelines concerning the data required for marketing authorization. The main issues of these guidelines are listed in Box 12.4-4 [6–8].

Only biosimilars of therapeutic proteins marketed within the EU are allowed. A biosimilar marketing authorization submission should contain the same

---

BOX 12.4-4. MAIN ELEMENTS CHMP GUIDELINES CONCERNING BIOSIMILARS

- The concept of similar biological product is applicable to any biological medicinal product, but it is more likely applied to highly purified products, which can be thoroughly characterized
- In order to support pharmacovigilance monitoring, the specific product given to the patient should be clearly identified
- The active substance of the biosimilar product must be similar in molecular and biological terms to the active substance of the reference medicinal product e. IFN alpha 2a is not similar to IFN alpha 2b
- The same reference product throughout the comparability program
- The pharmaceutical form, dose, and route of administration of the biosimilar and the reference product should be the same
- If the reference product has more than one indication, the safety and efficacy for all indications have to be justified or demonstrated for each indication separately
- The clinical safety must be monitored on an ungoing basis after marketing approval
- The issue of immunogenicity should always be addressed, and its long-term monitoring is necessary

extensive data on quality and safety as an innovative protein drug. In addition, the submission needs a supplement showing similarity in quality safety and efficacy between the biosimilar and the same reference product.

A biosimilar is defined as any therapeutic protein by its own manufacturing process that may influence the characteristics of the product itself but also introduce specific process-related impurities. The producers of biosimilars therefore need to demonstrate consistency and robustness of their production process. They need to do formulation studies showing stability and compatibility, even if the formulation is identical to the reference product.

The generation of comparative data concerning physico-chemical characteristics and preclinical testing will be complicated because unformulated bulk material of the innovator protein is unavailable to the biosimilar manufacturer. So the competitor can only use formulated marketed material. For some analytical techniques, isolation of the protein from the formulation is necessary. This may induce modifications that can hamper the comparisons. Comparing with the biosimilar isolated from the same formulation is being used to compensate for these possible modifications.

The biosimilar producer needs to do at least one comparative trial, usually a clinical trial with a sufficient number of patients to give the statistical power to find differences in efficacy and safety. The results may be extrapolated to other indications if the mode of action in other disease conditions is identical and the toxicity profile comparable with the population suffering from the other conditions. An example could be interferon alpha 2. If the biosimilar is proven to be similar in efficacy with the innovator protein in chronic hepatitis C virus infection, it will also be active in hepatitis B. However, an extension to cancer indications such as hairy cell leukemia would be blocked by a different mode of action in cancer and a different toxicity profile in tumor patients. For example, the immunogenicity of interferon alpha 2 is significantly different in cancer patients compared with patients with viral infections.

An important issue in the regulatory guidelines is immunogenicity, which is considered a major difference between protein drugs and low-molecular-weight drugs [9–11]. According to the CHMP guidelines on biosimilars, the issue of immunogenicity always needs to be addressed. There are only limited possibilities for the prediction of immunogenicity *in vitro* or in animal studies. This implies that the immunogenicity always needs to be monitored in clinical trials. The assay strategy and the assays employed need to be validated. Moreover, the assays need to meet specifications that allow the detection of immunogenicity differences between the biosimilar and the innovative protein therapeutic.

Sometimes the incidence of the induction of antibody formation in patients is too low to be evaluated in the clinical trials of the biosimilar, but the consequences are so severe that this side effect needs to be excluded for a biosimilar as is the case with epoetin-associated pure red cell aplasia. This can only be accomplished by strict postmarketing surveillance protocols.

The companies marketing biosimilars will need to show their capacity to manage postmarketing protocols, not only to monitor rare side effects but also because the biological and clinical characteristics of therapeutic proteins cannot be completely predicted by physico-chemical analyses, and possible batch-associated side effects need to be identified.

Another specific problem of protein therapeutics is their relative instability, which calls for strict regimens in storage and handling. In principle, protein drugs need a cold chain that is maintained from the manufacturing plant until administration to the patient. The CHMP has indicated that biosimilar producers need to show that they control the distribution of their products.

Nearly all innovative proteins are being administered parentally. They are mainly prescribed by hospital medical specialists for sometimes relatively rare, knowledge intensive conditions needing individualized treatment and specific diagnostic facilities. The marketing of biosimilars needs therefore to be accompanied by specific information to ensure their proper and safe use.

Considering the large investments manufacturers have to do to develop, produce, distribute, and market biosimilars, a large price drop by their introduction will be unlikely. Unlike with classic generics, the market will not be conquered by price alone, but the biosimilar manufacturers and distributors need to show quality and added value of their products.

Another consequence of the relative modest price drop by the introduction of biosimilars on the expensive biotechnology-derived protein drug treatments will be the attractiveness for racketeers to counterfeit protein drugs. Several biosimilars have been produced for many years in countries where patent protection or enforcement is lacking. The quality of these products varies (Figure 12.4-1). Proteins as epoetin, growth hormone, and insulin are also popular as performance enhancers in sports. This is an important part of the illegal demand. Both unbranded products as well as counterfeits, which look like authorized products, have been discovered. Counterfeits of therapeutic proteins have also been detected in the regular medical supply chains of these products. And products as growth hormone are also popular among the illegal drugs offered by the Internet (Figure 12.4-2).

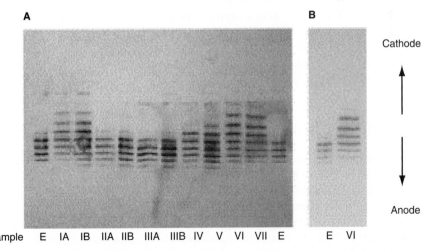

Figure 12.4-1. An example of biosimilar epoetins. This figure shows epoetin alpha's marketed in Asia and South America analyzed by iso-electrical focusing. The different products show large differences in isotype content. Difference can also be seen between different batches of the same manufacturer (IA and IB; IIA and IIB; IIIA and IIIB). E is the epoetin alpha marketed in Europe.

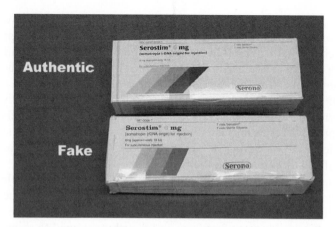

Figure 12.4-2. An example of a counterfeited growth hormone.

In Europe the first patents of protein drugs have expired and the legal conditions for launching biosimilars have been met, the general guidelines describing the regulatory demands have been implemented, and product-specific draft guidelines are under evaluation. The first marketing authorization requests have been submitted, and the marketing of these products will follow in 2006.

The biggest uncertainty concerning biosimilars is the level of similarity the regulatory bodies will find acceptable [12–13]. The only definition of a biosimilar that can be given is a protein that is similar in quality, safety, and efficacy to a reference product. Still, similarity is a matter of taste. Differences between products will always be found, and the question the regulatory agencies need to answer is how relevant these differences are. This is not an easy question because structurally comparable products can show significant differences in biological and clinical properties [14]. There are also examples of products with comparable clinical effects but considerably different structures. With room for interpretation about what a biosimilar is, legal procedures by the innovator or the generic industry challenging the decisions of the EU about marketing authorizations are to be expected. So, in the end the European Court of Justice will define what is a biosimilar.

REFERENCES

1. Crommelin D J, Storm G, Verrijk R, et al. (2003). Shifting paradigms: Biopharmaceuticals versus low molecular weight drugs. *Int. J. Pharm.* 266(1–2):3–16.

2. Crommelin D, Bermejo T, Bissig M, et al, (2005). Biosimilars, generic versions of the first generation of therapeutic proteins: Do they exist? *Contrib. Nephrol.* 149:287–294.

3. Schellekens H, Ryff J C (2002). "Biogenerics": The off-patent biotech products. *Trends Pharmacol. Sci.* 23(3):119–121.

4. Schellekens H (2005). Follow-on biologics: Challenges of the "next generation". *Nephrol. Dial. Transplant.* 20(suppl 4):iv31–36.

5. Schellekens H (2004). When biotech proteins go off-patent. *Trends Biotechnol.* 22(8):406–410.

6. Guidelines on comparability of medicinal products containing biotechnology-derived proteins as active substance: Quality issues. EMEA/CPMP/BWP/3207/00/Rev 1.

7. Guidelines on comparability of medicinal products containing biotechnology-derived proteins as active substance: Non-clinical and clinical issues. EMEA/CPMP/3097/02/Final.

8. Chamberlain P (2004). Biogenerics: Europe takes another step forward while the FDA dives for cover. *Drug Discov. Today.* 9(19):817–820.

9. Schellekens H (2002). Immunogenicity of therapeutic proteins: Clinical implications and future prospects. *Clin. Ther.* 24(11):1720–1740, discussion 1719.

10. Schellekens H (2002). Bioequivalence and the immunogenicity of biopharmaceuticals. *Nat. Rev. Drug Discov.* 1(6):457–462.

11. Schellekens H (2004). Biosimilar therapeutic agents: Issues with bioequivalence and immunogenicity. *Eur. J. Clin. Invest.* 34(12):797–799.

12. Schellekens H (2004). How similar do "biosimilars" need to be? *Nat. Biotechnol.* 22(11):1357–1359.

13. Combe C, Tredree R L, Schellekens H (2005). Biosimilar epoetins: An analysis based on recently implemented European medicines evaluation agency guidelines on comparability of biopharmaceutical proteins. *Pharmacother.* 25(7):954–962.

14. Nieminen O, Nordstrom K (2004). Regulation of biogenerics: A survey of viewpoints. *BioDrugs.* 18(6):399–406.

13.1

THE PROMISE OF INDIVIDUALIZED THERAPY

Michael Oettel

Jena, Germany

Chapter Contents

13.1.1	Background	1463
13.1.2	Drug Metabolizing Enzymes	1466
13.1.3	Drug Transporters	1469
13.1.4	Ion Channels	1471
13.1.5	Drug Receptors	1471
13.1.6	Cancer Drugs	1472
13.1.7	Cardiovascular Drugs	1473
13.1.8	Drugs Acting on the CNS	1474
13.1.9	Endocrinology	1475
13.1.10	Environmental Factors	1476
13.1.11	Ethnicity	1477
13.1.12	Technological Aspects	1478
13.1.13	Limitations	1480
13.1.14	Educational Aspects	1480
13.1.15	Concluding Remarks	1481
	References	1482

13.1.1 BACKGROUND

The rapid progress in molecular medicine has sought to understand the molecular basis of human disease with an ultimate goal of developing rationally designed therapies. The *gene discovery phase* has been largely driven by key technological advances including polymerase chain reaction (PCR) and other strategies and

Handbook of Pharmaceutical Biotechnology, Edited by Shayne Cox Gad.
Copyright © 2007 John Wiley & Sons, Inc.

methods, high-throughput sequencing, and bioinformatics. Now that the genome is sequenced, there are ongoing efforts to identify genetic polymorphisms (e.g., single nucleotide polymorphisms, SNPs) that may point to disease predisposition or unique responses to therapy such as untoward drug side effects.

Presently, physicians have to optimize a dosage regimen for an individual patient by a trail-and-error method. This kind of blind approach may cause adverse drug reactions (ADRs) in some patients. In fact, ADRs are found to occur in more than 2 million hospitalizations including approximate 100,000 deaths per year in the United States [1]. Similarly, according to a German study, about 6% of ADRs are attributed to new hospital admissions, and these ADRs were found to be preventable [2].

This interindividual variability in drug response could be caused by multiple factors such as disease determinants, genetic and environmental factors, and variability in drug target response (pharmacodynamic response) or idiosyncratic response. These factors affect drug absorption, distribution, metabolism, and excretion [3]. Drug concentrations in plasma can vary more than 600-fold between two individuals of the same weight on the same drug dosage [4]. An understanding of the variability in efficacy and toxicity of the same doses of medications in the human population, therefore, may provide safer and efficient drug therapy.

In general, genetic factors are estimated to account for 15–30% of interindividual differences in drug metabolism and response; but for certain drugs or classes of drugs, genetic factors are of the utmost importance and can account for up to 95% of interindividual variability in drug disposition and effects [4].

The idea that drug response is determined by genetic factors that alter pharmacokinetics and pharmacodynamics of compounds evolved in the late 1950s, when an inherited deficiency of glucose-6-phosphate dehydrogenase was shown to cause the severe hemolysis observed in some patients exposed to the antimalarial primaquine. This discovery explained why hemolysis was observed mainly in African-Americans, in whom the deficiency is common, and rarely in Caucasians of northern, western, and eastern descent [5]. In 1959, Vogel coined the term pharmacogenetics to describe inherited differences in drug responses [6].

Later in 1962, W. Kalow defined **pharmacogenetics** as ". . . the study of heredity and the response to drugs" [7]. It is a well-recognized fact that individuals respond differently to drug therapy; some drugs that are effective or well-tolerated by some people may be ineffective or toxic to others. This variability can often be traced by SNPs in genes encoding drug-metabolizing enzymes, transporters, ion channels, and drug receptors, all of which have been known to be associated with interindividual variation in drug response and have aroused considerable interest in recent years [8].

Taken together, pharmacogenetics is the study of genetic polymorphisms in drug-metabolizing enzymes and the translation of inherited differences to the variability in drug effects. Genes are described as "polymorphic" when allelic variants exist in the population, one or more of which alters the activity of the encoded protein compared with the wild-type sequence. Typically, the polymorphism leads to reduced activity of the encoded protein. Although the focus of pharmacogenetics is the study of drug-metabolizing enzymes like the CYP family, polymorphisms in drug transporters (such as P-glycoprotein) band drug targets (such as receptors) have also received attention.

The original task of pharmacogenetic research is to aid physicians in the prescription of the appropriate medicine in the right dosage in an attempt to attain maximum efficacy and minimum toxicity based on a genetic test, performed before the initiation of the therapy. The new paradigm moves toward the approach of screening for polymorphisms associated with a drug response and tailoring clinical and therapeutic decisions for an individual patient. This strategy of targeting drugs according to the patient's genetic constitution is termed "personalized medicine" [8, 9].

Toward the goal of personalized medicine, after completion of the human genome project, a haplotype map (HapMap) has been developed by the International HapMap Consortium with an intention of profiling DNA sequence variations across the human genome, which should provide a powerful tool to understand the genetic variants and drug responses (biomarkers). This knowledge may ultimately allow the development of personalized medications based on the genotype of each patient [10].

The term **"pharmacogenomics"** was introduced to reflect the transition from genetics to genomics and the use of genome-wide approaches to identify genes that contribute to a specific disease or drug response. A pharmacogenomics approach could allow a specific drug therapy to be targeted to genetically defined subsets of patients and could lead to a new disease and treatment classification on the molecular level.

Although the "blueprints" of human disease and for the drug action may be genetically encoded, the execution of the disease process and the pharmacodynamics of a given drug occurs through altered protein function. Researchers in molecular medicine are currently going from genomics to proteomics. One goal for clinical proteomics will be to characterize the information flow within single cells, tissues, or entire organisms under normal or disease conditions [11]. Therefore, identifying the genetic or epigenetic events leading to wanted or unwanted effects of drugs requires subsequent understanding of the proteomic consequences of these events. Therefore, **pharmacoproteomics** are focused on the influence of drugs on protein–protein interactions, their localization, or whether the encoded proteins are stable expressed, phosphorylated, cleaved, acetylated, glycosylated, or functionally "active." Mounting evidence confirms that the low-molecular-weight (LMW) range of circulatory proteome contains a rich source of information that may be able to detect or better characterize drug–protein interactions and stratify toxicological risk. Current mass spectrometry (MS) platforms can generate a rapid and high-resolution portrait of the LMW proteome. Emerging novel nanotechnology strategies to amplify and harvest these LMW biomarkers *in vivo* or *ex vivo* will greatly enhance our knowledge about pharmacoproteomics [12]. As an example for an pharmacoproteomic approach, it was found that in comparison with tamoxifen-sensitive breast tumor cells, in an tamoxifen-resistant line, 12 proteins were found upregulated, whereas nine were downregulated. Three of the identified proteins (AGL-2 interacting protein and two GDP-dissociation inhibitors) could be directly involved in the resistance phenomenon [13]. Curiously enough, the erythrocyte sedimentation reaction (ESR), which could be defined as a primitive forerunner of functional proteomics, has been used for a long time for clinical diagnostics and for monitoring of pharmacological therapy.

Although genetic variation is clearly important, it seems unlikely that personalized drug therapy will be enabled for a wide range of major diseases using genomic knowledge alone. Therefore, a third modern possibility for enabling personalized pharmacological therapy is the phenotyping by **pharmacometabonomics**. A major factor underlying interindividual variation in drug effects is variation in metabolic phenotype, which is influenced not only by genotype but also by different factors such as *in utero* effects, lifestyle, nutritional status, the gut microbiota, age, disease, environment, and the co- or pre-administration of other drugs. A new approach to personalized drug treatment is the examination of the metabolic profile. The profile, which is a measurement of small molecules such as sugars and amino acids, could be used to predict the pharmacodynamic or toxic response to drugs [14].

Pharmacometabonomics use a combination of pre-dose metabolite profiling and chemometrics to model and predict the responses of individual subjects. ^{1}H nuclear magnetic resonance (NMR) spectroscopy has been applied as a metabolite profiling tool for metabonomic studies, as it enables many endogenous metabolites to be quantified rapidly and reproducibly in biological fluids without derivatization or separation. Pharmacometabonomics has an important theoretical advantage over pharmacogenomics in that it can potentially take account of both genomic and environmental factors affecting drug-induced responses [15].

However, because the time of practical use of pharmacoproteomics and pharmacometabonomics is obviously more distant from today as such for the genetic approaches, we will focus this chapter more on pharmacogenetics and pharmacogenomics. The present knowledge certainly does not allow or recommend individualized therapy on the basis of pharmacoproteomics as well as pharmacometabonomics. They are not ready for profound optimization of clinical drug application or rationale drug development to become a reality in the near future. Intensification of research capacities is assumed, providing optimism for personalized medicine.

In the following sections, depending on the target respective, selected examples for the influence of genetic variability on the drug response are discussed. In most of the cases, we will cite publications from the first half of the year 2006, indicating the enormous progress in this field in the last months. For excellent reviews on individualized therapy solely from 2006, see Refs. 4, 8, 14, 16–25.

13.1.2   DRUG METABOLIZING ENZYMES

A considerable body of evidence suggests that SNP in genes encoding drug-metabolizing enzymes might determine drug efficacy and toxicity. The cytochrome P450 (CYP) enzymes consist of a superfamily of haem-containing monooxygenases, and multiple forms of CYP exist in all mammals. CYPs are responsible for the oxidation of many drugs, environmental chemicals, and endogenous substrates. They exist in the liver and in extrahepatic tissues such as the intestine, lung, and kidney. In humans, xenobiotics are metabolized primarily by three CYP subfamilies: CYP1, CYP2, and CYP3 [26, 27] (Table 13.1-1). Genetic polymorphisms of the genes for CYP2C9, CYP2C19, and CYP2D6 affect the metabolism of 20–30% of clinically used drugs. The frequency of variant alleles of CYP families varies among populations according to the race and ethnic background [28].

TABLE 13.1-1. Selected Drugs Metabolized by Various Cytochrome P450 (CYP) and Other Enzymes

Enzyme	Drug
CYP1A2	Imipramine, Tacrine, Propranolol
CYP1B1	17β-Estradiol, Estrone, Testosterone, Tamoxifen
CYP2B6	Carbamazepine, Efavirenz
CYP2C8	Paclitaxel, Carbamazepine
CYP2C9	Cyclosporine, Ifosfamide, Etoposide Nefazodone, Losartan, Warfarin, Tamoxifen, Phenytoin
CYP2C19	Omeprazole, Lansoprazole, Citalopram, Thalidomide
CYP2D6	Fluoxetine, Paroxetine, Desipramine, Amitryptiline, Imipramine, Nortryptilin, Maprotilin, Metoprolol, Propranolol, Tamoxifen
CYP 3A (responsible for the oxidative metabolism of more than 50% of clinically used drugs)	
CYP3A4	Amitryptiline, Imipramin, Cyclosporine, Docetaxel, Epipodophyllotoxin, Erythromycin, Losartan, Diazepam, Carbamazepine, Nefadozone, Nifedipine, Omeprazole, Terfenadine, Chlorpyrifos (Insecticide), Testosterone, Estradiol, Progesterone, and synthetic sexual steroids
CYP 3A5	Etoposide, Vincristine, Midazolam
CYP7A1	17α-Ethinylestradiol, Statins
CYP7B1	17α-Ethinylestradiol
CYP8B1	17α-Ethinylestradiol
Dihydropyrimidine Dehydrogenase (DPD)	Fluorouracil, Capecitabine, Nifedipine
Glucuronosyl transferase	Labetalol, Morphine, Naloxone
Glutathione—Transferase (GST)	
GSTA1	Cyclophosphamid
GSTM1	Anthracyclines (Doxorubicin, Idarubicin, Mitoxantrone)
GSTM3	Cisplatin
GSTP1	Etoposide, Doxorubicin, Cisplatin, Oxaliplatin, Carboplatin
GSTT1	Prednisolon
N-Acetyltransferase	Amonafide
S-Methyltransferase	Captopril
Sulfotransferase1A1 (SULT1A1)	Tamoxifen
Thiopurine Methyltransferase (TPMT)	6-Mercaptopurine, Azathioprine, Methotrexate
Uridine Diphosphate Glucuronosyltransferase (UGT)	Irinocetan, Etoposide, Epirubicine, Tipifarnib

For instance, there are 78 reported variants of CYP 2D6 that associated with adverse drug reactions. Many of these polymorphic genes encode inactive enzymes. However, these inactive enzymes may produce adverse drug reactions among patients because of their poor metabolic activity (e.g., the adverse effects of the neuroleptic risperidone) [29].

Similarly, several inactivating genetic polymorphisms have been reported in another member of the CYP family, namely CYP2C19 (CYP2C19*2 and CYP2C19*3), which is also associated with adverse drug reactions. This enzyme is responsible for the metabolism of proton pump inhibitors (e.g., omeprazole and lansoprazole) used for the treatment of gastric acid-related disorders such as peptic ulcer and gastroesophageal reflux disease. Approximately 2–4% of Caucasians and 4% of African-Americans have poor metabolism of these drugs [26]. Poor metabolizing patients of proton pump inhibitors carry two nonfunctional alleles, heterozygous extensive metabolizers have one nonfunctional and one wild-type allele, and extensive metabolizers are homozygous for the wild-type allele [4].

Another example is the coumarin warfarin, which is widely used for oral anticoagulation. Major side effects include bleeding complications. It was found that CYP2C9*2 and CYP2C9*3 alleles reduce the clearance of warfarin and increase the risk of bleeding [30]. The data indicate that patients carrying at least one variant CYP2C9 allele require lower maintenance doses and have a significantly higher risk of bleeding. However, SNPs in vitamin K epoxide reductase (VKORC1) may be more important. Recent studies have identified haplotype-dependent predictions for warfarin dosing. VKORC1 haplotype A predicted 21–25% of the required warfarin dose, and the inclusion of CYP2C9 genotypes reduced the required warfarin dose up to 31% in Caucasians. Combining nongenetic factors such as age, sex, body surface area, and drug interactions with the genotype information predicts up to 60% of warfarin dose. The remaining 40% of warfarin dosing variability remains unexplained [31].

CYP2C9*13 allele is associated with reduced metabolism of lornoxicam [32]. Similarly, CYP2C8 plays a role in the disposition of some therapeutic drugs [33].

The intestinal epithelium and liver contain the most abundant member of the CYP family, namely CAP3A, and these enzymes are responsible for the metabolism of more than half of the therapeutic drugs. Its activity also varies among members of a given population. In addition, this enzyme may undergo induction (rifamycins) and inhibition (calcium channel blockers) depending on the drug administration, which may account for its poor or higher metabolic activity. The interindividual variation in the immunosuppressive drugs cyclosporine and tacrolimus could be caused by interindividual differences in the expression of CYP 3A4 and 3A5 and the drug transporter P-glycoprotein. The most frequent allelic variant of CYP3A5 is CYP3A5*3, with a frequency of 87% of all alleles in a French population [34, 35].

An important P450 database in terms of genetic polymorphisms is the CYP alleles database (http://www.imm.ki.se/CYPalleles). However, genetic variants identified in the CYP3A4 and CYP3A5 genes have only a limited impact on the CYP3A-mediated drug metabolism [36], and hence the identification of the genotype for the ABCB1 transporter gene may provide further clues for the individualization of therapy with certain drugs (see below).

Apart from CYPs, some other enzymes involved in the biotransformation of drugs include N-acetyltransferase (NAT), glutathione-S-transferase, and uridine diphosphate-glucuronosyl transferase 1A1 (UGT1A1). One of the earlier discoveries of pharmacogenetics is the attribution of the neurological side effects of the antituberculosis drug isoniazid to genetic variability of NAT2. A 98.1% correlation between genotyping of NAT2 and acetylation phenotype has been demonstrated using an allele-specific PCR [37].

Gastric cancer patients treated with 5-FU and cisplatin and possessing the glutathione S-tranferase PI-105 Valine/Valine (GSTPI-105VV) genotype showed a response rate of 67% and medial survival time of 15 months compared with 21% and 6 months, respectively, in patients harboring one GSTPI-105 Isoleucine (GSTPI-1051) allele [38].

Pharmacogenetics of the UGT1A1 gene is known to influence irinotecan-induced diarrhea mediated via the glucuronidation of the active metabolite SN-38 [39]. Irinocetan is an inhibitor of topoisomerase used for the treatment of lung and colon cancer in adults and pediatric solid tumors such as rhabdomyosarcoma and neuroblastoma. The presence of seven repeats, instead of the wild-type number of six (UGT1A1*28) is associated with reduced UGT1A1 expression, leading to reduced SN-38 glucuronidation. It has been shown that the UGT1A1*28 allele leads to significantly increased amounts of SN-38 and a heightened risk of irinocetan-caused diarrhea and leukopenia. Patients homozygous or heterozygous for seven TA repeats have a sevenfold higher likelihood of diarrhea or leukopenia with irinocetan therapy than do patients with the wild-type genotype (six repeats). In a study conducted on Asians, UGT1A1*28 was found to be a common allele in Indians [40]. Thus, it is possible that UGT1A1 genotyping can be used to predict toxicity to irinocetan therapy. The UGT database contains data on genetic polymorphisms of the various UGT alleles (http://som.flinders.edu.au/FUSA/ClinPharm/UGT).

A good example for clinical reality of pharmacogenetic methods is the very frequently performed test for pseudocholinesterase. Inherited deficiency in pseudocholinesterase activity results in prolonged respiratory paralysis when deficient patients receive standard doses of the neuromuscular blockers suxamethonium or mivacurium [41].

13.1.3   DRUG TRANSPORTERS

Genetic variability in drug transporters also plays an important role for individual drug response (see Table 13.1-2). For instance, polymorphism in the ABC-binding cassette (ABC) gene may affect the function and expression of proteins [42] by inducing tumor cell resistance to anticancer therapy, altered disposition of chemotherapeutic agents, and associated chemotherapy toxicity, which may cause certain drug-induced side effects and uncertainty in treatment efficacy. ABCB1, also known as P-glycoprotein (Pgp) alias MDR1 for multidrug resistance, is a member of the ABC family that participates in the energy-dependent efflux of various substrates. A synonymous polymorphism in exon 26 (C345T) was found to influence drug response and exhibit interethnic variability [43]. The importance of haplotype analysis was shown by a meta-analysis of the influence of this SNP on digoxin pharmacokinetics and Pgp gene expression [44].

The ABCB1 genotype of the donor, but not of the recipient, may be a major risk factor for cyclosporine-related chronic nephrotoxicity in recipients of renal transplants. The ABCB1 3435TT genotype, which is associated with lower Pgp

TABLE 13.1-2. Selected Drug Transporters Associated with Individual Variations in Drug Response

Protein	Drug
P-Glycoprotein; Synonyms: Multidrug Resistance 1 (MDR1, ABCB1)	Digoxin, Vinblastine, Vincristine, Irinocetan
Multidrug Resistance-Associated Protein 2 (Canalicular-Multispecific Organic Anion Transporter, MRP2; ABCC2)	Doxorubicin, Vinblastine, Sulfinpyrazone, Irinocetan, SN38, Methotrexate
Breast Cancer Resistance Protein (BCRP, ABCBG2, ABCP)	Mitoxantrone, Methotrexate, Doxorubicin, Camptothecin-based drugs like Topocetan and SN-38
ABCG5/8	17α-Ethinylestradiol
Na$^+$/taurocholate cotransporting polypeptide (NTCP)	17α-Ethinylestradiol
Organic Anion Transporter Polypeptide-1B1 (OATP1B1; *SLCO1B1*)	Irinocetan
OATP1/2	17α-Ethinylestradiol
OATP1A2 (OATP-A)	Ajmalin, DHEA-Sulfate, 17β-Estradiol, Estrone, Microcystin, N-Methyl-quinidine, Ouabain, Prostaglandin E$_2$, Rocuronium, Thyroxin (T4), Triiodothyronine (T3)
OATP1B1 (OATP-C/OATP2)	Atorvastatin, Cerivastatin, Fluvastatin, Benzylpenicillin, Rifampicin
OATP1B3 (OATP8)	CCK-8, Digoxin, Fexofenadine, Paclitaxel

expression in renal parenchymal cells, is strongly associated with cyclosporine nephrotoxicity (odds ratio 13.4) [45]. On the other hand, the cyclosporin disposition in heart transplant patients may be influenced by Pgp haplotypes rather than genotypes [46].

Genetic polymorphisms in ABCB1 and ABCg2 may be important also in influencing the pharmacokinetics of irinotecan and its metabolites (see above).

Another notable example is that in certain patients the reduced rate of methotrexate metabolism produced a severe methotrexate overdosing and nephrotoxicity. This defect is attributed to the heterozygous mutation (R412G) in the highly conserved amino acid arginine [47] of the ABCC2 gene, which encodes the human multidrug resistant protein-2 (MRP-2). Interestingly, this mutated region is associated with substrate affinity and hence the mutant protein has a reduced rate of methotrexate elimination. In some other cases, a long-term use of methotrexate induces pancytopenia, which is determined by white blood cells and platelet counts [48]. However, it is also known that polymorphisms always need not have to produce functionally defective proteins. For example, in the multidrug resistant gene (MDR1), certain polymorphisms may not have any effect on the drug response [49]. However, this could be caused by nonsignificant statistical power.

Gwee et al. [50] have devised a rapid and robust assay to simultaneously screen SNPs of the MDR1 gene using a single-tube multiplex minisequencing strategy. Finally, there are several web-based transporter databases (http://www.tp-search.jp; http://www.gene.ucl.ac.uk/nomenclature/genefamily/abc.html and http://lab.digibench.net/transporter).

13.1.4   ION CHANNELS

Organic anion transporting polypeptides (OATPs) mediate the uptake of a broad range of compounds into cells. Substrates for members of the OATP family include bile salts, hormones, and steroid conjugates as well as drugs like the HMG-CoA-reductase inhibitors (statins), cardiac glycosides, anticancer agents like methotrexate, and antibiotics like rifampicin. The identification and functional characterization of naturally occurring variations in genes encoding human OATP family members is in the focus of transporter research [51]. There is a high degree of functional heterogeneity among OAT3 variants, with three variants (p.Arg149Ser, o. Gln239Stop, and p.Ile260 Arg) that results in complete loss of transporter function, and several other variants with significantly reduced function [52].

Common variation in the gene SCN1A affects the maximum dose of phenytoin and carbamazepine, which act on the sodium channel subunit encoded by this gene [53].

The etiology of drug-induced long QT syndrome (LQTS) could also be based on gene variability. A life-threatening form of cardiac arrhythmia has been associated with mutations in the ion channel genes [54]. Therefore, screening of LQTS-associated genes before the initiation of therapy with known QT-prolonging drugs could serve as a precautionary measure against life-threatening adverse effects by avoiding such drugs.

13.1.5   DRUG RECEPTORS

Genetic polymorphisms in drug receptors may alter pharmacological response. For example, variations in β-adrenoceptors, angiotensin-converting enzyme (ACE), and 5-hydroxytryptamine (5-HT) receptors alter drug response to β-adrenoceptor agonists and blockers, ACE inhibitors, and antipsychotic agents, respectively (see the sections cardiovascular dugs and drugs acting on the CNS in this chapter).

Genetic polymorphisms affecting amino acids at positions 16 and 27 within the β₂-adrenoceptor gene have been implicated in the asthma phenotypes and influence on the variability observed in response to use bronchodilator agents. It was found that Arg16 allele was slightly more frequent within the group with the unwanted tachyphylaxis phenomenon, whereas Gly16 allele carriers were overrepresented within the group of good responders (59.7%, $P = 0.0028$). On the other hand, the allele frequency of Gln27 and the proportion of Gln27 carriers was higher within the group with tachyphylaxis and Glu27 allele carriers were overrepresented within the group of good responders ($P = 0.026$) [55]. For asthma, see the section "Cardiovascular Drugs" in this chapter.

Concerning pharmacogenetic aspects of hormone receptors, see the "Endocrinology" section in this chapter.

13.1.6 CANCER DRUGS

Cancer chemotherapy is an area that requires continual monitoring and adjustment of antineoplastic agents to achieve optimal therapeutic outcome, which makes pharmacogenetic approaches attractive in this field.

Drugs such as azathioprines, mercaptopurines, and thioguanine have been used extensively to treat childhood acute lymphoblastic leukemia, rheumatic disease, inflammatory bowel disease, and used for solid organ transplantation. Thiopurine S-methyltransferase (TPMT) is a cytosolic enzyme that is involved in the metabolism of thiopurines (Table 13.1-1). It has been shown that polymorphisms in TPMT result in severe toxicity for patients prescribed with normal doses of the cytotoxic agents mercaptopurine or azathioprine. The variant enzyme was shown to misfold and subsequently form aggresome [56]. It has been reported that the TPMT genotype has a substantial impact on the mercaptopurine treatment response [57]. Previous studies also have shown that patients with homozygous mutant TPMT alleles exhibit very low enzyme activity and develop a severe hematopoietic toxicity after treatment with standard doses of thiopurines [58–60]. TPMT-deficient patients tend to accumulate excessive thioguanine nucleotide concentrations and are, therefore, at higher risk for hematological toxicity. TPMT deficiency can be largely attributed to three mutant alleles (TPMT*2, TPMT*3A, and TPMT*3C). Allele-specific PCR or PCR-restriction fragment length polymorphism strategies have been used to detect the three signature mutations, hence offering a rapid and affordable assay for identifying >90% of all mutant alleles [61]. Today, one of the most frequently performed pharmacogenetic tests are for TPMT. Patients who inherit two nonfunctional variant alleles should be given 6–10% of the standard dose of thiopurines. TPMT deficiency has also been associated with a high risk of irradiation-induced brain tumors in patients given thiopurines concomitantly with radiation therapy [62].

Similarly, the response rate of 5-fluorouracil (5-FU)-based treatment of advanced colorectal cancer is significantly linked to 677 C $\rightarrow$ T polymorphism in the methylenetetrahydrofolate reductase gene [63]. Additionally, polymorphisms in the thymidylate synthase gene promoter (TYMS enhancer region, TSER) has been linked to tumor downstaging in patients with rectal cancer who were treated preoperatively with 5-FU-based chemoradiation. In a study of 65 patients with stage T2–T4 rectal cancer, patients with at least one TSER*2 allele had a 38% increased frequency of tumor downstaging at the time of surgical resection compared with TSER*3/TSER*3 patients [64].

The first genotype-guided clinical trial in North America is based on TYMS TSER genotype. Rectal cancer patients (stage T3 and T4) with the "good risk" TSER*2 allele are treated in a phase II study consisting of standard therapy (radiation and 5FU). The sample size was calculated to detect a downstaging rate of 60%, compared with the historical downstaging rate of 45%. Patients homozygous for TSER*3 ("bad risk" genotype) are also enrolled in a phase II study, in which they receive the standard radiation and 5-FU along with additional irinocetan. The sample size was calculated to detect an improvement from the previously reported TSER*3/TSER*3 downstaging rate of 22–45%. Preliminary data implied an improved response rate in both treatment groups, suggesting an enrichment for positive response [22].

Genetic polymorphisms in the epidermal growth factor receptors (EGFR) impact on pharmacological response to gefitinib and erlotinib, tyrosine kinase inhibitors used as monotherapy in the treatment of metastatic non-small cell lung cancer. The drugs are effective in only 10–15% of patients. Responders were found to harbor activating mutations in the gene coding for EGFR [65]. The EGFR assay for tumor mutation analysis provides a platform in personalized therapy with the EGFR inhibitors gefitinib or erlotinib (http://www.dxsgenotyping.com).

The use of the other tyrosine kinase inhibitor imatinib, which blocks the enzymatic action of the BCR-ABL fusion protein, has represented a critical advance in chronic myeloid leukaemia (CML) treatment. However, a subset of patients initially fails to respond to this treatment. Use of complementary DNA (cDNA) microarray expression profiling, a set of 46 genes was differentially expressed in imatinib responders and non-responders. A six-gene prediction model was constructed, which was capable of distinguishing cytogenetic response with an accuracy of 80% [66].

Another case of already practical use of pharmacogenetic methods is the test for mutations in tumors that overexpress the human EGFR, HER2. Trastuzumab, a humanized monoclonal antibody, is effective in only 10–15% of breast cancer patients whose tumors overexpress HER2 [67]. Therefore, the pretreatment detection of HER2 is essential for the trastuzumab therapy.

13.1.7 CARDIOVASCULAR DRUGS

Variation in two genes encoding angiotensin-converting enzyme and endothelial nitric oxide synthase influence the effects of standard therapies [68]. In addition, polymorphism in the sodium channel gamma-subunit promoter region is significantly associated with blood pressure response to hydrochlorothiazide [69]. Similarly, SNPs in angiotensinogen (T1198C), apolipoprotein B (G10108A), and adrenoreceptor alpha 2A (A1817G) significantly predict the change in left ventricular mass during antihypertensive treatment [70]. Although common variants may influence the blood pressure response to a given class of antihypertensive medication, studies of polymorphisms have generally provided conflicting results [71]. For instance, polymorphisms in the alpha 2B adrenergic receptor does not show any association with azepexole hypertensive response [72]. However, patients with Gly 389 variant and Ser 49 homozygous of the beta-adrenergic receptor require increases in heart failure medication [73, 74].

One study has reported that the effect of statins in lowering low-density lipoprotein (LDL)-cholesterol levels was slightly greater in -204AA homozygotes of CYP7A1 [75].

Most of the studies to date have failed to demonstrate any link between polymorphism in tumor necrosis factor alpha and both cardiomyopathy and coronary artery disease [76].

In the case of asthma that causes substantial economic burden, morbidity, and mortality, patients exhibit an extensive interindividual variation in the response to beta-agonists acting at beta 2 adrenergic receptors, which could be caused by one nonsynonymous polymorphism [1772M9 of adenylyl cyclase type 9 (AC 9)] gene. This variation results in decreased catalytic activity (M772) and, therefore, alters albuterol (bronchodilator) responsiveness in the presence of a corticosteroid [77]. Additionally, in an Indian population, response to salbutamol treatment of asthmatic patients depends on polymorphism of the beta 2 adrenergic receptor [78].

13.1.8   DRUGS ACTING ON THE CNS

A meta-analysis of the quantitative contribution of CYP2D6 polymorphism to the interindividual variation in dosage of antidepressants has shown that the metabolism and dosage of imipramine, doxepin, maprotiline, trimipramine, desipramine, nortryptiline, clomipramine, and, partially, paroxetine depend on the CYP2D6 genotype and phenotype. Pharmacokinetic data suggest dose adjustments for these drugs that range from 28% to 60% of the normal dose for poor metabolizers and from 180% to 14% of the normal dose for ultrarapid metabolizers. In general, based on the impact of CYP2D6 on dosage adaption of antidepressants and antipsychotics, 40–50% of drugs may be subject to important pharmacokinetic alterations owing to CYP2D6 polymorphism [79].

A considerable variability also exists in efficiency and toxicity of other antipsychotic drugs. For instance, in the case of mood disorder, approximately 30–40% of patients do not completely respond to pharmacological treatment [80, 81]. However, serotonin transporter gene promoter (5HTTLPR) length polymorphisms has been implicated in the pathogenesis of mood disorders as well as in the therapeutic response to serotonergic drugs [82]. Reduction in the Liebowitz social anxiety scale and in the brief social phobia score during treatment with serotonin reuptake inhibitors (SSSRIs) was significantly associated with 5HTTLPR genotype [83]. When patients were treated with serotonin-blocking antidepressants, a significantly higher occurrence of side effects was found in patients with the HTTVNTR2.10/2.10 genotype (52.6%) than in patients with the 2.10/2.12 (12.5%) and 2.12/2.12 (0%) genotypes [84].

In patients with schizophrenia, Taq I polymorphism in the dopamine D2 receptor is associated with greater improvement of symptoms after treatment. Similarly, Gly 9 allele (Ser 9 Gly) of the dopamine D3 receptor and His 452 Tyr polymorphism in the 5-hydroxytryptamine 2A receptor (5-HT2A) are associated with response to clozapine. The side effects (weight gain) induced by antipsychotics seems to be associated with the -759C allele of the 5-HT2C receptor. Additionally, Gly 9-variant of dopamine D3, the 102C-variant of the 5-HT2A, and the Ser 23-variant of the 5-HT2C receptors (in females) seem to increase the susceptibility to tardive dyskinesia [85, 86].

Epilepsy is a difficult disease to treat because different patients require different ranges of doses, and some patients may even experience side effects such as increase in seizures, depression, and double vision. To control epilepsy, drugs such as phenytoin and carbamazepine have been extensively prescribed throughout the world. At present, evaluation of the allelic variation between individuals relies on the prior identification of candidate genes and their therapeutic effects of antiepileptic drugs [87]. Variants in the CYP2C9 and SCN1A (encodes a brain protein) genes are often found in patients treated with the highest doses of both phenytoin and carbamazepine [53]. Additionally, in Han Chinese, the carbamazepine side effects like Stevens–Johnson syndrome and toxic epidermal necrolysis are strongly associated with the HLA-B*1502 gene, which also means that genetic susceptibility to carbamazepine-induced cutaneous adverse drug reactions is phenotype-specific [88]. Pharmacoresistant epilepsy is still a major clinical problem in epilepsy, and it could be caused by multiple factors, but also multidrug transporters may play a key role in resistance phenotypes. However, studies on one variant in the ABCB1 gene provided inconclusive evidence so far [89].

Long-term treatment of Parkinson patients with L-Dopa exhibits L-Dopa-induced dyskinesis in some patients, which could be caused by genetic polymorphisms among patients. Therefore, pharmacogenetic studies may provide an explanation of neuronal plasticity among Parkinson patients [90].

Furthermore, drug addictions are major social and medical problems and therefore impose a significant burden on society. Epidemiological, linkage, and association studies have shown a significant contribution of genetic factors to the addictive diseases. Studies of polymorphisms in the mu-opioid receptors and transporter genes have contributed significantly to the knowledge of genetic influence on opioid and cocaine addiction and the efficacy of opioid therapy in pain management [91–94].

13.1.9 ENDOCRINOLOGY

At present, a lot of scientific efforts are being employed to use pharmacogenetics/pharmacogenomics for protein/peptidergic hormone therapy as well as for treatment with steroids. The main topics are polymorphisms in the membranous and nuclear receptors, including their subtypes and isoforms, and the genetics of steroid transforming enzymes (aromatases, 5α-reductases, sulfotransferases). It seems that, in comparison with other fields of pharmacological approaches, the pharmacogenetics in endocrinology are relatively advanced and, in certain parts, ready for clinical use.

Growth hormone receptor (GHR) transcripts exist in several isoforms in humans, among which is the retention (GHRfl) or exclusion (GHRd3) of exon 3, which encodes a 22-residue sequence in the extracellular domain of the membrane-located receptor. In Western Europe an populations, it has been estimated that 68–75% of alleles are GHRfl, whereas 25–32% are GHRd3. In short children, the homozygous or heterozygous presence of GHRd3 resulted in a significantly greater growth response in both year 1 and year 2 of GH therapy [95, 96]. Logically, patients who are homozygous for GHRd3 were less responsive to short-term and long-term hGH therapy [97].

Exogenous sexual hormones are used worldwide by women as oral contraceptives and hormonal replacement therapy. Some epidemiological studies have shown an increased risk of venous thromboembolism (VTE). It was found, that the risk/benefit ratio could be, in part, mediated by the genetic predisposition of women. Genetic thrombophilia might be implicated in the risk of VTE patients who use exogenous hormones.The most common causes of genetic hypercoagulability known today are factor V Leiden, G20210A prothrombin polymorphisms, and the genetic variant C677T of the methylenetetrahydrofolate reductase (MTHFR) gene. Therefore, an increasing number of kits for these two thrombophilic mutations are becoming commercially available, and screening for inherited thrombotic risk before giving the "pill" is among the most requested genetic tests in molecular diagnostic laboratories [98].

It seems that use of oral contraceptives or postmenopausal hormone replacement in women with the germ line mutations in the two genes BRCA1 and BRCA2 are more at risk for breast cancer than women also carrying these mutations but without hormonal interventions [99].

CYP1A2, CYP2C19, and CYP3A5 are responsible for estrone oxidation. These enzymes are all known to be genetically variant in the human population, and studies to asses the role of these CYP P450 enzymes in breast cancer risk are indicated [100].

The estrogen receptor-subtype ERα mediates the hepatotoxicity of 17α-ethinylestradiol (EE2). Upon EE2 treatment, ERα represses the expression of bile acid and cholesterol transporters (bile salt export pump, BSEP), Na^+/taurocholate cotransporting polypeptide (NTCP), OATP1, OATP2, ABCG5, and ABCG8 in the liver [101]. The genetic variability of some of these transporters is well known and could explain, at least in part, the interindividual differences for the tolerability of oral contraceptives.

Estrogen receptor α variations increase or decrease the action of estrogens; but, at present, no clinical studies are available that show that pharmacogenomic checking of ERα before estrogen-treatment (OCs or hormone replacement) can reduce the risk of adverse drug reactions.

The progesterone receptor 660L allele (PGR-12(rs1042638)V660L) may be associated with a moderately increased risk of breast cancer [102].

The polymorphism of UDP-glucuronosyl transferase (UGT2B17) is strongly associated with the bimodal distribution of the testosterone excretion [103].

Interestingly enough, besides the encouraging findings of genetics in the field of clinical endocrinology, proteomic approaches on the effects of estrogens, progestins, and androgens on the mammary gland are advancing step-by-step [13, 104, 105].

And what happens with diabetes mellitus? In 525 Caucasian type 2 diabetic patients, the common E23K variant of KCNJ11 encoding the pancreatic β-cell adenosine 5′-triphosphate-sensitive potassium channel subunit Kir6.2 was associated with increased risk for secondary failure to treatment with sulfonylurea-like glibenclamide [106].

13.1.10 ENVIRONMENTAL FACTORS

Needless to say, the genetic background and the gene variability is only one aspect of pharmacodynamics. Apart from potential gene–gene interactions, drug actions are also deeply affected by gene-environment interactions such that a particular genetic marker may present a variable pharmacological response in individuals with different nutritional states, lifestyle habits, and general well-being. Thus, the genetic information is not a reliable predictor of drug response, and therapeutic drug monitoring, with several notable exceptions, remains generally empirical. The sum and substance is as follows: Prescription genotyping serves more a predictive rather than a diagnostic role [8].

Here, we give only two examples for the role of the environment. Interindividual variability has been seen in liver UDP-glucuronosyltransferase 1A6 (UGT1A6) enzyme activity that glucuronates various drugs and toxins (Table 13.1-1). Its expression is associated with polymorphisms in the 5′-regulatory and exon 1 regions. The three most common non-synonymous polymorphisms are S7A, T181A, and R184S. However, it did not explain the interindividual variability in glucuronidation and alcohol consumption, which suggests that environmental factors may have a significant role in alcohol consumption [107, 108]. Similarly, alcohol dependence is not associated with single-nucleotide polymorphisms in the corticotrophin releasing hormone receptor 1 (CHRH 1) gene [109].

13.1.11 ETHNICITY

To use genomic knowledge to develop drugs and to improve health, we need to consider ethnical differences in different populations [110, 111]. There exists inter-ethnical differences in polymorphisms of genes encoding drug metabolizing enzymes, transporters, and disease-associated proteins [112, 113]. Meanwhile, a population genetics-based method to calculate the probability value for a variation in the gene is proposed [114]. Genetic differences are greater within socially defined racial groups than between other groups [115]. Additionally, it has been found that genetic diversity decreases in noncoding regions, whereas diversity of coding non-synomous SNPs is lower in regions containing a known protein sequence motif in individuals of European origin [116].

Drug treatment may be tailored for greater effect if important genetic variation exists between racial and ethnic groups. By knowing these variants, patients can be classified into low-, intermediate-, and high-dose groups [31, 117]. For instance, coumarins are characterized by a narrow therapeutic index and a wide interindividual variability in dose response; daily maintenance doses of warfarin range from less than 1 mg to over 20 mg [30] and, additionally, warfarin therapy shows a wide variation among patients of different ancestries. This variation could be caused by polymorphisms in the gene encoding vitamin K epoxide reductase complex 1. Accordingly, Chinese patients require lower dosages of heparin and warfarin than those usually recommended for white patients [118, 119]. Additionally, the combination of isosorbide dinitrate and hydralazine for treatment of heart failure in African-American heart patients reduced mortality by 43%, claiming that African-Americans and Caucasians differentially respond to the treatment, which is claimed to be because of genetic differences in the pathophysiology of heart failure between the two groups [120]. In other words, biological differences exist between the two racial groups. However, in this study, there is no comparison population and hence results should be interpreted cautiously. The distribution of haplotype profile of MDR1 (Pgp, ABCB1) has also been shown to exhibit inter-ethnic variabililty [121].

The ethical and moral concerns that develop in the midst of genetic testing are factors that hamper the development of personalized medicine. The deciphering of the genetic code may pose a threat to the protection of one's privacy. Moreover, some variants that predict drug response are also markers for disease predisposition. For example, the apolipoprotein E4 allele known to influence response to cholesterol-lowering (statin) is also associated with an increased risk of Alzheimer's disease, which may subsequently lead to medico-legal implications, such as the issue of data confidentiality and the possibility of stigmatization: whether employers and insurance companies should be given rights to asses the genetic data and the chance of the information falling into the hands of unauthorized parties. In addition, a positive result for a genetic determinant underlying therapeutic failure for a critical illness may inflict additional emotional trauma and dampen the willpower of the patient to combat the disease [8].

Nevertheless, the debate on the biological basis of race and ethnicity and pharmacogenetics may provide a useful understanding of ethnic and racial differences. Even in this case, however, we should not ignore several important parameters such as diet, economic, environmental, and psychosocial factors. However, pharmacogenetic studies on race and ethnicity are worthwhile because they are useful indicators of genetic variation. However, this kind of race and ethnicity classification for medical treatment could lead to discrimination [122].

13.1.12   TECHNOLOGICAL ASPECTS

Mutation screening technologies can generally be categorized into mass screening for novel variants and specific genotyping approaches. Although the former approach is more often adopted in academic research to uncover novel mutations and unravel functional consequences, the latter is more suited for practical diagnostic purposes. Rapid, precise, and cost-effective high-throughput technological platforms are essential for performing large-scale mutational analysis of genetic markers. However, genotyping techniques have generally been laborious in nature, rendering large-scale analysis time-consuming and inefficient from a cost perspective. However, SNP detection technologies have recently evolved to some of the most highly automated, robust, and affordable methods in biomedical research. Genotyping is often performed in conjunction with phenotyping (e.g., pharmacometabonomics) [8].

It is not the task of this chapter to discuss the advantages or disadvantages of the different technological platforms and bioinformatics tools in detail. Only a short presentation is possible (for review, see Ref. 8). Commercially provided services, such as Signature Genetics (http://www.signaturegenetics.com) are based on the analysis of integrated results from a detailed genetic test and comprehensive questionnaire. The report addresses the efficacy and toxicity of medications, potential drug interactions, and customized information on nutrition and recommended lifestyle modifications. Another example of a personalized medicine company (DxS, Manchester, U.K.) is focused on SNP testing, haplotyping, and clinical genotyping. DxS has applied the Amplification Refractory Mutation System and Scorpions (a homogenous fluorescent PCR detection system) technologies to the development of a highly sensitive oncology test panel [123].

The Roche AmpliChip CYP Genotyping test is FDA-approved and combines Roche's PCR amplification technology and Affymetrix high-density microarray technology to allow rapid, simultaneous analysis of multiple SNPs within the CYP family. This genotyping strategy relies on the hybridization of complementary fluorescent-tagged DNA sequences to an array of sequence-specific oligonucleotide probes   (http://www.roche-diagnostics.com/press_releases/archive/2003_06_25. html). Drug MEt is another microarray-based pharmacogenetic test used for simultaneous detection of 29 SNPs of CYP and phase II enzymes involved in drug metabolism (http://arrayit.com/Products/Microarray/DrugMEt/drugmet.html). Invader UGT1A1 Molecular Assay is an *in vitro* diagnostic test for genotyping UGT1A1 alleles and is FDA-approved. This assay appears to be an accurate method for the rapid detection of UGT1A1 polymorphisms (http://www.ons.org/fda/documents/FDA93995insert.pdf).

The TRUGENE Human Immunodeficiency Virus (HIV-1) Genotyping Kit and OpenGene DNA Sequencing System is yet another example illustrating the use of genetic testing in personalized medicine. TRUGENE is a sequence-based assay designed for detecting HIV genomic mutations (in the protease and part of the reverse transcriptase regions of HIV) that confer resistance to certain antiretroviral drugs. The assay has been shown to be robust, reproducible, and accurate and is considered a significant advance in the treatment of HIV infection [124].

New technologies like matrix-assisted laser desorption ionization-time-of-flight (MALDI-TOF) mass spectrometry (MS) and GOOD assay (requires no purification steps) have to bring in wider use. The next challenge is the demand for more accurate, economical, and large-scale technologies for SNP association studies. There are several review papers describing the technological platforms in pharmacogenetic research [8, 125–127].

The technological challenges for pharmacoproteomics are exceptional (see Table 13.1-3). Protein microarrays are an emerging class of nanotechnology for tracking many different proteins simultaneously. However, translation into the medical practice is very slow. On the other hand, proteomic changes in cultured cell lines might not fully reflect pharmacodynamic interactions because of the lack of the tissue microenvironment [11].

Although the molecular genotyping and phenotyping techniques are well established in major research institutes, the facilities for genetic testing and measurement of parent and metabolic concentrations are not always accessible in the diagnostic laboratory. Furthermore, mutational screening using the current state-of-the-art technology is still laborious and time-consuming. The hassle of having to courier samples to an external laboratory and the turnaround time for sample processing diminish the feasibility of adopting the approach in the fast-paced healthcare setting.

TABLE 13.1-3. The Hurdles for Protein Compared with Nucleic Acid Analysis in Human Tissue [11]

- More than 1 million proteins estimated (compared to 22.000 human genes)
- More than 300 posttranslational modifications known (phosphorylation, glycosylation)
- Wide dynamic range for protein abundances: 10^{10} (for mRNA:10^4)
- No protein amplification method available (comparable to PCR in genomics)
- Specific detection methods (only a few very good antibodies available)
- Senstivity (fM to aM needed, like ELISA)

PCR, polymerase chain reaction
ELISA, enzyme linked immunosorbent assay
fM, femtomolar (10^{-15} molar)
aM, attomolar (10^{-18} molar)

13.1.13 LIMITATIONS

A big problem limiting the progress of the pharmacogenetic approach is the fidelity of genotyping results and the confidence in associating SNPs with altered drug response. The ambiguity that sometimes develops in classifying an individual's genotype based on the laboratory results is another important contributory factor. In addition, the phenotyping method may give rise to false-positive results because the metabolic ratio can be influenced by other factors, such as epigenomic signaling, concomitantly administered drugs, nutritional state, and general health, hence affecting the validity of genotype-phenotype correlations [8].

The exact association between many SNPs of the drug target genes with therapeutic outcome is still unclear. However, few mutations have been characterized to ascertain their potential functional severity and limited definitive functional correlates established. As such, the functionality of these SNPs and their causative role remain largely speculative. In cases of functionally characterized SNPs, care should be exercised when translating research findings into an investigative tool, particularly in the extrapolation of observations from *in vitro* studies to a physiological effect. Even if the pharmacokinetic parameters are altered by genetic variants, the impact on pharmacodynamic or therapeutic effects may not be apparent. No standard guidelines exist on how the dosages of a drug should be adjusted, as a specific drug target may affect its panel of substrates to different extents, which increases the complexity of drug prescription because the dose adjustments differ among the various substrates in individuals carrying the same gene mutation [8].

Clinical use of pharmacogenetic testing has been severely limited by a lack of prospective clinical trials. Such trials are required to establish that pharmacogenetic testing benefits the selection of the appropriate drug and dose for the individual patient, thereby improving therapeutic responses or reducing ADRs. One key point that will affect the integration of pharmacogenetics into clinical practice will be the cost-effectiveness of these approaches, which may be influenced by several factors. Drugs with a narrow therapeutic index with more severer and expensive side effects are ideal candidates for pharmacogenetic testing. Drugs for which there are no established methods for monitoring adverse events (e.g., methotrexate) are also best-suited for pharmacogenomic analyses. However, for such approaches to be cost-effective, a well-established association should exist between genotype and clinical phenotype, and the frequency of the variant gene should be high. For example, if the frequency of a vriant allele is only 0.5%, then ~200 patients will have to be tested to identify one patient with the variant allele. Similarly, the strength of association between genotype and clinical phenotype will be important [128].

13.1.14 EDUCATIONAL ASPECTS

The resistance in the medical community to switching from the "trial-and-error" treatment approach to the gene-based approach is still prominently evident. Physicians in clinical practice, trainee physicians, and medical undergraduates have not been adequately educated in the field; the concept of pharmacogenetics has not been incorporated in the curriculum of medical courses worldwide. They are, thus, neither well-versed in the selection of target genes for ordering a genetic test nor equipped with the knowledge to interpret and analyze the report. Thus, the bridging of the gap between basic science and medicine requires the collaborative efforts of researchers and clinicians.

Professionals in medicine and the life sciences must be prepared to adapt to this new approach. However, systems-based pharmacogenomics is unlikely to be ready for clinical application in the near future. To benefit patients today, the already available options of pharmacogenetics should be carefully implemented in clinical practice as soon as possible. Teaching the current, continuously updated knowledge of pharmacogenomics should not be postponed until the new paradigm arrives [19].

13.1.15 CONCLUDING REMARKS

The well-known interindividual differences in drug response could be caused by genetic and environmental factors and by the dose-response curve of a given drug. Knowledge of the individual genetic variability in drug response is, therefore, clinically and economically very important. Pharmacogenetics (focus is on single genes) and pharmacogenomics (focus is on many genes) are the two recent developments to investigate interindividual variations of drug response. This type of genetic profiling of the population doubtless provides benefits for future medical care by predicting the individual drug response.

The field of pharmacogenetics/pharmacogenomics has seen exciting advances in the recent past. The Human Genome Project and International HapMap projects have uncovered a wealth of information for researchers. The discovery of clinically predictive genotypes (e.g., UGT1A1*28 for irinocetan therapy; TPMT alleles for avoiding severe ADRs if patients receive standard doses of mercaptopurine and azathioprine; TYMS TSER for treatment of rectal cancer; HER2 for optimizing the trastuzumab treatment in mammary Ca patients; BCA1, BRCA2, and factor V Leiden for improvement of oral contraception as well as hormone replacement), haplotypes (e.g., VKORC1 haplotype A for individualization of warfarin therapy), and somatic mutations (e.g., epidermal growth factor receptor for tailoring of tyrosine kinase inhibitors), along with the introduction of FDA-approved pharmacogenetic tests (UGT1A1*28) and the initiation of a genotype-guided clinical trial for cancer therapy (TYMS TSER in rectal cancer) have provided the first steps toward the integration of pharmacogenomics into clinical practice [22].

The translation from population-based (one dose fits all) to personalized medicine in the clinical setting is progressing at an incredibly slow pace. But why?

Several issues and problems need to be considered and solved before pharmacogenetics can be fully integrated into clinical practice (and also into drug development in the pharmaceutical industry). The ideal pharmacogenetic assay would quickly, accurately, and inexpensively provide composite genotypes for an individual patient to allow selection of the most suitable drug for the patient. Today, some approaches including suitable assays are very successful and have reached the level of clinical routine methods. Most other approaches (some are presently under investigation) are not mature. However, ongoing research is sure to bring one of the promises of the human genome project to fruition soon, that being individualized drug therapy. We should always keep in mind that, although in some cases polymorphism in a gene is associated with poor efficacy and adverse drug reactions, in many cases the clinical relevance remains to be understood or is irrelevant. Therefore, pharmacogenetics/pharmacogenomics may not be applicable to all diseases and all treatments [129]. However, the many pharmacogenomic complexities, and particularly time-dependent changes of gene expression, will never allow personalized medicine to become an error-free entity.

The suggestion that individuals will be genotyped at birth and their "HapMap genotype" carried lifelong as an implantable identity may be too Orwellian for some. However, it may be close to the reality of clinical pharmacology practice in decades to come. By the way, we will have the data needed to extract the genetic risk from a genome and effectively model dosages and the risks of adverse reactions based on an individual's genotype, which is further than most imagined electronic prescription would go, but it is an attractive prospect. Pharmacogenomics has a long way to go to achieve this goal, but its potential is clear enough [130].

However, at present, pharmacogenomics' practical impact on medicine is more or less minimal, and the greatest challenge is to understand the genotype-environmental factor interactions, extensive geographic variations in genes (ethnicity) and to optimize study design for the accuracy, high level of quality, and consistency of technologies [131].

In comparison with pharmacogenomics, the successful transition of proteomic technologies (including novel nanotechnology strategies) from research tools to integrated diagnostic platforms will require much more effort and much more time.

After the great enthusiasm about mastering of the Human Genome Project, at present, we are in the post-genetics skepticism, which should not block our efforts for bringing pharmacogenomics and pharmacoproteomics into the clinical practice, which, however, is a long and interesting road. Like other methods in medicine, pharmacogenomics as well as pharmacoproteomics will optimize only certain parts of pharmacological therapy, not the whole field of clinical pharmacology. Taken together, prescription genotyping is only intended to aid the doctor in making individualized therapeutic decisions and is not a substitute for a physician's judgment and clinical experience.

REFERENCES

1. Lazarou J, Pomeranz B H, Corey P N (1998). Incidence of adverse drug reactions in hospitalized patients: a meta-analysis of prospective studies. *JAMA.* 279:1200–1205.

2. Dorman H, Neubert A, Criegee-Rieck M, et al. (2004). Readmissions and adverse drug reactions in internal medicine: the economic impact. *J. Int. Med.* 255:653–663.

3. Zheng C J, Sun L Z, Han L Y, et al. (2004). Drug ADME associated protein data base as a resource for facilitating pharmacogenomics research. *Drug. Dev. Res.* 62:134–142.

4. Eichelbaum M, Ingelman-Sundberg M, Evans W E (2006). Pharmacogenomics and individualized drug therapy. *Annu. Rev. Med.* 57:119–137.

5. Beutler E (1969). Drug induced haemolytic anaemia. *Pharmacol. Rev.* 21:73–103.

6. Vogel F (1959). Moderne Probleme der Humangenetik. *Ergebn Inn Med Kinderheilk.* 12:52–125.

7. Kalow W (1962). *Pharmacogenetics, Heredity and the Response to Drugs.* WB Saunders, Philadelphia, PA.

8. Koo S H, Lee E D (2006). Pharmacogenetics approach to therapeutics. *Clin. Exp. Pharmacol. Physiol.* 33:525–532.

9. Webster A, Martin P, Lewis G, et al. (2004). Integrating pharmacogenetics into society: In search of a model. *Nat. Rev. Genet.* 5:663–669.

10. Lin M, Aquilante C, Johnson J A, et al. (2005). Sequencing drug response with HapMap. *Pharmacogenom. J.* 5:149–156.

11. Becker K-F, Metzger V, Hipp S, et al. (2006). Clinical proteomics: New trends for protein microarrays. *Curr. Med. Chem.* 13:1831–1837.

12. Calvo K R, Liotta L A, Petricoin E F (2005). Clinical proteomics: From biomarker and cell signalling profiles to individualized personal therapy. *Biosci. Reports.* 25:107–125.

13. Besada V, Diaz M, Becker M, et al. (2006). Proteomics of xenografted human breast cancer indicates novel targets related to tamoxifen resistance. *Proteom.* 6:1038–1048.

14. Haselden J N, Nicholls A W (2006). Personalized medicine in progress. *Nat. Med.* 12:510–511.

15. Clayton T A, Lindon J C, Cloarec O, et al. (2006). Pharmaco-metabonomic phenotyping and personalized drug treatment. *Nature.* 440:1073–1077.

16. Arnold H P, McHale D (2006). Pharmacogenetics: Developmental issues and solutions for safe and effective medicines. *Pharmacogen.* 7:149–155.

17. Bosch T M, Meijerman I, Beijnen J H, et al. (2006). Genetic polymorphisms of drug-metabolising enzymes and drug transporters in the chemotherapeutic treatment of cancer. *Clin. Pharmacokinet.* 45:253–285.

18. Kalow W (2006). Pharmacogenetics and pharmacogenomics: Origin, status, and the hope for personalized medicine. *Pharmacogen. J.* 6:162–165.

19. Lunshof J (2006). Teaching and practicing pharmacogenomics: A complex matter. *Pharmacogen.* 7:243–246.

20. Lunshof E J, Pirmohamed M, Gurwitz D (2006). Personalized medicine: Decades away? *Pharmacogen.* 7:237–241.

21. Mahlknecht U (2006). Pharmakogenomik: Prinzip and Perspektive. *Dtsch Med Wschr.* 131:310–313.

22. Marsh S, McLeod H L (2006). Pharmacogenomics: From bediside to clinical practice. *Human Molec. Genet.* 15:R89–R93.

23. Mendrick D L, Brazell C, Mansfield E A, et al. (2006). Pharmacogenomics and regulatory decision making: an international perspective. *Pharmacogen. J.* 6:154–157.

24. Shabo A (2006). Clinical genomics data standards for pharmacogenetics and pharmacogenomics. *Pharmacogen.* 7:247–253.

25. Shastry B S (2006). Pharmacogenetics and the concept of individualized medicine. *Pharmacogen. J.* 6:16–21.

26. Wilkinson G R (2005). Drug metabolism and variability among patients in drug response. *N. Engl. J. Med.* 352:2211–2221.

27. Emoto C, Iwasaki K I (2006). Enzymatic characteristics of CYP3A5 and CYP3A4: A comparison of *in vitro* kinetic and drug-drug interaction patterns. *Xenobio.* 36:219–233.

28. Xie H-G, Kim R B, Wood A J J, et al. (2001). Molecular basis of ethnic differences in drug disposition and response. *Ann. Rev. Pharmacol. Toxicol.* 41:815–850.

29. De Leon J, Susce M T, Pan R M, et al. (2005). The CYP2D6 poor metabolizer phenotype may be associated with risperidone adverse drug reactions and discontinuation. *J. Clin. Psychiatry.* 66:15–27.

30. Voora D, Eby C, Linder M W, et al. (2005). Prospective dosing of warfarin based on cytochrome P450 2C9 genotype. *Thromb. Haemost.* 93:7000–7005.

31. Rieder M J, Reiner A P, Gage B F, et al. (2005). Effect of VKORC1 haplotypes on transcriptional regulation and warfarin dose. *N. Engl. J. Med.* 352:1185–2293.

32. Guo Y, Zhang Y, Wang Y, et al. (2005). Role of CYP2C9 and its variants (CYP2C9*3 and CYP2C9*13) in the metabolism of lornoxicam in humans. *Drug. Metab.* 33:749–753.

33. Totah R A, Rettie A E (2005). Cytochrome P450 2C8: substrates, inhibitors, pharmacogenetics and clinical relevance. *Clin. Pharmacol. Ther.* 77:341–352.

34. Hesselink D A, van Gelder T, van Schaik R H (2005). The pharmacogenetics of calcineurin inhibitors: One step closer toward individualized immunosuppression? *Pharmacogen.* 6:323–337.

35. Rogausch A, Brockmoller J, Himmel W (2005). Pharmacogenetics in future medical care—implications for patients and physicians. *Gesundheitswesen.* 67:257–263.

36. He P, Court M H, Greenblatt D J, et al. (2005). Genotype-phenotype associations of cytochrome P450 3A4 and 3A5 polymorphism with midazolam clearance *in vivo. Clin. Pharmacol. Ther.* 77:373–378.

37. Zhao B, Seow A, Lee E J, et al. (2000). Correlation between acetylation phenotype and genotype in Chinese women. *Eur. J. Clin. Pharmacol.* 56:689–692.

38. Goekkurt E, Hoehn S, Wolschke C, et al. (2006). Polymorphisms of glutathione S-transferase (GST) and thymidylate synthase (TS)—novel predictors for response and survival in gastric cancer patients. *Br. J. Canc.* 94:281–286.

39. Iyer L, Das S, Janisch L, et al. (2002). UGT1A1*28 polymorphism as a determinant of irinocetan disposition and toxicity. *Pharmacogen. J.* 2:43–47.

40. Balram C, Sabapathy K, Frei G, et al. (2002). Genetic olymorphisms of UDP-glucuronosyltransferase in Asians: UGT1A1*28 is a common allele in Indians. *Pharmacogen.* 12:81–83.

41. Kalow W, Grant D M. Pharmacogenetics. In C R Scriver, A L Beaudet, W S Sly, (eds.), *The Metabolic and Molecular Basis of Inherited Disease.* McGraw-Hill, New York, pp. 293–326.

42. Lepper E R, Nooter K, Verweij J, et al. (2005). Mechanism of resistance to anticancer drugs: The role of the polymorphic ABS transporters ABCB1 and ABCG2. *Pharmacogenom.* 6:115–138.

43. Kafka A, Sauer G, Jaeger C, et al. (2003). Polymorphism C3435T of the MDR-1 gene predicts response to properative chemotherapy in locally advanced breast cancer. *Int. J. Oncol.* 22:1117–1121.

44. Chowbay B, Li H, David M, et al. (2005). Meta-analysis of the influence of MDR1 C3435T polymorphism on digoxin pharmacokinetics and MDR1 gene expression. *Br. J. Clin. Pharmacol.* 60:159–171.

45. Hauser I A, Schaeffeler E, Gauer S, et al. (2005). ABCB1 genotype of the donor but not of the recipient is a major risk factor for cyclosporine-related nephrotoxicity after renal transplantation. *J. Am. Soc. Nephrol.* 16:1501–1511.

46. Chowbay B, Cumaraswamy S, Cheung Y B, et al. (2003). Genetic polymorphisms in MDR1 and CYP3A4 genes in Asians and the influence of MDR1 haplotypes on cyclosporine disposition in heart transplant recipients. *Pharmacogen.* 13:89–95.

47. Hulot J S, Villard E, Maguy A, et al. (2005). A mutation in the drug transporter gene ABCC2 associated with impaired methotrexate elimination. *Pharmacogenet. Genomics.* 15:277–285.

48. Lim A Y, Gaffney K, Scott D G (2005). Methotrexate-induced pancytopenia: serious and under reported? Our experience of 25 cases in 5 years. *Rheumatol.* 44:1051–1055.

49. Sills G J, Mohanraj R, Butler E, et al. (2005). Lack of association between the C3435T polymorphism in the human multidrug resistance (MDR1) gene and response to antiepileptic drug treatment. *Epilepsia.* 46:643–647.

50. Gwee P C, Tang K, Chua J M, et al. (2003). Simultaneous genotyping of seven single-nucleotide polymorphisms in the MDR1 gene by single-tube multiplex minisequencing. *Clin. Chem.* 49:672–676.

51. König J, Seithel A, Gradhand U, et al. (2006). Pharmacogenomics of human OATP transporters. *Naunyn-Schmiedeberg's Arch. Pharmacol.* 372:432–443.

52. Erdman A R, Mangravite L M, Urban T J, et al. (2006). The human organic anion transporter 3 (OAT3; SLC22A8): Genetic variation and functional genomics. *A. J. Physiol. Renal. Physiol.* 290:F905–F912.

53. Tate S K, Depondt C, Sisodiya S M, et al. (2005). Genetic predictors of the maximum doses patients receive during clinical use of the anti-epileptic drugs carbamazepine and phenytoin. *Proc. Natl. Acad. Sci. USA.* 102:5507–5512.

54. Paulussen A D, Gilissen R A, Armstrong M, et al. (2004). Genetic variations of KCNQ1, KCNH2, SCN5A, KCNE1, and KCNE2 in drug-induced long QT syndrome patients. *J. Mol. Med.* 82:182–188.

55. Tellería J J, Blanco-Quirós A, Muntión S, et al. (2006). Tachyphylaxis to β_2-agonists in Spanish asthmatic patients could be modulated by β_2-adrenoceptor gene polymorphisms. *Reg. Med.* 100:1072–1078.

56. Wang L, Nguyen T V, McLaughlin R W, et al. (2005). Human thiopurine S-methyltransferase pharmacogenetics: Variant allozyme misfolding and aggresome formation. *Proc. Natl. Acad. Sci. USA.* 102:9394–9399.

57. Stanulla M, Schaeffeler E, Flohr T, et al. (2005). Thiopurine methyltransferase (TPMT) genotype and early treatment response to mercaptopurine in childhood acute lymphoblastic leukaemia. *JAMA.* 293:1485–1489.

58. McLeod H L, Krynetski E Y, Relling M V, et al. (2000). Genetic polymorphism of thiopurine methyltransferase and its clinical relevance for childhood acute lymphoblastic leukaemia. *Leukemia.* 14:567–572.

59. Schaeffeler E, Fischer C, Brockmeier D, et al. (2004). Comprehensive analysis of thiopurine S-methyltransferase (TPMT) phenotype-genotype correlation in a large population of German-Caucasians and identification of novel TPMT variants. *Pharmacogen.* 14:407–417.

60. Gearry R B, Barclay M L (2005). Azathioprine and 6-mercaptopurine pharmacogenetics and metabolite monitoring in inflammatory bowel disease. *Gastroenterol. Hepatol.* 20:1149–1157.

61. Yates C R, Krynetski E Y, Loennechen T, et al. (1997). Molecular diagnosis of thiopurine S-methyltransferase deficiency: Genetic basis for azathioprine and mercaptopurine intolerance. *Ann. Intern. Med.* 126:608–614.

62. Relling M V, Rubnitz J E, Rivera G K, et al. (1999). High incidence of secondary brain tumors related to radiation and antimetabolite therapy. *Lancet.* 354:34–39.

63. Etienne M C, Formento J L, Chazal M, et al. (2004). Methylenetetrahydrofolate reductase gene polymorphisms and response to fluorouracil-based treatment in advanced colorectal cancer patients. *Pharmacogen.* 14:785–792.

64. Villafranca E, Okruzhnov Y, Dominguez M A, et al. (2001). Polymorphisms of the repeated sequences in the anhancer region of the thymidilate synthase gene promoter may predict downstaging after properative chemoradiation in rectal cancer. *J. Clin. Oncol.* 19:1779–1786.

65. Lynch T J, Bell D W, Sordella R, et al. (2004). Activating mutations in the epidermal growth factor receptor underlying responsiveness of non-small-cell lung cancer to gefitinib. *N. Engl. J. Med.* 350:2129–2139.

66. Villuendes R, Steegmann J L, Pollán M, et al. (2006). Identification of genes involved in imatinib resistance in CML: a gene-expression profiling approach. *Leukemia.* 20:1047–1054.

67. Couzin J (2004). Pharmacogenomics: Cancer sharpshooters rely on DNA tests for a better aim. *Science.* 305:1222–1223.

68. McNamara D M (2004). Pharmacogenetics in heart failure: Genomic markers of endothelial and neurohumoral function. *Congest. Heart Failure.* 10:302–308.

69. Maitland-van der Zee A H, Turner S T, Schwartz G L, et al. (2005). A multilocus approach to the antihypertensive pharmacogenetics of hydrochlorothiazide. *Pharmacogenet. Genomics.* 15:287–293.

70. Liljedahl U, Kahan T, Malmqvist K, et al. (2005). Single nucleotide polymorphisms predict the change in left ventricular mass in response to antihypertensive treatment. *J. Hyperten.* 22:2273–2275.

71. Mellen P B, Herrington D M (2005). Pharmacogenomics of blood pressure response to antihypertensive treatment. *J. Hyperten.* 23:1311–1325.

72. King D, Etzel J P, Chopra S, et al. (2005). Human response to alpha 2-adrenergic agonist stimulation studied in an isolated vascular bed *in vivo*: Biphasic influence of dose, age, gender and receptor genotype. *Clin. Pharmacol.* 77:388–403.

73. Terra S G, Pauly D F, Lee C R, et al. (2005). Beta-adrenergic receptor polymorphisms and responses during titration of metoprolol controlled release/extended release in heart failure. *Clin. Pharmacol. Ther.* 77:127–137.

74. Taylor M R, Bristow M R (2004). The emerging pharmacogenomics of the beta-adrenergic receptors. *Congest. Heart Failure.* 10:281–288.

75. Hubacek J A, Bobkova D (2006). Role of cholesterol 7α-hydroxylase (CYP7A1) in nutrigenetics and pharmacogenetics of cholesterol lowering. *Mol. Diag. Ther.* 10:93–100.

76. Vadlamani L, Lyenger S (2004). Tumor necrosis factor alpha polymorphism in heart failure/cardiomyopathy. *Congest. Heart Failure.* 10:289–292.

77. Tantisira K G, Small K M, Litonjua A A, et al. (2005). Molecular properties and pharmacogenetics of a polymorphism of adenylyl cyclase type 9 in asthma: interaction between beta-agonist and corticosteroid pathways. *Hum. Mol. Genet.* 14:1671–1677.

78. Kukreti R, Bhatnagar P, B-Rao C, et al. (2005). Beta (2)-adrenergic receptor polymorphisms and response to salbutamol among Indian asthmatics. *Pharmacogen.* 6:399–410.

79. Kirchheiner J, Nickchen K, Bauer M, et al. (2004). Pharmacogenetics of antidepressants and antipsychotics: The contribution of allelic variations to the phenotype of drug response. *Mol. Psychiat.* 9:442–473.

80. Serrette A, Artioli P, Quartesan R (2005). Pharmacogenetics in the treatment of depression: pharmacodynamic studies. *Pharmacogenet. Genomics.* 15:61–67.

81. Sink K M, Holden K F, Yaffe K (2005). Pharmacological treatment of neuropsychiatric symptoms of dementia. *JAMA.* 293:596–608.

82. Rybakowski J K, Suwalska A, Czerski P M, et al. (2005). Prophylactic effect of lithium in bipolar affective illness may be related to serotonin transporter genotype. *Pharmacol. Rep.* 57:124–127.

83. Stein M S, Seedat S, Gelernter J (2006). Serotonin transporter gene promotor polymorphism predicts SSRI response in generalized social anxiety disorder. *Psychopharmacol.* 187:68–72.

84. Popp J, Leucht S, Heres S, Steimer W (2006). Serotonin transporter polymorphisms and side effects in antidepressant therapy—a pilot study. *Pharmacogen.* 7:159–166.

85. Reynolds G P, Yao Z, Zhang X, et al. (2005). Pharmacogenetics of treatment in first-episode schizophrenia: D3 and 5-HT2C receptor polymorphisms separately associate with positive and negative symptom response. *Eur. Neuropsychopharmacol.* 15: 143–151.

86. Wilffert B, Zaal R, Brouwers J R (2005). Pharmacogenetics as a tool in the therapy of schizophrenia. *Pharm. World Sci.* 27:20–30.

87. Ferraro T N, Buono R J (2005). The relationship between pharmacology of antiepileptic drugs and human gene variation: An overview. *Epilepsy Behav.* 7:18–36.

88. Hung S L, Chung W-H, Jee S-H, et al. (2006). Genetic susceptibilty to carbamazepine-induced cutaneous adverse drug rections. *Pharmacogen. Genomics.* 16:297–306.

89. Soranzo N, Goldstein D B, Sisodiya S M (2005). The role of common variation in drug transporter gene in refractory epilepsy. *Expert. Opin. Pharmacother.* 6:1305–1312.

90. Linazasoro G (2005). New ideas on the origin of l-Dopa-induced dyskinesias: Age, genes and neural plasticity. *Trends Pharmacol. Sci.* 26:391–397.

91. Stamer U M, Bayerer B, Stuber F (2005). Genetics and variability in opioid response. *Eur. J. Pain.* 9:101–104.

92. Kreek M J, Bart G, Lilly C, et al. (2005). Pharmacogenetics and human molecular genetics of opiate and cocaine addictions and their treatments. *Pharmacol. Rev.* 57:1–26.

93. Gourlay G K (2005). Advances in opioid pharmacology. *Support Care Cancer.* 13:153–159.

94. Mogil J S, Ritchie J, Smith S B, et al. (2005). Melanocortin-1 receptor gene variants affect pain and (micro)-opioid analgesin in mice and humans. *J. Med. Genet.* 42:583–587.

95. Rosenfeld R G (2006). Editorial: The Phasrmacogenomics of Human Growth. *J. Clin. Endocrinol. Metab.* 91:795–796.

96. Dos Santos C, Essioux L, Teinturier C, et al. (2004). A common polymorphism of the growth hormone receptor is associated with increased responsiveness to growth hormone. *Nat. Genet.* 36:720–724.

97. Jorge A A L, Marchisotti F G, Montenegro L R, et al. (2006). Growth hormone (GH) pharmacogenetics: Influence of GH receptor exon 3 retention or deletion on first-year growth response and final height in patients with severe GH deficiency. *J. Clin. Endocrinol. Metab.* 91:1076–1080.

98. Andreassi M G, Botto N, Maffei S (2006). Factor V Leiden, prothrombin G20210A substitution and hormone therapy: Indications for molecular screening. *Clin. Chem. Lab. Med.* 44:514–521.

99. Sade R B B, Chetrit A, Figer A, et al. (2006). Hormone replacement therapy is more prevalent among Jewish BRCA1/2 mutation carriers. *Eur. J. Cancer.* 42:650–655.

100. Cribb A E, Knight M J, Dryer D, et al. (2006). Role of polymorphic human cytochrome P450 enzymes in estrone oxidation. *Cancer Epidemiol. Biomarkers Prev.* 15:551–558.

101. Yamamoto Y, Moore R, Hess H A, et al. (2006). Estrogen receptor α mediates 17α-ethynylestradiol causing hepatotoxicity. *J. Biol. Chem.* 281:16625–16631.

102. Pooley K A, Healey C S, Smith P L, et al. (2006). Association of the progesterone receptor gene with breast cancer risk: A single-nucleotide polymorphism tagging approach. *Cancer Epidemiol. Biomarkers Prev.* 15:675–682.

103. Jakobsson J, Ekström L, Inotsume N, et al. (2006). Large differences in testosterone excretion in Korean and Swedish men are strongly associated with a UDP-glucurono-syl transferase 2B17 polymorphism. *J. Clin. Endocrinol. Metab.* 91:687–693.

104. Traub F, Hess R, Schorn K, et al. (2006). Peptidomic analysis of breast cancer reveals a putative surrogate marker for estrogen receptor-negative carcinomas. *Lab. Invest.* 86:246–253.

105. Zhang L, Liu X, Zhang J, et al. (2006). Proteome analysis of combined effects of androgen and estrogen on the mouse mammary gland. *Proteomics.* 6:487–497.

106. Sesti G, Laratta E, Cardellini M, et al. (2006). The E23K variant of KCNJ11 encoding the pancreatic β-cell adenosine 5'-triphosphate-sensitive potassiuj channel subunit Kir&.2 is associated with an increased risk of secondary failure to sulfonylurea in patients with type 2 diabetes. *J. Clin. Endocrinol. Metab.* 91:2334–2339.

107. Krishnaswamy S, Hao Q, Al-Rohaimi A, et al. (2005). UDP Glucuronosyltransferase (UGT) 1A6 pharmacogenetics: identification of polymorphisms in the 5'-regulatory and exon 1 regions, and association with human liver UGT1A6 gene expression and glucuronidation. *Pharmacol. Exp. Ther.* 313:1331–1339.

108. Krishnaswamy S, Hao Q, Al-Rohaimi A, et al. (2005). UDP glucuronosyltransferase (UGT) 1A6 pharmacogenetics: functional impact of the three most common non-synonymous UGT1A6 polymorphisms (S7A, T181A, and R184S) *Pharmacol. Exp. Ther.* 313:1340–1346.

109. Dahl J P, Doyle G A, Oslin D W, et al. (2005). Lack of association between single nucleotide polymorphisms in the corticotrophin releasing hormone receptor 1 (CRHR 1) gene and alcohol dependence. *J. Psychiatr. Res.* 39:475–479.

110. Daar A S, Singer P A (2005). Pharmacogenetics and geographical ancestry: Implications for drug development and global health. *Nat. Rev. Genet.* 6:241–246.

111. Rahemtulla T, Bhopal R (2005). Pharmacogenetics and ethnically targeted therapies. *BMJ.* 330:1036–1037.

112. Chowbay B, Zhou S, Lee E J (2005). An interethnic comparison of polymorphisms of the genes encoding drug-metabolizing enzymes and drug transporters: Experience in Singapore. *Drug Metab. Rev.* 37:327–378.

113. Mori M, Yamada R, KLobayashi K, et al. (2005). Ethnic differences in allele frequency of autoimmune disease-associated SNPs. *J. Hum. Genet.* 50:264–266.

114. Mitchell A A, Chakravarti A, Cutler D J (2005). On the probability that a novel variant is a disease-causing mutation. *Genome Res.* 15:960–966.

115. Kaplan J B, Bennett T (2003). Use of race and ethnicity in biomedical publication. *JAMA.* 289:2709–2716.

116. Freudenberg-Hua Y, Freudenberg J, Winantea J, et al. (2005). Systematic investigation of genetic variability in 111 human genes—implication for studying variable drug response. *Pharmacogen. J.* 5:183–192.

117. Hall A M, Wilkins M R (2005). Warfarin: A case history in pharmacogenetics. *Heart.* 9:563–564.

118. Yu C M, Chan T Y K, Critchley J A J H, et al. (1996). Factors determining the maintenance dose of warfarin in Chinese patients. *Q. J. Med.* 89:127–135.

119. Yu C M, Chan T Y, Tsoi W C, et al. (1997). Heparin therapy in the Chinese—lower doses are required. *Q. J. Med.* 90:535–543.

120. Taylor A L, Zieshe S, Yancy C, et al. (2004). Combination of isosorbide dinitrate and hydralazine in blacks with heart failure. *N. Engl. J. Med.* 351:2049–2057.

121. Tang K, Ngoi S M, Gwee P C, et al. (2002). Distinct haplotype profiles and strong linkage disequilibrium at the MDR1 multidrug transporter gene locus in three ethnic Asian populations. *Pharmacogen.* 12:437–450.

122. Harty L, Johnson K, Power A (2006). Race and ethnicity in the era of emerging pharmacogenomics. *J. Clin. Pharmacol.* 46:405–407.

123. Little S (2003). DxS Ltd. *Pharmacogen.* 4:97–101.

124. Jagodzinski L L, Colley J D, Weber M, et al. (2003). Performance characteristics of human immunodeficiency virus type 1 (HIV-1) genotyping systems in sequence-based analysis of subtypes other than HIV-1 subtype B. *J. Clin. Microbiol.* 41:998–1003.

125. Shi M M (2001). Enabling large-scale pharmacogenetic studies by high-throughput mutation detection and genotyping technologies. *Clin. Chem.* 47:164–172.

126. Jannetto P J, Laleli-Sahin E, Wong S H (2004). Pharmacogenomic genotyping methodologies. *Clin. Chem. Lab.* 42:1256–1264.

127. Koch W H (2004). Technology platforms for pharmacogenomic diagnostic assays. *Nat. Rev. Drug Discov.* 3:749–761.

128. Ranganathan P, McLeod H L (2006). *Arthritis Rheumat.* 54:1366–1377.

129. Lindpaintner K (2005). Pharmacogenetics and pharmacogenomics. *Methods Mol. Med.* 108:235–260.

130. O'Shaughnessy (2006). HapMap, pharmacogenomics, and the goal of personalized prescribing. *Br. J. Clin. Pharmacol.* 61:783–786.

131. Stoughton R B, Friend S H (2005). How molecular profiling could revolutionize drug discovery. *Nat. Rev. Drug Discov.* 4:345–350.

13.2

ENHANCED PROTEOMIC ANALYSIS BY HPLC PREFRACTIONATION

PIERRE C. HAVUGIMANA, PETER WONG, AND ANDREW EMILI
University of Toronto, Toronto, Ontario, Canada

Chapter Contents

13.2.1	Introduction	1491
13.2.2	Materials	1493
	13.2.2.1 Chemicals	1493
	13.2.2.2 Yeast Protein Extraction	1493
	13.2.2.3 IEX–HPLC Analysis	1494
	13.2.2.4 Precipitations and Proteolysis of Proteins in IEX–HPLC Fractions	1494
	13.2.2.5 1D-LC–MS Shotgun Tandem Mass Spectrometry Proteomic Analysis	1494
13.2.3	Method	1495
	13.2.3.1 HPLC Sample Preparation	1496
	13.2.3.2 HPLC Sample Prefractionation	1497
	13.2.3.3 LC–MS Sample and Data Analysis	1499
	References	1500

13.2.1 INTRODUCTION

Proteomics is often cited as the global (genome-wide) study of protein expression in a given living cell, tissue, or organism. Over the past decade, tandem mass spectrometry (MS/MS) has emerged as an essential analytical tool for large-scale

Handbook of Pharmaceutical Biotechnology, Edited by Shayne Cox Gad.
Copyright © 2007 John Wiley & Sons, Inc.

proteomic analysis [1, 2]. The critical challenge is to accurately and reproducibly identify and quantify large numbers of proteins in a given sample with good overall sensitivity and dynamic range. Currently, two shotgun (peptide level) tandem mass spectrometry-based proteomic approaches are widely employed to survey protein levels in complex biological samples. The first, known as MudPIT (for Multidimensional Protein Identification Technology), was developed by Yates and colleagues as an effective brute-force method for the comprehensive shotgun sequencing of large collections of peptide products of proteolytic digests of entire proteomes [3–5], whereas the second approach, prefractionation, requires sample separations to alleviate protein complexity before MS/MS analysis. In the latter approach, two avenues have generally been explored: gel electrophoresis (GE) and high-performance liquid chromatography (HPLC) pre-fractionations [6–8].

GE is a well-documented method [9] and is systematically used around the globe even though it is tedious, time-consuming, and often difficult to achieve reproducible results. On the other hand, HPLC prefractionation techniques [10, 11] such as size exclusion chromatography, reverse phase chromatography, affinity chromatography, and anion exchange chromatography represent more robust, simple-to-execute, rapid, and ultimately more reproducible sample prefractionation platforms that are also amenable to automation. These different HPLC methods can be characterized in terms of resolution as well as whether the method is denaturing. Choice of HPLC method will depend on the separation goals, with the primary goal usually being of obtaining the maximum nonoverlapping polypeptide fractions. Reverse phase HPLC, a popular and particularly effective separation method that fractionates proteins according to hydrophobicity, is typically used if sample denaturation (and the subsequent loss in protein activity) is not a consideration [12]. Alternatively, size exclusion chromatography, which separates proteins according to molecular size, is an excellent method to recover the proteins in their native conformation, albeit with relatively low resolution. Affinity chromatography is a powerful way to isolate proteins for which a biospecific interaction to a suitable selective adsorbent is available, but it is not suitable for the study of a whole proteome. The most flexible, convenient, and "widely used" method, however, for protein fractionation in biochemical studies is ion-exchange chromatography, or IEX–HPLC. Similar to size exclusion and affinity chromatography, this method also allows for the collection of proteins in their native conformations, without significant sample loss, but maintains the maximum possible resolution suitable for effective separations of complex biological samples. Surprisingly, despite its exciting potential, only limited use of IEX–HPLC in combination with proteomic analysis has been reported to date.

In IEX–HPLC, the resolution of a mixture of proteins is performed according to the differential retention due to inherent variations in the surface charge properties of the globular proteins in solution [13]. It involves the selective binding of target proteins to a suitably charged solid phase material or adsorbent, followed by sequential elution by increasing the mobile phase salt concentration (or, alternatively, under denaturing conditions, by altering the pH) to disrupt ionic interactions. Many such resins or prepacked columns are commercially available and can be used to efficiently resolve protein mixtures in a reasonable time frame. Anion exchange columns, which consist of positively charged functional groups, can be used to separate negatively charged proteins (which represent most typical cellular soluble proteins under near physiological pH), whereas cation exchange columns resolve positively (e.g., histidine, lysine, and arginine rich) proteins. In principle, the development of a method that combines the two types of columns should provide a more effective analytical separation tool for protein fractionations before proteomic analysis using MS/MS that could be applied to different types of biological samples.

In this chapter, we described an easily implemented, effective, and highly reproducible dual-column HPLC prefractionation method that we have developed for improving routine proteomic analyses. The approach results in multifold increases in the numbers of proteins that can be confidently identified by LC–MS of whole cell lysates without fractionation. We outline the key steps in the overall procedure, from sample preparation through to MS/MS and attendant data analysis, using yeast soluble protein extract as a test mixture.

13.2.2 MATERIALS

The chemicals, biologicals, equipment, and instruments listed are those routinely used in our laboratory for proteomics analyses. Substitutions could be made with equivalent items as available.

13.2.2.1 Chemicals

1. Tris-(hydroxymethyl)-aminomethane (TRIS base), Sodium chloride (NaCl), Trichloroacetic acid (TCA), Sodium azide (NaN$_3$), Ammonium bicarbonate (NH$_4$HCO$_3$), HPLC-grade water, Calcium chloride (CaCl$_2$), HPLC grade acetonitrile (ACN), and HPLC-grade glacial acetic acid (AA) were all purchased from Fisher Scientific (Whitby, ON, Canada).
2. Hydrochloric acid (6N HCl) was from VWR International (Mississauga, ON, Canada).
3. HPLC grade acetone was obtained from Sigma-Aldrich (Oakville, ON, Canada).
4. Bio-Rad dye reagent (Bio-Rad, Mississauga, ON, Canada).
5. Heptafluorobutyric acid (HFBA) was purchased from Pierce (Rockford, IL).

13.2.2.2 Yeast Protein Extraction

1. Standard laboratory *Saccharomyces cerevisiae* yeast strain W303 (Mata ade2-1, ura3-1, his3-11,15, trp1-1, leu2-3,112, can1-100) (Open Biosystems, Huntsville, AL).
2. Lysis Solution: 10-mM Tris-HCl, 30-mM NaCl, and 3-mM NaN$_3$ (pH 7.8) to minimize microbial growth in the buffer.
3. Protease cocktail inhibitor (Roche Diagnostic, Laval, QC, Canada).
4. Bovine serum albumin (BSA) (Sigma-Aldrich, Oakville, ON, Canada).
5. Silica beads (Fisher scientific, Whitby, ON, Canada).
6. Ice bucket.

7. Microtube vortex adaptor (placed in cold room).

8. Eppendorf centrifuge (cooled at 4°C) or placed in cold room.

9. 1.5-mL Eppendorf microcentrifuge tubes cooled at 4°C.

10. 1.0-mL plastic cuvettes (Bio-Rad, Mississauga, ON, Canada).

11. Ultrospec1100 pro UV/VIS Spectrophotometer (Biochrom, Cambridge, U.K.) or an equivalent instrument.

13.2.2.3   IEX-HPLC Analysis

1. 80–100-μg total soluble protein from yeast (W303) crude soluble cell lysate as measured by Bradford (approximate concentration: $[10 \mu g \cdot \mu L^{-1}]$).

2. Buffer A (binding buffer): 10-mM Tris-HCl, pH 7.5 with 3-mM NaN_3.

3. Buffer B (eluting buffer): Buffer A with 600-mM NaCl.

4. 0.22-μm membrane vacuum-driven disposable bottle top filter (Millipore, Cambridge, ON, Canada).

5. Anion exchange column (PolyWAX LP, 50 × 2.1 mm i.d., 5 μm, 1000 Å), cation exchange column (PolyCAT A, 50 × 2.1 mm i.d., 5 μm, 1000 Å), anion exchange guard cartridge (PolyWaX LP, 10 × 2.1 mm i.d., 5 μm, 1000 Å), and pre-guard column sieves with high flow, 2 μm (IT9085-20-10) were from Canadian Life Science (Peterborough, ON, Canada).

6. Agilent 1100 quaternary HPLC pump system (Agilent Technologies, Mississauga, ON, Canada).

7. Microcentrifuge tubes (precooled at 4°C).

13.2.2.4   Precipitations and Proteolysis of Proteins in IEX–HPLC Fractions

1. 100% TCA in HPLC-grade water: dissolve 20 g of solid TCA in 9.0 mL of water (prepare fresh solution).

2. Applied Biosystems immobilized trypsin beads (Applied Biosystems, Foster City, CA).

3. Pierce immobilized TPCK-trypsin beads (Pierce, Rockford, IL).

4. Digestion buffer: 50-mM NH_4HCO_3, 1-mM $CaCl_2$, pH 8.0.

5. 1-M $CaCl_2$ in HPLC-grade water.

6. Ice cold HPLC-grade acetone (keep at –20°C until use).

7. Eppendorf centrifuge 5417C (14,000 rpm capacity) or its equivalent.

13.2.2.5   1D-LC-MS Shotgun Tandem Mass Spectrometry Proteomic Analysis

1. An LTQ linear ion trap tandem mass spectrometer (Thermo Finnigan Corp, San Jose, CA) is routinely used in our laboratory. XCalibur Software is used to automate the acquisition of tandem mass spectra over a 400–1600-m/z range and to control precursor ion selection by the instrument in a data-dependent, data-acquisition mode, with dynamic exclusion activated.

2. 100-μm capillary microcolumn silica tubing (Polymicro Technologies, Phoenix, AZ).

3. 5-μm pore size C18 reverse-phase packing material (Zorbax eclipse XDB-C_{18} resin) (Agilent Technologies, Mississauga, ON, Canada).

4. Solvent A: 5% ACN, 0.5% AA, and 0.02% HFBA.

5. Solvent B: 100% ACN.

6. HPLC quaternary gradient pump, with flow splitter.

13.2.3   METHOD

The following method is designed to allow the routine prefractionation, leading to enhanced identification of proteins from complex sample matrices, such as a representative crude soluble cell extract prepared from a standard laboratory *S. cerevisiae* yeast strain, by IEX–HPLC and LC–MS techniques. This method can be readily customized to investigate other biological samples. Specific details regarding sample preparation, chromatography prefractionation, precipitation, and tryptic proteolysis of the protein samples followed by proteomic LC–MS analysis are included. The major steps of the entire procedure, from protein sample mixture prefractionation to subsequent analysis by LC–MS, are summarized schematically in Figure 13.2-1.

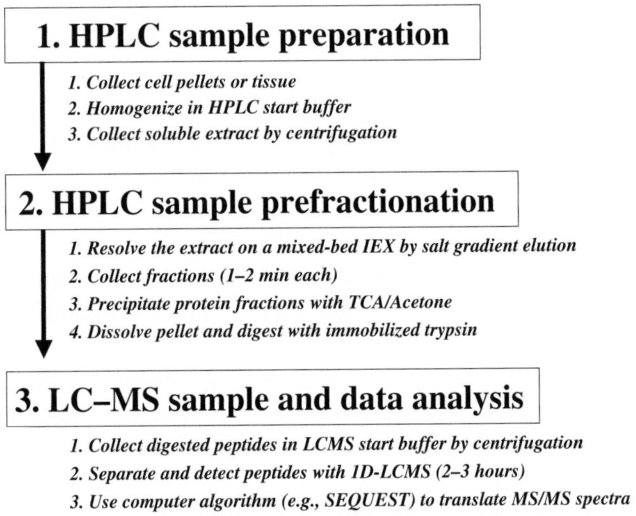

1. HPLC sample preparation

1. Collect cell pellets or tissue
2. Homogenize in HPLC start buffer
3. Collect soluble extract by centrifugation

2. HPLC sample prefractionation

1. Resolve the extract on a mixed-bed IEX by salt gradient elution
2. Collect fractions (1–2 min each)
3. Precipitate protein fractions with TCA/Acetone
4. Dissolve pellet and digest with immobilized trypsin

3. LC–MS sample and data analysis

1. Collect digested peptides in LCMS start buffer by centrifugation
2. Separate and detect peptides with 1D-LCMS (2–3 hours)
3. Use computer algorithm (e.g., SEQUEST) to translate MS/MS spectra

Figure 13.2-1. Representative workflow for IEX–HPLC prefractionation of a proteome sample (e.g., the yeast *S. cerevisiae*) before proteomic analysis by LC–MS. (1) Sample preparation is performed in IEX–HPLC starting buffer (pH 7.8) containing protease inhibitors. The soluble extract is recovered and clarified by centrifugation. (2) The sample is prefractionated using IEX–HPLC salt gradient, and a total of 27 fractions (1.3 minutes each) is collected using a dual-column system (tandem anion and cation-exchange columns placed in series) and a standard HPLC binary pump system. The collected protein fractions and a crude soluble sample lysate, as a reference control, are subjected to precipitation and tryptic digestion. (3) The peptide mixtures are separated and detected by single-dimension liquid chromatography coupled to automated tandem mass spectrometry. The resulting tandem mass spectra are computationally translated to a short amino acid sequence, and their corresponding cognate protein identity is determined using a suitable database algorithm searching against a relevant protein sequence database. High confidence protein identifications ($\geq$90% probability) are parsed into an in-house database, and diverse data clustering and mining strategies used to find interesting patterns of protein expression for biological validation and detailed analysis. (This figure is available in full color at ftp://ftp.wiley.com/public/sci_tech_med/pharmaceutical_biotech/.)

13.2.3.1 HPLC Sample Preparation

To facilitate the process and to maintain reproducible results, minimal sample handling is highly recommended for success of an IEX–HPLC experiment. To meet this requirement, a one-step sample pretreatment approach is employed. This approach allows recovering the sample in a low salt buffer compatible with HPLC while preserving the native conformation of the proteins. Here, a simple Tris-HCl-based buffer system at or near the physiological pH is used to solubilize the yeast cell pellet and to collect a soluble fraction by high-speed centrifugation. The yield of this crude extract is approximately 10-mg protein/mL. The protocol below allows for collection of a protein sample mixture ready for the direct injection onto a standard analytical grade HPLC column without further treatment.

1. To approximately 500 μL of *S. cerevisiae* cell pellets on ice, add an equal volume of the fresh Lysis Buffer Solution (described in Section 13.2.2.2, prefiltered through a 0.22-μm membrane) containing protease inhibitors (follow the vendor's instructions for preparation) and vortex extensively to fully resuspend the pellet.
2. Allow the mixture to stand for 1 minute on ice and add a half volume of glass beads (stored at 4°C) for cell disruption.
3. Insert tube into a vortex adaptor placed in the cold room and vortex vigorously at maximum speed for a period of 1 minute; repeat 10 times, equilibrating the tubes on ice for 2 minutes between bursts.
4. Allow to stand for 10 minutes and sediment the cell debris and the beads by microcentrifugation at 14,000 rpm for 10 minutes.
5. Collect the supernatant in a new prechilled microtube with a micropipette[1] and clarify the sample by centrifugation 10 minutes at 14,000 rpm in an eppendorf centrifuge.[2]
6. Measure protein concentration before injection onto the column.[3] Take 1- to 10-μL aliquots to measure the protein concentration, following the standard procedure Bio-Rad Assay for microtiter plate instructions, using BSA as a standard reference protein.
7. Dispense sample by dividing into 20-μL aliquots in the 1.5-mL microtubes for injection and immediately freeze the remainder at −80°C (if possible, freeze the sample in nitrogen liquid first).

---

[1] The tube contains two layers; take only the top layer that contains soluble protein, and estimate the concentration of protein for the subsequent step.
[2] The sample to be loaded onto the HPLC system must be free of the particulates susceptible to clog its valves, lines, and columns. The particulates may be removed either by filtration with 0.22–0.45-μm membrane or by centrifugation. Yeast crude cell lysate solution can be difficult to filter due to the presence of glass beads so centrifugation is the method of choice.
[3] Failure to measure the amount of the protein in the sample may lead to the injection of an excessively concentrated sample, which may cause the protein to precipitate out in the HPLC tubing or within the column itself, causing a pressure rise. The shelf lifetime of the column is also shortened by loading the protein in excess of the maximum capacity of the column.

13.2.3.2 HPLC Sample Prefractionation

To achieve the high degree of resolution while maintaining results reproducibility and getting fractionation of the protein in a reasonable time frame, a binary or quaternary eluent strategy is required. Our basic binary strategy involves the separation of protein mixtures using a binary eluent IEX–HPLC method. This approach uses a combination of two mobile phase buffers differing only in salt concentration, either 10-mM Tris-HCl, pH 7.5, no salt (*buffer A*), or 10-mM Tris-HCl + 600-mM NaCl, pH 7.5. The fractionation is accomplished by applying the gradient of salt during a 30–90-minute chromatographic run to resolve and separate the proteins.

The protocol described in this section is an example of the one currently performed in our laboratory for routine proteomic analyses. All of our fractionation experiments are carried out on the Agilent 1100 HPLC system equipped with a vacuum degasser, a binary pump, a refrigerated autosampler with a 100-µL injector loop, 2-D column compartments with a thermostat, a multiwavelength detector (MWD), a cooled autocollection fraction module, and an Agilent ChemStation for chromatograms acquisition and instrument control. As a test case, soluble yeast proteins were fractionated using two combined column chromatography, an anion-exchange column (PolyWAX LP 50 × 2.1 mm i.d., 5 µm, 1000 Å) connected upstream of the cation-exchange column (PolyCAT A, 50 × 2.1 mm i.d., 5 µm, 1000 Å) and protected with a precolumn guard cartridge.[4]

Using the above-described HPLC system and columns together with the salt gradient and the appropriate instrument preparation, 27 HPLC fractions may be collected in less than 35 minutes. Figure 13.2-2 depicts the typical chromatograms obtained when the following protocol was applied to the yeast soluble cell lysate:

1. Flush and equilibrate the entire HPLC system with 40 mL of the binding buffer (Buffer A) at 2 mL.min^{-1} using an empty union in place of the column.

2. Stop the flow, remove the union, and install the guard cartridge and the columns (make sure your columns are compatible with the buffer systems); use a 10-cm-long Peek tubing (0.13 mm i.d.) to connect the anion exchanger to a cation-exchange column.

3. Set the column temperature at 17°C to enhance protein integrity during the run, and turn on the HPLC system, flush columns with 80% of Buffer B at a flow rate of 250 µL.min^{-1} for 20 minutes. Follow the manufacturer's instructions when using new columns, and carefully monitor the column pressure for any backpressure indicating possible clogging.

4. Condition columns with pure Buffer A (10-mM Tris-HCl, 3-mM NaN$_3$, pH 7.5) at the above flow rate and time or until you get a flat Ultraviolet baseline at 280 nm.

---

[4] As discussed, crude cell lysate can contaminate the anion-exchange column resulting in increased backpressure and decreased efficiency. Poly WAX LP (10 × 2.1 mm, 5 µm, 1000 Å) precolumn is used before the PolyWAX LP column to act as a prefilter because the material is the same as the anion exchanger but is inexpensive. The use of a preguard column extends the lifetime of the anion-exchange column and so can save both time and money.

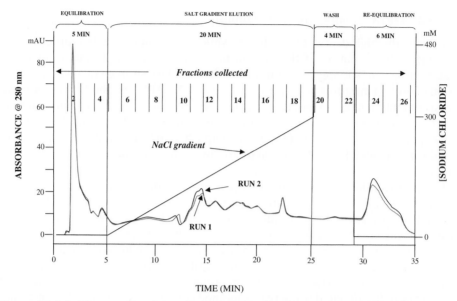

Figure 13.2-2. Repeat ultraviolet-traces (280 nm) recorded for a representative yeast whole cell lysate separated by IEX–HPLC showing the reproducibility of the procedure. Chromatography was performed at 17°C using a mixed bed of ion-exchange columns, PolyWAX LP (50 × 2.1 mm i.d., 5 μm, 1000 Å) and PolyCAT A (50 × 2.1 mm i.d., 5 μm, 1000 Å). The injection of 80-μg total protein onto the column was followed by 5 minutes of column equilibration with 100% of buffer A, and then 20 minutes of protein fractionation with a linear NaCl gradient from 0% to 50% of buffer B. After protein separation, the column was flushed with 80% buffer B for 4 minutes and then re-equilibrated for 6 minutes with the starting buffer A. (This figure is available in full color at ftp://ftp.wiley.com/public/sci_tech_med/pharmaceutical_biotech/.)

5. Apply 5–10 μL of the centrifuged crude soluble cell extracts (80–100-μg total protein) onto the column using automated (autosampler) injector that has been kept at 4°C.

6. Refer to Figure 13.2-2 for protein fractionation with the given salt gradient condition.

7. Monitor protein separation with a MWD set at 280 nm.

8. Collect HPLC fractions based on time (e.g., 1–2 minutes per fraction) by use of an automated sample collector, kept at 4°C, with 1.5-mL Agilent Well Plate fraction collector (325-μL fractions are typically collected per 1.5-mL microtubes).

PROTEOLYSIS AND LC–MS SAMPLE PREPARATION. The following procedure allows for digestion of the proteins in collected HPLC fractions or the control unfractionated crude cell lysate to enable their analysis by LC–MS. After fractionation with HPLC, fractions are precipitated and digested as follows:

1. To each HPLC fraction, add one-tenth cold TCA (100%, fresh solution); vortex the tubes and incubate 30 minutes on ice or overnight at 4°C (cold room).

2. Spin the tubes for 30 minutes at maximum speed (14,000 rpm) in a cooled or refrigerated microcentrifuge; aspirate half of the supernatant (be careful not to disturb the pellet).

3. To each tube, add 600-µL ice-cold HPLC-grade acetone and incubate at −20°C for 10–20 min.

4. Spin tubes as in step 2 and aspirate the total supernatant out of the tube without breaking the pellet.

5. Leave the cap open for air drying (in the fume-hood) for 20 minutes with a protective tissue (e.g., Kimwipes) cover held loosely over it (check the sample frequently to prevent overdrying).

6. Dissolve the pellet in 30 µL of the immobilized trypsin solution (600 µL of the digestion buffer, 60 µL of Pierce immobilized trypsin beads, 30 µL of the Applied Biosciences immobilized trypsin beads, and 2 µL of 1-M $CaCl_2$); check the pH (should be ~8.0) with the pH strip; and neutralize with 5 µL of 1-M TRIS (pH 8.0) if necessary.

7. Incubate the sample for two days at 30°C with rotation or agitation.

8. Stop digestion by adding 30 µL of the LC–MS solvent A, and spin the peptides mixtures for 5 minutes at maximum speed (room temperature) in a microcentrifuge to sediment the trypsin beads.

9. Carefully, recover 20 µL of the peptide mixtures into small disposable (e.g., PCR) tubes, and either immediately analyze by LC–MS or store at −20°C.

13.2.3.3 LC-MS Sample and Data Analysis

A single capillary scale reverse phase high-performance liquid chromatography system coupled online to automated electrospray ion trap tandem mass spectrometry (1D–LCMS) is usually performed to characterize the protein fractions. The following protocol is an example of the implementation of the method to identify the peptide components of tryptic digests of the yeast HPLC fractions analysis using the ion trap tandem mass spectrometer instrument described in Section 13.2.2.5, and typical results are shown in Figure 13.2-3.

1. Pack a silica capillary-scale microcolumn (150 µm i.d.) 7.5 cm of C18 reverse phase packing material (Zorbax eclipse XDB-C_{18} resin).

2. Place the packed column in line with the LC-MS buffer pump and ion source power supply.

3. Load the peptide mixture onto the column using an autosampler.

4. The digested proteome is chromatographically resolved and the peptides subject to automated precursor ion selection. The following gradient may be increased or decreased according to sample complexity: 100% of solvent A for 1 minute, 0% to 5% of solvent B in 1 minute, 5% to 30% of solvent B in 58 minutes, 30% to 80% of solvent B in 10 minutes, 80% of solvent B for 5 minutes, 80% to 30% of solvent B during 5 minutes, and then step the gradient back at 100% of solvent A. The flow rate at the tip of the needle is set to 150 µL.min^{-1} for 2 minutes, rise to 175 µL.min^{-1} in 58 minutes, and maintain the flow constant for 25 minutes and then reduce to 5 µL.min^{-1} in 5 minutes.

A

HPLC Fractions

B

LC–MS Coverage

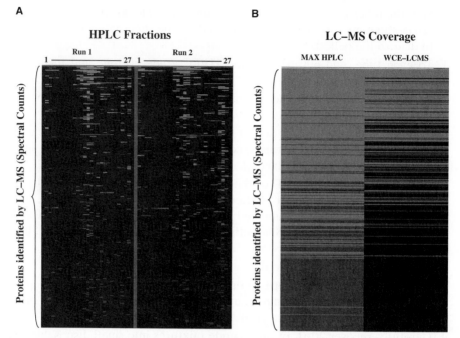

Figure 13.2-3. Yeast cell extract HPLC fractionation and LC–MS analysis. (*A*) Comparison of representative proteomic patterns recorded for two consecutive HPLC prefractionations. (*B*) Graphical comparison of HPLC prefractionation and straight analysis of the whole cell extract (WCE) by LC–MS (a maximum protein spectral count was recorded for two successive HPLC runs and compared with the corresponding value for the WCE). Black-to-red color shading indicates increasing spectral counts. (This figure is available in full color at ftp://ftp.wiley.com/public/sci_tech_med/pharmaceutical_biotech/.)

5. The mass spectrometer cycles through a full mass precursor ion scan, followed by 2–10 successive tandem mass scans using data-dependent ion isolation and fragmentation of the 10 most intense ions as the gradient progresses. Dynamic exclusion is implemented to prevent repeated fragmentations of the same peptide.

6. Candidate peptide sequences from the mixture are later identified using a computerized database search algorithm (e.g., SEQUEST) [14] and validated using the STATQUEST probabilistic scoring program [15].

7. Comparison of relative protein expression levels between chromatographic runs can be performed using the clustering algorithm Cluster 3.0 and the patterns visualized in a "heat map" format using Java Treeview freeware software.

REFERENCES

1. Domon B, Aebersold R (2006). Mass spectrometry and protein analysis. *Science.* 312:212–217.

2. Meyers R A (2005). *Encyclopedia of Molecular Cell Biology and Molecular Medicine.* Wiley-VCH Verlag, Weinheim, Germany, pp. 1–43.

3. Kislinger T, Emili A (2003). Going global: Protein expression profiling using shotgun mass spectrometry. *Curr. Opin. Mol. Ther.* 5:285–293.

4. Link A J, Eng J, Schieltz D M, et al. (1999). Direct analysis of protein complexes using mass spectrometry. *Nat. Biotechnol.* 17:676–682.

5. Washburn M P, Wolters D, Yates J R, 3rd. (2001). Large-scale analysis of the yeast proteome by multidimensional protein identification technology. *Nat. Biotechnol.* 19:242–247.

6. Badock V, Steinhusen U, Bommert K, et al. (2001). Prefractionation of protein samples for proteome analysis using reversed-phase high-performance liquid chromatography. *Electrophor.* 22:2856–2864.

7. Gao M-X, Hong J, Yang P-Y, et al. (2005). Chromatographic prefractionation prior to two-dimensional electrophoresis and mass spectrometry identifies: Application to the complex proteome analysis in rat liver. *Anal. Chim. Acta.* 553:83–92.

8. Issaq H J, Conrads T P, Janini G M, et al. (2002). Methods for fractionation, separation and profiling of proteins and peptides. *Electrophor.* 23:3048–3061.

9. Righetti P G, Castagna A, Antonioli P, et al. (2005). Prefractionation techniques in proteome analysis: the mining tools of the third millennium. *Electrophor.* 26:297–319.

10. Barnea E, Sorkin R, Ziv T, et al. (2005). Evaluation of prefractionation methods as a preparatory step for multidimensional based chromatography of serum proteins. *Proteom.* 5:3367–3375.

11. Zhang Z, Smith D L, Smith J B (2001). Multiple separations facilitate identification of protein variants by mass spectrometry. *Proteom.* 1:1001–1009.

12. Snyder L R, Glajch J L, Kirkland J J (1997). Practical HPLC method development. John Wiley & Sons, New York.

13. Lescuyer P, Hochstrasser D F, Sanchez J C (2004). Comprehensive proteome analysis by chromatographic protein prefractionation. *Electrophor.* 25:1125–1135.

14. Eng J K, McCormack A L, Yates I I I J R (1994). An approach to correlate tandem mass spectral data of peptides with amino acid sequences in a protein database. *J. Amer. Soc. Mass Spectrom.* 5:976–989.

15. Kislinger T, Rahman K, Radulovic D, et al. (2003). PRISM, a generic large scale proteomic investigation strategy for mammals. *Mol. Cell Proteom.* 2:96–106.

13.3

AN OVERVIEW OF METABONOMICS TECHNIQUES AND APPLICATIONS

JOHN C. LINDON

Imperial College London, South Kensington, London, United Kingdom

Chapter Contents

13.3.1 Introduction   1503
13.3.2 Metabonomics Analytical Technologies   1505
     13.3.2.1 NMR Spectroscopy   1505
     13.3.2.2 Mass Spectrometry   1510
     13.3.2.3 Other Technologies   1511
13.3.3 Data Analysis using Chemometrics   1512
     13.3.3.1 Chemometrics Methods   1512
     13.3.3.2 Biomarker Identification using Chemometrics   1514
13.3.4 Selected Applications of Metabonomics   1515
     13.3.4.1 Phenotypic and Physiological Effects   1515
     13.3.4.2 PreClinical Drug Candidate Safety Assessment   1515
     13.3.4.3 Disease Diagnosis and Therapeutic Efficacy   1517
13.3.5 Integration of -Omics Results   1518
13.3.6 Future Prospects   1519
     References   1521

13.3.1 INTRODUCTION

Metabonomics is the comprehensive and simultaneous systematic profiling of metabolite levels and their temporal changes in whole organisms through the study of biofluids, tissues, and tissue extracts [1, 2]. A parallel approach mainly from plant science and from the study of *in vitro* systems has led to the term *metabolomics* also

Handbook of Pharmaceutical Biotechnology, Edited by Shayne Cox Gad.

being coined [3], and the methods and approaches used in the two disciplines are highly convergent. For the pharmaceutical and medical communities, this multivariate approach holds out the promise of a means by which real disease and drug effect endpoints can be obtained. In this monograph, the main technologies used in metabonomics are summarized, brief details of the types of samples used are given, and the current applications of metabonomics are described. Some prospects for the future are then discussed.

One main problem with integrating information at the three main levels of biomolecular organization and control, transcriptomic, proteomic, and metabonomic, is because these are highly interdependent and can have very different time scales of change. As a result, difficulties can occur in correlation of effects seen by the different omics approaches because some time courses can be very rapid (gene switching), some require much longer time scales (protein synthesis), or some encompass enormous ranges of time scales (metabolite levels). Additionally, biochemical changes do not always occur in the order that would intuitively be expected, i.e., transcriptomic, proteomic, and metabolic, because, for example, pharmacological or toxicological effects at the metabolic level can induce subsequent adaptation effects at the proteomic or transcriptomic levels. One important potential role for metabonomics, therefore, could be to direct the timing of proteomic and genomic analyses in order to maximize the probability of observing "omic" biological changes that are relevant to functional outcomes.

In addition, overlaid with this complexity, is the fact that environmental and lifestyle effects have a large effect on gene and protein expression and on metabolite levels and these have to be considered as part of intersample and interindividual variation. Interpretation of genomic data, in terms of real biological endpoints, is therefore a major challenge because of the modification of gene expression levels by environmental factors. The modeling of such diverse information sets poses significant challenges in terms of bioinformatics. Highly complex animals such as man can be considered as "superorganisms" with an internal ecosystem of diverse symbiotic gut microbiota and parasites that have interactive metabolic processes and for which, in many cases, the genome is not known. The many levels of complexity of the mammalian system and the diverse features that need to be measured to allow "omic" data to be fully used have been reviewed recently [4]. In addition, novel approaches will continue to be required to measure and model metabolic processes in various compartments in different interacting cell types, with genomes that are connected by cometabolic processes in such a global mammalian system [5].

Typically, mammalian metabonomics studies of relevance to the pharmaceutical industry are carried out on biofluids because these are often easy to obtain and can provide an integrated view of the whole system's biology. The biochemical profiles of the main diagnostic fluids, blood plasma, cerebrospinal fluid (CSF), and urine, can reflect both normal variation and the impact of disease and drug toxicity or efficacy on single or multiple organ systems. Urine and plasma are obtained essentially noninvasively, and hence, are most appropriate for clinical trials monitoring and disease diagnosis. However, there is a wide range of fluids that can be, and have been, studied, including seminal fluids, amniotic fluid, synovial fluid, digestive fluids, blister and cyst fluids, lung aspirates, and dialysis fluids. In addition, several metabonomics studies have used analysis of tissue biopsy samples and their lipid and aqueous extracts. This is particularly true for studies in fields other than mammalian systems, such as for plants, or for model organisms such as yeast and cell culture studies. Additionally, the approach can be used to characterize *in vitro* cell systems such as Caco-2 cells commonly used for cell uptake studies [6] or tissue spheroids, which can be used, for example, as model systems for liver or tumor investigations [7].

As described in more detail in Section 13.3.2, the main analytical techniques that are employed for metabonomic studies are based on nuclear magnetic resonance (NMR) spectroscopy and mass spectrometry (MS). The latter technique requires a preseparation of the metabolic components using either gas chromatography (GC) after chemical derivatization or liquid chromatography (LC), with the newer method of ultra-high-pressure LC (UPLC) being used increasingly. The use of capillary electrophoresis (CE) coupled to MS has also shown promise. Other more specialized techniques such as Fourier transform infrared spectroscopy and arrayed electrochemical detection have been used in some cases.

All metabonomics studies result in complex multivariate datasets that require a variety of chemometric, bioinformatic, and visualization tools for effective interpretation. The aim of these procedures is to produce biochemically based fingerprints that are of diagnostic or other classification value. A second stage, crucial in such studies, is to identify the substances causing the diagnosis or classification, and these become the combination of biomarkers that reflects actual biological events. Thus, metabonomics studies allow real-world or biomedical endpoint observations to be obtained.

There have been several reviews of metabonomics and metabolomics recently that describe the various techniques used and that summarize the main areas of application. These provide more detail than can be given here and serve to act as pointers to the original literature studies [8–11].

13.3.2 METABONOMICS ANALYTICAL TECHNOLOGIES

13.3.2.1 NMR Spectroscopy

NMR spectroscopy is a nondestructive technique, widely used in chemistry, that provides detailed information on molecular structure, both for pure compounds and in complex mixtures [12]. NMR spectroscopic methods can also be used to probe metabolite molecular dynamics and mobility as well as substance concentrations through the interpretation of NMR spin relaxation times and by the determination of molecular diffusion coefficients [13].

Standard pulse-Fourier transform NMR spectra typically take only a few minutes to acquire, often using robotic flow-injection methods, with automatic sample preparation involving buffering and addition of D_2O as a magnetic field lock signal for the spectrometer. For large-scale studies, bar-coded vials containing the biofluid are used and the contents of these can be transferred and prepared for analysis using robotic liquid handling technology into 96-well plates with the whole process under LIMS system control. Using NMR flow probes, the capacity for NMR analysis has increased enormously recently, and around 200 samples per day can be measured on one spectrometer. Alternatively, for more precious samples or for those of limited volume, conventional 5-mm NMR tubes are usually used, either individually or using a commercial sample tube changer and automatic data acquisition. Typical NMR spectra of human biofluids are shown in Figure 13.3-1.

A ^{1}H NMR spectrum of urine acquired as seen in Figure 13.3-1, typically contains thousands of sharp lines from predominantly low-molecular-weight metabolites. This spectrum is the result of 64 co-added scans each requiring about 5 s, thus yielding a total data acquisition time of around 5 minutes. The large interfering NMR signal arising from water in all biofluids is easily eliminated by use of appropriate standard NMR solvent suppression methods, either by secondary radio-frequency (RF) irradiation at the water peak chemical shift or by use of a specialized NMR pulse sequence that does not excite the water resonance. The position of each spectral band (known as its chemical shift and measured in frequency terms, in ppm, from that of an added standard reference substance) gives information on molecular group identity and its molecular environment, e.g., methyl group on an aromatic ring, olefinic hydrogen, pyridyl ring proton, or aldehyde. The reference compound used in aqueous media is usually the sodium salt of 3-trimethylsilylpropionic acid (TSP) with the methylene groups deuterated to avoid giving rise to peaks in the ^{1}H NMR spectrum. The multiplicity of the splitting pattern on each band, and the magnitudes of the splittings (caused by a nuclear spin–spin interaction mediated through the electrons of the chemical bonds, and known as

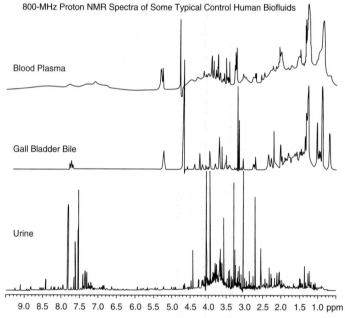

Figure 13.3-1. Shown are 800-MHz ^{1}H NMR spectra of control human biofluids. The peaks arise from different chemical types of hydrogen in the biochemicals present. The peak areas are related to molar concentrations, and the peak positions and splittings allow information to be obtained, after expert interpretation, of the molecules responsible for the peaks. The signal from water has been suppressed by an NMR procedure to avoid problems of dynamic range in the detection process.

J-coupling), provide knowledge about nearby protons, their through-bond connectivities, the relative orientation of nearby C–H bonds, and hence molecular conformations. The band areas relate directly to the number of protons, giving rise to the peak and hence to the relative concentrations of the substances in the sample. Absolute concentrations can be obtained if the sample contains an added internal standard of known concentration, if a standard addition of the analyte of interest is added to the sample, or if the concentration of a substance is known by independent means (e.g., glucose in plasma can be quantified by a conventional biochemical assay).

Blood plasma and serum contain both low- and high-molecular-weight components, and these give a wide range of signal line widths. Broad bands from protein and lipoprotein signals contribute strongly to the ^1H NMR spectra, with sharp peaks from small molecules superimposed on them. Standard NMR pulse sequences, where the observed peak intensities are edited on the basis of molecular diffusion coefficients or on NMR relaxation times (known as T_1, $T_{1\rho}$, and T_2), can be used to select only the contributions from proteins, and other macromolecules and micelles, or alternatively to select only the signals from the small-molecule metabolites, respectively [12]. A typical ^1H NMR spectrum from human blood serum is shown in Figure 13.3-1, and a series of edited NMR spectra, from rat blood serum, based on the approaches described above are given in Figure 13.3-2.

The spin-echo spectrum has the broader peaks from nuclei with shorter spin relaxation times attenuated (these are macromolecules and substances involved in

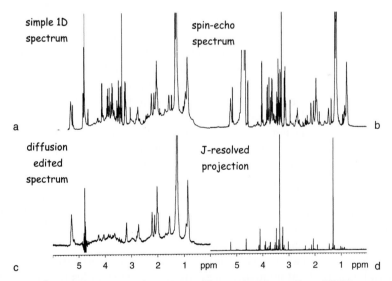

Figure 13.3-2. ^1H NMR spectra of rat serum illustrating the various NMR responses that are possible through the use of different pulse sequences, which edit the spectral intensities: (a) standard water suppressed spectrum, showing all metabolites; (b) CPMG spin-echo spectrum, with attenuation of peaks from fast relaxing components such as macromolecules and lipoproteins; (c) diffusion-edited spectrum, with attenuation of peaks from fast diffusing components such as small molecules; and (d) a projection of a 2D J-resolved spectrum on to the chemical shift axis, showing removal of all spin–spin coupling and peaks from fast relaxing species.

chemical exchange). The diffusion-edited spectrum, on the other hand, has peaks only from macromolecules and other slow-moving species present, with the peaks from highly mobile small molecules attenuated. Thus, it is also possible to use these approaches to investigate molecular mobility and flexibility, and to study inter-molecular interactions such as the reversible binding between small molecules and proteins [14].

Two recent improvements in NMR detector technology have resulted in major improvements in sample quantity requirements for NMR spectroscopy. The first is the commercialization of miniaturized detectors. Now it is possible to study meta-bolic profiles by NMR using as little as 2–20 μL of sample, and examples have been published using CSF and blood plasma [15]. Secondly, the availability of cryogenic NMR probe technology where the detector coil and preamplifier are cooled to around 20K has provided an improvement in spectral signal-to-noise ratios of up to a factor of 5 by reducing the thermal noise in the electronics of the spectrometer. This allows use of smaller samples, and when this approach is combined with the use of miniaturized probes as described above, optimum sensitivity and sample requirements are achieved. Conversely, because the NMR signal-to-noise ratio is proportional to the square root of the number of co-added scans, shorter data acquisition times by up to a factor of 20–25 become possible for the same amount of sample. NMR spectroscopy of biofluids detecting the much less sensitive ^{13}C nuclei, which also only have a natural abundance (1.1%), also becomes possible because of the increase in signal-to-noise ratio [16].

Within the last few years, the development of a technique called high-resolution ^{1}H magic angle spinning (MAS) NMR spectroscopy has made feasible the acquisi-tion of high-resolution NMR data on small pieces of intact tissues with no pretreat-ment [17–21]. Rapid spinning of the sample (typically at ~4–6 kHz) at an angle of 54.7° relative to the applied magnetic field serves to reduce the loss of information caused by line broadening effects observed in nonliquid samples such as tissues. These broadenings are caused by sample heterogeneity and residual anisotropic NMR parameters that are normally averaged out in free solution where molecules can tumble isotropically and rapidly. NMR spectroscopy on a tissue matrix in an MAS experiment is the same as solution-state NMR, and all common pulse tech-niques can be employed to study metabolic changes and to perform molecular structure elucidation. Typical ^{1}H NMR spectra from a range of tissue types are shown in Figure 13.3-3. In most cases, a standard set of one-dimensional (1D) sequences is used to describe the biochemical changes in the pool of low-molecular-weight metabolites and lipids. The different spin properties of macromolecules and small molecules can also be exploited using spectral editing techniques to filter out certain subgroups of peaks, as in solution-state NMR spectroscopy.

MAS NMR spectroscopy has straightforward sample preparation, although this still has to be carried out manually. Snap-frozen tissue samples (as little as 10 mg), which have been stored at −80°C, are defrosted and cut to select the region of interest. The original samples should be frozen rapidly using small specimens in liquid nitrogen to avoid microcrystallization of the water in the cells and conse-quent cell damage. Then, the tissue is rinsed with 0.9% D_2O/saline to wash off remaining blood and the specimen is transferred into a 4-mm-diameter zirconia rotor and a Teflon spacer is used to restrict the sample volume, to eliminate trapped air bubbles, and to increase sample homogeneity. As soon as the sample has been

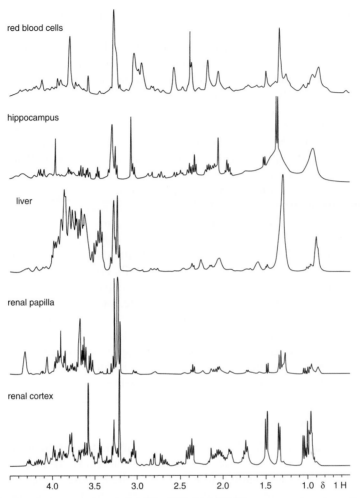

Figure 13.3-3. High-resolution 400-MHz ^{1}H MAS NMR spectra of various tissues, with sample spinning at 4.2 kHz.

transferred into the rotor, a trace volume of 0.9% D_2O/saline is added to provide a field-frequency lock for the ^{1}H MAS NMR data acquisitions. In cases where the chemical shifts of signals have not been identified previously, it is necessary to add a reference standard such as TSP in the saline solution, although sometimes it is possible to use a well-characterized peak of an easily assigned compound, such as the anomeric proton resonance of α-glucose, as a secondary chemical shift standard. Fast rotation of the sample can cause frictional heating, and this is prevented by a controlled cooling of the input gas to compensate.

Identification of biomarkers can involve the application of a range of techniques, including two-dimensional (2D) NMR experiments [12]. Although, all of the armory of the usual analytical physical chemistry can and should be used, including mass spectrometry, ^{1}H NMR spectra of urine and other biofluids, even though they are very complex, allow many resonances to be assigned directly based on their chemical shifts, signal multiplicities, and by adding authentic material, and further information can be obtained by using spectral editing techniques, as described above.

Two-dimensional NMR spectroscopy can be useful for increasing signal dispersion and for elucidating the connectivities between signals, thereby enhancing the information content and helping to identify biochemical substances. These include the ^{1}H-^{1}H 2-D J-resolved experiment, which attenuates the peaks from macromolecules and yields information on the multiplicity and coupling patterns of resonances, which is a good aid to molecule identification. The projection of such a spectrum on to the chemical shift axis yields a fingerprint of peaks from only the most highly mobile small molecules, with the added benefit that all spin-coupling peak multiplicities have been removed (see Figure 13.3-2). Other 2D experiments known as COSY and TOCSY provide ^{1}H-^{1}H spin–spin coupling connectivities, thus giving information on which hydrogens in a molecule are close in chemical bond terms. Use of other types of nuclei, such as naturally abundant ^{13}C or ^{15}N or where present ^{31}P, can be important to help assign NMR peaks, and here heteronuclear correlation NMR experiments can be obtained by use of appropriate NMR pulse sequences. These benefit by the use of so-called inverse detection, where the lower sensitivity or less abundant nucleus NMR spectrum (such as ^{13}C) is detected indirectly using the more sensitive/abundant nucleus (^{1}H) by making use of spin–spin interactions such as the one-bond ^{13}C-^{1}H spin–spin coupling between the nuclei to effect the connection. These yield both ^{1}H and ^{13}C NMR chemical shifts of CH, CH$_2$, and CH$_3$ groups, which are useful again for identification purposes. There is also a sequence that allows correlation of protons to quaternary carbons based on long-range ^{13}C-^{1}H spin–spin coupling between the nuclei.

13.3.2.2 Mass Spectrometry

Mass spectrometry has also been widely used in metabolic fingerprinting and metabolite identification, with most studies to date on plant extracts and model cell system extracts, although its application to mammalian studies is increasing. In general, with the exception of some studies using Fourier transform, or ion-cyclotron-MS, a prior separation of the complex mixture sample using chromatography is required. MS is inherently considerably more sensitive than NMR spectroscopy, but in complex mixtures of very variable composition such as biofluids, it is necessary generally to employ different separation techniques (e.g., different LC column packings) for different classes of substances. MS is also a mainstay technique for molecular identification purposes, especially through the use of MS^n methods for fragment ion studies. Analyte quantitation by MS in complex mixtures of highly variable composition can also be impaired by variable ionization and ion suppression effects. Chemical derivatization might be necessary to ensure volatility and analytical reproducibility, in complex mixtures such as biofluids.

For both profiling and metabolite identification, most published metabonomics studies on mammalian biological systems have used NMR spectroscopy, but HPLC–MS techniques are increasing in usage, particularly using electrospray ionization. Thus, for metabonomics applications on biofluids such as urine, an HPLC chromatogram is generated with MS detection, and usually both positive and negative ion chromatograms are measured. At each sampling point in the chromatgram, there is a full mass spectrum and so the data are three-dimensional in nature, i.e., retention time, mass-to-charge ratio, and intensity. Given this very high resolution, it is easy to cut out any mass peaks from interfering substances such as drug metabolites, without compromising the dataset.

For plant metabonomics studies, the principal approach has been to make extracts of the samples followed by chemical derivatization, and then to employ separation using GC with molecular identification using MS by comparing spectra against publically available and single-laboratory databases [10, 11].

UPLC is a new combination of a 1.7-μm reversed phase packing material and a chromatographic system, operating at around 12,000 psi. This has enabled a marked improvement in chromatographic performance to be obtained for complex mixture separation, with better peak resolution and increased speed and sensitivity. UPLC gives more than doubling of peak capacity compared with HPLC, an almost 10-fold increase in speed, and a 3- to 5-fold increase in sensitivity compared with that generated with a conventional stationary phase. Because of the much improved chromatographic resolution of UPLC, the problem of ion suppression from co-eluting peaks is greatly reduced. UPLC–MS has been used for metabolic profiling of urines from males and females of two groups of phenotypically normal mouse strains and a nude mouse strain [22].

Recently, CE coupled to mass spectrometry has also been explored as a suitable technology for metabonomics studies [23]. Charged metabolites are first separated by CE based on charge and size and then selectively detected using MS by monitoring over a large range of m/z values. This method was used to measure 352 metabolic standards and then employed for the analysis of 1692 metabolites from *Bacillus subtilis* extracts, revealing significant changes in metabolites during the bacterial sporulation.

For biomarker identification, it is also possible to separate out substances of interest from a complex biofluid sample using techniques such as solid phase extraction or HPLC. For metabolite identification, directly coupled chromatography–NMR spectroscopy methods can be used. The most powerful of these "hyphenated" approaches is HPLC–NMR–MS [24] in which the eluting HPLC peak is split with parallel analysis by directly coupled NMR and MS techniques. This can be operated in on-flow, stopped-flow, and loop-storage modes and thus can provide the full array of NMR and MS-based molecular identification tools. These include MS–MS for identification of fragment ions and FT–MS or TOF–MS for accurate mass measurement and hence derivation of molecular empirical formulae.

In summary NMR and MS approaches are highly complementary, and use of both is often necessary for full molecular characterization. MS can be more sensitive with lower detection limits provided the substance of interest can be ionized, but NMR spectroscopy is particularly useful for distinguishing isomers, for obtaining molecular conformation information, and for studies of molecular dynamics and compartmentation.

13.3.2.3   Other Technologies

Although most metabonomics applications have used either NMR spectroscopy or chromatography coupled to mass spectrometry, other techniques have been explored. Fourier transform infrared spectroscopy has been applied to a few metabolomics investigations including the evaluation of different *E. coli* mutants by analysis of their secreted metabolites [25]. Although in principal, different responses can be achieved from different sample classes, and analyzed using pattern recognition methods, the main limitation of this technique is the low level of detailed molecular identification that can be achieved, and indeed in the case quoted above, MS was also employed for metabolite identification.

An alternative approach has been pioneered by Lewitt et al. [26]. This uses an array of coulometric detectors following HPLC separation to detect redox-active compounds in a complex mixture such as in CSF or tissue extracts. Although this approach does not identify compounds directly, the combination of retention time and redox properties can serve as a basis for database searching of libraries of standard compounds. The separation output can also be directed to a mass spectrometer for additional identification experiments.

13.3.3   DATA ANALYSIS USING CHEMOMETRICS

13.3.3.1   Chemometrics Methods

The complex data that arise from, for example, an NMR spectrum of a sample can be thought of as an object with a multidimensional set of metabolic coordinates, the values of which are the spectral intensities at each data point. Thus, each spectrum becomes a point in a multidimensional metabolic hyperspace. In chemistry, the term *chemometrics* is generally applied to describe the use of both parametric and nonparametric multivariate statistical approaches to chemical numerical data. The general aim is to classify an object based on identification of inherent patterns in a set of experimental measurements or descriptors and to identify those descriptors responsible for the classification. The approach can also be used for reducing the dimensionality of complex datasets, for example, by 2D or 3D mapping procedures, to enable easy visualization of any clustering or similarity of the various samples. Alternatively, in what are known as "supervised" methods, multiparametric datasets can be modeled so that the class of separate samples (a "validation set") can be predicted based on a series of mathematical models derived from the original data or "training set" [27].

Principal components analysis (PCA) is one of the simplest techniques that has been used extensively in metabonomics, and this expresses most of the variance within a dataset using a smaller number of factors or principal components. Each PC is a linear combination of the original data parameters whereby each successive PC explains the maximum amount of variance possible, not accounted for by the previous PCs. Each PC is orthogonal and therefore independent of the other PCs. Thus, the variation in the spectral set is usually described by many fewer PCs than the original data point values because the less important PCs describe the noise variation in the spectra. Conversion of the data matrix to PCs results in two matrices known as scores and loadings. Scores, the linear combinations of the original variables, are the coordinates for the samples in the established model and may be regarded as the new variables. In a scores plot, each point represents a single sample spectrum. The PC loadings define the orientation of the computed PC with respect to the original variables and thus indicate which variables carry the greatest weight in transforming the position of the original samples from the data matrix into their new position in the scores matrix. In the loadings plot, each point represents a different spectral intensity. Thus, the cause of any spectral clustering observed in a PC scores plot is interpreted by examination of the loadings that cause any cluster separation. In addition, there are many other visualization methods such as non-linear mapping and hierarchical cluster analysis.

The upper part of Figure 13.3-4 shows a PC scores plot where each point is based on the ^{1}H NMR spectrum of a rat urine sample. In this case, the open and filled symbols represent samples of rat urine received from two different sites, from animals dosed with a model liver toxin, hydrazine, and sampled at various time points in hours after dosing as shown [28].

One of many widely used supervised methods (i.e., using a training set of data with known outcomes) is partial least squares (PLS). This method relates a data matrix containing independent variables from samples, such as spectral intensity values (an X matrix), to a matrix containing dependent variables (or measurements of response) for those samples (a Y matrix). PLS can also be used to examine the

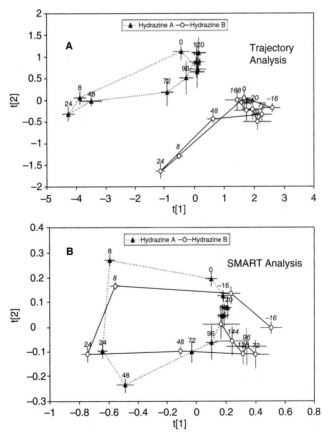

Figure 13.3-4. Principal components scores plots (PC1 vs. PC2) based on 600-MHz ^{1}H NMR spectra of rat urine. (a) Standard trajectory analysis of urinary NMR data obtained after treatment with hydrazine from identical studies repeated at different sites, open and closed symbols. (b) The same data after application of the trajectory matching technique, SMART. Error bars represent the standard error for each time point average [28].

influence of time on a dataset, which is particularly useful for biofluid NMR data collected from samples taken over a time course of the progression of a pathological effect. PLS can also be combined with discriminant analysis (DA) to establish the optimal position to place a discriminant surface that best separates classes. It is possible to use such supervised models to provide classification probabilities or even quantitative response factors for a wide range of sample types, but given the strong possibility of chance correlations when the number of descriptors is large, it is important to build and test such chemometric models using independent training data and validation datasets. Extensions of this approach allow the evaluation of those descriptors that are completely independent (orthogonal) to the Y matrix of endpoint data. This orthogonal signal correction (OSC) can be used to remove irrelevant and confusing parameters.

Apart from the methods described above that use linear combinations of parameters for dimension reduction, classification, or regression, other methods exist that are not limited in this way. For example, neural networks comprise a widely used nonlinear approach for modeling data. A training set of data is used to develop algorithms, which "learn" the structure of the data and can cope with complex functions. The basic software network consists of three or more layers, including an input level of neurons (spectral descriptors or other variables), one or more hidden layers of neurons that adjust the weighting functions for each variable, and an output layer that designates the class of the object or sample. Recently, probabilistic neural networks, which represent an extension to the approach, have shown promise for metabonomics applications in toxicity [29]. Other approaches that are currently being tested include genetic algorithms, machine learning, and Bayesian modeling [30].

13.3.3.2 Biomarker Identification using Chemometrics

Recently, a new method for identifying multiple NMR peaks from the same molecule in a complex mixture, hence providing a new approach to molecular identification, has been introduced. This is based on the concept of statistical total correlation spectroscopy and has been termed *STOCSY* [31]. This takes advantage of the multicollinearity of the intensity variables in a set of spectra (e.g., ^{1}H NMR spectra) to generate a pseudo-two-dimensional NMR spectrum that displays the correlation among the intensities of the various peaks across the whole sample. This method is not limited to the usual connectivities that are deducible from more standard two-dimensional NMR spectroscopic methods, such as TOCSY. Added information is available by examining lower correlation coefficients or even negative correlations, because this leads to a connection between two or more molecules involved in the same biochemical pathway. In an extension of the method, the combination of STOCSY with supervised chemometrics methods offers a new framework for analysis of metabonomic data. In a first step, a supervised multivariate discriminant analysis can be used to extract the parts of NMR spectra related to discrimination between two sample classes. This information is then cross-combined with the STOCSY results to help identify the molecules responsible for the metabolic variation. To illustrate the applicability of the method, it has been applied to 114 ^{1}H NMR spectra of urine from a metabonomic study of a model of insulin resistance based on the administration of a carbohydrate diet to three different mice strains in which a series of metabolites of biological importance could be conclusively assigned and identified by use of the STOCSY approach [31].

13.3.4   SELECTED APPLICATIONS OF METABONOMICS

13.3.4.1   Phenotypic and Physiological Effects

To determine therapeutic or toxic effects or to understand the biochemical alterations caused by disease, it is necessary first to understand any underlying physiological sources of variation. To this end, metabonomics has been used to separate classes of experimental animals such as mice and rats according to several inherent and external factors based on the endogenous metabolite patterns in their biofluids [32]. Such differences may help explain differential toxicity of drugs between strains and interanimal variation within a study. Many other effects can be distinguished using metabonomics, including male/female differences, age-related changes, estrus cycle effects in females, diet, diurnal effects, and interspecies differences and similarities [32].

Metabonomics has also been used for the phenotyping of mutant or transgenic animals and the investigation of the consequences of transgenesis such as the transfection process itself [33]. Genetic modifications used in the development of genetically engineered animal models of disease are often made using such transfection procedures, and it is important to differentiate often-seen unintended consequences of this process from the intended result. Metabonomic approaches can give insight into the metabolic similarities or differences between mutant or transgenic animals and the human disease processes that they are intended to simulate and hence their appropriateness for monitoring the efficacy of novel therapeutic agents. This suggests the method may be appropriate for following treatment regimes such as gene therapy.

The importance of gut microfloral populations on urine composition has been highlighted by a study in which axenic (germ free) rats were allowed to acclimatize in normal laboratory conditions and their urine biochemical makeup was monitored for 21 days [34]. The combined influence of gut microflora and parasitic infections on urinary metabolite profiles has also been elucidated [35].

13.3.4.2   PreClinical Drug Candidate Safety Assessment

Despite the regulatory requirements and the huge investment by pharmaceutical companies in safety testing in animals, unexpected results are sometimes observed in the clinic and drugs occasionally still have to be withdrawn from the marketplace. The selection of robust candidate drugs for development based on mimimization of the occurrence of drug adverse effects is therefore one of the most important aims of pharmaceutical R&D, and the pharmaceutical industry is now embracing metabonomics for evaluating the adverse effects of candidate drugs. The National Center for Toxicological Research, a part of the U.S. Food and Drug Administration, is also investigating the usefulness of the approach.

In this application, NMR-based metabonomics can be used for (1) definition of the metabolic hyperspace occupied by normal samples; (2) the consequential rapid classification of a biofluid sample as normal or abnormal; (3) if abnormal, classification of the target organ or region of toxicity; (4) biochemical mechanism of that toxin; (5) identification of combination biomarkers of toxic effect; and (6) evaluation of the time course of the effect, e.g., the onset, evolution, and regression of toxicity. An example of a metabolic trajectory is shown in Figure 13.3-4(a), where each point represents an NMR spectrum of a rat urine sample at various time points after dosing with the model liver toxin, hydrazine [28]. There have been many studies using ^{1}H NMR spectroscopy of biofluids to characterize drug toxicity going back to the 1980s [36], and the role of metabonomics in general, and magnetic resonance in particular, in toxicological evaluation of drugs has been comprehensively reviewed recently [37].

Metabonomics has already been applied in fields outside human and other mammalian systems. For example, studies in the environmental pollution field have highlighted the potential benefits of this approach by studies of caterpillar hemolymph [38] and earthworm biochemical changes as a result of soil pollution by model toxic substances [39]. In addition, a study of heavy metal toxicity (As^{3+} and Cd^{2+}) in wild rodents living on polluted sites has been concluded successfully [40]. In terms of monitoring water quality, one study has evaluated adverse effects in abalone using NMR-based metabonomics [41].

The usefulness of metabonomics for the evaluation of xenobiotic toxicity effects has recently been comprehensively and successfully explored by the Consortium for Metabonomic Toxicology (COMET). This was formed by five pharmaceutical companies and Imperial College, London, United Kingdom [42], with the aim of developing methodologies for the acquisition and evaluation of metabonomic data generated using ^{1}H NMR spectroscopy of urine and blood serum from rats and mice for preclinical toxicological screening of candidate drugs. The successful outcome is evidenced by the generated databases of spectral and conventional results for a wide range of model toxins (147 in total) that served as the basis for computer-based expert systems for toxicity prediction. The project goals of the generation of comprehensive metabonomic databases (now around 35,000 NMR spectra) and multivariate statistical models (expert systems) for prediction of toxicity, initially for liver and kidney toxicity in the rat and mouse, have now been achieved, and the predictive systems and databases have been transferred to the sponsoring companies.

A feasibility study was carried out at the start of the project, using the same detailed protocol and using the same model toxin, over seven sites in the companies and their appointed contract research organizations. This was used to evaluate the levels of analytical and biological variation that could both arise through the use of metabonomics on a multisite basis. The intersite NMR analytical reproducibility revealed the high degree of robustness expected for this technique when the same samples were analyzed both at Imperial College and at various company sites. This gave a multivariate coefficient of regression between paired samples of only about 1.6% [43].

Additionally, the biological variability was evaluated by a detailed comparison of the ability of the companies to provide consistent urine and serum samples for an in-life study of the same toxin, with all samples measured at Imperial College. There was a high degree of consistency between samples from the various companies, and dose-related effects could be distinguished from intersite variation [43].

As a precursor to developing the final predictive expert systems, metabonomic models were constructed for urine from control rats and mice, enabling identification of outlier samples and the metabolic reasons for the deviation. To achieve the project goals, new methodologies for analyzing and classifying the complex datasets were developed. For example, as the expert system takes into account the metabolic trajectory over time, a new way of comparing and scaling these multivariate trajectories was developed (called SMART), and this is illustrated in Figure 13.3-4 [28]. Additionally, a novel classification method for identifying the class of toxicity based on all NMR data for a given study has been generated. This has been termed "Classification Of Unknowns by Density Superposition (CLOUDS)" and is a novel non-neural implementation of a classification technique developed from probabilistic neural networks [44]. Modeling the urinary NMR data according to organ of effect (control, liver, kidney, or other organ), using a model training set of 50% of the samples and predicting the other 50%, over 90% of the test samples were classified as belonging to the correct group with only a 2% misclassification rate between these classes. This work showed that it is possible to construct predictive and informative models of metabonomic data, delineating the whole time course of toxicity, the ultimate goal of the COMET project.

13.3.4.3 Disease Diagnosis and Therapeutic Efficacy

Many examples exist in the literature on the use of NMR-based metabolic profiling to aid human disease diagnosis, including the use of plasma to study diabetes, CSF for investigating Alzheimer's disease, synovial fluid for osteoarthritis, seminal fluid for male infertility, and urine in the investigation of drug overdose, renal transplantation, and various renal diseases. Most of the earlier studies using NMR spectroscopy have been reviewed [45]. For example, a promising use of NMR spectroscopy of urine and plasma, as evidenced by the number of publications on the subject, is in the diagnosis of inborn errors of metabolism in children [46].

Some studies have been undertaken in the area of cancer diagnosis using perchloric acid extracts of various types of human brain tumor tissue [47], and the spectra could be classified using neural network software giving ~85% correct classification. Tissues can be studied by metabonomics through the MAS NMR technique, and published examples include prostate cancer [16, 48], renal cell carcinoma [18], breast cancer [19, 49], and various brain tumors [50]. Other recent studies include an NMR-based urinary metabonomic study of multiple sclerosis in humans and non-human primates [51].

Recently metabonomics has been applied to provide a method for diagnosis of coronary artery disease noninvasively through analysis of a blood serum sample using NMR spectroscopy [52]. Patients were classified, based on angiography, into two groups, those with normal coronary arteries and those with triple coronary vessel disease. Around 80% of the NMR spectra were used as a training set to provide a two-class model after appropriate data filtering techniques had been applied, and the samples from the two classes were easily distinguished. The remaining 20% of the samples were used as a test set, and their class was then predicted based on the derived model with a sensitivity of 92% and a specificity of 93% based on a 99% confidence limit for class membership.

It was also possible to diagnose the severity of the disease that was present by employing serum samples from patients with stenosis of one, two, or three of the coronary arteries. Although this is a simplistic indicator of disease severity, separation of the three sample classes was evident even though none of the wide range

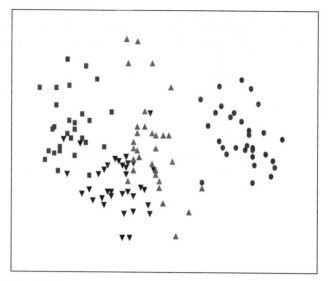

Figure 13.3-5. Partial least-squares discriminant analysis model for the classification of blood plasma samples in terms of coronary artery disease, based on their ^{1}H NMR spectra with visualization of the degree of coronary artery occlusion. Each point is based on data from a ^{1}H NMR spectrum of human blood plasma from subjects with different degrees of coronary artery occlusion. Circles—no stenosis, triangles—stenosis of one artery, inverted triangles—stenosis of two arteries, and squares—stenosis of three arteries. (This figure is available in full color at ftp://ftp.wiley.com/public/sci_tech_med/pharmaceutical_biotech/.)

of conventional clinical risk factors that had been measured was significantly different between the classes. The visualization of the separation of patients with zero, one, two, or three coronary arteries occluded is shown in Figure 13.3-5.

13.3.5 INTEGRATION OF -OMICS RESULTS

The value of obtaining multiple datasets from various biofluid samples and tissues of the same animals collected at different time points has been demonstrated. This procedure has been termed "integrated metabonomics" [2] and can be used to describe the changes in metabolic chemistry in different body compartments affected by exposure to toxic drugs. An illustration of the types of information that can be obtained is shown in Figure 13.3-6, from an NMR spectroscopic study of the acute toxicity of α-naphthylisothiocyanate, a model liver toxin [53]. Such timed profiles in multiple compartments are characteristic of particular types and mechanisms of pathology and can be used to give a more complete description of the biochemical consequences than can be obtained from one fluid or tissue alone [54].

Integration of metabonomics data with that from other multivariate techniques in molecular biology such as from gene array experiments or proteomics is also feasible. Thus, it has also been possible to integrate data from transcriptomics and metabonomics to find, after acetaminophen administration to mice, common metabolic pathways implicated by both gene expression changes and changes in metabolism [55]. In a similar fashion, changes in gene expression detected in microarray

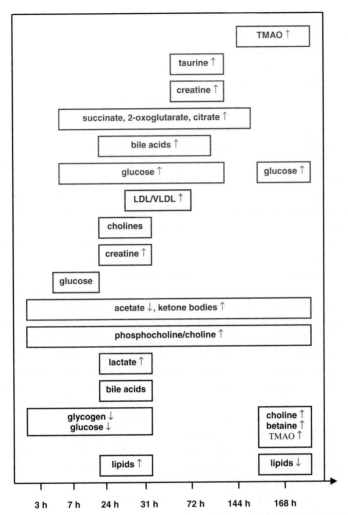

Figure 13.3-6. Representation of the information that can be obtained from an integrated metabonomics study. This shows the metabolites that are changed in a range of biofluids and tissues as a function of time after a single acute administration of the liver toxin α-naphthylisothiocyanate [53]. (This figure is available in full color at ftp://ftp.wiley.com/public/sci_tech_med/pharmaceutical_biotech/.)

experiments can lead to the identification of changed enzyme activity, and this can also be achieved by analysis of metabolic perturbations [56].

13.3.6 FUTURE PROSPECTS

It has become accepted that the main pharmaceutical areas where metabonomics is impacting include validation of animal models of disease, including genetically modified animals; preclinical evaluation of drug safety studies; allowing ranking of candidate compounds; assessment of safety in humans in clinical trials; after product launch, quantitation, or ranking of the beneficial effects of pharmaceuticals, both in development and clinically; improved understanding of the causes of highly sporadic idiosyncratic toxicity of marketed drugs; and patient stratification for clinical trials and drug treatment (pharmaco-metabonomics).

In addition, in terms of disease studies, metabonomics is playing a role in improved, differential diagnosis and prognosis of human diseases, particularly for chronic and degenerative diseases, and for diseases caused by genetic effects. A better understanding of large-scale human population differences through epidemiological studies is also being achieved.

Other applications include nutritional studies, sports medicine, and lifestyle studies, including the effects of diet, exercise, and stress, and evaluation of the effects of interactions between drugs, and between drugs and diet. Additionally, the metabolic profiles of plants, including genetically modified species, and quality control studies (essential equivalence) are now being achieved together with applications in microbiological characterization; forensic science, including the biochemical effects of drug overdose; and environmental effects of pollutants monitored using marker species, both terrestrial and aquatic.

One long-term goal of using pharmacogenomic approaches is to understand the genetic makeup of different individuals (their genetic polymorphism) and their varying abilities to handle pharmaceuticals both for their beneficial effects and for identifying adverse effects. If personalized health care is to become a reality, an individual's drug treatments must be balanced so as to achieve maximal efficacy and avoid adverse drug reactions. Very recently, for the first time, an alternative approach has been proposed for understanding intersubject variability in response to drug treatment using a combination of multivariate metabolic profiling and chemometrics to predict the metabolism and toxicity of a dosed substance, based solely on the analysis and modeling of a predose metabolic profile [57]. Unlike pharmacogenomics, this approach, which has been termed *pharmaco-metabonomics*, is sensitive to both the genetic and the modifying environmental influences that determine the basal metabolic fingerprint of an individual, because these will also influence the outcome of a chemical intervention. This new approach has been illustrated with studies of the toxicity and metabolism of acetaminophen (paracetamol) administered to rats.

A major initiative has been under way to investigate the reporting needs and to consider recommendations for standardizing reporting arrangements for metabonomics studies, and to this end, a Standard Metabolic Reporting Structures (SMRS) group was formed (see www.smrsgroup.org). This group has produced a draft policy document that covers all of those aspects of a metabolic study that are recommended for recording, from the origin of a biological sample, the analysis of material from that sample, and chemometric and statistical approaches to retrieve information from the sample data, and a summary publication has ensued [58]. The various levels and consequent detail for reporting needs, including journal submissions, public databases, and regulatory submissions, have also been addressed. This has been followed up with a workshop and discussion meeting sponsored by the U.S. National Institutes of Health, from which firm plans are being developed to define standards in a number of areas relevant to metabonomics, including characterization of sample-related meta-data, technical standards, and related data; meta-data and QC matters for the analytical instrumentation; data transfer methodologies and schema for implementation of such activities; and development of standard vocabularies to enable transparent exchange of data.

NMR- and MS-based metabonomics are now recognized as independent and widely used techniques for evaluating the toxicity of drug candidate compounds, and they have been adopted by several pharmaceutical companies into their drug development protocols. For drug safety studies, it is possible to identify the target organ of toxicity, derive the biochemical mechanism of the toxicity, and determine the combination of biochemical biomarkers for the onset, progression, and regression of the lesion. Additionally, the technique has been shown to be able to provide a metabolic fingerprint of an organism ("metabotyping") as an adjunct to functional genomics and hence has applications in design of drug clinical trials and for evaluation of genetically modified animals as disease models.

Using metabonomics, it has proved possible to derive new biochemically based assays for disease diagnosis and to identify combination biomarkers for disease, which can then be used to monitor the efficacy of drugs in clinical trials. Thus, based on differences observed in metabonomic databases from control animals and from animal models of disease, diagnostic methods and biomarker combinations might be derivable in a preclinical setting. Similarly, the use of databases to derive predictive expert systems for human disease diagnosis and the effects of therapy require compilations from both normal human populations and patients before, during, and after therapy.

It is expected that further integration of the techniques into pharmaceutical R&D activities will continue. One potential advantage of this approach is the transferability in principle of metabolic biomarkers between species and ultimately into humans.

The ultimate goal of systems biology must be the integration of data acquired from living organisms at the genomic, protein, and metabolite levels. In this respect, transcriptomics, proteomics, and metabonomics will all play an important role. Through the combination of these, and related approaches, will come an improved understanding of an organism's total biology, and with this, better understanding of the causes and progression of human diseases and, given the twenty-first century goal of personalized health care, the improved design and development of new and better targeted pharmaceuticals.

REFERENCES

1. Nicholson J K, Lindon J C, Holmes E (1999). "Metabonomics": Understanding the metabolic responses of living systems to pathophysiological stimuli via multivariate statistical analysis of biological NMR spectroscopic data. *Xenobiot.* 29:1181–1189.

2. Nicholson J K, Connelly J, Lindon J C, et al. (2002). Metabonomics: A platform for studying drug toxicity and gene function. *Nat. Rev. Drug Disc.* 1:153–162.

3. Fiehn O (2002). Metabolomics—the link between genotypes and phenotypes. *Plant Molec. Biol.* 48:155–171.

4. Nicholson J K, Wilson I D (2003). Understanding "global" systems biology: Metabonomics and the continuum of metabolism. *Nat. Rev. Drug Disc.* 2:668–676.

5. Nicholson J K, Holmes E, Lindon J C, et al. (2004). The challenges of modeling mammalian biocomplexity. *Nat. Biotechnol.* 22:1268–1274.

6. Lamers R J A N, Wessels E C H H, van der Sandt J J M, et al. (2003). A pilot study to investigate effects of inulin on Caco-2 cells through *in vitro* metabolic fingerprinting. *J. Nutr.* 133:3080–3084.

7. Bollard M E, Xu J S, Purcell W, Griffin J L, et al. (2002). Metabolic profiling of the effects of D-galactosamine in liver spheroids using ¹H NMR and MAS NMR spectroscopy. *Chem. Res. Toxicol.* 15:1351–1359.

8. Lindon J C, Nicholson J K, Holmes E, et al. (2000). Metabonomics: Metabolic processes studied by NMR spectroscopy of biofluids. *Concepts Magn. Reson.* 12: 289–320.

9. Lindon J C, Holmes E, Bollard M E, et al. (2004). Metabonomics technologies and their applications in physiological monitoring, drug safety assessment and disease diagnosis. *Biomarkers.* 9:1–31.

10. Bino R J, Hall R D, Fiehn O, et al. (2004). Potential of metabolomics as a functional genomics tool. *Trends Plant Sci.* 9:418–425.

11. Sumner L W, Mendes P, Dixon R A (2003). Plant metabolomics: Large-scale phytochemistry in the functional genomics era. *Phytochem.* 62:817–836.

12. Claridge, T D W (1999). *High-Resolution NMR Techniques in Organic Chemistry.* Elsevier Science, Oxford, UK, pp. 384.

13. Liu M, Nicholson J K, Lindon J C (1996). High resolution diffusion and relaxation edited one- and two-dimensional ¹H NMR spectroscopy of biological fluids. *Analyt. Chem.* 68:3370–3376.

14. Nicholson J K, Foxall P J D, Spraul M, et al. (1995). 750MHz ¹H and ¹H-¹³C NMR spectroscopy of human blood plasma. *Analyt. Chem.* 67:793–811.

15. Khandelwal P, Beyer C E, Lin Q, et al. (2004). Nanoprobe NMR spectroscopy and *in vivo* microdialysis: New analytical methods to study brain neurochemistry. *J. Neurosci. Meth.* 133:181–189.

16. Keun H C, Beckonert O, Griffin J L, et al. (2002). Cryogenic probe ¹³C NMR spectroscopy of urine for metabonomic studies. *Analyt. Chem.* 74:4588–4593.

17. Tomlins A, Foxall P J D, Lindon J C, et al. (1998). High resolution magic angle spinning ¹H nuclear magnetic resonance analysis of intact prostatic hyperplastic and tumour tissues. *Analyt. Communicat.* 35:113–115.

18. Garrod S L, Humpfer E, Spraul M, et al. (1999). High-resolution magic angle spinning ¹H NMR spectroscopic studies on intact rat renal cortex and medulla. *Magn. Reson. Med.* 41:1108–1118.

19. Moka D, Vorreuther R, Schicha H, et al. (1998). Biochemical classification of kidney carcinoma biopsy samples using magic-angle-spinning ¹H nuclear magnetic resonance spectroscopy. *J. Pharmaceut. Biomed. Anal.* 17:125–132.

20. Cheng L L, Chang I W, Smith B L, et al. (1998). Evaluating human breast ductal carcinomas with high-resolution magic-angle spinning proton magnetic resonance spectroscopy. *J. Magn. Reson.* 135:194–202.

21. Cheng L L, Chang I W, Louis D N, et al. (1998). Correlation of high-resolution magic angle spinning proton magnetic resonance spectroscopy with histopathology of intact human brain tumor specimens. *Cancer Res.* 58:1825–1832.

22. Wilson I D, Nicholson J K, Castro-Perez J, et al. (2005). High resolution "ultra performance" liquid chromatography coupled to oa-TOF mass spectrometry as a tool for differential metabolic pathway profiling in functional genomic studies. *J. Proteome Res.* 4:591–598.

23. Soga T, Ohashi Y, Ueno Y, et al. (2003). Quantitative metabolome analysis using capillary electrophoresis mass spectrometry. *J. Proteome Res.* 2:488–494.

24. Lindon J C, Nicholson J K, Wilson I D (2000). Directly-coupled HPLC-NMR and HPLC-NMR-MS in pharmaceutical research and development. *J. Chromatogr.* B. 748:233–258.

25. Kaderbhai N N, Broadhurst D I, Ellis D I, et al. (2003). Functional genomics via meta-bolic footprinting: Monitoring metabolite secretion by *Escherichia coli* tryptophan metabolism mutants using FT-IR and direct injection electrospray mass spectrometry. *Compar. Funct. Genom.* 4:376–391.

26. Lewitt P A, Galloway M P, Matson W, et al. (1992). Markers of dopamine metabolism in Parkinsons disease. *Neurol.* 42:2111–2117.

27. Lindon J C, Holmes E, Nicholson J K (2001). Pattern recognition methods and applica-tions in biomedical magnetic resonance. *Progr. NMR Spectros.* 39:1–40.

28. Keun H C, Ebbels T M D, Bollard M E, et al. (2004). Geometric trajectory analysis of metabolic responses to toxicity can define treatment specific profiles. *Chem. Res. Toxicol.* 17:579–587.

29. Holmes E, Nicholson J K, Tranter G (2001). Metabonomic characterization of genetic variations in toxicological and metabolic responses using probabilistic neural networks. *Chem. Res. Toxicol.* 14:182–191.

30. Kell D B (2002). Metabolomics and machine learning: Explanatory analysis of complex metabolome data using genetic programming to produce simple, robust rules. *Molec. Biol. Rep.* 29:237–241.

31. Cloarec O, Dumas M E, Craig A, et al. (2005). Statistical total correlation spectroscopy: An exploratory approach for latent biomarker identification from metabolic ^{1}H NMR data sets. *Analyt. Chem.* 77:1282–1289.

32. Bollard M E, Stanley E G, Lindon J C, et al. (2005). NMR-based metabonomics approaches for evaluating physiological influences on biofluid composition. *NMR Biomed.* 18:143–162.

33. Lehtimaki K K, Valonen P K, Griffin J L, et al. (2003). Metabolite changes in BT4C rat gliomas undergoing ganciclovir-thymidine kinase gene therapy-induced programmed cell death as studied by H-1 NMR spectroscopy in vivo, ex vivo, and in vitro. *J. Biolog. Chem.* 278:45915–45923.

34. Nicholls A W, Mortishire-Smith R J, Nicholson J K (2003). NMR spectroscopic based metabonomic studies of urinary metabolite variation in acclimatizing germ-free rats. *Chem. Res. Toxicol.* 16:1395–1404.

35. Wang Y, Holmes E, Nicholson J K, et al. (2004). Metabonomic investigations in mice infected with *Schistosoma mansoni*: An approach for biomarker identification. *Proc. Nat. Acad. Sci. USA.* 101:12676–12681.

36. Nicholson J K, Wilson I D (1989). High-resolution proton magnetic resonance spec-troscopy of biological fluids. *Prog. Nucl. Magn. Reson. Spectros.* 21:449–501.

37. Lindon J C, Holmes E, Nicholson J K (2004). Toxicological applications of magnetic resonance. *Prog. Nucl. Mag. Reson. Spectros.* 45:109–143.

38. Phalaraksh C, Lenz E M, Nicholson J K, et al. (1999). NMR spectroscopic studies on the haemolymph of the tobacco hornworm, *Manduca sexta*: Assignment of ^{1}H and ^{13}C NMR spectra. *Insect Biochem. Molec. Biol.* 29:795–805.

39. Bundy J G, Lenz E M, Bailey N J, et al. (2002). Metabonomic investigation into the toxicity of 4-fluoroaniline, 3,5-difluoroaniline and 2-fluoro-4-methylaniline to the earthworm *Eisenia veneta (Rosa)*: Identification of novel endogenous biomarkers. *Environ. Toxicol. Chem.* 21:1966–1972.

40. Griffin J L, Walker L A, Shore R F, et al. (2001). High-resolution magic angle spinning ^{1}H NMR spectroscopy studies on the renal biochemistry in the bank vole (*Clethriono-mys glareolus*) and the effects of arsenic (As^{3+}) toxicity. *Xenobiot.* 31:377–385.

41. Viant M R, Rosenblum E S, Tjeerdema R S (2003). NMR based metabolomics: a power-ful approach for characterizing the effects of environmental stressors on organism health. *Environ. Sci. Technol.* 37:4982–4989.

42. Lindon J C, Nicholson J K, Holmes E, Breau A P, Cantor G H, Bible R H, et al. (2003). Contemporary issues in toxicology: the role of metabonomics in toxicology and its evaluation by the COMET project. *Toxicol. Appl. Pharmacol.* 187:137–146.

43. Keun H C, Ebbels T M D, Antti H, et al. (2002). Analytical reproducibility in ^{1}H NMR-based metabonomic urinalysis. *Chem. Res. Toxicol.* 15:1380–1386.

44. Ebbels T, Keun H, Beckonert O, et al. (2003). Toxicity classification from metabonomic data using a density superposition approach: "CLOUDS". *Analyt. Chim. Acta.* 490: 109–122.

45. Lindon J C, Nicholson J K, Everett J R (1999). NMR spectroscopy of biofluids. In G A Webb (ed.), *Annual Reports on NMR Spectroscopy*, Vol. 38. Academic Press, Oxford, UK, pp. 1–88.

46. Moolenaar S H, Engelke U F H, Wevers R A (2003). Proton nuclear magnetic resonance spectroscopy of body fluids in the field of inborn errors of metabolism. *Annals Clin. Biochem.* 40:16–24.

47. Maxwell R J, Martinez-Perez I, Cerdan S, et al. (1998). Pattern recognition analysis of ^{1}H NMR spectra from perchloric acid extracts of human brain tumor biopsies. *Magn. Reson. Med.* 39:869–877.

48. Swanson M G, Vigneron D B, Tabatabai Z L, et al. (2003). Proton HR-MAS spectroscopy and quantitative pathologic analysis of MRI/3D-MRSI-targeted postsurgical prostate tissues. *Magn. Reson. Med.* 50:944–954.

49. Sitter B, Sonnewald U, Spraul M, et al. (2002). High-resolution magic angle spinning MRS of breast cancer tissue. *NMR in Biomed.* 15:327–337.

50. Barton S J, Howe F A, Tomlins A M, et al. (1999). Comparison of in vivo ^{1}H MRS of human brain tumors with ^{1}H HR-MAS spectroscopy of intact biopsy samples in vitro. *Magma.* 8:121–128.

51. 't Hart B A, Vogels J T W E, Spijksma G, et al. (2003). NMR spectroscopy combined with pattern recognition analysis reveals characteristic chemical patterns in urines of MS patients and non-human primates with MS-like disease. *J. Neurolog. Sci.* 212: 21–30.

52. Brindle J T, Antti H, Holmes E, et al. (2002). Rapid and noninvasive diagnosis of the presence and severity of coronary heart disease using H-1 NMR-based metabonomics. *Nat. Med.* 8:1439–1445.

53. Waters N J, Holmes E, Williams A, et al. (2001). NMR and pattern recognition studies on the time-related metabolic effects of α-naphthylisothiocyanate on liver, urine, and plasma in the rat: An integrative metabonomic approach. *Chem. Res. Toxicol.* 14:1401–1412.

54. Coen M, Lenz E M, Nicholson J K, et al. (2003). An integrated metabonomic investigation of acetaminophen toxicity in the mouse using NMR spectroscopy. *Chem. Res. Toxicol.* 16:295–303.

55. Coen M, Ruepp S U, Lindon J C, et al. (2004). Integrated application of transcriptomics and metabonomics yields new insight into the toxicity due to paracetamol in the mouse. *J. Pharmaceut. Biomed. Anal.* 35:93–105.

56. Griffin J L, Bonney S A, Mann C, et al. (2004). An integrated reverse functional genomic and metabolic approach to understanding orotic acid-induced fatty liver. *Physiolog. Genom.* 17:140–149.

57. Clayton T A, Lindon J C, Antti H, et al. (2005). Pharmaco-metabonomic phenotyping: A new paradigm for personalized healthcare? *Nat. Biotechnol.* submitted.

58. Lindon J C,, Nicholson J K, Holmes E, et al. (2005). Summary recommendations for standardization and reporting of metabolic analyses. *Nat. Biotechnol.* 23:833–838.

13.4

BIOTERRORISM

DANY SHOHAM

Begin-Sadat Center for Strategic Studies, Bar Ilan University, Ramat-Gan, Israel

Chapter Contents

13.4.1	Preface	1526
13.4.2	Definition and Essence of Bioterrorism	1528
13.4.3	Concept and Incentives	1530
13.4.4	Bioterrorism-Patterned Natural Epidemics	1532
	13.4.4.1 Ebola Hemorrhagic Fever	1533
	13.4.4.2 AIDS	1534
	13.4.4.3 SARS	1535
	13.4.4.4 West-Nile Encephalitis	1536
	13.4.4.5 Avian Influenza	1537
	13.4.4.6 Cholera	1539
13.4.5	Significance of Bioterrorism-Patterned Natural Epidemics	1540
13.4.6	Spectrum of Applicable Pathogens and Toxins	1543
13.4.7	Bacterial Agents	1545
	13.4.7.1 Plague	1545
	13.4.7.2 Legionnaires Disease (Legionellosis)	1546
	13.4.7.3 Brucellosis	1547
	13.4.7.4 Tuberculosis (TB)	1548
13.4.8	Viral Agents	1549
	13.4.8.1 Rift Valley Fever (RVF)	1549
	13.4.8.2 Yellow Fever	1549
	13.4.8.3 Norwalk	1550
	13.4.8.4 Hepatitis A	1551
13.4.9	Toxins	1551
	13.4.9.1 Botulism Toxin	1552
	13.4.9.2 Staphylococcal Enterotoxin B (SEB)	1554
	13.4.9.3 T-2 Toxin	1554
	13.4.9.4 Aflatoxin	1555
	13.4.9.5 Aconitine	1557

Handbook of Pharmaceutical Biotechnology, Edited by Shayne Cox Gad.
Copyright © 2007 John Wiley & Sons, Inc.

13.4.10	Availability of Infective Agents and Toxins	1557
13.4.11	Technical Feasibility of Biosabotage Acts	1560
13.4.12	Operational Modes	1563
13.4.13	Spatially and Temporally Varying Impacts	1567
13.4.14	History and Evolution	1569
13.4.15	State-Sponsored Bioterrorism	1577
	13.4.15.1 Germany	1577
	13.4.15.2 Japan	1578
	13.4.15.3 United States	1578
	13.4.15.4 South Africa	1579
	13.4.15.5 USSR and Russia	1581
	13.4.15.6 Iraq	1584
	13.4.15.7 Iran	1585
13.4.16	Nonstate-Sponsored Bioterrorism	1585
	13.4.16.1 al-Qaeda	1585
	13.4.16.2 Jemaah Islamiah	1587
	13.4.16.3 Hezbollah	1588
	13.4.16.4 Radical Palestinian Terror Organizations	1588
	13.4.16.5 Aum Shinrikyo Cult	1589
	13.4.16.6 Bhagwan Shri Rajneesh Cult—Typhoid Epidemic	1589
	13.4.16.7 Shigella Dysenteriae Type 2 Outbreak	1591
	13.4.16.8 Hepatitis A Outbreak	1591
	13.4.16.9 Ascaris Suum Outbreak	1591
13.4.17	The Anthrax Letters	1592
13.4.18	The Ricin Course	1597
13.4.19	The Smallpox Complex	1600
13.4.20	Agroterrorism	1606
13.4.21	Narcoterrorism	1610
13.4.22	Novel Agents	1612
	13.4.22.1 Anti-Organism Power Multipliers	1613
	13.4.22.2 Supremely Engineered Bacteria and Viruses	1614
	13.4.22.3 Anti-Inanimate Object Power Multipliers	1614
	13.4.22.4 Bioengineering Availability	1615
13.4.23	Exploitability of Biotechnology and Biomedicine	1617
13.4.24	Preventative and Countermeasures	1620
13.4.25	Projections and Prognosis	1630
	References	1635

13.4.1   PREFACE

The very fundamental characteristics of bioterrorism are presented, alongside with the distinct multidisciplinary configuration of the term and phenomenon. Some unique attributes are outlined that make bioterrorism an entity nourished by different essentials and bearing various, in part far-reaching, implications.

Infectious diseases—the most effective extraneous regulator controlling the size of many animal populations in nature—are largely restrained in relation to mankind, thanks to the advancement of preventive and therapeutic medicine. The balance thus created is at any rate fragile, especially as human population density is steadily increasing, overall, whereas pathogens persist in acquiring drug-resistance. Pharmaceutical biotechnology is one major contributing arena to medicine, in that context; bioterrorism constitutes a counter-arena that unfortunately may amplify or substitute for the natural control mechanisms signified for by infectious diseases. Paradoxically, yet, pharmaceutical biotechnology in itself represents a potential key resource for bioterrorism. This is the case because pharmaceutical biotechnology is becoming an unparalleled front area, bearing a melting pot of remarkably sophisticated and multidisciplinary know-how, substances, and appliances. Large portions of which are dual use and may readily be exploited decently or viciously.

This built-in, intriguing duality is an essential feature, in that bioterrorism may imbibe extra potency from the very capacity destined to prevent and cope with bioterrorism. Although a very old way of sabotage, bioterrorism is presently, or soon to be, immeasurably more powerful, potentially, than ever, due to marked breakthroughs achieved in biopharmacology and biomedicine. Scientifically, this evolving, dichotomic course is inevitable, albeit significantly augmenting bioterrorism resources. The comprehension and practice allowing for combating pathogens and toxins may serve then, as well, to upgrade bioterrorism. It is becoming but a matter of their degree of availability of those bioagents. The same principle applies for pathogens of farm animals and plants, which form an additional dimension of bioterrorism—agricultural bioterrorism (agroterrorism). Also, the increasing availability of recreational drugs—mostly a sort of phytotoxins, in their essence—propels another terrorism variant—narcoterrorism.

In a sense, then, bioterrorism may embody the golem that rose against its creator. It can trigger a confined outbreak, an extensive epidemic, or a colossal pandemic, intentionally engined by man himself. The perpetrators may not even figure the anticipated impact—medically, psychologically, and logistically—which may immensely exceed what they want to achieve. Inversely, the in-effect impact may be but marginal, much less than that quested for. This uncertainty stems from various unknowns still marking the behavioral profile of infectious diseases. The interface between pathogens and man retains its enigmatic identity, although largely deciphered. At any rate, as James Woolsey, former CIA director has put it: "Germ terrorism is the single most dangerous threat to our national security in the foreseeable future." The magnitude of bioterrorism, yet, extends far beyond the boundaries of the United States.

Man only can eradicate, either locally or globally, long-lasting formidable pathogens—as was the case with smallpox—by means of extraneous, far-reaching intervention, namely vaccination; antibiotics are not less important for wiping out certain bacterial pathogens. Yet man is, at the same time, the prime messenger of their possible reemergence, particularly through bioterrorism. Conceivably, his intelligence and impulse both shape, thus, a delicate balance, one that might branch unthinkable ramifications. It is, then, his own and absolute responsibility to pursue a steady state that would eliminate or minimize the risk of epidemic self-destruction. In general, calculated self-destruction is unsupprisingly, antagonistic to the very nature of most species.

The menace of terrorism at large, in its broad sense, is an unparalleled phenomenon that may be considered, basically, to be a sort of a limited self-destruction mechanism for mankind. In a way, it is even worse then wars, because of its inherent unpredictability, irrationalism, and wide scope, as well as the built-in, unapparent, vitality, and durability marking it. When the dimension of biological warfare is added to this featuring, it turns out to pose uncontrollable potentiality.

Addressing the complicated domain of bioterrorism has to necessarily be conducted through a multidisciplinary approach. Therefore, the following chapter covers varitype aspects altogether, namely, microbiological, epidemiological, medical, technological, demographical, political, and strategic; the integration of which is essential, so as to comprehend the substance, complexity, and implications of bioterrorism. The entire issue of bioterrorism may thus gain a typical multidimensional structure and meaning.

13.4.2 DEFINITION AND ESSENCE OF BIOTERRORISM

Integrating two spheres that are innately unconnected to each other—terrorism and pathogens—the entity of bioterrorism reflects a complex, although in a sense fairly old, phenomenon of mankind. Given this starting point, the definition and essence of bioterrorism, together with its meaningfulness, are complicated. At the same time, the protean nature of bioterrorism endows it with uncommon attractiveness and advantageousness that may meet various courses.

The two elements comprising bioterrorism, namely terrorizing by means of biological agents, represent two distinct spheres, each bearing multiple contents. Those two spheres form, thereby, a singular, mighty conjunction between strategic studies and life sciences.

Terrorism at large, can be variedly featured, according to the following parameters:

- Conducted by a country, organization, group, or individual.
- Against a country, organization, group, or individual.
- By threatening and/or attacking.
- Covertly or overtly.
- Terror may be the goal or a by-product of an in-effect damage.
- The direct impact may be the ultimate objective or propel the occurrence of the ultimate objective.
- The impact is intended to form in the short, medium, or long run.
- With or without taking responsibility by the perpetrator.

Biological agents include, basically, pathogens and toxins, which may be classified in the following ways:

- Live—hence reproducing—agents (pathogens) or nonviable agents (toxins). Prions, as reproducing molecules, constitute a notable, exceptional intermediate.

- Lethal or sublethal (not a clear-cut division).
- Transmissible—hence epidemic—pathogens or nontransmissible pathogens.
- Natural or modified/engineered pathogens and toxins.
- Abruptly or gradually affecting the target.
- Affecting humans, husbandry, crops, or materials.
- The vehicle may be natural (infected arthropod, animal, or human being) or artificial (man-made disseminator, i.e., sprayers and envelopes).
- The route of penetrating the body is the respiratory system, alimentary tract, or skin.

Remarkably, almost any combination of the various mentioned parameters featuring terrorism and biological agents is feasible. That is one main attribute underlying the attractiveness and potential might of bioterrorism. Another one is the outstanding ratio between the amount of biological agent to be used and the resulting impact, especially with reference being made to epidemic pathogens. At the extreme, which is certainly feasible, an individual saboteur can disperse a scarcely existent, unapparent amount of a fully epidemic, virulent pathogen, be it pestilence, smallpox, or influenza, and give rise to a transgressing, questionably controllable, lethal plague.

Biocrimes, biosabotage, and biowarfare are terms appreciably dovetailing with bioterrorism, in different manners. Although biological crimes accentuate the illegal dimension of bioterrorism, they pertain, as well, to a variety of acts, such as the very holding or transferring of a certain microorganism, which are in violation of national or international rules and conventions. Biological sabotage is the in-effect employment of biological agents for whatever operational sabotage purpose, whether or not terrorism-oriented, usually by means of guerilla warfare. Biological warfare basically reflects a military confrontation in which biological weaponry is used; yet broadly, it may be waged against civilian target populations, thus explicitly having the quality of bioterrorism.

Bioterrorism may be perceived, overall, as the calculated, deliberate use of pathogens or toxins (or threat of using them) against civilian populations or economic/logistical infrastructures, in order to attain goals that are political, social, religious, financial, ideological, or personal in nature; this is done through intimidation or coercion or instilling fear. It may still be subclassified, then, as follows:

- The unlawful release of biological agents or toxins with the intent to intimidate or coerce a government or civilian population to further political or social objectives. Humans, farm animals, and cultivars are often targets.
- Use of microorganisms or toxins to kill, sicken, or cause other malfunction in people, animals, plants, or useful materials.

In its broad scope, thus, bioterrorism includes, beyond its classic targeting against humans:

- Agroterrorism, which may involve anti-plant pests, in addition to anti-plant and anti-animal pathogens.

- Narcoterrorism, namely the constant on-purpose input of recreational drugs onto a target population.
- Any other type of sabotage achievable by means of microbes, such as fuel-eating bacteria, asphalt eating-bacteria, and alike.

In a way, the core of bioterrorism, both objectively and subjectively, is horrifying, whereas the involved weapon in itself consists in, absurdly, some of the most power-ful natural foes of humanity, and can in practice bring about a very significant direct impact, far beyond horror. Hence, it may regretfully be regarded, in a sense, as the potential promoter of various infectious diseases that are currently being suppressed, pharmacologically and medically, just waiting for their opportunity to come.

One chief character of bioterrorism is, nonetheless, the conceivably immeasur-able disproportion between effort and outcome, particularly when the latter is at its maximum. It thus poses a strong temptation. Any other weapons, terror means—and apparently any other context—are not competent in sustaining such dispropor-tion. Technically, then, this attribute mostly represents the essence of bioterrorism, such that nears a legendary identity. The distance from reality is at any rate second-ary, especially as it can readily be bridged over, in affect, quite soon. In that concern, recent breakthroughs made in the field of genetic engineering amplify at the same time, paradoxically, both imagination and practicality.

All in all, within the sphere of terrorism at large, three distinctions typify the essence of bioterrorism:

1. The distinction between bioterrorism and conventional terrorism.
2. Between bioterrorism and other unconventional terrorism (chemical, radio-logical, nuclear).
3. Between bioterrorism and biological warfare.

The projection of those three distinctions altogether reflects the whole substance of bioterrorism, as aimed, among other things, to be shown in this chapter.

13.4.3   CONCEPT AND INCENTIVES

Several common denominators are outlined, which mark both terrorism-oriented states and terror organizations, on their way to resorting to bioterrorism. Other elements of their concept and incentives regarding bioterrorism comparatively differ, naturally, owing to their dissimilar essences. Various motives and objectives may then underlie their encouraged attitude toward bioterrorism. The roots are at any rate historical, appreciably shaping during time.

Intuitively, not having any concrete knowledge in terms of toxinology and bacteriology until the nineteenth century, man adopted poisonous and infective substances for attacking rivals. It was mainly practical common sense that guided him. Much later, paving its way toward scientific horizons, man gradually mastered the related technical fundamentals. The recognition of infective microorganisms and naturally produced toxic molecules accumulated and ripened. It reached the level of having the full capacity to operate them as warfare agents.

But, anthropologically, the concept underlying bioterrorism extends far beyond technical and operational features, whether state or non-state sponsored. The very notion of afflicting passive civilian populations with infectious diseases seems to be, objectively, profoundly evil, even if accompanied by different considerations and calculations. In its simplest form, then, bioterrorism reflects but an extreme way of malevolence. Variably, the dimensions of governance and insidiousness may certainly play a roll, in conjunction. *A priori*, the power to generate epidemics was possessed only by God and nature. Thus, biblically, God waged, initially, the epidemics included within the Ten Plagues of Egypt. Later on, comprehending the principal mechanism and impact of epidemics, man tried to emulate and immobilize that route against adversaries. As a matter of fact, it was around 1500 BC, when the Hittites sent plague victims into the lands of their enemies [1].

The fundamental incentive and concept of bioterrorism did not substantially change, up to these days. What did alter, obviously, are the know-how, the technical tools, and the strategic approach. Strategically, there are three basic alternatives shaping the concept of bioterrorism acts:

- The perpetrator (either state or non-state sponsored) would seek to remain unidentified.
- The perpetrator would be indifferent as to its identity.
- The perpetrator would seek to be identified.

The first and third alternatives have each two variants, respectively:

- Anonymity would rely on objective inability to epidemiologically determine whether the event was an act of bioterrorism.
- Anonymity would rely on objective inability to trace the perpetrator by means of investigation or intelligence (even if epidemiologically affirmed to be an act of bioterrorism).
- Identification would rely on direct announcement, either before or subsequent to the act of bioterrorism.
- Identification would rely on investigative and intelligence findings—either evidential or, deliberately, merely circumstantial (hence less valid)—without taking responsibility.

Several technological, political, and psychological explanations underly the quest of terrorists and terrorism-oriented states to opt for acquiring weapons of mass destruction (WMD), particularly biological agents, among them are the cheap cost of such weapons and the sense of prestige and security they grant to the owner. Parallel to offensive considerations, the inverse dimension of deterring capacity also plays a roll. All in all, terrorists or terrorism-oriented states may choose bioterrorism among many options, both conventional and unconventional. Preference may be given to bioterrorism due to various reasons. In terms of a pragmatic, clinical feasibility study, bioterrorism is, at least potentially, the most efficient form of sabotage, particularly when conducted indistinguishably from a natural event. Moreover, it may have a fully strategic impact; hence, it is often reckoned to be the ultimate mode of terrorism. Notably, most biological agents are not contagious; yet they might generate extensive outbreaks, if not epidemics, due to their transmissibility in other ways.

A terrorist group or terrorism-oriented state, intending to employ biological agents, will likely be attracted to their mass killing potential [2]. In addition, their delayed impact may also enable the saboteur to escape detection. Also, an attacker might reckon the resemblance between a natural and an unnatural epidemic to be close enough to divert suspicion, or at least make it extremely difficult to trace [3]. That would make it difficult to retaliate with a massive punitive strike, the expectation of which ordinarily serves as deterrent to an unconventional assault. Thus, the subsequent intimidation and chaos resulting from the inherently unknowable may make biological agents the terrorist weapon of choice [4]. On the whole, the related line of terms accentuating the significance and potential impact of bioterrorism at large include, then: high consequence terrorism; catastrophic terrorism; total terrorism; asymmetric terrorism; superterrorism; ultraterrorism, and so on.

Beyond, terrorists and terrorism-oriented states are certain to decide for themselves whether the conduction of bioterrorism serves their cause. Moreover, it is believed that biological weapons are preferable for terrorist groups, and that the allure for terrorists of biological weapons is currently intensifying due to the increase in availability of biotechnologists and sophistication of manufacturing methods [5].

Overall, one fundamental motive of bioterrorism (within the context of WMD at large), then, is the will to generate what is currently called high consequence terrorism, total terrorism, or catastrophic terrorism. Under certain circumstances it may be regarded as pure asymmetric terrorism, the purpose of which is not to take lives or destroy property. It is rather the mechanism by which the attacker seeks to achieve weakening of the sense of cohesion that binds communities together, to reduce its social capital, and to sow distrust, fear, and insecurity. Within this narrow course, it is but a method of waging a sort of social and psychological warfare.

The availability of pathogenic agents, alongside with technical feasibility and operational modality of bioterrorism at large, are basically rather in favor of the saboteur or terrorism-oriented state. Still, the bioterrorism events that already took place overall, seemingly exhibit but a portion of its potentiality, whereas the essence of the related threshold of proliferation is vague. Resultantly, built-in limitations interfere with thoroughly comprehending and foreseeing the course of bioterrorism.

13.4.4 BIOTERRORISM-PATTERNED NATURAL EPIDEMICS

The apparent resemblance between natural epidemics and induced ones mostly leads to recognizing the relevance of bioterrorism-patterned natural epidemics as prime demonstrative occurrences. Let alone, that various infectious diseases, chiefly viral, are emerging or reemerging in an unpredictable, often overwhelming, manner. It would therefore be advisable to diligently observe some prominent natural epidemics that took place during the recent decade.

The most distinct bioterrorism-patterned natural epidemics are typically those concerned with water- and food-borne infectious gastroenteritis, stemming from the contamination of a current collective source. Such epidemics would include, principally, two phases:

- The first wave of infections is generated directly by the consumption of the collective source.
- An additional wave is brought about, dependent on the degree of ongoing communicability or contagiousness of the causative agent toward potential secondary victims that are prone to be exposed, irrespective of the initial collective source.

This type of epidemic locally occurs very often, worldwide, in the form of relatively restricted innocent outbreaks. They bear the potency to expand, however. And they are quite easy to be produced deliberately, in practice. But, at any rate, they represent the simplest bioterrorism-patterned natural epidemics.

Bioterrorism-patterned natural epidemics might be by far more complex and hazardous, nonetheless. It so happened that the past 10 years were marked by some outstanding epidemic events perfectly illustrating the tremendous potentiality of biological agents at large, as well as the candidature of certain pathogens for acts of bioterrorism, in particular. Those events are presented here in detail, then, so as to demonstrate the general effectuality of various bioterrorism-patterned scenarios. They include six virus- and germ-propelled catastrophic epidemics.

13.4.4.1 Ebola Hemorrhagic Fever

The 14th of April, 1995, Zaire (now the Democratic Republic of Congo). A 36-year-old laboratory technician checked into the medical clinic in Kikwit, complaining of a severe headache, stomach pains, fever, dizziness, weakness, and exhaustion [6].

Surgeons did an exploratory operation to try to find the cause of his illness. To their horror, they found his entire gastrointestinal tract was necrotic and putrefying. He bled uncontrollably and within hours was dead. By the next day, the five medical workers who cared for him, including an Italian nun who assisted in the operation began to show similar symptoms, including high fevers, fatigue, bloody diarrhea, rashes, red and itchy eyes, vomiting, and bleeding from every body orifice. Less than 48 hours later, they, too, were dead, and the disease turned to form a deadly, uncontrollable epidemic.

As panicked residents fled into the bush, government officials responded to calls for help by closing off all travel, including humanitarian aid into or out of Kikwit, about 400 km from Kinshasa, the national capital. Fearful neighboring villages felled trees across the roads to seal off the pestilent city. No one dared enter houses where dead corpses rotted in the intense tropical heat. Boats plying the adjacent Kwilu River refused to stop to take on or discharge passengers or cargo. Food and clean water became scarce. Hos-pitals could hardly function as medications and medical personnel became scarce.

Deadly tropical fevers are an unfortunate fact of life in Central Africa but rarely are they this contagious or lethal. The plague that afflicted Kikwit was Ebola hemorrhagic fever, a viral disease for which there is no known treatment. Within a few weeks, about 400 people in Kikwit had contracted the virus and 350 were dead.

Where a 10% mortality rate is considered high for most infectious diseases, Ebola kills up to 90% of its victims, usually within only a few days after exposure. Still, contagiousness seems to be low.

The Kikwit Ebola outbreak was neither the first nor the last appearance of this dread disease. The first recognized Ebola epidemic occurred in 1976 in Yambuku, Zaire (near the Ebola River, after which the virus was named), where at least 280 people died. Three years later, 22 patients died in Sudan from a slightly different and less virulent form of Ebola. In 1996, about 100 people were killed by Ebola in two separate episodes in Gabon, and in 1999, an outbreak in the gold mining town of Durba in the Democratic Republic of Congo killed at least 63 people.

Ebola is one of two members of a family of RNA viruses called the Filoviridae (the other one being Marburg fever). No host or vector is known for Ebola, but it has been observed that monkeys and other primates can contract related diseases. Why viruses remain peacefully in their hosts for many years without causing much more trouble than a common cold, but then erupt sporadically and unpredictably into terrible human epidemics, is a new and growing question in environmental health and risk assessment. Although certainly a formidable pathogen, this virus exhibits changing rates of virulence (which is at times extremely high, more than any other pathogen), communicability, and, hence, epidemicity, the latter being rather restrained, in effect (400 infected and diseased people, out of 600,000 residents in Kikwit). Therefore, a shift in the transmissibility level of Ebola virus may be crucial and is immensely feared.

13.4.4.2 AIDS

AIDS, or acquired immunodeficiency syndrome, is a fatal disease caused by a rapidly mutating retrovirus that attacks the immune system and leaves the victim vulnerable to infections, malignancies, and neurological disorders. It was first recognized as a disease in 1981. The virus was isolated for the first time not long ago, in 1983. It is steadily spreading. During 2004, as a representation, around five million adults and children became infected with HIV (human immunodeficiency virus), the virus that causes AIDS. By the end of the year, an estimated 39.4 million people worldwide were living with HIV/AIDS. The year also saw more than three million deaths from AIDS, despite the availability of HIV antiretroviral therapy, which reduced the number of deaths in high-income countries. African and South-American tropical regions are heavily afflicted in particular for long periods of time.

But in certain regions the disease has risen remarkably swiftly during recent years. One of those arenas is Leningrad, were a notably exponential AIDS epidemic started in 1997. Beginning in that year and proceeding up to 2001, all patients in the Narcology Hospital of Leningrad Regional Center of Addictions (LRCA), were tested for HIV antibody [7]. These clinical records (i.e., serostatus, gender, age, and addiction) and data from the HIV/AIDS Center in the Leningrad Region were reviewed.

It has thus been shown that HIV prevalence at the LRCA increased from 0% to 12.7% overall, 33.4% among drug-dependent patients, and 1.2% among alcohol-dependent patients. During the same 5-year period (1997–2001), 2826 persons were registered at the HIV/AIDS Center: 6, 6, 51, 780, and 1983 persons in 1997, 1998, 1999, 2000, and 2001, respectively.

So, HIV infection is exploding in the Leningrad Region, currently in injection drug users, but potentially more broadly. The known high-per-capita alcohol intake in Russia heightens concern regarding the sexual transmission of HIV. Overall, the number of HIV-infected people in the Leningrad Region has risen to 2929 by the beginning of 2001, which is 2.3 times more than in 2000. Young people aged from 14 to 30, mostly drug addicts, account for 85% of the overall number of HIV-infected people. The number of teenagers sick with AIDS has risen by 2.5 times since last year. Particular concern was caused by a 10-fold increase in the number of AIDS victims among pregnant women. A total of 350 births given by HIV-infected mothers has been registered in the past year, and in 10 cases, parents gave up their babies [8].

However, the AIDS pandemic, in its broad essence, is actually a global long-term eroding factor toward mankind at large. Caused by a type of attritional bioterrorism agent, analogicolly, its ultimate real impact is expected sometime in the future, unforeseeably. For now, and in relation to the magnitude of humanity, it is just crawling, slowly yet steadily. It may be regarded as an inherent, low-rate—although undefeatable, for the time being—cryptic, ostensibly overt pandemic, accumulating its critical mass.

13.4.4.3 SARS

The SARS (sever acute respiratory syndrome) virus emerged—and possibly rather formed, somehow, little earlier, in reality—on the 16 of November 2002. In Foshan, southern China, on the Pearl River Delta, the first case of SARS involved a man, a "super-infector" who subsequently infected four others [9].

It soon exploded. By November 31, 2002, a deadly outbreak of SARS carried on in China and has killed at least 34 people in the south and three in Beijing. Hundreds have been infected. Clearly an infectious disease, the illness was named and registered only clinically, owing to the unidentified nature of its casual agent. It was certainly recognized to be a virulent virus, at any rate. At that stage, the new virus seemed to be fairly confined. But during December, the disease "hitch-hiked" 200 km to Heyuan city, the provincial capital of Guangzhou. Virus spread to five hospital workers. Panic ensued in Heyuan, and the local authorities tried to calm the people with news that there was not epidemic. Within days, the alarm subsided. Somewhere during this time, provincial party leaders curbed the media from publishing details of the disease, hindering further efforts to contain the virus. By the end of December 2002, Guangdong had reported at least 300 cases.

Although showing to be a new virus, a comprehensive genomic analysis has established that the relationship between SARS Coronavirus and group 2 CoVs is monophyletic [10]. Apparently, then, the SARS virus evolved as a strange zoonotic offshoot of group 2 CoVs, bearing greatly enhanced infectivity toward man.

Actually, SARS is the first major novel infectious disease to hit the international community in the twenty-first century. It originated in southern China in November 2002 and spread rapidly thereafter to 29 countries/regions on five continents. At the end of the epidemic, the global cumulative total was 8098 with 774 deaths. Seven Asian countries/regions were among the top 10 on the list. It has alarmed the world with its infectivity and significant morbidity and mortality, its lack of a rapid, reliable diagnostic test, and lack of effective specific treatment and vaccination. The adverse impact on travel and business around the world, particularly in Asia, has been enormous [11].

The consequences of the SARS epidemic were remarkable. In Taiwan, as an example, the impact of the SARS epidemic on the utilization of medical services was studied [12]. Using interrupted time-series analysis and National Health Insurance data between January 2000 and August 2003, this study assessed the impact of SARS epidemic on medical service utilization in Taiwan. At the peak of the SARS epidemic, significant reductions in ambulatory care (23.9%), inpatient care (35.2%), and dental care (16.7%) were observed. People's fears of SARS appear to have had a strong impact on access to care. Adverse health outcomes resulting from accessibility barriers posed by the fear of SARS are significant.

13.4.4.4 West-Nile Encephalitis

In late August of 1999, an unusual cluster of severe cases of encephalitis was noted in an area around Queens in New York City [13]. An epidemiologic investigation by the New York City Department of Health identified eight such cases and revealed that all of the patients had been previously healthy, had resided within the same 16 square mile area, and had recently engaged in outdoor activities. All but one had developed severe acute flaccid paralysis in the setting of encephalitis. At that point, the ethological agent had not yet been deciphered, though it seemed to be a basic virulent virus.

Both before and during the human encephalitis investigation, an epizootic among birds associated with a high fatality rate had been noted in and around New York City [14]. Pathologic assessment of the dead birds displayed involvement of multiple organs, including evidence of encephalitis; however, common avian pathogens were not detected [15]. The diseased birds were initially felt to be unrelated to the human epidemic. But genomic analyses using polymerase chain reaction and genome sequencing with specimens from New York City birds, infected mosquitoes collected in Connecticut, and human brain tissue from a fatal case of encephalitis, as well as expanded serological testing of specimens from suspected human cases identified WNE virus, a mosquito-borne pathogen, as the etiologic agent of this outbreak—4 weeks after the outbreak in humans was first reported to New York City public health officials [16–19]! By the end of the summer of 1999, 62 patients with serologic evidence of acute WNE virus infection, including 59 hospitalized patients, had been identified. Seven died. Similar to more recent outbreaks of the virus, the epidemic seemed to be associated with a high rate of CNS involvement and a preponderance of cases in patients older than 60 years [20].

The introduction of WNE virus in North America was followed by progressive spread throughout the United States. During the summer of 2000, 21 cases of human WNE virus illnes, were identified, 2 died occurred among 10 counties in northeastern states [21]. The following year, 66 cases were detected among a much more widespread geographic area, involving 38 counties in 10 states. The spread of human cases seemed to follow avian deaths; thus, avian death surveillance and, to a lesser extent, mosquito pool surveillance became important parts of public health efforts to track the virus and predict potential human cases [21]. In addition to avian and human illness, a substantial number of equine cases were documented throughout the United States, a pattern also being observed in other parts of the world.

An eco-epidemiological critical mass has been neared in 2002. During the summer of 2002 the number of WNV cases in North America ascended, and was the largest outbreak of arboviral meningoencephalitis ever documented in the Western Hemisphere. The WNE virus expanded its geographic range from the Mississippi River area at the conclusion of the 2001 season to the Pacific Coast by the end of 2002. Virus activity considerably increased and then found expression in the following figures: 2002—1460 cases; 66 deaths; 2003—9862 cases; 264 deaths; 2004—2470 cases; 88 deaths.

13.4.4.5 Avian Influenza

During 1997, an epidemic in Hong Kong of avian influenza in chickens and ducks occurred and was—merely marginally, then—transmitted to humans. Fortunately, the virus did not move from person to person and seemingly died out after the reservoir of domestic birds was completely depleted by the killing of approximately 1.6 million chickens, ducks, and geese by Hong Kong authorities. A variant of the type A H5N1 flu virus has been identified to be the cause of that event [22].

It was but an ostensible defeat of that epidemic H5N1 variant. In actuality it silently proliferated spatially, and then catastrophic outbreaks of influenza H5N1 took place among poultry in many countries in Asia during late 2003 and early 2004. At that time, more than 100 million birds in the affected countries either died from the disease or were killed to try to control the outbreak. (Thailand 36 million, Vietnam 36 million, Indonesia 15 million, China 5 million, Pakistan 4 million) (Cambodia, Laos South Korea, and Japan have been less affected, so far.)

By March 2004, the outbreak was reported to be under control. Beginning in late June 2004, however, new deadly outbreaks of influenza H5N1 among poultry were reported by several countries in Asia (Cambodia, China, Indonesia, Malaysia [first-time reports], Thailand, and Vietnam). These outbreaks—since extended— have been ongoing, and in 2005 reached Europe. They reached Africa in 2006.

The H5N1 virus does not usually infect humans. In 1997, however, the first case of spread from a bird to a human being was observed during the outbreak of bird flu in poultry in Hong Kong. The virus caused severe respiratory illness in 18 people, 6 of whom died. Since that time, there have been other cases of H5N1 infection among humans. Most recently, human cases of H5N1 infection have occurred in Thailand, Vietnam, and Cambodia during large H5N1 outbreaks in poultry. The death rate for these reported cases has been about 50%. Most of these cases occurred from contact with infected poultry or contaminated surfaces. Remarkably, however, it is thought that a few cases of human-to-human spread of H5N1 have occurred.

So far, spread of H5N1 virus from person to person has been rare, and proliferation has not continued beyond one person. Yet, because all influenza viruses have the ability to change, there is a great concern that the H5N1 virus could one day be able to become contagious and spread easily from one person to another. Because these viruses do not commonly infect humans, there is little or no immune protection against them in the human population. If the H5N1 virus were able to infect people and spread easily from person to person, a new flu pandemic could probably begin. It is impossible to foresee when a pandemic might occur. However, experts from around the world are watching the H5N1 situation in Asia very closely and are preparing for the possibility that the virus may begin to spread more easily and widely from person to person. Still, the H5N1 strain is but one variant within an immeasurable natural gallery of avian flu viruses. A few of them have already infected man, whereas the majority are taking their time, for now. A highly communicable derivative of them or of the H5N1 type will most probably generate the next colossal pandemic.

Until now, the most demanding flu pandemic that has been documented was the "Spanish Flu" (also known as "Swine Flu"), which killed 40–90 million people during just two years (1918–1919). The virus was of the H1N1 subtype. Later on, in 1957 a new serotype, H2N2, emerged and generated another pandemic, substituting the H1N1 virus. The next pandemic was given rise to in 1968 by the serotype H3N2, which replaced the H2N2. In 1977, a mild H1N1 virus resurfaced, bringing about another pandemic, but it did not substituted the H3N2 virus, and they are both prevailing currently. As a matter of fact, all pandemic and any other influenza virus strains are derived from avian viral strains. Moreover, at least some of them are likely preserved in perennial lake ice, mostly in the Northern Hemisphere, and may reintroduce—subsequent to ice thawing—genes or entire genomes onto the current genetic pool of that protean virus [23]. In regard to the current H5N1 strain, which is considerably virulent but noncontagious, for now, it may attain transmissibility by even a minor genetic alteration, the full essence of which is not yet clear.

In case this virus would retain its ongoing high communicability and virulence toward farm birds (currently chickens and ducks mainly) concomitantly with such genetic alteration, then an unparalleled pathogen, both pandemic and epizootic, may rise. Theoretically, it could disease pigs and horses as well, as the influenza A virus is in general infective and at times virulent toward those domestic mammals. But apparently this cannot occur in actuality, because there is, in all likelihood, some inherent viral incompetence to have such amplified affinity toward more then one host simultaneously. The remarkable potentiality of influenza A virus to attack various farm animals signifies, at any rate, its distinct candidature in the context of agroterrorism (see below). Yet even in terms of merely an anti-human bioterrorism weapon, this varitype virus has been reckoned to be an ultimate pathogen, in some respects even exceeding the formidability of the notorious smallpox virus [24].

Pandemic spread would presumably bring about 60–80 million deaths worldwide. According to US Homeland Security Council's estimates, 90 million people could be infected in the US alone in case a bird flu pandemic breaks out, of which 45 million will require medical care, 10 million will require hospital admission and 2 million people could die [25].

A British government report warns that The health service will be plunged into chaos if Britain is struck by a bird flu pandemic. Faced with a possible 4.5 million victims, demand for hospital beds would outstrip supply and doctors might have to deny treatment to the sick and elderly to save younger, fitter patients [26].

Not casually, all of the above-detailed illustrative epidemics represent, basically, bioterrorism-patterned scenarios specifically propelled by viruses. Antibiotics, the main combat-tool against germs, are of no relevance against viruses. Antiviral drugs are very limited in their scope. Vaccines are efficient against most viruses, but they cannot cope with viruses that are currently dynamically altering. Antisera against viruses are of marginal pharmacological contribution, for now. Viruses pose, then, in a sense, the ultimate biological weapon, and can be used for bioterrorism purposes. Still, the cultivation and handling of viruses is complicated in comparison with bacteria, due to the necessity of host tissue, whereby viruses obligatorily propagate.

13.4.4.6 Cholera

But, beyond viruses, other classes of pathogens and toxins are not at all to be depreciated, certainly. The germ causing cholera—mostly a water-borne disease—is nowadays the most epidemic and demanding bacterial pathogen, with reference being made to acute illness (parallel to the inversely lingering course of illness of tuberculosis, another remarkably demanding bacterial disease). In contrast to viruses, the cardinal characteristic marking pathogenic bacteria is their interface with various antibiotics, which is a very tangled and dynamic one. Thus, during March and April 2002, a resurgence of the causative agent *Vibrio cholerae* O139 occurred in Dhaka and adjoining areas of Bangladesh with an estimated 30,000 cases of cholera [27]. The reemerged O139 strains belong to a single ribotype corresponding to one of two ribotypes that caused the initial O139 outbreak in 1993. Unlike the strains of 1993, however, the recent strains are susceptible to trimethoprim, streptomycin, and sulphamethoxazole, but resistant to nalidixic acid. This alternating profile exemplifies a predominant intensifying property of various bacterial pathogens resisting antibacterial drugs in an unpredictable fashion.

In general, cholera is an acute intestinal infection caused by ingestion of contaminated water or food. It has a short incubation period, from less than 1 day to 5 days, and produces an enterotoxin that causes copious, painless, watery diarrhea that can quickly lead to severe dehydration and death if treatment is not promptly given. Vomiting also occurs in most patients. Man-made and natural disasters can intensify the risk of epidemics considerably, as can conditions in crowded refugee camps. Explosive outbreaks with high case-fatality rates are often the result. For example, in the aftermath of the Rwanda crisis in 1994, outbreaks of cholera caused at least 48,000 cases and 23,800 deaths within one month in the refugee camps in Goma, the Congo [28]. Although rarely so deadly, cholera outbreaks continue to be a major public health concern, causing considerable socioeconomic disruption as well as loss of life. In 2001 alone, the WHO and its partners in the Global Outbreak Alert and Response Network participated in the verification of 41 cholera outbreaks in 28 countries.

Throughout history, populations all over the world have sporadically been affected by devastating outbreaks of cholera. Records from Hippocrates (460–377 BC) and Galen (129–216 AD) already described an illness that might well have been cholera, and numerous hints indicate that a cholera-like malady has also been known in the plains of the Ganges River since antiquity.

Modern knowledge about cholera, however, dates only from the beginning of the nineteenth century when researchers began to make progress toward a better understanding of the causes of the disease and its appropriate treatment. The first cholera pandemic, or global epidemic, started in 1817 from its endemic area in Southeast Asia and subsequently spread to other parts of the world. The first and subsequent pandemics inflicted a heavy toll, spreading all over the world before receding.

In 1961, the seventh cholera pandemic wave began in Indonesia and spread rapidly to other countries in Asia, Europe, Africa, and finally in 1991 to Latin America, which had been free of cholera for more than one century. The disease spread rapidly in Latin America, causing nearly 400,000 reported cases and over 4000 deaths in 16 countries of the Americas that year [29]. In 1992, a new sero-group—a genetic derivative of the EI-Tor biotype—emerged in Bangladesh and caused an extensive epidemic. Designated *V. cholerae* 0139 Bengal (as mentioned above), the new serogroup has now been detected in 11 countries and likewise warrants close surveillance. Although no evidence is available to gauge the significance of these developments, the possibility of a new pandemic cannot be excluded. EI-Tor, for example, was originally isolated as an avirulent strain in 1905 and subsequently acquired sufficient virulence to cause the current pandemic.

The economic and social impacts of cholera are immense. In addition to human suffering caused by cholera, cholera outbreaks generate panic, disrupt the social and economic structure, and can impede development in the affected communities. Unjustified panic-induced reactions by other countries include curtailing or restricting travel from countries where a cholera outbreak is occurring or imposing import restrictions on certain foods. For example, the cholera outbreak in Peru in 1991 cost the country US\$ 770 million due to food trade embargos and adverse effects on tourism.

The germ *Vibrio cholerae* is indeed a current unrestrained foe. Prominently transmissible among humans, this pathogen may readily be used for bioterrorism purposes.

13.4.5   SIGNIFICANCE OF BIOTERRORISM-PATTERNED NATURAL EPIDEMICS

The principles and practicality of differentiating between epidemics stemming from bioterrorism acts and natural epidemics are outlined, to the extent attainable. The comprehension of bioterrorism-patterned natural epidemics thus turns out to be essential, and the related factors are presented in details, so as to facilitate the analysis of abrupt outbreaks.

The occurrence of epidemics, or outbreaks, at large, is the outcome of three alternative courses:

- A natural biological process or move, such as the emergence of a new pathogenic variant, the spread of an infected vector onto a new biotope, and so on.
- An innocent, either passive or active, artificial intervention, such as the arrival of an infected person at an epidemiologically virgin area/place, with relation to the pathogen harbored by him.
- An act of bioterrorism (or a military biological war fare operation).

The significance of bioterrorism-patterned natural epidemics appears to be contributive in two major senses. It helps to appraise the profile and impact of bioterrorism scenarios, and it may facilitate the differentiation of an epidemic (or an outbreak), being a natural or man-made—either intentional or unintentional—occurrence. Hence, the recognition of a bioterrorism act is thus facilitated as well. This paradigmatic axis is vital for the very basic classification and handling of any epidemic or outbreak of an infectious disease or poisoning. In certain cases of diseased individuals, it ought to be applied as well, so as to arrive at the respecting differentiation. Thereby, the field of forensic microbiology plays an important roll.

In practice, classification and handling of various occurrences and events of concern ought to rely on the following elements:

- Biotlogical and chemical analysis of specific markers related to the pathogen/toxin involved, so as to trace its provenance.
- Epidemiological analysis through which gauging of the event being a natural, innocently man-made, or deliberately man-made one, is allowed for.
- Information obtained or assessment made by intelligence sources regarding identified or possible perpetrators, if any.

One essential trait of bioterrorism, then, is the degree of resemblance of a natural event. The perpetrator might be trying to emulate a natural occurrence—so as to mask his act—or might be indifferent in that respect. Being an artificial intervention, bioterrorism can rarely or scarcely initiate an infectious course that would otherwise take place naturally in the same way of appearance. Yet this incriminating disadvantage would in most cases remain uncertain owing to ambiguity, let alone events of innocent human intervention. To minimize that gray area, the mechanisms underlying natural epidemiological phenomena ought to be meticulously and thoroughly recognized. In general, they are rather gradually evolving, mostly unapparent, although their outcome may be abrupt and drastic, eventually. Acts of bioterrorism, even if looking similar in their resultant impact, lack the preceding evolving links that mark the natural apparatus. Of paramount importance, then, is the formation of a formulated, multifactorial epidemiological profile with respect to every infectious disease or toxin of special concern. This is particularly important regarding the deciphering of unexplained biological invasions of new pathogens, toxins or infected vectors/carriers.

A proper case study is the totally unexpected appearance of West Nile encephalitis (WNE) in New York City, which reflected, actually, a much broader and fundamental phenomenon, namely the introduction of a pathogen prevailing for long in the Eastern Hemisphere onto the Western Hemisphere (or vice versa). In that case, inherent difficulty emerged in both identifying the etiological agent and deciphering, later on, its provenance. Although a well-known disease in the Eastern Hemisphere, it was impossible, subjectively, to clinically identify, or even suspect, WNE. This means, instantly, that the inventory of diseases being looked into ought to be, *a priori*, perfectly global, even in terms of initial symptomatology. The term "exotic" (disease or pathogen)—hence less likely, ostensibly—largely loses its essence, therefore. As the primary suspicion was that of an arthropod-borne virus (arbovirus) encephalitis, early testing was directed at common eastern North-American arboviruses. The synchronic mortality of collocated birds could have been instrumental for postulating that the concerned virus is basically an avian pathogen. Early serologic testing displayed IgM antibodies against the St. Louis encephalitis virus by enzyme-linked immunosorbent assay, a basically correct finding that later turned to be misleading, showing a built-in limitation in identifying a pathogen new to a given locality.

Misdiagnosis is at times worse than lack of diagnosis, particularly with respect to infectious diseases, let alone in the context of epidemiological tracing intended to gauge bioterrorism-patterned events. Deliberate introduction of an encephalitic virus has been considered, among other possibilities, and has emerged to be more specific when it became clear that the etiological agent is WNE virus. It was a conjunction of three elements that propelled such assessment:

- Information indicating that Iraq cultivated WNE virus as a biological agents [30].
- An Iraqi intention to strike the United States by whatever means [30].
- Ostensible infeasibility of WNE virus natural transfer from the Western to the Eastern Hemisphere.

Sensibly, however, the possibility of bioterrorism was not regarded to be a prime one, in this case. The probability of a natural or accidental event still appeared to be higher. Although the mechanism of the introduction of the WNE virus into North America remains unknown, it seems clear that the source of the WNE virus strain detected in New York City originated in the Middle East. A similar avian epizootic among domestic geese in Israel during 1997 and 1998 had been attributed to WNE virus [16, 31]. Human cases of WNE virus occurred simultaneously in Israel and New York in August 1999, and when the genomic sequences of WNE virus isolates or infected human brain tissue from the New York City outbreak were compared with various non-U.S. strains, the greatest homology was found with a WNE virus strain isolated from a goose from the Israeli 1998 epizootic and subsequently with a strain detected in the brain tissue of an Israeli patient who died of WNE in 1999 [16]. In addition, both the pattern of high avian mortality previously not associated with WNE virus outbreaks as well as the severity of human CNS disease seen in New York City and Israel were similar during the 1999 outbreaks. In between 1998 and 1999, this specific strain of WNE virus could have probably been harbored by certain migrating avian host species (transfected by mosquitoes) along an axis extending from the Western Hemisphere, through Greenland, and, later, onto New York [32].

Deliberate introduction or release of infected birds or mosquitos compatible with the above-described natural apparatus could be equally feasible, conceivably; yet certainly extremely sophisticated, somewhat beyond the expected capacity of a terror organization or even a terrorism-sponsoring state. On the other hand, the employment of a current epidemic strain of cholera, for instance, through introducing it into food supplies or water systems, even if conducted far away from an ongoing epidemic, could generate a wide outbreak that by all means may seem natural, resembling an infected traveler that innocently took a transcontinental flight. The chain of transmission in such cases is much simpler; hence, they are liable to be followed, untraceably, in the form of an act of bioterrorism. Furthermore, concentrating on AIDS and Ebola, an elaborated analysis inquired into those aspects, and aroused certain viewpoints that illustrate the significance of bioterrorism-patterned natural epidemics [33].

Notably, a recent outbreak of mumps (June 2006, Kansas)—a viral, often epidemic disease—is expected to help health officials prepare for an act of bioterrorism or a flu pandemic [34]. The Kansas Health and Environment Department had found 761 confirmed or suspected cases of mumps around the state. In dealing with the outbreak, the department for the first time used its incident command system to spread information to agencies around the state. The same system would be used in the event of a natural or intentional disease outbreak.

Once the bioterrorism-employed pathogen has been introduced into the targeted site, the subsequent epidemiological featuring of the resultant events may expectedly be entirely futile, in terms of pointing to an act of bioterrorism, because those events would be shaped by whatever environmental, medical, and demographic circumstances prevailing at any rate in the targeted site. In the absence of direct evidence witnessing an act of bioterrorism, the only potentially incriminating indications can stem from the very specificity of the pathogenic strain involved, the particular manner of introduction, and the existence or lack of collateral factors such as natural vectors, carriers, or reservoirs, including natural abiotic vehicles, like rivers, of the pathogen in concern. Currently, the only fully reliable way to establish specificity is sequence analysis of both genomic and extragenomic DNA/RNA. This would be a necessary, yet usually insufficient, precondition/prerequisite for deciphering the provenance of a pathogen used in an act of bioterrorism.

Therefore, diligent modeling is vital. Evaluating surveillance systems for the early detection of bioterrorism is particularly challenging when systems are designed to detect events for which there are few or no historical examples [35]. One approach to benchmarking outbreak detection performance is to create semisynthetic datasets containing authentic baseline patient data (noise) and injected artificial patient clusters, as signal. This useful pattern has very recently been developed.

It has thereby been pointed out, indeed, that doctors in training had difficulties diagnosing diseases associated with bioterrorism. The study involved 631 doctors, mostly medical residents—doctors still in training—in 30 internal medicine residency programs in 16 states and Washington, D.C.; it was found that half of the doctors surveyed misdiagnosed botulism, 84% misdiagnosed plague, and chickenpox was misdiagnosed as smallpox 42% of the time [36].

Yet beyond the clinical domain, various, at times crucial, gaps still prevailing within our overall knowledge about the biology of pathogens may play a significantly negative roll in that they can possibly bring about false incrimination or, inversely, false discrimination, referring to bioterrorism-resembling events. That is but one reason, even if a major one, for deepening our research and understanding of the essence and pronunciations of the complex mechanisms underlying the emergence and prognosis of infectious diseases. Mankind, at large, would expectedly endeavor to be most familiar with its prime extraneous foes, especially as their might is currently and steadily increasing.

13.4.6 SPECTRUM OF APPLICABLE PATHOGENS AND TOXINS

A general featuring of the spectrum of pathogens and toxins applicable for bioterrorism is given. Basic distinctions among various bioterrorism warfare agents, together with the implications they bear, are thereby presented.

Basically, then, any epidemic pathogen may be regarded, potentially, as a candidate for carrying out a bioterrorism act, the germ causing plague, for example. To those agents should at any rate be added many nonepidemic—namely, incommunicable—pathogens, such as anthrax, and a large variety of toxins. The spectrum thus formed is in a way misleading in its amplitude.

In actuality, a much narrower range of pathogens and toxins that may be used, practically, as weapons is notable, although but a marginal fraction, quantitatively, of the entire class of bacteria, viruses, fungi, and other infectious agents plus numerous different toxins found in nature. This being the case, their characteristics should meet, one way or another, at least part of the following fundamental criteria:

- Infective/toxic and virulent.
- Procurable or reproducible.
- Conveyable and stable.
- Introducible into the targeted area and people.

Still, the spectrum of agents applicable for bioterrorism acts is broader than those applicable for armed conflicts, due to the greater vulnerability and wider variety of civilian infrastructures that may be targeted. Qualitatively, however, this specific offense-oriented range of pathogens and toxins is marked by very potent and aggressive biological agents, some of which are extremely contagious and medically untreatable.

Preferring, mostly justifiably, a comprehensive approach, the United Kingdom recently increased the number of biological agents that must be secured to ensure they are not obtained and used for acts of terrorism. The government boosted, thus, the number of restricted agents listed in the 2001 Antiterrorism, Crime and Security Act from 47 to 103. The list of controlled bioagents includes 45 viruses, 21 bacteria, 2 fungi, 13 toxins and 18 animal pathogens [37].

It seems, then, as if the most significant distinctions can be made between epidemic and nonepidemic agents, on the one hand, and, independently, between treatable and untreatable agents, on the other hand. Although the former distinction relates equally to bacterial and viral pathogens, the latter reflects a fundamental difference between those two major classes. Regardless of antisera, antiviral preparations are of limited efficacy, although they are expectedly being upgraded. Connectedly, within the sphere of toxins, the related spectrum may basically be divided into specifically treatable—protein toxins, hence may be coped by antitoxins—and specifically untreatable ones, which are otherwise chemically structured, like alkaloids. Vaccines, as prophylactic measures, are in principle efficient against viruses, bacteria, and protein toxins.

At the same time, it would be insensible to refer to those distinctions as absolute categorizations, because some pathogens represent intermediaries, to a certain degree. Other distinctions such as between lethal and sublethal pathogens are rather artificial, although of course not meaningless. Influenza, as one example, is generally regarded as an incapacitating disease, while potentially much more virulent, in reality. A genuine distinction, obviously, is the one categorizing self-reproducing agents—hence a multiplying mass of weapon—as against unviable agents, meaning toxins. The former, if effectively applied, constitute an infectious apparatus that intensify itself; the latter rather resemble, in that sense, chemical warfare agents, bearing their own immense capacity.

13.4.7 BACTERIAL AGENTS

Varitype pathogenic germs may be used for bioterrorism purposes through different modes, and they are here represented by four entities: plague, brucellosis, Legionnaires disease, and tuberculosis. Their heterogeneity with respect to source of infection, impact, as well as short-, medium-, and long-term endurance is thereby accentuated.

Within the context of bioterrorism, bacterial pathogens are signified for, foremost, by the germs responsible for the diseases of plague, anthrax, and alimentary tract infections (the latter group already typified above in the form of cholera). Other bacterial pathogens, such as those causing tularemia, brucellosis, glanders, Legionnaires disease, Klebsiella-associated pneumonia, and Q-fever, are of secondary—certainly not negligible—importance. A third category of potential microbial agents may be added, in case the perpetrator chooses to employ a slow-acting, scarcely traceable type of germ. Those three categories are here visited, then, through four representatives, namely, the causative agents of plague, Legionnaires disease, brucellosis, and tuberculosis. Anthrax is separately covered.

13.4.7.1 Plague

The plague bacterium is an old, salient foe to mankind. Posing three significantly different forms of an aggressive disease—bubonic, pneumonic, and septicemic plague—this germ is chiefly marked by two hosts, prior to man infection, namely rat and flee; hence, it can in principal be employed as a bioterrorism agent through each of those three modes, namely, inhalable air-borne bacteria, infected rats, or infected flees. Naturally, the humane plague cycle is usually initiated by the bite of a rodent flea or occasionally by the human flea.

The bubonic plague is the common form, so called because of the buboes or swellings, which are the obvious visible symptom, communicated by flea bites and attacking the human lymphatic system [38]. Simple bubonic plague may kill within 5 days, if antibiotic is not applied. It is regarded as the "true" or classic form of the disease. Pneumonic plague occurs when a victim develops pneumonia as a consequence of first getting the bubonic plague. The pneumonic variant occurs when the bacilli colonize the alveoli and cause latent suppurative pneumonia. This results in hemorrhagic sputum that can spread the disease via aerosolized droplets. In that case, the infection is then communicated by exhalation and affects the bronchial system of the next host directly. The pneumonic variant kills within 1–3 days. The third variant, the septicemic plague, is extremely rare and occurs where the bloodstream is directly affected raising the concentration of bacilli in the bloodstream to a level at which it can be transmitted directly by the human flea without passing through the intermediary host of the rat. This plague variant can kill within hours and explains contemporary accounts of people who went to bed apparently well and when they awoke in the morning they were, like the biblical Assyrians (also victims of a god-sent plague), all dead men.

The bubonic form of the plague is the one most widely described by chroniclers and, contrary to popular belief, can be survived, especially if the buboes burst or are lanced, although this was not understood at the time. Both the pneumonic and the septicemic varieties were fulminant and 100% lethal. The pneumonic variety, however, could be transmitted only by close human contact, in the same way as influenza today, but it accounted for the high mortality among monks, priests, and others tending to the dying, putting their faces close to the patient. As those infected in this way were unlikely to get far in the period between infection and death, the pneumonic plague remained localized within families and communities, for example, monasteries where it could spread like wildfire. Henry of Knighton, a regular Canon in the Abbey at Leicester and opponent of the Franciscan Order, reported the death of the entire population of 150 Friars in a Franciscan monastery in Marseilles, adding as a malicious afterthought, a good thing too. The bubonic version also kills its hosts, but more slowly, especially in the case of the flea that can survive dormant for weeks, even months, in the goods of a traveling merchant. The causative agent, *Yersinia pestis*, will, however, die off when it has itself killed off all accessible and susceptible hosts, whether human, rat, or flea. To sustain itself the bacillus must find a host that is not affected by it or that reproduces faster than the bacillus can kill the host.

13.4.7.2 Legionnaires Disease (Legionellosis)

The particular significance of *Legionalla* sp. as a human pathogen was not recognized until 1976, when a mysterious epidemic of pneumonia struck members of the Pennsylvania American Legion. Indeed, the most common presentation of *Legionella pneumophila* is acute pneumonia (legionellosis); potentially any species (about 40, allover) of *Legionella* may cause the disease. Extrapulmonary disease (e.g., pericarditis and endocarditis) is rare. Less often, the disease presents as a nonpneumonic epidemic, influenza-like illness called Pontiac fever [39].

Remarkably, *Legionella* bacilli are primarily waterborne. They reside in surface and drinking water and are usually transmitted to humans in watery aerosols. The pathogenesis of *Legionella* infections begins with a supply of water containing virulent bacteria and with a means for dissemination to humans. Person-to-person transmission has never been observed, and *Legionella* is not a member of the bacterial flora of humans. Infection starts in the lower respiratory tract. Alveolar macrophages, which are the primary defense against bacterial infection of the lungs, engulf the bacteria; however, *Legionella* is, unlike most bacterial pathogens, a facultative intracellular parasite and multiplies freely in alveolar macrophages. Recruited neutrophils and monocytes, as well as bacterial enzymes, produce destructive alveolar inflammation. Direct inoculation of surgical wounds by contaminated tap water has been described.

In most cases, the initial infective link is an aerosol of water contaminated with the organisms. Evaporative condensers and cooling towers are proven sources of outdoor infection. Indoors, nebulizers, and humidifiers filled with contaminated drinking water have disseminated *Legionella* to susceptible patients. The automatic misting devices that keep supermarket produce fresh have even been fingered as culprits in outbreaks of pneumonia. Notably, aerosols are produced in numerous ways in our environment, from taking a shower to flushing the toilet. An epidemic of Pontiac fever caused by *Legionella anisa* was associated with an ornamental fountain in a public place. Clusters of *Legionella* pneumonia have occurred after exposure to whirlpool spas in hotels or cruise ships. In most cases, the source of the infection remains unknown. Direct infection of surgical wounds has been linked to washing of patients with tap water that harbored pathogenic *Legionella* organisms.

Legionella infections may be sporadic or epidemic, community acquired, or nosocomial. There is great geographic variation in the frequency of infection even within communities, presumably reflecting the presence of suitable aquatic environments and susceptible subjects. Both sporadic and epidemic cases are more common during summer than winter months, apparently because of increased use of air-cooling equipment that generates aerosols. As a bioterrorism agent, this prominent bacterium is prone to be applied, then, in both water sources and air circulating systems.

13.4.7.3 Brucellosis

Brucellosis represents a narrow, uniquely significant category of diseases combining proximate and lingering effects altogether. This clinical profile may meet, then, certain needs of both short- and long-term warfare. In the short run, it is featured by an acute phase. After a relatively long and variable incubation period (1–8 weeks), it most often manifests as an acute bacteremic disease, and it may be complicated by secondary hematogenous localization in almost every organ [40, 41]. Inversely, chronic brucellosis is defined as a disease evolving for more than a year, without evident secondary localization. During asymptomatic infection, the pathogen is sequestered in lymphoid tissues, bone, and liver. It would tend to persist for many years, at times in a steadily debilitating incurable state. Four bacterial species are responsible for human brucellosis: *Brucella suis* (from pigs; high pathogenicity), *Brucella melitensis* (from sheep; highest pathogenicity), *Brucella abortus* (from cattle; moderate pathogenicity), and *Brucella canis* (from dogs; moderate pathogenicity). Many animals are potential reservoirs for various *Brucella* species [40, 41].

Epidemiologically, brucellosis is a zoonotic infection transmitted from animals to humans by ingestion of infected food products (milk and dairy products are the main sources for human contamination), direct contact with an infected animal, or inhalation of aerosols. This last mode of transmission is remarkably efficient given the relatively low concentration of organisms (as few as 10–100 bacteria) needed to establish infection in humans and has brought renewed attention to this old disease. Descriptions of the disease date back to the days of Hippocrates, although the organism was not isolated until 1887, when British Army physician David Bruce isolated the organism that bears his name from the spleens of five patients with fatal cases on Malta. The disease gets its additional names from both its course (undulant fever) and location (Malta fever, Crimean fever).

Given the ease of effectiveness of aerosol transmission of *Brucella* species, researchers attempted to develop it into a biological weapon beginning in 1942. In 1954, it became the first agent weaponized by the old U.S. offensive biological weapons program. Field testing on animals soon followed. By 1955, the United States was producing *B suis*-filled cluster bombs for the U.S. Air Force at the Pine

Bluff Arsenal in Arkansas. Not much later, it was weaponized and standardizes by the USSR. The pathogen may meet both large-scale and guerilla warfare purposes.

13.4.7.4 Tuberculosis (TB)

Although AIDS represents an attritional viral disease, as described, tuberculosis is, in parallel, a bacterial disease of a long-lasting debilitating course, both individually and demographically. Migration of contagion within community and the course of the disease in itself are very clumsy, yet consistent, aided by fortified stability toward antibacterial drugs. A staggering 1.9 million around the globe die of tuberculosis each year—another 1.9 billion are infected with the causative agent, *Mycobacterium tuberculosis*, and are at risk for active disease [42]. It's sobering to realize that this ancient scourge, which has been found in 2000-year-old mummies (but much earlier evolved), remains a severe global health threat despite modern medicine. In fact, TB is a leading infectious disease killer in the world, alongside AIDS and malaria.

Millions of people—fully one third of the world's population—are infected with TB. Five to 10% of those will develop the active disease, for reasons only partially understood. This is, certainly, a key trait. The bacteria can lie dormant for many years and become active when the host is weakened by factors such as stress, poor nutrition, diabetes, or HIV infection. Once active, TB destroys tissue in the lungs and becomes contagious. Because it is airborne, there is little that can be done to protect against infection.

Antibiotics and public health measures failed to wipe out TB, although until the mid-1980s its rates had been brought very low, even in the developed world. Then the TB bacteria acquired immunity to many drugs, spreading rapidly among homeless people, AIDS patients, and other vulnerable groups worldwide. Several drugs can cure TB, but only if they are taken for many months. With abbreviated treatment, bacteria that have evolved ways of evading the drugs can escape eradication and proliferate as a resistant strain. Then, even when treated with the strongest drugs for 2 years, the resistant TB is fatal about 60% of the time.

Russian prisons are a major source of multidrug-resistant TB. As tens of thousands of infected inmates reenter the civilian world every year, TB spreads when they cough and sneeze. Some experts fear that with the speed and frequency of modern travel, the resistant epidemic could spread to Europe and the United States. To counter the threat of a worldwide epidemic, the costly second-line drugs are supplied, and prison health workers are being helped to ensure that infected inmates take their entire course of drugs—even if it means the prisoners have to remain in confinement for up to 2 years past the start of treatment.

In 2005, in Japan, the Tuberculosis Prevention Law was set to be abolished so that bacteria that cause the disease can be included in counterterrorism regulations in an updated Infectious Disease [43]. The Tuberculosis Prevention Law (passed in Japan in 1951) does not contain counterterrorism provisions or regulations requiring the quick investigation of an epidemic. The change is aimed at restricting transfer of the disease (alongside with other ones), allowing inspections, and requiring registration of ownership of samples. The bacterium causing TB is regarded, in a sense, as an attritional bioterrorism agent.

13.4.8 VIRAL AGENTS

In addition to the viruses mentioned above, some other viral pathogens are notice-
able, of which several are profiled herewith as varitype representatives: Rift Valley
fever, yellow fever, Norwalk, and hepatitis A. The smallpox virus is presented later.
Notably, the handling of viruses is much more complicated then that of bacteria;
hence, it is less likely to be mastered by terrorist organizations, individuals, or per-
petrators, with reference being made to the viruses mentioned below as well as
others.

13.4.8.1 Rift Valley Fever (RVF)

The RVF virus typically represents a zoonotic hemorrhagic pathogen equally
potent against man as well as various mammalian farm animals (such as cattle,
buffalo, sheep, goats, and camels). For viruses, this wide spectrum featuring is rare,
making this pathogen a biological agent of appreciable uniqueness. This trait char-
acteristically found expression in the late 1970s, in Egypt. After an initial epizootic
of RVF in Egypt in 1971, the first recorded epidemic of RVF in Egypt, in 1977–
1980, inflicted an estimated 200,000 human cases, with some 600 reported deaths,
plus enormous losses of farm animals [44]. An epidemic of the disease occurred
again in Egypt in 1993. RVF virus is naturally a mosquito-borne pathogen, but it
is adequately infective in the form of an air-borne aerosol, making it a biological
agent applicable through those two modes. It is found in the Old World only, principally in Africa. In September 2000, an RVF outbreak was reported in Saudi
Arabia and subsequently Yemen. These cases represent the first RVF cases identified outside Africa. Fortunately, it is not carried by migrating birds, as is the case
with many other arboviruses (including the parallel mosquito-borne West Nile
virus) and with influenza A virus. RVF virus, then, is basically a geographically
slowly crawling virus of notable potentiality, particularly with respect to tropical
and subtropical regions. It may be employed as a bioterrorism agent in the
context of both antihuman bioterrorism as well as agroterrorism. Basically a
mosquito-dependent tropical virus, it could doubtfully colonize temperate regions,
however.

RVF is generally observed during years in which unusually heavy rainfall and
localized flooding occur [45]. The excessive rainfall allows mosquito eggs, usually
of the genus *Aedes*, to hatch. The mosquito eggs are naturally infected with the
RVF virus, and the resulting mosquitoes transfer the virus to the livestock on which
they feed. Once the livestock is infected, other species of mosquitoes can become
infected from the animals and can spread the disease among other hosts, including
man.

13.4.8.2 Yellow Fever

The Yellow Fever virus is another mosquito-borne pathogen, in that case affecting
both man and primates. In the past, it was mass-accumulated as a biological
weapon, in the form of infected mosquitoes, by the U.S. Army. Although somewhat
complex, this is an effective way of employing this virus as a weapon either through
guerilla or apparent warfare. It so happened that this standardized weapon-form

resembled and reestablished the first human virus ever isolated. It was in 1901 that the pathogen causing yellow fever was proved to be a filterable virus carried by mosquitoes.

Yellow fever is a viral hemorrhagic fever. There are three types of transmission cycle: sylvatic (or jungle), intermediate, and urban. All three cycles exist in Africa, but in South America, only sylvatic and urban yellow fever occur [46]. Sylvatic yellow fever occurs in tropical rainforests where monkeys, infected by sylvatic mosquitoes, pass the virus onto other mosquitoes that feed on them; these mosquitoes, in turn, bite and infect humans entering the forest. This produces sporadic cases, the majority of which are often young men working in the forest, e.g., logging.

On occasion, the virus spreads beyond the affected individual. The intermediate cycle of yellow fever transmission occurs in humid or semi-humid savannahs of Africa and can produce small-scale epidemics in rural villages. Semi-domestic mosquitoes infect both monkey and human hosts and increased contact between man and infected mosquito leads to disease. This is the most common type of outbreak observed in recent decades in Africa. Urban yellow fever results in large explosive epidemics when travelers from rural areas introduce the virus into areas with high human population density. Domestic mosquitoes, most notably *Aedes aegypti*, carry the virus from person to person. These outbreaks tend to spread outward from one source to cover a wide area.

13.4.8.3 Norwalk

The Norwalk virus is an enterovirus with pronounced eco-epidemiological traits. It causes acute gastroenteritis with nausea, vomiting, fever, and myalgia that lasts 24–48 hours. The virus is transmitted through fecal-oral contact. The Norwalk virus is well established as the chief cause of viral gastroenteritis epidemics. The disease occurs throughout the year without a seasonal predominance.

Norwalk virus was first associated with gastroenteritis in 1972. For a long time, it was an unnoticed, although extremely important, pathogen. It was identified by electron microscopy of stool samples that had been saved from a 1968 gastroenteritis epidemic that occurred in Norwalk, Ohio [47]. In a 2-day period, acute gastroenteritis developed in 50% of 232 students or teachers in an elementary school. The virus initially was labeled as a small, round, structured virus, and it was named after the city in which the outbreak occurred. The virus is transmitted through contaminated food, water, or infected contacts. After ingestion, the virus infects the mucosa of the proximal small intestine, damages microvilli, and causes malabsorption. Although no histopathological lesions can be found in the stomach mucosa, the virus causes abnormal gastric motility and delayed gastric emptying.

In the United States, the Norwalk virus is estimated to cause approximately 40% of cases of nonbacterial gastroenteritis. The Norwalk virus is, then, the leading cause of viral gastroenteritis in the United States. January 1, 2002, to December 2, 2002: Norwalk virus was attributed to 9 of the 21 outbreaks of acute gastroenteritis on cruise ships reported to the U.S. Centers for Disease Control and Prevention's Vessel Sanitation Program in this time period. The Norwalk virus brings about approximately 23 million cases of acute gastroenteritis each year and is the major cause of outbreaks of gastroenteritis. An extensive outbreak has been recorded in Europe (Finland). As a bioterrorism weapon, this enterovirus is a typical incapacitating agent.

13.4.8.4 Hepatitis A

Inversely a water/food-borne virus of a relatively long incubation period (several weeks) and a causative agent of a lingering incapacitating infectious disease, hepatitis A virus (HAV), has been intended mainly for attrition warfare purposes and has apparently been employed by the Soviet army in Afghanistan, through contaminating water or food sources. Also, hepatitis A virus is reported to have been studied as a possible bioterrorism agent by South Africa.

Infectious hepatitis caused by HAV is common worldwide and is usually spread as a result of poor personal hygiene and can be a problem in day care centers, refugee camps, and among troops in the field. The virus is hardy and can withstand relatively harsh chemical treatments and can survive outside the body for days. It attacks humans only, invading the liver and generating, hence, a systemic debilitating illness.

The disease has an incubation period that is relatively long for a biological weapon at 3–4 weeks, but the onset of the disease is abrupt and victims become exhausted and develop jaundice as their livers lose function [48]. Victims typically take about 10 weeks to recover. The disease is rarely fatal, and victims have lifelong immunity. Overall, hepatitis A virus may be regarded, then, as a representative of a class of water/food borne antihuman viruses, in that case, a pathogen of a rather attritional nature. Lasting for more than 3 months, the entire course of the virus since contracted by a victim produces a fairly chronic infection.

13.4.9 TOXINS

Overall, about 400 toxins have been identified, which are certainly but a small proportion of what is actually found in nature. Out of this remarkable inventory, several are reckoned as typical bioterrorism agents. Presented here are few prominent toxins: two protein toxins (botulinum and SEB)—hence detectible and treatable by anti-toxins—and three non-protein toxins (T-2 toxin, aflatoxin and aconitine), hence hardly detectable or treatable by antidotes. Etiological diagnosis of toxins is often extremely complicated.

The realm of toxins is a very wide one. Those microbial, fungal, plant, and animal biomolecules exhibit an unthinkably diversified inventory of natural substances, all poisonous toward man, by definition. They are extremely variable chemically and physically. But dosage plays a cardinal role, both in terms of its great differential among toxins, as well as the delicate transition onto the tangential domain of toxin-based pharmacological applications. Moreover, most usable for man—as proteins, chiefly—in preparing toxoids and antisera, toxins are inwardly inert—sometimes rather immensely beneficial—within their own toxinogenic bio-system, or extremely poisonous, outwardly, as natural toxicants. Fundamentally, they constitute the pristine form of biochemical warfare in nature, much earlier to the emergence of mankind. Thus, for instance, Hydrogen cyanide, reckoned as a classical chemical warfare agent, is found in various toxic plants, thereby equipping them with a powerful poison.

Overall, about 400 toxins have been identified, which are certainly but a small proportion of what is actually found in nature. They include, however, bacterial toxins—49; plant—36; fungal—26; algal—22; (other) marine organism—111; snake—124; insect—22; amphibian—5; and in total—395. Out of this remarkable inventory, several are reckoned as typical bioterrorism warfare agents. Presented here are few prominent toxins: two protein toxins (botulinum and SEB)—hence detectible and treatable by antitoxins—and three nonprotein toxins (T-2 toxin, aflatoxin, and aconitine), which are hardly detectable and treatable by antidotes. Ricin, an outstanding bioterrorism toxin on its own, is covered separately. Other notable toxins, such as tetradotoxin, saxitoxin, mamba toxin, and many others are not discussed, although certainly potential candidates for bioterrorism tasks, particularly state-sponsored ones. Mamba toxin, as one example, has been developed as a weapon in South Africa (see below). Also, various plant and fungal psycho-toxins—part of which constitute important psychopharmacological agens—have in some countries been exploited experimentally, to the least, as a major means to attain compelled mind control over normal human beings— an act of governance regardable as bioterrorism.

13.4.9.1   Botulinum Toxin

Botulism toxin, a bacterial protein, is reputed as the most known toxic molecule. It is produced by *Clostridium botulinum*, a gram-positive spore-forming anaerobe germ that dwells in the soil. There are 7 different antigenic types of this toxin: A through G. Disease in humans is caused by types A, B, E, or F. Aerosol exposure studies in rhesus monkeys indicate that type F is the most toxic and that it is 60× more toxic than type B, the least toxic. Extrapolations from primate studies predict that the fatal dose of Botulinum toxin in humans is on the order of:

- 70 µg by mouth
- 0.7–0.9 µg via the respiratory route
- 0.09 µg by intravenous or intramuscular routes

The toxin is a 150-kDa zinc-dependent metalloproteinase that cleaves proteins involved in the docking and fusion of synaptic vesicles to the membrane at the neuromuscular junction. The toxin consists of two chains: a heavy chain (100 kDa) and a light chain (50 kDa) linked by a disulfide bond. The crystal structure of the molecule has been solved to 3.3-A resolution [49]. The protein structure revealed a 50-residue belt that partially obscures the active site access channel; the authors note that this unusual feature makes rational inhibitor design more difficult.

The deadly toxin irreversibly blocks acetylcholine release from peripheral nerves resulting in muscle paralysis. It has been developed as a biological weapon by many government research programs, including the United States, Russia, and Iraq. Currently, public health services classifies botulinum toxin as a Category A Bioterrorism Agent, indicating that it is a high-priority risk to public health. As a bioterrorism weapon, it is a potent substance that is easy to produce and transport. A large population of victims requiring intensive care could easily overwhelm health-care systems. It can be introduced into the food and water supply or be aerosolized. The toxin is unstable after a few days in surface water, and it cannot withstand chlorine treatment. Although it is more stable in packaged foods, it is inactivated by normal cooking practices (heat > 85°). Aerosolization is technically difficult but would bring about the poisoning of a large number of people.

At the conclusion of the Gulf War, the Iraqis reported that they produced 19,000 L of pure botulinum toxin to a UN team. A total of 19,000 L of botulinum toxin is estimated, just arithmetically, to be sufficient to kill three times the current human population by inhalation. Overall, 10,000 L of the toxin had been loaded into weapons. They loaded botulinum toxin into thirteen 600-km missiles and one hundred 400-lb bombs. Some of the toxin has not been accounted for. The JAMA consensus statement points out that it is "noteworthy that Iraq chose to weaponize more botulinum toxin than any other of its known biological agents."

Incubation times vary according to the amount of toxin ingested and the toxin type. The range reported from natural infections is between 6 hours and 10 days. Symptoms most commonly present between 12 and 36 hours of eating contaminated food. Inhalational exposures are predicted to progress more rapidly. Regardless of the route of intoxication, the neurological symptoms are similar. Gastrointestinal cases may also display nausea and vomiting. Botulinum toxicity presents as a febrile, descending paralysis. In cases of severe exposure, it is possible that respiratory failure will occur suddenly. Without treatment, the cause of death is respiratory failure due to obstruction from the pharyngeal muscles and failure of the diaphragm to move adequate volumes of air.

The standardized antitoxin is equine in origin, and of high efficacy, provided that it is administered quite early. When treatment is administered promptly, the risk of death is significantly decreased. In the past 50 years, the mortality has decreased from 50% to 8%. It is important to note that the antitoxin prevents worsening of the condition, but recovery may not occur until after several weeks to months of intensive care. Complications such as long-lasting weakness, aspiration pneumonia, and nervous system dysfunction may occur. This would at any rate make botulinum toxin a deadly and/or medium-term debilitating agent. The toxoid is highly efficient as prophylaxis.

Most experts predict that a terrorist attack with botulinum toxin would attempt to distribute the agent via aerosolization to intoxicate the largest number of people. The toxin does not survive long in the water supply. In fact, there are no reported cases of water-transmitted botulism. The food supply is at some risk, but the agent can be inactivated by food preparation. Weapons programs, such as the one in Iraq, have demonstrated that it is not difficult to produce and store large amounts of concentrated toxin.

13.4.9.2   Staphylococcal Enterotoxin B (SEB)

One of seven staphylococcal enterotoxins, this bacterial protein is the best studied potential bioterrorism weapon of a group of molecules described as superantigens; others include two streptococcal pyrogenic toxins, and the toxic shock syndrome toxin [50]. Superantigens create stable bonds across class II molecules of major histocompatibility complex (MHC) and specific variable (Vß) T-cell receptors on T_H cells, causing a release of disease-producing cytokines. Naturally, SEB intoxication occurs due to consumption of contaminated foods or drinks. SEB may be aerosolized or used to sabotage food supplies. As a weapon it would induce substantial, prolonged morbidity (up to 2 weeks) with little mortality.

From 1 to 6 hours after aerosol exposure to SEB, a distinct syndrome of high fever, chills, myalgias, nonproductive cough, dyspnea, and severe substernal chest pain may appear. Headache is also common, and nausea, vomiting, and anorexia may occur. Illness usually lasts for a few days, but it is rarely fatal. Physical examination may be unremarkable or reveal inspiratory and/or expiratory rales. After SEB ingestion, symptoms are primarily vomiting and diarrhea; fever only occurs in about one quarter of ill persons and respiratory involvement is absent. Although symptoms may be quite severe and lead to dehydration or even shock, illness usually lasts less than 12 hours. Physical examination may be unremarkable or reveal inspiratory and/or expiratory rales. Were Toxic shock syndrome toxin-1 (TSST-1) or streptococcal pyrogenic exotins to be used, they would likely cause acute erythroderma followed by desquamation and multiorgan failure. Differential diagnosis is complicated by many other diseases. Yet although toxin is transient in serum, it accumulates in the urine and is detectable for several hours postexposure. Antibodies generally develop within 6 days of exposure.

SEB, a very potent and enduring protein toxin, was included in the past U.S. BW program and standardized as a powerful biological warfare agent. Although reckoned to be an effectual incapacitating agent, it has not been connected, so far, with terror organizations.

13.4.9.3   T-2 Toxin

The Soviet military discovered—and later on materialiged—the potential use of the fungal trichothecene toxins—chiefly T-2 toxin—as aggressive bioterrorism agents shortly after World War II, when many Russian civilians ate bread baked from flour, naturally contaminated with extremely toxinogenic species of *Fusarium* mold. During this vast epidemic, numerous victims developed a protracted lethal illness characterized by initial symptoms of abdominal pain, diarrhea, vomiting, and exhaustion followed within days by fever, chills, muscular pain, and an imbalance of the red and white blood cells accompanied by pus-forming or other disease-causing organisms or their toxins in the blood or tissues [51]. It was then named Alimentary Toxic Aleucia.

The trichothecene mycotoxins are nonvolatile compounds produced by molds. They are very stable and resist heat- and ultraviolet light-induced inactivation. Only after heating at 500°F for 30 minutes will the toxins inactivate.

This discussion focuses on the hemorrhagic T-2 mycotoxin, a highly toxic agent that causes several illnesses in humans and animals, as described. From the 1970s and 1980s, trichothecene mycotoxins surfaced in the press as bioterrorism warfare agents in incidents labeled "yellow rain" attacks against civilians in Southeast Asia [52]. Although such attacks have not been verified officially, it has been observed that trichothecene mycotoxins, T-2 in particular, have been produced in the USSR, weaponized, transferred to North Vietnam, and employed by North Vietnamese airplanes. The impact was vast and terrorizing, owing to many civilians that were vulnerably exposed, and then suffered a prolonged course of conspicuous pathological effects, with high mortality rate. Somehow, the evident employment of trichothecene mycotoxins in Southeast Asia, destined for bioterrorism—as well as live human field experimentation—purposes, turned into a saga involving alternative, natural explanations for the "yellow rain" phenomenon, which are certainly legitimate but do not refute the coexistence of the two—vicious and naïve—occurrences.

Reportedly, the Egyptian Air Force employed trichothecene mycotoxins during the Yemen War (in the 1960s) against Yemenis civilians—the main impact being terrorizing [53]. Later on, the Iraqi Air Force allegedly employed trichothecene mycotoxins against Iranian soldiers during the Iraq–Iran War in 1984 [54, 55]. According to UNSCOM, the Iraqis researched trichothecene mycotoxins, including T-2 [56]; they probably weaponized the T-2 toxin.

Unlike most biotoxins and microorganisms that do not affect the skin, T-2 toxin is an aggresive dermal irritant and can severely harm an unprotected person's skin and eyes. The pain associated with the exposure occurs within seconds to minutes. Larger doses produce incapacitation and death within minutes to hours. A larger amount of T-2 toxin is required for a lethal dose than of the chemical warfare agents VX, soman, or sarin. Comparisons with blister agents such as sulfur mustard show the T-2 toxin is about 400 times more efficient in producing blisters: It takes approximately 50 ng of T-2 toxin to produce the same injury to the skin as 20 µg (20,000 ng) of mustard. The T-2 toxin has a diverse effect depending on the manner and amount of exposure with vomiting and diarrhea noted at exposure doses one fifth to one tenth the lethal dose [52].

Exposure causes skin pain, itching, redness, blisters, and sloughing (shedding) of dead skin. Effects on the airway include nose and throat pain, nasal discharge, itching and sneezing, cough, shortness of breath, wheezing, and chest pain; the victim spits blood as a result of pulmonary or bronchial hemorrhage. The T-2 toxin also produces effects after ingestion or eye contact. Severe poisoning results in prostration, weakness, jerky movement, collapse, shock, and death [57]. The only protection against T-2 toxin effects is the individual protective mask and chemical protective overgarment. No chemotherapy, vaccine, or specific antidote is available [57].

Altogether, the trichothecene mycotoxins are reckoned as primarily blister agents that, at lower exposure concentrations, would cause extreme skin and eye irritation, and at larger doses would produce considerable incapacitation and death within minutes to hours [52].

13.4.9.4 Aflatoxin

As a public health hazard, the major routes for aflatoxin exposure are inhalation, pulmonary mycotoxicosis, especially of grain dusts, and ingestion as a result of eating food made with contaminated grains. The aflatoxin problem was first recognized in 1960, when there was a severe outbreak of a disease referred to as "Turkey 'X' Disease" in the United Kingdom, in which over 100,000 turkey poults died. The cause of the disease was due to toxins in peanut meal infected with *Aspergillus flavus*, and the toxins were called aflatoxins. The potentialities of aflatoxins as toxic carcinogens, mutagens, teratogens, and immunosuppressive agents are well documented [58]. They are produced as secondary metabolites by the fungus *Aspergillus flavus* and *A. parasiticus* on a variety of food products. Chemically, aflatoxins normally refer to the group of difuranocoumarins.

Two forms of illness have been observed:

1. (Primary) Acute aflatoxicosis is produced when moderate-to-high levels of aflatoxins are consumed. Specific, acute episodes of disease may include hemorrhage, acute liver damage, edema, alteration in digestion, absorption and/or metabolism of nutrients, and possibly death.

2. (Primary) Chronic aflatoxicosis results from ingestion of low-to-moderate levels of aflatoxins. The effects are usually subclinical and difficult to recognize. Some common symptoms are impaired food conversion and slower rates of growth with or without the production of an overt aflatoxin syndrome. In the long term, exposure to aflatoxin is thought to explain the high rates of primary liver cancer in Africa and parts of Asia. Pooled data from Kenya, Mozambique, Swaziland, and Thailand show a positive correlation between daily dietary aflatoxin intake (in the range of 3.5- to 222.4-ng/kg body weight per day) and the crude incidence rate of primary liver cancer (ranging from 1.2 to 13.0 cases per 100,000 people per year).

Iraq developed and weaponized aflatoxins, chiefly as a bioterrorism agent. It has then been noted that "The discovery (by UNSCOM, during the 90s) that Iraq was researching aflatoxin, not a traditional BW candidate, was a cause for some surprise. It is a carcinogen, the effects of which manifest themselves only after many years, and several Western experts have rationalized this Iraqi program only in terms of genocidal goals. If aflatoxin were used against the Kurds, for instance, it would be impossible definitively to prove the use of BW once the symptoms emerged. Another possible explanation is its potential use as an immune suppressant, making victims more susceptible to other agents. However, the aflatoxin declaration may also hide other aspects of Iraq's BW program: according to Iraq's depositions, the production program never encountered any mishap (as other parts of the BW program had) and, to judge from the declared time-frame for the total amount produced, production could never have stopped, even for cleaning of the equipment. This raises the suspicion that Iraq declared an excessive amount of aflatoxin in order to disguise the fact that other, more destructive agents had been produced in greater quantities [59]."

Iraq developed and produced aflatoxin as a long-term debilitating bioterrorism agent, intended to be used against the Kurds. There is documentary evidence and statements obtained by UNSCOM that Iraq was mixing aflatoxin with riot-control gas [60]. Although seemingly peculiar, some rational can be figured out regarding this type of weaponization (much more detail is given in Section 13.4.15.6).

13.4.9.5 Aconitine

This relatively unnoticed plant toxin is an intensely poisonous alkaloid obtained from aconite root. It occurs in colorless crystals. Aconite tubers are most toxic wild plants distributed from Asia to western Europe [61–64]. It achieved some notoriety in the nineteenth century as an agent for homicides and suicides. In modern times, aconite has been used as Chinese herbal medicine, which is freely purchased from herb shops and consumed as a doctation by a herbal practitioner for pain control in the Northern Hemisphere [65]. In Japan, some cases of aconite poisoning appeared as a result of committing suicide or accidental ingestion, which were mistaken for edible grass. However, aconite alkaloids have the potential for serious and even fatal cardiotoxity, which management has continued to fail save patients with therapeutic resistant fatal arrhythmia [66, 67].

Serious aconitine poisoning is characterized by hypotension, palpitations, shock, delay in myocardial conduction, and dysrhythmia beginning within 6 hours of ingestion. Respiration is progressively depressed by the effect of aconitine on the bulbar respiratory center. Tingling of the tongue, mouth, and skin, followed by numbness and anesthesia, are characteristic signs of aconitine poisoning. Excessive salivation is a characteristic sign of poisoning. The skin becomes cold, clammy, sweaty, and pale. Initial bradycardia is due to vagal stimulation. Soon after ingestion, aconite causes a tingling, burning sensation on the lips, tongue, mouth, and throat, which is followed by numbness and constriction of the throat. When applied to the skin, aconite causes a tingling sensation and then numbness. Blurring of vision occurs. Initial miosis is followed by mydriasis. There is a feeling of constriction in the throat [68].

According to some reports, Egyptian Armed Forces Commander-in-Chief Field Marshall Mohammed Abd el Hakim Amer, who headed the defeated Egyptian Army in the 1967 Six Days War, was later poisoned, institutionally, by this substance. Allegedly, he committed suicide by means of aconitine that had been introduced to him while arrested [69]. An unidentified toxicant, possibly the same plant toxin, was used to replace the antidote with which Egyptian autoinjectors were filled, for personal assassinations [70].

Another trivial, yet peculiar episode, involved an English doctor who used aconitine to murder his crippled 18-year-old nephew for his inheritance. Percy Malcolm John was a resident at Blenheim School in Wimbledon and owned a small property that would go to his uncle on his death. Lamson visited the boy, bringing a cake and a capsule that he said was medicine. A few hours later he died in agony from aconitine poisoning [71]. Reportedly, aconitine toxin has been adopted by some states for assassination purposes.

13.4.10 AVAILABILITY OF INFECTIVE AGENTS AND TOXINS

A broad spectrum of availability of pathogens and toxins is discussed, especially pertaining to potential resources worldwide. Still, the spectrum is much narrower regarding weaponized agents, making the course of the attacker complicated, unless "raw" infective agents or toxins are to be employed. Different episodes illustrating that variability are presented.

The common nonstate-sponsored scenario of a just-graduated microbiologist cultivating a pathogenic germ in a rudimentary household laboratory, so as to produce it as a weapon for bioterrorism acts, is partially feasible. It is much less likely with regard to weaponizing or engineering bioterrorism agents. A significant advantage emerges concerning state-sponsored bioterrorism, because the relevant pathogens and toxins are appreciably more procurable, reproducible, and conveyable for a state, generally speaking. Therefore, the inventory of agents, their handling and the feasibility of conducting bioterrorism are scaled up when a state is involved, either intending to carry out an act of bioterrorism on its own or to assist a terror organization. Any step made by a country may basically be institutionalized, hence be more legitimate compared with an unaffiliated or noninstitutionalized body, element, or person.

Overall, the wide range of bioterrorism agents is reflected by considerably varying degrees of availability. Ricin toxin can readily be produced, roughly, by unqualified saboteurs. Conversely, an expedient way to meet the need for highleveled know-how often faced by terror organizations would be in the form of "scientific mercenaries." At large, varitype availability of pathogens and toxins is demonstrated though the following episodes (and further ones mentioned in Sections 13.4.14, 13.4.15 and 13.4.16).

In Japan, in 1965, outbreaks of typhoid and dysentery were intentionally induced by a local bacteriologist, with professional direct access to those germs [72].

In 1972 two men affiliated with the U.S. group "Order of the Rising Sun," who eventually fled to Cuba, had conspired to contaminate the water supplies of some large U.S. Midwestern cities with stocks of typhoid fever germs cultivated by one of them. Up to 40 kg of bacteria cultures were found in a college laboratory [73].

In 1973, the American leftist terrorist group, the Weather Underground, reportedly attempted to blackmail a homosexual officer at USAMRIID into supplying organisms that would be used to contaminate municipal water supplies in the United States. The plot was discovered when the officer requested several items "unrelated to his work."

In 1975, the Symbionese Liberation Army was found in possession of technical manuals on how to produce bioterrorism weapons.

Moving to Europe, in 1980, police raided a German Red Army Faction apartment in Paris and found a miniature laboratory containing a culture medium of the germ that produces botulinum toxin. Notes about bacteria-induced diseases were found in the apartment as well [74]. Once again in Europe, several Muslim terrorists affiliated with the Algerian "Armed Muslim Group" were arrested in Belgium, holding a lot of information about biological (and chemical) weapons and about the World Cup football games [75].

Outstandingly, protesters claimed to have taken infected soil from the Scottish island of Gruinard and placed it at the Microbiological Defence Establishment at Porton Down, Britain. The island has been closed to the public since germ warfare experiments on sheep were conducted there in 1941. The anthrax spores used in the experiments can remain dangerous for decades.

The American Type Culture Collection (ATCC) has often been approached. Two Canadians attempted to procure botulinum and tetanus cultures from the ATCC. Reportedly the first phone order, of less deadly cultures, was fulfilled, and it was not until the second order that ATCC employees became sufficiently suspicious to notify authorities [76].

A member of the white supremacist Aryan Nation acquired freeze-dried bubonic plague bacteria from the ATCC [77]. In May 1995, in the United States, Larry Wayne Harris was arrested for illegally obtaining the plague bacteria *Yersinia pestis*. Using his previous employer's certification, Harris obtained the samples through the mail from the ATCC. He was sentenced to 18 months probation and 200 hours of community service. Harris was again arrested in 1998 when he and another individual were found allegedly in possession of anthrax cultures, which were later determined to be anthrax vaccine. Due to the ease of obtaining dangerous pathogens, the CDC established rigorous guidelines for shipment of specific pathogens that may be used as bioterrorism agents.

Yet, as a matter of fact, a wide variety of pathogens and toxins serving the Iraqi BW program during the 1980s has been freely procured from different Western and Eastern sources, mostly the ATCC. The Iraqi case study provides remarkable demonstration, indeed, of repeated procurements of highly dangerous pathogens and toxins that readily took place albeit an integral part of the past Iraqi BW program [78]. And beyond, paradoxically, anti-bioterrorism programs worldwide bring about academic, commercial, and industrial institutionalization of numerous new pathogens-holdings labs which constitute potential resources for pathogens seekers. It so happened, that 245 facilities are now authorized to work with live anthrax in the US, and about 100 actually do so , compared to roughly 12 prior to the 2001 anthrax mailings [79].

Inadequately secured state-held stockpiles of BW constitute a potential resource. Reportedly, bin Laden's associates managed to receive anthrax and plague cultures or weaponized agents, from former Soviet facilities in Kazakhstan, or elsewhere. This has not been evidenced, although attempts to obtain those and other germs have certainly been made by al-Qaeda, perhaps fruitlessly. Still, assuming that al-Qaeda is responsible (together with Iraq) for the Sept. 2001 anthrax letter attacks, its members were at least involved in the final step of the envelope posting, thus possessing, temporarily, this pathogen [80]. Also, aerosol-released bioterrorism weapons were often in the possession of Spetsnaz operatives (Soviet Special Forces) who were believed to be involved with terrorists, especially those associated with bin Laden.

A typical example demonstrating an endeavor to exploit an institutionalized researcher as a 'scientific mercenary' was the case in which al-Qaeda had direct connections with a senior Pakistani microbiologist, Dr. Abdur Rauf, who attempted to support al-Qaeda's pursue of BW [81].

Adriana Stujit, an acknowledged Dutch journalist, contended that radical Muslims have gotten control over South Africa's biological weapons stockpiles. Quantities of anthrax, Ebola, Congo fever, and other agents are missing from South Africa, with the only explanation being theft. There has been no explanation or accurate documentation of the amounts, locations, and contents of the stockpiles [82]. This is not at all surprising, considering the picture described below regarding South Africa.

Obtaining pathogens and toxins from their natural environment is another sensible, though skill-demanding mode. In 1993, the Japanese cult Aum Shinrikyo sent a group of 16 cult doctors and nurses to Zaire, on a supposed medical mission. The actual purpose of the trip to Central Africa was to learn as much as possible about and, ideally, to bring back samples of Ebola virus. In early 1994, cult doctors were quoted on Russian radio as discussing the possibility of using Ebola as a bioterrorism weapon. The attempt was fruitless. Also, they attempted to purchase a Q-fever culture from a Japanese academic researcher but were rebuffed. But they did obtain botulinum germs from earth in Japan, anthrax germs, and cholera germs. The anthrax strain was consistent with strain Sterne 34F2, which is used in Japan for animal prophylaxis against anthrax [83].

Domestic naive facilities may readily be contributive. In Israel, much concern was raised that regular microbiological laboratories—whether private or institutionalized, and particularly those located in the Palestinian-occupied territories—may be used as minifactories for producing biological agents in quantities sufficient for bioterrorism purposes [84].

A broad spectrum of resources of pathogens and toxins is notable, then, especially pertaining to potential resources worldwide. Still, it is much narrower regarding weaponized agents, making the course of the attacker complicated, unless "raw" infective agents or toxins are to be employed in an act of bioterrorism. But, on the other hand, dozens of biotech firms now offer to synthesize complete genes from the chemical components of DNA, making it feasible, seemingly, that a bioterrorist, armed with only a fake e-mail address, could probably order such deadly biological components online and receive them by mail [85]. He would doubtfully be able to transform, however, those components into a weapon, if not substantially assisted.

13.4.11   TECHNICAL FEASIBILITY OF BIOSABOTAGE ACTS

Pragmatically, biosabotage acts are featured by high technical feasibility, making them attractive, in that sense. Various operational options allowing for such feasibility are presented and discussed. Still, an appreciable level of know-how is needed for all preparatory stages preceding the in-effect act of bioterrorism.

The most plain, nearly self-evident, act of bioterrorism is demonstrated by one person emptying a tube into a water source so as to contaminate it with a waterborne pathogen. Against such an act stands mainly the level of water chlorination, which does not interfere, practically, with the distinct technical feasibility of this deed. It is, indeed, a simple thing to do.

But the technical feasibility of bioterrorism may extend much further. A fairly sophisticated practice involves the preparation and dispersing of a toxin or pathogen in an inhalable aerosol form, either dry or wet. Various advanced aerosol technology-based devices structured to dispersing inhalable aerosols of pharmacological quality are available and may likewise release a pathogen or toxin. Alternatively, dissemination may be achieved by infected insects. During an infamous biowarfare attack in 1941, the Japanese Military released an estimated 150-million plague-infected fleas from airplanes over villages in China and Manchuria, resulting in several plague outbreaks in those villages [86]. The victims were, deliberately, merely civilians. The same principle may be applied for disseminating plague by the release of infected rats in an enemy's settlements. Another example, similarly complex, yet plainly feasible act of bioterrorism, rabid, pre-symptomatic dogs or

other canines infiltrated covertly, steadily, and untraceably in the wild through bor-
derlines, may then bring about the spread of rabies onto local dogs and farm animals
plus humans, gradually elevating the overall level of rabies endemicity.

Currently, effective worldwide dissemination is achievable through postal systems. The following recent incident, although certainly a naïve one, illustrates the
technical feasibility of biosabotage acts at large, as one, outstanding example:

Thousands of scientists were scrambling at the urging of global health authorities
to destroy vials of a pandemic flu strain sent to laboratories in 18 countries as part
of routine testing. The rush, urged by the World Health Organization, was sparked
by a slim, but real, risk that the samples could spark a global flu epidemic. The
vials of virus sent by a U.S. company went to nearly 5000 laboratories, mostly in
the United States [87].

Referring to the incident, WHO's influenza chief, Klaus Stohr contended that
"The risk is relatively low that a lab worker will get sick, but a large number of labs
got it and if someone does get infected, the risk of severe illness is high and this
virus has shown to be fully transmissible. The risk is low but things can go wrong
as long as these samples are out there and there are some still out there."

The 1957 pandemic strain, which killed between 1 million and 4 million people,
was in the proficiency test kits routinely sent to laboratories. This strain has not
been included in the flu vaccine since 1968, and anyone born after that date has no
immunity to it.

Most samples were sent at the request of the College of American Pathologists,
which helps laboratories do proficiency testing. A private company, Meridian Bioscience Inc. of Cincinnati, Ohio, is paid to prepare the samples. The firm was told
to pick an influenza type A virus sample and chose from its stockpile the deadly
1957 H2N2 strain. The reason for choosing this past, mighty strain, and not a
current strain, remains unclear [88]. Still, some other test kit providers besides the
college also used the 1957 pandemic strain in samples sent to laboratories in
the United States.

The majority of the laboratories that got the test kits are in the United States.
Fourteen were in Canada, and 61 samples went to laboratories in 16 other
countries in Europe, Asia, the Middle East, and South America, according to the
WHO. The test kits are used for internal quality control checks to demonstrate that
a laboratory is able to correctly identify viruses or as a way for laboratories to get
certified by the College of American Pathologists. The kits involve blind samples.
The laboratory then has to correctly identify the pathogen in the vial in order to
pass the test. Usually, the influenza virus included in these kits is one that is currently circulating, or at least one that has recently been in circulation. The WHO
then notified the health authorities in all countries that received the kits and recommended that all samples be destroyed immediately. That same day, the College
of American Pathologists faxed the laboratories asking them to immediately incinerate the samples and to confirm in writing that the operation had been
completed.

Unintentionally, this demonstration of the technical feasibility of biosabotage
operations marks both simultaneous, wide-range distributable shipment of a virulent—in that case, rather recoverably pandemic—pathogen in a stable infective
form, as well as specifically, on-spot targeted sites intended to be attacked biologically. The two options are equally feasible.

Actually, two events of unintentional postal distribution of smallpox—yet in that case generating epidemics in effect—happened to occur already in 1901. Smallpox had developed in a woman in Saginaw, Michigan, after she received a letter form her sweetheart, a soldier in Alaska.

He had written it while recovering from this disease. The infection subsequently spread to 33 other persons in Saginaw [89]. Also in 1901, an outbreak of 5 cases of smallpox was recorded at the Mormon headquarters in Nottingham, England, apparently after receipt of "letters or other fomites" from Salt Lake City, Utah, where smallpox was widespread [90].

Heavily crowded facilities single out the technical feasibility of bioterrorism. Thus, the British army tested model bacilli spore powder released (1963) from a window of a tube train traveling in the London Underground [91]. The trial concluded that the spores can be carried for several miles on the tube system, and locally can persist as an aerosol of high concentration for a considerable period. Also, widespread dispersal of bacteria was found in a May 1965 secret release of innocent bacilli at Washington's National Airport and its Greyhound bus terminal, according to declassified military reports. More than 130 passengers who had been exposed to the bacteria traveled to 39 cities in seven states in the two weeks following the mock attack [92]. By 1966, in a similar trail in New-York Subway, bacilli powder-carrying light bulbs were dropped [93].

Water is an easy target, basically. The feasibility of water contamination has been pointed out to be concrete in Israel: Deliberate influx of sewage from Judea and Samaria hills onto the Israel Coast Plain was been observed during the 1990s, according to the Israel Minister for Environment Quality, causing the contamination of Israeli water systems [94, 95]. Objectively, this may become much more feasible as a result of the 2005 Israeli disengagement. It should be mentioned that, in the past, outbreaks of cholera occurred in Judea and Samaria as a result of sewage being routinely and innocently used for the irrigation of vegetables. Connectedly, the factual situation is that most streams found in Israel stem from the mountainous aquifers, which are mostly located in the Judean and Samarian Hills, parts of which are now outside of Israeli territorial control. The other Israeli water sources stem from the Golan Heights and Southern Lebanon.

Furthermore, rather unusual modes illustrating the technical feasibility of bio-sabotage acts have been noted in Israel:

In the body of one person who survived suicide bomber sabotage, bone fragments were discovered that are believed to come from a suicide bomber. The bone fragments tested positive for hepatitis B. Medical experts believe "This is possibly the first report of human bone fragments acting as foreign bodies in a blast injury, and consequently all survivors of these attacks in Israel are now vaccinated for hepatitis B." The experts suggested that bone fragments embedded in attack victims should be routinely tested [96]. Theoretically, bone fragments might also spread other diseases including HIV, dengue fever, syphilis, and Creutz–Jakob disease [97].

In August 2004, an Iranian-made unmanned drone launched from southern Lebanon by the Iranian-supported Shi'lte militia, the Hezbollah flew for about 15 minutes along Israel's northern Mediterranean coast until it reached the coastal resort of Nahariyah. "Iran has not only supplied Hezbollah with these UAVs but has also trained 30 of the group's members to operate them," an official (a senior commander in the Iranian Revolutionary Guards) told the London-based Arab daily *al-Sharq al-Awsat* 11 Nov 2004 [98]. The efficiency of UAVs as aerosol disseminators of biological agents is well known.

Drifted saboteurs may even be full residents of the country to be attacked. A terrorist state or organization might use such residents, particularly extremists, to carry out bioterrorism acts, thus considerably elevating the technical feasibility of those operations. The attacked might catch the extremists, but not understand they were minor figures, and present them instead as major figures. All in all, the practicability of bioterrorism is fairly plain, although a degree of professional knowledge is needed for planning and preparations.

It is also possible to disseminate infected insects or infected animals (like rats, for instance). Also, deliberate infection of contacts by an infected individual harboring a contagious disease (AIDS, for example—a plot already carried out not infrequently) is, in a sense, an act of bioterrorism. Water and ventilation systems may be targeted. During the First Gulf War, there were serious concerns in the United States, that the ventilation systems of buildings might be attacked by terrorists using BW agents [99].

According to Norqvist, terrorists might effectively and easily disseminate biological agents through water systems, resulting in a high number of casualties, the terrorism focusing either on large cities or military facilities. It has thereupon been pointed out that chlorination can be neutralized by using naturally chlorine-resistant microorganisms or constructing bacteria resistant to chlorine concentrations normally used in the municipal drinking water systems [100]. Simple devices might be used to deliver biological agents into the ventilation systems of buildings. Open-air experiments carried out in the 1960s demonstrated that throwing a light bulb filled with biological agent before an incoming subway train during rush hour is sufficient to infect tens of thousands of people [101]. Plausibly, state-supported terrorist groups would be the most likely of all terrorist groups to get a hold of biological weapons by the supporting government transferring entire systems from their national programs [5].

13.4.12   OPERATIONAL MODES

A wide variety of modes that may serve for operating bioterrorism is described. Essentially aimed at attaining advantage through asymmetric warfare, they may lead, at most, to an impact equaling catastrophic terrorism. They might be limited, inversely, to merely assassination. Hoax acts are visited as well. One-man operational mode can bring about any of the mentioned outcomes.

The operational modes marking bioterrorism are closely related to asymmetric and catastrophic warfare. Asymmetric warfare is a military term to describe warfare in which the two belligerents are mismatched in their total capabilities or accustomed methods of engagement such that the inferior side must press its special advantages or effectively exploit its enemy's particular weaknesses if they are to have any hope of *prevailing* [102–104]. Thus, asymmetric warfare is a recipe for engaging an opponent who has superior military power at his disposal and where the main target is not the armed forces but the fabric of society itself. Catastrophic terrorism differs from the more traditional forms of terrorism because it has no closely defined political aim. It is based on the belief that the current world order is dominated by certain "universal values" that the terrorists despise and that this order must be destroyed.

Consequence management roles vary depending on whether the attack is a terrorist event or an act of war. Yet, the distinction between terrorist and military use of WMDs is increasingly problematic. State adversaries, perhaps acting through terrorist surrogates, may be inclined to use WMDs early, unconventionally, and if possible, anonymously. They may use WMDs against such military targets as ports, airfields, staging areas, and overseas bases to prepare the military battlefield by slowing logistics and power projection. Similarly, they may psychologically undermine public support or politically divide a coalition for an operation by attacking domestic or allied civilian targets. Eager to deter regional involvement while avoiding an overwhelming retaliatory response, perpetrators may try to obscure their identity. Under these circumstances, the line between terrorism and symmetric warfare may then vanish.

The stealthy qualities of BW further complicate the distinction between terrorism and war. An adversary with effective agent dissemination capabilities could employ BW as part of a covert attack nearly impossible to detect until casualties appear. Depending on the agent used, the attacked might not know whether an outbreak was natural, a terrorist attack, or the opening assault of a war. For all the legitimate concern about bioterrorism, systemized weapons are mostly still difficult to employ effectively without access to state-developed technology. Rogue states and well-financed terrorists with access to state-developed technology remain the most serious danger from the effective, large-scale use of biological agents through guerilla warfare.

The repertoire of bioterrorism modes is potentially an incredible one, extending from ricin-based assassination of one person to a worldwide smallpox pandemic, even if not aimed, a priori, to reach such magnitude. Those two edges may each take shape in actuality due to a seemingly slight act carried out by a single perpetrator. The former mode has practically been materialized through the murder of Markov, the Bulgarian journalist in London, whereas instead of ricin the lethal toxin used could variedly be of another type, as detailed above. The latter mode is a continuously floating menace, the possible realization of which is attentively being faced, in terms of both preventive measures and preparedness for handling such occurrence. The one-man mode was exemplified in 1998, when the British government issued a warning to all ports about an Iraqi attempt to bring large quantities of the deadly germs of anthrax into Britain (and other countries) inside cosmetics bottles, cigarette lighters, and perfume sprays, disguised as duty-free goods [104].

A bioterrorism attack could consist, then, of nothing more than a person deliberately coughing on people and surfaces after being infected with a pathogen, as correctly pointed out in a study conducted by the Australian Strategic Policy Institute. This could happen by a terrorist knowingly infected and incubating an infection flying to Australia and then walking around a crowded shopping center for some hours, coughing near people and over surfaces. The study found that even without their knowledge, infected individuals could likewise be used as carriers for bioterrorism acts [105].

One major operational mode of massive bioterrorism may focus on endpoints of collective source reservoirs right before their physical division into much smaller to individually consumed portions. Such reservoirs may include blood, sera, or other preparations intended to be introduced into the body as well as foods, drinks, and water. Thus, certain collective infusion fluids may be contaminated before being packed: AIDS, hepatitis B, hepatitis C, arboviruses, and other blood-adapted pathogens.

Milk supplies, as another example, meet that very same biosabotage principle and have aroused much worry. About a third of an ounce of botulism toxin poured by bioterrorists into a milk truck en route from a dairy farm to a processing plant could cause hundreds of thousands of deaths and billions of dollars in economic losses, according to a scientific analysis [106].

The analysis considered what might happen if terrorists poured into a milk tanker truck a couple of gallons of concentrated sludge containing as much as 10 g of botulinum toxin. Because milk from many sources is combined in huge tanks holding hundreds of thousands of gallons, the toxin would get widely distributed in low, but potentially lethal, concentrations and within days be consumed by about 568,000 people, the report concludes. The researchers acknowledge that their numbers are very rough. But depending on how thoroughly the milk was pasteurized (which partially inactivates toxins) and how promptly the outbreak was detected and supplies recalled, about 400,000 people would be likely to fall ill, they conclude. Although only 6% of victims would generally be expected to die, the death rate could easily hit 60%, they conclude, because there would not be nearly enough mechanical ventilators or doses of antitoxin to treat so many victims.

A different mode of bioterrorism may aim to bring about the outward leakage of infectious microorganisms by physically damaging facilities containing them. A variety of installations may serve for such a mode, including P-3 and P-4 laboratories, vaccine factories, and institutionalized depositories. Such sabotage may questionably be regarded as bioterrorism, particularly if directed toward a facility engaged in biological warfare, even if but defensive. Yet the consequences and ultimate impact may be very severe. Hardly accusable may as well be bioterrorists choosing a mode of amplifying an ongoing outbreak or epidemic, especially if the very same causative pathogenic strain is concurrently employed by them for that purpose.

The various options of bioterrorism conducted as a sole operative mode or in conjunction with other sabotage acts add a further dimension. Bioterrorism acts may on the spot be accompanied by simultaneous conventional and/or unconventional—mostly chemical or radiological—sabotage. Such synchronism may form camouflage and be misleading, making the bioterrorism component scarcely thinkable or detectable. Spatial diversity may concomitantly increase the blurring effect. Combined employment of WMDs indeed drew the Pentagon's attention. The Pentagon has implemented, thus, a plan to expand its support of civilian authorities in the event of multiple domestic attacks involving chemical, biological, radiological, or nuclear weapons, The Virginia-based Joint Task Force-Civil Support was previously the Pentagon's only force dedicated to such a mission [107].

"We have identified capabilities within our force structure—beyond Joint Task Force-Civil Support—in order to ensure that we could respond not simply to a domestic attack involving a WMD, but to multiple attacks at diverse locations, several cities perhaps at once where terrorists might have employed weapons of mass destruction," Paul McHale, deputy assistant secretary of defense for homeland defense, said. "It is now the established policy of the Department of Defense that we will train and equip for the mission requirement of multiple WMD response," said McHale [107].

Two overt modes are notable:

1. Suicide by a person carrying a detectable contagion.
2. Bioterrorism combined with conventional terrorism, simultaneously serving for two advantages:
   a. Temporary misleading (until infected cases present).
   b. Increasing vulnerability to bioterrorism agents by conventional wounding.

Opposite in polarity to the extreme mode involving in-effect bioterrorism in conjunction with multi-WMDs, are bioterrorism hoaxes. Bioterrorism hoaxes—whether consisting of unviable or viable substances that are benign—constitute an increasing mode of bioterrorism. The anonymously angry of the world are creating serious trouble today with just a stamp, an envelope, and a household product. It takes only moments for someone to send a package containing a suspicious but ultimately harmless powder. Moreover, hoaxers often see the powder-filled letters as a way to send a message, without considering that they are committing a crime. The act, however, can force entire buildings to be locked down and tie up emergency personnel for hours. Each incident must be taken seriously in case the contents turn out to be anthrax rather than crushed aspirin.

It so happened that the uncertainty prevailing around a hoax caused Australian Prime Minister John Howard to say in the early hours of a crisis thereupon created due to an powder containing envelope received in the Indonesian Embassy (June 2005): "This is a very serious development for our country, and I can't overstate the sense of concern I feel that such a recklessly criminal act should have been committed." He added he believed the threat was meant as retaliation for the 20-year prison sentence Australian national Schapelle Corby received in Indonesia for smuggling drugs. Such acts could damage relations between the two nations and invite retaliations against Australians in Indonesia [108].

Actually, however, people were putting powder in the mail to scare others years before someone sent envelopes laced with anthrax that killed five American people in the fall of 2001. There were 22 incidents reported worldwide in 2000 involving faked biological agents, according to the Monterey Institute's Center for Nonproliferation Studies. The count jumped to about 730 in 2001, due largely to Clayton Lee Waagner's campaign against abortion providers [109].

Waagner was arrested late that year after sending roughly 550 powder-filled letters, and the number of bioterrorism hoaxes dropped to 70 in 2002. However, excluding his count, both 2001 and 2002 saw significant increases in such incidents from 2000, spurred by the anthrax mailings and the resulting media coverage, according to Sundara Vadlamudi, research associate for the center's WMD Terrorism Research Program.

Although certainly the most relentless, Waagner was not the first anti-abortion bioterrorism hoaxer or the last. More than 20 letters containing fake anthrax were mailed to abortion providers and abortion rights organizations in January 2002, several weeks after his arrest. Currently, however, the trend is increasing; notable receiving powder-filled envelopes since April 2005: the Israeli Embassy in Washington, D.C., the Danish embassies in Stockholm and Vienna, the New Mexico state capitol, a Slovenian government office, the office of Quebec Premier Jean Charest, a Vermont multimedia company, Israel Army Radio, the Bank of Israel, and NATO's Joint Warfare Center, Norway.

13.4.13 SPATIALLY AND TEMPORALLY VARYING IMPACTS

The expected impact of bioterrorism acts, together with the featuring of their spatial and temporal variance, are configured. A typical multifactorial system is thus formed, the handling of which is remarkably complicated, demanding, and vital. It includes a far-reaching range of bioterrorism scenarios, some evident, some tentative, beginning with hoaxes scenarios, and ending in the form of a pandemic.

The impact of bioterrorism acts is appreciably varied, both spatially and temporally. Its variability is shaped by the following factors:

- Initial area coverage (via air, water, food, or animal vectors/carriers)
- Duration of pathogenetic course
- Curability
- Environmental stability of the pathogen/toxin
- Contagiousness
- Demographic conditions
- Climatic conditions
- Conduction and effectiveness of preventing measures (before and after the act of bioterrorism)

The multifactorial integral thus formed is complex. For automating the testing and validation of spatial and temporal cluster detection algorithms, and enabling the ready creation of datasets for benchmarking outbreak detection systems, a specific software tool has been developed [35]. It allows for the creation of simulated clusters with controlled feature sets, varying the desired cluster radius, density, distance, relative location from a reference point, and temporal epidemiological growth pattern. This tool does not require the use of an external geographical information system program for cluster creation. Based on user-specified parameters describing the location, properties, and temporal pattern of simulated clusters, it creates clusters accurately and uniformly.

The impact is pronounced within several domains:

- Personal illness that may lead to death
- Paralyzed manpower
- Logistical efforts needed to medically support and isolate the infected/sickened victims
- Meticulous, extremely demanding managing of the apparently uninfected population

- Demoralization that may ascend to total panic
- Overall instability

The remarkable psychological responses subsequent to a bioterrorism attack were summarized as follows: horror, wrath, panic, paranoia, demoralization, magical thinking about germs and viruses, fear of invisible pathogenic agents, fright of contagion, anger at terrorists, government, or both, attribution of arousal symptoms to infection, scapegoating, social isolation, and loss of faith in social institutions [110]. Panic is a power multiplier in itself. Some plead that the use of biological agents can cause severe panic and hysteria among the civic population. For example, during the Sept. 2001 anthrax letter attack, thousands of civilians came to hospitals, although only few of them were found to be suffering from exposure to the anthrax powder. When BW agents are used, there is a great fear that the number of civilians turning to get medical help will overload the health service systems and those who really need treatment will not be able to get it. This overload, besides keeping treatment from those who need it, will also cause mass hysteria. Mass hysteria is very significant. In case of contagious diseases, there is a grave concern that the mass hysteria and the urge to flee the scene will only cause the disease to spread. That is contrary to the proper way of fighting it, namely, closing and isolating the contaminated area.

A typically acute bioterrorism episode would be featured by a toxin or pathogen of a short incubation period, pronounced illness, recovery or death, absence of secondary epidemic waves, and the creation of long-lasting immunity of the surviving population, which would inhibit relapsing occurrences. Such a prototypic episode may at times not exceed a week, just as it had been designed by the perpetrator. But, contrastingly, attritional or demographically debilitating impact may result from a delayed, relapsing, or slowly progressing pathogenetic course that may prevail for many years in a given population.

Spatially, however, any combination with the above-mentioned temporal variations is possible, basically. This means that in a relatively confined area a several-weeks transient bioterrorism episode or a lingering impact stemming from a bioterrorism act lasting for years, may take place. Equally, a vast territory may be afflicted by a swiftly moving epidemic or, alternatively, by a clumsily evolving pathogen, in terms of epidemic rate and virulence, whether it is an incapacitating or deadly agent. The dynamics underlying this complex variability is foremost shaped by the nature of the toxin or pathogen employed in-effect, and the countermeasures to be specifically taken against it. Fundamentally, this fluctuating dynamics is well illustrated, both spatially and temporally, through the varitype bacteria, viruses, and toxins presented within the above corresponding sections. It prevails, thereby, for both the dimension of the individual victim as well as the demographic dimension at large.

On the other hand are the bioterrorism hoax incidents, of which the spatial and temporal impacts, though, are still significant. Even an absolutely empty post envelope bearing but one word, like "biohazard," would bring about those undesirable impacts. If not empty, it may contain but paper and/or innocent chemical, toxin, non-infective microorganism, infective yet attenuated microorganism, or pathogenic microorganism. In this order of possibilities, there is an ascending line of spatial and temporal impacts. (The envelope content may as well comprise insects, infected or naïve.)

Three recent examples are herewith mentioned, then, resembling the spatial and temporal variance of impacts:

After a harmless bacillus powder was found in an post envelope sent to the Indonesian Embassy in Australia, the Australian capital of Canberra sustained a barrage of bioterrorism hoax mailings beginning 1 June 2005 to its Parliament building, Prime Minister and Cabinet Department, and the embassies of Indonesia, the United States, United Kingdom, Japan, Italy, and South Korea [111]. Emergency personnel in Canberra had to decontaminate 46 staffers at the Indonesian Embassy when it appeared some form of powered toxin had been sent to the building, according to Australian media reports. Tests later indicated the substance was not dangerous [112].

Some days later, continuation took place far away. On 10 June 2005, an employee in the collections department of Imperial Parking in Vancouver, British Columbia, opened an envelope containing a similarly suspicious white powder. Vancouver emergency services were quickly alerted [113]. The affected section of the building was locked down—the air-conditioning system and elevators were shut off, and no one was allowed to enter or leave. Twenty-five workers were quarantined for several hours until tests determined the substance was not dangerous. Forty-eight emergency workers and 18 pieces of equipment were sent to the powder scare at the Imperial Parking office. Firefighters suited up in protective gear to retrieve a sample, whereas others managed the quarantine and organized decontamination for the hazardous materials workers. Emergency medical personnel examined firefighters before they entered the building and after they exited, and waited to see whether they would have to take anyone to the hospital. Police officers handled crowd control and traffic and provided an escort when the sample was taken for further testing at a laboratory. This occurred, unfortunately, while the city's emergency services were also handling a hotel fire, an injured person, and the nearby appearance of the Aga Khan, spiritual leader of Shia Imami Ismaili Muslims.

A third episode followed soon. An Israeli embassy employee in the United States checking mail at 5 p.m. 16 June 2005 opened one envelope that contained a small amount of a white powder. A section of the building was closed, and some employees remained inside until tests indicated at 11 p.m. that the substance was benign. The letter, which reportedly contained anti-Semitic language, was traced to a man jailed in North Carolina, according to the FBI [113]. The logistical, mental (and medical) impacts would be much worse, certainly, both temporally and spatially, in case a mailed envelope contains a living pathogen, as was the 2001 anthrax letter attack (presented below).

13.4.14 HISTORY AND EVOLUTION

An old tool of mankind, the chronology of bioterrorism from 1500 BC to now is presented. It demonstrates the gradual, consistent transition from intuitive bioterrorism to knowledge-based bioterrorism. It shows, as well, the past, recent, and present trends of bioterrorism as an evolving course, the extrapolative featuring of which is troublesome.

The evolutionary course of biological warfare at large is fascinatingly long. Almost as soon as humans figured out how to make arrows, they were dipping them in animal feces or sick persons so as to contaminate them. Practically, bioterrorism agents were the earliest weapons of mass destruction ever used by mankind. Long before having any idea regarding the in-effect nature of poisonous and infectious substances, man used toxins, feces, carcasses, and the like as effective weapons, in order to disease, kill, and terrorize opponents and enemies, thus following, from the very beginning, the three very basic fundamentals: incapacitating, killing, and intimidating. Rather a sort of intuitive bioterrorism, primarily.

Still, most of the ancient world liked to believe it fought with a code of honor. According to Adrienne Mayor, "Archers were disdained because they shot safely from afar: long range missiles implied unwillingness to face the enemy at close range. And long range missiles daubed with poisons seemed even more cowardly." Yet Odysseus, hardly a coward, returns home to kill his wife's suitors with poison arrows [1].

The roots of repugnance toward contaminated weapons are deep. Such weapons were singled out for disdain more than 2000 years ago in Greek and Roman codes of conduct, as well as in Hindu writings. But the course of using pathogens and toxins for terrorism continued. The rulers of ancient India were no less conflicted than the Greeks about arrows "barbed, poisoned or blazing with flame." These instruments violated the "traditional Hindu laws of conduct for Brahmans and high castes, the Laws of Manu." But in the Arthashastra, the Brahman military strategist Kautilya advised his king to use whatever means necessary to attain his military goals, including poisons.

In the Near East, it was after the mythic Greek hero Hercules slew the multi-headed Hydra that he developed a technology at the heart of today's most pressing international issue. In her illuminating history of warfare, the acknowledged folklorist Adrienne Mayor argues, in a sense, that "by steeping his arrows in the monster's venom, Hercules created the first biological weapon [1]."

Amazingly, then, the ancient world already made an important distinction between using disease weapons for purely defensive purposes as opposed to "first strikes." Although this constraint was rooted partially in ethics, Mayor contends it also reflected a shrewd understanding of epidemiology: "The principle of summoning plague for self-defense may be related to the reality that invaders are 'immunologically naïve' and therefore more vulnerable to epidemic diseases in foreign lands than the local population." Mayor is intrigued with the often fantastic devices the ancients used in war, but "Greek Fire" also excavates ancient attitudes toward biological arms and terrorism tools that are startlingly relevant today. Poisonous arrows were the Bronze Age's terror weapons. "Almost as soon as they were created," Mayor writes, "poison weapons set in motion a relentless train of tragedies for Hercules and the Greeks—not to mention the Greeks' enemies, the Trojans."

Similar, unrecorded biotterrorism-like events followed, most probably. Yet, beyond Greek mythology, even earlier, the ancient world also contained examples of bioterrorism warfare against settled populations stretching back to 1500 BC, when the Hittites sent plague victims into the lands of their enemies. Later, it was in about 600 BC that Solon of Athens put hellebore roots in the drinking water of Kirrha to kill the inhabitants. At about the same time, the Assyrians used to poison enemy wells with a fungus that would make the enemy delusional. Around 500 BC, Socrates was poisoned by the juice of Conium, which was the state poison of the Athenians. Necrotizing germs were intuitively employed in 400 BC, whereas Scythian archers systematically used arrows dipped in blood and manure or decomposing bodies to prevent wounds from healing. There are accounts from 300 BC of parts of the dead bodies of humans or animals being used by the Romans to poison water supplies. There have also been instances of the dead bodies of those who died from plague being catapulted into besieged cities. Later on, by 200 BC, Carthaginians used Mandrake root left in wine to sedate the enemy. Furthermore, in 184 BC, Carthaginian leader Hannibal is credited with an interesting use of bioterrorism weapons. In anticipation of a naval battle with the Pergamenes at Eurymedon, he ordered his troops to fill clay pots with snakes. During the battle, Hannibal sent the pots crashing down on the deck on the Pergamene ship. The confused Pergamenes lost the battle, having to fight both Hannibal's forces and a ship full of snakes. The act brought about panic and injured enemy sailors. Pergamenes, a king in Asia Minor, remarked that "he did not think any general would want to obtain a victory by the use of means which might in turn be directed against himself [1]." Honey bees are known to have been used as weapons since Roman times, as was then the case of Lucullus, who defeated Mithridates (74 BC) via bees.

As man expended onto the Far East and, later, the Americas, tropical plants became a common source for toxins. Thus, poisoned arrows are used widely throughout the jungles of Burma, Malaysia, and Assam. The principal sources of arrow poison are varieties of *Antiaris, Strychnos*, and *Strophanthus* [114]. *Antiaris toxicaria*, for example, is a tree of the mulberry and breadfruit family, common in Java and the neighboring islands. The active agents in all of these are either contained within the milky sap or the juice of crushed seeds. This is smeared behind the arrow point on its own, or mixed with another plant latex. When introduced into the bloodstream, the active ingredient (either antiarin, strychnine, or strophanthin depending on the species) acts quickly, attacking the central nervous system causing paralysis, convulsions, and cardiac failure.

The story of the *Strychnos*-derived toxin Curare is remarkable, in that it perfectly embodies a complete bivalent evolutionary line, starting with a very old arrow-poison—still in its very same traditional use presently—and ending, for the time being, as an important muscle relaxant. Contemporarily, curare is usable for selected assassinations just like "modern" toxins. In the sixteenth century, a group of Spanish explorers traveled the Amazon River. During the voyage, one explorer was hit in the hand by an arrow and died soon after. The culprit was curare, used widely as an arrow poison by many Amazon Indian groups (as it is still used by a few today). The complex processes used to make curare were a guarded secret. Often 30 or more ingredients could be found in one recipe. Indigenous Amazonians often mixed plants of different genera to concoct their potent toxins; their skill and knowledge in safely preparing these poisons is a testimony to their incredible ingenuity. Amazonian curares are divided into two groups based on the container the plant is stored in: pots or tubes. Pot curare in the East Amazon is predominately from the species *Strychnos guianensis*. Tube curare in the West Amazon is from *Chrondrodendron tomentosum*. (The curare in modern medicine is made from the latter species, therefore, its name: tubocurarine.)

For many centuries, the exact content of curare remained a mystery to Western observers; not until 1800 did Alexander Von Humboldt witness and document the preparation of curare by the Indians from the Orinco River. In 1814, an explorer named Charles Waterton injected a donkey with curare. Within 10 minutes, the donkey appeared dead. Waterton cut a small hole in her throat and inserted a pair of bellows, and then pumped to inflate the lungs. The donkey held her head up and looked around. Waterton continued artificial respiration for 2 hours until the effects of curare had worn off. Curare was found to block the transmission of nerve impulses to muscle, including the diaphragm muscle, which controls breathing.

Back to Europe, the Middle Ages, the use of infectious diseases to break sieges of castles and fortified towns is widespread. The most common method is to use catapults to hurl dead human or animal bodies over walls to spread disease. This same method is used to poison water sources. Plagued rats and infected flies were employed as well for introducing and disseminating the contagion. This was the case in 1155, when Barbarossa uses dead bodies to spread pathogens among the enemy during the battle of Tortona [115]. About 200 years later, in 1340 attackers hurled dead horses and other animals by catapult at the castle of Thun L'Eveque in Hainault, in what is now northern France. The defenders reported that "the stink and the air were so abominable . . . they could not long endure" and negotiated a truce.

In 1346, Tartar forces led by Khan Janibeg attacked the city of Kaffa (now Feodossia, Ukraine), catapulting the plague-infected bodies of their own men over the city's walls, and forcing the defending Genoese to abandon it when plague spread. Connectedly, ships carrying plague infected refugees (and possibly rats) sailed to Constantinople, Genoa, Venice, and other Mediterranean ports and are thought to have contributed to the second plague pandemic. (The first plague pandemic in 541 AD spread from Egypt to other parts of the world and killed 50–60% of the world population). It was perhaps the trigger of a subsequent outbreak of Bubonic plague that swept medieval Europe, causing 25 million deaths. Using dead bodies and excrement as weapons continued in Europe during the Black Plague of the fourteenth and fifteenth centuries. Even as late as 1710, Russian troops fighting Sweden resorted to catapulting plagued bodies over the city walls of Reval.

During that era, three peculiar events took place, as well:

In 1422, at Karlstein in Bohemia, attacking forces launched the decaying cadavers of men killed in battle over the castle walls. They also stockpiled animal manure in the hope of spreading illness. Yet the defense held fast, and the siege was abandoned after 5 months. In 1495, the Spanish tried wine infected with leprosy patients' blood against the French near Naples, and, in 1650, Polish artillery General put saliva from rabid dogs into hollow spheres for firing against his enemies.

In between, a sort of bioterrorism came to the New World in the fifteenth century, aimed to defeat the Indians. Spanish conquistador Pizarro gave clothing contaminated with the smallpox virus to natives in South America. During the French and Indian War (1754–1767) Sir Jeffrey Amherst, commander of British forces in North America, suggested the deliberate use of smallpox to "reduce" Native American tribes hostile to the British [116]. An outbreak of smallpox at Fort Pitt results in the opportunity to execute Amherst's plan. On June 24, 1763, Captain Ecuyer, Amherst's subordinate, gives blankets and a handkerchief from the smallpox hospital to the Native Americans and records in his journal, "I hope it will have the desired effect." This was followed by an epidemic of smallpox among Native American tribes in the Ohio River valley, which may also have been spread by contact with settlers. Transmission of smallpox by fomites (on blankets) is inefficient compared with respiratory droplet transmission.

Still, although the ancient world's arsenal of primitive bioterrorism weapons was trivial compared with the horrors of the modern world, those weapons raised the same terrifying moral and political dilemmas then as now. Thus, the absence of brought out events whereby BW were employed throughout the nineteenth century was possibly the latency preceding proliferation. Let alone that the nineteenth century marked the first isolation of germs in science.

During the twentieth century, then, an ascending course took place. In modern times, BW was used first for sabotage by Germany during WWI. German forces reportedly spread glanders and anthrax to debilitate enemy cavalries (detailed above).

In 1918, the Japanese formed a biological weapons section in the Japanese Army (Unit 731). Later on, in 1931, Japan expanded its territory into Manchuria and made available "an endless supply of human experiment materials" (prisoners of war, mostly civilians) for Unit 731. Biological weapons experiments in Harbin, Manchuria, continued until 1945. A post-World War II autopsy investigation of 1000 victims revealed that most were exposed to aerosolized anthrax. It is estimated that up to 3000 more prisoners and Chinese nationals may have died in this facility. During an infamous biowarfare attack in 1941, the Japanese Military released an estimated 150-million plague-infected fleas from airplanes over villages in China and Manchuria, resulting in several plague outbreaks in those villages. Overall, the Japanese Imperial Army experimented with and operated about 16 biological agents as tools of warfare and terrorism between 1932 and 1945. This took place in numerous locations in Asia, and it has been estimated that a total of 10,000 Chinese prisoners, U.S. prisoners of war, and British detainees were killed by some of the most gruesome human experimentation in history. The Japanese used BW agents such as anthrax, plague, tularemia, and smallpox [117, 118].

During World War II, another effort, taken by Britain, was the production of over 5 million anthrax infected cattle cakes, which would have been dropped over Germany in an attempt to decrease meat stocks by some 30%. Events overtook plans to put these into operation [119].

Ken Alibek, formerly a chief scientist of the Soviet offensive biological warfare program, has alleged that the Soviets employed the germs causing tularemia during World War II. In his book *Biohazard*, he states that there is evidence tularemia was used by the Soviet troops to help stop the German panzer troops in the Battle of Stalingrad. The resulting tularemia outbreak may have halted the Nazi advance, but the Soviet troops also developed the disease because of what Alibek suspects was a sudden change in wind direction. Over 100,000 cases of tularemia were reported in the Soviet Union in 1942, a 10-fold increase in incidence experienced in 1941 and 1943. Seventy percent of the cases were the respiratory form of the disease, which is the form that would have been expected from a BW rather than a natural outbreak of the disease [120]. Nonetheless, this episode may constitute a good example of biological warfare, rather than bioterrorism.

Avner Cohen, an expert on unconventional weapons proliferation, has catalogued reported uses of bioterrorism weapons by Jewish forces during the 1948 War of Independence in Palestine. Connectedly, the Israeli historian Uri Milstein alleged that "in many conquered Arab villages, the water supply was poisoned to prevent the inhabitants from coming back." Milstein states that one of the largest of such covert operations caused the typhoid outbreak in Acre in May 1948 [121]. Within that context, the Palestinian Arab Higher Committee reported in July 1948 that there was some evidence that Jewish forces were responsible for a cholera outbreak in Egypt in November 1947 and in Syrian villages near the Palestinian–Syrian border in February 1948. Furthermore, in May 1948, the Egyptian ministry of defense stated that four "Zionists" had been captured while trying to contaminate artesian wells in Gaza with "a liquid which was discovered to contain germs of dysentery and typhoid."

A pause took place for almost two decades (apart from minor or marginal incidents), but since 1965, the trend significantly changed. Multiple bioterrorism occurrences have been observed, as follows.

1965—In Japan, outbreaks of typhoid and dysentery were deliberately induced by a local bacteriologist [72].

1970—In Canada, several students became badly ill after eating food deliberately contaminated with the eggs of parasitic ringworm [122].

    During the 1970s, the Soviets used mycotoxins in Laos and Cambodia and hepatitis A virus in Afganistan.

1972—Two men affiliated with the U.S. group "Order of the Rising Sun," who eventually fled to Cuba, had conspired to contaminate the water supplies of some large Midwestern cities with stocks of typhoid fever germs cultivated by one of them. Up to 40 kg of bacteria cultures were found in a college laboratory [73].

1975—The Symbionese Liberation Army was found in possession of technical manuals on how to produce bioweapons.

1980—Assassination of CIA agent Boris Korczak in McLean, Virginia, Tyson's Corner, using a ricin weapon, possibly in umbrella configuration [75].

    Police raided a German Red Army Faction apartment in Paris and found a miniature laboratory containing a culture medium of the germ that produces botulinum toxin. Notes about bacteria-induced diseases were found in the apartment as well [74].

    Protesters claimed to have taken infected soil from the Hebridean island of Gruinard and placed it at the chemical defense establishment at Porton Down. The island has been closed to the public since germ warfare experiments on sheep were conducted there in 1941. The anthrax spores used in the experiments can remain dangerous for decades [75].

1983—The FBI obtained one ounce of ricin in a 35-mm film canister from an individual in Springfield, Massachusetts, who had manufactured it himself. This is believed to be one of several confiscations of ricin [75].

    Some 750 people were sickened due to typhoid fever, consequent to bacterial contamination of restaurant salad bars in Oregon, conducted by a local cult attempting to affect the outcome of a local election [73].

1984—Australian authorities received an anonymous threat warning that foot-and-mouth disease virus would be released among livestock if reforms in Queensland Prison were not implemented [123].

    A Cuban expert defected and testified that one third of the United States could have been contaminated if a stockpile of toxins held by Cuba were to be "strategically placed in the Mississippi River [75]".

Two Canadians attempt to procure tetanus and botulism cultures from ATCC. Reportedly the first phone order, of less deadly cultures, was fulfilled, and it was not until the second order that ATCC employees become sufficiently suspicious to notify authorities [71].

1990–1995—Japanese cult Aum Shinriky attempted at least three times to disperse aerosolized botulinum toxin and anthrax in downtown Tokyo and at U.S. military installations in Japan. All attempts failed.

1990—In Scotland, a limited outbreak of giardiasis occurred as result of deliberate water contamination [124].

1991—During the Gulf War, there were serious concerns in the United States, that the ventilation systems of buildings might be attacked by terrorists using BW agents [98].

1993—An Arkansas man with survivalist group connections attempted to smuggle 130 g of ricin from Alaska into Canada to use as a weapon [125].

1994—An Iraqi scientist specializing in genetic engineering and implanted in New York by Saddam Hussein's regime intended to conduct an act of bioterrorism, due to having access to various local laboratories [126].

1995—Two members of the Minnesota Patriots Council were convicted of conspiracy to assassinate a deputy U.S. Marshal and International Revenue Service agents by ricin [127].

A member of the white supremacist Aryan Nation acquired freeze-dried bubonic plague bacteria from the ATCC [76].

1996—In Texas, 12 laboratory workers at a medical center became ill as a result of eating muffins and doughnuts intentionally contaminated by dysentery germs type 2 [185].

1998—An Iraqi terrorist network was maintained in the United States, intending to conduct acts of bioterrorism and reportedly furnished with BW agents by Iraqi women that smuggle agents filled vials into the United States within their bodies [128].

The British government has issued a warning to all ports about an Iraqi attempt to bring large quantities of the deadly germs of anthrax into Britain (and other countries) inside cosmetics bottles, cigarette lighters, and perfume sprays, disguised as duty-free goods [104].

Several Muslim terrorists affiliated with the Algerian "Armed Muslim Group" were arrested in Belgium, holding a lot of information about biological and chemical weapons and about the World Cup football games [129].

1999—A group of medical workers, mostly Bulgarian, employed in the Benghazi children's hospital, Libya, were detained, subsequent to an explosive AIDS epidemic that apparently started in 1998 in that hospital, and involved, thereupon, hundreds of children. By the year 2000, the medics were charged with deliberately infecting 393 Libyan children in their care with HIV, by injecting them with infected products. The HIV type was the same in each case—a rare and previously unrecorded strain that originated in West Africa when two existing viruses combined [130]. An analysis conducted by European experts suggested that the outbreak has the hallmarks of accidental cross-contamination, where poor hygiene and ineffective sterilization procedures allowed contaminated blood to be spread between patients from a single infected child. Highly involved in the affair, the charged medics were then sentenced to death, however. Libyan leader Muammar Gadhafi said he believes the medics are guilty.

2001—The anthrax letters were sent in the USA, containing highly sophisticated militory-grade spore powder (details are given bellow).

2003—Former Texas Tech professor Thomas Butler said that 30 vials of plague bacteria were missing from the university. In fact, he stole, smuggled, and illicitly transported the bacteria to Tanzania [131].

Further episodes have been discussed above. All in all, the various above-described episodes are at any rate but a small segment of the entire picture. By 1998, Carus undertook a comprehensive inventory and assessment of bioterrorism (together with biocrimes) in the twentieth century. He has documented 222 cases, categorizing the cases and number of reported cases [132]:

- Confirmed use of bioterrorism agents—24
- Probable or possible use—28
- Threatened use (probable or confirmed possession)—11
- Threatened use (no confirmed possession)—121
- Confirmed possession (no known attempts or threats to use)—5
- Probable or possible possession—6
- Possible interest in acquisition (no known possession)—13
- False cases and hoaxes—14

Surprisingly, there have been only 222 bioterrorism-related incidents in a 100-year period and in only 24 cases have there been confirmed attacks—an average of 1 every 4 years worldwide. Most were abortive. Fourteen of the 24 confirmed cases of bioterrorism or biocrimes are food or agriculture-related; of these cases, 11 involved food poisoning and only 3 targeted commercial animals or plants. Of the 222 documented incidents, only 6 appear to be clearly linked to attacks on commercial plants and animals.

Significantly, yet, the survey made by Carus points at 144 incidents that occurred in the 1990s, meaning nearly two thirds of the total. This may reflect better incident tracking and record keeping in recent years, or it may indicate a dramatic increase in the propensity of terrorists or criminals to employ bioterrorism agents. Available evidence supports the latter premise. For example, FBI statistics indicate that U. S. incidents involving weapons of mass destruction using chemical, biological, radiological, or nuclear materials have soared from 37 in 1996 to over 200 in 1999, with three fourths of the cases involving bioterrorism agents—usually the threatened release of anthrax. Notably, the vast majority of incidents have been directed against individuals or small groups, not mass populations. On the whole, if any sort of extrapolation can be made, then the next decade is supposed to be a disturbing one, at the least.

13.4.15   STATE-SPONSORED BIOTERRORISM

State-sponsored bioterrorism may be carried out by saboteurs either affiliated with the concerned state or acting on their own but institutionally assisted by some country specifically aware of the ultimate outcome. Biosabotage programs and projects, in part realized, have been identified mainly in Germany, Japan, USSR, the United States, South Africa, and Iraq. They are here presented in detail, illustrating the conceptual and practical paradigm of state-sponsored bioterrorism.

States running a methodical biological weapons program would usually adopt a collateral subsystem dealing with bioterrorism means and operations. But the opposite equation remains open, in that the absence of a methodical program may not at all impair an effectual framework in charge of developing, manufacturing, and deploying of terrorism-oriented biological agents. The perpetrators tasked for carrying out the biosabotage acts may equally be affiliated with the involved state, or with another state, be it a friendly or a hostile state, including the target country itself. State-sponsored bioterrorism may as well be directed against domestic opponents and carried out internally. Another variance lies between bioterrorism initiated by a state and such that is initiated by a terror organization (or a second state) but is crucially and knowingly assisted by the concerned state. All those modes may be categorized as state-sponsored bioterrorism and are looked into through the following case studies.

13.4.15.1   Germany

From 1915 to 1918, Germany waged an ambitious campaign of covert biosabotage on animals being shipped from neutral countries to the Allies. The program used the germs causing glanders and anthrax, and employed secret agents to administer the bacterial cultures to animals penned for shipment. The cultures were sometimes injected using needles dipped into the cultures, sometimes poured onto feed, or (later in the war) contained in capillary tubes embedded in lumps of sugar that were fed to the animals. Horses and mules were the main targets, but in some cases sheep and cattle appear to have been targeted as well.

The programs were initiated nearly simultaneously in Romania and the United States. The Romanian campaign was administered by Major Nodolny of the German General Staff, through his Military Attaché in Bucharest. The agents disseminating the cultures were Bulgarian, run by the Bulgarian embassy. Cultures were shipped from Berlin. The program lasted until the August 1916 Romanian declaration of war against Austro-Hungary and the expulsion of German diplomats.

The campaign in the United States was operated by a U.S.-born, German-raised, physician, Anton Dilger. Dr. Dilger brought seed cultures with him to the United States in 1915 and set up a culture facility in the suburban Washington, D.C. home that he rented. He supplied cultures to the German merchant-ship captain Hinsch, stranded in the United States by the British naval blockade, who ran the agents, largely stevedores. Dr. Dilger returned to Germany in early 1916, and the campaign came to a halt a few months later [133, 134]. Similar campaigns were conducted concomitantly in Argentina, Mesopotamia, and Norway. During WWII, the only known conduction of biosabotage by Germany was the contamination of a large reservoir in Bohemia with sewage, in 1945. Throughout that war, however, the Nazis routinely used non-German-captured civilians for induced infection experimentations.

13.4.15.2   Japan

Civilians were the main object of mass terror attacks launched by the Japanese bacteriological Unit 731 on multiple Chinese targets during the 1930s and 1940s, various bombs and other disseminating devices being thereupon applied. Furthermore, if terrorism at large is a term of relevance with respect to enemy civilian prisoners, then the vast biological experimentations conducted by Unit 731 on thousands of arrested civilian Chinese during the 1930s and 1940s are certainly a sort of an extremely brutal bioterrorism. First-hand accounts testify the Japanese infected civilians through the distribution of contaminated foodstuffs, such as dumplings and vegetables. There are also reports of contaminated water supplies. Such estimates report over 580,000 victims, largely due to plague, anthrax, tularemia, smallpox, and cholera outbreaks [135]. Plague was often induced through infected flees, as well. In addition, repeated seasonal outbreaks after the conclusion of the war brought the death toll to much higher. On one occasion at least (1939), the Japanese military contaminated Soviet water sources with typhoid bacteria at the former Mongolian border.

13.4.15.3   United States

Declassified documents reveal the past existence of at least two bioterrorism operations worked out, if not conducted, by the CIA against Cuba:

Operation FULL-UP, the objective of which was to destroy confidence in fuel supplied by the Soviet Bloc by indicating it is contaminated. The operation was to be accomplished by introducing a known biological agent into jet fuel storage facilities. This agent flourishes in jet fuel and grows until it consumes all the space inside the tank [136].

Operation MONGOOSE: A document entitled "Project Cuba", dated 18 January 1962, sets forth the aims and the 32 original tasks of what subsequently became known as "Operation MONGOOSE [137]." For task 21, it states that, on 15 February 1962, the CIA would submit a plan to disrupt the harvest of food crops in Cuba. The following two sections of the declassified text of this document, which might be expected to clarify the method to be used in pursuit of this objective, appear to be censored: Clearly their content was so repugnant that even the officials responsible for declassifying the document saw fit to keep that part of it secret. However, in materials provided to the U.S. Senate Select Committee on Intelligence Activities by the CIA during the mid-1970s, the CIA acknowledged that it had developed methods and systems for carrying out a covert attack against crops and causing severe crop loss. The CIA denied that it had ever employed such systems [138].

The United States has repeatedly been accused by Cuba of biosabotage acts that allegedly occurred. Reportedly, in 1977, a U.S. intelligence source admitted that the United States used the swine-fever virus as biological warfare against Cuba. The agent told U.S. media he was ordered to transport the virus from a U.S. Army Base and CIA training center in the Panama Canal Zone to a group of right-wing

Cuban exiles who in turn delivered it to operatives inside Cuba in March 1971 [139]. This may seemingly be connected to the outbreak of swine-fever virus in Cuba on 6 May, 1971—its first appearance in the Western Hemisphere. The highly contagious virus is lethal to pigs. Six weeks into the epidemic, the Cubans were forced to slaughter a half-million pigs to stem the spread of the epidemic. Also, the *New York Times* reported in 1983 how the head of a Miami-based anti-Cuban terrorist group admitted in a U.S. court that he had taken germs to Cuba in 1980.

Furthermore, Project MKNAOMI is notable as well. In the 1970s, the CIA publicly revealed that a U.S. Army team called the Special Operations Division (SOD) at Fort Detrick, Maryland, developed biological and chemical weapons for the CIA under a Top Secret project that would last almost 20 years [140]. This project, MKNAOMI, was practically unknown at the CIA due to the extreme sensitivity of its mission. Few written records were kept. CIA personnel working at Fort Detrick used the cover of Special Support Staff of the Department of Defense. Indeed, on 23 October 1962, the U.S. Patent Office granted patent 3,060,165 to four persons "as represented by the Secretary of the Army." The patent was first filed 3 July 1952, Serial Number 297,142 for the use of ricin as a biological weapon. The strikingly honest descriptive language used to apply for this U.S. patent, in 1952, is very revealing: "Ricin is a protoplasmic poison prepared from castor beans after the extraction of castor oil therefrom. It is most effective as a poison when injected intravenously or inhaled. A very fine particle size was necessary so that the product might be used as a toxic weapon." Also, in the early 50s the SOD conducted—as revealed in a document partially declassified by the FBI— a biosabotage experiment, using the Pentagon building as a model target, so as to prove that all persons present therein would be unknowingly exposed within a relatively short period of time [141]. Additional model targets where selected to simulate biosabotage operations, for example the Washington's National Airport and the New-York Subway during the 60s (details given above). The CIA and possibly other elements within the U.S. intelligence community apparently persisted in holding bioterrorism agents after the United States destroyed its biological weapons arsenal. Besides, an increasing debate has aroused as to the shear innocence of the US outstandingly growing research of virulent pathogens, referring to the fragile borderline separating between defensive- and offensive-oriented implications. However, the very same debate may equally pertain, in principle, to other states.

13.4.15.4 South Africa

Rather unexpectedly, an extremely meticulous biosabotage program has been uncovered in South Africa. Apparently, it constitutes an outstanding case of a state-sponsored bioterrorism paradigm that has been fairly brought out in details [142].

The most characteristic feature of the South African biological (and chemical) terrorism program was undoubtedly the development, testing, and utilization of a wide array of hard-to-trace agents to assassinate "enemies of the state [143]." As insider testimony and the notorious "sales list" of 1989 (Telemedicine Research Center document 52) indicate, several of the highly poisonous substances produced at both Delta G and Roodeplaat Research Laboratories (ad hoc front companies) were actually deployed by clandestine units ADF and SAP "death squads," above all the Security Branch's C[ounterinsurgency]1 section (later renamed C10), housed at the Vlakplaas base, in covert assassination operations [144].

There is no doubt whatsoever that high-ranking officers within the SADF and SAP and other "securocrats" within the government were generally aware of these activities, many of which they in fact authorized. Some civilian Afrikaner paramilitary groups, whose pro-apartheid members remain violently opposed to black majority rule, have even publicly threatened to attack their enemies with biological and chemical agents [145, 146].

In the mid-1980s, a higher level and more formalized assassination program formed when the Teen-Rewolusionêre Inligting Taakspan (TREWITS: Counter-Revolutionary Intelligence Task Force) was created. Then, on the verbal instructions of a senior cardiologist, Colonel Dr. Wouter Basson—head of Project Coast: Apartheid's Chemical and Biological Warfare Program—a host of freeze-dried pathogens and highly toxic substances that had been produced either at Delta G or RRL—was secretly transferred and thereafter stored in a refrigerator inside a fireproof and bombproof walk-in safe in his own office to military and police personnel through various channels. The deadly agents were passed on, either to the aforementioned persons in innocuous public places like restaurants or to Basson himself in the latter's office at South African Medical Services (SAMS) headquarters in Centurion. The specific recipients of these lethal substances and contaminated items were operatives of the "death squads," officers who either deployed some of them personally or later distributed them to the so-called "hit team" members; an ex-psychologist who in 1988 assumed control over Systems Research and Development; a bioengineering company set up in part to manufacture special "applicators," i.e., arcane assassination devices such as rings, screwdrivers, walking sticks, and umbrellas that had been transformed into weapons by means of the addition of poison compartments and injectors or firing mechanisms for poisoned pellets; and Basson himself. Furthermore, holes were drilled in cans of Game orange soda, into which some substance was injected, and then closed by means of soldering so that they were no longer visible.

The actual substances included the viruses of Ebola, Congo, and Marburg; lethal toxins such as mamba toxin, botulinum, and ricin; bacterial agents such as plague, anthrax, brucellosis, salmonellosis, and bottles of cholera bacteria; and a wide variety of foodstuffs, beverages, household items, and cigarettes that had been contaminated with these biological agents [147]. There can be little doubt that several of these materials, items, or devices were subsequently often used to murder or sicken opponents of the apartheid regime. If one excludes the hundreds of drugged and secretly disposed guerrilla opponents, the total number of victims appears to have been in the dozens.

International ties formed as well within that dark context. Dr. Wouter Basson, or other Coast personnel, may have transferred dangerous biological warfare materials and know-how to elements of a loose international network of right-wing extremists. Fears have been expressed that Basson and other Coast scientists were associated with an even broader international right-wing network, purportedly known as Die Organisasie (The Organization), among whose members are said to be expatriate Rhodesians and South Africans who emigrated to other countries both during the apartheid era and as the apartheid system was collapsing [148]. Basson also had strong bonds with Arabic elements, particularly Iraqi and Libyan.

Connectedly, according to a pair of Federal Bureau of Investigation (FBI) informants, in the mid-1980s, the American doctor Larry Ford transferred a suitcase full of dangerous "kaffir-killing" pathogens to Surgeon-General Knobel at the Los Angeles residence of the South African trade attaché, Gideon Bouwer [149]. It has also emerged that at Knobel's request Ford lectured Coast scientists about the contamination of household items with biological agents [150–153]. The whole apparatus was dismantled during the 1990s. Still, such a complex state-sponsored bioterrorism program, operating both inside South Africa and beyond, all over, has not elsewhere been uncovered publicly, yet certainly not an isolated one, worldwide; the Soviet, and later Russian parallel system did not—and apparently does not—lag far behind.

13.4.15.5 USSR and Russia

Two bioterrorism affairs, one momentary—the planned assassination of an opponent by ricin toxin in London—and the other one lasting for years—the relapsing employment of fungal toxins against civilians in South East Asia—were sponsored, in effect, by USSR, whereas Bulgaria and North Vietnam were involved as partners, respectively. Not singular (referring to the Soviet apparatus at large), those two affairs attracted much attention. Although the latter has been surreptitiously—yet massively—conducted by the Soviet MOD (details given in Section 13.4.9.3), the former was a fine product of the KGB.

Various modes of biosabotage have been specifically encountered with regard to the USSR and Russia, in the book titled *Biological Espionage*, written by former KGB officer Alexander Kouzminov who worked for the so-called "Department 12, Directorate S (the KGB Operational Technical Support Directorate) [154]." This directorate oversees Moscow's "illegals" (i.e., Russian agents posing as Westerners, operating under deep cover). Department 12 (and Department 8) of the KGB were tasked with preparing "clandestine acts of biological sabotage against 'potential strike targets' on the enemy's territory." These potential targets include military research laboratories, combat units, weapon stockpiles, public drinking water, food stores, vaccine repositories, pharmaceutical plants, and the overall economy of the target country. These two departments would also assassinate, incapacitate, or kidnap foreign officials, political enemies, and "important persons" (where "important" is determined by the exigencies of war).

Department 12 was referred to in intelligence circles as the "Chamber" or "Kamera." KGB General Viktor Chebrikov, Andropov's closest subordinate in the KGB, was then the Director of this department. The "Chamber" developed, among many technical devices and substances various toxins, such as ricin, which is but one example. The work of Department 12 "has grown" since the collapse of the Soviet Union. Genetic engineering has brought forth new bioterrorism horrors. And these are to be unleashed in the event of something called "Day X," which signifies, conceptually, the beginning of the next world war—a large-scale war. Soviet military thinkers believe that such a war will tentatively involve the mass use of nuclear as well as biological weapons, concomitant with multiple bioterrorism operations. The KGB thus had its own R&D centers for special toxic matters including bioterrorism toxins and pathogens for use in espionage aims and sabotage (so-called "Fleita" program). In parallel, aerosol-released bioterrorism weapons were often in the possession of Spetsnaz operatives (Soviet military Special Forces).

Connectedly, the manners of the KGB are well illustrated in detail through the Soviet–Bulgarian ricin plot. Georgi Ivanov Markov was born 01.03.1929 in Sofia, Bulgaria and died in London due to induced ricin intoxication. He was executed by DS (Durzhavna Sigornost—Bulgarian State Security) in London on 11.09.1978, thanks to vital assistance knowingly afforded by the KGB. A totally independent journalist, Markov was Bulgaria's most revered dissident and Bulgarian Communism's arch enemy. Bulgarian Communist dictator, Todor Zhivkov, was very well informed about Markov's activities by the DS [155]. In 1977, Zhivkov asked the KGB to help him silence Markov. The Russians did not hesitate. Both President Yuri Andropov and Vladimir Krutchkov, head of KGB, personally approved and ordered General Sergei Golubev (Chief of the Security Service and specialist on "murder") to cooperate with DS. The KGB granted DS access to the resources in the "Chamber."

Golubev received instructions in the KGB headquarter relevant "Chamber," and the next week he flew to Sofia with Ivan Surov. Surov's job was to transmit the Bulgarian Intelligence Service practical know-how in the use of special poisons, which could not be traced after the victim's death. Golubev and Surov discussed with the Bulgarian's intelligence officers the various options of killing Markov. They worked out one plan to use a poison that could be surreptitiously dissolved in tea, coffee, or any liquid that Markov might drink. In 1978, Gen. Golubev traveled to Sofia three or four times to help DS with planning of the secret operation. Three attempts to assassinate Markov followed. The first attempt was made in Munich in the spring of 1978 when Markov was visiting friends and colleagues at Radio Free Europe. Someone put a toxin into Markov's drink at a dinner in his honor. The attempt to kill him failed. The second assassination effort occurred on the Italian island of Sardinia, where Markov was on vacation with his family. The plan also failed for reasons unknown.

Golubev returned to Sofia to work out a new plan to kill Markov. The KGB decided to use a camouflaged weapon. A folding umbrella was adapted with a firing mechanism and silencer to shoot a small pellet at close range, one and a half to two meters. Golubev requested that the KGB Residency in Washington purchase several U.S.-manufactured umbrellas and send them to the Center. The KGB head resident in Washington bought several umbrellas and sent them to the KGB, and an OTU operational technical unit rebuilt the umbrellas. The "Chamber" then adapted the umbrella tip to enable it to shoot the victim with a tiny metal pellet containing ricin. Golubev then took the converted umbrellas to Sofia to instruct the assassin on how to use this weapon. The pellet was supposed to penetrate the clothing and be lodged in the upper skin layer. Consequently, the final, and successful, attempt was staged in London 07.09.1978, on Bulgarian's Communist dictator Todor Zhivkov's sixty-seventh birthday. Before his assassination, Markov received a threatening anonymous phone call: "Not this time," said anonymous caller. "This time you will not become a martyr. You will simply die of natural causes. You will be killed by a poison that the West cannot detect nor treat." The tasked ricin toxin did the job, indeed.

On the day of his assassination, Markov worked a double shift at the BBC. After finishing the early morning shift, he went home for rest and lunch. Returning to work by car, he drove to a parking lot on the south side of Waterloo Bridge. It was his habit to take a bus across the half-mile Waterloo Bridge to the BBC headquarters in the Bush House. After parking his car in a parking lot near the Waterloo Bridge, Markov climbed the stairs to the bus stop. As he neared the queue of people waiting for the bus, he experienced a sudden stinging pain in the back of his right thigh. He turned and saw a man bending to pick up a dropped umbrella. The man who was facing away from Markov, apologized. The assassin then hailed a taxi and departed. Although in pain, Markov boarded the bus to work. But the pain continued.

Markov noticed a small blood spot on his jeans. He told colleagues at the BBC what happened and showed one friend a pimple-like red swelling on his thigh. By evening, he had developed a high fever. He was hospitalized and treated for an undetermined form of blood poisoning. His condition fast worsened. He was not responding to doctor's efforts. The next day he went into shock, and after three days of agony and delirium, he died on 11.09.1978.

An autopsy was performed at Wandsworth Public Mortuary. The doctors found a tiny metal sphere the size of a pinhead in the wound. When they attempted to extract the "pin," a tiny pellet fell on the table. The police took the pellet to the Chemical and Microbiological Warfare Establishment at Porton Down, commonly called the "Germ Warfare Center." There, a team of the England's foremost specialists in forensic medicine, and, reportedly, Dr. Christopher Green of the CIA examined the pellet. The pellet was 1.52 mm in diameter, embedded in his calf, and composed of 90% platinum and 10% iridium [156]. They found that two 0.34-mm holes had been drilled in the pellet, possibly using a high-technology laser at right angles to each other, producing an X-shaped cavity. The holes were empty.

This prevented investigators from establishing the type of substance that had been used, but it was sufficient to determine that Markov had "not died of natural causes." BATS (British Anti-Terrorist Squad), detectives then joined the Scotland Yard investigating team. After weeks of research and experimentation, in January 1979, a Coroner's Inquest in London Gavin Thurston ruled that Georgi Markov had been killed via ricin toxin. Traces of ricin were possibly found later thereupon.

But several years later, two former top KGB officers, Oleg Gordievsky and Oleg Kalugin, publicly admitted Soviet complicity in Markov's murder by means of ricin toxin. The case was dormant until after the arresting of Communist dictator Todor Zhivkov and the fall of the Communist government in Bulgaria in 1989. Although the DS act about the case Markov is destroyed or has been sent to Moscow, there is no doubt that some DS and KGB officers know the true about his death.

In 1991, former chief of Bulgarian Foreign Intelligence Vasil Kotsev, who was identified as the person in charge of the Markov operation, died in a questionable and mysterious automobile accident. A second suspect, General Stoyan Savov, preferred to commit suicide on 09.01.1992 rather than face trial also for destroying the documents. In 1994 the British Parliament asked Russia to help them find 15 past KGB agents who might have been involved in or had knowledge of the murder. The request remains unanswered. The Markov murder case remains officially unsolved. No one has been brought to justice for the murder of Markov, although *prima facie* evidence points fairly clearly at the involvement of Soviet and Bulgarian elements, chiefly the KGB.

The Russian KGB officially ceased to exist in November 1991, but its successor organization, the FSB, is functionally extremely similar to the KGB. Even after many details about the related bioterrorism mechanisms were revealed, during the 1990s, and although many alterations took place in Russia, this system did not substantially change.

13.4.15.6 Iraq

In the past, Iraq conducted bioterrorism in the following cases:

In 1988, the Iraqi army deliberately introduced typhoid bacteria into the water supply of the Kurdish city of Sulaimaniyah, bringing about an outbreak [157].

In 1989, outbreaks of cholera were generated as a result of BWA being experimented with on Kurdish populations by Iraq [158].

In 1990, an apparent, severe malaria outbreak that exceptionally occurred in the Kurdish Biharka concentration camp in Iraq has been attributed to Iraqi experimental employment of a BWA [159].

Also, Iraq developed and produced aflatoxin (at least 2200 L) as a long-term debilitating bioterrorism agent intended to be used against the Kurds. It has thereupon been noted that:

> The discovery (by UNSCOM, during the 1990s) that Iraq was researching aflatoxin, not a traditional BW candidate, was a cause for some surprise. It is a carcinogen, the effects of which manifest themselves only after many years, and several Western experts have rationalized this Iraqi program only in terms of genocidal goals. If aflatoxin were used against the Kurds, for instance, it would be impossible definitively to prove the use of BW once the symptoms emerged. Another possible explanation is its potential use as an immune suppressant, making victims more susceptible to other agents. However, the aflatoxin declaration may also hide other aspects of Iraq's BW program: according to Iraq's depositions, the production program never encountered any mishap (as other parts of the BW program had) and, to judge from the declared time-frame for the total amount produced, production could never have stopped, even for cleaning of the equipment. This raises the suspicion that Iraq declared an excessive amount of aflatoxin in order to disguise the fact that other, more destructive agents had been produced in greater quantities [59].

Being a very potent, although typically slow-acting hepatotoxin and nephrotoxin, aflatoxin may have been developed as a BWA by the Iraqis for purposes of long-term terrorism, or some short-term acute impact of yet unknown nature [5]. Iraq was unable to justify the weaponization of aflatoxin from the research data obtained from its own experimentation. One bomb that by Iraq's account should have contained aflatoxin instead tested positive for botulinum toxin and negative for aflatoxin [160]; this indicated another mode of weaponization. And indeed, there is documentary evidence and statements obtained by UNSCOM that Iraq was mixing aflatoxin with riot-control gas. They note that it would not be unthinkable for a leader who has used chemical weapons on part of his population. "It is a great way to keep colonels from becoming generals," said an UNSCOM inspector familiar with the searches. "Saddam Hussein hasn't made any weapon that he hasn't used on his own population, with the exception of Scud missiles [60]." Doubtfully, the entire, tainted past Iraqi–Kurdish interface has been revealed by him since being in prison.

Notably, the Iraqi Intelligence Service provided the BW program with security and participated in bioterrorism research, probably for its own purposes, from the beginning of Iraq's BW effort in the early 1970s until the final days of Saddam Hussein's regime. The Iraqi Intelligence Service had a series of laboratories that conducted bioterrorism work including research into BW agents for assassination purposes until the mid-1990s. The Iraqi Survey Group found that the Iraqi Intelligence service produced ricin and tested it on political prisoners. ISG has not been able to establish the scope and nature of the work at these laboratories or determine whether any of the work was related to military development of a BW agent [161]. But they could certainly suffice to sponsor and support the 2001 anthrax letters' sabotage, together with al-Qaeda, as has previously been suggested [80]. All in all, the option for bioterrorism operations was persistently maintained by Iraq until its collapse in 2003 [162].

13.4.15.7 Iran

According to Eisenstadt, Iran has probably deployed BWs, which it could deliver via terrorist saboteurs, aerosol tanks mounted on aircraft or ships, or via missiles [163]. Taking care, thus, of the entire spectrum of BWs, Tehran did not ignore their importance in the context of terrorist actions; it equipped itself with micro-warfare means destined to employ bioterrorism agents by on-spot spraying and by contaminating water systems. Iran, as well as Muslim BW possessors like Sudan and Syria, can readily be assisted by some Muslim terror organization, in case they want to carry out bioterrorism acts through nonresidential saboteurs. More concretely, such cooperation between Iran and its terror organization ally—the Hezbollah—has been feared.

Other states runing BW programs that may potentially support of sponsor bioterrorism acts include, mainly, Syria, North Korea and Cuba.

13.4.16 NONSTATE-SPONSORED BIOTERRORISM

Persisting organizations, temporary communes, or sporadic individuals may initiate and conduct bioterrorism acts, without there being any institutionalized assistance from any country. The following section deals with those three types of elements, practically or potentially involved in bioterrorism-related activities. Also, it presents in detail prominent episodes of in-effect bioterrorism (further episodes are included in Section 13.4.14, 13.4.17 and 13.4.18).

13.4.16.1 al-Qaeda

This terror organization has been, and still is, the most significant one worldwide, in terms of bioterrorism, especially that it was involved, most probably, in the Sept. 2001 anthrax letter attack. In the mid-1990s, while being sheltered in Sudan, its head, Osama bin Laden, financed, in part, the construction of BW (and CW) facilities in Sudan. Later on, members of al-Qaeda were apparently trained in Iraq (by its intelligence apparatus) for BW (and other nonconventional weapons) employment. In 1998, relationship with Iraqi intelligence was established, so as to obtain poisons—appatently toxins—and gases training. After the USS Cole bombing in 2000, two al-Qaeda operatives were sent to Iraq for CBW-related training beginning in Dec. 2000. Iraqi intelligence was "encouraged" after the embassy and USS Cole bombings to provide this training [164].

Moreover, in August 2002, reports have emerged that Ansar al-Islam, an al-Qaeda affiliate active in Iraqi Kurdistan since September 2001 as a militant group, has been involved in testing various poisons including ricin [165]. According to an ABC News report, the group tested ricin powder as an aerosol on animals such as donkeys and chickens, and perhaps even an unwitting human subject. The experiments were ordered and financed by a senior al-Qaeda official, who was providing money and guidance from elsewhere in the region.

Also, selected al-Qaeda terrorists were guided as to the methods of cultivating germs and preparing toxins to be used as bioterrorism agents and how to convert them into weapons, exploiting easily available equipment and materials. Those who plotted in the caves of Afghanistan have left behind diagrams of American cities and landmarks. Also, manufacturing instructions have been posted on the Internet. One of the simulated operations in Kandahar, Afghanistan, was contaminating a water main of a European city (apparently London) using equipment that could fit inside a backpack. Several al-Qaeda cells have been trained in Afghanistan, where they have learned to use bioterrorism agents, including anthrax, ricin, and botulism toxins. Later, after the fall of the Taliban regime, those groups continued their experiments in the Pankisi Gorge, on the territory of Georgia, bordering Chechnya [166, 167]. Not too far away, in Fallugah, Iraq, an improvised laboratory belonging to Iraqi al-Qaeda-directed rebels was revealed in 2004. It was found to contain materials for making chemical blood agents, as well as a "cookbook" on how to produce a deadly form of anthrax [168]. Two months earlier, notably, the very same Abu Musan al-Zarkawi's derivative group of al-Qaedat ("Tahwid and Jihad"—"Oneness of God and Holy War") tried hard to take care of two women known as the Iraqi anthrax masterminds. It kidnapped two Americans and a British man, threatening to kill them if Iraqi women prisoners were not released. The two Iraqi masterly women were not set free, and the three kidnapped men were then killed [169].

Another training base was located at Zenica, Bosnia. It is not clear, however, whether any experimentation practically included, beyond toxins, in-effect living germs. Reportedly, bin Laden's associates managed to receive anthrax and plague biological weapons from former Soviet facilities in Kazakhstan. Records and operations manuals captured in Afghanistan and elsewhere disclose that bin Laden has devoted money and personnel to pursue smallpox, among other biological weapons. Furthermore, uncorroborated testimony in a high-profile Egyptian trial in 1999 indicated that al-Qaeda had equired Ebola virus and Salmonella germs.

al-Qaeda, and particularly its second in command, Ayman al-Zawahiri, a physician, had long been eager to acquire biological agents, particularly anthrax, according to Khalid Shaikh Mohammed, one of Osama bin Laden's top lieutenants. Also, al-Qaeda agents had inquired about renting crop-dusters to spread pathogens, especially anthrax. During interrogations with terror suspects in custody, it has come to light that the al-Qaeda terror group has been actively seeking weapons-grade anthrax [170]. According to an article by Milton Leitenberg, computer hard drives and handwritten notes seized at the home where Mohammed was arrested included an order to buy anthrax, along with other evidence of an interest in acquiring anthrax and other dangerous germs [171]. Until the American invasion of Afghanistan, launched after the Sept. 11, 2001 attacks, al-Qaeda's anthrax program was based in Kandahar, Afghanistan, and was led by two men: Riduan Isamuddin, known as Hambali, and Yazid Sufaat, a Malaysian member of Jemaah Islamiyah, an al-Qaeda-affiliated group. Although Sufaat tried to acquire anthrax, there is no evidence that he was able to procure the appropriate strain used for the 2001 attacks (Ames strain). This does not mean, yet, that al-Qaeda did not obtain that deadly strain through other channels. Hambali had been trying to open a new BW project for al-Qaeda in the Far-East, when he was arrested. Reportedly, anthrax powder was found, eventually, in March 2006 in a house occupied by the Taliban in Afghanistan [172].

A London-based offshoot was engaged in preparing ricin and botulinum toxins. In January 2003, British authorities arrested six Arab men that intended to produce ricin in their north London apartment, led by Kamel Bourgass, who had been trained in al-Qaeda terror camps in Afghanistan and was specially selected for instruction in making poisons [173]. Accurate recipes and ingredients for poisons including botulinum and ricin (plus cyanide), and the blueprint for a bomb were found in the apartment, in addition to castor beans.

Nonetheless, after the 2001 anthrax letter attack (covered bellow in special subchapter) and the U.S. military operations against al-Qaeda in Afghanistan, on many occasions repeated bioterrorism warnings emerged, stemming from intelligence information and from various announcements made by al-Qaeda. Just as one example, six flights bound for America, including four from Britain, were cancelled after intelligence suggested that al-Qaeda was planning a spectacular attack using a hijacked airliner and weapons of mass destruction, particularly BWs [174]. Alongside with cases of in-effect bioterrorism-oriented activities, like the ricin affair in London, various bioterrorism hoaxes that could not be identified may probably be attributed to al-Qaeda. At any rate, al-Qaeda certainly wants to use biological weapons and widen the scope. The British M15 recently reported, thereby, that al-Qaeda operatives are training in germ warfare, and trying to recruit university students with access to microbiological laboratories, so as to steal virulent pathogens [175].

13.4.16.2 Jemaah Islamiah

A Southeast Asian Islamic militant group affiliated with al-Qaeda, this organization does not lag far behind the latter. Connectedly, Dr Rohan Gunaratna, who heads the terror unit at Singapore's Institute of Defence and Strategic Studies, has warned Australia of a new, global generation of Islamist terrorists "armed, trained financed and ideologicised" to use biological weapons. In declaring that part of this new global terror wave would come out of Southeast Asia, he added he found the Jemaah Islamiah group had come extremely close to developing bioterrorism-chemical weapons, recounting an alarming analysis of a Jemaah Islamiah biological-terror rudimentary training manual that was taken from a Philippines safe house in late 2003. His further generalization is notable, saying "It is only a question of time that a group that has these intentions will have the capability to develop them. We are seeing a new generation of terrorists being trained in the use of biotlogical weapons. We rarely saw this in the 1990s but today we are seeing increasingly this kind of training. Groups are being trained to use these agents, though the probability of attack is still low [176]."

13.4.16.3 Hezbollah

Former CIA director James Wolsey described the Hezbollah as a potential tool for bioterrorism. That threat has been accentuated and is particularly realistic, because the Hezbollah is a radical terror organization, directly supported and encouraged by Iran, which possesses biological weapons. The evident acquisition and successful employment of unmanned drones by the Hezbollah, as well as its access to the sources of the Jordan River, may increase a potential bioterrorism threat posed by this dangerous organization. In 1998 the Hezbollah attempted to obtain biological and chemical weapons through two businessmen situated in Switzerland [177]. Elements of Hezbollah and al-Qaeda joined forces to plan a series of attacks using toxic substances, apparently chemicals and toxins. Groups of al-Qaeda and Hezbollah activists from the Sidon area of Lebanon met in Africa in December 2002, to set up the needed preparations there or in Sidon [178]. The 2006 confrontation between Hezbollah and Israel, demonstrated effective operational capacities of the Hezbollah to employ a variety of Iranian rockets and missiles, some of which are tipped—in Iran—with biological warheads.

13.4.16.4 Radical Palestinian Terror Organizations

An article appeared on the 13 August 2001 edition of the Lebanon-based Palestinian weekly, *Al-Manar*, stating that there is "serious thinking" among the Palestinians about obtaining BWs. It says regarding BWs that "obtaining its primary components is possible without too much effort, let alone the fact that there are hundreds of experts who are capable of handling them and use them as weapons of deterrence, thus creating a balance of horror. Anyone who is capable, with complete self-control, of turning his body into shrapnel and scattered organs, is also capable of carrying a small device that cannot be traced and throw it in the targeted location [179]." It has been pointed out that Palestinian organizations are inclined to resort to the acquisition and introduction of nonconventional capabilities—in particular biological and chemical—mainly for deterrence purposes. According to that source, many Palestinians believe that biological and chemical weapons are a legitimate and desirable means in the struggle against Israel [180]. Basically, a radical terror organization, like the Hamas, is apt to acquire BWs, although so far it is not known to have obtained them, in that case.

Notably, a novel mode of infection with hepatitis B virus has been suggested, following penetrating human bone fragments due to the explosion of a Palestinian suicide bomber [96]. Connectedly, perhaps, the Tanzim planned to use a suicide-bomber arrested with explosives bearing AIDS-infected blood, the thinking being that anyone who survived would get AIDS. The operation was foiled [181]. Another Palestinian body, the al-Aqsa Martyrs' Brigades (affiliated with the Palestinian Fatah political party) sent an announcement in 2006 to the Ramattan News Agency, stating: "we are pleased to say that we succeeded in developing some 20 different types of biological and chemical weapons, this after a three-year effort", adding that those agents might be employed under certain circumstances [182].

13.4.16.5 Aum Shinrikyo Cult

The Aum Shinrikyo Cult ("religion of supreme truth"), a Japan-based apocalyptic religious sect, produced biological agents and tried to use them. The Japanese police discovered that the Aum included among its members skilled scientists and technicians, including some with training in microbiology, who attempted to generate weapons using anthrax, cholera, botulinum toxin, Q-fever, and even Ebola virus [183]. These accounts also suggest that there were four separate attempts to use biological agents, including anthrax once and botulinum toxin three times:

- In April 1990, the Aum Shinrikyo outfitted an automobile to disseminate botulinum toxin through the engine's exhaust. The car was then driven around Japan's parliament building.
- In early June 1993, the cult attempted to disrupt the planned wedding of Prince Naruhito, Japan's Crown Prince, by spreading botulinum toxin in downtown Tokyo using a specially equipped automobile.
- In late June 1993, the cult attempted to spread anthrax in Tokyo using a sprayer system on the roof of an Aum-owned building in eastern Tokyo. The anthrax was disseminated for four days.
- On March 15, 1995, the Aum planted in the Tokyo subway three briefcases designed to release botulinum toxin. Apparently, the individual responsible for filling the botulinum toxin had qualms about the planned attack and substituted a nontoxic substance. The failure of this attack led the cult to use sarin in its March 20, 1995 subway attack. A helicopter and two UAVs were purchased by the cult and intended to be equiped with sprayers.

Fortunately, the Aum scientists apparently made mistakes in either the way they produced or disseminated the agents, and no one became ill or died from the attacks.

13.4.16.6 Bhagwan Shri Rajneesh Cult—Typhoid Epidemic

As the summer of 1984 waned, a cult led by the Bhagwan Shri Rajneesh near the town of The Dalles, Oregon, used a biological agent to sicken hundreds in an apparent dress rehearsal to sway the outcome of a local election. Three years earlier, the Bhagwan brought followers with him from India when he immigrated in 1981. The hard-working Rajneeshees were isolationist, with some 150 armed people to keep outsiders away from their ranch. When the Bhagwan decided to enlarge his ranch and his flock, he took over the small town of Antelope, christening the new town Rajneesheepuram.

Oregon's attorney general stated, however, that the municipality was unconstitutional because it did not separate church and state. To outmaneuver the attorney general, the cult's hierarchy hatched a plan to make the Wasco county residents too sick to vote in November 1984, enabling the Rajneeshees to seat their favored candidate on the county court. The cult's nurse was the scientific brain behind attempts to put a bioterrorism plan into action [184].

Although the cultists considered additional organisms (AIDS and *Salmonella typhi*), the Rajneeshees decided upon *Salmonella typhimurium*, so as to bring about a massive food contamination. The cult bought bactrol disks from a Seattle medical supply company under false pretenses. A trio of cult members worked in a laboratory equipped with an incubator and freeze dryer to brew what they called a "salsa." Several more Rajneeshees were involved in distributing the agent on various occasions. Starting on 29 August 1984, the Rajneeshees began sprinkling their *S. typhimurium* in personal drinking glasses, on doorknobs, and urinal handles; on produce at the local supermarket; and on salad bars in 11 restaurants.

Soon, a steady stream of patients was reporting to local physicians and hospitals, with symptoms ranging from nausea and diarrhea to headache and fever. In total, 751 fell ill. Wasco County commissioners and ordinary citizens were among the victims. Within 4 days, local health-care providers were able to identify the *S. typhimurium* as the source, but over a year passed before there was confirmation that a single strain caused all of the illnesses and the Centers for Disease Control filed its report. No one died in this test to see whether a ballot box could be fixed, but the Bhagwan reportedly observed that one should not worry if a few perished. Law enforcement authorities thought that the Rajneeshees were practicing to poison the water system of The Dalles. Cult members had already put dead rodents and perhaps raw sewage and salmonella salsa into The Dalles' water supply.

Several cult members were involved in the planning and execution of the salmonella attacks, but only two of the Baghwan's chief lieutenants were prosecuted. This pair received multiple concurrent 20 year sentences, among other penalties. According to one analyst who studied the Rajneeshees carefully, the cult did not work its way up the ladder of violence to bioterrorism. Rather, the Rajneeshees appear to have abruptly embarked on their salmonella spree as a means to a specific end.

An epidemiologic analysis was then done. Cohort and case-control investigations were conducted among groups of restaurant patrons and employees to identify exposures associated with illness. All 751 persons with Salmonella gastroenteritis were associated with eating or working at area restaurants. Most patients were identified through passive surveillance; active surveillance was conducted for selected groups. A case was defined either by clinical criteria or by a stool culture yielding *S. typhimurium*. It has been found that the outbreak occurred in two waves, September 9 through 18 and September 19 through October 10. Most cases were associated with 10 restaurants, and epidemiologic studies of customers at 4 restaurants and of employees at all 10 restaurants implicated eating from salad bars as the major risk factor for infection. Eight (80%) of 10 affected restaurants compared with only 3 (11%) of the 28 other restaurants in The Dalles operated salad bars (relative risk, 7.5; 95% confidence interval, 2.4–22.7; $P < 0.001$). The implicated food items on the salad bars differed from one restaurant to another. The investigation did not identify any water supply, food item, supplier, or distributor common to all affected restaurants, nor were employees exposed to any single common source. In some instances, infected employees may have contributed to the spread of illness by inadvertently contaminating foods. However, no evidence was found linking ill employees to initiation of the outbreak. Errors in food rotation and inadequate refrigeration on ice-chilled salad bars may have facilitated growth of the *S. typhimurium* but could not have caused the outbreak. A subsequent criminal investigation revealed that members of a religious commune had deliberately contaminated the salad bars. An *S. typhimurium* strain found in a laboratory at the commune was indistinguishable from the outbreak strain [184].

Although the above information pertains to organizational or cultist bioterrorism, the following episodes represent individual bioterrorism.

13.4.16.7   Shigella Dysenteriae Type 2 Outbreak

Another case of intentional contamination of food with a bacterial pathogen occurred in 1996, involving clinical laboratory workers infected with *S. dysenteriae* type 2 [185]. *S. dysenteriae* type 2 is a rare organism, and outbreaks are seldom seen in the general population. After consumption of muffins and donuts anonymously placed in the break room of a Texas laboratory, 12 of 45 laboratory workers developed severe, acute diarrheal illness. Rapid onset of symptoms allowed investigators to retrieve a muffin sample, later cultured and found to contain *S. dysenteriae* type 2, matching those cultured from infected workers. Laboratory reference cultures were found in disarray, and cultures were missing, indicating laboratory cultures had been taken, grown, and used to contaminate foodstuffs. All case patients reported having eaten muffins or doughnuts placed in the staff break room on October 29. Isolates from nine case patients were indistinguishable from *S. dysenteriae* type 2 recovered from an uneaten muffin and from the laboratory's stock strain, a portion of which was missing.

13.4.16.8   Hepatitis A Outbreak

Another instance of apparent intentional contamination of food with hepatitis A virus was documented in 1965. Twenty-three cases of hepatitis were reported among personnel at a Naval Air Station in 1961 [186]. The course of disease among those infected was moderately severe with no deaths. At the time, hepatitis A (infectious hepatitis) and hepatitis B (serum hepatitis) diagnoses were made on clinical and epidemiological grounds. To determine the cause of the infections, an extensive investigation of risk factors was conducted. Dining hall records and food questionnaires identified potato salad as the most likely contaminated food source. Questioning of the sole salad cook revealed that he had experienced symptoms resembling hepatitis, and a background investigation revealed aberrant social behavior. Connectedly, two incidents of inappropriate urination gave investigators a possible route of contamination of the potato salad when coupled with the individuals' symptomology.

13.4.16.9   Ascaris Suum Outbreak

The last instance of intentional contamination involved four college students who were peculiarly exposed to a massive dose of embryonated *Ascaris suum* eggs (a large ringworm infecting pigs) while attending a Winter Carnival in Canada during February 1970. The most likely source of infection was a meal served to the students during the Winter Carnival [187].

More episodes, mostly of secondary importance, are mentioned below in Section 13.14.4.

13.4.17 THE ANTHRAX LETTERS

A colossal event generated by means of a minute amount of anthrax powder, the September 2001 anthrax letter attack, is discussed in detail. The remarkable potentiality of anthrax as a bioterrorism agent in general, and within the context of the letter attack in particular, is demonstrated. Resultant short- and long-term impacts are accentuated, alongside with the severe inability to point at the provenance of the anthrax powder contained in the letters.

The germ *Bacillus anthracis* is the etiological agent of anthrax, a disease primarily of herbivores, but one that humans occasionally acquire through contact with infected animals or with contaminated animal products. The mode of infection may be cutaneous, intestinal, or pulmonary. The latter represents, as well, a distinct form of employing anthrax germs as an inhalable bioterrorism agent, either sprayed (wet aerosol) or powdered (dry aerosol). The efficacy of the aerosol thus generated is shaped by a variety of inherent attributes largely depending on the way the aerosol material has thereby been structured. In general, the aerosol material contains anthrax germs spores—a resilient form of life typifying all species of the genus *Bacillus* sp.—plus various, most vital, additives.

Under optimal meteorological conditions, producing 50% fatalities over a one-square-mile area would require about just 10 g of anthrax spores, in case there is no medical intervention, whatsoever [188]. Additional data show the impact of anthrax in terms of some 20,000 to 80,000 casualties afflicted by 30 kg of this biological agent, as compared with 400 to 6000 casualties afflicted by 300 kg of sarin nerve gas, and 80,000 casualties afflicted by 20 kiloton of a nuclear weapon [189]. Moreover, official U.S. estimates for casualties produced by an airplane, flying upwind of a city, releasing successfully a cloud of anthrax germs, range from 100,000 to three million dead. An individual driving a car around a medium-sized city spewing anthrax out the tailpipe would cause some 70,000 fatalities, and two individuals in two cars in two cities, 140,000 (referring to an unprotected population). These are, of course, traditional theoretical estimates; yet they illustrate the proportional impact of a biological agent, in this case, anthrax—a noncontagious bacterial pathogen. Still, even by 2003, it was held that a little more than two pounds of anthrax spores efficiently spilled into the air over a city the size of New York could be expected to kill more than 120,000 people unless state and federal officials respond much more aggressively than they currently plan to, according to the first comprehensive computer model of such a terrorist act [190].

The September 2001 anthrax letter attack marked an outstanding, most significant milestone in the course of bioterrorism. It was a quantum leap, scarcely expected in practical terms, albeit with a lot of indicative intelligence preceding the event. Overall, this bioterrorism campaign has been, operationally, complicated, assuming that the spore powder was produced outside the United States. A most comprehensive picture has been presented by the School of Public Health, Department of Epidemiology, University of California. It meticulously covers the technical, epidemiological, medical, and logistical aspects altogether. Seven letters containing anthrax spores were probably mailed, with similar written messages from Trenton, NJ. Five letters were sent on September 18 (postal facilities [191, 192] [probable cross-contamination]), one going to American Media in Boca Raton, Florida [193, 194] (not recovered); a second to the *New York Post* (recovered) [195–197]; a third to Tom Brokaw of NBC News (recovered) [198, 199]; a fourth to ABC News (not recovered) [200]; and a fifth to Dan Rather of CBS News (not recovered) [201]. On October 9, two more letters were sent from Trenton, NJ. (probably cross-contamination) [202–206] via Brentwood mail processing facility [207–210], one to Senator Tom Daschle (recovered), and the other to Senator Patrick Leahy (recovered) [211]. Letter(s) cross-contaminated with the Daschle and Leahy letters were sent from Trenton, NJ to Wallingford, CT, with at least one letter probably going to Oxford, CT [212]. The final anthrax case in the outbreak remains a mystery, but possibly arose from contact with the September 18 letters or cross-contamination with the October 9 Leahy letter in Trenton, NJ. To do so, spores from the Leahy letter would need to have adhered to an envelope of another letter destined for the Bronx, New York City [213]. In summary, the 22 cases (one was removed by the CDC) that comprised the American Anthrax Outbreak of 2001 likely had contact with one or more of seven spore-laden envelopes.

Four letters laden with anthrax spores were discovered, all dated by an unknown author as "09-11-01," and all sent from Trenton, NJ. Two letters were postmarked September 18, 2001 in Trenton, one of which was sent to the *New York Post* where it was handled by several staff members, and the other to Tom Brokaw of NBC, opened September 19–25 but not found until October 12, 2001 in case 2's file drawer [198]. The *New York Post* letter, handled but not opened, was found on October 19, 2001. It was dampened before being discovered, turning the spore contents into a granular or clumped state.

The second two letters were postmarked on October 9, 2001 and mailed to the Washington, D.C. offices of Senator Tom Daschle of North Carolina, Majority Leader, and Senator Patrick Leahy of Vermont, Chair of the Judiciary Committee. Both letters went though Washington, D.C.'s Brentwood mail processing facility, which handles all incoming federal government mail. Both letters contained the same anthrax strain and were of the same potency.

The Daschle letter was opened in the sixth floor office at 9:45 a.m. by an aide in the Senator's Hart Senate Office Building suite on October 15, 2001. It was believed to contain about 2 g of powder comprising 200 billion to 2 trillion spores. Based on nasal swabs, all 18 persons who were in the area of Daschle's sixth floor office tested positive for anthrax exposure, as did 7 of 25 (i.e, 28%) in the area of the Senator's fifth floor office (an open staircase connected the two offices).

The Leahy letter never arrived at his office. Instead an optical reader misread the handwritten 20510 ZIP code for the Capitol as 20520, which serves the State Department. As a result, the letter was routed to the State Department, where it arrived on October 15, infecting a State Department postal worker (case 20) [211]. Shortly thereafter, all mail was isolated and sealed in plastic bags for a later search.

On November 16, 2001, the Leahy letter was found; then after special preparation, it was opened on December 6 in a laboratory setting. It contained about 1 g of anthrax, made fresh no more than 2 years before it was sent. The contents of the enclosed letter were identical to the wording of the Daschle letter. The anthrax spores in the Daschle and Leahy envelopes were uniformly between 1 and 3 μm in size, and were coated with fine particles of frothy silica glass. More investigation is underway.

Also processed at the Trenton, NJ postal facility was a small number of letters sent to the Southern Connecticut Processing and Distribution Center in Wallingford, CT. Here letters arrived on October 11 that had been cross-contaminated with anthrax spores from the October 9 Daschle or Leahy envelops. Anthrax spores were found on mail-sorting equipment in Wallingford. One letter that went through the Wallingford distribution center was found in Seymour, nearby to Oxford where case 23 resided. Likely case 23 was infected via a similar cross-contaminated letter that came in contact with mail in the distribution center in Wallingford (not discovered) [212].

Most uncertain is the origin of case 22, although there is a connection between case 22's neighborhood and the Trenton, NJ post office [213]. A printout from the post office showed that an unrelated letter went to a shop around the corner from case 22's home. This unrelated letter was processed 2 minutes after the Leahy letter and 18 minutes before the Daschle letter. Thus the mail sent to case 22 might also have been cross-contaminated with spores from the Leahy or Daschle envelopes. Alternatively, case 22 might have had contact with one or more of the unrecovered September 18th letters following their disposal in Manhattan, similar to case 19 and case 21 [196].

But the anthrax letters affair was not limited to the US. The American embassy in Vilnius, Lithuania, was likewise concurrently targeted. Also, Three businesses in Karachi, Pakistan, were targeted, and received anthrax letters. Anthrax spores were found in a letter sent from Europe to a doctor's surgery in Chile, according to Chilean officials [214].

For the time being, the culmination of bioterrorism worldwide has been this act of distributing mail envelopes containing anthrax spore powder. It reflected noticeable supremacy of a simple act of bioterrorism (irrespective of preparing the anthrax powder in itself, which was very sophisticated), in several senses:

a. Uncontrollable preparing of the postal envelopes containing the anthrax powder
b. Uncontrollable, repeated mailings
c. Undetectable conveying of the mailed envelopes until reaching their various destinies
d. Untraceable footprints of the perpetrators
e. Inability to even postulate whether the sabotage was state or nonstate sponsored

Despite or, rather, because of those points of supremacy, the leading agencies responsible for foreseeing or, at the least, deciphering in retrospect such strikes— FBI, CIA, DIA, and the like—were after nearly 2 years still reluctant to grant public access to an unclassified report on lessons learned from the anthrax letter attacks of 2001; the Pentagon has finally released a redacted version of the document [215]. "The anthrax attacks revealed weaknesses in almost every aspect of U.S. bioterrorism-preparedness and response," the report said. "As simple as these attacks were, their impact was far-reaching." The report, authored by David Heyman of the Center for Strategic and International Studies, is based on a day-long forum convened by CSIS under contract to the Defense Threat Reduction Agency in December 2001.

It provides a detailed and informative but hardly unsuspected inventory of shortcomings in emergency preparedness and response. Ironically, the Defense Department's two-year denial of repeated requests for release of the document exemplifies one of the central problems identified in the report: "The failure to communicate a clear message to the public was one of the greatest problems observed during the anthrax attacks . . . [including] failure to provide timely and accurate information." The Department's inability to efficiently process requests for this document suggests that it still has a long way to go to remedy this particular failure. The redacted version of the report was finally released in response to a Freedom of Information Act (FOIA) appeal [215].

Also, the Pentagon did not contend that the withheld portions of the anthrax report are classified (exemption 1), even though they are said to concern "vulnerabilities and capabilities of the US Government to respond to another . . . attack." Rather, in an expansive interpretation of the law, it said that release of the redacted portions would make it possible to "circumvent Department of Defense rules and practices. . . ." (exemption 2). See the March 15 transmittal letter from H.J. McIntyre of the DoD FOIA Policy Office here: Mike Denny Senior Force Protection Consultant Guardian Group International (GGI) [216].

In the long run, the impact of the anthrax letter attack still prevails in several respects:

- Individually, those who survived the infection have not been fully cured.
- Globally, the short-run impact encouraged terror-oriented entities to persistently explore or increase bioterrorism options.
- Practically, numerous consequent hoaxes of bioterrorism threats through powder containing post envelopes extremely proved burdensome to wide logistical systems.
- Environmentally, even as of 2005, rooms contaminated in 2001 are repeatedly disinfected.
- Publicly, the interface with institutional classified information holders sharpened.
- Strategically, the provenance of the anthrax powder has not been deciphered, although somewhat traced [217].

Thus, since the anthrax attacks in 2001, work at the nation's post offices has been disrupted by more than 20,000 incidents of suspicious powder leaking from envelopes and packages. All but a few have turned out to be nothing more than soap, dust, talc, or other nonlethal substances. Sand was the culprit in one incident included in wedding invitations for a beach ceremony. Still, the scares have taken a toll in nerves, lost time, and money. Postal workers have been instructed, if they see something like that to consider it dangerous. The area then is sealed off, and local hazardous materials teams and the Postal Inspection Service are called in. Leaking suspicious powders that turned out to be harmless include powdered Alfredo sauce, ground lentils, pudding mix, and coffee creamer, officials said.

Other cases have included leaking samples of detergent, sugar, or baking soda [218].

In November 2003, a secret cabinet-level "tabletop" exercise, designed to be very stressful to the system was thus conducted, which simulated the simultaneous release of anthrax in different types of aerosols in several American cities. The drill, code named Scarlet Cloud, found that the country was better able to detect an anthrax attack than it was 2 years ago, said officials knowledgeable about the exercise. But they said the exercise also showed that antibiotics in some cities could not be distributed and administered quickly enough and that a widespread attack could kill thousands; particularly, enormous difficulties stopping the spread of contamination through the country and into Canada were observed [219].

The drill was notable for the top-level attention it drew and the gaps it showed in the effort to protect against bioterrorism. About three dozen senior officials involved in domestic defense, including two cabinet officers—the secretary of homeland security, and the secretary of transportation—as well as the head of the White House's Homeland Security Council, participated in the exercise at the Pentagon's National Defense University.

Moreover, the contents of the former American Media Inc. building in Boca Raton, FL—namely, thousands of boxes—exposed to anthrax in 2001, were set to be decontaminated for a second time to ensure all of the pathogen is eliminated, given the approval of the Environmental Protection Agency and the Palm Beach Country Health Department [220].

Overall, the impact of the anthrax letter attack was colossal. In a sense, the very fact that the anthrax letter attack has not been deciphered since 2001 is no less meaningful and important than this unprecedented act of bioterrorism itself. In contrast to the wealth of empirical data collected and published with respect to the specific Ames strain of the 2001 anthrax letter attack and the resultant medical cases, a relative paucity of information has been brought out regarding the chemical and physical properties of the substance contained in the mail envelopes, namely the anthrax spore powder. Primarily its provenance has not yet been deciphered. This peculiar powder certainly reflects an outstanding assortment of different disciplines: microbiological, technological, medical, political, and strategic. Connectedly, it still poses, in parallel, an enigmatic, extremely complex intelligence issue that ought to be elucidated, for numerous reasons. Although the anthrax germs used for this act of bioterrorism have been fully identified to be indistinguishable from the strain Ames—a remarkably virulent strain initially isolated in 1981 in Texas but later shared with laboratories outside the United States—it is still not even known whether the anthrax powder contained in the envelopes has been structured in the United States or elsewhere [221]. Various indications point at Al-Qaeda together with Iraq as being the perpetrators [80].

Overall, dozens of buildings were contaminated with anthrax as a result of the five mailings, which contained, altogether, about 18 gr. of the sabotage spore powder. The decontamination of the Brentwood postal facility took 26 months (and cost US$130 million). The Hamilton, NJ postal facility remained closed for 41 months (its cleanup cost US$65 million). The Environmental Protection Agency spent US$41.7 million to clean up government buildings in Washington, D.C. [222]. One FBI document said the total damage exceeded US$1 billion [223].

13.4.18   THE RICIN COURSE

Various traits making ricin the toxin of choice for bioterrorism purposes are discussed. The actual outcome of that appeal, as shown, has for long been relapsing episodes involving ricin, worldwide. Clear tendency emerges of al-Qaeda and affiliated groups, at the least, to persist in deploying ricin, and is doubtfully controllable.

Ricin molecules contain a heterodimeric type 2 ribosome-inactivating enzyme (32 kDa), also known as the A chain. It is a very potent cytotoxic RNA N-glycosidase that inactivates eukaryotic ribosomes. It is linked by a disulfide bond to the galactose/N-acetylgalactosamine-binding lectin (34 kDa), also called the B chain. Weakness, fever, cough and pulmonary edema occur 18–24 hours after inhalation exposure, followed by severe respiratory distress and death from hypoxemia in 36–72 hours. Direct administration into the blood system brings about a similar picture, whereas food poisoning is basically slower, certainly dose-dependent [224].

This protein plant toxin may be regarded, foremost, as the prime toxin for bioterrorism, owing to the integral formed by several different characteristics, altogether:

a. High toxicity
b. Atypical symptomatology
c. Effective contraction through respiratory and alimentary tract
d. Effectual administration through blood stream
e. Availability of the raw material (castor beans)
f. Typical dual usability of the raw material
g. Easiness of manufacturing

Ricin's properties have been known since ancient times, and it has a long history of accidental and intentional intoxication [225]. Ricin was studied as a possible weapon in World War I, and the United States, Canada, and the United Kingdom developed it as a field weapon during World War II, also known as Agent W [226]. The World War II effort led to refined ricin in a crystalline form, although it was the amorphous form that was used in high-explosive bombs and shells and in more specialized delivery systems, such as plastic containers and cluster bombs. Attaining effective particle sizes with ricin powder is difficult, and use of the material in water or suspended in glycerol or carbon tetrachloride is more effective [227]. The agent is difficult to detect. It is fairly stable in clear, dry weather, persisting in the soil or environment for up to 3 days [228].

But the perfect usability of ricin-oriented for sabotage acts, in parallel, has not at all been ignored. The course of ricin oriented sabotage began long before 1978; yet its in-effect emergence took place in that year. It was the successful ricin-based assassination of the Bulgarian anticommunist reporter Georgi Markov that then occurred in London (full details given above). Two years later, an assassination of CIA agent Boris Korczak occurred in McLean, Virginia, Tyson's Corner, ricin being used as weapon, possibly repeatedly in umbrella configuration [76].

The attractiveness of ricin did not diminish because it found new grounds in the United States. In 1983, the FBI obtained one ounce of ricin oriented in a 35-mm film canister from an individual in Springfield, Massachusetts, who had manufactured it himself. This is believed to be one of several confiscations of ricin [76]. In 1993, an Arkansas man with survivalist group connections attempted to smuggle 130 g of ricin from Alaska into Canada to use as a weapon [164]. Later, in 1995 two members of the Minnesota Patriots Council were convicted of conspiracy to assassinate a deputy U.S. Marshal and International Revenue Service agents by ricin [127].

Syria and Iran acquired ricin as a weapon during the 1990s. Also, the Iraqi Survey Group concluded that the Iraqi Intelligence Service produced ricin contemporarily, and tested it on political prisoners. The toxin subsequently became an agent of increasing interest within some Muslim terrorist organizations. Thus, in August 2001, the FSB (Russian Federal Security Service) claimed it had intercepted a recorded conversation between two Chechen field commanders about instructions on the "homemade production of poison" for use against Russian soldiers. Russian authorities reportedly then seized materials, including confiscated papers also containing instructions on how to produce ricin from castor beans. The Russian response did not take place too late, employing the lethal castor molecule. In March 2002 the FSB apparently continued its long tradition of poisoning, while killing Chechen rebel Amir Khattab, by sending him a letter impregnated with ricin [229].

al-Qaeda got into action. Traces of ricin and instructions on its use were discovered in 2002 in an al-Qaeda house in Afghanistan [230]. The head of an al-Qaeda affiliate network, Abu Massab al-Zarqawi, was trained in the use of ricin in Afghanistan before he relocated to Iraq in 2002. It was regarded to be an ideal bioterrorism weapon for terrorists. An al-Qaeda suspect arrested in Italy before the 2003 Iraq War told his interrogators that members of the al-Zarqawi network had purchased toxins from Iraq [231]. Ricin, botulinum, and gas-gangrene toxins were the main toxins possessed by Iraq.

And, indeed, in August 2002 reports have emerged that Ansar al-Islam, an al-Qaeda affiliate active in Iraqi, has been involved in testing poisons and chemicals including ricin. The toxin powder has been tested as an aerosol on animals such as donkeys and chickens, perhaps even on an unwitting human subject.

The outcome has first surfaced in London, marking the ricin route from Asia onto Europe: London, 5 January 2003. British authorities arrested six men suspected of producing ricin in their north London apartment. All six, including two teenaged asylum seekers and four individuals in their twenties and thirties, are believed to be Arabs from Algeria or other North African countries. On 8 January, a seventh man, age 33, was also arrested in connection with the case [232]. The next day sources in Whitehall indicated that at least one of these seven men had attended an al-Qaeda training camp in Afghanistan, whereas others appear to have received terrorist training in Chechnya and the Pankisi Gorge region of the Newly Independent State of Georgia [233–235].

In the apartment where the six original arrestees resided, the authorities found several castor oil beans and equipment that could be used to process those beans. The toxin has not been detected. Five other locales were subsequently searched in conjunction with this incident, and on 13 January, Scotland Yard officials arrested five more men and a woman in Bournemouth. One day later, another Islamist who was being arrested by Manchester police attacked them with a knife, killing one officer and wounding four others [236]. British antiterrorist investigators soon identified this same 27-year-old Algerian, "Kamel Bourgass" (also known as Nadir Habra), as "a very senior player" in the network thought to be behind the ricin plot [237].

These events occurred after a series of arrests of other Islamist radicals in Rome, Paris, and London. Because some of these individuals, particularly those in the Rome case, were allegedly planning to carry out terrorist attacks using poisons, the most recent incident in London raised several potentially worrisome questions [238]. One is whether the arrestees actually intended to employ the deadly toxin as an assassination or mass casualty weapon. A second is whether they were linked to components of the al-Qaeda network, such as the Groupe Salafiste pour la Predication et le Combat (GSPC: Salafist Group for Preaching and Fighting—see footnote) in Algeria, or were instead affiliated with some other Islamist terrorist organization such as the Algerian Groupe Islamique Armée (GIA: Armed Islamic Group). A third is whether certain states that have previously produced and tested ricin as a potential weapon played any role at all in transmitting their technical expertise or unused stocks of ricin to violent anti-Western Islamist groups. Iran, Syria, and Iraq (until occupied, if not later) are known to possess ricin.

Still in Europe, ricin was then discovered in Paris, two months later. Traces of the toxin have been found at a railway station in Paris, according to the French interior ministry. Two vials of the potentially deadly substance were found inside a locker at the Gare de Lyon, according to ministry officials. The locker contained two vials with a powder, a bottle filled with a liquid and two smaller bottles also containing a liquid. The two smaller bottles contained traces of ricin in a mixture that turned out to be a highly toxic [218].

The United States was repeatedly the next station of the ricin course, chronologically, although apparently involving but domestic American perpetrators questing to achieve influence. In October 2003, a metallic container was discovered at a Greenville, SC postal facility with ricin in it. The small container was in an envelope along with a threatening note. Authorities did not believe this was a terrorism-related incident. The note expressed anger against regulations overseeing the trucking industry. But one month later, the target was the White House.

The U.S. Secret Service intercepted a letter addressed to the White House in November 2003 that contained sparkled ricin powder, but it never revealed the incident publicly and delayed telling the FBI and other agencies; the affair was disclosed in February 2004. It was addressed to the U.S. Department of Transportation. The letter—signed by "Fallen Angel" and containing complaints about trucking regulations—was nearly identical to one discovered in October 2003 at a Greenville, SC, mail-sorting facility. But the existence of a similar letter sent to the White House was not disclosed until yesterday, and then only by law enforcement

---

[1] The name *Salafi*, which derives from the Arabic verb *salafa* (to precede), refers to the original companions of Muhammad, who are collectively known as al-Salaf al-Salih (the virtuous forefathers of the faith). In this context, a Salafi is a traditionalist who demands that all Muslims follow the exemplary, pious, and uncorrupted behavior of Muhammad and his trusted original companions. Note also that the GSPC is a breakaway faction of the GIA that later allied with al-Qaeda.

officials who asked not to be identified by name. Until then, the whole thing was absolutely kept under wraps on a national security basis.

Although seemingly—at the least—purely domestic, Secret Service spokeswoman declined to comment on details of the case or why it was kept secret, citing the ongoing investigation. He also declined to comment as to whether workers at the mail facility were tested or underwent decontamination procedures, and said the facility's location was kept secret for security reasons. Officials have previously noted that one mail facility used by the White House is located at Bolling Air Force Base. The letter is believed to have been sent from the Chattanooga area as it passed through a mail facility there [240]. The subsequent and last link in the United States, for now, has been a ricin-containing postal enveloped mailed in Feb. 2004 to the U.S. Senate office of the Majority Leader. Federal investigators have examined about 20,000 pieces of mail in hopes of finding the source of the ricin discovered on Capitol Hill, but so far they have turned up nothing to lead them to a suspect in the case [241].

One month later, possibly by sheer coincidence, FBI agents searching Miron Tereshchuk's home in March 2004 found firecrackers, chemicals, recipes for ricin, and a postcard with the message "Here is your poison—enjoy," according to documents filed in the Virginia court. Back to Iraq, the same month, the U.S. Army staged raids in Baghdad, uncovering evidence of an attempt to produce ricin, so as to put some in mortar rounds [242]. And again, in the United States, in November 2005, plans for producing ricin, along with bomb materials and diagrams were found in the home of Sergio Maldonado, a 20-year-old man known as a criminal [243].

In a markedly similar event (June 2006), a jar containing ricin derived form castor beans, along with ricin residue in a bowl, gun silencers, pipe bombs and bomb-making materials, were found in a shed in Nashville, Tennessee [244].

Most probably, the ricin course did not reach an endpoint.

13.4.19 THE SMALLPOX COMPLEX

The unique properties and particular importance of smallpox virus are discussed, both as a bioterrorism threat and as a major biohazard in general. It appears to pose, potentially, a colossal menace, which ought to be intensely coped with. The issues of renewed vaccination and ongoing research are then addressed, showing the complexity related to that singular pathogen.

The virus causing smallpox appears to be, objectively, the most devastating pathogen ever faced—as far as known—by humanity. Presumably, many billions of people lost their life due to smallpox during human history (an estimated 300-million people worldwide in the twentieth century alone). The virus has been effectively used for bioterrorism purposes (details given above). It has successfully been eradicated worldwide, primarily thanks to the fact that it is infective toward man only, hence not having the capacity to endure within any other creature. This plainly means that the absence of this pathogen from human populations signifies for its final elimination, except for persisting in the frozen state in laboratories or, perhaps, in nature. Moreover, the lacking of alternative hosts, apart from man, considerably restricts—both quantitatively and qualitatively—the overall gene pool of this virus and, therefore, its ability to markedly evolve and give rise to novel variants capable of evading human herd immunity, which has extremely been fortified in the past by mass inoculations. Resultantly, the magnitude of susceptible unvaccinated segments of human populations was, at the time, too small to sustain the virus. Consequently, the virus vanished.

Furthermore, the genetic distance between *Variola major*, the virus that generates smallpox, and the monkey-pox virus, which is sometimes regarded as a possible future reproducer of a new variant of human pox, is still appreciable. But the genetic distance from certain viral strains held in some laboratories over the world is zero. And, at the same time, human herd immunity against smallpox is steadily diminishing, since inoculations were stopped (around 1980). The temporal equation thus formed seems to be formidable. Moreover, the sound, natural equilibrium between the spatial distribution of immunized and nonimmunized segments of human populations worldwide has substantially been impaired as a result of vaccinations, particularly since the elimination of the virus (as a natural selection factor) from the wild. Objectively, favoring for a moment an authentic natural perspective, this seems to be, then, a sort of anomaly. Such circumstances may immensely increase the vulnerability of mankind to the impact of an act of bioterrorism involving the smallpox virus.

If not held in some laboratories until now, it would apparently be, then, a non-revivable creature. But in actuality, things are different. The story begins in that there are no clear data as to the overall local attempts to isolate—and, hence, preserve—the virus during the occurrences of natural smallpox epidemics over many countries worldwide throughout the twentieth century, until the last 1972 epidemic. Still, even if assuming the total and complete destruction of this virus worldwide, it is apt to be recreated, based on the knowledge attained on its full genome.

Smallpox (also known by the Latin names *Variola* or *Variola vera*) is a highly contagious disease unique to humans. The causative virus has two variants called *Variola major* and *Variola minor*. *V. major* is the more deadly form, with a typical mortality of 20–40% of those infected. The other type, *V. minor*, only kills 1% of its victims. Many survivors are left blind in one or both eyes from corneal ulcerations, and persistent skin scarring—pockmarks—is nearly universal. Smallpox was responsible for an estimated 300–500 million deaths in the twentieth century. In 1967, for instance, still prevailing worldwide in spite of massive inoculations, the WHO estimated that 15 million people contracted the disease and that 2 million died in that year [245].

Transmission is by droplets, and infection in the natural disease will be via the lungs. The incubation period for obvious disease is around 12 days. In the initial growth phase, the virus seems to move from cell to cell, but around the twelfth day, lysis of many infected cells occurs and the virus will be found in the bloodstream in large numbers. The initial or prodromal symptoms are essentially similar to other viral diseases such as influenza and the common cold—fevers, muscle pain, stomach aches, etc. The digestive tract is commonly involved, leading to vomiting. Most cases will be prostrated.

The smallpox virus preferentially attacks skin cells, and by days 14–15, smallpox infection becomes obvious. The attack on skin cells causes the characteristic pimples associated with the disease. The pimples tend to erupt first in the mouth, then the arms and the hands, and later the rest of the body. At that point, the pimples, called macules, should still be fairly small. At this stage, the victim is most contagious.

By days 15–16 the condition worsens; at this point, the disease can take two vastly different courses. The first form is of classic ordinary smallpox, in which the pimples grow into pauples, and then fill up with pus (turning them into pustules). Ordinary smallpox generally takes one of two basic courses. In *discrete* ordinary smallpox, the pustules stand out on the skin separately—there is a greater chance of surviving this form. In *confluent* ordinary smallpox, the blisters merge together into sheets that begin to detach the outer layers of skin from the underlying flesh—this form is usually fatal. If a victim of ordinary smallpox survives for the course of the disease, the pustules will deflate in time (the duration is variable), and will start to dry up, usually beginning on day 28. Eventually the pustules will completely dry and start to flake off. Once all of the pustules flake off, the patient is considered cured.

In the other form of smallpox, known as hemorrhagic smallpox, an entirely different set of symptoms starts to develop. The skin does not blister, but it remains smooth. Instead, bleeding will occur under the skin, making the skin look charred and black (this is known as black pox). The eyes will also hemorrhage, making the whites of the eyes turn deep red (and, if the victim lives long enough, eventually black). At the same time, bleeding begins in the organs. Death may occur from bleeding (fatal loss of blood or by other causes such as brain hemorrhage), or from loss of fluid. The entry of other infectious organisms, because the skin and the intestine are no longer a barrier, can also lead to multiorgan failure. This form of smallpox occurs in anywhere from 3% to 25% of fatal cases (depending on the virulence of the smallpox strain).

Edward Jenner developed a smallpox vaccine by using cowpox fluid (hence the name vaccination, from the Latin *vacca*, cow); his first inoculation occurred on May 14, 1796. After independent confirmation, this practice of vaccination against smallpox spread quickly in Europe. The first smallpox vaccination in North America occurred on June 2, 1800. National laws requiring vaccination began appearing as early as 1805 [245]. The cowpox virus is still the immunogenic tool, dismissing, fortunately, the need to use the smallpox virus itself for vaccine preparation, as is the case with many pathogens. Compulsory vaccinations were increased, and the virus has been wiped out globally.

The last case of wild smallpox occurred on September 11, 1977. One last victim was claimed by the disease in the United Kingdom in September 1978, when Janet Parker, a medical photographer in the University of Birmingham Medical School, contracted the disease and died, and the Professor responsible for the unit killed himself. A research project on smallpox was being conducted in the building at the time, although the exact route by which Ms. Parker became infected has never been fully elucidated.

After successful vaccination campaigns, the WHO in 1979 declared the eradication of smallpox, although cultures of the virus were kept by the Centers for Disease Control and Prevention (CDC) in the United States and at the Vector Institute in Koltsovo, Novosibirsk in Siberia, Russia, where a regiment of troops guard it. Under such tight control, smallpox would, it was thought, never be let out again. All other known stocks of smallpox were ostensibly destroyed. Smallpox vaccination was discontinued in most countries in the late 1970s; the risks of vaccination include death (~1 per million), among other serious side effects.

Nonetheless, after the 2001 anthrax attacks took place in the United States, concerns about smallpox have resurfaced as a possible agent for bioterrorism. As a result, there has been increased concern about the availability of vaccine stocks. Moreover, President George W. Bush has ordered all American military personnel to be vaccinated against smallpox and has implemented a voluntary program for vaccinating emergency medical personnel. Vaccination is seriously being reconsidered over many countries. It is also feared that additional stocks of the virus may exist in research collections, the product of the accumulatory nature of microbiologists. Additional collections of the virus almost certainly exist as the result of certain military and biologicol warfare programs, particularly in Russia. Iran, Syria, Israel, China, North Korea, and Cuba have been mentioned as current possessors, as well. The possibility that the virus is currently being held by a terrorist organization has not been excluded.

In light of those circumstances, and considering the extremely undesirable possession of this virus by various countries—hence its expected migration to terror elements, as well—the final and complete destruction of virus stocks held anywhere, and particularly in the CDC and Vector, was ordered in 1993, 1994, 1995, and 1996, but they have not yet been destroyed in those two facilities. Yet the continuation of this actuality has been approved. The World Health Assembly has passed resolutions (WHA52.10) (WHA 55.15) authorizing temporary retention of the existing stocks of variola virus for the purpose of further essential research [246].

The program of research is overseen by the WHO Advisory Committee on Variola Virus Research, composed of members from all WHO regions and advised by some 10 scientific academic experts from such areas as public health, fundamental applied research, and regulatory agencies. Current reports are submitted to the World Health Assembly and are worth detailing due to the uniqueness of the topic at large, as follows.

Articles summarizing recent research overseen by the WHO Advisory Committee on Variola Virus Research

- Exploring the potential of variola virus infection of cynomolgus macaques as a model for human smallpox [pdf 564 kb] [247]
- The host response to smallpox: Analysis of the gene expression program in peripheral blood cells in a nonhuman primate model [pdf 897 kb] [248]

Abstracts summarizing recent research overseen by the WHO Advisory Committee on Variola Virus Research—2004

- Diagnostic Development at the CDC: Update 2004
- Inferring the Phylogeny of Variola from Genomic SNPs Analysis
- Variola Morphogenesis effected by an erb-B Tyrosine Kinase Inhibitor: Possible therapeutic functions
- The WHO Collaborating Center for Smallpox and other Poxviruses at the Centers for Disease Control and Prevention Atlanta: 2004 report on the variola repository [249]

Abstracts summarizing recent research overseen by the WHO Advisory Committee on Variola Virus Research—2003

- Analysis of Nucleotide Sequences of Individual Orthopoxvirus Genes
- Comparative restriction analysis of genomic DNAs of the variola virus strains from the Russian collection
- Human Combinatorial Antibodies against Orthopoxviruses
- Update on variola and orthopoxvirus diagnostic development and use
- Update on search for antivirals against human-pathogenic orthopoxviruses
- Viability estimation of variola virus isolates from the Russian collection [250]

Abstracts summarizing recent research overseen by the WHO Advisory Committee on Variola Virus Research—2002

- Capillary-electrophoresis restriction fragment length polymorphism. A new method for poxvirus fingerprinting [251]
- Cidofovir Treatment of Variola (Smallpox) in the Hemorrhagic Smallpox Primate Model and the IV Monkeypox Primate Model [252]
- Genome-wide analysis of the host response to variola infection [253, 254]
- Lethal infection of primates with variola virus as a model for human smallpox [255]
- New PCR assays for identification of smallpox virus [232]
- Variola virus genomics [256]

So far, so reasonable, perhaps. But a dangerous shift might have taken place. American scientists are now awaiting World Health Assembly approval to begin experiments to genetically modify the smallpox virus. Researchers have already been given the go-ahead by a technical committee (WHO Advisory Committee on Variola Virus Research), which accepts the argument that the research could bring new vaccines and treatments for smallpox closer. Still, in the debate during the full assembly of the WHO, whose representatives from 192 member states met from 16 to 25 May 2005, WHO said it will ensure that any research will only be conducted after detailed proposals have been thoroughly examined on a case-by-case basis by the Advisory Committee, paying particular attention to biosafety and bio-security issues [257].

Some 200 tons of smallpox virus have been produced by the USSR as a weapon and inherited by Russia. Their fate is unclear. However, details of the development of smallpox as a weapon by the Soviets became available. A report was elicited from General Prof. Peter Burgasov, former Chief Sanitary Physician of the Soviet Army and a senior researcher within the BWP. Admitting that development of BW by the Soviets did take place, in the form of live field tests, he described a "smallpox incident" that happened in the 1970s, and was then hashed up: "On Vozrazhdenie Island in the Aral Sea, the strongest recipes of smallpox were tested. Suddenly I was informed that there were mysterious cases of mortalities in Aralsk. A research ship of the Aral fleet came 15 km away from the island (it was forbidden to come any closer than 40 km). The lab technician of this ship took samples of plankton twice a day from the top deck. The smallpox formulation—400 gr. of which was exploded on the island—'got her' and she became infected. After returning home to Aralsk, she infected several people including children. All of them died. I suspected the reason for this and called the Chief of General Staff of Ministry of Defense and requested to forbid the stop of the Alma-Ata—Moscow train in Aralsk. As a result, the epidemic around the country was prevented. I called Andropov, who at that time was Chief of KGB, and informed him of the exclusive recipe of smallpox obtained on Vozrazhdenie Island" [258].

Actually, it was a new lethal strain of smallpox that traveled far and away from the island-based BW testing facility in the Aral Sea, to infect people downwind on a ship. Most of the adults exposed to the strain contracted smallpox despite being immunized. Promptly, remarkably extensive countermeasures, including vaccination, disinfection, and quarantining were conducted [259].

In March 2003, smallpox scabs were found tucked inside an envelope in a book on Civil War medicine in Santa Fe, New Mexico [260]. The envelope was labeled as containing the scabs and listed the names of the patients that were vaccinated with them. Assuming the contents could be dangerous, the librarian who found them did not open the envelope. This was fortunate, as unlike bacteria (with the exception of those that produce spores), viruses can theoretically survive for many years. The scabs ended up with employees from the National Center for Disease Control, who responded quickly once in-formed of the discovery. The discovery raised concerns that smallpox DNA could be extracted from these and other scabs and used for a bioterrorism attack.

Nevertheless, the chances of successfully doing that seem to be slim, for now. Notably, Ramses V, who lived some 3000 years ago, is the earliest known victim of smallpox, based on an analysis of his mummy [261]. Many virus-like particles were revealed by electron microscopy and identified serologically as smallpox virus in a 400-year-old mummy from Italy. The antigenic structure of the particles was well preserved [262]. The virions in the mummy's skin had lost their viability. Viral antigen was not be detected by EIA or RPHA, and its DNA was not detected by molecular hybridization [263]. Attempts to recover the virus from the frozen bodies of persons who died of smallpox continue in Arctic sites in Siberia [264], and in Canada [265].

The *Variola major* virus is very stable and survives in exudates from patients for many months. The virus is unlikely to survive in dried crusts for more than a year [266]. Yet, it can be preserved in sealed ampules at 4°C for many years, and indefinitely by freeze-drying. This leads to the presumption that the virus still may be preserved alive within bodies of victims buried in permafrost. Much more feasible, however, the practical employment of smallpox virus for bioterrorism purposes is remarkably worrisome and reckoned to be a prime threat.

The current, perpetual phase along the smallpox virus course is facilitated, in effect, by an informal Russian–American status quo, one that has virtually been established lately, acknowledging the nontermination of BWs, for the time being. In a sense, the cardinal, unsolved issue of the smallpox virus constitutes an illustrative reflection of that much broader, not as yet untied, tangle. Sheltered by seemingly acceptable justifications of essential ongoing research into the most dreadful pathogen of mankind is thus still being kept in Russia, the United States, and most likely, some other places. It perfectly symbolizes the bivalent potential of further, extremely ominous pathogens held, militarily, by Russia and additional states. In Russia, it is retained, probably, as a stockpiled weapon, alongside with an arsenal of various pathogens and toxins. The argument that full acquaintances with pathogens reckoned as BWA—and, likewise, with their aggressive attributes, in terms of bioterrorism threats—are necessary for achieving effective protection, is a very reasonable one, and may hardly be debated. It is therefore, objectively, a salient dual-purpose factor; particularly as engineered pathogens and bioterrorism menaces are becoming common. Concurrently, and perhaps genuinely, this attitude is applied within the U.S. Army as defensively oriented "bioprofiling" [202]. At any rate, the U.S. Army Medical Research Institute of Infectious Diseases, at Fort Detrick, has accumulated "credible evidence" that several terrorist groups and nations have obtained, or are trying to obtain, clandestine stocks of smallpox and are actively trying to produce weaponized armaments based on the virus [267].

Connectedly, an international exercise in dealing with a series of international terrorist attacks involving smallpox virus has shown that the impact would be catastrophic for almost the entire world, and that the lives of millions of people could be lost. This outcome of the "Atlantic Storm" session seems all the more alarming given all the measures taken to combat bioterrorism since the 2001 anthrax letters attacks [268]. The fictitious scenario of "Atlantic Storm" centered on a wave of attacks with the smallpox virus. The exercise was held at a hotel in Washington, D.C., with some 150 observers present to see how 11 former government officials and leaders of inter-governmental organizations, all with the necessary experience, would deal with such a calamity.

13.4.20   AGROTERRORISM

The essence and characteristics of agricultural terrorism (agroterrorism) are presented. Reference is here made to infective agents only, meaning pathogens of farm animals and cultivars. Mention is not made here of bioterrorism warfare agents common to man and animals, such as anthrax (those are discussed above). Accordingly, various episodes of agroterrorism that took place since WWII are reveiwed.

Agroterrorism is the deliberate introduction of pathogens or chemicals (toxic/radioactive), either against livestock or crops, for the purpose of causing economic losses, undermining stability and/or generating fear. The direct impacts are economical, logistical, and demographical (up to hunger). Agroterrorism may be conducted in the form of state-sponsored or nonstate-sponsored acts. The outcomes show in the form of epizootics (animal epidemics) or epiphytotics (plant epidemics).

The United States, the USSR, Germany, Iraq, and Iran. foremost, are known to have had programs that included potential agroterrorism pathogens (in Russia and Iran, at least, this inheritance presently persists), as follows:

United States: The viruses causing fowl plague (epizootic avian influenza), rinderpest (cattle plague), hog cholera (swine contagion), and Newcastle disease (poultry contagion), plus late blight of potato fungus, were explored by the United States as candidates for agroterrorism. Weaponized have been the fungal causative agents of wheat rust and rice blast [269], making the former, explored ones, of no less importance. A wealth of British scientific experience has thereupon been exploited by the US. One salient finding has been in that 3 g of the rice blast fungi per hectare could infect between 50% and 90% of the crops exposed [270]. Other fungal agents evaluated—in that case against drug crops, mainly poppy and coca fields—were *Fusarium oxysporum* and *Pleospora papaveracea*.

Germany: Explored: foot and mouth disease, potato beetle, potato stalk rot, potato tuber decay, wheat fungus, turnip weevils and antler moths.

USSR: Explored: African swine fever virus, rinderpest virus, and the fungi casing wheat stem rust and rice blast [271]. An uncertain part of those pathogens have been weaponized, plus, apparently, cow pox and sheep pox viruses [272].

Iraq: Explored: sheep pox, goat pox, camel pox, and foot-and-mouth disease viruses; wheat smut and cereal rust fungi. Weaponization of the fungus causing weat smut has been evidenced [30].

Iran: Rinderpest and foot-and-mouth disease viruses, as well as crop fungal pathogens apparently have been explored and possibly weaponized [273].

Those lists exhibit, basically, the severity of agroterrorism, although principally the spectrum of potential agents is much broader. Thus, mention should be made, in addition, of some further examples. For instance, a 2005 GAO report accentuates the special significance of Asian soybean rust [274]. In an experts meeting "Biosystems and Bioterrorism: Can Arthropods Be Used as Agents of Destruction," the immense importance of arthropods as disease disseminators has been pointed at, with reference being made, among other, to the glassy-winged sharpshooter scare, the Mediterranean fruit fly and the potential for Tsetse and Trypanosomosis reinfestation or establishment in free areas [275].

Illustratively, the devastating New World screwworm fly, regarded to be the most serious insect pest of cattle in the New World, appeared for the first time in the Old World in Libya, during 1988. It aroused, justifiably, great fear, and the countermeasure conducted was an act of biological warfare in itself, namely the sterile male technique: from December 1990 and until October 1991, sterile male flies at the rates of 3.5 building up to 40 million per week were released so as to suppress the fertile male invaders [276]. The release of sterile flies was terminated 6 months after the last detected case of screwworm myiasis in Libya. Notably, Libya accused the United States of intentionally delivering the initial, reproductive pest, although it was the Mexican-American Commission for Eradication of Screwworms that thereafter carried out the eradication campaign.

Peculiarly, an outstanding agroterrorism-related interface evolved between Cuba and the United States. Cuba has blamed the United States for attacking Cuban crops and livestock on as many as 21 different occasions. According to Raymond Zilinskas, of the few incidents for which information is available, agents include Newcastle Disease (1962), African wine sever (1971, 1979–1980), Tobacco Blue Mold Disease (1979–1980), Sugarcane Rust Disease (1978), and Thrips insect infestation (1997). Only in the case of the Thrips did Cuba make a formal complaint. According to the Cuban account, the United States flew a crop-duster operated by the State Department over Cuba and released the insects. The United States has denied the allegation, and while U.S. agriculture experts discount Cuban claims, the United Nations has undertaken an investigation. Considering the above-mentioned information on the U.S. programs as presented in this section and the Section 13.4.15.3, there is support for some Cuban allegations. Yet, according to Zilinskas, the most likely explanation for all these incidents was nature or accidental human transmittal through commerce [5]. Connectedly, a Florida university professor informed the CIA that a Florida citrus canker outbreak was the result of a Cuban bioterrorism weapons program. Although the CIA could not substantiate the claim, it did investigate the case. During that same time period, Cuba claimed that an outbreak of Thrips Palmi disease on the island was bioterrorism warfare introduced by the United States [277]. Apparently, no final conclusion can be reached concerning those issues.

At any rate, the Cuban file was not the only one within the context of agroterrorism affairs. Since 1915, a variety of agroterrorism-related incidents have taken place, worldwide [278], and included, mostly, the following events involving animal and plant pathogens/toxins:

Richard Ford, a prominent British naturalist, has made the accusation that Germany dropped Colorado Potato Beetles on the United Kingdom during WWII, accounting for their unusual appearance in parts of the United Kingdom. According to Ford the bombs were made of cardboard and contained 50 to 100 beetles. The allegation has not been verified or refuted.

In a Ministry of Forestry report dated June 15, 1950, the East German government accused the United States of scattering Colorado Potato beetles over potato crops in May and June of 1950 [279, 280].

The Mau Mau, a nationalist liberation movement, poisoned some 33 steers at a British mission station, using what is believed to be a local toxic plant known as "African milk bush" [281–285].

Ken Alibek, First Deputy Chief of the Russian Biopreparat system, alleged he was informed by a senior Soviet military officer that the Soviet Union attacked the Afghan Mujaheddin with glanders on at least one occasion. According to Alibek, this would have the dual effect of sickening the Mujaheddin and killing their horses, their main mode of transportation [120, 280, 286].

Sometime from 1983 to 1987, a Tamil militant group threatened to use bioterrorism agents against Sinhalese and crops in Sri Lanka. The communiqu' threatened to introduce foreign diseases into the local tea crop and to use Leaf Curl to infect rubber trees [287–290].

Queensland's State Premier received a letter threatening to infect wild pigs with foot-and-mouth disease (FMD), which was feared might spread to cattle and sheep, unless prison reforms were implemented within 12 weeks. Ultimately, this incident proved to be a hoax as the perpetrator turned out to be a 37-year-old murderer serving a life sentence in a local jail. In December 1984, Queensland's premier received a similar letter from an unidentified individual [120, 280, 286].

An outstanding pattern emerged regarding the spread of the Mediterranean fruit fly, a major threat to agriculture in California, in 1989. Despite heroic attempts to eradicate this insect, new infestations repeatedly appeared in odd and unexpected places. Concomitantly, the Mayor of Los Angeles received several letters from a group calling itself The Breeders, which claimed to be spreading the insect to protest California's agricultural practices [287–290].

In 2003, Israel has been blamed for spreading foot-and-mouth disease in the Palestinian Authority territory [291, 292].

It so happened that the majority of antihuman bioterrorism agents are basically and currently pathogens of animals, making them zoonotic. The sense lying in that phenomenon is in that, naturally, those pathogens rarely infect humans; hence human herd immunity against them is marginal, bringing about considerable vulnerability. Some of those zoonotic pathogens are important as both antihuman and anti-animal agents, like Rift Valley fever, epiornithic avian influenza ("fowl-plague"), glanders, anthrax, and brucellosis. They are potentially usable for both agro-and bioterrorism. But another class of pathogens includes specifically livestock attackers. The FMD virus is seemingly their best representative. It is, indeed a formidable agroterrorism agent.

It was again, thus, a viral epidemic—rather an epizootic—this time belonging to the category of livestock attacking agents, which exhibited the aggressiveness of an agroterrorism-patterned course, in that case—the FMD virus. Although very rarely infective toward humans, it is extremely contagious and virulent for mammalian farm animals. It is therefore a prime candidate for agricultural bioterrorism and a very effective one.

Great Britain was the arena, 2001. Despite a single outbreak of FMD in cattle on the Isle of Wight, Hampshire in 1981, there had been no outbreaks of FMD in Great Britain since 1968, until 2001. On 20 February 2001 an outbreak of FMD caused by the O1 Pan Asia strain of virus was confirmed in pigs in an abattoir in Essex. The source of the infection was traced to a pig unit in northeast England where disease was believed to have been introduced at the beginning of February. The provenance of the causative agent could not be identified. Deliberate introduction has not been ruled out. Sheep on a neighboring holding became infected by airborne spread from the pig unit. These sheep were subsequently moved through markets in Northumberland and Cumbria between 13 and 22 February 2001. Disease was then disseminated to many other parts of Great Britain and Northern Ireland as a result of movements through markets and dealers before the existence of disease in the country was recognized. Notably, between 20 February and 30 September 2001, a total of 2025 apparently resultant FMD outbreaks were confirmed in Great Britain; a further 4 were confirmed in Northern Ireland. Over 4 million animals were killed as part of the FMD control program. The epidemic was essentially sheep based, other classes of livestock becoming infected through direct contact with infected sheep or through the movements of people, vehicles, and fomites [293].

The impact of the epidemic was colossal. The overall economic costs in the food and farming sectors of the U.K. economy totaled an estimated £5 billion, roughly $10 billion [9]. This figure compares with an annual gross output of the entire U.K. agricultural sector of £25 billion [10]. The U.K. economy suffered other economic costs associated with the outbreak of FMD. Outside the agricultural sector costs, the United Kingdom endured additional losses to the leisure and tourism sector of the economy. The U.K. Department of Environmental Food and Rural Affairs estimated that the leisure and tourism sector of the economy lost £5–6 billion pounds as a result of the outbreak of FMD in 2001 [11].

Finally, an image of a list of livestock diseases found in a cave in Afghanistan was used as evidence that terrorists are considering attacks on agriculture. It has been accentuated, in connection, that there are many operatives that were planning such acts [180].

13.4.21 NARCOTERRORISM

The nature and specific features of narcoterrorism are discussed. Various connotations of narcoterrorism, which are compatible or concerned with biosabotage-related activities, are presented. The main terrorist organizations involved in narcoterrorism worldwide are outlined.

Primarily, narcoterrorism is a term coined by former President Belaunde Terry of Peru in 1983, when describing terrorist-type attacks against his nation's antinarcotics police. In the original context, narcoterrorism is understood to mean the attempts of narcotics—actually, recreational drugs—traffickers to influence the policies of government, the enforcement of the law, and the administration of justice by the systematic threat or use of violence. Pablo Escobar's ruthless dealings with the Colombian government is probably the best known and best documented example of narcoterrorism [296].

The term has become a subject of controversy, and it is being increasingly used for known terrorist organizations that engage in drug trafficking activity to fund their operations and gain recruits and expertise. Such organizations include, mainly, FARC, ELN, and AUC in Colombia, Hezbollah in Lebanon, and al-Qaeda throughout the Middle East, Europe, and Central Asia. Although al-Qaeda is often said to finance its activities through drug trafficking, the 9/11 Commission Report notes that "While the drug trade was a source of income for the Taliban, it did not serve the same purpose for al Qaeda, and there is no reliable evidence that Bin Laden was involved in or made his money through drug trafficking."

Although financial considerations seem to be the leading ones, narcoterrorism is often marked, substantially, by a constant attempt to sustain and enlarge the proportion of drug consumers in a given target population, hence weakening the latter physically, mentally, and economically. In that sense, it is indeed a sort of terrorism. But beyond the direct effects of narcoterrorism, it is very meaningful, although secondarily, in that it greatly facilitates the ongoing spread of blood-transmitted pathogens, such as AIDS and hepatitis, having their own significant and demographic attritional impact.

Moreover, usually the drugs consumed by narcomans are in fact psychotropic plant toxins: heroin, cocaine, opium, hashish, marijuana, and so on. Narcoterrorism is often based on massive cultivars of the related plants that are currently being overseen by the concerned terror organization, namely self-production of the involved hallucinogenic phytotoxins thereafter consumed by narcomans. In addition, the terrorists engaged in producing those phytotoxins accumulate the know-how and experience needed, basically, for the manufacturing of typical plant toxins, such as ricin, aconitine, and other potent ones. Chemical refinement is often essential. Also, the financial, technical, and logistical infrastructures involved in drug trading may be appreciably supportive of narcoterrorism, bioterrorism, and agroterrorism altogether. Like agroterrorism, narcoterrorism may be regarded, thus, as an additional variant of bioterrorism, although rather of a different sphere.

Most (more than half, certainly) of drug consumption worldwide is fueled by directed narcoterrorism. Some terrorist groups, like Colombia's FARC, collect taxes from people who cultivate or process illicit drugs on lands that it controls; others, including Hezbollah and Colombia's AUC, traffic in drugs themselves.

Moreover, some terrorist groups are supported by states funded by the drug trade; Afghanistan's former Taliban rulers, for instance, earned an estimated $40 million to $50 million per year from taxes related to opium. The drug trade is also a significant part of the economies of Syria—which has funded terrorist organizations such as Hezbollah, the Popular Front for the Liberation of Palestine-General Command, and Palestinian Islamic Jihad—and Lebanon, a haven for numerous terrorist groups including Hezbollah and Hamas [297].

The drug trade is extremely lucrative. Heroin, cocaine, and marijuana are uncomplicated and cheap to produce, but because they are illegal and therefore risky to supply, they can earn more than their weight in gold on the vast international black market. The United Nations estimated that the illicit drug business generates about $400 billion per year. Also, because the drug trade is secretive, terrorists can amass large sums of cash without being detected by authorities.

The terror organizations involved in narcoterrorism at large are mostly the following ones [298]:

- The Revolutionary Armed Forces of Colombia (FARC), a Colombian leftist group, which raises funds by taxing coca farmers in the Switzerland-sized zone of the country it controls. Experts say FARC may force peasant farmers to grow the coca used to make cocaine. It also makes money by protecting cocaine laboratories and clandestine airstrips and by trafficking in drugs locally.
- The National Liberation Army (ELN), another Colombian leftist group, taxes growers of marijuana and opium poppies and protects drug-lab operations. But it generates far less of its funding from drugs than does FARC.
- The United Self-Defense Forces of Colombia (AUC), which includes several right-wing paramilitary groups, says it gets 70% of its income from processing and exporting cocaine. It claims to be leaving the drug business, but experts doubt that all of its members will comply.
- Remnants of Shining Path, a Peruvian leftist group, finance some operations by "protecting" cocaine smugglers in jungle areas under its control and by taxing the coca trade.
- Some members of the Liberation Tigers of Tamil Eelam, a Sri Lankan separatist group, traffic in heroin, and the group reportedly has close ties to drug-trafficking networks in nearby Burma.
- Hezbollah smuggles Latin American cocaine to Europe and the Middle East and has smuggled opiates out of Lebanon's Bekaa Valley, although poppy cultivation there is declining.
- The Kurdistan Workers' Party (PKK), a Marxist separatist group based in Turkey, taxes ethnic Kurdish drug traffickers, and individual PKK cells traffic in heroin.
- The Real IRA, an Irish Republican Army (IRA) splinter group that opposes the peace process in Northern Ireland, is suspected of trafficking drugs, although the extent of its involvement is unclear.
- Basque Fatherland and Liberty (ETA), a separatist group in Spain, is reportedly involved in drug trafficking.

• al-Qaeda does not appear to have direct links to the drug trade. But its former protector in Afghanistan, the Taliban, supported itself in part through opium poppy production and trafficking.

Chaotic countries constitute warm nests for narcoterrorism. Drug traffickers and terrorists tend to thrive in failed states with ineffective governments that have been destabilized by war and internal conflict, experts say. For example, Colombia—a large, fragmented country in the throes of a decades-long conflict over power and resources—produces 80% of the world's cocaine and 70% of the U.S. heroin supply. Lebanon has been plagued by drug traffickers and terrorist groups since its own harrowing 15-year civil war began in 1975. In Afghanistan, the 1990 retreat by occupying Soviet troops left the economically devastated country vulnerable to control by warlords and Islamist extremists. Furthermore, by promoting violence, tax evasion, and lawlessness, terrorists and drug traffickers make it harder for a weakened state to form a stable central government. All in all, the magnitude of narcoterrorism is a remarkable one, and coping with it is difficult and complicated. Narcoterrorism is, in a sense, a long-term attritional mode of terrorism. It is unlikely to diminish, and the main way to effectively struggle against it is expediently to remove the drug producing cultivars themselves. Significant achievements were thus reached by mass employment of fungi that specifically attack those cultivars: cocaine plantations in Columbia and opium plantations in Uzbekistan—a sensible mode of biological warfare by itself.

Paradoxically—yet sensibly, by all means—vast narcoterrorism cultivars were eliminated by agroterrorism-like warfare launched in the last decade, including the effective use of specific fungi capable of destroying poppy fields in Uzbekistan and coca fields in Columbia.

13.4.22   NOVEL AGENTS

Novel biological agents constitute a category of enormous interest, in many senses. It is apt to chiefly be pronounced in terms of paradigmatic combat power multipliers on the one hand, and plain bioterrorism parameters, on the other hand. Reference is made, however, to the inability of terrorist organizations to develop or acquire novel bioterrorism agents, unless extraneously assisted.

A combat power multiplier is defined, primarily, as a warfare element sufficient enough to balance a quantitative inferiority of troops. This concept bears, however, a broader meaning in the context of strategy and military affaires. In these areas, the power multipliers have greater and more complex implications, starting with terror attacks and ending in an overall geopolitical power balance at its highest level, such as international pacts as we saw in the rivalry between NATO and the Warsaw pact.

The future battle field is absorbing, tentatively, a variety of multifunctional technologies. So, the sheer importance of these power multipliers is less in balancing the quantity disadvantage and more as a pragmatic method to decrease the number of the fighting corps (order of battle). Connectedly, a unique range of action is dedicated to biological, chemical, and radiological guerilla warfare technologies, both terrorism-oriented and military-oriented. This section will discuss these technologies and their ramifications at the macrolevel, within the biological sphere, synthetic biology being, thereby, the chief platform.

13.4.22.1 Anti-Organism Power Multipliers

Anti-organism power multipliers are based on technologies designed to harm humans, farm animals, or cultivars, whether during offensive, defensive, or calm situations. Although the new generation of these power multipliers is characterized by nonlethal (or sublethal) weaponry, some of them are distinctively lethal.

This new generation include:

- Novel toxins
- Bacteria and viruses in a supremely engineered condition

Toxins are being upgraded as power multipliers. This being the case, through the following biotechnological featuring:

- Molecular design, avoiding or decreasing the body ability to activate an immunological response
- Molecular design that reinforces their toxicity
- Embedding coded genes (regarding protein toxins) in noncontagious bacteria and turning these into "production lines"
- Embedding coded genes in contagious bacteria and viruses and making these even more harmful

The range of toxins designed to perform as power multipliers is vast. Here are two examples, representing both ends of the spectrum: The Pepper spray and the group of agents called "Prions."

The Oleoresin Capsicum (OC) called "the Pepper spray" can be naturally found in hot chilies. The OC has been recently found as more efficient than the common police incapacitating agents, because it causes a faster and longer lasting reaction. The spray can be used for on-spot or wider purposes. Simultaneously, the "Pepper gel" was also developed. This gel is launched from its container by air-pressure; when contact occurs, the gel clings to any surface (if it touches the face, it may cause temporary blindness). The gel formula contains 10% of OC mixed with the gel. The gel is not flammable [299].

Prions, on the other hand, are extremely destructive protein molecules. They are the cause of "Mad Cow Disease" (Creutzfeldt–Jacob Disease). The prions are infectious molecules, which means their penetration to the body brings about normally functioning proteins that turn into abnormal proteins like themselves. The infection-like condition they cause is developing slowly but deadly. They are host-specific. The research of prions, chemically and biologically, is in progressive process and some see them as one of the future measures designed to target a specific civil population for long-term influence. Some prions harm a variety of farm animals [300].

13.4.22.2   Supremely Engineered Bacteria and Viruses

Genetic engineering is intruding this field, in two main ways:

- The probability of finding gene fragments of highly virulent extinct pathogens is more likely than finding the complete genome and succeeding to revive it. At any rate, an intervention of genetic engineering is necessary to reconstruct and assemble the genome so it can imitate the natural pathogen as a living-reproducing system. In that manner, the polio virus has been recently produced, out of the blue, in laboratories. Similarly, the devastating 1918 "Swine Flu" virus has lately been reconstructed and resurrected.
- The ability of the defender to immunize its population even against the most virulent variants forces the offender to engineer them so immunity will turn out as ineffective.

Genetic engineering is covering far more than those two just mentioned, by creating different and sophisticated power multipliers. The current front in genetic research is promising great achievements for the medical treatments of humans, animals, and plants disease by what is referred to as gene therapy. But, a military aspect of this research also exists. This aspect permits a new form of bioterrorism warfare to take place—genetic warfare. An unavoidable outcome is the discussion on how to use these methods of genetic treatment as a weapon: how to deploy these weapons of genetic warfare and how it is possible to expose the deployment of these weapons [301].

In his futuristic–realistic article, Mark Willis predicts the appearance of the biological genetic-engineered warfare agents according to the following criteria [302]:

- Synthetic viruses
- Cellular pathogens bearing unusual virulence
- Cell-like synthetic entities that cannot reproduce, as vectors of biochemical warfare agents
- Dormant and secret pathogens
- Pathogens specific to farm animals and agricultural vegetation
- Pathogens specific to ethnic groups
- Long-lasting or delayed-effect biological warfare agents
- Ethnic pathogens that can cause autoimmune diseases such as infertility

In practice, at the last decade, we saw the use of specific fungus able to destroy poppy fields in Uzbekistan and coca fields in Columbia. In future perspective, there is no doubt that this is just the tip of the iceberg.

13.4.22.3   Anti-Inanimate Object Power Multipliers

Anti-inanimate object power multipliers are derived from technologies aimed to harm, disrupt, or jam weapon systems or other valuable assets. They can be implemented during times of offense, defense, and calm. In this context, power multipliers designed to contaminate water or food sources will not be regarded as an anti-inanimate object power multiplier because water and food are used, in that case, to attack humans, serving as vehicles. To our discussion, mention is made, then, of bacteria that can be fed from key materials, naturally or genetically engineered. That includes [303]:

- Plastic, like polyurethane (for example, digesting and removing the special cover enables specific aircrafts to reduce their radar signatures, and by that allowing their detection)
- Rubber, steel, and paint used in strategic and logistic facilities
- Asphalt used in runways
- Crude oil and petrol

13.4.22.4 Bioengineering Availability

This ethical–technological entanglement has begat a discussion, both locally and globally, concerning increasing or reducing bioengineering availability of the power multipliers being discussed. That, while the freedom of information principle, universal communication interfaces, and the natural curiosity of different scientific circles are taking their place and diverting the reality toward a constant increasing technology availability at large. The fact that a lot of these novel agents are defined as nonlethal (ethical, so to speak) and antiriot countermeasures is contributing to this trend. This trend is also encouraged by the involvement of a university's laboratories in the military R&D process. Those universities, especially in Western countries, mostly get the publication they want.

The latest developments in the fields of political and strategic thinking regarding the use of existing or developing weapon systems (including nonlethal biological weapons) and the ramifications over the operational doctrine in the context of the contemporary trends in the international relations are very significant [304]. This last publication maintains that the alleged revolution of the military-technology affairs has to be considered in light of discussions on military use of nonlethal weapons in existing and future conflicts. The issue is presented from a political-strategic point of view, criticizing the revolution in military affairs as a starting point in the discussion on the use of nonlethal weapons during conflicts. These issues are presented to politicians and strategic analysts asked to formulate a technology-based policy that may be used in future political/social conflicts. The danger of neglecting other important dimensions in politics and strategy, emphasizing the "force to force" aspect, are considered.

Most of the scientific and technological developments mentioned here are openly reported and documented in the United States, which is probably the world leader on these subjects. Russia is, probably, not far behind, but seems to lessen the publishing of this domain. The Russian BWs are very advanced [305], and Russia also puts a lot of effort in biological espionage, trying to get its hands on relevant knowledge and technological means [306].

At large, the global research for upgrading biological power multipliers is striving to develop:

- New measures: both new categories and new means included in already known categories.
- Measures that although they are known, the enemy will not be able to detect, identify, and handle. That may happen due to objective-inherent technological limitations (in such case, then, measures that have to be activated from a far, with no offender presence) or by using a known technological weakness the enemy suffers from.
- Measures that will complicate the enemy's ability—military demogrophical, or economical—significantly and even decisive, although known by the enemy.
- Specific measures designed against a given rival, affecting his military forces, population, or assets uniquely.

The range covered by these new agents is amazing by both its extent and its intensity, starting with incapacitating agents like Pepper spray and ending, hypothetically for now, with ethnical weapons, targeting specific races only. Not in vain, these two examples represent, even if superficially, derivatives of two powerful, inexhaustible technologies—molecular biology and genetic engineering. So, the Pepper spray is becoming a more favorable substitute to the regular tear-gases, being a molecular copy of the biological extract in the hot Mexican chile. So, the biological-ethnic weapons—whose mechanism is using specific genetic elements in the target race or nationality—are conceivably prone to be the unavoidable outcome of highly complex genetic mechanisms, being deciphered by man.

Toxins can be produced in three alternative methods:

- Extracting it from a natural source (the classic mode)
- Chemically synthesizing the biological molecule, after its full and accurate identification
- Producing it by a gene able to code for it (when it is a protein molecule)

The range of scenarios, where the novel agents discussed can be used is vast, in any parameter:

- Activation by state or terror organization, if only by an individual (that can also be sent by state)
- Activation in a concealed or open manner
- Activation in order to achieve an affect in the short, medium, or long range
- Activation during a military conflict or during an international nonmilitary conflict
- Activation against humans, animals, plants, or still objects
- Activation at times of offense, defense, or at calm

Any combination of the parameters or the subparameters is possible.

Today and in the near future it is not expected that the availability of sophisticated biologically engineered power multipliers will proliferate. It will probably be an asset of few supremely qualified laboratories. However, with time they will become more and more accessible, presumably. With no connection to the BW agents that the Iraqi government has produced in its time (undoubtedly, industrial amounts), it is appropriate to mention two agents reflecting an unusual way of thinking suiting aspects aforementioned, although on an embryonic level [30]:

- Camel poxvirus—a pathogen that can attack humans, but not those who live next to camels and so are naturally immune. The Iraqis have also used this virus as a model to the small pox virus.
- The fungal toxin aflatoxin, which generates terrible diseases, but only on the long term (was apparently destined to be used against the Kurds in Iraq).

In Russia, or elsewhere, chimeras of VEE-smallpox virus, Ebola-smallpox virus, and snake-toxin-gene-bearing influenza virus have possibly been developed successfully, but this has not been verified or refuted.

Furthermore, it was recently argued that advances in nanotechnology could be used in the development of new types of biological (and chemical) weapons, based on the dual-use application of nanotechnology in the fields of biotechnology and medicine, which can be used for offensive purposes. Such a potential military application of nanotechnolgoy could undermine existing international laws that ban biological weapons [307].

To sum, the technological absorbability (of a country) is indeed the cardinal factor, but it is not sufficient for these frontline technologies to be adopted (even more so—for their upgrading) by one. But mention is made of the option that hostile countries or organizations can get their hands on standardized, instantly usable, weapons. The discussed novel agents are very sophisticated, and so transferring, maintaining, and operating them may be very complicated. However, it is possible that hostile elements will get extraneous assistance at these aspects as well. Particularly, a great deal of these weapons includes, so-called, nonlethal and anti-inanimate agents, hence more obtainable.

Referring to the increasing threat of novel bioagents, Steven Block, former president of the Biophysical Society, was quite pessimistic: "The biological weapons threat is multiplying and will do so regardless of the countermeasures we try to take. You can't stop it, any more than you can stop the progress of mankind. You just have to hope that your collective brainpower can muster more resources than your adversaries [308]."

13.4.23   EXPLOITABILITY OF BIOTECHNOLOGY AND BIOMEDICINE

The vast space inevitably formed by biotechnology and biomedicine, in terms of dual usability, is briefly presented. Its affinity to bioterrorism-oriented activities is shown, with regard to both pathogens and toxins. Two prominent toxins—botulinum and ricin—are thereby visited, to accentuate the scarcely observable borderline lying in between legitimacy and illegitimacy.

Apparently, the dual usability marking the fields of biotechnology and biomedicine, namely their capacity to serve for desirable and undesirable purposes at the same time, is the most extreme one, comparing with almost any other scientific domain. A relatively simple instance of this: the undisputable need to thoroughly understand the mechanisms underlying the above-described alternating profile of drug resistance of *Vibrio cholerae* potentially enables, in parallel, the creation of a multistable pathogen. The shift is fairly slight, if any.

Fundamentally, all categories of biomedicine targeted at infectious diseases—antibacterial and antiviral drugs, vaccines, and antisera—as well as gene therapy are apt to be exploited for improper development, parallel to their impressive, invaluable advancements. Moreover, regretfully, a principle of "amplified reversibility" underlies the undesired, yet inevitable, usability of biomedicine and bioengineering resources for the purpose of bioterrorism. That is to say, although most pathogens and toxins have found a way to become harmless and beneficial, thanks to the scientific comprehension gained from their usefulness in various respects, that very paved course may readily be reciprocal; in particular their road back to pathogenicity and virulence—often appreciably amplified, artificially—may take place in the form of accidental leakage or bioterrorism acts.

In pragmatic terms, two dimensions of exploitability can be observed: One is the general modularity of equipment and materials that can equally be employed for innocent or bioterrorism-oriented activities. The other one is the various untainted activities currently engaging specific pathogens and toxins that are reckoned, at the same time, as potential agents for bioterrorism; hence, they are prone to undergo, in principle, a turnover at the level of intentions and practice of their possessors. Thus, the harvest of a seed virus propagated for subsequent production of a vaccine by attenuation may equally be kept as is, before attenuation takes place, serving as a weapon. In actuality, the two courses are not in contrast with each other; hence, they may expediently be applied in parallel, in whatever proportions the possessor chooses, so as to attain effective camouflage.

The principle is fairly plain, even an old one. In 1939, the active ingredient of curare—an ancient plant toxin weapon still in use by Indians—was isolated for the first time. In 1943, it was introduced successfully into anesthesiology. Curare provided adequate muscle relaxation without the depressant effect of deep anesthesia induced by ether or chloroform. Over the last 20 years, physicians have used curare to ease the stiffened muscles caused by polio and to treat such diverse conditions as lockjaw, epilepsy, and cholea (a nervous disorder characterized by uncontrollable muscle movements). Eventually, more effective treatments were found for these illnesses, but the active ingredient of curare, d-tubocurarine, led to the skeletal muscle relaxant Intocostrin, which has been used in surgery ever since. Synthetic analogs of d-tubocurarine are used tens of thousands of times per day in the operating room [309].

Tubocurarines reflect, yet, but one, relatively marginal example of the dual exploitability of biotechnology and biomedicine. Ricin toxin and botulinum toxin are more actual agents, as shown. Global castor seed production for civilian purposes is around 1 million tons per year. The beans are widely used for the production of castor oil, making their control impossible, practically. The castor oil manufacturer may readily use the remains of the beans for obtaining ricin. Moreover, ricin in itself is typically a dual-use substance.

Because of its cytotoxic potency, modified ricin is being used for the selective killing of unwanted cells and for the toxigenic ablation of cell lineages in transgenic organisms. Ricin is the most commonly used toxin in conjugates for selective cell killing, so-called targeted toxins. Beneficial uses of ricin include, potentially the treatment of cancer and AIDS. Recently, there is, indeed, increasing interest in the medicinal applications of ricin as immunotoxin, which have been successfully applied in several human diseases. In addition, the mole-cule of ricin in itself is the precursor for the preparation of toxoid and anti-serum.

Ricin can be targeted to specific cells, such as cancer cells, by conjugating the RTA subunit to antibodies or growth factors that preferentially bind the unwanted cells. These immunotoxins have worked very well for *in vitro* applications, e.g., bone marrow transplants. Although they have not worked very well in many *in vivo* situations, progress in this area of research shows promise for the future. In bone marrow transplant procedures, RTA-immunotoxins have been used success-fully to destroy T lymphocytes in bone marrow taken from histocompatible donors. This reduces rejection of the donor bone marrow, a problem called "graft-vs-host disease" (GVHD). In steroid-resistant, acute GVDH situations, RTA-immunotoxins helped alleviate the condition. Also, in autologous bone marrow transplantation, a sample of the patients own bone marrow is treated with anti-T-cell immunotoxins to destroy malignant T cells in T-cell leukemias and lymphomas [310]:

For the *in vivo* treatment of solid tumors, considerable problems can arise due to poor access of the immunotoxin to the tumor mass, lack of immunotoxin speci-ficity, and further disadvantages. Still, research efforts to expand and develop immunotoxins and therapies for clinical use in cancer and AIDS are continuing with strategies using recombinant DNA technology [311]. On the whole, the fol-lowing examples of recent studies may be noted: Oncologic applications: Immuno-toxin treatment of brain tumors; cytotoxins directed at Interleukin-4 receptors as therapy for human brain tumors; bispecific monoclonal antibodies for the targeting of type I ribosome-inactivating proteins against hematological malignancies; tyro-sine kinase inhibitors against EGF receptor-positive malignancies; targeting tumor vasculature using VEGF-toxin conjugates; and gene therapy with immunotoxins. Alternative applications: effects of selective immunotoxic lesions on learning and memory; targeting toxins to neural antigens and receptors; *in vivo* testing of anti-HIV immunotoxins. All in all promising, apparently, but yet legitimating the access to purified ricin.

Botulinum toxin constitutes another notable case. The most poisonous molecule ever identified, compared with any other natural or synthetic molecule, could not gain, apparently, a title better then "A Bug with Beauty and Weapon [312]."

The germ *Clostridium botulinum*, a gram-positive, anaerobic spore-forming bacterium, is distinguished by its significant clinical applications as well as its potential to be used as a producer of a singular bioterror agent. Growing cells secrete botulinum neurotoxin, a multitype protein bearing exceptional toxicity. Although botulinum toxin is the causative agent of deadly neuroparalytic botulism, it also permits a remarkably effective treatment for involuntary muscle disorders such as torticollus, dystonia blepharospasm, strabismus, hemifacial spasm, certain types of spasticity in children, and other ailments. It is also used for "off label" indications such as migraine headaches. Furthermore, this extraordinarily potent toxin is also used in cosmetology for the treatment of glabellar lines and is well-known as the active component of the anti-aging medications Botox and Dysport. In addition, recent reports show that botulinum neurotoxin can be used as a tool for pharmaceutical drug delivery. However, botulinum toxin remains the deadliest of all toxins and is a potent agent of bioterrorism. Among seven serotypes, *C. botulinum* type A is responsible for the highest mortality rate in botulism, and thus it has the greatest potential to act as a bioterrorism weapon. Genome sequencing of *C. botulinum* type A Hall strain (ATCC 3502) has been completed. It may readily serve to dually engineer the related molecule in terms of toxicity, antigenicity, and stability.

The availability of medicinal Botox is not likely to contribute to a terrorist attack as the vials contain dilute toxin: approximately 0.3% of the lethal inhalational dose estimate and 0.005% of the lethal oral dose estimate. Yet if available right before being diluted, it would certainly be an attractive object. Obviously, the toxin is needed for producing a toxoid and an antiserum.

Finally, the next two examples may tentatively illustrate an entire scope of the exploitability of biomedicine and biotechnology through its two edges, namely the rudimentary one as against the sophisticated one. The former can be demonstrated, then, by crude fermentation of various fungi for the manufacturing of antibiotics or mycotoxins (a dual-use biotechnology largely adopted by the then USSR), whereas the latter one is the usage—for now experimental—recently made of viruses as drug delivery systems, owing to their unique capacity to inwardly parasitize host cells. This biomedical horizon appears to be fascinating, and a far reaching one, in both a desirable and an undesirable fashion.

13.4.24   PREVENTIVE AND COUNTERMEASURE

Having been recognized as a prime threat with global potentiality, different approaches and schools configured to fight bioterrorism are here discussed, rather exhibiting the absence of adequate preparedness. Moreover, various factors, some inherent, may hinder the effectual establishment of future preparedness and are considered. Much progress has though been achieved and is reviewed, particularly in the United States. An endeavor is taking place in Europe and Russia. All over, the efforts are doctrinal, scientific, and logistical, both preventive and reactive.

Aiming to defy any sort of assistance lent by (or from) states to terrorists within the entire context of biological, chemical, or nuclear weapons, the UN Security Council formulated in 2004 a resolution saying that "... All States shall refrain from providing any form of support to non-state actors that attempt to develop, acquire, manufacture, possess, transport, transfer or use nuclear, chemical or biological weapons and their means of delivery. All States, in accordance with their national procedures, shall adopt and enforce appropriate effective laws which prohibit any non-state actor to manufacture, acquire, possess, develop, transport, transfer or use nuclear, chemical or biological weapons and their means of delivery, in particular for terrorist purposes, as well as attempts to engage in any of the foregoing activities, participate in them as an accomplice, assist or finance them ..." [313].

Globally, apart from many national frameworks, various international initiatives have been founded, aimed at the prevention of, and coping with, possible bioterrorism scenarios. In 2006, the U.N. General Assembly released its new counterterrorism strategy. It recommends development of a "biological incidents" database by the United Nations and its member countries, in order to fight the threat of bioterrorism [314].

In Europe, much progress was achieved during 2004–2006, owing to the New Dedence Agenda (NDA) (later on renamed as Security and Defence Angenda—SDA)—a regular professional discussion forum involving NATO, the EU, and the World Health Organization plus national ministries, industry figures, and journalists, to debate safety and defense issues. It aims to raise awareness in Europe of the bioterrorism threat and to define a set of recommendations for EU policy makers to prevent and protect against attacks.

The 2004 NDA conference outlined that security awareness at epidemiology research laboratories across the industrialized world is lagging behind the growing threat of bioterrorism. Tighter cooperation is needed among laboratories, international customs, and transport authorities, and NATO could provide important logistical and communications support in the event of a bioterror attack in Europe. Notably, Roger Roffey, research director of the Swedish Defense Ministry's Department of International and Security Affairs, accentuated that "The risk of bioterrorism is on the rise, and one of the deficiencies we all face concerns the control of bioterrorism agents within laboratories and their transfer beyond, especially at labs across the former Soviet Union. We should expect bioterrorists to go beyond an attack by aerosol delivery into product-tampering of crops and animals, into our very food and water supplies. We need new multilateral initiatives to improve biosecurity facilities [315]."

The 2005 NDA conference was an example of excellent EU–U.S. collaboration in the context of bioterrorism. If there was one conclusion that could be drawn from the conference, it was that such teamwork had to be duplicated in the actual fight against bioterrorism. Not that the event lacked ideas, as these were ever present. But there were few signs of real cooperation, and the specter of different "threat perceptions" was forever hovering in the background [316]. This line strengthened in the 2006 SDA conference.

The Interpol and the European Homeland Security Association pay much attention to bioterrorism threats. Interpol President Jackie Selebi recently observed that "Major panic, temporary paralysis of government functions and private businesses and even civil disorder are all likely outcomes of a bioterrorism attack. In fact, bioterrorism appears particularly well suited to the small, well-informed groups. A bioterrorist's lab could well be the size of a household kitchen and the weapon built there could be smaller than a toaster, and the range of options available to terrorists will continue to grow." Furthermore, he warned of bioterror attacks on livestock or the food chain [317].

Interpol Secretary General Ronald Noble added that "There is no criminal threat with greater potential danger to all countries, regions and people in the world than the threat of bioterrorism. There is no crime area where police have as little training than in preventing—or responding to—bioterrorist attacks. Terrorists do want to use biological weapons. The threat is worthy of immediate preparation" [318].

French Interior Minister Dominique de Villepin called for creation of a UN-affiliated organization to track potential biowarfare agents and keep them away from terrorists. He did not propose giving inspection powers to such an agency, but said that biotechnology companies, laboratories, hospitals, and universities need to better monitor themselves on issues of hiring, pathogen work, and access to sensitive areas [317]. Connectedly, in Britain, MI5 officials recently warned British laboratories that Islamist terrorists—chiefly al-Qaeda, through recruited university students—may try to steal deadly pathogens. Scientists and lab staff handing biological agents such as samples of avian flu, tuberculosis, rabies and polio, have been told their security measures will be vetted by police [319]. In consequence, police is supposed to conduct background checks on scientists and others who work with restricted materials at universities, hospitals and pharmaceutical firms [320]. Government officials are intended to inspect such laboratories, and regular audits are planned of agent inventories. The idea is that it will be up to scientists to prove they have good reason for using the materials in their work, according to a senior of the British National Institute for Biological Standards and Controls.

During 2006, cooperation between prominent European states and Russia considerably strengthened. Thus, in Kyiv, Russia, the highly qualified Interpol Workshop on Preventing Bioterrorism took place. Also, a fairly thorough "Strategic Study on Bioterrorism" produced by 20 high level bio-experts from the Russian Federation and other European countries, was released by the Center for Strategic and International Studies.

In Israel, much concern was raised that regular microbiological laboratories—particularly those located in the Palestinian-occupied territories—may be used as minifactories for producing bioterrorism agents. It has therefore been decided that the National Security Council—in coordination with the General Security Service, Ministry of Health, academic institutions, and industrial bodies—will oversee such laboratories [84]. The Israeli Parliament building and facilities are expected to be reviewed for potential vulnerabilities to biological (and chemical) terrorism. Air conditioning and other ducts are expected to be examined to determine any entry point for airborne agents [321].

In some Muslim states, professional forums intended to fight bioterrorism have been formed. In Iran, parallel to its ongoing, progressing BW program—which includes, among other things, guerilla means for applying bioterrorism warfare—The First Conference on [the] Campaign Against Bioterrorism was held in Tehran. The conference, which was a joint project of the Iranian Red Crescent Society (IRCS) and the infectious diseases department of the Tehran Medical Science University examined the health risks posed by bioterrorism and attempts to develop an action plan for use by relief workers in the event of a bioterror event. Over 300 IRCS relief instructors attended the event along with IRCS managers and students from the University [322].

In Saudi Arabia, where there are still elements in support of al-Qaeda, the Center of Studies and Research at Naif Arab University for Security Sciences in Saudi Arabia held a seminar on bioterrorism. Experts participating in the seminar were from Saudi Arabia, Jordan, Bahrain, Comoros, Sudan, Syria, Palestine, Qatar, Iraq, Kuwait, Lebanon, and Egypt [323].

Beyond comparison, however, is the effort against bioterrorism made by the United States. The preventive and countermeasures taken by the United States against bioterrorism are indeed the most impressive ones, in terms of concept, thoroughness, and scope. They are nourished by fundamentals that preceded the 2001 anthrax attack, which means, dually, that on the one hand the need for preemptive preparedness was soberly recognized in advance, and on the other hand, that the saboteurs succeeded to carry out their attack plan despite the defensive preparedness already established, then, by the United States.

In principle, a variety of ways in which the U.S. agencies were implementing President George W. Bush's April 2004 Biodefense for the 21st Century initiative, organized around the four "pillars" of awareness, prevention, detection, and response [324]. Project Bioshield Presidential law reflects the genuine national American incentive to get prepared toward the bioterrorism threat. In this fashion, lately (2004), the President of the United States signed a new law to implement a $7.8-billion (estimated) project aimed to develop advanced vaccines. This law permits the government to use, in case of national emergency, medications and treatment not yet approved by the FDA. President Bush said this step will help the country to be better prepared in case of a terror attack. Also, said Bush: "It sends a message about our direction in the war on terror. We refuse to remain idle while modern technology might be turned against us." The President also added: "We will rally the great promise of American science and innovation to confront the greatest danger of our time [325]."

The U.S. Homeland Security Subcommittee on Prevention of Nuclear and Bioterrorism Attack handles the bioterrorism threat. It pointed out that the United States must seek the proper balance between agility of response and countermeasure stockpiling in defending against bioterrorism. Moreover, the United States is pursuing multiple avenues of defense against a possible terrorist attack using bioengineered pathogens [326]. Top leaders such as the Homeland Security Secretary have stepped up calls for greater use of terrorist threat information in setting a hierarchy of planning and spending priorities. Witnesses warned the Government Reform National Security, Emerging Threats and International Relations Subcommittee, though, about limits to what can be known about potential engineered threats. They stressed the need for maintaining a broad, flexible array of countermeasures [327]. Homeland Security and the Health and Human Services Department have developed a strategy to address the potential for a bioengineered attack, a document that highlights monitoring of scientific research around the world, as well as broad countermeasure development to give the United States flexibility when confronting an unknown pathogen. Complementarily, two annexes of the new U.S. National Response Plan are particularly relevant to bioterrorism: Emergency Support Function 8 (Public Health and Medical Services) and the Biological Incident Annex. These annexes describe specialized application of the NRP to the delivery of public health and medical services and biological incidents [328].

Within that colossal program, the Bush administration's fiscal 2006 budget plans for civilian bioterrorism defense measures total at least $5.1 billion, according to a new nongovernmental analysis. The amount brings the total requested since 2001 to at least $27.7 billion. The budget request includes increases for protecting national food and water supplies and a significant decrease for funding state and local public health departments, compared with requested spending last year [329].

Further budgetary and organizational details are worth mentioning, then, as presented in that publication [330]. The fiscal 2006 budgeting, an aggregate of spending across numerous agencies, reflects as much as a $2.5 billion drop from what was sought for fiscal 2005, but the decrease results mostly from the absence of a one-time appropriation last year for drug and vaccine purchases through 2008 as part of the Project Bioshield law. "Civilian biodefense spending, not including the Bioshield bill, has reached a consistent level of about $5 billion from fiscal 2003 to fiscal 2006," the article says. The fiscal 2006 total in the study does not include Defense Department budgeting for civilian bioterrorism defenses, as do previous years' figures, because the Pentagon "was unable to furnish numbers for the requisite programs," the study says.

Military bioterrorism defense spending for civilians in the previous two fiscal years averaged about $200 million. The analysis says, though, that those figures do not truly account for all Pentagon funding for civilian bioterrorism defense measures. "Some DOD research has direct civilian benefit, but because the majority of these funds are primarily military in application, these lines were excluded from calculation of total DOD expenditures," it says.

As in previous years, the largest amount of money—$4.1 billion according to the article—was budgeted for the Health and Human Services Department, which funds the National Institutes of Health, the Food and Drug Administration, and the Centers for Disease Control and Prevention. The article describes a planned $130 million cut to CDC funding for state and local public health departments, bringing the total CDC budget request down to $797 million. Another substantial cut is a $119 million reduction from the National Institute of Allergy and Infectious Diseases' budget for research facility construction, down to $30 million, and intended to "offset the increase in research funding," the analysis says. The Homeland Security Department's $362 million budget is roughly the same as for fiscal 2005—except for the $2.5 billion drop reflecting the advanced purchase for fiscal 2005 under the Bioshield law.

The Agriculture Department was budgeted $354 million, a 26% increase, for bioterrorism defense activities, the State Department $71.8 million, and the National Science Foundation $31.3 million, particularly, in accordance with Homeland Security Presidential Directive 9 (HSPD-9), Defense of United States agriculture and food, January 30, 2004. Finally, the Environmental Protection Agency received an 87% increase to $185 million, primarily for decontamination capabilities, to protect water and food supplies, and for training, the study says.

Apart from doctrinal and laboratory studies conducted by those agencies, field studies are being carried out as well. Scientists from several agencies released two tracer gases in a 2-km^2 area of midtown Manhattan. In a series of experiments, they are tracking how harmful particles might disperse through street canyons, subway tunnels, and buildings. Dubbed the "Urban Dispersion Project," this Department of Homeland Security-sponsored effort aims to produce a computer model of airflow patterns that could help state and local officials better respond to an emergency or to a bioterrorism act [331]. The studies are intended to conclude in 2007.

Overall, a possible counterproductive outcome has been assessed, naturally, as some scientists are concerned that the increase in the number of researchers working in the United States to counter bioterrorism elevates, potentially, the risk of an attack and the accidental release of biological agents [332]. More than 300 institutions and 12,000 individuals have access to weaponizable biological agents, said Richard Ebright, a Rutgers University molecular biologist and critic of the expansion in biodefense since the anthrax attacks of 2001. Biodefense watchdog, the Sunshine Project, claims that 97% of principal investigators who received grants from the U.S. National Institute for Allergy and Infectious Diseases from

2001 to 2005 to study six biological agents had not previously conducted similar work. The explosion of "NIAID newbies" increases the likelihood of accidents, said Edward Hammond, U.S. director for the Sunshine Project. Jeanne Guillemin, a senior fellow at the Security Studies Program at the Massachusetts Institute of Technology, has the view that increasing access to pathogens heightens the chances that rogue scientists could use them in an attack. "What [NIAID Director Anthony Fauci] and others haven't thought through is the particular kind of expertise, from basic bench work to aerosolization, that comes with defensive biological weapons programs" [332].

Furthermore, the extreme, objective complexity of implementing bioterrorism-defense measures aroused, unavoidably, critics saying that despite substantial funding on bioterrorism defenses since the Sept. 11 terror attacks, the country remains substantially unprepared for a mass-casualty bioterrorism attack [330]. A variety of concrete steps and moves has been conducted, however, as follows.

Nonetheless, a large bioterrorism research network has been established in the United States. The U.S. National Institute of Allergy and Infectious Diseases supports 11 academic institutions throughout the United States, which operate as Regional Centers of Excellence for Biodefense and Emerging Infectious Diseases Research. The new network is working diligently to uncover new knowledge and create preventive, therapeutic, and diagnostic tools that will leave us far less vulnerable to bioterrorism, and is regarded as a key element of a U.S. strategic plan to counter bioterrorism and emerging infectious diseases. Each Center for Excellence leads a group of local universities to conduct research on next-generation treatments for anthrax, smallpox, plague, and other diseases. The consortiums encourage bioterrorism research, train personnel, maintain support resources, push for research on the development of new countermeasures, open facilities to researchers from academia and the business world, and provide support for first responders [333]. Within that framework, an 83,154-square foot Biosafety Level 3 laboratory is to be constructed as part of George Mason University National Center for Biodefense and Infectious Diseases. Personnel at the facility will conduct research on development of techniques and products to combat bioterrorism and treat natural infectious disease outbreaks. Researchers will focus on bioterror threats identified by the U.S. government, such as anthrax, tularemia, and plague, along with diseases such as SARS, West Nile virus, and influenza [334]. In addition, Army Medical Research and Materiel Commander Eric Schoomaker focused on the coming benefits of interagency cooperation at the planned National Interagency Biodefense Campus at Fort Detrick, MD. The Defense, Health and Human Services and Homeland Security departments are participating in the campus project.

At the same time, the multiplicity of pathogen holding research facilities is disadvantageous in that they are vulnerable to technical intelligence assaults, thefts and sabotage acts which may serve bioterrorism. This applies, in principle, to a variety of biotechnological resources at large, and thereby propelled, for example, the Congressional Seminar on Preventing Terrorist Exploitation of the Biotechnology Revolution, (June 5, 2006). Connectedly, perhaps, an FBI secrecy initiative formed with key bio-facilities such as the Public Health Research Institute (PHRI) in Newark, New Jersey, so as to refrain from "publicly disclose which specific select agent pathogens and/or strains are stored at PHRI [335]."

In practical terms, early detection systems of various pathogens gained high priority. BioWatch—a Department of Homeland Security's effort to collect air samples form dozens of cities around the country to detect biological agents—bio-sensors have been deployed in more than 30 urban areas (at a cost of $79 million for 2006). Air is thus monitored air 24 hours a day and samples are collected daily and taken to labs that are part of the Centers for Disease Control and Prevention Laboratory Response Network. The results are provided within 12 to 36 hours [336].

Connectedly, the U.S. Postal Service has installed more than 1000 biological agents—anthrax plus two undisclosed agents—detectors at 271 mail processing facilities since the 2001 anthrax mailings. Moreover, President Bush has staked out the right of the U.S. government to open citizens' mail without first obtaining a warrant [337].

A further methodology, mass spectrometry based "finger print" data base, has been developed by the Food and Drug Administration (FDA) National Center of Toxicological Research (NCTR), for immediate differentiation between biological contaminants, toxins and other, nontoxic substances. Although the concerned device cannot distinguish between living and dead cells, it is highly advantageous in terms of instant findings [338].

Also, nanotechnology could significantly increase the speed of detection of an act of bioterrorism or a natural outbreak. Using a diagnostic test, researchers can measure the frequency change of a nearinfrared laser while it scatters a virus' DNA or RNA. The change in frequency is distinct [339].

Beyond, the most advanced biodetectors are the Triangulation Identification for Genetic Evaluation of Risks—the first multi-purpose diagnostic system that can work with samples of various materials. It can simultaneously identify all pathogens in a soil, water, air or blood samples within four hours, and roughly five such devices are now being used in the US by the US Army and Agriculture Department [340]. In connection, a comprehensive national surveillance network—Project Tripwire—aimed at detecting diseases in wildlife that may be linked to bioterrorism, is being developed by the Wildlife Center of Virginia [341].

Coping with specific pathogens is complicated, naturally, though the very fact that a certain pathogen is being referred to removes a lot of vagueness. Illustratively, the optional adoption of a pre-attack line of defense, as against a post-attack one, is indeed an intriguing one. Researchers from United States and Canada said post-attack anthrax immunization and antibiotic therapy is more effective than pre-attack vaccination [342]. Still, this is doubtfully the correct strategy concerning smallpox, for instance, particularly considering the irrelevance of antibiotic drugs.

Currently, cities and counties throughout the United States are creating stockpiles of anthrax countermeasures that mirror the drugs available in the Strategic National Stockpile [343]. The distributed stockpiles are necessary because it would take too long to distribute countermeasures from the National Stockpile. The drugs, given to cities to prepare for a terrorist attack, are meant to immediately treat first responders in the event of an anthrax attack without having to wait for drugs from the National Stockpile.

The 2001 anthrax letters sabotage was at any rate a cardinal turning point. Using those attacks as a case study, a group of U.S. researchers have developed a risk-based decision-making system that they believe will allow governments to make better judgments following a bioterrorism attack [344]. In "Bayes, Bugs and Bioterrorists: Lesson Learned from the Anthrax Attacks," Kimberly Thompson, Robert Armstrong, and Donald Thompson argue that governments must develop "decision trees," or methods for evaluating several courses of action, following a bioterrorism attack. They claim this would improve the evaluation of costs, risks, and benefits and would create more effective policy development [344].

"Using this type of approach, the government can better characterize the costs, risks and benefits of different policy options and ensure the integration of policy development," the report states. "Additionally, confirmed use and refinement of decision trees during exercises will provide analysis of the long-term consequences of decisions made during an event and give policymakers insights to improve initial decisions." The study, dated April 2005 and published by the National Defense University's Center for Technology and National Security Policy, says poor planning prior to the fall 2001 bioterrorism attacks in which anthrax was sent to U.S. Senate office buildings through the mail led to improper allocation of resources after the attack [345]. The report estimates the direct costs to the Postal Service could exceed $3 billion, with additional expenditures of over $1 billion for unnecessary countermeasures.

Despite the attack, coordination between U.S. agencies has made development of a comprehensive response plan difficult. To remedy the problem, the report urges establishing a decision tree so that multiple paths of action can be evaluated at once, making individual agency's responsibility more clear. "With this approach, analysts can quickly communicate with decision makers about the implications of combinations of options," the report says. "We emphasize that this approach of focusing on decisions provides a means to cross interdisciplinary and other boundaries . . . and consequently it provides a useful organization and communication tool to promote effective management." The report adds that decisions should be separated into different categories, including who should be immunized, what response should be, how to allocate Strategic National Stockpile resources, how to contain biological agents, and how much information should be made public. The Health and Human Services Department should head the effort to form the decision tree, drawing on the expertise of other agencies when necessary.

The 2001 anthrax attacks clearly demonstrated the need for a better decision-making process, the report says. Lack of investment in the public health infrastructure, a poor understanding of the disease caused by anthrax, inadequate training of first responders, and poor communication are just a few of the problems that plagued the response to the attack. The report argues that a decision tree would have vastly improved response, allowing for a better understanding of who should be vaccinated, how relevant agencies should have responded, a better plan for containment, and improved information management. The report is careful to say that it is not critical of the 2001 response. However, researchers believe a decision tree would have prevented panic and yielded a more appropriate response. Researchers hope decision trees can guide policy discussions as a comprehensive response plan is formulated. "Given our recent experience with anthrax, the specific decision trees for anthrax are offered, as an analytical tool to aid future policy decisions," the report says. "Indeed, the lessons learned from the 2001 attack should facilitate the use of these trees."

Plague has been visited as well, when a wide-scale drill took place in the United States in November 2003. The drill was an effort to follow up on weaknesses in federal emergency response plans identified in a simulated pneumonic plague bioterrorism attack. That exercise, called Top Off 2, was organized by the Department of Homeland Security and involved 8000 local, state, and federal officials. It simulated a pneumonic plague attack on Chicago (and a radiological attack on Seattle) [219].

As for smallpox, the United States intends to acquire up to 80 million doses of the modified virus Ankara smallpox vaccine, as part of its defense against the threat of smallpox virus being used for bioterrorism purposes. It would be useful for people who could suffer reactions to the existing vaccine. That includes patients with suppressed immune systems or those with the skin condition eczema. The vaccine contract could be worth more than $1 billion [346]. President George W. Bush has ordered all American military personnel to be vaccinated against smallpox and has implemented a voluntary program for vaccinating emergency medical personnel. Stockpiling and vaccination is seriously being reconsidered over many countries. Japan is a recent example; after announcing in 2001 that it would begin stockpiling enough vaccine for 3 million people, a governmental panel of experts recommended—in July 2005—the stockpiling of smallpox vaccine for 56 million people in preparation for a possible bioterrorism attack [347]. Russia, with its efficient oral—plus aerosol—vaccines, does not lag far behind. Lately, an international exercise—the "Atlantic Storm Scenario" in dealing with a series of international terrorist attacks involving smallpox virus has shown that the impact would be catastrophic for almost the entire world, and that the lives of millions of people could be lost. This outcome of the "Atlantic Storm" session seems all the more alarming given all the measures taken to combat bioterrorism since 11 September 2001 [268]. The Atlantic Storm Scenario has been preceded by "Exercise Global Mercury," which was the first worldwide simulation project modeling a smallpox bioterrorism attack [348].

Considering influenza virus to be the most powerful potential bioterrorism weapon, Madjid et al. proposed several steps to address the threat of influenza as a terrorist weapon, including the following [349]:

- The Centers for Disease Control and Prevention should classify influenza virus as a "critical agent" for bioterrorism.
- Immunization should be expanded, possibly by requiring it for all medical personnel.
- Laboratories that work with influenza virus should strengthen their security.
- Antiviral drugs should be stockpiled, and vaccine-making capacity should be increased.
- The government should consider a gene-sequencing and vaccine development program.
- Surveillance efforts should be increased and should include incentives for reporting of clinical cases.
- The fitting of ventilation systems with virus detection and inactivation.

A super-flu virus, particularly a pandemic derivative of the current H5N1 avian flu, would presumably kill 200,000 and up to 1.9 million Americans, plus many millions sickened, according to recent estimates. A government's plan to fight such contagion designates not just who cares for the diseased, but who will keep the country running amid the chaos. Specific countermeasures include stockpiling of both the anti-flu drug Tamiflu as well as avian flu vaccine [350]. Still, the protean nature of influenza type A virus at large may be defying. Experts in Hong Kong observed that the human H5N1 strain that surfaced in northern Vietnam during 2005 had proved to be resistant to Tamiflu [351]. Also, the present antigenic quality of the H5N1 may undesirably alter. Also, the present intact strain of avian flu, as isolated from its human victims, and although not yet a contagious pathogen, is a putative bioterrorism agent itself, due to its respiratory infectivity, lethality, and panic-causing nature.

In general, then, ideally—although seemingly impractically—a complete genetic inventory of all pathogenic strains found worldwide could provide the proper basis for initial, rather essential, inquiry into bioterrorism acts. It may facilitate, as well, controlling the registration of culture collection so as to hamper theft of dangerous pathogens. Thus oriented, the United States is making an effort to fully catalog and selectively to obtain prioritized pathogenic strains such as those then included in the Soviet BW program. Within that context, Kazakhstan, Azerbaijan, Ukraine, and Georgia are appreciably cooperative.

Consequently, more than 60 pathogens—including anthrax and plague strains—from the former Soviet Union's Azerbaijan-based biological weapons program plus epidemiological system arrived at Dover Air Force Base in Delaware [352]. In Azerbaijan—largely a Muslim country—several facilities—including field-test site—were engaged in the then USSR BW development program plus epidemic control. The samples were transferred as part of the Cooperative Threat Reduction program with Azerbaijan, aimed at getting the U.S. help to improve security for pathogens.

Moving to Russia, a different picture emerges with regard to a country fully acknowledging the essentiality of anti-bioterrorism moves, while still possessing an inventory of BWs and an active BW program [305]. The anti-bioterrorism outline and legislative base that were established in Russia—so as to deny possible access by terrorist states and terrorist groups to dual-use biotechnologies, plus materials—have been described by Vorobiev. He thereupon pointed at potential international cooperative efforts to be implemented, accordingly [353]. Eventually, in 2001, a commitment was achieved between Russian and U.S. Presidents, Vladimir V. Putin and George W. Bush, to pursue cooperation to counter the threat of bioterrorism, including a focus on health-related measures.

American assistance to Russia within the "Reduce the Common (WMD) Threats" program fared much. In the bioterrorism sphere, a chief course of that program for conversion has been the anti-bioterrorism one. The Novosibirsk-based Biopreparat-affiliated Vector microbiological center has been a pioneering facility, within that context. A long time before others, Vector's Director, Academician Sandakhchiev, started to cooperate with international institutions, and already in the early 1990s focused on issues related to monitoring and prevention of bioterrorism [354]. In addition, there are plenty of microbiological laboratories all over Russia eager—for financial, professional, or even ideological reasons—to save forbidden strains, toxins, equipment, and information. On the other hand, with time the Americans became more attentive to Russian arguments about development of anti-terror measures.

Consequently, The Russian-U.S. BioIndustry Initiative (BII) began in 2002, constituting the newest proliferation threat reduction program. It aims to counter bioterrorism through targeted reconfiguration of large-scale, formerly Soviet BW research, development, and production facilities for civilian purposes, by creating Russian-U.S. research partnerships [355]. Collaboration has then been established in 2003 between the International Science and Technology Center (ISTC), Moscow, and the Boston-based Center for Integration of Medicine and Innovative Technology (CIMIT), to implement the BII. Conversion is thus intended to take place through formation of systems in Russia to link scientists, physicians, and engineers to solve medical and scientific problems, and to identify innovative technologies and commercialization opportunities [356]. Also, the U.S. Defense Department has been increasingly engaged in efforts to secure from proliferation dozens of former Soviet pathogen collection and research facilities included within the "anti-plague system," widespread across 11 former Soviet states (excluding Russia). Apparently, many of those facilities lack sufficient safety and security and their scientists on average are poorly paid [357]. The concerned, so called Anti-Plague Institutes, are thus being turned into controllable "central reference laboratories" [358].

13.4.25   PROJECTIONS AND PROGNOSIS

Built-in limitations, which interfere with comprehending and foreseeing the course of bioterrorism, are outlined and discussed, together with objective constrains that restrict the scope of countering it. Nonetheless, the conjunction of methodological intelligence monitoring, systematic biosecurity globalization, and creative scientific research may furnish a crucial tool for effectively fighting bioterrorism.

The very supremacy of mankind, in terms of know-how and technology, may possibly make it prone, paradoxically, to self-destruction, in various manners. Bioterrorism is one main way to materialize that undesirable predisposition. Apparently, an outstanding conjunction marked the recent decade, bringing about the prominence of bioterrorism as a colossal issue, including:

- Internet-contained unclassified and declassified information.
- The outcomes of the birth of the formerly Soviet republics—particularly the Muslim and semi-Muslim ones—with their Soviet BW inheritance, namely highly qualified, equipped facilities, culture collections, and many unemployed experts, plus partial Islamic orientation.
- The accelerated augmentation of international terrorism, at large.
- The rise and persistence of al-Qaeda with its radical philosophy and unlimited financial resources.
- The global strengthening of Islam and its fundamentalistic inclination.
- Extremely meaningful breakthroughs made in life sciences and apparatus engineering.
- The increasing emergence and reemergence of infectious diseases worldwide.

Altogether, the ongoing integration of those various, interacting factors will probably shape, both conceptually and practically, the future of bioterrorism, one way or another. If a new breed of anarchistic multinational terrorists—not necessarily connected with al-Qaeda—is currently sprouting up, as at times claimed, the outcome may be unforeseeably catastrophic.

The outlining—let alone structuring—of a defensive alignment that would fully address the bioterrorism threat is, objectively, an impossible mission. Even while referring, primarily, to natural (nonengineered) pathogens and toxins, and assuming a consensus prioritizing anthrax, smallpox, plague, influenza, ricin, and botulinum, one can hardly face the two main resultant questions:

a. How to be best prepared toward each of those agents?
b. Should other potential bioterrorism agents be totally ignored, and if not—what ought to be the respecting mode of preparedness?

And, beyond, of course—what about potential engineered pathogens and toxins? Is it unfeasible, for instance, to form an anthrax germ resistant to the antibiotic Ciproflaxin? Overall, the immeasurable complexity thus formed seems to be insoluble even if budgetary aspects are ignored. Although frustrating, those issues can and should in part be practically administered by bioterrorism-strategic planners, provided they are fully aware of and acknowledging the concurrent constraints and limitations.

Increasing success of acts of so-called conventional terrorism is not regarded to decrease the potentiality and potency of bioterrorism (or other WMD-terrorism). Recurrent operations of conventional sabotage—even if fully effective—may intensify a perpetrator's daring quest for escalation and resonance through varitype means and modes of terrorism. Perhaps the most significant feature of bioterrorism, the ratio between the low probability of threat realization on the one hand, as against the high impact generated in case of threat realization on the other hand, is an extremely problematic factor for the defender. Let alone, that bioterrorism act that took place—the enthrax letters—in effect looks, in part, successful. In a sense, the very fact that the anthrax letter attack—apparently the most significant bioterrorism act ever conducted—has not been deciphered since 2001 is no less meaningful and important than this unprecedented act of bioterrorism itself.

Apparently, the prime and ultimate apparatus to cope with the bioterrorism threat is effective monitoring by intelligence sources, at least in that if adequately structured it would expediently shape the real and concrete needs and deeds of defense, in whatever sense. Therefore, intelligence ought to be supremely prioritized. All defensive measures have to actually form as derivatives of the intelligence picture, which may certainly be dynamic, though, in itself. The defense planner would not like that dynamics, in case the latter is not a plainly evolving one, but this is an inherent disadvantage. Conversely, yet, or rather complementarily, the defense planner may rely on his own considerations, as long as they can objectively be figured out. The significance of intelligence has been accentuated by the U.S. House of Representatives Homeland Security Subcommittee on Prevention of Nuclear and Biological Attack Chairman John Linder, who called for better coordination between intelligence and disease-fighting agencies. "Science, tools, reagents and technology may be ubiquitous. Scientists, however, are not," Linder said. "We have to do a better job of keeping track of those individuals with skill sets that are attractive to potential terrorists [359]."

In conjunction, it has been accentuated that a harmonized international regime that enhances biosecurity is essential for reducing the risk of bioterrorism [360]. Presumably, like other security regimes, this will entail mutually reinforcing strands, which need to include enactment of legally binding control of access to dangerous pathogens, transparency for sanctioned biodefense programs, technology transfer, and assistance to developing countries to jointly advance biosafety and biosecurity, global awareness of the dual-use dilemma and the potential misuse of science by terrorists, and development of a global ethic of compliance. To work, this effort must be undertaken collectively, using the international and regional institutions that already have a role to play in providing safety and security.

To those two essential elements—methodological intelligence monitoring and systematic international endeavor—should be added the dimension of creative scientific research. Apparently, the prospects for attaining a breakthrough in upgrading preparedness lie in the integral formed, then, by those three different spheres (beyond the self-evident readiness achieved by means of doctrine-structuring and medical stockpiling). The scientific sphere seems to bear enormous potential. Specific immunoglobulins may constitute a good example of a not yet adequately developed defensive tool, which potentially can be very efficient against pathogens and protein toxins, both for instant prophylaxis and treatment for as well as detection. This tool may be valid against both recognized and future engineered agents, provided that biotechnological processes will allow for at once production of fully specific immunoglobulins against a given agent.

Seemingly, to the least, this direction is worth elaborating on. Unlike vaccines and antibiotics, which are largely prioritized, antisera are less noticed, apparently unjustifiably. Antisera could constitute an optimal means for treatment, provided that instant and specific ethiological diagnosis is available. Studies and case reports evaluating convalescent plasma as therapy (or prophylaxis) of infectious diseases showed to be fairly promising. In the event of a bioterrorism attack, the usefulness of active immunization may be limited. Vaccine efficacy often requires time, multiple doses, and a competent immune system. Prophylactic immunization might provide an effective defense against bioterrorism agents but has the disadvantage that many individuals would have to be vaccinated to protect against an attack that might never occur. In this situation, even a small number of vaccine-related side effects would be unacceptable. In addition, vaccines do not induce protective immunity in all recipients, especially immunocompromised individuals [361]. Such inadequacies should carefully be considered. Yet in contrast, passive immunization involves the administration of preformed antibody to provide a state of immediate immunity. The two modes of immunization are used together in certain circumstances, such as rabies prophylaxis following possible exposure, where a passive antibody provides immediate protection and a vaccine elicits a protective immune response.

A protective antibody suitable for passive immunization could be used in concert with vaccines and drugs to provide a multilayered defense against attacks with biological agents. It should be possible to create a strategic reserve of immunoglobulins against the major biological warfare agents that can be rapidly administered to exposed individuals in the event of an attack. The availability of a strategic reserve of specific antibody preparations would have a significant deterrent value, as aggressors would be aware that the lethality of their weapon could potentially be counteracted by prompt administration of antibody to susceptible individuals. As antibody can be administered intramuscularly, immunoglobulin preparations could be packaged in self-injectable, disposable, single-use containers. Self-administration of antidote would avoid taxing the health-care system with the need for intravenous administration. Given the stability of immunoglobulin preparations, it should be possible to store antibody for many years. Developing, producing, and stockpiling antibody reagents for defense against bioterrorism agents is a sensible strategy that needs to be considered as steps are taken to prepare against the threat of bioterrorism [336]. On the whole, so it seems, the practicality of antisera should rather equalize that of vaccines and drugs, if not beyond. Connectedly, it has been highlighted that, in the U.S. National Institute of Allergy and Infectious Diseases work on boosting the human innate immune system, a strategy is formed that could lead to countermeasures that would be useful against a wide variety of different agents [359].

Such strategy may become vital, owing to the considerably diminishing usefulness of antibiotics against bacterial pathogens, as predicted by Prof George Poste, Director of the Biodesign Institute at Arizona State University and an advisor to the U.S. President: "Frankly, most governments are asleep at the switch"; he predicts that from 2010 to 2015 will be a "window of vulnerability" when the toll of the super-bug will reach its peak as a result of naturally augmenting antibiotic resistance [362]. Antibiotics may therefore be appreciably replaced by other protective agents, such as antibodies, receptor decoys, dominant-negative inhibitors of translocation, small-molecule inhibitors, and substrate analogues [363].

A new horizon may be gained through the Neugene antisense technology. Neugene antisense "rapid response therapeutics" are synthetic compounds that mirror a critical portion of a disease-causing organism's genetic code and bind to specific portions of the target genetic sequence. Like a key in a lock, Neugene antisense compounds are designed to match up perfectly with a specific gene sequence, blocking the function of the target gene [364]. Resultant antisense preparations have been able to block the cellular mechanisms used by the causative agents of anthrax, Ebola and Marburg, during pathogenesis. Another variant has been reported to block ricin toxin as well [365].

Relying on a similar principle, a novel drug has lately been developed that could expectedly be used to treat people exposed to anthrax bacteria specifically engineered to overcome antibiotics, due to an increasing concern that therapeutics developed for bioterrorism agents may be rendered ineffective if the microbial target is altered intentionally. This problem could be overcome, however, by designing inhibitors that block host proteins used by pathogens or their toxins to cause disease [366].

Thus, although the bioterrorism threat stemming from nonengineered pathogens and toxins is more or less characterized, in terms of expected impacts (although not in terms of likelihood, timing, and locality), the vagueness marking the threat posed by engineered pathogens and toxins is intriguing. Synthetic biology may be a key arena. Security experts fear that scientists building synthetic biochemical compounds could create a pathogen with no natural countermeasures, hence the possibility of making an unprecedented deadly bioterrorism weapon [367].

Synthetic biologists are combining existing chemical components of DNA or RNA taken from cells and viruses to form compounds that do not occur naturally. Scientists hope to use this technology to produce computers, medicines, and energy sources. The field is attracting attention from investors and prominent biologists; yet some biologists and security experts warn that these new compounds could be used to develop a biological weapon. "There are certainly a lot of national security implications with synthetic biology."

Concerns were played down that a new agent could exterminate the human race but warned that the threat of new, engineered pathogens remains serious [368]. Scientists are already exploring ways to self-police synthetic biological research, including requiring reports of sales of materials that could be used to create a tailored biological-weapon. "There are now tens of thousands of people, worldwide, who could engineer drug-resistant anthrax," said Professor Kenneth Alibek, who as a consultant to the U.S. government has received numerous briefings on U.S. and Soviet biological weapon programs.

A testimony prepared to the House of Representatives Homeland Security Subcommittee on Prevention of Nuclear and Biological Attack may well illustrate the issue of bioengineered agents. Addressing it, National Institute of Allergy and Infectious Diseases Director Anthony Fauci contented that agents could be made more virulent through "resistance to one or more antibiotic or antiviral drugs, increased infectiousness or pathogenicity or, in the somewhat longer term, a new virulent pathogen made by combining genes from more than one organism. As the power of biological science and technology continues to grow, it will become increasingly possible that we will face an attack with a pathogen that has been deliberately engineered for increased virulence." Fauci added, however, that creating an agent whose transmissibility could be sustained on such a scale, even as authorities worked to counter it, would be a daunting task. "Would you end up with a microbe that functionally will . . . essentially wipe out everyone from the face of the Earth? . . . It would be very, very difficult to do that," he said.

Centers for Disease Control and Prevention Director Julie Gerberding said a deadly agent could be engineered with relative ease that could spread throughout the world if left unchecked, but that the outbreak would be unlikely to defeat countries' detection and response systems. "The technical obstacles are really trivial," Gerberding said. "What's difficult is the distribution of agents in ways that would bypass our capacity to recognize and intervene effectively."

Today, the CDC's list of pathogens that must be reviewed comprises more than 60 agents. Each of them may completely be explored, so as to become fully manageable. But the catch is in that the existent pathogens are recognized—hence controllable—yet currently available for the potential attacker, whereas the unknown future engineered pathogens will scarcely be available, yet hardly controllable. The dichotomic menace thus emerging is an extremely challenging one. Possibly, as mentioned, immunoglobulins, polyvalent inhibitors, and other opproaches, may furnish a solution.

All in all, the threat of bioterrorism is indeed a challenging one and bears, potentially, overwhelming, somewhat enigmatic impacts. Henry Crumpton, the US State Department Coordinator for Counter-Terrorism, described a biological attack on West "simply a matter of time", adding that such an attack could pose an even greater menace to security than a nuclear strike [369].

U.N. Secretary General Kofi Annan, in the General Assembly, laid emphasis on the hazardous conjunction of bioterrorism and biotechnology:

"The international community must give a higher priority to developing new strategies against bioterrorism. Biotechnology has value in the fight against disease, but scientific advances could also "bring incalculable harm if put to destructive use by those who seek to develop designer drugs and pathogens. Soon, tens of thousands of laboratories worldwide will be operating in a multibillion-dollar industry," Annan said, according to the Associated Press. "Even students working in small laboratories will be able to carry out gene manipulation. U.N. nations should consider organizing of forum of governmental officials and science and public health experts to develop a strategy "to ensure that biotechnology's advances are used for the public good and that the benefits are shared equitably around the world [370]."

Finally, perhaps most plainly forwarded by the head of the Interpol, Ronald Noble, the bioterrorism treat has been featured as a global, severe, unprecedented menace: "The world is ill prepared for the looming threat of a bioterrorism attack. The danger of an al-Qaeda attack has not diminished since the 9/11 strikes on the US. The potential cost of a bioterrorism attack left no room for complacency. When you talk about bioterrorism, that's one crime we can't try to solve after it happens because the harm will be too great. How could we ever forgive ourselves if millions or hundreds . . . or tens of thousands of people were killed simply because our priorities did not include bioterrorism?" Noble acknowledged that governments and security agencies were better organized against the threat than ever before, but "none of us can let our guards down and assume that the problem has been addressed". Were al-Qaeda to launch a "spectacular bioterrorism attack which could cause contagious disease to be spread, no entity in the world is prepared for it," he said. "Not the US, not Europe, not Asia, not Africa [371]." He may possibly be right.

REFERENCES

1. Currie J. The earliest weapons of mass destruction. *Christ. Sci. Mon.* Oct. 9, 2003 Available: http://www.csmonitor.com/2003/1009/p17s01-bogn.html.
2. Simon J D (1997). Biological terrorism. Preparing to meet the threat. *J. Amer. Med. Assoc.* 278(5):428–430.
3. SIPRI (1993). The Problem of Chemical and Bioterrorism Warfare, Vol. 2. Humanities Press, New York, p. 143.
4. Zelicoff A P (1996). Preparing for bioterrorism terrorism: First, do not harm. *Politics Life Sci.* 15(2):235–236.
5. Zilinskas R (1999). Cuban allegations of biological warfare by the United States: Assessing the evidence. *Crit. Rev. Microbiol.* 25(3):173–227.
6. Cunningham W, Cunningham M (1998). Principles of Environmental Science. Available: http://highered.mcgraw-hill.com/sites/0072452706/student_view0/chapter8/additional_case_studies.html#ddt.
7. Krupitsky E, et al. (2004). The onset of HIV infection in the Leningrad region of Russia: A focus on drug and alcohol dependence. *HIV Med.* 5(1):30.
8. TASS, Online, Moscow, 2 April 2002.

9. SARS- Origins of the Virus. Available: http://www.bikesutra.com/sars/origins_old. html.

10. Zhu G, Chen H-W (2004). Monophyletic relationship between Severe Acute Respiratory Syndrome Coronavirus and Group 2 Coronaviruses. *J. Infec. Dis.* 189:1676.

11. Lam W K, et al. (2003). Overview on SARS in Asia and the world. *Respirol.* 8(s1): S2.

12. Chang H-J, et al. (2004). The impact of the SARS epidemic on the utilization of medical services: SARS and the fear of SARS. *Amer. J. Pub. Health.* 94(4):562–564.

13. Nash D, Mostashari F, Fine A, et al. (2001). The Outbreak of West Nile virus infection in the New York City area in 1999. *N. Engl. J. Med.* 344:1807–1814.

14. Komar N, Panella N A, Burns J E, et al. (2001). Serologic evidence for West Nile virus infection in birds in the New York City vicinity during an Outbreak in 1999. *Emerg. Infect. Dis.* 7:621–625.

15. Steele K E, Linn M J, Schoepp R J, et al. (2000). Pathology of fatal West Nile virus infections in native and exotic birds during the 1999 Outbreak in New York City, New York, *Vet. Pathol.* 37:208–224.

16. Giladi M, Metzkor-Cotter E, Martin D A, et al. (2001). West Nile encephalitis in Isael, 1999: The New York connection. *Emerg. Infect. Dis.* 7:659–661.

17. Briese T, Jia X Y, Huang C, et al. (1999). Identification of a Kunjin/West Nile-like flavivirus in brains of patients with New York encephalitis. *Lancet.* 354: 1261–1262.

18. Jia X Y, Briese T, Jordan I, et al. (1999). Genetic analysis of West Nile New York 1999 encephalitis virus. *Lancet.* 354:1971–1972.

19. Lanciotti R S, Roehrig J T, Deubel V, et al. (1999). Origin of the West Nile virus responsible for an Outbreak of encephalitis in the northeastern United States. *Science.* 286:2333–2337.

20. Weiss D, Carr D, Kellachan J, et al. (2000). Clinical findings of West Nile virus infection in hospitalized patients, New York and New Jersey, 2000. *Emerg. Infect. Dis.* 7:654–658.

21. Marfin A A, Petersen L R, Eidson M, et al. (2000). Widespread West Nile virus activity, eastern United States, 2000. *Emerg. Infec. Dis.* 7:730–735.

22. WHO. Avian Influenza. Available: www.who.int/csr/disease/avian_influenza/ en/.

23. Shoham D (1993). Biotic-abiotic mechanisms for long-term preservation and reemergence of influenza type A virus genes. *Prog. Med. Virol.* 40:178–192.

24. Madjid M, Lillibridge S, Mirhaji P, et al. (2003). Influenza as a bioweapon. *J. R. Soc. Med.* 96:345–346.

25. US ready with guide on pandemics, 8 Jan. 2007, Times of India, New Delhi.

26 NHS "meltdown" predicted by Government bird flu report, UK Telegraph, Aug. 28, 2006.

27. Faruque S M, Chowdhury N, Kamruzzaman M, et al. (2003). Reemergence of epidemic Vibrio cholerae O139, Bangladesh. *Emerg. Infect. Dis.* 9:1116–1122.

28. Global Epidemics and Impact of Cholera. Available: http://www.who.int/topics/ cholera/impact/en/.

29. Cloete T E, Nevondo T S (2001). The Global Cholera Pandemic. Available: http:// www.scienceinafrica.co.za/2001/September/cholera.htm.

30. Shoham D (2000). Iraq's biological warfare agents: A comprehensive analysis. *Crit. Rev. Microbiol.* 26:179–204.

31. Malkinson M, Banet C, Weisman Y, et al. (2002). Introduction of West Nile virus in the Middle East by migrating white storks. *Emerg. Infect. Dis.* 8:392–397.

32. Shoham D (2005). *Viral Pathogenes of Humans Likely to be Preserved in Natural Ice.* Princeton University Press, Princeton, NJ, p. 221.

33. Horowitz G L (1996). *Emerging Viruses: AIDS and Ebola-Nature, Accident or Intentional?* Tetrahedron Inc., Sandpoint, ID.

34. Mumps Outbreak Aids Kansas Biological Preparedness, Associated Press/The Wichita Eagle, June 16, 2006.

35. Cassa C A, Iancu K, Olson K L, et al. (2005). A software tool for creating simulated Outbreaks to benchmark surveillance systems. *BMC Med. Informat. Decis. Making.* 5:22.

36. Cosgrove S E, Perl T M, Song X, et al. (2005). Ability of Physicians to diagnose and manage illness due to Category A bioterrorism agents. *Arch. Intern. Med.* 165:2002–2006.

37. U.K. Increases Biological Security, Tendler/McGrory, The Times, Jan. 25, 2007.

38. The Black Death Available: http://www.studentcentral.co.uk/black_death_9886.

39. Winn Jr. W C (1996). Medical microbiology. In B Samuel (ed.), *Legionella.* The University of Texas Medical Branch at Galveston, Galveston, TX.

40. Corbel M J (1997). Brucellosis: An overview. *Emerg. Infect. Dis.* 3:213–221.

41. Young E J (1995). An overview of human brucellosis. *Clin. Infect. Dis.* 21:283– 290.

42. Dye C. Scheele S, Dolin P, et al. (1999). Consensus statement. Global burden of tuberculosis: Estimated incidence, prevalence, and mortality by country. WHO Global Surveillance and Monitoring Project. *JAMA.* 282:677–686.

43. Anonymous (2005). TB to be Subject to Parameters of Bioterrorism Law. *The Japan Times.*, Sept. 20, 2005 Available: http://www.japantimes.co.jp/cgi-bin/makeprfy.pl5?nn20050920b1.htm.

44. Meegan J M, Hoogstraal H, Moussa M I (1979). An epizootic of Rift Valley fever in Egypt in 1977. *Vet. Rec.* 105:124–125.

45. CDC. Rift Valley Fever Fact Sheet. Available: http://www.cdc.gov/ncidod/dvrd/spd/mnpages/dispages/Fact_Sheets/Rift_Valley_Fever_Fact_Sheet.pdf.

46. WHO (2001). Yellow Fever Fact Sheet. Available: http://www.cdc.gov/ncidod/dvrd/spd/mnpages/dispages/Fact_Sheets/Rift_Valley_Fever_Fact_Sheet.pdf.

47. Jaworski M (2003). Norwalk Virus. Available: http://www.emedicine.com/med/topic1648.htm#section~author_information.

48. Rubins K H, Hensley L E, Jahrling P B, et al. (2004). The host response to smallpox: Analysis of the gene expression program in peripheral blood cells in nonhuman primate model. *Proc. Natl. Acad. Sci. USA.* 101:15190– 15195.

49. Lacy D B, Tepp W, Cohen A C, et al. (1998). Crystal structure of botulinum neurotoxin type A and implications for toxicity. *Nat. Struct. Biol.* 5:898–902.

50. McCormick J B, Steele J H, Hendricks K (2005). Staphylococcal Enterotoxin B-Factsheet. Available: http://www.dshs.state.tx.us/idcu/disease/staphylococcal/seb/factsheet.

51. *US Army Medical Research Insitute of Infectious Diseases*, (1998) Medical Management of Biological Causalties Handbook 3rd ed.: Frederick, Fort Detrick, MD, p. 108, 109.

52. Wannemacher Jr. R W, Wiener S L (1997). Trichothecene Mycotoxins, Medical aspects of chemical and biological warfare, F R Sidell, COL E T Takafuji, COL D R Franz (eds.). In: B G R Zajtchuk, COL R F Bellamy, *Part I, Warfare, Weapons, and the Casualty, Textbook of Military Medicine: Medical Aspects of Chemical and Biological Warfare*, Office of the Surgeon General, Walter Reed Army Medical Center, Washington, D.C., p. 656.

53. Seagrave, S (1981). *Yellow Rain*. Evans & Co., New York, P. 134.

54. Kadivar H, Adams S (1991). Treatment of chemical and bioterrorism warfare injuries: insights derived from the 1984 Iraqi attack on Majnoon Island. *Military Med*. 156: 170.

55. Heyndrickx A, et al. (1989). Detection of trichothecene mycotoxins in Iranian soldiers treated as victims of a gas attack. *Rivista Toxicol. Speriment*. 19:7.

56. United Nations, Department of Public Information, JY (1996). The United Nations and the Iraq-Kuwait Conflict, 1990–1996, In: *The United Nations Blue Book Series*, p. 784.

57. US Army Medical Research Institute of Infectious Diseases (1998). *Medical Management of Biological Casualties Handbook*, 3rd ed.: Frederick, Fort Detrick, MD, p. 107.

58. Yourtee D M, Raj H G, Prasanna H R, et al. (eds) (1989). Proceedings of the international symposium on agricultural and biological aspects of aflatoxin related health hazards. *J. Toxicol. Toxin Rev*. 8:3–375.

59. SIPRI Fact Sheet (1998). Iraq: The UNSCOM Experience. www.sipri.org/contents/webmaster/publications.

60. UNSCOM Periodical Report (1999). Current e-mail notifications on Iraq; by Lautie Mylroie sam11@erols.com, January 25.

61. French G (1958). Aconitine-induced cardiac arrhythmia. *Br. Heart J*. 20:140–142.

62. Tai Y T, But P P, Young K, et al. (1992). Cardiotoxity after accidental herb-induced Aconite poisoning. *Lancet*. 340:1254–1256.

63. Chan T Y (1994). Aconitine poisoning: A global perspective. *Vet. Hum. Toxicolo*. 36:326–328.

64. Chan T Y, Tomlinson B, Tse L K, et al. (1994). Aconitine poisoning due to Chinese herbal medicines: A review. *Vet. Hum. Toxicolo*. 36:452–455.

65. Fatovich D M (1992). Aconite: A lethal Chinese herb. *Ann. Emerg. Med*. 21:309–311.

66. Kolev S T, Leman P, Kite G C, et al. (1996). Toxicity following accidental ingestion of Acotinum containing Chinese remedy. *Hum. Exp. Toxicol*. 15:839–842.

67. Honerjager P, Meissner A (1983). The positive inotropic effect of Aconitine. *Naunyn Schmiedebergs Arch. Pharmacol*. 322:49–58.

68. H'ctor A, de Landoni H J (1990). Aconitum Napellus spp. Available: http://www.inchem.org/documents/pims/plant/aconitum.htm#SectionTitle: 12.3%20Other.

69. El–Siad (Lebanon), The Death of Marsall Amer, Editorial 14 August 1975.

70. Anonymous (1976). Latest dirty trick is fake nerve as antidote. *New Scientist*. 20 May, p. 423.

71. Adam H L (1953). Trial of Dr. Lamson (The Blenheim School Murder). *J. Crim. Law, Criminol. Police Sci*. 44:361–362.

72. Anonymous (1966). Deliberated spreading of Typhoid in Japan. *Sci. J*. 2:11–12.

73. U.S. Congress Office of Technology Assessment (1992). *Technology against terrorism: structuring security*, OTA-ISC-511: US Government Printing Office, Washington, D.C.

74. Bishop S (1980). GRA Seeks Botulin *International Herald Tribune*, November 8–9.

75. Shalev R, Bioterrorists Arrested in Belgium. Ma'ariv, March 23, 1998, p. 3.

76. Douglass J D (1990). America the Vulnerable, Lexingtom Books, p. 196.

77. CNN, Plague Plot, Feb 19, 1998.

78. Shoham D (2000). *Crit. Rev. Microb*. 26:(3):179–204.

79. Dave Altimari, Hartford Courant, Oct. 9, 2006.

80. Shoham D (2003). The anthrax evidence points to Iraq. *Intern. J. Intell. CounterIntell.* 16(1):39–68.

81. Joby Warrick, Suspect and A Setback In Al-Qaeda Anthrax Case, Washington Post, October 31, 2006; A01.

82. September 19, 2001. Available: Worldnetdaily.com. Biological Weapons in South Africa.

83. Keim P, Smith K L, Keys C, et al. (2001). Molecular investigation of the Aum Shinri-kyo anthrax release in Kameido, Japan. *J. Clin. Microbiol.* 39:4566–4567.

84. Alon B 19 April 2005. Ma'ariv (Israel). Palestinian Labs Worry, p. 4.

85. Anonymous (2005). Your bioweapon is in the mail. UPI, Washington, Nov 9.

86. Bureau of Emergency Preparedness and Response: Arizona Department of Health Service. History of biowarfare and bioterrorism. Available: http://www.azdhs.gov/phs/edc/edrp/es/bthistor2.htm.

87. Ross E (2005). Labs urged to destroy pandemic flu strain. *Associated Press*, London, U.K., April 12.

88. WHO Memorandum (2005). Missing Samples of Deadly Flu Strain, April 19, 2005.

89. Anonymous, A letter blamed for an epidemic of smallpox, *N Y Med. J.* 1901;73:600.

90. Boobbyer P (1901). Smallpox in Nottingham, *Br. Med. J.* 1901;1:1054.

91. Audrey Woods (2002). 1960s germ tests carried out in London Underground, Associated Press February 26, 2002, Available from: URL http://www.nti.org/d_newswire/issues/2002/2/27/13s.html.

92. Carlton, Jim, Of Microbes and Mock Attacks: Years Ago, The Military Sprayed Germs on U.S. Cities, Wall Street Journal, October 22, 2001.

93. BBC Radio 4 Documentary, Hotel Anthrax, February 13, 2006.

94. Alon B Maariv, (Israel). 3 November, 1996. Sewage Influx Causes Concerns, p. 6.

95. Alon B Maariv, (Israel). 1 December, 1996. Sewage Contamination Increases, p. 5.

96. Braveman I, Wexler D, Oren M (2002). A novel mode of infection with hepatitis B: Penetrating bone fragments due to the explosion of a suicide bomber. *Isr. Med. Assoc. J.* 4:528–529.

97. MacKenzie D (2002). Suicide Bombers May Spread Disease. Available: http://www.newscientist.com/article.ns?id=dn2588.

98. Abdel-Aziz G Arab daily "al-Sharq al-Awsat" 11 Nov 2004. Iranian UAVs Operated by Hezbollah, p. 2.

99. Anonymous Janes Defence Weekly, 13 August, 1997. Ventilation Systems Vulnerable to Bioterrorism, p. 14.

100. Norqvist A (1995). Covert use of bioterrorism and toxin weapons—For what purpose? *Proc. 5th International Symposium Protection Against CB Warfare Agents*, Stockholm, Sweden, pp. 367–372.

101. Geissler E (1996). Joint international action is necessary to counter the threat of C/B terrorism. *Polit. Life Sci.* 15(2):205–207.

102. Grange D L (2001). Asymmetric warfare. *The Officer.* Available: http://www.findarticles.com/p/articles/mi_m0IBY/is_2_77/ai_73326816#continue.

103. McKenzie K (2001). The Rise of Asymmetric Threats: Priorities for Defense Planning, Institute for National Strategic Studies. Available: http://www.ndu.edu/inss/press/QDR_2001/sdcasch03.html.

104. (1998). BBC, March 23.

105. Blenkin, Max, Australian Associated Press/The Age, Dec. 1, 2005.

106. Faruque S, Chowdhury N, Kamruzzaman M, et al. Reemergence of epidemic-Vibrio cholerae O139, Bangladesh. Emerging Infections Diseases, 9(9):1116–1122, 2003.

107. McHale P (2005). Pentagon preparing for multiple WMD attacks. *Inside Missile Defense*, May 25.

108. Brissenden M (2005). Biological agent sent to Indonesian Embassy. *The 7:30 Report* ABC Australia.

109. Associated Press, July 8, 2005. Man Sentenced to 19 years for Anthrax Hoaxes.

110. Holloway H C, et al. (1997). The threat of bioterrorism weapons. *JAMA*. 278(5): 425–427.

111. McPhedran I (2005). Suspicious Powder letter, Daily Telegraph June, 4.

112. Australian AP, June 2, 2005. Osborne P Powder Letter Found to be Harmless.

113. Schneidmiller C. Years after Anthrax Attacks, Hoaxes Persist. Global Security Newswire. Available: http://www.nti.org/d_newswire/issues/2005/7/19/9c68a408-b230-431e-9769-6805f6855b99.html.

114. Poisoned Arrows, V&A. Available: http://www.vam.ac.uk/collections/asia/object_stories/arrows/.

115. Above Top Secret—War On Terrorism. Available: http://www.abovetopsecret.com/forum/viewthread.php?tid=18930&s.

116. B-WAR. Available: www.junglelogic.tv/artf/b_war.html.

117. McGovern T W, Christopher G W (2000). Biological warfare and its cutaneous manifestations. In: Dr. R Drugge, H A Dunn (eds.), *The Electronic Textbook of Dermatology*. Available: http://www.telemedicine.org/BioterrorismWar/bioterrorismlogic.htm.

118. Harris S H (1994). *Factories of Death: Japanese Biological Warfare, 1932–1945, and the American Cover-Up*. Routledge, New York, pp. 1–147.

119. Spencer, R C, Bacillus anthracis, Journal of Clinical Pathology, 2003, 56:182–187.

120. Alibek K, Handelman S (1999). *Biohazard*. Random House, New York, pp. 29–31.

121. US Politics Online: A Political Discussion Forum—what about Israel?! Available: www.uspoliticsonline.com/forums/archive/index.php/t-18141.htm.

122. Phills J A, et al. (1972). Pulmonary ringworm infestation in man. *N. Eng. J. Med.* 2:11–12.

123. Simon J D (1997). Biological terrorism. Preparing to meet the threat. *J. Amer. Med. Assoc.* 278(5):8.

124. Robert S (1991). Root-Bernstein, Infectious Terrorism, The Atlantic Monthly, 267: 44–50.

125. Kifner, J (1995), Man Arrested in a Case Involving Ricin Ny Times, Dec. 23.

126. Shemer P (1994), Yediot Ahronot, 10 October, 1994. Iragi Scientist Implanted in NY, p. 8.

127. Herbert B (1995). Militia Madness. *The New York Times*, June 7.

128. Rapaport A (1998), Ma'ariv, 22 February, 1998. Iragi Women Intend to Smuggle Bio-agents, p. 3.

129. Alon B (1998). Muslim Bioterrorists in Belgium, p. 2. Ma'ariv, 23 March, 1998.

130. Sarrar S (2005). Libya Acts in HIV Row with Bulgaria. *The Guardian*, April 13.

131. Blaney B (2005) Everything is Under Consideration. *Associated Press/Star-telegram*, October 25.

132. Carus W S (1998). *Bioterrorism and Biocrimes: The Illicit Use of Bioterrorism Agents in the 20 Century*. Center for Counterproliferation Research, National Defense University, Washington, D.C.

133. Wheelis M (2004). *A short History of Biological Warfare and Weapons*. ISO Press, Amsterdam, The Netherlands.

134. Wheelis M (1999). Bioterrorism sabotage in World War I. In: E Geissler, J Ellis, van Courtland Moon (eds.), *Biological and Toxin Weapons: Research, Development, and Use from the Middle Ages to 1945*: Oxford University Press, Oxford, UK, pp. 35–62.

135. Biological Warfare. Available: http://www.nutrition-report.com/articles/Biological_warfare.

136. Nelson A K (1998). JFK assassination review board releases top secret documents. *Organization of American Historians Newsletter*. Available: http://www.indiana.edu/%7Eoah/nl/98feb/jfk.html.

137. Ad Hoc Group of the States Parties to the BW Convention (2000). University of Bradford. Available: http://www.brad.ac.uk/acad/sbtwc/ahg52wp/wp417.pdf.

138. Livingstone N C, Douglass J D (1984). CBW: The poor man's atomic bomb. *National Security Papers*. Vol 1. Institute for Foreign Policy Analysis, Inc., Cambridge, MA.

139. Anonymous Book review: Swine fever warfare (1997). *Hartford Web Publishing*. Available: http://www.hartford-hwp.com/archives/43b/069.html.

140. Cummings R H (1996). Code Name Piccadilly—The Murder of Georgi Markov. *Balkan Info*. Available: http://www.b-info.com/places/Bulgaria/news/96–02/feb09.scb.

141. FBI File: Bacteriological Warfare in the United States (released May 2006) www.thememoryhole.org/fbi/biowar.htm.

142. South Africa Profile Available: http://www.nti.org/e_research/profiles/SAfrica/Biological/index.html#fnB2.

143. Truth and Reconciliation Commission, South Africa (1998). Chemical and Biological Warfare Hearings. Available: http://www.totse.com/en/politics/the_world_beyond_the_usa/camera.html.

144. Burger M, Gould C (2002). *Secrets and Lies: Wouter Basson and South Africa's Chemical and Biological Warfare Program*. Zebra, Cape Town, South Africa, pp. 34–35.

145. Kemp A (1990). *Victory or Violence: The Story of the AWB*. Forma, Pretoria, South Africa, p. 47.

146. Van Rooyen J (1994). *Hard Right: The New White Power in South Africa*. I.B.Tauris, New York, pp. 96, 97, 199.

147. NTI: Country Overviews: South Africa: Chemical Overview. Available: www.nti.org/e_research/profiles/SAfrica/Chemical.

148. Mangold T, Goldberg J (1999). *Plague Wars: The Terrifying Reality of Biological Warfare*. St. Martin's, New York, pp. 250, 254, 277–279.

149. Humes E (2001). The Medicine Man. *Los Angeles Magazine*, July:95–96.

150. CBS News (2002). "Dr. Death" and his Accomplice. *60 Minutes*, November 7.

151. Collins J (2000). Ford advised S. Africa on Warfare Devices. *Orange County Register*, 15 March.

152. Allen A (2002). Mad Scientist. *Salon Magazine*, June 26. Available: http://archive.salon.com/health/feature/2000/06/26/biofem.

153. Bell R (2004). Court TV's Crime Library. Available: www.crimelibrary.com/terrorists_spies/larry_ford/1.html.

154. Kouzminov A (2005). *Biological Espionage: Special Operations of the Soviet and Russian Foreign Intelligence Services in the West*. Greenhill, London.

155. Markov G. Available: http://www.videofact.com/english/defectors_33_en.html.

156. Ridley Y (2005). Don't Let Facts Spoil a Good Story. Available: http://www.ukwatch. net/article/450.

157. Miller J (1993). Evidence Grows on Bioterrorism Weapons. *New York Times Magazine*, January 3, p. 33.

158. Anonymous (1990). A Case of Finding Evidence, Jane's Defence Weekly, July 14, p. 54.

159. Anonymous (1990). On the Street. *The ASA Newsletter*, August issue.

160. UNSCOM Periodical Report (1998) Current e-mail notifications on Iraq; by Laurie Mylroie sam11@erols.com, June 3.

161. Duelfer C (2004). Comprehensive Report of the Special Advisor to the DCI on Iraq's WMD. Available: http: //www.cia.gov/cia/reports/iraq_wmd_2004/index.html.

162. Shoham D (2006). An antithesis on the fate of Iraq's chemical and biological weapons. *Internat J. Intell. Counterintell.* 19(1):59–83.

163. Eisenstadt M (1996). Iranian *Military Power: Capabilities and Intentions.* Washington Inst. for Near East Policy, Washington, D.C., p. 26.

164. Gaffney Jr. F J (2004). Terror-tied by Memo. *The Washington Times*, May 9.

165. Schanzer J (2003). *A. Al-Islam: Iraq's Al-Qaeda Connection.* The Washington Institute for Near East Policy, Washington, D.C. Available: http://www.freerepublic. com/focus/news/824263/posts.

166. Warrick J, al-Qaeda's Bio-lab, Washington Post May 5, 2004.

167. Parry D. Interfax/Mosnews.com, March 1, 2005. al-Qaeda Explores Bioagents.

168. Gertz B (2004). Iraqi Bomb Labs Signal Attacks in the Works. *The Washington Times*, November 30.

169. Anonymous (2004). Iraqi Women Not Being Released. *BBC News World Edition*, September 22.

170. McMahon M. NewsMax.com Monday, December 29, 2003. al-Qaeda Seeks Anthrax; no page.

171. Leitenberg M (2002). Biological weapons and bioterrorism in the first years of the twenty-first century. *Politics Life Sci.* 21(2):3–27.

172. Taliban Official Said Found With Anthrax, Agence France-Presse/The Nation, Jan. 17, 2007.

173. Associated Press (2005). 8 of 9 Acquitted in U.K Ricin Trial. April 13.

174. Uhlig R, Fleming N (2004). Flights Axed in New Terror Alert. Available: http://www. telegraph.co.uk/news/main.jhtml?xml=/news/2004/02/02/nterr02.xml.

175. Anonymous, MI5 warns UK labs of threat from al Qaeda, London Daily Mail, January 25, 2007.

176. Bockmann M W (2004). Biochemical Threat Next Terror Wave. *The Australian.*

177. Anonymous *The Times*, 3 March, 1998. Hezbollah Trying to Obtain Biochemicals page unknown.

178. *Reuters*, 9 Feb. 2003.

179. Al-Manar (2001). Will We Reach the Option of Bioterrorism Deterrence? August 13.

180. Luft, G (2003). The Palestinian security forces and the second intifada. *Institute for the Analysis of Global Security, Presentation given at the National Press Club.* Washington, D.C.

181. Rapaport A, Ma'ariv, 13 April 2004. Tanzim Plants to Use Aids Virus, p. 7.

182. Palestinians Claim Biological and Chemical Weapons Capability, Israel Today, June 26, 2006.

183. Kaplan D E, Marshall A (1996). *The Cult at the End of the World*. Crown Publishers, Inc., New York.

184. Torok T J, Tauxe R V, Wise R P, et al. (1997). A large community outbreak of salmonellosis caused by intentional contamination of restaurant salad bars. *JAMA*. 278: 389–395.

185. Kolavic S A, Kimura A, Simons S L, et al. (1997). An outbreak of Shigella dysenteriae type 2 among laboratory workers due to intentional food contamination. *JAMA*. 278:396–398.

186. Joseph P R, Millar J D, Henderson D A (1965). An outbreak of hepatitis traced to food contamination. *N. Engl. J. Med*. 273:188–194.

187. Phills J A, Harrold A J, Whiteman G V, et al. (1972). Pulmonary infiltrates, asthma and eosinophilia due to Ascaris suum infestation in man. *N. Engl. J. Med*. 286:965–970.

188. Chester C V, Zimmerman G P (1994). Civil Defense Implications of Bioterrorism Weapons. *J. Civil Def*. 16(6):6–12.

189. Fetter S (1991). International Security, Summer 1991, p. 27. Ballistic Missiles and WMD.

190. MSNBC, 13 March 2003. The Impact of Anthrax Available: http://www.msnbc.com/news/886761.asp?0cv=CB20&cp1=1.

191. American Anthrax Outbreak of 2001, Case 3. UCLA School of Public Health, Los Angeles, CA. Available: http://www.ph.ucla.edu/epi/bioter/detect/antdetect_case3.html.

192. American Anthrax Outbreak of 2001, Case 4. UCLA School of Public Health, Los Angeles, CA. Available: http://www.ph.ucla.edu/epi/bioter/detect/antdetect_case4.html.

193. American Anthrax Outbreak of 2001, Case 5. UCLA School of Public Health, Los Angeles, CA. Available: http://www.ph.ucla.edu/epi/bioter/detect/antdetect_case5.html.

194. American Anthrax Outbreak of 2001, Case 7. UCLA School of Public Health. Available: http://www.ph.ucla.edu/epi/bioter/detect/antdetect_case7.html.

195. American Anthrax Outbreak of 2001, Case 1. UCLA School of Public Health, Los Angeles, CA. Available: http://www.ph.ucla.edu/epi/bioter/detect/antdetect_case1.html.

196. American Anthrax Outbreak of 2001, Case 19. UCLA School of Public Health, Los Angeles, CA. Available: http://www.ph.ucla.edu/epi/bioter/detect/antdetect_case19.html.

197. American Anthrax Outbreak of 2001, Case 21. UCLA School of Public Health, Los Angeles, CA. Available: http://www.ph.ucla.edu/epi/bioter/detect/antdetect_case21.html.

198. American Anthrax Outbreak of 2001, Case 2. UCLA School of Public Health, Los Angeles, CA. Available: http://www.ph.ucla.edu/epi/bioter/detect/antdetect_case2.html.

199. American Anthrax outbreak of 2001, Case 6. UCLA School of Public Health, Los Angeles, CA. Available: http://www.ph.ucla.edu/epi/bioter/detect/antdetect_case6.html.

200. American Anthrax Outbreak of 2001, Case 8. UCLA School of Public Health, Los Angeles, CA. Available: http://www.ph.ucla.edu/epi/bioter/detect/antdetect_case8.html.

201. American Anthrax outbreak of 2001, Case 9. UCLA School of Public Health, Los Angeles, CA. Available: http://www.ph.ucla.edu/epi/bioter/detect/antdetect_case9.html.

202. American Anthrax Outbreak of 2001, Case 10. UCLA School of Public Health, Los Angeles. Available: http://www.ph.ucla.edu/epi/bioter/detect/antdetect_case10.html.

203. American Anthrax Outbreak of 2001, Case 11. UCLA School of Public Health, Los Angeles, CA. Available: http://www.ph.ucla.edu/epi/bioter/detect/antdetect_case11.html.

204. American Anthrax Outbreak of 2001, Case 12. UCLA School of Public Health, Los Angeles, CA. Available: http://www.ph.ucla.edu/epi/bioter/detect/antdetect_case12.html.

205. American Anthrax Outbreak of 2001, Case 13. UCLA School of Public Health, Los Angeles, CA. Available: http://www.ph.ucla.edu/epi/bioter/detect/antdetect_case13.html.

206. American Anthrax Outbreak of 2001, Case 18. UCLA School of Public Health, Los Angeles, CA. Available: http://www.ph.ucla.edu/epi/bioter/detect/antdetect_case18.html.

207. American Anthrax Outbreak of 2001, Case 14. UCLA School of Public Health, Los Angeles, CA. Available: http://www.ph.ucla.edu/epi/bioter/detect/antdetect_case14.html.

208. American Anthrax Outbreak of 2001, Case 15. UCLA School of Public Health, Los Angeles, CA. Available: http://www.ph.ucla.edu/epi/bioter/detect/antdetect_case15.html.

209. American Anthrax Outbreak of 2001, Case 16. UCLA School of Public Health, Los Angeles, CA. Available: http://www.ph.ucla.edu/epi/bioter/detect/antdetect_case16.html.

210. American Anthrax Outbreak of 2001, Case 17. UCLA School of Public Health, Los Angeles, CA. Available: http://www.ph.ucla.edu/epi/bioter/detect/antdetect_case17.html.

211. American Anthrax outbreak of 2001, Case 20. UCLA School of Public Health, Los Angeles, CA. Available: http://www.ph.ucla.edu/epi/bioter/detect/antdetect_case20.html.

212. American Anthrax Outbreak of 2001, Case 23. UCLA School of Public Health, Los Angeles, CA. Available: http://www.ph.ucla.edu/epi/bioter/detect/antdetect_case23.html.

213. American Anthrax Outbreak of 2001, Case 22. UCLA School of Public Health, Los Angeles, CA. Available: http://www.ph.ucla.edu/epi/bioter/detect/antdetect_case22.html.

214. BBC News, November 22, 2001.

215. Heyman D (2002). Lessons from the anthrax attacks. Intelligence Resource Program. Available: http://www.fas.org/irp/threat/cbw/dtra02.pdf.

216. Mcintyre H J (2004). Project on Government Secrecy. Ref: 04-F-0521(A), DoD FOIA Pollicy Office.

217. Shoham D, and Jacobsen S T (2007). Technical intelligence in retrospect: the 2001 anthrax letters powder, *International J. of Intelligence and Counterintelligence*, 20(1):79–105, 2007.

218. Associated Press, Washington D.C. 11 March 2004, Powder Letters Hoaxes.

219. Miller J (2003). U.S. Has New Concerns about Anthrax Readiness. December 28.

220. Valdemoro T (2005). Cleaning of Anthrax Contents Continues. *Cox News Service-Herald Today*, July 7. Available: http://www.bradenton.com/mld/bradenton/news/local/12072828.htm.

221. Matsumoto G (2003). Anthrax powder: State of the art? *Science*. 302:1492–1497.

222. 2001 anthrax attacks—Wikipedia, the free encyclopedia, en.wikipedia.org/wiki/ Website.

223. Allan Lengel, Little Progress In FBI Probe of Anthrax Attacks, Washington Post, September 16, 2005; A01.

224. Ricin Toxin from Castor Bean Plant. Cornell University, Ithaca, NY. Poisonous Plants Informational Database Available: http://www.ansci.cornell.edu/plants/toxicagents/ ricin/ricin.html.

225. Klain G J, Jaeger J J (1990). *Castor Seed Poisoning In Humans: A Review*. Letterman Army Inst. of Research, San Francisco, CA.

226. Historical records from Army Office of Research and Development (OSRD), 1946, Agent W-Ricin.

227. Franke S (1976). *Bacterial, Animal and Plant Toxins as Combat Agents, Manual of Military Chemistry*, Vol. 2. Militaerverlag der DDR., Berlin, Germany, pp. 484, 485, 488–496.

228. Sanches M L, Russell C R, Randolph C L (1993). *Chemical Weapons Convention (CWC) Signatures Analysis: Technical Analysis*. Defense Nuclear Agency, Alexandria, VA, pp. B171–788.

229. Usama bin Laden. Available: www.nmhschool.org/tthornton/mehistorydatabase/ afterseptembereleven.htm.

230. Frankel G (2003). *The Washington Post*. Jan. 8, Ricin Found in Afghomistan.

231. Gold D (2004). *Tower of Babble*. Crown Forum, New York, p. 129.

232. Seventh Arrest in Ricin Case. *BBC News Online*, 8 January 2003.

233. Norton-Taylor R, Hopkins N, Henley J (2003). Poison Suspect Trained at al-Qaida Camp. *The Guardian* (London), January 10.

234. Sergeyev V (2003). London Poisoners Came from Chechnya. *Gazeta* (Moscow), January 10.

235. Ivanov S (2003) France, UK Security Follow Trail of New Terrorist Structures with Chechen cell. *ITAR-TASS News Agency* (Moscow), January 13.

236. Gray C, Herbert I, Bennetto J (2003). Policeman Stabbed to Death in Terror Raid. *The Independent* (London), January 15.

237. Carter H, Ward D, Hopkins N (2003). Murder Suspect "is Senior Player" in Ricin Plot Network. *The Guardian* (London), January 16.

238. Chemical Terrorist Plot in Rome? Center for Nonproliferation Studies. Available: http://cns.miis.edu/pubs/week/020311.htm.

239. Anonymous (2003). Ricin found in Paris. *BBC News*, March 21.

240. Eggen D (2004). Letter with Ricin Vial Sent To White House, November Discovery Was Kept Quiet. *The Washington Post*, February 4, A:07.

241. Lengel A (2004). Ricin Investigation Still Wide Open. *The Washington Post*, April 6, B:01.

242. Barry R (2004). Rebels Tried to Make Ricin. *Associated Press*, Washington, D.C., June 10.

243. Woodward C (2005). Ricin Intended to be Produced *Associated Press/RedOrbit.com*, November 21.

244. Jar of Ricin Found in Tennessee Shed, Associated Press/San Diego Union-Tribune, June 2, 2006.

245. What Is Smallpox? Available: http://skin-care.health-cares.net/smallpox.php.

246. Smallpox: Destruction of variola virus stocks. World Health Organization. Available: http://www.who.int/gb/ebwha/pdf_files/WHA58/A58_10-en.pdf.

247. Jahrling P B, Hensley L E, Martinez M J, et al. (2004). Exploring the potential of variola virus infection of cynomolgus macaques as a model for human smallpox. *Proc. Natl. Acad. Sci. USA.* 101:15196–15200.337.

248. Rubins K H, Hensley L E, Jahrling P B, et al. (2004). The host response to smallpox: analysis of the gene expression program in peripheral blood cells in a nonhuman primate model. *Proc. Natl. Acad. Sci. USA.* 101:15190–15195.

249. WHO (2004). Abstracts summarizing recent research overseen by the WHO Advisory Committee on Variola Virus Research. Available: http://www.who.int/csr/disease/smallpox/abstracts2004/en/index1.html.

250. WHO (2003). Abstracts summarizing recent research over seen by the WHO Advisory Committee on Variola Virus Research. Available: http://www.who.int/csr/disease/smallpox/abstracts2003/en/index1.html.

251. Li Y, Kline R, Regnery R. (2002). Capillary-Electrophoresis Restriction Fragment Length Polymorphism: A New Method for Poxvirus Fingerprinting. Available: http://www.who.int/csr/disease/smallpox/capillary/en/index.html.

252. Huggins J W, Zwiers S H, Baker R O, et al. Cidofovir Treatment of Variola (Smallpox) in the Hemorrhagic Smallpox Primate Model and the IV Monkeypox Primate Model. Available: http://www.who.int/csr/disease/smallpox/cidofovirtreatment/en/index.html.

253. Rubins K, Whitney A, Jahrling P, et al. Genome-Wide Analysis of the Host Response to Variola Infection. Available: http://www.who.int/csr/disease/smallpox/genomewide/en/index.html.

254. Pulford D J, Damon I, Ulaeto D. New PCR Assays for Identification of Smallpox Virus. Available: http://www.who.int/csr/disease/smallpox/newpcr/en/index.html.

255. Jahrling P B, Hensley L, Huggins J, et al. Lethal Infection of Primates with Variola Virus as a Model for Human Smallpox. Available: http://www.who.int/csr/disease/smallpox/lethalinfection/en/index.html.

256. Esposito J J, Sammons S, Frace M, et al. Variola Virus Genomics. Available: http://www.who.int/csr/disease/smallpox/virusgenomics/en/index.html.

257. Boseley S, Borger J (2005). US Scientists Push for Go-Ahead to Genetically Modify Smallpox Virus. *The Guardian,* May 16.

258. Burgasov P (2001). Smallpox Field Tested By USSR *Courier,* November 24.

259. Anonynmous (2003). Soviet Smallpox Most Lethal *Vedomosty* (Moscow), March 27.

260. Smallpox: Facts and Information. Available: http://www.absoluteastronomy.com/encyclopedia/s/sm/smallpox.htm.

261. Hopkins D R, Ramses V (1980). Earliest Known Victim? *World Health,* p. 220.

262. Fornaciari G, Marcetti A (1986). Intact smallpox virus particles in an Italian mummy of 16th century. *The Lancet.* 2(8507):625.

263. Marennikova S S, Shelukhina E M, Zhukova O A, et al. (1990). Smallpox diagnosed 400 years later: Results of skin lesions examination of 16th century Italian mummy. *J. Hyg. Epidemiol. Microbiol. Immunol.* 34:227–231.

264. Orent W (1998). Escape from Moscow. *The Sciences,* pp. 29–31, May 22.

265. Lewin P K (1985). Mummy riddles unraveled. *Microsc. Soc. Can. Bull.* 12:4–8.

266. Arita I (1980). Can we stop smallpox vaccination? *World Health.* May:27–29.

267. Thomas, Gordon (2003). Ethnic Bomb to Kill Whites Only Tied to North Korea. *American Free Press,* October, 17.

268. Sombre Conclusions follow "Atlantic Storm". Radio Netherlands. Available: www2. rnw.nl/rnw/en/currentaffairs/region/internationalorganisations/ter050117.

269. National Security Archive (1970). The September 11th Source Books: National Security Archive Online Readers on Terrorism, Intelligence and the Next War. Volume III: BIOWAR: The Nixon Administration's Decision to End U.S. Biological Warfare Programs. Available: http://www.gwu.edu/~nsarchiv/NSAEBB/NSAEBB58/RNCBW22.pdf.

270. Beckett B (1983). *Weapons of Tomorrow*. Plenum Press, New York.

271. Leitenberg M (2001). Biological Weapons in the Twentieth Century: A Review and Analysis. Available: http://www.fas.org/bwc/papers/bw20th.htm.

272. Anonymous Times, Sept. 14, 1999, Soviet Bioagents Against Livestock.

273. Shoham D (2005). Image versus reality of Iranian chemical and biological weapons. *Internat. J. Intell. Counterintell.* 18(1):89–141.

274. Robinson R A. USDA's Preparation for Asian Soybean Rust. Available: http://www. gao.gov/new.items/d05668r.pdf.

275. Vargas R I (2002). Biosystems and Bioterrorism: Can Arthropods Be Used as Agents of Destruction. Entomological Society of America Meeting, Fort Lauderolale, FL.

276. Lindquist D A, Abusowa M, Hall MJ (1992). The New World screwworm fly in Libya: A review of its introduction and eradication. *Med. Vet. Entomol.* 6:2–8.

277. Windrem R (1999). U.S. to launch war on "agro-terror." *MSNBC.com*, September 22. Available: http://www.msnbc.com/news/314627.asp.

278. Treble A (2005). Agro-Terrorism. Available: http://cns.miis.edu/research/cbw/agromain.htm.

279. Stockholm International Peace Research Institute (1971). *The Problem of Chemical and Biological Warfare*, Vol. I. Humanities Press, New York, p. 224.

280. Treble A (2002). Chemical and Biological Weapons: Possession and Programs Past and Present. Available: http://cns.miis.edu/research/cbw/possess.htm#US.

281. Thorold P W (1953). Suspected malicious poisoning. *J. South African Vet. Med. Assoc.* 24:215–217.

282. Morton J F (1977). Poisonous and injurious higher plants and fungi. In: C C Tedeschi, W G Eckert, L G Tedeschi (eds.), *Forensic Medicine: A study in trauma and environment, Volume III: Environmental Hazards*. W.B. Saunders, Philadelphia, PA, p. 1504.

283. Verdcourt B, Trump E C (1969). *Common Poisonous Plants of East Africa*. Collins, London, p. 62.

284. Maloba W O (1993). Mau *Mau and Kenya: An Analysis of a Peasant Revolt*. Indiana University Press, Bloomington, IN.

285. Edgerton R B (1989). *Mau Mau: An African Crucible*. Free Press, New York.

286. Alibek K (1999). The Soviet Union's anti-agricultural biological weapons. *Annals N. Y. Acad. Sci.* 894:18–19.

287. Narayan Swamy M R (1994). *Tigers of Lanka, From Boys to Guerrillas*. Konark Publishers, Delhi.

288. Hellmann-Rajanayagam D (1994). *The Tamil Tigers: Armed Struggle for Identity*. Franz Steiner Verlag, Stuttgart.

289. O'Balance E (1989). *The Cyanide War: Tamil Insurrection in Sri Lanka 1973–1988*. Brassey's, Washington, D.C.

290. Gunaratna R (1987). *War and Peace in Sri Lanka*. Institute of Fundamental Studies, Sri Lanka, pp. 51, 52.

291. Duboudin T (1984). Australian Livestock Threatened. *The Times* (London), January 21, p. 5.

292. Duboudin T (1984). Murderer in Court Over Virus Threat. *The Times* (London), February 22, p. 5.

293. Root R S (1991). Infectious terrorism. *Atlantic Monthly.* May, p. 4450.

294. Office International des Eepizootics, Headquarter, Paris (2002). Foot and mouth disease in the United Kingdom. 2001 Report to the OIE Part I: Great Britain Department for Environment, Food & Rural Affairs 14 January, 2002.

295. Anonymous (2005). Putative Agroterrorism Bioagents in Afghanistan Associated Press, 17 July, 2005.

296. Narcoterrorism. Available: http://www.sciencedaily.com/encyclopedia/narcoterrorism.

297. What is Narcoterrorism? Terrorism: Questions & Answers. Available: http://cfrterrorism.org/terrorism/narcoterrorism_print.html.

298. Narco-Terrorismus. Zentraler Informatikdienst- Universität Wien. Available: http://gerda.univie.ac.at/ie/ws03/fischer/wiki/index.php/Narco-Terrorismus.

299. Department of the Navy (1998). Legal Review of Oleoresin Capsicum (OC) Pepper Spray, (Ser 103/353). Office of the Judge Advocate General, Alexandria, Virginia.

300. Common and New Types of Biological Weapons. Available: whalonlab.msu.edu/2003_EC_Projects/Biological_Weapons/ISB%20Extra%20Credit%20Project/Common%20Types.html.

301. Black J L (2003). Genome projects and gene therapy: Gateways to next generation biological weapons. *Military Med.* 168(11):864–871.

302. Wheelis A M (2004). Will the new biology lead to new weapons? *Arms Control Today.* 34(6):6–13.

303. van Aken J, Hammond E (2003). Genetic engineering and biological weapons: New technologies, desires and threats from biological research. *Euro. Molec. Biol. Org. Rep.* 4 (suppl 1):S57–S60.

304. Quille G (2001). The revolution in military affairs debate and non-lethal weapons, medicine, conflicts and survival. *Internat. Security Informat. Serv. London, UK.* 17(3):207–220.

305. Shoham D, Wolfson Z (2004). The Russian biological weapons program: Vanished or disappeared? *Crit. Rev. Microbiol.* 30(4):241–261.

306. Former KGB spy, living in New Zealand, soon to publish a book on Russian biological warfare. Available: http://www.freerepublic.com/focus/f-news/1347344/posts.

307. Oppenheimer A (2005). Nanotechnology paves way for new weapons. Jane's Chem-Bio Web. www.janes.com/defence/news/jcbw.

308. Joby Warrick, Engineered Microbes Pose New Bioterror Threat, Washington Post, July 31, 2006.

309. King S R (1992). Medicines that changed the world. Available: http://www.accessexcellence.org/RC/Ethnobotany/page4.html. Ethnobotany Index.

310. Ricin toxin from castor bean plant. Available: http://7plus7.com/MedicalNews/News/Ricin.htm.

311. Lord J M, Roberts L M, Robertus J D (1994). Ricin: Structure, mode of action and some current applications. *FASEB J.* 8(2):201–208.

312. Shukla H, Sharma S (2005). Clostridium botulinum: A bug with beauty and weapon. *Crit. Rev. Microbiol.* 31(1):11–18.

313. UN Security Council Resolution 1540 (2004), United Nations, New York City, April 28.

314. U.N. Releases Counterterrorism Strategy, Associated Press, Sep. 4, 2006.

315. Tigner B (2004). European group urges bioterror defense. *Defense News*, June 29.

316. Bioterrorism Reporting Group Report (2005). Countering bioterrorism: How can the EU and US work together? New Defence Agenda *NDA* July, 18.

317. Trevelyan M (2005). *Reuters*, March 1. Outcomes of a Biotterrorism Attack.

318. Ceran F (2005). *Associated Press*, March 1, 2005. The Threat of Bioterrorism.

319. MI5 warns UK labs of threat from al-Qaeda, London Daily Mail, January 25, 2007.

320. U.K. Increases Biological Security, Tendler/McGrory, The Times, January 25, 2007.

321. Rozen K (2005). NIS 500,000 to Find Knesset Vulnerabilities to Non-Conventional Weapons. *Israel National News*, September 22.

322. Anonymous (2003). Iran Red Crescent holds first conference on campaign against bioterrorism, May 29. Available: http://www.fbis.gov. FBIS doc IAP 20030529000082.

323. Anonymous (2005). Seminar on Bioterrorism Opens in Saudi Capital. *Financial Times Information*, March 7.

324. Biodefense for the 21st century. Available: http://www.whitehouse.gov/homeland/20040430.html.

325. Remarks by the President at the Signing of S.15-Project BioShield Act of 2004, The White House, July 21, 2004.

326. Fiorill J (2005). Experts paint dire picture of bioterrorism threat. *Global Security Newswire*, July 13.

327. Fiorill J (2005). U.S. plans to defend against engineered bioattack. *Global Security Newswire*, June 15.

328. Bioterrorism and the National Response Plan 2005. Available: apha.confex.com/apha/133am/techprogram/paper_110332.htm.

329. Schuler A (2005). Billions for Biodefense: Federal Agency Biodefense Budgeting, FY2005-FY2006. *Biosecur. Bioterror.* 3(2):91–101.

330. Ruppe D (2005). *Global Security Newswire*, July 14, 2005. Bush Budget Holds Steady for Civiliam Biodefences.

331. Ember L (2005). Manhattan experiments track gases for terror response biomonitoring. *Chem. Engin. News.* 83(33):10.

332. Schwellenbach N (2005). Biodefense: A Plague of Researchers Bulletin of the Atomic Scientists, May–June. pp. 8–10.

333. National Institute of Allergy and Infectious Diseases release, June 1, 2005. US to Complete Bioterrorism Research Network.

334. Anonymous. *Associated Press*, 8 September 2005. New Biocontainment Lab.

335. FBI Wants Biodefense Work Secret says Newark, New Jersey Lab. The Sunshine Project—Biosafety Bites (v.2) p. 17 September 20, 2006.

336. Jennifer Lebovich, BioWatch Air Samplers Used to Detect Deadly Germs, jlebovich@MiamiHerald.com Nov. 01, 2006.

337. Hall/Jackson, Bush Asserts Mail-Opening Authority, USA Today, Jan. 5, 2007.

338. FDA's National Center for Toxicological Research Publishes Study on Distinguishing Potential Hoax Materials from Bioterror Agents, FDA New, PO6-109, August 3, 2006.

339. Nanotechnology Offers Fast Virus Detection, United Press International, RedOrbit.com, Nov. 15, 2006.

340. Alison Walker-Baird, U.S. Researchers Develop Tool Against Agroterrorism, The Frederick News-Post, October 31, 2006.

341. Network Would Help Detect Bioterrorism, The Richmond Times Dispatch, May 23, 2006.

342. Anonymous *Biotech Week*, June 8, 2005. Post-Attack Anthrax Medical Measures.

343. Barrett G (2005). *Baltimore Sun*, June 10, 2005. Stockpiling Bioterror Countermeasures.

344. Francis D. Researchers suggest risk-based decision making model to combat bioterrorism attacks. *Global Security Newswire.* June 3, 2005.

345. Thompson K M, Armstrong R E, Thompson D F (2005). Bayes, Bugs, and Bioterrorists. National Defense University, Washington, D.C. Available: http://www.ndu.edu/ctnsp/Def_Tech/DTP14%20Bayes%20Bugs%20Bioterrorists.pdf.

346. Harrison P (2005). Smallpox Vaccine to be Acquired *Reuters*, August 16, 2005.

347. *Daily Yomiuri*, July 12, 2005. Japan Intends to Stockpile Smallpox Vaccine.

348. Exercise Global Mercury: Post Exercise Report. Available: www.dh.gov.uk . . . /ChiefMedicalOfficer/Features/FeaturesArticle/fs/en.

349. Madjid M, Lillibridge S, Parsa M, et al. (2003). Influenza as a bioweapon. *J. R. Soc. Med.* 96:345, 346.

350. Neergaard L (2005). U.S. working on plan for flu pandemic. *Associated Press*, Washington D.C., October 8.

351. Anonymous (2005). Avian Flu Resistant to Tamiflu. CNN (*Reuters*), September 30.

352. Zeleny J (2005). Pathogens Transferred from Formen USSR to USA. *Chicago Tribune*, September 3.

353. Vorobiev A (1996). Countering chemical/biological terrorism in the former Soviet Union: The need for cooperative efforts. *Politics Life Sci.* 15:233–235.

354. Vector Outline (2003). Available: http://www.vector.nsc.ru/art1-r.htm.

355. Available: www.state.gov/documents/organization/28781.

356. Gulobev E (2003). New Russian-American Collaboration Against Bioterrorism: Moscow's International Science and Technology Center (ISTC) participating, Pravda, 17 September 2003. Pravda Online English Version.

357. Ruppe D (2005). U.S. helps secure former soviet antiplague sites. *Global Security Newswire*, November 23.

358. David Francis, U.S. Program to Secure Former Soviet Pathogens Grows, Global Security Newswire, March 29, 2006.

359. Fiorill J (2005). Top U.S. disease fighters warn of new engineered pathogens but call bioweapons doomsday unlikely. *Global Security Newswire*, July 29.

360. Atlas R M, Reppy J (2005). Globalizing biosecurity, *Biosecurity and Bioterrorism*. 3(1):51–60.

361. Casadevall A (2002). Antibodies for defense against biological attack. *Nat. Biotechnol.* 20:114.

362. Anonymous *Associated Press* (2005). Antibiotics Less Useful March 3.

363. Rainey G J, Young J A (2004). Antitoxins: Novel strategies to target agents of bioterrorism. *Nat. Rev. Microbiol.* 2(9):721–726.

364. AVI BioPharma Receives Notice of Allowance for Key Patent Application Covering NEUGENE Antisense Technology for RNA Viruses, BioExchange Newsroom. Available: http://www.bioexchange.com/news/news_page.cfm?id=20709.

365. AVI BioPharma Info (2005). Promising new technology may block biological agent effects, *GSN*, October 18. Available: http://www.mmrs.fema.gov/news/publichealth/2005/oct/nph2005-10-18a.aspx.

366. Basha, Saleem et al, Polyvalent inhibitors of anthrax toxin that target host receptors, PNAS, September 5, 2006 vol. 103, no. 36, 13509–13513.

367. Elias P (2005). *Associated Press/Journal-Gazette*, August 19. Super Pathogens.

368. Fiorill J (2005). *GSN*, July 13. Experts Paint Dire Picture of Bioterrorism Threat.

369. Biological attack on West "simply a question of time": US official (AFP) Jan 17, 2006.

370. Edith Lederer, Annan Calls for Bioterror Defense Strategies, Associated Press/phillyBrubs.com, May 3, 2006.

371. Anonymous (2005). Interpol sounds bioterrorism alarm. *BBC*, February 23. Available: http://news.bbc.co.uk/go/pr/fr/-/1/hi/world/europe/4289485.stm.

INDEX

4-6-20 Rule, 582

Absorption, 100
Accuracy, 573–575, 615*t*, 617–618
Aconitine, 1557
Actinium radionuclides, 918
  See also radionuclides
Active (vector), 951
Adenoviruses, 1152
  gene therapy of, 1289
  recombinant vectors, 1267–1268
Adhesion formation, 536
Adjuvants, 992–993
ADMET, 100
Adoptive transfer of *in vitro* transfected
  APCS, 991
Adsorption, 390–391
Aging, 435
Aggregation, 386–387
AIDS, 1534–1535
Alphatoxin,
Altered foci assay, 530
Alzheimer's disease, 1125
Amplification,
Animal lectins, 135–136*t*
Anti-drug antibody assay, 625–626
Anti-HEV antibody, 526

Anti-infectious agents, 139
Anti-infectious diseases vaccines, 993
Anti-inflammatory assay, 536
Antibodies
  anti-drug antibody assay, 625–626
  anti-HEV antibody, 526
  antibody arrays, 647
  advantages and disadvantages of the
    most commonly used assays to detect
    binding anti-drug antibodies, 821–822*t*
  delivery of, 869
  FDA-approved antibody
    pharmaceuticals, 716–720*t*
  monoclonal antibodies, 494, 496*t*
  panning antibody libraries, 857
  pathogens detected by recombinant
    antibodies, 863*t*
Anticancer
  activity, 139–140, 523
  assay, 523
Antiestrogen assay, 525
Antimalarial activity assay, 556
Antimalarial assay, 555
Antiplatelet aggregation assay, 544
Antisense, 976
Antisense chemistries, 1064
Antitubercular assay, 525

Antitumor vaccines, 995–996
Antivenoms, 692–693
Antiviral assay, 524
Arterial thrombosis assay, 545
Arthritis, 1133
Artificial cells, 1360–1370
Artificial noses, 27
Assays
    advantages and disadvantages of the
        most commonly used assays to detect
        binding anti-drug antibodies,
        821–822t
    altered foci assay, 530
    anti-drug antibody assay, 625–626
    anti-inflammatory assay, 536
    anticancer assay, 524
    antiestrogen assay, 525
    antimalarial activity assay, 556
    antimalarial assay, 555
    antiplatelet aggregation assay, 544
    antitubercular assay, 525
    antiviral assay, 524
    binding assays, 821–822
    blood pressure-related assay, 558
    body weight-based liver tumor incidence
        assay, 530, 833
    cerebral artery occlusion assay (MCAO),
        558, 561
    chloroquine assay, 557
    dendritic cell assay, 546
    ear edema assay, 544
    estrogen assay, 551
    fibrinolytic area assay, 546
    fluorescent protein expression assay,
        554
    hepatoprotective assay, 530–531
    IFN-γ assay, 546
    immunomodulating assay, 546
    lymphoid organ assay, 548
    neutralizing assays, 822
    PBMC proliferation assay, 536
    pulmonary hypertension assay, 558–559
    radial assay of chemotaxis, 550
    rat mast cell assay, 546
    thread-induced thrombosis assay
    thrombisis assay, 541–542
    thrombolytic assay, 546
    yeast oestrogen assay, 551–552
Astatine radionuclides, 900
    See also radionuclides
Autoimmune diseases, 435, 1133
Avian influenza, 1537–1538

Baby hamster kidney (BHK), 11
Baculovirus, 10–12
B-Lactam, 34–35
    compounds, 34
Binding assays, 919–922
    advantages and disadvantages of the
        most commonly used assays to detect
        binding anti-drug antibodies, 821t, 822
Binding properties of antibodies, 926
Biocatalysts, 37
Biogenerics, 362–363
Bioluminescence, 1306–1307
Biomarker, 622–625
Biosimilar products, 115–116
    CHMP guidelines concerning
        biosimilars, 1457
    regulatory criteria, 115t, 1420–1422
Biosynthetic pathways
    characterization of, 13, 15t, 19
Biotechnology industry organization, 694
Bismuth radionuclides, 919
    See also radionuclides
Blood pressure-related assay, 558–559
Body weight-based liver tumor incidence
    assay, 530, 533
Botulinum toxin, 1552–1553
Bovine spongiform encephalopathy (BSE),
    350
Bromine radionuclides, 896
    See also radionuclides
Brucellosis, 1547

Cancer, 1116, 1118–1120, 1122–1123,
    1130–1131, 1133, 1137, 1140–1142,
    1145–1146, 1149–1150, 1153, 1163,
    1165–1168, 1170–1172, 1204, 1211,
    1219, 1223, 1262, 1268–1272
    drugs, 1471–1473
Cardiovascular diseases, 435
Cardiovascular drugs, 1471, 1473
Cationic dendrimers, 1027
Cationic lipids, 1028–1029, 1031, 1075
Cationic polymers, 1016, 1019, 1155
Cavitation, 304–305
Cell-based delivery of therapeutic genes,
    949–950
Cell penetrating peptides (CRPs),
    1156–1157
Cellular senescence and immortalization,
    1362–1363
Cerebral artery occlusion assay (MCAO),
    561

Chaperones, 5
Chemokines, 1198–1214
Chitosan, 130, 133, 140, 144
Chloroquine assay, 557
CHMP guidelines concerning biosimilars, 1458
Cholera, 1539–1540
Classification of vaccines, 1379*t*
Claude, Albert, 973
Clinical candidates,
 selection of 98, 102
Collagens, 1035
Combinatorial biosynthesis, 19–20
Contaminant clearance, 337
Convention on biological diversity (CBD), 1405–1406
Copper radionuclides, 916–917
 See also radionuclides
Copyright, 1395
CYP, 1466–1468, 1476, 1478, 1467*t*
 CYP1, 1466
 CYP2, 1466
 CYP3, 1466
 CYP 2D6, 1466
 CYP 3A4, 1468

Database and material deposit rights, 1395
Deamidation, 372–375, 404–405
Deamination, 380
Delivery devices, 110–111, 401, 777
Delivery of antibodies, 869
Denaturation, 387–392, 394, 396, 399, 740–744
Dendritic cell assay, 546
Depurination, 378–379, 384
Depyrimidation, 378–379, 384
Diabetes, 435
Direct encapsulation, 1031, 1033, 1037
Disease targets for gene therapy, 953, 955*t*
Distribution, 255
DNA microarray genotyping, 1103
DNA vaccines, 870–871, 1034, 1262
Drug delivery vehicles, 750, 1150
Drug receptors, 1464, 1471
Drugs acting on the CNS, 1471, 1474
Drugs involving growth factors, cytokines, and chemokines, 1211, 1212*t*

Ear edema assay, 512, 536, 544
Ebola hemorrhagic fever, 1525, 1533
Electronic tongues, 27

ELISA validation, 835, 845
Embryonic stem cells, 1327–1330, 1335, 1348, 1369
 properties of, 1313, 1315
Endocrine disruptors, 511, 523
Endodermal diffentiation, 1321
Endosomal escape, 1042–1044
Endotoxin removal, 36
Escherichia coli, 3–6, 4*t*, 11–12, 30–31, 35, 60, 106, 296, 393, 445, 524, 673
Estrogen assay, 551
Ethnicity, 1477–1478
Evaluation of homogeneity, 919
Excipients, 392–393, 399*t*
Excretion, 100
Experimental joint pain, 1182
Expiration of biologics patents, 1456*t*
Expression-related functions, 976
Extracellular limitations to nonviral gene delivery, 1019*t*
Exubera, 400, 1213

Fabry disease, 728
Failure mode and effect, 346
Fault tree analysis (FTA), 346
FDA, 444, 320, 359, 400, 402, 450
FDA-approved antibody pharmaceuticals, 716*t*
FDA-approved enzyme pharmaceuticals, 725–727
FDA-approved protein/peptide pharmaceuticals, 700–707*t*
Fermentation, 332–333, 353*t*
Fertility hormones, 710
Fibrinolytic area assay, 546
Filamentous fungi, 10
First inhaled version of insulin (exubera), 1213
 See also exubera and insulin
Fluorescence microscopy, 1302
Fluorescent protein expression assay, 554
Fluorine radionuclides, 898–900
 See also radionuclides
Foldases, 5
Food and Drug Administration (FDA), 320, 611
Formulation, 108–113, 111*t*, 393
Formulation of growth factors, cytokines, and chemokines, 1213*t*
Fusarium, 511, 513
 mycotoxins, 513

Gallium radionuclides, 915–916
  See also radionuclides
Gannt chart for production scheduling of
    one batch of adenoviral vector, 1286*f*
Gelsinger, Jesse, 966, 1399
Gene amplification, 8
Gene delivery targeting, 951
Gene delivery vectors, 946–948, 947*t*
Gene expression
  control systems, 952
  enhancement of, 18–19
Gene-Gun, 991, 1005
Gene therapy, 998–1002, 1262
  in adenoviruses, 1264–1265
  milestones, 1262*t*
Genetic diseases, 1270–1271
Genomics, 421
Genotoxicity, 518
Glycosylation, 819
Gold standard, 94
Growth regulatory hormones, 710–712

Hematopoietic stem cells (HSCs),
    1334–1339
Hemophilia, 692
Hepatitis A, 1551
Hepatitis B, 994–995
Hepatocyte primary culture, 595–596
Hepatoprotective assay, 534–536
Hepatotoxicity, 590
Herpes simplex virus, 1153
High-throughput screening for targets,
    1117–1119
Hirudin, 1227–1229, 1238
Homeostatic hormones, 699
Hormones
  fertility hormones, 710
  growth regulatory hormones, 710–712
  homeostatic hormones, 699
  immunomodulatory hormones, 712
Human gene therapy (HGT), 943
Human genome, 53–87
  approaches to sequence, 69–73
  mapping of, 65–68, 67*f*, 68*f*
  human genome program (HGP),
    55–57
Human insulin, 386
Huntington's disease, 1134
Hyaluronic acid, 131
Hybridization-based approaches, 1066
Hybridization probes, 1090–1092
Hydrodynamic degradation, 301, 304

Hydrogels, 1036–1037
Hydrolysis probes, 1090–1091

IFN-γ assay, 548–549
Inflammation, 1214
Imaging of gene transfer, 1297–1299
Immortalization of human cells, 1361–1366
Immune mechanisms, 817–818
Immune tolerance, 818
  of animals, 824
Immune tolerant animals, 824
Immunocamouflage, 500
Immunodetection of pathogens, 861
Immunogenicity, 444–448
  of proteins, 745
Immunomodulating assay, 546–547
Immunomodulatory hormones, 712
Impurities, 349
Impurity clearance, 337–338
Indium radionuclides, 908
  See also radionuclides
Inhaled insulin, 763–764
  See also insulin
Insect cell, 11
Insulin, 393, 489
  background, 759
  first inhaled version of insulin (exubera),
    1213
  human insulin, 387
  inhaled, 763
  intranasal, 764
  rapid-acting insulin, 762
Intellectual property, 114
Interchain disulphide bridges, 109
Interferon, 778–783
  background, 778
  formulation, 781–783
  pharmacokinetics in animals, 866
  pharmacokinetics in humans, 868
International Conference on
    harmonization (ICH), 321, 364, 395,
    612
Intracellular limitations, 1040
  to nonviral gene delivery, 1019–1020*t*
  toxicity of, 1041–1042
Intranasal insulin, 764–765
Intranasal siRNA injection, 1073
Iodine radionuclides, 892
  See also radionuclides
Ion-exchange separation of proteins, 602
Ion channels, 1471
Iraq, 1584

Jenner, Edward, 692, 1602
"Junk" DNA, 82

Labeling mabs and peptides, 892–896
Labeling with radiometals, 914–919
Lambda-based vectors, 57–61
Laser-induced Fluorescence (LIF), 480
Lead identification, 94–99
  technologies, 95–96
Lead optimization, 100–104, 101*t*
Lead radionuclides, 918
  See also radionuclides
Legionnaires disease (Legionellosis), 1546
Lentiviruses, 1153
Lipoplexes, 1158–1161
Liposomes, 455
Low-throughput screening for targets, 1116
Luciferase, 553
Lutetium radionuclides, 918
  See also radionuclides
Lymphoid organ assay, 548

MAbs, 494
Magnetic resonance imaging (MRI), 1299
Major functions of growth factors,
  cytokines, and chemokines, 1200*t*
Malaria, 994
Mass spectrometry, 1510
Master cell bandk, 959
Matrix-based delivery, 1031
Matrix-tethered delivery, 1037
Measurement of radiochemical purity,
  (RCP), 921
Mesodermal differentiation, 1323
Metabolic shifts, 22
Metabolism, 94
Method transfer, 580
Microcystin toxicity, 516
Monoclonal antibodies, 494, 496*t*
Mononuclear phagocyte system (MPS), 129
MPS, 131
Mucosa surface, 145
Multidimensional liquid chromatography
  of proteins, 603
Multidrug resistance (MDR), 141

Nanocarriers for vascular drug delivery,
  1151*f*
Nanoparticles, 454
Nanoparticular drug carriers, 128
Narcoterrorism, 1610–1612
Natural polymers, 1037

Nerve growth factor (NGF), 1174
Neurodegenerative diseases, 435
Neurological disorders, 1125
Neuropathic pain syndromes, 1174
Neutralizing assays, 822
NMR spectroscopy, 1505–1510
Nonlipidic polycation gene delivery
  systems, 1155
Nonviral vectors, 948–949
Northern blotting, 1089
Norwalk, 1550–1551
Nuclear localization, 947
Nucleic acid delivery, 944

Occular disorders, 1134
Oligonucleotides, 243
Ovarian toxicity, 521–522
Oxygen mass transfer coefficient, 29

p53, 1365*f*
Packaging of nucleic acids for therapy,
  980–982
Pancreatic islets, 454
Panning antibody libraries, 857
Parasite infections, 1124
Parkinson's disease, 1117
Passive immunotherapy, 866
Passive targeting, 951
Pasteur, Louis, 692
Patent application, 1412–1413
Patent criteria, 1409–1411
Patent expiration of biologics, 1456*t*
Patent filing, 1413
Pathogens detected by recombinant
  antibodies, 863*t*
PBMC proliferation assay, 602
PEGylated, 696, 706, 711–712, 726–730,
  746, 748
Pegylation, 749–751, 754–755, 759, 764
Peroxisome proliferators activator receptor
  (PPAR) agonists, 166
pH sensitivity, 281
Pharmaceutical production and
  regulations, 1002–1004
Pharmacogenetics, 1464
Pharmacogenomics, 1465
Pharmacometabonomics, 1466
Pharmacophore, 97*t*, 98
Pharmacoproteomics, 1465
"Pharming", 1434
Phase ½ trial, 963
Phase 2 trial, 963

Physical scale-up parameters, 29
Plague, 1545–1546
Plant breeders' rights, 1395
Plant-made pharmaceuticals (PMPs), 1434–1447
  threates posed, 1434–1437
Plasmid DNA-based anti-tumor vaccines, 995
PLGA microspheres, 751
Pluripotency, 1315–1319
Polymerase chain reaction (PCR), 1089
Polyplexes as cellular transfection agents, 1025
Positron emission tomography (PET), 1307–1308
Posttranslational gene silencing, 17
Posttranslational modification, 11
Powdermed gene-gun, 1005
Poxviruses, 1152
Precipitation, 386–387
Precision, 574–575, 615t, 619–620, 627
Preclinical biodistribution, 924–925
Preclinical drug candidate safety assessment, 1515–1517
Prions, 693
Process behavior
  improvement of, 18–19
Process characterization, 26–28
Process optimization, 28–29
Product potency tests, 962
Product purification, 30–31
Product recovery, 31–32
Proteomics, 422, 501–502
Pulmonary hypertension assay, 559–560
Pulmozyme, 695

QSAR, 97t, 98, 100
Quality Issues, 37

Radial assay of chemotaxis, 550–551
Radiochemical purity (RCP), 921–923
Radioisotopic imaging, 1304
Radionuclides, 887–888t
  actinium radionuclides, 918
  astatine radionuclides, 900
  bismuth radionuclides, 919
  bromine radionuclides, 896–898
  copper radionuclides, 916–918
  for tumor imaging, 886–889
  fluorine radionuclides, 898–900
  gallium radionuclides, 915–916
  iodine radionuclides, 892–896

indium radionuclides, 908–913
  lead radionuclides, 941
  lutetium radionuclides, 918
  radionuclides suitable for labeling biomolecules, 888t
  rhenium radionuclides, 906–907
  technetium radionuclides, 902–906
  yttrium radionuclides, 913–915
Rapid-acting insulin, 762–763
  See also insulin
Rat mast cell assay, 608–609
Recognition of nucleic acids by immune cells, 978
Recombinant adenovirus vectors, 1267, 1291–1294
Recombinant expression systems, 3, 3t, 11, 13, 39
  characterisitics of, 15t
Red blood cells (RBCs), 452
Reduced plasmid stability, 26
Regulation of biotechnology products, 1380
Regulation of small molecules, 1377
Regulatory criteria
  biosimilar products, 115t
Respiratory diseases, 720t, 1116, 1132
Ricin, 1558, 1564, 1574–1575, 1579, 1597–1598, 1618–1619, 1640, 1642, 1645
Rift valley fever (RVF), 1549
Risk assessment, 345–346
Rhenium radionuclides, 906
  See also radionuclides
RNA interference RNAi, 1069
  companies developing RNAi drugs, 1120t
  delivery of RNAi compounds, 1129
  shRNA, 943, 945
Robustness, 573, 580, 615, 622
Routes of administration, 1066, 1214
  of growth factors, cytokines, and chemokines, 1199t
RU486, 952

Saccharomyces cerevisiae, 3, 7, 41, 48, 50, 56
Safety pharmacology, 103
SARS, 1625, 1636
SCID trial in France, 966
Secreters, 11–12
Secretion signal, 7
Secretory expression systems, 5, 13
Selected drug transporters, 1470t

Selected drugs metabolized by various cytochrome P450, 1467*t*
Selection of clinical candidates, 102, 111
Selectivity, 576, 616–617, 626, 628, 630, 695
Serial analysis of gene expression (SAGE), 1092
Sialic acids, 133
Single nucleotide polymorphisms (SNPs), 670
Small interfering RNA (siRNA), 1013, 1016, 1146
  in clinical trials, 1076, 1162
  therapy, 1128
Small-pox
  complex, 1600
  vaccine, 692
Specificity, 573–574, 623*t*, 625–626, 641
Stability, 495, 579
Staphylococcal enterotoxin B (SEB), 1525, 1554
StarLink Bt Corn, 1443, 1438–1439
Stem cells, 1497–1514
  general classification, 1498–1500
  hematopoietic stem cells (HSCs), 1313, 1333
Streptomyces, 139
Structure-activity relationships, 1029, 1046
Swine, 511, 526
Syngenta Bt corn, 1433, 1440

T-2 toxin, 1551–1552, 1554–1555
Target product profile, 94
Target selection, 89–90, 92
TCDD, 511, 517, 520–521
Technetium radionuclides, 883, 902
  See also radionuclides
Telomerase, 1318
Telomeres, 1362–1364
Therapeutic proteins
  characteristics, 1222
Thread-induced thrombosis assay, 545
Thrombisis assay, 512, 545
Thrombolytic assay, 512, 546
Total error, 573, 578, 582, 584
Toxicology, 103
Toxicology and biodistribution of Ad5FGF-4, 956–969
Trichothecene mycotoxins, 513–514
Trade mark, 1396
Trade related aspects of intellectual property rights (TRIPs), 1401, 1428

Trade secret, 1394
Transfer RNA (tRNA)
  "wobbling" of, 26
Transmission electron microscopy (TEM), 1300
Tuberculosis, 1525, 1548
Tumor necrosis factor α (TNFα), 1117–1174, 1193
Tumor targeting, 1052
Twin arginine translocation (TAT), 6

Urotoxicity, 511, 522

Vaccination, 997
Vaccine classification, 1549*t*
Vaccine formulations, 1149–1157
Vaccines
  active immunotherapy, 851, 870
  anti-infectious diseases vaccines, 993
  antitumor vaccines, 995–997
  classification, 1379*t*
  DNA vaccine, 871
  formulations, 1162–1163
  plasmid DNA-based anti-tumor vaccines, 995–997
  smallpox vaccine, 692
  subunit vaccines, 738–739
Vasopressin, 709
Vector construction, 958–959
Vector decomplexation, 1044
Vectors, 1263
Viral clearance studies, 333–336
Viral diseases, 1132–1133
Viral infections, 1123–1124
Virosomes, 1157–1158, 1158*f*
Virus Removal, 36

West-Nile encephalitis, 1536–1537
Working cell bank, 959

Yeasts, 6–10, 8*t*, 9*t*
  yeast cells as artificial chromosomes (YACs), 59–60, 61*f*, 63
  yeast display, 859–860
  yeast protein extraction, 1493
Yeast oestrogen assay, 551–552
Yellow fever, 1549–1550
Yield, 697
Yttrium radionuclides, 913
  See also radionuclides